List of Applications

BUSINESS AND ECONOMICS

401(k) investors, 314
Accumulated value of an income stream, 506
Accumulation years of baby boomers, 197, 541
Advertising, 86, 150, 239, 278, 329, 541, 573
Ailing financial institutions, 129, 147
Aircraft structural integrity, 259
Airplane arrival times, 671
Air travel, 390
Alternative energy sources, 461
Alternative minimum tax, 281, 337
Amusement park attendance, 196, 505
Annual retail sales, 95, 154
Annuities, 359, 391, 475
Approximating changes in revenue, 595
Assembly time of workers, 278, 383, 434
Authentication technology, 578
Auto financing, 445
Auto replacement parts market, 89
Average age of cars in U.S., 300
Average daily TV-viewing time, 617
Banking, 56, 147, 405
Blackberry subscribers, 87
Black Monday, 285
Book design, 91, 323
Box-office receipts, 114, 181, 296
Broadband Internet households, 61
Budget deficit and surplus, 68, 252
Business cycles, 791, 795, 803
Business spending on technology, 281
Cable ad revenue, 96
Cable TV subscription, 244, 407, 418, 448, 664
Calling cards, 577
Capacity of man-made lake, 485
Capital value, 529, 726
Cargo volume, 266
Cash reserves at Blue Cross and Blue Shield, 282
CDs, 351, 489
Cellular phone subscription, 262
Charter-flight revenue, 84, 92
Chip sales, 87
City planning, 118, 182
Coal production, 447, 489
Cobb–Douglas production function, 550, 555, 588, 595, 617
COLAs, 61
Commissions, 130
Commodity prices, 130, 158, 368, 497
Common stock transactions, 4, 358
Commuter airlines, 496
Compact disc sales, 497
Comparison of bank rates, 359
Complementary commodities, 551, 556
Computer game sales, 532
Computer resale value, 488
Construction jobs, 76, 103, 184
Consumer decisions, 30
Consumer demand, 168, 181, 245, 394
Consumer price index, 169, 273, 394
Consumers' surplus, 467, 473, 474, 476, 477, 489, 505, 510, 519, 532
Consumption function, 61
Consumption of electricity, 435
Consumption of petroleum, 519
Cost of laying cable, 26, 30
Cost of producing calculators, 329
Cost of producing guitars, 406
Cost of producing loudspeakers, 503
Cost of producing PDAs, 75
Cost of producing solar cell panels, 414
Cost of producing surfboards, 150
Cost of removing toxic waste, 181, 296
Cost of wireless phone calls, 245
Creation of new jobs, 195
Credit card debt, 88, 407
Crop yield, 148, 371
Cruise ship bookings, 195
Demand for agricultural commodities, 239
Demand for butter, 551, 679
Demand for CD players, 616
Demand for commodities, 650
Demand for computer software, 532
Demand for digital camcorder tapes, 489
Demand for DVDs, 574
Demand for electricity, 580
Demand for perfume, 367
Demand for personal computers, 195, 387
Demand for RNs, 279
Demand for wine, 368
Demand for wristwatches, 181, 195, 757, 766
Depletion of Social Security funds, 315
Depreciation, 88, 365, 448
Designing a cruise ship pool, 587
Determining the optimal site, 568
Digital camera sales, 168
Digital TV sales, 280
Digital TV services, 44
Digital TV shipments, 95
Digital versus film cameras, 88
Disability benefits, 217
Disposable annual incomes, 86
Document management, 87
Driving costs, 81, 114, 154
Driving range of an automobile, 10
Drug spending, 281
DVD sales, 176, 407
Effect of advertising on bank deposits, 278
Effect of advertising on hotel revenue, 281
Effect of advertising on profit, 150, 239
Effect of advertising on sales, 86, 169, 235, 273, 387, 461
Effect of capital and labor on productivity, 598
Effect of housing starts on jobs, 195
Effect of inflation on salaries, 359
Effect of inventory and floor space on profit, 597
Effect of luxury tax on consumption, 194
Effect of mortgage rates on housing starts, 75, 239
Effect of price increase on quantity demanded, 239, 242
Effect of production on profit, 597
Effect of speed on operating cost of a truck, 235
Effect of tax cut on spending, 729, 770
Effect of TV advertising on car sales, 461
Efficiency studies, 169, 280, 438
Elasticity of demand, 205, 208, 210, 211, 230
E-mail usage, 87
Energy conservation, 452, 460
Energy consumption and productivity, 130, 358
Energy efficiency of appliances, 367
Establishing a trust fund, 529, 770
Estimating changes in profit, 617
Expected demand, 394
Expressway tollbooths, 664
Factory worker wages, 690, 691, 757
Federal budget deficit, 68, 252
Federal debt, 96, 314
Female self-employed workforce, 309
Financing a college education, 359, 475
Financing a home, 239, 241
Forecasting commodity prices, 239
Forecasting profits, 239, 281
Forecasting sales, 158, 416
Franchises, 475, 506
Fuel capacity of a space shuttle, 483
Fuel consumption of domestic cars, 518
Fuel economy of cars, 172, 248
Gasoline prices, 291
Gasoline self-service sales, 57
Gas station sales, 679
Gender gap, 60
Google's revenue, 282
Gross domestic product, 150, 166, 217, 239, 276, 311
Growth of bank deposits, 56
Growth of HMOs, 173, 498
Growth of managed services, 261
Growth of service industries, 520, 648, 708
Growth of Web sites, 336
Health-care costs, 170, 407
Health club membership, 158, 189
Home mortgages, 538, 543, 544, 766
Home sales, 173
Home-shopping industry, 135
Hotel occupancy rate, 75, 89, 194, 808
Households with microwaves, 389
Housing prices, 358, 447
Housing starts, 76, 195, 225
Illegal ivory trade, 88
Income distribution of a country, 490
Incomes of American families, 370
Income streams, 468, 529
Indian gaming industry, 94
Inflation, 215
Information security software sales, 576
Installment contract sales, 489
Internal rate of return, 762, 765
Inventory control and planning, 129, 321, 322, 325, 329, 330
Investment analysis, 359, 469, 475
Investment groups, 766
Investment options, 358
Investment returns, 240, 358, 394
IRAs, 470
Keogh accounts, 240, 489
Land prices, 555, 568, 613
Life span of car batteries, 694
Life span of color television tubes, 664, 679
Life span of light bulbs, 658, 669
Loan amortization, 370, 544
Loan consolidation, 358
Loans at Japanese banks, 367
Locating a TV relay station, 566
Lorentz curves, 472, 475, 506
Magazine circulation, 403
Mail-order phone sales, 664
Management decisions, 281, 469
Manufacturing capacity, 67, 173, 266, 284
Manufacturing capacity operating rate, 307
Marginal average cost function, 200, 201, 209, 210
Marginal cost function, 198, 199, 209, 210, 437, 488, 489
Marginal productivity of labor and capital, 550
Marginal productivity of money, 589
Marginal profit, 203, 209, 210
Marginal propensity to consume, 210
Marginal propensity to save, 210
Marginal revenue, 203, 209, 210, 311, 367, 480
Market equilibrium, 83, 91, 95, 157, 158, 466, 757
Market for cholesterol reducing drugs, 78
Market for drugs, 577, 581
Market share, 148, 404
Markup on a car, 10

(*continued*)

List of Applications (*continued*)

Mass transit subsidies, 576
Maximizing crop yield, 323
Maximizing oil production, 368
Maximizing production, 591
Maximizing profit, 303, 309, 310, 328, 329, 565, 568, 574, 586, 590
Maximizing revenue, 310, 324, 367, 617
Maximizing sales, 591
Meeting profit goals, 10
Meeting sales targets, 10
Metal fabrication, 322
Minimizing construction costs, 322, 329, 590
Minimizing container costs, 319, 323, 329, 591
Minimizing costs of laying cable, 324
Minimizing fencing costs, 617
Minimizing heating and cooling costs, 569
Minimizing packaging costs, 323, 329
Minimizing production costs, 310
Minimizing shipping costs, 29
Morning traffic rush, 267
Mortgage rates, 504
MP3 player reliability, 757
Multimedia sales, 220, 285
Multiplier effect, 726
Navigation systems, 44, 48
Net investment flow, 448
Net sales, 576
New construction jobs, 184
Newsmagazine shows, 418
Nielsen television polls, 134, 147
Office rents, 311
Oil production, 448, 460, 489, 495
Oil spills, 230, 514, 532
Online ad sales, 407
Online banking, 366, 390, 577
Online buyers, 168, 377
Online hotel reservations, 328
Online retail sales, 358
Online sales of used autos, 577
Online shopping, 96
Online spending, 96, 577
Online travel, 581
Operating costs of a truck, 235
Operating rates of factories, mines, and utilities, 307
Optimal charter flight fare, 324
Optimal market price, 364
Optimal selling price, 368
Optimal speed of a truck, 325
Optimal subway fare, 318
Outpatient service companies, 408
Outsourcing of jobs, 87, 193, 281
Ownership of portable phones, 168
Packaging, 52, 91, 317, 319, 329, 568, 569, 688
PC shipments, 281
Pensions, 358, 359
Perpetual net income stream, 529
Perpetuities, 529, 532, 729
Personal consumption expenditure, 210
Petroleum consumption, 517
Portable phone services, 168, 578
Present value of a franchise, 490, 498
Present value of an income stream, 469, 527
Price earnings ratio, 598
Price of replacement automobile parts, 89
Prime interest rate, 130
Producers' surplus, 467, 473, 474, 476, 505, 519, 532
Product design, 323
Production costs, 208, 209, 433
Production function, 543
Production of steam coal, 497
Productivity fueled by oil, 368
Productivity of a country, 555
Product reliability, 664, 691
Profit of a vineyard, 92, 325
Projected retirement funds, 262
Projection TV sales, 488
Purchasing power, 358
Quality control, 10, 406, 715
Racetrack design, 325
Rate of bank failures, 220, 266, 314
Rate of change of DVD sales, 176
Rate of change of housing starts, 225
Rate of net investment, 625
Rate of return on investment, 358, 498
Real estate, 355, 359, 428, 447, 518
Real estate investment groups, 766
Reliability of computer chips, 387
Reliability of microprocessors, 664, 717
Reliability of robots, 664
Resale value, 387, 650
Restaurant revenue, 795, 803
Retirement planning, 358, 359, 489
Revenue growth of a home theater business, 358
Revenue of a charter yacht, 324
Revenue of a travel agency, 598
Reverse annuity mortgage, 475
Sales forecasts, 51
Sales growth and decay, 44
Sales of a sporting good store, 38
Sales of digital signal processors, 95, 169
Sales of digital TVs, 86
Sales of drugs, 577
Sales of DVD players versus VCRs, 89
Sales of functional food products, 262
Sales of GPS equipment, 577
Sales of loudspeakers, 488
Sales of mobile processors, 281
Sales of pocket computers, 438
Sales of prerecorded music, 60
Sales promotions, 367
Sales tax, 61
Satellite radio subscriptions, 406
Selling price of DVD recorders, 87, 193
Shopping habits, 679
Sickouts, 314
Sinking funds, 471, 641
Social Security beneficiaries, 136
Social Security contributions, 43
Social Security wage base, 577
Solvency of the Social Security system, 299, 315
Spending on Medicare, 169
Starbucks' annual sales, 580
Starbucks' store count, 576
Stock prices, 794, 801, 803, 804
Stock purchase, 4
Substitute commodities, 551, 556
Supply and demand, 83, 90, 168, 226, 230, 418, 625, 632
Tax planning, 358
Television pilots, 666
Testing new products, 217
Time on the market, 285, 314
Training personnel, 641
Tread lives of tires, 520
Truck leasing, 61
Trust funds, 529
TV viewing patterns, 134, 193
U.S. daily oil consumption, 519
U.S. drug sales, 577
Use of diesel engines, 314
U.S. nutritional supplements market, 88
U.S. online banking households, 577
U.S. strategic petroleum reserves, 519
Value of an art object, 39
Value of an investment, 74
VCR ownership, 505
Venture-capital investment, 311
Wages, 145
Waiting times, 664, 666
Web hosting, 262
Wilson lot size formula, 544
Worker efficiency, 62, 86, 169, 280, 329
World production of coal, 447, 489
Worldwide production of vehicles, 197
Yahoo! in Europe, 377

SOCIAL SCIENCES

Age of drivers in crash fatalities, 263
Aging drivers, 86
Aging population, 193, 218, 815
Air pollution, 194, 262, 263, 267, 282, 408, 519
Air purification, 217
Alcohol-related traffic accidents, 497
Annual college costs, 581
Arson for profit, 543
Bursts of knowledge, 124
Civil service exams, 691
Closing the gender gap in education, 61
College admissions, 43, 576
Commuter trends, 488
Continuing education enrollment, 194, 708
Cost of removing toxic waste, 114, 178, 181, 296
Crime, 217, 239, 257, 311
Cube rule, 62
Curbing population growth, 170
Decline of union membership, 67
Demographics, 388
Dependency ratio, 282
Disability benefits, 217
Disability rates, 336
Dissemination of information, 388
Distribution of families by size, 666
Distribution of incomes, 10, 360, 473, 475
Driving age requirement, 681
Educational level of senior citizens, 40, 575
Effect of budget cuts on crime rate, 280
Effect of immigration on population growth, 641
Effect of smoking bans, 280
Elderly workforce, 262
Endowments, 527, 529, 729
Energy conservation, 456
Energy needs, 435
Families' annual income, 360
Female life expectancy, 192, 418, 617
Food stamp recipients, 315
Foreign-born residents, 311
Gender gap, 60
Global epidemic, 440
Global supply of plutonium, 75
Grade distributions, 691
Growth of HMOs, 173, 284
Health-care spending, 73, 170, 407, 578
HMOs, 79
Immigration, 89, 386
Income distributions, 473
Increase in juvenile offenders, 371
Intervals between phone calls, 532
Lay teachers at Roman Catholic schools, 388, 391
Learning curves, 124, 129, 181, 239, 387, 418, 641, 650
Logistic curves, 385
Male life expectancy, 245, 578
Marijuana arrests, 96, 440
Married households, 336
Married households with children, 168
Mass transit, 318, 576
Medical school applicants, 262
Membership in credit unions, 448
Narrowing gender gap, 44
Nuclear plant utilization, 43
Oil spills, 158, 194, 230, 418, 514, 532
Overcrowding of prisons, 75, 263
Ozone pollution, 408

(*continued on back endpaper*)

Applied Calculus

for the Managerial, Life, and Social Sciences

Seventh Edition

S. T. TAN

STONEHILL COLLEGE

Australia • Brazil • Canada • Mexico • Singapore • Spain
United Kingdom • United States

Applied Calculus
for the Managerial, Life, and Social Sciences, 7e
S. T. Tan

Acquisitions Editor: Carolyn Crocket
Development Editor: Danielle Derbenti
Assistant Editor: Beth Gershman
Editorial Assistant: Ashley Summers
Technology Project Manager: Donna Kelley
Marketing Manager: Joe Rogove
Marketing Assistant: Jennifer Liang
Marketing Communications Manager: Jessica Perry
Project Manager, Editorial Production: Janet Hill
Creative Director: Rob Hugel
Print Buyer: Becky Cross
Permissions Editor: Bob Kauser
Production Service: Newgen–Austin—Jamie Armstrong
Text Designer: Diane Beasley
Photo Researcher: Kathleen Olson
Cover Designer: Irene Morris
Cover Image: Portrait of Navin Khaneja by Peter Kiar
Cover Printer: Coral Graphic Services
Compositor: Newgen
Printer: Quebecor World—Taunton

Printed in Canada
2 3 4 5 6 7 11 10 09 08 07

Library of Congress Control Number: 2006931576

Student Edition
ISBN-13: 978-0-495-01582-6
ISBN-10: 0-495-01582-2

Thomson Higher Education
10 Davis Drive
Belmont, CA 94002-3098
USA

TO PAT, BILL, AND MICHAEL

Contents

Preface x

CHAPTER 1

Preliminaries 1

1.1 Precalculus Review I 2
1.2 Precalculus Review II 11
1.3 The Cartesian Coordinate System 23
1.4 Straight Lines 31
Chapter 1 Summary of Principal Formulas and Terms 46
Chapter 1 Concept Review Questions 46
Chapter 1 Review Exercises 47
Chapter 1 Before Moving On 48

CHAPTER 2

Functions, Limits, and the Derivative 49

2.1 Functions and Their Graphs 50
Using Technology: Graphing a Function 64
2.2 The Algebra of Functions 68
2.3 Functions and Mathematical Models 76
Using Technology: Finding the Points of Intersection of Two Graphs and Modeling 93
2.4 Limits 97
Using Technology: Finding the Limit of a Function 116
2.5 One-Sided Limits and Continuity 119
Using Technology: Finding the Points of Discontinuity of a Function 132
2.6 The Derivative 135
Using Technology: Graphing a Function and Its Tangent Line 152
Chapter 2 Summary of Principal Formulas and Terms 155
Chapter 2 Concept Review Questions 155
Chapter 2 Review Exercises 156
Chapter 2 Before Moving On 158

CHAPTER 3

Differentiation 159

3.1 Basic Rules of Differentiation 160
Using Technology: Finding the Rate of Change of a Function 171

Note: Sections marked with an asterisk are not prerequisites for later material.

3.2 The Product and Quotient Rules 174
Using Technology: The Product and Quotient Rules 183
3.3 The Chain Rule 185
Using Technology: Finding the Derivative of a Composite Function 196
3.4 Marginal Functions in Economics 197
3.5 Higher-Order Derivatives 212
PORTFOLIO: Steve Regenstreif 213
Using Technology: Finding the Second Derivative of a Function at a Given Point 219
*3.6 Implicit Differentiation and Related Rates 221
3.7 Differentials 232
Using Technology: Finding the Differential of a Function 240
Chapter 3 Summary of Principal Formulas and Terms 242
Chapter 3 Concept Review Questions 243
Chapter 3 Review Exercises 243
Chapter 3 Before Moving On 245

CHAPTER 4

Applications of the Derivative 247

4.1 Applications of the First Derivative 248
Using Technology: Using the First Derivative to Analyze a Function 264
4.2 Applications of the Second Derivative 267
Using Technology: Finding the Inflection Points of a Function 283
4.3 Curve Sketching 285
Using Technology: Analyzing the Properties of a Function 298
4.4 Optimization I 300
Using Technology: Finding the Absolute Extrema of a Function 313
4.5 Optimization II 315
Chapter 4 Summary of Principal Terms 327
Chapter 4 Concept Review Questions 327
Chapter 4 Review Exercises 328
Chapter 4 Before Moving On 330

CHAPTER 5

Exponential and Logarithmic Functions 331

5.1 Exponential Functions 332
Using Technology 338
5.2 Logarithmic Functions 339
5.3 Compound Interest 347
5.4 Differentiation of Exponential Functions 360
PORTFOLIO: Robert Derbenti 361
Using Technology 370
5.5 Differentiation of Logarithmic Functions 371
*5.6 Exponential Functions as Mathematical Models 379
Using Technology: Analyzing Mathematical Models 389
Chapter 5 Summary of Principal Formulas and Terms 392
Chapter 5 Concept Review Questions 392

Chapter 5 Review Exercises *393*
Chapter 5 Before Moving On *394*

CHAPTER 6

Integration 395

6.1 Antiderivatives and the Rules of Integration 396
6.2 Integration by Substitution 410
6.3 Area and the Definite Integral 420
6.4 The Fundamental Theorem of Calculus 429
Using Technology: Evaluating Definite Integrals 440
6.5 Evaluating Definite Integrals 441
Using Technology: Evaluating Definite Integrals for Piecewise-Defined Functions 450
6.6 Area between Two Curves 452
Using Technology: Finding the Area between Two Curves 463
*6.7 Applications of the Definite Integral to Business and Economics 464
Using Technology: Business and Economic Applications 476
*6.8 Volumes of Solids of Revolution 478
Chapter 6 Summary of Principal Formulas and Terms *485*
Chapter 6 Concept Review Questions *487*
Chapter 6 Review Exercises *487*
Chapter 6 Before Moving On *490*

CHAPTER 7

Additional Topics in Integration 491

7.1 Integration by Parts 492
*7.2 Integration Using Tables of Integrals 499
7.3 Numerical Integration 506
7.4 Improper Integrals 521
Chapter 7 Summary of Principal Formulas and Terms *530*
Chapter 7 Concept Review Questions *531*
Chapter 7 Review Exercises *531*
Chapter 7 Before Moving On *533*

CHAPTER 8

Calculus of Several Variables 535

8.1 Functions of Several Variables 536
8.2 Partial Derivatives 545
Using Technology: Finding Partial Derivatives at a Given Point 558
8.3 Maxima and Minima of Functions of Several Variables 559
PORTFOLIO: Kirk Hoiberg 562
8.4 The Method of Least Squares 570
Using Technology: Finding an Equation of a Least-Squares Line 579
8.5 Constrained Maxima and Minima and the Method of Lagrange Multipliers 581
*8.6 Total Differentials 592
*8.7 Double Integrals 599
*8.8 Applications of Double Integrals 606

Chapter 8 Summary of Principal Terms 614
Chapter 8 Concept Review Questions 615
Chapter 8 Review Exercises 615
Chapter 8 Before Moving On 617

CHAPTER 9
Differential Equations 619

9.1 Differential Equations 620
9.2 Separation of Variables 627
9.3 Applications of Separable Differential Equations 633
9.4 Approximate Solutions of Differential Equations 642
Chapter 9 Summary of Principal Terms 649
Chapter 9 Concept Review Questions 649
Chapter 9 Review Exercises 650
Chapter 9 Before Moving On 651

CHAPTER 10
Probability and Calculus 653

10.1 Probability Distributions of Random Variables 654
Using Technology: Graphing a Histogram 665
10.2 Expected Value and Standard Deviation 666
PORTFOLIO: Gary Li 672
Using Technology: Finding the Mean and Standard Deviation 680
10.3 Normal Distributions 682
Chapter 10 Summary of Principal Formulas and Terms 692
Chapter 10 Concept Review Questions 693
Chapter 10 Review Exercises 694
Chapter 10 Before Moving On 695

CHAPTER 11
Taylor Polynomials and Infinite Series 697

11.1 Taylor Polynomials 698
11.2 Infinite Sequences 710
11.3 Infinite Series 718
11.4 Series with Positive Terms 730
11.5 Power Series and Taylor Series 741
11.6 More on Taylor Series 751
***11.7** Newton's Method 758
Chapter 11 Summary of Principal Formulas and Terms 767
Chapter 11 Concept Review Questions 768
Chapter 11 Review Exercises 769
Chapter 11 Before Moving On 770

CHAPTER 12
Trigonometric Functions 771

12.1 Measurement of Angles 772
12.2 The Trigonometric Functions 777

12.3 Differentiation of Trigonometric Functions 785
Using Technology: Analyzing Trigonometric Functions 797
12.4 Integration of Trigonometric Functions 799
Using Technology: Evaluating Integrals of Trigonometric Functions 805
Chapter 12 Summary of Principal Formulas and Terms 806
Chapter 12 Concept Review Questions 807
Chapter 12 Review Exercises 807
Chapter 12 Before Moving On 808

APPENDIX A **Inverse Functions 809**

A.1 The Inverse of a Function 810
A.2 The Graphs of Inverse Functions 812
A.3 Functions That Have Inverses 812
A.4 Finding the Inverse of a Function 813

APPENDIX B **Table 817**

The Standard Normal Distribution 817

Answers to Odd-Numbered Exercises 819
Index 865

Preface

Math is an integral part of our daily life. *Applied Calculus for the Managerial, Life, and Social Sciences, Seventh Edition*, attempts to illustrate this point with its applied approach to mathematics. This text is appropriate for use in a two-semester or three-quarter introductory calculus course for students in the managerial, life, and social sciences. Our objective for this Seventh Edition is twofold: (1) to write an applied text that motivates students and (2) to make the book a useful tool for instructors. We hope that with this present edition we have come closer to realizing our goal.

General Approach

- **Level of Presentation** Our approach is intuitive, and we state the results informally. However, we have taken special care to ensure that this approach does not compromise the mathematical content and accuracy.

- **Approach** A problem-solving approach is stressed throughout the book. Numerous examples and applications are used to illustrate each new concept and result in order to help the students comprehend the material presented. An emphasis is placed on helping the students formulate, solve, and interpret the results of the problems involving applications. Very early on in the text, students are given practice in setting up word problems (Section 2.3) and developing modeling skills. As another example, when optimization problems are covered, the problems are presented in two sections. First students are asked to solve optimization problems in which the objective function to be optimized is given (Section 4.4) and then students are asked to solve problems where they have to formulate the optimization problems to be solved (Section 4.5).

Guidelines for Constructing Mathematical Models

1. Assign a letter to each variable mentioned in the problem. If appropriate, draw and label a figure.
2. Find an expression for the quantity sought.
3. Use the conditions given in the problem to write the quantity sought as a function f of one variable. Note any restrictions to be placed on the domain of f from physical considerations of the problem.

APPLIED EXAMPLE 5 Enclosing an Area The owner of the Rancho Los Feliz has 3000 yards of fencing with which to enclose a rectangular piece of grazing land along the straight portion of a river. Fencing is not required along the river. Letting x denote the width of the rectangle, find a function f in the variable x giving the area of the grazing land if she uses all of the fencing (Figure 23).

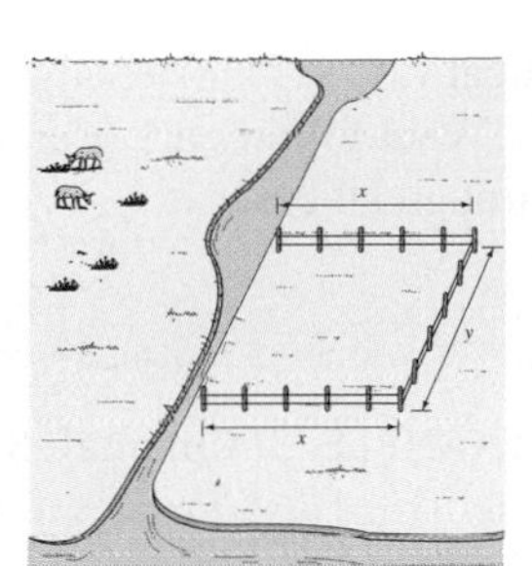

FIGURE 23
The rectangular grazing land has width x and length y.

Solution

1. This information was given.
2. The area of the rectangular grazing land is $A = xy$. Next, observe that the amount of fencing is $2x + y$ and this must be equal to 3000 since all the fencing is used; that is,
$$2x + y = 3000$$
3. From the equation we see that $y = 3000 - 2x$. Substituting this value of y into the expression for A gives
$$A = xy = x(3000 - 2x) = 3000x - 2x^2$$
Finally, observe that both x and y must be nonnegative since they represent the width and length of a rectangle, respectively. Thus, $x \geq 0$ and $y \geq 0$. But the latter is equivalent to $3000 - 2x \geq 0$, or $x \leq 1500$. So the required function is $f(x) = 3000x - 2x^2$ with domain $0 \leq x \leq 1500$. ■

- **Intuitive Introduction to Concepts** Mathematical concepts are introduced with concrete real-life examples, wherever appropriate. My goal here is to capture students' interest and show the relevance of mathematics to their everyday life. For example, curve-sketching (Section 4.3) is introduced via a "Black Monday" example.

Consider, for example, the graph of the function giving the Dow-Jones Industrial Average (DJIA) on Black Monday, October 19, 1987 (Figure 45). Here, $t = 0$ corresponds to 8:30 a.m., when the market was open for business, and $t = 7.5$ corresponds to 4 p.m., the closing time. The following information may be gleaned from studying the graph.

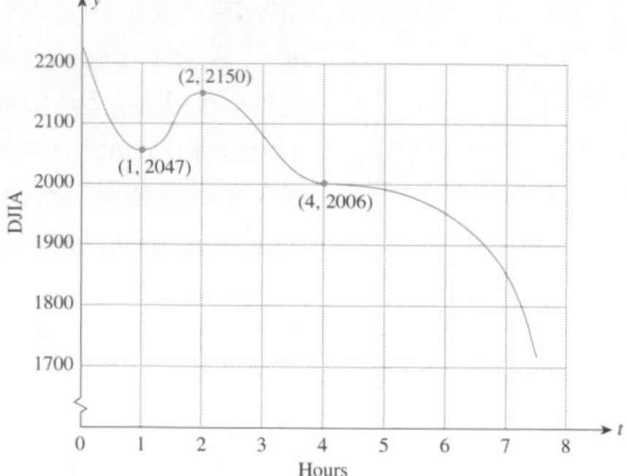

FIGURE 45
The Dow-Jones Industrial Average on Black Monday

Source: Wall Street Journal

The graph is *decreasing* rapidly from $t = 0$ to $t = 1$, reflecting the sharp drop in the index in the first hour of trading. The point (1, 2047) is a *relative minimum* point of the function, and this turning point coincides with the start of an aborted recovery. The short-lived rally, represented by the portion of the graph that is *increasing* on the interval (1, 2), quickly fizzled out at $t = 2$ (10:30 a.m.). The *relative maximum* point (2, 2150) marks the highest point of the recovery. The function is decreasing in the rest of the interval. The point (4, 2006) is an *inflection point* of the function; it shows that there was a temporary respite at $t = 4$ (12:30 p.m.). However, selling pressure continued unabated, and the DJIA continued to fall until the closing bell. Finally, the graph also shows that the index opened at the high of the day [$f(0) = 2247$ is the *absolute maximum* of the function] and closed at the low of the day [$f(\frac{15}{2}) = 1739$ is the *absolute minimum* of the function], a drop of 508 points!*

Motivation

Illustrating the practical value of mathematics in applied areas is an important objective of our approach. What follows are examples of how we have implemented this relevant approach throughout the text.

- **Real-life Applications** Current and relevant examples and exercises are drawn from the fields of business, economics, social and behavioral sciences, life sciences, physical sciences, and other fields of interest. In the examples, these are highlighted with new icons that illustrate the various applications.

APPLIED EXAMPLE 3 Optimal Subway Fare A city's Metropolitan Transit Authority (MTA) operates a subway line for commuters from a certain suburb to the downtown metropolitan area. Currently, an average of 6000 passengers a day take the trains, paying a fare of \$3.00 per ride. The board of the MTA, contemplating raising the fare to \$3.50 per ride in order to generate a larger revenue, engages the services of a consulting firm. The firm's study reveals that for each \$.50 increase in fare, the ridership will be reduced by an average of 1000 passengers a day. Thus, the consulting firm recommends that MTA stick to the current fare of \$3.00 per ride, which already yields a maximum revenue. Show that the consultants are correct.

Developing Modeling Skills We believe that one of the most important skills a student can learn is the ability to translate a real problem into a real model that can provide insight into the problem. Many of the applications are based on mathematical models (functions) that the author has constructed using data drawn from various sources, including current newspapers, magazines, and data obtained through the Internet. Sources are given in the text for these applied problems. In Functions and Mathematical Models (Section 2.3), the modeling process is discussed and students are asked to use models (functions) constructed from real-life data to answer questions about the Market for Cholesterol-Reducing Drugs, HMO Membership, and the Driving Costs for a Ford Taurus.

29. BLACKBERRY SUBSCRIBERS According to a study conducted in 2004, the number of subscribers of BlackBerry, the handheld e-mail devices manufactured by Research in Motion Ltd., is expected to be

$$N(t) = -0.0675t^4 + 0.5083t^3 - 0.893t^2 + 0.66t + 0.32 \quad (0 \le t \le 4)$$

where $N(t)$ is measured in millions and t in years, with $t = 0$ corresponding to the beginning of 2002.

a. How many BlackBerry subscribers were there at the beginning of 2002?

b. What is the projected number of BlackBerry subscribers at the beginning of 2006?

Source: ThinkEquity Partners

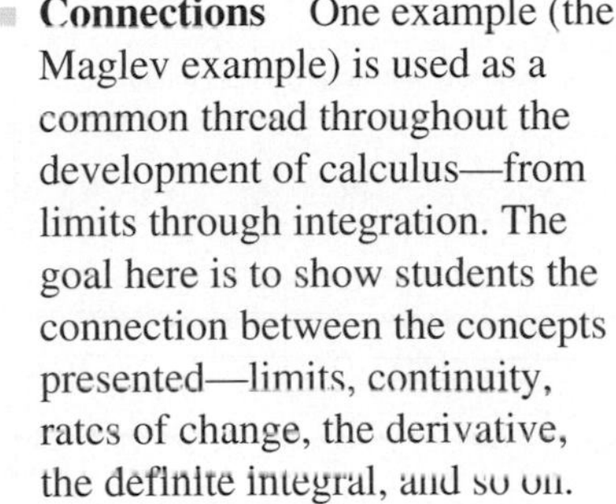

Connections One example (the Maglev example) is used as a common thread throughout the development of calculus—from limits through integration. The goal here is to show students the connection between the concepts presented—limits, continuity, rates of change, the derivative, the definite integral, and so on.

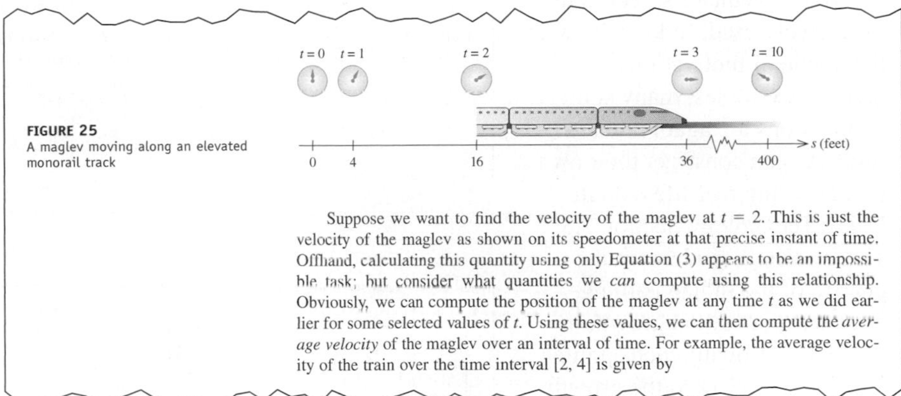

FIGURE 25
A maglev moving along an elevated monorail track

Suppose we want to find the velocity of the maglev at $t = 2$. This is just the velocity of the maglev as shown on its speedometer at that precise instant of time. Offhand, calculating this quantity using only Equation (3) appears to be an impossible task; but consider what quantities we *can* compute using this relationship. Obviously, we can compute the position of the maglev at any time t as we did earlier for some selected values of t. Using these values, we can then compute the *average velocity* of the maglev over an interval of time. For example, the average velocity of the train over the time interval [2, 4] is given by

Utilizing Tools Students Use

Technology Technology is used to explore mathematical ideas and as a tool to solve problems throughout the text.

Exploring with Technology Questions Here technology is used to explore mathematical concepts and to shed further light on examples in the text. These optional questions appear throughout the main body of the text and serve to enhance the student's understanding of the concepts and theory presented. Often the solution of an example in the text is augmented with a graphical or numerical solution. Complete solutions to these exercises are given in the *Instructor's Solution Manual.*

EXPLORING WITH TECHNOLOGY

In the opening paragraph of Section 5.1, we pointed out that the accumulated amount of an account earning interest *compounded continuously* will eventually outgrow by far the accumulated amount of an account earning interest at the same nominal rate but earning simple interest. Illustrate this fact using the following example.

Suppose you deposit \$1000 in account I, earning interest at the rate of 10% per year compounded continuously so that the accumulated amount at the end of t years is $A_1(t) = 1000e^{0.1t}$. Suppose you also deposit \$1000 in account II, earning simple interest at the rate of 10% per year so that the accumulated amount at the end of t years is $A_2(t) = 1000(1 + 0.1t)$. Use a graphing utility to sketch the graphs of the functions A_1 and A_2 in the viewing window $[0, 20] \times [0, 10{,}000]$ to see the accumulated amounts $A_1(t)$ and $A_2(t)$ over a 20-year period.

- **Using Technology** These are optional subsections that appear after the exercises. They can be used in the classroom if desired or as material for self-study by the student. Here the graphing calculator and Excel spreadsheets are used as a tool to solve problems. The subsections are written in the traditional example-exercise format with answers given at the back of the book. Illustrations showing graphing calculator screens are extensively used. In keeping with the theme of motivation through real-life examples, many sourced applications are again included. Students can construct their own models using real-life data in Using Technology Section 2.3. These include models for the growth of the Indian gaming industry, population growth in the fastest growing metropolitan area in the U.S., and the growth in online spending, among others. In Using Technology Section 4.3, students are asked to predict when the assets of the Social Security "trust fund" (unless changes are made) will be exhausted.

APPLIED EXAMPLE 3 Indian Gaming Industry The following data gives the estimated gross revenues (in billions of dollars) from the Indian gaming industries from 1990 ($t = 0$) to 1997 ($t = 7$).

Year	0	1	2	3	4	5	6	7
Revenue	0.5	0.7	1.6	2.6	3.4	4.8	5.6	6.8

a. Use a graphing utility to find a polynomial function f of degree 4 that models the data.
b. Plot the graph of the function f, using the viewing window $[0, 8] \times [0, 10]$.
c. Use the function evaluation capability of the graphing utility to compute $f(0)$, $f(1), \ldots, f(7)$ and compare these values with the original data.
Source: Christiansen/Cummings Associates

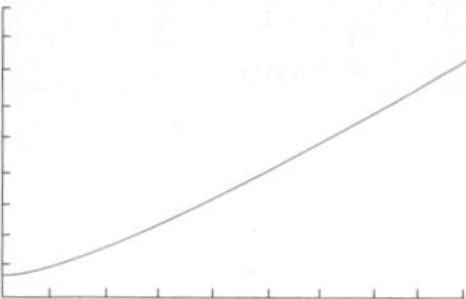

FIGURE T3
The graph of f in the viewing window $[0, 8] \times [0, 10]$

Solution

a. Choosing P4REG (fourth-order polynomial regression) from the STAT CALC (statistical calculations) menu of a graphing utility, we find

$$f(t) = 0.00379t^4 - 0.06616t^3 + 0.41667t^2 - 0.07291t + 0.48333$$

b. The graph of f is shown in Figure T3.
c. The required values, which compare favorably with the given data, follow:

t	0	1	2	3	4	5	6	7
$f(t)$	0.5	0.8	1.5	2.5	3.6	4.6	5.7	6.8

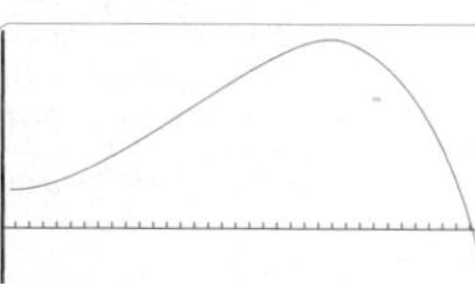

FIGURE T5
The graph of $f(t)$

APPLIED EXAMPLE 3 Solvency of Social Security Fund Unless payroll taxes are increased significantly and/or benefits are scaled back drastically, it is a matter of time before the current Social Security system goes broke. Data show that the assets of the system—the Social Security "trust fund"—may be approximated by

$$f(t) = -0.0129t^4 + 0.3087t^3 + 2.1760t^2 + 62.8466t + 506.2955 \quad (0 \le t \le 35)$$

where $f(t)$ is measured in millions of dollars and t is measured in years, with $t = 0$ corresponding to 1995.

a. Use a graphing calculator to sketch the graph of f.
b. Based on this model, when can the Social Security system be expected to go broke?
Source: Social Security Administration

Solution

a. The graph of f in the window $[0, 35] \times [-1000, 3500]$ is shown in Figure T5.
b. Using the function for finding the roots on a graphing utility, we find that $y = 0$ when $t \approx 34.1$, and this tells us that the system is expected to go broke around 2029.

Exercise Sets The exercise sets are designed to help students understand and apply the concepts developed in each section. Three types of exercises are included in these sets.

- **Self-Check Exercises** offer students immediate feedback on key concepts with worked-out solutions following the section exercises.

2.6 Self-Check Exercises

1. Let $f(x) = -x^2 - 2x + 3$.
   a. Find the derivative f' of f, using the definition of the derivative.
   b. Find the slope of the tangent line to the graph of f at the point (0, 3).
   c. Find the rate of change of f when $x = 0$.
   d. Find an equation of the tangent line to the graph of f at the point (0, 3).
   e. Sketch the graph of f and the tangent line to the curve at the point (0, 3).

2. The losses (in millions of dollars) due to bad loans extended chiefly in agriculture, real estate, shipping, and energy by the Franklin Bank are estimated to be

$$A = f(t) = -t^2 + 10t + 30 \qquad (0 \le t \le 10)$$

where t is the time in years ($t = 0$ corresponds to the beginning of 1994). How fast were the losses mounting at the beginning of 1997? At the beginning of 1999? At the beginning of 2001?

Solutions to Self-Check Exercises 2.6 can be found on page 152.

- **New Concept Questions** are designed to test students' understanding of the basic concepts discussed in the section and at the same time encourage students to explain these concepts in their own words.

2.6 Concept Questions

1. Let $P(2, f(2))$ and $Q(2 + h, f(2 + h))$ be points on the graph of a function f.
   a. Find an expression for the slope of the secant line passing through P and Q.
   b. Find an expression for the slope of the tangent line passing through P.
   c. Compare your answers for part (a) and (b) with those of Exercise 1.

2. Refer to Question 1.
   a. Find an expression for the average rate of change of f over the interval $[2, 2 + h]$.
   b. Find an expression for the instantaneous rate of change of f at 2.

3. a. Give a geometric and a physical interpretation of the expression

$$\frac{f(x + h) - f(x)}{h}$$

   b. Give a geometric and a physical interpretation of the expression

$$\lim_{h \to 0} \frac{f(x + h) - f(x)}{h}$$

4. Under what conditions does a function fail to have a derivative at a number? Illustrate your answer with sketches.

- **Exercises** provide an ample set of problems of a routine computational nature followed by an extensive set of application-oriented problems.

2.6 Exercises

1. Average Weight of an Infant The following graph shows the weight measurements of the average infant from the time of birth ($t = 0$) through age 2 ($t = 24$). By computing the slopes of the respective tangent lines, estimate the rate of change of the average infant's weight when $t = 3$ and when $t = 18$. What is the average rate of change in the average infant's weight over the first year of life?

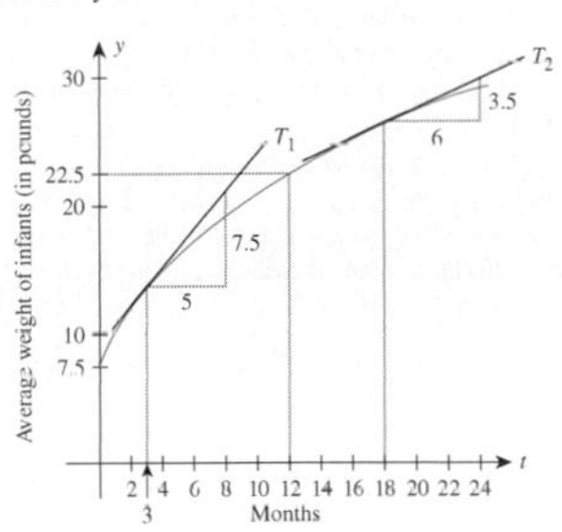

2. Forestry The following graph shows the volume of wood produced in a single-species forest. Here $f(t)$ is measured in cubic meters/hectare and t is measured in years. By computing the slopes of the respective tangent lines, estimate the rate at which the wood grown is changing at the beginning of year 10 and at the beginning of year 30.

Source: *The Random House Encyclopedia*

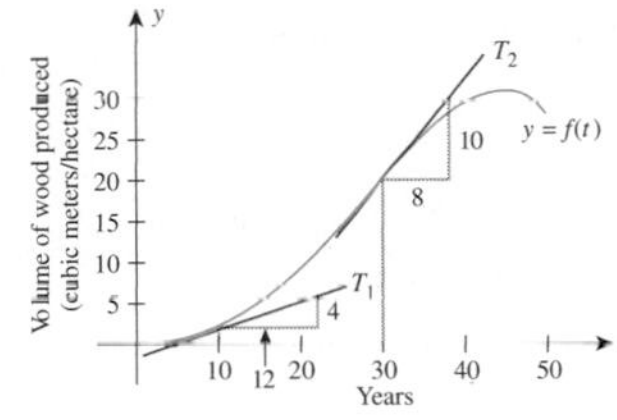

Review Sections These sections are designed to help students review the material in each section and assess their understanding of basic concepts as well as problem-solving skills.

- **Summary of Principal Formulas and Terms** highlights important equations and terms with page numbers given for quick review.

CHAPTER 2 **Summary of Principal Formulas and Terms**

FORMULAS

1. Average rate of change of f over $[x, x + h]$ or Slope of the secant line to the graph of f through $(x, f(x))$ and $(x + h, f(x + h))$ or Difference quotient $\dfrac{f(x+h) - f(x)}{h}$

TERMS

function (50)
domain (50)
range (50)
independent variable (51)
dependent variable (51)
polynomial function (80)
linear function (80)
quadratic function (80)
cubic function (80)
rational function (80)
limit of a function (100)
indeterminate form (103)
limit of a function at infinity (107)
right-hand limit of a function (119)
left-hand limit of a function (119)

- **New Concept Review Questions** give students a chance to check their knowledge of the basic definitions and concepts given in each chapter.

CHAPTER 2 **Concept Review Questions**

Fill in the blanks.

1. If f is a function from the set A to the set B, then A is called the _____ of f, and the set of all values of $f(x)$ as x takes on all possible values in A is called the _____ of f. The range of f is contained in the set _____.
2. The graph of a function is the set of all points (x, y) in the xy-plane such that x is in the _____ of f and $y =$ _____. The vertical-line test states that a curve in the xy-plane is the graph of a function $y = f(x)$ if and only if each _____ line intersects it in at most one _____.
4. The composition of g and f is the function with rule $(g \circ f)(x) =$ _____. Its domain is the set of all x in the domain of _____ such that _____ lies in the domain of _____.
5. **a.** A polynomial function of degree n is a function of the form _____.
   b. A polynomial function of degree 1 is called a _____ function; one of degree 2 is called a _____ function; one of degree 3 is called a _____ function.
   c. A rational function is a/an _____ of two _____.
   d. A power function has the form $f(x) =$ _____.

- **Review Exercises** offer routine computational exercises followed by applied problems.

CHAPTER 2 **Review Exercises**

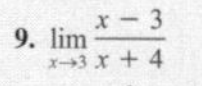

1. Find the domain of each function:
   a. $f(x) = \sqrt{9 - x}$ **b.** $f(x) = \dfrac{x + 3}{2x^2 - x - 3}$
2. Let $f(x) = 3x^2 + 5x - 2$. Find:
   a. $f(-2)$ **b.** $f(a + 2)$
   c. $f(2a)$ **d.** $f(a + h)$
8. $\lim\limits_{x \to -1} (3x^2 + 4)(2x - 1)$
9. $\lim\limits_{x \to 3} \dfrac{x - 3}{x + 4}$
10. $\lim\limits_{x \to 2} \dfrac{x + 3}{x^2 - 9}$
11. $\lim\limits_{x \to -2} \dfrac{x^2 - 2x - 3}{x^2 + 5x + 6}$
12. $\lim\limits_{x \to 3} \sqrt{2x^3 - 5}$

- **New Before Moving On . . . Exercises** give students a chance to see if they have mastered the basic computational skills developed in each chapter. If they solve a problem incorrectly, they can go to the companion website and try again. In fact, they can keep on trying until they get it right. If students need step-by-step help, they can utilize the ThomsonNOW Tutorials that are keyed to the text and work out similar problems at their own pace.

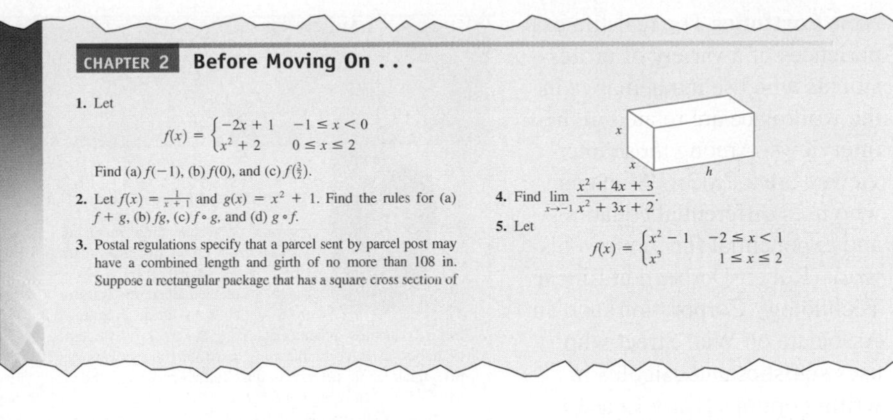

CHAPTER 2 **Before Moving On . . .**

1. Let

$$f(x) = \begin{cases} -2x + 1 & -1 \le x < 0 \\ x^2 + 2 & 0 \le x \le 2 \end{cases}$$

Find (a) $f(-1)$, (b) $f(0)$, and (c) $f(\frac{3}{2})$.

2. Let $f(x) = \frac{1}{x+1}$ and $g(x) = x^2 + 1$. Find the rules for (a) $f + g$, (b) fg, (c) $f \circ g$, and (d) $g \circ f$.

3. Postal regulations specify that a parcel sent by parcel post may have a combined length and girth of no more than 108 in. Suppose a rectangular package that has a square cross section of

4. Find $\lim_{x \to -1} \frac{x^2 + 4x + 3}{x^2 + 3x + 2}$.

5. Let

$$f(x) = \begin{cases} x^2 - 1 & -2 \le x < 1 \\ x^3 & 1 \le x \le 2 \end{cases}$$

- **Explore & Discuss** are optional questions appearing throughout the main body of the text that can be discussed in class or assigned as homework. These questions generally require more thought and effort than the usual exercises. They may also be used to add a writing component to the class or as team projects. Complete solutions to these exercises are given in the *Instructor's Solutions Manual*.

EXPLORE & DISCUSS

The average price of gasoline at the pump over a 3-month period, during which there was a temporary shortage of oil, is described by the function f defined on the interval $[0, 3]$. During the first month, the price was increasing at an increasing rate. Starting with the second month, the good news was that the rate of increase was slowing down, although the price of gas was still increasing. This pattern continued until the end of the second month. The price of gas peaked at the end of $t = 2$ and began to fall at an increasing rate until $t = 3$.

1. Describe the signs of $f'(t)$ and $f''(t)$ over each of the intervals $(0, 1)$, $(1, 2)$, and $(2, 3)$.
2. Make a sketch showing a plausible graph of f over $[0, 3]$.

- **New Portfolios** The real-life experiences of a variety of professionals who use mathematics in the workplace are related in these interviews. Among those interviewed are a Process Manager who uses differential equations and exponential functions in his work (Robert Derbenti at Linear Technology Corporation) and an Associate on Wall Street who uses statistics and calculus in writing options (Gary Li at J P Morgan Chase & Co.).

PORTFOLIO Gary Li

TITLE Associate
INSTITUTION JPMorgan Chase

As one of the leading financial institutions in the world, JPMorgan Chase & Co. depends on a wide range of mathematical disciplines from statistics to linear programming to calculus. Whether assessing the credit worthiness of a borrower, recommending portfolio investments or pricing an exotic derivative, quantitative understanding is a critical tool in serving the financial needs of clients.

I work in the Fixed-Income Derivatives Strategy group. A derivative in finance is an instrument whose value depends on the price of some other underlying instrument. A simple type of derivative is the forward contract, where two parties agree to a future trade at a specified price. In agriculture, for instance, farmers will often pledge their crops for sale to buyers at an agreed price before even planting the harvest. Depending on the weather, demand and other factors, the actual price may turn out higher or lower. Either the buyer or seller of the with interest rates. With trillions of dollars in this form, especially government bonds and mortgages, fixed-income derivatives are vital to the economy. As a strategy group, our job is to track and anticipate key drivers and developments in the market using, in significant part, quantitative analysis. Some of the derivatives we look at are of the forward kind, such as interest-rate swaps, where over time you receive fixed-rate payments in exchange for paying a floating-rate or vice-versa. A whole other class of derivatives where statistics and calculus are especially relevant are options.

Whereas forward contracts bind both parties to a future trade, options give the holder the right but not the obligation to trade at a specified time and price. Similar to an insurance policy, the holder of the option pays an upfront premium in exchange for potential gain. Solving this pricing problem requires statistics, stochastic calculus and enough insight to win a Nobel prize. Fortunately for us, this was taken care of by Fischer Black, Myron

- **New Skillbuilder Videos**, available through ThomsonNOW and Enhanced WebAssign, offer hours of video instruction from award-winning teacher Deborah Upton of Stonehill College. Watch as she walks students through key examples from the text, step by step—giving them a foundation in the skills that they need to know. Each example available online is identified by the video icon located in the margin.

VIDEO

APPLIED EXAMPLE 7 Oxygen-Restoration Rate in a Pond When organic waste is dumped into a pond, the oxidation process that takes place reduces the pond's oxygen content. However, given time, nature will restore the oxygen content to its natural level. Suppose the oxygen content t days after organic waste has been dumped into the pond is given by

$$f(t) = 100\left[\frac{t^2 + 10t + 100}{t^2 + 20t + 100}\right] \qquad (0 < t < \infty)$$

percent of its normal level.

Other Changes in the Seventh Edition

- **New Applications** More than 100 new real-life applications have been introduced. Among these applications are Sales of GPS Equipment, Broadband Internet Households, Cancer Survivors, Global Supply of Plutonium, Testosterone Use, Blackberry Subscribers, Outsourcing of Jobs, Spending on Medicare, Obesity in America, U.S. Nursing Shortage, Effects of Smoking Bans, Google's Revenue, Computer Security, Yahoo! In Europe, Satellite Radio Subscriptions, Gastric Bypass Surgeries, and the Surface Area of the New York Central Park Reservoir.

- **A Revised and Expanded Student Solutions Manual** Problem-solving strategies and additional algebra steps and review for selected problems have been added to this supplement.
- **Other Changes** In Functions and Mathematical Models (Section 2.3), a new model describing the membership of HMOs is now discussed by using a scatter plot of the real-life data and the graph of a function that describes that data. Another model describing the driving costs of a Ford Taurus is also presented in this same fashion. In Section 3.6, an additional applied example illustrating the solution of related-rates problems has been added. In Section 4.2, an example calling for the interpretation of the first and second derivatives to help sketch the graph of a function has been added. In Section 6.4, the definite integral as a measure of net change is now discussed along with a new example giving the Population Growth in Clark County. In Section 8.6, the introduction to total differentials has been rewritten. In Section 9.4, the explanation of Euler's method has been simplified and rewritten. Also, a new intuitive example is used to introduce Infinite Sequences (Section 11.1) and Infinite Series (Section 11.3).

Teaching Aids

- **Instructor's Solutions Manual** includes complete solutions for all exercises in the text, as well as answers to the *Exploring with Technology* and *Explore & Discuss* questions.
- **Instructor's Suite CD** includes the **Instructor's Solutions Manual** and **Test Bank** in formats compatible with Microsoft Office®.
- **Printed Test Bank,** which includes test questions (including multiple-choice) and sample tests for each chapter of the text, is available to adopters of the text.
- **Enhanced WebAssign** offers an easy way for instructors to deliver, collect, grade, and record assignments via the web. Within **WebAssign** you will find:
  - 1500 problems that match the text's end-of-section exercises.
  - Active examples integrated into each problem that allow students to work step-by-step at their own pace.
  - Links to Skillbuilder Videos that provide further instruction on each problem.
  - Portable PDFs of the textbook that match the assigned section.
- **ExamView® Computerized Testing** allows instructors to create, deliver, and customize tests and study guides (both print and online) in minutes with this easy-to-use assessment and tutorial system, which contains all questions from the **Test Bank** in electronic format.
- **JoinIn™ on Turning Point®** offers instructors text-specific JoinIn content for electronic response systems. Instructors can transform their classroom and assess students' progress with instant in-class quizzes and polls. Turning Point software lets instructors pose book-specific questions and display students' answers seamlessly within Microsoft PowerPoint lecture slides, in conjunction with a choice of "clicker" hardware. Enhance how your students interact with you, your lecture, and one another.

Learning Aids

- **Student Solutions Manual** contains complete solutions for all odd-numbered exercises in the text, plus problem-solving strategies and additional algebra steps and review for selected problems. 0-495-11936-9

- **ThomsonNOW™ for Tan's *Applied Calculus for the Managerial, Life, and Social Sciences,* Seventh Edition**, designed for self-study, offers text-specific tutorials that require no setup or involvement by instructors. (If desired, instructors can assign the tutorials online and track students' progress via an instructor gradebook.) Students can explore active examples from the text as well as Skillbuilder Videos that provide additional reinforcement. Along the way, they can check their comprehension by taking quizzes and receiving immediate feedback.
- **vMentor™** allows students to talk (using their own computer microphones) to tutors who will skillfully guide them through a problem using an interactive whiteboard for illustration. Up to 40 hours of live tutoring a week is available and can be accessed through http://www.thomsonedu.com via the access card shrink-wrapped to every new book.

Acknowledgments

I wish to express my personal appreciation to each of the following reviewers, whose many suggestions have helped make a much improved book.

Faiz Al-Rubaee
University of North Florida

James V. Balch
Middle Tennessee State University

Albert Bronstein
Purdue University

Kimberly Jordan Burch
Montclair State University

Michael Button
San Diego City College

Peter Casazza
University of Missouri–Columbia

Matthew P. Coleman
Fairfield University

William Coppage
Wright State University

Lisa Cox
Texas A&M University

Frank Deutsch
Penn State University

Carl Droms
James Madison University

Bruce Edwards
University of Florida at Gainesville

Janice Epstein
Texas A&M University

Gary J. Etgen
University of Houston

Charles S. Frady
Georgia State University

Howard Frisinger
Colorado State University

Larry Gerstein
University of California at Santa Barbara

Matthew Gould
Vanderbilt University

Harvey Greenwald
California Polytechnic State University–San Luis Obispo

John Haverhals
Bradley University

Yvette Hester
Texas A&M University

Frank Jenkins
John Carroll University

David E. Joyce
Clark University

Mohammed Kazemi
University of North Carolina–Charlotte

James H. Liu
James Madison University

Norman R. Martin
Northern Arizona University

Sandra Wray McAfee
University of Michigan

Maurice Monahan
South Dakota State University

Dean Moore
Florida Community College at Jacksonville

Ralph J. Neuhaus
University of Idaho

Gertrude Okhuysen
Mississippi State University

James Olsen
North Dakota State University

Lloyd Olson
North Dakota State University

Wesley Orser
Clark College

Richard Porter
Northeastern University

Virginia Puckett
Miami Dade College

Richard Quindley
Bridgewater State College

Mary E. Rerick
University of North Dakota

Thomas N. Roe
South Dakota State University

Donald R. Sherbert
University of Illinois

Anne Siswanto
East Los Angeles College

Jane Smith
University of Florida

Devki Talwar
Indiana University of Pennsylvania

Larry Taylor
North Dakota State University

Giovanni Viglino
Ramapo College of New Jersey

Hiroko K. Warshauer
Texas State University–San Marcos

Lawrence V. Welch
Western Illinois University

Jennifer Whitfield
Texas A&M University

I also wish to thank my colleague, Deborah Upton, who did a great job preparing the videos that now accompany the text and who helped with the accuracy check of the text. I also wish to thank Kevin Charlwood and Tao Guo for their many helpful suggestions for improving the text. My thanks also go to the editorial, production, and marketing staffs of Brooks/Cole: Carolyn Crockett, Danielle Derbenti, Beth Gershman, Janet Hill, Joe Rogove, Becky Cross, Donna Kelley, Jennifer Liang, and Ashley Summers for all of their help and support during the development and production of this edition. I also thank Jamie Armstrong and the staff at Newgen for their production services. Finally, a special thanks to the mathematicians—Chris Shannon and Mark van der Lann at Berkeley, Peter Blair Henry at Stanford, Jonathan D. Farley at MIT, and Navin Khaneja at Harvard for taking time off from their busy schedules to describe how mathematics is used in their research. Their pictures appear on the covers of my applied mathematics series.

S. T. Tan

About the Author

SOO T. TAN received his S.B. degree from Massachusetts Institute of Technology, his M.S. degree from the University of Wisconsin–Madison, and his Ph.D. from the University of California at Los Angeles. He has published numerous papers in Optimal Control Theory, Numerical Analysis, and Mathematics of Finance. He is currently a Professor of Mathematics at Stonehill College.

"By the time I started writing the first of what turned out to be a series of textbooks in mathematics for students in the managerial, life, and social sciences, I had quite a few years of experience teaching mathematics to non-mathematics majors. One of the most important lessons I learned from my early experience teaching these courses is that many of the students come into these courses with some degree of apprehension. This awareness led to the intuitive approach I have adopted in all of my texts. As you will see, I try to introduce each abstract mathematical concept through an example drawn from a common, real-life experience. Once the idea has been conveyed, I then proceed to make it precise, thereby assuring that no mathematical rigor is lost in this intuitive treatment of the subject. Another lesson I learned from my students is that they have a much greater appreciation of the material if the applications are drawn from their fields of interest and from situations that occur in the real world. This is one reason you will see so many exercises in my texts that are modeled on data gathered from newspapers, magazines, journals, and other media. Whether it be the market for cholesterol-reducing drugs, financing a home, bidding for cable rights, broadband Internet households, or Starbucks' annual sales, I weave topics of current interest into my examples and exercises to keep the book relevant to all of my readers."

1 Preliminaries

What sales figure can be predicted for next year? In Example 10, page 38, you will see how the manager of a local sporting goods store used sales figures from the previous years to predict the sales level for next year.

© Taxi/Getty Images

THE FIRST TWO sections of this chapter contain a brief review of algebra. We then introduce the Cartesian coordinate system, which allows us to represent points in the plane in terms of ordered pairs of real numbers. This in turn enables us to compute the distance between two points algebraically. This chapter also covers straight lines. The slope of a straight line plays an important role in the study of calculus.

1.1 Precalculus Review I

Sections 1.1 and 1.2 review some basic concepts and techniques of algebra that are essential in the study of calculus. The material in this review will help you work through the examples and exercises in this book. You can read through this material now and do the exercises in areas where you feel a little "rusty," or you can review the material on an as-needed basis as you study the text. We begin our review with a discussion of real numbers.

The Real Number Line

The real number system is made up of the set of real numbers together with the usual operations of addition, subtraction, multiplication, and division.

We can represent real numbers geometrically by points on a **real number**, or **coordinate, line**. This line can be constructed as follows. Arbitrarily select a point on a straight line to represent the number 0. This point is called the **origin**. If the line is horizontal, then a point at a convenient distance to the right of the origin is chosen to represent the number 1. This determines the scale for the number line. Each positive real number lies at an appropriate distance to the right of the origin, and each negative real number lies at an appropriate distance to the left of the origin (Figure 1).

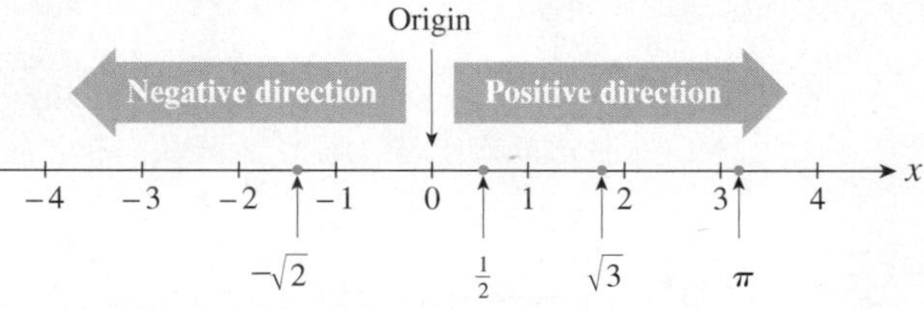

FIGURE 1
The real number line

A *one-to-one correspondence* is set up between the set of all real numbers and the set of points on the number line; that is, exactly one point on the line is associated with each real number. Conversely, exactly one real number is associated with each point on the line. The real number that is associated with a point on the real number line is called the **coordinate** of that point.

Intervals

Throughout this book, we will often restrict our attention to subsets of the set of real numbers. For example, if x denotes the number of cars rolling off a plant assembly line each day, then x must be nonnegative—that is, $x \geq 0$. Further, suppose management decides that the daily production must not exceed 200 cars. Then, x must satisfy the inequality $0 \leq x \leq 200$.

More generally, we will be interested in the following subsets of real numbers: open intervals, closed intervals, and half-open intervals. The set of all real numbers that lie *strictly* between two fixed numbers a and b is called an **open interval** (a, b). It consists of all real numbers x that satisfy the inequalities $a < x < b$, and it is called "open" because neither of its endpoints is included in the interval. A **closed interval** contains *both* of its endpoints. Thus, the set of all real numbers x that satisfy the inequalities $a \leq x \leq b$ is the closed interval $[a, b]$. Notice that square brackets are used to indicate that the endpoints are included in this interval. **Half-open intervals** contain only *one* of their endpoints. Thus, the interval $[a, b)$ is the set of all real numbers x that satisfy $a \leq x < b$, whereas the interval $(a, b]$ is described by the inequalities $a < x \leq b$. Examples of these **finite intervals** are illustrated in Table 1.

TABLE 1

Finite Intervals

Interval		Graph	Example	
Open	(a, b)	a b x	$(-2, 1)$	-3 -2 -1 0 1 2 3 x
Closed	$[a, b]$	a b x	$[-1, 2]$	-1 0 1 2 x
Half-open	$(a, b]$	a b x	$(\frac{1}{2}, 3]$	0 $\frac{1}{2}$ 1 2 3 x
Half-open	$[a, b)$	a b x	$[-\frac{1}{2}, 3)$	$-\frac{1}{2}$ 0 1 2 3 x

In addition to finite intervals, we will encounter **infinite intervals**. Examples of infinite intervals are the half lines (a, ∞), $[a, \infty)$, $(-\infty, a)$, and $(-\infty, a]$ defined by the set of all real numbers that satisfy $x > a$, $x \geq a$, $x < a$, and $x \leq a$, respectively. The symbol ∞, called *infinity*, is not a real number. It is used here only for notational purposes in conjunction with the definition of infinite intervals. The notation $(-\infty, \infty)$ is used for the set of all real numbers x since, by definition, the inequalities $-\infty < x < \infty$ hold for any real number x. Infinite intervals are illustrated in Table 2.

TABLE 2

Infinite Intervals

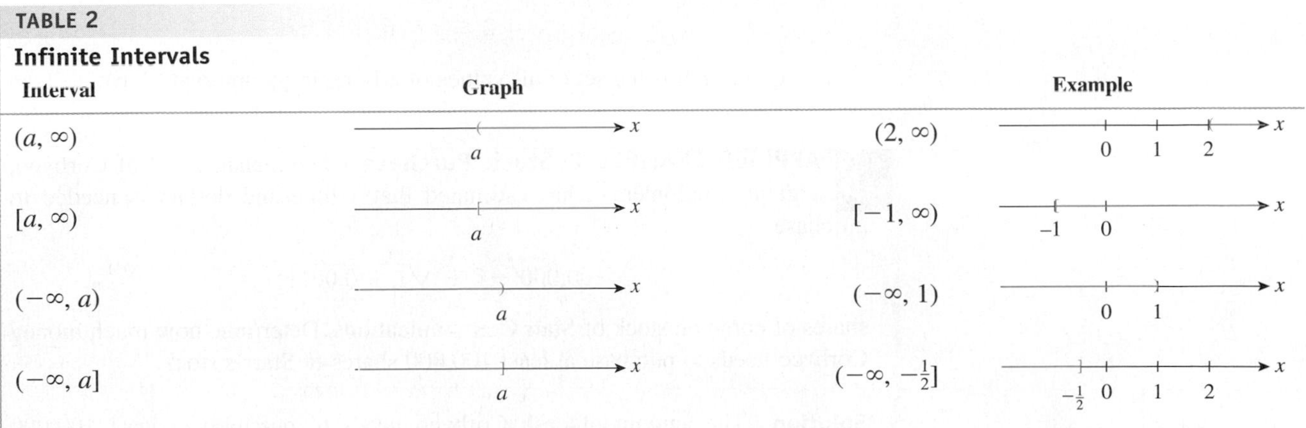

Interval	Graph	Example	
(a, ∞)	a x	$(2, \infty)$	0 1 2 x
$[a, \infty)$	a x	$[-1, \infty)$	-1 0 x
$(-\infty, a)$	a x	$(-\infty, 1)$	0 1 x
$(-\infty, a]$	a x	$(-\infty, -\frac{1}{2}]$	$-\frac{1}{2}$ 0 1 2 x

Properties of Inequalities

The following properties may be used to solve one or more inequalities involving a variable.

Properties of Inequalities

If a, b, and c, are any real numbers, then

		Example
Property 1	If $a < b$ and $b < c$, then $a < c$.	$2 < 3$ and $3 < 8$, so $2 < 8$.
Property 2	If $a < b$, then $a + c < b + c$.	$-5 < -3$, so $-5 + 2 < -3 + 2$; that is, $-3 < -1$.
Property 3	If $a < b$ and $c > 0$, then $ac < bc$.	$-5 < -3$, and since $2 > 0$, we have $(-5)(2) < (-3)(2)$; that is, $-10 < -6$.
Property 4	If $a < b$ and $c < 0$, then $ac > bc$.	$-2 < 4$, and since $-3 < 0$, we have $(-2)(-3) > (4)(-3)$; that is, $6 > -12$.

Similar properties hold if each inequality sign, $<$, between a and b and between b and c is replaced by $\geq$, $>$, or $\leq$. Note that Property 4 says that an inequality sign is reversed if the inequality is multiplied by a negative number.

A real number is a *solution of an inequality* involving a variable if a true statement is obtained when the variable is replaced by that number. The set of all real numbers satisfying the inequality is called the *solution set*. We often use interval notation to describe the solution set.

EXAMPLE 1 Find the set of real numbers that satisfy $-1 \leq 2x - 5 < 7$.

Solution Add 5 to each member of the given double inequality, obtaining

$$4 \leq 2x < 12$$

Next, multiply each member of the resulting double inequality by $\frac{1}{2}$, yielding

$$2 \leq x < 6$$

Thus, the solution is the set of all values of x lying in the interval $[2, 6)$. ■

APPLIED EXAMPLE 2 Stock Purchase The management of Corbyco, a giant conglomerate, has estimated that x thousand dollars is needed to purchase

$$100{,}000(-1 + \sqrt{1 + 0.001x})$$

shares of common stock of Starr Communications. Determine how much money Corbyco needs to purchase at least 100,000 shares of Starr's stock.

Solution The amount of cash Corbyco needs to purchase at least 100,000 shares is found by solving the inequality

$$100{,}000(-1 + \sqrt{1 + 0.001x}) \geq 100{,}000$$

Proceeding, we find

$$\begin{aligned} -1 + \sqrt{1 + 0.001x} &\geq 1 \\ \sqrt{1 + 0.001x} &\geq 2 \\ 1 + 0.001x &\geq 4 \qquad \text{Square both sides.} \\ 0.001x &\geq 3 \\ x &\geq 3000 \end{aligned}$$

so Corbyco needs at least \$3,000,000. (Recall that x is measured in thousands of dollars.) ■

Absolute Value

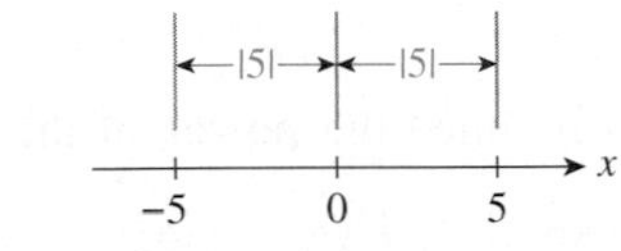

(a)

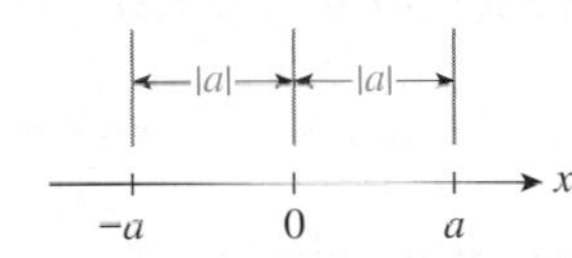

(b)

FIGURE 2
The absolute value of a number

Absolute Value
The **absolute value** of a number a is denoted by $|a|$ and is defined by

$$|a| = \begin{cases} a & \text{if } a \geq 0 \\ -a & \text{if } a < 0 \end{cases}$$

Since $-a$ is a positive number when a is negative, it follows that the absolute value of a number is always nonnegative. For example, $|5| = 5$ and $|-5| = -(-5) = 5$. Geometrically, $|a|$ is the distance between the origin and the point on the number line that represents the number a (Figure 2).

Absolute Value Properties
If a and b are any real numbers, then

		Example
Property 5	$\lvert -a\rvert = \lvert a\rvert$	$\lvert -3\rvert = -(-3) = 3 = \lvert 3\rvert$
Property 6	$\lvert ab\rvert = \lvert a\rvert\,\lvert b\rvert$	$\lvert (2)(-3)\rvert = \lvert -6\rvert = 6 = (2)(3) = \lvert 2\rvert\,\lvert -3\rvert$
Property 7	$\left\lvert \dfrac{a}{b} \right\rvert = \dfrac{\lvert a\rvert}{\lvert b\rvert} \quad (b \neq 0)$	$\left\lvert \dfrac{(-3)}{(-4)} \right\rvert = \left\lvert \dfrac{3}{4} \right\rvert = \dfrac{3}{4} = \dfrac{\lvert -3\rvert}{\lvert -4\rvert}$
Property 8	$\lvert a + b\rvert \leq \lvert a\rvert + \lvert b\rvert$	$\lvert 8 + (-5)\rvert = \lvert 3\rvert = 3 \leq \lvert 8\rvert + \lvert -5\rvert = 13$

Property 8 is called the **triangle inequality**.

EXAMPLE 3 Evaluate each of the following expressions:

a. $|\pi - 5| + 3$ **b.** $|\sqrt{3} - 2| + |2 - \sqrt{3}|$

Solution

a. Since $\pi - 5 < 0$, we see that $|\pi - 5| = -(\pi - 5)$. Therefore,

$$|\pi - 5| + 3 = -(\pi - 5) + 3 = 8 - \pi$$

b. Since $\sqrt{3} - 2 < 0$, we see that $|\sqrt{3} - 2| = -(\sqrt{3} - 2)$. Next, observe that $2 - \sqrt{3} > 0$, so $|2 - \sqrt{3}| = 2 - \sqrt{3}$. Therefore,

$$\begin{aligned}|\sqrt{3} - 2| + |2 - \sqrt{3}| &= -(\sqrt{3} - 2) + (2 - \sqrt{3})\\ &= 4 - 2\sqrt{3} = 2(2 - \sqrt{3})\end{aligned}$$

Exponents and Radicals

Recall that if b is any real number and n is a positive integer, then the expression b^n (read "b to the power n") is defined as the number

$$b^n = \underbrace{b \cdot b \cdot b \cdot \cdots \cdot b}_{n \text{ factors}}$$

The number b is called the **base**, and the superscript n is called the **power** of the exponential expression b^n. For example,

$$2^5 = 2 \cdot 2 \cdot 2 \cdot 2 \cdot 2 = 32 \quad \text{and} \quad \left(\frac{2}{3}\right)^3 = \left(\frac{2}{3}\right)\left(\frac{2}{3}\right)\left(\frac{2}{3}\right) = \frac{8}{27}$$

If $b \neq 0$, we define

$$b^0 = 1$$

For example, $2^0 = 1$ and $(-\pi)^0 = 1$, but the expression 0^0 is undefined.

Next, recall that if n is a positive integer, then the expression $b^{1/n}$ is defined to be the number that, when raised to the nth power, is equal to b. Thus,

$$(b^{1/n})^n = b$$

Such a number, if it exists, is called the ***n*th root of *b***, also written $\sqrt[n]{b}$.

If n is even, the nth root of a negative number is not defined. For example, the square root of -2 ($n = 2$) is not defined because there is no real number b such that $b^2 = -2$. Also, given a number b, more than one number might satisfy our definition of the nth root. For example, both 3 and -3 squared equal 9, and each is a square root of 9. So, to avoid ambiguity, we define $b^{1/n}$ to be the positive nth root of b whenever it exists. Thus, $\sqrt{9} = 9^{1/2} = 3$. That's why your calculator will give the answer 3 when you use it to evaluate $\sqrt{9}$.

Next, recall that if p/q (p, q, positive integers with $q \neq 0$) is a rational number in lowest terms, then the expression $b^{p/q}$ is defined as the number $(b^{1/q})^p$ or, equivalently, $\sqrt[q]{b^p}$, whenever it exists. Expressions involving negative rational exponents are taken care of by the definition

$$b^{-p/q} = \frac{1}{b^{p/q}}$$

Examples are

$$2^{3/2} = (2^{1/2})^3 \approx (1.4142)^3 \approx 2.8283$$

and

$$4^{-5/2} = \frac{1}{4^{5/2}} = \frac{1}{(4^{1/2})^5} = \frac{1}{2^5} = \frac{1}{32}$$

The rules defining the exponential expression a^n, where $a > 0$ for all rational values of n, are given in Table 3.

TABLE 3

Rules for Defining a^n

Definition of a^n $(a > 0)$	Example	Definition of a^n $(a > 0)$	Example
Integer exponent: If n is a positive integer, then $a^n = a \cdot a \cdot a \cdot \cdots \cdot a$ (n factors of a)	$2^5 = 2 \cdot 2 \cdot 2 \cdot 2 \cdot 2$ (5 factors) $= 32$	**Fractional exponent:** **a.** If n is a positive integer, then $a^{1/n}$ or $\sqrt[n]{a}$ denotes the nth root of a.	$16^{1/2} = \sqrt{16}$ $= 4$
Zero exponent: If n is equal to zero, then $a^0 = 1$ (0^0 is not defined.)	$7^0 = 1$	**b.** If m and n are positive integers, then $a^{m/n} = \sqrt[n]{a^m} = (\sqrt[n]{a})^m$	$8^{2/3} = (\sqrt[3]{8})^2$ $= 4$
Negative exponent: If n is a positive integer, then $a^{-n} = \dfrac{1}{a^n}$ $(a \neq 0)$	$6^{-2} = \dfrac{1}{6^2}$ $= \dfrac{1}{36}$	**c.** If m and n are positive integers, then $a^{-m/n} = \dfrac{1}{a^{m/n}}$ $(a \neq 0)$	$9^{-3/2} = \dfrac{1}{9^{3/2}}$ $= \dfrac{1}{27}$

The first three definitions in Table 3 are also valid for negative values of a. The fourth definition holds for all values of a if n is odd, but only for nonnegative values of a if n is even. Thus,

$$(-8)^{1/3} = \sqrt[3]{-8} = -2 \qquad n \text{ is odd.}$$
$$(-8)^{1/2} \text{ has no real value} \qquad n \text{ is even.}$$

Finally, note that it can be shown that a^n has meaning for *all* real numbers n. For example, using a pocket calculator with a $\boxed{y^x}$ key, we see that $2^{\sqrt{2}} \approx 2.665144$.

The five laws of exponents are listed in Table 4.

TABLE 4

Laws of Exponents

Law		Example
1. $a^m \cdot a^n = a^{m+n}$		$x^2 \cdot x^3 = x^{2+3} = x^5$
2. $\dfrac{a^m}{a^n} = a^{m-n}$	$(a \neq 0)$	$\dfrac{x^7}{x^4} = x^{7-4} = x^3$
3. $(a^m)^n = a^{m \cdot n}$		$(x^4)^3 = x^{4 \cdot 3} = x^{12}$
4. $(ab)^n = a^n \cdot b^n$		$(2x)^4 = 2^4 \cdot x^4 = 16x^4$
5. $\left(\dfrac{a}{b}\right)^n = \dfrac{a^n}{b^n}$	$(b \neq 0)$	$\left(\dfrac{x}{2}\right)^3 = \dfrac{x^3}{2^3} = \dfrac{x^3}{8}$

These laws are valid for any real numbers a, b, m, and n whenever the quantities are defined.

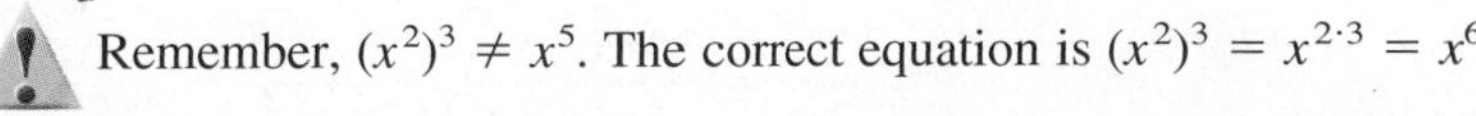

Remember, $(x^2)^3 \neq x^5$. The correct equation is $(x^2)^3 = x^{2 \cdot 3} = x^6$.

The next several examples illustrate the use of the laws of exponents.

EXAMPLE 4 Simplify the expressions:

a. $(3x^2)(4x^3)$ **b.** $\dfrac{16^{5/4}}{16^{1/2}}$ **c.** $(6^{2/3})^3$ **d.** $(x^3y^{-2})^{-2}$ **e.** $\left(\dfrac{y^{3/2}}{x^{1/4}}\right)^{-2}$

Solution

a. $(3x^2)(4x^3) = 12x^{2+3} = 12x^5$ Law 1

b. $\dfrac{16^{5/4}}{16^{1/2}} = 16^{5/4-1/2} = 16^{3/4} = (\sqrt[4]{16})^3 = 2^3 = 8$ Law 2

c. $(6^{2/3})^3 = 6^{(2/3)(3)} = 6^{6/3} = 6^2 = 36$ Law 3

d. $(x^3y^{-2})^{-2} = (x^3)^{-2}(y^{-2})^{-2} = x^{(3)(-2)}y^{(-2)(-2)} = x^{-6}y^4 = \dfrac{y^4}{x^6}$ Law 4

e. $\left(\dfrac{y^{3/2}}{x^{1/4}}\right)^{-2} = \dfrac{y^{(3/2)(-2)}}{x^{(1/4)(-2)}} = \dfrac{y^{-3}}{x^{-1/2}} = \dfrac{x^{1/2}}{y^3}$ Law 5 ■

We can also use the laws of exponents to simplify expressions involving radicals, as illustrated in the next example.

EXAMPLE 5 Simplify the expressions. (Assume that x, y, m, and n are positive.)

a. $\sqrt[4]{16x^4y^8}$ **b.** $\sqrt{12m^3n}\cdot\sqrt{3m^5n}$ **c.** $\dfrac{\sqrt[3]{-27x^6}}{\sqrt[3]{8y^3}}$

Solution

a. $\sqrt[4]{16x^4y^8} = (16x^4y^8)^{1/4} = 16^{1/4}\cdot x^{4/4}y^{8/4} = 2xy^2$

b. $\sqrt{12m^3n}\cdot\sqrt{3m^5n} = \sqrt{36m^8n^2} = (36m^8n^2)^{1/2} = 36^{1/2}\cdot m^{8/2}n^{2/2} = 6m^4n$

c. $\dfrac{\sqrt[3]{-27x^6}}{\sqrt[3]{8y^3}} = \dfrac{(-27x^6)^{1/3}}{(8y^3)^{1/3}} = \dfrac{-27^{1/3}x^{6/3}}{8^{1/3}y^{3/3}} = -\dfrac{3x^2}{2y}$ ■

If a radical appears in the numerator or denominator of an algebraic expression, we often try to simplify the expression by eliminating the radical from the numerator or denominator. This process, called **rationalization**, is illustrated in the next two examples.

EXAMPLE 6 Rationalize the denominator of the expression $\dfrac{3x}{2\sqrt{x}}$.

Solution

$$\frac{3x}{2\sqrt{x}} = \frac{3x}{2\sqrt{x}}\cdot\frac{\sqrt{x}}{\sqrt{x}} = \frac{3x\sqrt{x}}{2\sqrt{x^2}} = \frac{3x\sqrt{x}}{2x} = \frac{3}{2}\sqrt{x}$$ ■

EXAMPLE 7 Rationalize the numerator of the expression $\dfrac{3\sqrt{x}}{2x}$.

Solution

$$\frac{3\sqrt{x}}{2x} = \frac{3\sqrt{x}}{2x}\cdot\frac{\sqrt{x}}{\sqrt{x}} = \frac{3\sqrt{x^2}}{2x\sqrt{x}} = \frac{3x}{2x\sqrt{x}} = \frac{3}{2\sqrt{x}}$$ ■

1.1 Exercises

In Exercises 1–4, determine whether the statement is true or false.

1. $-3 < -20$
2. $-5 \le -5$
3. $\frac{2}{3} > \frac{5}{6}$
4. $-\frac{5}{6} < -\frac{11}{12}$

In Exercises 5–10, show the interval on a number line.

5. $(3, 6)$
6. $(-2, 5]$
7. $[-1, 4)$
8. $\left[-\frac{6}{5}, -\frac{1}{2}\right]$
9. $(0, \infty)$
10. $(-\infty, 5]$

In Exercises 11–20, find the values of x that satisfy the inequality (inequalities).

11. $2x + 4 < 8$
12. $-6 > 4 + 5x$
13. $-4x \ge 20$
14. $-12 \le -3x$
15. $-6 < x - 2 < 4$
16. $0 \le x + 1 \le 4$
17. $x + 1 > 4$ or $x + 2 < -1$
18. $x + 1 > 2$ or $x - 1 < -2$
19. $x + 3 > 1$ and $x - 2 < 1$
20. $x - 4 \le 1$ and $x + 3 > 2$

In Exercises 21–30, evaluate the expression.

21. $|-6 + 2|$
22. $4 + |-4|$
23. $\frac{|-12 + 4|}{|16 - 12|}$
24. $\left|\frac{0.2 - 1.4}{1.6 - 2.4}\right|$
25. $\sqrt{3}|-2| + 3|-\sqrt{3}|$
26. $|-1| + \sqrt{2}|-2|$
27. $|\pi - 1| + 2$
28. $|\pi - 6| - 3$
29. $|\sqrt{2} - 1| + |3 - \sqrt{2}|$
30. $|2\sqrt{3} - 3| - |\sqrt{3} - 4|$

In Exercises 31–36, suppose a and b are real numbers other than zero and that $a > b$. State whether the inequality is true or false.

31. $b - a > 0$
32. $\frac{a}{b} > 1$
33. $a^2 > b^2$
34. $\frac{1}{a} > \frac{1}{b}$
35. $a^3 > b^3$
36. $-a < -b$

In Exercises 37–42, determine whether the statement is true for all real numbers a and b.

37. $|-a| = a$
38. $|b^2| = b^2$
39. $|a - 4| = |4 - a|$
40. $|a + 1| = |a| + 1$
41. $|a + b| = |a| + |b|$
42. $|a - b| = |a| - |b|$

In Exercises 43–58, evaluate the expression.

43. $27^{2/3}$
44. $8^{-4/3}$
45. $\left(\frac{1}{\sqrt{3}}\right)^0$
46. $(7^{1/2})^4$
47. $\left[\left(\frac{1}{8}\right)^{1/3}\right]^{-2}$
48. $\left[\left(-\frac{1}{3}\right)^2\right]^{-3}$
49. $\left(\frac{7^{-5} \cdot 7^2}{7^{-2}}\right)^{-1}$
50. $\left(\frac{9}{16}\right)^{-1/2}$
51. $(125^{2/3})^{-1/2}$
52. $\sqrt[3]{2^6}$
53. $\frac{\sqrt{32}}{\sqrt{8}}$
54. $\sqrt[3]{\frac{-8}{27}}$
55. $\frac{16^{5/8}16^{1/2}}{16^{7/8}}$
56. $\left(\frac{9^{-3} \cdot 9^5}{9^{-2}}\right)^{-1/2}$
57. $16^{1/4} \cdot 8^{-1/3}$
58. $\frac{6^{2.5} \cdot 6^{-1.9}}{6^{-1.4}}$

In Exercises 59–68, determine whether the statement is true or false. Give a reason for your choice.

59. $x^4 + 2x^4 = 3x^4$
60. $3^2 \cdot 2^2 = 6^2$
61. $x^3 \cdot 2x^2 = 2x^6$
62. $3^3 + 3 = 3^4$
63. $\frac{2^{4x}}{1^{3x}} = 2^{4x-3x}$
64. $(2^2 \cdot 3^2)^2 = 6^4$
65. $\frac{1}{4^{-3}} = \frac{1}{64}$
66. $\frac{4^{3/2}}{2^4} = \frac{1}{2}$
67. $(1.2^{1/2})^{-1/2} = 1$
68. $5^{2/3} \cdot (25)^{2/3} = 25$

In Exercises 69–74, rewrite the expression using positive exponents only.

69. $(xy)^{-2}$
70. $3s^{1/3} \cdot s^{-7/3}$
71. $\frac{x^{-1/3}}{x^{1/2}}$
72. $\sqrt{x^{-1}} \cdot \sqrt{9x^{-3}}$
73. $12^0(s + t)^{-3}$
74. $(x - y)(x^{-1} + y^{-1})$

In Exercises 75–90, simplify the expression. (Assume that *x*, *y*, *r*, *s*, and *t* are positive.)

75. $\dfrac{x^{7/3}}{x^{-2}}$

76. $(49x^{-2})^{-1/2}$

77. $(x^2y^{-3})(x^{-5}y^3)$

78. $\dfrac{5x^6y^3}{2x^2y^7}$

79. $\dfrac{x^{3/4}}{x^{-1/4}}$

80. $\left(\dfrac{x^3y^2}{z^2}\right)^2$

81. $\left(\dfrac{x^3}{-27y^{-6}}\right)^{-2/3}$

82. $\left(\dfrac{e^x}{e^{x-2}}\right)^{-1/2}$

83. $\left(\dfrac{x^{-3}}{y^{-2}}\right)^2\left(\dfrac{y}{x}\right)^4$

84. $\dfrac{(r^n)^4}{r^{5-2n}}$

85. $\sqrt[3]{x^{-2}}\cdot\sqrt{4x^5}$

86. $\sqrt{81x^6y^{-4}}$

87. $-\sqrt[4]{16x^4y^8}$

88. $\sqrt[3]{x^{3a+b}}$

89. $\sqrt[6]{64x^8y^3}$

90. $\sqrt[3]{27r^6}\cdot\sqrt{s^2t^4}$

In Exercises 91–94, use the fact that $2^{1/2} \approx 1.414$ and $3^{1/2} \approx 1.732$ to evaluate the expression without using a calculator.

91. $2^{3/2}$

92. $8^{1/2}$

93. $9^{3/4}$

94. $6^{1/2}$

In Exercises 95–98, use the fact that $10^{1/2} \approx 3.162$ and $10^{1/3} \approx 2.154$ to evaluate the expression without using a calculator.

95. $10^{3/2}$

96. $1000^{3/2}$

97. $10^{2.5}$

98. $(0.0001)^{-1/3}$

In Exercises 99–104, rationalize the denominator of the expression.

99. $\dfrac{3}{2\sqrt{x}}$

100. $\dfrac{3}{\sqrt{xy}}$

101. $\dfrac{2y}{\sqrt{3y}}$

102. $\dfrac{5x^2}{\sqrt{3x}}$

103. $\dfrac{1}{\sqrt[3]{x}}$

104. $\sqrt{\dfrac{2x}{y}}$

In Exercises 105–110, rationalize the numerator of the expression.

105. $\dfrac{2\sqrt{x}}{3}$

106. $\dfrac{\sqrt[3]{x}}{24}$

107. $\sqrt{\dfrac{2y}{x}}$

108. $\sqrt[3]{\dfrac{2x}{3y}}$

109. $\dfrac{\sqrt[3]{x^2z}}{y}$

110. $\dfrac{\sqrt[3]{x^2y}}{2x}$

111. DRIVING RANGE OF A CAR An advertisement for a certain car states that the EPA fuel economy is 20 mpg city and 27 mpg highway and that the car's fuel-tank capacity is 18.1 gal. Assuming ideal driving conditions, determine the driving range for the car from the foregoing data.

112. Find the minimum cost C (in dollars), given that

$$5(C - 25) \geq 1.75 + 2.5C$$

113. Find the maximum profit P (in dollars) given that

$$6(P - 2500) \leq 4(P + 2400)$$

114. CELSIUS AND FAHRENHEIT TEMPERATURES The relationship between Celsius (°C) and Fahrenheit (°F) temperatures is given by the formula

$$C = \frac{5}{9}(F - 32)$$

a. If the temperature range for Montreal during the month of January is $-15° < °C < -5°$, find the range in degrees Fahrenheit in Montreal for the same period.

b. If the temperature range for New York City during the month of June is $63° < °F < 80°$, find the range in degrees Celsius in New York City for the same period.

115. MEETING SALES TARGETS A salesman's monthly commission is 15% on all sales over \$12,000. If his goal is to make a commission of at least \$3000/month, what minimum monthly sales figures must he attain?

116. MARKUP ON A CAR The markup on a used car was at least 30% of its current wholesale price. If the car was sold for \$5600, what was the maximum wholesale price?

117. QUALITY CONTROL PAR Manufacturing manufactures steel rods. Suppose the rods ordered by a customer are manufactured to a specification of 0.5 in. and are acceptable only if they are within the *tolerance limits* of 0.49 in. and 0.51 in. Letting x denote the diameter of a rod, write an inequality using absolute value signs to express a criterion involving x that must be satisfied in order for a rod to be acceptable.

118. QUALITY CONTROL The diameter x (in inches) of a batch of ball bearings manufactured by PAR Manufacturing satisfies the inequality

$$|x - 0.1| \leq 0.01$$

What is the smallest diameter a ball bearing in the batch can have? The largest diameter?

119. MEETING PROFIT GOALS A manufacturer of a certain commodity has estimated that her profit in thousands of dollars is given by the expression

$$-6x^2 + 30x - 10$$

where x (in thousands) is the number of units produced. What production range will enable the manufacturer to realize a profit of at least \$14,000 on the commodity?

120. DISTRIBUTION OF INCOMES The distribution of income in a certain city can be described by the exponential model

$y = (2.8 \cdot 10^{11})(x)^{-1.5}$, where y is the number of families with an income of x or more dollars.

a. How many families in this city have an income of $30,000 or more?

b. How many families have an income of $60,000 or more?

c. How many families have an income of $150,000 or more?

In Exercises 121–124, determine whether the statement is true or false. If it is true, explain why it is true. If it is false, give an example to show why it is false.

121. If $a < b$, then $a - c > b - c$.

122. $|a - b| = |b - a|$

123. $|a - b| \le |b| + |a|$

124. $\sqrt{a^2 - b^2} = |a| - |b|$

1.2 Precalculus Review II

Operations with Algebraic Expressions

In calculus, we often work with algebraic expressions such as

$$2x^{4/3} - x^{1/3} + 1 \qquad 2x^2 - x - \frac{2}{\sqrt{x}} \qquad \frac{3xy + 2}{x + 1} \qquad 2x^3 + 2x + 1$$

An algebraic expression of the form ax^n, where the coefficient a is a real number and n is a nonnegative integer, is called a **monomial**, meaning it consists of one term. For example, $7x^2$ is a monomial. A **polynomial** is a monomial or the sum of two or more monomials. For example,

$$x^2 + 4x + 4 \qquad x^3 + 5 \qquad x^4 + 3x^2 + 3 \qquad x^2y + xy + y$$

are all polynomials.

Constant terms and terms containing the same variable factor are called **like**, or **similar, terms**. Like terms may be combined by adding or subtracting their numerical coefficients. For example,

$$3x + 7x = 10x \quad \text{and} \quad \frac{1}{2}xy + 3xy = \frac{7}{2}xy$$

The distributive property of the real number system,

$$ab + ac = a(b + c)$$

is used to justify this procedure.

To add or subtract two or more algebraic expressions, first remove the parentheses and then combine like terms. The resulting expression is written in order of decreasing degree from left to right.

*

EXAMPLE 1

a. $(2x^4 + 3x^3 + 4x + 6) - (3x^4 + 9x^3 + 3x^2)$

$= 2x^4 + 3x^3 + 4x + 6 - 3x^4 - 9x^3 - 3x^2$ Remove parentheses.

$= 2x^4 - 3x^4 + 3x^3 - 9x^3 - 3x^2 + 4x + 6$

$= -x^4 - 6x^3 - 3x^2 + 4x + 6$ Combine like terms.

*The symbol indicates that these examples were selected from the calculus portion of the text in order to help you review the algebraic computations you will *actually* be using in calculus.

b. $2t^3 - \{t^2 - [t - (2t - 1)] + 4\}$
$= 2t^3 - \{t^2 - [t - 2t + 1] + 4\}$
$= 2t^3 - \{t^2 - [-t + 1] + 4\}$ Remove parentheses and combine like terms within brackets.
$= 2t^3 - \{t^2 + t - 1 + 4\}$ Remove brackets.
$= 2t^3 - \{t^2 + t + 3\}$ Combine like terms within braces.
$= 2t^3 - t^2 - t - 3$ Remove braces. ■

An algebraic expression is said to be **simplified** if none of its terms are similar. Observe that when the algebraic expression in Example 1b was simplified, the innermost grouping symbols were removed first; that is, the parentheses ( ) were removed first, the brackets [ ] second, and the braces { } third.

When algebraic expressions are multiplied, each term of one algebraic expression is multiplied by each term of the other. The resulting algebraic expression is then simplified.

EXAMPLE 2 Perform the indicated operations:

a. $(x^2 + 1)(3x^2 + 10x + 3)$ **b.** $(e^t + e^{-t})e^t - e^t(e^t - e^{-t})$

Solution

a. $(x^2 + 1)(3x^2 + 10x + 3) = x^2(3x^2 + 10x + 3) + 1(3x^2 + 10x + 3)$
$= 3x^4 + 10x^3 + 3x^2 + 3x^2 + 10x + 3$
$= 3x^4 + 10x^3 + 6x^2 + 10x + 3$

b. $(e^t + e^{-t})e^t - e^t(e^t - e^{-t}) = e^{2t} + e^0 - e^{2t} + e^0$
$= e^{2t} - e^{2t} + e^0 + e^0$
$= 1 + 1$ Recall that $e^0 = 1$.
$= 2$ ■

Certain product formulas that are frequently used in algebraic computations are given in Table 5.

TABLE 5
Some Useful Product Formulas

Formula	Example
$(a + b)^2 = a^2 + 2ab + b^2$	$(2x + 3y)^2 = (2x)^2 + 2(2x)(3y) + (3y)^2$ $= 4x^2 + 12xy + 9y^2$
$(a - b)^2 = a^2 - 2ab + b^2$	$(4x - 2y)^2 = (4x)^2 - 2(4x)(2y) + (2y)^2$ $= 16x^2 - 16xy + 4y^2$
$(a + b)(a - b) = a^2 - b^2$	$(2x + y)(2x - y) = (2x)^2 - (y)^2$ $= 4x^2 - y^2$

Factoring

Factoring is the process of expressing an algebraic expression as a product of other algebraic expressions. For example, by applying the distributive property, we may write

$$3x^2 - x = x(3x - 1)$$

To factor an algebraic expression, first check to see if it contains any common terms. If it does, then factor out the greatest common term. For example, the common factor of the algebraic expression $2a^2x + 4ax + 6a$ is $2a$, because

$$2a^2x + 4ax + 6a = 2a \cdot ax + 2a \cdot 2x + 2a \cdot 3 = 2a(ax + 2x + 3)$$

EXAMPLE 3 Factor out the greatest common factor in each expression:

a. $-0.3t^2 + 3t$ **b.** $2x^{3/2} - 3x^{1/2}$ **c.** $2ye^{xy^2} + 2xy^3e^{xy^2}$

d. $4x(x + 1)^{1/2} - 2x^2\left(\frac{1}{2}\right)(x + 1)^{-1/2}$

Solution

a. $-0.3t^2 + 3t = -0.3t(t - 10)$

b. $2x^{3/2} - 3x^{1/2} = x^{1/2}(2x - 3)$

c. $2ye^{xy^2} + 2xy^3e^{xy^2} = 2ye^{xy^2}(1 + xy^2)$

d. $$\begin{aligned} 4x(x + 1)^{1/2} - 2x^2\left(\frac{1}{2}\right)(x + 1)^{-1/2} &= 4x(x + 1)^{1/2} - x^2(x + 1)^{-1/2} \\ &= x(x + 1)^{-1/2}[4(x + 1)^{1/2}(x + 1)^{1/2} - x] \\ &= x(x + 1)^{-1/2}[4(x + 1) - x] \\ &= x(x + 1)^{-1/2}(4x + 4 - x) = x(x + 1)^{-1/2}(3x + 4) \end{aligned}$$

Here we select $(x + 1)^{-1/2}$ as the common factor because it is the lowest power of $(x + 1)$ in each algebraic term. In particular, observe that

$$(x + 1)^{-1/2}(x + 1)^{1/2}(x + 1)^{1/2} = (x + 1)^{-1/2+1/2+1/2} = (x + 1)^{1/2}$$ ■

Sometimes an algebraic expression may be factored by regrouping and rearranging its terms so that a common term can be factored out. This technique is illustrated in Example 4.

EXAMPLE 4 Factor:

a. $2ax + 2ay + bx + by$ **b.** $3x\sqrt{y} - 4 - 2\sqrt{y} + 6x$

Solution

a. First, factor the common term $2a$ from the first two terms and the common term b from the last two terms. Thus,

$$2ax + 2ay + bx + by = 2a(x + y) + b(x + y)$$

Since $(x + y)$ is common to both terms of the polynomial, we may factor it out. Hence,

$$2a(x + y) + b(x + y) = (2a + b)(x + y)$$

b. $$\begin{aligned} 3x\sqrt{y} - 4 - 2\sqrt{y} + 6x &= 3x\sqrt{y} - 2\sqrt{y} + 6x - 4 && \text{Regroup terms.} \\ &= \sqrt{y}(3x - 2) + 2(3x - 2) && \text{Factor out common terms.} \\ &= (3x - 2)(\sqrt{y} + 2) \end{aligned}$$ ■

As we have seen, the first step in factoring a polynomial is to find the common factors. The next step is to express the polynomial as the product of a constant and/or one or more prime polynomials.

Certain product formulas that are useful in factoring binomials and trinomials are listed in Table 6.

TABLE 6

Product Formulas Used in Factoring

Formula	Example
Difference of two squares	
$x^2 - y^2 = (x + y)(x - y)$	$x^2 - 36 = (x + 6)(x - 6)$
	$8x^2 - 2y^2 = 2(4x^2 - y^2)$
	$= 2(2x + y)(2x - y)$
	$9 - a^6 = (3 + a^3)(3 - a^3)$
Perfect-square trinomial	
$x^2 + 2xy + y^2 = (x + y)^2$	$x^2 + 8x + 16 = (x + 4)^2$
$x^2 - 2xy + y^2 = (x - y)^2$	$4x^2 - 4xy + y^2 = (2x - y)^2$
Sum of two cubes	
$x^3 + y^3 = (x + y)(x^2 - xy + y^2)$	$z^3 + 27 = z^3 + (3)^3$
	$= (z + 3)(z^2 - 3z + 9)$
Difference of two cubes	
$x^3 - y^3 = (x - y)(x^2 + xy + y^2)$	$8x^3 - y^6 = (2x)^3 - (y^2)^3$
	$= (2x - y^2)(4x^2 + 2xy^2 + y^4)$

The factors of the second-degree polynomial with integral coefficients

$$px^2 + qx + r$$

are $(ax + b)(cx + d)$, where $ac = p$, $ad + bc = q$, and $bd = r$. Since only a limited number of choices are possible, we use a trial-and-error method to factor polynomials having this form.

For example, to factor $x^2 - 2x - 3$, we first observe that the only possible first-degree terms are

$$(x \quad)(x \quad) \qquad \text{Since the coefficient of } x^2 \text{ is 1}$$

Next, we observe that the product of the constant term is (-3). This gives us the following possible factors:

$$(x - 1)(x + 3)$$
$$(x + 1)(x - 3)$$

Looking once again at the polynomial $x^2 - 2x - 3$, we see that the coefficient of x is -2. Checking to see which set of factors yields -2 for the coefficient of x, we find that

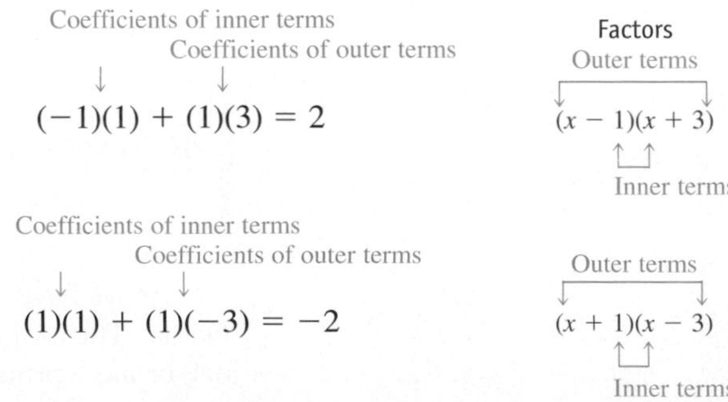

and we conclude that the correct factorization is

$$x^2 - 2x - 3 = (x + 1)(x - 3)$$

With practice, you will soon find that you can perform many of these steps mentally and the need to write out each step will be eliminated.

EXAMPLE 5 Factor:

a. $3x^2 + 4x - 4$ **b.** $3x^2 - 6x - 24$

Solution

a. Using trial and error, we find that the correct factorization is

$$3x^2 + 4x - 4 = (3x - 2)(x + 2)$$

b. Since each term has the common factor 3, we have

$$3x^2 - 6x - 24 = 3(x^2 - 2x - 8)$$

Using the trial-and-error method of factorization, we find that

$$x^2 - 2x - 8 = (x - 4)(x + 2)$$

Thus, we have

$$3x^2 - 6x - 24 = 3(x - 4)(x + 2)$$

■

Roots of Polynomial Equations

A polynomial equation of degree n in the variable x is an equation of the form

$$a_n x^n + a_{n-1}x^{n-1} + \cdots + a_0 = 0$$

where n is a nonnegative integer and $a_0, a_1, \ldots, a_n$ are real numbers with $a_n \neq 0$. For example, the equation

$$-2x^5 + 8x^3 - 6x^2 + 3x + 1 = 0$$

is a polynomial equation of degree 5 in x.

The **roots of a polynomial equation** are precisely the values of x that satisfy the given equation.* One way of finding the roots of a polynomial equation is to first factor the polynomial and then solve the resulting equation. For example, the polynomial equation

$$x^3 - 3x^2 + 2x = 0$$

may be rewritten in the form

$$x(x^2 - 3x + 2) = 0 \quad \text{or} \quad x(x - 1)(x - 2) = 0$$

Since the product of two real numbers can be equal to zero if and only if one (or both) of the factors is equal to zero, we have

$$x = 0 \qquad x - 1 = 0 \qquad \text{or} \qquad x - 2 = 0$$

from which we see that the desired roots are $x = 0$, 1, and 2.

---

*In this book, we are interested only in the *real* roots of an equation.

The Quadratic Formula

In general, the problem of finding the roots of a polynomial equation is a difficult one. But the roots of a quadratic equation (a polynomial equation of degree 2) are easily found either by factoring or by using the following quadratic formula.

Quadratic Formula

The solutions of the equation $ax^2 + bx + c = 0$ $(a \neq 0)$ are given by

$$x = \frac{-b \pm \sqrt{b^2 - 4ac}}{2a}$$

Note If you use the quadratic formula to solve a quadratic equation, first make sure that the equation is in the *standard form* $ax^2 + bx + c = 0$. ■

EXAMPLE 6 Solve each of the following quadratic equations:

a. $2x^2 + 5x - 12 = 0$ **b.** $x^2 = -3x + 8$

Solution

a. The equation is in standard form, with $a = 2$, $b = 5$, and $c = -12$. Using the quadratic formula, we find

$$\begin{aligned} x &= \frac{-b \pm \sqrt{b^2 - 4ac}}{2a} = \frac{-5 \pm \sqrt{5^2 - 4(2)(-12)}}{2(2)} \\ &= \frac{-5 \pm \sqrt{121}}{4} = \frac{-5 \pm 11}{4} \\ &= -4 \text{ or } \frac{3}{2} \end{aligned}$$

This equation can also be solved by factoring. Thus,

$$2x^2 + 5x - 12 = (2x - 3)(x + 4) = 0$$

from which we see that the desired roots are $x = \frac{3}{2}$ or $x = -4$, as obtained earlier.

b. We first rewrite the given equation in the standard form $x^2 + 3x - 8 = 0$, from which we see that $a = 1$, $b = 3$, and $c = -8$. Using the quadratic formula, we find

$$\begin{aligned} x &= \frac{-b \pm \sqrt{b^2 - 4ac}}{2a} = \frac{-3 \pm \sqrt{3^2 - 4(1)(-8)}}{2(1)} \\ &= \frac{-3 \pm \sqrt{41}}{2} \end{aligned}$$

That is, the solutions are

$$\frac{-3 + \sqrt{41}}{2} \approx 1.7 \quad \text{and} \quad \frac{-3 - \sqrt{41}}{2} \approx -4.7$$

In this case, the quadratic formula proves quite handy! ■

Rational Expressions

Quotients of polynomials are called **rational expressions.** Examples of rational expressions are

$$\frac{6x - 1}{2x + 3} \qquad \frac{3x^2y^3 - 2xy}{4x} \qquad \frac{2}{5ab}$$

Since rational expressions are quotients in which the variables represent real numbers, the properties of the real numbers apply to rational expressions as well, and operations with rational fractions are performed in the same manner as operations with arithmetic fractions. For example, using the properties of the real number system, we may write

$$\frac{ac}{bc} = \frac{a}{b} \cdot \frac{c}{c} = \frac{a}{b} \cdot 1 = \frac{a}{b}$$

where a, b, and c are any real numbers and b and c are not zero.

Similarly, using the same properties of real numbers, we may write

$$\frac{(x + 2)(x - 3)}{(x - 2)(x - 3)} = \frac{x + 2}{x - 2} \qquad (x \neq 2, 3)$$

after "canceling" the common factors.

An example of incorrect cancellation is

$$\frac{\not{3} + 4x}{\not{3}} \neq 1 + 4x$$

because 3 is not a factor of the numerator. Instead, we need to write

$$\frac{3 + 4x}{3} = \frac{3}{3} + \frac{4x}{3} = 1 + \frac{4x}{3}$$

A rational expression is simplified, or in lowest terms, when the numerator and denominator have no common factors other than 1 and -1 and the expression contains no negative exponents.

EXAMPLE 7 Simplify the following expressions:

a. $\dfrac{x^2 + 2x - 3}{x^2 + 4x + 3}$ **b.** $\dfrac{[(t^2 + 4)(2t - 4) - (t^2 - 4t + 4)(2t)]}{(t^2 + 4)^2}$

Solution

a. $\dfrac{x^2 + 2x - 3}{x^2 + 4x + 3} = \dfrac{(x + 3)(x - 1)}{(x + 3)(x + 1)} = \dfrac{x - 1}{x + 1}$

b. $\dfrac{[(t^2 + 4)(2t - 4) - (t^2 - 4t + 4)(2t)]}{(t^2 + 4)^2}$

$= \dfrac{2t^3 - 4t^2 + 8t - 16 - 2t^3 + 8t^2 - 8t}{(t^2 + 4)^2}$ Carry out the indicated multiplication.

$= \dfrac{4t^2 - 16}{(t^2 + 4)^2}$ Combine like terms.

$= \dfrac{4(t^2 - 4)}{(t^2 + 4)^2}$ Factor.

The operations of multiplication and division are performed with algebraic fractions in the same manner as with arithmetic fractions (Table 7).

TABLE 7

Rules of Multiplication and Division: Algebraic Fractions

Operation	Example
If P, Q, R, and S are polynomials, then	
Multiplication	
$\frac{P}{Q} \cdot \frac{R}{S} = \frac{PR}{QS} \quad (Q, S \neq 0)$	$\frac{2x}{y} \cdot \frac{(x+1)}{(y-1)} = \frac{2x(x+1)}{y(y-1)} = \frac{2x^2+2x}{y^2-y}$
Division	
$\frac{P}{Q} \div \frac{R}{S} = \frac{P}{Q} \cdot \frac{S}{R} = \frac{PS}{QR} \quad (Q, R, S \neq 0)$	$\frac{x^2+3}{y} \div \frac{y^2+1}{x} = \frac{x^2+3}{y} \cdot \frac{x}{y^2+1} = \frac{x^3+3x}{y^3+y}$

When rational expressions are multiplied and divided, the resulting expressions should be simplified if possible.

EXAMPLE 8 Perform the indicated operations and simplify:

$$\frac{2x-8}{x+2} \cdot \frac{x^2+4x+4}{x^2-16}$$

Solution

$$\begin{aligned}\frac{2x-8}{x+2} \cdot \frac{x^2+4x+4}{x^2-16} &= \frac{2(x-4)}{x+2} \cdot \frac{(x+2)^2}{(x+4)(x-4)} \\ &= \frac{2(x-4)(x+2)(x+2)}{(x+2)(x+4)(x-4)} \\ &= \frac{2(x+2)}{x+4}\end{aligned}$$

Cancel the common factors $(x+2)(x-4)$.

For rational expressions, the operations of addition and subtraction are performed by finding a common denominator of the fractions and then adding or subtracting the fractions. Table 8 shows the rules for fractions with equal denominators.

TABLE 8

Rules of Addition and Subtraction: Fractions with Equal Denominators

Operation	Example
If P, Q, and R are polynomials, then	
Addition	
$\frac{P}{R} + \frac{Q}{R} = \frac{P+Q}{R} \quad (R \neq 0)$	$\frac{2x}{x+2} + \frac{6x}{x+2} = \frac{2x+6x}{x+2} = \frac{8x}{x+2}$
Subtraction	
$\frac{P}{R} - \frac{Q}{R} = \frac{P-Q}{R} \quad (R \neq 0)$	$\frac{3y}{y-x} - \frac{y}{y-x} = \frac{3y-y}{y-x} = \frac{2y}{y-x}$

To add or subtract fractions that have different denominators, first find a common denominator, preferably the least common denominator (LCD). Then carry out the indicated operations following the procedure described in Table 8.

To find the LCD of two or more rational expressions:

1. *Find the prime factors of each denominator.*
2. *Form the product of the different prime factors that occur in the denominators.* Each prime factor in this product should be raised to the highest power of that factor appearing in the denominators.

$$\frac{x}{2+y} \neq \frac{x}{2} + \frac{x}{y}$$

EXAMPLE 9 Simplify:

a. $\dfrac{2x}{x^2+1} + \dfrac{6(3x^2)}{x^3+2}$ **b.** $\dfrac{1}{x+h} - \dfrac{1}{x}$

Solution

a.
$$\begin{aligned}\frac{2x}{x^2+1} + \frac{6(3x^2)}{x^3+2} &= \frac{2x(x^3+2)+6(3x^2)(x^2+1)}{(x^2+1)(x^3+2)} && \text{LCD} = (x^2+1)(x^3+2)\\ &= \frac{2x^4+4x+18x^4+18x^2}{(x^2+1)(x^3+2)} && \text{Carry out the indicated multiplication.}\\ &= \frac{20x^4+18x^2+4x}{(x^2+1)(x^3+2)} && \text{Combine like terms.}\\ &= \frac{2x(10x^3+9x+2)}{(x^2+1)(x^3+2)} && \text{Factor.}\end{aligned}$$

b.
$$\begin{aligned}\frac{1}{x+h} - \frac{1}{x} &= \frac{x-(x+h)}{x(x+h)} && \text{LCD} = x(x+h)\\ &= \frac{x-x-h}{x(x+h)} && \text{Remove parentheses.}\\ &= \frac{-h}{x(x+h)} && \text{Combine like terms.}\end{aligned}$$

Other Algebraic Fractions

The techniques used to simplify rational expressions may also be used to simplify algebraic fractions in which the numerator and denominator are not polynomials, as illustrated in Example 10.

EXAMPLE 10 Simplify:

a. $\dfrac{1+\dfrac{1}{x+1}}{x-\dfrac{4}{x}}$ **b.** $\dfrac{x^{-1}+y^{-1}}{x^{-2}-y^{-2}}$

Solution

a. $$\frac{1+\dfrac{1}{x+1}}{x-\dfrac{4}{x}} = \frac{\dfrac{x+1+1}{x+1}}{\dfrac{x^2-4}{x}}$$ LCD for numerator is $x + 1$ and LCD for denominator is x.

$$= \frac{x+2}{x+1}\cdot\frac{x}{x^2-4} = \frac{x+2}{x+1}\cdot\frac{x}{(x+2)(x-2)}$$

$$= \frac{x}{(x+1)(x-2)}$$

b. $$\frac{x^{-1}+y^{-1}}{x^{-2}-y^{-2}} = \frac{\dfrac{1}{x}+\dfrac{1}{y}}{\dfrac{1}{x^2}-\dfrac{1}{y^2}} = \frac{\dfrac{y+x}{xy}}{\dfrac{y^2-x^2}{x^2y^2}}$$ $x^{-n} = \dfrac{1}{x^n}$

$$= \frac{y+x}{xy}\cdot\frac{x^2y^2}{y^2-x^2} = \frac{y+x}{xy}\cdot\frac{(xy)^2}{(y+x)(y-x)}$$

$$= \frac{xy}{y-x}$$

EXAMPLE 11 Perform the given operations and simplify:

a. $\dfrac{x^2(2x^2+1)^{1/2}}{x-1}\cdot\dfrac{4x^3-6x^2+x-2}{x(x-1)(2x^2+1)}$ **b.** $\dfrac{12x^2}{\sqrt{2x^2+3}} + 6\sqrt{2x^2+3}$

Solution

a. $$\frac{x^2(2x^2+1)^{1/2}}{x-1}\cdot\frac{4x^3-6x^2+x-2}{x(x-1)(2x^2+1)} = \frac{x(4x^3-6x^2+x-2)}{(x-1)^2(2x^2+1)^{1-1/2}}$$

$$= \frac{x(4x^3-6x^2+x-2)}{(x-1)^2(2x^2+1)^{1/2}}$$

b. $$\frac{12x^2}{\sqrt{2x^2+3}} + 6\sqrt{2x^2+3} = \frac{12x^2}{(2x^2+3)^{1/2}} + 6(2x^2+3)^{1/2}$$ Write radicals in exponential form.

$$= \frac{12x^2 + 6(2x^2+3)^{1/2}(2x^2+3)^{1/2}}{(2x^2+3)^{1/2}}$$ LCD is $(2x^2+3)^{1/2}$.

$$= \frac{12x^2+6(2x^2+3)}{(2x^2+3)^{1/2}}$$

$$= \frac{24x^2+18}{(2x^2+3)^{1/2}} = \frac{6(4x^2+3)}{\sqrt{2x^2+3}}$$

Rationalizing Algebraic Fractions

When the denominator of an algebraic fraction contains sums or differences involving radicals, we may **rationalize the denominator**—that is, transform the fraction into an equivalent one with a denominator that does not contain radicals. In doing so, we make use of the fact that

$$(\sqrt{a}+\sqrt{b})(\sqrt{a}-\sqrt{b}) = (\sqrt{a})^2 - (\sqrt{b})^2$$
$$= a - b$$

This procedure is illustrated in Example 12.

EXAMPLE 12 Rationalize the denominator: $\dfrac{1}{1+\sqrt{x}}$.

Solution Upon multiplying the numerator and the denominator by $(1-\sqrt{x})$, we obtain

$$\frac{1}{1+\sqrt{x}} = \frac{1}{1+\sqrt{x}}\cdot\frac{1-\sqrt{x}}{1-\sqrt{x}} \qquad \frac{1-\sqrt{x}}{1-\sqrt{x}}=1$$

$$= \frac{1-\sqrt{x}}{1-(\sqrt{x})^2}$$

$$= \frac{1-\sqrt{x}}{1-x}$$

In other situations, it may be necessary to rationalize the numerator of an algebraic expression. In calculus, for example, one encounters the following problem.

EXAMPLE 13 Rationalize the numerator: $\dfrac{\sqrt{1+h}-1}{h}$.

Solution

$$\frac{\sqrt{1+h}-1}{h} = \frac{\sqrt{1+h}-1}{h}\cdot\frac{\sqrt{1+h}+1}{\sqrt{1+h}+1}$$

$$= \frac{(\sqrt{1+h})^2-(1)^2}{h(\sqrt{1+h}+1)}$$

$$= \frac{1+h-1}{h(\sqrt{1+h}+1)} \qquad (\sqrt{1+h})^2 = \sqrt{1+h}\cdot\sqrt{1+h} = 1+h$$

$$= \frac{h}{h(\sqrt{1+h}+1)}$$

$$= \frac{1}{\sqrt{1+h}+1}$$

1.2 Exercises

In Exercises 1–22, perform the indicated operations and simplify each expression.

1. $(7x^2-2x+5)+(2x^2+5x-4)$
2. $(3x^2+5xy+2y)+(4-3xy-2x^2)$
3. $(5y^2-2y+1)-(y^2-3y-7)$
4. $3(2a-b)-4(b-2a)$
5. $x-\{2x-[-x-(1-x)]\}$
6. $3x^2-\{x^2+1-x[x-(2x-1)]\}+2$
7. $\left(\frac{1}{3}-1+e\right)-\left(-\frac{1}{3}-1+e^{-1}\right)$
8. $-\frac{3}{4}y-\frac{1}{4}x+100+\frac{1}{2}x+\frac{1}{4}y-120$
9. $3\sqrt{8}+8-2\sqrt{y}+\frac{1}{2}\sqrt{x}-\frac{3}{4}\sqrt{y}$
10. $\frac{8}{9}x^2+\frac{2}{3}x+\frac{16}{3}x^2-\frac{16}{3}x-2x+2$
11. $(x+8)(x-2)$
12. $(5x+2)(3x-4)$
13. $(a+5)^2$
14. $(3a-4b)^2$
15. $(x+2y)^2$
16. $(6-3x)^2$
17. $(2x+y)(2x-y)$
18. $(3x+2)(2-3x)$
19. $(x^2-1)(2x)-x^2(2x)$

20. $(x^{1/2} + 1)\left(\frac{1}{2}x^{-1/2}\right) - (x^{1/2} - 1)\left(\frac{1}{2}x^{-1/2}\right)$

21. $2(t + \sqrt{t})^2 - 2t^2$ **22.** $2x^2 + (-x + 1)^2$

In Exercises 23–30, factor out the greatest common factor from each expression.

23. $4x^5 - 12x^4 - 6x^3$

24. $4x^2y^2z - 2x^5y^2 + 6x^3y^2z^2$

25. $7a^4 - 42a^2b^2 + 49a^3b$

26. $3x^{2/3} - 2x^{1/3}$ **27.** $e^{-x} - xe^{-x}$

28. $2ye^{xy^2} + 2xy^3e^{xy^2}$ **29.** $2x^{-5/2} - \frac{3}{2}x^{-3/2}$

30. $\frac{1}{2}\left(\frac{2}{3}u^{3/2} - 2u^{1/2}\right)$

In Exercises 31–44, factor each expression.

31. $6ac + 3bc - 4ad - 2bd$

32. $3x^3 - x^2 + 3x - 1$

33. $4a^2 - b^2$ **34.** $12x^2 - 3y^2$

35. $10 - 14x - 12x^2$ **36.** $x^2 - 2x - 15$

37. $3x^2 - 6x - 24$ **38.** $3x^2 - 4x - 4$

39. $12x^2 - 2x - 30$ **40.** $(x + y)^2 - 1$

41. $9x^2 - 16y^2$ **42.** $8a^2 - 2ab - 6b^2$

43. $x^6 + 125$ **44.** $x^3 - 27$

In Exercises 45–52, perform the indicated operations and simplify each expression.

45. $(x^2 + y^2)x - xy(2y)$ **46.** $2kr(R - r) - kr^2$

47. $2(x - 1)(2x + 2)^3[4(x - 1) + (2x + 2)]$

48. $5x^2(3x^2 + 1)^4(6x) + (3x^2 + 1)^5(2x)$

49. $4(x - 1)^2(2x + 2)^3(2) + (2x + 2)^4(2)(x - 1)$

50. $(x^2 + 1)(4x^3 - 3x^2 + 2x) - (x^4 - x^3 + x^2)(2x)$

51. $(x^2 + 2)^2[5(x^2 + 2)^2 - 3](2x)$

52. $(x^2 - 4)(x^2 + 4)(2x + 8) - (x^2 + 8x - 4)(4x^3)$

In Exercises 53–58, find the real roots of each equation by factoring.

53. $x^2 + x - 12 = 0$ **54.** $3x^2 - x - 4 = 0$

55. $4t^2 + 2t - 2 = 0$ **56.** $-6x^2 + x + 12 = 0$

57. $\frac{1}{4}x^2 - x + 1 = 0$ **58.** $\frac{1}{2}a^2 + a - 12 = 0$

In Exercises 59–64, solve the equation by using the quadratic formula.

59. $4x^2 + 5x - 6 = 0$ **60.** $3x^2 - 4x + 1 = 0$

61. $8x^2 - 8x - 3 = 0$ **62.** $x^2 - 6x + 6 = 0$

63. $2x^2 + 4x - 3 = 0$ **64.** $2x^2 + 7x - 15 = 0$

In Exercises 65–70, simplify the expression.

65. $\dfrac{x^2 + x - 2}{x^2 - 4}$ **66.** $\dfrac{2a^2 - 3ab - 9b^2}{2ab^2 + 3b^3}$

67. $\dfrac{12t^2 + 12t + 3}{4t^2 - 1}$ **68.** $\dfrac{x^3 + 2x^2 - 3x}{-2x^2 - x + 3}$

69. $\dfrac{(4x - 1)(3) - (3x + 1)(4)}{(4x - 1)^2}$

70. $\dfrac{(1 + x^2)^2(2) - 2x(2)(1 + x^2)(2x)}{(1 + x^2)^4}$

In Exercises 71–88, perform the indicated operations and simplify each expression.

71. $\dfrac{2a^2 - 2b^2}{b - a} \cdot \dfrac{4a + 4b}{a^2 + 2ab + b^2}$

72. $\dfrac{x^2 - 6x + 9}{x^2 - x - 6} \cdot \dfrac{3x + 6}{2x^2 - 7x + 3}$

73. $\dfrac{3x^2 + 2x - 1}{2x + 6} \div \dfrac{x^2 - 1}{x^2 + 2x - 3}$

74. $\dfrac{3x^2 - 4xy - 4y^2}{x^2y} \div \dfrac{(2y - x)^2}{x^3y}$

75. $\dfrac{58}{3(3t + 2)} + \dfrac{1}{3}$ **76.** $\dfrac{a + 1}{3a} + \dfrac{b - 2}{5b}$

77. $\dfrac{2x}{2x - 1} - \dfrac{3x}{2x + 5}$ **78.** $\dfrac{-xe^x}{x + 1} + e^x$

79. $\dfrac{4}{x^2 - 9} - \dfrac{5}{x^2 - 6x + 9}$

80. $\dfrac{x}{1 - x} + \dfrac{2x + 3}{x^2 - 1}$

81. $\dfrac{1 + \frac{1}{x}}{1 - \frac{1}{x}}$ **82.** $\dfrac{\frac{1}{x} + \frac{1}{y}}{1 - \frac{1}{xy}}$

83. $\dfrac{4x^2}{2\sqrt{2x^2 + 7}} + \sqrt{2x^2 + 7}$

84. $6(2x + 1)^2\sqrt{x^2 + x} + \dfrac{(2x + 1)^4}{2\sqrt{x^2 + x}}$

85. $\dfrac{2x(x + 1)^{-1/2} - (x + 1)^{1/2}}{x^2}$

86. $\dfrac{(x^2+1)^{1/2} - 2x^2(x^2+1)^{-1/2}}{1-x^2}$

87. $\dfrac{(2x+1)^{1/2} - (x+2)(2x+1)^{-1/2}}{2x+1}$

88. $\dfrac{2(2x-3)^{1/3} - (x-1)(2x-3)^{-2/3}}{(2x-3)^{2/3}}$

In Exercises 89–94, rationalize the denominator of each expression.

89. $\dfrac{1}{\sqrt{3}-1}$

90. $\dfrac{1}{\sqrt{x}+5}$

91. $\dfrac{1}{\sqrt{x}-\sqrt{y}}$

92. $\dfrac{a}{1-\sqrt{a}}$

93. $\dfrac{\sqrt{a}+\sqrt{b}}{\sqrt{a}-\sqrt{b}}$

94. $\dfrac{2\sqrt{a}+\sqrt{b}}{2\sqrt{a}-\sqrt{b}}$

In Exercises 95–100, rationalize the numerator of each expression.

95. $\dfrac{\sqrt{x}}{3}$

96. $\dfrac{\sqrt[3]{y}}{x}$

97. $\dfrac{1-\sqrt{3}}{3}$

98. $\dfrac{\sqrt{x}-1}{x}$

99. $\dfrac{1+\sqrt{x+2}}{\sqrt{x+2}}$

100. $\dfrac{\sqrt{x+3}-\sqrt{x}}{3}$

In Exercises 101–104, determine whether the statement is true or false. If it is true, explain why it is true. If it is false, give an example to show why it is false.

101. If $b^2 - 4ac > 0$, then $ax^2 + bx + c = 0$ $(a \neq 0)$ has two real roots.

102. If $b^2 - 4ac < 0$, then $ax^2 + bx + c = 0$ $(a \neq 0)$ has no real roots.

103. $\dfrac{a}{b+c} = \left(\dfrac{a}{b} + \dfrac{a}{c}\right)$

104. $\sqrt{(a+b)(b-a)} = \sqrt{b^2 - a^2}$ for all real numbers a and b.

1.3 The Cartesian Coordinate System

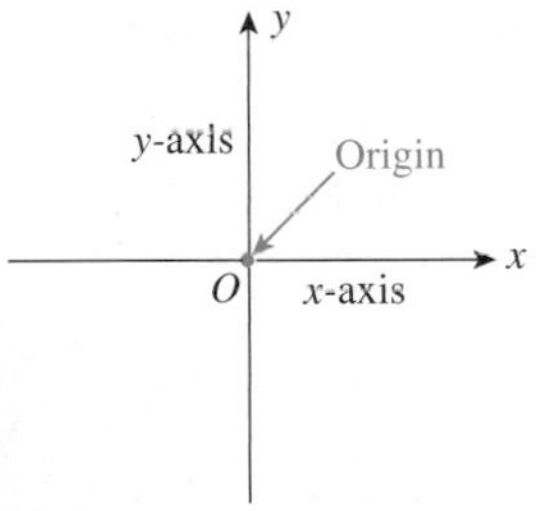

FIGURE 3
The Cartesian coordinate system

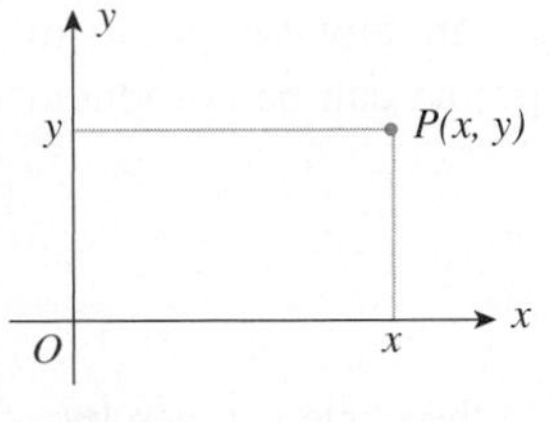

FIGURE 4
An ordered pair (x, y)

The Cartesian Coordinate System

In Section 1.1, we saw how a one-to-one correspondence between the set of real numbers and the points on a straight line leads to a coordinate system on a line (a one-dimensional space).

A similar representation for points in a plane (a two-dimensional space) is realized through the **Cartesian coordinate system**, which is constructed as follows: Take two perpendicular lines, one of which is normally chosen to be horizontal. These lines intersect at a point O, called the **origin** (Figure 3). The horizontal line is called the ***x*-axis**, and the vertical line is called the ***y*-axis**. A number scale is set up along the x-axis, with the positive numbers lying to the right of the origin and the negative numbers lying to the left of it. Similarly, a number scale is set up along the y-axis, with the positive numbers lying above the origin and the negative numbers lying below it.

The number scales on the two axes need not be the same. Indeed, in many applications different quantities are represented by x and y. For example, x may represent the number of typewriters sold and y the total revenue resulting from the sales. In such cases, it is often desirable to choose different number scales to represent the different quantities. Note, however, that the zeros of both number scales coincide at the origin of the two-dimensional coordinate system.

A point in the plane can now be represented uniquely in this coordinate system by an **ordered pair** of numbers—that is, a pair (x, y), where x is the first number and y the second. To see this, let P be any point in the plane (Figure 4). Draw perpendiculars from P to the x-axis and y-axis, respectively. Then the number x is precisely the number that corresponds to the point on the x-axis at which the perpendicular through P hits the x-axis. Similarly, y is the number that corresponds to the point on the y-axis at which the perpendicular through P crosses the y-axis.

Conversely, given an ordered pair (x, y) with x as the first number and y the second, a point P in the plane is uniquely determined as follows: Locate the point on the x-axis represented by the number x and draw a line through that point parallel to the y-axis. Next, locate the point on the y-axis represented by the number y and draw a line through that point parallel to the x-axis. The point of intersection of these two lines is the point P (see Figure 4).

In the ordered pair (x, y), x is called the **abscissa**, or ***x*-coordinate**, y is called the **ordinate**, or ***y*-coordinate**, and x and y together are referred to as the **coordinates** of the point P.

Letting $P(a, b)$ denote the point with x-coordinate a and y-coordinate b, the points $A(2, 3)$, $B(-2, 3)$, $C(-2, -3)$, $D(2, -3)$, $E(3, 2)$, $F(4, 0)$, and $G(0, -5)$ are plotted in Figure 5. The fact that, in general, $P(x, y) \neq P(y, x)$ is clearly illustrated by points A and E.

The axes divide the plane into four quadrants. Quadrant I consists of the points $P(x, y)$ that satisfy $x > 0$ and $y > 0$; Quadrant II, the points $P(x, y)$, where $x < 0$ and $y > 0$; Quadrant III, the points $P(x, y)$, where $x < 0$ and $y < 0$; and Quadrant IV, the points $P(x, y)$, where $x > 0$ and $y < 0$ (Figure 6).

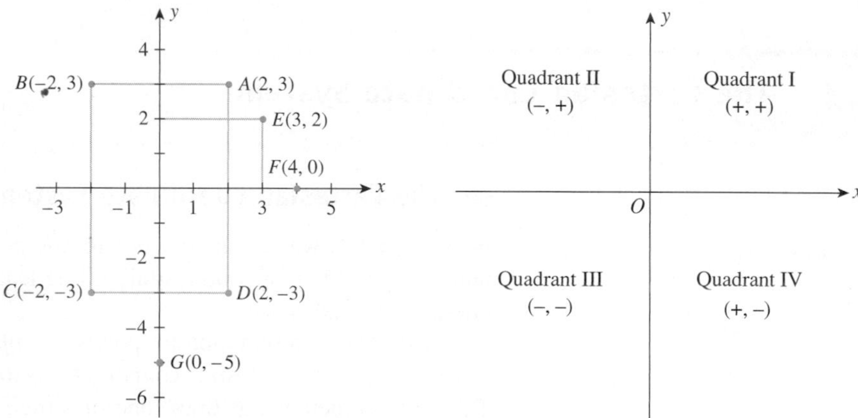

FIGURE 5
Several points in the Cartesian plane

FIGURE 6
The four quadrants in the Cartesian plane

The Distance Formula

One immediate benefit that arises from using the Cartesian coordinate system is that the distance between any two points in the plane may be expressed solely in terms of their coordinates. Suppose, for example, (x_1, y_1) and (x_2, y_2) are any two points in the plane (Figure 7). Then the distance between these two points can be computed using the following formula.

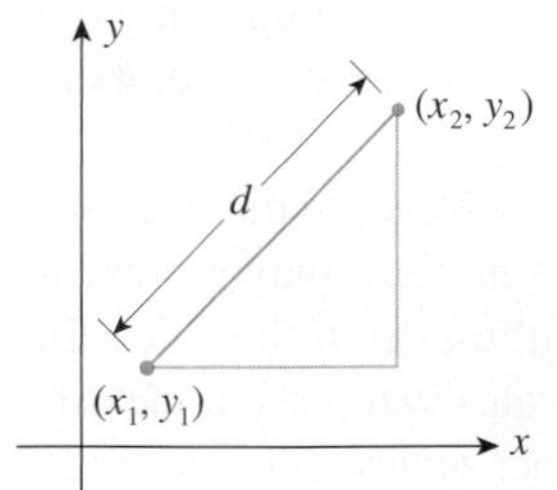

FIGURE 7
The distance d between the points (x_1, y_1) and (x_2, y_2)

Distance Formula

The distance d between two points $P_1(x_1, y_1)$ and $P_2(x_2, y_2)$ in the plane is given by

$$d = \sqrt{(x_2 - x_1)^2 + (y_2 - y_1)^2} \tag{1}$$

For a proof of this result, see Exercise 47, page 30.

In what follows, we give several applications of the distance formula.

EXAMPLE 1 Find the distance between the points $(-4, 3)$ and $(2, 6)$.

Solution Let $P_1(-4, 3)$ and $P_2(2, 6)$ be points in the plane. Then, we have

$$x_1 = -4 \qquad y_1 = 3 \qquad x_2 = 2 \qquad y_2 = 6$$

Using Formula (1), we have

$$\begin{aligned} d &= \sqrt{[2-(-4)]^2 + (6-3)^2} \\ &= \sqrt{6^2 + 3^2} \\ &= \sqrt{45} = 3\sqrt{5} \end{aligned}$$

EXPLORE & DISCUSS

Refer to Example 1. Suppose we label the point $(2, 6)$ as P_1 and the point $(-4, 3)$ as P_2. (1) Show that the distance d between the two points is the same as that obtained earlier. (2) Prove that, in general, the distance d in Formula (1) is independent of the way we label the two points.

EXAMPLE 2 Let $P(x, y)$ denote a point lying on the circle with radius r and center $C(h, k)$ (Figure 8). Find a relationship between x and y.

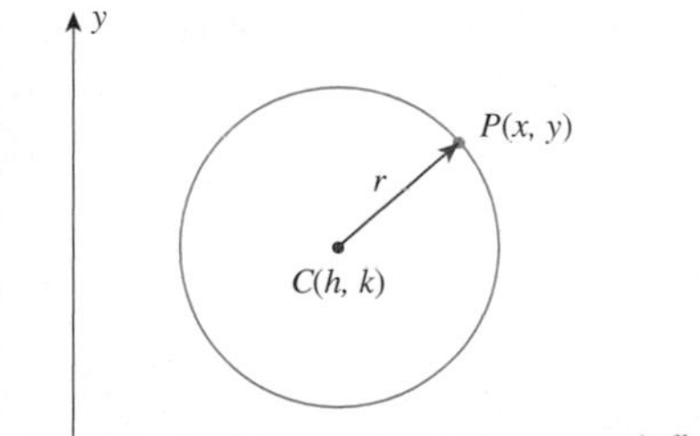

FIGURE 8
A circle with radius r and center $C(h, k)$

Solution By the definition of a circle, the distance between $C(h, k)$ and $P(x, y)$ is r. Using Formula (1), we have

$$\sqrt{(x-h)^2 + (y-k)^2} = r$$

which, upon squaring both sides, gives the equation

$$(x-h)^2 + (y-k)^2 = r^2$$

that must be satisfied by the variables x and y.

A summary of the result obtained in Example 2 follows.

Equation of a Circle

An equation of the circle with center $C(h, k)$ and radius r is given by

$$(x-h)^2 + (y-k)^2 = r^2 \qquad \textbf{(2)}$$

EXAMPLE 3 Find an equation of the circle with

a. Radius 2 and center $(-1, 3)$.

b. Radius 3 and center located at the origin.

Solution

a. We use Formula (2) with $r = 2$, $h = -1$, and $k = 3$, obtaining

$$[x - (-1)]^2 + (y - 3)^2 = 2^2 \quad \text{or} \quad (x + 1)^2 + (y - 3)^2 = 4$$

(Figure 9a).

b. Using Formula (2) with $r = 3$ and $h = k = 0$, we obtain

$$x^2 + y^2 = 3^2 \quad \text{or} \quad x^2 + y^2 = 9$$

(Figure 9b).

FIGURE 9

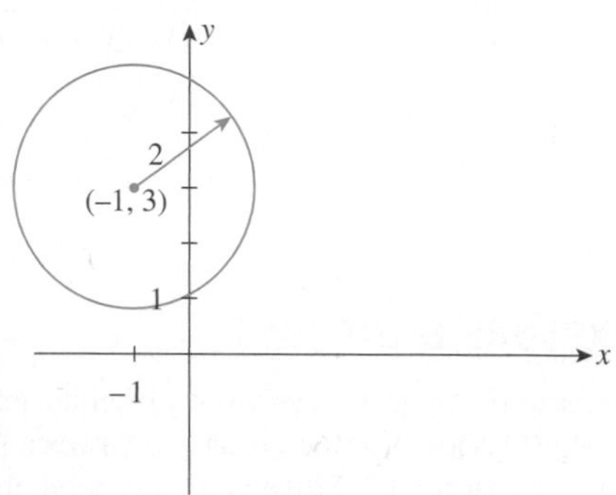

(a) The circle with radius 2 and center $(-1, 3)$

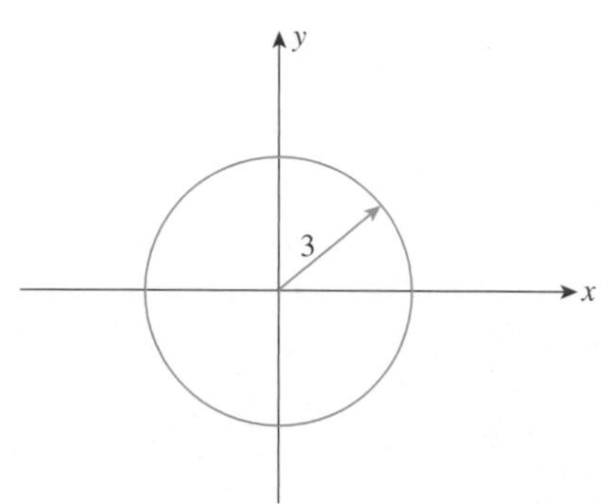

(b) The circle with radius 3 and center $(0, 0)$

EXPLORE & DISCUSS

1. Use the distance formula to help you describe the set of points in the xy-plane satisfying each of the following inequalities.

a. $(x - h)^2 + (y - k)^2 \leq r^2$ **b.** $(x - h)^2 + (y - k)^2 < r^2$

c. $(x - h)^2 + (y - k)^2 \geq r^2$ **d.** $(x - h)^2 + (y - k)^2 > r^2$

2. Consider the equation $x^2 + y^2 = 4$.

a. Show that $y = \pm\sqrt{4 - x^2}$.

b. Describe the set of points (x, y) in the xy-plane satisfying the following equations:

(i) $y = \sqrt{4 - x^2}$ **(ii)** $y = -\sqrt{4 - x^2}$

APPLIED EXAMPLE 4 Cost of Laying Cable In Figure 10, S represents the position of a power relay station located on a straight coastal highway, and M shows the location of a marine biology experimental station on an island. A cable is to be laid connecting the relay station with the experimental station. If the cost of running the cable on land is \$1.50 per running foot and the cost of running the cable under water is \$2.50 per running foot, find the total cost for laying the cable.

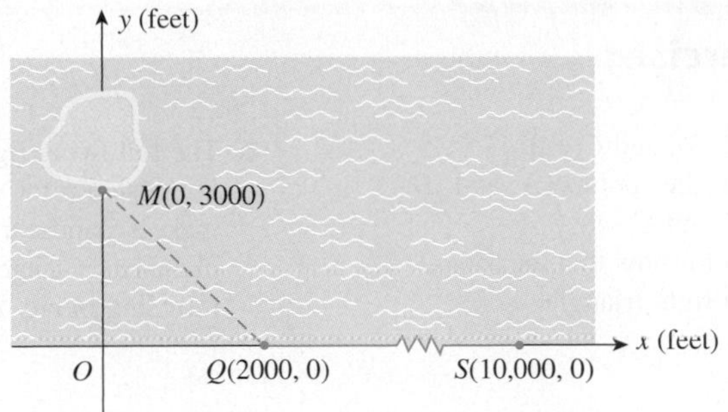

FIGURE 10
Cable connecting relay station S to experimental station M

Solution The length of cable required on land is given by the distance from S to Q. This distance is (10,000 − 2000), or 8000 feet. Next, we see that the length of cable required underwater is given by the distance from M to Q. This distance is

$$\begin{aligned}\sqrt{(0-2000)^2+(3000-0)^2} &= \sqrt{2000^2+3000^2}\\ &= \sqrt{13{,}000{,}000}\\ &\approx 3605.55\end{aligned}$$

or approximately 3605.55 feet. Therefore, the total cost for laying the cable is

$$1.5(8000) + 2.5(3605.55) = 21{,}013.875$$

or approximately \$21,014.

EXPLORE & DISCUSS

In the Cartesian coordinate system, the two axes are perpendicular to each other. Consider a coordinate system in which the x- and y-axes are not collinear and are not perpendicular to each other (see the accompanying figure).

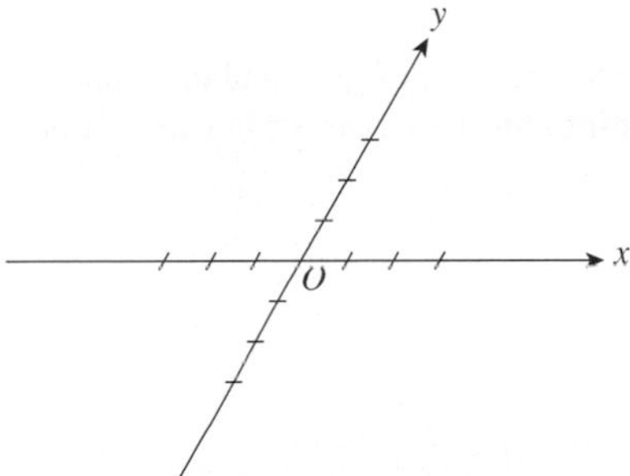

1. Describe how a point is represented in this coordinate system by an ordered pair (x, y) of real numbers. Conversely, show how an ordered pair (x, y) of real numbers uniquely determines a point in the plane.

2. Suppose you want to find a formula for the distance between two points $P_1(x_1, y_1)$ and $P_2(x_2, y_2)$ in the plane. What is the advantage that the Cartesian coordinate system has over the coordinate system under consideration? Comment on your answer.

1.3 Self-Check Exercises

1. **a.** Plot the points $A(4, -2)$, $B(2, 3)$, and $C(-3, 1)$.
   b. Find the distance between the points A and B; between B and C; between A and C.
   c. Use the Pythagorean theorem to show that the triangle with vertices A, B, and C is a right triangle.

2. The following figure shows the location of cities A, B, and C. Suppose a pilot wishes to fly from city A to city C but must make a mandatory stopover in city B. If the single-engine light plane has a range of 650 miles, can she make the trip without refueling in city B?

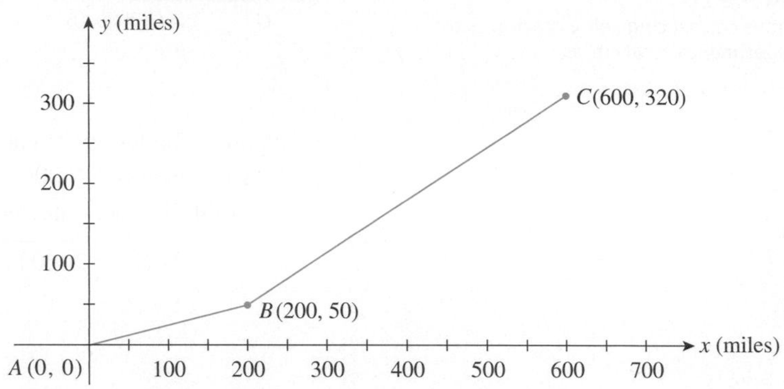

Solutions to Self-Check Exercises 1.3 can be found on page 31.

1.3 Concept Questions

1. What can you say about the signs of a and b if the point $P(a,b)$ lies in (a) the second quadrant? (b) the third quadrant? (c) the fourth quadrant?

2. **a.** What is the distance between $P_1(x_1, y_1)$ and $P_2(x_2, y_2)$?
   b. When you use the distance formula, does it matter which point is labeled P_1 and which point is labeled P_2? Explain.

1.3 Exercises

In Exercises 1–6, refer to the following figure and determine the coordinates of each point and the quadrant in which it is located.

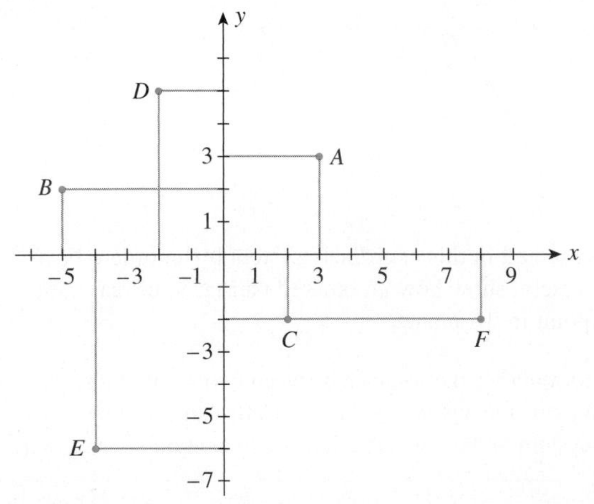

1. A 2. B 3. C
4. D 5. E 6. F

In Exercises 7–12, refer to the following figure.

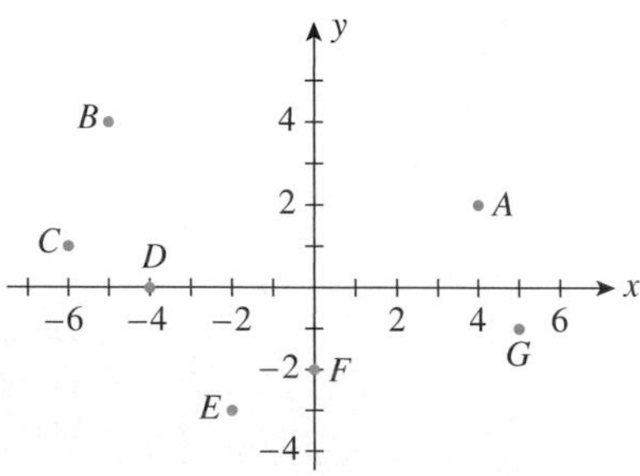

7. Which point has coordinates (4, 2)?
8. What are the coordinates of point B?
9. Which points have negative y-coordinates?

10. Which point has a negative x-coordinate and a negative y-coordinate?

11. Which point has an x-coordinate that is equal to zero?

12. Which point has a y-coordinate that is equal to zero?

In Exercises 13–20, sketch a set of coordinate axes and plot each point.

13. $(-2, 5)$

14. $(1, 3)$

15. $(3, -1)$

16. $(3, -4)$

17. $\left(8, -\frac{7}{2}\right)$

18. $\left(-\frac{5}{2}, \frac{3}{2}\right)$

19. $(4.5, -4.5)$

20. $(1.2, -3.4)$

In Exercises 21–24, find the distance between the given points.

21. $(1, 3)$ and $(4, 7)$

22. $(1, 0)$ and $(4, 4)$

23. $(-1, 3)$ and $(4, 9)$

24. $(-2, 1)$ and $(10, 6)$

25. Find the coordinates of the points that are 10 units away from the origin and have a y-coordinate equal to -6.

26. Find the coordinates of the points that are 5 units away from the origin and have an x-coordinate equal to 3.

27. Show that the points $(3, 4)$, $(-3, 7)$, $(-6, 1)$, and $(0, -2)$ form the vertices of a square.

28. Show that the triangle with vertices $(-5, 2)$, $(-2, 5)$, and $(5, -2)$ is a right triangle.

In Exercises 29–34, find an equation of the circle that satisfies the given conditions.

29. Radius 5 and center $(2, -3)$

30. Radius 3 and center $(-2, -4)$

31. Radius 5 and center at the origin

32. Center at the origin and passes through $(2, 3)$

33. Center $(2, -3)$ and passes through $(5, 2)$

34. Center $(-a, a)$ and radius $2a$

35. Distance Traveled A grand tour of four cities begins at city A and makes successive stops at cities B, C, and D before returning to city A. If the cities are located as shown in the following figure, find the total distance covered on the tour.

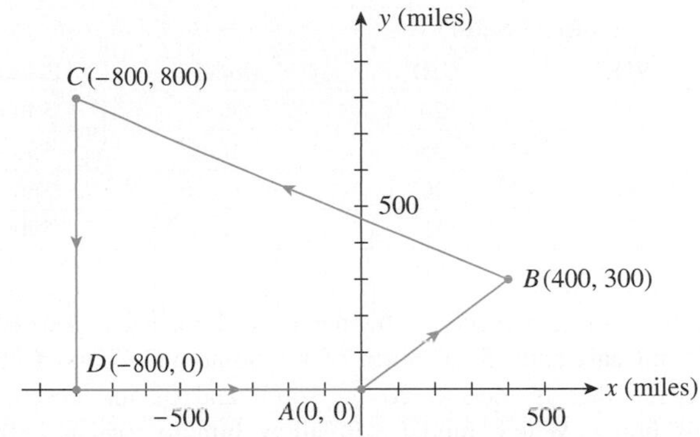

36. Delivery Charges A furniture store offers free setup and delivery services to all points within a 25-mi radius of its warehouse distribution center. If you live 20 mi east and 14 mi south of the warehouse, will you incur a delivery charge? Justify your answer.

37. Optimizing Travel Time Towns A, B, C, and D are located as shown in the following figure. Two highways link town A to town D. Route 1 runs from town A to town D via town B, and Route 2 runs from town A to town D via town C. If a salesman wishes to drive from town A to town D and traffic conditions are such that he could expect to average the same speed on either route, which highway should he take to arrive in the shortest time?

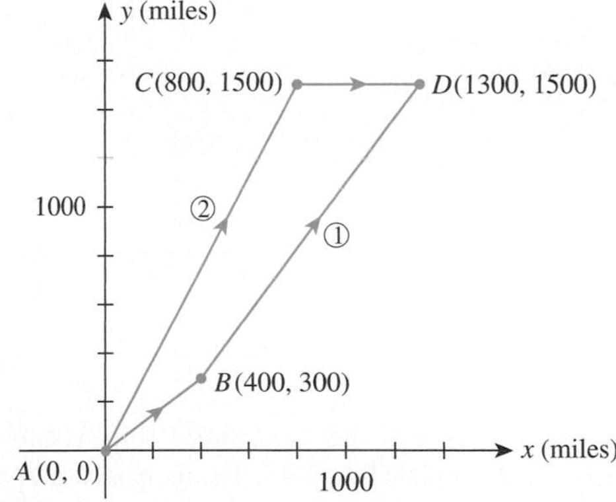

38. Minimizing Shipping Costs Refer to the figure for Exercise 37. Suppose a fleet of 100 automobiles are to be shipped from an assembly plant in town A to town D. They may be shipped either by freight train along Route 1 at a cost of 22¢/mile per automobile or by truck along Route 2 at a cost

of 21¢/mile per automobile. Which means of transportation minimizes the shipping cost? What is the net savings?

39. CONSUMER DECISIONS Ivan wishes to determine which antenna he should purchase for his home. The TV store has supplied him with the following information:

Range in Miles			
VHF	UHF	Model	Price
30	20	A	$40
45	35	B	$50
60	40	C	$60
75	55	D	$70

Ivan wishes to receive Channel 17 (VHF) that is located 25 mi east and 35 mi north of his home and Channel 38 (UHF) that is located 20 mi south and 32 mi west of his home. Which model will allow him to receive both channels at the least cost? (Assume that the terrain between Ivan's home and both broadcasting stations is flat.)

40. COST OF LAYING CABLE In the following diagram, S represents the position of a power relay station located on a straight coastal highway, and M shows the location of a marine biology experimental station on an island. A cable is to be laid connecting the relay station with the experimental station. If the cost of running the cable on land is \$1.50/running foot and the cost of running cable under water is \$2.50/running foot, find an expression in terms of x that gives the total cost for laying the cable. What is the total cost when $x = 2500$? when $x = 3000$?

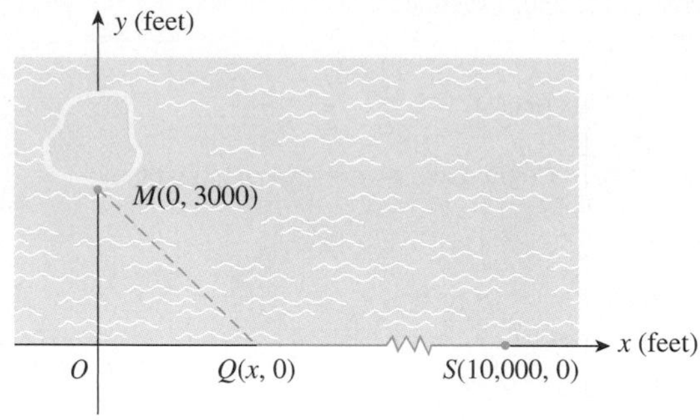

41. Two ships leave port at the same time. Ship A sails north at a speed of 20 mph while ship B sails east at a speed of 30 mph.
   a. Find an expression in terms of the time t (in hours) giving the distance between the two ships.
   b. Using the expression obtained in part (a), find the distance between the two ships 2 hr after leaving port.

42. Ship A leaves port sailing north at a speed of 25 mph. A half hour later, ship B leaves the same port sailing east at a speed of 20 mph. Let t (in hours) denote the time ship B has been at sea.
   a. Find an expression in terms of t giving the distance between the two ships.
   b. Use the expression obtained in part (a) to find the distance between the two ships 2 hr after ship A has left port.

In Exercises 43–46, determine whether the statement is true or false. If it is true, explain why it is true. If it is false, give an example to show why it is false.

43. The point $(-a, b)$ is symmetric to the point (a, b) with respect to the y-axis.

44. The point $(-a, -b)$ is symmetric to the point (a, b) with respect to the origin.

45. If the distance between the points $P_1(a, b)$ and $P_2(c, d)$ is D, then the distance between the points $P_1(a, b)$ and $P_3(kc, kd)$, $(k \neq 0)$, is given by $|k|D$.

46. The circle with equation $kx^2 + ky^2 = a^2$ lies inside the circle with equation $x^2 + y^2 = a^2$, provided $k > 1$.

47. Let (x_1, y_1) and (x_2, y_2) be two points lying in the xy-plane. Show that the distance between the two points is given by

$$d = \sqrt{(x_2 - x_1)^2 + (y_2 - y_1)^2}$$

Hint: Refer to the accompanying figure and use the Pythagorean theorem.

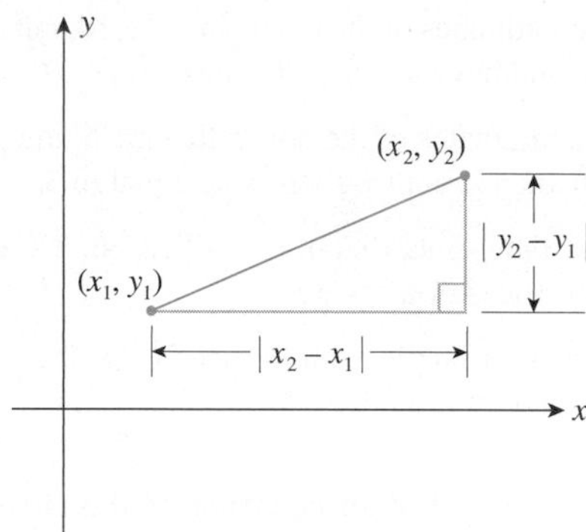

48. Show that an equation of a circle can be written in the form

$$x^2 + y^2 + Cx + Dy + E = 0$$

where C, D, and E are constants. This is called the general form of an equation of a circle.

1.3 Solutions to Self-Check Exercises

1. a. The points are plotted in the following figure:

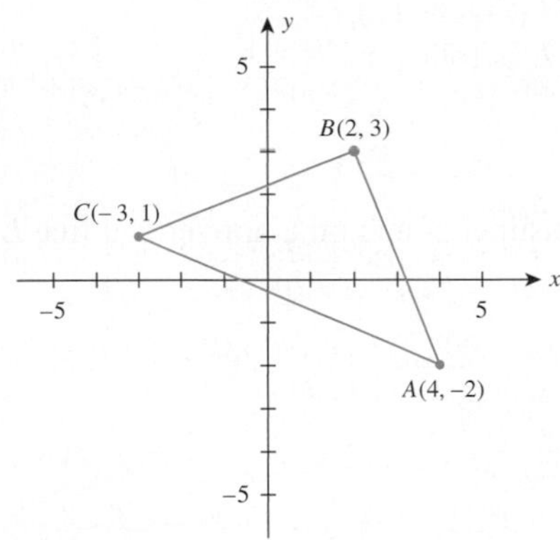

b. The distance between A and B is

$$d(A, B) = \sqrt{(2-4)^2 + [3-(-2)]^2} = \sqrt{(-2)^2 + 5^2} = \sqrt{4 + 25} = \sqrt{29}$$

The distance between B and C is

$$d(B, C) = \sqrt{(-3-2)^2 + (1-3)^2} = \sqrt{(-5)^2 + (-2)^2} = \sqrt{25 + 4} = \sqrt{29}$$

The distance between A and C is

$$d(A, C) = \sqrt{(-3-4)^2 + [1-(-2)]^2} = \sqrt{(-7)^2 + 3^2} = \sqrt{49 + 9} = \sqrt{58}$$

c. We will show that

$$[d(A, C)]^2 = [d(A, B)]^2 + [d(B, C)]^2$$

From part (b), we see that $[d(A, B)]^2 = 29$, $[d(B, C)]^2 = 29$, and $[d(A, C)]^2 = 58$, and the desired result follows.

2. The distance between city A and city B is

$$d(A, B) = \sqrt{200^2 + 50^2} \approx 206$$

or 206 mi. The distance between city B and city C is

$$d(B, C) = \sqrt{[600 - 200]^2 + [320 - 50]^2} = \sqrt{400^2 + 270^2} \approx 483$$

or 483 mi. Therefore, the total distance the pilot would have to cover is 689 mi, so she must refuel in city B.

1.4 Straight Lines

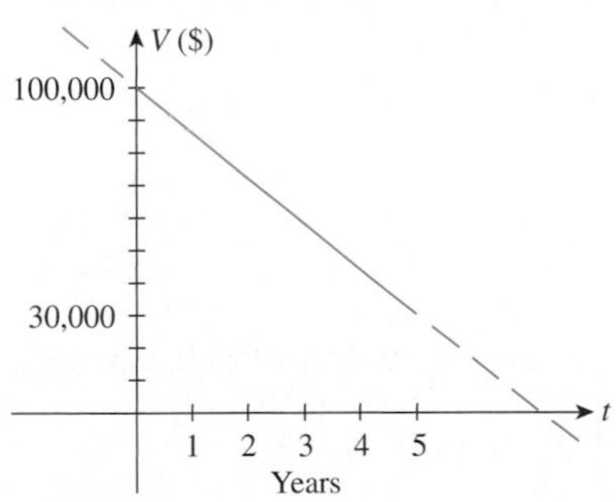

FIGURE 11
Linear depreciation of an asset

In computing income tax, business firms are allowed by law to depreciate certain assets such as buildings, machines, furniture, automobiles, and so on, over a period of time. Linear depreciation, or the straight-line method, is often used for this purpose. The graph of the straight line shown in Figure 11 describes the book value V of a computer that has an initial value of \$100,000 and that is being depreciated linearly over 5 years with a scrap value of \$30,000. Note that only the solid portion of the straight line is of interest here.

The book value of the computer at the end of year t, where t lies between 0 and 5, can be read directly from the graph. But there is one shortcoming in this approach: The result depends on how accurately you draw and read the graph. A better and more accurate method is based on finding an *algebraic* representation of the depreciation line.

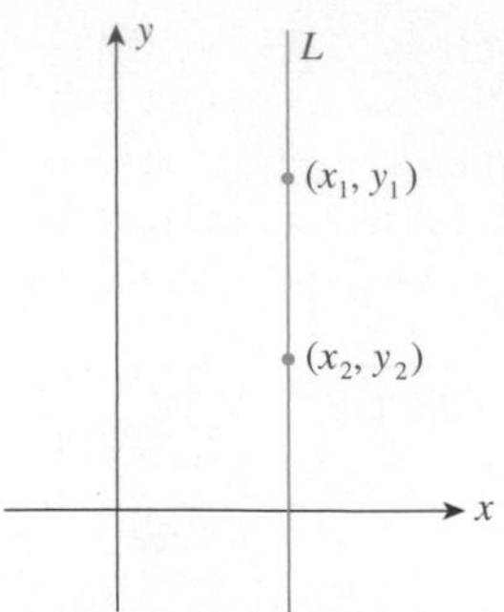

FIGURE 12
m is undefined.

Slope of a Line

To see how a straight line in the xy-plane may be described algebraically, we need to first recall certain properties of straight lines. Let L denote the unique straight line that passes through the two distinct points (x_1, y_1) and (x_2, y_2). If $x_1 = x_2$, then L is a vertical line, and the slope is undefined (Figure 12).

If $x_1 \neq x_2$, we define the slope of L as follows:

Slope of a Nonvertical Line

If (x_1, y_1) and (x_2, y_2) are any two distinct points on a nonvertical line L, then the slope m of L is given by

$$m = \frac{\Delta y}{\Delta x} = \frac{y_2 - y_1}{x_2 - x_1} \tag{3}$$

See Figure 13.

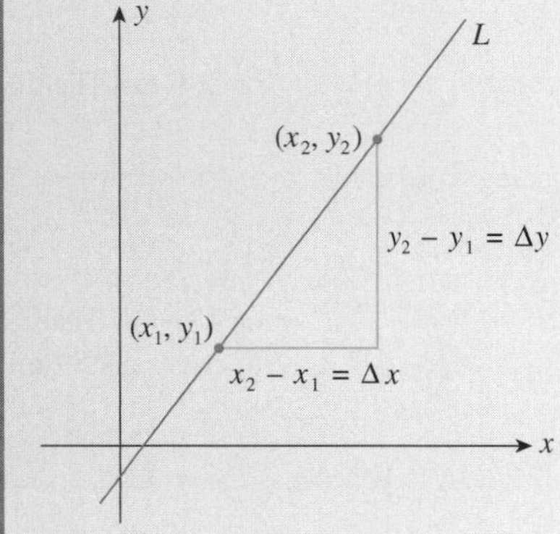

FIGURE 13

Observe that the slope of a straight line is a constant whenever it is defined. The number $\Delta y = y_2 - y_1$ (Δy is read "delta y") is a measure of the vertical change in y, and $\Delta x = x_2 - x_1$ is a measure of the horizontal change in x, as shown in Figure 13. From this figure, we can see that the slope m of a straight line L is a measure of the *rate of change of y with respect to x.*

Figure 14a shows a straight line L_1 with slope 2. Observe that L_1 has the property that a 1-unit increase in x results in a 2-unit increase in y. To see this, let

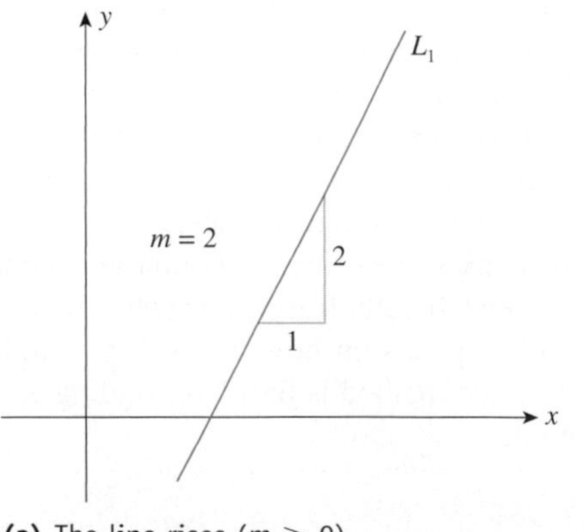

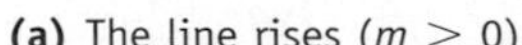
(a) The line rises ($m > 0$).

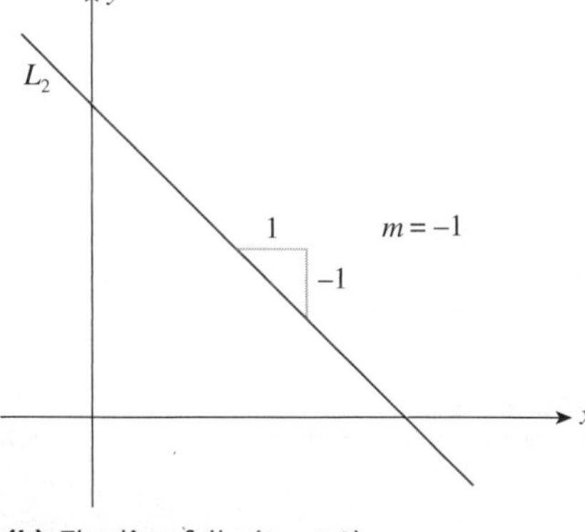

(b) The line falls ($m < 0$).

FIGURE 14

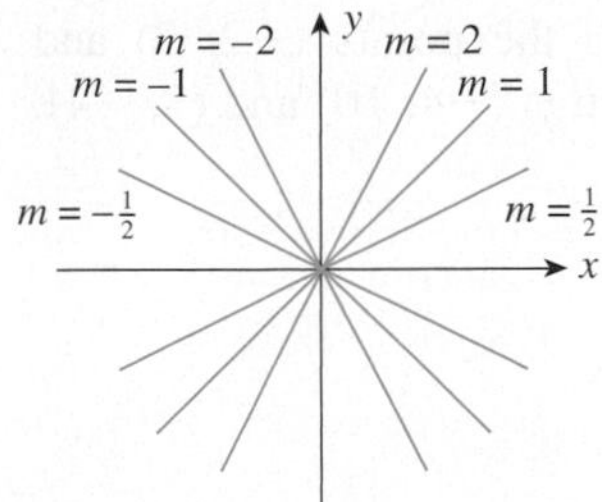

FIGURE 15
A family of straight lines

$\Delta x = 1$ in Formula (3) so that $m = \Delta y$. Since $m = 2$, we conclude that $\Delta y = 2$. Similarly, Figure 14b shows a line L_2 with slope -1. Observe that a straight line with positive slope slants upward from left to right (y increases as x increases), whereas a line with negative slope slants downward from left to right (y decreases as x increases). Finally, Figure 15 shows a family of straight lines passing through the origin with indicated slopes.

EXAMPLE 1 Sketch the straight line that passes through the point $(-2, 5)$ and has slope $-\frac{4}{3}$.

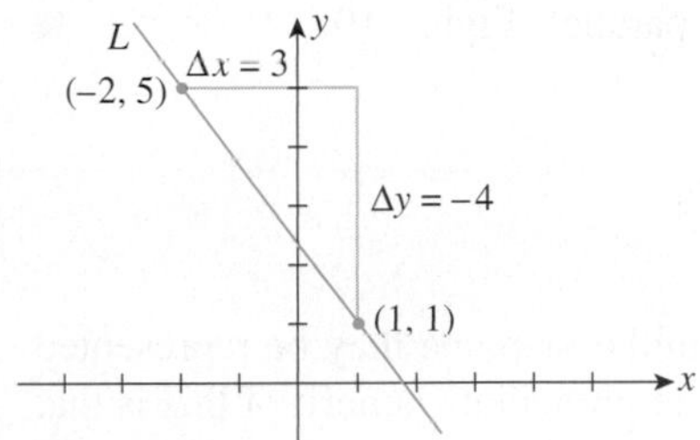

FIGURE 16
L has slope $-\frac{4}{3}$ and passes through $(-2, 5)$.

Solution First, plot the point $(-2, 5)$ (Figure 16). Next, recall that a slope of $-\frac{4}{3}$ indicates that an increase of 1 unit in the x-direction produces a decrease of $\frac{4}{3}$ units in the y-direction, or equivalently, a 3-unit increase in the x-direction produces a $3(\frac{4}{3})$, or 4-unit, decrease in the y-direction. Using this information, we plot the point $(1, 1)$ and draw the line through the two points. ■

EXPLORE & DISCUSS

Show that the slope of a nonvertical line is independent of the two distinct points $P_1(x_1, y_1)$ and $P_2(x_2, y_2)$ used to compute it.

Hint: Suppose we pick two other distinct points, $P_3(x_3, y_3)$ and $P_4(x_4, y_4)$ lying on L. Draw a picture and use similar triangles to demonstrate that using P_3 and P_4 gives the same value as that obtained using P_1 and P_2.

EXAMPLE 2 Find the slope m of the line that passes through the points $(-1, 1)$ and $(5, 3)$.

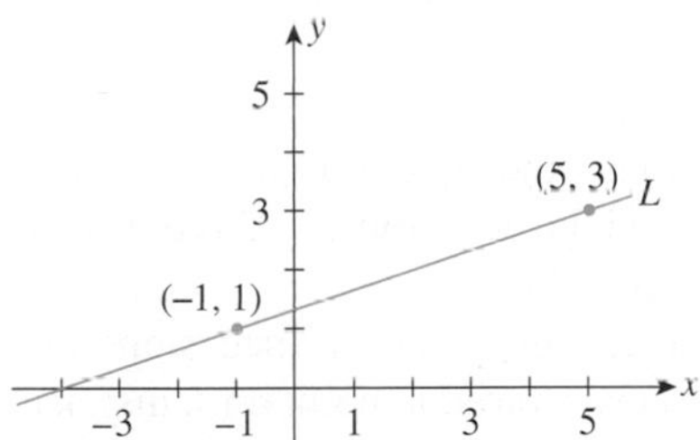

FIGURE 17
L passes through $(5, 3)$ and $(-1, 1)$.

Solution Choose (x_1, y_1) to be the point $(-1, 1)$ and (x_2, y_2) to be the point $(5, 3)$. Then, with $x_1 = -1$, $y_1 = 1$, $x_2 = 5$, and $y_2 = 3$, we find

$$m = \frac{y_2 - y_1}{x_2 - x_1} = \frac{3 - 1}{5 - (-1)} = \frac{1}{3} \quad \text{Use Formula (3).}$$

(Figure 17). Try to verify that the result obtained would have been the same had we chosen the point $(-1, 1)$ to be (x_2, y_2) and the point $(5, 3)$ to be (x_1, y_1). ■

EXAMPLE 3 Find the slope of the line that passes through the points $(-2, 5)$ and $(3, 5)$.

Solution The slope of the required line is given by

$$m = \frac{5 - 5}{3 - (-2)} = \frac{0}{5} = 0$$

(Figure 18). ■

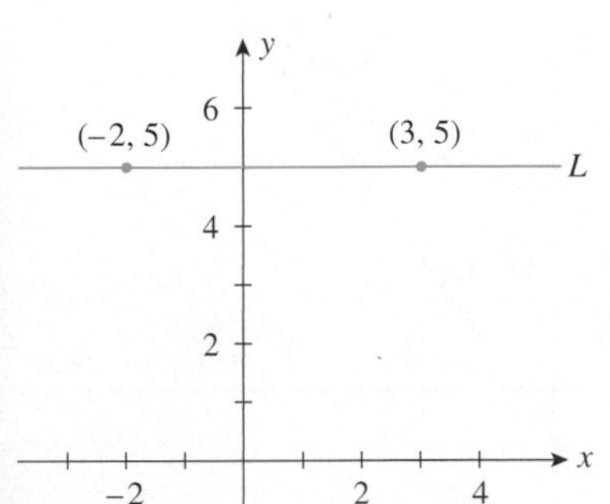

FIGURE 18
The slope of the horizontal line L is 0.

Note In general, the slope of a horizontal line is zero. ■

We can use the slope of a straight line to determine whether a line is parallel to another line.

Parallel Lines
Two distinct lines are **parallel** if and only if their slopes are equal or their slopes are undefined.

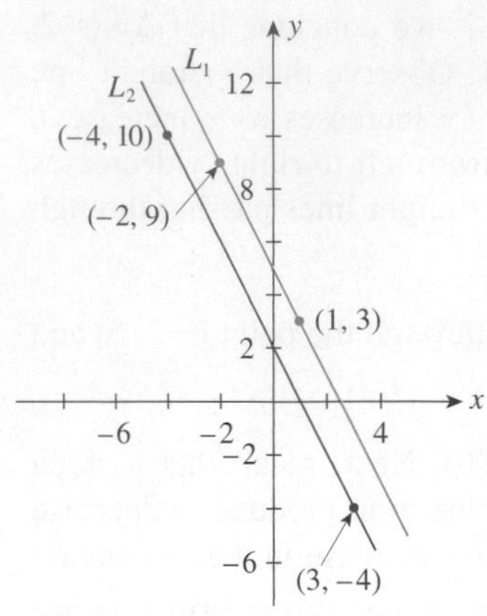

FIGURE 19
L_1 and L_2 have the same slope and hence are parallel.

EXAMPLE 4 Let L_1 be a line that passes through the points $(-2, 9)$ and $(1, 3)$ and let L_2 be the line that passes through the points $(-4, 10)$ and $(3, -4)$. Determine whether L_1 and L_2 are parallel.

Solution The slope m_1 of L_1 is given by

$$m_1 = \frac{3 - 9}{1 - (-2)} = -2$$

The slope m_2 of L_2 is given by

$$m_2 = \frac{-4 - 10}{3 - (-4)} = -2$$

Since $m_1 = m_2$, the lines L_1 and L_2 are in fact parallel (Figure 19). ■

Equations of Lines

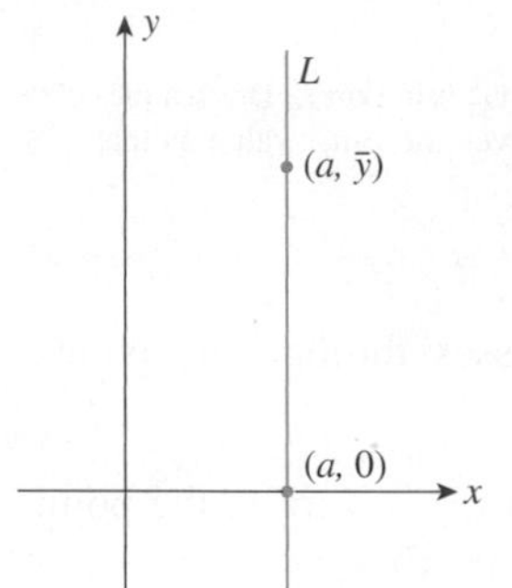

FIGURE 20
The vertical line $x = a$

We will now show that every straight line lying in the xy-plane may be represented by an equation involving the variables x and y. One immediate benefit of this is that problems involving straight lines may be solved algebraically.

Let L be a straight line parallel to the y-axis (perpendicular to the x-axis) (Figure 20). Then, L crosses the x-axis at some point $(a, 0)$ with the x-coordinate given by $x = a$, where a is some real number. Any other point on L has the form $(a, \bar{y})$, where $\bar{y}$ is an appropriate number. Therefore, the vertical line L is described by the sole condition

$$x = a$$

and this is, accordingly, the equation of L. For example, the equation $x = -2$ represents a vertical line 2 units to the left of the y-axis, and the equation $x = 3$ represents a vertical line 3 units to the right of the y-axis (Figure 21).

Next, suppose L is a nonvertical line so that it has a well-defined slope m. Suppose (x_1, y_1) is a fixed point lying on L and (x, y) is a variable point on L distinct from (x_1, y_1) (Figure 22).

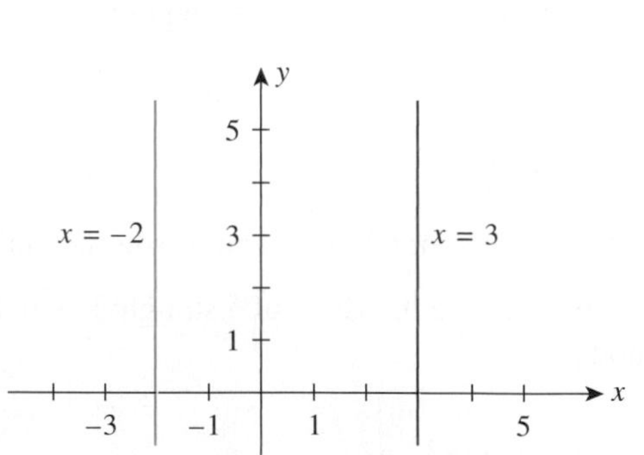

FIGURE 21
The vertical lines $x = -2$ and $x = 3$

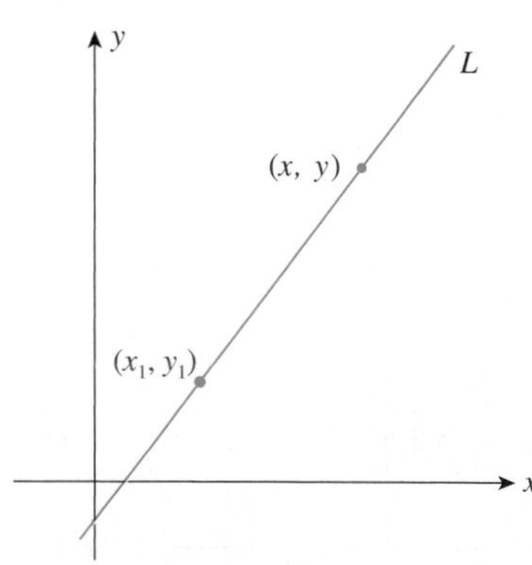

FIGURE 22
L passes through (x_1, y_1) and has slope m.

Using Formula (3) with the point $(x_2, y_2) = (x, y)$, we find that the slope of L is given by

$$m = \frac{y - y_1}{x - x_1}$$

Upon multiplying both sides of the equation by $x - x_1$, we obtain Formula (4).

Point-Slope Form

An equation of the line that has slope m and passes through the point (x_1, y_1) is given by

$$y - y_1 = m(x - x_1) \tag{4}$$

Equation (4) is called the **point-slope form of the equation of a line** since it utilizes a given point (x_1, y_1) on a line and the slope m of the line.

EXAMPLE 5 Find an equation of the line that passes through the point (1, 3) and has slope 2.

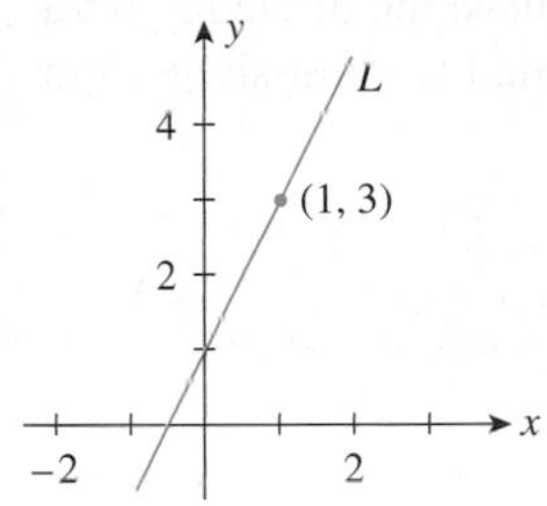

FIGURE 23
L passes through (1, 3) and has slope 2.

Solution Using the point-slope form of the equation of a line with the point (1, 3) and $m = 2$, we obtain

$$y - 3 = 2(x - 1) \qquad y - y_1 = m(x - x_1)$$

which, when simplified, becomes

$$2x - y + 1 = 0$$

(Figure 23).

VIDEO

EXAMPLE 6 Find an equation of the line that passes through the points (−3, 2) and (4, −1).

Solution The slope of the line is given by

$$m = \frac{-1 - 2}{4 - (-3)} = -\frac{3}{7}$$

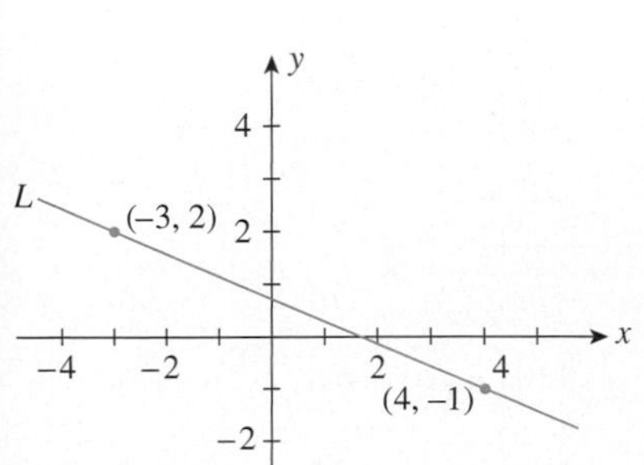

FIGURE 24
L passes through (−3, 2) and (4, −1).

Using the point-slope form of an equation of a line with the point (4, −1) and the slope $m = -\frac{3}{7}$, we have

$$\begin{aligned} y + 1 &= -\frac{3}{7}(x - 4) \qquad y - y_1 = m(x - x_1) \\ 7y + 7 &= -3x + 12 \\ 3x + 7y - 5 &= 0 \end{aligned}$$

(Figure 24).

We can use the slope of a straight line to determine whether a line is perpendicular to another line.

Perpendicular Lines

If L_1 and L_2 are two distinct nonvertical lines that have slopes m_1 and m_2, respectively, then L_1 is **perpendicular** to L_2 (written $L_1 \perp L_2$) if and only if

$$m_1 = -\frac{1}{m_2}$$

If the line L_1 is vertical (so that its slope is undefined), then L_1 is perpendicular to another line, L_2, if and only if L_2 is horizontal (so that its slope is zero). For a proof of these results, see Exercise 84, page 44.

EXAMPLE 7 Find an equation of the line that passes through the point (3, 1) and is perpendicular to the line of Example 5.

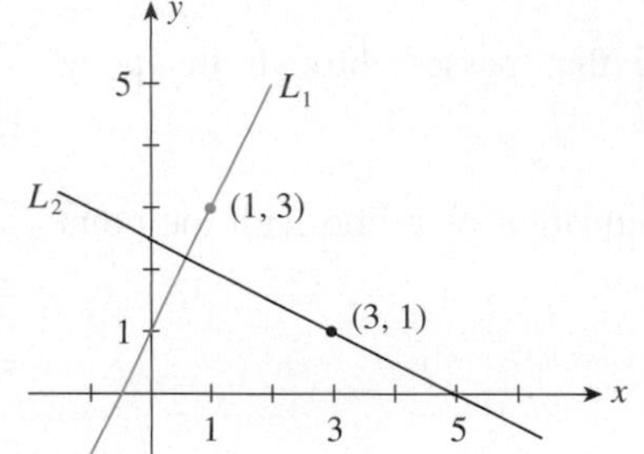

FIGURE 25
L_2 is perpendicular to L_1 and passes through (3, 1).

Solution Since the slope of the line in Example 5 is 2, the slope of the required line is given by $m = -\frac{1}{2}$, the negative reciprocal of 2. Using the point-slope form of the equation of a line, we obtain

$$y - 1 = -\frac{1}{2}(x - 3) \qquad y - y_1 = m(x - x_1)$$

$$2y - 2 = -x + 3$$

$$x + 2y - 5 = 0$$

(See Figure 25). ■

A straight line L that is neither horizontal nor vertical cuts the x-axis and the y-axis at, say, points $(a, 0)$ and $(0, b)$, respectively (Figure 26). The numbers a and b are called the ***x*-intercept** and ***y*-intercept**, respectively, of L.

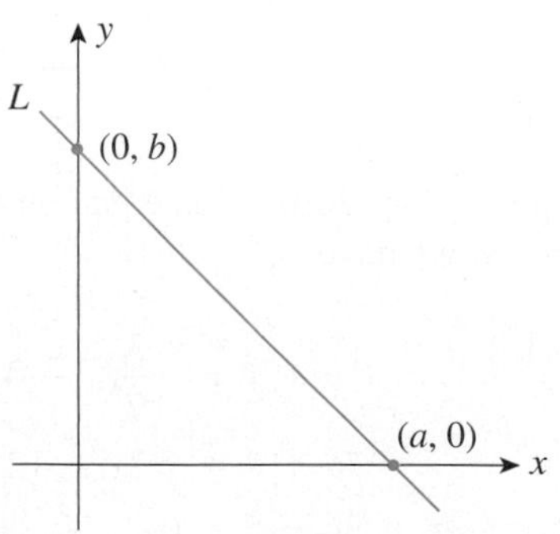

FIGURE 26
The line L has x-intercept a and y-intercept b.

EXPLORING WITH TECHNOLOGY

1. Use a graphing utility to plot the straight lines L_1 and L_2 with equations $2x + y - 5 = 0$ and $41x + 20y - 11 = 0$ on the same set of axes, using the standard viewing window.
   a. Can you tell if the lines L_1 and L_2 are parallel to each other?
   b. Verify your observations by computing the slopes of L_1 and L_2 algebraically.
2. Use a graphing utility to plot the straight lines L_1 and L_2 with equations $x + 2y - 5 = 0$ and $5x - y + 5 = 0$ on the same set of axes, using the standard viewing window.
   a. Can you tell if the lines L_1 and L_2 are perpendicular to each other?
   b. Verify your observation by computing the slopes of L_1 and L_2 algebraically.

Now, let L be a line with slope m and y-intercept b. Using Formula (4), the point-slope form of the equation of a line, with the point $(0, b)$ and slope m, we have

$$y - b = m(x - 0)$$
$$y = mx + b$$

Slope-Intercept Form

An equation of the line that has slope m and intersects the y-axis at the point $(0, b)$ is given by

$$y = mx + b \qquad (5)$$

EXAMPLE 8 Find an equation of the line that has slope 3 and y-intercept -4.

Solution Using Equation (5) with $m = 3$ and $b = -4$, we obtain the required equation

$$y = 3x - 4$$

■

EXAMPLE 9 Determine the slope and y-intercept of the line whose equation is $3x - 4y = 8$.

Solution Rewrite the given equation in the slope-intercept form. Thus,

$$3x - 4y = 8$$
$$-4y = 8 - 3x$$
$$y = \frac{3}{4}x - 2$$

Comparing this result with Equation (5), we find $m = \frac{3}{4}$ and $b = -2$, and we conclude that the slope and y-intercept of the given line are $\frac{3}{4}$ and -2, respectively. ■

EXPLORE & DISCUSS

Consider the slope-intercept form of an equation of a straight line $y = mx + b$. Describe the family of straight lines obtained by keeping

1. The value of m fixed and allowing the value of b to vary.
2. The value of b fixed and allowing the value of m to vary.

EXPLORING WITH TECHNOLOGY

1. Use a graphing utility to plot the straight lines with equations $y = -2x + 3$, $y = -x + 3$, $y = x + 3$, and $y = 2.5x + 3$ on the same set of axes, using the standard viewing window. What effect does changing the coefficient m of x in the equation $y = mx + b$ have on its graph?
2. Use a graphing utility to plot the straight lines with equations $y = 2x - 2$, $y = 2x - 1$, $y = 2x$, $y = 2x + 1$, and $y = 2x + 4$ on the same set of axes, using the standard viewing window. What effect does changing the constant b in the equation $y = mx + b$ have on its graph?
3. Describe in words the effect of changing both m and b in the equation $y = mx + b$.

APPLIED EXAMPLE 10 Sales of a Sporting Goods Store The sales manager of a local sporting goods store plotted sales versus time for the last 5 years and found the points to lie approximately along a straight line (Figure 27). By using the points corresponding to the first and fifth years, find an equation of the trend line. What sales figure can be predicted for the sixth year?

Solution Using Formula (3) with the points (1, 20) and (5, 60), we find that the slope of the required line is given by

$$m = \frac{60 - 20}{5 - 1} = 10$$

Next, using the point-slope form of the equation of a line with the point (1, 20) and $m = 10$, we obtain

$$\begin{aligned} y - 20 &= 10(x - 1) \qquad y - y_1 = m(x - x_1) \\ y &= 10x + 10 \end{aligned}$$

as the required equation.

The sales figure for the sixth year is obtained by letting $x = 6$ in the last equation, giving

$$y = 70$$

or $70,000. ■

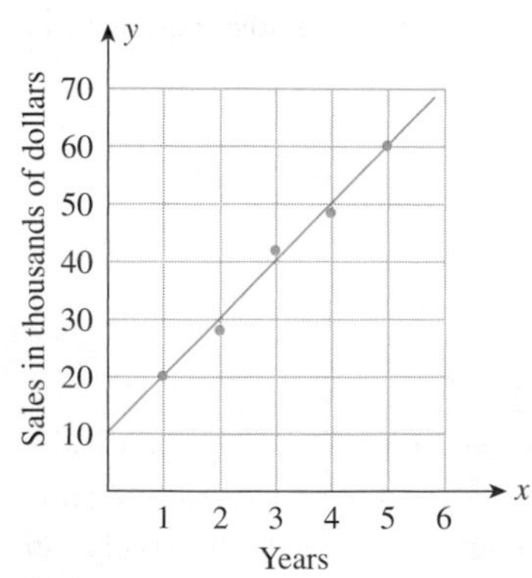

FIGURE 27
Sales of a sporting goods store

APPLIED EXAMPLE 11 Appreciation in Value of an Art Object Suppose an art object purchased for \$50,000 is expected to appreciate in value at a constant rate of \$5000 per year for the next 5 years. Use Formula (5) to write an equation predicting the value of the art object in the next several years. What will be its value 3 years from the date of purchase?

EXPLORE & DISCUSS

Refer to Example 11. Can the equation predicting the value of the art object be used to predict long-term growth?

Solution Let x denote the time (in years) that has elapsed since the date the object was purchased and let y denote the object's value (in dollars). Then, $y = 50,000$ when $x = 0$. Furthermore, the slope of the required equation is given by $m = 5000$, since each unit increase in x (1 year) implies an increase of 5000 units (dollars) in y. Using (5) with $m = 5000$ and $b = 50,000$, we obtain

$$y = 5000x + 50,000 \qquad y = mx + b$$

Three years from the date of purchase, the value of the object will be given by

$$y = 5000(3) + 50,000$$

or \$65,000. ■

General Form of an Equation of a Line

We have considered several forms of an equation of a straight line in the plane. These different forms of the equation are equivalent to each other. In fact, each is a special case of the following equation.

General Form of a Linear Equation

The equation

$$Ax + By + C = 0 \qquad (6)$$

where A, B, and C are constants and A and B are not both zero, is called the general form of a linear equation in the variables x and y.

We will now state (without proof) an important result concerning the algebraic representation of straight lines in the plane.

THEOREM 1

An equation of a straight line is a linear equation; conversely, every linear equation represents a straight line.

This result justifies the use of the adjective *linear* describing Equation (6).

EXAMPLE 12 Sketch the straight line represented by the equation

$$3x - 4y - 12 = 0$$

Solution Since every straight line is uniquely determined by two distinct points, we need find only two such points through which the line passes in order to sketch it. For convenience, let's compute the x- and y-intercepts. Setting $y = 0$, we find $x = 4$; thus, the x-intercept is 4. Setting $x = 0$ gives $y = -3$, and the y-intercept is -3. A sketch of the line appears in Figure 28. ■

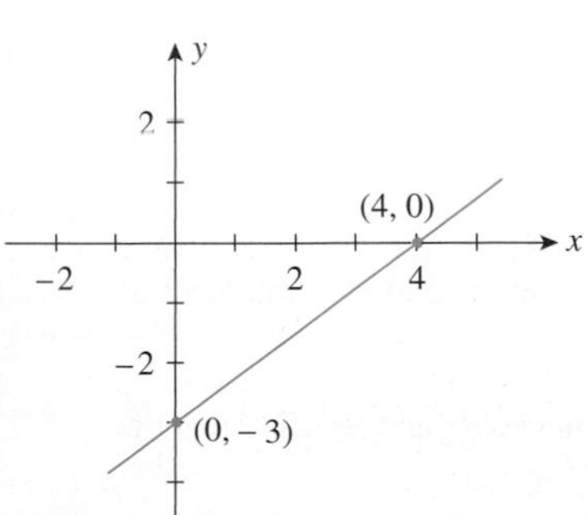

FIGURE 28
To sketch $3x - 4y - 12 = 0$, first find the x-intercept, 4, and the y-intercept, -3.

Following is a summary of the common forms of the equations of straight lines discussed in this section.

Equations of Straight Lines

Vertical line:	$x = a$
Horizontal line:	$y = b$
Point-slope form:	$y - y_1 = m(x - x_1)$
Slope-intercept form:	$y = mx + b$
General form:	$Ax + By + C = 0$

1.4 Self-Check Exercises

1. Determine the number a so that the line passing through the points $(a, 2)$ and $(3, 6)$ is parallel to a line with slope 4.

2. Find an equation of the line that passes through the point $(3, -1)$ and is perpendicular to a line with slope $-\frac{1}{2}$.

3. Does the point $(3, -3)$ lie on the line with equation $2x - 3y - 12 = 0$? Sketch the graph of the line.

4. The percent of people over age 65 who have high school diplomas is summarized in the following table:

Year, x	1960	1965	1970	1975	1980	1985	1990
Percent with Diplomas, y	20	25	30	36	42	47	52

Source: The World Almanac

   a. Plot the percent of people over age 65 who have high school diplomas (y) versus the year (x).
   b. Draw the straight line L through the points (1960, 20) and (1990, 52).
   c. Find an equation of the line L.
   d. Assuming this trend continued, estimate the percent of people over age 65 who obtained high school diplomas by 1995.

Solutions to Self-Check Exercises 1.4 can be found on page 45.

1.4 Concept Questions

1. What is the slope of a nonvertical line? What can you say about the slope of a vertical line?

2. Give (a) the point-slope form, (b) the slope-intercept form, and (c) the general form of the equation of a line.

3. Let L_1 have slope m_1 and L_2 have slope m_2. State the conditions on m_1 and m_2 if (a) L_1 is parallel to L_2 and (b) L_1 is perpendicular to L_2.

1.4 Exercises

In Exercises 1–6, match the statement with one of the graphs (a)–(f).

1. The slope of the line is zero.

2. The slope of the line is undefined.

3. The slope of the line is positive, and its y-intercept is positive.

4. The slope of the line is positive, and its y-intercept is negative.

5. The slope of the line is negative, and its x-intercept is negative.

6. The slope of the line is negative, and its x-intercept is positive.

(a)

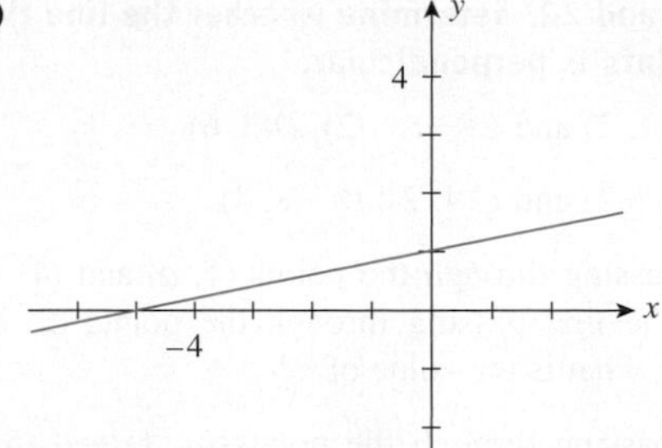

(b)

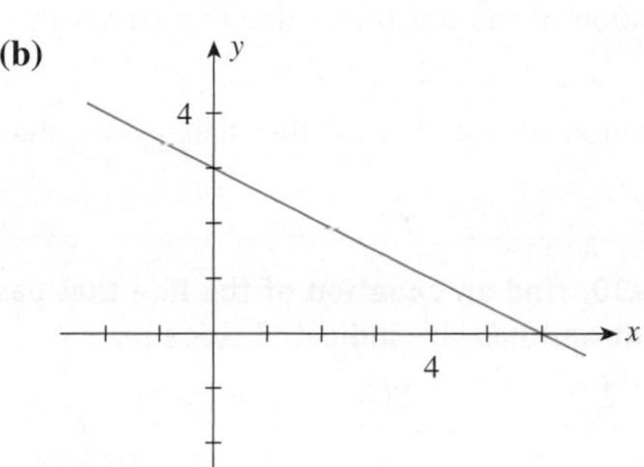

(c)

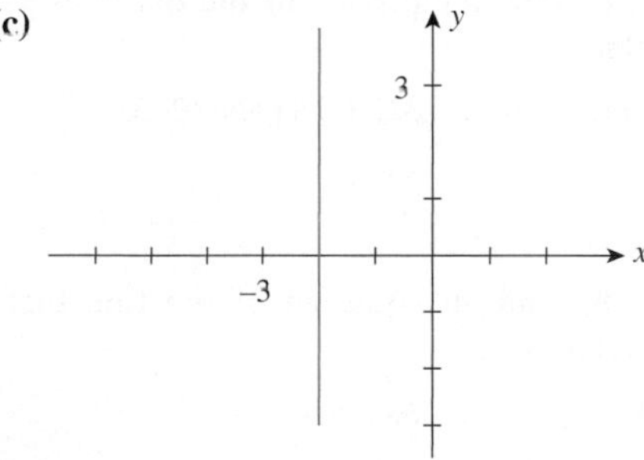

(d)

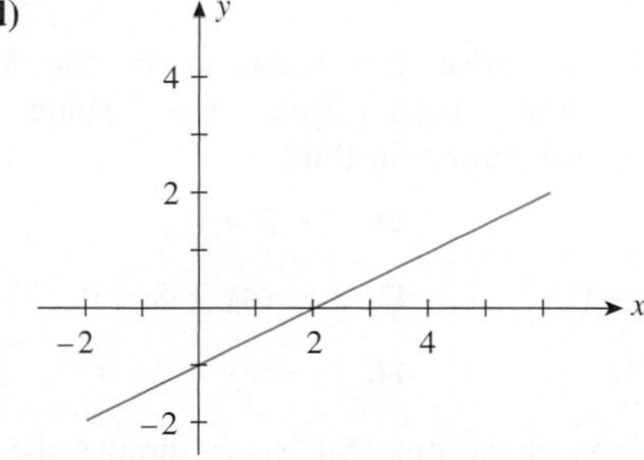

(e)

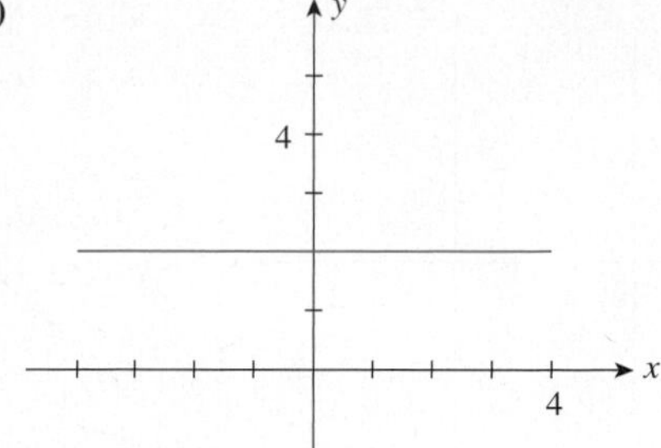

(f)

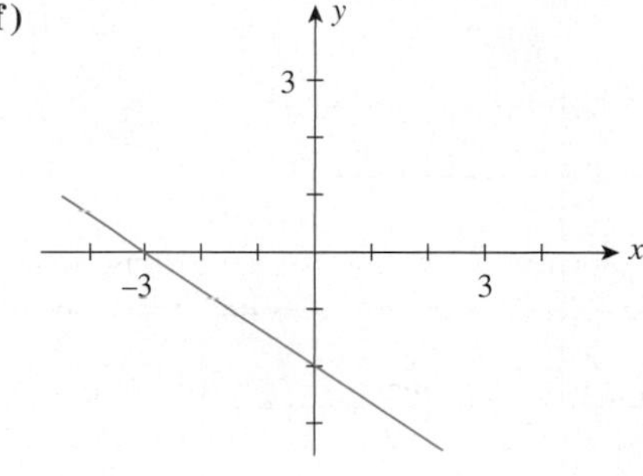

In Exercises 7–10, find the slope of the line shown in each figure.

7.

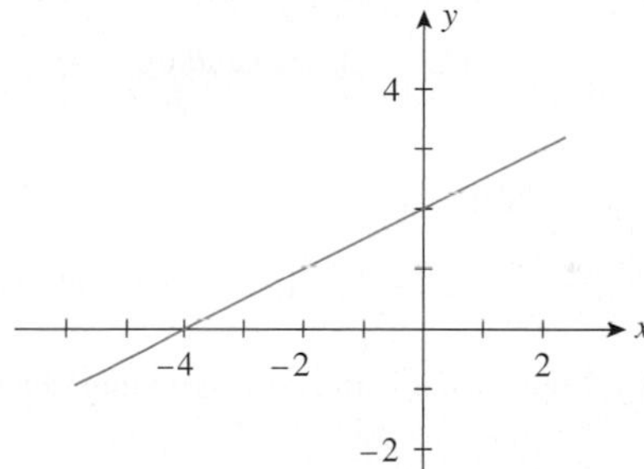

8.

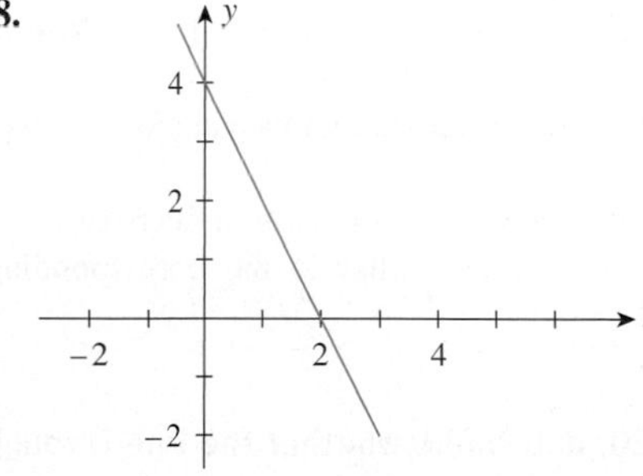

9.

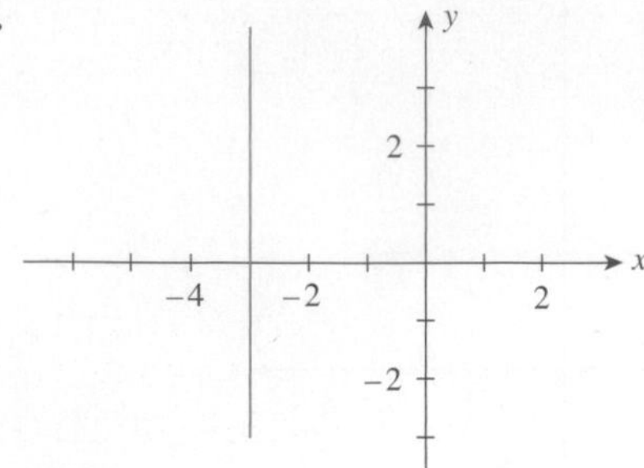

10.

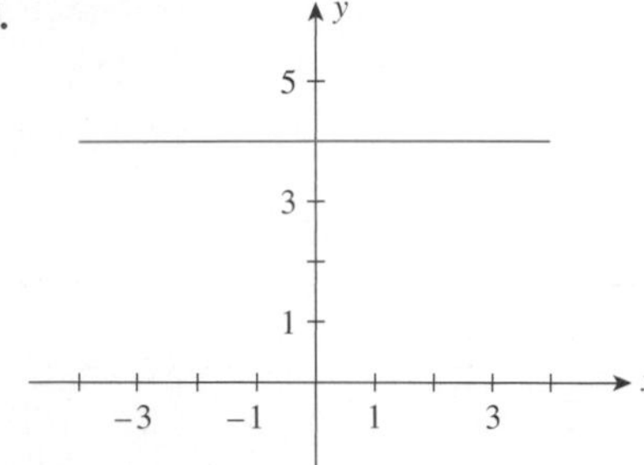

In Exercises 11–16, find the slope of the line that passes through each pair of points.

11. (4, 3) and (5, 8)

12. (4, 5) and (3, 8)

13. (−2, 3) and (4, 8)

14. (−2, −2) and (4, −4)

15. (a, b) and (c, d)

16. $(-a + 1, b - 1)$ and $(a + 1, -b)$

17. Given the equation $y = 4x - 3$, answer the following questions:
   a. If x increases by 1 unit, what is the corresponding change in y?
   b. If x decreases by 2 units, what is the corresponding change in y?

18. Given the equation $2x + 3y = 4$, answer the following questions:
   a. Is the slope of the line described by this equation positive or negative?
   b. As x increases in value, does y increase or decrease?
   c. If x decreases by 2 units, what is the corresponding change in y?

In Exercises 19 and 20, determine whether the line through each pair of points is parallel.

19. $A(1, -2)$, $B(-3, -10)$ and $C(1, 5)$, $D(-1, 1)$

20. $A(2, 3)$, $B(2, -2)$ and $C(-2, 4)$, $D(-2, 5)$

In Exercises 21 and 22, determine whether the line through each pair of points is perpendicular.

21. $A(-2, 5)$, $B(4, 2)$ and $C(-1, -2)$, $D(3, 6)$

22. $A(2, 0)$, $B(1, -2)$ and $C(4, 2)$, $D(-8, 4)$

23. If the line passing through the points $(1, a)$ and $(4, -2)$ is parallel to the line passing through the points (2, 8) and $(-7, a + 4)$, what is the value of a?

24. If the line passing through the points $(a, 1)$ and (5, 8) is parallel to the line passing through the points (4, 9) and $(a + 2, 1)$, what is the value of a?

25. Find an equation of the horizontal line that passes through (−4, −3).

26. Find an equation of the vertical line that passes through (0, 5).

In Exercises 27–30, find an equation of the line that passes through the point and has the indicated slope *m*.

27. (3, −4); $m = 2$

28. (2, 4); $m = -1$

29. (−3, 2); $m = 0$

30. (1, 2); $m = -\frac{1}{2}$

In Exercises 31–34, find an equation of the line that passes through the points.

31. (2, 4) and (3, 7)

32. (2, 1) and (2, 5)

33. (1, 2) and (−3, −2)

34. (−1, −2) and (3, −4)

In Exercises 35–38, find an equation of the line that has slope *m* and *y*-intercept *b*.

35. $m = 3$; $b = 4$

36. $m = -2$; $b = -1$

37. $m = 0$; $b = 5$

38. $m = -\frac{1}{2}$; $b = \frac{3}{4}$

In Exercises 39–44, write the equation in the slope-intercept form and then find the slope and *y*-intercept of the corresponding line.

39. $x - 2y = 0$

40. $y - 2 = 0$

41. $2x - 3y - 9 = 0$

42. $3x - 4y + 8 = 0$

43. $2x + 4y = 14$

44. $5x + 8y - 24 = 0$

45. Find an equation of the line that passes through the point (−2, 2) and is parallel to the line $2x - 4y - 8 = 0$.

46. Find an equation of the line that passes through the point (2, 4) and is perpendicular to the line $3x + 4y - 22 = 0$.

In Exercises 47–52, find an equation of the line that satisfies the given condition.

47. The line parallel to the x-axis and 6 units below it

48. The line passing through the origin and parallel to the line joining the points (2, 4) and (4, 7)

49. The line passing through the point (a, b) with slope equal to zero

50. The line passing through $(-3, 4)$ and parallel to the x-axis

51. The line passing through $(-5, -4)$ and parallel to the line joining $(-3, 2)$ and $(6, 8)$

52. The line passing through (a, b) with undefined slope

53. Given that the point $P(-3, 5)$ lies on the line $kx + 3y + 9 = 0$, find k.

54. Given that the point $P(2, -3)$ lies on the line $-2x + ky + 10 = 0$, find k.

In Exercises 55–60, sketch the straight line defined by the given linear equation by finding the x- and y-intercepts.

Hint: See Example 12, page 39.

55. $3x - 2y + 6 = 0$

56. $2x - 5y + 10 = 0$

57. $x + 2y - 4 = 0$

58. $2x + 3y - 15 = 0$

59. $y + 5 = 0$

60. $-2x - 8y + 24 = 0$

61. Show that an equation of a line through the points $(a, 0)$ and $(0, b)$ with $a \neq 0$ and $b \neq 0$ can be written in the form

$$\frac{x}{a} + \frac{y}{b} = 1$$

(Recall that the numbers a and b are the x- and y-intercepts, respectively, of the line. This form of an equation of a line is called the *intercept form.*)

In Exercises 62–65, use the results of Exercise 61 to find an equation of a line with the given x- and y-intercepts.

62. x-intercept 3; y-intercept 4

63. x-intercept -2; y-intercept -4

64. x-intercept $-\frac{1}{2}$; y-intercept $\frac{3}{4}$

65. x-intercept 4; y-intercept $-\frac{1}{2}$

In Exercises 66 and 67, determine whether the given points lie on a straight line.

66. $A(-1, 7)$, $B(2, -2)$, and $C(5, -9)$

67. $A(-2, 1)$, $B(1, 7)$, and $C(4, 13)$

68. TEMPERATURE CONVERSION The relationship between the temperature in degrees Fahrenheit (°F) and the temperature in degrees Celsius (°C) is

$$F = \frac{9}{5}C + 32$$

a. Sketch the line with the given equation.
b. What is the slope of the line? What does it represent?
c. What is the F-intercept of the line? What does it represent?

69. NUCLEAR PLANT UTILIZATION The United States is not building many nuclear plants, but the ones it has are running full tilt. The output (as a percent of total capacity) of nuclear plants is described by the equation

$$y = 1.9467t + 70.082$$

where t is measured in years, with $t = 0$ corresponding to the beginning of 1990.
a. Sketch the line with the given equation.
b. What are the slope and the y-intercept of the line found in part (a)?
c. Give an interpretation of the slope and the y-intercept of the line found in part (a).
d. If the utilization of nuclear power continues to grow at the same rate and the total capacity of nuclear plants in the United States remains constant, by what year can the plants be expected to be generating at maximum capacity?

Source: Nuclear Energy Institute

70. SOCIAL SECURITY CONTRIBUTIONS For wages less than the maximum taxable wage base, Social Security contributions by employees are 7.65% of the employee's wages.
a. Find an equation that expresses the relationship between the wages earned (x) and the Social Security taxes paid (y) by an employee who earns less than the maximum taxable wage base.
b. For each additional dollar that an employee earns, by how much is his or her Social Security contribution increased? (Assume that the employee's wages are less than the maximum taxable wage base.)
c. What Social Security contributions will an employee who earns \$35,000 (which is less than the maximum taxable wage base) be required to make?

Source: Social Security Administration

71. COLLEGE ADMISSIONS Using data compiled by the Admissions Office at Faber University, college admissions officers estimate that 55% of the students who are offered admission to the freshman class at the university will actually enroll.
a. Find an equation that expresses the relationship between the number of students who actually enroll (y) and the number of students who are offered admission to the university (x).
b. If the desired freshman class size for the upcoming academic year is 1100 students, how many students should be admitted?

72. WEIGHT OF WHALES The equation $W = 3.51L - 192$, expressing the relationship between the length L (in feet) and the expected weight W (in British tons) of adult blue whales, was adopted in the late 1960s by the International Whaling Commission.
   a. What is the expected weight of an 80-ft blue whale?
   b. Sketch the straight line that represents the equation.

73. THE NARROWING GENDER GAP Since the founding of the Equal Employment Opportunity Commission and the passage of equal-pay laws, the gulf between men's and women's earnings has continued to close gradually. At the beginning of 1990 ($t = 0$), women's wages were 68% of men's wages, and by the beginning of 2000 ($t = 10$), women's wages were projected to be 80% of men's wages. If this gap between women's and men's wages continued to narrow *linearly*, what percent of men's wages are women's wages expected to be at the beginning of 2008?
   Source: Journal of Economic Perspectives

74. SALES GROWTH Metro Department Store's annual sales (in millions of dollars) during the past 5 yr were

Annual Sales, y	5.8	6.2	7.2	8.4	9.0
Year, x	1	2	3	4	5

   a. Plot the annual sales (y) versus the year (x).
   b. Draw a straight line L through the points corresponding to the first and fifth years.
   c. Derive an equation of the line L.
   d. Using the equation found in part (c), estimate Metro's annual sales 4 yr from now ($x = 9$).

75. SALES OF GPS EQUIPMENT The annual sales (in billions of dollars) of global positioning system (GPS) equipment from the year 2000 through 2006 follow. (Sales for 2004–2006 were projections.) Here $x = 0$ corresponds to 2000.

Annual Sales, y	7.9	9.6	11.5	13.3	15.2	17	18.8
Year, x	0	1	2	3	4	5	6

   a. Plot the annual sales (y) versus the year (x).
   b. Draw a straight line L through the points corresponding to 2000 and 2006.
   c. Derive an equation of the line L.
   d. Use the equation found in part (c) to estimate the annual sales of GPS equipment for 2005. Compare this figure with the projected sales for that year.
   Source: ABI Research

76. DIGITAL TV SERVICES The percent of homes with digital TV services, which stood at 5% at the beginning of 1999 ($t = 0$) was projected to grow linearly so that at the beginning of 2003 ($t = 4$) the percent of such homes was projected to be 25%.
   a. Derive an equation of the line passing through the points $A(0, 5)$ and $B(4, 25)$.
   b. Plot the line with the equation found in part (a).
   c. Using the equation found in part (a), find the percent of homes with digital TV services at the beginning of 2001.
   Source: Paul Kagan Associates

In Exercises 77–81, determine whether the statement is true or false. If it is true, explain why it is true. If it is false, give an example to show why it is false.

77. Suppose the slope of a line L is $-\frac{1}{2}$ and P is a given point on L. If Q is the point on L lying 4 units to the left of P, then Q is situated 2 units above P.

78. The line with equation $Ax + By + C = 0$, $(B \neq 0)$, and the line with equation $ax + by + c = 0$, $(b \neq 0)$, are parallel if $Ab - aB = 0$.

79. If the slope of the line L_1 is positive, then the slope of a line L_2 perpendicular to L_1 may be positive or negative.

80. The lines with equations $ax + by + c_1 = 0$ and $bx - ay + c_2 = 0$, where $a \neq 0$ and $b \neq 0$, are perpendicular to each other.

81. If L is the line with equation $Ax + By + C = 0$, where $A \neq 0$, then L crosses the x-axis at the point $(-C/A, 0)$.

82. Is there a difference between the statements "The slope of a straight line is zero" and "The slope of a straight line does not exist (is not defined)"? Explain your answer.

83. Show that two distinct lines with equations $a_1x + b_1y + c_1 = 0$ and $a_2x + b_2y + c_2 = 0$, respectively, are parallel if and only if $a_1b_2 - b_1a_2 = 0$.
   Hint: Write each equation in the slope-intercept form and compare.

84. Prove that if a line L_1 with slope m_1 is perpendicular to a line L_2 with slope m_2, then $m_1m_2 = -1$.
   Hint: Refer to the following figure. Show that $m_1 = b$ and $m_2 = c$. Next, apply the Pythagorean theorem to triangles OAC, OCB, and OBA to show that $1 = -bc$.

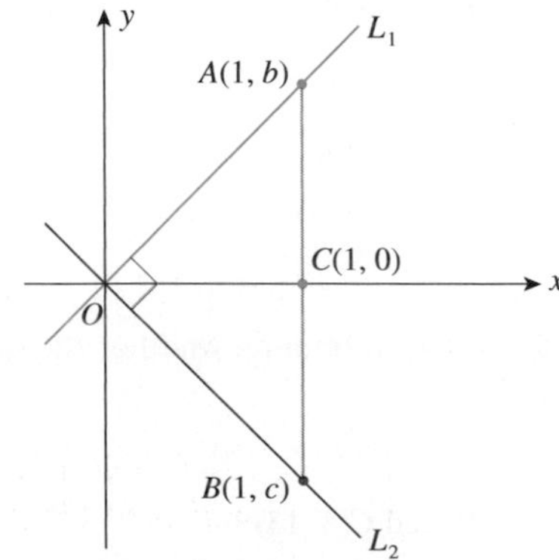

1.4 Solutions to Self-Check Exercises

1. The slope of the line that passes through the points $(a, 2)$ and $(3, 6)$ is

$$m = \frac{6 - 2}{3 - a} = \frac{4}{3 - a}$$

Since this line is parallel to a line with slope 4, m must be equal to 4; that is,

$$\frac{4}{3 - a} = 4$$

or, upon multiplying both sides of the equation by $3 - a$,

$$\begin{aligned} 4 &= 4(3 - a) \\ 4 &= 12 - 4a \\ 4a &= 8 \\ a &= 2 \end{aligned}$$

2. Since the required line L is perpendicular to a line with slope $-1/2$, the slope of L is

$$m = \frac{-1}{-\frac{1}{2}} = 2$$

Next, using the point-slope form of the equation of a line, we have

$$\begin{aligned} y - (-1) &= 2(x - 3) \\ y + 1 &= 2x - 6 \\ y &= 2x - 7 \end{aligned}$$

3. Substituting $x = 3$ and $y = -3$ into the left-hand side of the given equation, we find

$$2(3) - 3(-3) - 12 = 3$$

which is not equal to zero (the right-hand side). Therefore, $(3, -3)$ does not lie on the line with equation $2x - 3y - 12 = 0$. (See the accompanying figure.)

Setting $x = 0$, we find $y = -4$, the y-intercept. Next, setting $y = 0$ gives $x = 6$, the x-intercept. We now draw the line passing through the points $(0, -4)$ and $(6, 0)$ as shown.

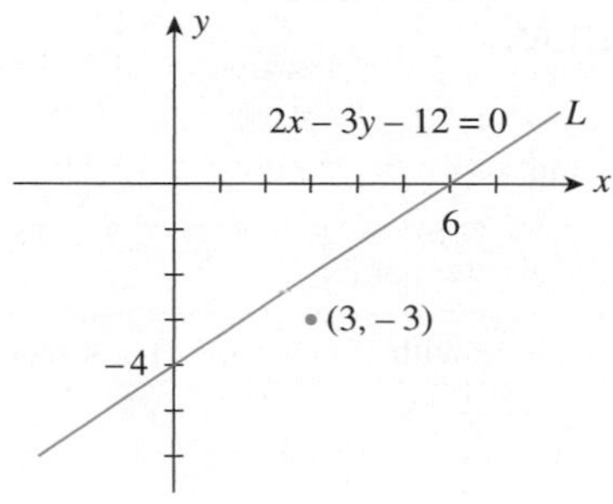

4. **a** and **b.** See the accompanying figure.

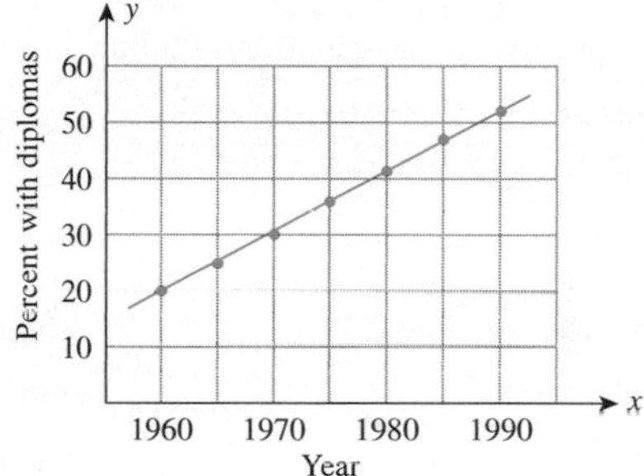

c. The slope of L is

$$m = \frac{52 - 20}{1990 - 1960} = \frac{32}{30} = \frac{16}{15}$$

Using the point-slope form of the equation of a line with the point (1960, 20), we find

$$\begin{aligned} y - 20 &= \frac{16}{15}(x - 1960) = \frac{16}{15}x - \frac{(16)(1960)}{15} \\ y &= \frac{16}{15}x - \frac{6272}{3} + 20 \\ &= \frac{16}{15}x - \frac{6212}{3} \end{aligned}$$

d. To estimate the percent of people over age 65 who obtained high school diplomas by 1995, let $x = 1995$ in the equation obtained in part (c). Thus, the required estimate is

$$y = \frac{16}{15}(1995) - \frac{6212}{3} \approx 57.33$$

or approximately 57%.

CHAPTER 1 Summary of Principal Formulas and Terms

FORMULAS

1. Quadratic formula	$x = \dfrac{-b \pm \sqrt{b^2 - 4ac}}{2a}$
2. Distance between two points	$d = \sqrt{(x_2 - x_1)^2 + (y_2 - y_1)^2}$
3. Equation of a circle with center $C(h, k)$ and radius r	$(x - h)^2 + (y - k)^2 = r^2$
4. Slope of a line	$m = \dfrac{y_2 - y_1}{x_2 - x_1}$
5. Equation of a vertical line	$x = a$
6. Equation of a horizontal line	$y = b$
7. Point-slope form of the equation of a line	$y - y_1 = m(x - x_1)$
8. Slope-intercept form of the equation of a line	$y = mx + b$
9. General equation of a line	$Ax + By + C = 0$

TERMS

real number (coordinate) line (2)
open interval (2)
closed interval (3)
half-open interval (3)
finite interval (3)
infinite interval (3)
absolute value (5)
triangle inequality (5)
polynomial (11)
roots of a polynomial equation (15)
Cartesian coordinate system (23)
ordered pair (23)
parallel lines (33)
perpendicular lines (36)

CHAPTER 1 Concept Review Questions

Fill in the blanks.

1. A point in the plane can be represented uniquely by a/an _____ pair of numbers. The first number of the pair is called the _____, and the second number of the pair is called the _____.

2. a. The point $P(a, 0)$ lies on the _____-axis, and the point $P(0, b)$ lies on the _____-axis.

   b. If the point $P(a, b)$ lies in the fourth quadrant, then the point $P(-a, b)$ lies in the _____ quadrant.

3. The distance between two points $P(a, b)$ and $P(c, d)$ is _____.

4. An equation of a circle with center $C(a, b)$ and radius r is given by _____.

5. a. If $P_1(x_1, y_1)$ and $P_2(x_2, y_2)$ are any two distinct points on a nonvertical line L, then the slope of L is $m =$ _____.

   b. The slope of a vertical line is _____.

   c. The slope of a horizontal line is _____.

   d. The slope of a line that slants upward is _____.

6. If L_1 and L_2 are nonvertical lines with slopes m_1 and m_2, respectively, then L_1 is parallel to L_2 if and only if _____ and L_1 is perpendicular to L_2 if and only if _____.

7. a. An equation of the line passing through the point $P(x_1, y_1)$ and having slope m is _____. This form of the equation of a line is called the _____ _____.

   b. An equation of the line that has slope m and y-intercept b is _____. It is called the _____ form of an equation of a line.

8. a. The general form of an equation of a line is _____.

   b. If a line has equation $ax + by + c = 0$ ($b \neq 0$), then its slope is _____.

CHAPTER 1 Review Exercises

In Exercises 1–4, find the values of x that satisfy the inequality (inequalities).

1. $-x + 3 \le 2x + 9$
2. $-2 \le 3x + 1 \le 7$
3. $x - 3 > 2$ or $x + 3 < -1$
4. $2x^2 > 50$

In Exercises 5–8, evaluate the expression.

5. $|-5 + 7| + |-2|$
6. $\left|\dfrac{5 - 12}{-4 - 3}\right|$
7. $|2\pi - 6| - \pi$
8. $|\sqrt{3} - 4| + |4 - 2\sqrt{3}|$

In Exercises 9–14, evaluate the expression.

9. $\left(\dfrac{9}{4}\right)^{3/2}$
10. $\dfrac{5^6}{5^4}$
11. $(3 \cdot 4)^{-2}$
12. $(-8)^{5/3}$
13. $\dfrac{(3 \cdot 2^{-3})(4 \cdot 3^5)}{2 \cdot 9^3}$
14. $\dfrac{3\sqrt[3]{54}}{\sqrt[3]{18}}$

In Exercises 15–20, simplify the expression.

15. $\dfrac{4(x^2 + y)^3}{x^2 + y}$
16. $\dfrac{a^6b^{-5}}{(a^3b^{-2})^{-3}}$
17. $\dfrac{\sqrt[4]{16x^5yz}}{\sqrt[4]{81xyz^5}}$
18. $(2x^3)(-3x^{-2})\left(\dfrac{1}{6}x^{-1/2}\right)$
19. $\left(\dfrac{3xy^2}{4x^3y}\right)^{-2}\left(\dfrac{3xy^3}{2x^2}\right)^3$
20. $\sqrt[3]{81x^5y^{10}}\sqrt[3]{9xy^2}$

In Exercises 21–24, factor the expression.

21. $-2\pi^2r^3 + 100\pi r^2$
22. $2v^3w + 2vw^3 + 2u^2vw$
23. $16 - x^2$
24. $12t^3 - 6t^2 - 18t$

In Exercises 25–28, solve the equation by factoring.

25. $8x^2 + 2x - 3 = 0$
26. $-6x^2 - 10x + 4 = 0$
27. $-x^3 - 2x^2 + 3x = 0$
28. $2x^4 + x^2 = 1$

In Exercises 29 and 30, use the quadratic formula to solve the quadratic equation.

29. $x^2 - 2x - 5 = 0$
30. $2x^2 + 8x + 7 = 0$

In Exercises 31–34, perform the indicated operations and simplify the expression.

31. $\dfrac{(t + 6)(60) - (60t + 180)}{(t + 6)^2}$
32. $\dfrac{6x}{2(3x^2 + 2)} + \dfrac{1}{4(x + 2)}$
33. $\dfrac{2}{3}\left(\dfrac{4x}{2x^2 - 1}\right) + 3\left(\dfrac{3}{3x - 1}\right)$
34. $\dfrac{-2x}{\sqrt{x + 1}} + 4\sqrt{x + 1}$
35. Rationalize the numerator: $\dfrac{\sqrt{x} - 1}{x - 1}$.
36. Rationalize the denominator: $\dfrac{\sqrt{x} - 1}{2\sqrt{x}}$.

In Exercises 37 and 38, find the distance between the two points.

37. $(-2, -3)$ and $(1, -7)$
38. $\left(\dfrac{1}{2}, \sqrt{3}\right)$ and $\left(-\dfrac{1}{2}, 2\sqrt{3}\right)$

In Exercises 39–44, find an equation of the line L that passes through the point $(-2, 4)$ and satisfies the condition.

39. L is a vertical line.
40. L is a horizontal line.
41. L passes through the point $\left(3, \dfrac{7}{2}\right)$.
42. The x-intercept of L is 3.
43. L is parallel to the line $5x - 2y = 6$.
44. L is perpendicular to the line $4x + 3y = 6$.
45. Find an equation of the straight line that passes through the point $(2, 3)$ and is parallel to the line with equation $3x + 4y - 8 = 0$.
46. Find an equation of the straight line that passes through the point $(-1, 3)$ and is parallel to the line passing through the points $(-3, 4)$ and $(2, 1)$.
47. Find an equation of the line that passes through the point $(-3, -2)$ and is parallel to the line passing through the points $(-2, -4)$ and $(1, 5)$.
48. Find an equation of the line that passes through the point $(-2, -4)$ and is perpendicular to the line with equation $2x - 3y - 24 = 0$.

49. Sketch the graph of the equation $3x - 4y = 24$.

50. Sketch the graph of the line that passes through the point (3, 2) and has slope $-2/3$.

51. Find the minimum cost C (in dollars) given that

$$2(1.5C + 80) \leq 2(2.5C - 20)$$

52. Find the maximum revenue R (in dollars) given that

$$12(2R - 320) \leq 4(3R + 240)$$

53. **Sales of Navigation Systems** The projected number of navigation systems (in millions) installed in vehicles in North America, Europe, and Japan from 2002 through 2006 follow. Here $x = 0$ corresponds to 2002.

Systems Installed, y	3.9	4.7	5.8	6.8	7.8
Year, x	0	1	2	3	4

a. Plot the annual sales (y) versus the year (x).
b. Draw a straight line L through the points corresponding to 2002 and 2006.
c. Derive an equation of the line L.
d. Use the equation found in part (c) to estimate the number of navigation systems installed for 2005. Compare this figure with the projected sales for that year.

Source: ABI Research

The problem-solving skills that you learn in each chapter are building blocks for the rest of the course. Therefore, it is a good idea to make sure that you have mastered these skills before moving on to the next chapter. The Before Moving On exercises that follow are designed for that purpose. After taking this test, you can see where your weaknesses, if any, are. Then you can go to http://www.thomsonedu.com/math/tan where you will find a link to our Companion Website. Here, you can click on the Before Moving On button, which will lead you to other versions of these tests. There you can re-test yourself on those exercises that you solved incorrectly. (You can also test yourself on these basic skills before taking your course quizzes and exams.)

If you feel that you need additional help with these exercises, visit http://www.thomsonedu.com/login where you will find ThomsonNow tutorials with video examples, as well as vMentor for live help from a tutor.

CHAPTER 1 Before Moving On . . .

1. Evaluate:

   a. $|\pi - 2\sqrt{3}| - |\sqrt{3} - \sqrt{2}|$ b. $\left[\left(-\frac{1}{3}\right)^{-3}\right]^{1/3}$

2. Simplify:

   a. $\sqrt[3]{64x^6} \cdot \sqrt{9y^2x^6}$ b. $\left(\frac{a^{-3}}{b^{-4}}\right)^2\left(\frac{b}{a}\right)^{-3}$

3. Rationalize the denominator:

   a. $\frac{2x}{3\sqrt{y}}$ b. $\frac{x}{\sqrt{x} - 4}$

4. Perform each operation and simplify:

   a. $\frac{(x^2 + 1)(\frac{1}{2}x^{-1/2}) - x^{1/2}(2x)}{(x^2 + 1)^2}$

   b. $-\frac{3x}{\sqrt{x + 2}} + 3\sqrt{x + 2}$

5. Rationalize the numerator: $\frac{\sqrt{x} + \sqrt{y}}{\sqrt{x} - \sqrt{y}}$.

6. Factor:

   a. $12x^3 - 10x^2 - 12x$ b. $2bx - 2by + 3cx - 3cy$

7. Solve each equation:

   a. $12x^2 - 9x - 3 = 0$ b. $3x^2 - 5x + 1 = 0$

8. Find the distance between $(-2, 4)$ and $(6, 8)$.

9. Find an equation of the line that passes through $(-1, -2)$ and $(4, 5)$.

10. Find an equation of the line that has slope $-\frac{1}{3}$ and y-intercept $\frac{4}{3}$.

2 Functions, Limits, and the Derivative

How does the change in the demand for a certain make of tires affect the unit price of the tires? The management of the Titan Tire Company has determined the demand function that relates the unit price of its Super Titan tires to the quantity demanded. In Example 7, page 143, you will see how this function can be used to compute the rate of change of the unit price of the Super Titan tires with respect to the quantity demanded.

IN THIS CHAPTER, we define a *function*, a special relationship between two variables. The concept of a function enables us to describe many relationships that exist in applications. We also begin the study of differential calculus. Historically, differential calculus was developed in response to the problem of finding the tangent line to an arbitrary curve. But it quickly became apparent that solving this problem provided mathematicians with a method for solving many practical problems involving the rate of change of one quantity with respect to another. The basic tool used in differential calculus is the *derivative* of a function. The concept of the derivative is based, in turn, on a more fundamental notion—that of the *limit* of a function.

2.1 Functions and Their Graphs

Functions

A manufacturer would like to know how his company's profit is related to its production level; a biologist would like to know how the size of the population of a certain culture of bacteria will change over time; a psychologist would like to know the relationship between the learning time of an individual and the length of a vocabulary list; and a chemist would like to know how the initial speed of a chemical reaction is related to the amount of substrate used. In each instance, we are concerned with the same question: How does one quantity depend upon another? The relationship between two quantities is conveniently described in mathematics by using the concept of a function.

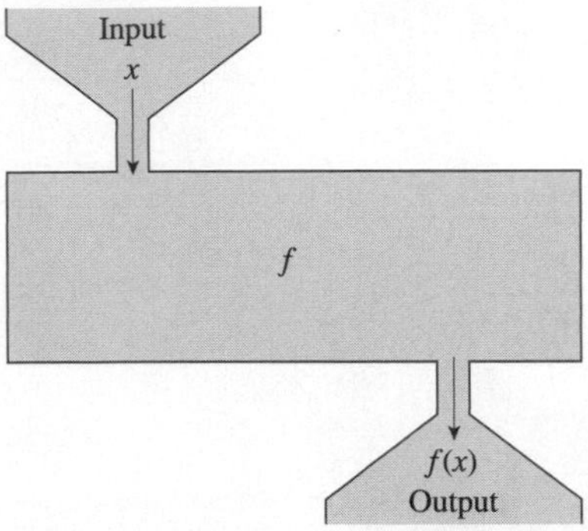

FIGURE 1
A function machine

> **Function**
> A **function** is a rule that assigns to each element in a set A one and only one element in a set B.

The set A is called the **domain** of the function. It is customary to denote a function by a letter of the alphabet, such as the letter f. If x is an element in the domain of a function f, then the element in B that f associates with x is written $f(x)$ (read "f of x") and is called the value of f at x. The set comprising all the values assumed by $y = f(x)$ as x takes on all possible values in its domain is called the **range** of the function f.

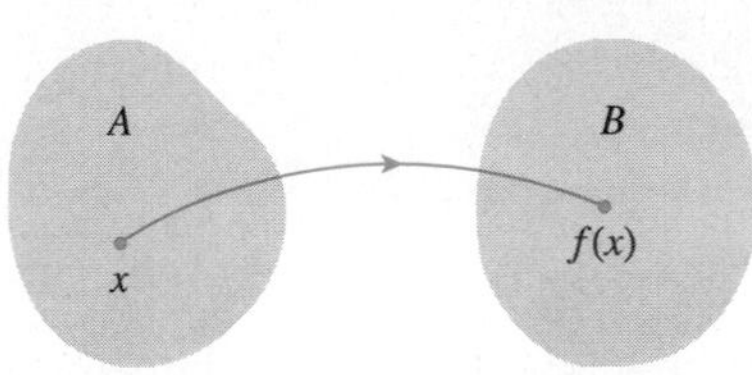

FIGURE 2
The function f viewed as a mapping

We can think of a function f as a machine. The domain is the set of inputs (raw material) for the machine, the rule describes how the input is to be processed, and the value(s) of the function are the outputs of the machine (Figure 1).

We can also think of a function f as a mapping in which an element x in the domain of f is mapped onto a unique element $f(x)$ in B (Figure 2).

Notes

1. The output $f(x)$ associated with an input x is unique. To appreciate the importance of this uniqueness property, consider a rule that associates with each item x in a department store its selling price y. Then, each x must correspond to *one and only one* y. Notice, however, that different x's may be associated with the same y. In the context of the present example, this says that different items may have the same price.
2. Although the sets A and B that appear in the definition of a function may be quite arbitrary, in this book they will denote sets of real numbers. ■

An example of a function may be taken from the familiar relationship between the area of a circle and its radius. Letting x and y denote the radius and area of a circle, respectively, we have, from elementary geometry,

$$y = \pi x^2 \tag{1}$$

Equation (1) defines y as a function of x since for each admissible value of x (that is, for each nonnegative number representing the radius of a certain circle) there corresponds precisely one number $y = \pi x^2$ that gives the area of the circle. The rule defining this "area function" may be written as

$$f(x) = \pi x^2 \tag{2}$$

To compute the area of a circle of radius 5 inches, we simply replace x in Equation (2) with the number 5. Thus, the area of the circle is

$$f(5) = \pi 5^2 = 25\pi$$

or 25π square inches.

In general, to evaluate a function at a specific value of x, we replace x with that value, as illustrated in Examples 1 and 2.

EXAMPLE 1 Let the function f be defined by the rule $f(x) = 2x^2 - x + 1$. Find:

a. $f(1)$ **b.** $f(-2)$ **c.** $f(a)$ **d.** $f(a + h)$

Solution

a. $f(1) = 2(1)^2 - (1) + 1 = 2 - 1 + 1 = 2$
b. $f(-2) = 2(-2)^2 - (-2) + 1 = 8 + 2 + 1 = 11$
c. $f(a) = 2(a)^2 - (a) + 1 = 2a^2 - a + 1$
d. $f(a + h) = 2(a + h)^2 - (a + h) + 1 = 2a^2 + 4ah + 2h^2 - a - h + 1$ ■

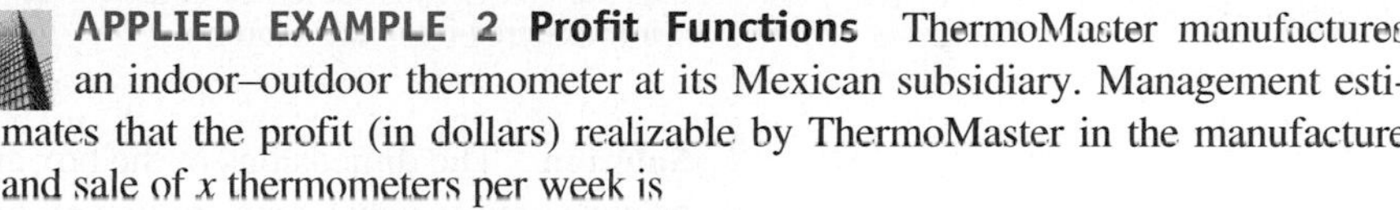

APPLIED EXAMPLE 2 Profit Functions ThermoMaster manufactures an indoor–outdoor thermometer at its Mexican subsidiary. Management estimates that the profit (in dollars) realizable by ThermoMaster in the manufacture and sale of x thermometers per week is

$$P(x) = -0.001x^2 + 8x - 5000$$

Find ThermoMaster's weekly profit if its level of production is (a) 1000 thermometers per week and (b) 2000 thermometers per week.

Solution

a. The weekly profit when the level of production is 1000 units per week is found by evaluating the profit function P at $x = 1000$. Thus,

$$P(1000) = -0.001(1000)^2 + 8(1000) - 5000 = 2000$$

or \$2000.

b. When the level of production is 2000 units per week, the weekly profit is given by

$$P(2000) = -0.001(2000)^2 + 8(2000) - 5000 = 7000$$

or \$7000. ■

Determining the Domain of a Function

Suppose we are given the function $y = f(x)$.* Then, the variable x is called the **independent variable**. The variable y, whose value depends on x, is called the **dependent variable**.

---

*It is customary to refer to a function f as $f(x)$ or by the equation $y = f(x)$ defining it.

To determine the domain of a function, we need to find what restrictions, if any, are to be placed on the independent variable x. In many practical applications, the domain of a function is dictated by the nature of the problem, as illustrated in Example 3.

APPLIED EXAMPLE 3 Packaging An open box is to be made from a rectangular piece of cardboard 16 inches long and 10 inches wide by cutting away identical squares (x inches by x inches) from each corner and folding up the resulting flaps (Figure 3). Find an expression that gives the volume V of the box as a function of x. What is the domain of the function?

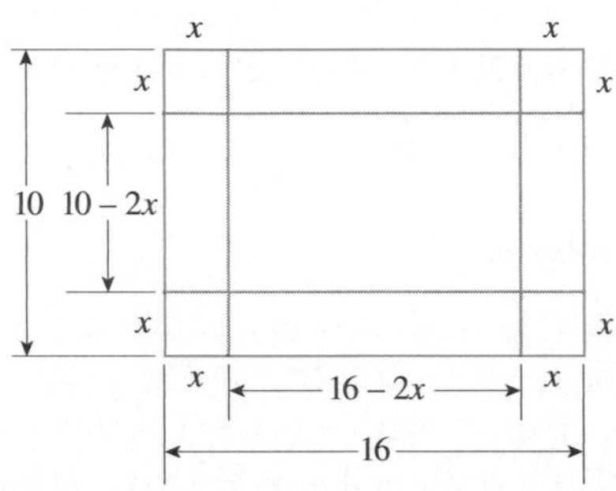

(a) The box is constructed by cutting x-by-x-inch squares from each corner.

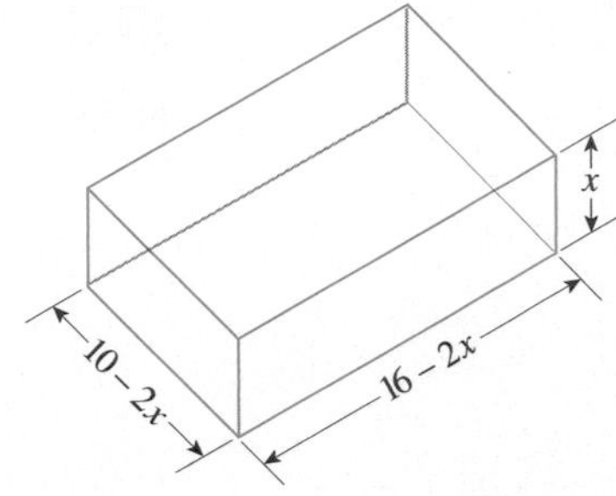

(b) The dimensions of the resulting box are $(10 - 2x)''$ by $(16 - 2x)''$ by x''.

FIGURE 3

Solution The dimensions of the box are $(16 - 2x)$ inches long, $(10 - 2x)$ inches wide, and x inches high, so its volume (in cubic inches) is given by

$$\begin{aligned} V = f(x) &= (16 - 2x)(10 - 2x)x \qquad \text{Length} \cdot \text{width} \cdot \text{height} \\ &= (160 - 52x + 4x^2)x \\ &= 4x^3 - 52x^2 + 160x \end{aligned}$$

Since the length of each side of the box must be greater than or equal to zero, we see that

$$16 - 2x \geq 0 \qquad 10 - 2x \geq 0 \qquad x \geq 0$$

simultaneously; that is,

$$x \leq 8 \qquad x \leq 5 \qquad x \geq 0$$

All three inequalities are satisfied simultaneously provided that $0 \leq x \leq 5$. Thus, the domain of the function f is the interval $[0, 5]$. ■

In general, if a function is defined by a rule relating x to $f(x)$ without specific mention of its domain, it is understood that the domain will consist of all values of x for which $f(x)$ is a real number. In this connection, you should keep in mind that (1) division by zero is not permitted and (2) the square root of a negative number is not defined.

EXAMPLE 4 Find the domain of each function.

a. $f(x) = \sqrt{x - 1}$ **b.** $f(x) = \dfrac{1}{x^2 - 4}$ **c.** $f(x) = x^2 + 3$

Solution

a. Since the square root of a negative number is undefined, it is necessary that $x - 1 \geq 0$. The inequality is satisfied by the set of real numbers $x \geq 1$. Thus, the domain of f is the interval $[1, \infty)$.

b. The only restriction on x is that $x^2 - 4$ be different from zero since division by zero is not allowed. But $(x^2 - 4) = (x + 2)(x - 2) = 0$ if $x = -2$ or $x = 2$. Thus, the domain of f in this case consists of the intervals $(-\infty, -2)$, $(-2, 2)$, and $(2, \infty)$.

c. Here, any real number satisfies the equation, so the domain of f is the set of all real numbers. ■

Graphs of Functions

If f is a function with domain A, then corresponding to each real number x in A there is precisely one real number $f(x)$. We can also express this fact by using **ordered pairs** of real numbers. Write each number x in A as the first member of an ordered pair and each number $f(x)$ corresponding to x as the second member of the ordered pair. This gives exactly one ordered pair $(x, f(x))$ for each x in A. This observation leads to an **alternative definition of a function** f:

Function (Alternative Definition)
A function f with domain A is the set of all ordered pairs $(x, f(x))$ where x belongs to A.

Observe that the condition that there be one and only one number $f(x)$ corresponding to each number x in A translates into the requirement that *no two ordered pairs have the same first number.*

Since ordered pairs of real numbers correspond to points in the plane, we have found a way to exhibit a function graphically.

Graph of a Function of One Variable
The **graph of a function** f is the set of all points (x, y) in the xy-plane such that x is in the domain of f and $y = f(x)$.

Figure 4 shows the graph of a function f. Observe that the y-coordinate of the point (x, y) on the graph of f gives the height of that point (the distance above the x-axis), if $f(x)$ is positive. If $f(x)$ is negative, then $-f(x)$ gives the depth of the point (x, y) (the distance below the x-axis). Also, observe that the domain of f is a set of real numbers lying on the x-axis, whereas the range of f lies on the y-axis.

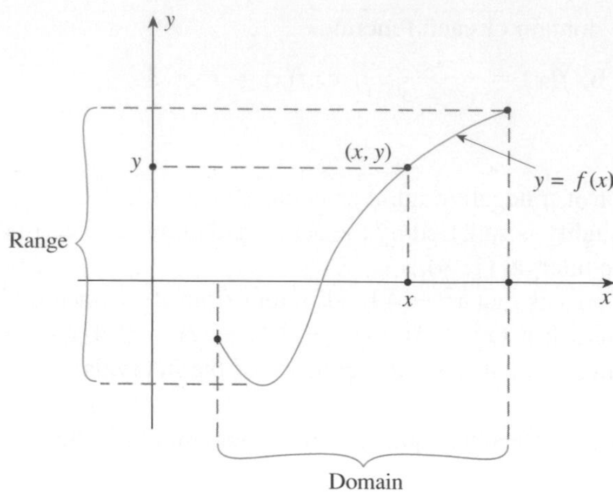

FIGURE 4
The graph of f

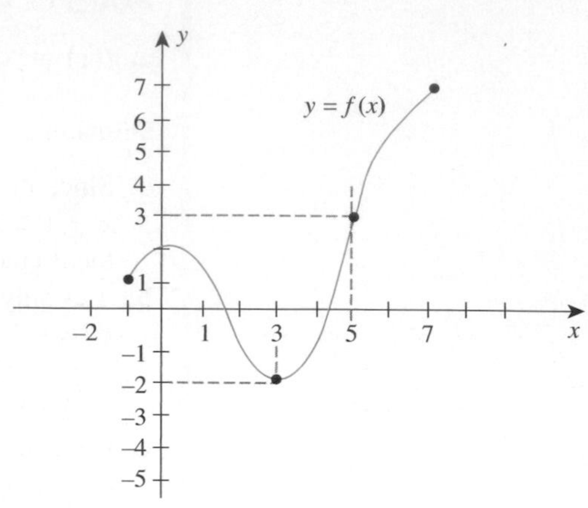

FIGURE 5

EXAMPLE 5 The graph of a function f is shown in Figure 5.

a. What is the value of $f(3)$? The value of $f(5)$?
b. What is the height or depth of the point $(3, f(3))$ from the x-axis? The point $(5, f(5))$ from the x-axis?
c. What is the domain of f? The range of f?

Solution

a. From the graph of f, we see that $y = -2$ when $x = 3$ and conclude that $f(3) = -2$. Similarly, we see that $f(5) = 3$.
b. Since the point $(3, -2)$ lies below the x-axis, we see that the depth of the point $(3, f(3))$ is $-f(3) = -(-2) = 2$ units below the x-axis. The point $(5, f(5))$ lies above the x-axis and is located at a height of $f(5)$, or 3 units above the x-axis.
c. Observe that x may take on all values between $x = -1$ and $x = 7$, inclusive, and so the domain of f is $[-1, 7]$. Next, observe that as x takes on all values in the domain of f, $f(x)$ takes on all values between -2 and 7, inclusive. (You can easily see this by running your index finger along the x-axis from $x = -1$ to $x = 7$ and observing the corresponding values assumed by the y-coordinate of each point of the graph of f.) Therefore, the range of f is $[-2, 7]$. ■

Much information about the graph of a function can be gained by plotting a few points on its graph. Later on we will develop more systematic and sophisticated techniques for graphing functions.

EXAMPLE 6 Sketch the graph of the function defined by the equation $y = x^2 + 1$. What is the range of f?

Solution The domain of the function is the set of all real numbers. By assigning several values to the variable x and computing the corresponding values for y, we obtain the following solutions to the equation $y = x^2 + 1$:

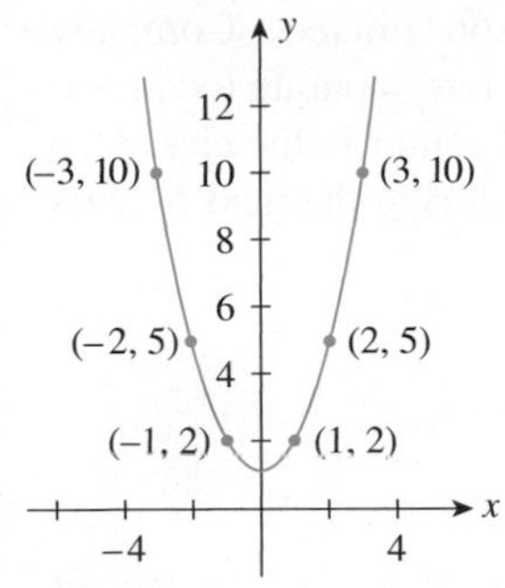

FIGURE 6
The graph of $y = x^2 + 1$ is a parabola.

x	-3	-2	-1	0	1	2	3
y	10	5	2	1	2	5	10

By plotting these points and then connecting them with a smooth curve, we obtain the graph of $y = f(x)$, which is a parabola (Figure 6). To determine the range of f, we observe that $x^2 \geq 0$, if x is any real number and so $x^2 + 1 \geq 1$ for all real numbers x. We conclude that the range of f is $[1, \infty)$. The graph of f confirms this result visually. ■

EXPLORING WITH TECHNOLOGY

Let $f(x) = x^2$.

1. Plot the graphs of $F(x) = x^2 + c$ on the same set of axes for $c = -2, -1, -\frac{1}{2}, 0, \frac{1}{2}, 1, 2$.
2. Plot the graphs of $G(x) = (x + c)^2$ on the same set of axes for $c = -2, -1, -\frac{1}{2}, 0, \frac{1}{2}, 1, 2$.
3. Plot the graphs of $H(x) = cx^2$ on the same set of axes for $c = -2, -1, -\frac{1}{2}, -\frac{1}{4}, 0, \frac{1}{4}, \frac{1}{2}, 1, 2$.
4. Study the family of graphs in parts 1–3 and describe the relationship between the graph of a function f and the graphs of the functions defined by (a) $y = f(x) + c$, (b) $y = f(x + c)$, and (c) $y = cf(x)$, where c is a constant.

A function that is defined by more than one rule is called a **piecewise-defined function**.

EXAMPLE 7 Sketch the graph of the function f defined by

$$f(x) = \begin{cases} -x & \text{if } x < 0 \\ \sqrt{x} & \text{if } x \geq 0 \end{cases}$$

Solution The function f is defined in a piecewise fashion on the set of all real numbers. In the subdomain $(-\infty, 0)$, the rule for f is given by $f(x) = -x$. The equation $y = -x$ is a linear equation in the slope-intercept form (with slope -1 and intercept 0). Therefore, the graph of f corresponding to the subdomain $(-\infty, 0)$ is the half line shown in Figure 7. Next, in the subdomain $[0, \infty)$, the rule for f is given by $f(x) = \sqrt{x}$. The values of $f(x)$ corresponding to $x = 0, 1, 2, 3, 4, 9$, and 16 are shown in the following table:

x	0	1	2	3	4	9	16
$f(x)$	0	1	$\sqrt{2}$	$\sqrt{3}$	2	3	4

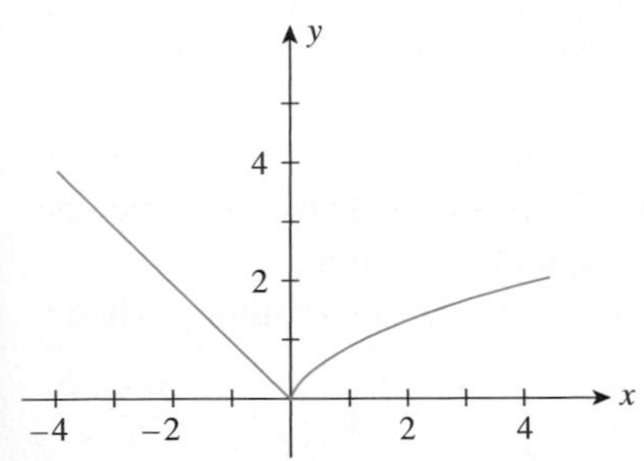

FIGURE 7
The graph of $y = f(x)$ is obtained by graphing $y = -x$ over $(-\infty, 0)$ and $y = \sqrt{x}$ over $[0, \infty)$.

Using these values, we sketch the graph of the function f as shown in Figure 7. ■

APPLIED EXAMPLE 8 Bank Deposits Madison Finance Company plans to open two branch offices 2 years from now in two separate locations: an industrial complex and a newly developed commercial center in the city. As a result of these expansion plans, Madison's total deposits during the next 5 years are expected to grow in accordance with the rule

$$f(x) = \begin{cases} \sqrt{2x} + 20 & \text{if } 0 \le x \le 2 \\ \frac{1}{2}x^2 + 20 & \text{if } 2 < x \le 5 \end{cases}$$

where $y = f(x)$ gives the total amount of money (in millions of dollars) on deposit with Madison in year x ($x = 0$ corresponds to the present). Sketch the graph of the function f.

Solution The function f is defined in a piecewise fashion on the interval [0, 5]. In the subdomain [0, 2], the rule for f is given by $f(x) = \sqrt{2x} + 20$. The values of $f(x)$ corresponding to $x = 0$, 1, and 2 may be tabulated as follows:

x	0	1	2
$f(x)$	20	21.4	22

Next, in the subdomain (2, 5], the rule for f is given by $f(x) = \frac{1}{2}x^2 + 20$. The values of $f(x)$ corresponding to $x = 3$, 4, and 5 are shown in the following table:

x	3	4	5
$f(x)$	24.5	28	32.5

Using the values of $f(x)$ in this table, we sketch the graph of the function f as shown in Figure 8. ■

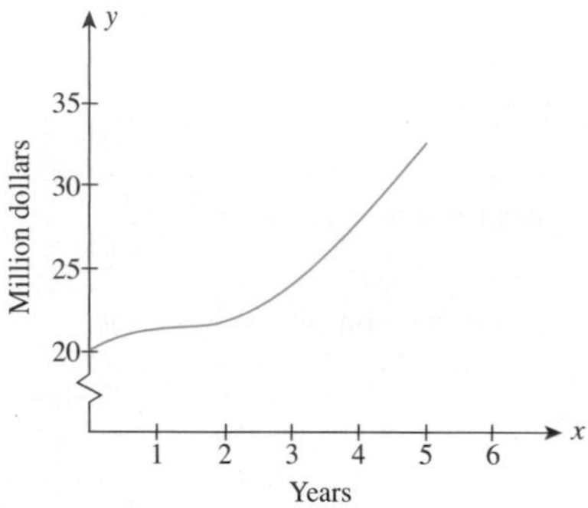

FIGURE 8
We obtain the graph of the function $y = f(x)$ by graphing $y = \sqrt{2x} + 20$ over [0, 2] and $y = \frac{1}{2}x^2 + 20$ over (2, 5].

The Vertical-Line Test

Although it is true that every function f of a variable x has a graph in the xy-plane, it is not true that every curve in the xy-plane is the graph of a function. For example, consider the curve depicted in Figure 9. This is the graph of the equation $y^2 = x$. In general, the **graph of an equation** is the set of all ordered pairs (x, y) that satisfy the given equation. Observe that the points $(9, -3)$ and $(9, 3)$ both lie on the curve. This implies that the number $x = 9$ is associated with *two* numbers: $y = -3$ and $y = 3$. But this clearly violates the uniqueness property of a function. Thus, we conclude that the curve under consideration cannot be the graph of a function.

This example suggests the following **vertical-line test** for determining when a curve is the graph of a function.

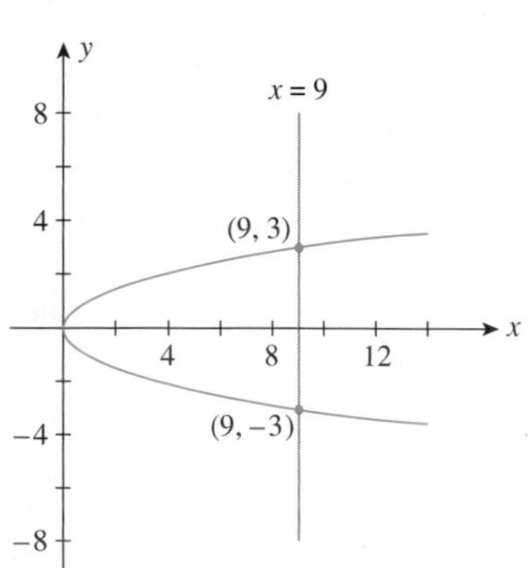

FIGURE 9
Since a vertical line passes through the curve at more than one point, we deduce that it is *not* the graph of a function.

Vertical-Line Test

A curve in the xy-plane is the graph of a function $y = f(x)$ if and only if each vertical line intersects it in at most one point.

EXAMPLE 9 Determine which of the curves shown in Figure 10 are the graphs of functions of x.

Solution The curves depicted in Figure 10a, c, and d are graphs of functions because each curve satisfies the requirement that each vertical line intersects the curve in at most one point. Note that the vertical line shown in Figure 10c does *not* intersect the graph because the point on the x-axis through which this line passes does not lie in the domain of the function. The curve depicted in Figure 10b is *not* the graph of a function because the vertical line shown there intersects the graph at three points.

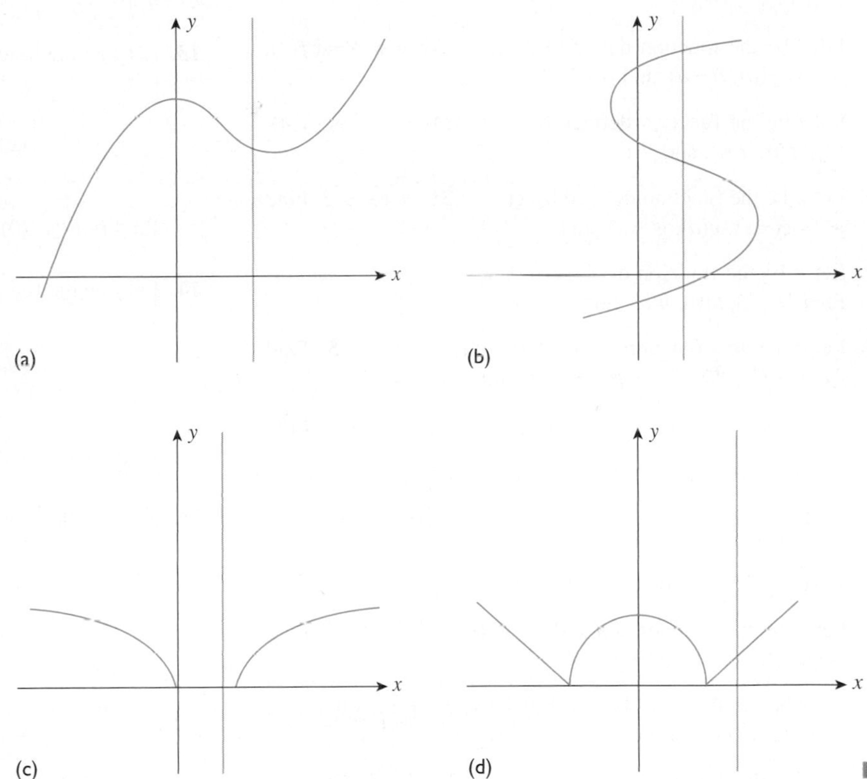

FIGURE 10
The vertical-line test can be used to determine which of these curves are graphs of functions.

2.1 Self-Check Exercises

1. Let f be the function defined by

$$f(x) = \frac{\sqrt{x+1}}{x}$$

   a. Find the domain of f.
   b. Compute $f(3)$.
   c. Compute $f(a + h)$.

2. Statistics show that more and more motorists are pumping their own gas. The following function gives self-serve sales as a percent of all U.S. gas sales:

$$f(t) = \begin{cases} 6t + 17 & \text{if } 0 \le t \le 6 \\ 15.98(t - 6)^{1/4} + 53 & \text{if } 6 < t \le 20 \end{cases}$$

   Here t is measured in years, with $t = 0$ corresponding to the beginning of 1974.

   a. Sketch the graph of the function f.
   b. What percent of all gas sales at the beginning of 1978 were self-serve? At the beginning of 1994?

   Source: Amoco Corporation

3. Let $f(x) = \sqrt{2x + 1} + 2$. Determine whether the point (4, 6) lies on the graph of f.

Solutions to Self-Check Exercises 2.1 can be found on page 63.

2.1 Concept Questions

1. a. What is a function?
   b. What is the domain of a function? The range of a function?
   c. What is an independent variable? A dependent variable?

2. a. What is the graph of a function? Use a drawing to illustrate the graph, the domain, and the range of a function.
   b. If you are given a curve in the xy-plane, how can you tell if the graph is that of a function f defined by $y = f(x)$?

2.1 Exercises

1. Let f be the function defined by $f(x) = 5x + 6$. Find $f(3)$, $f(-3)$, $f(a)$, $f(-a)$, and $f(a + 3)$.

2. Let f be the function defined by $f(x) = 4x - 3$. Find $f(4)$, $f(\frac{1}{4})$, $f(0)$, $f(a)$, and $f(a + 1)$.

3. Let g be the function defined by $g(x) = 3x^2 - 6x - 3$. Find $g(0)$, $g(-1)$, $g(a)$, $g(-a)$, and $g(x + 1)$.

4. Let h be the function defined by $h(x) = x^3 - x^2 + x + 1$. Find $h(-5)$, $h(0)$, $h(a)$, and $h(-a)$.

5. Let f be the function defined by $f(x) = 2x + 5$. Find $f(a + h)$, $f(-a)$, $f(a^2)$, $f(a - 2h)$, and $f(2a - h)$.

6. Let g be the function defined by $g(x) = -x^2 + 2x$. Find $g(a + h)$, $g(-a)$, $g(\sqrt{a})$, $a + g(a)$, and $\dfrac{1}{g(a)}$.

7. Let s be the function defined by $s(t) = \dfrac{2t}{t^2 - 1}$. Find $s(4)$, $s(0)$, $s(a)$, $s(2 + a)$, and $s(t + 1)$.

8. Let g be the function defined by $g(u) = (3u - 2)^{3/2}$. Find $g(1)$, $g(6)$, $g(\frac{11}{3})$, and $g(u + 1)$.

9. Let f be the function defined by $f(t) = \dfrac{2t^2}{\sqrt{t - 1}}$. Find $f(2)$, $f(a)$, $f(x + 1)$, and $f(x - 1)$.

10. Let f be the function defined by $f(x) = 2 + 2\sqrt{5 - x}$. Find $f(-4)$, $f(1)$, $f(\frac{11}{4})$, and $f(x + 5)$.

11. Let f be the function defined by

$$f(x) = \begin{cases} x^2 + 1 & \text{if } x \le 0 \\ \sqrt{x} & \text{if } x > 0 \end{cases}$$

Find $f(-2)$, $f(0)$, and $f(1)$.

12. Let g be the function defined by

$$g(x) = \begin{cases} -\dfrac{1}{2}x + 1 & \text{if } x < 2 \\ \sqrt{x - 2} & \text{if } x \ge 2 \end{cases}$$

Find $g(-2)$, $g(0)$, $g(2)$, and $g(4)$.

13. Let f be the function defined by

$$f(x) = \begin{cases} -\dfrac{1}{2}x^2 + 3 & \text{if } x < 1 \\ 2x^2 + 1 & \text{if } x \ge 1 \end{cases}$$

Find $f(-1)$, $f(0)$, $f(1)$, and $f(2)$.

14. Let f be the function defined by

$$f(x) = \begin{cases} 2 + \sqrt{1 - x} & \text{if } x \le 1 \\ \dfrac{1}{1 - x} & \text{if } x > 1 \end{cases}$$

Find $f(0)$, $f(1)$, and $f(2)$.

15. Refer to the graph of the function f in the following figure.

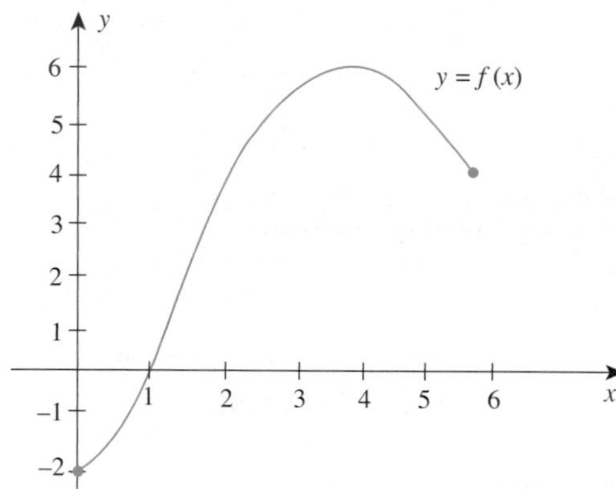

a. Find the value of $f(0)$.
b. Find the value of x for which (i) $f(x) = 3$ and (ii) $f(x) = 0$.
c. Find the domain of f.
d. Find the range of f.

16. Refer to the graph of the function f in the following figure.

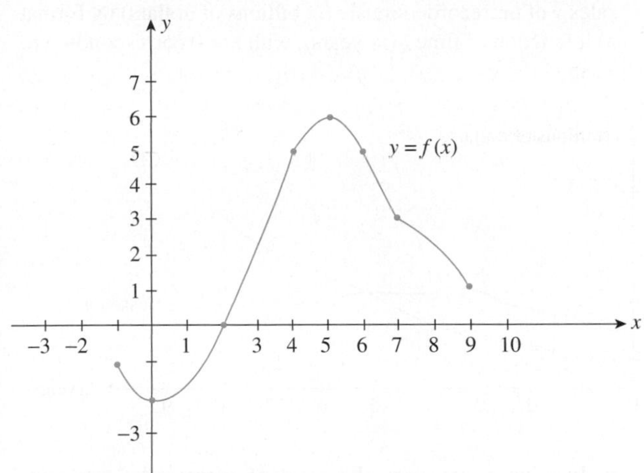

a. Find the value of $f(7)$.
b. Find the values of x corresponding to the point(s) on the graph of f located at a height of 5 units from the x-axis.
c. Find the point on the x-axis at which the graph of f crosses it. What is the value of $f(x)$ at this point?
d. Find the domain and range of f.

In Exercises 17–20, determine whether the point lies on the graph of the function.

17. $(2, \sqrt{3})$; $g(x) = \sqrt{x^2 - 1}$

18. $(3, 3)$; $f(x) = \dfrac{x + 1}{\sqrt{x^2 + 7}} + 2$

19. $(-2, -3)$; $f(t) = \dfrac{|t - 1|}{t + 1}$

20. $\left(-3, -\dfrac{1}{13}\right)$; $h(t) = \dfrac{|t + 1|}{t^3 + 1}$

In Exercises 21–34, find the domain of the function.

21. $f(x) = x^2 + 3$

22. $f(x) = 7 - x^2$

23. $f(x) = \dfrac{3x + 1}{x^2}$

24. $g(x) = \dfrac{2x + 1}{x - 1}$

25. $f(x) = \sqrt{x^2 + 1}$

26. $f(x) = \sqrt{x - 5}$

27. $f(x) = \sqrt{5 - x}$

28. $g(x) = \sqrt{2x^2 + 3}$

29. $f(x) = \dfrac{x}{x^2 - 1}$

30. $f(x) = \dfrac{1}{x^2 + x - 2}$

31. $f(x) = (x + 3)^{3/2}$

32. $g(x) = 2(x - 1)^{5/2}$

33. $f(x) = \dfrac{\sqrt{1 - x}}{x^2 - 4}$

34. $f(x) = \dfrac{\sqrt{x - 1}}{(x + 2)(x - 3)}$

35. Let f be a function defined by the rule $f(x) = x^2 - x - 6$.
a. Find the domain of f.
b. Compute $f(x)$ for $x = -3, -2, -1, 0, \frac{1}{2}, 1, 2, 3$.
c. Use the results obtained in parts (a) and (b) to sketch the graph of f.

36. Let f be a function defined by the rule $f(x) = 2x^2 + x - 3$.
a. Find the domain of f.
b. Compute $f(x)$ for $x = -3, -2, -1, -\frac{1}{2}, 0, 1, 2, 3$.
c. Use the results obtained in parts (a) and (b) to sketch the graph of f.

In Exercises 37–48, sketch the graph of the function with the given rule. Find the domain and range of the function.

37. $f(x) = 2x^2 + 1$

38. $f(x) = 9 - x^2$

39. $f(x) = 2 + \sqrt{x}$

40. $g(x) = 4 - \sqrt{x}$

41. $f(x) = \sqrt{1 - x}$

42. $f(x) = \sqrt{x - 1}$

43. $f(x) = |x| - 1$

44. $f(x) = |x| + 1$

45. $f(x) = \begin{cases} x & \text{if } x < 0 \\ 2x + 1 & \text{if } x \geq 0 \end{cases}$

46. $f(x) = \begin{cases} 4 - x & \text{if } x < 2 \\ 2x - 2 & \text{if } x \geq 2 \end{cases}$

47. $f(x) = \begin{cases} -x + 1 & \text{if } x \leq 1 \\ x^2 - 1 & \text{if } x > 1 \end{cases}$

48. $f(x) = \begin{cases} -x - 1 & \text{if } x < -1 \\ 0 & \text{if } -1 \leq x \leq 1 \\ x + 1 & \text{if } x > 1 \end{cases}$

In Exercises 49–56, use the vertical-line test to determine whether the graph represents y as a function of x.

49. **50.**

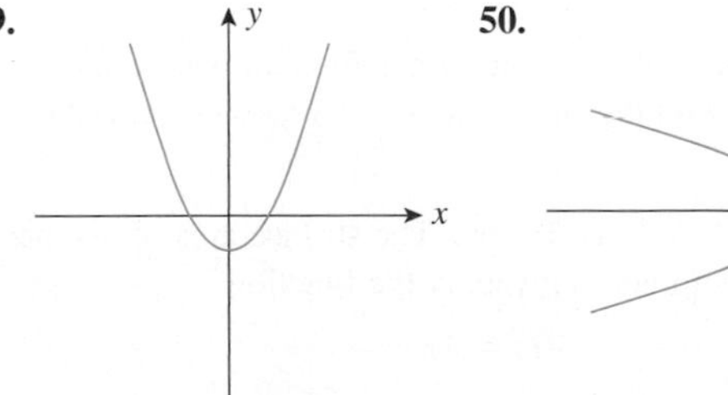

51. 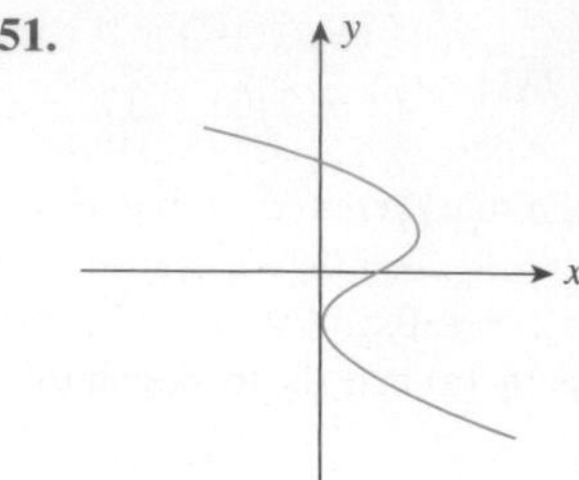

52.

53. 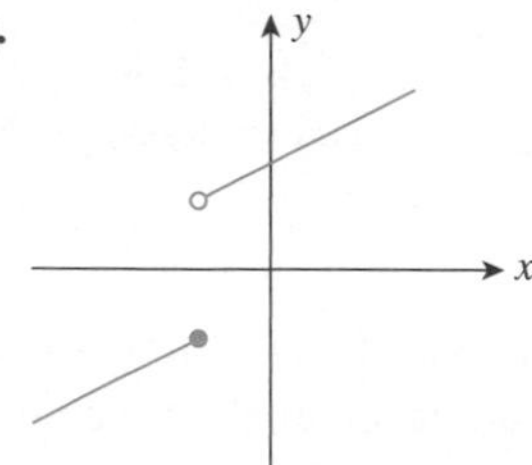

54.

55.

56.

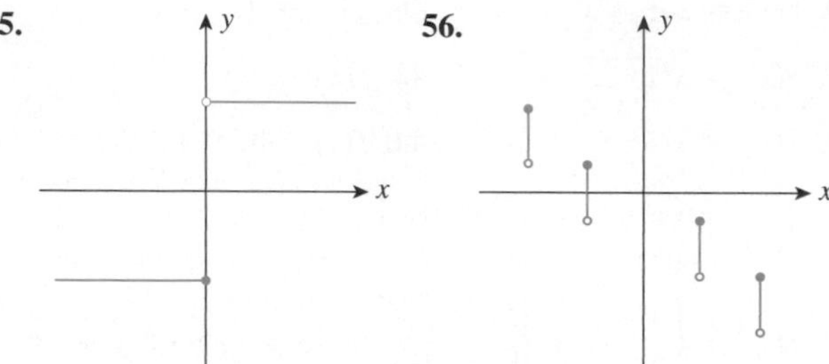

57. The circumference of a circle is given by $C(r) = 2\pi r$, where r is the radius of the circle. What is the circumference of a circle with a 5-in. radius?

58. The volume of a sphere of radius r is given by $V(r) = \frac{4}{3}\pi r^3$. Compute $V(2.1)$ and $V(2)$. What does the quantity $V(2.1) - V(2)$ measure?

59. **Growth of a Cancerous Tumor** The volume of a spherical cancerous tumor is given by the function

$$V(r) = \frac{4}{3}\pi r^3$$

where r is the radius of the tumor in centimeters. By what factor is the volume of the tumor increased if its radius is doubled?

60. **Growth of a Cancerous Tumor** The surface area of a spherical cancerous tumor is given by the function

$$S(r) = 4\pi r^2$$

where r is the radius of the tumor in centimeters. After extensive chemotherapy treatment, the surface area of the tumor is reduced by 75%. What is the radius of the tumor after treatment?

61. **Sales of Prerecorded Music** The following graphs show the sales y of prerecorded music (in billions of dollars) by format as a function of time t (in years), with $t = 0$ corresponding to 1985.

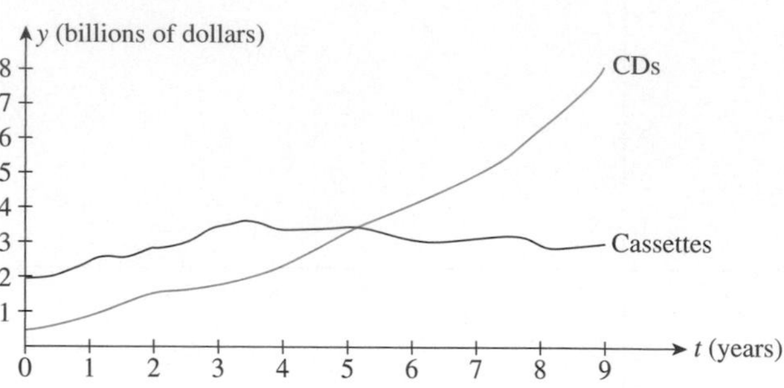

a. In what years were the sales of prerecorded cassettes greater than those of prerecorded CDs?
b. In what years were the sales of prerecorded CDs greater than those of prerecorded cassettes?
c. In what year were the sales of prerecorded cassettes the same as those of prerecorded CDs? Estimate the level of sales in each format at that time.

Source: Recording Industry Association of America

62. **The Gender Gap** The following graph shows the ratio of women's earnings to men's from 1960 through 2000.

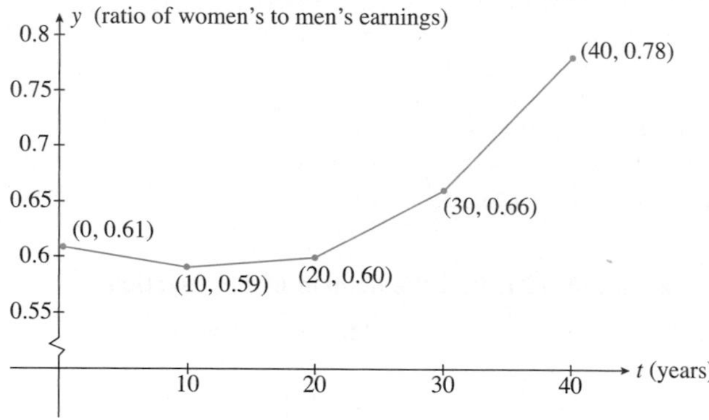

a. Write the rule for the function f giving the ratio of women's earnings to men's in year t, with $t = 0$ corresponding to 1960.
Hint: The function f is defined piecewise and is linear over each of four subintervals.
b. In what decade(s) was the gender gap expanding? Shrinking?
c. Refer to part (b). How fast was the gender gap (the ratio/year) expanding or shrinking in each of these decades?

Source: U.S. Bureau of Labor Statistics

63. **Closing the Gender Gap in Education** The following graph shows the ratio of bachelor's degrees earned by women to men from 1960 through 1990.

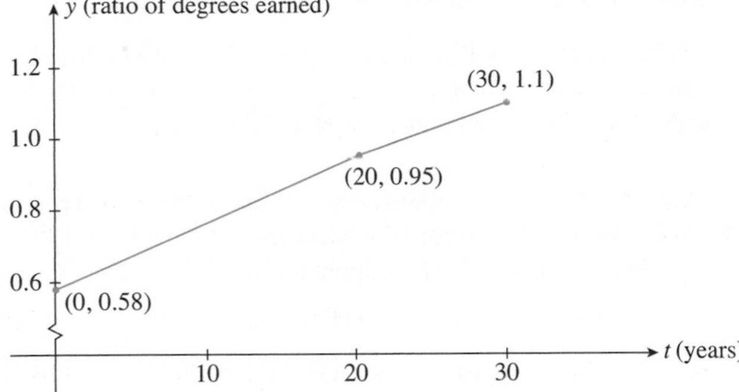

a. Write the rule for the function f giving the ratio of bachelor's degrees earned by women to men in year t, with $t = 0$ corresponding to 1960.
Hint: The function f is defined piecewise and is linear over each of two subintervals.

b. How fast was the ratio changing in the period from 1960 to 1980? From 1980 to 1990?

c. In what year (approximately) was the number of bachelor's degrees earned by women equal for the first time to that earned by men?

Source: Department of Education

64. **Consumption Function** The consumption function in a certain economy is given by the equation

$$C(y) = 0.75y + 6$$

where $C(y)$ is the personal consumption expenditure, y is the disposable personal income, and both $C(y)$ and y are measured in billions of dollars. Find $C(0)$, $C(50)$, and $C(100)$.

65. **Sales Taxes** In a certain state, the sales tax T on the amount of taxable goods is 6% of the value of the goods purchased (x), where both T and x are measured in dollars.

a. Express T as a function of x.

b. Find $T(200)$ and $T(5.65)$.

66. **Surface Area of a Single-Celled Organism** The surface area S of a single-celled organism may be found by multiplying 4π times the square of the radius r of the cell. Express S as a function of r.

67. **Friend's Rule** Friend's rule, a method for calculating pediatric drug dosages, is based on a child's age. If a denotes the adult dosage (in milligrams) and if t is the age of the child (in years), then the child's dosage is given by

$$D(t) = \frac{2}{25}ta$$

If the adult dose of a substance is 500 mg, how much should a 4-yr-old child receive?

68. **COLAs** Social Security recipients receive an automatic cost-of-living adjustment (COLA) once each year. Their monthly benefit is increased by the amount that consumer prices increased during the preceding year. Suppose that consumer prices increased by 5.3% during the preceding year.

a. Express the adjusted monthly benefit of a Social Security recipient as a function of his or her current monthly benefit.

b. If Harrington's monthly Social Security benefit is now \$620, what will be his adjusted monthly benefit?

69. **Broadband Internet Households** The number of U.S. broadband Internet households stood at 20 million at the beginning of 2002 and is projected to grow at the rate of 7.5 million households per year for the next 6 yr.

a. Find a function $f(t)$ giving the projected U.S. broadband Internet households (in millions) in year t, where $t = 0$ corresponds to the beginning of 2002.
Hint: The graph of f is a straight line.

b. What is the projected size of U.S. broadband Internet households at the beginning of 2008?

Source: Strategy Analytics Inc.

70. **Cost of Renting a Truck** Ace Truck leases its 10-ft box truck at \$30/day and \$.45/mi, whereas Acme Truck leases a similar truck at \$25/day and \$.50/mi.

a. Find the daily cost of leasing from each company as a function of the number of miles driven.

b. Sketch the graphs of the two functions on the same set of axes.

c. Which company should a customer rent a truck from for 1 day if she plans to drive at most 70 mi and wishes to minimize her cost?

71. **Linear Depreciation** A new machine was purchased by National Textile for \$120,000. For income tax purposes, the machine is depreciated linearly over 10 yr; that is, the book value of the machine decreases at a constant rate, so that at the end of 10 yr the book value is zero.

a. Express the book value of the machine (V) as a function of the age, in years, of the machine (n).

b. Sketch the graph of the function in part (a).

c. Find the book value of the machine at the end of the sixth year.

d. Find the rate at which the machine is being depreciated each year.

72. **Linear Depreciation** Refer to Exercise 71. An office building worth \$1 million when completed in 1986 was depreciated linearly over 50 yr. What was the book value of the building in 2001? What will be the book value in 2005? In 2009? (Assume that the book value of the building will be zero at the end of the 50th year.)

73. **Boyle's Law** As a consequence of Boyle's law, the pressure P of a fixed sample of gas held at a constant temperature is related to the volume V of the gas by the rule

$$P = f(V) = \frac{k}{V}$$

where k is a constant. What is the domain of the function f? Sketch the graph of the function f.

74. **Poiseuille's Law** According to a law discovered by the 19th-century physician Poiseuille, the velocity (in centimeters/second) of blood r cm from the central axis of an artery is given by

$$v(r) = k(R^2 - r^2)$$

where k is a constant and R is the radius of the artery. Suppose that for a certain artery, $k = 1000$ and $R = 0.2$ so that $v(r) = 1000(0.04 - r^2)$.

a. What is the domain of the function $v(r)$?
b. Compute $v(0)$, $v(0.1)$, and $v(0.2)$ and interpret your results.
c. Sketch the graph of the function v on the interval $[0, 0.2]$.
d. What can you say about the velocity of blood as we move away from the central axis toward the artery wall?

75. **Cancer Survivors** The number of living Americans who have had a cancer diagnosis has increased drastically since 1971. In part, this is due to more testing for cancer and better treatment for some cancers. In part, it is because the population is older, and cancer is largely a disease of the elderly. The number of cancer survivors (in millions) between 1975 ($t = 0$) and 2000 ($t = 25$) is approximately

$$N(t) = 0.0031t^2 + 0.16t + 3.6 \qquad (0 \le t \le 25)$$

a. How many living Americans had a cancer diagnosis in 1975? In 2000?
b. Assuming the trend continued, how many cancer survivors were there in 2005?

Source: National Cancer Institute

76. **Worker Efficiency** An efficiency study conducted for Elektra Electronics showed that the number of "Space Commander" walkie-talkies assembled by the average worker t hr after starting work at 8:00 a.m. is given by

$$N(t) = -t^3 + 6t^2 + 15t \qquad (0 \le t \le 4)$$

How many walkie-talkies can an average worker be expected to assemble between 8:00 and 9:00 a.m.? Between 9:00 and 10:00 a.m.?

77. **Politics** Political scientists have discovered the following empirical rule, known as the "cube rule," which gives the relationship between the proportion of seats in the House of Representatives won by Democratic candidates $s(x)$ and the proportion of popular votes x received by the Democratic presidential candidate:

$$s(x) = \frac{x^3}{x^3 + (1 - x)^3} \qquad (0 \le x \le 1)$$

Compute $s(0.6)$ and interpret your result.

78. **Prevalence of Alzheimer's Patients** Based on a study conducted in 1997, the percent of the U.S. population by age afflicted with Alzheimer's disease is given by the function

$$P(x) = 0.0726x^2 + 0.7902x + 4.9623 \qquad (0 \le x \le 25)$$

where x is measured in years, with $x = 0$ corresponding to age 65. What percent of the U.S. population at age 65 is expected to have Alzheimer's disease? At age 90?

Source: Alzheimer's Association

79. **Registered Vehicles in Massachusetts** The number of registered vehicles (in millions) in Massachusetts between 1991 ($t = 0$) and 2003 ($t = 12$) is approximately

$$N(t) = -0.0014t^3 + 0.027t^2 - 0.008t + 4.1 \qquad (0 \le t \le 12)$$

a. How many registered vehicles were there in Massachusetts in 1991?
b. How many registered vehicles where there in Massachusetts in 2003?

Source: Mass. Registry of Motor Vehicles

80. **Postal Regulations** In 2006 the postage for first-class mail was raised to 39¢ for the first ounce or fraction thereof and 24¢ for each additional ounce or fraction thereof. Any parcel not exceeding 12 oz may be sent by first-class mail. Letting x denote the weight of a parcel in ounces and $f(x)$ the postage in cents, complete the following description of the "postage function" f:

$$f(x) = \begin{cases} 39 & \text{if } 0 < x \le 1 \\ 63 & \text{if } 1 < x \le 2 \\ \cdots & \\ ? & \text{if } 11 < x \le 12 \end{cases}$$

a. What is the domain of f?
b. Sketch the graph of f.

81. **Harbor Cleanup** The amount of solids discharged from the MWRA (Massachusetts Water Resources Authority) sewage treatment plant on Deer Island (near Boston Harbor) is given by the function

$$f(t) = \begin{cases} 130 & \text{if } 0 \le t \le 1 \\ -30t + 160 & \text{if } 1 < t \le 2 \\ 100 & \text{if } 2 < t \le 4 \\ -5t^2 + 25t + 80 & \text{if } 4 < t \le 6 \\ 1.25t^2 - 26.25t + 162.5 & \text{if } 6 < t \le 10 \end{cases}$$

where $f(t)$ is measured in tons/day and t is measured in years, with $t = 0$ corresponding to 1989.

a. What amount of solids were discharged per day in 1989? In 1992? In 1996?
b. Sketch the graph of f.

Source: Metropolitan District Commission

82. **Rising Median Age** Increased longevity and the aging of the baby boom generation—those born between 1946 and 1965—are the primary reasons for a rising median age. The median age (in years) of the U.S. population from 1900 through 2000 is approximated by the function

$$f(t) = \begin{cases} 1.3t + 22.9 & \text{if } 0 \le t \le 3 \\ -0.7t^2 + 7.2t + 11.5 & \text{if } 3 < t \le 7 \\ 2.6t + 9.4 & \text{if } 7 < t \le 10 \end{cases}$$

where t is measured in decades, with $t = 0$ corresponding to the beginning of 1900.

a. What was the median age of the U.S. population at the beginning of 1900? At the beginning of 1950? At the beginning of 1990?

b. Sketch the graph of f.

Source: U.S. Census Bureau

In Exercises 83–86, determine whether the statement is true or false. If it is true, explain why it is true. If it is false, give an example to show why it is false.

83. If $a = b$, then $f(a) = f(b)$.

84. If $f(a) = f(b)$, then $a = b$.

85. If f is a function, then $f(a + b) = f(a) + f(b)$.

86. A vertical line must intersect the graph of $y = f(x)$ at exactly one point.

2.1 Solutions to Self-Check Exercises

1. a. The expression under the radical sign must be nonnegative, so $x + 1 \ge 0$ or $x \ge -1$. Also, $x \ne 0$ because division by zero is not permitted. Therefore, the domain of f is $[-1, 0) \cup (0, \infty)$.

b. $f(3) = \dfrac{\sqrt{3 + 1}}{3} = \dfrac{\sqrt{4}}{3} = \dfrac{2}{3}$

c. $f(a + h) = \dfrac{\sqrt{(a + h) + 1}}{a + h} = \dfrac{\sqrt{a + h + 1}}{a + h}$

2. a. For t in the subdomain $[0, 6]$, the rule for f is given by $f(t) = 6t + 17$. The equation $y = 6t + 17$ is a linear equation, so that portion of the graph of f is the line segment joining the points (0, 17) and (6, 53). Next, in the subdomain $(6, 20]$, the rule for f is given by $f(t) = 15.98(t - 6)^{1/4} + 53$. Using a calculator, we construct the following table of values of $f(t)$ for selected values of t.

t	6	8	10	12	14	16	18	20
$f(t)$	53	72	75.6	78	79.9	81.4	82.7	83.9

We have included $t = 6$ in the table, although it does not lie in the subdomain of the function under consideration, in order to help us obtain a better sketch of that portion of the graph of f in the subdomain $(6, 20]$. The graph of f follows:

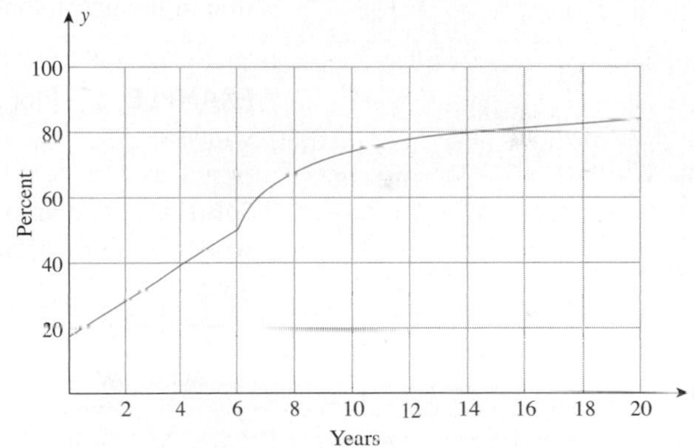

b. The percent of all self-serve gas sales at the beginning of 1978 is found by evaluating f at $t = 4$. Since this point lies in the interval $[0, 6]$, we use the rule $f(t) = 6t + 17$ and find

$$f(4) = 6(4) + 17 = 41$$

giving 41% as the required figure. The percent of all self-serve gas sales at the beginning of 1994 is given by

$$f(20) = 15.98(20 - 6)^{1/4} + 53 \approx 83.9$$

or approximately 83.9%.

3. A point (x, y) lies on the graph of the function f if and only if the coordinates satisfy the equation $y = f(x)$. Now,

$$f(4) = \sqrt{2(4) + 1} + 2 = \sqrt{9} + 2 = 5 \ne 6$$

and we conclude that the given point does *not* lie on the graph of f.

USING TECHNOLOGY

Graphing a Function

Most of the graphs of functions in this book can be plotted with the help of a graphing utility. Furthermore, a graphing utility can be used to analyze the nature of a function. However, the amount and accuracy of the information obtained using a graphing utility depend on the experience and sophistication of the user. As you progress through this book, you will see that the more knowledge of calculus you gain, the more effective the graphing utility will prove to be as a tool in problem solving.

Finding a Suitable Viewing Window

The first step in plotting the graph of a function with a graphing utility is to select a suitable viewing window. We usually do this by experimenting. For example, you might first plot the graph using the *standard viewing window* $[-10, 10]$ by $[-10, 10]$. If necessary, you then might adjust the viewing window by enlarging it or reducing it to obtain a sufficiently complete view of the graph or at least the portion of the graph that is of interest.

EXAMPLE 1 Plot the graph of $f(x) = 2x^2 - 4x - 5$ in the standard viewing window.

Solution The graph of f, shown in Figure T1a, is a parabola. From our previous work (Example 6, Section 2.1), we know that the figure does give a good view of the graph.

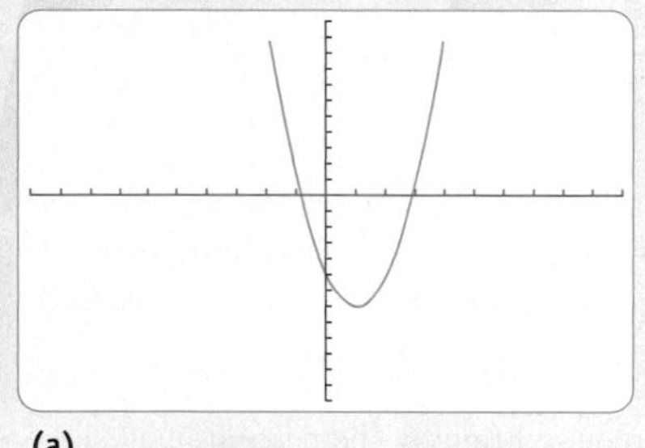

(a)

WINDOW
X_{min} = −10
X_{max} = 10
X_{scl} = 1
Y_{min} = −10
Y_{max} = 10
Y_{scl} = 1
X_{res} = 1

(b)

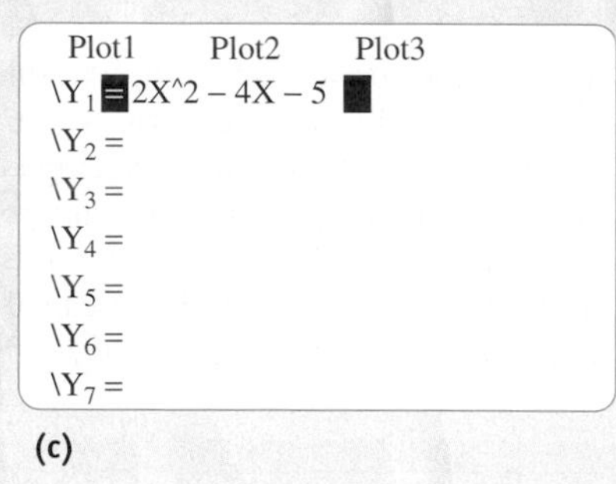

(c)

FIGURE T1
(a) The graph of $f(x) = 2x^2 - 4x - 5$ on $[-10, 10] \times [-10, 10]$; (b) the TI-83 window screen for (a); (c) the TI-83 equation screen

EXAMPLE 2 Let $f(x) = x^3(x - 3)^4$.

a. Plot the graph of f in the standard viewing window.
b. Plot the graph of f in the window $[-1, 5] \times [-40, 40]$.

Solution

a. The graph of f in the standard viewing window is shown in Figure T2a. Since the graph does not appear to be complete, we need to adjust the viewing window.

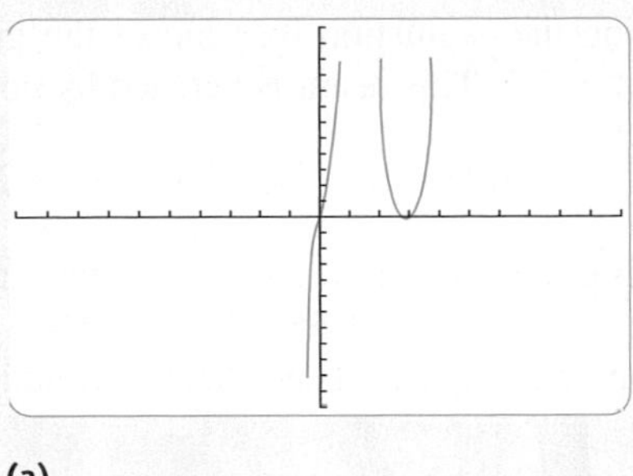

Plot1 Plot2 Plot3
\Y₁ = X^3(X − 3)^4
\Y₂ =
\Y₃ =
\Y₄ =
\Y₅ =
\Y₆ =
\Y₇ =

(a) (b)

FIGURE T2
(a) An incomplete sketch of $f(x) = x^3(x - 3)^4$ on $[-10, 10] \times [-10, 10]$; (b) the TI-83 equation screen

b. The graph of f in the window $[-1, 5] \times [-40, 40]$, shown in Figure T3a, is an improvement over the previous graph. (Later we will be able to show that the figure does in fact give a rather complete view of the graph of f.)

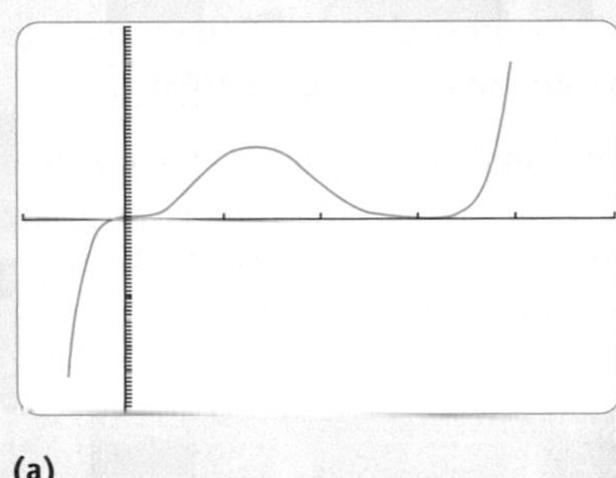

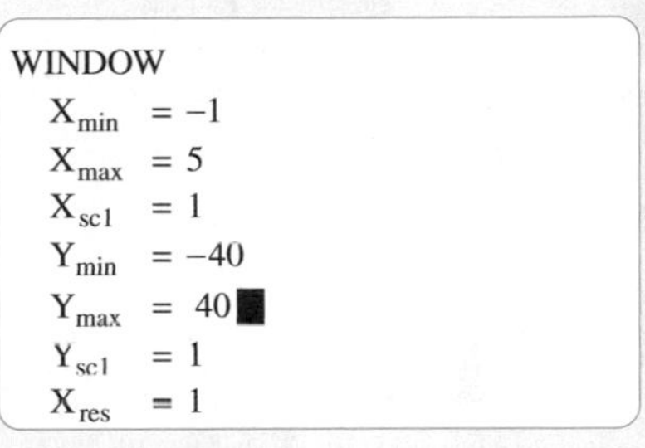

(a) (b)

FIGURE T3
(a) A complete sketch of $f(x) = x^3(x - 3)^4$ is shown using the window $[-1, 5] \times [-40, 40]$; (b) the TI-83 window screen

Evaluating a Function

A graphing utility can be used to find the value of a function with minimal effort, as the next example shows.

EXAMPLE 3 Let $f(x) = x^3 - 4x^2 + 4x + 2$.

a. Plot the graph of f in the standard viewing window.
b. Find $f(3)$ and verify your result by direct computation.
c. Find $f(4.215)$.

Solution

a. The graph of f is shown in Figure T4a.

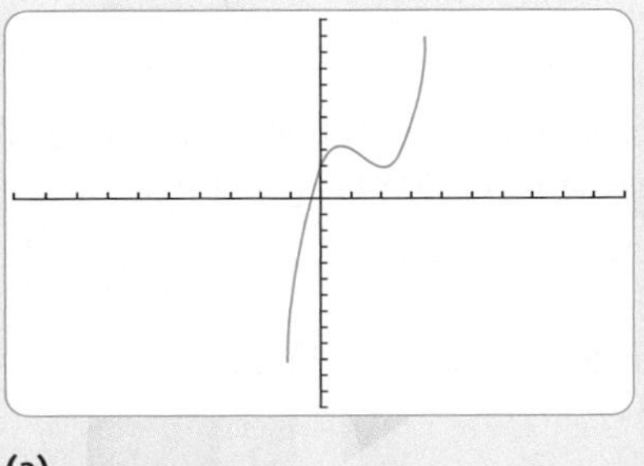

Plot1 Plot2 Plot3
\Y₁ = X^3 − 4X^2 + 4X + 2
\Y₂ =
\Y₃ =
\Y₄ =
\Y₅ =
\Y₆ =

(a) (b)

FIGURE T4
(a) The graph of $f(x) = x^3 - 4x^2 + 4x + 2$ in the standard viewing window; (b) the TI-83 equation screen

(continued)

b. Using the evaluation function of the graphing utility and the value 3 for x, we find $y = 5$. This result is verified by computing

$$f(3) = 3^3 - 4(3^2) + 4(3) + 2 = 27 - 36 + 12 + 2 = 5$$

c. Using the evaluation function of the graphing utility and the value 4.215 for x, we find $y = 22.679738375$. Thus, $f(4.215) = 22.679738375$. The efficacy of the graphing utility is clearly demonstrated here! ■

APPLIED EXAMPLE 4 Number of Alzheimer's Patients The number of Alzheimer's patients in the United States is given by

$$f(t) = -0.0277t^4 + 0.3346t^3 - 1.1261t^2 + 1.7575t + 3.7745 \qquad (0 \le t \le 6)$$

where $f(t)$ is measured in millions and t is measured in decades, with $t = 0$ corresponding to the beginning of 1990.

a. Use a graphing utility to plot the graph of f in the viewing window $[0, 7] \times [0, 12]$.
b. What is the anticipated number of Alzheimer's patients in the United States at the beginning of 2010 ($t = 2$)? At the beginning of 2030 ($t = 4$)?

Source: Alzheimer's Association

Solution

a. The graph of f in the viewing rectangle $[0, 7] \times [0, 12]$ is shown in Figure T5a.

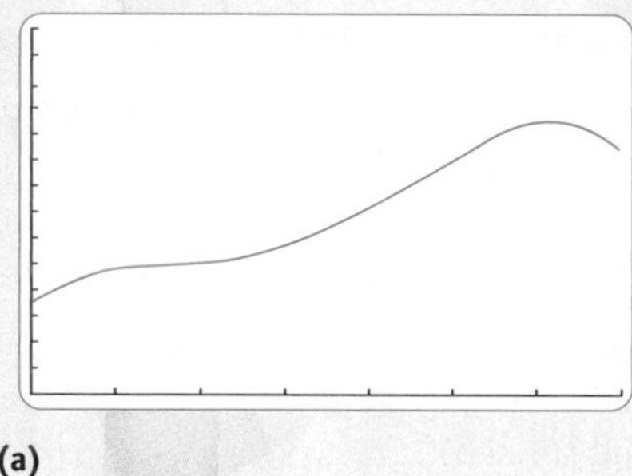

(a)

Plot1 Plot2 Plot3
\Y1 = –.0277X^4 +
.3346X^3 – 1.1261X^2 +
1.7575X + 3.7745
\Y2 =
\Y3 =
\Y4 =
\Y5 =

(b)

FIGURE T5
(a) The graph of f in the viewing window $[0, 7] \times [0, 12]$; (b) the TI-83 equation screen

b. Using the evaluation function of the graphing utility and the value 2 for x, we see that the anticipated number of Alzheimer's patients at the beginning of 2010 is given by

$$f(2) = 5.0187$$

or approximately 5 million. The anticipated number of Alzheimer's patients at the beginning of 2030 is given by

$$f(4) = 7.1101$$

or approximately 7.1 million. ■

TECHNOLOGY EXERCISES

In Exercises 1–4, plot the graph of the function f in the standard viewing window.

1. $f(x) = 2x^2 - 16x + 29$

2. $f(x) = -2.01x^3 + 1.21x^2 - 0.78x + 1$

3. $f(x) = 0.2x^4 - 2.1x^2 + 1$

4. $f(x) = \dfrac{\sqrt{x} + 1}{\sqrt{x} - 1}$

In Exercises 5–10, plot the graph of the function f in (a) the standard viewing window and (b) the indicated window.

5. $f(x) = 2x^2 - 32x + 125$; $[5, 15] \times [-5, 10]$

6. $f(x) = x^3 - 20x^2 + 8x - 10$; $[-20, 20] \times [-1200, 100]$

7. $f(x) = x^4 - 2x^2 + 8$; $[-2, 2] \times [6, 10]$

8. $f(x) = \dfrac{4}{x^2 - 8}$; $[-5, 5] \times [-5, 5]$

9. $f(x) = x\sqrt{4 - x^2}$; $[-3, 3] \times [-2, 2]$

10. $f(x) = x - 2\sqrt{x}$; $[0, 20] \times [-2, 10]$

In Exercises 11–16, plot the graph of the function f in an appropriate viewing window. (*Note:* The answer is *not* unique.)

11. $f(x) = 2x^3 - 10x^2 + 5x - 10$

12. $f(x) = -x^3 + 5x^2 - 14x + 20$

13. $f(x) = 2x^4 - 3x^3 + 5x^2 - 20x + 40$

14. $f(x) = -2x^4 + 5x^2 - 4$

15. $f(x) = \dfrac{x^3}{x^3 + 1}$

16. $f(x) = \dfrac{2x^4 - 3x}{x^2 - 1}$

In Exercises 17–20, use the evaluation function of your graphing utility to find the value of f at the given value of x. Verify your result by direct computation.

17. $f(x) = -3x^3 + 5x^2 - 2x + 8$; $x = -1$

18. $f(x) = 2x^4 - 3x^3 + 2x^2 + x - 5$; $x = 2$

19. $f(x) = \dfrac{x^4 - 3x^2}{x - 2}$; $x = 1$

20. $f(x) = \dfrac{\sqrt{x^2 - 1}}{3x + 4}$; $x = 2$

In Exercises 21–24, use the evaluation function of your graphing utility to find the value of f at the indicated value of x. Express your answer accurate to four decimal places.

21. $f(x) = 3x^3 - 2x^2 + x - 4$; $x = 2.145$

22. $f(x) = 5x^4 - 2x^2 + 8x - 3$; $x = 1.28$

23. $f(x) = \dfrac{2x^3 - 3x + 1}{3x - 2}$; $x = 2.41$

24. $f(x) = \sqrt{2x^2 + 1} + \sqrt{3x^2 - 1}$; $x = 0.62$

25. **Manufacturing Capacity** Data show that the annual increase in manufacturing capacity between 1994 and 2000 is given by

$$f(t) = 0.0094t^3 - 0.4266t^2 + 2.7489t + 5.54 \qquad (0 \le t \le 6)$$

where $f(t)$ is expressed as a percent and t is measured in years, with $t = 0$ corresponding to the beginning of 1994.

a. Plot the graph of f in the viewing window $[0, 8] \times [0, 12]$.
b. What was the annual increase in manufacturing capacity at the beginning of 1996 ($t = 2$)? At the beginning of 2000 ($t = 6$)?

Source: Federal Reserve

26. **Decline of Union Membership** The total union membership as a percent of the private workforce is given by

$$f(t) = 0.00017t^4 - 0.00921t^3 + 0.15437t^2 - 1.360723t + 16.8028 \qquad (0 \le t \le 10)$$

where t is measured in years, with $t = 0$ corresponding to the beginning of 1983.

a. Plot the graph of f in the viewing window $[0, 11] \times [8, 20]$.
b. What was the total union membership as a percent of the private force at the beginning of 1986? At the beginning of 1993?

Source: American Federation of Labor and Congress of Industrial Organizations

27. **Keeping with the Traffic Flow** By driving at a speed to match the prevailing traffic speed, you decrease the chances of an accident. According to data obtained in a university study, the number of accidents/100 million vehicle

(*continued*)

miles, y, is related to the deviation from the mean speed, x, in mph by

$$y = 1.05x^3 - 21.95x^2 + 155.9x - 327.3 \qquad (6 \le x \le 11)$$

a. Plot the graph of y in the viewing window $[6, 11] \times [20, 150]$.
b. What is the number of accidents/100 million vehicle miles if the deviation from the mean speed is 6 mph, 8 mph, and 11 mph?

Source: University of Virginia School of Engineering and Applied Science

28. SAFE DRIVERS The fatality rate in the United States (per 100 million miles traveled) by age of driver (in years) is given by the function

$$f(x) = 0.00000304x^4 - 0.0005764x^3 + 0.04105x^2 - 1.30366x + 16.579 \qquad (18 \le x \le 82)$$

a. Plot the graph of f in the viewing window $[18, 82] \times [0, 8]$.
b. What is the fatality rate for 18-yr-old drivers? For 50-yr-old drivers? For 80-yr-old drivers?

Source: National Highway Traffic Safety Administration

2.2 The Algebra of Functions

The Sum, Difference, Product, and Quotient of Functions

Let $S(t)$ and $R(t)$ denote, respectively, the federal government's spending and revenue at any time t, measured in billions of dollars. The graphs of these functions for the period between 1990 and 2000 are shown in Figure 11.

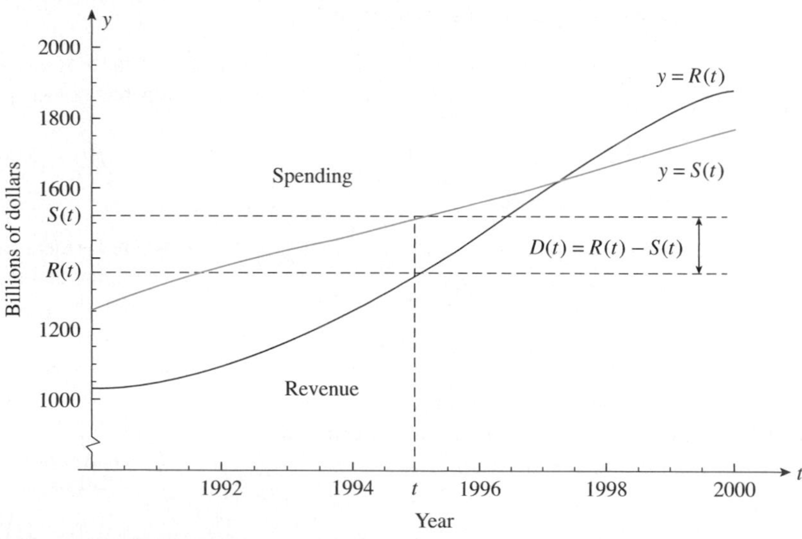

Source: Office of Management and Budget

FIGURE 11
$R(t) - S(t)$ gives the federal budget deficit (surplus) at any time t.

The difference $R(t) - S(t)$ gives the deficit (surplus) in billions of dollars at any time t if $R(t) - S(t)$ is negative (positive). This observation suggests that we can define a function D whose value at any time t is given by $R(t) - S(t)$. The function D, the *difference* of the two functions R and S, is written $D = R - S$ and may be called the "deficit (surplus) function" since it gives the budget deficit or surplus at any time t. It has the same domain as the functions S and R. The graph of the function D is shown in Figure 12.

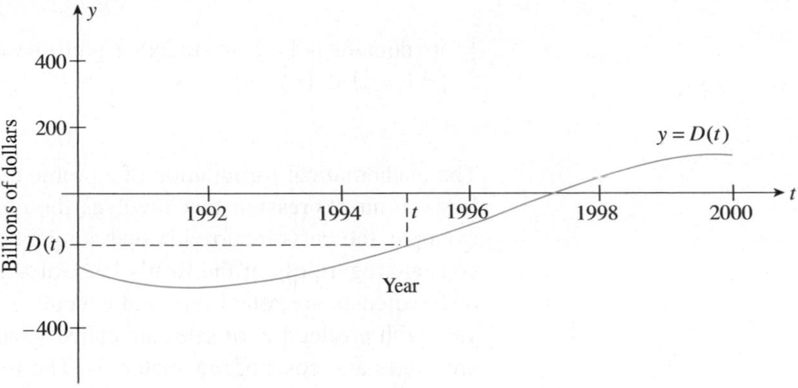

Source: Office of Management and Budget

FIGURE 12
The graph of $D(t)$

Most functions are built up from other, generally simpler functions. For example, we may view the function $f(x) = 2x + 4$ as the sum of the two functions $g(x) = 2x$ and $h(x) = 4$. The function $g(x) = 2x$ may in turn be viewed as the product of the functions $p(x) = 2$ and $q(x) = x$.

In general, given the functions f and g, we define the sum $f + g$, the difference $f - g$, the product fg, and the quotient f/g of f and g as follows.

The Sum, Difference, Product, and Quotient of Functions

Let f and g be functions with domains A and B, respectively. Then the **sum** $f + g$, **difference** $f - g$, and **product** fg of f and g are functions with domain $A \cap B$* and rule given by

$$(f + g)(x) = f(x) + g(x) \qquad \text{Sum}$$

$$(f - g)(x) = f(x) - g(x) \qquad \text{Difference}$$

$$(fg)(x) = f(x)g(x) \qquad \text{Product}$$

The **quotient** f/g of f and g has domain $A \cap B$ excluding all numbers x such that $g(x) = 0$ and rule given by

$$\left(\frac{f}{g}\right)(x) = \frac{f(x)}{g(x)} \qquad \text{Quotient}$$

*$A \cap B$ is read "A intersected with B" and denotes the set of all points common to both A and B.

EXAMPLE 1 Let $f(x) = \sqrt{x + 1}$ and $g(x) = 2x + 1$. Find the sum s, the difference d, the product p, and the quotient q of the functions f and g.

Solution Since the domain of f is $A = [-1, \infty)$ and the domain of g is $B = (-\infty, \infty)$, we see that the domain of s, d, and p is $A \cap B = [-1, \infty)$. The rules follow.

$$s(x) = (f + g)(x) = f(x) + g(x) = \sqrt{x + 1} + 2x + 1$$

$$d(x) = (f - g)(x) = f(x) - g(x) = \sqrt{x + 1} - (2x + 1) = \sqrt{x + 1} - 2x - 1$$

$$p(x) = (fg)(x) = f(x)g(x) = \sqrt{x + 1}(2x + 1) = (2x + 1)\sqrt{x + 1}$$

The quotient function q has rule

$$q(x) = \left(\frac{f}{g}\right)(x) = \frac{f(x)}{g(x)} = \frac{\sqrt{x+1}}{2x+1}$$

Its domain is $[-1, \infty)$ together with the restriction $x \neq -\frac{1}{2}$. We denote this by $[-1, -\frac{1}{2}) \cup (-\frac{1}{2}, \infty)$. ■

The mathematical formulation of a problem arising from a practical situation often leads to an expression that involves the combination of functions. Consider, for example, the costs incurred in operating a business. Costs that remain more or less constant regardless of the firm's level of activity are called **fixed costs**. Examples of fixed costs are rental fees and executive salaries. On the other hand, costs that vary with production or sales are called **variable costs**. Examples of variable costs are wages and costs of raw materials. The **total cost** of operating a business is thus given by the *sum* of the variable costs and the fixed costs, as illustrated in the next example.

APPLIED EXAMPLE 2 Cost Functions Suppose Puritron, a manufacturer of water filters, has a monthly fixed cost of \$10,000 and a variable cost of

$$-0.0001x^2 + 10x \qquad (0 \leq x \leq 40{,}000)$$

dollars, where x denotes the number of filters manufactured per month. Find a function C that gives the total monthly cost incurred by Puritron in the manufacture of x filters.

Solution Puritron's monthly fixed cost is always \$10,000, regardless of the level of production, and it is described by the constant function $F(x) = 10{,}000$. Next, the variable cost is described by the function $V(x) = -0.0001x^2 + 10x$. Since the total cost incurred by Puritron at any level of production is the sum of the variable cost and the fixed cost, we see that the required total cost function is given by

$$\begin{aligned} C(x) &= V(x) + F(x) \\ &= -0.0001x^2 + 10x + 10{,}000 \qquad (0 \leq x \leq 40{,}000) \end{aligned}$$ ■

Next, the **total profit** realized by a firm in operating a business is the *difference* between the total revenue realized and the total cost incurred; that is,

$$P(x) = R(x) - C(x)$$

APPLIED EXAMPLE 3 Profit Functions Refer to Example 2. Suppose the total revenue realized by Puritron from the sale of x water filters is given by the total revenue function

$$R(x) = -0.0005x^2 + 20x \qquad (0 \leq x \leq 40{,}000)$$

a. Find the total profit function—that is, the function that describes the total profit Puritron realizes in manufacturing and selling x water filters per month.
b. What is the profit when the level of production is 10,000 filters per month?

Solution

a. The total profit realized by Puritron in manufacturing and selling x water filters per month is the difference between the total revenue realized and the total cost incurred. Thus, the required total profit function is given by

$$\begin{aligned} P(x) &= R(x) - C(x) \\ &= (-0.0005x^2 + 20x) - (-0.0001x^2 + 10x + 10{,}000) \\ &= -0.0004x^2 + 10x - 10{,}000 \end{aligned}$$

b. The profit realized by Puritron when the level of production is 10,000 filters per month is

$$P(10{,}000) = -0.0004(10{,}000)^2 + 10(10{,}000) - 10{,}000 = 50{,}000$$

or \$50,000 per month. ■

Composition of Functions

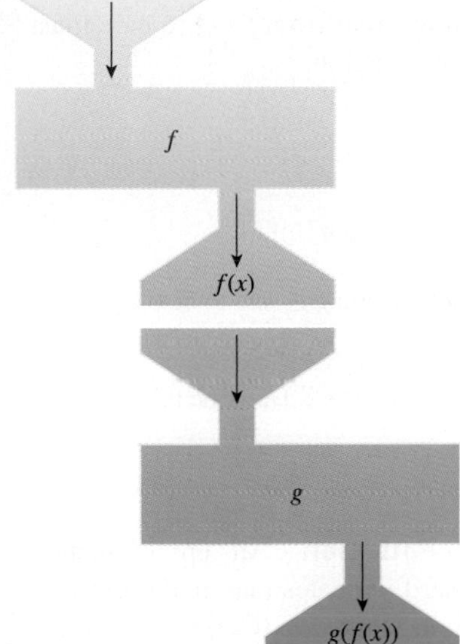

FIGURE 13
The composite function $h = g \circ f$ viewed as a machine

Another way to build up a function from other functions is through a process known as the *composition of functions*. Consider, for example, the function h, whose rule is given by $h(x) = \sqrt{x^2 - 1}$. Let f and g be functions defined by the rules $f(x) = x^2 - 1$ and $g(x) = \sqrt{x}$. Evaluating the function g at the point $f(x)$ [remember that for each real number x in the domain of f, $f(x)$ is simply a real number], we find that

$$g(f(x)) = \sqrt{f(x)} = \sqrt{x^2 - 1}$$

which is just the rule defining the function h!

In general, the composition of a function g with a function f is defined as follows.

The Composition of Two Functions

Let f and g be functions. Then the composition of g and f is the function $g \circ f$ defined by

$$(g \circ f)(x) = g(f(x))$$

The domain of $g \circ f$ is the set of all x in the domain of f such that $f(x)$ lies in the domain of g.

The function $g \circ f$ (read "g circle f") is also called a **composite function**. The interpretation of the function $h = g \circ f$ as a machine is illustrated in Figure 13 and its interpretation as a mapping is shown in Figure 14.

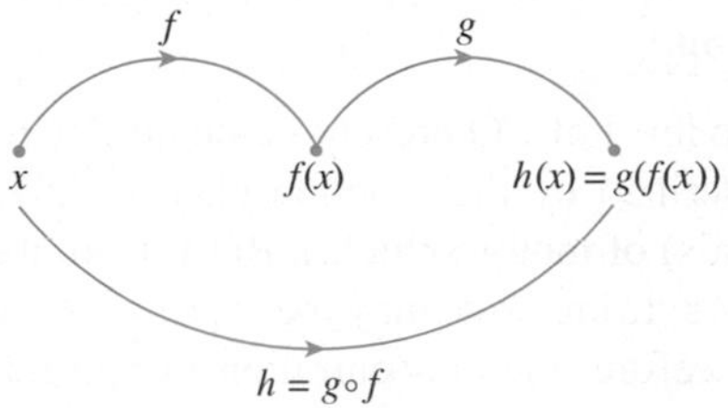

FIGURE 14
The function $h = g \circ f$ viewed as a mapping

EXAMPLE 4 Let $f(x) = x^2 - 1$ and $g(x) = \sqrt{x} + 1$. Find:

a. The rule for the composite function $g \circ f$.
b. The rule for the composite function $f \circ g$.

Solution

a. To find the rule for the composite function $g \circ f$, evaluate the function g at $f(x)$. We obtain

$$(g \circ f)(x) = g(f(x)) = \sqrt{f(x)} + 1 = \sqrt{x^2 - 1} + 1$$

b. To find the rule for the composite function $f \circ g$, evaluate the function f at $g(x)$. Thus,

$$\begin{aligned}(f \circ g)(x) &= f(g(x)) = (g(x))^2 - 1 = (\sqrt{x} + 1)^2 - 1 \\ &= x + 2\sqrt{x} + 1 - 1 = x + 2\sqrt{x}\end{aligned}$$

Example 4 reminds us that in general $g \circ f$ is different from $f \circ g$, so care must be taken when finding the rule for a composite function.

EXPLORE & DISCUSS

Let $f(x) = \sqrt{x} + 1$ for $x \geq 0$ and let $g(x) = (x - 1)^2$ for $x \geq 1$.

1. Show that $(g \circ f)(x)$ and $(f \circ g)(x) = x$. (*Note:* The function g is said to be the *inverse* of f and vice versa.)

2. Plot the graphs of f and g together with the straight line $y = x$. Describe the relationship between the graphs of f and g.

APPLIED EXAMPLE 5 Automobile Pollution An environmental impact study conducted for the city of Oxnard indicates that, under existing environmental protection laws, the level of carbon monoxide (CO) present in the air due to pollution from automobile exhaust will be $0.01x^{2/3}$ parts per million when the number of motor vehicles is x thousand. A separate study conducted by a state government agency estimates that t years from now the number of motor vehicles in Oxnard will be $0.2t^2 + 4t + 64$ thousand.

a. Find an expression for the concentration of CO in the air due to automobile exhaust t years from now.
b. What will be the level of concentration 5 years from now?

Solution

a. The level of CO present in the air due to pollution from automobile exhaust is described by the function $g(x) = 0.01x^{2/3}$, where x is the number (in thousands) of motor vehicles. But the number of motor vehicles x (in thousands) t years from now may be estimated by the rule $f(t) = 0.2t^2 + 4t + 64$. Therefore, the concentration of CO due to automobile exhaust t years from now is given by

$$C(t) = (g \circ f)(t) = g(f(t)) = 0.01(0.2t^2 + 4t + 64)^{2/3}$$

parts per million.

b. The level of concentration 5 years from now will be

$$C(5) = 0.01[0.2(5)^2 + 4(5) + 64]^{2/3}$$
$$= (0.01)89^{2/3} \approx 0.20$$

or approximately 0.20 parts per million. ■

2.2 Self-Check Exercises

1. Let f and g be functions defined by the rules

$$f(x) = \sqrt{x} + 1 \quad \text{and} \quad g(x) = \frac{x}{1 + x}$$

respectively. Find the rules for

a. The sum s, the difference d, the product p, and the quotient q of f and g.

b. The composite functions $f \circ g$ and $g \circ f$.

2. Health-care spending per person by the private sector includes payments by individuals, corporations, and their insurance companies and is approximated by the function

$$f(t) = 2.48t^2 + 18.47t + 509 \qquad (0 \leq t \leq 6)$$

where $f(t)$ is measured in dollars and t is measured in years, with $t = 0$ corresponding to the beginning of 1994. The corresponding government spending—including expenditures for Medicaid, Medicare, and other federal, state, and local government public health care—is

$$g(t) = -1.12t^2 + 29.09t + 429 \qquad (0 \leq t \leq 6)$$

where t has the same meaning as before.

a. Find a function that gives the difference between private and government health-care spending per person at any time t.

b. What was the difference between private and government expenditures per person at the beginning of 1995? At the beginning of 2000?

Source: Health Care Financing Administration

Solutions to Self-Check Exercises 2.2 can be found on page 76.

2.2 Concept Questions

1. a. Explain what is meant by the sum, difference, product, and quotient of the functions f and g with domains A and B, respectively.

b. If $f(2) = 3$ and $g(2) = -2$, what is $(f + g)(2)$? $(f - g)(2)$? $(fg)(2)$? $(f/g)(2)$?

2. a. What is the composition of the functions f and g? The functions g and f?

b. If $f(2) = 3$ and $g(3) = 8$, what is $(g \circ f)(2)$? Can you conclude from the given information what $(f \circ g)(3)$ is? Explain.

2.2 Exercises

In Exercises 1–8, let $f(x) = x^3 + 5$, $g(x) = x^2 - 2$, and $h(x) = 2x + 4$. Find the rule for each function.

1. $f + g$ **2.** $f - g$ **3.** fg **4.** gf

5. $\frac{f}{g}$ **6.** $\frac{f - g}{h}$ **7.** $\frac{fg}{h}$ **8.** fgh

In Exercises 9–18, let $f(x) = x - 1$, $g(x) = \sqrt{x + 1}$, and $h(x) = 2x^3 - 1$. Find the rule for each function.

9. $f + g$ **10.** $g - f$ **11.** fg **12.** gf

13. $\frac{g}{h}$ **14.** $\frac{h}{g}$ **15.** $\frac{fg}{h}$ **16.** $\frac{fh}{g}$

17. $\frac{f - h}{g}$ **18.** $\frac{gh}{g - f}$

In Exercises 19–24, find the functions $f + g$, $f - g$, fg, and f/g.

19. $f(x) = x^2 + 5$; $g(x) = \sqrt{x} - 2$

20. $f(x) = \sqrt{x - 1}$; $g(x) = x^3 + 1$

21. $f(x) = \sqrt{x + 3}$; $g(x) = \dfrac{1}{x - 1}$

22. $f(x) = \dfrac{1}{x^2 + 1}$; $g(x) = \dfrac{1}{x^2 - 1}$

23. $f(x) = \dfrac{x + 1}{x - 1}$; $g(x) = \dfrac{x + 2}{x - 2}$

24. $f(x) = x^2 + 1$; $g(x) = \sqrt{x + 1}$

In Exercises 25–30, find the rules for the composite functions $f \circ g$ and $g \circ f$.

25. $f(x) = x^2 + x + 1$; $g(x) = x^2$

26. $f(x) = 3x^2 + 2x + 1$; $g(x) = x + 3$

27. $f(x) = \sqrt{x} + 1$; $g(x) = x^2 - 1$

28. $f(x) = 2\sqrt{x} + 3$; $g(x) = x^2 + 1$

29. $f(x) = \dfrac{x}{x^2 + 1}$; $g(x) = \dfrac{1}{x}$

30. $f(x) = \sqrt{x + 1}$; $g(x) = \dfrac{1}{x - 1}$

In Exercises 31–34, evaluate $h(2)$, where $h = g \circ f$.

31. $f(x) = x^2 + x + 1$; $g(x) = x^2$

32. $f(x) = \sqrt[3]{x^2 - 1}$; $g(x) = 3x^3 + 1$

33. $f(x) = \dfrac{1}{2x + 1}$; $g(x) = \sqrt{x}$

34. $f(x) = \dfrac{1}{x - 1}$; $g(x) = x^2 + 1$

In Exercises 35–42, find functions f and g such that $h = g \circ f$. (*Note:* The answer is *not* unique.)

35. $h(x) = (2x^3 + x^2 + 1)^5$

36. $h(x) = (3x^2 - 4)^{-3}$

37. $h(x) = \sqrt{x^2 - 1}$

38. $h(x) = (2x - 3)^{3/2}$

39. $h(x) = \dfrac{1}{x^2 - 1}$

40. $h(x) = \dfrac{1}{\sqrt{x^2 - 4}}$

41. $h(x) = \dfrac{1}{(3x^2 + 2)^{3/2}}$

42. $h(x) = \dfrac{1}{\sqrt{2x + 1}} + \sqrt{2x + 1}$

In Exercises 43–46, find $f(a + h) - f(a)$ for each function. Simplify your answer.

43. $f(x) = 3x + 4$

44. $f(x) = -\dfrac{1}{2}x + 3$

45. $f(x) = 4 - x^2$

46. $f(x) = x^2 - 2x + 1$

In Exercises 47–52, find and simplify

$$\frac{f(a + h) - f(a)}{h} \qquad (h \neq 0)$$

for each function.

47. $f(x) = x^2 + 1$

48. $f(x) = 2x^2 - x + 1$

49. $f(x) = x^3 - x$

50. $f(x) = 2x^3 - x^2 + 1$

51. $f(x) = \dfrac{1}{x}$

52. $f(x) = \sqrt{x}$

53. RESTAURANT REVENUE Nicole owns and operates two restaurants. The revenue of the first restaurant at time t is $f(t)$ dollars, and the revenue of the second restaurant at time t is $g(t)$ dollars. What does the function $F(t) = f(t) + g(t)$ represent?

54. BIRTHRATE OF ENDANGERED SPECIES The birthrate of an endangered species of whales in year t is $f(t)$ whales/year. This species of whales is dying at the rate of $g(t)$ whales/year in year t. What does the function $F(t) = f(t) - g(t)$ represent?

55. VALUE OF AN INVESTMENT The number of IBM shares that Nancy owns is given by $f(t)$. The price per share of the stock of IBM at time t is $g(t)$ dollars. What does the function $f(t)g(t)$ represent?

56. PRODUCTION COSTS The total cost incurred by time t in the production of a certain commodity is $f(t)$ dollars. The number of products produced by time t is $g(t)$ units. What does the function $f(t)/g(t)$ represent?

57. CARBON MONOXIDE POLLUTION The number of cars running in the business district of a town at time t is given by $f(t)$. Carbon monoxide pollution coming from these cars is given by $g(x)$ parts per million, where x is the number of cars being operated in the district. What does the function $g \circ f$ represent?

58. EFFECT OF ADVERTISING ON REVENUE The revenue of Leisure Travel is given by $f(x)$ dollars, where x is the dollar amount spent by the company on advertising. The amount spent by Leisure at time t on advertising is given by $g(t)$ dollars. What does the function $f \circ g$ represent?

59. MANUFACTURING COSTS TMI, a manufacturer of blank audiocassette tapes, has a monthly fixed cost of \$12,100 and a variable cost of \$.60/tape. Find a function C that gives the total cost incurred by TMI in the manufacture of x tapes/month.

60. SPAM MESSAGES The total number of email messages per day (in billions) between 2003 and 2007 is forecast to be

$$f(t) = 1.54t^2 + 7.1t + 31.4 \qquad (0 \leq t \leq 4)$$

where t is measured in years, with $t = 0$ corresponding to 2003. Over the same period, the total number of spam messages per day (in billions) is forecast to be

$$g(t) = 1.21t^2 + 6t + 14.5 \qquad (0 \leq t \leq 4)$$

a. Find the rule for the function $D = f - g$. Compute $D(4)$ and explain what it measures.

b. Find the rule for the function $P = f/g$. Compute $P(4)$ and explain what it means.

Source: Technology Review

61. GLOBAL SUPPLY OF PLUTONIUM The global stockpile of plutonium for military applications between 1990 ($t = 0$) and 2003 ($t = 13$) stood at a constant 267 tons. On the other hand, the global stockpile of plutonium for civilian use was

$$2t^2 + 46t + 733$$

tons in year t over the same period.

a. Find the function f giving the global stockpile of plutonium for military use from 1990 through 2003 and the function g giving the global stockpile of plutonium for civilian use over the same period.

b. Find the function h giving the total global stockpile of plutonium between 1990 and 2003.

c. What was the total global stockpile of plutonium in 2003?

Source: Institute for Science and International Security

62. COST OF PRODUCING PDAS Apollo manufactures PDAs at a variable cost of

$$V(x) = 0.000003x^3 - 0.03x^2 + 200x$$

dollars, where x denotes the number of units manufactured per month. The monthly fixed cost attributable to the division that produces these PDAs is \$100,000. Find a function C that gives the total cost incurred by the manufacture of x PDAs. What is the total cost incurred in producing 2000 units/month?

63. PROFIT FROM SALE OF PDAS Refer to Exercise 62. Suppose the total revenue realized by Apollo from the sale of x PDAs is given by the total revenue function

$$R(x) = -0.1x^2 + 500x \qquad (0 \le x \le 5000)$$

where $R(x)$ is measured in dollars.

a. Find the total profit function.

b. What is the profit when 1500 units are produced and sold each month?

64. PROFIT FROM SALE OF PAGERS A division of Chapman Corporation manufactures a pager. The weekly fixed cost for the division is \$20,000, and the variable cost for producing x pagers/week is

$$V(x) = 0.000001x^3 - 0.01x^2 + 50x$$

dollars. The company realizes a revenue of

$$R(x) = -0.02x^2 + 150x \qquad (0 \le x \le 7500)$$

dollars from the sale of x pagers/week.

a. Find the total cost function.

b. Find the total profit function.

c. What is the profit for the company if 2000 units are produced and sold each week?

65. OVERCROWDING OF PRISONS The 1980s saw a trend toward old-fashioned punitive deterrence as opposed to the more liberal penal policies and community-based corrections popular in the 1960s and early 1970s. As a result, prisons became more crowded, and the gap between the number of people in prison and the prison capacity widened. The number of prisoners (in thousands) in federal and state prisons is approximated by the function

$$N(t) = 3.5t^2 + 26.7t + 436.2 \qquad (0 \le t \le 10)$$

where t is measured in years, with $t = 0$ corresponding to 1983. The number of inmates for which prisons were designed is given by

$$C(t) = 24.3t + 365 \qquad (0 \le t \le 10)$$

where $C(t)$ is measured in thousands and t has the same meaning as before.

a. Find an expression that shows the gap between the number of prisoners and the number of inmates for which the prisons were designed at any time t.

b. Find the gap at the beginning of 1983 and at the beginning of 1986.

Source: U.S. Department of Justice

66. EFFECT OF MORTGAGE RATES ON HOUSING STARTS A study prepared for the National Association of Realtors estimated that the number of housing starts per year over the next 5 yr will be

$$N(r) = \frac{7}{1 + 0.02r^2}$$

million units, where r (percent) is the mortgage rate. Suppose the mortgage rate over the next t mo is

$$r(t) = \frac{10t + 150}{t + 10} \qquad (0 \le t \le 24)$$

percent/year.

a. Find an expression for the number of housing starts per year as a function of t, t mo from now.

b. Using the result from part (a), determine the number of housing starts at present, 12 mo from now, and 18 mo from now.

67. HOTEL OCCUPANCY RATE The occupancy rate of the all-suite Wonderland Hotel, located near an amusement park, is given by the function

$$r(t) = \frac{10}{81}t^3 - \frac{10}{3}t^2 + \frac{200}{9}t + 55 \qquad (0 \le t \le 11)$$

where t is measured in months and $t = 0$ corresponds to the beginning of January. Management has estimated that the monthly revenue (in thousands of dollars) is approximated by the function

$$R(r) = -\frac{3}{5000}r^3 + \frac{9}{50}r^2 \qquad (0 \le r \le 100)$$

where r (percent) is the occupancy rate.

a. What is the hotel's occupancy rate at the beginning of January? At the beginning of June?
b. What is the hotel's monthly revenue at the beginning of January? At the beginning of June?
Hint: Compute $R(r(0))$ and $R(r(5))$.

68. **Housing Starts and Construction Jobs** The president of a major housing construction firm reports that the number of construction jobs (in millions) created is given by

$$N(x) = 1.42x$$

where x denotes the number of housing starts. Suppose the number of housing starts in the next t mo is expected to be

$$x(t) = \frac{7(t+10)^2}{(t+10)^2 + 2(t+15)^2}$$

million units/year. Find an expression for the number of jobs created per month in the next t mo. How many jobs will have been created 6 mo and 12 mo from now?

In Exercises 69–72, determine whether the statement is true or false. If it is true, explain why it is true. If it is false, give an example to show why it is false.

69. If f and g are functions with domain D, then $f + g = g + f$.

70. If $g \circ f$ is defined at $x = a$, then $f \circ g$ must also be defined at $x = a$.

71. If f and g are functions, then $f \circ g = g \circ f$.

72. If f is a function, then $f \circ f = f^2$.

2.2 Solutions to Self-Check Exercises

1. a. $s(x) = f(x) + g(x) = \sqrt{x} + 1 + \dfrac{x}{1+x}$

$$d(x) = f(x) - g(x) = \sqrt{x} + 1 - \frac{x}{1+x}$$

$$p(x) = f(x)g(x) = (\sqrt{x} + 1)\cdot\frac{x}{1+x} = \frac{x(\sqrt{x}+1)}{1+x}$$

$$q(x) = \frac{f(x)}{g(x)} = \frac{\sqrt{x}+1}{\dfrac{x}{1+x}} = \frac{(\sqrt{x}+1)(1+x)}{x}$$

b. $(f \circ g)(x) = f(g(x)) = \sqrt{\dfrac{x}{1+x}} + 1$

$$(g \circ f)(x) = g(f(x)) = \frac{\sqrt{x}+1}{1 + (\sqrt{x}+1)} = \frac{\sqrt{x}+1}{\sqrt{x}+2}$$

2. a. The difference between private and government health-care spending per person at any time t is given by the function d with the rule

$$\begin{aligned} d(t) = f(t) - g(t) &= (2.48t^2 + 18.47t + 509) \\ &\quad - (-1.12t^2 + 29.09t + 429) \\ &= 3.6t^2 - 10.62t + 80 \end{aligned}$$

b. The difference between private and government expenditures per person at the beginning of 1995 is given by

$$d(1) = 3.6(1)^2 - 10.62(1) + 80$$

or \$72.98/person.

The difference between private and government expenditures per person at the beginning of 2000 is given by

$$d(6) = 3.6(6)^2 - 10.62(6) + 80$$

or \$145.88/person.

2.3 Functions and Mathematical Models

Mathematical Models

One of the fundamental goals in this book is to show how mathematics and, in particular, calculus can be used to solve real-world problems such as those arising from the world of business and the social, life, and physical sciences. You have already seen some of these problems earlier. Here are a few more examples of real-world phenomena that we will analyze in this and ensuing chapters.

- The average U.S. credit card debt (p. 88).
- The population growth in the fastest-growing metropolitan area in the United States (p. 96).

- The increase in the revenue collected by the IRS from the alternative minimum tax (p. 281).
- The prevalence of Alzheimer's patients in the United States (p. 62).
- The Social Security retirement and disability trust funds (p. 315).
- The concentration of glucose in the bloodstream (p. 389).

Regardless of the field from which the real-world problem is drawn, the problem is analyzed using a process called **mathematical modeling**. The four steps in this process, as illustrated in Figure 15, follow.

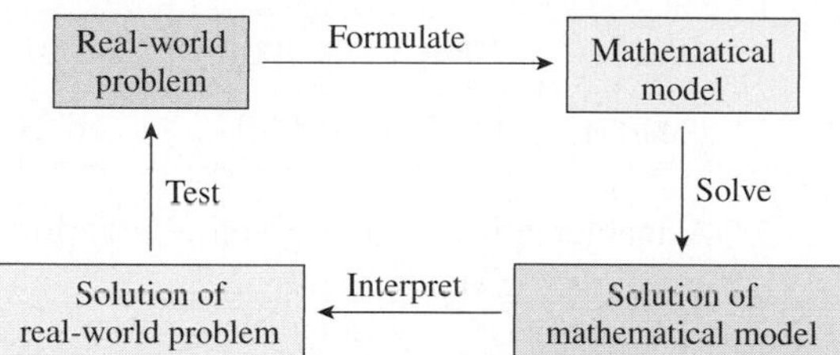

FIGURE 15

1. **Formulate** Given a real-world problem, our first task is to formulate the problem, using the language of mathematics. The many techniques used in constructing mathematical models range from theoretical consideration of the problem on the one extreme to an interpretation of data associated with the problem on the other. For example, the mathematical model giving the accumulated amount at any time when a certain sum of money is deposited in the bank can be derived theoretically (see Chapter 5). On the other hand, many of the mathematical models in this book are constructed by studying the data associated with the problem (see Using Technology, pages 93–96). In calculus, we are primarily concerned with how one (dependent) variable depends on one or more (independent) variables. Consequently, most of our mathematical models will involve functions of one or more variables or equations defining these functions (implicitly).
2. **Solve** Once a mathematical model has been constructed, we can use the appropriate mathematical techniques, which we will develop throughout the book, to solve the problem.
3. **Interpret** Bearing in mind that the solution obtained in step 3 is just the solution of the mathematical model, we need to interpret these results in the context of the original real-world problem.
4. **Test** Some mathematical models of real-world applications describe the situations with complete accuracy. For example, the model describing a deposit in a bank account gives the exact accumulated amount in the account at any time. But other mathematical models give, at best, an approximate description of the real-world problem. In this case we need to test the accuracy of the model by observing how well it describes the original real-world problem and how well it predicts past and/or future behavior. If the results are unsatisfactory, then we may have to reconsider the assumptions made in the construction of the model or, in the worst case, return to step 1.

Before going on, let's look at two mathematical models. The first one is used to estimate the U.S. market for cholesterol-reducing drugs from 1999 through 2004, and the second is used to project the growth of the number of people enrolled in health maintenance organizations (HMOs). These models are derived from data using the least-squares technique. In Using Technology on pages 93–96, you can see more examples of mathematical models that are constructed from raw data. In Section 8.4 we will show how the model in Example 1 is constructed using the method of least squares.

VIDEO

APPLIED EXAMPLE 1 Market for Cholesterol-Reducing Drugs In a study conducted in early 2000, experts projected a rise in the market for cholesterol-reducing drugs. The U.S. market (in billions of dollars) for such drugs from 1999 through 2004 is given in the following table:

Year	1999	2000	2001	2002	2003	2004
Market	12.07	14.07	16.21	18.28	20	21.72

A mathematical model giving the approximate U.S. market over the period in question is given by

$$M(t) = 1.95t + 12.19$$

where t is measured in years, with $t = 0$ corresponding to 1999.

a. Sketch the graph of the function M and the given data on the same set of axes.
b. Assuming that the projection held and the trend continued, what was the market for cholesterol-reducing drugs in 2005 ($t = 6$)?
c. What was the rate of increase of the market for cholesterol-reducing drugs over the period in question?

Source: S. G. Cowen

Solution

a. The graph of M is shown in Figure 16.

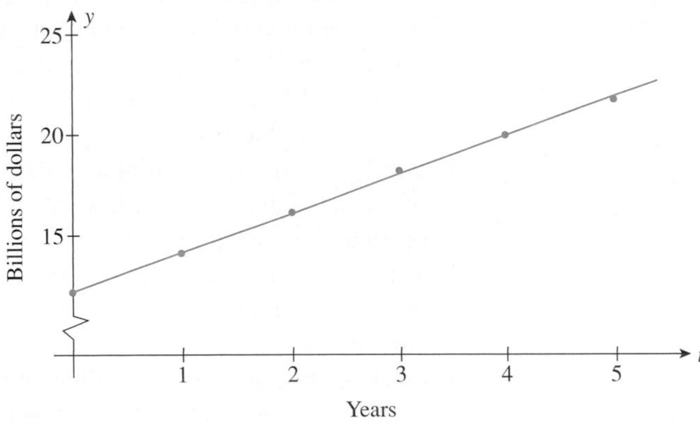

FIGURE 16
Projected U.S. market for cholesterol-reducing drugs

b. The projected market in 2005 for cholesterol-reducing drugs was

$$\begin{aligned} M(6) &= 1.95(6) + 12.19 \\ &= 23.89 \end{aligned}$$

or $23.89 billion.

c. The function M is linear, and so we see that the rate of increase of the market for cholesterol-reducing drugs is given by the slope of the straight line represented by M, which is approximately $1.95 billion per year. ■

The next example utilizes a polynomial of degree 4 to help us construct a model giving the membership in health maintenance organizations (HMOs).

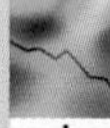

APPLIED EXAMPLE 2 HMO Membership The number of people (in millions) enrolled in HMOs from 1994 through 2002 is given in the following table:

Year	1994	1995	1996	1997	1998	1999	2000	2001	2002
People	45.4	50.6	58.7	67.0	76.4	81.3	80.9	80.0	74.2

The **scatter plot** (graph of the data) associated with this data is shown in Figure 17a, where $t = 0$ corresponds to 1994. A mathematical model giving the approximate number of people, $N(t)$ (in millions), enrolled in HMOs during this period is

$$N(t) = 0.030915t^4 - 0.67974t^3 + 3.704t^2 + 1.63t + 45.5 \qquad (0 \leq t \leq 8)$$

The graph of N is shown in Figure 17b.

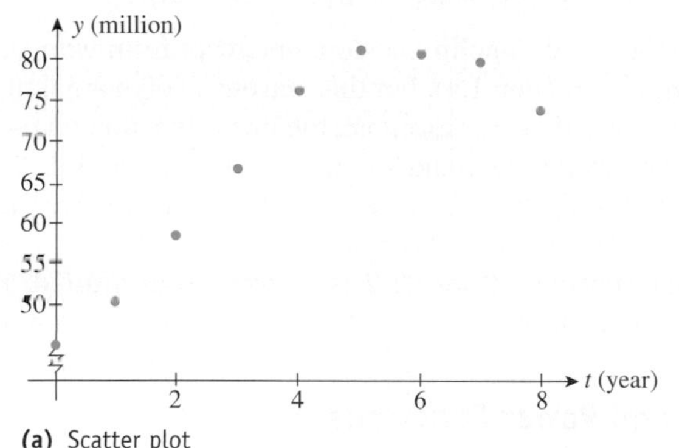

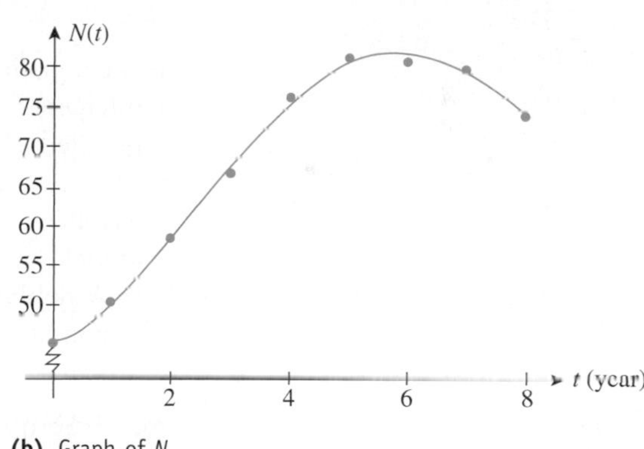

FIGURE 17 (a) Scatter plot (b) Graph of N

a. Use the model to estimate the number of people enrolled in HMOs in 2000. How does this number compare with the actual number?

b. Assume that the trend continued and use the model to predict the number of people enrolled in HMOs in 2003.

Source: Group Health Association of America

Solution

a. The number of people that were enrolled in HMOs at the beginning of 2000 ($t = 6$) is given by

$$N(6) = 0.030915(6)^4 - 0.67974(6)^3 + 3.704(6)^2 + 1.63(6) + 45.5 \approx 81.87$$

or approximately 81.9 million. This number is very close to the actual number of 80.9 million people.

b. Assuming that the trend continued, the number of people enrolled in HMOs in 2003 ($t = 9$) was given by

$$N(9) = 0.030915(9)^4 - 0.67974(9)^3 + 3.704(9)^2 + 1.63(9) + 45.5 \approx 67.50$$

or approximately 67.5 million. ■

Polynomial Functions

We begin by recalling a special class of functions, polynomial functions.

Polynomial Function

A **polynomial function** of degree n is a function of the form

$$f(x) = a_0x^n + a_1x^{n-1} + \cdots + a_{n-1}x + a_n \qquad (a_0 \neq 0)$$

where $a_0, a_1, \ldots, a_n$ are constants and n is a nonnegative integer.

For example, the functions

$$f(x) = 4x^5 - 3x^4 + x^2 - x + 8$$
$$g(x) = 0.001x^3 - 2x^2 + 20x + 400$$

are polynomial functions of degrees 5 and 3, respectively. Observe that a polynomial function is defined everywhere so that it has domain $(-\infty, \infty)$.

A polynomial function of degree 1 ($n = 1$)

$$f(x) = a_0x + a_1 \qquad (a_0 \neq 0)$$

is the equation of a straight line in the slope-intercept form with slope $m = a_0$ and y-intercept $b = a_1$ (see Section 1.4). For this reason, a polynomial function of degree 1 is called a **linear function**. For example, the linear function $f(x) = 2x + 3$ may be written as a linear equation in x and y—namely, $y = 2x + 3$ or $2x - y + 3 = 0$. Conversely, the linear equation $2x - 3y + 4 = 0$ can be solved for y in terms of x to yield the linear function $y = f(x) = \frac{2}{3}x + \frac{4}{3}$.

A polynomial function of degree 2 is referred to as a **quadratic function**. A polynomial function of degree 3 is called a **cubic function**, and so on.

Rational and Power Functions

Another important class of functions is rational functions. A **rational function** is simply the quotient of two polynomials. Examples of rational functions are

$$F(x) = \frac{3x^3 + x^2 - x + 1}{x - 2}$$
$$G(x) = \frac{x^2 + 1}{x^2 - 1}$$

In general, a rational function has the form

$$R(x) = \frac{f(x)}{g(x)}$$

where $f(x)$ and $g(x)$ are polynomial functions. Since division by zero is not allowed, we conclude that the domain of a rational function is the set of all real numbers except the zeros of g—that is, the roots of the equation $g(x) = 0$. Thus, the domain of the function F is the set of all numbers except $x = 2$, whereas the domain of the function G is the set of all numbers except those that satisfy $x^2 - 1 = 0$, or $x = \pm 1$.

Functions of the form

$$f(x) = x^r$$

where r is any real number, are called **power functions**. We encountered examples of power functions earlier in our work. For example, the functions

$$f(x) = \sqrt{x} = x^{1/2} \quad \text{and} \quad g(x) = \frac{1}{x^2} = x^{-2}$$

are power functions.

Many of the functions we will encounter later will involve combinations of the functions introduced here. For example, the following functions may be viewed as combinations of such functions:

$$f(x) = \sqrt{\frac{1 - x^2}{1 + x^2}}$$

$$g(x) = \sqrt{x^2 - 3x + 4}$$

$$h(x) = (1 + 2x)^{1/2} + \frac{1}{(x^2 + 2)^{3/2}}$$

As with polynomials of degree 3 or greater, analyzing the properties of these functions is facilitated by using the tools of calculus, to be developed later.

In the next example, we use a power function to construct a model that describes the driving costs of a car.

APPLIED EXAMPLE 3 Driving Costs A study of driving costs based on the 2002 Ford Taurus SEL found the following average costs (car payments, gas, insurance, upkeep, and depreciation), measured in cents per mile.

Miles/year, x	5000	10,000	15,000	20,000
Cost/mile, y (¢)	80.0	60.0	49.8	44.9

A mathematical model (using least squares techniques) giving the average cost in cents per mile is

$$C(x) = \frac{157.6}{x^{0.421}}$$

where x (in thousands) denotes the number of miles the car is driven in 1 year. The scatter plot associated with this data and the graph of C are shown in Figure 18. Using this model, estimate the average cost of driving the 2002 Ford Taurus SEL 8000 miles per year and 18,000 miles per year.

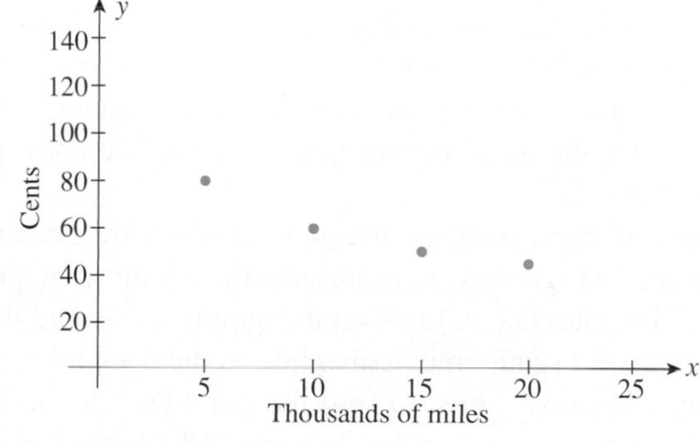

(a) Scatter plot

Source: American Automobile Association

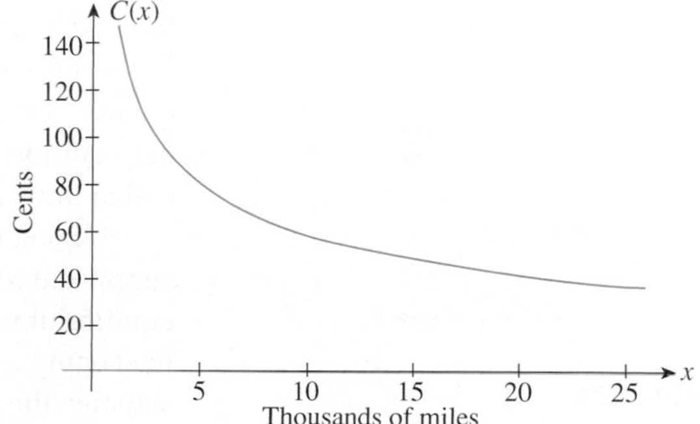

(b) Graph of C

FIGURE 18
(a) The scatter plot and (b) the graph of the model for driving costs

Solution The average cost for driving the car 8000 miles per year is

$$C(8) = \frac{157.6}{8^{0.421}} \approx 65.67$$

or approximately 65.7¢/mile. The average cost for driving it 18,000 miles per year is

$$C(18) = \frac{157.6}{18^{0.421}} \approx 46.68$$

or approximately 46.7¢/mile. ■

Some Economic Models

In the remainder of this section, we look at some economic models.

In a free-market economy, consumer demand for a particular commodity depends on the commodity's unit price. A **demand equation** expresses the relationship between the unit price and the quantity demanded. The graph of the demand equation is called a **demand curve**. In general, the quantity demanded of a commodity decreases as the commodity's unit price increases, and vice versa. Accordingly, a **demand function** defined by $p = f(x)$, where p measures the unit price and x measures the number of units of the commodity in question, is generally characterized as a decreasing function of x; that is, $p = f(x)$ decreases as x increases. Since both x and p assume only nonnegative values, the demand curve is that part of the graph of $f(x)$ that lies in the first quadrant (Figure 19).

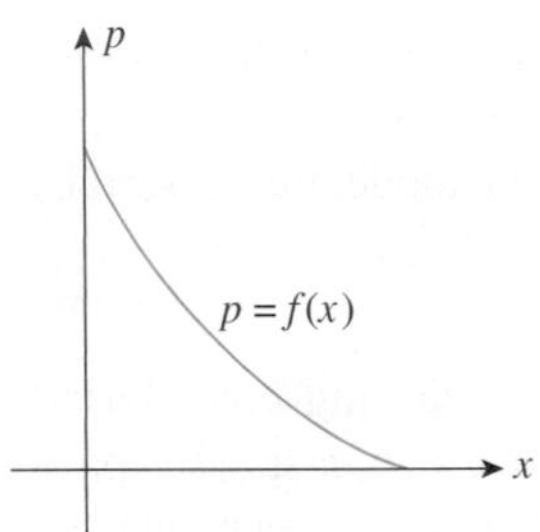

FIGURE 19
A demand curve

In a competitive market, a relationship also exists between the unit price of a commodity and the commodity's availability in the market. In general, an increase in the commodity's unit price induces the producer to increase the supply of the commodity. Conversely, a decrease in the unit price generally leads to a drop in the supply. The equation that expresses the relation between the unit price and the quantity supplied is called a **supply equation**, and its graph is called a **supply curve**. A **supply function** defined by $p = f(x)$ is generally characterized as an increasing function of x; that is, $p = f(x)$ increases as x increases. Since both x and p assume only nonnegative values, the supply curve is that part of the graph of $f(x)$ that lies in the first quadrant (Figure 20).

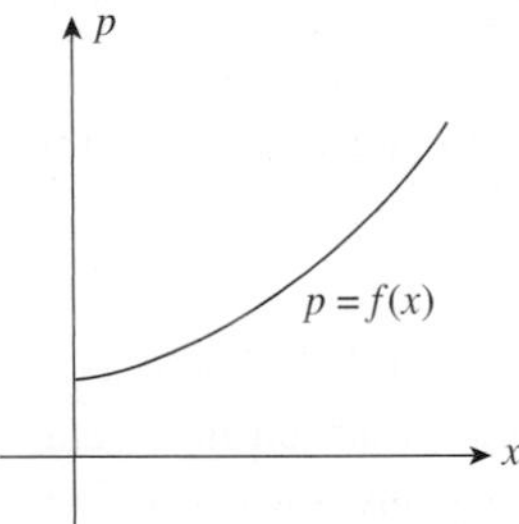

FIGURE 20
A supply curve

Under pure competition, the price of a commodity will eventually settle at a level dictated by the following condition: The supply of the commodity will be equal to the demand for it. If the price is too high, the consumer will not buy; if the price is too low, the supplier will not produce. **Market equilibrium** prevails when the quantity produced is equal to the quantity demanded. The quantity produced at market equilibrium is called the **equilibrium quantity**, and the corresponding price is called the **equilibrium price**.

Market equilibrium corresponds to the point at which the demand curve and the supply curve intersect. In Figure 21, x_0 represents the equilibrium quantity and p_0 the equilibrium price. The point (x_0, p_0) lies on the supply curve and therefore satisfies the supply equation. At the same time, it also lies on the demand curve and therefore satisfies the demand equation. Thus, to find the point (x_0, p_0), and hence the equilibrium quantity and price, we solve the demand and supply equations simultaneously for x and p. For meaningful solutions, x and p must both be positive.

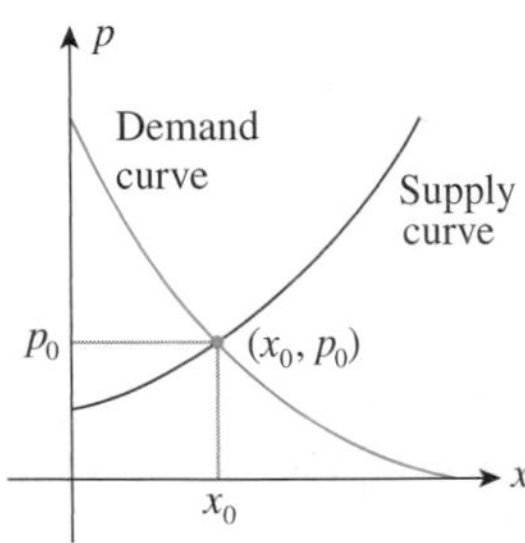

FIGURE 21
Market equilibrium corresponds to (x_0, p_0), the point at which the supply and demand curves intersect.

APPLIED EXAMPLE 4 Supply-Demand The demand function for a certain brand of videocassette is given by

$$p = d(x) = -0.01x^2 - 0.2x + 8$$

and the corresponding supply function is given by

$$p = s(x) = 0.01x^2 + 0.1x + 3$$

where p is expressed in dollars and x is measured in units of a thousand. Find the equilibrium quantity and price.

Solution We solve the following system of equations:

$$\begin{aligned} p &= -0.01x^2 - 0.2x + 8 \\ p &= 0.01x^2 + 0.1x + 3 \end{aligned}$$

Substituting the first equation into the second yields

$$-0.01x^2 - 0.2x + 8 = 0.01x^2 + 0.1x + 3$$

which is equivalent to

$$\begin{aligned} 0.02x^2 + 0.3x - 5 &= 0 \\ 2x^2 + 30x - 500 &= 0 \\ x^2 + 15x - 250 &= 0 \\ (x + 25)(x - 10) &= 0 \end{aligned}$$

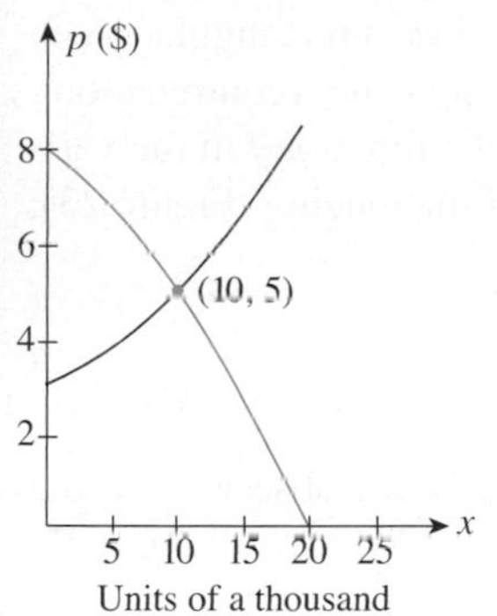

FIGURE 22
The supply curve and the demand curve intersect at the point (10, 5).

Thus, $x = -25$ or $x = 10$. Since x must be nonnegative, the root $x = -25$ is rejected. Therefore, the equilibrium quantity is 10,000 videocassettes. The equilibrium price is given by

$$p = 0.01(10)^2 + 0.1(10) + 3 = 5$$

or \$5 per videocassette (Figure 22). ■

EXPLORING WITH TECHNOLOGY

1. **a.** Use a graphing utility to plot the straight lines L_1 and L_2 with equations $y = 2x - 1$ and $y = 2.1x + 3$, respectively, on the same set of axes, using the standard viewing window. Do the lines appear to intersect?
   b. Plot the straight lines L_1 and L_2, using the viewing window $[-100, 100] \times [-100, 100]$. Do the lines appear to intersect? Can you find the point of intersection using TRACE and ZOOM? Using the "intersection" function of your graphing utility?
   c. Find the point of intersection of L_1 and L_2 algebraically.
   d. Comment on the effectiveness of the methods of solutions in parts (b) and (c).
2. **a.** Use a graphing utility to plot the straight lines L_1 and L_2 with equations $y = 3x - 2$ and $y = -2x + 3$, respectively, on the same set of axes, using the standard viewing window. Then use TRACE and ZOOM to find the point of intersection of L_1 and L_2. Repeat using the "intersection" function of your graphing utility.
   b. Find the point of intersection of L_1 and L_2 algebraically.
   c. Comment on the effectiveness of the methods.

Constructing Mathematical Models

We close this section by showing how some mathematical models can be constructed using elementary geometric and algebraic arguments.

The following guidelines can be used to construct mathematical models.

Guidelines for Constructing Mathematical Models

1. Assign a letter to each variable mentioned in the problem. If appropriate, draw and label a figure.
2. Find an expression for the quantity sought.
3. Use the conditions given in the problem to write the quantity sought as a function f of one variable. Note any restrictions to be placed on the domain of f from physical considerations of the problem.

APPLIED EXAMPLE 5 Enclosing an Area The owner of the Rancho Los Feliz has 3000 yards of fencing with which to enclose a rectangular piece of grazing land along the straight portion of a river. Fencing is not required along the river. Letting x denote the width of the rectangle, find a function f in the variable x giving the area of the grazing land if she uses all of the fencing (Figure 23).

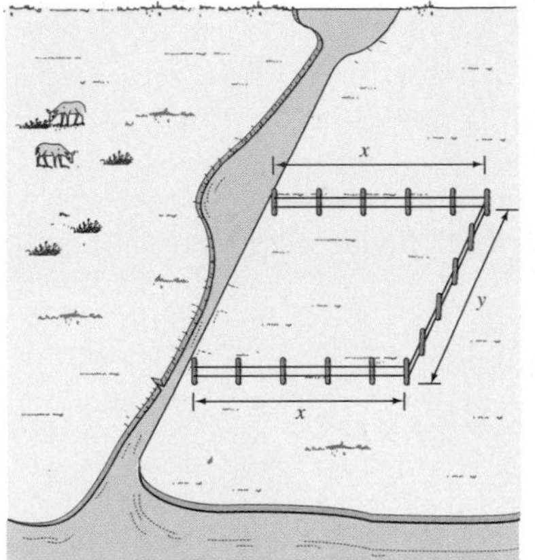

FIGURE 23
The rectangular grazing land has width x and length y.

Solution

1. This information was given.
2. The area of the rectangular grazing land is $A = xy$. Next, observe that the amount of fencing is $2x + y$ and this must be equal to 3000 since all the fencing is used; that is,
$$2x + y = 3000$$
3. From the equation we see that $y = 3000 - 2x$. Substituting this value of y into the expression for A gives
$$A = xy = x(3000 - 2x) = 3000x - 2x^2$$
Finally, observe that both x and y must be nonnegative since they represent the width and length of a rectangle, respectively. Thus, $x \geq 0$ and $y \geq 0$. But the latter is equivalent to $3000 - 2x \geq 0$, or $x \leq 1500$. So the required function is $f(x) = 3000x - 2x^2$ with domain $0 \leq x \leq 1500$. ■

Note Observe that if we view the function $f(x) = 3000x - 2x^2$ strictly as a mathematical entity, then its domain is the set of all real numbers. But physical considerations dictate that its domain should be restricted to the interval $[0, 1500]$. ■

APPLIED EXAMPLE 6 Charter-Flight Revenue If exactly 200 people sign up for a charter flight, Leisure World Travel Agency charges \$300 per person. However, if more than 200 people sign up for the flight (assume this is the case), then each fare is reduced by \$1 for each additional person. Letting x denote the number of passengers above 200, find a function giving the revenue realized by the company.

Solution

1. This information was given.
2. If there are x passengers above 200, then the number of passengers signing up for the flight is $200 + x$. Furthermore, the fare will be $(300 - x)$ dollars per passenger.
3. The revenue will be

$$R = (200 + x)(300 - x) \quad \text{Number of passengers} \times \text{the fare per passenger}$$
$$= -x^2 + 100x + 60{,}000$$

Clearly, x must be nonnegative, and $300 - x \geq 0$, or $x \leq 300$. So the required function is $f(x) = -x^2 + 100x + 60{,}000$ with domain $[0, 300]$. ■

2.3 Self-Check Exercises

1. Thomas Young has suggested the following rule for calculating the dosage of medicine for children from ages 1 to 12 yr. If a denotes the adult dosage (in milligrams) and t is the age of the child (in years), then the child's dosage is given by

$$D(t) = \frac{at}{t + 12}$$

If the adult dose of a substance is 500 mg, how much should a 4-yr-old child receive?

2. The demand function for Mrs. Baker's cookies is given by

$$d(x) = -\frac{2}{15}x + 4$$

where $d(x)$ is the wholesale price in dollars/pound and x is the quantity demanded each week, measured in thousands of pounds. The supply function for the cookies is given by

$$s(x) = \frac{1}{75}x^2 + \frac{1}{10}x + \frac{3}{2}$$

where $s(x)$ is the wholesale price in dollars/pound and x is the quantity, in thousands of pounds, that will be made available in the market each week by the supplier.

a. Sketch the graphs of the functions d and s.

b. Find the equilibrium quantity and price.

Solutions to Self-Check Exercises 2.3 can be found on page 92.

2.3 Concept Questions

1. Describe mathematical modeling in your own words.
2. Define (a) a polynomial function and (b) a rational function. Give an example of each.
3. a. What is a demand function? A supply function?

   b. What is market equilibrium? Describe how you would go about finding the equilibrium quantity and equilibrium price given the demand and supply equations associated with a commodity.

2.3 Exercises

In Exercises 1–8, determine whether the equation defines y as a linear function of x. If so, write it in the form $y = mx + b$.

1. $2x + 3y = 6$
2. $-2x + 4y = 7$
3. $x = 2y - 4$
4. $2x = 3y + 8$
5. $2x - 4y + 9 = 0$
6. $3x - 6y + 7 = 0$
7. $2x^2 - 8y + 4 = 0$
8. $3\sqrt{x} + 4y = 0$

In Exercises 9–14, determine whether the given function is a polynomial function, a rational function, or some other function. State the degree of each polynomial function.

9. $f(x) = 3x^6 - 2x^2 + 1$
10. $f(x) = \dfrac{x^2 - 9}{x - 3}$
11. $G(x) = 2(x^2 - 3)^3$
12. $H(x) = 2x^{-3} + 5x^{-2} + 6$
13. $f(t) = 2t^2 + 3\sqrt{t}$
14. $f(r) = \dfrac{6r}{r^3 - 8}$

15. Find the constants m and b in the linear function $f(x) = mx + b$ so that $f(0) = 2$ and $f(3) = -1$.

16. Find the constants m and b in the linear function $f(x) = mx + b$ so that $f(2) = 4$ and the straight line represented by f has slope -1.

17. A manufacturer has a monthly fixed cost of $40,000 and a production cost of $8 for each unit produced. The product sells for $12/unit.

a. What is the cost function?
b. What is the revenue function?
c. What is the profit function?
d. Compute the profit (loss) corresponding to production levels of 8000 and 12,000 units.

18. A manufacturer has a monthly fixed cost of $100,000 and a production cost of $14 for each unit produced. The product sells for $20/unit.

a. What is the cost function?
b. What is the revenue function?
c. What is the profit function?
d. Compute the profit (loss) corresponding to production levels of 12,000 and 20,000 units.

19. DISPOSABLE INCOME Economists define the *disposable annual income* for an individual by the equation $D = (1 - r)T$, where T is the individual's total income and r is the net rate at which he or she is taxed. What is the disposable income for an individual whose income is $60,000 and whose net tax rate is 28%?

20. DRUG DOSAGES A method sometimes used by pediatricians to calculate the dosage of medicine for children is based on the child's surface area. If a denotes the adult dosage (in milligrams) and S is the surface area of the child (in square meters), then the child's dosage is given by

$$D(S) = \frac{Sa}{1.7}$$

If the adult dose of a substance is 500 mg, how much should a child whose surface area is 0.4 m² receive?

21. COWLING'S RULE Cowling's rule is a method for calculating pediatric drug dosages. If a denotes the adult dosage (in milligrams) and t is the age of the child (in years), then the child's dosage is given by

$$D(t) = \left(\frac{t+1}{24}\right)a$$

If the adult dose of a substance is 500 mg, how much should a 4-yr-old child receive?

22. WORKER EFFICIENCY An efficiency study showed that the average worker at Delphi Electronics assembled cordless telephones at the rate of

$$f(t) = -\frac{3}{2}t^2 + 6t + 10 \qquad (0 \le t \le 4)$$

phones/hour, t hr after starting work during the morning shift. At what rate does the average worker assemble telephones 2 hr after starting work?

23. EFFECT OF ADVERTISING ON SALES The quarterly profit of Cunningham Realty depends on the amount of money x spent on advertising/quarter according to the rule

$$p(x) = -\frac{1}{8}x^2 + 7x + 30 \qquad (0 \le x \le 50)$$

where $P(x)$ and x are measured in thousands of dollars. What is Cunningham's profit when its quarterly advertising budget is $28,000?

24. SPENDING ON MEDICAL DEVICES The U.S. market size for medical devices (in billions of dollars) in year t, from 1999 ($t = 0$) through 2005 ($t = 6$) is approximated by the function

$$S(t) = 0.288t^2 + 3.03t + 45.9 \qquad (0 \le t \le 6)$$

What was the U.S. market size for medical devices in 1999? In 2005?

Source: Frost and Sullivan

25. SOLAR POWER More and more businesses and homeowners are installing solar panels on their roofs to draw energy from the Sun's rays. According to the U.S. Department of Energy, the solar cell kilowatt-hour used in the United States (in millions) is projected to be

$$S(t) = 0.73t^2 + 15.8t + 2.7 \qquad (0 \le t \le 8)$$

in year t, with $t = 0$ corresponding to 2000. What is the projected solar cell kilowatt-hour use in the United States for 2006? For 2008?

Source: U.S. Department of Energy

26. SALES OF DIGITAL TVS The number of homes with digital TVs is expected to grow according to the function

$$f(t) = 0.1714t^2 + 0.6657t + 0.7143 \qquad (0 \le t \le 6)$$

where t is measured in years, with $t = 0$ corresponding to the beginning of 2000, and $f(t)$ is measured in millions of homes.

a. How many homes had digital TVs at the beginning of 2000?
b. How many homes had digital TVs at the beginning of 2005?

Source: Consumer Electronics Manufacturers Association

27. AGING DRIVERS The number of fatalities due to car crashes, based on the number of miles driven, begins to climb after the driver is past age 65. Aside from declining ability as one ages, the older driver is more fragile. The number of fatalities per 100 million vehicle miles driven is approximately

$$N(x) = 0.0336x^3 - 0.118x^2 + 0.215x + 0.7 \qquad (0 \le x \le 7)$$

where x denotes the age group of drivers, with $x = 0$ corresponding to those aged 50–54, $x = 1$ corresponding to those aged 55–59, $x = 2$ corresponding to those aged 60–64, . . . , and $x = 7$ corresponding to those aged 85–89. What is the fatality rate per 100 million vehicle miles driven for an average driver in the 50–54 age group? In the 85–89 age group?

Source: U.S. Department of Transportation

28. **Testosterone Use** Fueled by the promotion of testosterone as an antiaging elixir, use of the hormone by middle-age and older men grew dramatically. The total number of prescriptions for testosterone from 1999 through 2002 is given by

$$N(t) = -35.8t^3 + 202t^2 + 87.8t + 648 \qquad (0 \le t \le 3)$$

where $N(t)$ is measured in thousands and t is measured in years, with $t = 0$ corresponding to 1999. Find the total number of prescriptions for testosterone in 1999, 2000, 2001, and 2002.

Source: IMS Health

29. **BlackBerry Subscribers** According to a study conducted in 2004, the number of subscribers of BlackBerry, the handheld e-mail devices manufactured by Research in Motion Ltd., is expected to be

$$N(t) = -0.0675t^4 + 0.5083t^3 - 0.893t^2 + 0.66t + 0.32 \qquad (0 \le t \le 4)$$

where $N(t)$ is measured in millions and t in years, with $t = 0$ corresponding to the beginning of 2002.

a. How many BlackBerry subscribers were there at the beginning of 2002?

b. What is the projected number of BlackBerry subscribers at the beginning of 2006?

Source: ThinkEquity Partners

30. **Chip Sales** The worldwide flash memory chip sales (in billions of dollars) is projected to be

$$S(t) = 4.3(t + 2)^{0.94} \qquad (0 \le t \le 6)$$

where t is measured in years, with $t = 0$ corresponding to 2002. Flash chips are used in cell phones, digital cameras, and other products.

a. What were the worldwide flash memory chip sales in 2002?

b. What are the projected sales for 2008?

Source: Web-Feet Research Inc.

31. **Outsourcing of Jobs** According to a study conducted in 2003, the total number of U.S. jobs (in millions) that are projected to leave the country by year t, where $t = 0$ corresponds to 2000, is

$$N(t) = 0.0018425(t + 5)^{2.5} \qquad (0 \le t \le 15)$$

How many jobs will be outsourced by 2005? By 2010?

Source: Forrester Research

32. **Selling Price of DVD Recorders** The rise of digital music and the improvement to the DVD format are part of the reasons why the average selling price of standalone DVD recorders will drop in the coming years. The function

$$A(t) = \frac{699}{(t + 1)^{0.94}} \qquad (0 \le t \le 5)$$

gives the projected average selling price (in dollars) of standalone DVD recorders in year t, where $t = 0$ corresponds to the beginning of 2002. What was the average selling price of standalone DVD recorders at the beginning of 2002? What will it be at the beginning of 2007?

Source: Consumer Electronics Association

33. **E-mail Usage** The number of international e-mailings per day (in millions) is approximated by the function

$$f(t) = 38.57t^2 - 24.29t + 79.14 \qquad (0 \le t \le 4)$$

where t is measured in years, with $t = 0$ corresponding to the beginning of 1998.

a. Sketch the graph of f.

b. How many international e-mailings per day were there at the beginning of 2002?

Source: Pioneer Consulting

34. **Document Management** The size (measured in millions of dollars) of the document-management business is described by the function

$$f(t) = 0.22t^2 + 1.4t + 3.77 \qquad (0 \le t \le 6)$$

where t is measured in years, with $t = 0$ corresponding to the beginning of 1996.

a. Sketch the graph of f.

b. What was the size of the document-management business at the beginning of 2002?

Source: Sun Trust Equitable Securities

35. **Reaction of a Frog to a Drug** Experiments conducted by A. J. Clark suggest that the response $R(x)$ of a frog's heart muscle to the injection of x units of acetylcholine (as a percent of the maximum possible effect of the drug) may be approximated by the rational function

$$R(x) = \frac{100x}{b + x} \qquad (x \ge 0)$$

where b is a positive constant that depends on the particular frog.

a. If a concentration of 40 units of acetylcholine produces a response of 50% for a certain frog, find the "response function" for this frog.

b. Using the model found in part (a), find the response of the frog's heart muscle when 60 units of acetylcholine are administered.

36. **Walking versus Running** The oxygen consumption (in milliliter/pound/minute) for a person walking at x mph is approximated by the function

$$f(x) = \frac{5}{3}x^2 + \frac{5}{3}x + 10 \qquad (0 \le x \le 9)$$

whereas the oxygen consumption for a runner at x mph is approximated by the function

$$g(x) = 11x + 10 \qquad (4 \le x \le 9)$$

a. Sketch the graphs of f and g.
b. At what speed is the oxygen consumption the same for a walker as it is for a runner? What is the level of oxygen consumption at that speed?
c. What happens to the oxygen consumption of the walker and the runner at speeds beyond that found in part (b)?

Source: William McArdley, Frank Katch, and Victor Katch, *Exercise Physiology*

37. **Digital versus Film Cameras** The sales of digital cameras (in millions of units) in year t is given by the function

$$f(t) = 3.05t + 6.85 \qquad (0 \le t \le 3)$$

where $t = 0$ corresponds to 2001. Over that same period, the sales of film cameras (in millions of units) is given by

$$g(t) = -1.85t + 16.58 \qquad (0 \le t \le 3)$$

a. Show that more film cameras than digital cameras were sold in 2001.
b. When did the sales of digital cameras first exceed those of film cameras?

Source: Popular Science

38. **Cricket Chirping and Temperature** Entomologists have discovered that a linear relationship exists between the number of chirps of crickets of a certain species and the air temperature. When the temperature is 70°F, the crickets chirp at the rate of 120 times/minute, and when the temperature is 80°F, they chirp at the rate of 160 times/minute.
a. Find an equation giving the relationship between the air temperature T and the number of chirps/minute, N, of the crickets.
b. Find N as a function of T and use this formula to determine the rate at which the crickets chirp when the temperature is 102°F.

39. **Linear Depreciation** In computing income tax, businesses are allowed by law to depreciate certain assets such as buildings, machines, furniture, automobiles, and so on, over a period of time. The linear depreciation, or straight-line method, is often used for this purpose. Suppose an asset has an initial value of \$$C$ and is to be depreciated linearly over n yr with a scrap value of \$$S$. Show that the book value of the asset at any time t $(0 \le t \le n)$ is given by the linear function

$$V(t) = C - \frac{C - S}{n}t$$

Hint: Find an equation of the straight line that passes through the points $(0, C)$ and (n, S). Then rewrite the equation in the slope-intercept form.

40. **Linear Depreciation** Using the linear depreciation model of Exercise 39, find the book value of a printing machine at the end of the second year if its initial value is \$100,000 and it is depreciated linearly over 5 yr with a scrap value of \$30,000.

41. **Credit Card Debt** Following the introduction in 1950 of the nation's first credit card, the Diners Club Card, credit cards have proliferated over the years. More than 720 different cards are now used at more than 4 million locations in the United States. The average U.S. credit card debt (per household) in thousands of dollars is approximately given by

$$D(t) = \begin{cases} 4.77(1 + t)^{0.2676} & \text{if } 0 \le t \le 2 \\ 5.6423t^{0.1818} & \text{if } 2 < t \le 6 \end{cases}$$

where t is measured in years, with $t = 0$ corresponding to the beginning of 1994. What was the average U.S. credit card debt (per household) at the beginning of 1994? At the beginning of 1996? At the beginning of 1999?

Source: David Evans and Richard Schmalensee, *Paying with Plastic: The Digital Revolution in Buying and Borrowing*

42. **Obese Children in the United States** The percent of obese children aged 12–19 in the United States is approximately

$$P(t) = \begin{cases} 0.04t + 4.6 & \text{if } 0 \le t < 10 \\ -0.01005t^2 + 0.945t - 3.4 & \text{if } 10 \le t \le 30 \end{cases}$$

where t is measured in years, with $t = 0$ corresponding to the beginning of 1970. What was the percent of obese children aged 12–19 at the beginning of 1970? At the beginning of 1985? At the beginning of 2000?

Source: Centers for Disease Control

43. **U.S. Nutritional Supplements Market** The size of the U.S. nutritional supplements market from 1999 through 2003 is approximated by the function

$$A(t) = 16.4(t + 1)^{0.1} \qquad (0 \le t \le 4)$$

where $A(t)$ is measured in billions of dollars and t is measured in years, with $t = 0$ corresponding to 1999.
a. Compute $A(0)$, $A(1)$, $A(2)$, $A(3)$, and $A(4)$. Interpret your results.
b. Use the results of part (a) to sketch the graph of A.

Source: Nutrition Business Journal

44. **Price of Ivory** According to the World Wildlife Fund, a group in the forefront of the fight against illegal ivory trade, the price of ivory (in dollars/kilo) compiled from a variety of legal and black market sources is approximated by the function

$$f(t) = \begin{cases} 8.37t + 7.44 & \text{if } 0 \le t \le 8 \\ 2.84t + 51.68 & \text{if } 8 < t \le 30 \end{cases}$$

where t is measured in years, with $t = 0$ corresponding to the beginning of 1970.
a. Sketch the graph of the function f.
b. What was the price of ivory at the beginning of 1970? At the beginning of 1990?

Source: World Wildlife Fund

45. **Senior Citizens' Health Care** According to a study, the out-of-pocket cost to senior citizens for health care, $f(t)$ (as a percent of income), in year t where $t = 0$ corresponds to 1977, is given by

$$y = \begin{cases} \frac{2}{7}t + 12 & \text{if } 0 \le t \le 7 \\ t + 7 & \text{if } 7 < t \le 10 \\ \frac{3}{5}t + 11 & \text{if } 10 < t \le 20 \end{cases}$$

a. Sketch the graph of f.
b. What was the out-of-pocket cost, as a percent of income, to senior citizens for health care in 1982? In 1992?

Source: Senate Select Committee on Aging, AARP

46. **WORKING-AGE POPULATION** The ratio of working-age population to the elderly in the United States (including projections after 2000) is given by

$$f(t) = \begin{cases} 4.1 & \text{if } 0 \le t < 5 \\ -0.03t + 4.25 & \text{if } 5 \le t < 15 \\ -0.075t + 4.925 & \text{if } 15 \le t \le 35 \end{cases}$$

with $t = 0$ corresponding to the beginning of 1995.
a. Sketch the graph of f.
b. What was the ratio at the beginning of 2005? What will be the ratio at the beginning of 2020?
c. Over what years is the ratio constant?
d. Over what years is the decline of the ratio greatest?

Source: U.S. Census Bureau

47. **PRICE OF AUTOMOBILE PARTS** For years, automobile manufacturers had a monopoly on the replacement-parts market, particularly for sheet metal parts such as fenders, doors, and hoods, the parts most often damaged in a crash. Beginning in the late 1970s, however, competition appeared on the scene. In a report conducted by an insurance company to study the effects of the competition, the price of an OEM (original equipment manufacturer) fender for a particular 1983 model car was found to be

$$f(t) = \frac{110}{\frac{1}{2}t + 1} \qquad (0 \le t \le 2)$$

where $f(t)$ is measured in dollars and t is in years. Over the same period of time, the price of a non-OEM fender for the car was found to be

$$g(t) = 26\left(\frac{1}{4}t^2 - 1\right)^2 + 52 \qquad (0 \le t \le 2)$$

where $g(t)$ is also measured in dollars. Find a function $h(t)$ that gives the difference in price between an OEM fender and a non-OEM fender. Compute $h(0)$, $h(1)$, and $h(2)$. What does the result of your computation seem to say about the price gap between OEM and non-OEM fenders over the 2 yr?

48. **SALES OF DVD PLAYERS VERSUS VCRS** The sales of DVD players in year t (in millions of units) is given by the function

$$f(t) = 5.6(1 + t) \qquad (0 \le t \le 3)$$

where $t = 0$ corresponds to 2001. Over the same period, the sales of VCRs (in millions of units) is given by

$$g(t) = \begin{cases} -9.6t + 22.5 & \text{if } 0 \le t \le 1 \\ -0.5t + 13.4 & \text{if } 1 < t \le 2 \\ -7.8t + 28 & \text{if } 2 < t \le 3 \end{cases}$$

a. Show that more VCRs than DVD players were sold in 2001.
b. When did the sales of DVD players first exceed those of VCRs?

Source: Popular Science

49. **HOTEL OCCUPANCY RATE** A forecast released by PricewaterhouseCoopers in June of 2004 predicted the occupancy rate of U.S. hotels between 2001 ($t = 0$) and 2005 ($t = 4$) to be

$$P(t) = \begin{cases} -0.9t + 59.8 & \text{if } 0 \le t < 1 \\ 0.3t + 58.6 & \text{if } 1 \le t < 2 \\ 56.79t^{0.06} & \text{if } 2 \le t \le 4 \end{cases}$$

percent.
a. Compute $P(0)$, $P(1)$, $P(2)$, $P(3)$, and $P(4)$.
b. Sketch the graph of P.
c. What was the predicted occupancy rate of hotels for 2004?

Source: PricewaterhouseCoopers LLP Hospitality & Leisure Research

50. **IMMIGRATION TO THE UNITED STATES** The immigration to the United States from Europe, as a percent of the total immigration, is approximately

$$P(t) = 0.767t^3 - 0.636t^2 - 19.17t + 52.7 \qquad (0 \le t \le 4)$$

where t is measured in decades, with $t = 0$ corresponding to the decade of the 1950s.
a. Complete the table:

t	0	1	2	3	4
$P(t)$					

b. Use the result of part (a) to sketch the graph of P.
c. Use the result of part (b) to estimate the decade when the immigration, as a percent of the total immigration, was the greatest and the smallest.

Source: Jeffrey Williamson, Harvard University

For the demand equations in Exercises 51–54, where x represents the quantity demanded in units of a thousand and p is the unit price in dollars, (a) sketch the demand curve and (b) determine the quantity demanded when the unit price is set at \$$p$.

51. $p = -x^2 + 16;\ p = 7$ **52.** $p = -x^2 + 36;\ p = 11$

53. $p = \sqrt{18 - x^2};\ p = 3$ **54.** $p = \sqrt{9 - x^2};\ p = 2$

For the supply equations in Exercises 55–58, where x is the quantity supplied in units of a thousand and p is the unit price in dollars, (a) sketch the supply curve and (b) determine the price at which the supplier will make 2000 units of the commodity available in the market.

55. $p = x^2 + 16x + 40$ **56.** $p = 2x^2 + 18$

57. $p = x^3 + 2x + 3$ **58.** $p = x^3 + x + 10$

59. **DEMAND FOR CLOCK RADIOS** In the accompanying figure, L_1 is the demand curve for the model A clock radios manufactured by Ace Radio, and L_2 is the demand curve for their model B clock radios. Which line has the greater slope? Interpret your results.

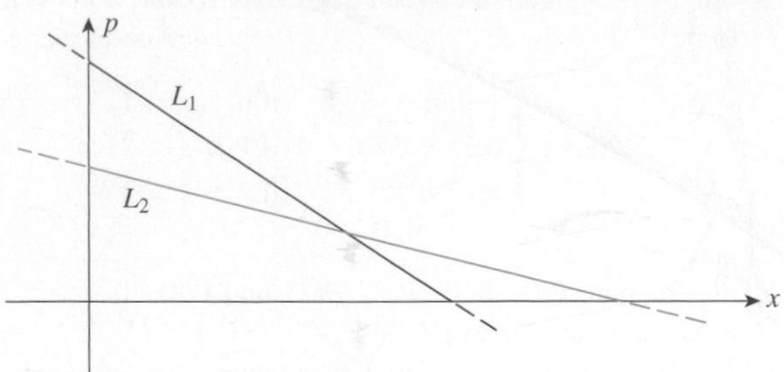

60. **Supply of Clock Radios** In the accompanying figure, L_1 is the supply curve for the model A clock radios manufactured by Ace Radio, and L_2 is the supply curve for their model B clock radios. Which line has the greater slope? Interpret your results.

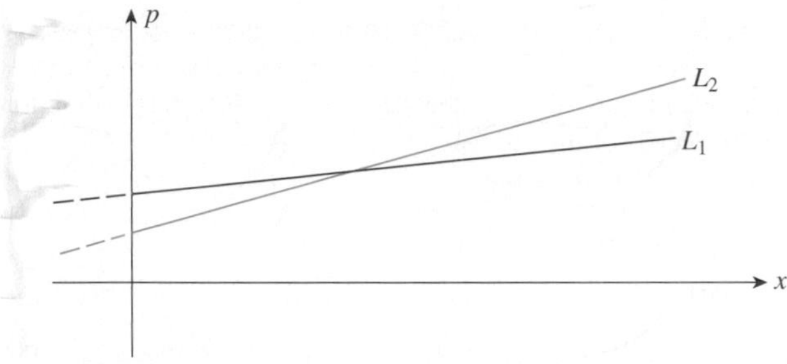

61. **Demand for Smoke Alarms** The demand function for the Sentinel smoke alarm is given by

$$p = \frac{30}{0.02x^2 + 1} \qquad (0 \le x \le 10)$$

where x (measured in units of a thousand) is the quantity demanded per week and p is the unit price in dollars.

a. Sketch the graph of the demand function.

b. What is the unit price that corresponds to a quantity demanded of 10,000 units?

62. **Demand for Commodities** Assume that the demand function for a certain commodity has the form

$$p = \sqrt{-ax^2 + b} \qquad (a \ge 0, b \ge 0)$$

where x is the quantity demanded, measured in units of a thousand and p is the unit price in dollars. Suppose the quantity demanded is 6000 ($x = 6$) when the unit price is \$8.00 and 8000 ($x = 8$) when the unit price is \$6.00. Determine the demand equation. What is the quantity demanded when the unit price is set at \$7.50?

63. **Supply Functions** The supply function for the Luminar desk lamp is given by

$$p = 0.1x^2 + 0.5x + 15$$

where x is the quantity supplied (in thousands) and p is the unit price in dollars.

a. Sketch the graph of the supply function.

b. What unit price will induce the supplier to make 5000 lamps available in the marketplace?

64. **Supply Functions** Suppliers of transistor radios will market 10,000 units when the unit price is \$20 and 62,500 units when the unit price is \$35. Determine the supply function if it is known to have the form

$$p = a\sqrt{x} + b \qquad (a \ge 0, b \ge 0)$$

where x is the quantity supplied and p is the unit price in dollars. Sketch the graph of the supply function. What unit price will induce the supplier to make 40,000 transistor radios available in the marketplace?

65. Suppose the demand and supply equations for a certain commodity are given by $p = ax + b$ and $p = cx + d$, respectively, where $a < 0$, $c > 0$, and $b > d > 0$ (see the accompanying figure).

a. Find the equilibrium quantity and equilibrium price in terms of a, b, c, and d.

b. Use part (a) to determine what happens to the market equilibrium if c is increased, while a, b, and d remain fixed. Interpret your answer in economic terms.

c. Use part (a) to determine what happens to the market equilibrium if b is decreased while a, c, and d remain fixed. Interpret your answer in economic terms.

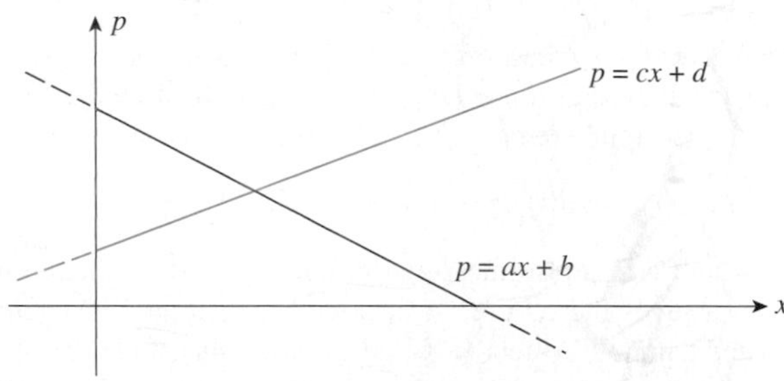

For each pair of supply and demand equations in Exercises 66–69, where x represents the quantity demanded in units of a thousand and p the unit price in dollars, find the equilibrium quantity and the equilibrium price.

66. $p = -x^2 - 2x + 100$ and $p = 8x + 25$

67. $p = -2x^2 + 80$ and $p = 15x + 30$

68. $p = 60 - 2x^2$ and $p = x^2 + 9x + 30$

69. $11p + 3x - 66 = 0$ and $2p^2 + p - x = 10$

70. **Market Equilibrium** The weekly demand and supply functions for Sportsman 5×7 tents are given by

$$p = -0.1x^2 - x + 40$$
$$p = 0.1x^2 + 2x + 20$$

respectively, where p is measured in dollars and x is measured in units of a hundred. Find the equilibrium quantity and price.

71. **Market Equilibrium** The management of Titan Tire Company has determined that the weekly demand and supply functions for their Super Titan tires are given by

$$p = 144 - x^2$$
$$p = 48 + \frac{1}{2}x^2$$

respectively, where p is measured in dollars and x is measured in units of a thousand. Find the equilibrium quantity and price.

72. **Enclosing an Area** Patricia wishes to have a rectangular-shaped garden in her backyard. She has 80 ft of fencing with which to enclose her garden. Letting x denote the width of the garden, find a function f in the variable x giving the area of the garden. What is its domain?

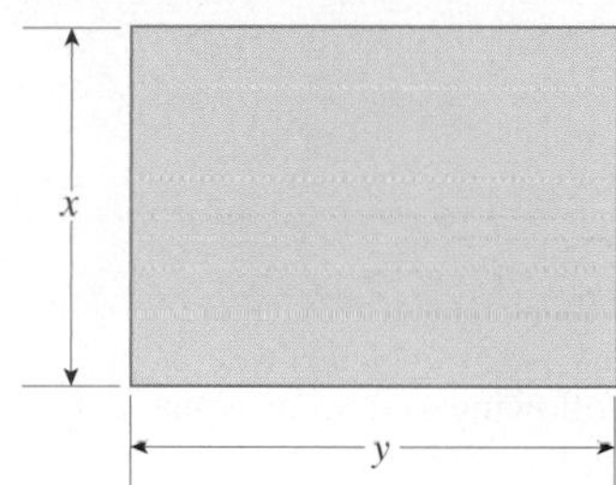

73. **Enclosing an Area** Patricia's neighbor, Juanita, also wishes to have a rectangular-shaped garden in her backyard. But Juanita wants her garden to have an area of 250 ft^2. Letting x denote the width of the garden, find a function f in the variable x giving the length of the fencing required to construct the garden. What is the domain of the function?

Hint: Refer to the figure for Exercise 72. The amount of fencing required is equal to the perimeter of the rectangle, which is twice the width plus twice the length of the rectangle.

74. **Packaging** By cutting away identical squares from each corner of a rectangular piece of cardboard and folding up the resulting flaps, an open box may be made. If the cardboard is 15 in. long and 8 in. wide and the square cutaways have dimensions of x in. by x in., find a function giving the volume of the resulting box.

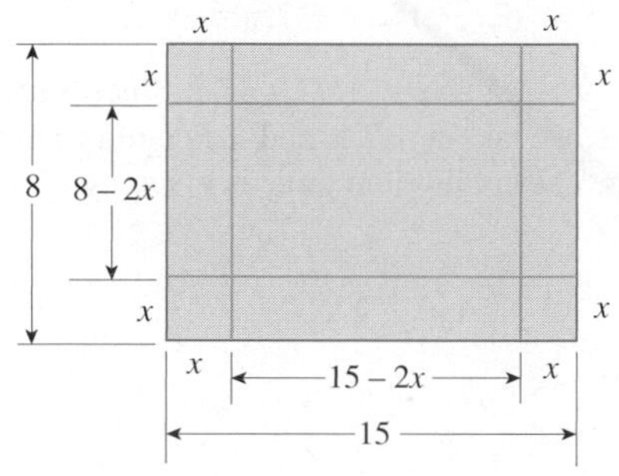

75. **Construction Costs** A rectangular box is to have a square base and a volume of 20 ft^3. The material for the base costs 30¢/ft^2, the material for the sides costs 10¢/ft^2, and the material for the top costs 20¢/ft^2. Letting x denote the length of one side of the base, find a function in the variable x giving the cost of constructing the box.

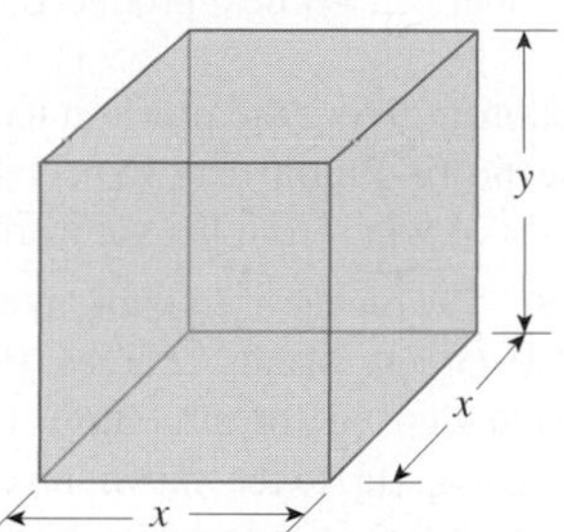

76. **Area of a Norman Window** A Norman window has the shape of a rectangle surmounted by a semicircle (see the accompanying figure). Suppose a Norman window is to have a perimeter of 28 ft; find a function in the variable x giving the area of the window.

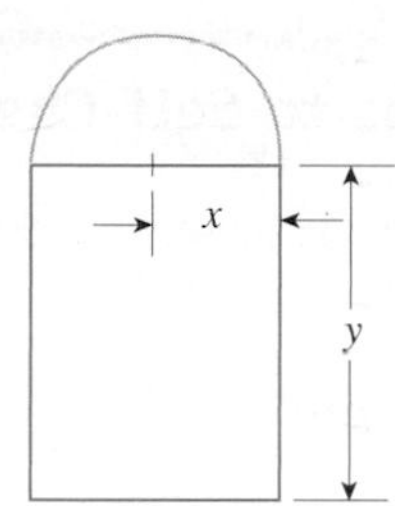

77. **Yield of an Apple Orchard** An apple orchard has an average yield of 36 bushels of apples/tree if tree density is 22 trees/acre. For each unit increase in tree density, the yield decreases by 2 bushels/tree. Letting x denote the number of trees beyond 22/acre, find a function in x that gives the yield of apples.

78. **Book Design** A book designer has decided that the pages of a book should have 1-in. margins at the top and bottom and $\frac{1}{2}$-in. margins on the sides. She further stipulated that each page should have an area of 50 in.^{2}. Find a function in the variable x, giving the area of the printed page. What is the domain of the function?

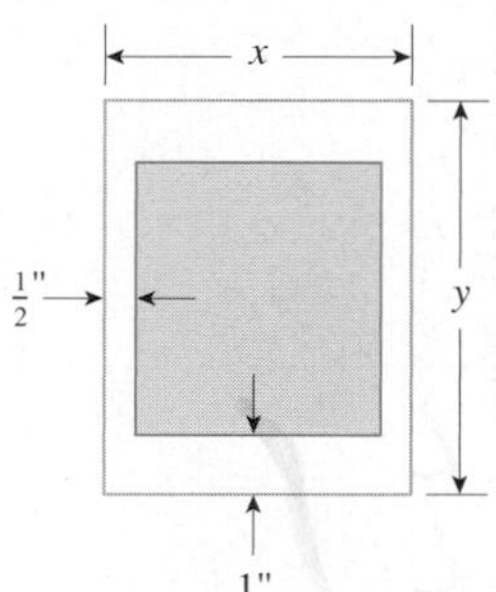

79. **Profit of a Vineyard** Phillip, the proprietor of a vineyard, estimates that if 10,000 bottles of wine were produced this season, then the profit would be \$5/bottle. But if more than 10,000 bottles were produced, then the profit/bottle would drop by \$0.0002 for each additional bottle sold. Assume at least 10,000 bottles of wine are produced and sold and let x denote the number of bottles produced and sold above 10,000.
   a. Find a function P giving the profit in terms of x.
   b. What is the profit Phillip can expect from the sale of 16,000 bottles of wine from his vineyard?

80. **Charter Revenue** The owner of a luxury motor yacht that sails among the 4000 Greek islands charges \$600/person/day if exactly 20 people sign up for the cruise. However, if more than 20 people sign up (up to the maximum capacity of 90) for the cruise, then each fare is reduced by \$4 for each additional passenger. Assume at least 20 people sign up for the cruise and let x denote the number of passengers above 20.
   a. Find a function R giving the revenue/day realized from the charter.
   b. What is the revenue/day if 60 people sign up for the cruise?
   c. What is the revenue/day if 80 people sign up for the cruise?

In Exercises 81–84, determine whether the statement is true or false. If it is true, explain why it is true. If it is false, give an example to show why it is false.

81. A polynomial function is a sum of constant multiples of power functions.

82. A polynomial function is a rational function, but the converse is false.

83. If $r > 0$, then the power function $f(x) = x^r$ is defined for all values of x.

84. The function $f(x) = 2^x$ is a power function.

2.3 Solutions to Self-Check Exercises

1. Since the adult dose of the substance is 500 mg, $a = 500$; thus, the rule in this case is

$$D(t) = \frac{500t}{t + 12}$$

A 4-yr-old should receive

$$D(4) = \frac{500(4)}{4 + 12}$$

or 125 mg of the substance.

2. a. The graphs of the functions d and s are shown in the following figure:

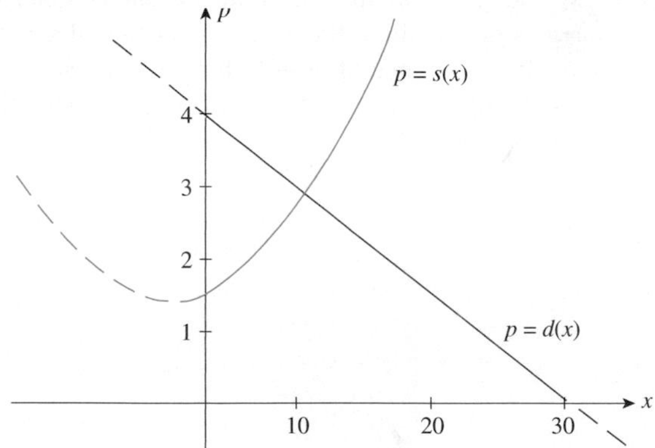

b. Solve the following system of equations:

$$p = -\frac{2}{15}x + 4$$
$$p = \frac{1}{75}x^2 + \frac{1}{10}x + \frac{3}{2}$$

Substituting the first equation into the second yields

$$\frac{1}{75}x^2 + \frac{1}{10}x + \frac{3}{2} = -\frac{2}{15}x + 4$$
$$\frac{1}{75}x^2 + \left(\frac{1}{10} + \frac{2}{15}\right)x - \frac{5}{2} = 0$$
$$\frac{1}{75}x^2 + \frac{7}{30}x - \frac{5}{2} = 0$$

Multiplying both sides of the last equation by 150, we have

$$2x^2 + 35x - 375 = 0$$
$$(2x - 15)(x + 25) = 0$$

Thus, $x = -25$ or $x = 15/2 = 7.5$. Since x must be nonnegative, we take $x = 7.5$, and the equilibrium quantity is 7500 lb. The equilibrium price is given by

$$p = -\frac{2}{15}\left(\frac{15}{2}\right) + 4$$

or \$3/lb.

USING TECHNOLOGY

Finding the Points of Intersection of Two Graphs and Modeling

A graphing utility can be used to find the point(s) of intersection of the graphs of two functions.

EXAMPLE 1 Find the points of intersection of the graphs of

$$f(x) = 0.3x^2 - 1.4x - 3 \quad \text{and} \quad g(x) = -0.4x^2 + 0.8x + 6.4$$

Solution The graphs of both f and g in the standard viewing window are shown in Figure T1a. Using TRACE and ZOOM or the function for finding the points of intersection of two graphs on your graphing utility, we find the point(s) of intersection, accurate to four decimal places, to be $(-2.4158, 2.1329)$ (Figure T1b) and $(5.5587, -1.5125)$ (Figure T1c).

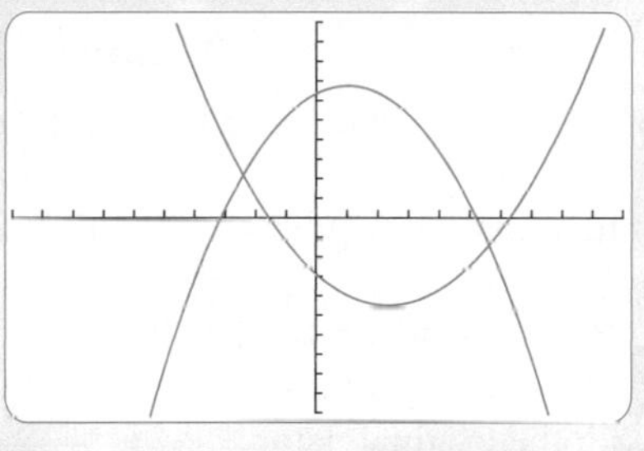

(a)

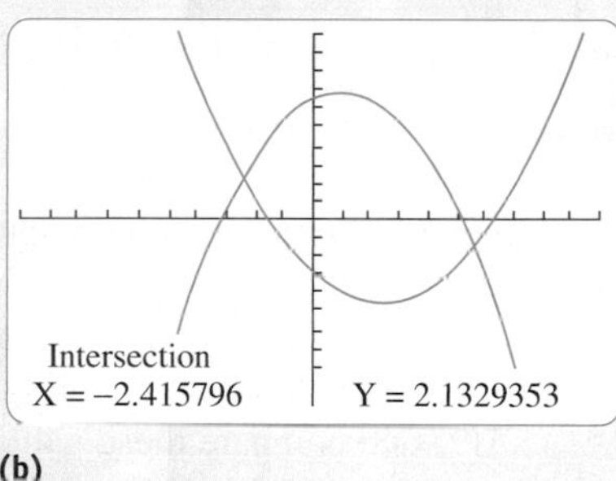

(b)

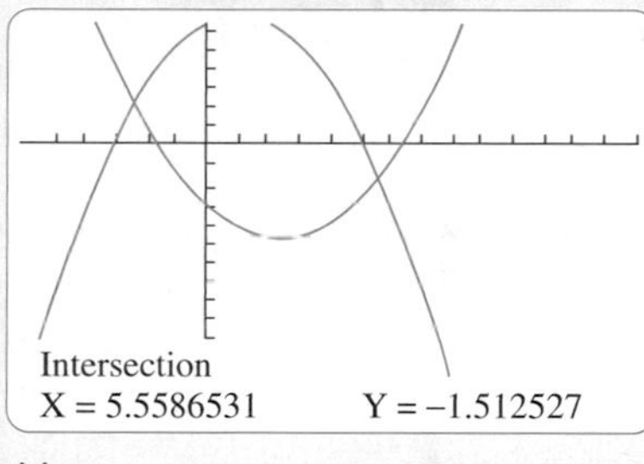

(c)

FIGURE T1
(a) The graphs of f and g in the standard viewing window; (b) and (c) the TI-83 intersection screens

EXAMPLE 2 Consider the demand and supply functions

$$p = d(x) = -0.01x^2 - 0.2x + 8 \quad \text{and} \quad p = s(x) = 0.01x^2 + 0.1x + 3$$

of Example 4 in Section 2.3.

a. Plot the graphs of d and s in the viewing window $[0, 15] \times [0, 10]$.
b. Verify that the equilibrium point is $(10, 5)$, as obtained in Example 4.

Solution

a. The graphs of d and s are shown in Figure T2a.

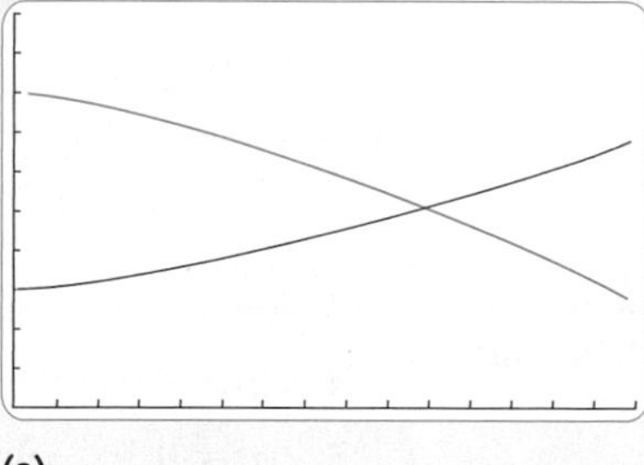

(a)

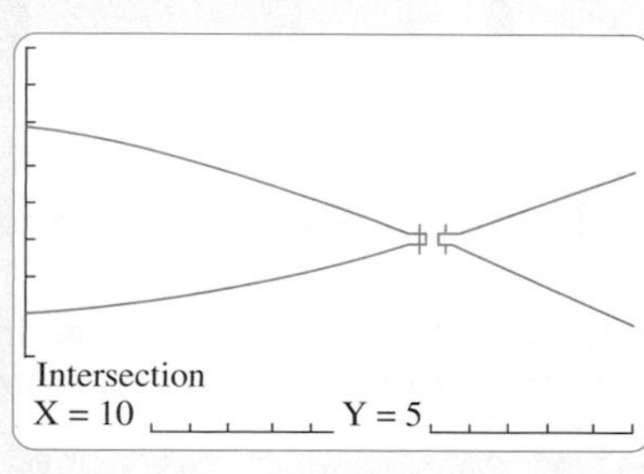

(b)

FIGURE T2
(a) The graphs of d and s in the window $[0, 15] \times [0, 10]$; (b) the TI-83 intersection screen

(continued)

b. Using **TRACE** and **ZOOM** or the function for finding the point of intersection of two graphs, we see that $x = 10$ and $y = 5$ (Figure T2b), so the equilibrium point is (10, 5), as obtained before. ■

Constructing Mathematical Models from Raw Data

A graphing utility can sometimes be used to construct mathematical models from sets of data. For example, if the points corresponding to the given data are scattered about a straight line, then use **LINR** (linear regression) from the **STAT CALC** (statistical calculation) menu of the graphing utility to obtain a function (model) that approximates the data at hand. If the points seem to be scattered along a parabola (the graph of a quadratic function), then use **P2REG** (second-order polynomial regression), and so on. (These are functions on the TI-83 calculator.)

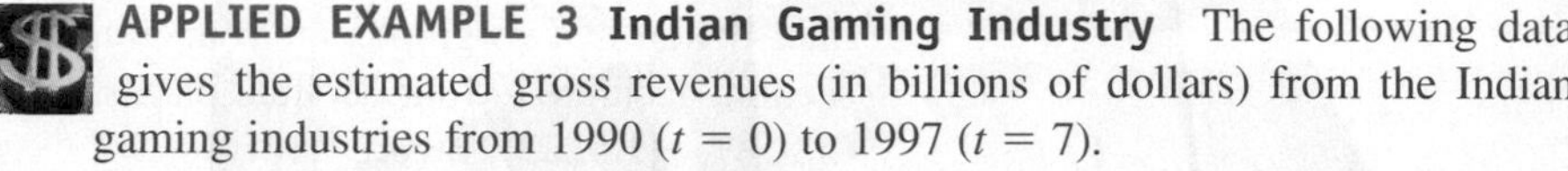

APPLIED EXAMPLE 3 Indian Gaming Industry The following data gives the estimated gross revenues (in billions of dollars) from the Indian gaming industries from 1990 ($t = 0$) to 1997 ($t = 7$).

Year	0	1	2	3	4	5	6	7
Revenue	0.5	0.7	1.6	2.6	3.4	4.8	5.6	6.8

a. Use a graphing utility to find a polynomial function f of degree 4 that models the data.

b. Plot the graph of the function f, using the viewing window $[0, 8] \times [0, 10]$.

c. Use the function evaluation capability of the graphing utility to compute $f(0)$, $f(1), \ldots, f(7)$ and compare these values with the original data.

Source: Christiansen/Cummings Associates

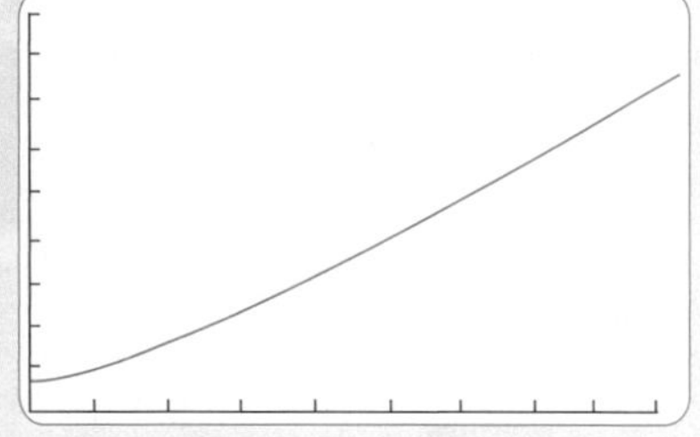

FIGURE T3
The graph of f in the viewing window $[0, 8] \times [0, 10]$

Solution

a. Choosing **P4REG** (fourth-order polynomial regression) from the **STAT CALC** (statistical calculations) menu of a graphing utility, we find

$$f(t) = 0.00379t^4 - 0.06616t^3 + 0.41667t^2 - 0.07291t + 0.48333$$

b. The graph of f is shown in Figure T3.

c. The required values, which compare favorably with the given data, follow:

t	0	1	2	3	4	5	6	7
$f(t)$	0.5	0.8	1.5	2.5	3.6	4.6	5.7	6.8

■

TECHNOLOGY EXERCISES

In Exercises 1–6, find the points of intersection of the graphs of the functions. Express your answer accurate to four decimal places.

1. $f(x) = 1.2x + 3.8$; $g(x) = -0.4x^2 + 1.2x + 7.5$

2. $f(x) = 0.2x^2 - 1.3x - 3$; $g(x) = -1.3x + 2.8$

3. $f(x) = 0.3x^2 - 1.7x - 3.2$; $g(x) = -0.4x^2 + 0.9x + 6.7$

4. $f(x) = -0.3x^2 + 0.6x + 3.2$; $g(x) = 0.2x^2 - 1.2x - 4.8$

5. $f(x) = 0.3x^3 - 1.8x^2 + 2.1x - 2$; $g(x) = 2.1x - 4.2$

6. $f(x) = -0.2x^3 + 1.2x^2 - 1.2x + 2$; $g(x) = -0.2x^2 + 0.8x + 2.1$

7. **Market Equilibrium** The monthly demand and supply functions for a certain brand of wall clock are given by

$$p = -0.2x^2 - 1.2x + 50$$
$$p = 0.1x^2 + 3.2x + 25$$

respectively, where p is measured in dollars and x is measured in units of a hundred.
 a. Plot the graphs of both functions in an appropriate viewing window.
 b. Find the equilibrium quantity and price.

8. **Market Equilibrium** The quantity demanded x (in units of a hundred) of Mikado miniature cameras/week is related to the unit price p (in dollars) by

$$p = -0.2x^2 + 80$$

The quantity x (in units of a hundred) that the supplier is willing to make available in the market is related to the unit price p (in dollars) by

$$p = 0.1x^2 + x + 40$$

 a. Plot the graphs of both functions in an appropriate viewing window.
 b. Find the equilibrium quantity and price.

In Exercises 9–18, use the STAT CALC menu to construct a mathematical model associated with the given data.

9. **Sales of DSPs** The projected sales (in billions of dollars) of digital signal processors (DSPs) and the scatter plot for these data follow:

Year	1997	1998	1999	2000	2001	2002
Sales	3.1	4	5	6.2	8	10

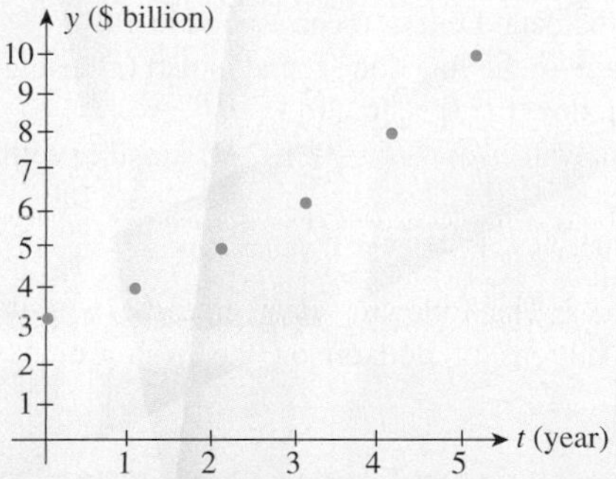

 a. Use **P2REG** to find a second-degree polynomial regression model for the data. Let $t = 0$ correspond to 1997.
 b. Plot the graph of the function f found in part (a), using the viewing window $[0, 5] \times [0, 12]$.
 c. Compute the values of $f(t)$ for $t = 0, 1, 2, 3, 4$, and 5. How does your model compare with the given data?

Source: A. G. Edwards & Sons, Inc.

10. **Annual Retail Sales** Annual retail sales in the United States from 1990 through the year 2000 (in billions of dollars) are given in the following table:

Year	1990	1991	1992	1993	1994	1995
Sales	471.6	485.4	519.2	553.4	595	625.5

Year	1996	1997	1998	1999	2000
Sales	656.6	685.6	727.2	781.7	877.7

 a. Let $t = 0$ correspond to 1990 and use **P2REG** to find a second-degree polynomial regression model based on the given data.
 b. Plot the graph of the function f found in part (a) using the viewing window $[0, 10] \times [0, 1000]$.
 c. Compute $f(0)$, $f(5)$, and $f(10)$. Compare these values with the given data.

Source: National Retail Federation

11. **Digital TV Shipments** The estimated number of digital TV shipments between the year 2000 and 2006 (in millions of units) and the scatter plot for these data follow:

Year	2000	2001	2002	2003	2004	2005	2006
Units Shipped	0.63	1.43	2.57	4.1	6	8.1	10

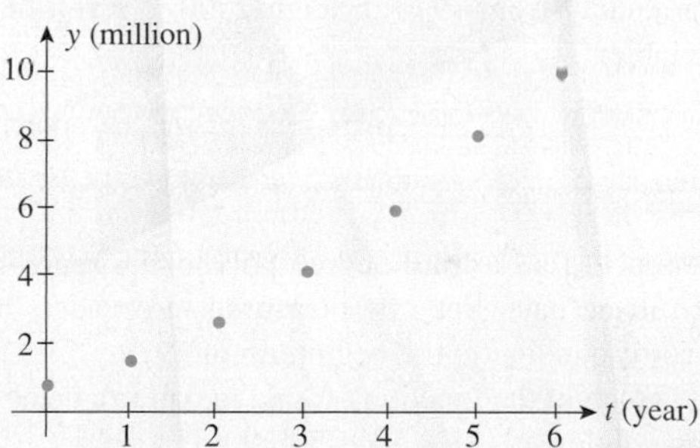

 a. Use **P3REG** to find a third-degree polynomial regression model for the data. Let $t = 0$ correspond to the year 2000.
 b. Plot the graph of the function f found in part (a), using the viewing window $[0, 6] \times [0, 11]$.
 c. Compute the values of $f(t)$ for $t = 0, 1, 2, 3, 4, 5$, and 6.

Source: Consumer Electronics Manufacturers Association

12. **Prison Population** The following data give the past, present, and projected U.S. prison population (in millions) from 1980 through 2005:

Year	1980	1985	1990	1995	2000	2005
Population	0.52	0.77	1.18	1.64	2.23	3.20

 a. Let $t = 0$ correspond to the beginning of 1980, where t is measured in 5-yr intervals, and use **P2REG** to find a

(continued)

second-degree polynomial regression model based on the given data.

b. Plot the graph of the function f found in part (a), using the viewing window $[0, 5] \times [0, 3.5]$.

c. Compute $f(0)$, $f(1)$, $f(2)$, $f(3)$, $f(4)$, and $f(5)$. Compare these values with the given data.

13. FEDERAL DEBT According to data obtained from the Congressional Budget Office, the federal debt (in trillions of dollars) for selected years from 1990 through 2010, with figures after 2000 being estimates, is given in the following table:

Year	1990	1997	2000	2010
Debt	2.55	3.84	3.41	1.70

a. Use P3REG to find a third-degree polynomial regression model for the data, letting $t = 0$ correspond to the beginning of 1990.

b. Plot the graph of the function f found in part (a).

c. Compare the values of f at $t = 0, 7, 10$, and 20, with the given data.

Source: Congressional Budget Office

14. POPULATION GROWTH IN CLARK COUNTY Clark County in Nevada—dominated by greater Las Vegas—is the fastest-growing metropolitan area in the United States. The population of the county from 1970 through 2000 is given in the following table:

Year	1970	1980	1990	2000
Population	273,288	463,087	741,459	1,375,765

a. Use P3REG to find a third-degree polynomial regression model for the data. Let t be measured in decades, with $t = 0$ corresponding to the beginning of 1970.

b. Plot the graph of the function f found in part (a), using the viewing window $[0, 3] \times [0, 1{,}500{,}000]$.

c. Compare the values of f at $t = 0, 1, 2$, and 3, with the given data.

Source: U.S. Census Bureau

15. CABLE AD REVENUE The revenues (in billions of dollars) from cable advertisement for the years 1995 through 2000 follow:

Year	1995	1996	1997	1998	1999	2000
Revenue	5.1	6.6	8.1	9.4	11.1	13.7

a. Use P3REG to find a third-degree polynomial regression model for the data. Let $t = 0$ correspond to 1995.

b. Plot the graph of the function f found in part (a), using the viewing window $[0, 6] \times [0, 14]$.

c. Compare the values of f at $t = 1, 2, 3, 4$, and 5, with the given data.

Source: National Cable Television Association

16. ONLINE SHOPPING The following data give the revenue per year (in billions of dollars) from Internet shopping:

Year	1997	1998	1999	2000	2001
Revenue	2.4	5	8	12	17.4

a. Use P3REG to find a third-degree polynomial regression model for the data. Let $t = 0$ correspond to 1997.

b. Plot the graph of the function f found in part (a), using the viewing window $[0, 4] \times [0, 20]$.

c. Compare the values of f at $t = 0, 1, 2, 3$, and 4, with the given data.

Source: Forrester Research, Inc.

17. MARIJUANA ARRESTS The number of arrests (in thousands) for marijuana sales and possession in New York City from 1992 through 1997 and the scatter plot for these data follow:

Year	1992	1993	1994	1995	1996	1997
Arrests	5.0	5.8	8.8	11.7	18.5	27.5

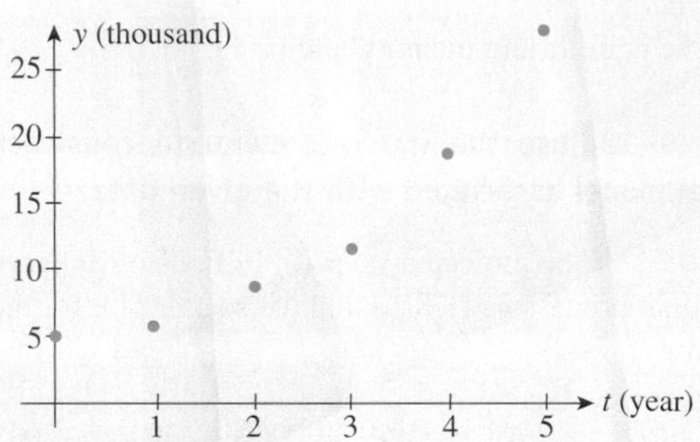

a. Use P4REG to find a fourth-degree polynomial regression model for the data. Let $t = 0$ correspond to 1992.

b. Plot the graph of the function f found in part (a), using the viewing window $[0, 5] \times [0, 30]$.

c. Compare the values of f at $t = 0, 1, 2, 3, 4$, and 5, with the given data.

Source: New York State Division of Criminal Justice Services

18. ONLINE SPENDING The following data give the worldwide spending (in billions of dollars) on the Web from 1997 through 2002.

Year	1997	1998	1999	2000	2001	2002
Spending	5.0	10.5	20.5	37.5	60	95

a. Use P4REG to find a fourth-degree polynomial regression model for the data. Let $t = 0$ correspond to 1997.

b. Plot the graph of the function f found in part (a), using the viewing window $[0, 5] \times [0, 100]$.

Source: International Data Corporation

2.4 Limits

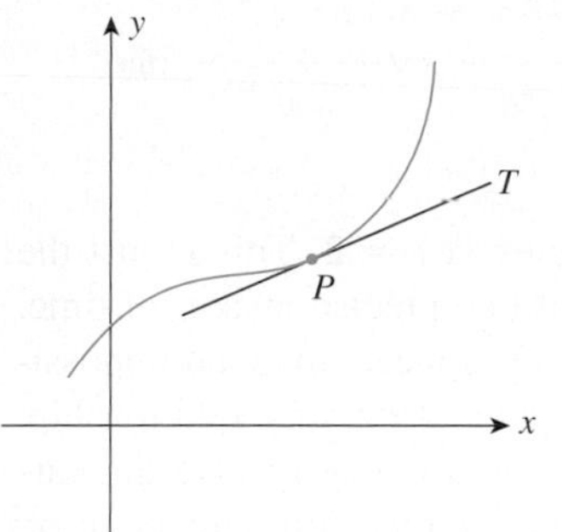

(a) What is the slope of the tangent line T at point P?

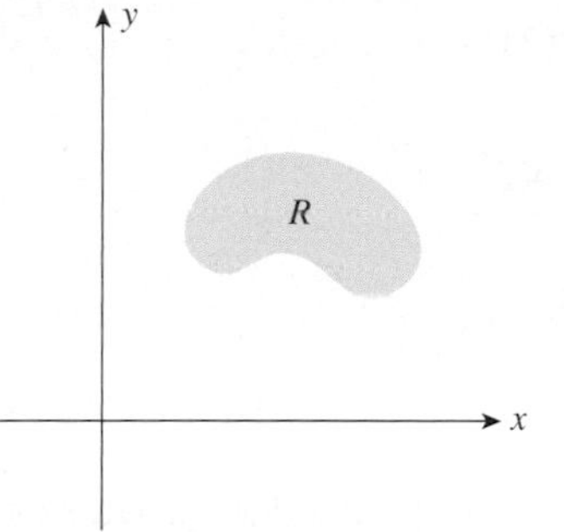

(b) What is the area of the region R?

FIGURE 24

Introduction to Calculus

Historically, the development of calculus by Isaac Newton (1642–1727) and Gottfried Wilhelm Leibniz (1646–1716) resulted from the investigation of the following problems:

1. Finding the tangent line to a curve at a given point on the curve (Figure 24a)
2. Finding the area of a planar region bounded by an arbitrary curve (Figure 24b)

The tangent-line problem might appear to be unrelated to any practical applications of mathematics, but as you will see later, the problem of finding the *rate of change* of one quantity with respect to another is mathematically equivalent to the geometric problem of finding the slope of the *tangent line* to a curve at a given point on the curve. It is precisely the discovery of the relationship between these two problems that spurred the development of calculus in the 17th century and made it such an indispensable tool for solving practical problems. The following are a few examples of such problems:

- Finding the velocity of an object
- Finding the rate of change of a bacteria population with respect to time
- Finding the rate of change of a company's profit with respect to time
- Finding the rate of change of a travel agency's revenue with respect to the agency's expenditure for advertising

The study of the tangent-line problem led to the creation of *differential calculus*, which relies on the concept of the *derivative* of a function. The study of the area problem led to the creation of *integral calculus*, which relies on the concept of the *antiderivative*, or *integral*, of a function. (The derivative of a function and the integral of a function are intimately related, as you will see in Section 6.4.) Both the derivative of a function and the integral of a function are defined in terms of a more fundamental concept—the limit—our next topic.

A Real-Life Example

From data obtained in a test run conducted on a prototype of a maglev (magnetic levitation train), which moves along a straight monorail track, engineers have determined that the position of the maglev (in feet) from the origin at time t is given by

$$s = f(t) = 4t^2 \qquad (0 \leq t \leq 30) \tag{3}$$

where f is called the **position function** of the maglev. The position of the maglev at time $t = 0, 1, 2, 3, \ldots, 10$, measured from its initial position, is

$$f(0) = 0 \qquad f(1) = 4 \qquad f(2) = 16 \qquad f(3) = 36, \ldots \qquad f(10) = 400$$

feet (Figure 25).

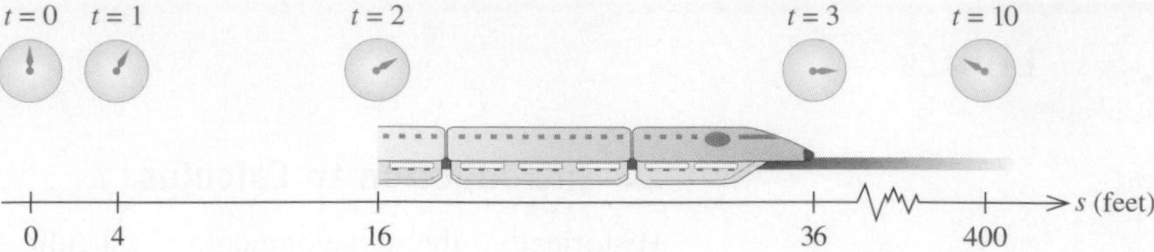

FIGURE 25
A maglev moving along an elevated monorail track

Suppose we want to find the velocity of the maglev at $t = 2$. This is just the velocity of the maglev as shown on its speedometer at that precise instant of time. Offhand, calculating this quantity using only Equation (3) appears to be an impossible task; but consider what quantities we *can* compute using this relationship. Obviously, we can compute the position of the maglev at any time t as we did earlier for some selected values of t. Using these values, we can then compute the *average velocity* of the maglev over an interval of time. For example, the average velocity of the train over the time interval [2, 4] is given by

$$\begin{aligned}\frac{\text{Distance covered}}{\text{Time elapsed}} &= \frac{f(4) - f(2)}{4 - 2} \\ &= \frac{4(4^2) - 4(2^2)}{2} \\ &= \frac{64 - 16}{2} = 24\end{aligned}$$

or 24 feet/second.

Although this is not quite the velocity of the maglev at $t = 2$, it does provide us with an approximation of its velocity at that time.

Can we do better? Intuitively, the smaller the time interval we pick (with $t = 2$ as the left endpoint), the better the average velocity over that time interval will approximate the actual velocity of the maglev at $t = 2$.*

Now, let's describe this process in general terms. Let $t > 2$. Then, the average velocity of the maglev over the time interval $[2, t]$ is given by

$$\frac{f(t) - f(2)}{t - 2} = \frac{4t^2 - 4(2^2)}{t - 2} = \frac{4(t^2 - 4)}{t - 2} \tag{4}$$

By choosing the values of t closer and closer to 2, we obtain a sequence of numbers that give the average velocities of the maglev over smaller and smaller time intervals. As we observed earlier, this sequence of numbers should approach the *instantaneous velocity* of the train at $t = 2$.

Let's try some sample calculations. Using Equation (4) and taking the sequence $t = 2.5, 2.1, 2.01, 2.001$, and 2.0001, which approaches 2, we find

The average velocity over [2, 2.5] is $\dfrac{4(2.5^2 - 4)}{2.5 - 2} = 18$, or 18 feet/second.

The average velocity over [2, 2.1] is $\dfrac{4(2.1^2 - 4)}{2.1 - 2} = 16.4$, or 16.4 feet/second.

and so forth. These results are summarized in Table 1.

*Actually, any interval containing $t = 2$ will do.

TABLE 1

		t approaches 2 from the right.			
t	2.5	2.1	2.01	2.001	2.0001
Average Velocity over [2, t]	18	16.4	16.04	16.004	16.0004

Average velocity approaches 16 from the right.

From Table 1, we see that the average velocity of the maglev seems to approach the number 16 as it is computed over smaller and smaller time intervals. These computations suggest that the instantaneous velocity of the train at $t = 2$ is 16 feet/second.

Note Notice that we cannot obtain the instantaneous velocity for the maglev at $t = 2$ by substituting $t = 2$ into Equation (4) because this value of t is not in the domain of the average velocity function. ■

Intuitive Definition of a Limit

Consider the function g defined by

$$g(t) = \frac{4(t^2 - 4)}{t - 2}$$

which gives the average velocity of the maglev [see Equation (4)]. Suppose we are required to determine the value that $g(t)$ approaches as t approaches the (fixed) number 2. If we take the sequence of values of t approaching 2 from the right-hand side, as we did earlier, we see that $g(t)$ approaches the number 16. Similarly, if we take a sequence of values of t approaching 2 from the left, such as $t = 1.5, 1.9, 1.99, 1.999$, and 1.9999, we obtain the results shown in Table 2.

TABLE 2

		t approaches 2 from the left.			
t	1.5	1.9	1.99	1.999	1.9999
$g(t)$	14	15.6	15.96	15.996	15.9996

$g(t)$ approaches 16 from the left.

Observe that $g(t)$ approaches the number 16 as t approaches 2—this time from the left-hand side. In other words, as t approaches 2 from *either* side of 2, $g(t)$ approaches 16. In this situation, we say that the limit of $g(t)$ as t approaches 2 is 16, written

$$\lim_{t \to 2} g(t) = \lim_{t \to 2} \frac{4(t^2 - 4)}{t - 2} = 16$$

The graph of the function g, shown in Figure 26, confirms this observation.

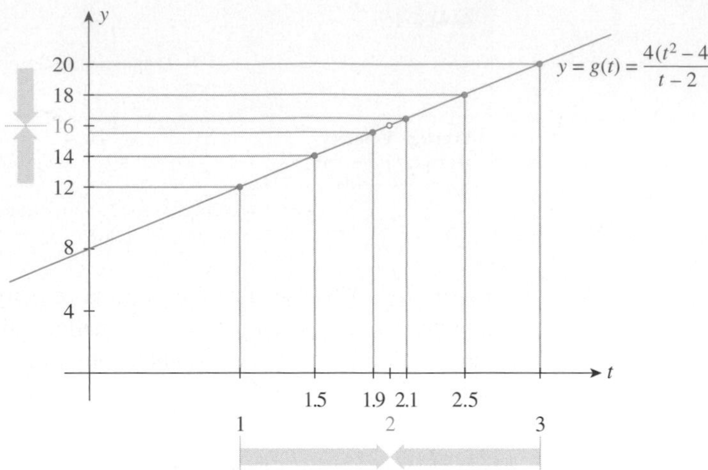

FIGURE 26
As t approaches $t = 2$ from either direction, $g(t)$ approaches $y = 16$.

Observe that the point $t = 2$ is not in the domain of the function g [for this reason, the point (2, 16) is missing from the graph of g]. This, however, is inconsequential because the value, if any, of $g(t)$ at $t = 2$ plays no role in computing the limit.

This example leads to the following informal definition.

Limit of a Function

The function f has the **limit** L as x approaches a, written

$$\lim_{x \to a} f(x) = L$$

if the value of $f(x)$ can be made as close to the number L as we please by taking x sufficiently close to (but not equal to) a.

EXPLORING WITH TECHNOLOGY

1. Use a graphing utility to plot the graph of
$$g(x) = \frac{4(x^2 - 4)}{x - 2}$$
in the viewing window $[0, 3] \times [0, 20]$.
2. Use ZOOM and TRACE to describe what happens to the values of $g(x)$ as x approaches 2, first from the right and then from the left.
3. What happens to the y-value when you try to evaluate $g(x)$ at $x = 2$? Explain.
4. Reconcile your results with those of the preceding example.

Evaluating the Limit of a Function

Let's now consider some examples involving the computation of limits.

EXAMPLE 1 Let $f(x) = x^3$ and evaluate $\lim_{x \to 2} f(x)$.

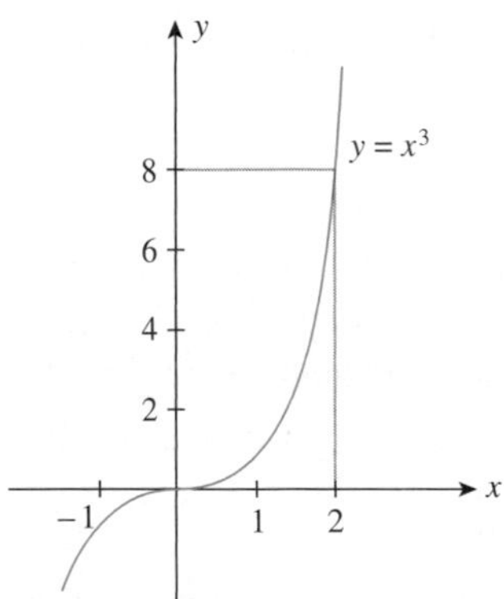

FIGURE 27
$f(x)$ is close to 8 whenever x is close to 2.

Solution The graph of f is shown in Figure 27. You can see that $f(x)$ can be made as close to the number 8 as we please by taking x sufficiently close to 2.

Therefore,

$$\lim_{x \to 2} x^3 = 8$$

■

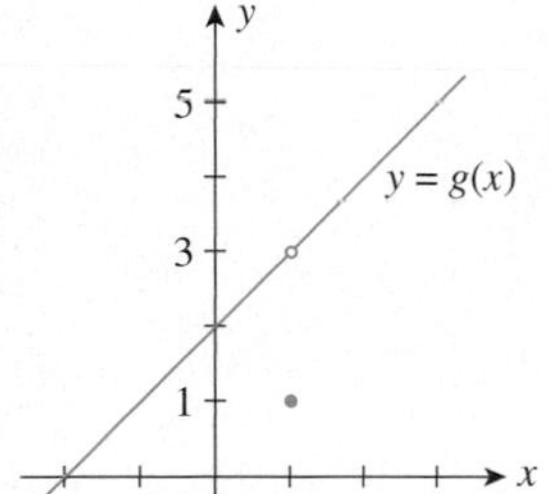

FIGURE 28
$\lim_{x \to 1} g(x) = 3$

EXAMPLE 2 Let

$$g(x) = \begin{cases} x + 2 & \text{if } x \neq 1 \\ 1 & \text{if } x = 1 \end{cases}$$

Evaluate $\lim_{x \to 1} g(x)$.

Solution The domain of g is the set of all real numbers. From the graph of g shown in Figure 28, we see that $g(x)$ can be made as close to 3 as we please by taking x sufficiently close to 1. Therefore,

$$\lim_{x \to 1} g(x) = 3$$

Observe that $g(1) = 1$, which is not equal to the limit of the function g as x approaches 1. [Once again, the value of $g(x)$ at $x = 1$ has no bearing on the existence or value of the limit of g as x approaches 1.] ■

EXAMPLE 3 Evaluate the limit of the following functions as x approaches the indicated point.

a. $f(x) = \begin{cases} -1 & \text{if } x < 0 \\ 1 & \text{if } x \geq 0 \end{cases}; \; x = 0$ **b.** $g(x) = \dfrac{1}{x^2}; \; x = 0$

Solution The graphs of the functions f and g are shown in Figure 29.

FIGURE 29

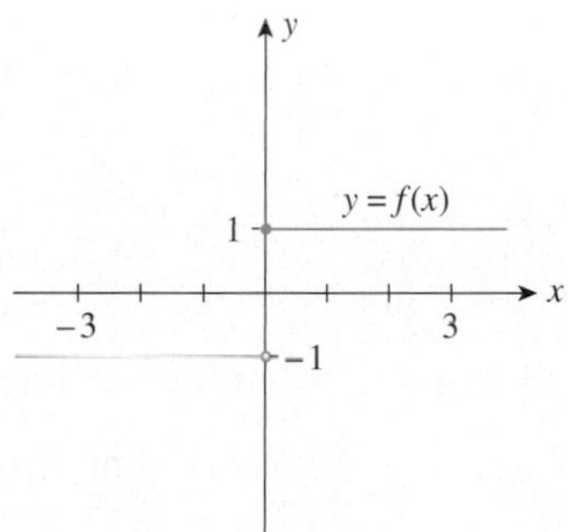

(a) $\lim_{x \to 0} f(x)$ does not exist.

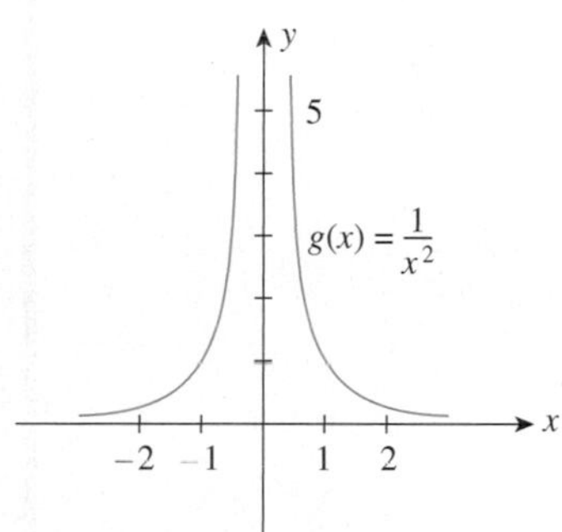

(b) $\lim_{x \to 0} g(x)$ does not exist.

a. Referring to Figure 29a, we see that no matter how close x is to zero, $f(x)$ takes on the values 1 or -1, depending on whether x is positive or negative. Thus, there is no *single* real number L that $f(x)$ approaches as x approaches zero. We conclude that the limit of $f(x)$ does *not* exist as x approaches zero.

b. Referring to Figure 29b, we see that as x approaches zero (from either side), $g(x)$ increases without bound and thus does not approach any specific real number. We conclude, accordingly, that the limit of $g(x)$ does *not* exist as x approaches zero. ■

EXPLORE & DISCUSS

Consider the graph of the function h whose graph is depicted in the following figure.

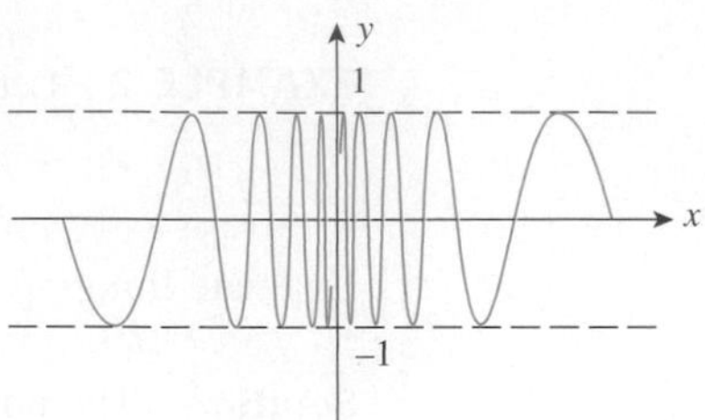

It has the property that as x approaches zero from either the right or the left, the curve oscillates more and more frequently between the lines $y = -1$ and $y = 1$.

1. Explain why $\lim_{x \to 0} h(x)$ does not exist.
2. Compare this function with those in Example 3. More specifically, discuss the different ways the functions fail to have a limit at $x = 0$.

Until now, we have relied on knowing the actual values of a function or the graph of a function near $x = a$ to help us evaluate the limit of the function $f(x)$ as x approaches a. The following properties of limits, which we list without proof, enable us to evaluate limits of functions algebraically.

THEOREM 1

Properties of Limits

Suppose

$$\lim_{x \to a} f(x) = L \quad \text{and} \quad \lim_{x \to a} g(x) = M$$

Then,

1. $\lim_{x \to a} [f(x)]^r = [\lim_{x \to a} f(x)]^r = L^r$ r, a real number
2. $\lim_{x \to a} cf(x) = c \lim_{x \to a} f(x) = cL$ c, a real number
3. $\lim_{x \to a} [f(x) \pm g(x)] = \lim_{x \to a} f(x) \pm \lim_{x \to a} g(x) = L \pm M$
4. $\lim_{x \to a} [f(x)g(x)] = [\lim_{x \to a} f(x)][\lim_{x \to a} g(x)] = LM$
5. $\lim_{x \to a} \dfrac{f(x)}{g(x)} = \dfrac{\lim_{x \to a} f(x)}{\lim_{x \to a} g(x)} = \dfrac{L}{M}$ Provided that $M \neq 0$

EXAMPLE 4 Use Theorem 1 to evaluate the following limits.

a. $\lim_{x \to 2} x^3$ **b.** $\lim_{x \to 4} 5x^{3/2}$ **c.** $\lim_{x \to 1} (5x^4 - 2)$

d. $\lim_{x \to 3} 2x^3 \sqrt{x^2 + 7}$ **e.** $\lim_{x \to 2} \dfrac{2x^2 + 1}{x + 1}$

Solution

a. $\lim_{x\to 2} x^3 = [\lim_{x\to 2} x]^3$ Property 1

$= 2^3 = 8$ $\quad \lim_{x\to 2} x = 2$

b. $\lim_{x\to 4} 5x^{3/2} = 5[\lim_{x\to 4} x^{3/2}]$ Property 2

$= 5(4)^{3/2} = 40$

c. $\lim_{x\to 1}(5x^4 - 2) = \lim_{x\to 1} 5x^4 - \lim_{x\to 1} 2$ Property 3

To evaluate $\lim_{x\to 1} 2$, observe that the constant function $g(x) = 2$ has value 2 for all values of x. Therefore, $g(x)$ must approach the limit 2 as x approaches $x = 1$ (or any other point for that matter!). Therefore,

$$\lim_{x\to 1}(5x^4 - 2) = 5(1)^4 - 2 = 3$$

d. $\lim_{x\to 3} 2x^3 \sqrt{x^2 + 7} = 2 \lim_{x\to 3} x^3 \sqrt{x^2 + 7}$ Property 2

$= 2 \lim_{x\to 3} x^3 \lim_{x\to 3} \sqrt{x^2 + 7}$ Property 4

$= 2(3)^3 \sqrt{3^2 + 7}$ Property 1

$= 2(27)\sqrt{16} = 216$

e. $\lim_{x\to 2} \dfrac{2x^2 + 1}{x + 1} = \dfrac{\lim_{x\to 2}(2x^2 + 1)}{\lim_{x\to 2}(x + 1)}$ Property 5

$= \dfrac{2(2)^2 + 1}{2 + 1} = \dfrac{9}{3} = 3$

Indeterminate Forms

Let's emphasize once again that Property 5 of limits is valid only when the limit of the function that appears in the denominator is not equal to zero at the number in question.

If the numerator has a limit different from zero and the denominator has a limit equal to zero, then the limit of the quotient does not exist at the number in question. This is the case with the function $g(x) = 1/x^2$ in Example 3b. Here, as x approaches zero, the numerator approaches 1 but the denominator approaches zero, so the quotient becomes arbitrarily large. Thus, as observed earlier, the limit does not exist.

Next, consider

$$\lim_{x\to 2} \frac{4(x^2 - 4)}{x - 2}$$

which we evaluated earlier by looking at the values of the function for x near $x = 2$. If we attempt to evaluate this expression by applying Property 5 of limits, we see that both the numerator and denominator of the function

$$\frac{4(x^2 - 4)}{x - 2}$$

approach zero as x approaches 2; that is, we obtain an expression of the form 0/0. In this event, we say that the limit of the quotient $f(x)/g(x)$ as x approaches 2 has the **indeterminate form 0/0**.

We need to evaluate limits of this type when we discuss the derivative of a function, a fundamental concept in the study of calculus. As the name suggests, the meaningless expression 0/0 does not provide us with a solution to our problem. One strategy that can be used to solve this type of problem follows.

Strategy for Evaluating Indeterminate Forms

1. Replace the given function with an appropriate one that takes on the same values as the original function everywhere except at $x = a$.
2. Evaluate the limit of this function as x approaches a.

Examples 5 and 6 illustrate this strategy.

EXAMPLE 5 Evaluate:

$$\lim_{x \to 2} \frac{4(x^2 - 4)}{x - 2}$$

Solution Since both the numerator and the denominator of this expression approach zero as x approaches 2, we have the indeterminate form 0/0. We rewrite

$$\frac{4(x^2 - 4)}{x - 2} = \frac{4(x - 2)(x + 2)}{(x - 2)}$$

which, upon canceling the common factors, is equivalent to $4(x + 2)$, provided $x \neq 2$. Next, we replace $4(x^2 - 4)/(x - 2)$ with $4(x + 2)$ and find that

$$\lim_{x \to 2} \frac{4(x^2 - 4)}{x - 2} = \lim_{x \to 2} 4(x + 2) = 16$$

The graphs of the functions

$$f(x) = \frac{4(x^2 - 4)}{x - 2} \quad \text{and} \quad g(x) = 4(x + 2)$$

are shown in Figure 30. Observe that the graphs are identical except when $x = 2$. The function g is defined for all values of x and, in particular, its value at $x = 2$ is $g(2) = 4(2 + 2) = 16$. Thus, the point (2, 16) is on the graph of g. However, the function f is not defined at $x = 2$. Since $f(x) = g(x)$ for all values of x except $x = 2$, it follows that the graph of f must look exactly like the graph of g, with the exception that the point (2, 16) is missing from the graph of f. This illustrates graphically why we can evaluate the limit of f by evaluating the limit of the "equivalent" function g.

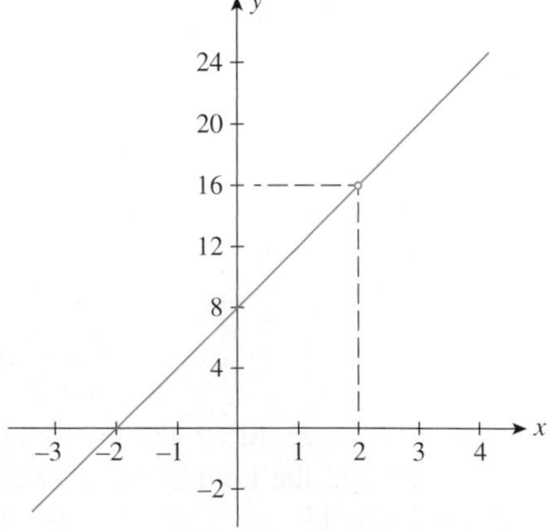

(a) $f(x) = \frac{4(x^2 - 4)}{x - 2}$

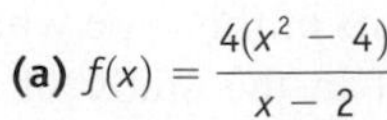

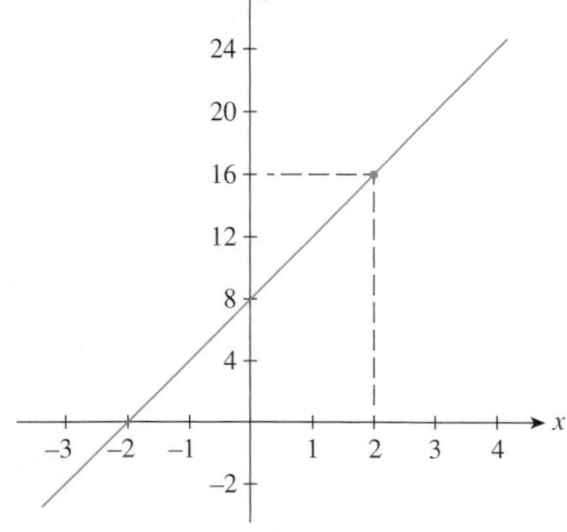

(b) $g(x) = 4(x + 2)$

FIGURE 30
The graphs of $f(x)$ and $g(x)$ are identical except at the point (2, 16).

Note Notice that the limit in Example 5 is the same limit that we evaluated earlier when we discussed the instantaneous velocity of a maglev at a specified time. ■

EXPLORING WITH TECHNOLOGY

1. Use a graphing utility to plot the graph of

$$f(x) = \frac{4(x^2 - 4)}{x - 2}$$

in the viewing window $[0, 3] \times [0, 20]$. Then use ZOOM and TRACE to find

$$\lim_{x \to 2} \frac{4(x^2 - 4)}{x - 2}$$

2. Use a graphing utility to plot the graph of $g(x) = 4(x + 2)$ in the viewing window $[0, 3] \times [0, 20]$. Then use ZOOM and TRACE to find $\lim_{x \to 2} 4(x + 2)$. What happens to the y-value when you try to evaluate $f(x)$ at $x = 2$? Explain.

3. Can you distinguish between the graphs of f and g?

4. Reconcile your results with those of Example 5.

EXAMPLE 6 Evaluate:

$$\lim_{h \to 0} \frac{\sqrt{1 + h} - 1}{h}$$

Solution Letting h approach zero, we obtain the indeterminate form 0/0. Next, we rationalize the numerator of the quotient (see page 21) by multiplying both the numerator and the denominator by the expression $(\sqrt{1 + h} + 1)$, obtaining

$$\begin{aligned} \frac{\sqrt{1 + h} - 1}{h} &= \frac{(\sqrt{1 + h} - 1)(\sqrt{1 + h} + 1)}{h(\sqrt{1 + h} + 1)} \\ &= \frac{1 + h - 1}{h(\sqrt{1 + h} + 1)} \qquad (\sqrt{a} - \sqrt{b})(\sqrt{a} + \sqrt{b}) = a - b \\ &= \frac{h}{h(\sqrt{1 + h} + 1)} \\ &= \frac{1}{\sqrt{1 + h} + 1} \end{aligned}$$

Therefore,

$$\begin{aligned} \lim_{h \to 0} \frac{\sqrt{1 + h} - 1}{h} &= \lim_{h \to 0} \frac{1}{\sqrt{1 + h} + 1} \\ &= \frac{1}{\sqrt{1} + 1} = \frac{1}{2} \end{aligned}$$

■

EXPLORING WITH TECHNOLOGY

1. Use a graphing utility to plot the graph of
$$g(x) = \frac{\sqrt{1+x} - 1}{x}$$ in the viewing window $[-1, 2] \times [0, 1]$. Then use **ZOOM** and **TRACE** to find
$$\lim_{x \to 0} \frac{\sqrt{1+x} - 1}{x}$$ by observing the values of $g(x)$ as x approaches zero from the left and from the right.
2. Use a graphing utility to plot the graph of
$$f(x) = \frac{1}{\sqrt{1+x} + 1}$$ in the viewing window $[-1, 2] \times [0, 1]$. Then use **ZOOM** and **TRACE** to find $\lim_{x \to 0} \frac{1}{\sqrt{1+x} + 1}$. What happens to the y-value when x takes on the value zero? Explain.
3. Can you distinguish between the graphs of f and g?
4. Reconcile your results with those of Example 6.

Limits at Infinity

Up to now we have studied the limit of a function as x approaches a (finite) number a. There are occasions, however, when we want to know whether $f(x)$ approaches a unique number as x increases without bound. Consider, for example, the function P, giving the number of fruit flies (*Drosophila*) in a container under controlled laboratory conditions, as a function of a time t. The graph of P is shown in Figure 31. You can see from the graph of P that, as t increases without bound (gets larger and larger), $P(t)$ approaches the number 400. This number, called the *carrying capacity* of the environment, is determined by the amount of living space and food available, as well as other environmental factors.

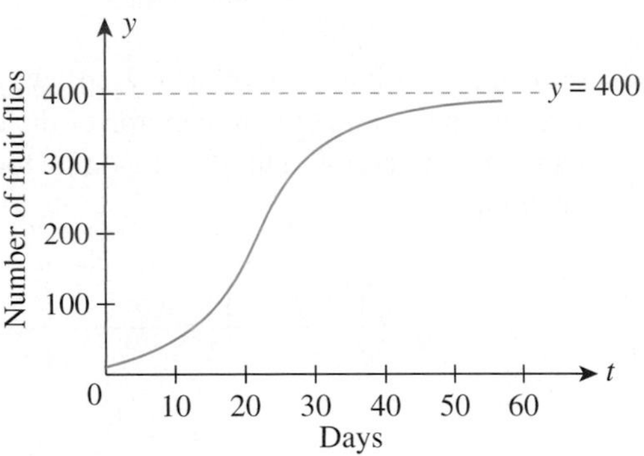

FIGURE 31
The graph of $P(t)$ gives the population of fruit flies in a laboratory experiment.

As another example, suppose we are given the function

$$f(x) = \frac{2x^2}{1 + x^2}$$

and we want to determine what happens to $f(x)$ as x gets larger and larger. Picking the sequence of numbers 1, 2, 5, 10, 100, and 1000 and computing the corresponding values of $f(x)$, we obtain the following table of values:

x	1	2	5	10	100	1000
$f(x)$	1	1.6	1.92	1.98	1.9998	1.999998

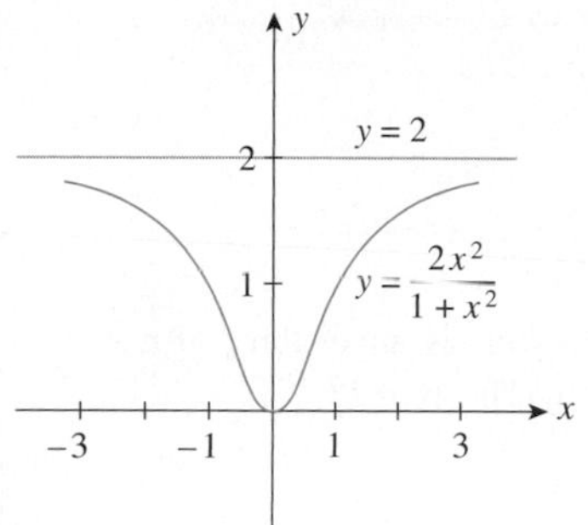

FIGURE 32

The graph of $y = \frac{2x^2}{1 + x^2}$ has a horizontal asymptote at $y = 2$.

From the table, we see that as x gets larger and larger, $f(x)$ gets closer and closer to 2. The graph of the function f shown in Figure 32 confirms this observation. We call the line $y = 2$ a **horizontal asymptote.*** In this situation, we say that the limit of the function $f(x)$ as x increases without bound is 2, written

$$\lim_{x\to\infty} \frac{2x^2}{1 + x^2} = 2$$

In the general case, the following definition for a **limit of a function at infinity** is applicable.

Limit of a Function at Infinity

The function f has the limit L as x increases without bound (or, as x approaches infinity), written

$$\lim_{x\to\infty} f(x) = L$$

if $f(x)$ can be made arbitrarily close to L by taking x large enough.

Similarly, the function f has the limit M as x decreases without bound (or as x approaches negative infinity), written

$$\lim_{x\to-\infty} f(x) = M$$

if $f(x)$ can be made arbitrarily close to M by taking x to be negative and sufficiently large in absolute value.

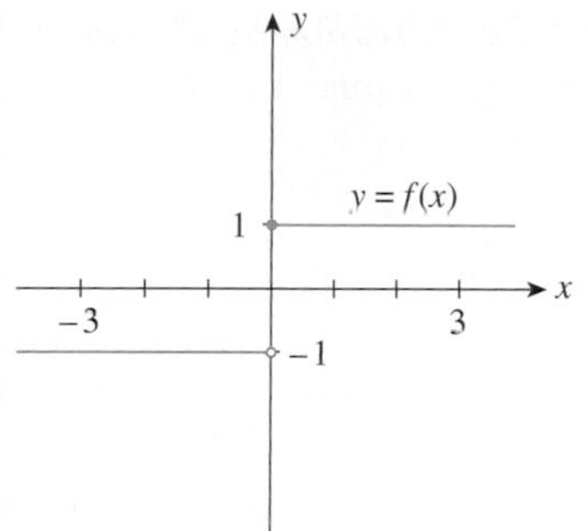

(a) $\lim_{x\to\infty} f(x) = 1$ and $\lim_{x\to-\infty} f(x) = -1$

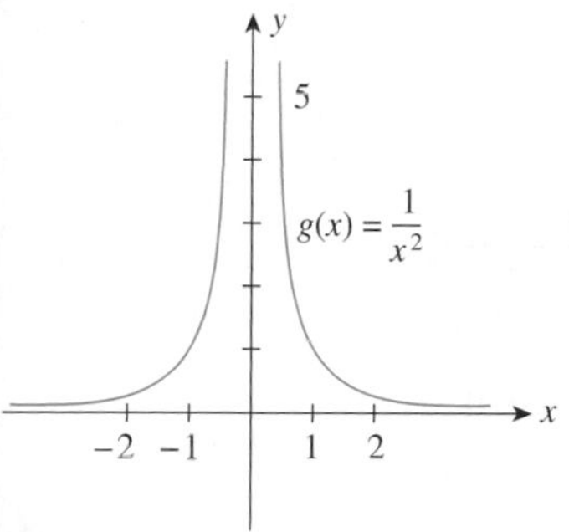

(b) $\lim_{x\to\infty} g(x) = 0$ and $\lim_{x\to-\infty} g(x) = 0$

FIGURE 33

EXAMPLE 7 Let f and g be the functions

$$f(x) = \begin{cases} -1 & \text{if } x < 0 \\ 1 & \text{if } x \geq 0 \end{cases} \qquad \text{and} \qquad g(x) = \frac{1}{x^2}$$

Evaluate:

a. $\lim_{x\to\infty} f(x)$ and $\lim_{x\to-\infty} f(x)$ **b.** $\lim_{x\to\infty} g(x)$ and $\lim_{x\to-\infty} g(x)$

Solution The graphs of $f(x)$ and $g(x)$ are shown in Figure 33. Referring to the graphs of the respective functions, we see that

a. $\lim_{x\to\infty} f(x) = 1$ and $\lim_{x\to-\infty} f(x) = -1$ **b.** $\lim_{x\to\infty} \frac{1}{x^2} = 0$ and $\lim_{x\to-\infty} \frac{1}{x^2} = 0$

All the properties of limits listed in Theorem 1 are valid when a is replaced by ∞ or $-\infty$. In addition, we have the following property for the limit at infinity.

THEOREM 2

For all $n > 0$, $\lim_{x\to\infty} \frac{1}{x^n} = 0$ and $\lim_{x\to-\infty} \frac{1}{x^n} = 0$

provided that $\frac{1}{x^n}$ is defined.

*We will discuss asymptotes in greater detail in Section 4.3.

EXPLORING WITH TECHNOLOGY

1. Use a graphing utility to plot the graphs of

$$y_1 = \frac{1}{x^{0.5}} \qquad y_2 = \frac{1}{x} \qquad y_3 = \frac{1}{x^{1.5}}$$

in the viewing window $[0, 200] \times [0, 0.5]$. What can you say about $\lim_{x\to\infty} \frac{1}{x^n}$ if $n = 0.5$, $n = 1$, and $n = 1.5$? Are these results predicted by Theorem 2?

2. Use a graphing utility to plot the graphs of

$$y_1 = \frac{1}{x} \qquad \text{and} \qquad y_2 = \frac{1}{x^{5/3}}$$

in the viewing window $[-50, 0] \times [-0.5, 0]$. What can you say about $\lim_{x\to-\infty} \frac{1}{x^n}$ if $n = 1$ and $n = \frac{5}{3}$? Are these results predicted by Theorem 2?

Hint: To graph y_2, write it in the form y2 = 1/(x^(1/3))^5.

We often use the following technique to evaluate the limit at infinity of a rational function: *Divide the numerator and denominator of the expression by x^n, where n is the highest power present in the denominator of the expression.*

EXAMPLE 8 Evaluate

$$\lim_{x\to\infty} \frac{x^2 - x + 3}{2x^3 + 1}$$

Solution Since the limits of both the numerator and the denominator do not exist as x approaches infinity, the property pertaining to the limit of a quotient (Property 5) is not applicable. Let's divide the numerator and denominator of the rational expression by x^3, obtaining

$$\begin{aligned}\lim_{x\to\infty} \frac{x^2 - x + 3}{2x^3 + 1} &= \lim_{x\to\infty} \frac{\frac{1}{x} - \frac{1}{x^2} + \frac{3}{x^3}}{2 + \frac{1}{x^3}} \\ &= \frac{0 - 0 + 0}{2 + 0} = \frac{0}{2} \qquad \text{Use Theorem 2.} \\ &= 0\end{aligned}$$

■

EXAMPLE 9 Let

$$f(x) = \frac{3x^2 + 8x - 4}{2x^2 + 4x - 5}$$

Compute $\lim_{x\to\infty} f(x)$ if it exists.

Solution Again, we see that Property 5 is not applicable. Dividing the numerator and the denominator by x^2, we obtain

$$\lim_{x\to\infty} \frac{3x^2 + 8x - 4}{2x^2 + 4x - 5} = \lim_{x\to\infty} \frac{3 + \frac{8}{x} - \frac{4}{x^2}}{2 + \frac{4}{x} - \frac{5}{x^2}}$$

$$= \frac{\lim_{x\to\infty} 3 + 8 \lim_{x\to\infty} \frac{1}{x} - 4 \lim_{x\to\infty} \frac{1}{x^2}}{\lim_{x\to\infty} 2 + 4 \lim_{x\to\infty} \frac{1}{x} - 5 \lim_{x\to\infty} \frac{1}{x^2}}$$

$$= \frac{3 + 0 - 0}{2 + 0 - 0} \qquad \text{Use Theorem 2.}$$

$$= \frac{3}{2}$$

■

EXAMPLE 10 Let $f(x) = \dfrac{2x^3 - 3x^2 + 1}{x^2 + 2x + 4}$ and evaluate:

a. $\lim_{x\to\infty} f(x)$ **b.** $\lim_{x\to -\infty} f(x)$

Solution

a. Dividing the numerator and the denominator of the rational expression by x^2, we obtain

$$\lim_{x\to\infty} \frac{2x^3 - 3x^2 + 1}{x^2 + 2x + 4} = \lim_{x\to\infty} \frac{2x - 3 + \frac{1}{x^2}}{1 + \frac{2}{x} + \frac{4}{x^2}}$$

Since the numerator becomes arbitrarily large whereas the denominator approaches 1 as x approaches infinity, we see that the quotient $f(x)$ gets larger and larger as x approaches infinity. In other words, the limit does not exist. We indicate this by writing

$$\lim_{x\to\infty} \frac{2x^3 - 3x^2 + 1}{x^2 + 2x + 4} = \infty$$

b. Once again, dividing both the numerator and the denominator by x^2, we obtain

$$\lim_{x\to -\infty} \frac{2x^3 - 3x^2 + 1}{x^2 + 2x + 4} = \lim_{x\to -\infty} \frac{2x - 3 + \frac{1}{x^2}}{1 + \frac{2}{x} + \frac{4}{x^2}}$$

In this case, the numerator becomes arbitrarily large in magnitude but negative in sign, whereas the denominator approaches 1 as x approaches negative infinity. Therefore, the quotient $f(x)$ decreases without bound, and the limit does not exist. We indicate this by writing

$$\lim_{x\to -\infty} \frac{2x^3 - 3x^2 + 1}{x^2 + 2x + 4} = -\infty$$

■

Example 11 gives an application of the concept of the limit of a function at infinity.

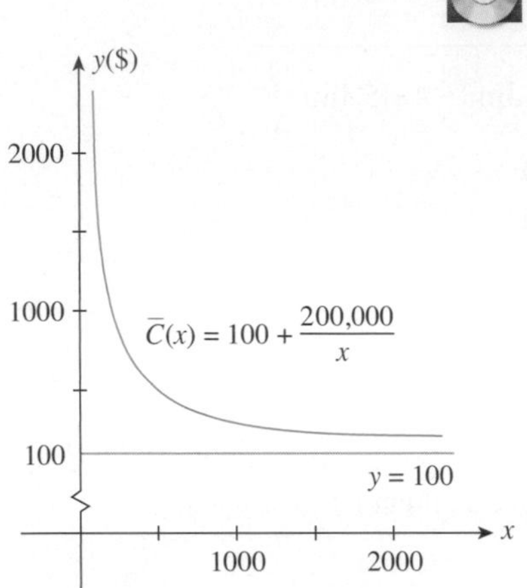

FIGURE 34
As the level of production increases, the average cost approaches $100 per desk.

VIDEO

APPLIED EXAMPLE 11 Average Cost Functions Custom Office makes a line of executive desks. It is estimated that the total cost of making x Senior Executive Model desks is $C(x) = 100x + 200,000$ dollars per year, so the average cost of making x desks is given by

$$\overline{C}(x) = \frac{C(x)}{x} = \frac{100x + 200,000}{x} = 100 + \frac{200,000}{x}$$

dollars per desk. Evaluate $\lim_{x\to\infty} \overline{C}(x)$ and interpret your results.

Solution

$$\lim_{x\to\infty} \overline{C}(x) = \lim_{x\to\infty}\left(100 + \frac{200,000}{x}\right)$$

$$= \lim_{x\to\infty} 100 + \lim_{x\to\infty} \frac{200,000}{x} = 100$$

A sketch of the graph of the function $\overline{C}(x)$ appears in Figure 34. The result we obtained is fully expected if we consider its economic implications. Note that as the level of production increases, the fixed cost per desk produced, represented by the term $(200,000/x)$, drops steadily. The average cost should approach a constant unit cost of production—$100 in this case.

EXPLORE & DISCUSS

Consider the graph of the function f depicted in the following figure:

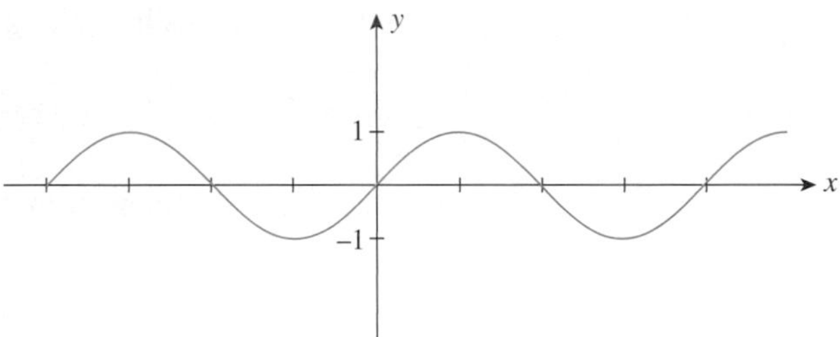

It has the property that the curve oscillates between $y = -1$ and $y = 1$ indefinitely in either direction.

1. Explain why $\lim_{x\to-\infty} f(x)$ and $\lim_{x\to\infty} f(x)$ do not exist.
2. Compare this function with those of Example 10. More specifically, discuss the different ways each function fails to have a limit at infinity or minus infinity.

2.4 Self-Check Exercises

1. Find the indicated limit if it exists.

   a. $\lim_{x \to 3} \dfrac{\sqrt{x^2 + 7} + \sqrt{3x - 5}}{x + 2}$

   b. $\lim_{x \to -1} \dfrac{x^2 - x - 2}{2x^2 - x - 3}$

2. The average cost per disc (in dollars) incurred by Herald Records in pressing x compact discs (CDs) is given by the average cost function

$$\overline{C}(x) = 1.8 + \frac{3000}{x}$$

Evaluate $\lim_{x \to \infty} \overline{C}(x)$ and interpret your result.

Solutions to Self-Check Exercises 2.4 can be found on page 115.

2.4 Concept Questions

1. Explain what is meant by the statement $\lim_{x \to 2} f(x) = 3$.

2. a. If $\lim_{x \to 3} f(x) = 5$, what can you say about $f(3)$? Explain.

   b. If $f(2) = 6$, what can you say about $\lim_{x \to 2} f(x)$? Explain.

3. Evaluate and state the property of limits that you use at each step.

   a. $\lim_{x \to 4} \sqrt{x}(2x^2 + 1)$ b. $\lim_{x \to 1} \left(\dfrac{2x^2 + x + 5}{x^4 + 1} \right)^{3/2}$

4. What is an indeterminate form? Illustrate with an example.

5. Explain in your own words the meaning of $\lim_{x \to \infty} f(x) = L$ and $\lim_{x \to -\infty} f(x) = M$.

2.4 Exercises

In Exercises 1–8, use the graph of the given function f to determine $\lim_{x \to a} f(x)$ at the indicated value of a, if it exists.

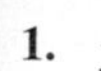

1\. 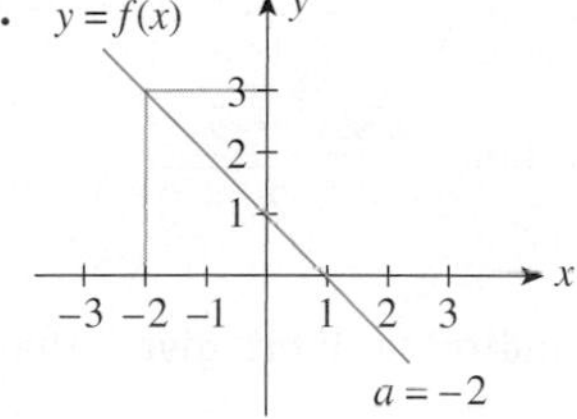

2\. 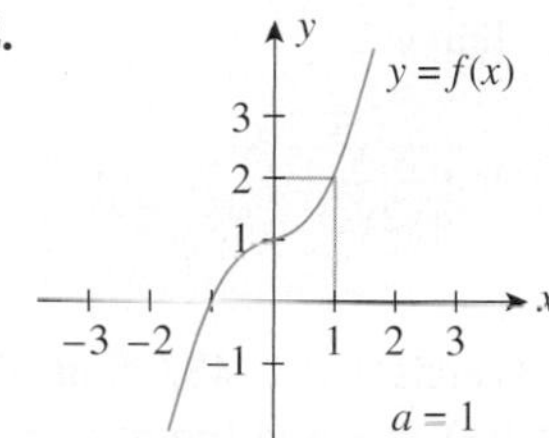

3\. 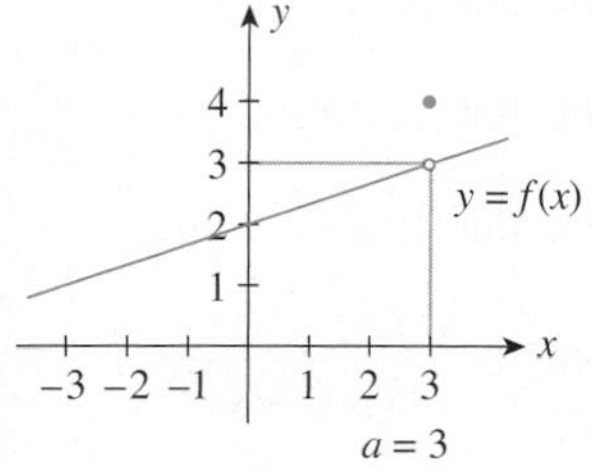

4\. 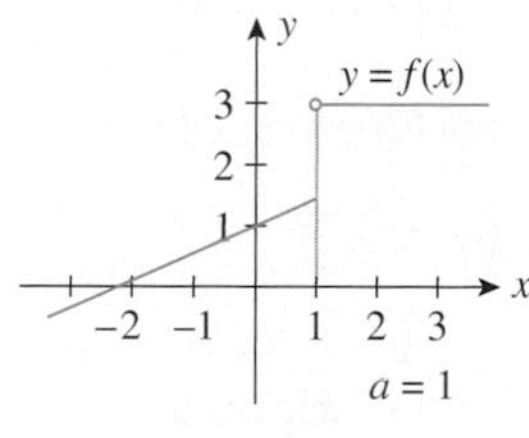

5\. 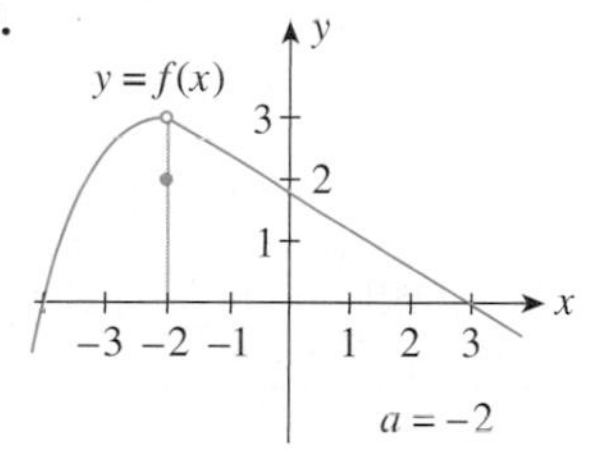

6\. 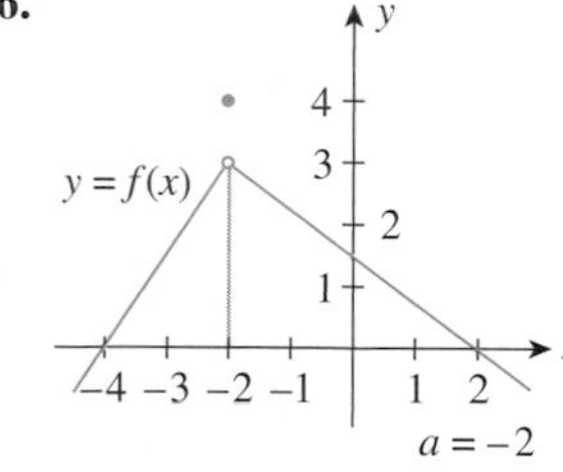

7\. 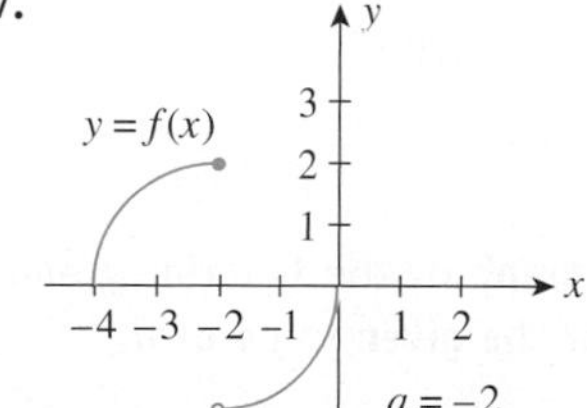

8\. 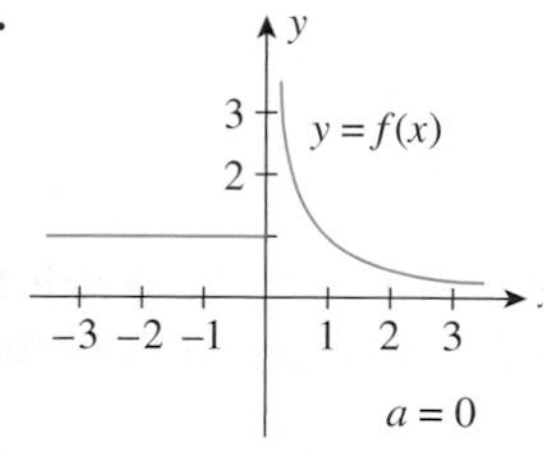

In Exercises 9–16, complete the table by computing $f(x)$ at the given values of x. Use these results to estimate the indicated limit (if it exists).

9. $f(x) = x^2 + 1$; $\lim_{x \to 2} f(x)$

x	1.9	1.99	1.999	2.001	2.01	2.1
$f(x)$						

10. $f(x) = 2x^2 - 1$; $\lim_{x \to 1} f(x)$

x	0.9	0.99	0.999	1.001	1.01	1.1
$f(x)$						

11. $f(x) = \dfrac{|x|}{x}$; $\lim_{x \to 0} f(x)$

x	−0.1	−0.01	−0.001	0.001	0.01	0.1
$f(x)$						

12. $f(x) = \dfrac{|x-1|}{x-1}$; $\lim_{x \to 1} f(x)$

x	0.9	0.99	0.999	1.001	1.01	1.1
$f(x)$						

13. $f(x) = \dfrac{1}{(x-1)^2}$; $\lim_{x \to 1} f(x)$

x	0.9	0.99	0.999	1.001	1.01	1.1
$f(x)$						

14. $f(x) = \dfrac{1}{x-2}$; $\lim_{x \to 2} f(x)$

x	1.9	1.99	1.999	2.001	2.01	2.1
$f(x)$						

15. $f(x) = \dfrac{x^2 + x - 2}{x - 1}$; $\lim_{x \to 1} f(x)$

x	0.9	0.99	0.999	1.001	1.01	1.1
$f(x)$						

16. $f(x) = \dfrac{x-1}{x-1}$; $\lim_{x \to 1} f(x)$

x	0.9	0.99	0.999	1.001	1.01	1.1
$f(x)$						

In Exercises 17–22, sketch the graph of the function f and evaluate $\lim_{x \to a} f(x)$, if it exists, for the given value of a.

17. $f(x) = \begin{cases} x - 1 & \text{if } x \leq 0 \\ -1 & \text{if } x > 0 \end{cases} \quad (a = 0)$

18. $f(x) = \begin{cases} x - 1 & \text{if } x \leq 3 \\ -2x + 8 & \text{if } x > 3 \end{cases} \quad (a = 3)$

19. $f(x) = \begin{cases} x & \text{if } x < 1 \\ 0 & \text{if } x = 1 \\ -x + 2 & \text{if } x > 1 \end{cases} \quad (a = 1)$

20. $f(x) = \begin{cases} -2x + 4 & \text{if } x < 1 \\ 4 & \text{if } x = 1 \\ x^2 + 1 & \text{if } x > 1 \end{cases} \quad (a = 1)$

21. $f(x) = \begin{cases} |x| & \text{if } x \neq 0 \\ 1 & \text{if } x = 0 \end{cases} \quad (a = 0)$

22. $f(x) = \begin{cases} |x - 1| & \text{if } x \neq 1 \\ 0 & \text{if } x = 1 \end{cases} \quad (a = 1)$

In Exercises 23–40, find the indicated limit.

23. $\lim_{x \to 2} 3$

24. $\lim_{x \to -2} -3$

25. $\lim_{x \to 3} x$

26. $\lim_{x \to -2} -3x$

27. $\lim_{x \to 1} (1 - 2x^2)$

28. $\lim_{t \to 3} (4t^2 - 2t + 1)$

29. $\lim_{x \to 1} (2x^3 - 3x^2 + x + 2)$

30. $\lim_{x \to 0} (4x^5 - 20x^2 + 2x + 1)$

31. $\lim_{s \to 0} (2s^2 - 1)(2s + 4)$

32. $\lim_{x \to 2} (x^2 + 1)(x^2 - 4)$

33. $\lim_{x \to 2} \dfrac{2x + 1}{x + 2}$

34. $\lim_{x \to 1} \dfrac{x^3 + 1}{2x^3 + 2}$

35. $\lim_{x \to 2} \sqrt{x + 2}$

36. $\lim_{x \to -2} \sqrt[3]{5x + 2}$

37. $\lim_{x \to -3} \sqrt{2x^4 + x^2}$

38. $\lim_{x \to 2} \sqrt{\dfrac{2x^3 + 4}{x^2 + 1}}$

39. $\lim_{x \to -1} \dfrac{\sqrt{x^2 + 8}}{2x + 4}$

40. $\lim_{x \to 3} \dfrac{x\sqrt{x^2 + 7}}{2x - \sqrt{2x + 3}}$

In Exercises 41–48, find the indicated limit given that $\lim_{x \to a} f(x) = 3$ and $\lim_{x \to a} g(x) = 4$.

41. $\lim_{x \to a} [f(x) - g(x)]$

42. $\lim_{x \to a} 2f(x)$

43. $\lim_{x \to a} [2f(x) - 3g(x)]$

44. $\lim_{x \to a} [f(x)g(x)]$

45. $\lim_{x \to a} \sqrt{g(x)}$

46. $\lim_{x \to a} \sqrt[3]{5f(x) + 3g(x)}$

47. $\lim_{x \to a} \dfrac{2f(x) - g(x)}{f(x)g(x)}$

48. $\lim_{x \to a} \dfrac{g(x) - f(x)}{f(x) + \sqrt{g(x)}}$

In Exercises 49–62, find the indicated limit, if it exists.

49. $\lim_{x\to 1} \dfrac{x^2-1}{x-1}$

50. $\lim_{x\to -2} \dfrac{x^2-4}{x+2}$

51. $\lim_{x\to 0} \dfrac{x^2-x}{x}$

52. $\lim_{x\to 0} \dfrac{2x^2-3x}{x}$

53. $\lim_{x\to -5} \dfrac{x^2-25}{x+5}$

54. $\lim_{b\to -3} \dfrac{b+1}{b+3}$

55. $\lim_{x\to 1} \dfrac{x}{x-1}$

56. $\lim_{x\to 2} \dfrac{x+2}{x-2}$

57. $\lim_{x\to -2} \dfrac{x^2-x-6}{x^2+x-2}$

58. $\lim_{z\to 2} \dfrac{z^3-8}{z-2}$

59. $\lim_{x\to 1} \dfrac{\sqrt{x}-1}{x-1}$

Hint: Multiply by $\dfrac{\sqrt{x}+1}{\sqrt{x}+1}$.

60. $\lim_{x\to 4} \dfrac{x-4}{\sqrt{x}-2}$

Hint: See Exercise 59.

61. $\lim_{x\to 1} \dfrac{x-1}{x^3+x^2-2x}$

62. $\lim_{x\to -2} \dfrac{4-x^2}{2x^2+x^3}$

In Exercises 63–68, use the graph of the function f to determine $\lim_{x\to\infty} f(x)$ and $\lim_{x\to -\infty} f(x)$, if they exist.

63.

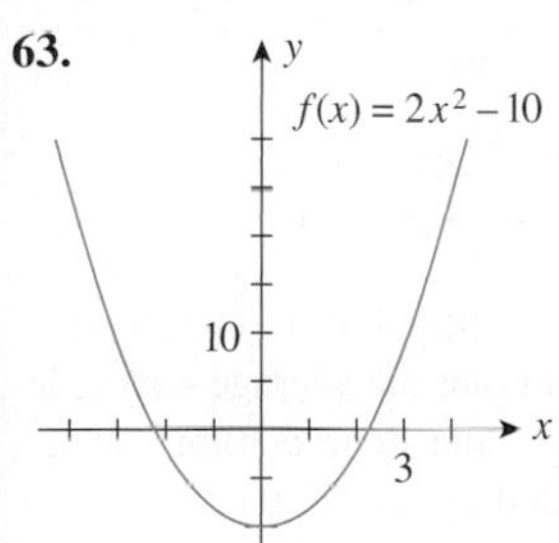

64.

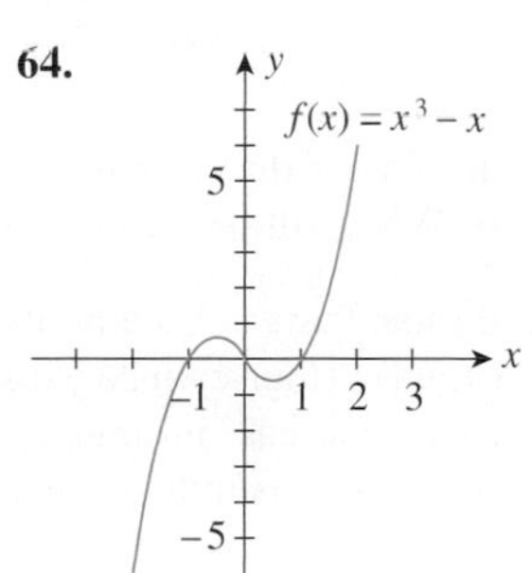

65.

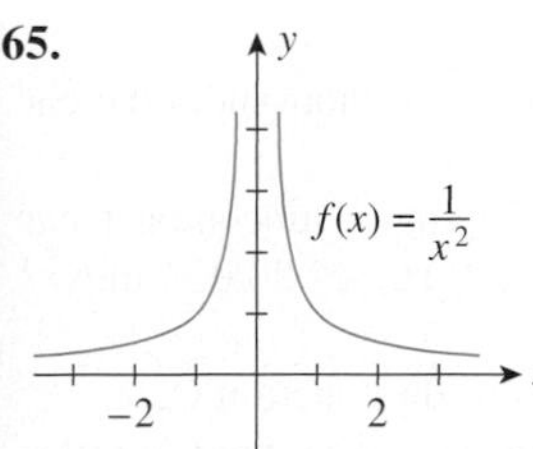

66.

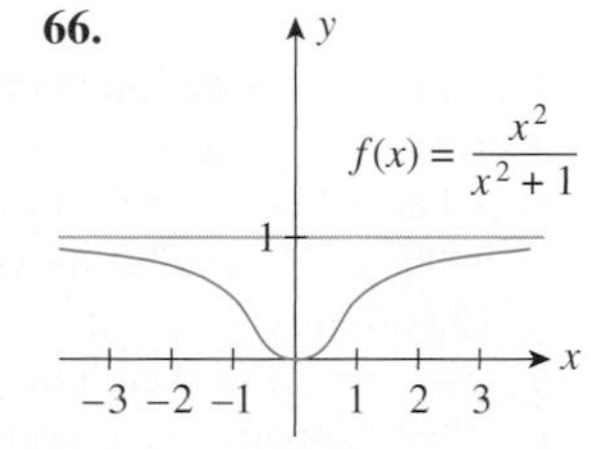

67.

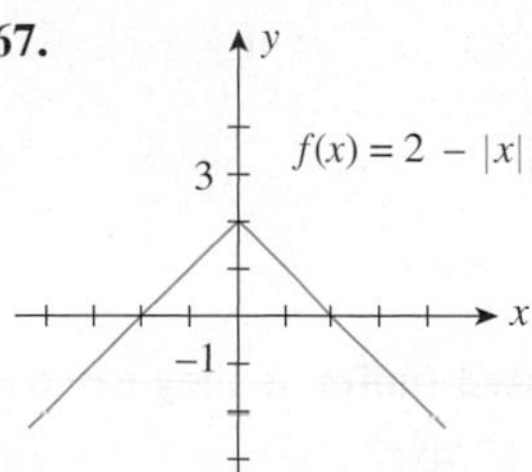

68.

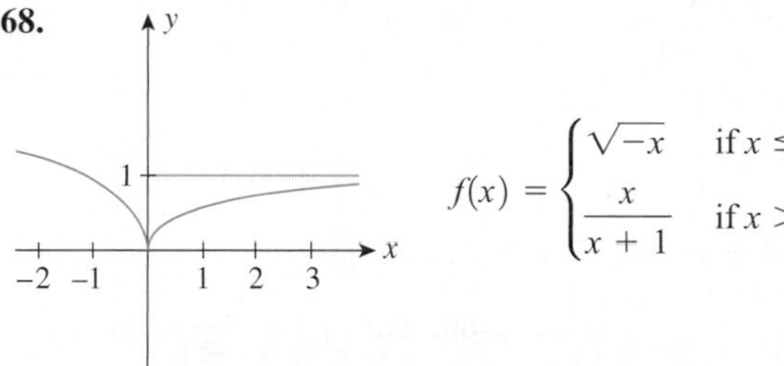

$$f(x) = \begin{cases} \sqrt{-x} & \text{if } x \le 0 \\ \dfrac{x}{x+1} & \text{if } x > 0 \end{cases}$$

In Exercises 69–72, complete the table by computing $f(x)$ at the given values of x. Use the results to guess at the indicated limits, if they exist.

69. $f(x) = \dfrac{1}{x^2+1}$; $\lim_{x\to\infty} f(x)$ and $\lim_{x\to -\infty} f(x)$

x	1	10	100	1000
$f(x)$				

x	−1	−10	−100	−1000
$f(x)$				

70. $f(x) = \dfrac{2x}{x+1}$; $\lim_{x\to\infty} f(x)$ and $\lim_{x\to -\infty} f(x)$

x	1	10	100	1000
$f(x)$				

x	−5	−10	−100	−1000
$f(x)$				

71. $f(x) = 3x^3 - x^2 + 10$; $\lim_{x\to\infty} f(x)$ and $\lim_{x\to -\infty} f(x)$

x	1	5	10	100	1000
$f(x)$					

x	−1	−5	−10	−100	−1000
$f(x)$					

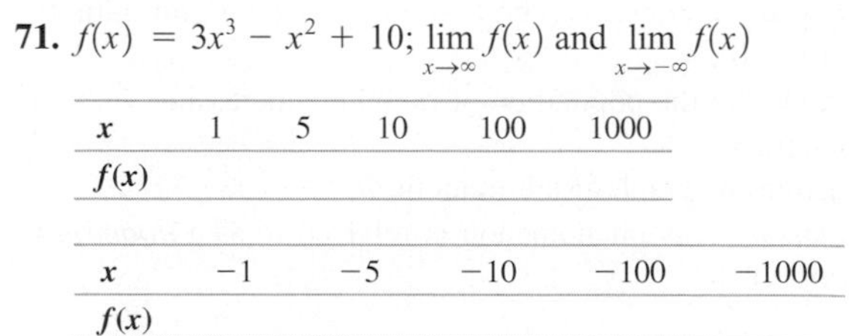

72. $f(x) = \dfrac{|x|}{x}$; $\lim_{x\to\infty} f(x)$ and $\lim_{x\to-\infty} f(x)$

x	1	10	100	-1	-10	-100
$f(x)$						

In Exercises 73–80, find the indicated limits, if they exist.

73. $\lim_{x\to\infty} \dfrac{3x+2}{x-5}$

74. $\lim_{x\to-\infty} \dfrac{4x^2-1}{x+2}$

75. $\lim_{x\to-\infty} \dfrac{3x^3+x^2+1}{x^3+1}$

76. $\lim_{x\to\infty} \dfrac{2x^2+3x+1}{x^4-x^2}$

77. $\lim_{x\to-\infty} \dfrac{x^4+1}{x^3-1}$

78. $\lim_{x\to\infty} \dfrac{4x^4-3x^2+1}{2x^4+x^3+x^2+x+1}$

79. $\lim_{x\to\infty} \dfrac{x^5-x^3+x-1}{x^6+2x^2+1}$

80. $\lim_{x\to\infty} \dfrac{2x^2-1}{x^3+x^2+1}$

81. TOXIC WASTE A city's main well was recently found to be contaminated with trichloroethylene, a cancer-causing chemical, as a result of an abandoned chemical dump leaching chemicals into the water. A proposal submitted to city council members indicates that the cost, measured in millions of dollars, of removing $x\%$ of the toxic pollutant is given by

$$C(x) = \frac{0.5x}{100-x} \qquad (0 < x < 100)$$

a. Find the cost of removing 50%, 60%, 70%, 80%, 90%, and 95% of the pollutant.

b. Evaluate

$$\lim_{x\to 100} \frac{0.5x}{100-x}$$

and interpret your result.

82. A DOOMSDAY SITUATION The population of a certain breed of rabbits introduced into an isolated island is given by

$$P(t) = \frac{72}{9-t} \qquad (0 \le t < 9)$$

where t is measured in months.

a. Find the number of rabbits present in the island initially (at $t = 0$).

b. Show that the population of rabbits is increasing without bound.

c. Sketch the graph of the function P.

(*Comment:* This phenomenon is referred to as a *doomsday situation.*)

83. AVERAGE COST The average cost/disc in dollars incurred by Herald Records in pressing x DVDs is given by the average cost function

$$\overline{C}(x) = 2.2 + \frac{2500}{x}$$

Evaluate $\lim_{x\to\infty} \overline{C}(x)$ and interpret your result.

84. CONCENTRATION OF A DRUG IN THE BLOODSTREAM The concentration of a certain drug in a patient's bloodstream t hr after injection is given by

$$C(t) = \frac{0.2t}{t^2+1}$$

mg/cm^3. Evaluate $\lim_{t\to\infty} C(t)$ and interpret your result.

85. BOX-OFFICE RECEIPTS The total worldwide box-office receipts for a long-running blockbuster movie are approximated by the function

$$T(x) = \frac{120x^2}{x^2+4}$$

where $T(x)$ is measured in millions of dollars and x is the number of months since the movie's release.

a. What are the total box-office receipts after the first month? The second month? The third month?

b. What will the movie gross in the long run (when x is very large)?

86. POPULATION GROWTH A major corporation is building a 4325-acre complex of homes, offices, stores, schools, and churches in the rural community of Glen Cove. As a result of this development, the planners have estimated that Glen Cove's population (in thousands) t yr from now will be given by

$$P(t) = \frac{25t^2+125t+200}{t^2+5t+40}$$

a. What is the current population of Glen Cove?

b. What will be the population in the long run?

87. DRIVING COSTS A study of driving costs of 1992 model subcompact (four-cylinder) cars found that the average cost (car payments, gas, insurance, upkeep, and depreciation), measured in cents/mile, is approximated by the function

$$C(x) = \frac{2010}{x^{2.2}} + 17.80$$

where x denotes the number of miles (in thousands) the car is driven in a year.

a. What is the average cost of driving a subcompact car 5000 mi/yr? 10,000 mi/yr? 15,000 mi/yr? 20,000 mi/yr? 25,000 mi/yr?

b. Use part (a) to sketch the graph of the function C.

c. What happens to the average cost as the number of miles driven increases without bound?

Source: American Automobile Association

88. PHOTOSYNTHESIS The rate of production R in photosynthesis is related to the light intensity I by the function

$$R(I) = \frac{aI}{b + I^2}$$

where a and b are positive constants.

a. Taking $a = b = 1$, compute $R(I)$ for $I = 0, 1, 2, 3, 4$, and 5.

b. Evaluate $\lim_{I \to \infty} R(I)$.

c. Use the results of parts (a) and (b) to sketch the graph of R. Interpret your results.

In Exercises 89–94, determine whether the statement is true or false. If it is true, explain why it is true. If it is false, give an example to show why it is false.

89. If $\lim_{x \to a} f(x)$ exists, then f is defined at $x = a$.

90. If $\lim_{x \to 0} f(x) = 4$ and $\lim_{x \to 0} g(x) = 0$, then $\lim_{x \to 0} f(x)g(x) = 0$.

91. If $\lim_{x \to 2} f(x) = 3$ and $\lim_{x \to 2} g(x) = 0$, then $\lim_{x \to 2} [f(x)]/[g(x)]$ does not exist.

92. If $\lim_{x \to 3} f(x) = 0$ and $\lim_{x \to 3} g(x) = 0$, then $\lim_{x \to 3} [f(x)]/[g(x)]$ does not exist.

93. $\lim_{x \to 2} \left(\frac{x}{x + 1} + \frac{3}{x - 1} \right) = \lim_{x \to 2} \frac{x}{x + 1} + \lim_{x \to 2} \frac{3}{x - 1}$

94. $\lim_{x \to 1} \left(\frac{2x}{x - 1} - \frac{2}{x - 1} \right) = \lim_{x \to 1} \frac{2x}{x - 1} - \lim_{x \to 1} \frac{2}{x - 1}$

95. SPEED OF A CHEMICAL REACTION Certain proteins, known as enzymes, serve as catalysts for chemical reactions in living things. In 1913 Leonor Michaelis and L. M. Menten discovered the following formula giving the initial speed V (in moles/liter/second) at which the reaction begins in terms of the amount of substrate x (the substance being acted upon, measured in moles/liters) present:

$$V = \frac{ax}{x + b}$$

where a and b are positive constants. Evaluate

$$\lim_{x \to \infty} \frac{ax}{x + b}$$

and interpret your result.

96. Show by means of an example that $\lim_{x \to a} [f(x) + g(x)]$ may exist even though neither $\lim_{x \to a} f(x)$ nor $\lim_{x \to a} g(x)$ exists. Does this example contradict Theorem 1?

97. Show by means of an example that $\lim_{x \to a} [f(x)g(x)]$ may exist even though neither $\lim_{x \to a} f(x)$ nor $\lim_{x \to a} g(x)$ exists. Does this example contradict Theorem 1?

98. Show by means of an example that $\lim_{x \to a} f(x)/g(x)$ may exist even though neither $\lim_{x \to a} f(x)$ nor $\lim_{x \to a} g(x)$ exists. Does this example contradict Theorem 1?

2.4 Solutions to Self-Check Exercises

1. a. $$\lim_{x \to 3} \frac{\sqrt{x^2 + 7} + \sqrt{3x - 5}}{x + 2} = \frac{\sqrt{9 + 7} + \sqrt{3(3) - 5}}{3 + 2} = \frac{\sqrt{16} + \sqrt{4}}{5} = \frac{6}{5}$$

b. Letting x approach -1 leads to the indeterminate form 0/0. Thus, we proceed as follows:

$$\lim_{x \to -1} \frac{x^2 - x - 2}{2x^2 - x - 3} = \lim_{x \to -1} \frac{(x + 1)(x - 2)}{(x + 1)(2x - 3)}$$

$$= \lim_{x \to -1} \frac{x - 2}{2x - 3} \quad \text{Cancel the common factors.}$$

$$= \frac{-1 - 2}{2(-1) - 3}$$

$$= \frac{3}{5}$$

2. $$\lim_{x \to \infty} \overline{C}(x) = \lim_{x \to \infty} \left(1.8 + \frac{3000}{x} \right) = \lim_{x \to \infty} 1.8 + \lim_{x \to \infty} \frac{3000}{x} = 1.8$$

Our computation reveals that, as the production of CDs increases "without bound," the average cost drops and approaches a unit cost of \$1.80/disc.

USING TECHNOLOGY

Finding the Limit of a Function

A graphing utility can be used to help us find the limit of a function, if it exists, as illustrated in the following examples.

EXAMPLE 1 Let $f(x) = \dfrac{x^3 - 1}{x - 1}$.

a. Plot the graph of f in the viewing window $[-2, 2] \times [0, 4]$.

b. Use ZOOM to find $\lim_{x \to 1} \dfrac{x^3 - 1}{x - 1}$.

c. Verify your result by evaluating the limit algebraically.

Solution

a. The graph of f in the viewing window $[-2, 2] \times [0, 4]$ is shown in Figure T1a.

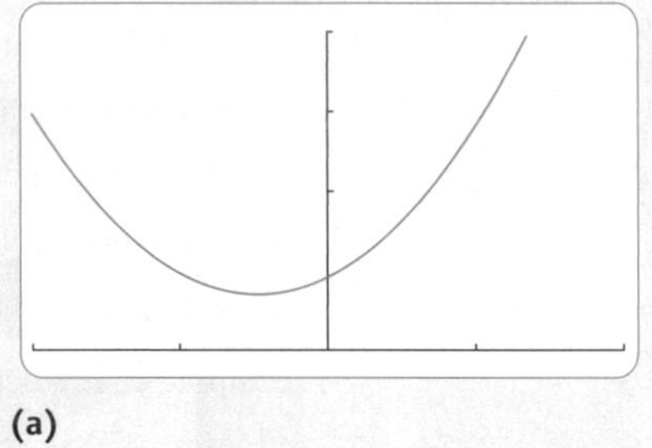

(a)

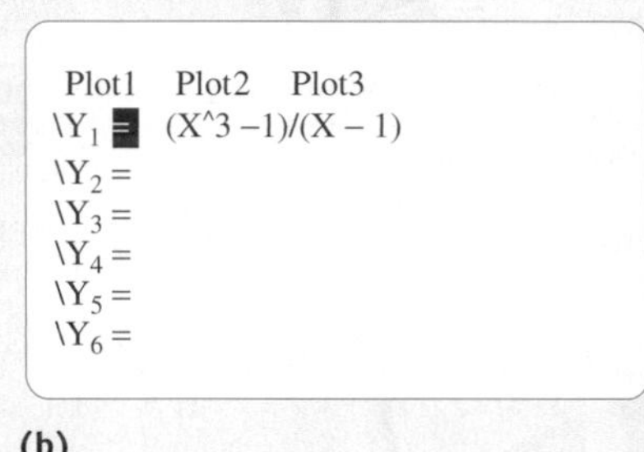

(b)

FIGURE T1
(a) The graph of $f(x) = \dfrac{x^3 - 1}{x - 1}$ in the viewing window $[-2, 2] \times [0, 4]$; (b) the TI-83 equation screen

b. Using ZOOM-IN repeatedly, we see that the y-value approaches 3 as the x-value approaches 1. We conclude, accordingly, that

$$\lim_{x \to 1} \frac{x^3 - 1}{x - 1} = 3$$

c. We compute

$$\lim_{x \to 1} \frac{x^3 - 1}{x - 1} = \lim_{x \to 1} \frac{(x - 1)(x^2 + x + 1)}{x - 1}$$

$$= \lim_{x \to 1} (x^2 + x + 1) = 3$$

Note If you attempt to find the limit in Example 1 by using the evaluation function of your graphing utility to find the value of $f(x)$ when $x = 1$, you will see that the graphing utility does not display the y-value. This happens because $x = 1$ is not in the domain of f.

EXAMPLE 2 Use ZOOM to find $\lim_{x \to 0} (1 + x)^{1/x}$.

Solution We first plot the graph of $f(x) = (1 + x)^{1/x}$ in a suitable viewing window. Figure T2a shows a plot of f in the window $[-1, 1] \times [0, 4]$. Using ZOOM-IN repeatedly, we see that $\lim_{x \to 0} (1 + x)^{1/x} \approx 2.71828$.

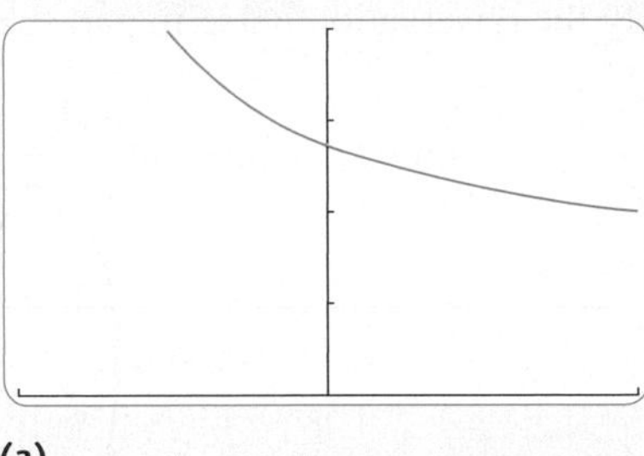

(a)

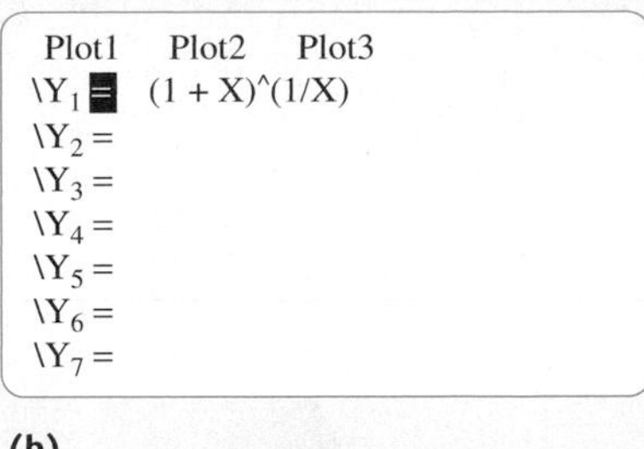

(b)

FIGURE T2
(a) The graph of $f(x) = (1 + x)^{1/x}$ in the viewing window $[-1, 1] \times [0, 4]$; (b) the T1-83 equation screen

The limit of $f(x) = (1 + x)^{1/x}$ as x approaches zero, denoted by the letter e, plays a very important role in the study of mathematics and its applications (see Section 5.6). Thus,

$$\lim_{x \to 0} (1 + x)^{1/x} = e$$

where, as we have just seen, $e \approx 2.71828$. ■

APPLIED EXAMPLE 3 Oxygen Content of a Pond When organic waste is dumped into a pond, the oxidation process that takes place reduces the pond's oxygen content. However, given time, nature will restore the oxygen content to its natural level. Suppose the oxygen content t days after the organic waste has been dumped into the pond is given by

$$f(t) = 100\left(\frac{t^2 + 10t + 100}{t^2 + 20t + 100}\right)$$

percent of its normal level.

a. Plot the graph of f in the viewing window $[0, 200] \times [70, 100]$.
b. What can you say about $f(t)$ when t is very large?
c. Verify your observation in part (b) by evaluating $\lim_{t \to \infty} f(t)$.

Solution

a. The graph of f is shown in Figure T3a.
b. From the graph of f, it appears that $f(t)$ approaches 100 steadily as t gets larger and larger. This observation tells us that eventually the oxygen content of the pond will be restored to its natural level.

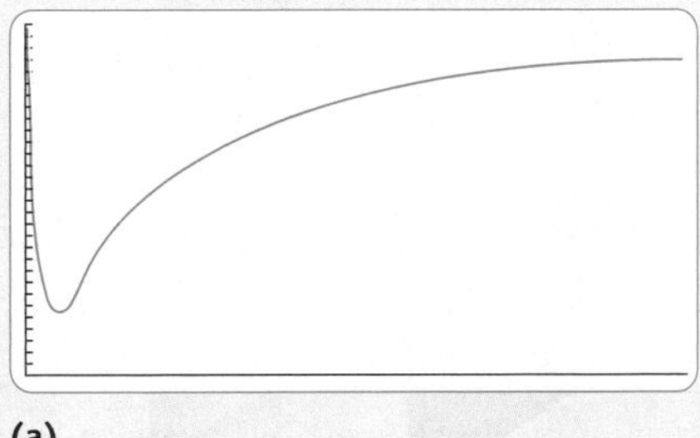

(a)

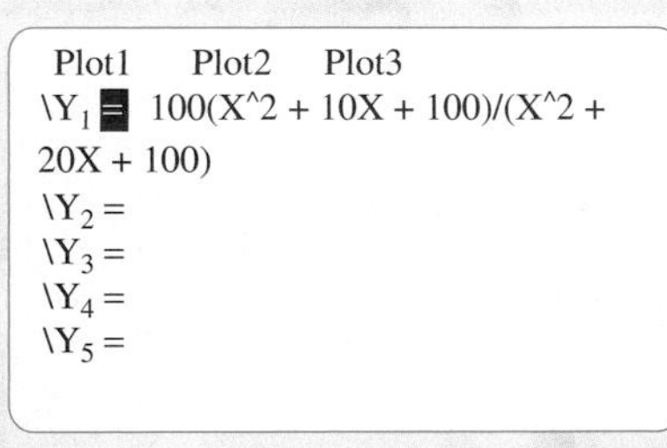

(b)

FIGURE T3
(a) The graph of f in the viewing window $[0, 200] \times [70, 100]$; (b) the TI-83 equation screen

(continued)

c. To verify the observation made in part (b), we compute

$$\lim_{t\to\infty} f(t) = \lim_{t\to\infty} 100\left(\frac{t^2 + 10t + 100}{t^2 + 20t + 100}\right)$$

$$= 100 \lim_{t\to\infty}\left(\frac{1 + \frac{10}{t} + \frac{100}{t^2}}{1 + \frac{20}{t} + \frac{100}{t^2}}\right) = 100$$

TECHNOLOGY EXERCISES

In Exercises 1–10, find the indicated limit by first plotting the graph of the function in a suitable viewing window and then using the ZOOM-IN feature of the calculator.

1. $\lim_{x\to 1} \frac{2x^3 - 2x^2 + 3x - 3}{x - 1}$

2. $\lim_{x\to -2} \frac{2x^3 + 3x^2 - x + 2}{x + 2}$

3. $\lim_{x\to -1} \frac{x^3 + 1}{x + 1}$

4. $\lim_{x\to -1} \frac{x^4 - 1}{x - 1}$

5. $\lim_{x\to 1} \frac{x^3 - x^2 - x + 1}{x^3 - 3x + 2}$

6. $\lim_{x\to 2} \frac{x^3 + 2x^2 - 16}{2x^3 - x^2 + 2x - 16}$

7. $\lim_{x\to 0} \frac{\sqrt{x + 1} - 1}{x}$

8. $\lim_{x\to 0} \frac{(x + 4)^{3/2} - 8}{x}$

9. $\lim_{x\to 0} (1 + 2x)^{1/x}$

10. $\lim_{x\to 0} \frac{2^x - 1}{x}$

11. Show that $\lim_{x\to 3} \frac{2}{x - 3}$ does not exist.

12. Show that $\lim_{x\to 2} \frac{x^3 - 2x + 1}{x - 2}$ does not exist.

13. CITY PLANNING A major developer is building a 5000-acre complex of homes, offices, stores, schools, and churches in the rural community of Marlboro. As a result of this development, the planners have estimated that Marlboro's population (in thousands) t yr from now will be given by

$$P(t) = \frac{25t^2 + 125t + 200}{t^2 + 5t + 40}$$

a. Plot the graph of P in the viewing window $[0, 50] \times [0, 30]$.

b. What will be the population of Marlboro in the long run?
Hint: Find $\lim_{t\to\infty} P(t)$.

14. AMOUNT OF RAINFALL The total amount of rain (in inches) after t hr during a rainfall is given by

$$T(t) = \frac{0.8t}{t + 4.1}$$

a. Plot the graph of T in the viewing window $[0, 30] \times [0, 0.8]$.

b. What is the total amount of rain during this rainfall?
Hint: Find $\lim_{t\to\infty} T(t)$.

2.5 One-Sided Limits and Continuity

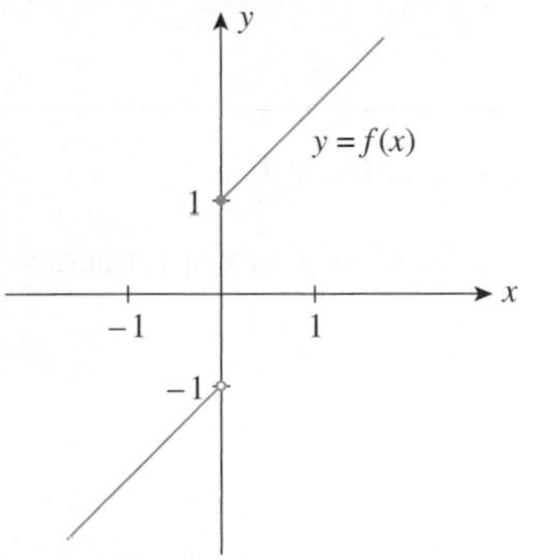

FIGURE 35
The function f does not have a limit as x approaches zero.

One-Sided Limits

Consider the function f defined by

$$f(x) = \begin{cases} x - 1 & \text{if } x < 0 \\ x + 1 & \text{if } x \geq 0 \end{cases}$$

From the graph of f shown in Figure 35, we see that the function f does not have a limit as x approaches zero because, no matter how close x is to zero, $f(x)$ takes on values that are close to 1 if x is positive and values that are close to -1 if x is negative. Therefore, $f(x)$ cannot be close to a single number L—no matter how close x is to zero. Now, if we restrict x to be greater than zero (to the right of zero), then we see that $f(x)$ can be made as close to 1 as we please by taking x sufficiently close to zero. In this situation we say that the right-hand limit of f as x approaches zero (from the right) is 1, written

$$\lim_{x \to 0^+} f(x) = 1$$

Similarly, we see that $f(x)$ can be made as close to -1 as we please by taking x sufficiently close to, but to the left of, zero. In this situation we say that the left-hand limit of f as x approaches zero (from the left) is -1, written

$$\lim_{x \to 0^-} f(x) = -1$$

These limits are called **one-sided limits.** More generally, we have the following definitions.

> **One-Sided Limits**
> The function f has the **right-hand limit** L as x approaches a from the right, written
>
> $$\lim_{x \to a^+} f(x) = L$$
>
> if the values of $f(x)$ can be made as close to L as we please by taking x sufficiently close to (but not equal to) a and to the right of a.
>
> Similarly, the function f has the **left-hand limit** M as x approaches a from the left, written
>
> $$\lim_{x \to a^-} f(x) = M$$
>
> if the values of $f(x)$ can be made as close to M as we please by taking x sufficiently close to (but not equal to) a and to the left of a.

The connection between one-sided limits and the two-sided limit defined earlier is given by the following theorem.

> **THEOREM 3**
> Let f be a function that is defined for all values of x close to $x = a$ with the possible exception of a itself. Then
>
> $$\lim_{x \to a} f(x) = L \qquad \text{if and only if} \qquad \lim_{x \to a^+} f(x) = \lim_{x \to a^-} f(x) = L$$

Thus, the two-sided limit exists if and only if the one-sided limits exist and are equal.

EXAMPLE 1 Let

$$f(x) = \begin{cases} \sqrt{x} & \text{if } x > 0 \\ -x & \text{if } x \le 0 \end{cases} \quad \text{and} \quad g(x) = \begin{cases} -1 & \text{if } x < 0 \\ 1 & \text{if } x \ge 0 \end{cases}$$

a. Show that $\lim\limits_{x\to 0} f(x)$ exists by studying the one-sided limits of f as x approaches $x = 0$.

b. Show that $\lim\limits_{x\to 0} g(x)$ does not exist.

Solution

a. For $x > 0$, we find

$$\lim_{x\to 0^+} f(x) = \lim_{x\to 0^+} \sqrt{x} = 0$$

and for $x \le 0$

$$\lim_{x\to 0^-} f(x) = \lim_{x\to 0^-} (-x) = 0$$

Thus,

$$\lim_{x\to 0} f(x) = 0$$

(Figure 36a).

b. We have

$$\lim_{x\to 0^-} g(x) = -1 \quad \text{and} \quad \lim_{x\to 0^+} g(x) = 1$$

and since these one-sided limits are not equal, we conclude that $\lim\limits_{x\to 0} g(x)$ does not exist (Figure 36b).

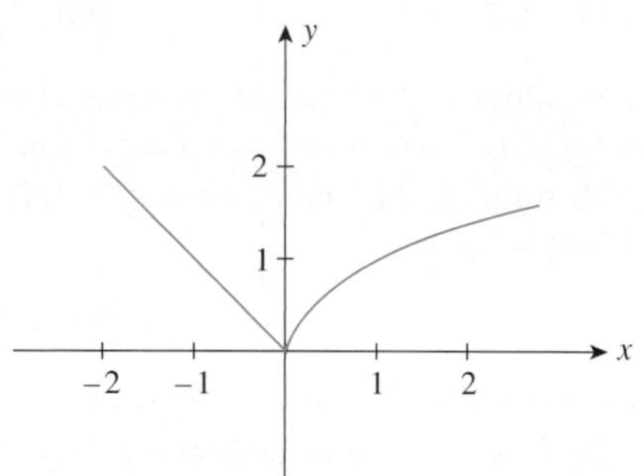

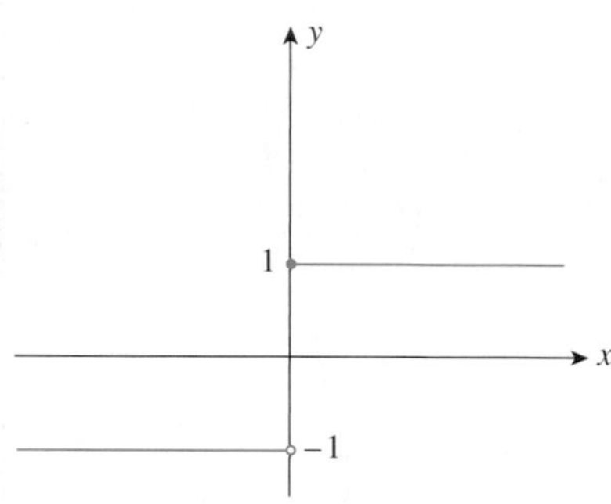

FIGURE 36

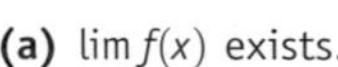
(a) $\lim\limits_{x\to 0} f(x)$ exists.

(b) $\lim\limits_{x\to 0} g(x)$ does not exist.

Continuous Functions

Continuous functions will play an important role throughout most of our study of calculus. Loosely speaking, a function is continuous at a point if the graph of the function at that point is devoid of holes, gaps, jumps, or breaks. Consider, for example, the graph of the function f depicted in Figure 37.

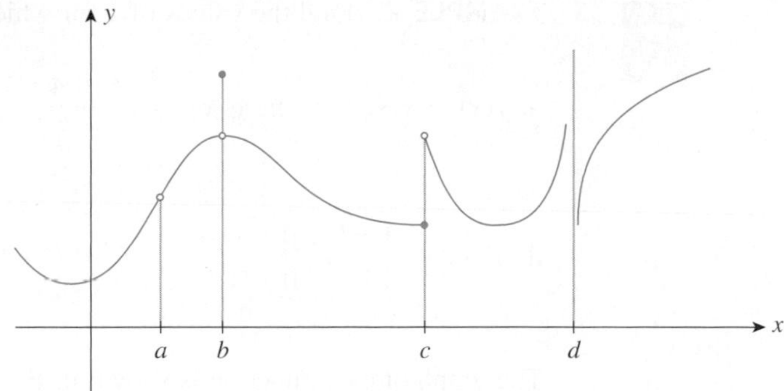

FIGURE 37
The graph of this function is not continuous at $x = a$, $x = b$, $x = c$, and $x = d$.

Let's take a closer look at the behavior of f at or near $x = a$, $x = b$, $x = c$, and $x = d$. First, note that f is not defined at $x = a$; that is, $x = a$ is not in the domain of f, thereby resulting in a "hole" in the graph of f. Next, observe that the value of f at b, $f(b)$, is not equal to the limit of $f(x)$ as x approaches b, resulting in a "jump" in the graph of f at $x = b$. The function f does not have a limit at $x = c$ since the left-hand and right-hand limits of $f(x)$ are not equal, also resulting in a jump in the graph of f at $x = c$. Finally, the limit of f does not exist at $x = d$, resulting in a break in the graph of f. The function f is *discontinuous* at each of these numbers. It is *continuous* everywhere else.

Continuity of a Function at a Number

A function f is **continuous at a number** $x = a$ if the following conditions are satisfied.

1. $f(a)$ is defined. **2.** $\lim_{x \to a} f(x)$ exists. **3.** $\lim_{x \to a} f(x) = f(a)$

Thus, a function f is continuous at $x = a$ if the limit of f at $x = a$ exists and has the value $f(a)$. Geometrically, f is continuous at $x = a$ if the proximity of x to a implies the proximity of $f(x)$ to $f(a)$.

If f is not continuous at $x = a$, then f is said to be **discontinuous** at $x = a$. Also, f is **continuous on an interval** if f is continuous at every number in the interval.

Figure 38 depicts the graph of a continuous function on the interval (a, b). Notice that the graph of the function over the stated interval can be sketched without lifting one's pencil from the paper.

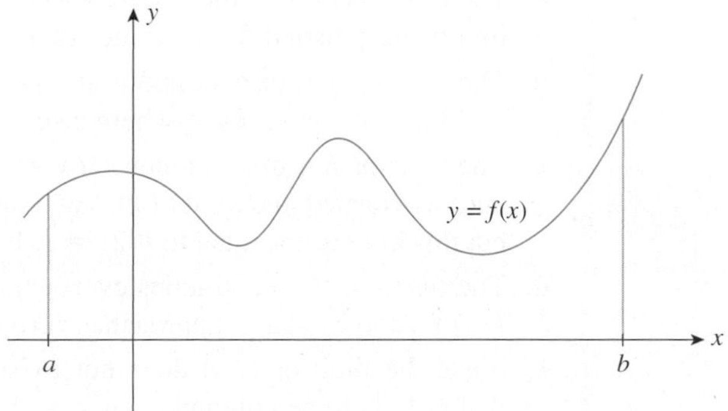

FIGURE 38
The graph of f is continuous on the interval (a, b).

VIDEO

EXAMPLE 2 Find the values of x for which each function is continuous.

a. $f(x) = x + 2$ **b.** $g(x) = \dfrac{x^2 - 4}{x - 2}$ **c.** $h(x) = \begin{cases} x + 2 & \text{if } x \neq 2 \\ 1 & \text{if } x = 2 \end{cases}$

d. $F(x) = \begin{cases} -1 & \text{if } x < 0 \\ 1 & \text{if } x \geq 0 \end{cases}$ **e.** $G(x) = \begin{cases} \dfrac{1}{x} & \text{if } x > 0 \\ -1 & \text{if } x \leq 0 \end{cases}$

The graph of each function is shown in Figure 39.

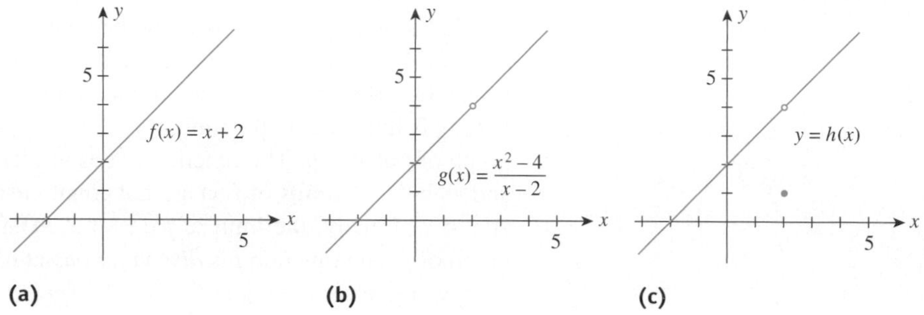

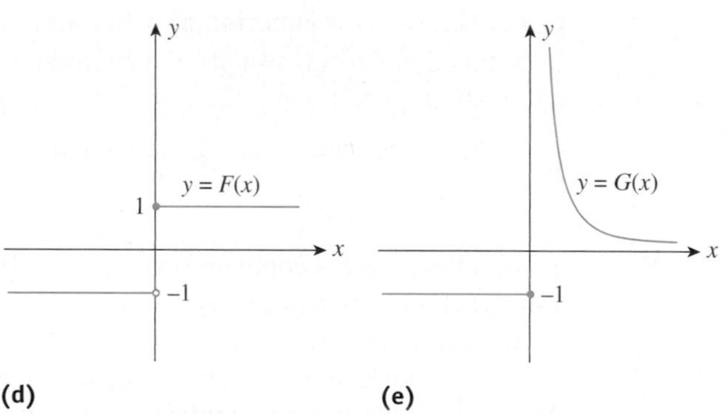

FIGURE 39

Solution

a. The function f is continuous everywhere because the three conditions for continuity are satisfied for all values of x.

b. The function g is discontinuous at $x = 2$ because g is not defined at that number. It is continuous everywhere else.

c. The function h is discontinuous at $x = 2$ because the third condition for continuity is violated; the limit of $h(x)$ as x approaches 2 exists and has the value 4, but this limit is not equal to $h(2) = 1$. It is continuous for all other values of x.

d. The function F is continuous everywhere except at $x = 0$, where the limit of $F(x)$ fails to exist as x approaches zero (see Example 3a, Section 2.4).

e. Since the limit of $G(x)$ does not exist as x approaches zero, we conclude that G fails to be continuous at $x = 0$. The function G is continuous everywhere else.

Properties of Continuous Functions

The following properties of continuous functions follow directly from the definition of continuity and the corresponding properties of limits. They are stated without proof.

Properties of Continuous Functions

1. The constant function $f(x) = c$ is continuous everywhere.
2. The identity function $f(x) = x$ is continuous everywhere.

If f and g are continuous at $x = a$, then

3. $[f(x)]^n$, where n is a real number, is continuous at $x = a$ whenever it is defined at that number.
4. $f \pm g$ is continuous at $x = a$.
5. fg is continuous at $x = a$.
6. f/g is continuous at $x = a$ provided $g(a) \neq 0$.

Using these properties of continuous functions, we can prove the following results. (A proof is sketched in Exercise 97, pages 131–132.)

Continuity of Polynomial and Rational Functions

1. A polynomial function $y = P(x)$ is continuous at every value of x.
2. A rational function $R(x) = p(x)/q(x)$ is continuous at every value of x where $q(x) \neq 0$.

EXAMPLE 3 Find the values of x for which each function is continuous.

a. $f(x) = 3x^3 + 2x^2 - x + 10$ **b.** $g(x) = \dfrac{8x^{10} - 4x + 1}{x^2 + 1}$

c. $h(x) = \dfrac{4x^3 - 3x^2 + 1}{x^2 - 3x + 2}$

Solution

a. The function f is a polynomial function of degree 3, so $f(x)$ is continuous for all values of x.

b. The function g is a rational function. Observe that the denominator of g—namely, $x^2 + 1$—is never equal to zero. Therefore, we conclude that g is continuous for all values of x.

c. The function h is a rational function. In this case, however, the denominator of h is equal to zero at $x = 1$ and $x = 2$, which can be seen by factoring it. Thus,

$$x^2 - 3x + 2 = (x - 2)(x - 1)$$

We therefore conclude that h is continuous everywhere except at $x = 1$ and $x = 2$, where it is discontinuous. ■

Up to this point, most of the applications we have discussed involved functions that are continuous everywhere. In Example 4, we consider an application from the field of educational psychology that involves a discontinuous function.

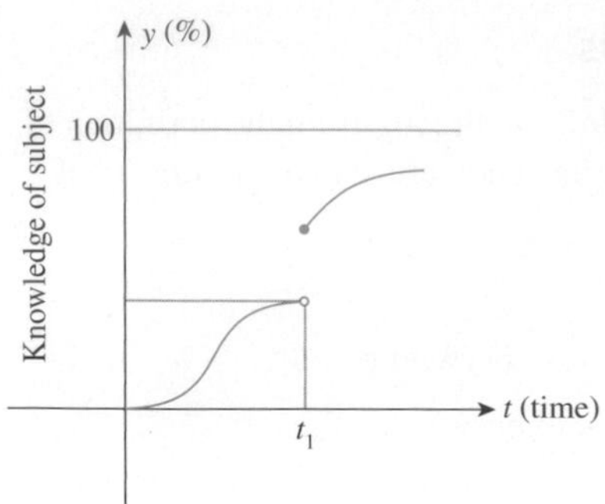

FIGURE 40
A learning curve that is discontinuous at $t = t_1$

APPLIED EXAMPLE 4 Learning Curves Figure 40 depicts the learning curve associated with a certain individual. Beginning with no knowledge of the subject being taught, the individual makes steady progress toward understanding it over the time interval $0 \leq t < t_1$. In this instance, the individual's progress slows as we approach time t_1 because he fails to grasp a particularly difficult concept. All of a sudden, a breakthrough occurs at time t_1, propelling his knowledge of the subject to a higher level. The curve is discontinuous at t_1. ■

Intermediate Value Theorem

Let's look again at our model of the motion of the maglev on a straight stretch of track. We know that the train cannot vanish at any instant of time and it cannot skip portions of the track and reappear someplace else. To put it another way, the train cannot occupy the positions s_1 and s_2 without at least, at some time, occupying an intermediate position (Figure 41).

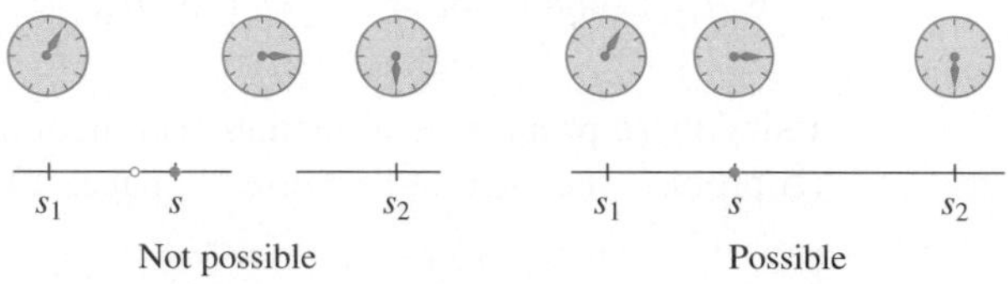

FIGURE 41
The position of the maglev

To state this fact mathematically, recall that the position of the maglev as a function of time is described by

$$f(t) = 4t^2 \qquad (0 \leq t \leq 10)$$

Suppose the position of the maglev is s_1 at some time t_1 and its position is s_2 at some time t_2 (Figure 42). Then, if s_3 is any number between s_1 and s_2 giving an intermediate position of the maglev, there must be at least one t_3 between t_1 and t_2 giving the time at which the train is at s_3—that is, $f(t_3) = s_3$.

This discussion carries the gist of the intermediate value theorem. The proof of this theorem can be found in most advanced calculus texts.

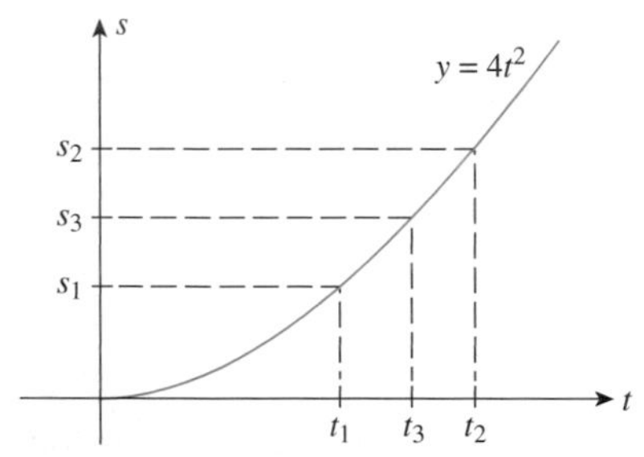

FIGURE 42
If $s_1 \leq s_3 \leq s_2$, then there must be at least one t_3 $(t_1 \leq t_3 \leq t_2)$ such that $f(t_3) = s_3$.

THEOREM 4

The Intermediate Value Theorem

If f is a continuous function on a closed interval $[a, b]$ and M is any number between $f(a)$ and $f(b)$, then there is at least one number c in $[a, b]$ such that $f(c) = M$ (Figure 43).

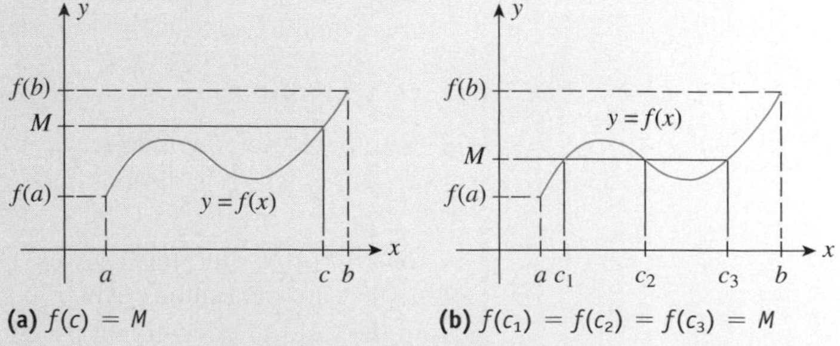

(a) $f(c) = M$

(b) $f(c_1) = f(c_2) = f(c_3) = M$

FIGURE 43

To illustrate the intermediate value theorem, let's look at the example involving the motion of the maglev again (see Figure 25, page 98). Notice that the initial position of the train is $f(0) = 0$ and the position at the end of its test run is $f(10) = 400$. Furthermore, the function f is continuous on $[0, 10]$. So, the intermediate value theorem guarantees that if we arbitrarily pick a number between 0 and 400—say, 100—giving the position of the maglev, there must be a $\bar{t}$ (read "t bar") between 0 and 10 at which time the train is at the position $s = 100$. To find the value of $\bar{t}$, we solve the equation $f(\bar{t}) = s$, or

$$4\bar{t}^2 = 100$$

giving $\bar{t} = 5$ (t must lie between 0 and 10).

It is important to remember when we use Theorem 4 that the function f must be continuous. The conclusion of the intermediate value theorem may not hold if f is not continuous (see Exercise 98, page 132).

The next theorem is an immediate consequence of the intermediate value theorem. It not only tells us when a zero of a function f [root of the equation $f(x) = 0$] exists but also provides the basis for a method of approximating it.

THEOREM 5

Existence of Zeros of a Continuous Function

If f is a continuous function on a closed interval $[a, b]$, and if $f(a)$ and $f(b)$ have opposite signs, then there is at least one solution of the equation $f(x) = 0$ in the interval (a, b) (Figure 44).

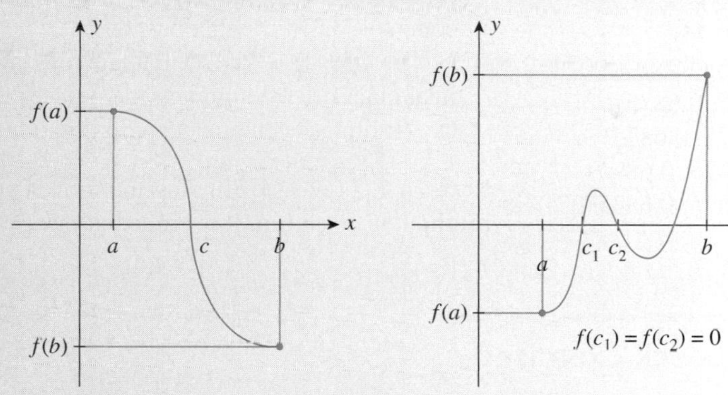

FIGURE 44
If $f(a)$ and $f(b)$ have opposite signs, there must be at least one number c $(a < c < b)$ such that $f(c) = 0$.

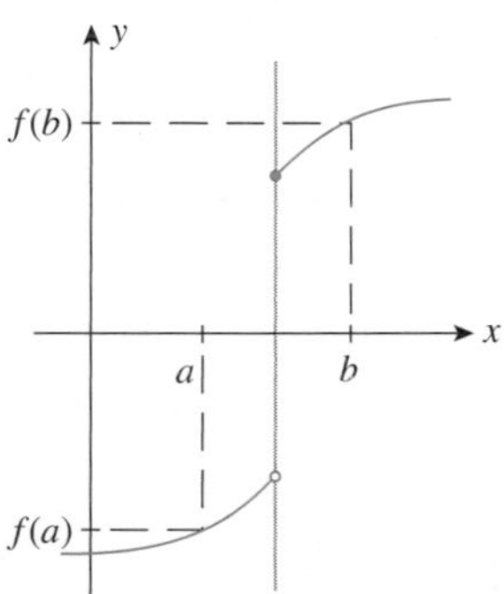

FIGURE 45
$f(a) < 0$ and $f(b) > 0$, but the graph of f does not cross the x-axis between a and b because f is discontinuous.

Geometrically, this property states that if the graph of a continuous function goes from above the x-axis to below the x-axis, or vice versa, it must *cross* the x-axis. This is not necessarily true if the function is discontinuous (Figure 45).

EXAMPLE 5 Let $f(x) = x^3 + x + 1$.

a. Show that f is continuous for all values of x.

b. Compute $f(-1)$ and $f(1)$ and use the results to deduce that there must be at least one number $x = c$, where c lies in the interval $(-1, 1)$ and $f(c) = 0$.

Solution

a. The function f is a polynomial function of degree 3 and is therefore continuous everywhere.

b. $f(-1) = (-1)^3 + (-1) + 1 = -1$ and $f(1) = 1^3 + 1 + 1 = 3$
Since $f(-1)$ and $f(1)$ have opposite signs, Theorem 5 tells us that there must be at least one number $x = c$ with $-1 < c < 1$ such that $f(c) = 0$. ■

The next example shows how the intermediate value theorem can be used to help us find the zero of a function.

EXAMPLE 6 Let $f(x) = x^3 + x - 1$. Since f is a polynomial function, it is continuous everywhere. Observe that $f(0) = -1$ and $f(1) = 1$ so that Theorem 5 guarantees the existence of at least one root of the equation $f(x) = 0$ in $(0, 1)$.*

We can locate the root more precisely by using Theorem 5 once again as follows: Evaluate $f(x)$ at the midpoint of $[0, 1]$, obtaining

$$f(0.5) = -0.375$$

Because $f(0.5) < 0$ and $f(1) > 0$, Theorem 5 now tells us that the root must lie in $(0.5, 1)$.

Repeat the process: Evaluate $f(x)$ at the midpoint of $[0.5, 1]$, which is

$$\frac{0.5 + 1}{2} = 0.75$$

Thus,

$$f(0.75) = 0.1719$$

Because $f(0.5) < 0$ and $f(0.75) > 0$, Theorem 5 tells us that the root is in $(0.5, 0.75)$. This process can be continued. Table 3 summarizes the results of our computations through nine steps.

From Table 3 we see that the root is approximately 0.68, accurate to two decimal places. By continuing the process through a sufficient number of steps, we can obtain as accurate an approximation to the root as we please. ■

TABLE 3

Step	Root of $f(x) = 0$ Lies in
1	(0, 1)
2	(0.5, 1)
3	(0.5, 0.75)
4	(0.625, 0.75)
5	(0.625, 0.6875)
6	(0.65625, 0.6875)
7	(0.671875, 0.6875)
8	(0.6796875, 0.6875)
9	(0.6796875, 0.6835937)

Note The process of finding the root of $f(x) = 0$ used in Example 6 is called the **method of bisection**. It is crude but effective. ■

2.5 Self-Check Exercises

1. Evaluate $\lim_{x \to -1^-} f(x)$ and $\lim_{x \to -1^+} f(x)$, where

$$f(x) = \begin{cases} 1 & \text{if } x < -1 \\ 1 + \sqrt{x + 1} & \text{if } x \geq -1 \end{cases}$$

Does $\lim_{x \to -1} f(x)$ exist?

2. Determine the values of x for which the given function is discontinuous. At each number where f is discontinuous, indicate which condition(s) for continuity are violated. Sketch the graph of the function.

a. $f(x) = \begin{cases} -x^2 + 1 & \text{if } x \leq 1 \\ x - 1 & \text{if } x > 1 \end{cases}$

b. $g(x) = \begin{cases} -x + 1 & \text{if } x < -1 \\ 2 & \text{if } -1 < x \leq 1 \\ -x + 3 & \text{if } x > 1 \end{cases}$

Solutions to Self-Check Exercises 2.5 can be found on page 132.

*It can be shown that f has precisely one zero in (0, 1) (see Exercise 103, Section 4.1).

2.5 Concept Questions

1. Explain what is meant by the statement $\lim_{x\to 3^-} f(x) = 2$ and $\lim_{x\to 3^+} f(x) = 4$.

2. Suppose $\lim_{x\to 1^-} f(x) = 3$ and $\lim_{x\to 1^+} f(x) = 4$.
   a. What can you say about $\lim_{x\to 1} f(x)$? Explain.
   b. What can you say about $f(1)$? Explain.

3. Explain what it means for a function f to be continuous (a) at a number a and (b) on an interval I.

4. Determine whether each function f is continuous or discontinuous. Explain your answer.
   a. $f(t)$ gives the altitude of an airplane at time t.
   b. $f(t)$ measures the total amount of rainfall at time t at the Municipal Airport.
   c. $f(s)$ measures the fare as a function of the distance s for taking a cab from Kennedy Airport to downtown Manhattan.
   d. $f(t)$ gives the interest rate charged by a financial institution at time t.

5. Explain the intermediate value theorem in your own words.

2.5 Exercises

In Exercises 1–8, use the graph of the function f to find $\lim_{x\to a^-} f(x)$, $\lim_{x\to a^+} f(x)$, and $\lim_{x\to a} f(x)$ at the indicated value of a, if the limit exists.

1.
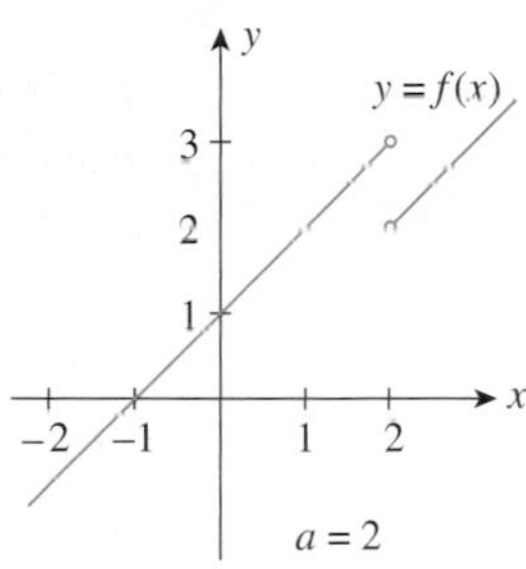

2.
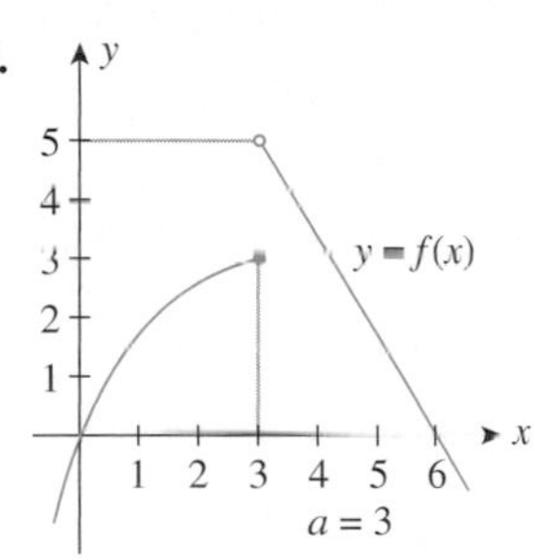

3.
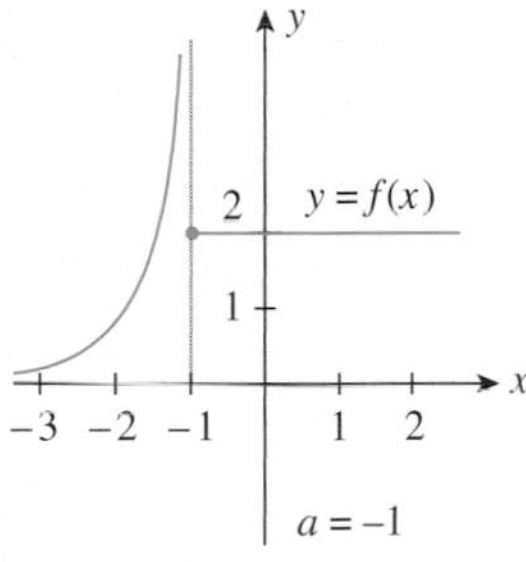

4.
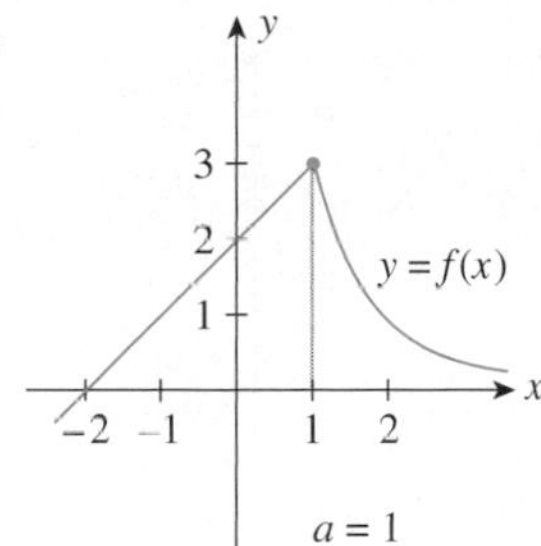

5.
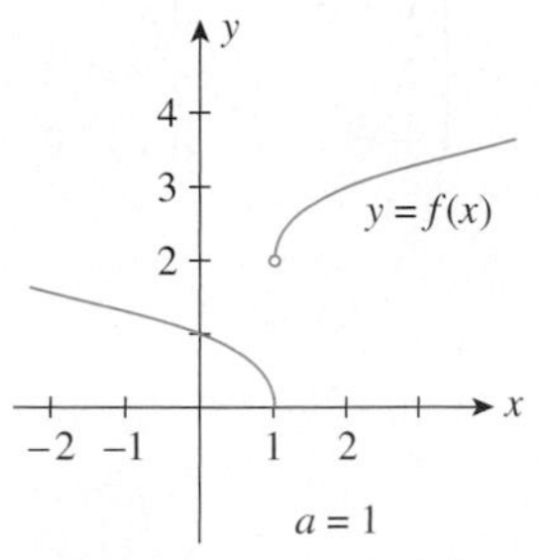

6.
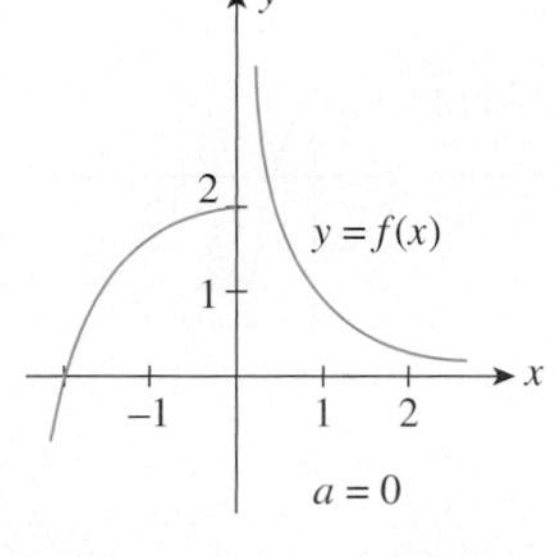

7.
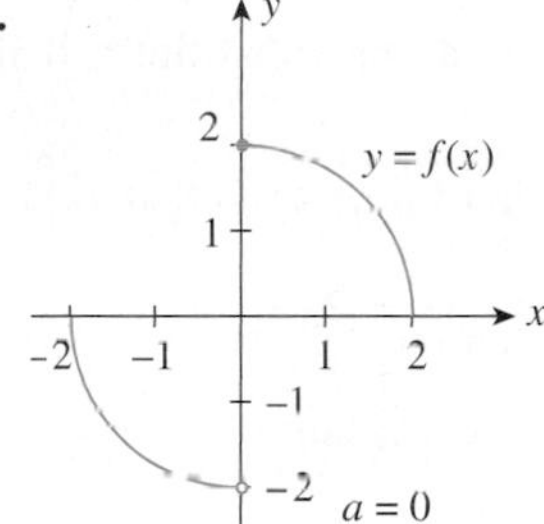

8.
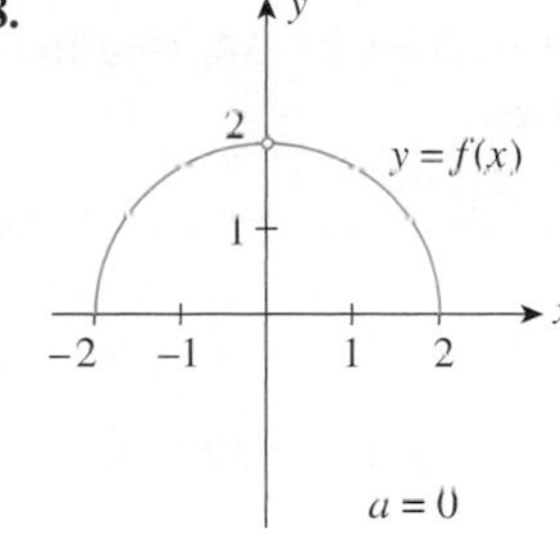

In Exercises 9–14, refer to the graph of the function f and determine whether each statement is true or false.

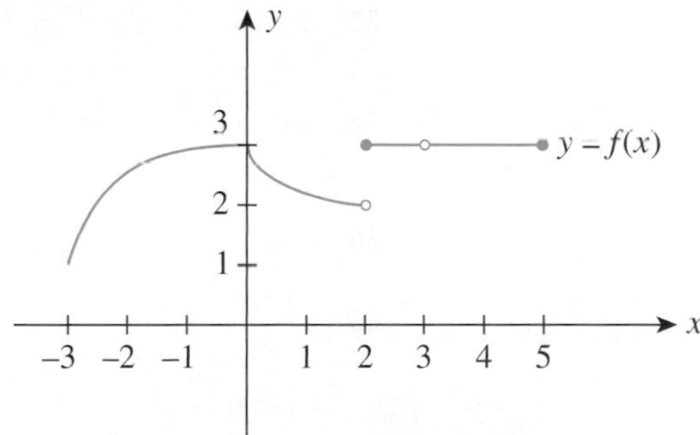

9. $\lim_{x\to -3^+} f(x) = 1$

10. $\lim_{x\to 0} f(x) = f(0)$

11. $\lim_{x\to 2^-} f(x) = 2$

12. $\lim_{x\to 2^+} f(x) = 3$

13. $\lim_{x\to 3} f(x)$ does not exist.

14. $\lim_{x\to 5^-} f(x) = 3$

In Exercises 15–20, refer to the graph of the function f and determine whether each statement is true or false.

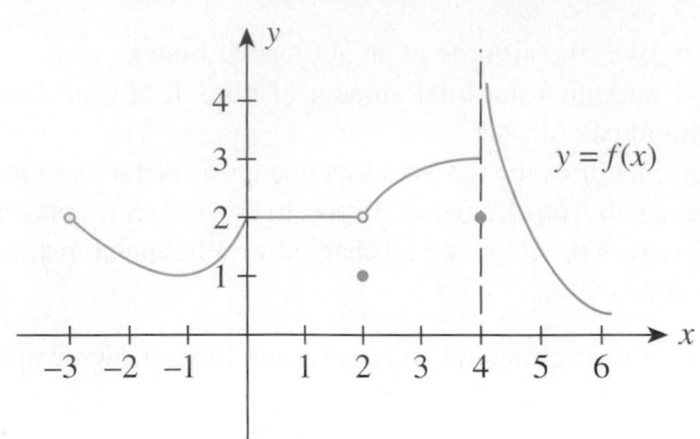

15. $\lim_{x \to -3^+} f(x) = 2$

16. $\lim_{x \to 0} f(x) = 2$

17. $\lim_{x \to 2} f(x) = 1$

18. $\lim_{x \to 4^-} f(x) = 3$

19. $\lim_{x \to 4^+} f(x)$ does not exist.

20. $\lim_{x \to 4} f(x) = 2$

In Exercises 21–38, find the indicated one-sided limit, if it exists.

21. $\lim_{x \to 1^+} (2x + 4)$

22. $\lim_{x \to 1^-} (3x - 4)$

23. $\lim_{x \to 2^-} \dfrac{x - 3}{x + 2}$

24. $\lim_{x \to 1^+} \dfrac{x + 2}{x + 1}$

25. $\lim_{x \to 0^+} \dfrac{1}{x}$

26. $\lim_{x \to 0^-} \dfrac{1}{x}$

27. $\lim_{x \to 0^+} \dfrac{x - 1}{x^2 + 1}$

28. $\lim_{x \to 2^+} \dfrac{x + 1}{x^2 - 2x + 3}$

29. $\lim_{x \to 0^+} \sqrt{x}$

30. $\lim_{x \to 2^+} 2\sqrt{x - 2}$

31. $\lim_{x \to -2^+} (2x + \sqrt{2 + x})$

32. $\lim_{x \to -5^+} x(1 + \sqrt{5 + x})$

33. $\lim_{x \to 1^-} \dfrac{1 + x}{1 - x}$

34. $\lim_{x \to 1^+} \dfrac{1 + x}{1 - x}$

35. $\lim_{x \to 2^-} \dfrac{x^2 - 4}{x - 2}$

36. $\lim_{x \to -3^+} \dfrac{\sqrt{x + 3}}{x^2 + 1}$

37. $\lim_{x \to 0^+} f(x)$ and $\lim_{x \to 0^-} f(x)$, where

$$f(x) = \begin{cases} 2x & \text{if } x < 0 \\ x^2 & \text{if } x \geq 0 \end{cases}$$

38. $\lim_{x \to 0^+} f(x)$ and $\lim_{x \to 0^-} f(x)$, where

$$f(x) = \begin{cases} -x + 1 & \text{if } x \leq 0 \\ 2x + 3 & \text{if } x > 0 \end{cases}$$

In Exercises 39–44, determine the values of x, if any, at which each function is discontinuous. At each number where f is discontinuous, state the condition(s) for continuity that are violated.

39. $f(x) = \begin{cases} 2x - 4 & \text{if } x \leq 0 \\ 1 & \text{if } x > 0 \end{cases}$

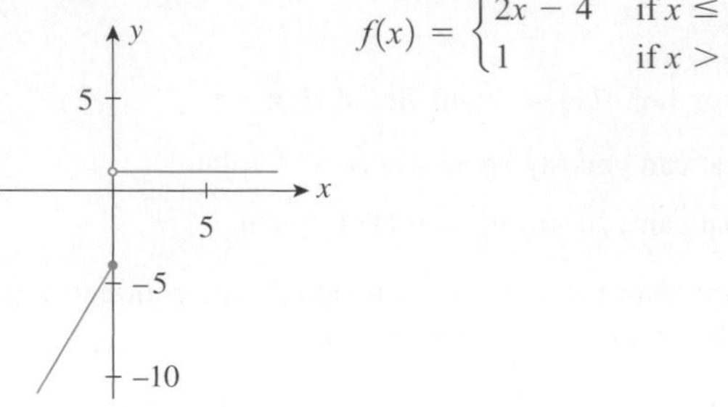

40. $f(x) = \begin{cases} x^2 + 1 & \text{if } x \neq 0 \\ 0 & \text{if } x = 0 \end{cases}$

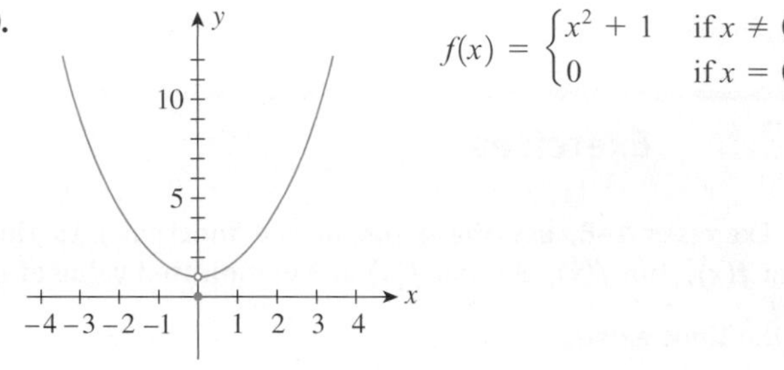

41. $f(x) = \begin{cases} x + 5 & \text{if } x \leq 0 \\ -x^2 + 5 & \text{if } x > 0 \end{cases}$

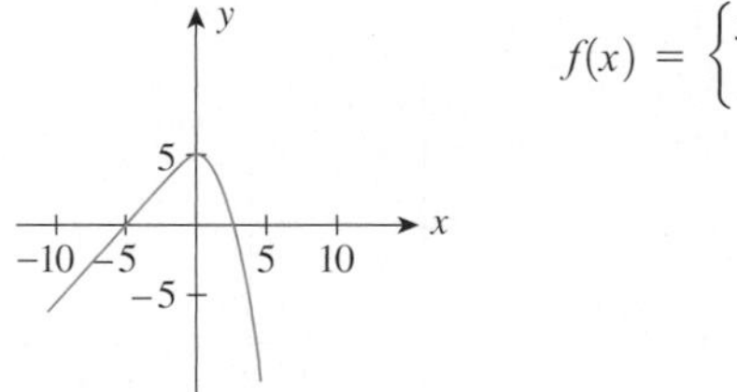

42. $f(x) = |x - 1|$

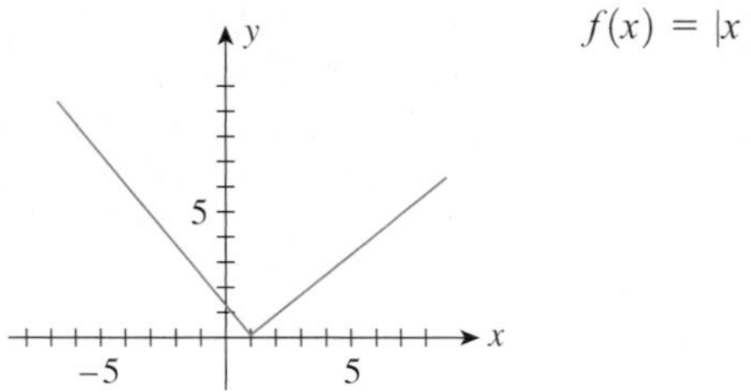

43. $f(x) = \begin{cases} x + 5 & \text{if } x < 0 \\ 2 & \text{if } x = 0 \\ -x^2 + 5 & \text{if } x > 0 \end{cases}$

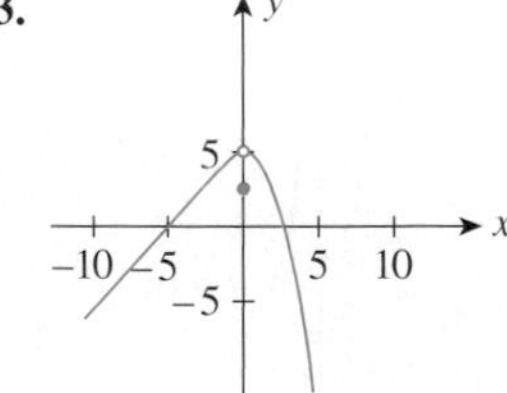

44.

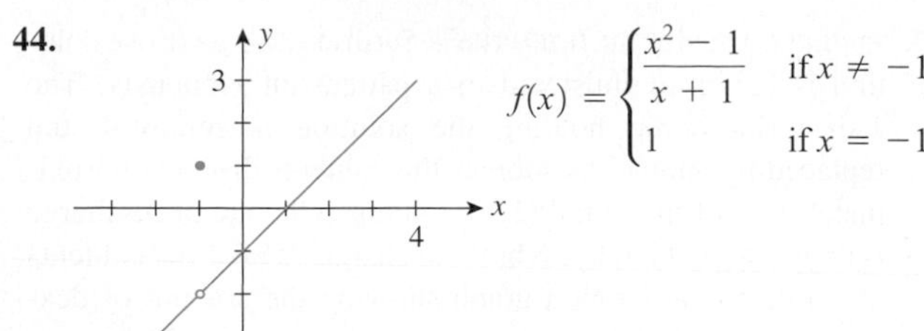

$$f(x) = \begin{cases} \dfrac{x^2 - 1}{x + 1} & \text{if } x \neq -1 \\ 1 & \text{if } x = -1 \end{cases}$$

In Exercises 45–56, find the values of x for which each function is continuous.

45. $f(x) = 2x^2 + x - 1$

46. $f(x) = x^3 - 2x^2 + x - 1$

47. $f(x) = \dfrac{2}{x^2 + 1}$

48. $f(x) = \dfrac{x}{2x^2 + 1}$

49. $f(x) = \dfrac{2}{2x - 1}$

50. $f(x) = \dfrac{x + 1}{x - 1}$

51. $f(x) = \dfrac{2x + 1}{x^2 + x - 2}$

52. $f(x) = \dfrac{x - 1}{x^2 + 2x - 3}$

53. $f(x) = \begin{cases} x & \text{if } x \leq 1 \\ 2x - 1 & \text{if } x > 1 \end{cases}$

54. $f(x) = \begin{cases} -2x + 1 & \text{if } x < 0 \\ x^2 + 1 & \text{if } x \geq 0 \end{cases}$

55. $f(x) = |x + 1|$

56. $f(x) = \dfrac{|x - 1|}{x - 1}$

In Exercises 57–60, determine all values of x at which the function is discontinuous.

57. $f(x) = \dfrac{2x}{x^2 - 1}$

58. $f(x) = \dfrac{1}{(x - 1)(x - 2)}$

59. $f(x) = \dfrac{x^2 - 2x}{x^2 - 3x + 2}$

60. $f(x) = \dfrac{x^2 - 3x + 2}{x^2 - 2x}$

61. The Postage Function The graph of the "postage function" for 2006,

$$f(x) = \begin{cases} 39 & \text{if } 0 < x \leq 1 \\ 63 & \text{if } 1 < x \leq 2 \\ \vdots & \\ 303 & \text{if } 11 < x \leq 12 \end{cases}$$

where x denotes the weight of a parcel in ounces and $f(x)$ the postage in cents, is shown in the accompanying figure. Determine the values of x for which f is discontinuous.

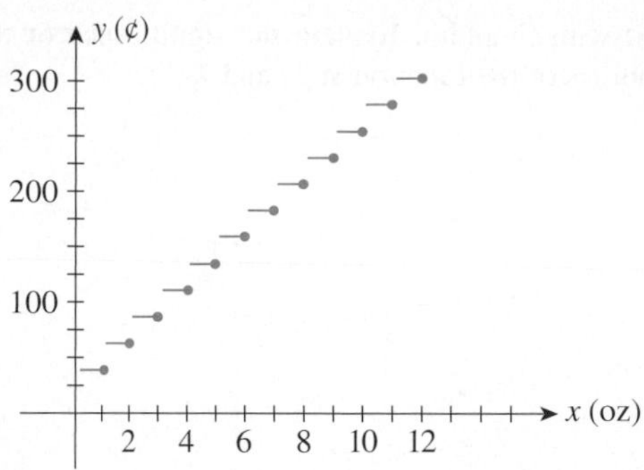

62. Inventory Control As part of an optimal inventory policy, the manager of an office supply company orders 500 reams of photocopy paper every 20 days. The accompanying graph shows the *actual* inventory level of paper in an office supply store during the first 60 business days of 2005. Determine the values of t for which the "inventory function" is discontinuous and give an interpretation of the graph.

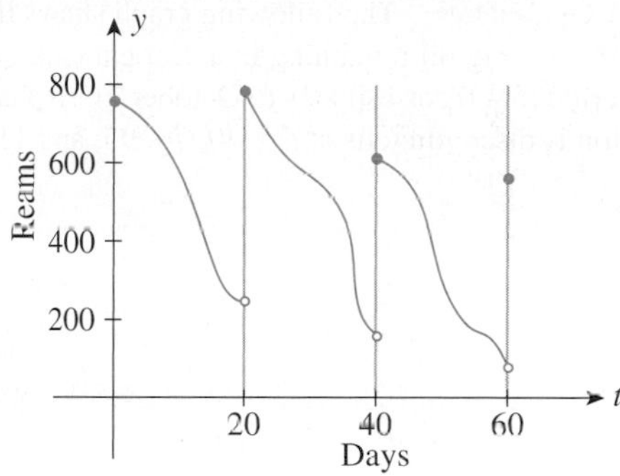

63. Learning Curves The following graph describes the progress Michael made in solving a problem correctly during a mathematics quiz. Here, y denotes the percent of work completed, and x is measured in minutes. Give an interpretation of the graph.

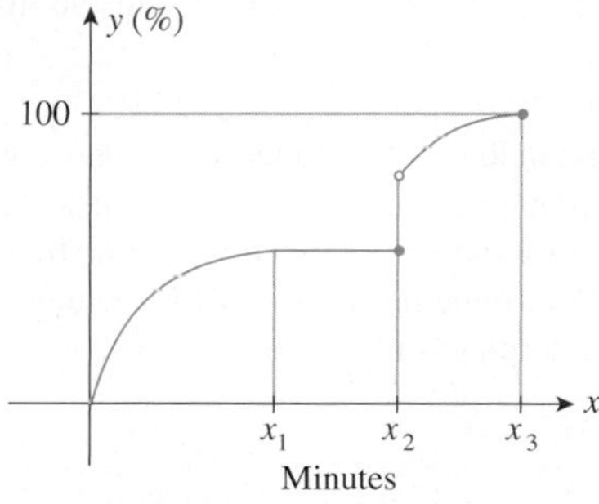

64. Ailing Financial Institutions Franklin Savings and Loan acquired two ailing financial institutions in 2006. One of them was acquired at time $t = T_1$, and the other was acquired at time $t = T_2$ ($t = 0$ corresponds to the beginning of 2006). The following graph shows the total amount of money on

deposit with Franklin. Explain the significance of the discontinuities of the function at T_1 and T_2.

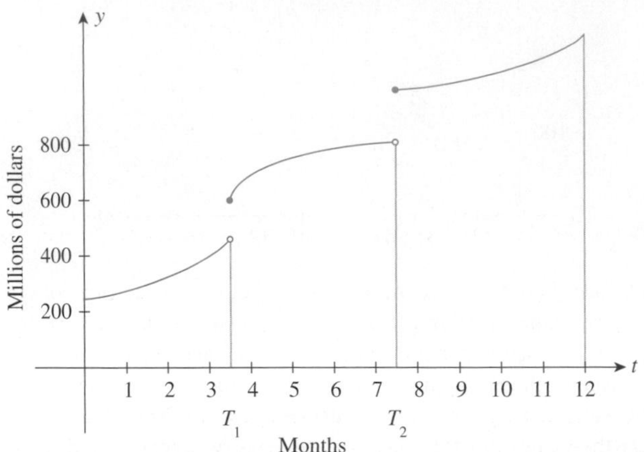

65. **Energy Consumption** The following graph shows the amount of home heating oil remaining in a 200-gal tank over a 120-day period ($t = 0$ corresponds to October 1). Explain why the function is discontinuous at $t = 40$, 70, 95, and 110.

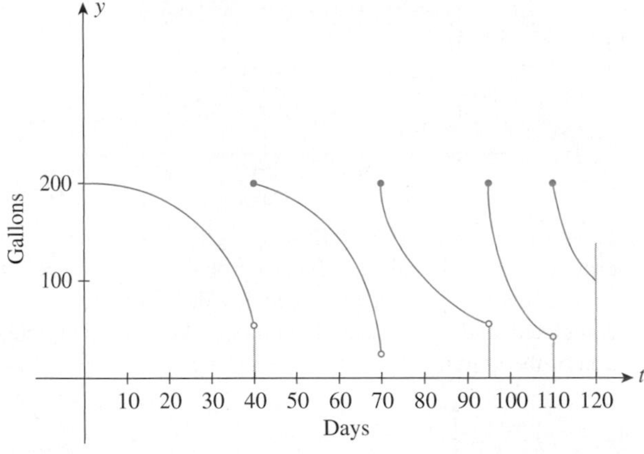

66. **Prime Interest Rate** The function P, whose graph follows, gives the prime rate (the interest rate banks charge their best corporate customers) as a function of time for the first 32 wk in 1989. Determine the values of t for which P is discontinuous and interpret your results.

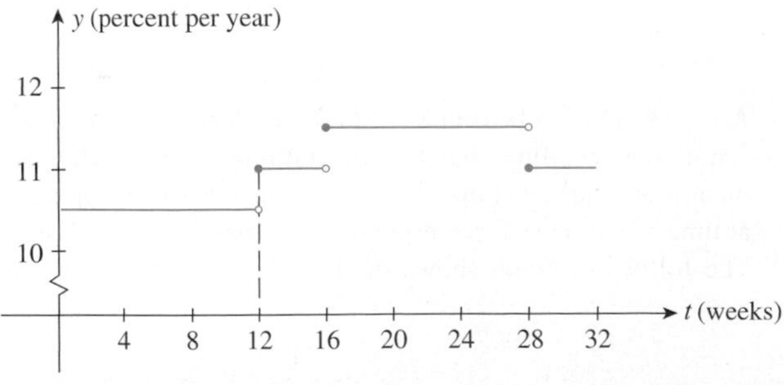

67. **Administration of an Intravenous Solution** A dextrose solution is being administered to a patient intravenously. The 1-liter (L) bottle holding the solution is removed and replaced by another as soon as the contents drop to approximately 5% of the initial (1-L) amount. The rate of discharge is constant, and it takes 6 hr to discharge 95% of the contents of a full bottle. Draw a graph showing the amount of dextrose solution in a bottle in the IV system over a 24-hr period, assuming that we started with a full bottle.

68. **Commissions** The base monthly salary of a salesman working on commission is \$12,000. For each \$50,000 of sales beyond \$100,000, he is paid a \$1000 commission. Sketch a graph showing his earnings as a function of the level of his sales x. Determine the values of x for which the function f is discontinuous.

69. **Parking Fees** The fee charged per car in a downtown parking lot is \$2.00 for the first half hour and \$1.00 for each additional half hour or part thereof, subject to a maximum of \$10.00. Derive a function f relating the parking fee to the length of time a car is left in the lot. Sketch the graph of f and determine the values of x for which the function f is discontinuous.

70. **Commodity Prices** The function that gives the cost of a certain commodity is defined by

$$C(x) = \begin{cases} 5x & \text{if } 0 < x < 10 \\ 4x & \text{if } 10 \le x < 30 \\ 3.5x & \text{if } 30 \le x < 60 \\ 3.25x & \text{if } x \ge 60 \end{cases}$$

where x is the number of pounds of a certain commodity sold and $C(x)$ is measured in dollars. Sketch the graph of the function C and determine the values of x for which the function C is discontinuous.

71. **Weiss's Law** According to Weiss's law of excitation of tissue, the strength S of an electric current is related to the time t the current takes to excite tissue by the formula

$$S(t) = \frac{a}{t} + b \qquad (t > 0)$$

where a and b are positive constants.

a. Evaluate $\lim_{t \to 0^+} S(t)$ and interpret your result.

b. Evaluate $\lim_{t \to \infty} S(t)$ and interpret your result.

(*Note:* The limit in part (b) is called the threshold strength of the current. Why?)

72. **Energy Expended by a Fish** Suppose a fish swimming a distance of L ft at a speed of v ft/sec relative to the water and against a current flowing at the rate of u ft/sec ($u < v$) expends a total energy given by

$$E(v) = \frac{aLv^3}{v - u}$$

where E is measured in foot-pounds (ft-lb) and a is a constant.

a. Evaluate $\lim_{v \to u^+} E(v)$ and interpret your result.

b. Evaluate $\lim_{v \to \infty} E(v)$ and interpret your result.

73. Let

$$f(x) = \begin{cases} x + 2 & \text{if } x \leq 1 \\ kx^2 & \text{if } x > 1 \end{cases}$$

Find the value of k that will make f continuous on $(-\infty, \infty)$.

74. Let

$$f(x) = \begin{cases} \dfrac{x^2 - 4}{x + 2} & \text{if } x \neq -2 \\ k & \text{if } x = -2 \end{cases}$$

For what value of k will f be continuous on $(-\infty, \infty)$?

75. a. Suppose f is continuous at a and g is discontinuous at a. Is the sum $f + g$ discontinuous at a? Explain.

b. Suppose f and g are both discontinuous at a. Is the sum $f + g$ necessarily discontinuous at a? Explain.

76. a. Suppose f is continuous at a and g is discontinuous at a. Is the product fg necessarily discontinuous at a? Explain.

b. Suppose f and g are both discontinuous at a. Is the product fg necessarily discontinuous at a? Explain.

In Exercises 77–80, (a) show that the function f is continuous for all values of x in the interval $[a, b]$ and (b) prove that f must have at least one zero in the interval (a, b) by showing that $f(a)$ and $f(b)$ have opposite signs.

77. $f(x) = x^2 - 6x + 8$; $a = 1$, $b = 3$

78. $f(x) = 2x^3 - 3x^2 - 36x + 14$; $a = 0$, $b = 1$

79. $f(x) = x^3 - 2x^2 + 3x + 2$; $a = -1$, $b = 1$

80. $f(x) = 2x^{5/3} - 5x^{4/3}$; $a = 14$, $b = 16$

In Exercises 81–82, use the intermediate value theorem to find the value of c such that $f(c) = M$.

81. $f(x) = x^2 - 4x + 6$ on $[0, 3]$; $M = 4$

82. $f(x) = x^2 - x + 1$ on $[-1, 4]$; $M = 7$

83. Use the method of bisection (see Example 6) to find the root of the equation $x^5 + 2x - 7 = 0$ accurate to two decimal places.

84. Use the method of bisection (see Example 6) to find the root of the equation $x^3 - x + 1 = 0$ accurate to two decimal places.

85. Falling Object Joan is looking straight out a window of an apartment building at a height of 32 ft from the ground. A boy throws a tennis ball straight up by the side of the building where the window is located. Suppose the height of the ball (measured in feet) from the ground at time t is $h(t) = 4 + 64t - 16t^2$.

a. Show that $h(0) = 4$ and $h(2) = 68$.

b. Use the intermediate value theorem to conclude that the ball must cross Joan's line of sight at least once.

c. At what time(s) does the ball cross Joan's line of sight? Interpret your results.

86. Oxygen Content of a Pond The oxygen content t days after organic waste has been dumped into a pond is given by

$$f(t) = 100\left(\frac{t^2 + 10t + 100}{t^2 + 20t + 100}\right)$$

percent of its normal level.

a. Show that $f(0) = 100$ and $f(10) = 75$.

b. Use the intermediate value theorem to conclude that the oxygen content of the pond must have been at a level of 80% at some time.

c. At what time(s) is the oxygen content at the 80% level? Hint: Use the quadratic formula.

In Exercises 87–93, determine whether the statement is true or false. If it is true, explain why it is true. If it is false, give an example to show why it is false.

87. If $f(2) = 4$, then $\lim_{x \to 2} f(x) = 4$.

88. If $\lim_{x \to 0} f(x) = 3$, then $f(0) = 3$.

89. Suppose the function f is defined on the interval $[a, b]$. If $f(a)$ and $f(b)$ have the same sign, then f has no zero in $[a, b]$.

90. If $\lim_{x \to a} f(x) = L$, then $\lim_{x \to a^+} f(x) - \lim_{x \to a^-} f(x) \neq 0$.

91. If $\lim_{x \to a^-} f(x) = L$ and $\lim_{x \to a^+} f(x) = L$, then $f(a) = L$.

92. If $\lim_{x \to a} f(x) = L$ and $g(a) = M$, then $\lim_{x \to a} f(x)g(x) = LM$.

93. If f is continuous for all $x \neq 0$ and $f(0) = 0$, then $\lim_{x \to 0} f(x) = 0$.

94. Suppose f is continuous on $[a, b]$ and $f(a) < f(b)$. If M is a number that lies outside the interval $[f(a), f(b)]$, then there does not exist a number $a < c < b$ such that $f(c) = M$. Does this contradict the intermediate value theorem?

95. Let $f(x) = x - \sqrt{1 - x^2}$.

a. Show that f is continuous for all values of x in the interval $[-1, 1]$.

b. Show that f has at least one zero in $[-1, 1]$.

c. Find the zeros of f in $[-1, 1]$ by solving the equation $f(x) = 0$.

96. Let $f(x) = \dfrac{x^2}{x^2 + 1}$.

a. Show that f is continuous for all values of x.

b. Show that $f(x)$ is nonnegative for all values of x.

c. Show that f has a zero at $x = 0$. Does this contradict Theorem 5?

97. a. Prove that a polynomial function $y = P(x)$ is continuous at every number x. Follow these steps:

(1) Use Properties 2 and 3 of continuous functions to establish that the function $g(x) = x^n$, where n is a positive integer, is continuous everywhere.

(2) Use Properties 1 and 5 to show that $f(x) = cx^n$, where c is a constant and n is a positive integer, is continuous everywhere.

(3) Use Property 4 to complete the proof of the result.

b. Prove that a rational function $R(x) = p(x)/q(x)$ is continuous at every point x, where $q(x) \neq 0$.
Hint: Use the result of part (a) and Property 6.

98. Show that the conclusion of the intermediate value theorem does not hold if f is discontinuous on $[a, b]$.

2.5 Solutions to Self-Check Exercises

1. For $x < -1$, $f(x) = 1$, and so

$$\lim_{x\to-1^-} f(x) = \lim_{x\to-1^-} 1 = 1$$

For $x \geq -1$, $f(x) = 1 + \sqrt{x+1}$, and so

$$\lim_{x\to-1^+} f(x) = \lim_{x\to-1^+} (1 + \sqrt{x+1}) = 1$$

Since the left-hand and right-hand limits of f exist as x approaches $x = -1$ and both are equal to 1, we conclude that

$$\lim_{x\to-1} f(x) = 1$$

2. a. The graph of f follows:

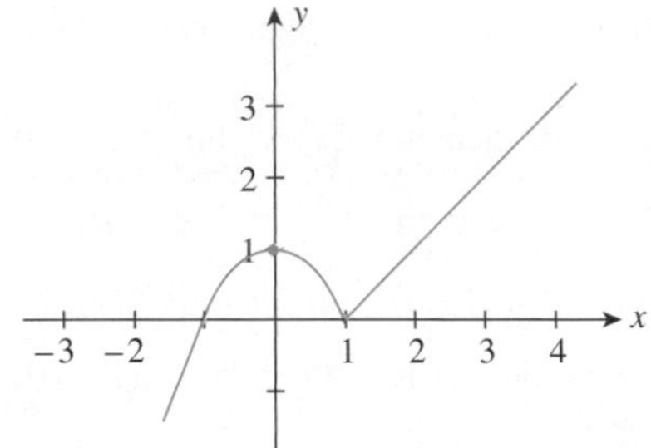

We see that f is continuous everywhere.

b. The graph of g follows:

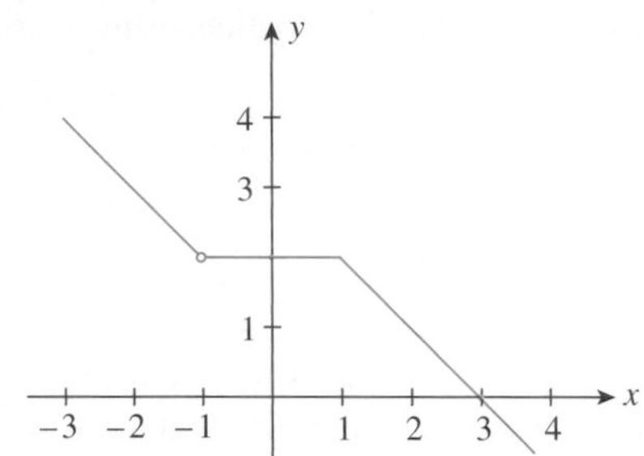

Since g is not defined at $x = -1$, it is discontinuous there. It is continuous everywhere else.

USING TECHNOLOGY

Finding the Points of Discontinuity of a Function

You can very often recognize the points of discontinuity of a function f by examining its graph. For example, Figure T1a shows the graph of $f(x) = x/(x^2 - 1)$ obtained using a graphing utility. It is evident that f is discontinuous at $x = -1$ and $x = 1$. This observation is also borne out by the fact that both these points are not in the domain of f.

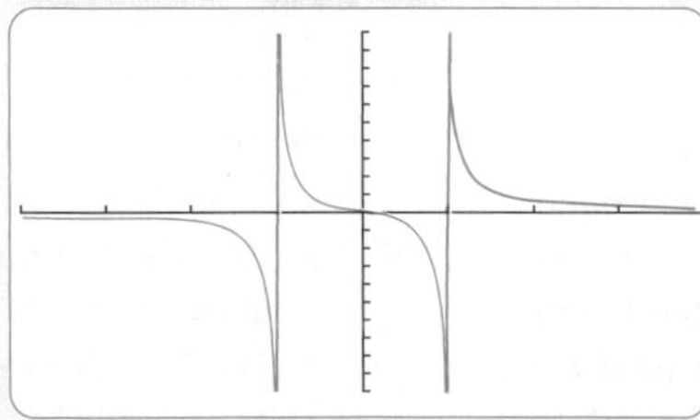

(a)

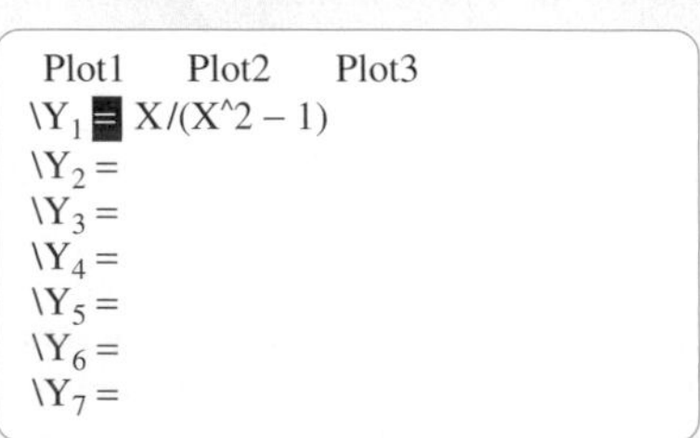

(b)

FIGURE T1
(a) The graph of $f(x) = \dfrac{x}{x^2 - 1}$ in the viewing window $[-4, 4] \times [-10, 10]$; (b) the TI-83 equation screen

Consider the function

$$g(x) = \frac{2x^3 + x^2 - 7x - 6}{x^2 - x - 2}$$

Using a graphing utility, we obtain the graph of g shown in Figure T2a.

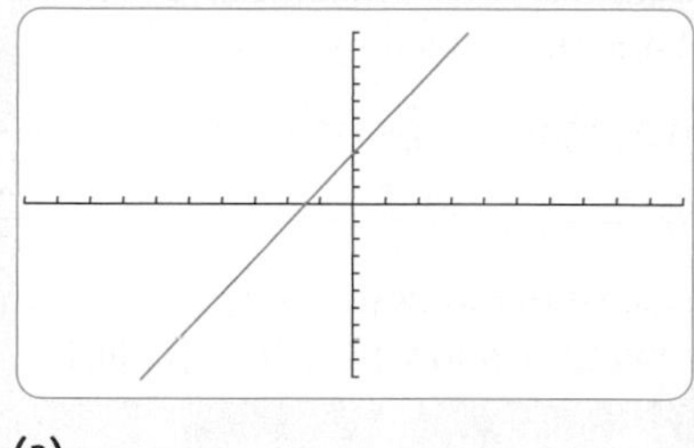

(a)

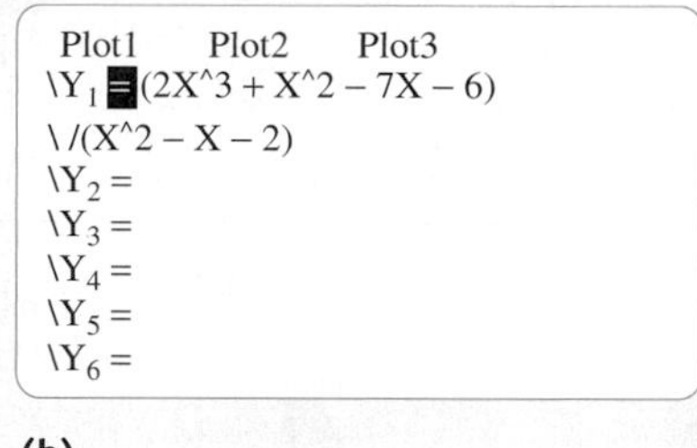

(b)

FIGURE T2
(a) The graph of $g(x) = \dfrac{2x^3 + x^2 - 7x - 6}{x^2 - x - 2}$ in the standard viewing window; (b) the TI-83 equation screen.

An examination of this graph does not reveal any points of discontinuity. However, if we factor both the numerator and the denominator of the rational expression, we see that

$$\begin{aligned} g(x) &= \frac{(x + 1)(x - 2)(2x + 3)}{(x + 1)(x - 2)} \\ &= 2x + 3 \end{aligned}$$

provided $x \neq -1$ and $x \neq 2$, so that its graph in fact looks like that shown in Figure T3.

This example shows the limitation of the graphing utility and reminds us of the importance of studying functions analytically!

FIGURE T3
The graph of g has holes at $(-1, 1)$ and $(2, 7)$.

Graphing Functions Defined Piecewise

The following example illustrates how to plot the graphs of functions defined in a piecewise manner on a graphing utility.

EXAMPLE 1 Plot the graph of

$$f(x) = \begin{cases} x + 1 & \text{if } x \leq 1 \\ \dfrac{2}{x} & \text{if } x > 1 \end{cases}$$

Solution We enter the function

$$\text{y1} = (\text{x} + 1)(\text{x} \leq 1) + (2/\text{x})(\text{x} > 1)$$

Figure T4 shows the graph of the function in the viewing window $[-5, 5] \times [-2, 4]$.

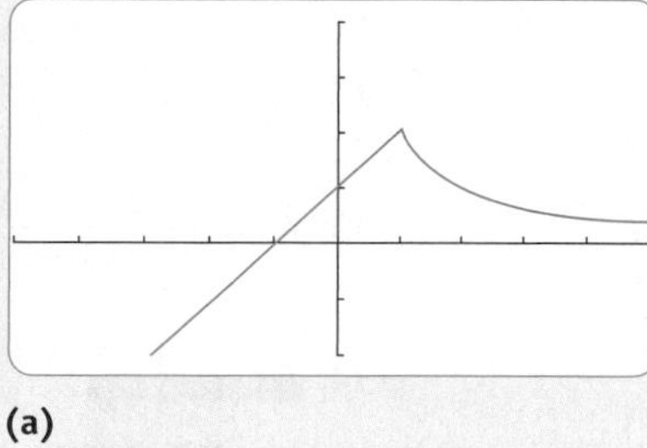

(a)

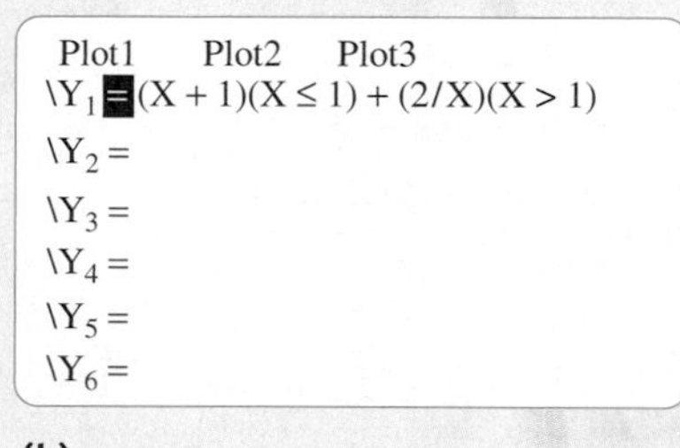

(b)

FIGURE T4
(a) The graph of f in the viewing window $[-5, 5] \times [-2, 4]$; (b) the TI-83 equation screen.

(*continued*)

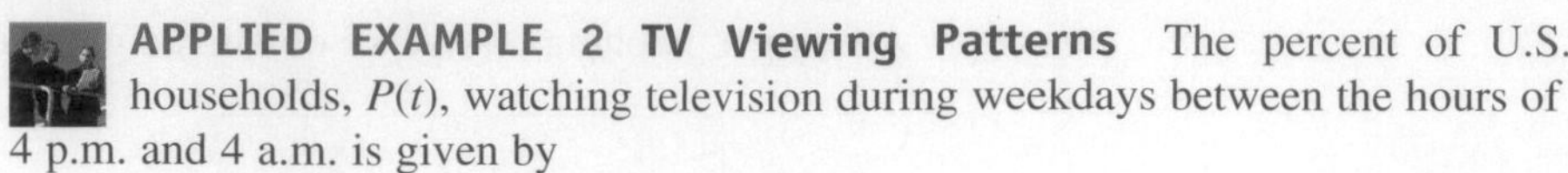

APPLIED EXAMPLE 2 TV Viewing Patterns The percent of U.S. households, $P(t)$, watching television during weekdays between the hours of 4 p.m. and 4 a.m. is given by

$$P(t) = \begin{cases} 0.01354t^4 - 0.49375t^3 + 2.58333t^2 + 3.8t + 31.60704 & \text{if } 0 \leq t \leq 8 \\ 1.35t^2 - 33.05t + 208 & \text{if } 8 < t \leq 12 \end{cases}$$

where t is measured in hours, with $t = 0$ corresponding to 4 p.m. Plot the graph of P in the viewing window $[0, 12] \times [0, 80]$.

Source: A. C. Nielsen Co.

Solution We enter the function

$$\text{y1} = (.01354x\text{^}4 - .49375x\text{^}3 + 2.58333x\text{^}2 + 3.8x + 31.60704)(x \geq 0)(x \leq 8) + (1.35x\text{^}2 - 33.05x + 208)(x > 8)(x \leq 12)$$

Figure T5a shows the graph of P.

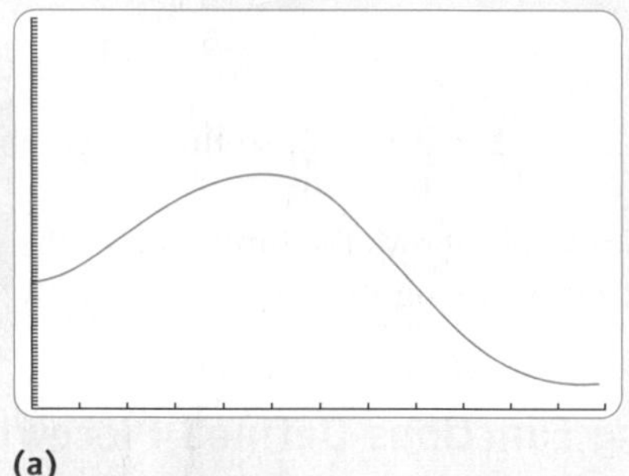

(a)

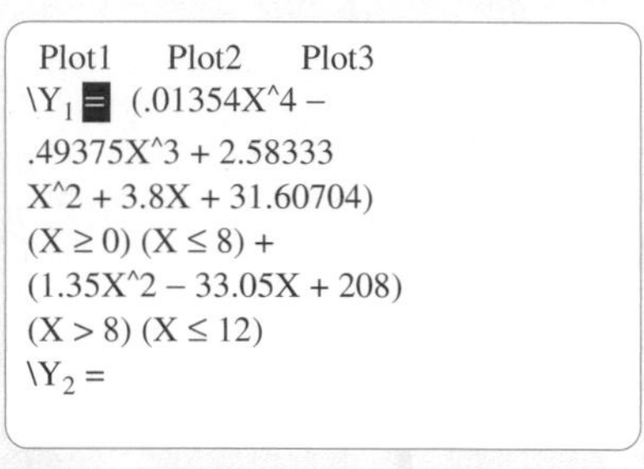

(b)

FIGURE T5
(a) The graph of P in the viewing window $[0, 12] \times [0, 80]$; (b) the TI-83 equation screen

TECHNOLOGY EXERCISES

In Exercises 1–10, plot the graph of f and find the points of discontinuity of f. Then use analytical means to verify your observation and find all numbers where f is discontinuous.

1. $f(x) = \dfrac{2}{x^2 - x}$

2. $f(x) = \dfrac{2x + 1}{x^2 + x - 2}$

3. $f(x) = \dfrac{\sqrt{x}}{x^2 - x - 2}$

4. $f(x) = \dfrac{3}{\sqrt{x}(x + 1)}$

5. $f(x) = \dfrac{6x^3 + x^2 - 2x}{2x^2 - x}$

6. $f(x) = \dfrac{2x^3 - x^2 - 13x - 6}{2x^2 - 5x - 3}$

7. $f(x) = \dfrac{2x^4 - 3x^3 - 2x^2}{2x^2 - 3x - 2}$

8. $f(x) = \dfrac{6x^4 - x^3 + 5x^2 - 1}{6x^2 - x - 1}$

9. $f(x) = \dfrac{x^3 + x^2 - 2x}{x^4 + 2x^3 - x - 2}$
   Hint: $x^4 + 2x^3 - x - 2 = (x^3 - 1)(x + 2)$

10. $f(x) = \dfrac{x^3 - x}{x^{4/3} - x + x^{1/3} - 1}$
    Hint: $x^{4/3} - x + x^{1/3} - 1 = (x^{1/3} - 1)(x + 1)$
    Can you explain why part of the graph is missing?

In Exercises 11–14, plot the graph of f in the indicated viewing window.

11. $f(x) = \begin{cases} -1 & \text{if } x \leq 1 \\ x + 1 & \text{if } x > 1 \end{cases}$; $[-5, 5] \times [-2, 8]$

12. $f(x) = \begin{cases} \frac{1}{3}x^2 - 2x & \text{if } x \leq 3 \\ -x + 6 & \text{if } x > 3 \end{cases}$; $[0, 7] \times [-5, 5]$

13. $f(x) = \begin{cases} 2 & \text{if } x \leq 0 \\ \sqrt{4 - x^2} & \text{if } x > 0 \end{cases}$; $[-2, 2] \times [-4, 4]$

14. $f(x) = \begin{cases} -x^2 + x + 2 & \text{if } x \leq 1 \\ 2x^3 - x^2 - 4 & \text{if } x > 1 \end{cases}$; $[-4, 4] \times [-5, 5]$

15. **Flight Path of a Plane** The function

$$f(x) = \begin{cases} 0 & \text{if } 0 \le x < 1 \\ -0.00411523x^3 + 0.0679012x^2 \\ \quad -0.123457x + 0.0596708 & \text{if } 1 \le x < 10 \\ 1.5 & \text{if } 10 \le x \le 100 \end{cases}$$

where both x and $f(x)$ are measured in units of 1000 ft, describes the flight path of a plane taking off from the origin and climbing to an altitude of 15,000 ft. Plot the graph of f to visualize the trajectory of the plane.

16. **Home Shopping Industry** According to industry sources, revenue from the home shopping industry for the years since its inception may be approximated by the function

$$R(t) = \begin{cases} -0.03t^3 + 0.25t^2 - 0.12t & \text{if } 0 \le t \le 3 \\ 0.57t - 0.63 & \text{if } 3 < t \le 11 \end{cases}$$

where $R(t)$ measures the revenue in billions of dollars and t is measured in years, with $t = 0$ corresponding to the beginning of 1984. Plot the graph of R.

Source: Paul Kagan Associates

2.6 The Derivative

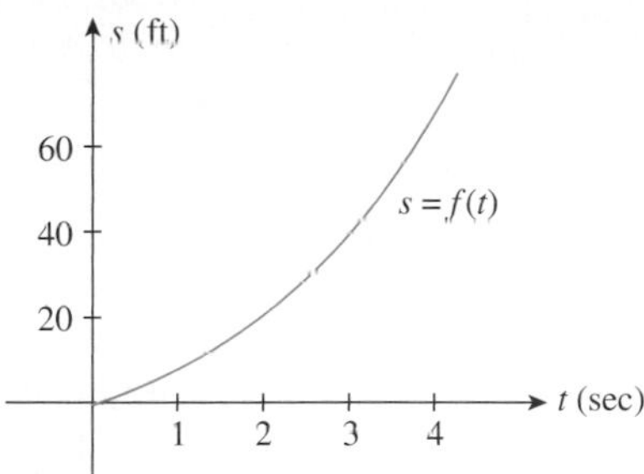

FIGURE 46
Graph showing the position s of a maglev at time t

An Intuitive Example

We mentioned in Section 2.4 that the problem of finding the *rate of change* of one quantity with respect to another is mathematically equivalent to the problem of finding the *slope of the tangent line* to a curve at a given point on the curve. Before going on to establish this relationship, let's show its plausibility by looking at it from an intuitive point of view.

Consider the motion of the maglev discussed in Section 2.4. Recall that the position of the maglev at any time t is given by

$$s = f(t) = 4t^2 \qquad (0 \le t \le 30)$$

where s is measured in feet and t in seconds. The graph of the function f is sketched in Figure 46.

Observe that the graph of f rises slowly at first but more rapidly as t increases, reflecting the fact that the speed of the maglev is increasing with time. This observation suggests a relationship between the speed of the maglev at any time t and the *steepness* of the curve at the point corresponding to this value of t. Thus, it would appear that we can solve the problem of finding the speed of the maglev at any time if we can find a way to measure the steepness of the curve at any point on the curve.

To discover a yardstick that will measure the steepness of a curve, consider the graph of a function f such as the one shown in Figure 47a. Think of the curve as representing a stretch of roller coaster track (Figure 47b). When the car is at the point P on the curve, a passenger sitting erect in the car and looking straight ahead will have a line of sight that is parallel to the line T, the tangent to the curve at P.

As Figure 47a suggests, the steepness of the curve—that is, the rate at which y is increasing or decreasing with respect to x—is given by the slope of the tangent line to the graph of f at the point $P(x, f(x))$. But for now we will show how this relationship can be used to estimate the rate of change of a function from its graph.

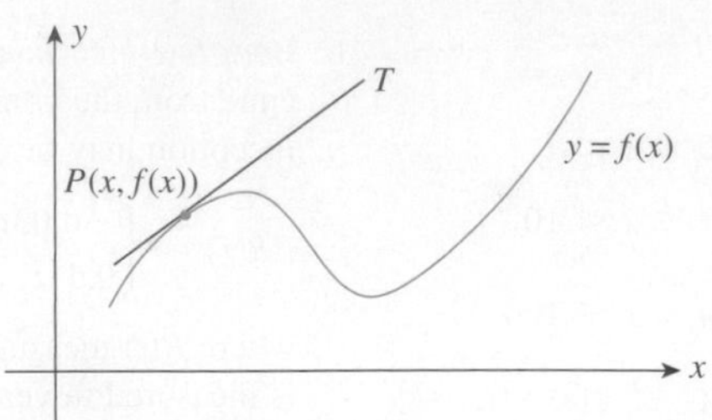

(a) T is the tangent line to the curve at P.

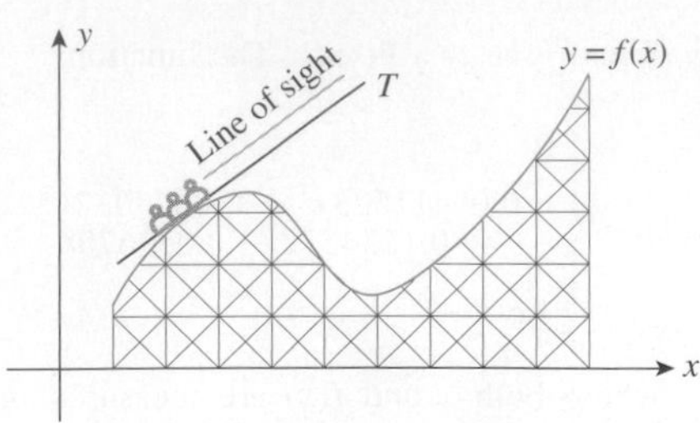

(b) T is parallel to the line of sight.

FIGURE 47

APPLIED EXAMPLE 1 Social Security Beneficiaries The graph of the function $y = N(t)$, shown in Figure 48, gives the number of Social Security beneficiaries from the beginning of 1990 ($t = 0$) through the year 2045 ($t = 55$).

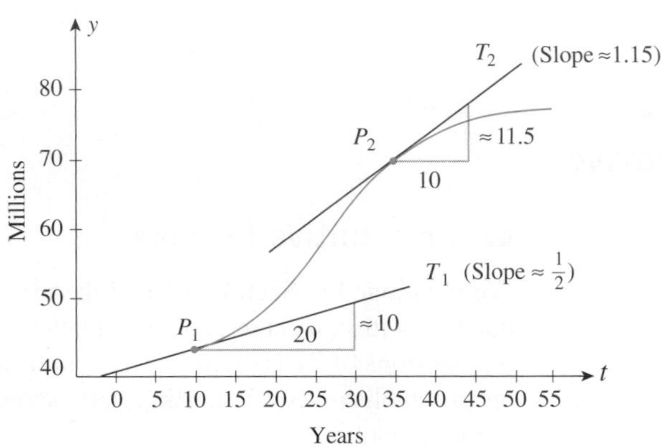

FIGURE 48
The number of Social Security beneficiaries from 1990 through 2045. We can use the slope of the tangent line at the indicated points to estimate the rate at which the number of Social Security beneficiaries will be changing.

Use the graph of $y = N(t)$ to estimate the rate at which the number of Social Security beneficiaries was growing at the beginning of the year 2000 ($t = 10$). How fast will the number be growing at the beginning of 2025 ($t = 35$)? [Assume that the rate of change of the function N at any value of t is given by the slope of the tangent line at the point $P(t, N(t))$.]
Source: Social Security Administration

Solution From the figure, we see that the slope of the tangent line T_1 to the graph of $y = N(t)$ at $P_1(10, 44.7)$ is approximately 0.5. This tells us that the quantity y is increasing at the rate of $\frac{1}{2}$ unit per unit increase in t, when $t = 10$. In other words, at the beginning of the year 2000, the number of Social Security beneficiaries was increasing at the rate of approximately 0.5 million, or 500,000, per year.

The slope of the tangent line T_2 at $P_2(35, 71.9)$ is approximately 1.15. This tells us that at the beginning of 2025 the number of Social Security beneficiaries will be growing at the rate of approximately 1.15 million, or 1,150,000, per year. ■

Slope of a Tangent Line

In Example 1 we answered the questions raised by drawing the graph of the function N and estimating the position of the tangent lines. Ideally, however, we would like to solve a problem analytically whenever possible. To do this we need a precise definition of the slope of a tangent line to a curve.

To define the tangent line to a curve C at a point P on the curve, fix P and let Q be any point on C distinct from P (Figure 49). The straight line passing through P and Q is called a **secant line**.

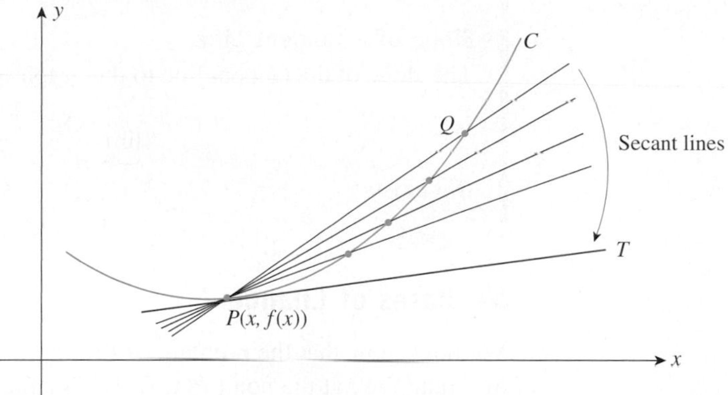

FIGURE 49
As Q approaches P along the curve C, the secant lines approach the tangent line T.

Now, as the point Q is allowed to move toward P along the curve, the secant line through P and Q rotates about the fixed point P and approaches a fixed line through P. This fixed line, which is the limiting position of the secant lines through P and Q as Q approaches P, is the **tangent line to the graph of f** at the point P.

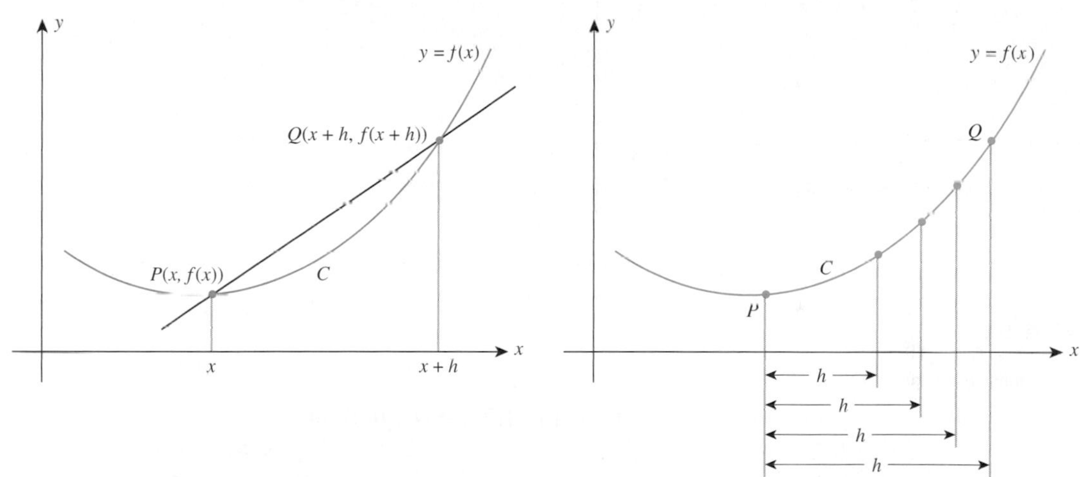

FIGURE 50
(a) The points $P(x, f(x))$ and $Q(x + h, f(x + h))$
(b) As h approaches zero, Q approaches P.

We can describe the process more precisely as follows. Suppose the curve C is the graph of a function f defined by $y = f(x)$. Then the point P is described by $P(x, f(x))$ and the point Q by $Q(x + h, f(x + h))$, where h is some appropriate nonzero number (Figure 50a). Observe that we can make Q approach P along the curve C by letting h approach zero (Figure 50b).

Next, using the formula for the slope of a line, we can write the slope of the secant line passing through $P(x, f(x))$ and $Q(x + h, f(x + h))$ as

$$\frac{f(x + h) - f(x)}{(x + h) - x} = \frac{f(x + h) - f(x)}{h} \tag{5}$$

As observed earlier, Q approaches P, and therefore the secant line through P and Q approaches the tangent line T as h approaches zero. Consequently, we might expect that the slope of the secant line would approach the slope of the tangent line T as h approaches zero. This leads to the following definition.

Slope of a Tangent Line

The slope of the tangent line to the graph of f at the point $P(x, f(x))$ is given by

$$\lim_{h \to 0} \frac{f(x + h) - f(x)}{h} \tag{6}$$

if it exists.

Rates of Change

We now show that the problem of finding the slope of the tangent line to the graph of a function f at the point $P(x, f(x))$ is mathematically equivalent to the problem of finding the rate of change of f at x. To see this, suppose we are given a function f that describes the relationship between the two quantities x and y—that is, $y = f(x)$. The number $f(x + h) - f(x)$ measures the change in y that corresponds to a change h in x (Figure 51).

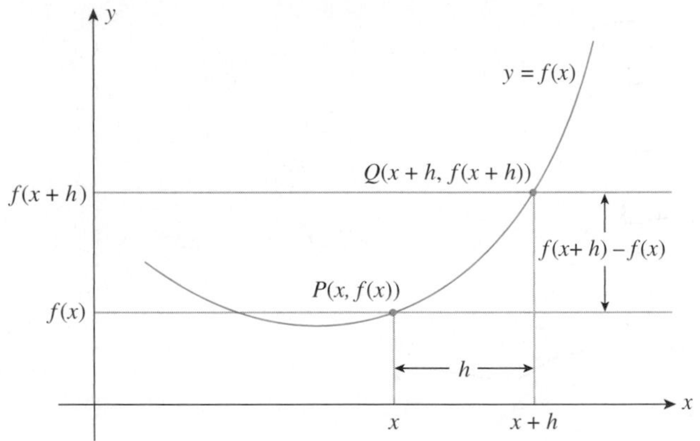

FIGURE 51
$f(x + h) - f(x)$ is the change in y that corresponds to a change h in x.

Then, the **difference quotient**

$$\frac{f(x + h) - f(x)}{h} \tag{7}$$

measures the **average rate of change of y with respect to x** over the interval $[x, x + h]$. For example, if y measures the position of a car at time x, then quotient (7) gives the average velocity of the car over the time interval $[x, x + h]$.

Observe that the difference quotient (7) is the same as (5). We conclude that the difference quotient (7) also measures the slope of the secant line that passes through the two points $P(x, f(x))$ and $Q(x + h, f(x + h))$ lying on the graph of $y = f(x)$. Next, by taking the limit of the difference quotient (7) as h goes to zero—that is, by evaluating

$$\lim_{h \to 0} \frac{f(x + h) - f(x)}{h} \tag{8}$$

we obtain the **rate of change of f at x**. For example, if y measures the position of a car at time x, then the limit (8) gives the velocity of the car at time x. For emphasis, the rate of change of a function f at x is often called the **instantaneous rate of**

change of f at x. This distinguishes it from the average rate of change of f, which is computed over an *interval* $[x, x + h]$ rather than at a *number* x.

Observe that the limit (8) is the same as (6). Therefore, the limit of the difference quotient also measures the slope of the tangent line to the graph of $y = f(x)$ at the point $(x, f(x))$. The following summarizes this discussion.

Average and Instantaneous Rates of Change

The **average rate of change** of f over the interval $[x, x + h]$ or **slope of the secant line** to the graph of f through the points $(x, f(x))$ and $(x + h, f(x + h))$ is

$$\frac{f(x+h) - f(x)}{h} \tag{9}$$

The **instantaneous rate of change** of f at x or **slope of the tangent line** to the graph of f at $(x, f(x))$ is

$$\lim_{h \to 0} \frac{f(x+h) - f(x)}{h} \tag{10}$$

EXPLORE & DISCUSS

Explain the difference between the average rate of change of a function and the instantaneous rate of change of a function.

The Derivative

The limit (6) or (10), which measures both the slope of the tangent line to the graph of $y = f(x)$ at the point $P(x, f(x))$ and the (instantaneous) rate of change of f at x, is given a special name: the **derivative of f at x**.

Derivative of a Function

The derivative of a function f with respect to x is the function f' (read "f prime"),

$$f'(x) = \lim_{h \to 0} \frac{f(x+h) - f(x)}{h} \tag{11}$$

The domain of f' is the set of all x where the limit exists.

Thus, the derivative of a function f is a function f' that gives the slope of the tangent line to the graph of f at *any* point $(x, f(x))$ and also the rate of change of f at x (Figure 52).

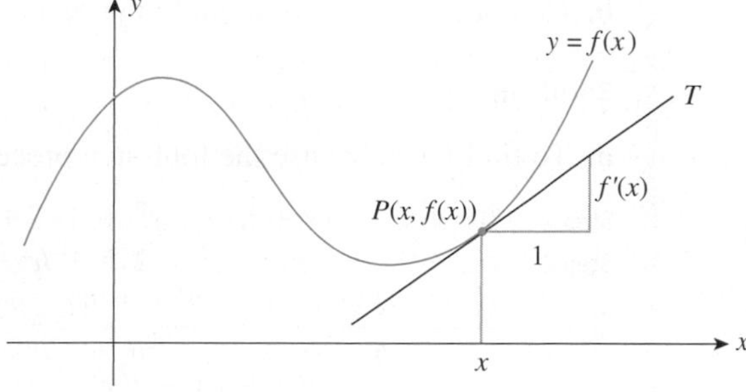

FIGURE 52
The slope of the tangent line at $P(x, f(x))$ is $f'(x)$; f changes at the rate of $f'(x)$ units per unit change in x at x.

Other notations for the derivative of f include:

$D_x f(x)$ Read "d sub x of f of x"

$\dfrac{dy}{dx}$ Read "d y d x"

y' Read "y prime"

The last two are used when the rule for f is written in the form $y = f(x)$.

The calculation of the derivative of f is facilitated using the following four-step process.

Four-Step Process for Finding $f'(x)$

1. Compute $f(x + h)$.

2. Form the difference $f(x + h) - f(x)$.

3. Form the quotient $\dfrac{f(x + h) - f(x)}{h}$.

4. Compute $f'(x) = \lim_{h \to 0} \dfrac{f(x + h) - f(x)}{h}$.

EXAMPLE 2 Find the slope of the tangent line to the graph of $f(x) = 3x + 5$ at any point $(x, f(x))$.

Solution The slope of the tangent line at any point on the graph of f is given by the derivative of f at x. To find the derivative, we use the four-step process:

Step 1 $f(x + h) = 3(x + h) + 5 = 3x + 3h + 5$

Step 2 $f(x + h) - f(x) = (3x + 3h + 5) - (3x + 5) = 3h$

Step 3 $\dfrac{f(x + h) - f(x)}{h} = \dfrac{3h}{h} = 3$

Step 4 $f'(x) = \lim_{h \to 0} \dfrac{f(x + h) - f(x)}{h} = \lim_{h \to 0} 3 = 3$

We expect this result since the tangent line to any point on a straight line must coincide with the line itself and therefore must have the same slope as the line. In this case, the graph of f is a straight line with slope 3. ■

EXAMPLE 3 Let $f(x) = x^2$.

a. Find $f'(x)$.
b. Compute $f'(2)$ and interpret your result.

Solution

a. To find $f'(x)$, we use the four-step process:

Step 1 $f(x + h) = (x + h)^2 = x^2 + 2xh + h^2$

Step 2 $f(x + h) - f(x) = x^2 + 2xh + h^2 - x^2 = 2xh + h^2 = h(2x + h)$

Step 3 $\dfrac{f(x + h) - f(x)}{h} = \dfrac{h(2x + h)}{h} = 2x + h$

Step 4 $f'(x) = \lim_{h \to 0} \dfrac{f(x + h) - f(x)}{h} = \lim_{h \to 0} (2x + h) = 2x$

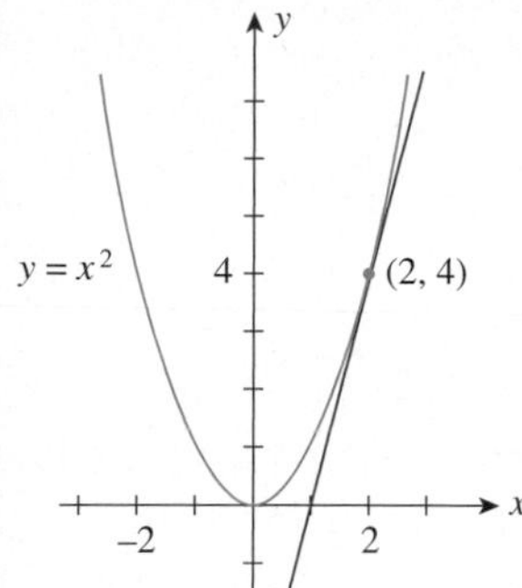

FIGURE 53
The tangent line to the graph of $f(x) = x^2$ at (2, 4)

b. $f'(2) = 2(2) = 4$. This result tells us that the slope of the tangent line to the graph of f at the point (2, 4) is 4. It also tells us that the function f is changing at the rate of 4 units per unit change in x at $x = 2$. The graph of f and the tangent line at (2, 4) are shown in Figure 53. ■

EXPLORING WITH TECHNOLOGY

1. Consider the function $f(x) = x^2$ of Example 3. Suppose we want to compute $f'(2)$, using Equation (11). Thus,

$$f'(2) = \lim_{h \to 0} \frac{f(2 + h) - f(2)}{h} = \lim_{h \to 0} \frac{(2 + h)^2 - 2^2}{h}$$

Use a graphing utility to plot the graph of

$$g(x) = \frac{(2 + x)^2 - 4}{x}$$

in the viewing window $[-3, 3] \times [-2, 6]$.

2. Use **ZOOM** and **TRACE** to find $\lim_{h \to 0} g(x)$.
3. Explain why the limit found in part 2 is $f'(2)$.

EXAMPLE 4 Let $f(x) = x^2 - 4x$.

a. Find $f'(x)$.
b. Find the point on the graph of f where the tangent line to the curve is horizontal.
c. Sketch the graph of f and the tangent line to the curve at the point found in part (b).
d. What is the rate of change of f at this point?

Solution

a. To find $f'(x)$, we use the four-step process:

Step 1 $f(x + h) = (x + h)^2 - 4(x + h) = x^2 + 2xh + h^2 - 4x - 4h$

Step 2 $f(x + h) - f(x) = x^2 + 2xh + h^2 - 4x - 4h - (x^2 - 4x)$
$= 2xh + h^2 - 4h = h(2x + h - 4)$

Step 3 $\dfrac{f(x + h) - f(x)}{h} = \dfrac{h(2x + h - 4)}{h} = 2x + h - 4$

Step 4 $f'(x) = \lim_{h \to 0} \dfrac{f(x + h) - f(x)}{h} = \lim_{h \to 0} (2x + h - 4) = 2x - 4$

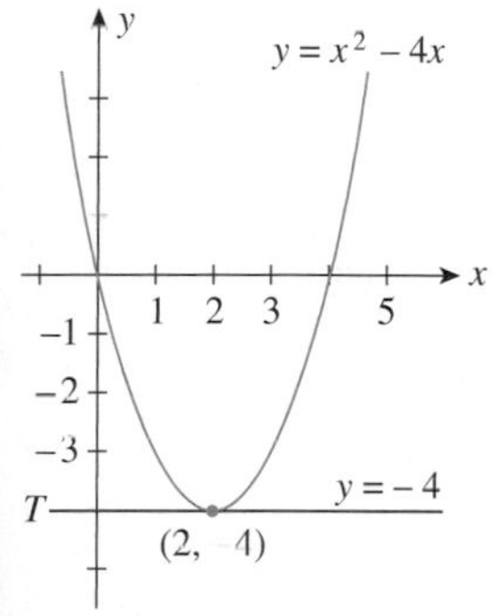

b. At a point on the graph of f where the tangent line to the curve is horizontal and hence has slope zero, the derivative f' of f is zero. Accordingly, to find such point(s) we set $f'(x) = 0$, which gives $2x - 4 = 0$, or $x = 2$. The corresponding value of y is given by $y = f(2) = -4$, and the required point is $(2, -4)$.
c. The graph of f and the tangent line are shown in Figure 54.
d. The rate of change of f at $x = 2$ is zero. ■

FIGURE 54
The tangent line to the graph of $y = x^2 - 4x$ at $(2, -4)$ is $y = -4$.

> **EXPLORE & DISCUSS**
> Can the tangent line to the graph of a function intersect the graph at more than one point? Explain your answer using illustrations.

EXAMPLE 5 Let $f(x) = \dfrac{1}{x}$.

a. Find $f'(x)$.
b. Find the slope of the tangent line T to the graph of f at the point where $x = 1$.
c. Find an equation of the tangent line T in part (b).

Solution

a. To find $f'(x)$, we use the four-step process:

Step 1 $f(x+h) = \dfrac{1}{x+h}$

Step 2 $f(x+h) - f(x) = \dfrac{1}{x+h} - \dfrac{1}{x} = \dfrac{x-(x+h)}{x(x+h)} = -\dfrac{h}{x(x+h)}$

Step 3 $\dfrac{f(x+h) - f(x)}{h} = -\dfrac{h}{x(x+h)} \cdot \dfrac{1}{h} = -\dfrac{1}{x(x+h)}$

Step 4 $f'(x) = \lim\limits_{h \to 0} \dfrac{f(x+h) - f(x)}{h} = \lim\limits_{h \to 0} -\dfrac{1}{x(x+h)} = -\dfrac{1}{x^2}$

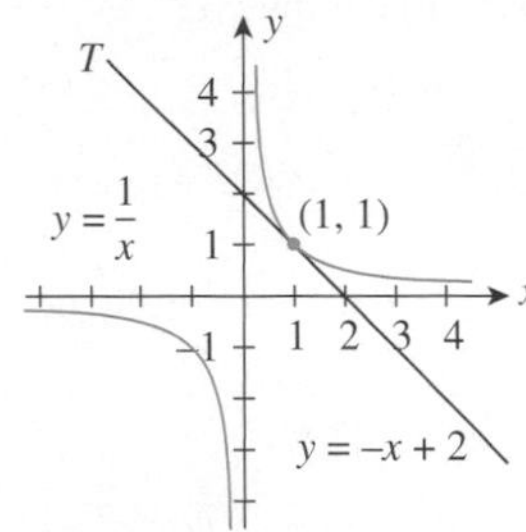

FIGURE 55
The tangent line to the graph of $f(x) = 1/x$ at (1, 1)

b. The slope of the tangent line T to the graph of f where $x = 1$ is given by $f'(1) = -1$.

c. When $x = 1$, $y = f(1) = 1$ and T is tangent to the graph of f at the point (1, 1). From part (b), we know that the slope of T is -1. Thus, an equation of T is

$$y - 1 = -1(x - 1)$$
$$y = -x + 2$$

(Figure 55).

EXPLORING WITH TECHNOLOGY

1. Use the results of Example 5 to draw the graph of $f(x) = 1/x$ and its tangent line at the point (1, 1) by plotting the graphs of $y_1 = 1/x$ and $y_2 = -x + 2$ in the viewing window $[-4, 4] \times [-4, 4]$.
2. Some graphing utilities draw the tangent line to the graph of a function at a given point automatically—you need only specify the function and give the x-coordinate of the point of tangency. If your graphing utility has this feature, verify the result of part 1 without finding an equation of the tangent line.

EXPLORE & DISCUSS

Consider the following alternative approach to the definition of the derivative of a function: Let h be a positive number and suppose $P(x - h, f(x - h))$ and $Q(x + h, f(x + h))$ are two points on the graph of f.

1. Give a geometric and a physical interpretation of the quotient
$$\frac{f(x+h) - f(x-h)}{2h}$$
Make a sketch to illustrate your answer.
2. Give a geometric and a physical interpretation of the limit
$$\lim_{h \to 0} \frac{f(x+h) - f(x-h)}{2h}$$
Make a sketch to illustrate your answer.
3. Explain why it makes sense to define
$$f'(x) = \lim_{h \to 0} \frac{f(x+h) - f(x-h)}{2h}$$
4. Using the definition given in part 3, formulate a four-step process for finding $f'(x)$ similar to that given on page 140 and use it to find the derivative of $f(x) = x^2$. Compare your answer with that obtained in Example 3 on page 141.

APPLIED EXAMPLE 6 Average Velocity of a Car Suppose the distance (in feet) covered by a car moving along a straight road t seconds after starting from rest is given by the function $f(t) = 2t^2$ $(0 \leq t \leq 30)$.

a. Calculate the average velocity of the car over the time intervals [22, 23], [22, 22.1], and [22, 22.01].
b. Calculate the (instantaneous) velocity of the car when $t = 22$.
c. Compare the results obtained in part (a) with that obtained in part (b).

Solution

a. We first compute the average velocity (average rate of change of f) over the interval $[t, t + h]$ using Formula (9). We find

$$\begin{aligned}\frac{f(t+h) - f(t)}{h} &= \frac{2(t+h)^2 - 2t^2}{h} \\ &= \frac{2t^2 + 4th + 2h^2 - 2t^2}{h} \\ &= 4t + 2h\end{aligned}$$

Next, using $t = 22$ and $h = 1$, we find that the average velocity of the car over the time interval [22, 23] is

$$4(22) + 2(1) = 90$$

or 90 feet per second. Similarly, using $t = 22$, $h = 0.1$, and $h = 0.01$, we find that its average velocities over the time intervals [22, 22.1] and [22, 22.01] are 88.2 and 88.02 feet per second, respectively.

b. Using the limit (10), we see that the instantaneous velocity of the car at any time t is given by

$$\begin{aligned}\lim_{h \to 0} \frac{f(t+h) - f(t)}{h} &= \lim_{h \to 0} (4t + 2h) \qquad \text{Use the results from part (a).} \\ &= 4t\end{aligned}$$

In particular, the velocity of the car 22 seconds from rest $(t = 22)$ is given by

$$v = 4(22)$$

or 88 feet per second.

c. The computations in part (a) show that, as the time intervals over which the average velocity of the car are computed become smaller and smaller, the average velocities over these intervals do approach 88 feet per second, the instantaneous velocity of the car at $t = 22$. ■

APPLIED EXAMPLE 7 Demand for Tires The management of Titan Tire Company has determined that the weekly demand function of their Super Titan tires is given by

$$p = f(x) = 144 - x^2$$

where p is measured in dollars and x is measured in units of a thousand (Figure 56).

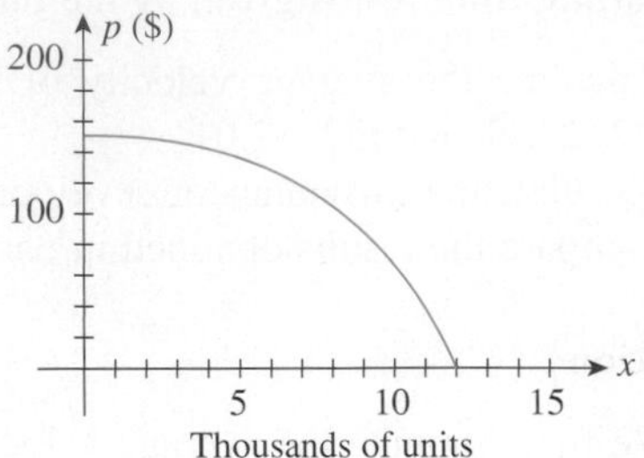

FIGURE 56
The graph of the demand function $p = 144 - x^2$

a. Find the average rate of change in the unit price of a tire if the quantity demanded is between 5000 and 6000 tires, between 5000 and 5100 tires, and between 5000 and 5010 tires.

b. What is the instantaneous rate of change of the unit price when the quantity demanded is 5000 units?

Solution

a. The average rate of change of the unit price of a tire if the quantity demanded is between x and $x + h$ is

$$\begin{aligned} \frac{f(x+h) - f(x)}{h} &= \frac{[144 - (x+h)^2] - (144 - x^2)}{h} \\ &= \frac{144 - x^2 - 2xh - h^2 - 144 + x^2}{h} \\ &= -2x - h \end{aligned}$$

To find the average rate of change of the unit price of a tire when the quantity demanded is between 5000 and 6000 tires (that is, over the interval [5, 6]), we take $x = 5$ and $h = 1$, obtaining

$$-2(5) - 1 = -11$$

or $-\$11$ per 1000 tires. (Remember, x is measured in units of a thousand.) Similarly, taking $h = 0.1$ and $h = 0.01$ with $x = 5$, we find that the average rates of change of the unit price when the quantities demanded are between 5000 and 5100 and between 5000 and 5010 are $-\$10.10$ and $-\$10.01$ per 1000 tires, respectively.

b. The instantaneous rate of change of the unit price of a tire when the quantity demanded is x units is given by

$$\begin{aligned} \lim_{h \to 0} \frac{f(x+h) - f(x)}{h} &= \lim_{h \to 0} (-2x - h) \qquad \text{Use the results from part (a).} \\ &= -2x \end{aligned}$$

In particular, the instantaneous rate of change of the unit price per tire when the quantity demanded is 5000 is given by $-2(5)$, or $-\$10$ per 1000 tires. ■

The derivative of a function provides us with a tool for measuring the rate of change of one quantity with respect to another. Table 4 lists several other applications involving this limit.

TABLE 4

Applications Involving Rate of Change

x Stands for	y Stands for	$\frac{f(a+h)-f(a)}{h}$ Measures	$\lim_{h\to 0}\frac{f(a+h)-f(a)}{h}$ Measures
Time	**Concentration of a drug** in the bloodstream at time x	Average rate of change in the concentration of the drug over the time interval $[a, a+h]$	Instantaneous rate of change in the concentration of the drug in the bloodstream at time $x = a$
Number of items sold	**Revenue** at a sales level of x units	Average rate of change in the revenue when the sales level is between $x = a$ and $x = a + h$	Instantaneous rate of change in the revenue when the sales level is a units
Time	**Volume of sales** at time x	Average rate of change in the volume of sales over the time interval $[a, a+h]$	Instantaneous rate of change in the volume of sales at time $x = a$
Time	**Population** of *Drosophila* (fruit flies) at time x	Average rate of growth of the fruit fly population over the time interval $[a, a+h]$	Instantaneous rate of change of the fruit fly population at time $x = a$
Temperature in a chemical reaction	**Amount of product formed in the chemical reaction** when the temperature is x degrees	Average rate of formation of chemical product over the temperature range $[a, a+h]$	Instantaneous rate of formation of chemical product when the temperature is a degrees

Differentiability and Continuity

In practical applications, one encounters continuous functions that fail to be **differentiable**—that is, do not have a derivative—at certain values in the domain of the function f. It can be shown that a continuous function f fails to be differentiable at $x = a$ when the graph of f makes an abrupt change of direction at $(a, f(a))$. We call such a point a "corner." A function also fails to be differentiable at a point where the tangent line is vertical since the slope of a vertical line is undefined. These cases are illustrated in Figure 57.

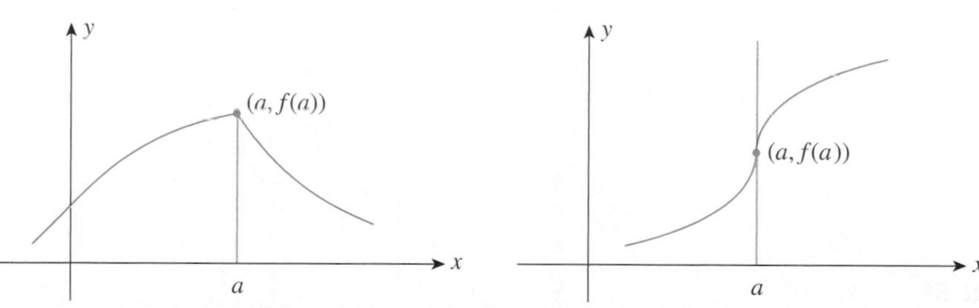

FIGURE 57

(a) The graph makes an abrupt change of direction at $x = a$.

(b) The slope at $x = a$ is undefined.

The next example illustrates a function that is not differentiable at a point.

APPLIED EXAMPLE 8 Wages Mary works at the B&O department store, where, on a weekday, she is paid \$8 an hour for the first 8 hours and \$12 an hour for overtime. The function

$$f(x) = \begin{cases} 8x & \text{if } 0 \le x \le 8 \\ 12x - 32 & \text{if } 8 < x \end{cases}$$

gives Mary's earnings on a weekday in which she worked x hours. Sketch the graph of the function f and explain why it is not differentiable at $x = 8$.

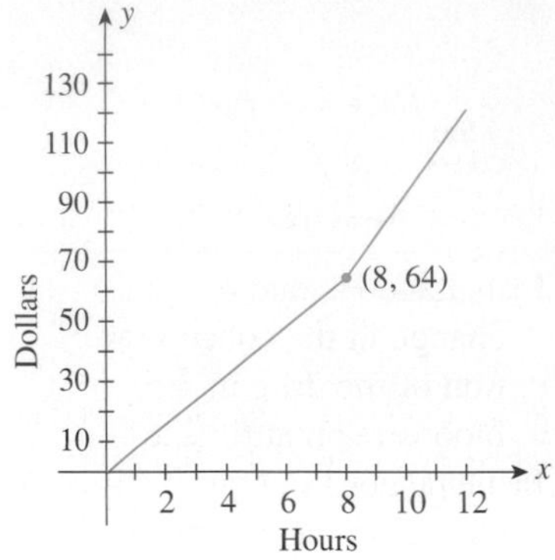

FIGURE 58
The function f is not differentiable at (8, 64).

Solution The graph of f is shown in Figure 58. Observe that the graph of f has a corner at $x = 8$ and consequently is not differentiable at $x = 8$. ■

We close this section by mentioning the connection between the continuity and the differentiability of a function at a given value $x = a$ in the domain of f. By reexamining the function of Example 8, it becomes clear that f is continuous everywhere and, in particular, when $x = 8$. This shows that in general the continuity of a function at $x = a$ does not necessarily imply the differentiability of the function at that number. The converse, however, is true: If a function f is differentiable at $x = a$, then it is continuous there.

Differentiability and Continuity
If a function is differentiable at $x = a$, then it is continuous at $x = a$.

For a proof of this result, see Exercise 60, page 151.

EXPLORE & DISCUSS

Suppose a function f is differentiable at $x = a$. Can there be two tangent lines to the graphs of f at the point $(a, f(a))$? Explain your answer.

EXPLORING WITH TECHNOLOGY

1. Use a graphing utility to plot the graph of $f(x) = x^{1/3}$ in the viewing window $[-2, 2] \times [-2, 2]$.
2. Use a graphing utility to draw the tangent line to the graph of f at the point (0, 0). Can you explain why the process breaks down?

EXAMPLE 9 Figure 59 depicts a portion of the graph of a function. Explain why the function fails to be differentiable at each of the numbers $x = a, b, c, d, e, f$, and g.

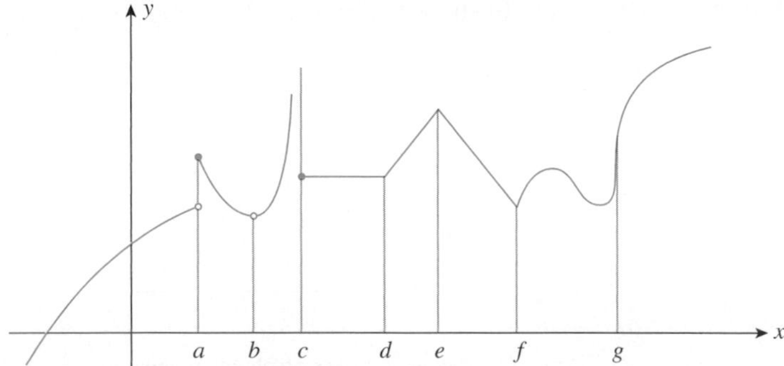

FIGURE 59
The graph of this function is not differentiable at the numbers a–g.

Solution The function fails to be differentiable at $x = a, b$, and c because it is discontinuous at each of these numbers. The derivative of the function does not exist at $x = d, e$, and f because it has a kink at each point on the graph corresponding to these numbers. Finally, the function is not differentiable at $x = g$ because the tangent line is vertical at $(g, f(g))$. ■

2.6 Self-Check Exercises

1. Let $f(x) = -x^2 - 2x + 3$.
   a. Find the derivative f' of f, using the definition of the derivative.
   b. Find the slope of the tangent line to the graph of f at the point (0, 3).
   c. Find the rate of change of f when $x = 0$.
   d. Find an equation of the tangent line to the graph of f at the point (0, 3).
   e. Sketch the graph of f and the tangent line to the curve at the point (0, 3).

2. The losses (in millions of dollars) due to bad loans extended chiefly in agriculture, real estate, shipping, and energy by the Franklin Bank are estimated to be

$$A = f(t) = -t^2 + 10t + 30 \qquad (0 \le t \le 10)$$

where t is the time in years ($t = 0$ corresponds to the beginning of 1994). How fast were the losses mounting at the beginning of 1997? At the beginning of 1999? At the beginning of 2001?

Solutions to Self-Check Exercises 2.6 can be found on page 152.

2.6 Concept Questions

1. Let $P(2, f(2))$ and $Q(2 + h, f(2 + h))$ be points on the graph of a function f.
   a. Find an expression for the slope of the secant line passing through P and Q.
   b. Find an expression for the slope of the tangent line passing through P.

2. Refer to Question 1.
   a. Find an expression for the average rate of change of f over the interval $[2, 2 + h]$.
   b. Find an expression for the instantaneous rate of change of f at 2.
   c. Compare your answers for part (a) and (b) with those of Exercise 1.

3. a. Give a geometric and a physical interpretation of the expression
$$\frac{f(x+h) - f(x)}{h}$$
   b. Give a geometric and a physical interpretation of the expression
$$\lim_{h \to 0} \frac{f(x+h) - f(x)}{h}$$

4. Under what conditions does a function fail to have a derivative at a number? Illustrate your answer with sketches.

2.6 Exercises

1. Average Weight of an Infant The following graph shows the weight measurements of the average infant from the time of birth ($t = 0$) through age 2 ($t = 24$). By computing the slopes of the respective tangent lines, estimate the rate of change of the average infant's weight when $t = 3$ and when $t = 18$. What is the average rate of change in the average infant's weight over the first year of life?

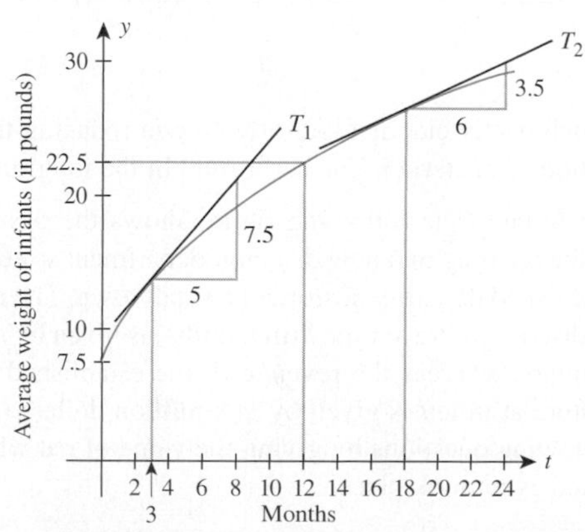

2. Forestry The following graph shows the volume of wood produced in a single-species forest. Here $f(t)$ is measured in cubic meters/hectare and t is measured in years. By computing the slopes of the respective tangent lines, estimate the rate at which the wood grown is changing at the beginning of year 10 and at the beginning of year 30.

Source: The Random House Encyclopedia

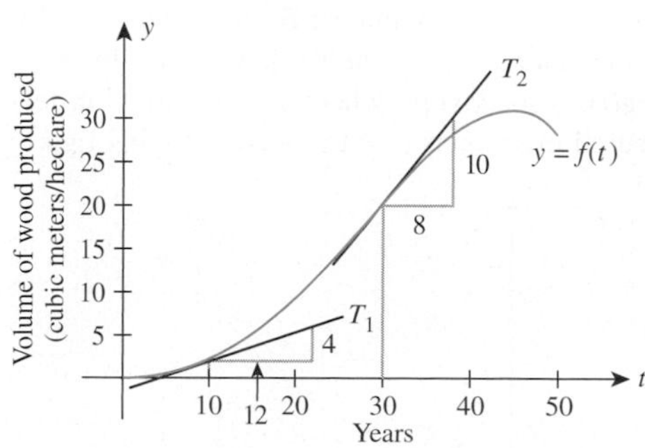

3. **TV-Viewing Patterns** The following graph shows the percent of U.S. households watching television during a 24-hr period on a weekday ($t = 0$ corresponds to 6 a.m.). By computing the slopes of the respective tangent lines, estimate the rate of change of the percent of households watching television at 4 p.m. and 11 p.m.

Source: A. C. Nielsen Company

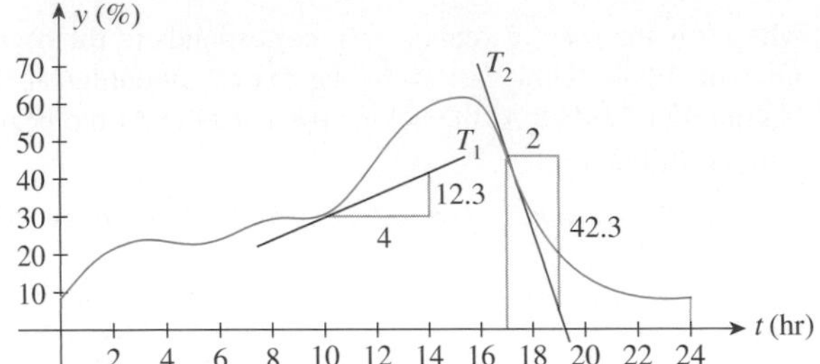

4. **Crop Yield** Productivity and yield of cultivated crops are often reduced by insect pests. The following graph shows the relationship between the yield of a certain crop, $f(x)$, as a function of the density of aphids x. (Aphids are small insects that suck plant juices.) Here, $f(x)$ is measured in kilograms/4000 square meters, and x is measured in hundreds of aphids/bean stem. By computing the slopes of the respective tangent lines, estimate the rate of change of the crop yield with respect to the density of aphids when that density is 200 aphids/bean stem and when it is 800 aphids/bean stem.

Source: The Random House Encyclopedia

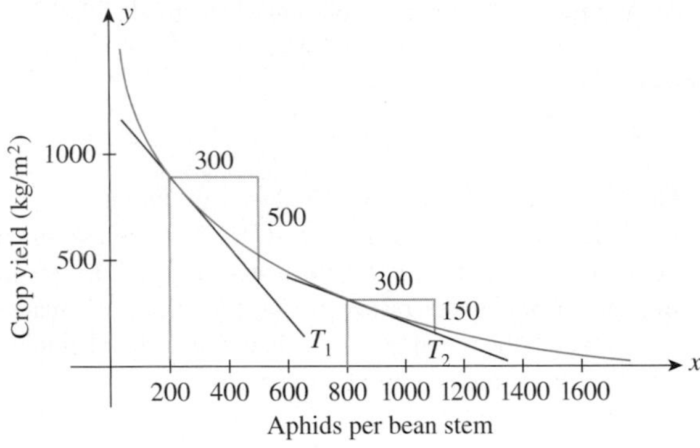

5. The position of car A and car B, starting out side by side and traveling along a straight road, is given by $s = f(t)$ and $s = g(t)$, respectively, where s is measured in feet and t is measured in seconds (see the accompanying figure).

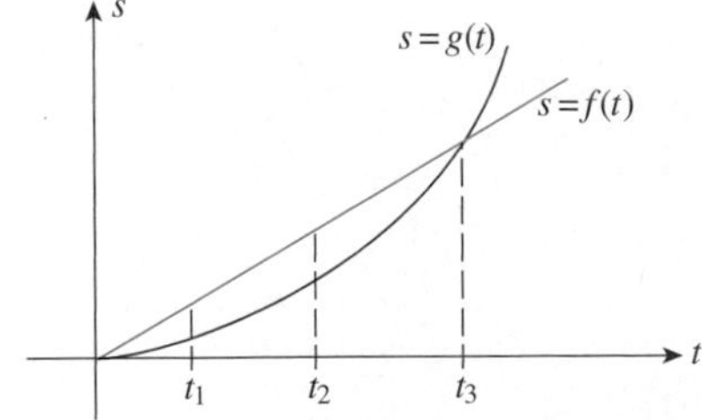

a. Which car is traveling faster at t_1?
b. What can you say about the speed of the cars at t_2? Hint: Compare tangent lines.
c. Which car is traveling faster at t_3?
d. What can you say about the positions of the cars at t_3?

6. The velocity of car A and car B, starting out side by side and traveling along a straight road, is given by $v = f(t)$ and $v = g(t)$, respectively, where v is measured in feet/second and t is measured in seconds (see the accompanying figure).

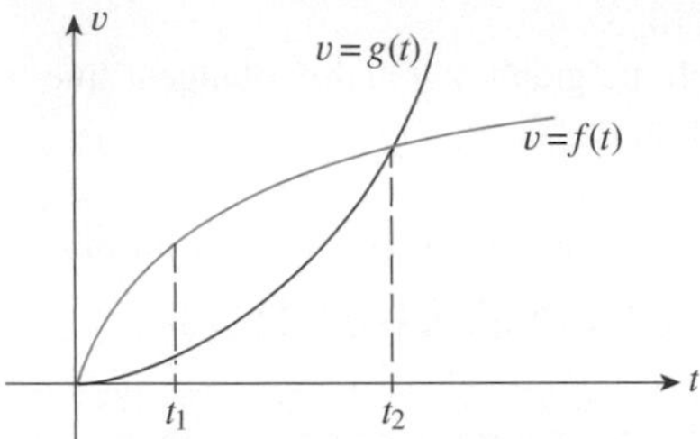

a. What can you say about the velocity and acceleration of the two cars at t_1? (Acceleration is the rate of change of velocity.)
b. What can you say about the velocity and acceleration of the two cars at t_2?

7. **Effect of a Bactericide on Bacteria** In the following figure, $f(t)$ gives the population P_1 of a certain bacteria culture at time t after a portion of bactericide A was introduced into the population at $t = 0$. The graph of g gives the population P_2 of a similar bacteria culture at time t after a portion of bactericide B was introduced into the population at $t = 0$.
a. Which population is decreasing faster at t_1?
b. Which population is decreasing faster at t_2?

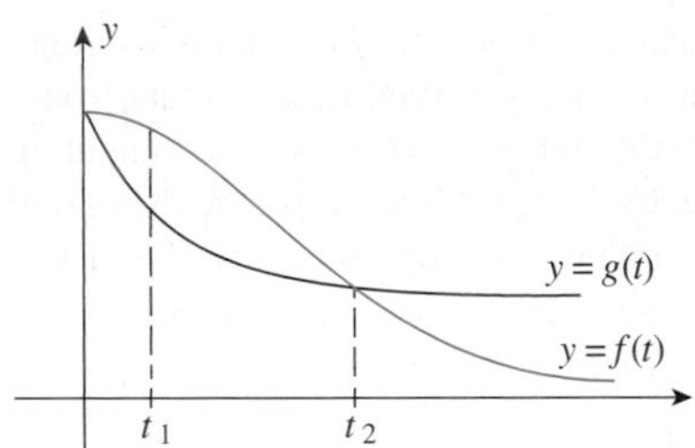

c. Which bactericide is more effective in reducing the population of bacteria in the short run? In the long run?

8. **Market Share** The following figure shows the devastating effect the opening of a new discount department store had on an established department store in a small town. The revenue of the discount store at time t (in months) is given by $f(t)$ million dollars, whereas the revenue of the established department store at time t is given by $g(t)$ million dollars. Answer the following questions by giving the value of t at which the specified event took place.

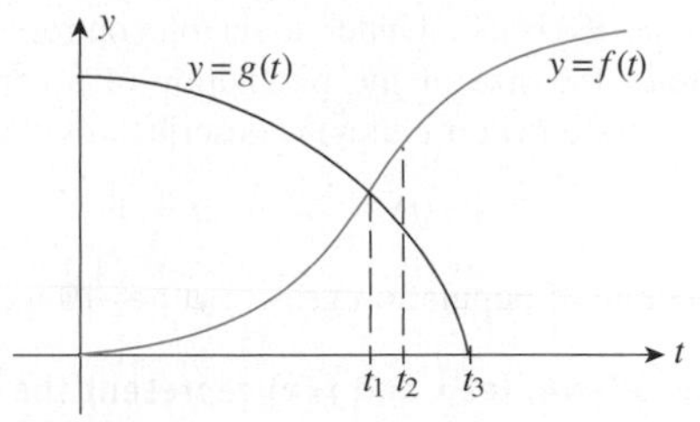

a. The revenue of the established department store is decreasing at the slowest rate.
b. The revenue of the established department store is decreasing at the fastest rate.
c. The revenue of the discount store first overtakes that of the established store.
d. The revenue of the discount store is increasing at the fastest rate.

In Exercises 9–16, use the four-step process to find the slope of the tangent line to the graph of the given function at any point.

9. $f(x) = 13$
10. $f(x) = -6$
11. $f(x) = 2x + 7$
12. $f(x) = 8 - 4x$
13. $f(x) = 3x^2$
14. $f(x) = -\frac{1}{2}x^2$
15. $f(x) = -x^2 + 3x$
16. $f(x) = 2x^2 + 5x$

In Exercises 17–22, find the slope of the tangent line to the graph of each function at the given point and determine an equation of the tangent line.

17. $f(x) = 2x + 7$ at $(2, 11)$
18. $f(x) = -3x + 4$ at $(-1, 7)$
19. $f(x) = 3x^2$ at $(1, 3)$
20. $f(x) = 3x - x^2$ at $(-2, -10)$
21. $f(x) = -\frac{1}{x}$ at $\left(3, -\frac{1}{3}\right)$
22. $f(x) = \frac{3}{2x}$ at $\left(1, \frac{3}{2}\right)$
23. Let $f(x) = 2x^2 + 1$.
    a. Find the derivative f' of f.
    b. Find an equation of the tangent line to the curve at the point $(1, 3)$.
    c. Sketch the graph of f.
24. Let $f(x) = x^2 + 6x$.
    a. Find the derivative f' of f.
    b. Find the point on the graph of f where the tangent line to the curve is horizontal.
    Hint: Find the value of x for which $f'(x) = 0$.
    c. Sketch the graph of f and the tangent line to the curve at the point found in part (b).
25. Let $f(x) = x^2 - 2x + 1$.
    a. Find the derivative f' of f.
    b. Find the point on the graph of f where the tangent line to the curve is horizontal.
    c. Sketch the graph of f and the tangent line to the curve at the point found in part (b).
    d. What is the rate of change of f at this point?
26. Let $f(x) = \dfrac{1}{x - 1}$.
    a. Find the derivative f' of f.
    b. Find an equation of the tangent line to the curve at the point $(-1, -\frac{1}{2})$.
    c. Sketch the graph of f and the tangent line to the curve at $(-1, -\frac{1}{2})$.
27. Let $y = f(x) = x^2 + x$.
    a. Find the average rate of change of y with respect to x in the interval from $x = 2$ to $x = 3$, from $x = 2$ to $x = 2.5$, and from $x = 2$ to $x = 2.1$.
    b. Find the (instantaneous) rate of change of y at $x = 2$.
    c. Compare the results obtained in part (a) with that of part (b).
28. Let $y = f(x) = x^2 - 4x$.
    a. Find the average rate of change of y with respect to x in the interval from $x = 3$ to $x = 4$, from $x = 3$ to $x = 3.5$, and from $x = 3$ to $x = 3.1$.
    b. Find the (instantaneous) rate of change of y at $x = 3$.
    c. Compare the results obtained in part (a) with that of part (b).
29. Velocity of a Car Suppose the distance s (in feet) covered by a car moving along a straight road after t sec is given by the function $f(t) = 2t^2 + 48t$.
    a. Calculate the average velocity of the car over the time intervals [20, 21], [20, 20.1], and [20, 20.01].
    b. Calculate the (instantaneous) velocity of the car when $t = 20$.
    c. Compare the results of part (a) with that of part (b).
30. Velocity of a Ball Thrown into the Air A ball is thrown straight up with an initial velocity of 128 ft/sec, so that its height (in feet) after t sec is given by $s(t) = 128t - 16t^2$.
    a. What is the average velocity of the ball over the time intervals [2, 3], [2, 2.5], and [2, 2.1]?
    b. What is the instantaneous velocity at time $t = 2$?
    c. What is the instantaneous velocity at time $t = 5$? Is the ball rising or falling at this time?
    d. When will the ball hit the ground?
31. During the construction of a high-rise building, a worker accidentally dropped his portable electric screwdriver from a height of 400 ft. After t sec, the screwdriver had fallen a distance of $s = 16t^2$ ft.
    a. How long did it take the screwdriver to reach the ground?
    b. What was the average velocity of the screwdriver between the time it was dropped and the time it hit the ground?

c. What was the velocity of the screwdriver at the time it hit the ground?

32. A hot-air balloon rises vertically from the ground so that its height after t sec is $h = \frac{1}{2}t^2 + \frac{1}{2}t$ ft $(0 \le t \le 60)$.
 a. What is the height of the balloon at the end of 40 sec?
 b. What is the average velocity of the balloon between $t = 0$ and $t = 40$?
 c. What is the velocity of the balloon at the end of 40 sec?

33. At a temperature of 20°C, the volume V (in liters) of 1.33 g of O_2 is related to its pressure p (in atmospheres) by the formula $V = 1/p$.
 a. What is the average rate of change of V with respect to p as p increases from $p = 2$ to $p = 3$?
 b. What is the rate of change of V with respect to p when $p = 2$?

34. **Cost of Producing Surfboards** The total cost $C(x)$ (in dollars) incurred by Aloha Company in manufacturing x surfboards a day is given by

$$C(x) = -10x^2 + 300x + 130 \qquad (0 \le x \le 15)$$

 a. Find $C'(x)$.
 b. What is the rate of change of the total cost when the level of production is ten surfboards a day?

35. **Effect of Advertising on Profit** The quarterly profit (in thousands of dollars) of Cunningham Realty is given by

$$P(x) = -\frac{1}{3}x^2 + 7x + 30 \qquad (0 \le x \le 50)$$

where x (in thousands of dollars) is the amount of money Cunningham spends on advertising per quarter.
 a. Find $P'(x)$.
 b. What is the rate of change of Cunningham's quarterly profit if the amount it spends on advertising is \$10,000/quarter ($x = 10$) and \$30,000/quarter ($x = 30$)?

36. **Demand for Tents** The demand function for Sportsman 5 × 7 tents is given by

$$p = f(x) = -0.1x^2 - x + 40$$

where p is measured in dollars and x is measured in units of a thousand.
 a. Find the average rate of change in the unit price of a tent if the quantity demanded is between 5000 and 5050 tents; between 5000 and 5010 tents.
 b. What is the rate of change of the unit price if the quantity demanded is 5000?

37. **A Country's GDP** The gross domestic product (GDP) of a certain country is projected to be

$$N(t) = t^2 + 2t + 50 \qquad (0 \le t \le 5)$$

billion dollars t yr from now. What will be the rate of change of the country's GDP 2 yr and 4 yr from now?

38. **Growth of Bacteria** Under a set of controlled laboratory conditions, the size of the population of a certain bacteria culture at time t (in minutes) is described by the function

$$P = f(t) = 3t^2 + 2t + 1$$

Find the rate of population growth at $t = 10$ min.

In Exercises 39–44, let x and $f(x)$ represent the given quantities. Fix $x = a$ and let h be a small positive number. Give an interpretation of the quantities

$$\frac{f(a+h) - f(a)}{h} \quad \text{and} \quad \lim_{h \to 0} \frac{f(a+h) - f(a)}{h}$$

39. x denotes time and $f(x)$ denotes the population of seals at time x.

40. x denotes time and $f(x)$ denotes the prime interest rate at time x.

41. x denotes time and $f(x)$ denotes a country's industrial production.

42. x denotes the level of production of a certain commodity, and $f(x)$ denotes the total cost incurred in producing x units of the commodity.

43. x denotes altitude and $f(x)$ denotes atmospheric pressure.

44. x denotes the speed of a car (in mph), and $f(x)$ denotes the fuel economy of the car measured in miles per gallon (mpg).

In each of Exercises 45–50, the graph of a function is shown. For each function, state whether or not (a) $f(x)$ has a limit at $x = a$, (b) $f(x)$ is continuous at $x = a$, and (c) $f(x)$ is differentiable at $x = a$. Justify your answers.

45. y x a

46. y x a

47.

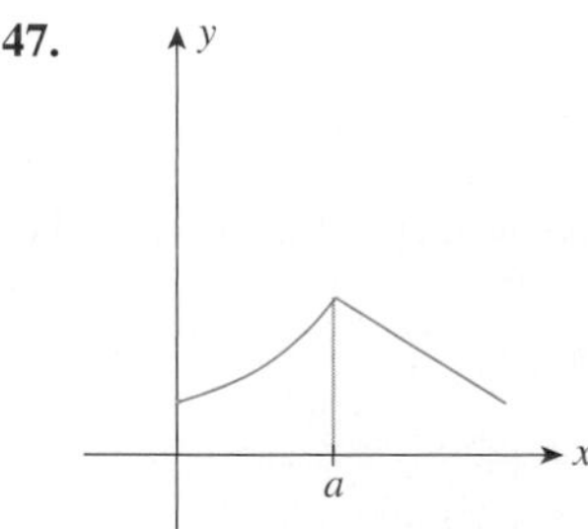

48.

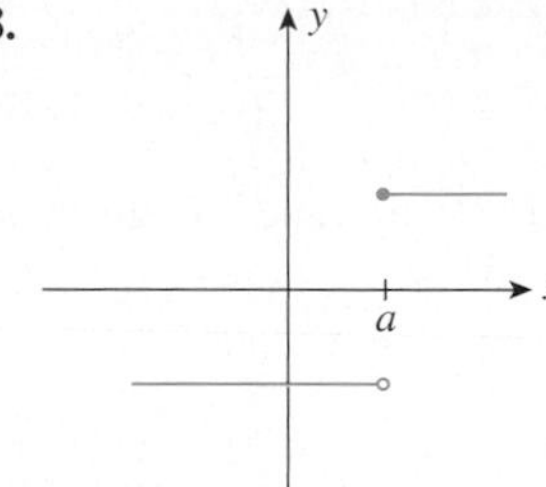

49.

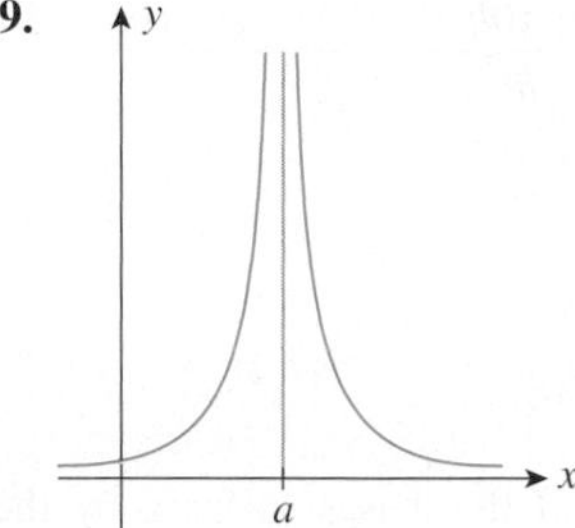

50.

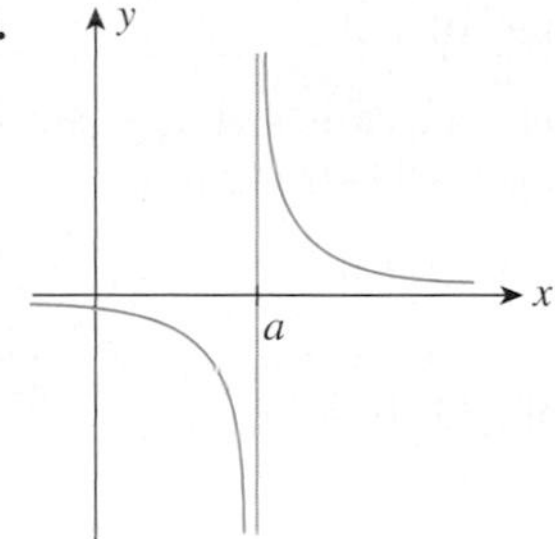

51. The distance s (in feet) covered by a motorcycle traveling in a straight line and starting from rest in t sec is given by the function

$$s(t) = -0.1t^3 + 2t^2 + 24t$$

Calculate the motorcycle's average velocity over the time interval $[2, 2 + h]$ for $h = 1, 0.1, 0.01, 0.001, 0.0001$, and 0.00001 and use your results to guess at the motorcycle's instantaneous velocity at $t = 2$.

52. The daily total cost $C(x)$ incurred by Trappee and Sons for producing x cases of TexaPep hot sauce is given by

$$C(x) = 0.000002x^3 + 5x + 400$$

Calculate

$$\frac{C(100 + h) - C(100)}{h}$$

for $h = 1, 0.1, 0.01, 0.001$, and 0.0001 and use your results to estimate the rate of change of the total cost function when the level of production is 100 cases/day.

In Exercises 53 and 54, determine whether the statement is true or false. If it is true, explain why it is true. If it is false, give an example to show why it is false.

53. If f is continuous at $x = a$, then f is differentiable at $x = a$.

54. If f is continuous at $x = a$ and g is differentiable at $x = a$, then $\lim_{x \to a} f(x)g(x) = f(a)g(a)$.

55. Sketch the graph of the function $f(x) = |x + 1|$ and show that the function does not have a derivative at $x = -1$.

56. Sketch the graph of the function $f(x) = 1/(x - 1)$ and show that the function does not have a derivative at $x = 1$.

57. Let

$$f(x) = \begin{cases} x^2 & \text{if } x \le 1 \\ ax + b & \text{if } x > 1 \end{cases}$$

Find the values of a and b so that f is continuous and has a derivative at $x = 1$. Sketch the graph of f.

58. Sketch the graph of the function $f(x) = x^{2/3}$. Is the function continuous at $x = 0$? Does $f'(0)$ exist? Why or why not?

59. Prove that the derivative of the function $f(x) = |x|$ for $x \ne 0$ is given by

$$f'(x) = \begin{cases} 1 & \text{if } x > 0 \\ -1 & \text{if } x < 0 \end{cases}$$

Hint: Recall the definition of the absolute value of a number.

60. Show that if a function f is differentiable at $x = a$, then f must be continuous at that number.
Hint: Write

$$f(x) - f(a) = \left[\frac{f(x) - f(a)}{x - a}\right](x - a)$$

Use the product rule for limits and the definition of the derivative to show that

$$\lim_{x \to a} [f(x) - f(a)] = 0$$

2.6 Solutions to Self-Check Exercises

1. a. $f'(x)$

$$= \lim_{h \to 0} \frac{f(x+h) - f(x)}{h}$$
$$= \lim_{h \to 0} \frac{[-(x+h)^2 - 2(x+h) + 3] - (-x^2 - 2x + 3)}{h}$$
$$= \lim_{h \to 0} \frac{-x^2 - 2xh - h^2 - 2x - 2h + 3 + x^2 + 2x - 3}{h}$$
$$= \lim_{h \to 0} \frac{h(-2x - h - 2)}{h}$$
$$= \lim_{h \to 0} (-2x - h - 2) = -2x - 2$$

b. From the result of part (a), we see that the slope of the tangent line to the graph of f at any point $(x, f(x))$ is given by

$$f'(x) = -2x - 2$$

In particular, the slope of the tangent line to the graph of f at $(0, 3)$ is

$$f'(0) = -2$$

c. The rate of change of f when $x = 0$ is given by $f'(0) = -2$, or -2 units/unit change in x.

d. Using the result from part (b), we see that an equation of the required tangent line is

$$y - 3 = -2(x - 0)$$
$$y = -2x + 3$$

e.

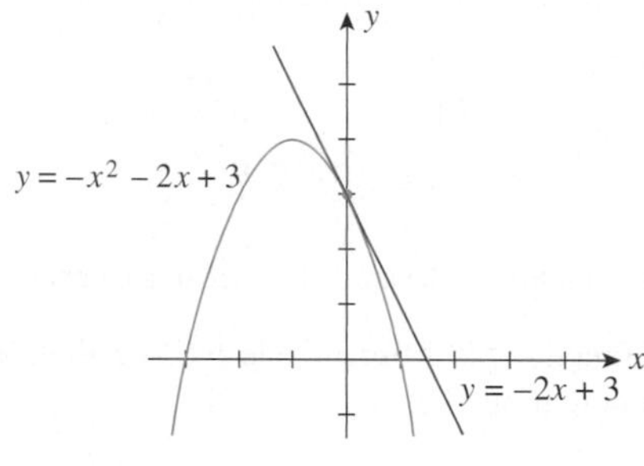

2. The rate of change of the losses at any time t is given by $f'(t)$

$$= \lim_{h \to 0} \frac{f(t+h) - f(t)}{h}$$
$$= \lim_{h \to 0} \frac{[-(t+h)^2 + 10(t+h) + 30] - (-t^2 + 10t + 30)}{h}$$
$$= \lim_{h \to 0} \frac{-t^2 - 2th - h^2 + 10t + 10h + 30 + t^2 - 10t - 30}{h}$$
$$= \lim_{h \to 0} \frac{h(-2t - h + 10)}{h}$$
$$= \lim_{h \to 0} (-2t - h + 10)$$
$$= -2t + 10$$

Therefore, the rate of change of the losses suffered by the bank at the beginning of 1997 ($t = 3$) was

$$f'(3) = -2(3) + 10 = 4$$

In other words, the losses were increasing at the rate of \$4 million/year. At the beginning of 1999 ($t = 5$),

$$f'(5) = -2(5) + 10 = 0$$

and we see that the growth in losses due to bad loans was zero at this point. At the beginning of 2001 ($t = 7$),

$$f'(7) = -2(7) + 10 = -4$$

and we conclude that the losses were decreasing at the rate of \$4 million/year.

USING TECHNOLOGY

Graphing a Function and Its Tangent Line

We can use a graphing utility to plot the graph of a function f and the tangent line at any point on the graph.

EXAMPLE 1 Let $f(x) = x^2 - 4x$.

a. Find an equation of the tangent line to the graph of f at the point $(3, -3)$.
b. Plot both the graph of f and the tangent line found in part (a) on the same set of axes.

Solution

a. The slope of the tangent line at any point on the graph of f is given by $f'(x)$. But from Example 4 (page 141) we find $f'(x) = 2x - 4$. Using this result, we see that the slope of the required tangent line is

$$f'(3) = 2(3) - 4 = 2$$

Finally, using the point-slope form of the equation of a line, we find that an equation of the tangent line is

$$\begin{aligned} y - (-3) &= 2(x - 3) \\ y + 3 &= 2x - 6 \\ y &= 2x - 9 \end{aligned}$$

b. Figure T1a shows the graph of f in the standard viewing window and the tangent line of interest.

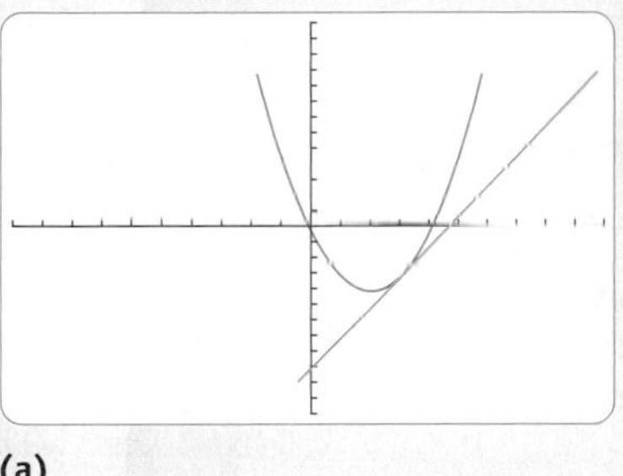

(a)

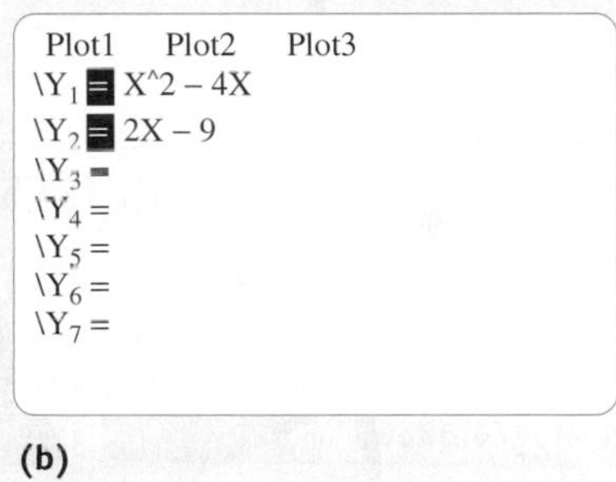

(b)

FIGURE T1
(a) The graph of $f(x) = x^2 - 4x$ and the tangent line $y = 2x - 9$ in the standard viewing window; (b) the TI-83 equation screen.

Note Some graphing utilities will draw both the graph of a function f and the tangent line to the graph of f at a specified point when the function and the specified value of x are entered.

Finding the Derivative of a Function at a Given Point

The numerical derivative operation of a graphing utility can be used to give an approximate value of the derivative of a function for a given value of x.

EXAMPLE 2 Let $f(x) = \sqrt{x}$.

a. Use the numerical derivative operation of a graphing utility to find the derivative of f at (4, 2).

b. Find an equation of the tangent line to the graph of f at (4, 2).

c. Plot the graph of f and the tangent line on the same set of axes.

Solution

a. Using the numerical derivative operation of a graphing utility, we find that

$$f'(4) = \frac{1}{4}$$

(Figure T2).

```
nDeriv(X^.5, X, 4)
          .250000002
```

FIGURE T2
The TI-83 numerical derivative screen.

(continued)

b. An equation of the required tangent line is

$$y - 2 = \frac{1}{4}(x - 4)$$

$$y = \frac{1}{4}x + 1$$

c. Figure T3a shows the graph of f and the tangent line in the viewing window $[0, 15] \times [0, 4]$.

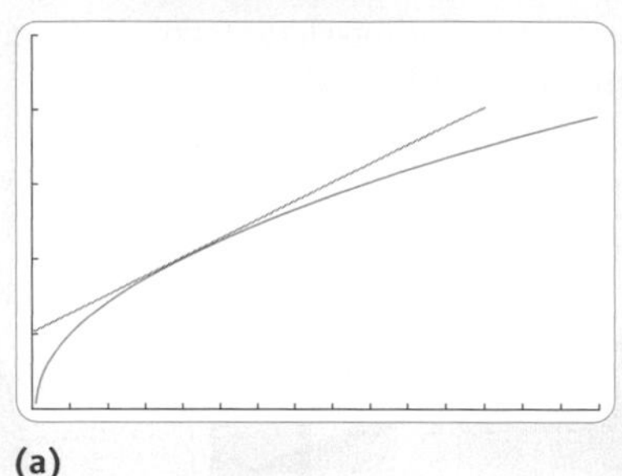

(a)

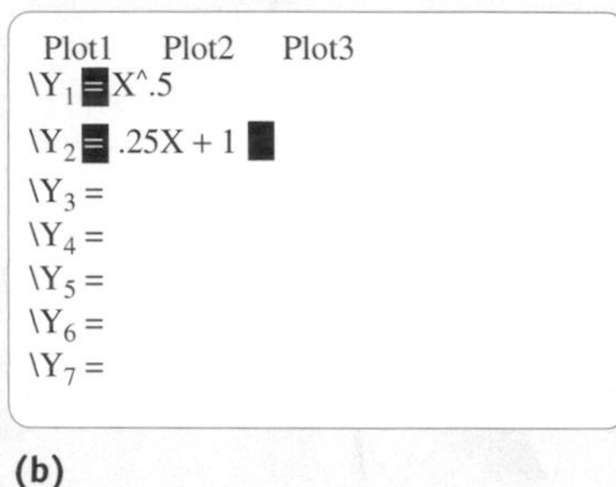

(b)

FIGURE T3
(a) The graph of $f(x) = \sqrt{x}$ and the tangent line $y = \frac{1}{4}x + 1$ in the viewing window $[0, 15] \times [0, 4]$; (b) the TI-83 equation screen

TECHNOLOGY EXERCISES

In Exercises 1–6, (a) find an equation of the tangent line to the graph of f at the indicated point and (b) plot the graph of f and the tangent line on the same set of axes. Use a suitable viewing window.

1. $f(x) = 4x - 3$; $(2, 5)$
2. $f(x) = 2x^2 + x$; $(-2, 6)$
3. $f(x) = 2x^2 + x - 3$; $(2, 7)$
4. $f(x) = x + \dfrac{1}{x}$; $(1, 2)$
5. $f(x) = \sqrt{x}$; $(4, 2)$
6. $f(x) = \dfrac{1}{\sqrt{x}}$; $\left(4, \dfrac{1}{2}\right)$

In Exercises 7–12, (a) use the numerical derivative operation to find the derivative of f for the given value of x (to two desired places of accuracy), (b) find an equation of the tangent line to the graph of f at the indicated point, and (c) plot the graph of f and the tangent line on the same set of axes. Use a suitable viewing window.

7. $f(x) = x^3 + x + 1$; $x = 1$; $(1, 3)$
8. $f(x) = x^4 - 3x^2 + 1$; $x = 2$; $(2, 5)$
9. $f(x) = x - \sqrt{x}$; $x = 4$; $(4, 2)$
10. $f(x) = \dfrac{1}{x + 1}$; $x = 1$; $\left(1, \dfrac{1}{2}\right)$
11. $f(x) = x\sqrt{x^2 + 1}$; $x = 2$; $(2, 2\sqrt{5})$
12. $f(x) = \dfrac{x}{\sqrt{x^2 + 1}}$; $x = 1$; $\left(1, \dfrac{\sqrt{2}}{2}\right)$

13. **Driving Costs** The average cost of owning and operating a car in the United States from 1991 through 2001 is approximated by the function

$$C(t) = 0.06t^2 + 0.74t + 37.3 \qquad (0 \le t \le 11)$$

where $C(t)$ is measured in cents/mile and t is measured in years, with $t = 0$ corresponding to the beginning of 1991.
a. Plot the graph of C in the viewing window $[0, 10] \times [35, 52]$.
b. What was the average cost of driving a car at the beginning of 1995?
c. How fast was the average cost of driving a car changing at the beginning of 1995?
Source: Automobile Association of America

14. **Annual Retail Sales** The annual retail sales in the United States from 1990 through the year 2000 (in millions of dollars) are approximated by the function

$$S(t) = 1.9152t^2 + 18.7176t + 473.8 \qquad (0 \le t \le 11)$$

where t is measured in years, with $t = 0$ corresponding to 1990.
a. Plot the graph of S in the viewing window $[0, 10] \times [0, 1000]$.
b. What were the annual retail sales in the United States in 1999 ($t = 9$)?
c. Approximately, how fast were the retail sales changing in 1999 ($t = 9$)?
Source: U.S. Census Bureau

CHAPTER 2 Summary of Principal Formulas and Terms

FORMULAS

1. Average rate of change of f over $[x, x + h]$ or Slope of the secant line to the graph of f through $(x, f(x))$ and $(x + h, f(x + h))$ or Difference quotient	$\dfrac{f(x + h) - f(x)}{h}$
2. Instantaneous rate of change of f at $(x, f(x))$ or Slope of tangent line to the graph of f at $(x, f(x))$ at x or Derivative of f	$\lim_{h \to 0} \dfrac{f(x + h) - f(x)}{h}$

TERMS

function (50)
domain (50)
range (50)
independent variable (51)
dependent variable (51)
ordered pairs (53)
function (alternative definition) (53)
graph of a function (53)
graph of an equation (56)
vertical-line test (56)
composite function (71)
polynomial function (80)
linear function (80)
quadratic function (80)
cubic function (80)
rational function (80)
power function (81)
demand function (82)
supply function (82)
market equilibrium (82)
equilibrium quantity (82)
equilibrium price (82)
limit of a function (100)
indeterminate form (103)
limit of a function at infinity (107)
right-hand limit of a function (119)
left-hand limit of a function (119)
continuity of a function at a number (121)
secant line (137)
tangent line to the graph of f (137)
differentiable function (145)

CHAPTER 2 Concept Review Questions

Fill in the blanks.

1. If f is a function from the set A to the set B, then A is called the _____ of f, and the set of all values of $f(x)$ as x takes on all possible values in A is called the _____ of f. The range of f is contained in the set _____.

2. The graph of a function is the set of all points (x, y) in the xy-plane such that x is in the _____ of f and y = _____. The vertical-line test states that a curve in the xy-plane is the graph of a function $y = f(x)$ if and only if each _____ line intersects it in at most one _____.

3. If f and g are functions with domains A and B, respectively, then (a) $(f \pm g)(x)$ = _____, (b) $(fg)(x)$ = _____, and (c) $\left(\dfrac{f}{g}\right)(x)$ = _____. The domain of $f + g$ is _____. The domain of $\dfrac{f}{g}$ is _____ with the additional condition that $g(x)$ is never _____.

4. The composition of g and f is the function with rule $(g \circ f)(x)$ = _____. Its domain is the set of all x in the domain of _____ such that _____ lies in the domain of _____.

5. a. A polynomial function of degree n is a function of the form _____.

b. A polynomial function of degree 1 is called a _____ function; one of degree 2 is called a _____ function; one of degree 3 is called a _____ function.

c. A rational function is a/an _____ of two _____.

d. A power function has the form $f(x)$ = _____.

6. The statement $\lim_{x \to a} f(x) = L$ means that there is a number _____ such that the values of _____ can be made as close to _____ as we please by taking x sufficiently close to _____.

7. If $\lim_{x\to a} f(x) = L$ and $\lim_{x\to a} g(x) = M$, then

a. $\lim_{x\to a} [f(x)]^r =$ ____, where r is a real number.

b. $\lim_{x\to a} [f(x) \pm g(x)] =$ ____.

c. $\lim_{x\to a} [f(x)g(x)] =$ ____.

d. $\lim_{x\to a} \dfrac{f(x)}{g(x)} =$ ____ provided that ____.

8. a. The statement $\lim_{x\to\infty} f(x) = L$ means that $f(x)$ can be made arbitrarily close to ____ by taking ____ large enough.

b. The statement $\lim_{x\to-\infty} f(x) = M$ means that $f(x)$ can be made arbitrarily close to ____ by taking x to be ____ and sufficiently large in ____ value.

9. a. The statement $\lim_{x\to a^+} f(x) = L$ is similar to the statement $\lim_{x\to a} f(x) = L$, but here x is required to lie to the ____ of a.

b. The statement $\lim_{x\to a^-} f(x) = L$ is similar to the statement $\lim_{x\to a} f(x) = L$, but here x is required to lie to the ____ of a.

c. $\lim_{x\to a} f(x) = L$ if and only if both $\lim_{x\to a^-} f(x) =$ ____ and $\lim_{x\to a^+} f(x) =$ ____.

10. a. If $f(a)$ is defined, $\lim_{x\to a} f(x)$ exists and $\lim_{x\to a} f(x) = f(a)$, then f is ____ at a.

b. If f is not continuous at a, then it is ____ at a.

c. f is continuous on an interval I if f is continuous at ____ number in the interval.

11. a. If f and g are continuous at a, then $f \pm g$ and fg are continuous at ____. Also, $\frac{f}{g}$ is continuous at ____, provided ____ $\neq 0$.

b. A polynomial function is continuous ____.

c. A rational function $R = \frac{P}{Q}$ is continuous everywhere except at values of x where ____ $= 0$.

12. a. Suppose f is continuous on $[a, b]$ and $f(a) < M < f(b)$. Then the intermediate value theorem guarantees the existence of at least one number c in ____ such that ____.

b. If f is continuous on $[a, b]$ and $f(a)f(b) < 0$, then there must be at least one solution of the equation ____ in the interval ____.

13. a. The tangent line at $P(a, f(a))$ to the graph of f is the line passing through P and having slope ____.

b. If the slope of the tangent line at $P(a, f(a))$ is m, then an equation of the tangent line at P is ____.

14. a. The slope of the secant line passing through $P(a, f(a))$ and $Q(a + h, f(a + h))$ and the average rate of change of f over the interval $[a, a + h]$ are both given by ____.

b. The slope of the tangent line at $P(a, f(a))$ and the instantaneous rate of change of f at a are both given by ____.

CHAPTER 2 Review Exercises

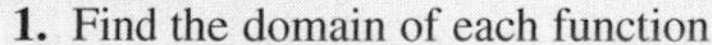

1. Find the domain of each function:

a. $f(x) = \sqrt{9 - x}$ **b.** $f(x) = \dfrac{x + 3}{2x^2 - x - 3}$

2. Let $f(x) = 3x^2 + 5x - 2$. Find:

a. $f(-2)$ **b.** $f(a + 2)$

c. $f(2a)$ **d.** $f(a + h)$

3. Let $y^2 = 2x + 1$.

a. Sketch the graph of this equation.

b. Is y a function of x? Why?

c. Is x a function of y? Why?

4. Sketch the graph of the function defined by

$$f(x) = \begin{cases} x + 1 & \text{if } x < 1 \\ -x^2 + 4x - 1 & \text{if } x \geq 1 \end{cases}$$

5. Let $f(x) = 1/x$ and $g(x) = 2x + 3$. Find:

a. $f(x)g(x)$ **b.** $f(x)/g(x)$

c. $f(g(x))$ **d.** $g(f(x))$

In Exercises 6–19, find the indicated limits, if they exist.

6. $\lim_{x\to 0} (5x - 3)$ **7.** $\lim_{x\to 1} (x^2 + 1)$

8. $\lim_{x\to -1} (3x^2 + 4)(2x - 1)$

9. $\lim_{x\to 3} \dfrac{x - 3}{x + 4}$ **10.** $\lim_{x\to 2} \dfrac{x + 3}{x^2 - 9}$

11. $\lim_{x\to -2} \dfrac{x^2 - 2x - 3}{x^2 + 5x + 6}$ **12.** $\lim_{x\to 3} \sqrt{2x^3 - 5}$

13. $\lim_{x\to 3} \dfrac{4x - 3}{\sqrt{x + 1}}$ **14.** $\lim_{x\to 1^+} \dfrac{x - 1}{x(x - 1)}$

15. $\lim_{x\to 1^-} \dfrac{\sqrt{x} - 1}{x - 1}$ **16.** $\lim_{x\to\infty} \dfrac{x^2}{x^2 - 1}$

17. $\lim_{x\to -\infty} \dfrac{x + 1}{x}$ **18.** $\lim_{x\to\infty} \dfrac{3x^2 + 2x + 4}{2x^2 - 3x + 1}$

19. $\lim_{x\to -\infty} \dfrac{x^2}{x + 1}$

20. Sketch the graph of the function

$$f(x) = \begin{cases} 2x - 3 & \text{if } x \leq 2 \\ -x + 3 & \text{if } x > 2 \end{cases}$$

and evaluate $\lim_{x\to a^+} f(x)$, $\lim_{x\to a^-} f(x)$, and $\lim_{x\to a} f(x)$ at the point $a = 2$, if the limits exist.

21. Sketch the graph of the function

$$f(x) = \begin{cases} 4 - x & \text{if } x \leq 2 \\ x + 2 & \text{if } x > 2 \end{cases}$$

and evaluate $\lim_{x \to a^+} f(x)$, $\lim_{x \to a^-} f(x)$, and $\lim_{x \to a} f(x)$ at the point $a = 2$, if the limits exist.

In Exercises 22–25, determine all values of *x* for which each function is discontinuous.

22. $g(x) = \begin{cases} x + 3 & \text{if } x \neq 2 \\ 0 & \text{if } x = 2 \end{cases}$

23. $f(x) = \dfrac{3x + 4}{4x^2 - 2x - 2}$

24. $f(x) = \begin{cases} \dfrac{1}{(x + 1)^2} & \text{if } x \neq -1 \\ 2 & \text{if } x = -1 \end{cases}$

25. $f(x) = \dfrac{|2x|}{x}$

26. Let $y = x^2 + 2$.

a. Find the average rate of change of y with respect to x in the intervals $[1, 2]$, $[1, 1.5]$, and $[1, 1.1]$.

b. Find the (instantaneous) rate of change of y at $x = 1$.

27. Use the definition of the derivative to find the slope of the tangent line to the graph of the function $f(x) = 3x + 5$ at any point $P(x, f(x))$ on the graph.

28. Use the definition of the derivative to find the slope of the tangent line to the graph of the function $f(x) = -1/x$ at any point $P(x, f(x))$ on the graph.

29. Use the definition of the derivative to find the slope of the tangent line to the graph of the function $f(x) = \frac{3}{2}x + 5$ at the point $(-2, 2)$ and determine an equation of the tangent line.

30. Use the definition of the derivative to find the slope of the tangent line to the graph of the function $f(x) = -x^2$ at the point $(2, -4)$ and determine an equation of the tangent line.

31. The graph of the function f is shown in the accompanying figure.

a. Is f continuous at $x = a$? Why?

b. Is f differentiable at $x = a$? Justify your answers.

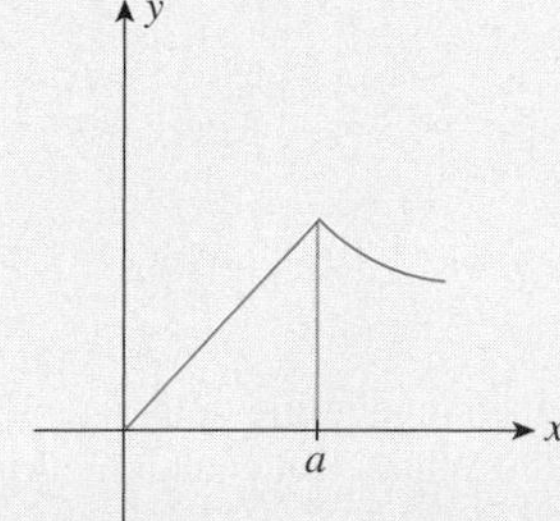

32. SALES OF CLOCK RADIOS Sales of a certain stereo clock radio are approximated by the relationship $S(x) = 6000x + 30{,}000$ $(0 \leq x \leq 5)$, where $S(x)$ denotes the number of clock radios sold in year x ($x = 0$ corresponds to the year 2002). Find the number of clock radios expected to be sold in 2006.

33. SALES OF A COMPANY A company's total sales (in millions of dollars) are approximately linear as a function of time (in years). Sales in 2001 were \$2.4 million, whereas sales in 2006 amounted to \$7.4 million.

a. Find an equation that gives the company's sales as a function of time.

b. What were the sales in 2004?

34. PROFIT FUNCTIONS A company has a fixed cost of \$30,000 and a production cost of \$6 for each unit it manufactures. A unit sells for \$10.

a. What is the cost function?

b. What is the revenue function?

c. What is the profit function?

d. Compute the profit (loss) corresponding to production levels of 6000, 8000, and 12,000 units, respectively.

35. Find the point of intersection of the two straight lines having the equations $y = \frac{3}{4}x + 6$ and $3x - 2y + 3 = 0$.

36. The cost and revenue functions for a certain firm are given by $C(x) = 12x + 20{,}000$ and $R(x) = 20x$, respectively. Find the company's profit function.

37. MARKET EQUILIBRIUM Given the demand equation $3x + p - 40 = 0$ and the supply equation $2x - p + 10 = 0$, where p is the unit price in dollars and x represents the quantity in units of a thousand, determine the equilibrium quantity and the equilibrium price.

38. CLARK'S RULE Clark's rule is a method for calculating pediatric drug dosages based on a child's weight. If a denotes the adult dosage (in milligrams) and if w is the weight of the child (in pounds), then the child's dosage is given by

$$D(w) = \frac{aw}{150}$$

If the adult dose of a substance is 500 mg, how much should a child who weighs 35 lb receive?

39. REVENUE FUNCTIONS The revenue (in dollars) realized by Apollo from the sale of its ink-jet printers is given by

$$R(x) = -0.1x^2 + 500x$$

where x denotes the number of units manufactured each month. What is Apollo's revenue when 1000 units are produced?

40. REVENUE FUNCTIONS The monthly revenue R (in hundreds of dollars) realized in the sale of Royal electric shavers is related to the unit price p (in dollars) by the equation

$$R(p) = -\frac{1}{2}p^2 + 30p$$

Find the revenue when an electric shaver is priced at \$30.

41. Health Club Membership The membership of the newly opened Venus Health Club is approximated by the function

$$N(x) = 200(4 + x)^{1/2} \qquad (1 \leq x \leq 24)$$

where $N(x)$ denotes the number of members x mo after the club's grand opening. Find $N(0)$ and $N(12)$ and interpret your results.

42. Population Growth A study prepared for a Sunbelt town's Chamber of Commerce projected that the population of the town in the next 3 yr will grow according to the rule

$$P(x) = 50{,}000 + 30x^{3/2} + 20x$$

where $P(x)$ denotes the population x mo from now. By how much will the population increase during the next 9 mo? During the next 16 mo?

43. Thurstone Learning Curve Psychologist L. L. Thurstone discovered the following model for the relationship between the learning time T and the length of a list n:

$$T = f(n) = An\sqrt{n - b}$$

where A and b are constants that depend on the person and the task. Suppose that, for a certain person and a certain task, $A = 4$ and $b = 4$. Compute $f(4), f(5), \ldots, f(12)$ and use this information to sketch the graph of the function f. Interpret your results.

44. Forecasting Sales The annual sales of Crimson Drug Store are expected to be given by

$$S_1(t) = 2.3 + 0.4t$$

million dollars t yr from now, whereas the annual sales of Cambridge Drug Store are expected to be given by

$$S_2(t) = 1.2 + 0.6t$$

million dollars t yr from now. When will the annual sales of Cambridge first surpass the annual sales of Crimson?

45. Market Equilibrium The monthly demand and supply functions for the Luminar desk lamp are given by

$$p = d(x) = -1.1x^2 + 1.5x + 40$$
$$p = s(x) = 0.1x^2 + 0.5x + 15$$

respectively, where p is measured in dollars and x in units of a thousand. Find the equilibrium quantity and price.

46. Oil Spills The oil spilling from the ruptured hull of a grounded tanker spreads in all directions in calm waters. Suppose the area polluted is a circle of radius r and the radius is increasing at the rate of 2 ft/sec.
a. Find a function f giving the area polluted in terms of r.
b. Find a function g giving the radius of the polluted area in terms of t.
c. Find a function h giving the area polluted in terms of t.
d. What is the size of the polluted area 30 sec after the hull was ruptured?

47. Film Conversion Prices PhotoMart transfers movie films to CDs. The fees charged for this service are shown in the following table. Find a function C relating the cost $C(x)$ to the number of feet x of film transferred. Sketch the graph of the function C and discuss its continuity.

Length of Film in Feet, x	Price (\$) for Conversion
$1 \leq x \leq 100$	5.00
$100 < x \leq 200$	9.00
$200 < x \leq 300$	12.50
$300 < x \leq 400$	15.00
$x > 400$	$7 + 0.02x$

48. Average Price of a Commodity The average cost (in dollars) of producing x units of a certain commodity is given by

$$\overline{C}(x) = 20 + \frac{400}{x}$$

Evaluate $\lim_{x \to \infty} \overline{C}(x)$ and interpret your results.

CHAPTER 2 Before Moving On . . .

1. Let

$$f(x) = \begin{cases} -2x + 1 & -1 \leq x < 0 \\ x^2 + 2 & 0 \leq x \leq 2 \end{cases}$$

Find (a) $f(-1)$, (b) $f(0)$, and (c) $f(\frac{3}{2})$.

2. Let $f(x) = \frac{1}{x+1}$ and $g(x) = x^2 + 1$. Find the rules for (a) $f + g$, (b) fg, (c) $f \circ g$, and (d) $g \circ f$.

3. Postal regulations specify that a parcel sent by parcel post may have a combined length and girth of no more than 108 in. Suppose a rectangular package that has a square cross section of x in. $\times$ x in. is to have a combined length and girth of exactly 108 in. Find a function in terms of x giving the volume of the package.
Hint: The length plus the girth is $4x + h$ (see the accompanying figure).

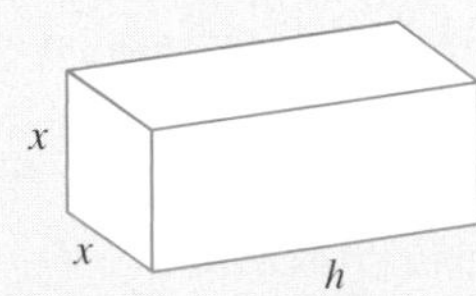

4. Find $\lim_{x \to -1} \frac{x^2 + 4x + 3}{x^2 + 3x + 2}$.

5. Let

$$f(x) = \begin{cases} x^2 - 1 & -2 \leq x < 1 \\ x^3 & 1 \leq x \leq 2 \end{cases}$$

Find (a) $\lim_{x \to 1^-} f(x)$ and (b) $\lim_{x \to 1^+} f(x)$. Is f continuous at $x = 1$? Explain.

6. Find the slope of the tangent line to the graph of $x^2 - 3x + 1$ at the point $(1, -1)$. What is an equation of the tangent line?

3 Differentiation

How is a pond's oxygen content affected by organic waste? In Example 7, page 178, you will see how to find the rate at which oxygen is being restored to the pond after organic waste has been dumped into it.

© Theo Allofs/Corbis

THIS CHAPTER GIVES several rules that will greatly simplify the task of finding the derivative of a function, thus enabling us to study how fast one quantity is changing with respect to another in many real-world situations. For example, we will be able to find how fast the population of an endangered species of whales grows after certain conservation measures have been implemented, how fast an economy's consumer price index (CPI) is changing at any time, and how fast the time taken to learn the items on a list changes with respect to the length of a list. We also see how these rules of differentiation facilitate the study of marginal analysis, the study of the rate of change of economic quantities. Finally, we introduce the notion of the differential of a function. Using differentials is a relatively easy way of approximating the change in one quantity due to a small change in a related quantity.

3.1 Basic Rules of Differentiation

Four Basic Rules

The method used in Chapter 2 for computing the derivative of a function is based on a faithful interpretation of the definition of the derivative as the limit of a quotient. Thus, to find the rule for the derivative f' of a function f, we first computed the difference quotient

$$\frac{f(x+h)-f(x)}{h}$$

and then evaluated its limit as h approached zero. As you have probably observed, this method is tedious even for relatively simple functions.

The main purpose of this chapter is to derive certain rules that will simplify the process of finding the derivative of a function. Throughout this book, we will use the notation

$$\frac{d}{dx}[f(x)] \qquad \text{Read "}d, d\,x \text{ of } f \text{ of } x\text{"}$$

to mean "the derivative of f with respect to x at x." In stating the rules of differentiation, we assume that the functions f and g are differentiable.

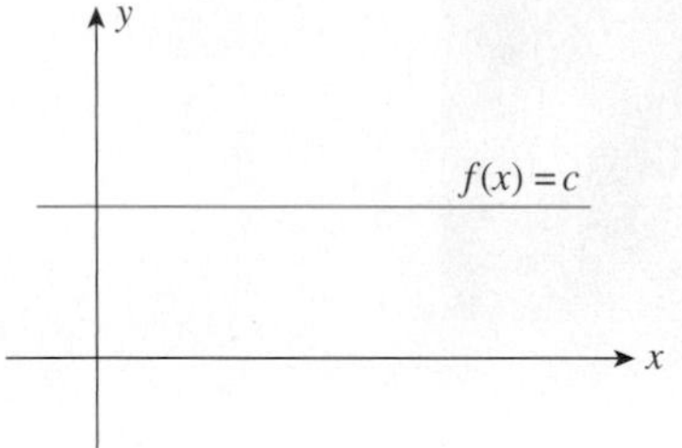

FIGURE 1
The slope of the tangent line to the graph of $f(x) = c$, where c is a constant, is zero.

Rule 1: Derivative of a Constant

$$\frac{d}{dx}(c) = 0 \qquad (c, \text{ a constant})$$

The derivative of a constant function is equal to zero.

We can see this from a geometric viewpoint by recalling that the graph of a constant function is a straight line parallel to the x-axis (Figure 1). Since the tangent line to a straight line at any point on the line coincides with the straight line itself, its slope [as given by the derivative of $f(x) = c$] must be zero. We can also use the definition of the derivative to prove this result by computing

$$\begin{aligned} f'(x) &= \lim_{h\to 0}\frac{f(x+h)-f(x)}{h} \\ &= \lim_{h\to 0}\frac{c-c}{h} \\ &= \lim_{h\to 0} 0 = 0 \end{aligned}$$

EXAMPLE 1

a. If $f(x) = 28$, then

$$f'(x) = \frac{d}{dx}(28) = 0$$

b. If $f(x) = -2$, then

$$f'(x) = \frac{d}{dx}(-2) = 0$$

Rule 2: The Power Rule

If n is any real number, then $\dfrac{d}{dx}(x^n) = nx^{n-1}$.

Let's verify the power rule for the special case $n = 2$. If $f(x) = x^2$, then

$$\begin{aligned} f'(x) = \frac{d}{dx}(x^2) &= \lim_{h \to 0} \frac{f(x + h) - f(x)}{h} \\ &= \lim_{h \to 0} \frac{(x + h)^2 - x^2}{h} \\ &= \lim_{h \to 0} \frac{x^2 + 2xh + h^2 - x^2}{h} \\ &= \lim_{h \to 0} \frac{2xh + h^2}{h} = \lim_{h \to 0} (2x + h) = 2x \end{aligned}$$

as we set out to show.

The proof of the power rule for the general case is not easy to prove and will be omitted. However, you will be asked to prove the rule for the special case $n = 3$ in Exercise 77, page 170.

EXAMPLE 2

a. If $f(x) = x$, then

$$f'(x) = \frac{d}{dx}(x) = 1 \cdot x^{1-1} = x^0 = 1$$

b. If $f(x) = x^8$, then

$$f'(x) = \frac{d}{dx}(x^8) = 8x^7$$

c. If $f(x) = x^{5/2}$, then

$$f'(x) = \frac{d}{dx}(x^{5/2}) = \frac{5}{2}x^{3/2}$$

■

To differentiate a function whose rule involves a radical, we first rewrite the rule using fractional powers. The resulting expression can then be differentiated using the power rule.

EXAMPLE 3 Find the derivative of the following functions:

a. $f(x) = \sqrt{x}$ **b.** $g(x) = \dfrac{1}{\sqrt[3]{x}}$

Solution

a. Rewriting $\sqrt{x}$ in the form $x^{1/2}$, we obtain

$$\begin{aligned} f'(x) &= \frac{d}{dx}(x^{1/2}) \\ &= \frac{1}{2}x^{-1/2} = \frac{1}{2x^{1/2}} = \frac{1}{2\sqrt{x}} \end{aligned}$$

b. Rewriting $\dfrac{1}{\sqrt[3]{x}}$ in the form $x^{-1/3}$, we obtain

$$\begin{aligned} g'(x) &= \frac{d}{dx}(x^{-1/3}) \\ &= -\frac{1}{3}x^{-4/3} = -\frac{1}{3x^{4/3}} \end{aligned}$$

Rule 3: Derivative of a Constant Multiple of a Function

$$\frac{d}{dx}[cf(x)] = c\frac{d}{dx}[f(x)] \qquad (c\text{, a constant})$$

The derivative of a constant times a differentiable function is equal to the constant times the derivative of the function.

This result follows from the following computations.

If $g(x) = cf(x)$, then

$$\begin{aligned} g'(x) &= \lim_{h\to 0}\frac{g(x+h) - g(x)}{h} = \lim_{h\to 0}\frac{cf(x+h) - cf(x)}{h} \\ &= c\lim_{h\to 0}\frac{f(x+h) - f(x)}{h} \\ &= cf'(x) \end{aligned}$$

EXAMPLE 4

a. If $f(x) = 5x^3$, then

$$\begin{aligned} f'(x) &= \frac{d}{dx}(5x^3) = 5\frac{d}{dx}(x^3) \\ &= 5(3x^2) = 15x^2 \end{aligned}$$

b. If $f(x) = \dfrac{3}{\sqrt{x}}$, then

$$\begin{aligned} f'(x) &= \frac{d}{dx}(3x^{-1/2}) \\ &= 3\left(-\frac{1}{2}x^{-3/2}\right) = -\frac{3}{2x^{3/2}} \end{aligned}$$

Rule 4: The Sum Rule

$$\frac{d}{dx}[f(x) \pm g(x)] = \frac{d}{dx}[f(x)] \pm \frac{d}{dx}[g(x)]$$

The derivative of the sum (difference) of two differentiable functions is equal to the sum (difference) of their derivatives.

This result may be extended to the sum and difference of any finite number of differentiable functions. Let's verify the rule for a sum of two functions.

If $s(x) = f(x) + g(x)$, then

$$
\begin{aligned}
s'(x) &= \lim_{h\to 0} \frac{s(x+h) - s(x)}{h} \\
&= \lim_{h\to 0} \frac{[f(x+h) + g(x+h)] - [f(x) + g(x)]}{h} \\
&= \lim_{h\to 0} \frac{[f(x+h) - f(x)] + [g(x+h) - g(x)]}{h} \\
&= \lim_{h\to 0} \frac{f(x+h) - f(x)}{h} + \lim_{h\to 0} \frac{g(x+h) - g(x)}{h} \\
&= f'(x) + g'(x)
\end{aligned}
$$

EXAMPLE 5 Find the derivatives of the following functions:

a. $f(x) = 4x^5 + 3x^4 - 8x^2 + x + 3$ **b.** $g(t) = \dfrac{t^2}{5} + \dfrac{5}{t^3}$

Solution

a.

$$
\begin{aligned}
f'(x) &= \frac{d}{dx}(4x^5 + 3x^4 - 8x^2 + x + 3) \\
&= \frac{d}{dx}(4x^5) + \frac{d}{dx}(3x^4) - \frac{d}{dx}(8x^2) + \frac{d}{dx}(x) + \frac{d}{dx}(3) \\
&= 20x^4 + 12x^3 - 16x + 1
\end{aligned}
$$

b. Here, the independent variable is t instead of x, so we differentiate with respect to t. Thus,

$$g'(t) = \frac{d}{dt}\left(\frac{1}{5}t^2 + 5t^{-3}\right) \qquad \text{Rewrite } \frac{1}{t^3} \text{ as } t^{-3}.$$

$$= \frac{2}{5}t - 15t^{-4}$$

$$= \frac{2t^5 - 75}{5t^4} \qquad \text{Rewrite } t^{-4} \text{ as } \frac{1}{t^4} \text{ and simplify.}$$

■

EXAMPLE 6 Find the slope and an equation of the tangent line to the graph of $f(x) = 2x + 1/\sqrt{x}$ at the point $(1, 3)$.

Solution The slope of the tangent line at any point on the graph of f is given by

$$f'(x) = \frac{d}{dx}\left(2x + \frac{1}{\sqrt{x}}\right)$$

$$= \frac{d}{dx}(2x + x^{-1/2}) \qquad \text{Rewrite } \frac{1}{\sqrt{x}} \text{ as } \frac{1}{x^{1/2}} = x^{-1/2}.$$

$$= 2 - \frac{1}{2}x^{-3/2} \qquad \text{Use the sum rule.}$$

$$= 2 - \frac{1}{2x^{3/2}}$$

In particular, the slope of the tangent line to the graph of f at (1, 3) (where $x = 1$) is

$$f'(1) = 2 - \frac{1}{2(1^{3/2})} = 2 - \frac{1}{2} = \frac{3}{2}$$

Using the point-slope form of the equation of a line with slope $\frac{3}{2}$ and the point (1, 3), we see that an equation of the tangent line is

$$y - 3 = \frac{3}{2}(x - 1) \qquad y - y_1 = m(x - x_1)$$

or, upon simplification,

$$y = \frac{3}{2}x + \frac{3}{2}$$

(see Figure 2).

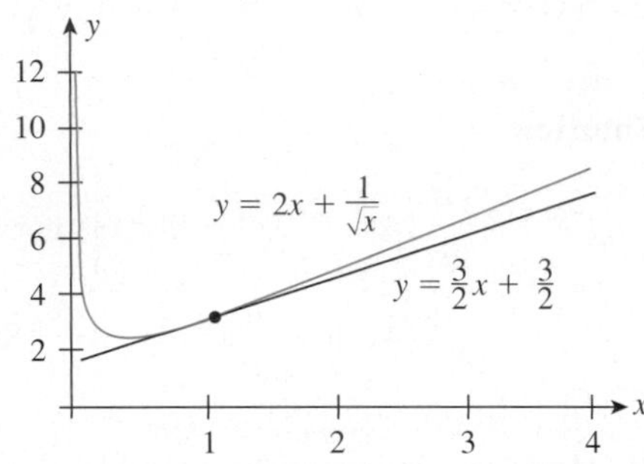

FIGURE 2
The tangent line to the graph of $f(x) = 2x + 1/\sqrt{x}$ at (1, 3).

APPLIED EXAMPLE 7 Conservation of a Species A group of marine biologists at the Neptune Institute of Oceanography recommended that a series of conservation measures be carried out over the next decade to save a certain species of whale from extinction. After implementing the conservation measures, the population of this species is expected to be

$$N(t) = 3t^3 + 2t^2 - 10t + 600 \qquad (0 \le t \le 10)$$

where $N(t)$ denotes the population at the end of year t. Find the rate of growth of the whale population when $t = 2$ and $t = 6$. How large will the whale population be 8 years after implementing the conservation measures?

Solution The rate of growth of the whale population at any time t is given by

$$N'(t) = 9t^2 + 4t - 10$$

In particular, when $t = 2$ and $t = 6$, we have

$$\begin{aligned} N'(2) &= 9(2)^2 + 4(2) - 10 \\ &= 34 \\ N'(6) &= 9(6)^2 + 4(6) - 10 \\ &= 338 \end{aligned}$$

Thus, the whale population's rate of growth will be 34 whales per year after 2 years and 338 per year after 6 years.

The whale population at the end of the eighth year will be

$$\begin{aligned} N(8) &= 3(8)^3 + 2(8)^2 - 10(8) + 600 \\ &= 2184 \text{ whales} \end{aligned}$$

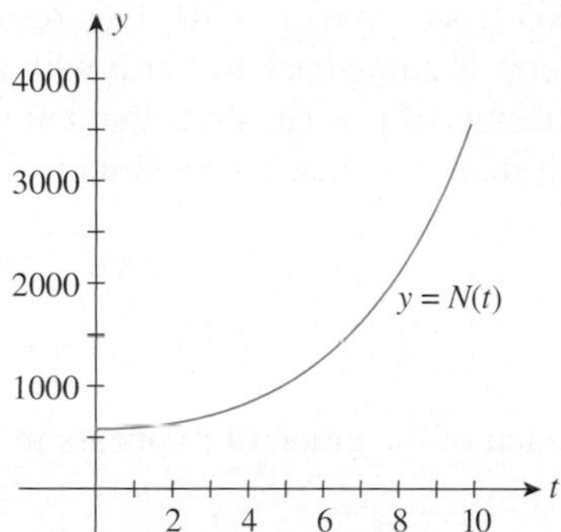

FIGURE 3
The whale population after year t is given by $N(t)$.

The graph of the function N appears in Figure 3. Note the rapid growth of the population in the later years, as the conservation measures begin to pay off, compared with the growth in the early years. ■

APPLIED EXAMPLE 8 Altitude of a Rocket The altitude of a rocket (in feet) t seconds into flight is given by

$$s = f(t) = -t^3 + 96t^2 + 195t + 5 \qquad (t \geq 0)$$

a. Find an expression v for the rocket's velocity at any time t.

b. Compute the rocket's velocity when $t = 0, 30, 50, 65$, and 70. Interpret your results.

c. Using the results from the solution to part (b) and the observation that at the highest point in its trajectory the rocket's velocity is zero, find the maximum altitude attained by the rocket.

Solution

a. The rocket's velocity at any time t is given by

$$v = f'(t) = -3t^2 + 192t + 195$$

b. The rocket's velocity when $t = 0, 30, 50, 65$, and 70 is given by

$$\begin{aligned} f'(0) &= -3(0)^2 + 192(0) + 195 = 195 \\ f'(30) &= -3(30)^2 + 192(30) + 195 = 3255 \\ f'(50) &= -3(50)^2 + 192(50) + 195 = 2295 \\ f'(65) &= -3(65)^2 + 192(65) + 195 = 0 \\ f'(70) &= -3(70)^2 + 192(70) + 195 = -1065 \end{aligned}$$

or 195, 3255, 2295, 0, and -1065 feet per second (ft/sec).

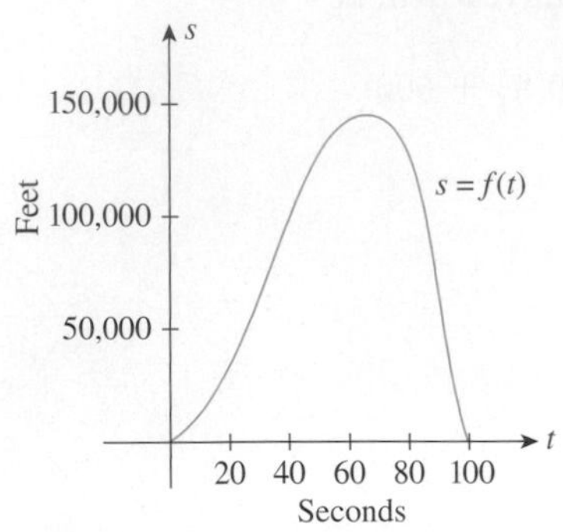

FIGURE 4
The rocket's altitude t seconds into flight is given by $f(t)$.

Thus, the rocket has an initial velocity of 195 ft/sec at $t = 0$ and accelerates to a velocity of 3255 ft/sec at $t = 30$. Fifty seconds into the flight, the rocket's velocity is 2295 ft/sec, which is less than the velocity at $t = 30$. This means that the rocket begins to decelerate after an initial period of acceleration. (Later on we will learn how to determine the rocket's maximum velocity.)

The deceleration continues: The velocity is 0 ft/sec at $t = 65$ and -1065 ft/sec when $t = 70$. This result tells us that 70 seconds into flight the rocket is heading back to Earth with a speed of 1065 ft/sec.

c. The results of part (b) show that the rocket's velocity is zero when $t = 65$. At this instant, the rocket's maximum altitude is

$$\begin{aligned} s = f(65) &= -(65)^3 + 96(65)^2 + 195(65) + 5 \\ &= 143{,}655 \text{ feet} \end{aligned}$$

A sketch of the graph of f appears in Figure 4. ■

EXPLORING WITH TECHNOLOGY

Refer to Example 8.

1. Use a graphing utility to plot the graph of the velocity function

$$v = f'(t) = -3t^2 + 192t + 195$$

using the viewing window $[0, 120] \times [-5000, 5000]$. Then, using **ZOOM** and **TRACE** or the root-finding capability of your graphing utility, verify that $f'(65) = 0$.

2. Plot the graph of the position function of the rocket

$$s = f(t) = -t^3 + 96t^2 + 195t + 5$$

using the viewing window $[0, 120] \times [0, 150{,}000]$. Then, using **ZOOM** and **TRACE** repeatedly, verify that the maximum altitude of the rocket is 143,655 feet.

3. Use **ZOOM** and **TRACE** or the root-finding capability of your graphing utility to find when the rocket returns to Earth.

3.1 Self-Check Exercises

1. Find the derivative of each function using the rules of differentiation.
   a. $f(x) = 1.5x^2 + 2x^{1.5}$
   b. $g(x) = 2\sqrt{x} + \dfrac{3}{\sqrt{x}}$

2. Let $f(x) = 2x^3 - 3x^2 + 2x - 1$.
   a. Compute $f'(x)$.
   b. What is the slope of the tangent line to the graph of f when $x = 2$?
   c. What is the rate of change of the function f at $x = 2$?

3. A certain country's gross domestic product (GDP) (in millions of dollars) is described by the function

$$G(t) = -2t^3 + 45t^2 + 20t + 6000 \qquad (0 \le t \le 11)$$

where $t = 0$ corresponds to the beginning of 1995.
   a. At what rate was the GDP changing at the beginning of 2000? At the beginning of 2002? At the beginning of 2005?
   b. What was the average rate of growth of the GDP over the period 2000–2005?

Solutions to Self-Check Exercises 3.1 can be found on page 171.

3.1 Concept Questions

1. State the following rules of differentiation in your own words.
   a. The rule for differentiating a constant function
   b. The power rule
   c. The constant multiple rule
   d. The sum rule

2. If $f'(2) = 3$ and $g'(2) = -2$, find
   a. $h'(2)$ if $h(x) = 2f(x)$
   b. $F'(2)$ if $F(x) = 3f(x) - 4g(x)$

3.1 Exercises

In Exercises 1–34, find the derivative of the function f by using the rules of differentiation.

1. $f(x) = -3$
2. $f(x) = 365$
3. $f(x) = x^5$
4. $f(x) = x^7$
5. $f(x) = x^{2.1}$
6. $f(x) = x^{0.8}$
7. $f(x) = 3x^2$
8. $f(x) = -2x^3$
9. $f(r) = \pi r^2$
10. $f(r) = \frac{4}{3}\pi r^3$
11. $f(x) = 9x^{1/3}$
12. $f(x) = \frac{5}{4}x^{4/5}$
13. $f(x) = 3\sqrt{x}$
14. $f(u) = \frac{2}{\sqrt{u}}$
15. $f(x) = 7x^{-12}$
16. $f(x) = 0.3x^{-1.2}$
17. $f(x) = 5x^2 - 3x + 7$
18. $f(x) = x^3 - 3x^2 + 1$
19. $f(x) = -x^3 + 2x^2 - 6$
20. $f(x) = x^4 - 2x^2 + 5$
21. $f(x) = 0.03x^2 - 0.4x + 10$
22. $f(x) = 0.002x^3 - 0.05x^2 + 0.1x - 20$
23. $f(x) = \dfrac{x^3 - 4x^2 + 3}{x}$
24. $f(x) = \dfrac{x^3 + 2x^2 + x - 1}{x}$
25. $f(x) = 4x^4 - 3x^{5/2} + 2$
26. $f(x) = 5x^{4/3} - \frac{2}{3}x^{3/2} + x^2 - 3x + 1$
27. $f(x) = 3x^{-1} + 4x^{-2}$
28. $f(x) = -\frac{1}{3}(x^{-3} - x^6)$
29. $f(t) = \dfrac{4}{t^4} - \dfrac{3}{t^3} + \dfrac{2}{t}$
30. $f(x) = \dfrac{5}{x^3} - \dfrac{2}{x^2} - \dfrac{1}{x} + 200$
31. $f(x) = 2x - 5\sqrt{x}$
32. $f(t) = 2t^2 + \sqrt{t^3}$
33. $f(x) = \dfrac{2}{x^2} - \dfrac{3}{x^{1/3}}$
34. $f(x) = \dfrac{3}{x^3} + \dfrac{4}{\sqrt{x}} + 1$
35. Let $f(x) = 2x^3 - 4x$. Find:
   a. $f'(-2)$ b. $f'(0)$ c. $f'(2)$
36. Let $f(x) = 4x^{5/4} + 2x^{3/2} + x$. Find:
   a. $f'(0)$ b. $f'(16)$

In Exercises 37–40, find each limit by evaluating the derivative of a suitable function at an appropriate point.
Hint: Look at the definition of the derivative.

37. $\lim\limits_{h \to 0} \dfrac{(1 + h)^3 - 1}{h}$
38. $\lim\limits_{x \to 1} \dfrac{x^5 - 1}{x - 1}$
    Hint: Let $h = x - 1$.
39. $\lim\limits_{h \to 0} \dfrac{3(2 + h)^2 - (2 + h) - 10}{h}$
40. $\lim\limits_{t \to 0} \dfrac{1 - (1 + t)^2}{t(1 + t)^2}$

In Exercises 41–44, find the slope and an equation of the tangent line to the graph of the function f at the specified point.

41. $f(x) = 2x^2 - 3x + 4$; $(2, 6)$
42. $f(x) = -\frac{5}{3}x^2 + 2x + 2$; $\left(-1, -\frac{5}{3}\right)$
43. $f(x) = x^4 - 3x^3 + 2x^2 - x + 1$; $(1, 0)$
44. $f(x) = \sqrt{x} + \dfrac{1}{\sqrt{x}}$; $\left(4, \frac{5}{2}\right)$
45. Let $f(x) = x^3$.
   a. Find the point on the graph of f where the tangent line is horizontal.
   b. Sketch the graph of f and draw the horizontal tangent line.
46. Let $f(x) = x^3 - 4x^2$. Find the point(s) on the graph of f where the tangent line is horizontal.

47. Let $f(x) = x^3 + 1$.
   a. Find the point(s) on the graph of f where the slope of the tangent line is equal to 12.
   b. Find the equation(s) of the tangent line(s) of part (a).
   c. Sketch the graph of f showing the tangent line(s).

48. Let $f(x) = \frac{2}{3}x^3 + x^2 - 12x + 6$. Find the values of x for which:
   a. $f'(x) = -12$ b. $f'(x) = 0$
   c. $f'(x) = 12$

49. Let $f(x) = \frac{1}{4}x^4 - \frac{1}{3}x^3 - x^2$. Find the point(s) on the graph of f where the slope of the tangent line is equal to:
   a. $-2x$ b. 0 c. $10x$

50. A straight line perpendicular to and passing through the point of tangency of the tangent line is called the *normal* to the curve. Find an equation of the tangent line and the normal to the curve $y = x^3 - 3x + 1$ at the point (2, 3).

51. Growth of a Cancerous Tumor The volume of a spherical cancerous tumor is given by the function

$$V(r) = \frac{4}{3}\pi r^3$$

where r is the radius of the tumor in centimeters. Find the rate of change in the volume of the tumor when
   a. $r = \frac{2}{3}$ cm b. $r = \frac{5}{4}$ cm

52. Velocity of Blood in an Artery The velocity (in centimeters/second) of blood r cm from the central axis of an artery is given by

$$v(r) = k(R^2 - r^2)$$

where k is a constant and R is the radius of the artery (see the accompanying figure). Suppose $k = 1000$ and $R = 0.2$ cm. Find $v(0.1)$ and $v'(0.1)$ and interpret your results.

Blood vessel

53. Sales of Digital Cameras According to projections made in 2004, the worldwide shipments of digital point-and-shoot cameras are expected to grow in accordance with the rule

$$N(t) = 16.3t^{0.8766} \qquad (1 \le t \le 6)$$

where $N(t)$ is measured in millions and t is measured in years, with $t = 1$ corresponding to 2001.
   a. How many digital cameras were sold in 2001 ($t = 1$)?
   b. How fast were sales increasing in 2001?
   c. What were the projected sales in 2005?
   d. How fast were the sales projected to grow in 2005?

Source: International Data Corp.

54. Online Buyers As use of the Internet grows, so does the number of consumers who shop online. The number of online buyers, as a percent of net users, is expected to be

$$P(t) = 53t^{0.12} \qquad (1 \le t \le 7)$$

where t is measured in years, with $t = 1$ corresponding to the beginning of 2002.
   a. How many online buyers, as a percent of net users, are there expected to be at the beginning of 2007?
   b. How fast is the number of online buyers, as a percent of net users, expected to be changing at the beginning of 2007?

Source: Strategy Analytics

55. Married Households with Children The percent of families that were married households with children between 1970 and 2000 is approximately

$$P(t) = \frac{49.6}{t^{0.27}} \qquad (1 \le t \le 4)$$

where t is measured in decades, with $t = 1$ corresponding to 1970.
   a. What percent of families were married households with children in 1970? In 1980? In 1990? In 2000?
   b. How fast was the percent of families that were married households with children changing in 1980? In 1990?

Source: U.S. Census Bureau

56. Effect of Stopping on Average Speed According to data from a study, the average speed of your trip A (in mph) is related to the number of stops/mile you make on the trip x by the equation

$$A = \frac{26.5}{x^{0.45}}$$

Compute dA/dx for $x = 0.25$ and $x = 2$. How is the rate of change of the average speed of your trip affected by the number of stops/mile?

Source: General Motors

57. Portable Phones The percent of the U.S. population with portable phones is projected to be

$$P(t) = 24.4t^{0.34} \qquad (1 \le t \le 10)$$

where t is measured in years, with $t = 1$ corresponding to the beginning of 1998.
   a. What percent of the U.S. population is expected to have portable phones by the beginning of 2006?
   b. How fast is the percent of the U.S. population with portable phones expected to be changing at the beginning of 2006?

Source: BancAmerica Robertson Stephens

58. Demand Functions The demand function for the Luminar desk lamp is given by

$$p = f(x) = -0.1x^2 - 0.4x + 35$$

where x is the quantity demanded (measured in thousands) and p is the unit price in dollars.

a. Find $f'(x)$.

b. What is the rate of change of the unit price when the quantity demanded is 10,000 units ($x = 10$)? What is the unit price at that level of demand?

59. STOPPING DISTANCE OF A RACING CAR During a test by the editors of an auto magazine, the stopping distance s (in feet) of the MacPherson X-2 racing car conformed to the rule

$$s = f(t) = 120t - 15t^2 \qquad (t \geq 0)$$

where t was the time (in seconds) after the brakes were applied.

a. Find an expression for the car's velocity v at any time t.

b. What was the car's velocity when the brakes were first applied?

c. What was the car's stopping distance for that particular test?

Hint: The stopping time is found by setting $v = 0$.

60. SALES OF DSPS The sales of digital signal processors (DSPs) in billions of dollars is projected to be

$$S(t) = 0.14t^2 + 0.68t + 3.1 \qquad (0 \leq t \leq 6)$$

where t is measured in years, with $t = 0$ corresponding to the beginning of 1997.

a. What were the sales of DSPs at the beginning of 1997? What were the sales at the beginning of 2002?

b. How fast was the level of sales increasing at the beginning of 1997? How fast were sales increasing at the beginning of 2002?

Source: World Semiconductor Trade Statistics

61. CHILD OBESITY The percent of obese children, ages 12–19, in the United States has grown dramatically in recent years. The percent of obese children from 1980 through the year 2000 is approximated by the function

$$P(t) = -0.0105t^2 + 0.735t + 5 \qquad (0 \leq t \leq 20)$$

where t is measured in years, with $t = 0$ corresponding to the beginning of 1980.

a. What percent of children were obese at the beginning of 1980? At the beginning of 1990? At the beginning of the year 2000?

b. How fast was the percent of obese children changing at the beginning of 1985? At the beginning of 1990?

Source: Centers for Disease Control and Prevention

62. SPENDING ON MEDICARE Based on the current eligibility requirement, a study conducted in 2004 showed that federal spending on entitlement programs, particularly Medicare, would grow enormously in the future. The study predicted that spending on Medicare, as a percent of the gross domestic product (GDP), will be

$$P(t) = 0.27t^2 + 1.4t + 2.2 \qquad (0 \leq t \leq 5)$$

percent in year t, where t is measured in decades, with $t = 0$ corresponding to 2000.

a. How fast will the spending on Medicare, as a percent of the GDP, be growing in 2010? In 2020?

b. What will the predicted spending on Medicare be in 2010? In 2020?

Source: Congressional Budget Office

63. FISHERIES The total groundfish population on Georges Bank in New England between 1989 and 1999 is approximated by the function

$$f(t) = 5.303t^2 - 53.977t + 253.8 \qquad (0 \leq t \leq 10)$$

where $f(t)$ is measured in thousands of metric tons and t is measured in years, with $t = 0$ corresponding to the beginning of 1989.

a. What was the rate of change of the groundfish population at the beginning of 1994? At the beginning of 1996?

b. Fishing restrictions were imposed on Dec. 7, 1994. Were the conservation measures effective?

Source: New England Fishery Management Council

64. WORKER EFFICIENCY An efficiency study conducted for Elektra Electronics showed that the number of Space Commander walkie-talkies assembled by the average worker t hr after starting work at 8 a.m. is given by

$$N(t) = -t^3 + 6t^2 + 15t$$

a. Find the rate at which the average worker will be assembling walkie-talkies t hr after starting work.

b. At what rate will the average worker be assembling walkie-talkies at 10 a.m.? At 11 a.m.?

c. How many walkie-talkies will the average worker assemble between 10 a.m. and 11 a.m.?

65. CONSUMER PRICE INDEX An economy's consumer price index (CPI) is described by the function

$$I(t) = -0.2t^3 + 3t^2 + 100 \qquad (0 \leq t \leq 10)$$

where $t = 0$ corresponds to 1997.

a. At what rate was the CPI changing in 2002? In 2004? In 2007?

b. What was the average rate of increase in the CPI over the period from 2002 to 2007?

66. EFFECT OF ADVERTISING ON SALES The relationship between the amount of money x that Cannon Precision Instruments spends on advertising and the company's total sales $S(x)$ is given by the function

$$S(x) = -0.002x^3 + 0.6x^2 + x + 500 \qquad (0 \leq x \leq 200)$$

where x is measured in thousands of dollars. Find the rate of change of the sales with respect to the amount of money spent on advertising. Are Cannon's total sales increasing at a faster rate when the amount of money spent on advertising is (a) \$100,000 or (b) \$150,000?

67. SUPPLY FUNCTIONS The supply function for a certain make of transistor radio is given by

$$p = f(x) = 0.0001x^{5/4} + 10$$

where x is the quantity supplied and p is the unit price in dollars.

a. Find $f'(x)$.
b. What is the rate of change of the unit price if the quantity supplied is 10,000 transistor radios?

68. POPULATION GROWTH A study prepared for a Sunbelt town's chamber of commerce projected that the town's population in the next 3 yr will grow according to the rule

$$P(t) = 50{,}000 + 30t^{3/2} + 20t$$

where $P(t)$ denotes the population t mo from now. How fast will the population be increasing 9 mo and 16 mo from now?

69. AVERAGE SPEED OF A VEHICLE ON A HIGHWAY The average speed of a vehicle on a stretch of Route 134 between 6 a.m. and 10 a.m. on a typical weekday is approximated by the function

$$f(t) = 20t - 40\sqrt{t} + 50 \qquad (0 \leq t \leq 4)$$

where $f(t)$ is measured in mph and t is measured in hours, with $t = 0$ corresponding to 6 a.m.
a. Compute $f'(t)$.
b. What is the average speed of a vehicle on that stretch of Route 134 at 6 a.m.? At 7 a.m.? At 8 a.m.?
c. How fast is the average speed of a vehicle on that stretch of Route 134 changing at 6:30 a.m.? At 7 a.m.? At 8 a.m.?

70. CURBING POPULATION GROWTH Five years ago, the government of a Pacific Island state launched an extensive propaganda campaign toward curbing the country's population growth. According to the Census Department, the population (measured in thousands of people) for the following 4 yr was

$$P(t) = -\frac{1}{3}t^3 + 64t + 3000$$

where t is measured in years and $t = 0$ corresponds to the start of the campaign. Find the rate of change of the population at the end of years 1, 2, 3, and 4. Was the plan working?

71. CONSERVATION OF SPECIES A certain species of turtle faces extinction because dealers collect truckloads of turtle eggs to be sold as aphrodisiacs. After severe conservation measures are implemented, it is hoped that the turtle population will grow according to the rule

$$N(t) = 2t^3 + 3t^2 - 4t + 1000 \qquad (0 \leq t \leq 10)$$

where $N(t)$ denotes the population at the end of year t. Find the rate of growth of the turtle population when $t = 2$ and $t = 8$. What will be the population 10 yr after the conservation measures are implemented?

72. FLIGHT OF A ROCKET The altitude (in feet) of a rocket t sec into flight is given by

$$s = f(t) = -2t^3 + 114t^2 + 480t + 1 \qquad (t \geq 0)$$

a. Find an expression v for the rocket's velocity at any time t.
b. Compute the rocket's velocity when $t = 0$, 20, 40, and 60. Interpret your results.
c. Using the results from the solution to part (b), find the maximum altitude attained by the rocket.
Hint: At its highest point, the velocity of the rocket is zero.

73. OBESITY IN AMERICA The body mass index (BMI) measures body weight in relation to height. A BMI of 25 to 29.9 is considered overweight, a BMI of 30 or more is considered obese, and a BMI of 40 or more is morbidly obese. The percent of the U.S. population that is obese is approximated by the function

$$P(t) = 0.0004t^3 + 0.0036t^2 + 0.8t + 12 \qquad (0 \leq t \leq 13)$$

where t is measured in years, with $t = 0$ corresponding to the beginning of 1991.
a. What percent of the U.S. population was deemed obese at the beginning of 1991? At the beginning of 2004?
b. How fast was the percent of the U.S. population that is deemed obese changing at the beginning of 1991? At the beginning of 2004?

(*Note:* A formula for calculating the BMI of a person is given in Exercise 27, page 544.)
Source: Centers for Disease Control and Prevention

74. HEALTH-CARE SPENDING Despite efforts at cost containment, the cost of the Medicare program is increasing. Two major reasons for this increase are an aging population and extensive use by physicians of new technologies. Based on data from the Health Care Financing Administration and the U.S. Census Bureau, health-care spending through the year 2000 may be approximated by the function

$$S(t) = 0.02836t^3 - 0.05167t^2 + 9.60881t + 41.9 \qquad (0 \leq t \leq 35)$$

where $S(t)$ is the spending in billions of dollars and t is measured in years, with $t = 0$ corresponding to the beginning of 1965.
a. Find an expression for the rate of change of health-care spending at any time t.
b. How fast was health-care spending changing at the beginning of 1980? At the beginning of 2000?
c. What was the amount of health-care spending at the beginning of 1980? At the beginning of 2000?
Source: Health Care Financing Administration and U.S. Census Bureau

In Exercises 75 and 76, determine whether the statement is true or false. If it is true, explain why it is true. If it is false, give an example to show why it is false.

75. If f and g are differentiable, then

$$\frac{d}{dx}[2f(x) - 5g(x)] = 2f'(x) - 5g'(x)$$

76. If $f(x) = \pi^x$, then $f'(x) = x\pi^{x-1}$.

77. Prove the power rule (Rule 2) for the special case $n = 3$.
Hint: Compute $\lim_{h \to 0}\left[\frac{(x+h)^3 - x^3}{h}\right]$.

3.1 Solutions to Self-Check Exercises

1. a. $$f'(x) = \frac{d}{dx}(1.5x^2) + \frac{d}{dx}(2x^{1.5})$$
$$= (1.5)(2x) + (2)(1.5x^{0.5})$$
$$= 3x + 3\sqrt{x} = 3(x + \sqrt{x})$$

b. $$g'(x) = \frac{d}{dx}(2x^{1/2}) + \frac{d}{dx}(3x^{-1/2})$$
$$= (2)\left(\frac{1}{2}x^{-1/2}\right) + (3)\left(-\frac{1}{2}x^{-3/2}\right)$$
$$= x^{-1/2} - \frac{3}{2}x^{-3/2}$$
$$= \frac{1}{2}x^{-3/2}(2x - 3) = \frac{2x - 3}{2x^{3/2}}$$

2. a. $$f'(x) = \frac{d}{dx}(2x^3) - \frac{d}{dx}(3x^2) + \frac{d}{dx}(2x) - \frac{d}{dx}(1)$$
$$= (2)(3x^2) - (3)(2x) + 2$$
$$= 6x^2 - 6x + 2$$

b. The slope of the tangent line to the graph of f when $x = 2$ is given by

$$f'(2) = 6(2)^2 - 6(2) + 2 = 14$$

c. The rate of change of f at $x = 2$ is given by $f'(2)$. Using the results of part (b), we see that the required rate of change is 14 units/unit change in x.

3. a. The rate at which the GDP was changing at any time t $(0 < t < 11)$ is given by

$$G'(t) = -6t^2 + 90t + 20$$

In particular, the rates of change of the GDP at the beginning of the years 2000 $(t = 5)$, 2002 $(t = 7)$, and 2005 $(t = 10)$ are given by

$$G'(5) = 320 \qquad G'(7) = 356 \qquad G'(10) = 320$$

respectively—that is, by \$320 million/year, \$356 million/year, and \$320 million/year, respectively.

b. The average rate of growth of the GDP over the period from the beginning of 2000 $(t = 5)$ to the beginning of 2005 $(t = 10)$ is given by

$$\frac{G(10) - G(5)}{10 - 5} = \frac{[-2(10)^3 + 45(10)^2 + 20(10) + 6000]}{5}$$
$$- \frac{[-2(5)^3 + 45(5)^2 + 20(5) + 6000]}{5}$$
$$= \frac{8700 - 6975}{5}$$

or \$345 million/year.

USING TECHNOLOGY

Finding the Rate of Change of a Function

We can use the numerical derivative operation of a graphing utility to obtain the value of the derivative at a given value of x. Since the derivative of a function $f(x)$ measures the rate of change of the function with respect to x, the numerical derivative operation can be used to answer questions pertaining to the rate of change of one quantity y with respect to another quantity x, where $y = f(x)$, for a specific value of x.

EXAMPLE 1 Let $y = 3t^3 + 2\sqrt{t}$.

a. Use the numerical derivative operation of a graphing utility to find how fast y is changing with respect to t when $t = 1$.

b. Verify the result of part (a), using the rules of differentiation of this section.

Solution

```
nDeriv((3X^3+2X^
.5), X, 1)
        10.00000313
```

FIGURE T1
The TI-83 numerical derivative screen for computing $f'(1)$

a. Write $f(t) = 3t^3 + 2\sqrt{t}$. Using the numerical derivative operation of a graphing utility, we find that the rate of change of y with respect to t when $t = 1$ is given by $f'(1) = 10$ (Figure T1).

(continued)

b. Here, $f(t) = 3t^3 + 2t^{1/2}$ and

$$f'(t) = 9t^2 + 2\left(\frac{1}{2}t^{-1/2}\right) = 9t^2 + \frac{1}{\sqrt{t}}$$

Using this result, we see that when $t = 1$, y is changing at the rate of

$$f'(1) = 9(1^2) + \frac{1}{\sqrt{1}} = 10$$

units per unit change in t, as obtained earlier. ■

APPLIED EXAMPLE 2 Fuel Economy of Cars According to data obtained from the U.S. Department of Energy and the Shell Development Company, a typical car's fuel economy depends on the speed it is driven and is approximated by the function

$$\begin{aligned} f(x) &= 0.00000310315x^4 - 0.000455174x^3 \\ &\quad + 0.00287869x^2 + 1.25986x \qquad (0 \le x \le 75) \end{aligned}$$

where x is measured in mph and $f(x)$ is measured in miles per gallon (mpg).

a. Use a graphing utility to graph the function f on the interval [0, 75].

b. Find the rate of change of f when $x = 20$ and when $x = 50$.

c. Interpret your results.

Source: U.S. Department of Energy and the Shell Development Company

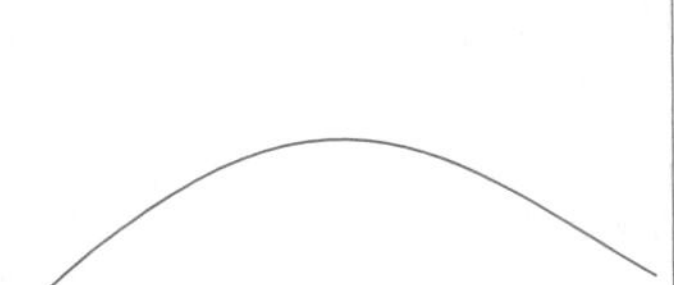

FIGURE T2
The graph of the function f on the interval [0, 75]

Solution

a. The graph is shown in Figure T2.

b. Using the numerical derivative operation of a graphing utility, we see that $f'(20) = .9280996$. The rate of change of f when $x = 50$ is given by $f'(50) = -.3145009995$. (See Figure T3a and T3b.)

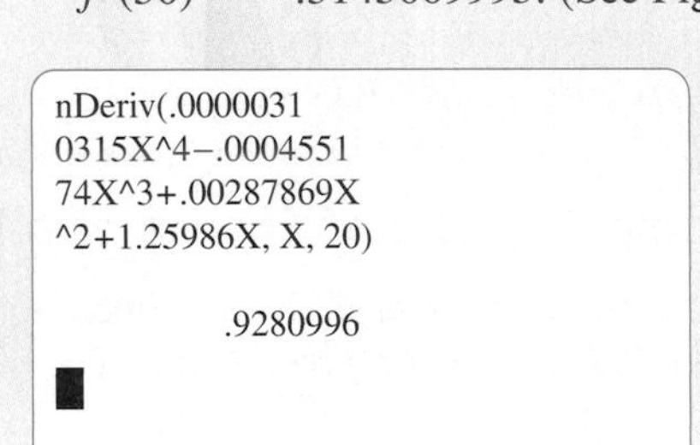

(a)

nDeriv(.0000031
0315X^4−.0004551
74X^3+.00287869X
^2+1.25986X, X, 5
0)
−.3145009995

(b)

FIGURE T3
The TI-83 numerical derivative screen for computing (a) $f'(20)$ and (b) $f'(50)$

c. The results of part (b) tell us that when a typical car is being driven at 20 mph, its fuel economy increases at the rate of approximately 0.9 mpg per 1 mph increase in its speed. At a speed of 50 mph, its fuel economy decreases at the rate of approximately 0.3 mpg per 1 mph increase in its speed. ■

TECHNOLOGY EXERCISES

In Exercises 1–6, use the numerical derivative operation to find the rate of change of $f(x)$ at the given value of x. Give your answer accurate to four decimal places.

1. $f(x) = 4x^5 - 3x^3 + 2x^2 + 1$; $x = 0.5$

2. $f(x) = -x^5 + 4x^2 + 3$; $x = 0.4$

3. $f(x) = x - 2\sqrt{x}$; $x = 3$

4. $f(x) = \dfrac{\sqrt{x} - 1}{x}$; $x = 2$

5. $f(x) = x^{1/2} - x^{1/3}; x = 1.2$

6. $f(x) = 2x^{5/4} + x; x = 2$

7. **Carbon Monoxide in the Atmosphere** The projected average global atmospheric concentration of carbon monoxide is approximated by the function

$$f(t) = 0.881443t^4 - 1.45533t^3 + 0.695876t^2 + 2.87801t + 293 \qquad (0 \leq t \leq 4)$$

where t is measured in 40-yr intervals with $t = 0$ corresponding to the beginning of 1860, and $f(t)$ is measured in parts per million by volume.

a. Plot the graph of f in the viewing window $[0, 4] \times [280, 400]$.

b. Use a graphing utility to estimate how fast the projected average global atmospheric concentration of carbon monoxide was changing at the beginning of the year 1900 $(t = 1)$ and at the beginning of 2000 $(t = 3.5)$.

Source: Meadows et al. "Beyond the Limits"

8. **Growth of HMOs** Based on data compiled by the Group Health Association of America, the number of people receiving their care in an HMO (health maintenance organization) from the beginning of 1984 through 1994 is approximated by the function

$$f(t) = 0.0514t^3 - 0.853t^2 + 6.8147t + 15.6524 \qquad (0 \leq t \leq 11)$$

where $f(t)$ gives the number of people in millions and t is measured in years, with $t = 0$ corresponding to the beginning of 1984.

a. Plot the graph of f in the viewing window $[0, 11] \times [0, 80]$.

b. How fast was the number of people receiving their care in an HMO changing at the beginning of 1992?

Source: Group Health Association of America

9. **Home Sales** The average number of days a single-family home remains for sale from listing to accepted offer (in the greater Boston area) is approximated by the function

$$f(t) = 0.0171911t^4 - 0.662121t^3 + 6.18083t^2 - 8.97086t + 53.3357 \qquad (0 \leq t \leq 12)$$

where t is measured in years, with $t = 0$ corresponding to the beginning of 1984.

a. Plot the graph of f in the viewing window $[0, 12] \times [0, 120]$.

b. How fast was the average number of days a single-family home remained for sale from listing to accepted offer changing at the beginning of 1984 $(t = 0)$? At the beginning of 1988 $(t = 4)$?

Source: Greater Boston Real Estate Board—Multiple Listing Service

10. **Spread of HIV** The estimated number of children newly infected with HIV through mother-to-child contact worldwide is given by

$$f(t) = -0.2083t^3 + 3.0357t^2 + 44.0476t + 200.2857 \qquad (0 \leq t \leq 12)$$

where $f(t)$ is measured in thousands and t is measured in years, with $t = 0$ corresponding to the beginning of 1990.

a. Plot the graph of f in the viewing window $[0, 12] \times [0, 800]$.

b. How fast was the estimated number of children newly infected with HIV through mother-to-child contact worldwide increasing at the beginning of the year 2000?

Source: United Nations

11. **Manufacturing Capacity** Data show that the annual change in manufacturing capacity between 1994 and 2000 is given by

$$f(t) = 0.009417t^3 - 0.426571t^2 + 2.74894t + 5.54 \qquad (0 \leq t \leq 6)$$

percent, where t is measured in years, with $t = 0$ corresponding to the beginning of 1994.

a. Plot the graph of f in the viewing window $[0, 8] \times [0, 10]$.

b. How fast was $f(t)$ changing at the beginning of 1996 $(t = 2)$? At the beginning of 1998 $(t = 4)$?

Source: Federal Reserve

12. **Fisheries** The total groundfish population on Georges Bank in New England between 1989 and 1999 is approximated by the function

$$f(t) = 5.303t^2 - 53.977t + 253.8 \qquad (0 \leq t \leq 10)$$

where $f(t)$ is measured in thousands of metric tons and t is measured in years, with $t = 0$ corresponding to the beginning of 1989.

a. Plot the graph of f in the viewing window $[0, 10] \times [100, 250]$.

b. How fast was the total groundfish population changing at the beginning of 1990? At the beginning of 1996?

(*Note:* Fishing restrictions were imposed on December 7, 1994.)

Source: New England Fishery Management Council

3.2 The Product and Quotient Rules

In this section we study two more rules of differentiation: the **product rule** and the **quotient rule**.

The Product Rule

The derivative of the product of two differentiable functions is given by the following rule:

Rule 5: The Product Rule

$$\frac{d}{dx}[f(x)g(x)] = f(x)g'(x) + g(x)f'(x)$$

The derivative of the product of two functions is the first function times the derivative of the second plus the second function times the derivative of the first.

The product rule may be extended to the case involving the product of any finite number of functions (see Exercise 63, p. 182). We prove the product rule at the end of this section.

The derivative of the product of two functions is *not* given by the product of the derivatives of the functions; that is, in general

$$\frac{d}{dx}[f(x)g(x)] \neq f'(x)g'(x)$$

EXAMPLE 1 Find the derivative of the function

$$f(x) = (2x^2 - 1)(x^3 + 3)$$

Solution By the product rule,

$$\begin{aligned} f'(x) &= (2x^2 - 1)\frac{d}{dx}(x^3 + 3) + (x^3 + 3)\frac{d}{dx}(2x^2 - 1) \\ &= (2x^2 - 1)(3x^2) + (x^3 + 3)(4x) \\ &= 6x^4 - 3x^2 + 4x^4 + 12x \\ &= 10x^4 - 3x^2 + 12x \\ &= x(10x^3 - 3x + 12) \end{aligned}$$

EXAMPLE 2 Differentiate (that is, find the derivative of) the function

$$f(x) = x^3(\sqrt{x} + 1)$$

Solution First, we express the function in exponential form, obtaining

$$f(x) = x^3(x^{1/2} + 1)$$

By the product rule,

$$\begin{aligned} f'(x) &= x^3\frac{d}{dx}(x^{1/2}+1) + (x^{1/2}+1)\frac{d}{dx}x^3 \\ &= x^3\left(\frac{1}{2}x^{-1/2}\right) + (x^{1/2}+1)(3x^2) \\ &= \frac{1}{2}x^{5/2} + 3x^{5/2} + 3x^2 \\ &= \frac{7}{2}x^{5/2} + 3x^2 \end{aligned}$$

Note We can also solve the problem by first expanding the product before differentiating *f*. Examples for which this is not possible will be considered in Section 3.3, where the true value of the product rule will be appreciated.

The Quotient Rule

The derivative of the quotient of two differentiable functions is given by the following rule:

Rule 6: The Quotient Rule

$$\frac{d}{dx}\left[\frac{f(x)}{g(x)}\right] = \frac{g(x)f'(x) - f(x)g'(x)}{[g(x)]^2} \qquad (g(x) \neq 0)$$

As an aid to remembering this expression, observe that it has the following form:

$$\frac{d}{dx}\left[\frac{f(x)}{g(x)}\right] = \frac{(\text{Denominator})\begin{pmatrix}\text{Derivative of}\\ \text{numerator}\end{pmatrix} - (\text{Numerator})\begin{pmatrix}\text{Derivative of}\\ \text{denominator}\end{pmatrix}}{(\text{Square of denominator})}$$

For a proof of the quotient rule, see Exercise 64, page 182.

The derivative of a quotient is *not* equal to the quotient of the derivatives; that is,

$$\frac{d}{dx}\left[\frac{f(x)}{g(x)}\right] \neq \frac{f'(x)}{g'(x)}$$

For example, if $f(x) = x^3$ and $g(x) = x^2$, then

$$\frac{d}{dx}\left[\frac{f(x)}{g(x)}\right] = \frac{d}{dx}\left(\frac{x^3}{x^2}\right) = \frac{d}{dx}(x) = 1$$

which is *not* equal to

$$\frac{f'(x)}{g'(x)} = \frac{\frac{d}{dx}(x^3)}{\frac{d}{dx}(x^2)} = \frac{3x^2}{2x} = \frac{3}{2}x$$

EXAMPLE 3 Find $f'(x)$ if $f(x) = \dfrac{x}{2x - 4}$.

Solution Using the quotient rule, we obtain

$$\begin{aligned} f'(x) &= \frac{(2x - 4)\dfrac{d}{dx}(x) - x\dfrac{d}{dx}(2x - 4)}{(2x - 4)^2} \\ &= \frac{(2x - 4)(1) - x(2)}{(2x - 4)^2} \\ &= \frac{2x - 4 - 2x}{(2x - 4)^2} = -\frac{4}{(2x - 4)^2} \end{aligned}$$

EXAMPLE 4 Find $f'(x)$ if $f(x) = \dfrac{x^2 + 1}{x^2 - 1}$.

Solution By the quotient rule,

$$\begin{aligned} f'(x) &= \frac{(x^2 - 1)\dfrac{d}{dx}(x^2 + 1) - (x^2 + 1)\dfrac{d}{dx}(x^2 - 1)}{(x^2 - 1)^2} \\ &= \frac{(x^2 - 1)(2x) - (x^2 + 1)(2x)}{(x^2 - 1)^2} \\ &= \frac{2x^3 - 2x - 2x^3 - 2x}{(x^2 - 1)^2} \\ &= -\frac{4x}{(x^2 - 1)^2} \end{aligned}$$

EXAMPLE 5 Find $h'(x)$ if $h(x) = \dfrac{\sqrt{x}}{x^2 + 1}$.

Solution Rewrite $h(x)$ in the form $h(x) = \dfrac{x^{1/2}}{x^2 + 1}$. By the quotient rule, we find

$$\begin{aligned} h'(x) &= \frac{(x^2 + 1)\dfrac{d}{dx}(x^{1/2}) - x^{1/2}\dfrac{d}{dx}(x^2 + 1)}{(x^2 + 1)^2} \\ &= \frac{(x^2 + 1)(\frac{1}{2}x^{-1/2}) - x^{1/2}(2x)}{(x^2 + 1)^2} \\ &= \frac{\frac{1}{2}x^{-1/2}(x^2 + 1 - 4x^2)}{(x^2 + 1)^2} && \text{Factor out } \tfrac{1}{2}x^{-1/2} \text{ from the numerator.} \\ &= \frac{1 - 3x^2}{2\sqrt{x}(x^2 + 1)^2} \end{aligned}$$

VIDEO

APPLIED EXAMPLE 6 Rate of Change of DVD Sales The sales (in millions of dollars) of a DVD recording of a hit movie t years from the date of release is given by

$$S(t) = \frac{5t}{t^2 + 1}$$

a. Find the rate at which the sales are changing at time t.

b. How fast are the sales changing at the time the DVDs are released ($t = 0$)? Two years from the date of release?

Solution

a. The rate at which the sales are changing at time t is given by $S'(t)$. Using the quotient rule, we obtain

$$\begin{aligned} S'(t) &= \frac{d}{dt}\left[\frac{5t}{t^2+1}\right] = 5\frac{d}{dt}\left[\frac{t}{t^2+1}\right] \\ &= 5\left[\frac{(t^2+1)(1) - t(2t)}{(t^2+1)^2}\right] \\ &= 5\left[\frac{t^2+1-2t^2}{(t^2+1)^2}\right] = \frac{5(1-t^2)}{(t^2+1)^2} \end{aligned}$$

b. The rate at which the sales are changing at the time the DVDs are released is given by

$$S'(0) = \frac{5(1-0)}{(0+1)^2} = 5$$

That is, they are increasing at the rate of \$5 million per year.

Two years from the date of release, the sales are changing at the rate of

$$S'(2) = \frac{5(1-4)}{(4+1)^2} = -\frac{3}{5} = -0.6$$

That is, they are decreasing at the rate of \$600,000 per year.

The graph of the function S is shown in Figure 5.

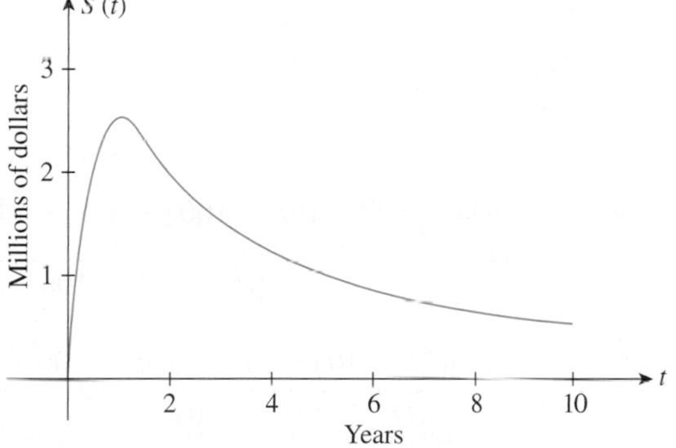

FIGURE 5
After a spectacular rise, the sales begin to taper off.

EXPLORING WITH TECHNOLOGY

Refer to Example 6.

1. Use a graphing utility to plot the graph of the function S, using the viewing window $[0, 10] \times [0, 3]$.
2. Use **TRACE** and **ZOOM** to determine the coordinates of the highest point on the graph of S in the interval $[0, 10]$. Interpret your results.

EXPLORE & DISCUSS

Suppose the revenue of a company is given by $R(x) = xp(x)$, where x is the number of units of the product sold at a unit price of $p(x)$ dollars.

1. Compute $R'(x)$ and explain, in words, the relationship between $R'(x)$ and $p(x)$ and/or its derivative.
2. What can you say about $R'(x)$ if $p(x)$ is constant? Is this expected?

APPLIED EXAMPLE 7 Oxygen-Restoration Rate in a Pond When organic waste is dumped into a pond, the oxidation process that takes place reduces the pond's oxygen content. However, given time, nature will restore the oxygen content to its natural level. Suppose the oxygen content t days after organic waste has been dumped into the pond is given by

$$f(t) = 100\left[\frac{t^2 + 10t + 100}{t^2 + 20t + 100}\right] \qquad (0 < t < \infty)$$

percent of its normal level.

a. Derive a general expression that gives the rate of change of the pond's oxygen level at any time t.
b. How fast is the pond's oxygen content changing 1 day, 10 days, and 20 days after the organic waste has been dumped?

Solution

a. The rate of change of the pond's oxygen level at any time t is given by the derivative of the function f. Thus, the required expression is

$$\begin{aligned}
f'(t) &= 100\frac{d}{dt}\left[\frac{t^2 + 10t + 100}{t^2 + 20t + 100}\right] \\
&= 100\left[\frac{(t^2 + 20t + 100)\frac{d}{dt}(t^2 + 10t + 100) - (t^2 + 10t + 100)\frac{d}{dt}(t^2 + 20t + 100)}{(t^2 + 20t + 100)^2}\right] \\
&= 100\left[\frac{(t^2 + 20t + 100)(2t + 10) - (t^2 + 10t + 100)(2t + 20)}{(t^2 + 20t + 100)^2}\right] \\
&= 100\left[\frac{2t^3 + 10t^2 + 40t^2 + 200t + 200t + 1000 - 2t^3 - 20t^2 - 20t^2 - 200t - 200t - 2000}{(t^2 + 20t + 100)^2}\right] \\
&= 100\left[\frac{10t^2 - 1000}{(t^2 + 20t + 100)^2}\right]
\end{aligned}$$

b. The rate at which the pond's oxygen content is changing 1 day after the organic waste has been dumped is given by

$$f'(1) = 100\left[\frac{10 - 1000}{(1 + 20 + 100)^2}\right] \approx -6.76$$

That is, it is dropping at the rate of 6.8% per day. After 10 days, the rate is

$$f'(10) = 100\left[\frac{10(10)^2 - 1000}{(10^2 + 20(10) + 100)^2}\right] = 0$$

That is, it is neither increasing nor decreasing. After 20 days, the rate is

$$f'(20) = 100\left[\frac{10(20)^2 - 1000}{(20^2 + 20(20) + 100)^2}\right] \approx 0.37$$

That is, the oxygen content is increasing at the rate of 0.37% per day, and the restoration process has indeed begun.

Verification of the Product Rule

We will now verify the product rule. If $p(x) = f(x)g(x)$, then

$$\begin{aligned} p'(x) &= \lim_{h\to 0} \frac{p(x+h) - p(x)}{h} \\ &= \lim_{h\to 0} \frac{f(x+h)g(x+h) - f(x)g(x)}{h} \end{aligned}$$

By adding $-f(x + h)g(x) + f(x + h)g(x)$ (which is zero!) to the numerator and factoring, we have

$$\begin{aligned} p'(x) &= \lim_{h\to 0} \frac{f(x+h)[g(x+h) - g(x)] + g(x)[f(x+h) - f(x)]}{h} \\ &= \lim_{h\to 0} \left\{ f(x+h)\left[\frac{g(x+h) - g(x)}{h}\right] + g(x)\left[\frac{f(x+h) - f(x)}{h}\right] \right\} \\ &= \lim_{h\to 0} f(x+h)\left[\frac{g(x+h) - g(x)}{h}\right] \\ &\quad + \lim_{h\to 0} g(x)\left[\frac{f(x+h) - f(x)}{h}\right] && \text{By Property 3 of limits} \\ &= \lim_{h\to 0} f(x+h) \cdot \lim_{h\to 0} \frac{g(x+h) - g(x)}{h} \\ &\quad + \lim_{h\to 0} g(x) \cdot \lim_{h\to 0} \frac{f(x+h) - f(x)}{h} && \text{By Property 4 of limits} \\ &= f(x)g'(x) + g(x)f'(x) \end{aligned}$$

Observe that in the second from the last link in the chain of equalities, we have used the fact that $\lim_{h\to 0} f(x + h) = f(x)$ because f is continuous at x.

3.2 Self-Check Exercises

1. Find the derivative of $f(x) = \dfrac{2x + 1}{x^2 - 1}$.
2. What is the slope of the tangent line to the graph of

$$f(x) = (x^2 + 1)(2x^3 - 3x^2 + 1)$$

at the point (2, 25)? How fast is the function f changing when $x = 2$?
3. The total sales of Security Products in its first 2 yr of operation are given by

$$S = f(t) = \frac{0.3t^3}{1 + 0.4t^2} \qquad (0 \le t \le 2)$$

where S is measured in millions of dollars and $t = 0$ corresponds to the date Security Products began operations. How fast were the sales increasing at the beginning of the company's second year of operation?

Solutions to Self-Check Exercises 3.2 can be found on page 182.

3.2 Concept Questions

1. State the rule of differentiation in your own words.
   a. Product rule   b. Quotient rule
2. If $f(1) = 3$, $g(1) = 2$, $f'(1) = -1$, and $g'(1) = 4$, find:
   a. $h'(1)$ if $h(x) = f(x)g(x)$ b. $F'(1)$ if $F(x) = \dfrac{f(x)}{g(x)}$

3.2 Exercises

In Exercises 1–30, find the derivative of each function.

1. $f(x) = 2x(x^2 + 1)$
2. $f(x) = 3x^2(x - 1)$
3. $f(t) = (t - 1)(2t + 1)$
4. $f(x) = (2x + 3)(3x - 4)$
5. $f(x) = (3x + 1)(x^2 - 2)$
6. $f(x) = (x + 1)(2x^2 - 3x + 1)$
7. $f(x) = (x^3 - 1)(x + 1)$
8. $f(x) = (x^3 - 12x)(3x^2 + 2x)$
9. $f(w) = (w^3 - w^2 + w - 1)(w^2 + 2)$
10. $f(x) = \frac{1}{5}x^5 + (x^2 + 1)(x^2 - x - 1) + 28$
11. $f(x) = (5x^2 + 1)(2\sqrt{x} - 1)$
12. $f(t) = (1 + \sqrt{t})(2t^2 - 3)$
13. $f(x) = (x^2 - 5x + 2)\left(x - \frac{2}{x}\right)$
14. $f(x) = (x^3 + 2x + 1)\left(2 + \frac{1}{x^2}\right)$
15. $f(x) = \dfrac{1}{x - 2}$
16. $g(x) = \dfrac{3}{2x + 4}$
17. $f(x) = \dfrac{x - 1}{2x + 1}$
18. $f(t) = \dfrac{1 - 2t}{1 + 3t}$
19. $f(x) = \dfrac{1}{x^2 + 1}$
20. $f(u) = \dfrac{u}{u^2 + 1}$
21. $f(s) = \dfrac{s^2 - 4}{s + 1}$
22. $f(x) = \dfrac{x^3 - 2}{x^2 + 1}$
23. $f(x) = \dfrac{\sqrt{x}}{x^2 + 1}$
24. $f(x) = \dfrac{x^2 + 1}{\sqrt{x}}$
25. $f(x) = \dfrac{x^2 + 2}{x^2 + x + 1}$
26. $f(x) = \dfrac{x + 1}{2x^2 + 2x + 3}$
27. $f(x) = \dfrac{(x + 1)(x^2 + 1)}{x - 2}$
28. $f(x) = (3x^2 - 1)\left(x^2 - \frac{1}{x}\right)$
29. $f(x) = \dfrac{x}{x^2 - 4} - \dfrac{x - 1}{x^2 + 4}$
30. $f(x) = \dfrac{x + \sqrt{3x}}{3x - 1}$

In Exercises 31–34, suppose f and g are functions that are differentiable at $x = 1$ and that $f(1) = 2$, $f'(1) = -1$, $g(1) = -2$, and $g'(1) = 3$. Find the value of $h'(1)$.

31. $h(x) = f(x)g(x)$
32. $h(x) = (x^2 + 1)g(x)$
33. $h(x) = \dfrac{xf(x)}{x + g(x)}$
34. $h(x) = \dfrac{f(x)g(x)}{f(x) - g(x)}$

In Exercises 35–38, find the derivative of each function and evaluate $f'(x)$ at the given value of x.

35. $f(x) = (2x - 1)(x^2 + 3)$; $x = 1$
36. $f(x) = \dfrac{2x + 1}{2x - 1}$; $x = 2$
37. $f(x) = \dfrac{x}{x^4 - 2x^2 - 1}$; $x = -1$
38. $f(x) = (\sqrt{x} + 2x)(x^{3/2} - x)$; $x = 4$

In Exercises 39–42, find the slope and an equation of the tangent line to the graph of the function f at the specified point.

39. $f(x) = (x^3 + 1)(x^2 - 2)$; $(2, 18)$
40. $f(x) = \dfrac{x^2}{x + 1}$; $\left(2, \frac{4}{3}\right)$
41. $f(x) = \dfrac{x + 1}{x^2 + 1}$; $(1, 1)$

42. $f(x) = \dfrac{1 + 2x^{1/2}}{1 + x^{3/2}}$; $\left(4, \dfrac{5}{9}\right)$

43. Find an equation of the tangent line to the graph of the function $f(x) = (x^3 + 1)(3x^2 - 4x + 2)$ at the point (1, 2).

44. Find an equation of the tangent line to the graph of the function $f(x) = \dfrac{3x}{x^2 - 2}$ at the point (2, 3).

45. Let $f(x) = (x^2 + 1)(2 - x)$. Find the point(s) on the graph of f where the tangent line is horizontal.

46. Let $f(x) = \dfrac{x}{x^2 + 1}$. Find the point(s) on the graph of f where the tangent line is horizontal.

47. Find the point(s) on the graph of the function $f(x) = (x^2 + 6)(x - 5)$ where the slope of the tangent line is equal to -2.

48. Find the point(s) on the graph of the function $f(x) = \dfrac{x + 1}{x - 1}$ where the slope of the tangent line is equal to $-\dfrac{1}{2}$.

49. A straight line perpendicular to and passing through the point of tangency of the tangent line is called the *normal* to the curve. Find the equation of the tangent line and the normal to the curve $y = \dfrac{1}{1 + x^2}$ at the point $(1, \frac{1}{2})$.

50. **Concentration of a Drug in the Bloodstream** The concentration of a certain drug in a patient's bloodstream t hr after injection is given by

$$C(t) = \frac{0.2t}{t^2 + 1}$$

a. Find the rate at which the concentration of the drug is changing with respect to time.
b. How fast is the concentration changing $\frac{1}{2}$ hr, 1 hr, and 2 hr after the injection?

51. **Cost of Removing Toxic Waste** A city's main well was recently found to be contaminated with trichloroethylene, a cancer-causing chemical, as a result of an abandoned chemical dump leaching chemicals into the water. A proposal submitted to the city's council members indicates that the cost, measured in millions of dollars, of removing x% of the toxic pollutant is given by

$$C(x) = \frac{0.5x}{100 - x}$$

Find $C'(80)$, $C'(90)$, $C'(95)$, and $C'(99)$. What does your result tell you about the cost of removing *all* of the pollutant?

52. **Drug Dosages** Thomas Young has suggested the following rule for calculating the dosage of medicine for children 1 to 12 yr old. If a denotes the adult dosage (in milligrams) and if t is the child's age (in years), then the child's dosage is given by

$$D(t) = \frac{at}{t + 12}$$

Suppose the adult dosage of a substance is 500 mg. Find an expression that gives the rate of change of a child's dosage with respect to the child's age. What is the rate of change of a child's dosage with respect to his or her age for a 6-yr-old child? A 10-yr-old child?

53. **Effect of Bactericide** The number of bacteria $N(t)$ in a certain culture t min after an experimental bactericide is introduced obeys the rule

$$N(t) = \frac{10{,}000}{1 + t^2} + 2000$$

Find the rate of change of the number of bacteria in the culture 1 min and 2 min after the bactericide is introduced. What is the population of the bacteria in the culture 1 min and 2 min after the bactericide is introduced?

54. **Demand Functions** The demand function for the Sicard wristwatch is given by

$$d(x) = \frac{50}{0.01x^2 + 1} \qquad (0 \le x \le 20)$$

where x (measured in units of a thousand) is the quantity demanded per week and $d(x)$ is the unit price in dollars.
a. Find $d'(x)$.
b. Find $d'(5)$, $d'(10)$, and $d'(15)$ and interpret your results.

55. **Learning Curves** From experience, Emory Secretarial School knows that the average student taking Advanced Typing will progress according to the rule

$$N(t) = \frac{60t + 180}{t + 6} \qquad (t \ge 0)$$

where $N(t)$ measures the number of words/minute the student can type after t wk in the course.
a. Find an expression for $N'(t)$.
b. Compute $N'(t)$ for $t = 1$, 3, 4, and 7 and interpret your results.
c. Sketch the graph of the function N. Does it confirm the results obtained in part (b)?
d. What will be the average student's typing speed at the end of the 12-wk course?

56. **Box-Office Receipts** The total worldwide box-office receipts for a long-running movie are approximated by the function

$$T(x) = \frac{120x^2}{x^2 + 4}$$

where $T(x)$ is measured in millions of dollars and x is the number of years since the movie's release. How fast are the total receipts changing 1 yr, 3 yr, and 5 yr after its release?

57. **Formaldehyde Levels** A study on formaldehyde levels in 900 homes indicates that emissions of various chemicals can decrease over time. The formaldehyde level (parts per million) in an average home in the study is given by

$$f(t) = \frac{0.055t + 0.26}{t + 2} \qquad (0 \le t \le 12)$$

where t is the age of the house in years. How fast is the formaldehyde level of the average house dropping when it is new? At the beginning of its fourth year?

Source: Bonneville Power Administration

58. **POPULATION GROWTH** A major corporation is building a 4325-acre complex of homes, offices, stores, schools, and churches in the rural community of Glen Cove. As a result of this development, the planners have estimated that Glen Cove's population (in thousands) t yr from now will be given by

$$P(t) = \frac{25t^2 + 125t + 200}{t^2 + 5t + 40}$$

a. Find the rate at which Glen Cove's population is changing with respect to time.
b. What will be the population after 10 yr? At what rate will the population be increasing when $t = 10$?

In Exercises 59–62, determine whether the statement is true or false. If it is true, explain why it is true. If it is false, give an example to show why it is false.

59. If f and g are differentiable, then

$$\frac{d}{dx}[f(x)g(x)] = f'(x)g'(x)$$

60. If f is differentiable, then

$$\frac{d}{dx}[xf(x)] = f(x) + xf'(x)$$

61. If f is differentiable, then

$$\frac{d}{dx}\left[\frac{f(x)}{x^2}\right] = \frac{f'(x)}{2x}$$

62. If f, g, and h are differentiable, then

$$\frac{d}{dx}\left[\frac{f(x)g(x)}{h(x)}\right] = \frac{f'(x)g(x)h(x) + f(x)g'(x)h(x) - f(x)g(x)h'(x)}{[h(x)]^2}$$

63. Extend the product rule for differentiation to the following case involving the product of three differentiable functions: Let $h(x) = u(x)v(x)w(x)$ and show that $h'(x) = u(x)v(x)w'(x) + u(x)v'(x)w(x) + u'(x)v(x)w(x)$.
Hint: Let $f(x) = u(x)v(x)$, $g(x) = w(x)$, and $h(x) = f(x)g(x)$ and apply the product rule to the function h.

64. Prove the quotient rule for differentiation (Rule 6).
Hint: Let $k(x) = f(x)/g(x)$ and verify the following steps:

a. $$\frac{k(x+h) - k(x)}{h} = \frac{f(x+h)g(x) - f(x)g(x+h)}{hg(x+h)g(x)}$$

b. By adding $[-f(x)g(x) + f(x)g(x)]$ to the numerator and simplifying, show that

$$\frac{k(x+h) - k(x)}{h} = \frac{1}{g(x+h)g(x)} \times \left\{\left[\frac{f(x+h) - f(x)}{h}\right]\cdot g(x) - \left[\frac{g(x+h) - g(x)}{h}\right]\cdot f(x)\right\}$$

c. $$k'(x) = \lim_{h \to 0} \frac{k(x+h) - k(x)}{h} = \frac{g(x)f'(x) - f(x)g'(x)}{[g(x)]^2}$$

3.2 Solutions to Self-Check Exercises

1. We use the quotient rule to obtain

$$\begin{aligned} f'(x) &= \frac{(x^2-1)\frac{d}{dx}(2x+1) - (2x+1)\frac{d}{dx}(x^2-1)}{(x^2-1)^2} \\ &= \frac{(x^2-1)(2) - (2x+1)(2x)}{(x^2-1)^2} \\ &= \frac{2x^2 - 2 - 4x^2 - 2x}{(x^2-1)^2} \\ &= \frac{-2x^2 - 2x - 2}{(x^2-1)^2} \\ &= \frac{-2(x^2 + x + 1)}{(x^2-1)^2} \end{aligned}$$

2. The slope of the tangent line to the graph of f at any point is given by

$$\begin{aligned} f'(x) &= (x^2+1)\frac{d}{dx}(2x^3 - 3x^2 + 1) \\ &\quad + (2x^3 - 3x^2 + 1)\frac{d}{dx}(x^2+1) \\ &= (x^2+1)(6x^2 - 6x) + (2x^3 - 3x^2 + 1)(2x) \end{aligned}$$

In particular, the slope of the tangent line to the graph of f when $x = 2$ is

$$\begin{aligned} f'(2) &= (2^2+1)[6(2^2) - 6(2)] \\ &\quad + [2(2^3) - 3(2^2) + 1][2(2)] \\ &= 60 + 20 = 80 \end{aligned}$$

Note that it is not necessary to simplify the expression for $f'(x)$ since we are required only to evaluate the expression at $x = 2$. We also conclude, from this result, that the function f is changing at the rate of 80 units/unit change in x when $x = 2$.

3. The rate at which the company's total sales are changing at any time t is given by

$$S'(t) = \frac{(1 + 0.4t^2)\frac{d}{dt}(0.3t^3) - (0.3t^3)\frac{d}{dt}(1 + 0.4t^2)}{(1 + 0.4t^2)^2}$$

$$= \frac{(1 + 0.4t^2)(0.9t^2) - (0.3t^3)(0.8t)}{(1 + 0.4t^2)^2}$$

Therefore, at the beginning of the second year of operation, Security Products' sales were increasing at the rate of

$$S'(1) = \frac{(1 + 0.4)(0.9) - (0.3)(0.8)}{(1 + 0.4)^2} = 0.520408$$

or $520,408/year.

USING TECHNOLOGY

The Product and Quotient Rules

EXAMPLE 1 Let $f(x) = (2\sqrt{x} + 0.5x)\left(0.3x^3 + 2x - \frac{0.3}{x}\right)$. Find $f'(0.2)$.

Solution Using the numerical derivative operation of a graphing utility, we find

$$f'(0.2) = 6.4797499802$$

See Figure T1.

```
nDeriv((2X^.5+.5
X)(.3X^3+2X-.3/X),
X,.2)
        6.4797499802
```

FIGURE T1
The TI-83 numerical derivative screen for computing $f'(0.2)$

APPLIED EXAMPLE 2 Importance of Time in Treating Heart Attacks According to the American Heart Association, the treatment benefit for heart attacks depends on the time until treatment and is described by the function

$$f(t) = \frac{-16.94t + 203.28}{t + 2.0328} \qquad (0 \le t \le 12)$$

where t is measured in hours and $f(t)$ is expressed as a percent.

a. Use a graphing utility to graph the function f using the viewing window $[0, 13] \times [0, 120]$.
b. Use a graphing utility to find the derivative of f when $t = 0$ and $t = 2$.
c. Interpret the results obtained in part (b).

Source: American Heart Association

Solution

a. The graph of f is shown in Figure T2.

FIGURE T2

(continued)

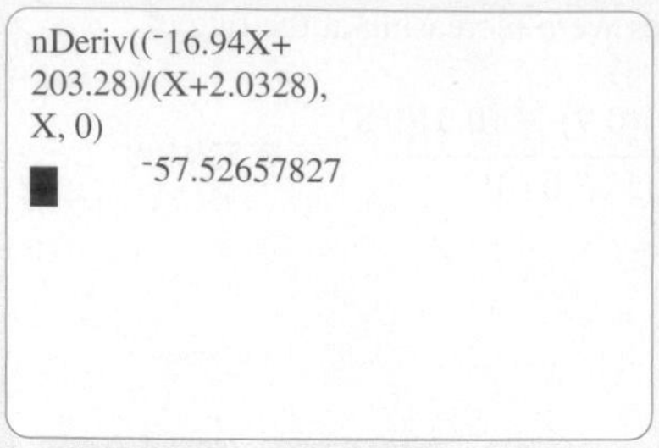

```
nDeriv((-16.94X+
203.28)/(X+2.0328),
X, 0)
        -57.52657827
```

FIGURE T3
The TI-83 numerical derivative screen for computing $f'(0)$

b. Using the numerical derivative operation of a graphing utility, we find

$$f'(0) \approx -57.5266$$
$$f'(2) \approx -14.6165$$

(see Figure T3).

c. The results of part (b) show that the treatment benefit drops off at the rate of 58% per hour at the time when the heart attack first occurs and falls off at the rate of 15% per hour when the time to treatment is 2 hours. Thus, it is extremely urgent that a patient suffering a heart attack receive medical attention as soon as possible. ■

TECHNOLOGY EXERCISES

In Exercises 1–6, use the numerical derivative operation to find the rate of change of $f(x)$ at the given value of x. Give your answer accurate to four decimal places.

1. $f(x) = (2x^2 + 1)(x^3 + 3x + 4)$; $x = -0.5$
2. $f(x) = (\sqrt{x} + 1)(2x^2 + x - 3)$; $x = 1.5$
3. $f(x) = \dfrac{\sqrt{x} - 1}{\sqrt{x} + 1}$; $x = 3$
4. $f(x) = \dfrac{\sqrt{x}(x^2 + 4)}{x^3 + 1}$; $x = 4$
5. $f(x) = \dfrac{\sqrt{x}(1 + x^{-1})}{x + 1}$; $x = 1$
6. $f(x) = \dfrac{x^2(2 + \sqrt{x})}{1 + \sqrt{x}}$; $x = 1$
7. New Construction Jobs The president of a major housing construction company claims that the number of construction jobs created in the next t mo is given by

$$f(t) = 1.42\left(\frac{7t^2 + 140t + 700}{3t^2 + 80t + 550}\right)$$

where $f(t)$ is measured in millions of jobs/year. At what rate will construction jobs be created 1 yr from now, assuming her projection is correct?

8. Population Growth A major corporation is building a 4325-acre complex of homes, offices, stores, schools, and churches in the rural community of Glen Cove. As a result of this development, the planners have estimated that Glen Cove's population (in thousands) t yr from now will be given by

$$P(t) = \frac{25t^2 + 125t + 200}{t^2 + 5t + 40}$$

a. What will be the population 10 yr from now?
b. At what rate will the population be increasing 10 yr from now?

3.3 The Chain Rule

This section introduces another rule of differentiation called the **chain rule.** When used in conjunction with the rules of differentiation developed in the last two sections, the chain rule enables us to greatly enlarge the class of functions that we are able to differentiate.

The Chain Rule

Consider the function $h(x) = (x^2 + x + 1)^2$. If we were to compute $h'(x)$ using only the rules of differentiation from the previous sections, then our approach might be to expand $h(x)$. Thus,

$$\begin{aligned} h(x) &= (x^2 + x + 1)^2 = (x^2 + x + 1)(x^2 + x + 1) \\ &= x^4 + 2x^3 + 3x^2 + 2x + 1 \end{aligned}$$

from which we find

$$h'(x) = 4x^3 + 6x^2 + 6x + 2$$

But what about the function $H(x) = (x^2 + x + 1)^{100}$? The same technique may be used to find the derivative of the function H, but the amount of work involved in this case would be prodigious! Consider, also, the function $G(x) = \sqrt{x^2 + 1}$. For each of the two functions H and G, the rules of differentiation of the previous sections cannot be applied directly to compute the derivatives H' and G'.

Observe that both H and G are **composite functions**; that is, each is composed of, or built up from, simpler functions. For example, the function H is composed of the two simpler functions $f(x) = x^2 + x + 1$ and $g(x) = x^{100}$ as follows:

$$\begin{aligned} H(x) &= g[f(x)] = [f(x)]^{100} \\ &= (x^2 + x + 1)^{100} \end{aligned}$$

In a similar manner, we see that the function G is composed of the two simpler functions $f(x) = x^2 + 1$ and $g(x) = \sqrt{x}$. Thus,

$$\begin{aligned} G(x) &= g[f(x)] = \sqrt{f(x)} \\ &= \sqrt{x^2 + 1} \end{aligned}$$

As a first step toward finding the derivative h' of a composite function $h = g \circ f$ defined by $h(x) = g[f(x)]$, we write

$$u = f(x) \quad \text{and} \quad y = g[f(x)] = g(u)$$

The dependency of h on g and f is illustrated in Figure 6. Since u is a function of x, we may compute the derivative of u with respect to x, if f is a differentiable function, obtaining $du/dx = f'(x)$. Next, if g is a differentiable function of u, we may compute the derivative of g with respect to u, obtaining $dy/du = g'(u)$. Now, since the function h is composed of the function g and the function f, we might suspect that the rule $h'(x)$ for the derivative h' of h will be given by an expression that involves the rules for the derivatives of f and g. But how do we combine these derivatives to yield h'?

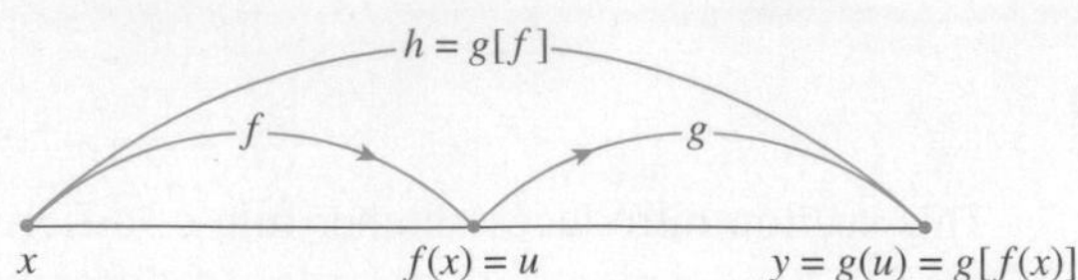

FIGURE 6
The composite function $h(x) = g[f(x)]$

This question can be answered by interpreting the derivative of each function as giving the rate of change of that function. For example, suppose $u = f(x)$ changes three times as fast as x—that is,

$$f'(x) = \frac{du}{dx} = 3$$

And suppose $y = g(u)$ changes twice as fast as u—that is,

$$g'(u) = \frac{dy}{du} = 2$$

Then, we would expect $y = h(x)$ to change six times as fast as x—that is,

$$h'(x) = g'(u)f'(x) = (2)(3) = 6$$

or equivalently,

$$\frac{dy}{dx} = \frac{dy}{du} \cdot \frac{du}{dx} = (2)(3) = 6$$

This observation suggests the following result, which we state without proof.

Rule 7: The Chain Rule
If $h(x) = g[f(x)]$, then

$$h'(x) = \frac{d}{dx}g(f(x)) = g'(f(x))f'(x) \quad \textbf{(1)}$$

Equivalently, if we write $y = h(x) = g(u)$, where $u = f(x)$, then

$$\frac{dy}{dx} = \frac{dy}{du} \cdot \frac{du}{dx} \quad \textbf{(2)}$$

Notes

1. If we label the composite function h in the following manner

   Inside function ↓
   $$h(x) = g[f(x)]$$
   ↑ Outside function

   then $h'(x)$ is just the *derivative* of the "outside function" *evaluated at* the "inside function" times the *derivative* of the "inside function."
2. Equation (2) can be remembered by observing that if we "cancel" the du's, then

$$\frac{dy}{dx} = \frac{dy}{\cancel{du}} \cdot \frac{\cancel{du}}{dx} = \frac{dy}{dx}$$

■

The Chain Rule for Powers of Functions

Many composite functions have the special form $h(x) = g(f(x))$, where g is defined by the rule $g(x) = x^n$ (n, a real number)—that is,

$$h(x) = [f(x)]^n$$

In other words, the function h is given by the power of a function f. The functions

$$h(x) = (x^2 + x + 1)^2 \qquad H = (x^2 + x + 1)^{100} \qquad G = \sqrt{x^2 + 1}$$

discussed earlier are examples of this type of composite function. By using the following corollary of the chain rule, the general power rule, we can find the derivative of this type of function much more easily than by using the chain rule directly.

The General Power Rule

If the function f is differentiable and $h(x) = [f(x)]^n$ (n, a real number), then

$$h'(x) = \frac{d}{dx}[f(x)]^n = n[f(x)]^{n-1}f'(x) \tag{3}$$

To see this, we observe that $h(x) = g(f(x))$, where $g(x) = x^n$, so that, by virtue of the chain rule, we have

$$\begin{aligned} h'(x) &= g'(f(x))f'(x) \\ &= n[f(x)]^{n-1}f'(x) \end{aligned}$$

since $g'(x) = nx^{n-1}$.

EXAMPLE 1 Let $F(x) = (3x + 1)^2$.

a. Find $F'(x)$, using the general power rule.
b. Verify your result without the benefit of the general power rule.

Solution

a. Using the general power rule, we obtain

$$\begin{aligned} F'(x) &= 2(3x + 1)^1\frac{d}{dx}(3x + 1) \\ &= 2(3x + 1)(3) \\ &= 6(3x + 1) \end{aligned}$$

b. We first expand $F(x)$. Thus,

$$F(x) = (3x + 1)^2 = 9x^2 + 6x + 1$$

Next, differentiating, we have

$$\begin{aligned} F'(x) &= \frac{d}{dx}(9x^2 + 6x + 1) \\ &= 18x + 6 \\ &= 6(3x + 1) \end{aligned}$$

as before. ■

EXAMPLE 2 Differentiate the function $G(x) = \sqrt{x^2 + 1}$.

Solution We rewrite the function $G(x)$ as

$$G(x) = (x^2 + 1)^{1/2}$$

and apply the general power rule, obtaining

$$\begin{aligned} G'(x) &= \frac{1}{2}(x^2 + 1)^{-1/2}\frac{d}{dx}(x^2 + 1) \\ &= \frac{1}{2}(x^2 + 1)^{-1/2} \cdot 2x = \frac{x}{\sqrt{x^2 + 1}} \end{aligned}$$

VIDEO

EXAMPLE 3 Differentiate the function $f(x) = x^2(2x + 3)^5$.

Solution Applying the product rule followed by the general power rule, we obtain

$$\begin{aligned} f'(x) &= x^2\frac{d}{dx}(2x + 3)^5 + (2x + 3)^5\frac{d}{dx}(x^2) \\ &= (x^2)5(2x + 3)^4 \cdot \frac{d}{dx}(2x + 3) + (2x + 3)^5(2x) \\ &= 5x^2(2x + 3)^4(2) + 2x(2x + 3)^5 \\ &= 2x(2x + 3)^4(5x + 2x + 3) = 2x(7x + 3)(2x + 3)^4 \end{aligned}$$

EXAMPLE 4 Find $f'(x)$ if $f(x) = (2x^2 + 3)^4(3x - 1)^5$.

Solution Applying the product rule, we have

$$f'(x) = (2x^2 + 3)^4\frac{d}{dx}(3x - 1)^5 + (3x - 1)^5\frac{d}{dx}(2x^2 + 3)^4$$

Next, we apply the general power rule to each term, obtaining

$$\begin{aligned} f'(x) &= (2x^2 + 3)^4 \cdot 5(3x - 1)^4\frac{d}{dx}(3x - 1) + (3x - 1)^5 \cdot 4(2x^2 + 3)^3\frac{d}{dx}(2x^2 + 3) \\ &= 5(2x^2 + 3)^4(3x - 1)^4 \cdot 3 + 4(3x - 1)^5(2x^2 + 3)^3(4x) \end{aligned}$$

Finally, observing that $(2x^2 + 3)^3(3x - 1)^4$ is common to both terms, we can factor and simplify as follows:

$$\begin{aligned} f'(x) &= (2x^2 + 3)^3(3x - 1)^4[15(2x^2 + 3) + 16x(3x - 1)] \\ &= (2x^2 + 3)^3(3x - 1)^4(30x^2 + 45 + 48x^2 - 16x) \\ &= (2x^2 + 3)^3(3x - 1)^4(78x^2 - 16x + 45) \end{aligned}$$

EXAMPLE 5 Find $f'(x)$ if $f(x) = \dfrac{1}{(4x^2 - 7)^2}$.

Solution Rewriting $f(x)$ and then applying the general power rule, we obtain

$$\begin{aligned} f'(x) &= \frac{d}{dx}\left[\frac{1}{(4x^2 - 7)^2}\right] = \frac{d}{dx}(4x^2 - 7)^{-2} \\ &= -2(4x^2 - 7)^{-3}\frac{d}{dx}(4x^2 - 7) \\ &= -2(4x^2 - 7)^{-3}(8x) = -\frac{16x}{(4x^2 - 7)^3} \end{aligned}$$

EXAMPLE 6 Find the slope of the tangent line to the graph of the function

$$f(x) = \left(\frac{2x + 1}{3x + 2}\right)^3$$

at the point $\left(0, \frac{1}{8}\right)$.

Solution The slope of the tangent line to the graph of f at any point is given by $f'(x)$. To compute $f'(x)$, we use the general power rule followed by the quotient rule, obtaining

$$\begin{aligned} f'(x) &= 3\left(\frac{2x + 1}{3x + 2}\right)^2 \frac{d}{dx}\left(\frac{2x + 1}{3x + 2}\right) \\ &= 3\left(\frac{2x + 1}{3x + 2}\right)^2 \left[\frac{(3x + 2)(2) - (2x + 1)(3)}{(3x + 2)^2}\right] \\ &= 3\left(\frac{2x + 1}{3x + 2}\right)^2 \left[\frac{6x + 4 - 6x - 3}{(3x + 2)^2}\right] \\ &= \frac{3(2x + 1)^2}{(3x + 2)^4} \end{aligned}$$

In particular, the slope of the tangent line to the graph of f at $\left(0, \frac{1}{8}\right)$ is given by

$$f'(0) = \frac{3(0 + 1)^2}{(0 + 2)^4} = \frac{3}{16}$$

EXPLORING WITH TECHNOLOGY

Refer to Example 6.

1. Use a graphing utility to plot the graph of the function f, using the viewing window $[-2, 1] \times [-1, 2]$. Then draw the tangent line to the graph of f at the point $(0, \frac{1}{8})$.
2. For a better picture, repeat part 1 using the viewing window $[-1, 1] \times [-0.1, 0.3]$.
3. Use the numerical differentiation capability of the graphing utility to verify that the slope of the tangent line at $(0, \frac{1}{8})$ is $\frac{3}{16}$.

VIDEO

APPLIED EXAMPLE 7 Growth in a Health Club Membership The membership of The Fitness Center, which opened a few years ago, is approximated by the function

$$N(t) = 100(64 + 4t)^{2/3} \qquad (0 \le t \le 52)$$

where $N(t)$ gives the number of members at the beginning of week t.

a. Find $N'(t)$.
b. How fast was the center's membership increasing initially ($t = 0$)?
c. How fast was the membership increasing at the beginning of the 40th week?
d. What was the membership when the center first opened? At the beginning of the 40th week?

Solution

a. Using the general power rule, we obtain

$$\begin{aligned}
N'(t) &= \frac{d}{dt}[100(64 + 4t)^{2/3}] \\
&= 100\frac{d}{dt}(64 + 4t)^{2/3} \\
&= 100\left(\frac{2}{3}\right)(64 + 4t)^{-1/3}\frac{d}{dt}(64 + 4t) \\
&= \frac{200}{3}(64 + 4t)^{-1/3}(4) \\
&= \frac{800}{3(64 + 4t)^{1/3}}
\end{aligned}$$

b. The rate at which the membership was increasing when the center first opened is given by

$$N'(0) = \frac{800}{3(64)^{1/3}} \approx 66.7$$

or approximately 67 people per week.

c. The rate at which the membership was increasing at the beginning of the 40th week is given by

$$N'(40) = \frac{800}{3(64 + 160)^{1/3}} \approx 43.9$$

or approximately 44 people per week.

d. The membership when the center first opened is given by

$$N(0) = 100(64)^{2/3} = 100(16)$$

or approximately 1600 people. The membership at the beginning of the 40th week is given by

$$N(40) = 100(64 + 160)^{2/3} \approx 3688.3$$

or approximately 3688 people. ■

EXPLORE & DISCUSS

The profit P of a one-product software manufacturer depends on the number of units of its products sold. The manufacturer estimates that it will sell x units of its product per week. Suppose $P = g(x)$ and $x = f(t)$, where g and f are differentiable functions.

1. Write an expression giving the rate of change of the profit with respect to the number of units sold.
2. Write an expression giving the rate of change of the number of units sold per week.
3. Write an expression giving the rate of change of the profit per week.

APPLIED EXAMPLE 8 Arteriosclerosis Arteriosclerosis begins during childhood when plaque (soft masses of fatty material) forms in the arterial walls, blocking the flow of blood through the arteries and leading to heart attacks, strokes, and gangrene. Suppose the idealized cross section of the aorta is

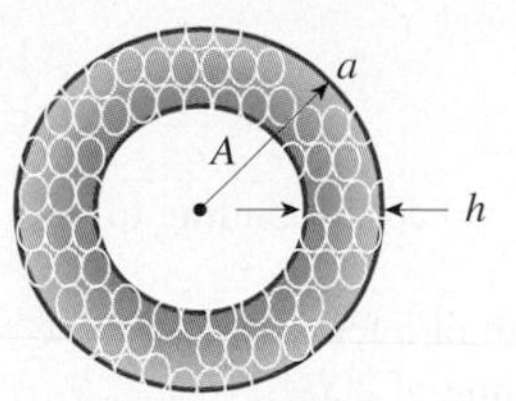

FIGURE 7
Cross section of the aorta

circular with radius a cm and by year t the thickness of the plaque (assume it is uniform) is $h = g(t)$ cm (Figure 7). Then the area of the opening is given by $A = \pi(a - h)^2$ square centimeters (cm^2).

Suppose the radius of an individual's artery is 1 cm ($a = 1$) and the thickness of the plaque in year t is given by

$$h = g(t) = 1 - 0.01(10{,}000 - t^2)^{1/2} \text{ cm}$$

Since the area of the arterial opening is given by

$$A = f(h) = \pi(1 - h)^2$$

the rate at which A is changing with respect to time is given by

$$\begin{aligned}
\frac{dA}{dt} &= \frac{dA}{dh} \cdot \frac{dh}{dt} = f'(h) \cdot g'(t) && \text{By the chain rule} \\
&= 2\pi(1 - h)(-1)\left[-0.01\left(\frac{1}{2}\right)(10{,}000 - t^2)^{-1/2}(-2t)\right] && \text{Use the chain rule twice.} \\
&= -2\pi(1 - h)\left[\frac{0.01t}{(10{,}000 - t^2)^{1/2}}\right] \\
&= -\frac{0.02\pi(1 - h)t}{\sqrt{10{,}000 - t^2}}
\end{aligned}$$

For example, when $t = 50$,

$$h = g(50) = 1 - 0.01(10{,}000 - 2500)^{1/2} \approx 0.134$$

so that

$$\frac{dA}{dt} = -\frac{0.02\pi(1 - 0.134)50}{\sqrt{10{,}000 - 2500}} \approx -0.03$$

That is, the area of the arterial opening is decreasing at the rate of 0.03 cm^2 per year. ■

EXPLORE & DISCUSS

Suppose the population P of a certain bacteria culture is given by $P = f(T)$, where T is the temperature of the medium. Further, suppose the temperature T is a function of time t in seconds—that is, $T = g(t)$. Give an interpretation of each of the following quantities:

1. $\dfrac{dP}{dT}$ **2.** $\dfrac{dT}{dt}$ **3.** $\dfrac{dP}{dt}$ **4.** $(f \circ g)(t)$ **5.** $f'(g(t))g'(t)$

3.3 Self-Check Exercises

1. Find the derivative of

$$f(x) = -\frac{1}{\sqrt{2x^2 - 1}}$$

2. Suppose the life expectancy at birth (in years) of a female in a certain country is described by the function

$$g(t) = 50.02(1 + 1.09t)^{0.1} \qquad (0 \le t \le 150)$$

where t is measured in years, with $t = 0$ corresponding to the beginning of 1900.

a. What is the life expectancy at birth of a female born at the beginning of 1980? At the beginning of 2000?

b. How fast is the life expectancy at birth of a female born at any time t changing?

Solutions to Self-Check Exercises 3.3 can be found on page 196.

3.3 Concept Questions

1. In your own words, state the chain rule for differentiating the composite function $h(x) = g[f(x)]$.

2. State the general power rule for differentiating the function $h(x) = [f(x)]^n$, where n is a real number.

3.3 Exercises

In Exercises 1–48, find the derivative of each function.

1. $f(x) = (2x - 1)^4$
2. $f(x) = (1 - x)^3$
3. $f(x) = (x^2 + 2)^5$
4. $f(t) = 2(t^3 - 1)^5$
5. $f(x) = (2x - x^2)^3$
6. $f(x) = 3(x^3 - x)^4$
7. $f(x) = (2x + 1)^{-2}$
8. $f(t) = \frac{1}{2}(2t^2 + t)^{-3}$
9. $f(x) = (x^2 - 4)^{3/2}$
10. $f(t) = (3t^2 - 2t + 1)^{3/2}$
11. $f(x) = \sqrt{3x - 2}$
12. $f(t) = \sqrt{3t^2 - t}$
13. $f(x) = \sqrt[3]{1 - x^2}$
14. $f(x) = \sqrt{2x^2 - 2x + 3}$
15. $f(x) = \dfrac{1}{(2x + 3)^3}$
16. $f(x) = \dfrac{2}{(x^2 - 1)^4}$
17. $f(t) = \dfrac{1}{\sqrt{2t - 3}}$
18. $f(x) = \dfrac{1}{\sqrt{2x^2 - 1}}$
19. $y = \dfrac{1}{(4x^4 + x)^{3/2}}$
20. $f(t) = \dfrac{4}{\sqrt[3]{2t^2 + t}}$
21. $f(x) = (3x^2 + 2x + 1)^{-2}$
22. $f(t) = (5t^3 + 2t^2 - t + 4)^{-3}$
23. $f(x) = (x^2 + 1)^3 - (x^3 + 1)^2$
24. $f(t) = (2t - 1)^4 + (2t + 1)^4$
25. $f(t) = (t^{-1} - t^{-2})^3$
26. $f(v) = (v^{-3} + 4v^{-2})^3$
27. $f(x) = \sqrt{x + 1} + \sqrt{x - 1}$
28. $f(u) = (2u + 1)^{3/2} + (u^2 - 1)^{-3/2}$
29. $f(x) = 2x^2(3 - 4x)^4$
30. $h(t) = t^2(3t + 4)^3$
31. $f(x) = (x - 1)^2(2x + 1)^4$
32. $g(u) = (1 + u^2)^5(1 - 2u^2)^8$
33. $f(x) = \left(\dfrac{x + 3}{x - 2}\right)^3$
34. $f(x) = \left(\dfrac{x + 1}{x - 1}\right)^5$
35. $s(t) = \left(\dfrac{t}{2t + 1}\right)^{3/2}$
36. $g(s) = \left(s^2 + \dfrac{1}{s}\right)^{3/2}$
37. $g(u) = \sqrt{\dfrac{u + 1}{3u + 2}}$
38. $g(x) = \sqrt{\dfrac{2x + 1}{2x - 1}}$
39. $f(x) = \dfrac{x^2}{(x^2 - 1)^4}$
40. $g(u) = \dfrac{2u^2}{(u^2 + u)^3}$
41. $h(x) = \dfrac{(3x^2 + 1)^3}{(x^2 - 1)^4}$
42. $g(t) = \dfrac{(2t - 1)^2}{(3t + 2)^4}$
43. $f(x) = \dfrac{\sqrt{2x + 1}}{x^2 - 1}$
44. $f(t) = \dfrac{4t^2}{\sqrt{2t^2 + 2t - 1}}$
45. $g(t) = \dfrac{\sqrt{t + 1}}{\sqrt{t^2 + 1}}$
46. $f(x) = \dfrac{\sqrt{x^2 + 1}}{\sqrt{x^2 - 1}}$
47. $f(x) = (3x + 1)^4(x^2 - x + 1)^3$
48. $g(t) = (2t + 3)^2(3t^2 - 1)^{-3}$

In Exercises 49–54, find $\dfrac{dy}{du}$, $\dfrac{du}{dx}$, and $\dfrac{dy}{dx}$.

49. $y = u^{4/3}$ and $u = 3x^2 - 1$
50. $y = \sqrt{u}$ and $u = 7x - 2x^2$

51. $y = u^{-2/3}$ and $u = 2x^3 - x + 1$

52. $y = 2u^2 + 1$ and $u = x^2 + 1$

53. $y = \sqrt{u} + \dfrac{1}{\sqrt{u}}$ and $u = x^3 - x$

54. $y = \dfrac{1}{u}$ and $u = \sqrt{x} + 1$

55. Suppose $F(x) = g(f(x))$ and $f(2) = 3, f'(2) = -3, g(3) = 5$, and $g'(3) = 4$. Find $F'(2)$.

56. Suppose $h = f \circ g$. Find $h'(0)$ given that $f(0) = 6, f'(5) = -2$, $g(0) = 5$, and $g'(0) = 3$.

57. Suppose $F(x) = f(x^2 + 1)$. Find $F'(1)$ if $f'(2) = 3$.

58. Let $F(x) = f(f(x))$. Does it follow that $F'(x) = [f'(x)]^2$?
Hint: Let $f(x) = x^2$.

59. Suppose $h = g \circ f$. Does it follow that $h' = g' \circ f'$?
Hint: Let $f(x) = x$ and $g(x) = x^2$.

60. Suppose $h = f \circ g$. Show that $h' = (f' \circ g)g'$.

In Exercises 61–64, find an equation of the tangent line to the graph of the function at the given point.

61. $f(x) = (1 - x)(x^2 - 1)^2$; $(2, -9)$

62. $f(x) = \left(\dfrac{x + 1}{x - 1}\right)^2$, $(3, 4)$

63. $f(x) = x\sqrt{2x^2 + 7}$; $(3, 15)$

64. $f(x) = \dfrac{8}{\sqrt{x^2 + 6x}}$; $(2, 2)$

65. TELEVISION VIEWING The number of viewers of a television series introduced several years ago is approximated by the function

$$N(t) = (60 + 2t)^{2/3} \qquad (1 \le t \le 26)$$

where $N(t)$ (measured in millions) denotes the number of weekly viewers of the series in the tth week. Find the rate of increase of the weekly audience at the end of week 2 and at the end of week 12. How many viewers were there in week 2? In week 24?

66. OUTSOURCING OF JOBS According to a study conducted in 2003, the total number of U.S. jobs that are projected to leave the country by year t, where $t = 0$ corresponds to 2000, is

$$N(t) = 0.0018425(t + 5)^{2.5} \qquad (0 \le t \le 15)$$

where $N(t)$ is measured in millions. How fast will the number of U.S. jobs that are outsourced be changing in 2005? In 2010 ($t = 10$)?
Source: Forrester Research

67. WORKING MOTHERS The percent of mothers who work outside the home and have children younger than age 6 yr is approximated by the function

$$P(t) = 33.55(t + 5)^{0.205} \qquad (0 \le t \le 21)$$

where t is measured in years, with $t = 0$ corresponding to the beginning of 1980. Compute $P'(t)$. At what rate was the percent of these mothers changing at the beginning of 2000? What was the percent of these mothers at the beginning of 2000?
Source: U.S. Bureau of Labor Statistics

68. SELLING PRICE OF DVD RECORDERS The rise of digital music and the improvement to the DVD format are some of the reasons why the average selling price of standalone DVD recorders will drop in the coming years. The function

$$A(t) = \frac{699}{(t + 1)^{0.94}} \qquad (0 \le t \le 5)$$

gives the projected average selling price (in dollars) of standalone DVD recorders in year t, where $t = 0$ corresponds to the beginning of 2002. How fast was the average selling price of standalone DVD recorders falling at the beginning of 2002? How fast was it falling at the beginning of 2006?
Source: Consumer Electronics Association

69. SOCIALLY RESPONSIBLE FUNDS Since its inception in 1971, socially responsible investments, or SRIs, have yielded returns to investors on par with investments in general. The assets of socially responsible funds (in billions of dollars) from 1991 through 2001 is given by

$$f(t) = 23.7(0.2t + 1)^{1.32} \qquad (0 \le t \le 11)$$

where $t = 0$ corresponds to the beginning of 1991.

a. Find the rate at which the assets of SRIs were changing at the beginning of 2000.
b. What were the assets of SRIs at the beginning of 2000?

Source: Thomson Financial Wiesenberger

70. AGING POPULATION The population of Americans age 55 and over as a percent of the total population is approximated by the function

$$f(t) = 10.72(0.9t + 10)^{0.3} \qquad (0 \le t \le 20)$$

where t is measured in years, with $t = 0$ corresponding to the year 2000. At what rate was the percent of Americans age 55 and over changing at the beginning of 2000? At what rate will the percent of Americans age 55 and over be changing in 2010? What will be the percent of the population of Americans age 55 and over in 2010?
Source: U.S. Census Bureau

71. CONCENTRATION OF CARBON MONOXIDE (CO) IN THE AIR According to a joint study conducted by Oxnard's Environmental Management Department and a state government agency, the concentration of CO in the air due to automobile exhaust t yr from now is given by

$$C(t) = 0.01(0.2t^2 + 4t + 64)^{2/3}$$

parts per million.

a. Find the rate at which the level of CO is changing with respect to time.
b. Find the rate at which the level of CO will be changing 5 yr from now.

72. CONTINUING EDUCATION ENROLLMENT The registrar of Kellogg University estimates that the total student enrollment in the Continuing Education division will be given by

$$N(t) = -\frac{20,000}{\sqrt{1 + 0.2t}} + 21,000$$

where $N(t)$ denotes the number of students enrolled in the division t yr from now. Find an expression for $N'(t)$. How fast is the student enrollment increasing currently? How fast will it be increasing 5 yr from now?

73. AIR POLLUTION According to the South Coast Air Quality Management District, the level of nitrogen dioxide, a brown gas that impairs breathing, present in the atmosphere on a certain May day in downtown Los Angeles is approximated by

$$A(t) = 0.03t^3(t - 7)^4 + 60.2 \qquad (0 \leq t \leq 7)$$

where $A(t)$ is measured in pollutant standard index and t is measured in hours, with $t = 0$ corresponding to 7 a.m.
a. Find $A'(t)$.
b. Find $A'(1)$, $A'(3)$, and $A'(4)$ and interpret your results.

74. EFFECT OF LUXURY TAX ON CONSUMPTION Government economists of a developing country determined that the purchase of imported perfume is related to a proposed "luxury tax" by the formula

$$N(x) = \sqrt{10,000 - 40x - 0.02x^2} \qquad (0 \leq x \leq 200)$$

where $N(x)$ measures the percentage of normal consumption of perfume when a "luxury tax" of x% is imposed on it. Find the rate of change of $N(x)$ for taxes of 10%, 100%, and 150%.

75. PULSE RATE OF AN ATHLETE The pulse rate (the number of heartbeats/minute) of a long-distance runner t sec after leaving the starting line is given by

$$P(t) = \frac{300\sqrt{\frac{1}{2}t^2 + 2t + 25}}{t + 25} \qquad (t \geq 0)$$

Compute $P'(t)$. How fast is the athlete's pulse rate increasing 10 sec, 60 sec, and 2 min into the run? What is her pulse rate 2 min into the run?

76. THURSTONE LEARNING MODEL Psychologist L. L. Thurstone suggested the following relationship between learning time T and the length of a list n:

$$T = f(n) = An\sqrt{n - b}$$

where A and b are constants that depend on the person and the task.

a. Compute dT/dn and interpret your result.
b. For a certain person and a certain task, suppose $A = 4$ and $b = 4$. Compute $f'(13)$ and $f'(29)$ and interpret your results.

77. OIL SPILLS In calm waters, the oil spilling from the ruptured hull of a grounded tanker spreads in all directions. Assuming that the area polluted is a circle and that its radius is increasing at a rate of 2 ft/sec, determine how fast the area is increasing when the radius of the circle is 40 ft.

78. ARTERIOSCLEROSIS Refer to Example 8, page 190. Suppose the radius of an individual's artery is 1 cm and the thickness of the plaque (in centimeters) t yr from now is given by

$$h = g(t) = \frac{0.5t^2}{t^2 + 10} \qquad (0 \leq t \leq 10)$$

How fast will the arterial opening be decreasing 5 yr from now?

79. TRAFFIC FLOW Opened in the late 1950s, the Central Artery in downtown Boston was designed to move 75,000 vehicles a day. The number of vehicles moved per day is approximated by the function

$$x = f(t) = 6.25t^2 + 19.75t + 74.75 \qquad (0 \leq t \leq 5)$$

where x is measured in thousands and t in decades, with $t = 0$ corresponding to the beginning of 1959. Suppose the average speed of traffic flow in mph is given by

$$S = g(x) = -0.00075x^2 + 67.5 \qquad (75 \leq x \leq 350)$$

where x has the same meaning as before. What was the rate of change of the average speed of traffic flow at the beginning of 1999? What was the average speed of traffic flow at that time?
Hint: $S = g[f(t)]$.

80. HOTEL OCCUPANCY RATES The occupancy rate of the all-suite Wonderland Hotel, located near an amusement park, is given by the function

$$r(t) = \frac{10}{81}t^3 - \frac{10}{3}t^2 + \frac{200}{9}t + 60 \qquad (0 \leq t \leq 12)$$

where t is measured in months, with $t = 0$ corresponding to the beginning of January. Management has estimated that the monthly revenue (in thousands of dollars/month) is approximated by the function

$$R(r) = -\frac{3}{5000}r^3 + \frac{9}{50}r^2 \qquad (0 \leq r \leq 100)$$

where r is the occupancy rate.
a. Find an expression that gives the rate of change of Wonderland's occupancy rate with respect to time.
b. Find an expression that gives the rate of change of Wonderland's monthly revenue with respect to the occupancy rate.

c. What is the rate of change of Wonderland's monthly revenue with respect to time at the beginning of January? At the beginning of July?
Hint: Use the chain rule to find $R'(r(0))r'(0)$ and $R'(r(6))r'(6)$.

81. **Effect of Housing Starts on Jobs** The president of a major housing construction firm claims that the number of construction jobs created is given by

$$N(x) = 1.42x$$

where x denotes the number of housing starts. Suppose the number of housing starts in the next t mo is expected to be

$$x(t) = \frac{7t^2 + 140t + 700}{3t^2 + 80t + 550}$$

million units/year. Find an expression that gives the rate at which the number of construction jobs will be created t mo from now. At what rate will construction jobs be created 1 yr from now?

82. **Demand for PCs** The quantity demanded per month, x, of a certain make of personal computer (PC) is related to the average unit price, p (in dollars), of PCs by the equation

$$x = f(p) = \frac{100}{9}\sqrt{810{,}000 - p^2}$$

It is estimated that t mo from now, the average price of a PC will be given by

$$p(t) = \frac{400}{1 + \frac{1}{8}\sqrt{t}} + 200 \qquad (0 \le t \le 60)$$

dollars. Find the rate at which the quantity demanded per month of the PCs will be changing 16 mo from now.

83. **Demand for Watches** The demand equation for the Sicard wristwatch is given by

$$x = f(p) = 10\sqrt{\frac{50 - p}{p}} \qquad (0 < p \le 50)$$

where x (measured in units of a thousand) is the quantity demanded each week and p is the unit price in dollars. Find the rate of change of the quantity demanded of the wristwatches with respect to the unit price when the unit price is \$25.

84. **Cruise Ship Bookings** The management of Cruise World, operators of Caribbean luxury cruises, expects that the percent of young adults booking passage on their cruises in the years ahead will rise dramatically. They have constructed the following model, which gives the percent of young adult passengers in year t:

$$p = f(t) = 50\left(\frac{t^2 + 2t + 4}{t^2 + 4t + 8}\right) \qquad (0 \le t \le 5)$$

Young adults normally pick shorter cruises and generally spend less on their passage. The following model gives an approximation of the average amount of money R (in dollars) spent per passenger on a cruise when the percent of young adults is p:

$$R(p) = 1000\left(\frac{p + 4}{p + 2}\right)$$

Find the rate at which the price of the average passage will be changing 2 yr from now.

In Exercises 85–88, determine whether the statement is true or false. If it is true, explain why it is true. If it is false, give an example to show why it is false.

85. If f and g are differentiable and $h = f \circ g$, then $h'(x) = f'[g(x)]g'(x)$.

86. If f is differentiable and c is a constant, then

$$\frac{d}{dx}[f(cx)] = cf'(cx).$$

87. If f is differentiable, then

$$\frac{d}{dx}\sqrt{f(x)} = \frac{f'(x)}{2\sqrt{f(x)}}$$

88. If f is differentiable, then

$$\frac{d}{dx}\left[f\left(\frac{1}{x}\right)\right] = f'\left(\frac{1}{x}\right)$$

89. In Section 3.1, we proved that

$$\frac{d}{dx}(x^n) = nx^{n-1}$$

for the special case when $n = 2$. Use the chain rule to show that

$$\frac{d}{dx}(x^{1/n}) = \frac{1}{n}x^{1/n - 1}$$

for any nonzero integer n, assuming that $f(x) = x^{1/n}$ is differentiable.
Hint: Let $f(x) = x^{1/n}$ so that $[f(x)]^n = x$. Differentiate both sides with respect to x.

90. With the aid of Exercise 89, prove that

$$\frac{d}{dx}(x^r) = rx^{r-1}$$

for any rational number r.
Hint: Let $r = m/n$, where m and n are integers, with $n \neq 0$, and write $x^r = (x^m)^{1/n}$.

3.3 Solutions to Self-Check Exercises

1. Rewriting, we have

$$f(x) = -(2x^2 - 1)^{-1/2}$$

Using the general power rule, we find

$$\begin{aligned} f'(x) &= -\frac{d}{dx}(2x^2 - 1)^{-1/2} \\ &= -\left(-\frac{1}{2}\right)(2x^2 - 1)^{-3/2}\frac{d}{dx}(2x^2 - 1) \\ &= \frac{1}{2}(2x^2 - 1)^{-3/2}(4x) \\ &= \frac{2x}{(2x^2 - 1)^{3/2}} \end{aligned}$$

2. a. The life expectancy at birth of a female born at the beginning of 1980 is given by

$$g(80) = 50.02[1 + 1.09(80)]^{0.1} \approx 78.29$$

or approximately 78 yr. Similarly, the life expectancy at birth of a female born at the beginning of the year 2000 is given by

$$g(100) = 50.02[1 + 1.09(100)]^{0.1} \approx 80.04$$

or approximately 80 yr.

b. The rate of change of the life expectancy at birth of a female born at any time t is given by $g'(t)$. Using the general power rule, we have

$$\begin{aligned} g'(t) &= 50.02\frac{d}{dt}(1 + 1.09t)^{0.1} \\ &= (50.02)(0.1)(1 + 1.09t)^{-0.9}\frac{d}{dt}(1 + 1.09t) \\ &= (50.02)(0.1)(1.09)(1 + 1.09t)^{-0.9} \\ &= 5.45218(1 + 1.09t)^{-0.9} \\ &= \frac{5.45218}{(1 + 1.09t)^{0.9}} \end{aligned}$$

USING TECHNOLOGY

Finding the Derivative of a Composite Function

EXAMPLE 1 Find the rate of change of $f(x) = \sqrt{x}(1 + 0.02x^2)^{3/2}$ when $x = 2.1$.

Solution Using the numerical derivative operation of a graphing utility, we find

$$f'(2.1) = 0.5821463392$$

or approximately 0.58 unit per unit change in x. (See Figure T1.)

```
nDeriv(X^.5(1+.0
2X^2)^1.5, X, 2.1)

        .5821463392
```

FIGURE T1
The TI-83 numerical derivative screen for computing $f'(2.1)$

APPLIED EXAMPLE 2 Amusement Park Attendance The management of AstroWorld ("The Amusement Park of the Future") estimates that the total number of visitors (in thousands) to the amusement park t hours after opening time at 9 a.m. is given by

$$N(t) = \frac{30t}{\sqrt{2 + t^2}}$$

What is the rate at which visitors are admitted to the amusement park at 10:30 a.m.?

Solution Using the numerical derivative operation of a graphing utility, we find

$$N'(1.5) \approx 6.8481$$

or approximately 6848 visitors per hour. (See Figure T2.)

```
nDeriv((30X)/(2+
X^2)^.5,X,1.5)
        6.848066034
```

FIGURE T2
The TI-83 numerical derivative screen for computing $N'(1.5)$

TECHNOLOGY EXERCISES

In Exercises 1–6, use the numerical derivative operation to find the rate of change of $f(x)$ at the given value of x. Give your answer accurate to four decimal places.

1. $f(x) = \sqrt{x^2 - x^4}$; $x = 0.5$
2. $f(x) = x - \sqrt{1 - x^2}$; $x = 0.4$
3. $f(x) = x\sqrt{1 - x^2}$; $x = 0.2$
4. $f(x) = (x + \sqrt{x^2 + 4})^{3/2}$; $x = 1$
5. $f(x) = \dfrac{\sqrt{1 + x^2}}{x^3 + 2}$; $x = -1$
6. $f(x) = \dfrac{x^3}{1 + (1 + x^2)^{3/2}}$; $x = 3$
7. Worldwide Production of Vehicles The worldwide production of vehicles between 1960 and 1990 is given by the function

$$f(t) = 16.5\sqrt{1 + 2.2t} \qquad (0 \le t \le 3)$$

where $f(t)$ is measured in units of a million and t is measured in decades, with $t = 0$ corresponding to the beginning of 1960. What was the rate of change of the worldwide production of vehicles at the beginning of 1970? At the beginning of 1980?
Source: Automotive News

8. Accumulation Years People from their mid-40s to their mid-50s are in the prime investing years. Demographic studies of this type are of particular importance to financial institutions. The function

$$N(t) = 34.4(1 + 0.32125t)^{0.15} \qquad (0 \le t \le 12)$$

gives the projected number of people in this age group in the United States (in millions) in year t, where $t = 0$ corresponds to the beginning of 1996.
   a. How large is this segment of the population projected to be at the beginning of 2005?
   b. How fast will this segment of the population be growing at the beginning of 2005?

Source: U.S. Census Bureau

3.4 Marginal Functions in Economics

Marginal analysis is the study of the rate of change of economic quantities. For example, an economist is not merely concerned with the value of an economy's gross domestic product (GDP) at a given time but is equally concerned with the rate at which it is growing or declining. In the same vein, a manufacturer is not only interested in the total cost corresponding to a certain level of production of a commodity but also is interested in the rate of change of the total cost with respect to the level of production, and so on. Let's begin with an example to explain the meaning of the adjective *marginal*, as used by economists.

Cost Functions

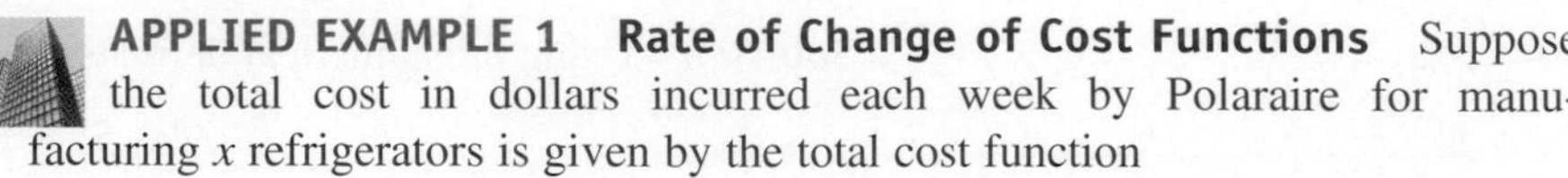

APPLIED EXAMPLE 1 Rate of Change of Cost Functions Suppose the total cost in dollars incurred each week by Polaraire for manufacturing x refrigerators is given by the total cost function

$$C(x) = 8000 + 200x - 0.2x^2 \qquad (0 \le x \le 400)$$

a. What is the actual cost incurred for manufacturing the 251st refrigerator?

b. Find the rate of change of the total cost function with respect to x when $x = 250$.

c. Compare the results obtained in parts (a) and (b).

Solution

a. The actual cost incurred in producing the 251st refrigerator is the difference between the total cost incurred in producing the first 251 refrigerators and the total cost of producing the first 250 refrigerators:

$$\begin{aligned} C(251) - C(250) &= [8000 + 200(251) - 0.2(251)^2] \\ &\quad - [8000 + 200(250) - 0.2(250)^2] \\ &= 45{,}599.8 - 45{,}500 \\ &= 99.80 \end{aligned}$$

or \$99.80.

b. The rate of change of the total cost function C with respect to x is given by the derivative of C—that is, $C'(x) = 200 - 0.4x$. Thus, when the level of production is 250 refrigerators, the rate of change of the total cost with respect to x is given by

$$\begin{aligned} C'(250) &= 200 - 0.4(250) \\ &= 100 \end{aligned}$$

or \$100.

c. From the solution to part (a), we know that the actual cost for producing the 251st refrigerator is \$99.80. This answer is very closely approximated by the answer to part (b), \$100. To see why this is so, observe that the difference $C(251) - C(250)$ may be written in the form

$$\frac{C(251) - C(250)}{1} = \frac{C(250 + 1) - C(250)}{1} = \frac{C(250 + h) - C(250)}{h}$$

where $h = 1$. In other words, the difference $C(251) - C(250)$ is precisely the average rate of change of the total cost function C over the interval [250, 251], or, equivalently, the slope of the secant line through the points (250, 45,500) and (251, 45,599.8). However, the number $C'(250) = 100$ is the instantaneous rate of change of the total cost function C at $x = 250$, or, equivalently, the slope of the tangent line to the graph of C at $x = 250$.

Now when h is small, the average rate of change of the function C is a good approximation to the instantaneous rate of change of the function C, or, equivalently, the slope of the secant line through the points in question is a good approximation to the slope of the tangent line through the point in question. Thus, we may expect

$$\begin{aligned} C(251) - C(250) &= \frac{C(251) - C(250)}{1} = \frac{C(250 + h) - C(250)}{h} \\ &\approx \lim_{h \to 0} \frac{C(250 + h) - C(250)}{h} = C'(250) \end{aligned}$$

which is precisely the case in this example. ■

The actual cost incurred in producing an additional unit of a certain commodity given that a plant is already at a certain level of operation is called the **marginal cost.** Knowing this cost is very important to management. As we saw in Example 1, the marginal cost is approximated by the rate of change of the total cost function evaluated at the appropriate point. For this reason, economists have defined the **marginal cost function** to be the derivative of the corresponding total cost function. In other words, if C is a total cost function, then the marginal cost function is defined to be its derivative C'. Thus, the adjective *marginal* is synonymous with *derivative of.*

APPLIED EXAMPLE 2 Marginal Cost Functions A subsidiary of Elektra Electronics manufactures a programmable pocket calculator. Management determined that the daily total cost of producing these calculators (in dollars) is given by

$$C(x) = 0.0001x^3 - 0.08x^2 + 40x + 5000$$

where x stands for the number of calculators produced.

a. Find the marginal cost function.
b. What is the marginal cost when $x = 200, 300, 400$, and 600?
c. Interpret your results.

Solution

a. The marginal cost function C' is given by the derivative of the total cost function C. Thus,

$$C'(x) = 0.0003x^2 - 0.16x + 40$$

b. The marginal cost when $x = 200, 300, 400$, and 600 is given by

$$\begin{aligned} C'(200) &= 0.0003(200)^2 - 0.16(200) + 40 = 20 \\ C'(300) &= 0.0003(300)^2 - 0.16(300) + 40 = 19 \\ C'(400) &= 0.0003(400)^2 - 0.16(400) + 40 = 24 \\ C'(600) &= 0.0003(600)^2 - 0.16(600) + 40 = 52 \end{aligned}$$

or \$20, \$19, \$24, and \$52, respectively.

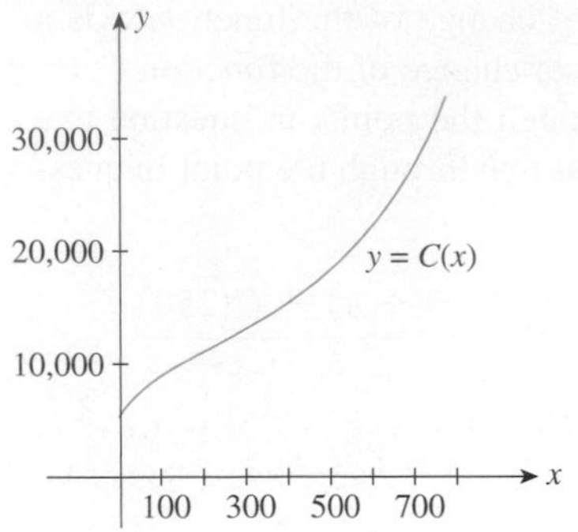

FIGURE 8
The cost of producing x calculators is given by $C(x)$.

c. From the results of part (b), we see that Elektra's actual cost for producing the 201st calculator is approximately \$20. The actual cost incurred for producing one additional calculator when the level of production is already 300 calculators is approximately \$19, and so on. Observe that when the level of production is already 600 units, the actual cost of producing one additional unit is approximately \$52. The higher cost for producing this additional unit when the level of production is 600 units may be the result of several factors, among them excessive costs incurred because of overtime or higher maintenance, production breakdown caused by greater stress and strain on the equipment, and so on. The graph of the total cost function appears in Figure 8. ■

Average Cost Functions

Let's now introduce another marginal concept closely related to the marginal cost. Let $C(x)$ denote the total cost incurred in producing x units of a certain commodity. Then the **average cost** of producing x units of the commodity is obtained by dividing the total production cost by the number of units produced. This leads to the following definition:

Average Cost Function

Suppose $C(x)$ is a total cost function. Then the **average cost function**, denoted by $\overline{C}(x)$ (read "C bar of x"), is

$$\frac{C(x)}{x} \tag{4}$$

The derivative $\overline{C}'(x)$ of the average cost function, called the **marginal average cost function**, measures the rate of change of the average cost function with respect to the number of units produced.

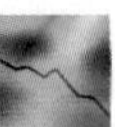

APPLIED EXAMPLE 3 Marginal Average Cost Functions The total cost of producing x units of a certain commodity is given by

$$C(x) = 400 + 20x$$

dollars.

a. Find the average cost function $\overline{C}$.
b. Find the marginal average cost function $\overline{C}'$.
c. What are the economic implications of your results?

Solution

a. The average cost function is given by

$$\overline{C}(x) = \frac{C(x)}{x} = \frac{400 + 20x}{x} = 20 + \frac{400}{x}$$

b. The marginal average cost function is

$$\overline{C}'(x) = -\frac{400}{x^2}$$

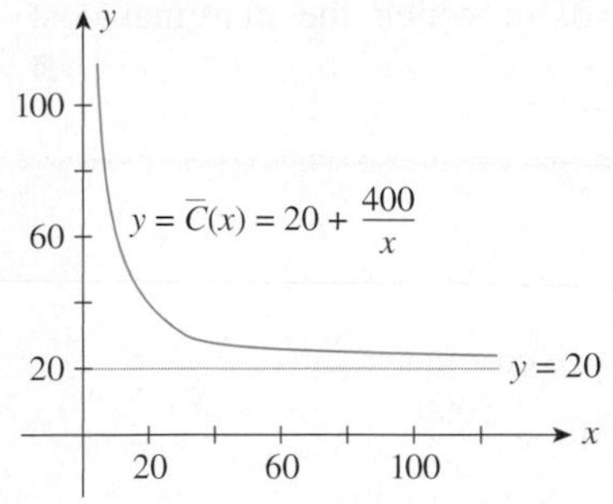

FIGURE 9
As the level of production increases, the average cost approaches $20.

c. Since the marginal average cost function is negative for all admissible values of x, the rate of change of the average cost function is negative for all $x > 0$; that is, $\overline{C}(x)$ decreases as x increases. However, the graph of $\overline{C}$ always lies above the horizontal line $y = 20$, but it approaches the line since

$$\lim_{x \to \infty} \overline{C}(x) = \lim_{x \to \infty} \left(20 + \frac{400}{x}\right) = 20$$

A sketch of the graph of the function $\overline{C}(x)$ appears in Figure 9. This result is fully expected if we consider the economic implications. Note that as the level of production increases, the fixed cost per unit of production, represented by the term $(400/x)$, drops steadily. The average cost approaches the constant unit of production, which is \$20 in this case. ■

APPLIED EXAMPLE 4 Marginal Average Cost Functions Once again consider the subsidiary of Elektra Electronics. The daily total cost for producing its programmable calculators is given by

$$C(x) = 0.0001x^3 - 0.08x^2 + 40x + 5000$$

dollars, where x stands for the number of calculators produced (see Example 2).

a. Find the average cost function $\overline{C}$.
b. Find the marginal average cost function $\overline{C}'$. Compute $\overline{C}'(500)$.
c. Sketch the graph of the function $\overline{C}$ and interpret the results obtained in parts (a) and (b).

Solution

a. The average cost function is given by

$$\overline{C}(x) = \frac{C(x)}{x} = 0.0001x^2 - 0.08x + 40 + \frac{5000}{x}$$

b. The marginal average cost function is given by

$$\overline{C}'(x) = 0.0002x - 0.08 - \frac{5000}{x^2}$$

Also,

$$\overline{C}'(500) = 0.0002(500) - 0.08 - \frac{5000}{(500)^2} = 0$$

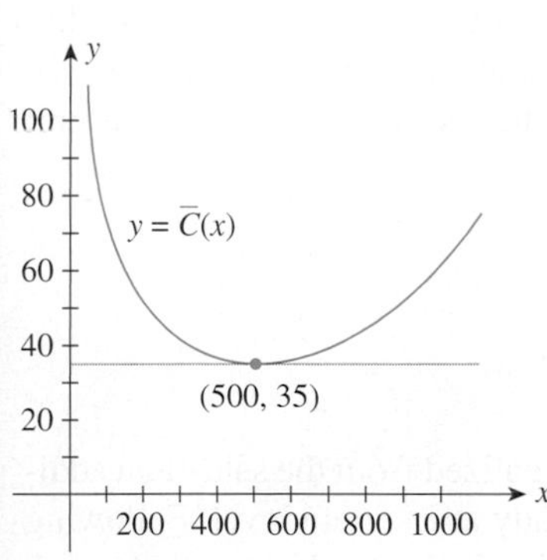

FIGURE 10
The average cost reaches a minimum of $35 when 500 calculators are produced.

c. To sketch the graph of the function $\overline{C}$, observe that if x is a small positive number, then $\overline{C}(x) > 0$. Furthermore, $\overline{C}(x)$ becomes arbitrarily large as x approaches zero from the right, since the term $(5000/x)$ becomes arbitrarily large as x approaches zero. Next, the result $\overline{C}'(500) = 0$ obtained in part (b) tells us that the tangent line to the graph of the function $\overline{C}$ is horizontal at the point (500, 35) on the graph. Finally, plotting the points on the graph corresponding to, say, $x = 100, 200, 300, \ldots, 900$, we obtain the sketch in Figure 10. As expected, the average cost drops as the level of production increases. But in this case, as opposed to the case in Example 3, the average cost reaches a minimum value of \$35, corresponding to a production level of 500, and *increases* thereafter.

This phenomenon is typical in situations where the marginal cost increases from some point on as production increases, as in Example 2. This situation is in contrast to that of Example 3, in which the marginal cost remains constant at any level of production. ■

EXPLORING WITH TECHNOLOGY

Refer to Example 4.

1. Use a graphing utility to plot the graph of the average cost function
$$\overline{C}(x) = 0.0001x^2 - 0.08x + 40 + \frac{5000}{x}$$
using the viewing window $[0, 1000] \times [0, 100]$. Then, using **ZOOM** and **TRACE**, show that the lowest point on the graph of $\overline{C}$ is (500, 35).
2. Draw the tangent line to the graph of $\overline{C}$ at (500, 35). What is its slope? Is this expected?
3. Plot the graph of the marginal average cost function
$$\overline{C}'(x) = 0.0002x - 0.08 - \frac{5000}{x^2}$$
using the viewing window $[0, 2000] \times [-1, 1]$. Then use **ZOOM** and **TRACE** to show that the zero of the function $\overline{C}'$ occurs at $x = 500$. Verify this result using the root-finding capability of your graphing utility. Is this result compatible with that obtained in part 2? Explain your answer.

Revenue Functions

Recall that a revenue function $R(x)$ gives the revenue realized by a company from the sale of x units of a certain commodity. If the company charges p dollars per unit, then

$$R(x) = px \tag{5}$$

However, the price that a company can command for the product depends on the market in which it operates. If the company is one of many—none of which is able to dictate the price of the commodity—then in this competitive market environment the price is determined by market equilibrium (see Section 2.3). On the other hand, if the company is the sole supplier of the product, then under this monopolistic situation it can manipulate the price of the commodity by controlling the supply. The unit selling price p of the commodity is related to the quantity x of the commodity demanded. This relationship between p and x is called a *demand equation* (see Section 2.3). Solving the demand equation for p in terms of x, we obtain the unit price function f. Thus,

$$p = f(x)$$

and the revenue function R is given by

$$R(x) = px = xf(x)$$

The **marginal revenue** gives the actual revenue realized from the sale of an additional unit of the commodity given that sales are already at a certain level. Following an argument parallel to that applied to the cost function in Example 1, you can convince yourself that the marginal revenue is approximated by $R'(x)$. Thus, we define the **marginal revenue function** to be $R'(x)$, where R is the revenue function. The derivative R' of the function R measures the rate of change of the revenue function.

APPLIED EXAMPLE 5 Marginal Revenue Functions Suppose the relationship between the unit price p in dollars and the quantity demanded x of the Acrosonic model F loudspeaker system is given by the equation

$$p = -0.02x + 400 \qquad (0 \le x \le 20{,}000)$$

a. Find the revenue function R.
b. Find the marginal revenue function R'.
c. Compute $R'(2000)$ and interpret your result.

Solution

a. The revenue function R is given by

$$\begin{aligned} R(x) &= px \\ &= x(-0.02x + 400) \\ &= -0.02x^2 + 400x \qquad (0 \le x \le 20{,}000) \end{aligned}$$

b. The marginal revenue function R' is given by

$$R'(x) = -0.04x + 400$$

c.
$$R'(2000) = -0.04(2000) + 400 = 320$$

Thus, the actual revenue to be realized from the sale of the 2001st loudspeaker system is approximately \$320.

Profit Functions

Our final example of a marginal function involves the profit function. The profit function P is given by

$$P(x) = R(x) - C(x) \qquad \textbf{(6)}$$

where R and C are the revenue and cost functions and x is the number of units of a commodity produced and sold. The **marginal profit function** $P'(x)$ measures the rate of change of the profit function P and provides us with a good approximation of the actual profit or loss realized from the sale of the $(x + 1)$st unit of the commodity (assuming the xth unit has been sold).

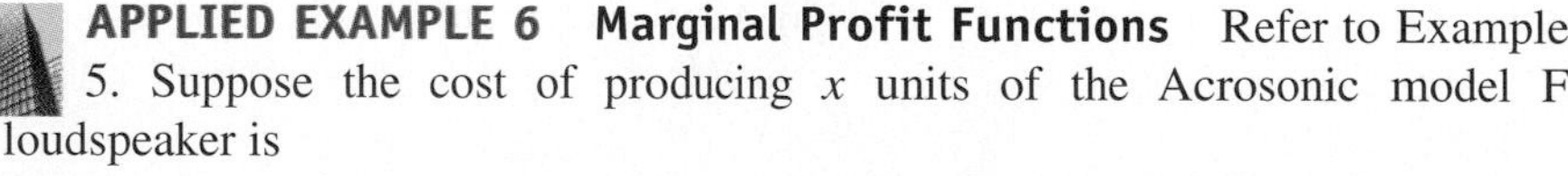

APPLIED EXAMPLE 6 Marginal Profit Functions Refer to Example 5. Suppose the cost of producing x units of the Acrosonic model F loudspeaker is

$$C(x) = 100x + 200{,}000$$

dollars.
a. Find the profit function P.
b. Find the marginal profit function P'.
c. Compute $P'(2000)$ and interpret your result.
d. Sketch the graph of the profit function P.

Solution

a. From the solution to Example 5a, we have

$$R(x) = -0.02x^2 + 400x$$

Thus, the required profit function P is given by

$$\begin{aligned} P(x) &= R(x) - C(x) \\ &= (-0.02x^2 + 400x) - (100x + 200{,}000) \\ &= -0.02x^2 + 300x - 200{,}000 \end{aligned}$$

b. The marginal profit function P' is given by

$$P'(x) = -0.04x + 300$$

c.

$$P'(2000) = -0.04(2000) + 300 = 220$$

Thus, the actual profit realized from the sale of the 2001st loudspeaker system is approximately \$220.

d. The graph of the profit function P appears in Figure 11.

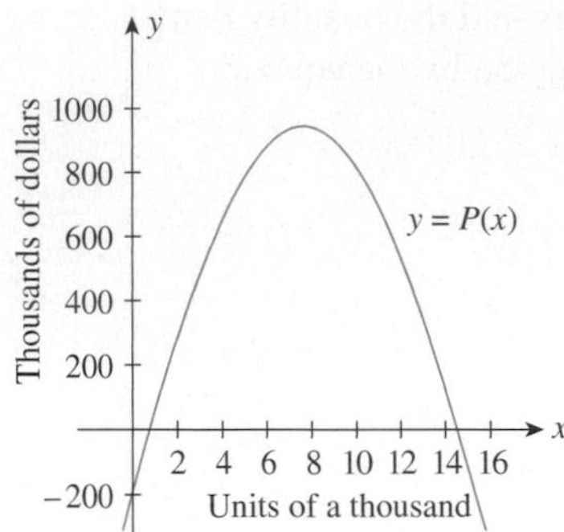

FIGURE 11
The total profit made when x loudspeakers are produced is given by $P(x)$.

Elasticity of Demand

Finally, let's use the marginal concepts introduced in this section to derive an important criterion used by economists to analyze the demand function: elasticity of demand.

In what follows, it will be convenient to write the demand function f in the form $x = f(p)$; that is, we will think of the quantity demanded of a certain commodity as a function of its unit price. Since the quantity demanded of a commodity usually decreases as its unit price increases, the function f is typically a decreasing function of p (Figure 12a).

FIGURE 12

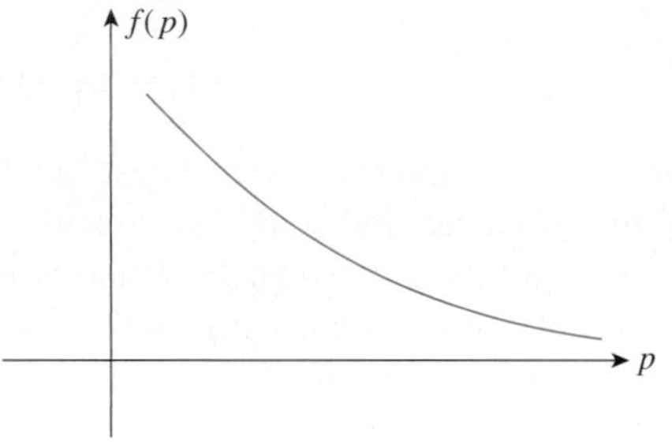

(a) A demand function

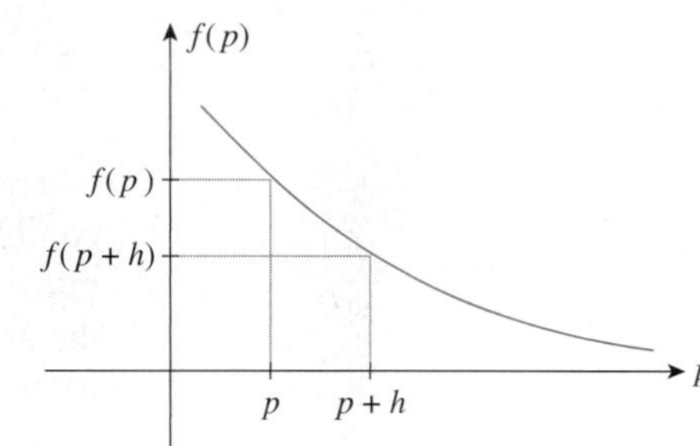

(b) $f(p + h)$ is the quantity demanded when the unit price increases from p to $p + h$ dollars.

Suppose the unit price of a commodity is increased by h dollars from p dollars to $(p + h)$ dollars (Figure 12b). Then the quantity demanded drops from $f(p)$ units to $f(p + h)$ units, a change of $[f(p + h) - f(p)]$ units. The percentage change in the unit price is

$$\frac{h}{p}(100) \qquad \frac{\text{Change in unit price}}{\text{Price } p}(100)$$

and the corresponding percentage change in the quantity demanded is

$$100\left[\frac{f(p+h)-f(p)}{f(p)}\right] \quad \frac{\text{Change in quantity demanded}}{\text{Quantity demanded at price } p}(100)$$

Now, one good way to measure the effect that a percentage change in price has on the percentage change in the quantity demanded is to look at the ratio of the latter to the former. We find

$$\frac{\text{Percentage change in the quantity demanded}}{\text{Percentage change in the unit price}} = \frac{100\left[\dfrac{f(p+h)-f(p)}{f(p)}\right]}{100\left(\dfrac{h}{p}\right)} = \frac{\dfrac{f(p+h)-f(p)}{h}}{\dfrac{f(p)}{p}}$$

If f is differentiable at p, then

$$\frac{f(p+h)-f(p)}{h} \approx f'(p)$$

when h is small. Therefore, if h is small, then the ratio is approximately equal to

$$\frac{f'(p)}{\dfrac{f(p)}{p}} = \frac{pf'(p)}{f(p)}$$

Economists call the negative of this quantity the elasticity of demand.

Elasticity of Demand

If f is a differentiable demand function defined by $x = f(p)$, then the **elasticity of demand** at price p is given by

$$E(p) = -\frac{pf'(p)}{f(p)} \tag{7}$$

Note It will be shown later (Section 4.1) that if f is decreasing on an interval, then $f'(p) < 0$ for p in that interval. In light of this, we see that since both p and $f(p)$ are positive, the quantity $\dfrac{pf'(p)}{f(p)}$ is negative. Because economists would rather work with a positive value, the elasticity of demand $E(p)$ is defined to be the negative of this quantity. ■

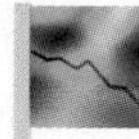

APPLIED EXAMPLE 7 Elasticity of Demand Consider the demand equation

$$p = -0.02x + 400 \qquad (0 \le x \le 20{,}000)$$

which describes the relationship between the unit price in dollars and the quantity demanded x of the Acrosonic model F loudspeaker systems.

a. Find the elasticity of demand $E(p)$.
b. Compute $E(100)$ and interpret your result.
c. Compute $E(300)$ and interpret your result.

Solution

a. Solving the given demand equation for x in terms of p, we find

$$x = f(p) = -50p + 20{,}000$$

from which we see that

$$f'(p) = -50$$

Therefore,

$$E(p) = -\frac{pf'(p)}{f(p)} = -\frac{p(-50)}{-50p + 20{,}000} = \frac{p}{400 - p}$$

b. $$E(100) = \frac{100}{400 - 100} = \frac{1}{3}$$

which is the elasticity of demand when $p = 100$. To interpret this result, recall that $E(100)$ is the negative of the ratio of the percentage change in the quantity demanded to the percentage change in the unit price when $p = 100$. Therefore, our result tells us that when the unit price p is set at \$100 per speaker, an increase of 1% in the unit price will cause a decrease of approximately 0.33% in the quantity demanded.

c. $$E(300) = \frac{300}{400 - 300} = 3$$

which is the elasticity of demand when $p = 300$. It tells us that when the unit price is set at \$300 per speaker, an increase of 1% in the unit price will cause a decrease of approximately 3% in the quantity demanded. ■

Economists often use the following terminology to describe demand in terms of elasticity.

Elasticity of Demand

The demand is said to be **elastic** if $E(p) > 1$.

The demand is said to be **unitary** if $E(p) = 1$.

The demand is said to be **inelastic** if $E(p) < 1$.

As an illustration, our computations in Example 7 revealed that demand for Acrosonic loudspeakers is elastic when $p = 300$ but inelastic when $p = 100$. These computations confirm that when demand is elastic, a small percentage change in the unit price will result in a greater percentage change in the quantity demanded; and when demand is inelastic, a small percentage change in the unit price will cause a smaller percentage change in the quantity demanded. Finally, when demand is unitary, a small percentage change in the unit price will result in the same percentage change in the quantity demanded.

We can describe the way revenue responds to changes in the unit price using the notion of elasticity. If the quantity demanded of a certain commodity is related to its unit price by the equation $x = f(p)$, then the revenue realized through the sale of x units of the commodity at a price of p dollars each is

$$R(p) = px = pf(p)$$

The rate of change of the revenue with respect to the unit price p is given by

$$\begin{aligned} R'(p) &= f(p) + pf'(p) \\ &= f(p)\left[1 + \frac{pf'(p)}{f(p)}\right] \\ &= f(p)[1 - E(p)] \end{aligned}$$

Now, suppose demand is elastic when the unit price is set at a dollars. Then $E(a) > 1$, and so $1 - E(a) < 0$. Since $f(p)$ is positive for all values of p, we see that

$$R'(a) = f(a)[1 - E(a)] < 0$$

and so $R(p)$ is decreasing at $p = a$. This implies that a small increase in the unit price when $p = a$ results in a decrease in the revenue, whereas a small decrease in the unit price will result in an increase in the revenue. Similarly, you can show that if the demand is inelastic when the unit price is set at a dollars, then a small increase in the unit price will cause the revenue to increase, and a small decrease in the unit price will cause the revenue to decrease. Finally, if the demand is unitary when the unit price is set at a dollars, then $E(a) = 1$ and $R'(a) = 0$. This implies that a small increase or decrease in the unit price will not result in a change in the revenue. The following statements summarize this discussion.

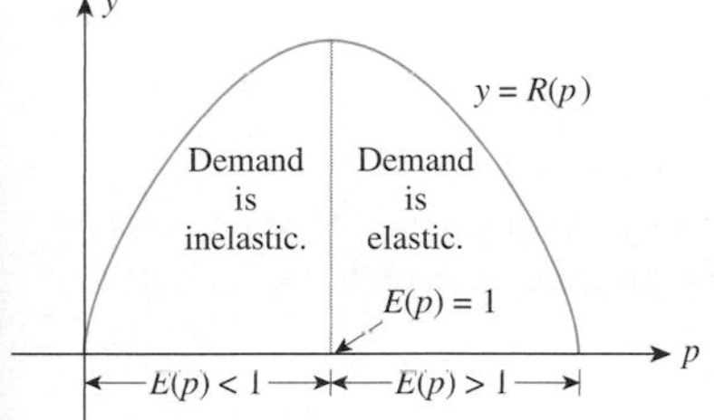

FIGURE 13
The revenue is increasing on an interval where the demand is inelastic, decreasing on an interval where the demand is elastic, and stationary at the point where the demand is unitary.

1. If the demand is elastic at p $[E(p) > 1]$, then an increase in the unit price will cause the revenue to decrease, whereas a decrease in the unit price will cause the revenue to increase.
2. If the demand is inelastic at p $[E(p) < 1]$, then an increase in the unit price will cause the revenue to increase, and a decrease in the unit price will cause the revenue to decrease.
3. If the demand is unitary at p $[E(p) = 1]$, then an increase in the unit price will cause the revenue to stay about the same.

These results are illustrated in Figure 13.

Note As an aid to remembering this, note the following:

1. If demand is elastic, then the change in revenue and the change in the unit price move in opposite directions.
2. If demand is inelastic, then they move in the same direction. ■

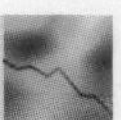

APPLIED EXAMPLE 8 Elasticity of Demand Refer to Example 7.

a. Is demand elastic, unitary, or inelastic when $p = 100$? When $p = 300$?
b. If the price is \$100, will raising the unit price slightly cause the revenue to increase or decrease?

Solution

a. From the results of Example 7, we see that $E(100) = \frac{1}{3} < 1$ and $E(300) = 3 > 1$. We conclude accordingly that demand is inelastic when $p = 100$ and elastic when $p = 300$.
b. Since demand is inelastic when $p = 100$, raising the unit price slightly will cause the revenue to increase. ■

3.4 Self-Check Exercises

1. The weekly demand for Pulsar DVD recorders is given by the demand equation

$$p = -0.02x + 300 \qquad (0 \le x \le 15{,}000)$$

where p denotes the wholesale unit price in dollars and x denotes the quantity demanded. The weekly total cost function associated with manufacturing these recorders is

$$C(x) = 0.000003x^3 - 0.04x^2 + 200x + 70{,}000$$

dollars.

a. Find the revenue function R and the profit function P.
b. Find the marginal cost function C', the marginal revenue function R', and the marginal profit function P'.
c. Find the marginal average cost function $\overline{C}'$.
d. Compute $C'(3000)$, $R'(3000)$, and $P'(3000)$ and interpret your results.

2. Refer to the preceding exercise. Determine whether the demand is elastic, unitary, or inelastic when $p = 100$ and when $p = 200$.

Solutions to Self-Check Exercises 3.4 can be found on page 211.

3.4 Concept Questions

1. Explain each term in your own words:
a. Marginal cost function
b. Average cost function
c. Marginal average cost function
d. Marginal revenue function
e. Marginal profit function

2. **a.** Define the elasticity of demand.
b. When is the elasticity of demand elastic? Unitary? Inelastic? Explain the meaning of each term.

3.4 Exercises

1. Production Costs The graph of a typical total cost function $C(x)$ associated with the manufacture of x units of a certain commodity is shown in the following figure.
a. Explain why the function C is always increasing.
b. As the level of production x increases, the cost/unit drops so that $C(x)$ increases but at a slower pace. However, a level of production is soon reached at which the cost/unit begins to increase dramatically (due to a shortage of raw material, overtime, breakdown of machinery due to excessive stress and strain) so that $C(x)$ continues to increase at a faster pace. Use the graph of C to find the approximate level of production x_0 where this occurs.

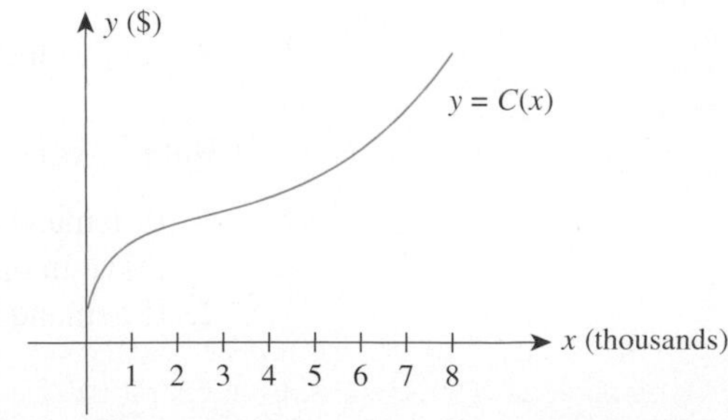

2. **PRODUCTION COSTS** The graph of a typical average cost function $A(x) = C(x)/x$, where $C(x)$ is a total cost function associated with the manufacture of x units of a certain commodity is shown in the following figure.
   a. Explain in economic terms why $A(x)$ is large if x is small and why $A(x)$ is large if x is large.
   b. What is the significance of the numbers x_0 and y_0, the x- and y-coordinates of the lowest point on the graph of the function A?

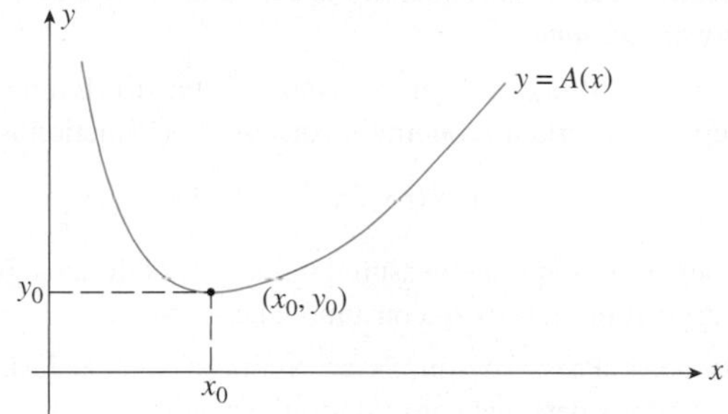

3. **MARGINAL COST** The total weekly cost (in dollars) incurred by Lincoln Records in pressing x compact discs is

$$C(x) = 2000 + 2x - 0.0001x^2 \qquad (0 \leq x \leq 6000)$$

   a. What is the actual cost incurred in producing the 1001st and the 2001st disc?
   b. What is the marginal cost when $x = 1000$ and 2000?

4. **MARGINAL COST** A division of Ditton Industries manufactures the Futura model microwave oven. The daily cost (in dollars) of producing these microwave ovens is

$$C(x) = 0.0002x^3 - 0.06x^2 + 120x + 5000$$

where x stands for the number of units produced.
   a. What is the actual cost incurred in manufacturing the 101st oven? The 201st oven? The 301st oven?
   b. What is the marginal cost when $x = 100$, 200, and 300?

5. **MARGINAL AVERAGE COST** Custom Office makes a line of executive desks. It is estimated that the total cost for making x units of their Senior Executive model is

$$C(x) = 100x + 200{,}000$$

dollars/year.
   a. Find the average cost function $\overline{C}$.
   b. Find the marginal average cost function $\overline{C}'$.
   c. What happens to $\overline{C}(x)$ when x is very large? Interpret your results.

6. **MARGINAL AVERAGE COST** The management of ThermoMaster Company, whose Mexican subsidiary manufactures an indoor–outdoor thermometer, has estimated that the total weekly cost (in dollars) for producing x thermometers is

$$C(x) = 5000 + 2x$$

   a. Find the average cost function $\overline{C}$.
   b. Find the marginal average cost function $\overline{C}'$.
   c. Interpret your results.

7. Find the average cost function $\overline{C}$ and the marginal average cost function $\overline{C}'$ associated with the total cost function C of Exercise 3.

8. Find the average cost function $\overline{C}$ and the marginal average cost function $\overline{C}'$ associated with the total cost function C of Exercise 4.

9. **MARGINAL REVENUE** Williams Commuter Air Service realizes a monthly revenue of

$$R(x) = 8000x - 100x^2$$

dollars when the price charged per passenger is x dollars.
   a. Find the marginal revenue R'.
   b. Compute $R'(39)$, $R'(40)$, and $R'(41)$.
   c. Based on the results of part (b), what price should the airline charge in order to maximize their revenue?

10. **MARGINAL REVENUE** The management of Acrosonic plans to market the ElectroStat, an electrostatic speaker system. The marketing department has determined that the demand for these speakers is

$$p = -0.04x + 800 \qquad (0 \leq x \leq 20{,}000)$$

where p denotes the speaker's unit price (in dollars) and x denotes the quantity demanded.
   a. Find the revenue function R.
   b. Find the marginal revenue function R'.
   c. Compute $R'(5000)$ and interpret your results.

11. **MARGINAL PROFIT** Refer to Exercise 10. Acrosonic's production department estimates that the total cost (in dollars) incurred in manufacturing x ElectroStat speaker systems in the first year of production will be

$$C(x) = 200x + 300{,}000$$

   a. Find the profit function P.
   b. Find the marginal profit function P'.
   c. Compute $P'(5000)$ and $P'(8000)$.
   d. Sketch the graph of the profit function and interpret your results.

12. **MARGINAL PROFIT** Lynbrook West, an apartment complex, has 100 two-bedroom units. The monthly profit (in dollars) realized from renting x apartments is

$$P(x) = -10x^2 + 1760x - 50{,}000$$

   a. What is the actual profit realized from renting the 51st unit, assuming that 50 units have already been rented?

b. Compute the marginal profit when $x = 50$ and compare your results with that obtained in part (a).

13. MARGINAL COST, REVENUE, AND PROFIT The weekly demand for the Pulsar 25 color LED television is

$$p = 600 - 0.05x \qquad (0 \le x \le 12{,}000)$$

where p denotes the wholesale unit price in dollars and x denotes the quantity demanded. The weekly total cost function associated with manufacturing the Pulsar 25 is given by

$$C(x) = 0.000002x^3 - 0.03x^2 + 400x + 80{,}000$$

where $C(x)$ denotes the total cost incurred in producing x sets.

a. Find the revenue function R and the profit function P.
b. Find the marginal cost function C', the marginal revenue function R', and the marginal profit function P'.
c. Compute $C'(2000)$, $R'(2000)$, and $P'(2000)$ and interpret your results.
d. Sketch the graphs of the functions C, R, and P and interpret parts (b) and (c), using the graphs obtained.

14. MARGINAL COST, REVENUE, AND PROFIT Pulsar also manufactures a series of 19-in. color television sets. The quantity x of these sets demanded each week is related to the wholesale unit price p by the equation

$$p = -0.006x + 180$$

The weekly total cost incurred by Pulsar for producing x sets is

$$C(x) = 0.000002x^3 - 0.02x^2 + 120x + 60{,}000$$

dollars. Answer the questions in Exercise 13 for these data.

15. MARGINAL AVERAGE COST Find the average cost function $\overline{C}$ associated with the total cost function C of Exercise 13.

a. What is the marginal average cost function $\overline{C}'$?
b. Compute $\overline{C}'(5000)$ and $\overline{C}'(10{,}000)$ and interpret your results.
c. Sketch the graph of $\overline{C}$.

16. MARGINAL AVERAGE COST Find the average cost function $\overline{C}$ associated with the total cost function C of Exercise 14.

a. What is the marginal average cost function $\overline{C}'$?
b. Compute $\overline{C}'(5000)$ and $\overline{C}'(10{,}000)$ and interpret your results.

17. MARGINAL REVENUE The quantity of Sicard wristwatches demanded each month is related to the unit price by the equation

$$p = \frac{50}{0.01x^2 + 1} \qquad (0 \le x \le 20)$$

where p is measured in dollars and x in units of a thousand.

a. Find the revenue function R.
b. Find the marginal revenue function R'.
c. Compute $R'(2)$ and interpret your result.

18. MARGINAL PROPENSITY TO CONSUME The consumption function of the U.S. economy from 1929 to 1941 is

$$C(x) = 0.712x + 95.05$$

where $C(x)$ is the personal consumption expenditure and x is the personal income, both measured in billions of dollars. Find the rate of change of consumption with respect to income, dC/dx. This quantity is called the *marginal propensity to consume.*

19. MARGINAL PROPENSITY TO CONSUME Refer to Exercise 18. Suppose a certain economy's consumption function is

$$C(x) = 0.873x^{1.1} + 20.34$$

where $C(x)$ and x are measured in billions of dollars. Find the marginal propensity to consume when $x = 10$.

20. MARGINAL PROPENSITY TO SAVE Suppose $C(x)$ measures an economy's personal consumption expenditure and x the personal income, both in billions of dollars. Then,

$$S(x) = x - C(x) \qquad \text{Income minus consumption}$$

measures the economy's savings corresponding to an income of x billion dollars. Show that

$$\frac{dS}{dx} = 1 - \frac{dC}{dx}$$

The quantity dS/dx is called the *marginal propensity to save.*

21. Refer to Exercise 20. For the consumption function of Exercise 18, find the marginal propensity to save.

22. Refer to Exercise 20. For the consumption function of Exercise 19, find the marginal propensity to save when $x = 10$.

For each demand equation in Exercises 23–28, compute the elasticity of demand and determine whether the demand is elastic, unitary, or inelastic at the indicated price.

23. $x = -\frac{5}{4}p + 20; p = 10$

24. $x = -\frac{3}{2}p + 9; p = 2$

25. $x + \frac{1}{3}p - 20 = 0; p = 30$

26. $0.4x + p - 20 = 0; p = 10$

27. $p = 169 - x^2; p = 29$ **28.** $p = 144 - x^2; p = 96$

29. ELASTICITY OF DEMAND The demand equation for the Roland portable hair dryer is given by

$$x = \frac{1}{5}(225 - p^2) \qquad (0 \le p \le 15)$$

where x (measured in units of a hundred) is the quantity demanded per week and p is the unit price in dollars.

a. Is the demand elastic or inelastic when $p = 8$ and when $p = 10$?

b. When is the demand unitary?
Hint: Solve $E(p) = 1$ for p.

c. If the unit price is lowered slightly from \$10, will the revenue increase or decrease?

d. If the unit price is increased slightly from \$8, will the revenue increase or decrease?

30. Elasticity of Demand The management of Titan Tire Company has determined that the quantity demanded x of their Super Titan tires per week is related to the unit price p by the equation

$$x = \sqrt{144 - p} \qquad (0 \le p \le 144)$$

where p is measured in dollars and x in units of a thousand.

a. Compute the elasticity of demand when $p = 63$, 96, and 108.

b. Interpret the results obtained in part (a).

c. Is the demand elastic, unitary, or inelastic when $p = 63$, 96, and 108?

31. Elasticity of Demand The proprietor of the Showplace, a video store, has estimated that the rental price p (in dollars) of prerecorded videodiscs is related to the quantity x rented/day by the demand equation

$$x = \frac{2}{3}\sqrt{36 - p^2} \qquad (0 \le p \le 6)$$

Currently, the rental price is \$2/disc.

a. Is the demand elastic or inelastic at this rental price?

b. If the rental price is increased, will the revenue increase or decrease?

32. Elasticity of Demand The quantity demanded each week x (in units of a hundred) of the Mikado digital camera is related to the unit price p (in dollars) by the demand equation

$$x = \sqrt{400 - 5p} \qquad (0 \le p \le 80)$$

a. Is the demand elastic or inelastic when $p = 40$? When $p = 60$?

b. When is the demand unitary?

c. If the unit price is lowered slightly from \$60, will the revenue increase or decrease?

d. If the unit price is increased slightly from \$40, will the revenue increase or decrease?

33. Elasticity of Demand The demand function for a certain make of exercise bicycle sold exclusively through cable television is

$$p = \sqrt{9 - 0.02x} \qquad (0 \le x \le 450)$$

where p is the unit price in hundreds of dollars and x is the quantity demanded/week. Compute the elasticity of demand and determine the range of prices corresponding to inelastic, unitary, and elastic demand.
Hint: Solve the equation $E(p) = 1$.

34. Elasticity of Demand The demand equation for the Sicard wristwatch is given by

$$x = 10\sqrt{\frac{50 - p}{p}} \qquad (0 < p \le 50)$$

where x (measured in units of a thousand) is the quantity demanded/week and p is the unit price in dollars. Compute the elasticity of demand and determine the range of prices corresponding to inelastic, unitary, and elastic demand.

In Exercises 35 and 36, determine whether the statement is true or false. If it is true, explain why it is true. If it is false, give an example to show why it is false.

35. If C is a differentiable total cost function, then the marginal average cost function is

$$\overline{C}'(x) = \frac{xC'(x) - C(x)}{x^2}$$

36. If the marginal profit function is positive at $x = a$, then it makes sense to decrease the level of production.

3.4 Solutions to Self-Check Exercises

1. a.
$$\begin{aligned} R(x) &= px \\ &= x(-0.02x + 300) \\ &= -0.02x^2 + 300x \qquad (0 \le x \le 15{,}000) \end{aligned}$$

$$\begin{aligned} P(x) &= R(x) - C(x) \\ &= -0.02x^2 + 300x \\ &\quad - (0.000003x^3 - 0.04x^2 + 200x + 70{,}000) \\ &= -0.000003x^3 + 0.02x^2 + 100x - 70{,}000 \end{aligned}$$

b.
$$\begin{aligned} C'(x) &= 0.000009x^2 - 0.08x + 200 \\ R'(x) &= -0.04x + 300 \\ P'(x) &= -0.000009x^2 + 0.04x + 100 \end{aligned}$$

c. The average cost function is

$$\begin{aligned} \overline{C}(x) &= \frac{C(x)}{x} \\ &= \frac{0.000003x^3 - 0.04x^2 + 200x + 70{,}000}{x} \\ &= 0.000003x^2 - 0.04x + 200 + \frac{70{,}000}{x} \end{aligned}$$

Therefore, the marginal average cost function is

$$\overline{C}'(x) = 0.000006x - 0.04 - \frac{70{,}000}{x^2}$$

d. Using the results from part (b), we find

$$C'(3000) = 0.000009(3000)^2 - 0.08(3000) + 200$$
$$= 41$$

That is, when the level of production is already 3000 recorders, the actual cost of producing one additional recorder is approximately \$41. Next,

$$R'(3000) = -0.04(3000) + 300 = 180$$

That is, the actual revenue to be realized from selling the 3001st recorder is approximately \$180. Finally,

$$P'(3000) = -0.000009(3000)^2 + 0.04(3000) + 100$$
$$= 139$$

That is, the actual profit realized from selling the 3001st DVD recorder is approximately \$139.

2. We first solve the given demand equation for x in terms of p, obtaining

$$x = f(p) = -50p + 15{,}000$$
$$f'(p) = -50$$

Therefore,

$$E(p) = -\frac{pf'(p)}{f(p)} = -\frac{p}{-50p + 15{,}000}(-50)$$
$$= \frac{p}{300 - p} \qquad (0 \le p < 300)$$

Next, we compute

$$E(100) = \frac{100}{300 - 100} = \frac{1}{2} < 1$$

and we conclude that demand is inelastic when $p = 100$. Also,

$$E(200) = \frac{200}{300 - 200} = 2 > 1$$

and we see that demand is elastic when $p = 200$.

3.5 Higher-Order Derivatives

Higher-Order Derivatives

The derivative f' of a function f is also a function. As such, the differentiability of f' may be considered. Thus, the function f' has a derivative f'' at a point x in the domain of f' if the limit of the quotient

$$\frac{f'(x + h) - f'(x)}{h}$$

exists as h approaches zero. In other words, it is the derivative of the first derivative.

The function f'' obtained in this manner is called the **second derivative of** the function f, just as the derivative f' of f is often called the first derivative of f. Continuing in this fashion, we are led to considering the third, fourth, and higher-order derivatives of f whenever they exist. Notations for the first, second, third, and, in general, nth derivatives of a function f at a point x are

$$f'(x), f''(x), f'''(x), \ldots, f^{(n)}(x)$$

or

$$D^1f(x), D^2f(x), D^3f(x), \ldots, D^nf(x)$$

If f is written in the form $y = f(x)$, then the notations for its derivatives are

$$y', y'', y''', \ldots, y^{(n)}$$
$$\frac{dy}{dx}, \frac{d^2y}{dx^2}, \frac{d^3y}{dx^3}, \ldots, \frac{d^ny}{dx^n}$$

or

$$D^1y, D^2y, D^3y, \ldots, D^ny$$

respectively.

PORTFOLIO Steve Regenstreif

TITLE Director, Retiree Program
INSTITUTION American Federation of State, County and Municipal Employees

One of the biggest problems American Federation of State, County and Municipal Employees (AFSCME) and other retirees face in planning retirement finances is price inflation, the continuous increase in the cost of living over the years. Many wonder how financial planning is possible when future costs and income levels seem so unpredictable. Even relatively low rates of inflation can cause prices to more than double in the 15–20 years most of us can expect to be retired.

For example, if the annual inflation rate is 4%, the purchasing power of your income is reduced by 50% in 16 years. In 30 years, the purchasing power of that income is reduced by 75%. Preserving or building purchasing power for retirement means that savings and investments should produce an after-tax annual return or growth that equals or exceeds the annual rate of inflation.

The majority of pension plans provide a fixed payment—that is, the same amount of money each month for the life of the retiree. But a number of public retirement systems also agree to increase pension amounts as the cost of living rises—either through periodic adjustments (requiring legislative action each time) or automatic annual cost-of-living adjustments (COLAs). Even the latter, however, are almost never full COLAs. They are generally capped at a certain percentage (commonly 3%) regardless of how high inflation may rise in a year.

AFSCME RETIREES

EXAMPLE 1 Find the derivatives of all orders of the polynomial function

$$f(x) = x^5 - 3x^4 + 4x^3 - 2x^2 + x - 8$$

Solution We have

$$\begin{aligned}
f'(x) &= 5x^4 - 12x^3 + 12x^2 - 4x + 1 \\
f''(x) &= \frac{d}{dx}f'(x) = 20x^3 - 36x^2 + 24x - 4 \\
f'''(x) &= \frac{d}{dx}f''(x) = 60x^2 - 72x + 24 \\
f^{(4)}(x) &= \frac{d}{dx}f'''(x) = 120x - 72 \\
f^{(5)}(x) &= \frac{d}{dx}f^{(4)}(x) = 120
\end{aligned}$$

and, in general,

$$f^{(n)}(x) = 0 \qquad (\text{for } n > 5)$$

EXAMPLE 2 Find the third derivative of the function f defined by $y = x^{2/3}$. What is its domain?

Solution We have

$$\begin{aligned}
y' &= \frac{2}{3}x^{-1/3} \\
y'' &= \left(\frac{2}{3}\right)\left(-\frac{1}{3}\right)x^{-4/3} = -\frac{2}{9}x^{-4/3}
\end{aligned}$$

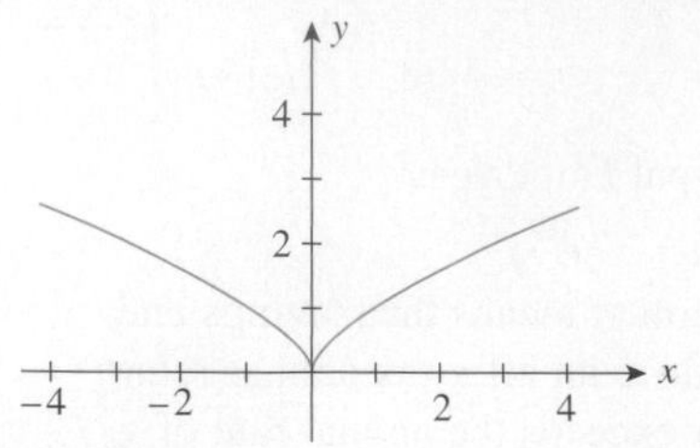

FIGURE 14
The graph of the function $y = x^{2/3}$

so the required derivative is

$$y''' = \left(-\frac{2}{9}\right)\left(-\frac{4}{3}\right)x^{-7/3} = \frac{8}{27}x^{-7/3} = \frac{8}{27x^{7/3}}$$

The common domain of the functions f', f'', and f''' is the set of all real numbers except $x = 0$. The domain of $y = x^{2/3}$ is the set of all real numbers. The graph of the function $y = x^{2/3}$ appears in Figure 14. ■

Note Always simplify an expression before differentiating it to obtain the next order derivative. ■

EXAMPLE 3 Find the second derivative of the function $y = (2x^2 + 3)^{3/2}$.

Solution We have, using the general power rule,

$$y' = \frac{3}{2}(2x^2 + 3)^{1/2}(4x) = 6x(2x^2 + 3)^{1/2}$$

Next, using the product rule and then the chain rule, we find

$$\begin{aligned} y'' &= (6x)\cdot\frac{d}{dx}(2x^2 + 3)^{1/2} + \left[\frac{d}{dx}(6x)\right](2x^2 + 3)^{1/2} \\ &= (6x)\left(\frac{1}{2}\right)(2x^2 + 3)^{-1/2}(4x) + 6(2x^2 + 3)^{1/2} \\ &= 12x^2(2x^2 + 3)^{-1/2} + 6(2x^2 + 3)^{1/2} \\ &= 6(2x^2 + 3)^{-1/2}[2x^2 + (2x^2 + 3)] \\ &= \frac{6(4x^2 + 3)}{\sqrt{2x^2 + 3}} \end{aligned}$$

■

Just as the derivative of a function f at a point x measures the rate of change of the function f at that point, the second derivative of f (the derivative of f') measures the rate of change of the derivative f' of the function f. The third derivative of the function f, f''', measures the rate of change of f'', and so on.

In Chapter 4, we will discuss applications involving the geometric interpretation of the second derivative of a function. The following example gives an interpretation of the second derivative in a familiar role.

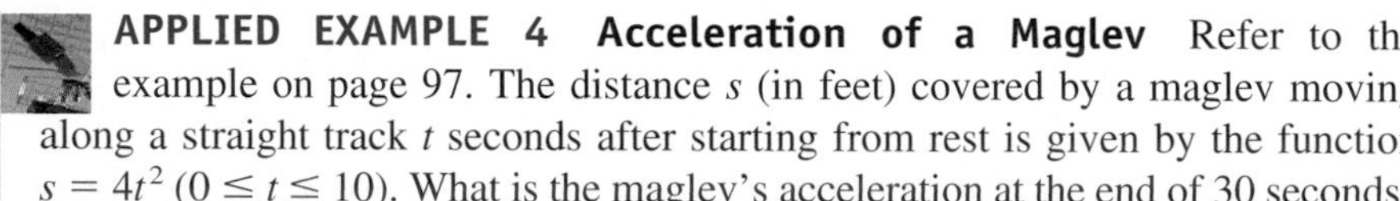

APPLIED EXAMPLE 4 Acceleration of a Maglev Refer to the example on page 97. The distance s (in feet) covered by a maglev moving along a straight track t seconds after starting from rest is given by the function $s = 4t^2$ $(0 \le t \le 10)$. What is the maglev's acceleration at the end of 30 seconds?

Solution The velocity of the maglev t seconds from rest is given by

$$v = \frac{ds}{dt} = \frac{d}{dt}(4t^2) = 8t$$

The acceleration of the maglev t seconds from rest is given by the rate of change of the velocity of t—that is,

$$a = \frac{d}{dt}v = \frac{d}{dt}\left(\frac{ds}{dt}\right) = \frac{d^2s}{dt^2} = \frac{d}{dt}(8t) = 8$$

or 8 feet per second per second, normally abbreviated 8 ft/sec^2. ■

APPLIED EXAMPLE 5 Acceleration and Velocity of a Falling Object A ball is thrown straight up into the air from the roof of a building. The height of the ball as measured from the ground is given by

$$s = -16t^2 + 24t + 120$$

where s is measured in feet and t in seconds. Find the velocity and acceleration of the ball 3 seconds after it is thrown into the air.

Solution The velocity v and acceleration a of the ball at any time t are given by

$$v = \frac{ds}{dt} = \frac{d}{dt}(-16t^2 + 24t + 120) = -32t + 24$$

and

$$a = \frac{d^2s}{dt^2} = \frac{d}{dt}\left(\frac{ds}{dt}\right) = \frac{d}{dt}(-32t + 24) = -32$$

Therefore, the velocity of the ball 3 seconds after it is thrown into the air is

$$v = -32(3) + 24 = -72$$

That is, the ball is falling downward at a speed of 72 ft/sec. The acceleration of the ball is 32 ft/sec^2 downward at any time during the motion. ■

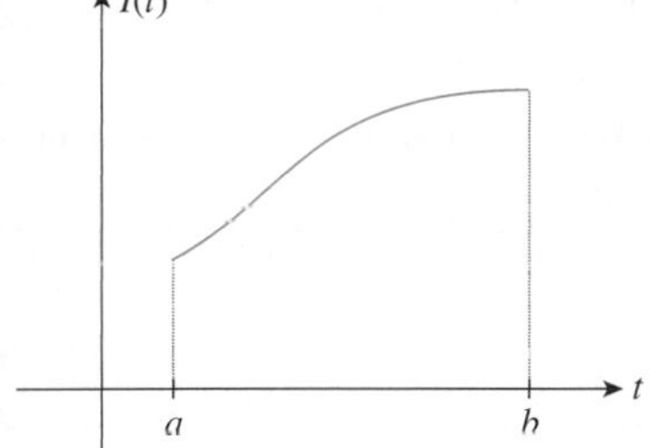

FIGURE 15
The CPI of a certain economy from year a to year b is given by $I(t)$.

Another interpretation of the second derivative of a function—this time from the field of economics—follows. Suppose the consumer price index (CPI) of an economy between the years a and b is described by the function $I(t)$ $(a \leq t \leq b)$ (Figure 15). Then the first derivative of I at $t = c$, $I'(c)$, where $a < c < b$, gives the rate of change of I at c. The quantity

$$\frac{I'(c)}{I(c)}$$

called the *relative rate of change of* $I(t)$ with respect to t at $t = c$, measures the *inflation rate* of the economy at $t = c$. The second derivative of I at $t = c$, $I''(c)$, gives the rate of change of I' at $t = c$. Now, it is possible for $I'(t)$ to be positive and $I''(t)$ to be negative at $t = c$ (see Example 6). This tells us that at $t = c$ the economy is experiencing inflation (the CPI is increasing) but the rate at which inflation is growing is in fact decreasing. This is precisely the situation described by an economist or a politician when she claims that "inflation is slowing." One may not jump to the conclusion from the aforementioned quote that prices of goods and services are about to drop!

APPLIED EXAMPLE 6 Inflation Rate of an Economy The function

$$I(t) = -0.2t^3 + 3t^2 + 100 \qquad (0 \leq t \leq 9)$$

gives the CPI of an economy, where $t = 0$ corresponds to the beginning of 1995.

a. Find the inflation rate at the beginning of 2001 ($t = 6$).
b. Show that inflation was moderating at that time.

Solution

a. We find $I'(t) = -0.6t^2 + 6t$. Next, we compute

$$I'(6) = -0.6(6)^2 + 6(6) = 14.4 \quad \text{and} \quad I(6) = -0.2(6)^3 + 3(6)^2 + 100 = 164.8$$

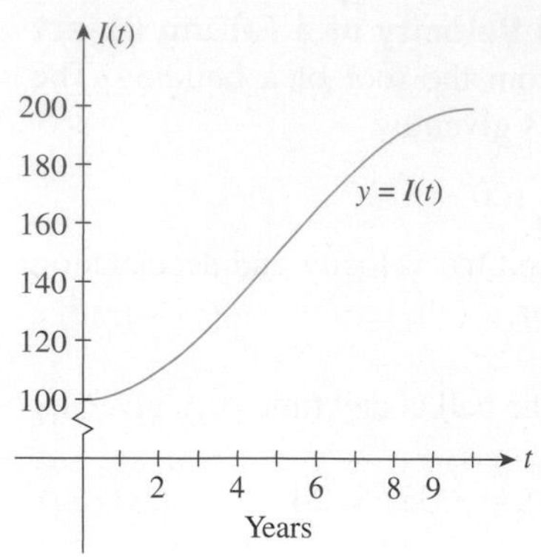

FIGURE 16
The CPI of an economy is given by $I(t)$.

from which we see that the inflation rate is

$$\frac{I'(6)}{I(6)} = \frac{14.4}{164.8} \approx 0.0874$$

or approximately 8.7%.

b. We find

$$I''(t) = \frac{d}{dt}(-0.6t^2 + 6t) = -1.2t + 6$$

Since

$$I''(6) = -1.2(6) + 6 = -1.2$$

we see that I' is indeed decreasing at $t = 6$ and conclude that inflation was moderating at that time (Figure 16).

3.5 Self-Check Exercises

1. Find the third derivative of

$$f(x) = 2x^5 - 3x^3 + x^2 - 6x + 10$$

2. Let

$$f(x) = \frac{1}{1 + x}$$

Find $f'(x)$, $f''(x)$, and $f'''(x)$.

3. A certain species of turtles faces extinction because dealers collect truckloads of turtle eggs to be sold as aphrodisiacs. After severe conservation measures are implemented, it is hoped that the turtle population will grow according to the rule

$$N(t) = 2t^3 + 3t^2 - 4t + 1000 \qquad (0 \le t \le 10)$$

where $N(t)$ denotes the population at the end of year t. Compute $N''(2)$ and $N''(8)$. What do your results tell you about the effectiveness of the program?

Solutions to Self-Check Exercises 3.5 can be found on page 218.

3.5 Concept Questions

1. **a.** What is the second derivative of a function f?
   b. How do you find the second derivative of a function f, assuming that it exists?

2. If $s = f(t)$ gives the position of an object moving on the coordinate line, what do $f'(t)$ and $f''(t)$ measure?

3. If $I(t)$ gives the CPI of an economy, how do you compute the relative rate of change of I with respect to t at time $t = c$?

3.5 Exercises

In Exercises 1–20, find the first and second derivatives of the function.

1. $f(x) = 4x^2 - 2x + 1$
2. $f(x) = -0.2x^2 + 0.3x + 4$
3. $f(x) = 2x^3 - 3x^2 + 1$
4. $g(x) = -3x^3 + 24x^2 + 6x - 64$
5. $h(t) = t^4 - 2t^3 + 6t^2 - 3t + 10$
6. $f(x) = x^5 - x^4 + x^3 - x^2 + x - 1$
7. $f(x) = (x^2 + 2)^5$
8. $g(t) = t^2(3t + 1)^4$
9. $g(t) = (2t^2 - 1)^2(3t^2)$
10. $h(x) = (x^2 + 1)^2(x - 1)$
11. $f(x) = (2x^2 + 2)^{7/2}$
12. $h(w) = (w^2 + 2w + 4)^{5/2}$

13. $f(x) = x(x^2 + 1)^2$

14. $g(u) = u(2u - 1)^3$

15. $f(x) = \dfrac{x}{2x + 1}$

16. $g(t) = \dfrac{t^2}{t - 1}$

17. $f(s) = \dfrac{s - 1}{s + 1}$

18. $f(u) = \dfrac{u}{u^2 + 1}$

19. $f(u) = \sqrt{4 - 3u}$

20. $f(x) = \sqrt{2x - 1}$

In Exercises 21–28, find the third derivative of the given function.

21. $f(x) = 3x^4 - 4x^3$

22. $f(x) = 3x^5 - 6x^4 + 2x^2 - 8x + 12$

23. $f(x) = \dfrac{1}{x}$

24. $f(x) = \dfrac{2}{x^2}$

25. $g(s) = \sqrt{3s - 2}$

26. $g(t) = \sqrt{2t + 3}$

27. $f(x) = (2x - 3)^4$

28. $g(t) = \left(\frac{1}{2}t^2 - 1\right)^5$

29. **Acceleration of a Falling Object** During the construction of an office building, a hammer is accidentally dropped from a height of 256 ft. The distance (in feet) the hammer falls in t sec is $s = 16t^2$. What is the hammer's velocity when it strikes the ground? What is its acceleration?

30. **Acceleration of a Car** The distance s (in feet) covered by a car after t sec is given by

$$s = -t^3 + 8t^2 + 20t \qquad (0 \le t \le 6)$$

Find a general expression for the car's acceleration at any time t $(0 \le t \le 6)$. Show that the car is decelerating after $2\frac{2}{3}$ sec.

31. **Crime Rates** The number of major crimes committed in Bronxville between 1988 and 1995 is approximated by the function

$$N(t) = -0.1t^3 + 1.5t^2 + 100 \qquad (0 \le t \le 7)$$

where $N(t)$ denotes the number of crimes committed in year t, with $t = 0$ corresponding to 1988. Enraged by the dramatic increase in the crime rate, Bronxville's citizens, with the help of the local police, organized "Neighborhood Crime Watch" groups in early 1992 to combat this menace.

a. Verify that the crime rate was increasing from 1988 through 1995.
Hint: Compute $N'(0), N'(1), \ldots, N'(7)$.

b. Show that the Neighborhood Crime Watch program was working by computing $N''(4)$, $N''(5)$, $N''(6)$, and $N''(7)$.

32. **GDP of a Developing Country** A developing country's gross domestic product (GDP) from 1996 to 2004 is approximated by the function

$$G(t) = -0.2t^3 + 2.4t^2 + 60 \qquad (0 \le t \le 8)$$

where $G(t)$ is measured in billions of dollars, with $t = 0$ corresponding to 1996.

a. Compute $G'(0), G'(1), \ldots, G'(8)$.

b. Compute $G''(0), G''(1), \ldots, G''(8)$.

c. Using the results obtained in parts (a) and (b), show that after a spectacular growth rate in the early years, the growth of the GDP cooled off.

33. **Disability Benefits** The number of persons aged 18–64 receiving disability benefits through Social Security, the Supplemental Security income, or both, from 1990 through 2000 is approximated by the function

$$N(t) = 0.00037t^3 - 0.0242t^2 + 0.52t + 5.3 \qquad (0 \le t \le 10)$$

where $f(t)$ is measured in units of a million and t is measured in years, with $t = 0$ corresponding to the beginning of 1990. Compute $N(8)$, $N'(8)$, and $N''(8)$ and interpret your results.

Source: Social Security Administration

34. **Obesity in America** The body mass index (BMI) measures body weight in relation to height. A BMI of 25 to 29.9 is considered overweight, a BMI of 30 or more is considered obese, and a BMI of 40 or more is morbidly obese. The percent of the U.S. population that is obese is approximated by the function

$$P(t) = 0.0004t^3 + 0.0036t^2 + 0.8t + 12 \qquad (0 \le t \le 13)$$

where t is measured in years, with $t = 0$ corresponding to the beginning of 1991. Show that the rate of the rate of change of the percent of the U.S. population that is deemed obese was positive from 1991 to 2004. What does this mean?

Source: Centers for Disease Control and Prevention

35. **Test Flight of a VTOL** In a test flight of the McCord Terrier, McCord Aviation's experimental VTOL (vertical takeoff and landing) aircraft, it was determined that t sec after liftoff, when the craft was operated in the vertical takeoff mode, its altitude (in feet) was

$$h(t) = \frac{1}{16}t^4 - t^3 + 4t^2 \qquad (0 \le t \le 8)$$

a. Find an expression for the craft's velocity at time t.

b. Find the craft's velocity when $t = 0$ (the initial velocity), $t = 4$, and $t = 8$.

c. Find an expression for the craft's acceleration at time t.

d. Find the craft's acceleration when $t = 0$, 4, and 8.

e. Find the craft's height when $t = 0$, 4, and 8.

36. **Air Purification** During testing of a certain brand of air purifier, the amount of smoke remaining t min after the start of the test was

$$A(t) = -0.00006t^5 + 0.00468t^4 - 0.1316t^3 + 1.915t^2 - 17.63t + 100$$

percent of the original amount. Compute $A'(10)$ and $A''(10)$ and interpret your results.
Source: Consumer Reports

37. **Aging Population** The population of Americans age 55 yr and over as a percent of the total population is approximated by the function

$$f(t) = 10.72(0.9t + 10)^{0.3} \qquad (0 \le t \le 20)$$

where t is measured in years, with $t = 0$ corresponding to 2000. Compute $f''(10)$ and interpret your result.
Source: U.S. Census Bureau

38. **Working Mothers** The percent of mothers who work outside the home and have children younger than age 6 yr is approximated by the function

$$P(t) = 33.55(t + 5)^{0.205} \qquad (0 \le t \le 21)$$

where t is measured in years, with $t = 0$ corresponding to the beginning of 1980. Compute $P''(20)$ and interpret your result.
Source: U.S. Bureau of Labor Statistics

In Exercises 39–43, determine whether the statement is true or false. If it is true, explain why it is true. If it is false, give an example to show why it is false.

39. If the second derivative of f exists at $x = a$, then $f''(a) = [f'(a)]^2$.

40. If $h = fg$ where f and g have second-order derivatives, then

$$h''(x) = f''(x)g(x) + 2f'(x)g'(x) + f(x)g''(x)$$

41. If $f(x)$ is a polynomial function of degree n, then $f^{(n+1)}(x) = 0$.

42. Suppose $P(t)$ represents the population of bacteria at time t and suppose $P'(t) > 0$ and $P''(t) < 0$; then the population is increasing at time t but at a decreasing rate.

43. If $h(x) = f(2x)$, then $h''(x) = 4f''(2x)$.

44. Let f be the function defined by the rule $f(x) = x^{7/3}$. Show that f has first- and second-order derivatives at all points x, and in particular at $x = 0$. Also show that the third derivative of f does *not* exist at $x = 0$.

45. Construct a function f that has derivatives of order up through and including n at a point a but fails to have the $(n + 1)$st derivative there.
Hint: See Exercise 44.

46. Show that a polynomial function has derivatives of all orders.
Hint: Let $P(x) = a_0x^n + a_1x^{n-1} + a_2x^{n-2} + \cdots + a_n$ be a polynomial of degree n, where n is a positive integer and $a_0, a_1, \ldots, a_n$ are constants with $a_0 \ne 0$. Compute $P'(x), P''(x), \ldots$.

3.5 Solutions to Self-Check Exercises

1. $f'(x) = 10x^4 - 9x^2 + 2x - 6$
$f''(x) = 40x^3 - 18x + 2$
$f'''(x) = 120x^2 - 18$

2. We write $f(x) = (1 + x)^{-1}$ and use the general power rule, obtaining

$$f'(x) = (-1)(1 + x)^{-2}\frac{d}{dx}(1 + x) = -(1 + x)^{-2}(1)$$
$$= -(1 + x)^{-2} = -\frac{1}{(1 + x)^2}$$

Continuing, we find

$$f''(x) = -(-2)(1 + x)^{-3}$$
$$= 2(1 + x)^{-3} = \frac{2}{(1 + x)^3}$$

$$f'''(x) = 2(-3)(1 + x)^{-4}$$
$$= -6(1 + x)^{-4}$$
$$= -\frac{6}{(1 + x)^4}$$

3. $N'(t) = 6t^2 + 6t - 4$ and $N''(t) = 12t + 6 = 6(2t + 1)$

Therefore, $N''(2) = 30$ and $N''(8) = 102$. The results of our computations reveal that at the end of year 2, the *rate* of growth of the turtle population is increasing at the rate of 30 turtles/year/year. At the end of year 8, the rate is increasing at the rate of 102 turtles/year/year. Clearly, the conservation measures are paying off handsomely.

USING TECHNOLOGY

Finding the Second Derivative of a Function at a Given Point

Some graphing utilities have the capability of numerically computing the second derivative of a function at a point. If your graphing utility has this capability, use it to work through the examples and exercises of this section.

EXAMPLE 1 Use the (second) numerical derivative operation of a graphing utility to find the second derivative of $f(x) = \sqrt{x}$ when $x = 4$.

Solution Using the (second) numerical derivative operation, we find

$$f''(4) = \text{der2}(\text{x}^\wedge.5, \text{x}, 4) = -.03125$$

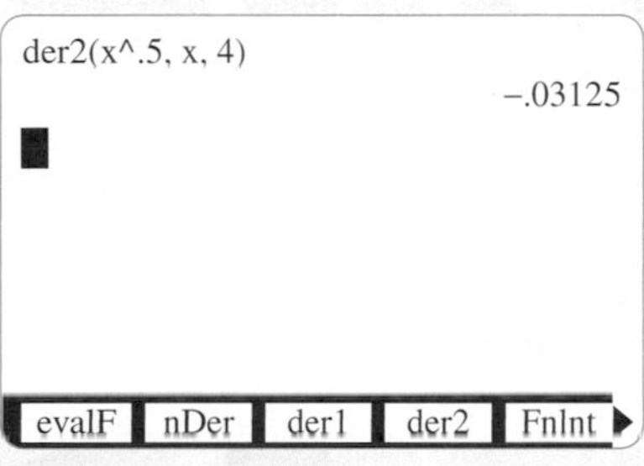

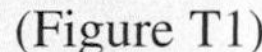

FIGURE T1
The TI-86 second derivative screen for computing $f''(4)$

(Figure T1).

APPLIED EXAMPLE 2 Prevalence of Alzheimer's Patients The number of Alzheimer's patients in the United States is given by

$$f(t) = -0.02765t^4 + 0.3346t^3 - 1.1261t^2 + 1.7575t + 3.7745 \qquad (0 \le t \le 6)$$

where $f(t)$ is measured in millions and t is measured in decades, with $t = 0$ corresponding to the beginning of 1990.

a. How fast is the number of Alzheimer's patients in the United States anticipated to be changing at the beginning of 2030?
b. How fast is the rate of change of the number of Alzheimer's patients in the United States anticipated to be changing at the beginning of 2030?
c. Plot the graph of f in the viewing window $[0, 7] \times [0, 12]$.

Source: Alzheimer's Association

Solution

a. Using the numerical derivative operation of a graphing utility, we find that the number of Alzheimer's patients at the beginning of 2030 can be anticipated to be changing at the rate of

$$f'(4) = 1.7311$$

That is, the number is increasing at the rate of approximately 1.7 million patients per decade.

(continued)

b. Using the (second) numerical derivative operation of a graphing utility, we find that

$$f''(4) = .4694$$

```
der2(-.02765 x^4+.334
6 x^3-1.1261 x^2+1.75
75 x+3.7745, x, 4)
                .4694
```

evalF | nDer | der1 | der2 | FnInt

FIGURE T2
The TI-86 second derivative screen for computing $f''(4)$

(Figure T2); that is, the rate of change of the number of Alzheimer's patients is increasing at the rate of approximately 0.5 million patients per decade per decade.

c. Figure T3 shows the graph.

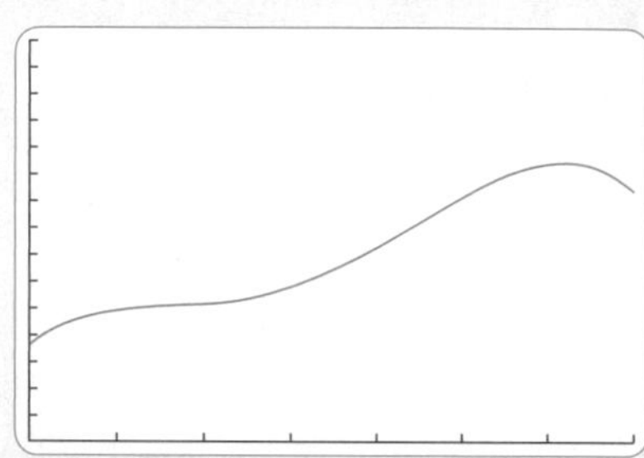

FIGURE T3
The graph of f in the viewing window $[0, 7] \times [0, 12]$

TECHNOLOGY EXERCISES

In Exercises 1–8, find the value of the second derivative of f at the given value of x. Express your answer correct to four decimal places.

1. $f(x) = 2x^3 - 3x^2 + 1;\ x = -1$

2. $f(x) = 2.5x^5 - 3x^3 + 1.5x + 4;\ x = 2.1$

3. $f(x) = 2.1x^{3.1} - 4.2x^{1.7} + 4.2;\ x = 1.4$

4. $f(x) = 1.7x^{4.2} - 3.2x^{1.3} + 4.2x - 3.2;\ x = 2.2$

5. $f(x) = \dfrac{x^2 + 2x - 5}{x^3 + 1};\ x = 2.1$

6. $f(x) = \dfrac{x^3 + x + 2}{2x^2 - 5x + 4};\ x = 1.2$

7. $f(x) = \dfrac{x^{1/2} + 2x^{3/2} + 1}{2x^{1/2} + 3};\ x = 0.5$

8. $f(x) = \dfrac{\sqrt{x} - 1}{2x + \sqrt{x} + 4};\ x = 2.3$

9. **Rate of Bank Failures** The rate at which banks were failing between 1982 and 1994 is given by

$$f(t) = -0.063447t^4 - 1.953283t^3 + 14.632576t^2 - 6.684704t + 47.458874 \qquad (0 \le t \le 12)$$

where $f(t)$ is measured in the number of banks/year and t is measured in years, with $t = 0$ corresponding to the beginning of 1982. Compute $f''(6)$ and interpret your results.
Source: Federal Deposit Insurance Corporation

10. **Multimedia Sales** Sales in the multimedia market (hardware and software) are given by

$$S(t) = -0.0094t^4 + 0.1204t^3 - 0.0868t^2 + 0.0195t + 3.3325 \qquad (0 \le t \le 10)$$

where $S(t)$ is measured in billions of dollars and t is measured in years, with $t = 0$ corresponding to 1990. Compute $S''(7)$ and interpret your results.
Source: Electronics Industries Association

3.6 Implicit Differentiation and Related Rates

Differentiating Implicitly

Up to now we have dealt with functions expressed in the form $y = f(x)$; that is, the dependent variable y is expressed *explicitly* in terms of the independent variable x. However, not all functions are expressed in this form. Consider, for example, the equation

$$x^2y + y - x^2 + 1 = 0 \tag{8}$$

This equation does express y *implicitly* as a function of x. In fact, solving (8) for y in terms of x, we obtain

$$(x^2 + 1)y = x^2 - 1 \qquad \text{Implicit equation}$$

$$y = f(x) = \frac{x^2 - 1}{x^2 + 1} \qquad \text{Explicit equation}$$

which gives an explicit representation of f.

Next, consider the equation

$$y^4 - y^3 - y + 2x^3 - x = 8$$

When certain restrictions are placed on x and y, this equation defines y as a function of x. But in this instance, we would be hard pressed to find y explicitly in terms of x. The following question arises naturally: How does one go about computing dy/dx in this case?

As it turns out, thanks to the chain rule, a method *does* exist for computing the derivative of a function directly from the implicit equation defining the function. This method is called **implicit differentiation** and is demonstrated in the next several examples.

EXAMPLE 1 Given the equation $y^2 = x$, find $\dfrac{dy}{dx}$.

Solution Differentiating both sides of the equation with respect to x, we obtain

$$\frac{d}{dx}(y^2) = \frac{d}{dx}(x)$$

To carry out the differentiation of the term $\dfrac{d}{dx}(y^2)$, we note that y is a function of x.

Writing $y = f(x)$ to remind us of this fact, we find that

$$\begin{aligned} \frac{d}{dx}(y^2) &= \frac{d}{dx}[f(x)]^2 && \text{Write } y = f(x). \\ &= 2f(x)f'(x) && \text{Use the chain rule.} \\ &= 2y\frac{dy}{dx} && \text{Return to using } y \text{ instead of } f(x). \end{aligned}$$

Therefore, the equation

$$\frac{d}{dx}(y^2) = \frac{d}{dx}(x)$$

is equivalent to

$$2y\frac{dy}{dx} = 1$$

Solving for $\frac{dy}{dx}$ yields

$$\frac{dy}{dx} = \frac{1}{2y}$$

■

Before considering other examples, let's summarize the important steps involved in implicit differentiation. (Here we assume that dy/dx exists.)

Finding $\frac{dy}{dx}$ by Implicit Differentiation

1. Differentiate both sides of the equation *with respect to x*. (Make sure that the derivative of any term involving y includes the factor dy/dx).
2. Solve the resulting equation for dy/dx in terms of x and y.

EXAMPLE 2 Find $\frac{dy}{dx}$ given the equation

$$y^3 - y + 2x^3 - x = 8$$

Solution Differentiating both sides of the given equation with respect to x, we obtain

$$\frac{d}{dx}(y^3 - y + 2x^3 - x) = \frac{d}{dx}(8)$$

$$\frac{d}{dx}(y^3) - \frac{d}{dx}(y) + \frac{d}{dx}(2x^3) - \frac{d}{dx}(x) = 0$$

Now, recalling that y is a function of x, we apply the chain rule to the first two terms on the left. Thus,

$$3y^2\frac{dy}{dx} - \frac{dy}{dx} + 6x^2 - 1 = 0$$

$$(3y^2 - 1)\frac{dy}{dx} = 1 - 6x^2$$

$$\frac{dy}{dx} = \frac{1 - 6x^2}{3y^2 - 1}$$

■

EXPLORE AND DISCUSS

Refer to Example 2. Suppose we think of the equation $y^3 - y + 2x^3 - x = 8$ as defining x implicitly as a function of y. Find dx/dy and justify your method of solution.

EXAMPLE 3 Consider the equation $x^2 + y^2 = 4$.

a. Find dy/dx by implicit differentiation.
b. Find the slope of the tangent line to the graph of the function $y = f(x)$ at the point $(1, \sqrt{3})$.
c. Find an equation of the tangent line of part (b).

Solution

a. Differentiating both sides of the equation with respect to x, we obtain

$$\frac{d}{dx}(x^2 + y^2) = \frac{d}{dx}(4)$$

$$\frac{d}{dx}(x^2) + \frac{d}{dx}(y^2) = 0$$

$$2x + 2y\frac{dy}{dx} = 0$$

$$\frac{dy}{dx} = -\frac{x}{y} \qquad (y \neq 0)$$

b. The slope of the tangent line to the graph of the function at the point $(1, \sqrt{3})$ is given by

$$\left.\frac{dy}{dx}\right|_{(1, \sqrt{3})} = \left.-\frac{x}{y}\right|_{(1, \sqrt{3})} = -\frac{1}{\sqrt{3}}$$

(*Note:* This notation is read "dy/dx evaluated at the point $(1, \sqrt{3})$.")

c. An equation of the tangent line in question is found by using the point-slope form of the equation of a line with the slope $m = -1/\sqrt{3}$ and the point $(1, \sqrt{3})$. Thus,

$$y - \sqrt{3} = -\frac{1}{\sqrt{3}}(x - 1)$$

$$\sqrt{3}y - 3 = -x + 1$$

$$x + \sqrt{3}y - 4 = 0$$

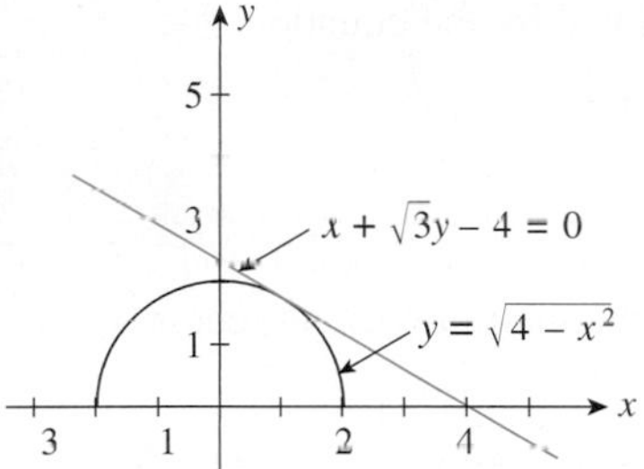

FIGURE 17
The line $x + \sqrt{3}y - 4 = 0$ is tangent to the graph of the function $y = f(x)$.

A sketch of this tangent line is shown in Figure 17.

We can also solve the equation $x^2 + y^2 = 4$ explicitly for y in terms of x. If we do this, we obtain

$$y = \pm\sqrt{4 - x^2}$$

From this, we see that the equation $x^2 + y^2 = 4$ defines the two functions

$$y = f(x) = \sqrt{4 - x^2}$$

$$y = g(x) = -\sqrt{4 - x^2}$$

Since the point $(1, \sqrt{3})$ does not lie on the graph of $y = g(x)$, we conclude that

$$y = f(x) = \sqrt{4 - x^2}$$

is the required function. The graph of f is the upper semicircle shown in Figure 17. ■

Note The notation

$$\left.\frac{dy}{dx}\right|_{(a, b)}$$

is used to denote the value of dy/dx at the point (a, b). ■

EXPLORE AND DISCUSS

Refer to Example 3. Yet another function defined implicitly by the equation $x^2 + y^2 = 4$ is the function

$$y = h(x) = \begin{cases} \sqrt{4 - x^2} & \text{if } -2 \le x < 0 \\ -\sqrt{4 - x^2} & \text{if } 0 \le x \le 2 \end{cases}$$

1. Sketch the graph of h.
2. Show that $h'(x) = -x/y$.
3. Find an equation of the tangent line to the graph of h at the point $(1, -\sqrt{3})$.

To find dy/dx at a *specific point* (a, b), differentiate the given equation implicitly with respect to x and then replace x and y by a and b, respectively, *before* solving the equation for dy/dx. This often simplifies the amount of algebra involved.

EXAMPLE 4 Find $\dfrac{dy}{dx}$ given that x and y are related by the equation

$$x^2y^3 + 6x^2 = y + 12$$

and that $y = 2$ when $x = 1$.

Solution Differentiating both sides of the given equation with respect to x, we obtain

$$\frac{d}{dx}(x^2y^3) + \frac{d}{dx}(6x^2) = \frac{d}{dx}(y) + \frac{d}{dx}(12)$$

$$x^2 \cdot \frac{d}{dx}(y^3) + y^3 \cdot \frac{d}{dx}(x^2) + 12x = \frac{dy}{dx} \qquad \text{Use the product rule on } \frac{d}{dx}(x^2y^3).$$

$$3x^2y^2\frac{dy}{dx} + 2xy^3 + 12x = \frac{dy}{dx}$$

Substituting $x = 1$ and $y = 2$ into this equation gives

$$3(1)^2(2)^2\frac{dy}{dx} + 2(1)(2)^3 + 12(1) = \frac{dy}{dx}$$

$$12\frac{dy}{dx} + 16 + 12 = \frac{dy}{dx}$$

and, solving for $\dfrac{dy}{dx}$,

$$\frac{dy}{dx} = -\frac{28}{11}$$

Note that it is not necessary to find an explicit expression for dy/dx. ■

Note In Examples 3 and 4, you can verify that the points at which we evaluated dy/dx actually lie on the curve in question by showing that the coordinates of the points satisfy the given equations. ■

EXAMPLE 5 Find $\frac{dy}{dx}$ given that x and y are related by the equation

$$\sqrt{x^2 + y^2} - x^2 = 5$$

Solution Differentiating both sides of the given equation with respect to x, we obtain

$$\frac{d}{dx}(x^2 + y^2)^{1/2} - \frac{d}{dx}(x^2) = \frac{d}{dx}(5)$$

Write $\sqrt{x^2 + y^2} = (x^2 + y^2)^{1/2}$.

$$\frac{1}{2}(x^2 + y^2)^{-1/2}\frac{d}{dx}(x^2 + y^2) - 2x = 0$$

Use the general power rule on the first term.

$$\frac{1}{2}(x^2 + y^2)^{-1/2}\left(2x + 2y\frac{dy}{dx}\right) - 2x = 0$$

$$2x + 2y\frac{dy}{dx} = 4x(x^2 + y^2)^{1/2}$$

Transpose $2x$ and multiply both sides by $2(x^2 + y^2)^{1/2}$.

$$2y\frac{dy}{dx} = 4x(x^2 + y^2)^{1/2} - 2x$$

$$\frac{dy}{dx} = \frac{2x\sqrt{x^2 + y^2} - x}{y}$$

■

Related Rates

Implicit differentiation is a useful technique for solving a class of problems known as **related-rates** problems. The following is a typical related-rates problem: Suppose x and y are two quantities that depend on a third quantity t and we know the relationship between x and y in the form of an equation. Can we find a relationship between dx/dt and dy/dt? In particular, if we know one of the rates of change at a specific value of t—say, dx/dt—can we find the other rate, dy/dt, at that value of t?

APPLIED EXAMPLE 6 Rate of Change of Housing Starts A study prepared for the National Association of Realtors estimates that the number of housing starts in the southwest, $N(t)$ (in units of a million), over the next 5 years is related to the mortgage rate $r(t)$ (percent per year) by the equation

$$9N^2 + r = 36$$

What is the rate of change of the number of housing starts with respect to time when the mortgage rate is 11% per year and is increasing at the rate of 1.5% per year?

Solution We are given that

$$r = 11 \quad \text{and} \quad \frac{dr}{dt} = 1.5$$

at a certain instant of time, and we are required to find dN/dt. First, by substituting $r = 11$ into the given equation, we find

$$9N^2 + 11 = 36$$

$$N^2 = \frac{25}{9}$$

or $N = 5/3$ (we reject the negative root). Next, differentiating the given equation implicitly on both sides with respect to t, we obtain

$$\frac{d}{dt}(9N^2) + \frac{d}{dt}(r) = \frac{d}{dt}(36)$$

$$18N\frac{dN}{dt} + \frac{dr}{dt} = 0 \qquad \text{Use the chain rule on the first term.}$$

Then, substituting $N = 5/3$ and $dr/dt = 1.5$ into this equation gives

$$18\left(\frac{5}{3}\right)\frac{dN}{dt} + 1.5 = 0$$

Solving this equation for dN/dt then gives

$$\frac{dN}{dt} = -\frac{1.5}{30} \approx -0.05$$

Thus, at the instant of time under consideration, the number of housing starts is decreasing at the rate of 50,000 units per year. ■

APPLIED EXAMPLE 7 Supply-Demand A major audiotape manufacturer is willing to make x thousand ten-packs of metal alloy audiocassette tapes available in the marketplace each week when the wholesale price is $\$p$ per ten-pack. It is known that the relationship between x and p is governed by the supply equation

$$x^2 - 3xp + p^2 = 5$$

How fast is the supply of tapes changing when the price per ten-pack is \$11, the quantity supplied is 4000 ten-packs, and the wholesale price per ten-pack is increasing at the rate of \$.10 per ten-pack each week?

Solution We are given that

$$p = 11, \qquad x = 4, \qquad \frac{dp}{dt} = 0.1$$

at a certain instant of time, and we are required to find dx/dt. Differentiating the given equation on both sides with respect to t, we obtain

$$\frac{d}{dt}(x^2) - \frac{d}{dt}(3xp) + \frac{d}{dt}(p^2) = \frac{d}{dt}(5)$$

$$2x\frac{dx}{dt} - 3\left(p\frac{dx}{dt} + x\frac{dp}{dt}\right) + 2p\frac{dp}{dt} = 0 \qquad \text{Use the product rule on the second term.}$$

Substituting the given values of p, x, and dp/dt into the last equation, we have

$$2(4)\frac{dx}{dt} - 3\left[(11)\frac{dx}{dt} + 4(0.1)\right] + 2(11)(0.1) = 0$$

$$8\frac{dx}{dt} - 33\frac{dx}{dt} - 1.2 + 2.2 = 0$$

$$25\frac{dx}{dt} = 1$$

$$\frac{dx}{dt} = 0.04$$

Thus, at the instant of time under consideration the supply of ten-pack audiocassettes is increasing at the rate of (0.04)(1000), or 40, ten-packs per week. ■

In certain related-rates problems, we need to formulate the problem mathematically before analyzing it. The following guidelines can be used to help solve problems of this type.

Solving Related-Rates Problems

1. Assign a variable to each quantity. Draw a diagram if needed.
2. Write the *given* values of the variables and their rates of change with respect to t.
3. Find an equation giving the relationship between the variables.
4. Differentiate both sides of this equation implicitly with respect to t.
5. Replace the variables and their derivatives by the numerical data found in step 2 and solve the equation for the required rate of change.

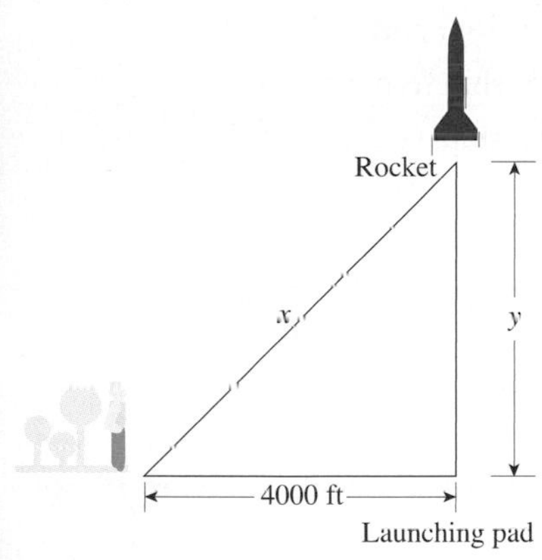

FIGURE 18
The rate at which x is changing with respect to time is related to the rate of change of y with respect to time.

APPLIED EXAMPLE 8 Watching a Rocket Launch At a distance of 4000 feet from the launch site, a spectator is observing a rocket being launched. If the rocket lifts off vertically and is rising at a speed of 600 feet/second when it is at an altitude of 3000 feet, how fast is the distance between the rocket and the spectator changing at that instant?

Solution

Step 1 Let

y = altitude of the rocket

x = distance between the rocket and the spectator

at any time t (Figure 18).

Step 2 We are given that at a certain instant of time

$$y = 3000 \quad \text{and} \quad \frac{dy}{dt} = 600$$

and are asked to find dx/dt at that instant.

Step 3 Applying the Pythagorean theorem to the right triangle in Figure 18, we find that

$$x^2 = y^2 + 4000^2$$

Therefore, when $y = 3000$,

$$x = \sqrt{3000^2 + 4000^2} = 5000$$

Step 4 Next, we differentiate the equation $x^2 = y^2 + 4000^2$ with respect to t, obtaining

$$2x\frac{dx}{dt} = 2y\frac{dy}{dt}$$

(Remember, both x and y are functions of t.)

Step 5 Substituting $x = 5000$, $y = 3000$, and $dy/dt = 600$, we find

$$2(5000)\frac{dx}{dt} = 2(3000)(600)$$

$$\frac{dx}{dt} = 360$$

Therefore, the distance between the rocket and the spectator is changing at a rate of 360 feet/second. ■

Be sure that you do *not* replace the variables in the equation found in Step 3 by their numerical values before differentiating the equation.

EXAMPLE 9 A passenger ship and an oil tanker left port sometime in the morning; the former headed north, and the latter headed east. At noon, the passenger ship was 40 miles from port and sailing at 30 mph, while the oil tanker was 30 miles from port and sailing at 20 mph. How fast was the distance between the two ships changing at that time?

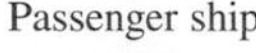
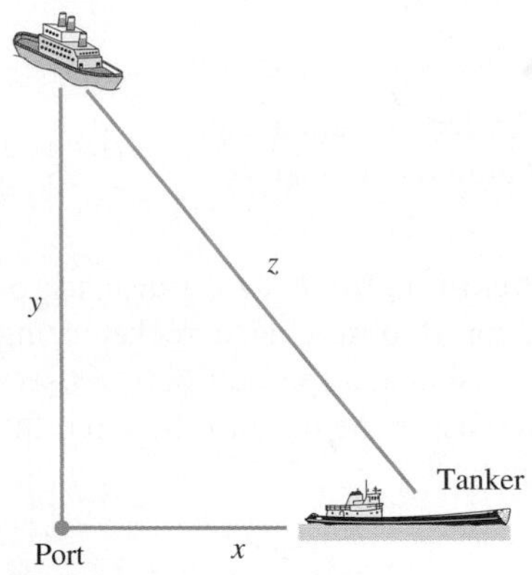

FIGURE 19
We want to find dz/dt, the rate at which the distance between the two ships is changing at a certain instant of time.

Solution

Step 1 Let

x = distance of the oil tanker from port
y = distance of the passenger ship from port
z = distance between the two ships

See Figure 19.

Step 2 We are given that at noon

$$x = 30 \qquad y = 40 \qquad \frac{dx}{dt} = 20 \qquad \frac{dy}{dt} = 30$$

and we are required to find $\frac{dz}{dt}$ at that time.

Step 3 Applying the Pythagorean theorem to the right triangle in Figure 19, we find that

$$z^2 = x^2 + y^2 \tag{9}$$

In particular, when $x = 30$ and $y = 40$, we have

$$z^2 = 30^2 + 40^2 = 2500 \quad \text{or} \quad z = 50.$$

Step 4 Differentiating (9) implicitly with respect to t, we obtain

$$2z\frac{dz}{dt} = 2x\frac{dx}{dt} + 2y\frac{dy}{dt}$$

$$z\frac{dz}{dt} = x\frac{dx}{dt} + y\frac{dy}{dt}$$

Step 5 Finally, substituting $x = 30$, $y = 40$, $z = 50$, $dx/dt = 20$, and $dy/dt = 30$ into the last equation, we find

$$50\frac{dz}{dt} = (30)(20) + (40)(30) \quad \text{and} \quad \frac{dz}{dt} = 36$$

Therefore, at noon on the day in question, the ships are moving apart at the rate of 36 mph. ■

3.6 Self-Check Exercises

1. Given the equation $x^3 + 3xy + y^3 = 4$, find dy/dx by implicit differentiation.

2. Find an equation of the tangent line to the graph of
$$16x^2 + 9y^2 = 144$$
at the point
$$\left(2, -\frac{4\sqrt{5}}{3}\right)$$

Solutions to Self-Check Exercises 3.6 can be found on page 232.

3.6 Concept Questions

1. **a.** Suppose the equation $F(x, y) = 0$ defines y as a function of x. Explain how implicit differentiation can be used to find dy/dx.
   b. What is the role of the chain rule in implicit differentiation?

2. Suppose the equation $xg(y) + yf(x) = 0$, where f and g are differentiable functions, defines y as a function of x. Find an expression for dy/dx.

3. In your own words, describe what a related-rates problem is.

4. Give the steps that you would use to solve a related-rates problem.

3.6 Exercises

In Exercises 1–8, find the derivative *dy/dx* (a) by solving each of the implicit equations for *y* explicitly in terms of *x* and (b) by differentiating each of the equations implicitly. Show that, in each case, the results are equivalent.

1. $x + 2y = 5$
2. $3x + 4y = 6$
3. $xy = 1$
4. $xy - y - 1 = 0$
5. $x^3 - x^2 - xy = 4$
6. $x^2y - x^2 + y - 1 = 0$
7. $\dfrac{x}{y} - x^2 = 1$
8. $\dfrac{y}{x} - 2x^3 = 4$

In Exercises 9–30, find *dy/dx* by implicit differentiation.

9. $x^2 + y^2 = 16$
10. $2x^2 + y^2 = 16$
11. $x^2 - 2y^2 = 16$
12. $x^3 + y^3 + y - 4 = 0$
13. $x^2 - 2xy = 6$
14. $x^2 + 5xy + y^2 = 10$
15. $x^2y^2 - xy = 8$
16. $x^2y^3 - 2xy^2 = 5$
17. $x^{1/2} + y^{1/2} = 1$
18. $x^{1/3} + y^{1/3} = 1$
19. $\sqrt{x + y} = x$
20. $(2x + 3y)^{1/3} = x^2$
21. $\dfrac{1}{x^2} + \dfrac{1}{y^2} = 1$
22. $\dfrac{1}{x^3} + \dfrac{1}{y^3} = 5$
23. $\sqrt{xy} = x + y$
24. $\sqrt{xy} = 2x + y^2$
25. $\dfrac{x + y}{x - y} = 3x$
26. $\dfrac{x - y}{2x + 3y} = 2x$
27. $xy^{3/2} = x^2 + y^2$
28. $x^2y^{1/2} = x + 2y^3$
29. $(x + y)^3 + x^3 + y^3 = 0$
30. $(x + y^2)^{10} = x^2 + 25$

In Exercises 31–34, find an equation of the tangent line to the graph of the function *f* defined by the equation at the indicated point.

31. $4x^2 + 9y^2 = 36$; $(0, 2)$
32. $y^2 - x^2 = 16$; $(2, 2\sqrt{5})$
33. $x^2y^3 - y^2 + xy - 1 = 0$; $(1, 1)$
34. $(x - y - 1)^3 = x$; $(1, -1)$

In Exercises 35–38, find the second derivative *d²y/dx²* of each of the functions defined implicitly by the equation.

35. $xy = 1$
36. $x^3 + y^3 = 28$
37. $y^2 - xy = 8$
38. $x^{1/3} + y^{1/3} = 1$

39. The volume of a right-circular cylinder of radius r and height h is $V = \pi r^2 h$. Suppose the radius and height of the cylinder are changing with respect to time t.
   a. Find a relationship between dV/dt, dr/dt, and dh/dt.
   b. At a certain instant of time, the radius and height of the cylinder are 2 and 6 in. and are increasing at the rate of

0.1 and 0.3 in./sec, respectively. How fast is the volume of the cylinder increasing?

40. A car leaves an intersection traveling west. Its position 4 sec later is 20 ft from the intersection. At the same time, another car leaves the same intersection heading north so that its position 4 sec later is 28 ft from the intersection. If the speed of the cars at that instant of time is 9 ft/sec and 11 ft/sec, respectively, find the rate at which the distance between the two cars is changing.

41. PRICE-DEMAND Suppose the quantity demanded weekly of the Super Titan radial tires is related to its unit price by the equation

$$p + x^2 = 144$$

where p is measured in dollars and x is measured in units of a thousand. How fast is the quantity demanded changing when $x = 9$, $p = 63$, and the price/tire is increasing at the rate of \$2/week?

42. PRICE-SUPPLY Suppose the quantity x of Super Titan radial tires made available each week in the marketplace is related to the unit-selling price by the equation

$$p - \frac{1}{2}x^2 = 48$$

where x is measured in units of a thousand and p is in dollars. How fast is the weekly supply of Super Titan radial tires being introduced into the marketplace when $x = 6$, $p = 66$, and the price/tire is decreasing at the rate of \$3/week?

43. PRICE-DEMAND The demand equation for a certain brand of metal alloy audiotape is

$$100x^2 + 9p^2 = 3600$$

where x represents the number (in thousands) of ten-packs demanded each week when the unit price is \$$p$. How fast is the quantity demanded increasing when the unit price/ten-pack is \$14 and the selling price is dropping at the rate of \$.15/ten-pack/week?
Hint: To find the value of x when $p = 14$, solve the equation $100x^2 + 9p^2 = 3600$ for x when $p = 14$.

44. EFFECT OF PRICE ON SUPPLY Suppose the wholesale price of a certain brand of medium-sized eggs p (in dollars/carton) is related to the weekly supply x (in thousands of cartons) by the equation

$$625p^2 - x^2 = 100$$

If 25,000 cartons of eggs are available at the beginning of a certain week and the price is falling at the rate of 2¢/carton/week, at what rate is the supply falling?
Hint: To find the value of p when $x = 25$, solve the supply equation for p when $x = 25$.

45. SUPPLY-DEMAND Refer to Exercise 44. If 25,000 cartons of eggs are available at the beginning of a certain week and the supply is falling at the rate of 1000 cartons/week, at what rate is the wholesale price changing?

46. ELASTICITY OF DEMAND The demand function for a certain make of ink-jet cartridge is

$$p = -0.01x^2 - 0.1x + 6$$

where p is the unit price in dollars and x is the quantity demanded each week, measured in units of a thousand. Compute the elasticity of demand and determine whether the demand is inelastic, unitary, or elastic when $x = 10$.

47. ELASTICITY OF DEMAND The demand function for a certain brand of compact disc is

$$p = -0.01x^2 - 0.2x + 8$$

where p is the wholesale unit price in dollars and x is the quantity demanded each week, measured in units of a thousand. Compute the elasticity of demand and determine whether the demand is inelastic, unitary, or elastic when $x = 15$.

48. The volume V of a cube with sides of length x in. is changing with respect to time. At a certain instant of time, the sides of the cube are 5 in. long and increasing at the rate of 0.1 in./sec. How fast is the volume of the cube changing at that instant of time?

49. OIL SPILLS In calm waters, oil spilling from the ruptured hull of a grounded tanker spreads in all directions. If the area polluted is a circle and its radius is increasing at a rate of 2 ft/sec, determine how fast the area is increasing when the radius of the circle is 40 ft.

50. Two ships leave the same port at noon. Ship A sails north at 15 mph, and ship B sails east at 12 mph. How fast is the distance between them changing at 1 p.m.?

51. A car leaves an intersection traveling east. Its position t sec later is given by $x = t^2 + t$ ft. At the same time, another car leaves the same intersection heading north, traveling $y = t^2 + 3t$ ft in t sec. Find the rate at which the distance between the two cars will be changing 5 sec later.

52. At a distance of 50 ft from the pad, a man observes a helicopter taking off from a heliport. If the helicopter lifts off vertically and is rising at a speed of 44 ft/sec when it is at an altitude of 120 ft, how fast is the distance between the helicopter and the man changing at that instant?

53. A spectator watches a rowing race from the edge of a river bank. The lead boat is moving in a straight line that is 120 ft from the river bank. If the boat is moving at a constant speed of 20 ft/sec, how fast is the boat moving away from the spectator when it is 50 ft past her?

54. A boat is pulled toward a dock by means of a rope wound on a drum that is located 4 ft above the bow of the boat. If the rope is being pulled in at the rate of 3 ft/sec, how fast is the boat approaching the dock when it is 25 ft from the dock? (See figure on the next page.)

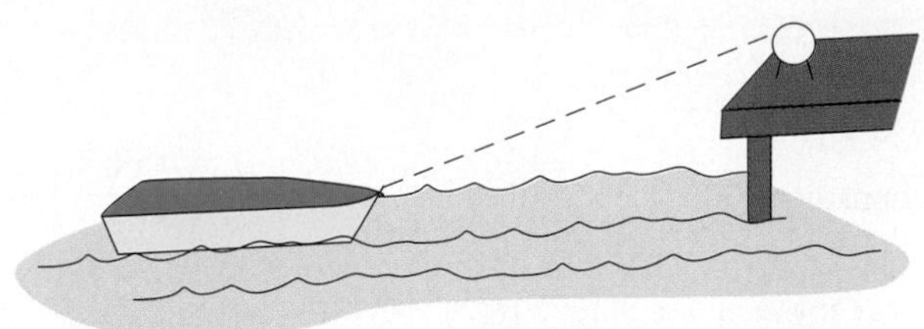

55. Assume that a snowball is in the shape of a sphere. If the snowball melts at a rate that is proportional to its surface area, show that its radius decreases at a constant rate.
Hint: Its volume is $V = (4/3)\pi r^3$, and its surface area is $S = 4\pi r^2$.

56. Blowing Soap Bubbles Carlos is blowing air into a soap bubble at the rate of 8 cm³/sec. Assuming that the bubble is spherical, how fast is its radius changing at the instant of time when the radius is 10 cm? How fast is the surface area of the bubble changing at that instant of time?

57. Coast Guard Patrol Search Mission The pilot of a Coast Guard patrol aircraft on a search mission had just spotted a disabled fishing trawler and decided to go in for a closer look. Flying in a straight line at a constant altitude of 1000 ft and at a steady speed of 264 ft/sec, the aircraft passed directly over the trawler. How fast was the aircraft receding from the trawler when it was 1500 ft from it?

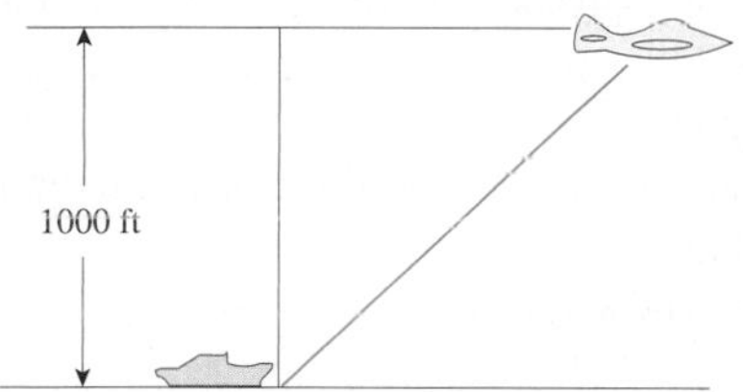

58. A coffee pot in the form of a circular cylinder of radius 4 in. is being filled with water flowing at a constant rate. If the water level is rising at the rate of 0.4 in./sec, what is the rate at which water is flowing into the coffee pot?

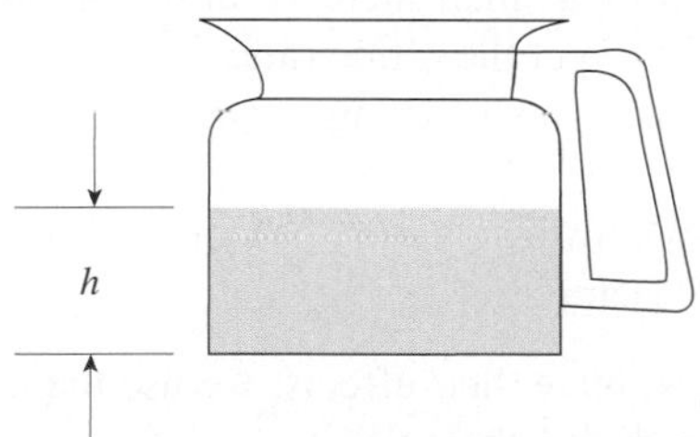

59. A 6-ft tall man is walking away from a street light 18 ft high at a speed of 6 ft/sec. How fast is the tip of his shadow moving along the ground?

60. A 20-ft ladder leaning against a wall begins to slide. How fast is the top of the ladder sliding down the wall at the instant of time when the bottom of the ladder is 12 ft from the wall and sliding away from the wall at the rate of 5 ft/sec?
Hint: Refer to the accompanying figure. By the Pythagorean theorem, $x^2 + y^2 = 400$. Find dy/dt when $x = 12$ and $dx/dt = 5$.

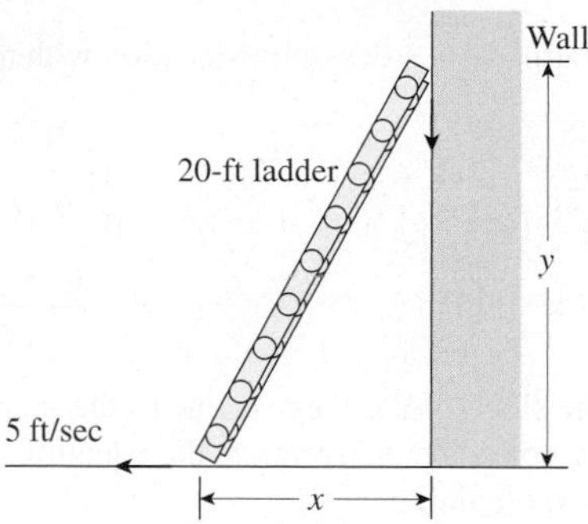

61. The base of a 13-ft ladder leaning against a wall begins to slide away from the wall. At the instant of time when the base is 12 ft from the wall, the base is moving at the rate of 8 ft/sec. How fast is the top of the ladder sliding down the wall at that instant of time?
Hint: Refer to the hint in Problem 60.

62. Water flows from a tank of constant cross-sectional area 50 ft² through an orifice of constant cross-sectional area 1.4 ft² located at the bottom of the tank (see the figure).

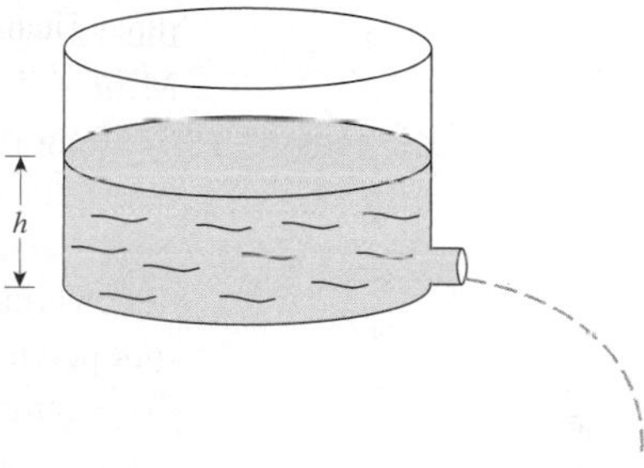

Initially, the height of the water in the tank was 20 ft and its height t sec later is given by the equation

$$2\sqrt{h} + \frac{1}{25}t - 2\sqrt{20} = 0 \qquad (0 \le t \le 50\sqrt{20})$$

How fast was the height of the water decreasing when its height was 8 ft?

In Exercises 63 and 64, determine whether the statement is true or false. If it is true, explain why it is true. If it is false, give an example to show why it is false.

63. If f and g are differentiable and $f(x)g(y) = 0$, then

$$\frac{dy}{dx} = -\frac{f'(x)g(y)}{f(x)g'(y)} \qquad (f(x) \neq 0 \text{ and } g'(y) \neq 0)$$

64. If f and g are differentiable and $f(x) + g(y) = 0$, then

$$\frac{dy}{dx} = -\frac{f'(x)}{g'(y)}$$

3.6 Solutions to Self-Check Exercises

1. Differentiating both sides of the equation with respect to x, we have

$$3x^2 + 3y + 3xy' + 3y^2y' = 0$$
$$(x^2 + y) + (x + y^2)y' = 0$$
$$y' = -\frac{x^2 + y}{x + y^2}$$

2. To find the slope of the tangent line to the graph of the function at any point, we differentiate the equation implicitly with respect to x, obtaining

$$32x + 18yy' = 0$$
$$y' = -\frac{16x}{9y}$$

In particular, the slope of the tangent line at $\left(2, -\frac{4\sqrt{5}}{3}\right)$ is

$$m = -\frac{16(2)}{9\left(-\frac{4\sqrt{5}}{3}\right)} = \frac{8}{3\sqrt{5}}$$

Using the point-slope form of the equation of a line, we find

$$y - \left(-\frac{4\sqrt{5}}{3}\right) = \frac{8}{3\sqrt{5}}(x - 2)$$
$$y = \frac{8\sqrt{5}}{15}x - \frac{36\sqrt{5}}{15} = \frac{8\sqrt{5}}{15}x - \frac{12\sqrt{5}}{5}$$

3.7 Differentials

The Millers are planning to buy a house in the near future and estimate that they will need a 30-year fixed-rate mortgage of $240,000. If the interest rate increases from the present rate of 7% per year to 7.4% per year between now and the time the Millers decide to secure the loan, approximately how much more per month will their mortgage be? (You will be asked to answer this question in Exercise 44, page 239.)

Questions such as this, in which one wishes to *estimate* the change in the dependent variable (monthly mortgage payment) corresponding to a small change in the independent variable (interest rate per year), occur in many real-life applications. For example:

- An economist would like to know how a small increase in a country's capital expenditure will affect the country's gross domestic output.
- A sociologist would like to know how a small increase in the amount of capital investment in a housing project will affect the crime rate.
- A businesswoman would like to know how raising a product's unit price by a small amount will affect her profit.
- A bacteriologist would like to know how a small increase in the amount of a bactericide will affect a population of bacteria.

To calculate these changes and estimate their effects, we use the *differential* of a function, a concept that will be introduced shortly.

Increments

Let x denote a variable quantity and suppose x changes from x_1 to x_2. This change in x is called the **increment in x** and is denoted by the symbol Δx (read "delta x"). Thus,

$$\Delta x = x_2 - x_1 \quad \text{Final value} - \text{initial value} \tag{10}$$

EXAMPLE 1 Find the increment in x as x changes (a) from 3 to 3.2 and (b) from 3 to 2.7.

Solution

a. Here, $x_1 = 3$ and $x_2 = 3.2$, so

$$\Delta x = x_2 - x_1 = 3.2 - 3 = 0.2$$

b. Here, $x_1 = 3$ and $x_2 = 2.7$. Therefore,

$$\Delta x = x_2 - x_1 = 2.7 - 3 = -0.3$$

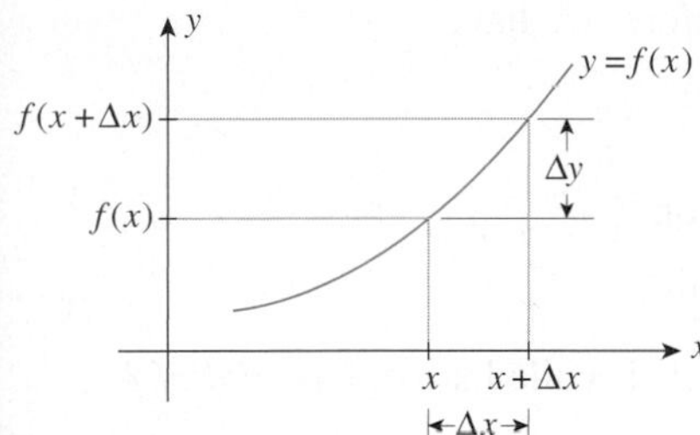

FIGURE 20
An increment of Δx in x induces an increment of $\Delta y = f(x + \Delta x) - f(x)$ in y.

Observe that Δx plays the same role that h played in Section 2.4.

Now, suppose two quantities, x and y, are related by an equation $y = f(x)$, where f is a function. If x changes from x to $x + \Delta x$, then the corresponding change in y is called the **increment in y**. It is denoted by Δy and is defined by

$$\Delta y = f(x + \Delta x) - f(x) \tag{11}$$

(see Figure 20).

EXAMPLE 2 Let $y = x^3$. Find Δx and Δy when x changes (a) from 2 to 2.01 and (b) from 2 to 1.98.

Solution Let $f(x) = x^3$.

a. Here, $\Delta x = 2.01 - 2 = 0.01$. Next,

$$\begin{aligned}\Delta y &= f(x + \Delta x) - f(x) = f(2.01) - f(2)\\ &= (2.01)^3 - 2^3 = 8.120601 - 8 = 0.120601\end{aligned}$$

b. Here, $\Delta x = 1.98 - 2 = -0.02$. Next,

$$\begin{aligned}\Delta y &= f(x + \Delta x) - f(x) = f(1.98) - f(2)\\ &= (1.98)^3 - 2^3 = 7.762392 - 8 = -0.237608\end{aligned}$$

Differentials

We can obtain a relatively quick and simple way of approximating Δy, the change in y due to a small change Δx, by examining the graph of the function f shown in Figure 21.

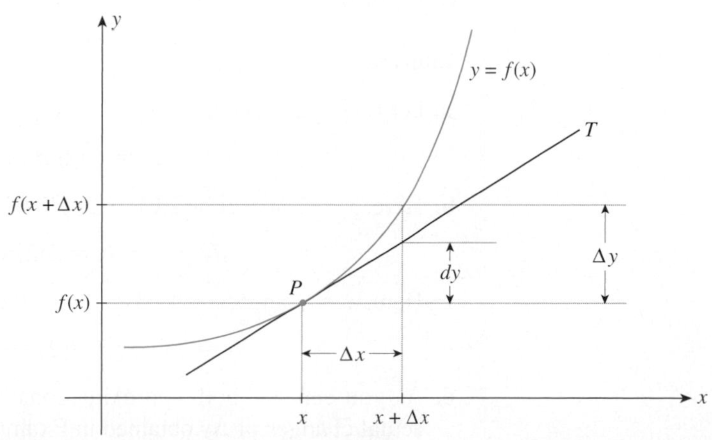

FIGURE 21
If Δx is small, dy is a good approximation of Δy.

Observe that near the point of tangency P, the tangent line T is close to the graph of f. Therefore, if Δx is small, then dy is a good approximation of Δy. We can find an expression for dy as follows: Notice that the slope of T is given by

$$\frac{dy}{\Delta x} \quad \text{Rise} \div \text{run}$$

However, the slope of T is given by $f'(x)$. Therefore, we have

$$\frac{dy}{\Delta x} = f'(x)$$

or $dy = f'(x)\Delta x$. Thus, we have the approximation

$$\Delta y \approx dy = f'(x)\Delta x$$

in terms of the derivative of f at x. The quantity dy is called the *differential of y*.

The Differential

Let $y = f(x)$ define a differentiable function of x. Then,

1. The **differential dx** of the independent variable x is $dx = \Delta x$.
2. The **differential dy** of the dependent variable y is

$$dy = f'(x)\Delta x = f'(x)\,dx \qquad \textbf{(12)}$$

Notes

1. For the independent variable x: There is no difference between Δx and dx—both measure the change in x from x to $x + \Delta x$.
2. For the dependent variable y: Δy measures the *actual* change in y as x changes from x to $x + \Delta x$, whereas dy measures the *approximate* change in y corresponding to the same change in x.
3. The differential dy depends on both x and dx, but for fixed x, dy is a linear function of dx. ■

EXAMPLE 3 Let $y = x^3$.

a. Find the differential dy of y.
b. Use dy to approximate Δy when x changes from 2 to 2.01.
c. Use dy to approximate Δy when x changes from 2 to 1.98.
d. Compare the results of part (b) with those of Example 2.

Solution

a. Let $f(x) = x^3$. Then,

$$dy = f'(x)\,dx = 3x^2\,dx$$

b. Here, $x = 2$ and $dx = 2.01 - 2 = 0.01$. Therefore,

$$dy = 3x^2\,dx = 3(2)^2(0.01) = 0.12$$

c. Here, $x = 2$ and $dx = 1.98 - 2 = -0.02$. Therefore,

$$dy = 3x^2\,dx = 3(2)^2(-0.02) = -0.24$$

d. As you can see, both approximations 0.12 and -0.24 are quite close to the actual changes of Δy obtained in Example 2: 0.120601 and -0.237608. ■

Observe how much easier it is to find an approximation to the exact change in a function with the help of the differential, rather than calculating the exact change in the function itself. In the following examples, we take advantage of this fact.

EXAMPLE 4 Approximate the value of $\sqrt{26.5}$ using differentials. Verify your result using the $\boxed{\sqrt{\ }}$ key on your calculator.

Solution Since we want to compute the square root of a number, let's consider the function $y = f(x) = \sqrt{x}$. Since 25 is the number nearest 26.5 whose square root is readily recognized, let's take $x = 25$. We want to know the change in y, Δy, as x changes from $x = 25$ to $x = 26.5$, an increase of $\Delta x = 1.5$ units. Using Equation (12), we find

$$\begin{aligned}\Delta y \approx dy &= f'(x)\Delta x\\ &= \left(\frac{1}{2\sqrt{x}}\bigg|_{x=25}\right)\cdot(1.5) = \left(\frac{1}{10}\right)(1.5) = 0.15\end{aligned}$$

Therefore,

$$\sqrt{26.5} - \sqrt{25} = \Delta y \approx 0.15$$
$$\sqrt{26.5} \approx \sqrt{25} + 0.15 = 5.15$$

The exact value of $\sqrt{26.5}$, rounded off to five decimal places, is 5.14782. Thus, the error incurred in the approximation is 0.00218. ■

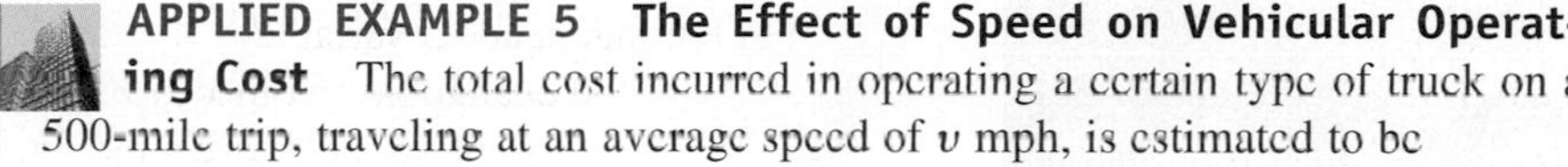

APPLIED EXAMPLE 5 The Effect of Speed on Vehicular Operating Cost The total cost incurred in operating a certain type of truck on a 500-mile trip, traveling at an average speed of v mph, is estimated to be

$$C(v) = 125 + v + \frac{4500}{v}$$

dollars. Find the approximate change in the total operating cost when the average speed is increased from 55 mph to 58 mph.

Solution With $v = 55$ and $\Delta v = dv = 3$, we find

$$\begin{aligned}\Delta C \approx dC = C'(v)dv &= \left(1 - \frac{4500}{v^2}\right)\bigg|_{v=55}(3)\\ &= \left(1 - \frac{4500}{3025}\right)(3) \approx -1.46\end{aligned}$$

so the total operating cost is found to decrease by \$1.46. This might explain why so many independent truckers often exceed the 55 mph speed limit. ■

APPLIED EXAMPLE 6 The Effect of Advertising on Sales The relationship between the amount of money x spent by Cannon Precision Instruments on advertising and Cannon's total sales $S(x)$ is given by the function

$$S(x) = -0.002x^3 + 0.6x^2 + x + 500 \qquad (0 \le x \le 200)$$

where x is measured in thousands of dollars. Use differentials to estimate the change in Cannon's total sales if advertising expenditures are increased from \$100,000 ($x = 100$) to \$105,000 ($x = 105$).

Solution The required change in sales is given by

$$\begin{aligned}\Delta S \approx dS &= S'(100)dx \\ &= -0.006x^2 + 1.2x + 1|_{x=100} \cdot (5) \qquad dx = 105 - 100 = 5 \\ &= (-60 + 120 + 1)(5) = 305\end{aligned}$$

—that is, an increase of \$305,000. ■

APPLIED EXAMPLE 7 The Rings of Neptune

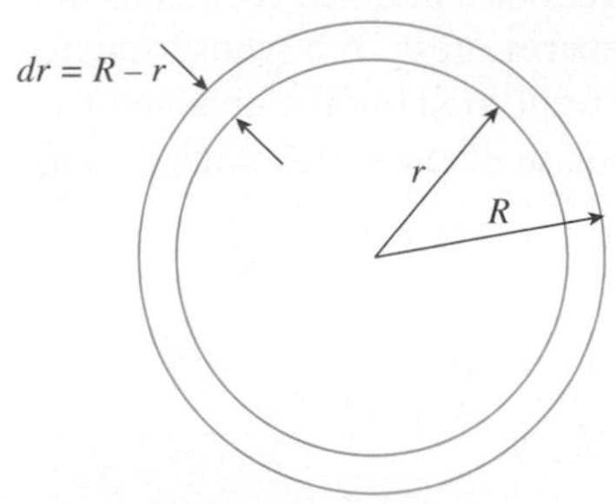

(a) The area of the ring is the circumference of the inner circle times the thickness.

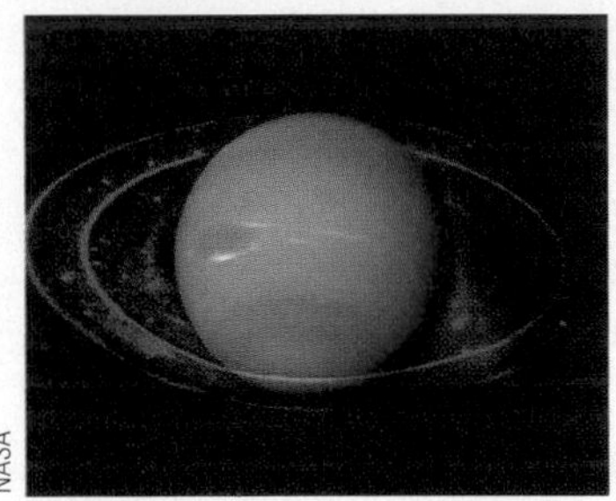

(b) Neptune and its rings

FIGURE 22

a. A ring has an inner radius of r units and an outer radius of R units, where $(R - r)$ is small in comparison to r (Figure 22a). Use differentials to estimate the area of the ring.

b. Recent observations, including those of *Voyager I* and *II*, showed that Neptune's ring system is considerably more complex than had been believed. For one thing, it is made up of a large number of distinguishable rings rather than one continuous great ring as previously thought (Figure 22b). The outermost ring, 1989N1R, has an inner radius of approximately 62,900 kilometers (measured from the center of the planet), and a radial width of approximately 50 kilometers. Using these data, estimate the area of the ring.

Solution

a. Since the area of a circle of radius x is $A = f(x) = \pi x^2$, we find

$$\begin{aligned}\pi R^2 - \pi r^2 &= f(R) - f(r) \\ &= \Delta A \qquad \text{Remember, } \Delta A = \text{change in } f \text{ when } x \text{ changes from } x = r \text{ to } x = R. \\ &\approx dA \\ &= f'(r)dr\end{aligned}$$

where $dr = R - r$. So, we see that the area of the ring is approximately $2\pi r(R - r)$ square units. In words, the area of the ring is approximately equal to

Circumference of the inner circle × Thickness of the ring

b. Applying the results of part (a) with $r = 62{,}900$ and $dr = 50$, we find that the area of the ring is approximately $2\pi(62{,}900)(50)$, or 19,760,618 square kilometers, which is roughly 4% of Earth's surface. ■

Before looking at the next example, we need to familiarize ourselves with some terminology. If a quantity with exact value q is measured or calculated with an error of Δq, then the quantity $\Delta q/q$ is called the *relative error* in the measurement or calculation of q. If the quantity $\Delta q/q$ is expressed as a percentage, it is then called the *percentage error*. Because Δq is approximated by dq, we normally approximate the relative error $\Delta q/q$ by dq/q.

APPLIED EXAMPLE 8 Estimating Errors in Measurement Suppose the radius of a ball-bearing is measured to be 0.5 inch, with a maximum error of ±0.0002 inch. Then, the relative error in r is

$$\frac{dr}{r} = \frac{\pm 0.0002}{0.5} = \pm 0.0004$$

and the percentage error is ±0.04%. ■

APPLIED EXAMPLE 9 Estimating Errors in Measurement Suppose the side of a cube is measured with a maximum percentage error of 2%. Use differentials to estimate the maximum percentage error in the calculated volume of the cube.

Solution Suppose the side of the cube is x, so its volume is

$$V = x^3$$

We are given that $\left|\dfrac{dx}{x}\right| \le 0.02$. Now,

$$dV = 3x^2dx$$

and so

$$\frac{dV}{V} = \frac{3x^2dx}{x^3} = 3\frac{dx}{x}$$

Therefore,

$$\left|\frac{dV}{V}\right| = 3\left|\frac{dx}{x}\right| \le 3(0.02) = 0.06$$

and we see that the maximum percentage error in the measurement of the volume of the cube is 6%. ■

Finally, if at some point in reading this section you have a sense of déjà vu, do not be surprised, because the notion of the differential was first used in Section 3.4 (see Example 1). There we took $\Delta x = 1$ since we were interested in finding the marginal cost when the level of production was increased from $x = 250$ to $x = 251$. If we had used differentials, we would have found

$$C(251) - C(250) \approx C'(250)\,dx$$

so that taking $dx = \Delta x = 1$, we have $C(251) - C(250) \approx C'(250)$, which agrees with the result obtained in Example 1. Thus, in Section 3.4, we touched upon the notion of the differential, albeit in the special case in which $dx = 1$.

3.7 Self-Check Exercises

1. Find the differential of $f(x) = \sqrt{x} + 1$.
2. A certain country's government economists have determined that the demand equation for corn in that country is given by
$$p = f(x) = \frac{125}{x^2 + 1}$$
where p is expressed in dollars/bushel and x, the quantity demanded each year, is measured in billions of bushels. The economists are forecasting a harvest of 6 billion bushels for the year. If the actual production of corn were 6.2 billion bushels for the year instead, what would be the approximate drop in the predicted price of corn/bushel?

Solutions to Self-Check Exercises 3.7 can be found on page 240.

3.7 Concept Questions

1. If $y = f(x)$, what is the differential of x? Write an expression for the differential dy.
2. Let $y = f(x)$. What is the relationship between the actual change in y, Δy, when x changes from x to $x + \Delta x$, and the differential dy of f at x? Illustrate this relationship graphically.

3.7 Exercises

In Exercises 1–14, find the differential of the function.

1. $f(x) = 2x^2$
2. $f(x) = 3x^2 + 1$
3. $f(x) = x^3 - x$
4. $f(x) = 2x^3 + x$
5. $f(x) = \sqrt{x + 1}$
6. $f(x) = \dfrac{3}{\sqrt{x}}$
7. $f(x) = 2x^{3/2} + x^{1/2}$
8. $f(x) = 3x^{5/6} + 7x^{2/3}$
9. $f(x) = x + \dfrac{2}{x}$
10. $f(x) = \dfrac{3}{x - 1}$
11. $f(x) = \dfrac{x - 1}{x^2 + 1}$
12. $f(x) = \dfrac{2x^2 + 1}{x + 1}$
13. $f(x) = \sqrt{3x^2 - x}$
14. $f(x) = (2x^2 + 3)^{1/3}$

15. Let f be the function defined by

$$y = f(x) = x^2 - 1$$

   a. Find the differential of f.
   b. Use your result from part (a) to find the approximate change in y if x changes from 1 to 1.02.
   c. Find the actual change in y if x changes from 1 to 1.02 and compare your result with that obtained in part (b).

16. Let f be the function defined by

$$y = f(x) = 3x^2 - 2x + 6$$

   a. Find the differential of f.
   b. Use your result from part (a) to find the approximate change in y if x changes from 2 to 1.97.
   c. Find the actual change in y if x changes from 2 to 1.97 and compare your result with that obtained in part (b).

17. Let f be the function defined by

$$y = f(x) = \frac{1}{x}$$

   a. Find the differential of f.
   b. Use your result from part (a) to find the approximate change in y if x changes from -1 to -0.95.
   c. Find the actual change in y if x changes from -1 to -0.95 and compare your result with that obtained in part (b).

18. Let f be the function defined by

$$y = f(x) = \sqrt{2x + 1}$$

   a. Find the differential of f.
   b. Use your result from part (a) to find the approximate change in y if x changes from 4 to 4.1.
   c. Find the actual change in y if x changes from 4 to 4.1 and compare your result with that obtained in part (b).

In Exercises 19–26, use differentials to approximate the quantity.

19. $\sqrt{10}$
20. $\sqrt{17}$
21. $\sqrt{49.5}$
22. $\sqrt{99.7}$
23. $\sqrt[3]{7.8}$
24. $\sqrt[4]{81.6}$
25. $\sqrt{0.089}$
26. $\sqrt[3]{0.00096}$

27. Use a differential to approximate $\sqrt{4.02} + \dfrac{1}{\sqrt{4.02}}$.

   Hint: Let $f(x) = \sqrt{x} + \dfrac{1}{\sqrt{x}}$ and compute dy with $x = 4$ and $dx = 0.02$.

28. Use a differential to approximate $\dfrac{2(4.98)}{(4.98)^2 + 1}$.

   Hint: Study the hint for Exercise 27.

29. Error Estimation The length of each edge of a cube is 12 cm, with a possible error in measurement of 0.02 cm. Use differentials to estimate the error that might occur when the volume of the cube is calculated.

30. Estimating the Amount of Paint Required A coat of paint of thickness 0.05 cm is to be applied uniformly to the faces of a cube of edge 30 cm. Use differentials to find the approximate amount of paint required for the job.

31. Error Estimation A hemisphere-shaped dome of radius 60 ft is to be coated with a layer of rust-proofer before painting. Use differentials to estimate the amount of rust-proofer needed if the coat is to be 0.01 in. thick.
   Hint: The volume of a hemisphere of radius r is $V = \frac{2}{3}\pi r^3$.

32. Growth of a Cancerous Tumor The volume of a spherical cancerous tumor is given by

$$V(r) = \frac{4}{3}\pi r^3$$

   If the radius of a tumor is estimated at 1.1 cm, with a maximum error in measurement of 0.005 cm, determine the error that might occur when the volume of the tumor is calculated.

33. Unclogging Arteries Research done in the 1930s by the French physiologist Jean Poiseuille showed that the resistance R of a blood vessel of length l and radius r is $R = kl/r^4$, where k is a constant. Suppose a dose of the drug TPA increases r by 10%. How will this affect the resistance R? Assume that l is constant.

34. GROSS DOMESTIC PRODUCT An economist has determined that a certain country's gross domestic product (GDP) is approximated by the function $f(x) = 640x^{1/5}$, where $f(x)$ is measured in billions of dollars and x is the capital outlay in billions of dollars. Use differentials to estimate the change in the country's GDP if the country's capital expenditure changes from \$243 billion to \$248 billion.

35. LEARNING CURVES The length of time (in seconds) a certain individual takes to learn a list of n items is approximated by

$$f(n) = 4n\sqrt{n - 4}$$

Use differentials to approximate the additional time it takes the individual to learn the items on a list when n is increased from 85 to 90 items.

36. EFFECT OF ADVERTISING ON PROFITS The relationship between Cunningham Realty's quarterly profits, $P(x)$, and the amount of money x spent on advertising per quarter is described by the function

$$P(x) = -\frac{1}{8}x^2 + 7x + 30 \qquad (0 \leq x \leq 50)$$

where both $P(x)$ and x are measured in thousands of dollars. Use differentials to estimate the increase in profits when advertising expenditure each quarter is increased from \$24,000 to \$26,000.

37. EFFECT OF MORTGAGE RATES ON HOUSING STARTS A study prepared for the National Association of Realtors estimates that the number of housing starts per year over the next 5 yr will be

$$N(r) = \frac{7}{1 + 0.02r^2}$$

million units, where r (percent) is the mortgage rate. Use differentials to estimate the decrease in the number of housing starts when the mortgage rate is increased from 12% to 12.5%.

38. SUPPLY-PRICE The supply equation for a certain brand of radio is given by

$$p = s(x) = 0.3\sqrt{x} + 10$$

where x is the quantity supplied and p is the unit price in dollars. Use differentials to approximate the change in price when the quantity supplied is increased from 10,000 units to 10,500 units.

39. DEMAND-PRICE The demand function for the Sentinel smoke alarm is given by

$$p = d(x) = \frac{30}{0.02x^2 + 1}$$

where x is the quantity demanded (in units of a thousand) and p is the unit price in dollars. Use differentials to estimate the change in the price p when the quantity demanded changes from 5000 to 5500 units/week.

40. SURFACE AREA OF AN ANIMAL Animal physiologists use the formula

$$S = kW^{2/3}$$

to calculate an animal's surface area (in square meters) from its weight W (in kilograms), where k is a constant that depends on the animal under consideration. Suppose a physiologist calculates the surface area of a horse ($k = 0.1$). If the horse's weight is estimated at 300 kg, with a maximum error in measurement of 0.6 kg, determine the percentage error in the calculation of the horse's surface area.

41. FORECASTING PROFITS The management of Trappee and Sons forecast that they will sell 200,000 cases of their TexaPep hot sauce next year. Their annual profit is described by

$$P(x) = -0.000032x^3 + 6x - 100$$

thousand dollars, where x is measured in thousands of cases. If the maximum error in the forecast is 15%, determine the corresponding error in Trappee's profits.

42. FORECASTING COMMODITY PRICES A certain country's government economists have determined that the demand equation for soybeans in that country is given by

$$p = f(x) = \frac{55}{2x^2 + 1}$$

where p is expressed in dollars/bushel and x, the quantity demanded each year, is measured in billions of bushels. The economists are forecasting a harvest of 1.8 billion bushels for the year, with a maximum error of 15% in their forecast. Determine the corresponding maximum error in the predicted price per bushel of soybeans.

43. CRIME STUDIES A sociologist has found that the number of serious crimes in a certain city each year is described by the function

$$N(x) = \frac{500(400 + 20x)^{1/2}}{(5 + 0.2x)^2}$$

where x (in cents/dollar deposited) is the level of reinvestment in the area in conventional mortgages by the city's ten largest banks. Use differentials to estimate the change in the number of crimes if the level of reinvestment changes from 20¢/dollar deposited to 22¢/dollar deposited.

44. FINANCING A HOME The Millers are planning to buy a home in the near future and estimate that they will need a 30-yr fixed-rate mortgage for \$240,000. Their monthly payment P (in dollars) can be computed using the formula

$$P = \frac{20{,}000r}{1 - \left(1 + \dfrac{r}{12}\right)^{-360}}$$

where r is the interest rate per year.

a. Find the differential of P.

b. If the interest rate increases from the present rate of 7%/year to 7.2%/year between now and the time the Millers decide to secure the loan, approximately how much more will their monthly mortgage payment be? How much more will it be if the interest rate increases to 7.3%/year? To 7.4%/year? To 7.5%/year?

45. **INVESTMENTS** Lupé deposits a sum of $10,000 into an account that pays interest at the rate of r/year compounded monthly. Her investment at the end of 10 yr is given by

$$A = 10{,}000\left(1 + \frac{r}{12}\right)^{120}$$

a. Find the differential of A.

b. Approximately how much more would Lupé's account be worth at the end of the term if her account paid 8.1%/year instead of 8%/year? 8.2%/year instead of 8%/year? 8.3%/year instead of 8%/year?

46. **KEOGH ACCOUNTS** Ian, who is self-employed, contributes $2000 a month into a Keogh account earning interest at the rate of r/year compounded monthly. At the end of 25 yr, his account will be worth

$$S = \frac{24{,}000\left[\left(1 + \frac{r}{12}\right)^{300} - 1\right]}{r}$$

dollars.

a. Find the differential of S.

b. Approximately how much more would Ian's account be worth at the end of 25 yr if his account earned 9.1%/year instead of 9%/year? 9.2%/year instead of 9%/year? 9.3%/year instead of 9%/year?

In Exercises 47 and 48, determine whether the statement is true or false. If it is true, explain why it is true. If it is false, give an example to show why it is false.

47. If $y = ax + b$ where a and b are constants, then $\Delta y = dy$.

48. If $A = f(x)$, then the percentage change in A is

$$\frac{100f'(x)}{f(x)}dx$$

3.7 Solutions to Self-Check Exercises

1. We find

$$f'(x) = \frac{1}{2}x^{-1/2} = \frac{1}{2\sqrt{x}}$$

Therefore, the required differential of f is

$$dy = \frac{1}{2\sqrt{x}}dx$$

2. We first compute the differential

$$dp = -\frac{250x}{(x^2 + 1)^2}dx$$

Next, using Equation (12) with $x = 6$ and $dx = 0.2$, we find

$$\Delta p \approx dp = -\frac{250(6)}{(36 + 1)^2}(0.2) = -0.22$$

or a drop in price of 22¢/bushel.

USING TECHNOLOGY

Finding the Differential of a Function

The calculation of the differential of f at a given value of x involves the evaluation of the derivative of f at that point and can be facilitated through the use of the numerical derivative function.

EXAMPLE 1 Use dy to approximate Δy if $y = x^2(2x^2 + x + 1)^{2/3}$ and x changes from 2 to 1.98.

Solution Let $f(x) = x^2(2x^2 + x + 1)^{2/3}$. Since $dx = 1.98 - 2 = -0.02$, we find the required approximation to be

$$dy = f'(2)(-0.02)$$

But using the numerical derivative operation, we find

$$f'(2) = 30.57581679$$

(see Figure T1). Thus,

$$dy = (-0.02)(30.57581679) = -0.6115163358$$

```
nDeriv(X^2(2X^2+
X+1)^(2/3), X, 2)
        30.57581679
```

FIGURE T1
The TI-83 numerical derivative screen for computing $f'(2)$

APPLIED EXAMPLE 2 Financing a Home The Meyers are considering the purchase of a house in the near future and estimate that they will need a loan of \$120,000. Based on a 30-year conventional mortgage with an annual interest rate of r, their monthly repayment will be

$$P = \frac{10{,}000r}{1 - \left(1 + \frac{r}{12}\right)^{-360}}$$

dollars. If the interest rate increases from 10% per year to 10.2% per year between now and the time the Meyers decide to secure the loan, approximately how much more will their monthly mortgage payment be?

Solution Let's write

$$P = f(r) = \frac{10{,}000r}{1 - \left(1 + \frac{r}{12}\right)^{-360}}$$

Then the increase in the mortgage payment will be approximately

$$\begin{aligned} dP &= f'(0.1)\,dr = f'(0.1)(0.002) && \text{Since } dr = 0.102 - 0.1 \\ &= (8867.543909)(0.002) \approx 17.7351 && \text{Use the numerical derivative operation.} \end{aligned}$$

or approximately \$17.74 per month. (See Figure T2.)

```
nDeriv((10000X)/
(1−(1+X/12)^ −360),
X, .1)
        8867.543909
```

FIGURE T2
The TI-83 numerical derivative screen for computing $f'(0.1)$

(continued)

TECHNOLOGY EXERCISES

In Exercises 1–6, use dy to approximate Δy for the function $y = f(x)$ when x changes from $x = a$ to $x = b$.

1. $f(x) = 0.21x^7 - 3.22x^4 + 5.43x^2 + 1.42x + 12.42$; $a = 3$, $b = 3.01$
2. $f(x) = \dfrac{0.2x^2 + 3.1}{1.2x + 1.3}$; $a = 2$, $b = 1.96$
3. $f(x) = \sqrt{2.2x^2 + 1.3x + 4}$; $a = 1$, $b = 1.03$
4. $f(x) = x\sqrt{2x^3 - x + 4}$; $a = 2$, $b = 1.98$
5. $f(x) = \dfrac{\sqrt{x^2 + 4}}{x - 1}$; $a = 4$, $b = 4.1$
6. $f(x) = 2.1x^2 + \dfrac{3}{\sqrt{x}} + 5$; $a = 3$, $b = 2.95$
7. Calculating Mortgage Payments Refer to Example 2. How much more will the Meyers' mortgage payment be each month if the interest rate increases from 10% to 10.3%/year? To 10.4%/year? To 10.5%/year?
8. Estimating the Area of a Ring of Neptune The ring 1989N2R of the planet Neptune has an inner radius of approximately 53,200 km (measured from the center of the planet) and a radial width of 15 km. Use differentials to estimate the area of the ring.
9. Effect of Price Increase on Quantity Demanded The quantity demanded each week of the Alpha Sports Watch, x (in thousands), is related to its unit price of p dollars by the equation
$$x = f(p) = 10\sqrt{\frac{50 - p}{p}} \qquad (0 \leq p \leq 50)$$
Use differentials to find the decrease in the quantity of the watches demanded each week if the unit price is increased from \$40 to \$42.
10. Period of a Communications Satellite According to Kepler's third law, the period T (in days) of a satellite moving in a circular orbit d mi above the surface of Earth is given by
$$T = 0.0588\left(1 + \frac{d}{3959}\right)^{3/2}$$
Suppose a communications satellite that was moving in a circular orbit 22,000 mi above Earth's surface at one time has, because of friction, dropped down to a new orbit that is 21,500 mi above Earth's surface. Estimate the decrease in the period of the satellite to the nearest $\frac{1}{100}$th hr.

CHAPTER 3 Summary of Principal Formulas and Terms

FORMULAS

1. Derivative of a constant	$\frac{d}{dx}(c) = 0 \qquad (c\text{, a constant})$
2. Power rule	$\frac{d}{dx}(x^n) = nx^{n-1}$
3. Constant multiple rule	$\frac{d}{dx}[cf(x)] = cf'(x)$
4. Sum rule	$\frac{d}{dx}[f(x) \pm g(x)] = f'(x) \pm g'(x)$
5. Product rule	$\frac{d}{dx}[f(x)g(x)] = f(x)g'(x) + g(x)f'(x)$
6. Quotient rule	$\frac{d}{dx}\left[\frac{f(x)}{g(x)}\right] = \frac{g(x)f'(x) - f(x)g'(x)}{[g(x)]^2}$
7. Chain rule	$\frac{d}{dx}g(f(x)) = g'(f(x))f'(x)$
8. General power rule	$\frac{d}{dx}[f(x)]^n = n[f(x)]^{n-1}f'(x)$
9. Average cost function	$\overline{C}(x) = \frac{C(x)}{x}$
10. Revenue function	$R(x) = px$
11. Profit function	$P(x) = R(x) - C(x)$
12. Elasticity of demand	$E(p) = -\frac{pf'(p)}{f(p)}$
13. Differential of y	$dy = f'(x)dx$

TERMS

marginal cost (199)
marginal cost function (199)
average cost (200)
marginal average cost function (200)
marginal revenue (202)
marginal revenue function (203)
marginal profit function (203)
elasticity of demand (205)
elastic demand (206)
unitary demand (206)
inelastic demand (206)
second derivative of f (212)
implicit differentiation (221)
related rates (225)

CHAPTER 3 Concept Review Questions

Fill in the blanks.

1. a. If c is a constant, then $\frac{d}{dx}(c) =$ _____.
   b. The power rule states that if n is any real number, then $\frac{d}{dx}(x^n) =$ _____.
   c. The constant multiple rule states that if c is a constant, then $\frac{d}{dx}[cf(x)] =$ _____.
   d. The sum rule states that $\frac{d}{dx}[f(x) \pm g(x)] =$ _____.
2. a. The product rule states that $\frac{d}{dx}[f(x)g(x)] =$ _____.
   b. The quotient rule states that $\frac{d}{dx}[f(x)/g(x)] =$ _____.
3. a. The chain rule states that if $h(x) = g[f(x)]$, then $h'(x) =$ _____.
   b. The general power rule states that if $h(x) = [f(x)]^n$, then $h'(x) =$ _____.
4. If C, R, P, and $\overline{C}$ denote the total cost function, the total revenue function, the profit function, and the average cost function, respectively, then C' denotes the _____ _____ function, R' denotes the _____ _____ function, P' denotes the _____ _____ function, and $\overline{C}'$ denotes the _____ _____ _____ function.
5. a. If f is a differentiable demand function defined by $x = f(p)$, then the elasticity of demand at price p is given by $E(p) =$ _____.
   b. The demand is _____ if $E(p) > 1$; it is _____ if $E(p) = 1$; it is _____ if $E(p) < 1$.
6. Suppose a function $y = f(x)$ is defined implicitly by an equation in x and y. To find $\frac{dy}{dx}$, we differentiate _____ _____ of the equation with respect to x and then solve the resulting equation for $\frac{dy}{dx}$. The derivative of a term involving y includes _____ as a factor.
7. In a related-rates problem, we are given a relationship between x and _____ that depends on a third variable t. Knowing the values of x, y, and $\frac{dx}{dt}$ at a, we want to find _____ at _____.
8. Let $y = f(t)$ and $x = g(t)$. If $x^2 + y^2 = 4$, then $\frac{dx}{dt} =$ _____. If $xy = 1$, then $\frac{dy}{dt} =$ _____.
9. a. If a variable quantity x changes from x_1 to x_2, then the increment in x is $\Delta x =$ _____.
   b. If $y = f(x)$ and x changes from x to $x + \Delta x$, then the increment in y is $\Delta y =$ _____.
10. If $y = f(x)$, where f is a differentiable function, then the differential dx of x is $dx =$ _____, where _____ is an increment in _____, and the differential dy of y is $dy =$ _____.

CHAPTER 3 Review Exercises

In Exercises 1–30, find the derivative of the function.

1. $f(x) = 3x^5 - 2x^4 + 3x^2 - 2x + 1$
2. $f(x) = 4x^6 + 2x^4 + 3x^2 - 2$
3. $g(x) = -2x^{-3} + 3x^{-1} + 2$
4. $f(t) = 2t^2 - 3t^3 - t^{-1/2}$
5. $g(t) = 2t^{-1/2} + 4t^{-3/2} + 2$
6. $h(x) = x^2 + \dfrac{2}{x}$
7. $f(t) = t + \dfrac{2}{t} + \dfrac{3}{t^2}$
8. $g(s) = 2s^2 - \dfrac{4}{s} + \dfrac{2}{\sqrt{s}}$
9. $h(x) = x^2 - \dfrac{2}{x^{3/2}}$
10. $f(x) = \dfrac{x + 1}{2x - 1}$
11. $g(t) = \dfrac{t^2}{2t^2 + 1}$
12. $h(t) = \dfrac{\sqrt{t}}{\sqrt{t} + 1}$
13. $f(x) = \dfrac{\sqrt{x} - 1}{\sqrt{x} + 1}$
14. $f(t) = \dfrac{t}{2t^2 + 1}$
15. $f(x) = \dfrac{x^2(x^2 + 1)}{x^2 - 1}$
16. $f(x) = (2x^2 + x)^3$
17. $f(x) = (3x^3 - 2)^8$

18. $h(x) = (\sqrt{x} + 2)^5$

19. $f(t) = \sqrt{2t^2 + 1}$

20. $g(t) = \sqrt[3]{1 - 2t^3}$

21. $s(t) = (3t^2 - 2t + 5)^{-2}$

22. $f(x) = (2x^3 - 3x^2 + 1)^{-3/2}$

23. $h(x) = \left(x + \dfrac{1}{x}\right)^2$

24. $h(x) = \dfrac{1 + x}{(2x^2 + 1)^2}$

25. $h(t) = (t^2 + t)^4(2t^2)$

26. $f(x) = (2x + 1)^3(x^2 + x)^2$

27. $g(x) = \sqrt{x}(x^2 - 1)^3$

28. $f(x) = \dfrac{x}{\sqrt{x^3 + 2}}$

29. $h(x) = \dfrac{\sqrt{3x + 2}}{4x - 3}$

30. $f(t) = \dfrac{\sqrt{2t + 1}}{(t + 1)^3}$

In Exercises 31–36, find the second derivative of the function.

31. $f(x) = 2x^4 - 3x^3 + 2x^2 + x + 4$

32. $g(x) = \sqrt{x} + \dfrac{1}{\sqrt{x}}$

33. $h(t) = \dfrac{t}{t^2 + 4}$

34. $f(x) = (x^3 + x + 1)^2$

35. $f(x) = \sqrt{2x^2 + 1}$

36. $f(t) = t(t^2 + 1)^3$

In Exercises 37–42, find *dy/dx* by implicit differentiation.

37. $6x^2 - 3y^2 = 9$

38. $2x^3 - 3xy = 4$

39. $y^3 + 3x^2 = 3y$

40. $x^2 + 2x^2y^2 + y^2 = 10$

41. $x^2 - 4xy - y^2 = 12$

42. $3x^2y - 4xy + x - 2y = 6$

43. Find the differential of $f(x) = x^2 + \dfrac{1}{x^2}$.

44. Find the differential of $f(x) = \dfrac{1}{\sqrt{x^3 + 1}}$.

45. Let f be the function defined by $f(x) = \sqrt{2x^2 + 4}$.
    a. Find the differential of f.
    b. Use your result from part (a) to find the approximate change in $y = f(x)$ if x changes from 4 to 4.1.
    c. Find the actual change in y if x changes from 4 to 4.1 and compare your result with that obtained in part (b).

46. Use a differential to approximate $\sqrt[3]{26.8}$.

47. Let $f(x) = 2x^3 - 3x^2 - 16x + 3$.
    a. Find the points on the graph of f at which the slope of the tangent line is equal to -4.
    b. Find the equation(s) of the tangent line(s) of part (a).

48. Let $f(x) = \frac{1}{3}x^3 + \frac{1}{2}x^2 - 4x + 1$.
    a. Find the points on the graph of f at which the slope of the tangent line is equal to -2.
    b. Find the equation(s) of the tangent line(s) of part (a).

49. Find an equation of the tangent line to the graph of $y = \sqrt{4 - x^2}$ at the point $(1, \sqrt{3})$.

50. Find an equation of the tangent line to the graph of $y = x(x + 1)^5$ at the point (1, 32).

51. Find the third derivative of the function

$$f(x) = \frac{1}{2x - 1}$$

What is its domain?

52. The demand equation for a certain product is $2x + 5p - 60 = 0$, where p is the unit price and x is the quantity demanded of the product. Find the elasticity of demand and determine whether the demand is elastic or inelastic, at the indicated prices.
    a. $p = 3$ b. $p = 6$ c. $p = 9$

53. The demand equation for a certain product is

$$x = \frac{25}{\sqrt{p}} - 1$$

where p is the unit price and x is the quantity demanded for the product. Compute the elasticity of demand and determine the range of prices corresponding to inelastic, unitary, and elastic demand.

54. The demand equation for a certain product is $x = 100 - 0.01p^2$.
    a. Is the demand elastic, unitary, or inelastic when $p = 40$?
    b. If the price is \$40, will raising the price slightly cause the revenue to increase or decrease?

55. The demand equation for a certain product is

$$p = 9\sqrt[3]{1000 - x}$$

    a. Is the demand elastic, unitary, or inelastic when $p = 60$?
    b. If the price is \$60, will raising the price slightly cause the revenue to increase or decrease?

56. ADULT OBESITY In the United States, the percent of adults (age 20–74) classified as obese held steady through the 1960s and 1970s at around 14% but began to rise rapidly during the 1980s and 1990s. This rise in adult obesity coincided with the period when an increasing number of Americans began eating more sugar and fats. The function

$$P(t) = 0.01484t^2 + 0.446t + 15 \qquad (0 \le t \le 22)$$

gives the percent of obese adults from 1978 ($t = 0$) through the year 2000 ($t = 22$).
    a. What percent of adults were obese in 1978? In 2000?
    b. How fast was the percent of obese adults increasing in 1980 ($t = 2$)? In 1998 ($t = 20$)?

Source: Journal of the American Medical Association

57. CABLE TV SUBSCRIBERS The number of subscribers to CNC Cable Television in the town of Randolph is approximated by the function

$$N(x) = 1000(1 + 2x)^{1/2} \qquad (1 \le x \le 30)$$

where $N(x)$ denotes the number of subscribers to the service in the xth week. Find the rate of increase in the number of subscribers at the end of the 12th week.

58. **Cost of Wireless Phone Calls** As cellular phone usage continues to soar, the airtime costs have dropped. The average price per minute of use (in cents) is projected to be

$$f(t) = 31.88(1 + t)^{-0.45} \qquad (0 \le t \le 6)$$

where t is measured in years and $t = 0$ corresponds to the beginning of 1998. Compute $f'(t)$. How fast was the average price/minute of use changing at the beginning of 2000? What was the average price/minute of use at the beginning of 2000?

Source: Cellular Telecommunications Industry Association

59. **Male Life Expectancy** Suppose the life expectancy of a male at birth in a certain country is described by the function

$$f(t) = 46.9(1 + 1.09t)^{0.1} \qquad (0 \le t \le 150)$$

where t is measured in years, with $t = 0$ corresponding to the beginning of 1900. How long can a male born at the beginning of 2000 in that country expect to live? What is the rate of change of the life expectancy of a male born in that country at the beginning of 2000?

60. **Cost of Producing DVDs** The total weekly cost in dollars incurred by Herald Media Corp. in producing x DVDs is given by the total cost function

$$C(x) = 2500 + 2.2x \qquad (0 \le x \le 8000)$$

a. What is the marginal cost when $x = 1000$ and 2000?
b. Find the average cost function $\overline{C}$ and the marginal average cost function $\overline{C}'$.
c. Using the results from part (b), show that the average cost incurred by Herald in pressing a DVD approaches \$2.20/disc when the level of production is high enough.

61. **Demand for Cordless Phones** The marketing department of Telecon has determined that the demand for their cordless phones obeys the relationship

$$p = -0.02x + 600 \qquad (0 \le x \le 30{,}000)$$

where p denotes the phone's unit price (in dollars) and x denotes the quantity demanded.
a. Find the revenue function R.
b. Find the marginal revenue function R'.
c. Compute $R'(10{,}000)$ and interpret your result.

62. **Demand for Photocopying Machines** The weekly demand for the LectroCopy photocopying machine is given by the demand equation

$$p = 2000 - 0.04x \qquad (0 \le x \le 50{,}000)$$

where p denotes the wholesale unit price in dollars and x denotes the quantity demanded. The weekly total cost function for manufacturing these copiers is given by

$$C(x) = 0.000002x^3 - 0.02x^2 + 1000x + 120{,}000$$

where $C(x)$ denotes the total cost incurred in producing x units.
a. Find the revenue function R, the profit function P, and the average cost function $\overline{C}$.
b. Find the marginal cost function C', the marginal revenue function R', the marginal profit function P', and the marginal average cost function $\overline{C}'$.
c. Compute $C'(3000)$, $R'(3000)$, and $P'(3000)$.
d. Compute $\overline{C}'(5000)$ and $\overline{C}'(8000)$ and interpret your results.

CHAPTER 3 Before Moving On . . .

1. Find the derivative of $f(x) = 2x^3 - 3x^{1/3} + 5x^{-2/3}$.
2. Differentiate $g(x) = x\sqrt{2x^2 - 1}$.
3. Find $\dfrac{dy}{dx}$ if $y = \dfrac{2x + 1}{x^2 + x + 1}$.
4. Find the first three derivatives of $f(x) = \dfrac{1}{\sqrt{x + 1}}$.
5. Find $\dfrac{dy}{dx}$ given that $xy^2 - x^2y + x^3 = 4$.
6. Let $y = x\sqrt{x^2 + 5}$.
   a. Find the differential of y.
   b. If x changes from $x = 2$ to $x = 2.01$, what is the approximate change in y?

4 Applications of the Derivative

What is the maximum altitude and the maximum velocity attained by the rocket? In Example 7, page 306, you will see how the techniques of calculus can be used to help answer these questions.

THIS CHAPTER FURTHER explores the power of the derivative as a tool to help analyze the properties of functions. The information obtained can then be used to accurately sketch graphs of functions. We also see how the derivative is used in solving a large class of optimization problems, including finding what level of production will yield a maximum profit for a company, finding what level of production will result in minimal cost to a company, finding the maximum height attained by a rocket, finding the maximum velocity at which air is expelled when a person coughs, and a host of other problems.

4.1 Applications of the First Derivative

Determining the Intervals Where a Function Is Increasing or Decreasing

According to a study by the U.S. Department of Energy and the Shell Development Company, a typical car's fuel economy as a function of its speed is described by the graph shown in Figure 1. Observe that the fuel economy $f(x)$ in miles per gallon (mpg) improves as x, the vehicle's speed in miles per hour (mph), increases from 0 to 42, and then drops as the speed increases beyond 42 mph. We use the terms *increasing* and *decreasing* to describe the behavior of a function as we move from left to right along its graph.

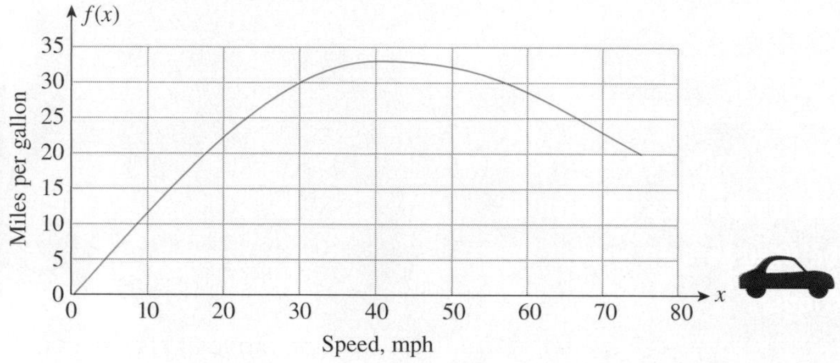

Source: U.S. Department of Energy and Shell Development Co.

FIGURE 1
A typical car's fuel economy improves as the speed at which it is driven increases from 0 mph to 42 mph and drops at speeds greater than 42 mph.

More precisely, we have the following definitions.

Increasing and Decreasing Functions

A function f is **increasing** on an interval (a, b) if for any two numbers x_1 and x_2 in (a, b), $f(x_1) < f(x_2)$ whenever $x_1 < x_2$ (Figure 2a).

A function f is **decreasing** on an interval (a, b) if for any two numbers x_1 and x_2 in (a, b), $f(x_1) > f(x_2)$ whenever $x_1 < x_2$ (Figure 2b).

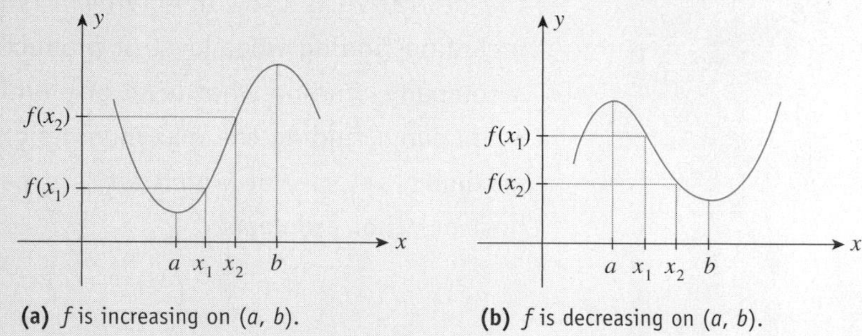

(a) f is increasing on (a, b). **(b)** f is decreasing on (a, b).

FIGURE 2

We say that f is *increasing at a number c* if there exists an interval (a, b) containing c such that f is increasing on (a, b). Similarly, we say that f is *decreasing at a number c* if there exists an interval (a, b) containing c such that f is decreasing on (a, b).

Since the rate of change of a function at $x = c$ is given by the derivative of the function at that number, the derivative lends itself naturally to being a tool for determining the intervals where a differentiable function is increasing or decreasing.

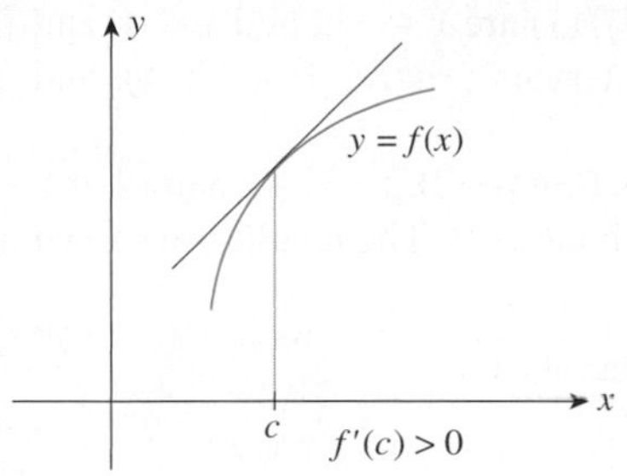

(a) f is increasing at $x = c$.

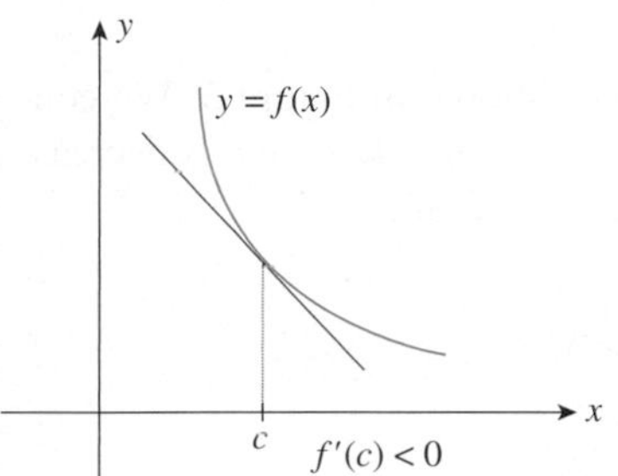

(b) f is decreasing at $x = c$.

FIGURE 3

Indeed, as we saw in Chapter 2, the derivative of a function at a number measures both the slope of the tangent line to the graph of the function at the point on the graph of f corresponding to that number and the rate of change of the function at that number. In fact, at a number where the derivative is positive, the slope of the tangent line to the graph is positive, and the function is increasing. At a number where the derivative is negative, the slope of the tangent line to the graph is negative, and the function is decreasing (Figure 3).

These observations lead to the following important theorem, which we state without proof.

THEOREM 1

a. If $f'(x) > 0$ for each value of x in an interval (a, b), then f is increasing on (a, b).

b. If $f'(x) < 0$ for each value of x in an interval (a, b), then f is decreasing on (a, b).

c. If $f'(x) = 0$ for each value of x in an interval (a, b), then f is constant on (a, b).

EXAMPLE 1 Find the interval where the function $f(x) = x^2$ is increasing and the interval where it is decreasing.

Solution The derivative of $f(x) = x^2$ is $f'(x) = 2x$. Since

$$f'(x) = 2x > 0 \quad \text{if } x > 0 \qquad \text{and} \qquad f'(x) = 2x < 0 \quad \text{if } x < 0$$

f is increasing on the interval $(0, \infty)$ and decreasing on the interval $(-\infty, 0)$ (Figure 4). ■

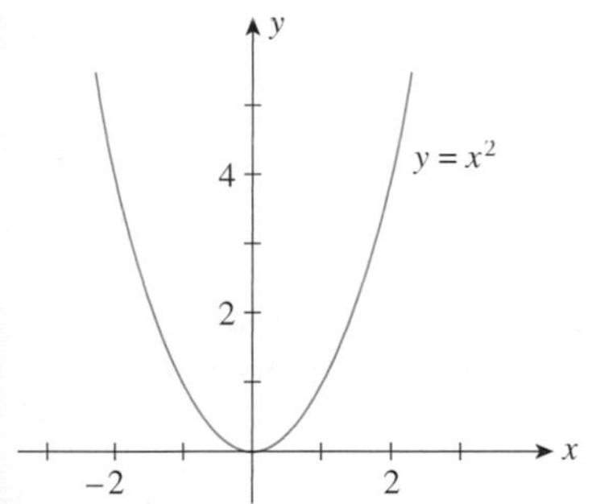

FIGURE 4
The graph of f falls on $(-\infty, 0)$ where $f'(x) < 0$ and rises on $(0, \infty)$ where $f'(x) > 0$.

Recall that the graph of a continuous function cannot have any breaks. As a consequence, a continuous function cannot change sign unless it equals zero for some value of x. (See Theorem 5, page 125.) This observation suggests the following procedure for determining the sign of the derivative f' of a function f, and hence the intervals where the function f is increasing and where it is decreasing.

Determining the Intervals Where a Function Is Increasing or Decreasing

1. Find all values of x for which $f'(x) = 0$ or f' is discontinuous and identify the open intervals determined by these numbers.
2. Select a test number c in each interval found in step 1 and determine the sign of $f'(c)$ in that interval.
   a. If $f'(c) > 0$, f is increasing on that interval.
   b. If $f'(c) < 0$, f is decreasing on that interval.

EXAMPLE 2 Determine the intervals where the function $f(x) = x^3 - 3x^2 - 24x + 32$ is increasing and where it is decreasing.

Solution

1. The derivative of f is

$$f'(x) = 3x^2 - 6x - 24 = 3(x + 2)(x - 4)$$

> **EXPLORE & DISCUSS**
>
> True or false? If f is continuous at c and f is increasing at c, then $f'(c) \neq 0$. Explain your answer.
>
> Hint: Consider $f(x) = x^3$ and $c = 0$.

and it is continuous everywhere. The zeros of $f'(x)$ are $x = -2$ and $x = 4$, and these numbers divide the real line into the intervals $(-\infty, -2)$, $(-2, 4)$, and $(4, \infty)$.

2. To determine the sign of $f'(x)$ in the intervals $(-\infty, -2)$, $(-2, 4)$, and $(4, \infty)$, compute $f'(x)$ at a convenient test point in each interval. The results are shown in the following table.

Interval	Test Point c	$f'(c)$	Sign of $f'(x)$
$(-\infty, -2)$	-3	21	+
$(-2, 4)$	0	-24	−
$(4, \infty)$	5	21	+

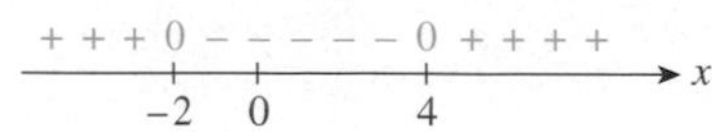

FIGURE 5
Sign diagram for f'

Using these results, we obtain the sign diagram shown in Figure 5. We conclude that f is increasing on the intervals $(-\infty, -2)$ and $(4, \infty)$ and is decreasing on the interval $(-2, 4)$. Figure 6 shows the graph of f.

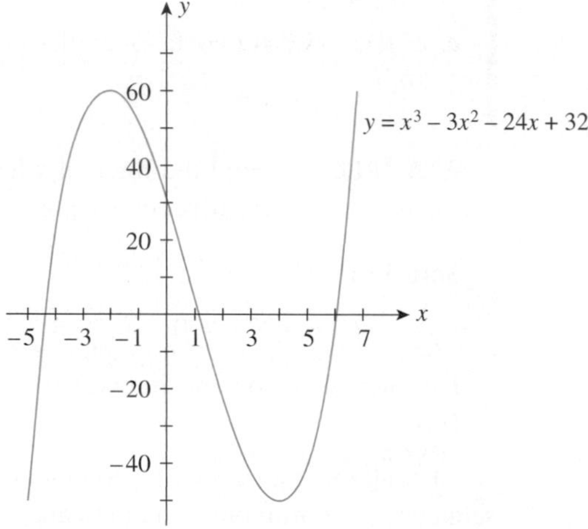

FIGURE 6
The graph of f rises on $(-\infty, -2)$, falls on $(-2, 4)$, and rises again on $(4, \infty)$.

Note Do not be concerned with how the graphs in this section are obtained. We will learn how to sketch these graphs later. However, if you are familiar with the use of a graphing utility, you may go ahead and verify each graph.

EXPLORING WITH TECHNOLOGY

Refer to Example 2.

1. Use a graphing utility to plot the graphs of
$$f(x) = x^3 - 3x^2 - 24x + 32$$
and its derivative function
$$f'(x) = 3x^2 - 6x - 24$$
using the viewing window $[-10, 10] \times [-50, 70]$.

2. By looking at the graph of f', determine the intervals where $f'(x) > 0$ and the intervals where $f'(x) < 0$. Next, look at the graph of f and determine the intervals where it is increasing and the intervals where it is decreasing. Describe the relationship. Is it what you expected?

EXAMPLE 3 Find the interval where the function $f(x) = x^{2/3}$ is increasing and the interval where it is decreasing.

Solution

1. The derivative of f is

$$f'(x) = \frac{2}{3}x^{-1/3} = \frac{2}{3x^{1/3}}$$

The function f' is not defined at $x = 0$, so f' is discontinuous there. It is continuous everywhere else. Furthermore, f' is not equal to zero anywhere. The number 0 divides the real line (the domain of f) into the intervals $(-\infty, 0)$ and $(0, \infty)$.

2. Pick a test point (say, $x = -1$) in the interval $(-\infty, 0)$ and compute

$$f'(-1) = -\frac{2}{3}$$

Since $f'(-1) < 0$, we see that $f'(x) < 0$ on $(-\infty, 0)$. Next, we pick a test point (say, $x = 1$) in the interval $(0, \infty)$ and compute

$$f'(1) = \frac{2}{3}$$

Since $f'(1) > 0$, we see that $f'(x) > 0$ on $(0, \infty)$. Figure 7 shows these results in the form of a sign diagram.

We conclude that f is decreasing on the interval $(-\infty, 0)$ and increasing on the interval $(0, \infty)$. The graph of f, shown in Figure 8, confirms these results. ■

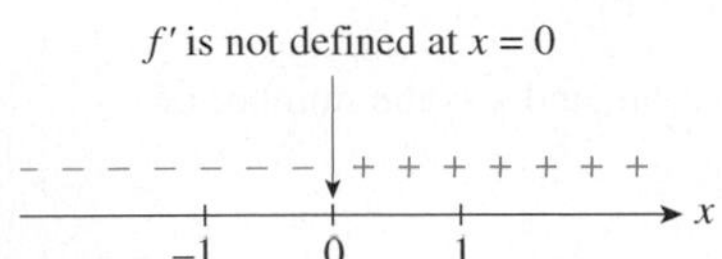

FIGURE 7
Sign diagram for f'

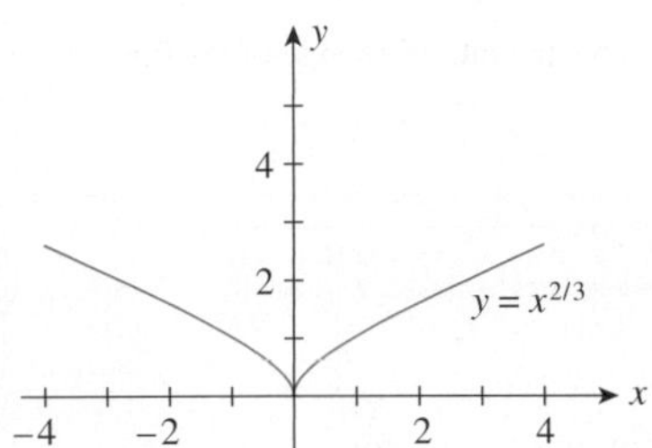

FIGURE 8
f decreases on $(-\infty, 0)$ and increases on $(0, \infty)$.

EXAMPLE 4 Find the intervals where the function $f(x) = x + \dfrac{1}{x}$ is increasing and where it is decreasing.

Solution

1. The derivative of f is

$$f'(x) = 1 - \frac{1}{x^2} = \frac{x^2 - 1}{x^2}$$

Since f' is not defined at $x = 0$, it is discontinuous there. Furthermore, $f'(x)$ is equal to zero when $x^2 - 1 = 0$ or $x = \pm 1$. These values of x partition the domain of f' into the open intervals $(-\infty, -1)$, $(-1, 0)$, $(0, 1)$, and $(1, \infty)$, where the sign of f' is different from zero.

2. To determine the sign of f' in each of these intervals, we compute $f'(x)$ at the test points $x = -2, -\frac{1}{2}, \frac{1}{2}$, and 2, respectively, obtaining $f'(-2) = \frac{3}{4}$, $f'(-\frac{1}{2}) = -3$, $f'(\frac{1}{2}) = -3$, and $f'(2) = \frac{3}{4}$. From the sign diagram for f' (Figure 9), we conclude that f is increasing on $(-\infty, -1)$ and $(1, \infty)$ and decreasing on $(-1, 0)$ and $(0, 1)$.

The graph of f appears in Figure 10. Note that f' does not change sign as we move across $x = 0$. (Compare this with Example 3.) ■

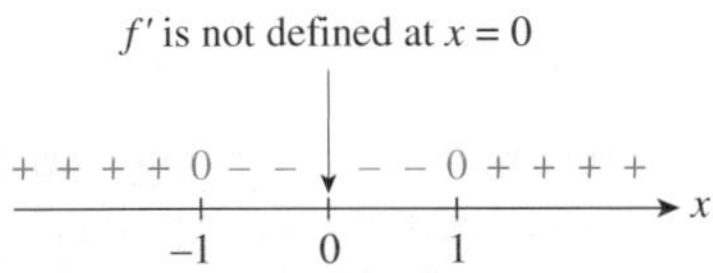

FIGURE 9
f' does not change sign as we move across $x = 0$.

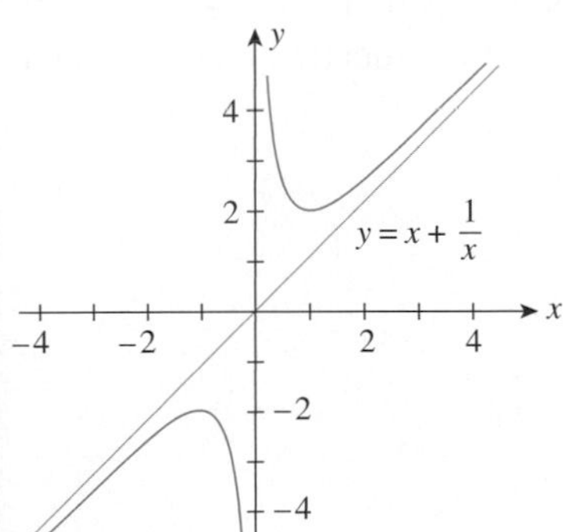

FIGURE 10
The graph of f rises on $(-\infty, -1)$, falls on $(-1, 0)$ and $(0, 1)$, and rises again on $(1, \infty)$.

! Example 4 reminds us that we must *not* automatically conclude that the derivative f' must change sign when we move across a number where f' is discontinous or a zero of f'.

EXPLORE & DISCUSS

Consider the profit function P associated with a certain commodity defined by

$$P(x) = R(x) - C(x) \qquad (x \geq 0)$$

where R is the revenue function, C is the total cost function, and x is the number of units of the product produced and sold.

1. Find an expression for $P'(x)$.
2. Find relationships in terms of the derivatives of R and C so that
   a. P is increasing at $x = a$.
   b. P is decreasing at $x = a$.
   c. P is neither increasing nor decreasing at $x = a$.
   Hint: Recall that the derivative of a function at $x = a$ measures the rate of change of the function at that number.
3. Explain the results of part 2 in economic terms.

EXPLORING WITH TECHNOLOGY

1. Use a graphing utility to sketch the graphs of $f(x) = x^3 - ax$ for $a = -2, -1, 0, 1$, and 2, using the viewing window $[-2, 2] \times [-2, 2]$.
2. Use the results of part 1 to guess at the values of a so that f is increasing on $(-\infty, \infty)$.
3. Prove your conjecture analytically.

Relative Extrema

Besides helping us determine where the graph of a function is increasing and decreasing, the first derivative may be used to help us locate certain "high points" and "low points" on the graph of f. Knowing these points is invaluable in sketching the graphs of functions and solving optimization problems. These "high points" and "low points" correspond to the *relative (local) maxima* and *relative minima* of a function. They are so called because they are the highest or the lowest points when compared with points nearby.

The graph shown in Figure 11 gives the U.S. budget deficit from 1980 ($t = 0$) to 1991. The relative maxima and the relative minima of the function f are indicated on the graph.

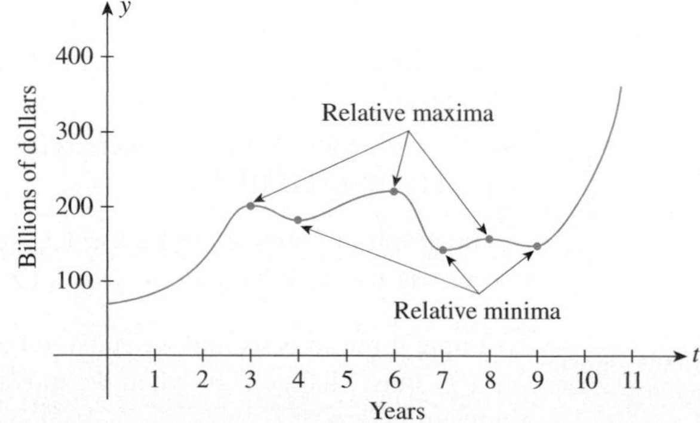

Source: Office of Management and Budget

FIGURE 11
U.S. budget deficit from 1980 to 1991

More generally, we have the following definition:

Relative Maximum

A function f has a **relative maximum** at $x = c$ if there exists an open interval (a, b) containing c such that $f(x) \leq f(c)$ for all x in (a, b).

Geometrically, this means that there is *some* interval containing $x = c$ such that no point on the graph of f with its x-coordinate in that interval can lie above the point $(c, f(c))$; that is, $f(c)$ is the largest value of $f(x)$ in some interval around $x = c$. Figure 12 depicts the graph of a function f that has a relative maximum at $x = x_1$ and another at $x = x_3$.

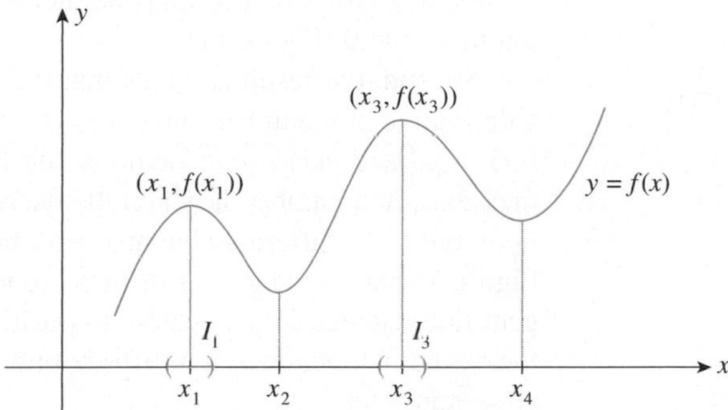

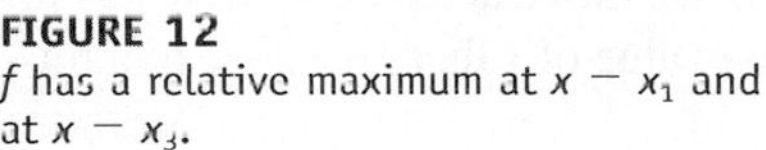
FIGURE 12
f has a relative maximum at $x = x_1$ and at $x = x_3$.

Observe that all the points on the graph of f with x-coordinates in the interval I_1 containing x (shown in blue) lie on or below the point $(x_1, f(x_1))$. This is also true for the point $(x_3, f(x_3))$ and the interval I_3. Thus, even though there are points on the graph of f that are "higher" than the points $(x_1, f(x_1))$ and $(x_3, f(x_3))$, the latter points are "highest" relative to points in their respective neighborhoods (intervals). Points on the graph of a function f that are "highest" and "lowest" with respect to *all* points in the domain of f will be studied in Section 4.4.

The definition of the relative minimum of a function parallels that of the relative maximum of a function.

Relative Minimum

A function f has a **relative minimum** at $x = c$ if there exists an open interval (a, b) containing c such that $f(x) \geq f(c)$ for all x in (a, b).

The graph of the function f, depicted in Figure 12, has a relative minimum at $x = x_2$ and another at $x = x_4$.

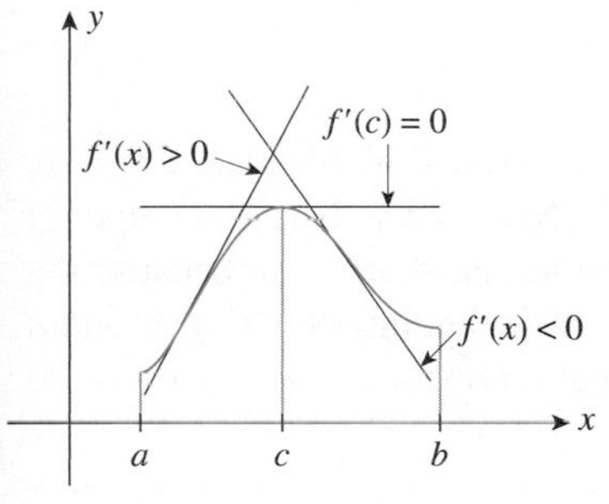

(a) f has a relative maximum at $x = c$.

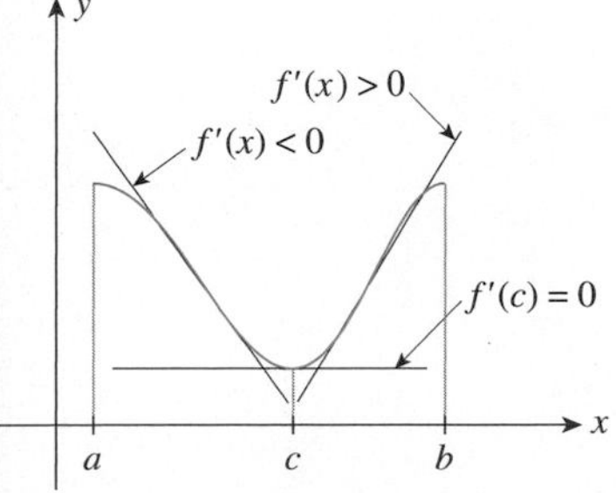

(b) f has a relative minimum at $x = c$.

FIGURE 13

Finding the Relative Extrema

We refer to the relative maximum and relative minimum of a function as the **relative extrema** of that function. As a first step in our quest to find the relative extrema of a function, we consider functions that have derivatives at such points. Suppose that f is a function that is differentiable in some interval (a, b) that contains a number c and that f has a relative maximum at $x = c$ (Figure 13a).

Observe that the slope of the tangent line to the graph of f must change from positive to negative as we move across $x = c$ from left to right. Therefore, the tangent line to the graph of f at the point $(c, f(c))$ must be horizontal; that is, $f'(c) = 0$ (Figure 13a).

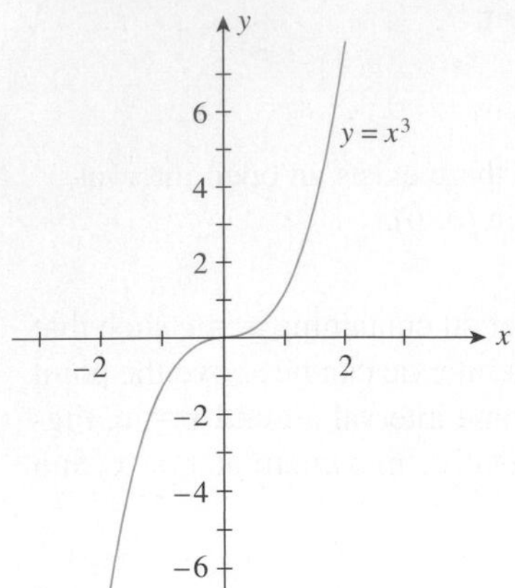

FIGURE 14
$f'(0) = 0$, but f does not have a relative extremum at (0, 0).

Using a similar argument, it may be shown that the derivative f' of a differentiable function f must also be equal to zero at $x = c$, where f has a relative minimum (Figure 13b).

This analysis reveals an important characteristic of the relative extrema of a differentiable function f: *At any number c where f has a relative extremum,* $f'(c) = 0$.

Before we develop a procedure for finding such numbers, a few words of caution are in order. First, this result tells us that if a differentiable function f has a relative extremum at a number $x = c$, then $f'(c) = 0$. The converse of this statement—if $f'(c) = 0$ at $x = c$, then f must have a relative extremum at that number—is *not* true. Consider, for example, the function $f(x) = x^3$. Here, $f'(x) = 3x^2$, so $f'(0) = 0$. Yet, f has neither a relative maximum nor a relative minimum at $x = 0$ (Figure 14).

Second, our result assumes that the function is differentiable and thus has a derivative at a number that gives rise to a relative extremum. The functions $f(x) = |x|$ and $g(x) = x^{2/3}$ demonstrate that a relative extremum of a function may exist at a number at which the derivative does not exist. Both these functions fail to be differentiable at $x = 0$, but each has a relative minimum there. Figure 15 shows the graphs of these functions. Note that the slopes of the tangent lines change from negative to positive as we move across $x = 0$, just as in the case of a function that is differentiable at a value of x that gives rise to a relative minimum.

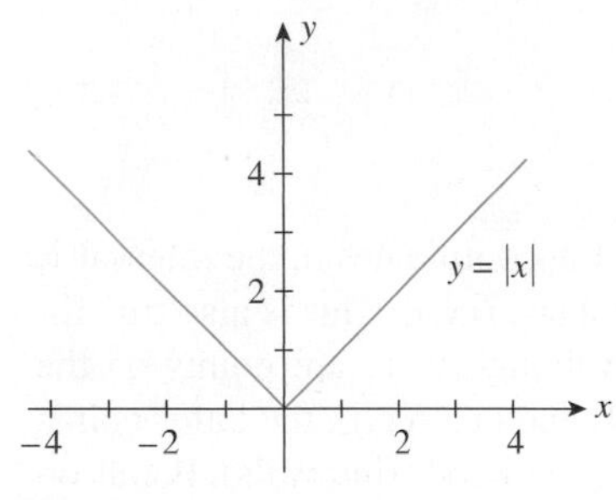

(a)

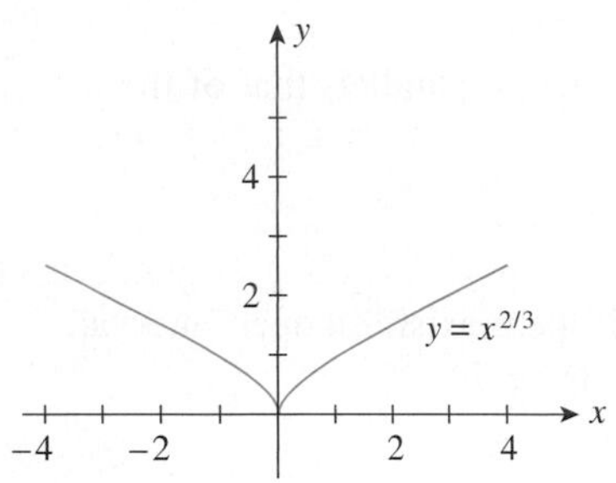

(b)

FIGURE 15
Each of these functions has a relative extremum at (0, 0), but the derivative does not exist there.

We refer to a number in the domain of f that *may* give rise to a relative extremum as a critical number.

Critical Number of *f*

A **critical number** of a function f is any number x in the domain of f such that $f'(x) = 0$ or $f'(x)$ does not exist.

Figure 16 depicts the graph of a function that has critical numbers at $x = a, b, c, d$, and e. Observe that $f'(x) = 0$ at $x = a, b$, and c. Next, since there is a corner at $x = d$, $f'(x)$ does not exist there. Finally, $f'(x)$ does not exist at $x = e$ because the tangent line there is vertical. Also, observe that the critical numbers $x = a, b$, and d give rise to relative extrema of f, whereas the critical numbers $x = c$ and $x = e$ do not.

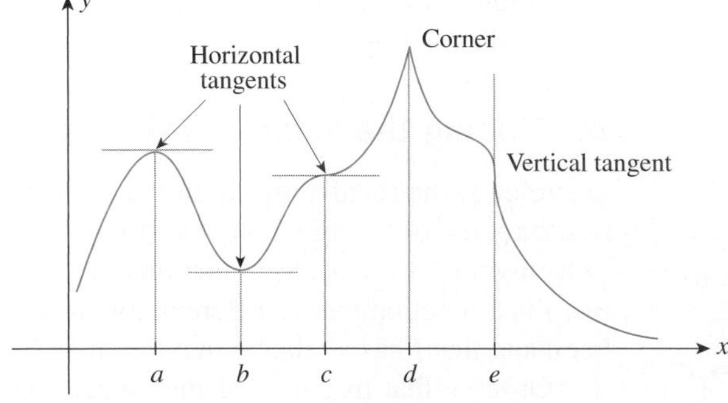

FIGURE 16
Critical numbers of f

Having defined what a critical number is, we can now state a formal procedure for finding the relative extrema of a continuous function that is differentiable everywhere except at isolated values of x. Incorporated into the procedure is the so-called **first derivative test,** which helps us determine whether a number gives rise to a relative maximum or a relative minimum of the function f.

The First Derivative Test

Procedure for Finding Relative Extrema of a Continuous Function f

1. Determine the critical numbers of f.
2. Determine the sign of $f'(x)$ to the left and right of each critical number.
   - **a.** If $f'(x)$ changes sign from *positive* to *negative* as we move across a critical number c, then $f(c)$ is a relative maximum.
   - **b.** If $f'(x)$ changes sign from *negative* to *positive* as we move across a critical number c, then $f(c)$ is a relative minimum.
   - **c.** If $f'(x)$ does not change sign as we move across a critical number c, then $f(c)$ is not a relative extremum.

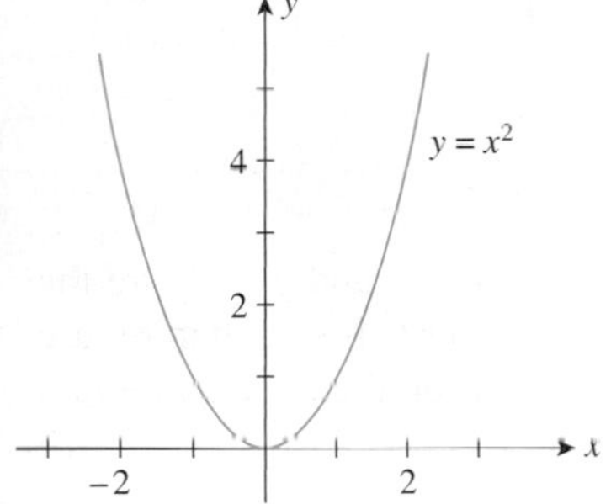

FIGURE 17
f has a relative minimum at $x = 0$.

EXAMPLE 5 Find the relative maxima and relative minima of the function $f(x) = x^2$.

Solution The derivative of $f(x) = x^2$ is given by $f'(x) = 2x$. Setting $f'(x) = 0$ yields $x = 0$ as the only critical number of f. Since

$$f'(x) < 0 \quad \text{if } x < 0 \qquad \text{and} \qquad f'(x) > 0 \quad \text{if } x > 0$$

we see that $f'(x)$ changes sign from negative to positive as we move across the critical number 0. Thus, we conclude that $f(0) = 0$ is a relative minimum of f (Figure 17). ■

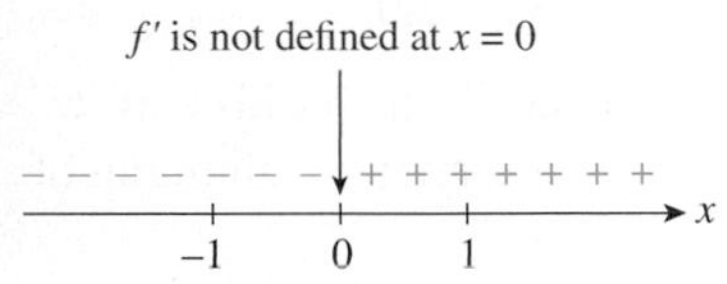

FIGURE 18
Sign diagram for f'

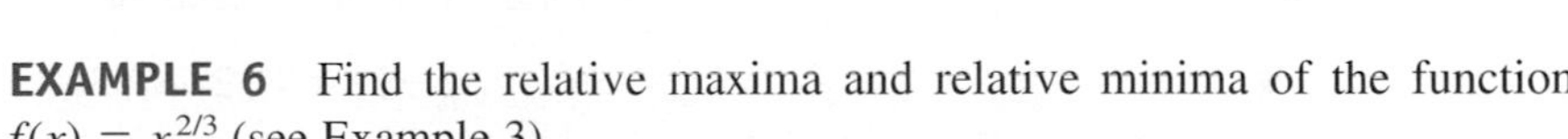

EXAMPLE 6 Find the relative maxima and relative minima of the function $f(x) = x^{2/3}$ (see Example 3).

Solution The derivative of f is $f'(x) = \frac{2}{3}x^{-1/3}$. As noted in Example 3, f' is not defined at $x = 0$, is continuous everywhere else, and is not equal to zero in its domain. Thus, $x = 0$ is the only critical number of the function f.

The sign diagram obtained in Example 3 is reproduced in Figure 18. We can see that the sign of $f'(x)$ changes from negative to positive as we move across $x = 0$ from left to right. Thus, an application of the first derivative test tells us that $f(0) = 0$ is a relative minimum of f (Figure 19).

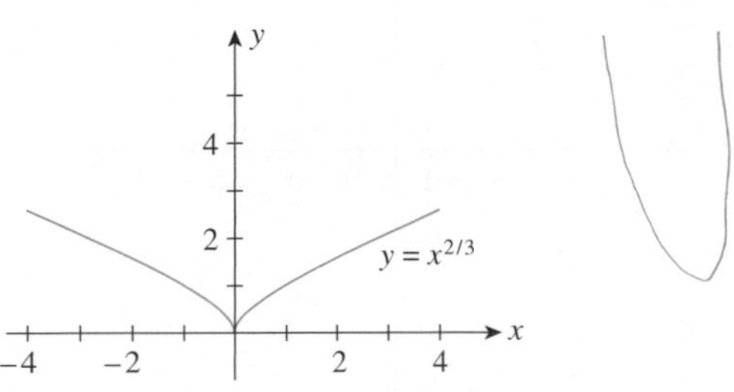

FIGURE 19
f has a relative minimum at $x = 0$. ■

EXPLORE & DISCUSS

Recall that the average cost function $\overline{C}$ is defined by

$$\overline{C} = \frac{C(x)}{x}$$

where $C(x)$ is the total cost function and x is the number of units of a commodity manufactured (see Section 3.4).

1. Show that

$$\overline{C}'(x) = \frac{C'(x) - \overline{C}(x)}{x} \qquad (x > 0)$$

2. Use the result of part 1 to conclude that $\overline{C}$ is decreasing for values of x at which $C'(x) < \overline{C}(x)$. Find similar conditions for which $\overline{C}$ is increasing and for which $\overline{C}$ is constant.

3. Explain the results of part 2 in economic terms.

EXAMPLE 7 Find the relative maxima and relative minima of the function

$$f(x) = x^3 - 3x^2 - 24x + 32$$

Solution The derivative of f is

$$f'(x) = 3x^2 - 6x - 24 = 3(x + 2)(x - 4)$$

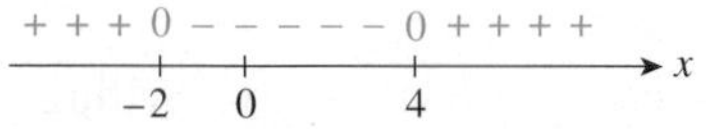

+ + + 0 − − − − − 0 + + + +
−2 0 4 x

FIGURE 20
Sign diagram for f'

and it is continuous everywhere. The zeros of $f'(x)$, $x = -2$ and $x = 4$, are the only critical numbers of the function f. The sign diagram for f' is shown in Figure 20. Examine the two critical numbers $x = -2$ and $x = 4$ for a relative extremum using the first derivative test and the sign diagram for f':

1. *The critical number* -2: Since the function $f'(x)$ changes sign from positive to negative as we move across $x = -2$ from left to right, we conclude that a relative maximum of f occurs at $x = -2$. The value of $f(x)$ when $x = -2$ is

$$f(-2) = (-2)^3 - 3(-2)^2 - 24(-2) + 32 = 60$$

2. *The critical number* 4: $f'(x)$ changes sign from negative to positive as we move across $x = 4$ from left to right, so $f(4) = -48$ is a relative minimum of f. The graph of f appears in Figure 21.

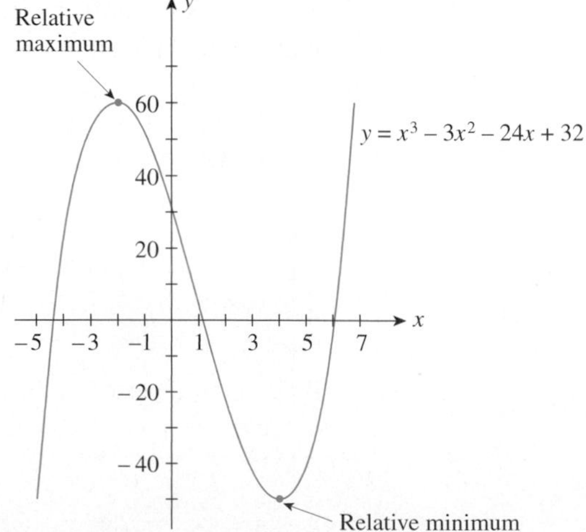

FIGURE 21
f has a relative maximum at $x = -2$ and a relative minimum at $x = 4$.

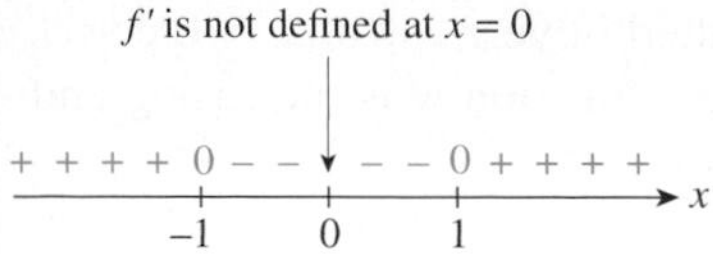

FIGURE 22
$x = 0$ is not a critical number because f is not defined at $x = 0$.

EXAMPLE 8 Find the relative maxima and the relative minima of the function

$$f(x) = x + \frac{1}{x}$$

Solution The derivative of f is

$$f'(x) = 1 - \frac{1}{x^2} = \frac{x^2 - 1}{x^2} = \frac{(x + 1)(x - 1)}{x^2}$$

Since f' is equal to zero at $x = -1$ and $x = 1$, these are critical numbers for the function f. Next, observe that f' is discontinuous at $x = 0$. However, because *f is not defined at that number, $x = 0$ does not qualify as a critical number of f*. Figure 22 shows the sign diagram for f'.

Since $f'(x)$ changes sign from positive to negative as we move across $x = -1$ from left to right, the first derivative test implies that $f(-1) = -2$ is a relative maximum of the function f. Next, $f'(x)$ changes sign from negative to positive as we move across $x = 1$ from left to right, so $f(1) = 2$ is a relative minimum of the function f. The graph of f appears in Figure 23. Note that this function has a relative maximum that lies below its relative minimum. ■

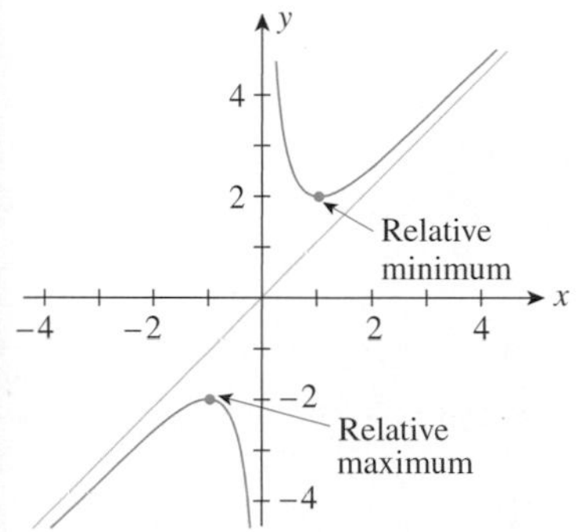

FIGURE 23
$f(x) = x + \frac{1}{x}$

EXPLORING WITH TECHNOLOGY

Refer to Example 8.

1. Use a graphing utility to plot the graphs of $f(x) = x + 1/x$ and its derivative function $f'(x) = 1 - 1/x^2$, using the viewing window $[-4, 4] \times [-8, 8]$.
2. By studying the graph of f', determine the critical numbers of f. Next, note the sign of $f'(x)$ immediately to the left and to the right of each critical number. What can you conclude about each critical number? Are your conclusions borne out by the graph of f?

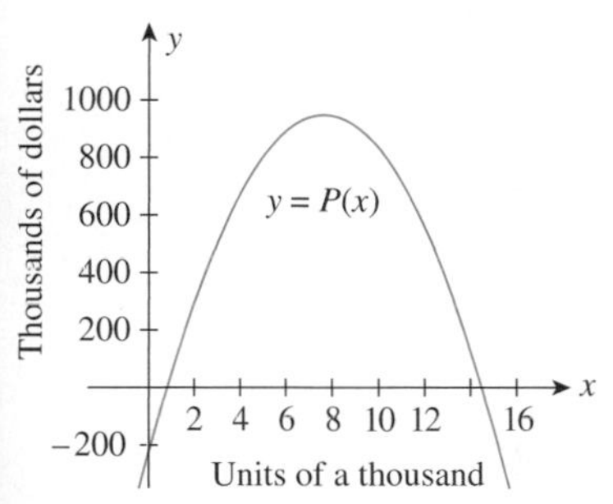

FIGURE 24
The profit function is increasing on (0, 7500) and decreasing on (7500, ∞).

APPLIED EXAMPLE 9 Profit Functions The profit function of Acrosonic Company is given by

$$P(x) = -0.02x^2 + 300x - 200{,}000$$

dollars, where x is the number of Acrosonic model F loudspeaker systems produced. Find where the function P is increasing and where it is decreasing.

Solution The derivative P' of the function P is

$$P'(x) = -0.04x + 300 = -0.04(x - 7500)$$

Thus, $P'(x) = 0$ when $x = 7500$. Furthermore, $P'(x) > 0$ for x in the interval $(0, 7500)$, and $P'(x) < 0$ for x in the interval $(7500, \infty)$. This means that the profit function P is increasing on $(0, 7500)$ and decreasing on $(7500, \infty)$ (Figure 24). ■

APPLIED EXAMPLE 10 Crime Rates The number of major crimes committed in the city of Bronxville from 1998 to 2005 is approximated by the function

$$N(t) = -0.1t^3 + 1.5t^2 + 100 \qquad (0 \le t \le 7)$$

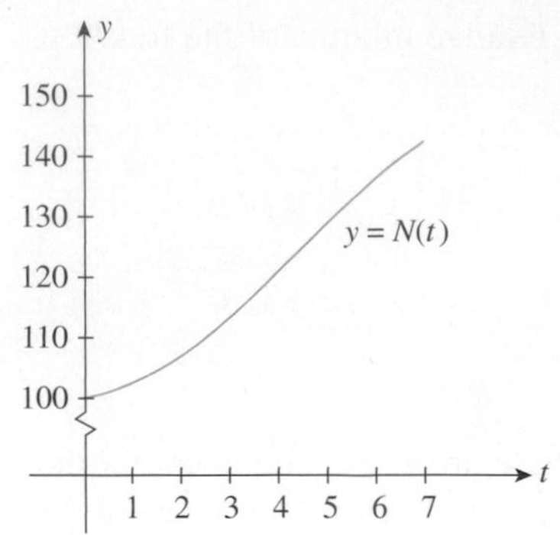

FIGURE 25
The number of crimes, $N(t)$, is increasing over the 7-year interval.

where $N(t)$ denotes the number of crimes committed in year t, with $t = 0$ corresponding to the beginning of 1998. Find where the function N is increasing and where it is decreasing.

Solution The derivative N' of the function N is

$$N'(t) = -0.3t^2 + 3t = -0.3t(t - 10)$$

Since $N'(t) > 0$ for t in the interval (0, 7), the function N is increasing throughout that interval (Figure 25). ■

4.1 Self-Check Exercises

1. Find the intervals where the function $f(x) = \frac{2}{3}x^3 - x^2 - 12x + 3$ is increasing and the intervals where it is decreasing.
2. Find the relative extrema of $f(x) = \dfrac{x^2}{1 - x^2}$.

Solutions to Self-Check Exercises 4.1 can be found on page 264.

4.1 Concept Questions

1. Explain each of the following:
   a. f is increasing on an interval I.
   b. f is decreasing on an interval I.
2. Describe a procedure for determining where a function is increasing and where it is decreasing.
3. Explain each term: **(a)** relative maximum and **(b)** relative minimum.
4. **a.** What is a critical number of a function f?
   b. Explain the role of a critical number in determining the relative extrema of a function.
5. Describe the first derivative test and describe a procedure for finding the relative extrema of a function.

4.1 Exercises

In Exercises 1–8, you are given the graph of a function f. Determine the intervals where f is increasing, constant, or decreasing.

1.

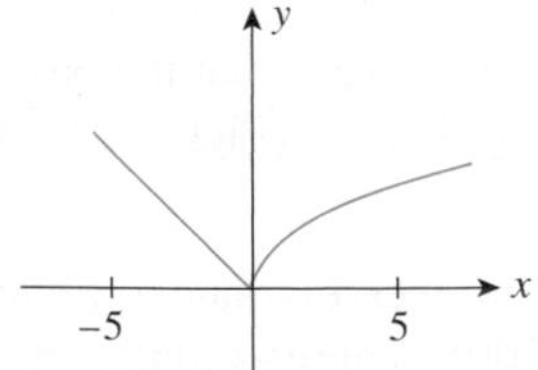

2.

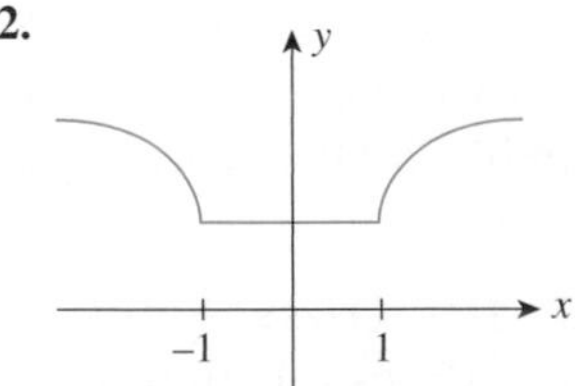

3.

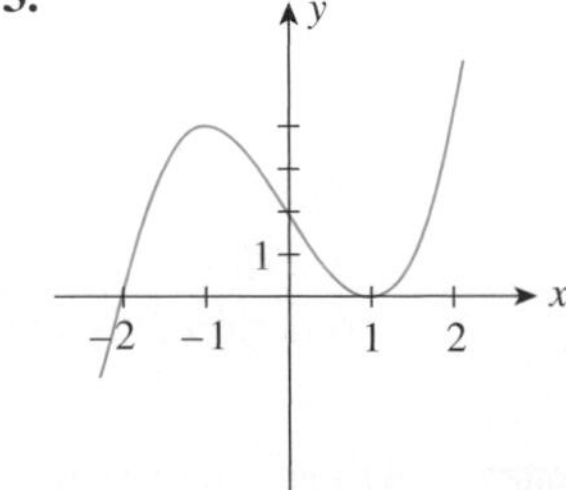

4.

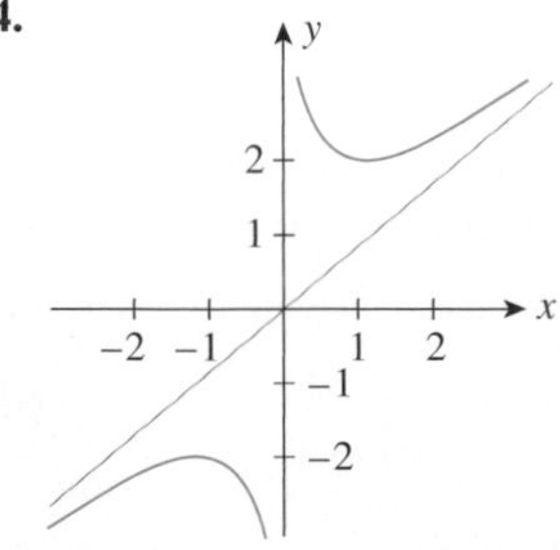

5.

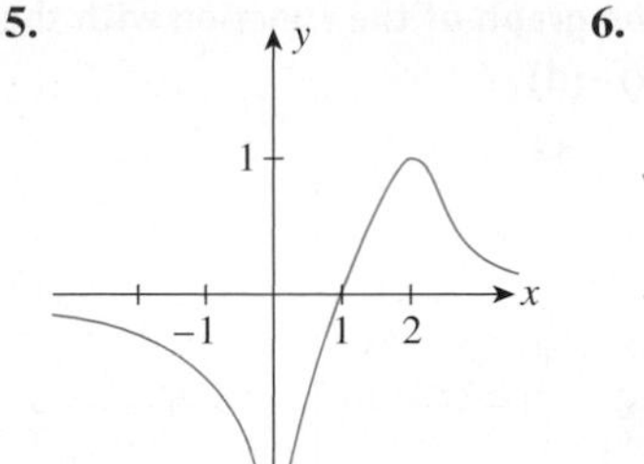

6.

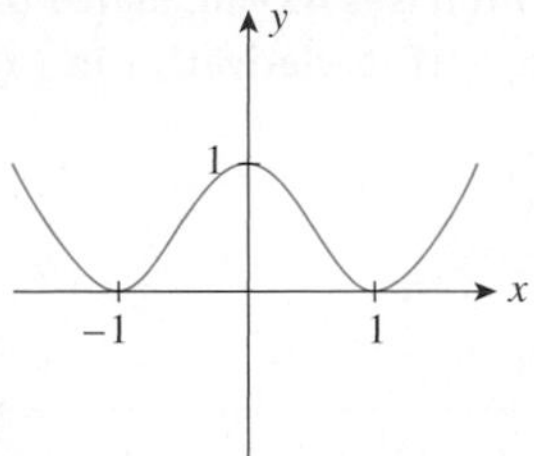

7.

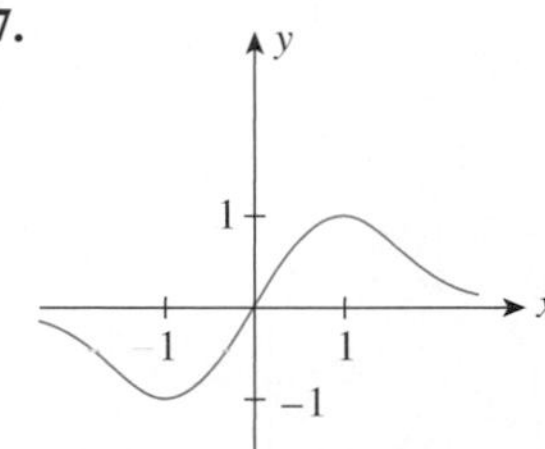

8.

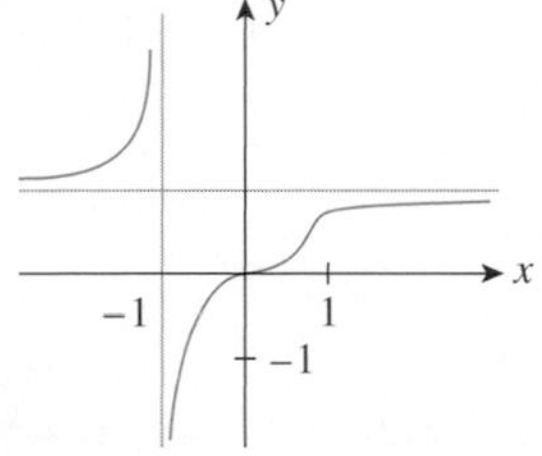

9. **The Boston Marathon** The graph of the function f shown in the accompanying figure gives the elevation of that part of the Boston Marathon course that includes the notorious Heartbreak Hill. Determine the intervals (stretches of the course) where the function f is increasing (the runner is laboring), where it is constant (the runner is taking a breather), and where it is decreasing (the runner is coasting).

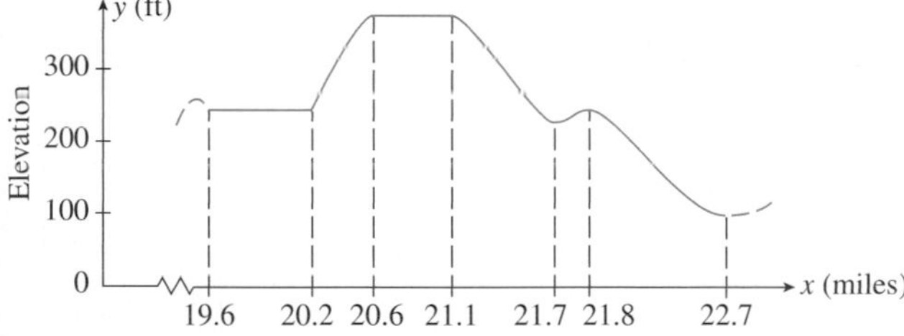

10. **Aircraft Structural Integrity** Among the important factors in determining the structural integrity of an aircraft is its age. Advancing age makes planes more likely to crack. The graph of the function f, shown in the accompanying figure, is referred to as a "bathtub curve" in the airline industry. It gives the fleet damage rate (damage due to corrosion, accident, and metal fatigue) of a typical fleet of commercial aircraft as a function of the number of years of service.

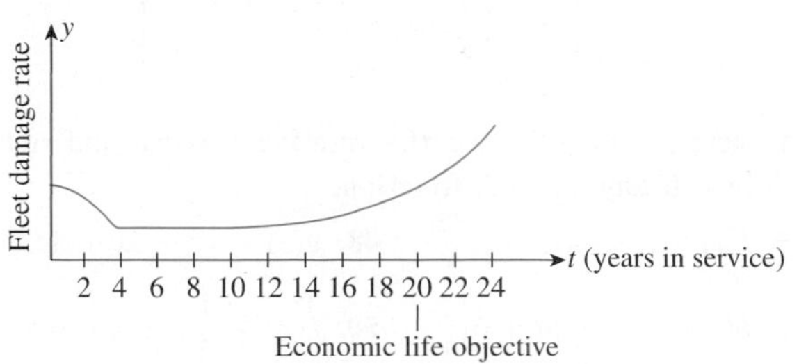

a. Determine the interval where f is decreasing. This corresponds to the time period when the fleet damage rate is dropping as problems are found and corrected during the initial "shakedown" period.

b. Determine the interval where f is constant. After the initial shakedown period, planes have few structural problems, and this is reflected by the fact that the function is constant on this interval.

c. Determine the interval where f is increasing. Beyond the time period mentioned in part (b), the function is increasing—reflecting an increase in structural defects due mainly to metal fatigue.

In Exercises 11–34, find the interval(s) where each function is increasing and the interval(s) where it is decreasing.

11. $f(x) = 3x + 5$

12. $f(x) = 4 - 5x$

13. $f(x) = x^2 - 3x$

14. $f(x) = 2x^2 + x + 1$

15. $g(x) = x - x^3$

16. $f(x) = x^3 - 3x^2$

17. $g(x) = x^3 + 3x^2 + 1$

18. $f(x) = x^3 - 3x + 4$

19. $f(x) = \frac{1}{3}x^3 - 3x^2 + 9x + 20$

20. $f(x) = \frac{2}{3}x^3 - 2x^2 - 6x - 2$

21. $h(x) = x^4 - 4x^3 + 10$

22. $g(x) = x^4 - 2x^2 + 4$

23. $f(x) = \dfrac{1}{x - 2}$

24. $h(x) = \dfrac{1}{2x + 3}$

25. $h(t) = \dfrac{t}{t - 1}$

26. $g(t) = \dfrac{2t}{t^2 + 1}$

27. $f(x) = x^{3/5}$

28. $f(x) = x^{2/3} + 5$

29. $f(x) = \sqrt{x + 1}$

30. $f(x) = (x - 5)^{2/3}$

31. $f(x) = \sqrt{16 - x^2}$

32. $g(x) = x\sqrt{x + 1}$

33. $f(x) = \dfrac{x^2 - 1}{x}$

34. $h(x) = \dfrac{x^2}{x - 1}$

In Exercises 35–42, you are given the graph of a function f. Determine the relative maxima and relative minima, if any.

35.

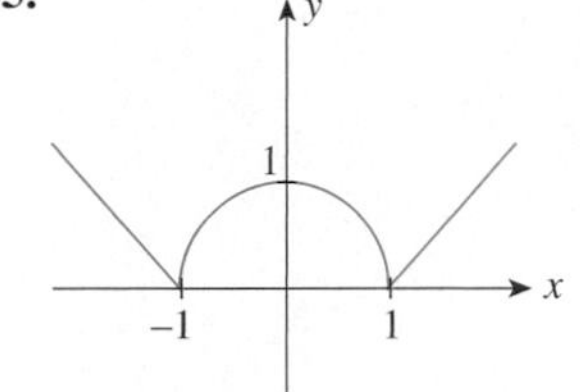

36.

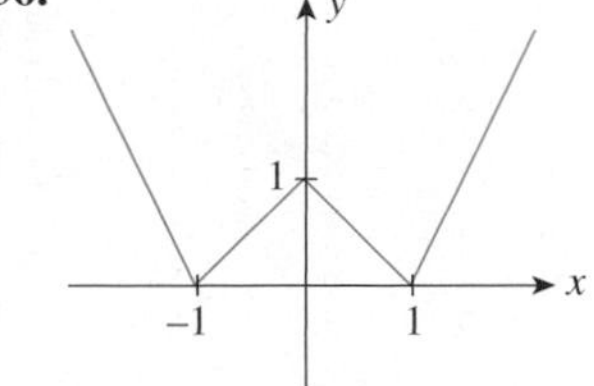

37.

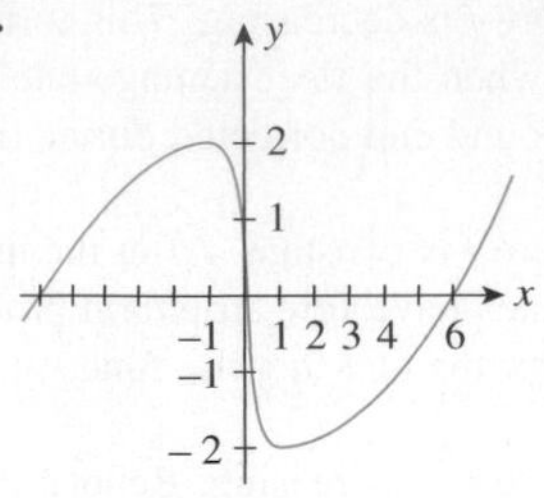

38.

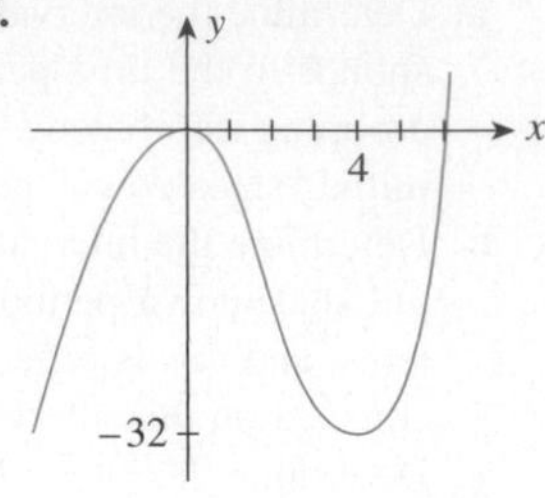

39.

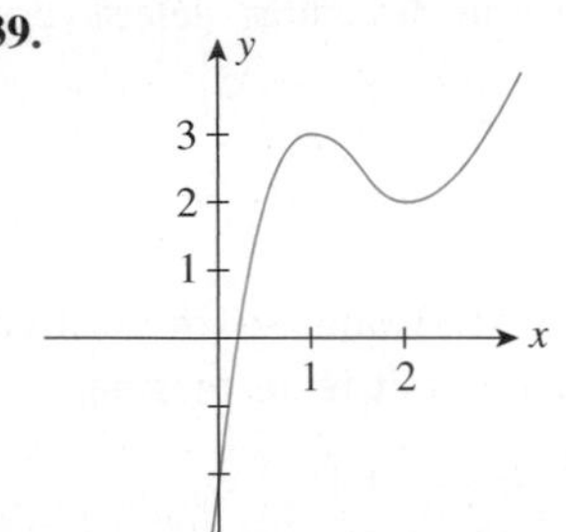

40.

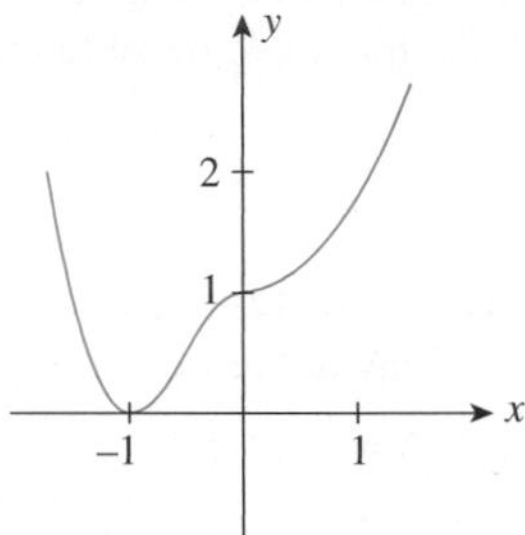

41.

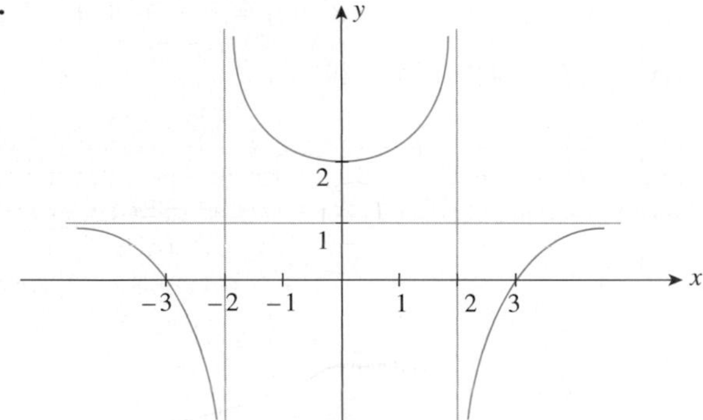

42.

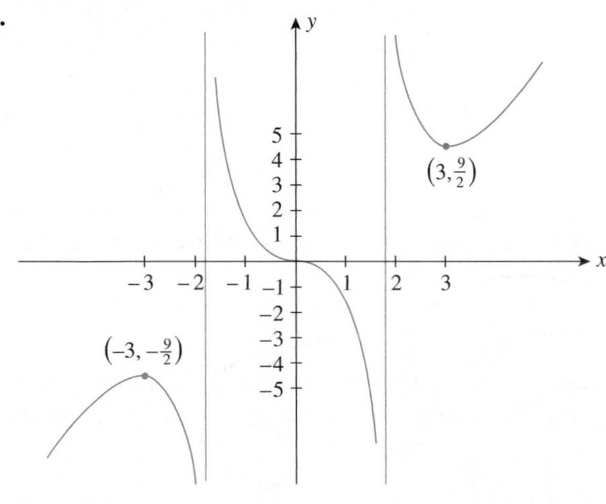

In Exercises 43–46, match the graph of the function with the graph of its derivative in (a)–(d).

43.

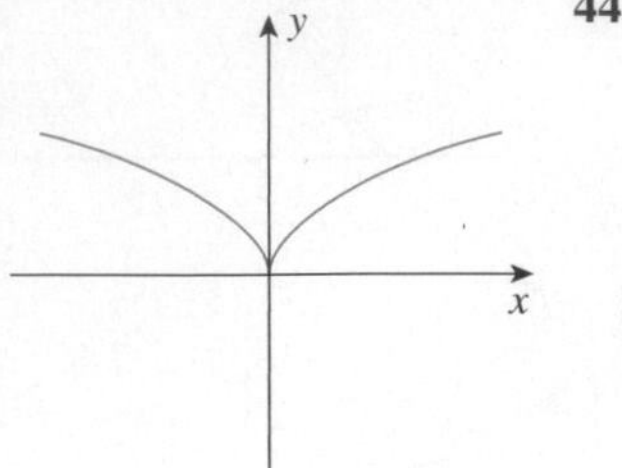

44.

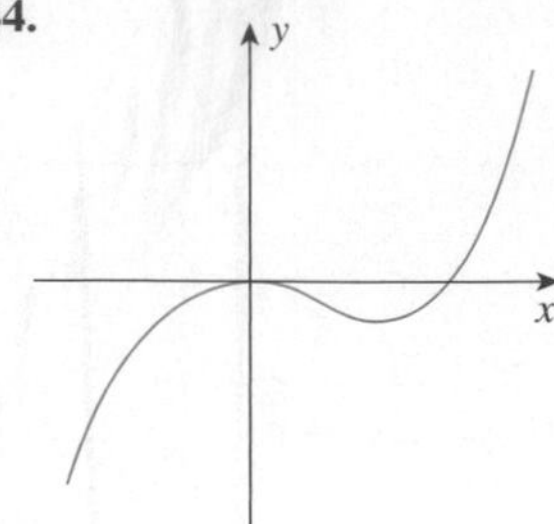

45.

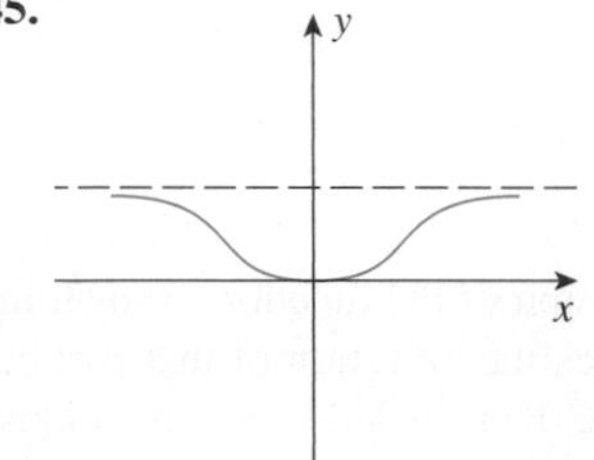

46.

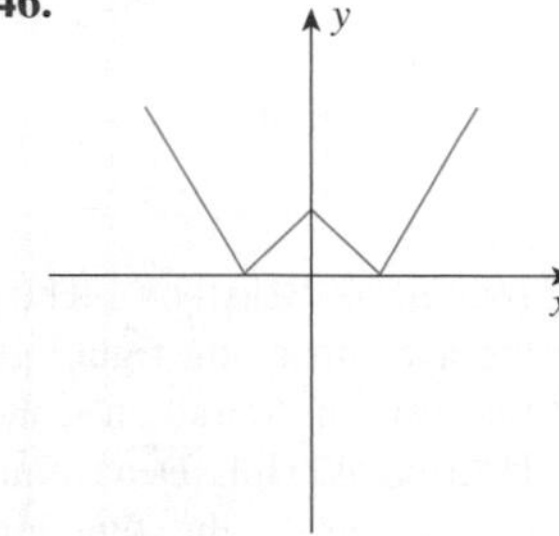

(a)

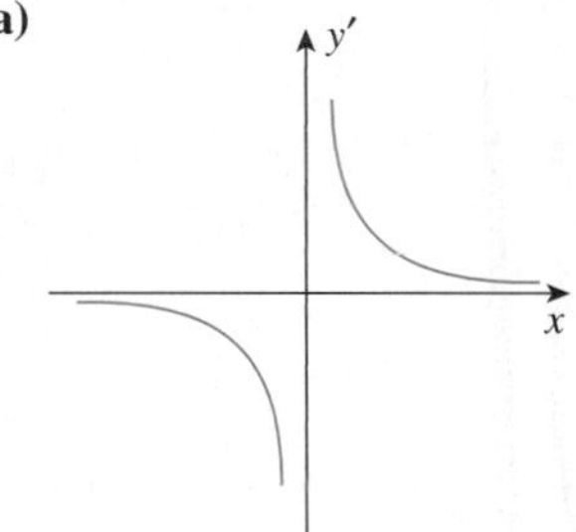

(b)

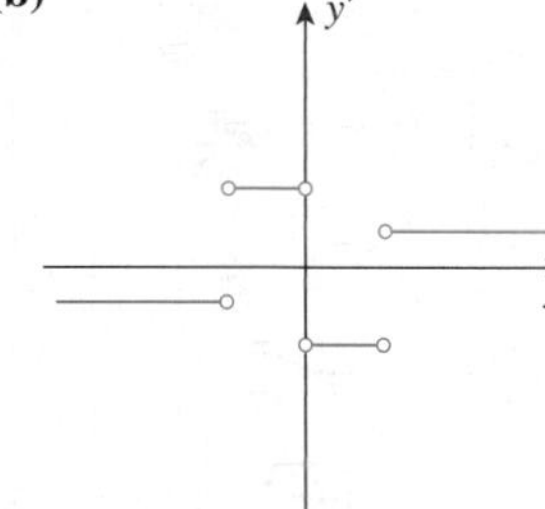

(c)

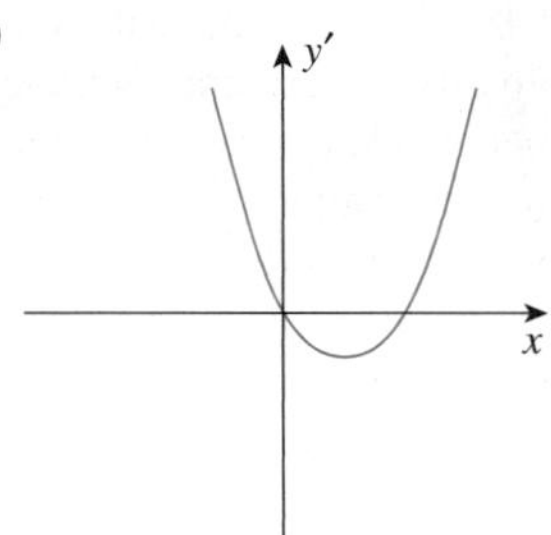

(d)

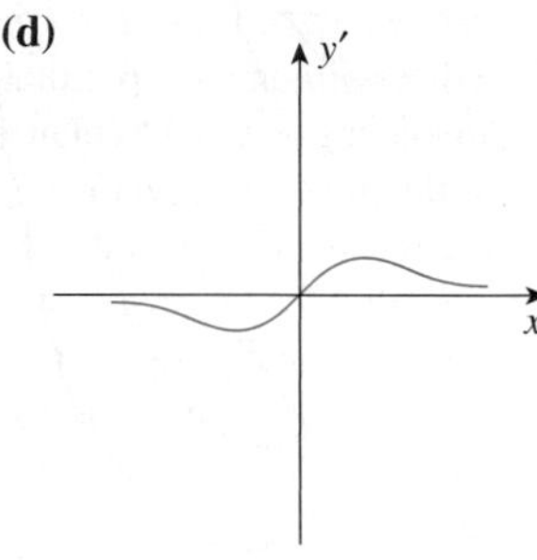

In Exercises 47–68, find the relative maxima and relative minima, if any, of each function.

47. $f(x) = x^2 - 4x$

48. $g(x) = x^2 + 3x + 8$

49. $h(t) = -t^2 + 6t + 6$

50. $f(x) = \frac{1}{2}x^2 - 2x + 4$

51. $f(x) = x^{5/3}$

52. $f(x) = x^{2/3} + 2$

53. $g(x) = x^3 - 3x^2 + 4$

54. $f(x) = x^3 - 3x + 6$

55. $f(x) = \frac{1}{2}x^4 - x^2$

56. $h(x) = \frac{1}{2}x^4 - 3x^2 + 4x - 8$

57. $F(x) = \frac{1}{3}x^3 - x^2 - 3x + 4$

58. $F(t) = 3t^5 - 20t^3 + 20$

59. $g(x) = x^4 - 4x^3 + 8$

60. $f(x) = 3x^4 - 2x^3 + 4$

61. $g(x) = \frac{x+1}{x}$

62. $h(x) = \frac{x}{x+1}$

63. $f(x) = x + \frac{9}{x} + 2$

64. $g(x) = 2x^2 + \frac{4000}{x} + 10$

65. $f(x) = \frac{x}{1+x^2}$

66. $g(x) = \frac{x}{x^2-1}$

67. $f(x) = (x-1)^{2/3}$

68. $g(x) = x\sqrt{x-4}$

69. A stone is thrown straight up from the roof of an 80-ft building. The distance (in feet) of the stone from the ground at any time t (in seconds) is given by

$$h(t) = -16t^2 + 64t + 80$$

When is the stone rising, and when is it falling? If the stone were to miss the building, when would it hit the ground? Sketch the graph of h.
Hint: The stone is on the ground when $h(t) = 0$.

70. PROFIT FUNCTIONS The Mexican subsidiary of ThermoMaster manufactures an indoor–outdoor thermometer. Management estimates that the profit (in dollars) realizable by the company for the manufacture and sale of x units of thermometers each week is

$$P(x) = -0.001x^2 + 8x - 5000$$

Find the intervals where the profit function P is increasing and the intervals where P is decreasing.

71. PREVALENCE OF ALZHEIMER'S PATIENTS Based on a study conducted in 1997, the percent of the U.S. population by age afflicted with Alzheimer's disease is given by the function

$$P(x) = 0.0726x^2 + 0.7902x + 4.9623 \qquad (0 \le x \le 25)$$

where x is measured in years, with $x = 0$ corresponding to age 65. Show that P is an increasing function of x on the interval (0, 25). What does your result tell you about the relationship between Alzheimer's disease and age for the population that is age 65 and over?
Source: Alzheimer's Association

72. GROWTH OF MANAGED SERVICES Almost half of companies let other firms manage some of their Web operations—a practice called Web hosting. Managed services—monitoring a customer's technology services—is the fastest growing part of Web hosting. Managed services sales are expected to grow in accordance with the function

$$f(t) = 0.469t^2 + 0.758t + 0.44 \qquad (0 \le t \le 6)$$

where $f(t)$ is measured in billions of dollars and t is measured in years, with $t = 0$ corresponding to 1999.
a. Find the interval where f is increasing and the interval where f is decreasing.
b. What does your result tell you about sales in managed services from 1999 through 2005?
Source: International Data Corp.

73. FLIGHT OF A ROCKET The height (in feet) attained by a rocket t sec into flight is given by the function

$$h(t) = -\frac{1}{3}t^3 + 16t^2 + 33t + 10 \qquad (t \ge 0)$$

When is the rocket rising, and when is it descending?

74. ENVIRONMENT OF FORESTS Following the lead of the National Wildlife Federation, the Department of the Interior of a South American country began to record an index of environmental quality that measured progress and decline in the environmental quality of its forests. The index for the years 1984 through 1994 is approximated by the function

$$I(t) = \frac{1}{3}t^3 - \frac{5}{2}t^2 + 80 \qquad (0 \le t \le 10)$$

where $t = 0$ corresponds to 1984. Find the intervals where the function I is increasing and the intervals where it is decreasing. Interpret your results.

75. AVERAGE SPEED OF A HIGHWAY VEHICLE The average speed of a vehicle on a stretch of Route 134 between 6 a.m. and 10 a.m. on a typical weekday is approximated by the function

$$f(t) = 20t - 40\sqrt{t} + 50 \qquad (0 \le t \le 4)$$

where $f(t)$ is measured in miles per hour and t is measured in hours, with $t = 0$ corresponding to 6 a.m. Find the interval where f is increasing and the interval where f is decreasing and interpret your results.

76. AVERAGE COST The average cost (in dollars) incurred by Lincoln Records each week in pressing x compact discs is given by

$$\overline{C}(x) = -0.0001x + 2 + \frac{2000}{x} \qquad (0 < x \le 6000)$$

Show that $\overline{C}(x)$ is always decreasing over the interval $(0, 6000)$.

77. **WEB HOSTING** Refer to Exercise 72. Sales in the Web-hosting industry are projected to grow in accordance with the function

$$f(t) = -0.05t^3 + 0.56t^2 + 5.47t + 7.5 \qquad (0 \le t \le 6)$$

where $f(t)$ is measured in billions of dollars and t is measured in years, with $t = 0$ corresponding to 1999.

a. Find the interval where f is increasing and the interval where f is decreasing.
Hint: Use the quadratic formula.

b. What does your result tell you about sales in the Web-hosting industry from 1999 through 2005?

Source: International Data Corp.

78. **CELLULAR PHONE REVENUE** According to a study conducted in 1997, the revenue (in millions of dollars) in the U.S. cellular phone market in the next 6 yr is approximated by the function

$$R(t) = 0.03056t^3 - 0.45357t^2 + 4.81111t + 31.7 \quad (0 \le t \le 6)$$

where t is measured in years, with $t = 0$ corresponding to 1997.

a. Find the interval where R is increasing and the interval where R is decreasing.

b. What does your result tell you about the revenue in the cellular phone market in the years under consideration?
Hint: Use the quadratic formula.

Source: Paul Kagan Associates, Inc.

79. **ELDERLY WORKFORCE** The percent of men 65 years and older in the workforce from 1970 through the year 2000 is approximated by the function

$$P(t) = 0.00093t^3 - 0.018t^2 - 0.51t + 25 \qquad (0 \le t \le 30)$$

where t is measured in years, with $t = 0$ corresponding to the beginning of 1970.

a. Find the interval where P is decreasing and the interval where P is increasing.

b. Interpret the results of part (a).

Source: U.S. Census Bureau

80. **MEDICAL SCHOOL APPLICANTS** According to a study from the American Medical Association, the number of medical school applicants from academic year 1997–1998 ($t = 0$) through the academic year 2002–2003 is approximated by the function

$$N(t) = -0.0333t^3 + 0.47t^2 - 3.8t + 47 \qquad (0 \le t \le 5)$$

where $N(t)$ measured in thousands.

a. Show that the number of medical school applicants had been declining over the period in question.
Hint: Use the quadratic formula.

b. What was the largest number of medical school applicants in any one academic year for the period in question? In what academic year did that occur?

Source: Journal of the American Medical Association

81. **SALES OF FUNCTIONAL FOOD PRODUCTS** The sales of functional food products—those that promise benefits beyond basic nutrition—have risen sharply in recent years. The sales (in billions of dollars) of foods and beverages with herbal and other additives is approximated by the function

$$S(t) = 0.46t^3 - 2.22t^2 + 6.21t + 17.25 \qquad (0 \le t \le 4)$$

where t is measured in years, with $t = 0$ corresponding to the beginning of 1997. Show that S is increasing on the interval $[0, 4]$.
Hint: Use the quadratic formula.

Source: Frost & Sullivan

82. **PROJECTED RETIREMENT FUNDS** Based on data from the Central Provident Fund of a certain country (a government agency similar to the Social Security Administration), the estimated cash in the fund in 1995 is given by

$$A(t) = -96.6t^4 + 403.6t^3 + 660.9t^2 + 250 \qquad (0 \le t \le 5)$$

where $A(t)$ is measured in billions of dollars and t is measured in decades, with $t = 0$ corresponding to 1995. Find the interval where A is increasing and the interval where A is decreasing and interpret your results.
Hint: Use the quadratic formula.

83. **SPENDING ON FIBER-OPTIC LINKS** U.S. telephone company spending on fiber-optic links to homes and businesses from 2001 to 2006 is projected to be

$$S(t) = -2.315t^3 + 34.325t^2 + 1.32t + 23 \qquad (0 \le t \le 5)$$

billion dollars in year t, where t is measured in years with $t = 0$ corresponding to 2001. Show that $S'(t) > 0$ for all t in the interval $[0, 5]$. What conclusion can you draw from this result?
Hint: Use the quadratic formula.

Source: RHK Inc.

84. **AIR POLLUTION** According to the South Coast Air Quality Management District, the level of nitrogen dioxide, a brown gas that impairs breathing, present in the atmosphere on a certain May day in downtown Los Angeles is approximated by

$$A(t) = 0.03t^3(t - 7)^4 + 60.2 \qquad (0 \le t \le 7)$$

where $A(t)$ is measured in pollutant standard index (PSI) and t is measured in hours, with $t = 0$ corresponding to 7 a.m. At what time of day is the air pollution increasing, and at what time is it decreasing?

85. **DRUG CONCENTRATION IN THE BLOOD** The concentration (in milligrams/cubic centimeter) of a certain drug in a patient's body t hr after injection is given by

$$C(t) = \frac{t^2}{2t^3 + 1} \qquad (0 \le t \le 4)$$

When is the concentration of the drug increasing, and when is it decreasing?

86. **Age of Drivers in Crash Fatalities** The number of crash fatalities per 100,000 vehicle miles of travel (based on 1994 data) is approximated by the model

$$f(x) = \frac{15}{0.08333x^2 + 1.91667x + 1} \qquad (0 \le x \le 11)$$

where x is the age of the driver in years, with $x = 0$ corresponding to age 16. Show that f is decreasing on $(0, 11)$ and interpret your result.

Source: National Highway Traffic Safety Administration

87. **Air Pollution** The amount of nitrogen dioxide, a brown gas that impairs breathing, present in the atmosphere on a certain May day in the city of Long Beach is approximated by

$$A(t) = \frac{136}{1 + 0.25(t - 4.5)^2} + 28 \qquad (0 \le t \le 11)$$

where $A(t)$ is measured in pollutant standard index (PSI) and t is measured in hours, with $t = 0$ corresponding to 7 a.m. Find the intervals where A is increasing and where A is decreasing and interpret your results.

88. **Prison Overcrowding** The 1980s saw a trend toward old-fashioned punitive deterrence as opposed to the more liberal penal policies and community based corrections popular in the 1960s and early 1970s. As a result, prisons became more crowded, and the gap between the number of people in prison and the prison capacity widened. The number of prisoners (in thousands) in federal and state prisons is approximated by the function

$$N(t) = 3.5t^2 + 26.7t + 436.2 \qquad (0 \le t \le 10)$$

where t is measured in years, with $t = 0$ corresponding to 1984. The number of inmates for which prisons were designed is given by

$$C(t) = 24.3t + 365 \qquad (0 \le t \le 10)$$

where $C(t)$ is measured in thousands and t has the same meaning as before. Show that the gap between the number of prisoners and the number for which the prisons were designed has been widening at any time t.

Hint: First, write a function G that gives the gap between the number of prisoners and the number for which the prisons were designed at any time t. Then show that $G'(t) > 0$ for all values of t in the interval $(0, 10)$.

Source: U.S. Department of Justice

89. **U.S. Nursing Shortage** The demand for nurses between 2000 and 2015 is estimated to be

$$D(t) = 0.0007t^2 + 0.0265t + 2 \qquad (0 \le t \le 15)$$

where $D(t)$ is measured in millions and $t = 0$ corresponds to the year 2000. The supply of nurses over the same time period is estimated to be

$$S(t) = -0.0014t^2 + 0.0326t + 1.9 \qquad (0 \le t \le 15)$$

where $S(t)$ is also measured in millions.

a. Find an expression $G(t)$ giving the gap between the demand and supply of nurses over the period in question.
b. Find the interval where G is decreasing and where it is increasing. Interpret your result.
c. Find the relative extrema of G. Interpret your result.

Source: U.S. Department of Health and Human Services

In Exercises 90–95, determine whether the statement is true or false. If it is true, explain why it is true. If it is false, give an example to show why it is false.

90. If f is decreasing on (a, b), then $f'(x) < 0$ for each x in (a, b).

91. If f and g are both increasing on (a, b), then $f + g$ is increasing on (a, b).

92. If f and g are both decreasing on (a, b), then $f - g$ is decreasing on (a, b).

93. If $f(x)$ and $g(x)$ are positive on (a, b) and both f and g are increasing on (a, b), then fg is increasing on (a, b).

94. If $f'(c) = 0$, then f has a relative maximum or a relative minimum at $x = c$.

95. If f has a relative minimum at $x = c$, then $f'(c) = 0$.

96. Using Theorem 1, verify that the linear function $f(x) = mx + b$ is (a) increasing everywhere if $m > 0$, (b) decreasing everywhere if $m < 0$, and (c) constant if $m = 0$.

97. Show that the function $f(x) = x^3 + x + 1$ has no relative extrema on $(-\infty, \infty)$.

98. Let

$$f(x) = \begin{cases} -3x & \text{if } x < 0 \\ 2x + 4 & \text{if } x \ge 0 \end{cases}$$

a. Compute $f'(x)$ and show that it changes sign from negative to positive as we move across $x = 0$.
b. Show that f does not have a relative minimum at $x = 0$. Does this contradict the first derivative test? Explain your answer.

99. Let

$$f(x) = \begin{cases} -x^2 + 3 & \text{if } x \ne 0 \\ 2 & \text{if } x = 0 \end{cases}$$

a. Compute $f'(x)$ and show that it changes sign from positive to negative as we move across $x = 0$.
b. Show that f does not have a relative maximum at $x = 0$. Does this contradict the first derivative test? Explain your answer.

100. Let

$$f(x) = \begin{cases} \dfrac{1}{x^2} & \text{if } x > 0 \\ x^2 & \text{if } x \le 0 \end{cases}$$

a. Compute $f'(x)$ and show that it does not change sign as we move across $x = 0$.
b. Show that f has a relative minimum at $x = 0$. Does this contradict the first derivative test? Explain your answer.

101. Show that the quadratic function

$$f(x) = ax^2 + bx + c \qquad (a \neq 0)$$

has a relative extremum when $x = -b/2a$. Also, show that the relative extremum is a relative maximum if $a < 0$ and a relative minimum if $a > 0$.

102. Show that the cubic function

$$f(x) = ax^3 + bx^2 + cx + d \qquad (a \neq 0)$$

has no relative extremum if and only if $b^2 - 3ac \leq 0$.

103. Refer to Example 6, page 126.
a. Show that f is increasing on the interval (0, 1).
b. Show that $f(0) = -1$ and $f(1) = 1$ and use the result of part (a) together with the intermediate value theorem to conclude that there is exactly one root of $f(x) = 0$ in (0, 1).

104. Show that the function

$$f(x) = \frac{ax + b}{cx + d}$$

does not have a relative extremum if $ad - bc \neq 0$. What can you say about f if $ad - bc = 0$?

4.1 Solutions to Self-Check Exercises

1. The derivative of f is

$$f'(x) = 2x^2 - 2x - 12 = 2(x + 2)(x - 3)$$

and it is continuous everywhere. The zeros of $f'(x)$ are $x = -2$ and $x = 3$. The sign diagram of f' is shown in the accompanying figure. We conclude that f is increasing on the intervals $(-\infty, -2)$ and $(3, \infty)$ and decreasing on the interval $(-2, 3)$.

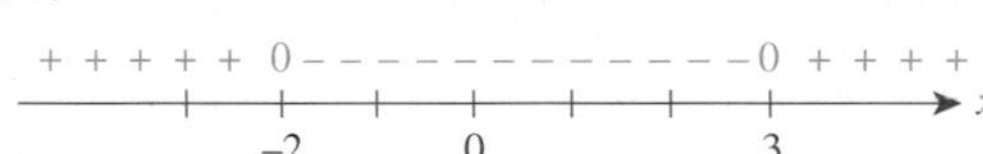

2. The derivative of f is

$$f'(x) = \frac{(1 - x^2)\dfrac{d}{dx}(x^2) - x^2\dfrac{d}{dx}(1 - x^2)}{(1 - x^2)^2}$$

$$= \frac{(1 - x^2)(2x) - x^2(-2x)}{(1 - x^2)^2} = \frac{2x}{(1 - x^2)^2}$$

and it is continuous everywhere except at $x = \pm 1$. Since $f'(x)$ is equal to zero at $x = 0$, $x = 0$ is a critical number of f. Next, observe that $f'(x)$ is discontinuous at $x = \pm 1$, but since these numbers are not in the domain of f, they do not qualify as critical numbers of f. Finally, from the sign diagram of f' shown in the accompanying figure, we conclude that $f(0) = 0$ is a relative minimum of f.

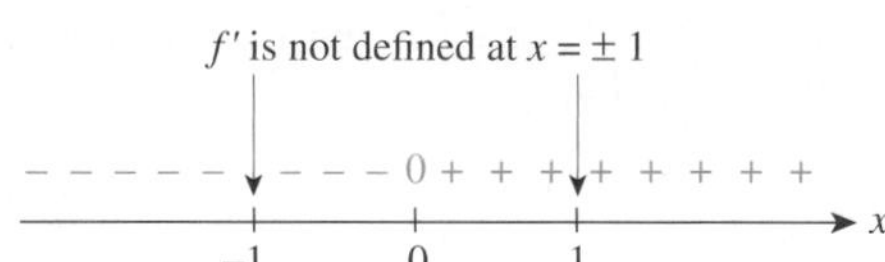

USING TECHNOLOGY

Using the First Derivative to Analyze a Function

A graphing utility is an effective tool for analyzing the properties of functions. This is especially true when we also bring into play the power of calculus, as the following examples show.

EXAMPLE 1 Let $f(x) = 2.4x^4 - 8.2x^3 + 2.7x^2 + 4x + 1$.

a. Use a graphing utility to plot the graph of f.
b. Find the intervals where f is increasing and the intervals where f is decreasing.
c. Find the relative extrema of f.

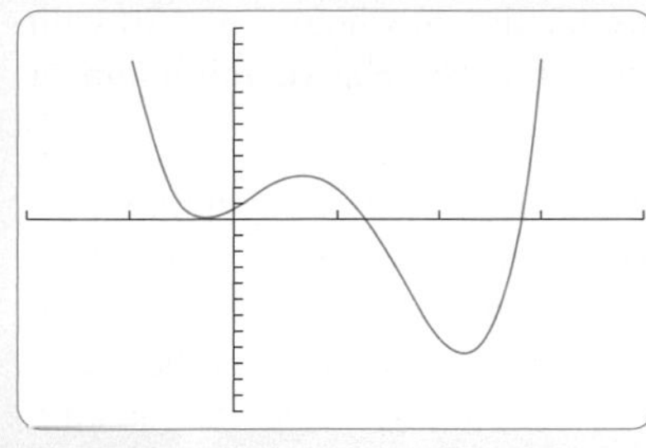

FIGURE T1
The graph of f in the viewing window $[-2, 4] \times [-10, 10]$

Solution

a. The graph of f in the viewing window $[-2, 4] \times [-10, 10]$ is shown in Figure T1.

b. We compute

$$f'(x) = 9.6x^3 - 24.6x^2 + 5.4x + 4$$

and observe that f' is continuous everywhere, so the critical numbers of f occur at values of x where $f'(x) = 0$. To solve this last equation, observe that $f'(x)$ is a *polynomial function* of degree 3. The easiest way to solve the polynomial equation

$$9.6x^3 - 24.6x^2 + 5.4x + 4 = 0$$

is to use the function on a graphing utility for solving polynomial equations. (Not all graphing utilities have this function.) You can also use TRACE and ZOOM, but this will not give the same accuracy without a much greater effort.

We find

$$x_1 \approx 2.22564943249 \qquad x_2 \approx 0.63272944121 \qquad x_3 \approx -0.295878873696$$

Referring to Figure T1, we conclude that f is decreasing on $(-\infty, -0.2959)$ and $(0.6327, 2.2256)$ (correct to four decimal places) and f is increasing on $(-0.2959, 0.6327)$ and $(2.2256, \infty)$.

c. Using the evaluation function of a graphing utility, we find the value of f at each of the critical numbers found in part (b). Upon referring to Figure T1 once again, we see that $f(x_3) \approx 0.2836$ and $f(x_1) \approx -8.2366$ are relative minimum values of f and $f(x_2) \approx 2.9194$ is a relative maximum value of f. ■

Note The equation $f'(x) = 0$ in Example 1 is a polynomial equation, and so it is easily solved using the function for solving polynomial equations. We could also solve the equation using the function for finding the roots of equations, but that would require much more work. For equations that are *not* polynomial equations, however, our only choice is to use the function for finding the roots of equations. ■

If the derivative of a function is difficult to compute or simplify and we do not require great precision in the solution, we can find the relative extrema of the function using a combination of ZOOM and TRACE. This technique, which does not require the use of the derivative of f, is illustrated in the following example.

EXAMPLE 2 Let $f(x) = x^{1/3}(x^2 + 1)^{-3/2}3^{-x}$.

a. Use a graphing utility to plot the graph of f.*

b. Find the relative extrema of f.

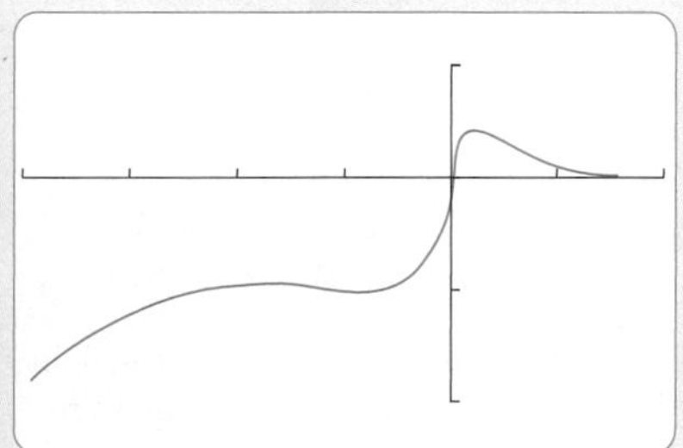

FIGURE T2
The graph of f in the viewing window $[-4, 2] \times [-2, 1]$

Solution

a. The graph of f in the viewing window $[-4, 2] \times [-2, 1]$ is shown in Figure T2.

b. From the graph of f in Figure T2, we see that f has relative maxima when $x \approx -2$ and $x \approx 0.25$ and a relative minimum when $x \approx -0.75$. To obtain a better approximation of the first relative maximum, we zoom-in with the cursor at approximately the point on the graph corresponding to $x \approx -2$. Then, using TRACE, we see that a relative maximum occurs when $x \approx -1.76$ with value

*Functions of the form $f(x) = 3^{-x}$ are called *exponential functions*, and we will study them in greater detail in Chapter 5.

(*continued*)

$y \approx -1.01$. Similarly, we find the other relative maximum where $x \approx 0.20$ with value $y \approx 0.44$. Repeating the procedure, we find the relative minimum at $x \approx -0.86$ and $y \approx -1.07$. ■

You can also use the "minimum" and "maximum" functions of a graphing utility to find the relative extrema of the function. See the Web site for the procedure.

Finally, we comment that if you have access to a computer and software such as Derive, Maple, or Mathematica, then symbolic differentiation will yield the derivative $f'(x)$ of any differentiable function. This software will also solve the equation $f'(x) = 0$ with ease. Thus, the use of a computer will simplify even more greatly the analysis of functions.

TECHNOLOGY EXERCISES

In Exercises 1–4, find (a) the intervals where *f* is increasing and the intervals where *f* is decreasing and (b) the relative extrema of *f*. Express your answers accurate to four decimal places.

1. $f(x) = 3.4x^4 - 6.2x^3 + 1.8x^2 + 3x - 2$
2. $f(x) = 1.8x^4 - 9.1x^3 + 5x - 4$
3. $f(x) = 2x^5 - 5x^3 + 8x^2 - 3x + 2$
4. $f(x) = 3x^5 - 4x^2 + 3x - 1$

In Exercises 5–8, use the ZOOM and TRACE features to find (a) the intervals where *f* is increasing and the intervals where *f* is decreasing and (b) the relative extrema of *f*. Express your answers accurate to two decimal places.

5. $f(x) = (2x + 1)^{1/3}(x^2 + 1)^{-2/3}$
6. $f(x) = [x^2(x^3 - 1)]^{1/3} + \dfrac{1}{x}$
7. $f(x) = x - \sqrt{1 - x^2}$
8. $f(x) = \dfrac{\sqrt{x}(x^2 - 1)^2}{x - 2}$
9. RATE OF BANK FAILURES The rate at which banks were failing between 1982 and 1994 was estimated to be

$$f(t) = 0.063447t^4 - 1.953283t^3 + 14.632576t^2 - 6.684704t + 47.458874 \quad (0 \le t \le 12)$$

where $f(t)$ is the number of banks failing each year and t is measured in years, with $t = 0$ corresponding to the beginning of 1982.
   a. Plot the graph of f in the viewing window $[0, 13] \times [0, 220]$.
   b. Determine the intervals where f is increasing and where f is decreasing and interpret your result.
   c. Find the relative maximum of f and interpret your result.

   Source: Federal Deposit Insurance Corporation

10. CARGO VOLUME The volume of cargo moved in the port of New York/New Jersey from 1991 through 2002 is modeled by the function

$$V(t) = 0.0094665t^3 - 0.052775t^2 + 0.39895t + 13.7 \quad (0 \le t \le 11)$$

where $V(t)$ is measured in millions of tons and t is measured in years, with $t = 0$ corresponding to 1991.
   a. Plot the graph of V in the viewing window $[0, 11] \times [0, 25]$.
   b. Where is V increasing? What does this tell us?
   c. Verify the result of part (b) analytically.

   Source: Port Authority of New York/New Jersey

11. MANUFACTURING CAPACITY Data show that the annual increase in manufacturing capacity between 1994 and 2000 is given by

$$f(t) = 0.009417t^3 - 0.426571t^2 + 2.74894t + 5.54 \quad (0 \le t \le 6)$$

percent where t is measured in years, with $t = 0$ corresponding to the beginning of 1994.
   a. Plot the graph of f in the viewing window $[0, 6] \times [0, 11]$.
   b. Determine the interval where f is increasing and the interval where f is decreasing and interpret your result.

   Source: Federal Reserve

12. SURGERIES IN PHYSICIANS' OFFICES Driven by technological advances and financial pressures, the number of surgeries performed in physicians' offices nationwide has been increasing over the years. The function

$$f(t) = -0.00447t^3 + 0.09864t^2 + 0.05192t + 0.8 \quad (0 \le t \le 15)$$

gives the number of surgeries (in millions) performed in physicians' offices in year t, with $t = 0$ corresponding to the beginning of 1986.
   a. Plot the graph of f in the viewing window $[0, 15] \times [0, 10]$.

b. Prove that f is increasing on the interval [0, 15].
Hint: Show that f' is positive on the interval.
Source: SMG Marketing Group

13. Morning Traffic Rush The speed of traffic flow on a certain stretch of Route 123 between 6 a.m. and 10 a.m. on a typical weekday is approximated by the function

$$f(t) = 20t - 40\sqrt{t} + 52 \qquad (0 \le t \le 4)$$

where $f(t)$ is measured in miles per hour and t is measured in hours, with $t = 0$ corresponding to 6 a.m. Find the interval where f is increasing, the interval where f is decreasing, and the relative extrema of f. Interpret your results.

14. Air Pollution The amount of nitrogen dioxide, a brown gas that impairs breathing, present in the atmosphere on a certain May day in the city of Long Beach, is approximated by

$$A(t) = \frac{136}{1 + 0.25(t - 4.5)^2} + 28 \qquad (0 \le t \le 11)$$

where $A(t)$ is measured in pollutant standard index (PSI) and t is measured in hours, with $t = 0$ corresponding to 7 a.m. When is the PSI increasing and when is it decreasing? At what time is the PSI highest, and what is its value at that time?

4.2 Applications of the Second Derivative

Determining the Intervals of Concavity

Consider the graphs shown in Figure 26, which give the estimated population of the world and of the United States through the year 2000. Both graphs are rising, indicating that both the U.S. population and the world population continued to increase through the year 2000. But observe that the graph in Figure 26a opens upward, whereas the graph in Figure 26b opens downward. What is the significance of this? To answer this question, let's look at the slopes of the tangent lines to various points on each graph (Figure 27).

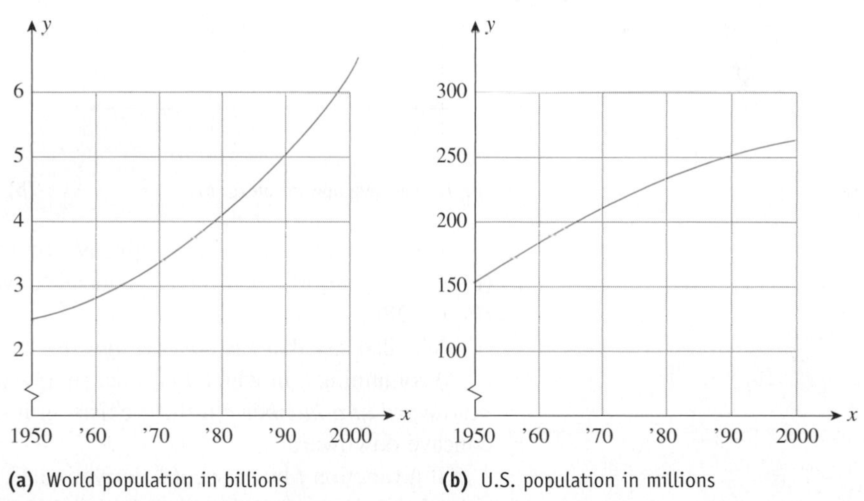

FIGURE 26

(a) World population in billions (b) U.S. population in millions

Source: U.S. Department of Commerce and Worldwatch Institute

In Figure 27a, we see that the slopes of the tangent lines to the graph are increasing as we move from left to right. Since the slope of the tangent line to the graph at a point on the graph measures the rate of change of the function at that point, we conclude that the world population was not only increasing through the year 2000 but was also increasing at an *increasing* pace. A similar analysis of Figure 27b reveals that the U.S. population was increasing, but at a *decreasing* pace.

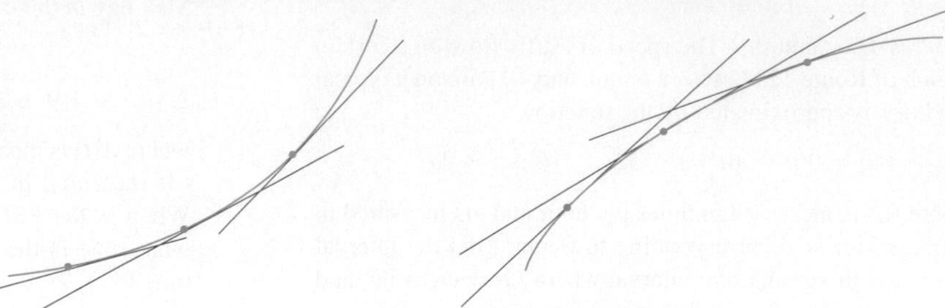

FIGURE 27

(a) Slopes of tangent lines are increasing. **(b)** Slopes of tangent lines are decreasing.

The shape of a curve can be described using the notion of concavity.

Concavity of a Function f

Let the function f be differentiable on an interval (a, b). Then,

1. f is **concave upward** on (a, b) if f' is increasing on (a, b).
2. f is **concave downward** on (a, b) if f' is decreasing on (a, b).

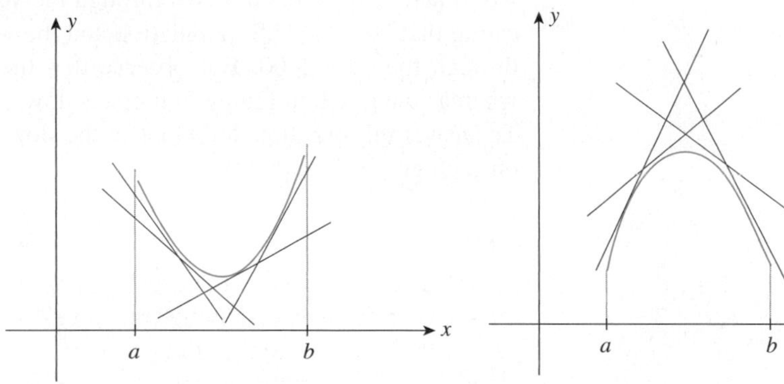

FIGURE 28

(a) f is concave upward on (a, b). **(b)** f is concave downward on (a, b).

Geometrically, a curve is concave upward if it lies above its tangent lines (Figure 28a). Similarly, a curve is concave downward if it lies below its tangent lines (Figure 28b).

We also say that f is *concave upward at a number c* if there exists an interval (a, b) containing c in which f is concave upward. Similarly, we say that f is *concave downward at a number c* if there exists an interval (a, b) containing c in which f is concave downward.

If a function f has a second derivative f'', we can use f'' to determine the intervals of concavity of the function. Recall that $f''(x)$ measures the rate of change of the slope $f'(x)$ of the tangent line to the graph of f at the point $(x, f(x))$. Thus, if $f''(x) > 0$ on an interval (a, b), then the slopes of the tangent lines to the graph of f are increasing on (a, b), and so f is concave upward on (a, b). Similarly, if $f''(x) < 0$ on (a, b), then f is concave downward on (a, b). These observations suggest the following theorem.

THEOREM 2

a. If $f''(x) > 0$ for each value of x in (a, b), then f is concave upward on (a, b).

b. If $f''(x) < 0$ for each value of x in (a, b), then f is concave downward on (a, b).

The following procedure, based on the conclusions of Theorem 2, may be used to determine the intervals of concavity of a function.

Determining the Intervals of Concavity of f

1. Determine the values of x for which f'' is zero or where f'' is not defined, and identify the open intervals determined by these numbers.
2. Determine the sign of f'' in each interval found in step 1. To do this, compute $f''(c)$, where c is any conveniently chosen test number in the interval.
   a. If $f''(c) > 0$, f is concave upward on that interval.
   b. If $f''(c) < 0$, f is concave downward on that interval.

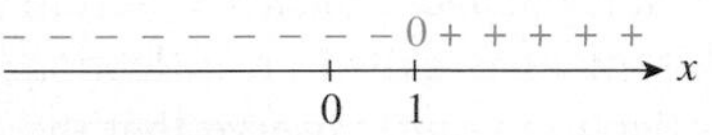

FIGURE 29
Sign diagram for f''

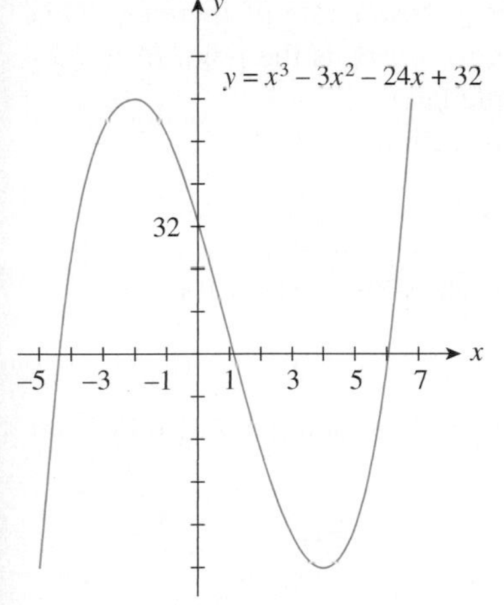

FIGURE 30
f is concave downward on $(-\infty, 1)$ and concave upward on $(1, \infty)$.

EXAMPLE 1 Determine where the function $f(x) = x^3 - 3x^2 - 24x + 32$ is concave upward and where it is concave downward.

Solution Here,

$$f'(x) = 3x^2 - 6x - 24$$
$$f''(x) = 6x - 6 = 6(x - 1)$$

and f'' is defined everywhere. Setting $f''(x) = 0$ gives $x = 1$. The sign diagram of f'' appears in Figure 29. We conclude that f is concave downward on the interval $(-\infty, 1)$ and is concave upward on the interval $(1, \infty)$. Figure 30 shows the graph of f. ■

EXPLORING WITH TECHNOLOGY

Refer to Example 1.

1. Use a graphing utility to plot the graph of $f(x) = x^3 - 3x^2 - 24x + 32$ and its second derivative $f''(x) = 6x - 6$ using the viewing window $[-10, 10] \times [-80, 90]$.
2. By studying the graph of f'', determine the intervals where $f''(x) > 0$ and the intervals where $f''(x) < 0$. Next, look at the graph of f and determine the intervals where the graph of f is concave upward and the intervals where the graph of f is concave downward. Are these observations what you might have expected?

EXAMPLE 2 Determine the intervals where the function $f(x) = x + \dfrac{1}{x}$ is concave upward and where it is concave downward.

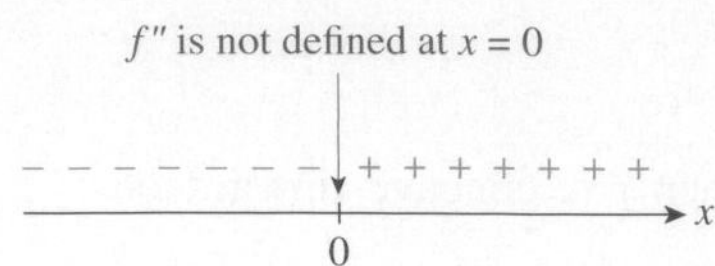

FIGURE 31
The sign diagram for f''

Solution We have

$$f'(x) = 1 - \frac{1}{x^2}$$

$$f''(x) = \frac{2}{x^3}$$

We deduce from the sign diagram for f'' (Figure 31) that the function f is concave downward on the interval $(-\infty, 0)$ and concave upward on the interval $(0, \infty)$. The graph of f is sketched in Figure 32. ■

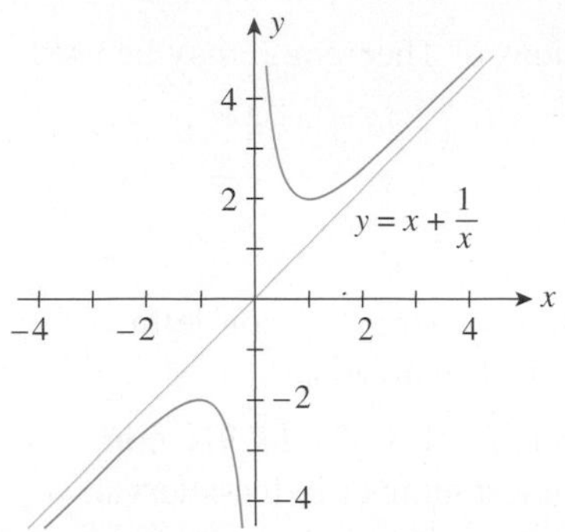

FIGURE 32
f is concave downward on $(-\infty, 0)$ and concave upward on $(0, \infty)$.

Inflection Points

Figure 33 shows the total sales S of a manufacturer of automobile air conditioners versus the amount of money x that the company spends on advertising its product. Notice that the graph of the continuous function $y = S(x)$ changes concavity—from upward to downward—at the point (50, 2700). This point is called an inflection point of S. To understand the significance of this inflection point, observe that the total sales increase rather slowly at first, but as more money is spent on advertising, the total sales increase rapidly. This rapid increase reflects the effectiveness of the company's ads. However, a point is soon reached after which any additional advertising expenditure results in increased sales but at a slower rate of increase. This point, commonly known as the *point of diminishing returns,* is the point of inflection of the function S. We will return to this example later.

Let's now state formally the definition of an inflection point.

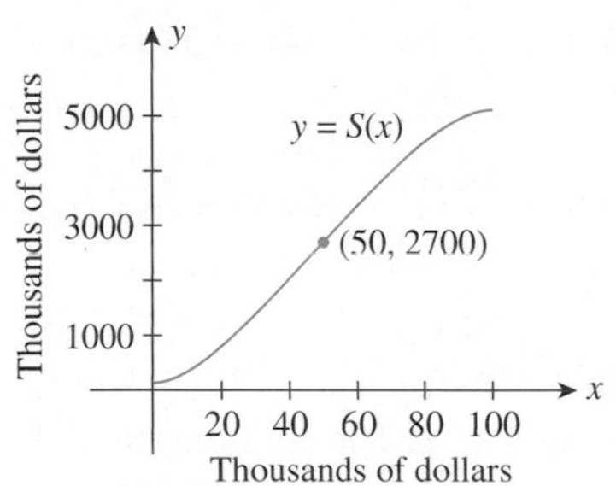

FIGURE 33
The graph of S has a point of inflection at (50, 2700).

Inflection Point

A point on the graph of a continuous function f where the tangent line exists and where the concavity changes is called an **inflection point**.

Observe that the graph of a function crosses its tangent line at a point of inflection (Figure 34).

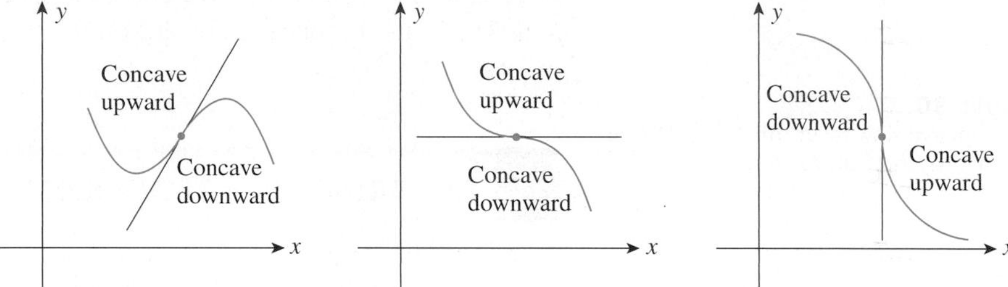

FIGURE 34
At each point of inflection, the graph of a function crosses its tangent line.

The following procedure may be used to find inflection points.

Finding Inflection Points

1. Compute $f''(x)$.
2. Determine the numbers in the domain of f for which $f''(x) = 0$ or $f''(x)$ does not exist.
3. Determine the sign of $f''(x)$ to the left and right of each number c found in step 2. If there is a change in the sign of $f''(x)$ as we move across $x = c$, then $(c, f(c))$ is an inflection point of f.

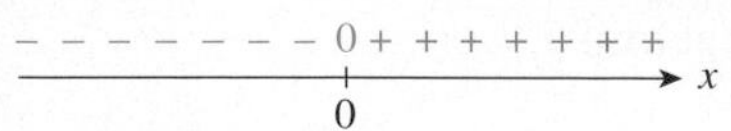

FIGURE 35
Sign diagram for f''

The numbers determined in step 2 are only *candidates* for the inflection points of f. For example, you can easily verify that $f''(0) = 0$ if $f(x) = x^4$, but a sketch of the graph of f will show that $(0, 0)$ is *not* an inflection point of f.

EXAMPLE 3 Find the points of inflection of the function $f(x) = x^3$.

Solution

$$f'(x) = 3x^2$$

$$f''(x) = 6x$$

Observe that f'' is continuous everywhere and is zero if $x = 0$. The sign diagram of f'' is shown in Figure 35. From this diagram, we see that $f''(x)$ changes sign as we move across $x = 0$. Thus, the point $(0, 0)$ is an inflection point of the function f (Figure 36).

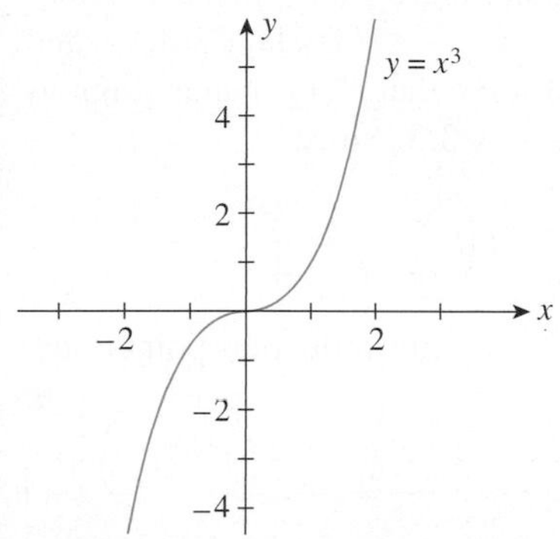

FIGURE 36
f has an inflection point at $(0, 0)$.

EXAMPLE 4 Determine the intervals where the function $f(x) = (x - 1)^{5/3}$ is concave upward and where it is concave downward and find the inflection points of f.

Solution The first derivative of f is

$$f'(x) = \frac{5}{3}(x - 1)^{2/3}$$

and the second derivative of f is

$$f''(x) = \frac{10}{9}(x - 1)^{-1/3} = \frac{10}{9(x - 1)^{1/3}}$$

We see that f'' is not defined at $x = 1$. Furthermore, $f''(x)$ is not equal to zero anywhere. The sign diagram of f'' is shown in Figure 37. From the sign diagram, we see that f is concave downward on $(-\infty, 1)$ and concave upward on $(1, \infty)$. Next, since $x = 1$ does lie in the domain of f, our computations also reveal that the point $(1, 0)$ is an inflection point of f (Figure 38).

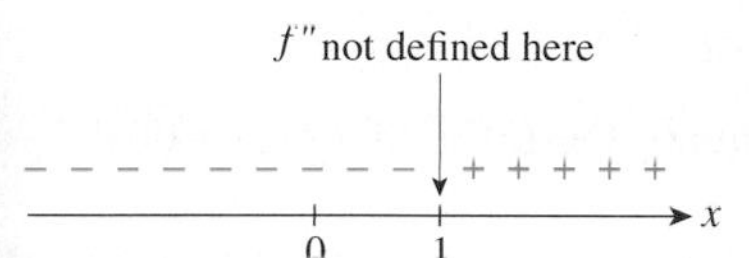

FIGURE 37
Sign diagram for f''

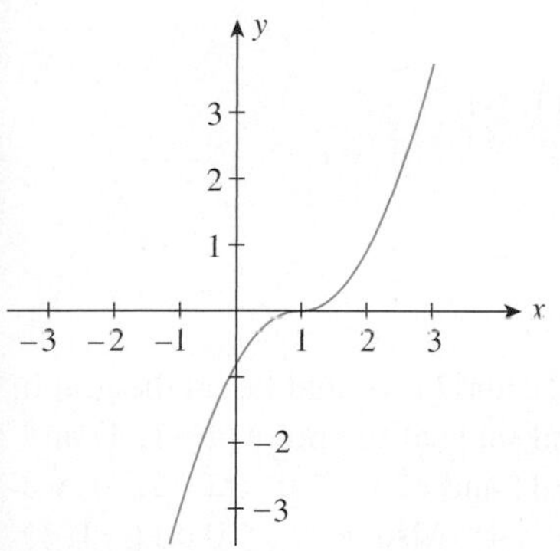

FIGURE 38
f has an inflection point at $(1, 0)$.

EXAMPLE 5 Determine the intervals where the function

$$f(x) = \frac{1}{x^2 + 1}$$

is concave upward and where it is concave downward and find the inflection points of f.

Solution The first derivative of f is

$$f'(x) = \frac{d}{dx}(x^2 + 1)^{-1} = -2x(x^2 + 1)^{-2} \quad \text{Rewrite the original function and use the general power rule.}$$

$$= -\frac{2x}{(x^2 + 1)^2}$$

Next, using the quotient rule, we find

$$f''(x) = \frac{(x^2 + 1)^2(-2) + (2x)2(x^2 + 1)(2x)}{(x^2 + 1)^4}$$

$$= \frac{(x^2 + 1)[-2(x^2 + 1) + 8x^2]}{(x^2 + 1)^4} = \frac{(x^2 + 1)(6x^2 - 2)}{(x^2 + 1)^4}$$

$$= \frac{2(3x^2 - 1)}{(x^2 + 1)^3} \quad \text{Cancel the common factors.}$$

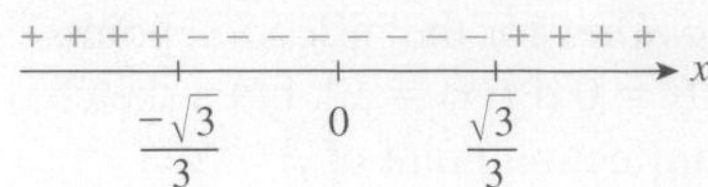

FIGURE 39
Sign diagram for f''

Observe that f'' is continuous everywhere and is zero if

$$3x^2 - 1 = 0$$
$$x^2 = \frac{1}{3}$$

or $x = \pm\sqrt{3}/3$. The sign diagram for f'' is shown in Figure 39. From the sign diagram for f'', we see that f is concave upward on $(-\infty, -\sqrt{3}/3) \cup (\sqrt{3}/3, \infty)$ and concave downward on $(-\sqrt{3}/3, \sqrt{3}/3)$. Also, observe that $f''(x)$ changes sign as we move across the numbers $x = -\sqrt{3}/3$ and $x = \sqrt{3}/3$. Since

$$f\left(-\frac{\sqrt{3}}{3}\right) = \frac{1}{\frac{1}{3} + 1} = \frac{3}{4} \quad \text{and} \quad f\left(\frac{\sqrt{3}}{3}\right) = \frac{3}{4}$$

we see that the points $(-\sqrt{3}/3, 3/4)$ and $(\sqrt{3}/3, 3/4)$ are inflection points of f. The graph of f is shown in Figure 40. ■

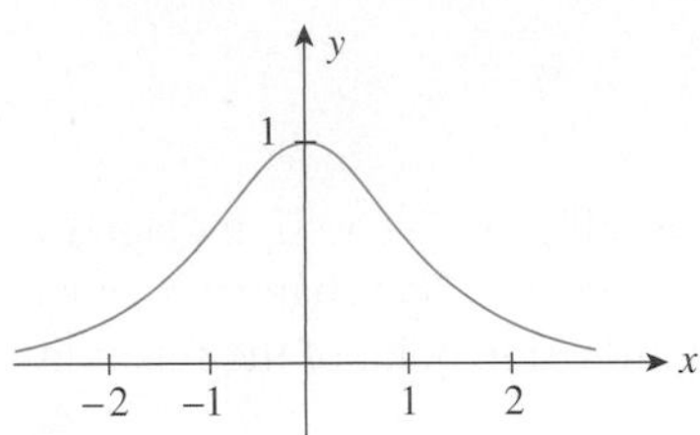

FIGURE 40
The graph of $f(x) = \frac{1}{x^2 + 1}$ is concave upward on $(-\infty, -\sqrt{3}/3) \cup (\sqrt{3}/3, \infty)$ and concave downward on $(-\sqrt{3}/3, \sqrt{3}/3)$.

EXPLORE & DISCUSS

1. Suppose $(c, f(c))$ is an inflection point of f. Can you conclude that f has no relative extremum at $x = c$? Explain your answer.
2. True or false: A polynomial function of degree 3 has exactly one inflection point. Hint: Study the function $f(x) = ax^3 + bx^2 + cx + d$ $(a \neq 0)$.

The next example uses an interpretation of the first and second derivatives to help us sketch a graph of a function.

EXAMPLE 6 Sketch the graph of a function having the following properties:

$$f(-1) = 4$$
$$f(0) = 2$$
$$f(1) = 0$$
$$f'(-1) = 0$$
$$f'(1) = 0$$
$$f'(x) > 0 \quad \text{on } (-\infty, -1) \cup (1, \infty)$$
$$f'(x) < 0 \quad \text{on } (-1, 1)$$
$$f''(x) < 0 \quad \text{on } (-\infty, 0)$$
$$f''(x) > 0 \quad \text{on } (0, \infty)$$

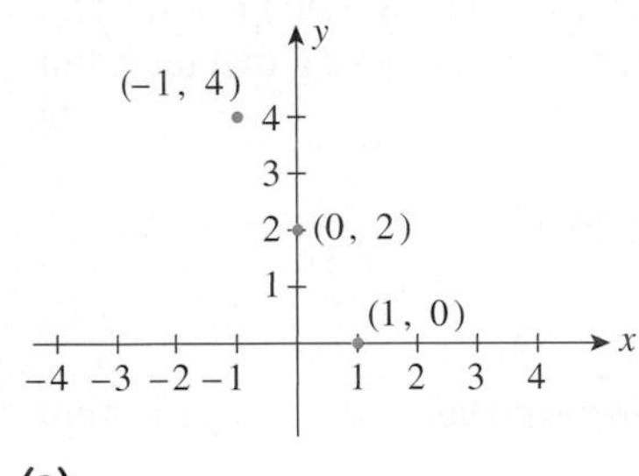

(a)

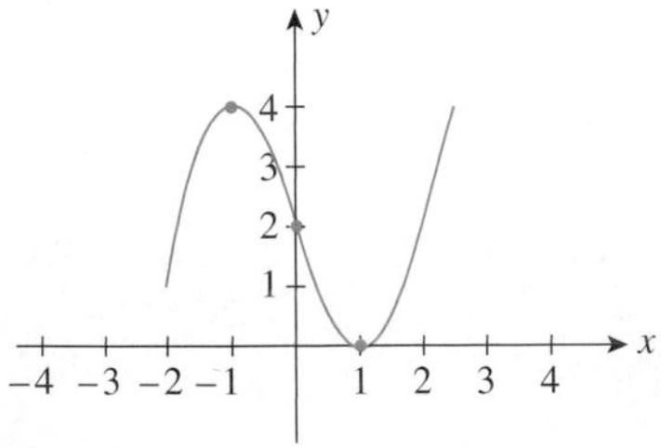

(b)
FIGURE 41

Solution First, we plot the points $(-1, 4)$, $(0, 2)$, and $(1, 0)$ that lie on the graph of f. Since $f'(-1) = 0$ and $f'(1) = 0$, the tangent lines at the points $(-1, 4)$ and $(1, 0)$ are horizontal. Since $f'(x) > 0$ on $(-\infty, -1)$ and $f'(x) < 0$ on $(-1, 1)$, we see that f has a relative maximum at the point $(-1, 4)$. Also, $f'(x) < 0$ on $(-1, 1)$ and $f'(x) > 0$ on $(1, \infty)$ implies that f has a relative minimum at the point $(1, 0)$ (Figure 41a).

Since $f''(x) < 0$ on $(-\infty, 0)$ and $f''(x) > 0$ on $(0, \infty)$, we see that the point $(0, 2)$ is an inflection point. Finally, we complete the graph making use of the fact that f is increasing on $(-\infty, -1) \cup (1, \infty)$, where it is given that $f'(x) > 0$, and f is decreasing on $(-1, 1)$, where $f'(x) < 0$. Also, make sure that f is concave downward on $(-\infty, 0)$ and concave upward on $(0, \infty)$ (Figure 41b). ■

Examples 7 and 8 illustrate familiar interpretations of the significance of the inflection point of a function.

APPLIED EXAMPLE 7 Effect of Advertising on Sales The total sales S (in thousands of dollars) of Arctic Air Corporation, a manufacturer of automobile air conditioners, is related to the amount of money x (in thousands of dollars) the company spends on advertising its products by the formula

$$S = -0.01x^3 + 1.5x^2 + 200 \qquad (0 \le x \le 100)$$

Find the inflection point of the function S.

Solution The first two derivatives of S are given by

$$S' = -0.03x^2 + 3x$$
$$S'' = -0.06x + 3$$

Setting $S'' = 0$ gives $x = 50$. So $(50, S(50))$ is the only candidate for an inflection point of S. Moreover, since

$$S'' > 0 \quad \text{for} \quad x < 50$$

and

$$S'' < 0 \quad \text{for} \quad x > 50$$

the point (50, 2700) is an inflection point of the function S. The graph of S appears in Figure 42. Notice that this is the graph of the function we discussed earlier. ■

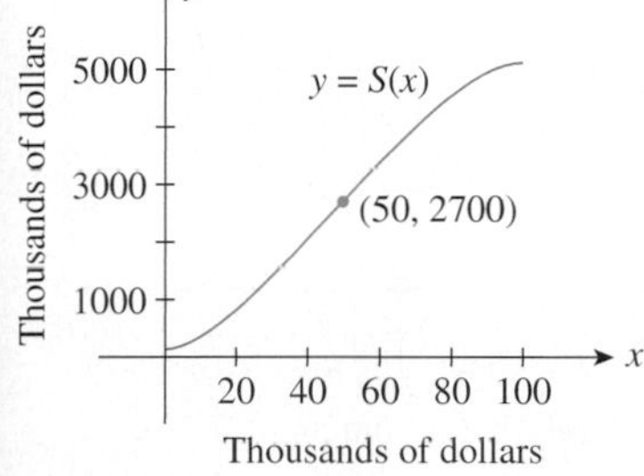

FIGURE 42
The graph of $S(x)$ has a point of inflection at (50, 2700).

APPLIED EXAMPLE 8 Consumer Price Index An economy's consumer price index (CPI) is described by the function

$$I(t) = -0.2t^3 + 3t^2 + 100 \qquad (0 \le t \le 9)$$

where $t = 0$ corresponds to the year 1995. Find the point of inflection of the function I and discuss its significance.

Solution The first two derivatives of I are given by

$$I'(t) = -0.6t^2 + 6t$$
$$I''(t) = -1.2t + 6 = -1.2(t - 5)$$

Setting $I''(t) = 0$ gives $t = 5$. So $(5, I(5))$ is the only candidate for an inflection point of I. Next, we observe that

$$I'' > 0 \quad \text{for} \quad t < 5$$
$$I'' < 0 \quad \text{for} \quad t > 5$$

so the point (5, 150) is an inflection point of I. The graph of I is sketched in Figure 43.

Since the second derivative of I measures the rate of change of the inflation rate, our computations reveal that the rate of inflation had in fact peaked at $t = 5$. Thus, relief actually began at the beginning of 2000. ■

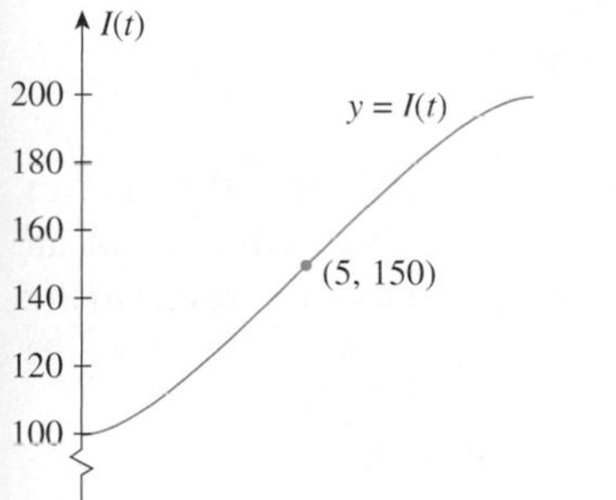

FIGURE 43
The graph of $I(t)$ has a point of inflection at (5, 150).

The Second Derivative Test

We now show how the second derivative f'' of a function f can be used to help us determine whether a critical number of f gives rise to a relative extremum of f.

Figure 44a shows the graph of a function that has a relative maximum at $x = c$. Observe that f is concave downward at that number. Similarly, Figure 44b shows that at a relative minimum of f the graph is concave upward. But from our previous work, we know that f is concave downward at $x = c$ if $f''(c) < 0$ and f is concave upward at $x = c$ if $f''(c) > 0$. These observations suggest the following alternative procedure for determining whether a critical number of f gives rise to a relative extremum of f. This result is called the **second derivative test** and is applicable when f'' exists.

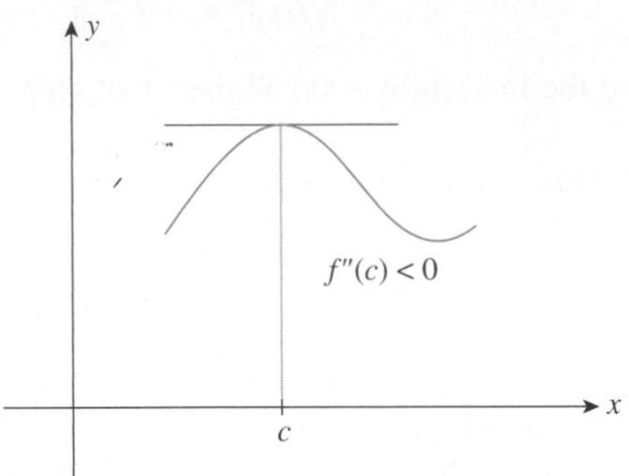

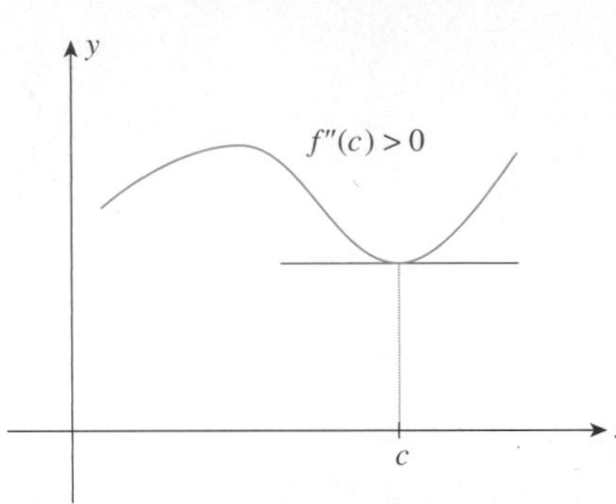

FIGURE 44

(a) f has a relative maximum at $x = c$. **(b)** f has a relative minimum at $x = c$.

The Second Derivative Test

1. Compute $f'(x)$ and $f''(x)$.
2. Find all the critical numbers of f at which $f'(x) = 0$.
3. Compute $f''(c)$ for each such critical number c.
   a. If $f''(c) < 0$, then f has a relative maximum at c.
   b. If $f''(c) > 0$, then f has a relative minimum at c.
   c. If $f''(c) = 0$, the test fails; that is, it is inconclusive.

Note The second derivative test does not yield a conclusion if $f''(c) = 0$ or if $f''(c)$ does not exist. In other words, $x = c$ may give rise to a relative extremum or an inflection point (see Exercise 102, page 283). In such cases, you should revert to the first derivative test. ■

EXAMPLE 9 Determine the relative extrema of the function

$$f(x) = x^3 - 3x^2 - 24x + 32$$

using the second derivative test. (See Example 7, Section 4.1.)

Solution We have

$$f'(x) = 3x^2 - 6x - 24 = 3(x + 2)(x - 4)$$

so $f'(x) = 0$ gives $x = -2$ and $x = 4$, the critical numbers of f, as in Example 7. Next, we compute

$$f''(x) = 6x - 6 = 6(x - 1)$$

Since

$$f''(-2) = 6(-2 - 1) = -18 < 0$$

the second derivative test implies that $f(-2) = 60$ is a relative maximum of f. Also,

$$f''(4) = 6(4 - 1) = 18 > 0$$

and the second derivative test implies that $f(4) = -48$ is a relative minimum of f, which confirms the results obtained earlier. ■

EXPLORE & DISCUSS

Suppose a function f has the following properties:

1. $f''(x) > 0$ for all x in an interval (a, b).

2. There is a number c between a and b such that $f'(c) = 0$.

What special property can you ascribe to the point $(c, f(c))$? Answer the question if Property 1 is replaced by the property that $f''(x) < 0$ for all x in (a, b).

Comparing the First and Second Derivative Tests

Notice that both the first derivative test and the second derivative test are used to classify the critical numbers of f. What are the pros and cons of the two tests? Since the second derivative test is applicable only when f'' exists, it is less versatile than the first derivative test. For example, it cannot be used to locate the relative minimum $f(0) = 0$ of the function $f(x) = x^{2/3}$.

Furthermore, the second derivative test is inconclusive when f'' is equal to zero at a critical number of f, whereas the first derivative test always yields positive conclusions. The second derivative test is also inconvenient to use when f'' is difficult to compute. On the plus side, if f'' is computed easily, then we use the second derivative test since it involves just the evaluation of f'' at the critical number(s) of f. Also, the conclusions of the second derivative test are important in theoretical work.

We close this section by summarizing the different roles played by the first derivative f' and the second derivative f'' of a function f in determining the properties of the graph of f. The first derivative f' tells us where f is increasing and where f is decreasing, whereas the second derivative f'' tells us where f is concave upward and where f is concave downward. These different properties of f are reflected by the signs of f' and f'' in the interval of interest. The following table shows the general characteristics of the function f for various possible combinations of the signs of f' and f'' in the interval (a, b).

Signs of f' and f''	Properties of the Graph of f	General Shape of the Graph of f
$f'(x) > 0$ $f''(x) > 0$	f increasing f concave upward	
$f'(x) > 0$ $f''(x) < 0$	f increasing f concave downward	
$f'(x) < 0$ $f''(x) > 0$	f decreasing f concave upward	
$f'(x) < 0$ $f''(x) < 0$	f decreasing f concave downward	

4.2 Self-Check Exercises

1. Determine where the function $f(x) = 4x^3 - 3x^2 + 6$ is concave upward and where it is concave downward.

2. Using the second derivative test, if applicable, find the relative extrema of the function $f(x) = 2x^3 - \frac{1}{2}x^2 - 12x - 10$.

3. A certain country's gross domestic product (GDP) (in millions of dollars) in year t is described by the function

$$G(t) = -2t^3 + 45t^2 + 20t + 6000 \qquad (0 \le t \le 11)$$

where $t = 0$ corresponds to the beginning of 1995. Find the inflection point of the function G and discuss its significance.

Solutions to Self-Check Exercises 4.2 can be found on page 283.

4.2 Concept Questions

1. Explain what it means for a function f to be (a) concave upward and (b) concave downward on an open interval I. Given that f has a second derivative on I (except at isolated numbers), how do you determine where the graph of f is concave upward and where it is concave downward?

2. What is an inflection point of the graph of a function f? How do you find the inflection point(s) of the graph of a function f whose rule is given?

3. State the second derivative test. What are the pros and cons of using the first derivative test and the second derivative test?

4.2 Exercises

In Exercises 1–8, you are given the graph of a function f. Determine the intervals where f is concave upward and where it is concave downward. Also, find all inflection points of f, if any.

1.

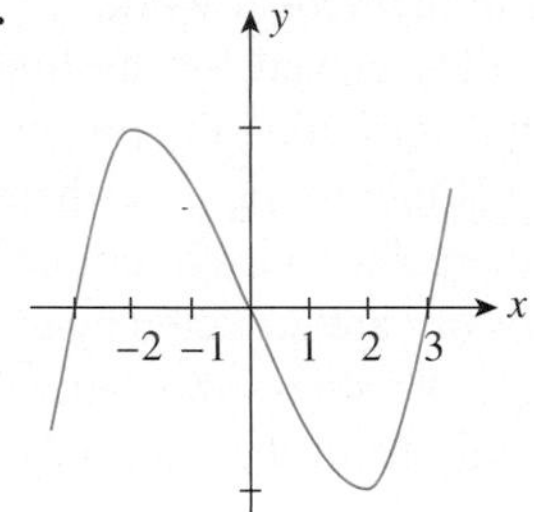

2.

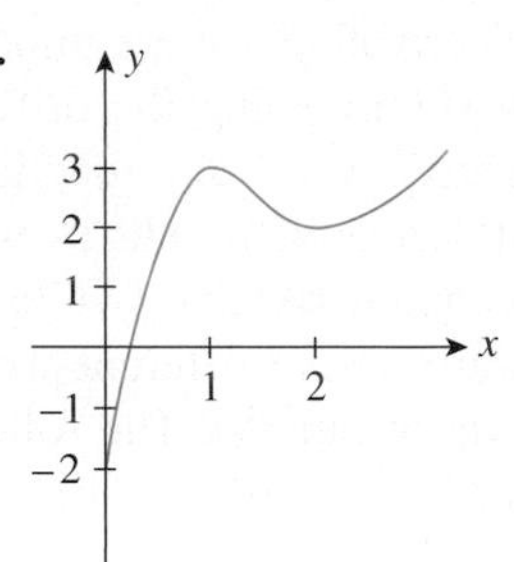

3.

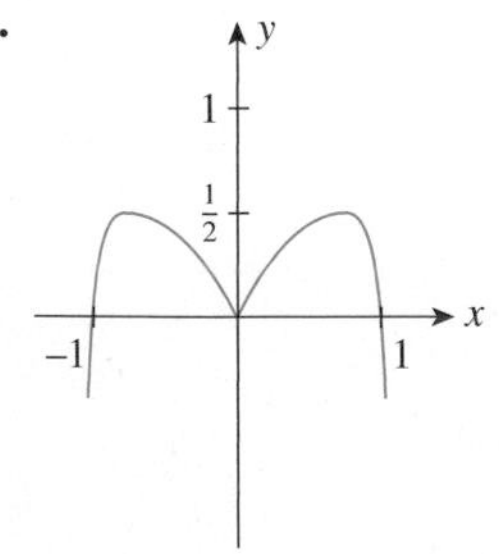

4.

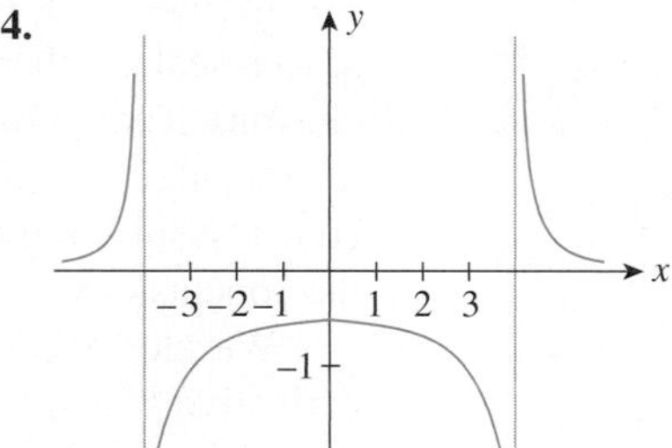

5.

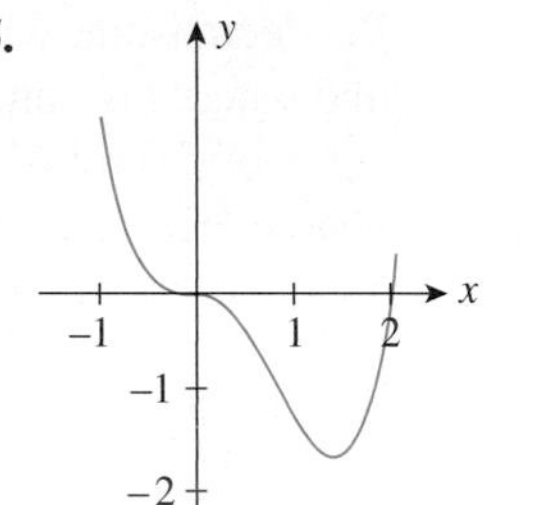

6.

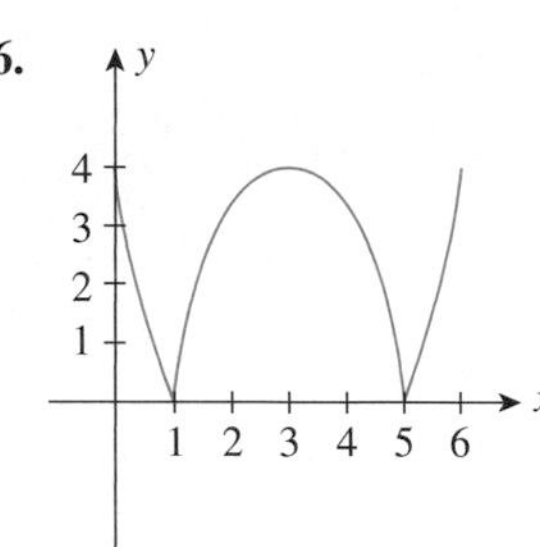

7.

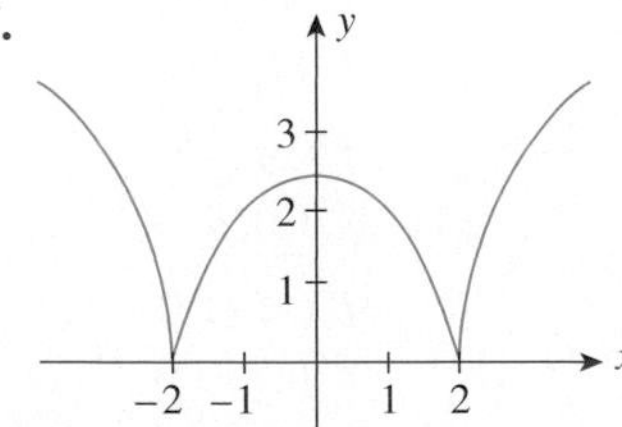

8.

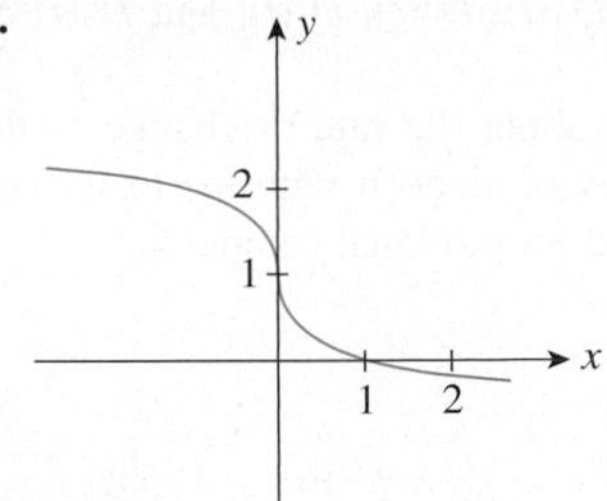

In Exercises 9–12, determine which graph—(a), (b), or (c)—is the graph of the function f with the specified properties.

9. $f(2) = 1, f'(2) > 0$, and $f''(2) < 0$

(a)

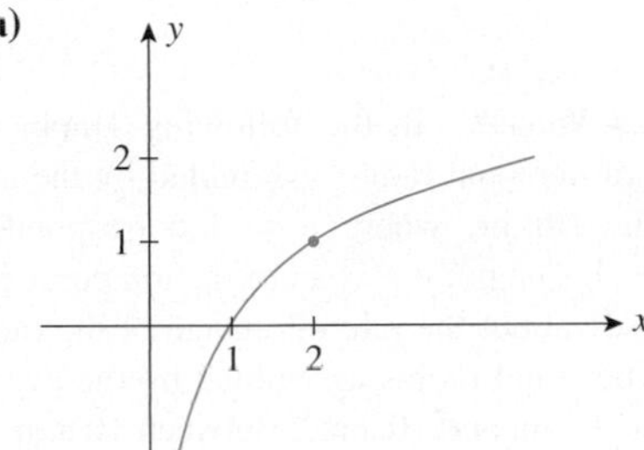

(b)

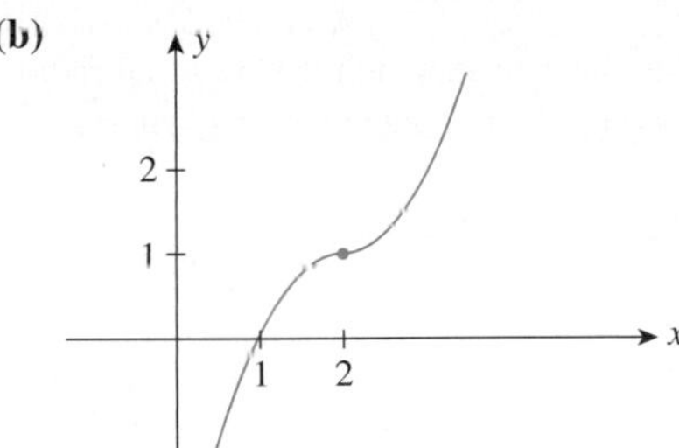

(c)

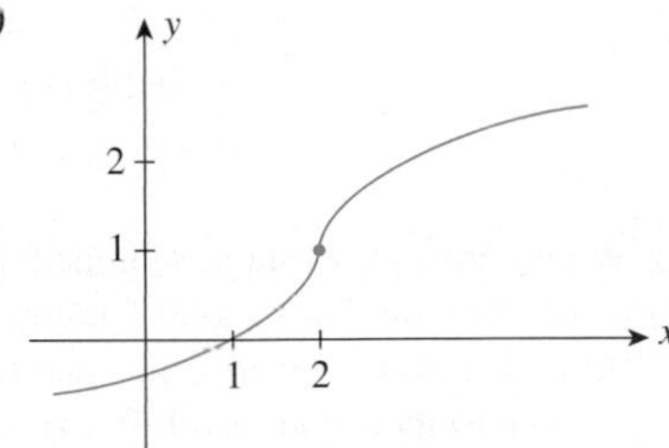

10. $f(1) = 2, f'(x) > 0$ on $(-\infty, 1) \cup (1, \infty)$, and $f''(1) = 0$

(a)

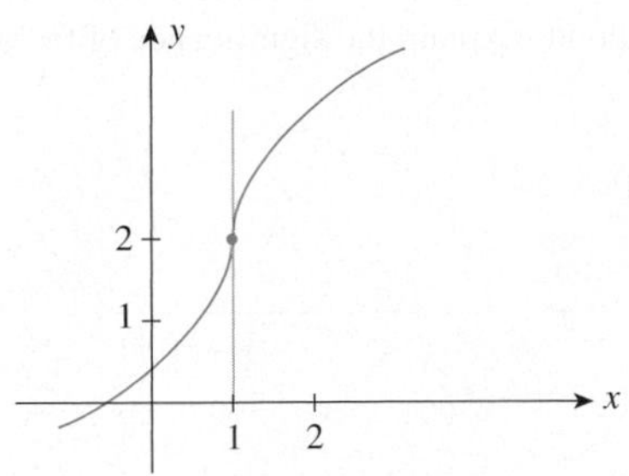

(b)

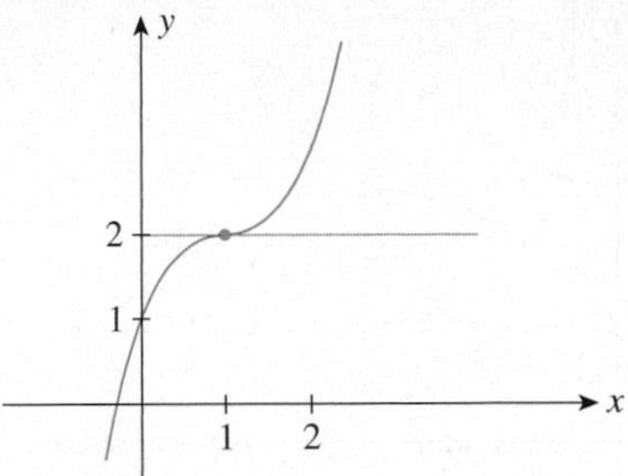

(c)

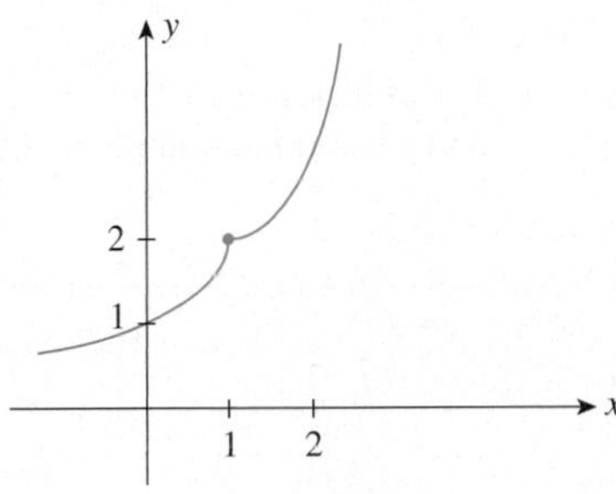

11. $f'(0)$ is undefined, f is decreasing on $(-\infty, 0)$, f is concave downward on $(0, 3)$, and f has an inflection point at $x = 3$.

(a)

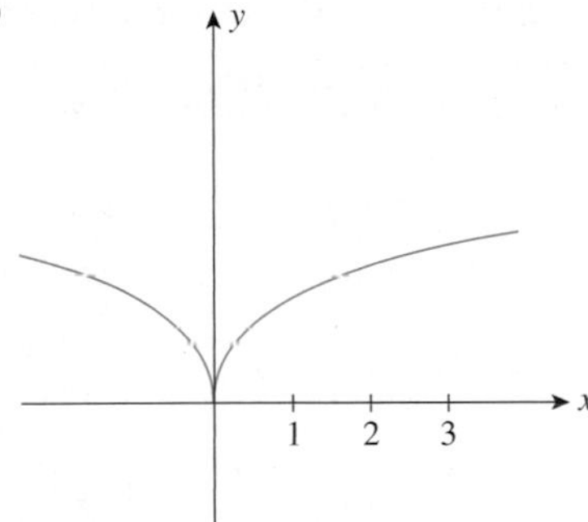

(b)

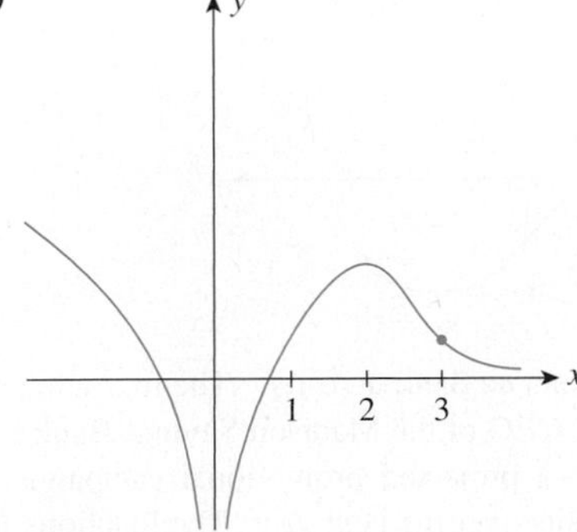

(c)

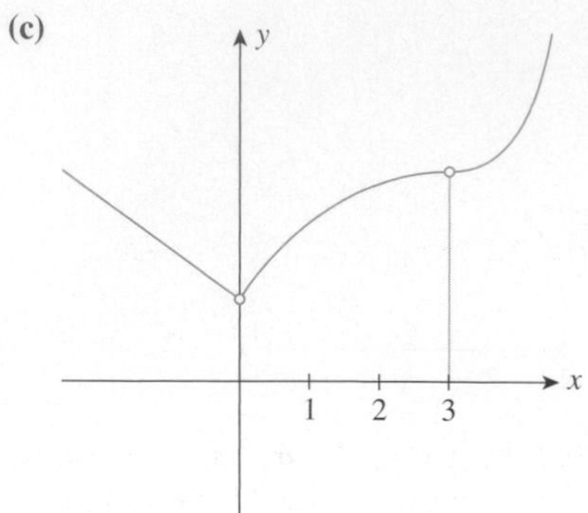

12. f is decreasing on $(-\infty, 2)$ and increasing on $(2, \infty)$, f is concave upward on $(1, \infty)$, and f has inflection points at $x = 0$ and $x = 1$.

(a)

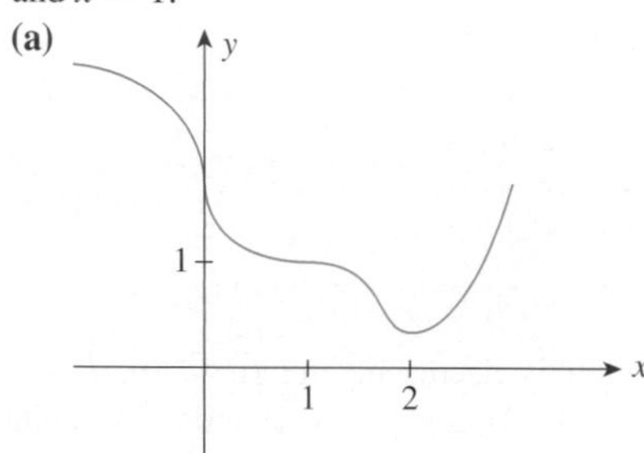

(b)

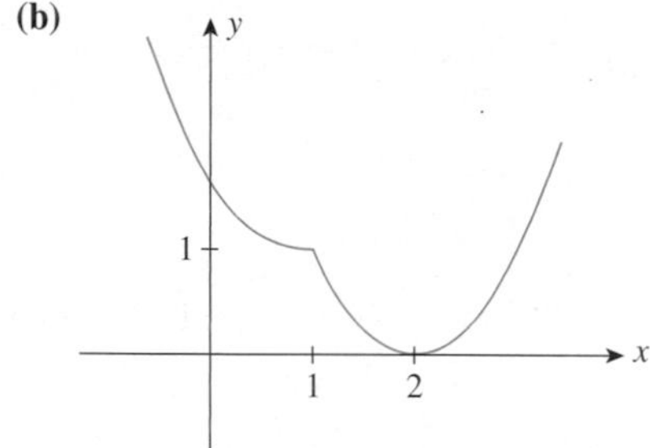

(c)

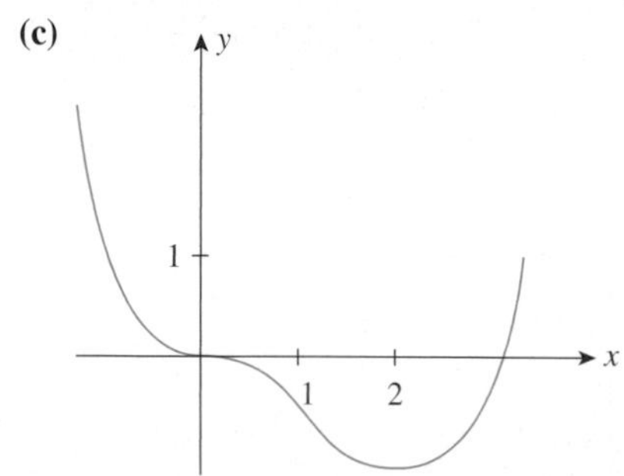

13. Effect of Advertising on Bank Deposits The following graphs were used by the CEO of the Madison Savings Bank to illustrate what effect a projected promotional campaign would have on its deposits over the next year. The functions D_1 and D_2 give the projected amount of money on deposit with the bank over the next 12 mo with and without the proposed promotional campaign, respectively.

a. Determine the signs of $D_1'(t)$, $D_2'(t)$, $D_1''(t)$, and $D_2''(t)$ on the interval (0, 12).

b. What can you conclude about the rate of change of the growth rate of the money on deposit with the bank with and without the proposed promotional campaign?

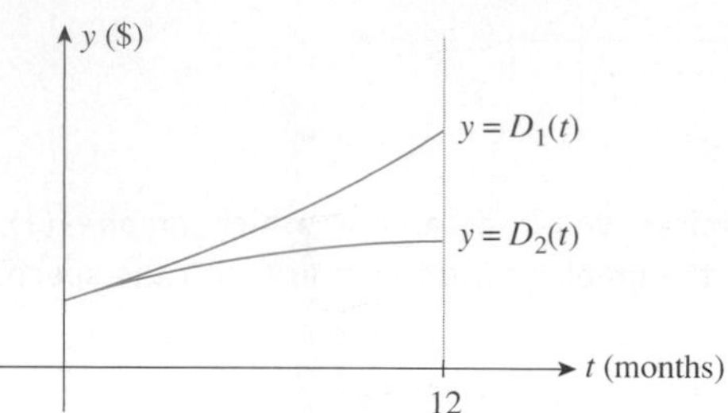

14. Assembly Time of a Worker In the following graph, $N(t)$ gives the number of personal radios assembled by the average worker by the tth hr, where $t = 0$ corresponds to 8 a.m. and $0 \leq t \leq 4$. The point P is an inflection point of N.

a. What can you say about the rate of change of the rate of the number of personal radios assembled by the average worker between 8 a.m. and 10 a.m.? Between 10 a.m. and 12 a.m.?

b. At what time is the rate at which the personal radios are being assembled by the average worker greatest?

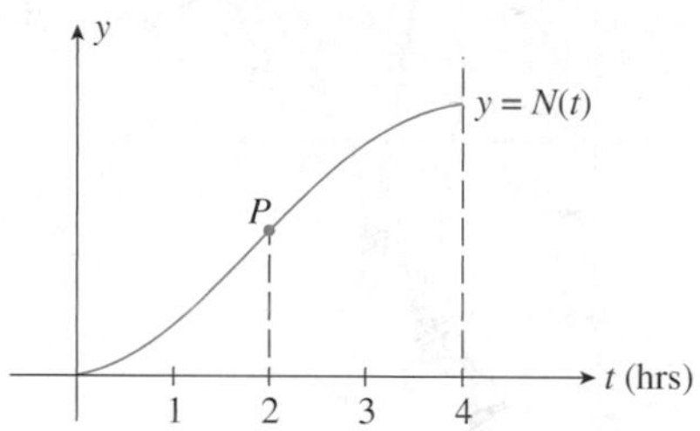

15. Water Pollution When organic waste is dumped into a pond, the oxidation process that takes place reduces the pond's oxygen content. However, given time, nature will restore the oxygen content to its natural level. In the following graph, $P(t)$ gives the oxygen content (as a percent of its normal level) t days after organic waste has been dumped into the pond. Explain the significance of the inflection point Q.

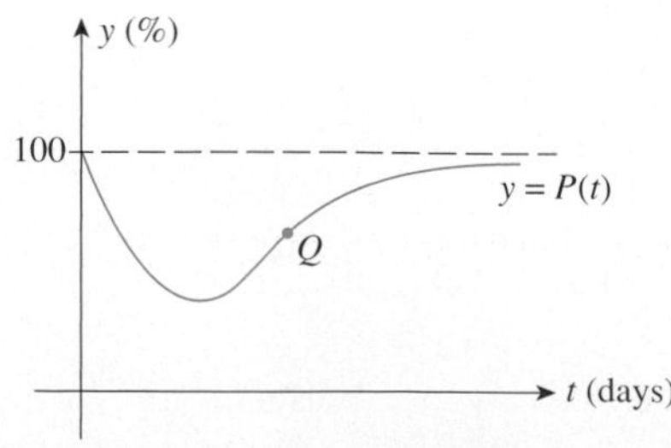

16. **Spread of a Rumor** Initially, a handful of students heard a rumor on campus. The rumor spread and, after t hr, the number had grown to $N(t)$. The graph of the function N is shown in the following figure:

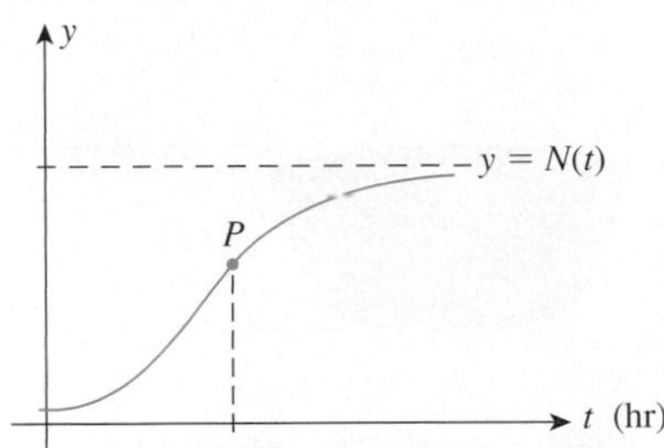

Describe the spread of the rumor in terms of the speed it was spread. In particular, explain the significance of the inflection point P of the graph of N.

In Exercises 17–20, show that the function is concave upward wherever it is defined.

17. $f(x) = 4x^2 - 12x + 7$

18. $g(x) = x^4 + \frac{1}{2}x^2 + 6x + 10$

19. $f(x) = \frac{1}{x^4}$

20. $g(x) = -\sqrt{4 - x^2}$

In Exercises 21–40, determine where the function is concave upward and where it is concave downward.

21. $f(x) = 2x^2 - 3x + 4$

22. $g(x) = -x^2 + 3x + 4$

23. $f(x) = x^3 - 1$

24. $g(x) = x^3 - x$

25. $f(x) = x^4 - 6x^3 + 2x + 8$

26. $f(x) = 3x^4 - 6x^3 + x - 8$

27. $f(x) = x^{4/7}$

28. $f(x) = \sqrt[3]{x}$

29. $f(x) = \sqrt{4 - x}$

30. $g(x) = \sqrt{x - 2}$

31. $f(x) = \frac{1}{x - 2}$

32. $g(x) = \frac{x}{x + 1}$

33. $f(x) = \frac{1}{2 + x^2}$

34. $g(x) = \frac{x}{1 + x^2}$

35. $h(t) = \frac{t^2}{t - 1}$

36. $f(x) = \frac{x + 1}{x - 1}$

37. $g(x) = x + \frac{1}{x^2}$

38. $h(r) = -\frac{1}{(r - 2)^2}$

39. $g(t) = (2t - 4)^{1/3}$

40. $f(x) = (x - 2)^{2/3}$

In Exercises 41–52, find the inflection point(s), if any, of each function.

41. $f(x) = x^3 - 2$

42. $g(x) = x^3 - 6x$

43. $f(x) = 6x^3 - 18x^2 + 12x - 15$

44. $g(x) = 2x^3 - 3x^2 + 18x - 8$

45. $f(x) = 3x^4 - 4x^3 + 1$

46. $f(x) = x^4 - 2x^3 + 6$

47. $g(t) = \sqrt[3]{t}$

48. $f(x) = \sqrt[5]{x}$

49. $f(x) = (x - 1)^3 + 2$

50. $f(x) = (x - 2)^{4/3}$

51. $f(x) = \frac{2}{1 + x^2}$

52. $f(x) = 2 + \frac{3}{x}$

In Exercises 53–68, find the relative extrema, if any, of each function. Use the second derivative test, if applicable.

53. $f(x) = -x^2 + 2x + 4$

54. $g(x) = 2x^2 + 3x + 7$

55. $f(x) = 2x^3 + 1$

56. $g(x) = x^3 - 6x$

57. $f(x) = \frac{1}{3}x^3 - 2x^2 - 5x - 10$

58. $f(x) = 2x^3 + 3x^2 - 12x - 4$

59. $g(t) = t + \frac{9}{t}$

60. $f(t) = 2t + \frac{3}{t}$

61. $f(x) = \frac{x}{1 - x}$

62. $f(x) = \frac{2x}{x^2 + 1}$

63. $f(t) = t^2 - \frac{16}{t}$

64. $g(x) = x^2 + \frac{2}{x}$

65. $g(s) = \frac{s}{1 + s^2}$

66. $g(x) = \frac{1}{1 + x^2}$

67. $f(x) = \frac{x^4}{x - 1}$

68. $f(x) = \frac{x^2}{x^2 + 1}$

In Exercises 69–74, sketch the graph of a function having the given properties.

69. $f(2) = 4, f'(2) = 0, f''(x) < 0$ on $(-\infty, \infty)$

70. $f(2) = 2, f'(2) = 0, f'(x) > 0$ on $(-\infty, 2), f'(x) > 0$ on $(2, \infty), f''(x) < 0$ on $(-\infty, 2), f''(x) > 0$ on $(2, \infty)$

71. $f(-2) = 4, f(3) = -2, f'(-2) = 0, f'(3) = 0, f'(x) > 0$ on $(-\infty, -2) \cup (3, \infty), f'(x) < 0$ on $(-2, 3)$, inflection point at $(1, 1)$

72. $f(0) = 0, f'(0)$ does not exist, $f''(x) < 0$ if $x \neq 0$

73. $f(0) = 1, f'(0) = 0, f(x) > 0$ on $(-\infty, \infty), f''(x) < 0$ on $(-\sqrt{2}/2, \sqrt{2}/2), f''(x) > 0$ on $(-\infty, -\sqrt{2}/2) \cup (\sqrt{2}/2, \infty)$

74. f has domain $[-1, 1], f(-1) = -1, f(-\frac{1}{2}) = -2, f'(-\frac{1}{2}) = 0, f''(x) > 0$ on $(-1, 1)$

75. **Demand for RNs** The following graph gives the total number of help-wanted ads for RNs (registered nurses) in 22 cities over the last 12 mo as a function of time t (t measured in months).
   a. Explain why $N'(t)$ is positive on the interval $(0, 12)$.
   b. Determine the signs of $N''(t)$ on the interval $(0, 6)$ and the interval $(6, 12)$.

c. Interpret the results of part (b).

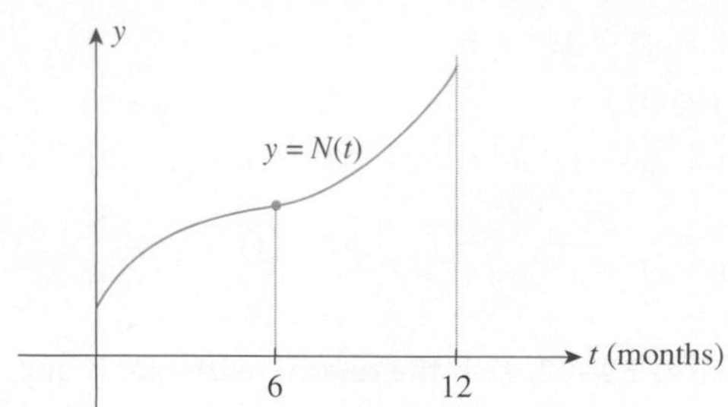

76. Effect of Budget Cuts on Drug-Related Crimes The graphs below were used by a police commissioner to illustrate what effect a budget cut would have on crime in the city. The number $N_1(t)$ gives the projected number of drug-related crimes in the next 12 mo. The number $N_2(t)$ gives the projected number of drug-related crimes in the same time frame if next year's budget is cut.

a. Explain why $N_1'(t)$ and $N_2'(t)$ are both positive on the interval (0, 12).

b. What are the signs of $N_1''(t)$ and $N_2''(t)$ on the interval (0, 12)?

c. Interpret the results of part (b).

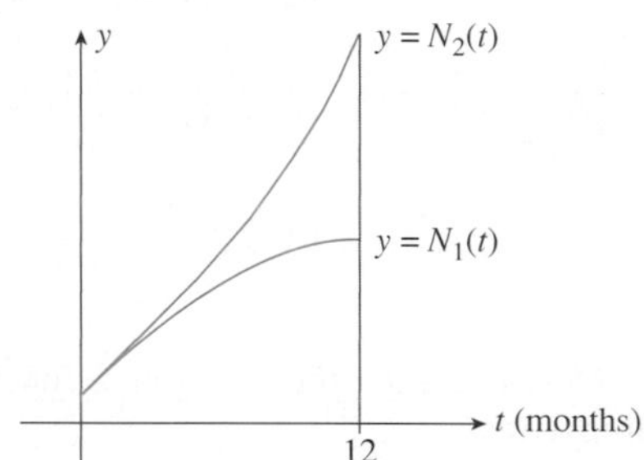

77. In the following figure, water is poured into the vase at a constant rate (in appropriate units), and the water level rises to a height of $f(t)$ units at time t as measured from the base of the vase. The graph of f follows. Explain the shape of the curve in terms of its concavity. What is the significance of the inflection point?

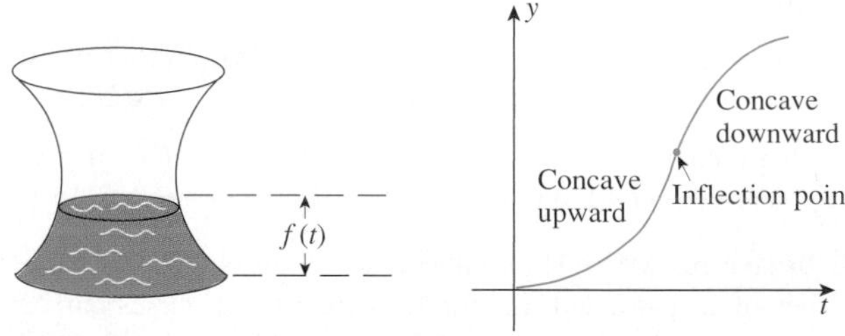

78. In the following figure, water is poured into an urn at a constant rate (in appropriate units), and the water level rises to a height of $f(t)$ units at time t as measured from the base of the urn. Sketch the graph of f and explain its shape, indicating where it is concave upward and concave downward. Indicate the inflection point on the graph and explain its significance.
Hint: Study Exercise 77.

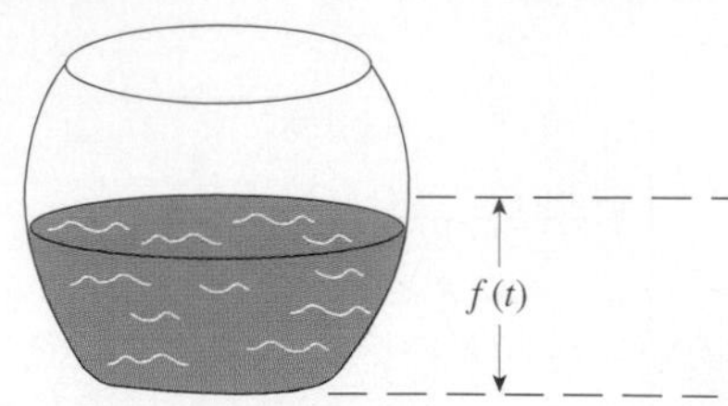

79. Effect of Smoking Bans The sales (in billions of dollars) in restaurants and bars in California from 1993 ($t = 0$) through 2000 ($t = 7$) are approximated by the function

$$S(t) = 0.195t^2 + 0.32t + 23.7 \qquad (0 \leq t \leq 7)$$

a. Show that the sales in restaurants and bars continued to rise after smoking bans were implemented in restaurants in 1995 and in bars in 1998.
Hint: Show that S is increasing in the interval (2, 7).

b. What can you say about the rate at which the sales were rising after smoking bans were implemented?

Source: California Board of Equalization

80. Digital Television Sales Since their introduction into the market in the late 1990s, the sales of digital televisions, including high-definition television sets, have slowly gathered momentum. The model

$$S(t) = 0.164t^2 + 0.85t + 0.3 \qquad (0 \leq t \leq 4)$$

describes the sales of digital television sets (in billions of dollars) between 1999 ($t = 0$) and 2003 ($t = 4$).

a. Find $S'(t)$ and $S''(t)$.

b. Use the results of part (a) to conclude that the sales of digital TVs were increasing between 1999 and 2003 and that the sales were increasing at an increasing rate over that time interval.

Source: Consumer Electronics Association

81. Worker Efficiency An efficiency study conducted for Elektra Electronics showed that the number of Space Commander walkie-talkies assembled by the average worker t hr after starting work at 8 a.m. is given by

$$N(t) = -t^3 + 6t^2 + 15t \qquad (0 \leq t \leq 4)$$

At what time during the morning shift is the average worker performing at peak efficiency?

82. Flight of a Rocket The altitude (in feet) of a rocket t sec into flight is given by

$$s = f(t) = -t^3 + 54t^2 + 480t + 6 \qquad (t \geq 0)$$

Find the point of inflection of the function f and interpret your result. What is the maximum velocity attained by the rocket?

83. **Business Spending on Technology** In a study conducted in 2003, business spending on technology (in billions of dollars) from 2000 through 2005 was projected to be

$$S(t) = -1.88t^3 + 30.33t^2 - 76.14t + 474 \qquad (0 \leq t \leq 5)$$

where t is measured in years, with $t = 0$ corresponding to 2000. Show that the graph of S is concave upward on the interval (0, 5). What does this result tell you about the rate of business spending on technology over the period in question?

Source: Quantit Economic Group

84. **Alternative Minimum Tax** Congress created the alternative minimum tax (AMT) in the late 1970s to ensure that wealthy people paid their fair share of taxes. But because of quirks in the law, even middle-income taxpayers have started to get hit with the tax. The AMT (in billions of dollars) projected to be collected by the IRS from 2001 through 2010 is

$$f(t) = 0.0117t^3 + 0.0037t^2 + 0.7563t + 4.1 \qquad (0 \leq t \leq 9)$$

where t is measured in years, with $t = 0$ corresponding to 2001.

a. Show that f is increasing on the interval (0, 9). What does this result tell you about the projected amount of AMT paid over the years in question?

b. Show that f' is increasing on the interval (0, 9). What conclusion can you draw from this result concerning the rate of growth at which the AMT is paid over the years in question?

Source: U.S. Congress Joint Economic Committee

85. **Effect of Advertising on Hotel Revenue** The total annual revenue R of the Miramar Resorts Hotel is related to the amount of money x the hotel spends on advertising its services by the function

$$R(x) = -0.003x^3 + 1.35x^2 + 2x + 8000 \qquad (0 \leq x \leq 400)$$

where both R and x are measured in thousands of dollars.

a. Find the interval where the graph of R is concave upward and the interval where the graph of R is concave downward. What is the inflection point of R?

b. Would it be more beneficial for the hotel to increase its advertising budget slightly when the budget is \$140,000 or when it is \$160,000?

86. **Forecasting Profits** As a result of increasing energy costs, the growth rate of the profit of the 4-yr old Venice Glassblowing Company has begun to decline. Venice's management, after consulting with energy experts, decides to implement certain energy-conservation measures aimed at cutting energy bills. The general manager reports that, according to his calculations, the growth rate of Venice's profit should be on the increase again within 4 yr. If Venice's profit (in hundreds of dollars) t yr from now is given by the function

$$P(t) = t^3 - 9t^2 + 40t + 50 \qquad (0 \leq t \leq 8)$$

determine whether the general manager's forecast will be accurate.

Hint: Find the inflection point of the function P and study the concavity of P.

87. **Outsourcing** The amount (in billions of dollars) spent by the top 15 U.S. financial institutions on IT (information technology) offshore outsourcing is projected to be

$$A(t) = 0.92(t + 1)^{0.61} \qquad (0 \leq t \leq 4)$$

where t is measured in years, with $t = 0$ corresponding to 2004.

a. Show that A is increasing on (0, 4) and interpret your result.

b. Show that A is concave downward on (0, 4). Interpret your result.

Source: Tower Group

88. **Sales of Mobile Processors** The rising popularity of notebook computers is fueling the sales of mobile PC processors. In a study conducted in 2003, the sales of these chips (in billions of dollars) was projected to be

$$S(t) = 6.8(t + 1.03)^{0.49} \qquad (0 \leq t \leq 4)$$

where t is measured in years, with $t = 0$ corresponding to 2003.

a. Show that S is increasing on the interval (0, 4) and interpret your result.

b. Show that the graph of S is concave downward on the interval (0, 4). Interpret your result.

Source: International Data Corp.

89. **Drug Spending** Medicaid spending on drugs in Massachusetts started slowing down in part after the state demanded that patients use more generic drugs and limited the range of drugs available to the program. The annual pharmacy spending (in millions of dollars) from 1999 through 2004 is given by

$$S(t) = -1.806t^3 + 10.238t^2 + 93.35t + 583 \qquad (0 \leq t \leq 5)$$

where t is measured in years with $t = 0$ corresponding to 1999. Find the inflection point of S and interpret your result.

Source: MassHealth

90. **PC Shipments** In a study conducted in 2003, it was projected that worldwide PC shipments (in millions) through 2005 will be given by

$$N(t) = 0.75t^3 - 1.5t^2 + 8.25t + 133 \qquad (0 \leq t \leq 4)$$

where t is measured in years, with $t = 0$ corresponding to 2001.

a. Determine the intervals where N is concave upward and where it is concave downward.

b. Find the inflection point of N and interpret your result.

Source: International Data Corporation

91. **Google's Revenue** The revenue for Google from 1999 ($t = 0$) through 2003 ($t = 4$) is approximated by the function

$$R(t) = 24.975t^3 - 49.81t^2 + 41.25t + 0.2 \qquad (0 \le t \le 4)$$

where $R(t)$ is measured in millions of dollars.

a. Find $R'(t)$ and $R''(t)$.

b. Show that $R'(t) > 0$ for all t in the interval $(0, 4)$ and interpret your result.
Hint: Use the quadratic formula.

c. Find the inflection point of R and interpret your result.

Source: Company Report

92. **Population Growth in Clark County** Clark County in Nevada—dominated by greater Las Vegas—is the fastest-growing metropolitan area in the United States. The population of the county from 1970 through 2000 is approximated by the function

$$P(t) = 44560t^3 - 89394t^2 + 234633t + 273288 \qquad (0 \le t \le 4)$$

where t is measured in decades, with $t = 0$ corresponding to the beginning of 1970.

a. Show that the population of Clark County was always increasing over the time period in question.
Hint: Show that $P'(t) > 0$ for all t in the interval $(0, 4)$.

b. Show that the population of Clark County was increasing at the slowest pace some time toward the middle of August 1976.
Hint: Find the inflection point of P in the interval $(0, 4)$.

Source: U.S. Census Bureau

93. **Air Pollution** The level of ozone, an invisible gas that irritates and impairs breathing, present in the atmosphere on a certain May day in the city of Riverside was approximated by

$$A(t) = 1.0974t^3 - 0.0915t^4 \qquad (0 \le t \le 11)$$

where $A(t)$ is measured in pollutant standard index (PSI) and t is measured in hours, with $t = 0$ corresponding to 7 a.m. Use the second derivative test to show that the function A has a relative maximum at approximately $t = 9$. Interpret your results.

94. **Cash Reserves at Blue Cross and Blue Shield** Based on company financial reports, the cash reserves of Blue Cross and Blue Shield as of the beginning of year t is approximated by the function

$$R(t) = -1.5t^4 + 14t^3 - 25.4t^2 + 64t + 290 \qquad (0 \le t \le 6)$$

where $R(t)$ is measured in millions of dollars and t is measured in years, with $t = 0$ corresponding to the beginning of 1998.

a. Find the inflection points of R.
Hint: Use the quadratic formula.

b. Use the result of part (a) to show that the cash reserves of the company was growing at the greatest rate at the beginning of 2002.

Source: Blue Cross and Blue Shield

95. **Women's Soccer** Starting with the youth movement that took hold in the 1970s and buoyed by the success of the U.S. national women's team in international competition in recent years, girls and women have taken to soccer in ever-growing numbers. The function

$$N(t) = -0.9307t^3 + 74.04t^2 + 46.8667t + 3967 \qquad (0 \le t \le 16)$$

gives the number of participants in women's soccer in year t, with $t = 0$ corresponding to the beginning of 1985.

a. Verify that the number of participants in women's soccer had been increasing from 1985 through 2000.
Hint: Use the quadratic formula.

b. Show that the number of participants in women's soccer had been increasing at an increasing rate from 1985 through 2000.
Hint: Show that the sign of N'' is positive on the interval in question.

Source: NCCA News

96. **Dependency Ratio** The share of the world population that is over 60 years of age compared to the rest of the working population in the world is of concern to economists. An increasing dependency ratio means that there will be fewer workers to support an aging population. The dependency ratio over the next century is forecast to be

$$R(t) = 0.00731t^4 - 0.174t^3 + 1.528t^2 + 0.48t + 19.3 \qquad (0 \le t \le 10)$$

in year t, where t is measured in decades with $t = 0$ corresponding to 2000.

a. Show that the dependency ratio will be increasing at the fastest pace around 2052.
Hint: Use the quadratic formula.

b. What will the dependency ratio be at that time?

Source: International Institute for Applied Systems Analysis

In Exercises 97–100, determine whether the statement is true or false. If it is true, explain why it is true. If it is false, give an example to show why it is false.

97. If the graph of f is concave upward on (a, b), then the graph of $-f$ is concave downward on (a, b).

98. If the graph of f is concave upward on (a, c) and concave downward on (c, b), where $a < c < b$, then f has an inflection point at $(c, f(c))$.

99. If c is a critical number of f where $a < c < b$ and $f''(x) < 0$ on (a, b), then f has a relative maximum at $x = c$.

100. A polynomial function of degree n ($n \ge 3$) can have at most $(n - 2)$ inflection points.

101. Show that the quadratic function

$$f(x) = ax^2 + bx + c \qquad (a \ne 0)$$

is concave upward if $a > 0$ and concave downward if $a < 0$. Thus, by examining the sign of the coefficient of x^2, one can tell immediately whether the parabola opens upward or downward.

102. Consider the functions $f(x) = x^3$, $g(x) = x^4$, and $h(x) = -x^4$.

a. Show that $x = 0$ is a critical number of each of the functions f, g, and h.
b. Show that the second derivative of each of the functions f, g, and h equals zero at $x = 0$.
c. Show that f has neither a relative maximum nor a relative minimum at $x = 0$, that g has a relative minimum at $x = 0$, and that h has a relative maximum at $x = 0$.

4.2 Solutions to Self-Check Exercises

1. We first compute

$$f'(x) = 12x^2 - 6x$$
$$f''(x) = 24x - 6 = 6(4x - 1)$$

Observe that f'' is continuous everywhere and has a zero at $x = \frac{1}{4}$. The sign diagram of f'' is shown in the accompanying figure.

$$\begin{array}{c} - - - - - - 0 + + + + + + \\ \longrightarrow x \\ 0 \quad \tfrac{1}{4} \end{array}$$

From the sign diagram for f'', we see that f is concave upward on $(\frac{1}{4}, \infty)$ and concave downward on $(-\infty, \frac{1}{4})$.

2. First, we find the critical numbers of f by solving the equation

$$f'(x) = 6x^2 - x - 12 = 0$$

That is,

$$(3x + 4)(2x - 3) = 0$$

giving $x = -\frac{4}{3}$ and $x = \frac{3}{2}$. Next, we compute

$$f''(x) = 12x - 1$$

Since

$$f''\left(-\frac{4}{3}\right) = 12\left(-\frac{4}{3}\right) - 1 = -17 < 0$$

the second derivative test implies that $f(-\frac{4}{3}) = \frac{10}{27}$ is a relative maximum of f. Also,

$$f''\left(\frac{3}{2}\right) = 12\left(\frac{3}{2}\right) - 1 = 17 > 0$$

and we see that $f(\frac{3}{2}) = -\frac{179}{8}$ is a relative minimum.

3. We compute the second derivative of G. Thus,

$$G'(t) = -6t^2 + 90t + 20$$
$$G''(t) = -12t + 90$$

Now, G'' is continuous everywhere, and $G''(t) = 0$, where $t = \frac{15}{2}$, giving $t = \frac{15}{2}$ as the only candidate for an inflection point of G. Since $G''(t) > 0$ for $t < \frac{15}{2}$ and $G''(t) < 0$ for $t > \frac{15}{2}$, we see that $(\frac{15}{2}, \frac{15{,}675}{2})$ is an inflection point of G. The results of our computations tell us that the country's GDP was increasing most rapidly at the beginning of July 2002.

USING TECHNOLOGY

Finding the Inflection Points of a Function

A graphing utility can be used to find the inflection points of a function and hence the intervals where the graph of the function is concave upward and the intervals where it is concave downward. Some graphing utilities have an operation for finding inflection points directly. For example, both the TI-85 and TI-86 graphing calculators have this capability. If your graphing utility has this capability, use it to work through the example and exercises in this section.

EXAMPLE 1 Let $f(x) = 2.5x^5 - 12.4x^3 + 4.2x^2 - 5.2x + 4$.

a. Use a graphing utility to plot the graph of f.
b. Find the inflection points of f.
c. Find the intervals where f is concave upward and where it is concave downward.

(continued)

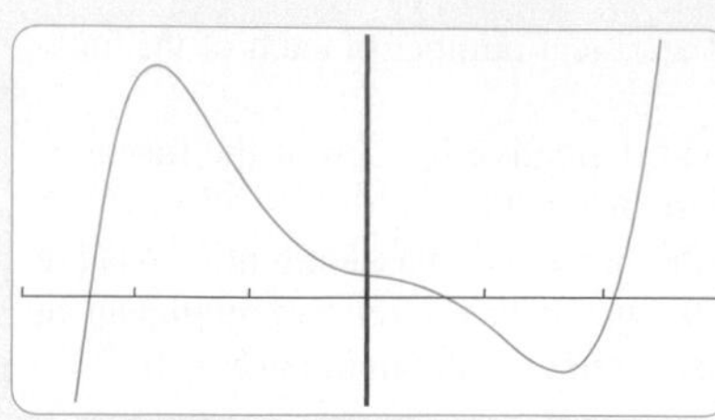

FIGURE T1
The graph of f in the viewing window $[-3, 3] \times [-25, 60]$

Solution

a. The graph of f, using the viewing window $[-3, 3] \times [-25, 60]$, is shown in Figure T1.

b. We describe here the procedure using the TI-85. See the Web site for instructions for using the TI-86. From Figure T1 we see that f has three inflection points—one occurring at the point where the x-coordinate is approximately -1, another at the point where $x \approx 0$, and the third at the point where $x \approx 1$. To find the first inflection point, we use the inflection operation, moving the cursor to the point on the graph of f where $x \approx -1$. We obtain the point $(-1.2728, 34.6395)$ (accurate to four decimal places). Next, setting the cursor near $x = 0$ yields the inflection point $(0.1139, 3.4440)$. Finally, with the cursor set at $x = 1$, we obtain the third inflection point $(1.1589, -10.4594)$. (See Figure T2a–c.)

c. From the results of part (b), we see that f is concave upward on the intervals $(-1.2728, 0.1139)$ and $(1.1589, \infty)$ and concave downward on $(-\infty, -1.2728)$ and $(0.1139, 1.1589)$.

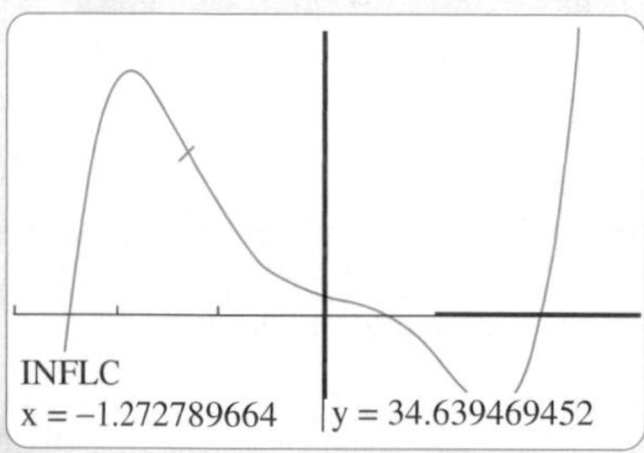

(a)

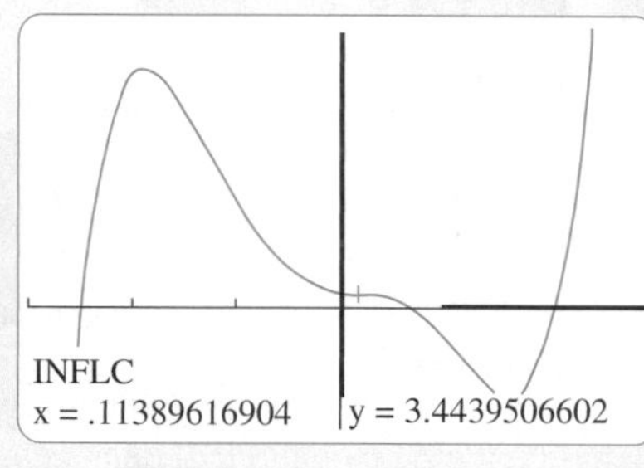

(b)

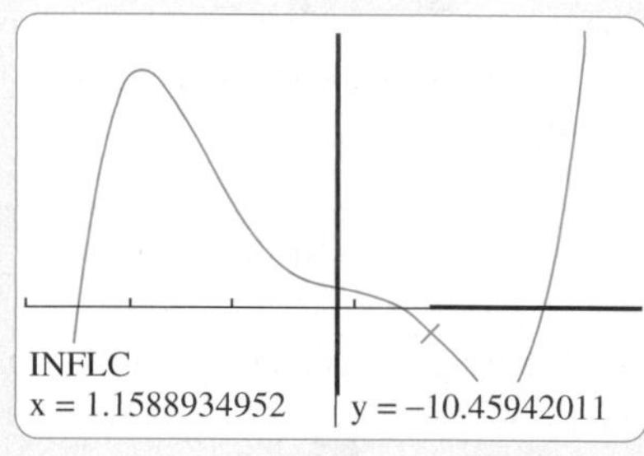

(c)

FIGURE T2
The TI-85 inflection point screens showing the points (a) $(-1.2728, 34.6395)$, (b) $(0.1139, 3.4440)$, and (c) $(1.1589, -10.4594)$

TECHNOLOGY EXERCISES

In Exercises 1–8, find (a) the intervals where f is concave upward and the intervals where f is concave downward and (b) the inflection points of f. Express your answers accurate to four decimal places.

1. $f(x) = 1.8x^4 - 4.2x^3 + 2.1x + 2$
2. $f(x) = -2.1x^4 + 3.1x^3 + 2x^2 - x + 1.2$
3. $f(x) = 1.2x^5 - 2x^4 + 3.2x^3 - 4x + 2$
4. $f(x) = -2.1x^5 + 3.2x^3 - 2.2x^2 + 4.2x - 4$
5. $f(x) = x^3(x^2 + 1)^{-1/3}$
6. $f(x) = x^2(x^3 - 1)^3$
7. $f(x) = \dfrac{x^2 - 1}{x^3}$
8. $f(x) = \dfrac{x + 1}{\sqrt{x}}$
9. **Growth of HMOs** Data show that the number of people receiving their care in an HMO (health maintenance organization) from the beginning of 1984 through 1994 is approximated by the function

$$f(t) = 0.0514t^3 - 0.853t^2 + 6.8147t + 15.6524 \quad (0 < t \leq 11)$$

where $f(t)$ gives the number of people in millions and t is measured in years, with $t = 0$ corresponding to the beginning of 1984.

a. Plot the graph of f in the viewing window $[0, 12] \times [0, 120]$.
b. Find the points of inflection of f.
c. At what time in the given time interval was the number of people receiving their care at an HMO increasing at the slowest rate?

Source: Group Health Association of America

10. **Manufacturing Capacity** Data show that the annual increase in manufacturing capacity between 1988 and 1994 is given by

$$f(t) = 0.0388889t^3 - 0.283333t^2 + 0.477778t + 2.04286 \quad (0 \leq t \leq 6)$$

percent where t is measured in years, with $t = 0$ corresponding to the beginning of 1988.

a. Plot the graph of f in the viewing window $[0, 8] \times [0, 4]$.
b. Find the point of inflection and interpret your result.

Source: Federal Reserve

11. **Time on the Market** The average number of days a single-family home remains for sale from listing to accepted offer (in the greater Boston area) is approximated by the function

$$f(t) = 0.0171911t^4 - 0.662121t^3 + 6.18083t^2 - 8.97086t + 53.3357 \qquad (0 \le t \le 10)$$

where t is measured in years, with $t = 0$ corresponding to the beginning of 1984.
a. Plot the graph of f in the viewing window $[0, 12] \times [0, 120]$.
b. Find the points of inflection and interpret your result.

Source: Greater Boston Real Estate Board—Multiple Listing Service

12. **Multimedia Sales** Sales in the multimedia market (hardware and software) are approximated by the function

$$S(t) = -0.0094t^4 + 0.1204t^3 - 0.0868t^2 + 0.0195t + 3.3325 \qquad (0 \le t \le 10)$$

where $S(t)$ is measured in billions of dollars and t is measured in years, with $t = 0$ corresponding to 1990.
a. Plot the graph of S in the viewing window $[0, 12] \times [0, 25]$.
b. Find the inflection point of S and interpret your result.

Source: Electronic Industries Association

13. **Surgeries in Physicians' Offices** Driven by technological advances and financial pressures, the number of surgeries performed in physicians' offices nationwide has been increasing over the years. The function

$$f(t) = -0.00447t^3 + 0.09864t^2 + 0.05192t + 0.8 \qquad (0 \le t \le 15)$$

gives the number of surgeries (in millions) performed in physicians' offices in year t, with $t = 0$ corresponding to the beginning of 1986.
a. Plot the graph of f in the viewing window $[0, 15] \times [0, 10]$.
b. At what time in the period under consideration is the number of surgeries performed in physicians' offices increasing at the fastest rate?

Source: SMG Marketing Group

14. **Computer Security** The number of computer-security incidents, including computer viruses and intrusions, in which the same tool or exploit is used by an intruder, from 1999 through 2003 is approximated by the function

$$N(t) = 2.0417t^4 - 16.083t^3 + 44.46t^2 - 19.42t + 10 \qquad (0 \le t \le 4)$$

where $N(t)$ is measured in thousands and t is measured in years, with $t = 0$ corresponding to 1999.
a. Show that the number of computer-security incidents was always increasing between 2000 and 2003.
Hint: Plot the graph of N' and show that it always lies above the t-axis for $1 \le t \le 4$.
b. Show that between 2000 and 2001, the number of computer-security incidents was increasing at the fastest rate in the middle of 2000 and that between 2001 and 2003 the number of incidents was increasing at the slowest rate in the middle of 2001.
Hint: Study the nature of the inflection points of N.

Source: CERT Coordination Center

4.3 Curve Sketching

A Real-Life Example

As we have seen on numerous occasions, the graph of a function is a useful aid for visualizing the function's properties. From a practical point of view, the graph of a function also gives, at one glance, a complete summary of all the information captured by the function.

Consider, for example, the graph of the function giving the Dow-Jones Industrial Average (DJIA) on Black Monday, October 19, 1987 (Figure 45). Here, $t = 0$ corresponds to 8:30 a.m., when the market was open for business, and $t = 7.5$ corresponds to 4 p.m., the closing time. The following information may be gleaned from studying the graph.

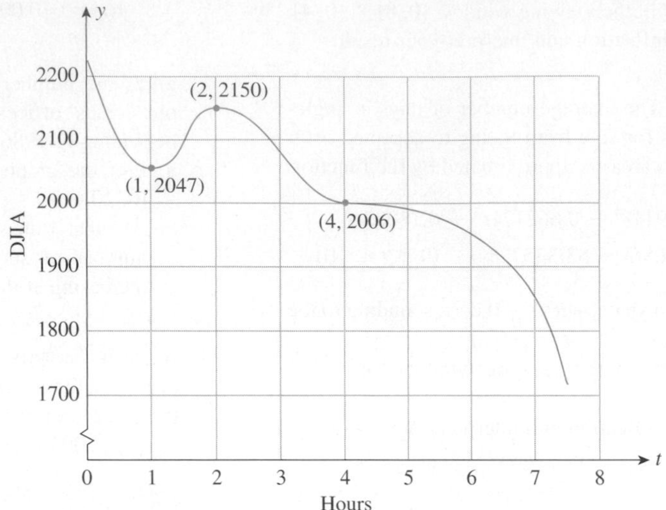

FIGURE 45
The Dow-Jones Industrial Average on Black Monday.

Source: Wall Street Journal

The graph is *decreasing* rapidly from $t = 0$ to $t = 1$, reflecting the sharp drop in the index in the first hour of trading. The point (1, 2047) is a *relative minimum* point of the function, and this turning point coincides with the start of an aborted recovery. The short-lived rally, represented by the portion of the graph that is *increasing* on the interval (1, 2), quickly fizzled out at $t = 2$ (10:30 a.m.). The *relative maximum* point (2, 2150) marks the highest point of the recovery. The function is decreasing in the rest of the interval. The point (4, 2006) is an *inflection point* of the function; it shows that there was a temporary respite at $t = 4$ (12:30 p.m.). However, selling pressure continued unabated, and the DJIA continued to fall until the closing bell. Finally, the graph also shows that the index opened at the high of the day [$f(0) = 2247$ is the *absolute maximum* of the function] and closed at the low of the day [$f(\frac{15}{2}) = 1739$ is the *absolute minimum* of the function], a drop of 508 points!*

Before we turn our attention to the actual task of sketching the graph of a function, let's look at some properties of graphs that will be helpful in this connection.

Vertical Asymptotes

Before going on, you might want to review the material on one-sided limits and the limit at infinity of a function (Sections 2.4 and 2.5).

Consider the graph of the function

$$f(x) = \frac{x + 1}{x - 1}$$

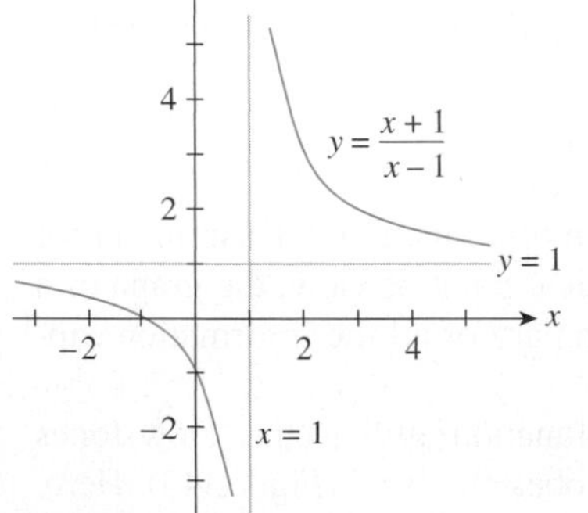

FIGURE 46
The graph of f has a vertical asymptote at $x = 1$.

shown in Figure 46. Observe that $f(x)$ increases without bound (tends to infinity) as x approaches $x = 1$ from the right; that is,

$$\lim_{x \to 1^+} \frac{x + 1}{x - 1} = \infty$$

You can verify this by taking a sequence of values of x approaching $x = 1$ from the right and looking at the corresponding values of $f(x)$.

*Absolute maxima and absolute minima of functions are covered in Section 4.4.

Here is another way of looking at the situation: Observe that if x is a number that is a little larger than 1, then both $(x + 1)$ and $(x - 1)$ are positive, so $(x + 1)/(x - 1)$ is also positive. As x approaches $x = 1$, the numerator $(x + 1)$ approaches the number 2, but the denominator $(x - 1)$ approaches zero, so the quotient $(x + 1)/(x - 1)$ approaches infinity, as observed earlier. The line $x = 1$ is called a vertical asymptote of the graph of f.

For the function $f(x) = (x + 1)/(x - 1)$, you can show that

$$\lim_{x\to 1^-} \frac{x + 1}{x - 1} = -\infty$$

and this tells us how $f(x)$ approaches the asymptote $x = 1$ from the left.

More generally, we have the following definition:

Vertical Asymptote

The line $x = a$ is a **vertical asymptote** of the graph of a function f if either

$$\lim_{x\to a^+} f(x) = \infty \quad \text{or} \quad -\infty$$

or

$$\lim_{x\to a^-} f(x) = \infty \quad \text{or} \quad -\infty$$

Note Although a vertical asymptote of a graph is not part of the graph, it serves as a useful aid for sketching the graph. ■

For rational functions

$$f(x) = \frac{P(x)}{Q(x)}$$

there is a simple criterion for determining whether the graph of f has any vertical asymptotes.

Finding Vertical Asymptotes of Rational Functions

Suppose f is a rational function

$$f(x) = \frac{P(x)}{Q(x)}$$

where P and Q are polynomial functions. Then, the line $x = a$ is a vertical asymptote of the graph of f if $Q(a) = 0$ but $P(a) \neq 0$.

For the function

$$f(x) = \frac{x + 1}{x - 1}$$

considered earlier, $P(x) = x + 1$ and $Q(x) = x - 1$. Observe that $Q(1) = 0$ but $P(1) = 2 \neq 0$, so $x = 1$ is a vertical asymptote of the graph of f.

EXAMPLE 1 Find the vertical asymptotes of the graph of the function

$$f(x) = \frac{x^2}{4 - x^2}$$

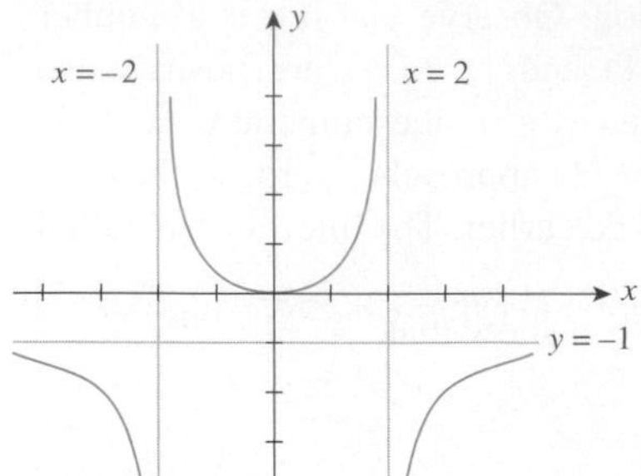

FIGURE 47
$x = -2$ and $x = 2$ are vertical asymptotes of the graph of f.

Solution The function f is a rational function with $P(x) = x^2$ and $Q(x) = 4 - x^2$. The zeros of Q are found by solving

$$4 - x^2 = 0$$

—that is,

$$(2 - x)(2 + x) = 0$$

giving $x = -2$ and $x = 2$. These are candidates for the vertical asymptotes of the graph of f. Examining $x = -2$, we compute $P(-2) = (-2)^2 = 4 \neq 0$, and we see that $x = -2$ is indeed a vertical asymptote of the graph of f. Similarly, we find $P(2) = 2^2 = 4 \neq 0$, and so $x = 2$ is also a vertical asymptote of the graph of f. The graph of f sketched in Figure 47 confirms these results. ■

Recall that in order for the line $x = a$ to be a vertical asymptote of the graph of a rational function f, *only* the denominator of $f(x)$ must be equal to zero at $x = a$. If *both* $P(a)$ and $Q(a)$ are equal to zero, then $x = a$ need *not* be a vertical asymptote. For example, look at the function

$$f(x) = \frac{4(x^2 - 4)}{x - 2}$$

whose graph appears in Figure 30a, page 104.

Horizontal Asymptotes

Let's return to the function f defined by

$$f(x) = \frac{x + 1}{x - 1}$$

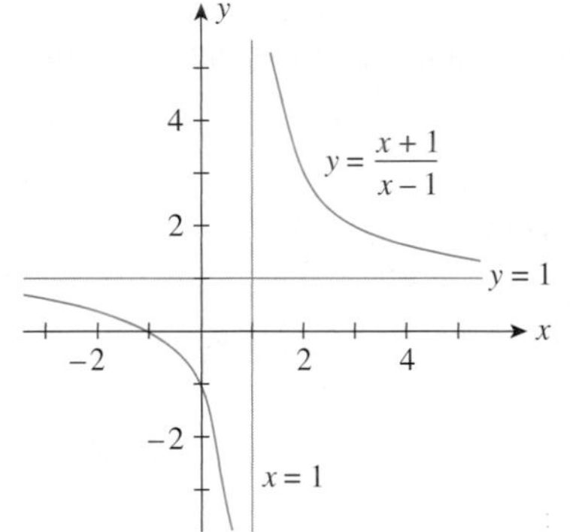

FIGURE 48
The graph of f has a horizontal asymptote at $y = 1$.

(Figure 48).

Observe that $f(x)$ approaches the horizontal line $y = 1$ as x approaches infinity, and, in this case, $f(x)$ approaches $y = 1$ as x approaches minus infinity as well. The line $y = 1$ is called a horizontal asymptote of the graph of f. More generally, we have the following definition:

Horizontal Asymptote

The line $y = b$ is a **horizontal asymptote** of the graph of a function f if either

$$\lim_{x \to \infty} f(x) = b \quad \text{or} \quad \lim_{x \to -\infty} f(x) = b$$

For the function

$$f(x) = \frac{x + 1}{x - 1}$$

we see that

$$\lim_{x \to \infty} \frac{x + 1}{x - 1} = \lim_{x \to \infty} \frac{1 + \frac{1}{x}}{1 - \frac{1}{x}} \qquad \text{Divide numerator and denominator by } x.$$

$$= 1$$

Also,

$$\lim_{x \to -\infty} \frac{x+1}{x-1} = \lim_{x \to -\infty} \frac{1+\frac{1}{x}}{1-\frac{1}{x}}$$
$$= 1$$

In either case, we conclude that $y = 1$ is a horizontal asymptote of the graph of f, as observed earlier.

EXAMPLE 2 Find the horizontal asymptotes of the graph of the function

$$f(x) = \frac{x^2}{4 - x^2}$$

Solution We compute

$$\lim_{x \to \infty} \frac{x^2}{4 - x^2} = \lim_{x \to \infty} \frac{1}{\frac{4}{x^2} - 1} \qquad \text{Divide numerator and denominator by } x^2.$$
$$= -1$$

and so $y = -1$ is a horizontal asymptote, as before. (Similarly, $\lim_{x \to -\infty} f(x) = -1$, as well.) The graph of f sketched in Figure 49 confirms this result.

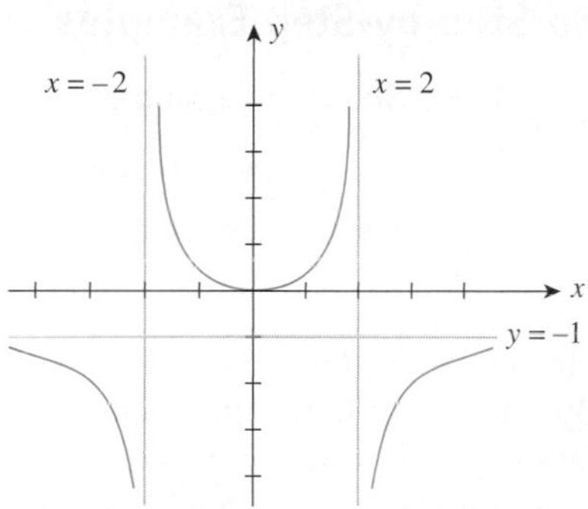

FIGURE 49
The graph of f has a horizontal asymptote at $y = -1$.

We next state an important property of polynomial functions.

> A polynomial function has no vertical or horizontal asymptotes.

To see this, note that a polynomial function $P(x)$ can be written as a rational function with denominator equal to 1. Thus,

$$P(x) = \frac{P(x)}{1}$$

Since the denominator is never equal to zero, P has no vertical asymptotes. Next, if P is a polynomial of degree greater than or equal to 1, then

$$\lim_{x \to \infty} P(x) \quad \text{and} \quad \lim_{x \to -\infty} P(x)$$

are either infinity or minus infinity; that is, they do not exist. Therefore, P has no horizontal asymptotes.

In the last two sections, we saw how the first and second derivatives of a function are used to reveal various properties of the graph of a function f. We now show how this information can be used to help us sketch the graph of f. We begin by giving a general procedure for curve sketching.

A Guide to Curve Sketching

1. Determine the domain of f.
2. Find the x- and y-intercepts of f.*
3. Determine the behavior of f for large absolute values of x.
4. Find all horizontal and vertical asymptotes of f.
5. Determine the intervals where f is increasing and where f is decreasing.
6. Find the relative extrema of f.
7. Determine the concavity of f.
8. Find the inflection points of f.
9. Plot a few additional points to help further identify the shape of the graph of f and sketch the graph.

*The equation $f(x) = 0$ may be difficult to solve, in which case one may decide against finding the x-intercepts or to use technology, if available, for assistance.

We now illustrate the techniques of curve sketching with several examples.

Two Step-by-Step Examples

EXAMPLE 3 Sketch the graph of the function

$$y = f(x) = x^3 - 6x^2 + 9x + 2$$

Solution Obtain the following information on the graph of f.

1. The domain of f is the interval $(-\infty, \infty)$.
2. By setting $x = 0$, we find that the y-intercept is 2. The x-intercept is found by setting $y = 0$, which in this case leads to a cubic equation. Since the solution is not readily found, we will not use this information.
3. Since

$$\lim_{x\to-\infty} f(x) = \lim_{x\to-\infty} (x^3 - 6x^2 + 9x + 2) = -\infty$$
$$\lim_{x\to\infty} f(x) = \lim_{x\to\infty} (x^3 - 6x^2 + 9x + 2) = \infty$$

we see that f decreases without bound as x decreases without bound and that f increases without bound as x increases without bound.

4. Since f is a polynomial function, there are no asymptotes.
5. $$f'(x) = 3x^2 - 12x + 9 = 3(x^2 - 4x + 3) = 3(x - 3)(x - 1)$$

Setting $f'(x) = 0$ gives $x = 1$ or $x = 3$. The sign diagram for f' shows that f is increasing on the intervals $(-\infty, 1)$ and $(3, \infty)$ and decreasing on the interval $(1, 3)$ (Figure 50).

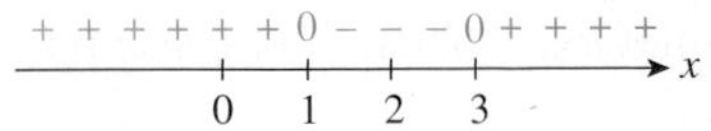

FIGURE 50
Sign diagram for f'

6. From the results of step 5, we see that $x = 1$ and $x = 3$ are critical numbers of f. Furthermore, f' changes sign from positive to negative as we move across $x = 1$, so a relative maximum of f occurs at $x = 1$. Similarly, we see that a

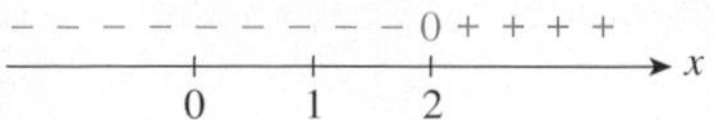

FIGURE 51
Sign diagram for f''

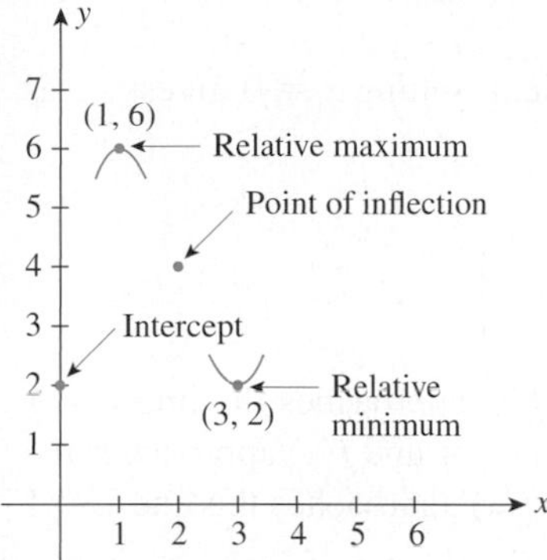

FIGURE 52
We first plot the intercept, the relative extrema, and the inflection point.

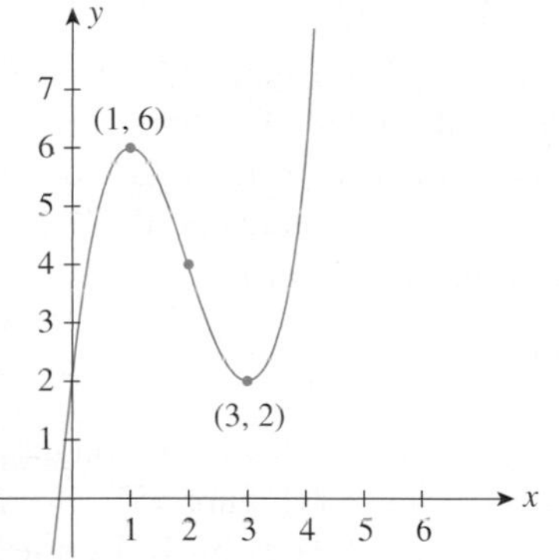

FIGURE 53
The graph of $y = x^3 - 6x^2 + 9x + 2$

relative minimum of f occurs at $x = 3$. Now,

$$f(1) = 1 - 6 + 9 + 2 = 6$$
$$f(3) = 3^3 - 6(3)^2 + 9(3) + 2 = 2$$

so $f(1) = 6$ is a relative maximum of f and $f(3) = 2$ is a relative minimum of f.

7. $$f''(x) = 6x - 12 = 6(x - 2)$$

which is equal to zero when $x = 2$. The sign diagram of f'' shows that f is concave downward on the interval $(-\infty, 2)$ and concave upward on the interval $(2, \infty)$ (Figure 51).

8. From the results of step 7, we see that f'' changes sign as we move across $x = 2$. Next,

$$f(2) = 2^3 - 6(2)^2 + 9(2) + 2 = 4$$

and so the required inflection point of f is (2, 4).

Summarizing, we have the following:

Domain: $(-\infty, \infty)$
Intercept: (0, 2)
$\lim_{x\to-\infty} f(x)$; $\lim_{x\to\infty} f(x)$: $-\infty$; ∞
Asymptotes: None
Intervals where f is ↗ or ↘: ↗ on $(-\infty, 1) \cup (3, \infty)$; ↘ on (1, 3)
Relative extrema: Relative maximum at (1, 6); relative minimum at (3, 2)
Concavity: Downward on $(-\infty, 2)$; upward on $(2, \infty)$
Point of inflection: (2, 4)

In general, it is a good idea to start graphing by plotting the intercept(s), relative extrema, and inflection point(s) (Figure 52). Then, using the rest of the information, we complete the graph of f, as sketched in Figure 53. ■

EXPLORE & DISCUSS

The average price of gasoline at the pump over a 3-month period, during which there was a temporary shortage of oil, is described by the function f defined on the interval [0, 3]. During the first month, the price was increasing at an increasing rate. Starting with the second month, the good news was that the rate of increase was slowing down, although the price of gas was still increasing. This pattern continued until the end of the second month. The price of gas peaked at the end of $t = 2$ and began to fall at an increasing rate until $t = 3$.

1. Describe the signs of $f'(t)$ and $f''(t)$ over each of the intervals (0, 1), (1, 2), and (2, 3).
2. Make a sketch showing a plausible graph of f over [0, 3].

EXAMPLE 4 Sketch the graph of the function

$$y = f(x) = \frac{x + 1}{x - 1}$$

Solution Obtain the following information:

1. f is undefined when $x = 1$, so the domain of f is the set of all real numbers other than $x = 1$.
2. Setting $y = 0$ gives -1, the x-intercept of f. Next, setting $x = 0$ gives -1 as the y-intercept of f.
3. Earlier we found that

$$\lim_{x \to \infty} \frac{x + 1}{x - 1} = 1 \quad \text{and} \quad \lim_{x \to -\infty} \frac{x + 1}{x - 1} = 1$$

(see pp. 288–289). Consequently, we see that $f(x)$ approaches the line $y = 1$ as $|x|$ becomes arbitrarily large. For $x > 1$, $f(x) > 1$ and $f(x)$ approaches the line $y = 1$ from above. For $x < 1$, $f(x) < 1$, so $f(x)$ approaches the line $y = 1$ from below.

4. The straight line $x = 1$ is a vertical asymptote of the graph of f. Also, from the results of step 3, we conclude that $y = 1$ is a horizontal asymptote of the graph of f.
5. $$f'(x) = \frac{(x - 1)(1) - (x + 1)(1)}{(x - 1)^2} = -\frac{2}{(x - 1)^2}$$

and is discontinuous at $x = 1$. The sign diagram of f' shows that $f'(x) < 0$ whenever it is defined. Thus, f is decreasing on the intervals $(-\infty, 1)$ and $(1, \infty)$ (Figure 54).

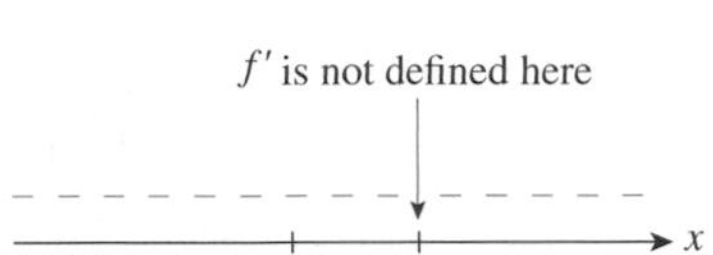

FIGURE 54
The sign diagram for f'

6. From the results of step 5, we see that there are no critical numbers of f since $f'(x)$ is never equal to zero for any value of x in the domain of f.
7. $$f''(x) = \frac{d}{dx}[-2(x - 1)^{-2}] = 4(x - 1)^{-3} = \frac{4}{(x - 1)^3}$$

The sign diagram of f'' shows immediately that f is concave downward on the interval $(-\infty, 1)$ and concave upward on the interval $(1, \infty)$ (Figure 55).

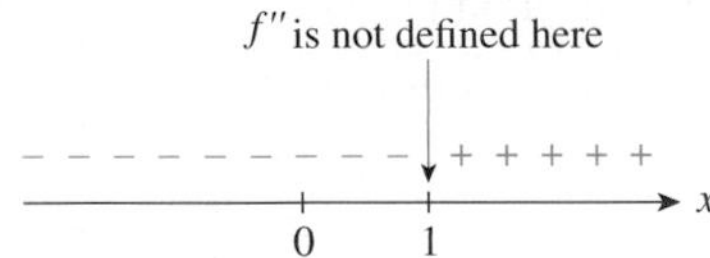

FIGURE 55
The sign diagram for f''

8. From the results of step 7, we see that there are no candidates for inflection points of f since $f''(x)$ is never equal to zero for any value of x in the domain of f. Hence, f has no inflection points.

Summarizing, we have the following:

Domain: $(-\infty, 1) \cup (1, \infty)$
Intercepts: $(0, -1)$; $(-1, 0)$
$\lim_{x \to -\infty} f(x)$; $\lim_{x \to \infty} f(x)$: 1; 1
Asymptotes: $x = 1$ is a vertical asymptote $y = 1$ is a horizontal asymptote
Intervals where f is ↗ or ↘: ↘ on $(-\infty, 1) \cup (1, \infty)$
Relative extrema: None
Concavity: Downward on $(-\infty, 1)$; upward on $(1, \infty)$
Points of inflection: None

The graph of f is sketched in Figure 56. ■

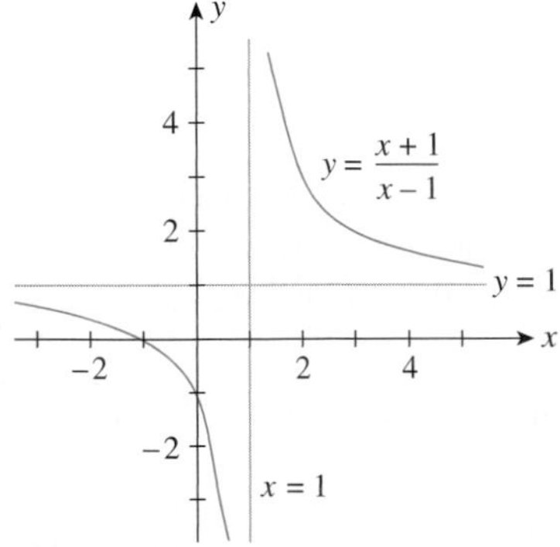

FIGURE 56
The graph of f has a horizontal asymptote at $y = 1$ and a vertical asymptote at $x = 1$.

4.3 Self-Check Exercises

1. Find the horizontal and vertical asymptotes of the graph of the function

$$f(x) = \frac{2x^2}{x^2 - 1}$$

2. Sketch the graph of the function

$$f(x) = \frac{2}{3}x^3 - 2x^2 - 6x + 4$$

Solutions to Self-Check Exercises 4.3 can be found on page 297.

4.3 Concept Questions

1. Explain the following terms in your own words:
   a. Vertical asymptote **b.** Horizontal asymptote

2. **a.** How many vertical asymptotes can the graph of a function f have? Explain using graphs.
   b. How many horizontal asymptotes can the graph of a function f have? Explain using graphs.

3. How do you find the vertical asymptotes of a rational function?

4. Give a procedure for sketching the graph of a function.

4.3 Exercises

In Exercises 1–10, find the horizontal and vertical asymptotes of the graph.

1.

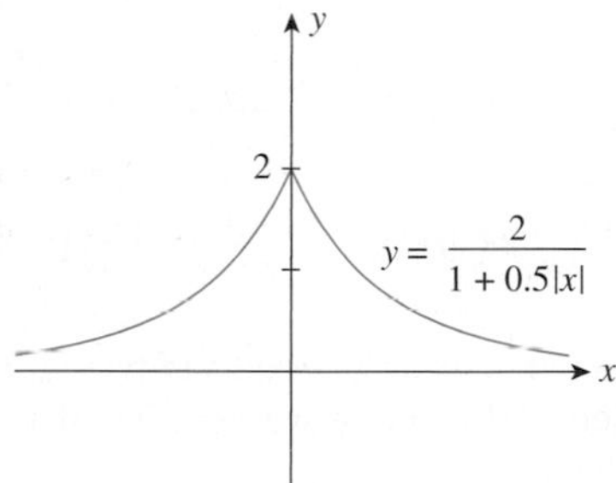

2.

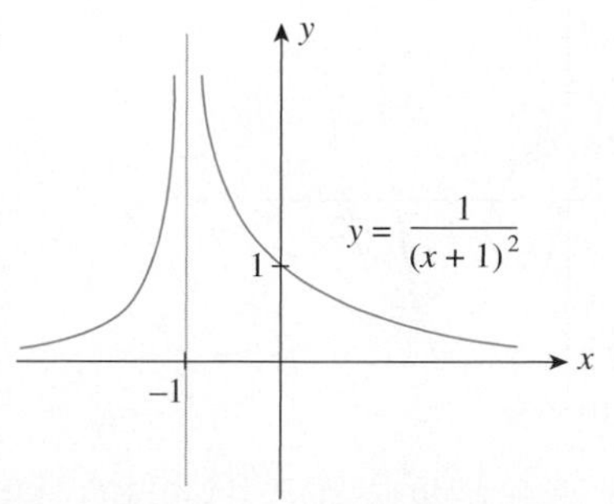

3.

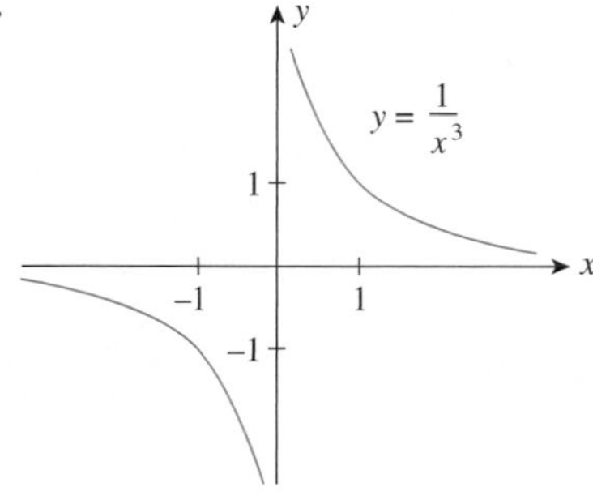

4.

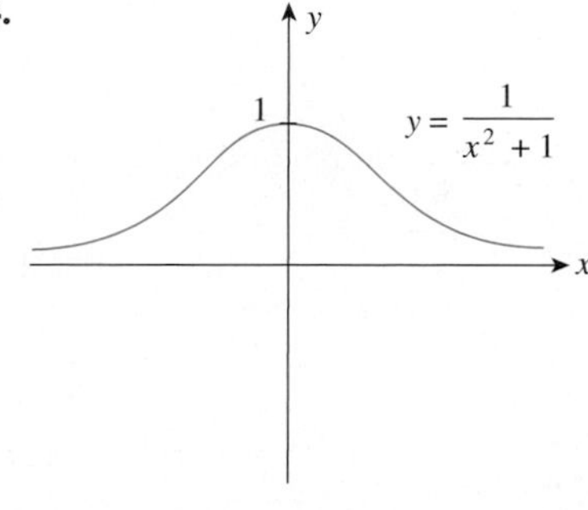

5.

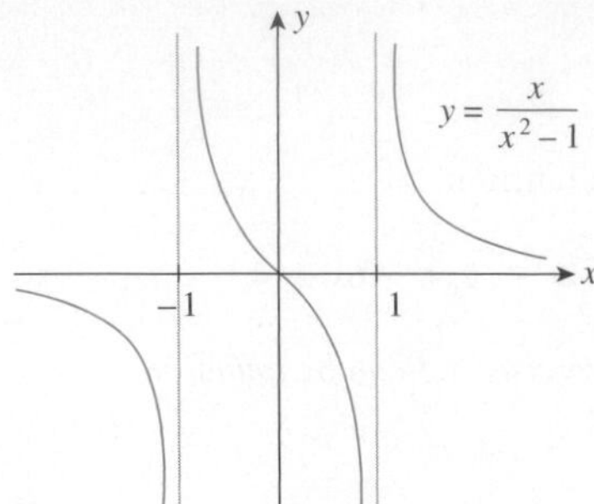

6.

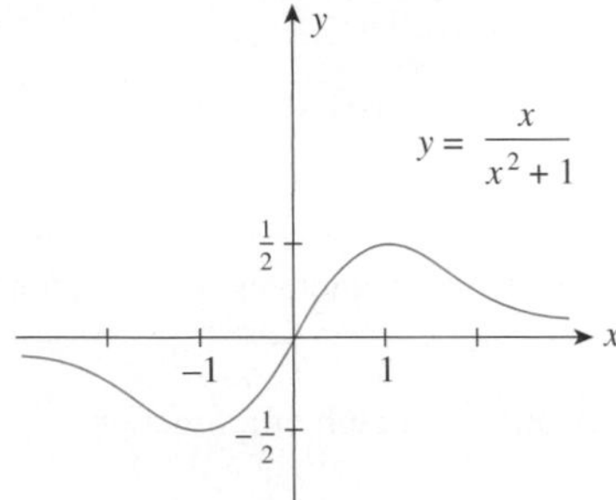

7.

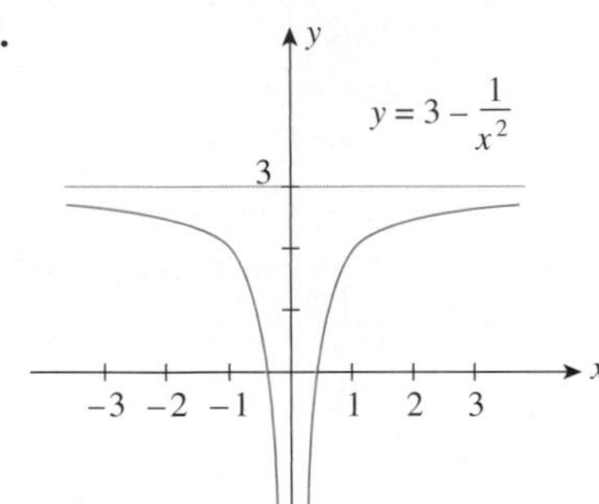

8.

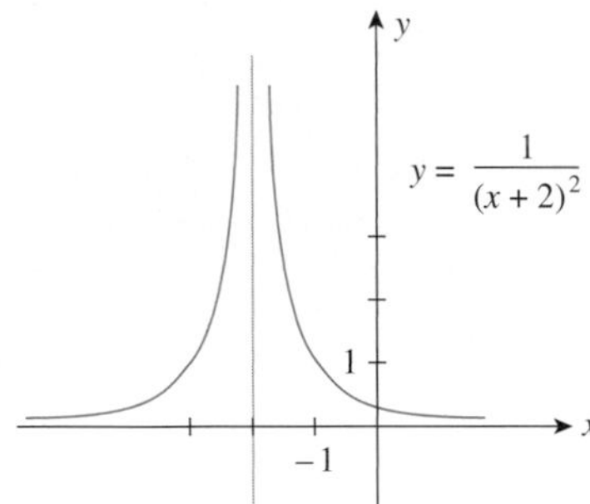

9.

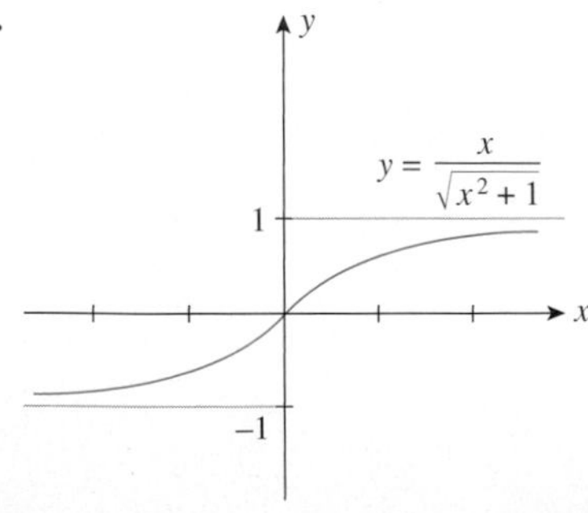

10.

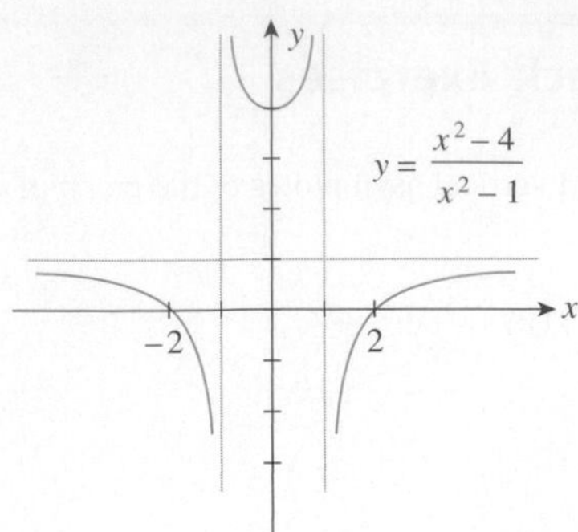

In Exercises 11–28, find the horizontal and vertical asymptotes of the graph of the function. (You need not sketch the graph.)

11. $f(x) = \frac{1}{x}$

12. $f(x) = \frac{1}{x + 2}$

13. $f(x) = -\frac{2}{x^2}$

14. $g(x) = \frac{1}{1 + 2x^2}$

15. $f(x) = \frac{x - 1}{x + 1}$

16. $g(t) = \frac{t + 1}{2t - 1}$

17. $h(x) = x^3 - 3x^2 + x + 1$

18. $g(x) = 2x^3 + x^2 + 1$

19. $f(t) = \frac{t^2}{t^2 - 9}$

20. $g(x) = \frac{x^3}{x^2 - 4}$

21. $f(x) = \frac{3x}{x^2 - x - 6}$

22. $g(x) = \frac{2x}{x^2 + x - 2}$

23. $g(t) = 2 + \frac{5}{(t - 2)^2}$

24. $f(x) = 1 + \frac{2}{x - 3}$

25. $f(x) = \frac{x^2 - 2}{x^2 - 4}$

26. $h(x) = \frac{2 - x^2}{x^2 + x}$

27. $g(x) = \frac{x^3 - x}{x(x + 1)}$

28. $f(x) = \frac{x^4 - x^2}{x(x - 1)(x + 2)}$

In Exercises 29 and 30, you are given the graphs of two functions f and g. One function is the derivative function of the other. Identify each of them.

29.

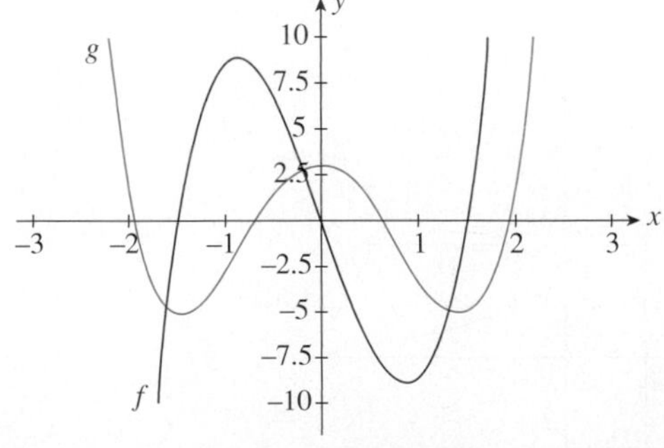

30.

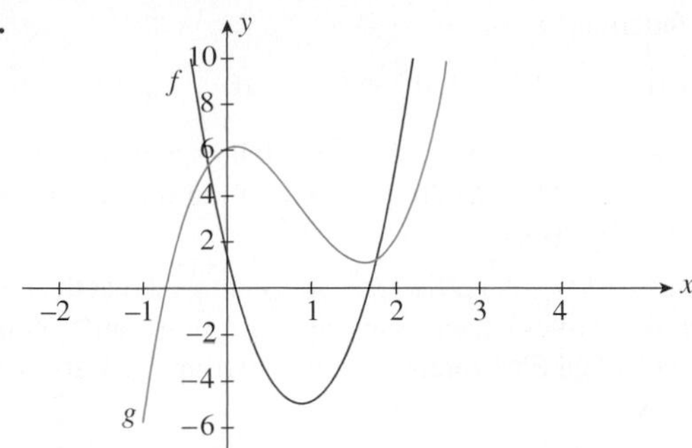

31. Terminal Velocity A skydiver leaps from the gondola of a hot-air balloon. As she free-falls, air resistance, which is proportional to her velocity, builds up to a point where it balances the force due to gravity. The resulting motion may be described in terms of her velocity as follows: Starting at rest (zero velocity), her velocity increases and approaches a constant velocity, called the *terminal velocity*. Sketch a graph of her velocity v versus time t.

32. Spread of a Flu Epidemic Initially, 10 students at a junior high school contracted influenza. The flu spread over time, and the total number of students who eventually contracted the flu approached but never exceeded 200. Let $P(t)$ denote the number of students who had contracted the flu after t days, where P is an appropriate function.

a. Make a sketch of the graph of P. (Your answer will *not* be unique.)
b. Where is the function increasing?
c. Does P have a horizontal asymptote? If so, what is it?
d. Discuss the concavity of P. Explain its significance.
e. Is there an inflection point on the graph of P? If so, explain its significance.

In Exercises 33–36, use the information summarized in the table to sketch the graph of f.

33. $f(x) = x^3 - 3x^2 + 1$

Domain: $(-\infty, \infty)$
Intercept: y-intercept: 1
Asymptotes: None
Intervals where f is ↗ and ↘: ↗ on $(-\infty, 0) \cup (2, \infty)$;
↘ on $(0, 2)$
Relative extrema: Rel. max. at $(0, 1)$; rel. min. at $(2, -3)$
Concavity: Downward on $(-\infty, 1)$; upward on $(1, \infty)$
Point of inflection: $(1, -1)$

34. $f(x) = \dfrac{1}{9}(x^4 - 4x^3)$

Domain: $(-\infty, \infty)$
Intercepts: x-intercepts: 0, 4; y-intercept: 0
Asymptotes: None
Intervals where f is ↗ and ↘: ↗ on $(3, \infty)$;
↘ on $(-\infty, 3)$
Relative extrema: Rel. min. at $(3, -3)$
Concavity: Downward on $(0, 2)$;
upward on $(-\infty, 0) \cup (2, \infty)$
Points of inflection: $(0, 0)$ and $(2, -\frac{16}{9})$

35. $f(x) = \dfrac{4x - 4}{x^2}$

Domain: $(-\infty, 0) \cup (0, \infty)$
Intercept: x-intercept: 1
Asymptotes: x-axis and y-axis
Intervals where f is ↗ and ↘: ↗ on $(0, 2)$;
↘ on $(-\infty, 0) \cup (2, \infty)$
Relative extrema: Rel. max. at $(2, 1)$
Concavity: Downward on $(-\infty, 0) \cup (0, 3)$;
upward on $(3, \infty)$
Point of inflection: $(3, \frac{8}{9})$

36. $f(x) = x - 3x^{1/3}$

Domain: $(-\infty, \infty)$
Intercepts: x-intercepts: $\pm 3\sqrt{3}, 0$
Asymptotes: None
Intervals where f is ↗ and ↘: ↗ on $(-\infty, -1) \cup (1, \infty)$;
↘ on $(-1, 1)$
Relative extrema: Rel. max. at $(-1, 2)$; rel. min. at $(1, -2)$
Concavity: Downward on $(-\infty, 0)$; upward on $(0, \infty)$
Point of inflection: $(0, 0)$

In Exercises 37–60, sketch the graph of the function, using the curve-sketching guide of this section.

37. $g(x) = 4 - 3x - 2x^3$

38. $f(x) = x^2 - 2x + 3$

39. $h(x) = x^3 - 3x + 1$

40. $f(x) = 2x^3 + 1$

41. $f(x) = -2x^3 + 3x^2 + 12x + 2$

42. $f(t) = 2t^3 - 15t^2 + 36t - 20$

43. $h(x) = \dfrac{3}{2}x^4 - 2x^3 - 6x^2 + 8$

44. $f(t) = 3t^4 + 4t^3$

45. $f(t) = \sqrt{t^2 - 4}$

46. $f(x) = \sqrt{x^2 + 5}$

47. $g(x) = \dfrac{1}{2}x - \sqrt{x}$

48. $f(x) = \sqrt[3]{x^2}$

49. $g(x) = \dfrac{2}{x - 1}$

50. $f(x) = \dfrac{1}{x + 1}$

51. $h(x) = \dfrac{x + 2}{x - 2}$

52. $g(x) = \dfrac{x}{x - 1}$

53. $f(t) = \dfrac{t^2}{1 + t^2}$

54. $g(x) = \dfrac{x}{x^2 - 4}$

55. $g(t) = -\dfrac{t^2 - 2}{t - 1}$

56. $f(x) = \dfrac{x^2 - 9}{x^2 - 4}$

57. $g(t) = \dfrac{t^2}{t^2 - 1}$

58. $h(x) = \dfrac{1}{x^2 - x - 2}$

59. $h(x) = (x - 1)^{2/3} + 1$ **60.** $g(x) = (x + 2)^{3/2} + 1$

61. Cost of Removing Toxic Pollutants A city's main well was recently found to be contaminated with trichloroethylene (a cancer-causing chemical) as a result of an abandoned chemical dump that leached chemicals into the water. A proposal submitted to the city council indicated that the cost, measured in millions of dollars, of removing $x\%$ of the toxic pollutants is given by

$$C(x) = \frac{0.5x}{100 - x}$$

a. Find the vertical asymptote of $C(x)$.
b. Is it possible to remove 100% of the toxic pollutant from the water?

62. Average Cost of Producing DVDs The average cost per disc (in dollars) incurred by Herald Media Corporation in pressing x DVDs is given by the average cost function

$$\overline{C}(x) = 2.2 + \frac{2500}{x}$$

a. Find the horizontal asymptote of $\overline{C}(x)$.
b. What is the limiting value of the average cost?

63. Concentration of a Drug in the Bloodstream The concentration (in milligrams/cubic centimeter) of a certain drug in a patient's bloodstream t hr after injection is given by

$$C(t) = \frac{0.2t}{t^2 + 1}$$

a. Find the horizontal asymptote of $C(t)$.
b. Interpret your result.

64. Effect of Enzymes on Chemical Reactions Certain proteins, known as enzymes, serve as catalysts for chemical reactions in living things. In 1913 Leonor Michaelis and L. M. Menten discovered the following formula giving the initial speed V (in moles/liter/second) at which the reaction begins in terms of the amount of substrate x (the substance that is being acted upon, measured in moles/liter):

$$V = \frac{ax}{x + b}$$

where a and b are positive constants.
a. Find the horizontal asymptote of V.
b. What does the result of part (a) tell you about the initial speed at which the reaction begins, if the amount of substrate is very large?

65. GDP of a Developing Country A developing country's gross domestic product (GDP) from 1997 to 2005 is approximated by the function

$$G(t) = -0.2t^3 + 2.4t^2 + 60 \qquad (0 \le t \le 8)$$

where $G(t)$ is measured in billions of dollars, with $t = 0$ corresponding to 1997. Sketch the graph of the function G and interpret your results.

66. Worker Efficiency An efficiency study showed that the total number of cordless telephones assembled by an average worker at Delphi Electronics t hr after starting work at 8 a.m. is given by

$$N(t) = -\frac{1}{2}t^3 + 3t^2 + 10t \qquad (0 \le t \le 4)$$

Sketch the graph of the function N and interpret your results.

67. Concentration of a Drug in the Bloodstream The concentration (in millimeters/cubic centimeter) of a certain drug in a patient's bloodstream t hr after injection is given by

$$C(t) = \frac{0.2t}{t^2 + 1}$$

Sketch the graph of the function C and interpret your results.

68. Oxygen Content of a Pond When organic waste is dumped into a pond, the oxidation process that takes place reduces the pond's oxygen content. However, given time, nature will restore the oxygen content to its natural level. Suppose the oxygen content t days after organic waste has been dumped into the pond is given by

$$f(t) = 100\left(\frac{t^2 - 4t + 4}{t^2 + 4}\right) \qquad (0 \le t < \infty)$$

percent of its normal level. Sketch the graph of the function f and interpret your results.

69. Box-Office Receipts The total worldwide box-office receipts for a long-running movie are approximated by the function

$$T(x) = \frac{120x^2}{x^2 + 4}$$

where $T(x)$ is measured in millions of dollars and x is the number of years since the movie's release. Sketch the graph of the function T and interpret your results.

70. Cost of Removing Toxic Pollutants Refer to Exercise 61. The cost, measured in millions of dollars, of removing $x\%$ of a toxic pollutant is given by

$$C(x) = \frac{0.5x}{100 - x}$$

Sketch the graph of the function C and interpret your results.

4.3 Solutions to Self-Check Exercises

1. Since

$$\lim_{x\to\infty}\frac{2x^2}{x^2-1} = \lim_{x\to\infty}\frac{2}{1-\frac{1}{x^2}} \qquad \text{Divide the numerator and denominator by } x^2.$$
$$= 2$$

we see that $y = 2$ is a horizontal asymptote. Next, since

$$x^2 - 1 = (x + 1)(x - 1) = 0$$

implies $x = -1$ or $x = 1$, these are candidates for the vertical asymptotes of f. Since the numerator of f is not equal to zero for $x = -1$ or $x = 1$, we conclude that $x = -1$ and $x = 1$ are vertical asymptotes of the graph of f.

2. We obtain the following information on the graph of f.
(1) The domain of f is the interval $(-\infty, \infty)$.
(2) By setting $x = 0$, we find the y-intercept is 4.
(3) Since

$$\lim_{x\to-\infty} f(x) = \lim_{x\to-\infty}\left(\frac{2}{3}x^3 - 2x^2 - 6x + 4\right) = -\infty$$

$$\lim_{x\to\infty} f(x) = \lim_{x\to\infty}\left(\frac{2}{3}x^3 - 2x^2 - 6x + 4\right) = \infty$$

we see that $f(x)$ decreases without bound as x decreases without bound and that $f(x)$ increases without bound as x increases without bound.
(4) Since f is a polynomial function, there are no asymptotes.

(5) $$f'(x) = 2x^2 - 4x - 6 = 2(x^2 - 2x - 3)$$
$$= 2(x + 1)(x - 3)$$

Setting $f'(x) = 0$ gives $x = -1$ or $x = 3$. The accompanying sign diagram for f' shows that f is increasing on the intervals $(-\infty, -1)$ and $(3, \infty)$ and decreasing on $(-1, 3)$.

+ + + + 0 – – – – – – – 0 + + + +
-1 0 3 x

Sign diagram for f'

(6) From the results of step 5, we see that $x = -1$ and $x = 3$ are critical numbers of f. Furthermore, the sign diagram of f' tells us that $x = -1$ gives rise to a relative maximum of f and $x = 3$ gives rise to a relative minimum of f. Now,

$$f(-1) = \frac{2}{3}(-1)^3 - 2(-1)^2 - 6(-1) + 4 = \frac{22}{3}$$

$$f(3) = \frac{2}{3}(3)^3 - 2(3)^2 - 6(3) + 4 = -14$$

so $f(-1) = \frac{22}{3}$ is a relative maximum of f and $f(3) = -14$ is a relative minimum of f.

(7) $$f''(x) = 4x - 4 = 4(x - 1)$$

which is equal to zero when $x = 1$. The accompanying sign diagram of f'' shows that f is concave downward on the interval $(-\infty, 1)$ and concave upward on the interval $(1, \infty)$.

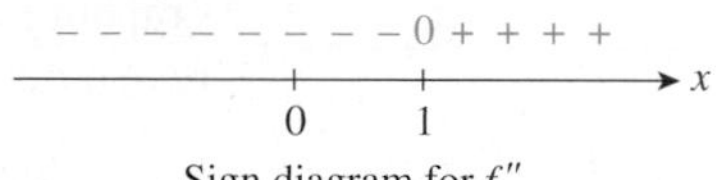

Sign diagram for f''

(8) From the results of step 7, we see that $x = 1$ is the only candidate for an inflection point of f. Since $f''(x)$ changes sign as we move across the point $x = 1$ and

$$f(1) = \frac{2}{3}(1)^3 - 2(1)^2 - 6(1) + 4 = -\frac{10}{3}$$

we see that the required inflection point is $(1, -\frac{10}{3})$.
(9) Summarizing this information, we have the following:

Domain: $(-\infty, \infty)$

Intercept: $(0, 4)$
$\lim_{x\to-\infty} f(x)$; $\lim_{x\to\infty} f(x)$: $-\infty$; ∞
Asymptotes: None

Intervals where f is ↗ or ↘: ↗ on $(-\infty, -1) \cup (3, \infty)$;
↘ on $(-1, 3)$

Relative extrema: Rel. max. at $(-1, \frac{22}{3})$; rel. min. at $(3, -14)$

Concavity: Downward on $(-\infty, 1)$; upward on $(1, \infty)$

Point of inflection: $(1, -\frac{10}{3})$

The graph of f is sketched in the accompanying figure.

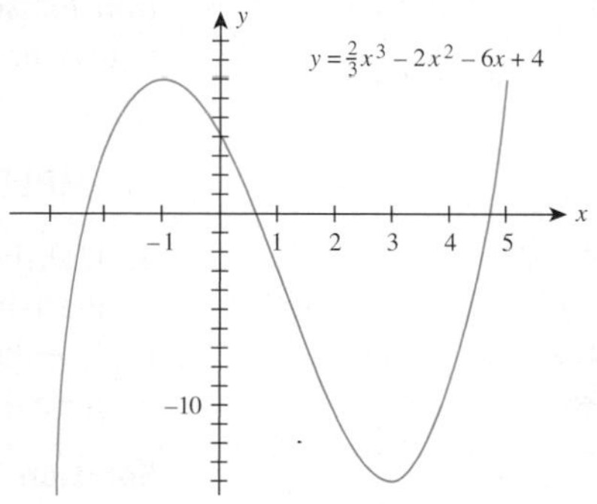

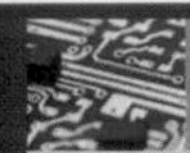

USING TECHNOLOGY

Analyzing the Properties of a Function

One of the main purposes of studying Section 4.3 is to see how the many concepts of calculus come together to paint a picture of a function. The techniques of graphing also play a very practical role. For example, using the techniques of graphing developed in Section 4.3, you can tell if the graph of a function generated by a graphing utility is reasonably complete. Furthermore, these techniques can often reveal details that are missing from a graph.

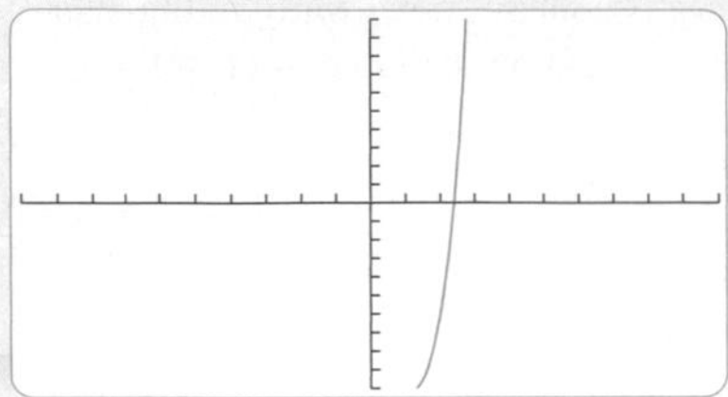

FIGURE T1
The graph of f in the standard viewing window

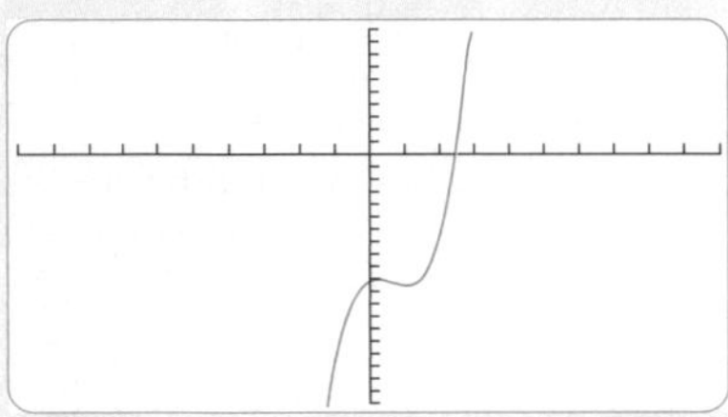

FIGURE T2
The graph of f in the viewing window $[-10, 10] \times [-20, 10]$

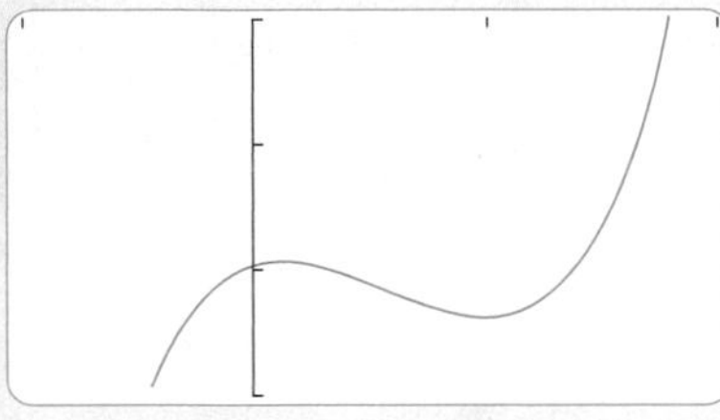

FIGURE T3
The graph of f in the viewing window $[-1, 2] \times [-11, -9]$

EXAMPLE 1 Consider the function $f(x) = 2x^3 - 3.5x^2 + x - 10$. A plot of the graph of f in the standard viewing window is shown in Figure T1. Since the domain of f is the interval $(-\infty, \infty)$, we see that Figure T1 does not reveal the part of the graph to the left of the y-axis. This suggests that we enlarge the viewing window accordingly. Figure T2 shows the graph of f in the viewing window $[-10, 10] \times [-20, 10]$.

The behavior of f for large values of f

$$\lim_{x \to -\infty} f(x) = -\infty \quad \text{and} \quad \lim_{x \to \infty} f(x) = \infty$$

suggests that this viewing window has captured a sufficiently complete picture of f. Next, an analysis of the first derivative of f,

$$f'(x) = 6x^2 - 7x + 1 = (6x - 1)(x - 1)$$

reveals that f has critical values at $x = \frac{1}{6}$ and $x = 1$. In fact, a sign diagram of f' shows that f has a relative maximum at $x = \frac{1}{6}$ and a relative minimum at $x = 1$, details that are not revealed in the graph of f shown in Figure T2. To examine this portion of the graph of f, we use, say, the viewing window $[-1, 2] \times [-11, -9]$. The resulting graph of f is shown in Figure T3, which certainly reveals the hitherto missing details! Thus, through an interaction of calculus and a graphing utility, we are able to obtain a good picture of the properties of f. ■

Finding *x*-Intercepts

As noted in Section 4.3, it is not always easy to find the x-intercepts of the graph of a function. But this information is very important in applications. By using the function for solving polynomial equations or the function for finding the roots of an equation, we can solve the equation $f(x) = 0$ quite easily and hence yield the x-intercepts of the graph of a function.

EXAMPLE 2 Let $f(x) = x^3 - 3x^2 + x + 1.5$.

a. Use the function for solving polynomial equations on a graphing utility to find the x-intercepts of the graph of f.

b. Use the function for finding the roots of an equation on a graphing utility to find the x-intercepts of the graph of f.

Solution

a. Observe that f is a polynomial function of degree 3, and so we may use the function for solving polynomial equations to solve the equation $x^3 - 3x^2 + x + 1.5 = 0$ $[f(x) = 0]$. We find that the solutions (x-intercepts) are

$$x_1 \approx -0.525687120865 \qquad x_2 \approx 1.2586520225 \qquad x_3 \approx 2.26703509836$$

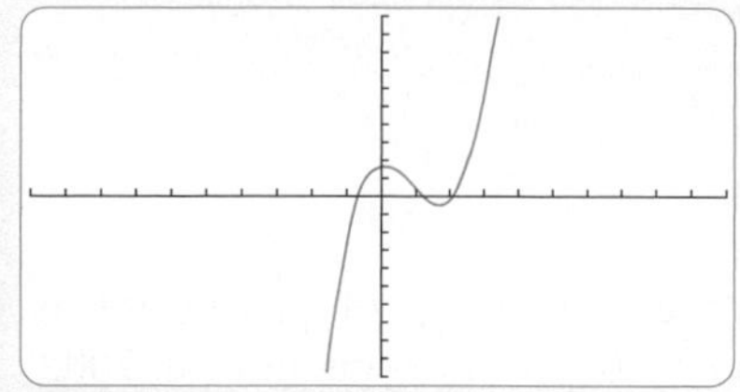

FIGURE T4
The graph of $f(x) = x^3 - 3x^2 + x + 1.5$

b. Using the graph of f (Figure T4), we see that $x_1 \approx -0.5$, $x_2 \approx 1$, and $x_3 \approx 2$. Using the function for finding the roots of an equation on a graphing utility, and these values of x as initial guesses, we find

$$x_1 \approx -0.5256871209 \qquad x_2 \approx 1.2586520225 \qquad x_3 \approx 2.2670350984$$

Note The function for solving polynomial equations on a graphing utility will solve a polynomial equation $f(x) = 0$, where f is a polynomial function. The function for finding the roots of a polynomial, however, will solve equations $f(x) = 0$ even if f is not a polynomial.

APPLIED EXAMPLE 3 Solvency of Social Security Fund Unless payroll taxes are increased significantly and/or benefits are scaled back drastically, it is a matter of time before the current Social Security system goes broke. Data show that the assets of the system—the Social Security "trust fund"—may be approximated by

$$f(t) = -0.0129t^4 + 0.3087t^3 + 2.1760t^2 + 62.8466t + 506.2955 \qquad (0 \le t \le 35)$$

where $f(t)$ is measured in millions of dollars and t is measured in years, with $t = 0$ corresponding to 1995.

a. Use a graphing calculator to sketch the graph of f.
b. Based on this model, when can the Social Security system be expected to go broke?

Source: Social Security Administration

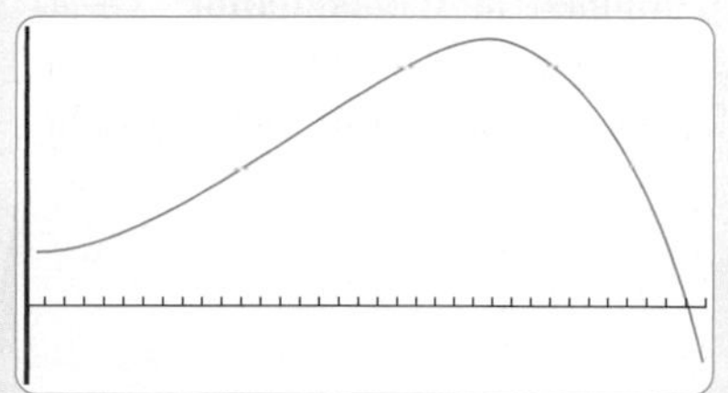

FIGURE T5
The graph of $f(t)$

Solution

a. The graph of f in the window $[0, 35] \times [-1000, 3500]$ is shown in Figure T5.
b. Using the function for finding the roots on a graphing utility, we find that $y = 0$ when $t \approx 34.1$, and this tells us that the system is expected to go broke around 2029.

TECHNOLOGY EXERCISES

In Exercises 1–4, use the method of Example 1 to analyze the function. (*Note:* Your answers will *not* be unique.)

1. $f(x) = 4x^3 - 4x^2 + x + 10$
2. $f(x) = x^3 + 2x^2 + x - 12$
3. $f(x) = \frac{1}{2}x^4 + x^3 + \frac{1}{2}x^2 - 10$
4. $f(x) = 2.25x^4 - 4x^3 + 2x^2 + 2$

In Exercises 5–10, find the *x*-intercepts of the graph of *f*. Give your answer accurate to four decimal places.

5. $f(x) = 0.2x^3 - 1.2x^2 + 0.8x + 2.1$
6. $f(x) = -0.5x^3 + 1.7x^2 - 1.2$
7. $f(x) = 0.3x^4 - 1.2x^3 + 0.8x^2 + 1.1x - 2$
8. $f(x) = -0.2x^4 + 0.8x^3 - 2.1x + 1.2$
9. $f(x) = 2x^2 - \sqrt{x + 1} - 3$
10. $f(x) = x - \sqrt{1 - x^2}$

4.4 Optimization I

Absolute Extrema

The graph of the function f in Figure 57 shows the average age of cars in use in the United States from the beginning of 1946 ($t = 0$) to the beginning of 2002 ($t = 56$). Observe that the highest average age of cars in use during this period is 9 years, whereas the lowest average age of cars in use during the same period is $5\frac{1}{2}$ years. The number 9, the largest value of $f(t)$ for all values of t in the interval $[0, 56]$ (the domain of f), is called the *absolute maximum value of f* on that interval. The number $5\frac{1}{2}$, the smallest value of $f(t)$ for all values of t in $[0, 56]$, is called the *absolute minimum value of f* on that interval. Notice, too, that the absolute maximum value of f is attained at the endpoint $t = 0$ of the interval, whereas the absolute minimum value of f is attained at the points $t = 12$ (corresponding to 1958) and $t = 23$ (corresponding to 1969) that lie within the interval $(0, 56)$.

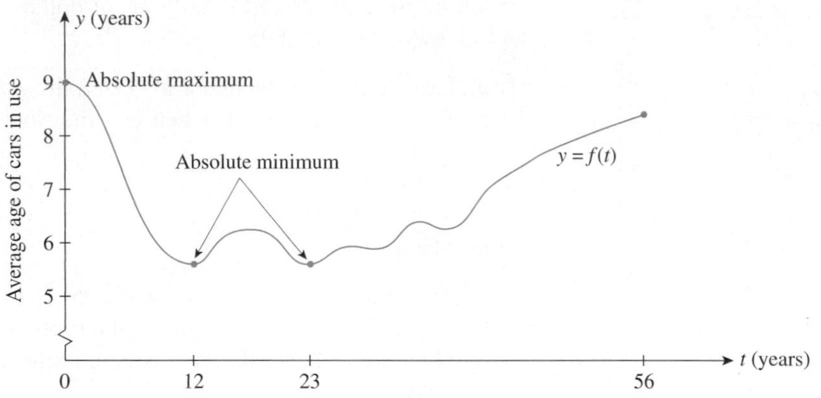

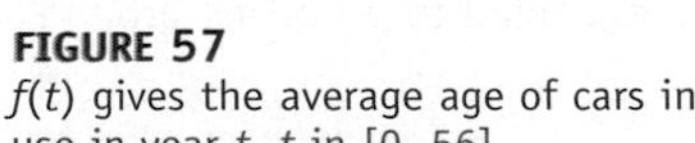
FIGURE 57
$f(t)$ gives the average age of cars in use in year t, t in $[0, 56]$.

Source: American Automobile Association

(Incidentally, it is interesting to note that 1946 marked the first year of peace following World War II, and the two years, 1958 and 1969, marked the end of two periods of prosperity in recent U.S. history!)

A precise definition of the **absolute extrema** (absolute maximum or absolute minimum) of a function follows.

The Absolute Extrema of a Function f

If $f(x) \leq f(c)$ for all x in the domain of f, then $f(c)$ is called the **absolute maximum value** of f.

If $f(x) \geq f(c)$ for all x in the domain of f, then $f(c)$ is called the **absolute minimum value** of f.

Figure 58 shows the graphs of several functions and gives the absolute maximum and absolute minimum of each function, if they exist.

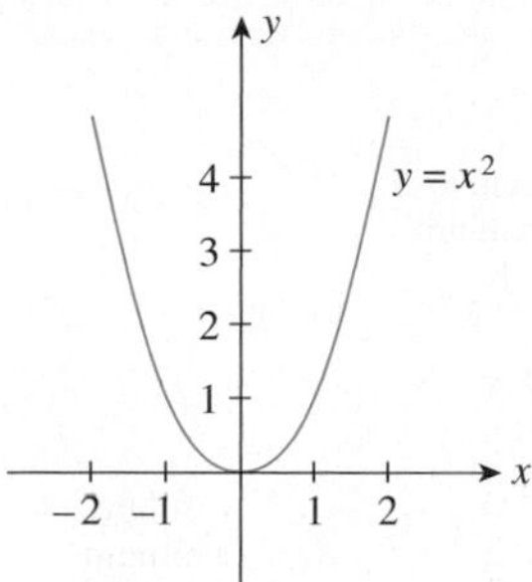

(a) $f(0) = 0$ is the absolute minimum of f; f has no absolute maximum.

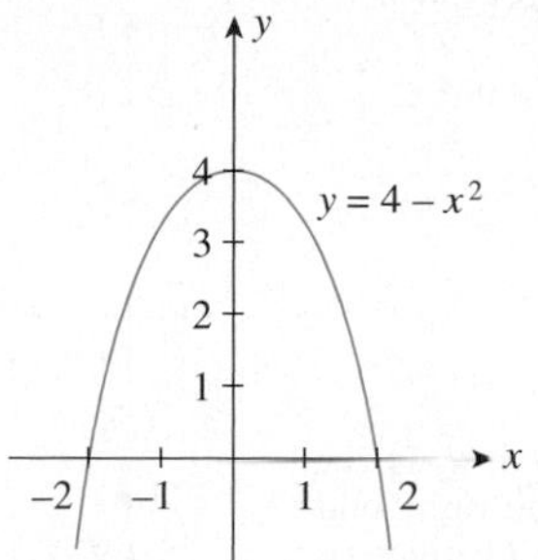

(b) $f(0) = 4$ is the absolute maximum of f; f has no absolute minimum.

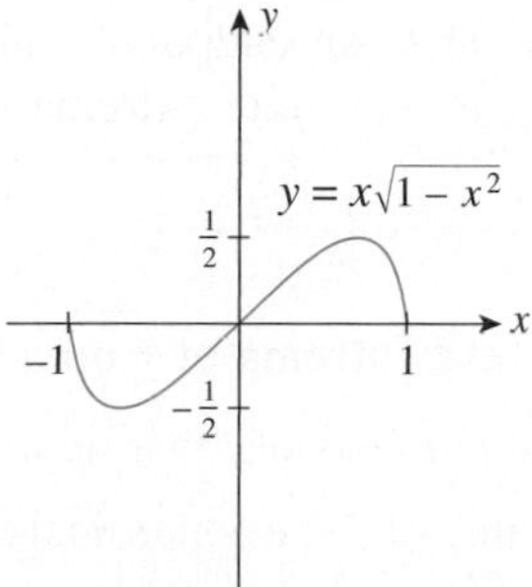

(c) $f(\sqrt{2}/2) = 1/2$ is the absolute maximum of f; $f(-\sqrt{2}/2) = -1/2$ is the absolute minimum of f.

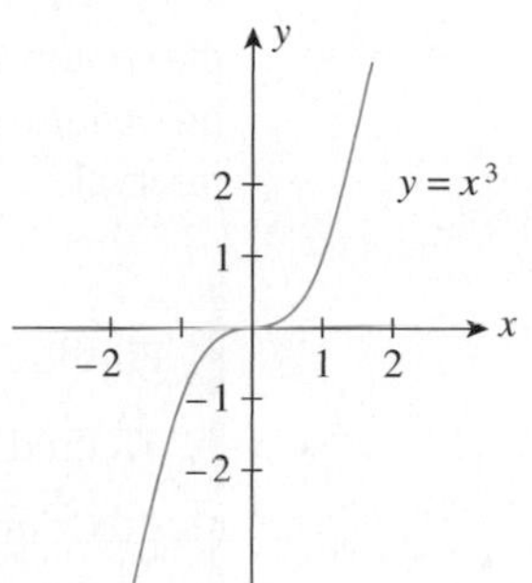

(d) f has no absolute extrema.

FIGURE 58

Absolute Extrema on a Closed Interval

As the preceding examples show, a continuous function defined on an arbitrary interval does not always have an absolute maximum or an absolute minimum. But an important case arises often in practical applications in which both the absolute maximum and the absolute minimum of a function are guaranteed to exist. This occurs where a continuous function is defined on a *closed* interval. Let's state this important result in the form of a theorem, whose proof we will omit.

> **THEOREM 3**
>
> If a function f is continuous on a closed interval $[a, b]$, then f has both an absolute maximum value and an absolute minimum value on $[a, b]$.

Observe that if an absolute extremum of a continuous function f occurs at a point in an open interval (a, b), then it must be a relative extremum of f and hence its x-coordinate must be a critical number of f. Otherwise, the absolute extremum of f must occur at one or both of the endpoints of the interval $[a, b]$. A typical situation is illustrated in Figure 59.

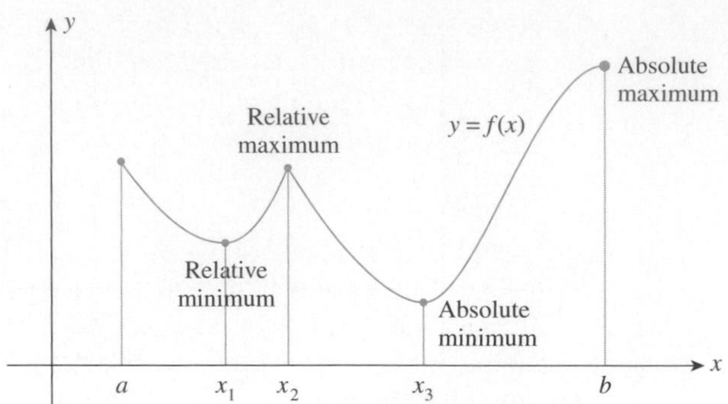

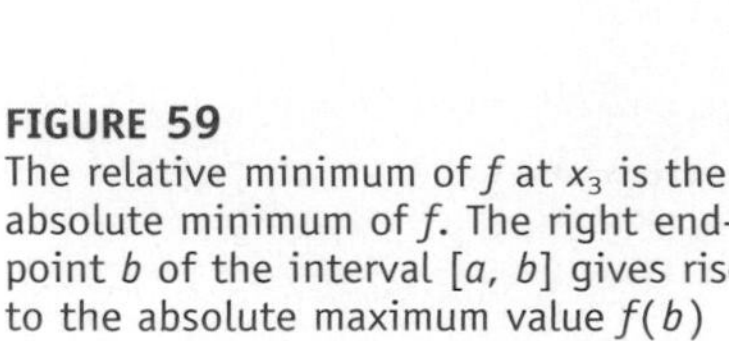
FIGURE 59
The relative minimum of f at x_3 is the absolute minimum of f. The right endpoint b of the interval $[a, b]$ gives rise to the absolute maximum value $f(b)$ of f.

Here x_1, x_2, and x_3 are critical numbers of f. The absolute minimum of f occurs at x_3, which lies in the open interval (a, b) and is a critical number of f. The absolute maximum of f occurs at b, an endpoint. This observation suggests the following procedure for finding the absolute extrema of a continuous function on a closed interval.

Finding the Absolute Extrema of f on a Closed Interval

1. Find the critical numbers of f that lie in (a, b).
2. Compute the value of f at each critical number found in step 1 and compute $f(a)$ and $f(b)$.
3. The absolute maximum value and absolute minimum value of f will correspond to the largest and smallest numbers, respectively, found in step 2.

EXAMPLE 1 Find the absolute extrema of the function $F(x) = x^2$ defined on the interval $[-1, 2]$.

Solution The function F is continuous on the closed interval $[-1, 2]$ and differentiable on the open interval $(-1, 2)$. The derivative of F is

$$F'(x) = 2x$$

so 0 is the only critical number of F. Next, evaluate $F(x)$ at $x = -1$, $x = 0$, and $x = 2$. Thus,

$$F(-1) = 1 \qquad F(0) = 0 \qquad F(2) = 4$$

It follows that 0 is the absolute minimum value of F and 4 is the absolute maximum value of F. The graph of F, in Figure 60, confirms our results. ■

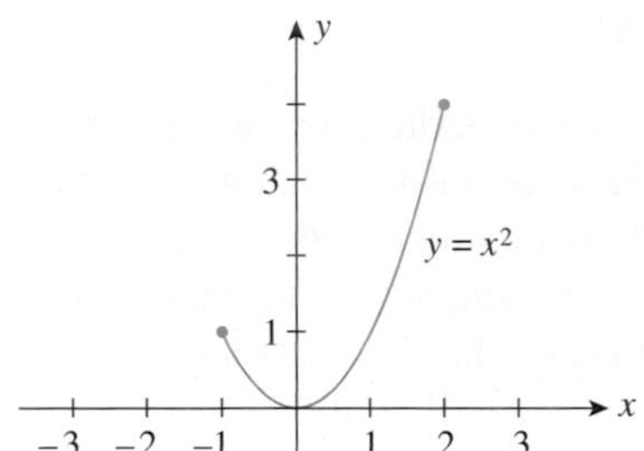

FIGURE 60
F has an absolute minimum value of 0 and an absolute maximum value of 4.

EXAMPLE 2 Find the absolute extrema of the function

$$f(x) = x^3 - 2x^2 - 4x + 4$$

defined on the interval $[0, 3]$.

Solution The function f is continuous on the closed interval $[0, 3]$ and differentiable on the open interval $(0, 3)$. The derivative of f is

$$f'(x) = 3x^2 - 4x - 4 = (3x + 2)(x - 2)$$

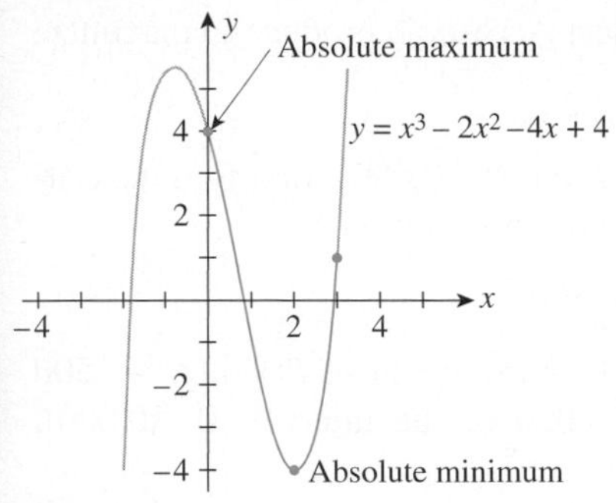

FIGURE 61
f has an absolute maximum value of 4 and an absolute minimum value of -4.

and it is equal to zero when $x = -\frac{2}{3}$ and $x = 2$. Since $x = -\frac{2}{3}$ lies outside the interval [0, 3], it is dropped from further consideration, and $x = 2$ is seen to be the sole critical number of f. Next, we evaluate $f(x)$ at the critical number of f as well as the endpoints of f, obtaining

$$f(0) = 4 \qquad f(2) = -4 \qquad f(3) = 1$$

From these results, we conclude that -4 is the absolute minimum value of f and 4 is the absolute maximum value of f. The graph of f, which appears in Figure 61, confirms our results. Observe that the absolute maximum of f occurs at the endpoint $x = 0$ of the interval [0, 3], while the absolute minimum of f occurs at $x = 2$, which lies in the interval (0, 3). ■

EXPLORING WITH TECHNOLOGY

Let $f(x) = x^3 - 2x^2 - 4x + 4$. (This is the function of Example 2.)

1. Use a graphing utility to plot the graph of f, using the viewing window $[0, 3] \times [-5, 5]$. Use TRACE to find the absolute extrema of f on the interval [0, 3] and thus verify the results obtained analytically in Example 2.
2. Plot the graph of f, using the viewing window $[-2, 1] \times [-5, 6]$. Use ZOOM and TRACE to find the absolute extrema of f on the interval $[-2, 1]$. Verify your results analytically.

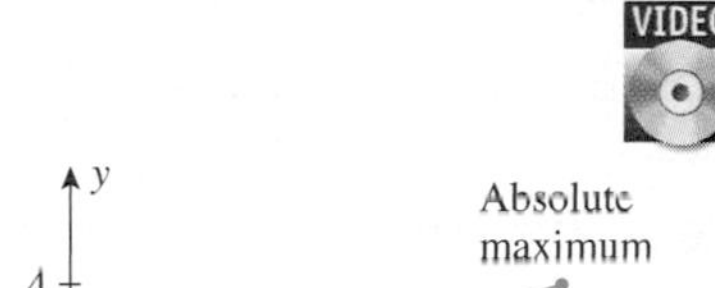

EXAMPLE 3 Find the absolute maximum and absolute minimum values of the function $f(x) = x^{2/3}$ on the interval $[-1, 8]$.

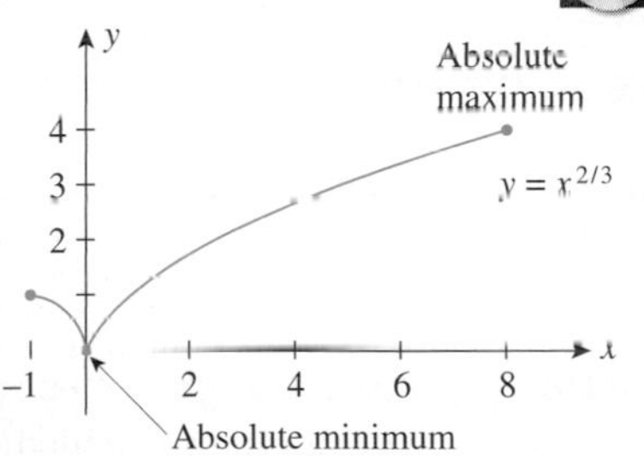

FIGURE 62
f has an absolute minimum value of $f(0) = 0$ and an absolute maximum value of $f(8) = 4$.

Solution The derivative of f is

$$f'(x) = \frac{2}{3}x^{-1/3} = \frac{2}{3x^{1/3}}$$

Note that f' is not defined at $x = 0$, is continuous everywhere else, and does not equal zero for all x. Therefore, 0 is the only critical number of f. Evaluating $f(x)$ at $x = -1$, 0, and 8, we obtain

$$f(-1) = 1 \qquad f(0) = 0 \qquad f(8) = 4$$

We conclude that the absolute minimum value of f is 0, attained at $x = 0$, and the absolute maximum value of f is 4, attained at $x = 8$ (Figure 62). ■

Many real-world applications call for finding the absolute maximum value or the absolute minimum value of a given function. For example, management is interested in finding what level of production will yield the maximum profit for a company; a farmer is interested in finding the right amount of fertilizer to maximize crop yield; a doctor is interested in finding the maximum concentration of a drug in a patient's body and the time at which it occurs; and an engineer is interested in finding the dimension of a container with a specified shape and volume that can be constructed at a minimum cost.

APPLIED EXAMPLE 4 Maximizing Profits Acrosonic's total profit (in dollars) from manufacturing and selling x units of their model F loudspeaker systems is given by

$$P(x) = -0.02x^2 + 300x - 200{,}000 \qquad (0 \leq x \leq 20{,}000)$$

How many units of the loudspeaker system must Acrosonic produce to maximize its profits?

Solution To find the absolute maximum of P on [0, 20,000], first find the critical points of P on the interval (0, 20,000). To do this, compute

$$P'(x) = -0.04x + 300$$

Solving the equation $P'(x) = 0$ gives $x = 7500$. Next, evaluate $P(x)$ at $x = 7500$ as well as the endpoints $x = 0$ and $x = 20{,}000$ of the interval [0, 20,000], obtaining

$$P(0) = -200{,}000$$
$$P(7500) = 925{,}000$$
$$P(20{,}000) = -2{,}200{,}000$$

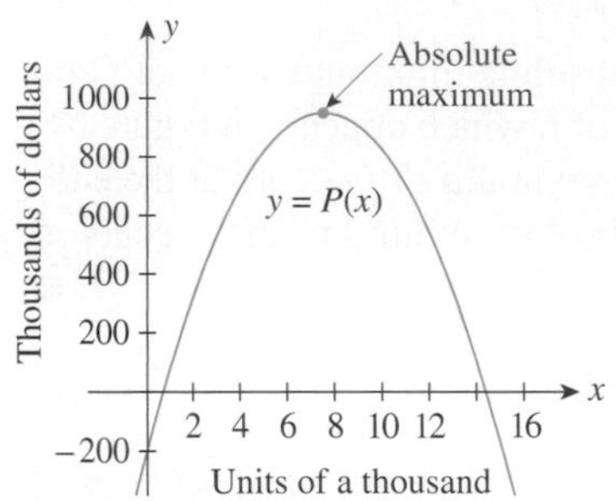

FIGURE 63
P has an absolute maximum at (7500, 925,000).

From these computations we see that the absolute maximum value of the function P is 925,000. Thus, by producing 7500 units, Acrosonic will realize a maximum profit of \$925,000. The graph of P is sketched in Figure 63. ■

EXPLORE & DISCUSS

Recall that the total profit function P is defined as $P(x) = R(x) - C(x)$, where R is the total revenue function, C is the total cost function, and x is the number of units of a product produced and sold. (Assume all derivatives exist.)

1. Show that at the level of production x_0 that yields the maximum profit for the company, the following two conditions are satisfied:

$$R'(x_0) = C'(x_0) \quad \text{and} \quad R''(x_0) < C''(x_0)$$

2. Interpret the two conditions in part 1 in economic terms and explain why they make sense.

APPLIED EXAMPLE 5 Trachea Contraction during a Cough When a person coughs, the trachea (windpipe) contracts, allowing air to be expelled at a maximum velocity. It can be shown that during a cough the velocity v of airflow is given by the function

$$v = f(r) = kr^2(R - r)$$

where r is the trachea's radius (in centimeters) during a cough, R is the trachea's normal radius (in centimeters), and k is a positive constant that depends on the length of the trachea. Find the radius r for which the velocity of airflow is greatest.

Solution To find the absolute maximum of f on $[0, R]$, first find the critical numbers of f on the interval $(0, R)$. We compute

$$\begin{aligned} f'(r) &= 2kr(R - r) - kr^2 \qquad \text{Use the product rule.} \\ &= -3kr^2 + 2kRr = kr(-3r + 2R) \end{aligned}$$

Setting $f'(r) = 0$ gives $r = 0$ or $r = \frac{2}{3}R$, and so $\frac{2}{3}R$ is the sole critical number of f ($r = 0$ is an endpoint). Evaluating $f(r)$ at $r = \frac{2}{3}R$, as well as at the endpoints

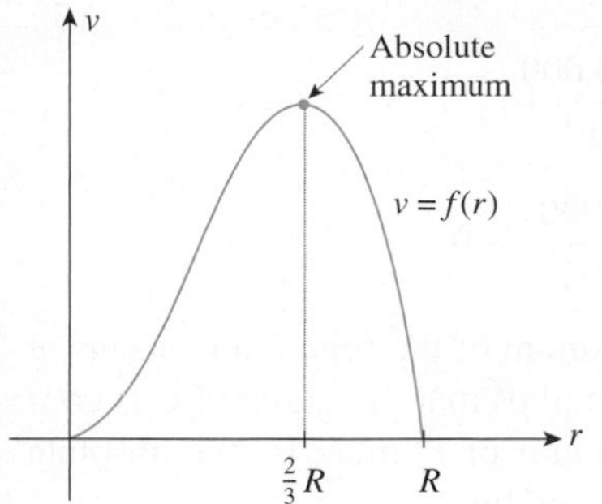

FIGURE 64
The velocity of airflow is greatest when the radius of the contracted trachea is $\frac{2}{3}R$.

$r = 0$ and $r = R$, we obtain

$$f(0) = 0$$
$$f\left(\frac{2}{3}R\right) = \frac{4k}{27}R^3$$
$$f(R) = 0$$

from which we deduce that the velocity of airflow is greatest when the radius of the contracted trachea is $\frac{2}{3}R$—that is, when the radius is contracted by approximately 33%. The graph of the function f is shown in Figure 64. ■

EXPLORE & DISCUSS

Prove that if a cost function $C(x)$ is concave upward $[C''(x) > 0]$, then the level of production that will result in the smallest average production cost occurs when

$$\overline{C}(x) = C'(x)$$

—that is, when the average cost $\overline{C}(x)$ is equal to the marginal cost $C'(x)$.

Hints:

1. Show that

$$\overline{C}'(x) = \frac{xC'(x) - C(x)}{x^2}$$

so that the critical number of the function $\overline{C}$ occurs when

$$xC'(x) - C(x) = 0$$

2. Show that at a critical number of $\overline{C}$

$$\overline{C}''(x) = \frac{C''(x)}{x}$$

Use the second derivative test to reach the desired conclusion.

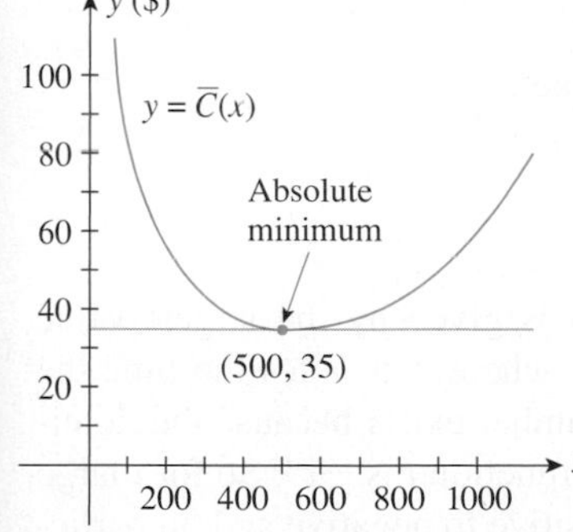

FIGURE 65
The minimum average cost is $35 per unit.

APPLIED EXAMPLE 6 Minimizing Average Cost The daily average cost function (in dollars per unit) of Elektra Electronics is given by

$$\overline{C}(x) = 0.0001x^2 - 0.08x + 40 + \frac{5000}{x} \qquad (x > 0)$$

where x stands for the number of programmable calculators that Elektra produces. Show that a production level of 500 units per day results in a minimum average cost for the company.

Solution The domain of the function $\overline{C}$ is the interval $(0, \infty)$, which is not closed. To solve the problem, we resort to the graphical method. Using the techniques of graphing from the last section, we sketch the graph of $\overline{C}$ (Figure 65). Now,

$$\overline{C}'(x) = 0.0002x - 0.08 - \frac{5000}{x^2}$$

Substituting the given value of x, 500, into $\overline{C}'(x)$ gives $\overline{C}'(500) = 0$, so 500 is a critical number of $\overline{C}$. Next,

$$\overline{C}''(x) = 0.0002 + \frac{10{,}000}{x^3}$$

Thus,

$$\overline{C}''(500) = 0.0002 + \frac{10{,}000}{(500)^3} > 0$$

and by the second derivative test, a relative minimum of the function $\overline{C}$ occurs at 500. Furthermore, $\overline{C}''(x) > 0$ for $x > 0$, which implies that the graph of $\overline{C}$ is concave upward everywhere, so the relative minimum of $\overline{C}$ must be the absolute minimum of $\overline{C}$. The minimum average cost is given by

$$\begin{aligned}\overline{C}(500) &= 0.0001(500)^2 - 0.08(500) + 40 + \frac{5000}{500} \\ &= 35\end{aligned}$$

or \$35 per unit. ■

EXPLORING WITH TECHNOLOGY

Refer to the preceding Explore & Discuss and Example 6.

1. Using a graphing utility, plot the graphs of

$$\overline{C}(x) = 0.0001x^2 - 0.08x + 40 + \frac{5000}{x}$$

$$C'(x) = 0.0003x^2 - 0.16x + 40$$

using the viewing window $[0, 1000] \times [0, 150]$.

Note: $C(x) = 0.0001x^3 - 0.08x^2 + 40x + 5000$ (Why?)

2. Find the point of intersection of the graphs of $\overline{C}$ and C' and thus verify the assertion in the Explore & Discuss for the special case studied in Example 6.

APPLIED EXAMPLE 7 Flight of a Rocket The altitude (in feet) of a rocket t seconds into flight is given by

$$s = f(t) = -t^3 + 96t^2 + 195t + 5 \qquad (t \geq 0)$$

a. Find the maximum altitude attained by the rocket.
b. Find the maximum velocity attained by the rocket.

Solution

a. The maximum altitude attained by the rocket is given by the largest value of the function f in the closed interval $[0, T]$, where T denotes the time the rocket impacts Earth. We know that such a number exists because the dominant term in the expression for the continuous function f is $-t^3$. So for t large enough, the value of $f(t)$ must change from positive to negative and, in particular, it must attain the value 0 for some T.

To find the absolute maximum of f, compute

$$\begin{aligned}f'(t) &= -3t^2 + 192t + 195 \\ &= -3(t - 65)(t + 1)\end{aligned}$$

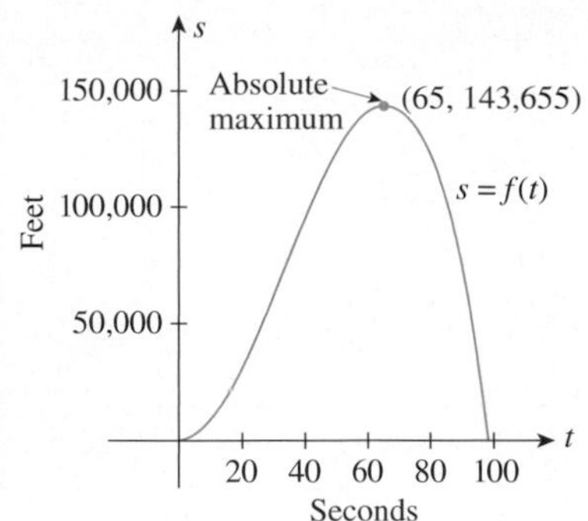

FIGURE 66
The maximum altitude of the rocket is 143,655 feet.

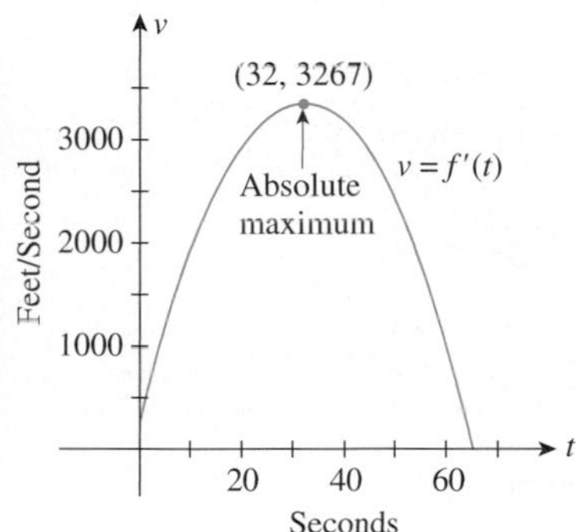

FIGURE 67
The maximum velocity of the rocket is 3267 feet per second.

and solve the equation $f'(t) = 0$, obtaining $t = -1$ and $t = 65$. Ignore $t = -1$ since it lies outside the interval $[0, T]$. This leaves the critical number 65 of f. Continuing, we compute

$$f(0) = 5 \qquad f(65) = 143{,}655 \qquad f(T) = 0$$

and conclude, accordingly, that the absolute maximum value of f is 143,655. Thus, the maximum altitude of the rocket is 143,655 feet, attained 65 seconds into flight. The graph of f is sketched in Figure 66.

b. To find the maximum velocity attained by the rocket, find the largest value of the function that describes the rocket's velocity at any time t—namely,

$$v = f'(t) = -3t^2 + 192t + 195 \qquad (t \geq 0)$$

We find the critical number of v by setting $v' = 0$. But

$$v' = -6t + 192$$

and the critical number of v is 32. Since

$$v'' = -6 < 0$$

the second derivative test implies that a relative maximum of v occurs at $t = 32$. Our computation has in fact clarified the property of the "velocity curve." Since $v'' < 0$ everywhere, the velocity curve is concave downward everywhere. With this observation, we assert that the relative maximum must in fact be the absolute maximum of v. The maximum velocity of the rocket is given by evaluating v at $t = 32$,

$$f'(32) = -3(32)^2 + 192(32) + 195$$

or 3267 feet per second. The graph of the velocity function v is sketched in Figure 67. ■

4.4 Self-Check Exercises

1. Let $f(x) = x - 2\sqrt{x}$
   a. Find the absolute extrema of f on the interval [0, 9].
   b. Find the absolute extrema of f.

2. Find the absolute extrema of $f(x) = 3x^4 + 4x^3 + 1$ on $[-2, 1]$.

3. The operating rate (expressed as a percent) of factories, mines, and utilities in a certain region of the country on the tth day of 2006 is given by the function

$$f(t) = 80 + \frac{1200t}{t^2 + 40{,}000} \qquad (0 \leq t \leq 250)$$

On which day of the first 250 days of 2006 was the manufacturing capacity operating rate highest?

Solutions to Self-Check Exercises 4.4 can be found on page 312.

4.4 Concept Questions

1. Explain the following terms: (a) absolute maximum and (b) absolute minimum.

2. Describe the procedure for finding the absolute extrema of a continuous function on a closed interval.

4.4 Exercises

In Exercises 1–8, you are given the graph of a function f defined on the indicated interval. Find the absolute maximum and the absolute minimum of f, if they exist.

1.

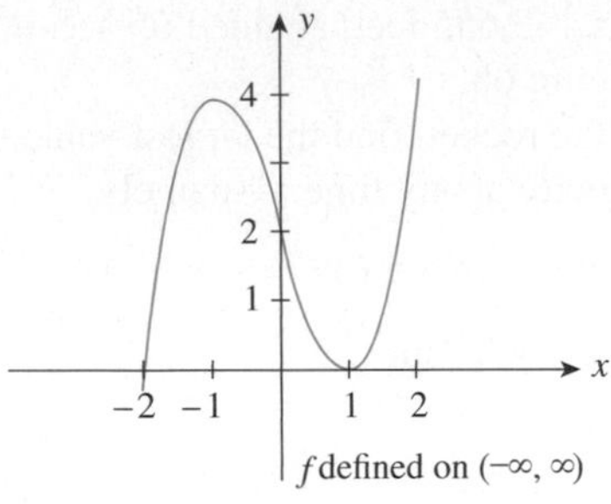

f defined on $(-\infty, \infty)$

2.

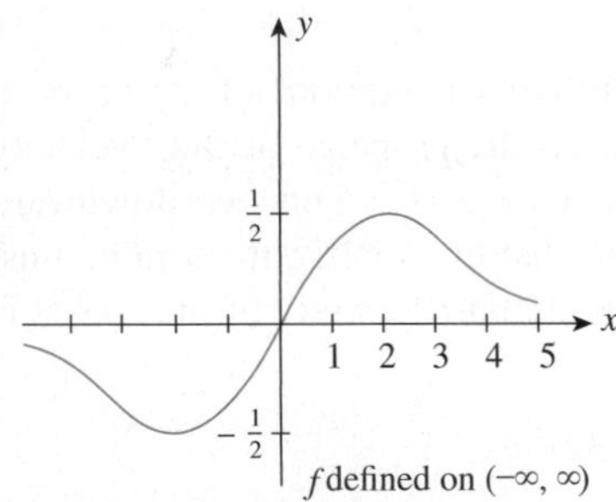

f defined on $(-\infty, \infty)$

3.

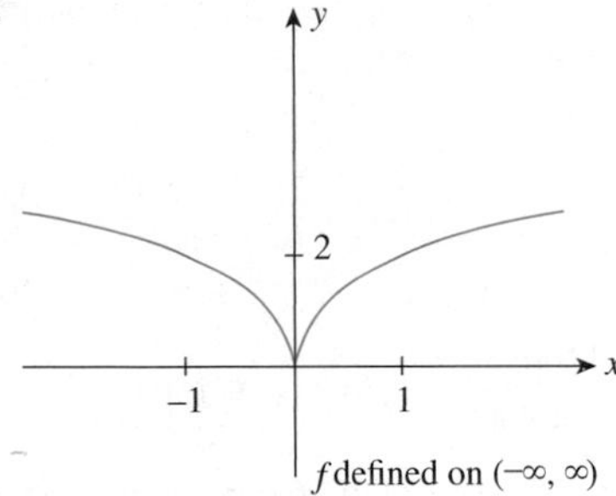

f defined on $(-\infty, \infty)$

4.

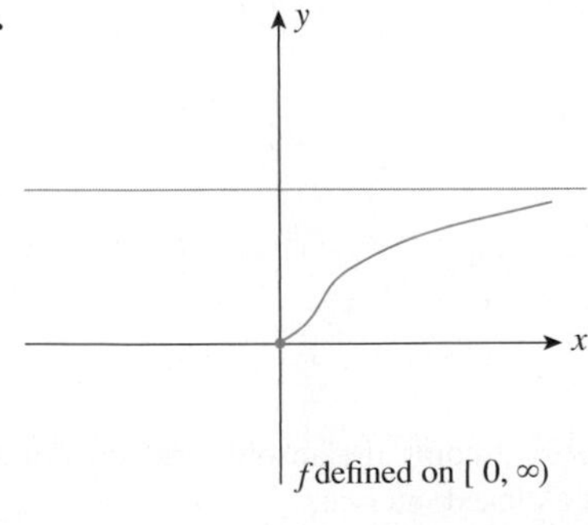

f defined on $[0, \infty)$

5.

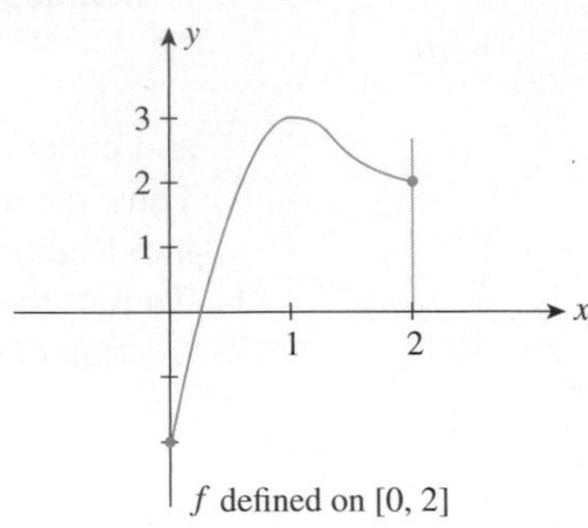

f defined on $[0, 2]$

6.

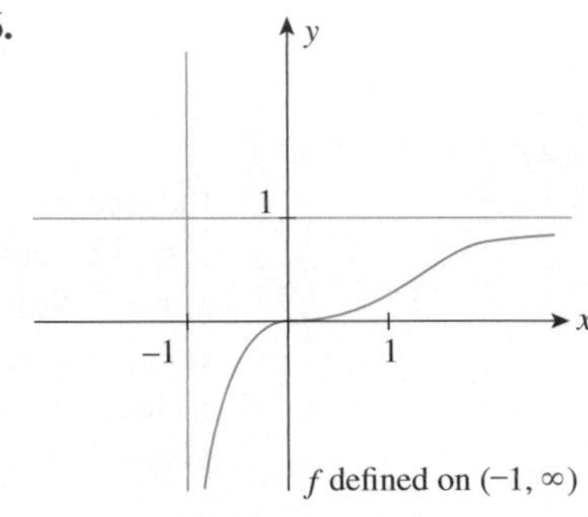

f defined on $(-1, \infty)$

7.

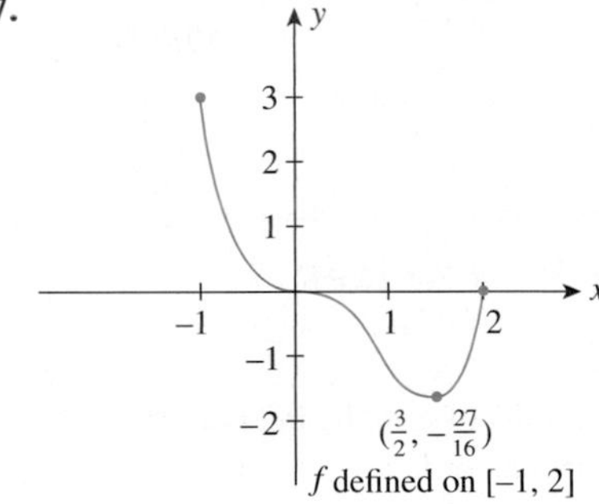

f defined on $[-1, 2]$

8.

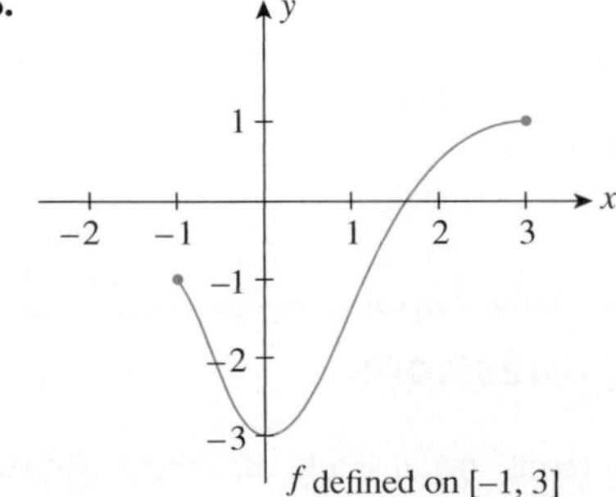

f defined on $[-1, 3]$

In Exercises 9–38, find the absolute maximum value and the absolute minimum value, if any, of each function.

9. $f(x) = 2x^2 + 3x - 4$
10. $g(x) = -x^2 + 4x + 3$
11. $h(x) = x^{1/3}$
12. $f(x) = x^{2/3}$
13. $f(x) = \dfrac{1}{1 + x^2}$
14. $f(x) = \dfrac{x}{1 + x^2}$
15. $f(x) = x^2 - 2x - 3$ on $[-2, 3]$
16. $g(x) = x^2 - 2x - 3$ on $[0, 4]$
17. $f(x) = -x^2 + 4x + 6$ on $[0, 5]$
18. $f(x) = -x^2 + 4x + 6$ on $[3, 6]$
19. $f(x) = x^3 + 3x^2 - 1$ on $[-3, 2]$
20. $g(x) = x^3 + 3x^2 - 1$ on $[-3, 1]$
21. $g(x) = 3x^4 + 4x^3$ on $[-2, 1]$
22. $f(x) = \frac{1}{2}x^4 - \frac{2}{3}x^3 - 2x^2 + 3$ on $[-2, 3]$
23. $f(x) = \dfrac{x + 1}{x - 1}$ on $[2, 4]$
24. $g(t) = \dfrac{t}{t - 1}$ on $[2, 4]$
25. $f(x) = 4x + \dfrac{1}{x}$ on $[1, 3]$
26. $f(x) = 9x - \dfrac{1}{x}$ on $[1, 3]$
27. $f(x) = \frac{1}{2}x^2 - 2\sqrt{x}$ on $[0, 3]$
28. $g(x) = \frac{1}{8}x^2 - 4\sqrt{x}$ on $[0, 9]$
29. $f(x) = \dfrac{1}{x}$ on $(0, \infty)$
30. $g(x) = \dfrac{1}{x + 1}$ on $(0, \infty)$
31. $f(x) = 3x^{2/3} - 2x$ on $[0, 3]$
32. $g(x) = x^2 + 2x^{2/3}$ on $[-2, 2]$
33. $f(x) = x^{2/3}(x^2 - 4)$ on $[-1, 2]$
34. $f(x) = x^{2/3}(x^2 - 4)$ on $[-1, 3]$
35. $f(x) = \dfrac{x}{x^2 + 2}$ on $[-1, 2]$
36. $f(x) = \dfrac{1}{x^2 + 2x + 5}$ on $[-2, 1]$
37. $f(x) = \dfrac{x}{\sqrt{x^2 + 1}}$ on $[-1, 1]$
38. $g(x) = x\sqrt{4 - x^2}$ on $[0, 2]$

39. A stone is thrown straight up from the roof of an 80-ft building. The height (in feet) of the stone at any time t (in seconds), measured from the ground, is given by

$$h(t) = -16t^2 + 64t + 80$$

What is the maximum height the stone reaches?

40. Maximizing Profits Lynbrook West, an apartment complex, has 100 two-bedroom units. The monthly profit (in dollars) realized from renting out x apartments is given by

$$P(x) = -10x^2 + 1760x - 50{,}000$$

To maximize the monthly rental profit, how many units should be rented out? What is the maximum monthly profit realizable?

41. Seniors in the Workforce The percent of men, age 65 and above, in the workforce from 1950 ($t = 0$) through 2000 ($t = 50$) is approximately

$$P(t) = 0.0135t^2 - 1.126t + 41.2 \qquad (0 \leq t \leq 50)$$

Show that the percent of men, age 65 and above, in the workforce in the period of time under consideration was smallest around mid-September 1991. What is that percent?
Source: U.S. Census Bureau

42. Flight of a Rocket The altitude (in feet) attained by a model rocket t sec into flight is given by the function

$$h(t) = -\frac{1}{3}t^3 + 4t^2 + 20t + 2 \qquad (t \geq 0)$$

Find the maximum altitude attained by the rocket.

43. Female Self-Employed Workforce Data show that the number of nonfarm, full-time, self-employed women can be approximated by

$$N(t) = 0.81t - 1.14\sqrt{t} + 1.53 \qquad (0 \leq t \leq 6)$$

where $N(t)$ is measured in millions and t is measured in 5-yr intervals, with $t = 0$ corresponding to the beginning of 1963. Determine the absolute extrema of the function N on the interval $[0, 6]$. Interpret your results.
Source: U.S. Department of Labor

44. Average Speed of a Vehicle on a Highway The average speed of a vehicle on a stretch of Route 134 between 6 a.m. and 10 a.m. on a typical weekday is approximated by the function

$$f(t) = 20t - 40\sqrt{t} + 50 \qquad (0 \leq t \leq 4)$$

where $f(t)$ is measured in miles per hour and t is measured in hours, with $t = 0$ corresponding to 6 a.m. At what time of the morning commute is the traffic moving at the slowest rate? What is the average speed of a vehicle at that time?

45. Maximizing Profits The management of Trappee and Sons, producers of the famous TexaPep hot sauce, estimate that their profit (in dollars) from the daily production and sale of x cases (each case consisting of 24 bottles) of the hot sauce is given by

$$P(x) = -0.000002x^3 + 6x - 400$$

What is the largest possible profit Trappee can make in 1 day?

46. **Maximizing Profits** The quantity demanded each month of the Walter Serkin recording of Beethoven's *Moonlight Sonata*, manufactured by Phonola Record Industries, is related to the price/compact disc. The equation

$$p = -0.00042x + 6 \qquad (0 \le x \le 12{,}000)$$

where p denotes the unit price in dollars and x is the number of discs demanded, relates the demand to the price. The total monthly cost (in dollars) for pressing and packaging x copies of this classical recording is given by

$$C(x) = 600 + 2x - 0.00002x^2 \qquad (0 \le x \le 20{,}000)$$

To maximize its profits, how many copies should Phonola produce each month?
Hint: The revenue is $R(x) = px$, and the profit is $P(x) = R(x) - C(x)$.

47. **Maximizing Profit** A manufacturer of tennis rackets finds that the total cost $C(x)$ (in dollars) of manufacturing x rackets/day is given by $C(x) = 400 + 4x + 0.0001x^2$. Each racket can be sold at a price of p dollars, where p is related to x by the demand equation $p = 10 - 0.0004x$. If all rackets that are manufactured can be sold, find the daily level of production that will yield a maximum profit for the manufacturer.

48. **Maximizing Profit** The weekly demand for the Pulsar 25-in. color console television is given by the demand equation

$$p = -0.05x + 600 \qquad (0 \le x \le 12{,}000)$$

where p denotes the wholesale unit price in dollars and x denotes the quantity demanded. The weekly total cost function associated with manufacturing these sets is given by

$$C(x) = 0.000002x^3 - 0.03x^2 + 400x + 80{,}000$$

where $C(x)$ denotes the total cost incurred in producing x sets. Find the level of production that will yield a maximum profit for the manufacturer.
Hint: Use the quadratic formula.

49. **Maximizing Profit** A division of Chapman Corporation manufactures a pager. The weekly fixed cost for the division is \$20,000, and the variable cost for producing x pagers/week is

$$V(x) = 0.000001x^3 - 0.01x^2 + 50x$$

dollars. The company realizes a revenue of

$$R(x) = -0.02x^2 + 150x \qquad (0 \le x \le 7500)$$

dollars from the sale of x pagers/week. Find the level of production that will yield a maximum profit for the manufacturer.
Hint: Use the quadratic formula.

50. **Minimizing Average Cost** Suppose the total cost function for manufacturing a certain product is $C(x) = 0.2(0.01x^2 + 120)$ dollars, where x represents the number of units produced. Find the level of production that will minimize the average cost.

51. **Minimizing Production Costs** The total monthly cost (in dollars) incurred by Cannon Precision Instruments for manufacturing x units of the model M1 camera is given by the function

$$C(x) = 0.0025x^2 + 80x + 10{,}000$$

a. Find the average cost function $\overline{C}$.
b. Find the level of production that results in the smallest average production cost.
c. Find the level of production for which the average cost is equal to the marginal cost.
d. Compare the result of part (c) with that of part (b).

52. **Minimizing Production Costs** The daily total cost (in dollars) incurred by Trappee and Sons for producing x cases of TexaPep hot sauce is given by the function

$$C(x) = 0.000002x^3 + 5x + 400$$

Using this function, answer the questions posed in Exercise 51.

53. **Maximizing Revenue** Suppose the quantity demanded per week of a certain dress is related to the unit price p by the demand equation $p = \sqrt{800 - x}$, where p is in dollars and x is the number of dresses made. To maximize the revenue, how many dresses should be made and sold each week?
Hint: $R(x) = px$.

54. **Maximizing Revenue** The quantity demanded each month of the Sicard wristwatch is related to the unit price by the equation

$$p = \frac{50}{0.01x^2 + 1} \qquad (0 \le x \le 20)$$

where p is measured in dollars and x is measured in units of a thousand. To yield a maximum revenue, how many watches must be sold?

55. **Oxygen Content of a Pond** When organic waste is dumped into a pond, the oxidation process that takes place reduces the pond's oxygen content. However, given time, nature will restore the oxygen content to its natural level. Suppose the oxygen content t days after organic waste has been dumped into the pond is given by

$$f(t) = 100\left[\frac{t^2 - 4t + 4}{t^2 + 4}\right] \qquad (0 \le t \le \infty)$$

percent of its normal level.
a. When is the level of oxygen content lowest?
b. When is the rate of oxygen regeneration greatest?

56. **Air Pollution** The amount of nitrogen dioxide, a brown gas that impairs breathing, present in the atmosphere on a certain May day in the city of Long Beach is approximated by

$$A(t) = \frac{136}{1 + 0.25(t - 4.5)^2} + 28 \qquad (0 \le t \le 11)$$

where $A(t)$ is measured in pollutant standard index (PSI) and t is measured in hours, with $t = 0$ corresponding to 7 a.m. Determine the time of day when the pollution is at its highest level.

57. **Maximizing Revenue** The average revenue is defined as the function

$$\overline{R}(x) = \frac{R(x)}{x} \qquad (x > 0)$$

Prove that if a revenue function $R(x)$ is concave downward $[R''(x) < 0]$, then the level of sales that will result in the largest average revenue occurs when $\overline{R}(x) = R'(x)$.

58. **Velocity of Blood** According to a law discovered by the 19th-century physician Jean Louis Marie Poiseuille, the velocity (in centimeters/second) of blood r cm from the central axis of an artery is given by

$$v(r) = k(R^2 - r^2)$$

where k is a constant and R is the radius of the artery. Show that the velocity of blood is greatest along the central axis.

59. **GDP of a Developing Country** A developing country's gross domestic product (GDP) from 1997 to 2005 is approximated by the function

$$G(t) = -0.2t^3 + 2.4t^2 + 60 \qquad (0 \le t \le 8)$$

where $G(t)$ is measured in billions of dollars and $t = 0$ corresponds to 1997. Show that the growth rate of the country's GDP was maximal in 2001.

60. **Crime Rates** The number of major crimes committed in the city of Bronxville between 1997 and 2004 is approximated by the function

$$N(t) = -0.1t^3 + 1.5t^2 + 100 \qquad (0 \le t \le 7)$$

where $N(t)$ denotes the number of crimes committed in year t ($t = 0$ corresponds to 1997). Enraged by the dramatic increase in the crime rate, the citizens of Bronxville, with the help of the local police, organized "Neighborhood Crime Watch" groups in early 2001 to combat this menace. Show that the growth in the crime rate was maximal in 2002, giving credence to the claim that the Neighborhood Crime Watch program was working.

61. **Foreign-Born Residents** The percent of foreign-born residents in the United States from 1910 through 2000 is approximated by the function

$$P(t) = 0.04363t^3 - 0.267t^2 - 1.59t + 14.7 \qquad (0 \le t \le 9)$$

where t is measured in decades, with $t = 0$ corresponding to 1910. Show that the percentage of foreign-born residents was lowest in early 1970.
Hint: Use the quadratic formula.
Source: Journal of American Medical Association

62. **Average Prices of Homes** The average annual price of single-family homes in Massachusetts between 1990 and 2002 is approximated by the function

$$P(t) = -0.183t^3 + 4.65t^2 - 17.3t + 200 \qquad (0 \le t \le 12)$$

where $P(t)$ is measured in thousands of dollars and t is measured in years, with $t = 0$ corresponding to 1990. In what year was the average annual price of single-family homes in Massachusetts lowest? What was the approximate lowest average annual price?
Hint: Use the quadratic formula.
Source: Massachusetts Association of Realtors

63. **Office Rents** After the economy softened, the sky-high office space rents of the late 1990s started to come down to earth. The function R gives the approximate price per square foot in dollars, $R(t)$, of prime space in Boston's Back Bay and Financial District from 1997 ($t = 0$) through 2002, where

$$R(t) = -0.711t^3 + 3.76t^2 + 0.2t + 36.5 \qquad (0 \le t \le 5)$$

Show that the office space rents peaked at about the middle of 2000. What was the highest office space rent during the period in question?
Hint: Use the quadratic formula.
Source: Meredith & Grew Inc./Oncor

64. **World Population** The total world population is forecast to be

$$P(t) = 0.00074t^3 - 0.0704t^2 + 0.89t + 6.04 \qquad (0 \le t \le 10)$$

in year t, where t is measured in decades with $t = 0$ corresponding to 2000 and $P(t)$ is measured in billions.

a. Show that the world population is forecast to peak around 2071.
Hint: Use the quadratic formula.

b. What will the population peak at?

Source: International Institute for Applied Systems Analysis

65. **Venture-Capital Investment** Venture-capital investment increased dramatically in the late 1990s but came to a screeching halt after the dot-com bust. The venture-capital investment (in billions of dollars) from 1995 ($t = 0$) through 2003 ($t = 8$) is approximated by the function

$$C(t) = \begin{cases} 0.6t^2 + 2.4t + 7.6 & 0 \le t < 3 \\ 3t^2 + 18.8t - 63.2 & 3 \le t < 5 \\ -3.3167t^3 + 80.1t^2 - 642.583t + 1730.8025 & 5 \le t \le 8 \end{cases}$$

a. In what year did venture-capital investment peak over the period under consideration? What was the amount of that investment?

b. In what year was the venture-capital investment lowest over this period? What was the amount of that investment?
Hint: Find the absolute extrema of C on each of the closed intervals [0, 3], [3, 5], and [5, 8].

Sources: Venture One; Ernst & Young

66. Energy Expended by a Fish It has been conjectured that a fish swimming a distance of L ft at a speed of v ft/sec relative to the water and against a current flowing at the rate of u ft/sec ($u < v$) expends a total energy given by

$$E(v) = \frac{aLv^3}{v - u}$$

where E is measured in foot-pounds (ft-lb) and a is a constant. Find the speed v at which the fish must swim in order to minimize the total energy expended. (*Note:* This result has been verified by biologists.)

67. Reaction to a Drug The strength of a human body's reaction R to a dosage D of a certain drug is given by

$$R = D^2\left(\frac{k}{2} - \frac{D}{3}\right)$$

where k is a positive constant. Show that the maximum reaction is achieved if the dosage is k units.

68. Refer to Exercise 67. Show that the rate of change in the reaction R with respect to the dosage D is maximal if $D = k/2$.

In Exercises 69–72, determine whether the statement is true or false. If it is true, explain why it is true. If it is false, give an example to show why it is false.

69. If f is defined on a closed interval $[a, b]$, then f has an absolute maximum value.

70. If f is continuous on an open interval (a, b), then f does not have an absolute minimum value.

71. If f is not continuous on the closed interval $[a, b]$, then f cannot have an absolute maximum value.

72. If $f''(x) < 0$ on (a, b) and $f'(c) = 0$ where $a < c < b$, then $f(c)$ is the absolute maximum value of f on $[a, b]$.

73. Let f be a constant function—that is, let $f(x) = c$, where c is some real number. Show that every number a gives rise to an absolute maximum and, at the same time, an absolute minimum of f.

74. Show that a polynomial function defined on the interval $(-\infty, \infty)$ cannot have both an absolute maximum and an absolute minimum unless it is a constant function.

75. One condition that must be satisfied before Theorem 3 (p. 301) is applicable is that the function f must be continuous on the closed interval $[a, b]$. Define a function f on the closed interval $[-1, 1]$ by

$$f(x) = \begin{cases} \frac{1}{x} & \text{if } x \in [-1, 1] \quad (x \neq 0) \\ 0 & \text{if } x = 0 \end{cases}$$

a. Show that f is not continuous at $x = 0$.
b. Show that $f(x)$ does not attain an absolute maximum or an absolute minimum on the interval $[-1, 1]$.
c. Confirm your results by sketching the function f.

76. One condition that must be satisfied before Theorem 3 (page 301) is applicable is that the interval on which f is defined must be a closed interval $[a, b]$. Define a function f on the *open* interval $(-1, 1)$ by $f(x) = x$. Show that f does not attain an absolute maximum or an absolute minimum on the interval $(-1, 1)$.
Hint: What happens to $f(x)$ if x is close to but not equal to $x = -1$? If x is close to but not equal to $x = 1$?

4.4 Solutions to Self-Check Exercises

1. a. The function f is continuous in its domain and differentiable in the interval $(0, 9)$. The derivative of f is

$$f'(x) = 1 - x^{-1/2} = \frac{x^{1/2} - 1}{x^{1/2}}$$

and it is equal to zero when $x = 1$. Evaluating $f(x)$ at the endpoints $x = 0$ and $x = 9$ and at the critical number 1 of f, we have

$$f(0) = 0 \qquad f(1) = -1 \qquad f(9) = 3$$

From these results, we see that -1 is the absolute minimum value of f and 3 is the absolute maximum value of f.

b. In this case, the domain of f is the interval $[0, \infty)$, which is not closed. Therefore, we resort to the graphic method. Using the techniques of graphing, we sketch in the accompanying figure the graph of f.

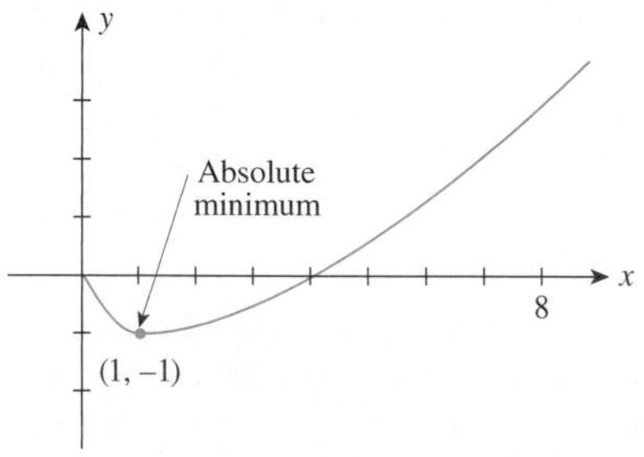

The graph of f shows that -1 is the absolute minimum value of f, but f has no absolute maximum since $f(x)$ increases without bound as x increases without bound.

2. The function f is continuous on the interval $[-2, 1]$. It is also differentiable on the open interval $(-2, 1)$. The derivative of f is

$$f'(x) = 12x^3 + 12x^2 = 12x^2(x + 1)$$

and it is continuous on $(-2, 1)$. Setting $f'(x) = 0$ gives -1 and 0 as critical numbers of f. Evaluating $f(x)$ at these critical numbers of f as well as at the endpoints of the interval $[-2, 1]$, we obtain

$$f(-2) = 17 \qquad f(-1) = 0 \qquad f(0) = 1 \qquad f(1) = 8$$

From these results, we see that 0 is the absolute minimum value of f and 17 is the absolute maximum value of f.

3. The problem is solved by finding the absolute maximum of the function f on $[0, 250]$. Differentiating $f(t)$, we obtain

$$f'(t) = \frac{(t^2 + 40{,}000)(1200) - 1200t(2t)}{(t^2 + 40{,}000)^2} = \frac{-1200(t^2 - 40{,}000)}{(t^2 + 40{,}000)^2}$$

Upon setting $f'(t) = 0$ and solving the resulting equation, we obtain $t = -200$ or 200. Since -200 lies outside the interval $[0, 250]$, we are interested only in the critical number 200 of f. Evaluating $f(t)$ at $t = 0$, $t = 200$, and $t = 250$, we find

$$f(0) = 80 \qquad f(200) = 83 \qquad f(250) = 82.93$$

We conclude that the manufacturing capacity operating rate was the highest on the 200th day of 2006—that is, a little past the middle of July 2006.

USING TECHNOLOGY

Finding the Absolute Extrema of a Function

Some graphing utilities have a function for finding the absolute maximum and the absolute minimum values of a continuous function on a closed interval. If your graphing utility has this capability, use it to work through the example and exercises of this section.

EXAMPLE 1 Let $f(x) = \dfrac{2x + 4}{(x^2 + 1)^{3/2}}$.

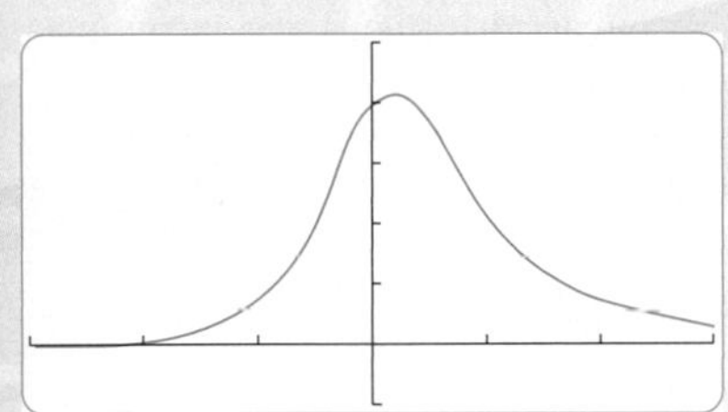

FIGURE T1
The graph of f in the viewing window $[-3, 3] \times [-1, 5]$

a. Use a graphing utility to plot the graph of f in the viewing window $[-3, 3] \times [-1, 5]$.

b. Find the absolute maximum and absolute minimum values of f on the interval $[-3, 3]$. Express your answers accurate to four decimal places.

Solution

a. The graph of f is shown in Figure T1.

b. Using the function on a graphing utility for finding the absolute minimum value of a continuous function on a closed interval, we find the absolute minimum value of f to be -0.0632. Similarly, using the function for finding the absolute maximum value, we find the absolute maximum value to be 4.1593. ■

Note Some graphing utilities will enable you to find the absolute minimum and absolute maximum values of a continuous function on a closed interval without having to graph the function. ■

(continued)

TECHNOLOGY EXERCISES

In Exercises 1–6, find the absolute maximum and the absolute minimum values of f in the given interval using the method of Example 1. Express your answers accurate to four decimal places.

1. $f(x) = 3x^4 - 4.2x^3 + 6.1x - 2$; $[-2, 3]$

2. $f(x) = 2.1x^4 - 3.2x^3 + 4.1x^2 + 3x - 4$; $[-1, 2]$

3. $f(x) = \dfrac{2x^3 - 3x^2 + 1}{x^2 + 2x - 8}$; $[-3, 1]$

4. $f(x) = \sqrt{x}(x^3 - 4)^2$; $[0.5, 1]$

5. $f(x) = \dfrac{x^3 - 1}{x^2}$; $[1, 3]$

6. $f(x) = \dfrac{x^3 - x^2 + 1}{x - 2}$; $[1, 3]$

7. Rate of Bank Failures The estimated rate at which banks were failing between 1982 and 1994 is given by

$$f(t) = 0.063447t^4 - 1.953283t^3 + 14.632576t^2 - 6.684704t + 47.458874 \qquad (0 \le t \le 12)$$

where $f(t)$ is the number of banks per year and t is measured in years, with $t = 0$ corresponding to the beginning of 1982.

a. Plot the graph of f in the viewing window $[0, 12] \times [0, 220]$.

b. What is the highest rate of bank failures during the period in question?

Source: Federal Deposit Insurance Corporation

8. Time on the Market The average number of days a single-family home remains for sale from listing to accepted offer (in the greater Boston area) is approximated by the function

$$f(t) = 0.0171911t^4 - 0.662121t^3 + 6.18083t^2 - 8.97086t + 53.3357 \qquad (0 \le t \le 10)$$

where t is measured in years, with $t = 0$ corresponding to the beginning of 1984.

a. Plot the graph of f in the viewing window $[0, 12] \times [0, 120]$.

b. Find the absolute maximum value and the absolute minimum value of f in the interval $[0, 12]$. Interpret your results.

Source: Greater Boston Real Estate Board—Multiple Listing Service

9. Use of Diesel Engines Diesel engines are popular in cars in Europe, where fuel prices are high. The percent of new vehicles in Western Europe equipped with diesel engines is approximated by the function

$$f(t) = 0.3t^4 - 2.58t^3 + 8.11t^2 - 7.71t + 23.75 \qquad (0 \le t \le 4)$$

where t is measured in years, with $t = 0$ corresponding to the beginning of 1996.

a. Use a graphing utility to sketch the graph of f on $[0, 4] \times [0, 40]$.

b. What was the lowest percent of new vehicles equipped with diesel engines for the period in question?

Source: German Automobile Industry Association

10. Federal Debt According to data obtained from the Congressional Budget Office, the national debt (in trillions of dollars) is given by the function

$$f(t) = 0.001532t^3 - 0.0588t^2 + 0.5208t + 2.55 \qquad (0 \le t \le 20)$$

where t is measured in years, with $t = 0$ corresponding to the beginning of 1990.

a. Sketch the graph of f, using the viewing window $[0, 20] \times [0, 4]$.

b. Estimate (to the nearest year) when the federal debt was at the highest level over the period under consideration. What was that level?

Source: Congressional Budget Office

11. Sickouts In a sickout by pilots of American Airlines in February 1999, the number of canceled flights from February 6 ($t = 0$) through February 14 ($t = 8$) is approximated by the function

$$N(t) = 1.2576t^4 - 26.357t^3 + 127.98t^2 + 82.3t + 43 \qquad (0 \le t \le 8)$$

where t is measured in days. The sickout ended after the union was threatened with millions of dollars of fines.

a. Show that the number of canceled flights was increasing at the fastest rate on February 8.

b. Estimate the maximum number of canceled flights in a day during the sickout.

Source: Associated Press

12. 401(k) Investors The average account balance of a 401(k) investor from 1996 through 2002 is approximately

$$A(t) = 0.28636t^4 - 3.4864t^3 + 11.689t^2 - 6.08t + 37.6 \qquad (0 \le t \le 6)$$

where $A(t)$ is measured in thousands of dollars and t is measured in years with $t = 0$ corresponding to the beginning of 1996.

a. Plot the graph of A, using the viewing window $[0, 6] \times [0, 60]$.

b. When was the average account balance lowest in the period under consideration? When was it highest?

c. What were the lowest average account balance and the highest average account balance during the period under consideration?

Source: Investment Company Institute

13. WHY SS BENEFITS MAY EXCEED PAYROLL TAXES Unless payroll taxes are increased significantly and/or benefits are scaled back drastically, it is only a matter of time before the current Social Security system goes broke. Data show that the assets of the system—the Social Security "trust fund"—may be approximated by

$$f(t) = -0.0129t^4 + 0.3087t^3 + 2.1760t^2 + 62.8466t + 506.2955 \qquad (0 \le t \le 35)$$

where $f(t)$ is measured in millions of dollars and t is measured in years, with $t = 0$ corresponding to the beginning of 1995.

a. Use a graphing utility to sketch the graph of f.

b. Based on this model, when will the Social Security system start to pay out more benefits than it gets in payroll taxes?

Source: Board of Trustees, Social Security Administration

14. FOOD STAMP RECIPIENTS The number of people in the United States receiving food stamps from 1988 through 1998 is approximated by the function

$$f(t) = 0.00944t^4 - 0.2311t^3 + 1.516t^2 - 1.35t + 18.6 \qquad (0 \le t \le 10)$$

where $f(t)$ is measured in millions of people and t is measured in years, with $t = 0$ corresponding to the fiscal year ending in September 1988.

a. Plot the graph of f, using the viewing window $[0, 10] \times [0, 28]$.

b. Based on this model, in what year did the number of food stamps recipients peak?

Source: U.S. Department of Agriculture

4.5 Optimization II

Section 4.4 outlined how to find the solution to certain optimization problems in which the objective function is given. In this section, we consider problems in which we are required to first find the appropriate function to be optimized. The following guidelines will be useful for solving these problems.

Guidelines for Solving Optimization Problems

1. Assign a letter to each variable mentioned in the problem. If appropriate, draw and label a figure.
2. Find an expression for the quantity to be optimized.
3. Use the conditions given in the problem to write the quantity to be optimized as a function f of *one* variable. Note any restrictions to be placed on the domain of f from physical considerations of the problem.
4. Optimize the function f over its domain using the methods of Section 4.4.

Note In carrying out step 4, remember that if the function f to be optimized is continuous on a closed interval, then the absolute maximum and absolute minimum of f are, respectively, the largest and smallest values of $f(x)$ on the set composed of the critical numbers of f and the endpoints of the interval. If the domain of f is not a closed interval, then we resort to the graphical method. ■

Maximization Problems

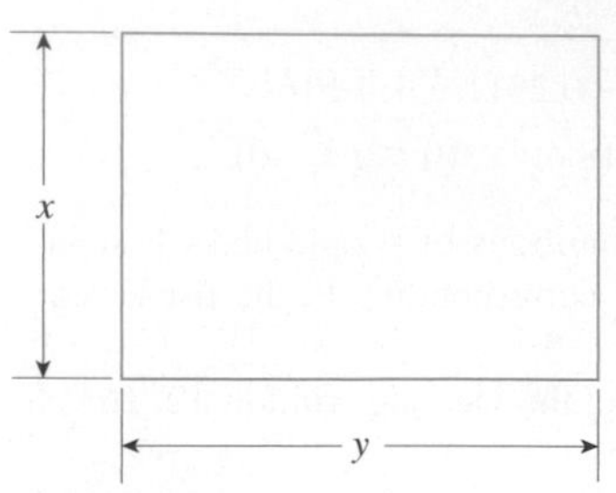

FIGURE 68
What is the maximum rectangular area that can be enclosed with 50 feet of fencing?

APPLIED EXAMPLE 1 Fencing a Garden A man wishes to have a rectangular-shaped garden in his backyard. He has 50 feet of fencing with which to enclose his garden. Find the dimensions for the largest garden he can have if he uses all of the fencing.

Solution

Step 1 Let x and y denote the dimensions (in feet) of two adjacent sides of the garden (Figure 68) and let A denote its area.

Step 2 The area of the garden

$$A = xy \tag{1}$$

is the quantity to be maximized.

Step 3 The perimeter of the rectangle, $(2x + 2y)$ feet, must equal 50 feet. Therefore, we have the equation

$$2x + 2y = 50$$

Next, solving this equation for y in terms of x yields

$$y = 25 - x \tag{2}$$

which, when substituted into Equation (1), gives

$$\begin{aligned} A &= x(25 - x) \\ &= -x^2 + 25x \end{aligned}$$

(Remember, the function to be optimized must involve just one variable.) Since the sides of the rectangle must be nonnegative, we must have $x \geq 0$ and $y = 25 - x \geq 0$; that is, we must have $0 \leq x \leq 25$. Thus, the problem is reduced to that of finding the absolute maximum of $A = f(x) = -x^2 + 25x$ on the closed interval $[0, 25]$.

Step 4 Observe that f is continuous on $[0, 25]$, so the absolute maximum value of f must occur at the endpoint(s) of the interval or at the critical number(s) of f. The derivative of the function A is given by

$$A' = f'(x) = -2x + 25$$

Setting $A' = 0$ gives

$$-2x + 25 = 0$$

or 12.5, as the critical number of A. Next, we evaluate the function $A = f(x)$ at $x = 12.5$ and at the endpoints $x = 0$ and $x = 25$ of the interval $[0, 25]$, obtaining

$$f(0) = 0 \qquad f(12.5) = 156.25 \qquad f(25) = 0$$

We see that the absolute maximum value of the function f is 156.25. From Equation (2) we see that $y = 12.5$ when $x = 12.5$. Thus, the garden of maximum area (156.25 square feet) is a square with sides of length 12.5 feet. ■

APPLIED EXAMPLE 2 Packaging By cutting away identical squares from each corner of a rectangular piece of cardboard and folding up the resulting flaps, the cardboard may be turned into an open box. If the cardboard is 16 inches long and 10 inches wide, find the dimensions of the box that will yield the maximum volume.

Solution

Step 1 Let x denote the length (in inches) of one side of each of the identical squares to be cut out of the cardboard (Figure 69) and let V denote the volume of the resulting box.

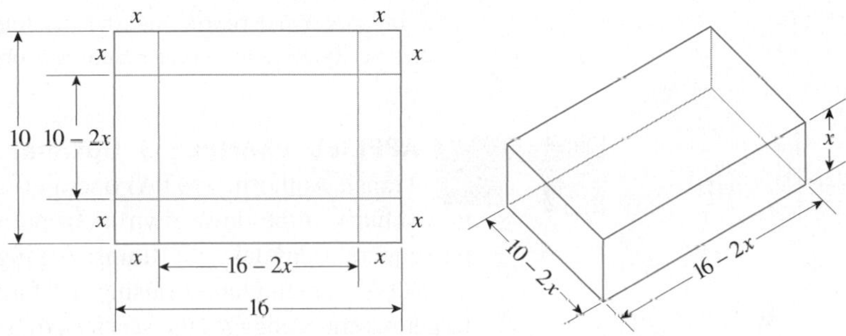

FIGURE 69
The dimensions of the open box are $(16 - 2x)''$ by $(10 - 2x)''$ by x''.

Step 2 The dimensions of the box are $(16 - 2x)$ inches by $(10 - 2x)$ inches by x inches. Therefore, its volume (in cubic inches),

$$V = (16 - 2x)(10 - 2x)x$$
$$= 4(x^3 - 13x^2 + 40x) \quad \text{Expand the expression.}$$

is the quantity to be maximized.

Step 3 Since each side of the box must be nonnegative, x must satisfy the inequalities $x \geq 0$, $16 - 2x \geq 0$, and $10 - 2x \geq 0$. This set of inequalities is satisfied if $0 \leq x \leq 5$. Thus, the problem at hand is equivalent to that of finding the absolute maximum of

$$V = f(x) = 4(x^3 - 13x^2 + 40x)$$

on the closed interval $[0, 5]$.

Step 4 Observe that f is continuous on $[0, 5]$, so the absolute maximum value of f must be attained at the endpoint(s) or at the critical number(s) of f.

Differentiating $f(x)$, we obtain

$$f'(x) = 4(3x^2 - 26x + 40)$$
$$= 4(3x - 20)(x - 2)$$

Upon setting $f'(x) = 0$ and solving the resulting equation for x, we obtain $x = \frac{20}{3}$ or $x = 2$. Since $\frac{20}{3}$ lies outside the interval $[0, 5]$, it is no longer considered, and we are interested only in the critical number 2 of f. Next, evaluating $f(x)$ at $x = 0$, $x = 5$ (the endpoints of the interval $[0, 5]$), and $x = 2$, we obtain

$$f(0) = 0 \qquad f(2) = 144 \qquad f(5) = 0$$

Thus, the volume of the box is maximized by taking $x = 2$. The dimensions of the box are $12'' \times 6'' \times 2''$, and the volume is 144 cubic inches. ■

EXPLORING WITH TECHNOLOGY

Refer to Example 2.

1. Use a graphing utility to plot the graph of

$$f(x) = 4(x^3 - 13x^2 + 40x)$$

using the viewing window $[0, 5] \times [0, 150]$. Explain what happens to $f(x)$ as x increases from $x = 0$ to $x = 5$ and give a physical interpretation.

2. Using ZOOM and TRACE, find the absolute maximum of f on the interval $[0, 5]$ and thus verify the solution for Example 2 obtained analytically.

APPLIED EXAMPLE 3 Optimal Subway Fare A city's Metropolitan Transit Authority (MTA) operates a subway line for commuters from a certain suburb to the downtown metropolitan area. Currently, an average of 6000 passengers a day take the trains, paying a fare of \$3.00 per ride. The board of the MTA, contemplating raising the fare to \$3.50 per ride in order to generate a larger revenue, engages the services of a consulting firm. The firm's study reveals that for each \$.50 increase in fare, the ridership will be reduced by an average of 1000 passengers a day. Thus, the consulting firm recommends that MTA stick to the current fare of \$3.00 per ride, which already yields a maximum revenue. Show that the consultants are correct.

Solution

Step 1 Let x denote the number of passengers per day, p denote the fare per ride, and R be MTA's revenue.

Step 2 To find a relationship between x and p, observe that the given data imply that when $x = 6000$, $p = 3$, and when $x = 5000$, $p = 3.50$. Therefore, the points (6000, 3) and (5000, 3.50) lie on a straight line. (Why?) To find the linear relationship between p and x, use the point-slope form of the equation of a straight line. Now, the slope of the line is

$$m = \frac{3.50 - 3}{5000 - 6000} = -0.0005$$

Therefore, the required equation is

$$\begin{aligned} p - 3 &= -0.0005(x - 6000) \\ &= -0.0005x + 3 \\ p &= -0.0005x + 6 \end{aligned}$$

Therefore, the revenue

$$R = f(x) = xp = -0.0005x^2 + 6x \qquad \text{Number of riders} \times \text{unit fare}$$

is the quantity to be maximized.

Step 3 Since both p and x must be nonnegative, we see that $0 \leq x \leq 12{,}000$, and the problem is that of finding the absolute maximum of the function f on the closed interval $[0, 12{,}000]$.

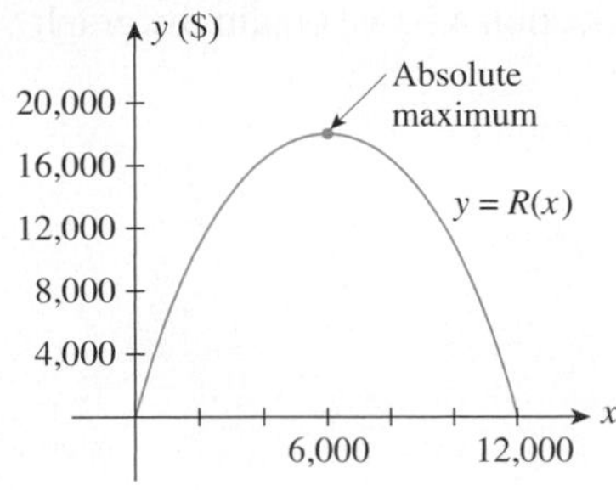

FIGURE 70
f has an absolute maximum of 18,000 when $x = 6000$.

Step 4 Observe that f is continuous on $[0, 12{,}000]$. To find the critical number of R, we compute

$$f'(x) = -0.001x + 6$$

and set it equal to zero, giving $x = 6000$. Evaluating the function f at $x = 6000$, as well as at the endpoints $x = 0$ and $x = 12{,}000$, yields

$$f(0) = 0$$
$$f(6000) = 18{,}000$$
$$f(12{,}000) = 0$$

We conclude that a maximum revenue of \$18,000 per day is realized when the ridership is 6000 per day. The optimum price of the fare per ride is therefore \$3.00, as recommended by the consultants. The graph of the revenue function R is shown in Figure 70. ■

Minimization Problems

APPLIED EXAMPLE 4 Packaging Betty Moore Company requires that its corned beef hash containers have a capacity of 54 cubic inches, have the shape of right circular cylinders, and be made of aluminum. Determine the radius and height of the container that requires the least amount of metal.

Solution

Step 1 Let the radius and height of the container be r and h inches, respectively, and let S denote the surface area of the container (Figure 71).

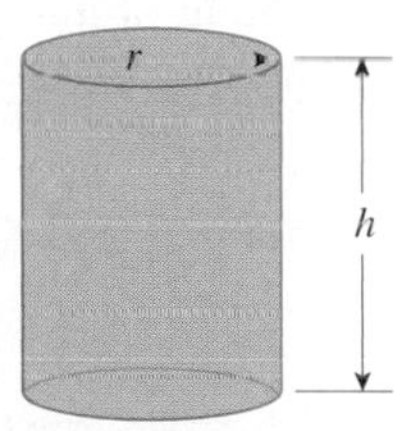

FIGURE 71
We want to minimize the amount of material used to construct the container.

Step 2 The amount of aluminum used to construct the container is given by the total surface area of the cylinder. Now, the area of the base and the top of the cylinder are each πr^2 square inches and the area of the side is $2\pi rh$ square inches. Therefore,

$$S = 2\pi r^2 + 2\pi rh \tag{3}$$

is the quantity to be minimized.

Step 3 The requirement that the volume of a container be 54 cubic inches implies that

$$\pi r^2 h = 54 \tag{4}$$

Solving Equation (4) for h, we obtain

$$h = \frac{54}{\pi r^2} \tag{5}$$

which, when substituted into (3), yields

$$S = 2\pi r^2 + 2\pi r\left(\frac{54}{\pi r^2}\right)$$
$$= 2\pi r^2 + \frac{108}{r}$$

Clearly, the radius r of the container must satisfy the inequality $r > 0$. The problem now is reduced to finding the absolute minimum of the function $S = f(r)$ on the interval $(0, \infty)$.

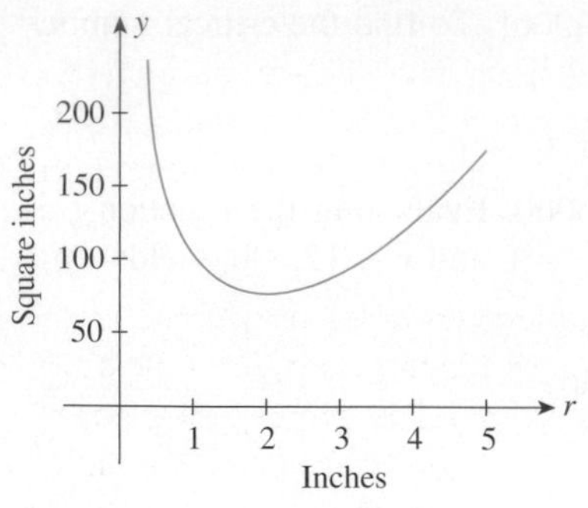

FIGURE 72
The total surface area of the right cylindrical container is graphed as a function of r.

Step 4 Using the curve-sketching techniques of Section 4.3, we obtain the graph of f in Figure 72.

To find the critical number of f, we compute

$$S' = 4\pi r - \frac{108}{r^2}$$

and solve the equation $S' = 0$ for r:

$$4\pi r - \frac{108}{r^2} = 0$$

$$4\pi r^3 - 108 = 0$$

$$r^3 = \frac{27}{\pi}$$

$$r = \frac{3}{\sqrt[3]{\pi}} \approx 2 \tag{6}$$

Next, let's show that this value of r gives rise to the absolute minimum of f. To show this, we first compute

$$S'' = 4\pi + \frac{216}{r^3}$$

Since $S'' > 0$ for $r = 3/\sqrt[3]{\pi}$, the second derivative test implies that the value of r in Equation (6) gives rise to a relative minimum of f. Finally, this relative minimum of f is also the absolute minimum of f since f is always concave upward ($S'' > 0$ for all $r > 0$). To find the height of the given container, we substitute the value of r given in (6) into (5). Thus,

$$h = \frac{54}{\pi r^2} = \frac{54}{\pi\left(\frac{3}{\pi^{1/3}}\right)^2}$$

$$= \frac{54\pi^{2/3}}{(\pi)9}$$

$$= \frac{6}{\pi^{1/3}} = \frac{6}{\sqrt[3]{\pi}}$$

$$= 2r$$

We conclude that the required container has a radius of approximately 2 inches and a height of approximately 4 inches, or twice the size of the radius. ■

An Inventory Problem

One problem faced by many companies is that of controlling the inventory of goods carried. Ideally, the manager must ensure that the company has sufficient stock to meet customer demand at all times. At the same time, she must make sure that this is accomplished without overstocking (incurring unnecessary storage costs) and also without having to place orders too frequently (incurring reordering costs).

APPLIED EXAMPLE 5 Inventory Control and Planning Dixie Import-Export is the sole agent for the Excalibur 250-cc motorcycle. Management estimates that the demand for these motorcycles is 10,000 per year and that they will sell at a uniform rate throughout the year. The cost incurred in ordering each shipment of motorcycles is \$10,000, and the cost per year of storing each motorcycle is \$200.

Dixie's management faces the following problem: Ordering too many motorcycles at one time ties up valuable storage space and increases the storage cost. On the other hand, placing orders too frequently increases the ordering costs. How large should each order be, and how often should orders be placed, to minimize ordering and storage costs?

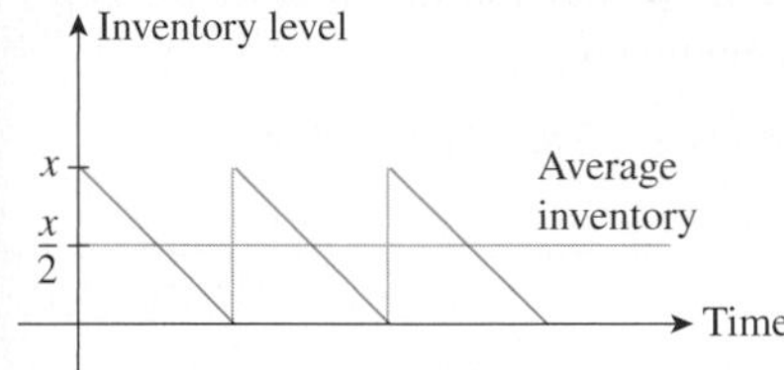

FIGURE 73
As each lot is depleted, the new lot arrives. The average inventory level is $x/2$ if x is the lot size.

Solution Let x denote the number of motorcycles in each order (the lot size). Then, assuming that each shipment arrives just as the previous shipment has been sold, the average number of motorcycles in storage during the year is $x/2$. You can see that this is the case by examining Figure 73. Thus, Dixie's storage cost for the year is given by $200(x/2)$, or $100x$ dollars.

Next, since the company requires 10,000 motorcycles for the year and since each order is for x motorcycles, the number of orders required is

$$\frac{10{,}000}{x}$$

This gives an ordering cost of

$$10{,}000\left(\frac{10{,}000}{x}\right) = \frac{100{,}000{,}000}{x}$$

dollars for the year. Thus, the total yearly cost incurred by Dixie, which includes the ordering and storage costs attributed to the sale of these motorcycles, is given by

$$C(x) = 100x + \frac{100{,}000{,}000}{x}$$

The problem is reduced to finding the absolute minimum of the function C in the interval $(0, 10{,}000]$. To accomplish this, we compute

$$C'(x) = 100 - \frac{100{,}000{,}000}{x^2}$$

Setting $C'(x) = 0$ and solving the resulting equation, we obtain $x = \pm 1000$. Since the number -1000 is outside the domain of the function C, it is rejected, leaving 1000 as the only critical number of C. Next, we find

$$C''(x) = \frac{200{,}000{,}000}{x^3}$$

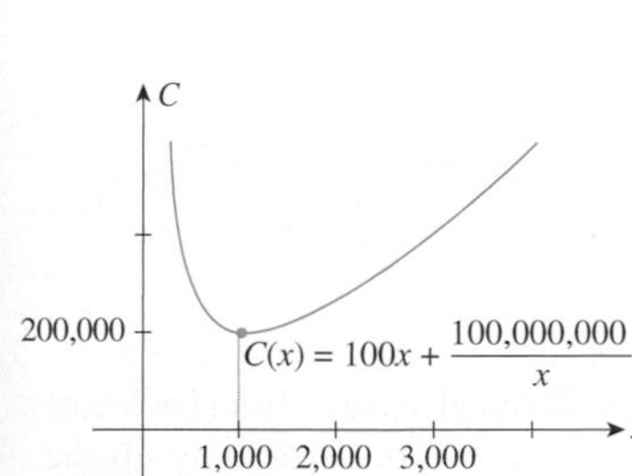

FIGURE 74
C has an absolute minimum at (1000, 200,000).

Since $C''(1000) > 0$, the second derivative test implies that the critical number 1000 is a relative minimum of the function C (Figure 74). Also, since $C''(x) > 0$ for all x in $(0, 10{,}000]$, the function C is concave upward everywhere so that $x = 1000$ also gives the absolute minimum of C. Thus, to minimize the ordering and storage costs, Dixie should place 10,000/1000, or 10, orders a year, each for a shipment of 1000 motorcycles. ■

4.5 Self-Check Exercises

1. A man wishes to have an enclosed vegetable garden in his backyard. If the garden is to be a rectangular area of 300 ft^2, find the dimensions of the garden that will minimize the amount of fencing material needed.

2. The demand for the Super Titan tires is 1,000,000/year. The setup cost for each production run is \$4000, and the manufacturing cost is \$20/tire. The cost of storing each tire over the year is \$2. Assuming uniformity of demand throughout the year and instantaneous production, determine how many tires should be manufactured per production run in order to keep the production cost to a minimum.

Solutions to Self-Check Exercises 4.5 can be found on page 326.

4.5 Concept Questions

1. If the domain of a function f is not a closed interval, how would you find the absolute extrema of f, if they exist?

2. Refer to Example 4 (page 319). In the solution given in the example, we solved for h in terms of r, resulting in a function of r, which we then optimized with respect to r. Write S in terms of h and re-solve the problem. Which choice is better?

4.5 Exercises

1. Enclosing the Largest Area The owner of the Rancho Los Feliz has 3000 yd of fencing material with which to enclose a rectangular piece of grazing land along the straight portion of a river. If fencing is not required along the river, what are the dimensions of the largest area that he can enclose? What is this area?

2. Enclosing the Largest Area Refer to Exercise 1. As an alternative plan, the owner of the Rancho Los Feliz might use the 3000 yd of fencing material to enclose the rectangular piece of grazing land along the straight portion of the river and then subdivide it by means of a fence running parallel to the sides. Again, no fencing is required along the river. What are the dimensions of the largest area that can be enclosed? What is this area? (See the accompanying figure.)

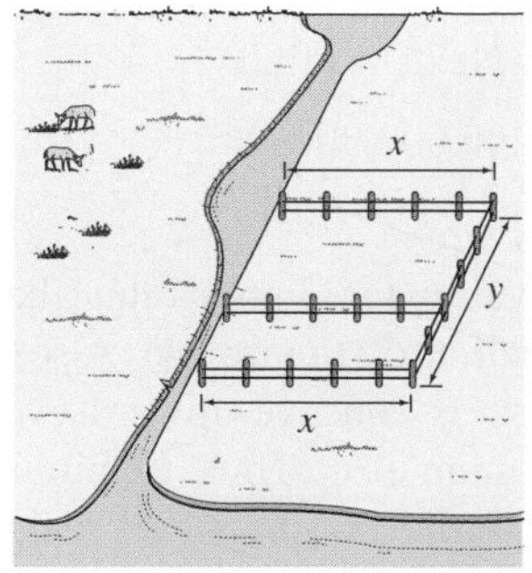

3. Minimizing Construction Costs The management of the UNICO department store has decided to enclose an 800-ft^2 area outside the building for displaying potted plants and flowers. One side will be formed by the external wall of the store, two sides will be constructed of pine boards, and the fourth side will be made of galvanized steel fencing material. If the pine board fencing costs \$6/running foot and the steel fencing costs \$3/running foot, determine the dimensions of the enclosure that can be erected at minimum cost.

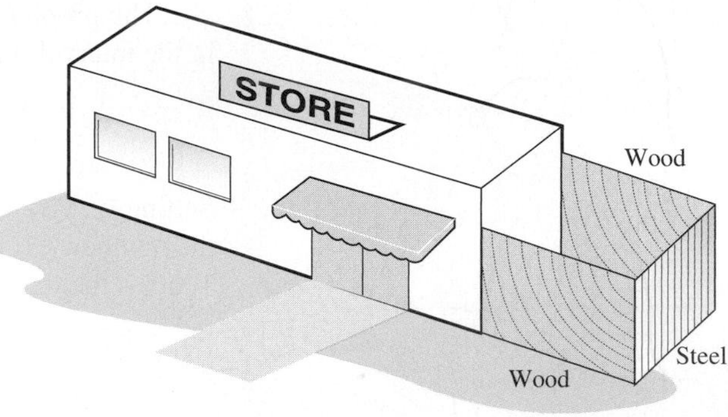

4. Packaging By cutting away identical squares from each corner of a rectangular piece of cardboard and folding up the resulting flaps, an open box may be made. If the cardboard is 15 in. long and 8 in. wide, find the dimensions of the box that will yield the maximum volume.

5. Metal Fabrication If an open box is made from a tin sheet 8 in. square by cutting out identical squares from each corner

and bending up the resulting flaps, determine the dimensions of the largest box that can be made.

6. **Minimizing Packaging Costs** If an open box has a square base and a volume of 108 in.^{3} and is constructed from a tin sheet, find the dimensions of the box, assuming a minimum amount of material is used in its construction.

7. **Minimizing Packaging Costs** What are the dimensions of a closed rectangular box that has a square cross section, a capacity of 128 in.^{3}, and is constructed using the least amount of material?

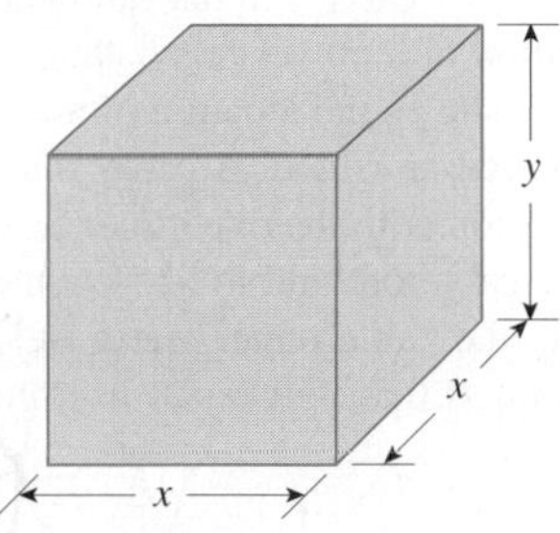

8. **Minimizing Packaging Costs** A rectangular box is to have a square base and a volume of 20 ft^3. If the material for the base costs 30¢/square foot, the material for the sides costs 10¢/square foot, and the material for the top costs 20¢/square foot, determine the dimensions of the box that can be constructed at minimum cost. (Refer to the figure for Exercise 7.)

9. **Parcel Post Regulations** Postal regulations specify that a parcel sent by parcel post may have a combined length and girth of no more than 108 in. Find the dimensions of a rectangular package that has a square cross section and the largest volume that may be sent through the mail. What is the volume of such a package?
Hint: The length plus the girth is $4x + h$ (see the accompanying figure).

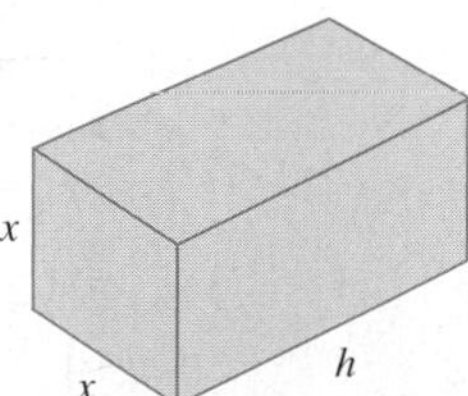

10. **Book Design** A book designer has decided that the pages of a book should have 1-in. margins at the top and bottom and $\frac{1}{2}$-in. margins on the sides. She further stipulated that each page should have an area of 50 in.^{2} (see the accompanying figure). Determine the page dimensions that will result in the maximum printed area on the page.

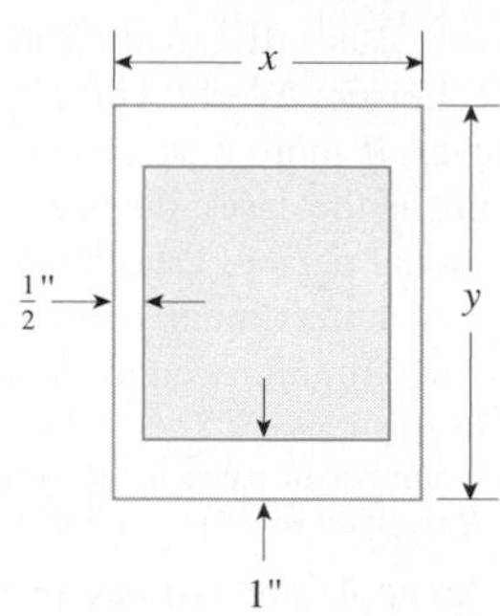

11. **Parcel Post Regulations** Postal regulations specify that a parcel sent by parcel post may have a combined length and girth of no more than 108 in. Find the dimensions of the cylindrical package of greatest volume that may be sent through the mail. What is the volume of such a package? Compare with Exercise 9.
Hint: The length plus the girth is $2\pi r + l$.

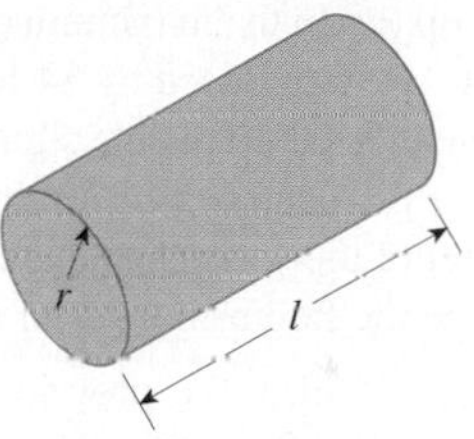

12. **Minimizing Costs** For its beef stew, Betty Moore Company uses aluminum containers that have the form of right circular cylinders. Find the radius and height of a container if it has a capacity of 36 in.^{3} and is constructed using the least amount of metal.

13. **Product Design** The cabinet that will enclose the Acrosonic model D loudspeaker system will be rectangular and will have an internal volume of 2.4 ft^3. For aesthetic reasons, it has been decided that the height of the cabinet is to be 1.5 times its width. If the top, bottom, and sides of the cabinet are constructed of veneer costing 40¢/square foot and the front (ignore the cutouts in the baffle) and rear are constructed of particle board costing 20¢/square foot, what are the dimensions of the enclosure that can be constructed at a minimum cost?

14. **Designing a Norman Window** A Norman window has the shape of a rectangle surmounted by a semicircle (see the accompanying figure). If a Norman window is to have a perimeter of 28 ft, what should its dimensions be in order to allow the maximum amount of light through the window?

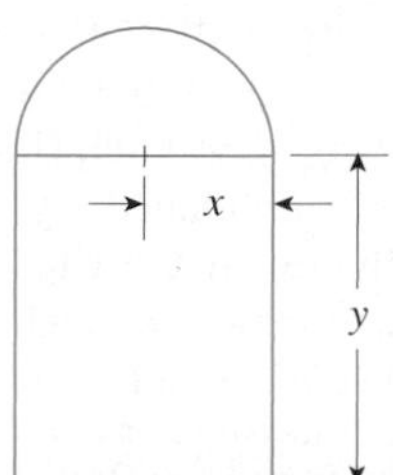

15. **Optimal Charter-Flight Fare** If exactly 200 people sign up for a charter flight, Leisure World Travel Agency charges \$300/person. However, if more than 200 people sign up for the flight (assume this is the case), then each fare is reduced by \$1 for each additional person. Determine how many passengers will result in a maximum revenue for the travel agency. What is the maximum revenue? What would be the fare per passenger in this case?
Hint: Let x denote the number of passengers above 200. Show that the revenue function R is given by $R(x) = (200 + x)(300 - x)$.

16. **Maximizing Yield** An apple orchard has an average yield of 36 bushels of apples/tree if tree density is 22 trees/acre. For each unit increase in tree density, the yield decreases by 2 bushels/tree. How many trees should be planted in order to maximize the yield?

17. **Charter Revenue** The owner of a luxury motor yacht that sails among the 4000 Greek islands charges \$600/person/day if exactly 20 people sign up for the cruise. However, if more than 20 people sign up (up to the maximum capacity of 90) for the cruise, then each fare is reduced by \$4 for each additional passenger. Assuming at least 20 people sign up for the cruise, determine how many passengers will result in the maximum revenue for the owner of the yacht. What is the maximum revenue? What would be the fare/passenger in this case?

18. **Profit of a Vineyard** Phillip, the proprietor of a vineyard, estimates that the first 10,000 bottles of wine produced this season will fetch a profit of \$5/bottle. However, the profit from each bottle beyond 10,000 drops by \$0.0002 for each additional bottle sold. Assuming at least 10,000 bottles of wine are produced and sold, what is the maximum profit? What would be the price/bottle in this case?

19. **Strength of a Beam** A wooden beam has a rectangular cross section of height h in. and width w in. (see the accompanying figure). The strength S of the beam is directly proportional to its width and the square of its height. What are the dimensions of the cross section of the strongest beam that can be cut from a round log of diameter 24 in.?
Hint: $S = kh^2w$, where k is a constant of proportionality.

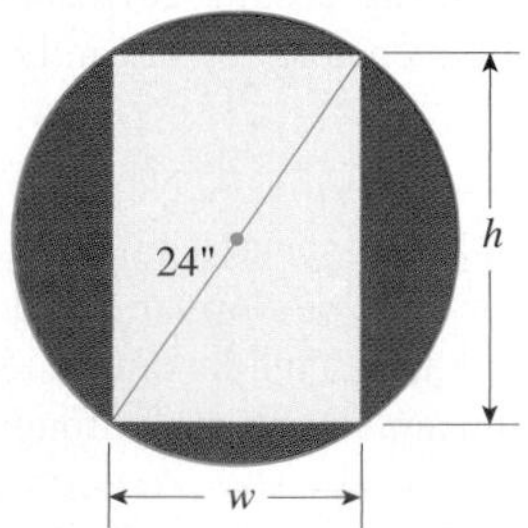

20. **Designing a Grain Silo** A grain silo has the shape of a right circular cylinder surmounted by a hemisphere (see the accompanying figure). If the silo is to have a capacity of 504π ft^3, find the radius and height of the silo that requires the least amount of material to construct.
Hint: The volume of the silo is $\pi r^2 h + \frac{2}{3}\pi r^3$, and the surface area (including the floor) is $\pi(3r^2 + 2rh)$.

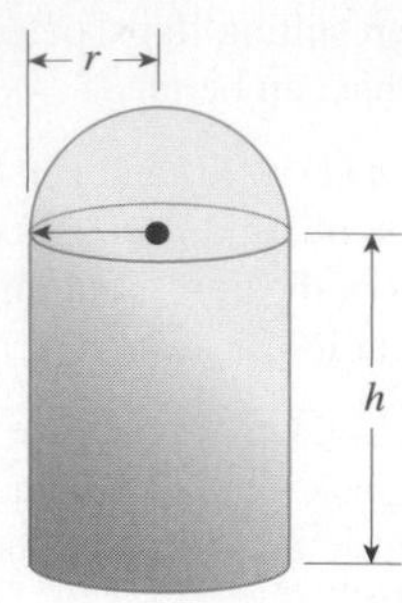

21. **Minimizing Cost of Laying Cable** In the following diagram, S represents the position of a power relay station located on a straight coast, and E shows the location of a marine biology experimental station on an island. A cable is to be laid connecting the relay station with the experimental station. If the cost of running the cable on land is \$1.50/running foot and the cost of running the cable under water is \$2.50/running foot, locate the point P that will result in a minimum cost (solve for x).

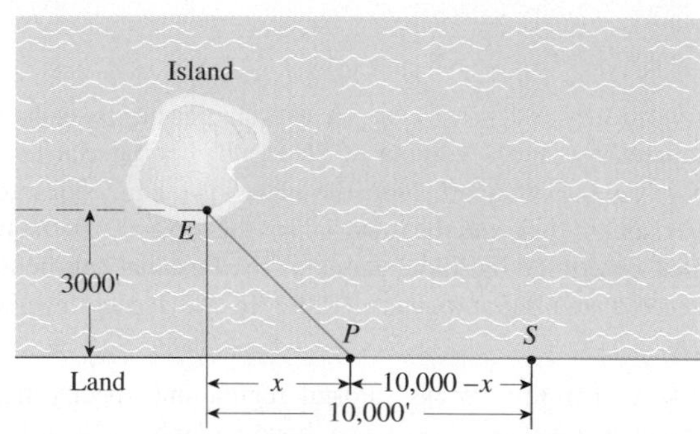

22. **Storing Radioactive Waste** A cylindrical container for storing radioactive waste is to be constructed from lead and have a thickness of 6 in. (see the accompanying figure). If the volume of the outside cylinder is to be 16π ft^3, find the radius and the height of the inside cylinder that will result in a container of maximum storage capacity.

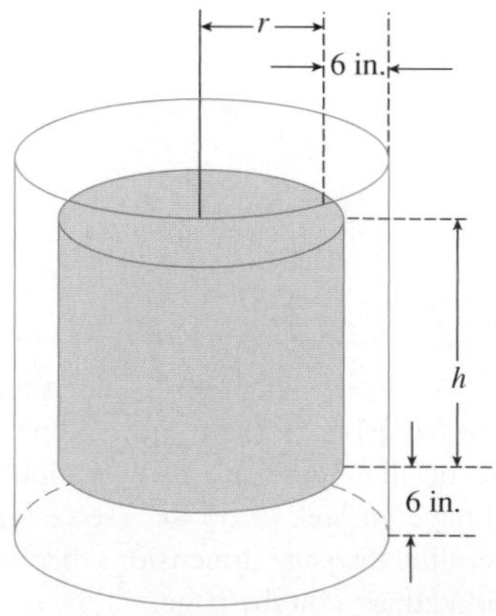

Hint: Show that the storage capacity (inside volume) is given by

$$V(r) = \pi r^2\left[\frac{16}{(r+\frac{1}{2})^2} - 1\right] \qquad \left(0 \le r \le \frac{7}{2}\right)$$

23. FLIGHTS OF BIRDS During daylight hours, some birds fly more slowly over water than over land because some of their energy is expended in overcoming the downdrafts of air over open bodies of water. Suppose a bird that flies at a constant speed of 4 mph over water and 6 mph over land starts its journey at the point E on an island and ends at its nest N on the shore of the mainland, as shown in the accompanying figure. Find the location of the point P that allows the bird to complete its journey in the minimum time (solve for x).

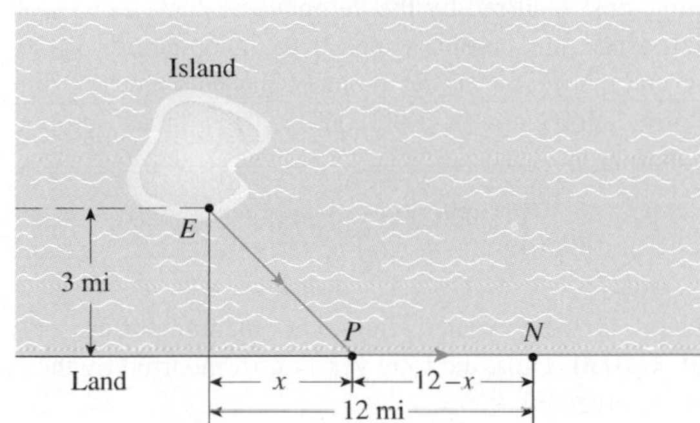

24. OPTIMAL SPEED OF A TRUCK A truck gets $600/x$ mpg when driven at a constant speed of x mph (between 50 and 70 mph). If the price of fuel is \$3/gallon and the driver is paid \$18/hour, at what speed between 50 and 70 mph is it most economical to drive?

25. RACETRACK DESIGN The accompanying figure depicts a racetrack with ends that are semicircular in shape. The length of the track is 1760 ft ($\frac{1}{3}$ mi). Find l and r so that the area enclosed by the rectangular region of the racetrack is as large as possible. What is the area enclosed by the track in this case?

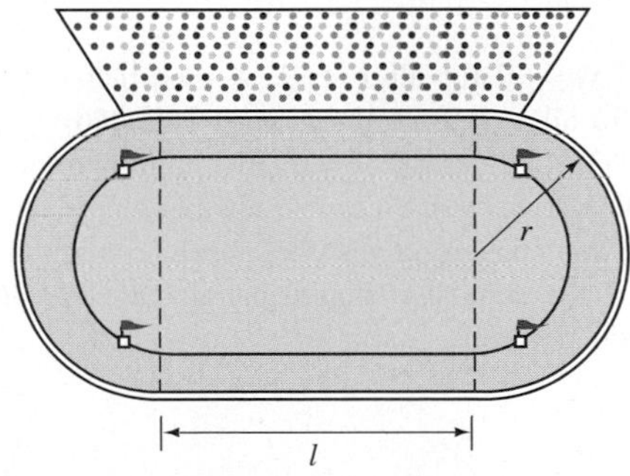

26. INVENTORY CONTROL AND PLANNING The demand for motorcycle tires imported by Dixie Import-Export is 40,000/year and may be assumed to be uniform throughout the year. The cost of ordering a shipment of tires is \$400, and the cost of storing each tire for a year is \$2. Determine how many tires should be in each shipment if the ordering and storage costs are to be minimized. (Assume that each shipment arrives just as the previous one has been sold.)

27. INVENTORY CONTROL AND PLANNING McDuff Preserves expects to bottle and sell 2,000,000 32-oz jars of jam. The company orders its containers from Consolidated Bottle Company. The cost of ordering a shipment of bottles is \$200, and the cost of storing each empty bottle for a year is \$.40. How many orders should McDuff place per year and how many bottles should be in each shipment if the ordering and storage costs are to be minimized? (Assume that each shipment of bottles is used up before the next shipment arrives.)

28. INVENTORY CONTROL AND PLANNING Neilsen Cookie Company sells its assorted butter cookies in containers that have a net content of 1 lb. The estimated demand for the cookies is 1,000,000 1-lb containers. The setup cost for each production run is \$500, and the manufacturing cost is \$.50 for each container of cookies. The cost of storing each container of cookies over the year is \$.40. Assuming uniformity of demand throughout the year and instantaneous production, how many containers of cookies should Neilsen produce per production run in order to minimize the production cost?
Hint: Following the method of Example 5, show that the total production cost is given by the function

$$C(x) = \frac{500{,}000{,}000}{x} + 0.2x + 500{,}000$$

Then minimize the function C on the interval (0, 1,000,000).

29. INVENTORY CONTROL AND PLANNING A company expects to sell D units of a certain product per year. Sales are assumed to be at a steady rate with no shortages allowed. Each time an order for the product is placed, an ordering cost of K dollars is incurred. Each item costs p dollars, and the holding cost is h dollars per item per year.

a. Show that the inventory cost (the combined ordering cost, purchasing cost, and holding cost) is

$$C(x) = \frac{KD}{x} + pD + \frac{hx}{2} \qquad (x > 0)$$

where x is the order quantity (the number of items in each order).

b. Use the result of part (a) to show that the inventory cost is minimized if

$$x = \sqrt{\frac{2KD}{h}}$$

This quantity is called the *economic order quantity* (EOQ).

30. INVENTORY CONTROL AND PLANNING Refer to Exercise 29. The Camera Store sells 960 Yamaha A35 digital cameras per year. Each time an order for cameras is placed with the manufacturer, an ordering cost of \$10 is incurred. The store pays \$80 for each camera, and the cost for holding a camera (mainly due to the opportunity cost incurred in tying up capital in inventory) is \$12/yr. Assume that the cameras sell at a uniform rate and no shortages are allowed.

a. What is the EOQ?

b. How many orders will be placed each year?

c. What is the interval between orders?

4.5 Solutions to Self-Check Exercises

1. Let x and y (measured in feet) denote the length and width of the rectangular garden.

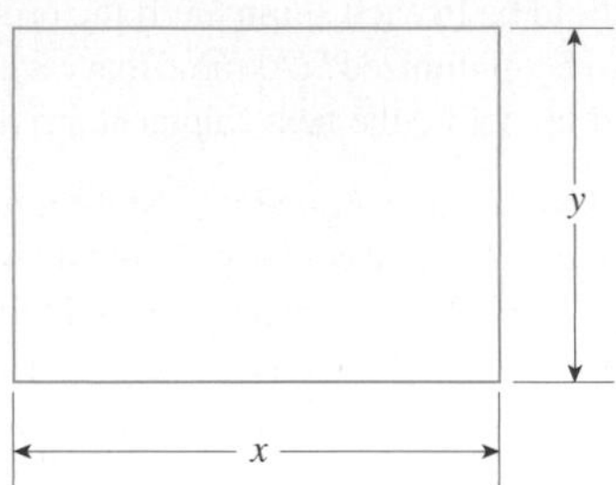

Since the area is to be 300 ft^2, we have

$$xy = 300$$

Next, the amount of fencing to be used is given by the perimeter, and this quantity is to be minimized. Thus, we want to minimize

$$2x + 2y$$

or, since $y = 300/x$ (obtained by solving for y in the first equation), we see that the expression to be minimized is

$$\begin{aligned} f(x) &= 2x + 2\left(\frac{300}{x}\right) \\ &= 2x + \frac{600}{x} \end{aligned}$$

for positive values of x. Now

$$f'(x) = 2 - \frac{600}{x^2}$$

Setting $f'(x) = 0$ yields $x = -\sqrt{300}$ or $x = \sqrt{300}$. We consider only the critical number $\sqrt{300}$ since $-\sqrt{300}$ lies outside the interval $(0, \infty)$. We then compute

$$f''(x) = \frac{1200}{x^3}$$

Since

$$f''(300) > 0$$

the second derivative test implies that a relative minimum of f occurs at $x = \sqrt{300}$. In fact, since $f''(x) > 0$ for all x in $(0, \infty)$, we conclude that $x = \sqrt{300}$ gives rise to the absolute minimum of f. The corresponding value of y, obtained by substituting this value of x into the equation $xy = 300$, is $y = \sqrt{300}$. Therefore, the required dimensions of the vegetable garden are approximately 17.3 ft × 17.3 ft.

2. Let x denote the number of tires in each production run. Then, the average number of tires in storage is $x/2$, so the storage cost incurred by the company is $2(x/2)$, or x dollars. Next, since the company needs to manufacture 1,000,000 tires for the year in order to meet the demand, the number of production runs is $1{,}000{,}000/x$. This gives setup costs amounting to

$$4000\left(\frac{1{,}000{,}000}{x}\right) = \frac{4{,}000{,}000{,}000}{x}$$

dollars for the year. The total manufacturing cost is \$20,000,000. Thus, the total yearly cost incurred by the company is given by

$$C(x) = x + \frac{4{,}000{,}000{,}000}{x} + 20{,}000{,}000$$

Differentiating $C(x)$, we find

$$C'(x) = 1 - \frac{4{,}000{,}000{,}000}{x^2}$$

Setting $C'(x) = 0$ gives 63,246 as the critical number in the interval (0, 1,000,000). Next, we find

$$C''(x) = \frac{8{,}000{,}000{,}000}{x^3}$$

Since $C''(x) > 0$ for all $x > 0$, we see that C is concave upward for all $x > 0$. Furthermore, $C''(63{,}246) > 0$ implies that $x = 63{,}246$ gives rise to a relative minimum of C (by the second derivative test). Since C is always concave upward for $x > 0$, $x = 63{,}246$ gives the absolute minimum of C. Therefore, the company should manufacture 63,246 tires in each production run.

CHAPTER 4 Summary of Principal Terms

TERMS

increasing function (248)
decreasing function (248)
relative maximum (253)
relative minimum (253)
relative extrema (253)
critical number (254)
first derivative test (255)
concave upward (268)
concave downward (268)
inflection point (270)
second derivative test (274)
vertical asymptote (287)
horizontal asymptote (288)
absolute extrema (300)
absolute maximum value (300)
absolute minimum value (300)

CHAPTER 4 Concept Review Questions

Fill in the blanks.

1. a. A function f is increasing on an interval I, if for any two numbers x_1 and x_2 in I, $x_1 < x_2$ implies that _____.

b. A function f is decreasing on an interval I, if for any two numbers x_1 and x_2 in I, $x_1 < x_2$ implies that _____.

2. a. If f is differentiable on an open interval (a, b) and $f'(x) > 0$ on (a, b), then f is _____ on (a, b).

b. If f is differentiable on an open interval (a, b) and _____ on (a, b), then f is decreasing on (a, b).

c. If $f'(x) = 0$ for each value of x in the interval (a, b), then f is _____ on (a, b).

3. a. A function f has a relative maximum at c if there exists an open interval (a, b) containing c such that _____ for all x in (a, b).

b. A function f has a relative minimum at c if there exists an open interval (a, b) containing c such that _____ for all x in (a, b).

4. a. A critical number of a function f is any number in the _____ of f at which $f'(c)$ _____ or $f'(c)$ does *not* _____.

b. If f has a relative extremum at c, then c must be a _____ _____ of f.

c. If c is a critical number of f, then f may or may not have a _____ _____ at c.

5. a. A differentiable function f is concave upward on an interval I if _____ is increasing on I.

b. If f has a second derivative on an open interval I and $f''(x)$ _____ on I, then the graph of f is concave upward on I.

c. If the graph of a continuous function f has a tangent line at $P(c, f(c))$ and the graph of f changes _____ at P, then P is called an inflection point of the graph of f.

d. Suppose f has a continuous second derivative on an interval (a, b), containing a critical number c of f. If $f''(c) < 0$, then f has a _____ _____ at c. If $f''(c) = 0$, then f may or may not have a _____ _____ at c.

6. The line $x = a$ is a vertical asymptote of the graph f if at least one of the following is true: $\lim_{x \to a^+} f(x) =$ _____ or $\lim_{x \to a^-} f(x) =$ _____.

7. For a rational function $f(x) = \dfrac{P(x)}{Q(x)}$, the line $x = a$ is a vertical asymptote of the graph of f if $Q(a) =$ _____ but $P(a) \neq$ _____.

8. The line $y = b$ is a horizontal asymptote of the graph of a function f if either $\lim_{x \to \infty} f(x) =$ _____ or $\lim_{x \to -\infty} f(x) =$ _____.

9. a. A function f has an absolute maximum at c if _____ for all x in the domain D of f. The number $f(c)$ is called the _____ _____ _____ of f on D.

b. A function f has a relative minimum at c if _____ for all values of x in some _____ _____ containing c.

10. The extreme value theorem states that if f is _____ on the closed interval $[a, b]$, then f has both an _____ maximum value and an _____ minimum value on $[a, b]$.

CHAPTER 4 Review Exercises

In Exercises 1–10, (a) find the intervals where the function f is increasing and where it is decreasing, (b) find the relative extrema of f, (c) find the intervals where f is concave upward and where it is concave downward, and (d) find the inflection points, if any, of f.

1. $f(x) = \frac{1}{3}x^3 - x^2 + x - 6$
2. $f(x) = (x - 2)^3$
3. $f(x) = x^4 - 2x^2$
4. $f(x) = x + \frac{4}{x}$
5. $f(x) = \frac{x^2}{x - 1}$
6. $f(x) = \sqrt{x - 1}$
7. $f(x) = (1 - x)^{1/3}$
8. $f(x) = x\sqrt{x - 1}$
9. $f(x) = \frac{2x}{x + 1}$
10. $f(x) = \frac{-1}{1 + x^2}$

In Exercises 11–18, obtain as much information as possible on each function. Then use this information to sketch the graph of the function.

11. $f(x) = x^2 - 5x + 5$
12. $f(x) = -2x^2 - x + 1$
13. $g(x) = 2x^3 - 6x^2 + 6x + 1$
14. $g(x) = \frac{1}{3}x^3 - x^2 + x - 3$
15. $h(x) = x\sqrt{x - 2}$
16. $h(x) = \frac{2x}{1 + x^2}$
17. $f(x) = \frac{x - 2}{x + 2}$
18. $f(x) = x - \frac{1}{x}$

In Exercises 19–22, find the horizontal and vertical asymptotes of the graph of each function. Do not sketch the graph.

19. $f(x) = \frac{1}{2x + 3}$
20. $f(x) = \frac{2x}{x + 1}$
21. $f(x) = \frac{5x}{x^2 - 2x - 8}$
22. $f(x) = \frac{x^2 + x}{x(x - 1)}$

In Exercises 23–32, find the absolute maximum value and the absolute minimum value, if any, of the function.

23. $f(x) = 2x^2 + 3x - 2$
24. $g(x) = x^{2/3}$
25. $g(t) = \sqrt{25 - t^2}$
26. $f(x) = \frac{1}{3}x^3 - x^2 + x + 1$ on $[0, 2]$
27. $h(t) = t^3 - 6t^2$ on $[2, 5]$
28. $g(x) = \frac{x}{x^2 + 1}$ on $[0, 5]$
29. $f(x) = x - \frac{1}{x}$ on $[1, 3]$
30. $h(t) = 8t - \frac{1}{t^2}$ on $[1, 3]$
31. $f(s) = s\sqrt{1 - s^2}$ on $[-1, 1]$
32. $f(x) = \frac{x^2}{x - 1}$ on $[-1, 3]$

33. **Maximizing Profits** Odyssey Travel Agency's monthly profit (in thousands of dollars) depends on the amount of money x (in thousands of dollars) spent on advertising each month according to the rule

$$P(x) = -x^2 + 8x + 20$$

To maximize its monthly profits, what should be Odyssey's monthly advertising budget?

34. **Online Hotel Reservations** The online lodging industry is expected to grow dramatically. In a study conducted in 1999, analysts projected the U.S. online travel spending for lodging to be approximately

$$f(t) = 0.157t^2 + 1.175t + 2.03 \qquad (0 \le t \le 6)$$

billion dollars, where t is measured in years, with $t = 0$ corresponding to 1999.
   a. Show that f is increasing on the interval (0, 6).
   b. Show that the graph of f is concave upward on (0, 6).
   c. What do your results from parts (a) and (b) tell you about the growth of online travel spending over the years in question?

Source: International Data Corp.

35. **Sales of Camera Phones** Camera phones, virtually nonexistent a few years ago, are quickly gaining in popularity. The function

$$N(t) = 8.125t^2 + 24.625t + 18.375 \qquad (0 \le t \le 3)$$

gives the projected worldwide shipments of camera phones (in millions of units) in year t, with $t = 0$ corresponding to 2002.
   a. Find $N'(t)$. What does this say about the sales of camera phones between 2002 and 2005?
   b. Find $N''(t)$. What does this say about the rate of the rate of sales of camera phones between 2002 and 2005?

Source: In-Stat/MDR

36. Effect of Advertising on Sales The total sales S of Cannon Precision Instruments is related to the amount of money x that Cannon spends on advertising its products by the function

$$S(x) = -0.002x^3 + 0.6x^2 + x + 500 \qquad (0 \leq x \leq 200)$$

where S and x are measured in thousands of dollars. Find the inflection point of the function S and discuss its significance.

37. Cost of Producing Calculators A subsidiary of Elektra Electronics manufactures graphing calculators. Management determines that the daily cost $C(x)$ (in dollars) of producing these calculators is

$$C(x) = 0.0001x^3 - 0.08x^2 + 40x + 5000$$

where x is the number of calculators produced. Find the inflection point of the function C and interpret your result.

38. Index of Environmental Quality The Department of the Interior of an African country began to record an index of environmental quality to measure progress or decline in the environmental quality of its wildlife. The index for the years 1984 through 1994 is approximated by the function

$$I(t) = \frac{50t^2 + 600}{t^2 + 10} \qquad (0 \leq t \leq 10)$$

a. Compute $I'(t)$ and show that $I(t)$ is decreasing on the interval (0, 10).
b. Compute $I''(t)$. Study the concavity of the graph of I.
c. Sketch the graph of I.
d. Interpret your results.

39. Maximizing Profits The weekly demand for DVDs manufactured by Herald Media Corporation is given by

$$p = -0.0005x^2 + 60$$

where p denotes the unit price in dollars and x denotes the quantity demanded. The weekly total cost function associated with producing these discs is given by

$$C(x) = -0.001x^2 + 18x + 4000$$

where $C(x)$ denotes the total cost (in dollars) incurred in pressing x discs. Find the production level that will yield a maximum profit for the manufacturer.
Hint: Use the quadratic formula.

40. Maximizing Profits The estimated monthly profit (in dollars) realizable by Cannon Precision Instruments for manufacturing and selling x units of its model M1 digital camera is

$$P(x) = -0.04x^2 + 240x - 10{,}000$$

To maximize its profits, how many cameras should Cannon produce each month?

41. Minimizing Average Cost The total monthly cost (in dollars) incurred by Carlota Music in manufacturing x units of its Professional Series guitars is given by the function

$$C(x) = 0.001x^2 + 100x + 4000$$

a. Find the average cost function $\overline{C}$.
b. Determine the production level that will result in the smallest average production cost.

42. Worker Efficiency The average worker at Wakefield Avionics can assemble

$$N(t) = -2t^3 + 12t^2 + 2t \qquad (0 \leq t \leq 4)$$

ready-to-fly radio-controlled model airplanes t hr into the 8 a.m. to 12 noon morning shift. At what time during this shift is the average worker performing at peak efficiency?

43. Senior Workforce The percent of women 65 years and older in the workforce from 1970 through the year 2000 is approximated by the function

$$P(t) = -0.0002t^3 + 0.018t^2 - 0.36t + 10 \qquad (0 \leq t \leq 30)$$

where t is measured in years, with $t = 0$ corresponding to the beginning of 1970.
a. Find the interval where P is decreasing and the interval where P is increasing.
b. Find the absolute minimum of P.
c. Interpret the results of parts (a) and (b).
Source: U.S. Census Bureau

44. Spread of a Contagious Disease The incidence (number of new cases/day) of a contagious disease spreading in a population of M people is given by

$$R(x) = kx(M - x)$$

where k is a positive constant and x denotes the number of people already infected. Show that the incidence R is greatest when half the population is infected.

45. Maximizing the Volume of a Box A box with an open top is to be constructed from a square piece of cardboard, 10 in. wide, by cutting out a square from each of the four corners and bending up the sides. What is the maximum volume of such a box?

46. Minimizing Construction Costs A man wishes to construct a cylindrical barrel with a capacity of 32π ft^3. The cost/square foot of the material for the side of the barrel is half that of the cost/square foot for the top and bottom. Help him find the dimensions of the barrel that can be constructed at a minimum cost in terms of material used.

47. Packaging You wish to construct a closed rectangular box that has a volume of 4 ft^3. The length of the base of the box will be twice as long as its width. The material for the top

and bottom of the box costs 30¢/square foot. The material for the sides of the box costs 20¢/square foot. Find the dimensions of the least expensive box that can be constructed.

48. **Inventory Control and Planning** Lehen Vinters imports a certain brand of beer. The demand, which may be assumed to be uniform, is 800,000 cases/year. The cost of ordering a shipment of beer is $500, and the cost of storing each case of beer for a year is $2. Determine how many cases of beer should be in each shipment if the ordering and storage costs are to be kept at a minimum. (Assume that each shipment of beer arrives just as the previous one has been sold.)

49. In what interval is the quadratic function

$$f(x) = ax^2 + bx + c \qquad (a \neq 0)$$

increasing? In what interval is f decreasing?

50. Let

$$f(x) = \begin{cases} x^3 + 1 & \text{if } x \neq 0 \\ 2 & \text{if } x = 0 \end{cases}$$

a. Compute $f'(x)$ and show that it does not change sign as we move across $x = 0$.

b. Show that f has a relative maximum at $x = 0$. Does this contradict the first derivative test? Explain your answer.

CHAPTER 4 Before Moving On . . .

1. Find the interval(s) where $f(x) = \dfrac{x^2}{1 - x}$ is increasing and where it is decreasing.

2. Find the relative maxima and relative minima, if any, of $f(x) = 2x^2 - 12x^{1/3}$.

3. Find the intervals where $f(x) = \frac{1}{3}x^3 - \frac{1}{4}x^2 - \frac{1}{2}x + 1$ is concave upward, the intervals where f is concave downward, and the inflection point(s) of f.

4. Sketch the graph of $f(x) = 2x^3 - 9x^2 + 12x - 1$.

5. Find the absolute maximum and absolute minimum values of $f(x) = 2x^3 + 3x^2 - 1$ on the interval $[-2, 3]$.

6. An open bucket in the form of a right circular cylinder is to be constructed with a capacity of 1 ft^3. Find the radius and height of the cylinder if the amount of material used is minimal.

5 Exponential and Logarithmic Functions

How many bacteria will there be in a culture at the end of a certain period of time? How fast will the bacteria population be growing at the end of that time? Example 1, page 380, answers these questions.

THE EXPONENTIAL FUNCTION is, without doubt, the most important function in mathematics and its applications. After a brief introduction to the exponential function and its *inverse*, the logarithmic function, we learn how to differentiate such functions. This lays the foundation for exploring the many applications involving exponential functions. For example, we look at the role played by exponential functions in computing earned interest on a bank account and in studying the growth of a bacteria population in the laboratory, the way radioactive matter decays, the rate at which a factory worker learns a certain process, and the rate at which a communicable disease is spread over time.

5.1 Exponential Functions

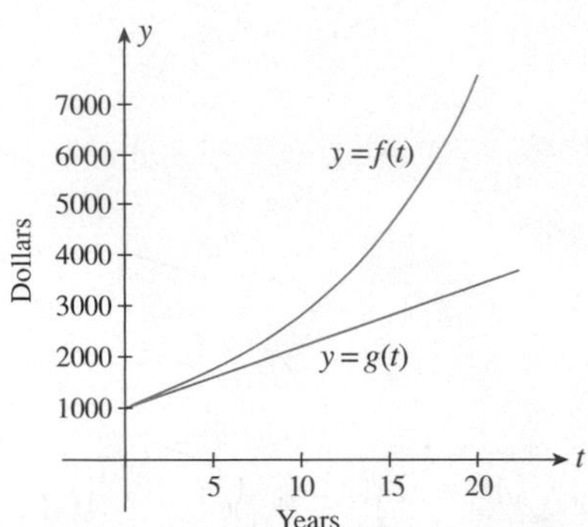

FIGURE 1
Under continuous compounding, a sum of money grows exponentially.

Exponential Functions and Their Graphs

Suppose you deposit a sum of \$1000 in an account earning interest at the rate of 10% per year *compounded continuously* (the way most financial institutions compute interest). Then, the accumulated amount at the end of t years $(0 \le t \le 20)$ is described by the function f, whose graph appears in Figure 1.* This function is called an *exponential function*. Observe that the graph of f rises rather slowly at first but very rapidly as time goes by. For purposes of comparison, we have also shown the graph of the function $y = g(t) = 1000(1 + 0.10t)$, giving the accumulated amount for the same principal (\$1000) but earning *simple* interest at the rate of 10% per year. The moral of the story: It is never too early to save.

Exponential functions play an important role in many real-world applications, as you will see throughout this chapter.

Observe that whenever b is a positive number and n is any real number, the expression b^n is a real number. This enables us to define an exponential function as follows:

Exponential Function

The function defined by

$$f(x) = b^x \qquad (b > 0, b \neq 1)$$

is called an **exponential function with base *b* and exponent *x***. The domain of f is the set of all real numbers.

For example, the exponential function with base 2 is the function

$$f(x) = 2^x$$

with domain $(-\infty, \infty)$. The values of $f(x)$ for selected values of x follow:

$$f(3) = 2^3 = 8 \qquad f\left(\frac{3}{2}\right) = 2^{3/2} = 2\cdot 2^{1/2} = 2\sqrt{2} \qquad f(0) = 2^0 = 1$$

$$f(-1) = 2^{-1} = \frac{1}{2} \qquad f\left(-\frac{2}{3}\right) = 2^{-2/3} = \frac{1}{2^{2/3}} = \frac{1}{\sqrt[3]{4}}$$

Computations involving exponentials are facilitated by the laws of exponents. These laws were stated in Section 1.1, and you might want to review the material there. For convenience, however, we will restate these laws.

Laws of Exponents

Let a and b be positive numbers and let x and y be real numbers. Then,

1. $b^x \cdot b^y = b^{x+y}$

2. $\dfrac{b^x}{b^y} = b^{x-y}$

3. $(b^x)^y = b^{xy}$

4. $(ab)^x = a^x b^x$

5. $\left(\dfrac{a}{b}\right)^x = \dfrac{a^x}{b^x}$

*We will derive the rule for f in Section 5.3.

The use of the laws of exponents is illustrated in the next example.

EXAMPLE 1

a. $16^{7/4} \cdot 16^{-1/2} = 16^{7/4-1/2} = 16^{5/4} = 2^5 = 32$ Law 1

b. $\dfrac{8^{5/3}}{8^{-1/3}} = 8^{5/3-(-1/3)} = 8^2 = 64$ Law 2

c. $(64^{4/3})^{-1/2} = 64^{(4/3)(-1/2)} = 64^{-2/3}$

$$= \frac{1}{64^{2/3}} = \frac{1}{(64^{1/3})^2} = \frac{1}{4^2} = \frac{1}{16}$$ Law 3

d. $(16 \cdot 81)^{-1/4} = 16^{-1/4} \cdot 81^{-1/4} = \dfrac{1}{16^{1/4}} \cdot \dfrac{1}{81^{1/4}} = \dfrac{1}{2} \cdot \dfrac{1}{3} = \dfrac{1}{6}$ Law 4

e. $\left(\dfrac{3^{1/2}}{2^{1/3}}\right)^4 = \dfrac{3^{4/2}}{2^{4/3}} = \dfrac{9}{2^{4/3}}$ Law 5 ■

EXAMPLE 2 Let $f(x) = 2^{2x-1}$. Find the value of x for which $f(x) = 16$.

Solution We want to solve the equation

$$2^{2x-1} = 16 = 2^4$$

But this equation holds if and only if

$$2x - 1 = 4 \qquad b^m = b^n \Rightarrow m = n$$

giving $x = \dfrac{5}{2}$. ■

Exponential functions play an important role in mathematical analysis. Because of their special characteristics, they are some of the most useful functions and are found in virtually every field where mathematics is applied. To mention a few examples: Under ideal conditions the number of bacteria present at any time t in a culture may be described by an exponential function of t; radioactive substances decay over time in accordance with an "exponential" law of decay; money left on fixed deposit and earning compound interest grows exponentially; and some of the most important distribution functions encountered in statistics are exponential.

Let's begin our investigation into the properties of exponential functions by studying their graphs.

EXAMPLE 3 Sketch the graph of the exponential function $y = 2^x$.

Solution First, as discussed earlier, the domain of the exponential function $y = f(x) = 2^x$ is the set of real numbers. Next, putting $x = 0$ gives $y = 2^0 = 1$, the y-intercept of f. There is no x-intercept since there is no value of x for which $y = 0$. To find the range of f, consider the following table of values:

x	-5	-4	-3	-2	-1	0	1	2	3	4	5
y	$\frac{1}{32}$	$\frac{1}{16}$	$\frac{1}{8}$	$\frac{1}{4}$	$\frac{1}{2}$	1	2	4	8	16	32

We see from these computations that 2^x decreases and approaches zero as x decreases without bound and that 2^x increases without bound as x increases without bound. Thus, the range of f is the interval $(0, \infty)$—that is, the set of positive real numbers. Finally, we sketch the graph of $y = f(x) = 2^x$ in Figure 2. ■

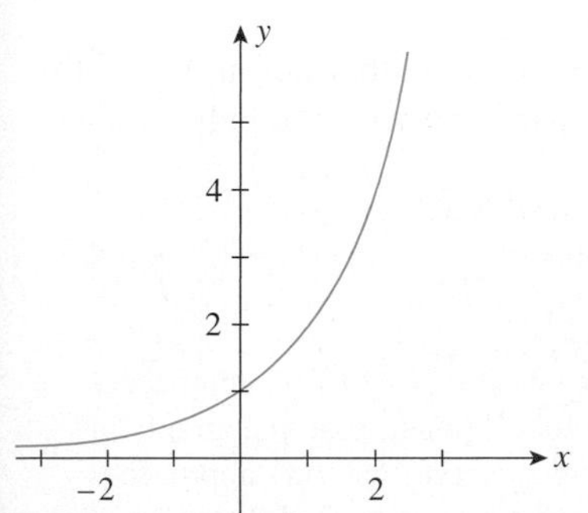

FIGURE 2
The graph of $y = 2^x$

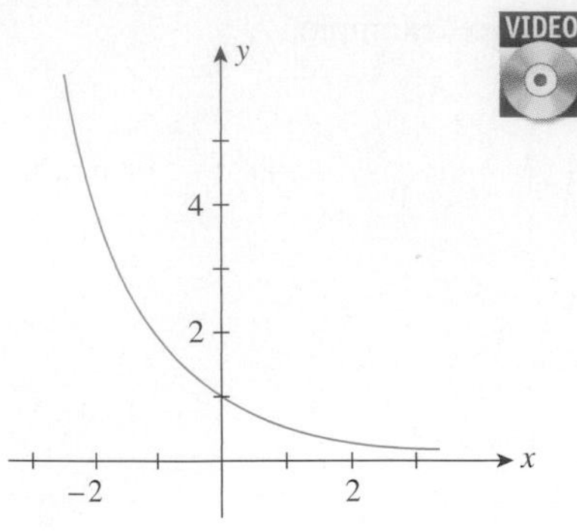

FIGURE 3
The graph of $y = \left(\frac{1}{2}\right)^x$

VIDEO

EXAMPLE 4 Sketch the graph of the exponential function $y = (1/2)^x$.

Solution The domain of the exponential function $y = (1/2)^x$ is the set of all real numbers. The y-intercept is $(1/2)^0 = 1$; there is no x-intercept since there is no value of x for which $y = 0$. From the following table of values

x	−5	−4	−3	−2	−1	0	1	2	3	4	5
y	32	16	8	4	2	1	$\frac{1}{2}$	$\frac{1}{4}$	$\frac{1}{8}$	$\frac{1}{16}$	$\frac{1}{32}$

we deduce that $(1/2)^x = 1/2^x$ increases without bound as x decreases without bound and that $(1/2)^x$ decreases and approaches zero as x increases without bound. Thus, the range of f is the interval $(0, \infty)$. The graph of $y = f(x) = (1/2)^x$ is sketched in Figure 3. ■

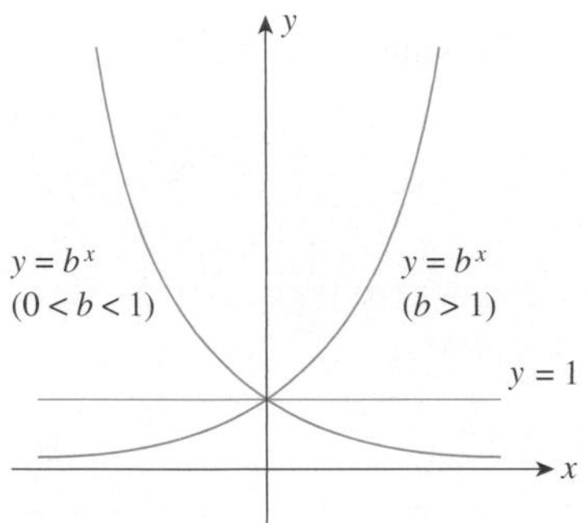

FIGURE 4
$y = b^x$ is an increasing function of x if $b > 1$, a constant function if $b = 1$, and a decreasing function if $0 < b < 1$.

The functions $y = 2^x$ and $y = (1/2)^x$, whose graphs you studied in Examples 3 and 4, are special cases of the exponential function $y = f(x) = b^x$, obtained by setting $b = 2$ and $b = 1/2$, respectively. In general, the exponential function $y = b^x$ with $b > 1$ has a graph similar to $y = 2^x$, whereas the graph of $y = b^x$ for $0 < b < 1$ is similar to that of $y = (1/2)^x$ (Exercises 27 and 28 on page 336). When $b = 1$, the function $y = b^x$ reduces to the constant function $y = 1$. For comparison, the graphs of all three functions are sketched in Figure 4.

Properties of the Exponential Function

The exponential function $y = b^x$ ($b > 0, b \neq 1$) has the following properties:

1. Its domain is $(-\infty, \infty)$.
2. Its range is $(0, \infty)$.
3. Its graph passes through the point $(0, 1)$.
4. It is continuous on $(-\infty, \infty)$.
5. It is increasing on $(-\infty, \infty)$ if $b > 1$ and decreasing on $(-\infty, \infty)$ if $b < 1$.

The Base *e*

Exponential functions to the base e, where e is an irrational number whose value is 2.7182818. . . , play an important role in both theoretical and applied problems. It can be shown, although we will not do so here, that

$$e = \lim_{m\to\infty}\left(1 + \frac{1}{m}\right)^m \tag{1}$$

However, you may convince yourself of the plausibility of this definition of the number e by examining Table 1, which may be constructed with the help of a calculator.

TABLE 1

m	$\left(1 + \frac{1}{m}\right)^m$
10	2.59374
100	2.70481
1000	2.71692
10,000	2.71815
100,000	2.71827
1,000,000	2.71828

EXPLORING WITH TECHNOLOGY

To obtain a visual confirmation of the fact that the expression $(1 + 1/m)^m$ approaches the number $e = 2.71828\ldots$ as m increases without bound, plot the graph of $f(x) = (1 + 1/x)^x$ in a suitable viewing window and observe that $f(x)$ approaches 2.71828. . . as x increases without bound. Use ZOOM and TRACE to find the value of $f(x)$ for large values of x.

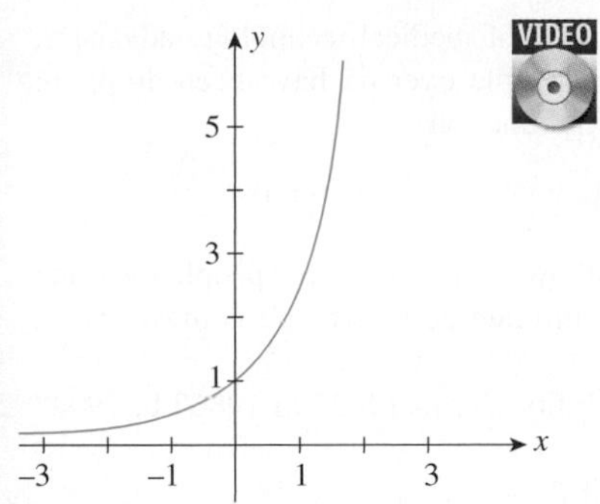

FIGURE 5
The graph of $y = e^x$

VIDEO

EXAMPLE 5 Sketch the graph of the function $y = e^x$.

Solution Since $e > 1$, it follows from our previous discussion that the graph of $y = e^x$ is similar to the graph of $y = 2^x$ (see Figure 2). With the aid of a calculator, we obtain the following table:

x	-3	-2	-1	0	1	2	3
y	0.05	0.14	0.37	1	2.72	7.39	20.09

The graph of $y = e^x$ is sketched in Figure 5. ■

Next, we consider another exponential function to the base e that is closely related to the previous function and is particularly useful in constructing models that describe "exponential decay."

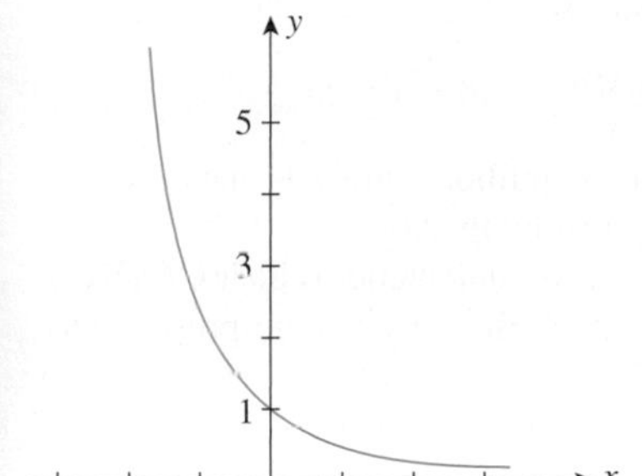

FIGURE 6
The graph of $y = e^{-x}$

EXAMPLE 6 Sketch the graph of the function $y = e^{-x}$.

Solution Since $e > 1$, it follows that $0 < 1/e < 1$, so $f(x) = e^{-x} = 1/e^x = (1/e)^x$ is an exponential function with base less than 1. Therefore, it has a graph similar to that of the exponential function $y = (1/2)^x$. As before, we construct the following table of values of $y = e^{-x}$ for selected values of x:

x	-3	-2	-1	0	1	2	3
y	20.09	7.39	2.72	1	0.37	0.14	0.05

Using this table, we sketch the graph of $y = e^{-x}$ in Figure 6. ■

5.1 Self-Check Exercises

1. Solve the equation $2^{2x+1} \cdot 2^{-3} = 2^{x-1}$.
2. Sketch the graph of $y = e^{0.4x}$.

Solutions to Self-Check Exercises 5.1 can be found on page 337.

5.1 Concept Questions

1. Define the exponential function f with base b and exponent x. What restrictions, if any, are placed on b?
2. For the exponential function $y = b^x$ ($b > 0$, $b \neq 1$), state (a) its domain and range, (b) its y-intercept, (c) where it is continuous, and (d) where it is increasing and where it is decreasing for the case $b > 1$ and the case $b < 1$.

5.1 Exercises

In Exercises 1–8, evaluate the expression.

1. **a.** $4^{-3} \cdot 4^5$ **b.** $3^{-3} \cdot 3^6$
2. **a.** $(2^{-1})^3$ **b.** $(3^{-2})^3$
3. **a.** $9(9)^{-1/2}$ **b.** $5(5)^{-1/2}$
4. **a.** $\left[\left(-\frac{1}{2}\right)^3\right]^{-2}$ **b.** $\left[\left(-\frac{1}{3}\right)^2\right]^{-3}$

5. **a.** $\dfrac{(-3)^4(-3)^5}{(-3)^8}$ **b.** $\dfrac{(2^{-4})(2^6)}{2^{-1}}$

6. **a.** $3^{1/4} \cdot 9^{-5/8}$ **b.** $2^{3/4} \cdot 4^{-3/2}$

7. **a.** $\dfrac{5^{3.3} \cdot 5^{-1.6}}{5^{-0.3}}$ **b.** $\dfrac{4^{2.7} \cdot 4^{-1.3}}{4^{-0.4}}$

8. **a.** $\left(\dfrac{1}{16}\right)^{-1/4}\left(\dfrac{27}{64}\right)^{-1/3}$ **b.** $\left(\dfrac{8}{27}\right)^{-1/3}\left(\dfrac{81}{256}\right)^{-1/4}$

In Exercises 9–16, simplify the expression.

9. **a.** $(64x^9)^{1/3}$ **b.** $(25x^3y^4)^{1/2}$

10. **a.** $(2x^3)(-4x^{-2})$ **b.** $(4x^{-2})(-3x^5)$

11. **a.** $\dfrac{6a^{-5}}{3a^{-3}}$ **b.** $\dfrac{4b^{-4}}{12b^{-6}}$

12. **a.** $y^{-3/2}y^{5/3}$ **b.** $x^{-3/5}x^{8/3}$

13. **a.** $(2x^3y^2)^3$ **b.** $(4x^2y^2z^3)^2$

14. **a.** $(x^{r/s})^{s/r}$ **b.** $(x^{-b/a})^{-a/b}$

15. **a.** $\dfrac{5^0}{(2^{-3}x^{-3}y^2)^2}$ **b.** $\dfrac{(x+y)(x-y)}{(x-y)^0}$

16. **a.** $\dfrac{(a^m \cdot a^{-n})^{-2}}{(a^{m+n})^2}$ **b.** $\left(\dfrac{x^{2n-2}y^{2n}}{x^{5n+1}y^{-n}}\right)^{1/3}$

In Exercises 17–26, solve the equation for *x*.

17. $6^{2x} = 6^4$ **18.** $5^{-x} = 5^3$

19. $3^{3x-4} = 3^5$ **20.** $10^{2x-1} = 10^{x+3}$

21. $(2.1)^{x+2} = (2.1)^5$ **22.** $(-1.3)^{x-2} = (-1.3)^{2x+1}$

23. $8^x = \left(\dfrac{1}{32}\right)^{x-2}$ **24.** $3^{x-x^2} = \dfrac{1}{9^x}$

25. $3^{2x} - 12 \cdot 3^x + 27 = 0$

26. $2^{2x} - 4 \cdot 2^x + 4 = 0$

In Exercises 27–36, sketch the graphs of the given functions on the same axes.

27. $y = 2^x$, $y = 3^x$, and $y = 4^x$

28. $y = \left(\dfrac{1}{2}\right)^x$, $y = \left(\dfrac{1}{3}\right)^x$, and $y = \left(\dfrac{1}{4}\right)^x$

29. $y = 2^{-x}$, $y = 3^{-x}$, and $y = 4^{-x}$

30. $y = 4^{0.5x}$ and $y = 4^{-0.5x}$

31. $y = 4^{0.5x}$, $y = 4^x$, and $y = 4^{2x}$

32. $y = e^x$, $y = 2e^x$, and $y = 3e^x$

33. $y = e^{0.5x}$, $y = e^x$, and $y = e^{1.5x}$

34. $y = e^{-0.5x}$, $y = e^{-x}$, and $y = e^{-1.5x}$

35. $y = 0.5e^{-x}$, $y = e^{-x}$, and $y = 2e^{-x}$

36. $y = 1 - e^{-x}$ and $y = 1 - e^{-0.5x}$

37. **Disability Rates** Because of medical technology advances, the disability rates for people over 65 have been dropping rather dramatically. The function

$$R(t) = 26.3e^{-0.016t} \qquad (0 \leq t \leq 18)$$

gives the disability rate $R(t)$, in percent, for people over age 65 from 1982 ($t = 0$) through 2000, where t is measured in years.

a. What was the disability rate in 1982? In 1986? In 1994? In 2000?

b. Sketch the graph of R.

Source: Frost and Sullivan

38. **Tracking with GPS** Employers are increasingly turning to GPS (global positioning system) technology to keep track of their fleet vehicles. In a study conducted in 2004, the estimated number of automatic vehicle trackers installed on fleet vehicles in the United States is

$$N(t) = 0.6e^{0.17t} \qquad (0 \leq t \leq 5)$$

where $N(t)$ is measured in millions and t is measured in years, with $t = 0$ corresponding to 2000.

a. What was the number of automatic vehicles trackers installed in the year 2000? How many were projected to be installed in 2005?

b. Sketch the graph of N.

Source: C. J. Driscoll Associates

39. **Growth of Web Sites** According to a study conducted in 2000, the projected number of Web addresses (in billions) is approximated by the function

$$N(t) = 0.45e^{0.5696t} \qquad (0 \leq t \leq 5)$$

where t is measured in years, with $t = 0$ corresponding to 1997.

a. Complete the following table by finding the number of Web addresses in each year:

Year	0	1	2	3	4	5
Number of Web Addresses (billions)						

b. Sketch the graph of N.

40. **Married Households** The percent of families that were married households between 1970 and 2000 is approximately

$$P(t) = 86.9e^{-0.05t} \qquad (0 \leq t \leq 3)$$

where t is measured in decades, with $t = 0$ corresponding to 1970.

a. What percent of families were married households in 1970? In 1980? In 1990? In 2000?

b. Sketch the graph of P.

Source: U.S. Census Bureau

41. **Alternative Minimum Tax** The alternative minimum tax was created in 1969 to prevent the very wealthy from using creative deductions and shelters to avoid having to pay anything to the Internal Revenue Service. But it has increasingly hit the middle class. The number of taxpayers subject to an alternative minimum tax is projected to be

$$N(t) = \frac{35.5}{1 + 6.89e^{-0.8674t}} \qquad (0 \le t \le 6)$$

where $N(t)$ is measured in millions and t is measured in years, with $t = 0$ corresponding to 2004. What is the projected number of taxpayers subjected to an alternative minimum tax in 2010?

Source: Brookings Institution

42. **Absorption of Drugs** The concentration of a drug in an organ at any time t (in seconds) is given by

$$C(t) = \begin{cases} 0.3t - 18(1 - e^{-t/60}) & \text{if } 0 \le t \le 20 \\ 18e^{-t/60} - 12e^{-(t-20)/60} & \text{if } t > 20 \end{cases}$$

where $C(t)$ is measured in grams/cubic centimeter (g/cm^3).

a. What is the initial concentration of the drug in the organ?

b. What is the concentration of the drug in the organ after 10 sec?

c. What is the concentration of the drug in the organ after 30 sec?

d. What will be the concentration of the drug in the long run?

43. **Absorption of Drugs** The concentration of a drug in an organ at any time t (in seconds) is given by

$$x(t) = 0.08 + 0.12(1 - e^{-0.02t})$$

where $x(t)$ is measured in grams/cubic centimeter (g/cm^3).

a. What is the initial concentration of the drug in the organ?

b. What is the concentration of the drug in the organ after 20 sec?

c. What will be the concentration of the drug in the organ in the long run?

d. Sketch the graph of x.

44. **Absorption of Drugs** Jane took 100 mg of a drug in the morning and another 100 mg of the same drug at the same time the following morning. The amount of the drug in her body t days after the first dosage was taken is given by

$$A(t) = \begin{cases} 100e^{-1.4t} & \text{if } 0 \le t < 1 \\ 100(1 + e^{1.4})e^{-1.4t} & \text{if } t \ge 1 \end{cases}$$

a. What was the amount of drug in Jane's body immediately after taking the second dose? After 2 days? In the long run?

b. Sketch the graph of A.

In Exercises 45–48, determine whether the statement is true or false. If it is true, explain why it is true. If it is false, give an example to show why it is false.

45. $(x^2 + 1)^3 = x^6 + 1$

46. $e^{xy} = e^x e^y$

47. If $x < y$, then $e^x < e^y$.

48. If $0 < b < 1$ and $x < y$, then $b^x > b^y$.

5.1 Solutions to Self-Check Exercises

1. $2^{2x+1} \cdot 2^{-3} = 2^{x-1}$

$$\frac{2^{2x+1}}{2^{x-1}} \cdot 2^{-3} = 1 \qquad \text{Divide both sides by } 2^{x-1}.$$

$$2^{(2x+1)-(x-1)-3} = 1$$

$$2^{x-1} = 1$$

This is true if and only if $x - 1 = 0$ or $x = 1$.

2. We first construct the following table of values:

x	-3	-2	-1	0	1	2	3	4
$y = e^{0.4x}$	0.3	0.4	0.7	1	1.5	2.2	3.3	5

Next, we plot these points and join them by a smooth curve to obtain the graph of f shown in the accompanying figure.

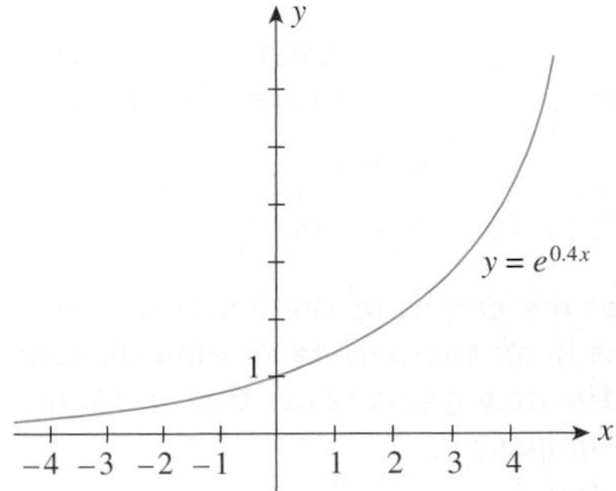

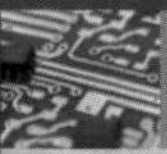

USING TECHNOLOGY

Although the proof is outside the scope of this book, it can be proved that an exponential function of the form $f(x) = b^x$, where $b > 1$, will ultimately grow faster than the power function $g(x) = x^n$ for *any* positive real number n. To give a visual demonstration of this result for the special case of the exponential function $f(x) = e^x$, we can use a graphing utility to plot the graphs of both f and g (for selected values of n) on the same set of axes in an appropriate viewing window and observe that the graph of f ultimately lies above that of g.

EXAMPLE 1 Use a graphing utility to plot the graphs of (a) $f(x) = e^x$ and $g(x) = x^3$ on the same set of axes in the viewing window $[0, 6] \times [0, 250]$ and (b) $f(x) = e^x$ and $g(x) = x^5$ in the viewing window $[0, 20] \times [0, 1{,}000{,}000]$.

Solution

a. The graphs of $f(x) = e^x$ and $g(x) = x^3$ in the viewing window $[0, 6] \times [0, 250]$ are shown in Figure T1a.

b. The graphs of $f(x) = e^x$ and $g(x) = x^5$ in the viewing window $[0, 20] \times [0, 1{,}000{,}000]$ are shown in Figure T1b.

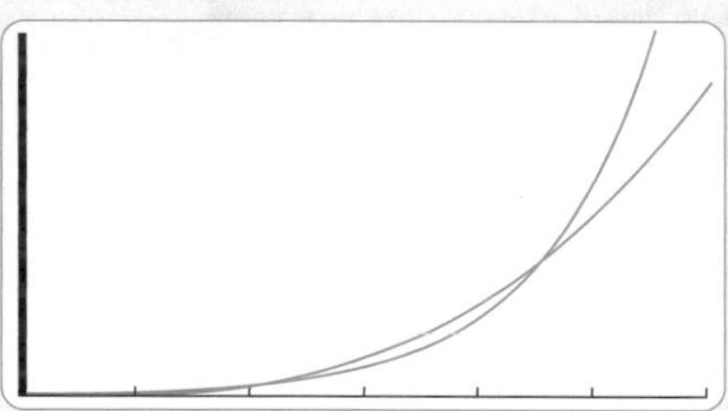

(a) The graphs of $f(x) = e^x$ and $g(x) = x^3$ in the viewing window $[0, 6] \times [0, 250]$

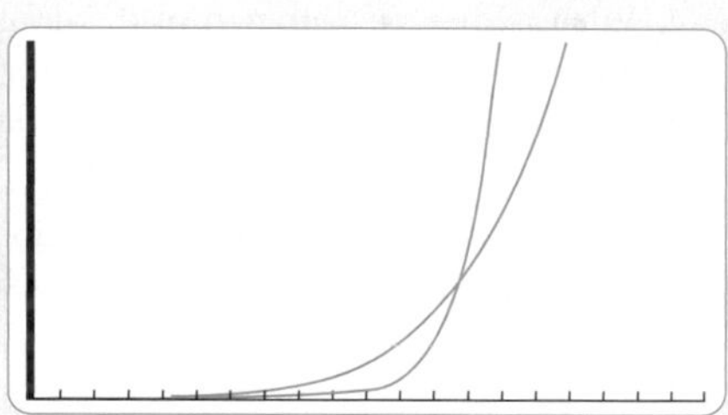

(b) The graphs of $f(x) = e^x$ and $g(x) = x^5$ in the viewing window $[0, 20] \times [0, 1{,}000{,}000]$

FIGURE T1

In the exercises that follow, you are asked to use a graphing utility to reveal the properties of exponential functions.

TECHNOLOGY EXERCISES

In Exercises 1 and 2, plot the graphs of the functions *f* and *g* on the same set of axes in the specified viewing window.

1. $f(x) = e^x$ and $g(x) = x^2$; $[0, 4] \times [0, 30]$
2. $f(x) = e^x$ and $g(x) = x^4$; $[0, 15] \times [0, 20{,}000]$

In Exercises 3 and 4, plot the graphs of the functions *f* and *g* on the same set of axes in an appropriate viewing window to demonstrate that *f* ultimately grows faster than *g*. (*Note:* Your answer will *not* be unique.)

3. $f(x) = 2^x$ and $g(x) = x^{2.5}$
4. $f(x) = 3^x$ and $g(x) = x^3$
5. Plot the graphs of $f(x) = 2^x$, $g(x) = 3^x$, and $h(x) = 4^x$ on the same set of axes in the viewing window $[0, 5] \times [0, 100]$. Comment on the relationship between the base b and the growth of the function $f(x) = b^x$.
6. Plot the graphs of $f(x) = (1/2)^x$, $g(x) = (1/3)^x$, and $h(x) = (1/4)^x$ on the same set of axes in the viewing window $[0, 4] \times [0, 1]$. Comment on the relationship between the base b and the growth of the function $f(x) = b^x$.
7. Plot the graphs of $f(x) = e^x$, $g(x) = 2e^x$, and $h(x) = 3e^x$ on the same set of axes in the viewing window $[-3, 3] \times [0, 10]$. Comment on the role played by the constant k in the graph of $f(x) = ke^x$.

8. Plot the graphs of $f(x) = -e^x$, $g(x) = -2e^x$, and $h(x) = -3e^x$ on the same set of axes in the viewing window $[-3, 3] \times [-10, 0]$. Comment on the role played by the constant k in the graph of $f(x) = ke^x$.

9. Plot the graphs of $f(x) = e^{0.5x}$, $g(x) = e^x$, and $h(x) = e^{1.5x}$ on the same set of axes in the viewing window $[-2, 2] \times [0, 4]$. Comment on the role played by the constant k in the graph of $f(x) = e^{kx}$.

10. Plot the graphs of $f(x) = e^{-0.5x}$, $g(x) = e^{-x}$, and $h(x) = e^{-1.5x}$ on the same set of axes in the viewing window $[-2, 2] \times [0, 4]$. Comment on the role played by the constant k in the graph of $f(x) = e^{kx}$.

11. Absorption of Drugs The concentration of a drug in an organ at any time t (in seconds) is given by

$$x(t) = 0.08 + 0.12(1 - e^{-0.02t})$$

where $x(t)$ is measured in grams/cubic centimeter (g/cm^3).

a. Plot the graph of the function x in the viewing window $[0, 200] \times [0, 0.2]$.
b. What is the initial concentration of the drug in the organ?
c. What is the concentration of the drug in the organ after 20 sec?
d. What will be the concentration of the drug in the organ in the long run?

12. Absorption of Drugs Jane took 100 mg of a drug in the morning and another 100 mg of the same drug at the same time the following morning. The amount of the drug in her body t days after the first dosage was taken is given by

$$A(t) = \begin{cases} 100e^{-1.4t} & \text{if } 0 \le t < 1 \\ 100(1 + e^{1.4})e^{-1.4t} & \text{if } t \ge 1 \end{cases}$$

a. Plot the graph of the function A in the viewing window $[0, 5] \times [0, 140]$.
b. Verify the results of Exercise 44, page 337.

13. Absorption of Drugs The concentration of a drug in an organ at any time t (in seconds) is given by

$$C(t) = \begin{cases} 0.3t - 18(1 - e^{-t/60}) & \text{if } 0 \le t \le 20 \\ 18e^{-t/60} - 12e^{-(t-20)/60} & \text{if } t > 20 \end{cases}$$

where $C(t)$ is measured in grams/cubic centimeter (g/cm^3).

a. Plot the graph of the function C in the viewing window $[0, 120] \times [0, 1]$.
b. How long after the drug is first introduced will it take for the concentration of the drug to reach a peak?
c. How long after the concentration of the drug has peaked will it take for the concentration of the drug to fall back to 0.5 g/cm^3?
Hint: Plot the graphs of $y_1 = C(x)$ and $y_2 = 0.5$ and use the ISECT function of your graphing utility.

5.2 Logarithmic Functions

Logarithms

You are already familiar with exponential equations of the form

$$b^y = x \qquad (b > 0, b \neq 1)$$

where the variable x is expressed in terms of a real number b and a variable y. But what about solving this same equation for y? You may recall from your study of algebra that the number y is called the **logarithm of x to the base b** and is denoted by $\log_b x$. It is the power to which the base b must be raised in order to obtain the number x.

Logarithm of x to the Base b

$$y = \log_b x \quad \text{if and only if} \quad x = b^y \qquad (x > 0)$$

Observe that the logarithm $\log_b x$ is defined only for positive values of x.

EXAMPLE 1

a. $\log_{10} 100 = 2$ since $100 = 10^2$
b. $\log_5 125 = 3$ since $125 = 5^3$

c. $\log_3 \frac{1}{27} = -3$ since $\frac{1}{27} = \frac{1}{3^3} = 3^{-3}$

d. $\log_{20} 20 = 1$ since $20 = 20^1$ ■

EXAMPLE 2 Solve each of the following equations for x.

a. $\log_3 x = 4$ **b.** $\log_{16} 4 = x$ **c.** $\log_x 8 = 3$

Solution

a. By definition, $\log_3 x = 4$ implies $x = 3^4 = 81$.

b. $\log_{16} 4 = x$ is equivalent to $4 = 16^x = (4^2)^x = 4^{2x}$, or $4^1 = 4^{2x}$, from which we deduce that

$$2x = 1 \qquad b^m = b^n \Rightarrow m = n$$

$$x = \frac{1}{2}$$

c. Referring once again to the definition, we see that the equation $\log_x 8 = 3$ is equivalent to

$$8 = 2^3 = x^3$$

$$x = 2 \qquad a^m = b^m \Rightarrow a = b$$ ■

The two widely used systems of logarithms are the system of **common logarithms**, which uses the number 10 as the base, and the system of **natural logarithms**, which uses the irrational number $e = 2.71828\ldots$ as the base. Also, it is standard practice to write **log** for $\log_{10}$ and **ln** for $\log_e$.

Logarithmic Notation

$$\log x = \log_{10} x \qquad \text{Common logarithm}$$

$$\ln x = \log_e x \qquad \text{Natural logarithm}$$

The system of natural logarithms is widely used in theoretical work. Using natural logarithms rather than logarithms to other bases often leads to simpler expressions.

Laws of Logarithms

Computations involving logarithms are facilitated by the following **laws of logarithms**.

Laws of Logarithms

If m and n are positive numbers, then

1. $\log_b mn = \log_b m + \log_b n$
2. $\log_b \frac{m}{n} = \log_b m - \log_b n$
3. $\log_b m^n = n \log_b m$
4. $\log_b 1 = 0$
5. $\log_b b = 1$

Do not confuse the expression log m/n (Law 2) with the expression log m/log n. For example,

$$\log\frac{100}{10} = \log 100 - \log 10 = 2 - 1 = 1 \neq \frac{\log 100}{\log 10} = \frac{2}{1} = 2$$

You will be asked to prove these laws in Exercises 60–62 on page 347. Their derivations are based on the definition of a logarithm and the corresponding laws of exponents. The following examples illustrate the properties of logarithms.

EXAMPLE 3

a. $\log(2 \cdot 3) = \log 2 + \log 3$ **b.** $\ln\frac{5}{3} = \ln 5 - \ln 3$

c. $\log\sqrt{7} = \log 7^{1/2} = \frac{1}{2}\log 7$ **d.** $\log_5 1 = 0$

e. $\log_{45} 45 = 1$ ■

EXAMPLE 4 Given that $\log 2 \approx 0.3010$, $\log 3 \approx 0.4771$, and $\log 5 \approx 0.6990$, use the laws of logarithms to find

a. log 15 **b.** log 7.5 **c.** log 81 **d.** log 50

Solution

a. Note that $15 = 3 \cdot 5$, so by Law 1 for logarithms,

$$\begin{aligned}\log 15 &= \log 3 \cdot 5\\ &= \log 3 + \log 5\\ &\approx 0.4771 + 0.6990\\ &= 1.1761\end{aligned}$$

b. Observing that $7.5 = 15/2 = (3 \cdot 5)/2$, we apply Laws 1 and 2, obtaining

$$\begin{aligned}\log 7.5 &= \log\frac{(3)(5)}{2}\\ &= \log 3 + \log 5 - \log 2\\ &\approx 0.4771 + 0.6990 - 0.3010\\ &= 0.8751\end{aligned}$$

c. Since $81 = 3^4$, we apply Law 3 to obtain

$$\begin{aligned}\log 81 &= \log 3^4\\ &= 4\log 3\\ &\approx 4(0.4771)\\ &= 1.9084\end{aligned}$$

d. We write $50 = 5 \cdot 10$ and find

$$\begin{aligned}\log 50 &= \log(5)(10)\\ &= \log 5 + \log 10\\ &\approx 0.6990 + 1 \qquad \text{Use Law 5}\\ &= 1.6990\end{aligned}$$

■

EXAMPLE 5 Expand and simplify the following expressions:

a. $\log_3 x^2y^3$ **b.** $\log_2 \dfrac{x^2+1}{2^x}$ **c.** $\ln \dfrac{x^2\sqrt{x^2-1}}{e^x}$

Solution

a. $\log_3 x^2y^3 = \log_3 x^2 + \log_3 y^3$ Law 1

$= 2\log_3 x + 3\log_3 y$ Law 3

b. $\log_2 \dfrac{x^2+1}{2^x} = \log_2(x^2+1) - \log_2 2^x$ Law 2

$= \log_2(x^2+1) - x\log_2 2$ Law 3

$= \log_2(x^2+1) - x$ Law 5

c. $\ln \dfrac{x^2\sqrt{x^2-1}}{e^x} = \ln \dfrac{x^2(x^2-1)^{1/2}}{e^x}$ Rewrite

$= \ln x^2 + \ln(x^2-1)^{1/2} - \ln e^x$ Laws 1 and 2

$= 2\ln x + \dfrac{1}{2}\ln(x^2-1) - x\ln e$ Law 3

$= 2\ln x + \dfrac{1}{2}\ln(x^2-1) - x$ Law 5

Logarithmic Functions and Their Graphs

The definition of a logarithm implies that if b and n are positive numbers and b is different from 1, then the expression $\log_b n$ is a real number. This enables us to define a logarithmic function as follows:

Logarithmic Function

The function defined by

$$f(x) = \log_b x \qquad (b > 0, b \neq 1)$$

is called the **logarithmic function with base *b*.** The domain of f is the set of all positive numbers.

One easy way to obtain the graph of the logarithmic function $y = \log_b x$ is to construct a table of values of the logarithm (base b). However, another method—and a more instructive one—is based on exploiting the intimate relationship between logarithmic and exponential functions.

If a point (u, v) lies on the graph of $y = \log_b x$, then

$$v = \log_b u$$

But we can also write this equation in exponential form as

$$u = b^v$$

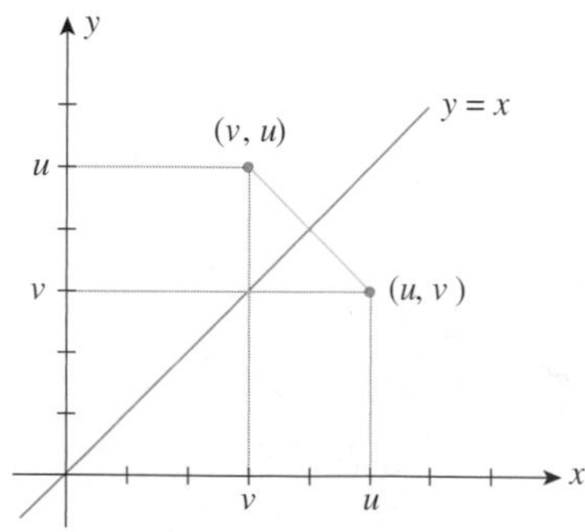

FIGURE 7
The points (u, v) and (v, u) are mirror reflections of each other.

So the point (v, u) also lies on the graph of the function $y = b^x$. Let's look at the relationship between the points (u, v) and (v, u) and the line $y = x$ (Figure 7). If we think of the line $y = x$ as a mirror, then the point (v, u) is the mirror reflection of the point (u, v). Similarly, the point (u, v) is a mirror reflection of the point (v, u). We can take

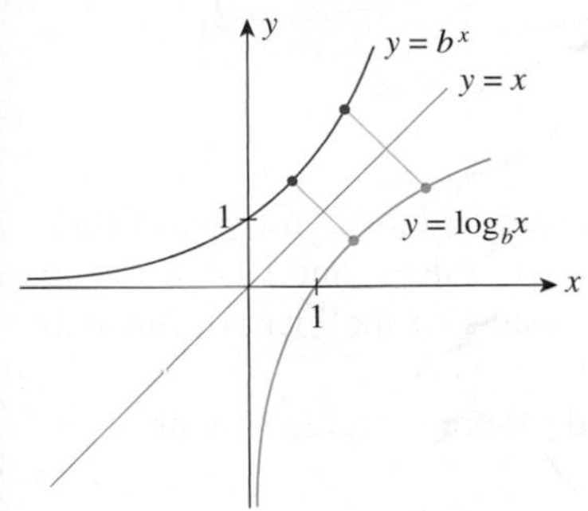

FIGURE 8
The graphs of $y = b^x$ and $y = \log_b x$ are mirror reflections of each other.

advantage of this relationship to help us draw the graph of logarithmic functions. For example, if we wish to draw the graph of $y = \log_b x$, where $b > 1$, then we need only draw the mirror reflection of the graph of $y = b^x$ with respect to the line $y = x$ (Figure 8).

You may discover the following properties of the logarithmic function by taking the reflection of the graph of an appropriate exponential function (Exercises 33 and 34 on page 345).

Properties of the Logarithmic Function

The logarithmic function $y = \log_b x$ $(b > 0, b \neq 1)$ has the following properties:

1. Its domain is $(0, \infty)$.
2. Its range is $(-\infty, \infty)$.
3. Its graph passes through the point $(1, 0)$.
4. It is continuous on $(0, \infty)$.
5. It is increasing on $(0, \infty)$ if $b > 1$ and decreasing on $(0, \infty)$ if $b < 1$.

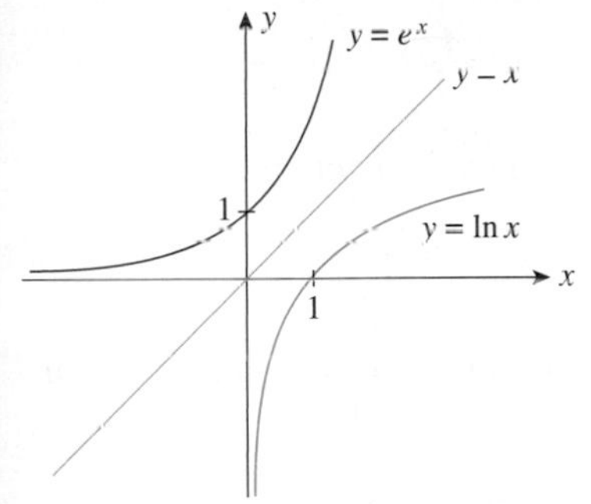

FIGURE 9
The graph of $y = \ln x$ is the mirror reflection of the graph of $y = e^x$.

EXAMPLE 6 Sketch the graph of the function $y = \ln x$.

Solution We first sketch the graph of $y = e^x$. Then, the required graph is obtained by tracing the mirror reflection of the graph of $y = e^x$ with respect to the line $y = x$ (Figure 9). ■

Properties Relating the Exponential and Logarithmic Functions

We made use of the relationship that exists between the exponential function $f(x) = e^x$ and the logarithmic function $g(x) = \ln x$ when we sketched the graph of g in Example 6. This relationship is further described by the following properties, which are an immediate consequence of the definition of the logarithm of a number.

Properties Relating e^x and $\ln x$

$$e^{\ln x} = x \qquad (x > 0) \qquad \textbf{(2)}$$

$$\ln e^x = x \qquad (\text{for any real number } x) \qquad \textbf{(3)}$$

(Try to verify these properties.)

From Properties 2 and 3, we conclude that the composite function

$$\begin{aligned}(f \circ g)(x) &= f[g(x)] \\ &= e^{\ln x} = x \\ (g \circ f)(x) &= g[f(x)] \\ &= \ln e^x = x\end{aligned}$$

Thus,

$$f[g(x)] = g[f(x)]$$
$$= x$$

Any two functions f and g that satisfy this relationship are said to be **inverses** of each other. Note that the function f undoes what the function g does, and vice versa, so the composition of the two functions in any order results in the identity function $F(x) = x$.*

The relationships expressed in Equations (2) and (3) are useful in solving equations that involve exponentials and logarithms.

EXPLORING WITH TECHNOLOGY

You can demonstrate the validity of Properties 2 and 3, which state that the exponential function $f(x) = e^x$ and the logarithmic function $g(x) = \ln x$ are inverses of each other as follows:

1. Sketch the graph of $(f \circ g)(x) = e^{\ln x}$, using the viewing window $[0, 10] \times [0, 10]$. Interpret the result.
2. Sketch the graph of $(g \circ f)(x) = \ln e^x$, using the standard viewing window. Interpret the result.

EXAMPLE 7 Solve the equation $2e^{x+2} = 5$.

Solution We first divide both sides of the equation by 2 to obtain

$$e^{x+2} = \frac{5}{2} = 2.5$$

Next, taking the natural logarithm of each side of the equation and using Equation (3), we have

$$\ln e^{x+2} = \ln 2.5$$
$$x + 2 = \ln 2.5$$
$$x = -2 + \ln 2.5$$
$$\approx -1.08$$

■

EXPLORE & DISCUSS

Consider the equation $y = y_0 b^{kx}$, where y_0 and k are positive constants and $b > 0$, $b \neq 1$. Suppose we want to express y in the form $y = y_0 e^{px}$. Use the laws of logarithms to show that $p = k \ln b$ and hence that $y = y_0 e^{(k \ln b)x}$ is an alternative form of $y = y_0 b^{kx}$ using the base e.

EXAMPLE 8 Solve the equation $5 \ln x + 3 = 0$.

Solution Adding -3 to both sides of the equation leads to

$$5 \ln x = -3$$
$$\ln x = -\frac{3}{5} = -0.6$$

and so

$$e^{\ln x} = e^{-0.6}$$

Using Equation (2), we conclude that

$$x = e^{-0.6}$$
$$\approx 0.55$$

■

*For a more extensive treatment of inverse functions, see the appendix.

5.2 Self-Check Exercises

1. Sketch the graph of $y = 3^x$ and $y = \log_3 x$ on the same set of axes.

2. Solve the equation $3e^{x+1} - 2 = 4$.

Solutions to Self-Check Exercises 5.2 can be found on page 347.

5.2 Concept Questions

1. a. Define $y = \log_b x$.
   b. Define the logarithmic function f with base b. What restrictions, if any, are placed on b?

2. For the logarithmic function $y = \log_b x$ $(b > 0, b \neq 1)$, state (a) its domain and range, (b) its x-intercept, (c) where it is continuous, and (d) where it is increasing and where it is decreasing for the case $b > 1$ and the case $b < 1$.

3. a. If $x > 0$, what is $e^{\ln x}$?
   b. If x is any real number, what is $\ln e^x$?

5.2 Exercises

In Exercises 1–10, express each equation in logarithmic form.

1. $2^6 = 64$
2. $3^5 = 243$
3. $3^{-2} = \dfrac{1}{9}$
4. $5^{-3} = \dfrac{1}{125}$
5. $\left(\dfrac{1}{3}\right)^1 = \dfrac{1}{3}$
6. $\left(\dfrac{1}{2}\right)^{-4} = 16$
7. $32^{3/5} = 8$
8. $81^{3/4} = 27$
9. $10^{-3} = 0.001$
10. $16^{-1/4} = 0.5$

In Exercises 11–16, use the facts that log 3 = 0.4771 and log 4 = 0.6021 to find the value of each logarithm.

11. $\log 12$
12. $\log \dfrac{3}{4}$
13. $\log 16$
14. $\log \sqrt{3}$
15. $\log 48$
16. $\log \dfrac{1}{300}$

In Exercises 17–20, write the expression as the logarithm of a single quantity.

17. $2 \ln a + 3 \ln b$
18. $\dfrac{1}{2} \ln x + 2 \ln y - 3 \ln z$
19. $\ln 3 + \dfrac{1}{2} \ln x + \ln y - \dfrac{1}{3} \ln z$
20. $\ln 2 + \dfrac{1}{2} \ln(x + 1) - 2 \ln(1 + \sqrt{x})$

In Exercises 21–28, use the laws of logarithms to expand and simplify the expression.

21. $\log x(x + 1)^4$
22. $\log x(x^2 + 1)^{-1/2}$
23. $\log \dfrac{\sqrt{x + 1}}{x^2 + 1}$
24. $\ln \dfrac{e^x}{1 + e^x}$
25. $\ln xe^{-x^2}$
26. $\ln x(x + 1)(x + 2)$
27. $\ln \dfrac{x^{1/2}}{x^2\sqrt{1 + x^2}}$
28. $\ln \dfrac{x^2}{\sqrt{x}(1 + x)^2}$

In Exercises 29–32, sketch the graph of the equation.

29. $y = \log_3 x$
30. $y = \log_{1/3} x$
31. $y = \ln 2x$
32. $y = \ln \dfrac{1}{2}x$

In Exercises 33 and 34, sketch the graphs of the equations on the same coordinate axes.

33. $y = 2^x$ and $y = \log_2 x$
34. $y = e^{3x}$ and $y = \ln 3x$

In Exercises 35–44, use logarithms to solve the equation for *t*.

35. $e^{0.4t} = 8$
36. $\dfrac{1}{3}e^{-3t} = 0.9$
37. $5e^{-2t} = 6$
38. $4e^{t-1} = 4$
39. $2e^{-0.2t} - 4 = 6$
40. $12 - e^{0.4t} = 3$
41. $\dfrac{50}{1 + 4e^{0.2t}} = 20$
42. $\dfrac{200}{1 + 3e^{-0.3t}} = 100$
43. $A = Be^{-t/2}$
44. $\dfrac{A}{1 + Be^{t/2}} = C$

45. Blood Pressure A normal child's systolic blood pressure may be approximated by the function

$$p(x) = m(\ln x) + b$$

where $p(x)$ is measured in millimeters of mercury, x is

measured in pounds, and m and b are constants. Given that $m = 19.4$ and $b = 18$, determine the systolic blood pressure of a child who weighs 92 lb.

46. MAGNITUDE OF EARTHQUAKES On the Richter scale, the magnitude R of an earthquake is given by the formula

$$R = \log \frac{I}{I_0}$$

where I is the intensity of the earthquake being measured and I_0 is the standard reference intensity.

a. Express the intensity I of an earthquake of magnitude $R = 5$ in terms of the standard intensity I_0.

b. Express the intensity I of an earthquake of magnitude $R = 8$ in terms of the standard intensity I_0. How many times greater is the intensity of an earthquake of magnitude 8 than one of magnitude 5?

c. In modern times, the greatest loss of life attributable to an earthquake occurred in eastern China in 1976. Known as the Tangshan earthquake, it registered 8.2 on the Richter scale. How does the intensity of this earthquake compare with the intensity of an earthquake of magnitude $R = 5$?

47. SOUND INTENSITY The relative loudness of a sound D of intensity I is measured in decibels (db), where

$$D = 10 \log \frac{I}{I_0}$$

and I_0 is the standard threshold of audibility.

a. Express the intensity I of a 30-db sound (the sound level of normal conversation) in terms of I_0.

b. Determine how many times greater the intensity of an 80-db sound (rock music) is than that of a 30-db sound.

c. Prolonged noise above 150 db causes immediate and permanent deafness. How does the intensity of a 150-db sound compare with the intensity of an 80-db sound?

48. BAROMETRIC PRESSURE Halley's law states that the barometric pressure (in inches of mercury) at an altitude of x mi above sea level is approximated by the equation

$$p(x) = 29.92e^{-0.2x} \qquad (x \geq 0)$$

If the barometric pressure as measured by a hot-air balloonist is 20 in. of mercury, what is the balloonist's altitude?

49. HEIGHT OF TREES The height (in feet) of a certain kind of tree is approximated by

$$h(t) = \frac{160}{1 + 240e^{-0.2t}}$$

where t is the age of the tree in years. Estimate the age of an 80-ft tree.

50. NEWTON'S LAW OF COOLING The temperature of a cup of coffee t min after it is poured is given by

$$T = 70 + 100e^{-0.0446t}$$

where T is measured in degrees Fahrenheit.

a. What is the temperature of the coffee when it was poured?

b. When will the coffee be cool enough to drink (say, 120°F)?

51. LENGTHS OF FISH The length (in centimeters) of a typical Pacific halibut t yr old is approximately

$$f(t) = 200(1 - 0.956e^{-0.18t})$$

Suppose a Pacific halibut caught by Mike measures 140 cm. What is its approximate age?

52. ABSORPTION OF DRUGS The concentration of a drug in an organ at any time t (in seconds) is given by

$$x(t) = 0.08(1 - e^{-0.02t})$$

where $x(t)$ is measured in grams/cubic centimeter (g/cm^3).

a. How long would it take for the concentration of the drug in the organ to reach 0.02 g/cm^3?

b. How long would it take for the concentration of the drug in the organ to reach 0.04 g/cm^3?

53. ABSORPTION OF DRUGS The concentration of a drug in an organ at any time t (in seconds) is given by

$$x(t) = 0.08 + 0.12e^{-0.02t}$$

where $x(t)$ is measured in grams/cubic centimeter (g/cm^3).

a. How long would it take for the concentration of the drug in the organ to reach 0.18 g/cm^3?

b. How long would it take for the concentration of the drug in the organ to reach 0.16 g/cm^3?

54. FORENSIC SCIENCE Forensic scientists use the following law to determine the time of death of accident or murder victims. If T denotes the temperature of a body t hr after death, then

$$T = T_0 + (T_1 - T_0)(0.97)^t$$

where T_0 is the air temperature and T_1 is the body temperature at the time of death. John Doe was found murdered at midnight in his house, when the room temperature was 70°F and his body temperature was 80°F. When was he killed? Assume that the normal body temperature is 98.6°F.

In Exercises 55–58, determine whether the statement is true or false. If it is true, explain why it is true. If it is false, give an example to show why it is false.

55. $(\ln x)^3 = 3 \ln x$ for all x in $(0, \infty)$.

56. $\ln a - \ln b = \ln(a - b)$ for all positive real numbers a and b.

57. The function $f(x) = \dfrac{1}{\ln x}$ is continuous on $(1, \infty)$.

58. The function $f(x) = \ln|x|$ is continuous for all $x \neq 0$.

59. a. Given that $2^x = e^{kx}$, find k.

b. Show that, in general, if b is a nonnegative real number, then any equation of the form $y = b^x$ may be written in the form $y = e^{kx}$, for some real number k.

60. Use the definition of a logarithm to prove

a. $\log_b mn = \log_b m + \log_b n$

b. $\log_b \frac{m}{n} = \log_b m - \log_b n$

Hint: Let $\log_b m = p$ and $\log_b n = q$. Then, $b^p = m$ and $b^q = n$.

61. Use the definition of a logarithm to prove

$$\log_b m^n = n \log_b m$$

62. Use the definition of a logarithm to prove

a. $\log_b 1 = 0$

b. $\log_b b = 1$

5.2 Solutions to Self-Check Exercises

1. First, sketch the graph of $y = 3^x$ with the help of the following table of values:

x	−3	−2	−1	0	1	2	3
$y = 3^x$	1/27	1/9	1/3	0	3	9	27

Next, take the mirror reflection of this graph with respect to the line $y = x$ to obtain the graph of $y = \log_3 x$.

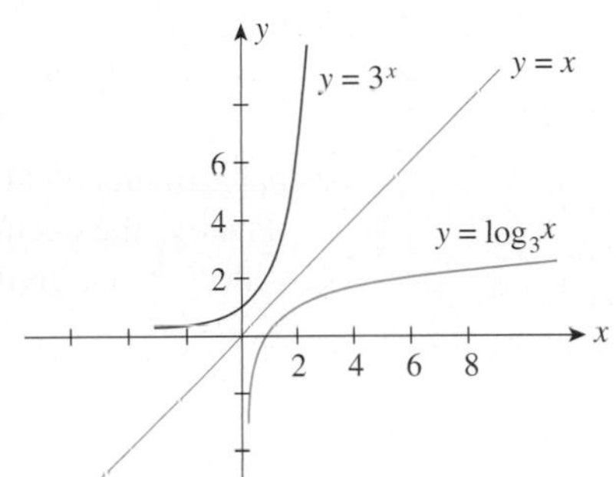

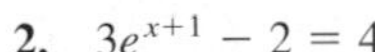

2. $$\begin{aligned} 3e^{x+1} - 2 &= 4 \\ 3e^{x+1} &= 6 \\ e^{x+1} &= 2 \\ \ln e^{x+1} &= \ln 2 && \text{Take the logarithm of both sides.} \\ (x+1)\ln e &= \ln 2 && \text{Law 3} \\ x + 1 &= \ln 2 && \text{Law 5} \\ x &= \ln 2 - 1 \\ &\approx -0.3069 \end{aligned}$$

5.3 Compound Interest

Compound Interest

Compound interest is a natural application of the exponential function to the business world. We begin by recalling that simple interest is interest that is computed only on the original principal. Thus, if I denotes the interest on a principal P (in dollars) at an interest rate of r per year for t years, then we have

$$I = Prt$$

The **accumulated amount** A, the sum of the principal and interest after t years, is given by

$$\begin{aligned} A &= P + I = P + Prt \\ &= P(1 + rt) && \text{Simple interest formula} \qquad (4) \end{aligned}$$

Frequently, interest earned is periodically added to the principal and thereafter earns interest itself at the same rate. This is called **compound interest**. To find a formula for the accumulated amount, let's consider a numerical example. Suppose

\$1000 (the principal) is deposited in a bank for a **term** of 3 years, earning interest at the rate of 8% per year (called the **nominal,** or **stated, rate**) compounded annually. Then, using Formula (4) with $P = 1000$, $r = 0.08$, and $t = 1$, we see that the accumulated amount at the end of the first year is

$$\begin{aligned} A_1 &= P(1 + rt) \\ &= 1000[1 + 0.08(1)] = 1000(1.08) = 1080 \end{aligned}$$

or \$1080.

To find the accumulated amount A_2 at the end of the second year, we use (4) once again, this time with $P = A_1$. (Remember, the principal *and* interest now earn interest over the second year.) We obtain

$$\begin{aligned} A_2 &= P(1 + rt) = A_1(1 + rt) \\ &= 1000[1 + 0.08(1)][1 + 0.08(1)] \\ &= 1000(1 + 0.08)^2 = 1000(1.08)^2 \approx 1166.40 \end{aligned}$$

or approximately \$1166.40.

Finally, the accumulated amount A_3 at the end of the third year is found using (4) with $P = A_2$, giving

$$\begin{aligned} A_3 &= P(1 + rt) = A_2(1 + rt) \\ &= 1000[1 + 0.08(1)]^2[1 + 0.08(1)] \\ &= 1000(1 + 0.08)^3 = 1000(1.08)^3 \approx 1259.71 \end{aligned}$$

or approximately \$1259.71.

If you reexamine our calculations in this example, you will see that the accumulated amounts at the end of each year have the following form:

$$\begin{array}{llll} \text{First year:} & A_1 = 1000(1 + 0.08) & \text{or} & A_1 = P(1 + r) \\ \text{Second year:} & A_2 = 1000(1 + 0.08)^2 & \text{or} & A_2 = P(1 + r)^2 \\ \text{Third year:} & A_3 = 1000(1 + 0.08)^3 & \text{or} & A_3 = P(1 + r)^3 \end{array}$$

These observations suggest the following general result: If P dollars are invested over a term of t years earning interest at the rate of r per year compounded annually, then the accumulated amount is

$$A = P(1 + r)^t \tag{5}$$

Formula (5) was derived under the assumption that interest was compounded *annually*. In practice, however, interest is usually compounded more than once a year. The interval of time between successive interest calculations is called the **conversion period**.

If interest at a nominal rate of r per year is compounded m times a year on a principal of P dollars, then the simple interest rate per conversion period is

$$i = \frac{r}{m} \qquad \frac{\text{Annual interest rate}}{\text{Periods per year}}$$

For example, if the nominal interest rate is 8% per year ($r = 0.08$) and interest is compounded quarterly ($m = 4$), then

$$i = \frac{r}{m} = \frac{0.08}{4} = 0.02$$

or 2% per period.

To find a general formula for the accumulated amount when a principal of P dollars is deposited in a bank for a term of t years and earns interest at the (nominal) rate of r per year compounded m times per year, we proceed as before using Formula (5) repeatedly with the interest rate $i = r/m$. We see that the accumulated amount at the end of each period is as follows:

$$\begin{aligned}
&\text{First period:} && A_1 = P(1+i) \\
&\text{Second period:} && A_2 = A_1(1+i) = [P(1+i)](1+i) = P(1+i)^2 \\
&\text{Third period:} && A_3 = A_2(1+i) = [P(1+i)^2](1+i) = P(1+i)^3 \\
&\quad\vdots && \quad\vdots \\
&n\text{th period:} && A_n = A_{n-1}(1+i) = [P(1+i)^{n-1}](1+i) = P(1+i)^n
\end{aligned}$$

But there are $n = mt$ periods in t years (number of conversion periods times the term). Therefore, the accumulated amount at the end of t years is given by

Compound Interest Formula

$$A = P\left(1 + \frac{r}{m}\right)^{mt} \tag{6}$$

where

A = Accumulated amount at the end of t years

P = Principal

r = Nominal interest rate per year

m = Number of conversion periods per year

t = Term (number of years)

EXAMPLE 1 Find the accumulated amount after 3 years if \$1000 is invested at 8% per year compounded (a) annually, (b) semiannually, (c) quarterly, (d) monthly, and (e) daily.

Solution

a. Here, $P = 1000$, $r = 0.08$, $m = 1$, and $t = 3$, so Formula (6) gives

$$\begin{aligned} A &= 1000(1 + 0.08)^3 \\ &= 1259.71 \end{aligned}$$

or \$1259.71

b. Here, $P = 1000$, $r = 0.08$, $m = 2$, and $t = 3$, so (6) gives

$$\begin{aligned} A &= 1000\left(1 + \frac{0.08}{2}\right)^{(2)(3)} \\ &= 1265.32 \end{aligned}$$

or \$1265.32.

c. In this case, $P = 1000$, $r = 0.08$, $m = 4$, and $t = 3$, so (6) gives

$$\begin{aligned} A &= 1000\left(1 + \frac{0.08}{4}\right)^{(4)(3)} \\ &= 1268.24 \end{aligned}$$

or \$1268.24

d. Here, $P = 1000$, $r = 0.08$, $m = 12$, and $t = 3$, so (6) gives

$$A = 1000\left(1 + \frac{0.08}{12}\right)^{(12)(3)}$$
$$= 1270.24$$

or $1270.24.

e. Here, $P = 1000$, $r = 0.08$, $m = 365$, and $t = 3$, so (6) gives

$$A = 1000\left(1 + \frac{0.08}{365}\right)^{(365)(3)}$$
$$= 1271.22$$

or $1271.22. These results are summarized in Table 2.

TABLE 2

Nominal Rate, r	Conversion Period	Term in Years	Initial Investment	Accumulated Amount
8%	Annual ($m = 1$)	3	$1000	$1259.71
8	Semiannual ($m = 2$)	3	1000	1265.32
8	Quarterly ($m = 4$)	3	1000	1268.24
8	Monthly ($m = 12$)	3	1000	1270.24
8	Daily ($m = 365$)	3	1000	1271.22

Effective Rate of Interest

In the last example, we saw that the interest actually earned on an investment depends on the frequency with which the interest is compounded. Thus, the stated, or nominal, rate of 8% per year does not reflect the actual rate at which interest is earned. This suggests that we need to find a common basis for comparing interest rates. One such way of comparing interest rates is provided by using the effective rate. The **effective rate** is the *simple* interest rate that would produce the same accumulated amount in 1 year as the nominal rate compounded m times a year. The effective rate is also called the **true rate**.

To derive a relation between the nominal interest rate, r per year compounded m times, and its corresponding effective rate, r_{eff} per year, let's assume an initial investment of P dollars. Then, the accumulated amount after 1 year at a simple interest rate of r_{eff} per year is

$$A = P(1 + r_{\text{eff}})$$

Also, the accumulated amount after 1 year at an interest rate of r per year compounded m times a year is

$$A = P\left(1 + \frac{r}{m}\right)^{m} \qquad \text{Since } t = 1$$

Equating the two expressions gives

$$P(1 + r_{\text{eff}}) = P\left(1 + \frac{r}{m}\right)^{m}$$
$$1 + r_{\text{eff}} = \left(1 + \frac{r}{m}\right)^{m} \qquad \text{Divide both sides by } P.$$

or, upon solving for r_{eff}, we obtain the formula for computing the effective rate of interest:

Effective Rate of Interest Formula

$$r_{\text{eff}} = \left(1 + \frac{r}{m}\right)^m - 1 \tag{7}$$

where

r_{eff} = Effective rate of interest

r = Nominal interest rate per year

m = Number of conversion periods per year

EXAMPLE 2 Find the effective rate of interest corresponding to a nominal rate of 8% per year compounded (a) annually, (b) semiannually, (c) quarterly, (d) monthly, and (e) daily.

Solution

a. The effective rate of interest corresponding to a nominal rate of 8% per year compounded annually is of course given by 8% per year. This result is also confirmed by using Formula (7) with $r = 0.08$ and $m = 1$. Thus,

$$r_{\text{eff}} = (1 + 0.08) - 1 = 0.08$$

b. Let $r = 0.08$ and $m = 2$. Then, (7) yields

$$r_{\text{eff}} = \left(1 + \frac{0.08}{2}\right)^2 - 1 = 0.0816$$

so the required effective rate is 8.16% per year.

c. Let $r = 0.08$ and $m = 4$. Then, (7) yields

$$r_{\text{eff}} = \left(1 + \frac{0.08}{4}\right)^4 - 1 = 0.08243$$

so the corresponding effective rate in this case is 8.243% per year.

d. Let $r = 0.08$ and $m = 12$. Then, (7) yields

$$r_{\text{eff}} = \left(1 + \frac{0.08}{12}\right)^{12} - 1 = 0.08300$$

so the corresponding effective rate in this case is 8.3% per year.

e. Let $r = 0.08$ and $m = 365$. Then, (7) yields

$$r_{\text{eff}} = \left(1 + \frac{0.08}{365}\right)^{365} - 1 = 0.08328$$

so the corresponding effective rate in this case is 8.328% per year. ■

Now, if the effective rate of interest r_{eff} is known, then the accumulated amount after t years on an investment of P dollars may be more readily computed by using the formula

$$A = P(1 + r_{eff})^t$$

The 1968 Truth in Lending Act passed by Congress requires that the effective rate of interest be disclosed in all contracts involving interest charges. The passage of this act has benefited consumers because they now have a common basis for comparing the various nominal rates quoted by different financial institutions. Furthermore, knowing the effective rate enables consumers to compute the actual charges involved in a transaction. Thus, if the effective rates of interest found in Example 2 were known, the accumulated values of Example 1, shown in Table 3, could have been readily found.

TABLE 3

Nominal Rate, r	Frequency of Interest Payment	Effective Rate	Initial Investment	Accumulated Amount after 3 Years
8%	Annually	8%	\$1000	$1000(1 + 0.08)^3 = \$1259.71$
8	Semiannually	8.16	1000	$1000(1 + 0.0816)^3 = 1265.32$
8	Quarterly	8.243	1000	$1000(1 + 0.08243)^3 = 1268.23$
8	Monthly	8.300	1000	$1000(1 + 0.08300)^3 = 1270.24$
8	Daily	8.328	1000	$1000(1 + 0.08328)^3 = 1271.22$

Present Value

Let's return to the compound interest Formula (6), which expresses the accumulated amount at the end of t years when interest at the rate of r is compounded m times a year. The principal P in (6) is often referred to as the **present value**, and the accumulated value A is called the **future value** since it is realized at a future date. In certain instances an investor may wish to determine how much money he should invest now, at a fixed rate of interest, so that he will realize a certain sum at some future date. This problem may be solved by expressing P in terms of A. Thus, from (6) we find

Present Value Formula for Compound Interest

$$P = A\left(1 + \frac{r}{m}\right)^{-mt} \quad \textbf{(8)}$$

EXAMPLE 3 How much money should be deposited in a bank paying interest at the rate of 6% per year compounded monthly so that at the end of 3 years the accumulated amount will be \$20,000?

Solution Here, $A = 20{,}000$, $r = 0.06$, $m = 12$, and $t = 3$. Using Formula (8), we obtain

$$P = 20{,}000\left(1 + \frac{0.06}{12}\right)^{-(12)(3)}$$
$$\approx 16{,}713$$

or \$16,713. ■

EXAMPLE 4 Find the present value of \$49,158.60 due in 5 years at an interest rate of 10% per year compounded quarterly.

Solution Using Formula (8) with $A = 49{,}158.60$, $r = 0.1$, $m = 4$, and $t = 5$, we obtain

$$P = (49{,}158.6)\left(1 + \frac{0.1}{4}\right)^{-(4)(5)} \approx 30{,}000$$

or \$30,000. ■

Continuous Compounding of Interest

One question that arises naturally in the study of compound interest is: What happens to the accumulated amount over a fixed period of time if the interest is computed more and more frequently?

Intuition suggests that the more often interest is compounded, the larger the accumulated amount will be. This is confirmed by the results of Example 1, where we found that the accumulated amounts did in fact increase when we increased the number of conversion periods per year.

This leads us to another question: Does the accumulated amount approach a limit when the interest is computed more and more frequently over a fixed period of time?

To answer this question, let's look again at the compound interest formula:

$$A = P\left(1 + \frac{r}{m}\right)^{mt} \qquad \textbf{(9)}$$

Recall that m is the number of conversion periods per year. So to find an answer to our problem, we should let m get larger and larger (approach infinity) in (9). But first we will rewrite this equation in the form

$$A = P\left[\left(1 + \frac{r}{m}\right)^{m}\right]^{t} \qquad \text{Since } b^{xy} = (b^x)^y$$

Now, letting $m \to \infty$, we find that

$$\lim_{m\to\infty}\left[P\left(1 + \frac{r}{m}\right)^{m}\right]^{t} = P\left[\lim_{m\to\infty}\left(1 + \frac{r}{m}\right)^{m}\right]^{t} \qquad \text{Why?}$$

Next, upon making the substitution $u = m/r$ and observing that $u \to \infty$ as $m \to \infty$, the foregoing expression reduces to

$$P\left[\lim_{u\to\infty}\left(1 + \frac{1}{u}\right)^{ur}\right]^{t} = P\left[\lim_{u\to\infty}\left(1 + \frac{1}{u}\right)^{u}\right]^{rt}$$

But

$$\lim_{u\to\infty}\left(1 + \frac{1}{u}\right)^{u} = e \qquad \text{Use (1).}$$

so

$$\lim_{m\to\infty} P\left[\left(1 + \frac{r}{m}\right)^{m}\right]^{t} = Pe^{rt}$$

Our computations tell us that as the frequency with which interest is compounded increases without bound, the accumulated amount approaches Pe^{rt}. In this situation, we say that interest is *compounded continuously.* Let's summarize this important result.

Continuous Compound Interest Formula

$$A = Pe^{rt} \tag{10}$$

where

P = Principal

r = Annual interest rate compounded continuously

t = Time in years

A = Accumulated amount at the end of t years

EXAMPLE 5 Find the accumulated amount after 3 years if \$1000 is invested at 8% per year compounded (a) daily (assume a 365-day year) and (b) continuously.

Solution

a. Using Formula (6) with $P = 1000$, $r = 0.08$, $m = 365$, and $t = 3$, we find

$$A = 1000\left(1 + \frac{0.08}{365}\right)^{(365)(3)} \approx 1271.22$$

or \$1271.22.

b. Here we use Formula (10) with $P = 1000$, $r = 0.08$, and $t = 3$, obtaining

$$\begin{aligned} A &= 1000e^{(0.08)(3)} \\ &\approx 1271.25 \end{aligned}$$

or \$1271.25. ■

Observe that the accumulated amounts corresponding to interest compounded daily and interest compounded continuously differ by very little. The continuous compound interest formula is a very important tool in theoretical work in financial analysis.

EXPLORING WITH TECHNOLOGY

In the opening paragraph of Section 5.1, we pointed out that the accumulated amount of an account earning interest *compounded continuously* will eventually outgrow by far the accumulated amount of an account earning interest at the same nominal rate but earning simple interest. Illustrate this fact using the following example.

Suppose you deposit \$1000 in account I, earning interest at the rate of 10% per year compounded continuously so that the accumulated amount at the end of t years is $A_1(t) = 1000e^{0.1t}$. Suppose you also deposit \$1000 in account II, earning simple interest at the rate of 10% per year so that the accumulated amount at the end of t years is $A_2(t) = 1000(1 + 0.1t)$. Use a graphing utility to sketch the graphs of the functions A_1 and A_2 in the viewing window $[0, 20] \times [0, 10{,}000]$ to see the accumulated amounts $A_1(t)$ and $A_2(t)$ over a 20-year period.

If we solve Formula (10) for P, we obtain

$$P = Ae^{-rt} \tag{11}$$

which gives the present value in terms of the future (accumulated) value for the case of continuous compounding.

VIDEO

APPLIED EXAMPLE 6 Real Estate Investment Blakely Investment Company owns an office building located in the commercial district of a city. As a result of the continued success of an urban renewal program, local business is enjoying a miniboom. The market value of Blakely's property is

$$V(t) = 300{,}000e^{\sqrt{t}/2}$$

where $V(t)$ is measured in dollars and t is the time in years from the present. If the expected rate of appreciation is 9% compounded continuously for the next 10 years, find an expression for the present value $P(t)$ of the market price of the property valid for the next 10 years. Compute $P(7)$, $P(8)$, and $P(9)$, and interpret your results.

Solution Using Formula (11) with $A = V(t)$ and $r = 0.09$, we find that the present value of the market price of the property t years from now is

$$\begin{aligned} P(t) &= V(t)e^{-0.09t} \\ &= 300{,}000e^{-0.09t+\sqrt{t}/2} \qquad (0 \leq t \leq 10) \end{aligned}$$

Letting $t = 7$, 8, and 9, respectively, we find that

$$\begin{aligned} P(7) &= 300{,}000e^{-0.09(7)+\sqrt{7}/2} \approx 599{,}837 \quad \text{or} \quad \$599{,}837 \\ P(8) &= 300{,}000e^{-0.09(8)+\sqrt{8}/2} \approx 600{,}640 \quad \text{or} \quad \$600{,}640 \\ P(9) &= 300{,}000e^{-0.09(9)+\sqrt{9}/2} \approx 598{,}115 \quad \text{or} \quad \$598{,}115 \end{aligned}$$

From the results of these computations, we see that the present value of the property's market price seems to decrease after a certain period of growth. This suggests that there is an optimal time for the owners to sell. Later we will show that the highest present value of the property's market price is \$600,779, which occurs at time $t = 7.72$ years. ■

EXPLORING WITH TECHNOLOGY

The effective rate of interest is given by

$$r_{\text{eff}} = \left(1 + \frac{r}{m}\right)^m - 1$$

where the number of conversion periods per year is m. In Exercise 43 you will be asked to show that the effective rate of interest r_{eff} corresponding to a nominal interest rate r per year compounded continuously is given by

$$\hat{r}_{\text{eff}} = e^r - 1$$

To obtain a visual confirmation of this result, consider the special case where $r = 0.1$ (10% per year).

1. Use a graphing utility to plot the graph of both

$$y_1 = \left(1 + \frac{0.1}{x}\right)^x - 1 \quad \text{and} \quad y_2 = e^{0.1} - 1$$

in the viewing window $[0, 3] \times [0, 0.12]$.

2. Does your result seem to imply that

$$\left(1 + \frac{r}{m}\right)^m - 1$$

approaches

$$\hat{r}_{\text{eff}} = e^r - 1$$

as m increases without bound for the special case $r = 0.1$?

The next two examples show how logarithms can be used to solve problems involving compound interest.

EXAMPLE 7 How long will it take \$10,000 to grow to \$15,000 if the investment earns an interest rate of 12% per year compounded quarterly?

Solution Using Formula (6) with $A = 15{,}000$, $P = 10{,}000$, $r = 0.12$, and $m = 4$, we obtain

$$15{,}000 = 10{,}000\left(1 + \frac{0.12}{4}\right)^{4t}$$

$$(1.03)^{4t} = \frac{15{,}000}{10{,}000} = 1.5$$

Taking the logarithm on each side of the equation gives

$$\ln(1.03)^{4t} = \ln 1.5$$

$$4t \ln 1.03 = \ln 1.5 \qquad \log_b m^n = n \log_b m$$

$$4t = \frac{\ln 1.5}{\ln 1.03}$$

$$t = \frac{\ln 1.5}{4 \ln 1.03} \approx 3.43$$

So it will take approximately 3.4 years for the investment to grow from \$10,000 to \$15,000. ■

EXAMPLE 8 Find the interest rate needed for an investment of \$10,000 to grow to an amount of \$18,000 in 5 years if the interest is compounded monthly.

Solution Using Formula (6) with $A = 18{,}000$, $P = 10{,}000$, $m = 12$, and $t = 5$, we obtain

$$18{,}000 = 10{,}000\left(1 + \frac{r}{12}\right)^{12(5)}$$

Dividing both sides of the equation by 10,000 gives

$$\frac{18{,}000}{10{,}000} = \left(1 + \frac{r}{12}\right)^{60}$$

or, upon simplification,

$$\left(1 + \frac{r}{12}\right)^{60} = 1.8$$

Now, we take the logarithm on each side of the equation, obtaining

$$\ln\left(1 + \frac{r}{12}\right)^{60} = \ln 1.8$$

$$60 \ln\left(1 + \frac{r}{12}\right) = \ln 1.8$$

$$\ln\left(1 + \frac{r}{12}\right) = \frac{\ln 1.8}{60} = 0.009796$$

$$\left(1 + \frac{r}{12}\right) = e^{0.009796} \qquad \text{By Property 2}$$

$$= 1.009844$$

and

$$\frac{r}{12} = 1.009844 - 1$$
$$r = 0.1181$$

or 11.81% per year. ■

5.3 Self-Check Exercises

1. Find the present value of $20,000 due in 3 yr at an interest rate of 12%/year compounded monthly.

2. Glen is a retiree living on Social Security and the income from his investment. Currently, his $100,000 investment in a 1-yr CD is yielding 6.6% interest compounded daily. If he reinvests the principal ($100,000) on the due date of the CD in another 1-yr CD paying 5.2% interest compounded daily, find the net decrease in his yearly income from his investment.

3. **a.** What is the accumulated amount after 5 yr if $10,000 is invested at 8%/year compounded continuously?
   b. Find the present value of $10,000 due in 5 yr at an interest rate of 8%/year compounded continuously.

Solutions to Self-Check Exercises 5.3 can be found on page 359.

5.3 Concept Questions

1. **a.** What is the difference between simple interest and compound interest?
   b. State the simple interest formula and the compound interest formula.

2. **a.** What is the effective rate of interest?
   b. State the formula for computing the effective rate of interest.

3. What is the present value formula for compound interest?

4. State the continuous compound interest formula.

5.3 Exercises

In Exercises 1–4, find the accumulated amount A if the principal P is invested at an interest rate of r per year for t years.

1. $P = \$2500$, $r = 7\%$, $t = 10$, compounded semiannually

2. $P = \$12{,}000$, $r = 8\%$, $t = 10$, compounded quarterly

3. $P = \$150{,}000$, $r = 10\%$, $t = 4$, compounded monthly

4. $P = \$150{,}000$, $r = 9\%$, $t = 3$, compounded daily

In Exercises 5 and 6, find the effective rate corresponding to the given nominal rate.

5. **a.** 10%/year compounded semiannually
   b. 9%/year compounded quarterly

6. **a.** 8%/year compounded monthly
   b. 8%/year compounded daily

In Exercises 7 and 8, find the present value of $40,000 due in 4 yr at the given rate of interest.

7. **a.** 8%/year compounded semiannually
   b. 8%/year compounded quarterly

8. **a.** 7%/year compounded monthly
   b. 9%/year compounded daily

9. Find the accumulated amount after 4 yr if $5000 is invested at 8%/year compounded continuously.

10. An amount of $25,000 is deposited in a bank that pays interest at the rate of 7%/year, compounded annually. What is the total amount on deposit at the end of 6 yr, assuming there are no deposits or withdrawals during those 6 yr? What is the interest earned in that period of time?

11. Find the interest rate needed for an investment of $5000 to grow to an amount of $7500 in 3 yr if interest is compounded monthly.

12. Find the interest rate needed for an investment of $5000 to grow to an amount of $7500 in 3 yr if interest is compounded quarterly.

13. Find the interest rate needed for an investment of $5000 to grow to an amount of $8000 in 4 yr if interest is compounded semiannually.

14. Find the interest rate needed for an investment of \$5000 to grow to an amount of \$5500 in 6 mo if interest is compounded monthly.

15. Find the interest rate needed for an investment of \$2000 to double in 5 yr if interest is compounded annually.

16. Find the interest rate needed for an investment of \$2000 to triple in 5 yr if interest is compounded monthly.

17. How long will it take \$5000 to grow to \$6500 if the investment earns interest at the rate of 12%/year compounded monthly?

18. How long will it take \$12,000 to grow to \$15,000 if the investment earns interest at the rate of 8%/year compounded monthly?

19. How long will it take an investment of \$2000 to double if the investment earns interest at the rate of 9%/year compounded monthly?

20. How long will it take an investment of \$5000 to triple if the investment earns interest at the rate of 8%/year compounded daily?

21. Find the interest rate needed for an investment of \$5000 to grow to an amount of \$6000 in 3 yr if interest is compounded continuously.

22. Find the interest rate needed for an investment of \$4000 to double in 5 yr if interest is compounded continuously.

23. How long will it take an investment of \$6000 to grow to \$7000 if the investment earns interest at the rate of $7\frac{1}{2}$% compounded continuously?

24. How long will it take an investment of \$8000 to double if the investment earns interest at the rate of 8% compounded continuously?

25. Housing Prices The Estradas are planning to buy a house 4 yr from now. Housing experts in their area have estimated that the cost of a home will increase at a rate of 9%/year during that 4-yr period. If this economic prediction holds true, how much can they expect to pay for a house that currently costs \$180,000?

26. Energy Consumption A metropolitan utility company in a western city of the United States expects the consumption of electricity to increase by 8%/year during the next decade, due mainly to the expected population increase. If consumption does increase at this rate, find the amount by which the utility company will have to increase its generating capacity in order to meet the area's needs at the end of the decade.

27. Pension Funds The managers of a pension fund have invested \$1.5 million in U.S. government certificates of deposit (CDs) that pay interest at the rate of 6.5%/year compounded semiannually over a period of 10 yr. At the end of this period, how much will the investment be worth?

28. Savings Accounts Bernie invested a sum of money 5 yr ago in a savings account, which has since paid interest at the rate of 8%/yr compounded quarterly. His investment is now worth \$22,289.22. How much did he originally invest?

29. Loan Consolidation The proprietors of the Coachmen Inn secured two loans from the Union Bank: one for \$8000 due in 3 yr and one for \$15,000 due in 6 yr, both at an interest rate of 10%/year compounded semiannually. The bank agreed to allow the two loans to be consolidated into one loan payable in 5 yr at the same interest rate. How much will the proprietors have to pay the bank at the end of 5 yr?

30. Tax-Deferred Annuities Kate is in the 28% tax bracket and has \$25,000 available for investment during her current tax year. Assume that she remains in the same tax bracket over the next 10 yr and determine the accumulated amount of her investment if she puts the \$25,000 into a

a. Tax-deferred annuity that pays 12%/year, tax deferred for 10 yr.

b. Taxable instrument that pays 12%/year for 10 yr.
Hint: In this case the yield after taxes is 8.64%/year.

31. Revenue Growth of a Home Theater Business Maxwell started a home theater business in 2002. The revenue of his company for that year was \$240,000. The revenue grew by 20% in 2003 and 30% in 2004. Maxwell projected that the revenue growth for his company in the next 3 yr will be at least 25%/year. How much does Maxwell expect his minimum revenue to be for 2007?

32. Online Retail Sales Online retail sales stood at \$23.5 billion for 2000. For the next 2 yr, they grew by 33.2% and 27.8% per year, respectively. For the next 6 yr, online retail sales are projected to grow at 30.5%, 19.9%, 24.3%, 14.0%, 17.6%, and 10.5% per year, respectively. What are the projected online sales for 2008?
Source: Jupiter Research

33. Purchasing Power The inflation rates in the U.S. economy in 2000 through 2003 were 3.4%, 2.8%, 1.6%, and 2.3%, respectively. What was the purchasing power of a dollar at the beginning of 2004 compared to that at the beginning of 2000?
Source: U.S. Census Bureau

34. Investment Returns Zoe purchased a house in 1999 for \$160,000. In 2005 she sold the house and made a net profit of \$56,000. Find the effective annual rate of return on her investment over the 6-yr period.

35. Investment Returns Julio purchased 1000 shares of a certain stock for \$25,250 (including commissions). He sold the shares 2 yr later and received \$32,100 after deducting commissions. Find the effective annual rate of return on his investment over the 2-yr period.

36. INVESTMENT OPTIONS Investment A offers a 10% return compounded semiannually, and investment B offers a 9.75% return compounded continuously. Which investment has a higher rate of return over a 4-yr period?

37. PRESENT VALUE Find the present value of $59,673 due in 5 yr at an interest rate of 8%/year compounded continuously.

38. REAL ESTATE INVESTMENTS A condominium complex was purchased by a group of private investors for $1.4 million and sold 6 yr later for $3.6 million. Find the annual rate of return (compounded continuously) on their investment.

39. SAVING FOR COLLEGE Having received a large inheritance, a child's parents wish to establish a trust for the child's college education. If 7 yr from now they need an estimated $70,000, how much should they set aside in trust now, if they invest the money at 10.5% compounded (a) quarterly? (b) continuously?

40. EFFECT OF INFLATION ON SALARIES Omar's current annual salary is $35,000. How much will he need to earn 10 yr from now in order to retain his present purchasing power if the rate of inflation over that period is 6%/year? Assume that inflation is continuously compounded.

41. PENSIONS Eleni, who is now 50 years old, is employed by a firm that guarantees her a pension of $40,000/year at age 65. What is the present value of her first year's pension if inflation over the next 15 yr is (a) 6%? (b) 8%? (c) 12%? Assume that inflation is continuously compounded.

42. REAL ESTATE INVESTMENTS An investor purchased a piece of waterfront property. Because of the development of a marina in the vicinity, the market value of the property is expected to increase according to the rule

$$V(t) = 80{,}000e^{\sqrt{t}/2}$$

where $V(t)$ is measured in dollars and t is the time in years from the present. If the inflation rate is expected to be 9% compounded continuously for the next 8 yr, find an expression for the present value $P(t)$ of the property's market price valid for the next 8 yr. What is $P(t)$ expected to be in 4 yr?

43. Show that the effective interest rate $\hat{r}_{\text{eff}}$ that corresponds to a nominal interest rate r per year compounded continuously is given by

$$\hat{r}_{\text{eff}} = e^r - 1$$

Hint: From Formula (7) we see that the effective rate $\hat{r}_{\text{eff}}$ corresponding to a nominal interest rate r per year compounded m times a year is given by

$$\hat{r}_{\text{eff}} = \left(1 + \frac{r}{m}\right)^m - 1$$

Let m tend to infinity in this expression.

44. Refer to Exercise 43. Find the effective interest rate that corresponds to a nominal rate of 10%/year compounded (a) quarterly, (b) monthly, and (c) continuously.

45. INVESTMENT ANALYSIS Refer to Exercise 43. Bank A pays interest on deposits at a 7% annual rate compounded quarterly, and Bank B pays interest on deposits at a $7\frac{1}{8}$% annual rate compounded continuously. Which bank has the higher effective rate of interest?

46. INVESTMENT ANALYSIS Find the nominal interest rate that, when compounded monthly, yields an effective interest rate of 10%/year.
Hint: Use Equation (7).

47. INVESTMENT ANALYSIS Find the nominal interest rate that, when compounded continuously, yields an effective interest rate of 10%/year.
Hint: See Exercise 43.

48. ANNUITIES An annuity is a sequence of payments made at regular time intervals. The future value of an annuity of n payments of R dollars each paid at the end of each investment period into an account that earns an interest rate of i/period is

$$S = R\left[\frac{(1 + i)^n - 1}{i}\right]$$

Determine

$$\lim_{i \to 0} R\left[\frac{(1 + i)^n - 1}{i}\right]$$

and interpret your result.
Hint: Use the definition of the derivative.

5.3 Solutions to Self-Check Exercises

1. Using Formula (8) with $A = 20{,}000$, $r = 0.12$, $m = 12$, and $t = 3$, we find the required present value to be

$$P = 20{,}000\left(1 + \frac{0.12}{12}\right)^{-(12)(3)} = 13{,}978.50$$

or $13,978.50.

2. The accumulated amount of Glen's current investment is found by using Formula (6) with $P = 100{,}000$, $r = 0.066$, and $m = 365$. Thus, the required accumulated amount is

$$A = 100{,}000\left(1 + \frac{0.066}{365}\right)^{365} = 106{,}822.03$$

or $106,822.03. Next, we compute the accumulated amount of Glen's reinvestment. Once again, using (6) with $P = 100{,}000$, $r = 0.052$, and $m = 365$, we find the required accumulated amount in this case to be

$$\bar{A} = 100{,}000\left(1 + \frac{0.052}{365}\right)^{365}$$

or $105,337.18. Therefore, Glen can expect to experience a net decrease in yearly income of

$$106{,}822.03 - 105{,}337.18$$

or $1,484.85.

3. **a.** Using Formula (10) with $P = 10{,}000$, $r = 0.08$, and $t = 5$, we find that the required accumulated amount is given by

$$A = 10{,}000e^{(0.08)(5)}$$
$$= 14{,}918.25$$

or $14,918.25.

b. Using Formula (11) with $A = 10{,}000$, $r = 0.08$, and $t = 5$, we see that the required present value is given by

$$P = 10{,}000e^{-(0.08)(5)}$$
$$= 6703.20$$

or $6703.20.

5.4 Differentiation of Exponential Functions

The Derivative of the Exponential Function

To study the effects of budget deficit-reduction plans at different income levels, it is important to know the income distribution of American families. Based on data from the House Budget Committee, the House Ways and Means Committee, and the U.S. Census Bureau, the graph of f shown in Figure 10 gives the number of American families y (in millions) as a function of their annual income x (in thousands of dollars) in 1990.

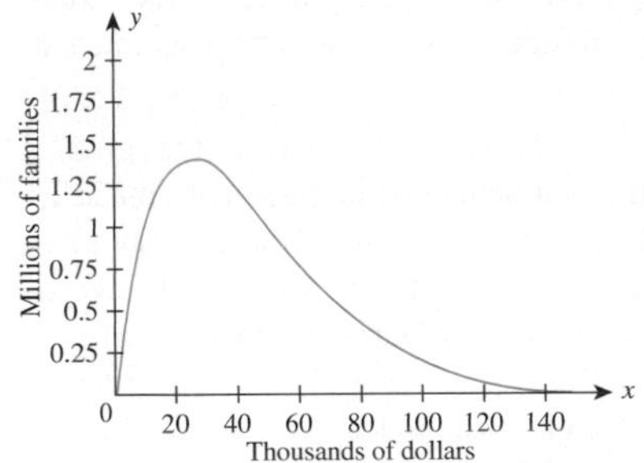

FIGURE 10
The graph of f shows the number of families versus their annual income.

Source: House Budget Committee, House Ways and Means Committee, and U.S. Census Bureau

Observe that the graph of f rises very quickly and then tapers off. From the graph of f, you can see that the bulk of American families earned less than $100,000 per year. In fact, 95% of U.S. families earned less than $102,358 per year in 1990. (We will refer to this model again in Using Technology at the end of this section.)

To analyze mathematical models involving exponential and logarithmic functions in greater detail, we need to develop rules for computing the derivative of these functions. We begin by looking at the rule for computing the derivative of the exponential function.

Rule 1: Derivative of the Exponential Function

$$\frac{d}{dx}(e^x) = e^x$$

Thus, the derivative of the exponential function with base e is equal to the function itself. To demonstrate the validity of this rule, we compute

$$f'(x) = \lim_{h \to 0} \frac{f(x+h) - f(x)}{h}$$
$$= \lim_{h \to 0} \frac{e^{x+h} - e^x}{h}$$
$$= \lim_{h \to 0} \frac{e^x(e^h - 1)}{h} \qquad \text{Write } e^{x+h} = e^x e^h \text{ and factor.}$$
$$= e^x \lim_{h \to 0} \frac{e^h - 1}{h} \qquad \text{Why?}$$

To evaluate

$$\lim_{h \to 0} \frac{e^h - 1}{h}$$

PORTFOLIO Robert Derbenti

TITLE Process Manager
INSTITUTION Linear Technology

When students are faced with the prospect of taking on the study of applied calculus, they are likely to ask if their pending challenge will have any useful application in the "real world." At a premiere high-tech company like Linear Technology, the answer is a definite, "Yes." Linear Technology Corporation designs, markets, and manufactures high-performance analog integrated circuits (analog "chips") that are used in such real-world systems as cell phones, disk drives, cable modems, and satellites, as well as numerous other scientific and industrial instruments. Behind every e-mail transmission and cell phone call, there is a lot of real science and the foundations of that science reside in the relevant uses of applied calculus.

The technology of manufacturing analog chips provides a concrete example of applying calculus to a real-world application. As the manager of the manufacturing process, I must coordinate a team of several process engineers who bring with them the knowledge of semiconductor (e.g., silicon) physics. It is our job to translate a high-performance design into a silicon chip through the precise placement of phosphorous, boron, and arsenic atoms within specific points of the silicon. How can such a precise manufacturing process be defined? The definition starts as the group discusses alternative possible experiments and decides upon a hypothetical process. Using software developed from differential equations and exponential functions to model and refine the hypothesis, we arrive at a final version. The hypothesis is tested and refined experimentally until a manufacturing process is stabilized and prototypes can be produced for performance evaluation.

Not surprisingly, this process closely resembles the scientific method approach to problem solving (i.e., hypothesis development, testability by experiment, conclusion). At each step, the appropriate differential equations and exponential functions inform the methodology of the process. Without exaggeration, one can say that "chip" technology, and all its derivative products, would not be possible without these fundamental uses of calculus.

TABLE 4

h	$\dfrac{e^h - 1}{h}$
0.1	1.0517
0.01	1.0050
0.001	1.0005
−0.1	0.9516
−0.01	0.9950
−0.001	0.9995

let's refer to Table 4, which is constructed with the aid of a calculator. From the table, we see that

$$\lim_{h \to 0} \frac{e^h - 1}{h} = 1$$

(Although a rigorous proof of this fact is possible, it is beyond the scope of this book. Also see Example 1, Using Technology, page 370.) Using this result, we conclude that

$$f'(x) = e^x \cdot 1 = e^x$$

as we set out to show.

EXAMPLE 1 Find the derivative of each of the following functions:

a. $f(x) = x^2 e^x$ **b.** $g(t) = (e^t + 2)^{3/2}$

Solution

a. The product rule gives

$$\begin{aligned} f'(x) &= \frac{d}{dx}(x^2 e^x) = x^2 \frac{d}{dx}(e^x) + e^x \frac{d}{dx}(x^2) \\ &= x^2 e^x + e^x(2x) = xe^x(x + 2) \end{aligned}$$

b. Using the general power rule, we find

$$g'(t) = \frac{3}{2}(e^t + 2)^{1/2}\frac{d}{dt}(e^t + 2) = \frac{3}{2}(e^t + 2)^{1/2}e^t = \frac{3}{2}e^t(e^t + 2)^{1/2}$$

EXPLORING WITH TECHNOLOGY

Consider the exponential function $f(x) = b^x$ $(b > 0, b \neq 1)$.

1. Use the definition of the derivative of a function to show that

$$f'(x) = b^x \cdot \lim_{h \to 0} \frac{b^h - 1}{h}$$

2. Use the result of part 1 to show that

$$\frac{d}{dx}(2^x) = 2^x \cdot \lim_{h \to 0} \frac{2^h - 1}{h}$$

$$\frac{d}{dx}(3^x) = 3^x \cdot \lim_{h \to 0} \frac{3^h - 1}{h}$$

3. Use the technique in Using Technology, page 370, to show that (to two decimal places)

$$\lim_{h \to 0} \frac{2^h - 1}{h} = 0.69 \quad \text{and} \quad \lim_{h \to 0} \frac{3^h - 1}{h} = 1.10$$

4. Conclude from the results of parts 2 and 3 that

$$\frac{d}{dx}(2^x) \approx (0.69)2^x \quad \text{and} \quad \frac{d}{dx}(3^x) \approx (1.10)3^x$$

Thus,

$$\frac{d}{dx}(b^x) = k \cdot b^x$$

where k is an appropriate constant.

5. The results of part 4 suggest that, for convenience, we pick the base b, where $2 < b < 3$, so that $k = 1$. This value of b is $e \approx 2.718281828. . . .$ Thus,

$$\frac{d}{dx}(e^x) = e^x$$

This is why we prefer to work with the exponential function $f(x) = e^x$.

Applying the Chain Rule to Exponential Functions

To enlarge the class of exponential functions to be differentiated, we appeal to the chain rule to obtain the following rule for differentiating composite functions of the form $h(x) = e^{f(x)}$. An example of such a function is $h(x) = e^{x^2-2x}$. Here, $f(x) = x^2 - 2x$.

Rule 2: Chain Rule for Exponential Functions

If $f(x)$ is a differentiable function, then

$$\frac{d}{dx}(e^{f(x)}) = e^{f(x)}f'(x)$$

To see this, observe that if $h(x) = g[f(x)]$, where $g(x) = e^x$, then by virtue of the chain rule,

$$h'(x) = g'(f(x))f'(x) = e^{f(x)}f'(x)$$

since $g'(x) = e^x$.

As an aid to remembering the chain rule for exponential functions, observe that it has the following form:

$$\frac{d}{dx}(e^{f(x)}) = e^{f(x)} \cdot \text{derivative of exponent}$$

(Same)

EXAMPLE 2 Find the derivative of each of the following functions:

a. $f(x) = e^{2x}$ **b.** $y = e^{-3x}$ **c.** $g(t) = e^{2t^2+t}$

Solution

a. $f'(x) = e^{2x}\dfrac{d}{dx}(2x) = e^{2x} \cdot 2 = 2e^{2x}$

b. $\dfrac{dy}{dx} = e^{-3x}\dfrac{d}{dx}(-3x) = -3e^{-3x}$

c. $g'(t) = e^{2t^2+t} \cdot \dfrac{d}{dt}(2t^2 + t) = (4t + 1)e^{2t^2+t}$ ■

EXAMPLE 3 Differentiate the function $y = xe^{-2x}$.

Solution Using the product rule, followed by the chain rule, we find

$$\begin{aligned}\frac{dy}{dx} &= x\frac{d}{dx}e^{-2x} + e^{-2x}\frac{d}{dx}(x)\\ &= xe^{-2x}\frac{d}{dx}(-2x) + e^{-2x} \qquad \text{Use the chain rule on the first term.}\\ &= -2xe^{-2x} + e^{-2x}\\ &= e^{-2x}(1 - 2x)\end{aligned}$$

■

EXAMPLE 4 Differentiate the function $g(t) = \dfrac{e^t}{e^t + e^{-t}}$.

Solution Using the quotient rule, followed by the chain rule, we find

$$\begin{aligned}g'(t) &= \frac{(e^t + e^{-t})\dfrac{d}{dt}(e^t) - e^t\dfrac{d}{dt}(e^t + e^{-t})}{(e^t + e^{-t})^2}\\ &= \frac{(e^t + e^{-t})e^t - e^t(e^t - e^{-t})}{(e^t + e^{-t})^2}\\ &= \frac{e^{2t} + 1 - e^{2t} + 1}{(e^t + e^{-t})^2} \qquad e^0 = 1\\ &= \frac{2}{(e^t + e^{-t})^2}\end{aligned}$$

■

EXAMPLE 5 In Section 5.6, we will discuss some practical applications of the exponential function

$$Q(t) = Q_0e^{kt}$$

where Q_0 and k are positive constants and $t \in [0, \infty)$. A quantity $Q(t)$ growing according to this law experiences exponential growth. Show that for a quantity $Q(t)$ experiencing exponential growth, the rate of growth of the quantity $Q'(t)$ at any time t is directly proportional to the amount of the quantity present.

Solution Using the chain rule for exponential functions, we compute the derivative Q' of the function Q. Thus,

$$\begin{aligned} Q'(t) &= Q_0e^{kt}\frac{d}{dt}(kt) \\ &= Q_0e^{kt}(k) \\ &= kQ_0e^{kt} \\ &= kQ(t) \qquad Q(t) = Q_0e^{kt} \end{aligned}$$

which is the desired conclusion. ■

EXAMPLE 6 Find the inflection points of the function $f(x) = e^{-x^2}$.

Solution The first derivative of f is

$$f'(x) = -2xe^{-x^2}$$

Differentiating $f'(x)$ with respect to x yields

$$\begin{aligned} f''(x) &= (-2x)(-2xe^{-x^2}) - 2e^{-x^2} \\ &= 2e^{-x^2}(2x^2 - 1) \end{aligned}$$

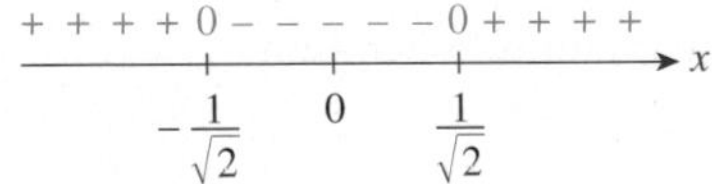

FIGURE 11
Sign diagram for f''

Setting $f''(x) = 0$ gives

$$2e^{-x^2}(2x^2 - 1) = 0$$

Since e^{-x^2} never equals zero for any real value of x, we see that $x = \pm 1/\sqrt{2}$ are the only candidates for inflection points of f. The sign diagram of f'', shown in Figure 11, tells us that both $x = -1/\sqrt{2}$ and $x = 1/\sqrt{2}$ give rise to inflection points of f.

Next,

$$f\left(-\frac{1}{\sqrt{2}}\right) = f\left(\frac{1}{\sqrt{2}}\right) = e^{-1/2}$$

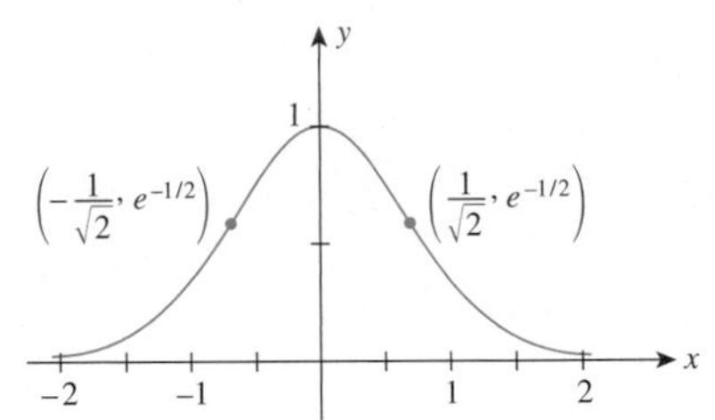

FIGURE 12
The graph of $y = e^{-x^2}$ has two inflection points.

and the inflection points of f are $(-1/\sqrt{2}, e^{-1/2})$ and $(1/\sqrt{2}, e^{-1/2})$. The graph of f appears in Figure 12. ■

Our final example involves finding the absolute maximum of an exponential function.

APPLIED EXAMPLE 7 Optimal Market Price Refer to Example 6, Section 5.3. The present value of the market price of the Blakely Office Building is given by

$$P(t) = 300{,}000e^{-0.09t+\sqrt{t}/2} \qquad (0 \le t \le 10)$$

Find the optimal present value of the building's market price.

Solution To find the maximum value of P over [0, 10], we compute

$$P'(t) = 300{,}000e^{-0.09t+\sqrt{t}/2}\frac{d}{dt}\left(-0.09t + \frac{1}{2}t^{1/2}\right)$$
$$= 300{,}000e^{-0.09t+\sqrt{t}/2}\left(-0.09 + \frac{1}{4}t^{-1/2}\right)$$

Setting $P'(t) = 0$ gives

$$-0.09 + \frac{1}{4t^{1/2}} = 0$$

since $e^{-0.09t+\sqrt{t}/2}$ is never zero for any value of t. Solving this equation, we find

$$\frac{1}{4t^{1/2}} = 0.09$$
$$t^{1/2} = \frac{1}{4(0.09)}$$
$$= \frac{1}{0.36}$$
$$t = \left(\frac{1}{0.36}\right)^2 \approx 7.72$$

the sole critical number of the function P. Finally, evaluating $P(t)$ at the critical number as well as at the endpoints of [0, 10], we have

t	0	7.72	10
$P(t)$	300,000	600,779	592,838

We conclude, accordingly, that the optimal present value of the property's market price is \$600,779 and that this will occur 7.72 years from now. ■

5.4 Self-Check Exercises

1. Let $f(x) = xe^{-x}$.
   a. Find the first and second derivatives of f.
   b. Find the relative extrema of f.
   c. Find the inflection points of f.

2. An industrial asset is being depreciated at a rate so that its book value t yr from now will be

$$V(t) = 50{,}000e^{-0.4t}$$

dollars. How fast will the book value of the asset be changing 3 yr from now?

Solutions to Self-Check Exercises 5.4 can be found on page 369.

5.4 Concept Questions

1. State the rule for differentiating (a) $f(x) = e^x$ and (b) $g(x) = e^{f(x)}$, where f is a differentiable function.

2. Let $f(x) = e^{kx}$.
   a. Compute $f'(x)$.
   b. Use the result to deduce the behavior of f for the case $k > 0$ and the case $k < 0$.

5.4 Exercises

In Exercises 1–28, find the derivative of the function.

1. $f(x) = e^{3x}$
2. $f(x) = 3e^{x}$
3. $g(t) = e^{-t}$
4. $f(x) = e^{-2x}$
5. $f(x) = e^{x} + x$
6. $f(x) = 2e^{x} - x^2$
7. $f(x) = x^3e^{x}$
8. $f(u) = u^2e^{-u}$
9. $f(x) = \dfrac{2e^{x}}{x}$
10. $f(x) = \dfrac{x}{e^{x}}$
11. $f(x) = 3(e^{x} + e^{-x})$
12. $f(x) = \dfrac{e^{x} + e^{-x}}{2}$
13. $f(w) = \dfrac{e^{w} + 1}{e^{w}}$
14. $f(x) = \dfrac{e^{x}}{e^{x} + 1}$
15. $f(x) = 2e^{3x-1}$
16. $f(t) = 4e^{3t+2}$
17. $h(x) = e^{-x^2}$
18. $f(x) = e^{x^2-1}$
19. $f(x) = 3e^{-1/x}$
20. $f(x) = e^{1/(2x)}$
21. $f(x) = (e^{x} + 1)^{25}$
22. $f(x) = (4 - e^{-3x})^3$
23. $f(x) = e^{\sqrt{x}}$
24. $f(x) = -e^{-\sqrt{2t}}$
25. $f(x) = (x - 1)e^{3x+2}$
26. $f(s) = (s^2 + 1)e^{-s^2}$
27. $f(x) = \dfrac{e^{x} - 1}{e^{x} + 1}$
28. $g(t) = \dfrac{e^{-t}}{1 + t^2}$

In Exercises 29–32, find the second derivative of the function.

29. $f(x) = e^{-4x} + 2e^{3x}$
30. $f(t) = 3e^{-2t} - 5e^{-t}$
31. $f(x) = 2xe^{3x}$
32. $f(t) = t^2e^{-2t}$

33. Find an equation of the tangent line to the graph of $y = e^{2x-3}$ at the point $(\frac{3}{2}, 1)$.

34. Find an equation of the tangent line to the graph of $y = e^{-x^2}$ at the point $(1, 1/e)$.

35. Determine the intervals where the function $f(x) = e^{-x^2/2}$ is increasing and where it is decreasing.

36. Determine the intervals where the function $f(x) = x^2e^{-x}$ is increasing and where it is decreasing.

37. Determine the intervals of concavity for the function $f(x) = \dfrac{e^{x} - e^{-x}}{2}$.

38. Determine the intervals of concavity for the function $f(x) = xe^{x}$.

39. Find the inflection point of the function $f(x) = xe^{-2x}$.

40. Find the inflection point(s) of the function $f(x) = 2e^{-x^2}$.

41. Find the equations of the tangent lines to the graph of $f(x) = e^{-x^2}$ at its inflection points.

42. Find an equation of the tangent line to the graph of $f(x) = xe^{-x}$ at its inflection point.

In Exercises 43–46, find the absolute extrema of the function.

43. $f(x) = e^{-x^2}$ on $[-1, 1]$
44. $h(x) = e^{x^2-4}$ on $[-2, 2]$
45. $g(x) = (2x - 1)e^{-x}$ on $[0, \infty)$
46. $f(x) = xe^{-x^2}$ on $[0, 2]$

In Exercises 47–50, use the curve-sketching guidelines of Chapter 4, page 290, to sketch the graph of the function.

47. $f(t) = e^{t} - t$
48. $h(x) = \dfrac{e^{x} + e^{-x}}{2}$
49. $f(x) = 2 - e^{-x}$
50. $f(x) = \dfrac{3}{1 + e^{-x}}$

51. **Percentage of Population Relocating** Based on data obtained from the Census Bureau, the manager of Plymouth Van Lines estimates that the percent of the total population relocating in year t ($t = 0$ corresponds to the year 1960) may be approximated by the formula

$$P(t) = 20.6e^{-0.009t} \qquad (0 \leq t \leq 35)$$

Compute $P'(10)$, $P'(20)$, and $P'(30)$ and interpret your results.

52. **Online Banking** In a study prepared in 2000, the percent of households using online banking was projected to be

$$f(t) = 1.5e^{0.78t} \qquad (0 \leq t \leq 4)$$

where t is measured in years, with $t = 0$ corresponding to the beginning of 2000.
  a. What was the projected percent of households using online banking at the beginning of 2003?
  b. How fast was the projected percent of households using online banking changing at the beginning of 2003?
  c. How fast was the rate of the projected percent of households using online banking changing at the beginning of 2003?
  Hint: We want $f''(3)$. Why?

Source: Online Banking Report

53. **Over-100 Population** Based on data obtained from the Census Bureau, the number of Americans over age 100 is expected to be

$$P(t) = 0.07e^{0.54t} \qquad (0 \leq t \leq 4)$$

where $P(t)$ is measured in millions and t is measured in decades, with $t = 0$ corresponding to the beginning of 2000.
  a. What was the population of Americans over age 100 at the beginning of 2000? What will it be at the beginning of 2030?

b. How fast was the population of Americans over age 100 changing at the beginning of 2000? How fast will it be changing at the beginning of 2030?
Source: U.S. Census Bureau

54. **World Population Growth** After its fastest rate of growth ever during the 1980s and 1990s, the rate of growth of world population is expected to slow dramatically in the 21st century. The function

$$G(t) = 1.58e^{-0.213t}$$

gives the projected annual average percent population growth/decade in the tth decade, with $t = 1$ corresponding to 2000.
a. What will the projected annual average population growth rate be in 2020 ($t = 3$)?
b. How fast will the projected annual average population growth rate be changing in 2020?
Source: U.S. Census Bureau

55. **Loans at Japanese Banks** The total loans outstanding at all Japanese banks have been declining in recent years. The function

$$L(t) = 4.6e^{-0.04t} \qquad (0 \le t \le 6)$$

gives the approximate total loans outstanding from 1998 ($t = 0$) through 2004 in trillions of dollars.
a. What were the total loans outstanding in 1998? In 2004?
b. How fast were the total loans outstanding declining in 1998? In 2004?
c. Show that the total loans outstanding were declining but at a slower rate over the years between 1998 and 2004.
Hint: Look at L'' on the interval (0, 6).
Source: Bank of Japan

56. **Energy Consumption of Appliances** The average energy consumption of the typical refrigerator/freezer manufactured by York Industries is approximately

$$C(t) = 1486e^{-0.073t} + 500 \qquad (0 \le t \le 20)$$

kilowatt-hours (kWh) per year, where t is measured in years, with $t = 0$ corresponding to 1972.
a. What was the average energy consumption of the York refrigerator/freezer at the beginning of 1972?
b. Prove that the average energy consumption of the York refrigerator/freezer is decreasing over the years in question.
c. All refrigerator/freezers manufactured as of January 1, 1990, must meet the 950-kWh/year maximum energy-consumption standard set by the National Appliance Conservation Act. Show that the York refrigerator/freezer satisfies this requirement.

57. **Sales Promotion** The Lady Bug, a women's clothing chain store, found that t days after the end of a sales promotion the volume of sales was given by

$$S(t) = 20{,}000(1 + e^{-0.5t}) \qquad (0 \le t \le 5)$$

dollars.
a. Find the rate of change of The Lady Bug's sales volume when $t = 1$, $t = 2$, $t = 3$, and $t = 4$.
b. After how many days will the sales volume drop below \$27,400?

58. **Blood Alcohol Level** The percentage of alcohol in a person's bloodstream t hr after drinking 8 fluid oz of whiskey is given by

$$A(t) = 0.23te^{-0.4t} \qquad (0 \le t \le 12)$$

a. What is the percentage of alcohol in a person's bloodstream after $\frac{1}{2}$ hr? After 8 hr?
b. How fast is the percentage of alcohol in a person's bloodstream changing after $\frac{1}{2}$ hr? After 8 hr?
Source: Encyclopedia Britannica

59. **Polio Immunization** Polio, a once-feared killer, declined markedly in the United States in the 1950s after Jonas Salk developed the inactivated polio vaccine and mass immunization of children took place. The number of polio cases in the United States from the beginning of 1959 to the beginning of 1963 is approximated by the function

$$N(t) = 5.3e^{0.095t^2-0.85t} \qquad (0 \le t \le 4)$$

where $N(t)$ gives the number of polio cases (in thousands) and t is measured in years with $t = 0$ corresponding to the beginning of 1959.
a. Show that the function N is decreasing over the time interval under consideration.
b. How fast was the number of polio cases decreasing at the beginning of 1959? At the beginning of 1962? (*Comment:* Following the introduction of the oral vaccine developed by Dr. Albert B. Sabin in 1963, polio in the United States has, for all practical purposes, been eliminated.)

60. **Marginal Revenue** The unit selling price p (in dollars) and the quantity demanded (in pairs) of a certain brand of women's gloves is given by the demand equation

$$p = 100e^{-0.0001x} \qquad (0 \le x \le 20{,}000)$$

a. Find the revenue function R.
Hint: $R(x) = px$.
b. Find the marginal revenue function R'.
c. What is the marginal revenue when $x = 10$?

61. **Maximizing Revenue** Refer to Exercise 60. How many pairs of the gloves must be sold to yield a maximum revenue? What will be the maximum revenue?

62. **Price of Perfume** The monthly demand for a certain brand of perfume is given by the demand equation

$$p = 100e^{-0.0002x} + 150$$

where p denotes the retail unit price (in dollars) and x denotes the quantity (in 1-oz bottles) demanded.

a. Find the rate of change of the price per bottle when $x = 1000$ and when $x = 2000$.

b. What is the price per bottle when $x = 1000$? When $x = 2000$?

63. PRICE OF WINE The monthly demand for a certain brand of table wine is given by the demand equation

$$p = 240\left(1 - \frac{3}{3 + e^{-0.0005x}}\right)$$

where p denotes the wholesale price per case (in dollars) and x denotes the number of cases demanded.

a. Find the rate of change of the price per case when $x = 1000$.

b. What is the price per case when $x = 1000$?

64. SPREAD OF AN EPIDEMIC During a flu epidemic, the total number of students on a state university campus who had contracted influenza by the xth day was given by

$$N(x) = \frac{3000}{1 + 99e^{-x}} \qquad (x \geq 0)$$

a. How many students had influenza initially?

b. Derive an expression for the rate at which the disease was being spread and prove that the function N is increasing on the interval $(0, \infty)$.

c. Sketch the graph of N. What was the total number of students who contracted influenza during that particular epidemic?

65. WEIGHTS OF CHILDREN The Ehrenberg equation

$$W = 2.4e^{1.84h}$$

gives the relationship between the height (in meters) and the average weight W (in kilograms) for children between 5 and 13 yr of age.

a. What is the average weight of a 10-yr-old child who stands at 1.6 m tall?

b. Use differentials to estimate the change in the average weight of a 10-yr-old child whose height increases from 1.6 m to 1.65 m.

66. MAXIMUM OIL PRODUCTION It has been estimated that the total production of oil from a certain oil well is given by

$$T(t) = -1000(t + 10)e^{-0.1t} + 10{,}000$$

thousand barrels t yr after production has begun. Determine the year when the oil well will be producing at maximum capacity.

67. OPTIMAL SELLING TIME Refer to Exercise 42, page 359. The present value of a piece of waterfront property purchased by an investor is given by the function

$$P(t) = 80{,}000e^{\sqrt{t}/2 - 0.09t} \qquad (0 \leq t \leq 8)$$

Determine the optimal time (based on present value) for the investor to sell the property. What is the property's optimal present value?

68. BLOOD ALCOHOL LEVEL Refer to Exercise 58, p. 367. At what time after drinking the alcohol is the percentage of alcohol in the person's bloodstream at its highest level? What is that level?

69. OIL USED TO FUEL PRODUCTIVITY A study on worldwide oil use was prepared for a major oil company. The study predicted that the amount of oil used to fuel productivity in a certain country is given by

$$f(t) = 1.5 + 1.8te^{-1.2t} \qquad (0 \leq t \leq 4)$$

where $f(t)$ denotes the number of barrels per \$1000 of economic output and t is measured in decades ($t = 0$ corresponds to 1965). Compute $f'(0)$, $f'(1)$, $f'(2)$, and $f'(3)$ and interpret your results.

70. PRICE OF A COMMODITY The price of a certain commodity in dollars per unit at time t (measured in weeks) is given by $p = 18 - 3e^{-2t} - 6e^{-t/3}$.

a. What is the price of the commodity at $t = 0$?

b. How fast is the price of the commodity changing at $t = 0$?

c. Find the equilibrium price of the commodity.
Hint: It is given by $\lim_{t \to \infty} p$.

71. PRICE OF A COMMODITY The price of a certain commodity in dollars per unit at time t (measured in weeks) is given by $p = 8 + 4e^{-2t} + te^{-2t}$.

a. What is the price of the commodity at $t = 0$?

b. How fast is the price of the commodity changing at $t = 0$?

c. Find the equilibrium price of the commodity.
Hint: It's given by $\lim_{t \to \infty} p$. Also, use the fact that $\lim_{t \to \infty} te^{-2t} = 0$.

72. ABSORPTION OF DRUGS A liquid carries a drug into an organ of volume V cm^3 at the rate of a cm^3/sec and leaves at the same rate. The concentration of the drug in the entering liquid is c g/cm^3. Letting $x(t)$ denote the concentration of the drug in the organ at any time t, we have $x(t) = c(1 - e^{-at/V})$.

a. Show that x is an increasing function on $(0, \infty)$.

b. Sketch the graph of x.

73. ABSORPTION OF DRUGS Refer to Exercise 72. Suppose the maximum concentration of the drug in the organ must *not* exceed m g/cm^3, where $m < c$. Show that the liquid must not be allowed to enter the organ for a time longer than

$$T = \left(\frac{V}{a}\right) \ln\left(\frac{c}{c - m}\right)$$

minutes.

74. CONCENTRATION OF A DRUG IN THE BLOODSTREAM The concentration of a drug in the bloodstream t sec after injection into a muscle is given by

$$y = c(e^{-bt} - e^{-at}) \qquad (\text{for } t \geq 0)$$

where a, b, and c are positive constants, with $a > b$.

a. Find the time at which the concentration is maximal.

b. Find the time at which the concentration of the drug in the bloodstream is decreasing most rapidly.

75. Absorption of Drugs The concentration of a drug in an organ at any time t (in seconds) is given by

$$C(t) = \begin{cases} 0.3t - 18(1 - e^{-t/60}) & \text{if } 0 \leq t \leq 20 \\ 18e^{-t/60} - 12e^{-(t-20)/60} & \text{if } t > 20 \end{cases}$$

where $C(t)$ is measured in grams/cubic centimeter (g/cm^3).

a. How fast is the concentration of the drug in the organ changing after 10 sec?

b. How fast is the concentration of the drug in the organ changing after 30 sec?

c. When will the concentration of the drug in the organ reach a maximum?

d. What is the maximum drug concentration in the organ?

76. Absorption of Drugs Jane took 100 mg of a drug in the morning and another 100 mg of the same drug at the same time the following morning. The amount of the drug in her body t days after the first dosage was taken is given by

$$A(t) = \begin{cases} 100e^{-1.4t} & \text{if } 0 \leq t < 1 \\ 100(1 + e^{1.4})e^{-1.4t} & \text{if } t \geq 1 \end{cases}$$

a. How fast was the amount of drug in Jane's body changing after 12 hr ($t = \frac{1}{2}$)? After 2 days?

b. When was the amount of drug in Jane's body a maximum?

c. What was the maximum amount of drug in Jane's body?

In Exercises 77–80, determine whether the statement is true or false. If it is true, explain why it is true. If it is false, give an example to show why it is false.

77. If $f(x) = 3^x$, then $f'(x) = x \cdot 3^{x-1}$.

78. If $f(x) = e^{\pi}$, then $f'(x) = e^{\pi}$.

79. If $f(x) = \pi^x$, then $f'(x) = \pi^x$.

80. If $x^2 + e^y = 10$, then $y' = \dfrac{-2x}{e^y}$.

5.4 Solutions to Self-Check Exercises

1. a. Using the product rule, we obtain

$$\begin{aligned} f'(x) &= x\frac{d}{dx}e^{-x} + e^{-x}\frac{d}{dx}x \\ &= -xe^{-x} + e^{-x} = (1 - x)e^{-x} \end{aligned}$$

Using the product rule once again, we obtain

$$\begin{aligned} f''(x) &= (1 - x)\frac{d}{dx}e^{-x} + e^{-x}\frac{d}{dx}(1 - x) \\ &= (1 - x)(-e^{-x}) + e^{-x}(-1) \\ &= -e^{-x} + xe^{-x} - e^{-x} = (x - 2)e^{-x} \end{aligned}$$

b. Setting $f'(x) = 0$ gives

$$(1 - x)e^{-x} = 0$$

Since $e^{-x} \neq 0$, we see that $1 - x = 0$, and this gives 1 as the only critical number of f. The sign diagram of f' shown in the accompanying figure tells us that the point $(1, e^{-1})$ is a relative maximum of f.

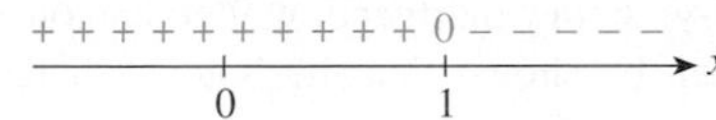

c. Setting $f''(x) = 0$ gives $x - 2 = 0$, so $x = 2$ is a candidate for an inflection point of f. The sign diagram of f'' (see the accompanying figure) shows that $(2, 2e^{-2})$ is an inflection point of f.

- - - - - - - - - - 0 + + + + +

x

0 2

2. The rate change of the book value of the asset t yr from now is

$$\begin{aligned} V'(t) &= 50{,}000\frac{d}{dt}e^{-0.4t} \\ &= 50{,}000(-0.4)e^{-0.4t} = -20{,}000e^{-0.4t} \end{aligned}$$

Therefore, 3 yr from now the book value of the asset will be changing at the rate of

$$V'(3) = -20{,}000e^{-0.4(3)} = -20{,}000e^{-1.2} \approx -6023.88$$

—that is, decreasing at the rate of approximately \$6024/year.

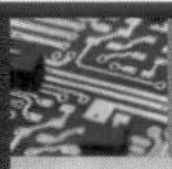

USING TECHNOLOGY

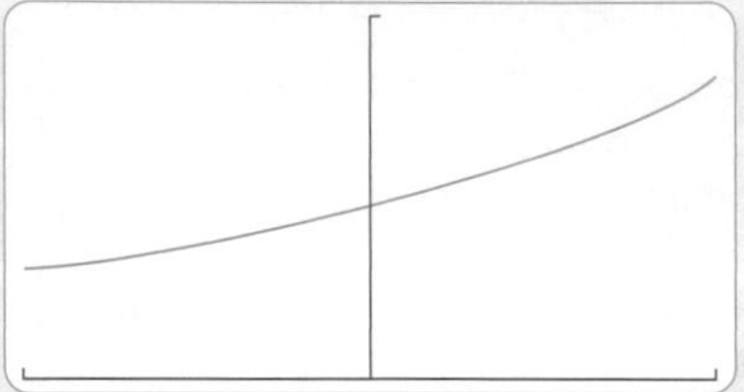

FIGURE T1
The graph of f in the viewing window $[-1, 1] \times [0, 2]$

EXAMPLE 1 At the beginning of Section 5.4, we demonstrated via a table of values of $(e^h - 1)/h$ for selected values of h the plausibility of the result

$$\lim_{h \to 0} \frac{e^h - 1}{h} = 1$$

To obtain a visual confirmation of this result, we plot the graph of

$$f(x) = \frac{e^x - 1}{x}$$

in the viewing window $[-1, 1] \times [0, 2]$ (Figure T1). From the graph of f, we see that $f(x)$ appears to approach 1 as x approaches 0. ■

The numerical derivative function of a graphing utility will yield the derivative of an exponential or logarithmic function for any value of x, just as it did for algebraic functions.*

*The rules for differentiating logarithmic functions will be covered in Section 5.5. However, the exercises given here can be done without using these rules.

TECHNOLOGY EXERCISES

In Exercises 1–6, use the numerical derivative operation to find the rate of change of $f(x)$ at the given value of x. Give your answer accurate to four decimal places.

1. $f(x) = x^3e^{-1/x}$; $x = -1$
2. $f(x) = (\sqrt{x} + 1)^{3/2}e^{-x}$; $x = 0.5$
3. $f(x) = x^3\sqrt{\ln x}$; $x = 2$
4. $f(x) = \dfrac{\sqrt{x}\ln x}{x + 1}$; $x = 3.2$
5. $f(x) = e^{-x}\ln(2x + 1)$; $x = 0.5$
6. $f(x) = \dfrac{e^{-\sqrt{x}}}{\ln(x^2 + 1)}$; $x = 1$
7. AN EXTINCTION SITUATION The number of saltwater crocodiles in a certain area of northern Australia is given by
$$P(t) = \frac{300e^{-0.024t}}{5e^{-0.024t} + 1}$$
 a. How many crocodiles were in the population initially?
 b. Show that $\lim_{t \to \infty} P(t) = 0$.
 c. Plot the graph of P in the viewing window $[0, 200] \times [0, 70]$.
 (*Comment:* This phenomenon is referred to as an *extinction situation.*)
8. INCOME OF AMERICAN FAMILIES Based on data, it is estimated that the number of American families y (in millions) who earned x thousand dollars in 1990 is related by the equation
$$y = 0.1584xe^{-0.0000016x^3 + 0.00011x^2 - 0.04491x} \qquad (x > 0)$$
 a. Plot the graph of the equation in the viewing window $[0, 150] \times [0, 2]$.
 b. How fast is y changing with respect to x when $x = 10$? When $x = 50$? Interpret your results.
 Source: House Budget Committee, House Ways and Means Committee, and U.S. Census Bureau
9. WORLD POPULATION GROWTH Based on data obtained in a study, the world population (in billions) is approximated by the function
$$f(t) = \frac{12}{1 + 3.74914e^{-1.42804t}} \qquad (0 \le t \le 4)$$
 where t is measured in half centuries, with $t = 0$ corresponding to the beginning of 1950.
 a. Plot the graph of f in the viewing window $[0, 5] \times [0, 14]$.
 b. How fast was the world population expected to increase at the beginning of 2000?
 Source: United Nations Population Division
10. LOAN AMORTIZATION The Sotos plan to secure a loan of \$160,000 to purchase a house. They are considering a conventional 30-yr home mortgage at 9%/year on the unpaid balance. It can be shown that the Sotos will have an outstanding principal of
$$B(x) = \frac{160{,}000(1.0075^{360} - 1.0075^x)}{1.0075^{360} - 1}$$
 dollars after making x monthly payments of \$1287.40.

a. Plot the graph of $B(x)$, using the viewing window $[0, 360] \times [0, 160{,}000]$.

b. Compute $B(0)$ and $B'(0)$ and interpret your results; compute $B(180)$ and $B'(180)$ and interpret your results.

11. **INCREASE IN JUVENILE OFFENDERS** The number of youths aged 15 to 19 increased by 21% between 1994 and 2005, pushing up the crime rate. According to the National Council on Crime and Delinquency, the number of violent crime arrests of juveniles under age 18 in year t is given by

$$f(t) = -0.438t^2 + 9.002t + 107 \qquad (0 \le t \le 13)$$

where $f(t)$ is measured in thousands and t in years, with $t = 0$ corresponding to 1989. According to the same source, if trends like inner-city drug use and wider availability of guns continues, then the number of violent crime arrests of juveniles under age 18 in year t is given by

$$g(t) = \begin{cases} -0.438t^2 + 9.002t + 107 & \text{if } 0 \le t < 4 \\ 99.456e^{0.07824t} & \text{if } 4 \le t \le 13 \end{cases}$$

where $g(t)$ is measured in thousands and $t = 0$ corresponds to 1989.

a. Compute $f(11)$ and $g(11)$ and interpret your results.

b. Compute $f'(11)$ and $g'(11)$ and interpret your results.

Source: National Council on Crime and Delinquency

12. **INCREASING CROP YIELDS** If left untreated on bean stems, aphids (small insects that suck plant juices) will multiply at an increasing rate during the summer months and reduce productivity and crop yield of cultivated crops. But if the aphids are treated in mid-June, the numbers decrease sharply to less than 100/bean stem, allowing for steep rises in crop yield. The function

$$F(t) = \begin{cases} 62e^{1.152t} & \text{if } 0 \le t < 1.5 \\ 349e^{-1.324(t-1.5)} & \text{if } 1.5 \le t \le 3 \end{cases}$$

gives the number of aphids on a typical bean stem at time t, where t is measured in months, with $t = 0$ corresponding to the beginning of May.

a. How many aphids are there on a typical bean stem at the beginning of June ($t = 1$)? At the beginning of July ($t = 2$)?

b. How fast is the population of aphids changing at the beginning of June? At the beginning of July?

Source: The Random House Encyclopedia

13. **PERCENT OF FEMALES IN THE LABOR FORCE** Based on data from the U.S. Census Bureau, the chief economist of Manpower, Inc., constructed the following formula giving the percent of the total female population in the civilian labor force, $P(t)$, at the beginning of the tth decade ($t = 0$ corresponds to the year 1900):

$$P(t) = \frac{74}{1 + 2.6e^{-0.166t + 0.04536t^2 - 0.0066t^3}} \qquad (0 \le t \le 11)$$

Assume this trend continued for the rest of the twentieth century.

a. What was the percent of the total female population in the civilian labor force at the beginning of 2000?

b. What was the growth rate of the percent of the total female population in the civilian labor force at the beginning of 2000?

Source: U.S. Census Bureau

5.5 Differentiation of Logarithmic Functions

The Derivative of ln *x*

Let's now turn our attention to the differentiation of logarithmic functions.

Rule 3: Derivative of ln *x*

$$\frac{d}{dx} \ln |x| = \frac{1}{x} \qquad (x \neq 0)$$

To derive Rule 3, suppose $x > 0$ and write $f(x) = \ln x$ in the equivalent form

$$x = e^{f(x)}$$

Differentiating both sides of the equation with respect to x, we find, using the chain rule,

$$1 = e^{f(x)} \cdot f'(x)$$

from which we see that

$$f'(x) = \frac{1}{e^{f(x)}}$$

or, since $e^{f(x)} = x$,

$$f'(x) = \frac{1}{x}$$

as we set out to show. You are asked to prove the rule for the case $x < 0$ in Exercise 71, page 378.

EXAMPLE 1 Find the derivative of each function:

a. $f(x) = x \ln x$ **b.** $g(x) = \dfrac{\ln x}{x}$

Solution

a. Using the product rule, we obtain

$$f'(x) = \frac{d}{dx}(x \ln x) = x \frac{d}{dx}(\ln x) + (\ln x)\frac{d}{dx}(x)$$

$$= x\left(\frac{1}{x}\right) + \ln x = 1 + \ln x$$

b. Using the quotient rule, we obtain

$$g'(x) = \frac{x \dfrac{d}{dx}(\ln x) - (\ln x)\dfrac{d}{dx}(x)}{x^2} = \frac{x\left(\dfrac{1}{x}\right) - \ln x}{x^2} = \frac{1 - \ln x}{x^2}$$

■

EXPLORE & DISCUSS

You can derive the formula for the derivative of $f(x) = \ln x$ directly from the definition of the derivative, as follows.

1. Show that

$$f'(x) = \lim_{h \to 0} \frac{f(x + h) - f(x)}{h} = \lim_{h \to 0} \ln\left(1 + \frac{h}{x}\right)^{1/h}$$

2. Put $m = x/h$ and note that $m \to \infty$ as $h \to 0$. Then, $f'(x)$ can be written in the form

$$f'(x) = \lim_{m \to \infty} \ln\left(1 + \frac{1}{m}\right)^{m/x}$$

3. Finally, use both the fact that the natural logarithmic function is continuous and the definition of the number e to show that

$$f'(x) = \frac{1}{x} \ln\left[\lim_{m \to \infty}\left(1 + \frac{1}{m}\right)^m\right] = \frac{1}{x}$$

The Chain Rule for Logarithmic Functions

To enlarge the class of logarithmic functions to be differentiated, we appeal once more to the chain rule to obtain the following rule for differentiating composite functions of the form $h(x) = \ln f(x)$, where $f(x)$ is assumed to be a positive differentiable function.

Rule 4: Chain Rule for Logarithmic Functions

If $f(x)$ is a differentiable function, then

$$\frac{d}{dx}[\ln f(x)] = \frac{f'(x)}{f(x)} \qquad [f(x) > 0]$$

To see this, observe that $h(x) = g[f(x)]$, where $g(x) = \ln x$ $(x > 0)$. Since $g'(x) = 1/x$, we have, using the chain rule,

$$\begin{aligned} h'(x) &= g'(f(x))f'(x) \\ &= \frac{1}{f(x)}f'(x) = \frac{f'(x)}{f(x)} \end{aligned}$$

Observe that in the special case $f(x) = x$, $h(x) = \ln x$, so the derivative of h is, by Rule 3, given by $h'(x) = 1/x$.

EXAMPLE 2 Find the derivative of the function $f(x) = \ln(x^2 + 1)$.

Solution Using Rule 4, we see immediately that

$$f'(x) = \frac{\frac{d}{dx}(x^2 + 1)}{x^2 + 1} = \frac{2x}{x^2 + 1}$$

When differentiating functions involving logarithms, the rules of logarithms may be used to advantage, as shown in Examples 3 and 4.

EXAMPLE 3 Differentiate the function $y = \ln[(x^2 + 1)(x^3 + 2)^6]$.

Solution We first rewrite the given function using the properties of logarithms:

$$\begin{aligned} y &= \ln[(x^2 + 1)(x^3 + 2)^6] \\ &= \ln(x^2 + 1) + \ln(x^3 + 2)^6 && \ln mn = \ln m + \ln n \\ &= \ln(x^2 + 1) + 6\ln(x^3 + 2) && \ln m^n = n \ln m \end{aligned}$$

Differentiating and using Rule 4, we obtain

$$\begin{aligned} y' &= \frac{\frac{d}{dx}(x^2 + 1)}{x^2 + 1} + \frac{6\frac{d}{dx}(x^3 + 2)}{x^3 + 2} \\ &= \frac{2x}{x^2 + 1} + \frac{6(3x^2)}{x^3 + 2} = \frac{2x}{x^2 + 1} + \frac{18x^2}{x^3 + 2} \end{aligned}$$

EXPLORING WITH TECHNOLOGY

Use a graphing utility to plot the graphs of $f(x) = \ln x$; its first derivative function, $f'(x) = 1/x$; and its second derivative function $f''(x) = -1/x^2$, using the same viewing window $[0, 4] \times [-3, 3]$.

1. Describe the properties of the graph of f revealed by studying the graph of $f'(x)$. What can you say about the rate of increase of f for large values of x?
2. Describe the properties of the graph of f revealed by studying the graph of $f''(x)$. What can you say about the concavity of f for large values of x?

EXAMPLE 4 Find the derivative of the function $g(t) = \ln(t^2 e^{-t^2})$.

Solution Here again, to save a lot of work, we first simplify the given expression using the properties of logarithms. We have

$$\begin{aligned} g(t) &= \ln(t^2 e^{-t^2}) \\ &= \ln t^2 + \ln e^{-t^2} && \ln mn = \ln m + \ln n \\ &= 2 \ln t - t^2 && \ln m^n = n \ln m \quad \text{and} \quad \ln e = 1 \end{aligned}$$

Therefore,

$$g'(t) = \frac{2}{t} - 2t = \frac{2(1 - t^2)}{t}$$

■

Logarithmic Differentiation

As we saw in the last two examples, the task of finding the derivative of a given function can be made easier by first applying the laws of logarithms to simplify the function. We now illustrate a process called **logarithmic differentiation**, which not only simplifies the calculation of the derivatives of certain functions but also enables us to compute the derivatives of functions we could not otherwise differentiate using the techniques developed thus far.

EXAMPLE 5 Differentiate $y = x(x + 1)(x^2 + 1)$, using logarithmic differentiation.

Solution First, we take the natural logarithm on both sides of the given equation, obtaining

$$\ln y = \ln x(x + 1)(x^2 + 1)$$

Next, we use the properties of logarithms to rewrite the right-hand side of this equation, obtaining

$$\ln y = \ln x + \ln(x + 1) + \ln(x^2 + 1)$$

If we differentiate both sides of this equation, we have

$$\begin{aligned} \frac{d}{dx} \ln y &= \frac{d}{dx} [\ln x + \ln(x + 1) + \ln(x^2 + 1)] \\ &= \frac{1}{x} + \frac{1}{x + 1} + \frac{2x}{x^2 + 1} && \text{Use Rule 4.} \end{aligned}$$

To evaluate the expression on the left-hand side, note that y is a function of x. Therefore, writing $y = f(x)$ to remind us of this fact, we have

$$\begin{aligned} \frac{d}{dx}\ln y &= \frac{d}{dx}\ln[f(x)] && \text{Write } y = f(x). \\ &= \frac{f'(x)}{f(x)} && \text{Use Rule 4.} \\ &= \frac{y'}{y} && \text{Return to using } y \text{ instead of } f(x). \end{aligned}$$

Therefore, we have

$$\frac{y'}{y} = \frac{1}{x} + \frac{1}{x+1} + \frac{2x}{x^2+1}$$

Finally, solving for y', we have

$$\begin{aligned} y' &= y\left(\frac{1}{x} + \frac{1}{x+1} + \frac{2x}{x^2+1}\right) \\ &= x(x+1)(x^2+1)\left(\frac{1}{x} + \frac{1}{x+1} + \frac{2x}{x^2+1}\right) \end{aligned}$$

■

Before considering other examples, let's summarize the important steps involved in logarithmic differentiation.

Finding $\dfrac{dy}{dx}$ by Logarithmic Differentiation

1. Take the natural logarithm on both sides of the equation and use the properties of logarithms to write any "complicated expression" as a sum of simpler terms.
2. Differentiate both sides of the equation with respect to x.
3. Solve the resulting equation for $\dfrac{dy}{dx}$.

EXAMPLE 6 Differentiate $y = x^2(x-1)(x^2+4)^3$.

Solution Taking the natural logarithm on both sides of the given equation and using the laws of logarithms, we obtain

$$\begin{aligned} \ln y &= \ln x^2(x-1)(x^2+4)^3 \\ &= \ln x^2 + \ln(x-1) + \ln(x^2+4)^3 \\ &= 2\ln x + \ln(x-1) + 3\ln(x^2+4) \end{aligned}$$

Differentiating both sides of the equation with respect to x, we have

$$\frac{d}{dx}\ln y = \frac{y'}{y} = \frac{2}{x} + \frac{1}{x-1} + 3\cdot\frac{2x}{x^2+4}$$

Finally, solving for y', we have

$$\begin{aligned} y' &= y\left(\frac{2}{x} + \frac{1}{x-1} + \frac{6x}{x^2+4}\right) \\ &= x^2(x-1)(x^2+4)^3\left(\frac{2}{x} + \frac{1}{x-1} + \frac{6x}{x^2+4}\right) \end{aligned}$$

■

EXAMPLE 7 Find the derivative of $f(x) = x^x$ $(x > 0)$.

Solution A word of caution! This function is neither a power function nor an exponential function. Taking the natural logarithm on both sides of the equation gives

$$\ln f(x) = \ln x^x = x \ln x$$

Differentiating both sides of the equation with respect to x, we obtain

$$\begin{aligned}\frac{f'(x)}{f(x)} &= x\frac{d}{dx}\ln x + (\ln x)\frac{d}{dx}x \\ &= x\left(\frac{1}{x}\right) + \ln x \\ &= 1 + \ln x\end{aligned}$$

Therefore,

$$f'(x) = f(x)(1 + \ln x) = x^x(1 + \ln x)$$

EXPLORING WITH TECHNOLOGY

Refer to Example 7.

1. Use a graphing utility to plot the graph of $f(x) = x^x$, using the viewing window $[0, 2] \times [0, 2]$. Then use ZOOM and TRACE to show that

$$\lim_{x \to 0^+} f(x) = 1$$

2. Use the results of part 1 and Example 7 to show that $\lim_{x \to 0^+} f'(x) = -\infty$. Justify your answer.

5.5 Self-Check Exercises

1. Find an equation of the tangent line to the graph of $f(x) = x \ln(2x + 3)$ at the point $(-1, 0)$.

2. Use logarithmic differentiation to compute y', given $y = (2x + 1)^3(3x + 4)^5$.

Solutions to Self-Check Exercises 5.5 can be found on page 378.

5.5 Concept Questions

1. State the rule for differentiating (a) $f(x) = \ln|x|$ $(x \neq 0)$, and $g(x) = \ln f(x)$ $[f(x) > 0]$, where f is a differentiable function.

2. Explain the technique of logarithmic differentiation.

5.5 Exercises

In Exercises 1–32, find the derivative of the function.

1. $f(x) = 5 \ln x$
2. $f(x) = \ln 5x$
3. $f(x) = \ln(x + 1)$
4. $g(x) = \ln(2x + 1)$
5. $f(x) = \ln x^8$
6. $h(t) = 2 \ln t^5$
7. $f(x) = \ln \sqrt{x}$
8. $f(x) = \ln(\sqrt{x} + 1)$

9. $f(x) = \ln \frac{1}{x^2}$

10. $f(x) = \ln \frac{1}{2x^3}$

11. $f(x) = \ln(4x^2 - 6x + 3)$

12. $f(x) = \ln(3x^2 - 2x + 1)$

13. $f(x) = \ln \frac{2x}{x + 1}$

14. $f(x) = \ln \frac{x + 1}{x - 1}$

15. $f(x) = x^2 \ln x$

16. $f(x) = 3x^2 \ln 2x$

17. $f(x) = \frac{2 \ln x}{x}$

18. $f(x) = \frac{3 \ln x}{x^2}$

19. $f(u) = \ln(u - 2)^3$

20. $f(x) = \ln(x^3 - 3)^4$

21. $f(x) = \sqrt{\ln x}$

22. $f(x) = \sqrt{\ln x + x}$

23. $f(x) = (\ln x)^3$

24. $f(x) = 2(\ln x)^{3/2}$

25. $f(x) = \ln(x^3 + 1)$

26. $f(x) = \ln \sqrt{x^2 - 4}$

27. $f(x) = e^x \ln x$

28. $f(x) = e^x \ln \sqrt{x + 3}$

29. $f(t) = e^{2t} \ln(t + 1)$

30. $g(t) = t^2 \ln(e^{2t} + 1)$

31. $f(x) = \frac{\ln x}{x}$

32. $g(t) = \frac{t}{\ln t}$

In Exercises 33–36, find the second derivative of the function.

33. $f(x) = \ln 2x$

34. $f(x) = \ln(x + 5)$

35. $f(x) = \ln(x^2 + 2)$

36. $f(x) = (\ln x)^2$

In Exercises 37–46, use logarithmic differentiation to find the derivative of the function.

37. $y = (x + 1)^2(x + 2)^3$

38. $y = (3x + 2)^4(5x - 1)^2$

39. $y = (x - 1)^2(x + 1)^3(x + 3)^4$

40. $y = \sqrt{3x + 5}(2x - 3)^4$

41. $y = \frac{(2x^2 - 1)^5}{\sqrt{x + 1}}$

42. $y = \frac{\sqrt{4 + 3x^2}}{\sqrt[3]{x^2 + 1}}$

43. $y = 3^x$

44. $y = x^{x+2}$

45. $y = (x^2 + 1)^x$

46. $y = x^{\ln x}$

47. Find an equation of the tangent line to the graph of $y = x \ln x$ at the point (1, 0).

48. Find an equation of the tangent line to the graph of $y = \ln x^2$ at the point (2, ln 4).

49. Determine the intervals where the function $f(x) = \ln x^2$ is increasing and where it is decreasing.

50. Determine the intervals where the function $f(x) = \frac{\ln x}{x}$ is increasing and where it is decreasing.

51. Determine the intervals of concavity for the function $f(x) = x^2 + \ln x^2$.

52. Determine the intervals of concavity for the function $f(x) = \frac{\ln x}{x}$.

53. Find the inflection points of the function $f(x) = \ln(x^2 + 1)$.

54. Find the inflection points of the function $f(x) = x^2 \ln x$.

55. Find an equation of the tangent line to the graph of $f(x) = x^2 + 2 \ln x$ at its inflection point.

56. Find an equation of the tangent line to the graph of $f(x) = e^{x/2} \ln x$ at its inflection point.
Hint: Show that (1, 0) is the only inflection point of f.

57. Find the absolute extrema of the function $f(x) = x - \ln x$ on $[\frac{1}{2}, 3]$.

58. Find the absolute extrema of the function $g(x) = \frac{x}{\ln x}$ on [2, 5].

59. **Strain on Vertebrae** The strain (percent of compression) on the lumbar vertebral disks in an adult human as a function of the load x (in kilograms) is given by

$$f(x) = 7.2956 \ln(0.0645012x^{0.95} + 1)$$

What is the rate of change of the strain with respect to the load when the load is 100 kg? When the load is 500 kg?
Source: Benedek and Villars, *Physics with Illustrative Examples from Medicine and Biology*

60. **Heights of Children** For children between the ages of 5 and 13 years old, the Ehrenberg equation

$$\ln W = \ln 2.4 + 1.84h$$

gives the relationship between the weight W (in kilograms) and the height h (in meters) of a child. Use differentials to estimate the change in the weight of a child who grows from 1 m to 1.1 m.

61. **Yahoo! in Europe** Yahoo! is putting more emphasis on Western Europe, where the number of online households is expected to grow steadily. In a study conducted in 2004, the number of online households (in millions) in Western Europe was projected to be

$$N(t) = 34.68 + 23.88 \ln(1.05t + 5.3) \qquad (0 \le t \le 2)$$

where $t = 0$ corresponds to the beginning of 2004.

a. What was the projected number of online households in Western Europe at the beginning of 2005?

b. How fast was the projected number of online households in Western Europe increasing at the beginning of 2005?

Source: Jupiter Research

62. **Online Buyers** The number of online buyers in Western Europe is expected to grow steadily in the coming years. The

function

$$P(t) = 28.5 + 14.42 \ln t \qquad (1 \le t \le 7)$$

gives the estimated online buyers as a percent of the total population, where t is measured in years with $t = 1$ corresponding to 2001.

a. What was the percent of online buyers in 2001 ($t = 1$)? How fast was it changing in 2001?

b. What was the percent of online buyers expected to be in 2006 ($t = 6$)? How fast was it expected to be changing in 2006?

Source: Jupiter Research

63. Absorption of Light When light passes through a window glass, some of it is absorbed. It can be shown that if $r\%$ of the light is absorbed by a glass of thickness w, then the percent of light absorbed by a piece of glass of thickness nw is

$$A(n) = 100\left[1 - \left(1 - \frac{r}{100}\right)^n\right] \qquad (0 \le r \le 100)$$

a. Show that A is an increasing function of n on $(0, \infty)$ if $0 < r < 100$.
Hint: Use logarithmic differentiation.

b. Sketch the graph of A for the special case where $r = 10$.

c. Evaluate $\lim_{n \to \infty} A(n)$ and interpret your result.

64. Lambert's Law of Absorption Lambert's law of absorption states that the light intensity $I(x)$ (in calories/square centimeter/second) at a depth of x m as measured from the surface of a material is given by $I = I_0 a^x$, where I_0 and a are positive constants.

a. Find the rate of change of the light intensity with respect to x at a depth of x m from the surface of the material.

b. Using the result of part (a), conclude that the rate of change $I'(x)$ at a depth of x m is proportional to $I(x)$. What is the constant of proportion?

65. Magnitude of Earthquakes On the Richter scale, the magnitude R of an earthquake is given by the formula

$$R = \log \frac{I}{I_0}$$

where I is the intensity of the earthquake being measured and I_0 is the standard reference intensity.

a. What is the magnitude of an earthquake that has intensity 1 million times that of I_0?

b. Suppose an earthquake is measured with a magnitude of 6 on the Richter scale with an error of at most 2%. Use differentials to find the error in the intensity of the earthquake.
Hint: Observe that $I = I_0 10^R$ and use logarithmic differentiation.

66. Weber–Fechner Law The Weber–Fechner law

$$R = k \ln \frac{S}{S_0}$$

where k is a positive constant, describes the relationship between a stimulus S and the resulting response R. Here, S_0, a positive constant, is the threshold level.

a. Show that $R = 0$ if the stimulus is at the threshold level S_0.

b. The derivative dR/dS is the *sensitivity* corresponding to the stimulus level S and measures the capability to detect small changes in the stimulus level. Show that dR/dS is inversely proportional to S and interpret your result.

In Exercises 67 and 68, use the guidelines on page 290 to sketch the graph of the given function.

67. $f(x) = \ln(x - 1)$ **68.** $f(x) = 2x - \ln x$

In Exercises 69 and 70, determine whether the statement is true or false. If it is true, explain why it is true. If it is false, give an example to show why it is false.

69. If $f(x) = \ln 5$, then $f'(x) = \frac{1}{5}$.

70. If $f(x) = \ln a^x$, then $f'(x) = \ln a$.

71. Prove that $\frac{d}{dx} \ln|x| = \frac{1}{x}$ $(x \ne 0)$ for the case $x < 0$.

72. Use the definition of the derivative to show that

$$\lim_{x \to 0} \frac{\ln(x + 1)}{x} = 1$$

5.5 Solutions to Self-Check Exercises

1. The slope of the tangent line to the graph of f at any point $(x, f(x))$ lying on the graph of f is given by $f'(x)$. Using the product rule, we find

$$\begin{aligned} f'(x) &= \frac{d}{dx}[x \ln(2x + 3)] \\ &= x \frac{d}{dx} \ln(2x + 3) + \ln(2x + 3) \cdot \frac{d}{dx}(x) \\ &= x\left(\frac{2}{2x + 3}\right) + \ln(2x + 3) \cdot 1 \\ &= \frac{2x}{2x + 3} + \ln(2x + 3) \end{aligned}$$

In particular, the slope of the tangent line to the graph of f at the point $(-1, 0)$ is

$$f'(-1) = \frac{-2}{-2 + 3} + \ln 1 = -2$$

Therefore, using the point-slope form of the equation of a line, we see that a required equation is

$$\begin{aligned} y - 0 &= -2(x + 1) \\ y &= -2x - 2 \end{aligned}$$

2. Taking the logarithm on both sides of the equation gives

$$\begin{aligned}\ln y &= \ln(2x+1)^3(3x+4)^5 \\ &= \ln(2x+1)^3 + \ln(3x+4)^5 \\ &= 3\ln(2x+1) + 5\ln(3x+4)\end{aligned}$$

Differentiating both sides of the equation with respect to x, keeping in mind that y is a function of x, we obtain

$$\begin{aligned}\frac{d}{dx}(\ln y) = \frac{y'}{y} &= 3 \cdot \frac{2}{2x+1} + 5 \cdot \frac{3}{3x+4} \\ &= 3\left(\frac{2}{2x+1} + \frac{5}{3x+4}\right)\end{aligned}$$

and

$$y' = 3(2x+1)^3(3x+4)^5\left(\frac{2}{2x+1} + \frac{5}{3x+4}\right)$$

5.6 Exponential Functions as Mathematical Models

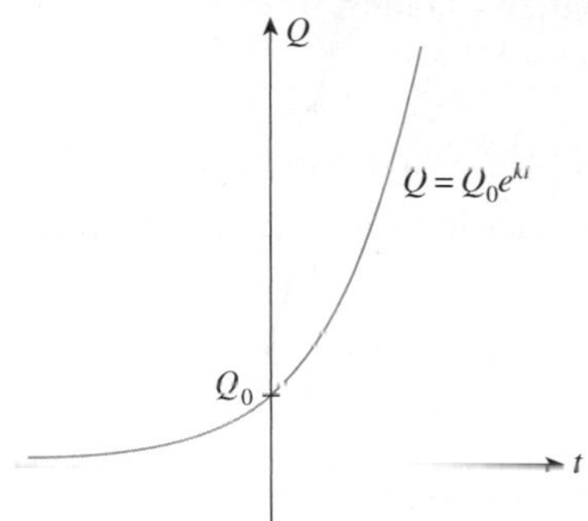

FIGURE 13
Exponential growth

Exponential Growth

Many problems arising from practical situations can be described mathematically in terms of exponential functions or functions closely related to the exponential function. In this section we look at some applications involving exponential functions from the fields of the life and social sciences.

In Section 5.1, we saw that the exponential function $f(x) = b^x$ is an increasing function when $b > 1$. In particular, the function $f(x) = e^x$ shares this property. From this result, one may deduce that the function $Q(t) = Q_0e^{kt}$, where Q_0 and k are positive constants, has the following properties:

1. $Q(0) = Q_0$
2. $Q(t)$ increases "rapidly" without bound as t increases without bound (Figure 13).

Property 1 follows from the computation

$$Q(0) = Q_0e^0 = Q_0$$

Next, to study the rate of change of the function $Q(t)$, we differentiate it with respect to t, obtaining

$$\begin{aligned}Q'(t) &= \frac{d}{dt}(Q_0e^{kt}) \\ &= Q_0\frac{d}{dt}(e^{kt}) \\ &= kQ_0e^{kt} \\ &= kQ(t) \end{aligned} \qquad \textbf{(12)}$$

Since $Q(t) > 0$ (because Q_0 is assumed to be positive) and $k > 0$, we see that $Q'(t) > 0$ and so $Q(t)$ is an increasing function of t. Our computation has in fact shed more light on an important property of the function $Q(t)$. Equation (12) says that the rate of increase of the function $Q(t)$ is proportional to the amount $Q(t)$ of the quantity present at time t. The implication is that as $Q(t)$ increases, so does the *rate of increase* of $Q(t)$, resulting in a very rapid increase in $Q(t)$ as t increases without bound.

Thus, the exponential function

$$Q(t) = Q_0 e^{kt} \qquad (0 \le t < \infty) \tag{13}$$

provides us with a mathematical model of a quantity $Q(t)$ that is initially present in the amount of $Q(0) = Q_0$ and whose rate of growth at any time t is directly proportional to the amount of the quantity present at time t. Such a quantity is said to exhibit **exponential growth**, and the constant k of proportionality is called the **growth constant**. Interest earned on a fixed deposit when compounded continuously exhibits exponential growth. Other examples of unrestricted exponential growth follow.

APPLIED EXAMPLE 1 Growth of Bacteria Under ideal laboratory conditions, the number of bacteria in a culture grows in accordance with the law $Q(t) = Q_0 e^{kt}$, where Q_0 denotes the number of bacteria initially present in the culture, k is a constant determined by the strain of bacteria under consideration, and t is the elapsed time measured in hours. Suppose 10,000 bacteria are present initially in the culture and 60,000 present 2 hours later.

a. How many bacteria will there be in the culture at the end of 4 hours?
b. What is the rate of growth of the population after 4 hours?

Solution

a. We are given that $Q(0) = Q_0 = 10{,}000$, so $Q(t) = 10{,}000e^{kt}$. Next, the fact that 60,000 bacteria are present 2 hours later translates into $Q(2) = 60{,}000$. Thus,

$$60{,}000 = 10{,}000e^{2k}$$
$$e^{2k} = 6$$

Taking the natural logarithm on both sides of the equation, we obtain

$$\ln e^{2k} = \ln 6$$
$$2k = \ln 6 \qquad \text{Since } \ln e = 1$$
$$k \approx 0.8959$$

Thus, the number of bacteria present at any time t is given by

$$Q(t) = 10{,}000e^{0.8959t}$$

In particular, the number of bacteria present in the culture at the end of 4 hours is given by

$$Q(4) = 10{,}000e^{0.8959(4)}$$
$$\approx 360{,}029$$

b. The rate of growth of the bacteria population at any time t is given by

$$Q'(t) = kQ(t)$$

Thus, using the result from part (a), we find that the rate at which the population is growing at the end of 4 hours is

$$Q'(4) = kQ(4)$$
$$\approx (0.8959)(360{,}029)$$
$$\approx 322{,}550$$

or approximately 322,550 bacteria per hour. ■

Exponential Decay

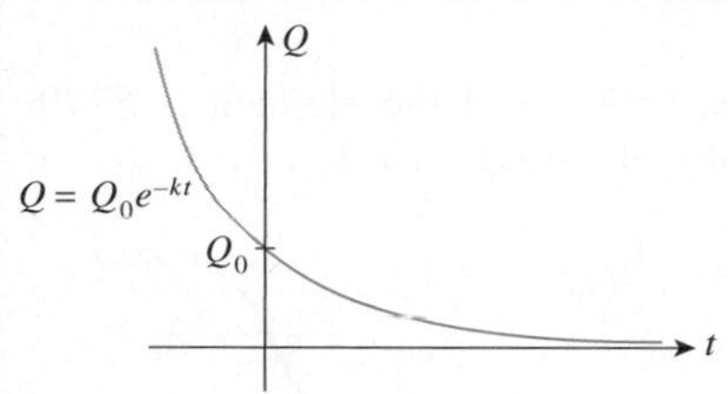

FIGURE 14
Exponential decay

In contrast to exponential growth, a quantity exhibits **exponential decay** if it decreases at a rate that is directly proportional to its size. Such a quantity may be described by the exponential function

$$Q(t) = Q_0e^{-kt} \qquad t \in [0, \infty) \tag{14}$$

where the positive constant Q_0 measures the amount present initially ($t = 0$) and k is some suitable positive number, called the **decay constant**. The choice of this number is determined by the nature of the substance under consideration. The graph of this function is sketched in Figure 14.

To verify the properties ascribed to the function $Q(t)$, we simply compute

$$\begin{aligned} Q(0) &= Q_0e^0 = Q_0 \\ Q'(t) &= \frac{d}{dt}(Q_0e^{-kt}) \\ &= Q_0\frac{d}{dt}(e^{-kt}) \\ &= -kQ_0e^{-kt} = -kQ(t). \end{aligned}$$

APPLIED EXAMPLE 2 Radioactive Decay Radioactive substances decay exponentially. For example, the amount of radium present at any time t obeys the law $Q(t) = Q_0e^{-kt}$, where Q_0 is the initial amount present and k is a suitable positive constant. The **half-life of a radioactive substance** is the time required for a given amount to be reduced by one-half. Now, it is known that the half-life of radium is approximately 1600 years. Suppose initially there are 200 milligrams of pure radium. Find the amount left after t years. What is the amount left after 800 years?

Solution The initial amount of radium present is 200 milligrams, so $Q(0) = Q_0 = 200$. Thus, $Q(t) = 200e^{-kt}$. Next, the datum concerning the half-life of radium implies that $Q(1600) = 100$, and this gives

$$\begin{aligned} 100 &= 200e^{-1600k} \\ e^{-1600k} &= \frac{1}{2} \end{aligned}$$

Taking the natural logarithm on both sides of this equation yields

$$\begin{aligned} -1600k \ln e &= \ln\frac{1}{2} \\ -1600k &= \ln\frac{1}{2} \qquad \ln e = 1 \\ k &= -\frac{1}{1600}\ln\left(\frac{1}{2}\right) = 0.0004332 \end{aligned}$$

Therefore, the amount of radium left after t years is

$$Q(t) = 200e^{-0.0004332t}$$

In particular, the amount of radium left after 800 years is

$$Q(800) = 200e^{-0.0004332(800)} \approx 141.42$$

or approximately 141 milligrams. ■

VIDEO

APPLIED EXAMPLE 3 Carbon-14 Decay Carbon 14, a radioactive isotope of carbon, has a half-life of 5770 years. What is its decay constant?

Solution We have $Q(t) = Q_0e^{-kt}$. Since the half-life of the element is 5770 years, half of the substance is left at the end of that period; that is,

$$Q(5770) = Q_0e^{-5770k} = \frac{1}{2}Q_0$$

$$e^{-5770k} = \frac{1}{2}$$

Taking the natural logarithm on both sides of this equation, we have

$$\ln e^{-5770k} = \ln\frac{1}{2}$$

$$-5770k = -0.693147$$

$$k \approx 0.00012$$

Carbon-14 dating is a well-known method used by anthropologists to establish the age of animal and plant fossils. This method assumes that the proportion of carbon 14 (C-14) present in the atmosphere has remained constant over the past 50,000 years. Professor Willard Libby, recipient of the Nobel Prize in chemistry in 1960, proposed this theory.

The amount of C-14 in the tissues of a living plant or animal is constant. However, when an organism dies, it stops absorbing new quantities of C-14, and the amount of C-14 in the remains diminishes because of the natural decay of the radioactive substance. Thus, the approximate age of a plant or animal fossil can be determined by measuring the amount of C-14 present in the remains.

APPLIED EXAMPLE 4 Carbon-14 Dating A skull from an archeological site has one-tenth the amount of C-14 that it originally contained. Determine the approximate age of the skull.

Solution Here,

$$Q(t) = Q_0e^{-kt}$$

$$= Q_0e^{-0.00012t}$$

where Q_0 is the amount of C-14 present originally and k, the decay constant, is equal to 0.00012 (see Example 3). Since $Q(t) = (1/10)Q_0$, we have

$$\frac{1}{10}Q_0 = Q_0e^{-0.00012t}$$

$$\ln\frac{1}{10} = -0.00012t \qquad \text{Take the natural logarithm on both sides.}$$

$$t = \frac{\ln\frac{1}{10}}{-0.00012}$$

$$\approx 19{,}200$$

or approximately 19,200 years.

Learning Curves

The next example shows how the exponential function may be applied to describe certain types of learning processes. Consider the function

$$Q(t) = C - Ae^{-kt}$$

where C, A, and k are positive constants. To sketch the graph of the function Q, observe that its y-intercept is given by $Q(0) = C - A$. Next, we compute

$$Q'(t) = kAe^{-kt}$$

Since both k and A are positive, we see that $Q'(t) > 0$ for all values of t. Thus, $Q(t)$ is an increasing function of t. Also,

$$\begin{aligned}\lim_{t\to\infty} Q(t) &= \lim_{t\to\infty} (C - Ae^{-kt}) \\ &= \lim_{t\to\infty} C - \lim_{t\to\infty} Ae^{-kt} \\ &= C\end{aligned}$$

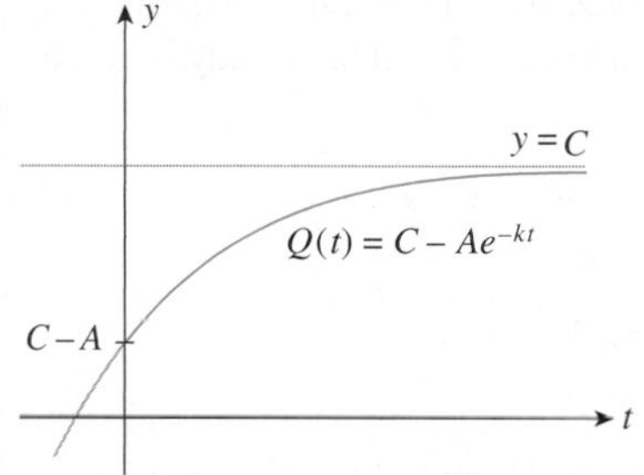

FIGURE 15
A learning curve

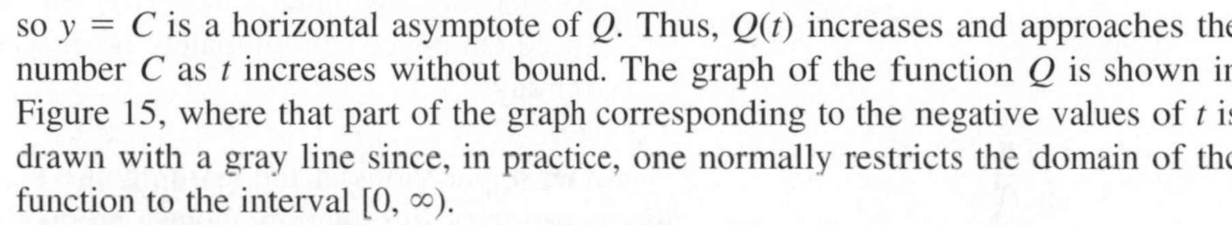

so $y = C$ is a horizontal asymptote of Q. Thus, $Q(t)$ increases and approaches the number C as t increases without bound. The graph of the function Q is shown in Figure 15, where that part of the graph corresponding to the negative values of t is drawn with a gray line since, in practice, one normally restricts the domain of the function to the interval $[0, \infty)$.

Observe that $Q(t)$ $(t > 0)$ increases rather rapidly initially but that the rate of increase slows down considerably after a while. To see this, we compute

$$\lim_{t\to\infty} Q'(t) = \lim_{t\to\infty} kAe^{-kt} = 0$$

This behavior of the graph of the function Q closely resembles the learning pattern experienced by workers engaged in highly repetitive work. For example, the productivity of an assembly-line worker increases very rapidly in the early stages of the training period. This productivity increase is a direct result of the worker's training and accumulated experience. But the rate of increase of productivity slows as time goes by, and the worker's productivity level approaches some fixed level due to the limitations of the worker and the machine. Because of this characteristic, the graph of the function $Q(t) = C - Ae^{-kt}$ is often called a **learning curve**.

APPLIED EXAMPLE 5 Assembly Time The Camera Division of Eastman Optical produces a 35-mm single-lens reflex camera. Eastman's training department determines that after completing the basic training program, a new, previously inexperienced employee will be able to assemble

$$Q(t) = 50 - 30e^{-0.5t}$$

model F cameras per day, t months after the employee starts work on the assembly line.

a. How many model F cameras can a new employee assemble per day after basic training?

b. How many model F cameras can an employee with 1 month of experience assemble per day? An employee with 2 months of experience? An employee with 6 months of experience?

c. How many model F cameras can the average experienced employee assemble per day?

Solution

a. The number of model F cameras a new employee can assemble is given by

$$Q(0) = 50 - 30 = 20$$

b. The number of model F cameras that an employee with 1 month of experience, 2 months of experience, and 6 months of experience can assemble per day is given by

$$Q(1) = 50 - 30e^{-0.5} \approx 31.80$$
$$Q(2) = 50 - 30e^{-1} \approx 38.96$$
$$Q(6) = 50 - 30e^{-3} \approx 48.51$$

or approximately 32, 39, and 49, respectively.

c. As t increases without bound, $Q(t)$ approaches 50. Hence, the average experienced employee can ultimately be expected to assemble 50 model F cameras per day.

Other applications of the learning curve are found in models that describe the dissemination of information about a product or the velocity of an object dropped into a viscous medium.

Logistic Growth Functions

Our last example of an application of exponential functions to the description of natural phenomena involves the **logistic** (also called the **S-shaped,** or **sigmoidal**) **curve**, which is the graph of the function

$$Q(t) = \frac{A}{1 + Be^{-kt}}$$

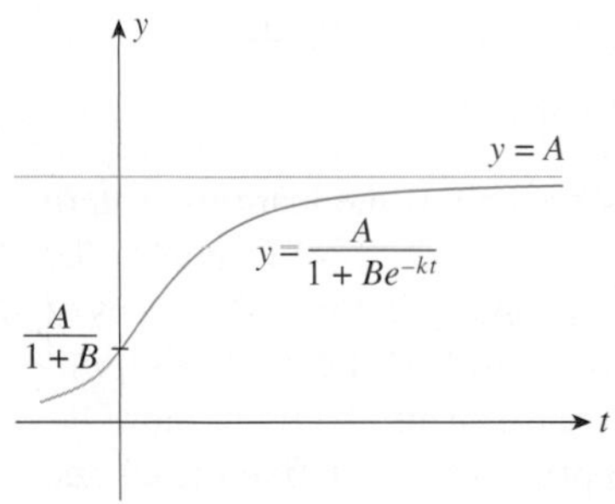

FIGURE 16
A logistic curve

where A, B, and k are positive constants. The function Q is called a **logistic growth function**, and the graph of the function Q is sketched in Figure 16.

Observe that $Q(t)$ increases rather rapidly for small values of t. In fact, for small values of t, the logistic curve resembles an exponential growth curve. However, the *rate of growth* of $Q(t)$ decreases quite rapidly as t increases and $Q(t)$ approaches the number A as t increases without bound.

Thus, the logistic curve exhibits both the property of rapid growth of the exponential growth curve as well as the "saturation" property of the learning curve. Because of these characteristics, the logistic curve serves as a suitable mathematical model for describing many natural phenomena. For example, if a small number of rabbits were introduced to a tiny island in the South Pacific, the rabbit population might be expected to grow very rapidly at first, but the growth rate would decrease quickly as overcrowding, scarcity of food, and other environmental factors affected it. The population would eventually stabilize at a level compatible with the life-support capacity of the environment. This level, given by A, is called the *carrying capacity* of the environment. Models describing the spread of rumors and epidemics are other examples of the application of the logistic curve.

VIDEO

APPLIED EXAMPLE 6 Spread of Flu The number of soldiers at Fort MacArthur who contracted influenza after t days during a flu epidemic is approximated by the exponential model

$$Q(t) = \frac{5000}{1 + 1249e^{-kt}}$$

If 40 soldiers contracted the flu by day 7, find how many soldiers contracted the flu by day 15.

Solution The given information implies that

$$Q(7) = \frac{5000}{1 + 1249e^{-7k}} = 40$$

Thus,

$$\begin{aligned} 40(1 + 1249e^{-7k}) &= 5000 \\ 1 + 1249e^{-7k} &= \frac{5000}{40} = 125 \\ e^{-7k} &= \frac{124}{1249} \\ -7k &= \ln \frac{124}{1249} \\ k &= -\frac{\ln \frac{124}{1249}}{7} \approx 0.33 \end{aligned}$$

Therefore, the number of soldiers who contracted the flu after t days is given by

$$Q(t) = \frac{5000}{1 + 1249e^{-0.33t}}$$

In particular, the number of soldiers who contracted the flu by day 15 is given by

$$\begin{aligned} Q(15) &= \frac{5000}{1 + 1249e^{-15(0.33)}} \\ &\approx 508 \end{aligned}$$

or approximately 508 soldiers.

EXPLORING WITH TECHNOLOGY

Refer to Example 6.

1. Use a graphing utility to plot the graph of the function Q, using the viewing window $[0, 40] \times [0, 5000]$.
2. Find how long it takes for the first 1000 soldiers to contract the flu.
   Hint: Plot the graphs of $y_1 = Q(t)$ and $y_2 = 1000$ and find the point of intersection of the two graphs.

5.6 Self-Check Exercise

Suppose the population (in millions) of a country at any time t grows in accordance with the rule

$$P = \left(P_0 + \frac{I}{k}\right)e^{kt} - \frac{I}{k}$$

where P denotes the population at any time t, k is a constant reflecting the natural growth rate of the population, I is a constant giving the (constant) rate of immigration into the country, and P_0 is the total population of the country at time $t = 0$. The population of the United States in 1980 ($t = 0$) was 226.5 million. If the natural growth rate is 0.8% annually ($k = 0.008$) and net immigration is allowed at the rate of half a million people per year ($I = 0.5$), what is the expected population of the United States in 2005?

Solution to Self-Check Exercise 5.6 can be found on page 389.

5.6 Concept Questions

1. Give the model for unrestricted exponential growth and the model for exponential decay. What effect does the magnitude of the growth (decay) constant have on the growth (decay) of a quantity?
2. What is the half-life of a radioactive substance?
3. What is the logistic growth function? What are its characteristics?

5.6 Exercises

1. **Exponential Growth** Given that a quantity $Q(t)$ is described by the exponential growth function

$$Q(t) = 400e^{0.05t}$$

where t is measured in minutes, answer the following questions:

a. What is the growth constant?
b. What quantity is present initially?
c. Complete the following table of values:

t	0	10	20	100	1000
Q					

2. **Exponential Decay** Given that a quantity $Q(t)$ exhibiting exponential decay is described by the function

$$Q(t) = 2000e^{-0.06t}$$

where t is measured in years, answer the following questions:

a. What is the decay constant?
b. What quantity is present initially?
c. Complete the following table of values:

t	0	5	10	20	100
Q					

3. **Growth of Bacteria** The growth rate of the bacterium *Escherichia coli*, a common bacterium found in the human intestine, is proportional to its size. Under ideal laboratory conditions, when this bacterium is grown in a nutrient broth medium, the number of cells in a culture doubles approximately every 20 min.

a. If the initial cell population is 100, determine the function $Q(t)$ that expresses the exponential growth of the number of cells of this bacterium as a function of time t (in minutes).
b. How long will it take for a colony of 100 cells to increase to a population of 1 million?
c. If the initial cell population were 1000, how would this alter our model?

4. **World Population** The world population at the beginning of 1990 was 5.3 billion. Assume that the population continues to grow at the rate of approximately 2%/year and find the function $Q(t)$ that expresses the world population (in billions) as a function of time t (in years), with $t = 0$ corresponding to the beginning of 1990.

a. Using this function, complete the following table of values and sketch the graph of the function Q.

Year	1990	1995	2000	2005
World Population				

Year	2010	2015	2020	2025
World Population				

b. Find the estimated rate of growth in 2010.

5. **World Population** Refer to Exercise 4.

a. If the world population continues to grow at the rate of approximately 2%/year, find the length of time t_0 required for the world population to triple in size.

b. Using the time t_0 found in part (a), what would be the world population if the growth rate were reduced to 1.8%/yr?

6. RESALE VALUE A certain piece of machinery was purchased 3 yr ago by Garland Mills for \$500,000. Its present resale value is \$320,000. Assuming that the machine's resale value decreases exponentially, what will it be 4 yr from now?

7. ATMOSPHERIC PRESSURE If the temperature is constant, then the atmospheric pressure P (in pounds/square inch) varies with the altitude above sea level h in accordance with the law

$$P = p_0 e^{-kh}$$

where p_0 is the atmospheric pressure at sea level and k is a constant. If the atmospheric pressure is 15 lb/in.^{2} at sea level and 12.5 lb/in.^{2} at 4000 ft, find the atmospheric pressure at an altitude of 12,000 ft. How fast is the atmospheric pressure changing with respect to altitude at an altitude of 12,000 ft?

8. RADIOACTIVE DECAY The radioactive element polonium decays according to the law

$$Q(t) = Q_0 \cdot 2^{-(t/140)}$$

where Q_0 is the initial amount and the time t is measured in days. If the amount of polonium left after 280 days is 20 mg, what was the initial amount present?

9. RADIOACTIVE DECAY Phosphorus 32 (P-32) has a half-life of 14.2 days. If 100 g of this substance are present initially, find the amount present after t days. What amount will be left after 7.1 days? How fast is P-32 decaying when $t = 7.1$?

10. NUCLEAR FALLOUT Strontium 90 (Sr-90), a radioactive isotope of strontium, is present in the fallout resulting from nuclear explosions. It is especially hazardous to animal life, including humans, because, upon ingestion of contaminated food, it is absorbed into the bone structure. Its half-life is 27 yr. If the amount of Sr-90 in a certain area is found to be four times the "safe" level, find how much time must elapse before an "acceptable level" is reached.

11. CARBON-14 DATING Wood deposits recovered from an archeological site contain 20% of the C-14 they originally contained. How long ago did the tree from which the wood was obtained die?

12. CARBON-14 DATING Skeletal remains of the so-called Pittsburgh Man, unearthed in Pennsylvania, had lost 82% of the C-14 they originally contained. Determine the approximate age of the bones.

13. LEARNING CURVES The American Court Reporting Institute finds that the average student taking Advanced Machine Shorthand, an intensive 20-wk course, progresses according to the function

$$Q(t) = 120(1 - e^{-0.05t}) + 60 \qquad (0 \le t \le 20)$$

where $Q(t)$ measures the number of words (per minute) of dictation that the student can take in machine shorthand after t wk in the course. Sketch the graph of the function Q and answer the following questions:

a. What is the beginning shorthand speed for the average student in this course?

b. What shorthand speed does the average student attain halfway through the course?

c. How many words per minute can the average student take after completing this course?

14. EFFECT OF ADVERTISING ON SALES Metro Department Store found that t wk after the end of a sales promotion the volume of sales was given by a function of the form

$$S(t) = B + Ae^{-kt} \qquad (0 \le t \le 4)$$

where $B = 50{,}000$ and is equal to the average weekly volume of sales before the promotion. The sales volumes at the end of the first and third weeks were \$83,515 and \$65,055, respectively. Assume that the sales volume is decreasing exponentially.

a. Find the decay constant k.

b. Find the sales volume at the end of the fourth week.

c. How fast is the sales volume dropping at the end of the fourth week?

15. DEMAND FOR COMPUTERS Universal Instruments found that the monthly demand for its new line of Galaxy Home Computers t mo after placing the line on the market was given by

$$D(t) = 2000 - 1500e^{-0.05t} \qquad (t > 0)$$

Graph this function and answer the following questions:

a. What is the demand after 1 mo? After 1 yr? After 2 yr? After 5 yr?

b. At what level is the demand expected to stabilize?

c. Find the rate of growth of the demand after the tenth month.

16. RELIABILITY OF COMPUTER CHIPS The percent of a certain brand of computer chips that will fail after t yr of use is estimated to be

$$P(t) = 100(1 - e^{-0.1t})$$

a. What percent of this brand of computer chips are expected to be usable after 3 yr?

b. Evaluate $\lim_{t \to \infty} P(t)$. Did you expect this result?

17. LENGTHS OF FISH The length (in centimeters) of a typical Pacific halibut t yr old is approximately

$$f(t) = 200(1 - 0.956e^{-0.18t})$$

a. What is the length of a typical 5-yr-old Pacific halibut?

b. How fast is the length of a typical 5-yr-old Pacific halibut increasing?

c. What is the maximum length a typical Pacific halibut can attain?

18. **Spread of an Epidemic** During a flu epidemic, the number of children in the Woodbridge Community School System who contracted influenza after t days was given by

$$Q(t) = \frac{1000}{1 + 199e^{-0.8t}}$$

a. How many children were stricken by the flu after the first day?
b. How many children had the flu after 10 days?
c. How many children eventually contracted the disease?

19. **Lay Teachers at Roman Catholic Schools** The change from religious to lay teachers at Roman Catholic schools has been partly attributed to the decline in the number of women and men entering religious orders. The percent of teachers who are lay teachers is given by

$$f(t) = \frac{98}{1 + 2.77e^{-t}} \qquad (0 \le t \le 4)$$

where t is measured in decades, with $t = 0$ corresponding to the beginning of 1960.
a. What percent of teachers were lay teachers at the beginning of 1990?
b. How fast was the percent of lay teachers changing at the beginning of 1990?
c. Find the year when the percent of lay teachers was increasing most rapidly.

Sources: National Catholic Education Association and the Department of Education

20. **Growth of a Fruit Fly Population** On the basis of data collected during an experiment, a biologist found that the growth of a fruit fly (*Drosophila*) with a limited food supply could be approximated by the exponential model

$$N(t) = \frac{400}{1 + 39e^{-0.16t}}$$

where t denotes the number of days since the beginning of the experiment.
a. What was the initial fruit fly population in the experiment?
b. What was the maximum fruit fly population that could be expected under this laboratory condition?
c. What was the population of the fruit fly colony on the 20th day?
d. How fast was the population changing on the 20th day?

21. **Demographics** The number of citizens aged 45–64 is projected to be

$$P(t) = \frac{197.9}{1 + 3.274e^{-0.0361t}} \qquad (0 \le t \le 20)$$

where $P(t)$ is measured in millions and t is measured in years, with $t = 0$ corresponding to the beginning of 1990. People belonging to this age group are the targets of insurance companies that want to sell them annuities. What is the projected population of citizens aged 45–64 in 2010?

Source: K. G. Securities

22. **Population Growth in the 21st Century** The U.S. population is approximated by the function

$$P(t) = \frac{616.5}{1 + 4.02e^{-0.5t}}$$

where $P(t)$ is measured in millions of people and t is measured in 30-yr intervals, with $t = 0$ corresponding to 1930. What is the expected population of the United States in 2020 ($t = 3$)?

23. **Dissemination of Information** Three hundred students attended the dedication ceremony of a new building on a college campus. The president of the traditionally female college announced a new expansion program, which included plans to make the college coeducational. The number of students who learned of the new program t hr later is given by the function

$$f(t) = \frac{3000}{1 + Be^{-kt}}$$

If 600 students on campus had heard about the new program 2 hr after the ceremony, how many students had heard about the policy after 4 hr? How fast was the news spreading 4 hr after the ceremony?

24. **Price of a Commodity** The unit price of a certain commodity is given by

$$p = f(t) = 6 + 4e^{-2t}$$

where p is measured in dollars and t is measured in months.
a. Show that f is decreasing on $(0, \infty)$.
b. Show that the graph of f is concave upward on $(0, \infty)$.
c. Evaluate $\lim_{t \to \infty} f(t)$. (*Note:* This value is called the *equilibrium price* of the commodity, and in this case, we have *price stability.*)
d. Sketch the graph of f.

25. **Chemical Mixtures** Two chemicals react to form another chemical. Suppose the amount of the chemical formed in time t (in hours) is given by

$$x(t) = \frac{15\left[1 - \left(\frac{2}{3}\right)^{3t}\right]}{1 - \frac{1}{4}\left(\frac{2}{3}\right)^{3t}}$$

where $x(t)$ is measured in pounds. How many pounds of the chemical are formed eventually?
Hint: You need to evaluate $\lim_{t \to \infty} x(t)$.

26. **Von Bertalanffy Growth Function** The length (in centimeters) of a common commercial fish is approximated by the von Bertalanffy growth function

$$f(t) = a(1 - be^{-kt})$$

where a, b, and k are positive constants.

a. Show that f is increasing on the interval $(0, \infty)$.
b. Show that the graph of f is concave downward on $(0, \infty)$.
c. Show that $\lim_{t \to \infty} f(t) = a$.
d. Use the results of parts (a)–(c) to sketch the graph of f.

27. Absorption of Drugs The concentration of a drug in grams/cubic centimeter (g/cm^3) t min after it has been injected into the bloodstream is given by

$$C(t) = \frac{k}{b - a}(e^{-at} - e^{-bt})$$

where a, b, and k are positive constants, with $b > a$.
a. At what time is the concentration of the drug the greatest?
b. What will be the concentration of the drug in the long run?

28. Concentration of Glucose in the Bloodstream A glucose solution is administered intravenously into the bloodstream at a constant rate of r mg/hr. As the glucose is being administered, it is converted into other substances and removed from the bloodstream. Suppose the concentration of the glucose solution at time t is given by

$$C(t) = \frac{r}{k} - \left[\left(\frac{r}{k}\right) - C_0\right]e^{-kt}$$

where C_0 is the concentration at time $t = 0$ and k is a positive constant. Assuming that $C_0 < r/k$, evaluate $\lim_{t \to \infty} C(t)$.
a. What does your result say about the concentration of the glucose solution in the long run?
b. Show that the function C is increasing on $(0, \infty)$.
c. Show that the graph of C is concave downward on $(0, \infty)$.
d. Sketch the graph of the function C.

29. Radioactive Decay A radioactive substance decays according to the formula

$$Q(t) = Q_0e^{-kt}$$

where $Q(t)$ denotes the amount of the substance present at time t (measured in years), Q_0 denotes the amount of the substance present initially, and k (a positive constant) is the decay constant.
a. Show that half-life of the substance is $\bar{t} = \ln 2/k$.
b. Suppose a radioactive substance decays according to the formula

$$Q(t) = 20e^{-0.0001238\bar{t}}$$

How long will it take for the substance to decay to half the original amount?

30. Gompertz Growth Curve Consider the function

$$Q(t) = Ce^{-Ae^{-kt}}$$

where $Q(t)$ is the size of a quantity at time t and A, C, and k are positive constants. The graph of this function, called the *Gompertz growth curve*, is used by biologists to describe restricted population growth.
a. Show that the function Q is always increasing.
b. Find the time t at which the growth rate $Q'(t)$ is increasing most rapidly.
Hint: Find the inflection point of Q.
c. Show that $\lim_{t \to \infty} Q(t) = C$ and interpret your result.

5.6 Solution to Self-Check Exercise

We are given that $P_0 = 226.5$, $k = 0.008$, and $I = 0.5$. So

$$P = \left(226.5 + \frac{0.5}{0.008}\right)e^{0.008t} - \frac{0.5}{0.008}$$
$$= 289e^{0.008t} - 62.5$$

Therefore, the expected population in 2005 is given by

$$P(25) = 289e^{0.2} - 62.5 \approx 290.5$$

or approximately 290.5 million.

USING TECHNOLOGY

Analyzing Mathematical Models

We can use a graphing utility to analyze the mathematical models encountered in this section.

APPLIED EXAMPLE 1 Households with Microwaves The number of households with microwave ovens increased greatly in the 1980s and 1990s.

(continued)

The percent of households with microwave ovens from 1981 through 1999 is given by

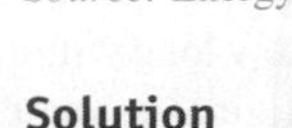

$$f(t) = \frac{87}{1 + 4.209e^{-0.3727t}} \qquad (0 \le t \le 18)$$

where t is measured in years, with $t = 0$ corresponding to the beginning of 1981.

a. Use a graphing utility to plot the graph of f on the interval [0, 18].
b. What percent of households owned microwave ovens at the beginning of 1984? At the beginning of 1994?
c. At what rate was the ownership of microwave ovens increasing at the beginning of 1984? At the beginning of 1994?
d. At what time was the increase in ownership of microwave ovens greatest?

Source: Energy Information Agency

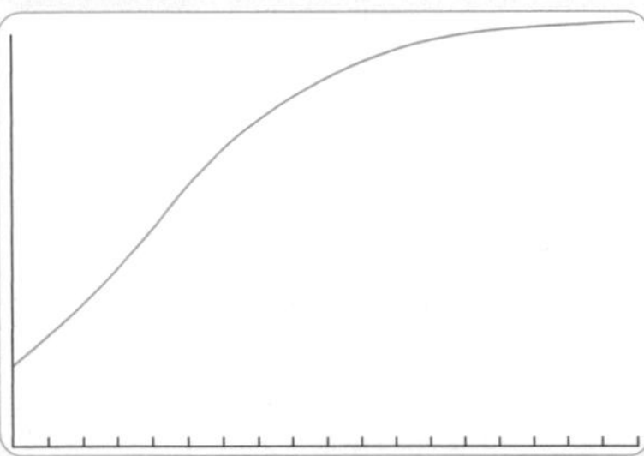

FIGURE T1
The graph of f in the viewing window $[0, 18] \times [0, 100]$

Solution

a. The graph of f using the viewing window $[0, 18] \times [0, 100]$ is shown in Figure T1.
b. Using the evaluation function, we find $f(3) \approx 36.6$ and $f(13) \approx 84.2$ and conclude that the percent of households owning microwave ovens in 1984 was 36.6 and that in 1994 was 84.2.
c. Using the numerical derivative operation, we find $f'(3) \approx 7.9$ and $f'(13) \approx 1.0$. We conclude that the ownership of microwave ovens was increasing at the rate of 7.9% per year in 1984 and at the rate of 1% per year in 1994.
d. We see that the inflection point of f occurs at (3.9, 43.5). Thus, the largest increase in ownership of microwave ovens occurred at the time when $t \approx 3.9$, or near the end of 1984. ■

TECHNOLOGY EXERCISES

1. **Online Banking** In a study prepared in 2000, the percent of households using online banking was projected to be

$$f(t) = 1.5e^{0.78t} \qquad (0 \le t \le 4)$$

where t is measured in years, with $t = 0$ corresponding to the beginning of 2000.
a. Plot the graph of f, using the viewing window $[0, 4] \times [0, 40]$.
b. How fast was the projected percent of households using online banking changing at the beginning of 2003?
c. How fast was the rate of the projected percent of households using online banking changing at the beginning of 2003?
Hint: We want $f''(3)$. Why?

Source: Online Banking Report

2. **Newton's Law of Cooling** The temperature of a cup of coffee t min after it is poured is given by

$$T = 70 + 100e^{-0.0446t}$$

where T is measured in degrees Fahrenheit.
a. Plot the graph of T, using the viewing window $[0, 30] \times [0, 200]$.
b. When will the coffee be cool enough to drink (say, 120°)?
Hint: Use the ISECT function.

3. **Air Travel** Air travel has been rising dramatically in the past 30 yr. In a study conducted in 2000, the FAA projected further exponential growth for air travel through 2010. The function

$$f(t) = 666e^{0.0413t} \qquad (0 \le t \le 10)$$

gives the number of passengers (in millions) in year t, with $t = 0$ corresponding to 2000.
a. Plot the graph of f, using the viewing window $[0, 10] \times [0, 1000]$.
b. How many air passengers were there in 2000? What was the projected number of air passengers for 2005?
c. What was the rate of change of the number of air passengers in 2005?

Source: Federal Aviation Administration

4. **Lengths of Fish** The length (in centimeters) of a typical Pacific halibut t yr old is approximately

$$f(t) = 200(1 - 0.956e^{-0.18t})$$

a. Plot the graph of f, using the viewing window $[0, 10] \times [0, 200]$.
b. Suppose a Pacific halibut caught by Mike measures 140 cm. What is its approximate age?

5. **Lay Teachers at Roman Catholic Schools** The change from religious to lay teachers at Roman Catholic schools has been partly attributed to the decline in the number of women and men entering religious orders. The percent of teachers who are lay teachers is given by

$$f(t) = \frac{98}{1 + 2.77e^{-t}} \qquad (0 \le t \le 4)$$

where t is measured in decades, with $t = 0$ corresponding to the beginning of 1960.
a. Plot the graph of f, using the viewing window $[0, 4] \times [0, 100]$.
b. What percent of teachers were lay teachers at the beginning of 1990?
c. How fast was the percent of lay teachers changing at the beginning of 1990?
d. Find the year when the percent of lay teachers was increasing most rapidly.

Sources: National Catholic Education Association and the Department of Education

6. **Growth of a Fruit Fly Population** On the basis of data collected during an experiment, a biologist found that the growth of a fruit fly (*Drosophila*) with a limited food supply could be approximated by the exponential model

$$N(t) = \frac{400}{1 + 39e^{-0.16t}}$$

where t denotes the number of days since the beginning of the experiment.
a. Plot the graph of f using the viewing window $[0, 50] \times [0, 400]$.
b. What was the initial fruit fly population in the experiment?
c. What was the maximum fruit fly population that could be expected under these laboratory conditions?
d. What was the population of the fruit fly colony on the 20th day?
e. How fast was the population changing on the 20th day?

7. **Population Growth in the 21st Century** The U.S. population is approximated by the function

$$P(t) = \frac{616.5}{1 + 4.02e^{-0.5t}}$$

where $P(t)$ is measured in millions of people and t is measured in 30-yr intervals, with $t = 0$ corresponding to 1930.
a. Plot the graph of f, using the viewing window $[0, 4] \times [0, 650]$.
b. What is the expected population of the United States in 2020 ($t = 3$)?
c. What is the expected rate of growth of the U.S. population in 2020?

8. **Time Rate of Growth of a Tumor** The rate at which a tumor grows, with respect to time, is given by

$$R = Ax \ln \frac{B}{x} \qquad (\text{for } 0 < x < B)$$

where A and B are positive constants and x is the radius of the tumor.
a. Plot the graph of R for the case $A = B = 10$.
b. Find the radius of the tumor when the tumor is growing most rapidly with respect to time.

9. **Absorption of Drugs** The concentration of a drug in an organ at any time t (in seconds) is given by

$$C(t) = \begin{cases} 0.3t - 18(1 - e^{-t/60}) & \text{if } 0 \le t \le 20 \\ 18e^{-t/60} - 12e^{-(t-20)/60} & \text{if } t > 20 \end{cases}$$

where $C(t)$ is measured in grams/cubic centimeter (g/cm^3).
a. Plot the graph of f, using the viewing window $[0, 120] \times [0, 1]$.
b. What is the initial concentration of the drug in the organ?
c. What is the concentration of the drug in the organ after 10 sec?
d. What is the concentration of the drug in the organ after 30 sec?
e. What will be the concentration of the drug in the long run?

10. **Annuities** At the time of retirement, Christine expects to have a sum of \$500,000 in her retirement account. Assuming that the account pays interest at the rate of 5%/year compounded continuously, her accountant pointed out to her that if she made withdrawals amounting to x dollars per year ($x > 25{,}000$), then the time required to deplete her savings would be T years, where

$$T = f(x) = 20 \ln\left(\frac{x}{x - 25{,}000}\right) \qquad (x > 25{,}000)$$

a. Plot the graph of f, using the viewing window $[25{,}000, 50{,}000] \times [0, 100]$.
b. How much should Christine plan to withdraw from her retirement account each year if she wants it to last for 25 yr?
c. Evaluate $\lim_{x \to 25{,}000^+} f(x)$. Is the result expected? Explain.
d. Evaluate $\lim_{x \to \infty} f(x)$. Is the result expected? Explain.

CHAPTER 5 Summary of Principal Formulas and Terms

FORMULAS

1. Exponential function with base b	$y = b^x$
2. The number e	$e = \lim_{m \to \infty} \left(1 + \frac{1}{m}\right)^m = 2.71828\ldots$
3. Exponential function with base e	$y = e^x$
4. Logarithmic function with base b	$y = \log_b x$
5. Logarithmic function with base e	$y = \ln x$
6. Inverse properties of $\ln x$ and e	$\ln e^x = x$ and $e^{\ln x} = x$
7. Compound interest (accumulated amount)	$A = P\left(1 + \frac{r}{m}\right)^{mt}$
8. Effective rate of interest	$r_{\text{eff}} = \left(1 + \frac{r}{m}\right)^m - 1$
9. Compound interest (present value)	$P = A\left(1 + \frac{r}{m}\right)^{-mt}$
10. Continuous compound interest	$A = Pe^{rt}$
11. Derivative of the exponential function	$\frac{d}{dx}(e^x) = e^x$
12. Chain rule for exponential functions	$\frac{d}{dx}(e^u) = e^u \frac{du}{dx}$
13. Derivative of the logarithmic function	$\frac{d}{dx} \ln \|x\| = \frac{1}{x}$
14. Chain rule for logarithmic functions	$\frac{d}{dx}(\ln u) = \frac{1}{u} \frac{du}{dx}$

TERMS

common logarithm (340)
natural logarithm (340)
compound interest (347)
logarithmic differentiation (374)
exponential growth (380)
growth constant (380)
exponential decay (381)
decay constant (381)
half-life of a radioactive substance (381)
logistic growth function (384)

CHAPTER 5 Concept Review Questions

Fill in the blanks.

1. The function $f(x) = x^b$ (b, a real number) is called a/an _____ function, whereas the function $g(x) = b^x$, where $b >$ _____ and $b \neq$ _____, is called a/an _____ function.
2. a. The domain of the function $y = 3^x$ is _____, and its range is _____.
   b. The graph of the function $y = 0.3^x$ passes through the point _____ and is decreasing on _____.
3. a. If $b > 0$ and $b \neq 1$, then the logarithmic function $y = \log_b x$ has domain _____ and range _____; its graph passes through the point _____.
   b. The graph of $y = \log_b x$ is decreasing if b _____ and increasing if b _____.
4. a. If $x > 0$, then $e^{\ln x} =$ _____.
   b. If x is any real number, then $\ln e^x =$ _____.

5. In the compound interest formula $A = P(1 + \frac{r}{m})^{mt}$, A stands for the _____ _____, P stands for the _____, r stands for the _____ _____ _____ per year, m stands for the _____ _____ _____ _____ per year, and t stands for the _____.

6. The effective rate r_{eff} is related to the nominal interest rate r per year and the number of conversion periods per year by $r_{\text{eff}} =$ _____.

7. If interest earned at the rate of r per year is compounded continuously over t years, then a principal of P dollars will have an accumulated value of $A =$ _____ dollars.

8. **a.** If $g(x) = e^{f(x)}$, where f is a differentiable function, then $g'(x) =$ _____.
b. If $g(x) = \ln f(x)$, where $f(x) > 0$ is differentiable, then $g'(x) =$ _____.

9. **a.** In the unrestricted exponential growth model $Q = Q_0e^{kt}$, Q_0 represents the quantity present _____, and k is called the _____ constant.
b. In the exponential decay model $Q = Q_0e^{-kt}$, k is called the _____ constant.
c. The half-life of a radioactive substance is the _____ required for a substance to decay to _____ _____ of its original amount.

10. **a.** For the model $Q(t) = C - Ae^{-kt}$ describing a learning curve, $y = C$ is a/an _____ _____ of the graph of Q. The value of $Q(t)$ never exceeds _____.
b. For the logistic growth model $Q(t) = \dfrac{A}{1 + Be^{-kt}}$, $y = A$ is a/an _____ _____ of the graph of Q. If the quantity $Q(t)$ is initially smaller than A, then $Q(t)$ will eventually approach _____ as t increases; the number A represents the life-support capacity of the environment and is called the _____ _____ of the environment.

CHAPTER 5 Review Exercises

1. Sketch on the same set of coordinate axes the graphs of the exponential functions defined by the equations.

a. $y = 2^{-x}$ **b.** $y = \left(\frac{1}{2}\right)^x$

In Exercises 2 and 3, express each equation in logarithmic form.

2. $\left(\frac{2}{3}\right)^3 = \frac{27}{8}$ **3.** $16^{-3/4} = 0.125$

In Exercises 4 and 5, solve each equation for x.

4. $\log_4(2x + 1) = 2$

5. $\ln(x - 1) + \ln 4 = \ln(2x + 4) - \ln 2$

In Exercises 6–8, given that ln 2 = x, ln 3 = y, and ln 5 = z, express each of the given logarithmic values in terms of x, y, and z.

6. $\ln 30$ **7.** $\ln 3.6$ **8.** $\ln 75$

9. Sketch the graph of the function $y = \log_2(x + 3)$.

10. Sketch the graph of the function $y = \log_3(x + 1)$.

In Exercises 11–28, find the derivative of the function.

11. $f(x) = xe^{2x}$ **12.** $f(t) = \sqrt{t}e^t + t$

13. $g(t) = \sqrt{t}e^{-2t}$ **14.** $g(x) = e^x\sqrt{1 + x^2}$

15. $y = \dfrac{e^{2x}}{1 + e^{-2x}}$ **16.** $f(x) = e^{2x^2-1}$

17. $f(x) = xe^{-x^2}$ **18.** $g(x) = (1 + e^{2x})^{3/2}$

19. $f(x) = x^2e^x + e^x$ **20.** $g(t) = t \ln t$

21. $f(x) = \ln(e^{x^2} + 1)$ **22.** $f(x) = \dfrac{x}{\ln x}$

23. $f(x) = \dfrac{\ln x}{x + 1}$ **24.** $y = (x + 1)e^x$

25. $y = \ln(e^{4x} + 3)$ **26.** $f(r) = \dfrac{re^r}{1 + r^2}$

27. $f(x) = \dfrac{\ln x}{1 + e^x}$ **28.** $g(x) = \dfrac{e^{x^2}}{1 + \ln x}$

29. Find the second derivative of the function $y = \ln(3x + 1)$.

30. Find the second derivative of the function $y = x \ln x$.

31. Find $h'(0)$ if $h(x) = g(f(x))$, $g(x) = x + \dfrac{1}{x}$, and $f(x) = e^x$.

32. Find $h'(1)$ if $h(x) = g(f(x))$, $g(x) = \dfrac{x + 1}{x - 1}$, and $f(x) = \ln x$.

33. Use logarithmic differentiation to find the derivative of $f(x) = (2x^3 + 1)(x^2 + 2)^3$.

34. Use logarithmic differentiation to find the derivative of $f(x) = \dfrac{x(x^2 - 2)^2}{(x - 1)}$.

35. Find an equation of the tangent line to the graph of $y = e^{-2x}$ at the point $(1, e^{-2})$.

36. Find an equation of the tangent line to the graph of $y = xe^{-x}$ at the point $(1, e^{-1})$.

37. Sketch the graph of the function $f(x) = xe^{-2x}$.

38. Sketch the graph of the function $f(x) = x^2 - \ln x$.

39. Find the absolute extrema of the function $f(t) = te^{-t}$.

40. Find the absolute extrema of the function

$$g(t) = \frac{\ln t}{t}$$

on [1, 2].

41. INVESTMENT RETURN A hotel was purchased by a conglomerate for \$4.5 million and sold 5 yr later for \$8.2 million. Find the annual rate of return (compounded continuously).

42. Find the present value of \$119,346 due in 4 yr at an interest rate of 10%/year compounded continuously.

43. CONSUMER PRICE INDEX At an annual inflation rate of 7.5%, how long will it take the Consumer Price Index (CPI) to double?

44. GROWTH OF BACTERIA A culture of bacteria that initially contained 2000 bacteria has a count of 18,000 bacteria after 2 hr.
   a. Determine the function $Q(t)$ that expresses the exponential growth of the number of cells of this bacterium as a function of time t (in minutes).
   b. Find the number of bacteria present after 4 hr.

45. RADIOACTIVE DECAY The radioactive element radium has a half-life of 1600 yr. What is its decay constant?

46. DEMAND FOR DVD PLAYERS VCA Television found that the monthly demand for its new line of DVD players t mo after placing the players on the market is given by

$$D(t) = 4000 - 3000e^{-0.06t} \qquad (t \geq 0)$$

Graph this function and answer the following questions:
   a. What was the demand after 1 mo? After 1 yr? After 2 yr?
   b. At what level is the demand expected to stabilize?

47. FLU EPIDEMIC During a flu epidemic, the number of students at a certain university who contracted influenza after t days could be approximated by the exponential model

$$Q(t) = \frac{3000}{1 + 499e^{-kt}}$$

If 90 students contracted the flu by day 10, how many students contracted the flu by day 20?

48. U.S. INFANT MORTALITY RATE The U.S. infant mortality rate (per 1000 live births) is approximated by the function

$$N(t) = 12.5e^{-0.0294t} \qquad (0 \leq t \leq 21)$$

where t is measured in years, with $t = 0$ corresponding to 1980.
   a. What was the mortality rate in 1980? In 1990? In 2000?
   b. Sketch the graph of N.

Source: U.S. Department of Health and Human Services

49. ABSORPTION OF DRUGS The concentration of a drug in an organ at any time t (in seconds) is given by

$$x(t) = 0.08(1 - e^{-0.02t})$$

where $x(t)$ is measured in grams/cubic centimeter (g/cm^3).
   a. What is the initial concentration of the drug in the organ?
   b. What is the concentration of the drug in the organ after 30 sec?
   c. What will be the concentration of the drug in the organ in the long run?
   d. Sketch the graph of x.

CHAPTER 5 Before Moving On . . .

1. Solve the equation $\dfrac{100}{1 + 2e^{0.3t}} = 40$ for t.

2. Find the accumulated amount after 4 yr if \$3000 is invested at 8%/year compounded weekly.

3. Find the slope of the tangent line to the graph of $f(x) = e^{\sqrt{x}}$.

4. Find the rate at which $y = x \ln(x^2 + 1)$ is changing at $x = 1$.

5. Find the second derivative of $y = e^{2x} \ln 3x$.

6. The temperature of a cup of coffee at time t (in minutes) is

$$T(t) = 70 + ce^{-kt}$$

Initially, the temperature of the coffee was 200°F. Three minutes later, it was 180°. When will the temperature of the coffee be 150°F?

6 Integration

How much will the solar cell panels cost? The head of Soloron Corporation's research and development department has projected that the cost of producing solar cell panels will drop at a certain rate in the next several years. In Example 7, page 414, you will see how this information can be used to predict the cost of solar cell panels in the coming years.

DIFFERENTIAL CALCULUS IS concerned with the problem of finding the rate of change of one quantity with respect to another. In this chapter, we begin the study of the other branch of calculus, known as integral calculus. Here we are interested in precisely the opposite problem: If we know the rate of change of one quantity with respect to another, can we find the relationship between the two quantities? The principal tool used in the study of integral calculus is the *antiderivative* of a function, and we develop rules for antidifferentiation, or *integration*, as the process of finding the antiderivative is called. We also show that a link is established between differential and integral calculus—via the fundamental theorem of calculus.

6.1 Antiderivatives and the Rules of Integration

Antiderivatives

Let's return, once again, to the example involving the motion of the maglev (Figure 1).

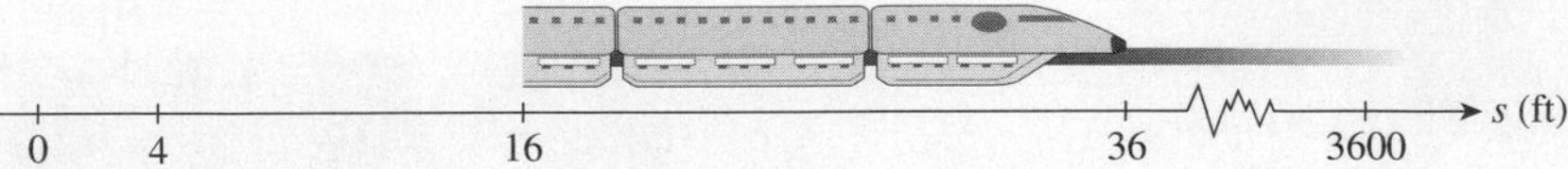

FIGURE 1
A maglev moving along an elevated monorail track

In Chapter 2, we discussed the following problem:

If we know the position of the maglev at any time t, can we find its velocity at time t?

As it turns out, if the position of the maglev is described by the position function f, then its velocity at any time t is given by $f'(t)$. Here f'—the velocity function of the maglev—is just the derivative of f.

Now, in Chapters 6 and 7, we will consider precisely the opposite problem:

If we know the velocity of the maglev at any time t, can we find its position at time t?

Stated another way, if we know the velocity function f' of the maglev, can we find its position function f?

To solve this problem, we need the concept of an antiderivative of a function.

Antiderivative

A function F is an **antiderivative** of f on an interval I if $F'(x) = f(x)$ for all x in I.

Thus, an antiderivative of a function f is a function F whose derivative is f. For example, $F(x) = x^2$ is an antiderivative of $f(x) = 2x$ because

$$F'(x) = \frac{d}{dx}(x^2) = 2x = f(x)$$

and $F(x) = x^3 + 2x + 1$ is an antiderivative of $f(x) = 3x^2 + 2$ because

$$F'(x) = \frac{d}{dx}(x^3 + 2x + 1) = 3x^2 + 2 = f(x)$$

EXAMPLE 1 Let $F(x) = \frac{1}{3}x^3 - 2x^2 + x - 1$. Show that F is an antiderivative of $f(x) = x^2 - 4x + 1$.

Solution Differentiating the function F, we obtain

$$F'(x) = x^2 - 4x + 1 = f(x)$$

and the desired result follows. ■

EXAMPLE 2 Let $F(x) = x$, $G(x) = x + 2$, and $H(x) = x + C$, where C is a constant. Show that F, G, and H are all antiderivatives of the function f defined by $f(x) = 1$.

Solution Since

$$F'(x) = \frac{d}{dx}(x) = 1 = f(x)$$

$$G'(x) = \frac{d}{dx}(x + 2) = 1 = f(x)$$

$$H'(x) = \frac{d}{dx}(x + C) = 1 = f(x)$$

we see that F, G, and H are indeed antiderivatives of f. ■

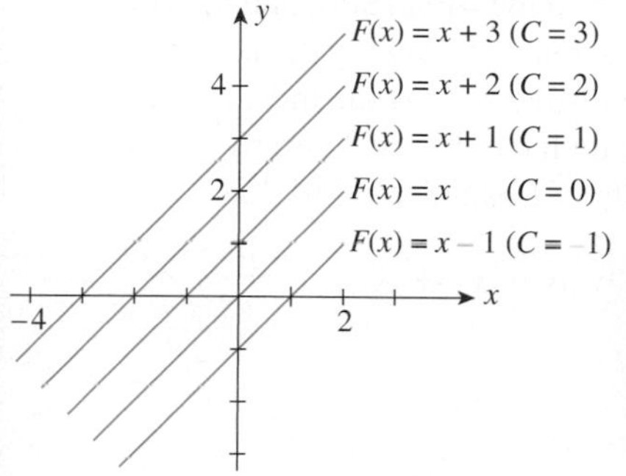

FIGURE 2
The graphs of some antiderivatives of $f(x) = 1$

Example 2 shows that once an antiderivative G of a function f is known, then another antiderivative of f may be found by adding an arbitrary constant to the function G. The following theorem states that no function other than one obtained in this manner can be an antiderivative of f. (We omit the proof.)

THEOREM 1

Let G be an antiderivative of a function f. Then, every antiderivative F of f must be of the form $F(x) = G(x) + C$, where C is a constant.

Returning to Example 2, we see that there are infinitely many antiderivatives of the function $f(x) = 1$. We obtain each one by specifying the constant C in the function $F(x) = x + C$. Figure 2 shows the graphs of some of these antiderivatives for selected values of C. These graphs constitute part of a family of infinitely many parallel straight lines, each having a slope equal to 1. This result is expected since there are infinitely many curves (straight lines) with a given slope equal to 1. The antiderivatives $F(x) = x + C$ (C, a constant) are precisely the functions representing this family of straight lines.

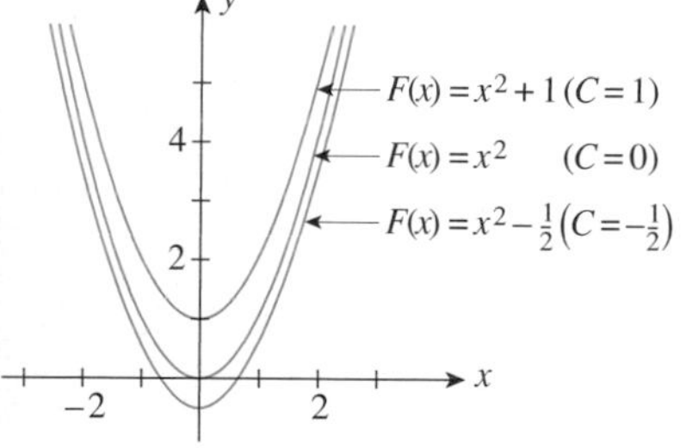

FIGURE 3
The graphs of some antiderivatives of $f(x) = 2x$

EXAMPLE 3 Prove that the function $G(x) = x^2$ is an antiderivative of the function $f(x) = 2x$. Write a general expression for the antiderivatives of f.

Solution Since $G'(x) = 2x = f(x)$, we have shown that $G(x) = x^2$ is an antiderivative of $f(x) = 2x$. By Theorem 1, every antiderivative of the function $f(x) = 2x$ has the form $F(x) = x^2 + C$, where C is some constant. The graphs of a few of the antiderivatives of f are shown in Figure 3. ■

EXPLORING WITH TECHNOLOGY

Let $f(x) = x^2 - 1$.

1. Show that $F(x) = \frac{1}{3}x^3 - x + C$, where C is an arbitrary constant, is an antiderivative of f.
2. Use a graphing utility to plot the graphs of the antiderivatives of f corresponding to $C = -2$, $C = -1$, $C = 0$, $C = 1$, and $C = 2$ on the same set of axes, using the viewing window $[-4, 4] \times [-4, 4]$.
3. If your graphing utility has the capability, draw the tangent line to each of the graphs in part 2 at the point whose x-coordinate is 2. What can you say about this family of tangent lines?
4. What is the slope of a tangent line in this family? Explain how you obtained your answer.

The Indefinite Integral

The process of finding all antiderivatives of a function is called **antidifferentiation**, or **integration**. We use the symbol $\int$, called an **integral sign**, to indicate that the operation of integration is to be performed on some function f. Thus,

$$\int f(x)\,dx = F(x) + C$$

[read "the indefinite integral of $f(x)$ with respect to x equals $F(x)$ plus C"] tells us that the **indefinite integral** of f is the family of functions given by $F(x) + C$, where $F'(x) = f(x)$. The function f to be integrated is called the **integrand**, and the constant C is called a **constant of integration**. The expression dx following the integrand $f(x)$ reminds us that the operation is performed with respect to x. If the independent variable is t, we write $\int f(t)\,dt$ instead. In this sense both t and x are "dummy variables."

Using this notation, we can write the results of Examples 2 and 3 as

$$\int 1\,dx = x + C \quad \text{and} \quad \int 2x\,dx = x^2 + K$$

where C and K are arbitrary constants.

Basic Integration Rules

Our next task is to develop some rules for finding the indefinite integral of a given function f. Because integration and differentiation are reverse operations, we discover many of the rules of integration by first making an "educated guess" at the antiderivative F of the function f to be integrated. Then this result is verified by demonstrating that $F' = f$.

Rule 1: The Indefinite Integral of a Constant

$$\int k\,dx = kx + C \qquad (k\text{, a constant})$$

To prove this result, observe that

$$F'(x) = \frac{d}{dx}(kx + C) = k$$

EXAMPLE 4 Find each of the following indefinite integrals:

a. $\int 2\,dx$ **b.** $\int \pi^2\,dx$

Solution Each of the integrands has the form $f(x) = k$, where k is a constant. Applying Rule 1 in each case yields

a. $\int 2\,dx = 2x + C$ **b.** $\int \pi^2\,dx = \pi^2 x + C$ ■

Next, from the rule of differentiation,

$$\frac{d}{dx}x^n = nx^{n-1}$$

we obtain the following rule of integration.

Rule 2: The Power Rule

$$\int x^n\,dx = \frac{1}{n+1}x^{n+1} + C \qquad (n \neq -1)$$

An antiderivative of a power function is another power function obtained from the integrand by increasing its power by 1 and dividing the resulting expression by the new power.

To prove this result, observe that

$$\begin{aligned} F'(x) &= \frac{d}{dx}\left(\frac{1}{n+1}x^{n+1} + C\right) \\ &= \frac{n+1}{n+1}x^n \\ &= x^n \\ &= f(x) \end{aligned}$$

EXAMPLE 5 Find each of the following indefinite integrals:

a. $\int x^3\,dx$ **b.** $\int x^{3/2}\,dx$ **c.** $\int \frac{1}{x^{3/2}}\,dx$

Solution Each integrand is a power function with exponent $n \neq -1$. Applying Rule 2 in each case yields the following results:

a. $\int x^3\,dx = \frac{1}{4}x^4 + C$

b. $\int x^{3/2}\,dx = \frac{1}{\frac{5}{2}}x^{5/2} + C = \frac{2}{5}x^{5/2} + C$

c. $\int \frac{1}{x^{3/2}}\,dx = \int x^{-3/2}\,dx = \frac{1}{-\frac{1}{2}}x^{-1/2} + C = -2x^{-1/2} + C = -\frac{2}{x^{1/2}} + C$

These results may be verified by differentiating each of the antiderivatives and showing that the result is equal to the corresponding integrand. ■

The next rule tells us that a constant factor may be moved through an integral sign.

Rule 3: The Indefinite Integral of a Constant Multiple of a Function

$$\int cf(x)\,dx = c\int f(x)\,dx \qquad (c\text{, a constant})$$

The indefinite integral of a constant multiple of a function is equal to the constant multiple of the indefinite integral of the function.

This result follows from the corresponding rule of differentiation (see Rule 3, Section 3.1).

Only a constant can be "moved out" of an integral sign. For example, it would be incorrect to write

$$\int x^2\,dx = x^2\int 1\,dx$$

In fact, $\int x^2\,dx = \frac{1}{3}x^3 + C$, whereas $x^2\int 1\,dx = x^2(x + C) = x^3 + Cx^2$.

EXAMPLE 6 Find each of the following indefinite integrals:

a. $\int 2t^3\, dt$ **b.** $\int -3x^{-2}\, dx$

Solution Each integrand has the form $cf(x)$, where c is a constant. Applying Rule 3, we obtain:

a. $\int 2t^3\, dt = 2\int t^3\, dt = 2\left(\frac{1}{4}t^4 + K\right) = \frac{1}{2}t^4 + 2K = \frac{1}{2}t^4 + C$

where $C = 2K$. From now on, we will write the constant of integration as C, since any nonzero multiple of an arbitrary constant is an arbitrary constant.

b. $\int -3x^{-2}\, dx = -3\int x^{-2}\, dx = (-3)(-1)x^{-1} + C = \frac{3}{x} + C$ ■

Rule 4: The Sum Rule

$$\int [f(x) + g(x)]\, dx = \int f(x)\, dx + \int g(x)\, dx$$

$$\int [f(x) - g(x)]\, dx = \int f(x)\, dx - \int g(x)\, dx$$

The indefinite integral of a sum (difference) of two integrable functions is equal to the sum (difference) of their indefinite integrals.

This result is easily extended to the case involving the sum and difference of any finite number of functions. As in Rule 3, the proof of Rule 4 follows from the corresponding rule of differentiation (see Rule 4, Section 3.1).

EXAMPLE 7 Find the indefinite integral

$$\int (3x^5 + 4x^{3/2} - 2x^{-1/2})\, dx$$

Solution Applying the extended version of Rule 4, we find that

$$\begin{aligned}
&\int (3x^5 + 4x^{3/2} - 2x^{-1/2})\, dx \\
&= \int 3x^5\, dx + \int 4x^{3/2}\, dx - \int 2x^{-1/2}\, dx \\
&= 3\int x^5\, dx + 4\int x^{3/2}\, dx - 2\int x^{-1/2}\, dx && \text{Rule 3} \\
&= (3)\left(\frac{1}{6}\right)x^6 + (4)\left(\frac{2}{5}\right)x^{5/2} - (2)(2)x^{1/2} + C && \text{Rule 2} \\
&= \frac{1}{2}x^6 + \frac{8}{5}x^{5/2} - 4x^{1/2} + C
\end{aligned}$$ ■

Observe that we have combined the three constants of integration, which arise from evaluating the three indefinite integrals, to obtain one constant C. After all, the sum of three arbitrary constants is also an arbitrary constant.

Rule 5: The Indefinite Integral of the Exponential Function

$$\int e^x\,dx = e^x + C$$

The indefinite integral of the exponential function with base e is equal to the function itself (except, of course, for the constant of integration).

EXAMPLE 8 Find the indefinite integral

$$\int (2e^x - x^3)\,dx$$

Solution We have

$$\begin{aligned}\int (2e^x - x^3)\,dx &= \int 2e^x\,dx - \int x^3\,dx\\ &= 2\int e^x\,dx - \int x^3\,dx\\ &= 2e^x - \frac{1}{4}x^4 + C\end{aligned}$$

The last rule of integration in this section covers the integration of the function $f(x) = x^{-1}$. Remember that this function constituted the only exceptional case in the integration of the power function $f(x) = x^n$ (see Rule 2).

Rule 6: The Indefinite Integral of the Function $f(x) = x^{-1}$

$$\int x^{-1}\,dx = \int \frac{1}{x}\,dx = \ln|x| + C \qquad (x \neq 0)$$

To prove Rule 6, observe that

$$\frac{d}{dx}\ln|x| = \frac{1}{x} \qquad \text{See Rule 3, Section 5.5.}$$

EXAMPLE 9 Find the indefinite integral

$$\int \left(2x + \frac{3}{x} + \frac{4}{x^2}\right) dx$$

Solution

$$\begin{aligned}\int \left(2x + \frac{3}{x} + \frac{4}{x^2}\right) dx &= \int 2x\,dx + \int \frac{3}{x}\,dx + \int \frac{4}{x^2}\,dx\\ &= 2\int x\,dx + 3\int \frac{1}{x}\,dx + 4\int x^{-2}\,dx\\ &= 2\left(\frac{1}{2}\right)x^2 + 3\ln|x| + 4(-1)x^{-1} + C\\ &= x^2 + 3\ln|x| - \frac{4}{x} + C\end{aligned}$$

Differential Equations

Let's return to the problem posed at the beginning of the section: *Given the derivative of a function, f', can we find the function f?* As an example, suppose we are given the function

$$f'(x) = 2x - 1 \tag{1}$$

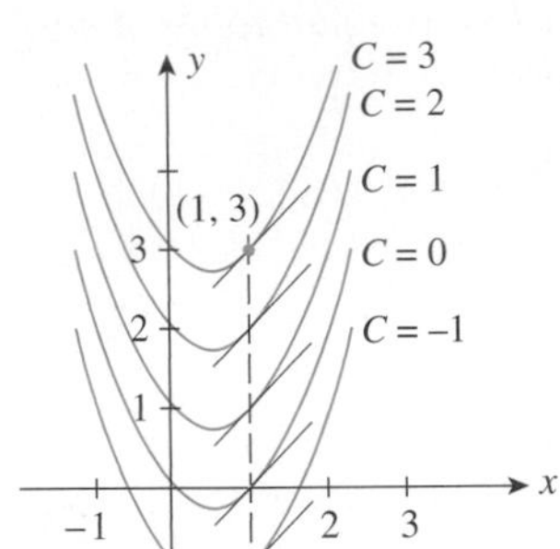

FIGURE 4
The graphs of some of the functions having the derivative $f'(x) = 2x - 1$. Observe that the slopes of the tangent lines to the graphs are the same for a fixed value of x.

and we wish to find $f(x)$. From what we now know, we can find f by integrating Equation (1). Thus,

$$f(x) = \int f'(x)\,dx = \int (2x - 1)\,dx = x^2 - x + C \tag{2}$$

where C is an arbitrary constant. Thus, infinitely many functions have the derivative f', each differing from the other by a constant.

Equation (1) is called a differential equation. In general, a **differential equation** is an equation that involves the derivative or differential of an unknown function. [In the case of Equation (1), the unknown function is f.] A **solution** of a differential equation is any function that satisfies the differential equation. Thus, Equation (2) gives *all* the solutions of the differential Equation (1), and it is, accordingly, called the **general solution** of the differential equation $f'(x) = 2x - 1$.

The graphs of $f(x) = x^2 - x + C$ for selected values of C are shown in Figure 4. These graphs have one property in common: For any fixed value of x, the tangent lines to these graphs have the same slope. This follows because any member of the family $f(x) = x^2 - x + C$ must have the same slope at x—namely, $2x - 1$!

Although there are infinitely many solutions to the differential equation $f'(x) = 2x - 1$, we can obtain a **particular solution** by specifying the value the function must assume at a certain value of x. For example, suppose we stipulate that the function f under consideration must satisfy the condition $f(1) = 3$ or, equivalently, the graph of f must pass through the point $(1, 3)$. Then, using the condition on the general solution $f(x) = x^2 - x + C$, we find that

$$f(1) = 1 - 1 + C = 3$$

and $C = 3$. Thus, the particular solution is $f(x) = x^2 - x + 3$ (see Figure 4).

The condition $f(1) = 3$ is an example of an initial condition. More generally, an **initial condition** is a condition imposed on the value of f at a point $x = a$.

Initial Value Problems

An **initial value problem** is one in which we are required to find a function satisfying (1) a differential equation and (2) one or more initial conditions. The following are examples of initial value problems.

EXAMPLE 10 Find the function f if it is known that

$$f'(x) = 3x^2 - 4x + 8 \quad \text{and} \quad f(1) = 9$$

Solution We are required to solve the initial value problem

$$\left.\begin{aligned} f'(x) &= 3x^2 - 4x + 8 \\ f(1) &= 9 \end{aligned}\right\}$$

Integrating the function f', we find

$$\begin{aligned} f(x) &= \int f'(x)\,dx \\ &= \int (3x^2 - 4x + 8)\,dx \\ &= x^3 - 2x^2 + 8x + C \end{aligned}$$

Using the condition $f(1) = 9$, we have

$$9 = f(1) = 1^3 - 2(1)^2 + 8(1) + C = 7 + C \quad \text{or} \quad C = 2$$

Therefore, the required function f is given by $f(x) = x^3 - 2x^2 + 8x + 2$. ■

APPLIED EXAMPLE 11 Velocity of a Maglev In a test run of a maglev along a straight elevated monorail track, data obtained from reading its speedometer indicate that the velocity of the maglev at time t can be described by the velocity function

$$v(t) = 8t \qquad (0 \le t \le 30)$$

Find the position function of the maglev. Assume that initially the maglev is located at the origin of a coordinate line.

Solution Let $s(t)$ denote the position of the maglev at any time t $(0 \le t \le 30)$. Then, $s'(t) = v(t)$. So, we have the initial value problem

$$\left.\begin{aligned} s'(t) &= 8t \\ s(0) &= 0 \end{aligned}\right\}$$

Integrating both sides of the differential equation $s'(t) = 8t$, we obtain

$$s(t) = \int s'(t)\,dt = \int 8t\,dt = 4t^2 + C$$

where C is an arbitrary constant. To evaluate C, we use the initial condition $s(0) = 0$ to write

$$s(0) = 4(0) + C = 0 \quad \text{or} \quad C = 0$$

Therefore, the required position function is $s(t) = 4t^2$ $(0 \le t \le 30)$. ■

APPLIED EXAMPLE 12 Magazine Circulation The current circulation of the *Investor's Digest* is 3000 copies per week. The managing editor of the weekly projects a growth rate of

$$4 + 5t^{2/3}$$

copies per week, t weeks from now, for the next 3 years. Based on her projection, what will the circulation of the digest be 125 weeks from now?

Solution Let $S(t)$ denote the circulation of the digest t weeks from now. Then $S'(t)$ is the rate of change in the circulation in the tth week and is given by

$$S'(t) = 4 + 5t^{2/3}$$

Furthermore, the current circulation of 3000 copies per week translates into the initial condition $S(0) = 3000$. Integrating the differential equation with respect to t gives

$$S(t) = \int S'(t)\,dt = \int (4 + 5t^{2/3})\,dt$$

$$= 4t + 5\left(\frac{t^{5/3}}{\frac{5}{3}}\right) + C = 4t + 3t^{5/3} + C$$

To determine the value of C, we use the condition $S(0) = 3000$ to write

$$S(0) = 4(0) + 3(0) + C = 3000$$

which gives $C = 3000$. Therefore, the circulation of the digest t weeks from now will be

$$S(t) = 4t + 3t^{5/3} + 3000$$

In particular, the circulation 125 weeks from now will be

$$S(125) = 4(125) + 3(125)^{5/3} + 3000 = 12{,}875$$

copies per week. ■

6.1 Self-Check Exercises

1. Evaluate $\displaystyle\int \left(\frac{1}{\sqrt{x}} - \frac{2}{x} + 3e^x\right) dx$.

2. Find the rule for the function f given that (1) the slope of the tangent line to the graph of f at any point $P(x, f(x))$ is given by the expression $3x^2 - 6x + 3$ and (2) the graph of f passes through the point (2, 9).

3. Suppose United Motors' share of the new cars sold in a certain country is changing at the rate of

$$f(t) = -0.01875t^2 + 0.15t - 1.2 \qquad (0 \le t \le 12)$$

percent at year t ($t = 0$ corresponds to the beginning of 1994). The company's market share at the beginning of 1994 was 48.4%. What was United Motors' market share at the beginning of 2006?

Solutions to Self-Check Exercises 6.1 can be found on page 409.

6.1 Concept Questions

1. What is an antiderivative? Give an example.

2. If $f'(x) = g'(x)$ for all x in an interval I, what is the relationship between f and g?

3. What is the difference between an antiderivative of f and the indefinite integral of f?

6.1 Exercises

In Exercises 1–4, verify directly that F is an antiderivative of f.

1. $F(x) = \frac{1}{3}x^3 + 2x^2 - x + 2$; $f(x) = x^2 + 4x - 1$

2. $F(x) = xe^x + \pi$; $f(x) = e^x(1 + x)$

3. $F(x) = \sqrt{2x^2 - 1}$; $f(x) = \dfrac{2x}{\sqrt{2x^2 - 1}}$

4. $F(x) = x \ln x - x$; $f(x) = \ln x$

In Exercises 5–8, (a) verify that G is an antiderivative of f, (b) find all antiderivatives of f, and (c) sketch the graphs of a few of the family of antiderivatives found in part (b).

5. $G(x) = 2x$; $f(x) = 2$

6. $G(x) = 2x^2$; $f(x) = 4x$

7. $G(x) = \frac{1}{3}x^3$; $f(x) = x^2$

8. $G(x) = e^x$; $f(x) = e^x$

In Exercises 9–50, find the indefinite integral.

9. $\int 6\,dx$

10. $\int \sqrt{2}\,dx$

11. $\int x^3\,dx$

12. $\int 2x^5\,dx$

13. $\int x^{-4}\,dx$

14. $\int 3t^{-7}\,dt$

15. $\int x^{2/3}\,dx$

16. $\int 2u^{3/4}\,du$

17. $\int x^{-5/4}\,dx$

18. $\int 3x^{-2/3}\,dx$

19. $\int \frac{2}{x^2}\,dx$

20. $\int \frac{1}{3x^5}\,dx$

21. $\int \pi\sqrt{t}\,dt$

22. $\int \frac{3}{\sqrt{t}}\,dt$

23. $\int (3 - 2x)\,dx$

24. $\int (1 + u + u^2)\,du$

25. $\int (x^2 + x + x^{-3})\,dx$

26. $\int (0.3t^2 + 0.02t + 2)\,dt$

27. $\int 4e^x\,dx$

28. $\int (1 + e^x)\,dx$

29. $\int (1 + x + e^x)\,dx$

30. $\int (2 + x + 2x^2 + e^x)\,dx$

31. $\int \left(4x^3 - \frac{2}{x^2} - 1\right)dx$

32. $\int \left(6x^3 + \frac{3}{x^2} - x\right)dx$

33. $\int (x^{5/2} + 2x^{3/2} - x)\,dx$

34. $\int (t^{3/2} + 2t^{1/2} - 4t^{-1/2})\,dt$

35. $\int \left(\sqrt{x} + \frac{3}{\sqrt{x}}\right)dx$

36. $\int \left(\sqrt[3]{x^2} - \frac{1}{x^2}\right)dx$

37. $\int \left(\frac{u^3 + 2u^2 - u}{3u}\right)du$

Hint: $\frac{u^3 + 2u^2 - u}{3u} = \frac{1}{3}u^2 + \frac{2}{3}u - \frac{1}{3}$

38. $\int \frac{x^4 - 1}{x^2}\,dx$

Hint: $\frac{x^4 - 1}{x^2} = x^2 - x^{-2}$

39. $\int (2t + 1)(t - 2)\,dt$

40. $\int u^{-2}(1 - u^2 + u^4)\,du$

41. $\int \frac{1}{x^2}(x^4 - 2x^2 + 1)\,dx$

42. $\int \sqrt{t}\,(t^2 + t - 1)\,dt$

43. $\int \frac{ds}{(s + 1)^{-2}}$

44. $\int \left(\sqrt{x} + \frac{3}{x} - 2e^x\right)dx$

45. $\int (e^t + t^e)\,dt$

46. $\int \left(\frac{1}{x^2} - \frac{1}{\sqrt[3]{x^2}} + \frac{1}{\sqrt{x}}\right)dx$

47. $\int \left(\frac{x^3 + x^2 - x + 1}{x^2}\right)dx$

Hint: Simplify the integrand first.

48. $\int \frac{t^3 + \sqrt[3]{t}}{t^2}\,dt$

Hint: Simplify the integrand first.

49. $\int \frac{(\sqrt{x} - 1)^2}{x^2}\,dx$

Hint: Simplify the integrand first.

50. $\int (x + 1)^2\left(1 - \frac{1}{x}\right)dx$

Hint: Simplify the integrand first.

In Exercises 51–58, find $f(x)$ by solving the initial value problem.

51. $f'(x) = 2x + 1;\ f(1) = 3$

52. $f'(x) = 3x^2 - 6x;\ f(2) = 4$

53. $f'(x) = 3x^2 + 4x - 1;\ f(2) = 9$

54. $f'(x) = \frac{1}{\sqrt{x}};\ f(4) = 2$

55. $f'(x) = 1 + \frac{1}{x^2};\ f(1) = 2$

56. $f'(x) = e^x - 2x;\ f(0) = 2$

57. $f'(x) = \frac{x + 1}{x};\ f(1) = 1$

58. $f'(x) = 1 + e^x + \frac{1}{x};\ f(1) = 3 + e$

In Exercises 59–62, find the function f given that the slope of the tangent line to the graph of f at any point $(x, f(x))$ is $f'(x)$ and that the graph of f passes through the given point.

59. $f'(x) = \frac{1}{2}x^{-1/2};\ (2, \sqrt{2})$

60. $f'(t) = t^2 - 2t + 3;\ (1, 2)$

61. $f'(x) = e^x + x;\ (0, 3)$

62. $f'(x) = \frac{2}{x} + 1;\ (1, 2)$

63. Bank Deposits Madison Finance opened two branches on September 1 ($t = 0$). Branch A is located in an established industrial park, and branch B is located in a fast-growing new development. The net rate at which money was deposited into branch A and branch B in the first 180 business days is given by the graphs of f and g, respectively (see the figure). Which branch has a larger amount on deposit at the end of 180 business days? Justify your answer.

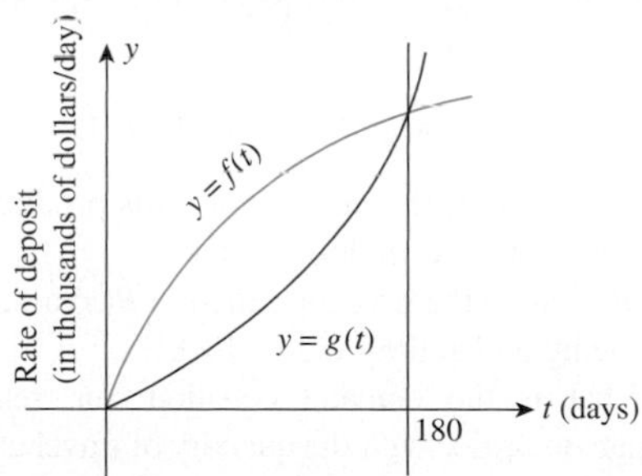

64. Velocity of a Car Two cars, side by side, start from rest and travel along a straight road. The velocity of car A is given by $v = f(t)$, and the velocity of car B is given by $v = g(t)$. The graphs of f and g are shown in the figure on the next page. Are the cars still side by side after T sec? If not, which car is ahead of the other? Justify your answer.

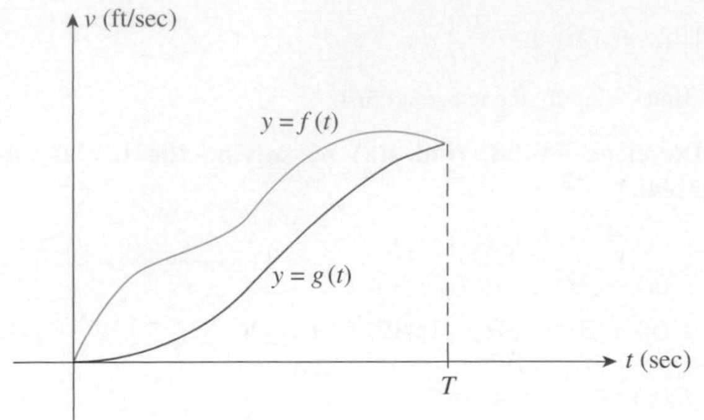

65. **Velocity of a Car** The velocity of a car (in feet/second) t sec after starting from rest is given by the function

$$f(t) = 2\sqrt{t} \qquad (0 \leq t \leq 30)$$

Find the car's position at any time t.

66. **Velocity of a Maglev** The velocity (in feet/second) of a maglev is

$$v(t) = 0.2t + 3 \qquad (0 \leq t \leq 120)$$

At $t = 0$, it is at the station. Find the function giving the position of the maglev at time t, assuming that the motion takes place along a straight stretch of track.

67. **Cost of Producing Clocks** Lorimar Watch Company manufactures travel clocks. The daily marginal cost function associated with producing these clocks is

$$C'(x) = 0.000009x^2 - 0.009x + 8$$

where $C'(x)$ is measured in dollars/unit and x denotes the number of units produced. Management has determined that the daily fixed cost incurred in producing these clocks is \$120. Find the total cost incurred by Lorimar in producing the first 500 travel clocks/day.

68. **Revenue Functions** The management of Lorimar Watch Company has determined that the daily marginal revenue function associated with producing and selling their travel clocks is given by

$$R'(x) = -0.009x + 12$$

where x denotes the number of units produced and sold and $R'(x)$ is measured in dollars/unit.

a. Determine the revenue function $R(x)$ associated with producing and selling these clocks.

b. What is the demand equation that relates the wholesale unit price with the quantity of travel clocks demanded?

69. **Profit Functions** Cannon Precision Instruments makes an automatic electronic flash with Thyrister circuitry. The estimated marginal profit associated with producing and selling these electronic flashes is

$$-0.004x + 20$$

dollars/unit/month when the production level is x units per month. Cannon's fixed cost for producing and selling these electronic flashes is \$16,000/month. At what level of production does Cannon realize a maximum profit? What is the maximum monthly profit?

70. **Cost of Producing Guitars** Carlota Music Company estimates that the marginal cost of manufacturing its Professional Series guitars is

$$C'(x) = 0.002x + 100$$

dollars/month when the level of production is x guitars/month. The fixed costs incurred by Carlota are \$4000/month. Find the total monthly cost incurred by Carlota in manufacturing x guitars/month.

71. **Quality Control** As part of a quality-control program, the chess sets manufactured by Jones Brothers are subjected to a final inspection before packing. The rate of increase in the number of sets checked per hour by an inspector t hr into the 8 a.m. to 12 noon morning shift is approximately

$$N'(t) = -3t^2 + 12t + 45 \qquad (0 \leq t \leq 4)$$

a. Find an expression $N(t)$ that approximates the number of sets inspected at the end of t hours.
Hint: $N(0) = 0$.

b. How many sets does the average inspector check during a morning shift?

72. **Measuring Temperature** The temperature on a certain day as measured at the airport of a city is changing at the rate of

$$T'(t) = 0.15t^2 - 3.6t + 14.4 \qquad (0 \leq t \leq 4)$$

°F/hr, where t is measured in hours, with $t = 0$ corresponding to 6 a.m. The temperature at 6 a.m. was 24°F.

a. Find an expression giving the temperature T at the airport at any time between 6 a.m. and 10 a.m.

b. What was the temperature at 10 a.m.?

73. **Satellite Radio Subscriptions** Based on data obtained by polling automobile buyers, the number of subscribers of satellite radios is expected to grow at the rate of

$$r(t) = -0.375t^2 + 2.1t + 2.45 \qquad (0 \leq t \leq 5)$$

million subscribers/year between 2003 ($t = 0$) and 2008 ($t = 5$). The number of satellite radio subscribers at the beginning of 2003 was 1.5 million.

a. Find an expression giving the number of satellite radio subscribers in year t ($0 \leq t \leq 5$).

b. Based on this model, what will be the number of satellite radio subscribers in 2008?

Source: Carmel Group

74. **Genetically Modified Crops** The total number of acres of genetically modified crops grown worldwide from 1997

through 2003 was changing at the rate of

$$R(t) = 2.718t^2 - 19.86t + 50.18 \qquad (0 \leq t \leq 6)$$

million acres/year. The total number of acres of such crops grown in 1997 ($t = 0$) was 27.2 million acres. How many acres of genetically modified crops were grown worldwide in 2003?

Source: International Services for the Acquisition of Agri-biotech Applications

75. **Credit Card Debt** The average credit card debt per U.S. household between 1990 ($t = 0$) and 2003 ($t = 13$) was growing at the rate of approximately

$$D(t) = -4.479t^2 + 69.8t + 279.5 \qquad (0 \leq t \leq 13)$$

dollars per year. The average credit card debt per U.S. household stood at $2917 in 1990.

a. Find an expression giving the approximate average credit card debt per U.S. household in year t ($0 \leq t \leq 13$).

b. Use the result of part (a) to estimate the average credit card debt per U.S. household in 2003.

Source: Encore Capital Group

76. **DVD Sales** The total number of DVDs sold to U.S. dealers for rental and sale from 1999 through 2003 grew at the rate of approximately

$$R(t) = -0.03t^2 + 0.218t - 0.032 \qquad (0 \leq t \leq 4)$$

billion units/year, where t is measured in years with $t = 0$ corresponding to 1999. The total number of DVDs sold as of 1999 was 0.1 billion units.

a. Find an expression giving the total number of DVDs sold by year t ($0 \leq t \leq 4$).

b. What was the total number of DVDs sold by 2003?

Source: Adams Media

77. **Gastric Bypass Surgeries** One method of weight loss gaining in popularity is stomach-reducing surgery. It is generally reserved for people at least 100 lb overweight because the procedure carries a serious risk of death or complications. According to the American Society of Bariatric Surgery, the number of morbidly obese patients undergoing the procedure was increasing at the rate of

$$R(t) = 9.399t^2 - 13.4t + 14.07 \qquad (0 \leq t \leq 3)$$

thousands/year, where $t = 0$ corresponds to 2000. The number of gastric bypass surgeries performed in 2000 was 36.7 thousand.

a. Find an expression giving the number of gastric bypass surgeries performed in year t ($0 \leq t \leq 3$).

b. Use the result of part (a) to find the number of gastric bypass surgeries performed in 2003.

Source: American Society for Bariatric Surgery

78. **Online Ad Sales** According to a study conducted in 2004, the share of online advertisement, worldwide, as a percent of the total ad market, is expected to grow at the rate of

$$R(t) = -0.033t^2 + 0.3428t + 0.07 \qquad (0 \leq t \leq 6)$$

percent/year at time t (in years), with $t = 0$ corresponding to the beginning of 2000. The online ad market at the beginning of 2000 was 2.9% of the total ad market.

a. What is the projected online ad market share at any time t?

b. What is the projected online ad market share at the beginning of 2005?

Source: Jupiter Media Metrix, Inc.

79. **Health-Care Costs** The average out-of-pocket costs for beneficiaries in traditional Medicare (including premiums, cost sharing, and prescription drugs not covered by Medicare) is projected to grow at the rate of

$$C'(t) = 12.288t^2 - 150.5594t + 695.23$$

dollars/year, where t is measured in 5-yr intervals, with $t = 0$ corresponding to 2000. The out-of-pocket costs for beneficiaries in 2000 were $3142.

a. Find an expression giving the average out-of-pocket costs for beneficiaries in year t.

b. What is the projected average out-of-pocket costs for beneficiaries in 2010?

Source: The Urban Institute

80. **Ballast Dropped from a Balloon** A ballast is dropped from a stationary hot-air balloon that is hovering at an altitude of 400 ft. Its velocity after t sec is $-32t$ ft/sec.

a. Find the height $h(t)$ of the ballast from the ground at time t.
Hint: $h'(t) = -32t$ and $h(0) = 400$.

b. When will the ballast strike the ground?

c. Find the velocity of the ballast when it hits the ground.

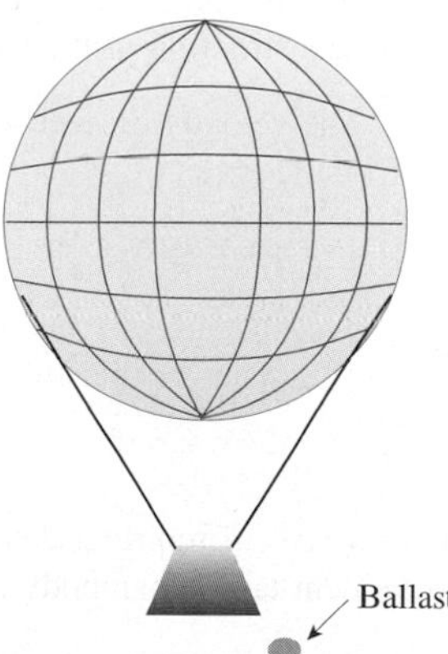

81. **Cable TV Subscribers** A study conducted by TeleCable estimates that the number of cable TV subscribers will grow at the rate of

$$100 + 210t^{3/4}$$

new subscribers/month t mo from the start date of the service. If 5000 subscribers signed up for the service before the starting date, how many subscribers will there be 16 mo from that date?

82. **Air Pollution** On an average summer day, the level of carbon monoxide (CO) in a city's air is 2 parts per million (ppm). An environmental protection agency's study predicts that, unless more stringent measures are taken to protect the city's atmosphere, the CO concentration present in the air will increase at the rate of

$$0.003t^2 + 0.06t + 0.1$$

ppm/year t yr from now. If no further pollution-control efforts are made, what will be the CO concentration on an average summer day 5 yr from now?

83. **Ozone Pollution** The rate of change of the level of ozone, an invisible gas that is an irritant and impairs breathing, present in the atmosphere on a certain May day in the city of Riverside is given by

$$R(t) = 3.2922t^2 - 0.366t^3 \qquad (0 < t < 11)$$

(measured in pollutant standard index/hour). Here, t is measured in hours, with $t = 0$ corresponding to 7 a.m. Find the ozone level $A(t)$ at any time t, assuming that at 7 a.m. it is zero.
Hint: $A'(t) = R(t)$ and $A(0) = 0$.
Source: Los Angeles Times

84. **Population Growth** The development of AstroWorld ("The Amusement Park of the Future") on the outskirts of a city will increase the city's population at the rate of

$$4500\sqrt{t} + 1000$$

people/year t yr from the start of construction. The population before construction is 30,000. Determine the projected population 9 yr after construction of the park has begun.

85. **Flight of a Rocket** The velocity, in feet/second, of a rocket t sec into vertical flight is given by

$$v(t) = -3t^2 + 192t + 120$$

Find an expression $h(t)$ that gives the rocket's altitude, in feet, t sec after liftoff. What is the altitude of the rocket 30 sec after liftoff?
Hint: $h'(t) = v(t)$; $h(0) = 0$.

86. **Surface Area of a Human** Empirical data suggest that the surface area of a 180-cm-tall human body changes at the rate of

$$S'(W) = 0.131773W^{-0.575}$$

square meters/kilogram, where W is the weight of the body in kilograms. If the surface area of a 180-cm-tall human body weighing 70 kg is 1.886277 m^2, what is the surface area of a human body of the same height weighing 75 kg?

87. **Outpatient Service Companies** The number of Medicare-certified home-health-care agencies (70% are freestanding, and 30% are owned by a hospital or other large facility) has been declining at the rate of

$$0.186e^{-0.02t} \qquad (0 \le t \le 14)$$

thousand agencies per year, between 1988 ($t = 0$) and 2002 ($t = 14$). The number of such agencies stood at 9.3 thousand units in 1988.

a. Find an expression giving the number of health-care agencies in year t.
b. What was the number of health-care agencies in 2002?
c. If this model held true through 2005, how many care agencies were there in 2005?

Source: Centers for Medicare and Medicaid Services

88. **Heights of Children** According to the Jenss model for predicting the height of preschool children, the rate of growth of a typical preschool child is

$$R(t) = 25.8931e^{-0.993t} + 6.39 \qquad \left(\frac{1}{4} \le t \le 6\right)$$

cm/yr, where t is measured in years. The height of a typical 3-mo-old preschool child is 60.2952 cm.

a. Find a model for predicting the height of a typical preschool child at age t.
b. Use the result of part (a) to estimate the height of a typical 1-yr-old child.

89. **Blood Flow in an Artery** Nineteenth-century physician Jean Louis Marie Poiseuille discovered that the rate of change of the velocity of blood r cm from the central axis of an artery (in centimeters/second/centimeter) is given by

$$a(r) = -kr$$

where k is a constant. If the radius of an artery is R cm, find an expression for the velocity of blood as a function of r (see the accompanying figure).
Hint: $v'(r) = a(r)$ and $v(R) = 0$. (Why?)

Blood vessel

90. **Acceleration of a Car** A car traveling along a straight road at 66 ft/sec accelerated to a speed of 88 ft/sec over a distance of 440 ft. What was the acceleration of the car, assuming it was constant?

91. **Deceleration of a Car** What constant deceleration would a car moving along a straight road have to be subjected to if it were brought to rest from a speed of 88 ft/sec in 9 sec? What would be the stopping distance?

92. **Carrier Landing** A pilot lands a fighter aircraft on an aircraft carrier. At the moment of touchdown, the speed of the aircraft is 160 mph. If the aircraft is brought to a complete stop in 1 sec and the deceleration is assumed to be constant, find the number of g's the pilot is subjected to during landing (1 g = 32 ft/sec^2).

93. **Crossing the Finish Line** After rounding the final turn in the bell lap, two runners emerged ahead of the pack. When runner A is 200 ft from the finish line, his speed is 22 ft/sec, a speed that he maintains until he crosses the line. At that instant of time, runner B, who is 20 ft behind runner A and running at a speed of 20 ft/sec, begins to spurt. Assuming that runner B sprints with a constant acceleration, what minimum acceleration will enable him to cross the finish line ahead of runner A?

94. **Draining a Tank** A tank has a constant cross-sectional area of 50 ft^2 and an orifice of constant cross-sectional area of $\frac{1}{2}$ ft^2 located at the bottom of the tank (see the accompanying figure).

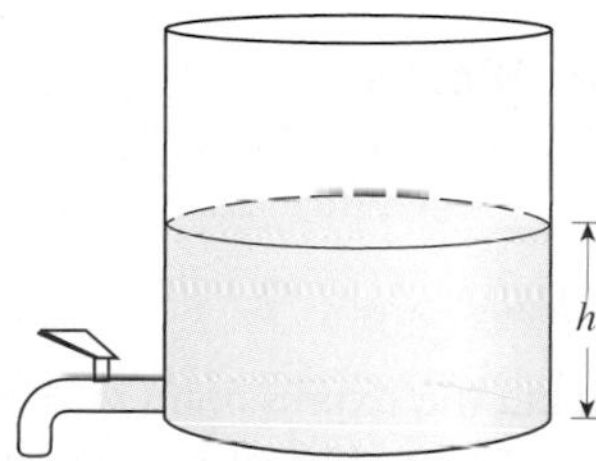

If the tank is filled with water to a height of h ft and allowed to drain, then the height of the water decreases at a rate that is described by the equation

$$\frac{dh}{dt} = -\frac{1}{25}\left(\sqrt{20} - \frac{t}{50}\right) \qquad (0 \le t \le 50\sqrt{20})$$

Find an expression for the height of the water at any time t if its height initially is 20 ft.

95. **Amount of Rainfall** During a thunderstorm, rain was falling at the rate of

$$\frac{8}{(t+4)^2} \qquad (0 \le t \le 2)$$

in./hr.

a. Find an expression giving the total amount of rainfall after t hr.
Hint: The total amount of rainfall at $t = 0$ is zero.

b. How much rain had fallen after 1 hr? After 2 hr?

96. **Launching a Fighter Aircraft** A fighter aircraft is launched from the deck of a Nimitz-class aircraft carrier with the help of a steam catapult. If the aircraft is to attain a takeoff speed of at least 240 ft/sec after traveling 800 ft along the flight deck, find the minimum acceleration it must be subjected to, assuming it is constant.

In Exercises 97–100, determine whether the statement is true or false. If it is true, explain why it is true. If it is false, give an example to show why it is false.

97. If F and G are antiderivatives of f on an interval I, then $F(x) = G(x) + C$ on I.

98. If F is an antiderivative of f on an interval I, then $\int f(x)\,dx = F(x)$.

99. If f and g are integrable, then $\int [2f(x) - 3g(x)]\,dx = 2\int f(x)\,dx - 3\int g(x)\,dx$.

100. If f and g are integrable, then $\int f(x)g(x)\,dx = [\int f(x)\,dx][\int g(x)\,dx]$.

6.1 Solutions to Self-Check Exercises

1. $$\int\left(\frac{1}{\sqrt{x}} - \frac{2}{x} + 3e^x\right)dx = \int\left(x^{-1/2} - \frac{2}{x} + 3e^x\right)dx$$
$$= \int x^{-1/2}\,dx - 2\int\frac{1}{x}\,dx + 3\int e^x\,dx$$
$$= 2x^{1/2} - 2\ln|x| + 3e^x + C$$
$$= 2\sqrt{x} - 2\ln|x| + 3e^x + C$$

2. The slope of the tangent line to the graph of the function f at any point $P(x, f(x))$ is given by the derivative f' of f. Thus, the first condition implies that

$$f'(x) = 3x^2 - 6x + 3$$

which, upon integration, yields

$$f(x) = \int (3x^2 - 6x + 3)\,dx$$
$$= x^3 - 3x^2 + 3x + k$$

where k is the constant of integration.

To evaluate k, we use the initial condition (2), which implies that $f(2) = 9$, or

$$9 = f(2) = 2^3 - 3(2)^2 + 3(2) + k$$

or $k = 7$. Hence, the required rule of definition of the

function f is

$$f(x) = x^3 - 3x^2 + 3x + 7$$

3. Let $M(t)$ denote United Motors' market share at year t. Then,

$$\begin{aligned} M(t) &= \int f(t)\, dt \\ &= \int (-0.01875t^2 + 0.15t - 1.2)\, dt \\ &= -0.00625t^3 + 0.075t^2 - 1.2t + C \end{aligned}$$

To determine the value of C, we use the initial condition $M(0) = 48.4$, obtaining $C = 48.4$. Therefore,

$$M(t) = -0.00625t^3 + 0.075t^2 - 1.2t + 48.4$$

In particular, United Motors' market share of new cars at the beginning of 2006 is given by

$$M(12) = -0.00625(12)^3 + 0.075(12)^2 - 1.2(12) + 48.4 = 34$$

or 34%.

6.2 Integration by Substitution

In Section 6.1, we developed certain rules of integration that are closely related to the corresponding rules of differentiation in Chapters 3 and 5. In this section, we introduce a method of integration called the **method of substitution**, which is related to the chain rule for differentiating functions. When used in conjunction with the rules of integration developed earlier, the method of substitution is a powerful tool for integrating a large class of functions.

How the Method of Substitution Works

Consider the indefinite integral

$$\int 2(2x + 4)^5\, dx \qquad (3)$$

One way of evaluating this integral is to expand the expression $(2x + 4)^5$ and then integrate the resulting integrand term by term. As an alternative approach, let's see if we can simplify the integral by making a change of variable. Write

$$u = 2x + 4$$

with differential

$$du = 2\, dx$$

If we formally substitute these quantities into Equation (3), we obtain

$$\int 2(2x + 4)^5\, dx = \underset{\uparrow\ \text{Rewrite}}{\int (2x + 4)^5 (2\, dx)} = \underset{\uparrow\ \left\{ u = 2x + 4,\ du = 2\,dx \right.}{\int u^5\, du}$$

Now, the last integral involves a power function and is easily evaluated using Rule 2 of Section 6.1. Thus,

$$\int u^5\, du = \frac{1}{6}u^6 + C$$

Therefore, using this result and replacing u by $u = 2x + 4$, we obtain

$$\int 2(2x + 4)^5\, dx = \frac{1}{6}(2x + 4)^6 + C$$

We can verify that the foregoing result is indeed correct by computing

$$\frac{d}{dx}\left[\frac{1}{6}(2x+4)^6\right] = \frac{1}{6}\cdot 6(2x+4)^5(2) \qquad \text{Use the chain rule.}$$
$$= 2(2x+4)^5$$

and observing that the last expression is just the integrand of (3).

The Method of Integration by Substitution

To see why the approach used in evaluating the integral in (3) is successful, write

$$f(x) = x^5 \quad \text{and} \quad g(x) = 2x + 4$$

Then, $g'(x) = 2\,dx$. Furthermore, the integrand of (3) is just the composition of f and g. Thus,

$$(f \circ g)(x) = f(g(x))$$
$$= [g(x)]^5 = (2x+4)^5$$

Therefore, (3) can be written as

$$\int f(g(x))g'(x)\,dx \tag{4}$$

Next, let's show that an integral having the form (4) can always be written as

$$\int f(u)\,du \tag{5}$$

Suppose F is an antiderivative of f. By the chain rule, we have

$$\frac{d}{dx}[F(g(x))] = F'(g(x))g'(x)$$

Therefore,

$$\int F'(g(x))g'(x)\,dx = F(g(x)) + C$$

Letting $F' = f$ and making the substitution $u = g(x)$, we have

$$\int f(g(x))g'(x)\,dx = F(u) + C = \int F'(u)\,du = \int f(u)\,du$$

as we wished to show. Thus, if the transformed integral is readily evaluated, as is the case with the integral (3), then the method of substitution will prove successful.

Before we look at more examples, let's summarize the steps involved in integration by substitution.

Integration by Substitution

Step 1 Let $u = g(x)$, where $g(x)$ is part of the integrand, usually the "inside function" of the composite function $f(g(x))$.

Step 2 Find $du = g'(x)\,dx$.

Step 3 Use the substitution $u = g(x)$ and $du = g'(x)\,dx$ to convert the *entire* integral into one involving *only* u.

Step 4 Evaluate the resulting integral.

Step 5 Replace u by $g(x)$ to obtain the final solution as a function of x.

Note Sometimes we need to consider different choices of g for the substitution $u = g(x)$ in order to carry out Step 3 and/or Step 4. ■

EXAMPLE 1 Find $\int 2x(x^2 + 3)^4\, dx$.

Solution

Step 1 Observe that the integrand involves the composite function $(x^2 + 3)^4$ with "inside function" $g(x) = x^2 + 3$. So, we choose $u = x^2 + 3$.

Step 2 Find $du = 2x\, dx$.

Step 3 Making the substitution $u = x^2 + 3$ and $du = 2x\, dx$, we obtain

$$\int 2x(x^2 + 3)^4\, dx = \underset{\substack{\uparrow \\ \text{Rewrite}}}{\int} (x^2 + 3)^4(2x\, dx) = \int u^4\, du$$

an integral involving only the variable u.

Step 4 Evaluate

$$\int u^4\, du = \frac{1}{5}u^5 + C$$

Step 5 Replacing u by $x^2 + 3$, we obtain

$$\int 2x(x^2 + 3)^4\, dx = \frac{1}{5}(x^2 + 3)^5 + C$$

■

EXAMPLE 2 Find $\int 3\sqrt{3x + 1}\, dx$.

Solution

Step 1 The integrand involves the composite function $\sqrt{3x + 1}$ with "inside function" $g(x) = 3x + 1$. So, let $u = 3x + 1$.

Step 2 Find $du = 3\, dx$.

Step 3 Making the substitution $u = 3x + 1$ and $du = 3\, dx$, we obtain

$$\int 3\sqrt{3x + 1}\, dx = \int \sqrt{3x + 1}(3\, dx) = \int \sqrt{u}\, du$$

an integral involving only the variable u.

Step 4 Evaluate

$$\int \sqrt{u}\, du = \int u^{1/2}\, du = \frac{2}{3}u^{3/2} + C$$

Step 5 Replacing u by $3x + 1$, we obtain

$$\int 3\sqrt{3x + 1}\, dx = \frac{2}{3}(3x + 1)^{3/2} + C$$

■

EXAMPLE 3 Find $\int x^2(x^3 + 1)^{3/2}\, dx$.

Solution

Step 1 The integrand contains the composite function $(x^3 + 1)^{3/2}$ with "inside function" $g(x) = x^3 + 1$. So, let $u = x^3 + 1$.

Step 2 Find $du = 3x^2\,dx$.

Step 3 Making the substitution $u = x^3 + 1$ and $du = 3x^2\,dx$, or $x^2\,dx = \frac{1}{3}\,du$, we obtain

$$\int x^2(x^3+1)^{3/2}\,dx = \int (x^3+1)^{3/2}(x^2\,dx)$$
$$= \int u^{3/2}\left(\frac{1}{3}\,du\right) = \frac{1}{3}\int u^{3/2}\,du$$

an integral involving only the variable u.

Step 4 We evaluate

$$\frac{1}{3}\int u^{3/2}\,du = \frac{1}{3}\cdot\frac{2}{5}u^{5/2} + C = \frac{2}{15}u^{5/2} + C$$

Step 5 Replacing u by $x^3 + 1$, we obtain

$$\int x^2(x^3+1)^{3/2}\,dx = \frac{2}{15}(x^3+1)^{5/2} + C$$

■

EXPLORE & DISCUSS

Let $f(x) = x^2(x^3+1)^{3/2}$. Using the result of Example 3, we see that an antiderivative of f is $F(x) = \frac{2}{15}(x^3+1)^{5/2}$. However, in terms of u (where $u = x^3 + 1$), an antiderivative of f is $G(u) = \frac{2}{15}u^{5/2}$. Compute $F(2)$. Next, suppose we want to compute $F(2)$ using the function G instead. At what value of u should you evaluate $G(u)$ in order to obtain the desired result? Explain your answer.

In the remaining examples, we drop the practice of labeling the steps involved in evaluating each integral.

EXAMPLE 4 Find $\int e^{-3x}\,dx$.

Solution Let $u = -3x$ so that $du = -3\,dx$, or $dx = -\frac{1}{3}\,du$. Then,

$$\int e^{-3x}\,dx = \int e^u\left(-\frac{1}{3}\,du\right) = -\frac{1}{3}\int e^u\,du$$
$$= -\frac{1}{3}e^u + C = -\frac{1}{3}e^{-3x} + C$$

■

EXAMPLE 5 Find $\displaystyle\int \frac{x}{3x^2+1}\,dx$.

Solution Let $u = 3x^2 + 1$. Then, $du = 6x\,dx$, or $x\,dx = \frac{1}{6}\,du$. Making the appropriate substitutions, we have

$$\int \frac{x}{3x^2+1}\,dx = \int \frac{\frac{1}{6}}{u}\,du$$
$$= \frac{1}{6}\int \frac{1}{u}\,du$$
$$= \frac{1}{6}\ln|u| + C$$
$$= \frac{1}{6}\ln(3x^2+1) + C \qquad \text{Since } 3x^2 + 1 > 0$$

■

EXAMPLE 6 Find $\displaystyle\int \frac{(\ln x)^2}{2x}\,dx$.

Solution Let $u = \ln x$. Then,

$$du = \frac{d}{dx}(\ln x)\,dx = \frac{1}{x}\,dx$$

$$\begin{aligned}\int \frac{(\ln x)^2}{2x}\,dx &= \frac{1}{2}\int \frac{(\ln x)^2}{x}\,dx \\ &= \frac{1}{2}\int u^2\,du \\ &= \frac{1}{6}u^3 + C \\ &= \frac{1}{6}(\ln x)^3 + C\end{aligned}$$

■

EXPLORE & DISCUSS

Suppose $\int f(u)\,du = F(u) + C$.

1. Show that $\int f(ax + b)\,dx = \dfrac{1}{a}F(ax + b) + C$.
2. How can you use this result to facilitate the evaluation of integrals such as $\int (2x + 3)^5\,dx$ and $\int e^{3x-2}\,dx$? Explain your answer.

Examples 7 and 8 show how the method of substitution can be used in practical situations.

APPLIED EXAMPLE 7 Cost of Producing Solar Cell Panels In 1990 the head of the research and development department of Soloron Corporation claimed that the cost of producing solar cell panels would drop at the rate of

$$\frac{58}{(3t + 2)^2} \qquad (0 \le t \le 10)$$

dollars per peak watt for the next t years, with $t = 0$ corresponding to the beginning of 1990. (A peak watt is the power produced at noon on a sunny day.) In 1990 the panels, which are used for photovoltaic power systems, cost \$10 per peak watt. Find an expression giving the cost per peak watt of producing solar cell panels at the beginning of year t. What was the cost at the beginning of 2000?

Solution Let $C(t)$ denote the cost per peak watt for producing solar cell panels at the beginning of year t. Then,

$$C'(t) = -\frac{58}{(3t + 2)^2}$$

Integrating, we find that

$$\begin{aligned}C(t) &= \int \frac{-58}{(3t + 2)^2}\,dt \\ &= -58\int (3t + 2)^{-2}\,dt\end{aligned}$$

Let $u = 3t + 2$ so that

$$du = 3\,dt \quad \text{or} \quad dt = \frac{1}{3}\,du$$

Then,

$$\begin{aligned} C(t) &= -58\left(\frac{1}{3}\right)\int u^{-2}\,du \\ &= -\frac{58}{3}(-1)u^{-1} + k \\ &= \frac{58}{3(3t+2)} + k \end{aligned}$$

where k is an arbitrary constant. To determine the value of k, note that the cost per peak watt of producing solar cell panels at the beginning of 1990 ($t = 0$) was 10, or $C(0) = 10$. This gives

$$C(0) = \frac{58}{3(2)} + k = 10$$

or $k = \frac{1}{3}$. Therefore, the required expression is given by

$$\begin{aligned} C(t) &= \frac{58}{3(3t+2)} + \frac{1}{3} \\ &= \frac{58 + (3t + 2)}{3(3t+2)} = \frac{3t + 60}{3(3t+2)} \\ &= \frac{t + 20}{3t + 2} \end{aligned}$$

The cost per peak watt for producing solar cell panels at the beginning of 2000 is given by

$$C(10) = \frac{10 + 20}{3(10) + 2} \approx 0.94$$

or approximately \$0.94 per peak watt. ■

EXPLORING WITH TECHNOLOGY

Refer to Example 7.

1. Use a graphing utility to plot the graph of

$$C(t) = \frac{t + 20}{3t + 2}$$

using the viewing window $[0, 10] \times [0, 5]$. Then, use the numerical differentiation capability of the graphing utility to compute $C'(10)$.

2. Plot the graph of

$$C'(t) = -\frac{58}{(3t+2)^2}$$

using the viewing window $[0, 10] \times [-10, 0]$. Then, use the evaluation capability of the graphing utility to find $C'(10)$. Is this value of $C'(10)$ the same as that obtained in part 1? Explain your answer.

APPLIED EXAMPLE 8 Computer Sales Projections A study prepared by the marketing department of Universal Instruments forecasts that, after its new line of Galaxy Home Computers is introduced into the market, sales will grow at the rate of

$$2000 - 1500e^{-0.05t} \qquad (0 \le t \le 60)$$

units per month. Find an expression that gives the total number of computers that will sell t months after they become available on the market. How many computers will Universal sell in the first year they are on the market?

Solution Let $N(t)$ denote the total number of computers that may be expected to be sold t months after their introduction in the market. Then, the rate of growth of sales is given by $N'(t)$ units per month. Thus,

$$N'(t) = 2000 - 1500e^{-0.05t}$$

so that

$$\begin{aligned} N(t) &= \int (2000 - 1500e^{-0.05t})\, dt \\ &= \int 2000\, dt - 1500 \int e^{-0.05t}\, dt \end{aligned}$$

Upon integrating the second integral by the method of substitution, we obtain

$$\begin{aligned} N(t) &= 2000t + \frac{1500}{0.05} e^{-0.05t} + C \qquad \text{Let } u = -0.05t, \text{ then } du = -0.05\, dt. \\ &= 2000t + 30{,}000e^{-0.05t} + C \end{aligned}$$

To determine the value of C, note that the number of computers sold at the end of month 0 is nil, so $N(0) = 0$. This gives

$$N(0) = 30{,}000 + C = 0 \qquad \text{Since } e^0 = 1$$

or $C = -30{,}000$. Therefore, the required expression is given by

$$\begin{aligned} N(t) &= 2000t + 30{,}000e^{-0.05t} - 30{,}000 \\ &= 2000t + 30{,}000(e^{-0.05t} - 1) \end{aligned}$$

The number of computers that Universal can expect to sell in the first year is given by

$$\begin{aligned} N(12) &= 2000(12) + 30{,}000(e^{-0.05(12)} - 1) \\ &\approx 10{,}464 \end{aligned}$$

6.2 Self-Check Exercises

1. Evaluate $\int \sqrt{2x + 5}\, dx$.

2. Evaluate $\int \frac{x^2}{(2x^3 + 1)^{3/2}}\, dx$.

3. Evaluate $\int xe^{2x^2 - 1}\, dx$.

4. According to a joint study conducted by Oxnard's Environmental Management Department and a state government agency, the concentration of carbon monoxide (CO) in the air due to automobile exhaust is increasing at the rate given by

$$f(t) = \frac{8(0.1t + 1)}{300(0.2t^2 + 4t + 64)^{1/3}}$$

parts per million (ppm) per year t. Currently, the CO concentration due to automobile exhaust is 0.16 ppm. Find an expression giving the CO concentration t yr from now.

Solutions to Self-Check Exercises 6.2 can be found on page 419.

6.2 Concept Questions

1. Explain how the method of substitution works by showing the steps used to find $\int f(g(x))g'(x)\,dx$.

2. Explain why the method of substitution works for the integral $\int xe^{-x^2}\,dx$, but not for the integral $\int e^{-x^2}\,dx$.

6.2 Exercises

In Exercises 1–50, find the indefinite integral.

1. $\int 4(4x+3)^4\,dx$
2. $\int 4x(2x^2+1)^7\,dx$
3. $\int (x^3-2x)^2(3x^2-2)\,dx$
4. $\int (3x^2-2x+1)(x^3-x^2+x)^4\,dx$
5. $\int \frac{4x}{(2x^2+3)^3}\,dx$
6. $\int \frac{3x^2+2}{(x^3+2x)^2}\,dx$
7. $\int 3t^2\sqrt{t^3+2}\,dt$
8. $\int 3t^2(t^3+2)^{3/2}\,dt$
9. $\int (x^2-1)^9x\,dx$
10. $\int x^2(2x^3+3)^4\,dx$
11. $\int \frac{x^4}{1-x^5}\,dx$
12. $\int \frac{x^2}{\sqrt{x^3-1}}\,dx$
13. $\int \frac{2}{x-2}\,dx$
14. $\int \frac{x^2}{x^3-3}\,dx$
15. $\int \frac{0.3x-0.2}{0.3x^2-0.4x+2}\,dx$
16. $\int \frac{2x^2+1}{0.2x^3+0.3x}\,dx$
17. $\int \frac{x}{3x^2-1}\,dx$
18. $\int \frac{x^2-1}{x^3-3x+1}\,dx$
19. $\int e^{-2x}\,dx$
20. $\int e^{-0.02x}\,dx$
21. $\int e^{2-x}\,dx$
22. $\int e^{2t+3}\,dt$
23. $\int xe^{-x^2}\,dx$
24. $\int x^2e^{x^3-1}\,dx$
25. $\int (e^x-e^{-x})\,dx$
26. $\int (e^{2x}+e^{-3x})\,dx$
27. $\int \frac{e^x}{1+e^x}\,dx$
28. $\int \frac{e^{2x}}{1+e^{2x}}\,dx$
29. $\int \frac{e^{\sqrt{x}}}{\sqrt{x}}\,dx$
30. $\int \frac{e^{-1/x}}{x^2}\,dx$
31. $\int \frac{e^{3x}+x^2}{(e^{3x}+x^3)^3}\,dx$
32. $\int \frac{e^x-e^{-x}}{(e^x+e^{-x})^{3/2}}\,dx$
33. $\int e^{2x}(e^{2x}+1)^3\,dx$
34. $\int e^{-x}(1+e^{-x})\,dx$
35. $\int \frac{\ln 5x}{x}\,dx$
36. $\int \frac{(\ln u)^3}{u}\,du$
37. $\int \frac{1}{x\ln x}\,dx$
38. $\int \frac{1}{x(\ln x)^2}\,dx$
39. $\int \frac{\sqrt{\ln x}}{x}\,dx$
40. $\int \frac{(\ln x)^{7/2}}{x}\,dx$
41. $\int \left(xe^{x^2}-\frac{x}{x^2+2}\right)dx$
42. $\int \left(xe^{-x^2}+\frac{e^x}{e^x+3}\right)dx$
43. $\int \frac{x+1}{\sqrt{x}-1}\,dx$
   Hint: Let $u=\sqrt{x}-1$.
44. $\int \frac{e^{-u}-1}{e^{-u}+u}\,du$
   Hint: Let $v=e^{-u}+u$.
45. $\int x(x-1)^5\,dx$
   Hint: $u=x-1$ implies $x=u+1$.
46. $\int \frac{t}{t+1}\,dt$
   Hint: $\frac{t}{t+1}=1-\frac{1}{t+1}$.
47. $\int \frac{1-\sqrt{x}}{1+\sqrt{x}}\,dx$
   Hint: Let $u=1+\sqrt{x}$.
48. $\int \frac{1+\sqrt{x}}{1-\sqrt{x}}\,dx$
   Hint: Let $u=1-\sqrt{x}$.
49. $\int v^2(1-v)^6\,dv$
   Hint: Let $u=1-v$.
50. $\int x^3(x^2+1)^{3/2}\,dx$
   Hint: Let $u=x^2+1$.

In Exercises 51–54, find the function f given that the slope of the tangent line to the graph of f at any point $(x, f(x))$ is $f'(x)$ and that the graph of f passes through the given point.

51. $f'(x)=5(2x-1)^4$; $(1, 3)$
52. $f'(x)=\frac{3x^2}{2\sqrt{x^3-1}}$; $(1, 1)$

53. $f'(x) = -2xe^{-x^2+1}$; $(1, 0)$

54. $f'(x) = 1 - \frac{2x}{x^2 + 1}$; $(0, 2)$

55. CABLE TELEPHONE SUBSCRIBERS The number of cable telephone subscribers stood at 3.2 million at the beginning of 2004 ($t = 0$). For the next 5 yr, the number was projected to grow at the rate of

$$R(t) = 3.36(t + 1)^{0.05} \qquad (0 \le t \le 4)$$

million subscribers/year. If the projection holds true, how many cable telephone subscribers will there be at the beginning of 2008 ($t = 4$)?

Source: Sanford C. Bernstein

56. TV VIEWERS: NEWSMAGAZINE SHOWS The number of viewers of a weekly TV newsmagazine show, introduced in the 2002 season, has been increasing at the rate of

$$3\left(2 + \frac{1}{2}t\right)^{-1/3} \qquad (1 \le t \le 6)$$

million viewers/year in its tth year on the air. The number of viewers of the program during its first year on the air is given by $9(5/2)^{2/3}$ million. Find how many viewers were expected in the 2007 season.

57. STUDENT ENROLLMENT The registrar of Kellogg University estimates that the total student enrollment in the Continuing Education division will grow at the rate of

$$N'(t) = 2000(1 + 0.2t)^{-3/2}$$

students/year t yr from now. If the current student enrollment is 1000, find an expression giving the total student enrollment t yr from now. What will be the student enrollment 5 yr from now?

58. SUPPLY: WOMEN'S BOOTS The rate of change of the unit price p (in dollars) of Apex women's boots is given by

$$p'(x) = \frac{240}{(5 - x)^2}$$

where x is the number of pairs in units of a hundred that the supplier will make available in the market daily when the unit price is \$$p$/pair. Find the supply equation for these boots if the quantity the supplier is willing to make available is 200 pairs daily ($x = 2$) when the unit price is \$50/pair.

59. DEMAND: WOMEN'S BOOTS The rate of change of the unit price p (in dollars) of Apex women's boots is given by

$$p'(x) = \frac{-250x}{(16 + x^2)^{3/2}}$$

where x is the quantity demanded daily in units of a hundred. Find the demand function for these boots if the quantity demanded daily is 300 pairs ($x = 3$) when the unit price is \$50/pair.

60. POPULATION GROWTH The population of a certain city is projected to grow at the rate of

$$r(t) = 400\left(1 + \frac{2t}{24 + t^2}\right) \qquad (0 \le t \le 5)$$

people/year, t years from now. The current population is 60,000. What will be the population 5 yr from now?

61. OIL SPILL In calm waters, the oil spilling from the ruptured hull of a grounded tanker forms an oil slick that is circular in shape. If the radius r of the circle is increasing at the rate of

$$r'(t) = \frac{30}{\sqrt{2t + 4}}$$

feet/minute t min after the rupture occurs, find an expression for the radius at any time t. How large is the polluted area 16 min after the rupture occurred?
Hint: $r(0) = 0$.

62. LIFE EXPECTANCY OF A FEMALE Suppose in a certain country the life expectancy at birth of a female is changing at the rate of

$$g'(t) = \frac{5.45218}{(1 + 1.09t)^{0.9}}$$

years/year. Here, t is measured in years, with $t = 0$ corresponding to the beginning of 1900. Find an expression $g(t)$ giving the life expectancy at birth (in years) of a female in that country if the life expectancy at the beginning of 1900 is 50.02 yr. What is the life expectancy at birth of a female born in 2000 in that country?

63. AVERAGE BIRTH HEIGHT OF BOYS Using data collected at Kaiser Hospital, pediatricians estimate that the average height of male children changes at the rate of

$$h'(t) = \frac{52.8706e^{-0.3277t}}{(1 + 2.449e^{-0.3277t})^2}$$

in./yr, where the child's height $h(t)$ is measured in inches and t, the child's age, is measured in years, with $t = 0$ corresponding to the age at birth. Find an expression $h(t)$ for the average height of a boy at age t if the height at birth of an average child is 19.4 in. What is the height of an average 8-yr-old boy?

64. LEARNING CURVES The average student enrolled in the 20-wk Court Reporting I course at the American Institute of Court Reporting progresses according to the rule

$$N'(t) = 6e^{-0.05t} \qquad (0 \le t \le 20)$$

where $N'(t)$ measures the rate of change in the number of words/minute of dictation the student takes in machine shorthand after t wk in the course. Assuming that the average student enrolled in this course begins with a dictation speed of 60 words/minute, find an expression $N(t)$ that gives the dictation speed of the student after t wk in the course.

65. AMOUNT OF GLUCOSE IN THE BLOODSTREAM Suppose a patient is given a continuous intravenous infusion of glucose at a constant rate of r mg/min. Then, the rate at which the amount of glucose in the bloodstream is changing at time t due to this infusion is given by

$$A'(t) = re^{-at}$$

mg/min, where a is a positive constant associated with the rate at which excess glucose is eliminated from the bloodstream and is dependent on the patient's metabolism rate. Derive an expression for the amount of glucose in the bloodstream at time t.
Hint: $A(0) = 0$.

66. **Concentration of a Drug in an Organ** A drug is carried into an organ of volume V cm^3 by a liquid that enters the organ at the rate of a cm^3/sec and leaves it at the rate of b cm^3/sec. The concentration of the drug in the liquid entering the organ is c g/cm^3. If the concentration of the drug in the organ at time t is increasing at the rate of

$$x'(t) = \frac{1}{V}(ac - bx_0)e^{-bt/V}$$

g/cm^3/sec, and the concentration of the drug in the organ initially is x_0 g/cm^3, show that the concentration of the drug in the organ at time t is given by

$$x(t) = \frac{ac}{b} + \left(x_0 - \frac{ac}{b}\right)e^{-bt/V}$$

6.2 Solutions to Self-Check Exercises

1. Let $u = 2x + 5$. Then, $du = 2\,dx$, or $dx = \frac{1}{2}\,du$. Making the appropriate substitutions, we have

$$\begin{aligned}\int \sqrt{2x+5}\,dx &= \int \sqrt{u}\left(\frac{1}{2}\,du\right) = \frac{1}{2}\int u^{1/2}\,du\\ &= \frac{1}{2}\left(\frac{2}{3}\right)u^{3/2} + C\\ &= \frac{1}{3}(2x+5)^{3/2} + C\end{aligned}$$

2. Let $u = 2x^3 + 1$, so that $du = 6x^2\,dx$, or $x^2\,dx = \frac{1}{6}\,du$. Making the appropriate substitutions, we have

$$\begin{aligned}\int \frac{x^2}{(2x^3+1)^{3/2}}\,dx &= \int \frac{\left(\frac{1}{6}\right)du}{u^{3/2}} = \frac{1}{6}\int u^{-3/2}\,du\\ &= \left(\frac{1}{6}\right)(-2)u^{-1/2} + C\\ &= -\frac{1}{3}(2x^3+1)^{-1/2} + C\\ &= -\frac{1}{3\sqrt{2x^3+1}} + C\end{aligned}$$

3. Let $u = 2x^2 - 1$, so that $du = 4x\,dx$, or $x\,dx = \frac{1}{4}\,du$. Then,

$$\begin{aligned}\int xe^{2x^2-1}\,dx &= \frac{1}{4}\int e^u\,du\\ &= \frac{1}{4}e^u + C\\ &= \frac{1}{4}e^{2x^2-1} + C\end{aligned}$$

4. Let $C(t)$ denote the CO concentration in the air due to automobile exhaust t yr from now. Then,

$$\begin{aligned}C'(t) = f(t) &= \frac{8(0.1t+1)}{300(0.2t^2+4t+64)^{1/3}}\\ &= \frac{8}{300}(0.1t+1)(0.2t^2+4t+64)^{-1/3}\end{aligned}$$

Integrating, we find

$$\begin{aligned}C(t) &= \int \frac{8}{300}(0.1t+1)(0.2t^2+4t+64)^{-1/3}\,dt\\ &= \frac{8}{300}\int (0.1t+1)(0.2t^2+4t+64)^{-1/3}\,dt\end{aligned}$$

Let $u = 0.2t^2 + 4t + 64$, so that $du = (0.4t + 4)\,dt = 4(0.1t + 1)\,dt$, or

$$(0.1t+1)\,dt = \frac{1}{4}\,du$$

Then,

$$\begin{aligned}C(t) &= \frac{8}{300}\left(\frac{1}{4}\right)\int u^{-1/3}\,du\\ &= \frac{1}{150}\left(\frac{3}{2}u^{2/3}\right) + k\\ &= 0.01(0.2t^2+4t+64)^{2/3} + k\end{aligned}$$

where k is an arbitrary constant. To determine the value of k, we use the condition $C(0) = 0.16$, obtaining

$$\begin{aligned}C(0) = 0.16 &= 0.01(64)^{2/3} + k\\ 0.16 &= 0.16 + k\\ k &= 0\end{aligned}$$

Therefore,

$$C(t) = 0.01(0.2t^2+4t+64)^{2/3}$$

6.3 Area and the Definite Integral

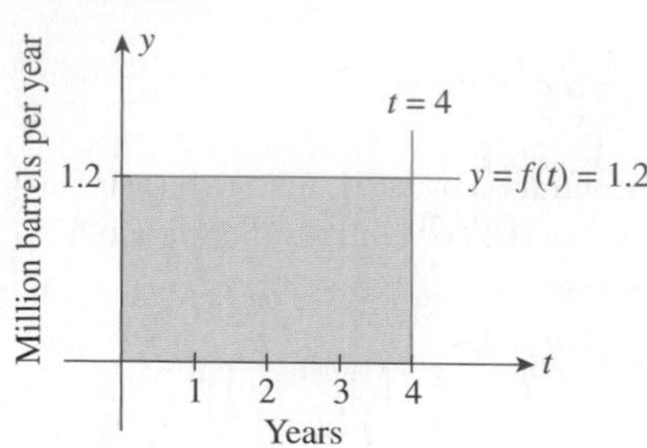

FIGURE 5
The total petroleum consumption is given by the area of the rectangular region.

An Intuitive Look

Suppose a certain state's annual rate of petroleum consumption over a 4-year period is constant and is given by the function

$$f(t) = 1.2 \qquad (0 \le t \le 4)$$

where t is measured in years and $f(t)$ in millions of barrels per year. Then, the state's total petroleum consumption over the period of time in question is

$$(1.2)(4 - 0) \qquad \text{Rate of consumption} \times \text{Time elapsed}$$

or 4.8 million barrels. If you examine the graph of f shown in Figure 5, you will see that this total is just the area of the rectangular region bounded above by the graph of f, below by the t-axis, and to the left and right by the vertical lines $t = 0$ (the y-axis) and $t = 4$, respectively.

Figure 6 shows the actual petroleum consumption of a certain New England state over a 4-year period from 1990 ($t = 0$) to 1994 ($t = 4$). Observe that the rate of consumption is not constant; that is, the function f is not a constant function. What is the state's total petroleum consumption over this 4-year period? It seems reasonable to conjecture that it is given by the "area" of the region bounded above by the graph of f, below by the t-axis, and to the left and right by the vertical lines $t = 0$ and $t = 4$, respectively.

This example raises two questions:

1. What is the "area" of the region shown in Figure 6?
2. How do we compute this area?

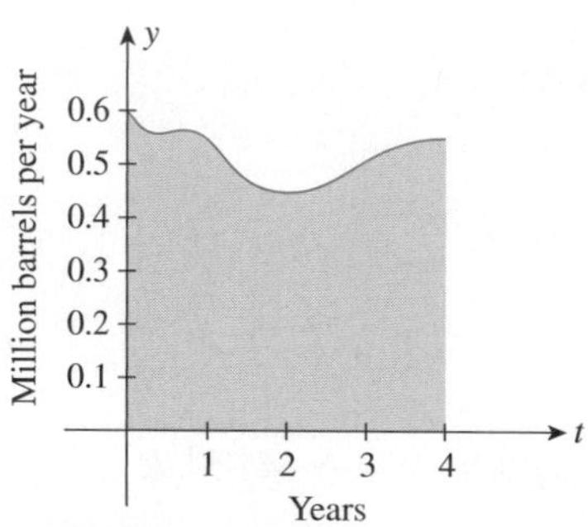

FIGURE 6
The daily petroleum consumption is given by the "area" of the shaded region.

The Area Problem

The preceding example touches on the second fundamental problem in calculus: Calculate the area of the region bounded by the graph of a nonnegative function f, the x-axis, and the vertical lines $x = a$ and $x = b$ (Figure 7). This area is called the **area under the graph of f** on the interval $[a, b]$, or from a to b.

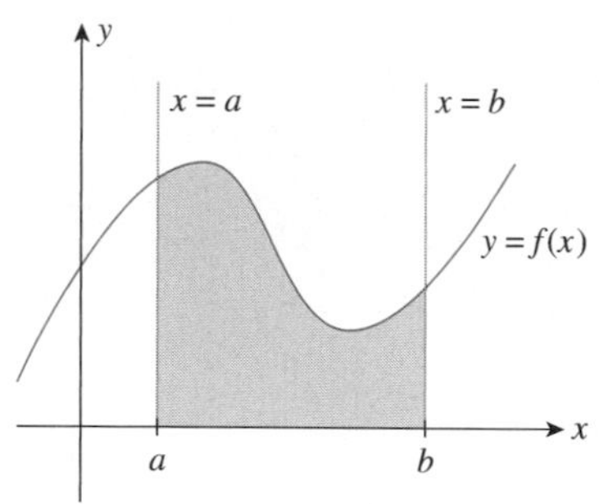

FIGURE 7
The area under the graph of f on $[a, b]$

Defining Area—Two Examples

Just as we used the slopes of secant lines (quantities that we could compute) to help us define the slope of the tangent line to a point on the graph of a function, we now adopt a parallel approach and use the areas of rectangles (quantities that we can compute) to help us define the area under the graph of a function. We begin by looking at a specific example.

EXAMPLE 1 Let $f(x) = x^2$ and consider the region R under the graph of f on the interval $[0, 1]$ (Figure 8a). To obtain an approximation of the area of R, let's construct four nonoverlapping rectangles as follows: Divide the interval $[0, 1]$ into four subintervals

$$\left[0, \frac{1}{4}\right], \quad \left[\frac{1}{4}, \frac{1}{2}\right], \quad \left[\frac{1}{2}, \frac{3}{4}\right], \quad \left[\frac{3}{4}, 1\right]$$

of equal length $\frac{1}{4}$. Next, construct four rectangles with these subintervals as bases

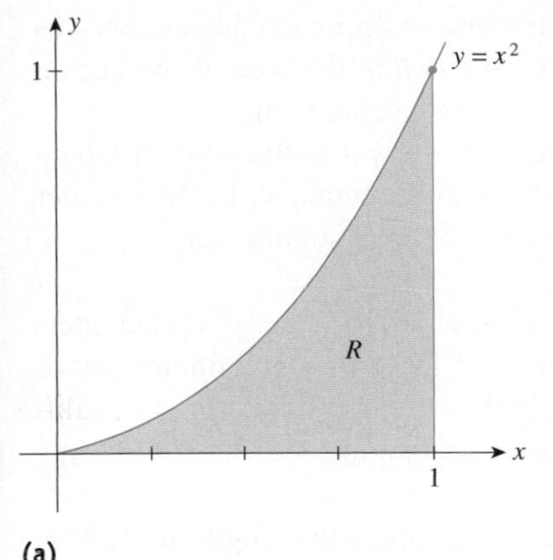

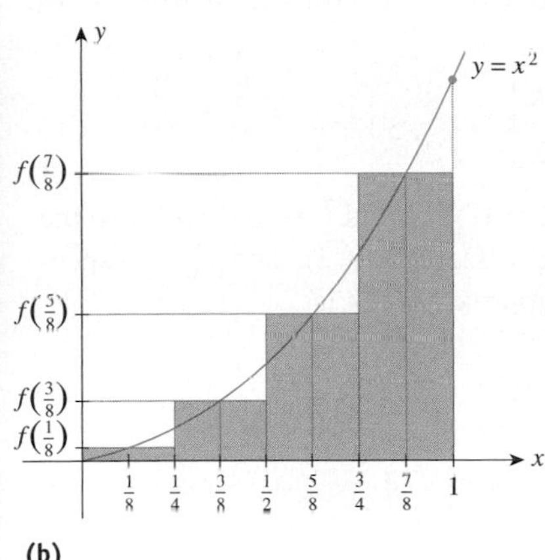

FIGURE 8
The area of the region under the graph of f on $[0, 1]$ in (a) is approximated by the sum of the areas of the four rectangles in (b).

and with heights given by the values of the function at the midpoints

$$\frac{1}{8}, \frac{3}{8}, \frac{5}{8}, \frac{7}{8}$$

of each subinterval. Then, each of these rectangles has width $\frac{1}{4}$ and height

$$f\left(\frac{1}{8}\right), \; f\left(\frac{3}{8}\right), \; f\left(\frac{5}{8}\right), \; f\left(\frac{7}{8}\right)$$

respectively (Figure 8b).

If we approximate the area A of R by the sum of the areas of the four rectangles, we obtain

$$\begin{aligned} A &\approx \frac{1}{4}f\left(\frac{1}{8}\right) + \frac{1}{4}f\left(\frac{3}{8}\right) + \frac{1}{4}f\left(\frac{5}{8}\right) + \frac{1}{4}f\left(\frac{7}{8}\right) \\ &= \frac{1}{4}\left[f\left(\frac{1}{8}\right) + f\left(\frac{3}{8}\right) + f\left(\frac{5}{8}\right) + f\left(\frac{7}{8}\right)\right] \\ &= \frac{1}{4}\left[\left(\frac{1}{8}\right)^2 + \left(\frac{3}{8}\right)^2 + \left(\frac{5}{8}\right)^2 + \left(\frac{7}{8}\right)^2\right] \qquad \text{Recall that } f(x) = x^2. \\ &= \frac{1}{4}\left(\frac{1}{64} + \frac{9}{64} + \frac{25}{64} + \frac{49}{64}\right) = \frac{21}{64} \end{aligned}$$

or approximately 0.328125 square unit. ■

Following the procedure of Example 1, we can obtain approximations of the area of the region R using any number n of rectangles ($n = 4$ in Example 1). Figure 9a shows the approximation of the area A of R using 8 rectangles ($n = 8$), and Figure 9b shows the approximation of the area A of R using 16 rectangles.

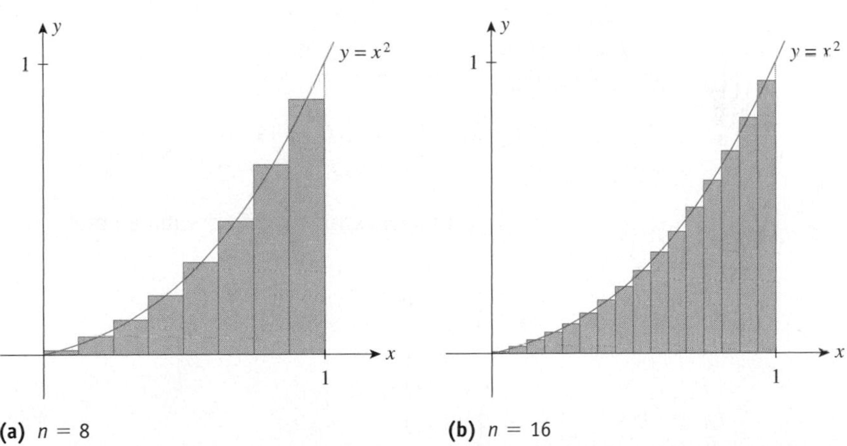

FIGURE 9
As n increases, the number of rectangles increases, and the approximation improves.

These figures suggest that the approximations seem to get better as n increases. This is borne out by the results given in Table 1, which were obtained using a computer.

TABLE 1

Number of Rectangles, n	4	8	16	32	64	100	200
Approximation of A	0.328125	0.332031	0.333008	0.333252	0.333313	0.333325	0.333331

Our computations seem to suggest that the approximations approach the number $\frac{1}{3}$ as n gets larger and larger. This result suggests that we *define* the area of the region under the graph of $f(x) = x^2$ on the interval $[0, 1]$ to be $\frac{1}{3}$ square unit.

In Example 1, we chose the *midpoint* of each subinterval as the point at which to evaluate $f(x)$ to obtain the height of the approximating rectangle. Let's consider another example, this time choosing the *left endpoint* of each subinterval.

EXAMPLE 2 Let R be the region under the graph of $f(x) = 16 - x^2$ on the interval $[1, 3]$. Find an approximation of the area A of R using four subintervals of $[1, 3]$ of equal length and picking the left endpoint of each subinterval to evaluate $f(x)$ to obtain the height of the approximating rectangle.

Solution The graph of f is sketched in Figure 10a. Since the length of $[1, 3]$ is 2, we see that the length of each subinterval is $\frac{2}{4}$, or $\frac{1}{2}$. Therefore, the four subintervals are

$$\left[1, \frac{3}{2}\right], \quad \left[\frac{3}{2}, 2\right], \quad \left[2, \frac{5}{2}\right], \quad \left[\frac{5}{2}, 3\right]$$

The left endpoints of these subintervals are $1, \frac{3}{2}, 2$, and $\frac{5}{2}$, respectively, so the heights of the approximating rectangles are $f(1)$, $f(\frac{3}{2})$, $f(2)$, and $f(\frac{5}{2})$, respectively (Figure 10b). Therefore, the required approximation is

$$\begin{aligned}
A &\approx \frac{1}{2}f(1) + \frac{1}{2}f\left(\frac{3}{2}\right) + \frac{1}{2}f(2) + \frac{1}{2}f\left(\frac{5}{2}\right) \\
&= \frac{1}{2}\left[f(1) + f\left(\frac{3}{2}\right) + f(2) + f\left(\frac{5}{2}\right)\right] \\
&= \frac{1}{2}\left\{[16 - (1)^2] + \left[16 - \left(\frac{3}{2}\right)^2\right]\right. \\
&\qquad \left. + [16 - (2)^2] + \left[16 - \left(\frac{5}{2}\right)^2\right]\right\} \qquad \text{Recall that } f(x) = 16 - x^2. \\
&= \frac{1}{2}\left(15 + \frac{55}{4} + 12 + \frac{39}{4}\right) = \frac{101}{4}
\end{aligned}$$

or approximately 25.25 square units.

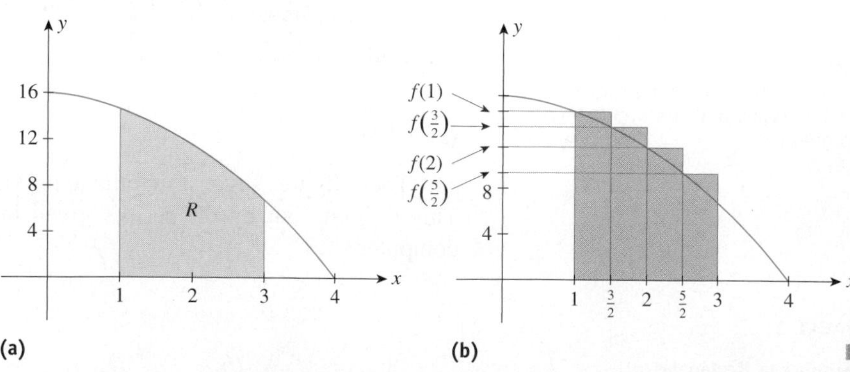

FIGURE 10
The area of R in (a) is approximated by the sum of the areas of the four rectangles in (b).

Table 2 shows the approximations of the area A of the region R of Example 2 when n rectangles are used for the approximation and the heights of the approximating rectangles are found by evaluating $f(x)$ at the left endpoints.

TABLE 2

Number of Rectangles, n	4	10	100	1,000	10,000	50,000	100,000
Approximation of A	25.2500	24.1200	23.4132	23.3413	23.3341	23.3335	23.3334

Once again, we see that the approximations seem to approach a unique number as n gets larger and larger—this time the number is $23\frac{1}{3}$. This result suggests that we *define* the area of the region under the graph of $f(x) = 16 - x^2$ on the interval [1, 3] to be $23\frac{1}{3}$ square units.

Defining Area—The General Case

Examples 1 and 2 point the way to defining the area A under the graph of an arbitrary but continuous and nonnegative function f on an interval $[a, b]$ (Figure 11a).

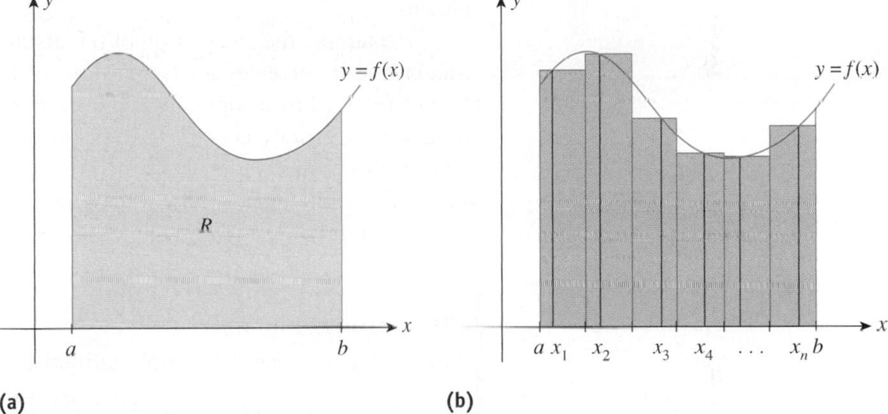

FIGURE 11
The area of the region under the graph of f on $[a, b]$ in (a) is approximated by the sum of the areas of the n rectangles shown in (b).

Divide the interval $[a, b]$ into n subintervals of equal length $\Delta x = (b - a)/n$. Next, pick n arbitrary points $x_1, x_2, \ldots, x_n$, called *representative points*, from the first, second, . . . , and nth subintervals, respectively (Figure 11b). Then, approximating the area A of the region R by the n rectangles of width Δx and heights $f(x_1)$, $f(x_2), \ldots, f(x_n)$, so that the areas of the rectangles are $f(x_1)\Delta x, f(x_2)\Delta x, \ldots, f(x_n)\Delta x$, we have

$$A \approx f(x_1)\Delta x + f(x_2)\Delta x + \cdots + f(x_n)\Delta x$$

The sum on the right-hand side of this expression is called a **Riemann sum** in honor of the German mathematician Bernhard Riemann (1826–1866). Now, as the earlier examples seem to suggest, the Riemann sum will approach a unique number as n becomes arbitrarily large.* We define this number to be the area A of the region R.

*Even though we chose the representative points to be the midpoints of the subintervals in Example 1 and the left endpoints in Example 2, it can be shown that each of the respective sums will always approach a unique number as n approaches infinity.

The Area under the Graph of a Function

Let f be a nonnegative continuous function on $[a, b]$. Then, the area of the region under the graph of f is

$$A = \lim_{n\to\infty} [f(x_1) + f(x_2) + \cdots + f(x_n)]\Delta x \tag{6}$$

where $x_1, x_2, \ldots, x_n$ are arbitrary points in the n subintervals of $[a, b]$ of equal width $\Delta x = (b - a)/n$.

The Definite Integral

As we have just seen, the area under the graph of a continuous *nonnegative* function f on an interval $[a, b]$ is defined by the limit of the Riemann sum

$$\lim_{n\to\infty} [f(x_1)\Delta x + f(x_2)\Delta x + \cdots + f(x_n)\Delta x]$$

We now turn our attention to the study of limits of Riemann sums involving functions that are not necessarily nonnegative. Such limits arise in many applications of calculus.

For example the calculation of the distance covered by a body traveling along a straight line involves evaluating a limit of this form. The computation of the total revenue realized by a company over a certain time period, the calculation of the total amount of electricity consumed in a typical home over a 24-hour period, the average concentration of a drug in a body over a certain interval of time, and the volume of a solid—all involve limits of this type.

We begin with the following definition.

The Definite Integral

Let f be a continuous function defined on $[a, b]$. If

$$\lim_{n\to\infty} [f(x_1)\Delta x + f(x_2)\Delta x + \cdots + f(x_n)\Delta x]$$

exists for all choices of representative points $x_1, x_2, \ldots, x_n$ in the n subintervals of $[a, b]$ of equal width $\Delta x = (b - a)/n$, then this limit is called the **definite integral** of f from a to b and is denoted by $\int_a^b f(x)\,dx$. Thus,

$$\int_a^b f(x)\,dx = \lim_{n\to\infty} [f(x_1)\Delta x + f(x_2)\Delta x + \cdots + f(x_n)\Delta x] \tag{7}$$

The number a is the **lower limit of integration**, and the number b is the **upper limit of integration**.

Notes

1. If f is nonnegative, then the limit in (7) is the same as the limit in (6); therefore, the definite integral gives the area under the graph of f on $[a, b]$.
2. The limit in (7) is denoted by the integral sign $\int$ because, as we will see later, the definite integral and the antiderivative of a function f are related.
3. It is important to realize that the definite integral $\int_a^b f(x)\,dx$ is a *number*, whereas

the indefinite integral $\int f(x)\,dx$ represents a *family of functions* (the antiderivatives of f).

4. If the limit in (7) exists, we say that f is **integrable** on the interval $[a, b]$. ■

When Is a Function Integrable?

The following theorem, which we state without proof, guarantees that a continuous function is integrable.

Integrability of a Function

Let f be continuous on $[a, b]$. Then, f is integrable on $[a, b]$; that is, the definite integral $\int_a^b f(x)\,dx$ exists.

Geometric Interpretation of the Definite Integral

If f is nonnegative and integrable on $[a, b]$, then we have the following geometric interpretation of the definite integral $\int_a^b f(x)\,dx$.

Geometric Interpretation of $\int_a^b f(x)\,dx$ for $f(x) \geq 0$ on $[a, b]$

If f is nonnegative and continuous on $[a, b]$, then

$$\int_a^b f(x)\,dx \tag{8}$$

is equal to the area of the region under the graph of f on $[a, b]$ (Figure 12).

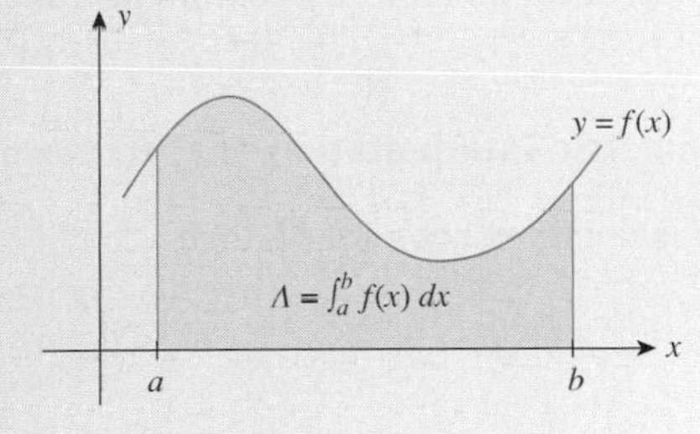

FIGURE 12
If $f(x) \geq 0$ on $[a, b]$, then $\int_a^b f(x)\,dx$ = area under the graph of f on $[a, b]$.

EXPLORE & DISCUSS

Suppose f is nonpositive [that is, $f(x) \leq 0$] and continuous on $[a, b]$. Explain why the area of the region below the x-axis and above the graph of f is given by $-\int_a^b f(x)\,dx$.

Next, let's extend our geometric interpretation of the definite integral to include the case where f assumes both positive as well as negative values on $[a, b]$. Consider a typical Riemann sum of the function f,

$$f(x_1)\Delta x + f(x_2)\Delta x + \cdots + f(x_n)\Delta x$$

corresponding to a partition of $[a, b]$ into n subintervals of equal width $(b - a)/n$, where $x_1, x_2, \ldots, x_n$ are representative points in the subintervals. The sum consists of n terms in which a positive term corresponds to the area of a rectangle of height $f(x_k)$ (for some positive integer k) lying above the x-axis and a negative term corresponds to the area of a rectangle of height $-f(x_k)$ lying below the x-axis. (See Figure 13, which depicts a situation with $n = 6$.)

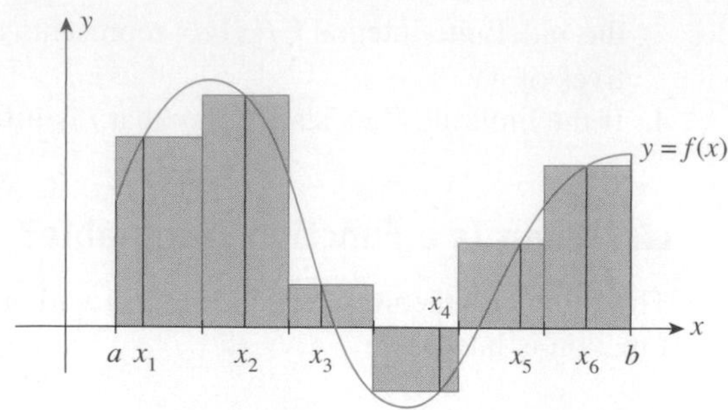

FIGURE 13
The positive terms in the Riemann sum are associated with the areas of the rectangles that lie above the x-axis, and the negative terms are associated with the areas of those that lie below the x-axis.

As n gets larger and larger, the sums of the areas of the rectangles lying above the x-axis seem to give a better and better approximation of the area of the region lying above the x-axis (Figure 14). Similarly, the sums of the areas of those rectangles lying below the x-axis seem to give a better and better approximation of the area of the region lying below the x-axis.

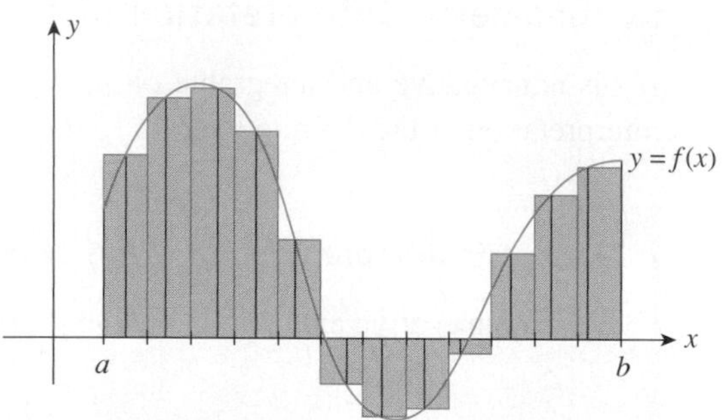

FIGURE 14
As n gets larger, the approximations get better. Here, $n = 12$ and we are approximating with twice as many rectangles as in Figure 13.

These observations suggest the following geometric interpretation of the definite integral for an arbitrary continuous function on an interval $[a, b]$.

Geometric Interpretation of $\int_a^b f(x)\,dx$ on $[a, b]$

If f is continuous on $[a, b]$, then

$$\int_a^b f(x)\,dx$$

is equal to the area of the region above $[a, b]$ minus the area of the region below $[a, b]$ (Figure 15).

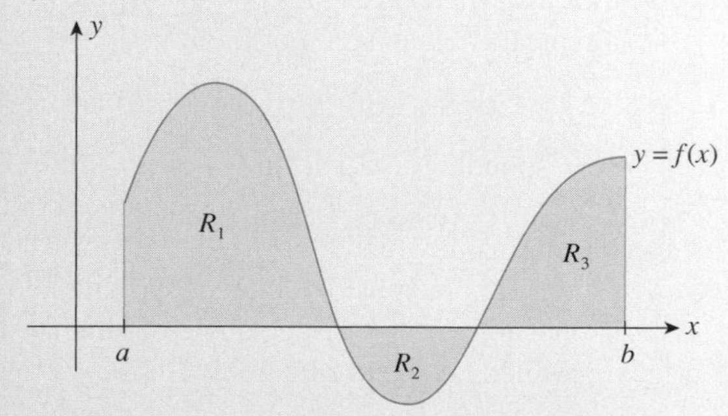

FIGURE 15
$\int_a^b f(x)\,dx$ = Area of R_1 − Area of R_2 + Area of R_3

6.3 Self-Check Exercise

Find an approximation of the area of the region R under the graph of $f(x) = 2x^2 + 1$ on the interval [0, 3], using four subintervals of [0, 3] of equal length and picking the midpoint of each subinterval as a representative point.

The solution to Self-Check Exercise 6.3 can be found on page 429.

6.3 Concept Questions

1. Explain how you would define the area of the region under the graph of a nonnegative continuous function f on the interval $[a, b]$.

2. Define the definite integral of a continuous function on the interval $[a, b]$. Give a geometric interpretation of $\int_a^b f(x)\,dx$ for the case where (a) f is nonnegative on $[a, b]$ and (b) f assumes both positive as well as negative values on $[a, b]$. Illustrate your answers graphically.

6.3 Exercises

In Exercises 1 and 2, find an approximation of the area of the region R under the graph of f by computing the Riemann sum of f corresponding to the partition of the interval into the subintervals shown in the accompanying figures. In each case, use the midpoints of the subintervals as the representative points.

1.
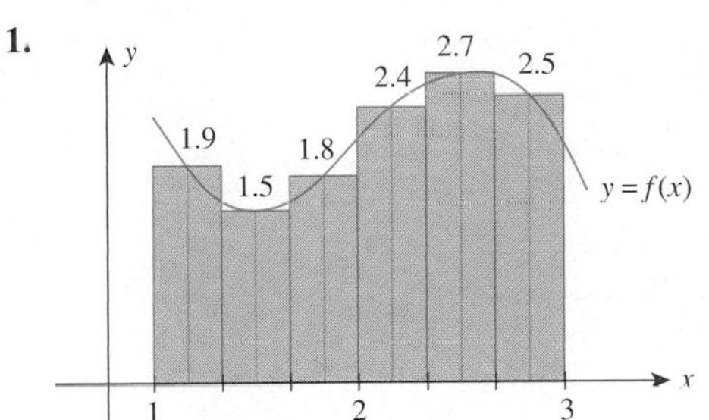

2.
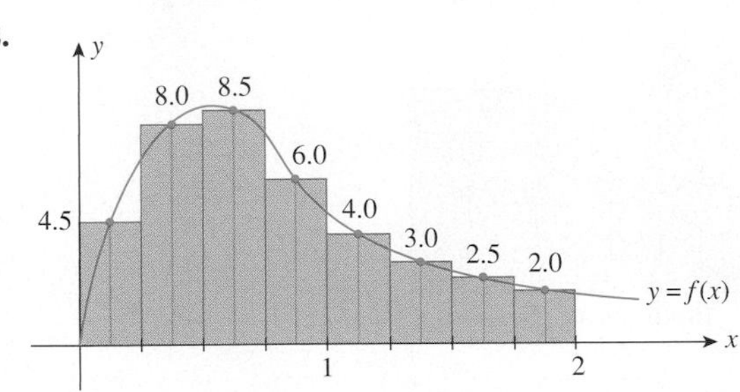

3. Let $f(x) = 3x$.
   a. Sketch the region R under the graph of f on the interval [0, 2] and find its exact area using geometry.
   b. Use a Riemann sum with four subintervals of equal length ($n = 4$) to approximate the area of R. Choose the representative points to be the left endpoints of the subintervals.
   c. Repeat part (b) with eight subintervals of equal length ($n = 8$).
   d. Compare the approximations obtained in parts (b) and (c) with the exact area found in part (a). Do the approximations improve with larger n?

4. Repeat Exercise 3, choosing the representative points to be the right endpoints of the subintervals.

5. Let $f(x) = 4 - 2x$.
   a. Sketch the region R under the graph of f on the interval [0, 2] and find its exact area using geometry.
   b. Use a Riemann sum with five subintervals of equal length ($n = 5$) to approximate the area of R. Choose the representative points to be the left endpoints of the subintervals.
   c. Repeat part (b) with ten subintervals of equal length ($n = 10$).
   d. Compare the approximations obtained in parts (b) and (c) with the exact area found in part (a). Do the approximations improve with larger n?

6. Repeat Exercise 5, choosing the representative points to be the right endpoints of the subintervals.

7. Let $f(x) = x^2$ and compute the Riemann sum of f over the interval [2, 4], using
   a. Two subintervals of equal length ($n = 2$).
   b. Five subintervals of equal length ($n = 5$).
   c. Ten subintervals of equal length ($n = 10$).

In each case, choose the representative points to be the midpoints of the subintervals.

d. Can you guess at the area of the region under the graph of f on the interval [2, 4]?

8. Repeat Exercise 7, choosing the representative points to be the left endpoints of the subintervals.

9. Repeat Exercise 7, choosing the representative points to be the right endpoints of the subintervals.

10. Let $f(x) = x^3$ and compute the Riemann sum of f over the interval [0, 1], using

a. Two subintervals of equal length ($n = 2$).

b. Five subintervals of equal length ($n = 5$).

c. Ten subintervals of equal length ($n = 10$).

In each case, choose the representative points to be the midpoints of the subintervals.

d. Can you guess at the area of the region under the graph of f on the interval [0, 1]?

11. Repeat Exercise 10, choosing the representative points to be the left endpoints of the subintervals.

12. Repeat Exercise 10, choosing the representative points to be the right endpoints of the subintervals.

In Exercises 13–16, find an approximation of the area of the region *R* under the graph of the function *f* on the interval [*a*, *b*]. In each case, use *n* subintervals and choose the representative points as indicated.

13. $f(x) = x^2 + 1$; [0, 2]; $n = 5$; midpoints

14. $f(x) = 4 - x^2$; [−1, 2]; $n = 6$; left endpoints

15. $f(x) = \dfrac{1}{x}$; [1, 3]; $n = 4$; right endpoints

16. $f(x) = e^x$; [0, 3]; $n = 5$; midpoints

17. REAL ESTATE Figure (a) shows a vacant lot with a 100-ft frontage in a development. To estimate its area, we introduce a coordinate system so that the x-axis coincides with the edge of the straight road forming the lower boundary of the property, as shown in Figure (b). Then, thinking of the upper boundary of the property as the graph of a continuous function f over the interval [0, 100], we see that the problem is mathematically equivalent to that of finding the area under the graph of f on [0, 100]. To estimate the area of the lot using a Riemann sum, we divide the interval [0, 100] into five equal subintervals of length 20 ft. Then, using surveyor's equipment, we measure the distance from the midpoint of each of these subintervals to the upper boundary of the property. These measurements give the values of $f(x)$ at $x = 10, 30, 50, 70$, and 90. What is the approximate area of the lot?

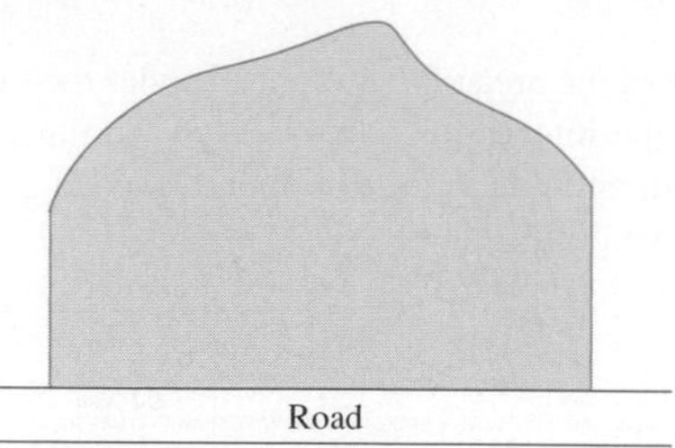

(a)

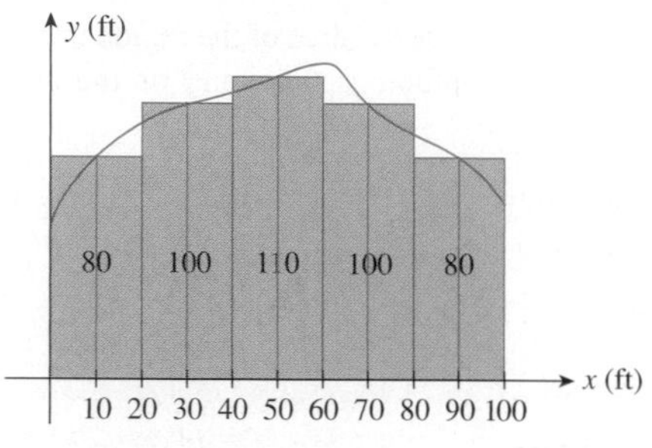

(b)

18. REAL ESTATE Use the technique of Exercise 17 to obtain an estimate of the area of the vacant lot shown in the accompanying figures.

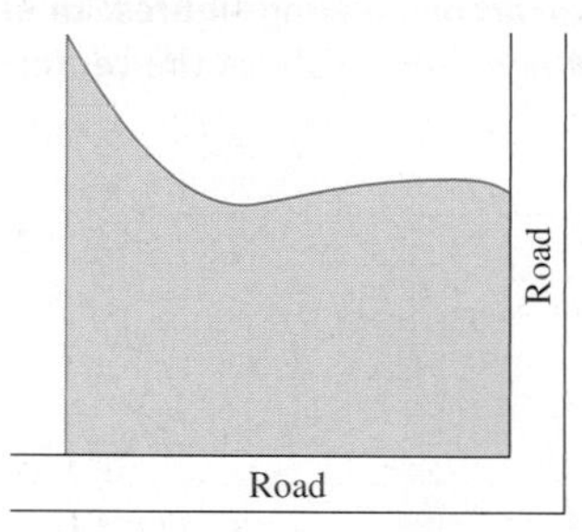

(a)

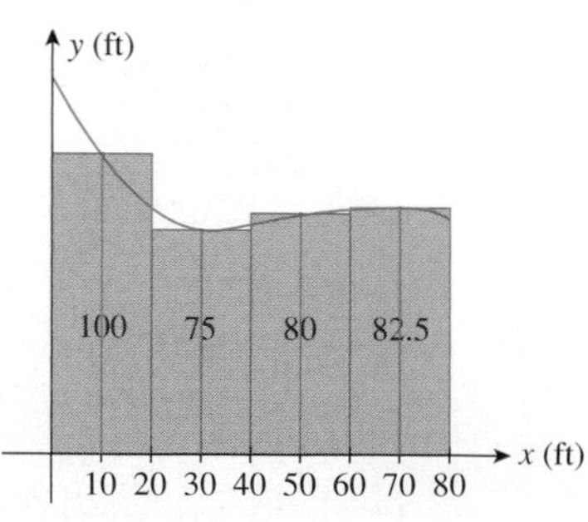

(b)

6.3 Solution to Self-Check Exercise

The length of each subinterval is $\frac{3}{4}$. Therefore, the four subintervals are

$$\left[0, \frac{3}{4}\right], \quad \left[\frac{3}{4}, \frac{3}{2}\right], \quad \left[\frac{3}{2}, \frac{9}{4}\right], \quad \left[\frac{9}{4}, 3\right]$$

The representative points are $\frac{3}{8}$, $\frac{9}{8}$, $\frac{15}{8}$, and $\frac{21}{8}$, respectively. Therefore, the required approximation is

$$\begin{aligned}
A &= \frac{3}{4}f\left(\frac{3}{8}\right) + \frac{3}{4}f\left(\frac{9}{8}\right) + \frac{3}{4}f\left(\frac{15}{8}\right) + \frac{3}{4}f\left(\frac{21}{8}\right) \\
&= \frac{3}{4}\left[f\left(\frac{3}{8}\right) + f\left(\frac{9}{8}\right) + f\left(\frac{15}{8}\right) + f\left(\frac{21}{8}\right)\right] \\
&= \frac{3}{4}\left\{\left[2\left(\frac{3}{8}\right)^2 + 1\right] + \left[2\left(\frac{9}{8}\right)^2 + 1\right] + \left[2\left(\frac{15}{8}\right)^2 + 1\right]\right. \\
&\qquad \left. + \left[2\left(\frac{21}{8}\right)^2 + 1\right]\right\} \\
&= \frac{3}{4}\left(\frac{41}{32} + \frac{113}{32} + \frac{257}{32} + \frac{473}{32}\right) = \frac{663}{32}
\end{aligned}$$

or approximately 20.72 square units.

6.4 The Fundamental Theorem of Calculus

The Fundamental Theorem of Calculus

In Section 6.3, we defined the definite integral of an arbitrary continuous function on an interval $[a, b]$ as a limit of Riemann sums. Calculating the value of a definite integral by actually taking the limit of such sums is tedious and in most cases impractical. It is important to realize that the numerical results we obtained in Examples 1 and 2 of Section 6.3 were *approximations* of the respective areas of the regions in question, even though these results enabled us to *conjecture* what the actual areas might be. Fortunately, there is a much better way of finding the exact value of a definite integral.

The following theorem shows how to evaluate the definite integral of a continuous function provided we can find an antiderivative of that function. Because of its importance in establishing the relationship between differentiation and integration, this theorem—discovered independently by Sir Isaac Newton (1642–1727) in England and Gottfied Wilhelm Leibniz (1646–1716) in Germany—is called the **fundamental theorem of calculus**.

THEOREM 2

The Fundamental Theorem of Calculus

Let f be continuous on $[a, b]$. Then,

$$\int_a^b f(x)\,dx = F(b) - F(a) \tag{9}$$

where F is any antiderivative of f; that is, $F'(x) = f(x)$.

We will explain why this theorem is true at the end of this section.

When applying the fundamental theorem of calculus, it is convenient to use the notation

$$F(x)\Big|_a^b = F(b) - F(a)$$

For example, using this notation, Equation (9) is written

$$\int_a^b f(x)\,dx = F(x)\Big|_a^b = F(b) - F(a)$$

EXAMPLE 1 Let R be the region under the graph of $f(x) = x$ on the interval $[1, 3]$. Use the fundamental theorem of calculus to find the area A of R and verify your result by elementary means.

Solution The region R is shown in Figure 16a. Since f is nonnegative on $[1, 3]$, the area of R is given by the definite integral of f from 1 to 3; that is,

$$A = \int_1^3 x\,dx$$

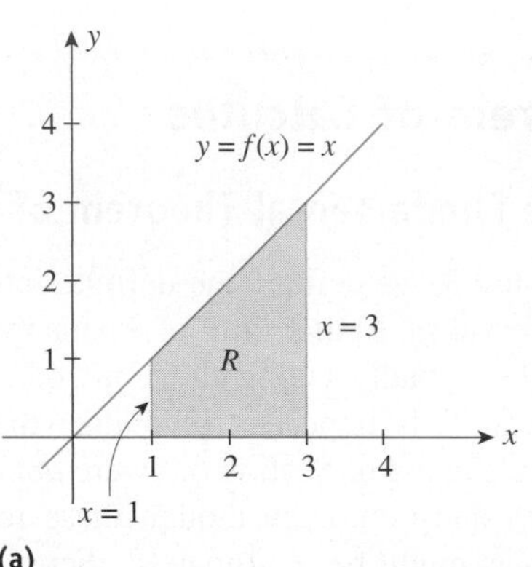

(a)

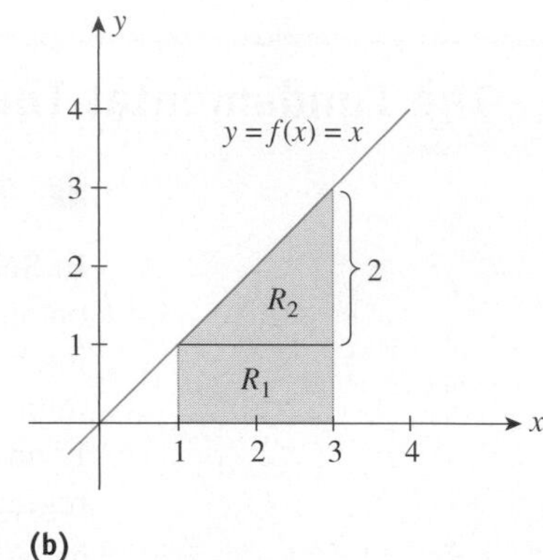

(b)

FIGURE 16
The area of R can be computed in two different ways.

To evaluate the definite integral, observe that an antiderivative of $f(x) = x$ is $F(x) = \frac{1}{2}x^2 + C$, where C is an arbitrary constant. Therefore, by the fundamental theorem of calculus, we have

$$\begin{aligned} A &= \int_1^3 x\,dx = \frac{1}{2}x^2 + C\Big|_1^3 \\ &= \left(\frac{9}{2} + C\right) - \left(\frac{1}{2} + C\right) = 4 \text{ square units} \end{aligned}$$

To verify this result by elementary means, observe that the area A is the area of the rectangle R_1 (width $\times$ height) plus the area of the triangle R_2 ($\frac{1}{2}$ base $\times$ height) (see Figure 16b); that is,

$$2(1) + \frac{1}{2}(2)(2) = 2 + 2 = 4$$

which agrees with the result obtained earlier. ■

Observe that in evaluating the definite integral in Example 1, the constant of integration "dropped out." This is true in general, for if $F(x) + C$ denotes an antiderivative of some function f, then

$$\begin{aligned} F(x) + C\Big|_a^b &= [F(b) + C] - [F(a) + C] \\ &= F(b) + C - F(a) - C \\ &= F(b) - F(a) \end{aligned}$$

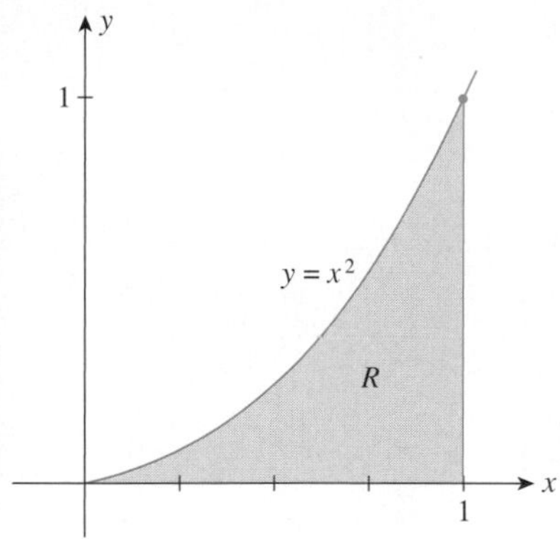

FIGURE 17

The area of R is $\int_0^1 x^2\,dx = \frac{1}{3}$.

With this fact in mind, we may, in all future computations involving the evaluation of a definite integral, drop the constant of integration from our calculations.

Finding the Area under a Curve

Having seen how effective the fundamental theorem of calculus is in helping us find the area of simple regions, we now use it to find the area of more complicated regions.

EXAMPLE 2 In Section 6.3, we conjectured that the area of the region R under the graph of $f(x) = x^2$ on the interval $[0, 1]$ was $\frac{1}{3}$ square unit. Use the fundamental theorem of calculus to verify this conjecture.

Solution The region R is reproduced in Figure 17. Observe that f is nonnegative on $[0, 1]$, so the area of R is given by $A = \int_0^1 x^2\,dx$. Since an antiderivative of $f(x) = x^2$ is $F(x) = \frac{1}{3}x^3$, we see, using the fundamental theorem of calculus, that

$$A = \int_0^1 x^2\,dx = \frac{1}{3}x^3\Big|_0^1 = \frac{1}{3}(1) - \frac{1}{3}(0) = \frac{1}{3} \text{ square unit}$$

as we wished to show. ■

Note It is important to realize that the value, $\frac{1}{3}$, is by definition the exact value of the area of R. ■

EXAMPLE 3 Find the area of the region R under the graph of $y = x^2 + 1$ from $x = -1$ to $x = 2$.

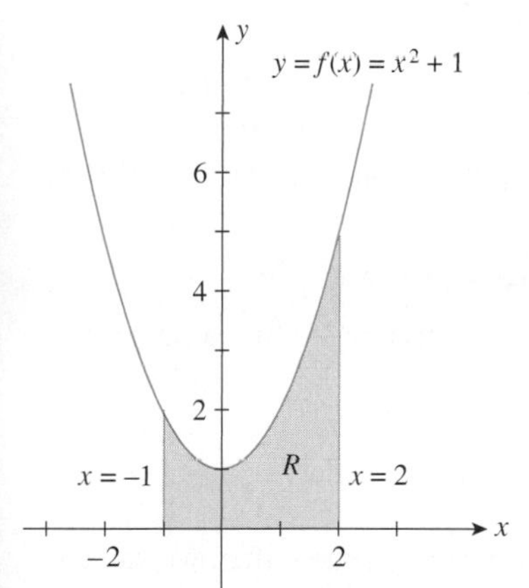

FIGURE 18
The area of R is $\int_{-1}^2 (x^2 + 1)\,dx$.

Solution The region R under consideration is shown in Figure 18. Using the fundamental theorem of calculus, we find that the required area is

$$\int_{-1}^2 (x^2 + 1)\,dx = \left(\frac{1}{3}x^3 + x\right)\Big|_{-1}^2$$
$$= \left[\frac{1}{3}(8) + 2\right] - \left[\frac{1}{3}(-1)^3 + (-1)\right] = 6$$

or 6 square units. ■

Evaluating Definite Integrals

In Examples 4 and 5, we use the rules of integration of Section 6.1 to help us evaluate the definite integrals.

EXAMPLE 4 Evaluate $\int_1^3 (3x^2 + e^x)\,dx$.

Solution

$$\int_1^3 (3x^2 + e^x)\,dx = x^3 + e^x\Big|_1^3$$
$$= (27 + e^3) - (1 + e) = 26 + e^3 - e$$ ■

EXAMPLE 5 Evaluate $\displaystyle\int_1^2 \left(\frac{1}{x} - \frac{1}{x^2}\right) dx.$

Solution

$$\begin{aligned}\int_1^2 \left(\frac{1}{x} - \frac{1}{x^2}\right) dx &= \int_1^2 \left(\frac{1}{x} - x^{-2}\right) dx \\ &= \ln|x| + \frac{1}{x}\bigg|_1^2 \\ &= \left(\ln 2 + \frac{1}{2}\right) - (\ln 1 + 1) \\ &= \ln 2 - \frac{1}{2} \qquad \text{Recall, } \ln 1 = 0.\end{aligned}$$

EXPLORE & DISCUSS

Consider the definite integral $\displaystyle\int_{-1}^1 \frac{1}{x^2}\, dx.$

1. Show that a formal application of Equation (9) leads to

$$\int_{-1}^1 \frac{1}{x^2}\, dx = -\frac{1}{x}\bigg|_{-1}^1 = -1 - 1 = -2$$

2. Observe that $f(x) = 1/x^2$ is positive at each value of x in $[-1, 1]$ where it is defined. Therefore, one might expect that the definite integral with integrand f has a positive value, if it exists.
3. Explain this apparent contradiction in the result (1) and the observation (2).

The Definite Integral as a Measure of Net Change

In real-world applications, we are often interested in the net change of a quantity over a period of time. For example, suppose P is a function giving the population, $P(t)$, of a city at time t. Then the *net change* in the population over the period from $t = a$ to $t = b$ is given by

$$P(b) - P(a) \qquad \text{Population at } t = b \text{ minus population at } t = a$$

If P has a continuous derivative P' in $[a, b]$, then we can invoke the fundamental theorem of calculus to write

$$P(b) - P(a) = \int_a^b P'(t)\, dt \qquad P \text{ is an antiderivative of } P'.$$

Thus, if we know the *rate of change* of the population at any time t, then we can calculate the net change in the population from $t = a$ to $t = b$ by evaluating an appropriate definite integral.

APPLIED EXAMPLE 6 Population Growth in Clark County Clark County in Nevada—dominated by Las Vegas—is the fastest-growing metropolitan area in the United States. From 1970 through 2000, the population was growing at the rate of

$$R(t) = 133680t^2 - 178788t + 234633 \qquad (0 \le t \le 3)$$

people per decade, where $t = 0$ corresponds to the beginning of 1970. What was the net change in the population over the decade from 1980 to 1990?

Source: U.S. Census Bureau

Solution The net change in the population over the decade from 1980 ($t = 1$) to 1990 ($t = 2$) is given by $P(2) - P(1)$, where P denotes the population in the county at time t. But $P' = R$, and so

$$\begin{aligned} P(2) - P(1) &= \int_1^2 P'(t)\,dt = \int_1^2 R(t)\,dt \\ &= \int_1^2 (133680t^2 - 178788t + 234633)\,dt \\ &= 44560t^3 - 89394t^2 + 234633t\Big|_1^2 \\ &= [44560(2)^3 - 89394(2)^2 + 234633(2)] \\ &\quad - [44560 - 89394 + 234633] \\ &= 278371 \end{aligned}$$

and so the net change is 278,371 people. ■

More generally, we have the following result. We assume that f has a continuous derivative, even though, the integrability of f' is sufficient.

Net Change Formula

The net change in a function f over an interval $[a, b]$ is given by

$$f(b) - f(a) = \int_a^b f'(x)\,dx \qquad \textbf{(10)}$$

provided f' is continuous on $[a, b]$.

As another example of the net change of a function, let's consider the following example.

APPLIED EXAMPLE 7 Production Costs The management of Staedtler Office Equipment has determined that the daily marginal cost function associated with producing battery-operated pencil sharpeners is given by

$$C'(x) = 0.000006x^2 - 0.006x + 4$$

where $C'(x)$ is measured in dollars per unit and x denotes the number of units produced. Management has also determined that the daily fixed cost incurred in producing these pencil sharpeners is \$100. Find Staedtler's daily total cost for producing (a) the first 500 units and (b) the 201st through 400th units.

Solution

a. Since $C'(x)$ is the marginal cost function, its antiderivative $C(x)$ is the total cost function. The daily fixed cost incurred in producing the pencil sharpeners is $C(0)$ dollars. Since the daily fixed cost is given as \$100, we have $C(0) = 100$. We are required to find $C(500)$. Let's compute $C(500) - C(0)$, the net change in the total cost function $C(x)$ over the interval $[0, 500]$. Using the fundamental theorem of calculus, we find

$$\begin{aligned}C(500) - C(0) &= \int_0^{500} C'(x)\,dx \\ &= \int_0^{500} (0.000006x^2 - 0.006x + 4)\,dx \\ &= 0.000002x^3 - 0.003x^2 + 4x \Big|_0^{500} \\ &= [0.000002(500)^3 - 0.003(500)^2 + 4(500)] \\ &\quad - [0.000002(0)^3 - 0.003(0)^2 + 4(0)] \\ &= 1500\end{aligned}$$

Therefore, $C(500) = 1500 + C(0) = 1500 + 100 = 1600$, so the total cost incurred daily by Staedtler in producing 500 pencil sharpeners is \$1600.

b. The daily total cost incurred by Staedtler in producing the 201st through 400th units of battery-operated pencil sharpeners is given by

$$\begin{aligned}C(400) - C(200) &= \int_{200}^{400} C'(x)\,dx \\ &= \int_{200}^{400} (0.000006x^2 - 0.006x + 4)\,dx \\ &= 0.000002x^3 - 0.003x^2 + 4x \Big|_{200}^{400} \\ &= [0.000002(400)^3 - 0.003(400)^2 + 4(400)] \\ &\quad - [0.000002(200)^3 - 0.003(200)^2 + 4(200)] \\ &= 552\end{aligned}$$

or \$552. ■

Since $C'(x)$ is nonnegative for x in the interval $(0, \infty)$, we have the following geometric interpretation of the two definite integrals in Example 7: $\int_0^{500} C'(x)\,dx$ is the area of the region under the graph of the function C' from $x = 0$ to $x = 500$, shown in Figure 19a, and $\int_{200}^{400} C'(x)\,dx$ is the area of the region from $x = 200$ to $x = 400$, shown in Figure 19b.

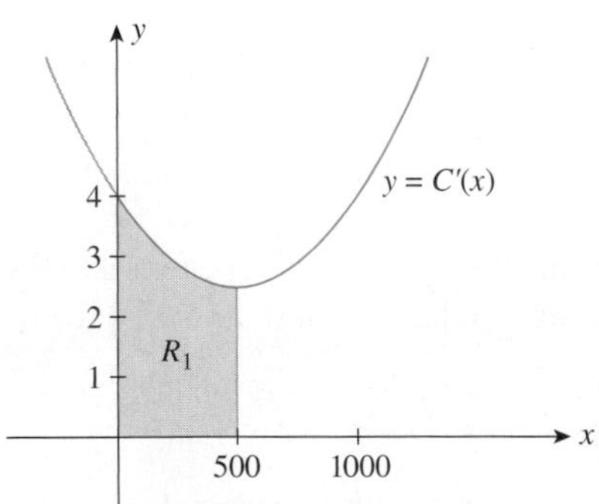

(a) Area of $R_1 = \int_0^{500} C'(x)\,dx$

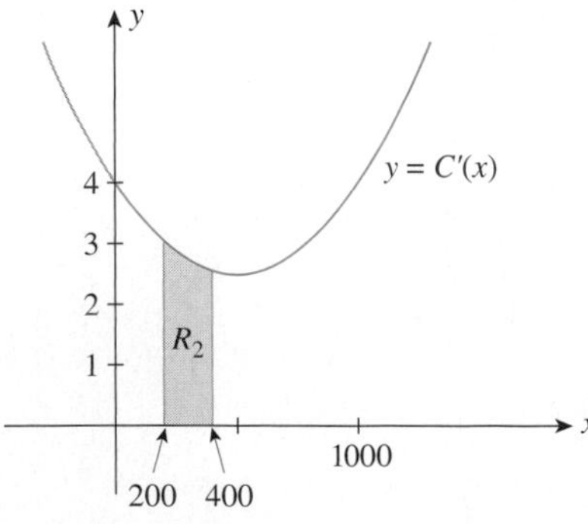

(b) Area of $R_2 = \int_{200}^{400} C'(x)\,dx$

FIGURE 19

VIDEO

APPLIED EXAMPLE 8 Assembly Time of Workers An efficiency study conducted for Elektra Electronics showed that the rate at which Space Commander walkie-talkies are assembled by the average worker t hours after starting work at 8 a.m. is given by the function

$$f(t) = -3t^2 + 12t + 15 \qquad (0 \leq t \leq 4)$$

Determine how many walkie-talkies can be assembled by the average worker in the first hour of the morning shift.

Solution Let $N(t)$ denote the number of walkie-talkies assembled by the average worker t hours after starting work in the morning shift. Then, we have

$$N'(t) = f(t) = -3t^2 + 12t + 15$$

Therefore, the number of units assembled by the average worker in the first hour of the morning shift is

$$\begin{aligned} N(1) - N(0) &= \int_0^1 N'(t)\,dt = \int_0^1 (-3t^2 + 12t + 15)\,dt \\ &= -t^3 + 6t^2 + 15t \Big|_0^1 = -1 + 6 + 15 \\ &= 20 \end{aligned}$$

or 20 units. ■

EXPLORING WITH TECHNOLOGY

You can demonstrate graphically that $\int_0^x t\,dt = \frac{1}{2}x^2$ as follows:

1. Plot the graphs of $y1 = \text{fnInt}(t, t, 0, x) = \int_0^x t\,dt$ and $y2 = \frac{1}{2}x^2$ on the same set of axes, using the viewing window $[-5, 5] \times [0, 10]$.
2. Compare the graphs of $y1$ and $y2$ and draw the desired conclusion.

APPLIED EXAMPLE 9 Projected Demand for Electricity A certain city's rate of electricity consumption is expected to grow exponentially with a growth constant of $k = 0.04$. If the present rate of consumption is 40 million kilowatt-hours (kWh) per year, what should be the total production of electricity over the next 3 years in order to meet the projected demand?

Solution If $R(t)$ denotes the expected rate of consumption of electricity t years from now, then

$$R(t) = 40e^{0.04t}$$

million kWh per year. Next, if $C(t)$ denotes the expected total consumption of electricity over a period of t years, then

$$C'(t) = R(t)$$

Therefore, the total consumption of electricity expected over the next 3 years is given by

EXPLORE & DISCUSS

The definite integral $\int_{-3}^{3} \sqrt{9 - x^2}\,dx$ cannot be evaluated using the fundamental theorem of calculus because the method of this section does not enable us to find an antiderivative of the integrand. But the integral can be evaluated by interpreting it as the area of a certain plane region. What is the region? And what is the value of the integral?

$$\int_0^3 C'(t)\,dt = \int_0^3 40e^{0.04t}\,dt$$
$$= \frac{40}{0.04}e^{0.04t}\Big|_0^3$$
$$= 1000(e^{0.12} - 1)$$
$$= 127.5$$

or 127.5 million kWh, the amount that must be produced over the next 3 years in order to meet the demand. ■

Validity of the Fundamental Theorem of Calculus

To demonstrate the plausibility of the fundamental theorem of calculus for the case where f is nonnegative on an interval $[a, b]$, let's define an "area function" A as follows. Let $A(t)$ denote the area of the region R under the graph of $y = f(x)$ from $x = a$ to $x = t$, where $a \le t \le b$ (Figure 20).

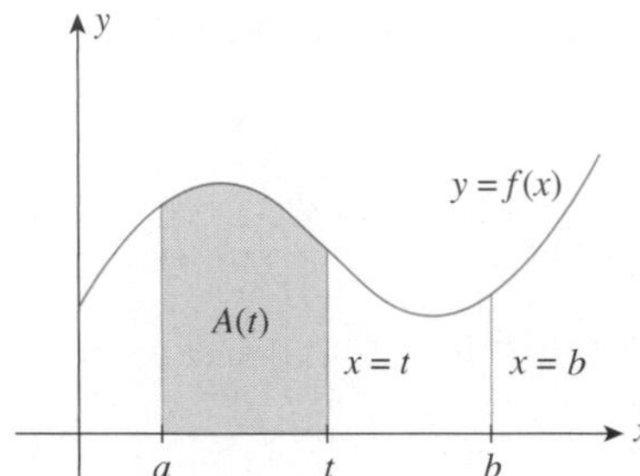

FIGURE 20
$A(t)$ = area under the graph of f from $x = a$ to $x = t$

If h is a small positive number, then $A(t + h)$ is the area of the region under the graph of $y = f(x)$ from $x = a$ to $x = t + h$. Therefore, the difference

$$A(t + h) - A(t)$$

is the area under the graph of $y = f(x)$ from $x = t$ to $x = t + h$ (Figure 21).

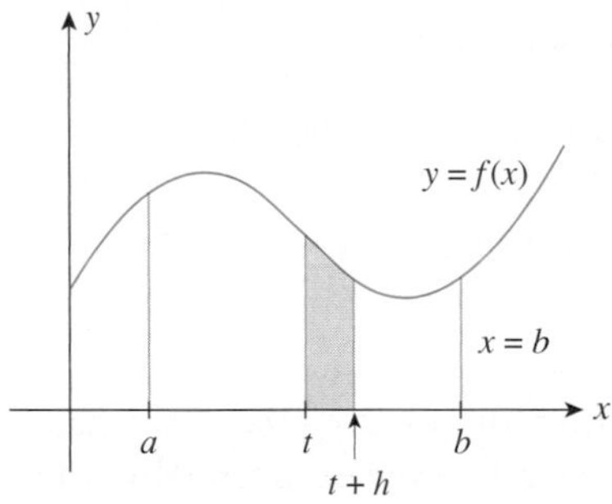

FIGURE 21
$A(t + h) - A(t)$ = area under the graph of f from $x = t$ to $x = t + h$

Now, the area of this last region can be approximated by the area of the rectangle of width h and height $f(t)$—that is, by the expression $h \cdot f(t)$ (Figure 22). Thus,

$$A(t + h) - A(t) \approx h \cdot f(t)$$

where the approximations improve as h is taken to be smaller and smaller.

Dividing both sides of the foregoing relationship by h, we obtain

$$\frac{A(t + h) - A(t)}{h} \approx f(t)$$

Taking the limit as h approaches zero, we find, by the definition of the derivative, that the left-hand side is

$$\lim_{h\to 0}\frac{A(t + h) - A(t)}{h} = A'(t)$$

The right-hand side, which is independent of h, remains constant throughout the limiting process. Because the approximation becomes exact as h approaches zero, we find that

$$A'(t) = f(t)$$

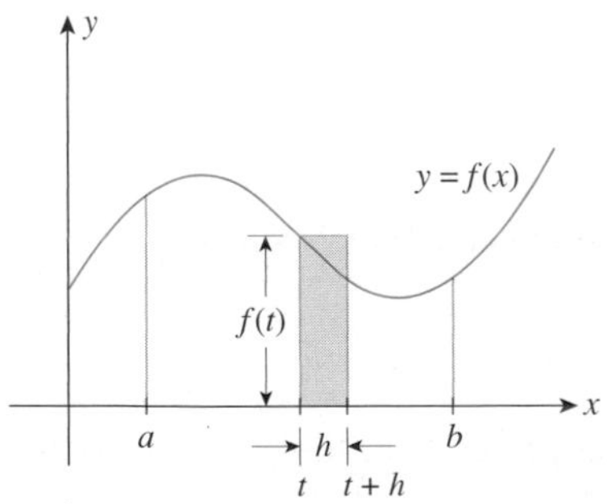

FIGURE 22
The area of the rectangle is $h \cdot f(t)$.

Since the foregoing equation holds for all values of t in the interval $[a, b]$, we have shown that the *area function* A is an antiderivative of the function $f(x)$. By Theorem 1 of Section 6.1, we conclude that $A(x)$ must have the form

$$A(x) = F(x) + C$$

where F is any antiderivative of f and C is a constant. To determine the value of C, observe that $A(a) = 0$. This condition implies that

$$A(a) = F(a) + C = 0$$

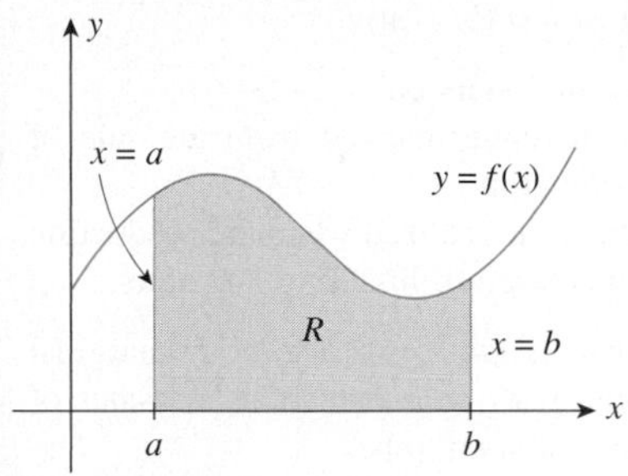

FIGURE 23
The area of R is given by $A(b)$.

or $C = -F(a)$. Next, since the area of the region R is $A(b)$ (Figure 23), we see that the required area is

$$\begin{aligned} A(b) &= F(b) + C \\ &= F(b) - F(a) \end{aligned}$$

Since the area of the region R is

$$\int_a^b f(x)\, dx$$

we have

$$\int_a^b f(x)\, dx = F(b) - F(a)$$

as we set out to show.

6.4 Self-Check Exercises

1. Evaluate $\int_0^2 (x + e^x)\, dx$.
2. The daily marginal profit function associated with producing and selling TexaPep hot sauce is

$$P'(x) = -0.000006x^2 + 6$$

where x denotes the number of cases (each case contains 24 bottles) produced and sold daily and $P'(x)$ is measured in dollars/units. The fixed cost is \$400.

a. What is the total profit realizable from producing and selling 1000 cases of TexaPep per day?
b. What is the additional profit realizable if the production and sale of TexaPep is increased from 1000 to 1200 cases/day?

Solutions to Self-Check Exercises 6.4 can be found on page 439.

6.4 Concept Questions

1. State the fundamental theorem of calculus.
2. State the net change formula and use it to answer the following questions:
   a. If a company generates income at the rate of R dollars/day, explain what $\int_a^b R(t)dt$ measures, where a and b are measured in days with $a < b$.
   b. If a private jet airplane consumes fuel at the rate of R gal/min, write an integral giving the net fuel consumption by the airplane between times $t = a$ and $t = b$ $(a < b)$, where t is measured in minutes.

6.4 Exercises

In Exercises 1–4, find the area of the region under the graph of the function f on the interval $[a, b]$, using the fundamental theorem of calculus. Then verify your result using geometry.

1. $f(x) = 2;\ [1, 4]$
2. $f(x) = 4;\ [-1, 2]$
3. $f(x) = 2x;\ [1, 3]$
4. $f(x) = -\frac{1}{4}x + 1;\ [1, 4]$

In Exercises 5–16, find the area of the region under the graph of the function f on the interval $[a, b]$.

5. $f(x) = 2x + 3;\ [-1, 2]$
6. $f(x) = 4x - 1;\ [2, 4]$
7. $f(x) = -x^2 + 4;\ [-1, 2]$
8. $f(x) = 4x - x^2;\ [0, 4]$
9. $f(x) = \frac{1}{x};\ [1, 2]$
10. $f(x) = \frac{1}{x^2};\ [2, 4]$

11. $f(x) = \sqrt{x}$; $[1, 9]$
12. $f(x) = x^3$; $[1, 3]$
13. $f(x) = 1 - \sqrt[3]{x}$; $[-8, -1]$
14. $f(x) = \dfrac{1}{\sqrt{x}}$; $[1, 9]$
15. $f(x) = e^x$; $[0, 2]$
16. $f(x) = e^x - x$; $[1, 2]$

In Exercises 17–40, evaluate the definite integral.

17. $\displaystyle\int_2^4 3\,dx$
18. $\displaystyle\int_{-1}^2 -2\,dx$
19. $\displaystyle\int_1^3 (2x + 3)\,dx$
20. $\displaystyle\int_{-1}^0 (4 - x)\,dx$
21. $\displaystyle\int_{-1}^3 2x^2\,dx$
22. $\displaystyle\int_0^2 8x^3\,dx$
23. $\displaystyle\int_{-2}^2 (x^2 - 1)\,dx$
24. $\displaystyle\int_1^4 \sqrt{u}\,du$
25. $\displaystyle\int_1^8 4x^{1/3}\,dx$
26. $\displaystyle\int_1^4 2x^{-3/2}\,dx$
27. $\displaystyle\int_0^1 (x^3 - 2x^2 + 1)\,dx$
28. $\displaystyle\int_1^2 (t^5 - t^3 + 1)\,dt$
29. $\displaystyle\int_2^4 \frac{1}{x}\,dx$
30. $\displaystyle\int_1^3 \frac{2}{x}\,dx$
31. $\displaystyle\int_0^4 x(x^2 - 1)\,dx$
32. $\displaystyle\int_0^2 (x - 4)(x - 1)\,dx$
33. $\displaystyle\int_1^3 (t^2 - t)^2\,dt$
34. $\displaystyle\int_{-1}^1 (x^2 - 1)^2\,dx$
35. $\displaystyle\int_{-3}^{-1} \frac{1}{x^2}\,dx$
36. $\displaystyle\int_1^2 \frac{2}{x^3}\,dx$
37. $\displaystyle\int_1^4 \left(\sqrt{x} - \frac{1}{\sqrt{x}}\right)dx$
38. $\displaystyle\int_0^1 \sqrt{2x}(\sqrt{x} + \sqrt{2})\,dx$
39. $\displaystyle\int_1^4 \frac{3x^3 - 2x^2 + 4}{x^2}\,dx$
40. $\displaystyle\int_1^2 \left(1 + \frac{1}{u} + \frac{1}{u^2}\right)du$

41. MARGINAL COST A division of Ditton Industries manufactures a deluxe toaster oven. Management has determined that the daily marginal cost function associated with producing these toaster ovens is given by

$$C'(x) = 0.0003x^2 - 0.12x + 20$$

where $C'(x)$ is measured in dollars/unit and x denotes the number of units produced. Management has also determined that the daily fixed cost incurred in the production is \$800.
   a. Find the total cost incurred by Ditton in producing the first 300 units of these toaster ovens per day.
   b. What is the total cost incurred by Ditton in producing the 201st through 300th units/day?

42. MARGINAL REVENUE The management of Ditton Industries has determined that the daily marginal revenue function associated with selling x units of their deluxe toaster ovens is given by

$$R'(x) = -0.1x + 40$$

where $R'(x)$ is measured in dollars/unit.
   a. Find the daily total revenue realized from the sale of 200 units of the toaster oven.
   b. Find the additional revenue realized when the production (and sales) level is increased from 200 to 300 units.

43. MARGINAL PROFIT Refer to Exercise 41. The daily marginal profit function associated with the production and sales of the deluxe toaster ovens is known to be

$$P'(x) = -0.0003x^2 + 0.02x + 20$$

where x denotes the number of units manufactured and sold daily and $P'(x)$ is measured in dollars/unit.
   a. Find the total profit realizable from the manufacture and sale of 200 units of the toaster ovens per day.
   Hint: $P(200) - P(0) = \int_0^{200} P'(x)\,dx$, $P(0) = -800$.
   b. What is the additional daily profit realizable if the production and sale of the toaster ovens are increased from 200 to 220 units/day?

44. EFFICIENCY STUDIES Tempco Electronics, a division of Tempco Toys, manufactures an electronic football game. An efficiency study showed that the rate at which the games are assembled by the average worker t hr after starting work at 8 a.m. is

$$-\frac{3}{2}t^2 + 6t + 20 \qquad (0 \le t \le 4)$$

units/hour.
   a. Find the total number of games the average worker can be expected to assemble in the 4-hr morning shift.
   b. How many units can the average worker be expected to assemble in the first hour of the morning shift? In the second hour of the morning shift?

45. SPEEDBOAT RACING In a recent pretrial run for the world water speed record, the velocity of the *Sea Falcon II* t sec after firing the booster rocket was given by

$$v(t) = -t^2 + 20t + 440 \qquad (0 \le t \le 20)$$

ft/sec. Find the distance covered by the boat over the 20-sec period after the booster rocket was activated.
Hint: The distance is given by $\int_0^{20} v(t)\,dt$.

46. POCKET COMPUTERS Annual sales (in millions of units) of pocket computers are expected to grow in accordance with the function

$$f(t) = 0.18t^2 + 0.16t + 2.64 \qquad (0 \le t \le 6)$$

where t is measured in years, with $t = 0$ corresponding to 1997. How many pocket computers were sold over the 6-yr period between the beginning of 1997 and the end of 2002?
Source: Dataquest, Inc.

47. SINGLE FEMALE-HEADED HOUSEHOLDS WITH CHILDREN The percent of families with children that are headed by single females grew at the rate of

$$R(t) = 0.8499t^2 - 3.872t + 5 \qquad (0 \le t \le 3)$$

households/decade between 1970 ($t = 0$) and 2000 ($t = 3$). The number of such households stood at 5.6% of all families in 1970.

a. Find an expression giving the percent of these households in the tth decade.

b. If the trend continued, estimate the percent of these households in 2010.

c. What was the net increase in the percent of these households from 1970 to 2000?

Source: U.S. Census Bureau

48. AIR PURIFICATION To test air purifiers, engineers ran a purifier in a smoke-filled 10-ft × 20-ft room. While conducting a test for a certain brand of air purifier, it was determined that the amount of smoke in the room was decreasing at the rate of

$$R(t) = 0.00032t^4 - 0.01872t^3 + 0.3948t^2 - 3.83t + 17.63 \qquad (0 \le t \le 20)$$

percent of the (original) amount of the smoke per minute, t min after the start of the test. How much smoke was left in the room 5 min after the start of the test? Ten minutes after the start of the test?

Source: Consumer Reports

49. SENIOR CITIZENS The population aged 65 years old and older (in millions) from 2000 to 2050 is projected to be

$$f(t) = \frac{85}{1 + 1.859e^{-0.66t}} \qquad (0 \le t \le 5)$$

where t is measured in decades, with $t = 0$ corresponding to 2000. What will be the average population aged 65 years and older over the years from 2000 to 2030?

Hint: The average population is given by $(1/3)\int_0^3 f(t)\,dt$. Multiply the integrand by $e^{0.66t}/e^{0.66t}$ and then use the method of substitution.

50. BLOOD FLOW Consider an artery of length L cm and radius R cm. Using Poiseuille's law (page 62), it can be shown that the rate at which blood flows through the artery (measured in cubic centimeters/second) is given by

$$V = \int_0^R \frac{k}{L}x(R^2 - x^2)\,dx$$

where k is a constant. Find an expression for V that does *not* involve an integral.

Hint: Use the substitution $u = R^2 - x^2$.

51. Find the area of the region bounded by the graph of the function $f(x) = x^4 - 2x^2 + 2$, the x-axis, and the lines $x = a$ and $x = b$, where $a < b$ and a and b are the x-coordinates of the relative maximum point and a relative minimum point of f, respectively.

52. Find the area of the region bounded by the graph of the function $f(x) = (x + 1)/\sqrt{x}$, the x axis, and the lines $x = a$ and $x = b$ where a and b are, respectively, the x-coordinates of the relative minimum point and the inflection point of f.

In Exercises 53–56, determine whether the statement is true or false. If it is true, explain why it is true. If it is false, give an example to show why it is false.

53. $\displaystyle\int_{-1}^{1} \frac{1}{x^3}\,dx = -\frac{1}{2x^2}\Big|_{-1}^{1} = -\frac{1}{2} - \left(-\frac{1}{2}\right) = 0$

54. $\displaystyle\int_{-1}^{1} \frac{1}{x}\,dx = \ln|x|\Big|_{-1}^{1} = \ln|1| - \ln|-1| = \ln 1 - \ln 1 = 0$

55. $\int_0^2 (1 - x)\,dx$ gives the area of the region under the graph of $f(x) = 1 - x$ on the interval $[0, 2]$.

56. The total revenue realized in selling the first 5000 units of a product is given by

$$\int_0^{500} R'(x)\,dx = R(500) - R(0)$$

where $R(x)$ is the total revenue.

6.4 Solutions to Self-Check Exercises

1.
$$\begin{aligned}\int_0^2 (x + e^x)\,dx &= \frac{1}{2}x^2 + e^x\Big|_0^2 \\ &= \left[\frac{1}{2}(2)^2 + e^2\right] - \left[\frac{1}{2}(0) + e^0\right] \\ &= 2 + e^2 - 1 \\ &= e^2 + 1\end{aligned}$$

2. a. We want $P(1000)$, but

$$\begin{aligned}P(1000) - P(0) &= \int_0^{1000} P'(x)\,dx = \int_0^{1000} (-0.000006x^2 + 6)\,dx \\ &= -0.000002x^3 + 6x\Big|_0^{1000} \\ &= -0.000002(1000)^3 + 6(1000) \\ &= 4000\end{aligned}$$

So, $P(1000) = 4000 + P(0) = 4000 - 400$, or \$3600/day $[P(0) = -C(0)]$.

b. The additional profit realizable is given by

$$\begin{aligned}\int_{1000}^{1200} P'(x)\,dx &= -0.000002x^3 + 6x\Big|_{1000}^{1200} \\ &= [-0.000002(1200)^3 + 6(1200)] \\ &\quad - [-0.000002(1000)^3 + 6(1000)] \\ &= 3744 - 4000 \\ &= -256\end{aligned}$$

That is, the company sustains a loss of \$256/day if production is increased to 1200 cases/day.

USING TECHNOLOGY

Evaluating Definite Integrals

Some graphing utilities have an operation for finding the definite integral of a function. If your graphing utility has this capability, use it to work through the example and exercises of this section.

EXAMPLE 1 Use the numerical integral operation of a graphing utility to evaluate

$$\int_{-1}^{2} \frac{2x + 4}{(x^2 + 1)^{3/2}}\, dx$$

Solution Using the numerical integral operation of a graphing utility, we find

$$\int_{-1}^{2} \frac{2x + 4}{(x^2 + 1)^{3/2}}\, dx = \text{fnInt}((2x + 4)/(x^2 + 1)\text{^}1.5, x, -1, 2) \approx 6.92592226$$ ■

TECHNOLOGY EXERCISES

In Exercises 1–4, find the area of the region under the graph of f on the interval $[a, b]$. Express your answer to four decimal places.

1. $f(x) = 0.002x^5 + 0.032x^4 - 0.2x^2 + 2$; $[-1.1, 2.2]$
2. $f(x) = x\sqrt{x^3 + 1}$; $[1, 2]$
3. $f(x) = \sqrt{x}e^{-x}$; $[0, 3]$
4. $f(x) = \dfrac{\ln x}{\sqrt{1 + x^2}}$; $[1, 2]$

In Exercises 5–10, evaluate the definite integral.

5. $\displaystyle\int_{-1.2}^{2.3} (0.2x^4 - 0.32x^3 + 1.2x - 1)\, dx$
6. $\displaystyle\int_{1}^{3} x(x^4 - 1)^{3.2}\, dx$
7. $\displaystyle\int_{0}^{2} \frac{3x^3 + 2x^2 + 1}{2x^2 + 3}\, dx$
8. $\displaystyle\int_{1}^{2} \frac{\sqrt{x} + 1}{2x^2 + 1}\, dx$
9. $\displaystyle\int_{0}^{2} \frac{e^x}{\sqrt{x^2 + 1}}\, dx$
10. $\displaystyle\int_{1}^{3} e^{-x} \ln(x^2 + 1)\, dx$
11. Rework Exercise 48, Exercises 6.4.
12. Rework Exercise 49, Exercises 6.4.
13. THE GLOBAL EPIDEMIC The number of AIDS-related deaths/year in the United States is given by the function

$$f(t) = -53.254t^4 + 673.7t^3 - 2801.07t^2 + 8833.379t + 20{,}000 \quad (0 \le t \le 9)$$

where $t = 0$ corresponds to the beginning of 1988. Find the total number of AIDS-related deaths in the United States between the beginning of 1988 and the end of 1996.
Source: Centers for Disease Control

14. MARIJUANA ARRESTS The number of arrests for marijuana sales and possession in New York City grew at the rate of approximately

$$f(t) = 0.0125t^4 - 0.01389t^3 + 0.55417t^2 + 0.53294t + 4.95238 \quad (0 \le t \le 5)$$

thousand/year, where t is measured in years, with $t = 0$ corresponding to the beginning of 1992. Find the approximate number of marijuana arrests in the city from the beginning of 1992 to the end of 1997.
Source: State Division of Criminal Justice Services

15. POPULATION GROWTH The population of a certain city is projected to grow at the rate of $18\sqrt{t + 1}\ln\sqrt{t + 1}$ thousand people/year t yr from now. If the current population is 800,000, what will be the population 45 yr from now?

6.5 Evaluating Definite Integrals

This section continues our discussion of the applications of the fundamental theorem of calculus.

Properties of the Definite Integral

Before going on, we list the following useful properties of the definite integral, some of which parallel the rules of integration of Section 6.1.

Properties of the Definite Integral

Let f and g be integrable functions; then,

1. $\int_a^a f(x)\,dx = 0$
2. $\int_a^b f(x)\,dx = -\int_b^a f(x)\,dx$
3. $\int_a^b cf(x)\,dx = c\int_a^b f(x)\,dx \quad (c, \text{ a constant})$
4. $\int_a^b [f(x) \pm g(x)]\,dx = \int_a^b f(x)\,dx \pm \int_a^b g(x)\,dx$
5. $\int_a^b f(x)\,dx = \int_a^c f(x)\,dx + \int_c^b f(x)\,dx \quad (a < c < b)$

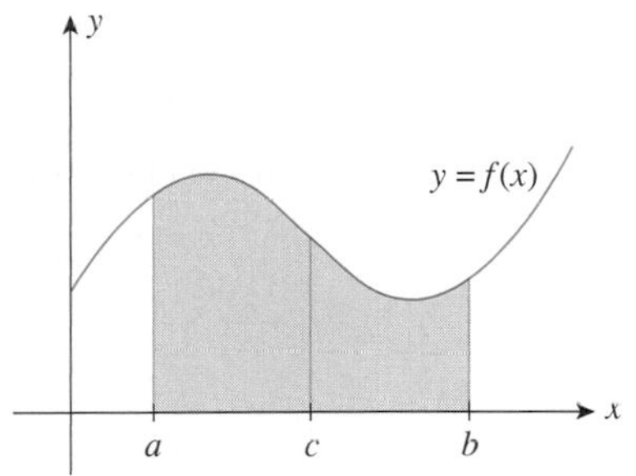

FIGURE 24
$\int_a^b f(x)\,dx = \int_a^c f(x)\,dx + \int_c^b f(x)\,dx$

Property 5 states that if c is a number lying between a and b so that the interval $[a, b]$ is divided into the intervals $[a, c]$ and $[c, b]$, then the integral of f over the interval $[a, b]$ may be expressed as the sum of the integral of f over the interval $[a, c]$ and the integral of f over the interval $[c, b]$.

Property 5 has the following geometric interpretation when f is nonnegative. By definition

$$\int_a^b f(x)\,dx$$

is the area of the region under the graph of $y = f(x)$ from $x = a$ to $x = b$ (Figure 24). Similarly, we interpret the definite integrals

$$\int_a^c f(x)\,dx \quad \text{and} \quad \int_c^b f(x)\,dx$$

as the areas of the regions under the graph of $y = f(x)$ from $x = a$ to $x = c$ and from $x = c$ to $x = b$, respectively. Since the two regions do not overlap, we see that

$$\int_a^b f(x)\,dx = \int_a^c f(x)\,dx + \int_c^b f(x)\,dx$$

The Method of Substitution for Definite Integrals

Our first example shows two approaches generally used when evaluating a definite integral using the method of substitution.

EXAMPLE 1 Evaluate $\int_0^4 x\sqrt{9 + x^2}\,dx$.

Solution

Method 1 We first find the corresponding indefinite integral:

$$I = \int x\sqrt{9 + x^2}\,dx$$

Make the substitution $u = 9 + x^2$ so that

$$\begin{aligned} du &= \frac{d}{dx}(9 + x^2)\,dx \\ &= 2x\,dx \\ x\,dx &= \frac{1}{2}\,du \qquad \text{Divide both sides by 2.} \end{aligned}$$

Then,

$$\begin{aligned} I &= \int \frac{1}{2}\sqrt{u}\,du = \frac{1}{2}\int u^{1/2}\,du \\ &= \frac{1}{3}u^{3/2} + C = \frac{1}{3}(9 + x^2)^{3/2} + C \qquad \begin{array}{l}\text{Substitute}\\ 9 + x^2 \text{ for } u.\end{array} \end{aligned}$$

Using this result, we now evaluate the given definite integral:

$$\begin{aligned} \int_0^4 x\sqrt{9 + x^2}\,dx &= \frac{1}{3}(9 + x^2)^{3/2}\Big|_0^4 \\ &= \frac{1}{3}[(9 + 16)^{3/2} - 9^{3/2}] \\ &= \frac{1}{3}(125 - 27) = \frac{98}{3} = 32\tfrac{2}{3} \end{aligned}$$

Method 2 *Changing the Limits of Integration.* As before, we make the substitution

$$u = 9 + x^2 \tag{11}$$

so that

$$\begin{aligned} du &= 2x\,dx \\ x\,dx &= \frac{1}{2}\,du \end{aligned}$$

Next, observe that the given definite integral is evaluated *with respect to x* with the range of integration given by the interval [0, 4]. If we perform the integration *with respect to u* via the substitution (11), then we must adjust the range of integration to reflect the fact that the integration is being performed with respect to the new variable u. To determine the proper range of integration, note that when $x = 0$, Equation (11) implies that

$$u = 9 + 0^2 = 9$$

which gives the required lower limit of integration with respect to u. Similarly, when $x = 4$,

$$u = 9 + 16 = 25$$

is the required upper limit of integration with respect to u. Thus, the range of integration when the integration is performed with respect to u is given by the interval [9, 25]. Therefore, we have

$$\begin{aligned}\int_0^4 x\sqrt{9 + x^2}\,dx &= \int_9^{25} \frac{1}{2}\sqrt{u}\,du = \frac{1}{2}\int_9^{25} u^{1/2}\,du \\ &= \frac{1}{3}u^{3/2}\Big|_9^{25} = \frac{1}{3}(25^{3/2} - 9^{3/2}) \\ &= \frac{1}{3}(125 - 27) = \frac{98}{3} = 32\tfrac{2}{3}\end{aligned}$$

which agrees with the result obtained using Method 1. ■

When you use the method of substitution, make sure you adjust the limits of integration to reflect integrating with respect to the new variable u.

EXPLORING WITH TECHNOLOGY

Refer to Example 1. You can confirm the results obtained there by using a graphing utility as follows:

1. Use the numerical integration operation of the graphing utility to evaluate

$$\int_0^4 x\sqrt{9 + x^2}\,dx$$

2. Evaluate $\dfrac{1}{2}\displaystyle\int_9^{25} \sqrt{u}\,du$.

3. Conclude that $\displaystyle\int_0^4 x\sqrt{9 + x^2}\,dx = \frac{1}{2}\int_9^{25} \sqrt{u}\,du$.

EXAMPLE 2 Evaluate $\int_0^2 xe^{2x^2}\,dx$.

Solution Let $u = 2x^2$ so that $du = 4x\,dx$, or $x\,dx = \frac{1}{4}\,du$. When $x = 0$, $u = 0$, and when $x = 2$, $u = 8$. This gives the lower and upper limits of integration with respect to u. Making the indicated substitutions, we find

$$\int_0^2 xe^{2x^2}\,dx = \int_0^8 \frac{1}{4}e^u\,du = \frac{1}{4}e^u\Big|_0^8 = \frac{1}{4}(e^8 - 1)$$ ■

EXAMPLE 3 Evaluate $\displaystyle\int_0^1 \frac{x^2}{x^3 + 1}\,dx$.

Solution Let $u = x^3 + 1$ so that $du = 3x^2\,dx$, or $x^2\,dx = \frac{1}{3}\,du$. When $x = 0$, $u = 1$, and when $x = 1$, $u = 2$. This gives the lower and upper limits of integration with respect to u. Making the indicated substitutions, we find

$$\begin{aligned}\int_0^1 \frac{x^2}{x^3 + 1}\,dx &= \frac{1}{3}\int_1^2 \frac{du}{u} = \frac{1}{3}\ln|u|\Big|_1^2 \\ &= \frac{1}{3}(\ln 2 - \ln 1) = \frac{1}{3}\ln 2\end{aligned}$$ ■

Finding the Area under a Curve

EXAMPLE 4 Find the area of the region R under the graph of $f(x) = e^{(1/2)x}$ from $x = -1$ to $x = 1$.

Solution The region R is shown in Figure 25. Its area is given by

$$A = \int_{-1}^{1} e^{(1/2)x}\, dx$$

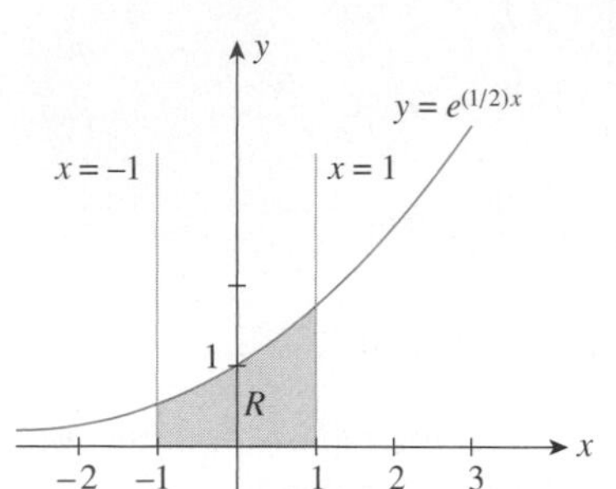

FIGURE 25
Area of $R = \int_{-1}^{1} e^{(1/2)x}\, dx$

To evaluate this integral, we make the substitution

$$u = \frac{1}{2}x$$

so that

$$du = \frac{1}{2}\, dx$$

$$dx = 2\, du$$

When $x = -1$, $u = -\frac{1}{2}$, and when $x = 1$, $u = \frac{1}{2}$. Making the indicated substitutions, we obtain

$$A = \int_{-1}^{1} e^{(1/2)x}\, dx = 2\int_{-1/2}^{1/2} e^u\, du$$

$$= 2e^u \Big|_{-1/2}^{1/2} = 2(e^{1/2} - e^{-1/2})$$

or approximately 2.08 square units. ■

EXPLORE & DISCUSS

Let f be a function defined piecewise by the rule

$$f(x) = \begin{cases} \sqrt{x} & \text{if } 0 \le x \le 1 \\ \dfrac{1}{x} & \text{if } 1 < x \le 2 \end{cases}$$

How would you use Property 5 of definite integrals to find the area of the region under the graph of f on $[0, 2]$? What is the area?

Average Value of a Function

The *average value* of a function over an interval provides us with an application of the definite integral. Recall that the average value of a set of n numbers is the number

$$\frac{y_1 + y_2 + \cdots + y_n}{n}$$

Now, suppose f is a continuous function defined on $[a, b]$. Let's divide the interval $[a, b]$ into n subintervals of equal length $(b - a)/n$. Choose points $x_1, x_2, \ldots, x_n$ in the first, second, . . . , and nth subintervals, respectively. Then, the average value of the numbers $f(x_1), f(x_2), \ldots, f(x_n)$, given by

$$\frac{f(x_1) + f(x_2) + \cdots + f(x_n)}{n}$$

is an approximation of the average of all the values of $f(x)$ on the interval $[a, b]$. This expression can be written in the form

$$\begin{aligned}
&\frac{(b-a)}{(b-a)}\left[f(x_1)\cdot\frac{1}{n}+f(x_2)\cdot\frac{1}{n}+\cdots+f(x_n)\cdot\frac{1}{n}\right]\\
&\quad=\frac{1}{b-a}\left[f(x_1)\cdot\frac{b-a}{n}+f(x_2)\cdot\frac{b-a}{n}+\cdots+f(x_n)\cdot\frac{b-a}{n}\right]\\
&\quad=\frac{1}{b-a}\left[f(x_1)\Delta x+f(x_2)\Delta x+\cdots+f(x_n)\Delta x\right] \qquad \textbf{(12)}
\end{aligned}$$

As n gets larger and larger, the expression (12) approximates the average value of $f(x)$ over $[a, b]$ with increasing accuracy. But the sum inside the brackets in (12) is a Riemann sum of the function f over $[a, b]$. In view of this, we have

$$\begin{aligned}
&\lim_{n\to\infty}\left[\frac{f(x_1)+f(x_2)+\cdots+f(x_n)}{n}\right]\\
&\quad=\frac{1}{b-a}\lim_{n\to\infty}\left[f(x_1)\Delta x+f(x_2)\Delta x+\cdots+f(x_n)\Delta x\right]\\
&\quad=\frac{1}{b-a}\int_a^b f(x)\,dx
\end{aligned}$$

This discussion motivates the following definition.

The Average Value of a Function

Suppose f is integrable on $[a, b]$. Then the **average value** of f over $[a, b]$ is

$$\frac{1}{b-a}\int_a^b f(x)\,dx$$

EXAMPLE 5 Find the average value of the function $f(x) = \sqrt{x}$ over the interval $[0, 4]$.

Solution The required average value is given by

$$\begin{aligned}
\frac{1}{4-0}\int_0^4 \sqrt{x}\,dx &= \frac{1}{4}\int_0^4 x^{1/2}\,dx\\
&= \frac{1}{6}x^{3/2}\Big|_0^4 = \frac{1}{6}(4^{3/2})\\
&= \frac{4}{3}
\end{aligned}$$

■

APPLIED EXAMPLE 6 Automobile Financing The interest rates charged by Madison Finance on auto loans for used cars over a certain 6-month period in 2000 are approximated by the function

$$r(t) = -\frac{1}{12}t^3 + \frac{7}{8}t^2 - 3t + 12 \qquad (0 \le t \le 6)$$

where t is measured in months and $r(t)$ is the annual percentage rate. What is the average rate on auto loans extended by Madison over the 6-month period?

Solution The average rate over the 6-month period in question is given by

$$\begin{aligned}\frac{1}{6-0}\int_0^6 \left(-\frac{1}{12}t^3+\frac{7}{8}t^2-3t+12\right)dt &= \frac{1}{6}\left(-\frac{1}{48}t^4+\frac{7}{24}t^3-\frac{3}{2}t^2+12t\right)\Bigg|_0^6 \\ &= \frac{1}{6}\left[-\frac{1}{48}(6^4)+\frac{7}{24}(6^3)-\frac{3}{2}(6^2)+12(6)\right] \\ &= 9\end{aligned}$$

or 9% per year. ■

APPLIED EXAMPLE 7 Drug Concentration in a Body The amount of a certain drug in a patient's body t days after it has been administered is

$$C(t) = 5e^{-0.2t}$$

units. Determine the average amount of the drug present in the patient's body for the first 4 days after the drug has been administered.

Solution The average amount of the drug present in the patient's body for the first 4 days after it has been administered is given by

$$\begin{aligned}\frac{1}{4-0}\int_0^4 5e^{-0.2t}\,dt &= \frac{5}{4}\int_0^4 e^{-0.2t}dt \\ &= \frac{5}{4}\left[\left(-\frac{1}{0.2}\right)e^{-0.2t}\Bigg|_0^4\right] \\ &= \frac{5}{4}(-5e^{-0.8}+5) \\ &\approx 3.44\end{aligned}$$

or approximately 3.44 units. ■

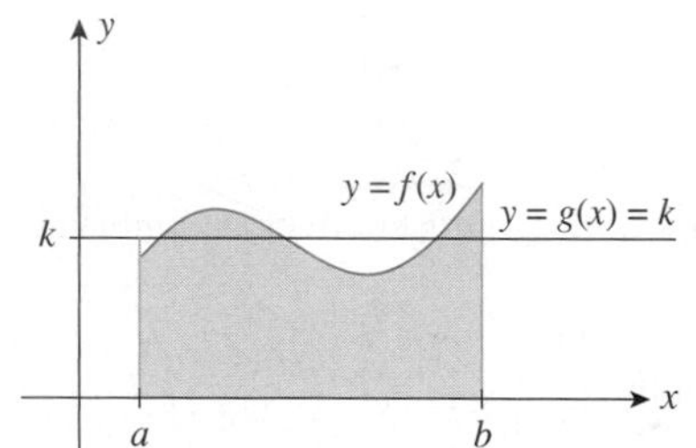

FIGURE 26
The average value of f over $[a, b]$ is k.

We now give a geometric interpretation of the average value of a function f over an interval $[a, b]$. Suppose $f(x)$ is nonnegative so that the definite integral

$$\int_a^b f(x)\,dx$$

gives the area under the graph of f from $x = a$ to $x = b$ (Figure 26). Observe that, in general, the "height" $f(x)$ varies from point to point. Can we replace $f(x)$ by a constant function $g(x) = k$ (which has constant height) such that the areas under each of the two functions f and g are the same? If so, since the area under the graph of g from $x = a$ to $x = b$ is $k(b - a)$, we have

$$k(b-a) = \int_a^b f(x)\,dx$$

$$k = \frac{1}{b-a}\int_a^b f(x)\,dx$$

so that k is the average value of f over $[a, b]$. Thus, the average value of a function f over an interval $[a, b]$ is the height of a rectangle with base of length $(b - a)$ that has the same area as that of the region under the graph of f from $x = a$ to $x = b$.

6.5 Self-Check Exercises

1. Evaluate $\int_0^2 \sqrt{2x + 5}\, dx$.

2. Find the average value of the function $f(x) = 1 - x^2$ over the interval $[-1, 2]$.

3. The median price of a house in a southwestern state between January 1, 2000, and January 1, 2005, is approximated by the function

$$f(t) = t^3 - 7t^2 + 17t + 280 \qquad (0 \le t \le 5)$$

where $f(t)$ is measured in thousands of dollars and t is expressed in years, with $t = 0$ corresponding to the beginning of 2000. Determine the average median price of a house over that time interval.

Solutions to Self-Check Exercises 6.5 can be found on page 450.

6.5 Concept Questions

1. Describe two approaches used to evaluate a definite integral using the method of substitution. Illustrate with the integral $\int_0^1 x^2(x^3 + 1)^2\, dx$.

2. Define the average value of a function f over an interval $[a, b]$. Give a geometric interpretation.

6.5 Exercises

In Exercises 1–28, evaluate the definite integral.

1. $\int_0^2 x(x^2 - 1)^3\, dx$

2. $\int_0^1 x^2(2x^3 - 1)^4\, dx$

3. $\int_0^1 x\sqrt{5x^2 + 4}\, dx$

4. $\int_1^3 x\sqrt{3x^2 - 2}\, dx$

5. $\int_0^2 x^2(x^3 + 1)^{3/2}\, dx$

6. $\int_1^5 (2x - 1)^{5/2}\, dx$

7. $\int_0^1 \frac{1}{\sqrt{2x + 1}}\, dx$

8. $\int_0^2 \frac{x}{\sqrt{x^2 + 5}}\, dx$

9. $\int_1^2 (2x - 1)^4\, dx$

10. $\int_1^2 (2x + 4)(x^2 + 4x - 8)^3\, dx$

11. $\int_{-1}^1 x^2(x^3 + 1)^4\, dx$

12. $\int_1^2 \left(x^3 + \frac{3}{4}\right)(x^4 + 3x)^{-2}\, dx$

13. $\int_1^5 x\sqrt{x - 1}\, dx$

14. $\int_1^4 x\sqrt{x + 1}\, dx$
Hint: Let $u = x + 1$.

15. $\int_0^2 xe^{x^2}\, dx$

16. $\int_0^1 e^{-x}\, dx$

17. $\int_0^1 (e^{2x} + x^2 + 1)\, dx$

18. $\int_0^2 (e^t - e^{-t})\, dt$

19. $\int_{-1}^1 xe^{x^2+1}\, dx$

20. $\int_0^4 \frac{e^{\sqrt{x}}}{\sqrt{x}}\, dx$

21. $\int_3^6 \frac{2}{x - 2}\, dx$

22. $\int_0^1 \frac{x}{1 + 2x^2}\, dx$

23. $\int_1^2 \frac{x^2 + 2x}{x^3 + 3x^2 - 1}\, dx$

24. $\int_0^1 \frac{e^x}{1 + e^x}\, dx$

25. $\int_1^2 \left(4e^{2u} - \frac{1}{u}\right) du$

26. $\int_1^2 \left(1 + \frac{1}{x} + e^x\right) dx$

27. $\int_1^2 \left(2e^{-4x} - \frac{1}{x^2}\right) dx$

28. $\int_1^2 \frac{\ln x}{x}\, dx$

In Exercises 29–38, find the average value of the function f over the indicated interval $[a, b]$.

29. $f(x) = 2x + 3$; $[0, 2]$

30. $f(x) = 8 - x$; $[1, 4]$

31. $f(x) = 2x^2 - 3$; $[1, 3]$

32. $f(x) = 4 - x^2$; $[-2, 3]$

33. $f(x) = x^2 + 2x - 3$; $[-1, 2]$

34. $f(x) = x^3$; $[-1, 1]$

35. $f(x) = \sqrt{2x + 1}$; $[0, 4]$

36. $f(x) = e^{-x}$; $[0, 4]$

37. $f(x) = xe^{x^2}$; $[0, 2]$

38. $f(x) = \frac{1}{x + 1}$; $[0, 2]$

39. World Production of Coal A study proposed in 1980 by researchers from the major producers and consumers of the world's coal concluded that coal could and must play an important role in fueling global economic growth over the next 20 yr. The world production of coal in 1980 was 3.5 bil-

lion metric tons. If output increased at the rate of $3.5e^{0.05t}$ billion metric tons/year in year t ($t = 0$ corresponding to 1980), determine how much coal was produced worldwide between 1980 and the end of the 20th century.

40. Newton's Law of Cooling A bottle of white wine at room temperature (68°F) is placed in a refrigerator at 4 p.m. Its temperature after t hr is changing at the rate of

$$-18e^{-0.6t}$$

°F/hr. By how many degrees will the temperature of the wine have dropped by 7 p.m.? What will the temperature of the wine be at 7 p.m.?

41. Net Investment Flow The net investment flow (rate of capital formation) of the giant conglomerate LTF incorporated is projected to be

$$t\sqrt{\frac{1}{2}t^2 + 1}$$

million dollars/year in year t. Find the accruement on the company's capital stock in the second year.
Hint: The amount is given by

$$\int_1^2 t\sqrt{\frac{1}{2}t^2 + 1}\,dt$$

42. Oil Production Based on a preliminary report by a geological survey team, it is estimated that a newly discovered oil field can be expected to produce oil at the rate of

$$R(t) = \frac{600t^2}{t^3 + 32} + 5 \qquad (0 \le t \le 20)$$

thousand barrels/year, t yr after production begins. Find the amount of oil that the field can be expected to yield during the first 5 yr of production, assuming that the projection holds true.

43. Depreciation: Double Declining-Balance Method Suppose a tractor purchased at a price of \$60,000 is to be depreciated by the *double declining-balance method* over a 10-yr period. It can be shown that the rate at which the book value will be decreasing is given by

$$R(t) = 13388.61e^{-0.22314t} \qquad (0 \le t \le 10)$$

dollars/year at year t. Find the amount by which the book value of the tractor will depreciate over the first 5 yr of its life.

44. Velocity of a Car A car moves along a straight road in such a way that its velocity (in feet/second) at any time t (in seconds) is given by

$$v(t) = 3t\sqrt{16 - t^2} \qquad (0 \le t \le 4)$$

Find the distance traveled by the car in the 4 sec from $t = 0$ to $t = 4$.

45. Average Velocity of a Truck A truck traveling along a straight road has a velocity (in feet/second) at time t (in seconds) given by

$$v(t) = \frac{1}{12}t^2 + 2t + 44 \qquad (0 \le t \le 5)$$

What is the average velocity of the truck over the time interval from $t = 0$ to $t = 5$?

46. Membership in Credit Unions Credit unions in Massachusetts have grown remarkably in recent years. Their tax-exempt status allows them to offer deposit and loan rates that are often more favorable than those offered by banks. The membership in Massachusetts credit unions grew at the rate of

$$R(t) = -0.0039t^2 + 0.0374t + 0.0046 \qquad (0 \le t \le 9)$$

million members/year between 1994 ($t = 0$) and 2003 ($t = 9$). Find the average rate of growth of membership in Massachusetts credit unions over the period in question.
Source: Massachusetts Credit Union League

47. Whale Population A group of marine biologists estimates that if certain conservation measures are implemented, the population of an endangered species of whale will be

$$N(t) = 3t^3 + 2t^2 - 10t + 600 \qquad (0 \le t \le 10)$$

where $N(t)$ denotes the population at the end of year t. Find the average population of the whales over the next 10 yr.

48. U.S. Citizens 65 and Older The number of U.S. citizens 65 and older from 1900 through 2050 is estimated to be growing at the rate of

$$R(t) = 0.063t^2 - 0.48t + 3.87 \qquad (0 \le t \le 15)$$

million people/decade, where t is measured in decades and $t = 0$ corresponds to 1900. Show that the average rate of growth of U.S. citizens 65 and older between 2000 and 2050 will be growing at twice the rate of that between 1950 and 2000.
Source: American Heart Association

49. Average Yearly Sales The sales of Universal Instruments in the first t yr of its operation are approximated by the function

$$S(t) = t\sqrt{0.2t^2 + 4}$$

where $S(t)$ is measured in millions of dollars. What were Universal's average yearly sales over its first 5 yr of operation?

50. Cable TV Subscribers The manager of TeleStar Cable Service estimates that the total number of subscribers to the service in a certain city t yr from now will be

$$N(t) = -\frac{40,000}{\sqrt{1 + 0.2t}} + 50,000$$

Find the average number of cable television subscribers over the next 5 yr if this prediction holds true.

51. Refer to Exercise 44. Find the average velocity of the car over the time interval [0, 4].

52. CONCENTRATION OF A DRUG IN THE BLOODSTREAM The concentration of a certain drug in a patient's bloodstream t hr after injection is

$$C(t) = \frac{0.2t}{t^2 + 1}$$

mg/cm^3. Determine the average concentration of the drug in the patient's bloodstream over the first 4 hr after the drug is injected.

53. AVERAGE PRICE OF A COMMODITY The price of a certain commodity in dollars/unit at time t (measured in weeks) is given by

$$p = 18 - 3e^{-2t} - 6e^{-t/3}$$

What is the average price of the commodity over the 5-wk period from $t = 0$ to $t = 5$?

54. FLOW OF BLOOD IN AN ARTERY According to a law discovered by 19th-century physician Jean Louis Marie Poiseuille, the velocity of blood (in centimeters/second) r cm from the central axis of an artery is given by

$$v(r) = k(R^2 - r^2)$$

where k is a constant and R is the radius of the artery. Find the average velocity of blood along a radius of the artery (see the accompanying figure).

Hint: Evaluate $\frac{1}{R}\int_0^R v(r)\,dr$.

Blood vessel

55. WASTE DISPOSAL When organic waste is dumped into a pond, the oxidization process that takes place reduces the pond's oxygen content. However, in time, nature will restore the oxygen content to its natural level. Suppose that the oxygen content t days after organic waste has been dumped into a pond is given by

$$f(t) = 100\left(\frac{t^2 + 10t + 100}{t^2 + 20t + 100}\right)$$

percent of its normal level. Find the average content of oxygen in the pond over the first 10 days after organic waste has been dumped into it.
Hint: Show that

$$\frac{t^2 + 10t + 100}{t^2 + 20t + 100} = 1 - \frac{10}{t + 10} + \frac{100}{(t + 10)^2}$$

56. VELOCITY OF A FALLING HAMMER During the construction of a high-rise apartment building, a construction worker accidently drops a hammer that falls vertically a distance of h ft. The velocity of the hammer after falling a distance of x ft is $v = \sqrt{2gx}$ ft/sec $(0 \le x \le h)$. Show that the average velocity of the hammer over this path is $\bar{v} = \frac{2}{3}\sqrt{2gh}$ ft/sec.

57. Prove Property 1 of the definite integral.
Hint: Let F be an antiderivative of f and use the definition of the definite integral.

58. Prove Property 2 of the definite integral.
Hint: See Exercise 57.

59. Verify by direct computation that

$$\int_1^3 x^2\,dx = -\int_3^1 x^2\,dx$$

60. Prove Property 3 of the definite integral.
Hint: See Exercise 57.

61. Verify by direct computation that

$$\int_1^9 2\sqrt{x}\,dx = 2\int_1^9 \sqrt{x}\,dx$$

62. Verify by direct computation that

$$\int_0^1 (1 + x - e^x)\,dx = \int_0^1 dx + \int_0^1 x\,dx - \int_0^1 e^x\,dx$$

What properties of the definite integral are demonstrated in this exercise?

63. Verify by direct computation that

$$\int_0^3 (1 + x^3)\,dx = \int_0^1 (1 + x^3)\,dx + \int_1^3 (1 + x^3)\,dx$$

What property of the definite integral is demonstrated here?

64. Verify by direct computation that

$$\int_0^3 (1 + x^3)\,dx$$

$$= \int_0^1 (1 + x^3)\,dx + \int_1^2 (1 + x^3)\,dx + \int_2^3 (1 + x^3)\,dx$$

hence showing that Property 5 may be extended.

65. Evaluate $\int_3^3 (1 + \sqrt{x})e^{-x}\,dx$.

66. Evaluate $\int_3^0 f(x)\,dx$, given that $\int_0^3 f(x)\,dx = 4$.

67. Given that $\int_{-1}^2 f(x)\,dx = -2$ and $\int_{-1}^2 g(x)\,dx = 3$, evaluate

a. $\int_{-1}^2 [2f(x) + g(x)]\,dx$

b. $\int_{-1}^2 [g(x) - f(x)]\,dx$

c. $\int_{-1}^2 [2f(x) - 3g(x)]\,dx$

68. Given that $\int_{-1}^2 f(x)\,dx = 2$ and $\int_0^2 f(x)\,dx = 3$, evaluate

a. $\int_{-1}^0 f(x)\,dx$ **b.** $\int_0^2 f(x)\,dx - \int_{-1}^0 f(x)\,dx$

In Exercises 69–74, determine whether the statement is true or false. If it is true, explain why it is true. If it is false, give an example to show why it is false.

69. $\displaystyle\int_2^2 \frac{e^x}{\sqrt{1+x}}\,dx = 0$

70. $\displaystyle\int_1^3 \frac{dx}{x-2} = -\int_3^1 \frac{dx}{x-2}$

71. $\displaystyle\int_0^1 x\sqrt{x+1}\,dx = \sqrt{x+1}\int_0^1 x\,dx = \frac{1}{2}x^2\sqrt{x+1}\,\Big|_0^1 = \frac{\sqrt{2}}{2}$

72. If f' is continuous on $[0, 2]$, then $\int_0^2 f'(x)\,dx = f(2) - f(0)$.

73. If f and g are continuous on $[a, b]$ and k is a constant, then

$$\int_a^b [kf(x) + g(x)]\,dx = k\int_a^b f(x)\,dx + \int_a^b g(x)\,dx$$

74. If f is continuous on $[a, b]$ and $a < c < b$, then

$$\int_b^c f(x)\,dx = \int_a^c f(x)\,dx - \int_a^b f(x)\,dx$$

6.5 Solutions to Self-Check Exercises

1. Let $u = 2x + 5$. Then, $du = 2\,dx$, or $dx = \frac{1}{2}\,du$. Also, when $x = 0$, $u = 5$, and when $x = 2$, $u = 9$. Therefore,

$$\begin{aligned}\int_0^2 \sqrt{2x+5}\,dx &= \int_0^2 (2x+5)^{1/2}\,dx \\ &= \frac{1}{2}\int_5^9 u^{1/2}\,du \\ &= \left(\frac{1}{2}\right)\left(\frac{2}{3}u^{3/2}\right)\Big|_5^9 \\ &= \frac{1}{3}[9^{3/2} - 5^{3/2}] \\ &= \frac{1}{3}(27 - 5\sqrt{5})\end{aligned}$$

2. The required average value is given by

$$\begin{aligned}\frac{1}{2-(-1)}\int_{-1}^2 (1-x^2)\,dx &= \frac{1}{3}\int_{-1}^2 (1-x^2)\,dx \\ &= \frac{1}{3}\left(x - \frac{1}{3}x^3\right)\Big|_{-1}^2 \\ &= \frac{1}{3}\left[\left(2-\frac{8}{3}\right) - \left(-1+\frac{1}{3}\right)\right] = 0\end{aligned}$$

3. The average median price of a house over the stated time interval is given by

$$\begin{aligned}\frac{1}{5-0}\int_0^5 (t^3 - 7t^2 + 17t + 280)\,dt &= \frac{1}{5}\left(\frac{1}{4}t^4 - \frac{7}{3}t^3 + \frac{17}{2}t^2 + 280t\right)\Big|_0^5 \\ &= \frac{1}{5}\left[\frac{1}{4}(5)^4 - \frac{7}{3}(5)^3 + \frac{17}{2}(5)^2 + 280(5)\right] \\ &= 295.417\end{aligned}$$

or \$295,417.

USING TECHNOLOGY

Evaluating Definite Integrals for Piecewise-Defined Functions

We continue using graphing utilities to find the definite integral of a function. But here we will make use of Property 5 of the properties of the definite integral (p. 441).

EXAMPLE 1 Use the numerical integral operation of a graphing utility to evaluate

$$\int_{-1}^2 f(x)\,dx$$

where

$$f(x) = \begin{cases} -x^2 & \text{if } x < 0 \\ \sqrt{x} & \text{if } x \geq 0 \end{cases}$$

Solution Using Property 5 of the definite integral, we can write

$$\int_{-1}^{2} f(x)\, dx = \int_{-1}^{0} -x^2\, dx + \int_{0}^{2} x^{1/2}\, dx$$

Using a graphing utility, we find

$$\begin{aligned} \int_{-1}^{2} f(x)\, dx &= \text{fnInt}(-x\text{^}2, x, -1, 0) + \text{fnInt}(x\text{^}0.5, x, 0, 2) \\ &\approx -0.333333 + 1.885618 \\ &= 1.552285 \end{aligned}$$

TECHNOLOGY EXERCISES

In Exercises 1–4, use Property 5 of the properties of the definite integral (page 441) to evaluate the definite integral accurate to six decimal places.

1. $\int_{-1}^{2} f(x)\, dx$, where

$$f(x) = \begin{cases} 2.3x^3 - 3.1x^2 + 2.7x + 3 & \text{if } x < 1 \\ -1.7x^2 + 2.3x + 4.3 & \text{if } x \geq 1 \end{cases}$$

2. $\int_{0}^{3} f(x)\, dx$, where $f(x) = \begin{cases} \dfrac{\sqrt{x}}{1 + x^2} & \text{if } 0 \leq x < 1 \\ 0.5e^{0.1x^2} & \text{if } x \geq 1 \end{cases}$

3. $\int_{-2}^{2} f(x)\, dx$, where $f(x) = \begin{cases} x^4 - 2x^2 + 4 & \text{if } x < 0 \\ 2\ln(x + e^2) & \text{if } x \geq 0 \end{cases}$

4. $\int_{-2}^{6} f(x)\, dx$, where

$$f(x) = \begin{cases} 2x^3 - 3x^2 + x + 2 & \text{if } x < -1 \\ \sqrt{3x + 4} - 5 & \text{if } -1 \leq x \leq 4 \\ x^2 - 3x - 5 & \text{if } x > 4 \end{cases}$$

5. **AIDS in Massachusetts** The rate of growth (and decline) of the number of AIDS cases diagnosed in Massachusetts from the beginning of 1989 ($t = 0$) through the end of 1997 ($t = 8$) is approximated by the function

$$f(t) = \begin{cases} 69.83333t^2 + 30.16667t + 1000 & \text{if } 0 \leq t < 3 \\ 1719 & \text{if } 3 \leq t < 4 \\ -28.79167t^3 + 491.37500t^2 & \text{if } 4 \leq t \leq 8 \\ \quad -2985.083333t + 7640 & \end{cases}$$

where $f(t)$ is measured in the number of cases/year. Estimate the total number of AIDS cases diagnosed in Massachusetts from the beginning of 1989 through the end of 1997.

Source: Massachusetts Department of Health

6. **Crop Yield** If left untreated on bean stems, aphids (small insects that suck plant juices) will multiply at an increasing rate during the summer months and reduce productivity and crop yield of cultivated crops. But if the aphids are treated in mid-June, the numbers decrease sharply to less than 100/bean stem, allowing for steep rises in crop yield. The function

$$F(t) = \begin{cases} 62e^{1.152t} & \text{if } 0 \leq t < 1.5 \\ 349e^{-1.324(t-1.5)} & \text{if } 1.5 \leq t \leq 3 \end{cases}$$

gives the number of aphids on a typical bean stem at time t, where t is measured in months, $t = 0$ corresponding to the beginning of May. Find the average number of aphids on a typical bean stem over the period from the beginning of May to the beginning of August.

7. **Absorption of Drugs** Jane took 100 mg of a drug in the morning and another 100 mg of the same drug at the same time the following morning. The amount of the drug in her body t days after the first dosage was taken is given by

$$A(t) = \begin{cases} 100e^{-1.4t} & \text{if } 0 \leq t < 1 \\ 100(1 + e^{1.4})e^{-1.4t} & \text{if } t \geq 1 \end{cases}$$

Find the average amount of the drug in Jane's body over the first 2 days.

8. **Absorption of Drugs** The concentration of a drug in an organ at any time t (in seconds) is given by

$$C(t) = \begin{cases} 0.3t - 18(1 - e^{-t/60}) & \text{if } 0 \leq t \leq 20 \\ 18e^{-t/60} - 12e^{-(t-20)/60} & \text{if } t > 20 \end{cases}$$

where $C(t)$ is measured in grams/cubic centimeter (g/cm^3). Find the average concentration of the drug in the organ over the first 30 sec after it is administered.

6.6 Area between Two Curves

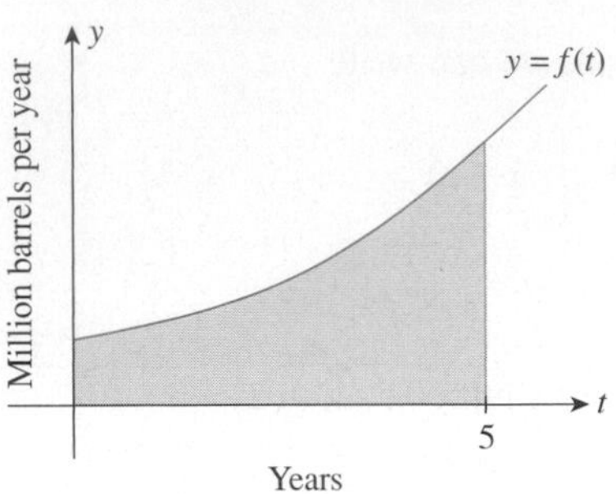

FIGURE 27
At a rate of consumption of $f(t)$ million barrels per year, the total petroleum consumption is given by the area of the region under the graph of f.

Suppose a certain country's petroleum consumption is expected to grow at the rate of $f(t)$ million barrels per year, t years from now, for the next 5 years. Then, the country's total petroleum consumption over the period of time in question is given by the area under the graph of f on the interval [0, 5] (Figure 27).

Next, suppose that because of the implementation of certain energy-conservation measures, the rate of growth of petroleum consumption is expected to be $g(t)$ million barrels per year instead. Then, the country's projected total petroleum consumption over the 5-year period is given by the area under the graph of g on the interval [0, 5] (Figure 28).

Therefore, the area of the shaded region S lying between the graphs of f and g on the interval [0, 5] (Figure 29) gives the amount of petroleum that would be saved over the 5-year period because of the conservation measures.

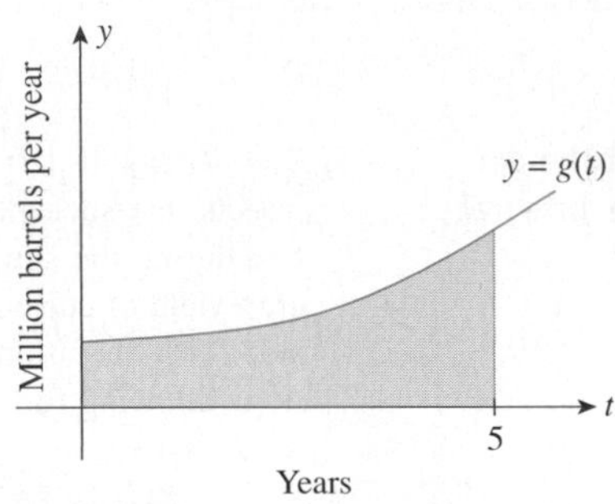

FIGURE 28
At a rate of consumption of $g(t)$ million barrels per year, the total petroleum consumption is given by the area of the region under the graph of g.

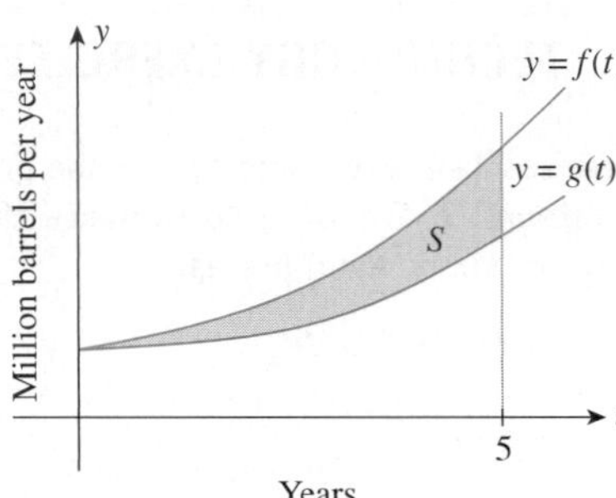

FIGURE 29
The area of S gives the amount of petroleum that would be saved over the 5-year period.

But the area of S is given by

$$\text{Area under the graph of } f \text{ on } [a, b] - \text{Area under the graph of } g \text{ on } [a, b]$$

$$= \int_0^5 f(t)\,dt - \int_0^5 g(t)\,dt$$

$$= \int_0^5 [f(t) - g(t)]\,dt \qquad \text{By Property 4, Section 6.5}$$

This example shows that some practical problems can be solved by finding the area of a region between two curves, which in turn can be found by evaluating an appropriate definite integral.

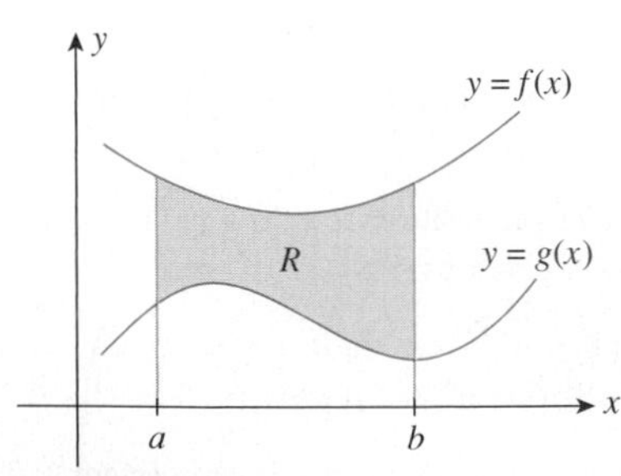

FIGURE 30
Area of $R = \int_a^b [f(x) - g(x)]\,dx$

Finding the Area between Two Curves

We now turn our attention to the general problem of finding the area of a plane region bounded both above and below by the graphs of functions. First, consider the situation in which the graph of one function lies above that of another. More specifically, let R be the region in the xy-plane (Figure 30) that is bounded above by the graph of a continuous function f, below by a continuous function g where $f(x) \geq g(x)$ on $[a, b]$, and to the left and right by the vertical lines $x = a$ and $x = b$, respectively. From the figure, we see that

$$\begin{aligned}\text{Area of } R &= \text{Area under } f(x) - \text{Area under } g(x)\\ &= \int_a^b f(x)\,dx - \int_a^b g(x)\,dx\\ &= \int_a^b [f(x) - g(x)]\,dx\end{aligned}$$

upon using Property 4 of the definite integral.

The Area between Two Curves

Let f and g be continuous functions such that $f(x) \geq g(x)$ on the interval $[a, b]$. Then, the area of the region bounded above by $y = f(x)$ and below by $y = g(x)$ on $[a, b]$ is given by

$$\int_a^b [f(x) - g(x)]\,dx \tag{13}$$

Even though we assumed that both f and g were nonnegative in the derivation of (13), it may be shown that this equation is valid if f and g are not nonnegative (see Exercise 55). Also, observe that if $g(x)$ is 0 for all x—that is, when the lower boundary of the region R is the x-axis—Equation (13) gives the area of the region under the curve $y = f(x)$ from $x = a$ to $x = b$, as we would expect.

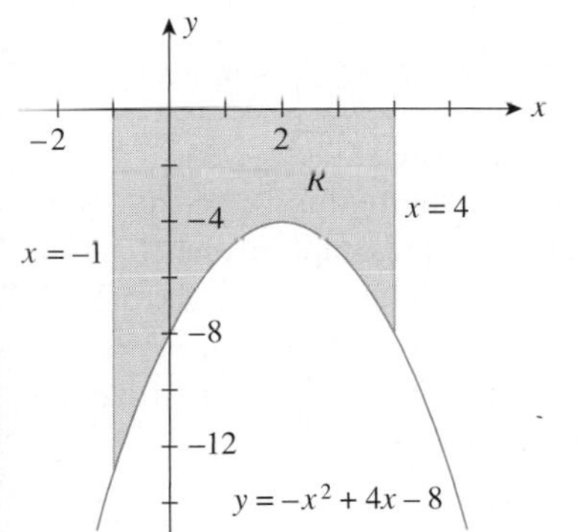

FIGURE 31
Area of $R = -\int_{-1}^{4} g(x)\,dx$

EXAMPLE 1 Find the area of the region bounded by the x-axis, the graph of $y = -x^2 + 4x - 8$, and the lines $x = -1$ and $x = 4$.

Solution The region R under consideration is shown in Figure 31. We can view R as the region bounded above by the graph of $f(x) = 0$ (the x-axis) and below by the graph of $g(x) = -x^2 + 4x - 8$ on $[-1, 4]$. Therefore, the area of R is given by

$$\begin{aligned}\int_a^b [f(x) - g(x)]\,dx &= \int_{-1}^{4} [0 - (-x^2 + 4x - 8)]\,dx\\ &= \int_{-1}^{4} (x^2 - 4x + 8)\,dx\\ &= \frac{1}{3}x^3 - 2x^2 + 8x\,\Big|_{-1}^{4}\\ &= \left[\frac{1}{3}(64) - 2(16) + 8(4)\right] - \left[\frac{1}{3}(-1) - 2(1) + 8(-1)\right]\\ &= 31\tfrac{2}{3}\end{aligned}$$

or $31\frac{2}{3}$ square units. ■

EXAMPLE 2 Find the area of the region R bounded by the graphs of

$$f(x) = 2x - 1 \quad \text{and} \quad g(x) = x^2 - 4$$

and the vertical lines $x = 1$ and $x = 2$.

Solution We first sketch the graphs of the functions $f(x) = 2x - 1$ and

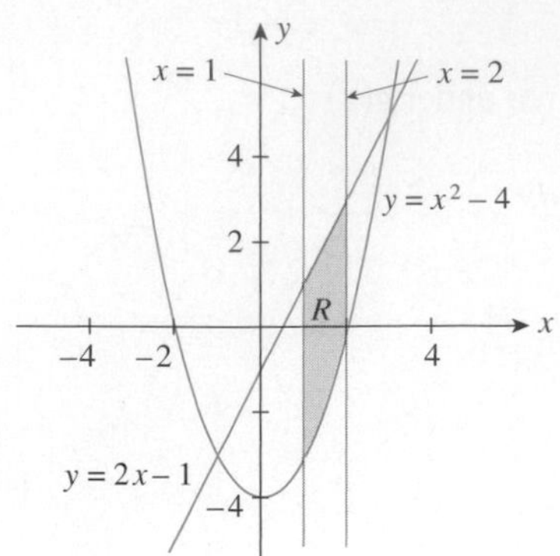

FIGURE 32
Area of $R = \int_1^2 [f(x) - g(x)]\, dx$

$g(x) = x^2 - 4$ and the vertical lines $x = 1$ and $x = 2$, and then we identify the region R whose area is to be calculated (Figure 32).

Since the graph of f always lies above that of g for x in the interval $[1, 2]$, we see by Equation (13) that the required area is given by

$$\begin{aligned}\int_1^2 [f(x) - g(x)]\, dx &= \int_1^2 [(2x - 1) - (x^2 - 4)]\, dx \\ &= \int_1^2 (-x^2 + 2x + 3)\, dx \\ &= -\frac{1}{3}x^3 + x^2 + 3x \Big|_1^2 \\ &= \left(-\frac{8}{3} + 4 + 6\right) - \left(-\frac{1}{3} + 1 + 3\right) = \frac{11}{3}\end{aligned}$$

or $\frac{11}{3}$ square units. ■

EXAMPLE 3 Find the area of the region R that is completely enclosed by the graphs of the functions

$$f(x) = 2x - 1 \quad \text{and} \quad g(x) = x^2 - 4$$

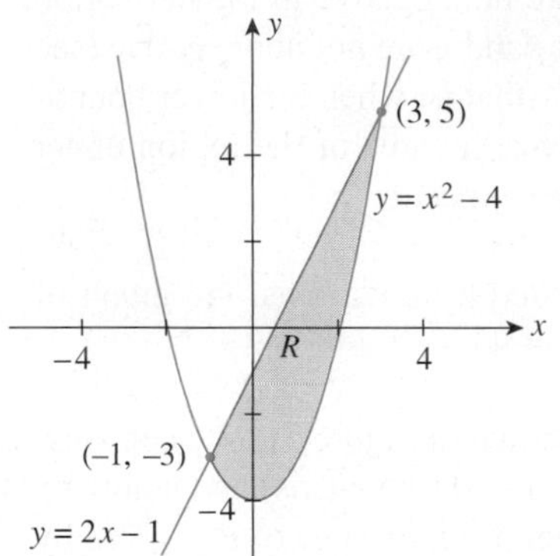

FIGURE 33
Area of $R = \int_{-1}^3 [f(x) - g(x)]\, dx$

Solution The region R is shown in Figure 33. First, we find the points of intersection of the two curves. To do this, we solve the system that consists of the two equations $y = 2x - 1$ and $y = x^2 - 4$. Equating the two values of y gives

$$\begin{aligned}x^2 - 4 &= 2x - 1 \\ x^2 - 2x - 3 &= 0 \\ (x + 1)(x - 3) &= 0\end{aligned}$$

so $x = -1$ or $x = 3$. That is, the two curves intersect when $x = -1$ and $x = 3$.

Observe that we could also view the region R as the region bounded above by the graph of the function $f(x) = 2x - 1$, below by the graph of the function $g(x) = x^2 - 4$, and to the left and right by the vertical lines $x = -1$ and $x = 3$, respectively.

Next, since the graph of the function f always lies above that of the function g on $[-1, 3]$, we can use (13) to compute the desired area:

$$\begin{aligned}\int_a^b [f(x) - g(x)]\, dx &= \int_{-1}^3 [(2x - 1) - (x^2 - 4)]\, dx \\ &= \int_{-1}^3 (-x^2 + 2x + 3)\, dx \\ &= -\frac{1}{3}x^3 + x^2 + 3x \Big|_{-1}^3 \\ &= (-9 + 9 + 9) - \left(\frac{1}{3} + 1 - 3\right) = \frac{32}{3} \\ &= 10\tfrac{2}{3}\end{aligned}$$

or $10\frac{2}{3}$ square units. ■

EXAMPLE 4 Find the area of the region R bounded by the graphs of the functions

$$f(x) = x^2 - 2x - 1 \quad \text{and} \quad g(x) = -e^x - 1$$

and the vertical lines $x = -1$ and $x = 1$.

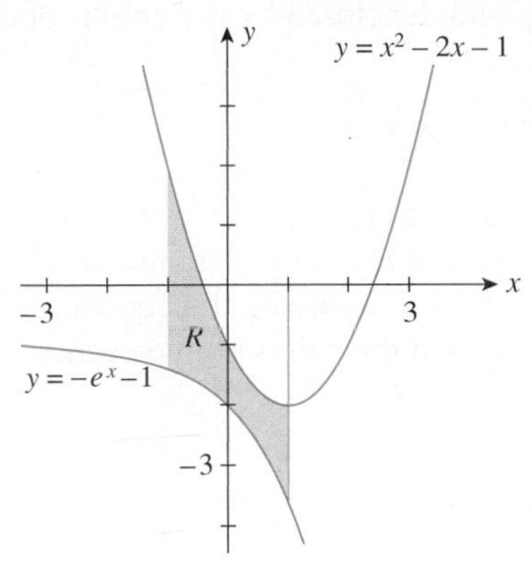

FIGURE 34
Area of $R = \int_{-1}^{1} [f(x) - g(x)]\, dx$

Solution The region R is shown in Figure 34. Since the graph of the function f always lies above that of the function g, the area of the region R is given by

$$\begin{aligned}\int_a^b [f(x) - g(x)]\, dx &= \int_{-1}^{1} [(x^2 - 2x - 1) - (-e^x - 1)]\, dx \\ &= \int_{-1}^{1} (x^2 - 2x + e^x)\, dx \\ &= \frac{1}{3}x^3 - x^2 + e^x \Big|_{-1}^{1} \\ &= \left(\frac{1}{3} - 1 + e\right) - \left(-\frac{1}{3} - 1 + e^{-1}\right) \\ &= \frac{2}{3} + e - \frac{1}{e}, \text{ or approximately 3.02 square units}\end{aligned}$$

Equation (13), which gives the area of the region between the curves $y = f(x)$ and $y = g(x)$ for $a \le x \le b$, is valid when the graph of the function f lies above that of the function g over the interval $[a, b]$. Example 5 shows how to use (13) to find the area of a region when the latter condition does not hold.

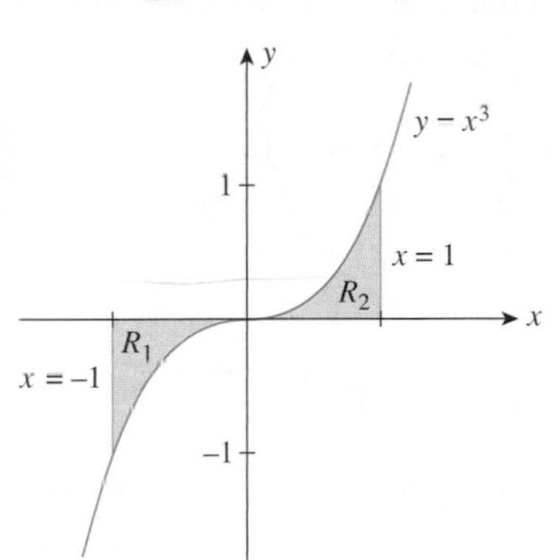

FIGURE 35
Area of R_1 = Area of R_2

EXAMPLE 5 Find the area of the region bounded by the graph of the function $f(x) = x^3$, the x-axis, and the lines $x = -1$ and $x = 1$.

Solution The region R under consideration can be thought of as composed of the two subregions R_1 and R_2, as shown in Figure 35.

Recall that the x-axis is represented by the function $g(x) = 0$. Since $g(x) \ge f(x)$ on $[-1, 0]$, we see that the area of R_1 is given by

$$\begin{aligned}\int_a^b [g(x) - f(x)]\, dx &= \int_{-1}^{0} (0 - x^3)\, dx = -\int_{-1}^{0} x^3\, dx \\ &= -\frac{1}{4}x^4 \Big|_{-1}^{0} = 0 - \left(-\frac{1}{4}\right) = \frac{1}{4}\end{aligned}$$

To find the area of R_2, we observe that $f(x) \ge g(x)$ on $[0, 1]$, so it is given by

$$\begin{aligned}\int_a^b [f(x) - g(x)]\, dx &= \int_0^1 (x^3 - 0)\, dx = \int_0^1 x^3\, dx \\ &= \frac{1}{4}x^4 \Big|_0^1 = \left(\frac{1}{4}\right) - 0 = \frac{1}{4}\end{aligned}$$

Therefore, the area of R is $\frac{1}{4} + \frac{1}{4}$, or $\frac{1}{2}$, square units.

By making use of symmetry, we could have obtained the same result by computing

$$-2\int_{-1}^{0} x^3\, dx \qquad \text{or} \qquad 2\int_0^1 x^3\, dx$$

as you may verify.

EXPLORE & DISCUSS

A function is *even* if it satisfies the condition $f(-x) = f(x)$, and it is *odd* if it satisfies the condition $f(-x) = -f(x)$. Show that the graph of an even function is symmetric with respect to the y-axis while the graph of an odd function is symmetric with respect to the origin. Explain why

$$\int_{-a}^{a} f(x)\, dx = 2\int_0^a f(x)\, dx \quad \text{if } f \text{ is even}$$

$$\int_{-a}^{a} f(x)\, dx = 0 \quad \text{if } f \text{ is odd}$$

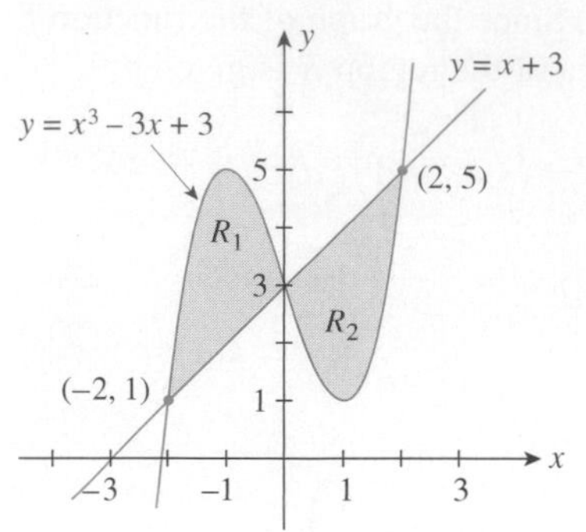

FIGURE 36
Area of R_1 + Area of R_2
$= \int_{-2}^{0} [f(x) - g(x)]\, dx + \int_{0}^{2} [g(x) - f(x)]\, dx$

EXAMPLE 6 Find the area of the region completely enclosed by the graphs of the functions

$$f(x) = x^3 - 3x + 3 \quad \text{and} \quad g(x) = x + 3$$

Solution First, sketch the graphs of $y = x^3 - 3x + 3$ and $y = x + 3$ and then identify the required region R. We can view the region R as being composed of the two subregions R_1 and R_2, as shown in Figure 36. By solving the equations $y = x + 3$ and $y = x^3 - 3x + 3$ simultaneously, we find the points of intersection of the two curves. Equating the two values of y, we have

$$\begin{aligned} x^3 - 3x + 3 &= x + 3 \\ x^3 - 4x &= 0 \\ x(x^2 - 4) &= 0 \\ x(x + 2)(x - 2) &= 0 \\ x &= 0, -2, 2 \end{aligned}$$

Hence, the points of intersection of the two curves are $(-2, 1)$, $(0, 3)$, and $(2, 5)$.

For $-2 \le x \le 0$, we see that the graph of the function f lies above that of the function g, so the area of the region R_1 is, by virtue of (13),

$$\begin{aligned} \int_{-2}^{0} [(x^3 - 3x + 3) - (x + 3)]\, dx &= \int_{-2}^{0} (x^3 - 4x)\, dx \\ &= \frac{1}{4}x^4 - 2x^2 \Big|_{-2}^{0} \\ &= -(4 - 8) \\ &= 4 \end{aligned}$$

or 4 square units. For $0 \le x \le 2$, the graph of the function g lies above that of the function f, and the area of R_2 is given by

$$\begin{aligned} \int_{0}^{2} [(x + 3) - (x^3 - 3x + 3)]\, dx &= \int_{0}^{2} (-x^3 + 4x)\, dx \\ &= -\frac{1}{4}x^4 + 2x^2 \Big|_{0}^{2} \\ &= -4 + 8 \\ &= 4 \end{aligned}$$

or 4 square units. Therefore, the required area is the sum of the area of the two regions $R_1 + R_2$—that is, $4 + 4$, or 8 square units. ■

APPLIED EXAMPLE 7 Conservation of Oil In a 1999 study for a developing country's Economic Development Board, government economists and energy experts concluded that if the Energy Conservation Bill were implemented in 2000, the country's oil consumption for the next 5 years would be expected to grow in accordance with the model

$$R(t) = 20e^{0.05t}$$

where t is measured in years ($t = 0$ corresponding to the year 2000) and $R(t)$ in millions of barrels per year. Without the government-imposed conservation measures, however, the expected rate of growth of oil consumption would be given by

$$R_1(t) = 20e^{0.08t}$$

millions of barrels per year. Using these models, determine how much oil would have been saved from 2000 through 2005 if the bill had been implemented.

Solution Under the Energy Conservation Bill, the total amount of oil that would have been consumed between 2000 and 2005 is given by

$$\int_0^5 R(t)\,dt = \int_0^5 20e^{0.05t}\,dt \qquad \textbf{(14)}$$

Without the bill, the total amount of oil that would have been consumed between 2000 and 2005 is given by

$$\int_0^5 R_1(t)\,dt = \int_0^5 20e^{0.08t}\,dt \qquad \textbf{(15)}$$

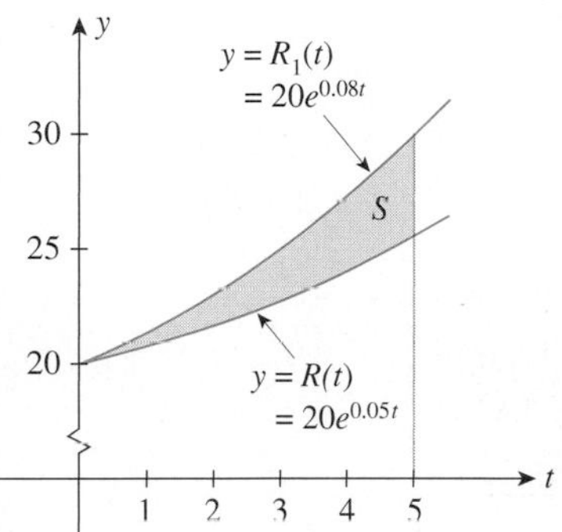

FIGURE 37
Area of $S = \int_0^5 [R_1(t) - R(t)]\,dt$

Equation (14) may be interpreted as the area of the region under the curve $y = R(t)$ from $t = 0$ to $t = 5$. Similarly, we interpret (15) as the area of the region under the curve $y = R_1(t)$ from $t = 0$ to $t = 5$. Furthermore, note that the graph of $y = R_1(t) = 20e^{0.08t}$ always lies on or above the graph of $y = R(t) = 20e^{0.05t}$ $(t \geq 0)$. Thus, the area of the shaded region S in Figure 37 shows the amount of oil that would have been saved from 2000 to 2005 if the Energy Conservation Bill had been implemented. But the area of the region S is given by

$$\begin{aligned}
\int_0^5 [R_1(t) - R(t)]\,dt &= \int_0^5 [20e^{0.08t} - 20e^{0.05t}]\,dt \\
&= 20\int_0^5 (e^{0.08t} - e^{0.05t})\,dt \\
&= 20\left(\frac{e^{0.08t}}{0.08} - \frac{e^{0.05t}}{0.05}\right)\Bigg|_0^5 \\
&= 20\left[\left(\frac{e^{0.4}}{0.08} - \frac{e^{0.25}}{0.05}\right) - \left(\frac{1}{0.08} - \frac{1}{0.05}\right)\right] \\
&\approx 9.3
\end{aligned}$$

or approximately 9.3 square units. Thus, the amount of oil that would have been saved is 9.3 million barrels. ■

EXPLORING WITH TECHNOLOGY

Refer to Example 7. Suppose we want to construct a mathematical model giving the amount of oil saved from 2000 through the year $2000 + x$, where $x \geq 0$. For example, in Example 7, $x = 5$.

1. Show that this model is given by

$$\begin{aligned}
F(x) &= \textstyle\int_0^x [R_1(t) - R(t)]\,dt \\
&= 250e^{0.08x} - 400e^{0.05x} + 150
\end{aligned}$$

   Hint: You may find it helpful to use some of the results of Example 7.
2. Use a graphing utility to plot the graph of F, using the viewing window $[0, 10] \times [0, 50]$.
3. Find $F(5)$ and thus confirm the result of Example 7.
4. What is the main advantage of this model?

6.6 Self-Check Exercises

1. Find the area of the region bounded by the graphs of $f(x) = x^2 + 2$ and $g(x) = 1 - x$ and the vertical lines $x = 0$ and $x = 1$.

2. Find the area of the region completely enclosed by the graphs of $f(x) = -x^2 + 6x + 5$ and $g(x) = x^2 + 5$.

3. The management of Kane Corporation, which operates a chain of hotels, expects its profits to grow at the rate of $1 + t^{2/3}$ million dollars/year t yr from now. However, with renovations and improvements of existing hotels and proposed acquisitions of new hotels, Kane's profits are expected to grow at the rate of $t - 2\sqrt{t} + 4$ million dollars/year in the next decade. What additional profits are expected over the next 10 yr if the group implements the proposed plans?

Solutions to Self-Check Exercises 6.6 can be found on page 462.

6.6 Concept Questions

1. Suppose f and g are continuous functions such that $f(x) \geq g(x)$ on the interval $[a, b]$. Write an integral giving the area of the region bounded above by the graph of f, below by the graph of g, and on the left and right by the lines $x = a$ and $x = b$.

2. Write an expression in terms of definite integrals giving the area of the shaded region in the following figure:

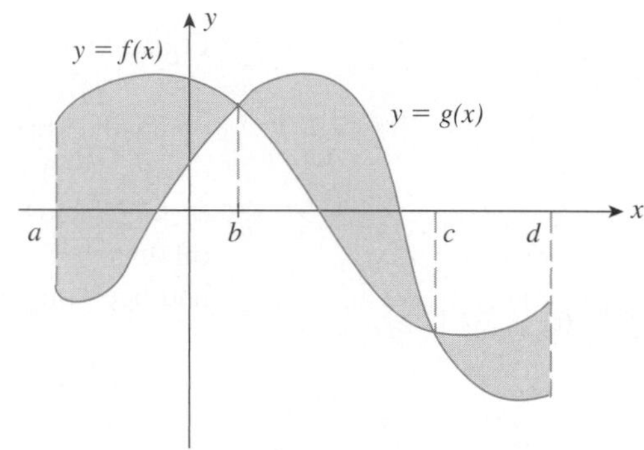

6.6 Exercises

In Exercises 1–8, find the area of the shaded region.

1.

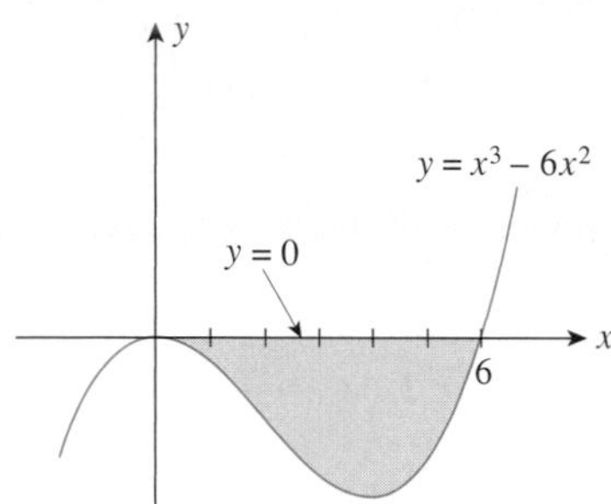

2.

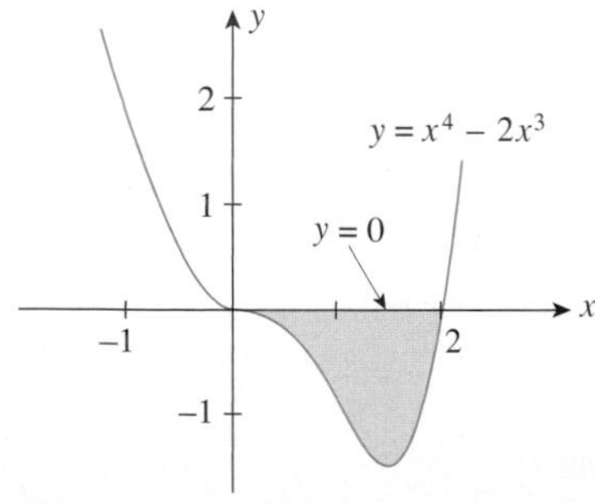

3.

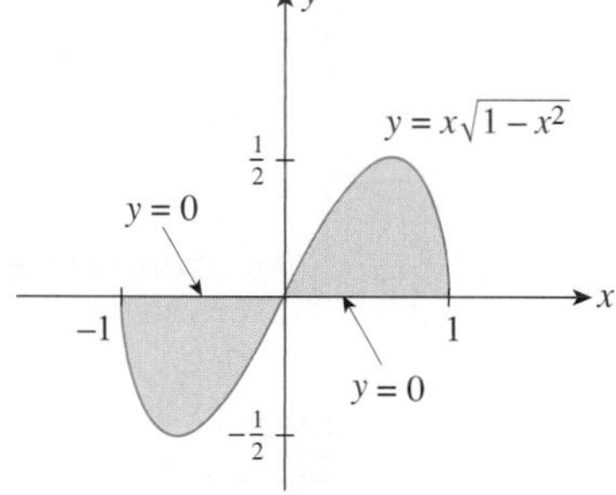

4.

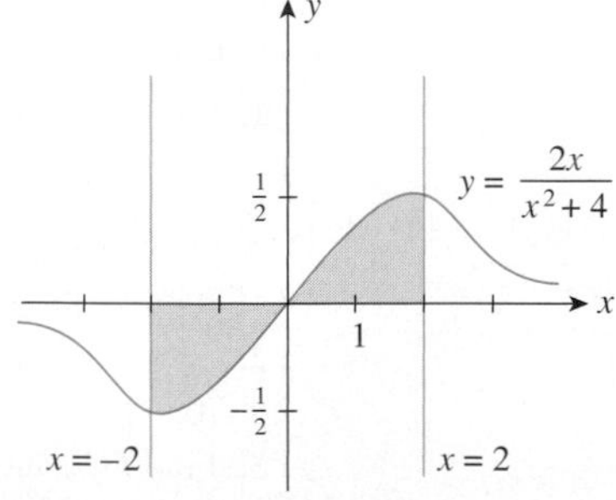

5.

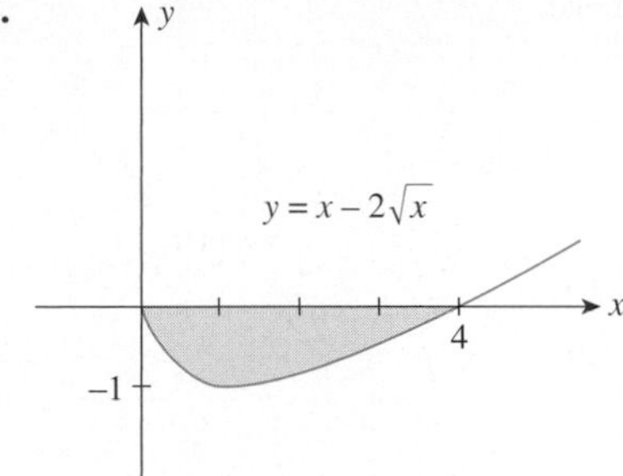

6.

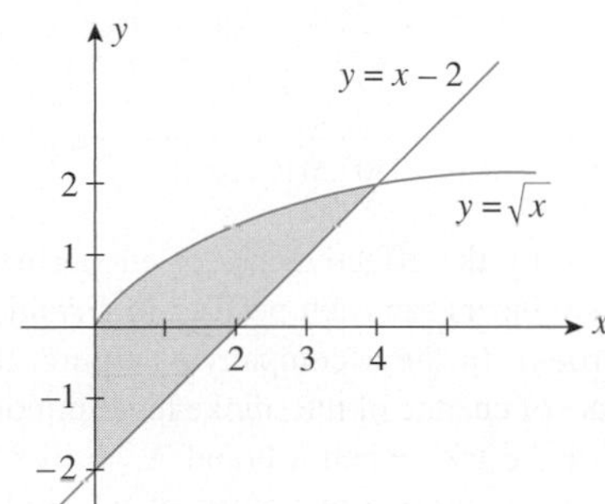

7.

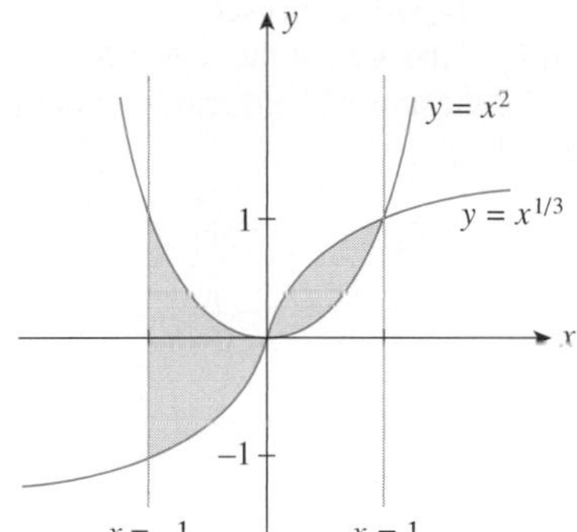

8.

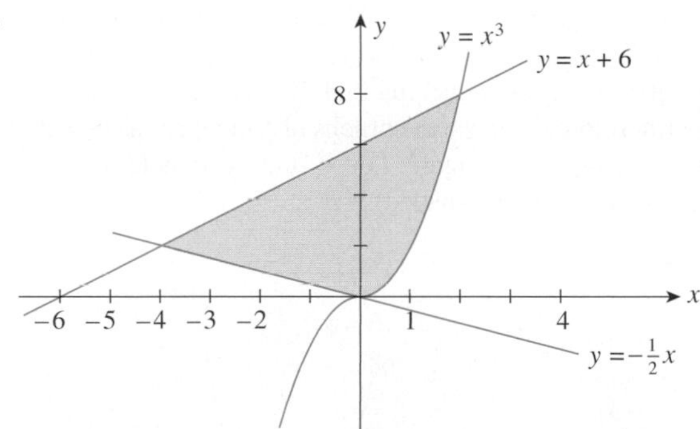

In Exercises 9–16, sketch the graph and find the area of the region bounded below by the graph of each function and above by the *x*-axis from $x = a$ to $x = b$.

9. $f(x) = -x^2;\ a = -1, b = 2$

10. $f(x) = x^2 - 4;\ a = -2, b = 2$

11. $f(x) = x^2 - 5x + 4;\ a = 1, b = 3$

12. $f(x) = x^3;\ a = -1, b = 0$

13. $f(x) = -1 - \sqrt{x};\ a = 0, b = 9$

14. $f(x) = \frac{1}{2}x - \sqrt{x};\ a = 0, b = 4$

15. $f(x) = -e^{(1/2)x};\ a = -2, b = 4$

16. $f(x) = -xe^{-x^2};\ a = 0, b = 1$

In Exercises 17–26, sketch the graphs of the functions *f* and *g* and find the area of the region enclosed by these graphs and the vertical lines $x = a$ and $x = b$.

17. $f(x) = x^2 + 3,\ g(x) = 1;\ a = 1, b = 3$

18. $f(x) = x + 2,\ g(x) = x^2 - 4;\ a = -1, b = 2$

19. $f(x) = -x^2 + 2x + 3,\ g(x) = -x + 3;\ a = 0, b = 2$

20. $f(x) = 9 - x^2,\ g(x) = 2x + 3;\ a = -1, b = 1$

21. $f(x) = x^2 + 1,\ g(x) = \frac{1}{3}x^3;\ a = -1, b = 2$

22. $f(x) = \sqrt{x},\ g(x) = -\frac{1}{2}x - 1;\ a = 1, b = 4$

23. $f(x) = \frac{1}{x},\ g(x) = 2x - 1;\ a = 1, b = 4$

24. $f(x) = x^2,\ g(x) = \frac{1}{x^2};\ a = 1, b = 3$

25. $f(x) = e^x,\ g(x) = \frac{1}{x};\ a = 1, b = 2$

26. $f(x) = x,\ g(x) = e^{2x};\ a = 1, b = 3$

In Exercises 27–34, sketch the graph and find the area of the region bounded by the graph of the function *f* and the lines $y = 0$, $x = a$, and $x = b$.

27. $f(x) = x;\ a = -1, b = 2$

28. $f(x) = x^2 - 2x;\ a = -1, b = 1$

29. $f(x) = -x^2 + 4x - 3;\ a = -1, b = 2$

30. $f(x) = x^3 - x^2;\ a = -1, b = 1$

31. $f(x) = x^3 - 4x^2 + 3x;\ a = 0, b = 2$

32. $f(x) = 4x^{1/3} + x^{4/3};\ a = -1, b = 8$

33. $f(x) = e^x - 1;\ a = -1, b = 3$

34. $f(x) = xe^{x^2};\ a = 0, b = 2$

In Exercises 35–42, sketch the graph and find the area of the region completely enclosed by the graphs of the given functions *f* and *g*.

35. $f(x) = x + 2$ and $g(x) = x^2 - 4$

36. $f(x) = -x^2 + 4x$ and $g(x) = 2x - 3$

37. $f(x) = x^2$ and $g(x) = x^3$

38. $f(x) = x^3 + 2x^2 - 3x$ and $g(x) = 0$

39. $f(x) = x^3 - 6x^2 + 9x$ and $g(x) = x^2 - 3x$

40. $f(x) = \sqrt{x}$ and $g(x) = x^2$

41. $f(x) = x\sqrt{9 - x^2}$ and $g(x) = 0$

42. $f(x) = 2x$ and $g(x) = x\sqrt{x + 1}$

43. Effect of Advertising on Revenue In the accompanying figure, the function f gives the rate of change of Odyssey Travel's revenue with respect to the amount x it spends on advertising with their current advertising agency. By engaging the services of a different advertising agency, it is expected that Odyssey's revenue will grow at the rate given by the function g. Give an interpretation of the area A of the region S and find an expression for A in terms of a definite integral involving f and g.

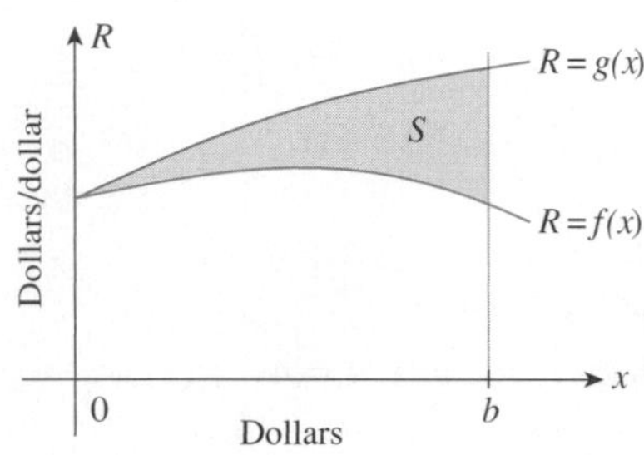

44. Pulse Rate During Exercise In the accompanying figure, the function f gives the rate of increase of an individual's pulse rate when he walked a prescribed course on a treadmill 6 mo ago. The function g gives the rate of increase of his pulse rate when he recently walked the same prescribed course. Give an interpretation of the area A of the region S and find an expression for A in terms of a definite integral involving f and g.

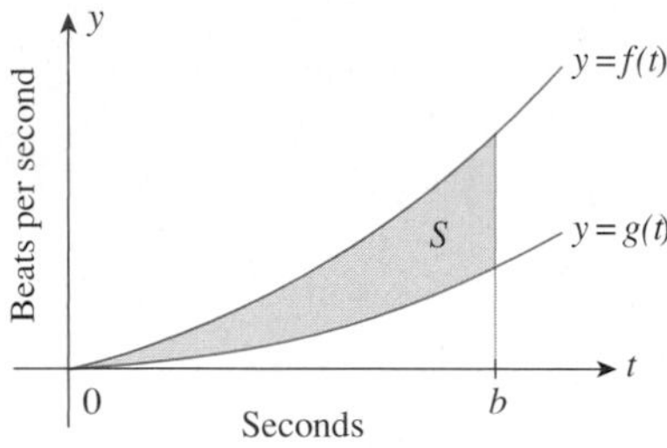

45. Oil Production Shortfall Energy experts disagree about when global oil production will begin to decline. In the following figure, the function f gives the annual world oil production in billions of barrels from 1980 to 2050, according to the Department of Energy projection. The function g gives the world oil production in billions of barrels per year over the same period, according to longtime petroleum geologist Colin Campbell. Find an expression in terms of the definite integrals involving f and g, giving the shortfall in the total oil production over the period in question heeding Campbell's dire warnings.

Source: U.S. Department of Energy; Colin Campbell

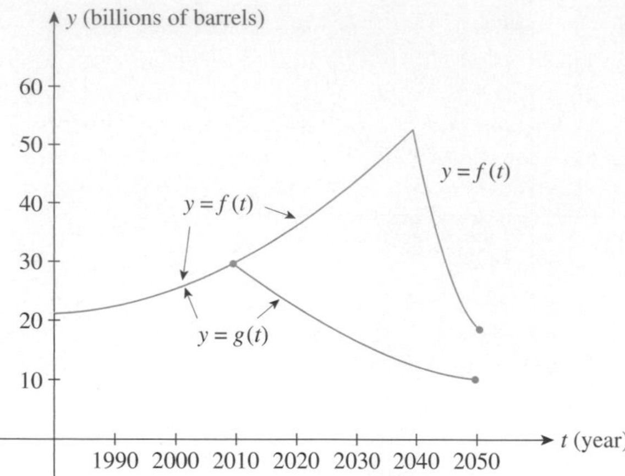

46. Air Purification To study the effectiveness of air purifiers in removing smoke, engineers ran each purifier in a smoke-filled 10-ft × 20-ft room. In the accompanying figure, the function f gives the rate of change of the smoke level/minute, t min after the start of the test, when a brand A purifier is used. The function g gives the rate of change of the smoke level/minute when a brand B purifier is used.

a. Give an interpretation of the area of the region S.

b. Find an expression for the area of S in terms of a definite integral involving f and g.

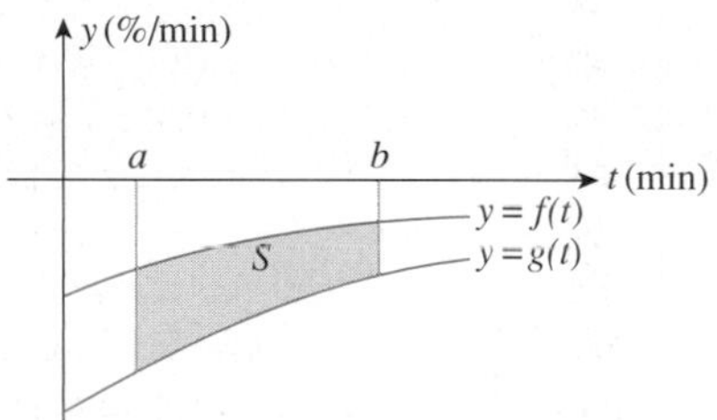

47. Two cars start out side by side and travel along a straight road. The velocity of car 1 is $f(t)$ ft/sec, the velocity of car 2 is $g(t)$ ft/sec over the interval $[0, T]$, and $0 < T_1 < T$. Furthermore, suppose the graphs of f and g are as depicted in the accompanying figure. Let A_1 and A_2 denote the areas of the regions (shown shaded).

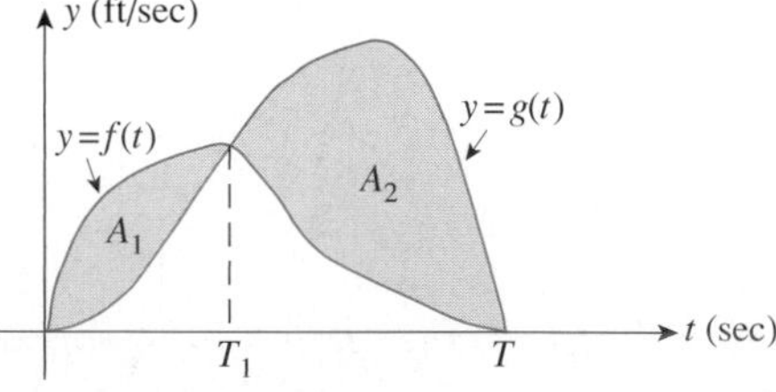

a. Write the number

$$\int_{T_1}^{T} [g(t) - f(t)]\, dt - \int_0^{T_1} [f(t) - g(t)]\, dt$$

in terms of A_1 and A_2.

b. What does the number obtained in part (a) represent?

48. The rate of change of the revenue of company A over the (time) interval $[0, T]$ is $f(t)$ dollars/week, whereas the rate of change of the revenue of company B over the same period is $g(t)$ dollars/week. The graphs of f and g are depicted in the accompanying figure. Find an expression in terms of definite integrals involving f and g giving the additional revenue that company B will have over company A in the period $[0, T]$.

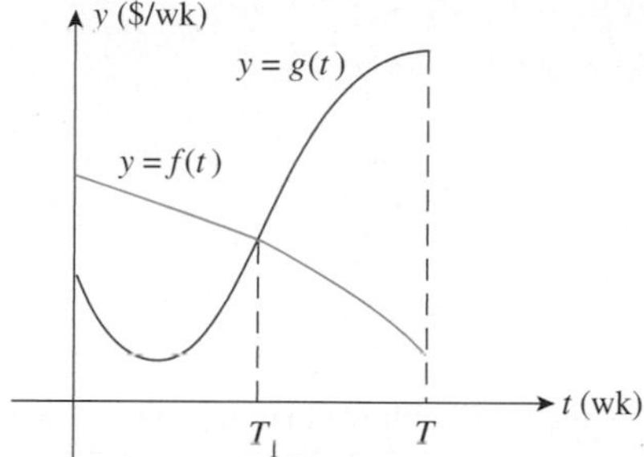

49. Turbo-Charged Engine vs. Standard Engine In tests conducted by *Auto Test Magazine* on two identical models of the Phoenix Elite—one equipped with a standard engine and the other with a turbo-charger—it was found that the acceleration of the former is given by

$$a = f(t) = 4 + 0.8t \qquad (0 \le t \le 12)$$

ft/sec/sec, t sec after starting from rest at full throttle, whereas the acceleration of the latter is given by

$$a = g(t) = 4 + 1.2t + 0.03t^2 \qquad (0 \le t \le 12)$$

ft/sec/sec. How much faster is the turbo-charged model moving than the model with the standard engine at the end of a 10-sec test run at full throttle?

50. Alternative Energy Sources Because of the increasingly important role played by coal as a viable alternative energy source, the production of coal has been growing at the rate of

$$3.5e^{0.05t}$$

billion metric tons/year t yr from 1980 (which corresponds to $t = 0$). Had it not been for the energy crisis, the rate of production of coal since 1980 might have been only

$$3.5e^{0.01t}$$

billion metric tons/year t yr from 1980. Determine how much additional coal was produced between 1980 and the end of the century as an alternate energy source.

51. Effect of TV Advertising on Car Sales Carl Williams, the proprietor of Carl Williams Auto Sales, estimates that with extensive television advertising, car sales over the next several years could be increasing at the rate of

$$5e^{0.3t}$$

thousand cars/year t yr from now, instead of at the current rate of

$$(5 + 0.5t^{3/2})$$

thousand cars/year t yr from now. Find how many more cars Carl expects to sell over the next 5 yr by implementing his advertising plans.

52. Population Growth In an endeavor to curb population growth in a Southeast Asian island state, the government has decided to launch an extensive propaganda campaign. Without curbs, the government expects the rate of population growth to have been

$$60e^{0.02t}$$

thousand people/year t yr from now, over the next 5 yr. However, successful implementation of the proposed campaign is expected to result in a population growth rate of

$$-t^2 + 60$$

thousand people/year t yr from now, over the next 5 yr. Assuming that the campaign is mounted, how many fewer people will there be in that country 5 yr from now then there would have been if no curbs had been imposed?

In Exercises 53 and 54, determine whether the statement is true or false. If it is true, explain why it is true. If it is false, give an example to show why it is false.

53. If f and g are continuous on $[a, b]$ and either $f(x) \ge g(x)$ for all x in $[a, b]$ or $f(x) \le g(x)$ for all x in $[a, b]$, then the area of the region bounded by the graphs of f and g and the vertical lines $x = a$ and $x = b$ is given by $\int_a^b |f(x) - g(x)|\, dx$.

54. The area of the region bounded by the graphs of $f(x) = 2 - x$ and $g(x) = 4 - x^2$ and the vertical lines $x = 0$ and $x = 2$ is given by $\int_0^2 [f(x) - g(x)]\, dx$.

55. Show that the area of a region R bounded above by the graph of a function f and below by the graph of a function g from $x = a$ to $x = b$ is given by

$$\int_a^b [f(x) - g(x)]\, dx$$

Hint: The validity of the formula was verified earlier for the case when both f and g were nonnegative. Now, let f and g be two functions such that $f(x) \ge g(x)$ for $a \le x \le b$. Then, there exists some nonnegative constant c such that the curves $y = f(x) + c$ and $y = g(x) + c$ are translated in the y-direction in such a way that the region R' has the same area as the region R (see the accompanying figures). Show that the area of R' is given by

$$\int_a^b \{[f(x) + c] - [g(x) + c]\}\, dx = \int_a^b [f(x) - g(x)]\, dx$$

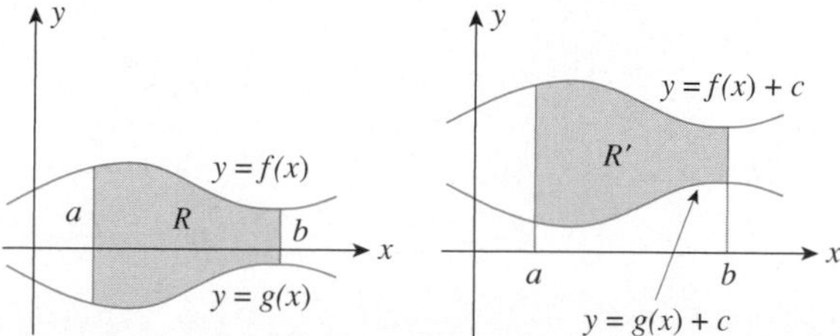

6.6 Solutions to Self-Check Exercises

1. The region in question is shown in the accompanying figure. Since the graph of the function f lies above that of the function g for $0 \le x \le 1$, we see that the required area is given by

$$\begin{aligned}\int_0^1 [(x^2 + 2) - (1 - x)]\, dx &= \int_0^1 (x^2 + x + 1)\, dx \\ &= \frac{1}{3}x^3 + \frac{1}{2}x^2 + x \Big|_0^1 \\ &= \frac{1}{3} + \frac{1}{2} + 1 \\ &= \frac{11}{6}\end{aligned}$$

or $\frac{11}{6}$ square units.

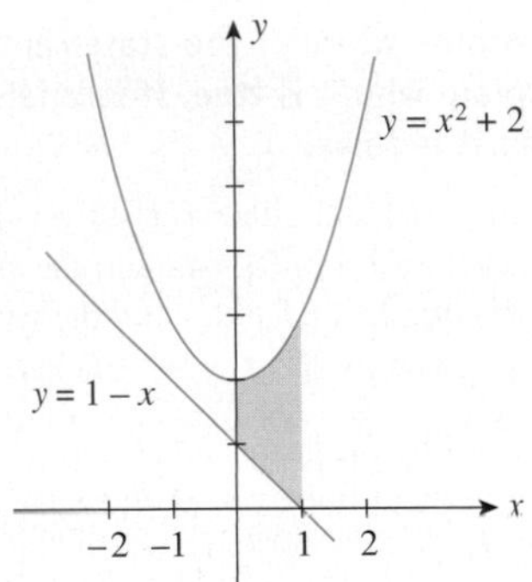

2. The region in question is shown in the accompanying figure. To find the points of intersection of the two curves, we solve the equations

$$\begin{aligned}-x^2 + 6x + 5 &= x^2 + 5 \\ 2x^2 - 6x &= 0 \\ 2x(x - 3) &= 0\end{aligned}$$

giving $x = 0$ or $x = 3$. Therefore, the points of intersection are (0, 5) and (3, 14).

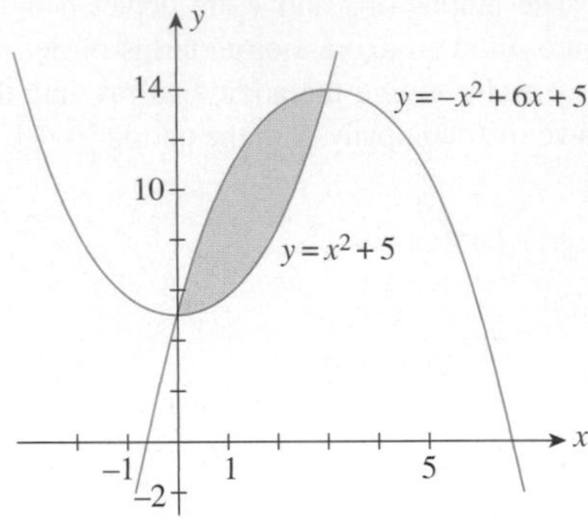

Since the graph of f always lies above that of g for $0 \le x \le 3$, we see that the required area is given by

$$\begin{aligned}\int_0^3 [(-x^2 + 6x + 5) - (x^2 + 5)]\, dx &= \int_0^3 (-2x^2 + 6x)\, dx \\ &= -\frac{2}{3}x^3 + 3x^2 \Big|_0^3 \\ &= -18 + 27 \\ &= 9\end{aligned}$$

or 9 square units.

3. The additional profits realizable over the next 10 yr are given by

$$\begin{aligned}&\int_0^{10} [(t - 2\sqrt{t} + 4) - (1 + t^{2/3})]\, dt \\ &= \int_0^{10} (t - 2t^{1/2} + 3 - t^{2/3})\, dt \\ &= \frac{1}{2}t^2 - \frac{4}{3}t^{3/2} + 3t - \frac{3}{5}t^{5/3} \Big|_0^{10} \\ &= \frac{1}{2}(10)^2 - \frac{4}{3}(10)^{3/2} + 3(10) - \frac{3}{5}(10)^{5/3} \\ &\approx 9.99\end{aligned}$$

or approximately \$10 million.

USING TECHNOLOGY

Finding the Area between Two Curves

The numerical integral operation can also be used to find the area between two curves. We do this by using the numerical integral operation to evaluate an appropriate definite integral or the sum (difference) of appropriate definite integrals. In the following example, the intersection operation is also used to advantage to help us find the limits of integration.

EXAMPLE 1 Use a graphing utility to find the area of the smaller region R that is completely enclosed by the graphs of the functions

$$f(x) = 2x^3 - 8x^2 + 4x - 3 \quad \text{and} \quad g(x) = 3x^2 + 10x - 11$$

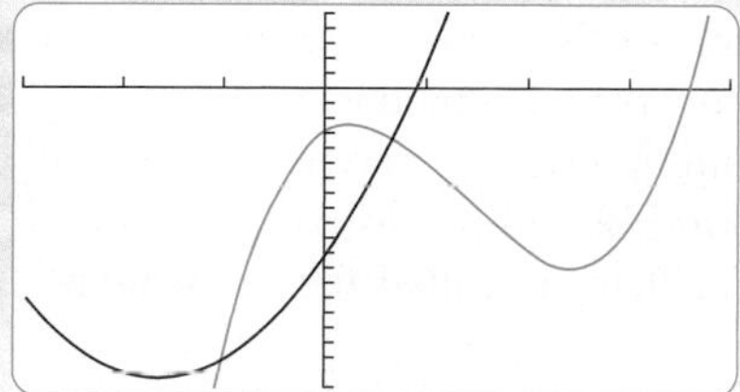

FIGURE T1
The region R is completely enclosed by the graphs of f and g.

Solution The graphs of f and g in the viewing window $[-3, 4] \times [-20, 5]$ are shown in Figure T1.

Using the intersection operation of a graphing utility, we find the x-coordinates of the points of intersection of the two graphs to be approximately -1.04 and 0.65, respectively. Since the graph of f lies above that of g on the interval $[-1.04, 0.65]$, we see that the area of R is given by

$$A = \int_{-1.04}^{0.65} [(2x^3 - 8x^2 + 4x - 3) - (3x^2 + 10x - 11)]\, dx$$

$$= \int_{-1.04}^{0.65} (2x^3 - 11x^2 - 6x + 8)\, dx$$

Using the numerical integral function of a graphing utility, we find $A \approx 9.87$, and so the area of R is approximately 9.87 square units. ■

TECHNOLOGY EXERCISES

In Exercises 1–6, (a) plot the graphs of the functions f and g and (b) find the area of the region enclosed by these graphs and the vertical lines $x = a$ and $x = b$. Express your answers accurate to four decimal places.

1. $f(x) = x^3(x - 5)^4$, $g(x) = 0$; $a = 1$, $b = 3$

2. $f(x) = x - \sqrt{1 - x^2}$, $g(x) = 0$; $a = -\frac{1}{2}$, $b = \frac{1}{2}$

3. $f(x) = x^{1/3}(x + 1)^{1/2}$, $g(x) = x^{-1}$; $a = 1.2$, $b = 2$

4. $f(x) = 2$, $g(x) = \ln(1 + x^2)$; $a = -1$, $b = 1$

5. $f(x) = \sqrt{x}$, $g(x) = \dfrac{x^2 - 3}{x^2 + 1}$; $a = 0$, $b = 3$

6. $f(x) = \dfrac{4}{x^2 + 1}$, $g(x) = x^4$; $a = -1$, $b = 1$

In Exercises 7–12, (a) plot the graphs of the functions f and g and (b) find the area of the region totally enclosed by the graphs of these functions.

7. $f(x) = 2x^3 - 8x^2 + 4x - 3$ and $g(x) = -3x^2 + 10x - 10$

8. $f(x) = x^4 - 2x^2 + 2$ and $g(x) = 4 - 2x^2$

9. $f(x) = 2x^3 - 3x^2 + x + 5$ and $g(x) = e^{2x} - 3$

10. $f(x) = \frac{1}{2}x^2 - 3$ and $g(x) = \ln x$

11. $f(x) = xe^{-x}$ and $g(x) = x - 2\sqrt{x}$

12. $f(x) = e^{-x^2}$ and $g(x) = x^4$

13. Refer to Example 1. Find the area of the larger region that is completely enclosed by the graphs of the functions f and g.

6.7 Applications of the Definite Integral to Business and Economics

In this section, we consider several applications of the definite integral in the fields of business and economics.

Consumers' and Producers' Surplus

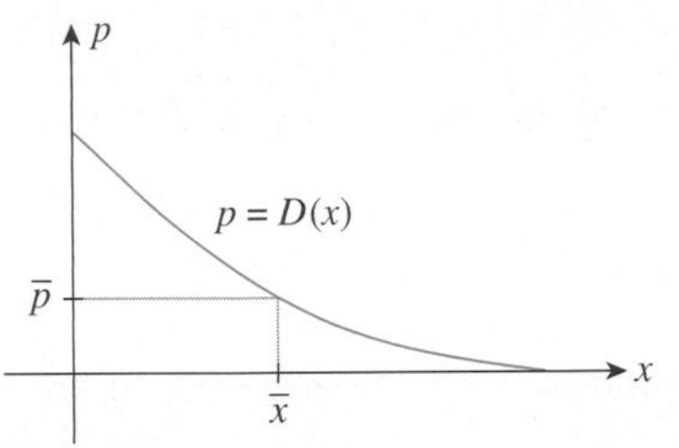

FIGURE 38
$D(x)$ is a demand function.

We begin by deriving a formula for computing the consumers' surplus. Suppose $p = D(x)$ is the demand function that relates the unit price p of a commodity to the quantity x demanded of it. Furthermore, suppose a fixed unit market price $\bar{p}$ has been established for the commodity and corresponding to this unit price the quantity demanded is $\bar{x}$ units (Figure 38). Then, those consumers who would be willing to pay a unit price higher than $\bar{p}$ for the commodity would in effect experience a savings. This difference between what the consumers *would* be willing to pay for $\bar{x}$ units of the commodity and what they *actually* pay for them is called the **consumers' surplus**.

To derive a formula for computing the consumers' surplus, divide the interval $[0, \bar{x}]$ into n subintervals, each of length $\Delta x = \bar{x}/n$, and denote the right endpoints of these subintervals by $x_1, x_2, \ldots, x_n = \bar{x}$ (Figure 39).

We observe in Figure 39 that there are consumers who would pay a unit price of at least $D(x_1)$ dollars for the first Δx units of the commodity instead of the market price of $\bar{p}$ dollars per unit. The savings to these consumers is approximated by

$$D(x_1)\Delta x - \bar{p}\Delta x = [D(x_1) - \bar{p}]\Delta x$$

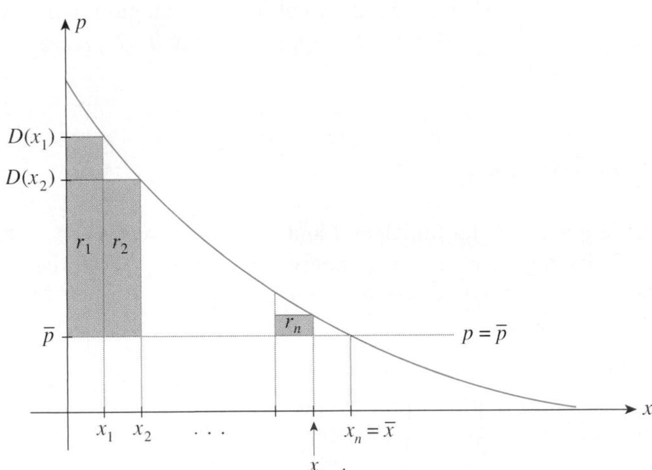

FIGURE 39
Approximating consumers' surplus by the sum of the rectangles $r_1, r_2, \ldots, r_n$

which is the area of the rectangle r_1. Pursuing the same line of reasoning, we find that the savings to the consumers who would be willing to pay a unit price of at least $D(x_2)$ dollars for the next Δx units (from x_1 through x_2) of the commodity, instead of the market price of $\bar{p}$ dollars per unit, is approximated by

$$D(x_2)\Delta x - \bar{p}\Delta x = [D(x_2) - \bar{p}]\Delta x$$

Continuing, we approximate the total savings to the consumers in purchasing $\bar{x}$ units of the commodity by the sum

$$\begin{aligned}&[D(x_1) - \bar{p}]\Delta x + [D(x_2) - \bar{p}]\Delta x + \cdots + [D(x_n) - \bar{p}]\Delta x \\ &\quad = [D(x_1) + D(x_2) + \cdots + D(x_n)]\Delta x - \underbrace{[\bar{p}\Delta x + \bar{p}\Delta x + \cdots + \bar{p}\Delta x]}_{n \text{ terms}} \\ &\quad = [D(x_1) + D(x_2) + \cdots + D(x_n)]\Delta x - n\bar{p}\Delta x \\ &\quad = [D(x_1) + D(x_2) + \cdots + D(x_n)]\Delta x - \bar{p}\,\bar{x}\end{aligned}$$

Now, the first term in the last expression is the Riemann sum of the demand function $p = D(x)$ over the interval $[0, \bar{x}]$ with representative points $x_1, x_2, \ldots, x_n$. Letting n approach infinity, we obtain the following formula for the consumers' surplus CS.

Consumers' Surplus

The consumers' surplus is given by

$$CS = \int_0^{\bar{x}} D(x)\,dx - \bar{p}\,\bar{x} \tag{16}$$

where D is the demand function, $\bar{p}$ is the unit market price, and $\bar{x}$ is the quantity sold.

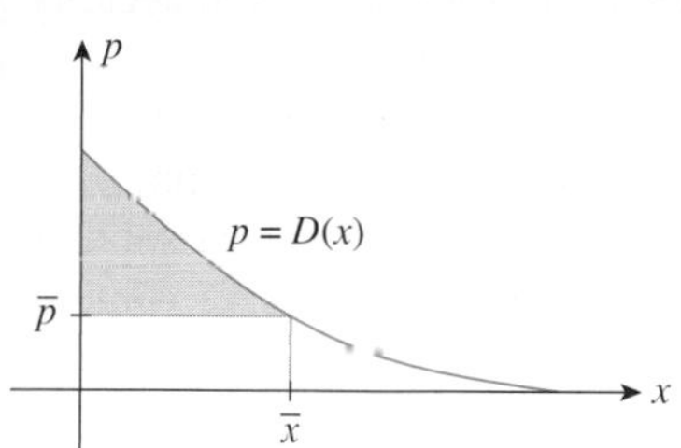

FIGURE 40
Consumers' surplus

The consumer's surplus is given by the area of the region bounded above by the demand curve $p = D(x)$ and below by the straight line $p = \bar{p}$ from $x = 0$ to $x = \bar{x}$ (Figure 40). We can also see this if we rewrite Equation (16) in the form

$$\int_0^{\bar{x}} [D(x) - \bar{p}]\,dx$$

and interpret the result geometrically.

Analogously, we can derive a formula for computing the producers' surplus. Suppose $p = S(x)$ is the supply equation that relates the unit price p of a certain commodity to the quantity x that the supplier will make available in the market at that price.

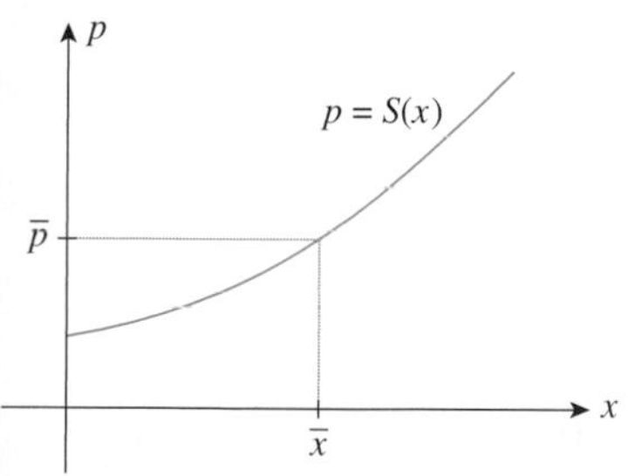

FIGURE 41
$S(x)$ is a supply function.

Again, suppose a fixed market price $\bar{p}$ has been established for the commodity and, corresponding to this unit price, a quantity of $\bar{x}$ units will be made available in the market by the supplier (Figure 41). Then, the suppliers who would be willing to make the commodity available at a lower price stand to gain from the fact that the market price is set as such. The difference between what the suppliers actually receive and what they would be willing to receive is called the **producers' surplus.** Proceeding in a manner similar to the derivation of the equation for computing the consumers' surplus, we find that the producers' surplus PS is defined as follows:

Producers' Surplus

The producers' surplus is given by

$$PS = \bar{p}\,\bar{x} - \int_0^{\bar{x}} S(x)\,dx \tag{17}$$

where $S(x)$ is the supply function, $\bar{p}$ is the unit market price, and $\bar{x}$ is the quantity supplied.

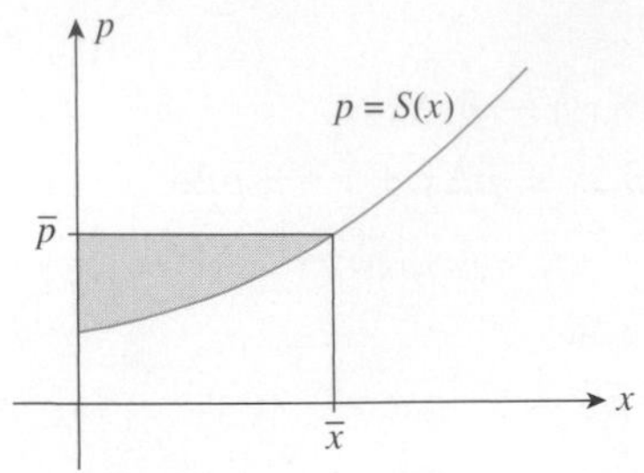

FIGURE 42
Producers' surplus

Geometrically, the producers' surplus is given by the area of the region bounded above by the straight line $p = \overline{p}$ and below by the supply curve $p = S(x)$ from $x = 0$ to $x = \overline{x}$ (Figure 42).

We can also show that the last statement is true by converting Equation (17) to the form

$$\int_0^{\overline{x}} [\overline{p} - S(x)]\, dx$$

and interpreting the definite integral geometrically.

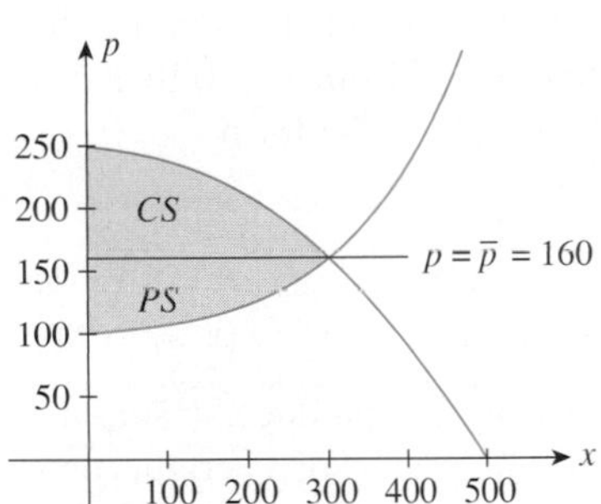

FIGURE 43
Consumers' surplus and producers' surplus when market price = equilibrium price

VIDEO

EXAMPLE 1 The demand function for a certain make of 10-speed bicycle is given by

$$p = D(x) = -0.001x^2 + 250$$

where p is the unit price in dollars and x is the quantity demanded in units of a thousand. The supply function for these bicycles is given by

$$p = S(x) = 0.0006x^2 + 0.02x + 100$$

where p stands for the unit price in dollars and x stands for the number of bicycles that the supplier will put on the market, in units of a thousand. Determine the consumers' surplus and the producers' surplus if the market price of a bicycle is set at the equilibrium price.

Solution Recall that the equilibrium price is the unit price of the commodity when market equilibrium occurs. We determine the equilibrium price by solving for the point of intersection of the demand curve and the supply curve (Figure 43). To solve the system of equations

$$\begin{aligned} p &= -0.001x^2 \qquad\qquad + 250 \\ p &= 0.0006x^2 + 0.02x + 100 \end{aligned}$$

we simply substitute the first equation into the second, obtaining

$$\begin{aligned} 0.0006x^2 + 0.02x + 100 &= -0.001x^2 + 250 \\ 0.0016x^2 + 0.02x - 150 &= 0 \\ 16x^2 + 200x - 1{,}500{,}000 &= 0 \\ 2x^2 + 25x - 187{,}500 &= 0 \end{aligned}$$

Factoring this last equation, we obtain

$$(2x + 625)(x - 300) = 0$$

Thus, $x = -625/2$ or $x = 300$. The first number lies outside the interval of interest, so we are left with the solution $x = 300$, with a corresponding value of

$$p = -0.001(300)^2 + 250 = 160$$

Thus, the equilibrium point is (300, 160); that is, the equilibrium quantity is 300,000, and the equilibrium price is \$160. Setting the market price at \$160 per unit and using Formula (16) with $\overline{p} = 160$ and $\overline{x} = 300$, we find that the consumers' surplus is given by

$$\begin{aligned} CS &= \int_0^{300} (-0.001x^2 + 250)\,dx - (160)(300) \\ &= \left(-\frac{1}{3000}x^3 + 250x\right)\Bigg|_0^{300} - 48{,}000 \\ &= -\frac{300^3}{3000} + (250)(300) - 48{,}000 \\ &= 18{,}000 \end{aligned}$$

or \$18,000,000.

(Recall that x is measured in units of a thousand.) Next, using (17), we find that the producers' surplus is given by

$$\begin{aligned} PS &= (160)(300) - \int_0^{300} (0.0006x^2 + 0.02x + 100)\,dx \\ &= 48{,}000 - (0.0002x^3 + 0.01x^2 + 100x)\Big|_0^{300} \\ &= 48{,}000 - [(0.0002)(300)^3 + (0.01)(300)^2 + 100(300)] \\ &= 11{,}700 \end{aligned}$$

or \$11,700,000. ■

The Future and Present Value of an Income Stream

Suppose a firm generates a stream of income over a period of time—for example, the revenue generated by a large chain of retail stores over a 5-year period. As the income is realized, it is reinvested and earns interest at a fixed rate. The **accumulated future income stream** over the 5-year period is the amount of money the firm ends up with at the end of that period.

The definite integral can be used to determine this accumulated, or total, future income stream over a period of time. The total future value of an income stream gives us a way to measure the value of such a stream. To find the **total future value of an income stream**, suppose

$R(t)$ = Rate of income generation at any time t — Dollars per year

r = Interest rate compounded continuously

T = Term — In years

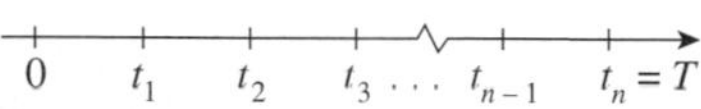

FIGURE 44
The time interval $[0, T]$ is partitioned into n subintervals.

Let's divide the time interval $[0, T]$ into n subintervals of equal length $\Delta t = T/n$ and denote the right endpoints of these intervals by $t_1, t_2, \ldots, t_n = T$, as shown in Figure 44.

If R is a continuous function on $[0, T]$, then $R(t)$ will not differ by much from $R(t_1)$ in the subinterval $[0, t_1]$ provided that the subinterval is small (which is true if n is large). Therefore, the income generated over the time interval $[0, t_1]$ is approximately

$$R(t_1)\Delta t \qquad \text{Constant rate of income} \times \text{Length of time}$$

dollars. The future value of this amount, T years from now, calculated as if it were earned at time t_1, is

$$[R(t_1)\Delta t]e^{r(T-t_1)} \qquad \text{Equation (10), Section 5.3}$$

dollars. Similarly, the income generated over the time interval $[t_1, t_2]$ is approximately $P(t_2)\Delta t$ dollars and has a future value, T years from now, of approximately

$$[R(t_2)\Delta t]e^{r(T-t_2)}$$

dollars. Therefore, the sum of the future values of the income stream generated over the time interval $[0, T]$ is approximately

$$\begin{aligned} &R(t_1)e^{r(T-t_1)}\Delta t + R(t_2)e^{r(T-t_2)}\Delta t + \cdots + R(t_n)e^{r(T-t_n)}\Delta t \\ &= e^{rT}[R(t_1)e^{-rt_1}\Delta t + R(t_2)e^{-rt_2}\Delta t + \cdots + R(t_n)e^{-rt_n}\Delta t] \end{aligned}$$

dollars. But this sum is just the Riemann sum of the function $e^{rT}R(t)e^{-rt}$ over the interval $[0, T]$ with representative points $t_1, t_2, \ldots, t_n$. Letting n approach infinity, we obtain the following result.

Accumulated or Total Future Value of an Income Stream

The accumulated, or total, future value after T years of an income stream of $R(t)$ dollars per year, earning interest at the rate of r per year compounded continuously, is given by

$$A = e^{rT}\int_0^T R(t)e^{-rt}\,dt \tag{18}$$

APPLIED EXAMPLE 2 Income Stream Crystal Car Wash recently bought an automatic car-washing machine that is expected to generate \$40,000 in revenue per year, t years from now, for the next 5 years. If the income is reinvested in a business earning interest at the rate of 12% per year compounded continuously, find the total accumulated value of this income stream at the end of 5 years.

Solution We are required to find the total future value of the given income stream after 5 years. Using Equation (18) with $R(t) = 40{,}000$, $r = 0.12$, and $T = 5$, we see that the required value is given by

$$\begin{aligned} &e^{0.12(5)}\int_0^5 40{,}000e^{-0.12t}\,dt \\ &= e^{0.6}\left[-\frac{40{,}000}{0.12}e^{-0.12t}\right]\Bigg|_0^5 \qquad \text{Integrate using the substitution } u = -0.12t. \\ &= -\frac{40{,}000e^{0.6}}{0.12}(e^{-0.6} - 1) \approx 274{,}039.60 \end{aligned}$$

or approximately \$274,040.

Another way of measuring the value of an income stream is by considering its present value. The **present value of an income stream** of $R(t)$ dollars per year over a term of T years, earning interest at the rate of r per year compounded continuously, is the principal P that will yield the same accumulated value as the income stream itself when P is invested today for a period of T years at the same rate of interest. In other words,

$$Pe^{rT} = e^{rT}\int_0^T R(t)e^{-rt}\,dt$$

Dividing both sides of the equation by e^{rT} gives the following result.

Present Value of an Income Stream

The present value of an income stream of $R(t)$ dollars per year, earning interest at the rate of r per year compounded continuously, is given by

$$PV = \int_0^T R(t)e^{-rt}\,dt \tag{19}$$

VIDEO

APPLIED EXAMPLE 3 Investment Analysis The owner of a local cinema is considering two alternative plans for renovating and improving the theater. Plan A calls for an immediate cash outlay of \$250,000, whereas plan B requires an immediate cash outlay of \$180,000. It has been estimated that adopting plan A would result in a net income stream generated at the rate of

$$f(t) = 630{,}000$$

dollars per year, whereas adopting plan B would result in a net income stream generated at the rate of

$$g(t) = 580{,}000$$

dollars per year for the next 3 years. If the prevailing interest rate for the next 5 years is 10% per year, which plan will generate a higher net income by the end of 3 years?

Solution Since the initial outlay is \$250,000, we find—using Equation (19) with $R(t) = 630{,}000$, $r = 0.1$, and $T = 3$—that the present value of the net income under plan A is given by

$$\begin{aligned}
&\int_0^3 630{,}000e^{-0.1t}\,dt - 250{,}000 \\
&\quad = \frac{630{,}000}{-0.1}e^{-0.1t}\Big|_0^3 - 250{,}000 && \text{Integrate using the substitution } u = -0.1t. \\
&\quad = -6{,}300{,}000e^{-0.3} + 6{,}300{,}000 - 250{,}000 \\
&\quad \approx 1{,}382{,}845
\end{aligned}$$

or approximately \$1,382,845.

To find the present value of the net income under plan B, we use (19) with $R(t) = 580{,}000$, $r = 0.1$, and $T = 3$, obtaining

$$\int_0^3 580{,}000e^{-0.1t}\,dt - 180{,}000$$

dollars. Proceeding as in the previous computation, we see that the required value is \$1,323,254 (see Exercise 8, page 474).

Comparing the present value of each plan, we conclude that plan A would generate a higher net income by the end of 3 years. ■

Note The function R in Example 3 is a constant function. If R is not a constant function, then we may need more sophisticated techniques of integration to evaluate the integral in (19). Exercises 7.1 and 7.2 contain problems of this type. ■

The Amount and Present Value of an Annuity

An annuity is a sequence of payments made at regular time intervals. The time period in which these payments are made is called the *term* of the annuity. Although the payments need not be equal in size, they are equal in many important applications, and we will assume that they are equal in our discussion. Examples of annuities are regular deposits to a savings account, monthly home mortgage payments, and monthly insurance payments.

The **amount of an annuity** is the sum of the payments plus the interest earned. A formula for computing the amount of an annuity A can be derived with the help of (18). Let

P = Size of each payment in the annuity
r = Interest rate compounded continuously
T = Term of the annuity (in years)
m = Number of payments per year

The payments into the annuity constitute a constant income stream of $R(t) = mP$ dollars per year. With this value of $R(t)$, (18) yields

$$\begin{aligned} A &= e^{rT}\int_0^T R(t)e^{-rt}\,dt = e^{rT}\int_0^T mPe^{-rt}\,dt \\ &= mPe^{rT}\left[-\frac{e^{-rt}}{r}\right]\Bigg|_0^T = mPe^{rT}\left[-\frac{e^{-rT}}{r} + \frac{1}{r}\right] \\ &= \frac{mP}{r}(e^{rT} - 1) \qquad \text{Since } e^{rT} \cdot e^{-rT} = 1 \end{aligned}$$

This leads us to the following formula.

Amount of an Annuity

The amount of an annuity is

$$A = \frac{mP}{r}(e^{rT} - 1) \tag{20}$$

where P, r, T, and m are as defined earlier.

APPLIED EXAMPLE 4 IRAs On January 1, 1990, Marcus Chapman deposited \$2000 into an Individual Retirement Account (IRA) paying interest at the rate of 10% per year compounded continuously. Assuming that he deposits \$2000 annually into the account, how much will he have in his IRA at the beginning of the year 2006?

Solution We use (20), with $P = 2000$, $r = 0.1$, $T = 16$, and $m = 1$, obtaining

$$\begin{aligned} A &= \frac{2000}{0.1}(e^{1.6} - 1) \\ &\approx 79{,}060.65 \end{aligned}$$

Thus, Marcus will have approximately \$79,061 in his account at the beginning of the year 2006. ■

EXPLORING WITH TECHNOLOGY

Refer to Example 4. Suppose Marcus wishes to know how much he will have in his IRA at any time in the future, not just at the beginning of 2006, as you were asked to compute in the example.

1. Using Formula (18) and the relevant data from Example 4, show that the required amount at any time x (x measured in years, $x > 0$) is given by

$$A = f(x) = 20{,}000(e^{0.1x} - 1)$$

2. Use a graphing utility to plot the graph of f, using the viewing window $[0, 30] \times [2000, 400{,}000]$.
3. Using ZOOM and TRACE, or using the function evaluation capability of your graphing utility, use the result of part 2 to verify the result obtained in Example 4. Comment on the advantage of the mathematical model found in part 1.

Using (19), we can derive the following formula for the present value of an annuity.

Present Value of an Annuity

The present value of an annuity is given by

$$PV = \frac{mP}{r}(1 - e^{-rT}) \tag{21}$$

where P, r, T, and m are as defined earlier.

APPLIED EXAMPLE 5 Sinking Funds Tomas Perez, the proprietor of a hardware store, wants to establish a fund from which he will withdraw \$1000 per month for the next 10 years. If the fund earns interest at the rate of 9% per year compounded continuously, how much money does he need to establish the fund?

Solution We want to find the present value of an annuity with $P = 1000$, $r = 0.09$, $T = 10$, and $m = 12$. Using Equation (21), we find

$$\begin{aligned} PV &= \frac{12{,}000}{0.09}(1 - e^{-(0.09)(10)}) \\ &\approx 79{,}124.05 \end{aligned}$$

Thus, Tomas needs approximately \$79,124 to establish the fund. ■

Lorentz Curves and Income Distributions

One method used by economists to study the distribution of income in a society is based on the **Lorentz curve**, named after American statistician M. D. Lorentz. To describe the Lorentz curve, let $f(x)$ denote the proportion of the total income received by the poorest $100x\%$ of the population for $0 \le x \le 1$. Using this terminology, $f(0.3) = 0.1$ simply states that the lowest 30% of the income recipients receive 10% of the total income.

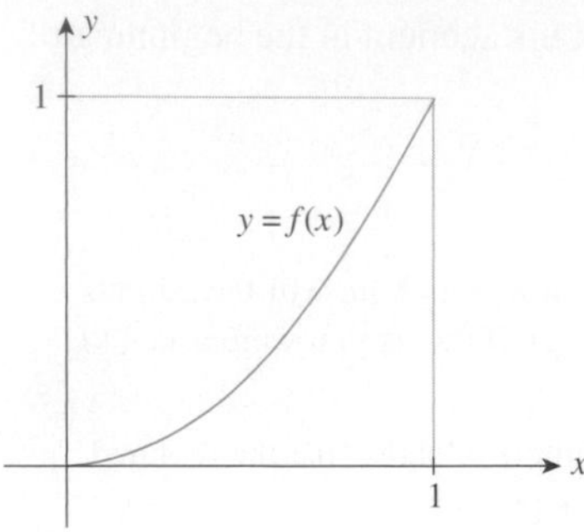

FIGURE 45
A Lorentz curve

The function f has the following properties:

1. The domain of f is $[0, 1]$.
2. The range of f is $[0, 1]$.
3. $f(0) = 0$ and $f(1) = 1$.
4. $f(x) \leq x$ for every x in $[0, 1]$.
5. f is increasing on $[0, 1]$.

The first two properties follow from the fact that both x and $f(x)$ are fractions of a whole. Property 3 is a statement that 0% of the income recipients receive 0% of the total income and 100% of the income recipients receive 100% of the total income. Property 4 follows from the fact that the lowest $100x\%$ of the income recipients cannot receive more than $100x\%$ of the total income. A typical Lorentz curve is shown in Figure 45.

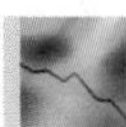

APPLIED EXAMPLE 6 Lorentz Curves A developing country's income distribution is described by the function

$$f(x) = \frac{19}{20}x^2 + \frac{1}{20}x$$

a. Sketch the Lorentz curve for the given function.
b. Compute $f(0.2)$ and $f(0.8)$ and interpret your results.

Solution

a. The Lorentz curve is shown in Figure 46.

b. $$f(0.2) = \frac{19}{20}(0.2)^2 + \frac{1}{20}(0.2) = 0.048$$

Thus, the lowest 20% of the people receive 4.8% of the total income.

$$f(0.8) = \frac{19}{20}(0.8)^2 + \frac{1}{20}(0.8) = 0.648$$

Thus, the lowest 80% of the people receive 64.8% of the total income.

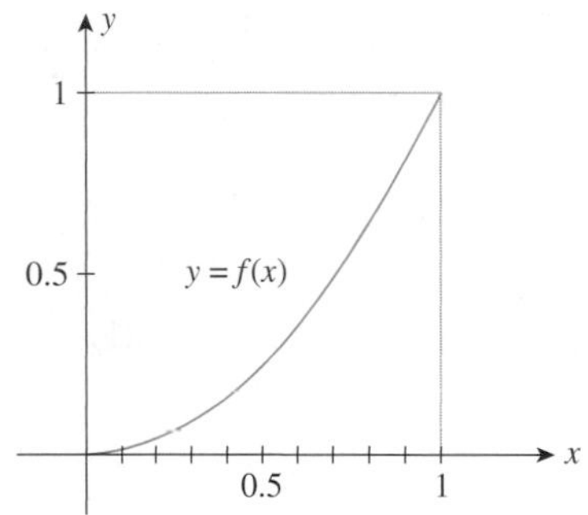

FIGURE 46
The Lorentz curve $f(x) = \frac{19}{20}x^2 + \frac{1}{20}x$

Next, let's consider the Lorentz curve described by the function $y = f(x) = x$. Since exactly $100x\%$ of the total income is received by the lowest $100x\%$ of income recipients, the line $y = x$ is called the **line of complete equality**. For example, 10% of the total income is received by the lowest 10% of income recipients, 20% of the total income is received by the lowest 20% of income recipients, and so on. Now, it is evident that the closer a Lorentz curve is to this line, the more equitable the income distribution is among the income recipients. But the proximity of a Lorentz curve to the line of complete equality is reflected by the area between the Lorentz curve and the line $y = x$ (Figure 47). The closer the curve is to the line, the smaller the enclosed area.

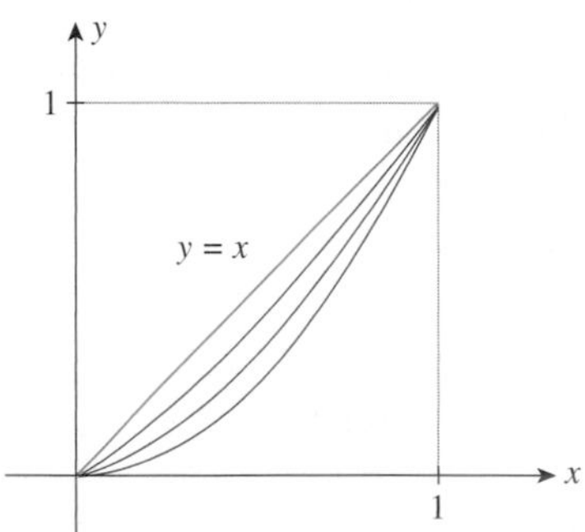

FIGURE 47
The closer the Lorentz curve is to the line, the more equitable the income distribution.

This observation suggests that we may define a number, called the coefficient of inequality of a Lorentz curve, as the ratio of the area between the line of complete equality and the Lorentz curve to the area under the line of complete equality. Since the area under the line of complete equality is $\frac{1}{2}$, we see that the coefficient of inequality is given by the following formula.

Coefficient of Inequality of a Lorentz Curve

The coefficient of inequality, or **Gini index**, of a Lorentz curve is

$$L = 2\int_0^1 [x - f(x)]\, dx \tag{22}$$

The coefficient of inequality is a number between 0 and 1. For example, a coefficient of zero implies that the income distribution is perfectly uniform.

APPLIED EXAMPLE 7 Income Distributions In a study conducted by a certain country's Economic Development Board with regard to the income distribution of certain segments of the country's workforce, it was found that the Lorentz curves for the distribution of income of medical doctors and of movie actors are described by the functions

$$f(x) = \frac{14}{15}x^2 + \frac{1}{15}x \quad \text{and} \quad g(x) = \frac{5}{8}x^4 + \frac{3}{8}x$$

respectively. Compute the coefficient of inequality for each Lorentz curve. Which profession has a more equitable income distribution?

Solution The required coefficients of inequality are, respectively,

$$\begin{aligned}
L_1 &= 2\int_0^1 \left[x - \left(\frac{14}{15}x^2 + \frac{1}{15}x\right)\right] dx = 2\int_0^1 \left(\frac{14}{15}x - \frac{14}{15}x^2\right) dx \\
&= \frac{28}{15}\int_0^1 (x - x^2)\, dx = \frac{28}{15}\left(\frac{1}{2}x^2 - \frac{1}{3}x^3\right)\Bigg|_0^1 \\
&= \frac{14}{45} \approx 0.311
\end{aligned}$$

$$\begin{aligned}
L_2 &= 2\int_0^1 \left[x - \left(\frac{5}{8}x^4 + \frac{3}{8}x\right)\right] dx = 2\int_0^1 \left(\frac{5}{8}x - \frac{5}{8}x^4\right) dx \\
&= \frac{5}{4}\int_0^1 (x - x^4)\, dx = \frac{5}{4}\left(\frac{1}{2}x^2 - \frac{1}{5}x^5\right)\Bigg|_0^1 \\
&= \frac{15}{40} \approx 0.375
\end{aligned}$$

We conclude that in this country the incomes of medical doctors are more evenly distributed than the incomes of movie actors. ■

6.7 Self-Check Exercise

The demand function for a certain make of exercise bicycle that is sold exclusively through cable television is

$$p = d(x) = \sqrt{9 - 0.02x}$$

where p is the unit price in hundreds of dollars and x is the quantity demanded/week. The corresponding supply function is given by

$$p = s(x) = \sqrt{1 + 0.02x}$$

where p has the same meaning as before and x is the number of exercise bicycles the supplier will make available at price p. Determine the consumers' surplus and the producers' surplus if the unit price is set at the equilibrium price.

The solution to Self-Check Exercise 6.7 can be found on page 476.

6.7 Concept Questions

1. **a.** Define consumers' surplus. Give a formula for computing it.
   b. Define producers' surplus. Give a formula for computing it.
2. **a.** Define the accumulated (future) value of an income stream. Give a formula for computing it.
   b. Define the present value of an income stream. Give a formula for computing it.
3. Define the amount of an annuity. Give a formula for computing it.
4. Explain the following terms: (a) Lorentz curve (b) Coefficient of inequality of a Lorentz curve.

6.7 Exercises

1. CONSUMERS' SURPLUS The demand function for a certain make of replacement cartridges for a water purifier is given by

$$p = -0.01x^2 - 0.1x + 6$$

where p is the unit price in dollars and x is the quantity demanded each week, measured in units of a thousand. Determine the consumers' surplus if the market price is set at \$4/cartridge.

2. CONSUMERS' SURPLUS The demand function for a certain brand of CD is given by

$$p = -0.01x^2 - 0.2x + 8$$

where p is the wholesale unit price in dollars and x is the quantity demanded each week, measured in units of a thousand. Determine the consumers' surplus if the wholesale market price is set at \$5/disc.

3. CONSUMERS' SURPLUS It is known that the quantity demanded of a certain make of portable hair dryer is x hundred units/week and the corresponding wholesale unit price is

$$p = \sqrt{225 - 5x}$$

dollars. Determine the consumers' surplus if the wholesale market price is set at \$10/unit.

4. PRODUCERS' SURPLUS The supplier of the portable hair dryers in Exercise 3 will make x hundred units of hair dryers available in the market when the wholesale unit price is

$$p = \sqrt{36 + 1.8x}$$

dollars. Determine the producers' surplus if the wholesale market price is set at \$9/unit.

5. PRODUCERS' SURPLUS The supply function for the CDs of Exercise 2 is given by

$$p = 0.01x^2 + 0.1x + 3$$

where p is the unit wholesale price in dollars and x stands for the quantity that will be made available in the market by the supplier, measured in units of a thousand. Determine the producers' surplus if the wholesale market price is set at the equilibrium price.

6. CONSUMERS' AND PRODUCERS' SURPLUS The management of the Titan Tire Company has determined that the quantity demanded x of their Super Titan tires/week is related to the unit price p by the relation

$$p = 144 - x^2$$

where p is measured in dollars and x is measured in units of a thousand. Titan will make x units of the tires available in the market if the unit price is

$$p = 48 + \frac{1}{2}x^2$$

dollars. Determine the consumers' surplus and the producers' surplus when the market unit price is set at the equilibrium price.

7. CONSUMERS' AND PRODUCERS' SURPLUS The quantity demanded x (in units of a hundred) of the Mikado miniature cameras/week is related to the unit price p (in dollars) by

$$p = -0.2x^2 + 80$$

and the quantity x (in units of a hundred) that the supplier is willing to make available in the market is related to the unit price p (in dollars) by

$$p = 0.1x^2 + x + 40$$

If the market price is set at the equilibrium price, find the consumers' surplus and the producers' surplus.

8. Refer to Example 3, page 469. Verify that

$$\int_0^3 580{,}000e^{-0.1t}\,dt - 180{,}000 \approx 1{,}323{,}254$$

9. **Present Value of an Investment** Suppose an investment is expected to generate income at the rate of

$$R(t) = 200{,}000$$

dollars/year for the next 5 yr. Find the present value of this investment if the prevailing interest rate is 8%/year compounded continuously.

10. **Franchises** Camille purchased a 15-yr franchise for a computer outlet store that is expected to generate income at the rate of

$$R(t) = 400{,}000$$

dollars/year. If the prevailing interest rate is 10%/year compounded continuously, find the present value of the franchise.

11. **The Amount of an Annuity** Find the amount of an annuity if \$250/month is paid into it for a period of 20 yr, earning interest at the rate of 8%/year compounded continuously.

12. **The Amount of an Annuity** Find the amount of an annuity if \$400/month is paid into it for a period of 20 yr, earning interest at the rate of 8%/year compounded continuously.

13. **The Amount of an Annuity** Aiso deposits \$150/month in a savings account paying 8%/year compounded continuously. Estimate the amount that will be in his account after 15 yr.

14. **Custodial Accounts** The Armstrongs wish to establish a custodial account to finance their children's education. If they deposit \$200 monthly for 10 yr in a savings account paying 9%/year compounded continuously, how much will their savings account be worth at the end of this period?

15. **IRA Accounts** Refer to Example 4, page 470. Suppose Marcus makes his IRA payment on April 1, 1990, and annually thereafter. If interest is paid at the same initial rate, approximately how much will Marcus have in his account at the beginning of the year 2006?

16. **Present Value of an Annuity** Estimate the present value of an annuity if payments are \$800 monthly for 12 yr and the account earns interest at the rate of 10%/year compounded continuously.

17. **Present Value of an Annuity** Estimate the present value of an annuity if payments are \$1200 monthly for 15 yr and the account earns interest at the rate of 10%/year compounded continuously.

18. **Lottery Payments** A state lottery commission pays the winner of the "Million Dollar" lottery 20 annual installments of \$50,000 each. If the prevailing interest rate is 8%/year compounded continuously, find the present value of the winning ticket.

19. **Reverse Annuity Mortgages** Sinclair wishes to supplement his retirement income by \$300/month for the next 10 yr. He plans to obtain a reverse annuity mortgage (RAM) on his home to meet this need. Estimate the amount of the mortgage he will require if the prevailing interest rate is 12%/year compounded continuously.

20. **Reverse Annuity Mortgage** Refer to Exercise 19. Leah wishes to supplement her retirement income by \$400/month for the next 15 yr by obtaining a RAM. Estimate the amount of the mortgage she will require if the prevailing interest rate is 9%/year compounded continuously.

21. **Lorentz Curves** A certain country's income distribution is described by the function

$$f(x) = \frac{15}{16}x^2 + \frac{1}{16}x$$

a. Sketch the Lorentz curve for this function.
b. Compute $f(0.4)$ and $f(0.9)$ and interpret your results.

22. **Lorentz Curves** In a study conducted by a certain country's Economic Development Board, it was found that the Lorentz curve for the distribution of income of college teachers was described by the function

$$f(x) = \frac{13}{14}x^2 + \frac{1}{14}x$$

and that of lawyers by the function

$$g(x) = \frac{9}{11}x^4 + \frac{2}{11}x$$

a. Compute the coefficient of inequality for each Lorentz curve.
b. Which profession has a more equitable income distribution?

23. **Lorentz Curves** A certain country's income distribution is described by the function

$$f(x) = \frac{14}{15}x^2 + \frac{1}{15}x$$

a. Sketch the Lorentz curve for this function.
b. Compute $f(0.3)$ and $f(0.7)$.

24. **Lorentz Curves** In a study conducted by a certain country's Economic Development Board, it was found that the Lorentz curve for the distribution of income of stockbrokers was described by the function

$$f(x) = \frac{11}{12}x^2 + \frac{1}{12}x$$

and that of high school teachers by the function

$$g(x) = \frac{5}{6}x^2 + \frac{1}{6}x$$

a. Compute the coefficient of inequality for each Lorentz curve.
b. Which profession has a more equitable income distribution?

6.7 Solution to Self-Check Exercise

We find the equilibrium price and equilibrium quantity by solving the system of equations

$$p = \sqrt{9 - 0.02x}$$
$$p = \sqrt{1 + 0.02x}$$

simultaneously. Substituting the first equation into the second, we have

$$\sqrt{9 - 0.02x} = \sqrt{1 + 0.02x}$$

Squaring both sides of the equation then leads to

$$9 - 0.02x = 1 + 0.02x$$
$$x = 200$$

Therefore,

$$\begin{aligned} p &= \sqrt{9 - 0.02(200)} \\ &= \sqrt{5} \approx 2.24 \end{aligned}$$

The equilibrium price is $224, and the equilibrium quantity is 200. The consumers' surplus is given by

$$\begin{aligned} CS &= \int_0^{200} \sqrt{9 - 0.02x}\, dx - (2.24)(200) \\ &= \int_0^{200} (9 - 0.02x)^{1/2}\, dx - 448 \\ &= -\frac{1}{0.02}\left(\frac{2}{3}\right)(9 - 0.02x)^{3/2}\Big|_0^{200} - 448 \qquad \text{Integrate by substitution.} \\ &= -\frac{1}{0.03}(5^{3/2} - 9^{3/2}) - 448 \\ &\approx 79.32 \end{aligned}$$

or approximately $7932.

Next, the producers' surplus is given by

$$\begin{aligned} PS &= (2.24)(200) - \int_0^{200} \sqrt{1 + 0.02x}\, dx \\ &= 448 - \int_0^{200} (1 + 0.02x)^{1/2}\, dx \\ &= 448 - \frac{1}{0.02}\left(\frac{2}{3}\right)(1 + 0.02x)^{3/2}\Big|_0^{200} \\ &= 448 - \frac{1}{0.03}(5^{3/2} - 1) \\ &\approx 108.66 \end{aligned}$$

or approximately $10,866.

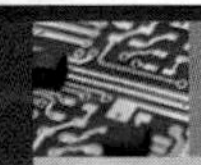

USING TECHNOLOGY

Business and Economic Applications/Technology Exercises

1. Re-solve Example 1, Section 6.7, using a graphing utility.
   Hint: Use the intersection operation to find the equilibrium quantity and the equilibrium price. Use the numerical integral operation to evaluate the definite integral.
2. Re-solve Exercise 7, Section 6.7, using a graphing utility.
   Hint: See Exercise 1.
3. The demand function for a certain brand of travel alarm clocks is given by

   $$p = -0.01x^2 - 0.3x + 10$$

   where p is the wholesale unit price in dollars and x is the quantity demanded each month, measured in units of a thousand. The supply function for this brand of clocks is given by

   $$p = -0.01x^2 + 0.2x + 4$$

   where p has the same meaning as before and x is the quantity, in thousands, the supplier will make available in the marketplace per month. Determine the consumers' surplus and the producers' surplus when the market unit price is set at the equilibrium price.

4. The quantity demanded of a certain make of compact disc organizer is x thousand units per week, and the corresponding wholesale unit price is

$$p = \sqrt{400 - 8x}$$

dollars. The supplier of the organizers will make x thousand units available in the market when the unit wholesale price is

$$p = 0.02x^2 + 0.04x + 5$$

dollars. Determine the consumers' surplus and the producers' surplus when the market unit price is set at the equilibrium price.

5. Investment A is expected to generate income at the rate of

$$R_1(t) = 50{,}000 + 10{,}000\sqrt{t}$$

dollars/year for the next 5 years and investment B is expected to generate income at the rate of

$$R_2(t) = 50{,}000 + 6000t$$

dollars/year over the same period of time. If the prevailing interest rate for the next 5 years is 10%/year, which investment will generate a higher net income by the end of 5 years?

6. Investment A is expected to generate income at the rate of

$$R_1(t) = 40{,}000 + 5000t + 100t^2$$

dollars/year for the next 10 years and investment B is expected to generate income at the rate of

$$R_2(t) = 60{,}000 + 2000t$$

dollars/year over the same period of time. If the prevailing interest rate for the next 10 years is 8%/year, which investment will generate a higher net income by the end of 10 years?

6.8 Volumes of Solids of Revolution (Optional)

Finding the Volume of a Solid of Revolution

In this section we use a Riemann sum to find a formula for computing the volume of a solid that results when a region is revolved about an axis. The solid is called a **solid of revolution**. To find the volume of a solid of revolution, suppose the plane region under the curve defined by a nonnegative continuous function $y = f(x)$ between $x = a$ and $x = b$ is revolved about the x-axis (Figure 48).

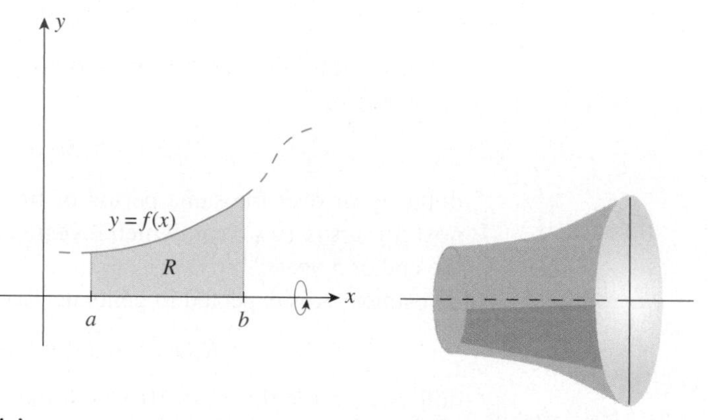

FIGURE 48

(a) Region R under the curve

(b) Solid obtained by revolving R about the x-axis

To derive a formula for finding the volume of the solid of revolution, we divide the interval $[a, b]$ into n subintervals, each of equal length, by means of the points $x_0 = a, x_1, x_2, \ldots, x_n = b$, so that the length of each subinterval is given by $\Delta x = (b - a)/n$. Also, let $p_1, p_2, \ldots, p_n$ be points in the subintervals $[x_0, x_1]$, $[x_1, x_2], \ldots, [x_{n-1}, x_n]$, respectively (Figure 49).

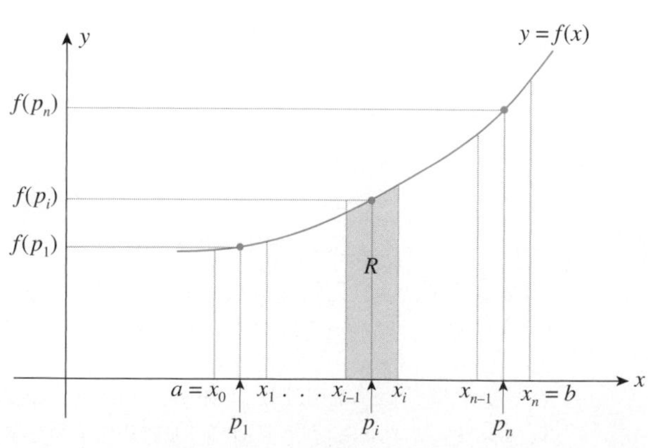

FIGURE 49
The region R is revolved about the x-axis.

Let's concentrate on that part of the solid of revolution swept out by revolving the region under the curve between $x = x_{i-1}$ and $x = x_i$ about the x-axis. The volume ΔV_i of this object, shown in Figure 50a, may be approximated by the volume of the disk shown in Figure 50b, of radius $f(p_i)$ and width Δx, obtained by revolving the rectangular region of height $f(p_i)$ and width Δx about the x-axis (Figure 50c).

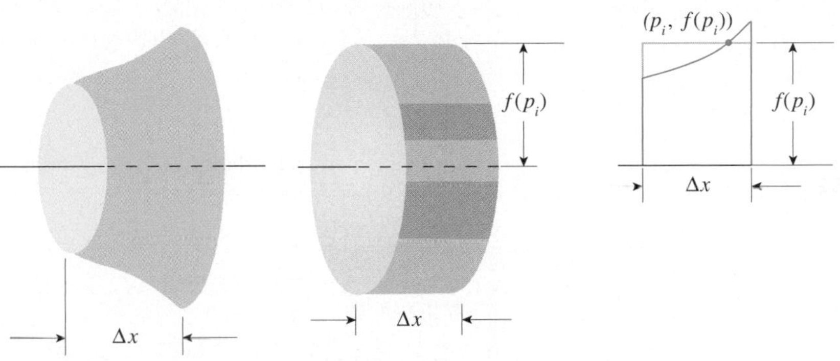

FIGURE 50

(a) Volume ΔV_i of object **(b)** Volume of disk **(c)** Approximating rectangle

Since the volume of a disk (cylinder) is equal to $\pi r^2 h$, we see that

$$\Delta V_i \approx \pi[f(p_i)]^2 \Delta x$$

This analysis suggests that the volume of the solid of revolution may be approximated by the sum of the volumes of n suitable disks—namely,

$$\pi[f(p_1)]^2 \Delta x + \pi[f(p_2)]^2 \Delta x + \cdots + \pi[f(p_n)]^2 \Delta x$$

(shown in Figure 51). This last expression is a Riemann sum of the function $g(x) = \pi[f(x)]^2$ over the interval $[a, b]$. Letting n approach infinity, we obtain the following formula.

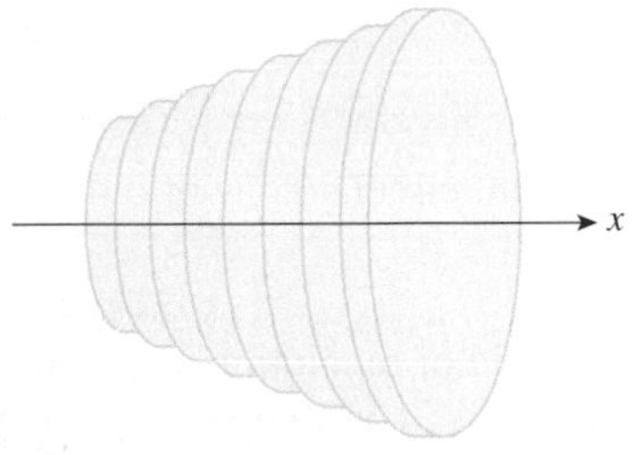

FIGURE 51
The solid of revolution is approximated by n disks.

Volume of a Solid of Revolution

The volume V of the solid of revolution obtained by revolving the region below the graph of a nonnegative function $y = f(x)$ from $x = a$ to $x = b$ about the x-axis is

$$V = \pi \int_a^b [f(x)]^2 \, dx \qquad \textbf{(23)}$$

EXAMPLE 1 Find the volume of the solid of revolution obtained by revolving the region under the curve $y = f(x) = e^{-x}$ from $x = 0$ to $x = 1$ about the x-axis.

Solution The region under the curve and the resulting solid of revolution are shown in Figure 52. Using Formula (23), we find that the required volume

FIGURE 52

(a) The region under the curve $y = e^{-x}$ from $x = 0$ to $x = 1$

(b) The solid of revolution obtained when R is revolved about the x-axis

is given by

$$\pi \int_0^1 (e^{-x})^2 \, dx = \pi \int_0^1 e^{-2x} \, dx$$

$$= -\frac{\pi}{2} e^{-2x} \Big|_0^1 = \frac{\pi}{2}(1 - e^{-2})$$

cubic units. ■

EXAMPLE 2 Derive a formula for the volume of a right circular cone of radius r and height h.

Solution The cone is shown in Figure 53a. It is the solid of revolution obtained by revolving the region R of Figure 53b about the x-axis. The region R is the region under the straight line $y = f(x) = (r/h)x$ between $x = 0$ and $x = h$.

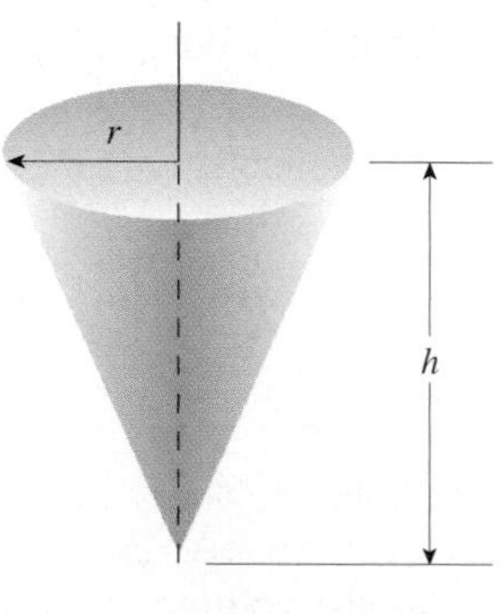

(a) A right circular cone of radius r and height h.

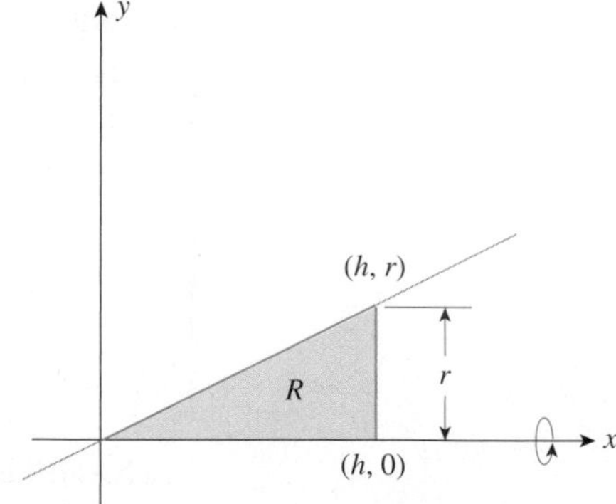

(b) R is the region under the straight line $y = \dfrac{r}{h}x$ from $x = 0$ to $x = h$.

FIGURE 53

Using Formula (23), we see that the required volume is given by

$$V = \pi \int_0^h \left[\left(\frac{r}{h}\right)x\right]^2 dx = \frac{\pi r^2}{h^2}\int_0^h x^2\,dx$$

$$= \left(\frac{\pi r^2}{3h^2}\right)x^3\Big|_0^h = \frac{1}{3}\pi r^2 h$$

cubic units. ■

Next, let's consider the solid of revolution obtained by revolving the region R bounded above by the graph of the nonnegative function $f(x)$ and below by the graph of the nonnegative function $g(x)$, from $x = a$ to $x = b$, about the x-axis (Figure 54).

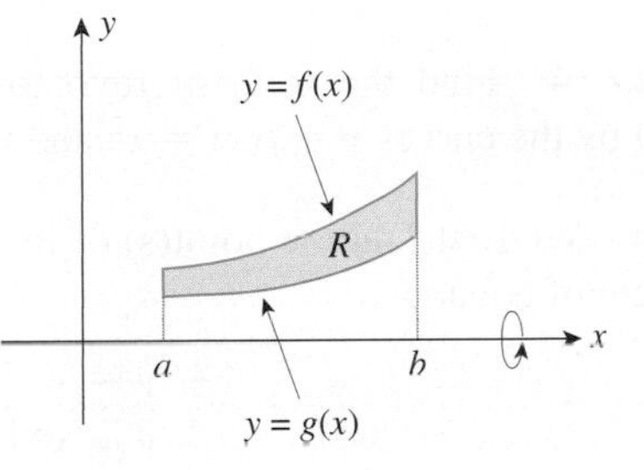

(a) R is the region bounded by the curves $y = f(x)$ and $y = g(x)$ from $x = a$ to $x = b$.

(b) The solid of revolution obtained by revolving R about the x-axis.

FIGURE 54

To derive a formula for computing the volume of this solid of revolution, observe that the required volume is the volume of the solid of revolution obtained by revolving the region under the curve $y = f(x)$ from $x = a$ to $x = b$ about the x-axis *minus* the volume of the solid of revolution obtained by revolving the region under the curve $y = g(x)$ from $x = a$ to $x = b$ about the x-axis. Thus, the required volume is given by

$$V = \pi \int_a^b [f(x)]^2\,dx - \pi \int_a^b [g(x)]^2\,dx \qquad \textbf{(24)}$$

or

$$V = \pi \int_a^b \{[f(x)]^2 - [g(x)]^2\}\,dx$$

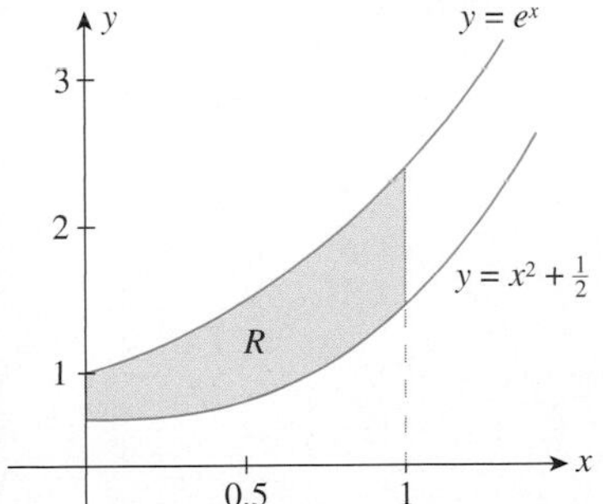

FIGURE 55
R is the region bounded by the curves $y = e^x$ and $y = x^2 + \frac{1}{2}$ from $x = 0$ to $x = 1$.

EXAMPLE 3 Find the volume of the solid of revolution obtained by revolving the region bounded by the curves $y = e^x$ and $y = x^2 + \frac{1}{2}$ from $x = 0$ to $x = 1$ about the x-axis (Figure 55).

Solution Using Formula (24) with $f(x) = e^x$ and $g(x) = x^2 + \frac{1}{2}$, we see that the required volume is given by

$$V = \pi \int_0^1 \left\{ [e^x]^2 - \left[x^2 + \frac{1}{2} \right]^2 \right\} dx$$

$$= \pi \int_0^1 \left(e^{2x} - x^4 - x^2 - \frac{1}{4} \right) dx$$

$$= \pi \left(\frac{1}{2} e^{2x} - \frac{1}{5} x^5 - \frac{1}{3} x^3 - \frac{1}{4} x \right) \Big|_0^1$$

$$= \pi \left[\left(\frac{1}{2} e^2 - \frac{1}{5} - \frac{1}{3} - \frac{1}{4} \right) - \frac{1}{2} \right]$$

$$= \frac{\pi}{60} (30e^2 - 77)$$

or approximately 7.57 cubic units. ■

EXAMPLE 4 Find the solid of revolution obtained by revolving the region bounded by the curves $y = f(x) = x^2$ and $y = g(x) = x^3$ about the x-axis.

Solution We first find the point(s) of intersection of the two curves by solving the system of equations

$$\left. \begin{aligned} y &= x^2 \\ y &= x^3 \end{aligned} \right\}$$

We have

$$\begin{aligned} x^3 &= x^2 \\ x^3 - x^2 &= 0 \\ x^2(x - 1) &= 0 \end{aligned}$$

so that $x = 0$ or $x = 1$, and the points of intersection are (0, 0) and (1, 1) (Figure 56).

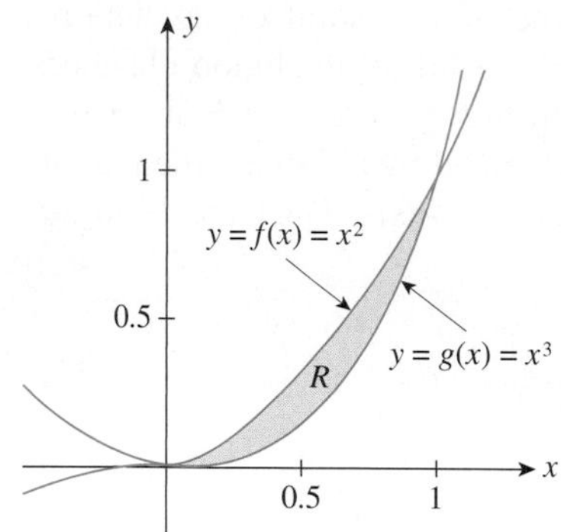

FIGURE 56
R is the region bounded by the curves $y = x^2$ and $y = x^3$.

Next, observe that $g(x) \le f(x)$ for all values of x between $x = 0$ and $x = 1$ so that, using Formula (24), we find the required volume to be

$$V = \pi \int_0^1 [(x^2)^2 - (x^3)^2] \, dx$$

$$= \pi \int_0^1 (x^4 - x^6) \, dx$$

$$= \pi \left(\frac{1}{5} x^5 - \frac{1}{7} x^7 \right) \Big|_0^1$$

$$= \pi \left(\frac{1}{5} - \frac{1}{7} \right)$$

$$= \frac{2\pi}{35}$$

cubic units. ■

APPLIED EXAMPLE 5 The external fuel tank for a space shuttle has a shape that may be obtained by revolving the region under the curve

$$f(x) = \begin{cases} 4\sqrt{10} & \text{if } -120 \le x \le 10 \\ \dfrac{1}{5}\sqrt{x}(30 - x) & \text{if } 10 < x \le 30 \end{cases}$$

from $x = -120$ to $x = 30$ about the x-axis (Figure 57) where all measurements are given in feet. The tank carries liquid hydrogen for fueling the shuttle's three main engines. Estimate the capacity of the tank (231 cubic inches = 1 gallon).

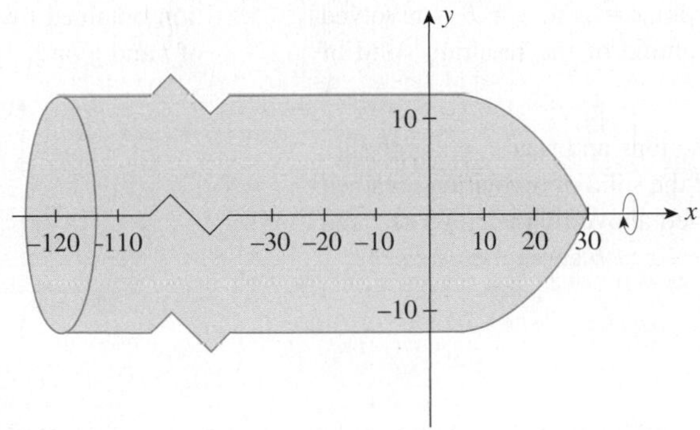

FIGURE 57
The solid of revolution obtained by revolving the region under the curve $y = f(x)$ about the x-axis.

Solution The volume of the tank is given by

$$\begin{aligned}
V &= \pi \int_{-120}^{30} [f(x)]^2\, dx \\
&= \pi \int_{-120}^{10} (4\sqrt{10})^2\, dx + \pi \int_{10}^{30} \left[\frac{1}{5}\sqrt{x}(30 - x)\right]^2 dx \\
&= 160\pi \int_{-120}^{10} dx + \frac{\pi}{25}\int_{10}^{30} x(30 - x)^2\, dx \\
&= 160\pi x \Big|_{-120}^{10} + \frac{\pi}{25}\int_{10}^{30} (900x - 60x^2 + x^3)\, dx \\
&= 160\pi(130) + \frac{\pi}{25}\left(450x^2 - 20x^3 + \frac{1}{4}x^4\right)\Big|_{10}^{30} \\
&= 20{,}800\pi + \frac{\pi}{25}\left\{\left[450(30)^2 - 20(30)^3 + \frac{1}{4}(30)^4\right]\right. \\
&\qquad \left. - \left[450(10)^2 - 20(10)^3 + \frac{1}{4}(10)^4\right]\right\} \\
&= 20{,}800\pi + 1600\pi = 22{,}400\pi
\end{aligned}$$

or approximately 70,372 cubic feet. Therefore, its capacity is approximately $(70{,}372)(12^3)/231$, or approximately 526,419 gallons. ■

6.8 Self-Check Exercise

Find the volume of the solid of revolution obtained by revolving the region bounded by the graphs of $f(x) = 5 - x^2$ and $g(x) = 1$ about the x-axis.

The solution to Self-Check Exercise 6.8 can be found on page 485.

6.8 Concept Questions

1. If f is a nonnegative function on the interval $[a, b]$ and the region under the graph of f from $x = a$ to $x = b$ is revolved about the axis, what is the volume of the resulting solid of revolution?

2. If f and g are nonnegative functions and $f(x) \geq g(x)$ for all x in $[a, b]$, what is the volume of the solid of revolution obtained by revolving the region bounded above by the graph of f and below by the graph of g for $a \leq x \leq b$ about the x-axis?

3. Write an expression giving the volume of the solid of revolution obtained by revolving the region bounded by the graphs of f and g on $[a, b]$ (shown below) about the x-axis.

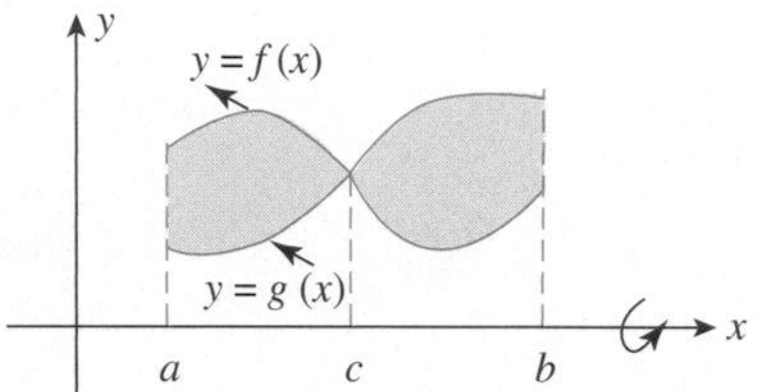

6.8 Exercises

In Exercises 1–10, find the volume of the solid of revolution obtained by revolving the region under the curve $y = f(x)$ from $x = a$ to $x = b$ about the x-axis.

1. $y = 3x$; $a = 0, b = 1$
2. $y = x + 1$; $a = 0, b = 2$
3. $y = \sqrt{x}$; $a = 1, b = 4$
4. $y = \sqrt{x + 1}$; $a = 0, b = 3$
5. $y = \sqrt{1 + x^2}$; $a = 0, b = 1$
6. $y = \sqrt{4 - x^2}$; $a = -2, b = 2$
7. $y = 1 - x^2$; $a = -1, b = 1$
8. $y = 2x - x^2$; $a = 0, b = 2$
9. $y = e^x$; $a = 0, b = 1$
10. $y = e^{-3x}$; $a = 0, b = 1$

In Exercises 11–18, find the volume of the solid of revolution obtained by revolving the region bounded above by the curve $y = f(x)$ and below by the curve $y = g(x)$ from $x = a$ to $x = b$ about the x-axis.

11. $f(x) = x$ and $g(x) = x^2$; $a = 0, b = 1$
12. $f(x) = 1$ and $g(x) = \sqrt{x}$; $a = 0, b = 1$
13. $f(x) = 4 - x^2$ and $g(x) = 3$; $a = -1, b = 1$
14. $f(x) = 8$ and $g(x) = x^{3/2}$; $a = 0, b = 4$
15. $f(x) = \sqrt{16 - x^2}$ and $g(x) = x$; $a = 0, b = 2\sqrt{2}$
16. $f(x) = e^x$ and $g(x) = \dfrac{1}{x}$; $a = 1, b = 2$
17. $f(x) = e^x$ and $g(x) = e^{-x}$; $a = 0, b = 1$
18. $f(x) = e^x$ and $g(x) = \sqrt{x}$; $a = 1, b = 2$

In Exercises 19–26, find the volume of the solid of revolution obtained by revolving the region bounded by the graphs of the functions about the x-axis.

19. $y = x$ and $y = \sqrt{x}$
20. $y = x^{1/3}$ and $y = x^2$
21. $y = \dfrac{1}{2}x + 3$ and $y = x^2$
22. $y = 2x$ and $y = x\sqrt{x + 1}$
23. $y = x^2$ and $y = 4 - x^2$
24. $y = x^2$ and $y = x(4 - x)$
25. $y = \dfrac{1}{x}$, $y = x$, and $y = 2x$
    Hint: You will need to evaluate two integrals.
26. $y = \sqrt{x - 1}$ and $y = (x - 1)^2$
27. By computing the volume of the solid of revolution obtained by revolving the region under the semicircle $y = \sqrt{r^2 - x^2}$ from $x = -r$ to $x = r$ about the x-axis, show that the volume of a sphere of radius r is $\frac{4}{3}\pi r^3$ cubic units.
28. Volume of a Football Find the volume of the prolate spheroid (a solid of revolution in the shape of a football) obtained

by revolving the region under the graph of the function $y = \frac{3}{5}\sqrt{25 - x^2}$ from $x = -5$ to $x = 5$ about the x-axis.

29. **Capacity of a Man-Made Lake** A man-made lake is approximately circular and has a cross section that is a region bounded above by the x-axis and below by the graph of

$$y = 10\left[\left(\frac{x}{100}\right)^2 - 1\right]$$

(see the accompanying figure). Find the approximate capacity of the lake.

Hint: Find the volume of the solid of revolution obtained by revolving the shaded region about the y-axis. Thus, $V = \pi \int_{-10}^{0} [f(y)]^2\,dy$, where $f(y)$ is obtained by solving the given equation for x in terms of y.

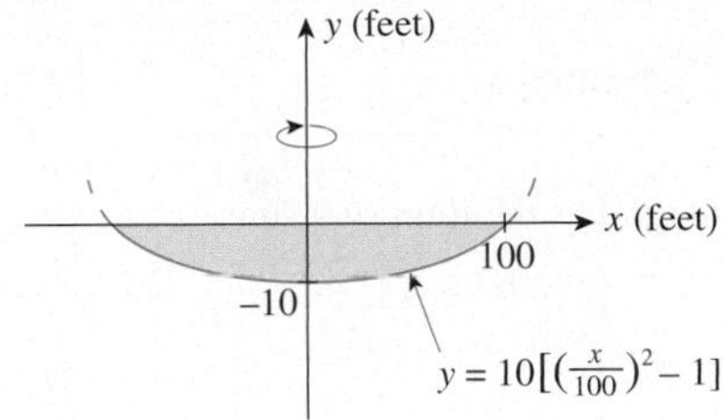

6.8 Solution to Self-Check Exercise

To find the points of intersection of the two curves, we solve the equation $5 - x^2 = 1$, giving $x = \pm 2$ (see the figure). Using Formula (24), we find the required volume:

$$\begin{aligned}
V &= \pi \int_{-2}^{2} [(5 - x^2)^2 - 1^2]\,dx \\
&= 2\pi \int_{0}^{2} (25 - 10x^2 + x^4 - 1)\,dx \qquad \text{By symmetry} \\
&= 2\pi \int_{0}^{2} (x^4 - 10x^2 + 24)\,dx \\
&= 2\pi \left(\frac{1}{5}x^5 - \frac{10}{3}x^3 + 24x\right)\Bigg|_0^2 \\
&= 2\pi \left(\frac{32}{5} - \frac{80}{3} + 48\right) \\
&= \frac{832\pi}{15} \text{ cubic units}
\end{aligned}$$

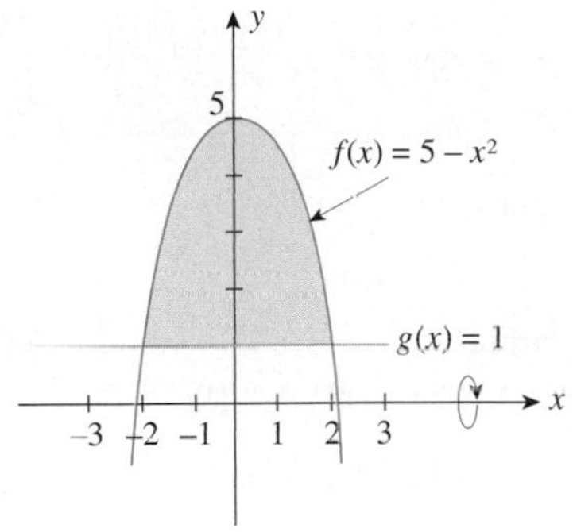

CHAPTER 6 Summary of Principal Formulas and Terms

FORMULAS

1. Indefinite integral of a constant	$\int k\,du = ku + C$
2. Power rule	$\int u^n\,du = \frac{u^{n+1}}{n+1} + C \quad (n \neq -1)$
3. Constant multiple rule	$\int kf(u)\,du = k\int f(u)\,du$ (k, a constant)
4. Sum rule	$\int [f(u) \pm g(u)]\,du = \int f(u)\,du \pm \int g(u)\,du$
5. Indefinite integral of the exponential function	$\int e^u\,du = e^u + C$

6. Indefinite integral of $f(u) = \frac{1}{u}$	$\int \frac{du}{u} = \ln\lvert u \rvert + C$
7. Method of substitution	$\int f'(g(x))g'(x)\, dx = \int f'(u)\, du$
8. Definite integral as the limit of a sum	$\int_a^b f(x)\, dx = \lim_{n\to\infty} S_n$, where S_n is a Riemann sum
9. Fundamental theorem of calculus	$\int_a^b f(x)\, dx = F(b) - F(a), F'(x) = f(x)$
10. Average value of f over $[a, b]$	$\frac{1}{b-a}\int_a^b f(x)\, dx$
11. Area between two curves	$\int_a^b [f(x) - g(x)]\, dx, f(x) \geq g(x)$
12. Consumers' surplus	$CS = \int_0^{\bar{x}} D(x)\, dx - \bar{p}\,\bar{x}$
13. Producers' surplus	$PS = \bar{p}\,\bar{x} - \int_0^{\bar{x}} S(x)\, dx$
14. Accumulated (future) value of an income stream	$A = e^{rT}\int_0^T R(t)e^{-rt}\, dt$
15. Present value of an income stream	$PV = \int_0^T R(t)e^{-rt}\, dt$
16. Amount of an annuity	$A = \frac{mP}{r}(e^{rT} - 1)$
17. Present value of an annuity	$PV = \frac{mP}{r}(1 - e^{-rT})$
18. Coefficient of inequality of a Lorentz curve	$L = 2\int_0^1 [x - f(x)]\, dx$
19. Volume of a solid of revolution	$V = \pi\int_a^b [f(x)]^2\, dx$

TERMS

antiderivative (396)
antidifferentiation (398)
integration (398)
indefinite integral (398)
integrand (398)
constant of integration (398)
differential equation (402)
initial value problem (402)
Riemann sum (423)
definite integral (424)
lower limit of integration (424)
upper limit of integration (424)
Lorentz curve (471)
line of complete equality (472)
solid of revolution (478)

CHAPTER 6 Concept Review Questions

Fill in the blanks.

1. **a.** A function F is an antiderivative of f on an interval, if _____ for all x in I.
   b. If F is an antiderivative of f on an interval I, then every antiderivative of f on I has the form _____.

2. **a.** $\int cf(x)\,dx =$ _____
   b. $\int [f(x) \pm g(x)]\,dx =$ _____

3. **a.** A differential equation is an equation that involves the derivative or differential of a/an _____ function.
   b. A solution of a differential equation on an interval I is any _____ that satisfies the differential equation.

4. If we let $u = g(x)$, then $du =$ _____, and the substitution transforms the integral $\int f(g(x))g'(x)\,dx$ into the integral _____ involving only u.

5. **a.** If f is continuous and nonnegative on an interval $[a, b]$, then the area of the region under the graph of f on $[a, b]$ is given by _____.
   b. If f is continuous on an interval $[a, b]$, then $\int_a^b f(x)\,dx$ is equal to the area(s) of the regions lying above the x-axis and bounded by the graph of f on $[a, b]$ _____ the area(s) of the regions lying below the x-axis and bounded by the graph of f on $[a, b]$.

6. **a.** The fundamental theorem of calculus states that if f is continuous on $[a, b]$, then $\int_a^b f(x)\,dx =$ _____, where F is a/an _____ of f.
   b. The net change in a function f over an interval $[a, b]$ is given by $f(b) - f(a) =$ _____, provided f' is continuous on $[a, b]$.

7. **a.** If f is continuous on $[a, b]$, then the average value of f over $[a, b]$ is the number _____.
   b. If f is a continuous and nonnegative function on $[a, b]$, then the average value of f over $[a, b]$ may be thought of as the _____ of the rectangle with base lying on the interval $[a, b]$ and having the same _____ as the region under the graph of f on $[a, b]$.

8. If f and g are continuous on $[a, b]$ and $f(x) \geq g(x)$ for all x in $[a, b]$, then the area of the region between the graphs of f and g and the vertical lines $x = a$ and $x = b$ is $A =$ _____.

9. **a.** The consumers' surplus is given by $CS =$ _____.
   b. The producers' surplus is given by $PS =$ _____.

10. **a.** The accumulated value after T years of an income stream of $R(t)$ dollars/year, earning interest of r/year compounded continuously, is given by $A =$ _____.
    b. The present value of an income stream is given by $PV =$ _____.

11. The amount of an annuity is $A =$ _____.

12. The coefficient of inequality of a Lorentz curve is $L =$ _____.

13. The volume V of the solid of revolution obtained by revolving the _____ below the graph of a nonnegative continuous function from _____ to _____ about the x-axis is _____.

CHAPTER 6 Review Exercises

In Exercises 1–20, find each indefinite integral.

1. $\int (x^3 + 2x^2 - x)\,dx$
2. $\int \left(\frac{1}{3}x^3 - 2x^2 + 8\right) dx$
3. $\int \left(x^4 - 2x^3 + \frac{1}{x^2}\right) dx$
4. $\int (x^{1/3} - \sqrt{x} + 4)\,dx$
5. $\int x(2x^2 + x^{1/2})\,dx$
6. $\int (x^2 + 1)(\sqrt{x} - 1)\,dx$
7. $\int \left(x^2 - x + \frac{2}{x} + 5\right) dx$
8. $\int \sqrt{2x + 1}\,dx$
9. $\int (3x - 1)(3x^2 - 2x + 1)^{1/3}\,dx$
10. $\int x^2(x^3 + 2)^{10}\,dx$
11. $\int \frac{x - 1}{x^2 - 2x + 5}\,dx$
12. $\int 2e^{-2x}\,dx$
13. $\int \left(x + \frac{1}{2}\right)e^{x^2+x+1}\,dx$
14. $\int \frac{e^{-x} - 1}{(e^{-x} + x)^2}\,dx$
15. $\int \frac{(\ln x)^5}{x}\,dx$
16. $\int \frac{\ln x^2}{x}\,dx$
17. $\int x^3(x^2 + 1)^{10}\,dx$
18. $\int x\sqrt{x + 1}\,dx$
19. $\int \frac{x}{\sqrt{x - 2}}\,dx$
20. $\int \frac{3x}{\sqrt{x + 1}}\,dx$

In Exercises 21–32, evaluate each definite integral.

21. $\int_0^1 (2x^3 - 3x^2 + 1)\,dx$

22. $\int_0^2 (4x^3 - 9x^2 + 2x - 1)\,dx$

23. $\int_1^4 (\sqrt{x} + x^{-3/2})\,dx$ 24. $\int_0^1 20x(2x^2 + 1)^4\,dx$

25. $\int_{-1}^0 12(x^2 - 2x)(x^3 - 3x^2 + 1)^3\,dx$

26. $\int_4^7 x\sqrt{x - 3}\,dx$ 27. $\int_0^2 \frac{x}{x^2 + 1}\,dx$

28. $\int_0^1 \frac{dx}{(5 - 2x)^2}$ 29. $\int_0^2 \frac{4x}{\sqrt{1 + 2x^2}}\,dx$

30. $\int_0^2 xe^{(-1/2)x^2}\,dx$ 31. $\int_{-1}^0 \frac{e^{-x}}{(1 + e^{-x})^2}\,dx$

32. $\int_1^e \frac{\ln x}{x}\,dx$

In Exercises 33–36, find the function f given that the slope of the tangent line to the graph at any point $(x, f(x))$ is $f'(x)$ and that the graph of f passes through the given point.

33. $f'(x) = 3x^2 - 4x + 1$; $(1, 1)$

34. $f'(x) = \frac{x}{\sqrt{x^2 + 1}}$; $(0, 1)$

35. $f'(x) = 1 - e^{-x}$; $(0, 2)$

36. $f'(x) = \frac{\ln x}{x}$; $(1, -2)$

37. Let $f(x) = -2x^2 + 1$ and compute the Riemann sum of f over the interval $[1, 2]$ by partitioning the interval into five subintervals of the same length ($n = 5$), where the points p_i ($1 \le i \le 5$) are taken to be the *right* endpoints of the respective subintervals.

38. Marginal Cost Functions The management of National Electric has determined that the daily marginal cost function associated with producing their automatic drip coffeemakers is given by

$$C'(x) = 0.00003x^2 - 0.03x + 20$$

where $C'(x)$ is measured in dollars/unit and x denotes the number of units produced. Management has also determined that the daily fixed cost incurred in producing these coffeemakers is \$500. What is the total cost incurred by National in producing the first 400 coffeemakers/day?

39. Marginal Revenue Functions Refer to Exercise 38. Management has also determined that the daily marginal revenue function associated with producing and selling their coffeemakers is given by

$$R'(x) = -0.03x + 60$$

where x denotes the number of units produced and sold and $R'(x)$ is measured in dollars/unit.

a. Determine the revenue function $R(x)$ associated with producing and selling these coffeemakers.

b. What is the demand equation relating the wholesale unit price to the quantity of coffeemakers demanded?

40. Computer Resale Value Franklin National Life Insurance Company purchased new computers for \$200,000. If the rate at which the computers' resale value changes is given by the function

$$V'(t) = 3800(t - 10)$$

where t is the length of time since the purchase date and $V'(t)$ is measured in dollars/year, find an expression $V(t)$ that gives the resale value of the computers after t yr. How much would the computers cost after 6 yr?

41. Projection TV Sales The marketing department of Vista Vision forecasts that sales of their new line of projection television systems will grow at the rate of

$$3000 - 2000e^{-0.04t} \qquad (0 \le t \le 24)$$

units/month once they are introduced into the market. Find an expression giving the total number of the projection television systems that Vista may expect to sell t mo from the time they are put on the market. How many units of the television systems can Vista expect to sell during the first year?

42. Commuter Trends Due to the increasing cost of fuel, the manager of the City Transit Authority estimates that the number of commuters using the city subway system will increase at the rate of

$$3000(1 + 0.4t)^{-1/2} \qquad (0 \le t \le 36)$$

per month t mo from now. If 100,000 commuters are currently using the system, find an expression giving the total number of commuters who will be using the subway t mo from now. How many commuters will be using the subway 6 mo from now?

43. Sales: Loudspeakers Sales of the Acrosonic model F loudspeaker systems have been growing at the rate of

$$f'(t) = 2000(3 - 2e^{-t})$$

units/yr, where t denotes the number of years these loudspeaker systems have been on the market. Determine the number of loudspeaker systems that were sold in the first 5 yr after they appeared on the market.

44. **Marginal Cost Functions** The management of a division of Ditton Industries has determined that the daily marginal cost function associated with producing their hot-air corn poppers is given by

$$C'(x) = 0.00003x^2 - 0.03x + 10$$

where $C'(x)$ is measured in dollars/unit and x denotes the number of units manufactured. Management has also determined that the daily fixed cost incurred in producing these corn poppers is $600. Find the total cost incurred by Ditton in producing the first 500 corn poppers.

45. **U.S. Census** The number of Americans aged 45–54 (which stood at 25 million at the beginning of 1990) grew at the rate of

$$R(t) = 0.00933t^3 + 0.019t^2 - 0.10833t + 1.3467$$

million people/year, t yr from the beginning of 1990. How many Americans aged 45 to 54 were added to the population between 1990 and the year 2000?

Source: U.S. Census Bureau

46. **World Coal Production** In 1980 the world produced 3.5 billion metric tons of coal. If output increased at the rate of

$$3.5e^{0.04t}$$

billion metric tons/year in year t ($t = 0$ corresponds to 1980), determine how much coal was produced worldwide between 1980 and the end of 1985.

47. Find the area of the region under the curve $y = 3x^2 + 2x + 1$ from $x = -1$ to $x = 2$.

48. Find the area of the region under the curve $y = e^{2x}$ from $x = 0$ to $x = 2$.

49. Find the area of the region bounded by the graph of the function $y = 1/x^2$, the x-axis, and the lines $x = 1$ and $x = 3$.

50. Find the area of the region bounded by the curve $y = -x^2 - x + 2$ and the x-axis.

51. Find the area of the region bounded by the graphs of the functions $f(x) = e^x$ and $g(x) = x$ and the vertical lines $x = 0$ and $x = 2$.

52. Find the area of the region that is completely enclosed by the graphs of $f(x) = x^4$ and $g(x) = x$.

53. Find the area of the region between the curve $y = x(x - 1)(x - 2)$ and the x-axis.

54. **Oil Production** Based on current production techniques, the rate of oil production from a certain oil well t yr from now is estimated to be

$$R_1(t) = 100e^{0.05t}$$

thousand barrels/year. Based on a new production technique, however, it is estimated that the rate of oil production from that oil well t yr from now will be

$$R_2(t) = 100e^{0.08t}$$

thousand barrels/year. Determine how much additional oil will be produced over the next 10 yr if the new technique is adopted.

55. Find the average value of the function

$$f(x) = \frac{x}{\sqrt{x^2 + 16}}$$

over the interval [0, 3].

56. **Average Temperature** The temperature (in °F) in Boston over a 12-hr period on a certain December day was given by

$$T = -0.05t^3 + 0.4t^2 + 3.8t + 5.6 \qquad (0 \leq t \leq 12)$$

where t is measured in hours, with $t = 0$ corresponding to 6 a.m. Determine the average temperature on that day over the 12-hr period from 6 a.m. to 6 p.m.

57. **Demand for Digital Camcorder Tapes** The demand function for a brand of blank digital camcorder tapes is given by

$$p = -0.01x^2 - 0.2x + 23$$

where p is the wholesale unit price in dollars and x is the quantity demanded each week, measured in units of a thousand. Determine the consumers' surplus if the wholesale unit price is $8/tape.

58. **Consumers' and Producers' Surplus** The quantity demanded x (in units of a hundred) of the Sportsman 5 × 7 tents, per week, is related to the unit price p (in dollars) by the relation

$$p = -0.1x^2 - x + 40$$

The quantity x (in units of a hundred) that the supplier is willing to make available in the market is related to the unit price by the relation

$$p = 0.1x^2 + 2x + 20$$

If the market price is set at the equilibrium price, find the consumers' surplus and the producers' surplus.

59. **Retirement Account Savings** Chi-Tai plans to deposit $4000/year in his Keogh Retirement Account. If interest is compounded continuously at the rate of 8%/year, how much will he have in his retirement account after 20 yr?

60. **Installment Contracts** Glenda sold her house under an installment contract whereby the buyer gave her a down payment of $9000 and agreed to make monthly payments of $925/month for 30 yr. If the prevailing interest rate is

12%/year compounded continuously, find the present value of the purchase price of the house.

61. **Present Value of a Franchise** Alicia purchased a 10-yr franchise for a health spa that is expected to generate income at the rate of

$$P(t) = 80{,}000$$

dollars/year. If the prevailing interest rate is 10%/year compounded continuously, find the present value of the franchise.

62. **Income Distribution of a Country** A certain country's income distribution is described by the function

$$f(x) = \frac{17}{18}x^2 + \frac{1}{18}x$$

a. Sketch the Lorentz curve for this function.

b. Compute $f(0.3)$ and $f(0.6)$ and interpret your results.

c. Compute the coefficient of inequality for this Lorentz curve.

63. **Population Growth** The population of a certain Sunbelt city, currently 80,000, is expected to grow exponentially in the next 5 yr with a growth constant of 0.05. If the prediction comes true, what will be the average population of the city over the next 5 yr?

64. Find the volume of the solid of revolution obtained by revolving the region under the curve $f(x) = 1/x$ from $x = 1$ to $x = 3$ about the x-axis.

65. Find the volume of the solid of revolution obtained by revolving the region bounded above by the curve $f(x) = \sqrt{x}$ and below by the curve $g(x) = x^2$ about the x-axis.

CHAPTER 6 Before Moving On . . .

1. Find $\int \left(2x^3 + \sqrt{x} + \frac{2}{x} - \frac{2}{\sqrt{x}}\right) dx$.

2. Find f if $f'(x) = e^x + x$ and $f(0) = 2$.

3. Find $\int \frac{x}{\sqrt{x^2 + 1}}\, dx$.

4. Evaluate $\int_0^1 x\sqrt{2 - x^2}\, dx$

5. Find the area of the region completely enclosed by the graphs of $y = x^2 - 1$ and $y = 1 - x$.

7 Additional Topics in Integration

What is the area of the oil spill caused by a grounded tanker? In Example 5, page 514, you will see how to determine the area of the oil spill.

© Lawson Wood/Corbis

BESIDES THE BASIC rules of integration developed in Chapter 6, there are more sophisticated techniques for finding the antiderivatives of functions. We begin this chapter by looking at the method of integration by parts. We then look at a technique of integration that involves using tables of integrals that have been compiled for this purpose. We also look at numerical methods of integration, which enable us to obtain approximate solutions to definite integrals, especially those whose exact value cannot be found otherwise. More specifically, we study the trapezoidal rule and Simpson's rule. Numerical integration methods are especially useful when the integrand is known only at discrete points. Finally, we learn how to evaluate integrals in which the intervals of integration are unbounded. Such integrals, called *improper integrals*, play an important role in the study of probability.

7.1 Integration by Parts

The Method of Integration by Parts

Integration by parts is another technique of integration that, like the method of substitution discussed in Chapter 6, is based on a corresponding rule of differentiation. In this case, the rule of differentiation is the product rule, which asserts that if f and g are differentiable functions, then

$$\frac{d}{dx}[f(x)g(x)] = f(x)g'(x) + g(x)f'(x) \tag{1}$$

If we integrate both sides of Equation (1) with respect to x, we obtain

$$\int \frac{d}{dx} f(x)g(x)\,dx = \int f(x)g'(x)\,dx + \int g(x)f'(x)\,dx$$

$$f(x)g(x) = \int f(x)g'(x)\,dx + \int g(x)f'(x)\,dx$$

This last equation, which may be written in the form

$$\int f(x)g'(x)\,dx = f(x)g(x) - \int g(x)f'(x)\,dx \tag{2}$$

is called the formula for **integration by parts**. This formula is useful since it enables us to express one indefinite integral in terms of another that may be easier to evaluate. Formula (2) may be simplified by letting

$$u = f(x) \qquad dv = g'(x)\,dx$$
$$du = f'(x)\,dx \qquad v = g(x)$$

giving the following version of the formula for integration by parts.

Integration by Parts Formula

$$\int u\,dv = uv - \int v\,du \tag{3}$$

EXAMPLE 1 Evaluate $\int xe^x\,dx$.

Solution No method of integration developed thus far enables us to evaluate the given indefinite integral in its present form. Therefore, we attempt to write it in terms of an indefinite integral that will be easier to evaluate. Let's use the integration by parts Formula (3) by letting

$$u = x \quad \text{and} \quad dv = e^x\,dx$$

so that

$$du = dx \quad \text{and} \quad v = e^x$$

Therefore,

$$\begin{aligned}\int xe^x\,dx &= \int u\,dv\\ &= uv - \int v\,du\\ &= xe^x - \int e^x\,dx\\ &= xe^x - e^x + C\\ &= (x-1)e^x + C\end{aligned}$$

The success of the method of integration by parts depends on the proper choice of u and dv. For example, if we had chosen

$$u = e^x \quad \text{and} \quad dv = x\,dx$$

in the last example, then

$$du = e^x\,dx \quad \text{and} \quad v = \frac{1}{2}x^2$$

Thus, (3) would have yielded

$$\begin{aligned}\int xe^x\,dx &= \int u\,dv\\ &= uv - \int v\,du\\ &= \frac{1}{2}x^2e^x - \int \frac{1}{2}x^2e^x\,dx\end{aligned}$$

Since the indefinite integral on the right-hand side of this equation is not readily evaluated (it is in fact more complicated than the original integral!), choosing u and dv as shown has not helped us evaluate the given indefinite integral.

In general, we can use the following guidelines.

Guidelines for Choosing u and dv

Choose u and dv so that

1. du is simpler than u.
2. dv is easy to integrate.

EXAMPLE 2 Evaluate $\int x \ln x\,dx$.

Solution Letting

$$u = \ln x \quad \text{and} \quad dv = x\,dx$$

we have

$$du = \frac{1}{x}\,dx \quad \text{and} \quad v = \frac{1}{2}x^2$$

Therefore,

$$\int x \ln x \, dx = \int u \, dv = uv - \int v \, du$$
$$= \frac{1}{2}x^2 \ln x - \int \frac{1}{2}x^2 \cdot \left(\frac{1}{x}\right) dx$$
$$= \frac{1}{2}x^2 \ln x - \frac{1}{2}\int x \, dx$$
$$= \frac{1}{2}x^2 \ln x - \frac{1}{4}x^2 + C$$
$$= \frac{1}{4}x^2(2 \ln x - 1) + C$$

■

EXAMPLE 3 Evaluate $\int \frac{xe^x}{(x+1)^2} \, dx$.

Solution Let

$$u = xe^x \quad \text{and} \quad dv = \frac{1}{(x+1)^2} \, dx$$

Then,

$$du = (xe^x + e^x) \, dx = e^x(x+1) \, dx \quad \text{and} \quad v = -\frac{1}{x+1}$$

Therefore,

$$\int \frac{xe^x}{(x+1)^2} \, dx = \int u \, dv = uv - \int v \, du$$
$$= xe^x\left(\frac{-1}{x+1}\right) - \int \left(-\frac{1}{x+1}\right) e^x(x+1) \, dx$$
$$= -\frac{xe^x}{x+1} + \int e^x \, dx$$
$$= -\frac{xe^x}{x+1} + e^x + C$$
$$= \frac{e^x}{x+1} + C$$

■

The next example shows that repeated applications of the technique of integration by parts is sometimes required to evaluate an integral.

EXAMPLE 4 Find $\int x^2 e^x \, dx$.

Solution Let

$$u = x^2 \quad \text{and} \quad dv = e^x \, dx$$

so that

$$du = 2x \, dx \quad \text{and} \quad v = e^x$$

Therefore,

$$\int x^2e^x dx = \int u\,dv = uv - \int v\,du$$
$$= x^2e^x - \int e^x(2x)\,dx = x^2e^x - 2\int xe^x\,dx$$

To complete the solution of the problem, we need to evaluate the integral

$$\int xe^x\,dx$$

But this integral may be found using integration by parts. In fact, you will recognize that this integral is precisely that of Example 1. Using the results obtained there, we now find

$$\int x^2e^x\,dx = x^2e^x - 2[(x-1)e^x] + C = e^x(x^2 - 2x + 2) + C$$

EXPLORE & DISCUSS

1. Use the method of integration by parts to derive the formula

$$\int x^n e^{ax}\,dx = \frac{1}{a}x^n e^{ax} - \frac{n}{a}\int x^{n-1}e^{ax}\,dx$$

where n is a positive integer and a is a real number.

2. Use the formula of part 1 to evaluate

$$\int x^3e^x\,dx.$$

Hint: You may find the results of Example 4 helpful.

VIDEO

APPLIED EXAMPLE 5 Oil Production The estimated rate at which oil will be produced from a certain oil well t years after production has begun is given by

$$R(t) = 100te^{-0.1t}$$

thousand barrels per year. Find an expression that describes the total production of oil at the end of year t.

Solution Let $T(t)$ denote the total production of oil from the well at the end of year t ($t \geq 0$). Then, the rate of oil production will be given by $T'(t)$ thousand barrels per year. Thus,

$$T'(t) = R(t) = 100te^{-0.1t}$$

so

$$T(t) = \int 100te^{-0.1t}\,dt$$
$$= 100\int te^{-0.1t}\,dt$$

We use the technique of integration by parts to evaluate this integral. Let

$$u = t \quad \text{and} \quad dv = e^{-0.1t}\,dt$$

so that

$$du = dt \quad \text{and} \quad v = -\frac{1}{0.1}e^{-0.1t} = -10e^{-0.1t}$$

Therefore,

$$\begin{aligned} T(t) &= 100\left[-10te^{-0.1t} + 10\int e^{-0.1t}\,dt\right] \\ &= 100[-10te^{-0.1t} - 100e^{-0.1t}] + C \\ &= -1000e^{-0.1t}(t + 10) + C \end{aligned}$$

To determine the value of C, note that the total quantity of oil produced at the end of year 0 is nil, so $T(0) = 0$. This gives

$$\begin{aligned} T(0) &= -1000(10) + C = 0 \\ C &= 10{,}000 \end{aligned}$$

Thus, the required production function is given by

$$T(t) = -1000e^{-0.1t}(t + 10) + 10{,}000$$

■

EXPLORING WITH TECHNOLOGY

Refer to Example 5.

1. Use a graphing utility to plot the graph of

$$T(t) = -1000e^{-0.1t}(t + 10) + 10{,}000$$

using the viewing window $[0, 100] \times [0, 12{,}000]$. Use **TRACE** to see what happens when t is very large.

2. Verify the result of part 1 by evaluating $\lim_{t\to\infty} T(t)$. Interpret your results.

7.1 Self-Check Exercises

1. Evaluate $\int x^2 \ln x\,dx$.
2. Since the inauguration of Ryan's Express at the beginning of 2002, the number of passengers (in millions) flying on this commuter airline has been growing at the rate of

$$R(t) = 0.1 + 0.2te^{-0.4t}$$

passengers/year ($t = 0$ corresponds to the beginning of 2002). Assuming that this trend continues through 2006, determine how many passengers will have flown on Ryan's Express by that time.

Solutions to Self-Check Exercises 7.1 can be found on page 499.

7.1 Concept Questions

1. Write the formula for integration by parts.
2. Explain how you would choose u and dv when using the integration by parts formula. Illustrate your answer with $\int x^2e^{-x}\,dx$. What happens if you reverse your choices of u and dv?

7.1 Exercises

In Exercises 1–26, find each indefinite integral.

1. $\int xe^{2x}\,dx$
2. $\int xe^{-x}\,dx$
3. $\int xe^{x/4}\,dx$
4. $\int 6xe^{3x}\,dx$
5. $\int (e^x - x)^2\,dx$
6. $\int (e^{-x} + x)^2\,dx$
7. $\int (x+1)e^x\,dx$
8. $\int (x-3)e^{3x}\,dx$
9. $\int x(x+1)^{-3/2}\,dx$
10. $\int x(x+4)^{-2}\,dx$
11. $\int x\sqrt{x-5}\,dx$
12. $\int \frac{x}{\sqrt{2x+3}}\,dx$
13. $\int x\ln 2x\,dx$
14. $\int x^2\ln 2x\,dx$
15. $\int x^3\ln x\,dx$
16. $\int \sqrt{x}\ln x\,dx$
17. $\int \sqrt{x}\ln\sqrt{x}\,dx$
18. $\int \frac{\ln x}{\sqrt{x}}\,dx$
19. $\int \frac{\ln x}{x^2}\,dx$
20. $\int \frac{\ln x}{x^3}\,dx$
21. $\int \ln x\,dx$

    Hint: Let $u = \ln x$ and $dv = dx$.
22. $\int \ln(x+1)\,dx$
23. $\int x^2e^{-x}\,dx$

    Hint: Integrate by parts twice.
24. $\int e^{-\sqrt{x}}\,dx$

    Hint: First, make the substitution $u = \sqrt{x}$; then, integrate by parts.
25. $\int x(\ln x)^2\,dx$

    Hint: Integrate by parts twice.
26. $\int x\ln(x+1)\,dx$

    Hint: First, make the substitution $u = x + 1$; then, integrate by parts.

In Exercises 27–32, evaluate each definite integral by using the method of integration by parts.

27. $\int_0^{\ln 2} xe^x\,dx$
28. $\int_0^2 xe^{-x}\,dx$
29. $\int_1^4 \ln x\,dx$
30. $\int_1^2 x\ln x\,dx$
31. $\int_0^2 xe^{2x}\,dx$
32. $\int_0^1 x^2e^{-x}\,dx$

33. Find the function f given that the slope of the tangent line to the graph of f at any point $(x, f(x))$ is xe^{-2x} and that the graph passes through the point $(0, 3)$.

34. Find the function f given that the slope of the tangent line to the graph of f at any point $(x, f(x))$ is $x\sqrt{x+1}$ and that the graph passes through the point $(3, 6)$.

35. Find the area of the region under the graph of $f(x) = \ln x$ from $x = 1$ to $x = 5$.

36. Find the area of the region under the graph of $f(x) = xe^{-x}$ from $x = 0$ to $x = 3$.

37. Velocity of a Dragster The velocity of a dragster t sec after leaving the starting line is

$$100te^{-0.2t}$$

ft/sec. What is the distance covered by the dragster in the first 10 sec of its run?

38. Production of Steam Coal In keeping with the projected increase in worldwide demand for steam coal, the boiler-firing fuel used for generating electricity, the management of Consolidated Mining has decided to step up its mining operations. Plans call for increasing the yearly production of steam coal by

$$2te^{-0.05t}$$

million metric tons/year for the next 20 yr. The current yearly production is 20 million metric tons. Find a function that describes Consolidated's total production of steam coal at the end of t yr. How much coal will Consolidated have produced over the next 20 yr if this plan is carried out?

39. Concentration of a Drug in the Bloodstream The concentration (in milligrams/milliliter) of a certain drug in a patient's bloodstream t hr after it has been administered is given by $C(t) = 3te^{-t/3}$ mg/mL. Find the average concentration of the drug in the patient's bloodstream over the first 12 hr after administration.

40. Alcohol-Related Traffic Accidents As a result of increasingly stiff laws aimed at reducing the number of alcohol-related traffic accidents in a certain state, preliminary data indicate that the number of such accidents has been changing at the rate of

$$R(t) = -10 - te^{0.1t}$$

accidents/month t mo after the laws took effect. There were 982 alcohol-related accidents for the year before the enactment of the laws. Determine how many alcohol-related accidents were expected during the first year the laws were in effect.

41. Compact Disc Sales Sales of the latest recording by Brittania, a British rock group, are currently $2te^{-0.1t}$ units/week (each unit representing 10,000 CDs), where t denotes the number of weeks since the recording's release. Find an expression that gives the total number of CDs sold as a function of t.

42. Average Price of a Commodity The price of a certain commodity in dollars/unit at time t (measured in weeks) is given by

$$p = 8 + 4e^{-2t} + te^{-2t}$$

What is the average price of the commodity over the 4-wk period from $t = 0$ to $t = 4$?

43. **Rate of Return on an Investment** Suppose an investment is expected to generate income at the rate of

$$P(t) = 30{,}000 + 800t$$

dollars/year for the next 5 yr. Find the present value of this investment if the prevailing interest rate is 8%/year compounded continuously.
Hint: Use Formula (19), Section 6.7 (page 469).

44. **Present Value of a Franchise** Tracy purchased a 15-yr franchise for a computer outlet store that is expected to generate income at the rate of

$$P(t) = 50{,}000 + 3000t$$

dollars/year. If the prevailing interest rate is 10%/year compounded continuously, find the present value of the franchise.
Hint: Use Formula (19), Section 6.7 (page 469).

45. **Growth of HMOs** The membership of the Cambridge Community Health Plan (a health maintenance organization) is projected to grow at the rate of $9\sqrt{t+1}\ln\sqrt{t+1}$ thousand people/year, t yr from now. If the HMO's current membership is 50,000, what will be the membership 5 yr from now?

46. **A Mixture Problem** Two tanks are connected in tandem as shown in the following figure. Each tank contains 60 gal of water. Starting at time $t = 0$, brine containing 3 lb/gal of salt flows into tank 1 at the rate of 2 gal/min. The mixture then enters and leaves tank 2 at the same rate. The mixtures in both tanks are stirred uniformly. It can be shown that the amount of salt in tank 2 after t min is given by

$$A(t) = 180(1 - e^{-t/30}) - 6te^{-t/30}$$

where $A(t)$ is measured in pounds.

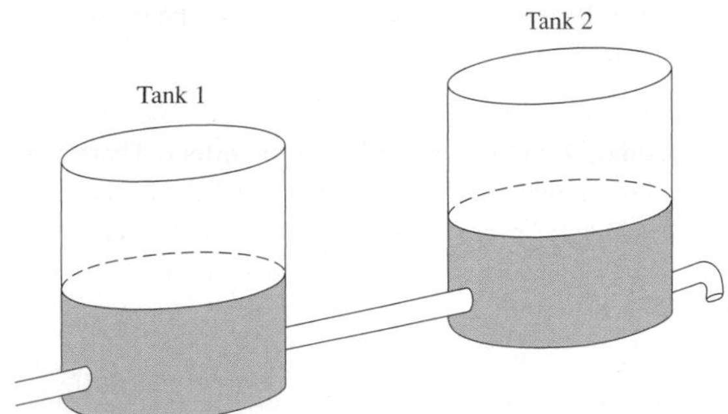

a. What is the initial amount of salt in tank 2?
b. What is the amount of salt in tank 2 after 3 hr (180 min)?
c. What is the average amount of salt in tank 2 over the first 3 hr?

47. **Diffusion** A cylindrical membrane with inner radius r_1 cm and outer radius r_2 cm containing a chemical solution is introduced into a salt bath with constant concentration c_2 moles/liter (see the accompanying figure).

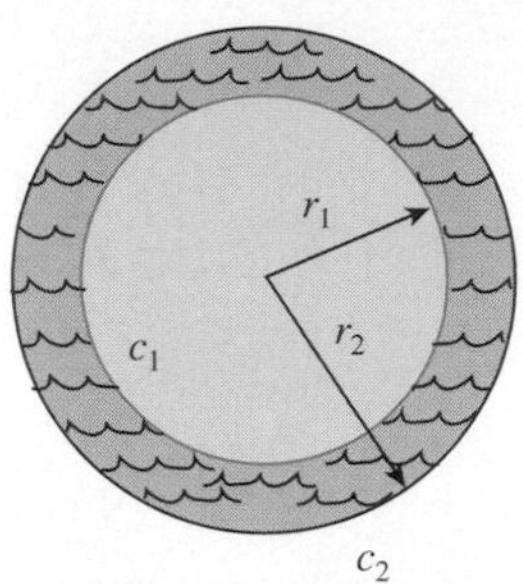

If the concentration of the chemical inside the membrane is kept constant at a different concentration of c_1 moles/liter, then the concentration of the chemical across the membrane will be given by

$$c(r) = \left(\frac{c_1 - c_2}{\ln r_1 - \ln r_2}\right)(\ln r - \ln r_2) + c_2 \qquad (r_1 < r < r_2)$$

moles/liter. Find the average concentration of the chemical across the membrane from $r = r_1$ to $r = r_2$.

48. Suppose f'' is continuous on [1, 3] and $f(1) = 2$, $f(3) = -1$, $f'(1) = 2$, and $f'(3) = 5$. Evaluate $\int_1^3 xf''(x)\,dx$.

In Exercises 49 and 50, determine whether the statement is true or false. If it is true, explain why it is true. If it is false, give an example to show why it is false.

49. $\int u\,dv + \int v\,du = uv$

50. $\int e^x g'(x)\,dx = e^x g(x) - \int e^x g(x)\,dx$

7.1 Solutions to Self-Check Exercises

1. Let $u = \ln x$ and $dv = x^2\,dx$ so that $du = \frac{1}{x}\,dx$ and $v = \frac{1}{3}x^3$. Therefore,

$$\begin{aligned}\int x^2 \ln x\,dx &= \int u\,dv = uv - \int v\,du \\ &= \frac{1}{3}x^3 \ln x - \int \frac{1}{3}x^2\,dx \\ &= \frac{1}{3}x^3 \ln x - \frac{1}{9}x^3 + C \\ &= \frac{1}{9}x^3(3 \ln x - 1) + C\end{aligned}$$

2. If $N(t)$ denote the total number of passengers who will have flown on Ryan's Express by the end of year t. Then $N'(t) = R(t)$, so that

$$\begin{aligned}N(t) &= \int R(t)\,dt \\ &= \int (0.1 + 0.2te^{-0.4t})\,dt \\ &= \int 0.1\,dt + 0.2\int te^{-0.4t}\,dt\end{aligned}$$

We now use the technique of integration by parts on the second integral. Letting $u = t$ and $dv = e^{-0.4t}\,dt$, we have

$$du = dt \quad \text{and} \quad v = -\frac{1}{0.4}e^{-0.4t} = -2.5e^{-0.4t}$$

Therefore,

$$\begin{aligned}N(t) &= 0.1t + 0.2\left(-2.5te^{-0.4t} + 2.5\int e^{-0.4t}\,dt\right) \\ &= 0.1t - 0.5te^{-0.4t} - \frac{0.5}{0.4}e^{-0.4t} + C \\ &= 0.1t - 0.5(t + 2.5)e^{-0.4t} + C\end{aligned}$$

To determine the value of C, note that $N(0) = 0$, which gives

$$\begin{aligned}N(0) &= -0.5(2.5) + C = 0 \\ C &= 1.25\end{aligned}$$

Therefore,

$$N(t) = 0.1t - 0.5(t + 2.5)e^{-0.4t} + 1.25$$

The number of passengers who will have flown on Ryan's Express by the end of 2006 is given by

$$\begin{aligned}N(5) &= 0.1(5) - 0.5(5 + 2.5)e^{-0.4(5)} + 1.25 \\ &= 1.242493\end{aligned}$$

—that is, 1,242,493 passengers.

7.2 Integration Using Tables of Integrals

A Table of Integrals

We have studied several techniques for finding an antiderivative of a function. However, useful as they are, these techniques are not always applicable. There are of course numerous other methods for finding an antiderivative of a function. Extensive lists of integration formulas have been compiled based on these methods.

A small sample of the integration formulas that can be found in many mathematical handbooks is given in the following table of integrals. The formulas are grouped according to the basic form of the integrand. Note that it may be necessary to modify the integrand of the integral to be evaluated in order to use one of these formulas.

TABLE OF INTEGRALS

Forms Involving $a + bu$

1. $\displaystyle\int \frac{u\,du}{a+bu} = \frac{1}{b^2}(a + bu - a\ln|a+bu|) + C$

2. $\displaystyle\int \frac{u^2\,du}{a+bu} = \frac{1}{2b^3}[(a+bu)^2 - 4a(a+bu) + 2a^2\ln|a+bu|] + C$

3. $\displaystyle\int \frac{u\,du}{(a+bu)^2} = \frac{1}{b^2}\left(\frac{a}{a+bu} + \ln|a+bu|\right) + C$

4. $\displaystyle\int u\sqrt{a+bu}\,du = \frac{2}{15b^2}(3bu - 2a)(a+bu)^{3/2} + C$

5. $\displaystyle\int \frac{u\,du}{\sqrt{a+bu}} = \frac{2}{3b^2}(bu-2a)\sqrt{a+bu} + C$

6. $\displaystyle\int \frac{du}{u\sqrt{a+bu}} = \frac{1}{\sqrt{a}}\ln\left|\frac{\sqrt{a+bu}-\sqrt{a}}{\sqrt{a+bu}+\sqrt{a}}\right| + C \qquad (\text{if } a > 0)$

Forms Involving $\sqrt{a^2 + u^2}$

7. $\displaystyle\int \sqrt{a^2+u^2}\,du = \frac{u}{2}\sqrt{a^2+u^2} + \frac{a^2}{2}\ln|u + \sqrt{a^2+u^2}| + C$

8. $\displaystyle\int u^2\sqrt{a^2+u^2}\,du = \frac{u}{8}(a^2+2u^2)\sqrt{a^2+u^2} - \frac{a^4}{8}\ln|u+\sqrt{a^2+u^2}| + C$

9. $\displaystyle\int \frac{du}{\sqrt{a^2+u^2}} = \ln|u+\sqrt{a^2+u^2}| + C$

10. $\displaystyle\int \frac{du}{u\sqrt{a^2+u^2}} = -\frac{1}{a}\ln\left|\frac{\sqrt{a^2+u^2}+a}{u}\right| + C$

11. $\displaystyle\int \frac{du}{u^2\sqrt{a^2+u^2}} = -\frac{\sqrt{a^2+u^2}}{a^2u} + C$

12. $\displaystyle\int \frac{du}{(a^2+u^2)^{3/2}} = \frac{u}{a^2\sqrt{a^2+u^2}} + C$

Forms Involving $\sqrt{u^2 - a^2}$

13. $\displaystyle\int \sqrt{u^2-a^2}\,du = \frac{u}{2}\sqrt{u^2-a^2} - \frac{a^2}{2}\ln|u+\sqrt{u^2-a^2}| + C$

14. $\displaystyle\int u^2\sqrt{u^2-a^2}\,du = \frac{u}{8}(2u^2-a^2)\sqrt{u^2+a^2} - \frac{a^4}{8}\ln|u+\sqrt{u^2-a^2}| + C$

15. $\displaystyle\int \frac{\sqrt{u^2-a^2}}{u^2}\,du = -\frac{\sqrt{u^2-a^2}}{u} + \ln|u+\sqrt{u^2-a^2}| + C$

16. $\displaystyle\int \frac{du}{\sqrt{u^2-a^2}} = \ln|u+\sqrt{u^2-a^2}| + C$

17. $\displaystyle\int \frac{du}{u^2\sqrt{u^2-a^2}} = \frac{\sqrt{u^2-a^2}}{a^2u} + C$

18. $\displaystyle\int \frac{du}{(u^2-a^2)^{3/2}} = -\frac{u}{a^2\sqrt{u^2-a^2}} + C$

TABLE OF INTEGRALS (*Continued*)

Forms Involving $\sqrt{a^2 - u^2}$

19. $\displaystyle\int \frac{\sqrt{a^2 - u^2}}{u}\,du = \sqrt{a^2 - u^2} - a \ln\left|\frac{a + \sqrt{a^2 - u^2}}{u}\right| + C$

20. $\displaystyle\int \frac{du}{u\sqrt{a^2 - u^2}} = -\frac{1}{a}\ln\left|\frac{a + \sqrt{a^2 - u^2}}{u}\right| + C$

21. $\displaystyle\int \frac{du}{u^2\sqrt{a^2 - u^2}} = -\frac{\sqrt{a^2 - u^2}}{a^2 u} + C$

22. $\displaystyle\int \frac{du}{(a^2 - u^2)^{3/2}} = \frac{u}{a^2\sqrt{a^2 - u^2}} + C$

Forms Involving e^{au} and $\ln u$

23. $\displaystyle\int ue^{au}\,du = \frac{1}{a^2}(au - 1)e^{au} + C$

24. $\displaystyle\int u^n e^{au}\,du = \frac{1}{a}u^n e^{au} - \frac{n}{a}\int u^{n-1}e^{au}\,du$

25. $\displaystyle\int \frac{du}{1 + be^{au}} = u - \frac{1}{a}\ln(1 + be^{au}) + C$

26. $\displaystyle\int \ln u\,du = u \ln u - u + C$

27. $\displaystyle\int u^n \ln u\,du = \frac{u^{n+1}}{(n + 1)^2}[(n + 1)\ln u - 1] + C \qquad (n \neq -1)$

28. $\displaystyle\int \frac{du}{u \ln u} = \ln|\ln u| + C$

29. $\displaystyle\int (\ln u)^n\,du = u(\ln u)^n - n\int (\ln u)^{n-1}\,du$

Using a Table of Integrals

We now consider several examples that illustrate how the table of integrals can be used to evaluate an integral.

VIDEO

EXAMPLE 1 Use the table of integrals to find $\displaystyle\int \frac{2x\,dx}{\sqrt{3 + x}}$.

Solution We first write

$$\int \frac{2x\,dx}{\sqrt{3 + x}} = 2\int \frac{x\,dx}{\sqrt{3 + x}}$$

Since $\sqrt{3 + x}$ is of the form $\sqrt{a + bu}$, with $a = 3$, $b = 1$, and $u = x$, we use Formula (5),

$$\int \frac{u\,du}{\sqrt{a + bu}} = \frac{2}{3b^2}(bu - 2a)\sqrt{a + bu} + C$$

obtaining

$$\begin{aligned} 2\int \frac{x}{\sqrt{3 + x}}\,dx &= 2\left[\frac{2}{3(1)}(x - 6)\sqrt{3 + x}\right] + C \\ &= \frac{4}{3}(x - 6)\sqrt{3 + x} + C \end{aligned}$$

EXPLORE & DISCUSS

All formulas given in the table of integrals can be verified by direct computation. Describe a method you would use and apply it to verify a formula of your choice.

EXAMPLE 2 Use the table of integrals to find $\int x^2\sqrt{3 + x^2}\,dx$.

Solution Observe that if we write 3 as $(\sqrt{3})^2$, then $3 + x^2$ has the form $\sqrt{a^2 + u^2}$, with $a = \sqrt{3}$ and $u = x$. Using Formula (8),

$$\int u^2\sqrt{a^2 + u^2}\,du = \frac{u}{8}(a^2 + 2u^2)\sqrt{a^2 + u^2} - \frac{a^4}{8}\ln|u + \sqrt{a^2 + u^2}| + C$$

we obtain

$$\int x^2\sqrt{3 + x^2}\,dx = \frac{x}{8}(3 + 2x^2)\sqrt{3 + x^2} - \frac{9}{8}\ln|x + \sqrt{3 + x^2}| + C$$

EXAMPLE 3 Use the table of integrals to evaluate

$$\int_3^4 \frac{dx}{x^2\sqrt{50 - 2x^2}}$$

Solution We first find the indefinite integral

$$I = \int \frac{dx}{x^2\sqrt{50 - 2x^2}}$$

Observe that $\sqrt{50 - 2x^2} = \sqrt{2(25 - x^2)} = \sqrt{2}\sqrt{25 - x^2}$, so we can write I as

$$I = \frac{1}{\sqrt{2}}\int \frac{dx}{x^2\sqrt{25 - x^2}} = \frac{\sqrt{2}}{2}\int \frac{dx}{x^2\sqrt{25 - x^2}} \qquad \frac{1}{\sqrt{2}} = \frac{\sqrt{2}}{2}$$

Next, using Formula (21),

$$\int \frac{du}{u^2\sqrt{a^2 - u^2}} = -\frac{\sqrt{a^2 - u^2}}{a^2u} + C$$

with $a = 5$ and $u = x$, we find

$$\begin{aligned} I &= \frac{\sqrt{2}}{2}\left[-\frac{\sqrt{25 - x^2}}{25x}\right] \\ &= -\left(\frac{\sqrt{2}}{50}\right)\frac{\sqrt{25 - x^2}}{x} \end{aligned}$$

Finally, using this result, we obtain

$$\begin{aligned} \int_3^4 \frac{dx}{x^2\sqrt{50 - 2x^2}} &= -\frac{\sqrt{2}}{50}\frac{\sqrt{25 - x^2}}{x}\bigg|_3^4 \\ &= -\frac{\sqrt{2}}{50}\frac{\sqrt{25 - 16}}{4} - \left(-\frac{\sqrt{2}}{50}\frac{\sqrt{25 - 9}}{3}\right) \\ &= -\frac{3\sqrt{2}}{200} + \frac{2\sqrt{2}}{75} = \frac{7\sqrt{2}}{600} \end{aligned}$$

EXAMPLE 4 Use the table of integrals to find $\int e^{2x}\sqrt{5 + 2e^x}\,dx$.

Solution Let $u = e^x$. Then $du = e^x\,dx$. Therefore, the given integral can be written

$$\int e^x\sqrt{5 + 2e^x}\,(e^x\,dx) = \int u\sqrt{5 + 2u}\,du$$

Using Formula (4),

$$\int u\sqrt{a + bu}\, du = \frac{2}{15b^2}(3bu - 2a)(a + bu)^{3/2} + C$$

with $a = 5$ and $b = 2$, we see that

$$\begin{aligned}\int u\sqrt{5 + 2u}\, du &= \frac{2}{15(4)}(6u - 10)(5 + 2u)^{3/2} + C \\ &= \frac{1}{15}(3u - 5)(5 + 2u)^{3/2} + C\end{aligned}$$

Finally, recalling the substitution $u = e^x$, we find

$$\int e^{2x}\sqrt{5 + 2e^x}\, dx = \frac{1}{15}(3e^x - 5)(5 + 2e^x)^{3/2} + C$$

■

EXPLORE & DISCUSS

The formulas given in the table of integrals were derived using various techniques, including the method of substitution and the method of integration by parts studied earlier. For example, Formula (1),

$$\int \frac{u\, du}{a + bu} = \frac{1}{b^2}[a + bu - a\ln|a + bu|] + C$$

can be derived using the method of substitution. Show how this is done.

As illustrated in the next example, we may need to apply a formula more than once in order to evaluate an integral.

EXAMPLE 5 Use the table of integrals to find $\int x^2 e^{(-1/2)x}\, dx$.

Solution Scanning the table of integrals for a formula involving e^{ax} in the integrand, we are led to Formula (24),

$$\int u^n e^{au}\, du = \frac{1}{a}u^n e^{au} - \frac{n}{a}\int u^{n-1}e^{au}\, du$$

With $n = 2$, $a = -\frac{1}{2}$, and $u = x$, we have

$$\begin{aligned}\int x^2 e^{(-1/2)x}\, dx &= \left(\frac{1}{-\frac{1}{2}}\right)x^2 e^{(-1/2)x} - \frac{1}{(-\frac{1}{2})}\int xe^{(-1/2)x}\, dx \\ &= -2x^2 e^{(-1/2)x} + 4\int xe^{(-1/2)x}\, dx\end{aligned}$$

If we use Formula (24) once again, with $n = 1$, $a = -\frac{1}{2}$, and $u = x$, to evaluate the integral on the right, we obtain

$$\begin{aligned}\int x^2 e^{(-1/2)x}\, dx &= -2x^2 e^{(-1/2)x} + 4\left[\left(\frac{1}{-\frac{1}{2}}\right)xe^{(-1/2)x} - \frac{2}{(-\frac{1}{2})}\int e^{(-1/2)x}\, dx\right] \\ &= -2x^2 e^{(-1/2)x} + 4\left[-2xe^{(-1/2)x} + 2\cdot\frac{1}{(-\frac{1}{2})}e^{(-1/2)x}\right] + C \\ &= -2e^{(-1/2)x}(x^2 + 4x + 8) + C\end{aligned}$$

■

VIDEO

APPLIED EXAMPLE 6 Mortgage Rates A study prepared for the National Association of Realtors estimated that the mortgage rate over the next t months will be

$$r(t) = \frac{8t + 100}{t + 10} \qquad (0 \le t \le 24)$$

percent per year. If the prediction holds true, what will be the average mortgage rate over the next 12 months?

Solution The average mortgage rate over the next 12 months will be given by

$$A = \frac{1}{12 - 0}\int_0^{12} \frac{8t + 100}{t + 10}\,dt = \frac{1}{12}\left[\int_0^{12} \frac{8t}{t + 10}\,dt + \int_0^{12} \frac{100}{t + 10}\,dt\right]$$

$$= \frac{8}{12}\int_0^{12} \frac{t}{t + 10}\,dt + \frac{100}{12}\int_0^{12} \frac{1}{t + 10}\,dt$$

Using Formula (1)

$$\int \frac{u\,du}{a + bu} = \frac{1}{b^2}[a + bu - a\ln|a + bu|] + C \qquad a = 10, b = 1, u = t$$

to evaluate the first integral, we have

$$A = \left(\frac{2}{3}\right)[10 + t - 10\ln(10 + t)]\Big|_0^{12} + \left(\frac{25}{3}\right)\ln(10 + t)\Big|_0^{12}$$

$$= \left(\frac{2}{3}\right)[(22 - 10\ln 22) - (10 - 10\ln 10)] + \left(\frac{25}{3}\right)[\ln 22 - \ln 10]$$

$$\approx 9.31$$

or approximately 9.31% per year.

7.2 Self-Check Exercises

1. Use the table of integrals to evaluate

$$\int_0^2 \frac{dx}{(5 - x^2)^{3/2}}$$

2. During a flu epidemic, the number of children in Easton Middle School who contracted influenza t days after the outbreak began was given by

$$N(t) = \frac{200}{1 + 9e^{-0.8t}}$$

Determine the average number of children who contracted the flu in the first 10 days of the epidemic.

Solutions to Self-Check Exercises 7.2 can be found on page 506.

7.2 Concept Questions

1. Consider the integral $\int \frac{\sqrt{2 - x^2}}{x}\,dx$.
   a. Which formula from the table of integrals would you choose to help you find the integral?
   b. Find the integral showing the appropriate substitutions you need to use to make the given integral conform to the formula.

2. Consider the integral $\int_2^3 \frac{dx}{\sqrt{2x^2 - 5}}$.
   a. Which formula from the table of integrals would you choose to help you evaluate the integral?
   b. Evaluate the integral.

7.2 Exercises

In Exercises 1–32, use the table of integrals in this section to find each integral.

1. $\int \frac{2x}{2+3x}\,dx$

2. $\int \frac{x}{(1+2x)^2}\,dx$

3. $\int \frac{3x^2}{2+4x}\,dx$

4. $\int \frac{x^2}{3+x}\,dx$

5. $\int x^2\sqrt{9+4x^2}\,dx$

6. $\int x^2\sqrt{4+x^2}\,dx$

7. $\int \frac{dx}{x\sqrt{1+4x}}$

8. $\int_0^2 \frac{x+1}{\sqrt{2+3x}}\,dx$

9. $\int_0^2 \frac{dx}{\sqrt{9+4x^2}}$

10. $\int \frac{dx}{x\sqrt{4+8x^2}}$

11. $\int \frac{dx}{(9-x^2)^{3/2}}$

12. $\int \frac{dx}{(2-x^2)^{3/2}}$

13. $\int x^2\sqrt{x^2-4}\,dx$

14. $\int_3^5 \frac{dx}{x^2\sqrt{x^2-9}}$

15. $\int \frac{\sqrt{4-x^2}}{x}\,dx$

16. $\int_0^1 \frac{dx}{(4-x^2)^{3/2}}$

17. $\int xe^{2x}\,dx$

18. $\int \frac{dx}{1+e^{-x}}$

19. $\int \frac{dx}{(x+1)\ln(1+x)}$

Hint: First use the substitution $u = x + 1$.

20. $\int \frac{x}{(x^2+1)\ln(x^2+1)}\,dx$

Hint: First use the substitution $u = x^2 + 1$.

21. $\int \frac{e^{2x}}{(1+3e^x)^2}\,dx$

22. $\int \frac{e^{2x}}{\sqrt{1+3e^x}}\,dx$

23. $\int \frac{3e^x}{1+e^{(1/2)x}}\,dx$

24. $\int \frac{dx}{1-2e^{-x}}$

25. $\int \frac{\ln x}{x(2+3\ln x)}\,dx$

26. $\int_1^e (\ln x)^2\,dx$

27. $\int_0^1 x^2e^x\,dx$

28. $\int x^3e^{2x}\,dx$

29. $\int x^2\ln x\,dx$

30. $\int x^3\ln x\,dx$

31. $\int (\ln x)^3\,dx$

32. $\int (\ln x)^4\,dx$

33. Consumers' Surplus Refer to Section 6.7. The demand function for Apex women's boots is

$$p = \frac{250}{\sqrt{16+x^2}}$$

where p is the wholesale unit price in dollars and x is the quantity demanded daily, in units of a hundred. Find the consumers' surplus if the wholesale price is set at \$50/pair.

34. Producers' Surplus Refer to Section 6.7. The supplier of Apex women's boots will make x hundred pairs of the boots available in the market daily when the wholesale unit price is

$$p = \frac{30x}{5-x}$$

dollars. Find the producers' surplus if the wholesale price is set at \$50/pair.

35. Amusement Park Attendance The management of AstroWorld ("The Amusement Park of the Future") estimates that the number of visitors (in thousands) entering the amusement park t hr after opening time at 9 a.m. is given by

$$R(t) = \frac{60}{(2+t^2)^{3/2}}$$

per hour. Determine the number of visitors admitted by noon.

36. Voter Registration The number of voters in a certain district of a city is expected to grow at the rate of

$$R(t) = \frac{3000}{\sqrt{4+t^2}}$$

people/year t yr from now. If the number of voters at present is 20,000, how many voters will be in the district 5 yr from now?

37. Growth of Fruit Flies Based on data collected during an experiment, a biologist found that the number of fruit flies (*Drosophila*) with a limited food supply could be approximated by the exponential model

$$N(t) = \frac{1000}{1+24e^{-0.02t}}$$

where t denotes the number of days since the beginning of the experiment. Find the average number of fruit flies in the colony in the first 10 days of the experiment and in the first 20 days.

38. VCR Ownership The percent of households that own VCRs is given by

$$P(t) = \frac{68}{1+21.67e^{-0.62t}} \quad (0 \le t \le 12)$$

where t is measured in years, with $t = 0$ corresponding to the beginning of 1981. Find the average percent of households owning VCRs from the beginning of 1981 to the beginning of 1993.

Source: Paul Kroger Associates

39. Recycling Programs The commissioner of the City of Newton Department of Public Works estimates that the number of people in the city who have been recycling their magazines in year t following the introduction of the recycling program at the beginning of 1990 is

$$N(t) = \frac{100{,}000}{2 + 3e^{-0.2t}}$$

Find the average number of people who will have recycled their magazines during the first 5 yr since the program was introduced.

40. FRANCHISES Elaine purchased a 10-yr franchise for a fast-food restaurant that is expected to generate income at the rate of $R(t) = 250{,}000 + 2000t^2$ dollars/year, t yr from now. If the prevailing interest rate is 10%/year compounded continuously, find the present value of the franchise.
Hint: Use Formula (19), Section 6.7.

41. ACCUMULATED VALUE OF AN INCOME STREAM The revenue of Virtual Reality, a video-game arcade, is generated at the rate of $R(t) = 20{,}000t$ dollars. If the revenue is invested t yr from now in a business earning interest at the rate of 15%/year compounded continuously, find the accumulated value of this stream of income at the end of 5 yr.
Hint: Use Formula (19), Section 6.7.

42. LORENTZ CURVES In a study conducted by a certain country's Economic Development Board regarding the income distribution of certain segments of the country's workforce, it was found that the Lorentz curve for the distribution of income of college professors is described by the function

$$g(x) = \frac{1}{3}x\sqrt{1 + 8x}$$

Compute the coefficient of inequality of the Lorentz curve.
Hint: Use Formula (22), Section 6.7.

7.2 Solutions to Self-Check Exercises

1. Using Formula (22), page 501, with $a^2 = 5$ and $u = x$, we see that

$$\int_0^2 \frac{dx}{(5 - x^2)^{3/2}} = \left.\frac{x}{5\sqrt{5 - x^2}}\right|_0^2 = \frac{2}{5\sqrt{5 - 4}} = \frac{2}{5}$$

2. The average number of children who contracted the flu in the first 10 days of the epidemic is given by

$$A = \frac{1}{10}\int_0^{10} \frac{200}{1 + 9e^{-0.8t}}\,dt = 20\int_0^{10} \frac{dt}{1 + 9e^{-0.8t}}$$

$$= 20\left[t + \frac{1}{0.8}\ln(1 + 9e^{-0.8t})\right]\Bigg|_0^{10} \qquad \text{Formula (25), } a = -0.8, b = 9, u = t$$

$$= 20\left[10 + \frac{1}{0.8}\ln(1 + 9e^{-8})\right] - 20\left(\frac{1}{0.8}\right)\ln 10$$

$$\approx 200.07537 - 57.56463$$

$$\approx 143$$

or 143 students.

7.3 Numerical Integration

Approximating Definite Integrals

One method of measuring cardiac output is to inject 5 to 10 milligrams (mg) of a dye into a vein leading to the heart. After making its way through the lungs, the dye returns to the heart and is pumped into the aorta, where its concentration is measured at equal time intervals. The graph of the function c in Figure 1 shows the concentration of dye in a person's aorta, measured at 2-second intervals after 5 mg of dye

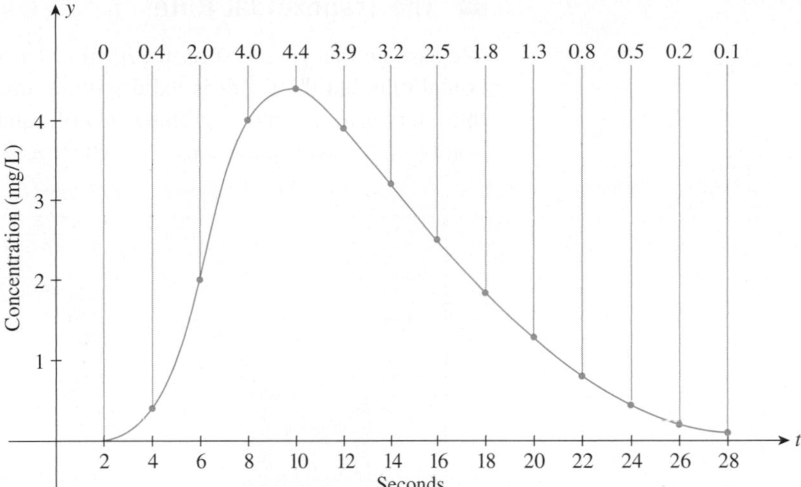

FIGURE 1
The function c gives the concentration of a dye measured at the aorta. The graph is constructed by drawing a smooth curve through a set of discrete points.

have been injected. The person's cardiac output, measured in liters per minute (L/min), is computed using the formula

$$R = \frac{60D}{\int_0^{28} c(t)\, dt} \tag{4}$$

where D is the quantity of dye injected (see Exercise 49, on page 520).

Now, to use Formula (4), we need to evaluate the definite integral

$$\int_0^{28} c(t)\, dt$$

But we do not have the algebraic rule defining the integrand c for all values of t in $[0, 28]$. In fact, we are given its values only at a set of discrete points in that interval. In situations such as this, the fundamental theorem of calculus proves useless because we cannot find an antiderivative of c. (We will complete the solution to this problem in Example 4.)

Other situations also arise in which an integrable function has an antiderivative that cannot be found in terms of elementary functions (functions that can be expressed as a finite combination of algebraic, exponential, logarithmic, and trigonometric functions). Examples of such functions are

$$f(x) = e^{x^2} \qquad g(x) = x^{-1/2}e^x \qquad h(x) = \frac{1}{\ln x}$$

Riemann sums provide us with a good approximation of a definite integral, provided the number of subintervals in the partitions is large enough. But there are better techniques and formulas, called *quadrature formulas*, that give a more efficient way of computing approximate values of definite integrals. In this section, we look at two rather simple but effective ways of approximating definite integrals.

The Trapezoidal Rule

We assume that $f(x) \geq 0$ on $[a, b]$ in order to simplify the derivation of the trapezoidal rule, but the result is valid without this restriction. We begin by subdividing the interval $[a, b]$ into n subintervals of equal length Δx, by means of the $(n + 1)$ points $x_0 = a, x_1, x_2, \ldots, x_n = b$, where n is a positive integer (Figure 2).

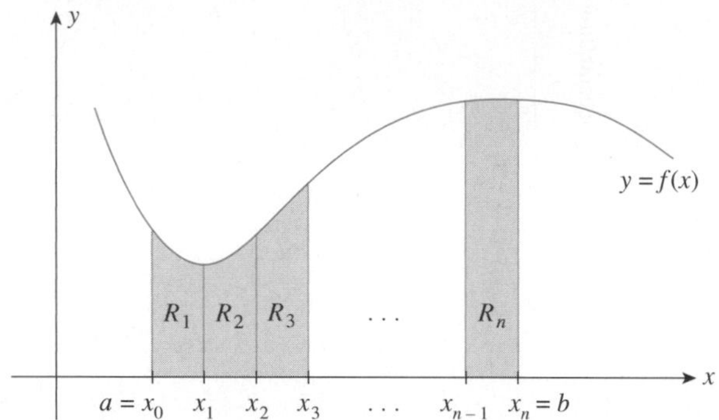

FIGURE 2
The area under the curve is equal to the sum of the n subregions $R_1, R_2, \ldots, R_n$.

Then, the length of each subinterval is given by

$$\Delta x = \frac{b - a}{n}$$

Furthermore, as we saw earlier, we may view the definite integral

$$\int_a^b f(x)\, dx$$

as the area of the region R under the curve $y = f(x)$ between $x = a$ and $x = b$. This area is given by the sum of the areas of the n nonoverlapping subregions $R_1, R_2, \ldots, R_n$, such that R_1 represents the region under the curve $y = f(x)$ from $x = x_0$ to $x = x_1$, and so on.

The basis for the trapezoidal rule lies in the approximation of each of the regions $R_1, R_2, \ldots, R_n$ by a suitable trapezoid. This often leads to a much better approximation than one obtained by means of rectangles (a Riemann sum).

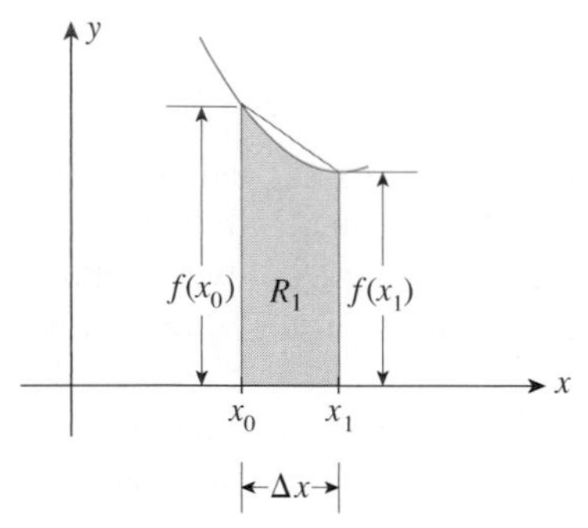

FIGURE 3
The area of R_1 is approximated by the area of the trapezoid.

Let's consider the subregion R_1, shown magnified for the sake of clarity in Figure 3. Observe that the area of the region R_1 may be approximated by the trapezoid of width Δx whose parallel sides are of lengths $f(x_0)$ and $f(x_1)$. The area of the trapezoid is given by

$$\left[\frac{f(x_0) + f(x_1)}{2}\right]\Delta x \qquad \text{Average of the lengths of the parallel sides} \times \text{Width}$$

square units. Similarly, the area of the region R_2 may be approximated by the trapezoid of width Δx and sides of lengths $f(x_1)$ and $f(x_2)$. The area of the trapezoid is given by

$$\left[\frac{f(x_1) + f(x_2)}{2}\right]\Delta x$$

Similarly, we see that the area of the last (nth) approximating trapezoid is given by

$$\left[\frac{f(x_{n-1}) + f(x_n)}{2}\right]\Delta x$$

Then, the area of the region R is approximated by the sum of the areas of the n trapezoids—that is,

$$\begin{aligned}
&\left[\frac{f(x_0) + f(x_1)}{2}\right]\Delta x + \left[\frac{f(x_1) + f(x_2)}{2}\right]\Delta x + \cdots + \left[\frac{f(x_{n-1}) + f(x_n)}{2}\right]\Delta x \\
&\quad = \frac{\Delta x}{2}[f(x_0) + f(x_1) + f(x_1) + f(x_2) + \cdots + f(x_{n-1}) + f(x_n)] \\
&\quad = \frac{\Delta x}{2}[f(x_0) + 2f(x_1) + 2f(x_2) + \cdots + 2f(x_{n-1}) + f(x_n)]
\end{aligned}$$

Since the area of the region R is given by the value of the definite integral we wished to approximate, we are led to the following approximation formula, which is called the **trapezoidal rule**.

Trapezoidal Rule

$$\int_a^b f(x)\,dx \approx \frac{\Delta x}{2}[f(x_0) + 2f(x_1) + 2f(x_2) + \cdots + 2f(x_{n-1}) + f(x_n)] \tag{5}$$

where $\Delta x = \dfrac{b-a}{n}$.

The approximation generally improves with larger values of n.

EXAMPLE 1 Approximate the value of

$$\int_1^2 \frac{1}{x}\,dx$$

using the trapezoidal rule with $n = 10$. Compare this result with the exact value of the integral.

Solution Here, $a = 1$, $b = 2$, and $n = 10$, so

$$\Delta x = \frac{b-a}{n} = \frac{1}{10} = 0.1$$

and

$$x_0 = 1 \qquad x_1 = 1.1 \qquad x_2 = 1.2 \qquad x_3 = 1.3, \ldots, x_9 = 1.9 \qquad x_{10} = 2$$

The trapezoidal rule yields

$$\begin{aligned}
\int_1^2 \frac{1}{x}\,dx &\approx \frac{0.1}{2}\left[1 + 2\left(\frac{1}{1.1}\right) + 2\left(\frac{1}{1.2}\right) + 2\left(\frac{1}{1.3}\right) + \cdots + 2\left(\frac{1}{1.9}\right) + \frac{1}{2}\right] \\
&\approx 0.693771
\end{aligned}$$

In this case, we can easily compute the actual value of the definite integral under consideration. In fact,

$$\int_1^2 \frac{1}{x}\,dx = \ln x \Big|_1^2 = \ln 2 - \ln 1 = \ln 2$$

$$\approx 0.693147$$

Thus, the trapezoidal rule with $n = 10$ yields a result with an error of 0.000624 to six decimal places. ■

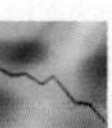

APPLIED EXAMPLE 2 Consumers' Surplus The demand function for a certain brand of perfume is given by

$$p = D(x) = \sqrt{10{,}000 - 0.01x^2}$$

where p is the unit price in dollars and x is the quantity demanded each week, measured in ounces. Find the consumers' surplus if the market price is set at \$60 per ounce.

Solution When $p = 60$, we have

$$\begin{aligned} \sqrt{10{,}000 - 0.01x^2} &= 60 \\ 10{,}000 - 0.01x^2 &= 3{,}600 \\ x^2 &= 640{,}000 \end{aligned}$$

or $x = 800$ since x must be nonnegative. Next, using the consumers' surplus formula (page 465) with $\bar{p} = 60$ and $\bar{x} = 800$, we see that the consumers' surplus is given by

$$CS = \int_0^{800} \sqrt{10{,}000 - 0.01x^2}\,dx - (60)(800)$$

It is not easy to evaluate this definite integral by finding an antiderivative of the integrand. Instead, let's use the trapezoidal rule with $n = 10$.

With $a = 0$ and $b = 800$, we find that

$$\Delta x = \frac{b - a}{n} = \frac{800}{10} = 80$$

and

$$x_0 = 0 \qquad x_1 = 80 \qquad x_2 = 160 \qquad x_3 = 240, \ldots, x_9 = 720 \qquad x_{10} = 800$$

so

$$\begin{aligned} &\int_0^{800} \sqrt{10{,}000 - 0.01x^2}\,dx \\ &\approx \frac{80}{2}\Big[100 + 2\sqrt{10{,}000 - (0.01)(80)^2} \\ &\quad + 2\sqrt{10{,}000 - (0.01)(160)^2} + \cdots + 2\sqrt{10{,}000 - (0.01)(720)^2} \\ &\quad + \sqrt{10{,}000 - (0.01)(800)^2}\Big] \\ &= 40(100 + 199.3590 + 197.4234 + 194.1546 + 189.4835 \\ &\quad + 183.3030 + 175.4537 + 165.6985 \\ &\quad + 153.6750 + 138.7948 + 60) \\ &\approx 70{,}293.82 \end{aligned}$$

Therefore, the consumers' surplus is approximately 70,294 − 48,000, or \$22,294. ■

EXPLORE & DISCUSS

Explain how you would approximate the value of $\int_0^2 f(x)\,dx$ using the trapezoidal rule with $n = 10$, where

$$f(x) = \begin{cases} \sqrt{1 + x^2} & \text{if } 0 \le x \le 1 \\ \dfrac{2}{\sqrt{1 + x^2}} & \text{if } 1 < x \le 2 \end{cases}$$

and find the value.

Simpson's Rule

Before stating Simpson's rule, let's review the two rules we have used in approximating a definite integral. Let f be a continuous nonnegative function defined on the interval $[a, b]$. Suppose the interval $[a, b]$ is partitioned by means of the $n + 1$ equally spaced points $x_0 = a, x_1, x_2, \ldots, x_n = b$, where n is a positive integer, so that the length of each subinterval is $\Delta x = (b - a)/n$ (Figure 4).

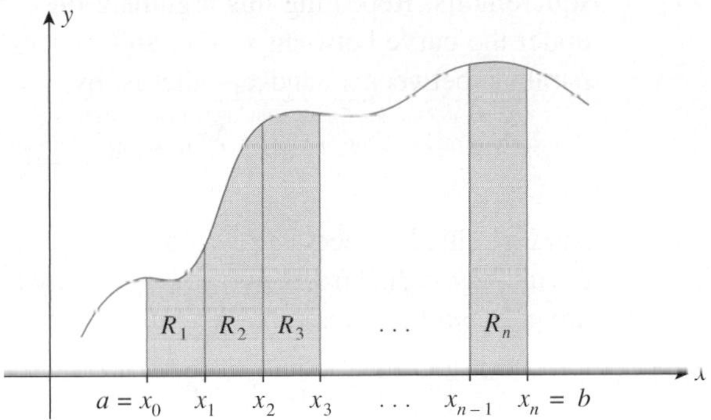

FIGURE 4
The area under the curve is equal to the sum of the n subregions $R_1, R_2, \ldots, R_n$.

Let's concentrate on the portion of the graph of $y = f(x)$ defined on the interval $[x_0, x_2]$. In using a Riemann sum to approximate the definite integral, we are in effect approximating the function $f(x)$ on $[x_0, x_1]$ by the *constant* function $y = f(p_1)$, where p_1 is chosen to be a point in $[x_0, x_1]$; the function $f(x)$ on $[x_1, x_2]$ by the constant function $y = f(p_2)$, where p_2 lies in $[x_1, x_2]$; and so on. Using a Riemann sum, we see that the area of the region under the curve $y = f(x)$ between $x = a$ and $x = b$ is approximated by the area under the approximating "step" function (Figure 5a).

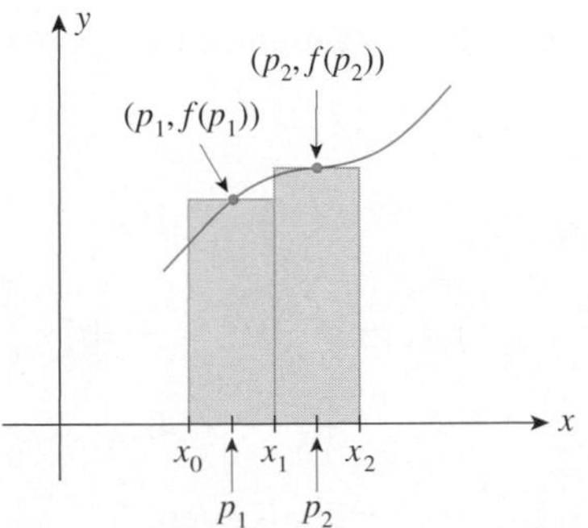

(a) The area under the curve is approximated by the area of the rectangles.

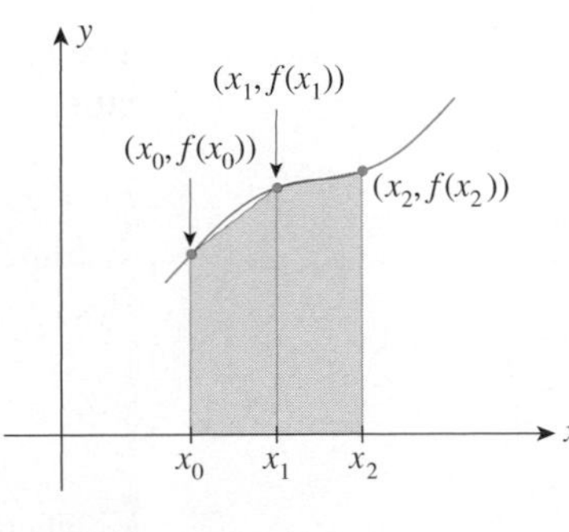

(b) The area under the curve is approximated by the area of the trapezoids.

FIGURE 5

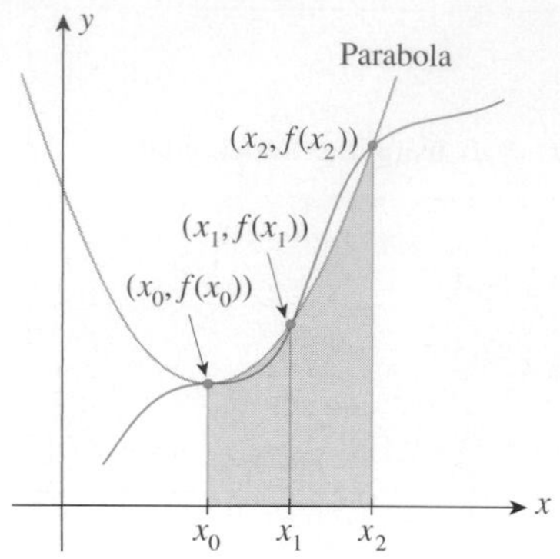

FIGURE 6
Simpson's rule approximates the area under the curve by the area under the parabola.

When we use the trapezoidal rule, we are in effect approximating the function $f(x)$ on the interval $[x_0, x_1]$ by a *linear* function through the two points $(x_0, f(x_0))$ and $(x_1, f(x_1))$; the function $f(x)$ on $[x_1, x_2]$ by a *linear* function through the two points $(x_1, f(x_1))$ and $(x_2, f(x_2))$; and so on. Thus, the trapezoidal rule simply approximates the actual area of the region under the curve $y = f(x)$ from $x = a$ to $x = b$ by the area under the approximating polygonal curve (Figure 5b).

A natural extension of the preceding idea is to approximate portions of the graph of $y = f(x)$ by means of portions of the graphs of second-degree polynomials (parts of parabolas). It can be shown that given any three noncollinear points there is a unique parabola that passes through the given points. Choose the points $(x_0, f(x_0))$, $(x_1, f(x_1))$, and $(x_2, f(x_2))$ corresponding to the first three points of the partition. Then, we can approximate the function $f(x)$ on $[x_0, x_2]$ by means of a quadratic function whose graph contains these three points (Figure 6).

Although we will not do so here, it can be shown that the area under the parabola between $x = x_0$ and $x = x_2$ is given by

$$\frac{\Delta x}{3}[f(x_0) + 4f(x_1) + f(x_2)]$$

square units. Repeating this argument on the interval $[x_2, x_4]$, we see that the area under the curve between $x = x_2$ and $x = x_4$ is approximated by the area under the parabola between x_2 and x_4—that is, by

$$\frac{\Delta x}{3}[f(x_2) + 4f(x_3) + f(x_4)]$$

square units. Proceeding, we conclude that if n is even (Why?), then the area under the curve $y = f(x)$ from $x = a$ to $x = b$ may be approximated by the sum of the areas under the $n/2$ approximating parabolas—that is,

$$\begin{aligned}
&\frac{\Delta x}{3}[f(x_0) + 4f(x_1) + f(x_2)] + \frac{\Delta x}{3}[f(x_2) + 4f(x_3) + f(x_4)] + \cdots \\
&\quad + \frac{\Delta x}{3}[f(x_{n-2}) + 4f(x_{n-1}) + f(x_n)] \\
&\quad = \frac{\Delta x}{3}[f(x_0) + 4f(x_1) + f(x_2) + f(x_2) + 4f(x_3) + f(x_4) + \cdots \\
&\quad\quad + f(x_{n-2}) + 4f(x_{n-1}) + f(x_n)] \\
&\quad = \frac{\Delta x}{3}[f(x_0) + 4f(x_1) + 2f(x_2) + 4f(x_3) \\
&\quad\quad + 2f(x_4) + \cdots + 4f(x_{n-1}) + f(x_n)]
\end{aligned}$$

The preceding is the derivation of the approximation formula known as **Simpson's rule**.

Simpson's Rule

$$\begin{aligned}
\int_a^b f(x)\,dx \approx \frac{\Delta x}{3}[&f(x_0) + 4f(x_1) + 2f(x_2) + 4f(x_3) + 2f(x_4) \\
&+ \cdots + 4f(x_{n-1}) + f(x_n)]
\end{aligned} \tag{6}$$

where $\Delta x = \dfrac{b-a}{n}$ and n is even.

In using this rule, remember that n must be even.

EXAMPLE 3 Find an approximation of

$$\int_1^2 \frac{1}{x}\,dx$$

using Simpson's rule with $n = 10$. Compare this result with that of Example 1 and also with the exact value of the integral.

Solution We have $a = 1$, $b = 2$, $f(x) = \frac{1}{x}$, and $n = 10$, so

$$\Delta x = \frac{b-a}{n} = \frac{1}{10} = 0.1$$

and

$$x_0 = 1 \qquad x_1 = 1.1 \qquad x_2 = 1.2 \qquad x_3 = 1.3, \ldots, x_9 = 1.9 \qquad x_{10} = 2$$

Simpson's rule yields

$$\begin{aligned}\int_1^2 \frac{1}{x}\,dx &\approx \frac{0.1}{3}[f(1) + 4f(1.1) + 2f(1.2) + \cdots + 4f(1.9) + f(2)] \\ &= \frac{0.1}{3}\left[1 + 4\left(\frac{1}{1.1}\right) + 2\left(\frac{1}{1.2}\right) + 4\left(\frac{1}{1.3}\right) + 2\left(\frac{1}{1.4}\right) + 4\left(\frac{1}{1.5}\right)\right. \\ &\quad \left. + 2\left(\frac{1}{1.6}\right) + 4\left(\frac{1}{1.7}\right) + 2\left(\frac{1}{1.8}\right) + 4\left(\frac{1}{1.9}\right) + \frac{1}{2}\right] \\ &\approx 0.693150\end{aligned}$$

The trapezoidal rule with $n = 10$ yielded an approximation of 0.693771, which is 0.000624 off the value of $\ln 2 \approx 0.693147$ to six decimal places. Simpson's rule yields an approximation with an error of 0.000003, a definite improvement over the trapezoidal rule. ■

APPLIED EXAMPLE 4 Cardiac Output Solve the problem posed at the beginning of this section. Recall that we wished to find a person's cardiac output by using the formula

$$R = \frac{60\,D}{\int_0^{28} c(t)\,dt}$$

where D (the quantity of dye injected) is equal to 5 milligrams and the function c has the graph shown in Figure 7. Use Simpson's rule with $n = 14$ to estimate the value of the integral.

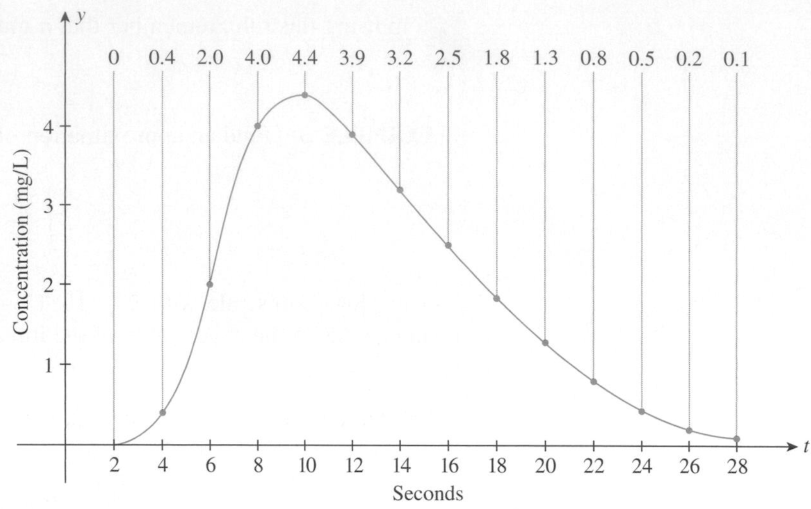

FIGURE 7
The function c gives the concentration of a dye measured at the aorta. The graph is constructed by drawing a smooth curve through a set of discrete points.

Solution Using Simpson's rule with $n = 14$ and $\Delta t = 2$ so that

$$t_0 = 0 \qquad t_1 = 2 \qquad t_2 = 4 \qquad t_3 = 6, \ldots, t_{14} = 28$$

we obtain

$$\begin{aligned}\int_0^{28} c(t)\,dt &\approx \frac{2}{3}[c(0) + 4c(2) + 2c(4) + 4c(6) + \cdots \\ &\quad + 4c(26) + c(28)] \\ &= \frac{2}{3}[0 + 4(0) + 2(0.4) + 4(2.0) + 2(4.0) \\ &\quad + 4(4.4) + 2(3.9) + 4(3.2) + 2(2.5) + 4(1.8) \\ &\quad + 2(1.3) + 4(0.8) + 2(0.5) + 4(0.2) + 0.1] \\ &\approx 49.9\end{aligned}$$

Therefore, the person's cardiac output is

$$R \approx \frac{60(5)}{49.9} \approx 6.0$$

or 6.0 L/min. ■

APPLIED EXAMPLE 5 Oil Spill An oil spill off the coastline was caused by a ruptured tank in a grounded oil tanker. Using aerial photographs, the Coast Guard was able to obtain the dimensions of the oil spill (Figure 8). Using Simpson's rule with $n = 10$, estimate the area of the oil spill.

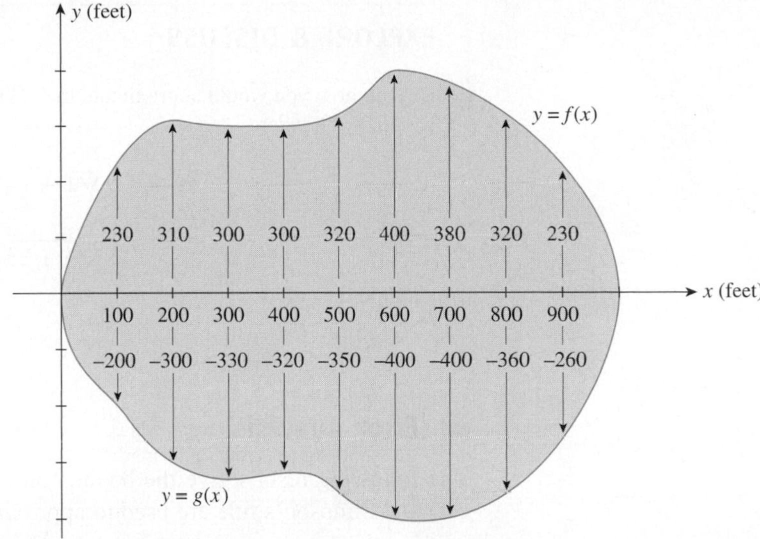

FIGURE 8
Simpson's rule can be used to calculate the area of the oil spill.

Solution We may think of the area affected by the oil spill as the area of the plane region bounded above by the graph of the function $f(x)$ and below by the graph of the function $g(x)$ between $x = 0$ and $x = 1000$ (Figure 8). Then, the required area is given by

$$A = \int_0^{1000} [f(x) - g(x)]\,dx$$

Using Simpson's rule with $n = 10$ and $\Delta x = 100$ so that

$$x_0 = 0 \qquad x_1 = 100 \qquad x_2 = 200, \ldots, x_{10} = 1000$$

we have

$$\begin{aligned}
A &= \int_0^{1000} [f(x) - g(x)]\,dx \\
&\approx \frac{\Delta x}{3}\{[f(x_0) - g(x_0)] + 4[f(x_1) - g(x_1)] + 2[f(x_2) - g(x_2)] \\
&\quad + \cdots + 4[f(x_9) - g(x_9)] + [f(x_{10}) - g(x_{10})]\} \\
&= \frac{100}{3}\{[0 - 0] + 4[230 - (-200)] + 2[310 - (-300)] \\
&\quad + 4[300 - (-330)] + 2[300 - (-320)] + 4[320 - (-350)] \\
&\quad + 2[400 - (-400)] + 4[380 - (-400)] + 2[320 - (-360)] \\
&\quad + 4[230 - (-260)] + [0 - 0]\} \\
&= \frac{100}{3}[0 + 4(430) + 2(610) + 4(630) + 2(620) + 4(670) \\
&\quad + 2(800) + 4(780) + 2(680) + 4(490) + 0] \\
&= \frac{100}{3}(17{,}420) \\
&\approx 580{,}667
\end{aligned}$$

or approximately 580,667 square feet. ■

EXPLORE & DISCUSS

Explain how you would approximate the value of $\int_0^2 f(x)\, dx$ using Simpson's rule with $n = 10$, where

$$f(x) = \begin{cases} \sqrt{1+x^2} & \text{if } 0 \le x \le 1 \\ \dfrac{2}{\sqrt{1+x^2}} & \text{if } 1 < x \le 2 \end{cases}$$

and find the value.

Error Analysis

The following results give the bounds on the errors incurred when the trapezoidal rule and Simpson's rule are used to approximate a definite integral (proof omitted).

Errors in the Trapezoidal and Simpson Approximations
Suppose the definite integral

$$\int_a^b f(x)\, dx$$

is approximated with n subintervals.

1. The *maximum* error incurred in using the trapezoidal rule is
$$\frac{M(b-a)^3}{12n^2} \qquad \textbf{(7)}$$
where M is a number such that $|f''(x)| \le M$ for all x in $[a, b]$.
2. The *maximum* error incurred in using Simpson's rule is
$$\frac{M(b-a)^5}{180n^4} \qquad \textbf{(8)}$$
where M is a number such that $|f^{(4)}(x)| \le M$ for all x in $[a, b]$.

Note In many instances, the actual error is less than the upper error bounds given.

EXAMPLE 6 Find bounds on the errors incurred when

$$\int_1^2 \frac{1}{x}\, dx$$

is approximated using (a) the trapezoidal rule and (b) Simpson's rule with $n = 10$. Compare these with the actual errors found in Examples 1 and 3.

Solution

a. Here, $a = 1$, $b = 2$, and $f(x) = 1/x$. Next, to find a value for M, we compute

$$f'(x) = -\frac{1}{x^2} \quad \text{and} \quad f''(x) = \frac{2}{x^3}$$

Since $f''(x)$ is positive and decreasing on (1, 2) (Why?), it attains its maximum value of 2 at $x = 1$, the left endpoint of the interval. Therefore, if we take $M = 2$, then $|f''(x)| \leq 2$. Using (7), we see that the maximum error incurred is

$$\frac{2(2-1)^3}{12(10)^2} = \frac{2}{1200} = 0.0016667$$

The actual error found in Example 1, 0.000624, is much less than the upper bound just found.

b. We compute

$$f'''(x) = \frac{-6}{x^4} \quad \text{and} \quad f^{(4)}(x) = \frac{24}{x^5}$$

Since $f^{(4)}(x)$ is positive and decreasing on (1, 2) (just look at $f^{(5)}$ to verify this fact), it attains its maximum at the left endpoint of [1, 2]. Now,

$$f^{(4)}(1) = 24$$

and so we may take $M = 24$. Using (8), we obtain the maximum error of

$$\frac{24(2-1)^5}{180(10)^4} = 0.0000133$$

The actual error is 0.000003 (see Example 3). ■

EXPLORE & DISCUSS

Refer to the Explore & Discuss on pages 511 and 516. Explain how you would find the maximum error incurred in using (1) the trapezoidal rule and (2) Simpson's rule with $n = 10$ to approximate $\int_0^2 f(x)\,dx$.

7.3 Self-Check Exercises

1. Use the trapezoidal rule and Simpson's rule with $n = 8$ to approximate the value of the definite integral

$$\int_0^2 \frac{2}{\sqrt{1+x^2}}\,dx$$

2. The graph in the accompanying figure shows the consumption of petroleum in the United States in quadrillion BTU, from 1976 to 1990. Using Simpson's rule with $n = 14$, estimate the average consumption during the 14-yr period.

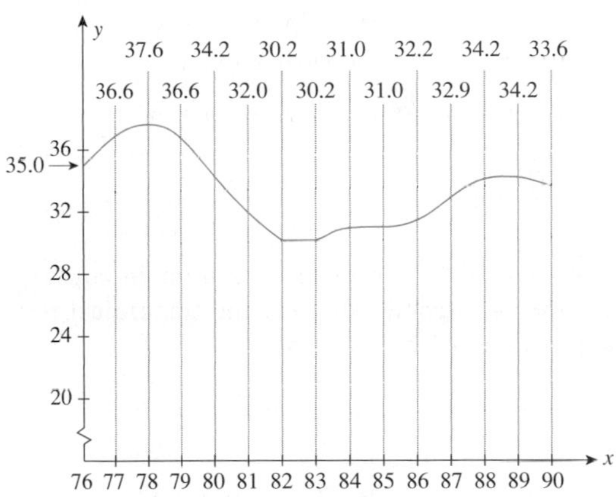

Source: The World Almanac

Solutions to Self-Check Exercises 7.3 can be found on page 521.

7.3 Concept Questions

1. Explain why n can be odd or even in the trapezoidal rule, but it must be even in Simpson's rule.
2. Explain, without alluding to the error formulas, why the trapezoidal rule gives the exact value of $\int_a^b f(x)\,dx$ if f is a linear function and why Simpson's rule gives the exact value of the integral if f is a quadratic function.
3. Refer to Concept Question 2 and answer the questions using the error formulas for the trapezoidal rule and Simpson's rule.

7.3 Exercises

In Exercises 1–14, use the trapezoidal rule and Simpson's rule to approximate the value of each definite integral. Compare your result with the exact value of the integral.

1. $\int_0^2 x^2\,dx;\ n = 6$

2. $\int_1^3 (x^2 - 1)\,dx;\ n = 4$

3. $\int_0^1 x^3\,dx;\ n = 4$

4. $\int_1^2 x^3\,dx;\ n = 6$

5. $\int_1^2 \frac{1}{x}\,dx;\ n = 4$

6. $\int_1^2 \frac{1}{x}\,dx;\ n = 8$

7. $\int_1^2 \frac{1}{x^2}\,dx;\ n = 4$

8. $\int_0^1 \frac{1}{1+x}\,dx;\ n = 4$

9. $\int_0^4 \sqrt{x}\,dx;\ n = 8$

10. $\int_0^2 x\sqrt{2x^2+1}\,dx;\ n = 6$

11. $\int_0^1 e^{-x}\,dx;\ n = 6$

12. $\int_0^1 xe^{-x^2}\,dx;\ n = 6$

13. $\int_1^2 \ln x\,dx;\ n = 4$

14. $\int_0^1 x\ln(x^2+1)\,dx;\ n = 8$

In Exercises 15–22, use the trapezoidal rule and Simpson's rule to approximate the value of each definite integral.

15. $\int_0^1 \sqrt{1+x^3}\,dx;\ n = 4$

16. $\int_0^2 x\sqrt{1+x^3}\,dx;\ n = 4$

17. $\int_0^2 \frac{1}{\sqrt{x^3+1}}\,dx;\ n = 4$

18. $\int_0^1 \sqrt{1-x^2}\,dx;\ n = 4$

19. $\int_0^2 e^{-x^2}\,dx;\ n = 4$

20. $\int_0^1 e^{x^2}\,dx;\ n = 6$

21. $\int_1^2 x^{-1/2}e^x\,dx;\ n = 4$

22. $\int_2^4 \frac{dx}{\ln x};\ n = 6$

In Exercises 23–28, find a bound on the error in approximating each definite integral using (a) the trapezoidal rule and (b) Simpson's rule with *n* intervals.

23. $\int_{-1}^2 x^5\,dx;\ n = 10$

24. $\int_0^1 e^{-x}\,dx;\ n = 8$

25. $\int_1^3 \frac{1}{x}\,dx;\ n = 10$

26. $\int_1^3 \frac{1}{x^2}\,dx;\ n = 8$

27. $\int_0^2 \frac{1}{\sqrt{1+x}}\,dx;\ n = 8$

28. $\int_1^3 \ln x\,dx;\ n = 10$

29. TRIAL RUN OF AN ATTACK SUBMARINE In a submerged trial run of an attack submarine, a reading of the sub's velocity was made every quarter hour, as shown in the accompanying table. Use the trapezoidal rule to estimate the distance traveled by the submarine during the 2-hr period.

Time, t (hr)	0	$\frac{1}{4}$	$\frac{1}{2}$	$\frac{3}{4}$	1	$\frac{5}{4}$	$\frac{3}{2}$	$\frac{7}{4}$	2
Velocity, $V(t)$ (mph)	19.5	24.3	34.2	40.5	38.4	26.2	18	16	8

30. REAL ESTATE Cooper Realty is considering development of a time-sharing condominium resort complex along the oceanfront property illustrated in the accompanying graph. To obtain an estimate of the area of this property, measurements of the distances from the edge of a straight road, which defines one boundary of the property, to the corresponding points on the shoreline are made at 100-ft intervals. Using Simpson's rule with $n = 10$, estimate the area of the oceanfront property.

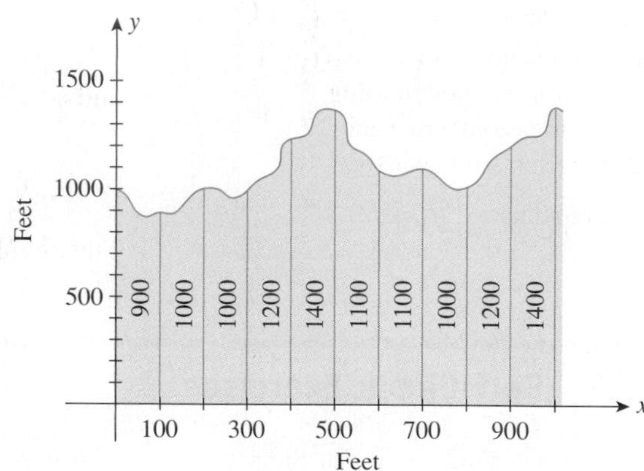

31. FUEL CONSUMPTION OF DOMESTIC CARS Thanks to smaller and more fuel-efficient models, American carmakers have doubled their average fuel economy over a 13-yr period, from 1974 to 1987. The following figure gives the average fuel consumption in miles per gallon (mpg) of domestic-built cars over the period under consideration ($t = 0$ corresponds to the beginning of 1974). Use the trapezoidal rule to estimate the

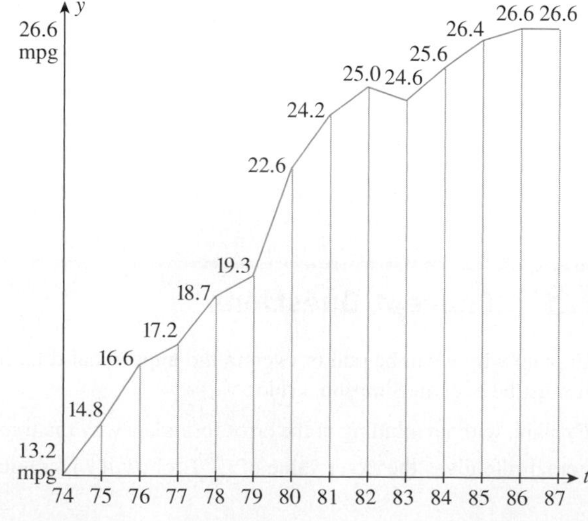

average fuel consumption of the domestic car built during this period.
Hint: Approximate the integral $\frac{1}{13}\int_0^{13} f(t)\,dt$.

32. **Average Temperature** The graph depicted in the following figure shows the daily mean temperatures recorded during one September in Cameron Highlands. Using (a) the trapezoidal rule and (b) Simpson's rule with $n = 10$, estimate the average temperature during that month.

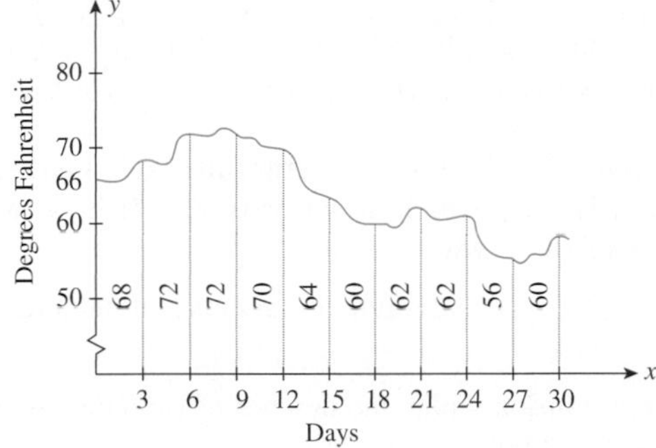

33. **U.S. Daily Oil Consumption** The following table gives the daily consumption of oil in the United States, in millions of barrels, measured in 2-yr intervals, from 1980 through 2000. Use Simpson's rule to estimate the average daily consumption of oil over the period in question.

Year	1980	1982	1984	1986	1988	1990
Consumption	17.1	15.3	15.7	16.3	17.3	17.0

Year	1992	1994	1996	1998	2000
Consumption	17.0	17.7	18.3	18.9	19.7

Source: Office of Transportation Technologies

34. **Surface Area of the Central Park Reservoir** The reservoir located in Central Park in New York City has the shape depicted in the following figure. The measurements shown were taken at 206-ft intervals. Use Simpson's rule with $n = 10$ to estimate the surface area of the lake.

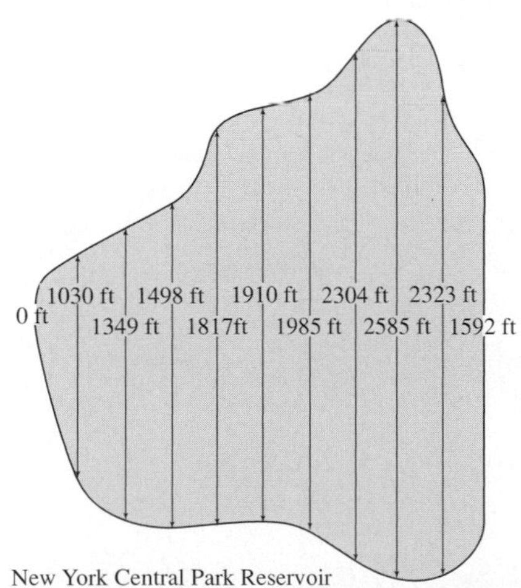

New York Central Park Reservoir

Source: Boston Globe

35. **Water Flow in a River** At a certain point, a river is 78 ft wide and its depth, measured at 6-ft intervals across the river, is recorded in the following table.

x (ft)	0	6	12	18	24	30	36
y (ft)	0.8	2.6	5.8	6.2	8.2	10.1	10.8

x (ft)	42	48	54	60	66	72	78
y (ft)	9.8	7.6	6.4	5.2	3.9	2.4	1.4

Here, x denotes the distance (in feet) from one bank of the river and y (in feet) is the corresponding depth. If the average rate of flow through this section of the river is 4 ft/sec, use the trapezoidal rule with $n = 13$ to find the rate of the volume of flow of water in the river.
Hint: Volume of flow = rate of flow × area of cross section.

36. **Consumers' Surplus** Refer to Section 6.7. The demand equation for the Sicard wristwatch is given by

$$p = S(x) = \frac{50}{0.01x^2 + 1} \qquad (0 \le x \le 20)$$

where x (measured in units of a thousand) is the quantity demanded per week and p is the unit price in dollars. Use (a) the trapezoidal rule and (b) Simpson's rule (take $n = 8$) to estimate the consumers' surplus if the market price is $25/watch.

37. **Producers' Surplus** Refer to Section 6.7. The supply function for the CD manufactured by Herald Records is given by

$$p = S(x) = \sqrt{0.01x^2 + 0.11x + 38}$$

where p is the unit wholesale price in dollars and x stands for the quantity that will be made available in the market by the supplier, measured in units of a thousand. Use (a) the trapezoidal rule and (b) Simpson's rule (take $n = 8$) to estimate the producers' surplus if the wholesale price is $8/CD.

38. **Air Pollution** The amount of nitrogen dioxide, a brown gas that impairs breathing, present in the atmosphere on a certain May day in the city of Long Beach has been approximated by

$$A(t) = \frac{136}{1 + 0.25(t - 4.5)^2} + 28 \qquad (0 \le t \le 11)$$

where $A(t)$ is measured in pollutant standard index (PSI), and t is measured in hours, with $t = 0$ corresponding to 7 a.m. Use the trapezoidal rule with $n = 10$ to estimate the average PSI between 7 a.m. and noon.
Hint: The average value is given by $\frac{1}{5}\int_0^5 A(t)\,dt$.

39. **U.S. Strategic Petroleum Reserves** According to data from the American Petroleum Institute, the U.S. strategic petroleum reserves from the beginning of 1981 through the beginning of 1990 can be approximated by the function

$$S(t) = \frac{613.7t^2 + 1449.1}{t^2 + 6.3} \qquad (0 \le t \le 9)$$

where $S(t)$ is measured in millions of barrels and t in years, with $t = 0$ corresponding to the beginning of 1981. Using the trapezoidal rule with $n = 9$, estimate the average petroleum reserves from the beginning of 1981 through the beginning of 1990.

Source: American Petroleum Institute

40. **Growth of Service Industries** It has been estimated that service industries, which currently make up 30% of the nonfarm workforce in a certain country, will continue to grow at the rate of

$$R(t) = 5e^{1/(t+1)}$$

percent/decade t decades from now. Estimate the percent of the nonfarm workforce in the service industries one decade from now.

Hint: (a) Show that the answer is given by $30 + \int_0^1 5e^{1/(t+1)}\,dt$ and (b) use Simpson's rule with $n = 10$ to approximate the definite integral.

41. **Tread Lives of Tires** Under normal driving conditions the percent of Super Titan radial tires expected to have a useful tread life of between 30,000 and 40,000 mi is given by

$$P = 100\int_{30,000}^{40,000} \frac{1}{2000\sqrt{2\pi}} e^{-(1/2)[(x-40,000)/2000]^2}\,dx$$

Use Simpson's rule with $n = 10$ to estimate P.

42. **Length of Infants at Birth** Medical records of infants delivered at Kaiser Memorial Hospital show that the percent of infants whose length at birth is between 19 and 21 in. is given by

$$P = 100\int_{19}^{21} \frac{1}{2.6\sqrt{2\pi}} e^{-(1/2)[(x-20)/2.6]^2}\,dx$$

Use Simpson's rule with $n = 10$ to estimate P.

43. **Measuring Cardiac Output** Eight milligrams of a dye are injected into a vein leading to an individual's heart. The concentration of the dye in the aorta (in mg/L) measured at 2-sec intervals is shown in the accompanying table. Use Simpson's rule and the formula of Example 4 to estimate the person's cardiac output.

t	0	2	4	6	8	10	12
$C(t)$	0	0	2.8	6.1	9.7	7.6	4.8

t	14	16	18	20	22	24
$C(t)$	3.7	1.9	0.8	0.3	0.1	0

44. **Estimating the Flow Rate of a River** A stream is 120 ft wide. The following table gives the depths of the river measured across a section of the river in intervals of 6 ft. Here, x denotes the distance from one bank of the river, and y denotes the corresponding depth (in feet). The average rate of flow of the river across this section of the river is 4.2 ft/sec. Use Simpson's rule with $n = 20$ to estimate the rate of the volume of flow of the river.

x (ft)	0	6	12	18	24	30	36	42	48	54	60
y (ft)	0.8	1.2	3.0	4.1	5.8	6.6	6.8	7.0	7.2	7.4	7.8

x (ft)	66	72	78	84	90	96	102	108	114	120
y (ft)	7.6	7.4	7.0	6.6	6.0	5.1	4.3	3.2	2.2	1.1

In Exercises 45–48, determine whether the statement is true or false. It it is true, explain why it is true. If it is false, give an example to show why it is false.

45. In using the trapezoidal rule, the number of subintervals n must be even.

46. In using Simpson's rule, the number of subintervals n may be chosen to be odd or even.

47. Simpson's rule is more accurate than the trapezoidal rule.

48. If f is a polynomial function of degree less than or equal to 3, then the approximation $\int_a^b f(x)\,dx$ using Simpson's rule is exact.

49. Derive the formula

$$R = \frac{60\,D}{\int_0^T c(t)\,dt}$$

for calculating the cardiac output of a person in L/min. Here, $c(t)$ is the concentration of dye in the aorta (in mg/L) at time t (in seconds) for t in $[0, T]$, and D is the amount of dye (in mg) injected into a vein leading to the heart.

Hint: Partition the interval $[0, T]$ into n subintervals of equal length Δt. The amount of dye that flows past the measuring point in the aorta during the time interval $[0, \Delta t]$ is approximately $c(t_i)(R\Delta t)/60$ (concentration times volume). Therefore, the total amount of dye measured at the aorta is

$$\frac{[c(t_1)R\Delta t + c(t_2)R\Delta t + \cdots + c(t_n)R\Delta t]}{60} = D$$

Take the limit of the Riemann sum to obtain

$$R = \frac{60\,D}{\int_0^T c(t)\,dt}$$

7.3 Solutions to Self-Check Exercises

1. We have $x = 0$, $b = 2$, and $n = 8$, so

$$\Delta x = \frac{b-a}{n} = \frac{2}{8} = 0.25$$

and $x_0 = 0$, $x_1 = 0.25$, $x_2 = 0.50$, $x_3 = 0.75, \ldots, x_7 = 1.75$, and $x_8 = 2$. The trapezoidal rule gives

$$\begin{aligned}\int_0^2 \frac{1}{\sqrt{1-x^2}}\,dx &\approx \frac{0.25}{2}\left[1 + \frac{2}{\sqrt{1+(0.25)^2}} + \frac{2}{\sqrt{1+(0.5)^2}} + \cdots + \frac{2}{\sqrt{1+(1.75)^2}} + \frac{1}{\sqrt{5}}\right]\\ &\approx 0.125(1 + 1.9403 + 1.7889 + 1.6000 + 1.4142\\ &\quad + 1.2494 + 1.1094 + 0.9923 + 0.4472)\\ &\approx 1.4427\end{aligned}$$

Using Simpson's rule with $n = 8$ gives

$$\begin{aligned}\int_0^2 \frac{1}{\sqrt{1-x^2}}\,dx &\approx \frac{0.25}{3}\left[1 + \frac{4}{\sqrt{1+(0.25)^2}} + \frac{2}{\sqrt{1+(0.5)^2}} + \frac{4}{\sqrt{1+(0.75)^2}} + \cdots + \frac{4}{\sqrt{1+(1.75)^2}} + \frac{1}{\sqrt{5}}\right]\\ &\approx \frac{0.25}{3}(1 + 3.8806 + 1.7889 + 3.2000 + 1.4142\\ &\quad + 2.4988 + 1.1094 + 1.9846 + 0.4472)\\ &\approx 1.4436\end{aligned}$$

2. The average consumption of petroleum during the 14-yr period is given by

$$\frac{1}{14}\int_0^{14} f(x)\,dx$$

where f is the function describing the given graph. Using Simpson's rule with $a = 0$, $b = 14$, and $n = 14$ so that $\Delta x = 1$ and

$$x_0 = 0 \qquad x_1 = 1 \qquad x_2 = 2, \ldots, x_{14} = 14$$

we have

$$\begin{aligned}\frac{1}{14}\int_0^{14} f(x)\,dx &\approx \left(\frac{1}{14}\right)\left(\frac{1}{3}\right)[f(x_0) + 4f(x_1) + 2f(x_2) + 4f(x_3) + \cdots\\ &\quad + 4f(x_{13}) + f(x_{14})]\\ &= \frac{1}{42}[35 + 4(36.6) + 2(37.6) + 4(36.6) + 2(34.2)\\ &\quad + 4(32.0) + 2(30.2) + 4(30.2) + 2(31.0) + 4(31.0)\\ &\quad + 2(32.2) + 4(32.9) + 2(34.2) + 4(34.2) + 33.6]\\ &\approx 33.4\end{aligned}$$

or approximately 33.4 quadrillion BTU/year.

7.4 Improper Integrals

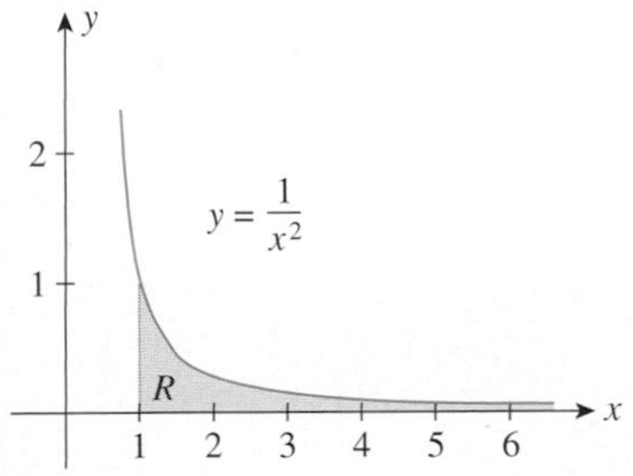

Improper Integrals

All the definite integrals we have encountered have had finite intervals of integration. In many applications, however, we are concerned with integrals that have unbounded intervals of integration. These integrals are called **improper integrals**.

To lead us to the definition of an improper integral of a function f over an infinite interval, consider the problem of finding the area of the region R under the curve $y = f(x) = 1/x^2$ and to the right of the vertical line $x = 1$, as shown in Figure 9. Because the interval over which the integration must be performed is unbounded, the method of integration presented previously cannot be applied directly in solving this problem. However, we can approximate the region R by the definite integral

$$\int_1^b \frac{1}{x^2}\,dx \tag{9}$$

FIGURE 9
The area of the unbounded region R can be approximated by a definite integral.

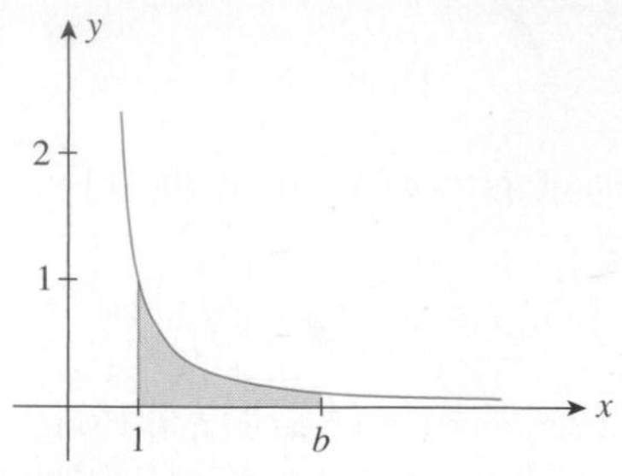

FIGURE 10

Area of shaded region $= \displaystyle\int_1^b \frac{1}{x^2}\,dx$

which gives the area of the region under the curve $y = f(x) = 1/x^2$ from $x = 1$ to $x = b$ (Figure 10). You can see that the approximation of the region R by the definite integral (9) improves as the upper limit of integration, b, becomes larger and larger. Figure 11 illustrates the situation for $b = 2$, 3, and 4, respectively.

This observation suggests that if we define a function $I(b)$ by

$$I(b) = \int_1^b \frac{1}{x^2}\,dx \tag{10}$$

then we can find the area of the required region R by evaluating the limit of $I(b)$ as b tends to infinity; that is, the area of R is given by

$$\lim_{b\to\infty} I(b) = \lim_{b\to\infty} \int_1^b \frac{1}{x^2}\,dx \tag{11}$$

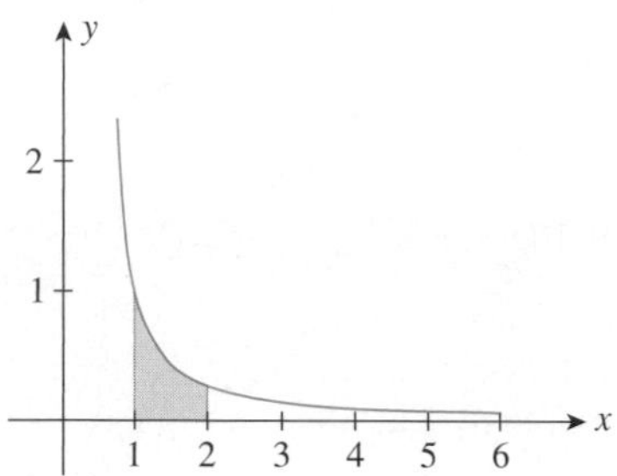

(a) Area of region under the graph of f on [1, 2]

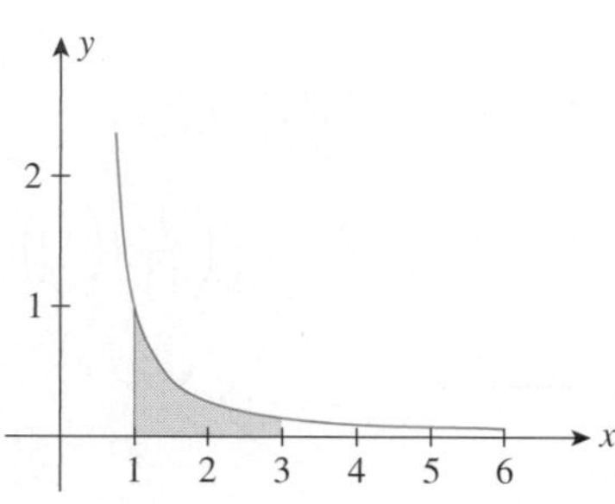

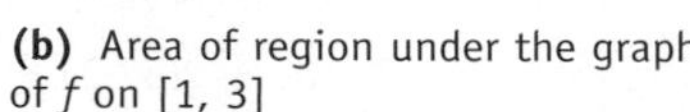

(b) Area of region under the graph of f on [1, 3]

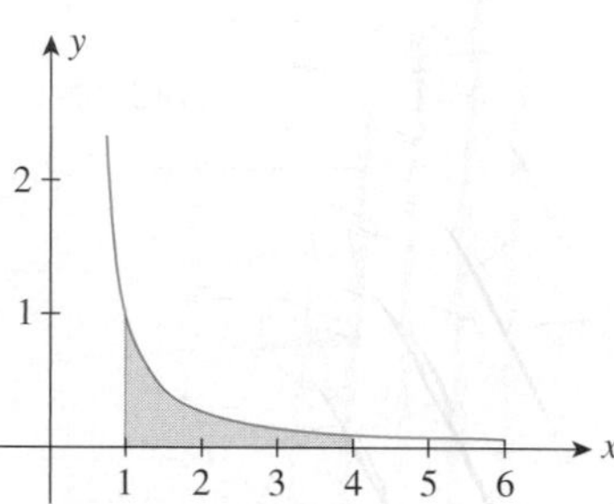

(c) Area of region under the graph of f on [1, 4]

FIGURE 11
As b increases, the approximation of R by the definite integral improves.

EXAMPLE 1

a. Evaluate the definite integral $I(b)$ in Equation (10).
b. Compute $I(b)$ for $b = 10$, 100, 1000, 10,000.
c. Evaluate the limit in Equation (11).
d. Interpret the results of parts (b) and (c).

Solution

a. $I(b) = \displaystyle\int_1^b \frac{1}{x^2}\,dx = -\frac{1}{x}\bigg|_1^b = -\frac{1}{b} + 1$

b. From the result of part (a),

$$I(b) = 1 - \frac{1}{b}$$

Therefore,

$$I(10) = 1 - \frac{1}{10} = 0.9$$
$$I(100) = 1 - \frac{1}{100} = 0.99$$
$$I(1000) = 1 - \frac{1}{1000} = 0.999$$
$$I(10{,}000) = 1 - \frac{1}{10{,}000} = 0.9999$$

c. Once again, using the result of part (a), we find

$$\begin{aligned}\lim_{b\to\infty} I(b) &= \lim_{b\to\infty}\int_1^b \frac{1}{x^2}\,dx\\ &= \lim_{b\to\infty}\left(1-\frac{1}{b}\right)\\ &= 1\end{aligned}$$

d. The result of part (c) tells us that the area of the region R is 1 square unit. The results of the computations performed in part (b) reinforce our expectation that $I(b)$ should approach 1, the area of the region R, as b approaches infinity. ■

The preceding discussion and the results of Example 1 suggest that we define the improper integral of a continuous function f over the unbounded interval $[a, \infty)$ as follows.

Improper Integral of f over $[a, \infty)$

Let f be a continuous function on the unbounded interval $[a, \infty)$. Then, the improper integral of f over $[a, \infty)$ is defined by

$$\int_a^\infty f(x)\,dx = \lim_{b\to\infty}\int_a^b f(x)\,dx \quad \textbf{(12)}$$

if the limit exists.

If the limit exists, the improper integral is said to be **convergent**. An improper integral for which the limit in Equation (12) fails to exist is said to be **divergent**.

EXAMPLE 2 Evaluate $\displaystyle\int_2^\infty \frac{1}{x}\,dx$ if it converges.

Solution

$$\begin{aligned}\int_2^\infty \frac{1}{x}\,dx &= \lim_{b\to\infty}\int_2^b \frac{1}{x}\,dx\\ &= \lim_{b\to\infty} \ln x\,\Big|_2^b\\ &= \lim_{b\to\infty}(\ln b - \ln 2)\end{aligned}$$

Since $\ln b \to \infty$ as $b \to \infty$, the limit does not exist, and we conclude that the given improper integral is divergent. ■

EXPLORE & DISCUSS

1. Suppose f is continuous and nonnegative on $[0, \infty)$. Furthermore, suppose $\lim_{x\to\infty} f(x) = L$, where L is a positive number. What can you say about the convergence of the improper integral $\int_0^\infty f(x)\,dx$? Explain your answer and illustrate with an example.
2. Suppose f is continuous and nonnegative on $[0, \infty)$ and satisfies the condition $\lim_{x\to\infty} f(x) = 0$. What can you say about $\int_0^\infty f(x)\,dx$? Explain and illustrate your answer with examples.

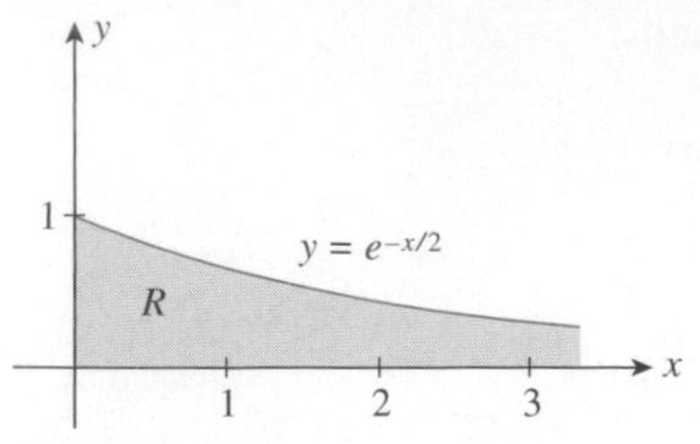

FIGURE 12

Area of $R = \int_0^\infty e^{-x/2}\,dx$

EXAMPLE 3 Find the area of the region R under the curve $y = e^{-x/2}$ for $x \geq 0$.

Solution The region R is shown in Figure 12. Taking $b > 0$, we compute the area of the region under the curve $y = e^{-x/2}$ from $x = 0$ to $x = b$—namely,

$$I(b) = \int_0^b e^{-x/2}\,dx = -2e^{-x/2}\Big|_0^b = -2e^{-b/2} + 2$$

Then, the area of the region R is given by

$$\begin{aligned}\lim_{b\to\infty} I(b) &= \lim_{b\to\infty}(2 - 2e^{-b/2}) = 2 - 2\lim_{b\to\infty}\frac{1}{e^{b/2}}\\ &= 2\end{aligned}$$

or 2 square units. ■

EXPLORING WITH TECHNOLOGY

You can see how fast the improper integral in Example 3 converges, as follows:

1. Use a graphing utility to plot the graph of $I(b) = 2 - 2e^{-b/2}$, using the viewing window $[0, 50] \times [0, 3]$.
2. Use TRACE to follow the values of y for increasing values of x, starting at the origin.

The improper integral defined in Equation (12) has an interval of integration that is unbounded on the right. Improper integrals with intervals of integration that are unbounded on the left also arise in practice and are defined in a similar manner.

Improper Integral of f over $(-\infty, b]$

Let f be a continuous function on the unbounded interval $(-\infty, b]$. Then, the improper integral of f over $(-\infty, b]$ is defined by

$$\int_{-\infty}^b f(x)\,dx = \lim_{a\to-\infty}\int_a^b f(x)\,dx \qquad \textbf{(13)}$$

if the limit exists.

In this case, the improper integral is said to be convergent. Otherwise, the improper integral is said to be divergent.

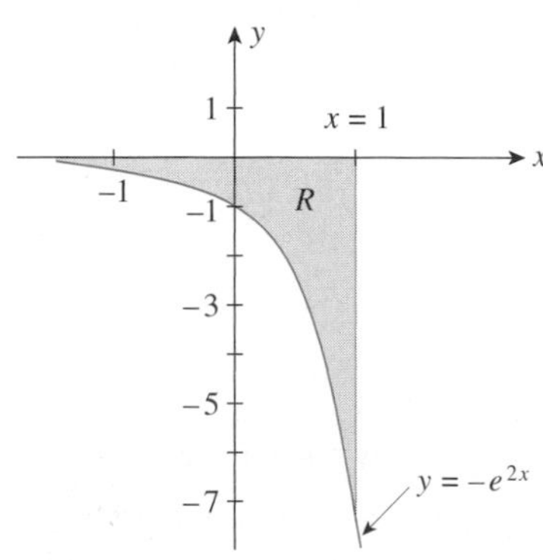

FIGURE 13

Area of $R = -\int_{-\infty}^1 -e^{2x}\,dx$

EXAMPLE 4 Find the area of the region R bounded above by the x-axis, below by the curve $y = -e^{2x}$, and on the right by the vertical line $x = 1$.

Solution The region R is shown in Figure 13. Taking $a < 1$, compute the area of the region bounded above by the x-axis ($y = 0$), and below by the curve

$y = -e^{2x}$ from $x = a$ to $x = 1$—namely,

$$\begin{aligned} I(a) &= \int_a^1 [0 - (-e^{2x})]\,dx = \int_a^1 e^{2x}\,dx \\ &= \frac{1}{2}e^{2x}\Big|_a^1 = \frac{1}{2}e^2 - \frac{1}{2}e^{2a} \end{aligned}$$

Then, the area of the required region is given by

$$\begin{aligned} \lim_{a \to -\infty} I(a) &= \lim_{a \to -\infty}\left(\frac{1}{2}e^2 - \frac{1}{2}e^{2a}\right) \\ &= \frac{1}{2}e^2 - \frac{1}{2}\lim_{a \to -\infty} e^{2a} \\ &= \frac{1}{2}e^2 \end{aligned}$$

Another improper integral found in practical applications involves the integration of a function f over the unbounded interval $(-\infty, \infty)$.

Improper Integral of f over $(-\infty, \infty)$

Let f be a continuous function over the unbounded interval $(-\infty, \infty)$. Let c be any real number and suppose both the improper integrals

$$\int_{-\infty}^{c} f(x)\,dx \quad \text{and} \quad \int_{c}^{\infty} f(x)\,dx$$

are convergent. Then, the improper integral of f over $(-\infty, \infty)$ is defined by

$$\int_{-\infty}^{\infty} f(x)\,dx = \int_{-\infty}^{c} f(x)\,dx + \int_{c}^{\infty} f(x)\,dx \qquad \textbf{(14)}$$

In this case, we say that the improper integral on the left in Equation (14) is convergent. If either one of the two improper integrals on the right in (14) is divergent, then the improper integral on the left is not defined.

Note Usually, we choose $c = 0$.

EXAMPLE 5 Evaluate the improper integral

$$\int_{-\infty}^{\infty} xe^{-x^2}\,dx$$

and give a geometric interpretation of the results.

Solution Take the number c in Equation (14) to be 0. Let's first evaluate

$$\begin{aligned} \int_{-\infty}^{0} xe^{-x^2}\,dx &= \lim_{a \to -\infty} \int_a^0 xe^{-x^2}\,dx \\ &= \lim_{a \to -\infty} -\frac{1}{2}e^{-x^2}\Big|_a^0 \\ &= \lim_{a \to -\infty}\left[-\frac{1}{2} + \frac{1}{2}e^{-a^2}\right] = -\frac{1}{2} \end{aligned}$$

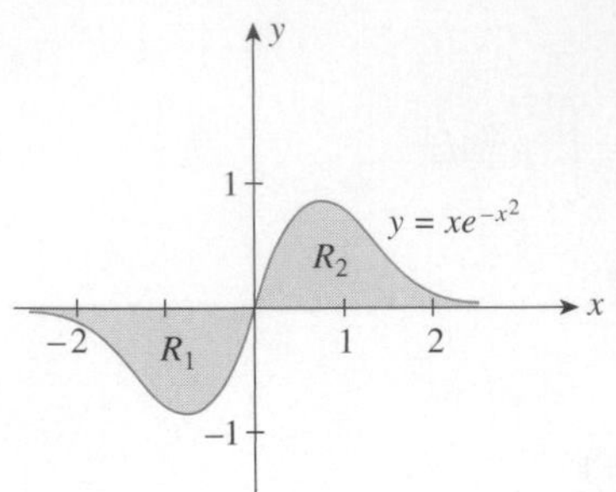

FIGURE 14

$$\int_{-\infty}^{\infty} xe^{-x^2}\,dx$$

$$= \int_{-\infty}^{0} xe^{-x^2}\,dx + \int_{0}^{\infty} xe^{-x^2}\,dx$$

Next, we evaluate

$$\begin{aligned}\int_{0}^{\infty} xe^{-x^2}\,dx &= \lim_{b\to\infty}\int_{0}^{b} xe^{-x^2}\,dx\\ &= \lim_{b\to\infty} -\frac{1}{2}e^{-x^2}\Big|_0^b\\ &= \lim_{b\to\infty}\left(-\frac{1}{2}e^{-b^2}+\frac{1}{2}\right) = \frac{1}{2}\end{aligned}$$

Therefore,

$$\begin{aligned}\int_{-\infty}^{\infty} xe^{-x^2}\,dx &= \int_{-\infty}^{0} xe^{-x^2}\,dx + \int_{0}^{\infty} xe^{-x^2}\,dx\\ &= -\frac{1}{2}+\frac{1}{2}\\ &= 0\end{aligned}$$

The graph of $y = xe^{-x^2}$ is sketched in Figure 14. A glance at the figure tells us that the improper integral

$$\int_{-\infty}^{0} xe^{-x^2}\,dx$$

gives the negative of the area of the region R_1, bounded above by the x-axis, below by the curve $y = xe^{-x^2}$, and on the right by the y-axis ($x = 0$).

However, the improper integral

$$\int_{0}^{\infty} xe^{-x^2}\,dx$$

gives the area of the region R_2 under the curve $y = xe^{-x^2}$ for $x \geq 0$. Since the graph of f is symmetric with respect to the origin, the area of R_1 is equal to the area of R_2. In other words,

$$\int_{-\infty}^{0} xe^{-x^2}\,dx = -\int_{0}^{\infty} xe^{-x^2}\,dx$$

Therefore,

$$\begin{aligned}\int_{-\infty}^{\infty} xe^{-x^2}\,dx &= \int_{-\infty}^{0} xe^{-x^2}\,dx + \int_{0}^{\infty} xe^{-x^2}\,dx\\ &= -\int_{0}^{\infty} xe^{-x^2}\,dx + \int_{0}^{\infty} xe^{-x^2}\,dx\\ &= 0\end{aligned}$$

as was shown earlier. ■

Perpetuities

Recall from Section 6.7 that the present value of an annuity is given by

$$PV \approx mP\int_{0}^{T} e^{-rt}\,dt = \frac{mP}{r}(1 - e^{-rT}) \tag{15}$$

Now, if the payments of an annuity are allowed to continue indefinitely, we have what is called a **perpetuity**. The present value of a perpetuity may be approximated by the improper integral

$$PV \approx mP\int_{0}^{\infty} e^{-rt}\,dt$$

obtained from Formula (15) by allowing the term of the annuity, T, to approach infinity. Thus,

$$\begin{aligned} mP\int_0^\infty e^{-rt}\,dt &= \lim_{b\to\infty} mP\int_0^b e^{-rt}\,dt \\ &= mP \lim_{b\to\infty}\int_0^b e^{-rt}\,dt \\ &= mP\lim_{b\to\infty}\left[-\frac{1}{r}e^{-rt}\Big|_0^b\right] \\ &= mP\lim_{b\to\infty}\left(-\frac{1}{r}e^{-rb}+\frac{1}{r}\right) = \frac{mP}{r} \end{aligned}$$

The Present Value of a Perpetuity

The **present value PV of a perpetuity** is given by

$$PV = \frac{mP}{r} \tag{16}$$

where m is the number of payments per year, P is the size of each payment, and r is the interest rate (compounded continuously).

APPLIED EXAMPLE 6 Endowments The Robinson family wishes to create a scholarship fund at a college. If a scholarship in the amount of \$5000 is awarded annually beginning 1 year from now, find the amount of the endowment they are required to make now. Assume that this fund will earn interest at a rate of 8% per year compounded continuously.

Solution The amount of the endowment, A, is given by the present value of a perpetuity, with $m = 1$, $P = 5000$, and $r = 0.08$. Using Formula (16), we find

$$\begin{aligned} A &= \frac{(1)(5000)}{0.08} \\ &= 62{,}500 \end{aligned}$$

or \$62,500. ■

The improper integral also plays an important role in the study of probability theory, as we will see in Chapter 10.

7.4 Self-Check Exercises

1. Evaluate $\displaystyle\int_{-\infty}^{\infty} \frac{x^3}{(1+x^4)^{3/2}}\,dx$.

2. Suppose an income stream is expected to continue indefinitely. Then, the present value of such a stream can be calculated from the formula for the present value of an income stream by letting T approach infinity. Thus, the required present value is given by

$$PV = \int_0^\infty P(t)e^{-rt}\,dt$$

Suppose Marcia has an oil well in her backyard that generates a stream of income given by

$$P(t) = 20e^{-0.02t}$$

where $P(t)$ is expressed in thousands of dollars per year and t

is the time in years from the present. Assuming that the prevailing interest rate in the foreseeable future is 10%/year compounded continuously, what is the present value of the income stream?

Solutions to Self-Check Exercises 7.4 can be found on page 529.

7.4 Concept Questions

1. a. Define $\int_a^{\infty} f(x)\,dx$, where f is continuous on $[a, \infty)$.

   b. Define $\int_{-\infty}^{b} f(x)\,dx$, where f is continuous on $(-\infty, b]$.

   c. Define $\int_{-\infty}^{\infty} f(x)\,dx$ where f is continuous on $(-\infty, \infty)$.

2. What is the present value of a perpetuity? Give a formula for computing it.

7.4 Exercises

In Exercises 1–10, find the area of the region under the curve $y = f(x)$ over the indicated interval.

1. $f(x) = \dfrac{2}{x^2}; x \geq 3$
2. $f(x) = \dfrac{2}{x^3}; x \geq 2$
3. $f(x) = \dfrac{1}{(x-2)^2}; x \geq 3$
4. $f(x) = \dfrac{2}{(x+1)^3}; x \geq 0$
5. $f(x) = \dfrac{1}{x^{3/2}}; x \geq 1$
6. $f(x) = \dfrac{3}{x^{5/2}}; x \geq 4$
7. $f(x) = \dfrac{1}{(x+1)^{5/2}}; x \geq 0$
8. $f(x) = \dfrac{1}{(1-x)^{3/2}}; x \leq 0$
9. $f(x) = e^{2x}; x \leq 2$
10. $f(x) = xe^{-x^2}; x \geq 0$
11. Find the area of the region bounded by the x-axis and the graph of the function
$$f(x) = \frac{x}{(1+x^2)^2}$$
12. Find the area of the region bounded by the x-axis and the graph of the function
$$f(x) = \frac{e^x}{(1+e^x)^2}$$
13. Consider the improper integral
$$\int_0^{\infty} \sqrt{x}\,dx$$
   a. Evaluate $I(b) = \int_0^b \sqrt{x}\,dx$.

   b. Show that
$$\lim_{b\to\infty} I(b) = \infty$$
   thus proving that the given improper integral is divergent.
14. Consider the improper integral
$$\int_1^{\infty} x^{-2/3}\,dx$$
   a. Evaluate $I(b) = \int_1^b x^{-2/3}\,dx$.

   b. Show that
$$\lim_{b\to\infty} I(b) = \infty$$
   thus proving that the given improper integral is divergent.

In Exercises 15–42, evaluate each improper integral whenever it is convergent.

15. $\int_1^{\infty} \dfrac{3}{x^4}\,dx$
16. $\int_1^{\infty} \dfrac{1}{x^3}\,dx$
17. $\int_4^{\infty} \dfrac{2}{x^{3/2}}\,dx$
18. $\int_1^{\infty} \dfrac{1}{\sqrt{x}}\,dx$
19. $\int_1^{\infty} \dfrac{4}{x}\,dx$
20. $\int_2^{\infty} \dfrac{3}{x}\,dx$
21. $\int_{-\infty}^{0} \dfrac{1}{(x-2)^3}\,dx$
22. $\int_2^{\infty} \dfrac{1}{(x+1)^2}\,dx$
23. $\int_1^{\infty} \dfrac{1}{(2x-1)^{3/2}}\,dx$
24. $\int_{-\infty}^{0} \dfrac{1}{(4-x)^{3/2}}\,dx$
25. $\int_0^{\infty} e^{-x}\,dx$
26. $\int_0^{\infty} e^{-x/2}\,dx$
27. $\int_{-\infty}^{0} e^{2x}\,dx$
28. $\int_{-\infty}^{0} e^{3x}\,dx$
29. $\int_1^{\infty} \dfrac{e^{\sqrt{x}}}{\sqrt{x}}\,dx$
30. $\int_1^{\infty} \dfrac{e^{-\sqrt{x}}}{\sqrt{x}}\,dx$
31. $\int_{-\infty}^{0} xe^x\,dx$
32. $\int_0^{\infty} xe^{-2x}\,dx$
33. $\int_{-\infty}^{\infty} x\,dx$
34. $\int_{-\infty}^{\infty} x^3\,dx$
35. $\int_{-\infty}^{\infty} x^3(1+x^4)^{-2}\,dx$
36. $\int_{-\infty}^{\infty} x(x^2+4)^{-3/2}\,dx$
37. $\int_{-\infty}^{\infty} xe^{1-x^2}\,dx$
38. $\int_{-\infty}^{\infty} \left(x - \dfrac{1}{2}\right)e^{-x^2+x-1}\,dx$

39. $\displaystyle\int_{-\infty}^{\infty} \frac{e^{-x}}{1+e^{-x}}\,dx$

40. $\displaystyle\int_{-\infty}^{\infty} \frac{xe^{-x^2}}{1+e^{-x^2}}\,dx$

41. $\displaystyle\int_{e}^{\infty} \frac{1}{x \ln^3 x}\,dx$

42. $\displaystyle\int_{e^2}^{\infty} \frac{1}{x \ln x}\,dx$

43. THE AMOUNT OF AN ENDOWMENT A university alumni group wishes to provide an annual scholarship in the amount of \$1500 beginning next year. If the scholarship fund will earn an interest rate of 8%/year compounded continuously, find the amount of the endowment the alumni are required to make now.

44. THE AMOUNT OF AN ENDOWMENT Mel Thompson wishes to establish a fund to provide a university medical center with an annual research grant of \$50,000 beginning next year. If the fund will earn an interest rate of 9%/year compounded continuously, find the amount of the endowment he is required to make now.

45. PERPETUAL NET INCOME STREAMS The present value of a perpetual stream of income that flows continually at the rate of $P(t)$ dollars/year is given by the formula

$$PV \approx \int_0^{\infty} P(t)e^{-rt}\,dt$$

where r is the interest rate compounded continuously. Using this formula, find the present value of a perpetual net income stream that is generated at the rate of

$$P(t) = 10{,}000 + 4000t$$

dollars/year.
Hint: $\lim_{b\to\infty} \frac{b}{e^{rb}} = 0$.

46. ESTABLISHING A TRUST FUND Becky Wilkinson wants to establish a trust fund that will provide her children and heirs with a perpetual annuity in the amount of

$$P(t) = 20 + t$$

thousand dollars/year beginning next year. If the trust fund will earn an interest rate of 10%/year compounded continuously, find the amount that she must place in the trust fund now.
Hint: Use the formula given in Exercise 45.

In Exercises 47–50, determine whether the statement is true or false. If it is true, explain why it is true. If it is false, give an example to show why it is false.

47. If $\int_a^{\infty} f(x)\,dx$ exists, then $\int_b^{\infty} f(x)\,dx$ exists for every real number $b > a$.

48. If f is continuous on $(-\infty, \infty)$, then
$$\int_{-\infty}^{\infty} f(x)\,dx = \lim_{t\to\infty} \int_{-t}^{t} f(x)\,dx.$$

49. If $\int_{-\infty}^{\infty} f(x)\,dx$ exists, then $\int_0^{\infty} f(x)\,dx$ exists, and
$$\int_{-\infty}^{\infty} f(x)\,dx = 2\int_0^{\infty} f(x)\,dx.$$

50. If $\int_a^{\infty} f(x)\,dx$ exists, then $\int_{-\infty}^{-a} f(x)\,dx$ exists, and
$$\int_a^{\infty} f(x)\,dx = -\int_{-\infty}^{-a} f(x)\,dx \text{ exists.}$$

51. CAPITAL VALUE The capital value (present sale value) CV of property that can be rented on a perpetual basis for R dollars annually is approximated by the formula

$$CV \approx \int_0^{\infty} Re^{-it}\,dt$$

where i is the prevailing continuous interest rate.
a. Show that $CV = R/i$.
b. Find the capital value of property that can be rented at \$10,000 annually when the prevailing continuous interest rate is 12%/year.

52. Show that an integral of the form $\int_a^{\infty} e^{-px}\,dx$ is convergent if $p > 0$ and divergent if $p < 0$.

53. Show that an integral of the form $\int_{-\infty}^{b} e^{px}\,dx$ is convergent if $p > 0$ and divergent if $p < 0$.

54. Find the values of p such that $\int_1^{\infty} \frac{1}{x^p}\,dx$ is convergent.

7.4 Solutions to Self-Check Exercises

1. Write

$$\int_{-\infty}^{\infty} \frac{x^3}{(1+x^4)^{3/2}}\,dx = \int_{-\infty}^{0} \frac{x^3}{(1+x^4)^{3/2}}\,dx + \int_0^{\infty} \frac{x^3}{(1+x^4)^{3/2}}\,dx$$

Now,

$$\begin{aligned}
\int_{-\infty}^{0} \frac{x^3}{(1+x^4)^{3/2}}\,dx &= \lim_{a\to-\infty} \int_a^0 x^3(1+x^4)^{-3/2}\,dx \\
&= \lim_{a\to-\infty} \frac{1}{4}(-2)(1+x^4)^{-1/2}\Big|_a^0 \qquad \text{Integrate by substitution.} \\
&= -\frac{1}{2}\lim_{a\to-\infty}\left[1 - \frac{1}{(1+a^4)^{1/2}}\right] \\
&= -\frac{1}{2}
\end{aligned}$$

Similarly, you can show that

$$\int_0^{\infty} \frac{x^3}{(1+x^4)^{3/2}}\,dx = \frac{1}{2}$$

Therefore,

$$\int_{-\infty}^{\infty} \frac{x^3}{(1+x^4)^{3/2}}\,dx = -\frac{1}{2} + \frac{1}{2} = 0$$

2. The required present value is given by

$$\begin{aligned} PV &= \int_0^{\infty} 20e^{-0.02t}e^{-0.10t}\,dt \\ &= 20\int_0^{\infty} e^{-0.12t}\,dt \\ &= 20\lim_{b\to\infty}\int_0^b e^{-0.12t}\,dt \\ &= -\frac{20}{0.12}\lim_{b\to\infty} e^{-0.12t}\Big|_0^b \\ &= -\frac{500}{3}\lim_{b\to\infty}(e^{-0.12b} - 1) \\ &= \frac{500}{3} \end{aligned}$$

or approximately \$166,667.

CHAPTER 7 Summary of Principal Formulas and Terms

FORMULAS

1. Integration by parts	$\int u\,dv = uv - \int v\,du$
2. Trapezoidal rule	$\int_a^b f(x)\,dx \approx \frac{\Delta x}{2}[f(x_0) + 2f(x_1) + 2f(x_2) + \cdots + 2f(x_{n-1}) + f(x_n)]$ where $\Delta x = \frac{b-a}{n}$
3. Simpson's rule	$\int_a^b f(x)\,dx \approx \frac{\Delta x}{3}[f(x_0) + 4f(x_1) + 2f(x_2) + 4f(x_3) + 2f(x_4) + \cdots + 4f(x_{n-1}) + f(x_n)]$ where $\Delta x = \frac{b-a}{n}$
4. Maximum error for trapezoidal rule	$\frac{M(b-a)^3}{12n^2}$, where $\lvert f''(x)\rvert \le M$ $(a \le x \le b)$
5. Maximum error for Simpson's rule	$\frac{M(b-a)^5}{180n^4}$, where $\lvert f^{(4)}(x)\rvert \le M$ $(a \le x \le b)$
6. Improper integral of f over $[a, \infty)$	$\int_a^{\infty} f(x)\,dx = \lim_{b\to\infty}\int_a^b f(x)\,dx$

7. Improper integral of f over $(-\infty, b]$	$\int_{-\infty}^{b} f(x)\,dx = \lim_{a\to-\infty} \int_{a}^{b} f(x)\,dx$
8. Improper integral of f over $(-\infty, \infty)$	$\int_{-\infty}^{\infty} f(x)\,dx = \int_{-\infty}^{c} f(x)\,dx + \int_{c}^{\infty} f(x)\,dx$
9. Present value of a perpetuity	$PV = \dfrac{mP}{r}$

TERMS

improper integral (521)

convergent integral (523)

divergent integral (523)

perpetuity (526)

CHAPTER 7 Concept Review Questions

Fill in the blanks.

1. The integration by parts formula is obtained by reversing the _____ rule. The formula for integration by parts is $\int u\,dv =$ _____. In choosing u and dv, we want du to be simpler than _____ and dv to be_____ _____ _____.

2. To find $I = \int x \ln(x^2 + 1)\,dx$ using the table of integrals, we need to first use the substitution with $u =$ _____ so that $du =$ _____ to transform I into the integral $I = \frac{1}{2}\int u \ln u\,du$. We then choose Formula _____ to evaluate this integral.

3. The trapezoidal rule states that $\int_a^b f(x)\,dx \approx$ _____ where $\Delta x = \frac{b-a}{n}$. The error E_n in approximating $\int_a^b f(x)\,dx$ by the trapezoidal rule satisfies $|E_n| \le$ _____.

4. Simpson's rule states that $\int_a^b f(x)\,dx \approx$ _____ where $\Delta x = \frac{b-a}{n}$ and n is _____. The error E_n in approximating $\int_a^b f(x)\,dx$ by Simpson's rule satisfies $|E_n| \le$ _____.

5. The improper integrals $\int_{-\infty}^{b} f(x)\,dx =$ _____, $\int_{a}^{\infty} f(x)\,dx =$ _____, and $\int_{-\infty}^{\infty} f(x)\,dx =$ _____.

CHAPTER 7 Review Exercises

In Exercises 1–6, evaluate the integral.

1. $\int 2xe^{-x}\,dx$

2. $\int xe^{4x}\,dx$

3. $\int \ln 5x\,dx$

4. $\int_1^4 \ln 2x\,dx$

5. $\int_0^1 xe^{-2x}\,dx$

6. $\int_0^2 xe^{2x}\,dx$

7. Find the function f given that the slope of the tangent line to the graph of f at any point $(x, f(x))$ is

$$f'(x) = \frac{\ln x}{\sqrt{x}}$$

and that the graph of f passes through the point $(1, -2)$.

8. Find the function f given that the slope of the tangent line to the graph of f at any point $(x, f(x))$ is

$$f'(x) = xe^{-3x}$$

and that the graph of f passes through the point $(0, 0)$.

In Exercises 9–14, use the table of integrals in Section 7.2 to evaluate the integral.

9. $\int \frac{x^2\,dx}{(3+2x)^2}$

10. $\int \frac{2x}{\sqrt{2x+3}}\,dx$

11. $\int x^2 e^{4x}\,dx$

12. $\int \frac{dx}{(x^2-25)^{3/2}}$

13. $\int \frac{dx}{x^2\sqrt{x^2-4}}$

14. $\int 8x^3 \ln 2x\,dx$

In Exercises 15–20, evaluate each improper integral whenever it is convergent.

15. $\int_0^\infty e^{-2x}\,dx$

16. $\int_{-\infty}^0 e^{3x}\,dx$

17. $\int_3^\infty \frac{2}{x}\,dx$

18. $\int_2^\infty \frac{1}{(x+2)^{3/2}}\,dx$

19. $\int_2^\infty \frac{dx}{(1+2x)^2}$

20. $\int_1^\infty 3e^{1-x}\,dx$

In Exercises 21–24, use the trapezoidal rule and Simpson's rule to approximate the value of the definite integral.

21. $\int_1^3 \frac{dx}{1+\sqrt{x}}$; $n = 4$

22. $\int_0^1 e^{x^2}\,dx$; $n = 4$

23. $\int_{-1}^1 \sqrt{1+x^4}\,dx$; $n = 4$

24. $\int_1^3 \frac{e^x}{x}\,dx$; $n = 4$

25. Find a bound on the error in approximating the integral $\int_0^1 \frac{dx}{x+1}$ with $n = 8$ using (a) the trapezoidal rule and (b) Simpson's rule.

26. Producers' Surplus The supply equation for the GTC Slim-Phone is given by

$$p = 2\sqrt{25 + x^2}$$

where p is the unit price in dollars and x is the quantity demanded per month in units of 10,000. Find the producers' surplus if the market price is \$26. Use the table of integrals in Section 7.2 to evaluate the definite integral.

27. Computer Game Sales The sales of Starr Communication's newest computer game, Laser Beams, are currently

$$te^{-0.05t}$$

units/month (each unit representing 1000 games), where t denotes the number of months since the release of the game. Find an expression that gives the total number of games sold as a function of t. How many games will be sold by the end of the first year?

28. Demand for Computer Software The demand equation for a computer software program is given by

$$p = 2\sqrt{325 + x^2}$$

where p is the unit price in dollars and x is the quantity demanded each month in units of a thousand. Find the consumers' surplus if the market price is \$30. Evaluate the definite integral using Simpson's rule with $n = 10$.

29. Oil Spills Using aerial photographs, the Coast Guard was able to determine the dimensions of an oil spill along an embankment on a coastline, as shown in the accompanying figure. Using (a) the trapezoidal rule and (b) Simpson's rule with $n = 10$, estimate the area of the oil spill.

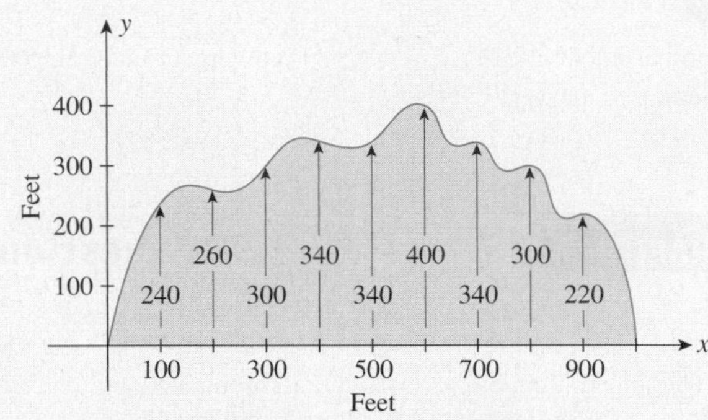

30. Surface Area of a Lake A manmade lake located in Lake View Condominiums has the shape depicted in the following figure. The measurements shown were taken at 15-ft intervals. Using Simpson's rule with $n = 10$, estimate the surface area of the lake.

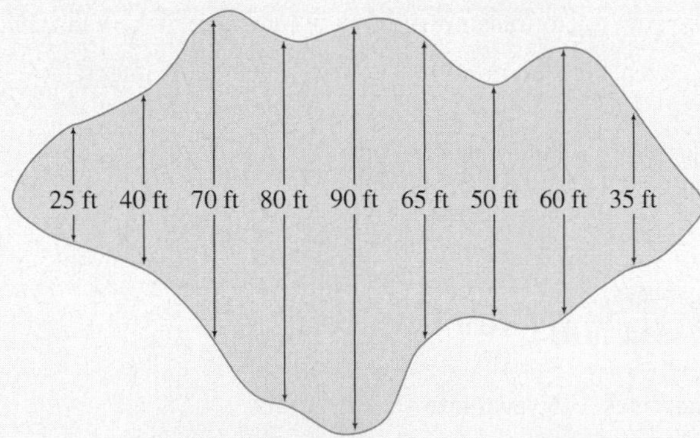

31. Perpetuities Lindsey wishes to establish a memorial fund at Newtown Hospital in the amount of \$10,000/year beginning next year. If the fund earns interest at a rate of 9%/year compounded continuously, find the amount of endowment that he is required to make now.

CHAPTER 7 Before Moving On . . .

1. Find $\int x^2 \ln x\, dx$.

2. Use the table of integrals to find $\int \dfrac{dx}{x^2\sqrt{8 + 2x^2}}$.

3. Evaluate $\int_2^4 \sqrt{x^2 + 1}\, dx$ using the trapezoidal rule with $n = 5$.

4. Evaluate $\int_1^3 e^{0.2x}\, dx$ using Simpson's rule with $n = 6$.

5. Evaluate $\int_1^\infty e^{-2x}\, dx$.

8 Calculus of Several Variables

What should the dimensions of the new swimming pool be? It will be built in an elliptical area located in the rear of the promenade deck. Subject to this constraint, what are the dimensions of the largest pool that can be built? See Example 5, page 587, to see how to solve this problem.

© Richard Cummins/Corbis

UP TO NOW, we have dealt with functions involving one variable. However, many situations involve two or more variables. For example, the Consumer Price Index (CPI) compiled by the Bureau of Labor Statistics depends on the price of more than 95,000 consumer items. To study such relationships, we need the notion of a function of several variables, the first topic in this chapter. Next, generalizing the concept of the derivative of a function of one variable, we study the *partial derivatives* of a function of two or more variables. Using partial derivatives, we study the rate of change of a function with respect to one variable while holding all other variables constant. We then learn how to find the extremum values of a function of several variables. As an application of optimization theory, we learn how to find an equation of the straight line that "best" fits a set of data points scattered about a straight line. Finally, we generalize the notion of the integral to the case involving a function of two variables.

8.1 Functions of Several Variables

Up to now, our study of calculus has been restricted to functions of one variable. In many practical situations, however, the formulation of a problem results in a mathematical model that involves a function of two or more variables. For example, suppose Ace Novelty determines that the profits are \$6, \$5, and \$4 for three types of souvenirs it produces. Let x, y, and z denote the number of type-A, type-B, and type-C souvenirs to be made; then the company's profit is given by

$$P = 6x + 5y + 4z$$

and P is a function of the three variables, x, y, and z.

Functions of Two Variables

Although this chapter deals with real-valued functions of several variables, most of our definitions and results are stated in terms of a function of two variables. One reason for adopting this approach, as you will soon see, is that there is a geometric interpretation for this special case, which serves as an important visual aid. We can then draw upon the experience gained from studying the two-variable case to help us understand the concepts and results connected with the more general case, which, by and large, is just a simple extension of the lower-dimensional case.

A Function of Two Variables

A real-valued **function of two variables** f, consists of

1. A set A of ordered pairs of real numbers (x, y) called the **domain** of the function.
2. A rule that associates with each ordered pair in the domain of f one and only one real number, denoted by $z = f(x, y)$.

The variables x and y are called **independent variables**, and the variable z, which is dependent on the values of x and y, is referred to as a **dependent variable**.

As in the case of a real-valued function of one real variable, the number $z = f(x, y)$ is called the **value of f** at the point (x, y). And, unless specified, the domain of the function f will be taken to be the largest possible set for which the rule defining f is meaningful.

EXAMPLE 1 Let f be the function defined by

$$f(x, y) = x + xy + y^2 + 2$$

Compute $f(0, 0)$, $f(1, 2)$, and $f(2, 1)$.

Solution We have

$$f(0, 0) = 0 + (0)(0) + 0^2 + 2 = 2$$
$$f(1, 2) = 1 + (1)(2) + 2^2 + 2 = 9$$
$$f(2, 1) = 2 + (2)(1) + 1^2 + 2 = 7$$

■

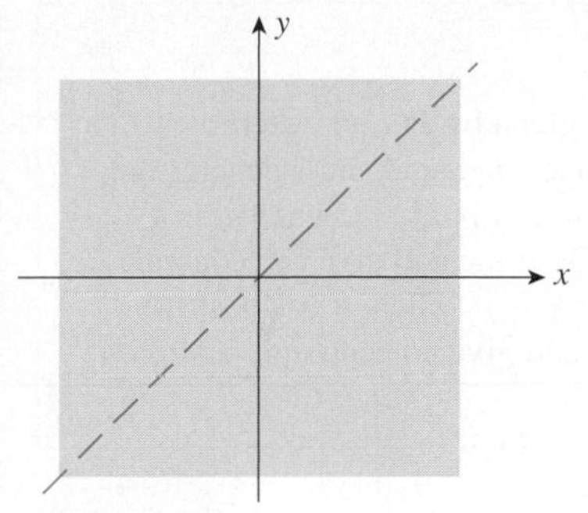

(a) Domain of g

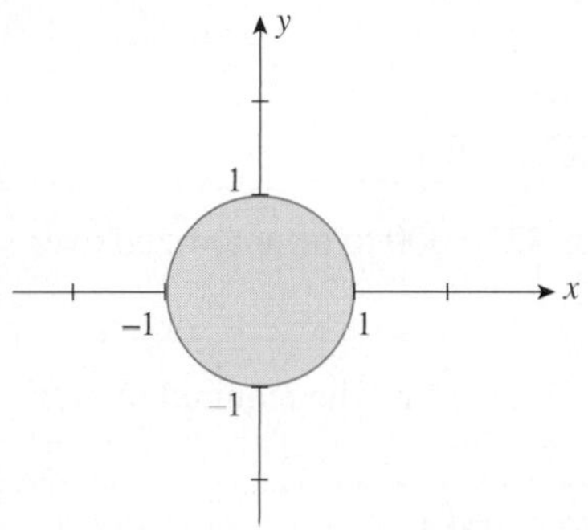

(b) Domain of h

FIGURE 1

The domain of a function of two variables $f(x, y)$ is a set of ordered pairs of real numbers and may therefore be viewed as a subset of the xy-plane.

EXAMPLE 2 Find the domain of each of the following functions.

a. $f(x, y) = x^2 + y^2$ **b.** $g(x, y) = \dfrac{2}{x - y}$ **c.** $h(x, y) = \sqrt{1 - x^2 - y^2}$

Solution

a. $f(x, y)$ is defined for all real values of x and y, so the domain of the function f is the set of all points (x, y) in the xy-plane.

b. $g(x, y)$ is defined for all $x \neq y$, so the domain of the function g is the set of all points in the xy-plane except those lying on the line $y = x$ (Figure 1a).

c. We require that $1 - x^2 - y^2 \geq 0$ or $x^2 + y^2 \leq 1$, which is just the set of all points (x, y) lying on and inside the circle of radius 1 with center at the origin (Figure 1b). ■

APPLIED EXAMPLE 3 Revenue Functions Acrosonic manufactures a bookshelf loudspeaker system that may be bought fully assembled or in a kit. The demand equations that relate the unit prices, p and q, to the quantities demanded weekly, x and y, of the assembled and kit versions of the loudspeaker systems are given by

$$p = 300 - \frac{1}{4}x - \frac{1}{8}y \quad \text{and} \quad q = 240 - \frac{1}{8}x - \frac{3}{8}y$$

a. What is the weekly total revenue function $R(x, y)$?

b. What is the domain of the function R?

Solution

a. The weekly revenue realizable from the sale of x units of the assembled speaker systems at p dollars per unit is given by xp dollars. Similarly, the weekly revenue realizable from the sale of y units of the kits at q dollars per unit is given by yq dollars. Therefore, the weekly total revenue function R is given by

$$\begin{aligned} R(x, y) &= xp + yq \\ &= x\left(300 - \frac{1}{4}x - \frac{1}{8}y\right) + y\left(240 - \frac{1}{8}x - \frac{3}{8}y\right) \\ &= -\frac{1}{4}x^2 - \frac{3}{8}y^2 - \frac{1}{4}xy + 300x + 240y \end{aligned}$$

b. To find the domain of the function R, let's observe that the quantities x, y, p, and q must be nonnegative. This observation leads to the following system of linear inequalities:

$$\begin{aligned} 300 - \frac{1}{4}x - \frac{1}{8}y &\geq 0 \\ 240 - \frac{1}{8}x - \frac{3}{8}y &\geq 0 \\ x &\geq 0 \\ y &\geq 0 \end{aligned}$$

The domain of the function R is sketched in Figure 2. ■

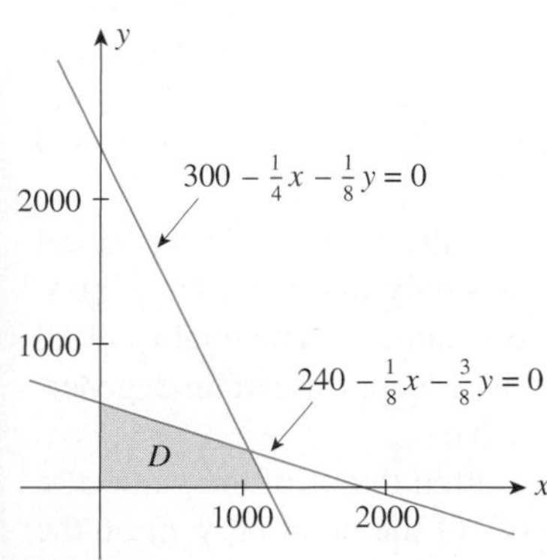

FIGURE 2
The domain of $R(x, y)$

EXPLORE & DISCUSS

Suppose the total profit of a two-product company is given by $P(x, y)$, where x denotes the number of units of the first product produced and sold and y denotes the number of units of the second product produced and sold. Fix $x = a$, where a is a positive number so that (a, y) is in the domain of P. Describe and give an economic interpretation of the function $f(y) = P(a, y)$. Next, fix $y = b$, where b is a positive number so that (x, b) is in the domain of P. Describe and give an economic interpretation of the function $g(x) = P(x, b)$.

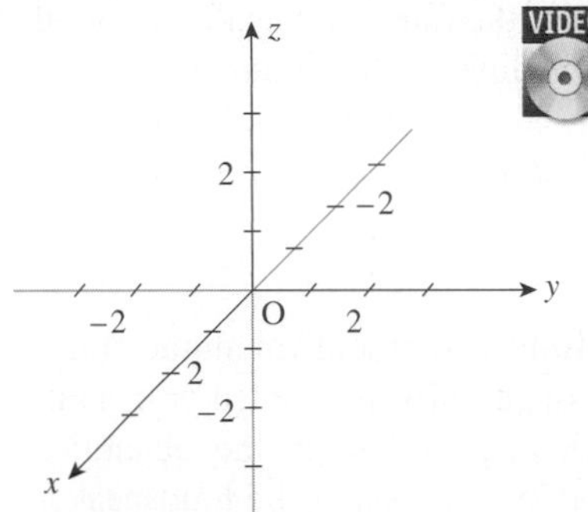

FIGURE 3
The three-dimensional Cartesian coordinate system

VIDEO

APPLIED EXAMPLE 4 Home Mortgage Payments The monthly payment that amortizes a loan of A dollars in t years when the interest rate is r per year is given by

$$P = f(A, r, t) = \frac{Ar}{12\left[1 - \left(1 + \frac{r}{12}\right)^{-12t}\right]}$$

Find the monthly payment for a home mortgage of \$270,000 to be amortized over 30 years when the interest rate is 10% per year.

Solution Letting $A = 270{,}000$, $r = 0.1$, and $t = 30$, we find the required monthly payment to be

$$P = f(270{,}000, 0.1, 30) = \frac{270{,}000(0.1)}{12\left[1 - \left(1 + \frac{0.1}{12}\right)^{-360}\right]}$$
$$\approx 2369.44$$

or approximately \$2369.44.

Graphs of Functions of Two Variables

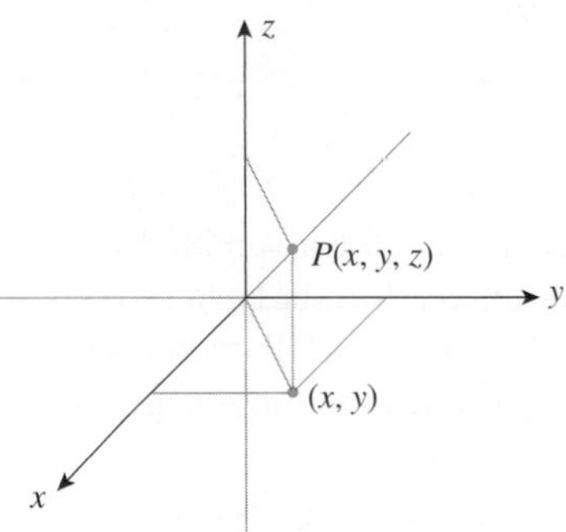

(a) A point in three-dimensional space

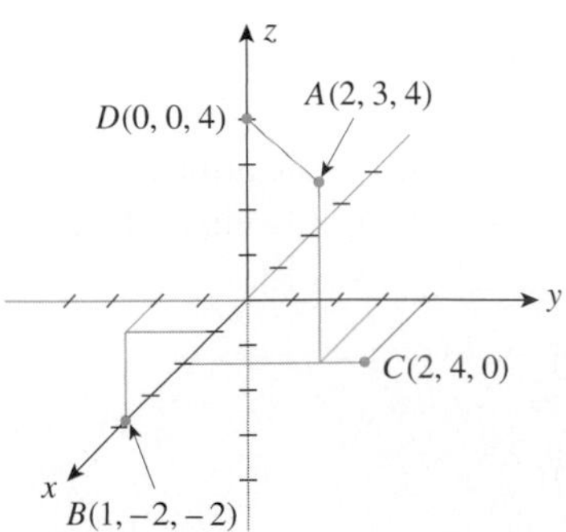

(b) Some sample points in three-dimensional space
FIGURE 4

To graph a function of two variables, we need a three-dimensional coordinate system. This is readily constructed by adding a third axis to the plane Cartesian coordinate system in such a way that the three resulting axes are mutually perpendicular and intersect at O. Observe that, by construction, the zeros of the three number scales coincide at the origin of the **three-dimensional Cartesian coordinate system** (Figure 3).

A point in three-dimensional space can now be represented uniquely in this coordinate system by an **ordered triple** of numbers (x, y, z), and, conversely, every ordered triple of real numbers (x, y, z) represents a point in three-dimensional space (Figure 4a). For example, the points $A(2, 3, 4)$, $B(1, -2, -2)$, $C(2, 4, 0)$, and $D(0, 0, 4)$ are shown in Figure 4b.

Now, if $f(x, y)$ is a function of two variables x and y, the domain of f is a subset of the xy-plane. Let $z = f(x, y)$ so that there is one and only one point $(x, y, z) \equiv (x, y, f(x, y))$ associated with each point (x, y) in the domain of f. The totality of all such points makes up the **graph** of the function f and is, except for certain degenerate cases, a surface in three-dimensional space (Figure 5).

In interpreting the graph of a function $f(x, y)$, one often thinks of the value $z = f(x, y)$ of the function at the point (x, y) as the "height" of the point (x, y, z) on the graph of f. If $f(x, y) > 0$, then the point (x, y, z) is $f(x, y)$ units above the xy-plane; if $f(x, y) < 0$, then the point (x, y, z) is $|f(x, y)|$ units below the xy-plane.

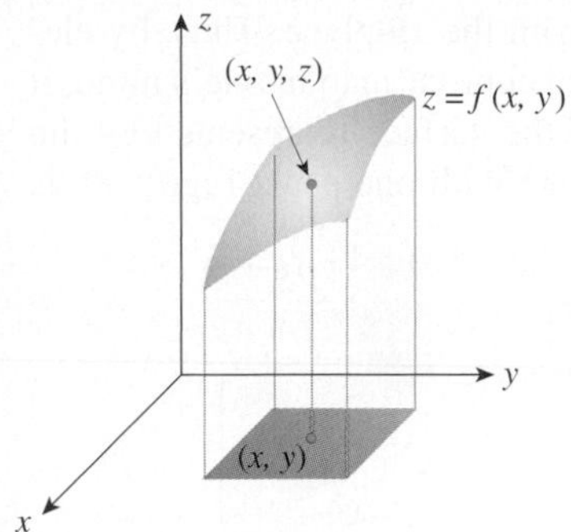

FIGURE 5
The graph of a function in three-dimensional space

In general, it is quite difficult to draw the graph of a function of two variables. But techniques have been developed that enable us to generate such graphs with minimum effort, using a computer. Figure 6 shows the computer-generated graphs of two functions.

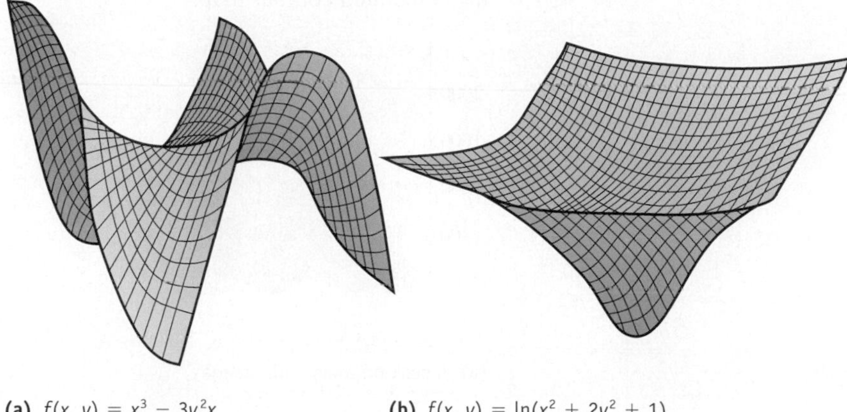

(a) $f(x, y) = x^3 - 3y^2x$ **(b)** $f(x, y) = \ln(x^2 + 2y^2 + 1)$

FIGURE 6
Two computer-generated graphs of functions of two variables

Level Curves

As mentioned earlier, the graph of a function of two variables is often difficult to sketch, and we will not develop a systematic procedure for sketching it. Instead, we describe a method that is used in constructing topographic maps. This method is relatively easy to apply and conveys sufficient information to enable one to obtain a feel for the graph of the function.

Suppose that $f(x, y)$ is a function of two variables x and y, with a graph as shown in Figure 7. If c is some value of the function f, then the equation $f(x, y) = c$ describes a curve lying on the plane $z = c$ called the **trace** of the graph of f in the plane $z = c$. If this trace is projected onto the xy-plane, the resulting curve in the xy-plane is called a **level curve**. By drawing the level curves corresponding to

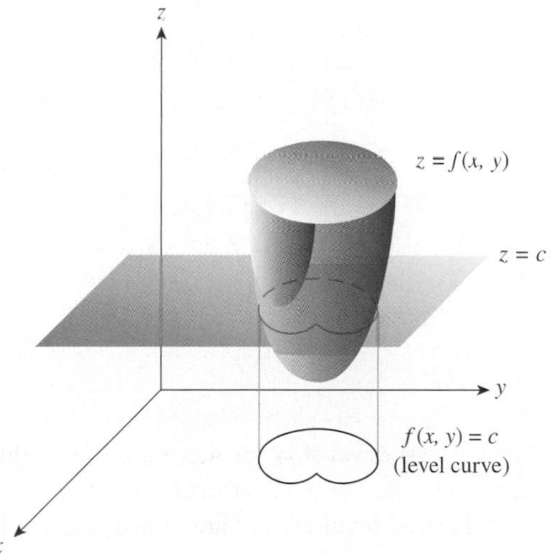

FIGURE 7
The graph of the function $z = f(x, y)$ and its intersection with the plane $z = c$

several admissible values of c, we obtain a **contour map**. Observe that, by construction, every point on a particular level curve corresponds to a point on the surface $z = f(x, y)$ that is a certain fixed distance from the xy-plane. Thus, by elevating or depressing the level curves that make up the contour map in one's mind, it is possible to obtain a feel for the general shape of the surface represented by the function f. Figure 8a shows a part of a mountain range with one peak; Figure 8b is the associated contour map.

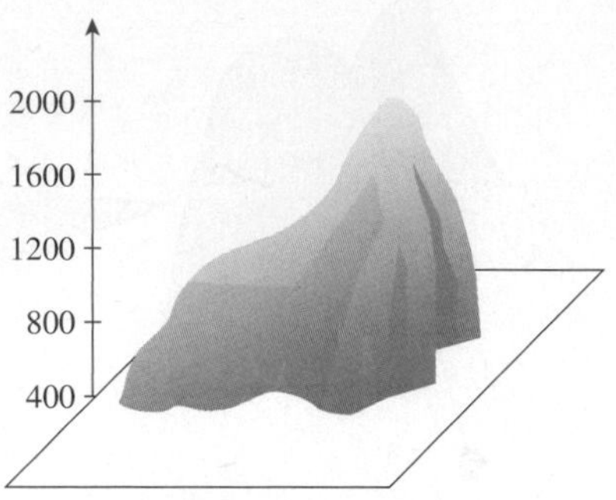

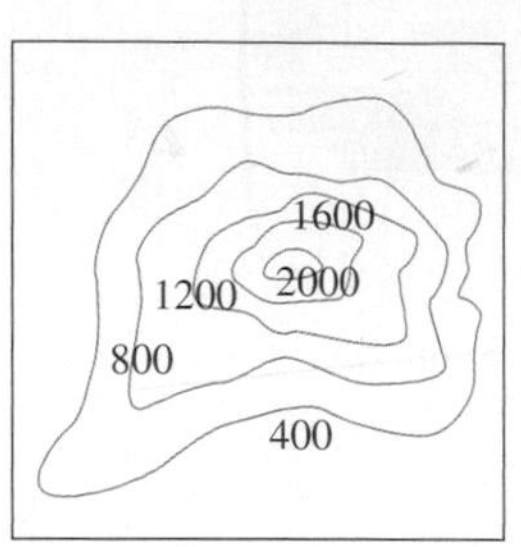

FIGURE 8

(a) A peak on a mountain range

(b) A contour map for the mountain peak

EXAMPLE 5 Sketch a contour map for the function $f(x, y) = x^2 + y^2$.

Solution The level curves are the graphs of the equation $x^2 + y^2 = c$ for nonnegative numbers c. Taking $c = 0, 1, 4, 9$, and 16, for example, we obtain

$$c = 0\colon x^2 + y^2 = 0$$
$$c = 1\colon x^2 + y^2 = 1$$
$$c = 4\colon x^2 + y^2 = 4 = 2^2$$
$$c = 9\colon x^2 + y^2 = 9 = 3^2$$
$$c = 16\colon x^2 + y^2 = 16 = 4^2$$

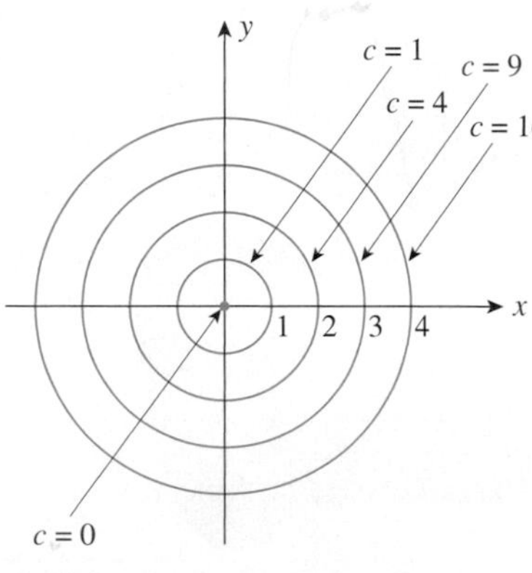

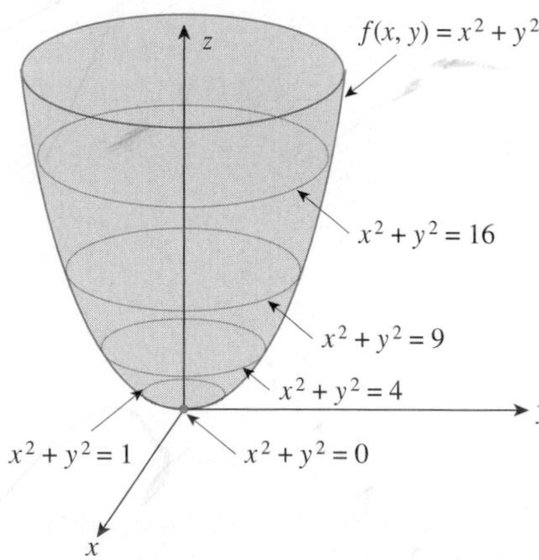

FIGURE 9

(a) Level curves of $f(x, y) = x^2 + y^2$

(b) The graph of $f(x, y) = x^2 + y^2$

The five level curves are concentric circles with center at the origin and radius given by $r = 0, 1, 2, 3$, and 4, respectively (Figure 9a). A sketch of the graph of $f(x, y) = x^2 + y^2$ is included for your reference in Figure 9b. ■

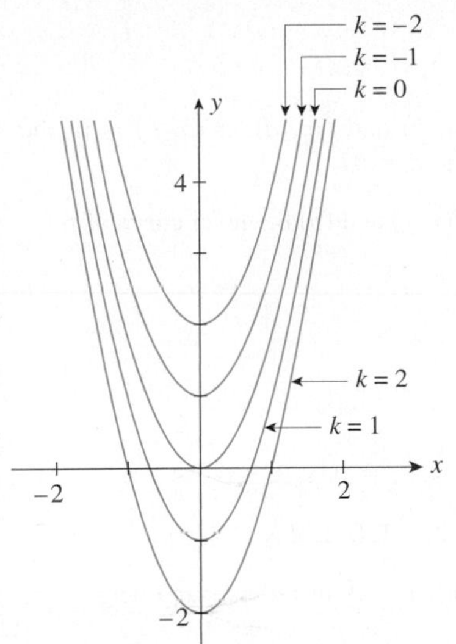

FIGURE 10
Level curves for $f(x, y) = 2x^2 - y$

EXAMPLE 6 Sketch the level curves for the function $f(x, y) = 2x^2 - y$ corresponding to $z = -2, -1, 0, 1$, and 2.

Solution The level curves are the graphs of the equation $2x^2 - y = k$ or $y = 2x^2 - k$ for $k = -2, -1, 0, 1$, and 2. The required level curves are shown in Figure 10. ■

Level curves of functions of two variables are found in many practical applications. For example, if $f(x, y)$ denotes the temperature at a location within the continental United States with longitude x and latitude y at a certain time of day, then the temperature at the point (x, y) is given by the "height" of the surface, represented by $z = f(x, y)$. In this situation the level curve $f(x, y) = k$ is a curve superimposed on a map of the United States, connecting points having the same temperature at a given time (Figure 11). These level curves are called **isotherms**.

Similarly, if $f(x, y)$ gives the barometric pressure at the location (x, y), then the level curves of the function f are called **isobars**, lines connecting points having the same barometric pressure at a given time.

As a final example, suppose $P(x, y, z)$ is a function of three variables x, y, and z giving the profit realized when x, y, and z units of three products, A, B, and C, respectively, are produced and sold. Then, the equation $P(x, y, z) = k$, where k is a constant, represents a surface in three-dimensional space called a **level surface** of P. In this situation, the level surface represented by $P(x, y, z) = k$ represents the product mix that results in a profit of exactly k dollars. Such a level surface is called an **isoprofit surface**.

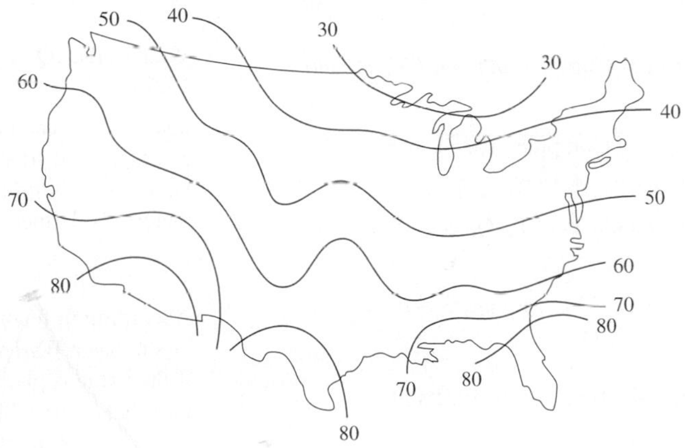

FIGURE 11
Isotherms: curves connecting points that have the same temperature

8.1 Self-Check Exercises

1. Let $f(x, y) = x^2 - 3xy + \sqrt{x + y}$. Compute $f(1, 3)$ and $f(-1, 1)$. Is the point $(-1, 0)$ in the domain of f?
2. Find the domain $f(x, y) = \dfrac{1}{x} + \dfrac{1}{x - y} - e^{x+y}$.
3. Odyssey Travel Agency has a monthly advertising budget of \$20,000. Odyssey's management estimates that if they spend x dollars on newspaper advertising and y dollars on television advertising, then the monthly revenue will be

$$f(x, y) = 30x^{1/4}y^{3/4}$$

dollars. What will be the monthly revenue if Odyssey spends \$5000/month on newspaper ads and \$15,000/month on television ads? If Odyssey spends \$4000/month on newspaper ads and \$16,000/month on television ads?

Solutions to Self-Check Exercises 8.1 can be found on page 544.

8.1 Concept Questions

1. What is a function of two variables? Give an example of a function of two variables and state its rule of definition and domain.

2. If f is a function of two variables, what can you say about the relationship between $f(a, b)$ and $f(c, d)$, if (a, b) is in the domain of f and $c = a$ and $d = b$?

3. Define (a) the graph of $f(x, y)$ and (b) the level curve of f.

8.1 Exercises

1. Let $f(x, y) = 2x + 3y - 4$. Compute $f(0, 0)$, $f(1, 0)$, $f(0, 1)$, $f(1, 2)$, and $f(2, -1)$.

2. Let $g(x, y) = 2x^2 - y^2$. Compute $g(1, 2)$, $g(2, 1)$, $g(1, 1)$, $g(-1, 1)$, and $g(2, -1)$.

3. Let $f(x, y) = x^2 + 2xy - x + 3$. Compute $f(1, 2)$, $f(2, 1)$, $f(-1, 2)$, and $f(2, -1)$.

4. Let $h(x, y) = (x + y)/(x - y)$. Compute $h(0, 1)$, $h(-1, 1)$, $h(2, 1)$, and $h(\pi, -\pi)$.

5. Let $g(s, t) = 3s\sqrt{t} + t\sqrt{s} + 2$. Compute $g(1, 2)$, $g(2, 1)$, $g(0, 4)$, and $g(4, 9)$.

6. Let $f(x, y) = xye^{x^2+y^2}$. Compute $f(0, 0)$, $f(0, 1)$, $f(1, 1)$, and $f(-1, -1)$.

7. Let $h(s, t) = s \ln t - t \ln s$. Compute $h(1, e)$, $h(e, 1)$, and $h(e, e)$.

8. Let $f(u, v) = (u^2 + v^2)e^{uv^2}$. Compute $f(0, 1)$, $f(-1, -1)$, $f(a, b)$, and $f(b, a)$.

9. Let $g(r, s, t) = re^{s/t}$. Compute $g(1, 1, 1)$, $g(1, 0, 1)$, and $g(-1, -1, -1)$.

10. Let $g(u, v, w) = (ue^{vw} + ve^{uw} + we^{uv})/(u^2 + v^2 + w^2)$. Compute $g(1, 2, 3)$ and $g(3, 2, 1)$.

In Exercises 11–18, find the domain of the function.

11. $f(x, y) = 2x + 3y$

12. $g(x, y, z) = x^2 + y^2 + z^2$

13. $h(u, v) = \dfrac{uv}{u - v}$

14. $f(s, t) = \sqrt{s^2 + t^2}$

15. $g(r, s) = \sqrt{rs}$

16. $f(x, y) = e^{-xy}$

17. $h(x, y) = \ln(x + y - 5)$

18. $h(u, v) = \sqrt{4 - u^2 - v^2}$

In Exercises 19–24, sketch the level curves of the function corresponding to each value of z.

19. $f(x, y) = 2x + 3y$; $z = -2, -1, 0, 1, 2$

20. $f(x, y) = -x^2 + y$; $z = -2, -1, 0, 1, 2$

21. $f(x, y) = 2x^2 + y$; $z = -2, -1, 0, 1, 2$

22. $f(x, y) = xy$; $z = -4, -2, 2, 4$

23. $f(x, y) = \sqrt{16 - x^2 - y^2}$; $z = 0, 1, 2, 3, 4$

24. $f(x, y) = e^x - y$; $z = -2, -1, 0, 1, 2$

25. The volume of a cylindrical tank of radius r and height h is given by

$$V = f(r, h) = \pi r^2 h$$

Find the volume of a cylindrical tank of radius 1.5 ft and height 4 ft.

26. IQs The IQ (intelligence quotient) of a person whose mental age is m yr and whose chronological age is c yr is defined as

$$f(m, c) = \frac{100m}{c}$$

What is the IQ of a 9-yr-old child who has a mental age of 13.5 yr?

27. Body Mass The body mass index (BMI) is used to identify, evaluate, and treat overweight and obese adults. The BMI value for an adult of weight w (in kilograms) and height h (in meters) is defined to be

$$M = f(w, h) = \frac{w}{h^2}$$

According to federal guidelines, an adult is overweight if he or she has a BMI value between 25 and 29.9 and is "obese" if the value is greater than or equal to 30.

a. What is the BMI of an adult who weighs in at 80 kg and stands 1.8 m tall?

b. What is the maximum weight for an adult of height 1.8 m, who is not classified as overweight or obese?

28. Poiseuille's Law Poiseuille's law states that the resistance R, measured in dynes, of blood flowing in a blood vessel of length l and radius r (both in centimeters) is given by

$$R = f(l, r) = \frac{kl}{r^4}$$

where k is the viscosity of blood (in dyne-sec/cm^2). What is the resistance, in terms of k, of blood flowing through an arteriole 4 cm long and of radius 0.1 cm?

29. Revenue Functions Country Workshop manufactures both finished and unfinished furniture for the home. The estimated quantities demanded each week of its rolltop desks in the

finished and unfinished versions are x and y units when the corresponding unit prices are

$$p = 200 - \frac{1}{5}x - \frac{1}{10}y$$

$$q = 160 - \frac{1}{10}x - \frac{1}{4}y$$

dollars, respectively.

a. What is the weekly total revenue function $R(x, y)$?

b. Find the domain of the function R.

30. For the total revenue function $R(x, y)$ of Exercise 29, compute $R(100, 60)$ and $R(60, 100)$. Interpret your results.

31. Revenue Functions Weston Publishing publishes a deluxe edition and a standard edition of its English language dictionary. Weston's management estimates that the number of deluxe editions demanded is x copies/day and the number of standard editions demanded is y copies/day when the unit prices are

$$p = 20 - 0.005x - 0.001y$$
$$q = 15 - 0.001x - 0.003y$$

dollars, respectively.

a. Find the daily total revenue function $R(x, y)$.

b. Find the domain of the function R.

32. For the total revenue function $R(x, y)$ of Exercise 31, compute $R(300, 200)$ and $R(200, 300)$. Interpret your results.

33. Volume of a Gas The volume of a certain mass of gas is related to its pressure and temperature by the formula

$$V = \frac{30.9T}{P}$$

where the volume V is measured in liters, the temperature T is measured in degrees Kelvin (obtained by adding 273° to the Celsius temperature), and the pressure P is measured in millimeters of mercury pressure.

a. Find the domain of the function V.

b. Calculate the volume of the gas at standard temperature and pressure—that is, when $T = 273$ K and $P = 760$ mm of mercury.

34. Surface Area of a Human Body An empirical formula by E. F. Dubois relates the surface area S of a human body (in square meters) to its weight W (in kilograms) and its height H (in centimeters). The formula, given by

$$S = 0.007184W^{0.425}H^{0.725}$$

is used by physiologists in metabolism studies.

a. Find the domain of the function S.

b. What is the surface area of a human body that weighs 70 kg and has a height of 178 cm?

35. Production Function Suppose the output of a certain country is given by

$$f(x, y) = 100x^{3/5}y^{2/5}$$

billion dollars if x billion dollars are spent for labor and y billion dollars are spent on capital. Find the output if the country spent \$32 billion on labor and \$243 billion on capital.

36. Production Function Economists have found that the output of a finished product, $f(x, y)$, is sometimes described by the function

$$f(x, y) = ax^by^{1-b}$$

where x stands for the amount of money expended for labor, y stands for the amount expended on capital, and a and b are positive constants with $0 < b < 1$.

a. If p is a positive number, show that $f(px, py) = pf(x, y)$.

b. Use the result of part (a) to show that if the amount of money expended for labor and capital are both increased by r percent, then the output is also increased by r percent.

37. Arson for Profit A study of arson for profit was conducted by a team of paid civilian experts and police detectives appointed by the mayor of a large city. It was found that the number of suspicious fires in that city in 2002 was very closely related to the concentration of tenants in the city's public housing and to the level of reinvestment in the area in conventional mortgages by the ten largest banks. In fact, the number of fires was closely approximated by the formula

$$N(x, y) = \frac{100(1000 + 0.03x^2y)^{1/2}}{(5 + 0.2y)^2} \quad (0 \leq x \leq 150; 5 \leq y \leq 35)$$

where x denotes the number of persons/census tract and y denotes the level of reinvestment in the area in cents/dollar deposited. Using this formula, estimate the total number of suspicious fires in the districts of the city where the concentration of public housing tenants was 100/census tract and the level of reinvestment was 20 cents/dollar deposited.

38. Continuously Compounded Interest If a principal of P dollars is deposited in an account earning interest at the rate of r/year compounded continuously, then the accumulated amount at the end of t yr is given by

$$A = f(P, r, t) = Pe^{rt}$$

dollars. Find the accumulated amount at the end of 3 yr if a sum of \$10,000 is deposited in an account earning interest at the rate of 10%/year.

39. Home Mortgages The monthly payment that amortizes a loan of A dollars in t yr when the interest rate is r per year is given by

$$P = f(A, r, t) = \frac{Ar}{12\left[1 - \left(1 + \frac{r}{12}\right)^{-12t}\right]}$$

a. What is the monthly payment for a home mortgage of \$100,000 that will be amortized over 30 yr with an interest rate of 8%/year? An interest rate of 10%/year?

b. Find the monthly payment for a home mortgage of \$100,000 that will be amortized over 20 yr with an interest rate of 8%/year.

40. HOME MORTGAGES Suppose a home buyer secures a bank loan of A dollars to purchase a house. If the interest rate charged is r/year and the loan is to be amortized in t yr, then the principal repayment at the end of i mo is given by

$$B = f(A, r, t, i) = A\left[\frac{(1 + \frac{r}{12})^i - 1}{(1 + \frac{r}{12})^{12t} - 1}\right] \qquad (0 \leq i \leq 12t)$$

Suppose the Blakelys borrow a sum of \$280,000 from a bank to help finance the purchase of a house and the bank charges interest at a rate of 6%/year. If the Blakelys agree to repay the loan in equal installments over 30 yr, how much will they owe the bank after the 60th payment (5 yr)? The 240th payment (20 yr)?

41. FORCE GENERATED BY A CENTRIFUGE A centrifuge is a machine designed for the specific purpose of subjecting materials to a sustained centrifugal force. The actual amount of centrifugal force, F, expressed in dynes (1 gram of force = 980 dynes) is given by

$$F = f(M, S, R) = \frac{\pi^2 S^2 MR}{900}$$

where S is in revolutions per minute (rpm), M is in grams, and R is in centimeters. Show that an object revolving at the rate of 600 rpm in a circle with radius of 10 cm generates a centrifugal force that is approximately 40 times gravity.

42. WILSON LOT-SIZE FORMULA The Wilson lot-size formula in economics states that the optimal quantity Q of goods for a store to order is given by

$$Q = f(C, N, h) = \sqrt{\frac{2\,CN}{h}}$$

where C is the cost of placing an order, N is the number of items the store sells per week, and h is the weekly holding cost for each item. Find the most economical quantity of 10-speed bicycles to order if it costs the store \$20 to place an order, \$5 to hold a bicycle for a week, and the store expects to sell 40 bicycles a week.

In Exercises 43–46, determine whether the statement is true or false. If it is true, explain why it is true. If it is false, give an example to show why it is false.

43. If h is a function of x and y, then there are functions f and g of one variable such that

$$h(x, y) = f(x) + g(y)$$

44. If f is a function of x and y and a is a real number, then

$$f(ax, ay) = af(x, y)$$

45. The domain of $f(x, y) = 1/(x^2 - y^2)$ is $\{(x, y) | y \neq x\}$.

46. Every point on the level curve $f(x, y) = c$ corresponds to a point on the graph of f that is c units above the xy-plane if $c > 0$ and $|c|$ units below the xy-plane if $c < 0$.

8.1 Solutions to Self-Check Exercises

1. $f(1, 3) = 1^2 - 3(1)(3) + \sqrt{1 + 3} = -6$
$f(-1, 1) = (-1)^2 - 3(-1)(1) + \sqrt{-1 + 1} = 4$

The point $(-1, 0)$ is not in the domain of f because the term $\sqrt{x + y}$ is not defined when $x = -1$ and $y = 0$. In fact, the domain of f consists of all real values of x and y that satisfy the inequality $x + y \geq 0$, the shaded half-plane shown in the accompanying figure.

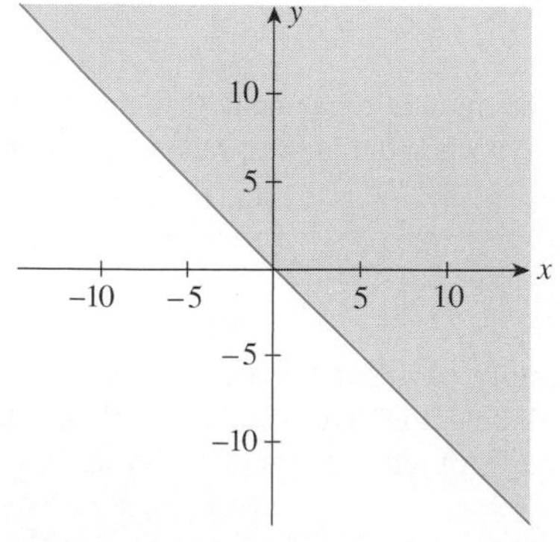

2. Since division by zero is not permitted, we see that $x \neq 0$ and $x - y \neq 0$. Therefore, the domain of f is the set of all points in the xy-plane not containing the y-axis ($x = 0$) and the straight line $x = y$.

3. If Odyssey spends \$5000/month on newspaper ads ($x = 5000$) and \$15,000/month on television ads ($y = 15{,}000$), then its monthly revenue will be given by

$$f(5000, 15{,}000) = 30(5000)^{1/4}(15{,}000)^{3/4} \approx 341{,}926.06$$

or approximately \$341,926. If the agency spends \$4000/month on newspaper ads and \$16,000/month on television ads, then its monthly revenue will be given by

$$f(4000, 16{,}000) = 30(4000)^{1/4}(16{,}000)^{3/4} \approx 339{,}411.26$$

or approximately \$339,411.

8.2 Partial Derivatives

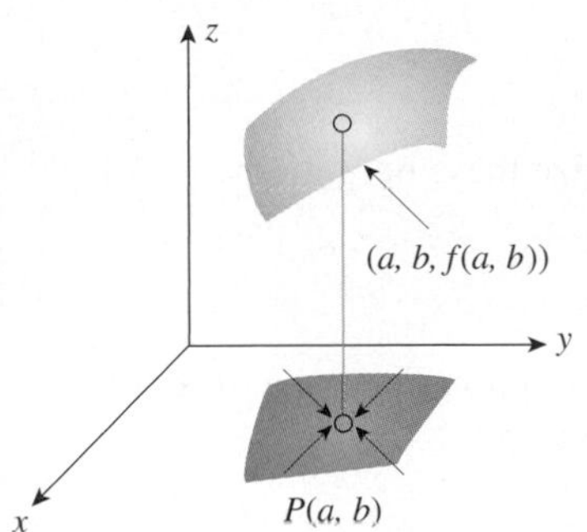

FIGURE 12
We can approach a point in the plane from infinitely many directions.

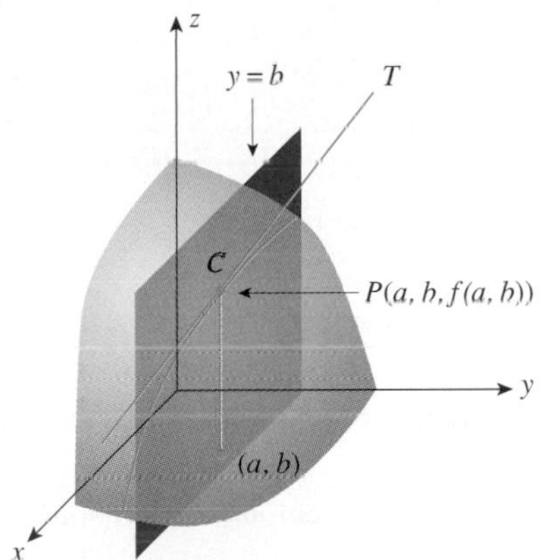

FIGURE 13
The curve C is formed by the intersection of the plane $y = b$ with the surface $z = f(x, y)$.

Partial Derivatives

For a function $f(x)$ of one variable x, there is no ambiguity when we speak about the rate of change of $f(x)$ with respect to x since x must be constrained to move along the x-axis. The situation becomes more complicated, however, when we study the rate of change of a function of two or more variables. For example, the domain D of a function of two variables $f(x, y)$ is a subset of the plane (Figure 12), so if $P(a, b)$ is any point in the domain of f, there are infinitely many directions from which one can approach the point P. We may therefore ask for the rate of change of f at P along any of these directions.

However, we will not deal with this general problem. Instead, we will restrict ourselves to studying the rate of change of the function $f(x, y)$ at a point $P(a, b)$ in each of two *preferred directions*—namely, the direction parallel to the x-axis and the direction parallel to the y-axis. Let $y = b$, where b is a constant, so that $f(x, b)$ is a function of the one variable x. Since the equation $z = f(x, y)$ is the equation of a surface, the equation $z = f(x, b)$ is the equation of the curve C on the surface formed by the intersection of the surface and the plane $y = b$ (Figure 13).

Because $f(x, b)$ is a function of one variable x, we may compute the derivative of f with respect to x at $x = a$. This derivative, obtained by keeping the variable y fixed and differentiating the resulting function $f(x, b)$ with respect to x, is called the **first partial derivative of f with respect to x** at (a, b), written

$$\frac{\partial z}{\partial x}(a, b) \quad \text{or} \quad \frac{\partial f}{\partial x}(a, b) \quad \text{or} \quad f_x(a, b)$$

Thus,

$$\frac{\partial z}{\partial x}(a, b) = \frac{\partial f}{\partial x}(a, b) = f_x(a, b) = \lim_{h \to 0} \frac{f(a + h, b) - f(a, b)}{h}$$

provided that the limit exists. The first partial derivative of f with respect to x at (a, b) measures both the slope of the tangent line T to the curve C and the rate of change of the function f in the x-direction when $x = a$ and $y = b$. We also write

$$\left.\frac{\partial f}{\partial x}\right|_{(a, b)} \equiv f_x(a, b)$$

Similarly, we define the **first partial derivative of f with respect to y** at (a, b), written

$$\frac{\partial z}{\partial y}(a, b) \quad \text{or} \quad \frac{\partial f}{\partial y}(a, b) \quad \text{or} \quad f_y(a, b)$$

as the derivative obtained by keeping the variable x fixed and differentiating the resulting function $f(a, y)$ with respect to y. That is,

$$\begin{aligned}\frac{\partial z}{\partial y}(a, b) &= \frac{\partial f}{\partial y}(a, b) = f_y(a, b)\\ &= \lim_{k \to 0} \frac{f(a, b + k) - f(a, b)}{k}\end{aligned}$$

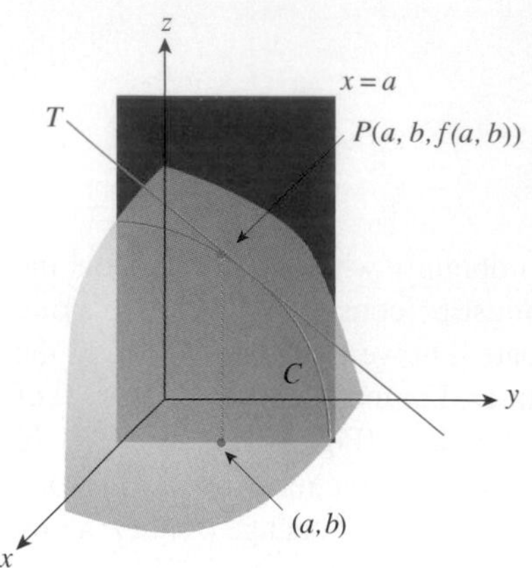

FIGURE 14
The first partial derivative of f with respect to y at (a, b) measures the slope of the tangent line T to the curve C with x held constant.

if the limit exists. The first partial derivative of f with respect to y at (a, b) measures both the slope of the tangent line T to the curve C, obtained by holding x constant (Figure 14), and the rate of change of the function f in the y-direction when $x = a$ and $y = b$. We write

$$\left.\frac{\partial f}{\partial y}\right|_{(a,\,b)} \equiv f_y(a, b)$$

Before looking at some examples, let's summarize these definitions.

First Partial Derivatives of $f(x, y)$

Suppose $f(x, y)$ is a function of the two variables x and y. Then, the **first partial derivative of f** with respect to x at the point (x, y) is

$$\frac{\partial f}{\partial x} = \lim_{h \to 0} \frac{f(x + h, y) - f(x, y)}{h}$$

provided the limit exists. The first partial derivative of f with respect to y at the point (x, y) is

$$\frac{\partial f}{\partial y} = \lim_{k \to 0} \frac{f(x, y + k) - f(x, y)}{k}$$

provided the limit exists.

EXAMPLE 1 Find the partial derivatives $\dfrac{\partial f}{\partial x}$ and $\dfrac{\partial f}{\partial y}$ of the function

$$f(x, y) = x^2 - xy^2 + y^3$$

What is the rate of change of the function f in the x-direction at the point $(1, 2)$? What is the rate of change of the function f in the y-direction at the point $(1, 2)$?

Solution To compute $\dfrac{\partial f}{\partial x}$, think of the variable y as a constant and differentiate the resulting function of x with respect to x. Let's write

$$f(x, y) = x^2 - xy^2 + y^3$$

where the variable y to be treated as a constant is shown in color. Then,

$$\frac{\partial f}{\partial x} = 2x - y^2$$

To compute $\dfrac{\partial f}{\partial y}$, think of the variable x as being fixed—that is, as a constant—and differentiate the resulting function of y with respect to y. In this case,

$$f(x, y) = x^2 - xy^2 + y^3$$

so that

$$\frac{\partial f}{\partial y} = -2xy + 3y^2$$

The rate of change of the function f in the x-direction at the point $(1, 2)$ is given by

$$f_x(1, 2) = \left.\frac{\partial f}{\partial x}\right|_{(1,\,2)} = 2(1) - 2^2 = -2$$

That is, f decreases 2 units for each unit increase in the x-direction, y being kept constant ($y = 2$). The rate of change of the function f in the y-direction at the point (1, 2) is given by

$$f_y(1, 2) = \left.\frac{\partial f}{\partial y}\right|_{(1,2)} = -2(1)(2) + 3(2)^2 = 8$$

That is, f increases 8 units for each unit increase in the y-direction, x being kept constant ($x = 1$).

EXPLORE & DISCUSS

Refer to the Explore & Discuss on page 538. Suppose management has decided that the projected sales of the first product is a units. Describe how you might help management decide how many units of the second product the company should produce and sell in order to maximize the company's total profit. Justify your method to management. Suppose, however, management feels that b units of the second product can be manufactured and sold. How would you help management decide how many units of the first product to manufacture in order to maximize the company's total profit?

EXAMPLE 2 Compute the first partial derivatives of each function.

a. $f(x, y) = \dfrac{xy}{x^2 + y^2}$ **b.** $g(s, t) = (s^2 - st + t^2)^5$

c. $h(u, v) = e^{u^2 - v^2}$ **d.** $f(x, y) = \ln(x^2 + 2y^2)$

Solution

a. To compute $\dfrac{\partial f}{\partial x}$, think of the variable y as a constant. Thus,

$$f(x, y) = \frac{xy}{x^2 + y^2}$$

so that, upon using the quotient rule, we have

$$\frac{\partial f}{\partial x} = \frac{(x^2 + y^2)y - xy(2x)}{(x^2 + y^2)^2} = \frac{x^2y + y^3 - 2x^2y}{(x^2 + y^2)^2}$$
$$= \frac{y(y^2 - x^2)}{(x^2 + y^2)^2}$$

upon simplification and factorization. To compute $\dfrac{\partial f}{\partial y}$, think of the variable x as a constant. Thus,

$$f(x, y) = \frac{xy}{x^2 + y^2}$$

so that, upon using the quotient rule once again, we obtain

$$\frac{\partial f}{\partial y} = \frac{(x^2 + y^2)x - xy(2y)}{(x^2 + y^2)^2} = \frac{x^3 + xy^2 - 2xy^2}{(x^2 + y^2)^2}$$
$$= \frac{x(x^2 - y^2)}{(x^2 + y^2)^2}$$

b. To compute $\partial g/\partial s$, we treat the variable t as if it were a constant. Thus,

$$g(s, t) = (s^2 - st + t^2)^5$$

Using the general power rule, we find

$$\begin{aligned}\frac{\partial g}{\partial s} &= 5(s^2 - st + t^2)^4 \cdot (2s - t)\\ &= 5(2s - t)(s^2 - st + t^2)^4\end{aligned}$$

To compute $\partial g/\partial t$, we treat the variable s as if it were a constant. Thus,

$$\begin{aligned}g(s, t) &= (s^2 - st + t^2)^5\\ \frac{\partial g}{\partial t} &= 5(s^2 - st + t^2)^4(-s + 2t)\\ &= 5(2t - s)(s^2 - st + t^2)^4\end{aligned}$$

c. To compute $\partial h/\partial u$, think of the variable v as a constant. Thus,

$$h(u, v) = e^{u^2 - v^2}$$

Using the chain rule for exponential functions, we have

$$\begin{aligned}\frac{\partial h}{\partial u} &= e^{u^2 - v^2} \cdot 2u\\ &= 2ue^{u^2 - v^2}\end{aligned}$$

Next, we treat the variable u as if it were a constant,

$$h(u, v) = e^{u^2 - v^2}$$

and we obtain

$$\begin{aligned}\frac{\partial h}{\partial v} &= e^{u^2 - v^2} \cdot (-2v)\\ &= -2ve^{u^2 - v^2}\end{aligned}$$

d. To compute $\partial f/\partial x$, think of the variable y as a constant. Thus,

$$f(x, y) = \ln(x^2 + 2y^2)$$

so that the chain rule for logarithmic functions gives

$$\frac{\partial f}{\partial x} = \frac{2x}{x^2 + 2y^2}$$

Next, treating the variable x as if it were a constant, we find

$$\begin{aligned}f(x, y) &= \ln(x^2 + 2y^2)\\ \frac{\partial f}{\partial y} &= \frac{4y}{x^2 + 2y^2}\end{aligned}$$

■

To compute the partial derivative of a function of several variables with respect to one variable—say, x—we think of the other variables as if they were constants and differentiate the resulting function with respect to x.

EXPLORE & DISCUSS

1. Let (a, b) be a point in the domain of $f(x, y)$. Put $g(x) = f(x, b)$ and suppose g is differentiable at $x = a$. Explain why you can find $f_x(a, b)$ by computing $g'(a)$. How would you go about calculating $f_y(a, b)$ using a similar technique? Give a geometric interpretation of these processes.
2. Let $f(x, y) = x^2y^3 - 3x^2y + 2$. Use the method of Problem 1 to find $f_x(1, 2)$ and $f_y(1, 2)$.

EXAMPLE 3 Compute the first partial derivatives of the function

$$w = f(x, y, z) = xyz - xe^{yz} + x \ln y$$

Solution Here we have a function of three variables, x, y, and z, and we are required to compute

$$\frac{\partial f}{\partial x}, \frac{\partial f}{\partial y}, \frac{\partial f}{\partial z}$$

To compute f_x, we think of the other two variables, y and z, as fixed, and we differentiate the resulting function of x with respect to x, thereby obtaining

$$f_x = yz - e^{yz} + \ln y$$

To compute f_y, we think of the other two variables, x and z, as constants, and we differentiate the resulting function of y with respect to y. We then obtain

$$f_y = xz - xze^{yz} + \frac{x}{y}$$

Finally, to compute f_z, we treat the variables x and y as constants and differentiate the function f with respect to z, obtaining

$$f_z = xy - xye^{yz}$$

EXPLORING WITH TECHNOLOGY

Refer to the Explore & Discuss on this page. Let

$$f(x, y) = \frac{e^{\sqrt{xy}}}{(1 + xy^2)^{3/2}}$$

1. Compute $g(x) = f(x, 1)$ and use a graphing utility to plot the graph of g in the viewing window $[0, 2] \times [0, 2]$.
2. Use the differentiation operation of your graphing utility to find $g'(1)$ and hence $f_x(1, 1)$.
3. Compute $h(y) = f(1, y)$ and use a graphing utility to plot the graph of g in the viewing window $[0, 2] \times [0, 2]$.
4. Use the differentiation operation of your graphing utility to find $h'(1)$ and hence $f_y(1, 1)$.

The Cobb–Douglas Production Function

For an economic interpretation of the first partial derivatives of a function of two variables, let's turn our attention to the function

$$f(x, y) = ax^b y^{1-b} \tag{1}$$

where a and b are positive constants with $0 < b < 1$. This function is called the **Cobb–Douglas production function**. Here, x stands for the amount of money expended for labor, y stands for the cost of capital equipment (buildings, machinery, and other tools of production), and the function f measures the output of the finished product (in suitable units) and is called, accordingly, the production function.

The partial derivative f_x is called the **marginal productivity of labor**. It measures the rate of change of production with respect to the amount of money expended for labor, with the level of capital expenditure held constant. Similarly, the partial derivative f_y, called the **marginal productivity of capital**, measures the rate of change of production with respect to the amount expended on capital, with the level of labor expenditure held fixed.

VIDEO

APPLIED EXAMPLE 4 Marginal Productivity A certain country's production in the early years following World War II is described by the function

$$f(x, y) = 30x^{2/3}y^{1/3}$$

units, when x units of labor and y units of capital were used.

a. Compute f_x and f_y.
b. What is the marginal productivity of labor and the marginal productivity of capital when the amounts expended on labor and capital are 125 units and 27 units, respectively?
c. Should the government have encouraged capital investment rather than increasing expenditure on labor to increase the country's productivity?

Solution

a. $f_x = 30 \cdot \frac{2}{3}x^{-1/3}y^{1/3} = 20\left(\frac{y}{x}\right)^{1/3}$

$f_y = 30x^{2/3} \cdot \frac{1}{3}y^{-2/3} = 10\left(\frac{x}{y}\right)^{2/3}$

b. The required marginal productivity of labor is given by

$$f_x(125, 27) = 20\left(\frac{27}{125}\right)^{1/3} = 20\left(\frac{3}{5}\right)$$

or 12 units per unit increase in labor expenditure (capital expenditure is held constant at 27 units). The required marginal productivity of capital is given by

$$f_y(125, 27) = 10\left(\frac{125}{27}\right)^{2/3} = 10\left(\frac{25}{9}\right)$$

or $27\frac{7}{9}$ units per unit increase in capital expenditure (labor outlay is held constant at 125 units).

c. From the results of part (b), we see that a unit increase in capital expenditure resulted in a much faster increase in productivity than a unit increase in labor expenditure would have. Therefore, the government should have encouraged increased spending on capital rather than on labor during the early years of reconstruction. ■

Substitute and Complementary Commodities

For another application of the first partial derivatives of a function of two variables in the field of economics, let's consider the relative demands of two commodities. We say that the two commodities are **substitute** (competitive) **commodities** if a decrease in the demand for one results in an increase in the demand for the other. Examples of substitute commodities are coffee and tea. Conversely, two commodities are referred to as **complementary commodities** if a decrease in the demand for one results in a decrease in the demand for the other as well. Examples of complementary commodities are automobiles and tires.

We now derive a criterion for determining whether two commodities A and B are substitute or complementary. Suppose the demand equations that relate the quantities demanded, x and y, to the unit prices, p and q, of the two commodities are given by

$$x = f(p, q) \quad \text{and} \quad y = g(p, q)$$

Let's consider the partial derivative $\partial f/\partial p$. Since f is the demand function for commodity A, we see that, for fixed q, f is typically a decreasing function of p—that is, $\partial f/\partial p < 0$. Now, if the two commodities were substitute commodities, then the quantity demanded of commodity B would increase with respect to p—that is, $\partial g/\partial p > 0$. A similar argument with p fixed shows that if A and B are substitute commodities, then $\partial f/\partial q > 0$. Thus, the two commodities A and B are substitute commodities if

$$\frac{\partial f}{\partial q} > 0 \quad \text{and} \quad \frac{\partial g}{\partial p} > 0$$

Similarly, A and B are complementary commodities if

$$\frac{\partial f}{\partial q} < 0 \quad \text{and} \quad \frac{\partial g}{\partial p} < 0$$

Substitute and Complementary Commodities

Two commodities A and B are substitute commodities if

$$\frac{\partial f}{\partial q} > 0 \quad \text{and} \quad \frac{\partial g}{\partial p} > 0 \tag{2}$$

Two commodities A and B are complementary commodities if

$$\frac{\partial f}{\partial q} < 0 \quad \text{and} \quad \frac{\partial g}{\partial p} < 0 \tag{3}$$

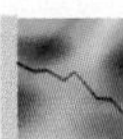

APPLIED EXAMPLE 5 Substitute and Complementary Commodities Suppose that the daily demand for butter is given by

$$x = f(p, q) = \frac{3q}{1 + p^2}$$

and the daily demand for margarine is given by

$$y = g(p, q) = \frac{2p}{1 + \sqrt{q}} \qquad (p > 0, q > 0)$$

where p and q denote the prices per pound (in dollars) of butter and margarine, respectively, and x and y are measured in millions of pounds. Determine whether these two commodities are substitute, complementary, or neither.

Solution We compute

$$\frac{\partial f}{\partial q} = \frac{3}{1 + p^2} \quad \text{and} \quad \frac{\partial g}{\partial p} = \frac{2}{1 + \sqrt{q}}$$

Since

$$\frac{\partial f}{\partial q} > 0 \quad \text{and} \quad \frac{\partial g}{\partial p} > 0$$

for all values of $p > 0$ and $q > 0$, we conclude that butter and margarine are substitute commodities.

Second-Order Partial Derivatives

The first partial derivatives $f_x(x, y)$ and $f_y(x, y)$ of a function $f(x, y)$ of the two variables x and y are also functions of x and y. As such, we may differentiate each of the functions f_x and f_y to obtain the **second-order partial derivatives of f** (Figure 15). Thus, differentiating the function f_x with respect to x leads to the second partial derivative

$$f_{xx} \equiv \frac{\partial^2 f}{\partial x^2} = \frac{\partial}{\partial x}(f_x)$$

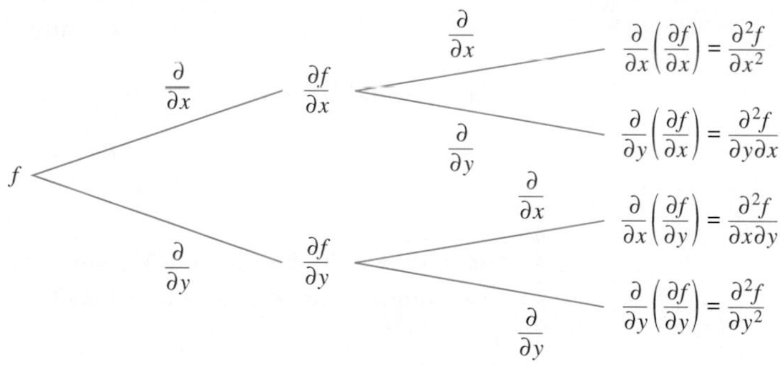

f —$\frac{\partial}{\partial x}$→ $\frac{\partial f}{\partial x}$ —$\frac{\partial}{\partial x}$→ $\frac{\partial}{\partial x}\left(\frac{\partial f}{\partial x}\right) = \frac{\partial^2 f}{\partial x^2}$

$\frac{\partial f}{\partial x}$ —$\frac{\partial}{\partial y}$→ $\frac{\partial}{\partial y}\left(\frac{\partial f}{\partial x}\right) = \frac{\partial^2 f}{\partial y \partial x}$

f —$\frac{\partial}{\partial y}$→ $\frac{\partial f}{\partial y}$ —$\frac{\partial}{\partial x}$→ $\frac{\partial}{\partial x}\left(\frac{\partial f}{\partial y}\right) = \frac{\partial^2 f}{\partial x \partial y}$

$\frac{\partial f}{\partial y}$ —$\frac{\partial}{\partial y}$→ $\frac{\partial}{\partial y}\left(\frac{\partial f}{\partial y}\right) = \frac{\partial^2 f}{\partial y^2}$

FIGURE 15
A schematic showing the four second-order partial derivatives of f

However, differentiation of f_x with respect to y leads to the second partial derivative

$$f_{xy} \equiv \frac{\partial^2 f}{\partial y \partial x} = \frac{\partial}{\partial y}(f_x)$$

Similarly, differentiation of the function f_y with respect to x and with respect to y leads to

$$f_{yx} \equiv \frac{\partial^2 f}{\partial x \partial y} = \frac{\partial}{\partial x}(f_y)$$

$$f_{yy} \equiv \frac{\partial^2 f}{\partial y^2} = \frac{\partial}{\partial y}(f_y)$$

respectively. Note that, in general, it is not true that $f_{xy} = f_{yx}$, but they are equal if both f_{xy} and f_{yx} are continuous. We might add that this is the case in most practical applications.

EXAMPLE 6 Find the second-order partial derivatives of the function

$$f(x, y) = x^3 - 3x^2y + 3xy^2 + y^2$$

Solution The first partial derivatives of f are

$$\begin{aligned} f_x &= \frac{\partial}{\partial x}(x^3 - 3x^2y + 3xy^2 + y^2) \\ &= 3x^2 - 6xy + 3y^2 \end{aligned}$$

$$\begin{aligned} f_y &= \frac{\partial}{\partial y}(x^3 - 3x^2y + 3xy^2 + y^2) \\ &= -3x^2 + 6xy + 2y \end{aligned}$$

Therefore,

$$\begin{aligned} f_{xx} &= \frac{\partial}{\partial x}(f_x) = \frac{\partial}{\partial x}(3x^2 - 6xy + 3y^2) \\ &= 6x - 6y = 6(x - y) \\ f_{xy} &= \frac{\partial}{\partial y}(f_x) = \frac{\partial}{\partial y}(3x^2 - 6xy + 3y^2) \\ &= -6x + 6y = 6(y - x) \\ f_{yx} &= \frac{\partial}{\partial x}(f_y) = \frac{\partial}{\partial x}(-3x^2 + 6xy + 2y) \\ &= -6x + 6y = 6(y - x) \\ f_{yy} &= \frac{\partial}{\partial y}(f_y) = \frac{\partial}{\partial y}(-3x^2 + 6xy + 2y) \\ &= 6x + 2 \end{aligned}$$

■

EXAMPLE 7 Find the second-order partial derivatives of the function

$$f(x, y) = e^{xy^2}$$

Solution We have

$$\begin{aligned} f_x &= \frac{\partial}{\partial x}(e^{xy^2}) \\ &= y^2e^{xy^2} \end{aligned}$$

$$\begin{aligned} f_y &= \frac{\partial}{\partial y}(e^{xy^2}) \\ &= 2xye^{xy^2} \end{aligned}$$

so the required second-order partial derivatives of f are

$$\begin{aligned}
f_{xx} &= \frac{\partial}{\partial x}(f_x) = \frac{\partial}{\partial x}(y^2e^{xy^2}) \\
&= y^4e^{xy^2} \\
f_{xy} &= \frac{\partial}{\partial y}(f_x) = \frac{\partial}{\partial y}(y^2e^{xy^2}) \\
&= 2ye^{xy^2} + 2xy^3e^{xy^2} \\
&= 2ye^{xy^2}(1 + xy^2) \\
f_{yx} &= \frac{\partial}{\partial x}(f_y) = \frac{\partial}{\partial x}(2xye^{xy^2}) \\
&= 2ye^{xy^2} + 2xy^3e^{xy^2} \\
&= 2ye^{xy^2}(1 + xy^2) \\
f_{yy} &= \frac{\partial}{\partial y}(f_y) = \frac{\partial}{\partial y}(2xye^{xy^2}) \\
&= 2xe^{xy^2} + (2xy)(2xy)e^{xy^2} \\
&= 2xe^{xy^2}(1 + 2xy^2)
\end{aligned}$$

■

8.2 Self-Check Exercises

1. Compute the first partial derivatives of $f(x, y) = x^3 - 2xy^2 + y^2 - 8$.

2. Find the first partial derivatives of $f(x, y) = x \ln y + ye^x - x^2$ at (0, 1) and interpret your results.

3. Find the second-order partial derivatives of the function of Self-Check Exercise 1.

4. A certain country's production is described by the function

$$f(x, y) = 60x^{1/3}y^{2/3}$$

when x units of labor and y units of capital are used.

a. What is the marginal productivity of labor and the marginal productivity of capital when the amounts expended on labor and capital are 125 units and 8 units, respectively?

b. Should the government encourage capital investment rather than increased expenditure on labor at this time in order to increase the country's productivity?

Solutions to Self-Check Exercises 8.2 can be found on page 557.

8.2 Concept Questions

1. **a.** What is the partial derivative of $f(x, y)$ with respect to x at (a, b)?
   b. Give a geometric interpretation of $f_x(a, b)$ and a physical interpretation of $f_x(a, b)$.

2. **a.** What are substitute commodities and complementary commodities? Give an example of each.
   b. Suppose $x = f(p, q)$ and $y = g(p, q)$ are demand functions for two commodities A and B, respectively. Give conditions for determining whether A and B are substitute or complementary commodities.

3. List all second-order partial derivatives of f.

8.2 Exercises

In Exercises 1–22, find the first partial derivatives of the function.

1. $f(x, y) = 2x + 3y + 5$
2. $f(x, y) = 2xy$
3. $g(x, y) = 2x^2 + 4y + 1$
4. $f(x, y) = 1 + x^2 + y^2$
5. $f(x, y) = \dfrac{2y}{x^2}$
6. $f(x, y) = \dfrac{x}{1 + y}$
7. $g(u, v) = \dfrac{u - v}{u + v}$
8. $f(x, y) = \dfrac{x^2 - y^2}{x^2 + y^2}$
9. $f(s, t) = (s^2 - st + t^2)^3$
10. $g(s, t) = s^2t + st^{-3}$
11. $f(x, y) = (x^2 + y^2)^{2/3}$
12. $f(x, y) = x\sqrt{1 + y^2}$
13. $f(x, y) = e^{xy+1}$
14. $f(x, y) = (e^x + e^y)^5$
15. $f(x, y) = x \ln y + y \ln x$
16. $f(x, y) = x^2e^{y^2}$
17. $g(u, v) = e^u \ln v$
18. $f(x, y) = \dfrac{e^{xy}}{x + y}$
19. $f(x, y, z) = xyz + xy^2 + yz^2 + zx^2$
20. $g(u, v, w) = \dfrac{2uvw}{u^2 + v^2 + w^2}$
21. $h(r, s, t) = e^{rst}$
22. $f(x, y, z) = xe^{y/z}$

In Exercises 23–32, evaluate the first partial derivatives of the function at the given point.

23. $f(x, y) = x^2y + xy^2$; $(1, 2)$
24. $f(x, y) = x^2 + xy + y^2 + 2x - y$; $(-1, 2)$
25. $f(x, y) = x\sqrt{y} + y^2$; $(2, 1)$
26. $g(x, y) = \sqrt{x^2 + y^2}$; $(3, 4)$
27. $f(x, y) = \dfrac{x}{y}$; $(1, 2)$
28. $f(x, y) = \dfrac{x + y}{x - y}$; $(1, -2)$
29. $f(x, y) = e^{xy}$; $(1, 1)$
30. $f(x, y) = e^x \ln y$; $(0, e)$
31. $f(x, y, z) = x^2yz^3$; $(1, 0, 2)$
32. $f(x, y, z) = x^2y^2 + z^2$; $(1, 1, 2)$

In Exercises 33–40, find the second-order partial derivatives of the function. In each case, show that the mixed partial derivatives f_{xy} and f_{yx} are equal.

33. $f(x, y) = x^2y + xy^3$
34. $f(x, y) = x^3 + x^2y + x + 4$
35. $f(x, y) = x^2 - 2xy + 2y^2 + x - 2y$
36. $f(x, y) = x^3 + x^2y^2 + y^3 + x + y$
37. $f(x, y) = \sqrt{x^2 + y^2}$
38. $f(x, y) = x\sqrt{y} + y\sqrt{x}$
39. $f(x, y) = e^{-x/y}$
40. $f(x, y) = \ln(1 + x^2y^2)$

41. **Productivity of a Country** The productivity of a South American country is given by the function

$$f(x, y) = 20x^{3/4}y^{1/4}$$

when x units of labor and y units of capital are used.
   a. What is the marginal productivity of labor and the marginal productivity of capital when the amounts expended on labor and capital are 256 units and 16 units, respectively?
   b. Should the government encourage capital investment rather than increased expenditure on labor at this time in order to increase the country's productivity?

42. **Productivity of a Country** The productivity of a country in Western Europe is given by the function

$$f(x, y) = 40x^{4/5}y^{1/5}$$

when x units of labor and y units of capital are used.
   a. What is the marginal productivity of labor and the marginal productivity of capital when the amounts expended on labor and capital are 32 units and 243 units, respectively?
   b. Should the government encourage capital investment rather than increased expenditure on labor at this time in order to increase the country's productivity?

43. **Land Prices** The rectangular region R shown in the figure on the next page represents a city's financial district. The price of land within the district is approximated by the function

$$p(x, y) = 200 - 10\left(x - \frac{1}{2}\right)^2 - 15(y - 1)^2$$

where $p(x, y)$ is the price of land at the point (x, y) in dollars per square foot and x and y are measured in miles. Compute

$$\frac{\partial p}{\partial x}(0, 1) \quad \text{and} \quad \frac{\partial p}{\partial y}(0, 1)$$

and interpret your results.

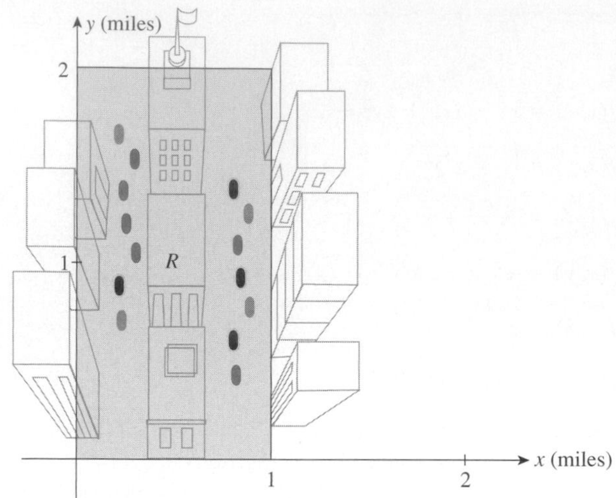

44. **Complementary and Substitute Commodities** In a survey conducted by *Home Entertainment* magazine, it was determined that the demand equation for VCRs is given by

$$x = f(p, q) = 10{,}000 - 10p + 0.2q^2$$

and the demand equation for DVD players is given by

$$y = g(p, q) = 5000 + 0.8p^2 - 20q$$

where p and q denote the unit prices (in dollars) for the VCRs and DVD players, respectively, and x and y denote the number of VCRs and DVD players demanded per week. Determine whether these two products are substitute, complementary, or neither.

45. **Complementary and Substitute Commodities** In a survey it was determined that the demand equation for VCRs is given by

$$x = f(p, q) = 10{,}000 - 10p - e^{0.5q}$$

The demand equation for blank VCR tapes is given by

$$y = g(p, q) = 50{,}000 - 4000q - 10p$$

where p and q denote the unit prices, respectively, and x and y denote the number of VCRs and the number of blank VCR tapes demanded each week. Determine whether these two products are substitute, complementary, or neither.

46. **Complementary and Substitute Commodities** Refer to Exercise 29, Exercises 8.1. Show that the finished and unfinished home furniture manufactured by Country Workshop are substitute commodities.
Hint: Solve the system of equations for x and y in terms of p and q.

47. **Revenue Functions** The total weekly revenue (in dollars) of Country Workshop associated with manufacturing and selling their rolltop desks is given by the function

$$R(x, y) = -0.2x^2 - 0.25y^2 - 0.2xy + 200x + 160y$$

where x denotes the number of finished units and y denotes the number of unfinished units manufactured and sold each week. Compute $\partial R/\partial x$ and $\partial R/\partial y$ when $x = 300$ and $y = 250$. Interpret your results.

48. **Profit Functions** The monthly profit (in dollars) of Bond and Barker Department Store depends on the level of inventory x (in thousands of dollars) and the floor space y (in thousands of square feet) available for display of the merchandise, as given by the equation

$$P(x, y) = -0.02x^2 - 15y^2 + xy + 39x + 25y - 20{,}000$$

Compute $\partial P/\partial x$ and $\partial P/\partial y$ when $x = 4000$ and $y = 150$. Interpret your results. Repeat with $x = 5000$ and $y = 150$.

49. **Wind Chill Factor** A formula used by meteorologists to calculate the wind chill temperature (the temperature that you feel in still air that is the same as the actual temperature when the presence of wind is taken into consideration) is

$$T = f(t, s) = 35.74 + 0.6215t - 35.75s^{0.16} + 0.4275ts^{0.16} \quad (s \geq 1)$$

where t is the actual air temperature in °F and s is the wind speed in mph.

a. What is the wind chill temperature when the actual air temperature is 32°F and the wind speed is 20 mph?

b. If the temperature is 32°F, by how much approximately will the wind chill temperature change if the wind speed increases from 20 mph to 21 mph?

50. **Engine Efficiency** The efficiency of an internal combustion engine is given by

$$E = \left(1 - \frac{v}{V}\right)^{0.4}$$

where V and v are the respective maximum and minimum volumes of air in each cylinder.

a. Show that $\partial E/\partial V > 0$ and interpret your result.

b. Show that $\partial E/\partial v < 0$ and interpret your result.

51. **Volume of a Gas** The volume V (in liters) of a certain mass of gas is related to its pressure P (in millimeters of mercury) and its temperature T (in degrees Kelvin) by the law

$$V = \frac{30.9T}{P}$$

Compute $\partial V/\partial T$ and $\partial V/\partial P$ when $T = 300$ and $P = 800$. Interpret your results.

52. **Surface Area of a Human Body** The formula

$$S = 0.007184W^{0.425}H^{0.725}$$

gives the surface area S of a human body (in square meters) in terms of its weight W (in kilograms) and its height H (in centimeters). Compute $\partial S/\partial W$ and $\partial S/\partial H$ when $W = 70$ kg and $H = 180$ cm. Interpret your results.

53. According to the *ideal gas law*, the volume V (in liters) of an ideal gas is related to its pressure P (in pascals) and temperature T (in degrees Kelvin) by the formula

$$V = \frac{kT}{P}$$

where k is a constant. Show that

$$\frac{\partial V}{\partial T} \cdot \frac{\partial T}{\partial P} \cdot \frac{\partial P}{\partial V} = -1$$

54. Kinetic Energy of a Body The kinetic energy K of a body of mass m and velocity v is given by

$$K = \frac{1}{2} mv^2$$

Show that $\dfrac{\partial K}{\partial m} \cdot \dfrac{\partial^2 K}{\partial v^2} = K.$

In Exercises 55–58, determine whether the statement is true or false. If it is true, explain why it is true. If it is false, give an example to show why it is false.

55. If $f_x(x, y)$ is defined at (a, b), then $f_y(x, y)$ must also be defined at (a, b).

56. If $f_x(a, b) < 0$, then f is decreasing with respect to x near (a, b).

57. If $f_{xy}(x, y)$ and $f_{yx}(x, y)$ are both continuous for all values of x and y, then $f_{xy} = f_{yx}$ for all values of x and y.

58. If both f_{xy} and f_{yx} are defined at (a, b), then f_{xx} and f_{yy} must be defined at (a, b).

8.2 Solutions to Self-Check Exercises

1. $f_x = \dfrac{\partial f}{\partial x} = 3x^2 - 2y^2$

$$f_y = \frac{\partial f}{\partial y} = 2x(2y) + 2y = 2y(1 - 2x)$$

2. $f_x = \ln y + ye^x - 2x; f_y = \dfrac{x}{y} + e^x$

In particular,

$$f_x(0, 1) = \ln 1 + 1e^0 - 2(0) = 1$$

$$f_y(0, 1) = \frac{0}{1} + e^0 = 1$$

The results tell us that at the point (0, 1), $f(x, y)$ increases 1 unit for each unit increase in the x-direction, y being kept constant; $f(x, y)$ also increases 1 unit for each unit increase in the y-direction, x being kept constant.

3. From the results of Self-Check Exercise 1,

$$f_x = 3x^2 - 2y^2$$

Therefore,

$$f_{xx} = \frac{\partial}{\partial x}(3x^2 - 2y^2) = 6x$$

$$f_{xy} = \frac{\partial}{\partial y}(3x^2 - 2y^2) = -4y$$

Also, from the results of Self-Check Exercise 1,

$$f_y = 2y(1 - 2x)$$

Thus,

$$f_{yx} = \frac{\partial}{\partial x}[2y(1 - 2x)] = -4y$$

$$f_{yy} = \frac{\partial}{\partial y}[2y(1 - 2x)] = 2(1 - 2x)$$

4. a. The marginal productivity of labor when the amounts expended on labor and capital are x and y units, respectively, is given by

$$f_x(x, y) = 60\left(\frac{1}{3}x^{-2/3}\right)y^{2/3} = 20\left(\frac{y}{x}\right)^{2/3}$$

In particular, the required marginal productivity of labor is given by

$$f_x(125, 8) = 20\left(\frac{8}{125}\right)^{2/3} = 20\left(\frac{4}{25}\right)$$

or 3.2 units/unit increase in labor expenditure, capital expenditure being held constant at 8 units. Next, we compute

$$f_y(x, y) = 60x^{1/3}\left(\frac{2}{3}y^{-1/3}\right) = 40\left(\frac{x}{y}\right)^{1/3}$$

and deduce that the required marginal productivity of capital is given by

$$f_y(125, 8) = 40\left(\frac{125}{8}\right)^{1/3} = 40\left(\frac{5}{2}\right)$$

or 100 units/unit increase in capital expenditure, labor expenditure being held constant at 125 units.

b. The results of part (a) tell us that the government should encourage increased spending on capital rather than on labor.

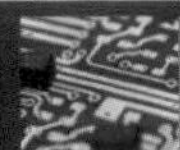

USING TECHNOLOGY

Finding Partial Derivatives at a Given Point

Suppose $f(x, y)$ is a function of two variables and we wish to compute

$$f_x(a, b) = \left.\frac{\partial f}{\partial x}\right|_{(a, b)}$$

Recall that in computing $\partial f/\partial x$, we think of y as being fixed. But in this situation, we are evaluating $\partial f/\partial x$ at (a, b). Therefore, we set y equal to b. Doing this leads to the function g of one variable, x, defined by

$$g(x) = f(x, b)$$

It follows from the definition of the partial derivative that

$$f_x(a, b) = g'(a)$$

Thus, the value of the partial derivative $\partial f/\partial x$ at a given point (a, b) can be found by evaluating the derivative of a function of one variable. In particular, the latter can be found by using the numerical derivative operation of a graphing utility. We find $f_y(a, b)$ in a similar manner.

EXAMPLE 1 Let $f(x, y) = (1 + xy^2)^{3/2}e^{x^2y}$. Find (a) $f_x(1, 2)$ and (b) $f_y(1, 2)$.

Solution

a. Define $g(x) = f(x, 2) = (1 + 4x)^{3/2}e^{2x^2}$. Using the numerical derivative operation to find $g'(1)$, we obtain

$$f_x(1, 2) = g'(1) \approx 429.585835$$

b. Define $h(y) = f(1, y) = (1 + y^2)^{3/2}e^y$. Using the numerical derivative operation to find $h'(2)$, we obtain

$$f_y(1, 2) = h'(2) \approx 181.7468642$$

TECHNOLOGY EXERCISES

Compute the following at the given point:

$$\frac{\partial f}{\partial x} \text{ and } \frac{\partial f}{\partial y}$$

1. $f(x, y) = \sqrt{x}(2 + xy^2)^{1/3}$; $(1, 2)$
2. $f(x, y) = \sqrt{xy}(1 + 2xy)^{2/3}$; $(1, 4)$
3. $f(x, y) = \dfrac{x + y^2}{1 + x^2y}$; $(1, 2)$
4. $f(x, y) = \dfrac{xy^2}{(\sqrt{x} + \sqrt{y})^2}$; $(4, 1)$
5. $f(x, y) = e^{-xy^2}(x + y)^{1/3}$; $(1, 1)$
6. $f(x, y) = \dfrac{\ln(\sqrt{x} + y^2)}{x^2 + y^2}$; $(4, 1)$

8.3 Maxima and Minima of Functions of Several Variables

Maxima and Minima

In Chapter 4, we saw that the solution of a problem often reduces to finding the extreme values of a function of one variable. In practice, however, situations also arise in which a problem is solved by finding the absolute maximum or absolute minimum value of a function of two or more variables.

For example, suppose Scandi Company manufactures computer desks in both assembled and unassembled versions. Its profit P is therefore a function of the number of assembled units, x, and the number of unassembled units, y, manufactured and sold per week; that is, $P = f(x, y)$. A question of paramount importance to the manufacturer is, How many assembled and unassembled desks should the company manufacture per week in order to maximize its weekly profit? Mathematically, the problem is solved by finding the values of x and y that will make $f(x, y)$ a maximum.

In this section we will focus our attention on finding the extrema of a function of two variables. As in the case of a function of one variable, we distinguish between the relative (or local) extrema and the absolute extrema of a function of two variables.

Relative Extrema of a Function of Two Variables

Let f be a function defined on a region R containing the point (a, b). Then, f has a **relative maximum** at (a, b) if $f(x, y) \leq f(a, b)$ for all points (x, y) that are sufficiently close to (a, b). The number $f(a, b)$ is called a **relative maximum value**. Similarly, f has a **relative minimum** at (a, b), with **relative minimum value** $f(a, b)$ if $f(x, y) \geq f(a, b)$ for all points (x, y) that are sufficiently close to (a, b).

Loosely speaking, f has a relative maximum at (a, b) if the point $(a, b, f(a, b))$ is the highest point on the graph of f when compared with all nearby points. A similar interpretation holds for a relative minimum.

If the inequalities in this last definition hold for *all* points (x, y) in the domain of f, then f has an **absolute maximum** (or **absolute minimum**) at (a, b) with **absolute maximum value** (or **absolute minimum value**) $f(a, b)$. Figure 16 shows

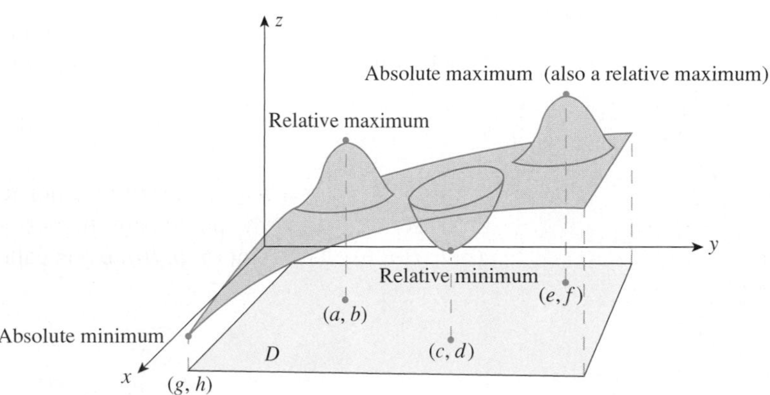

FIGURE 16

the graph of a function with relative maxima at (a, b) and (e, f) and a relative minimum at (c, d). The absolute maximum of f occurs at (e, f) and the absolute minimum of f occurs at (g, h).

Observe that in the case of a function of one variable, a relative extremum (relative maximum or relative minimum) may or may not be an absolute extremum.

Now let's turn our attention to the study of relative extrema of a function. Suppose that a differentiable function $f(x, y)$ of two variables has a relative maximum (relative minimum) at a point (a, b) in the domain of f. From Figure 17 it is clear that at the point (a, b) the slope of the "tangent lines" to the surface in any direction must be zero. In particular, this implies that both

$$\frac{\partial f}{\partial x}(a, b) \quad \text{and} \quad \frac{\partial f}{\partial y}(a, b)$$

must be zero.

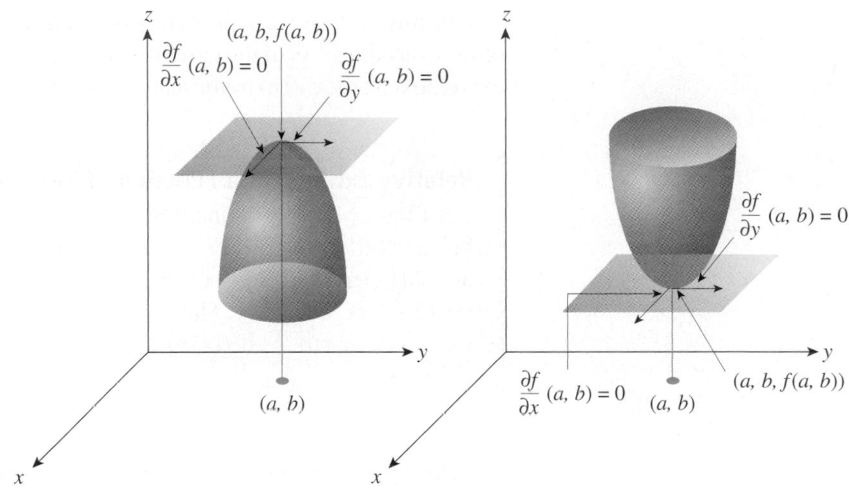

FIGURE 17

(a) f has a relative maximum at (a, b). **(b)** f has a relative minimum at (a, b).

Lest we are tempted to jump to the conclusion that a differentiable function f satisfying both the conditions

$$\frac{\partial f}{\partial x}(a, b) = 0 \quad \text{and} \quad \frac{\partial f}{\partial y}(a, b) = 0$$

at a point (a, b) must have a relative extremum at the point (a, b), let's examine the graph of the function f depicted in Figure 18. Here both

$$\frac{\partial f}{\partial x}(a, b) = 0 \quad \text{and} \quad \frac{\partial f}{\partial y}(a, b) = 0$$

but f has neither a relative maximum nor a relative minimum at the point (a, b) because some nearby points are higher and some are lower than the point $(a, b, f(a, b))$. The point $(a, b, f(a, b))$ is called a **saddle point**.

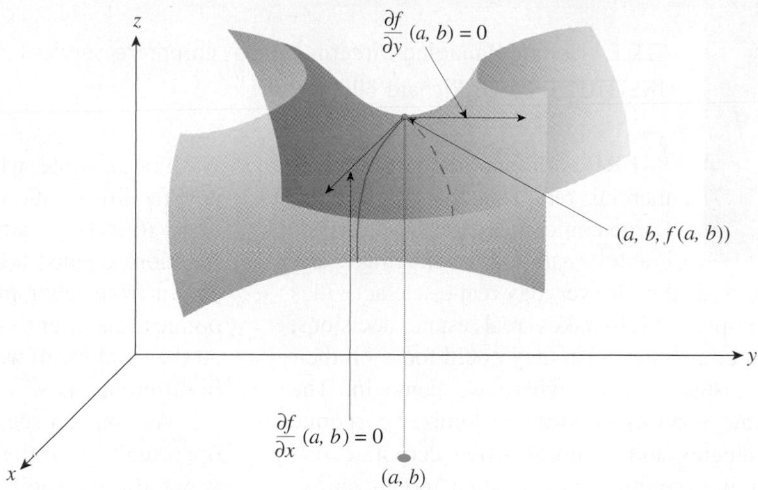

FIGURE 18
The point $(a, b, f(a, b))$ is called a saddle point.

Finally, an examination of the graph of the function f depicted in Figure 19 should convince you that f has a relative maximum at the point (a, b). But both $\partial f/\partial x$ and $\partial f/\partial y$ fail to be defined at (a, b).

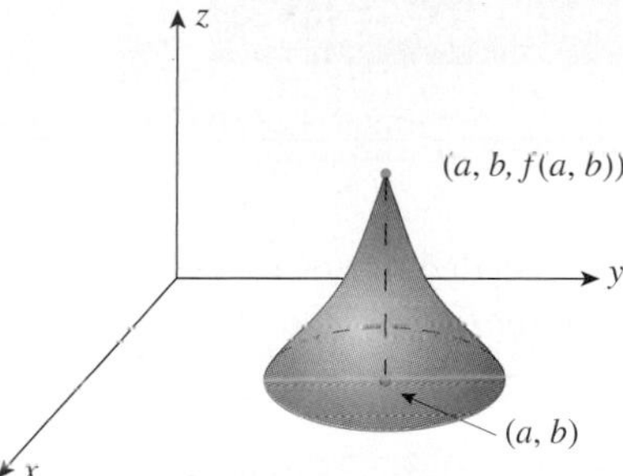

FIGURE 19
f has a relative maximum at (a, b), but neither $\partial f/\partial x$ nor $\partial f/\partial y$ exist at (a, b).

To summarize, a function f of two variables can only have a relative extremum at a point (a, b) in its domain where $\partial f/\partial x$ and $\partial f/\partial y$ both exist and are equal to zero at (a, b) or at least one of the partial derivatives does not exist. As in the case of one variable, we refer to a point in the domain of f that *may* give rise to a relative extremum as a critical point. The precise definition follows.

Critical Point of f

A **critical point** of f is a point (a, b) in the domain of f such that both

$$\frac{\partial f}{\partial x}(a, b) = 0 \quad \text{and} \quad \frac{\partial f}{\partial y}(a, b) = 0$$

or at least one of the partial derivatives does not exist.

To determine the nature of a critical point of a function $f(x, y)$ of two variables, we use the second partial derivatives of f. The resulting test, which helps us classify

PORTFOLIO Kirk Hoiberg

TITLE Senior Managing Director, Global Corporate Services
INSTITUTION CB Richard Ellis (CBRE)

CBRE is the world's largest commercial real estate services firm. Large companies often have considerable real estate holdings and inevitably have to deal with everyday real estate activities as well as complex, high-stakes real estate decisions. However, most companies wish they could focus on their core business instead. That's where we come in. The Global Corporate Services provides customized corporate real estate strategies and comprehensive corporate real estate outsourcing solutions to leading organizations worldwide. With over 40% of the Fortune 500 as clients, we deal with sophisticated businesses that demand rigorous analysis to support our recommendations and actions on their behalf. Doing this without regular use of quantitative and mathematical skills would be impossible.

For example, when a client needs to locate or acquire new facilities somewhere in the world, we use linear algebraic models to simultaneously optimize multivariable functions against a broad array of inputs and constraints drawn from labor markets, property markets, macroeconomics, the client's supply chain, and many other areas. At the local level, we regularly model the real estate cycle in different cities.

As you can see, the realm of real estate offers many opportunities for the use of applied mathematics. While it is not always readily apparent in college, these skills play a crucial role in the development of a successful business. In the future, the ability to understand and interpret the information generated by these techniques can be crucial to your own success.

these points, is called the **second derivative test** and is incorporated in the following procedure for finding and classifying the relative extrema of f.

Determining Relative Extrema

1. Find the critical points of $f(x, y)$ by solving the system of simultaneous equations

$$f_x = 0$$
$$f_y = 0$$

2. The second derivative test: Let

$$D(x, y) = f_{xx}f_{yy} - f_{xy}^2$$

Then,

a. $D(a, b) > 0$ and $f_{xx}(a, b) < 0$ implies that $f(x, y)$ has a **relative maximum** at the point (a, b).

b. $D(a, b) > 0$ and $f_{xx}(a, b) > 0$ implies that $f(x, y)$ has a **relative minimum** at the point (a, b).

c. $D(a, b) < 0$ implies that $f(x, y)$ has neither a relative maximum nor a relative minimum at the point (a, b).

d. $D(a, b) = 0$ implies that the test is inconclusive, so some other technique must be used to solve the problem.

EXAMPLE 1 Find the relative extrema of the function

$$f(x, y) = x^2 + y^2$$

Solution We have

$$f_x = 2x$$
$$f_y = 2y$$

To find the critical point(s) of f, we set $f_x = 0$ and $f_y = 0$ and solve the resulting system of simultaneous equations

$$2x = 0$$
$$2y = 0$$

obtaining $x = 0$, $y = 0$, or $(0, 0)$, as the sole critical point of f. Next, we apply the second derivative test to determine the nature of the critical point $(0, 0)$. We compute

$$f_{xx} = 2 \qquad f_{xy} = 0 \qquad f_{yy} = 2$$

and

$$D(x, y) = f_{xx}f_{yy} - f_{xy}^2 = (2)(2) - 0 = 4$$

In particular, $D(0, 0) = 4$. Since $D(0, 0) > 0$ and $f_{xx}(0, 0) = 2 > 0$, we conclude that $f(x, y)$ has a relative minimum at the point $(0, 0)$. The relative minimum value, 0, also happens to be the absolute minimum of f. The graph of the function f, shown in Figure 20, confirms these results. ■

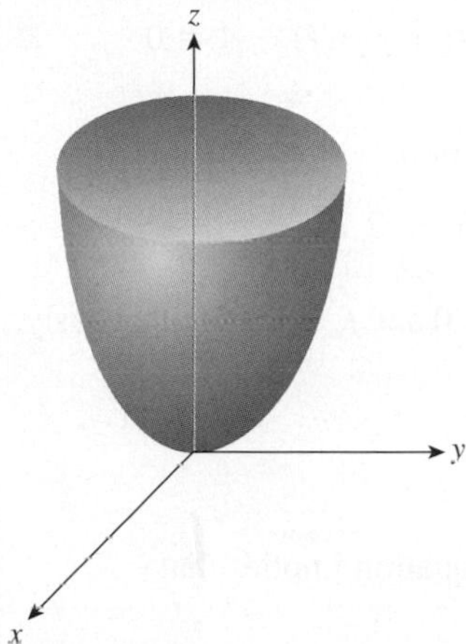

FIGURE 20
The graph of $f(x, y) = x^2 + y^2$

EXAMPLE 2 Find the relative extrema of the function

$$f(x, y) = 3x^2 - 4xy + 4y^2 - 4x + 8y + 4$$

Solution We have

$$f_x = 6x - 4y - 4$$
$$f_y = -4x + 8y + 8$$

To find the critical points of f, we set $f_x = 0$ and $f_y = 0$ and solve the resulting system of simultaneous equations

$$6x - 4y = 4$$
$$-4x + 8y = -8$$

Multiplying the first equation by 2 and the second equation by 3, we obtain the equivalent system

$$12x - 8y = 8$$
$$-12x + 24y = -24$$

Adding the two equations gives $16y = -16$, or $y = -1$. We substitute this value for y into either equation in the system to get $x = 0$. Thus, the only critical point of f is the point $(0, -1)$. Next, we apply the second derivative test to determine whether the point $(0, -1)$ gives rise to a relative extremum of f. We

> **EXPLORE & DISCUSS**
>
> Suppose $f(x, y)$ has a relative extremum (relative maximum or relative minimum) at a point (a, b). Let $g(x) = f(x, b)$ and $h(y) = f(a, y)$. Assuming that f and g are differentiable, explain why $g'(a) = 0$ and $h'(b) = 0$. Explain why these results are equivalent to the conditions $f_x(a, b) = 0$ and $f_y(a, b) = 0$.

compute

$$f_{xx} = 6 \qquad f_{xy} = -4 \qquad f_{yy} = 8$$

and

$$D(x, y) = f_{xx} f_{yy} - f_{xy}^2 = (6)(8) - (-4)^2 = 32$$

Since $D(0, -1) = 32 > 0$ and $f_{xx}(0, -1) = 6 > 0$, we conclude that $f(x, y)$ has a relative minimum at the point $(0, -1)$. The value of $f(x, y)$ at the point $(0, -1)$ is given by

$$f(0, -1) = 3(0)^2 - 4(0)(-1) + 4(-1)^2 - 4(0) + 8(-1) + 4 = 0$$

■

EXAMPLE 3 Find the relative extrema of the function

$$f(x, y) = 4y^3 + x^2 - 12y^2 - 36y + 2$$

Solution To find the critical points of f, we set $f_x = 0$ and $f_y = 0$ simultaneously, obtaining

$$f_x = 2x = 0$$
$$f_y = 12y^2 - 24y - 36 = 0$$

The first equation implies that $x = 0$. The second equation implies that

$$y^2 - 2y - 3 = 0$$
$$(y + 1)(y - 3) = 0$$

—that is, $y = -1$ or 3. Therefore, there are two critical points of the function f—namely, $(0, -1)$ and $(0, 3)$.

Next, we apply the second derivative test to determine the nature of each of the two critical points. We compute

$$f_{xx} = 2 \qquad f_{xy} = 0 \qquad f_{yy} = 24y - 24 = 24(y - 1)$$

Therefore,

$$D(x, y) = f_{xx} f_{yy} - f_{xy}^2 = 48(y - 1)$$

For the point $(0, -1)$,

$$D(0, -1) = 48(-1 - 1) = -96 < 0$$

Since $D(0, -1) < 0$, we conclude that the point $(0, -1)$ gives a saddle point of f. For the point $(0, 3)$,

$$D(0, 3) = 48(3 - 1) = 96 > 0$$

Since $D(0, 3) > 0$ and $f_{xx}(0, 3) > 0$, we conclude that the function f has a relative minimum at the point $(0, 3)$. Furthermore, since

$$f(0, 3) = 4(3)^3 + (0)^2 - 12(3)^2 - 36(3) + 2$$
$$= -106$$

we see that the relative minimum value of f is -106. ■

> **EXPLORE & DISCUSS**
>
> 1. Refer to the second derivative test. Can the condition $f_{xx}(a, b) < 0$ in part 2a be replaced by the condition $f_{yy}(a, b) < 0$? Explain your answer. How about the condition $f_{xx}(a, b) > 0$ in part 2b?
> 2. Let $f(x, y) = x^4 + y^4$.
>   **a.** Show that $(0, 0)$ is a critical point of f and that $D(0, 0) = 0$.
>   **b.** Explain why f has a relative (in fact, an absolute) minimum at $(0, 0)$. Does this contradict the second derivative test? Explain your answer.

As in the case of a practical optimization problem involving a function of one variable, the solution to an optimization problem involving a function of several variables calls for finding the *absolute* extremum of the function. Determining the absolute extremum of a function of several variables is more difficult than merely finding the relative extrema of the function. However, in many situations, the absolute extremum of a function actually coincides with the largest relative extremum of the function that occurs in the interior of its domain. We assume that the problems considered here belong to this category. Furthermore, the existence of the absolute extremum (solution) of a practical problem is often deduced from the geometric or physical nature of the problem.

VIDEO

APPLIED EXAMPLE 4 Maximizing Profits The total weekly revenue (in dollars) that Acrosonic realizes in producing and selling its bookshelf loudspeaker systems is given by

$$R(x, y) = -\frac{1}{4}x^2 - \frac{3}{8}y^2 - \frac{1}{4}xy + 300x + 240y$$

where x denotes the number of fully assembled units and y denotes the number of kits produced and sold each week. The total weekly cost attributable to the production of these loudspeakers is

$$C(x, y) = 180x + 140y + 5000$$

dollars, where x and y have the same meaning as before. Determine how many assembled units and how many kits Acrosonic should produce per week to maximize its profit.

Solution The contribution to Acrosonic's weekly profit stemming from the production and sale of the bookshelf loudspeaker systems is given by

$$\begin{aligned} P(x, y) &= R(x, y) - C(x, y) \\ &= \left(-\frac{1}{4}x^2 - \frac{3}{8}y^2 - \frac{1}{4}xy + 300x + 240y\right) - (180x + 140y + 5000) \\ &= -\frac{1}{4}x^2 - \frac{3}{8}y^2 - \frac{1}{4}xy + 120x + 100y - 5000 \end{aligned}$$

To find the relative maximum of the profit function $P(x, y)$, we first locate the critical point(s) of P. Setting $P_x(x, y)$ and $P_y(x, y)$ equal to zero, we obtain

$$\begin{aligned} P_x &= -\frac{1}{2}x - \frac{1}{4}y + 120 = 0 \\ P_y &= -\frac{3}{4}y - \frac{1}{4}x + 100 = 0 \end{aligned}$$

Solving the first of these equations for y yields

$$y = -2x + 480$$

which, upon substitution into the second equation, yields

$$\begin{aligned} -\frac{3}{4}(-2x + 480) - \frac{1}{4}x + 100 &= 0 \\ 6x - 1440 - x + 400 &= 0 \\ x &= 208 \end{aligned}$$

We substitute this value of x into the equation $y = -2x + 480$ to get

$$y = 64$$

Therefore, the function P has the sole critical point (208, 64). To show that the point (208, 64) is a solution to our problem, we use the second derivative test. We compute

$$P_{xx} = -\frac{1}{2} \qquad P_{xy} = -\frac{1}{4} \qquad P_{yy} = -\frac{3}{4}$$

So,

$$D(x, y) = \left(-\frac{1}{2}\right)\left(-\frac{3}{4}\right) - \left(-\frac{1}{4}\right)^2 = \frac{3}{8} - \frac{1}{16} = \frac{5}{16}$$

In particular, $D(208, 64) = 5/16 > 0$.

Since $D(208, 64) > 0$ and $P_{xx}(208, 64) < 0$, the point (208, 64) yields a relative maximum of P. This relative maximum is also the absolute maximum of P. We conclude that Acrosonic can maximize its weekly profit by manufacturing 208 assembled units and 64 kits of their bookshelf loudspeaker systems. The maximum weekly profit realizable from the production and sale of these loudspeaker systems is given by

$$\begin{aligned} P(208, 64) &= -\frac{1}{4}(208)^2 - \frac{3}{8}(64)^2 - \frac{1}{4}(208)(64) \\ &\quad + 120(208) + 100(64) - 5000 \\ &= 10{,}680 \end{aligned}$$

or \$10,680. ■

APPLIED EXAMPLE 5 Locating a Television Relay Station Site A television relay station will serve towns A, B, and C, whose relative locations are shown in Figure 21. Determine a site for the location of the station if the sum of the squares of the distances from each town to the site is minimized.

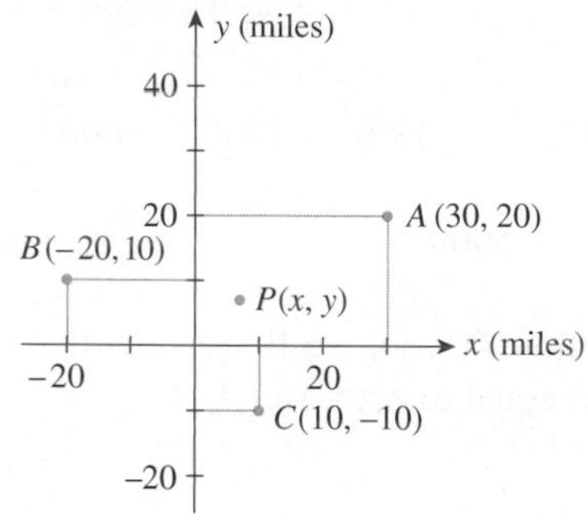

FIGURE 21
Locating a site for a television relay station

Solution Suppose the required site is located at the point $P(x, y)$. With the aid of the distance formula, we find that the square of the distance from town A to the site is

$$(x - 30)^2 + (y - 20)^2$$

The respective distances from towns B and C to the site are found in a similar manner, so the sum of the squares of the distances from each town to the site is given by

$$\begin{aligned} f(x, y) = (x - 30)^2 + (y - 20)^2 + (x + 20)^2 \\ + (y - 10)^2 + (x - 10)^2 + (y + 10)^2 \end{aligned}$$

To find the relative minimum of $f(x, y)$, we first find the critical point(s) of f. Using the chain rule to find $f_x(x, y)$ and $f_y(x, y)$ and setting each equal to zero, we obtain

$$f_x = 2(x - 30) + 2(x + 20) + 2(x - 10) = 6x - 40 = 0$$

$$f_y = 2(y - 20) + 2(y - 10) + 2(y + 10) = 6y - 40 = 0$$

from which we deduce that $(\frac{20}{3}, \frac{20}{3})$ is the sole critical point of f. Since

$$f_{xx} = 6 \qquad f_{xy} = 0 \qquad f_{yy} = 6$$

we have

$$D(x, y) = f_{xx} f_{yy} - f_{xy}^2 = (6)(6) - 0 = 36$$

Since $D(\frac{20}{3}, \frac{20}{3}) > 0$ and $f_{xx}(\frac{20}{3}, \frac{20}{3}) > 0$, we conclude that the point $(\frac{20}{3}, \frac{20}{3})$ yields a relative minimum of f. Thus, the required site has coordinates $x = \frac{20}{3}$ and $y = \frac{20}{3}$. ■

8.3 Self-Check Exercises

1. Let $f(x, y) = 2x^2 + 3y^2 - 4xy + 4x - 2y + 3$.
   a. Find the critical point of f.
   b. Use the second derivative test to classify the nature of the critical point.
   c. Find the relative extremum of f, if it exists.

2. Robertson Controls manufactures two basic models of setback thermostats: a standard mechanical thermostat and a deluxe electronic thermostat. Robertson's monthly revenue (in hundreds of dollars) is

$$R(x, y) = -\frac{1}{8}x^2 - \frac{1}{2}y^2 - \frac{1}{4}xy + 20x + 60y$$

where x (in units of a hundred) denotes the number of mechanical thermostats manufactured and y (in units of a hundred) denotes the number of electronic thermostats manufactured each month. The total monthly cost incurred in producing these thermostats is

$$C(x, y) = 7x + 20y + 280$$

hundred dollars. Find how many thermostats of each model Robertson should manufacture each month in order to maximize its profits. What is the maximum profit?

Solutions to Self-Check Exercises 8.3 can be found on page 569.

8.3 Concept Questions

1. Explain the terms (a) relative maximum of a function $f(x, y)$ and (b) absolute maximum of a function $f(x, y)$.
2. a. What is a critical point of a function $f(x, y)$?
   b. Explain the role of a critical point in determining the relative extrema of a function of two variables.
3. Explain how the second derivative test is used to determine the relative extrema of a function of two variables.

8.3 Exercises

In Exercises 1–20, find the critical point(s) of the function. Then use the second derivative test to classify the nature of each point, if possible. Finally, determine the relative extrema of the function.

1. $f(x, y) = 1 - 2x^2 - 3y^2$
2. $f(x, y) = x^2 - xy + y^2 + 1$
3. $f(x, y) = x^2 - y^2 - 2x + 4y + 1$
4. $f(x, y) = 2x^2 + y^2 - 4x + 6y + 3$
5. $f(x, y) = x^2 + 2xy + 2y^2 - 4x + 8y - 1$
6. $f(x, y) = x^2 - 4xy + 2y^2 + 4x + 8y - 1$
7. $f(x, y) = 2x^3 + y^2 - 9x^2 - 4y + 12x - 2$
8. $f(x, y) = 2x^3 + y^2 - 6x^2 - 4y + 12x - 2$
9. $f(x, y) = x^3 + y^2 - 2xy + 7x - 8y + 4$
10. $f(x, y) = 2y^3 - 3y^2 - 12y + 2x^2 - 6x + 2$
11. $f(x, y) = x^3 - 3xy + y^3 - 2$
12. $f(x, y) = x^3 - 2xy + y^2 + 5$
13. $f(x, y) = xy + \frac{4}{x} + \frac{2}{y}$
14. $f(x, y) = \frac{x}{y^2} + xy$
15. $f(x, y) = x^2 - e^{y^2}$
16. $f(x, y) = e^{x^2 - y^2}$
17. $f(x, y) = e^{x^2 + y^2}$
18. $f(x, y) = e^{xy}$
19. $f(x, y) = \ln(1 + x^2 + y^2)$
20. $f(x, y) = xy + \ln x + 2y^2$

21. **Maximizing Profit** The total weekly revenue (in dollars) of the Country Workshop realized in manufacturing and selling its rolltop desks is given by

$$R(x, y) = -0.2x^2 - 0.25y^2 - 0.2xy + 200x + 160y$$

where x denotes the number of finished units and y denotes the number of unfinished units manufactured and sold each week. The total weekly cost attributable to the manufacture of these desks is given by

$$C(x, y) = 100x + 70y + 4000$$

dollars. Determine how many finished units and how many unfinished units the company should manufacture each week in order to maximize its profit. What is the maximum profit realizable?

22. **Maximizing Profit** The total daily revenue (in dollars) that Weston Publishing realizes in publishing and selling its English-language dictionaries is given by

$$R(x, y) = -0.005x^2 - 0.003y^2 - 0.002xy + 20x + 15y$$

where x denotes the number of deluxe copies and y denotes the number of standard copies published and sold daily. The total daily cost of publishing these dictionaries is given by

$$C(x, y) = 6x + 3y + 200$$

dollars. Determine how many deluxe copies and how many standard copies Weston should publish each day to maximize its profits. What is the maximum profit realizable?

23. **Maximum Price** The rectangular region R shown in the accompanying figure represents the financial district of a city. The price of land within the district is approximated by the function

$$p(x, y) = 200 - 10\left(x - \frac{1}{2}\right)^2 - 15(y - 1)^2$$

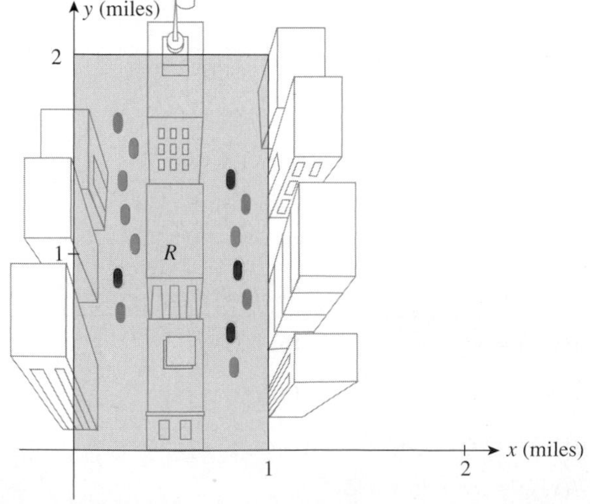

where $p(x, y)$ is the price of land at the point (x, y) in dollars/square foot and x and y are measured in miles. At what point within the financial district is the price of land highest?

24. **Maximizing Profit** C&G Imports imports two brands of white wine, one from Germany and the other from Italy. The German wine costs \$4/bottle, and the Italian wine costs \$3/bottle. It has been estimated that if the German wine retails at p dollars/bottle and the Italian wine is sold for q dollars/bottle, then

$$2000 - 150p + 100q$$

bottles of the German wine and

$$1000 + 80p - 120q$$

bottles of the Italian wine will be sold each week. Determine the unit price for each brand that will allow C&G to realize the largest possible weekly profit.

25. **Determining the Optimal Site** An auxiliary electric power station will serve three communities, A, B, and C, whose relative locations are shown in the accompanying figure. Determine where the power station should be located if the sum of the squares of the distances from each community to the site is minimized.

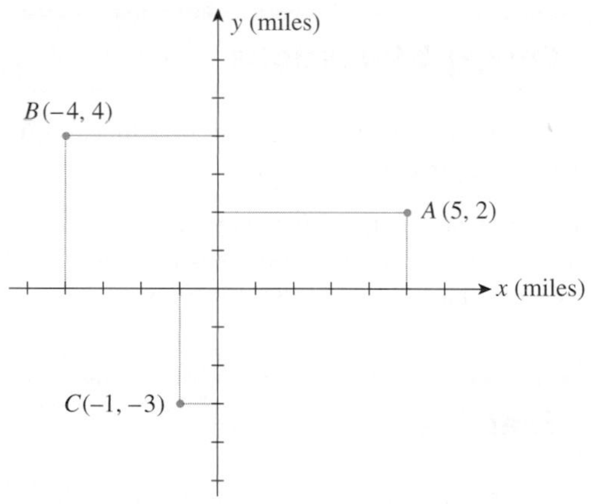

26. **Packaging** An open rectangular box having a volume of 108 in.^{3} is to be constructed from a tin sheet. Find the dimensions of such a box if the amount of material used in its construction is to be minimal.
Hint: Let the dimensions of the box be x'' by y'' by z''. Then, $xyz = 108$ and the amount of material used is given by $S = xy + 2yz + 2xz$. Show that

$$S = f(x, y) = xy + \frac{216}{x} + \frac{216}{y}$$

Minimize $f(x, y)$.

27. **Packaging** An open rectangular box having a surface area of 300 in.^{2} is to be constructed from a tin sheet. Find the dimen-

sions of the box if the volume of the box is to be as large as possible. What is the maximum volume?
Hint: Let the dimensions of the box be $x \times y \times z$ (see the figure that follows). Then the surface area is $xy + 2xz + 2yz$, and its volume is xyz.

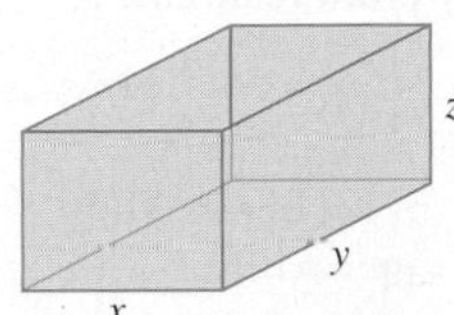

28. Packaging Postal regulations specify that the combined length and girth of a parcel sent by parcel post may not exceed 108 in. Find the dimensions of the rectangular package that would have the greatest possible volume under these regulations.
Hint: Let the dimensions of the box be x'' by y'' by z'' (see the figure below). Then, $2x + 2z + y = 108$, and the volume $V = xyz$. Show that

$$V = f(x, y) = 108xz - 2x^2z - 2xz^2$$

Maximize $f(x, z)$.

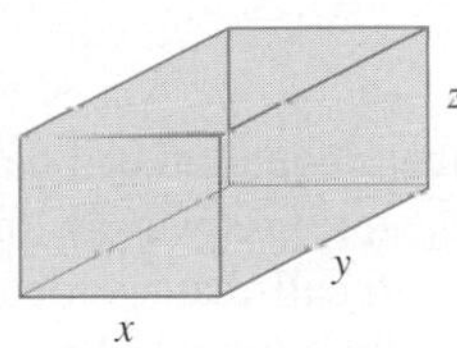

29. Minimizing Heating and Cooling Costs A building in the shape of a rectangular box is to have a volume of 12,000 ft^3 (see the figure). It is estimated that the annual heating and cooling costs will be \$2/square foot for the top, \$4/square foot for the front and back, and \$3/square foot for the sides. Find the dimensions of the building that will result in a minimal annual heating and cooling cost. What is the minimal annual heating and cooling cost?

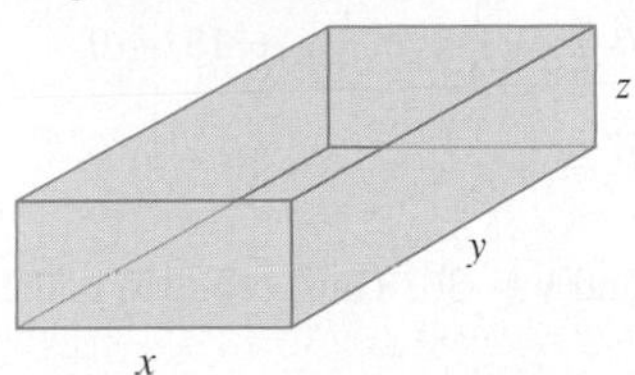

30. Packaging An open box having a volume of 48 in.^{3} is to be constructed. If the box is to include a partition that is parallel to a side of the box, as shown in the figure, and the amount of material used is to be minimal, what should be the dimensions of the box?

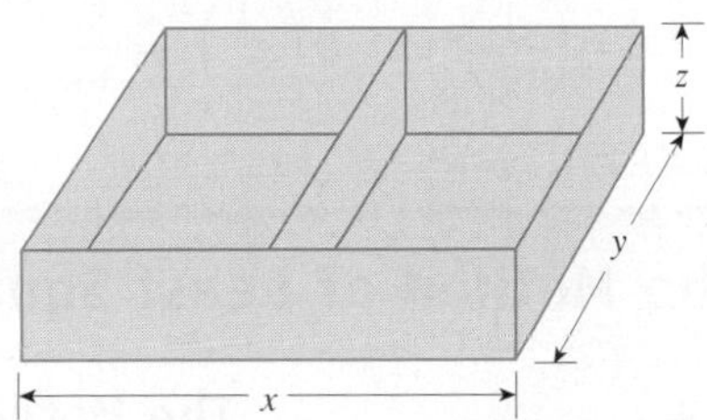

In Exercises 31 and 32, determine whether the statement is true or false. If it is true, explain why it is true. If it is false, give an example to show why it is false.

31. If $f_x(a, b) = 0$ and $f_y(a, b) = 0$, then f must have a relative extremum at (a, b).

32. If (a, b) is a critical point of f and both the conditions $f_{xx}(a, b) < 0$ and $f_{yy}(a, b) < 0$ hold, then f has a relative maximum at (a, b).

8.3 Solutions to Self-Check Exercises

1. a. To find the critical point(s) of f, we solve the system of equations

$$f_x = 4x - 4y + 4 = 0$$
$$f_y = -4x + 6y - 2 = 0$$

obtaining $x = -2$ and $y = -1$. Thus, the only critical point of f is the point $(-2, -1)$.

b. We have $f_{xx} = 4$, $f_{xy} = -4$, and $f_{yy} = 6$, so

$$D(x, y) = f_{xx}f_{yy} - f_{xy}^2 = (4)(6) - (-4)^2 = 8$$

Since $D(-2, -1) > 0$ and $f_{xx}(-2, -1) > 0$, we conclude that f has a relative minimum at the point $(-2, -1)$.

c. The relative minimum value of $f(x, y)$ at the point $(-2, -1)$ is

$$f(-2, -1) = 2(-2)^2 + 3(-1)^2 - 4(-2)(-1) + 4(-2) - 2(-1) + 3 = 0$$

2. Robertson's monthly profit is

$$P(x, y) = R(x, y) - C(x, y)$$
$$= \left(-\frac{1}{8}x^2 - \frac{1}{2}y^2 - \frac{1}{4}xy + 20x + 60y\right) - (7x + 20y + 280)$$
$$= -\frac{1}{8}x^2 - \frac{1}{2}y^2 - \frac{1}{4}xy + 13x + 40y - 280$$

The critical point of P is found by solving the system

$$P_x = -\frac{1}{4}x - \frac{1}{4}y + 13 = 0$$
$$P_y = -\frac{1}{4}x - y + 40 = 0$$

giving $x = 16$ and $y = 36$. Thus, $(16, 36)$ is the critical point of P. Next,

$$P_{xx} = -\frac{1}{4} \qquad P_{xy} = -\frac{1}{4} \qquad P_{yy} = -1$$

and

$$D(x, y) = f_{xx}f_{yy} - f_{xy}^2 = \left(-\frac{1}{4}\right)(-1) - \left(-\frac{1}{4}\right)^2 = \frac{3}{16}$$

Since $D(16, 36) > 0$ and $P_{xx}(16, 36) < 0$, the point $(16, 36)$ yields a relative maximum of P. We conclude that the monthly profit is maximized by manufacturing 1600 mechanical and 3600 electronic setback thermostats each month. The maximum monthly profit realizable is

$$P(16, 36) = -\frac{1}{8}(16)^2 - \frac{1}{2}(36)^2 - \frac{1}{4}(16)(36) + 13(16) + 40(36) - 280 = 544$$

or \$54,400.

8.4 The Method of Least Squares

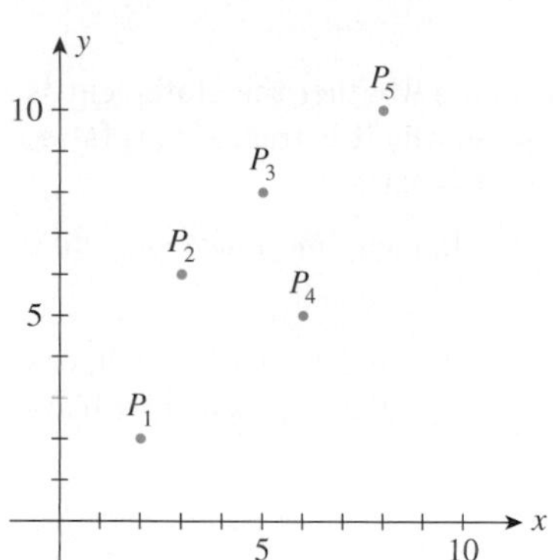

FIGURE 22
A scatter diagram

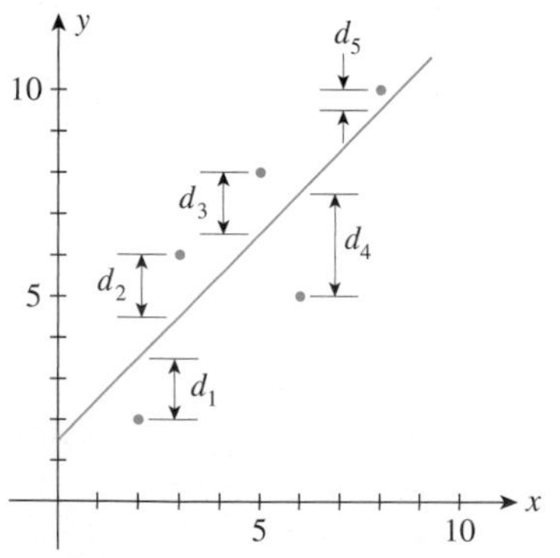

FIGURE 23
The approximating line misses each point by the amounts $d_1, d_2, \ldots, d_5$.

The Method of Least Squares

In Section 1.4, Example 10, we saw how a linear equation can be used to approximate the sales trend for a local sporting goods store. As we saw there, one use of a **trend line** is to predict a store's future sales. Recall that we obtained the line by requiring that it pass through two data points, the rationale being that such a line seems to *fit* the data reasonably well.

In this section, we describe a general method, known as the **method of least squares**, for determining a straight line that, in some sense, *best* fits a set of data points when the points are scattered about a straight line. To illustrate the principle behind the method of least squares, suppose, for simplicity, that we are given five data points,

$$P_1(x_1, y_1), \quad P_2(x_2, y_2), \quad P_3(x_3, y_3), \quad P_4(x_4, y_4), \quad P_5(x_5, y_5)$$

that describe the relationship between the two variables x and y. By plotting these data points, we obtain a graph called a **scatter diagram** (Figure 22).

If we try to fit a straight line to these data points, the line will miss the first, second, third, fourth, and fifth data points by the amounts d_1, d_2, d_3, d_4, and d_5, respectively (Figure 23).

The **principle of least squares** states that the straight line L that fits the data points best is the one chosen by requiring that the sum of the squares of $d_1, d_2, \ldots, d_5$—that is,

$$d_1^2 + d_2^2 + d_3^2 + d_4^2 + d_5^2$$

be made as small as possible. If we think of the amount d_1 as the error made when the value y_1 is approximated by the corresponding value of y lying on the straight line L, and d_2 as the error made when the value y_2 is approximated by the corresponding value of y, and so on, then it can be seen that the least-squares criterion calls for minimizing the sum of the squares of the errors. The line L obtained in this manner is called the **least-squares line**, or **regression line**.

To find a method for computing the regression line L, suppose L has representation $y = f(x) = mx + b$, where m and b are to be determined. Observe that

$$\begin{aligned} &d_1^2 + d_2^2 + d_3^2 + d_4^2 + d_5^2 \\ &\quad = [f(x_1) - y_1]^2 + [f(x_2) - y_2]^2 + [f(x_3) - y_3]^2 \\ &\qquad + [f(x_4) - y_4]^2 + [f(x_5) - y_5]^2 \\ &\quad = (mx_1 + b - y_1)^2 + (mx_2 + b - y_2)^2 + (mx_3 + b - y_3)^2 \\ &\qquad + (mx_4 + b - y_4)^2 + (mx_5 + b - y_5)^2 \end{aligned}$$

and may be viewed as a function of the two variables m and b. Thus, the least-squares criterion is equivalent to minimizing the function

$$\begin{aligned} f(m, b) = &(mx_1 + b - y_1)^2 + (mx_2 + b - y_2)^2 + (mx_3 + b - y_3)^2 \\ &+ (mx_4 + b - y_4)^2 + (mx_5 + b - y_5)^2 \end{aligned}$$

with respect to m and b. Using the chain rule, we compute

$$\begin{aligned} \frac{\partial f}{\partial m} &= 2(mx_1 + b - y_1)x_1 + 2(mx_2 + b - y_2)x_2 + 2(mx_3 + b - y_3)x_3 \\ &\quad + 2(mx_4 + b - y_4)x_4 + 2(mx_5 + b - y_5)x_5 \\ &= 2[mx_1^2 + bx_1 - x_1y_1 + mx_2^2 + bx_2 - x_2y_2 + mx_3^2 + bx_3 - x_3y_3 \\ &\quad + mx_4^2 + bx_4 - x_4y_4 + mx_5^2 + bx_5 - x_5y_5] \\ &= 2[(x_1^2 + x_2^2 + x_3^2 + x_4^2 + x_5^2)m + (x_1 + x_2 + x_3 + x_4 + x_5)b \\ &\quad - (x_1y_1 + x_2y_2 + x_3y_3 + x_4y_4 + x_5y_5)] \end{aligned}$$

and

$$\begin{aligned} \frac{\partial f}{\partial b} &= 2(mx_1 + b - y_1) + 2(mx_2 + b - y_2) + 2(mx_3 + b - y_3) \\ &\quad + 2(mx_4 + b - y_4) + 2(mx_5 + b - y_5) \\ &= 2[(x_1 + x_2 + x_3 + x_4 + x_5)m + 5b - (y_1 + y_2 + y_3 + y_4 + y_5)] \end{aligned}$$

Setting

$$\frac{\partial f}{\partial m} = 0 \quad \text{and} \quad \frac{\partial f}{\partial b} = 0$$

gives

$$\begin{aligned} &(x_1^2 + x_2^2 + x_3^2 + x_4^2 + x_5^2)m + (x_1 + x_2 + x_3 + x_4 + x_5)b \\ &\quad = x_1y_1 + x_2y_2 + x_3y_3 + x_4y_4 + x_5y_5 \end{aligned}$$

and

$$(x_1 + x_2 + x_3 + x_4 + x_5)m + 5b = y_1 + y_2 + y_3 + y_4 + y_5$$

Solving these two simultaneous equations for m and b then leads to an equation $y = mx + b$ of a straight line.

Before looking at an example, we state a more general result whose derivation is identical to the special case involving the five data points just discussed.

The Method of Least Squares

Suppose we are given n data points:

$$P_1(x_1, y_1), \quad P_2(x_2, y_2), \quad P_3(x_3, y_3), \ldots, P_n(x_n, y_n)$$

Then, the least-squares (regression) line for the data is given by the linear equation

$$y = f(x) = mx + b$$

where the constants m and b satisfy the equations

$$\begin{aligned}(x_1^2 + x_2^2 + x_3^2 + \cdots + x_n^2)m &+ (x_1 + x_2 + x_3 + \cdots + x_n)b \\ &= x_1y_1 + x_2y_2 + x_3y_3 + \cdots + x_ny_n\end{aligned} \tag{4}$$

and

$$\begin{aligned}(x_1 + x_2 + x_3 + \cdots + x_n)m &+ nb \\ &= y_1 + y_2 + y_3 + \cdots + y_n\end{aligned} \tag{5}$$

simultaneously. Equations (4) and (5) are called **normal equations.**

EXAMPLE 1 Find an equation of the least-squares line for the data

$$P_1(1, 1), \quad P_2(2, 3), \quad P_3(3, 4), \quad P_4(4, 3), \quad P_5(5, 6)$$

Solution Here, we have $n = 5$ and

$$\begin{array}{lllll} x_1 = 1 & x_2 = 2 & x_3 = 3 & x_4 = 4 & x_5 = 5 \\ y_1 = 1 & y_2 = 3 & y_3 = 4 & y_4 = 3 & y_5 = 6 \end{array}$$

so Equation (4) becomes

$$(1 + 4 + 9 + 16 + 25)m + (1 + 2 + 3 + 4 + 5)b = 1 + 6 + 12 + 12 + 30$$

or

$$55m + 15b = 61 \tag{6}$$

and (5) becomes

$$(1 + 2 + 3 + 4 + 5)m + 5b = 1 + 3 + 4 + 3 + 6$$

or

$$15m + 5b = 17 \tag{7}$$

Solving Equation (7) for b gives

$$b = -3m + \frac{17}{5} \tag{8}$$

which, upon substitution into (6), gives

$$\begin{aligned} 15\left(-3m + \frac{17}{5}\right) + 55m &= 61 \\ -45m + 51 + 55m &= 61 \\ 10m &= 10 \\ m &= 1 \end{aligned}$$

Substituting this value of m into (8) gives

$$b = -3 + \frac{17}{5} = \frac{2}{5} = 0.4$$

Therefore, the required equation of the least-squares line is

$$y = x + 0.4$$

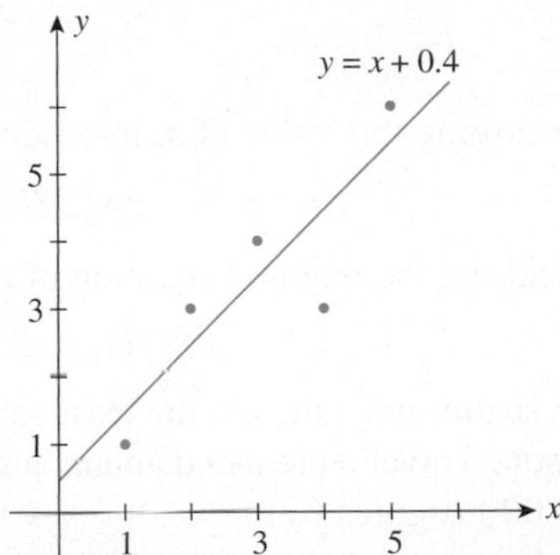

FIGURE 24
The scatter diagram and the least-squares line $y = x + 0.4$

The scatter diagram and the regression line are shown in Figure 24. ■

APPLIED EXAMPLE 2 Advertising Expense and a Firm's Profit The proprietor of Leisure Travel Service compiled the following data relating the firm's annual profit to its annual advertising expenditure (both measured in thousands of dollars).

Annual Advertising Expenditure, x	12	14	17	21	26	30
Annual Profit, y	60	70	90	100	100	120

a. Determine an equation of the least-squares line for these data.
b. Draw a scatter diagram and the least-squares line for these data.
c. Use the result obtained in part (a) to predict Leisure Travel's annual profit if the annual advertising budget is $20,000.

Solution

a. The calculations required for obtaining the normal equations may be summarized as follows:

	x	y	x^2	xy
	12	60	144	720
	14	70	196	980
	17	90	289	1,530
	21	100	441	2,100
	26	100	676	2,600
	30	120	900	3,600
Sum	120	540	2,646	11,530

The normal equations are

$$6b + 120m = 540 \tag{9}$$

$$120b + 2646m = 11,530 \tag{10}$$

Solving Equation (9) for b gives

$$b = -20m + 90 \tag{11}$$

which, upon substitution into Equation (10), gives

$$\begin{aligned} 120(-20m + 90) + 2646m &= 11{,}530 \\ -2400m + 10{,}800 + 2646m &= 11{,}530 \\ 246m &= 730 \\ m &\approx 2.97 \end{aligned}$$

Substituting this value of m into Equation (11) gives

$$b = -20(2.97) + 90 = 30.6$$

Therefore, the required equation of the least-squares line is given by

$$y = f(x) = 2.97x + 30.6$$

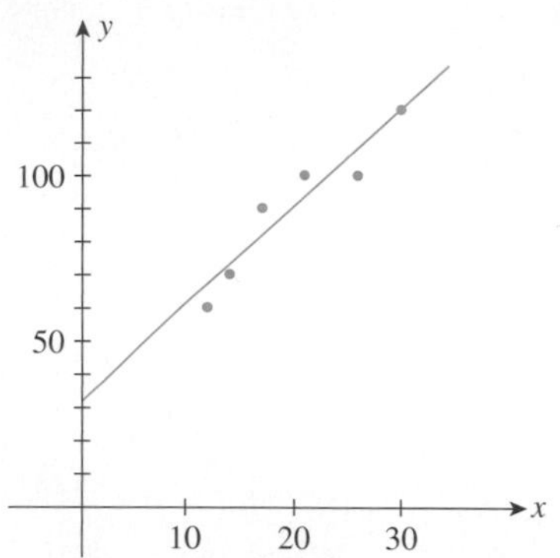

FIGURE 25
The scatter diagram and the least-squares line $y = 2.97x + 30.6$

b. The scatter diagram and the least-squares line are shown in Figure 25.

c. Leisure Travel's predicted annual profit corresponding to an annual budget of \$20,000 is given by

$$\begin{aligned} f(20) &= 2.97(20) + 30.6 \\ &= 90 \end{aligned}$$

or \$90,000. ■

APPLIED EXAMPLE 3 Maximizing Profit A market research study conducted for Century Communications provided the following data based on the projected monthly sales x (in thousands) of Century's DVD version of a box-office hit adventure movie with a proposed wholesale unit price of p dollars.

p	38	36	34.5	30	28.5
x	2.2	5.4	7.0	11.5	14.6

a. Find the demand equation if the demand curve is the least-squares line for these data.

b. The total monthly cost function associated with producing and distributing the DVD movies is given by

$$C(x) = 4x + 25$$

where x denotes the number of discs (in thousands) produced and sold and $C(x)$ is in thousands of dollars. Determine the unit wholesale price that will maximize Century's monthly profit.

Solution

a. The calculations required for obtaining the normal equations may be summarized as follows:

	x	p	x^2	xp
	2.2	38	4.84	83.6
	5.4	36	29.16	194.4
	7.0	34.5	49	241.5
	11.5	30	132.25	345
	14.6	28.5	213.16	416.1
Sum	40.7	167	428.41	1280.6

The normal equations are

$$5b + 40.7m = 167$$
$$40.7b + 428.41m = 1280.6$$

Solving this system of linear equations simultaneously, we find that

$$m \approx -0.81 \quad \text{and} \quad b \approx 39.99$$

Therefore, the required equation of the least-squares line is given by

$$p = f(x) = -0.81x + 39.99$$

which is the required demand equation, provided $0 \leq x \leq 49.37$.

b. The total revenue function in this case is given by

$$R(x) = xp = -0.81x^2 + 39.99x$$

and since the total cost function is

$$C(x) = 4x + 25$$

we see that the profit function is

$$P(x) = -0.81x^2 + 39.99x - (4x + 25)$$
$$= -0.81x^2 + 35.99x - 25$$

To find the absolute maximum of $P(x)$ over the closed interval [0, 49.37], we compute

$$P'(x) = -1.62x + 35.99$$

Since $P'(x) = 0$, we find $x \approx 22.22$ as the only critical point of P. Finally, from the table

x	0	22.22	49.37
$P(x)$	−25	374.78	−222.47

we see that the optimal wholesale price is

$$p = -0.81(22.22) + 39.99 = 21.99$$

or $21.99 per disc.

8.4 Self-Check Exercises

1. Find an equation of the least-squares line for the data

$$P_1(0, 3), \quad P_2(2, 6.5), \quad P_3(4, 10), \quad P_4(6, 16), \quad P_5(7, 16.5)$$

2. The following data give the percent of people over age 65 who have high school diplomas.

Year, x	0	6	11	16	22	26
Percent with Diplomas, y	19	25	30	35	44	48

Here, $x = 0$ corresponds to the beginning of the year 1959.

a. Find an equation of the least-squares line for this data.

b. Assuming that this trend continued, what percent of people over age 65 had high school diplomas at the beginning of the year 2003 ($x = 44$)?

Source: U.S. Department of Commerce

Solutions to Self-Check Exercises 8.4 can be found on page 578.

8.4 Concept Questions

1. Explain the terms (a) *scatter diagram* and (b) *least-squares line*.

2. Explain the *principle of least-squares* in your own words.

8.4 Exercises

In Exercises 1–6, (a) find an equation of the least-squares line for the data and (b) draw a scatter diagram for the data and graph the least-squares line.

1.

x	1	2	3	4
y	4	6	8	11

2.

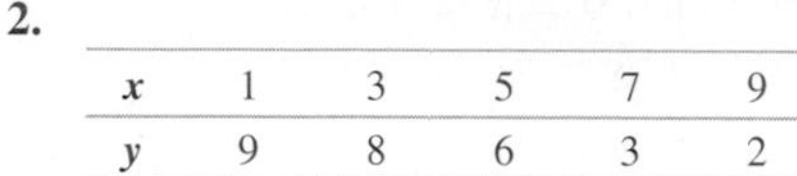

x	1	3	5	7	9
y	9	8	6	3	2

3.

x	1	2	3	4	4	6
y	4.5	5	3	2	3.5	1

4.

x	1	1	2	3	4	4	5
y	2	3	3	3.5	3.5	4	5

5. $P_1(1, 3)$, $P_2(2, 5)$, $P_3(3, 5)$, $P_4(4, 7)$, $P_5(5, 8)$

6. $P_1(1, 8)$, $P_2(2, 6)$, $P_3(5, 6)$, $P_4(7, 4)$, $P_5(10, 1)$

7. College Admissions The following data, compiled by the admissions office at Faber College during the past 5 yr, relate the number of college brochures and follow-up letters (x) sent to a preselected list of high school juniors who had taken the PSAT and the number of completed applications (y) received from these students (both measured in units of 1000):

x	4	4.5	5	5.5	6
y	0.5	0.6	0.8	0.9	1.2

a. Determine the equation of the least-squares line for these data.
b. Draw a scatter diagram and the least-squares line for these data.
c. Use the result obtained in part (a) to predict the number of completed applications that might be expected if 6400 brochures and follow-up letters are sent out during the next year.

8. Starbucks' Store Count According to company reports, the number of Starbucks stores worldwide between 1999 and 2003 are as follows ($x = 0$ corresponds to 1999):

Year, x	0	1	2	3	4
Stores, y	2135	3501	4709	5886	7225

a. Find an equation of the least-squares line for these data.
b. Use the result of part (a) to estimate the rate at which new stores were opened annually worldwide for the period in question.

Source: Company Reports

9. SAT Verbal Scores The following data, compiled by the superintendent of schools in a large metropolitan area, shows the average SAT verbal scores of high school seniors during the 5 yr since the district implemented the "back-to-basics" program:

Year, x	1	2	3	4	5
Average Score, y	436	438	428	430	426

a. Determine the equation of the least-squares line for these data.
b. Draw a scatter diagram and the least-squares line for these data.
c. Use the result obtained in part (a) to predict the average SAT verbal score of high school seniors 2 yr from now ($x = 7$).

10. Net Sales The management of Kaldor, a manufacturer of electric motors, submitted the following data in the annual report to its stockholders. The table shows the net sales (in millions of dollars) during the 5 yr that have elapsed since the new management team took over. (The first year the firm operated under the new management corresponds to the time period $x = 1$, and the four subsequent years correspond to $x = 2, 3, 4, 5$.)

Year, x	1	2	3	4	5
Net Sales, x	426	437	460	473	477

a. Determine the equation of the least-squares line for these data.
b. Draw a scatter diagram and the least-squares line for these data.
c. Use the result obtained in part (a) to predict the net sales for the upcoming year.

11. Mass Transit Subsidies The following table gives the projected state subsidies (in millions of dollars) to the Massachusetts Bay Transit Authority (MBTA) over a 5-yr period:

Year, x	1	2	3	4	5
Subsidy, y	20	24	26	28	32

a. Find an equation of the least-squares line for these data.
b. Use the result of part (a) to estimate the state subsidy to the MBTA for the eighth year ($x = 8$).

Source: Massachusetts Bay Transit Authority

12. Information Security Software Sales As online attacks persist, spending on information security software continues to rise. The following table gives the forecast for the worldwide sales (in billions of dollars) of information security software through 2007 ($x = 0$ corresponds to 2002):

Year, x	0	1	2	3	4	5
Spending, y	6.8	8.3	9.8	11.3	12.8	14.9

a. Find an equation of the least-squares line for these data.
b. Use the result of part (a) to forecast the spending on information security software in 2008, assuming the trend continues.

Source: International Data Corp.

13. U.S. DRUGS SALES The following table gives the total sales of drugs (in billions of dollars) in the United States from 1999 ($x = 0$) through 2003:

Year, x	0	1	2	3	4
Sales, y	126	144	171	191	216

a. Find an equation of the least-squares line for these data.
b. Use the result of part (a) to estimate the total sales of drugs in 2005, assuming that the trend continued.

Source: IMS Health

14. ONLINE SPENDING The convenience of shopping on the Web combined with high-speed broadband access services are spurring online spending. The following table gives the projected online spending per buyer (in dollars) from 2002 ($x = 0$) through 2008 ($x = 6$):

Year, x	0	1	2	3	4	5	6
Spending, y	501	540	585	631	680	728	779

a. Find an equation of the least-squares line for these data.
b. Use the result of part (a) to estimate the rate of change of spending per buyer between 2002 and 2008.

Source: Commerce Dept.

15. CALLING CARDS The market for prepaid calling cards is projected to grow steadily through 2008. The following table gives the projected sales of prepaid phone card sales, in billions of dollars, from 2002 through 2008:

Year, x	0	1	2	3	4	5	6
Sales, y	3.7	4.0	4.4	4.8	5.2	5.8	6.3

a. Find an equation of the least-squares line for these data.
b. Use the result of part (a) to estimate the rate at which the sales of prepaid phone cards will grow over the period in question.

Source: Atlantic-ACM

16. U.S. ONLINE BANKING HOUSEHOLDS The following table gives the projected U.S. online banking households as a percent of all U.S. banking households from 2001 ($x = 1$) through 2007 ($x = 7$):

Year, x	1	2	3	4	5	6	7
Percent of Households, y	21.2	26.7	32.2	37.7	43.2	48.7	54.2

a. Find an equation of the least-squares line for these data.
b. Use the result of part (a) to estimate the projected percent of U.S. online banking households in 2008.

Source: Jupiter Research

17. SALES OF GPS EQUIPMENT The annual sales (in billions of dollars) of global positioning system (GPS) equipment from 2000 through 2006 follow (sales in 2004 through 2006 are projections). Here, $x = 0$ corresponds to 2000.

Year, x	0	1	2	3	4	5	6
Annual sales, y	7.9	9.6	11.5	13.3	15.2	16	18.8

a. Find an equation of the least-squares line for these data.
b. Use the equation found in part (a) to estimate the annual sales of GPS equipment for 2005.

Source: ABI Research

18. ONLINE SALES OF USED AUTOS The amount (in millions of dollars) of used autos sold online in the United States is expected to grow in accordance with the figures given in the following table:

Year, x	0	1	2	3	4	5	6	7
Sales, y	1	1.4	2.2	2.8	3.6	4.2	5.0	5.8

(Here, $x = 0$ corresponds to 2000.)
a. Find an equation of the least-squares line for these data.
b. Use the result of part (a) to estimate the sales of used autos online in 2008, assuming that the predicted trend continues through that year.

Source: comScore Networks Inc.

19. SOCIAL SECURITY WAGE BASE The Social Security (FICA) wage base (in thousands of dollars) from 2001 to 2006 is given in the following table:

Year	2001	2002	2003	2004	2005	2006
Wage Base, y	80.4	84.9	87	87.9	90	94.2

a. Find an equation of the least-squares line for these data. (Let $x = 1$ represent 2001.)
b. Use your result of part (a) to estimate the FICA wage base in 2010.

Source: The World Almanac

20. MARKET FOR DRUGS Because of new, lower standards, experts in a study conducted in early 2000 projected a rise in the market for cholesterol-reducing drugs. The U.S. market (in billions of dollars) for such drugs from 1999 through 2004 is given in the following table ($x = 0$ represents 1999):

Year	1999	2000	2001	2002	2003	2004
Market, y	12.07	14.07	16.21	18.28	20	21.72

a. Find an equation of the least-squares line for these data.

b. Use the result of part (a) to estimate the U.S. market for cholesterol-reducing drugs in 2005, assuming the trend continued.

Source: S. G. Cowen

21. **Male Life Expectancy at 65** The projections of male life expectancy at age 65 in the United States are summarized in the following table ($x = 0$ corresponds to 2000):

Year, x	0	10	20	30	40	50
Years beyond 65, y	15.9	16.8	17.6	18.5	19.3	20.3

a. Find an equation of the least-squares line for these data.
b. Use the result of (a) to estimate the life expectancy at 65 of a male in 2040. How does this result compare with the given data for that year?
c. Use the result of (a) to estimate the life expectancy at 65 of a male in 2030.

Source: U.S. Census Bureau

22. **Authentication Technology** With computer security always a hot-button issue, demand is growing for technology that authenticates and authorizes computer users. The following table gives the authentication software sales (in billions of dollars), including projections, from 1999 through 2006 ($x = 0$ represents 1999):

Year, x	0	1	2	3	4	5
Sales, y	2.4	2.9	3.7	4.5	5.2	6.1

a. Find an equation of the least-squares line for these data.
b. Use the result of part (a) to estimate the sales for 2007, assuming the projection is accurate.

Source: International Data Corporation

23. **Portable-Phone Services** The projected number of wireless subscribers y (in millions) from 2000 through 2006 is summarized in the accompanying table. (Here, $x = 0$ corresponds to the beginning of 2000.)

Year, x	0	1	2	3	4	5	6
Subscribers, y	90.4	100.0	110.4	120.4	130.8	140.4	150.0

a. Find an equation of the least-squares line for these data.
b. Use the result of part (a) to estimate the projected number of wireless subscribers at the beginning of 2006. How does this result compare with the given data for that year?

Source: BancAmerica Robertson Stephens

24. **Outpatient Visits** With an aging population, the demand for health care, as measured by outpatient visits, is steadily growing. The number of outpatient visits (in millions) from 1991 through 2001 is recorded in the following table ($x = 0$ corresponds to 1991):

Year, x	0	1	2	3	4	5	6	7	8	9	10
Number of Visits, y	320	340	362	380	416	440	444	470	495	520	530

a. Find an equation of the least-squares line for these data.
b. Use the result of part (a) to estimate the number of outpatient visits in 2004, assuming the trend continued.

Source: PriceWaterhouse Coopers

In Exercises 25–28, determine whether the statement is true or false. If it is true, explain why it is true. If it is false, give an example to show why it is false.

25. The least-squares line must pass through at least one of the data points.

26. The sum of the squares of the errors incurred in approximating n data points using the least-squares linear function is zero if and only if the n data points lie along a straight line.

27. If the data consist of two distinct points, then the least-squares line is just the line that passes through the two points.

28. A data point lies on the least-squares line if and only if the vertical distance between the point and the line is equal to zero.

8.4 Solutions to Self-Check Exercises

1. We first construct the table:

	x	y	x^2	xy
	0	3	0	0
	2	6.5	4	13
	4	10	16	40
	6	16	36	96
	7	16.5	49	115.5
Sum	19	52	105	264.5

The normal equations are

$$5b + 19m = 52$$
$$19b + 105m = 264.5$$

Solving the first equation for b gives

$$b = -3.8m + 10.4$$

which, upon substitution into the second equation, gives

$$\begin{aligned} 19(-3.8m + 10.4) + 105m &= 264.5 \\ -72.2m + 197.6 + 105m &= 264.5 \\ 32.8m &= 66.9 \\ m &\approx 2.04 \end{aligned}$$

Substituting this value of m into the expression for b found earlier gives

$$b = -3.8(2.04) + 10.4 \approx 2.65$$

Therefore, the required least-squares line has the equation given by

$$y = 2.04x + 2.65$$

2. a. The calculations required for obtaining the normal equations may be summarized as follows:

	x	y	x^2	xy
	0	19	0	0
	6	25	36	150
	11	30	121	330
	16	35	256	560
	22	44	484	968
	26	48	676	1248
Sum	81	201	1573	3256

The normal equations are

$$\begin{aligned} 6b + 81m &= 201 \\ 81b + 1573m &= 3256 \end{aligned}$$

Solving this system of linear equations simultaneously, we find

$$m \approx 1.13 \quad \text{and} \quad b \approx 18.23$$

Therefore, the required least-squares line has the equation given by

$$y = f(x) = 1.13x + 18.23$$

b. The percent of people over the age of 65 who had high school diplomas at the beginning of the year 2003 is given by

$$\begin{aligned} f(44) &= 1.13(44) + 18.23 \\ &= 67.95 \end{aligned}$$

or approximately 68%.

USING TECHNOLOGY

Finding an Equation of a Least-Squares Line

A graphing utility is especially useful in calculating an equation of the least-squares line for a set of data. We simply enter the given data in the form of lists into the calculator and then use the linear regression function to obtain the coefficients of the required equation.

EXAMPLE 1 Find an equation of the least-squares line for the following data:

x	1.1	2.3	3.2	4.6	5.8	6.7	8
y	−5.8	−5.1	−4.8	−4.4	−3.7	−3.2	−2.5

Solution First, we enter the data as

$x_1 = 1.1$ $y_1 = -5.8$ $x_2 = 2.3$ $y_2 = -5.1$ $x_3 = 3.2$ $y_3 = -4.8$
$x_4 = 4.6$ $y_4 = -4.4$ $x_5 = 5.8$ $y_5 = -3.7$ $x_6 = 6.7$
$y_6 = -3.2$ $x_7 = 8$ $y_7 = -2.5$

Then, using the linear regression function from the statistics menu, we find

$a = -6.29996900666$ $b = 0.460560979389$ corr. $= .994488871079$ $n = 7$

(continued)

Therefore, an equation of the least-squares line ($y = a + bx$) is

$$y = -6.30 + 0.46x$$

The correlation coefficient of .99449 attests to the excellent fit of the regression line. ■

APPLIED EXAMPLE 2 Demand for Electricity According to Pacific Gas and Electric, the nation's largest utility company, the demand for electricity from 1990 through 2000 was as follows:

t	0	2	4	6	8	10
y	333	917	1500	2117	2667	3292

Here, $t = 0$ corresponds to 1990, and y gives the amount of electricity demanded in year t, measured in megawatts. Find an equation of the least-squares line for these data.

Source: Pacific Gas and Electric

Solution First, we enter the data as

$x_1 = 0$ $y_1 = 333$ $x_2 = 2$ $y_2 = 917$ $x_3 = 4$ $y_3 = 1500$
$x_4 = 6$ $y_4 = 2117$ $x_5 = 8$ $x_5 = 2667$ $x_6 = 10$ $y_6 = 3292$

Then, using the linear regression function from the statistic menu, we find

$$a = 328.476190476 \quad \text{and} \quad b = 295.171428571$$

Therefore, an equation of the least-squares line is

$$y = 328 + 295t$$

■

TECHNOLOGY EXERCISES

In Exercises 1–4, find an equation of the least-squares line for the data.

1.

x	2.1	3.4	4.7	5.6	6.8	7.2
y	8.8	12.1	14.8	16.9	19.8	21.1

2.

x	1.1	2.4	3.2	4.7	5.6	7.2
y	−0.5	1.2	2.4	4.4	5.7	8.1

3.

x	−2.1	−1.1	0.1	1.4	2.5	4.2	5.1
y	6.2	4.7	3.5	1.9	0.4	−1.4	−2.5

4.

x	−1.12	0.1	1.24	2.76	4.21	6.82
y	7.61	4.9	2.74	−0.47	−3.51	−8.94

5. STARBUCKS' ANNUAL SALES According to company reports, Starbucks' annual sales (in billions of dollars) for 1998 through 2003 are the following ($x = 0$ corresponds to 1998):

Year, x	0	1	2	3	4	5
Sales, y	1.28	1.73	2.18	2.65	3.29	4.08

a. Find an equation of the least-squares line for these data.
b. Use the result of part (a) to project Starbucks' sales for 2006, assuming that the trend continued.

Source: Company reports

6. SALES OF DRUGS Sales of drugs called analeptics, which are used to treat attention-deficit disorders, were rising even before some companies began advertising them to parents. The following table gives the sales of analeptics (in millions of dollars) from 1995 through 2000 ($x = 0$ represents 1995):

Year, x	0	1	2	3	4	5
Sales, y	382	455	536	618	664	758

a. Find an equation of the least-squares line for these data.
b. Use the result of part (a) to estimate the sales for 2002, assuming that the trend continued.

Source: IMS Health

7. WASTE GENERATION The amount of waste (in millions of tons/year) generated in the United States from 1960 to 1990 was

Year	1960	1965	1970	1975
Amount, y	81	100	120	124

Year	1980	1985	1990
Amount, y	140	152	164

a. Find an equation of the least-squares line for these data. (Let x be in units of 5 and let $x = 1$ represent 1960.)
b. Use the result of part (a) to estimate the amount of waste generated in 2000, assuming that the trend continued.

Source: Council on Environmental Quality

8. ONLINE TRAVEL More and more travelers are purchasing their tickets online. According to industry projections, the U.S. online travel revenue (in billions of dollars) from 2001 through 2005 are the following ($t = 0$ corresponds to 2001):

Year, t	0	1	2	3	4
Revenue, y	16.3	21.0	25.0	28.8	32.7

a. Find an equation of the least-squares line for these data.
b. Use the result of part (a) to estimate the U.S. online travel revenue for 2006, assuming that the trend continued.

Source: Forrester Research, Inc.

9. MARKET FOR DRUGS Because of new, lower standards, experts in a study conducted in early 2000 projected a rise in the market for cholesterol-reducing drugs. The following table gives the U.S. market (in billions of dollars) for such drugs from 1999 through 2004.

Year	1999	2000	2001	2002	2003	2004
Market, y	12.07	14.07	16.21	18.28	20	21.72

a. Find an equation of the least-squares line for these data. (Let $x = 0$ represent 1999.)
b. Use the result of part (a) to estimate the U.S. market for cholesterol-reducing drugs in 2005, assuming that the trend continued.

Source: S. G. Cowen

10. ANNUAL COLLEGE COSTS The following table gives the annual U.S. college costs (including tuition, room, and board) for academic years 1991–1992 through 2001–2002 ($x = 0$ corresponds to 1991–1992).

Academic Year, x	0	1	2	3	4
Cost ($), y	7077	7452	7931	8306	8800

Academic Year, x	5	6	7	8	9	10
Cost ($), y	9206	9588	10,040	10,444	10,818	11,454

a. Find an equation of the least-squares line for these data.
b. Use the result of part (a) to plot the least-squares line.
c. Use the result of part (a) to determine the approximate average rate of increase of college costs per year for the period in question.

Source: National Center for Education Statistics

8.5 Constrained Maxima and Minima and the Method of Lagrange Multipliers

Constrained Relative Extrema

In Section 8.3, we studied the problem of determining the relative extremum of a function $f(x, y)$ without placing any restrictions on the independent variables x and y—except, of course, that the point (x, y) lies in the domain of f. Such a relative extremum of a function f is referred to as an **unconstrained relative extremum** of f. However, in many practical optimization problems, we must maximize or minimize a function in which the independent variables are subjected to certain further constraints.

In this section, we discuss a powerful method for determining the relative extrema of a function $f(x, y)$ whose independent variables x and y are required to satisfy one or more constraints of the form $g(x, y) = 0$. Such a relative extremum of a function f is called a **constrained relative extremum** of f. We can see the difference between an unconstrained extremum of a function $f(x, y)$ of two variables and a constrained extremum of f, where the independent variables x and y are subjected to a constraint of the form $g(x, y) = 0$, by considering the geometry of the two cases. Figure 26a depicts the graph of a function $f(x, y)$ that has an unconstrained relative minimum at the point (0, 0). However, when the independent variables x and y are subjected to an equality constraint of the form $g(x, y) = 0$, the points (x, y, z) that satisfy both $z = f(x, y)$ and the constraint equation $g(x, y) = 0$ lie on a curve C. Therefore, the constrained relative minimum of f must also lie on C (Figure 26b).

Our first example involves an equality constraint $g(x, y) = 0$ in which we solve for the variable y explicitly in terms of x. In this case we may apply the technique used in Chapter 4 to find the relative extrema of a function of one variable.

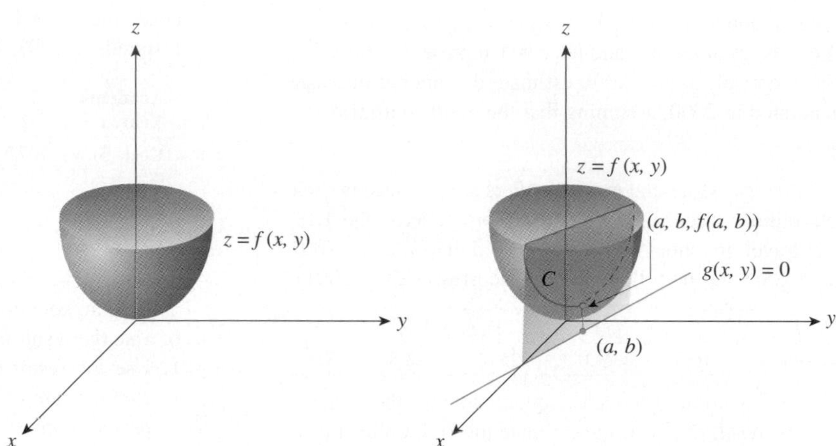

FIGURE 26

(a) $f(x, y)$ has an unconstrained relative extremum at (0, 0).

(b) $f(x, y)$ has a constrained relative extremum at $(a, b, f(a, b))$.

EXAMPLE 1 Find the relative minimum of the function

$$f(x, y) = 2x^2 + y^2$$

subject to the constraint $g(x, y) = x + y - 1 = 0$.

Solution Solving the constraint equation for y explicitly in terms of x, we obtain $y = -x + 1$. Substituting this value of y into the function $f(x, y) = 2x^2 + y^2$ results in a function of x,

$$h(x) = 2x^2 + (-x + 1)^2 = 3x^2 - 2x + 1$$

The function h describes the curve C lying on the graph of f on which the constrained relative minimum of f occurs. To find this point, use the technique developed in Chapter 4 to determine the relative extrema of a function of one variable:

$$h'(x) = 6x - 2 = 2(3x - 1)$$

Setting $h'(x) = 0$ gives $x = \frac{1}{3}$ as the sole critical point of the function h. Next, we find

$$h''(x) = 6$$

and, in particular,

$$h''\left(\frac{1}{3}\right) = 6 > 0$$

Therefore, by the second derivative test, the point $x = \frac{1}{3}$ gives rise to a relative minimum of h. Substitute this value of x into the constraint equation $x + y - 1 = 0$ to get $y = \frac{2}{3}$. Thus, the point $(\frac{1}{3}, \frac{2}{3})$ gives rise to the required constrained relative minimum of f. Since

$$f\left(\frac{1}{3}, \frac{2}{3}\right) = 2\left(\frac{1}{3}\right)^2 + \left(\frac{2}{3}\right)^2 = \frac{2}{3}$$

the required constrained relative minimum value of f is $\frac{2}{3}$ at the point $(\frac{1}{3}, \frac{2}{3})$. It may be shown that $\frac{2}{3}$ is in fact a constrained absolute minimum value of f (Figure 27).

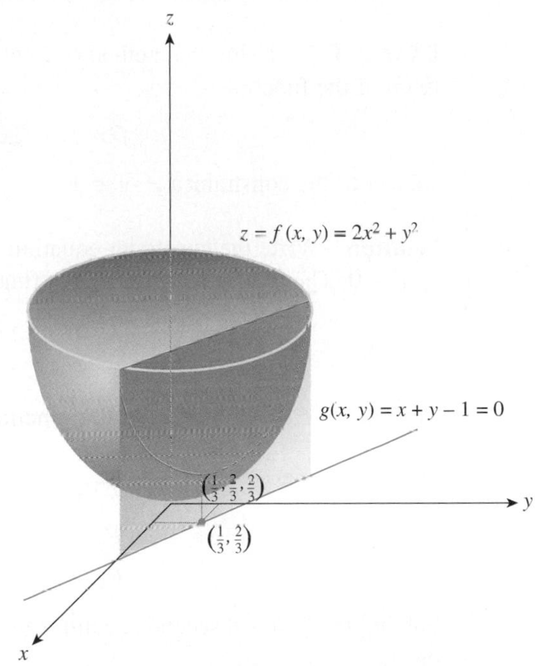

FIGURE 27
f has a constrained absolute minimum of $\frac{2}{3}$ at $(\frac{1}{3}, \frac{2}{3})$.

The Method of Lagrange Multipliers

The major drawback of the technique used in Example 1 is that it relies on our ability to solve the constraint equation $g(x, y) = 0$ for y explicitly in terms of x. This is not always an easy task. Moreover, even when we can solve the constraint equation $g(x, y) = 0$ for y explicitly in terms of x, the resulting function of one variable that is to be optimized may turn out to be unnecessarily complicated. Fortunately, an easier method exists. This method, called the **method of Lagrange multipliers** (Joseph Lagrange, 1736–1813), is as follows:

The Method of Lagrange Multipliers

To find the relative extrema of the function $f(x, y)$ subject to the constraint $g(x, y) = 0$ (assuming that these extreme values exist),

1. Form an auxiliary function

$$F(x, y, \lambda) = f(x, y) + \lambda g(x, y)$$

called the Lagrangian function (the variable λ is called the Lagrange multiplier).
2. Solve the system that consists of the equations

$$F_x = 0 \qquad F_y = 0 \qquad F_\lambda = 0$$

for all values of x, y, and λ.
3. The solutions found in step 2 are candidates for the extrema of f.

Let's re-solve Example 1 using the method of Lagrange multipliers.

EXAMPLE 2 Using the method of Lagrange multipliers, find the relative minimum of the function

$$f(x, y) = 2x^2 + y^2$$

subject to the constraint $x + y = 1$.

Solution Write the constraint equation $x + y = 1$ in the form $g(x, y) = x + y - 1 = 0$. Then, form the Lagrangian function

$$\begin{aligned} F(x, y, \lambda) &= f(x, y) + \lambda g(x, y) \\ &= 2x^2 + y^2 + \lambda(x + y - 1) \end{aligned}$$

To find the critical point(s) of the function F, solve the system composed of the equations

$$\begin{aligned} F_x &= 4x + \lambda = 0 \\ F_y &= 2y + \lambda = 0 \\ F_\lambda &= x + y - 1 = 0 \end{aligned}$$

Solving the first and second equations in this system for x and y in terms of λ, we obtain

$$x = -\frac{1}{4}\lambda \quad \text{and} \quad y = -\frac{1}{2}\lambda$$

which, upon substitution into the third equation, yields

$$-\frac{1}{4}\lambda - \frac{1}{2}\lambda - 1 = 0 \quad \text{or} \quad \lambda = -\frac{4}{3}$$

Therefore, $x = \frac{1}{3}$ and $y = \frac{2}{3}$, and $\left(\frac{1}{3}, \frac{2}{3}\right)$ affords a constrained minimum of the function f, in agreement with the result obtained earlier. ■

Note A disadvantage of the method of Lagrange multipliers is that there is no test analogous to the second derivative test mentioned in Section 8.3 for determining whether a critical point of a function of two or more variables leads to a relative maximum or relative minimum (and thus the absolute extrema) of the function. Here we have to rely on the geometric or physical nature of the problem to help us draw the necessary conclusions (see Example 2). ■

The method of Lagrange multipliers may be used to solve a problem involving a function of three or more variables, as illustrated in the next example.

EXAMPLE 3 Use the method of Lagrange multipliers to find the minimum of the function

$$f(x, y, z) = 2xy + 6yz + 8xz$$

subject to the constraint

$$xyz = 12{,}000$$

(*Note:* The existence of the minimum is suggested by the geometry of the problem.)

Solution Write the constraint equation $xyz = 12{,}000$ in the form $g(x, y, z) = xyz - 12{,}000$. Then, the Lagrangian function is

$$\begin{aligned} F(x, y, z, \lambda) &= f(x, y, z) + \lambda g(x, y, z) \\ &= 2xy + 6yz + 8xz + \lambda(xyz - 12{,}000) \end{aligned}$$

To find the critical point(s) of the function F, we solve the system composed of the equations

$$\begin{aligned} F_x &= 2y + 8z + \lambda yz = 0 \\ F_y &= 2x + 6z + \lambda xz = 0 \\ F_z &= 6y + 8x + \lambda xy = 0 \\ F_\lambda &= xyz - 12{,}000 = 0 \end{aligned}$$

Solving the first three equations of the system for λ in terms of x, y, and z, we have

$$\begin{aligned} \lambda &= -\frac{2y + 8z}{yz} \\ \lambda &= -\frac{2x + 6z}{xz} \\ \lambda &= -\frac{6y + 8x}{xy} \end{aligned}$$

Equating the first two expressions for λ leads to

$$\begin{aligned} \frac{2y + 8z}{yz} &= \frac{2x + 6z}{xz} \\ 2xy + 8xz &= 2xy + 6yz \\ x &= \frac{3}{4}y \end{aligned}$$

Next, equating the second and third expressions for λ in the same system yields

$$\begin{aligned} \frac{2x + 6z}{xz} &= \frac{6y + 8x}{xy} \\ 2xy + 6yz &= 6yz + 8xz \\ z &= \frac{1}{4}y \end{aligned}$$

Finally, substituting these values of x and z into the equation $xyz - 12{,}000 = 0$, the fourth equation of the first system of equations, we have

$$\left(\frac{3}{4}y\right)(y)\left(\frac{1}{4}y\right) - 12{,}000 = 0$$

$$y^3 = \frac{(12{,}000)(4)(4)}{3} = 64{,}000$$

$$y = 40$$

The corresponding values of x and z are given by $x = \frac{3}{4}(40) = 30$ and $z = \frac{1}{4}(40) = 10$. Therefore, we see that the point (30, 40, 10) gives the constrained minimum of f. The minimum value is

$$f(30, 40, 10) = 2(30)(40) + 6(40)(10) + 8(30)(10) = 7200$$

■

APPLIED EXAMPLE 4 Maximizing Profit Refer to Example 3, Section 8.1. The total weekly profit (in dollars) that Acrosonic realized in producing and selling its bookshelf loudspeaker systems is given by the profit function

$$P(x, y) = -\frac{1}{4}x^2 - \frac{3}{8}y^2 - \frac{1}{4}xy + 120x + 100y - 5000$$

where x denotes the number of fully assembled units and y denotes the number of kits produced and sold per week. Acrosonic's management decides that production of these loudspeaker systems should be restricted to a total of exactly 230 units each week. Under this condition, how many fully assembled units and how many kits should be produced each week to maximize Acrosonic's weekly profit?

Solution The problem is equivalent to the problem of maximizing the function

$$P(x, y) = -\frac{1}{4}x^2 - \frac{3}{8}y^2 - \frac{1}{4}xy + 120x + 100y - 5000$$

subject to the constraint

$$g(x, y) = x + y - 230 = 0$$

The Lagrangian function is

$$\begin{aligned} F(x, y, \lambda) &= P(x, y) + \lambda g(x, y) \\ &= -\frac{1}{4}x^2 - \frac{3}{8}y^2 - \frac{1}{4}xy + 120x + 100y \\ &\quad - 5000 + \lambda(x + y - 230) \end{aligned}$$

To find the critical point(s) of F, solve the following system of equations:

$$\begin{aligned} F_x &= -\frac{1}{2}x - \frac{1}{4}y + 120 + \lambda = 0 \\ F_y &= -\frac{3}{4}y - \frac{1}{4}x + 100 + \lambda = 0 \\ F_\lambda &= x + y - 230 = 0 \end{aligned}$$

Solving the first equation of this system for λ, we obtain

$$\lambda = \frac{1}{2}x + \frac{1}{4}y - 120$$

which, upon substitution into the second equation, yields

$$-\frac{3}{4}y - \frac{1}{4}x + 100 + \frac{1}{2}x + \frac{1}{4}y - 120 = 0$$
$$-\frac{1}{2}y + \frac{1}{4}x - 20 = 0$$

Solving the last equation for y gives

$$y = \frac{1}{2}x - 40$$

When we substitute this value of y into the third equation of the system, we have

$$x + \frac{1}{2}x - 40 - 230 = 0$$
$$x = 180$$

The corresponding value of y is $\frac{1}{2}(180) - 40$, or 50. Thus, the required constrained relative maximum of P occurs at the point (180, 50). Again, we can show that the point (180, 50) in fact yields a constrained absolute maximum for P. Thus, Acrosonic's profit is maximized by producing 180 assembled and 50 kit versions of their bookshelf loudspeaker systems. The maximum weekly profit realizable is given by

$$\begin{aligned} P(180, 50) &= -\frac{1}{4}(180)^2 - \frac{3}{8}(50)^2 - \frac{1}{4}(180)(50) \\ &\quad + 120(180) + 100(50) - 5000 \\ &= 10{,}312.5 \end{aligned}$$

or \$10,312.50.

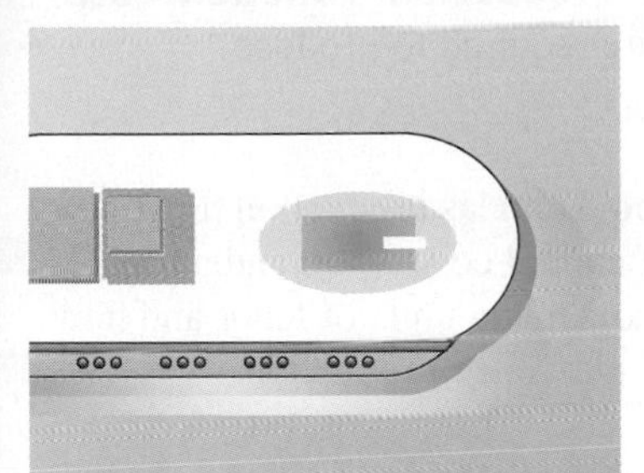

FIGURE 28
A rectangular-shaped pool will be built in the elliptical-shaped poolside area.

APPLIED EXAMPLE 5 Designing a Cruise-Ship Pool The operators of the *Viking Princess*, a luxury cruise liner, are contemplating the addition of another swimming pool to the ship. The chief engineer has suggested that an area in the form of an ellipse located in the rear of the promenade deck would be suitable for this purpose. This location would provide a poolside area with sufficient space for passenger movement and placement of deck chairs (Figure 28). It has been determined that the shape of the ellipse may be described by the equation $x^2 + 4y^2 = 3600$, where x and y are measured in feet. *Viking*'s operators would like to know the dimensions of the rectangular pool with the largest possible area that would meet these requirements.

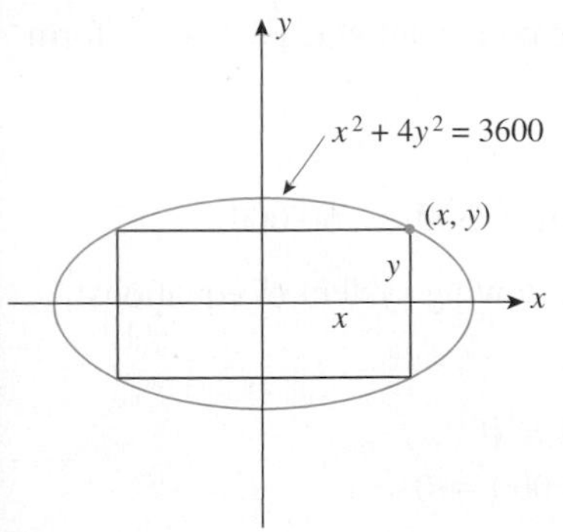

FIGURE 29
We want to find the largest rectangle that can be inscribed in the ellipse described by $x^2 + 4y^2 = 3600$.

Solution To solve this problem, we need to find the rectangle of largest area that can be inscribed in the ellipse with equation $x^2 + 4y^2 = 3600$. Letting the sides of the rectangle be $2x$ and $2y$ feet, we see that the area of the rectangle is $A = 4xy$ (Figure 29). Furthermore, the point (x, y) must be constrained to lie on the ellipse so that it satisfies the equation $x^2 + 4y^2 = 3600$. Thus, the problem is equivalent to the problem of maximizing the function

$$f(x, y) = 4xy$$

subject to the constraint $g(x, y) = x^2 + 4y^2 - 3600 = 0$. The Lagrangian function is

$$\begin{aligned} F(x, y, \lambda) &= f(x, y) + \lambda g(x, y) \\ &= 4xy + \lambda(x^2 + 4y^2 - 3600) \end{aligned}$$

To find the critical point(s) of F, we solve the following system of equations:

$$F_x = 4y + 2\lambda x = 0$$
$$F_y = 4x + 8\lambda y = 0$$
$$F_\lambda = x^2 + 4y^2 - 3600 = 0$$

Solving the first equation of this system for λ, we obtain

$$\lambda = -\frac{2y}{x}$$

which, upon substitution into the second equation, yields

$$4x + 8\left(-\frac{2y}{x}\right)y = 0 \quad \text{or} \quad x^2 - 4y^2 = 0$$

—that is, $x = \pm 2y$. Substituting these values of x into the third equation of the system, we have

$$4y^2 + 4y^2 - 3600 = 0$$

or, upon solving $y = \pm\sqrt{450} = \pm 15\sqrt{2}$. The corresponding values of x are $\pm 30\sqrt{2}$. Because both x and y must be nonnegative, we have $x = 30\sqrt{2}$ and $y = 15\sqrt{2}$. Thus, the dimensions of the pool with maximum area are $30\sqrt{2}$ feet $\times$ $60\sqrt{2}$ feet, or approximately 42 feet $\times$ 85 feet. ■

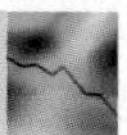

APPLIED EXAMPLE 6 Cobb–Douglas Production Function Suppose x units of labor and y units of capital are required to produce

$$f(x, y) = 100x^{3/4}y^{1/4}$$

units of a certain product (recall that this is a Cobb–Douglas production function). If each unit of labor costs \$200 and each unit of capital costs \$300 and a total of \$60,000 is available for production, determine how many units of labor and how many units of capital should be used in order to maximize production.

Solution The total cost of x units of labor at \$200 per unit and y units of capital at \$300 per unit is equal to $200x + 300y$ dollars. But \$60,000 is budgeted for production, so $200x + 300y = 60{,}000$, which we rewrite as

$$g(x, y) = 200x + 300y - 60{,}000 = 0$$

To maximize $f(x, y) = 100x^{3/4}y^{1/4}$ subject to the constraint $g(x, y) = 0$, we form the Lagrangian function

$$\begin{aligned} F(x, y, \lambda) &= f(x, y) + \lambda g(x, y) \\ &= 100x^{3/4}y^{1/4} + \lambda(200x + 300y - 60{,}000) \end{aligned}$$

To find the critical point(s) of F, we solve the following system of equations:

$$F_x = 75x^{-1/4}y^{1/4} + 200\lambda = 0$$
$$F_y = 25x^{3/4}y^{-3/4} + 300\lambda = 0$$
$$F_\lambda = 200x + 300y - 60{,}000 = 0$$

Solving the first equation for λ, we have

$$\lambda = -\frac{75x^{-1/4}y^{1/4}}{200} = -\frac{3}{8}\left(\frac{y}{x}\right)^{1/4}$$

which, when substituted into the second equation, yields

$$25\left(\frac{x}{y}\right)^{3/4} + 300\left(-\frac{3}{8}\right)\left(\frac{y}{x}\right)^{1/4} = 0$$

Multiplying the last equation by $\left(\frac{x}{y}\right)^{1/4}$ then gives

$$25\left(\frac{x}{y}\right) - \frac{900}{8} = 0$$

$$x = \left(\frac{900}{8}\right)\left(\frac{1}{25}\right)y = \frac{9}{2}y$$

Substituting this value of x into the third equation of the first system of equations, we have

$$200\left(\frac{9}{2}y\right) + 300y - 60,000 = 0$$

from which we deduce that $y = 50$. Hence, $x = 225$. Thus, maximum production is achieved when 225 units of labor and 50 units of capital are used. ■

When used in the context of Example 6, the negative of the Lagrange multiplier λ is called the **marginal productivity of money**. That is, if one additional dollar is available for production, then approximately $-\lambda$ units of a product can be produced. Here,

$$\lambda = -\frac{3}{8}\left(\frac{y}{x}\right)^{1/4} = -\frac{3}{8}\left(\frac{50}{225}\right)^{1/4} \approx -0.257$$

so, in this case, the marginal productivity of money is 0.257. For example, if \$65,000 is available for production instead of the originally budgeted figure of \$60,000, then the maximum production may be boosted from the original

$$f(225, 50) = 100(225)^{3/4}(50)^{1/4}$$

or 15,448 units, to

$$15,448 + 5000(0.257)$$

or 16,733 units.

8.5 Self-Check Exercises

1. Use the method of Lagrange multipliers to find the relative maximum of the function

$$f(x, y) = -2x^2 - y^2$$

subject to the constraint $3x + 4y = 12$.

2. The total monthly profit of Robertson Controls in manufacturing and selling x hundred of its standard mechanical setback thermostats and y hundred of its deluxe electronic setback thermostats each month is given by the total profit function

$$P(x, y) = -\frac{1}{8}x^2 - \frac{1}{2}y^2 - \frac{1}{4}xy + 13x + 40y - 280$$

where P is in hundreds of dollars. If the production of setback thermostats is to be restricted to a total of exactly 4000/month, how many of each model should Robertson manufacture in order to maximize its monthly profits? What is the maximum monthly profit?

Solutions to Self-Check Exercises 8.5 can be found on page 591.

8.5 Concept Questions

1. What is a constrained relative extremum of a function f?

2. Explain how the method of Lagrange multipliers is used to find the relative extrema of $f(x, y)$ subject to $g(x, y) = 0$.

8.5 Exercises

In Exercises 1–16, use the method of Lagrange multipliers to optimize the function subject to the given constraint.

1. Minimize the function $f(x, y) = x^2 + 3y^2$ subject to the constraint $x + y - 1 = 0$.

2. Minimize the function $f(x, y) = x^2 + y^2 - xy$ subject to the constraint $x + 2y - 14 = 0$.

3. Maximize the function $f(x, y) = 2x + 3y - x^2 - y^2$ subject to the constraint $x + 2y = 9$.

4. Maximize the function $f(x, y) = 16 - x^2 - y^2$ subject to the constraint $x + y - 6 = 0$.

5. Minimize the function $f(x, y) = x^2 + 4y^2$ subject to the constraint $xy = 1$.

6. Minimize the function $f(x, y) = xy$ subject to the constraint $x^2 + 4y^2 = 4$.

7. Maximize the function $f(x, y) = x + 5y - 2xy - x^2 - 2y^2$ subject to the constraint $2x + y = 4$.

8. Maximize the function $f(x, y) = xy$ subject to the constraint $2x + 3y - 6 = 0$.

9. Maximize the function $f(x, y) = xy^2$ subject to the constraint $9x^2 + y^2 = 9$.

10. Minimize the function $f(x, y) = \sqrt{y^2 - x^2}$ subject to the constraint $x + 2y - 5 = 0$.

11. Find the maximum and minimum values of the function $f(x, y) = xy$ subject to the constraint $x^2 + y^2 = 16$.

12. Find the maximum and minimum values of the function $f(x, y) = e^{xy}$ subject to the constraint $x^2 + y^2 = 8$.

13. Find the maximum and minimum values of the function $f(x, y) = xy^2$ subject to the constraint $x^2 + y^2 = 1$.

14. Maximize the function $f(x, y, z) = xyz$ subject to the constraint $2x + 2y + z = 84$.

15. Minimize the function $f(x, y, z) = x^2 + y^2 + z^2$ subject to the constraint $3x + 2y + z = 6$.

16. Find the maximum value of the function $f(x, y, z) = x + 2y - 3z$ subject to the constraint $z = 4x^2 + y^2$.

17. **Maximizing Profit** The total weekly profit (in dollars) realized by Country Workshop in manufacturing and selling its rolltop desks is given by the profit function

$$P(x, y) = -0.2x^2 - 0.25y^2 - 0.2xy + 100x + 90y - 4000$$

where x stands for the number of finished units and y denotes the number of unfinished units manufactured and sold each week. The company's management decides to restrict the manufacture of these desks to a total of exactly 200 units/week. How many finished and how many unfinished units should be manufactured each week to maximize the company's weekly profit?

18. **Maximizing Profit** The total daily profit (in dollars) realized by Weston Publishing in publishing and selling its dictionaries is given by the profit function

$$P(x, y) = -0.005x^2 - 0.003y^2 - 0.002xy + 14x + 12y - 200$$

where x stands for the number of deluxe editions and y denotes the number of standard editions sold daily. Weston's management decides that publication of these dictionaries should be restricted to a total of exactly 400 copies/day. How many deluxe copies and how many standard copies should be published each day to maximize Weston's daily profit?

19. **Minimizing Construction Costs** The management of UNICO Department Store decides to enclose an 800-ft² area outside their building to display potted plants. The enclosed area will be a rectangle, one side of which is provided by the external walls of the store. Two sides of the enclosure will be made of pine board, and the fourth side will be made of galvanized steel fencing material. If the pine board fencing costs \$6/running foot and the steel fencing costs \$3/running foot, determine the dimensions of the enclosure that will cost the least to erect.

20. Parcel Post Regulations Postal regulations specify that a parcel sent by parcel post may have a combined length and girth of no more than 108 in. Find the dimensions of the cylindrical package of greatest volume that may be sent through the mail. What is the volume of such a package?
Hint: The length plus the girth is $2\pi r + l$, and the volume is $\pi r^2 l$.

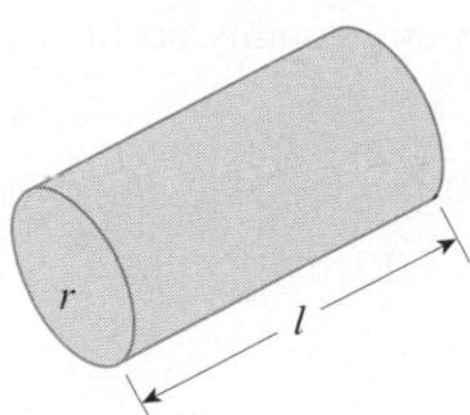

21. Minimizing Container Costs The Betty Moore Company requires that its corned beef hash containers have a capacity of 64 in.^{3}, be right circular cylinders, and be made of a tin alloy. Find the radius and height of the least expensive container that can be made if the metal for the side and bottom costs 4¢/in.^{2} and the metal for the pull-off lid costs \$2¢/in.^{2}.
Hint: Let the radius and height of the container be r and h in., respectively. Then, the volume of the container is $\pi r^2 h = 64$, and the cost is given by $C(r, h) = 8\pi rh + 6\pi r^2$.

22. Minimizing Construction Costs An open rectangular box is to be constructed from material that costs \$3/ft^2 for the bottom and \$1/ft^2 for its sides. Find the dimensions of the box of greatest volume that can be constructed for \$36.

23. Minimizing Construction Costs A closed rectangular box having a volume of 4 ft^3 is to be constructed. If the material for the sides costs \$1.00/ft^2 and the material for the top and bottom costs \$1.50/ft^2, find the dimensions of the box that can be constructed with minimum cost.

24. Maximizing Sales Ross–Simons Company has a monthly advertising budget of \$60,000. Their marketing department estimates that if they spend x dollars on newspaper advertising and y dollars on television advertising, then the monthly sales will be given by

$$z = f(x, y) = 90x^{1/4}y^{3/4}$$

dollars. Determine how much money Ross–Simons should spend on newspaper ads and on television ads each month to maximize its monthly sales.

25. Maximizing Production John Mills—the proprietor of Mills Engine Company, a manufacturer of model airplane engines—finds that it takes x units of labor and y units of capital to produce

$$f(x, y) = 100x^{3/4}y^{1/4}$$

units of the product. If a unit of labor costs \$100, a unit of capital costs \$200, and \$200,000 is budgeted for production, determine how many units should be expended on labor and how many units should be expended on capital in order to maximize production.

26. Use the method of Lagrange multipliers to solve Exercise 29, Exercises 8.3.

In Exercises 27 and 28, determine whether the statement is true or false. If it is true, explain why it is true. If it is false, give an example to show why it is false.

27. If (a, b) gives rise to a (constrained) relative extremum of f subject to the constraint $g(x, y) = 0$, then (a, b) also gives rise to the unconstrained relative extremum of f.

28. If (a, b) gives rise to a (constrained) relative extremum of f subject to the constraint $g(x, y) = 0$, then $f_x(a, b) = 0$ and $f_y(a, b) = 0$, simultaneously.

8.5 Solutions to Self-Check Exercises

1. Write the constraint equation in the form $g(x, y) = 3x + 4y - 12 = 0$. Then, the Lagrangian function is

$$F(x, y, \lambda) = -2x^2 - y^2 + \lambda(3x + 4y - 12)$$

To find the critical point(s) of F, we solve the system

$$F_x = -4x + 3\lambda = 0$$
$$F_y = -2y + 4\lambda = 0$$
$$F_\lambda = 3x + 4y - 12 = 0$$

Solving the first two equations for x and y in terms of λ, we find $x = \frac{3}{4}\lambda$ and $y = 2\lambda$. Substituting these values of x and y into the third equation of the system yields

$$3\left(\frac{3}{4}\lambda\right) + 4(2\lambda) - 12 = 0$$

or $\lambda = \frac{48}{41}$. Therefore, $x = \left(\frac{3}{4}\right)\left(\frac{48}{41}\right) = \frac{36}{41}$ and $y = 2\left(\frac{48}{41}\right) = \frac{96}{41}$, and we see that the point $\left(\frac{36}{41}, \frac{96}{41}\right)$ gives the constrained maximum of f. The maximum value is

$$f\left(\frac{36}{41}, \frac{96}{41}\right) = -2\left(\frac{36}{41}\right)^2 - \left(\frac{96}{41}\right)^2 = -\frac{11{,}808}{1681} = -\frac{288}{41}$$

2. We want to maximize

$$P(x, y) = -\frac{1}{8}x^2 - \frac{1}{2}y^2 - \frac{1}{4}xy + 13x + 40y - 280$$

subject to the constraint

$$g(x, y) = x + y - 40 = 0$$

The Lagrangian function is

$$F(x, y, \lambda) = P(x, y) + \lambda g(x, y)$$
$$= -\frac{1}{8}x^2 - \frac{1}{2}y^2 - \frac{1}{4}xy + 13x$$
$$+ 40y - 280 + \lambda(x + y - 40)$$

To find the critical points of F, solve the following system of equations:

$$F_x = -\frac{1}{4}x - \frac{1}{4}y + 13 + \lambda = 0$$
$$F_y = -\frac{1}{4}x - y + 40 + \lambda = 0$$
$$F_\lambda = x + y - 40 = 0$$

Subtracting the first equation from the second gives

$$-\frac{3}{4}y + 27 = 0 \quad \text{or} \quad y = 36$$

Substituting this value of y into the third equation yields $x = 4$. Therefore, to maximize its monthly profits, Robertson should manufacture 400 standard and 3600 deluxe thermostats. The maximum monthly profit is given by

$$P(4, 36) = -\frac{1}{8}(4)^2 - \frac{1}{2}(36)^2 - \frac{1}{4}(4)(36)$$
$$+ 13(4) + 40(36) - 280$$
$$= 526$$

or \$52,600.

8.6 Total Differentials (Optional)

Increments

Recall that if f is a function of one variable defined by $y = f(x)$, then the *increment* in y is defined to be

$$\Delta y = f(x + \Delta x) - f(x)$$

where Δx is an increment in x (Figure 30a). The increment of a function of two or more variables is defined in an analogous manner. For example, if z is a function of two variables defined by $z = f(x, y)$, then the **increment** in z is

$$\Delta z = f(x + \Delta x, y + \Delta y) - f(x, y) \qquad \textbf{(12)}$$

where Δx and Δy are the increments in the independent variables x and y, respectively. (See Figure 30b.)

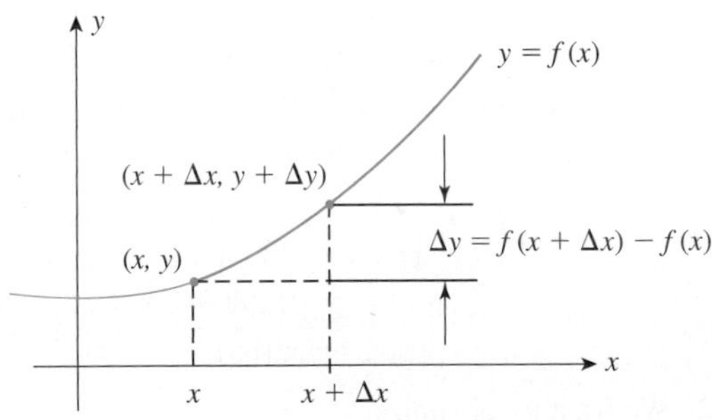

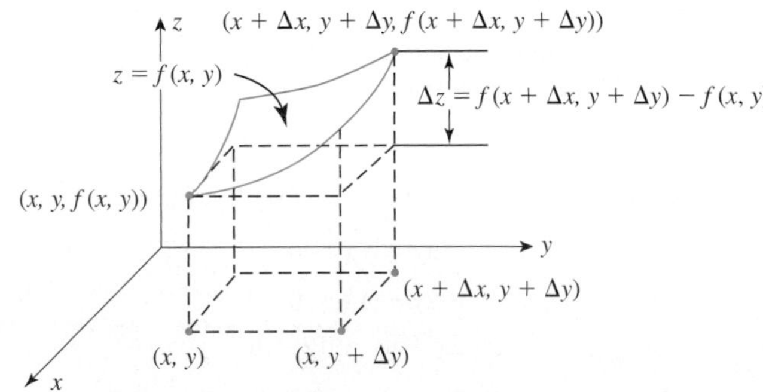

FIGURE 30 **a.** The increment Δy is the change in y as x changes from x to $x + \Delta x$.

b. The increment Δz is the change in z as x changes from x to $x + \Delta x$ and y changes from y to $y + \Delta y$.

EXAMPLE 1 Let $z = f(x, y) = 2x^2 - xy$. Find Δz. Then use your result to find the change in z if (x, y) changes from (1, 1) to (0.98, 1.03).

Solution Using (12), we obtain

$$\begin{aligned}\Delta z &= f(x + \Delta x, y + \Delta y) - f(x, y)\\ &= [2(x + \Delta x)^2 - (x + \Delta x)(y + \Delta y)] - (2x^2 - xy)\\ &= 2x^2 + 4x\Delta x + 2(\Delta x)^2 - xy - x\Delta y - y\Delta x - \Delta x\Delta y - 2x^2 + xy\\ &= (4x - y)\Delta x - x\Delta y + 2(\Delta x)^2 - \Delta x\Delta y\end{aligned}$$

Next, to find the increment in z if (x, y) changes from (1, 1) to (0.98, 1.03), we note that $\Delta x = 0.98 - 1 = -0.02$ and $\Delta y = 1.03 - 1 = 0.03$. Therefore, using the result obtained earlier with $x = 1$, $y = 1$, $\Delta x = -0.02$, and $\Delta y = 0.03$, we obtain

$$\begin{aligned}\Delta z &= [4(1) - 1](-0.02) - (1)(0.03) + 2(-0.02)^2 - (-0.02)(0.03)\\ &= -0.0886\end{aligned}$$

You can verify the correctness of this result by calculating the quantity $f(0.98, 1.03) - f(1, 1)$. ■

The Total Differential

Recall from Section 3.7 that if f is a function of one variable defined by $y = f(x)$, then the differential of f at x is defined by

$$dy = f'(x)\, dx$$

where $dx = \Delta x$ is the differential in x. Furthermore, we saw that

$$\Delta y \approx dy$$

if Δx is small (Figure 31a).

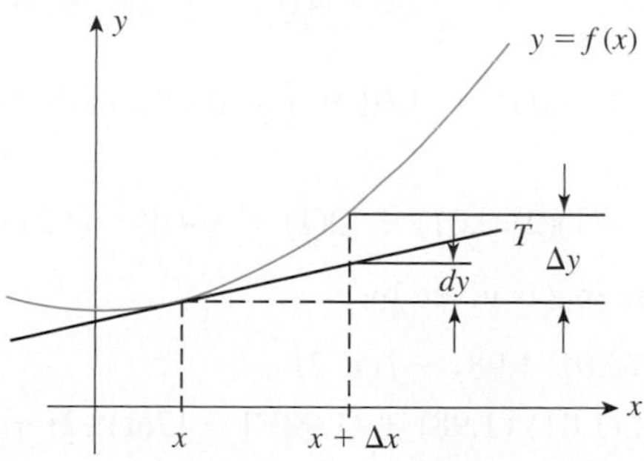

FIGURE 31

a. Relationship between dy and Δy

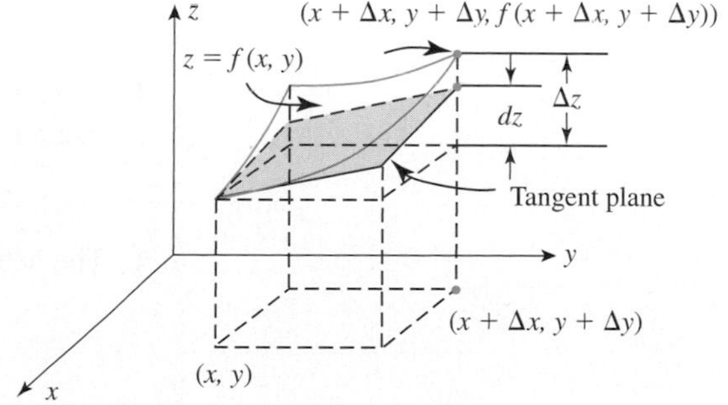

b. Relationship between dz and Δz. The tangent plane is the analog of tangent line T in the one-variable case.

The concept of the differential extends readily to a function of two or more variables.

Total Differential

Let $z = f(x, y)$ define a differentiable function of x and y.

1. The **differentials** of the independent variables x and y are $dx = \Delta x$ and $dy = \Delta y$.
2. The **differential** of the dependent variable z is

$$dz = \frac{\partial f}{\partial x}\,dx + \frac{\partial f}{\partial y}\,dy \tag{13}$$

Thus, analogous to the one-variable case, the total differential of z is a linear function of dx and dy. Furthermore, it provides us with an approximation of the exact change in z,

$$\Delta z = f(x + \Delta x, y + \Delta y) - f(x, y)$$

corresponding to a net change Δx in x from x to $x + \Delta x$ and a net change Δy in y from y to $y + \Delta y$; that is,

$$\Delta z \approx dz = \frac{\partial f}{\partial x}(x, y)\,dx + \frac{\partial f}{\partial y}(x, y)\,dy \tag{14}$$

provided $\Delta x = dx$ and $\Delta y = dy$ are sufficiently small.

EXAMPLE 2 Let $z = 2x^2y + y^3$.

a. Find the differential dz of z.
b. Find the approximate change in z when x changes from $x = 1$ to $x = 1.01$ and y changes from $y = 2$ to $y = 1.98$.
c. Find the actual change in z when x changes from $x = 1$ to $x = 1.01$ and y changes from $y = 2$ to $y = 1.98$. Compare the result with that obtained in part (b).

Solution

a. Let $f(x, y) = 2x^2y + y^3$. Then the required differential is

$$dz = \frac{\partial f}{\partial x}\,dx + \frac{\partial f}{\partial y}\,dy = 4xy\,dx + (2x^2 + 3y^2)\,dy$$

b. Here $x = 1$, $y = 2$, and $dx = 1.01 - 1 = 0.01$ and $dy = 1.98 - 2 = -0.02$. Therefore,

$$\Delta z \approx dz = 4(1)(2)(0.01) + [2(1) + 3(4)](-0.02) = -0.20$$

c. The actual change in z is given by

$$\begin{aligned}\Delta z &= f(1.01, 1.98) - f(1, 2)\\ &= [2(1.01)^2(1.98) + (1.98)^3] - [2(1)^2(2) + (2)^3]\\ &\approx 11.801988 - 12\\ &= -0.1980\end{aligned}$$

We see that $\Delta z \approx dz$, as expected. ■

VIDEO

APPLIED EXAMPLE 3 Approximating Changes in Revenue The weekly total revenue of Acrosonic Company resulting from the production and sales of x fully assembled bookshelf loudspeaker systems and y kit versions of the same loudspeaker system is

$$R(x, y) = -\frac{1}{4}x^2 - \frac{3}{8}y^2 - \frac{1}{4}xy + 300x + 240y$$

dollars. Determine the approximate change in Acrosonic's weekly total revenue when the level of production is increased from 200 assembled units and 60 kits per week to 206 assembled units and 64 kits per week.

Solution The approximate change in the weekly total revenue is given by the total differential R at $x = 200$ and $y = 60$, $dx = 206 - 200 = 6$ and $dy = 64 - 60 = 4$; that is, by

$$\begin{aligned} dR &= \frac{\partial R}{\partial x}dx + \frac{\partial R}{\partial y}dy \bigg|_{\substack{x=200,\, y=60 \\ dx=6,\, dy=4}} \\ &= \left(-\frac{1}{2}x - \frac{1}{4}y + 300\right)\bigg|_{(200,\, 60)} \cdot (6) \\ &\quad + \left(-\frac{3}{4}y - \frac{1}{4}x + 240\right)\bigg|_{(200,\, 60)} \cdot (4) \\ &= (-100 - 15 + 300)6 + (-45 - 50 + 240)4 \\ &= 1690 \end{aligned}$$

or \$1690. ■

APPLIED EXAMPLE 4 Cobb–Douglas Production Function The production for a certain country in the early years following World War II is described by the function

$$f(x, y) = 30x^{2/3}y^{1/3}$$

units, when x units of labor and y units of capital were utilized. Find the approximate change in output if the amount expended on labor had been decreased from 125 units to 123 units and the amount expended on capital had been increased from 27 to 29 units. Is your result as expected given the result of Example 4c, Section 8.2?

Solution The approximate change in output is given by the total differential of f at $x = 125$, $y = 27$, $dx = 123 - 125 = -2$, and $dy = 29 - 27 = 2$; that is, by

$$\begin{aligned} df &= \frac{\partial f}{\partial x}dx + \frac{\partial f}{\partial y}dy \bigg|_{\substack{x=125,\, y=27 \\ dx=-2,\, dy=2}} \\ &= 20x^{-1/3}y^{1/3}\bigg|_{(125,\, 27)} \cdot (-2) + 10x^{2/3}y^{-2/3}\bigg|_{(125,\, 27)} \cdot (2) \\ &= 20\left(\frac{27}{125}\right)^{1/3}(-2) + 10\left(\frac{125}{27}\right)^{2/3}(2) \\ &= -20\left(\frac{3}{5}\right)(2) + 10\left(\frac{25}{9}\right)(2) = \frac{284}{9} \end{aligned}$$

or $31\frac{5}{9}$ units. This result is fully compatible with the result of Example 4, where the recommendation was to encourage increased spending on capital rather than on labor. ■

If f is a function of the three variables x, y, and z, then the total differential of $w = f(x, y, z)$ is defined to be

$$dw = \frac{\partial f}{\partial x}\,dx + \frac{\partial f}{\partial y}\,dy + \frac{\partial f}{\partial z}\,dz$$

where $dx = \Delta x$, $dy = \Delta y$, and $dz = \Delta z$ are the actual changes in the independent variables x, y, and z as x changes from $x = a$ to $x = a + \Delta x$, y changes from $y = b$ to $y = b + \Delta y$, and z changes from $z = c$ to $z = c + \Delta z$, respectively.

APPLIED EXAMPLE 5 Error Analysis Find the maximum percentage error in calculating the volume of a rectangular box if an error of at most 1% is made in measuring the length, width, and height of the box.

Solution Let x, y, and z denote the length, width, and height, respectively, of the rectangular box. Then the volume of the box is given by $V = f(x, y, z) = xyz$ cubic units. Now suppose the true dimensions of the rectangular box are a, b, and c units, respectively. Since the error committed in measuring the length, width, and height of the box is at most 1%, we have

$$\begin{aligned} |\Delta x| &= |x - a| \leq 0.01a \\ |\Delta y| &= |y - b| \leq 0.01b \\ |\Delta z| &= |z - c| \leq 0.01c \end{aligned}$$

Therefore, the maximum error in calculating the volume of the box is

$$\begin{aligned} |\Delta V| \approx |dV| &= \left|\frac{\partial f}{\partial x}\,dx + \frac{\partial f}{\partial y}\,dy + \frac{\partial f}{\partial z}\,dz\right|\Bigg|_{x=a,\,y=b,\,z=c} \\ &= |yz\,dx + xz\,dy + xy\,dz|\,\Big|_{x=a,\,y=b,\,z=c} \\ &= |bc\,dx + ac\,dy + ab\,dz| \\ &\leq bc|dx| + ac|dy| + ab|dz| \\ &\leq bc(0.01a) + ac(0.01b) + ab(0.01c) \\ &= (0.03)abc \end{aligned}$$

Since the actual volume of the box is abc cubic units, we see that the maximum percentage error in calculating its volume is

$$\frac{|\Delta V|}{V|_{(a,b,c)}} \approx \frac{(0.03)abc}{abc} = 0.03$$

—that is, approximately 3%. ■

EXPLORE & DISCUSS

Refer to Example 5, where we found the maximum percentage error in calculating the volume of the rectangular box to be *approximately* 3%. What is the precise maximum percentage error?

8.6 Self-Check Exercise

Let f be a function defined by $z = f(x, y) = 3xy^2 - 4y$. Find the total differential of f at $(-1, 3)$. Then find the approximate change in z when x changes from $x = -1$ to $x = -0.98$ and y changes from $y = 3$ to $y = 3.01$.

The solution to Self-Check Exercise 8.6 can be found on page 598.

8.6 Concept Questions

1. If $z = f(x, y)$, what is the differential of x? The differential of y? What is the total differential of z?
2. Let $z = f(x, y)$. What is the relationship between the actual change, Δz, when x changes from x to $x + \Delta x$ and y changes from y to $y + \Delta y$, and the total differential, dz, of f at (x, y)?

8.6 Exercises

In Exercises 1–18, find the total differential of the function.

1. $f(x, y) = x^2 + 2y$
2. $f(x, y) = 2x^2 + 3y^2$
3. $f(x, y) = 2x^2 - 3xy + 4x$
4. $f(x, y) = xy^3 - x^2y^2$
5. $f(x, y) = \sqrt{x^2 + y^2}$
6. $f(x, y) = (x + 3y^2)^{1/3}$
7. $f(x, y) = \dfrac{5y}{x - y}$
8. $f(x, y) = \dfrac{x + y}{x - y}$
9. $f(x, y) = 2x^5 - ye^{-3x}$
10. $f(x, y) = xye^{x+y}$
11. $f(x, y) = x^2e^y + y \ln x$
12. $f(x, y) = \ln(x^2 + y^2)$
13. $f(x, y, z) = xy^2z^3$
14. $f(x, y, z) = x\sqrt{y} + y\sqrt{z}$
15. $f(x, y, z) = \dfrac{x}{y + z}$
16. $f(x, y, z) = \dfrac{x + y}{y + z}$
17. $f(x, y, z) = xyz + xe^{yz}$
18. $f(x, y, z) = \sqrt{e^x + e^y + ze^{xy}}$

In Exercises 19–30, find the approximate change in z when the point (x, y) changes from (x_0, y_0) to (x_1, y_1).

19. $f(x, y) = 4x^2 - xy$; from (1, 2) to (1.01, 2.02)
20. $f(x, y) = 2x^2 - 2x^3y^2 - y^3$; from (−1, 2) to (−0.98, 2.01)
21. $f(x, y) = x^{2/3}y^{1/2}$; from (8, 9) to (7.97, 9.03)
22. $f(x, y) = \sqrt{x^2 + y^2}$; from (1, 3) to (1.03, 3.03)
23. $f(x, y) = \dfrac{x}{x - y}$; from (−3, −2) to (−3.02, −1.98)
24. $f(x, y) = \dfrac{x - y}{x + y}$; from (−3, −2) to (−3.02, −1.98)
25. $f(x, y) = 2xe^{-y}$; from (4, 0) to (4.03, 0.03)
26. $f(x, y) = \sqrt{x}e^y$; from (1, 1) to (1.01, 0.98)
27. $f(x, y) = xe^{xy} - y^2$; from (−1, 0) to (−0.97, 0.03)
28. $f(x, y) = xe^{-y} + ye^{-x}$; from (1, 1) to (1.01, 0.90)
29. $f(x, y) = x \ln x + y \ln x$; from (2, 3) to (1.98, 2.89)
30. $f(x, y) = \ln(xy)^{1/2}$; from (5, 10) to (5.05, 9.95)
31. Effect of Inventory and Floor Space on Profit The monthly profit (in dollars) of Bond and Barker Department Store depends on the level of inventory x (in thousands of dollars) and the floor space y (in thousands of square feet) available for display of the merchandise, as given by the equation

$$P(x, y) = -0.02x^2 - 15y^2 + xy + 39x + 25y - 20{,}000$$

Currently, the level of inventory is \$4,000,000 ($x = 4000$), and the floor space is 150,000 square feet ($y = 150$). Find the anticipated change in monthly profit if management increases the level of inventory by \$500,000 and decreases the floor space for display of merchandise by 10,000 square feet.

32. Effect of Production on Profit The Country Workshop's total weekly profit (in dollars) realized in manufacturing and selling its rolltop desks is given by

$$P(x, y) = -0.2x^2 - 0.25y^2 - 0.2xy + 100x + 90y - 4000$$

where x stands for the number of finished units and y denotes the number of unfinished units manufactured and sold per week. Currently, the weekly output is 190 finished and 105 unfinished units. Determine the approximate change in the total weekly profit if the sole proprietor of the Country Workshop decides to increase the number of finished units to 200 per week and decrease the number of unfinished units to 100 per week.

33. **Revenue of a Travel Agency** The Odyssey Travel Agency's monthly revenue (in thousands of dollars) depends on the amount of money x (in thousands) spent on advertising per month and the number of agents y in its employ in accordance with the rule

$$R(x, y) = -x^2 - 0.5y^2 + xy + 8x + 3y + 20$$

Currently, the amount of money spent on advertising is \$10,000 per month, and there are 15 agents in the agency's employ. Estimate the change in revenue resulting from an increase of \$1000 per month in advertising expenditure and a decrease of 1 agent.

34. **Effect of Capital and Labor on Productivity** The production of a South American country is given by the function

$$f(x, y) = 20x^{3/4}y^{1/4}$$

when x units of labor and y units of capital are utilized. Find the approximate change in output if the amount expended on labor is decreased from 256 to 254 units and the amount expended on capital is increased from 16 to 18 units.

35. **Price–Earnings Ratio** The price–earnings ratio (PE ratio) of a stock is given by

$$R(x, y) = \frac{x}{y}$$

where x denotes the price per share of the stock and y denotes the earnings per share. Estimate the change in the PE ratio of a stock if its price increases from \$60/share to \$62/share while its earnings decrease from \$4/share to \$3.80/share.

36. **Error in Calculating the Surface Area of a Human** The formula

$$S = 0.007184W^{0.425}H^{0.725}$$

gives the surface area S of a human body (in square meters) in terms of its weight W in kilograms and its height H in centimeters. If an error of 1% is made in measuring the weight of a person and an error of 2% is made in measuring the height, what is the percentage error in the measurement of the person's surface area?

37. **Error in Calculating the Volume of a Cylinder** The radius and height of a right circular cylinder are measured with a maximum error of 0.1 cm in each measurement. Approximate the maximum error in calculating the volume of the cylinder if the measured dimensions $r = 8$ cm and $h = 20$ cm are used.

38. **Effect of Capital and Labor on Productivity** The production of a certain company is given by the function

$$f(x, y) = 50x^{1/3}y^{2/3}$$

when x units of labor and y units of capital are utilized. Find the approximate percentage change in the production of the company if labor is increased by 2% and capital is increased by 1%.

39. **Error in Calculating Total Resistance** The total resistance R of three resistors with resistance R_1, R_2, and R_3, connected in parallel, is given by the relationship

$$\frac{1}{R} = \frac{1}{R_1} + \frac{1}{R_2} + \frac{1}{R_3}$$

If R_1, R_2, and R_3 are measured at 100, 200, and 300 ohms, respectively, with a maximum error of 1% in each measurement, find the approximate maximum error in the calculated value of R.

40. **Error in Measuring Arterial Blood Flow** The flow of blood through an arteriole in cubic centimeters per second is given by

$$V = \frac{\pi p r^4}{8kl}$$

where l is the length (in cm) of the arteriole, r is its radius (in cm), p is the difference in pressure between the two ends of the arteriole (in dyne/cm^2), and k is the viscosity of blood (in dyne-sec/cm^2). Find the approximate percentage change in the flow of blood if an error of 2% is made in measuring the length of the arteriole and an error of 1% is made in measuring its radius. Assume that p and k are constant.

8.6 Solution to Self-Check Exercise

We find

$$\frac{\partial f}{\partial x} = 3y^2 \quad \text{and} \quad \frac{\partial f}{\partial y} = 6xy - 4$$

so that

$$\frac{\partial f}{\partial x}(-1, 3) = 3(3)^2 = 27$$

and

$$\frac{\partial f}{\partial y}(-1, 3) = 6(-1)(3) - 4 = -22$$

Therefore, the total differential is

$$\begin{aligned} dz &= \frac{\partial f}{\partial x}(-1, 3)\,dx + \frac{\partial f}{\partial y}(-1, 3)\,dy \\ &= 27\,dx - 22\,dy \end{aligned}$$

Now $dx = -0.98 - (-1) = 0.02$ and $dy = 3.01 - 3 = 0.01$, so the approximate change in z is

$$dz = 27(0.02) - 22(0.01) = 0.32$$

8.7 Double Integrals

A Geometric Interpretation of the Double Integral

To introduce the notion of the integral of a function of two variables, let's first recall the definition of the definite integral of a continuous function of one variable $y = f(x)$ over the interval $[a, b]$. We first divide the interval $[a, b]$ into n subintervals, each of equal length, by the points $x_0 = a < x_1 < x_2 < \cdots < x_n = b$ and define the **Reimann sum** by

$$S_n = f(p_1)h + f(p_2)h + \cdots + f(p_n)h$$

where $h = (b - a)/n$ and p_i is an arbitrary point in the interval $[x_{i-1}, x_i]$. The definite integral of f over $[a, b]$ is defined as the limit of the Riemann sum S_n as n tends to infinity, whenever it exists. Furthermore, recall that when f is a nonnegative continuous function on $[a, b]$, then the ith term of the Riemann sum, $f(p_i)h$, is an approximation (by the area of a rectangle) of the area under that part of the graph of $y = f(x)$ between $x = x_{i-1}$ and $x = x_i$, so that the Riemann sum S_n provides us with an approximation of the area under the curve $y = f(x)$ from $x = a$ to $x = b$. The integral

$$\int_a^b f(x)\,dx = \lim_{n\to\infty} S_n$$

gives the *actual* area under the curve from $x = a$ to $x = b$.

Now suppose $f(x, y)$ is a continuous function of two variables defined over a region R. For simplicity, we assume for the moment that R is a rectangular region in the plane (Figure 32). Let's construct a Riemann sum for this function over the rectangle R by following a procedure that parallels the case for a function of one variable over an interval I. We begin by observing that the analogue of a *partition* in the two-dimensional case is a rectangular **grid** composed of mn rectangles, each of length h and width k, as a result of partitioning the side of the rectangle R of length $(b - a)$ into m segments and the side of length $(d - c)$ into n segments. By construction

$$h = \frac{b - a}{m} \quad \text{and} \quad k = \frac{d - c}{n}$$

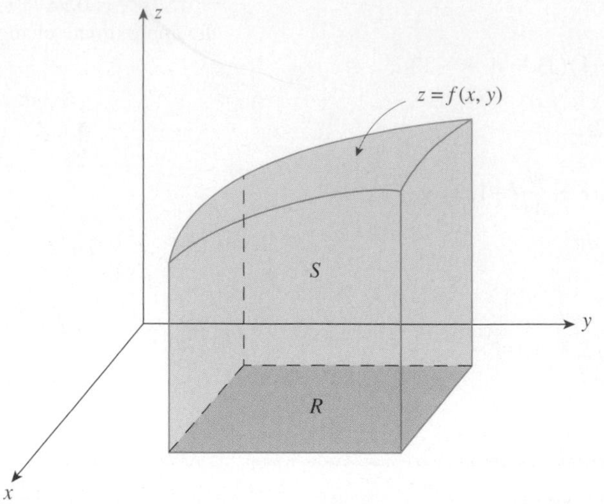

FIGURE 32
$f(x, y)$ is a function defined over a rectangular region R.

A sample grid with $m = 5$ and $n = 4$ is shown in Figure 33.

Let's label the rectangles $R_1, R_2, R_3, \ldots, R_{mn}$. If (x_i, y_i) is *any* point in R_i $(1 \le i \le mn)$, then the **Riemann sum of $f(x, y)$ over the region R** is defined as

$$S(m, n) = f(x_1, y_1)hk + f(x_2, y_2)hk + \cdots + f(x_{mn}, y_{mn})hk$$

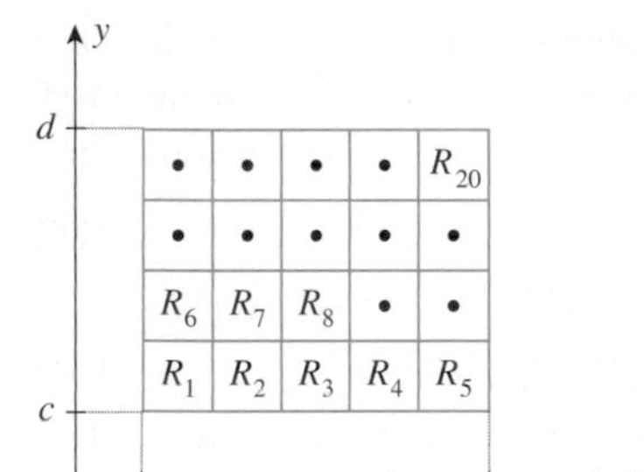

FIGURE 33
Grid with $m = 5$ and $n = 4$

If the limit of $S(m, n)$ exists as both m and n tend to infinity, we call this limit the value of the **double integral of $f(x, y)$ over the region R** and denote it by

$$\iint_R f(x, y)\, dA$$

EXPLORE & DISCUSS

Using a geometric interpretation, evaluate

$$\iint_R \sqrt{4 - x^2 - y^2}\, dA$$

where $R = \{(x, y) | x^2 + y^2 \le 4\}$.

If $f(x, y)$ is a nonnegative function, then it defines a solid S bounded above by the graph of f and below by the rectangular region R. Furthermore, the solid S is the union of the mn solids bounded above by the graph of f and below by the mn rectangular regions corresponding to the partition of R (Figure 34). The volume of a

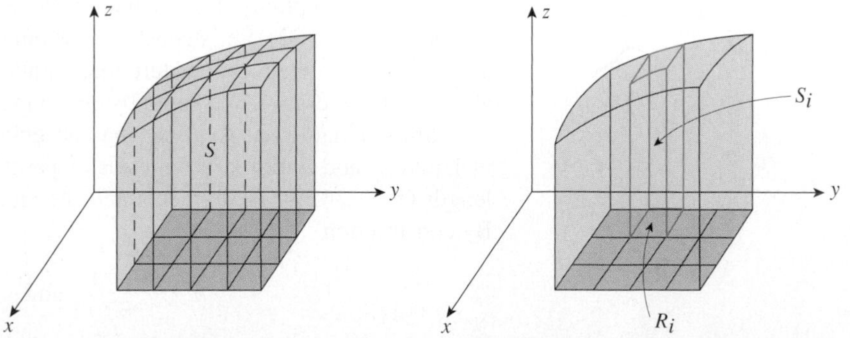

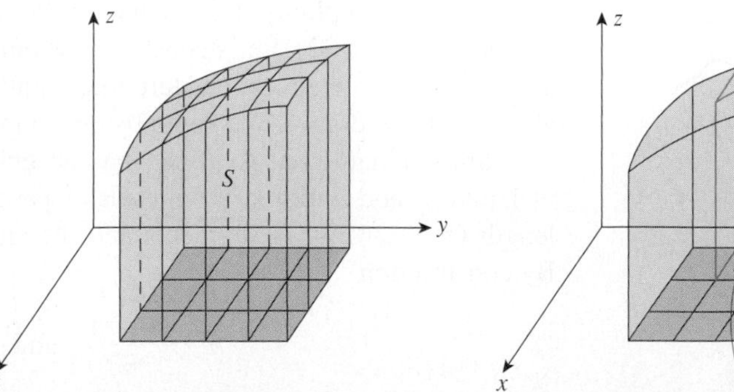

FIGURE 34

(a) The solid S is the union of mn solids (shown here with $m = 3$ and $n = 4$).

(b) A typical solid S_i is bounded above by the graph of f and lies above R_i.

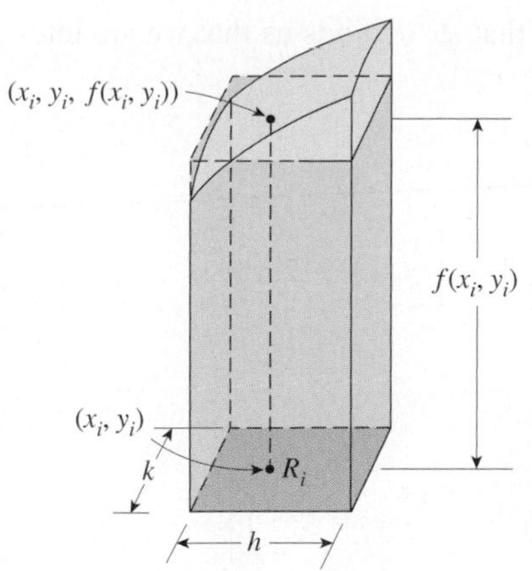

FIGURE 35
The volume of S_i is approximated by the parallelepiped with base R_i and height $f(x_i, y_i)$.

typical solid S_i can be approximated by a parallelepiped with base R_i and height $f(x_i, y_i)$ (Figure 35).

Therefore, the Riemann sum $S(m, n)$ gives us an approximation of the volume of the solid bounded above by the surface $z = f(x, y)$ and below by the plane region R. As both m and n tend to infinity, the Riemann sum $S(m, n)$ approaches the *actual* volume under the solid.

Evaluating a Double Integral over a Rectangular Region

Let's turn our attention to the evaluation of the double integral

$$\iint_R f(x, y)\, dA$$

where R is the rectangular region shown in Figure 32. As in the case of the definite integral of a function of one variable, it turns out that the double integral can be evaluated without our having to first find an appropriate Riemann sum and then take the limit of that sum. Instead, as we will now see, the technique calls for evaluating two single integrals—the so-called **iterated integrals**—in succession, using a process that might be called "antipartial differentiation." The technique is described in the following result, which we state without proof.

Let R be the rectangle defined by the inequalities $a \leq x \leq b$ and $c \leq y \leq d$ (see Figure 33). Then,

$$\iint_R f(x, y)\, dA = \int_c^d \left[\int_a^b f(x, y)\, dx \right] dy \tag{15}$$

where the iterated integrals on the right-hand side are evaluated as follows. We first compute the integral

$$\int_a^b f(x, y)\, dx$$

by treating y as if it were a constant and integrating the resulting function of x with respect to x (dx reminds us that we are integrating with respect to x). In this manner we obtain a value for the integral that may contain the variable y. Thus,

$$\int_a^b f(x, y)\, dx = g(y)$$

for some function g. Substituting this value into Equation (15) gives

$$\int_c^d g(y)\, dy$$

which may be integrated in the usual manner.

EXAMPLE 1 Evaluate $\iint_R f(x, y)\, dA$, where $f(x, y) = x + 2y$ and R is the rectangle defined by $1 \leq x \leq 4$ and $1 \leq y \leq 2$.

Solution Using Equation (15), we find

$$\iint_R f(x, y)\, dA = \int_1^2 \left[\int_1^4 (x + 2y)\, dx \right] dy$$

To compute

$$\int_1^4 (x + 2y)\, dx$$

we treat y as if it were a constant (remember that dx reminds us that we are integrating with respect to x). We obtain

$$\begin{aligned}\int_1^4 (x + 2y)\,dx &= \frac{1}{2}x^2 + 2xy\Big|_{x=1}^{x=4}\\ &= \left[\frac{1}{2}(16) + 2(4)y\right] - \left[\frac{1}{2}(1) + 2(1)y\right]\\ &= \frac{15}{2} + 6y\end{aligned}$$

Thus,

$$\begin{aligned}\iint_R f(x, y)\,dA &= \int_1^2 \left(\frac{15}{2} + 6y\right) dy = \left(\frac{15}{2}y + 3y^2\right)\Big|_1^2\\ &= (15 + 12) - \left(\frac{15}{2} + 3\right) = 16\tfrac{1}{2}\end{aligned}$$

Evaluating a Double Integral over a Plane Region

Up to now, we have assumed that the region over which a double integral is to be evaluated is rectangular. In fact, however, it is possible to compute the double integral of functions over rather arbitrary regions. The next theorem, which we state without proof, expands the number of types of regions over which we may integrate.

THEOREM 1

a. Suppose $g_1(x)$ and $g_2(x)$ are continuous functions on $[a, b]$ and the region R is defined by $R = \{(x, y) | g_1(x) \le y \le g_2(x);\ a \le x \le b\}$. Then,

$$\iint_R f(x, y)\,dA = \int_a^b \left[\int_{g_1(x)}^{g_2(x)} f(x, y)\,dy\right] dx \qquad \textbf{(16)}$$

(Figure 36a).

b. Suppose $h_1(y)$ and $h_2(y)$ are continuous functions on $[c, d]$ and the region R is defined by $R = \{(x, y) | h_1(y) \le x \le h_2(y);\ c \le y \le d\}$. Then,

$$\iint_R f(x, y)\,dA = \int_c^d \left[\int_{h_1(y)}^{h_2(y)} f(x, y)\,dx\right] dy \qquad \textbf{(17)}$$

(Figure 36b).

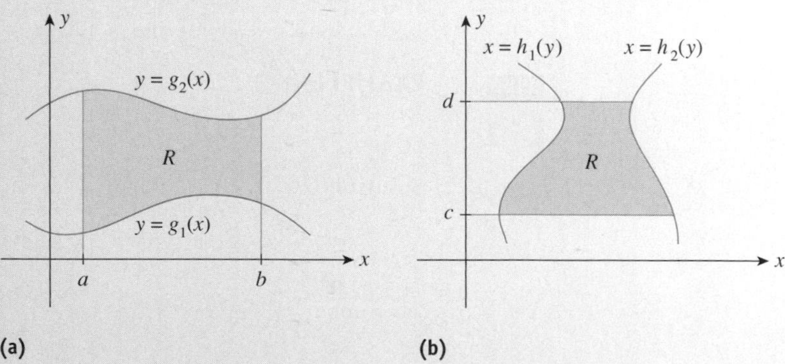

FIGURE 36

Notes

1. Observe that in (16) the lower and upper limits of integration with respect to y are given by $y = g_1(x)$ and $y = g_2(x)$. This is to be expected since, for a fixed value of x lying between $x = a$ and $x = b$, y runs between the lower curve defined by $y = g_1(x)$ and the upper curve defined by $y = g_2(x)$ (see Figure 36a). Observe, too, that in the special case when $g_1(x) = c$ and $g_2(x) = d$, the region R is rectangular, and (16) reduces to (15).
2. For a fixed value of y, x runs between $x = h_1(y)$ and $x = h_2(y)$, giving the indicated limits of integration with respect to x in (17) (see Figure 36b).
3. Note that the two curves in Figure 36b are not graphs of functions of x (use the vertical-line test), but they are graphs of functions of y. It is this observation that justifies the approach leading to (17). ■

We now look at several examples.

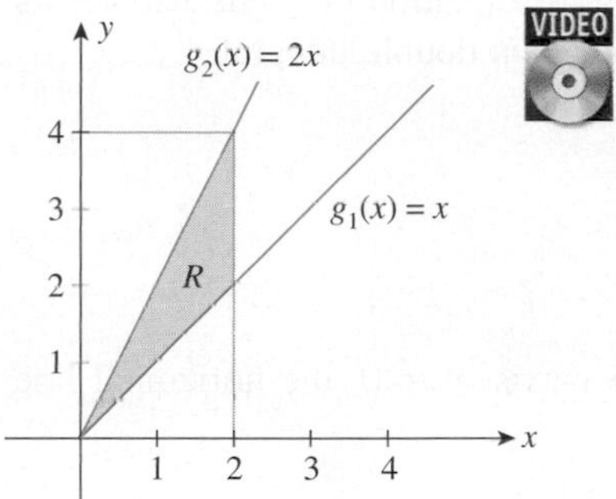

FIGURE 37
R is the region bounded by $g_1(x) = x$ and $g_2(x) = 2x$ for $0 \le x \le 2$.

VIDEO

EXAMPLE 2 Evaluate $\iint_R f(x, y)\, dA$ given that $f(x, y) = x^2 + y^2$ and R is the region bounded by the graphs of $g_1(x) = x$ and $g_2(x) = 2x$ for $0 \le x \le 2$.

Solution The region under consideration is shown in Figure 37. Using Equation (16), we find

$$\begin{aligned}\iint_R f(x, y)\, dA &= \int_0^2 \left[\int_x^{2x} (x^2 + y^2)\, dy\right] dx \\ &= \int_0^2 \left[\left(x^2y + \frac{1}{3}y^3\right)\Big|_x^{2x}\right] dx \\ &= \int_0^2 \left[\left(2x^3 + \frac{8}{3}x^3\right) - \left(x^3 + \frac{1}{3}x^3\right)\right] dx \\ &= \int_0^2 \frac{10}{3}x^3\, dx = \frac{5}{6}x^4\Big|_0^2 = 13\tfrac{1}{3}\end{aligned}$$

■

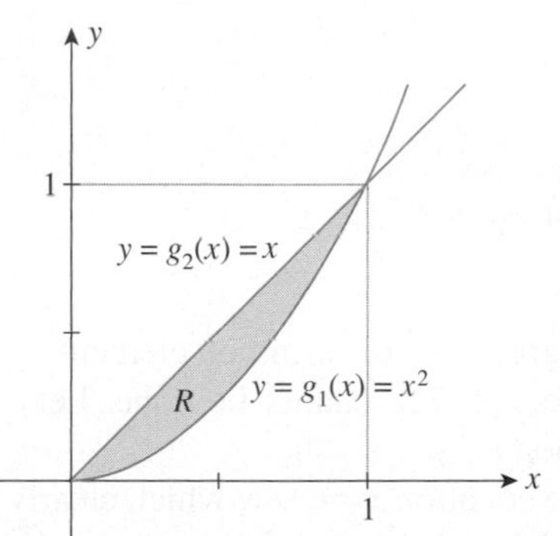

FIGURE 38
R is the region bounded by $y = x^2$ and $y = x$.

EXAMPLE 3 Evaluate $\iint_R f(x, y)\, dA$, where $f(x, y) = xe^y$ and R is the plane region bounded by the graphs of $y = x^2$ and $y = x$.

Solution The region in question is shown in Figure 38. The point of intersection of the two curves is found by solving the equation $x^2 = x$, giving $x = 0$ and $x = 1$. Using Equation (16), we find

$$\begin{aligned}\iint_R f(x, y)\, dA &= \int_0^1 \left[\int_{x^2}^{x} xe^y\, dy\right] dx = \int_0^1 \left[xe^y\Big|_{x^2}^{x}\right] dx \\ &= \int_0^1 (xe^x - xe^{x^2})\, dx = \int_0^1 xe^x\, dx - \int_0^1 xe^{x^2}\, dx\end{aligned}$$

and integrating the first integral on the right-hand side by parts,

$$\begin{aligned}&= \left[(x - 1)e^x - \frac{1}{2}e^{x^2}\right]\Big|_0^1 \\ &= -\frac{1}{2}e - \left(-1 - \frac{1}{2}\right) = \frac{1}{2}(3 - e)\end{aligned}$$

■

EXPLORE & DISCUSS

Refer to Example 3.

1. You can also view the region R as an example of the region shown in Figure 36b. Doing so, find the functions h_1 and h_2 and the numbers c and d.
2. Find an expression for $\iint_R f(x, y)\, dA$ in terms of iterated integrals using Formula (17).
3. Evaluate the iterated integrals of part 2 and hence verify the result of Example 3. Hint: Integrate by parts twice.
4. Does viewing the region R in two different ways make a difference?

The next example not only illustrates the use of Equation (17) but also shows that it may be the only viable way to evaluate the given double integral.

EXAMPLE 4 Evaluate

$$\iint_R xe^{y^2}\, dA$$

where R is the plane region bounded by the y-axis, $x = 0$, the horizontal line $y = 4$, and the graph of $y = x^2$.

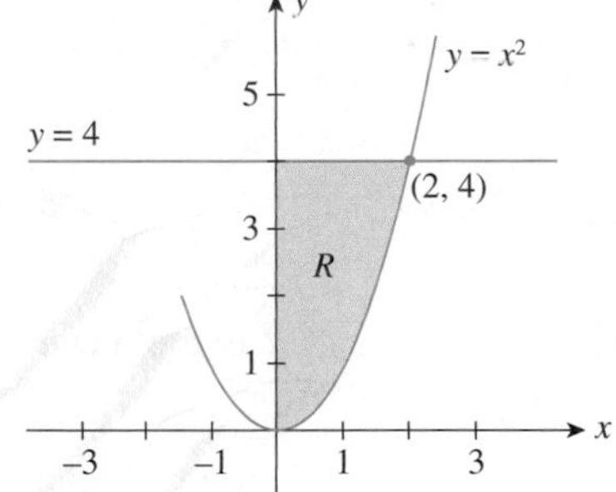

FIGURE 39
R is the region bounded by the y-axis, $x = 0$, $y = 4$, and $y = x^2$.

Solution The region R is shown in Figure 39. The point of intersection of the line $y = 4$ and the graph of $y = x^2$ is found by solving the equation $x^2 = 4$, giving $x = 2$ and the required point (2, 4). Using Equation (16) with $y = g_1(x) = x^2$ and $y = g_2(x) = 4$ leads to

$$\iint_R xe^{y^2}\, dA = \int_0^2 \left[\int_{x^2}^4 xe^{y^2}\, dy\right] dx$$

Now evaluation of the integral

$$\int_{x^2}^4 xe^{y^2}\, dy = x\int_{x^2}^4 e^{y^2}\, dy$$

calls for finding the antiderivative of the integrand e^{y^2} in terms of elementary functions, a task that, as was pointed out in Section 7.3, cannot be done. Let's begin afresh and attempt to make use of Equation (17).

Since the equation $y = x^2$ is equivalent to the equation $x = \sqrt{y}$, which clearly expresses x as a function of y, we may write, with $x = h_1(y) = 0$ and $h_2(y) = \sqrt{y}$,

$$\iint_R xe^{y^2}\, dA = \int_0^4 \left[\int_0^{\sqrt{y}} xe^{y^2}\, dx\right] dy = \int_0^4 \left[\frac{1}{2}x^2e^{y^2}\Big|_0^{\sqrt{y}}\right] dy$$

$$= \int_0^4 \frac{1}{2}ye^{y^2}\, dy = \frac{1}{4}e^{y^2}\Big|_0^4 = \frac{1}{4}(e^{16} - 1)$$

■

8.7 Self-Check Exercise

Evaluate $\iint_R (x + y)\, dA$, where R is the region bounded by the graphs of $g_1(x) = x$ and $g_2(x) = x^{1/3}$.

The solution to Self-Check Exercise 8.7 can be found on page 606.

8.7 Concept Questions

1. What is an iterated integral? How is $\iint_R f(x, y)\, dA$ evaluated in terms of iterated integrals, where R is the rectangular region defined by $a \leq x \leq b$ and $c \leq y \leq d$?

2. Suppose g_1 and g_2 are continuous on the interval $[a, b]$ and $R = \{(x, y) | g_1(x) \leq y \leq g_2(x), a \leq x \leq b\}$, what is $\iint_R f(x, y)\, dA$, where f is a continuous function defined on R?

3. Suppose h_1 and h_2 are continuous on the interval $[c, d]$ and $R = \{(x, y) | h_1(y) \leq x \leq h_2(y), c \leq y \leq d\}$, what is $\iint_R f(x, y)\, dA$, where f is a continuous function defined on R?

8.7 Exercises

In Exercises 1–25, evaluate the double integral $\iint_R f(x, y)\, dA$ for the function $f(x, y)$ and the region R.

1. $f(x, y) = y + 2x$; R is the rectangle defined by $1 \leq x \leq 2$ and $0 \leq y \leq 1$.

2. $f(x, y) = x + 2y$; R is the rectangle defined by $-1 \leq x \leq 2$ and $0 \leq y \leq 2$.

3. $f(x, y) = xy^2$; R is the rectangle defined by $-1 \leq x \leq 1$ and $0 \leq y \leq 1$.

4. $f(x, y) = 12xy^2 + 8y^3$; R is the rectangle defined by $0 \leq x \leq 1$ and $0 \leq y \leq 2$.

5. $f(x, y) = \dfrac{x}{y}$; R is the rectangle defined by $-1 \leq x \leq 2$ and $1 \leq y \leq e^3$.

6. $f(x, y) = \dfrac{xy}{1 + y^2}$; R is the rectangle defined by $-2 \leq x \leq 2$ and $0 \leq y \leq 1$.

7. $f(x, y) = 4xe^{2x^2+y}$; R is the rectangle defined by $0 \leq x \leq 1$ and $-2 \leq y \leq 0$.

8. $f(x, y) = \dfrac{y}{x^2} e^{y/x}$; R is the rectangle defined by $1 \leq x \leq 2$ and $0 \leq y \leq 1$.

9. $f(x, y) = \ln y$; R is the rectangle defined by $0 \leq x \leq 1$ and $1 \leq y \leq e$.

10. $f(x, y) = \dfrac{\ln y}{x}$; R is the rectangle defined by $1 \leq x \leq e^2$ and $1 \leq y \leq e$.

11. $f(x, y) = x + 2y$; R is bounded by $x = 0$, $x = 1$, $y = 0$, and $y = x$.

12. $f(x, y) = xy$; R is bounded by $x = 0$, $x = 1$, $y = 0$ and $y = x$.

13. $f(x, y) = 2x + 4y$; R is bounded by $x = 1$, $x = 3$, $y = 0$, and $y = x + 1$.

14. $f(x, y) = 2 - y$; R is bounded by $x = -1$, $x = 1 - y$, $y = 0$, and $y = 2$.

15. $f(x, y) = x + y$; R is bounded by $x = 0$, $x = \sqrt{y}$, $y = 0$, and $y = 4$.

16. $f(x, y) = x^2y^2$; R is bounded by $x = 0$, $x = 1$, $y = x^2$, and $y = x^3$.

17. $f(x, y) = y$; R is bounded by $x = 0, x = \sqrt{4 - y^2}, y = 0$, and $y = 2$.

18. $f(x, y) = \dfrac{y}{x^3 + 2}$; R is bounded by $x = 0, x = 1, y = 0$, and $y = x$.

19. $f(x, y) = 2xe^y$; R is bounded by $x = 0, x = 1, y = 0$, and $y = x$.

20. $f(x, y) = 2x$; R is bounded by $x = e^{2y}, x = y, y = 0$, and $y = 1$.

21. $f(x, y) = ye^x$; R is bounded by $y = \sqrt{x}$ and $y = x$.

22. $f(x, y) = xe^{-y^2}$; R is bounded by $x = 0, y = x^2$, and $y = 4$.

23. $f(x, y) = e^{y^2}$; R is bounded by $x = 0, x = 1, y = 2x$, and $y = 2$.

24. $f(x, y) = y$; R is bounded by $x = 1, x = e, y = 0$, and $y = \ln x$.

25. $f(x, y) = ye^{x^3}$; R is bounded by $x = \dfrac{y}{2}, x = 1, y = 0$, and $y = 2$.

In Exercises 26 and 27, determine whether the statement is true or false. If it is true, explain why it is true. If it is false, give an example to show why it is false.

26. If $h(x, y) = f(x)g(y)$, where f is continuous on $[a, b]$ and g is continuous on $[c, d]$, then $\iint_R h(x, y)\, dA = [\int_a^b f(x)\, dx][\int_c^d g(y)\, dy]$, where $R = \{(x, y) | a \leq x \leq b; c \leq y \leq d\}$.

27. If $\iint_{R_1} f(x, y)\, dA$ exists, where $R_1 = \{(x, y) | a \leq x \leq b; c \leq y \leq d\}$, then $\iint_{R_2} f(x, y)\, dA$ exists, where $R_2 = \{(x, y) | c \leq x \leq d; a \leq y \leq b\}$.

8.7 Solution to Self-Check Exercise

The region R is shown in the accompanying figure. The points of intersection of the two curves are found by solving the equation $x = x^{1/3}$, giving $x = 0$ and $x = 1$. Using Equation (16), we find

$$\begin{aligned}
\iint_R (x + y)\, dA &= \int_0^1 \left[\int_x^{x^{1/3}} (x + y)\, dy\right] dx \\
&= \int_0^1 \left[xy + \frac{1}{2}y^2 \Big|_x^{x^{1/3}} \right] dx \\
&= \int_0^1 \left[\left(x^{4/3} + \frac{1}{2}x^{2/3}\right) - \left(x^2 + \frac{1}{2}x^2\right) \right] dx \\
&= \int_0^1 \left(x^{4/3} + \frac{1}{2}x^{2/3} - \frac{3}{2}x^2 \right) dx \\
&= \frac{3}{7}x^{7/3} + \frac{3}{10}x^{5/3} - \frac{1}{2}x^3 \Big|_0^1 \\
&= \frac{3}{7} + \frac{3}{10} - \frac{1}{2} = \frac{8}{35}
\end{aligned}$$

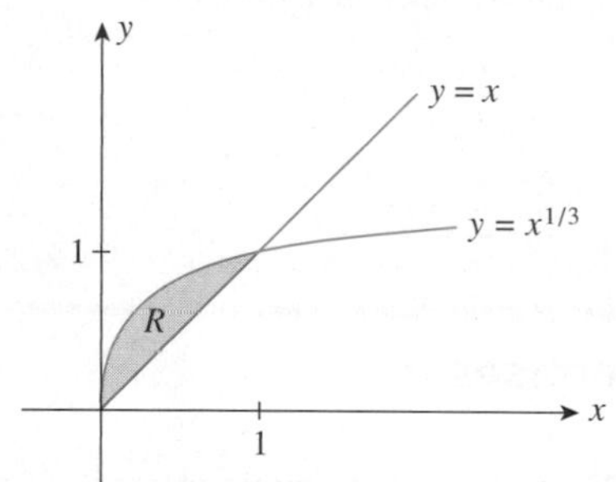

8.8 Applications of Double Integrals (Optional)

In this section we will give some sample applications involving the double integral.

Finding the Volume of a Solid by Double Integrals

As we saw earlier, the double integral

$$\iint_R f(x, y)\, dA$$

gives the volume of the solid bounded by the graph of $f(x, y)$ over the region R.

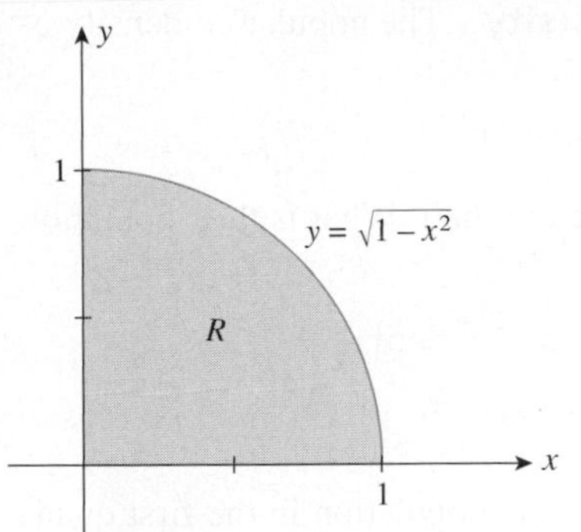

FIGURE 40
The plane region R defined by $y = \sqrt{1 - x^2}$ $(0 \le x \le 1)$

The Volume of a Solid under a Surface
Let R be a region in the xy-plane and let f be continuous and nonnegative on R. Then, the **volume of the solid under a surface** bounded above by $z = f(x, y)$ and below by R is given by

$$V = \iint_R f(x, y)\, dA$$

EXAMPLE 1 Find the volume of the solid bounded above by the plane $z = f(x, y) = y$ and below by the plane region R defined by $y = \sqrt{1 - x^2}$ $(0 \le x \le 1)$.

Solution The region R is sketched in Figure 40. Observe that $f(x, y) = y \ge 0$ for $(x, y) \in R$. Therefore, the required volume is given by

$$\iint_R y\, dA = \int_0^1 \left[\int_0^{\sqrt{1-x^2}} y\, dy\right] dx = \int_0^1 \left[\frac{1}{2}y^2 \Big|_0^{\sqrt{1-x^2}}\right] dx$$

$$= \int_0^1 \frac{1}{2}(1 - x^2)\, dx = \frac{1}{2}\left(x - \frac{1}{3}x^3\right)\Bigg|_0^1 = \frac{1}{3}$$

or $\frac{1}{3}$ cubic unit. The solid is shown in Figure 41. Note that it is not necessary to make a sketch of the solid in order to compute its volume.

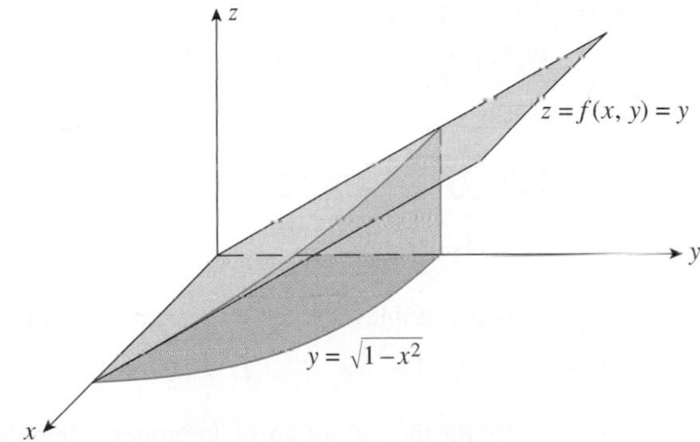

FIGURE 41
The solid bounded above by the plane $z = y$ and below by the plane region defined by $y = \sqrt{1 - x^2}$ $(0 \le x \le 1)$

Population of a City

Suppose the plane region R represents a certain district of a city and $f(x, y)$ gives the population density (the number of people per square mile) at any point (x, y) in R. Enclose the set R by a rectangle and construct a grid for it in the usual manner. In any rectangular region of the grid that has no point in common with R, set $f(x_i, y_i)hk = 0$ (Figure 42). Then, corresponding to any grid covering the set R, the general term of the Riemann sum $f(x_i, y_i)hk$ (population density times area) gives the number of people living in that part of the city corresponding to the rectangular region R_i. Therefore, the Riemann sum gives an approximation of the number of people living in the district represented by R and, in the limit, the double integral

$$\iint_R f(x, y)\, dA$$

gives the actual number of people living in the district under consideration.

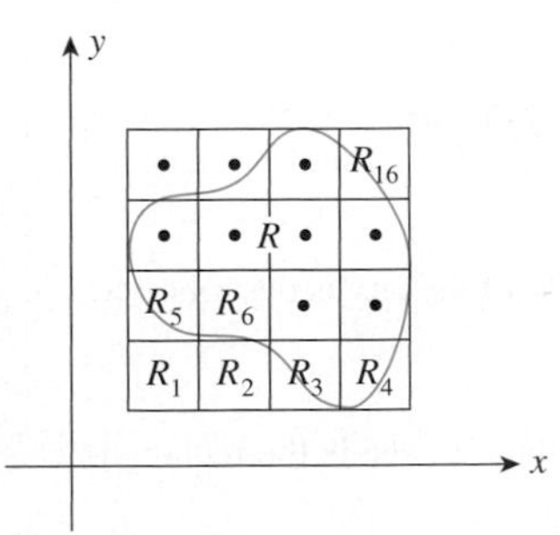

FIGURE 42
The rectangular region R representing a certain district of a city is enclosed by a rectangular grid.

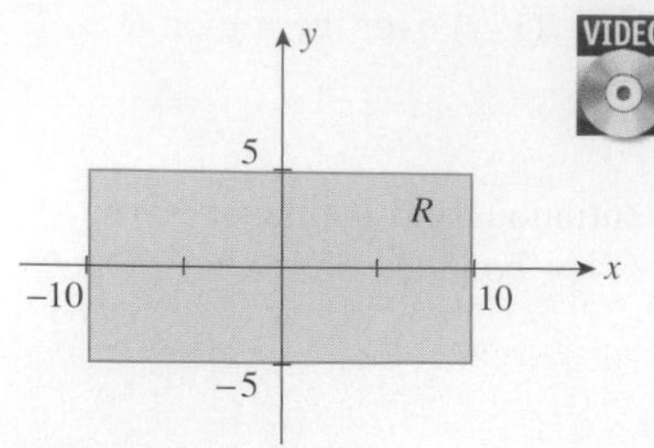

FIGURE 43
The rectangular region R represents a certain district of a city.

APPLIED EXAMPLE 2 Population Density The population density of a certain city is described by the function

$$f(x, y) = 10{,}000e^{-0.2|x|-0.1|y|}$$

where the origin (0, 0) gives the location of the city hall. What is the population inside the rectangular area described by

$$R = \{(x, y)|-10 \le x \le 10; -5 \le y \le 5\}$$

if x and y are in miles? (See Figure 43.)

Solution By symmetry, it suffices to compute the population in the first quadrant. (Why?) Then, upon observing that in this quadrant

$$f(x, y) = 10{,}000e^{-0.2x-0.1y} = 10{,}000e^{-0.2x}e^{-0.1y}$$

we see that the population in R is given by

$$\begin{aligned}\iint_R f(x, y)\, dA &= 4\int_0^{10}\left[\int_0^5 10{,}000e^{-0.2x}e^{-0.1y}\, dy\right] dx \\ &= 4\int_0^{10}\left[-100{,}000e^{-0.2x}e^{-0.1y}\Big|_0^5\right] dx \\ &= 400{,}000(1 - e^{-0.5})\int_0^{10} e^{-0.2x}\, dx \\ &= 2{,}000{,}000(1 - e^{-0.5})(1 - e^{-2})\end{aligned}$$

or approximately 680,438 people. ■

EXPLORE & DISCUSS

1. Consider the improper double integral $\iint_D f(x, y)\, dA$ of the continuous function f of two variables defined over the plane region

$$D = \{(x, y)|0 \le x < \infty; 0 \le y < \infty\}$$

Using the definition of improper integrals of functions of one variable (Section 7.4), explain why it makes sense to define

$$\begin{aligned}\iint_D f(x, y)\, dA &= \lim_{N\to\infty}\int_0^N\left[\lim_{M\to\infty}\int_0^M f(x, y)\, dx\right] dy \\ &= \lim_{M\to\infty}\int_0^M\left[\lim_{N\to\infty}\int_0^N f(x, y)\, dy\right] dx\end{aligned}$$

provided the limits exist.

2. Refer to Example 2. Assuming that the population density of the city is described by

$$f(x, y) = 10{,}000e^{-0.2|x|-0.1|y|}$$

for $-\infty < x < \infty$ and $-\infty < y < \infty$, show that the population outside the rectangular region

$$R = \{(x, y)|-10 < x < 10; -5 < y \le 5\}$$

of Example 2 is given by

$$4\iint_D f(x, y)\, dx\, dy - 680{,}438$$

(recall that 680,438 is the approximate population inside R).

3. Use the results of parts 1 and 2 to determine the population of the city outside the rectangular area R.

Average Value of a Function

In Section 6.5, we showed that the average value of a continuous function $f(x)$ over an interval $[a, b]$ is given by

$$\frac{1}{b-a}\int_a^b f(x)\, dx$$

That is, the average value of a function over $[a, b]$ is the integral of f over $[a, b]$ divided by the length of the interval. An analogous result holds for a function of two variables $f(x, y)$ over a plane region R. To see this, we enclose R by a rectangle and construct a rectangular grid. Let (x_i, y_i) be any point in the rectangle R_i of area hk. Now, the average value of the mn numbers $f(x_1, y_1), f(x_2, y_2), \ldots, f(x_{mn}, y_{mn})$ is given by

$$\frac{f(x_1, y_1) + f(x_2, y_2) + \cdots + f(x_{mn}, y_{mn})}{mn}$$

which can also be written as

$$\begin{aligned}\frac{hk}{hk}&\left[\frac{f(x_1, y_1) + f(x_2, y_2) + \cdots + f(x_{mn}, y_{mn})}{mn}\right]\\ &= \frac{1}{(mn)hk}[f(x_1, y_1) + f(x_2, y_2) + \cdots + f(x_{mn}, y_{mn})]hk\end{aligned}$$

Now the area of R is approximated by the sum of the mn rectangles (*omitting* those having no points in common with R), each of area hk. Note that this is the denominator of the previous expression. Therefore, taking the limit as m and n both tend to infinity, we obtain the following formula for the *average value of* $f(x, y)$ *over* R.

Average Value of $f(x, y)$ over the Region R

If f is integrable over the plane region R, then its average value over R is given by

$$\frac{\displaystyle\iint_R f(x, y)\, dA}{\text{Area of } R} \quad \text{or} \quad \frac{\displaystyle\iint_R f(x, y)\, dA}{\displaystyle\iint_R dA} \tag{18}$$

Note If we let $f(x, y) = 1$ for all (x, y) in R, then

$$\iint_R f(x, y)\, dA = \iint_R dA = \text{Area of } R$$

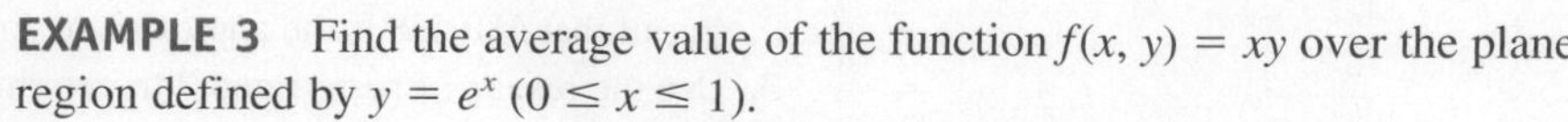

EXAMPLE 3 Find the average value of the function $f(x, y) = xy$ over the plane region defined by $y = e^x$ $(0 \le x \le 1)$.

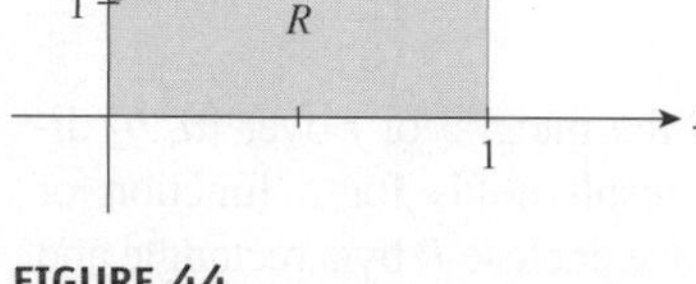

FIGURE 44
The plane region R defined by $y = e^x$ $(0 \le x \le 1)$

Solution The region R is shown in Figure 44. The area of the region R is given by

$$\int_0^1 \left[\int_0^{e^x} dy\right] dx = \int_0^1 \left[\, y \Big|_0^{e^x} \right] dx$$

$$= \int_0^1 e^x\, dx$$

$$= e^x \Big|_0^1$$

$$= e - 1$$

square units. We would obtain the same result had we viewed the area of this region as the area of the region under the curve $y = e^x$ from $x = 0$ to $x = 1$. Next, we compute

$$\iint_R f(x, y)\, dA = \int_0^1 \left[\int_0^{e^x} xy\, dy\right] dx$$

$$= \int_0^1 \left[\frac{1}{2}xy^2 \Big|_0^{e^x}\right] dx$$

$$= \int_0^1 \frac{1}{2}xe^{2x}\, dx$$

$$= \frac{1}{4}xe^{2x} - \frac{1}{8}e^{2x}\Big|_0^1 \qquad \text{Integrate by parts.}$$

$$= \left(\frac{1}{4}e^2 - \frac{1}{8}e^2\right) + \frac{1}{8}$$

$$= \frac{1}{8}(e^2 + 1)$$

square units. Therefore, the required average value is given by

$$\frac{\displaystyle\iint_R f(x, y)\, dA}{\displaystyle\iint_R dA} = \frac{\frac{1}{8}(e^2 + 1)}{e - 1} = \frac{e^2 + 1}{8(e - 1)}$$

APPLIED EXAMPLE 4 Population Density (Refer to Example 2.) The population density of a certain city (number of people per square mile) is described by the function

$$f(x, y) = 10{,}000e^{-0.2|x| - 0.1|y|}$$

where the origin gives the location of the city hall. What is the average population density inside the rectangular area described by

$$R = \{(x, y) | -10 \le x \le 10; -5 \le y \le 5\}$$

where x and y are measured in miles?

Solution From the results of Example 2, we know that

$$\iint_R f(x, y)\, dA \approx 680{,}438$$

From Figure 40, we see that the area of the plane rectangular region R is (20)(10), or 200, square miles. Therefore, the average population inside R is

$$\frac{\displaystyle\iint_R f(x, y)\, dA}{\displaystyle\iint_R dA} = \frac{680{,}438}{200} = 3402.19$$

or approximately 3402 people per square mile. ■

8.8 Self-Check Exercise

The population density of a coastal town located on an island is described by the function

$$f(x, y) = \frac{5000xe^y}{1 + 2x^2} \qquad (0 \le x \le 4; -2 \le y \le 0)$$

where x and y are measured in miles (see the accompanying figure).

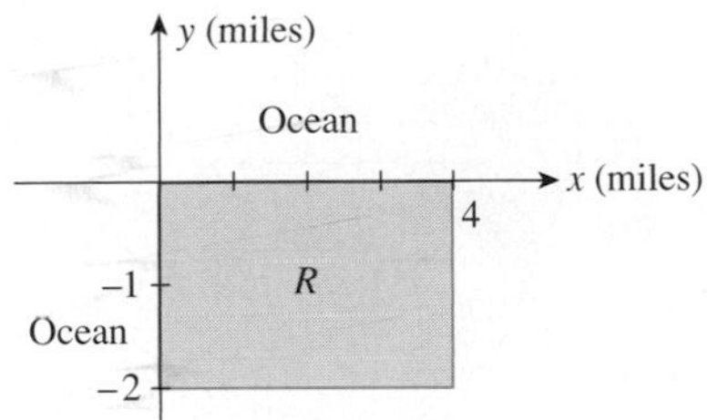

What is the population inside the rectangular area defined by $R = \{(x, y) | 0 \le x \le 4; -2 \le y \le 0\}$? What is the average population density in the area?

The solution to Self-Check Exercise 8.8 can be found on page 614.

8.8 Concept Questions

1. Give the formula for finding the volume of the solid bounded above by the graph of $z = f(x, y)$ and below by the region R in the xy-plane.
2. What is the average value of $f(x, y)$ over the region R?
3. Suppose a plane region R represents a certain district of a city, $f(x, y)$ gives the population density at any point (x, y) in R, and the set R is enclosed by a rectangle.
   a. Explain how a Riemann sum can be used to approximate the number of people living in the district represented by R.
   b. What does the general term of the Riemann sum $f(x_i, y_i)hk$ represent?
   c. What does $\displaystyle\iint_R f(x, y)\, dA$ represent?

8.8 Exercises

In Exercises 1–8, use a double integral to find the volume of the solid shown in the figure.

1.

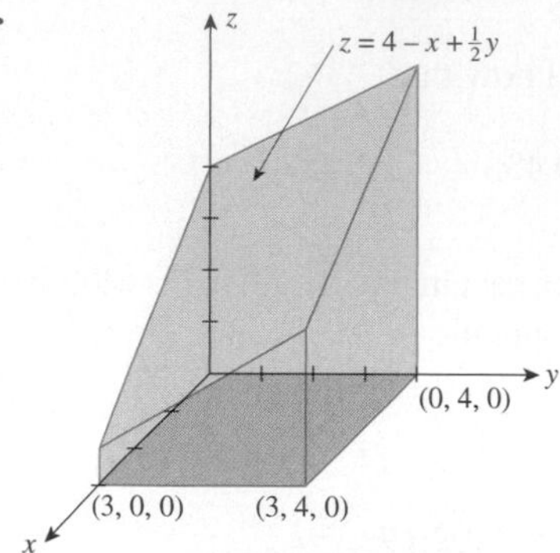

2.

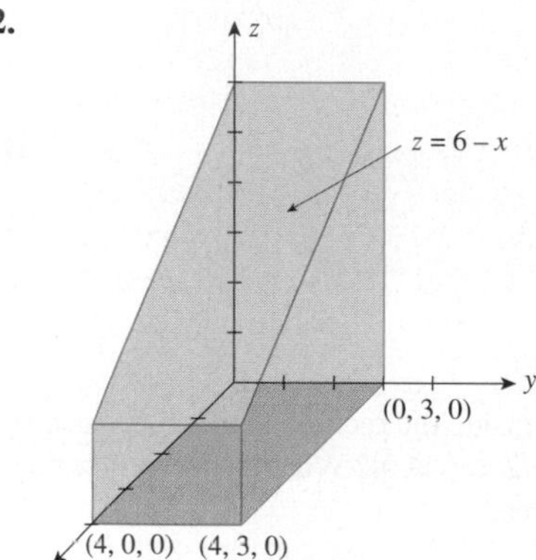

3.

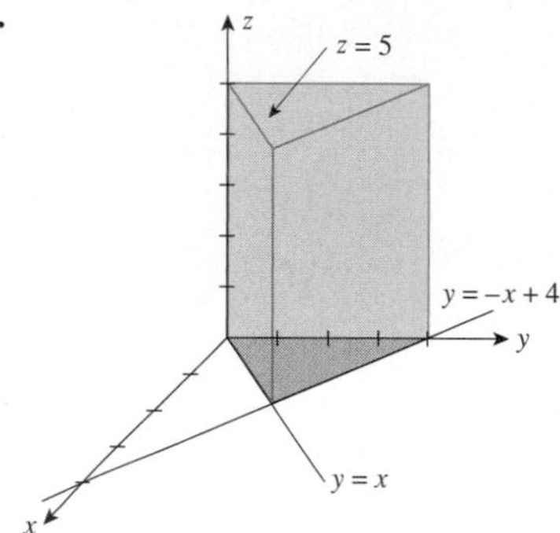

4.

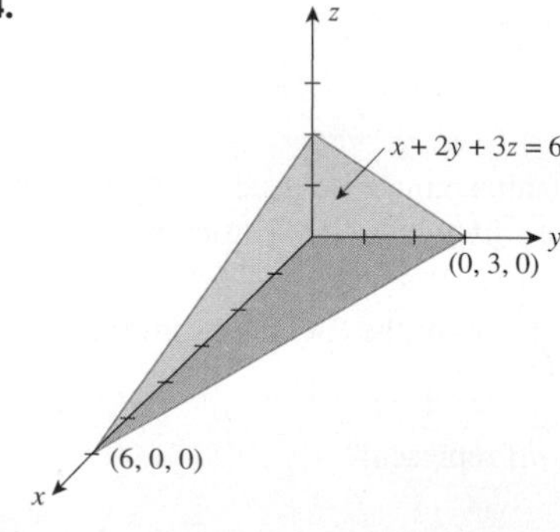

5.

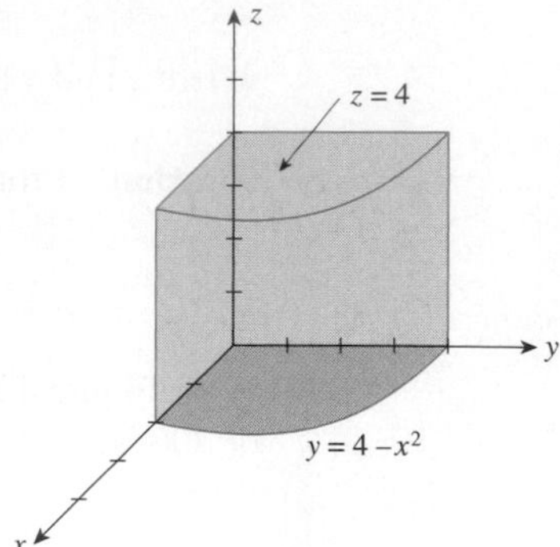

6.

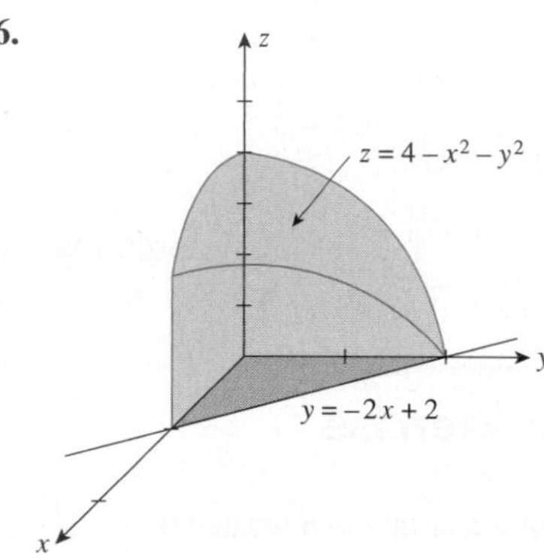

7.

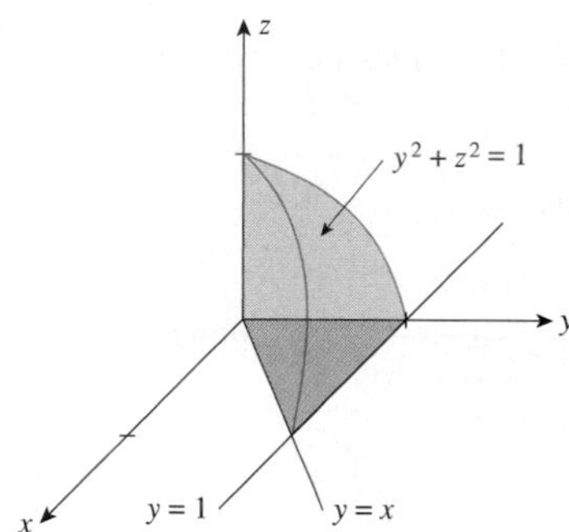

8.

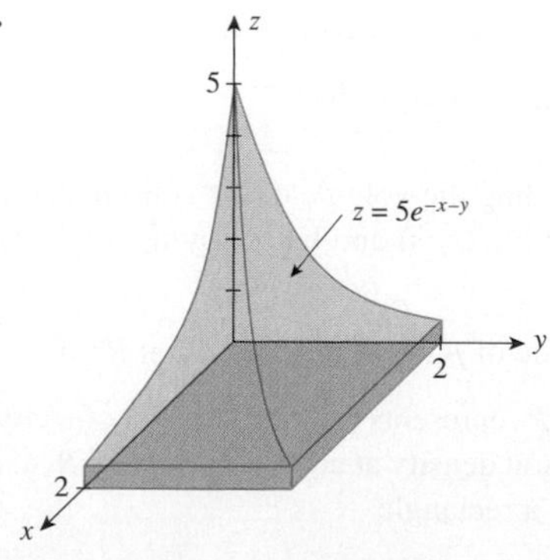

In Exercises 9–16, find the volume of the solid bounded above by the surface $z = f(x, y)$ and below by the plane region R.

9. $f(x, y) = 4 - 2x - y$; $R = \{(x, y) | 0 \leq x \leq 1; 0 \leq y \leq 2\}$

10. $f(x, y) = 2x + y$; R is the triangle bounded by $y = 2x$, $y = 0$, and $x = 2$.

11. $f(x, y) = x^2 + y^2$; R is the rectangle with vertices (0, 0), (1, 0), (1, 2), and (0, 2).

12. $f(x, y) = e^{x+2y}$; R is the triangle with vertices (0, 0), (1, 0), and (0, 1).

13. $f(x, y) = 2xe^y$; R is the triangle bounded by $y = x$, $y = 2$, and $x = 0$.

14. $f(x, y) = \dfrac{2y}{1 + x^2}$; R is the region bounded by $y = \sqrt{x}$, $y = 0$, and $x = 4$.

15. $f(x, y) = 2x^2y$; R is the region bounded by the graphs of $y = x$ and $y = x^2$.

16. $f(x, y) = x$; R is the region in the first quadrant bounded by the semicircle $y = \sqrt{16 - x^2}$, the x-axis, and the y-axis.

In Exercises 17–22, find the average value of the function $f(x, y)$ over the plane region R.

17. $f(x, y) = 6x^2y^3$; $R = \{(x, y) | 0 \leq x \leq 2; 0 \leq y \leq 3\}$

18. $f(x, y) = x + 2y$; R is the triangle with vertices (0, 0), (1, 0), and (1, 1).

19. $f(x, y) = xy$; R is the triangle bounded by $y = x$, $y = 2 - x$, and $y = 0$.

20. $f(x, y) = e^{-x^2}$; R is the triangle with vertices (0, 0), (1, 0), and (1, 1).

21. $f(x, y) = xe^y$; R is the triangle with vertices (0, 0), (1, 0), and (1, 1).

22. $f(x, y) = \ln x$; R is the region bounded by the graphs of $y = 2x$ and $y = 0$ from $x = 1$ to $x = 3$.
Hint: Use integration by parts.

23. **Population Density** The population density of a coastal town is described by the function

$$f(x, y) = \frac{10{,}000e^y}{1 + 0.5|x|} \qquad (-10 \leq x \leq 10; -4 \leq y \leq 0)$$

where x and y are measured in miles. Find the population inside the rectangular area described by

$$R = \{(x, y) | -5 \leq x \leq 5; -2 \leq y \leq 0\}$$

(See the accompanying figure.)

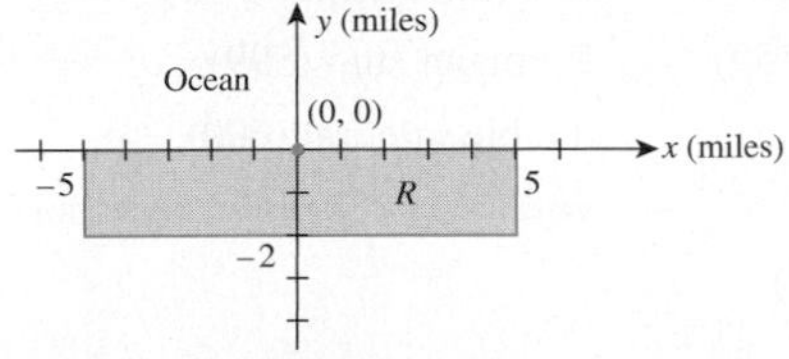

24. **Average Population Density** Refer to Exercise 23. Find the average population density inside the rectangular area R.

25. **Population Density** The population density of a certain city is given by the function

$$f(x, y) = \frac{50{,}000|xy|}{(x^2 + 20)(y^2 + 36)}$$

where the origin (0, 0) gives the location of the government center. Find the population inside the rectangular area described by

$$R = \{(x, y) | -15 \leq x \leq 15; -20 \leq y \leq 20\}$$

26. **Average Profit** The Country Workshop's total weekly profit (in dollars) realized in manufacturing and selling its rolltop desks is given by the profit function

$$P(x, y) = -0.2x^2 - 0.25y^2 - 0.2xy + 100x + 90y - 4000$$

where x stands for the number of finished units and y stands for the number of unfinished units manufactured and sold each week. Find the average weekly profit if the number of finished units manufactured and sold varies between 180 and 200 and the number of unfinished units varies between 100 and 120/week.

27. **Average Price of Land** The rectangular region R shown in the accompanying figure represents a city's financial district. The price of land in the district is approximated by the function

$$p(x, y) = 200 - 10\left(x - \frac{1}{2}\right)^2 - 15(y - 1)^2$$

where $p(x, y)$ is the price of land at the point (x, y) in dollars/square foot and x and y are measured in miles. What is the average price of land per square foot in the district?

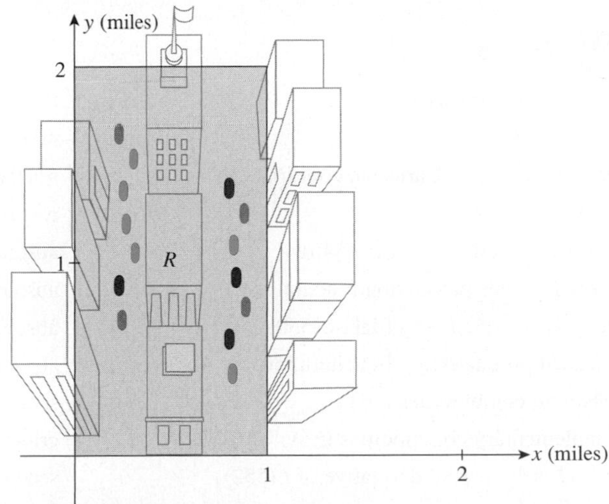

In Exercises 28 and 29, determine whether the statement is true or false. If it is true, explain why it is true. If it is false, give an example to show why it is false.

28. Let R be a region in the xy-plane and let f and g be continuous functions on R that satisfy the condition $f(x, y) \leq g(x, y)$ for all (x, y) in R. Then, $\iint_R [g(x, y) - f(x, y)]\, dA$ gives the volume of the solid bounded above by the surface $z = g(x, y)$ and below by the surface $z = f(x, y)$.

29. Suppose f is nonnegative and integrable over the plane region R. Then, the average value of f over R can be thought of as the (constant) height of the cylinder with base R and volume that is exactly equal to the volume of the solid under the graph of $z = f(x, y)$. (*Note:* The cylinder referred to here has sides perpendicular to R.)

8.8 Solution to Self-Check Exercise

The population in R is given by

$$\iint_R f(x, y)\, dA = \int_0^4 \left[\int_{-2}^0 \frac{5000xe^y}{1 + 2x^2}\, dy\right] dx$$

$$= \int_0^4 \left[\frac{5000xe^y}{1 + 2x^2}\bigg|_{-2}^{0}\right] dx$$

$$= 5000(1 - e^{-2}) \int_0^4 \frac{x}{1 + 2x^2}\, dx$$

$$= 5000(1 - e^{-2}) \left[\frac{1}{4} \ln(1 + 2x^2)\bigg|_0^4\right]$$

$$= 5000(1 - e^{-2}) \left(\frac{1}{4}\right) \ln 33$$

or approximately 3779 people. The average population density inside R is

$$\frac{\displaystyle\iint_R f(x, y)\, dA}{\displaystyle\iint_R dA} = \frac{3779}{(2)(4)}$$

or approximately 472 people/square mile.

CHAPTER 8 Summary of Principal Terms

TERMS

function of two variables (536)
domain (536)
three-dimensional Cartesian coordinate system (538)
level curve (539)
first partial derivatives of f (546)
Cobb–Douglas production function (550)
marginal productivity of labor (550)
marginal productivity of capital (550)
substitute commodities (551)
complementary commodities (551)
second-order partial derivative of f (552)
relative maximum (559)
relative maximum value (559)
relative minimum (559)
relative minimum value (559)
absolute maximum (559)
absolute minimum (559)
absolute maximum value (559)
absolute minimum value (559)
saddle point (560)
critical point (561)
second derivative test (562)
method of least squares (570)
scatter diagram (570)
least-squares line (regression line) (571)
normal equation (572)
constrained relative extremum (582)
method of Lagrange multipliers (583)
total differential (594)
Riemann sum (599)
double integral (600)
volume of a solid under a surface (607)

CHAPTER 8 Concept Review Questions

Fill in the blanks.

1. The domain of a function f of two variables is a subset of the ____-plane. The rule of f associates with each ____ ____ in the domain of f one and only one ____ ____, denoted by $z =$ ____.

2. If the function f has rule $z = f(x, y)$, then x and y are called ____ variables, and z is a/an ____ variable. The number z is also called the ____ of f.

3. The graph of a function f of two variables is the set of all points (x, y, z), where ____, and (x, y) is the domain of ____. The graph of a function of two variables is a ____ in three-dimensional space.

4. The trace of the graph of $f(x, y)$ in the plane $z = c$ is the curve with equation ____ lying in the plane $z = c$. The projection of the trace of f in the plane $z = c$ onto the xy-plane is called the ____ ____ of f. The contour map associated with f is obtained by drawing the ____ ____ of f corresponding to several admissible values of ____.

5. The partial derivative $\partial f/\partial x$ of f at (x, y) can be found by thinking of y as a/an ____ ____ in the expression for f, and differentiating this expression with respect to ____ as if it were a function of x alone.

6. The number $f_x(a, b)$ measures the ____ of the tangent line to the curve C obtained by the intersection of the graph of f and the plane $y = b$ at the point ____. It also measures the rate of change of f with respect to ____ at the point (a, b) with y held fixed with value ____.

7. A function $f(x, y)$ has a relative maximum at (a, b) if $f(x, y)$ ____ $f(a, b)$ for all points (x, y) that are sufficiently close to ____. The absolute maximum value of $f(x, y)$ is the number $f(a, b)$ such that $f(x, y)$ ____ $f(a, b)$ for all (x, y) in the ____ of f.

8. A critical point of $f(x, y)$ is a point (a, b) in the ____ of f such that ____ ____ ____ or at least one of the partial derivatives of f does not ____. A critical point of f is a ____ for a relative extremum of f.

9. By plotting the points associated with a set of data we obtain the ____ diagram for the data. The least-squares line is the line obtained by ____ the sum of the squares of the errors made when the y-values of the data points are approximated by the corresponding y-values of the ____ line. The least-squares equation is found by solving the ____ equations.

10. The method of Lagrange multipliers solves the problem of finding the relative extrema of a function $f(x, y)$ subject to the constraint ____. We first form the Lagrangian function $F(x, y, \lambda) =$ ____. Then we solve the system consisting of the three equations ____, ____, and ____ for x, y, and λ. These solutions give the critical points that give rise to the relative ____ of f.

11. For a differentiable function $z = f(x, y)$, the differentials of the ____ variables x and y are ____ and ____, and the differential of the dependent variable z is $dz =$ ____.

12. If $f(x, y)$ is continuous and nonnegative over a region R in the xy-plane and $\iint_R f(x, y)\, dA$ exists, then the double integral gives the____ of the ____ bounded by the graph of $f(x, y)$ over the region R.

13. The integral $\iint_R f(x, y)\, dA$ is evaluated using ____ integrals. For example, $\iint_R (2x + y^2)\, dA$ where $R = \{(x, y) | 0 \le x \le 1;\ 3 \le y \le 5\}$ is equal to $\int_0^1 \int_3^5 (2x + y^2)\, dy\, dx$ or the (iterated) integral ____.

CHAPTER 8 Review Exercises

1. Let $f(x, y) = \dfrac{xy}{x^2 + y^2}$. Compute $f(0, 1)$, $f(1, 0)$, and $f(1, 1)$. Does $f(0, 0)$ exist?

2. Let $f(x, y) = \dfrac{xe^y}{1 + \ln xy}$. Compute $f(1, 1)$, $f(1, 2)$, and $f(2, 1)$. Does $f(1, 0)$ exist?

3. Let $h(x, y, z) = xye^z + \dfrac{x}{y}$. Compute $h(1, 1, 0)$, $h(-1, 1, 1)$, and $h(1, -1, 1)$.

4. Find the domain of the function $f(u, v) = \dfrac{\sqrt{u}}{u - v}$.

5. Find the domain of the function $f(x, y) = \dfrac{x - y}{x + y}$.

6. Find the domain of the function $f(x, y) = x\sqrt{y} + y\sqrt{1 - x}$.

7. Find the domain of the function

$$f(x, y, z) = \frac{xy\sqrt{z}}{(1 - x)(1 - y)(1 - z)}$$

In Exercises 8–11, sketch the level curves of the function corresponding to each value of z.

8. $z = f(x, y) = 2x + 3y$; $z = -2, -1, 0, 1, 2$

9. $z = f(x, y) = y - x^2$; $z = -2, -1, 0, 1, 2$

10. $z = f(x, y) = \sqrt{x^2 + y^2}$; $z = 0, 1, 2, 3, 4$

11. $z = f(x, y) = e^{xy}$; $z = 1, 2, 3$

In Exercises 12–21, compute the first partial derivatives of the function.

12. $f(x, y) = x^2y^3 + 3xy^2 + \dfrac{x}{y}$

13. $f(x, y) = x\sqrt{y} + y\sqrt{x}$

14. $f(u, v) = \sqrt{uv^2 - 2u}$

15. $f(x, y) = \dfrac{x - y}{y + 2x}$

16. $g(x, y) = \dfrac{xy}{x^2 + y^2}$

17. $h(x, y) = (2xy + 3y^2)^5$

18. $f(x, y) = (xe^y + 1)^{1/2}$

19. $f(x, y) = (x^2 + y^2)e^{x^2+y^2}$

20. $f(x, y) = \ln(1 + 2x^2 + 4y^4)$

21. $f(x, y) = \ln\left(1 + \dfrac{x^2}{y^2}\right)$

In Exercises 22–27, compute the second-order partial derivatives of the function.

22. $f(x, y) = x^3 - 2x^2y + y^2 + x - 2y$

23. $f(x, y) = x^4 + 2x^2y^2 - y^4$

24. $f(x, y) = (2x^2 + 3y^2)^3$

25. $g(x, y) = \dfrac{x}{x + y^2}$

26. $g(x, y) = e^{x^2+y^2}$

27. $h(s, t) = \ln\left(\dfrac{s}{t}\right)$

28. Let $f(x, y, z) = x^3y^2z + xy^2z + 3xy - 4z$. Compute $f_x(1, 1, 0)$, $f_y(1, 1, 0)$, and $f_z(1, 1, 0)$ and interpret your results.

In Exercises 29–34, find the critical point(s) of the functions. Then use the second derivative test to classify the nature of each of these points, if possible. Finally, determine the relative extrema of each function.

29. $f(x, y) = 2x^2 + y^2 - 8x - 6y + 4$

30. $f(x, y) = x^2 + 3xy + y^2 - 10x - 20y + 12$

31. $f(x, y) = x^3 - 3xy + y^2$

32. $f(x, y) = x^3 + y^2 - 4xy + 17x - 10y + 8$

33. $f(x, y) = e^{2x^2+y^2}$

34. $f(x, y) = \ln(x^2 + y^2 - 2x - 2y + 4)$

In Exercises 35–38, use the method of Lagrange multipliers to optimize the function subject to the given constraints.

35. Maximize the function $f(x, y) = -3x^2 - y^2 + 2xy$ subject to the constraint $2x + y = 4$.

36. Minimize the function $f(x, y) = 2x^2 + 3y^2 - 6xy + 4x - 9y + 10$ subject to the constraint $x + y = 1$.

37. Find the maximum and minimum values of the function $f(x, y) = 2x - 3y + 1$ subject to the constraint $2x^2 + 3y^2 - 125 = 0$.

38. Find the maximum and minimum values of the function $f(x, y) = e^{x-y}$ subject to the constraint $x^2 + y^2 = 1$.

In Exercises 39 and 40, find the total differential of each function at the given point.

39. $f(x, y) = (x^2 + y^4)^{3/2}$; $(3, 2)$

40. $f(x, y) = xe^{x-y} + x \ln y$; $(1, 1)$

In Exercises 41 and 42, find the approximate change in z when the point (x, y) changes from (x_0, y_0) to (x_1, y_1).

41. $f(x, y) = 2x^2y^3 + 3y^2x^2 - 2xy$; from $(1, -1)$ to $(1.02, -0.98)$

42. $f(x, y) = 4x^{3/4}y^{1/4}$; from $(16, 81)$ to $(17, 80)$

In Exercises 43–46, evaluate the double integrals.

43. $f(x, y) = 3x - 2y$; R is the rectangle defined by $2 \le x \le 4$ and $-1 \le y \le 2$.

44. $f(x, y) = e^{-x-2y}$; R is the rectangle defined by $0 \le x \le 2$ and $0 \le y \le 1$.

45. $f(x, y) = 2x^2y$; R is bounded by $x = 0$, $x = 1$, $y = x^2$, and $y = x^3$.

46. $f(x, y) = \dfrac{y}{x}$, R is bounded by $x = 1$, $x = 2$, $y = 1$, and $y = x$.

In Exercises 47 and 48, find the volume of the solid bounded above by the surface $z = f(x, y)$ and below by the plane region R.

47. $f(x, y) = 4x^2 + y^2$; $R = \{0 \le x \le 2; 0 \le y \le 1\}$

48. $f(x, y) = x + y$; R is the region bounded by $y = x^2$, $y = 4x$, and $y = 4$.

49. Find the average value of the function

$$f(x, y) = xy + 1$$

over the plane region R bounded by $y = x^2$ and $y = 2x$.

50. Revenue Functions A division of Ditton Industries makes a 16-speed and a 10-speed electric blender. The company's management estimates that x units of the 16-speed model and y units of the 10-speed model are demanded daily when the unit prices are

$$p = 80 - 0.02x - 0.1y$$
$$q = 60 - 0.1x - 0.05y$$

dollars, respectively.

a. Find the daily total revenue function $R(x, y)$.
b. Find the domain of the function R.
c. Compute $R(100, 300)$ and interpret your result.

51. Demand for CD Players In a survey conducted by *Home Entertainment* magazine, it was determined that the demand equation for CD players is given by

$$x = f(p, q) = 900 - 9p - e^{0.4q}$$

whereas the demand equation for audio CDs is given by

$$y = g(p, q) = 20{,}000 - 3000q - 4p$$

where p and q denote the unit prices (in dollars) for the CD players and audio CDs, respectively, and x and y denote the number of CD players and audio CDs demanded per week. Determine whether these two products are substitute, complementary, or neither.

52. **Estimating Changes in Profit** The total daily profit function (in dollars) of Weston Publishing Company realized in publishing and selling its English language dictionaries is given by

$$P(x, y) = -0.0005x^2 - 0.003y^2 - 0.002xy + 14x + 12y - 200$$

where x denotes the number of deluxe copies and y denotes the number of standard copies published and sold daily. Currently the number of deluxe and standard copies of the dictionaries published and sold daily are 1000 and 1700, respectively. Determining the approximate daily change in the total daily profit if the number of deluxe copies is increased to 1050 and the number of standard copies is decreased to 1650 per day.

53. **Average Daily TV-Viewing Time** The following data were compiled by the Bureau of Television Advertising in a large metropolitan area, giving the average daily TV-viewing time per household in that area over the years 1996 to 2004.

Year	1996	1998
Daily Viewing Time, y	6 hr 9 min	6 hr 30 min

Year	2000	2002	2004
Daily Viewing Time, y	6 hr 36 min	7 hr	7 hr 16 min

a. Find the least-squares line for these data. (Let $x = 1$ represent 1996.)
b. Estimate the average daily TV-viewing time per household in 2006.

54. **Female Life Expectancy at 65** The projections of female life expectancy at age 65 in the United States are summarized in the following table ($x = 0$ corresponds to 2000):

Year, x	0	10	20	30	40	50
Years beyond 65, y	19.5	20.0	20.6	21.2	21.8	22.4

a. Find an equation of the least-squares line for these data.
b. Use the result of (a) to estimate life expectancy at 65 of a female in 2040. How does this result compare with the given data for that year?
c. Use the result of (a) to estimate the life expectancy at 65 of a female in 2030.

Source: U.S. Census Bureau

55. **Maximizing Revenue** Odyssey Travel Agency's monthly revenue depends on the amount of money x (in thousands of dollars) spent on advertising per month and the number of agents y in its employ in accordance with the rule

$$R(x, y) = -x^2 - 0.5y^2 + xy + 8x + 3y + 20$$

Determine the amount of money the agency should spend per month and the number of agents it should employ in order to maximize its monthly revenue.

56. **Minimizing Fencing Costs** The owner of the Rancho Grande wants to enclose a rectangular piece of grazing land along the straight portion of a river and then subdivide it using a fence running parallel to the sides. No fencing is required along the river. If the material for the sides costs \$3/running yard and the material for the divider costs \$2/running yard, what will be the dimensions of a 303,750-yd pasture if the cost of the fencing is kept to a minimum?

57. **Cobb–Douglas Production Functions** The production of Q units of a commodity is related to the amount of labor x and the amount of capital y (in suitable units) expended by the equation

$$Q = f(x, y) = x^{3/4}y^{1/4}$$

If an expenditure of 100 units is available for production, how should it be apportioned between labor and capital so that Q is maximized?
Hint: Use the method of Lagrange multipliers to maximize the function Q subject to the constraint $x + y = 100$.

CHAPTER 8 Before Moving On . . .

1. Find the domain of

$$f(x, y) = \frac{\sqrt{x} + \sqrt{y}}{(1 - x)(2 - y)}$$

2. Find the first- and second-order partial derivatives of $f(x, y) = x^2y + e^{xy}$.

3. Find the relative extrema, if any, of $f(x, y) = 2x^3 + 2y^3 - 6xy - 5$.

4. Find an equation of the least-squares line for the following data:

x	0	1	2	3	5
y	2.9	5.1	6.8	8.8	13.2

5. Use the method of Lagrange multipliers to find the minimum of $f(x, y) = 3x^2 + 3y^2 + 1$ subject to $x + y = 1$.

6. Let $z = 2x^2 - xy$.
a. Find the differential dz.
b. Compute the value of dz if (x, y) changes from (1, 1) to (0.98, 1.03).

7. Evaluate $\iint_R (1 - xy)\, dA$, where R is the region bounded by $x = 0$, $x = 1$, $y = x$, and $y = x^2$.

9 Differential Equations

If the amount of fertilizer used to cultivate land on a wheat farm is increased, will the crop yield increase substantially? Researchers at a midwestern university found that a new experimental fertilizer increased the wheat yield of the land at the university's experimental field station. In Example 2, page 635, you will see how they used a differential equation to estimate the crop yield.

Bob Stefko/Getty Images

AN EQUATION involving the derivative, or differential, of an unknown function is called a *differential equation*. In this chapter we show how differential equations are used to solve problems involving the growth of an amount of money earning interest compounded continuously, the growth of a population of bacteria, the decay of radioactive material, and the rate at which a person learns a new subject, to name just a few.

9.1 Differential Equations

Models Involving Differential Equations

We first encountered differential equations in Section 6.1. Recall that a **differential equation** is an equation that involves an unknown function and its derivative(s). Here are some examples of differential equations:

$$\frac{dy}{dx} = xe^x \qquad \frac{dy}{dx} + 2y = x^2 \qquad \frac{d^2y}{dt^2} + \left(\frac{dy}{dt}\right)^3 + ty - 8 = 0$$

Differential equations appear in practically every branch of applied mathematics, and the study of these equations remains one of the most active areas of research in mathematics. As you will see in the next few examples, models involving differential equations often arise from the mathematical formulation of practical problems.

Unrestricted Growth Models The unrestricted growth model was first discussed in Chapter 5. There we saw that the size of a population at any time t, $Q(t)$, increases at a rate that is proportional to $Q(t)$ itself. Thus,

$$\frac{dQ}{dt} = kQ \tag{1}$$

where k is a constant of proportionality. This is a differential equation involving the unknown function Q and its derivative Q'.

Restricted Growth Models In many applications the quantity $Q(t)$ does not exhibit unrestricted growth but approaches some definite upper bound. The learning curves and logistic functions we discussed in Chapter 5 are examples of restricted growth models. Let's derive the mathematical models that lead to these functions.

Suppose $Q(t)$ does not exceed some number C, called the *carrying capacity of the environment*. Furthermore, suppose the rate of growth of this quantity is *proportional* to the difference between its upper bound and its current size. The resulting differential equation is

$$\frac{dQ}{dt} = k(C - Q) \tag{2}$$

where k is a constant of proportionality. Observe that if the initial population is small relative to C, then the rate of growth of Q is relatively large. But as $Q(t)$ approaches C, the difference $C - Q(t)$ approaches zero, as does the rate of growth of Q. In Section 9.3 you will see that the solution of the differential Equation (2) is a function that describes a learning curve (Figure 1).

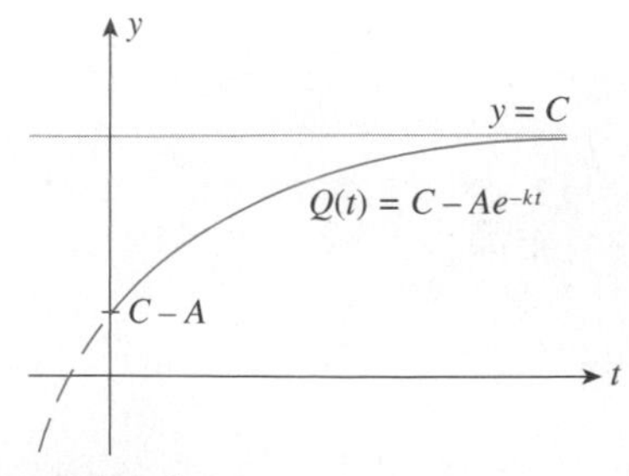

FIGURE 1
$Q(t)$ describes a learning curve.

Next, let's consider a restricted growth model in which the rate of growth of a quantity $Q(t)$ is *jointly proportional* to its current size and the difference between its upper bound and its current size; that is,

$$\frac{dQ}{dt} = kQ(C - Q) \tag{3}$$

where k is a constant of proportionality. Observe that when $Q(t)$ is small relative to C, the rate of growth of Q is approximately proportional to Q. But as $Q(t)$ approaches C, the growth rate slows down to zero. If $Q > C$, then $dQ/dt < 0$ and the

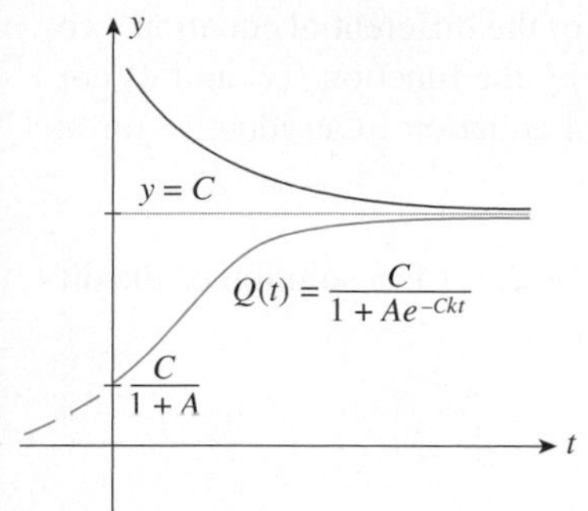

FIGURE 2
Two solutions of the logistic equation

quantity is decreasing with time, with the decay rate slowing down as Q approaches C. We will show later that the solution of the differential Equation (3) is just the logistic function we discussed in Chapter 5. Its graph is shown in Figure 2.

Stimulus Response In the quantitative theory of psychology, one model that describes the relationship between a stimulus S and the resulting response R is the Weber–Fechner law. This law asserts that the rate of change of a reaction R is inversely proportional to the stimulus S. Mathematically, this law may be expressed as

$$\frac{dR}{dS} = \frac{k}{S} \tag{4}$$

where k is a constant of proportionality. Furthermore, suppose that the threshold level, the lowest level of stimulation at which sensation is detected, is S_0. Then we have the condition $R = 0$ when $S = S_0$; that is, $R(S_0) = 0$. The graph of R versus S is shown in Figure 3.

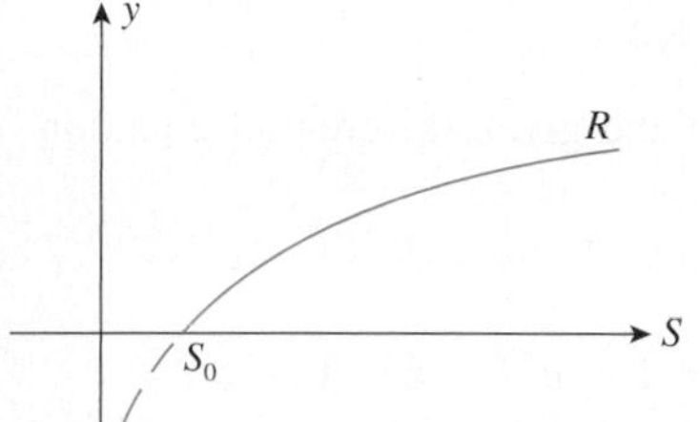

FIGURE 3
The solution to the differential equation (4) describes the response to a stimulus.

Mixture Problems Our next example is a typical mixture problem. Suppose a tank initially contains 10 gallons of pure water. Brine containing 3 pounds of salt per gallon flows into the tank at a rate of 2 gallons per minute, and the well-stirred mixture flows out of the tank at the same rate. How much salt is in the tank at any given time?

Let's formulate this problem mathematically. Suppose $A(t)$ denotes the amount of salt in the tank at any time t. Then the derivative dA/dt, the rate of change of the amount of salt at any time t, must satisfy the condition

$$\frac{dA}{dt} = (\text{rate of salt flowing in}) - (\text{rate of salt flowing out})$$

(Figure 4). But the rate at which salt flows into the tank is given by

$$(2 \text{ gal/min})(3 \text{ lb/gal}) \qquad \text{(rate of flow)} \times \text{(concentration)}$$

or 6 pounds per minute. Since the rate at which the solution leaves the tank is the same as the rate at which the brine is poured into it, that tank contains 10 gallons of the mixture at any time t. Since the salt content at any time t is A pounds, the concentration of the salt in the mixture is $(A/10)$ pounds per gallon. Therefore, the rate at which salt flows out of the tank is given by

$$(2 \text{ gal/min})\left(\frac{A}{10} \text{ lb/gal}\right)$$

or $(A/5)$ pounds per minute. Therefore, we are led to the differential equation

$$\frac{dA}{dt} = 6 - \frac{A}{5} \tag{5}$$

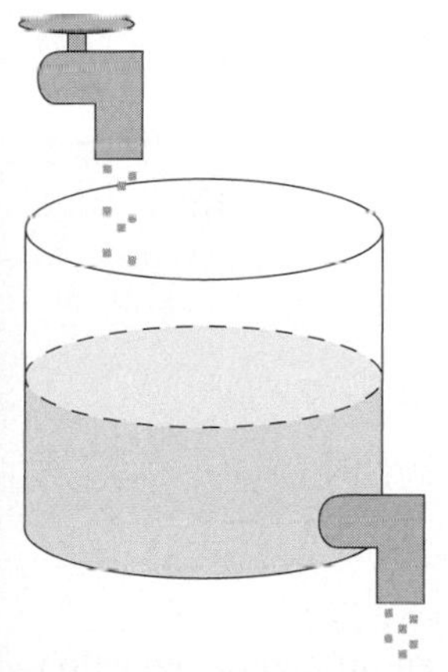

FIGURE 4
The rate of change of the amount of salt at time t = (rate of salt flowing in) − (rate of salt flowing out)

An additional condition arises from the fact that initially there is no salt in the solution. This condition may be expressed mathematically as $A = 0$ when $t = 0$ or, more concisely, $A(0) = 0$.

We will solve each of the differential equations we have introduced here in Section 9.3.

Solutions of Differential Equations

Suppose we are given a differential equation involving the derivative(s) of a function y. Recall that a **solution** to a differential equation is any function $f(x)$ that satisfies the differential equation. Thus, $y = f(x)$ is a solution of the differential equation, provided that the replacement of y and its derivative(s) by the function $f(x)$ and its corresponding derivatives reduces the given differential equation to an identity for all values of x.

EXAMPLE 1 Show that the function $f(x) = e^{-x} + x - 1$ is a solution of the differential equation

$$y' + y = x$$

Solution Let

$$y = f(x) = e^{-x} + x - 1$$

so that

$$y' = f'(x) = -e^{-x} + 1$$

Substituting these equations into the left side of the given differential equation yields

$$\overbrace{(-e^{-x} + 1)}^{y'} + \overbrace{(e^{-x} + x - 1)}^{y} = -e^{-x} + 1 + e^{-x} + x - 1 = x$$

which is equal to the right side of the given equation for all values of x. Therefore, $f(x) = e^{-x} + x - 1$ is a solution of the given differential equation. ■

In Example 1, we verified that $y = e^{-x} + x - 1$ is a solution of the differential equation $y' + y = x$. This is by no means the only solution of the differential equation, as the next example shows.

EXAMPLE 2 Show that any function of the form $f(x) = ce^{-x} + x - 1$, where c is a constant, is a solution of the differential equation

$$y' + y = x$$

Solution Let

$$y = f(x) = ce^{-x} + x - 1$$

so that

$$y' = f'(x) = -ce^{-x} + 1$$

Substituting these equations into the left side of the given differential equation yields

$$\overbrace{-ce^{-x} + 1}^{y'} + \overbrace{ce^{-x} + x - 1}^{y} = x$$

and we have verified the assertion. ■

It can be shown that *every* solution of the differential equation $y' + y = x$ must have the form $y = ce^{-x} + x - 1$, where c is a constant; therefore, this is the **general solution** of the differential equation $y' + y = x$. Figure 5 shows a family of solutions of this differential equation for selected values of c.

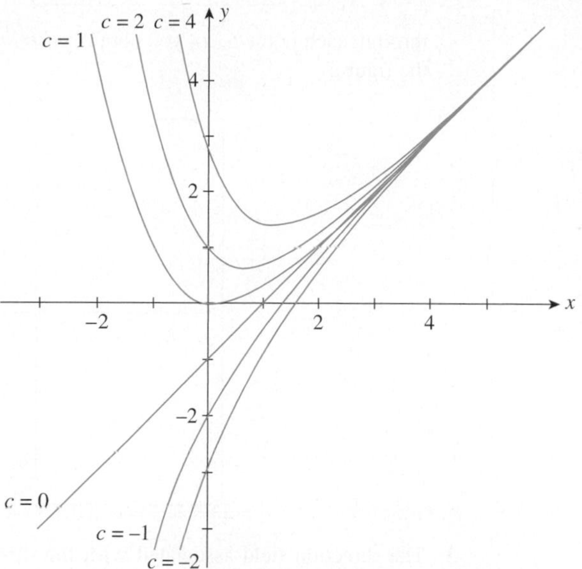

FIGURE 5
Some solutions of $y' + y = x$

Recall that the solution obtained by assigning a specific value to the constant c is called a **particular solution** of the differential equation. For example, the particular solution $y = e^{-x} + x - 1$ of Example 1 is obtained from the general solution by taking $c = 1$. In practice, a particular solution of a differential equation is obtained from the general solution of the differential equation by requiring that the solution and/or its derivative(s) satisfy certain conditions at one or more values of x.

EXAMPLE 3 Use the results of Example 2 to find the particular solution of the equation $y' + y = x$ that satisfies the condition $y(0) = 0$; that is, $f(0) = 0$, where f denotes the solution.

Solution From the results of Example 2, we see that the general solution of the given differential equation is given by

$$y = f(x) = ce^{-x} + x - 1$$

Using the given condition, we see that

$$f(0) = ce^0 + 0 - 1 = c - 1 = 0 \quad \text{or} \quad c = 1$$

Therefore, the required particular solution is $y = e^{-x} + x - 1$. ■

EXPLORE & DISCUSS

Consider the differential equation $dy/dx = F(x, y)$ and suppose $y = f(x)$ is a solution of the differential equation.

1. If (a, b) is a point in the domain of F, explain why $F(a, b)$ gives the slope of f at $x = a$.
2. For the differential equation $dy/dx = x/y$, compute $F(x, y)$ for selected integral values of x and y. (For example, try $x = 0, \pm 1, \pm 2$ and $y = \pm 1, \pm 2, \pm 3$.) Verify that if you draw a lineal element (a tiny line segment) having slope $F(x, y)$

through each point (x, y), you obtain a *direction field* similar to the one shown in the figure:

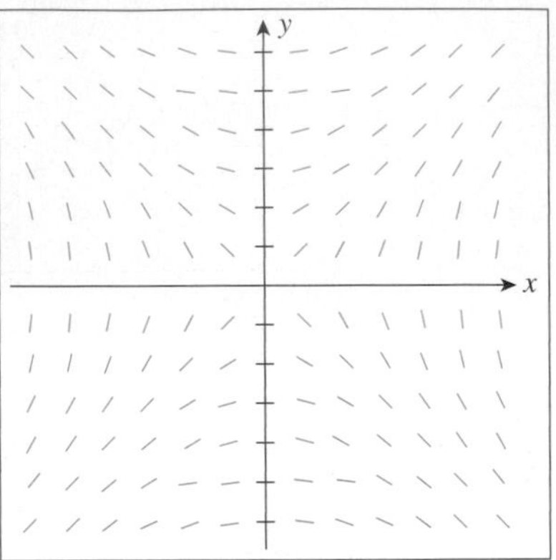

3. The direction field associated with the differential equation hints at the solution curves for the differential equation. Sketch a few solution curves for the differential equation. [You will be asked to verify your answer to part (3) in the next section.]

9.1 Self-Check Exercises

1. Consider the differential equation

$$xy' + 2y = 4x^2$$

   a. Show that $y = x^2 + (c/x^2)$ is the general solution of the differential equation.
   b. Find the particular solution of the differential equation that satisfies $y(1) = 4$.

2. The population of a certain species grows at a rate directly proportional to the square root of its size. If the initial population is N_0, find the population at any time t. Formulate but do not solve the problem.

Solutions to Self-Check Exercises 9.1 can be found on page 626.

9.1 Concept Questions

1. Define the following terms in your own words.
   a. A differential equation.
   b. The general solution of a differential equation.
   c. A particular solution of a differential equation.

2. Give a differential equation that describes the following situation.
   a. The size of a population at any time t, $Q(t)$, increases at a rate that is proportional to $Q(t)$ itself.
   b. The size of a population at any time t, $Q(t)$, does not exceed some number C, and the rate of growth of $Q(t)$ is proportional to the difference between its upper bound and its current size.

3. a. Does the equation of Question 2(a) describe restricted or unrestricted growth?
   b. Does the equation of Question 2(b) describe restricted or unrestricted growth?

4. Given a typical mixture problem where $A(t)$ denotes the amount of salt in the tank at any time t, what condition must the derivative dA/dt, the rate of change of the amount of salt at any time t, satisfy?

9.1 Exercises

In Exercises 1–12, verify that y is a solution of the differential equation.

1. $y = x^2$; $xy' + y = 3x^2$

2. $y = e^x$; $y' - y = 0$

3. $y = \frac{1}{2} + ce^{-x^2}$, c any constant; $y' + 2xy = x$

4. $y = Ce^{kx}$, C any constant; $\frac{dy}{dx} = ky$

5. $y = e^{-2x}$; $y'' + y' - 2y = 0$

6. $y = C_1e^x + C_2e^{2x}$; $y'' - 3y' + 2y = 0$

7. $y = C_1e^{-2x} + C_2xe^{-2x}$; $y'' + 4y' + 4y = 0$

8. $y = C_1 + C_2x^{1/3}$; $3xy'' + 2y' = 0$

9. $y = \frac{C_1}{x} + C_2\frac{\ln x}{x}$; $x^2y'' + 3xy' + y = 0$

10. $y = C_1e^x + C_2\,xe^x + C_3\,x^2e^x$; $y''' - 3y'' + 3y' - y = 0$

11. $y = C - Ae^{-kt}$, A and C constants; $\frac{dy}{dt} = k(C - y)$

12. $y = \frac{C}{1 + Ae^{-Ckt}}$, A and C constants; $\frac{dy}{dt} = ky(C - y)$

In Exercises 13–18, verify that y is a general solution of the differential equation. Then find a particular solution of the differential equation that satisfies the side condition.

13. $y = Cx^2 - 2x$; $y' - 2\left(\frac{y}{x}\right) = 2$; $y(1) = 10$

14. $y = Ce^{-x^2}$; $y' = -2xy$; $y(0) = y_0$

15. $y = \frac{C}{x}$; $y' + \left(\frac{1}{x}\right)y = 0$; $y(1) = 1$

16. $y = Ce^{2x} - 2x - 1$; $y' - 2y - 4x = 0$; $y(0) = 3$

17. $y = \frac{Ce^x}{x} + \frac{1}{2}xe^x$; $y' + \left(\frac{1 - x}{x}\right)y = e^x$; $y(1) = -\frac{1}{2}e$

18. $y = C_1x^3 + C_2x^2$; $x^2y'' - 4xy' + 6y = 0$; $y(2) = 0$ and $y'(2) = 4$

19. **Radioactive Decay** A radioactive substance decays at a rate directly proportional to the amount present. If the substance is present in the amount of Q_0 g initially ($t = 0$), find the amount present at any time t. Formulate the problem in terms of a differential equation with a side condition. Do not solve it.

20. **Supply and Demand** Let $S(t)$ denote the supply of a certain commodity as a function of time t. Suppose the rate of change of the supply is proportional to the difference between the demand $D(t)$ and the supply. Find a differential equation that describes this situation.

21. **Net Investment** The management of a company has decided that the level of investment should not exceed C dollars. Furthermore, management has decided that the rate of net investment (the rate of change of the total capital invested) should be proportional to the difference between C and the total capital invested. Formulate but do not solve the problem in terms of a differential equation.

22. **Lambert's Law of Absorption** Lambert's law of absorption states that the percentage of incident light L, absorbed in passing through a thin layer of material x, is proportional to the thickness of the material. If, for a certain material, x_0 in. of the material reduces the light to half its intensity, how much additional material is needed to reduce the intensity to a quarter of its initial value? Formulate but do not solve the problem in terms of a differential equation with a side condition.

23. **Concentration of a Drug in the Bloodstream** The rate at which the concentration of a drug in the bloodstream decreases is proportional to the concentration at any time t. Initially, the concentration of the drug in the bloodstream is C_0 g/mL. What is the concentration of the drug in the bloodstream at any time t? Formulate but do not solve the problem in terms of a differential equation with a side condition.

24. **Amount of Glucose in the Bloodstream** Suppose glucose is infused into the bloodstream at a constant rate of C g/min and, at the same time, the glucose is converted and removed from the bloodstream at a rate proportional to the amount of glucose present. Show that the amount of glucose $A(t)$ present in the bloodstream at any time t is governed by the differential equation.

$$A' = C - kA$$

where k is a constant.

25. **Newton's Law of Cooling** Newton's law of cooling states that the temperature of a body drops at a rate that is proportional to the difference between the temperature y of the body and the constant temperature C of the surrounding medium (assume that the temperature of the body is initially greater than C). Show that Newton's law of cooling may be expressed as the differential equation

$$\frac{dy}{dt} = -k(y - C) \qquad [y(0) = y_0]$$

where y_0 denotes the temperature of the body before immersion in the medium.

26. **Fisk's Law** Suppose a cell of volume V cc is surrounded by a homogeneous chemical solution of concentration C g/cc. Let y denote the concentration of the solute inside the cell at any time t and suppose that, initially, the concentration is y_0. Fisk's law, named after the German physiologist Adolf Fisk (1829–1901), states that the rate of change of the concentration of solute inside the cell at any time t is proportional to the difference between the concentration of the solute outside the cell and the concentration inside the cell and inversely proportional to the volume of the cell. Show that Fisk's law may be expressed as the differential equation

$$\frac{dy}{dt} = \frac{k}{V}(C - y) \qquad [y(0) = y_0]$$

where k is a constant. (*Note:* The constant of proportionality k depends on the area and permeability of the cell membrane.)

27. **Allometric Laws** Allometry is the study of the relative growth of a part in relation to an entire organism. Suppose $x(t)$ denotes the weight of an animal's organ at time t and $g(t)$ denotes the size of another organ in the same animal at the same time t. An allometric law states that the relative growth rate of one organ, $(dx/dt)/x$, is proportional to the relative growth rate of the other, $(dy/dt)/y$. Show that this allometric law may be stated in terms of the differential equation

$$\frac{1}{x}\frac{dx}{dt} - k\frac{1}{y}\frac{dy}{dt}$$

where k is a constant.

28. **Gompertz Growth Curve** Suppose a quantity $Q(t)$ does not exceed some number C; that is, $Q(t) \leq C$ for all t. Suppose further that the rate of growth of $Q(t)$ is jointly proportional to its current size and the difference between its upper bound and the natural logarithm of its current size. What is the size of the quantity $Q(t)$ at any time t? Show that the mathematical formulation of this problem leads to the differential equation

$$\frac{dQ}{dt} = kQ(C - \ln Q) \qquad [Q(0) = Q_0]$$

where Q_0 denotes the size of the quantity present initially. The graph of $Q(t)$ is called the *Gompertz growth curve*. This model, like the ones leading to the learning curve and the logistic curve, describes restricted growth.

In Exercises 29–34, determine whether the statement is true or false. If it is true, explain why it is true. If it is false, give an example to show why it is false.

29. The function $f(x) = x^2 + 2x + \frac{1}{x}$ is a solution of the differential equation $xy' + y = 3x^2 + 4x$.

30. The function $f(x) = \frac{1}{4}e^{3x} + ce^{-x}$ is a solution of the differential equation $y' + y = e^{3x}$.

31. The function $f(x) = 2 + ce^{-x^3}$ is a solution of the differential equation $y' + 3x^2y = x^2$.

32. The function $f(x) = 1 + cx^{-2}$ is a solution of the differential equation $xy' + 2y = 3$.

33. If $y = f(x)$ is a solution of a first-order differential equation, then $y = Cf(x)$ is also a solution.

34. If $y = f(x)$ is a solution of a first-order differential equation, then $y = f(x) + C$ is also a solution.

9.1 Solutions to Self-Check Exercises

1. **a.** We compute

$$y' = 2x - \frac{2c}{x^3}$$

Substituting this into the left side of the given differential equation gives

$$x\left(2x - \frac{2c}{x^3}\right) + 2\left(x^2 + \frac{c}{x^2}\right) = 2x^2 - \frac{2c}{x^2} + 2x^2 + \frac{2c}{x^2} = 4x^2$$

which equals the expression on the right side of the differential equation, and this verifies the assertion.

b. Using the given condition, we have

$$4 = 1^2 + \frac{c}{1^2} \qquad \text{or} \qquad c = 3$$

and the required particular solution is

$$y = x^2 + \frac{3}{x^2}$$

2. Let N denote the size of the population at any time t. Then the required differential equation is

$$\frac{dN}{dt} = kN^{1/2}$$

and the initial condition is $N(0) = N_0$.

9.2 Separation of Variables

The Method of Separation of Variables

Differential equations are classified according to their basic form. A compelling reason for this categorization is that different methods are used to solve different types of equations.

A differential equation may be classified by the order of its derivative. The **order** of a differential equation is the order of the highest derivative of the unknown function appearing in the equation. For example, the differential equations

$$y' = xe^x \qquad \text{and} \qquad y' + 2y = x^2$$

are **first-order equations**, whereas the differential equation

$$\frac{d^2y}{dt^2} + \left(\frac{dy}{dt}\right)^3 + ty - 8 = 0$$

is a second-order equation. For the remainder of this chapter, we restrict our study to first-order differential equations.

In this section we describe a method for solving an important class of first-order differential equations that can be written in the form

$$\frac{dy}{dx} = f(x)g(y)$$

where $f(x)$ is a function of x only and $g(y)$ is a function of y only. Such differential equations are said to be **separable** because the variables can be separated. Equations (1) through (5) are first-order separable differential equations. As another example, the equation

$$\frac{dQ}{dt} = kQ(C - Q)$$

has the form $dQ/dt = f(t)g(Q)$, where $f(t) = k$ and $g(Q) = Q(C - Q)$, and so it is separable. On the other hand, the differential equation

$$\frac{dy}{dx} = xy^2 + 2$$

is *not* separable.

Separable first-order equations can be solved using the **method of separation of variables.**

Method of Separation of Variables

Suppose we are given a first-order separable differential equation in the form

$$\frac{dy}{dx} = f(x)g(y) \tag{6}$$

Step 1 Write Equation (6) in the form

$$\frac{dy}{g(y)} = f(x)dx \tag{7}$$

When written in this form, the variables in (7) are said to be *separated.*

Step 2 Integrate each side of Equation (7) with respect to the appropriate variable.

We will justify this method at the end of this section.

Solving Separable Differential Equations

EXAMPLE 1 Find the general solution of the first-order differential equation

$$y' = \frac{xy}{x^2 + 1}$$

Solution

Step 1 Observe that the given differential equation has the form

$$\frac{dy}{dx} = \left(\frac{x}{x^2 + 1}\right)y = f(x)g(y)$$

where $f(x) = x/(x^2 + 1)$ and $g(y) = y$, and is therefore separable. Separating the variables, we obtain

$$\frac{dy}{y} = \left(\frac{x}{x^2 + 1}\right)dx$$

Step 2 Integrating each side of the last equation with respect to the appropriate variable, we have

$$\int \frac{dy}{y} = \int \frac{x}{x^2 + 1}\,dx$$

or

$$\ln|y| + C_1 = \frac{1}{2}\ln(x^2 + 1) + C_2$$

$$\ln|y| = \frac{1}{2}\ln(x^2 + 1) + C_2 - C_1$$

where C_1 and C_2 are arbitrary constants of integration. Letting C denote the constant such that $C_2 - C_1 = \ln|C|$, we have

$$\begin{aligned}\ln|y| &= \frac{1}{2}\ln(x^2 + 1) + \ln|C| \\ &= \ln\sqrt{x^2 + 1} + \ln|C| \\ &= \ln|C\sqrt{x^2 + 1}| \qquad \ln A + \ln B = \ln AB\end{aligned}$$

so the general solution is

$$y = C\sqrt{x^2 + 1}$$

EXPLORING WITH TECHNOLOGY

Refer to Example 1, where it was shown that the general solution of the given differential equation is $y = C\sqrt{x^2 + 1}$. Use a graphing utility to plot the graphs of the members of this family of solutions corresponding to $C = -3, -2, -1, 0, 1, 2,$ and 3. Use the standard viewing window.

EXAMPLE 2 Find the particular solution of the differential equation

$$ye^x + (y^2 - 1)y' = 0$$

that satisfies the condition $y(0) = 1$.

Solution

Step 1 Writing the given differential equation in the form

$$ye^x + (y^2 - 1)\frac{dy}{dx} = 0 \quad \text{or} \quad (y^2 - 1)\frac{dy}{dx} = -ye^x$$

and separating the variables, we obtain

$$\frac{y^2 - 1}{y}dy = -e^x dx$$

Step 2 Integrating each side of this equation with respect to the appropriate variable, we have

$$\int \frac{y^2 - 1}{y}dy = -\int e^x dx$$

$$\int \left(y - \frac{1}{y}\right) dy = -\int e^x dx$$

$$\frac{1}{2}y^2 - \ln|y| = -e^x + C_1$$

$$y^2 - \ln y^2 = -2e^x + C \qquad (C = 2C_1)$$

Using the condition $y(0) = 1$, we have

$$1 - \ln 1 = -2 + C \quad \text{or} \quad C = 3$$

Therefore, the required solution is

$$y^2 - \ln y^2 = -2e^x + 3$$

Example 2 is an initial value problem. In general, an **initial value problem** consists of a differential equation with one or more side conditions specified at a point. Also observe that the solution of Example 2 appeared as an implicit equation involving x and y. This often happens when we solve separable differential equations.

EXAMPLE 3 Find an equation describing f given that (1) the slope of the tangent line to the graph of f at any point $P(x, y)$ is given by the expression $-x/(2y)$ and (2) the graph of f passes through the point $P(1, 2)$.

Solution The slope of the tangent line to the graph of f at any point $P(x, y)$ is given by the derivative

$$y' = \frac{dy}{dx} = -\frac{x}{2y}$$

which is a separable first-order differential equation. Separating the variables, we obtain

$$2y\,dy = -x\,dx$$

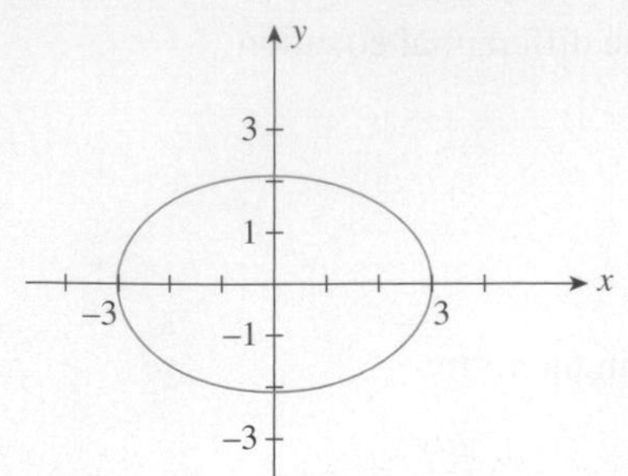

FIGURE 6
The graph of $x^2 + 2y^2 = 9$

which, upon integration, yields

$$y^2 = -\frac{1}{2}x^2 + C_1$$

or

$$x^2 + 2y^2 = C \qquad (C = 2C_1)$$

where C is an arbitrary constant.

To evaluate C, we use the second condition, which implies that when $x = 1$, $y = 2$. This gives

$$1^2 + 2(2^2) = C \qquad \text{or} \qquad C = 9$$

Hence, the required equation is

$$x^2 + 2y^2 = 9$$

The graph of the equation f appears in Figure 6.

EXPLORE & DISCUSS

Refer to the Explore & Discuss problem on page 623. Use the method of separation of variables to solve the differential equation $dy/dx = x/y$ and thus verify that your solution to part 3 in the problem is indeed correct.

Justification of the Method of Separation of Variables

To justify the method of separation of variables, let's consider the separable Equation (6) in its general form:

$$\frac{dy}{dx} = f(x)g(y)$$

If $g(y) \neq 0$, we may rewrite the equation in the form

$$\frac{1}{g(y)}\frac{dy}{dx} - f(x) = 0$$

Now, suppose that G is an antiderivative of $1/g$ and F is an antiderivative of f. Using the chain rule, we see that

$$\frac{d}{dx}[G(y) - F(x)] = G'(y)\frac{dy}{dx} - F'(x) = \frac{1}{g(y)}\frac{dy}{dx} - f(x)$$

Therefore,

$$\frac{d}{dx}[G(y) - F(x)] = 0$$

and so

$$G(y) - F(x) = C \qquad C\text{, a constant}$$

But the last equation is equivalent to

$$G(y) = F(x) + C \qquad \text{or} \qquad \int \frac{dy}{g(y)} = \int f(x)dx$$

which is precisely the result of step 2 in the method of separation of variables.

9.2 Self-Check Exercise

Find the solution of the differential equation $y' = 2x^2y + 2x^2$ that satisfies the condition $y(0) = 0$.

The solution to Self-Check Exercise 9.2 can be found on page 632.

9.2 Concept Questions

1. **a.** What is the order of a differential equation? Give an example.
   b. What is a separable equation? Give an example of a differential equation that is separable and of one that is *not* separable.
2. Explain how you would solve a separable equation.
3. What is an initial value problem?

9.2 Exercises

In Exercises 1–16, find the general solution of the first-order differential equation by separating variables.

1. $y' = \dfrac{x+1}{y^2}$
2. $y' = \dfrac{x^2}{y}$
3. $y' = \dfrac{e^x}{y^2}$
4. $y' = -\dfrac{x}{y}$
5. $y' = 2y$
6. $y' = 2(y-1)$
7. $y' = xy^2$
8. $y' = \dfrac{2y}{x+1}$
9. $y' = -2(3y+4)$
10. $y' = \dfrac{2y+3}{x^2}$
11. $y' = \dfrac{x^2+1}{3y^2}$
12. $y' = \dfrac{xe^x}{2y}$
13. $y' = \sqrt{\dfrac{y}{x}}$
14. $y' = \dfrac{xy^2}{\sqrt{1+x^2}}$
15. $y' = \dfrac{y \ln x}{x}$
16. $y' = \dfrac{(x-4)y^4}{x^3(y^2-3)}$

In Exercises 17–28, find the solution of the initial value problem.

17. $y' = \dfrac{2x}{y}$; $y(1) = -2$
18. $y' = xe^{-y}$; $y(0) = 1$
19. $y' = 2 - y$; $y(0) = 3$
20. $y' = \dfrac{y}{x}$; $y(1) = 1$
21. $y' = 3xy - 2x$; $y(0) = 1$
22. $y' = xe^{x^2}y$; $y(0) = 1$
23. $y' = \dfrac{xy}{x^2+1}$; $y(0) = 1$
24. $y' = x^2y^{-1/2}$; $y(1) = 1$
25. $y' = xye^x$; $y(1) = 1$
26. $y' = 2xe^{-y}$; $y(0) = 1$
27. $y' = 3x^2e^{-y}$; $y(0) = 1$
28. $y' = \dfrac{y^2}{x-2}$; $y(3) = 1$
29. Find an equation defining a function f given that (1) the slope of the tangent line to the graph of f at any point $P(x, y)$ is given by the expression $dy/dx = (3x^2)/(2y)$ and (2) the graph of f passes through the point (1, 3).
30. Find a function f given that (1) the slope of the tangent line to the graph of f at any point $P(x, y)$ is given by the expression $dy/dx = 3xy$ and (2) the graph of f passes through the point (0, 2).
31. **Exponential Decay** Use separation of variables to solve the differential equation
$$\frac{dQ}{dt} = -kQ \qquad [Q(0) = Q_0]$$
where k and Q_0 are positive constants, describing exponential decay.
32. **Fisk's Law** Refer to Exercise 26, Section 9.1. Use separation of variables to solve the differential equation
$$\frac{dy}{dt} = \frac{k}{V}(C - y) \qquad [y(0) = y_0]$$

where k, V, C, and y_0 are constants with $C - y > 0$. Find $\lim_{t \to \infty} y$ and interpret your result.

33. **Concentration of Glucose in the Bloodstream** Refer to Exercise 24, Section 9.1. Use separation of variables to solve the differential equation $A' = C - kA$, where C and k are positive constants.
Hint: Rewrite the given differential equation in the form

$$\frac{dA}{dt} = k\left(\frac{C}{k} - A\right)$$

34. **Allometric Laws** Refer to Exercise 27, Section 9.1. Use separation of variables to solve the differential equation

$$\frac{1}{x}\frac{dx}{dt} = k\frac{1}{y}\frac{dy}{dt}$$

where k is a constant.

35. **Supply and Demand** Assume that the rate of change of the supply of a commodity is proportional to the difference between the demand and the supply so that

$$\frac{dS}{dt} = k(D - S)$$

where k is a constant of proportionality. Suppose that D is constant and $S(0) = S_0$. Find a formula for $S(t)$.

36. **Supply and Demand** Assume that the rate of change of the unit price of a commodity is proportional to the difference between the demand and the supply so that

$$\frac{dp}{dt} = k(D - S)$$

where k is a constant of proportionality. Suppose that $D = 50 - 2p$, $S = 5 + 3p$, and $p(0) = 4$. Find a formula for $p(t)$.

In Exercises 37–42, determine whether the statement is true or false. If it is true, explain why it is true. If it is false, give an example to show why it is false.

37. The differential equation $y' = xy + 2x - y - 2$ is separable.

38. The differential equation $y' = x^2 - y^2$ is separable.

39. If the differentiable function $M(x, y)dx + N(x, y)dy = 0$ can be written so that $M(x, y) = f(x)g(y)$ and $N(x, y) = F(x)G(y)$ for functions f, g, F, and G, then it is separable.

40. The differential equation $(x^2 + 2)dx + (2x - 4xy)dy = 0$ is separable.

41. The differential equation $ydx - (y - xy^2)dy = 0$ is separable.

42. The differential equation $\dfrac{dy}{dx} = \dfrac{f(x)g(y)}{F(x) + G(y)}$ is separable.

9.2 Solution to Self-Check Exercise

Writing the differential equation in the form

$$\frac{dy}{dx} = 2x^2(y + 1)$$

and separating variables, we obtain

$$\frac{dy}{y + 1} = 2x^2dx$$

Integrating each side of the last equation with respect to the appropriate variable, we have

$$\int \frac{dy}{y + 1} = \int 2x^2dx \quad \text{or} \quad \ln|y + 1| = \frac{2}{3}x^3 + C$$

Using the initial condition $y(0) = 0$, we have

$$\ln 1 = C \quad \text{or} \quad C = 0$$

Therefore,

$$\begin{aligned} \ln|y + 1| &= \frac{2}{3}x^3 \\ y + 1 &= e^{(2/3)x^3} \\ y &= e^{(2/3)x^3} - 1 \end{aligned}$$

9.3 Applications of Separable Differential Equations

In this section we look at some applications of first-order separable differential equations. We begin by reexamining some of the applications discussed in Section 9.1.

Unrestricted Growth Models

The differential equation describing an unrestricted growth model is given by

$$\frac{dQ}{dt} = kQ$$

where $Q(t)$ represents the size of a certain population at any time t and k is a positive constant. Separating the variables in this differential equation, we have

$$\frac{dQ}{Q} = k\,dt$$

which, upon integration, yields

$$\int \frac{dQ}{Q} = \int k\,dt$$

$$\ln|Q| = kt + C_1$$

$$Q = e^{kt+C_1} = Ce^{kt}$$

where $C = e^{C_1}$ is an arbitrary positive constant. Thus, we may write the solution as

$$Q(t) = Ce^{kt}$$

Observe that if the quantity present initially is denoted by Q_0, then $Q(0) = Q_0$. Applying this condition yields the equation

$$Ce^0 = Q_0 \quad \text{or} \quad C = Q_0$$

Therefore, the model for unrestricted exponential growth with initial population Q_0 is given by

$$Q(t) = Q_0e^{kt} \tag{8}$$

(Figure 7).

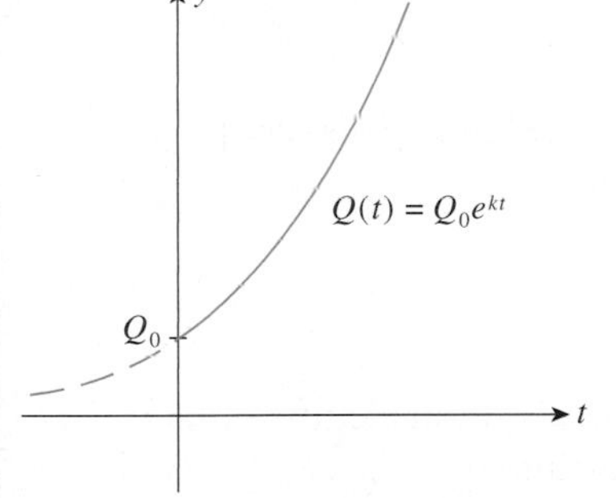

FIGURE 7
An unrestricted growth model

APPLIED EXAMPLE 1 Growth of Bacteria Under ideal laboratory conditions, the rate of growth of bacteria in a culture is proportional to the size of the culture at any time t. Suppose that 10,000 bacteria are present initially in a culture and 60,000 are present 2 hours later. How many bacteria will there be in the culture at the end of 4 hours?

Solution Let $Q(t)$ denote the number of bacteria present in the culture at time t. Then

$$\frac{dQ}{dt} = kQ$$

where k is a constant of proportionality. Solving this separable first-order differential equation, we obtain

$$Q(t) = Q_0 e^{kt} \qquad \text{Equation (8)}$$

where Q_0 denotes the initial bacteria population. Since $Q_0 = 10{,}000$, we have

$$Q(t) = 10{,}000e^{kt}$$

Next, the condition that 60,000 bacteria are present 2 hours later translates into $Q(2) = 60{,}000$, or

$$\begin{aligned} 60{,}000 &= 10{,}000e^{2k} \\ e^{2k} &= 6 \\ e^k &= 6^{1/2} \end{aligned}$$

Thus, the number of bacteria present at any time t is given by

$$\begin{aligned} Q(t) &= 10{,}000e^{kt} = 10{,}000(e^k)^t \\ &= (10{,}000)6^{(t/2)} \end{aligned}$$

In particular, the number of bacteria present in the culture at the end of 4 hours is given by

$$\begin{aligned} Q(4) &= 10{,}000(6^{4/2}) \\ &= 360{,}000 \end{aligned}$$

■

Restricted Growth Models

From Section 9.1 we see that a differential equation describing a restricted growth model is given by

$$\frac{dQ}{dt} = k(C - Q) \tag{9}$$

where both k and C are positive constants. To solve this separable first-order differential equation, we first separate the variables, obtaining

$$\frac{dQ}{C - Q} = k\,dt$$

Integrating each side with respect to the appropriate variable yields

$$\begin{aligned} \int \frac{dQ}{C - Q} &= \int k\,dt \\ -\ln|C - Q| &= kt + d \qquad d\text{, an arbitrary constant} \\ \ln|C - Q| &= -kt - d \\ C - Q &= e^{-kt-d} = e^{-kt}e^{-d} \\ Q(t) &= C - Ae^{-kt} \end{aligned} \tag{10}$$

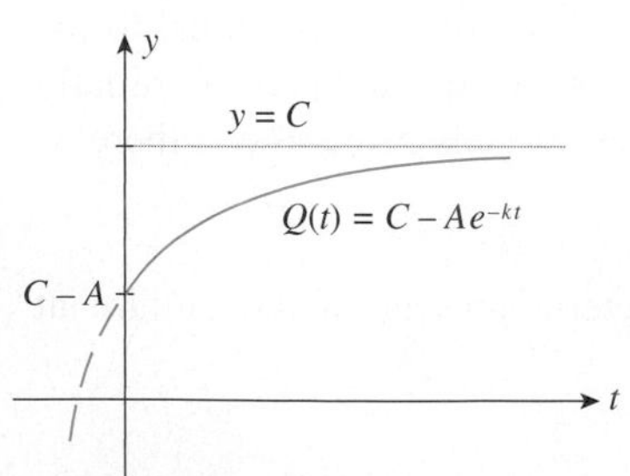

FIGURE 8
A restricted exponential growth model

where we have denoted the constant e^{-d} by A. This is the equation of the learning curve (Figure 8) studied in Chapter 5.

APPLIED EXAMPLE 2 Yield of a Wheat Field In an experiment conducted by researchers of the Agriculture Department of a midwestern university, it was found that the maximum yield of wheat in the university's experimental field station was 150 bushels per acre. Furthermore, the researchers discovered that the rate at which the yield of wheat increased was governed by the differential equation

$$\frac{dQ}{dx} = k(150 - Q)$$

where $Q(x)$ denotes the yield in bushels per acre and x is the amount in pounds of an experimental fertilizer used per acre of land. Data obtained in the experiment indicated that 10 pounds of fertilizer per acre of land would result in a yield of 80 bushels of wheat per acre, whereas 20 pounds of fertilizer per acre of land would result in a yield of 120 bushels of wheat per acre. Determine the yield if 30 pounds of fertilizer were used per acre.

Solution The given differential equation has the same form as Equation (9) with $C = 150$. Solving it directly or using the result obtained in the solution of Equation (9), we see that the yield per acre is given by

$$Q(x) = 150 - Ae^{-kx}$$

The first of the given conditions implies that $Q(10) = 80$; that is,

$$150 - Ae^{-10k} = 80$$

or $A = 70e^{10k}$. Therefore,

$$\begin{aligned} Q(x) &= 150 - 70e^{10k}e^{-kx} \\ &= 150 - 70e^{-k(x-10)} \end{aligned}$$

The second of the given conditions implies that $Q(20) = 120$, or

$$\begin{aligned} 150 - 70e^{-k(20-10)} &= 120 \\ 70e^{-10k} &= 30 \\ e^{-10k} &= \frac{3}{7} \end{aligned}$$

Taking the logarithm of each side of the equation, we find

$$\begin{aligned} \ln e^{-10k} &= \ln\left(\frac{3}{7}\right) \\ -10k &= \ln 3 - \ln 7 \approx -0.8473 \\ k &\approx 0.085 \end{aligned}$$

Therefore,

$$Q(x) = 150 - 70e^{-0.085(x-10)}$$

In particular, when $x = 30$, we have

$$\begin{aligned} Q(30) &= 150 - 70e^{-0.085(20)} \\ &= 150 - 70e^{-1.7} \\ &\approx 137 \end{aligned}$$

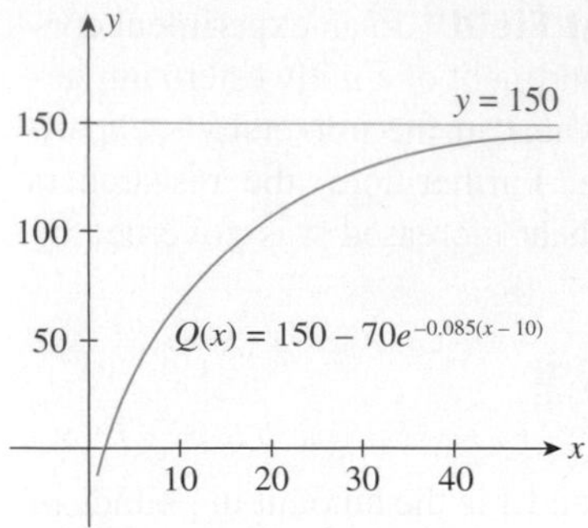

FIGURE 9
$Q(x)$ is a function relating crop yield to the amount of fertilizer used.

So the yield would be 137 bushels per acre if 30 pounds of fertilizer were used per acre. The graph of $Q(x)$ is shown in Figure 9. Note that it was not necessary to find the specific value of k in order to solve this problem. (Why?) ■

Next, let's consider a differential equation describing another type of restricted growth:

$$\frac{dQ}{dt} = kQ(C - Q)$$

where k and C are positive constants. Separating variables leads to

$$\frac{dQ}{Q(C - Q)} = k\,dt$$

Integrating each side of this equation with respect to the appropriate variable, we have

$$\int \frac{1}{Q(C - Q)}\,dQ = \int k\,dt$$

As it stands, the integrand on the left side of this equation is not in a form that can be easily integrated. However, observe that

$$\frac{1}{Q(C - Q)} = \frac{1}{C}\left[\frac{1}{Q} + \frac{1}{C - Q}\right]$$

as you may verify by adding the terms between the brackets on the right-hand side. Making use of this identity, we have

$$\int \frac{1}{C}\left[\frac{1}{Q} + \frac{1}{C - Q}\right]dQ = \int k\,dt$$

$$\int \frac{dQ}{Q} + \int \frac{dQ}{C - Q} = Ck\int dt$$

$$\ln|Q| - \ln|C - Q| = Ckt + b \qquad b\text{, an arbitrary constant}$$

$$\ln\left|\frac{Q}{C - Q}\right| = Ckt + b$$

$$\frac{Q}{C - Q} = e^{Ckt+b} = e^{b}e^{Ckt} = Be^{Ckt} \qquad (B = e^{b})$$

$$Q = CBe^{Ckt} - QBe^{Ckt}$$

$$(1 + Be^{Ckt})Q = CBe^{Ckt}$$

and

$$Q = \frac{CBe^{Ckt}}{1 + Be^{Ckt}}$$

or

$$Q(t) = \frac{C}{1 + Ae^{-Ckt}} \qquad \left(A = \frac{1}{B}\right) \tag{11}$$

(see Figure 2, page 621). In its final form, this function is equivalent to the logistic function encountered in Chapter 5.

APPLIED EXAMPLE 3 Spread of a Flu Epidemic During a flu epidemic, 5% of the 5000 army personnel stationed at Fort MacArthur had contracted influenza at time $t = 0$. Furthermore, the rate at which they were contracting influenza was jointly proportional to the number of personnel who had already contracted the disease and the noninfected population. If 20% of the personnel had contracted the flu by the 10th day, find the number of personnel who had contracted the flu by the 13th day.

Solution Let $Q(t)$ denote the number of army personnel who had contracted the flu after t days. Then

$$\frac{dQ}{dt} = kQ(5000 - Q)$$

We may solve this separable differential equation directly, or we may appeal to the result for the more general problem obtained earlier. Opting for the latter and noting that $C = 5000$, we find that, using Equation (11),

$$Q(t) = \frac{5000}{1 + Ae^{-5000kt}}$$

The condition that 5% of the population had contracted influenza at time $t = 0$ implies that

$$Q(0) = \frac{5000}{1 + A} = 250$$

from which we see that $A = 19$. Therefore,

$$Q(t) = \frac{5000}{1 + 19e^{-5000kt}}$$

Next, the condition that 20% of the population had contracted influenza by the tenth day implies that

$$Q(10) = \frac{5000}{1 + 19e^{-50,000k}} = 1000$$

or

$$\begin{aligned} 1 + 19e^{-50,000k} &= 5 \\ e^{-50,000k} &= \frac{4}{19} \\ -50,000k &= \ln 4 - \ln 19 \end{aligned}$$

and

$$\begin{aligned} k &= -\frac{1}{50,000}(\ln 4 - \ln 19) \\ &\approx 0.0000312 \end{aligned}$$

Therefore,

$$Q(t) = \frac{5000}{1 + 19e^{-0.156t}}$$

In particular, the number of army personnel who had contracted the flu by the 13th day is given by

$$Q(13) = \frac{5000}{1 + 19e^{-0.156(13)}} = \frac{5000}{1 + 19e^{-2.028}} \approx 1428$$

or approximately 1428, or 29%. The graph of $Q(t)$ is shown in Figure 10.

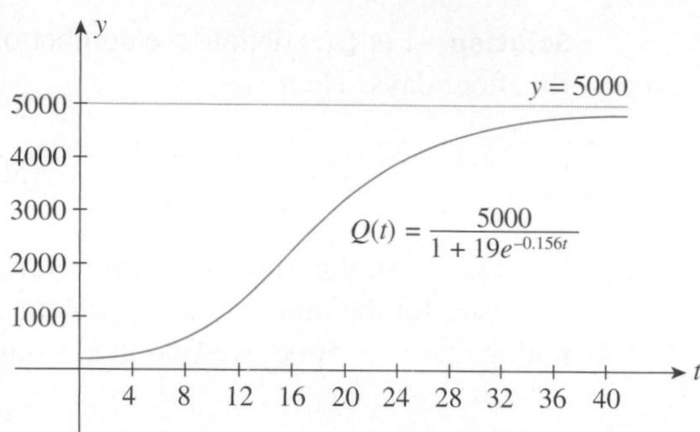

FIGURE 10
An epidemic model

EXPLORE & DISCUSS

Consider the model for restricted growth described by the differential equation $dQ/dt = kQ(C - Q)$ with the solution given in Equation (11).

1. Show that the rate of growth of Q is greatest at $t = (\ln A)/kC$.
   Hint: Use both the differential equation and Equation (11).
2. Refer to Example 3. At what time is the number of influenza cases increasing at the greatest rate?

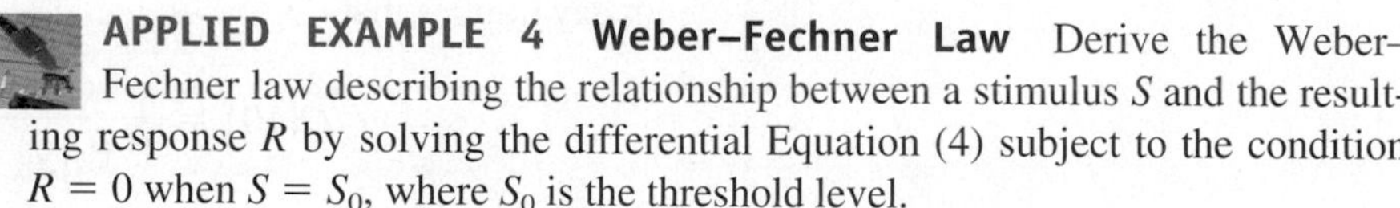

APPLIED EXAMPLE 4 Weber–Fechner Law Derive the Weber–Fechner law describing the relationship between a stimulus S and the resulting response R by solving the differential Equation (4) subject to the condition $R = 0$ when $S = S_0$, where S_0 is the threshold level.

Solution The differential equation under consideration is

$$\frac{dR}{dS} = \frac{k}{S}$$

and is separable. Separating the variables, we have

$$dR = k\frac{dS}{S}$$

which, upon integration, yields

$$\int dR = k\int \frac{dS}{S}$$

$$R = k \ln S + C$$

where C is an arbitrary constant. Using the condition $R = 0$ when $S = S_0$ gives

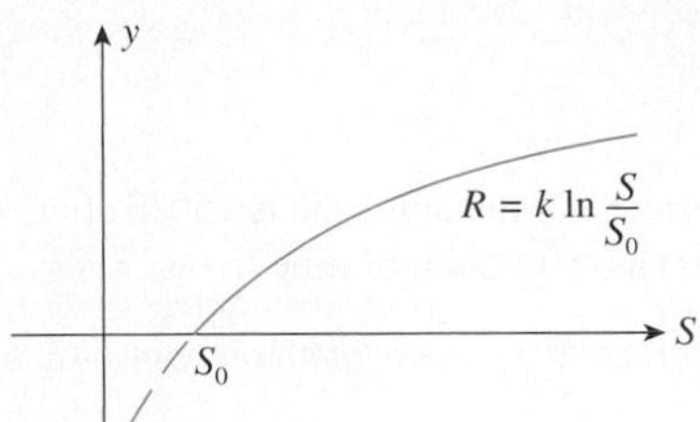

FIGURE 11
The Weber–Fechner law

$$0 = k \ln S_0 + C$$
$$C = -k \ln S_0$$

Substituting this value of C in the expression for R leads to

$$R = k \ln S - k \ln S_0$$
$$= k \ln \frac{S}{S_0}$$

the required relationship between R and S. The graph of R is shown in Figure 11. ■

APPLIED EXAMPLE 5 A Mixture Problem A tank initially contains 10 gallons of pure water. Brine containing 3 pounds of salt per gallon flows into the tank at a rate of 2 gallons per minute, and the well-stirred mixture flows out of the tank at the same rate. How much salt is present at the end of 10 minutes? How much salt is present in the long run?

Solution The problem was formulated mathematically on page 621, and we were led to the differential equation

$$\frac{dA}{dt} = 6 - \frac{A}{5}$$

subject to the condition $A(0) = 0$. The differential equation

$$\frac{dA}{dt} = 6 - \frac{A}{5} = \frac{30 - A}{5}$$

is separable. Separating the variables and integrating, we obtain

$$\int \frac{dA}{30 - A} = \int \frac{1}{5}\, dt$$

$$-\ln|30 - A| = \frac{1}{5}t + b \qquad b\text{, a constant}$$

$$\ln|30 - A| = -\frac{1}{5}t - b$$

$$30 - A = e^{-b}e^{-t/5}$$

$$A = 30 - Ce^{-t/5} \qquad (C = e^{-b})$$

The condition $A(0) = 0$ implies that

$$0 = 30 - C$$

giving $C = 30$, and so on

$$A(t) = 30(1 - e^{-t/5})$$

The amount of salt present after 10 minutes is given by

$$A(10) = 30(1 - e^{-2}) \approx 25.94$$

or 25.94 pounds. The amount of salt present in the long run is given by

$$\lim_{t\to\infty} A(t) = \lim_{t\to\infty} 30(1 - e^{-t/5}) = 30$$

or 30 pounds (Figure 12). ■

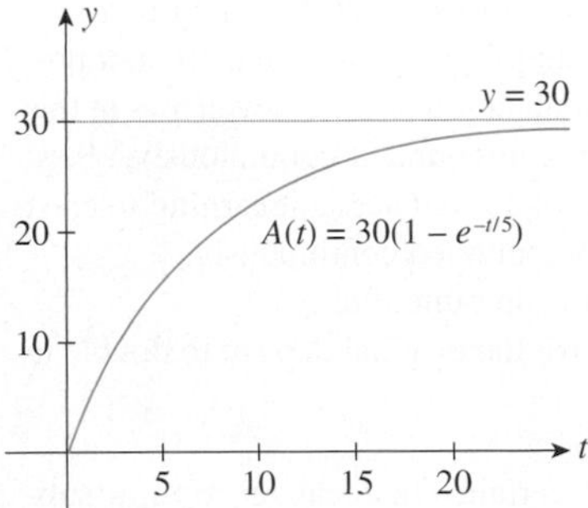

FIGURE 12
The solution of the differential equation $\frac{dA}{dt} = 6 - \frac{A}{5}$

9.3 Self-Check Exercise

Newton's law of cooling states that the temperature of an object drops at a rate that is proportional to the difference in the temperature between the object and that of the surrounding medium. Suppose that an apple pie is taken out of the oven at a temperature of 200°F and placed on the counter in a room where the temperature is 70°F. If the temperature of the apple pie is 150°F after 5 min, find its temperature $y(t)$ as a function of time t.

The solution to Self-Check Exercise 9.3 can be found on page 642.

9.3 Concept Questions

1. Consider the differential equation $\frac{dQ}{dt} = kQ$ $(k > 0)$ describing unrestricted growth.
   a. What does the equation say about the rate of growth of Q with respect to t? What happens to the rate of growth as t approaches infinity?
   b. Reconcile your answer to part (a) with the solution $Q(t) = Q_0e^{kt}$ of the differential equation.

2. Consider the differential equation $\frac{dR}{dS} = \frac{k}{S}$ $(k > 0)$ describing the relationship between a stimulus S and the resulting response R (the Weber–Fechner law).
   a. What can you say about the rate of change of R with respect to S? What happens to the rate of change as S approaches infinity?
   b. Reconcile your answers to part (a) with the solution $R = k \ln \frac{S}{S_0}$ of the differential equation.

9.3 Exercises

1. **Chemical Decomposition** The rate of decomposition of a certain chemical substance is directly proportional to the amount present at any time t. If y_0 g of the chemical are present at time $t = 0$, find an expression for the amount present at any time t.

2. **Growth of Bacteria** Under ideal laboratory conditions, the rate of growth of bacteria in a culture is proportional to the size of the culture at any time t. Suppose that 2000 bacteria are present initially in the culture and 5000 are present 1 hr later. How many bacteria will be in the culture at the end of 2 hr?

3. **World Population Growth** The world population at the beginning of 1980 was 4.5 billion. Assuming that the population continues to grow at its present rate of approximately 2%/year, find a function $Q(t)$ that expresses the world population (in billions) as a function of time t (in years). What will be the world population at the beginning of 2008?

4. **Population Growth** The population of a certain community is increasing at a rate directly proportional to the population at any time t. In the last 3 yr, the population has doubled. How long will it take for the population to triple?

5. **Lambert's Law of Absorption** According to Lambert's law of absorption, the percentage of incident light L, absorbed in passing through a thin layer of material x, is proportional to the thickness of the material. If $\frac{1}{2}$ in. of a certain material reduces the light to half of its intensity, how much additional material is needed to reduce the intensity to one-fourth of its initial value?

6. **Savings Accounts** An amount of money deposited in a savings account grows at a rate proportional to the amount present. (It can be shown that an amount of money grows in this manner if it earns interest compounded continuously.) Suppose $10,000 is deposited in a fixed account earning interest at the rate of 10%/year compounded continuously.
   a. What is the accumulated amount after 5 yr?
   b. How long does it take for the original deposit to double in value?

7. **Chemical Reactions** In a certain chemical reaction, a substance is converted into another substance at a rate proportional to the square of the amount of the first substance present at any time t. Initially ($t = 0$) 50 g of the first substance was present; 1 hr later, only 10 g of it remained. Find an ex-

pression that gives the amount of the first substance present at any time t. What is the amount present after 2 hr?

8. Newton's Law of Cooling Newton's law of cooling states that the rate at which the temperature of an object changes is directly proportional to the difference in temperature between the object and that of the surrounding medium. A horseshoe heated to a temperature of 100°C is immersed in a large tank of water at a (constant) temperature of 30°C at time $t = 0$. Three minutes later the temperature of the horseshoe is reduced to 70°C. Derive an expression that gives the temperature of the horseshoe at any time t. What is the temperature of the horseshoe 5 min after it has been immersed in the water?

9. Newton's Law of Cooling Newton's law of cooling states that the rate at which the temperature of an object changes is directly proportional to the difference in temperature between the object and that of the surrounding medium. A cup of coffee is prepared with boiling water (212°F) and left to cool on the counter in a room where the temperature is 72°F. If the temperature of the coffee is 140°F after 2 min, determine when the coffee will be cool enough to drink (say, 110°F).

10. Learning Curves The American Stenographic Institute finds that the average student taking Elementary Shorthand will progress at a rate given by

$$\frac{dQ}{dt} = k(80 - Q)$$

in a 20-wk course, where $Q(t)$ measures the number of words of dictation a student can take per minute after t wk in the course. If the average student can take 50 words of dictation per minute after 10 wk in the course, how many words per minute can the average student take after completing the course?

11. Training Personnel The personnel manager of Gibraltar Insurance Company estimates that the number of insurance claims an experienced clerk can process in a day is 40. Furthermore, the rate at which a clerk can process insurance claims during the tth wk of training is proportional to the difference between the maximum number possible (40) and the number he or she can process in the tth wk. If the number of claims the average trainee can process after 2 wk on the job is 10/day, determine how many claims the average trainee can process after 6 wk on the job.

12. Effect of Immigration on Population Growth Suppose a country's population at any time t grows in accordance with the rule

$$\frac{dP}{dt} = kP + I$$

where P denotes the population at any time t, k is a positive constant reflecting the natural growth rate of the population, and I is a constant giving the (constant) rate of immigration into the country. If the total population of the country at time $t = 0$ is P_0, find an expression for the population at any time t.

13. Effect of Immigration on Population Growth Refer to Exercise 12. The population of the United States in the year 1980 ($t = 0$) was 226.5 million. Suppose the natural growth rate is 0.8% annually ($k = 0.008$) and net immigration is allowed at the rate of .5 million people/year ($I = 0.5$). What will be the U.S. population in 2008?

14. Sinking Funds The proprietor of Carson Hardware Store has decided to set up a sinking fund for the purpose of purchasing a computer 2 yr from now. It is expected that the purchase will involve a sum of \$30,000. The fund grows at the rate of

$$\frac{dA}{dt} = rA + P$$

where A denotes the size of the fund at any time t, r is the annual interest rate earned by the fund compounded continuously, and P is the amount (in dollars) paid into the fund by the proprietor per year (assume this is done on a frequent basis in small deposits over the year so that it is essentially continuous). If the fund earns 10% interest per year compounded continuously, determine the size of the yearly investment the proprietor should pay into the fund.

15. Spread of a Rumor The rate at which a rumor spreads through an Alpine village of 400 residents is jointly proportional to the number of residents who have heard it and the number who have not. Initially, 10 residents heard the rumor, but 2 days later this number increased to 80. Find the number of people who will have heard the rumor after 1 wk.

16. Growth of a Fruit Fly Colony A biologist has determined that the maximum number of fruit flies that can be sustained in a carefully controlled environment (with a limited supply of space and food) is 400. Suppose that the rate at which the population of the colony increases obeys the rule

$$\frac{dQ}{dt} = kQ(C - Q)$$

where C is the carrying capacity (400) and Q denotes the number of fruit flies in the colony at any time t. If the initial population of fruit flies in the experiment is 10 and it grows to 45 after 10 days, determine the population of the colony of fruit flies on the 20th day.

17. Gompertz Growth Curves Refer to Exercise 28, Section 9.1. Consider the differential equation

$$\frac{dQ}{dt} = kQ(C - \ln Q)$$

with the side condition $Q(0) = Q_0$. The solution $Q(t)$ describes restricted growth and has a graph known as the Gompertz curve. Using separation of variables, solve this differential equation.

18. **Chemical Reaction Rates** Two chemical solutions, one containing N molecules of chemical A and another containing M molecules of chemical B, are mixed together at time $t = 0$. The molecules from the two chemicals combine to form another chemical solution containing $y(AB)$ molecules. The rate at which the AB molecules are formed, dy/dt, is called the *reaction rate* and is jointly proportional to $(N - y)$ and $(M - y)$. Thus,

$$\frac{dy}{dt} = k(N - y)(M - y)$$

where k is a constant. (We assume the temperature of the chemical mixture remains constant during the interaction.) Solve this differential equation with the side condition $y(0) = 0$ assuming that $N - y > 0$ and $M - y > 0$.
Hint: Use the identity

$$\frac{1}{(N - y)(M - y)} = \frac{1}{M - N}\left(\frac{1}{N - y} - \frac{1}{M - y}\right)$$

19. **Mixture Problems** A tank initially contains 20 gal of pure water. Brine containing 2 lb of salt per gallon flows into the tank at a rate of 3 gal/min, and the well-stirred mixture flows out of the tank at the same rate. How much salt is present in the tank at any time t? How much salt is present at the end of 20 min? How much salt is present in the long run?

20. **Mixture Problems** A tank initially contains 50 gal of brine, in which 10 lb of salt is dissolved. Brine containing 2 lb of dissolved salt per gallon flows into the tank at the rate of 2 gal/min, and the well-stirred mixture flows out of the tank at the same rate. How much salt is present in the tank at the end of 10 min?

9.3 Solution to Self-Check Exercise

We are required to solve the initial value problem

$$\frac{dy}{dt} = -k(y - 70) \qquad k\text{, a constant of proportionality}$$
$$y(0) = 200$$

To solve the differential equation, we separate variables and integrate, obtaining

$$\int \frac{dy}{y - 70} = \int -k\,dt$$
$$\ln|y - 70| = -kt + d \qquad d\text{, an arbitrary constant}$$
$$y - 70 = e^{-kt+d} = Ae^{-kt} \qquad (A = e^d)$$
$$y = 70 + Ae^{-kt}$$

Using the initial condition $y(0) = 200$, we find

$$200 = 70 + A$$
$$A = 130$$

Therefore,

$$y = 70 + 130e^{-kt}$$

To determine the value of k, we use the fact that $y(5) = 150$, obtaining

$$150 = 70 + 130e^{-5k}$$
$$130e^{-5k} = 80$$
$$e^{-5k} = \frac{80}{130}$$
$$-5k = \ln\frac{80}{130}$$
$$k = -\frac{1}{5}\ln\frac{80}{130}$$
$$= 0.097$$

Therefore, the required expression is

$$y(t) = 70 + 130e^{-0.097t}$$

9.4 Approximate Solutions of Differential Equations

Euler's Method

As in the case of definite integrals, there are many differential equations whose exact solutions cannot be found using any of the available methods. In such cases, we must once again resort to approximate solutions.

Many numerical methods have been developed for efficient computation of approximate solutions to differential equations. In this section we will look at a method for solving the problem

$$\frac{dy}{dx} = F(x, y) \qquad [y(x_0) = y_0] \tag{12}$$

Euler's method, named after Leonhard Euler (1707–1783), describes a way of finding an approximate solution of Equation (12). Basically, the technique calls for approximating the actual solution $y = f(x)$ at certain selected values of x. The values of f between two adjacent values of x are then found by linear interpolation. This situation is depicted geometrically in Figure 13. Thus, in Euler's method, the actual solution curve of the differential equation is approximated by a suitable polygonal curve.

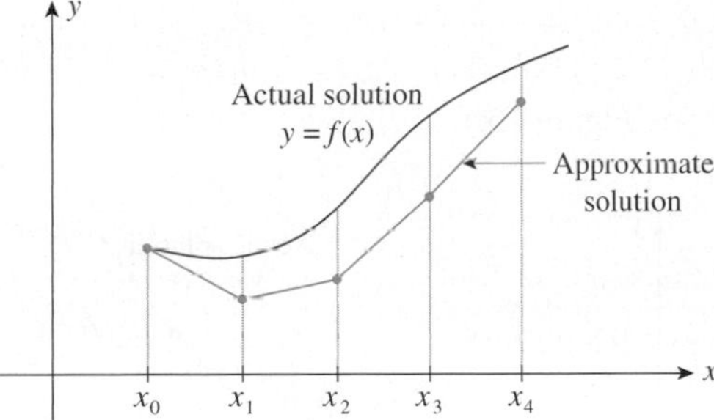

FIGURE 13
Using Euler's method, the actual solution curve of the differential equation is approximated by a polygonal curve.

To describe the method, let h be a small positive number and let $x_n = x_0 + nh$ $(n = 1, 2, 3, \ldots)$, that is,

$$x_1 = x_0 + h \qquad x_2 = x_0 + 2h \qquad x_3 = x_0 + 3h \qquad \ldots$$

Thus, the points $x_0, x_1, x_2, x_3, \ldots$ are spaced evenly apart, and the distance between any two adjacent points is h units.

We begin by finding an approximation y_1 to the value of the actual solution, $f(x_1)$ at $x = x_1$. Observe that the *initial* condition $y(x_0) = y_0$ of (12) tells us that the point (x_0, y_0) lies on the solution curve. Euler's method calls for approximating the part of the graph of f on the interval $[x_0, x_1]$ by the straight-line segment that is tangent to the graph of f at (x_0, y_0). To find an equation of this straight-line segment, observe that the slope of this line segment is equal to $F(x_0, y_0)$. So, using the point-slope form of an equation of a line, we see that the required equation is

$$y - y_0 = F(x_0, y_0)(x - x_0)$$

or

$$y = y_0 + F(x_0, y_0)(x - x_0)$$

Therefore, the approximation y_1 to $f(x_1)$ is obtained by replacing x by x_1. Thus,

$$\begin{aligned} y_1 &= y_0 + F(x_0, y_0)(x_1 - x_0) \\ &= y_0 + F(x_0, y_0)h \qquad \text{Since } x_1 - x_0 = h \end{aligned}$$

This situation is depicted in Figure 14.

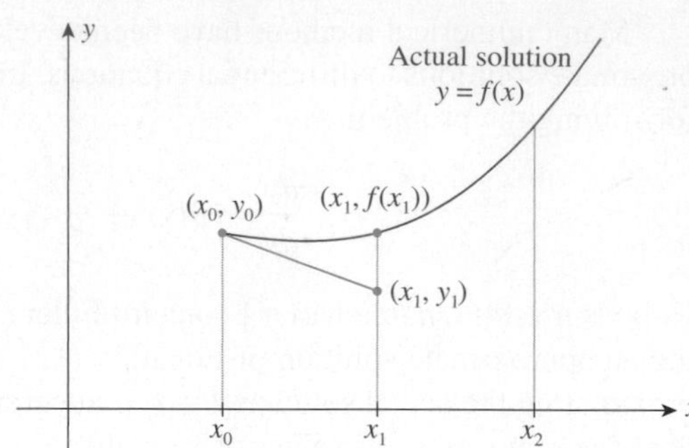

FIGURE 14
$y_1 = y_0 + hF(x_0, y_0)$ is the number used to approximate $f(x_1)$.

Next, to find an approximation y_2 to the value of the actual solution, $f(x_2)$, at $x = x_2$, we repeat the preceding argument but this time taking the slope of the straight-line segment on $[x_1, x_2]$ to be $F(x_1, y_1)$. We obtain

$$y_2 = y_1 + F(x_1, y_1)h$$

(See Figure 15.)

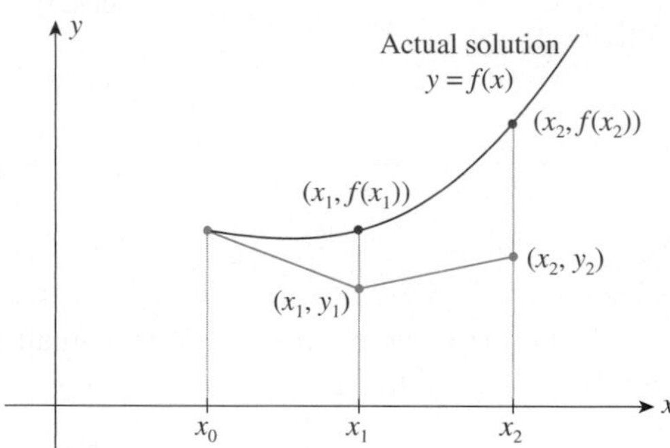

FIGURE 15
$y_2 = y_1 + hF(x_1, y_1)$ is the number used to approximate $f(x_2)$.

Continuing in this manner, we see that $y_1, y_2, \ldots, y_n$ can be found by the general formula

$$y_n = y_{n-1} + hF(x_{n-1}, y_{n-1}) \qquad (n = 1, 2, \ldots)$$

We now summarize this procedure.

Euler's Method

Suppose we are given the differential equation

$$\frac{dy}{dx} = F(x, y)$$

subject to the initial condition $y(x_0) = y_0$ and we wish to find an approximation of $y(b)$, where b is a number greater than x_0. Compute

$$h = \frac{b - x_0}{n}$$

$$x_1 = x_0 + h \qquad x_2 = x_0 + 2h \qquad x_3 = x_0 + 3h, \ldots \qquad x_n = x_0 + nh = b$$

and

$$\begin{aligned} y_0 &= y(x_0) \\ y_1 &= y_0 + hF(x_0, y_0) \\ y_2 &= y_1 + hF(x_1, y_1) \\ &\vdots \\ y_n &= y_{n-1} + hF(x_{n-1}, y_{n-1}) \end{aligned}$$

Then, y_n gives an approximation of the true value $y(b)$ of the solution to the initial value problem at $x = b$.

Solving Differential Equations with Euler's Method

VIDEO

EXAMPLE 1 Use Euler's method with $n = 8$ to obtain an approximation of the initial value problem

$$y' = x - y \qquad [y(0) = 1]$$

when $x = 2$.

Solution Here, $x_0 = 0$ and $b = 2$ so that, taking $n = 8$, we find

$$h = \frac{2 - 0}{8} = \frac{1}{4}$$

and

$$x_0 = 0 \quad x_1 = \frac{1}{4} \quad x_2 = \frac{1}{2} \quad x_3 = \frac{3}{4} \quad x_4 = 1$$

$$x_5 = \frac{5}{4} \quad x_6 = \frac{3}{2} \quad x_7 = \frac{7}{4} \quad x_8 = b = 2$$

Also,

$$F(x, y) = x - y \qquad \text{and} \qquad y_0 = y(0) = 1$$

Therefore, the approximations of the actual solution at the points $x_0, x_1, x_2, \ldots, x_n = b$ are

$$\begin{aligned} y_0 &= y(0) = 1 \\ y_1 &= y_0 + hF(x_0, y_0) = 1 + \frac{1}{4}(0 - 1) = \frac{3}{4} \\ y_2 &= y_1 + hF(x_1, y_1) = \frac{3}{4} + \frac{1}{4}\left(\frac{1}{4} - \frac{3}{4}\right) = \frac{5}{8} \\ y_3 &= y_2 + hF(x_2, y_2) = \frac{5}{8} + \frac{1}{4}\left(\frac{1}{2} - \frac{5}{8}\right) = \frac{19}{32} \\ y_4 &= y_3 + hF(x_3, y_3) = \frac{19}{32} + \frac{1}{4}\left(\frac{3}{4} - \frac{19}{32}\right) = \frac{81}{128} \\ y_5 &= y_4 + hF(x_4, y_4) = \frac{81}{128} + \frac{1}{4}\left(1 - \frac{81}{128}\right) = \frac{371}{512} \end{aligned}$$

$$y_6 = y_5 + hF(x_5, y_5) = \frac{371}{512} + \frac{1}{4}\left(\frac{5}{4} - \frac{371}{512}\right) = \frac{1753}{2048}$$

$$y_7 = y_6 + hF(x_6, y_6) = \frac{1753}{2048} + \frac{1}{4}\left(\frac{3}{2} - \frac{1753}{2048}\right) = \frac{8331}{8192}$$

$$y_8 = y_7 + hF(x_7, y_7) = \frac{8331}{8192} + \frac{1}{4}\left(\frac{7}{4} - \frac{8331}{8192}\right) = \frac{39{,}329}{32{,}768}$$

Thus, the approximate value of $y(2)$ is

$$\frac{39{,}329}{32{,}768} \approx 1.2002$$

■

EXAMPLE 2 Use Euler's method with (a) $n = 5$ and (b) $n = 10$ to approximate the solution of the initial value problem

$$y' = -2xy^2 \qquad [y(0) = 1]$$

on the interval $[0, 0.5]$. Find the actual solution of the initial value problem. Finally, sketch the graphs of the approximate solutions and the actual solution for $0 \le x \le 0.5$ on the same set of axes.

Solution

a. Here, $x_0 = 0$ and $b = 0.5$. Taking $n = 5$, we find

$$h = \frac{0.5 - 0}{5} = 0.1$$

and $x_0 = 0$, $x_1 = 0.1$, $x_2 = 0.2$, $x_3 = 0.3$, $x_4 = 0.4$, and $x_5 = b = 0.5$. Also,

$$F(x, y) = -2xy^2 \qquad \text{and} \qquad y_0 = y(0) = 1$$

Therefore,

$$y_0 = y(0) = 1$$
$$y_1 = y_0 + hF(x_0, y_0) = 1 + 0.1(-2)(0)(1)^2 = 1$$
$$y_2 = y_1 + hF(x_1, y_1) = 1 + 0.1(-2)(0.1)(1)^2 = 0.98$$
$$y_3 = y_2 + hF(x_2, y_2) = 0.98 + 0.1(-2)(0.2)(0.98)^2 = 0.9416$$
$$y_4 = y_3 + hF(x_3, y_3) = 0.9416 + 0.1(-2)(0.3)(0.9416)^2 = 0.8884$$
$$y_5 = y_4 + hF(x_4, y_4) = 0.8884 + 0.1(-2)(0.4)(0.8884)^2 = 0.8253$$

b. Here $x_0 = 0$ and $b = 0.5$. Taking $n = 10$, we find

$$h = \frac{0.5 - 0}{10}$$

and $x_0 = 0$, $x_1 = 0.05$, $x_2 = 0.10$, . . . , $x_9 = 0.45$, and $x_{10} = 0.5 = b$. Proceeding as in part (a), we obtain the approximate solutions listed in the following table:

x	0.00	0.05	0.10	0.15	0.20	0.25
y_n	1.0000	1.0000	0.9950	0.9851	0.9705	0.9517

x	0.30	0.35	0.40	0.45	0.50
y_n	0.9291	0.9032	0.8746	0.8440	0.8119

To obtain the actual solution of the differential equation, we separate variables, obtaining

$$\frac{dy}{y^2} = -2x\,dx$$

Integrating each side of the last equation with respect to the appropriate variables, we have

$$\int \frac{dy}{y^2} = -\int 2x\,dx$$

or

$$-\frac{1}{y} + C_1 = -x^2 + C_2$$

$$\frac{1}{y} = x^2 + C \qquad C = C_1 - C_2$$

$$y = \frac{1}{x^2 + C}$$

Using the condition $y(0) = 1$, we have

$$1 = \frac{1}{0 + C} \qquad \text{or} \qquad C = 1$$

Therefore, the required solution is given by

$$y = \frac{1}{x^2 + 1}$$

The graphs of the approximate solutions and the actual solution are sketched in Figure 16.

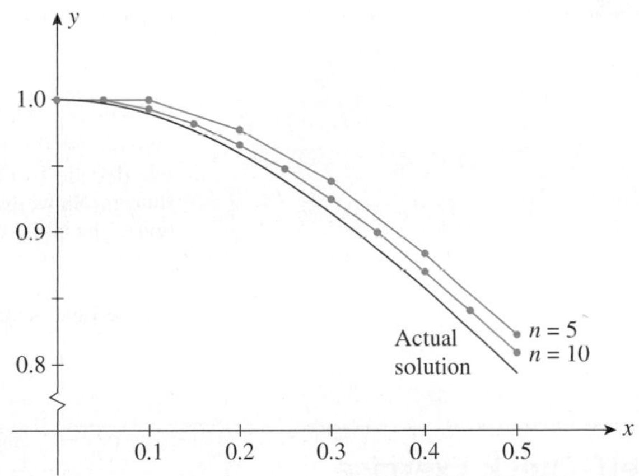

FIGURE 16
The approximate solutions and the actual solution to an initial value problem

9.4 Self-Check Exercise

Use Euler's method with $n = 5$ to obtain an approximation of the initial value problem

$$y' = 2x + y \qquad [y(0) = 1]$$

when $x = 1$.

The solution to Self-Check Exercise 9.4 can be found on page 648.

9.4 Concept Questions

1. Describe the ideas behind Euler's method for finding the approximate solution of the initial value problem

$$\frac{dy}{dx} = F(x, y) \qquad y(x_0) = y_0$$

in your own words.

2. Give Euler's method.

9.4 Exercises

A calculator is recommended for this exercise set. In Exercises 1–10, use Euler's method with (a) $n = 4$ and (b) $n = 6$ to obtain an approximation of the initial value problem when $x = b$.

1. $y' = x + y$, $y(0) = 1$; $b = 1$
2. $y' = x - 2y$, $y(0) = 1$; $b = 2$
3. $y' = 2x - y + 1$, $y(0) = 2$; $b = 2$
4. $y' = 2xy$, $y(0) = 1$; $b = 0.5$
5. $y' = -2xy^2$, $y(0) = 1$; $b = 0.5$
6. $y' = x^2 + y^2$, $y(0) = 1$; $b = 1.5$
7. $y' = \sqrt{x + y}$, $y(1) = 1$; $b = 1.5$
8. $y' = (1 + x^2)^{-1}$, $y(0) = 0$; $b = 1$
9. $y' = \dfrac{x}{y}$, $y(0) = 1$; $b = 1$
10. $y' = xy^{1/3}$, $y(0) = 1$; $b = 1$

In Exercises 11–15, use Euler's method with $n = 5$ to obtain an approximate solution to the initial value problem over the indicated interval.

11. $y' = \dfrac{1}{2}xy$, $y(0) = 1$; $0 \le x \le 1$
12. $y' = x^2y$, $y(0) = 2$; $0 \le x \le 0.6$
13. $y' = 2x - y + 1$, $y(0) = 2$; $0 \le x \le 1$
14. $y' = x + y^2$, $y(0) = 0$; $0 \le x \le 0.5$
15. $y' = x^2 + y$, $y(0) = 1$; $0 \le x \le 0.5$
16. Growth of Service Industries It has been estimated that service industries, which currently make up 30% of the nonfarm workforce in a certain country, will continue to grow at the rate of

$$R(t) = 5e^{1/(t+1)}$$

percent per decade t decades from now. Estimate the percentage of the nonfarm workforce in the service industries one decade from now.
Hint: (a) Show that the desired answer is $P(1)$, where P is the solution of the initial value problem

$$P' = 5e^{1/(t+1)} \qquad [P(0) = 30]$$

(b) Use Euler's method with $n = 10$ to approximate the solution.

9.4 Solution to Self-Check Exercise

Here, $x_0 = 0$ and $b = 1$ so that, taking $n = 5$, we find

$$h = \frac{1 - 0}{5} = \frac{1}{5}$$

and

$$x_0 = 0 \qquad x_1 = \frac{1}{5} \qquad x_2 = \frac{2}{5}$$

$$x_3 = \frac{3}{5} \qquad x_4 = \frac{4}{5} \qquad x_5 = b = 1$$

Also,

$$F(x, y) = 2x + y \qquad \text{and} \qquad y_0 = y(0) = 1$$

Therefore, the approximations of the actual solution at the points $x_0, x_1, x_2, \ldots, x_5 = 1$ are

$$y_0 = y(0) = 1$$

$$y_1 = y_0 + hF(x_0, y_0) = 1 + \frac{1}{5}(0 + 1) = \frac{6}{5}$$

$$y_2 = y_1 + hF(x_1, y_1) = \frac{6}{5} + \frac{1}{5}\left(\frac{2}{5} + \frac{6}{5}\right) = \frac{38}{25}$$

$$y_3 = y_2 + hF(x_2, y_2) = \frac{38}{25} + \frac{1}{5}\left(\frac{4}{5} + \frac{38}{25}\right) = \frac{248}{125}$$

$$y_4 = y_3 + hF(x_3, y_3) = \frac{248}{125} + \frac{1}{5}\left(\frac{6}{5} + \frac{248}{125}\right) = \frac{1638}{625}$$

$$y_5 = y_4 + hF(x_4, y_4) = \frac{1638}{625} + \frac{1}{5}\left(\frac{8}{5} + \frac{1638}{625}\right) = \frac{10{,}828}{3125}$$

Thus, the approximate value of $y(1)$ is

$$\frac{10{,}828}{3125} \approx 3.4650$$

CHAPTER 9 Summary of Principal Terms

TERMS

differential equation (620)

general solution of a differential equation (622)

particular solution of a differential equation (623)

first-order differential equation (627)

separable differential equation (627)

method of separation of variables (627)

initial value problem (629)

Euler's method (644)

CHAPTER 9 Concept Review Questions

Fill in the blanks.

1. a. An equation that involves an unknown function and its derivatives is called a _____ _____.

b. A solution of a differential equation is any function that _____ the differential system.

2. a. The solution of a differential equation that involves a constant c is called the _____ solution of the differential equation.

b. The solution obtained by assigning a specific value to c is called a _____ solution.

3. a. The order of a differential equation is the order of the _____ derivative of the _____ function in the equation.

b. A separable differential equation has the form $\frac{dy}{dx} =$ _____; the equation $y' = \frac{xy}{x^2 + y^2}$ is _____ _____; the equation $\frac{dy}{dx} = \frac{e^y}{1 + x^2}$ is _____.

c. To solve a separable equation, we first _____ the variables and then integrate each term with respect to the appropriate _____.

4. a. The differential equation describing an unrestricted growth model is given by $\frac{dQ}{dt} =$ _____.

b. The model for unrestricted exponential growth with initial population Q_0 is given by $Q(t) =$ _____.

c. The differential equations $\frac{dQ}{dt} = k(C - Q)$ and $\frac{dQ}{dt} = kQ(C - Q)$ describe _____ exponential growth.

5. Euler's method is used to find an _________ solution of an initial value problem. The _____ solution curve of a differential equation is approximated by a _______ curve.

CHAPTER 9 Review Exercises

In Exercises 1–3, verify that *y* is a solution of the differential equation.

1. $y = C_1e^{2x} + C_2e^{-3x}$; $y'' + y' - 6y = 0$

2. $y = 2e^{2x} + 3x - 2$; $y'' - y' - 2y = -6x + 1$

3. $y = Cx^{-4/3}$; $4xy^3\,dx + 3x^2y^2\,dy = 0$

In Exercises 4 and 5, verify that *y* is a general solution of the differential equation and find a particular solution of the differential equation satisfying the side condition.

4. $y = \dfrac{1}{x^2 - C}$; $\dfrac{dy}{dx} = -2xy^2$; $y(0) = 1$

5. $y = (9x + C)^{-1/3}$; $\dfrac{dy}{dx} = -3y^4$; $y(0) = \dfrac{1}{2}$

In Exercises 6–11, solve the differential equation.

6. $y' = \dfrac{x^3 + 1}{y^2}$

7. $\dfrac{dy}{dt} = 2(4 - y)$

8. $y' = \dfrac{y \ln x}{x}$

9. $y' = 3x^2y^2 + y^2$; $y(0) = -2$

10. $y' = x^2(1 - y)$; $y(0) = -2$

11. $\dfrac{dy}{dx} = -\dfrac{3}{2}x^2y$; $y(0) = 3$

12. Find a function *f* given that (1) the slope of the tangent line to the graph of *f* at any point $P(x, y)$ is given by the expression

$$y' = -\frac{4xy}{x^2 + 1}$$

and (2) the graph of *f* passes through the point (1, 1).

In Exercises 13–16, use Euler's method with (a) $n = 4$ and (b) $n = 6$ to obtain an approximation of the initial value problem when $x = b$.

13. $y' = x + y^2$, $y(0) = 0$; $b = 1$

14. $y' = x^2 + 2y^2$, $y(0) = 0$; $b = 1$

15. $y' = 1 + 2xy^2$, $y(0) = 0$; $b = 1$

16. $y' = e^x + y^2$, $y(0) = 0$; $b = 1$

In Exercises 17 and 18, use Euler's method with $n = 5$ to obtain an approximate solution to the initial value problem over the indicated interval.

17. $y' = 2xy$, $y(0) = 1$; $0 \le x \le 1$

18. $y' = x^2 + y^2$, $y(0) = 1$; $0 \le x \le 1$

19. Resale Value of a Machine The resale value of a certain machine decreases at a rate proportional to the machine's purchase price. The machine was purchased at \$50,000 and 2 yr later was worth \$32,000.
    a. Find an expression for the resale value of the machine at any time *t*.
    b. Find the value of the machine after 5 yr.

20. Brentano–Stevens Law The Brentano–Stevens law, which describes the rate of change of a response *R* to a stimulus *S*, is given by

$$\frac{dR}{dS} = k \cdot \frac{R}{S}$$

where *k* is a positive constant. Solve this differential equation.

21. Continuous Compound Interest HAL Corporation invests *P* dollars/year (assume this is done on a frequent basis in small deposits over the year so that it is essentially continuous) into a fund earning interest at the rate of *r*%/year compounded continuously. Then the size of the fund *A* grows at a rate given by

$$\frac{dA}{dt} = rA + P$$

Suppose $A = 0$ when $t = 0$. Determine the size of the fund after *t* yr. What is the size of the fund after 5 yr if $P =$ \$50,000 and $r = 12\%$/year?

22. Demand for a Commodity Suppose the demand *D* for a certain commodity is constant and the rate of change in the supply *S* over time *t* is proportional to the difference between the demand and the supply so that

$$\frac{dS}{dt} = k(D - S)$$

Find an expression for the supply at any time *t* if the supply at $t = 0$ is S_0.

23. Barbara placed an 11-lb rib roast, which had been standing at room temperature (68°F), into a 350°F oven at 4 P.M. At 6 P.M. the temperature of the roast was 118°F. At what time would the temperature of the roast have been 150°F (medium rare)?
Hint: Use Newton's law of cooling (or heating).

24. A Learning Curve The American Court Reporting Institute finds that the average student taking the Advanced Stenotype course will progress at a rate given by

$$\frac{dQ}{dt} = k(120 - Q)$$

in a 20-wk course, where $Q(t)$ measures the number of words of dictation the student can take per minute after t wk in the course. [Assume $Q(0) = 60$.] If the average student can take 90 words of dictation per minute after 10 wk in the course, how many words per minute can the average student take after completing the course?

25. **Spread of a Rumor** A rumor to the effect that a rental increase was imminent was first heard by four residents of the Chatham West Condominium Complex. The rumor spread through the complex of 200 single-family dwellings at a rate jointly proportional to the number of families who had heard it and the number who had not. Two days later, the number of families who had heard the rumor had increased to 40. Find how many families had heard the rumor after 5 days.

26. **A Mixture Problem** A tank initially contains 40 gal of pure water. Brine containing 3 lb of salt per gallon flows into the tank at a rate of 4 gal/min, and the well-stirred mixture flows out of the tank at the same rate. How much salt is in the tank at any time t? How much salt is in the tank in the long run?

CHAPTER 9 Before Moving On . . .

1. **a.** Show that $y = 2x^2 + cx$ is the general solution of the differential equation $xy' - y = 2x^2$.
   b. Find the particular solution of the differential equation of part (a) that satisfies $y(1) = 2$.

2. Find the solution of the differential equation

$$(1 + x)y\,dx + x\,dy = 0$$

that satisfies the condition $y(1) = 1$.

3. The population of a town grows at a rate that is proportional to the current population. If the initial population was 5000 and had doubled in size after 5 years, find an expression for the population P at year t. How long would it take for the population to reach 12,000?

4. Use Euler's method with $n = 5$ to approximate the solution of the initial value problem $y' = y^2 - x^2$, $y(0) = 1$, on the interval $[0, 0.5]$.

10 Probability and Calculus

Where do we go from here? Some of the top 10% of this senior class will further their education at one of the campuses of the state university system. In Example 6, page 688, we determine the minimum grade point average a senior needs to be eligible for admission to one of the state universities.

THE SYSTEMATIC STUDY of probability began in the seventeenth century, when certain aristocrats wanted to discover superior strategies to use in the gaming rooms of Europe. Some of the best mathematicians of the period were engaged in this pursuit. Since then, applications of probability have evolved in virtually every sphere of human endeavor that contains an element of uncertainty.

In this chapter we take a look at the role of calculus in the study of *probability* involving a *continuous random variable.* We see how probability can be used to find the average life span of a certain brand of color television tube, the average waiting time for patients in a health clinic, and the percentage of a current Mediterranean population who have serum cholesterol levels between 160 and 180 mg/dL—to name but a few applications.

10.1 Probability Distributions of Random Variables

Probability

We begin by mentioning some elementary notations important in the study of probability. For our purpose, an **experiment** is an activity with observable results called **outcomes**, or **sample points**. The totality of all outcomes is the **sample space** of the experiment. A subset of the sample space is called an **event** of the experiment.

Now, given an event associated with an experiment, our primary objective is to determine the likelihood that this event will occur. This likelihood, or **probability of the event**, is a number between 0 and 1 and may be viewed as the proportionate number of times that the event will occur if the experiment associated with the event is repeated indefinitely under independent and similar conditions. By way of example, let's consider the simple experiment of tossing an unbiased coin and observing whether it lands "heads" (H) or "tails" (T). Since the coin is *unbiased*, we see that the probability of each outcome is $\frac{1}{2}$, abbreviated

$$P(\text{H}) = \frac{1}{2} \quad \text{and} \quad P(\text{T}) = \frac{1}{2}$$

Discrete Random Variables

In many situations it is desirable to assign numerical values to the outcomes of an experiment. For example, suppose an experiment consists of casting a die and observing the face that lands up. If we let X denote the outcome of the experiment, then X assumes one of the values 1, 2, 3, 4, 5, or 6. Because the values assumed by X depend on the outcomes of a chance experiment, the outcome X is referred to as a **random variable**. In this case the random variable X is also said to be **finite discrete** since it can assume only a finite number of integer values.

The function P that associates with each value of a random variable its probability of occurrence is called a *probability function*. It is *discrete* since its domain consists of a finite set. In general, a discrete probability function is defined as follows.

Discrete Probability Function

A **discrete probability function** P with domain $\{x_1, x_2, \ldots, x_n\}$ satisfies these conditions:

1. $0 \leq P(x_i) \leq 1$
2. $P(x_1) + P(x_2) + \cdots + P(x_n) = 1$

Note These conditions imply that the probability assigned to an outcome must be nonnegative and less than or equal to 1 and that the sum of all probabilities must be 1.

Histograms

A discrete probability function or distribution may be exhibited graphically by means of a **histogram**. To construct a histogram of a probability distribution, first locate the values of the random variable on a number line. Then, above each such number, erect a rectangle with width 1 and height equal to the probability associated with that value of the random variable.

For example, consider the data in Table 1, the number of cars observed waiting in line at 2-minute intervals between 3 and 5 P.M. on a certain Friday at the drive-in teller of the Westwood Savings Bank and the corresponding frequency of occurrence. If we divide each number on the right of the table by 60 (the sum of these numbers), then we obtain the respective probabilities associated with the random variable X, when X assumes the values 0, 1, 2, . . . , 8. For example,

TABLE 1

Number of Cars	Frequency of Occurrence
0	2
1	9
2	16
3	12
4	8
5	6
6	4
7	2
8	1

$$P(X = 0) = \frac{2}{60} \approx .03$$

$$P(X = 1) = \frac{9}{60} = .15, \ldots$$

The resulting probability distribution is shown in Table 2.

TABLE 2

Probability Distribution for the Random Variable X

x	$P(X = x)$
0	.03
1	.15
2	.27
3	.20
4	.13
5	.10
6	.07
7	.03
8	.02

The histogram associated with this probability distribution is shown in Figure 1. Observe that the area of a rectangle in a histogram is associated with a value of a random variable X and that this area gives precisely the probability associated with that value of X. This follows since each rectangle, by construction, has width 1 and height corresponding to the probability associated with the value of the random variable.

Another consequence arising from the method of construction of a histogram is that the probability associated with more than one value of the random variable X is given by the sum of the areas of the rectangles associated with those values of X. For example, the probability that three or four cars are in line is given by

$$P(X - 3) + P(X - 4)$$

which may be obtained from the histogram by adding the areas of the rectangles associated with the values 3 and 4 of the random variable X. Thus, the required probability is

$$P(X = 3) + P(X = 4) = .20 + .13 = .33$$

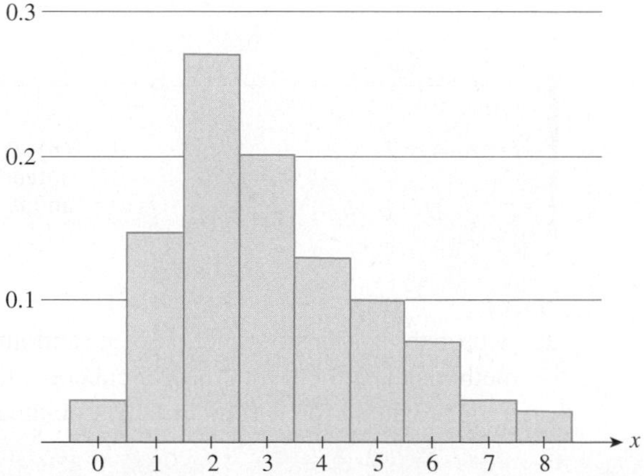

FIGURE 1
Probability distribution of the number of cars waiting in line.

Continuous Random Variables

A random variable x that can assume any value in an interval is called a **continuous random variable**. Examples of continuous random variables are the life span of a light bulb, the length of a telephone call, the length of an infant at birth, the daily amount of rainfall in Boston, and the life span of a certain plant species. For the remainder of this chapter, we will be interested primarily in continuous random variables.

Consider an experiment in which the associated random variable x has the interval $[a, b]$ as its sample space. Then an event of the experiment is any subset of $[a, b]$. For example, if x denotes the life span of a light bulb, then the sample space associated with the experiment is $[0, \infty)$, and the event that a light bulb selected at random has a life span between 500 and 600 hours inclusive is described by the interval [500, 600] or, equivalently, by the inequality $500 \leq x \leq 600$. The probability that the light bulb will have a life span of between 500 and 600 hours is denoted by $P(500 \leq x \leq 600)$.

In general, we will be interested in computing $P(a \leq x \leq b)$, the probability that a random variable x assumes a value in the interval $a \leq x \leq b$. This computation is based on the notion of a probability density function, which we now introduce.

Probability Density Function

A **probability density function** of a continuous random variable x in an interval I, where I may be bounded or unbounded, is a nonnegative function f having the following properties.

1. The total area of the region under the graph of f is equal to 1 (Figure 2a).
2. The probability that an observed value of the random variable x lies in the interval $[a, b]$ is given by

$$P(a \leq x \leq b) = \int_a^b f(x)\, dx$$

(Figure 2b).

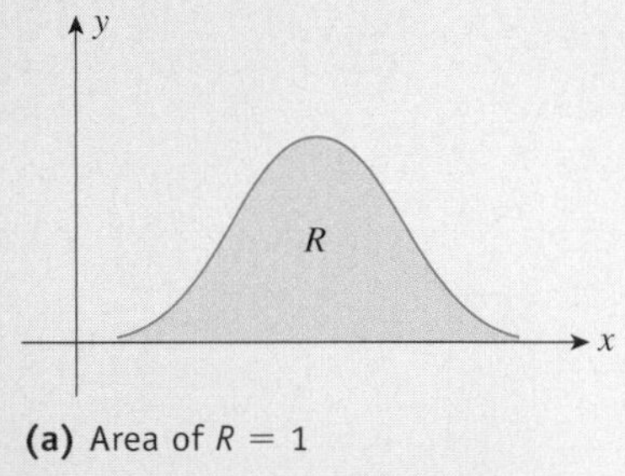

(a) Area of $R = 1$

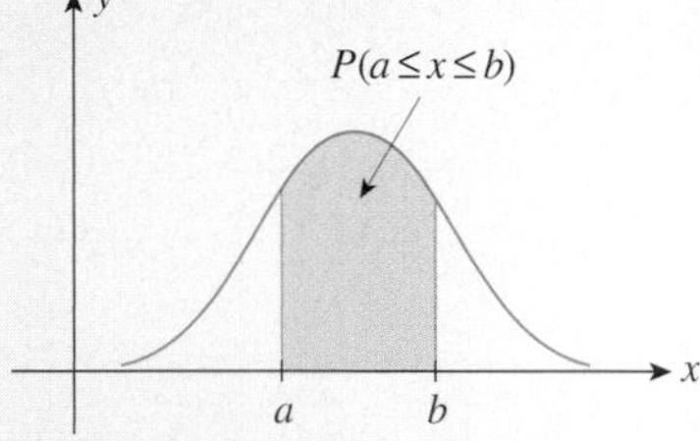

(b) $P(a \leq x \leq b)$ is the probability that an outcome of an experiment will lie between a and b.

FIGURE 2

Notes

1. A probability density function of a random variable x may be constructed using methods that range from theoretical considerations of the problem on the one extreme to an interpretation of data associated with the experiment on the other.

2. Property 1 states that the probability that a continuous random variable takes on a value lying in its range is 1, a certainty, which is expected.
3. Property 2 states that the probability that the random variable x assumes a value in an interval $a \leq x \leq b$ is given by the area of the region between the graph of f and the x-axis from $x = a$ to $x = b$. Because the area under one point of the graph of f is equal to zero, we see immediately that $P(a \leq x \leq b) = P(a < x \leq b) = P(a \leq x < b) = P(a < x < b)$.

EXAMPLE 1 Show that each of the following functions satisfies the nonnegativity condition and Property 1 of probability density functions.

a. $f(x) = \dfrac{2}{27}x(x - 1) \qquad (1 \leq x \leq 4)$

b. $f(x) = \dfrac{1}{3}e^{(-1/3)x} \qquad (0 \leq x < \infty)$

Solution

a. Since the factors x and $(x - 1)$ are both nonnegative, we see that $f(x) \geq 0$ on $[1, 4]$. Next, we compute

$$\begin{aligned}\int_1^4 \frac{2}{27}x(x-1)\,dx &= \frac{2}{27}\int_1^4 (x^2 - x)\,dx \\ &= \frac{2}{27}\left(\frac{1}{3}x^3 - \frac{1}{2}x^2\right)\Bigg|_1^4 \\ &= \frac{2}{27}\left[\left(\frac{64}{3} - 8\right) - \left(\frac{1}{3} - \frac{1}{2}\right)\right] \\ &= \frac{2}{27}\left(\frac{27}{2}\right) \\ &= 1\end{aligned}$$

showing that Property 1 of probability density functions holds as well.

b. First, $f(x) = \frac{1}{3}e^{(-1/3)x} \geq 0$ for all values of x in $[0, \infty)$. Next,

$$\begin{aligned}\int_0^\infty \frac{1}{3}e^{(-1/3)x}\,dx &= \lim_{b\to\infty}\int_0^b \frac{1}{3}e^{(-1/3)x}\,dx \\ &= \lim_{b\to\infty} -e^{(-1/3)x}\Bigg|_0^b \\ &= \lim_{b\to\infty}\left(-e^{(-1/3)b} + 1\right) \\ &= 1\end{aligned}$$

so the area under the graph of $f(x) = \frac{1}{3}e^{(-1/3)x}$ is equal to 1, as we set out to show.

EXAMPLE 2

a. Determine the value of the constant k such that the function $f(x) = kx^2$ is a probability density function on the interval $[0, 5]$.

b. If x is a continuous random variable with the probability density function given in part (a), compute the probability that x will assume a value between $x = 1$ and $x = 2$.

c. Find the probability that x will assume a value at $x = 3$.

Solution

a. We compute

$$\int_0^5 kx^2\,dx = k\int_0^5 x^2\,dx = \frac{k}{3}x^3\Big|_0^5 = \frac{125}{3}k$$

Since this value must be equal to 1, we find that $k = \frac{3}{125}$.

b. The required probability is given by

$$P(1 \le x \le 2) = \int_1^2 f(x)\,dx = \int_1^2 \frac{3}{125}x^2\,dx = \frac{1}{125}x^3\Big|_1^2 = \frac{1}{125}(8-1) = \frac{7}{125}$$

The graph of the probability density function f and the area corresponding to the probability $P(1 \le x \le 2)$ are shown in Figure 3.

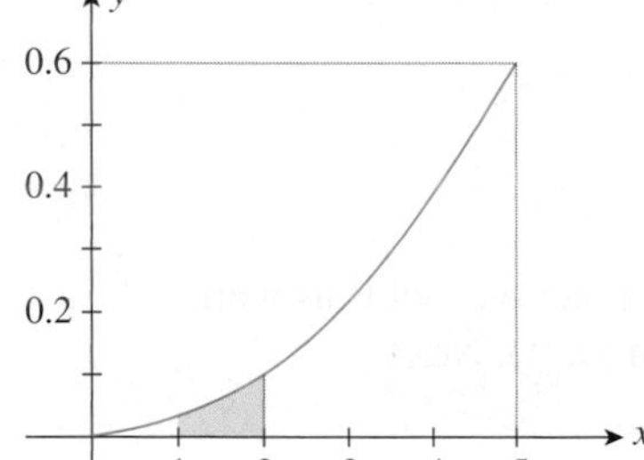

FIGURE 3
$P(1 \le x \le 2)$ for the probability density function $y = \frac{3}{125}x^2$

c. The required probability is given by

$$P(x = 3) = \int_3^3 f(x)\,dx = \int_3^3 \frac{3}{125}x^2\,dx = 0$$

Note Observe that, in general, the probability that x will assume a value at a point $x = a$ is zero since $P(x = a) = \int_a^a f(x)\,dx = 0$ by Property 1 of the definite integral (page 441).

APPLIED EXAMPLE 3 Life Span of Light Bulbs TKK Products manufactures a 200-watt electric light bulb. Laboratory tests show that the life spans of these light bulbs have a distribution described by the probability density function

$$f(x) = .001e^{-0.001x}$$

Determine the probability that a light bulb will have each of these life spans:

a. 500 hours or less
b. More than 500 hours
c. More than 1000 hours but less than 1500 hours

Solution Let x denote the life span of a light bulb.

a. The probability that a light bulb will have a life span of 500 hours or less is given by

$$\begin{aligned} P(0 \leq x \leq 500) &= \int_0^{500} 0.001e^{-0.001x}\,dx \\ &= -e^{-0.001x}\Big|_0^{500} \\ &= -e^{-0.5} + 1 \\ &\approx .3935 \end{aligned}$$

b. The probability that a light bulb will have a life span of more than 500 hours is given by

$$\begin{aligned} P(x > 500) &= \int_{500}^{\infty} 0.001e^{-0.001x}\,dx \\ &= \lim_{b\to\infty} \int_{500}^{b} 0.001e^{-0.001x}\,dx \\ &= \lim_{b\to\infty} -e^{-0.001x}\Big|_{500}^{b} \\ &= \lim_{b\to\infty}(-e^{-0.001b} + e^{-0.5}) \\ &= e^{-0.5} \approx .6065 \end{aligned}$$

This result may also be obtained by observing that

$$\begin{aligned} P(x > 500) &= 1 - P(x \leq 500) \\ &= 1 - .3935 \qquad \text{Using the result from part (a)} \\ &\approx .6065 \end{aligned}$$

c. The probability that a light bulb will have a life span of more than 1000 hours but less than 1500 hours is given by

$$\begin{aligned} P(1000 < x < 1500) &= \int_{1000}^{1500} .001e^{-0.001x}\,dx \\ &= -e^{-0.001x}\Big|_{1000}^{1500} \\ &= -e^{-1.5} + e^{-1} \\ &\approx -.2231 + .3679 \\ &= .1448 \end{aligned}$$

The probability density function of Example 3 has the form

$$f(x) = ke^{-kx}$$

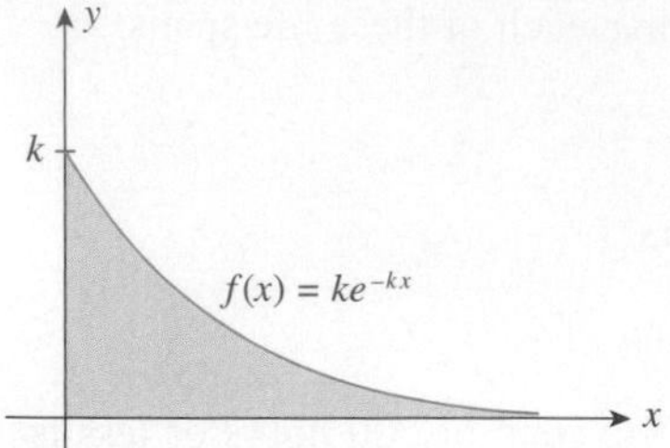

FIGURE 4
The area under the graph of the exponential density function is equal to 1.

where $x \geq 0$ and k is a positive constant. Its graph is shown in Figure 4. This probability function is called an **exponential density function**, and the random variable associated with it is said to be **exponentially distributed**. Exponential random variables are used to represent the life span of electronic components, the duration of telephone calls, the waiting time in a doctor's office, and the time between successive flight arrivals and departures in an airport, to mention but a few applications.

Joint Probability Density Functions

Sometimes the outcomes of an experiment are associated with more than one random variable. For example, we might be interested in the relationship between the weight and height of newborn infants or the price and tread life of automobile tires. To study such problems, we need to extend the concept of a probability density function of a random variable to functions of more than one variable. In the case involving two random variables, we have the following definition.

Joint Probability Density Function

A joint probability density function of the variables x and y on D is a nonnegative function $f(x, y)$ having the following properties:

1. $\displaystyle\iint_D f(x, y)\, dA = 1$

   Thus, the volume of the solid under the graph of f is equal to 1.
2. The probability that the observed values of the random variables x and y lie in a region $R \subset D$ is given by

$$P[(x, y) \text{ in } R] = \iint_R f(x, y)\, dA$$

EXAMPLE 4 Show that the function $f(x, y) = xy$ is a joint probability density function on $D = \{(x, y) \mid 0 \leq x \leq 1; 0 \leq y \leq 2\}$.

Solution First, observe that $f(x, y) = xy$ is nonnegative on D. Next, we compute

$$\begin{aligned}\iint_D f(x, y)\, dA &= \int_0^2 \int_0^1 xy\, dx\, dy \\ &= \int_0^2 \left(\frac{1}{2}x^2y\Big|_0^1\right) dy \\ &= \int_0^2 \frac{1}{2}y\, dy = \frac{1}{4}y^2\Big|_0^2 = 1\end{aligned}$$

Therefore, f is a joint probability density function on D. ■

EXAMPLE 5 Let $f(x, y) = xy$ be a joint probability density function on $D = \{(x, y) \mid 0 \leq x \leq 1; 0 \leq y \leq 2\}$. Find (a) $P(0 \leq x \leq \frac{1}{2}; 1 \leq y \leq 2)$ and (b) $P(\{(x, y) \mid x + y \leq 1\})$.

Solution

a. The required probability is given by

$$P\left(0 \le x \le \frac{1}{2};\, 1 \le y \le 2\right) = \int_1^2 \int_0^{1/2} xy\, dx\, dy$$
$$= \int_1^2 \left(\frac{1}{2}x^2 y \bigg|_0^{1/2}\right) dy$$
$$= \int_1^2 \frac{1}{8} y\, dy = \frac{1}{16} y^2 \bigg|_1^2 = \frac{3}{16}$$

b. The region $R = \{(x, y) \mid x + y \le 1\}$ is shown in Figure 5. The required probability is

$x + y = 1$

R

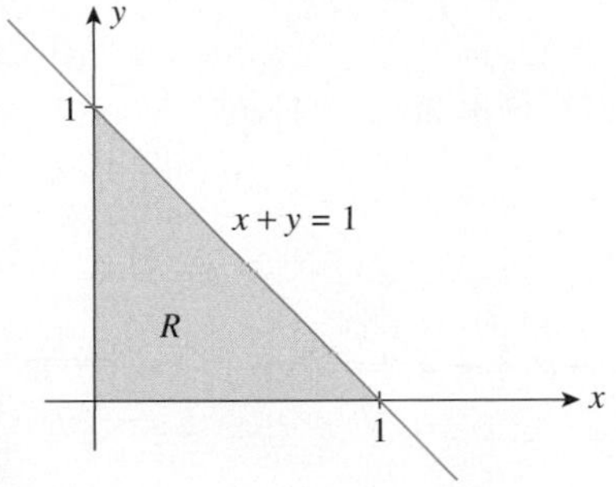

FIGURE 5

$$P(\{(x, y) \mid x + y \le 1\}) = \int_0^1 \int_0^{1-x} xy\, dy\, dx$$
$$= \int_0^1 \left(\frac{1}{2}xy^2 \bigg|_0^{1-x}\right) dx$$
$$= \int_0^1 \frac{1}{2}x(1 - x)^2\, dx$$
$$= \frac{1}{2}\int_0^1 (x - 2x^2 + x^3)\, dx$$
$$= \frac{1}{2}\left(\frac{1}{2}x^2 - \frac{2}{3}x^3 + \frac{1}{4}x^4\right)\bigg|_0^1$$
$$= \frac{1}{2}\left(\frac{1}{2} - \frac{2}{3} + \frac{1}{4}\right) = \frac{1}{24}$$

EXAMPLE 6 Let $f(x, y) = 2e^{-x-2y}$ be a joint probability density function on $D = \{(x, y) \mid x \ge 0 \text{ and } y \ge 0\}$.

a. Find the probability that x will assume a value between 0 and 1 and that y will assume a value between 1 and 2.

b. Find the probability that x will assume a value greater than 1 and that y will assume a value less than 2.

Solution

a. The required probability is given by

$$P(0 \le x \le 1;\, 1 \le y \le 2) = \int_1^2 \int_0^1 2e^{-x-2y}\, dx\, dy$$
$$= \int_1^2 \left(-2e^{-x-2y}\bigg|_0^1\right) dy$$
$$= \int_1^2 (-2e^{-1-2y} + 2e^{-2y})\, dy$$
$$= \int_1^2 2(1 - e^{-1})e^{-2y}\, dy \qquad \text{Simplify the integrand.}$$
$$= 2(1 - e^{-1})\left(-\frac{1}{2}e^{-2y}\right)\bigg|_1^2$$
$$= -(1 - e^{-1})e^{-2y}\bigg|_1^2$$
$$= -(1 - e^{-1})(e^{-4} - e^{-2}) \approx .0740$$

b. The required probability is given by

$$\begin{aligned} P(x \geq 1; 0 \leq y \leq 2) &= \int_0^2 \int_1^\infty 2e^{-x-2y}\, dx\, dy \\ &= \int_0^2 \left(\lim_{b\to\infty} \int_1^b 2e^{-x-2y}\, dx \right) dy \\ &= \int_0^2 \left(\lim_{b\to\infty} -2e^{-x-2y} \Big|_1^b \right) dy \\ &= \int_0^2 [\lim_{b\to\infty} (-2e^{-b-2y} + 2e^{-1-2y})]\, dy \\ &= \int_0^2 2e^{-1-2y}\, dy \\ &= -e^{-1-2y} \Big|_0^2 = -e^{-5} + e^{-1} \approx .3611 \end{aligned}$$

10.1 Self-Check Exercises

1. Determine the value of the constant k such that the function $f(x) = k(4x - x^2)$ is a probability density function on the interval [0, 4].

2. Suppose that x is a continuous random variable with the probability density function of Self-Check Exercise 1. Find the probability that x will assume a value between $x = 1$ and $x = 3$.

Solutions to Self-Check Exercises 10.1 can be found on page 665.

10.1 Concept Questions

1. Define the following terms in your own words:
   a. experiment
   b. sample point (outcome)
   c. event
   d. probability of an event

2. What is the difference between a finite discrete random variable and a continuous random variable?

3. **a.** What is a probability density function?
   b. What is a joint probability density function?

10.1 Exercises

In Exercises 1–12, show that the function is a probability density function on the indicated interval.

1. $f(x) = \frac{1}{3}$; $(3 \leq x \leq 6)$

2. $f(x) = \frac{1}{\pi}$; $(\pi \leq x \leq 2\pi)$

3. $f(x) = \frac{2}{32}x$; $(2 \leq x \leq 6)$

4. $f(x) = \frac{3}{8}x^2$; $(0 \leq x \leq 2)$

5. $f(x) = \frac{2}{9}(3x - x^2)$; $(0 \leq x \leq 3)$

6. $f(x) = \frac{3}{32}(x - 1)(5 - x)$; $(1 \leq x \leq 5)$

7. $f(x) = \frac{12 - x}{72}$; $(0 \leq x \leq 12)$

8. $f(x) = 20(x^3 - x^4)$; $(0 \leq x \leq 1)$

9. $f(x) = \frac{8}{7x^2}$; $(1 \leq x \leq 8)$

10. $f(x) = \frac{3}{14}\sqrt{x}$; $(1 \leq x \leq 4)$

11. $f(x) = \frac{x}{(x^2 + 1)^{3/2}}$; $(0 \leq x < \infty)$

12. $f(x) = 4xe^{-2x^2}$; $(0 \leq x < \infty)$

In Exercises 13–20, find the value of the constant k so that the function is a probability density function on the indicated interval.

13. $f(x) = k; [1, 4]$

14. $f(x) = kx; [0, 4]$

15. $f(x) = k(4 - x); [0, 4]$

16. $f(x) = kx^3; [0, 1]$

17. $f(x) = k\sqrt{x}; [0, 4]$

18. $f(x) = \frac{k}{x}; [1, 5]$

19. $f(x) = \frac{k}{x^3}; [1, \infty)$

20. $f(x) = ke^{-x/2}; [0, \infty)$

In Exercises 21–30, f is a probability density function defined on the given interval. Find the indicated probabilities.

21. $f(x) = \frac{1}{12}x; [1, 5]$
 a. $P(2 \le x \le 4)$ b. $P(1 \le x \le 4)$
 c. $P(x \ge 2)$ d. $P(x = 2)$

22. $f(x) = \frac{1}{9}x^2; [0, 3]$
 a. $P(1 \le x \le 2)$ b. $P(1 < x \le 3)$
 c. $P(x \le 2)$ d. $P(x = 1)$

23. $f(x) = \frac{3}{32}(4 - x^2); [-2, 2]$
 a. $P(-1 \le x \le 1)$ b. $P(x \le 0)$
 c. $P(x > -1)$ d. $P(x = 0)$

24. $f(x) = \frac{3}{16}\sqrt{x}; [0, 4]$
 a. $P(1 < x < 3)$ b. $P(x \le 2)$
 c. $P(x = 2)$ d. $P(x \ge 1)$

25. $f(x) = \frac{1}{4\sqrt{x}}; [1, 9]$
 a. $P(x \ge 4)$ b. $P(1 \le x < 8)$
 c. $P(x = 3)$ d. $P(x \le 4)$

26. $f(x) = \frac{1}{2}e^{-x/2}; [0, \infty)$
 a. $P(x \le 4)$ b. $P(1 < x < 2)$
 c. $P(x = 50)$ d. $P(x \ge 2)$

27. $f(x) = 4xe^{-2x^2}; [0, \infty)$
 a. $P(0 \le x \le 4)$ b. $P(x \ge 1)$

28. $f(x) = \frac{1}{9}xe^{-x/3}; [0, \infty)$
 a. $P(0 \le x \le 3)$ b. $P(x \ge 1)$

29. $f(x) = \begin{cases} x & \text{if } 0 \le x \le 1 \\ 2 - x & \text{if } 1 \le x \le 2 \end{cases}; [0, 2]$
 a. $P(\frac{1}{2} \le x \le 1)$ b. $P(\frac{1}{2} \le x \le \frac{3}{2})$
 c. $P(x \ge 1)$ d. $P(x \le \frac{3}{2})$

30. $f(x) = \begin{cases} \frac{3}{40}\sqrt{x} & 0 \le x \le 4 \\ \frac{12}{5x^2} & 4 < x < \infty \end{cases}; [0, \infty]$
 a. $P(1 \le x \le 4)$ b. $P(0 \le x \le 5)$

In Exercises 31–34, show that the function is a joint probability density function on D.

31. $f(x, y) = \frac{1}{4}; D = \{0 \le x \le 2; 1 \le y \le 3\}$

32. $f(x, y) = \frac{1}{4}(x + 2y); D = \{0 \le x \le 2; 0 \le y \le 1\}$

33. $f(x, y) = \frac{1}{3}xy; D = \{0 \le x \le 2; 1 \le y \le 2\}$

34. $f(x, y) = 4(1 - x)(2 - y); D = \{0 \le x \le 1; 1 \le y \le 2\}$

In Exercises 35–38, find the value of the constant k so that the function is a joint probability density function on D.

35. $f(x, y) = kx^2y; D = \{0 \le x \le 1; 1 \le y \le 2\}$

36. $f(x, y) = k\sqrt{x}(2 - y); D = \{1 \le x \le 2; 0 \le y \le 2\}$

37. $f(x, y) = k(x - x^2)e^{-2y}; D = \{0 \le x \le 1; 1 \le y < \infty\}$

38. $f(x, y) = kxye^{-(x^2+y^2)}; D = \{0 < x < \infty; 0 < y < \infty\}$

In Exercises 39–42, f is a joint probability density function on D. Find the indicated probabilities.

39. $f(x, y) = xy; D = \{(x, y) \mid 0 \le x \le 1; 0 \le y \le 2\}$
 a. $P(0 \le x \le 1; 0 \le y \le 1)$ b. $P(\{(x, y) \mid x + 2y \le 1\})$

40. $f(x, y) = \frac{1}{12}(x + y); D = \{(x, y) \mid 0 \le x \le 2; 1 \le y \le 3\}$
 a. $P(0 \le x \le 1; 1 \le y \le 2)$
 b. $P(\{x, y) \mid y \ge 1 \text{ and } x + y \le 3\})$

41. $f(x, y) = \frac{9}{224}\sqrt{xy}; D = \{(x, y) \mid 1 \le x \le 4; 0 \le y \le 4\}$
 a. $P(1 \le x \le 2; 0 \le y \le 1)$
 b. $P(\{(x, y) \mid 1 \le x \le 4; 0 \le y \le \sqrt{x}\})$

42. $f(x, y) = 3\sqrt{x}e^{-2y}; D = \{(x, y) \mid 0 \le x \le 1; 0 \le y < \infty\}$
 a. $P(0 \le x \le 1; 1 \le y < \infty)$
 b. $P(\{(x, y) \mid \frac{1}{2} \le x \le 1; 0 \le y < \infty\})$

43. **Life Span of a Plant** The life span of a certain plant species (in days) is described by the probability density function

$$f(x) = \frac{1}{100}e^{-x/100}$$

 a. Find the probability that a plant of this species will live for 100 days or less.
 b. Find the probability that a plant of this species will live longer than 120 days.

c. Find the probability that a plant of this species will live longer than 60 days but less than 140 days.

44. WAITING TIME AT A HEALTH CLINIC The average waiting time in minutes for patients arriving at the Newtown Health Clinic between 1 and 4 P.M. on a weekday is an exponentially distributed random variable x with associated probability density function $f(x) = \frac{1}{15} e^{-(1/15)x}$.

a. What is the probability that a patient arriving at the clinic between 1 and 4 P.M. will have to wait longer than 15 min?

b. What is the probability that a patient arriving at the clinic between 1 and 4 P.M. will have to wait between 10 and 12 min?

45. RELIABILITY OF ROBOTS National Welding Company uses industrial robots in some of its assembly-line operations. Management has determined that the lengths of time in hours between breakdowns are exponentially distributed with probability density function $f(t) = 0.001e^{-0.001t}$.

a. What is the probability that a robot selected at random will break down after between 600 and 800 hr of use?

b. What is the probability that a robot will break down after 1200 hr of use?

46. WAITING TIME AT AN EXPRESSWAY TOLLBOOTH Suppose the time intervals in seconds between arrivals of successive cars at an expressway tollbooth during rush hour are exponentially distributed with associated probability density function $f(t) = \frac{1}{8} e^{-(1/8)t}$. Find the probability that the average time interval between arrivals of successive cars is more than 8 sec.

47. MAIL-ORDER PHONE CALLS A study conducted by Uni-Mart, a mail-order department store, reveals that the time intervals in minutes between incoming telephone calls on its toll-free 800 line between 10 A.M. and 2 P.M. are exponentially distributed with probability density function $f(t) = \frac{1}{30} e^{-t/30}$. What is the probability that the time interval between successive calls is more than 2 min?

48. RELIABILITY OF A MICROPROCESSOR The microprocessors manufactured by United Motor Works, which are used in automobiles to regulate fuel consumption, are guaranteed against defects for 20,000 mi of use. Tests conducted in the laboratory under simulated driving conditions reveal that the distances driven in miles before the microprocessors break down are exponentially distributed with probability density function $f(x) = 0.00001e^{-0.00001t}$. What is the probability that a microprocessor selected at random will fail during the warranty period?

49. PROBABILITY OF SNOWFALL The amount of snowfall (in feet) in a remote region of Alaska in the month of January is a continuous random variable with probability density function $f(x) = \frac{2}{9} x(3 - x)$, $0 \leq x \leq 3$. Find the probability that the amount of snowfall will be between 1 and 2 ft; more than 1 ft.

50. NUMBER OF CHIPS IN CHOCOLATE CHIP COOKIES The number of chocolate chips in each cookie of a certain brand has a distribution described by the probability density function

$$f(x) = \frac{1}{36}(6x - x^2) \qquad (0 \leq x \leq 6)$$

Find the probability that the number of chocolate chips in a randomly chosen cookie is fewer than two.

51. LIFE EXPECTANCY OF COLOR TELEVISION TUBES The life expectancy (in years) of a certain brand of color television tube is a continuous random variable with probability density function

$$f(t) = 9(9 + t^2)^{-3/2} \qquad (0 \leq t < \infty)$$

Find the probability that a randomly chosen television tube will last more than 4 yr.
Hint: Integrate using a table of integrals.

52. PRODUCT RELIABILITY The tread life (in thousands of miles) of a certain make of tire is a continuous random variable with a probability density function

$$f(x) = 0.02e^{-0.02x} \qquad (0 \leq x < \infty)$$

a. Find the probability that a randomly selected tire of this make will have a tread life of at most 30,000 mi.

b. Find the probability that a randomly selected tire of this make will have a tread life between 40,000 and 60,000 mi.

c. Find the probability that a randomly selected tire of this make will have a tread life of at least 70,000 mi.

53. CABLE TV SUBSCRIBERS The management of Telstar Cable estimates that the number of new subscribers (in thousands) to their service next year in Foxborough, x, and the number of new subscribers (in thousands) next year in Sharon, y, have a distribution given by the joint probability density function

$$f(x, y) = \frac{9}{4000} xy \sqrt{25 - x^2}(4 - y);$$
$$D = \{(x, y) \mid 0 \leq x \leq 5; 0 \leq y \leq 4\}$$

Find the probability that the number of new subscribers next year in Foxborough will be between 2000 and 2500 and the number of new subscribers in Sharon will be between 1000 and 2000.

54. CABLE TV SUBSCRIBERS Refer to Exercise 53. What is the probability that the total number of new subscribers next year in Foxborough will be fewer than 3000 and the number of new subscribers next year in Sharon will be fewer than 2000?

In Exercises 55 and 56, determine whether the statement is true or false. If it is true, explain why it is true. If it is false, give an example to show why it is false.

55. If $\int_a^b f(x)\, dx = 1$, then f is a probability density function on $[a, b]$.

56. If f is a probability function on an interval $[a, b]$, then f is a probability function on $[c, d]$ for any real numbers c and d satisfying $a < c < d < b$.

10.1 Solutions to Self-Check Exercises

1. We compute

$$\int_0^4 k(4x - x^2)\,dx = k\left(2x^2 - \frac{1}{3}x^3\right)\Bigg|_0^4 = k\left(32 - \frac{64}{3}\right) = \frac{32}{3}k$$

Since this value must be equal to 1, we have $k = \frac{3}{32}$.

2. The required probability is given by

$$\begin{aligned} P(1 \le x \le 3) &= \int_1^3 f(x)\,dx \\ &= \int_1^3 \frac{3}{32}(4x - x^2)\,dx \\ &= \frac{3}{32}\left(2x^2 - \frac{1}{3}x^3\right)\Bigg|_1^3 \\ &= \frac{3}{32}\left[(18 - 9) - \left(2 - \frac{1}{3}\right)\right] = \frac{11}{16} \end{aligned}$$

USING TECHNOLOGY

Graphing a Histogram

Graphing Utility

A graphing utility can be used to plot the histogram for a given set of data, as illustrated by the following example.

APPLIED EXAMPLE 1 A survey of 90,000 households conducted in 1995 revealed the following percent of women who wear the given shoe size.

Shoe Size	<5	$5-5\frac{1}{2}$	$6-6\frac{1}{2}$	$7-7\frac{1}{2}$	$8-8\frac{1}{2}$	$9-9\frac{1}{2}$	$10-10\frac{1}{2}$	$>10\frac{1}{2}$
Women, %	1	5	15	27	29	14	7	2

Source: Footwear Market Insights survey

a. Plot a histogram for the given data.
b. What percent of women in the survey wear size $7-7\frac{1}{2}$ or $8-8\frac{1}{2}$ shoes?

Solution

a. Let X denote the random variable taking on the values 1 through 8, where 1 corresponds to a shoe size less than 5, 2 corresponds to a shoe size of $5-5\frac{1}{2}$, and so on. Enter the values of X as $x_1 = 1, x_2 = 2, \ldots, x_8 = 8$ and the corresponding values of Y as $y_1 = 1, y_2 = 5, \ldots, y_8 = 2$. Then use the DRAW function from the Statistics menu to draw the histogram shown in Figure T1.

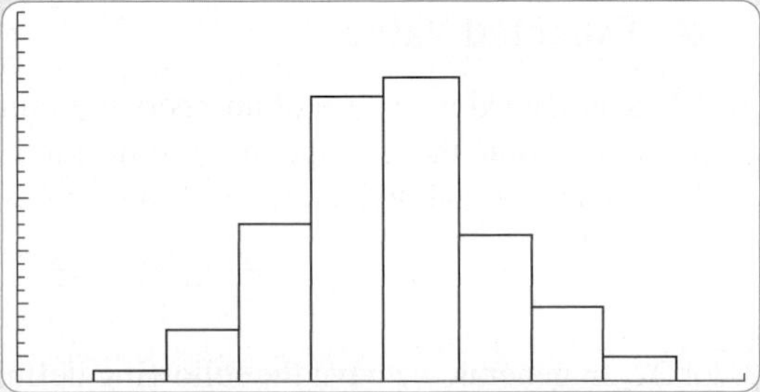

FIGURE T1
The histogram for the given data, using the viewing window $[0, 9] \times [0, 35]$

(continued)

b. The probability that a woman participating in the survey wears size $7\text{–}7\frac{1}{2}$ or $8\text{–}8\frac{1}{2}$ shoes is given by

$$P(X = 4) + P(X = 5) = .27 + .29 = .56$$

which tells us that 56% of the women wear either size $7\text{–}7\frac{1}{2}$ or $8\text{–}8\frac{1}{2}$ shoes. ■

TECHNOLOGY EXERCISES

1. Graph the histogram associated with the data given in Table 2, page 655. Compare your graph with that given in Figure 1, page 655.

2. DISTRIBUTION OF FAMILIES BY SIZE: A survey was conducted by the Public Housing Authority in a certain community among 1000 families to determine the distribution of families by size. The results follow.

Family Size	2	3	4	5	6	7	8
Frequency of Occurrence	350	200	245	125	66	10	4

   a. Find the probability distribution of the random variable X, where X denotes the number of persons in a randomly chosen family.
   b. Graph the histogram corresponding to the probability distribution found in part (a).

3. WAITING LINES The accompanying data were obtained in a study conducted by the manager of Sav-More Supermarket. In this study the number of customers waiting in line at the express checkout at the beginning of each 3-min interval between 9 A.M. and 12 noon on Saturday was observed.

Customers	0	1	2	3	4	5
Frequency of Occurrence	1	4	2	7	14	8

Customers	6	7	8	9	10
Frequency of Occurrence	10	6	3	4	1

   a. Find the probability distribution of the random variable X, where X denotes the number of customers observed waiting in line.
   b. Graph the histogram representing this probability distribution.

4. TELEVISION PILOTS After the private screening of a new television pilot, audience members were asked to rate the new show on a scale of 1 to 10 (10 being the highest rating). From a group of 140 people, these responses were obtained:

Rating	1	2	3	4	5
Frequency of Occurrence	1	4	3	11	23

Rating	6	7	8	9	10
Frequency of Occurrence	21	28	29	16	4

   Let the random variable X denote the rating given to the show by a randomly chosen audience member. Find the probability distribution and graph the histogram associated with these data.

10.2 Expected Value and Standard Deviation

Expected Value

The average value of a set of numbers is a familiar notion to most people. For example, to compute the average of the four numbers 12, 16, 23, and 37, we simply add these numbers and divide the resulting sum by 4, giving the average as

$$\frac{12 + 16 + 23 + 37}{4} = \frac{88}{4}$$

or 22. In general, we have the following definition.

Average, or Mean

The **average**, or **mean**, of the n numbers $x_1, x_2, \ldots, x_n$ is $\bar{x}$ (read "x bar"), where

$$\bar{x} = \frac{x_1 + x_2 + \cdots + x_n}{n}$$

APPLIED EXAMPLE 1 Waiting Times The number of cars observed waiting in line at the beginning of each 2-minute interval between 3 and 5 P.M. on a certain Friday at the drive-in teller of Westwood Savings Bank and the corresponding frequency of occurrence are shown in Table 3. Find the average number of cars observed waiting in line at the beginning of each 2-minute interval during the 2-hour period.

TABLE 3

Cars	0	1	2	3	4	5	6	7	8
Frequency of Occurrence	2	9	16	12	8	6	4	2	1

Solution Observe from Table 3 that the number 0 (of cars) occurs twice, the number 1 occurs nine times, and so on. There are altogether

$$2 + 9 + 16 + 12 + 8 + 6 + 4 + 2 + 1 = 60$$

numbers to be averaged. Therefore, the required average is given by

$$\frac{0 \cdot 2 + 1 \cdot 9 + 2 \cdot 16 + 3 \cdot 12 + 4 \cdot 8 + 5 \cdot 6 + 6 \cdot 4 + 7 \cdot 2 + 8 \cdot 1}{60} \approx 3.1 \quad \textbf{(1)}$$

or approximately 3.1 cars. ■

Let's reconsider the expression in Equation (1) that gives the average of the frequency distribution shown in Table 3. Dividing each term by the denominator, we can rewrite the expression in the form

$$0 \cdot \left(\frac{2}{60}\right) + 1 \cdot \left(\frac{9}{60}\right) + 2 \cdot \left(\frac{16}{60}\right) + 3 \cdot \left(\frac{12}{60}\right) + 4 \cdot \left(\frac{8}{60}\right) + 5 \cdot \left(\frac{6}{60}\right)$$
$$+ 6 \cdot \left(\frac{4}{60}\right) + 7 \cdot \left(\frac{2}{60}\right) + 8 \cdot \left(\frac{1}{60}\right)$$

Observe that each term in the sum is a product of two factors. The first factor is the value assumed by the random variable X, where X denotes the number of cars observed waiting in line, and the second factor is just the probability associated with that value of the random variable. This observation suggests the following general method for calculating the expected value (that is, the average, or mean) of a random variable X that assumes a finite number of values from the knowledge of its probability distribution.

Expected Value of a Discrete Random Variable X

Let X denote a random variable that assumes the values $x_1, x_2, \ldots, x_n$ with associated probabilities $p_1, p_2, \ldots, p_n$, respectively. Then the **expected value** of X, $E(X)$, is given by

$$E(X) = x_1p_1 + x_2p_2 + \cdots + x_np_n \quad (2)$$

TABLE 4

Probability Distribution for the Random Variable X

x	$P(X = x)$
0	.03
1	.15
2	.27
3	.20
4	.13
5	.10
6	.07
7	.03
8	.02

Note The numbers $x_1, x_2, \ldots, x_n$ may be positive, zero, or negative. For example, such a number will be positive if it represents a profit and negative if it represents a loss.

EXAMPLE 2 Refer to Example 1. Let the random variable X denote the number of cars observed waiting in line. Use the probability distribution for the random variable X (Table 4) to solve Example 1 again.

Solution Let X denote the number of cars observed waiting in line. Then, the average number of cars observed waiting in line is given by the expected value of X—that is, by

$$\begin{aligned} E(X) &= 0\cdot(.03) + 1\cdot(.15) + 2\cdot(.27) + 3\cdot(.20) + 4\cdot(.13) \\ &\quad + 5\cdot(.10) + 6\cdot(.07) + 7\cdot(.03) + 8\cdot(.02) \\ &= 3.1 \end{aligned}$$

or 3.1 cars, which agrees with the earlier results. ■

The expected value of a random variable X is a measure of the central tendency of the probability distribution associated with X. In repeated trials of an experiment with a random variable X, the average of the observed values of X gets closer and closer to the expected value of X as the number of trials gets larger and larger. Geometrically, the expected value of a random variable X has the following simple interpretation: If a laminate is made of the histogram of a probability distribution associated with a discrete random variable X, then the expected value of X corresponds to the point on the base of the laminate at which the latter will balance perfectly when the point is directly over a fulcrum (Figure 6).

Now suppose x is a continuous random variable and f is the probability density function associated with it. For simplicity, let's first assume that $a \leq x \leq b$. Divide

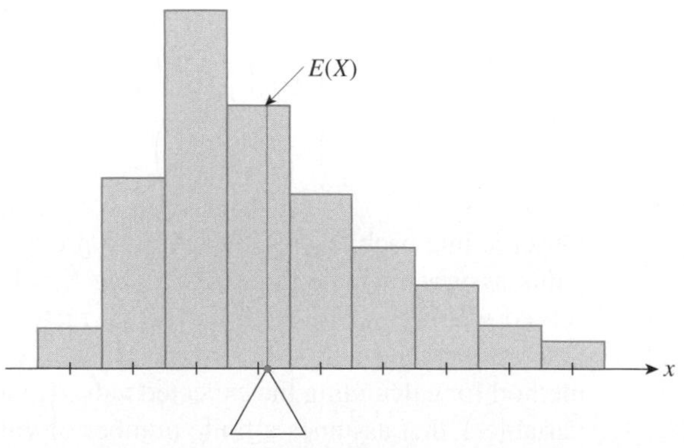

FIGURE 6
Expected value of a random variable X

the interval $[a, b]$ into n subintervals of equal length $\Delta x = (b - a)/n$ by means of the $(n + 1)$ points $x_0 = a, x_1, x_2, \ldots, x_n = b$ (Figure 7).

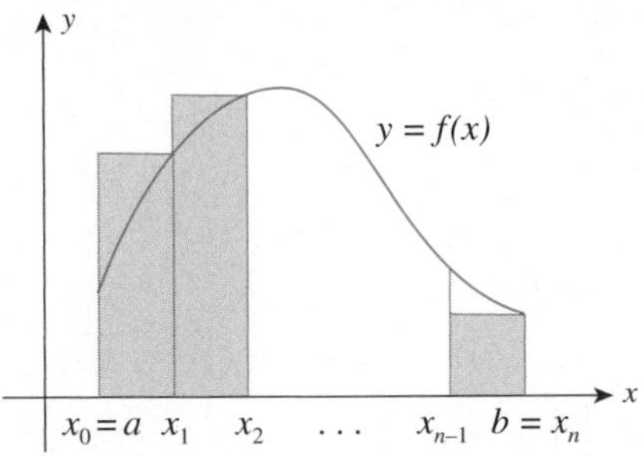

FIGURE 7
Approximating the expected value of a random variable x on $[a, b]$ by a Riemann sum

To find an approximation of the average value, or expected value, of x on the interval $[a, b]$, let's treat x as if it were a discrete random variable that takes on the values $x_1, x_2, \ldots, x_n$ with probabilities $p_1, p_2, \ldots, p_n$. Then

$$E(x) \approx x_1p_1 + x_2p_2 + \cdots + x_np_n$$

But p_1 is the probability that x is in the interval $[x_0, x_1]$, and this is just the area under the graph of f from $x = x_0$ to $x = x_1$, which may be approximated by $f(x_1)\,\Delta x$. The probabilities $p_2, \ldots, p_n$ may be approximated in a similar manner. Thus,

$$E(x) \approx x_1f(x_1)\Delta x + x_2f(x_2)\Delta x + \cdots + x_nf(x_n)\Delta x$$

which is seen to be the Riemann sum of the function $g(x) = xf(x)$ over the interval $[a, b]$. Letting n approach infinity, we obtain the following formula:

Expected Value of a Continuous Random Variable
Suppose the function f defined on the interval $[a, b]$ is the probability density function associated with a continuous random variable x. Thus, the **expected value** of x is

$$E(x) = \int_a^b xf(x)\,dx \tag{3}$$

If either $a = -\infty$ or $b = \infty$, then the integral in (3) becomes an improper integral.

The expected value of a random variable is used in many practical applications. For example, if x represents the life span of a certain brand of electronic component, then the expected value of x gives the average life span of these components. If x measures the waiting time in a doctor's office, then $E(x)$ gives the average waiting time, and so on.

APPLIED EXAMPLE 3 Average Life Span of a Light Bulb Show that if a continuous random variable x is exponentially distributed with the probability density function

$$f(x) = ke^{-kx}$$

then the expected value of x, $E(x)$, is equal to $1/k$. Using this result, determine the average life span of a 200-watt electric light bulb manufactured by TKK Products of Example 3, page 658.

Solution We compute

$$\begin{aligned} E(x) &= \int_0^\infty xf(x)\,dx \\ &= \int_0^\infty kxe^{-kx}\,dx \\ &= k\lim_{b\to\infty}\int_0^b xe^{-kx}\,dx \end{aligned}$$

Integrating by parts with

$$u = x \quad \text{and} \quad dv = e^{-kx}\,dx$$

so that

$$du = dx \quad \text{and} \quad v = -\frac{1}{k}e^{-kx}$$

we have

$$\begin{aligned} E(x) &= k \lim_{b\to\infty}\left(-\frac{1}{k}xe^{-kx}\Big|_0^b + \frac{1}{k}\int_0^b e^{-kx}\,dx\right) \\ &= k \lim_{b\to\infty}\left[-\left(\frac{1}{k}\right)be^{-kb} - \frac{1}{k^2}e^{-kx}\Big|_0^b\right] \\ &= k \lim_{b\to\infty}\left[-\left(\frac{1}{k}\right)be^{-kb} - \frac{1}{k^2}e^{-kb} + \frac{1}{k^2}\right] \\ &= -\lim_{b\to\infty}\frac{b}{e^{kb}} - \frac{1}{k}\lim_{b\to\infty}\frac{1}{e^{kb}} + \frac{1}{k}\lim_{b\to\infty} 1 \end{aligned}$$

Now, by taking a sequence of values of b that approaches infinity, for example, $b = 10, 100, 1000, 10{,}000, \ldots$, we see that, for a fixed k,

$$\lim_{b\to\infty}\frac{b}{e^{kb}} = 0$$

Therefore,

$$E(x) = \frac{1}{k}$$

as we set out to show. From our previous work, we know that $k = 0.001$. So the average life span of the TKK light bulb is $1/(0.001) = 1000$ hours. ■

Before considering another example, let's summarize the result obtained in Example 3.

Average Value of an Exponential Density Function

If a continuous random variable x is exponentially distributed with probability density function

$$f(x) = ke^{-kx}$$

then the **expected value** of x is given by

$$E = \frac{1}{k}$$

EXPLORING WITH TECHNOLOGY

Refer to Example 3. Plot the graphs of $f(x) = x/e^{kx}$ for different positive values of k, using the appropriate viewing windows to demonstrate that $\lim_{b\to\infty} b/e^{kb} = 0$ for a fixed $k > 0$. Repeat for the function $f(x) = 1/e^{kx}$ to demonstrate that $\lim_{b\to\infty} 1/e^{kb} = 0$.

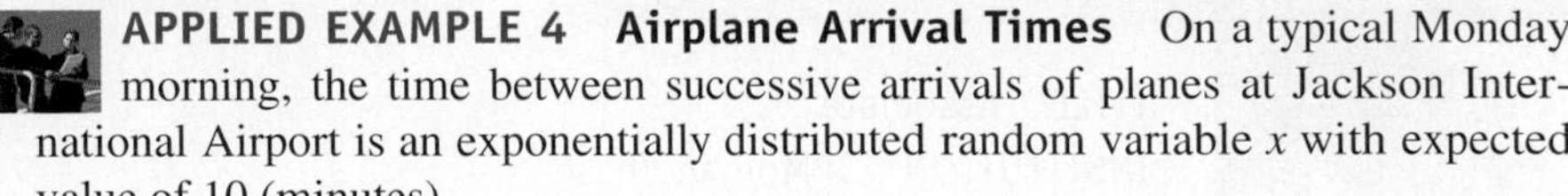

APPLIED EXAMPLE 4 Airplane Arrival Times On a typical Monday morning, the time between successive arrivals of planes at Jackson International Airport is an exponentially distributed random variable x with expected value of 10 (minutes).

a. Find the probability density function associated with x.
b. What is the probability that between 6 and 8 minutes will elapse between successive arrivals of planes?
c. What is the probability that the time between successive arrivals will be more than 15 minutes?

Solution

a. Since x is exponentially distributed, the associated probability density function has the form $f(x) = ke^{-kx}$. Next, since the expected value of x is 10, we see that

$$E(x) = \frac{1}{k} = 10$$

$$k = \frac{1}{10} = .1$$

so the required probability density function is

$$f(x) = .1e^{-0.1x}$$

b. The probability that between 6 and 8 minutes will elapse between successive arrivals is given by

$$\begin{aligned} P(6 \leq x \leq 8) &= \int_6^8 .1e^{-0.1x}\,dx = -e^{-0.1x}\Big|_6^8 \\ &= -e^{-0.8} + e^{-0.6} \\ &\approx .100 \end{aligned}$$

c. The probability that the time between successive arrivals will be more than 15 minutes is given by

$$\begin{aligned} P(x > 15) &= \int_{15}^{\infty} .1e^{-0.1x}\,dx \\ &= \lim_{b\to\infty} \int_{15}^{b} .1e^{-0.1x}\,dx \\ &= \lim_{b\to\infty} \left[-e^{-0.1x}\Big|_{15}^{b} \right] \\ &= \lim_{b\to\infty} (-e^{-0.1b} + e^{-1.5}) = e^{-1.5} \\ &\approx .22 \end{aligned}$$

Variance

The mean, or expected value, of a random variable enables us to express an important property of the probability distribution associated with the random variable in terms of a single number. But knowing the location, or central tendency, of a probability distribution alone is usually not enough to give a reasonably accurate picture

PORTFOLIO Gary Li

TITLE Associate
INSTITUTION JPMorgan Chase

As one of the leading financial institutions in the world, JPMorgan Chase & Co. depends on a wide range of mathematical disciplines from statistics to linear programming to calculus. Whether assessing the credit worthiness of a borrower, recommending portfolio investments or pricing an exotic derivative, quantitative understanding is a critical tool in serving the financial needs of clients.

I work in the Fixed-Income Derivatives Strategy group. A derivative in finance is an instrument whose value depends on the price of some other underlying instrument. A simple type of derivative is the forward contract, where two parties agree to a future trade at a specified price. In agriculture, for instance, farmers will often pledge their crops for sale to buyers at an agreed price before even planting the harvest. Depending on the weather, demand and other factors, the actual price may turn out higher or lower. Either the buyer or seller of the forward contract benefits accordingly. The value of the contract changes one-for-one with the actual price. In derivatives lingo, we borrow from calculus and say forward contracts have a delta of one.

Nowadays, the bulk of derivatives deal with interest-rate rather than agricultural risk. The value of any asset with fixed payments over time varies with interest rates. With trillions of dollars in this form, especially government bonds and mortgages, fixed-income derivatives are vital to the economy. As a strategy group, our job is to track and anticipate key drivers and developments in the market using, in significant part, quantitative analysis. Some of the derivatives we look at are of the forward kind, such as interest-rate swaps, where over time you receive fixed-rate payments in exchange for paying a floating-rate or vice-versa. A whole other class of derivatives where statistics and calculus are especially relevant are options.

Whereas forward contracts bind both parties to a future trade, options give the holder the right but not the obligation to trade at a specified time and price. Similar to an insurance policy, the holder of the option pays an upfront premium in exchange for potential gain. Solving this pricing problem requires statistics, stochastic calculus, and enough insight to win a Nobel prize. Fortunately for us, this was taken care of by Fischer Black, Myron Scholes, and Robert Merton in the early 1970s (including the 1997 Nobel Prize in Economics for Scholes and Merton). The Black-Scholes differential equation was the first accurate options pricing model, making possible the rapid growth of the derivatives market. Black-Scholes and the many derivatives of it, as it were, continue to be used to this day.

of the probability distribution. Consider, for example, the two probability distributions whose histograms appear in Figure 8. Both distributions have the same expected value, or mean, $\mu = 4$ (the Greek letter μ is read "mu"). Note that the probability distribution with the histogram shown in Figure 8a is closely concentrated about its mean μ, whereas the one with the histogram shown in Figure 8b is widely dispersed, or spread, about its mean.

As another example, suppose David Horowitz, host of the television show *The Consumer Advocate*, decides to demonstrate the accuracy of the weights of two popular brands of potato chips. Ten packages of potato chips of each brand are selected at random and weighed carefully. The weights in ounces are listed in Table 5.

We will verify in Example 7 that the mean weights for each of the two brands is 16 ounces. However, a cursory examination of the data now shows that the weights of the brand B packages exhibit much greater dispersion about the mean than those of brand A.

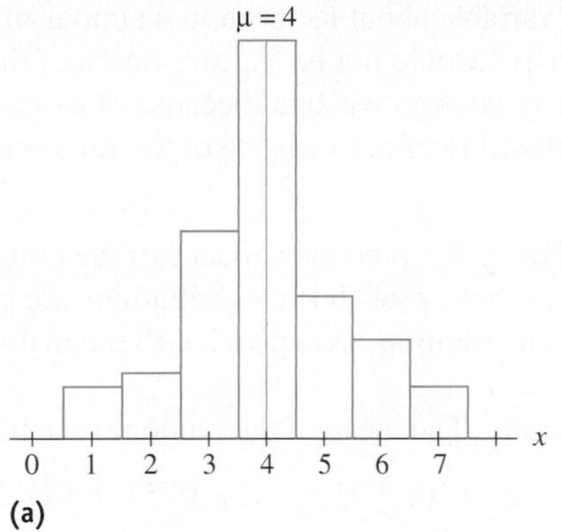

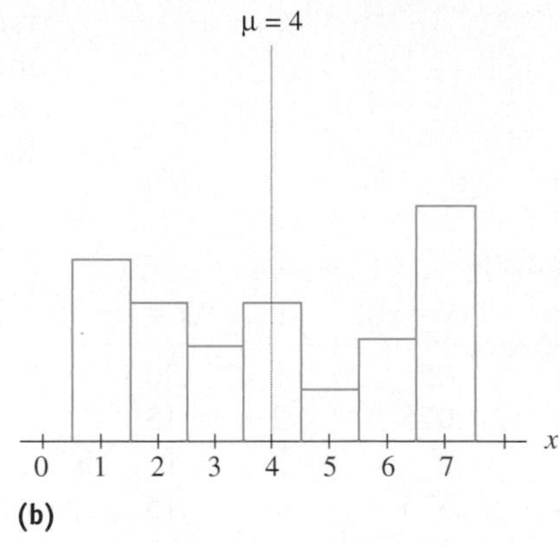

FIGURE 8
The histograms of two probability distributions

TABLE 5

Brand A	Brand B
16.1	16.3
16	15.7
15.8	15.8
16	16.2
15.9	15.9
16.1	16.1
15.9	15.7
16	16.2
16	16
16.2	16.1

One measure of the degree of dispersion, or spread, of a probability distribution about its mean is the variance of the random variable associated with the probability distribution. A probability distribution with a small spread about its mean will have a small variance, whereas one with a larger spread will have a larger variance. Thus, the variance of the random variable associated with the probability distribution whose histogram appears in Figure 8a is smaller than the variance of the random variable associated with the probability distribution whose histogram is shown in Figure 8b. Also, as we will see in Example 7, the variance of the random variable associated with the weights of the brand A potato chips is smaller than that of the random variable associated with the weights of the brand B potato chips.

Variance of a Discrete Random Variable *X*

Suppose a random variable has the probability distribution

x	x_1	x_2	x_3	$\cdots$	x_n
$P(X = x)$	p_1	p_2	p_3	$\cdots$	p_n

and expected value

$$E(X) = \mu$$

Then the **variance of a random variable** X is

$$\text{Var}(X) = p_1(x_1 - \mu)^2 + p_2(x_2 - \mu)^2 + \cdots + p_n(x_n - \mu)^2 \qquad \textbf{(4)}$$

Let's look a little closer at Formula (4). First, note that the numbers

$$x_1 - \mu, x_2 - \mu, \ldots, x_n - \mu \qquad \textbf{(5)}$$

measure the **deviations** of $x_1, x_2, \ldots, x_n$ from μ, respectively. Thus, the numbers

$$(x_1 - \mu)^2, (x_2 - \mu)^2, \ldots, (x_n - \mu)^2 \qquad \textbf{(6)}$$

measure the squares of the deviations of $x_1, x_2, \ldots, x_n$ from μ, respectively. Next, by multiplying each of the numbers in (6) by the probability associated with each value of the random variable X, we weight the numbers accordingly so that their sum is a measure of the variance of X about its mean. An attempt to define the variance of a random variable about its mean in a similar manner using the deviations rather than their squares would not be fruitful. Some of the deviations may be positive whereas others may be negative, and (because of cancellations) the sum will not give a satisfactory measure of the variance of the random variable.

TABLE 6

x	$P(X = x)$	y	$P(Y = y)$
1	.05	1	.20
2	.075	2	.15
3	.2	3	.1
4	.375	4	.15
5	.15	5	.05
6	.1	6	.1
7	.05	7	.25

EXAMPLE 5 Find the variance of the random variable X and of the random variable Y whose probability distributions are given in Table 6. These are the probability distributions associated with the histograms shown in Figures 8a and 8b.

Solution The mean of the random variable X is given by

$$\begin{aligned}\mu_X &= (1)(.05) + (2)(.075) + (3)(.2) + (4)(.375) + (5)(.15) \\ &\quad + (6)(.1) + (7)(.05) \\ &= 4\end{aligned}$$

Therefore, using Formula (4) and the data from the probability distribution of X, we find that the variance of X is given by

$$\begin{aligned}\text{Var}(X) &= (.05)(1 - 4)^2 + (.075)(2 - 4)^2 + (.2)(3 - 4)^2 \\ &\quad + (.375)(4 - 4)^2 + (.15)(5 - 4)^2 \\ &\quad + (.1)(6 - 4)^2 + (.05)(7 - 4)^2 \\ &= 1.95\end{aligned}$$

Next, we find that the mean of the random variable Y is given by

$$\begin{aligned}\mu_Y &= (1)(.2) + (2)(.15) + (3)(.1) + (4)(.15) + (5)(.05) \\ &\quad + (6)(.1) + (7)(.25) \\ &= 4\end{aligned}$$

and so the variance of Y is given by

$$\begin{aligned}\text{Var}(Y) &= (.2)(1 - 4)^2 + (.15)(2 - 4)^2 + (.1)(3 - 4)^2 \\ &\quad + (.15)(4 - 4)^2 + (.05)(5 - 4)^2 + (.1)(6 - 4)^2 \\ &\quad + (.25)(7 - 4)^2 \\ &= 5.2\end{aligned}$$

Note that Var(X) is smaller than Var(Y), which confirms the earlier observations about the spread, or dispersion, of the probability distribution of X and Y, respectively.

Standard Deviation

Because Formula (4), which gives the variance of the random variable X, involves the squares of the deviations, the unit of measurement of Var(X) is the square of the unit of measurement of the values of X. For example, if the values assumed by the random variable X are measured in units of a gram, then Var(X) will be measured in units involving the *square* of a gram. To remedy this situation, one normally works with the square root of Var(X) rather than Var(X) itself. The former is called the standard deviation of X.

Standard Deviation of a Discrete Random Variable X

The standard deviation of a random variable X, σ (pronounced "sigma"), is defined by

$$\begin{aligned}\sigma &= \sqrt{\text{Var}(X)}\\ &= \sqrt{p_1(x_1-\mu)^2 + p_2(x_2-\mu)^2 + \cdots + p_n(x_n-\mu)^2}\end{aligned} \tag{7}$$

where $x_1, x_2, \ldots, x_n$ denote the values assumed by the random variable X and $p_1 = P(X = x_1)$, $p_2 = P(X = x_2), \ldots, p_n = P(X = x_n)$.

EXAMPLE 6 Find the standard deviations of the random variables X and Y of Example 5.

Solution From the results of Example 5, we have $\text{Var}(X) = 1.95$ and $\text{Var}(Y) = 5.2$. Taking their respective square roots, we have

$$\begin{aligned}\sigma_X &= \sqrt{1.95}\\ &\approx 1.40\\ \sigma_Y &= \sqrt{5.2}\\ &\approx 2.28\end{aligned}$$

EXAMPLE 7 Let X and Y denote the random variables whose values are the weights of the brand A and brand B potato chips, respectively (see Table 5). Compute the means and standard deviations of X and Y and interpret your results.

Solution The probability distributions of X and Y may be computed from the data in Table 7.

TABLE 7

x	Relative Frequency of Occurrence	$P(X = x)$	y	Relative Frequency of Occurrence	$P(Y = y)$
15.8	1	.1	15.7	2	.2
15.9	2	.2	15.8	1	.1
16.0	4	.4	15.9	1	.1
16.1	2	.2	16.0	1	.1
16.2	1	.1	16.1	2	.2
			16.2	2	.2
			16.3	1	.1

The means of X and Y are given by

$$\begin{aligned}\mu_X &= (.1)(15.8) + (.2)(15.9) + (.4)(16.0) + (.2)(16.1)\\ &\quad + (.1)(16.2)\\ &= 16\\ \mu_Y &= (.2)(15.7) + (.1)(15.8) + (.1)(15.9) + (.1)(16.0)\\ &\quad + (.2)(16.1) + (.2)(16.2) + (.1)(16.3)\\ &= 16\end{aligned}$$

Therefore,

$$\begin{aligned}\text{Var}(X) &= (.1)(15.8 - 16)^2 + (.2)(15.9 - 16)^2 + (.4)(16 - 16)^2 \\ &\quad + (.2)(16.1 - 16)^2 + (.1)(16.2 - 16)^2 \\ &= 0.012 \\ \text{Var}(Y) &= (.2)(15.7 - 16)^2 + (.1)(15.8 - 16)^2 + (.1)(15.9 - 16)^2 \\ &\quad + (.1)(16 - 16)^2 + (.2)(16.1 - 16)^2 + (.2)(16.2 - 16)^2 \\ &\quad + (.1)(16.3 - 16)^2 \\ &= 0.042\end{aligned}$$

Thus, the required standard deviations are

$$\begin{aligned}\sigma_X &= \sqrt{\text{Var}(X)} \\ &= \sqrt{0.012} \\ &\approx 0.11 \\ \sigma_Y &= \sqrt{\text{Var}(Y)} \\ &= \sqrt{0.042} \\ &\approx 0.20\end{aligned}$$

The means of both X and Y are equal to 16. Therefore, the average weight of a package of potato chips of either brand is 16 ounces. However, the standard deviation of Y is greater than that of X. This tells us that the weights of the packages of brand B potato chips are more widely dispersed about the common mean of 16 than are those of brand A. ■

EXPLORE & DISCUSS

A useful alternative formula for the variance is

$$\sigma^2 = E(X^2) - \mu^2$$

where $E(X)$ is the expected value of X.

1. Establish the validity of the formula.
2. Use the formula to verify the calculations in Example 7.

Now suppose that x is a *continuous* random variable. Using an argument similar to that used earlier when we extended the result for the expected value from the discrete to the continuous case, we have the following definitions:

Variance and Standard Deviation of a Continuous Random Variable

Let x be a continuous random variable with probability density function $f(x)$ on $[a, b]$. Then the **variance** of x is

$$\text{Var}(x) = \int_a^b (x - \mu)^2 f(x)\, dx \qquad \textbf{(8)}$$

and the **standard deviation** of x is

$$\sigma = \sqrt{\text{Var}(x)} \qquad \textbf{(9)}$$

EXAMPLE 8 Find the expected value, variance, and standard deviation of the random variable x associated with the probability density function

$$f(x) = \frac{32}{15x^3}$$

on $[1, 4]$.

Solution Using (3), we find the mean of x:

$$\begin{aligned}
\mu &= \int_a^b xf(x)\,dx = \int_1^4 x \cdot \frac{32}{15x^3}\,dx \\
&= \frac{32}{15}\int_1^4 x^{-2}\,dx = \frac{32}{15}\left(-\frac{1}{x}\right)\Big|_1^4 \\
&= \frac{32}{15}\left(-\frac{1}{4} + 1\right) = \frac{8}{5}
\end{aligned}$$

Next, using (8), we find

$$\begin{aligned}
\text{Var}(x) &= \int_a^b (x - \mu)^2 f(x)\,dx = \int_1^4 \left(x - \frac{8}{5}\right)^2 \cdot \frac{32}{15x^3}\,dx \\
&= \frac{32}{15}\int_1^4 \left(x^2 - \frac{16}{5}x + \frac{64}{25}\right)\frac{1}{x^3}\,dx \\
&= \frac{32}{15}\int_1^4 \left(\frac{1}{x} - \frac{16}{5}x^{-2} + \frac{64}{25}x^{-3}\right)dx \\
&= \frac{32}{15}\left[\ln x + \frac{16}{5x} - \frac{32}{25x^2}\right]_1^4 \\
&= \frac{32}{15}\left[\left(\ln 4 + \frac{16}{5 \cdot 4} - \frac{32}{25 \cdot 16}\right) - \left(\ln 1 + \frac{16}{5} - \frac{32}{25}\right)\right] \\
&= \frac{32}{15}\left(\ln 4 + \frac{4}{5} - \frac{2}{25} - \frac{16}{5} + \frac{32}{25}\right) \\
&= \frac{32}{15}\left(\ln 4 - \frac{6}{5}\right) \approx 0.40
\end{aligned}$$

Finally, using (9), we find the required standard deviation to be

$$\sigma = \sqrt{\text{Var}(x)} \approx 0.63$$

■

Alternative Formula for Variance Using (8) to calculate the variance of a continuous random variable can be rather tedious. The following formula often makes this task easier:

$$\text{Var}(x) = \int_a^b x^2 f(x)\,dx - \mu^2 \qquad \textbf{(10)}$$

This equation follows from these computations:

$$\begin{aligned}
\operatorname{Var}(x) &= \int_a^b (x-\mu)^2 f(x)\,dx \\
&= \int_a^b (x^2 - 2\mu x + \mu^2) f(x)\,dx \\
&= \int_a^b x^2 f(x)\,dx - 2\mu\int_a^b xf(x)\,dx + \mu^2\int_a^b f(x)\,dx \\
&= \int_a^b x^2 f(x)\,dx - 2\mu\cdot\mu + \mu^2 \qquad \text{Since } \int_a^b xf(x)\,dx = \mu \text{ and } \int_a^b f(x)\,dx = 1 \\
&= \int_a^b x^2 f(x)\,dx - \mu^2
\end{aligned}$$

EXAMPLE 9 Use Formula (10) to calculate the variance of the random variable of Example 8.

Solution Using (10), we have

$$\begin{aligned}
\operatorname{Var}(x) &= \int_a^b x^2 f(x)\,dx - \mu^2 = \int_1^4 x^2\cdot\frac{32}{15x^3}\,dx - \left(\frac{8}{5}\right)^2 \\
&= \frac{32}{15}\int_1^4 \frac{1}{x}\,dx - \frac{64}{25} = \frac{32}{15}\ln x\Big|_1^4 - \frac{64}{25} \\
&= \frac{32}{15}\ln 4 - \frac{64}{25} = \frac{32}{15}\left(\ln 4 - \frac{6}{5}\right)
\end{aligned}$$

as obtained earlier. ■

10.2 Self-Check Exercise

Find the expected value, variance, and standard deviation of the random variable x associated with the probability density function $f(x) = 6(x - x^2)$ on $[0, 1]$.

The solution to Self-Check Exercise 10.2 can be found on page 680.

10.2 Concept Questions

1. a. What is the average or mean of a set of n numbers?
   b. What is the expected value of a discrete random variable?
   c. What is the expected value of a continuous random variable?

2. a. Define the variance $\operatorname{Var}(X)$ of a discrete random variable X.
   b. What is the standard deviation σ of a discrete random variable X?

3. What are the variance and standard deviation of a continuous random variable?

10.2 Exercises

In Exercises 1–14, find the mean, variance, and standard deviation of the random variable x associated with the probability density function over the indicated interval.

1. $f(x) = \frac{1}{3}$; $[3, 6]$

2. $f(x) = \frac{1}{4}$; $[2, 6]$

3. $f(x) = \frac{3}{125}x^2$; $[0, 5]$

4. $f(x) = \frac{3}{8}x^2$; $[0, 2]$

5. $f(x) = \frac{3}{32}(x - 1)(5 - x)$; $[1, 5]$

6. $f(x) = 20(x^3 - x^4)$; $[0, 1]$

7. $f(x) = \frac{8}{7x^2}$; $[1, 8]$

8. $f(x) = \frac{4}{3x^2}$; $[1, 4]$

9. $f(x) = \frac{3}{14}\sqrt{x}$; $[1, 4]$

10. $f(x) = \frac{5}{2}x^{3/2}$; $[0, 1]$

11. $f(x) = \frac{3}{x^4}$; $[1, \infty)$

12. $f(x) = 3.5x^{-4.5}$; $[1, \infty)$

13. $f(x) = \frac{1}{4}e^{-x/4}$; $[0, \infty)$

Hint: $\lim_{x\to\infty} x^n e^{kx} = 0, k < 0$

14. $f(x) = \frac{1}{9}xe^{-x/3}$; $[0, \infty)$

Hint: $\lim_{x\to\infty} x^n e^{kx} = 0, k < 0$

15. **Life Span of a Plant** The life span of a certain plant species (in days) is described by the probability density function

$$f(x) = \frac{1}{100}e^{-x/100}$$

If a plant of this species is selected at random, how long can the plant be expected to live?

16. **Number of Chips in Chocolate Chip Cookies** The number of chocolate chips in a certain brand of cookies has a distribution described by the probability density function

$$f(x) = \frac{1}{36}(6x - x^2) \qquad (0 \le x \le 6)$$

Find the expected number of chips in a cookie selected at random.

17. **Shopping Habits** The amount of time t (in minutes) a shopper spends browsing in the magazine section of a supermarket is a continuous random variable with probability density function

$$f(t) = \frac{2}{25}t \qquad (0 \le t \le 5)$$

How much time is a shopper chosen at random expected to spend in the magazine section?

18. **Reaction Time of a Motorist** The amount of time t (in seconds) it takes a motorist to react to a road emergency is a continuous random variable with probability density function

$$f(t) = \frac{9}{4t^3} \qquad (1 \le t \le 3)$$

What is the expected reaction time for a motorist chosen at random?

19. **Expected Snowfall** The amount of snowfall in feet in a remote region of Alaska in the month of January is a continuous random variable with probability density function

$$f(x) = \frac{2}{9}x(3 - x) \qquad (0 \le x \le 3)$$

Find the amount of snowfall one can expect in any given month of January in Alaska.

20. **Gas Station Sales** The amount of gas (in thousands of gallons) Al's Gas Station sells on a typical Monday is a continuous random variable with probability density function

$$f(x) = 4(x - 2)^3 \qquad (2 \le x \le 3)$$

How much gas can the gas station expect to sell each Monday?

21. **Demand for Butter** The quantity demanded x (in thousands of pounds) of a certain brand of butter per week is a continuous random variable with probability density function

$$f(x) = \frac{6}{125}x(5 - x) \qquad (0 \le x \le 5)$$

What is the expected demand for this brand of butter per week?

22. **Life Expectancy of Color Television Tubes** The life expectancy (in years) of a certain brand of color television tube is a continuous random variable with probability density function

$$f(t) = 9(9 + t^2)^{-3/2} \qquad (0 \le t < \infty)$$

How long is one of these color television tubes expected to last?

In Exercises 23–28, find the median of the random variable x with the probability density function defined on the indicated interval I. The median of x is defined to be the number m such that $P(x < m) = \frac{1}{2}$. Observe that half of the x-values lie below m and the other half lie above m.

23. $f(x) = \frac{1}{6}$; $[2, 8]$

24. $f(x) = \frac{2}{15}x$; $[1, 4]$

25. $f(x) = \frac{3}{16}\sqrt{x}$; $[0, 4]$

26. $f(x) = \frac{1}{6\sqrt{x}}$; $[1, 16]$

27. $f(x) = \frac{3}{x^2}$; $[1, \infty)$

28. $f(x) = \frac{1}{2}e^{-x/2}$; $[0, \infty)$

In Exercises 29 and 30, determine whether the statement is true or false. If it is true, explain why it is true. If it is false, give an example to show why it is false.

29. If f is a probability density function of a continuous random variable x in the interval $[a, b]$, then the expected value of x is given by $\int_a^b x^2 f(x)\,dx$.

30. If f is a probability density function of a continuous random variable x in the interval $[a, b]$, then

$$\text{Var}(x) = \int_a^b x^2 f(x)\,dx - \left[\int_a^b xf(x)\,dx\right]^2$$

10.2 Solution to Self-Check Exercise

Using (3), we find

$$\mu = \int_0^1 x f(x) = 6\int_0^1 (x^2 - x^3)\,dx$$
$$= 6\left[\frac{1}{3}x^3 - \frac{1}{4}x^4\right]\Bigg|_0^1$$
$$= 6\left(\frac{1}{3} - \frac{1}{4}\right) = \frac{1}{2}$$

Using (8), we find

$$\text{Var}(x) = \int_0^1 (x - \mu)^2 f(x)\,dx$$
$$= \int_0^1 \left(x - \frac{1}{2}\right)^2 6(x - x^2)\,dx$$
$$= 6\int_0^1 \left(x^2 - x + \frac{1}{4}\right)(x - x^2)\,dx$$
$$= 6\int_0^1 \left(-x^4 + 2x^3 - \frac{5}{4}x^2 + \frac{1}{4}x\right)dx$$
$$= 6\left(-\frac{1}{5}x^5 + \frac{1}{2}x^4 - \frac{5}{12}x^3 + \frac{1}{8}x^2\right)\Bigg|_0^1$$
$$= 6\left(-\frac{1}{5} + \frac{1}{2} - \frac{5}{12} + \frac{1}{8}\right) = 0.05$$

Using (9), we see that

$$\sigma = \sqrt{\text{Var}(x)} \approx 0.224$$

USING TECHNOLOGY

Finding the Mean and Standard Deviation

The calculation of the mean and standard deviation of a random variable is facilitated by the use of a graphing utility.

APPLIED EXAMPLE 1 A survey conducted in 1995 of the Fortune 1000 companies revealed the following age distribution of company directors:

Age	20–25	25–30	30–35	35–40	40–45	45–50	50–55
Directors	1	6	28	104	277	607	1142

Age	55–60	60–65	65–70	70–75	75–80	80–85	85–90
Directors	1413	1424	494	159	62	31	5

Source: Directorship

a. Plot a histogram for the given data.
b. Find the mean age and the standard deviation of the company directors.

Solution

a. Let X denote the random variable taking on the values 1 through 14, where 1 corresponds to the age bracket 20–25, 2 corresponds to the age bracket 25–30, and so on. Entering the values of X as $x_1 = 1, x_2 = 2, \ldots, x_{14} = 14$ and the corresponding values of Y as $y_1 = 1, y_2 = 6, \ldots, y_{14} = 5$, and then using the DRAW function from the Statistics menu, we obtain the histogram shown in Figure T1.

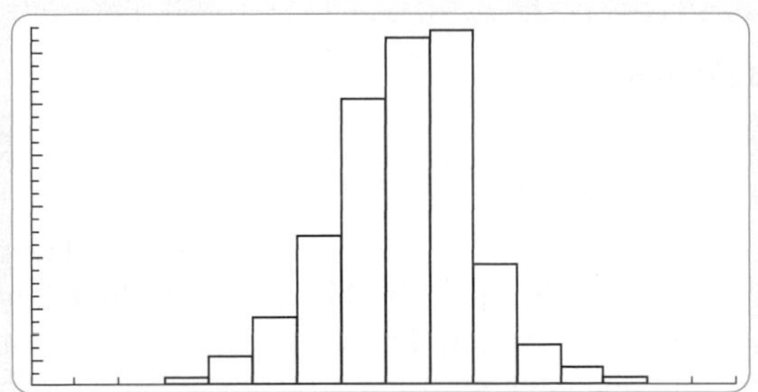

FIGURE T1
The histogram for the given data, using the viewing window $[0, 16] \times [0, 1500]$

b. Using the appropriate function from the Statistics menu, we find that $\bar{x} = 7.9193$ and $\sigma x = 1.6378$; that is, the mean of X is $\mu \approx 7.9$, and the standard deviation is $\sigma \approx 1.6$. Thus, the average age of the directors is in the 55- to 60-year-old bracket.

TECHNOLOGY EXERCISES

1. **a.** Graph the histogram associated with the random variable X in Example 5, page 674.
   b. Find the mean and the standard deviation for these data.
2. **a.** Graph the histogram associated with the random variable Y in Example 5, page 674.
   b. Find the mean and the standard deviation for these data.
3. **DRIVING AGE REQUIREMENTS** The minimum age requirement for a regular driver's license differs from state to state. The frequency distribution for this age requirement in the 50 states is given in the following table:

Minimum Age	15	16	17	18	19	21
Frequency of Occurrence	1	15	4	28	1	1

   a. Describe a random variable X that is associated with these data.
   b. Find the probability distribution for the random variable X.
   c. Graph the histogram associated with these data.
   d. Find the mean and the standard deviation for these data.
4. The distribution of the number of chocolate chips in a cookie is shown in the following table:

Chocolate Chips, x	0	1	2	3	4	5
$P(X = x)$	.01	.03	.05	.11	.13	.24

Chocolate Chips, x	6	7	8
$P(X = x)$	.22	.16	.05

   a. Describe a random variable X that is associated with these data.
   b. Find the probability distribution for the random variable X.
   c. Graph the histogram associated with these data.
   d. Find the mean and the standard deviation for these data.
5. A sugar refiner uses a machine to pack sugar in 5-lb cartons. In order to check the machine's accuracy, cartons are selected at random and weighed. The results follow:

4.98	5.02	4.96	4.97	5.03
4.96	4.98	5.01	5.02	5.06
4.97	5.04	5.04	5.01	4.99
4.98	5.04	5.01	5.03	5.05
4.96	4.97	5.02	5.04	4.97
5.03	5.01	5.00	5.01	4.98

   a. Describe a random variable X that is associated with these data.
   b. Find the probability distribution for the random variable X.
   c. Find the mean and the standard deviation for these data.
6. The scores of 25 students in a mathematics examination follow:

90	85	74	92	68	94	66
87	85	70	72	68	73	72
69	66	58	70	74	88	90
98	71	75	68			

   a. Describe a random variable X that is associated with these data.
   b. Find the probability distribution for the random variable X.
   c. Find the mean and the standard deviation for these data.

10.3 Normal Distributions

Normal Distributions

In Section 10.2 we saw the useful role played by exponential density functions in many applications. In this section we look at yet another class of continuous probability distributions known as **normal distributions**. The normal distribution is without doubt the most important of all the probability distributions. Many phenomena, such as the heights of people in a given population, the weights of newborn infants, the IQs of college students, and the actual weights of 16-ounce packages of cereals, have probability distributions that are normal. The normal distribution also provides us with an accurate approximation to the distributions of many random variables associated with random sampling problems.

The general **normal probability density function** with mean μ and standard deviation σ is defined to be

$$f(x) = \frac{e^{-(1/2)[(x-\mu)/\sigma]^2}}{\sigma\sqrt{2\pi}} \qquad (-\infty < x < \infty)$$

The graph of f, which is bell shaped, is called a **normal curve** (Figure 9).

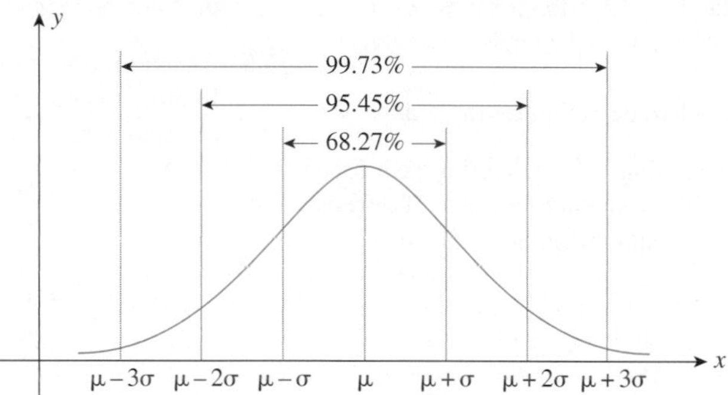

FIGURE 9
A normal curve

The normal curve (and therefore the corresponding normal distribution) is completely determined by its mean μ and standard deviation σ. In fact, the normal curve has the following characteristics, described in terms of these two parameters:

1. The curve has a peak at $x = \mu$.
2. The curve is symmetrical with respect to the vertical line $x = \mu$.
3. The curve always lies above the x-axis but approaches the x-axis as x extends indefinitely in either direction.
4. The area under the curve is 1.
5. For any normal curve, 68.27% of the area under the curve lies within 1 standard deviation of the mean (that is, between $\mu - \sigma$ and $\mu + \sigma$), 95.45% of the area lies within 2 standard deviations of the mean, and 99.73% of the area lies within 3 standard deviations of the mean.

Figure 10 shows two normal curves with different means μ_1 and μ_2 but the same deviation. Figure 11 shows two normal curves with the same mean but different standard deviations σ_1 and σ_2. (Which number is smaller?)

In general, the mean μ of a normal distribution determines where the center of the curve is located, whereas the standard deviation σ of a normal distribution determines the sharpness (or flatness) of the curve.

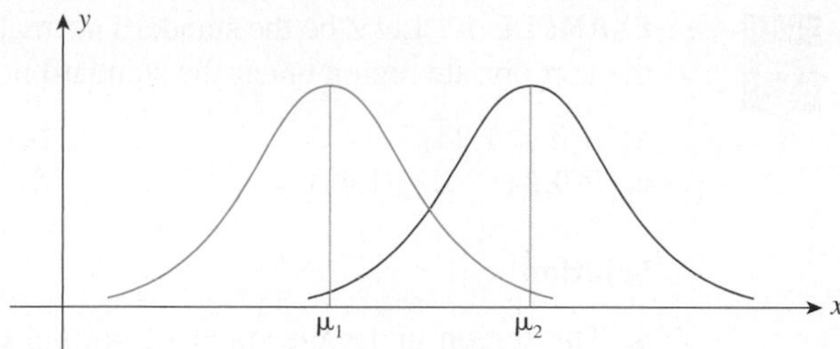

FIGURE 10
Two normal curves that have the same standard deviation but different means

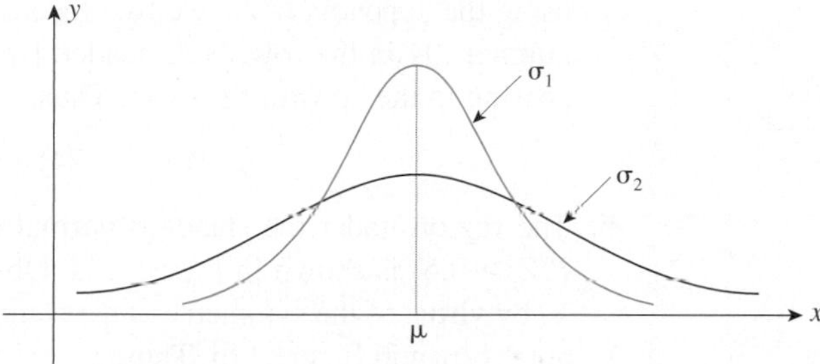

FIGURE 11
Two normal curves that have the same mean but different standard deviations

As this discussion reveals, infinitely many normal curves correspond to different choices of the parameters μ and σ, which characterize such curves. Fortunately, any normal curve may be transformed into any other normal curve (as we see later on), so in the study of normal curves it suffices to single out one such particular curve for special attention. The normal curve with mean $\mu = 0$ and standard deviation $\sigma = 1$ is called the **standard normal curve**. The corresponding distribution is called the **standard normal distribution**. The random variable itself is called the **standard normal variable** and is commonly denoted by Z.

EXPLORING WITH TECHNOLOGY

Consider the probability density function

$$f(x) = \frac{1}{\sqrt{2\pi}} e^{-x^2/2}$$

which is the formula given on page 682 with $\mu = 0$ and $\sigma = 1$.

1. Use a graphing utility to plot the graph of f using the viewing window $[-4, 4] \times [0, 0.5]$.

2. Use the numerical integration function of a graphing utility to find the area of the region under the graph of f on the intervals $[-1, 1]$, $[-2, 2]$, and $[-3, 3]$ and thus verify Property 5 of normal distributions for the special case where $\mu = 0$ and $\sigma = 1$.

Computations of Probabilities Associated with Normal Distributions

Areas under the standard normal curve have been extensively computed and tabulated. The appendix table gives the areas of the regions under the standard normal curve to the left of the number z; these areas correspond, of course, to probabilities of the form $P(Z < z)$ or $P(Z \leq z)$. The next several examples illustrate the use of this table in computations involving the probabilities associated with the standard normal variable.

EXAMPLE 1 Let Z be the standard normal variable. By first making a sketch of the appropriate region under the standard normal curve, find these values:

a. $P(Z < 1.24)$ **b.** $P(Z > 0.5)$
c. $P(0.24 < Z < 1.48)$ **d.** $P(-1.65 < Z < 2.02)$

Solution

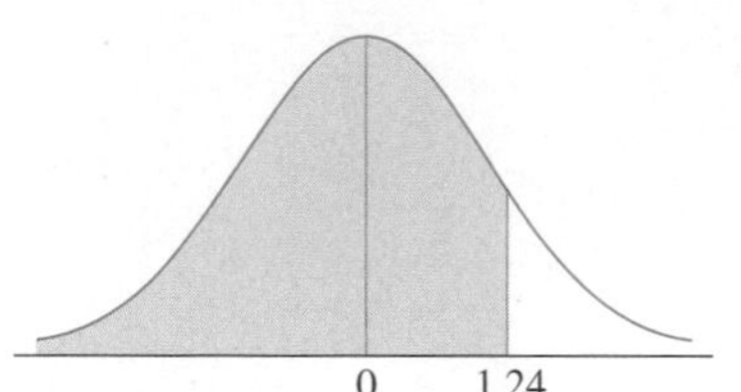

FIGURE 12
$P(Z < 1.24)$

a. The region under the standard normal curve associated with the probability $P(Z < 1.24)$ is shown in Figure 12. To find the area of the required region using the appendix table, we first locate the number 1.2 in the column and the number .04 in the row, both headed by z, and read off the number .8925 appearing in the body of the table. Thus,

$$P(Z < 1.24) = .8925$$

b. The region under the standard normal curve associated with the probability $P(Z > 0.5)$ is shown in Figure 13a. Observe, however, that the required area is, by virtue of the symmetry of the standard normal curve, equal to the shaded area shown in Figure 13b. Thus,

$$\begin{aligned} P(Z > 0.5) &= P(Z < -0.5) \\ &= .3085 \end{aligned}$$

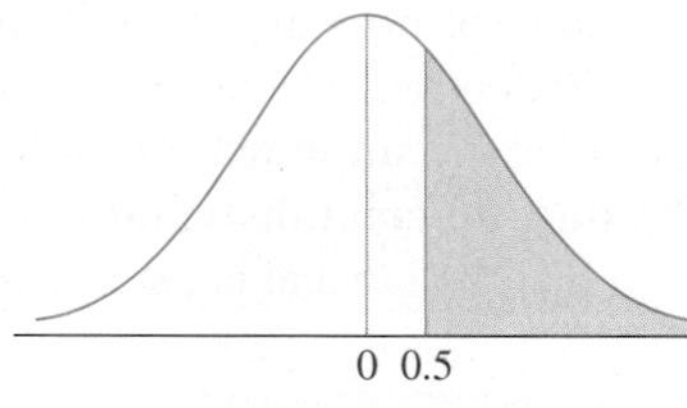

(a) $P(Z > 0.5)$

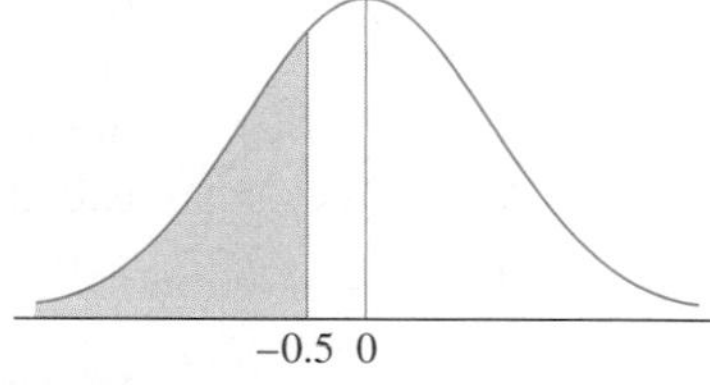

(b) $P(Z < -0.5)$

FIGURE 13

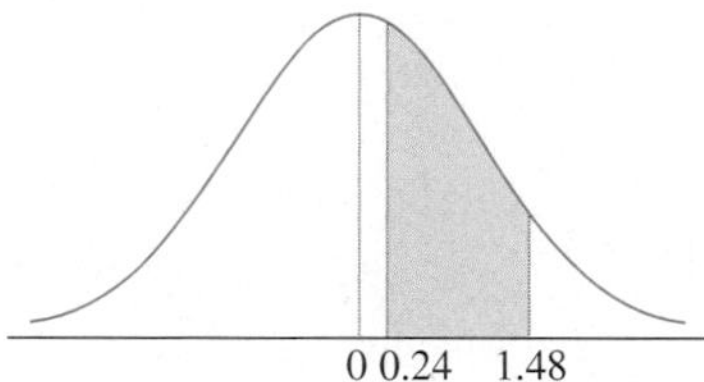

FIGURE 14
$P(0.24 < Z < 1.48)$

c. The probability $P(0.24 < Z < 1.48)$ is equal to the shaded area shown in Figure 14. This area is obtained by subtracting the area under the curve to the left of $z = 0.24$ from the area under the curve to the left of $z = 1.48$; that is

$$\begin{aligned} P(0.24 < Z < 1.48) &= P(Z < 1.48) - P(Z < 0.24) \\ &= .9306 - .5948 \\ &= .3358 \end{aligned}$$

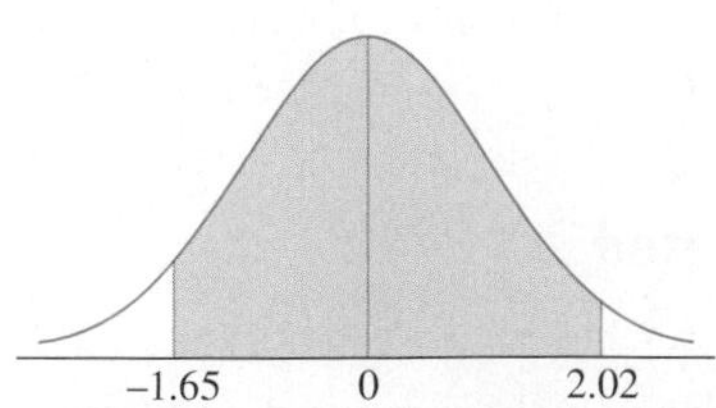

FIGURE 15
$P(-1.65 < Z < 2.02)$

d. The probability $P(-1.65 < Z < 2.02)$ is given by the shaded area in Figure 15. We have

$$\begin{aligned} P(-1.65 < Z < 2.02) &= P(Z < 2.02) - P(Z < -1.65) \\ &= .9783 - .0495 \\ &= .9288 \end{aligned}$$

■

EXAMPLE 2 Let Z be the standard normal variable. Find the value of z if z satisfies the following:

a. $P(Z < z) = .9474$ **b.** $P(Z > z) = .9115$ **c.** $P(-z < Z < z) = .7888$

Solution

a. Refer to Figure 16. We want the value of z such that the area of the region under the standard normal curve and to the left of $Z = z$ is .9474. Locating the number .9474 in the appendix table and reading back, we find that $z = 1.62$.

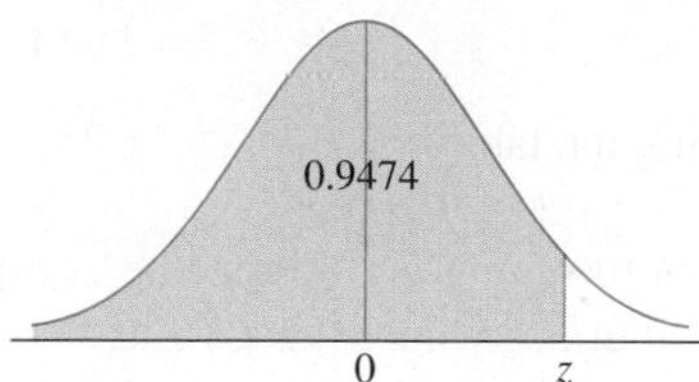

FIGURE 16
$P(Z < z) = .9474$

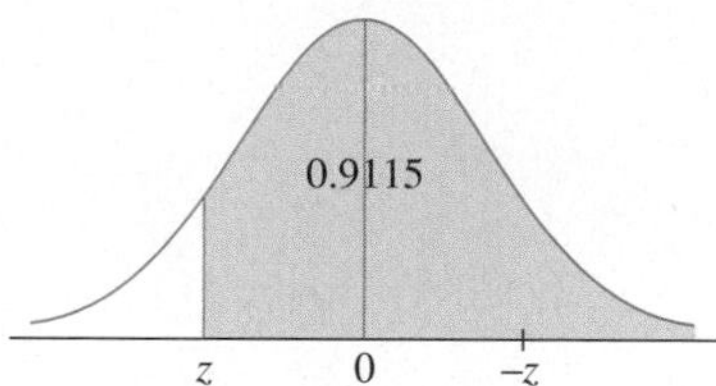

FIGURE 17
$P(Z > z) = .9115$

b. Since $P(Z > z)$ or, equivalently, the area of the region to the right of z is greater than 0.5, z must be negative (Figure 17). Therefore, $-z$ is positive. Furthermore, the area of the region to the right of z is the same as the area of the region to the left of $-z$:

$$\begin{aligned} P(Z > z) &= P(Z < -z) \\ &= .9115 \end{aligned}$$

Looking up the table, we find $-z = 1.35$, so $z = -1.35$.

c. The region associated with $P(-z < Z < z)$ is shown in Figure 18. Observe that by symmetry the area of this region is just double that of the area of the region between $Z = 0$ and $Z = z$; that is,

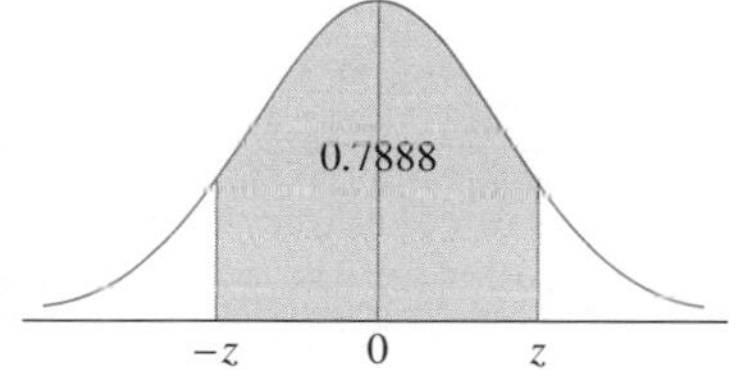

FIGURE 18
$P(-z < Z < z) = .7888$

$$P(-z < Z < z) = 2P(0 < Z < z)$$

or

$$P(0 < Z < z) = P(Z < z) - \frac{1}{2}$$

(see Figure 19). Therefore,

$$\frac{1}{2}P(-z < Z < z) = P(Z < z) - \frac{1}{2}$$

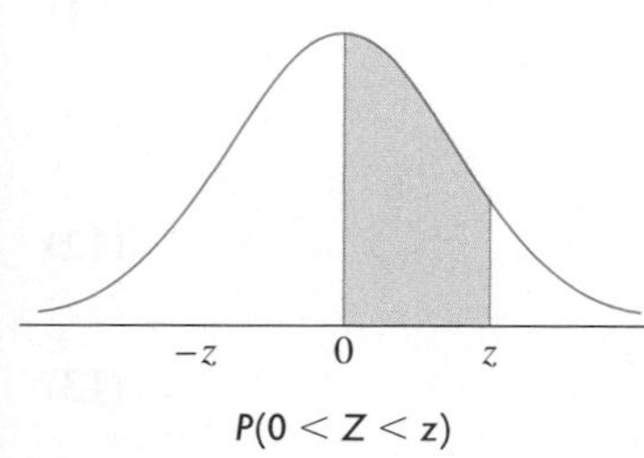

$P(0 < Z < z)$ =

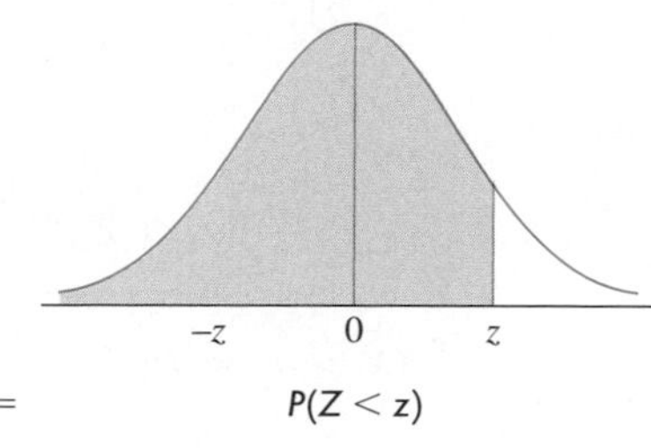

$P(Z < z)$ −

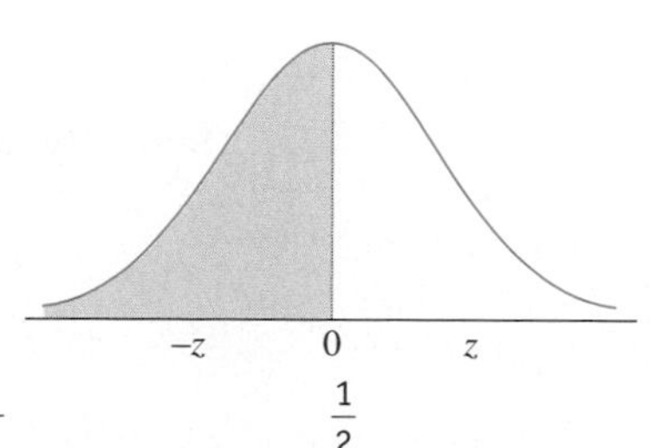

$\frac{1}{2}$

FIGURE 19

or, solving for $P(Z < z)$,

$$P(Z < z) = \frac{1}{2} + \frac{1}{2}P(-z < Z < z)$$
$$= \frac{1}{2}(1 + .7888)$$
$$= .8944$$

Consulting the table, we find $z = 1.25$. ■

We now turn our attention to the computation of probabilities associated with normal distributions whose means and standard deviations are not necessarily equal to 0 and 1, respectively. As mentioned earlier, any normal curve may be transformed into the standard normal curve. In particular, it may be shown that if X is a normal random variable with mean μ and standard deviation σ, then it can be transformed into the standard normal random variable Z by means of the substitution

$$Z = \frac{X - \mu}{\sigma}$$

(see Figure 20).

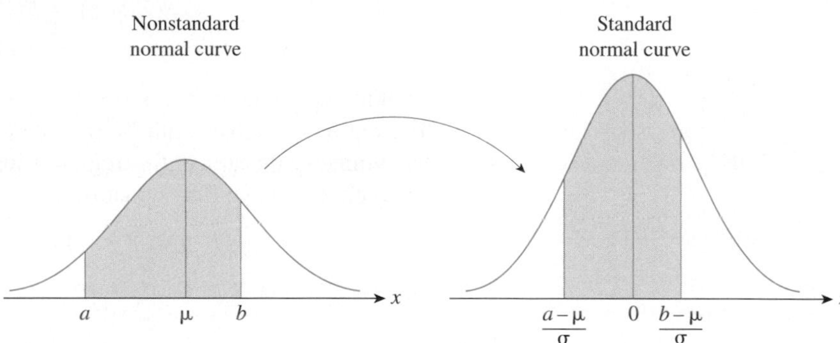

FIGURE 20 Area under the curve between a and b = Area under the curve between $\frac{a - \mu}{\sigma}$ and $\frac{b - \mu}{\sigma}$

The area of the region under the normal curve (with random variable X) between $x = a$ and $x = b$ is *equal* to the area of the region under the standard normal curve between $z = (a - \mu)/\sigma$ and $z = (b - \mu)/\sigma$. In terms of probabilities associated with these distributions, we have

$$P(a < X < b) = P\left(\frac{a - \mu}{\sigma} < Z < \frac{b - \mu}{\sigma}\right) \tag{11}$$

Similarly, we have

$$P(X < b) = P\left(Z < \frac{b - \mu}{\sigma}\right) \tag{12}$$

$$P(X > a) = P\left(Z > \frac{a - \mu}{\sigma}\right) \tag{13}$$

Thus, with the help of (11)–(13), computations of probabilities associated with any normal distribution may be reduced to computations of areas of regions under the standard normal curve.

EXAMPLE 3 Suppose X is a normal random variable with $\mu = 100$ and $\sigma = 20$. Find these values:

a. $P(X < 120)$ **b.** $P(X > 70)$ **c.** $P(75 < X < 110)$

Solution

a. Using (12) with $\mu = 100$, $\sigma = 20$, and $b = 120$, we have

$$P(X < 120) = P\left(Z < \frac{120 - 100}{20}\right)$$

$$= P(Z < 1) = .8413 \qquad \text{Using the table of values of } Z$$

b. Using (13) with $\mu = 100$, $\sigma = 20$, and $a = 70$, we have

$$P(X > 70) = P\left(Z > \frac{70 - 100}{20}\right)$$

$$= P(Z > -1.5)$$

$$= P(Z < 1.5) = .9332 \qquad \text{Using the table of values of } Z$$

c. Using (11) with $\mu = 100$, $\sigma = 20$, $a = 75$, and $b = 110$, we have

$$P(75 < X < 110)$$

$$= P\left(\frac{75 - 100}{20} < Z < \frac{110 - 100}{20}\right)$$

$$= P(-1.25 < Z < 0.5)$$

$$= P(Z < 0.5) - P(Z < -1.25) \qquad \text{See Figure 21.}$$

$$= .6915 - .1056 = .5859 \qquad \text{Using the table of values of } Z$$

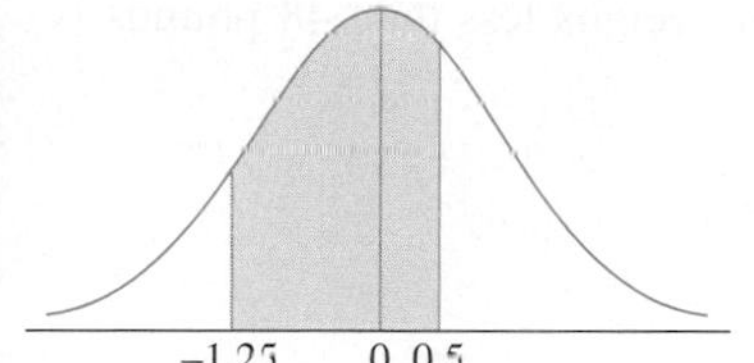

FIGURE 21

APPLIED EXAMPLE 4 Infants' Birth Weights The medical records of infants delivered at Kaiser Memorial Hospital show that the infants' birth weights in pounds are normally distributed with a mean of 7.4 and a standard deviation of 1.2. Find the probability that an infant selected at random from among those delivered at the hospital weighed more than 9.2 pounds at birth.

Solution Let X be the normal random variable denoting the birth weights of infants delivered at the hospital. Then the probability that an infant selected at random has a birth weight of more than 9.2 pounds is given by $P(X > 9.2)$. To compute $P(X > 9.2)$, we use Formula (13) with $\mu = 7.4$, $\sigma = 1.2$, and $a = 9.2$. We find

$$P(X > 9.2) = P\left(Z > \frac{9.2 - 7.4}{1.2}\right) \qquad P(X > a) = P\left(Z > \frac{a - \mu}{\sigma}\right)$$

$$= P(Z > 1.5)$$

$$= P(Z < -1.5)$$

$$= .0668$$

Thus, the probability that an infant delivered at the hospital weighed more than 9.2 pounds is .0668.

APPLIED EXAMPLE 5 Packaging Idaho Natural Produce ships potatoes to its distributors in bags with a mean weight of 50 pounds and a standard deviation of 0.5 pound. If a bag of potatoes is selected at random from a shipment, what are the probabilities of these events?

a. The bag weighs more than 51 pounds.
b. It weighs less than 48 pounds.
c. It weighs between 49 and 51 pounds.

Solution Let X denote the weight of a bag of potatoes packed by the company. Then, the mean and standard deviation of X are $\mu = 50$ and $\sigma = 0.5$, respectively.

a. The probability that a bag selected at random weighs more than 51 pounds is given by

$$\begin{aligned} P(X > 51) &= P\left(Z > \frac{51 - 50}{0.5}\right) \qquad P(X > a) = P\left(Z > \frac{a - \mu}{\sigma}\right) \\ &= P(Z > 2) \\ &= P(Z < -2) \\ &= .0228 \end{aligned}$$

b. The probability that a bag selected at random weighs less than 48 pounds is given by

$$\begin{aligned} P(X < 48) &= P\left(Z < \frac{48 - 50}{0.5}\right) \qquad P(X < b) = P\left(Z < \frac{b - \mu}{\sigma}\right) \\ &= P(Z < -4) \\ &= 0 \end{aligned}$$

c. The probability that a bag selected at random weighs between 49 and 51 pounds is given by

$$\begin{aligned} &P(49 < X < 51) \qquad P(a < X < b) = \\ &= P\left(\frac{49 - 50}{0.5} < Z < \frac{51 - 50}{0.5}\right) \qquad P\left(\frac{a - \mu}{\sigma} < Z < \frac{b - \mu}{\sigma}\right) \\ &= P(-2 < Z < 2) \\ &= P(Z < 2) - P(Z < -2) \\ &= .9772 - .0228 \\ &= .9544 \end{aligned}$$

■

APPLIED EXAMPLE 6 University Admission Eligibility The grade point average (GPA) of the senior class of Jefferson High School is normally distributed with a mean of 2.7 and a standard deviation of 0.4 point. If a senior in the top 10% of his or her class is eligible for admission to any of the nine campuses of the state university system, what is the minimum GPA that a senior should have to ensure eligibility for admission to the state university system?

Solution Let X denote the GPA of a randomly selected senior at Jefferson High School, and let x denote the minimum GPA to ensure his or her eligibility for admission to the university. Since only the top 10% are eligible for admission, x must satisfy the equation

$$P(X \geq x) = .1$$

Using Formula (13) with $\mu = 2.7$ and $\sigma = 0.4$, we find

$$P(X \geq x) = P\left(Z \geq \frac{x - 2.7}{0.4}\right) = .1 \qquad P(X > a) = P\left(Z > \frac{a - \mu}{\sigma}\right)$$

But, this is equivalent to the equation

$$P\left(Z \leq \frac{x - 2.7}{0.4}\right) = .9 \qquad \text{Why?}$$

Consulting the appendix table, we find

$$\frac{x - 2.7}{0.4} = 1.28$$

Upon solving for x, we obtain

$$\begin{aligned} x &= (1.28)(0.4) + 2.7 \\ &\approx 3.2 \end{aligned}$$

Thus, to ensure eligibility for admission to one of the nine campuses of the state university system, a senior at Jefferson High School should have a minimum 3.2 GPA. ■

10.3 Self-Check Exercises

1. Let Z be a standard normal variable.
   a. Find the value of $P(-1.2 < Z < 2.1)$ by first sketching the appropriate region under the standard normal curve.
   b. Find the value of z if z satisfies

$$P(-z < Z < z) = .8764$$

2. Let X be a normal random variable with $\mu = 80$ and $\sigma = 10$. Find these values:
   a. $P(X < 100)$
   b. $P(X > 60)$
   c. $P(70 < X < 90)$

3. The serum cholesterol levels (in mg/dL) in a current Mediterranean population are found to be normally distributed with a mean of 160 and a standard deviation of 50. Scientists at the National Heart, Lung, and Blood Institute consider this pattern ideal for a minimal risk of heart attacks. Find the percentage of the population who have blood cholesterol levels between 160 and 180 mg/dL.

Solutions to Self-Check Exercises 10.3 can be found on page 691.

10.3 Concept Questions

1. Suppose you are given a normal curve with mean μ and standard deviation σ.
   a. At what value of x does the curve peak?
   b. What is the equation of the vertical line that the curve is symmetric to?
   c. What is the value of the area under the curve?

2. Explain what is meant by a standard deviation of a normal curve.

3. a. What are the values of μ and σ for a standard normal curve?
   b. What is the standard normal distribution?

10.3 Exercises

In Exercises 1–6, find the value of the probability of the standard normal variable Z corresponding to the shaded area under the standard normal curve.

1. $P(Z < 1.45)$

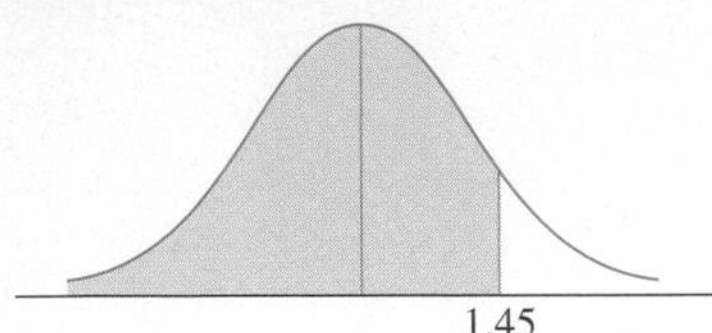

2. $P(Z > 1.11)$

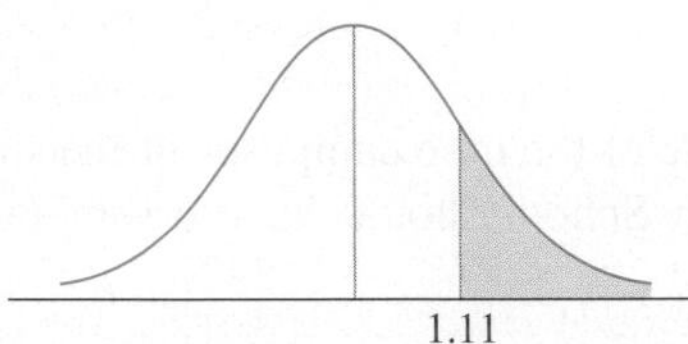

3. $P(Z < -1.75)$

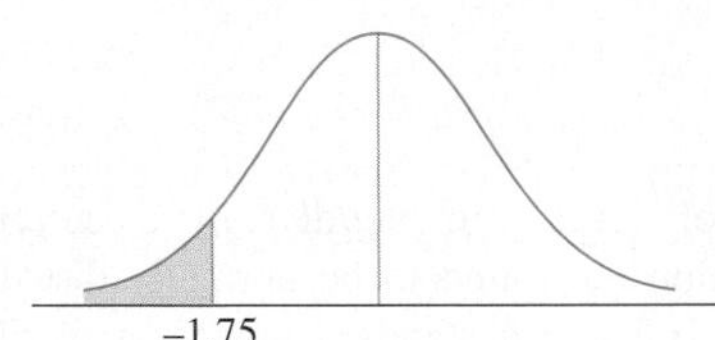

4. $P(0.3 < Z < 1.83)$

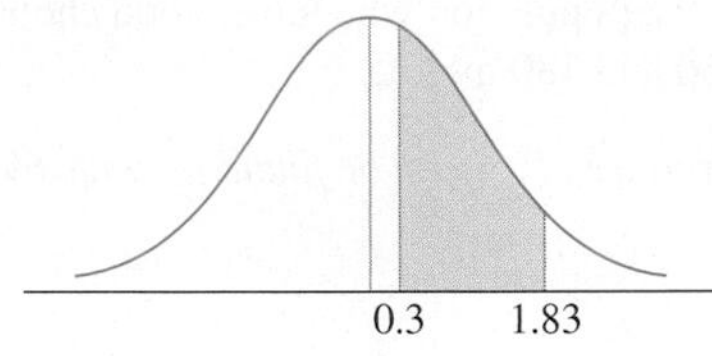

5. $P(-1.32 < Z < 1.74)$

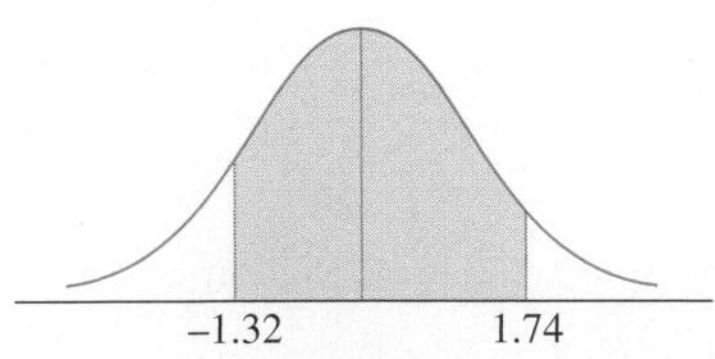

6. $P(-2.35 < Z < -0.51)$

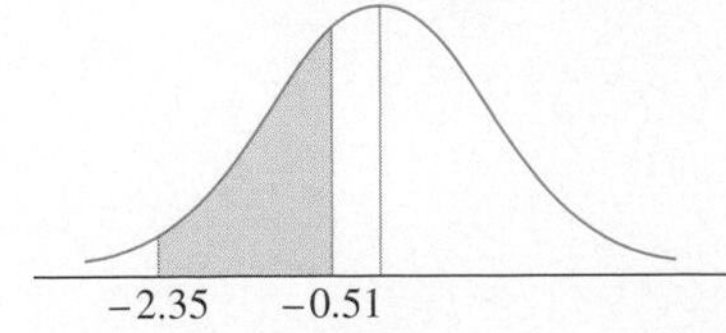

In Exercises 7–14, (a) sketch the area under the standard normal curve corresponding to the given probability and (b) find the value of the probability of the standard normal variable Z corresponding to this area.

7. $P(Z < 1.37)$
8. $P(Z > 2.24)$
9. $P(Z < -0.65)$
10. $P(0.45 < Z < 1.75)$
11. $P(Z > -1.25)$
12. $P(-1.48 < Z < 1.54)$
13. $P(0.68 < Z < 2.02)$
14. $P(-1.41 < Z < -0.24)$

15. Let Z be the standard normal variable. Find the values of z if z satisfies
    a. $P(Z < z) = .8907$ **b.** $P(Z < z) = .2090$

16. Let Z be the standard normal variable. Find the values of z if z satisfies
    a. $P(Z > z) = .9678$ **b.** $P(-z < Z < z) = .8354$

17. Let Z be the standard normal variable. Find the values of z if z satisfies
    a. $P(Z > -z) = .9713$ **b.** $P(Z < -z) = .9713$

18. Suppose X is a normal random variable with $\mu = 380$ and $\sigma = 20$. Find these values:
    a. $P(X < 405)$ **b.** $P(400 < X < 430)$
    c. $P(X > 400)$

19. Suppose X is a normal random variable with $\mu = 50$ and $\sigma = 5$. Find these values:
    a. $P(X < 60)$ **b.** $P(X > 43)$
    c. $P(46 < X < 58)$

20. Suppose X is a normal random variable with $\mu = 500$ and $\sigma = 75$. Find these values:
    a. $P(X < 750)$ **b.** $P(X > 350)$
    c. $P(400 < X < 600)$

21. MEDICAL RECORDS The medical records of infants delivered at Kaiser Memorial Hospital show that the infants' lengths at birth (in inches) are normally distributed with a mean of 20 and a standard deviation of 2.6. Find the probability that an infant selected at random from among those delivered at the hospital measures the following:
    a. More than 22 in. **b.** Less than 18 in.
    c. Between 19 and 21 in.

22. FACTORY WORKERS' WAGES According to the data released by a city's Chamber of Commerce, the weekly wages of factory workers are normally distributed with a mean of \$400 and a standard deviation of \$50. Find the probability that a worker selected at random from the city has a weekly wage of the following:
    a. Less than \$300 **b.** More than \$460
    c. Between \$350 and \$450

23. PRODUCT RELIABILITY TKK Products manufactures electric light bulbs in the 50-, 60-, 75-, and 100-W range. Laboratory tests show that the lives of these light bulbs are normally distributed with a mean of 750 hr and a standard deviation of 75 hr. Find the probability that a TKK light bulb selected at random will burn for the following hours:
    a. For more than 900 hr
    b. For less than 600 hr
    c. Between 750 and 900 hr
    d. Between 600 and 800 hr

24. EDUCATION On the average, a student takes 100 words/minute midway through an Advanced Shorthand course at the American Institute of Stenography. Assuming that the students' dictation speeds are normally distributed and that the standard deviation is 20 words/minute, find the probability that a student randomly selected from the course could take dictation at the following speeds:
    a. More than 120 words/minute
    b. Between 80 and 120 words/minute
    c. Less than 80 words/minute

25. IQS The IQs of students at Wilson Elementary School were measured recently and found to be normally distributed with a mean of 100 and a standard deviation of 15. Find the probability that a student selected at random will have an IQ of the following:
    a. 140 or higher
    b. 120 or higher
    c. Between 100 and 120
    d. 90 or less

26. PRODUCT RELIABILITY The tread lives of Super Titan radial tires under normal driving conditions are normally distributed with a mean of 40,000 mi and a standard deviation of 2000 mi. What is the probability that a tire selected at random will have a tread life of more than 35,000 mi? If four new tires are installed on a car and they experience even wear, determine the probability that all four tires still have useful tread lives after 35,000 mi of driving.

27. FEMALE FACTORY WORKERS' WAGES According to data released by a city's Chamber of Commerce, the weekly wages (in dollars) of female factory workers are normally distributed with a mean of 675 and a standard deviation of 50. Find the probability that a female factory worker selected at random from the city has a weekly wage of \$650 to \$750.

28. CIVIL SERVICE EXAMS To be eligible for further consideration, applicants for certain Civil Service positions must first pass a written qualifying examination on which a score of 70 or more must be obtained. In a recent examination, it was found that the scores were normally distributed with a mean of 60 points and a standard deviation of 10 points. Determine the percentage of applicants who passed the written qualifying examination.

29. WARRANTIES The general manager of the Service Department of MCA Television Company has estimated that the time that elapses between the dates of purchase and the dates on which the 19-in. sets manufactured by the company first require service is normally distributed with a mean of 22 mo and a standard deviation of 4 mo. If MCA gives a 1-yr warranty on parts and labor for these sets, determine what percentage of sets manufactured and sold may require service before the warranty period runs out.

30. GRADE DISTRIBUTIONS The scores on an economics examination are normally distributed with a mean of 72 and a standard deviation of 16. If the instructor assigns a grade of A to 10% of the class, what is the lowest score a student may have and still get an A?

31. GRADE DISTRIBUTIONS The scores on a sociology examination are normally distributed with a mean of 70 and a standard deviation of 10. If the instructor assigns A's to 15%, B's to 25%, C's to 40%, D's to 15%, and F's to 5% of the class, find the cutoff points for these grades.

10.3 Solutions to Self-Check Exercises

1. **a.** The probability $P(-1.2 < Z < 2.1)$ is given by the shaded area in the accompanying figure. We have

$$\begin{aligned} P(-1.2 < Z < 2.1) &= P(Z < 2.1) - P(Z < -1.2) \\ &= .9821 - .1151 \\ &= .867 \end{aligned}$$

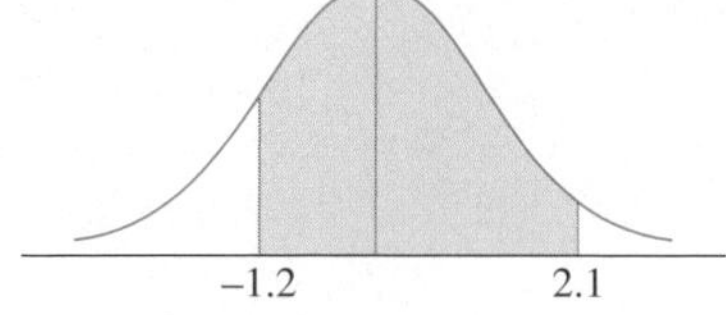

b. The region associated with $P(-z < Z < z)$ is shown in the accompanying figure. Observe that we have the following relationship:

$$P(Z < z) = \frac{1}{2}[1 + P(-z < Z < z)]$$

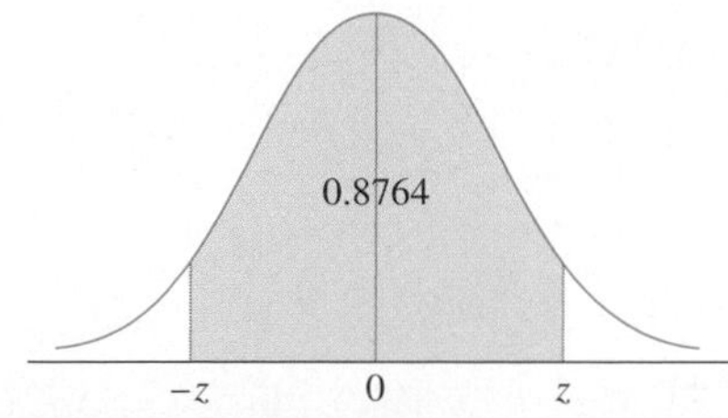

(see Example 2c). With $P(-z < Z < z) = .8764$, we find

$$P(Z < z) = \frac{1}{2}(1 + .8764) = .9382$$

Consulting the table, we find $z = 1.54$.

2. Using the transformations (11), (12), and (13) and the table of values of Z, we have

a. $$P(X < 100) = P\left(Z < \frac{100 - 80}{10}\right) = P(Z < 2) = .9772$$

b. $$P(X > 60) = P\left(Z > \frac{60 - 80}{10}\right) = P(Z > -2) = P(Z < 2) = .9772$$

c. $$P(70 < X < 90) = P\left(\frac{70 - 80}{10} < Z < \frac{90 - 80}{10}\right) = P(-1 < Z < 1) = P(Z < 1) - P(Z < -1) = .8413 - .1587 = .6826$$

3. Let X be the normal random variable denoting the serum cholesterol levels (in mg/dL) in the current Mediterranean population under consideration. Thus, the percentage of the population having blood cholesterol levels between 160 and 180 mg/dL is given by $P(160 < X < 180)$. To compute $P(160 < X < 180)$, we use (11), with $\mu = 160$, $\sigma = 50$, $a = 160$, and $b = 180$. We find

$$P(160 < X < 180) = P\left(\frac{160 - 160}{50} < Z < \frac{180 - 160}{50}\right) = P(0 < Z < 0.4) = P(Z < 0.4) - P(Z < 0) = .6554 - .5000 = .1554$$

so approximately 15.5% of the population has blood cholesterol levels between 160 and 180 mg/dL.

CHAPTER 10 Summary of Principal Formulas and Terms

FORMULAS

1. Expected value	$E(x) = x_1p_1 + x_2p_2 + \cdots + x_np_n$
2. Expected value of a continuous random variable	$E(x) = \int_a^b xf(x)\,dx$
3. Exponential density function	$f(x) = ke^{-kx}$
4. Expected value of an exponential density function	$E(x) = \frac{1}{k}$
5. Variance of a continuous random variable	$\text{Var}(x) = \int_a^b (x - \mu)^2 f(x)\,dx$
6. Standard deviation of a continuous random variable	$\sigma = \sqrt{\text{Var}(x)}$

TERMS

experiment (654)
outcomes (sample points) (654)
sample space (654)
event (654)
probability of the event (654)
random variable (654)
discrete probability function (654)
histogram (654)
continuous random variable (656)
probability density function (656)
average (mean) (667)
expected value (668)
variance of a random variable (673)
standard deviation of a random variable (675)
normal distribution (682)
normal curve (682)
standard normal variable (683)

CHAPTER 10 Concept Review Questions

Fill in the blanks.

1. **a.** An activity with observable results is called an ________.
   b. The observable results are called ________ or ________.
   c. The totality of all outcomes is called a ________ ________.
   d. The probability of an event is a number between ________ and ________.

2. **a.** The outcome of a chance experiment is called a ________ variable.
   b. A random variable that assumes only a finite number of integer values is said to be ________ ________.
   c. A discrete probability function P with domain $\{x_1, x_2, \ldots, x_n\}$ satisfies the two conditions ________ $\leq P(x_i) \leq$ ________ and $P(x_1) + P(x_2) + \cdots + P(x_n) =$ ________.

3. **a.** A random variable that can assume any value in an interval is called a ________ random variable.
   b. A probability density function of a continuous random variable is a ________ function such that the total area of the region under the graph of f is equal to ________, and the probability that an observed value of the random variable x lies in the interval $[a, b]$ is given by $P(a \leq x \leq b) =$ ________.

4. A joint probability density function of x and y on D is a ________ function $f(x, y)$ such that $\iint_D f(x, y)\, dA =$ ________, and the probability that the observed values of the random variables x and y lie in a region $R \subset D$ is given by $P[(x, y) \in R] =$ ________.

5. **a.** The average or mean of n numbers $x_1, x_2, \ldots, x_n$ is $\bar{x} =$ ________.
   b. The expected value of a random variable X is a measure of the ________ ________ of the probability distribution associated with ________.
   c. The expected value of a discrete random variable X is $E(X) =$ ________, where $x_1, x_2, \ldots, x_n$ are assumed by ________ and $p_1, p_2, \ldots, p_n$ are the associated ________.
   d. The expected value of a continuous random variable x is $E(x) =$ ________, where f is the ________ ________ function defined on ________ that is associated with x.

6. **a.** The variance of a random variable is a measure of the degree of ________ of a probability distribution about its mean.
   b. If X denotes a random variable that assumes the values $x_1, x_2, \ldots, x_n$ with associated probabilities $p_1, p_2, \ldots, p_n$ and $\mu = E(x)$, then the variance of the random variable X is $\text{Var}(X) =$ ________.
   c. The standard deviation of a random variable X is $\sigma =$ ________.

7. If x is a continuous random variable with probability density function $f(x)$ on $[a, b]$, then the variance of x is $\text{Var}(x) =$ ________ and the standard deviation of x is $\sigma =$ ________.

8. A normal curve has the following properties.
   a. The curve has a peak at ________.
   b. The curve is symmetric with respect to the line ________.
   c. The curve always lies ________ the x-axis and approaches ________ ________ as x extends indefinitely in either direction.
   d. The area under the curve is ________.
   e. For any normal curve, 68.27% of the area under the curve lies within 1 ________ ________ of the mean, 99.45% of the area lies within ________ standard deviations of the mean, and 99.73% of the area lies within ________ standard deviations of the mean.

CHAPTER 10 Review Exercises

In Exercises 1–4, show that the function is a probability density function on the given interval.

1. $f(x) = \frac{1}{28}(2x + 3)$; $[0, 4]$
2. $f(x) = \frac{3}{16}\sqrt{x}$; $[0, 4]$
3. $f(x) = \frac{1}{4}$; $[7, 11]$
4. $f(x) = \frac{4}{x^5}$; $[1, \infty)$

In Exercises 5–8, find the value of the constant k so that the function is a probability density function in the given interval.

5. $f(x) = kx^2$; $[0, 9]$
6. $f(x) = \frac{k}{\sqrt{x}}$; $[1, 16]$
7. $f(x) = \frac{k}{x^2}$; $[1, 3]$
8. $f(x) = \frac{k}{x^{2.5}}$; $[1, \infty)$

In Exercises 9–12, f is a probability density function defined on the given interval. Find the indicated probabilities.

9. $f(x) = \frac{2}{21}x$; $[2, 5]$
   a. $P(x \leq 4)$ b. $P(x = 4)$ c. $P(3 \leq x \leq 4)$
10. $f(x) = \frac{1}{4}$; $[1, 5]$
    a. $P(2 \leq x \leq 4)$ b. $P(x \leq 3)$ c. $P(x \geq 2)$
11. $f(x) = \frac{3}{16}\sqrt{x}$; $[0, 4]$
    a. $P(1 \leq x \leq 3)$ b. $P(x \leq 3)$ c. $P(x = 2)$
12. $f(x) = \frac{1}{x^2}$; $[1, \infty)$
    a. $P(x \leq 10)$ b. $P(2 \leq x \leq 4)$ c. $P(x \geq 2)$

In Exercises 13–16, find the mean, variance, and standard deviation of the random variable x associated with the probability density function f over the given interval.

13. $f(x) = \frac{1}{5}$; $[2, 7]$
14. $f(x) = \frac{1}{28}(2x + 3)$; $[0, 4]$
15. $f(x) = \frac{1}{4}(3x^2 + 1)$; $[-1, 1]$
16. $f(x) = \frac{4}{x^5}$; $[1, \infty)$

In Exercises 17–20, Z is the standard normal variable. Find the given probability.

17. $P(Z < 2.24)$
18. $P(Z > -1.24)$
19. $P(0.24 \leq Z \leq 1.28)$
20. $P(-1.37 \leq Z \leq 1.37)$

21. Show that the function

$$f(x, y) = \frac{1}{64}x^{1/2}y^{1/3};\ D = \{0 \leq x \leq 4;\ 0 \leq y \leq 8\}$$

is a joint probability density function on D.

22. Given that

$$f(x, y) = x(2 - y);\ D = \{(x, y) \mid 0 \leq x \leq 1;\ 0 \leq y \leq 2\}$$

is a joint probability density function on D, find the indicated probabilities.
   a. $P(0 \leq x \leq \frac{1}{2};\ 0 \leq y \leq 1)$
   b. $P(\{(x, y) \mid y \leq 2x\})$

23. Suppose X is a normal random variable with $\mu = 80$ and $\sigma = 8$. Find these values:
    a. $P(X \leq 84)$ b. $P(X \geq 70)$ c. $P(75 \leq X \leq 85)$

24. Suppose X is a normal random variable with $\mu = 45$ and $\sigma = 3$. Find these values:
    a. $P(X \leq 50)$ b. $P(X \geq 40)$ c. $P(40 \leq X \leq 50)$

25. Length of a Hospital Stay Records at Centerville Hospital indicate that the length of time in days that a maternity patient stays in the hospital has a probability density function given by

$$P(t) = \frac{1}{4}e^{-(1/4)t}$$

   a. What is the probability that a woman entering the maternity wing will be there longer than 6 days?
   b. What is the probability that a woman entering the maternity wing will be there less than 2 days?
   c. What is the average length of time that a woman entering the maternity wing stays in the hospital?

26. Life Span of a Car Battery The life span (in years) of a certain make of car battery is an exponentially distributed random variable with an expected value of 5. Find the probability that the life span of a battery is (a) less than 4 yr, (b) more than 6 yr, and (c) between 2 and 4 yr.

CHAPTER 10 Before Moving On . . .

1. Find the value of the constant k such that the function $f(x) = k(x + 2x^2)$ is a probability density function on the interval $[0, 1]$.

2. If $f(x) = 0.1e^{-0.1x}$ is a probability density function on the interval $[0, \infty)$, find $P(5 < x < 10)$.

3. Find the mean, variance, and standard deviation of the random variable x associated with the probability density function $f(x) = \dfrac{1}{4\sqrt{x}}$ over the interval $[1, 9]$.

4. Suppose X is a normal random variable with $\mu = 40$ and $\sigma = 4$. Find the following values.
   a. $P(X < 50)$ **b.** $P(X > 35)$ **c.** $P(30 < X < 50)$

11 Taylor Polynomials and Infinite Series

What percentage of the nonfarm workforce will be in the service industries one decade from now? In Example 4, page 703, you will see how a Taylor polynomial can be used to help answer this question.

Digital Vision/Getty Images

IN THIS CHAPTER we show how certain functions can be represented by a *power series*. A power series involves infinitely many terms, but when truncated it is just a polynomial. By approximating a function with a *Taylor polynomial*, we are often able to obtain approximate solutions to problems that we cannot otherwise solve.

We also look at *Newton's method* for finding the zeros of a function. For example, Newton's method can be used to find the critical points of a function, which, as you may recall, are candidates for the solution of the optimization problems considered in Chapter 4.

11.1 Taylor Polynomials

As we saw earlier, obtaining an exact solution to a problem is not always possible; in such cases we have to settle for an approximate solution. In this section we show how a function may be approximated near a given point by a polynomial. Polynomials, as we have seen time and again, are easy to work with; for example, they are easy to evaluate, differentiate, and integrate. Thus, by using polynomials rather than working with the original function itself, we can often obtain approximate solutions to a problem that we might otherwise not be able to solve.

Taylor Polynomials

Suppose we are given a differentiable function f and a number a in the domain of f. Then the polynomial of degree 0 that best approximates f *near* $x = a$ is the constant polynomial

$$P_0(x) = f(a)$$

which coincides with f at $x = a$ (Figure 1a).

Now, unless f itself is a constant function, it is possible in many cases to obtain a better approximation of f *near* $x = a$ by using a polynomial function of degree 1. Recall that the linear function

$$L(x) = f(a) + f'(a)(x - a)$$

is just an equation of the tangent line to the graph of the function f at the point $(a, f(a))$ (Figure 1b). As such, the value of the function L coincides with the value of f at $x = a$, and its slope coincides with the slope of f at $x = a$; that is,

$$L(a) = f(a) \qquad \text{and} \qquad L'(a) = f'(a)$$

Let's write

$$L(x) = P_1(x)$$

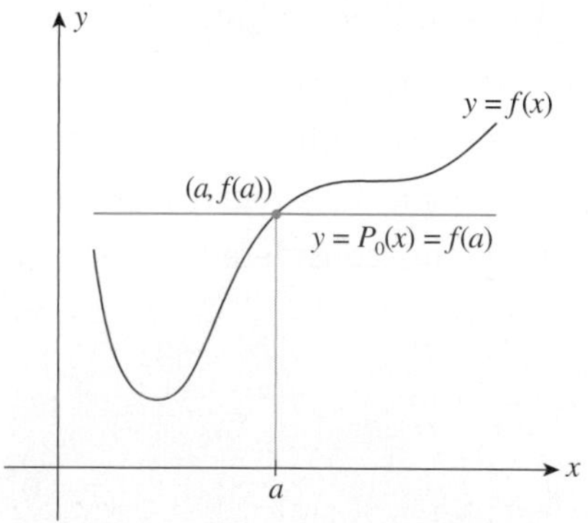

(a) $P_0(x) = f(a)$ is a zero-degree polynomial that approximates f near $x = a$.

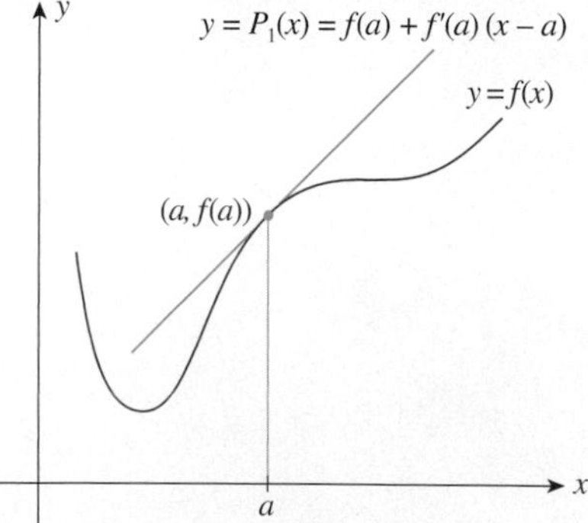

(b) $P_1(x) = f(a) + f'(a)(x - a)$ is a first-degree polynomial that approximates f near $x = a$.

FIGURE 1

Then

$$P_1(x) = f(a) + f'(a)(x - a)$$

is the required polynomial approximation (of degree 1) of f near $x = a$.

This discussion suggests that yet a better approximation to f at $x = a$ may be found by using a polynomial of degree 2, $P_2(x)$, and requiring that its value, slope, and concavity coincide with those of f at $x = a$. In other words, $P_2(x)$ should satisfy the three conditions

$$P_2(a) = f(a) \qquad P_2'(a) = f'(a) \qquad P_2''(a) = f''(a)$$

The third condition ensures that the graph of the polynomial bends in the right way, at least near $x = a$. Pursuing this line of reasoning, we are led to the search for a polynomial of degree n in $x - a$,

$$\begin{aligned} P_n(x) &= a_0 + a_1(x - a) + a_2(x - a)^2 \\ &\quad + a_3(x - a)^3 + \cdots + a_n(x - a)^n \end{aligned}$$

(where $a_0, a_1, \ldots, a_n$ are constants), that satisfies the conditions

$$P_n(a) = f(a), P_n'(a) = f'(a), P_n''(a) = f''(a), \ldots, P_n^{(n)}(a) = f^{(n)}(a) \tag{1}$$

To determine the required polynomial, we compute

$$\begin{aligned} P_n'(x) &= a_1 + 2a_2(x - a) + 3a_3(x - a)^2 + \cdots + na_n(x - a)^{n-1} \\ P_n''(x) &= 2a_2 + 3 \cdot 2a_3(x - a) + \cdots + n(n - 1)a_n(x - a)^{n-2} \\ P_n'''(x) &= 3 \cdot 2a_3 + 4 \cdot 3 \cdot 2a_4(x - a) + \cdots + n(n - 1)(n - 2)a_n(x - a)^{n-3} \\ &\vdots \\ P_n^{(n)}(x) &= n(n - 1)(n - 2) \cdots (1)a_n \end{aligned}$$

Setting $x = a$ in each of the expressions for $P_n(x), P_n'(x), P_n''(x), \ldots, P_n^{(n)}(x)$ in succession and using the conditions in (1), we find

$$\begin{aligned} P_n(a) &= a_0 = f(a) \\ P_n'(a) &= a_1 = f'(a) \\ P_n''(a) &= 2a_2 = f''(a) \\ P_n'''(a) &= 3 \cdot 2a_3 = f'''(a) \\ &\vdots \\ P_n^{(n)}(a) &= n(n - 1)(n - 2) \cdots (1)a_n = f^{(n)}(a) \end{aligned}$$

from which we deduce that

$$a_0 = f(a), a_1 = f'(a), a_2 = \frac{1}{2} f''(a), a_3 = \frac{1}{3 \cdot 2} f'''(a), \ldots,$$

$$a_n = \frac{1}{n(n - 1)(n - 2) \cdots (1)} f^{(n)}(a)$$

Let's introduce the expression $n!$ (read "n factorial"), defined by

$$\begin{aligned} n! &= n(n - 1)(n - 2)(n - 3) \cdots 3 \cdot 2 \cdot 1 \qquad (\text{for } n \geq 1) \\ 0! &= 1 \end{aligned}$$

Thus,

$$1! = 1 \qquad 4! = 4 \cdot 3 \cdot 2 \cdot 1 = 24$$
$$2! = 2 \cdot 1 = 2 \qquad 5! = 5 \cdot 4 \cdot 3 \cdot 2 \cdot 1 = 120$$
$$3! = 3 \cdot 2 \cdot 1 = 6$$

and so on. Using this notation, we may write the coefficients of $P_n(x)$ as

$$a_0 = f(a),\ a_1 = f'(a),\ a_2 = \frac{1}{2!}f''(a),\ a_3 = \frac{1}{3!}f'''(a), \ldots, a_n = \frac{1}{n!}f^{(n)}(a)$$

so that the required polynomial is

$$P_n(x) = f(a) + f'(a)(x-a) + \frac{f''(a)}{2!}(x-a)^2 + \cdots + \frac{f^{(n)}(a)}{n!}(x-a)^n$$

The *n*th Taylor Polynomial

Suppose that the function f and its first n derivatives are defined at $x = a$. Then the nth Taylor polynomial of f at $x = a$ is the polynomial

$$P_n(x) = f(a) + f'(a)(x-a) + \frac{f''(a)}{2!}(x-a)^2 + \cdots + \frac{f^{(n)}(a)}{n!}(x-a)^n \qquad \textbf{(2)}$$

which coincides with $f(x), f'(x), \ldots, f^{(n)}(x)$ at $x = a$; that is,

$$P_n(a) = f(a),\ P_n'(a) = f'(a), \ldots, P_n^{(n)}(a) = f^{(n)}(a)$$

Using a Taylor Polynomial to Approximate a Function

In many instances the Taylor polynomial $P_n(x)$ provides us with a good approximation of $f(x)$ near $x = a$.

EXAMPLE 1 Find the first four Taylor polynomials of $f(x) = e^x$ at $x = 0$ and sketch the graph of each polynomial superimposed upon the graph of $f(x) = e^x$.

Solution Here $a = 0$ and, since

$$f(x) = f'(x) = f''(x) = f'''(x) = f^{(4)}(x) = e^x$$

we find

$$f(0) = f'(0) = f''(0) = f'''(0) = f^{(4)}(0) = 1$$

Using Formula (2) with $n = 1, 2, 3$, and 4 in succession, we find that the first four Taylor polynomials are

$$P_1(x) = f(0) + f'(0)(x - 0) = 1 + x$$

$$P_2(x) = f(0) + f'(0)(x - 0) + \frac{f''(0)}{2!}(x - 0)^2 = 1 + x + \frac{1}{2}x^2$$

$$P_3(x) = f(0) + f'(0)(x - 0) + \frac{f''(0)}{2!}(x - 0)^2 + \frac{f'''(0)}{3!}(x - 0)^3$$

$$= 1 + x + \frac{1}{2}x^2 + \frac{1}{6}x^3$$

$$P_4(x) = f(0) + f'(0)(x - 0) + \frac{f''(0)}{2!}(x - 0)^2 + \frac{f'''(0)}{3!}(x - 0)^3$$

$$+ \frac{f^{(4)}(0)}{4!}(x - 0)^4 = 1 + x + \frac{1}{2}x^2 + \frac{1}{6}x^3 + \frac{1}{24}x^4$$

The graphs of these polynomials are shown in Figure 2. Observe that the approximation of $f(x)$ near $x = 0$ improves as the degree of the approximating Taylor polynomial increases.

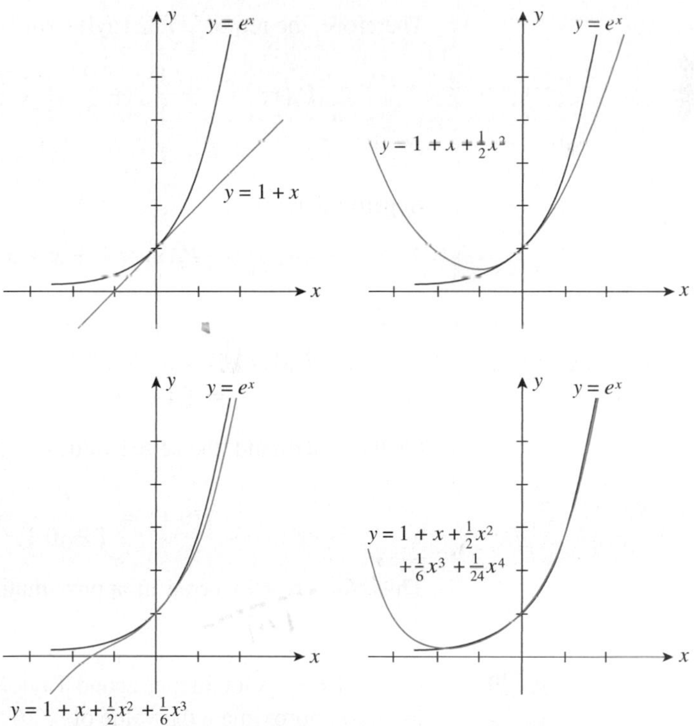

FIGURE 2
The graphs of the first four Taylor polynomials of $f(x) = e^x$ at $x = 0$ superimposed upon the graph of $f(x) = e^x$.

EXPLORING WITH TECHNOLOGY

Let $f(x) = xe^{-x}$.

1. Find the first four Taylor polynomials of f at $x = 0$.
2. Use a graphing utility to plot the graphs of f, P_1, P_2, P_3, and P_4 on the same set of axes in the viewing window $[-0.5, 1] \times [-0.5, 0.5]$.
3. Comment on the approximation of f by the polynomial P_n near $x = 0$ for $n = 1, 2, 3$, and 4. What happens to the approximation if x is "far" from the origin?

EXAMPLE 2 Find the nth Taylor polynomial of $f(x) = 1/(1 - x)$ at $x = 0$. Compute $P_5(0.1)$ and compare this result with $f(0.1)$.

Solution Here $a = 0$. Next, we find

$$\begin{aligned}
f(x) &= \frac{1}{1-x} = (1-x)^{-1} \\
f'(x) &= -(1-x)^{-2}(-1) = (1-x)^{-2} \\
f''(x) &= -2(1-x)^{-3}(-1) = 2(1-x)^{-3} \\
f'''(x) &= 3 \cdot 2(1-x)^{-4} \\
f^{(4)}(x) &= 4 \cdot 3 \cdot 2(1-x)^{-5} \\
&\vdots \\
f^{(n)}(x) &= n(n-1)(n-2) \cdots 1(1-x)^{-n-1}
\end{aligned}$$

so that

$$f(0) = 1, f'(0) = 1, f''(0) = 2, f'''(0) = 3!, f^{(4)}(0) = 4!, \ldots, f^{(n)}(0) = n!$$

Therefore, the required nth Taylor polynomial is

$$\begin{aligned}
P_n(x) &= 1 + \frac{1}{1}x + \frac{2}{2!}x^2 + \frac{3!}{3!}x^3 + \frac{4!}{4!}x^4 + \cdots + \frac{n!}{n!}x^n \\
&= 1 + x + x^2 + x^3 + x^4 + \cdots + x^n
\end{aligned}$$

In particular,

$$P_5(x) = 1 + x + x^2 + x^3 + x^4 + x^5$$

so that

$$\begin{aligned}
P_5(0.1) &= 1 + 0.1 + (0.1)^2 + (0.1)^3 + (0.1)^4 + (0.1)^5 \\
&= 1.11111
\end{aligned}$$

On the other hand, the actual value of $f(x)$ at $x = 0.1$ is

$$f(0.1) = \frac{1}{1-0.1} = \frac{1}{0.9} = 1.111\ldots$$

Thus, the error incurred in approximating $f(0.1)$ by $P_5(0.1)$ is 0.00000111. . . .

EXAMPLE 3 Obtain the second Taylor polynomial of $f(x) = \sqrt{x}$ at $x = 25$ and use it to approximate the value of $\sqrt{26.5}$.

Solution Here $a = 25$ and $n = 2$. Since

$$f(x) = x^{1/2} \qquad f'(x) = \frac{1}{2}x^{-1/2} = \frac{1}{2\sqrt{x}} \qquad f''(x) = -\frac{1}{4}x^{-3/2} = -\frac{1}{4x^{3/2}}$$

we find that

$$f(25) = 5 \qquad f'(25) = \frac{1}{10} \qquad f''(25) = -\frac{1}{500}$$

Using Formula (2), we obtain the required polynomial

$$P_2(x) = f(25) + f'(25)(x - 25) + \frac{f''(25)}{2!}(x - 25)^2$$
$$= 5 + \frac{1}{10}(x - 25) - \frac{1}{1000}(x - 25)^2$$

Next, using $P_2(x)$ as an approximation of $f(x)$ near $x = 25$, we find

$$\begin{aligned}\sqrt{26.5} &= f(26.5)\\ &\approx P_2(26.5)\\ &= 5 + \frac{1}{10}(26.5 - 25) - \frac{1}{1000}(26.5 - 25)^2\\ &\approx 5.14775\end{aligned}$$

The exact value of $\sqrt{26.5}$, rounded off to five decimal places, is 5.14782. Thus, the error incurred in the approximation is 0.00007. ■

EXPLORE & DISCUSS

Suppose you want to find an approximation to the value of $\sqrt[3]{26.5}$ using a second Taylor polynomial of some function f expanded about $x = a$.

1. How would you choose f and a?
2. Obtain an approximation of $\sqrt[3]{26.5}$ and compare your result with that obtained using a calculator.

The next example shows how a Taylor polynomial can be used to approximate an integral that involves an integrand whose antiderivative cannot be expressed as an elementary function.

VIDEO

APPLIED EXAMPLE 4 Growth of the Service Industries It has been estimated that service industries, which currently make up 30% of the nonfarm workforce in a certain country, will continue to grow at the rate of

$$R(t) = 5e^{1/(t+1)}$$

percent per decade, t decades from now. Estimate the percentage of the nonfarm workforce in the service industries one decade from now.

Solution The percentage of the nonfarm workforce in the service industries t decades from now will be given by

$$P(t) = \int 5e^{1/(t+1)}\,dt \qquad [P(0) = 30]$$

This integral cannot be expressed in terms of an elementary function. To obtain an approximate solution to the problem at hand, let's first make the substitution

$$u = \frac{1}{t+1}$$

so that

$$t + 1 = \frac{1}{u} \qquad \text{and} \qquad t = \frac{1}{u} - 1$$

giving

$$dt = -\frac{1}{u^2}\,du$$

The integral becomes

$$F(u) = 5\int e^u\left(-\frac{du}{u^2}\right) = -5\int \frac{e^u}{u^2}\,du$$

Next, let's approximate e^u at $u = 0$ by a fourth-degree Taylor polynomial. Proceeding as in Example 1, we find that

$$e^u \approx 1 + u + \frac{u^2}{2!} + \frac{u^3}{3!} + \frac{u^4}{4!}$$

Thus,

$$\begin{aligned} F(u) &\approx -5\int \frac{1}{u^2}\left(1 + u + \frac{u^2}{2} + \frac{u^3}{6} + \frac{u^4}{24}\right) du \\ &= -5\int\left(\frac{1}{u^2} + \frac{1}{u} + \frac{1}{2} + \frac{u}{6} + \frac{u^2}{24}\right) du \\ &= -5\left(-\frac{1}{u} + \ln u + \frac{1}{2}u + \frac{u^2}{12} + \frac{u^3}{72}\right) + C \end{aligned}$$

Therefore,

$$P(t) \approx -5\left[-(t+1) + \ln\left(\frac{1}{t+1}\right) + \frac{1}{2(t+1)} + \frac{1}{12(t+1)^2} + \frac{1}{72(t+1)^3}\right] + C$$

Using the condition $P(0) = 30$, we find

$$P(0) \approx -5\left(-1 + \ln 1 + \frac{1}{2} + \frac{1}{12} + \frac{1}{72}\right) + C = 30$$

or $C \approx 27.99$. So

$$P(t) \approx -5\left[-(t+1) + \ln\left(\frac{1}{t+1}\right) + \frac{1}{2(t+1)} + \frac{1}{12(t+1)^2} + \frac{1}{72(t+1)^3}\right] + 27.99$$

In particular, the percentage of the nonfarm workforce in the service industries one decade from now will be given by

$$\begin{aligned} P(1) &\approx -5\left[-2 + \ln\left(\frac{1}{2}\right) + \frac{1}{4} + \frac{1}{48} + \frac{1}{576}\right] + 27.99 \\ &\approx 40.09 \end{aligned}$$

or approximately 40.1%.

Errors in Taylor Polynomial Approximations

The **error** incurred in approximating $f(x)$ by its Taylor polynomial $P_N(x)$ at $x = a$ is

$$R_N(x) = f(x) - P_N(x)$$

A formula for $R_N(x)$, called the *remainder*, associated with $P_N(x)$ is stated without proof in what follows.

THEOREM 1

The Remainder Theorem

Suppose a function f has derivatives up to order $(N + 1)$ on an interval I containing the number a. Then for each x in I, there exists a number c lying between a and x such that

$$R_N(x) = \frac{f^{(N+1)}(c)}{(N+1)!}(x - a)^{N+1}$$

Usually the exact value of c is unknown. But from a practical point of view, this is not a serious drawback because what we need is just a bound on the error incurred when a Taylor polynomial of f is used to approximate $f(x)$. For this purpose, we have the following result, which is a consequence of the remainder theorem.

Error Bound for Taylor Polynomial Approximations

Suppose a function f has derivatives up to order $(N + 1)$ on an interval I containing the number a. Then for each x in I,

$$|R_N(x)| \leq \frac{M}{(N+1)!}|x - a|^{N+1} \tag{3}$$

where M is any number such that $|f^{(N+1)}(t)| \leq M$ for all t lying between a and x.

EXAMPLE 5 Refer to Example 2. Use Formula (3) to obtain an upper bound for the error incurred in approximating $f(0.1)$ by $P_5(0.1)$, where

$$f(x) = \frac{1}{1-x}$$

and $P_5(x)$ is the fifth Taylor polynomial of f at $x = 0$. Compare this result with the actual error incurred in the approximation.

Solution First, we need to find a number M such that

$$|f^{(6)}(t)| \leq M$$

for all t lying in the interval $[0, 0.1]$. From the solution of Example 2, we find that

$$\begin{aligned} f^{(6)}(t) &= (6)(5)(4)(3)(2)(1)(1-t)^{-6-1} \\ &= \frac{720}{(1-t)^7} \end{aligned}$$

Observe that

$$f^{(6)}(t) = \frac{720}{(1-t)^7}$$

is increasing on the interval $[0, 0.1]$. [Just look at $f^{(7)}(t)$.] Consequently, its maximum is attained at the right end point $t = 0.1$. Thus, we may take

$$M = \frac{720}{(1-0.1)^7} \approx 1505.3411$$

Next, with $a = 0$, $x = 0.1$, and $M = 1505.3411$, Formula (3) gives

$$\begin{aligned} |R_5(0.1)| &\leq \frac{1505.3411}{(5+1)!}(0.1)^{5+1} \\ &\approx 0.0000021 \end{aligned}$$

Finally, we refer once again to Example 2 to see that the actual error incurred in approximating $f(0.1)$ by $P_5(0.1)$ was 0.00000111. . . , and this is less than the error bound 0.0000021 just obtained. ■

EXPLORE & DISCUSS

Refer to the Explore & Discuss question on page 703. Obtain an upper bound for the error incurred in approximating $\sqrt[3]{26.5}$, using the Taylor polynomial you have chosen.

EXAMPLE 6 Use the fourth Taylor polynomial to obtain an approximation of $e^{0.5}$. Find a bound for the error in the approximation.

Solution The point $x = 0.5$ is near $x = 0$, so we use the fourth Taylor polynomial $P_4(x)$ at $x = 0$ to approximate $e^{0.5}$. From Example 1, we see that

$$P_4(x) = 1 + x + \frac{1}{2}x^2 + \frac{1}{6}x^3 + \frac{1}{24}x^4$$

Therefore,

$$\begin{aligned} e^{0.5} \approx P_4(0.5) &= 1 + 0.5 + \frac{1}{2}(0.5)^2 + \frac{1}{6}(0.5)^3 + \frac{1}{24}(0.5)^4 \\ &\approx 1.6484 \end{aligned}$$

To obtain an error bound for this approximation, we need to obtain an upper bound M for $|f^{(5)}(t)|$ for all t lying in the interval $[0, 0.5]$ (remember, $N = 4$). Now

$$f^{(5)}(t) = e^t$$

and since $e^t > 0$ for all values of t, $f^{(5)}(t) = |e^t| = e^t$. Next, we observe that e^t is an increasing function on $(-\infty, \infty)$ and in particular on $[0, 0.5]$. Therefore, e^t attains an absolute maximum at $x = 0.5$, the right end point of the interval in question; that is,

$$|f^{(5)}(t)| = |e^t| = e^t \le e^{0.5}$$

on $[0, 0.5]$. Now, $e < 3$, so certainly

$$|f^{(5)}(t)| \le e^{0.5} < 3^{0.5} = \sqrt{3} < 2$$

Thus, we may take $M = 2$. Finally, with $a = 0$, $x = 0.5$, and this value of M, Formula (3) gives

$$\begin{aligned} |R_4(0.5)| &\le \frac{2}{(4 + 1)!}|0.5 - 0|^{4+1} \\ &= \frac{2}{5!}(0.5)^5 \\ &\approx 0.00052 \end{aligned}$$

So our result, $e^{0.5} \approx 1.6484$, is guaranteed to be accurate to within 0.0005. ■

Notes

1. Using a calculator, we see that $e^{0.5} \approx 1.6487$, rounded to four decimal places. Therefore, the error incurred in approximating this number by $P_4(0.5)$ is $(1.6487 - 1.6484)$, or 0.0003, which is less than the upper error bound of 0.0005 obtained.
2. The error bound obtained in Example 6 is not the best possible. This stems partially from our choice of M. In practical applications, it is not essential that the error bound be the best possible as long as it is sharp enough. ■

11.1 Self-Check Exercises

1. Let $f(x) = \dfrac{1}{1 + x}$.
   a. Find the nth Taylor polynomial of f at $x = 0$.
   b. Use the third Taylor polynomial of f at $x = 0$ to obtain an approximation of $f(0.1)$. Compare this result with $f(0.1)$.

2. Refer to Exercise 1. Find a bound on the error in the approximation of $f(0.1)$.

Solutions to Self-Check Exercises 11.1 can be found on page 709.

11.1 Concept Questions

1. a. What is a Taylor polynomial?
   b. Suppose $P_n(x) = a_0 + a_1(x - a) + a_2(x - a)^2 + \cdots + a_n(x - a)^n$ is a Taylor polynomial. What is a_i $(0 \leq i \leq n)$ in terms of f and/or its derivatives?

2. a. What is the error incurred if $f(x)$ is approximated by its Nth-order Taylor polynomial $P_N(x)$? Give an expression for the error (remainder) $R_N(x)$.
   b. Write an expression giving the bound for the error when $f(x)$ is approximated by $P_N(x)$.

11.1 Exercises

In Exercises 1–10, find the first three Taylor polynomials of the function at the indicated number.

1. $f(x) = e^{-x}$; $x = 0$
2. $f(x) = e^{2x}$; $x = 0$
3. $f(x) = \dfrac{1}{x + 1}$; $x = 0$
4. $f(x) = \dfrac{1}{1 - x}$; $x = 2$
5. $f(x) = \dfrac{1}{x}$; $x = 1$
6. $f(x) = \dfrac{1}{x + 2}$; $x = 1$
7. $f(x) = \sqrt{1 - x}$; $x = 0$
8. $f(x) = \sqrt{x}$; $x = 4$
9. $f(x) = \ln(1 - x)$; $x = 0$
10. $f(x) = xe^{x}$; $x = 0$

In Exercises 11–20, find the nth Taylor polynomial of the function at the indicated number.

11. $f(x) = x^4$ at $x = 2$, $n = 2$
12. $f(x) = x^5$ at $x = -1$, $n = 3$
13. $f(x) = \ln x$ at $x = 1$, $n = 4$
14. $f(x) = \dfrac{1}{x}$ at $x = 2$, $n = 4$
15. $f(x) = e^x$ at $x = 1$, $n = 4$
16. $f(x) = e^{2x}$ at $x = 1$, $n = 4$
17. $f(x) = \sqrt{1 - x}$ at $x = 0$, $n = 3$
18. $f(x) = \sqrt{x + 1}$ at $x = 0$, $n = 3$
19. $f(x) = \dfrac{1}{2x + 3}$ at $x = 0$, $n = 3$
20. $f(x) = x^2e^x$ at $x = 0$, $n = 2$
21. Find the nth Taylor polynomial of $f(x) = \dfrac{1}{1 + x}$ at $x = 0$.

    Compute $P_4(0.1)$ and compare this result with $f(0.1)$.
22. Find the third Taylor polynomial of $f(x) = \sqrt{3x + 1}$ at $x = 0$. Compute $P_3(0.2)$ and compare this result with $f(0.2)$.
23. Find the fourth Taylor polynomial of $f(x) = e^{-x/2}$ at $x = 0$ and use it to estimate $e^{-0.1}$.
24. Find the third Taylor polynomial of $f(x) = \ln x$ at $x = 1$ and use it to estimate $\ln 1.1$.
25. Find the second Taylor polynomial of $f(x) = \sqrt{x}$ at $x = 16$ and use it to estimate the value of $\sqrt{15.6}$.
26. Find the second Taylor polynomial of $f(x) = x^{1/3}$ at $x = 8$ and use it to estimate the value of $\sqrt[3]{8.1}$.
27. Find the third Taylor polynomial of $f(x) = \ln(x + 1)$ at $x = 0$ and use it to estimate the value of

    $$\int_0^{1/2} \ln(x + 1)\, dx$$

    Compare your result with the exact value found using integration by parts.

28. Use the fourth Taylor polynomial of $f(x) = \ln(1 + x)$ to approximate

$$\int_{0.1}^{0.2} \frac{\ln(x + 1)}{x}\, dx$$

29. Use the second Taylor polynomial of $f(x) = \sqrt{x}$ at $x = 4$ to approximate $\sqrt{4.06}$ and find a bound for the error in the approximation.

30. Use the third Taylor polynomial of $f(x) = e^{-x}$ at $x = 0$ to approximate $e^{-0.2}$ and find a bound for the error in the approximation.

31. Use the third Taylor polynomial of $f(x) = \dfrac{1}{1 - x}$ at $x = 0$ to approximate $f(0.2)$. Find a bound for the error in the approximation and compare your results to the exact value of $f(0.2)$.

32. Use the second Taylor polynomial of $f(x) = \sqrt{x + 1}$ at $x = 0$ to approximate $f(0.1)$ and find a bound for the error in the approximation.

33. Use the second Taylor polynomial of $f(x) = \ln x$ at $x = 1$ to approximate ln 1.1 and find a bound for the error in the approximation.

34. Use the second Taylor polynomial of $f(x) = \ln x$ at $x = 1$ to approximate ln 0.9 and find a bound for the error in the approximation.

35. Let $f(x) = e^{-x}$.

a. Find a bound in the error incurred in approximating $f(x)$ by the third Taylor polynomial, $P_3(x)$, of f at $x = 0$ in the interval $[0, 1]$.

b. Estimate $\int_0^1 e^{-x}\, dx$ by computing $\int_0^1 P_3(x)\, dx$.

c. What is the bound on the error in the approximation in (b)?
Hint: Compute $\int_0^1 B\, dx$, where B is a bound on $|R_3(x)|$ for x in $[0, 1]$.

d. What is the actual error?

36. a. Find a bound on the error incurred in approximating $f(x) = \ln x$ by the fourth Taylor polynomial, $P_4(x)$, of f at $x = 1$ in the interval $[1, 1.5]$.

b. Approximate the area under the graph of f from $x = 1$ to $x = 1.5$ by computing

$$\int_1^{1.5} P_4(x)\, dx$$

c. What is the bound on the error in the approximation in part (b)?
Hint: Compute $\int_1^{1.5} B\, dx$, where B is a bound on $|R_4(x)|$ for x in $[1, 1.5]$.

d. What is the actual error?
Hint: Use integration by parts to evaluate $\int_1^{1.5} \ln x\, dx$.

37. Use the fourth Taylor polynomial at $x = 0$ to obtain an approximation of the area under the graph of $f(x) = e^{-x^2/2}$ from $x = 0$ to $x = 0.5$.
Hint: Use the substitution $u = -x^2/2$. See Example 4.

38. Use the fourth Taylor polynomial at $x = 0$ to obtain an approximation of the area under the graph of $f(x) = \ln(1 + x^2)$ from $x = -\frac{1}{2}$ to $x = \frac{1}{2}$.
Hint: First find the Taylor polynomial of $f(u) = \ln(1 + u)$; then use the substitution $u = x^2$.

39. GROWTH OF SERVICE INDUSTRIES It has been estimated that service industries, which currently make up 30% of the nonfarm workforce in a certain country, will continue to grow at the rate of

$$R(t) = 6e^{1/(2t+1)}$$

percent per decade t decades from now. Estimate the percentage of the nonfarm workforce in service industries two decades from now.

40. VACATION TRENDS The Travel Data Center of a European country has estimated that the percentage of vacations traditionally taken during the months of July, August, and September will continue to decline according to the rule

$$R(t) = -2e^{1/(t+1)}$$

percent per decade ($t = 0$ corresponds to the year 1996) as more and more vacationers switch to fall and winter vacations. There were 37% summer vacationers in that country in 1996. Estimate the percentage of vacations that were taken in the summer months in 2006.

41. CONTINUING EDUCATION ENROLLMENT The registrar of Kellogg University estimates that the total student enrollment in the Continuing Education division will be given by

$$N(t) = -\frac{20{,}000}{\sqrt{1 + 0.2t}} + 21{,}000$$

where $N(t)$ denotes the number of students enrolled in the division t yr from now. Using the second Taylor polynomial of N at $t = 0$, find an approximation of the average enrollment at Kellogg University between $t = 0$ and $t = 2$.

42. CONCENTRATION OF CARBON MONOXIDE IN THE AIR According to a joint study conducted by Oxnard's Environmental Management Department and a state government agency, the concentration of carbon monoxide (CO) in the air due to automobile exhaust t yr from now is given by

$$C(t) = 0.01(0.2t^2 + 4t + 64)^{2/3}$$

parts per million. Use the second Taylor polynomial of C at $t = 0$ to obtain an approximation of the average level of concentration of CO in the air between $t = 0$ and $t = 2$.

In Exercises 43–46, determine whether the statement is true or false. If it is true, explain why it is true. If it is false, give an example to show why it is false.

43. If f is a polynomial of degree four and $P_4(x)$ is the fourth Taylor polynomial of f at $x = a$, then $f(x) \neq P_4(x)$ for at least one value of x.

44. The function $f(x) = \dfrac{1}{x-2}$ has a Taylor polynomial at $x = a$ for any value of a except $a = 2$.

45. Suppose f is a polynomial of degree N and R_N is the remainder associated with P_N, the Nth Taylor polynomial of f at $x = a$. Then $R_N(x) = 0$ for every value of x.

46. Let P_N denote the Nth Taylor polynomial of the function $f(x) = e^{-x}$ at $x = a$. Then $R_N(x) = f(x) - P_N(x)$ is never equal to zero for all values of x.

11.1 Solutions to Self-Check Exercises

1. a. We rewrite $f(x)$ as

$$f(x) = (1 + x)^{-1}$$

and compute

$$\begin{aligned} f'(x) &= -(1+x)^{-2} \\ f''(x) &= (-1)(-2)(1+x)^{-3} = 2(1+x)^{-3} \\ f'''(x) &= 2(-3)(1+x)^{-4} = -3\cdot 2(1+x)^{-4} \\ f^{(4)}(x) &= -3\cdot 2(-4)(1+x)^{-5} = 4\cdot 3\cdot 2(1+x)^{-5} \\ &\vdots \\ f^{(n)}(x) &= (-1)^n n!(1+x)^{-(n+1)} \end{aligned}$$

Evaluating $f(x)$ and its derivatives at $x = 0$ gives

$$f(0) = 1, f'(0) = -1, f''(0) = 2!, f'''(0) = -3!$$
$$f^{(4)}(0) = 4!, \ldots, f^{(n)}(0) = (-1)^n n!$$

Therefore, the required nth Taylor polynomial is

$$\begin{aligned} P_n(x) &= 1 + \frac{(-1)}{1!}x + \frac{2}{2!}x^2 + \frac{-3!}{3!}x^3 \\ &\quad + \frac{4!}{4!}x^4 + \cdots + \frac{(-1)^n n!}{n!}x^n \\ &= 1 - x + x^2 - x^3 + x^4 - \cdots + (-1)^n x^n \end{aligned}$$

b. The third Taylor polynomial of f at $x = 0$ is

$$P_3(x) = 1 - x + x^2 - x^3$$

So

$$\begin{aligned} P_3(0.1) &= 1 - 0.1 + (0.1)^2 - (0.1)^3 \\ &= 0.909 \end{aligned}$$

The actual value of $f(x)$ at $x = 0.1$ is

$$f(0.1) = \frac{1}{1+0.1} = \frac{1}{1.1} = 0.909090\ldots$$

Thus, the error incurred in approximating $f(0.1)$ by $P_3(0.1)$ is 0.00009090. . . .

2. First, we need to find a number M such that

$$|f^{(4)}(t)| \le M$$

for all t lying in the interval [0, 0.1]. From the solution of Self-Check Exercise 1, we see that

$$f^{(5)}(x) = (-1)^5 5!(1+x)^{-6} = -5!(1+x)^{-6}$$

Since $f^{(5)}(x)$ is negative for x in [0, 0.1], $f^{(4)}(x)$ is decreasing on [0, 0.1]. Therefore, the maximum of the function

$$f^{(4)}(x) = 4!(1+x)^{-5}$$

is attained at $x = 0$, and its value is

$$f^{(4)}(0) = 4! = 24$$

So we can take $M = 24$. Next, with $a = 0, x = 0.1$, and $M = 24$, Formula (3) gives

$$\begin{aligned} |R_3(0.1)| &\le \frac{24}{(3+1)!}|0.1 - 0|^{3+1} \\ &= \frac{24}{24}(0.1)^4 = 0.0001 \end{aligned}$$

the required error bound.

11.2 Infinite Sequences

An idealized superball is dropped from a height of 1 meter onto a flat surface. Suppose that each time the ball hits the surface, it rebounds to two-thirds of its previous height. If we let a_1 denote the initial height of the ball, a_2 the maximum height attained on the first rebound, a_3 the maximum height attained on the second rebound, and so on, then we have

$$a_1 = 1, a_2 = \frac{2}{3}, a_3 = \frac{4}{9}, a_4 = \frac{8}{27}, \ldots$$

(see Figure 3).

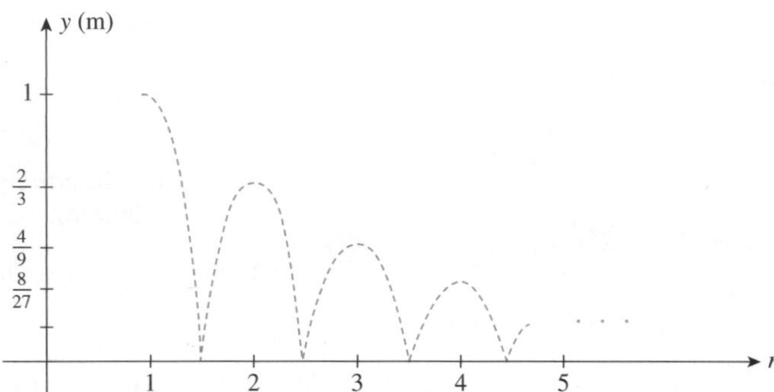

FIGURE 3
The ball rebounds to two-thirds of its previous height upon hitting the surface.

This array of numbers, $a_1, a_2, a_3, \ldots$, is an example of an *infinite sequence*, or simply a *sequence*. If we define the function f by $f(x) = (2/3)^{x-1}$ and allow x to take on the positive integral values $x = 1, 2, 3, \ldots, n, \ldots$, then we see that the sequence $a_1, a_2, a_3, \ldots$ may be viewed as the functional values of f at these numbers. Thus

$$\begin{array}{ccccc} f(1) = 1 & f(2) = \frac{2}{3} & f(3) = \frac{4}{9} & \ldots\ f(n) = \left(\frac{2}{3}\right)^{n-1} & \ldots \\ \downarrow & \downarrow & \downarrow & \downarrow & \\ a_1 & a_2 & a_3 & a_n & \ldots \end{array}$$

This discussion motivates the following definition.

Infinite Sequence

An **infinite sequence** $\{a_n\}$ is a function whose domain is the set of positive integers. The functional values $a_1, a_2, a_3, \ldots, a_n, \ldots$ are the **terms** of the sequence, and the term a_n is called the ***n*th term** of the sequence.

Notes

1. The sequence $\{a_n\}$ is also denoted by $\{a_n\}_{n=1}^{\infty}$.
2. Sometimes it is convenient to begin a sequence with a_k. In this case the sequence is $\{a_n\}_{n=k}^{\infty}$, and its terms are $a_k, a_{k+1}, a_{k+2}, \ldots, a_n, \ldots$.

EXAMPLE 1 List the terms of the sequence.

a. $\left\{\dfrac{n}{n+1}\right\}$ **b.** $\left\{\dfrac{\sqrt{n}}{2^{n-1}}\right\}$ **c.** $\{(-1)^n\sqrt{n-2}\}_{n=2}^{\infty}$

Solution

a. Here $a_n = f(n) = \dfrac{n}{n+1}$. Thus

$$a_1 = f(1) = \frac{1}{1+1} = \frac{1}{2},\ a_2 = f(2) = \frac{2}{2+1} = \frac{2}{3},\ a_3 = f(3) = \frac{3}{3+1} = \frac{3}{4},$$

and so forth and we see that the given sequence can be written as

$$\left\{\frac{n}{n+1}\right\} = \left\{\frac{1}{2}, \frac{2}{3}, \frac{3}{4}, \frac{4}{5}, \ldots, \frac{n}{n+1}, \ldots\right\}$$

b. $\left\{\dfrac{\sqrt{n}}{2^{n-1}}\right\} = \left\{\dfrac{\sqrt{1}}{2^0}, \dfrac{\sqrt{2}}{2}, \dfrac{\sqrt{3}}{2^2}, \dfrac{\sqrt{4}}{2^3}, \ldots, \dfrac{\sqrt{n}}{2^{n-1}}, \ldots\right\}$

c. $\{(-1)^n\sqrt{n-2}\}_{n=2}^{\infty}$

$$= \{(-1)^2\sqrt{0}, (-1)^3\sqrt{1}, (-1)^4\sqrt{2}, (-1)^5\sqrt{3}, \ldots, (-1)^n\sqrt{n-2}, \ldots\}$$
$$= \{0, -\sqrt{1}, \sqrt{2}, -\sqrt{3}, \ldots (-1)^n\sqrt{n-2}, \ldots\}$$

Notice that n starts from 2 in this example. (See Note 2, page 710.)

We can often determine the nth term of a sequence by studying the first few terms of the sequence and recognizing the pattern that emerges.

EXAMPLE 2 Find an expression for the nth term of each sequence.

a. $\left\{2, \dfrac{3}{\sqrt{2}}, \dfrac{4}{\sqrt{3}}, \dfrac{5}{\sqrt{4}}, \ldots\right\}$ **b.** $\left\{1, \dfrac{1}{8}, \dfrac{1}{27}, \dfrac{1}{64}, \ldots\right\}$ **c.** $\left\{1, -\dfrac{1}{2}, \dfrac{1}{3}, -\dfrac{1}{4}, \ldots\right\}$

Solution

a. The terms of the sequence may be written in the form

$$a_1 = \frac{1+1}{\sqrt{1}},\ a_2 = \frac{2+1}{\sqrt{2}},\ a_3 = \frac{3+1}{\sqrt{3}},\ a_4 = \frac{4+1}{\sqrt{4}}, \ldots$$

from which we see that $a_n = \dfrac{n+1}{\sqrt{n}}$.

b. Here

$$a_1 = \frac{1}{1^3},\ a_2 = \frac{1}{2^3},\ a_3 = \frac{1}{3^3},\ a_4 = \frac{1}{4^3}, \ldots$$

and so $a_n = \dfrac{1}{n^3}$.

c. Note that $(-1)^r$ is equal to 1 if r is an even integer and is equal to -1 if r is an odd integer. Using this result, we obtain

$$a_1 = \frac{(-1)^0}{1},\ a_2 = \frac{(-1)^1}{2},\ a_3 = \frac{(-1)^2}{3},\ a_4 = \frac{(-1)^3}{4}, \ldots$$

We conclude that the nth term is $a_n = (-1)^{n-1}/n$.

Graphs of Infinite Sequences

Since an infinite sequence is a function, we can sketch its graph. Because the domain of the function is the set of positive integers, the graph of a sequence consists of an infinite collection of points in the xy-plane.

EXAMPLE 3 Sketch the graph of the sequence.

a. $\{n + 1\}$ **b.** $\left\{\frac{n+1}{n}\right\}$ **c.** $\{(-1)^n\}$

Solution

a. The terms of the sequence $\{n + 1\}$ are 2, 3, 4, 5, ... Recalling that these numbers are precisely the functional values of $f(n) = n + 1$ for $n = 1, 2, 3, \ldots$, we obtain the following table of values of f:

n	1	2	3	4	5	$\cdots$
$f(n)$	2	3	4	5	6	$\cdots$

from which we construct the graph of $\{n + 1\}$ shown in Figure 4a.

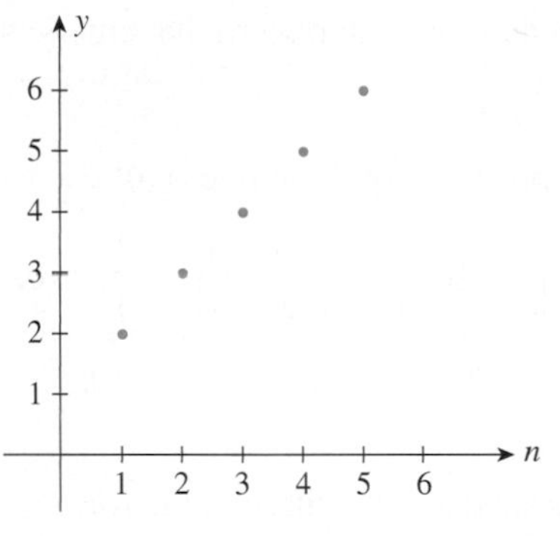

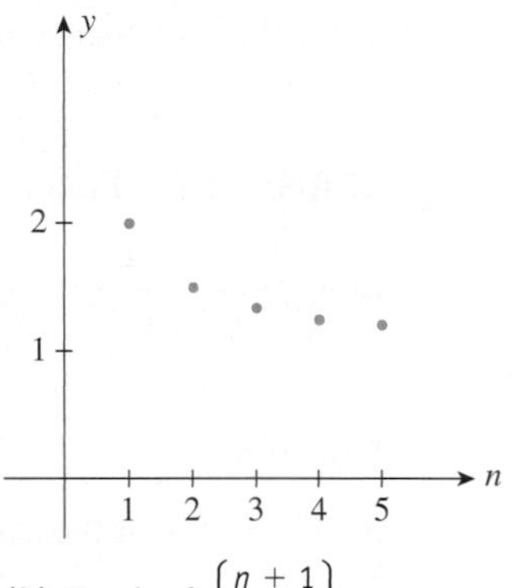

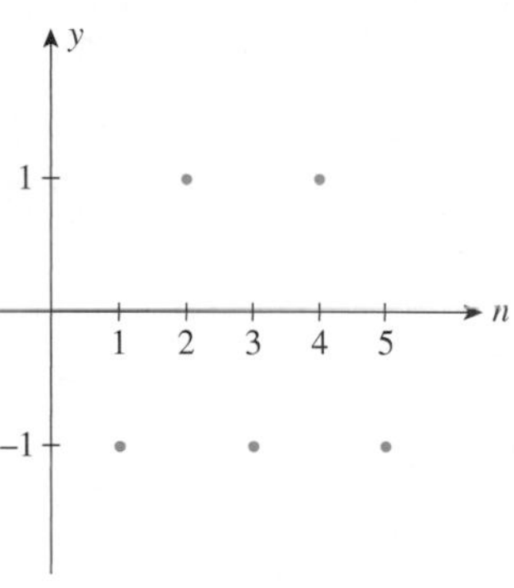

FIGURE 4 **(a)** Graph of $\{n + 1\}$ **(b)** Graph of $\left\{\frac{n+1}{n}\right\}$ **(c)** Graph of $\{(-1)^n\}$

b. From the following table of values of f,

n	1	2	3	4	5	$\cdots$
$f(n)$	2	$\frac{3}{2}$	$\frac{4}{3}$	$\frac{5}{4}$	$\frac{6}{5}$	$\cdots$

We sketch the graph shown in Figure 4b.

c. We use the following table of values for the sequence

n	1	2	3	4	5	$\cdots$
$f(n)$	-1	1	-1	1	-1	$\cdots$

to sketch the graph of $\{(-1)^n\}$ shown in Figure 4c. ■

Note Notice that the function $f(n) = n + 1$ defining the sequence $\{n + 1\}$ may be viewed as the function $f(x) = x + 1$ $(-\infty < x < \infty)$ with x *restricted* to the set of positive integers. Similarly, the function defining the sequence $\{(n + 1)/n\}$ is just the function $f(x) = (x + 1)/x$ with a similar restriction on its domain. Finally, the function defining the sequence $\{(-1)^n\}$ is the function $f(x) = \cos \pi x$ with x a positive integer.* These observations suggest that it is possible to study the properties of a sequence by analyzing the properties of a corresponding function defined for *all* values of x in some suitable interval. ■

The Limit of a Sequence

Given a sequence $\{a_n\}$, we may ask whether the terms a_n of the sequence approach some specific number L as n gets larger and larger. If they do, we say that the sequence a_n *converges* to L. More specifically, we have the following informal definition.

Limit of a Sequence

Let $\{a_n\}$ be a given sequence. We say that the sequence $\{a_n\}$ **converges** and has the limit L, written

$$\lim_{n \to \infty} a_n = L$$

if the terms of the sequence, a_n, can be made as close to L as we please by taking n sufficiently large. If a sequence is not convergent, it is said to be **divergent**.

EXAMPLE 4 Determine whether the infinite sequence converges or diverges.

a. $\left\{\dfrac{1}{n}\right\}$ **b.** $\{\sqrt{n}\}$ **c.** $\{(-1)^n\}$

Solution

a. The terms of the sequence,

$$1, \frac{1}{2}, \frac{1}{3}, \frac{1}{4}, \frac{1}{5}, \ldots, \frac{1}{n}, \ldots$$

approach the number $L = 0$ as n gets larger and larger. We conclude that

$$\lim_{n \to \infty} a_n = \lim_{n \to \infty} \frac{1}{n} = 0$$

Compare this with

$$\lim_{x \to \infty} \frac{1}{x} = 0$$

b. The terms of the sequence,

$$1, \sqrt{2}, \sqrt{3}, \sqrt{4}, \sqrt{5}, \ldots, \sqrt{n}, \ldots$$

*The trigonometric function $f(x) = \cos \pi x$ will be defined in Chapter 12.

get larger and larger as n gets larger and larger. Consequently, they do not approach any finite number L as n tends to infinity. Therefore, the sequence is divergent. Compare this with

$$\lim_{x\to\infty} \sqrt{x} = \infty$$

c. The terms of the sequence are

$$a_1 = -1, a_2 = 1, a_3 = -1, a_4 = 1, a_5 = -1, \ldots, a_n = (-1)^n, \ldots$$

Thus, no matter how large n is, there are terms that are equal to -1 (those with odd-numbered subscripts) and also terms that are equal to 1 (those with even-numbered subscripts). This implies that there cannot be a *unique* real number L such that a_n is arbitrarily close to L no matter how large n is. Therefore, the sequence is divergent.

EXPLORING WITH TECHNOLOGY

Refer to Example 4 and the Note on page 713.

1. Plot the graph of $f(x) = 1/x$, using the viewing window $[0, 10] \times [0, 3]$, and thus verify graphically that $\lim_{n\to\infty} (1/n) = 0$.
2. Plot the graph of $f(x) = \sqrt{x}$, using an appropriate viewing window, and thus verify graphically that $\lim_{n\to\infty} \sqrt{n}$ does not exist.
3. Plot the graph of $f(x) = \cos \pi x$, using the viewing window $[0, 10] \times [-2, 2]$, and thus verify graphically that $\lim_{n\to\infty} (-1)^n$ does not exist. (*Note:* Trigonometric functions will be studied in Chapter 12.)

The following properties of sequences, which parallel those of the limit of $f(x)$ at infinity, are helpful in computing limits of sequences.

Limit Properties of Sequences

Suppose

$$\lim_{n\to\infty} a_n = A \quad \text{and} \quad \lim_{n\to\infty} b_n = B$$

Then

1. $\lim_{n\to\infty} ca_n = c \lim_{n\to\infty} a_n = cA$ c, a constant
2. $\lim_{n\to\infty} (a_n \pm b_n) = \lim_{n\to\infty} a_n \pm \lim_{n\to\infty} b_n = A \pm B$ Sum rule
3. $\lim_{n\to\infty} a_n b_n = (\lim_{n\to\infty} a_n)(\lim_{n\to\infty} b_n) = AB$ Product rule
4. $\lim_{n\to\infty} \dfrac{a_n}{b_n} = \dfrac{\lim_{n\to\infty} a_n}{\lim_{n\to\infty} b_n} = \dfrac{A}{B}$ provided $B \neq 0$ Quotient rule

EXAMPLE 5 Evaluate

a. $\lim_{n\to\infty} \left(\dfrac{1}{2}\right)^n$ **b.** $\lim_{n\to\infty} \dfrac{2n^2 + 1}{3n^2 + n + 2}$

Solution

a. $\lim_{n\to\infty}\left(\frac{1}{2}\right)^n = \lim_{n\to\infty}\frac{1}{2^n}$

$$= \frac{1}{\lim_{n\to\infty} 2^n} = 0$$

b. Dividing the numerator and denominator by n^2, we find

$$\lim_{n\to\infty}\frac{2n^2+1}{3n^2+n+2} = \lim_{n\to\infty}\frac{2+\frac{1}{n^2}}{3+\frac{1}{n}+\frac{2}{n^2}}$$

$$= \frac{\lim_{n\to\infty}\left(2+\frac{1}{n^2}\right)}{\lim_{n\to\infty}\left(3+\frac{1}{n}+\frac{2}{n^2}\right)}$$

$$= \frac{2}{3}$$

EXPLORE & DISCUSS

Consider the sequences $\{a_n\}$ and $\{b_n\}$ defined by $a_n = (-1)^{n+1}$ and $b_n = (-1)^n$.

1. Show that $\lim_{n\to\infty} a_n$ and $\lim_{n\to\infty} b_n$ do not exist.
2. Show that $\lim_{n\to\infty} (a_n + b_n) = 0$.
3. Do the results of parts 1 and 2 contradict the limit properties of sequences listed on page 714? Explain your answer.

VIDEO

APPLIED EXAMPLE 6 Quality Control Of the spark plugs manufactured by the Parts Division of United Motors Corporation, 2% are defective. It can be shown that the probability of getting at least one defective plug in a random sample of n spark plugs is $f(n) = 1 - (0.98)^n$. Consider the sequence $\{a_n\}$ defined by $a_n = f(n)$.

a. Write down the terms a_5, a_{10}, a_{25}, a_{100}, and a_{200} of the sequence $\{a_n\}$.
b. Evaluate $\lim_{n\to\infty} a_n$ and interpret your results.

Solution

a. The required terms of the sequence are

0.10, 0.18, 0.40, 0.87, 0.98

For example, the probability of getting at least one defective plug in a random sample of 25 is .4—that is, a 40% chance.

b. $\lim_{n\to\infty} a_n = \lim_{n\to\infty} [1 - (0.98)^n]$

$= \lim_{n\to\infty} 1 - \lim_{n\to\infty} (0.98)^n$

$= 1 - 0$

$= 1$

The result tells us that if the sample is large enough, we will almost certainly pick at least one defective plug! ■

11.2 Self-Check Exercises

1. Determine whether the sequence $\left\{\dfrac{3n^2 + 2n + 1}{n^2 + 4}\right\}$ converges or diverges. If it converges, find its limit.

2. Consider the sequence $\left\{\dfrac{n}{n^2 + 1}\right\}$.
   a. Sketch the graph of the sequence.
   b. Show that the sequence is decreasing—that is, $a_1 > a_2 > a_3 > \cdots$.
   Hint: Consider $f(x) = x/(x^2 + 1)$ and show that f is decreasing by computing f'.

Solutions to Self-Check Exercises 11.2 can be found on page 717.

11.2 Concept Questions

Explain each of the following terms in your own words and give an example of each.

a. Sequence **b.** Convergent sequence
c. Divergent sequence **d.** Limit of a sequence

11.2 Exercises

In Exercises 1–9, write down the first five terms of the sequence.

1. $\{a_n\} = \{2^{n-1}\}$

2. $\{a_n\} = \left\{\dfrac{2n}{1 + n^2}\right\}$

3. $\{a_n\} = \left\{\dfrac{n - 1}{n + 1}\right\}$

4. $\{a_n\} = \left\{\left(-\dfrac{1}{3}\right)^n\right\}$

5. $\{a_n\} = \left\{\dfrac{2^{n-1}}{n!}\right\}$

6. $\{a_n\} = \left\{\dfrac{(-1)^n}{(2n)!}\right\}$

7. $\{a_n\} = \left\{\dfrac{e^n}{n^3}\right\}$

8. $\{a_n\} = \left\{\dfrac{\sqrt{n}}{\sqrt{n} + 1}\right\}$

9. $\{a_n\} = \left\{\dfrac{3n^2 - n + 1}{2n^2 + 1}\right\}$

In Exercises 10–21, find the general term of the sequence.

10. $\dfrac{1}{2}, \dfrac{1}{4}, \dfrac{1}{6}, \dfrac{1}{8}, \ldots$

11. $1, 4, 7, 10, \ldots$

12. $\dfrac{1}{3}, \dfrac{1}{7}, \dfrac{1}{11}, \dfrac{1}{15}, \ldots$

13. $1, \dfrac{1}{8}, \dfrac{1}{27}, \dfrac{1}{64}, \ldots$

14. $1, \dfrac{2}{3}, \dfrac{4}{9}, \dfrac{8}{27}, \ldots$

15. $2, \dfrac{8}{5}, \dfrac{32}{25}, \dfrac{128}{125}, \ldots$

16. $1, \dfrac{5}{4}, \dfrac{7}{5}, \dfrac{9}{6}, \ldots$

17. $1, -\dfrac{1}{2}, \dfrac{1}{4}, -\dfrac{1}{8}, \ldots$

18. $1 + \dfrac{1}{2}, 1 + \dfrac{1}{3}, 1 + \dfrac{1}{4}, 1 + \dfrac{1}{5}, \ldots$

19. $\dfrac{1}{2 \cdot 3}, \dfrac{2}{3 \cdot 4}, \dfrac{3}{4 \cdot 5}, \dfrac{4}{5 \cdot 6}, \ldots$

20. $1, \dfrac{2}{1 \cdot 3}, \dfrac{4}{1 \cdot 3 \cdot 5}, \dfrac{8}{1 \cdot 3 \cdot 5 \cdot 7}, \ldots$

21. $1, e, \dfrac{e^2}{2}, \dfrac{e^3}{6}, \dfrac{e^4}{24}, \dfrac{e^5}{120}, \ldots$

In Exercises 22–29, sketch the graph of the sequence.

22. $\{n^2\}$

23. $\left\{\dfrac{2n}{n + 1}\right\}$

24. $\left\{\dfrac{(-1)^n}{n}\right\}$

25. $\{\sqrt{n}\}$

26. $\{\ln n\}$

27. $\{e^n\}$

28. $\{ne^{-n}\}$

29. $\{n - \sqrt{n}\}$

In Exercises 30–44, determine the convergence or divergence of the sequence $\{a_n\}$. If the sequence converges, find its limit.

30. $a_n = \dfrac{n}{n^2 + 1}$

31. $a_n = \dfrac{n + 1}{2n}$

32. $a_n = \sqrt[3]{n}$

33. $a_n = \dfrac{(-1)^n}{\sqrt{n}}$

34. $a_n = \dfrac{1}{n + 1} - \dfrac{1}{n + 2}$

35. $a_n = \dfrac{\sqrt{n} - 1}{\sqrt{n} + 1}$

36. $a_n = \dfrac{3n^2 + n - 1}{6n^2 + n + 1}$

37. $a_n = \dfrac{2n^3 - 1}{n^3 + 2n + 1}$

38. $a_n = \dfrac{1 + (-1)^n}{3^n}$

39. $a_n = 2 - \dfrac{1}{2^n}$

40. $a_n = \dfrac{2n}{n!}$

41. $a_n = \dfrac{2^n}{3^n}$

42. $a_n = \dfrac{2^n - 1}{2^n}$

43. $a_n = \dfrac{n}{\sqrt{2n^2 + 3}}$

44. $a_n = \dfrac{2^n}{n!}$

45. AUTOMOBILE MICROPROCESSORS Of the microprocessors manufactured by a microelectronics firm for use in regulating fuel consumption in automobiles, $1\frac{1}{2}\%$ are defective. It can be shown that the probability of getting at least one defective microprocessor in a random sample of n microprocessors is $f(n) = 1 - (0.985)^n$. Consider the sequence $\{a_n\}$ defined by $a_n = f(n)$.
    a. Write down the terms a_1, a_{10}, a_{100}, and a_{1000} of the sequence $\{a_n\}$.
    b. Evaluate $\lim_{n\to\infty} a_n$ and interpret your results.

46. SAVINGS ACCOUNTS An amount of \$1000 is deposited in a bank that pays 8% interest/year compounded daily (take the number of days in a year to be 365). Let a_n denote the total amount on deposit after n days, assuming no deposits or withdrawals are made during the period in question.
    a. Find the formula for a_n.
    Hint:See Section 5.3.
    b. Compute a_1, a_{10}, a_{50}, and a_{100}.
    c. What amount is on deposit after 1 yr?

47. ACCUMULATED AMOUNT Suppose \$100 is deposited into an account earning interest at 12%/year compounded monthly. Let a_n denote the amount on deposit (called the accumulated amount or the future value) at the end of the nth mo.
    a. Show that $a_1 = 100(1.01)$, $a_2 = 100(1.01)^2$, and $a_3 = 100(1.01)^3$.
    b. Find the accumulated amount a_n.
    c. Find the 24th term of the sequence $\{a_n\}$ and interpret your result.

48. TRANSMISSION OF DISEASE In the early stages of an epidemic, the number of persons who have contracted the disease on the $(n + 1)$st day, a_{n+1}, is related to the number of persons who have the disease on the nth day, a_n, by the equation

$$a_{n+1} = (1 + aN - b)a_n$$

where N denotes the total population and a and b are positive constants that depend on the nature of the disease. Put $r = 1 + aN - b$.
    a. Write down the first n terms of the sequence $\{a_n\}$.
    b. Evaluate $\lim_{n\to\infty} a_n$ for each of the three cases $r < 1$, $r = 1$, and $r > 1$.
    c. Interpret the results of part (b).

In Exercises 49–52, determine whether the statement is true or false. If it is true, explain why it is true. If it is false, give an example to show why it is false.

49. If $\lim_{n\to\infty} a_n = L$ and $\lim_{n\to\infty} b_n = 0$, then $\lim_{n\to\infty} a_nb_n = 0$.

50. If $\{a_n\}$ and $\{b_n\}$ are sequences such that $\lim_{n\to\infty} (a_n + b_n)$ exists, then both $\lim_{n\to\infty} a_n$ and $\lim_{n\to\infty} b_n$ must exist.

51. If $\{a_n\}$ is bounded (that is $|a_n| \le M$ for some positive real numbers M and $n = 1, 2, 3, \ldots$) and $\{b_n\}$ converges, then $\lim_{n\to\infty} a_nb_n$ exists.

52. If $\lim_{n\to\infty} a_nb_n$ exists, then both $\lim_{n\to\infty} a_n$ and $\lim_{n\to\infty} b_n$ must exist.

11.2 Solutions to Self-Check Exercises

1. We compute

$$\lim_{n\to\infty} \frac{3n^2 + 2n + 1}{n^2 + 4} = \lim_{n\to\infty} \frac{3 + \frac{2}{n} + \frac{1}{n^2}}{1 + \frac{4}{n^2}} \qquad \text{Divide numerator and denominator by } n^2.$$

$$= \frac{3}{1} = 3$$

2. a. From the table

n	1	2	3	4	5	...
$f(n)$	$\frac{1}{2}$	$\frac{2}{5}$	$\frac{3}{10}$	$\frac{4}{17}$	$\frac{5}{26}$	...

we obtain the graph shown in the figure.

b. We compute

$$f'(x) = \frac{(x^2+1)(1) - x(2x)}{(x^2+1)^2} = \frac{1-x^2}{(x^2+1)^2} < 0$$

for $x > 1$, and so f is decreasing on $(1, \infty)$. We conclude that $\{a_n\}$ is decreasing.

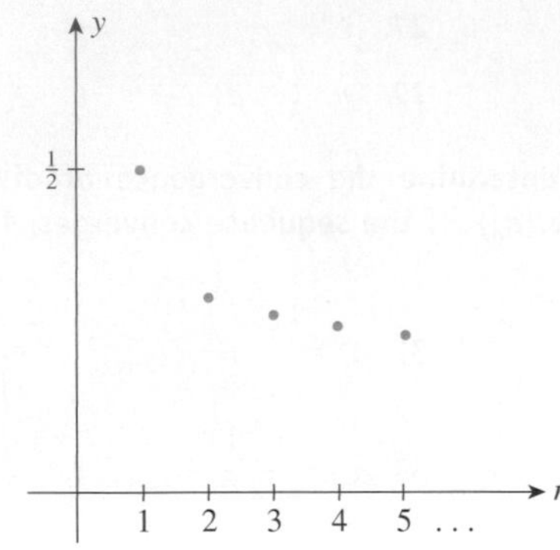

11.3 Infinite Series

The Sum of an Infinite Series

Consider again the example involving the bouncing ball. Earlier we found a sequence describing the maximum height traveled by the ball on each rebound after hitting a surface. The question that follows naturally is: How do we find the total distance traveled by the ball? To answer this question, recall that the initial height and the heights attained on each subsequent rebound are

$$1, \frac{2}{3}, \left(\frac{2}{3}\right)^2, \left(\frac{2}{3}\right)^3, \ldots$$

meters, respectively (Figure 5).

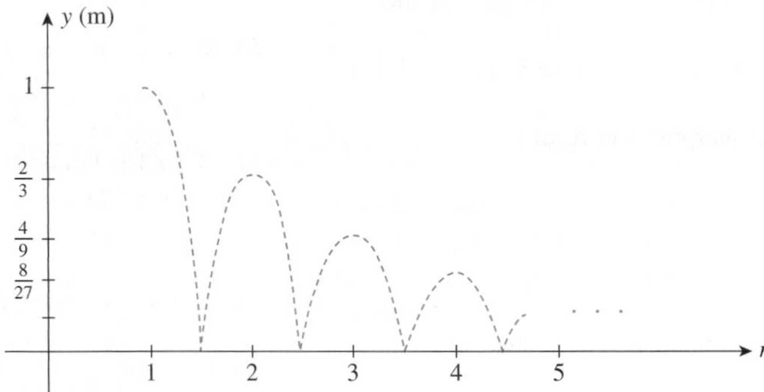

FIGURE 5

Observe that the distance traveled by the ball when it first hits the surface is 1 meter. When it hits the surface the second time, it will have traveled a total distance of

$$1 + 2\left(\frac{2}{3}\right) \quad \text{or} \quad 1 + \frac{4}{3}$$

meters. When it hits the surface the third time, it will have traveled a distance of

$$1 + 2\left(\frac{2}{3}\right) + 2\left(\frac{2}{3}\right)^2 \quad \text{or} \quad 1 + \frac{4}{3} + \frac{8}{3}$$

meters. Continuing in this fashion, we see that the total distance traveled by the ball is

$$1 + 2\left(\frac{2}{3}\right) + 2\left(\frac{2}{3}\right)^2 + 2\left(\frac{2}{3}\right)^3 + \cdots \tag{4}$$

meters. Observe that this last expression involves the sum of infinitely many terms. In general, an expression of the form

$$a_1 + a_2 + a_3 + \cdots + a_n + \cdots$$

is called an **infinite series** or, more simply, a **series**. The numbers $a_1, a_2, a_3, \ldots$ are called the **terms** of the series; a_n is called the ***n*th term**, or **general term**, of the series. The series itself is denoted by the symbol

$$\sum_{n=1}^{\infty} a_n$$

and is read "the sum of the numbers a_n for n running from 1 to infinity."

How do we define the "sum" of an infinite series, if it exists? To answer this question, we use the same technique we have employed several times before—using quantities we can compute in order to help us define new ones. For example, in defining the slope of the tangent line to the graph of a function, we take the limit of the slope of secant lines (quantities we can compute); and in defining the area under the graph of a function, we take the limit of the sum of the area of rectangles (again, quantities we can compute). Here, we define the sum of an infinite series as the limit of a sequence of *finite* sums (quantities that we can compute).

We can get an inkling of how this may be done by examining the series in (4), which gives the total distance traveled by the ball. Define the sequence $\{S_N\}$ by

$$\begin{aligned}
S_1 &= 1 \\
S_2 &= 1 + 2\left(\frac{2}{3}\right) \\
S_3 &= 1 + 2\left(\frac{2}{3}\right) + 2\left(\frac{2}{3}\right)^2 \\
&\vdots \\
S_N &= 1 + 2\left(\frac{2}{3}\right) + 2\left(\frac{2}{3}\right)^2 + \cdots + 2\left(\frac{2}{3}\right)^{N-1}
\end{aligned}$$

giving the total vertical distance traveled by the ball when it hits the surface the first time, the second time, the third time, . . . , and the Nth time, respectively. If the series in (4) has a sum S (the total distance traveled by the ball), then the terms of the sequence $\{S_N\}$ form a sequence of increasingly accurate approximations to S. This suggests that we define

$$S = \lim_{N \to \infty} S_N$$

We will complete the solution to this problem in Example 3.

Motivated by this discussion, we define the sum of an infinite series.

Sum of an Infinite Series

Given an infinite series $\sum_{n=1}^{\infty} a_n = a_1 + a_2 + a_3 + \cdots + a_n + \cdots$, the ***N*th partial sum** of the series is $S_N = \sum_{n=1}^{N} a_n = a_1 + a_2 + a_3 + \cdots + a_N$. If the sequence of partial sums $\{S_N\}$ **converges** to the number S, that is, if $\lim_{N\to\infty} S_N = S$, then the series $\sum a_n$ **converges** and has **sum** S, written

$$\sum_{n=1}^{\infty} a_n = a_1 + a_2 + a_3 + \cdots + a_n + \cdots = S$$

If $\{S_N\}$ diverges, then the series $\sum a_n$ **diverges.**

Be sure to note the difference between a sequence and a series. A sequence is a *succession* of terms, whereas a series is a *sum* of terms.

EXAMPLE 1

a. Show that the following infinite series diverges:

$$\sum_{n=1}^{\infty}(-1)^{n+1} = 1 - 1 + 1 - 1 + 1 - 1 + \cdots$$

b. Show that the following infinite series converges:

$$\sum_{n=1}^{\infty} \frac{1}{n(n+1)} = \frac{1}{1\cdot 2} + \frac{1}{2\cdot 3} + \frac{1}{3\cdot 4} + \cdots$$

Solution

a. The partial sums of the given infinite series are

$S_1 = 1, S_2 = 1 - 1 = 0, S_3 = 1 - 1 + 1 = 1, S_4 = 1 - 1 + 1 - 1 = 0, \ldots$

The sequence of partial sums $\{S_N\}$ evidently diverges, and so the given infinite series diverges.

b. Let's write

$$\frac{1}{n(n+1)} = \frac{1}{n} - \frac{1}{n+1}$$

an equality that is easily verified. The Nth partial sum of the given series is

$$\begin{aligned} S_N &= \sum_{n=1}^{N} \frac{1}{n(n+1)} = \sum_{n=1}^{N}\left(\frac{1}{n} - \frac{1}{n+1}\right) \\ &= \left(1 - \frac{1}{2}\right) + \left(\frac{1}{2} - \frac{1}{3}\right) + \left(\frac{1}{3} - \frac{1}{4}\right) + \cdots + \left(\frac{1}{N} - \frac{1}{N+1}\right) \\ &= 1 + \left(-\frac{1}{2} + \frac{1}{2}\right) + \left(-\frac{1}{3} + \frac{1}{3}\right) + \cdots + \left(-\frac{1}{N} + \frac{1}{N}\right) - \frac{1}{N+1} \\ &= 1 - \frac{1}{N+1} \end{aligned}$$

Since

$$\lim_{N\to\infty} S_N = \lim_{N\to\infty}\left(1 - \frac{1}{N+1}\right) = 1$$

we conclude that the given series converges and has a sum equal to 1; that is,

$$\sum_{n=1}^{\infty} \frac{1}{n(n+1)} = 1$$

The series in Example 1(b) is called a *telescoping series* because all the terms between the first and the last in the expression for S_N "collapse."

EXPLORING WITH TECHNOLOGY

Refer to Example 1b.

1. Verify graphically the results of Example 1b by plotting the graphs of the partial sums

$$S_1 = \frac{1}{1\cdot 2},\ S_2 = \frac{1}{1\cdot 2} + \frac{1}{2\cdot 3}, \ldots, S_6 = \frac{1}{1\cdot 2} + \frac{1}{2\cdot 3} + \cdots + \frac{1}{6\cdot 7}$$

and $S = 1$ in the viewing window $[0, 3] \times [0, 1]$.

2. Refer to the abbreviated form for S_N—namely, $S_N = \left(1 - \frac{1}{N+1}\right)$. By plotting the graphs of $y_1 = \left(1 - \frac{1}{x+1}\right)$ and $y_2 = 1$ in the viewing window $[0, 50] \times [0, 1.1]$, verify graphically that $\lim_{N\to\infty} S_N = 1$ and thus the result $\sum_{n=1}^{\infty} \frac{1}{n(n+1)} = 1$, as obtained in Example 1b.

EXPLORE & DISCUSS

1. An example of a divergent infinite series is

$$\sum_{n=0}^{\infty} 1 = 1 + 1 + 1 + \cdots$$

Show that the series is divergent by establishing the following:

a. The Nth partial sum of the infinite series is $S_N = N$.

b. $\lim_{N\to\infty} S_N = \infty$ so that $\{S_N\}$ is divergent and the desired result follows.

2. Since the *terms* of the infinite series in part 1 do not decrease, it is evident that the partial sums of the series must grow without bound. Therefore, it is not difficult to see that the infinite series cannot converge. Now consider the infinite series

$$\sum_{n=1}^{\infty} \frac{1}{n} = 1 + \frac{1}{2} + \frac{1}{3} + \cdots$$

Even though the terms of this series (called the *harmonic series*) approach zero as n goes to infinity, it can be shown that the harmonic series is divergent. Show that this result is intuitively true by establishing the following:

a. Observe that $S_2 = 1 + \frac{1}{2} > \frac{1}{2} + \frac{1}{2} = 1$

and

$$S_4 = 1 + \frac{1}{2} + \frac{1}{3} + \frac{1}{4} = S_2 + \frac{1}{3} + \frac{1}{4} > 1 + \left(\frac{1}{4} + \frac{1}{4}\right) = \frac{3}{2}$$

Using a similar argument, show that $S_8 > 4/2$ and $S_{16} > 5/2$. Conclude that in general

$$S_{2^n} > \frac{n+1}{2}$$

b. Explain why $S_1 < S_2 < S_3 < \cdots < S_n < \cdots$.

c. Use the results of parts (a) and (b) to explain why the harmonic series is divergent.

Geometric Series

In general, it is not an easy task to determine whether a given infinite series is convergent or divergent. It is an even more difficult problem to determine the sum of an infinite series that is known to be convergent. But there is an important and useful series whose sum, when it exists, is easy to find.

Geometric Series

A **geometric series** with ratio r is a series of the form

$$\sum_{n=0}^{\infty} ar^n = a + ar + ar^2 + ar^3 + \cdots + ar^n + \cdots \qquad \textbf{(5)}$$

The ratio here refers to the ratio of two consecutive terms.

Note that, for this series, we begin the summation with $n = 0$ instead of $n = 1$. To determine the conditions under which the geometric series (5) converges and find its sum, let's consider the Nth partial sum of the infinite series

$$S_N = a + ar + ar^2 + \cdots + ar^N$$

Multiplying both sides of the equation by r gives

$$rS_N = ar + ar^2 + ar^3 + \cdots + ar^{N+1}$$

Subtracting the second equation from the first yields

$$S_N - rS_N = a - ar^{N+1}$$
$$(1 - r)S_N = a(1 - r^{N+1})$$
$$S_N = \frac{a(1 - r^{N+1})}{1 - r}$$

provided $r \neq 1$. You are asked to show that if $|r| \geq 1$, then the series (5) diverges (Exercise 44). On the other hand, observe that if $|r| < 1$ (that is, if $-1 < r < 1$), then

$$\lim_{n \to \infty} r^{n+1} = 0$$

For example, if $r = \frac{1}{2}$, then

$$\lim_{n\to\infty} r^{n+1} = \lim_{n\to\infty}\left(\frac{1}{2}\right)^{n+1} = \lim_{n\to\infty}\frac{1}{2^{n+1}} = 0$$

Using this fact, together with the properties of limits stated earlier, we see that

$$\begin{aligned}\lim_{N\to\infty} S_N &= \lim_{N\to\infty}\frac{a(1 - r^{N+1})}{1 - r}\\ &= \frac{a}{1 - r}\lim_{N\to\infty}(1 - r^{N+1})\\ &= \frac{a}{1 - r}\end{aligned}$$

These results are summarized in Theorem 2.

THEOREM 2

The geometric series

$$\sum_{n=0}^{\infty} ar^n = a + ar + ar^2 + \cdots$$

converges and its sum is $\dfrac{a}{1 - r}$; that is,

$$\sum_{n=0}^{\infty} ar^n = a + ar + ar^2 + \cdots = \frac{a}{1 - r} \tag{6}$$

if $|r| < 1$. The series diverges if $|r| \geq 1$.

EXAMPLE 2 Show that the infinite series is a geometric series and find its sum if it is convergent.

a. $\displaystyle\sum_{n=0}^{\infty}\frac{1}{2^n}$ **b.** $\displaystyle\sum_{n=0}^{\infty} 3\left(\frac{5}{2}\right)^n$ **c.** $\displaystyle\sum_{n=1}^{\infty} 5\left(-\frac{3}{4}\right)^n$

Solution

a. Observe that

$$\frac{1}{2^n} = \left(\frac{1}{2}\right)^n$$

so that

$$\sum_{n=0}^{\infty}\frac{1}{2^n} = \sum_{n=0}^{\infty}\left(\frac{1}{2}\right)^n$$

is a geometric series with $a = 1$ and $r = \frac{1}{2}$. Since $|r| = \frac{1}{2} < 1$, we conclude that the series is convergent. Finally, using (6), we have

$$\sum_{n=0}^{\infty}\left(\frac{1}{2}\right)^n = 1 + \frac{1}{2} + \frac{1}{4} + \cdots = \frac{1}{1 - \frac{1}{2}} = 2$$

b. This is a geometric series with $a = 3$ and $r = \frac{5}{2}$. Since

$$|r| = \left|\frac{5}{2}\right| = \frac{5}{2} > 1$$

we deduce that the series is divergent.

c. The summation here begins with $n = 1$. However, we may rewrite the series as

$$\begin{aligned}\sum_{n=1}^{\infty} 5\left(-\frac{3}{4}\right)^n &= 5\left(-\frac{3}{4}\right) + 5\left(-\frac{3}{4}\right)^2 + 5\left(-\frac{3}{4}\right)^3 + \cdots \\ &= 5\left(-\frac{3}{4}\right)\left[1 + \left(-\frac{3}{4}\right) + \left(-\frac{3}{4}\right)^2 + \cdots\right] \\ &= \sum_{n=0}^{\infty}\left(\frac{-15}{4}\right)\left(-\frac{3}{4}\right)^n\end{aligned}$$

which is just a geometric series with $a = -\frac{15}{4}$ and $r = -\frac{3}{4}$. Since $|r| = |-\frac{3}{4}| = \frac{3}{4} < 1$, the series is convergent. Using (6), we find

$$\sum_{n=1}^{\infty} 5\left(-\frac{3}{4}\right)^n = \sum_{n=0}^{\infty}\left(-\frac{15}{4}\right)\left(-\frac{3}{4}\right)^n = \frac{-\frac{15}{4}}{1-(-\frac{3}{4})} = -\frac{\frac{15}{4}}{\frac{7}{4}} = -\frac{15}{7}$$

EXAMPLE 3 Complete the solution of the bouncing ball problem introduced at the beginning of this section. Recall that the total vertical distance traveled by the ball is given by

$$1 + 2\left(\frac{2}{3}\right) + 2\left(\frac{2}{3}\right)^2 + 2\left(\frac{2}{3}\right)^3 + \cdots$$

meters.

Solution If we denote the total vertical distance traveled by the ball by d, then

$$d = 1 + \sum_{n=1}^{\infty}\left(\frac{4}{3}\right)\left(\frac{2}{3}\right)^{n-1}$$

The expression after the first term is a geometric series with $a = \frac{4}{3}$ and $r = \frac{2}{3}$. Using Theorem 2, we obtain

$$d = 1 + \frac{\frac{4}{3}}{1-\frac{2}{3}} = 1 + 4 = 5$$

and conclude that the total distance traveled by the ball is 5 meters.

Properties of Infinite Series

The following properties of infinite series enable us to perform algebraic operations on convergent series.

Properties of Infinite Series

If $\sum_{n=1}^{\infty} a_n$ and $\sum_{n=1}^{\infty} b_n$ are convergent infinite series and c is a constant, then

1. $\sum_{n=1}^{\infty} ca_n = c\sum_{n=1}^{\infty} a_n$ **2.** $\sum_{n=1}^{\infty} (a_n \pm b_n) = \sum_{n=1}^{\infty} a_n \pm \sum_{n=1}^{\infty} b_n$

Thus, we may multiply each term of a convergent series by a constant c, which results in a convergent series whose sum is c times the sum of the original series. We may also add (subtract) the corresponding terms of two convergent series, which gives a convergent series whose sum is the sum (difference) of the sums of the original series.

EXAMPLE 4 Find the sum of the following series if it exists:

$$\sum_{n=0}^{\infty} \frac{2 \cdot 3^n - 2^n}{5^n} = 1 + \frac{4}{5} + \frac{14}{25} + \cdots$$

Solution Note that this series starts with $n = 0$. We can write

$$\sum_{n=0}^{\infty} \frac{2 \cdot 3^n - 2^n}{5^n} = \sum_{n=0}^{\infty} \left(\frac{2 \cdot 3^n}{5^n} - \frac{2^n}{5^n} \right)$$

Now observe that

$$\sum_{n=0}^{\infty} \frac{2 \cdot 3^n}{5^n} \quad \text{and} \quad \sum_{n=0}^{\infty} \frac{2^n}{5^n}$$

are convergent geometric series with ratios $r = \frac{3}{5} < 1$ and $r = \frac{2}{5} < 1$, respectively. Therefore, using Property 2 of infinite series, we have

$$\begin{aligned}
\sum_{n=0}^{\infty} \frac{2 \cdot 3^n - 2^n}{5^n} &= \sum_{n=0}^{\infty} \frac{2 \cdot 3^n}{5^n} - \sum_{n=0}^{\infty} \frac{2^n}{5^n} \\
&= 2 \sum_{n=0}^{\infty} \frac{3^n}{5^n} - \sum_{n=0}^{\infty} \frac{2^n}{5^n} \qquad \text{Using Property 1 on the first sum} \\
&= 2 \sum_{n=0}^{\infty} \left(\frac{3}{5}\right)^n - \sum_{n=0}^{\infty} \left(\frac{2}{5}\right)^n \\
&= 2\left(\frac{1}{1 - \frac{3}{5}}\right) - \left(\frac{1}{1 - \frac{2}{5}}\right) \\
&= 2\left(\frac{5}{2}\right) - \frac{5}{3} \\
&= \frac{10}{3}
\end{aligned}$$

We now consider some additional applications of geometric series.

EXAMPLE 5 Find the rational number that has the repeated decimal representation 0.222. . . .

Solution By definition, the decimal representation

$$\begin{aligned}
0.222\ldots &= \frac{2}{10} + \frac{2}{100} + \frac{2}{1000} + \cdots \\
&= \frac{2}{10}\left(1 + \frac{1}{10} + \frac{1}{100} + \cdots\right) \\
&= \sum_{n=0}^{\infty} \left(\frac{2}{10}\right)\left(\frac{1}{10}\right)^n
\end{aligned}$$

which is a geometric series with $a = \frac{2}{10}$ and $r = \frac{1}{10}$. Since $|r| = r = \frac{1}{10} < 1$, the series converges. In fact, using Formula (6), we have

$$0.222\ldots = \frac{\left(\frac{2}{10}\right)}{1 - \frac{1}{10}} = \frac{\frac{2}{10}}{\frac{9}{10}} = \frac{2}{9}$$

The following example illustrates a phenomenon in economics known as the **multiplier effect**.

APPLIED EXAMPLE 6 The Multiplier Effect Suppose the average wage earner saves 10% of her take-home pay and spends the other 90%. Estimate the impact that a proposed $20 billion tax cut will have on the economy over the long run in terms of the additional spending generated.

Solution Of the $20 billion received by the original beneficiaries of the proposed tax cut, (0.9)(20) billion dollars will be spent. Of the (0.9)(20) billion dollars reinjected into the economy, 90% of it, or (0.9)(0.9)(20) billion dollars, will find its way into the economy again. This process will go on ad infinitum, so this one-time proposed tax cut will result in additional spending over the years in the amount of

$$\begin{aligned}(0.9)(20) + (0.9)^2(20) + (0.9)^3(20) + \cdots &= (0.9)(20)[1 + 0.9 + 0.9^2 + 0.9^3 + \cdots]\\ &= 18\left[\frac{1}{1 - 0.9}\right]\\ &= 180\end{aligned}$$

or $180 billion.

A **perpetuity** is a sequence of payments made at regular time intervals and continuing on forever. The **capital value of a perpetuity** is the sum of the present values of all future payments. The following example illustrates these concepts.

APPLIED EXAMPLE 7 Establishing a Scholarship Fund The Robinson family wishes to establish a scholarship fund at a college. If a scholarship in the amount of $5000 is to be awarded on an annual basis beginning next year, find the amount of the endowment they are required to make now. Assume that this fund will earn interest at a rate of 10% per year compounded continuously.

Solution The amount of the endowment, A, is given by the sum of the present values of the amounts awarded annually in perpetuity. Now, the present value of the amount of the first award is equal to

$$5000e^{-0.1(1)}$$

(see Section 6.7). The present value of the amount of the second award is

$$5000e^{-0.1(2)}$$

and so on. Continuing, we see that the present value of the amount of the nth award is

$$5000e^{-0.1(n)}$$

Therefore, the amount of the endowment is

$$A = 5000e^{-0.1(1)} + 5000e^{-0.1(2)} + \cdots + 5000e^{-0.1(n)} + \cdots$$

To find the sum of the infinite series on the right-hand side, let

$$r = e^{-0.1}$$

Then

$$\begin{aligned} A &= 5000r^1 + 5000r^2 + \cdots + 5000r^n + \cdots \\ &= 5000r(1 + r + r^2 + \cdots + r^n + \cdots) \end{aligned}$$

The series inside the parentheses is a geometric series with $r = e^{-0.1} \approx 0.905 < 1$, so that, using (6), we find

$$A = 5000r\left(\frac{1}{1 - r}\right) = \frac{5000e^{-0.1}}{1 - e^{-0.1}} = 47{,}541.66$$

Thus, the amount of the endowment is \$47,541.66. ■

Our final example is an application of a geometric series in the field of medicine.

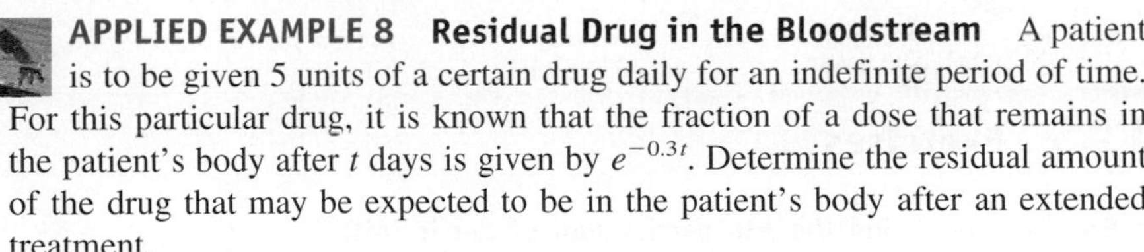

APPLIED EXAMPLE 8 Residual Drug in the Bloodstream A patient is to be given 5 units of a certain drug daily for an indefinite period of time. For this particular drug, it is known that the fraction of a dose that remains in the patient's body after t days is given by $e^{-0.3t}$. Determine the residual amount of the drug that may be expected to be in the patient's body after an extended treatment.

Solution The amount of the drug in the patient's body 1 day after the first dose is administered, and prior to administration of the second dose, is $5e^{-0.3}$ units. The amount of the drug in the patient's body 2 days later, and prior to administration of the third dose, consists of the residuals from the first two doses. Of the first dose, $5e^{-(0.3)2}$ units of the drug are left in the patient's body, and of the second dose, $5e^{-0.3}$ units of the drug are left. Thus, the amount of the drug 2 days later is given by

$$5e^{-0.3} + 5e^{-0.3(2)}$$

units. Continuing, we see that the amount of the drug left in the patient's body in the long run, and prior to administration of a fresh dose, is given by

$$R = 5e^{-0.3} + 5e^{-0.3(2)} + 5e^{-0.3(3)} + \cdots$$

To find the sum of the infinite series, we let

$$r = e^{-0.3}$$

Then

$$\begin{aligned} A &= 5r + 5r^2 + 5r^3 + \cdots = 5r(1 + r + r^2 + \cdots) \\ &= \frac{5r}{1 - r} = \frac{5e^{-0.3}}{1 - e^{-0.3}} \approx 14.29 \end{aligned}$$

Therefore, after an extended treatment, the residual amount of drug in the patient's body is approximately 14.29 units. ■

11.3 Self-Check Exercises

1. Determine whether the geometric series

$$\sum_{n=0}^{\infty} 5\left(-\frac{1}{3}\right)^n = 5 - 5\left(\frac{1}{3}\right) + 5\left(\frac{1}{9}\right) - \cdots$$

is convergent or divergent. If it is convergent, find its sum.

2. Suppose the average wage earner in a certain country saves 12% of his take-home pay and spends the other 88%. Estimate the impact that a proposed \$10 billion tax cut will have on the economy over the long run due to the additional spending generated.

Solutions to Self-Check Exercises 11.3 can be found on page 730.

11.3 Concept Questions

1. Explain the difference between:
a. a sequence and a series
b. a convergent sequence and a convergent series
c. a divergent sequence and a divergent series
d. the limit of a sequence and the sum of a series

2. Suppose $\sum_{n=1}^{\infty} a_n = 6$.
a. Evaluate $\lim_{N\to\infty} S_N$, where S_N is the Nth partial sum of $\sum_{n=1}^{\infty} a_n$.
b. Find $\sum_{n=2}^{\infty} a_n$ if it is known that $a_1 = \frac{1}{2}$.

11.3 Exercises

In Exercises 1–4, find the *N*th partial sum of the infinite series and evaluate its limit to determine whether the series converges or diverges. If the series is convergent, find its sum.

1. $\sum_{n=1}^{\infty} (-2)^n$

2. $\sum_{n=1}^{\infty} \left(\frac{1}{n+1} - \frac{1}{n+2}\right)$

3. $\sum_{n=1}^{\infty} \frac{1}{n^2+3n+2}$

Hint: $\frac{1}{n^2+3n+2} = \frac{1}{n+1} - \frac{1}{n+2}$

4. $\sum_{n=2}^{\infty} \left(\frac{1}{\ln n} - \frac{1}{\ln(n+1)}\right)$

In Exercises 5–16, determine whether the geometric series converges or diverges. If it converges, find its sum.

5. $\sum_{n=0}^{\infty} \left(\frac{1}{3}\right)^n$

6. $\sum_{n=0}^{\infty} 4\left(-\frac{2}{3}\right)^n$

7. $\sum_{n=0}^{\infty} 2(1.01)^n$

8. $\sum_{n=0}^{\infty} 3(0.9)^n$

9. $\sum_{n=0}^{\infty} \frac{(-2)^n}{3^n}$

10. $\sum_{n=0}^{\infty} \frac{3}{2^n}$

11. $\sum_{n=0}^{\infty} \frac{2^n}{3^{n+2}}$

12. $\sum_{n=0}^{\infty} \frac{3^{n+1}}{4^{n-1}}$

13. $\sum_{n=0}^{\infty} e^{-0.2n}$

14. $\sum_{n=0}^{\infty} 2e^{-0.1n}$

15. $\sum_{n=0}^{\infty} \left(-\frac{3}{\pi}\right)^n$

16. $\sum_{n=1}^{\infty} \frac{e^n}{3^{n+1}}$

In Exercises 17–26, determine whether the series converges or diverges. If it converges, find its sum.

17. $8 + 4 + \frac{1}{2} + \frac{1}{4} + \frac{1}{8} + \cdots$

18. $1 + 0.2 + 0.04 + 0.0016 + \cdots$

19. $3 - \frac{1}{3} + \frac{1}{9} - \frac{1}{27} + \cdots$

20. $5 - 1.01 + (1.01)^2 - (1.01)^3 + \cdots$

21. $\sum_{n=0}^{\infty} \frac{3+2^n}{3^n}$

22. $\sum_{n=0}^{\infty} \frac{2^n - 3^n}{4^n}$

23. $\sum_{n=0}^{\infty} \frac{3\cdot 2^n + 4^n}{3^n}$

24. $\sum_{n=0}^{\infty} \frac{2\cdot 3^n - 3\cdot 5^n}{7^n}$

25. $\sum_{n=1}^{\infty} \left[\left(\frac{e}{\pi}\right)^n + \left(\frac{\pi}{e^2}\right)^n\right]$

26. $\sum_{n=1}^{\infty} \left[\frac{1}{2^n} - \frac{1}{n(n+1)}\right]$

Hint: $\frac{1}{n(n+1)} = \frac{1}{n} - \frac{1}{n+1}$

In Exercises 27–30, express the decimal as a rational number.

27. 0.3333. . . **28.** 0.121212. . .

29. 1.213213213. . . **30.** 6.2314314314. . .

In Exercises 31–34, find the values of x for which the series converges and find the sum of the series.

Hint: First show that the series is a geometric series.

31. $\sum_{n=0}^{\infty} (-x)^n$ **32.** $\sum_{n=0}^{\infty} (x-2)^n$

33. $\sum_{n=1}^{\infty} 2^n(x-1)^n$ **34.** $\sum_{n=0}^{\infty} \frac{x^{2n}}{3^n}$

35. Effect of a Tax Cut on Spending Suppose the average wage earner saves 9% of her take-home pay and spends the other 91%. Estimate the impact that a proposed \$30 billion tax cut will have on the economy over the long run due to the additional spending generated.

36. A Bouncing Ball A ball is dropped from a height of 10 m. After hitting the ground, it rebounds to a height of 5 m and then continues to rebound at one-half of its former height thereafter. Find the total distance traveled by the ball before it comes to rest.

37. Winning a Toss Peter and Paul take turns tossing a pair of dice. The first to throw a 7 wins. If Peter starts the game, then it can be shown that his chances of winning are given by

$$p = \frac{1}{6} + \left(\frac{1}{6}\right)\left(\frac{5}{6}\right)^2 + \left(\frac{1}{6}\right)\left(\frac{5}{6}\right)^4 + \cdots$$

Find p.

38. Endowments Hal Corporation wants to establish a fund to provide the art center of a large metropolitan area with an annual grant of \$250,000 beginning next year. If the fund will earn interest at a rate of 10%/year compounded continuously, find the amount of endowment the corporation must make at this time.

39. Transmission of Disease Refer to Exercise 48, page 717. It can be shown that the total number of individuals, S_n, who have contracted the disease some time between the first and nth day is approximated by

$$S_n = a_1 + aNa_1 + aNa_2 + \cdots + aNa_{n-1}$$

Show that if $r < 1$, then the total number of persons who will have contracted the disease at some stage of the epidemic is no larger than $a_1b/(b - aN)$.

40. Capital Value of a Perpetuity Find a formula for the capital value of a perpetuity involving payments of P dollars each, paid at the end of each of m periods/year into a fund that earns interest at the nominal rate of r%/year compounded m times/year by verifying the following:

a. The present value of the nth payment is

$$P\left(1 + \frac{r}{m}\right)^{-n}$$

b. The capital value is

$$A = P\left(1 + \frac{r}{m}\right)^{-1} + P\left(1 + \frac{r}{m}\right)^{-2} + P\left(1 + \frac{r}{m}\right)^{-3} + \cdots$$

c. Using the fact that the series in part (b) is a geometric series, its sum is

$$A = \frac{mP}{r}$$

41. Capital Value of a Perpetuity Find a formula for the capital value of a perpetuity involving payments of P dollars paid at the end of each investment period into a fund that earns interest at the rate of r%/year compounded continuously.
Hint: Study Exercise 40.

42. Residual Drug in the Bloodstream Ten units of a certain drug are administered to a patient on a daily basis. The fraction of this drug that remains in the patient's bloodstream after t days is given by

$$f(t) = e^{(-1/4)t}$$

Determine the residual amount of the drug in the patient's bloodstream after a period of extended treatment with the drug.

43. Residual Drug in the Bloodstream Suppose a dose of C units of a certain drug is administered to a patient and the fraction of the dose remaining in the patient's bloodstream t hr after the dose is administered is given by Ce^{-kt}, where k is a positive constant.

a. Show that the residual concentration of the drug in the bloodstream after extended treatment when a dose of C units is administered at intervals of t hr is given by

$$R = \frac{Ce^{-kt}}{1 - e^{-kt}}$$

b. If the highest concentration of this particular drug that is considered safe is S units, find the minimal time that must exist between doses.
Hint: $C + R \leq S$

44. Let

$$\sum_{n=0}^{\infty} ar^n = a + ar + ar^2 + ar^3 + \cdots + ar^n + \cdots$$

be a geometric series with common ratio r. Show that if $|r| \geq 1$, then the series diverges.

In Exercises 45–48, determine whether the statement is true or false. If it is true, explain why it is true. If it is false, give an example to show why it is false.

45. If $\sum_{n=0}^{\infty} (a_n + b_n)$ converges, then both $\sum_{n=0}^{\infty} a_n$ and $\sum_{n=0}^{\infty} b_n$ must converge.

46. If $\sum_{n=0}^{\infty} a_n$ converges and $\sum_{n=0}^{\infty} b_n$ converges, then $\sum_{n=0}^{\infty} (ca_n + db_n)$ also converges, where c and d are constants.

47. If $|r| < 1$, then $\sum_{n=0}^{\infty} |r^n| = \dfrac{1}{1 - |r|}$.

48. If $|r| > 1$, then $\sum_{n=1}^{\infty} \dfrac{1}{r^n} = \dfrac{1}{r - 1}$.

11.3 Solutions to Self-Check Exercises

1. This is a geometric series with $r = -\frac{1}{3}$ and $a = 5$. Since $0 < \left|-\frac{1}{3}\right| < 1$, we see that the series is convergent. Its sum is

$$\frac{5}{1 - (-\frac{1}{3})} = \frac{5}{\frac{4}{3}} = \frac{15}{4} = 3\frac{3}{4}$$

2. Of the \$10 billion received by the original beneficiaries of the proposed cut, (0.88)(10) billion dollars will be spent. Of the (0.88)(10) billion dollars reinjected into the economy, 88% of it, or (0.88)(0.88)(10) billion dollars, will find its way into the economy again. This process will go on ad infinitum, so this one-time proposed tax cut will result in additional spending over the years in the amount of

$$\begin{aligned}
&(0.88)(10) + (0.88)^2(10) + (0.88)^3(10) + \cdots \\
&\quad = (0.88)(10)[1 + 0.88 + (0.88)^2 + \cdots] \\
&\quad = 8.8\left[\frac{1}{1 - 0.88}\right] \\
&\quad = 73.3
\end{aligned}$$

or \$73.3 billion.

11.4 Series with Positive Terms

The convergence or divergence of a telescoping series or a geometric series is relatively easy to determine because we can find a simple formula for the Nth partial sum S_N of these series. In fact, as we have seen earlier in this chapter, the knowledge of this formula helps us find the actual sum of a convergent series by simply evaluating $\lim_{N\to\infty} S_N$. More often than not, obtaining a simple formula for the Nth partial sum of an infinite series is very difficult or impossible, and we are forced to look for alternative ways to investigate the convergence or divergence of the series.

In what follows, we look at several tests for determining the convergence or divergence of an infinite series by examining the nth term a_n of the series. These tests will confirm the convergence of a series without yielding a value for its sum. From the practical point of view, however, this is all that is required. Once it has been ascertained that a series is convergent, we can approximate its sum to any degree of accuracy desired by adding up the terms of its Nth partial sum S_N, provided that N is chosen large enough.

Our first test tells us how to identify a divergent series.

The Test for Divergence

The following theorem tells us that the terms of a convergent series must ultimately approach zero.

THEOREM 3

If $\sum_{n=1}^{\infty} a_n$ converges, then $\lim_{n\to\infty} a_n = 0$.

To prove this result, let

$$S_n = a_1 + a_2 + \cdots + a_{n-1} + a_n = S_{n-1} + a_n$$

and so

$$a_n = S_n - S_{n-1}$$

Since $\sum_{n=1}^{\infty} a_n$ is convergent, the sequence $\{S_n\}$ is convergent. Let $\lim_{n\to\infty} S_n = S$. Then

$$\lim_{n\to\infty} a_n = \lim_{n\to\infty} (S_n - S_{n-1}) = \lim_{n\to\infty} S_n - \lim_{n\to\infty} S_{n-1} = S - S = 0$$

An important consequence of Theorem 3 is the following useful test for *divergence*.

THEOREM 4

The Test for Divergence

If $\lim_{n\to\infty} a_n$ does not exist or $\lim_{n\to\infty} a_n \neq 0$, then $\sum_{n=1}^{\infty} a_n$ diverges.

Observe that the test for divergence does *not* say

If $\lim_{n\to\infty} a_n = 0$, then $\sum_{n=1}^{\infty} a_n$ must converge.

In other words, the converse of Theorem 4 is not true in general. For example, $\lim_{n\to\infty} \frac{1}{n} = 0$, and yet, as we will see later, the **harmonic series** $\sum_{n=1}^{\infty} \frac{1}{n}$ is divergent. (See Example 3.) In short, the test for divergence rules out convergence for a series whose nth term does not approach zero but yields no information if the nth term of a series does approach zero—that is, the series may or may not converge.

EXAMPLE 1 Show that the following series are divergent:

a. $\sum_{n=1}^{\infty} (-1)^{n-1}$ **b.** $\sum_{n=1}^{\infty} \frac{2n^2 + 1}{3n^2 - 1}$

Solution

a. Here $a_n = (-1)^{n-1}$, and since

$$\lim_{n\to\infty} a_n = \lim_{n\to\infty} (-1)^{n-1}$$

does not exist, we conclude by the test for divergence that the series diverges.

b. Here

$$\lim_{n\to\infty} a_n = \lim_{n\to\infty} \frac{2n^2+1}{3n^2-1} = \lim_{n\to\infty} \frac{2+\frac{1}{n^2}}{3-\frac{1}{n^2}} = \frac{2}{3} \neq 0$$

and so, by the test for divergence, the series diverges. ■

We now look at several tests that tell us if a series is convergent. These tests apply only to series with positive terms.

The Integral Test

The integral test ties the convergence or divergence of an infinite series $\sum_{n=1}^{\infty} a_n$ to the convergence or divergence of the improper integral $\int_1^\infty f(x)\,dx$ where $f(n) = a_n$.

THEOREM 5

The Integral Test

Suppose f is a continuous, positive, and decreasing function on $[1, \infty)$. If $f(n) = a_n$ for $n \geq 1$, then

$$\sum_{n=1}^{\infty} a_n \quad \text{and} \quad \int_1^\infty f(x)\,dx$$

either both converge or both diverge.

Let's give an intuitive justification for this theorem. If you examine Figure 6a, you will see that the height of the first rectangle is $a_2 = f(2)$.

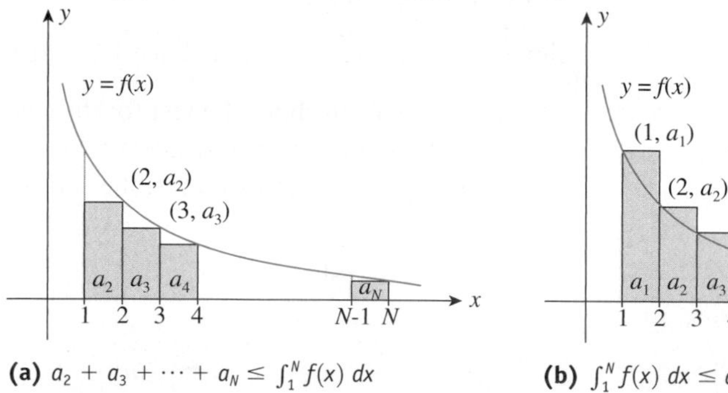

FIGURE 6

(a) $a_2 + a_3 + \cdots + a_N \leq \int_1^N f(x)\,dx$ **(b)** $\int_1^N f(x)\,dx \leq a_1 + a_2 + \cdots + a_{N-1}$

Since this rectangle has width one, the area of the rectangle is also $a_2 = f(2)$. Similarly, the area of the second is a_3, and so on. Comparing the sum of the areas of the first $(N-1)$ inscribed rectangles with the area under the graph of f over the interval $[1, N]$, we see that

$$a_2 + a_3 + \cdots + a_N \leq \int_1^N f(x)\,dx$$

which implies that

$$S_N = a_1 + a_2 + a_3 + \cdots + a_N \leq a_1 + \int_1^N f(x)\,dx$$

If $\int_1^\infty f(x)\,dx$ is convergent and has value L, then

$$S_N \leq a_1 + \int_1^N f(x)\,dx \leq a_1 + L$$

This shows that $\{S_N\}$ is bounded above. Also,

$$S_{N+1} = S_N + a_{N+1} \geq S_N \qquad \text{Because } a_{N+1} = f(N+1) \geq 0$$

shows that $\{S_N\}$ is increasing as well.

Now, intuitively, such a sequence must converge to a number no greater than its upper bound. (Although this result can be demonstrated with mathematical rigor, we will not do so here.) In other words $\sum_{n=1}^{\infty} a_n$ is convergent.

Next, by examining Figure 6b, you will see that

$$\int_1^N f(x)\,dx \leq a_1 + a_2 + \cdots + a_{N-1} = S_{N-1}$$

So, if

$$\int_1^\infty f(x)\,dx$$

diverges to infinity (because $f(x) \geq 0$), then $\lim_{N\to\infty} S_N = \infty$ and so $\sum_{n=1}^{\infty} a_n$ is divergent.

Notes

1. The integral test simply tells us whether a series converges or diverges. If it indicates that a series converges, we may not conclude that the (finite) value of the improper integral used in conjunction with the test is the *sum* of the convergent series.
2. Since the convergence of an infinite series is not affected by the omission or addition of a finite number of terms to the series, we sometimes study the series

$$\sum_{n=N}^{\infty} a_n = a_N + a_{N+1} + \cdots$$

rather than the series $\sum_{n=1}^{\infty} a_n$. In this case, the series is compared to the improper integral

$$\int_N^\infty f(x)\,dx$$

as we will see in Example 4. ■

EXAMPLE 2 Use the integral test to determine whether

$$\sum_{n=1}^{\infty} \frac{1}{n^2}$$

converges or diverges.

Solution Here

$$a_n = f(n) = \frac{1}{n^2}$$

and so we consider the function $f(x) = 1/x^2$. Since f is continuous, positive, and decreasing on $[1, \infty]$, we may use the integral test. Now

$$\int_1^\infty \frac{1}{x^2}\,dx = \lim_{b\to\infty}\int_1^b x^{-2}\,dx = \lim_{b\to\infty}\left[-\frac{1}{x}\Big|_1^b\right]$$
$$= \lim_{b\to\infty}\left(-\frac{1}{b}+1\right) = 1$$

Since

$$\int_1^\infty \frac{1}{x^2}\,dx$$

converges, we conclude that

$$\sum_{n=1}^\infty \frac{1}{n^2}$$

converges as well.

EXAMPLE 3 Use the integral test to determine whether the harmonic series $\sum_{n=1}^\infty \frac{1}{n}$ converges or diverges.

Solution Here $a_n = f(n) = 1/n$ and so we consider the function $f(x) = 1/x$. Since f is continuous, positive, and decreasing on $[1, \infty)$, we may use the integral test. But, as you may verify,

$$\int_1^\infty \frac{1}{x}\,dx = \infty$$

We conclude that $\sum_{n=1}^\infty \frac{1}{n}$ diverges.

EXAMPLE 4 Use the integral test to determine whether $\sum_{n=2}^\infty \frac{\ln n}{n}$ converges or diverges.

Solution Here $a_n = (\ln n)/n$ and so we consider the function $f(x) = (\ln x)/x$. Observe that f is continuous and positive on $[2, \infty)$. Next, we compute

$$f'(x) = \frac{x\left(\frac{1}{x}\right) - \ln x}{x^2} = \frac{1-\ln x}{x^2}$$

Note that $f'(x) < 0$ if $\ln x > 1$; that is, if $x > e$. This shows that f is decreasing on $[3, \infty)$. (See Note on page 733.) Therefore, we may use the integral test. Now,

$$\int_3^\infty \frac{\ln x}{x}\,dx = \lim_{b\to\infty}\int_3^b \frac{\ln x}{x}\,dx = \lim_{b\to\infty}\left[\frac{1}{2}(\ln x)^2\Big|_3^b\right]$$

$$= \lim_{b\to\infty}\frac{1}{2}[(\ln b)^2 - (\ln 3)^2] = \infty$$

and we conclude that $\sum_{n=2}^{\infty}\frac{\ln n}{n}$ diverges. ■

The p-Series

The following series will play an important role in our work later on.

p-Series

A ***p*-series** is a series of the form

$$\sum_{n=1}^{\infty}\frac{1}{n^p} = 1 + \frac{1}{2^p} + \frac{1}{3^p} + \cdots + \frac{1}{n^p} + \cdots$$

where p is a constant.

Observe that for $p = 1$, the p-series is just the harmonic series $\sum_{n=1}^{\infty}\frac{1}{n}$. The conditions for the convergence or divergence of the p-series can be found by applying the integral test to the series. We have the following result.

THEOREM 6

Convergence of p-Series

The p-series $\sum_{n=1}^{\infty}\frac{1}{n^p}$ converges if $p > 1$ and diverges if $p \le 1$.

Proof If $p < 0$, then

$$\lim_{n\to\infty}\frac{1}{n^p} = \infty$$

If $p = 0$, then

$$\lim_{n\to\infty}\frac{1}{n^p} = 1$$

In either case,

$$\lim_{n\to\infty}\frac{1}{n^p} \ne 0$$

and so the p-series diverges by the test for divergence. If $p > 0$, then the function $f(x) = 1/x^p$ is continuous, positive, and decreasing on $[1, \infty)$. It can be shown that

$$\int_1^\infty \frac{1}{x^p}\,dx \text{ converges if } p > 1 \text{ and diverges if } p \le 1$$

(see Exercise 53). Using this result and the integral test, we conclude that $\sum_{n=1}^{\infty} \frac{1}{n^p}$ converges if $p > 1$ and diverges if $0 < p \le 1$. Therefore, $\sum_{n=1}^{\infty} \frac{1}{n^p}$ converges if $p > 1$ and diverges if $p \le 1$.

EXAMPLE 5 Determine whether each series converges or diverges.

a. $\sum_{n=1}^{\infty} \frac{1}{n^2}$ **b.** $\sum_{n=1}^{\infty} \frac{1}{\sqrt{n}}$ **c.** $\sum_{n=1}^{\infty} n^{-1.001}$

Solution

a. This is a p-series with $p = 2 > 1$, and so by Theorem 6, the series converges.

b. Rewriting the series in the form $\sum_{n=1}^{\infty} \frac{1}{n^{1/2}}$, we see that the series is a p-series with $p = \frac{1}{2} < 1$, and so by Theorem 6, it diverges.

c. We rewrite the series in the form $\sum_{n=1}^{\infty} \frac{1}{n^{1.001}}$, which we recognize to be a p-series with $p = 1.001 > 1$ and conclude accordingly that the series converges. ■

The Comparison Test

The convergence or divergence of a given series $\Sigma\, a_n$ can be determined by comparing its terms with the terms of a *test series* that is known to be convergent or divergent. This is the basis for the comparison test for series that follows. In the rest of this section, we assume that all series under consideration have positive terms.

Suppose the terms of a series $\Sigma\, a_n$ are smaller than the corresponding terms of a series $\Sigma\, b_n$. This situation is illustrated in Figure 7, where the respective terms are represented by rectangles, each of width one and appropriate height.

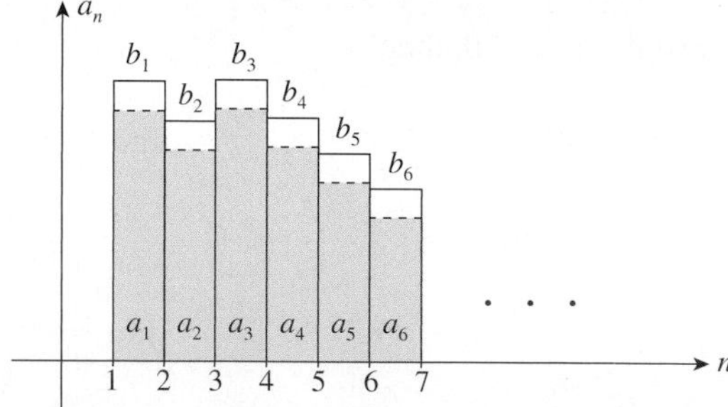

FIGURE 7
Each rectangle representing a_n is contained in the rectangle representing b_n.

If $\Sigma\, b_n$ is convergent, the total area of the rectangles representing this series is finite. Since each rectangle representing the terms of the series $\Sigma\, a_n$ is contained in a corresponding rectangle representing the terms of $\Sigma\, b_n$, the total area of the rectangles representing $\Sigma\, a_n$ must also be finite: that is, the series $\Sigma\, a_n$ must be convergent. A similar argument would seem to suggest that if all the terms of a series $\Sigma\, a_n$ are larger than the corresponding terms of a series $\Sigma\, b_n$ that is known to be divergent, then $\Sigma\, a_n$ must itself be divergent. These observations lead to the following theorem.

THEOREM 7

The Comparison Test

Suppose $\Sigma\, a_n$ and $\Sigma\, b_n$ are series with positive terms.

a. If $\Sigma\, b_n$ is convergent and $a_n \le b_n$ for all n, then $\Sigma\, a_n$ is also convergent.
b. If $\Sigma\, b_n$ is divergent and $a_n \ge b_n$ for all n, then $\Sigma\, a_n$ is also divergent.

Here is an intuitive justification for Theorem 7. Let

$$S_N = \sum_{n=1}^{N} a_n \qquad \text{and} \qquad T_N = \sum_{n=1}^{N} b_n$$

be the Nth terms of the sequence of partial sums of $\Sigma\, a_n$ and $\Sigma\, b_n$, respectively. Since both series have positive terms, $\{S_N\}$ and $\{T_N\}$ are increasing.

1. If $\sum_{n=1}^{\infty} b_n$ is convergent, then there exists a number L such that $\lim_{N\to\infty} T_N = L$ and $T_N \le L$ for all N. Since $a_n \le b_n$ for all n, we have $S_N \le T_N$, and this implies that $S_N \le L$ for all N. We have shown that $\{S_N\}$ is increasing and bounded above, and so, as before, we can argue intuitively that S_N and therefore $\Sigma\, a_n$ converges.
2. If $\Sigma\, b_n$ is divergent, then $\lim_{N\to\infty} T_N = \infty$ since $\{T_N\}$ is increasing. But $a_n \ge b_n$ for all n, and this implies that $S_N \ge T_N$, which in turn implies that $\lim_{N\to\infty} S_N = \infty$. Therefore, $\Sigma\, a_n$ diverges.

Note Since the convergence or divergence of a series is not affected by the omission of a finite number of terms of the series, the condition $a_n \le b_n$ $(a_n \ge b_n)$ for all n can be replaced by the condition that these inequalities hold for all $n \ge k$ for some integer k. ■

To use Theorem 7, we need a catalog of test series whose convergence and divergence are known. In what follows, we will use the geometric series and the p-series as test series.

EXAMPLE 6 Determine whether the series

$$\sum_{n=1}^{\infty} \frac{1}{n^2 + 2}$$

converges or diverges.

Solution Let

$$a_n = \frac{1}{n^2 + 2}$$

Observe that when n is large, $n^2 + 2$ behaves like n^2 and so a_n behaves like

$$b_n = \frac{1}{n^2}$$

This observation suggests that we compare $\Sigma\, a_n$ with the test series $\Sigma\, b_n$, which is a convergent p-series with $p = 2$. Now

$$0 < \frac{1}{n^2 + 2} < \frac{1}{n^2} \qquad (n \geq 1)$$

and the given series is indeed "smaller" than the test series

$$\sum_{n=1}^{\infty} \frac{1}{n^2}$$

Since the test series converges, we conclude by the comparison test that

$$\sum_{n=1}^{\infty} \frac{1}{n^2 + 2}$$

also converges. ■

EXAMPLE 7 Determine whether the series

$$\sum_{n=1}^{\infty} \frac{1}{3 + 2^n}$$

converges or diverges.

Solution Let

$$a_n = \frac{1}{3 + 2^n}$$

Observe that if n is large, $3 + 2^n$ behaves like 2^n, and so a_n behaves like $b_n = 1/2^n$. This observation suggests that we compare $\Sigma\, a_n$ with $\Sigma\, b_n$. Now the series $\Sigma\, 1/2^n = \Sigma\, (1/2)^n$ is a geometric series with $r = 1/2 < 1$ and so it is convergent. Since

$$a_n = \frac{1}{3 + 2^n} < \frac{1}{2^n} = b_n \qquad (n \geq 1)$$

the comparison test tells us that the given series is convergent. ■

EXAMPLE 8 Determine whether the series

$$\sum_{n=2}^{\infty} \frac{1}{\sqrt{n} - 1}$$

is convergent or divergent.

Solution Let

$$a_n = \frac{1}{\sqrt{n} - 1}$$

If n is large, $\sqrt{n} - 1$ behaves like $\sqrt{n}$, and so a_n behaves like

$$b_n = \frac{1}{\sqrt{n}}$$

Now the series

$$\sum_{n=2}^{\infty} b_n = \sum_{n=2}^{\infty} \frac{1}{\sqrt{n}} = \sum_{n=2}^{\infty} \frac{1}{n^{1/2}}$$

is a p-series with $p = 1/2 < 1$ and so it is divergent. Since

$$a_n = \frac{1}{\sqrt{n} - 1} > \frac{1}{\sqrt{n}} = b_n \qquad (\text{for } n \geq 2)$$

See the Note following Theorem 7.

the comparison test implies that the given series is divergent. ■

11.4 Self-Check Exercises

1. Show that the series $\sum_{n=1}^{\infty} \frac{2n^2 + 1}{3n^2 - 1}$ is divergent.

2. Use the integral test to determine whether the series $\sum_{n=1}^{\infty} \frac{\ln n}{n}$ converges or diverges.

Solutions to Self-Check Exercises 11.4 can be found on page 741.

11.4 Concept Questions

1. Consider the series $\sum_{n=1}^{\infty} \frac{1}{n} = \frac{1}{1} + \frac{1}{2} + \frac{1}{3} + \frac{1}{4} + \frac{1}{5} + \cdots$. Let $f(x) = \frac{1}{x}$.
   a. Sketch a figure similar to Figure 6b for this series and function and compute $a_1 = f(1)$, $a_2 = f(2)$, $a_3 = f(3), \ldots,$ $a_n = f(n)$.
   b. Explain why $S_{N-1} = \frac{1}{1} + \frac{1}{2} + \frac{1}{3} + \cdots + \frac{1}{N-1} \geq \int_1^N \frac{1}{x}\,dx$.
   c. Show that $\int_1^{\infty} \frac{1}{x}\,dx$ is divergent and conclude that $\sum_{n=1}^{\infty} \frac{1}{n}$ diverges.
   Note: This is the harmonic series that was shown to be divergent in Example 3.

2. Let $\Sigma\, a_n$ and $\Sigma\, b_n$ be series with positive terms.
   a. If $\Sigma\, b_n$ is convergent and $a_n \geq b_n$ for all n, what can you say about the convergence or divergence of $\Sigma\, a_n$? Give examples.
   b. If $\Sigma\, b_n$ is divergent and $a_n \leq b_n$ for all n, what can you say about the convergence or divergence of $\Sigma\, a_n$? Give examples.

3. Let $\Sigma\, a_n$, $\Sigma\, b_n$, and $\Sigma\, c_n$ be series with positive terms.
   a. If $\Sigma\, a_n$ is convergent and $b_n + c_n \leq a_n$ for all n, what can you say about the convergence or divergence of $\Sigma\, b_n$ and $\Sigma\, c_n$?
   b. If $\Sigma\, a_n$ is divergent and $b_n + c_n \geq a_n$ for all n, what can you say about the convergence or divergence of $\Sigma\, b_n$ and $\Sigma\, c_n$?

11.4 Exercises

In Exercises 1–10, show that the series is divergent.

1. $\frac{1}{2} + \frac{2}{3} + \frac{3}{4} + \cdots$

2. $1 - \frac{3}{2} + \frac{9}{4} - \frac{27}{8} + \cdots$

3. $\sum_{n=1}^{\infty} \frac{2n}{3n + 1}$

4. $\sum_{n=1}^{\infty} \frac{n^2}{2n^2 + 1}$

5. $\sum_{n=1}^{\infty} 2(1.5)^n$

6. $\sum_{n=0}^{\infty} \frac{(-1)^n 3^n}{2^{n-1}}$

7. $\sum_{n=1}^{\infty} \frac{1}{2 + 3^{-n}}$

8. $\sum_{n=1}^{\infty} \frac{n}{\sqrt{2n^2 + 1}}$

9. $\sum_{n=0}^{\infty} \left(-\frac{\pi}{3}\right)^n$

10. $\sum_{n=1}^{\infty} \frac{3^{n+1}}{e^n}$

In Exercises 11–20, use the integral test to determine whether the series is convergent or divergent.

11. $\sum_{n=1}^{\infty} \frac{1}{n + 1}$

12. $\sum_{n=1}^{\infty} \frac{3}{2n - 1}$

13. $\sum_{n=1}^{\infty} \frac{n}{2n^2 + 1}$

14. $\sum_{n=1}^{\infty} ne^{-n^2}$ **15.** $\sum_{n=1}^{\infty} ne^{-n}$ **16.** $\sum_{n=1}^{\infty} \frac{1}{n(2n-1)}$

17. $\sum_{n=1}^{\infty} \frac{n}{(n^2+1)^{3/2}}$ **18.** $\sum_{n=2}^{\infty} \frac{1}{n\sqrt{\ln n}}$ **19.** $\sum_{n=9}^{\infty} \frac{1}{n \ln^3 n}$

20. $\sum_{n=0}^{\infty} \frac{1}{e^n+1}$

In Exercises 21–26, determine whether the *p*-series is convergent or divergent.

21. $\sum_{n=1}^{\infty} \frac{1}{n^3}$ **22.** $\sum_{n=1}^{\infty} \frac{1}{n^{2/3}}$ **23.** $\sum_{n=1}^{\infty} \frac{1}{n^{1.01}}$

24. $\sum_{n=1}^{\infty} \frac{1}{n^e}$ **25.** $\sum_{n=1}^{\infty} n^{-\pi}$ **26.** $\sum_{n=1}^{\infty} n^{-0.98}$

In Exercises 27–36, use the comparison test to determine whether the series is convergent or divergent.

27. $\sum_{n=1}^{\infty} \frac{1}{2n^2+1}$ **28.** $\sum_{n=1}^{\infty} \frac{1}{n^2+2n}$

29. $\sum_{n=3}^{\infty} \frac{1}{n-2}$ **30.** $\sum_{n=2}^{\infty} \frac{1}{n^{2/3}-1}$

31. $\sum_{n=2}^{\infty} \frac{1}{\sqrt{n^2-1}}$ **32.** $\sum_{n=0}^{\infty} \frac{1}{\sqrt{n^3+1}}$

33. $\sum_{n=0}^{\infty} \frac{2^n}{3^n+1}$ **34.** $\sum_{n=3}^{\infty} \frac{3^n}{2^n-4}$

35. $\sum_{n=2}^{\infty} \frac{\ln n}{n}$ **36.** $\sum_{n=1}^{\infty} \frac{1}{n^n}$

In Exercises 37–48, determine whether the series is convergent or divergent.

37. $\sum_{n=0}^{\infty} \frac{1}{\sqrt{n+1}}$ **38.** $\sum_{n=1}^{\infty} \frac{n}{\sqrt{2n^2+1}}$

39. $\sum_{n=2}^{\infty} \frac{1}{n\sqrt{n^2+1}}$ **40.** $\sum_{n=2}^{\infty} \frac{\sqrt{n^2+1}}{n^2}$

41. $\sum_{n=1}^{\infty} \left(\frac{1}{n\sqrt{n}} + \frac{2}{n^2}\right)$ **42.** $\sum_{n=1}^{\infty} \left[\left(\frac{2}{3}\right)^n + \frac{1}{n^{3/2}}\right]$

43. $\sum_{n=2}^{\infty} \frac{\ln n}{\sqrt{n}}$ **44.** $\sum_{n=2}^{\infty} \frac{\ln n}{n^{2.1}}$

45. $\sum_{n=2}^{\infty} \frac{1}{n(\ln n)^2}$ **46.** $\sum_{n=1}^{\infty} \frac{e^{1/n}}{n^2}$

47. $\sum_{n=1}^{\infty} \frac{1}{\sqrt{n+4}}$

Hint: $\frac{1}{\sqrt{n+4}} > \frac{1}{3\sqrt{n}}$ if $n > 4$

48. $\sum_{n=1}^{\infty} \frac{1}{4n^2-1}$

Hint: $\frac{1}{4n^2-1} < \frac{1}{2n^2}$ if $n \geq 1$

In Exercises 49 and 50, find the value of *p* for which the series is convergent.

49. $\sum_{n=2}^{\infty} \frac{1}{n(\ln n)^p}$ **50.** $\sum_{n=1}^{\infty} \frac{\ln n}{n^p}$

51. Find the value(s) of a for which the series

$$\sum_{n=1}^{\infty} \left(\frac{a}{n+1} - \frac{1}{n+2}\right)$$

converges. Justify your answer.

52. Consider the series

$$\sum_{n=0}^{\infty} e^{-n}$$

a. Evaluate

$$\int_0^{\infty} e^{-x}\,dx$$

and deduce from the integral test that the given series is convergent.

b. Show that the given series is a geometric series and find its sum.

c. Conclude that although the convergence of

$$\int_0^{\infty} e^{-x}\,dx$$

implies convergence of the infinite series, its value does not give the sum of the infinite series.

53. Show that $\int_1^{\infty} \frac{1}{x^p}\,dx$ converges if $p > 1$ and diverges if $p \leq 1$.

54. Suppose

$$\sum_{n=1}^{\infty} a_n$$

is a convergent series with positive terms. Let $f(n) = a_n$, where f is a continuous and decreasing function for $x \geq N$, where N is some positive integer. Show that the error incurred in approximating the sum of the given series by the Nth partial sum of the series

$$S_N = \sum_{n=1}^{N} a_n$$

is less than $\int_N^{\infty} f(x)\,dx$.

In Exercises 55–60, determine whether the statement is true or false. If it is true, explain why it is true. If it is false, give an example to show why it is false.

55. True or false. The series

$$\sum_{n=1}^{\infty} \frac{x}{n}$$

converges only for $x = 0$.

56. If $\lim_{n\to\infty} a_n = 0$, then $\sum_{n=0}^{\infty} a_n$ converges.

57. If $\sum_{n=0}^{\infty} a_n$ diverges, then $\lim_{n\to\infty} a_n \neq 0$.

58. $\int_1^{\infty} \frac{2}{(x^2+1)^{1.1}}\,dx$ converges.

59. Suppose $\Sigma\, a_n$ and $\Sigma\, b_n$ are series with positive terms. If $\Sigma\, a_n$ is convergent and $b_n \geq a_n$ for all n, then $\Sigma\, b_n$ is divergent.

60. Suppose $\Sigma\, a_n$ and $\Sigma\, b_n$ are series with positive terms. If $\Sigma\, b_n$ is divergent and $a_n \leq b_n$, for all n, then $\Sigma\, a_n$ may or may not converge.

11.4 Solutions to Self-Check Exercises

1. Here

$$\begin{aligned}\lim_{n\to\infty} a_n &= \lim_{n\to\infty} \frac{2n^2+1}{3n^2-1}\\ &= \lim_{n\to\infty} \frac{2+\frac{1}{n^2}}{3-\frac{1}{n^2}}\\ &= \frac{2}{3} \neq 0\end{aligned}$$

and so by the divergence test, the series diverges.

2. Here $a_n = (\ln n)/n$ and so we consider the function $f(x) = (\ln x)/x$. Observe that f is continuous and positive on $[1, \infty)$. Next, we compute

$$\begin{aligned}f'(x) &= \frac{x\left(\frac{1}{x}\right) - \ln x}{x^2}\\ &= \frac{1-\ln x}{x^2}\end{aligned}$$

Note that $f'(x) < 0$ if $\ln x > 1$; that is, if $x > e$. This shows that f is decreasing on $[3, \infty)$. Therefore, we may use the integral test.

$$\begin{aligned}\int_3^{\infty} \frac{\ln x}{x}\,dx &= \lim_{b\to\infty} \int_3^b \frac{\ln x}{x}\,dx\\ &= \lim_{b\to\infty} \left[\frac{1}{2}(\ln x)^2\Big|_3^b\right]\\ &= \lim_{b\to\infty} \frac{1}{2}[(\ln b)^2 - (\ln 3)^2]\\ &= \infty\end{aligned}$$

and we conclude that $\sum_{n=1}^{\infty} (\ln n)/n$ diverges.

11.5 Power Series and Taylor Series

Power Series and Intervals of Convergence

One of our goals in this chapter is to see what happens when the number of terms of a Taylor polynomial is allowed to increase without bound. In other words, we wish to study the expression

$$f(a) + f'(a)(x-a) + \frac{f''(a)}{2!}(x-a)^2 + \cdots + \frac{f^{(n)}(a)}{n!}(x-a)^n + \cdots \quad (7)$$

called the **Taylor series of $f(x)$ at $x = a$**. If the series (7) is truncated after $(n + 1)$ terms, the result is a Taylor polynomial of degree n of $f(x)$ at $x = a$ (see Section 11.1). Note that when $a = 0$, (7) reduces to

$$f(0) + f'(0)x + \frac{f''(0)}{2!}x^2 + \cdots + \frac{f^{(n)}(0)}{n!}x^n + \cdots \tag{8}$$

which is referred to as the **Maclaurin series of $f(x)$**. Thus, the Maclaurin series is a special case of the Taylor series when $a = 0$.

Observe that the Taylor series (7) has the form

$$\sum_{n=0}^{\infty} a_n(x-a)^n = a_0 + a_1(x-a) + a_2(x-a)^2 + \cdots + a_n(x-a)^n + \cdots \tag{9}$$

In fact, comparing the Taylor series (7) with the infinite series (9), we see that

$$a_0 = f(a), a_1 = f'(a), a_2 = \frac{f''(a)}{2!}, \ldots, a_n = \frac{f^{(n)}(a)}{n!}, \ldots$$

The infinite series (9) is called a **power series centered at $x = a$**. Thus, the Taylor series is just a power series with coefficients that involve the values of some function f and its derivatives at $x = a$.

Let's examine some important properties of the power series given in (9). Observe that if x is assigned a value, then this power series becomes an infinite series with constant terms. Accordingly, we can determine, at least theoretically, whether this infinite series converges or diverges. Now the totality of all values of x for which this power series *converges* comprises a set of points in the real line called the **interval of convergence** of the power series. This observation suggests that we may view the power series (9) as a function f whose domain coincides with the interval of convergence of the series and whose functional values are the sums of the infinite series obtained by allowing x to take on all values in the interval of convergence. In this case, we also say the function f is **represented by the power series** (9).

Given such a power series, how does one determine its interval of convergence? The following theorem, whose proof we will omit, provides the answer to this question.

THEOREM 8

Suppose that we are given the power series.

$$\sum_{n=0}^{\infty} a_n(x-a)^n$$

Let

$$R = \lim_{n\to\infty}\left|\frac{a_n}{a_{n+1}}\right|$$

a. If $R = 0$, the series converges only for $x = a$.

b. If $0 < R < \infty$, the series converges for x in the interval $(a - R, a + R)$ and diverges for x outside this interval (Figure 8).

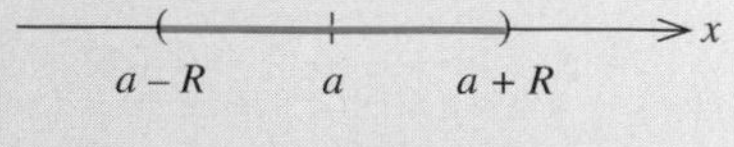

FIGURE 8

$a - R \quad a \quad a + R \quad x$

c. If $R = \infty$, the series converges for all x.

Notes

1. The domain of convergence of a power series is an interval. This **interval of convergence** is determined by R, the **radius of convergence**. Depending on whether $R = 0$, $0 < R < \infty$, or $R = \infty$, the interval of convergence may just be the degenerate interval consisting of the point a, a bona fide interval $(a - R, a + R)$, or the entire real line.
2. As mentioned earlier, a power series, inside its interval of convergence, represents a function f whose domain of definition coincides with the interval of convergence; that is,

$$f(x) = \sum_{n=0}^{\infty} a_n(x - a)^n \qquad x \in (a - R, a + R)$$

3. The power series may or may not converge at an endpoint. In general, it is difficult to determine whether a power series is convergent at an endpoint. For this reason, we restrict our attention to points inside the interval of convergence of a power series. ■

EXAMPLE 1 Find the radius of convergence and the interval of convergence of the power series

$$\sum_{n=0}^{\infty} \frac{(x - 1)^n}{2^n}$$

Show that $f(2)$ exists and find its value, where

$$f(x) = \sum_{n=0}^{\infty} \frac{(x - 1)^n}{2^n}$$

Solution Since $a_n = 1/2^n$, we have

$$R = \lim_{n \to \infty} \left| \frac{a_n}{a_{n+1}} \right| = \lim_{n \to \infty} \left| \frac{\frac{1}{2^n}}{\frac{1}{2^{n+1}}} \right|$$

$$= \lim_{n \to \infty} \frac{2^{n+1}}{2^n} = \lim_{n \to \infty} 2 = 2$$

Therefore, bearing in mind that $a = 1$, we see that the interval of convergence of the given series is $(-1, 3)$. Thus, in the interval $(-1, 3)$, the given power series defines a function

$$f(x) = \sum_{n=0}^{\infty} \frac{(x - 1)^n}{2^n} \qquad x \in (-1, 3)$$

Since $2 \in (-1, 3)$, $f(2)$ exists. In fact,

$$f(2) = \sum_{n=0}^{\infty} \frac{(2 - 1)^n}{2^n} = \sum_{n=0}^{\infty} \frac{1}{2^n}$$

which is a geometric series with $a = 1$ and $r = \frac{1}{2}$, so that

$$f(2) = \frac{1}{1 - \frac{1}{2}} = 2$$

■

EXAMPLE 2 Find the interval of convergence of each of the following power series:

a. $\sum_{n=0}^{\infty} n^3(x+2)^n$ **b.** $\sum_{n=0}^{\infty} n!(x-1)^n$ **c.** $\sum_{n=0}^{\infty} \frac{x^n}{n!}$

Solution

a. Here $a_n = n^3$, so

$$R = \lim_{n\to\infty}\left|\frac{a_n}{a_{n+1}}\right| = \lim_{n\to\infty}\frac{n^3}{(n+1)^3}$$

$$= \lim_{n\to\infty}\frac{1}{\left(1+\frac{1}{n}\right)^3} = 1 \qquad \text{Dividing numerator and denominator by } n^3$$

Since $a = -2$, we find that the interval of convergence of the series is $(-3, -1)$.

b. Here $a_n = n!$, so

$$R = \lim_{n\to\infty}\left|\frac{a_n}{a_{n+1}}\right| = \lim_{n\to\infty}\left|\frac{n!}{(n+1)!}\right|$$

$$= \lim_{n\to\infty}\frac{1}{n+1} = 0$$

Therefore, the series converges only at $a = 1$.

c. Here $a_n = 1/n!$ and

$$R = \lim_{n\to\infty}\left|\frac{a_n}{a_{n+1}}\right| = \lim_{n\to\infty}\left|\frac{\frac{1}{n!}}{\frac{1}{(n+1)!}}\right| = \lim_{n\to\infty}(n+1) = \infty$$

Therefore, the interval of convergence of the series is $(-\infty, \infty)$; that is, it converges for any value of x. ■

EXPLORE & DISCUSS

Suppose the power series $\sum_{n=0}^{\infty} a_n(x-a)^n$ has a radius of convergence R. What can you deduce about the radius of convergence of the series $\sum_{n=0}^{\infty} a_n(x-a)^{2n}$?

Finding a Taylor Series

Theorem 8 guarantees that a power series represents a function whose domain is precisely the interval of convergence of the series. We now show that if a function is defined in this manner, then the power series representing this function must be a Taylor series. To see this, we need the following theorem, which we state without proof.

THEOREM 9

Suppose the function f is defined by

$$f(x) = \sum_{n=0}^{\infty} a_n(x-a)^n = a_0 + a_1(x-a) + a_2(x-a)^2 + \cdots$$

with radius of convergence $R > 0$. Then

$$f'(x) = \sum_{n=1}^{\infty} na_n(x-a)^{n-1} = a_1 + 2a_2(x-a) + 3a_3(x-a)^2 + \cdots$$

on the interval $(a - R, a + R)$.

Thus, the derivative of f may be found by differentiating the power series term by term.

We now prove the statement asserted earlier. Suppose that f is represented by a power series centered about $x = a$; that is,

$$f(x) = a_0 + a_1(x - a) + a_2(x - a)^2 + a_3(x - a)^3 + \cdots + a_n(x - a)^n + \cdots$$

Then, applying Theorem 9 repeatedly, we find

$$\begin{aligned}
f'(x) &= a_1 + 2a_2(x - a) + 3a_3(x - a)^2 + \cdots + na_n(x - a)^{n-1} + \cdots \\
f''(x) &= 2a_2 + 3 \cdot 2a_3(x - a) + 4 \cdot 3a_4(x - a)^2 + \cdots \\
&\quad + n(n - 1)a_n(x - a)^{n-2} + \cdots \\
f'''(x) &= 3 \cdot 2a_3 + 4 \cdot 3 \cdot 2a_4(x - a) + \cdots \\
&\quad + n(n - 1)(n - 2)a_n(x - a)^{n-3} + \cdots \\
&= 3!a_3 + 4!a_4(x - a) + \cdots + n(n - 1)(n - 2)a_n(x - a)^{n-3} + \cdots \\
&\ \vdots \\
f^{(n)}(x) &= n!a_n + (n + 1)!a_{n+1}(x - a) + \cdots
\end{aligned}$$

Evaluating $f(x)$ and each of these derivatives at $x = a$ yields

$$\begin{aligned}
f(a) &= a_0 \\
f'(a) &= a_1 \\
f''(a) &= 2a_2 = 2!\, a_2 \\
f'''(a) &= 3 \cdot 2a_3 = 3!\, a_3 \\
&\ \vdots \\
f^{(n)}(a) &= n!\, a_n
\end{aligned}$$

Thus,

$$a_0 = f(a),\ a_1 = f'(a),\ a_2 = \frac{f''(a)}{2!},\ a_3 = \frac{f'''(a)}{3!}, \ldots, a_n = \frac{f^{(n)}(a)}{n!}, \ldots$$

as we set out to show.

Next we turn our attention to the converse problem. More precisely, suppose we are given a function f that has derivatives of *all* orders in an open interval I. Can we find a power series representation of f in that interval? Theorem 10 answers this question.

THEOREM 10

Taylor Series Representation of a Function

If a function f has derivatives of all orders in an open interval $I = (a - R, a + R)$ $(R > 0)$ centered at $x = a$, then

$$f(x) = \sum_{n=0}^{\infty} \frac{f^{(n)}(a)}{n!}(x - a)^n$$

if and only if

$$\lim_{N \to \infty} R_N(x) = 0$$

for all x in I, where $R_N(x) = f(x) - P_N(x)$ is as defined in Theorem 1.

Thus, f has a power series representation in the form of a Taylor series provided the error term associated with the Taylor polynomial tends to zero as the number of terms of the Taylor polynomial increases without bound. (A proof of this theorem is sketched in Exercise 33.) In what follows, we assume that the functions under consideration have Taylor series representations.

EXAMPLE 3 Find the Taylor series of the function $f(x) = \dfrac{1}{x-1}$ at $x = 2$.

Solution Here $a = 2$ and

$$\begin{aligned}
f(x) &= \frac{1}{x-1}\\
f'(x) &= -\frac{1}{(x-1)^2}\\
f''(x) &= \frac{2}{(x-1)^3}\\
f'''(x) &= -\frac{3\cdot 2}{(x-1)^4}\\
&\vdots\\
f^{(n)}(x) &= (-1)^n\frac{n!}{(x-1)^{n+1}}\\
&\vdots
\end{aligned}$$

so that

$$f(2) = 1, f'(2) = -1, f''(2) = 2,$$
$$f'''(2) = -6 = -3!, \ldots, f^{(n)}(2) = (-1)^n n!, \ldots$$

Therefore, the required Taylor series is

$$\begin{aligned}
\frac{1}{x-1} &= f(2) + f'(2)(x-2) + \frac{f''(2)}{2!}(x-2)^2 + \frac{f'''(2)}{3!}(x-2)^3 + \cdots\\
&\quad + \frac{f^{(n)}(2)}{n!}(x-2)^n + \cdots\\
&= 1 - (x-2) + \frac{2!}{2!}(x-2)^2 - \frac{3!}{3!}(x-2)^3 + \cdots\\
&\quad + (-1)^n\frac{n!}{n!}(x-2)^n + \cdots\\
&= 1 - (x-2) + (x-2)^2 - (x-2)^3 + \cdots + (-1)^n(x-2)^n + \cdots\\
&= \sum_{n=0}^{\infty}(-1)^n(x-2)^n
\end{aligned}$$

To find the radius of convergence of the series, we compute

$$\begin{aligned}
R &= \lim_{n\to\infty}\left|\frac{a_n}{a_{n+1}}\right| = \lim_{n\to\infty}\left|\frac{(-1)^n}{(-1)^{n+1}}\right|\\
&= \lim_{n\to\infty} 1 = 1
\end{aligned}$$

Thus, the series converges in the interval $(1, 3)$. By Theorem 10, we see that the function $f(x) = 1/(x - 1)$ is represented by the Taylor series in the interval $(1, 3)$. ■

The representation of the function $f(x) = 1/(x - 1)$ by its Taylor series at $x = 2$ is valid only in the interval $(1, 3)$, despite the fact that f itself has a domain that is the set of all real numbers except $x = 1$. This serves to remind us of the local nature of this representation.

EXPLORING WITH TECHNOLOGY

Refer to Example 3, where it was shown that the Taylor series at $x = 2$ representing $f(x) = 1/(x - 1)$ is

$$P(x) = \sum_{n=0}^{\infty} (-1)^n (x-2)^n = 1 - (x-2) + (x-2)^2 - (x-2)^3 + \cdots$$

for $1 < x < 3$. This means that if c is any number satisfying $1 < c < 3$, then $f(c) = 1/(c - 1) = P(c)$. In particular, this means that $\lim_{N\to\infty} | P_N(c) - f(c) = 0|$, where $\{P_N(x)\}$ is the sequence of partial sums of $P(x)$.

1. Plot the graphs of f, P_0, P_1, P_2, P_3, P_4, P_5, and P_6 on the same set of axes, using the viewing window $[2.5, 3.1] \times [-0.1, 1.1]$.
2. Do the results of part 1 give a visual confirmation of the statement $\lim_{N\to\infty} P_N(c) = f(c)$ for the special case where $c = 2.8$? Explain what happens when $c = 3$.

EXAMPLE 4 Find the Taylor series of the function $f(x) = \ln x$ at $x = 1$.

Solution Here $a = 1$ and

$$\begin{aligned}
f(x) &= \ln x \\
f'(x) &= \frac{1}{x} \\
f''(x) &= -\frac{1}{x^2} \\
f'''(x) &= \frac{2}{x^3} = \frac{2!}{x^3} \\
f^{(4)}(x) &= -\frac{3 \cdot 2}{x^4} = -\frac{3!}{x^4} \\
&\vdots \\
f^{(n)}(x) &= (-1)^{n+1}\frac{(n-1)!}{x^n}
\end{aligned}$$

so that

$$f(1) = 0, f'(1) = 1, f''(1) = -1, f'''(1) = 2!,$$
$$f^{(4)}(1) = -3!, \ldots, f^{(n)}(1) = (-1)^{n+1}(n-1)!, \ldots$$

Therefore, the required Taylor series is

$$
\begin{aligned}
f(x) = \ln x &= f(1) + f'(1)(x-1) + \frac{f''(1)}{2!}(x-1)^2 + \frac{f'''(1)}{3!}(x-1)^3 \\
&\quad + \cdots + \frac{f^{(n)}(1)}{n!}(x-1)^n + \cdots \\
&= (x-1) - \frac{1}{2!}(x-1)^2 + \frac{2!}{3!}(x-1)^3 - \frac{3!}{4!}(x-1)^4 + \cdots \\
&\quad + (-1)^{n+1}\frac{(n-1)!}{n!}(x-1)^n + \cdots \\
&= (x-1) - \frac{1}{2}(x-1)^2 + \frac{1}{3}(x-1)^3 - \frac{1}{4}(x-1)^4 + \cdots \\
&\quad + \frac{(-1)^{n+1}}{n}(x-1)^n + \cdots \\
&= \sum_{n=1}^{\infty} \frac{(-1)^{n+1}}{n}(x-1)^n
\end{aligned}
$$

Since

$$
\begin{aligned}
R = \lim_{n\to\infty} \left| \frac{a_n}{a_{n+1}} \right| &= \lim_{n\to\infty} \left| \frac{\dfrac{(-1)^{n+1}}{n}}{\dfrac{(-1)^{n+2}}{n+1}} \right| \\
&= \lim_{n\to\infty} \frac{n+1}{n} = \lim_{n\to\infty} \left(1 + \frac{1}{n}\right) = 1
\end{aligned}
$$

we see that the power series representation of $f(x) = \ln x$ is valid in the interval $(0, 2)$. It can be shown that the representation is valid at $x = 2$ as well. ■

EXAMPLE 5 Find the power series representation about $x = 0$ for the function $f(x) = e^x$.

Solution Here $a = 0$ and, since

$$f(x) = f'(x) = f''(x) = \cdots = e^x$$

we see that

$$f(0) = f'(0) = f''(0) = \cdots = e^0 = 1$$

Therefore, the required power series (Taylor series) is

$$
\begin{aligned}
f(x) = e^x &= f(0) + f'(0)(x-0) + \frac{f''(0)}{2!}(x-0)^2 + \cdots + \frac{f^{(n)}(0)}{n!}(x-0)^n + \cdots \\
&= 1 + x + \frac{x^2}{2!} + \frac{x^3}{3!} + \cdots + \frac{x^n}{n!} + \cdots \\
&= \sum_{n=0}^{\infty} \frac{x^n}{n!}
\end{aligned}
$$

Since

$$R = \lim_{n\to\infty}\left|\frac{a_n}{a_{n+1}}\right| = \lim_{n\to\infty}\left|\frac{\frac{1}{n!}}{\frac{1}{(n+1)!}}\right|$$
$$= \lim_{n\to\infty}(n+1) = \infty$$

we see that the series representation is valid for all x. ■

11.5 Self-Check Exercises

1. Find the interval of convergence of the power series $\sum_{n=0}^{\infty}\frac{x^n}{n^2+1}$ (disregard the end points).

2. Find the Taylor series of the function $f(x) = e^{-x}$ at $x = 1$ and determine its interval of convergence.

Solutions to Self-Check Exercises 11.5 can be found on page 750.

11.5 Concept Questions

1. a. Define a power series in x.
   b. Define a power series in $(x - a)$.

2. a. What is the radius of convergence of a power series?
   b. What is the interval of convergence of a power series?
   c. How do you find the radius and the interval of convergence of a power series?

3. Suppose that $\sum_{n=0}^{\infty} a_n x^n$ has radius of convergence 2. What can you say about the convergence or divergence of $\sum_{n=0}^{\infty} a_n(\frac{3}{2})^n$?

4. Suppose that $\sum_{n=0}^{\infty} a_n(x-2)^n$ diverges for $x = 0$. What can you say about the convergence or divergence of $\sum_{n=0}^{\infty} a_n 5^n$? What about $\sum_{n=0}^{\infty} a_n(\frac{1}{2})^n$?

5. a. What is a Taylor series?
   b. Suppose $f(x) = \sum_{n=0}^{\infty} a_n(x-c)^n$ for x in $(-R, R)$, where $R > 0$. What is $f^{(n)}(c)$?
   c. The Taylor series of $f(x)$ at $x = 1$ is $\sum_{n=0}^{\infty}(-1)^n\frac{(x-1)^{n+1}}{n+1}$. What is $f^{(5)}(1)$?

11.5 Exercises

In Exercises 1–20, find the radius of convergence and the interval of convergence of the power series.

1. $\sum_{n=0}^{\infty}(x-1)^n$

2. $\sum_{n=0}^{\infty}\left(\frac{x}{2}\right)^n$

3. $\sum_{n=1}^{\infty} n^2x^n$

4. $\sum_{n=0}^{\infty}\frac{(n+1)(x+2)^n}{2^n}$

5. $\sum_{n=0}^{\infty}\frac{(-1)^n x^n}{4^n}$

6. $\sum_{n=0}^{\infty}\frac{(2x)^n}{3^n}$

7. $\sum_{n=0}^{\infty}\frac{(x-1)^n}{n!\,2^n}$

8. $\sum_{n=0}^{\infty}(2n)!\,x^n$

9. $\sum_{n=0}^{\infty}\frac{(-1)^n n!(x+2)^n}{2^n}$

10. $\sum_{n=2}^{\infty}\frac{x^n}{n(n+1)}$

11. $\sum_{n=2}^{\infty}\frac{(x+3)^n}{(n+1)^2}$

12. $\sum_{n=0}^{\infty}\frac{n!(x+1)^n}{(3n)!}$

13. $\sum_{n=1}^{\infty}\frac{2n(x-3)^n}{(n+1)!}$

14. $\sum_{n=0}^{\infty}\frac{(-2x)^{2n}}{4^n(n+1)}$

15. $\sum_{n=1}^{\infty} \frac{n(-2x)^n}{n+1}$

16. $\sum_{n=0}^{\infty} \frac{x^{2n+1}}{(2n+1)!}$

17. $\sum_{n=0}^{\infty} \frac{n!(x+1)^n}{3^n}$

18. $\sum_{n=0}^{\infty} (n+1)(x+2)^n$

19. $\sum_{n=0}^{\infty} \frac{n^3(x-3)^n}{3^n}$

20. $\sum_{n=0}^{\infty} \frac{(-1)^{n+1}(x-2)^n}{n2^n}$

In Exercises 21–32, find the Taylor series of the function at the indicated number and give its radius and interval of convergence (disregard the end points).

21. $f(x) = \frac{1}{x}; x = 1$

22. $f(x) = \frac{1}{x+1}; x = 0$

23. $f(x) = \frac{1}{x+1}; x = 2$

24. $f(x) = \frac{1}{1-x}; x = 0$

25. $f(x) = \frac{1}{1-x}; x = 2$

26. $f(x) = \ln(x+1); x = 0$

27. $f(x) = \sqrt{x}; x = 1$

28. $f(x) = \sqrt{1-x}; x = 0$

29. $f(x) = e^{2x}; x = 0$

30. $f(x) = e^{2x}; x = 1$

31. $f(x) = \frac{1}{\sqrt{x+1}}; x = 0$

32. $f(x) = \sqrt{x+1}; x = 1$

33. Prove Theorem 10.
Hint: The Nth partial sum of the Taylor series is $S_N(x) = P_N(x)$ where $P_N(x)$ is the Nth Taylor polynomial. Use this fact to show that $\lim_{N\to\infty} S_N(x) = f(x)$ if and only if $\lim_{N\to\infty} R_N(x) = 0$.

In Exercises 34–36, determine whether the statement is true or false. If it is true, explain why it is true. If it is false, give an example to show why it is false.

34. If $\sum_{n=0}^{\infty} a_n(x-2)^n$ converges for $x = 4$, then it converges for $x = 1$.

35. If $\sum_{n=0}^{\infty} a_n(x-a)^n$ has radius of convergence R, then $\sum_{n=0}^{\infty} na_n(x-a)^n$ has radius of convergence R.

36. If $\sum_{n=0}^{\infty} a_n(x-a)^n$ has radius of convergence R, then $\sum_{n=0}^{\infty} a_n^2(x-a)^n$ has radius of convergence $\sqrt{R}$.

11.5 Solutions to Self-Check Exercises

1. We first find the radius of convergence of the power series. Since $a_n = \frac{1}{n^2+1}$, we have

$$R = \lim_{n\to\infty}\left|\frac{a_n}{a_{n+1}}\right| = \lim_{n\to\infty}\frac{\frac{1}{n^2+1}}{\frac{1}{(n+1)^2+1}}$$
$$= \lim_{n\to\infty}\frac{(n+1)^2+1}{n^2+1}$$
$$= \lim_{n\to\infty}\frac{\left(1+\frac{1}{n}\right)^2+\frac{1}{n^2}}{1+\frac{1}{n^2}} \qquad \text{Dividing numerator and denominator by } n^2$$
$$= 1$$

Therefore, the interval of convergence of the series is $(-1, 1)$.

2. Here $a = 1$ and

$$\begin{aligned} f(x) &= e^{-x} \\ f'(x) &= -e^{-x} \\ f''(x) &= e^{-x} \\ f'''(x) &= -e^{-x} \\ &\vdots \\ f^{(n)}(x) &= (-1)^n e^{-x} \end{aligned}$$

so that

$$f(1) = e^{-1}, f'(1) = -e^{-1}, f''(1) = e^{-1},$$
$$f'''(1) = -e^{-1}, \ldots, f^{(n)}(1) = (-1)^n e^{-1}$$

Therefore, the required Taylor series is

$$\begin{aligned} e^{-x} &= f(1) + f'(1)(x-1) + \frac{f''(1)}{2!}(x-1)^2 \\ &\quad + \frac{f'''(1)}{3!}(x-1)^3 + \cdots + \frac{f^{(n)}(1)}{n!}(x-1)^n + \cdots \\ &= e^{-1} - \frac{e^{-1}}{1!}(x-1) + \frac{e^{-1}}{2!}(x-1)^2 \\ &\quad - \frac{e^{-1}}{3!}(x-1)^3 + \cdots + \frac{(-1)^n e^{-1}}{n!}(x-1)^n + \cdots \\ &= \frac{1}{e} - \frac{1}{e}(x-1) + \frac{1}{2!\,e}(x-1)^2 - \frac{1}{3!\,e}(x-1)^3 + \cdots \\ &\quad + \frac{(-1)^n}{n!\,e}(x-1)^n + \cdots \\ &= \sum_{n=0}^{\infty}\frac{(-1)^n}{n!\,e}(x-1)^n \end{aligned}$$

11.6 More on Taylor Series

A Useful Technique for Finding Taylor Series

In the last section, we showed how to find the power series representation of certain functions. This representation turned out to be the Taylor series of f at $x = a$. The method we used to compute the series relies solely on our ability to find the higher-order derivatives of the function f. This method, however, is rather tedious.

In this section, we show how the Taylor series of a function can often be found by manipulating some well-known power series. For this purpose, we first catalog the power series of some of the most commonly used functions (Table 1). Recall that each of these representations was derived in Sections 11.1 and 11.5.

TABLE 1

Power Series Representations for Some Common Functions

1. $\dfrac{1}{1-x} = 1 + x + x^2 + \cdots + x^n + \cdots \qquad (-1 < x < 1)$

2. $e^x = 1 + x + \dfrac{1}{2!}x^2 + \dfrac{1}{3!}x^3 + \cdots + \dfrac{1}{n!}x^n + \cdots \qquad (-\infty < x < \infty)$

3. $\ln x = (x-1) - \dfrac{1}{2}(x-1)^2 + \dfrac{1}{3}(x-1)^3 - \cdots$
$$+ \frac{(-1)^{n+1}}{n}(x-1)^n + \cdots \qquad (0 < x \leq 2)$$

EXAMPLE 1 Find the Taylor series of each function at the indicated number.

a. $f(x) = \dfrac{1}{1+x}$; $x = 0$

b. $f(x) = \dfrac{1}{1+x}$; $x = 2$

c. $f(x) = \dfrac{1}{1-3x}$; $x = 0$

d. $f(x) = \dfrac{x}{1+x^2}$; $x = 0$

Solution

a. We write

$$f(x) = \frac{1}{1-(-x)}$$

and use the power series representation of $1/(1-x)$ with x replaced by $-x$ to obtain

$$\begin{aligned} f(x) &= \frac{1}{1+x} \\ &= 1 + (-x) + (-x)^2 + \cdots + (-x)^n + \cdots \\ &= 1 - x + x^2 - x^3 + \cdots + (-1)^n x^n + \cdots \qquad (-1 < x < 1) \end{aligned}$$

b. We write

$$f(x) = \frac{1}{1+x} = \frac{1}{3+(x-2)} = \frac{1}{3\left[1+\left(\frac{x-2}{3}\right)\right]} = \frac{1}{3}\left[\frac{1}{1+\left(\frac{x-2}{3}\right)}\right]$$

From the result of part (a), we have

$$\frac{1}{1+u} = 1 - u + u^2 - u^3 + \cdots + (-1)^n u^n + \cdots \qquad (-1 < u < 1)$$

Thus, with $u = \dfrac{x-2}{3}$, we find

$$\begin{aligned} f(x) &= \frac{1}{3}\left[\frac{1}{1+\left(\dfrac{x-2}{3}\right)}\right] \\ &= \frac{1}{3}\left[1 - \left(\frac{x-2}{3}\right) + \left(\frac{x-2}{3}\right)^2 - \left(\frac{x-2}{3}\right)^3 + \cdots + (-1)^n\left(\frac{x-2}{3}\right)^n + \cdots\right] \\ &= \frac{1}{3} - \frac{1}{3^2}(x-2) + \frac{1}{3^3}(x-2)^2 - \frac{1}{3^4}(x-2)^3 + \cdots + \frac{(-1)^n}{3^{n+1}}(x-2)^n + \cdots \end{aligned}$$

which converges when

$$-1 < \frac{x-2}{3} < 1 \qquad \text{or} \qquad -1 < x < 5$$

c. Here we simply replace x in the power series representation of $1/(1-x)$ by $3x$ to get

$$\begin{aligned} f(x) = \frac{1}{1-3x} &= 1 + 3x + (3x)^2 + (3x)^3 + \cdots + (3x)^n + \cdots \\ &= 1 + 3x + 3^2x^2 + 3^3x^3 + \cdots + 3^nx^n + \cdots \end{aligned}$$

The series converges for $-1 < 3x < 1$, or $-\dfrac{1}{3} < x < \dfrac{1}{3}$.

d. We write

$$\begin{aligned} f(x) = \frac{x}{1+x^2} &= x\left[\frac{1}{1-(-x^2)}\right] \\ &= x[1 + (-x^2) + (-x^2)^2 + (-x^2)^3 + \cdots + (-x^2)^n + \cdots] \\ &= x - x^3 + x^5 - x^7 + \cdots + (-1)^n x^{2n+1} + \cdots \end{aligned}$$

The series converges for $-1 < x < 1$. ■

EXAMPLE 2 Find the Taylor series of the function at the indicated number.

a. $f(x) = xe^{-x}$; $x = 0$ **b.** $f(x) = \ln(1+x)$; $x = 0$

Solution

a. First, we replace x with $-x$ in the expression

$$e^x = 1 + x + \frac{x^2}{2!} + \frac{x^3}{3!} + \cdots + \frac{x^n}{n!} + \cdots \qquad \text{See Table 1.}$$

to obtain

$$\begin{aligned} e^{-x} &= 1 + (-x) + \frac{(-x)^2}{2!} + \frac{(-x)^3}{3!} + \cdots + \frac{(-x)^n}{n!} + \cdots \\ &= 1 - x + \frac{x^2}{2!} - \frac{x^3}{3!} + \cdots + \frac{(-1)^n x^n}{n!} + \cdots \end{aligned}$$

Then, multiplying both sides of this expression by x gives the required expression

$$f(x) = xe^{-x} = x - x^2 + \frac{x^3}{2!} - \frac{x^4}{3!} + \cdots + \frac{(-1)^n x^{n+1}}{n!} + \cdots$$

b. From Table 1 we have

$$\ln x = (x-1) - \frac{1}{2}(x-1)^2 + \frac{1}{3}(x-1)^3 - \cdots + \frac{(-1)^{n+1}}{n}(x-1)^n + \cdots \qquad (0 < x \leq 2)$$

Replacing x with $1 + x$ in this expression, we obtain the required expression

$$f(x) = \ln(1+x) = x - \frac{1}{2}x^2 + \frac{1}{3}x^3 - \cdots + \frac{(-1)^{n+1}}{n}x^n + \cdots$$

valid for $-1 < x \leq 1$. (Why?) ■

EXPLORE & DISCUSS

The formula

$$\ln x = (x-1) - \frac{1}{2}(x-1)^2 + \frac{1}{3}(x-1)^3 - \cdots \qquad (0 < x \leq 2) \qquad \textbf{(A)}$$

from Table 1 can be used to compute the value of $\ln x$ for $0 < x \leq 2$. However, the restriction on x and the slow convergence of the series limit its effectiveness from the computational point of view. A more effective formula, first obtained by the Scottish mathematician James Gregory (1638–1675), follows:

1. Using Formula (A), derive the formulas

$$\ln(1+x) = x - \frac{x^2}{2} + \frac{x^3}{3} - \frac{x^4}{4} + \cdots \qquad (-1 < x \leq 1)$$

and

$$\ln(1-x) = -x - \frac{x^2}{2} - \frac{x^3}{3} - \frac{x^4}{4} - \cdots \qquad (-1 \leq x < 1)$$

2. Use part 1 to show that

$$\ln\left(\frac{1+x}{1-x}\right) = 2\left(x + \frac{x^3}{3} + \frac{x^5}{5} + \frac{x^7}{7} + \cdots\right) \qquad (-1 < x < 1)$$

3. To compute the natural logarithm of a positive number p, let

$$p = \frac{1+x}{1-x}$$

and show that

$$x = \frac{p-1}{p+1} \qquad (-1 < x < 1)$$

4. Use parts 2 and 3 to show that

$$\ln 2 = 2\left[\left(\frac{1}{3}\right) + \frac{(\frac{1}{3})^3}{3} + \frac{(\frac{1}{3})^5}{5} + \frac{(\frac{1}{3})^7}{7} + \cdots\right]$$

and this yields $\ln 2 \approx 0.6931$ when we add the first four terms of the series. This approximation of $\ln 2$ is accurate to four decimal places.

5. Compare this method of computing $\ln 2$ with that of using Formula (A) directly.

In Theorem 10 in the last section, we saw that a power series may be differentiated term by term to yield another power series whose interval of convergence coincides with that of the original series. The latter is of course the power series representation of the derivative f' of the function f represented by the first series. The following theorem tells us that we may integrate a power series term by term.

THEOREM 11

Suppose

$$\begin{aligned} f(x) &= \sum_{n=0}^{\infty} a_n(x-a)^n \\ &= a_0 + a_1(x-a) + a_2(x-a)^2 + a_3(x-a)^3 + \cdots \\ &\quad + a_n(x-a)^n + \cdots \qquad x \in (a-R, a+R) \end{aligned}$$

Then

$$\begin{aligned} \int f(x)\,dx &= \sum_{n=0}^{\infty} \frac{a_n}{n+1}(x-a)^{n+1} \\ &= a_0(x-a) + \frac{a_1}{2}(x-a)^2 + \frac{a_2}{3}(x-a)^3 + \cdots \\ &\quad + \frac{a_n}{n+1}(x-a)^{n+1} + \cdots \qquad x \in (a-R, a+R) \end{aligned}$$

Theorems 10 and 11 can also be used to help us find the power-series representation of a function starting from the series representation of some appropriate function, as the next two examples show.

EXAMPLE 3 Differentiate the power series for the function $1/(1-x)$ at $x = 0$ (see Table 1) to obtain a Taylor series representation of the function $f(x) = 1/(1-x)^2$ at $x = 0$.

Solution From Table 1, we have

$$\frac{1}{1-x} = 1 + x + x^2 + x^3 + \cdots + x^n + \cdots \qquad (-1 < x < 1)$$

Differentiating both sides of the equation with respect to x, and using Theorem 10, we obtain

$$\frac{d}{dx}\left(\frac{1}{1-x}\right) = \frac{d}{dx}(1 + x + x^2 + x^3 + \cdots + x^n + \cdots)$$

or

$$\begin{aligned} f(x) &= \frac{1}{(1-x)^2} \\ &= 1 + 2x + 3x^2 + \cdots + nx^{n-1} + \cdots \qquad (-1 < x < 1) \end{aligned}$$ ■

EXAMPLE 4 Integrate the Taylor series for the function $1/(1+x)$ at $x = 0$ (see Table 1) to obtain a power series representation for the function $f(x) = \ln(1+x)$ centered at $x = 0$. Compare your result with that of Example 2b.

Solution From Table 1 we see that

$$\frac{1}{1-x} = 1 + x + x^2 + x^3 + \cdots + x^n + \cdots \qquad (-1 < x < 1)$$

Replacing x by $-x$ gives

$$\frac{1}{1-(-x)} = 1 + (-x) + (-x)^2 + (-x)^3 + \cdots + (-x)^n + \cdots$$

or

$$\frac{1}{1+x} = 1 - x + x^2 - x^3 + \cdots + (-1)^n x^n + \cdots \qquad (-1 < x < 1)$$

Finally, integrating both sides of this equation with respect to x and using Theorem 11, we obtain

$$\int \frac{1}{1+x}\,dx = \int [1 - x + x^2 - x^3 + \cdots + (-1)^n x^n + \cdots]\,dx$$

or

$$\begin{aligned} f(x) = \ln(1+x) &= x - \frac{1}{2}x^2 + \frac{1}{3}x^3 - \cdots + \frac{(-1)^n}{n+1}x^{n+1} + \cdots \\ &= x - \frac{1}{2}x^2 + \frac{1}{3}x^3 - \cdots + \frac{(-1)^{n+1}}{n}x^n + \cdots \qquad (-1 < x < 1) \end{aligned}$$

This result is the same as that of Example 2b, as expected.

In many practical applications, all that is required is an approximation of the actual solution to a problem. Thus, rather than working with the Taylor series of a function at a point that is *equal* to $f(x)$ inside its interval of convergence, one often works with the truncated Taylor series. The truncated Taylor series is a Taylor polynomial that gives an acceptable approximation to the values of $f(x)$ in the neighborhood of the point about which the function is expanded, provided the degree of the polynomial or, equivalently, the number of terms of the Taylor series retained is large enough.

Before looking at an example, we want to point out the comparative ease with which one can obtain the Taylor polynomial approximation of a function using the method of this section, rather than obtaining it directly as was done in Section 11.1.

APPLIED EXAMPLE 5 Serum Cholesterol Levels The serum cholesterol levels (in mg/dL) in a current Mediterranean population are found to be normally distributed with a probability density function given by

$$f(x) = \frac{1}{50\sqrt{2\pi}} e^{-1/2[(x-160)/50]^2}$$

Scientists at the National Heart, Lung, and Blood Institute consider this pattern ideal for a minimal risk of heart attacks. Find the percentage of the population who have blood cholesterol levels between 160 and 180 mg/dL.

Solution The required probability is given by

$$P(160 \le x \le 180) = \frac{1}{50\sqrt{2\pi}} \int_{160}^{180} e^{-1/2[(x-160)/50]^2}\,dx$$

(see Section 10.3). Let's approximate the integrand by a sixth-degree Taylor polynomial about $x = 160$. From Table 1, we have

$$e^x = 1 + x + \frac{1}{2!}x^2 + \frac{1}{3!}x^3 + \cdots$$

Replacing x with $-\frac{1}{2}[(x - 160)/50]^2$, we obtain

$$\begin{aligned} e^{-1/2[(x-160)/50]^2} &\approx 1 - \frac{1}{2}\left(\frac{x-160}{50}\right)^2 + \frac{1}{2!}\left[-\frac{1}{2}\left(\frac{x-160}{50}\right)^2\right]^2 \\ &\quad + \frac{1}{3!}\left[-\frac{1}{2}\left(\frac{x-160}{50}\right)^2\right]^3 \\ &= 1 - \frac{(x-160)^2}{5000} + \frac{(x-160)^4}{5 \cdot 10^7} - \frac{(x-160)^6}{7.5 \cdot 10^{11}} \end{aligned}$$

Therefore,

$$\begin{aligned} &P(160 \le x \le 180) \\ &\quad \approx \frac{1}{50\sqrt{2\pi}} \int_{160}^{180} \left[1 - \frac{(x-160)^2}{5000} + \frac{(x-160)^4}{5 \cdot 10^7} - \frac{(x-160)^6}{7.5 \cdot 10^{11}}\right] dx \\ &\quad \approx \frac{1}{50\sqrt{2\pi}} \left[x - \frac{(x-160)^3}{15{,}000} + \frac{(x-160)^5}{2.5 \cdot 10^8} - \frac{(x-160)^7}{5.25 \cdot 10^{12}}\right]\Bigg|_{160}^{180} \\ &\quad \approx \frac{1}{50\sqrt{2\pi}}[(180 - 0.53333 + 0.0128 - 0.00024) - 160] \\ &\quad \approx 0.1554 \end{aligned}$$

and so approximately 15.5% of the population has blood cholesterol levels between 160 and 180 mg/dL. ■

11.6 Self-Check Exercises

1. Find the Taylor series of the function $f(x) = xe^{-x^2}$ at $x = 0$.
2. Use the result from Self-Check Exercise 1 to write the seventh Taylor polynomial of f about $x = 0$, and use this polynomial to approximate

$$\int_0^{0.5} xe^{-x^2}\, dx$$

Compare this result with the exact value of the integral.

Solutions to Self-Check Exercises 11.6 can be found on page 758.

11.6 Concept Questions

In Exercises 1–3, rewrite $f(x)$ in a form that will enable you to take advantage of the power series representations of functions in Table 1 to find the Taylor series of f at the indicated number. Do not find the series.

1. $f(x) = \dfrac{x}{1 + 2x}$; $x = 0$
2. $f(x) = \dfrac{1}{2x + 3}$; $x = 1$
3. $f(x) = \dfrac{2x}{4 + 3x^2}$; $x = 0$

11.6 Exercises

In Exercises 1–20, find the Taylor series of each function at the indicated number. Give the interval of convergence for each series.

1. $f(x) = \dfrac{1}{1 - x}; x = 2$

2. $f(x) = \dfrac{1}{1 + x}; x = 1$

3. $f(x) = \dfrac{1}{1 + 3x}; x = 0$

4. $f(x) = \dfrac{x}{1 - 2x}; x = 0$

5. $f(x) = \dfrac{1}{4 - 3x}; x = 0$

6. $f(x) = \dfrac{1}{4 - 3x}; x = 1$

7. $f(x) = \dfrac{1}{1 - x^2}; x = 0$

8. $f(x) = \dfrac{x^2}{1 + x^3}; x = 0$

9. $f(x) = e^{-x}; x = 0$

10. $f(x) = e^x; x = 1$

11. $f(x) = xe^{-x^2}; x = 0$

12. $f(x) = xe^{x/2}; x = 0$

13. $f(x) = \dfrac{1}{2}(e^x + e^{-x}); x = 0$

14. $f(x) = \dfrac{1}{2}(e^x - e^{-x}); x = 0$

15. $f(x) = \ln(1 + 2x); x = 0$

16. $f(x) = \ln\left(1 + \dfrac{x}{2}\right); x = 0$

17. $f(x) = \ln(1 + x^2); x = 0$

18. $f(x) = \ln(1 + 2x); x = 2$

19. $f(x) = (x - 2)\ln x; x = 2$

20. $f(x) = x^2 \ln\left(1 + \dfrac{x}{2}\right); x = 0$

21. Differentiate the power series for $\ln(1 + x)$ at $x = 0$ to obtain a series representation for the function $f(x) = 1/(1 + x)$.

22. Differentiate the power series for $1/(1 + x)$ at $x = 0$ to obtain a series representation for the function $f(x) = 1/(1 + x)^2$.

23. Integrate the power series for $1/(1 + x)$ to obtain a power series representation for the function $f(x) = \ln(1 + x)$.

24. Integrate the power series for $2x/(1 + x^2)$ to obtain a power series representation for the function $f(x) = \ln(1 + x^2)$.

25. Use the sixth-degree Taylor polynomial to approximate

$$\int_0^{0.5} \frac{1}{\sqrt{1 + x^2}}\, dx$$

Hint: $P_6(x) = 1 - \frac{1}{2}x^2 + \frac{3}{8}x^4 - \frac{5}{16}x^6$

26. Use the sixth-degree Taylor polynomial to approximate

$$\int_0^{0.4} \ln(1 + x^2)\, dx$$

27. Use the eighth-degree Taylor polynomial to approximate

$$\int_0^{1} e^{-x^2}\, dx$$

28. Use the sixth-degree Taylor polynomial to approximate

$$\int_0^{0.5} \frac{\ln(1 + x)}{x}\, dx$$

29. Use the eighth-degree Taylor polynomial of $f(x) = 1/(1 + x^2)$ at $x = 0$ and the relationship

$$\pi = 4\int_0^1 \frac{dx}{1 + x^2}$$

to obtain an approximation of π.

30. Factory Worker Wages According to data released by a city's Chamber of Commerce, the weekly wages of factory workers are normally distributed according to the probability density function

$$f(x) = \frac{1}{50\sqrt{2\pi}}\, e^{-1/2[(x-500)/50]^2}$$

Find the probability that a worker selected at random from the city has a weekly wage of \$450–\$550.

31. MP3 Player Reliability The General Manager of the Service Department of MCA Media Company has estimated that the time that elapses between the dates of purchase and the dates on which the MP3 players manufactured by the company first require service is normally distributed according to the probability density function

$$f(x) = \frac{1}{10\sqrt{2\pi}}\, e^{-1/2[(x-30)/10]^2}$$

where x is measured in months. Determine the percentage of players manufactured and sold by MCA that may require service 28–32 mo after purchase.

32. Demand for Wristwatches The demand equation for the "Tempus" quartz wristwatch is given by

$$p = 50e^{-0.1(x+1)^2}$$

where x (measured in units of a thousand) is the quantity demanded per week and p is the unit wholesale price in dollars.

National Importers, the supplier of the watches, will make x units (in thousands) available in the market if the unit wholesale price is

$$p = 10 + 5x^2$$

dollars. Use an appropriate Taylor polynomial approximation of the function

$$p = f(x) = 50e^{-0.1(x+1)^2}$$

to find the consumers' surplus.
Hint: The equilibrium quantity is 1689 units.

11.6 Solutions to Self-Check Exercises

1. First, we replace x in the expression

$$e^x = 1 + x + \frac{x^2}{2!} + \frac{x^3}{3!} + \cdots + \frac{x^n}{n!} + \cdots$$

(see Table 1) with $-x^2$ to obtain

$$\begin{aligned} e^{-x^2} &= 1 + (-x^2) + \frac{(-x^2)^2}{2!} \\ &\quad + \frac{(-x^2)^3}{3!} + \cdots + \frac{(-x^2)^n}{n!} + \cdots \\ &= 1 - x^2 + \frac{x^4}{2!} - \frac{x^6}{3!} + \cdots + \frac{(-1)^n x^{2n}}{n!} + \cdots \end{aligned}$$

Then multiplying both sides of this expression by x gives the required expression

$$\begin{aligned} f(x) = xe^{-x^2} &= x - x^3 + \frac{x^5}{2!} \\ &\quad - \frac{x^7}{3!} + \cdots + \frac{(-1)^n x^{2n+1}}{n!} + \cdots \\ &= \sum_{n=0}^{\infty} \frac{(-1)^n x^{2n+1}}{n!} \end{aligned}$$

2. Using the result from Exercise 1, we see that the seventh Taylor polynomial of f about $x = 0$ is

$$\begin{aligned} P_7(x) &= x - x^3 + \frac{x^5}{2!} - \frac{x^7}{3!} \\ &= x - x^3 + \frac{1}{2}x^5 - \frac{1}{6}x^7 \end{aligned}$$

Therefore,

$$\begin{aligned} \int_0^{0.5} xe^{-x^2}\,dx &\approx \int_0^{0.5} \left(x - x^3 + \frac{1}{2}x^5 - \frac{1}{6}x^7\right) dx \\ &= \frac{1}{2}x^2 - \frac{1}{4}x^4 + \frac{1}{12}x^6 - \frac{1}{48}x^8 \Big|_0^{0.5} \\ &\approx 0.1105957 \end{aligned}$$

The exact value of the integral is

$$\begin{aligned} \int_0^{0.5} xe^{-x^2}\,dx &= -\frac{1}{2}e^{-x^2}\Big|_0^{0.5} \qquad \text{Use the substitution } u = x^2. \\ &= -\frac{1}{2}(e^{-0.25} - 1) \end{aligned}$$

which is approximately 0.1105996. Thus, the error is 0.0000039.

11.7 Newton's Method

Newton's Method (Optional)

Previously we have had occasion to find the zeros of a function f or, equivalently, the **roots** of the equation $f(x) = 0$. For example, the x-intercepts of a function f are precisely the values of x that satisfy $f(x) = 0$; the critical points of f include the roots of the equation $f'(x) = 0$; and the candidates for the inflection points of f include the roots of the equation $f''(x) = 0$.

If $f(x)$ is a linear or quadratic function or $f(x)$ is a polynomial that is easily factored, the roots of f are readily found. In practice, however, we often encounter functions with zeros that cannot be found as readily. For example, the function

$$f(t) = -t^3 + 96t^2 + 195t + 5$$

of Example 8, page 165, that gives the altitude (in feet) of a rocket t seconds into flight, is not easily factored, so its zeros cannot be found by elementary algebraic

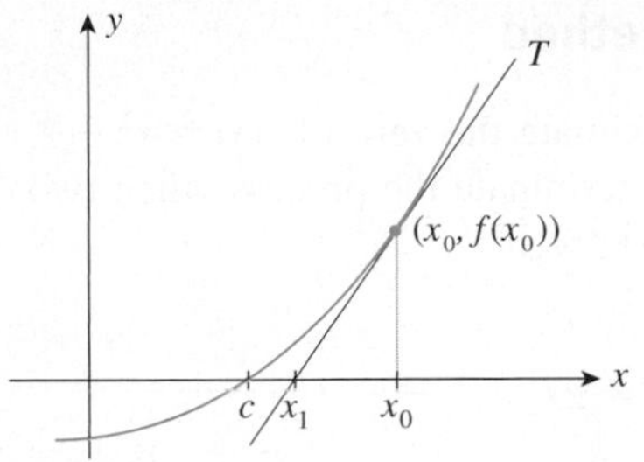

FIGURE 9
x_0 is our first guess in estimating the value of c, the zero of $f(x)$.

methods. (We will find the zeros of this function in Example 2.) Another example of a function with roots that are not easily found is

$$g(x) = e^{2x} - 3x - 2$$

In this section we develop an algorithm, based on the approximation of a function $f(x)$ by a first-degree Taylor polynomial, to approximate the value of a zero of $f(x)$ to any desired degree of accuracy.

Suppose the function f has a zero at $x = c$ (Figure 9). Let x_0 be an initial estimate of the actual zero c of $f(x)$. Now, the first-degree Taylor polynomial of $f(x)$ at $x = x_0$ is

$$P_1(x) = f(x_0) + f'(x_0)(x - x_0)$$

and, as observed earlier, is just the equation of the tangent line to the graph of $y = f(x)$ at the point $(x_0, f(x_0))$. Since the tangent line T is an approximation of the graph of $y = f(x)$ near $x = x_0$ (and x_0 is assumed to be close to c!), it may be expected that the zero of $P_1(x)$ is close to the zero $x = c$ of $f(x)$. But the zero of $P_1(x)$, a linear function, is found by setting $P_1(x) = 0$. Thus,

$$\begin{aligned} f(x_0) + f'(x_0)(x - x_0) &= 0 \\ f'(x_0)(x - x_0) &= -f(x_0) \\ x - x_0 &= -\frac{f(x_0)}{f'(x_0)} \\ x &= x_0 - \frac{f(x_0)}{f'(x_0)} \end{aligned}$$

This number provides us with another estimate of the zero of $f(x)$:

$$x_1 = x_0 - \frac{f(x_0)}{f'(x_0)}$$

In general, the estimate x_1 is better than the initial estimate x_0.

This process may be repeated with the initial estimate x_0 replaced by the recent estimate x_1. This leads to yet another estimate,

$$x_2 = x_1 - \frac{f(x_1)}{f'(x_1)}$$

which is usually better than x_1. In this manner, we generate a sequence of approximations $x_0, x_1, x_2, \ldots, x_n, x_{n+1}, \ldots,$ with

$$x_{n+1} = x_n - \frac{f(x_n)}{f'(x_n)}$$

which, in most instances, approaches the zero $x = c$ of $f(x)$.

A summary of this algorithm, which is called **Newton's method** (also called the **Newton–Raphson method**), follows.

Newton's Method

1. Pick an initial estimate x_0 of the root c.
2. Find a new estimate using the iterative formula

$$x_{n+1} = x_n - \frac{f(x_n)}{f'(x_n)} \qquad (n = 0, 1, 2, \ldots) \qquad \textbf{(10)}$$

3. Compute $|x_n - x_{n+1}|$. If this number is less than a prescribed positive number, stop. The required approximation to the root $x = c$ is $x = x_{n+1}$.

Solving Equations Using Newton's Method

EXAMPLE 1 Use Newton's method to approximate the zero of $f(x) = x^2 - 2$. Start the iteration with initial guess $x_0 = 1$ and terminate the process when two successive approximations differ by less than 0.00001.

Solution We have

$$f(x) = x^2 - 2 \qquad f'(x) = 2x$$

so that, by (10), the required iterative formula is

$$x_{n+1} = x_n - \frac{x_n^2 - 2}{2x_n} = \frac{x_n^2 + 2}{2x_n}$$

With $x_0 = 1$, we find

$$x_1 = \frac{1^2 + 2}{2(1)} = 1.5$$
$$x_2 = \frac{(1.5)^2 + 2}{2(1.5)} \approx 1.416667$$
$$x_3 = \frac{(1.416667)^2 + 2}{2(1.416667)} \approx 1.414216$$
$$x_4 = \frac{(1.414216)^2 + 2}{2(1.414216)} \approx 1.414214$$

Since $x_3 - x_4 = 0.000002 < 0.00001$, we terminate the process. The sequence generated converges to $\sqrt{2}$, which is one of the two roots of the equation $x^2 - 2 = 0$. Note that, to six places, $\sqrt{2} = 1.414214$!

APPLIED EXAMPLE 2 Flight of a Rocket Refer to Example 8, page 165. The altitude in feet of a rocket t seconds into flight is given by

$$s = f(t) = -t^3 + 96t^2 + 195t + 5 \qquad (t \geq 0)$$

Find the time T when the rocket hits Earth.

Solution The rocket hits Earth when the altitude is equal to zero. So we are required to solve the equation

$$s = f(t) = -t^3 + 96t^2 + 195t + 5 = 0$$

Let's use Newton's method with initial guess $t_0 = 100$ (Figure 10). Here

$$f'(t) = -3t^2 + 192t + 195$$

so the required iterative formula takes the form

$$t_{n+1} = t_n - \frac{-t_n^3 + 96t_n^2 + 195t_n + 5}{-3t_n^2 + 192t_n + 195}$$
$$= t_n - \frac{t_n^3 - 96t_n^2 - 195t_n - 5}{3t_n^2 - 192t_n - 195}$$
$$= \frac{3t_n^3 - 192t_n^2 - 195t_n - t_n^3 + 96t_n^2 + 195t_n + 5}{3t_n^2 - 192t_n - 195}$$
$$= \frac{2t_n^3 - 96t_n^2 + 5}{3t_n^2 - 192t_n - 195}$$

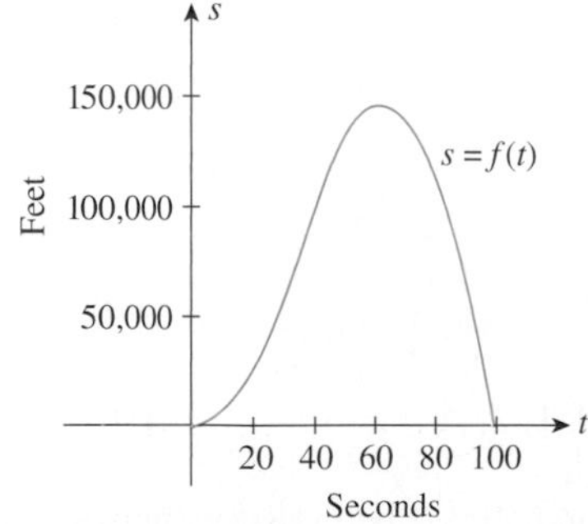

FIGURE 10
The rocket's altitude t seconds into flight is given by $f(t)$.

We have

$$t_1 = \frac{2(100)^3 - 96(100)^2 + 5}{3(100)^2 - 192(100) - 195} \approx 98.0674$$

$$t_2 = \frac{2(98.0674)^3 - 96(98.0674)^2 + 5}{3(98.0674)^2 - 192(98.0674) - 195} \approx 97.9906$$

$$t_3 = \frac{2(97.9906)^3 - 96(97.9906)^2 + 5}{3(97.9906)^2 - 192(97.9906) - 195} \approx 97.9905$$

Thus, we see that $T \approx 97.99$, so the rocket hits Earth approximately 98 seconds after liftoff. ■

The next example shows how Newton's method may be used to help determine a certain commodity's equilibrium quantity and price.

APPLIED EXAMPLE 3 Market Equilibrium The demand equation for the "Tempus" quartz wristwatch is given by

$$p = 50e^{-0.1(x+1)^2}$$

where x (measured in units of a thousand) is the quantity demanded per week and p is the unit wholesale price in dollars. National Importers, the supplier of the watches, will make x units (in thousands) available in the market if the unit wholesale price is

$$p = 10 + 5x^2$$

dollars. Find the equilibrium quantity and price.

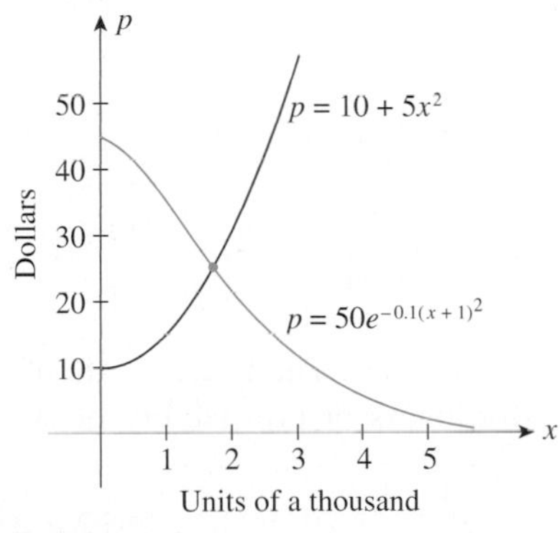

FIGURE 11
The equilibrium point is the point of intersection of the demand curve and the supply curve.

Solution We determine the equilibrium point by finding the point of intersection of the demand curve and the supply curve (Figure 11). To solve the system of equations

$$p = 50e^{-0.1(x+1)^2}$$
$$p = 10 + 5x^2$$

we substitute the second equation into the first, obtaining

$$10 + 5x^2 = 50e^{-0.1(x+1)^2}$$
$$10 + 5x^2 - 50e^{-0.1(x+1)^2} = 0$$

To solve the last equation, we use Newton's method with

$$\begin{aligned} f(x) &= 10 + 5x^2 - 50e^{-0.1(x+1)^2} \\ &= 5[2 + x^2 - 10e^{-0.1(x+1)^2}] \\ f'(x) &= 10x - 50e^{-0.1(x+1)^2} \cdot [-0.2(x+1)] \\ &= 10[x + (x+1)e^{-0.1(x+1)^2}] \end{aligned}$$

leading to the iterative formula

$$x_{n+1} = x_n - \frac{2 + x_n^2 - 10e^{-0.1(x_n+1)^2}}{2[x_n + (x_n+1)e^{-0.1(x_n+1)^2}]}$$

Referring to Figure 11, we see that a reasonable initial estimate is $x_0 = 2$. We have

$$x_1 = 2 - \frac{2 + 2^2 - 10e^{-0.1(3)^2}}{2[2 + 3e^{-0.1(3)^2}]} \approx 1.69962$$

$$x_2 = 1.69962 - \frac{2 + (1.69962)^2 - 10e^{-0.1(2.69962)^2}}{2[1.69962 + 2.69962e^{-0.1(2.69962)^2}]} \approx 1.68899$$

$$x_3 = 1.68899 - \frac{2 + (1.68899)^2 - 10e^{-0.1(2.68899)^2}}{2[1.68899 + 2.68899e^{-0.1(2.68899)^2}]} \approx 1.68898$$

Thus, the equilibrium quantity is approximately 1.689 units, and the equilibrium wholesale price is given by

$$p = 10 + 5(1.68898)^2 \approx 24.263$$

or approximately \$24.26 per watch. ■

The Internal Rate of Return on an Investment

Yet another use of Newton's method is to find the internal rate of return on an investment. Suppose a company has an initial outlay of C dollars in an investment that yields returns of $R_1, R_2, \ldots, R_n$ dollars at the end of the first, second, . . . , nth periods, respectively. Then the *net present value* of the investment is

$$\frac{R_1}{1 + r} + \frac{R_2}{(1 + r)^2} + \frac{R_3}{(1 + r)^3} + \cdots + \frac{R_n}{(1 + r)^n} - C$$

where r denotes the interest rate per period earned on the investment. Now the **internal rate of return** on the investment is defined as the rate of return for which the net present value of the investment is equal to zero; that is, it is the value of r that satisfies the equation

$$\frac{R_1}{1 + r} + \frac{R_2}{(1 + r)^2} + \frac{R_3}{(1 + r)^3} + \cdots + \frac{R_n}{(1 + r)^n} - C = 0$$

or, equivalently, the equation

$$C(1 + r)^n - R_1(1 + r)^{n-1} - R_2(1 + r)^{n-2} - R_3(1 + r)^{n-3} - \cdots - R_n = 0$$

obtained by multiplying both sides of the former by $(1 + r)^n$. The internal rate of return is used by management to decide on the worthiness or profitability of an investment.

APPLIED EXAMPLE 4 Internal Rate of Return The management of A-1 Rental—a tool and equipment rental service for industry, contractors, and homeowners—is contemplating purchasing new equipment. The initial outlay for the equipment, which has a useful life of 4 years, is \$45,000. It is expected that the investment will yield returns of \$15,000 at the end of the first year, \$18,000 at the end of the second year, \$14,000 at the end of the third year, and \$10,000 at the end of the fourth year. Find the internal rate of return on this investment.

Solution Here $n = 4$, $C = 45{,}000$, $R_1 = 15{,}000$, $R_2 = 18{,}000$, $R_3 = 14{,}000$, and $R_4 = 10{,}000$. So we are required to solve the equation

$$45{,}000(1+r)^4 - 15{,}000(1+r)^3 - 18{,}000(1+r)^2 - 14{,}000(1+r) - 10{,}000 = 0$$

for r. The equation may be written more simply by letting $x = 1 + r$. Thus,

$$f(x) = 45{,}000x^4 - 15{,}000x^3 - 18{,}000x^2 - 14{,}000x - 10{,}000 = 0$$

To solve the equation $f(x) = 0$ using Newton's method, we first compute

$$f'(x) = 180{,}000x^3 - 45{,}000x^2 - 36{,}000x - 14{,}000$$

The required iterative formula is

$$x_{n+1} = x_n - \frac{45{,}000x_n^4 - 15{,}000x_n^3 - 18{,}000x_n^2 - 14{,}000x_n - 10{,}000}{180{,}000x_n^3 - 45{,}000x_n^2 - 36{,}000x_n - 14{,}000}$$

Starting with the initial estimate $x_0 = 1.1$, we find

$$x_1 = 1.1 - \frac{45{,}000(1.1)^4 - 15{,}000(1.1)^3 - 18{,}000(1.1)^2 - 14{,}000(1.1) - 10{,}000}{180{,}000(1.1)^3 - 45{,}000(1.1)^2 - 36{,}000(1.1) - 14{,}000}$$

$$\approx 1.10958$$

$$x_2 \approx 1.10941$$

$$x_3 \approx 1.10941$$

Therefore, we may take $x \approx 1.1094$, in which case we see that $r \approx 0.1094$. Thus, the rate of return on the investment is approximately 10.94% per year. ■

Having seen how effective Newton's method can be in finding the zeros of a function, we want to close this section by reminding you that there are situations in which the method fails and that care must be exercised in applying it. Figure 12a illustrates a situation where $f'(x_n) = 0$ for some n (in this case, $n = 2$). Since the iterative Formula (11) involves division by $f'(x_n)$, it should be clear why the method fails to work in this case. However, if you choose a different initial estimate x_0, the situation may yet be salvaged (Figure 12b).

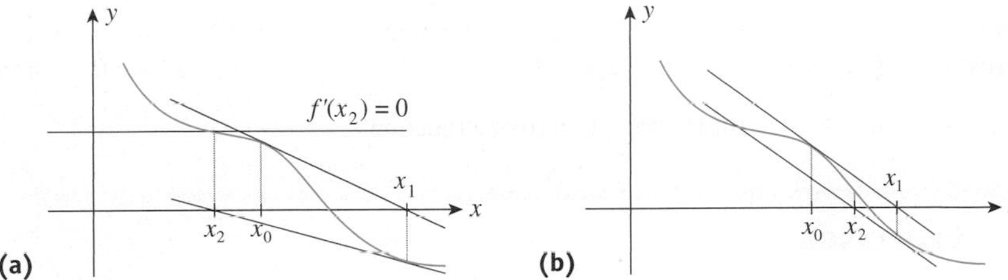

FIGURE 12
In (a), Newton's method fails to work because $f'(x_2) = 0$, but this situation is remedied in (b) by selecting a different initial estimate x_0.

EXPLORE & DISCUSS

For a concrete example of a situation similar to that depicted in Figure 12, consider the function

$$f(x) = x^3 - 1.5x^2 - 6x + 2$$

1. Show that Newton's method fails to work if we choose $x_0 = -1$ or $x_0 = 2$ for an initial estimate.
2. Using the initial estimates $x_0 = -2.5$, $x_0 = 1$, and $x_0 = 2.5$, show that the three roots of $f(x) = 0$ are -2, 0.313859, and 3.186141, respectively.
3. Using a graphing utility, plot the graph of f in the viewing window $[-3, 4] \times [-10, 7]$. Verify the results of part 2, using **TRACE** and **ZOOM** or the root-finding function of your calculator.

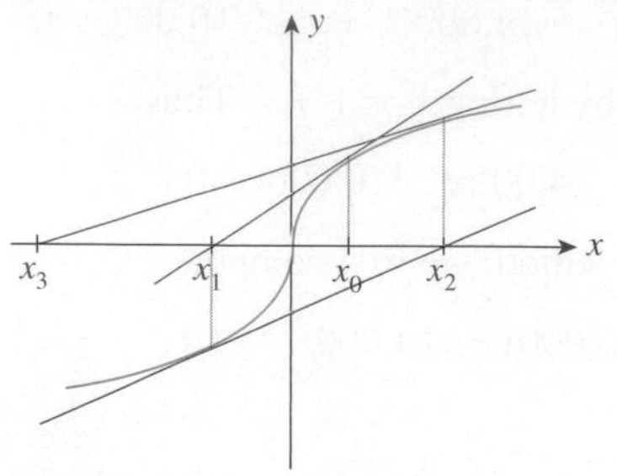

FIGURE 13
Newton's method fails here because the sequence of estimates diverges.

The next situation, shown in Figure 13, is more serious, and the method will not work for any choice of the initial estimate x_0 other than the actual zero of the function $f(x)$ because the sequence, $x_1, x_2, \ldots x_n$ diverges.

EXPLORE & DISCUSS

For a concrete example of a situation similar to that depicted in Figure 13, consider the function $f(x) = x^{1/3}$.

1. Show that Newton's iteration for solving the equation $f(x) = 0$ is $x_{n+1} = -2x_n$, with x_0 being an initial guess.
2. Show that the sequence $x_0, x_1, x_2, \ldots$ diverges for any choice of x_0 other than zero and thus Newton's method does not lead to the unique solution $x = 0$ for the problem under consideration.
3. Illustrate this situation geometrically.

11.7 Self-Check Exercises

1. Use three iterations of Newton's algorithm on an appropriate function $f(x)$ and initial guess x_0 to obtain an estimate of $\sqrt[3]{10}$.
2. A study prepared for a certain Sunbelt town's Chamber of Commerce projected that the town's population in the next 3 yr will grow according to the rule

$$P(t) = 50{,}000 + 30t^{3/2} + 20t$$

where $P(t)$ denotes the population t mo from now. Using this model, estimate the time when the population will reach 55,000.

Solutions to Self-Check Exercises 11.7 can be found on page 767.

11.7 Concept Questions

1. Give a geometric description of Newton's method for finding the zeros of the function f. Illustrate graphically.
2. Describe Newton's method for finding the zero of a function f.
3. Does Newton's method always work for any choice of the initial estimate x_0 of the root of $f(x) = 0$? Explain graphically.

11.7 Exercises

A calculator is recommended for this exercise set. In Exercises 1–6, estimate the value of each radical by using three iterations of Newton's method with the indicated initial guess for each function.

1. $\sqrt{3}; f(x) = x^2 - 3; x_0 = 1.5$
2. $\sqrt{5}; f(x) = x^2 - 5; x_0 = 2$
3. $\sqrt{7}; f(x) = x^2 - 7; x_0 = 2.5$
4. $\sqrt[3]{6}; f(x) = x^3 - 6; x_0 = 2$
5. $\sqrt[3]{14}; f(x) = x^3 - 14; x_0 = 2.5$
6. $\sqrt[4]{50}; f(x) = x^4 - 50; x_0 = 2.5$

In Exercises 7–10, use Newton's method to find the zero(s) of f to four decimal places by solving the equation $f(x) = 0$. Use the initial estimate(s) x_0.

7. $f(x) = -x^3 - 2x + 2; x_0 = 1$

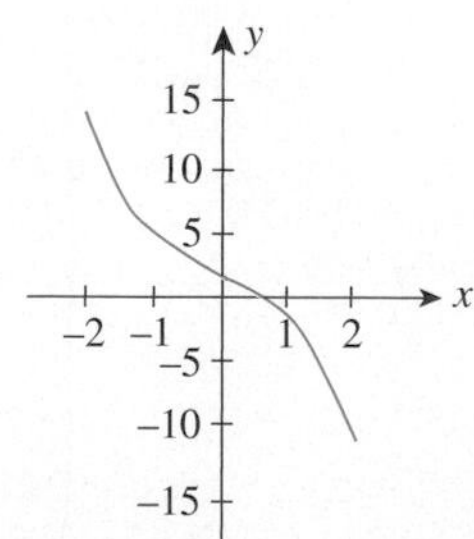

8. $f(x) = 2x^3 - 15x^2 + 36x - 20$; $x_0 = 1$

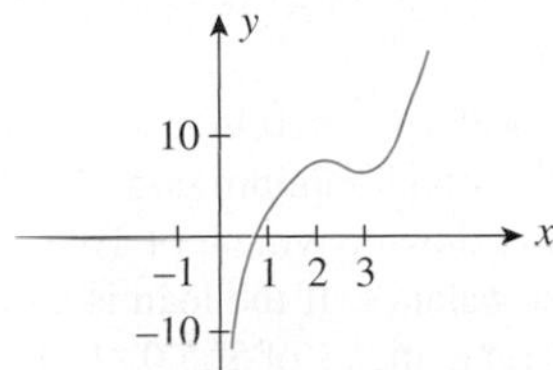

9. $f(x) = \frac{3}{2}x^4 - 2x^3 - 6x^2 + 8$; $x_0 = 1$ and $x_0 = 3$

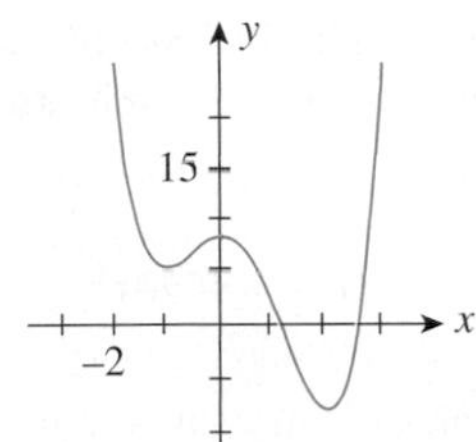

10. $f(x) = x - \sqrt{1 - x^2}$; $x_0 = 0.5$

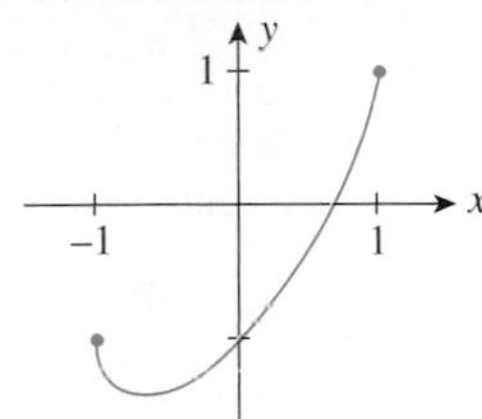

In Exercises 11–18, use Newton's method to approximate the indicated zero of each function. Continue with the iteration until two successive approximations differ by less than 0.0001.

11. The zero of $f(x) = x^2 - x - 3$ between $x = 2$ and $x = 3$

12. The zero of $f(x) = x^3 + x - 3$ between $x = 1$ and $x = 2$

13. The zero of $f(x) = x^3 + 2x^2 + x - 5$ between $x = 1$ and $x = 2$

14. The zero of $f(x) = x^5 + x - 1$ between $x = 0$ and $x = 1$

15. The zero of $f(x) = \sqrt{x + 1} - x$ between $x = 1$ and $x = 2$

16. The zero of $f(x) = e^{-x} - x$ between $x = 0$ and $x = 1$

17. The zero of $f(x) = e^x - (1/x)$ between $x = 0$ and $x = 1$

18. The zero of $f(x) = \ln x^2 - 0.7x + 1$ between $x = 6$ and $x = 7$

19. Let $f(x) = 2x^3 - 9x^2 + 12x - 2$.
   a. Show that $f(x) = 0$ has a root between $x = 0$ and $x = 1$.
   Hint: Compute $f(0)$ and $f(1)$ and use the fact that f is continuous.
   b. Use Newton's method to find the zero of f in the interval $(0, 1)$.

20. Let $f(x) = x^3 - x - 1$.
   a. Show that $f(x) = 0$ has a root between $x = 1$ and $x = 2$
   Hint: See Exercise 19.
   b. Use Newton's method to find the zero of f in the interval $(1, 2)$.

21. Let $f(x) = x^3 - 3x - 1$.
   a. Show that f has a zero between $x = 1$ and $x = 2$.
   Hint: See Exercise 19.
   b. Use Newton's method to find the zero of f.

22. Let $f(x) = x^4 - 4x^3 + 10$.
   a. Show that $f(x) = 0$ has a root between $x = 1$ and $x = 2$.
   b. Use Newton's method to find the zero of f.

In Exercises 23–28, make a rough sketch of the graphs of each of the given pairs of functions. Use your sketch to approximate the point(s) of intersection of the two graphs and then apply Newton's method to refine the approximation of the x-coordinate of the point of intersection.

23. $f(x) = 2\sqrt{x + 3}$; $g(x) = 2x - 1$

24. $f(x) = e^{-x^2}$; $g(x) = x^2$

25. $f(x) = e^{-x}$; $g(x) = x - 1$

26. $f(x) = \ln x$; $g(x) = 2 - x$

27. $f(x) = \sqrt{x}$; $g(x) = e^{-x}$

28. $f(x) = 2 - x^2$; $g(x) = \ln x$

29. Minimizing Average Cost A division of Ditton Industries manufactures the "Futura" model microwave oven. Given that the daily cost of producing these microwave ovens (in dollars) obeys the rule

$$C(x) = 0.0002x^3 - 0.06x^2 + 120x + 5000$$

where x stands for the number of units produced, find the level of production that minimizes the daily average cost per unit.

30. Altitude of a Rocket The altitude (in feet) of a rocket t sec into flight is given by

$$s = f(t) = -2t^3 + 114t^2 + 480t + 1 \qquad (t \geq 0)$$

Find the time T when the rocket hits Earth.

31. Temperature The temperature at 6 A.M. on a certain December day was measured at 15.6°F. In the next t hr, the temperature was given by the function

$$T = -0.05t^3 + 0.4t^2 + 3.8t + 15.6 \qquad (0 \leq t \leq 15)$$

where T is measured in degrees Fahrenheit. At what time was the temperature 0°F?

32. Internal Rate of Return on an Investment The proprietor of Qwik Film Lab recently purchased \$12,000 of new film-processing equipment. She expects that this investment, which has a useful life of 4 yr, will yield returns of \$4000 at the end of the first year, \$5000 at the end of the second year,

\$4000 at the end of the third year, and \$3000 at the end of the fourth year. Find the internal rate of return on the investment.

33. **INTERNAL RATE OF RETURN ON AN INVESTMENT** Executive Limousine Service recently acquired limousines worth \$120,000. The projected returns over the next 3 yr, the time period the limousines will be in service, are \$80,000 at the end of the first year, \$60,000 at the end of the second year, and \$40,000 at the end of the third year. Find the internal rate of return on the investment.

34. **INTERNAL RATE OF RETURN ON AN INVESTMENT** Suppose an initial outlay of \$$C$ in an investment yields returns of \$$R$ at the end of each period over N periods.

 a. Show that the internal rate of return on the investment, r, may be obtained by solving the equation

$$Cr + R[(1 + r)^{-N} - 1] = 0$$

 Hint: $1 + x + x^2 + \cdots + x^{N-1} = \dfrac{1 - x^{N+1}}{1 - x}$

 b. Show that r can be found by performing the iteration

$$r_{n+1} = r_n - \frac{Cr_n + R[(1 + r_n)^{-N} - 1]}{C - NR(1 + r_n)^{-N-1}}$$

 [r_0 (positive), an initial guess]

 Hint: Apply Newton's method to the function $f(r) = Cr + R[(1 + r)^{-N} - 1]$.

35. **HOME MORTGAGES** Refer to Exercise 34. The Flemings secured a loan of \$200,000 from a bank to finance the purchase of a house. They have agreed to repay the loan in equal monthly installments of \$2106 over 25 yr. The bank charges interest at the rate of $12r$/year on the unpaid balance, and interest computations are made at the end of each month. Find r.

36. **HOME MORTGAGES** The Blakelys borrowed a sum of \$240,000 from a bank to help finance the purchase of a house. The bank charges interest at the rate of $12r$/year on the unpaid balance, with interest being computed at the end of each month. The Blakelys have agreed to repay the loan in equal monthly installments of \$1931.10 over 30 yr. What is the true rate of interest charged by the bank?
 Hint: See Exercise 34.

37. **CAR LOANS** The price of a certain new car is \$16,000. Suppose an individual makes a down payment of 25% toward the purchase of the car and secures financing for the balance over 4 yr. If the monthly payment is \$304.36, what is the true rate of interest charged by the finance company?
 Hint: See Exercise 34.

38. **REAL ESTATE INVESTMENT GROUPS** Refer to Exercise 34. A group of private investors purchased a condominium complex for \$2 million. They made an initial down payment of 10% and have obtained financing for the balance. If the loan is amortized over 15 yr with quarterly repayments of \$65,039, determine the interest rate charged by the bank. Assume that interest is calculated at the end of each quarter and is based on the unpaid balance.

39. **DEMAND FOR WRISTWATCHES** The quantity of "Sicard" wristwatches demanded per month is related to the unit price by the equation

$$p = d(x) = \frac{50}{0.01x^2 + 1} \qquad (1 \le x \le 20)$$

 where p is measured in dollars and x is measured in units of a thousand. The supplier is willing to make x thousand wristwatches available per month when the price per watch is given by $p = s(x) = 0.1x + 20$ dollars. Find the equilibrium quantity and price.

40. **DEMAND FOR LCD TELEVISIONS** The weekly demand for the Pulsar 13-in. LCD television is given by the demand equation

$$p = -0.05x + 600 \qquad (0 \le x \le 12{,}000)$$

 where p denotes the wholesale unit price in dollars and x denotes the quantity demanded. The weekly total cost function associated with the manufacture of these televisions is given by

$$C = 0.000002x^3 - 0.03x^2 + 400x + 80{,}000$$

 where $C(x)$ denotes the total cost incurred in the production of x sets. Find the break-even level(s) of operation for the company.
 Hint: Solve the equation $P(x) = 0$, where P is the total profit function.

41. **a.** Show that using Newton's method for finding the nth root of a real number a leads to the iterative formula

$$x_{i+1} = \left(\frac{n-1}{n}\right)x_i + \frac{a}{nx_i^{n-1}}$$

 b. Use the result of part (a) to find $\sqrt[4]{42}$, accurate to three decimal places.

11.7 Solutions to Self-Check Exercises

1. Since $\sqrt[3]{10}$ is one of the cube roots of the equation $x^3 - 10 = 0$, let's take $f(x) = x^3 - 10$. We have

$$f'(x) = 3x^2$$

and so the required iteration formula is

$$\begin{aligned} x_{n+1} &= x_n - \frac{f(x_n)}{f'(x_n)} \\ &= x_n - \frac{x_n^3 - 10}{3x_n^2} \\ &= \frac{2x_n^3 + 10}{3x_n^2} \end{aligned}$$

Taking $x_0 = 2$, we find

$$\begin{aligned} x_1 &= \frac{2(2^3) + 10}{3(2^2)} \approx 2.166667 \\ x_2 &= \frac{2(2.166667)^3 + 10}{3(2.166667)^2} \approx 2.154504 \\ x_3 &= \frac{2(2.154504)^3 + 10}{3(2.154504)^2} \approx 2.154435 \end{aligned}$$

Therefore, $\sqrt[3]{10} \approx 2.1544$.

2. We need to solve the equation $P(t) = 55{,}000$ or

$$50{,}000 + 30t^{3/2} + 20t = 55{,}000$$

Rewriting, we have

$$\begin{aligned} 30t^{3/2} + 20t - 5000 &= 0 \\ 3t^{3/2} + 2t - 500 &= 0 \end{aligned}$$

Thus, the problem reduces to that of finding the zero of the function

$$f(t) = 3t^{3/2} + 2t - 500$$

Using Newton's method, we have

$$f'(t) = \frac{9}{2}t^{1/2} + 2 = \frac{9t^{1/2} + 4}{2}$$

So the required iteration formula is

$$\begin{aligned} t_{n+1} &= t_n - \frac{3t_n^{3/2} + 2t_n - 500}{\dfrac{9t_n^{1/2} + 4}{2}} \\ &= t_n - \frac{6t_n^{3/2} + 4t_n - 1000}{9t_n^{1/2} + 4} \\ &= \frac{9t_n^{3/2} + 4t_n - (6t_n^{3/2} + 4t_n - 1000)}{9t_n^{1/2} + 4} \\ &= \frac{3t_n^{3/2} + 1000}{9t_n^{1/2} + 4} \end{aligned}$$

Using the initial guess $t_0 = 24$, we find

$$\begin{aligned} t_1 &= \frac{3(24)^{3/2} + 1000}{9(24)^{1/2} + 4} \approx 28.128584 \\ t_2 &= \frac{3(28.128584)^{3/2} + 1000}{9(28.128584)^{1/2} + 4} \approx 27.981338 \\ t_3 &= \frac{3(27.981338)^{3/2} + 1000}{9(27.981338)^{1/2} + 4} \approx 27.981160 \end{aligned}$$

So we may take $t \approx 27.98$. Thus, the population will reach 55,000 approximately 28 mo from now.

CHAPTER 11 Summary of Principal Formulas and Terms

FORMULAS

1. nth Taylor polynomial of f at $x = a$	$P_n(x) = f(a) + f'(a)(x - a) + \dfrac{f''(a)}{2!}(x - a)^2 + \cdots + \dfrac{f^{n}(a)}{n!}(x - a)^n$
2. Infinite series	$\sum_{n=1}^{\infty} a_n = a_1 + a_2 + a_3 + \cdots$

3. Geometric series	$\sum_{n=0}^{\infty} ar^n = a + ar + \cdots$ $= \dfrac{a}{1-r} \quad (\lvert r\rvert < 1)$
4. Taylor series expansion of $f(x)$ at $x = a$	$f(x) = f(a) + f'(a)(x-a)$ $+ \dfrac{f''(a)}{2!}(x-a)^2$ $+ \dfrac{f'''(a)}{3!}(x-a)^3 + \cdots$
5. Iterative formula for Newton's method	$x_{n+1} = x_n - \dfrac{f(x_n)}{f'(x_n)}$ $(n = 0, 1, 2, \ldots)$

TERMS

error bound for Taylor polynomials (705)
infinite sequence (710)
limit of a sequence (713)
convergent sequence (713)
divergent sequence (713)
partial sum of an infinite series (720)
convergent series (720)
divergent series (720)
geometric series (722)
perpetuity (726)
capital value of a perpetuity (726)
p-series (735)
Taylor series of $f(x)$ (742)
Maclaurin series of $f(x)$ (742)
power series (742)
interval of convergence (742)
internal rate of return (762)

CHAPTER 11 Concept Review Questions

Fill in the blanks.

1. a. The nth Taylor polynomial of f at $x = a$ is a polynomial of the form $P_n(x) =$ ____.

b. In terms of f and its derivatives, we have $P_n(a) =$ ____, $P_n'(a) =$ ____, . . . , $P_n^{(n)}(a) =$ ____.

2. If f has derivatives up through order $N + 1$ in an interval I containing the number a, then for each x in ____ there exists a number c lying between ____ and ____ such that $R_N(x) =$ ____, where $R_N(x) = f(x) - P_N(x)$.

3. a. An infinite sequence is a ____ whose domain is the set of positive ____. The term $a_n =$ ____ is called the ____ ____ of the sequence.

b. If a_n can be made as close to the number L as we please by taking n sufficiently ____, then $\{a_n\}$ is said to ____ to L.

4. If $\lim_{n\to\infty} a_n = A$ and $\lim_{n\to\infty} b_n = B$ and c is any real number, then $\lim_{n\to\infty} ca_n =$ ____, $\lim_{n\to\infty}(a_n + b_n) =$ ____, $\lim_{n\to\infty} a_nb_n =$ ____, and $\lim_{n\to\infty} \dfrac{a_n}{b_n} =$ ____, provided ____.

5. a. A series $\sum_{n=1}^{\infty} a_n$ converges and has sum S if its sequence of ____ ____ converges to S.

b. The series $\sum_{n=0}^{\infty} ar^n$ is called a ____ series. It converges if $\lvert r\rvert <$ ____, and it diverges if $\lvert r\rvert \geq$ ____.

6. a. If $\sum_{n=1}^{\infty} a_n$ converges, then $\lim_{n\to\infty} a_n =$ ____. If $\lim_{n\to\infty} a_n$ does not exist or if $\lim_{n\to\infty} a_n \neq 0$, then $\sum_{n=1}^{\infty} a_n$ ____.

b. If $\sum_{n=1}^{\infty} a_n = A$ and $\sum_{n=1}^{\infty} b_n = B$ and c is any real number, then $\sum_{n=1}^{\infty}(ca_n + b_n) =$ ____.

7. a. If f is a continuous, positive, and decreasing function on $[1, \infty)$ and $f(n) = a_n$ for $n \geq 1$, then $\sum_{n=1}^{\infty} a_n$ and $\int_1^{\infty} f(x)\,dx$ are either both ____ or both ____.

b. The p-series has the form ____, and it converges if ____ and diverges if ____.

8. Let $\Sigma\, a_n$ and $\Sigma\, b_n$ be series with positive terms and with $a_n \le b_n$ for all n. If $\Sigma\, b_n$ converges, then $\Sigma\, a_n$ ____. If $\Sigma\, b_n$ diverges and ____ for all n, then $\Sigma\, a_n$ also diverges.

9. a. A power series in $(x - a)$ is a series of the form ____.
   b. For a power series in $(x - a)$, exactly one of the following is true: it converges only at ____; it converges for all ____; or it converges for x in the interval ____, where R is some positive number. In the last case, the series diverges for x ____ this interval.

10. a. The Taylor series of a function f at a has the form $\sum_{n=0}^{\infty} a_n(x - a)^n$, where $a_n =$ ____.
    b. If f has derivatives of all orders in I and $R_N(x) = f(x) - P_N(x)$, then $f(x) = \sum_{n=0}^{\infty} \frac{f^{(n)}}{n!}(x - a)^n$ if and only if ____. In this case, f is said to be represented by ____ ____ ____ of f at $x = a$.

CHAPTER 11 Review Examples

In Exercises 1–4, find the fourth Taylor polynomial of the function f at the indicated number.

1. $f(x) = \dfrac{1}{x + 2}$ at $x = -1$

2. $f(x) = e^{-x}$ at $x = 1$

3. $f(x) = \ln(1 + x^2)$ at $x = 0$

4. $f(x) = \dfrac{1}{(1 + x)^2}$ at $x = 0$

5. Find the second Taylor polynomial of $f(x) = \sqrt[3]{x}$ at $x = 8$ and use it to estimate the value of $\sqrt[3]{7.8}$.

6. Find the sixth Taylor polynomial of $f(x) = e^{-x^2}$ at $x = 0$ and use it to estimate the value of

$$\int_0^1 e^{-x^2}\,dx$$

7. Use the second Taylor polynomial of $f(x) = \sqrt[3]{x}$ at $x = 27$ to approximate $\sqrt[3]{26.98}$ and find a bound for the error in the approximation.

8. Use the third Taylor polynomial of $f(x) = 1/(1 + x)$ to approximate $f(0.1)$. Find a bound for the error in the approximation and compare your results to the exact value of $f(0.1)$.

9. Find an estimate of e^{-1} with an error less than 10^{-2}.

10. Estimate $\int_0^{0.2} e^{-(1/2)x^2}\,dx$ with an error less than 10^{-2}.

In Exercises 11–14, the nth term of a sequence is given. Determine whether the sequence converges or diverges. If the sequence converges, find its limit.

11. $a_n = \dfrac{2n^2 + 1}{3n^2 - 1}$

12. $a_n = \dfrac{(-1)^{n-1}n}{n + 1}$

13. $a_n = 1 - \dfrac{1}{2^n}$

14. $a_n = \dfrac{1 + \sqrt{n}}{1 - \sqrt{n}}$

In Exercises 15–18, find the sum of the geometric series if it converges.

15. $\sum_{n=1}^{\infty} \dfrac{2^n}{3^n}$

16. $\sum_{n=1}^{\infty} 2^{-n}3^{-n+1}$

17. $\sum_{n=1}^{\infty} (-1)^{n-1}\left(\dfrac{1}{\sqrt{2}}\right)^n$

18. $\sum_{n=1}^{\infty} \left(\dfrac{1}{e}\right)^n$

19. Express the repeating decimal 1.424242. . . as a rational number.

20. Express the repeating decimal 3.142142142. . . as a rational number.

In Exercises 21–24, determine whether the series is convergent or divergent.

21. $\sum_{n=1}^{\infty} \dfrac{n^2 + 1}{2n^2 - 1}$

22. $\sum_{n=1}^{\infty} \dfrac{n + 1}{2n^2 + 4n}$

23. $\sum_{n=1}^{\infty} \left(\dfrac{1}{n}\right)^{1.1}$

24. $\sum_{n=1}^{\infty} \dfrac{n^3}{n^5 + 2}$

In Exercises 25–28, find the radius of convergence and the interval of convergence of each power series.

25. $\sum_{n=0}^{\infty} \dfrac{x^n}{n^2 + 2}$

26. $\sum_{n=1}^{\infty} \dfrac{(-1)^{n-1}}{\sqrt{n}}x^n$

27. $\sum_{n=1}^{\infty} \dfrac{(x - 1)^n}{n(n + 1)}$

28. $\sum_{n=2}^{\infty} \dfrac{e^n}{n^2}(x - 2)^n$

In Exercises 29–32, find the Taylor series of the function at the indicated number. Give the interval of convergence for each series.

29. $f(x) = \dfrac{1}{2x - 1}$; $x = 0$

30. $f(x) = e^{-x}$; $x = 1$

31. $f(x) = \ln(1 + 2x)$; $x = 0$

32. $f(x) = x^2e^{-2x}$; $x = 0$

33. Apply Newton's method to an appropriate function $f(x)$ to obtain an estimate of $\sqrt[3]{12}$.

34. Use Newton's method to find the root of the equation $x^3 + x^2 - 1 = 0$ between $x = 0$ and $x = 1$.

35. Use Newton's method to find the point of intersection of the graphs of the functions $y = f(x) = 2x$ and $y = g(x) = e^{-x}$.

36. A suitcase released from rest at the top of a plane metal slide moves a distance of

$$x = f(t) = 27(t + 3e^{-t/3} - 3)$$

feet in t sec. If the metal slide is 24 ft long, how long does it take the suitcase to reach the bottom?
Hint: Use Newton's method.

37. ESTABLISHING A TRUST FUND Sam Simpson wishes to establish a trust fund that will provide each of his two children with an annual income of \$10,000 beginning next year and continuing throughout their lifetimes. Find the amount of money he is required to place in trust now if the fund will earn interest at the rate of 9%/year compounded continuously.

38. EFFECT OF A TAX CUT Suppose the average wage earner in a certain country saves 8% of her take-home pay and spends the other 92%.
   a. Estimate the impact that a proposed \$10 billion tax cut will have on the economy over the long run in terms of the additional spending generated.
   b. Estimate the impact that a \$10 billion tax cut will have on the economy over the long run if, at the same time, legislation is enacted to boost the rate of savings of the average taxpayer from 8% to 10%.

39. HEIGHTS OF WOMEN The heights of 4000 women who participated in a recent survey were found to be normally distributed according to the probability density function

$$f(x) = \frac{1}{2.5\sqrt{2\pi}}e^{-1/2[(x-64.5)/2.5]^2}$$

What percentage of these women have heights between 63.5 and 65.5 in.?

CHAPTER 11 Before Moving On . . .

1. Find the first three Taylor polynomials of $f(x) = xe^{-x}$ at $x = 0$.

2. Determine whether the sequence converges or diverges. If the sequence converges, find its limit.

   a. $\left\{\dfrac{2n^2}{3n^2 + 2n + 1}\right\}$ b. $\left\{\dfrac{2n}{\sqrt{n^2 + 1}}\right\}$

3. Determine whether the series converges or diverges. If it converges, find its sum.

   a. $\displaystyle\sum_{n=0}^{\infty} \frac{3 - 2^n}{5^n}$ b. $\displaystyle\sum_{n=1}^{\infty} \left(\frac{1}{n + 2} - \frac{1}{n + 3}\right)$

4. Determine whether the series is convergent or divergent.

   a. $\displaystyle\sum_{n=1}^{\infty} \frac{1}{2n^2 + n + 3}$ b. $\displaystyle\sum_{n=2}^{\infty} \frac{(\ln n)^2}{n}$

5. Find the Taylor series of $f(x) = \dfrac{1}{1 + x}$ at $x = -2$, and give its radius of convergence.

12 Trigonometric Functions

The revenue of McMenamy's Fish Shanty follows a cyclical pattern. When is the revenue of the restaurant increasing most rapidly? In Example 6, page 791, we attempt to answer this question.

Andre Jenny/Alamy

During the golden period of Greek civilization, Apollonius (262–200 B.C.) developed the trigonometric techniques necessary for calculating the radii of various circles. This was done in an effort to describe planetary motion, a physical process that exhibits a cyclical, or periodic, mode of behavior. Many other real-life phenomena, such as business cycles, earthquake vibrations, respiratory cycles, sales trends, and sound waves, exhibit cyclical behavior patterns.

In this chapter we extend our study of calculus to an important class of functions called the trigonometric functions. These functions are periodic and hence lend themselves readily to describing many natural phenomena.

12.1 Measurement of Angles

Angles

An **angle** consists of two rays that intersect at a common end point. If we rotate the ray l_1 in a *counterclockwise* direction about the point O, the angle generated is *positive* (Figure 1a). On the other hand, if we rotate the ray l_1 in a *clockwise* direction about the point O, the angle generated is *negative* (Figure 1b). We refer to the ray l_1 as the **initial ray**, the ray l_2 as the **terminal ray**, and the end point O as the **vertex** of the angle. If A and B are points on l_2 and l_1, respectively, then we refer to the angle as angle AOB (Figure 1c).

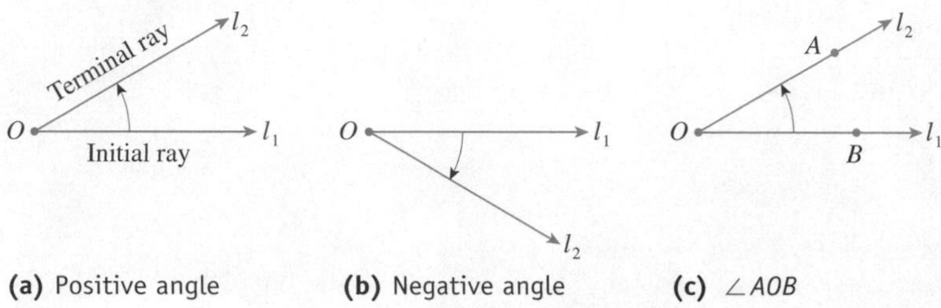

FIGURE 1 **(a)** Positive angle **(b)** Negative angle **(c)** $\angle AOB$

We can represent an angle in the rectangular coordinate system. An angle is in **standard position** if the vertex of the angle is centered at the origin and its initial side coincides with the positive x-axis (Figures 2a and b).

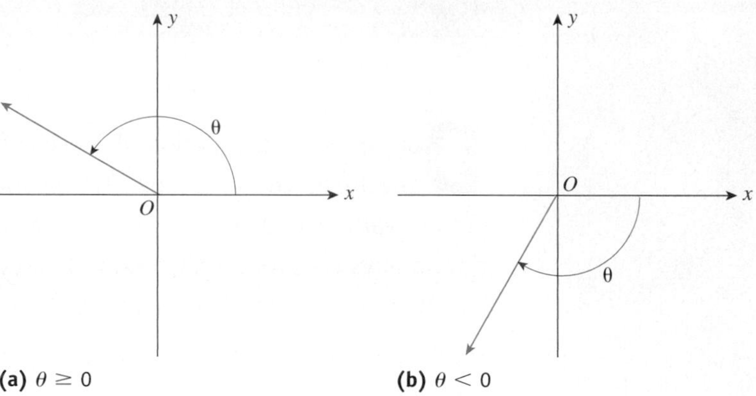

FIGURE 2 **(a)** $\theta \geq 0$ **(b)** $\theta < 0$

When we say that an angle lies in a certain quadrant, we are referring to the quadrant in which the terminal ray lies. The angle shown in Figure 3a lies in Quadrant II, and the angles shown in Figures 3b and 3c lie in Quadrant IV.

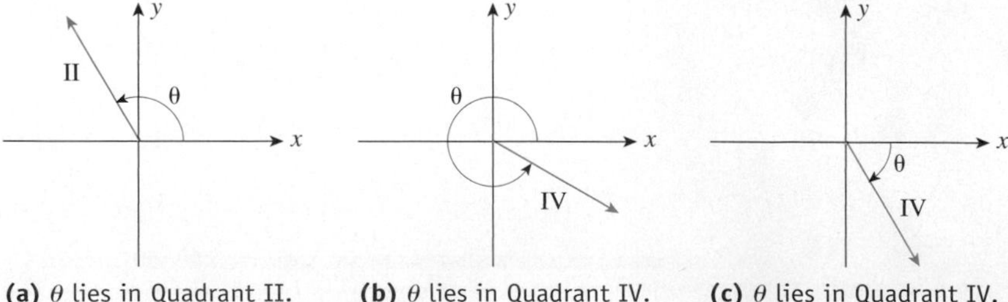

FIGURE 3 **(a)** θ lies in Quadrant II. **(b)** θ lies in Quadrant IV. **(c)** θ lies in Quadrant IV.

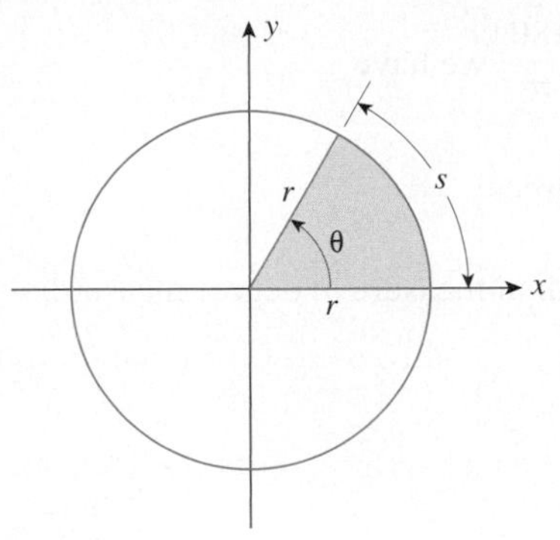

FIGURE 4
$\theta = \frac{s}{r}$ radians

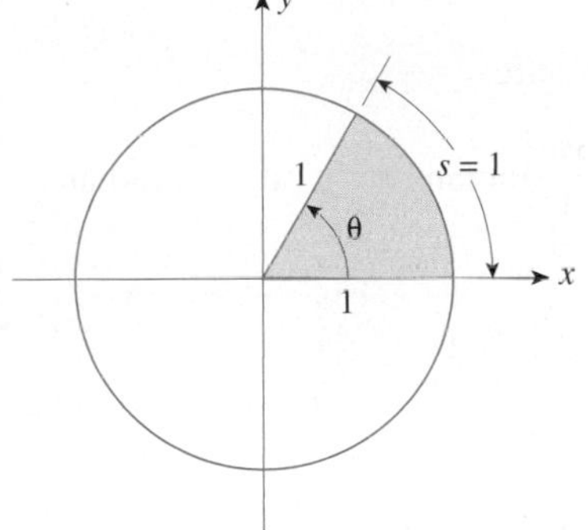

FIGURE 5
$\theta = \frac{s}{r} = 1$ radian

Degree and Radian Measure

An angle may be measured in either degrees or radians. A **degree** is the measure of the angle formed by $\frac{1}{360}$ of one complete revolution. If we rotate an initial ray in standard position through one complete revolution, we obtain an angle of 360°.

A **radian** is the measure of the central angle subtended by an arc equal in length to the radius of the circle. In Figure 4, if s is the length of the arc subtended by a central angle θ in a circle of radius r, then

$$\theta = \frac{s}{r} \text{ radians} \tag{1}$$

For convenience, we consider a circle of radius 1 centered at the origin. We refer to this circle as the **unit circle**. Then an angle of 1 radian is subtended by an arc of length 1 (Figure 5). The circumference of the unit circle is 2π, and, consequently, the angle subtended by one complete revolution is 2π radians. In terms of degrees, the angle subtended by one complete revolution is 360°. Therefore,

$$360 \text{ degrees} = 2\pi \text{ radians}$$

so that

$$1 \text{ degree} = \frac{\pi}{180} \text{ radians}$$

and

$$x \text{ degrees} = \frac{\pi}{180} x \text{ radians} \tag{2}$$

Since this relationship is linear, we may specify the rule for converting degree measure to radian measure in terms of the following linear function.

Converting Degrees to Radians

$$f(x) = \frac{\pi}{180} x \text{ radians} \tag{3}$$

where x is the number of degrees and $f(x)$ is the number of radians.

EXAMPLE 1 Convert each angle to radian measure:

a. 30° **b.** 45° **c.** 300° **d.** 450° **e.** −240°

Solution Using Formula (3), we have

a. $f(30) = \frac{\pi}{180}(30)$ radians, or $\frac{\pi}{6}$ radians

b. $f(45) = \frac{\pi}{180}(45)$ radians, or $\frac{\pi}{4}$ radians

c. $f(300) = \frac{\pi}{180}(300)$ radians, or $\frac{5\pi}{3}$ radians

d. $f(450) = \frac{\pi}{180}(450)$ radians, or $\frac{5\pi}{2}$ radians

e. $f(-240) = \frac{\pi}{180}(-240)$ radians, or $-\frac{4\pi}{3}$ radians ■

By multiplying both sides of Equation (2) by $\frac{180}{\pi}$, we have

$$x \text{ radians} = \frac{180}{\pi} x \text{ degrees}$$

Thus, we have the following rule for converting radian measure to degree measure.

Converting Radians to Degrees

$$g(x) = \frac{180}{\pi} x \text{ degrees} \tag{4}$$

where x is the number of radians and $g(x)$ is the number of degrees.

EXAMPLE 2 Convert each angle to degree measure:

a. $\frac{\pi}{2}$ radians **b.** $\frac{5\pi}{4}$ radians **c.** $-\frac{3\pi}{4}$ radians **d.** $\frac{7\pi}{2}$ radians

Solution Using Formula (4), we have

a. $g\left(\frac{\pi}{2}\right) = \frac{180}{\pi}\left(\frac{\pi}{2}\right)$ degrees, or 90 degrees

b. $g\left(\frac{5\pi}{4}\right) = \frac{180}{\pi}\left(\frac{5\pi}{4}\right)$ degrees, or 225 degrees

c. $g\left(-\frac{3\pi}{4}\right) = \frac{180}{\pi}\left(-\frac{3\pi}{4}\right)$ degrees, or -135 degrees

d. $g\left(\frac{7\pi}{2}\right) = \frac{180}{\pi}\left(\frac{7\pi}{2}\right)$ degrees, or 630 degrees ■

Take time to familiarize yourself with the radian and degree measures of the common angles given in Table 1.

TABLE 1

Degrees	0°	30°	45°	60°	90°	120°	135°	150°	180°	270°	360°
Radians	0	$\frac{\pi}{6}$	$\frac{\pi}{4}$	$\frac{\pi}{3}$	$\frac{\pi}{2}$	$\frac{2\pi}{3}$	$\frac{3\pi}{4}$	$\frac{5\pi}{6}$	π	$\frac{3\pi}{2}$	2π

You may have observed by now that more than one angle may be described by the same initial and terminal rays. The two angles in Figure 6a and b illustrate this case. The angle $\theta = 5\pi/4$ radians is generated by rotating the initial ray in a counterclockwise direction, and the angle $\theta = -3\pi/4$ radians is generated by rotating the initial ray in a clockwise direction. We refer to such angles as **coterminal angles**.

You may also have observed from the preceding examples that an angle may be greater than 2π radians. For example, an angle of 3π radians is generated by rotating a ray through 1.5 revolutions. As illustrated in Figure 7, an angle of 3π radians has

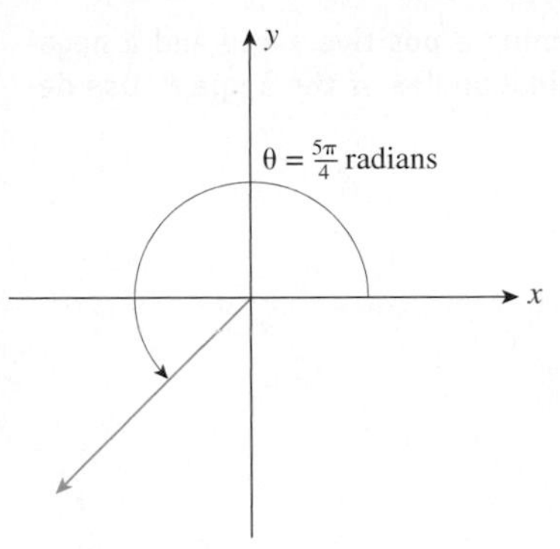

(a)

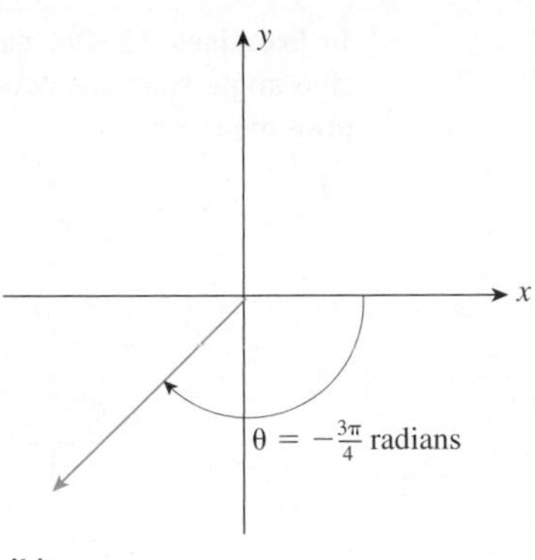

(b)

FIGURE 6
Coterminal angles

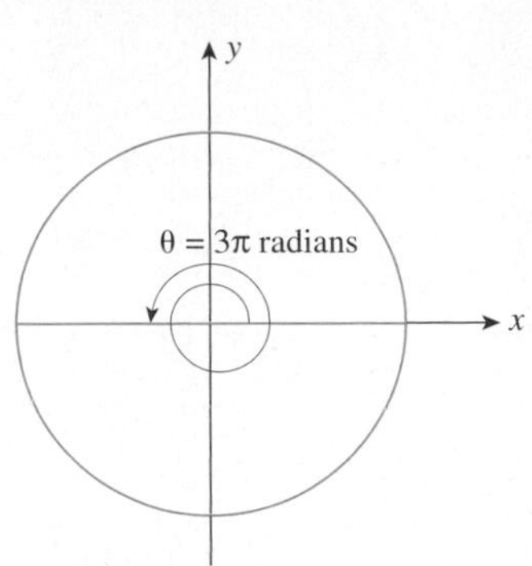

FIGURE 7
$\theta = 3\pi$ radians has its initial and terminal rays in the same position as $\theta = \pi$ radians.

its initial and terminal rays in the same position as an angle of π radians. A similar statement is true for the negative angles $-\pi$ and -3π radians. Thus, in identifying an angle, we must be careful to specify the *direction* of the rotation and the *number of revolutions* through which the angle has gone.

12.1 Self-Check Exercises

1. **a.** Convert 315° to radian measure.

   b. Convert $-\dfrac{5\pi}{4}$ radians to degree measure.

2. Make a sketch of the angle $-\dfrac{2\pi}{3}$ radians.

Solutions to Self-Check Exercises 12.1 can be found on page 777.

12.1 Concept Questions

1. Define the following terms in your own words: a. degree b. radian
2. When is an angle in standard position?
3. How do you convert degrees to radians and radians to degrees? Illustrate with examples.

12.1 Exercises

In Exercises 1–4, express the angle shown in the figure in radian measure.

1.

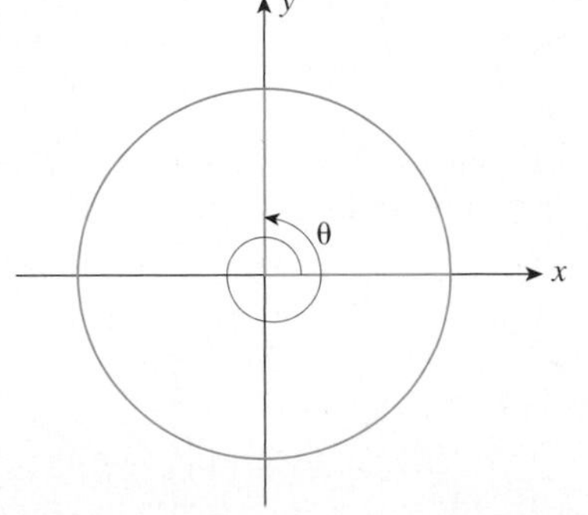

2.

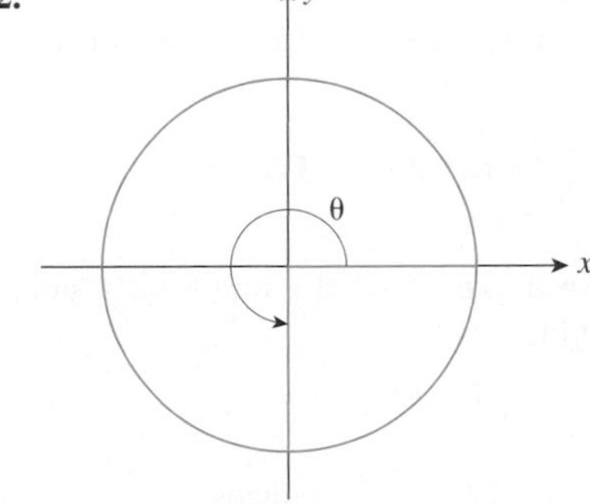

3.

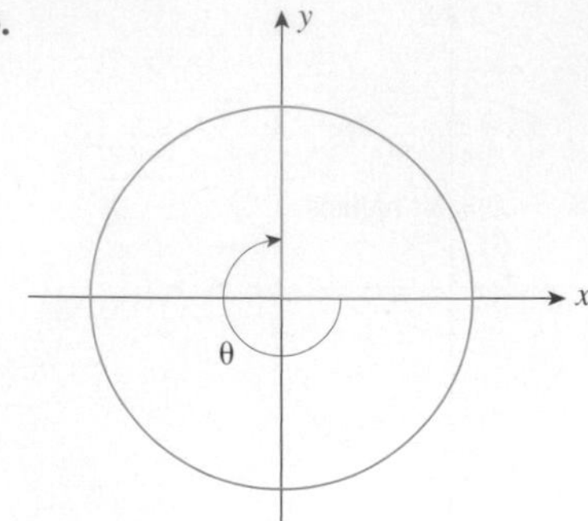

4.

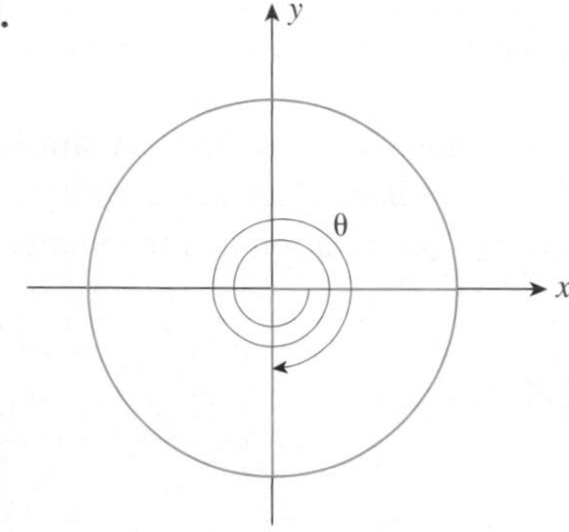

5. Identify the quadrant in which the angle lies.

a. 220° **b.** −110°
c. 460° **d.** −310°

6. Identify the quadrant in which the angle lies.

a. $\frac{13}{6}\pi$ radians **b.** $-\frac{11}{4}\pi$ radians
c. $\frac{17}{3}\pi$ radians **d.** $-\frac{25}{12}\pi$ radians

In Exercises 7–12, convert the angle to radian measure.

7. 75° **8.** 330° **9.** 160°

10. −210° **11.** 630° **12.** −420°

In Exercises 13–18, convert the angle to degree measure.

13. $\frac{2}{3}\pi$ radians **14.** $\frac{7}{6}\pi$ radians **15.** $-\frac{3}{2}\pi$ radians

16. $-\frac{13}{12}\pi$ radians **17.** $\frac{22}{18}\pi$ radians **18.** $-\frac{21}{6}\pi$ radians

In Exercises 19–22, make a sketch of the angle on a unit circle centered at the origin.

19. 225° **20.** −120°

21. $\frac{7}{3}\pi$ radians **22.** $-\frac{13}{6}\pi$ radians

In Exercises 23–26, determine a positive angle and a negative angle that are coterminal angles of the angle θ. Use degree measure.

23.

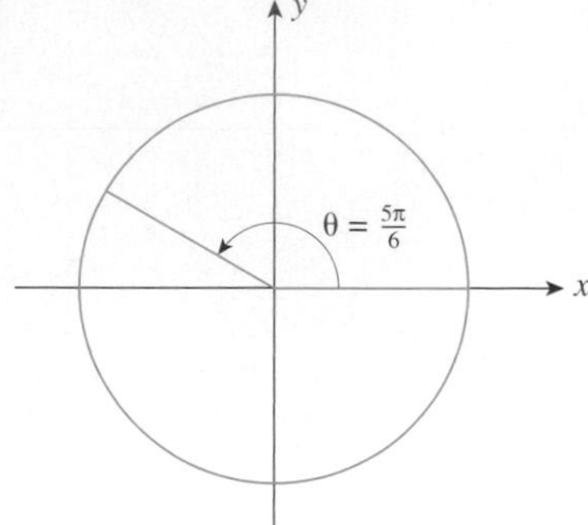

24.

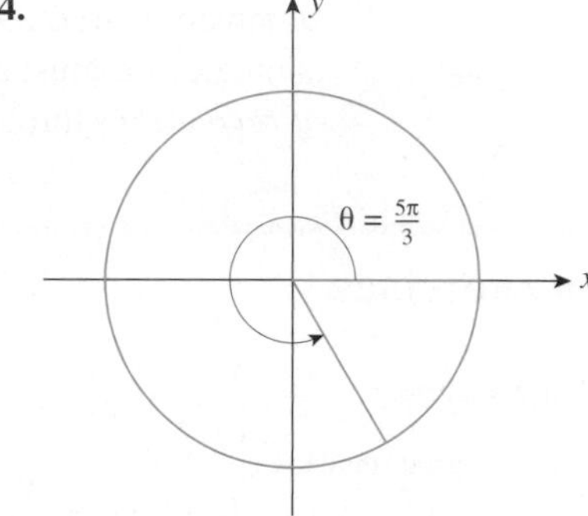

25.

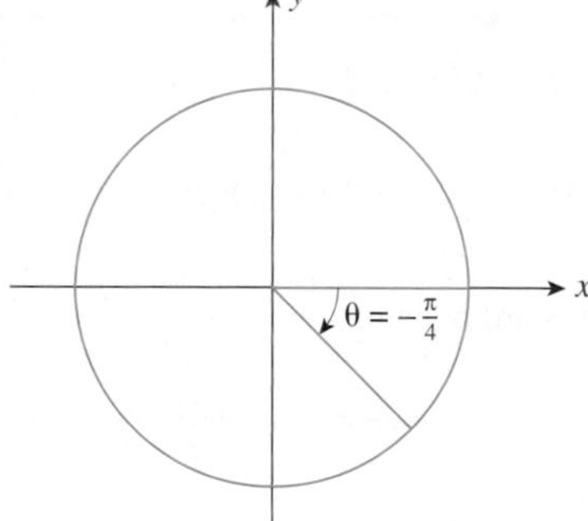

26.

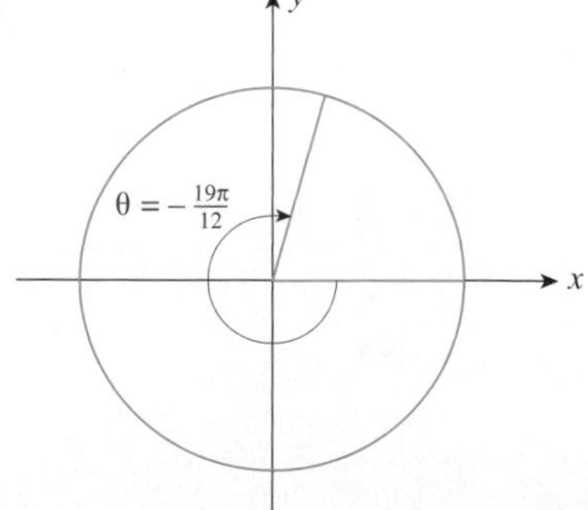

In Exercises 27–30, determine whether the statement is true or false. If it is true, explain why it is true. If it is false, give an example to show why it is false.

27. The angle 3630° lies in the first quadrant.

28. The angle $\dfrac{103\pi}{6}$ radians lies in the second quadrant.

29. If θ is any angle, then $\theta + n(360)$, where n is a nonzero integer, is coterminal with θ (all angles measured in degrees).

30. If x, y, and z are measured in degrees, then $(x + y + z)$ degrees is equal to $\dfrac{\pi}{180}(x + y + z)$ radians.

12.1 Solutions to Self-Check Exercises

1. a. Using Formula (3), we find that the required radian measure is

$$f(315) = \frac{\pi}{180}(315) = \frac{7\pi}{4}, \quad \text{or} \quad \frac{7\pi}{4} \text{ radians}$$

b. Using Formula (4), we find that the required degree measure is

$$g\left(-\frac{5\pi}{4}\right) = \frac{180}{\pi}\left(-\frac{5\pi}{4}\right) = -225, \quad \text{or} \quad -225°$$

2.

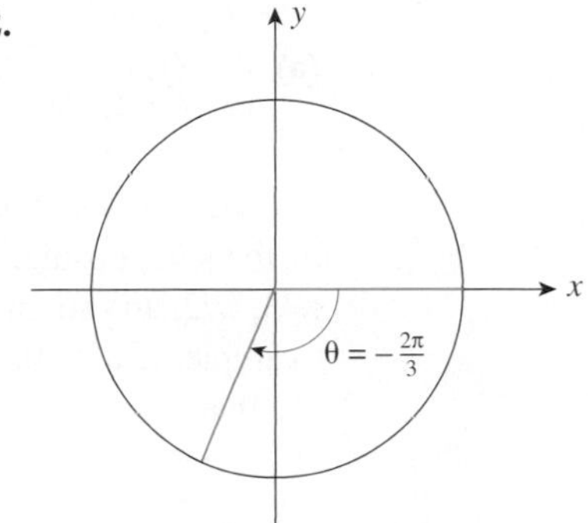

12.2 The Trigonometric Functions

The Trigonometric Functions and Their Graphs

Let $P(x, y)$ be a point on the unit circle so that the radius OP forms an angle of θ radians $(0 \leq \theta < 2\pi)$ with respect to the positive x-axis (see Figure 8).

We define the **sine** of the angle θ, written $\sin \theta$, to be the y-coordinate of P. Similarly, the **cosine** of the angle θ, written $\cos \theta$, is defined to be the x-coordinate of P. The other trigonometric functions, tangent, cosecant, secant, and cotangent of θ—written $\tan \theta$, $\csc \theta$, $\sec \theta$, and $\cot \theta$, respectively—are defined in terms of the sine and cosine functions.

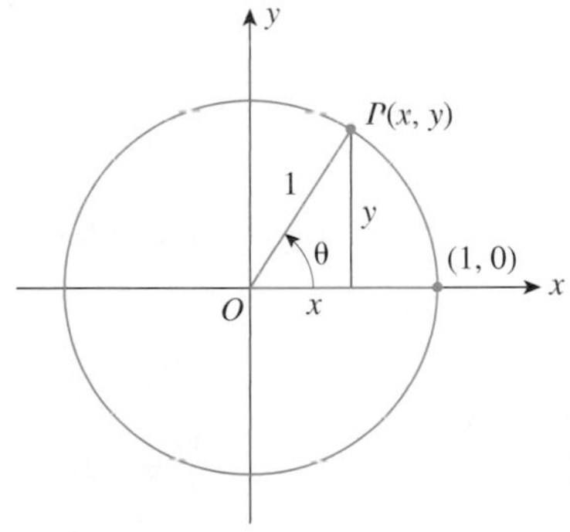

FIGURE 8
P is a point on the unit circle with coordinates $x = \cos \theta$ and $y = \sin \theta$.

Trigonometric Functions

If P is a point on the unit circle and the coordinates of P are (x, y), then

$$\cos \theta = x \quad \text{and} \quad \sin \theta = y$$

$$\tan \theta = \frac{y}{x} = \frac{\sin \theta}{\cos \theta} \qquad (x \neq 0)$$

$$\csc \theta = \frac{1}{y} = \frac{1}{\sin \theta} \qquad (y \neq 0)$$

$$\sec \theta = \frac{1}{x} = \frac{1}{\cos \theta} \qquad (x \neq 0)$$

$$\cot \theta = \frac{x}{y} = \frac{\cos \theta}{\sin \theta} \qquad (y \neq 0)$$

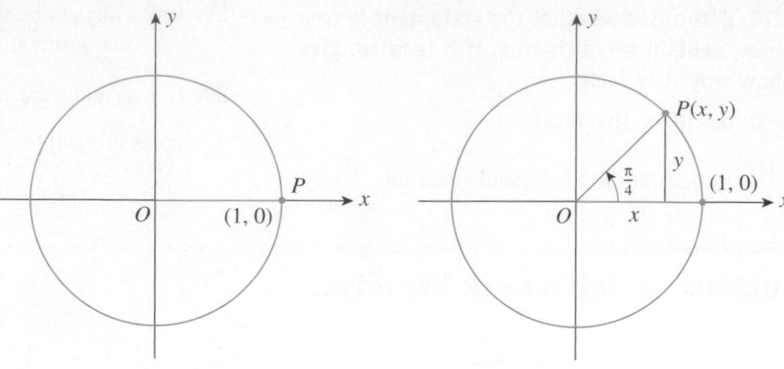

FIGURE 9

(a) $P = (1, 0)$, $\theta = 0$, and $\cos\theta = 1$, $\sin\theta = 0$

(b) $\theta = \pi/4$, $x = y$, and $\cos\theta = \sin\theta = \sqrt{2}/2$

As you work with trigonometric functions, it is useful to remember the values of the sine, cosine, and tangent of some important angles, such as $\theta = 0, \pi/6, \pi/4, \pi/3, \pi/2$, and so on. These values may be found using elementary trigonometry. For example, if $\theta = 0$, then the point P has coordinates $(1, 0)$ (see Figure 9a), and we see that

$$\sin 0 = y = 0 \qquad \cos 0 = x = 1 \qquad \tan 0 = \frac{y}{x} = 0$$

As another example, the diagram for $\theta = \pi/4$ (Figure 9b) suggests that $x = y$, so that, by the Pythagorean theorem, we have

$$x^2 + y^2 = 2x^2 = 1$$

and $x = y = \dfrac{\sqrt{2}}{2}$. Therefore,

$$\sin\frac{\pi}{4} = y = \frac{\sqrt{2}}{2} \qquad \cos\frac{\pi}{4} = x = \frac{\sqrt{2}}{2} \qquad \tan\frac{\pi}{4} = \frac{y}{x} = 1$$

The values of the sine, cosine, and tangent of some common angles are given in Table 2.

TABLE 2

θ in radians	0	$\frac{\pi}{6}$	$\frac{\pi}{4}$	$\frac{\pi}{3}$	$\frac{\pi}{2}$	$\frac{2}{3}\pi$	$\frac{3}{4}\pi$	$\frac{5}{6}\pi$	π	$\frac{7}{6}\pi$	$\frac{5}{4}\pi$	$\frac{4}{3}\pi$	$\frac{3}{2}\pi$	$\frac{5}{3}\pi$	$\frac{7}{4}\pi$	$\frac{11}{6}\pi$	2π
$\sin\theta$	0	$\frac{1}{2}$	$\frac{\sqrt{2}}{2}$	$\frac{\sqrt{3}}{2}$	1	$\frac{\sqrt{3}}{2}$	$\frac{\sqrt{2}}{2}$	$\frac{1}{2}$	0	$-\frac{1}{2}$	$-\frac{\sqrt{2}}{2}$	$-\frac{\sqrt{3}}{2}$	-1	$-\frac{\sqrt{3}}{2}$	$-\frac{\sqrt{2}}{2}$	$-\frac{1}{2}$	0
$\cos\theta$	1	$\frac{\sqrt{3}}{2}$	$\frac{\sqrt{2}}{2}$	$\frac{1}{2}$	0	$-\frac{1}{2}$	$-\frac{\sqrt{2}}{2}$	$-\frac{\sqrt{3}}{2}$	-1	$-\frac{\sqrt{3}}{2}$	$-\frac{\sqrt{2}}{2}$	$-\frac{1}{2}$	0	$\frac{1}{2}$	$\frac{\sqrt{2}}{2}$	$\frac{\sqrt{3}}{2}$	1
$\tan\theta$	0	$\frac{1}{\sqrt{3}}$	1	$\sqrt{3}$	∞	$-\sqrt{3}$	-1	$-\frac{1}{\sqrt{3}}$	0	$\frac{1}{\sqrt{3}}$	1	$\sqrt{3}$	$-\infty$	$-\sqrt{3}$	-1	$-\frac{1}{\sqrt{3}}$	0

Since the rotation of the radius OP by 2π radians leaves it in its original configuration (see Figure 10), we see that

$$\sin(\theta + 2\pi) = \sin\theta \qquad \text{and} \qquad \cos(\theta + 2\pi) = \cos\theta$$

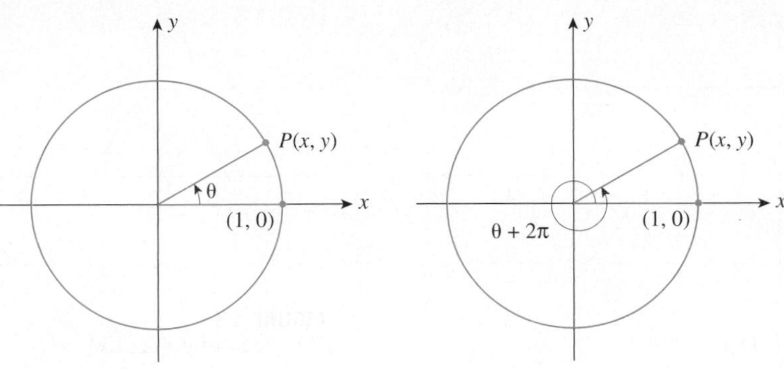

FIGURE 10 (a) The angle θ (b) The angle $(\theta + 2\pi)$

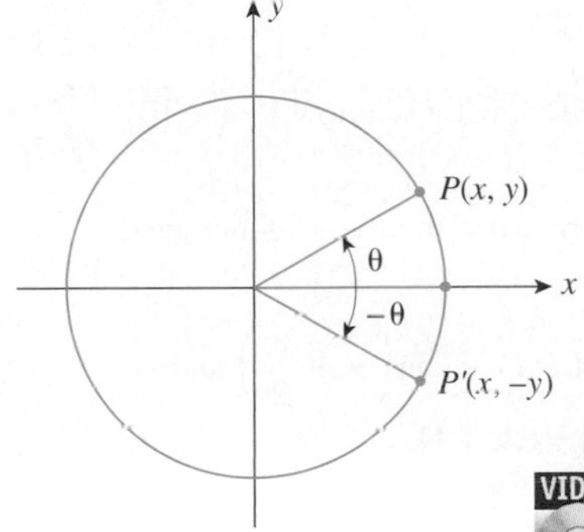

FIGURE 11
$\sin(-\theta) = -\sin\theta$ and $\cos(-\theta) = \cos\theta$.

It can be shown that 2π is the smallest positive number for which the preceding equations hold true. That is, the sine and cosine function are **periodic** with **period** $\mathbf{2\pi}$. It follows that

$$\sin(\theta + 2n\pi) = \sin\theta \quad \text{and} \quad \cos(\theta + 2n\pi) = \cos\theta \tag{5}$$

whenever n is an integer. Also, from Figure 11, we see that

$$\sin(-\theta) = -\sin\theta \quad \text{and} \quad \cos(-\theta) = \cos\theta \tag{6}$$

VIDEO

EXAMPLE 1 Evaluate

a. $\sin\dfrac{7\pi}{2}$ **b.** $\cos 5\pi$ **c.** $\sin\left(-\dfrac{5\pi}{2}\right)$ **d.** $\cos\left(-\dfrac{11\pi}{4}\right)$

Solution

a. Using (5) and Table 2, we find

$$\sin\left(\frac{7\pi}{2}\right) = \sin\left(2\pi + \frac{3\pi}{2}\right) = \sin\frac{3\pi}{2} = -1$$

b. Using (5) and Table 2, we have

$$\cos 5\pi = \cos(4\pi + \pi) = \cos\pi = -1$$

c. Using (5), (6), and Table 2, we have

$$\sin\left(-\frac{5\pi}{2}\right) = -\sin\left(\frac{5\pi}{2}\right) = -\sin\left(2\pi + \frac{\pi}{2}\right) = -\sin\frac{\pi}{2} = -1$$

d. Using (5), (6), and Table 2, we have

$$\cos\left(-\frac{11\pi}{4}\right) = \cos\frac{11\pi}{4} = \cos\left(2\pi + \frac{3\pi}{4}\right) = \cos\frac{3\pi}{4} = -\frac{\sqrt{2}}{2}$$

■

To draw the graph of the function $y = f(x) = \sin x$, we first note that $\sin x$ is defined for every real number x so that the domain of the sine function is $(-\infty, \infty)$. Next, since the sine function is periodic with period 2π, it suffices to concentrate on sketching that part of the graph of $y = \sin x$ on the interval $[0, 2\pi]$ and repeating it as necessary. With the help of Table 2, we sketch Figure 12, the graph of $y = \sin x$.

In a similar manner, we can sketch the graph of $y = \cos x$ (Figure 13).

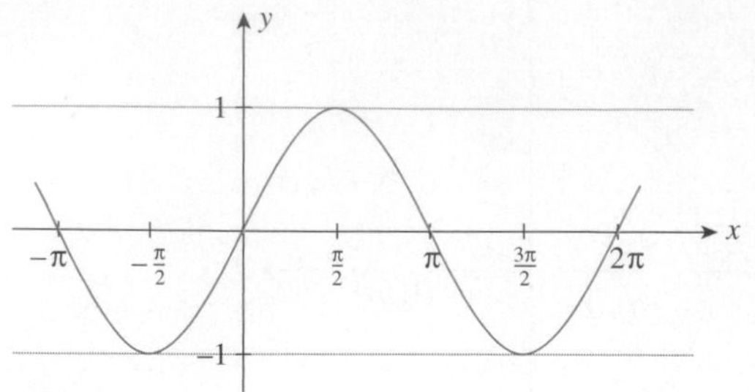

FIGURE 12
The graph of $y = \sin x$

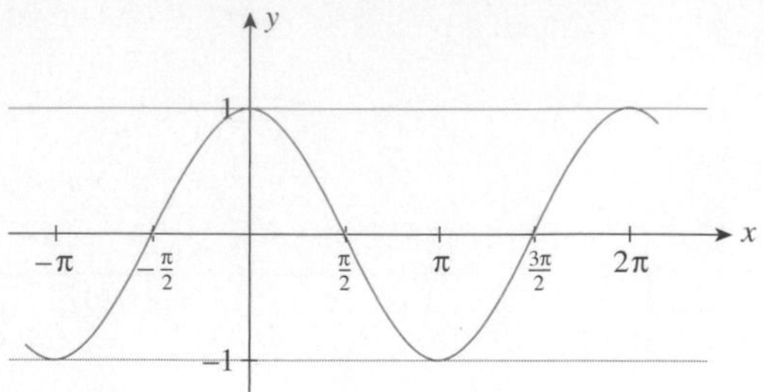

FIGURE 13
The graph of $y = \cos x$

EXPLORING WITH TECHNOLOGY

Plot the graphs of $f(x) = \sin x$ and $g(x) = \cos x$ in the viewing window $[-10, 10] \times [-1.1, 1.1]$.

1. What do the graphs suggest about the relationship between the two trigonometric functions?
2. Confirm your observation by plotting the graphs of $h(x) = \sin\left(x + \dfrac{\pi}{2}\right)$ and $g(x) = \cos x$ in the viewing window $[-10, 10] \times [-1.1, 1.1]$.

To sketch the graph of $y = \tan x$, note that $\tan x = \sin x/\cos x$, and so $\tan x$ is not defined when $\cos x = 0$—that is, when $x = (\pi/2) \pm n\pi (n = 0, 1, 2, 3, \ldots)$. The function is defined at all other points so that the domain of the tangent function is the set of all real numbers with the exception of the points just noted. Next, we can show that the vertical lines with equation $x = (\pi/2) \pm n\pi (n = 0, 1, 2, 3, \ldots)$ are vertical asymptotes of the function $f(x) = \tan x$. For example, since

$$\lim_{x \to \frac{\pi}{2}-} \tan x = \infty \qquad \text{and} \qquad \lim_{x \to \frac{\pi}{2}+} \tan x = -\infty$$

which we readily verify with the help of a calculator, we conclude that $x = \pi/2$ is a vertical asymptote of $y = \tan x$. Finally, using Table 2, we can sketch the graph of $y = \tan x$ (Figure 14).

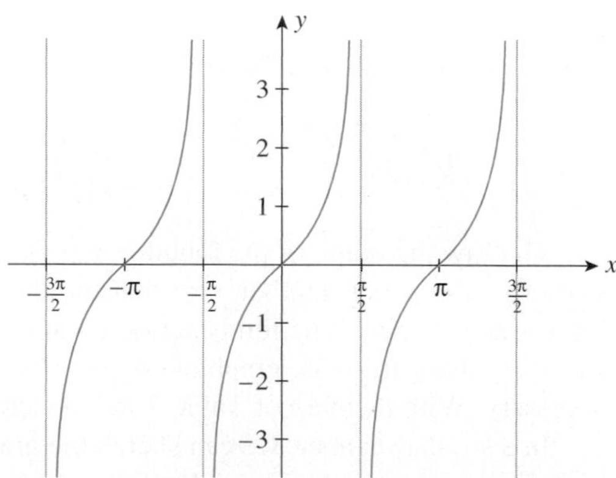

FIGURE 14
The graph of $y = \tan x$

Observe that the tangent function is periodic with period π. It follows that

$$\tan(x + n\pi) = \tan x \tag{7}$$

whenever n is an integer.

The graphs of $y = \sec x$, $y = \csc x$, and $y = \cot x$ may be sketched in a similar manner (see Exercises 23–25).

The Predator–Prey Population Model

We will now look at a specific mathematical model of a phenomenon exhibiting cyclical behavior—the so-called **predator–prey population model**.

APPLIED EXAMPLE 2 Predator–Prey Population The population of owls (predators) in a certain region over a 2-year period is estimated to be

$$P_1(t) = 1000 + 100 \sin\left(\frac{\pi t}{12}\right)$$

in month t, and the population of mice (prey) in the same area at time t is given by

$$P_2(t) = 20{,}000 + 4000 \cos\left(\frac{\pi t}{12}\right)$$

Sketch the graphs of these two functions and explain the relationship between the sizes of the two populations.

Solution We first observe that both of the given functions are periodic with period 24 (months). To see this, recall that both the sine and cosine functions are periodic with period 2π. Now the smallest value of $t > 0$ such that $\sin(\pi t/12) = 0$ is obtained by solving the equation

$$\frac{\pi t}{12} = 2\pi$$

giving $t = 24$ as the period of $\sin(\pi t/12)$. Since $P_1(t + 24) = P_1(t)$, we see that the function P_1 is periodic with period 24. Similarly, one verifies that the function P_2 is also periodic, with period 24, as asserted. Next, recall that both the sine and cosine functions oscillate between -1 and $+1$ so that $P_1(t)$ is seen to oscillate between $[1000 + 100(-1)]$, or 900, and $[1000 + 100(1)]$, or 1100, while $P_2(t)$ oscillates between $[20{,}000 + 4000(-1)]$, or 16,000, and $[20{,}000 + 4000(1)]$, or 24,000. Finally, plotting a few points on each graph for—say, $t = 0, 2, 3$, and so on—we obtain the graphs of the functions P_1 and P_2 as shown in Figure 15.

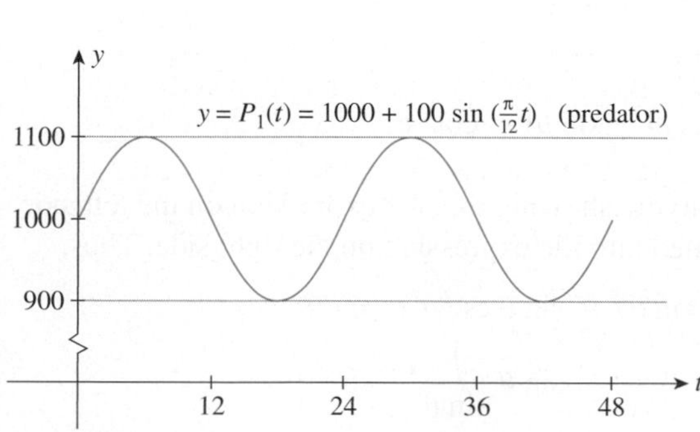

(a) The graph of the predator function $P_1(t)$

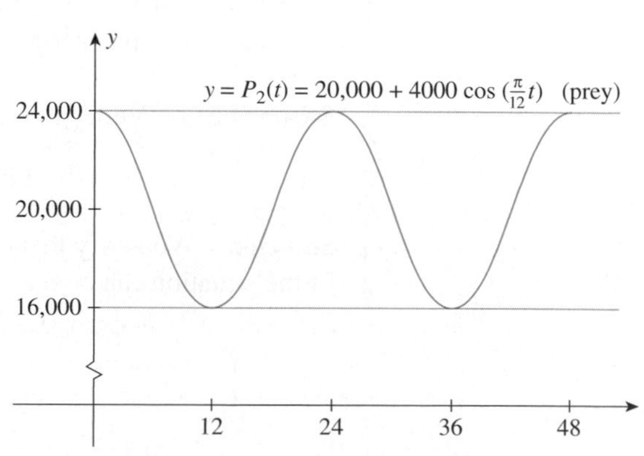

(b) The graph of the prey function $P_2(t)$

FIGURE 15

From the graphs, we see that at time $t = 0$ the predator population stands at 1000 owls. As it increases, the prey population decreases from 24,000 mice at that instant of time. Eventually, this decrease in the food supply causes the predator population to decrease, which in turn causes an increase in the prey population. But as the prey population increases, resulting in an increase in food supply, the predator population once again increases. The cycle is complete and starts all over again. ■

Trigonometric Identities

Equations expressing the relationships between trigonometric functions, such as

$$\sin(-\theta) = -\sin\theta \qquad \text{and} \qquad \cos(-\theta) = \cos\theta$$

are called **trigonometric identities**. Some other important trigonometric identities are listed in Table 3. Each identity holds true for every value of θ in the domain of the specified function. The proofs of these identities may be found in any elementary trigonometry book.

TABLE 3

Trigonometric Identities

Pythagorean Identities	Half-Angle Formulas	Sum and Difference Formulas
$\sin^2\theta + \cos^2\theta = 1$	$\cos^2\theta = \frac{1}{2}(1 + \cos 2\theta)$	$\sin(A \pm B) = \sin A\cos B \pm \cos A\sin B$
$\tan^2\theta + 1 = \sec^2\theta$	$\sin^2\theta = \frac{1}{2}(1 - \cos 2\theta)$	$\cos(A \pm B) = \cos A\cos B \mp \sin A\sin B$
$\cot^2\theta + 1 = \csc^2\theta$		

Double-Angle Formulas	Cofunctions of Complementary Angles
$\sin 2A = 2\sin A\cos A$	$\sin\theta = \cos\left(\frac{\pi}{2} - \theta\right)$
$\cos 2A = \cos^2 A - \sin^2 A$	$\cos\theta = \sin\left(\frac{\pi}{2} - \theta\right)$

As we will see later, these identities are useful in simplifying trigonometric expressions and equations and in deriving other trigonometric relationships. They are also used to verify other trigonometric identities, as illustrated in the next example.

EXAMPLE 3 Verify the identity

$$\sin\theta(\csc\theta - \sin\theta) = \cos^2\theta$$

Solution We verify this identity by showing that the expression on the left side of the equation can be transformed into the expression on the right side. Thus,

$$\begin{aligned}\sin\theta(\csc\theta - \sin\theta) &= \sin\theta\csc\theta - \sin^2\theta \\ &= \sin\theta\frac{1}{\sin\theta} - \sin^2\theta \\ &= 1 - \sin^2\theta \\ &= \cos^2\theta\end{aligned}$$

■

EXPLORING WITH TECHNOLOGY

1. You can confirm some of the trigonometric identities graphically in many ways. For example, explain why an identity given in the form $f(\theta) = g(\theta)$ over some domain D is equivalent to the following statements:
   a. The graphs of f and g coincide when viewed in any appropriate viewing window.
   b. The graphs of f and $-g$ are reflections of each other with respect to the θ-axis when viewed in any appropriate viewing window.
   c. The graph of $h = f - g$ is the graph of the zero function $h(\theta) = 0$ in the domain of f (and g).
2. Use the observations made in part 1 to verify graphically the identity $\sin^2\theta + \cos^2\theta = 1$ by taking $f(x) = \sin^2 x + \cos^2 x$ and $g(x) = 1$.
   Hint: Enter $f(x)$ as $y_1 = (\sin x)^2 + (\cos x)^2$ and use the viewing window $[-10, 10] \times [-1.1, 1.1]$.
3. Verify graphically the half-angle formula $\cos^2\theta = \frac{1}{2}(1 + \cos 2\theta)$.
4. Verify graphically the formula $\sin\theta = \cos\left(\frac{\pi}{2} - \theta\right)$.

12.2 Self-Check Exercises

1. Evaluate $\cos\left(-\frac{13\pi}{3}\right)$.
2. Solve the equation $\cos\theta = -\frac{\sqrt{2}}{2}$ for $0 \le \theta \le 2\pi$.
3. Sketch the graph of $y = 2\cos x$.

Solutions to Self-Check Exercises 12.2 can be found on page 785.

12.2 Concept Questions

1. Define $\sin\theta$, $\cos\theta$, $\tan\theta$, $\csc\theta$, $\sec\theta$, and $\cot\theta$ in terms of x and y, where (x, y) is a point on the unit circle.
2. Write the values of $\sin\theta$, $\cos\theta$, and $\tan\theta$ for $\theta = 0, \frac{\pi}{6}, \frac{\pi}{4}, \frac{\pi}{3}, \frac{\pi}{2}, \frac{2\pi}{3}, \frac{5\pi}{6}$, and π.
3. Complete the equation:
   a. $\sin^2\theta + \cos^2\theta =$ ______
   b. $\sec^2\theta =$ ______
   c. $\cot^2\theta + 1 =$ ______
4. Complete the equation:
   a. $\frac{1 + \cos 2\theta}{2} =$ ______
   b. $\frac{1 - \cos 2\theta}{2} =$ ______
   c. $\sin(A \pm B) =$ ______
5. Complete the equation:
   a. $\cos(A \pm B) =$ ______
   b. $\sin 2A =$ ______
   c. $\cos 2A =$ ______

12.2 Exercises

In Exercises 1–10, evaluate the trigonometric function.

1. $\sin 3\pi$
2. $\cos\left(-\frac{3}{2}\pi\right)$
3. $\sin\frac{9}{2}\pi$
4. $\cos\frac{13}{6}\pi$
5. $\sin\left(-\frac{4}{3}\pi\right)$
6. $\cos\left(-\frac{5}{4}\pi\right)$
7. $\tan\frac{\pi}{6}$
8. $\cot\left(-\frac{\pi}{3}\right)$
9. $\sec\left(-\frac{5}{8}\pi\right)$
10. $\csc\frac{9}{4}\pi$

In Exercises 11–14, find the six trigonometric functions of the angle.

11. $\frac{\pi}{2}$
12. $-\frac{\pi}{6}$
13. $\frac{5}{3}\pi$
14. $-\frac{3}{4}\pi$

In Exercises 15–22, find all values of θ that satisfy the equation over the interval $[0, 2\pi]$.

15. $\sin\theta = -\frac{1}{2}$

16. $\tan\theta = 1$

17. $\cot\theta = -\sqrt{3}$

18. $\csc\theta = \sqrt{2}$

19. $\sec\theta = -1$

20. $\cos\theta = \sin\theta$

21. $\sin\theta = \sin\left(-\frac{4}{3}\pi\right)$

22. $\cos\theta = \cos\left(-\frac{\pi}{6}\right)$

In Exercises 23–26, sketch the graph of the function over the interval $[0, 2\pi]$.

23. $y = \csc x$

24. $y = \sec x$

25. $y = \cot x$

26. $y = \tan 2x$

In Exercises 27–30, sketch the graph of the functions over the interval $[0, 2\pi]$ by comparing it to the graph of $y = \sin x$.

27. $y = \sin 2x$

28. $y = 2\sin x$

29. $y = -\sin x$

30. $y = -2\sin 2x$

In Exercises 31–38, verify each identity.

31. $\cos^2\theta - \sin^2\theta = 2\cos^2\theta - 1$

32. $1 - 2\sin^2\theta = 2\cos^2\theta - 1$

33. $(\sec\theta + \tan\theta)(1 - \sin\theta) = \cos\theta$

34. $\dfrac{\sin^2\theta}{1 + \cos^2\theta} = \dfrac{1 - \cos^2\theta}{2 - \sin^2\theta}$

35. $(1 + \cot^2\theta)\tan^2\theta = \sec^2\theta$

36. $\dfrac{\sec\theta - \cos\theta}{\tan\theta} = \sin\theta$

37. $\dfrac{\csc\theta}{\tan\theta + \cot\theta} = \cos\theta$

38. $\tan(A + B) = \dfrac{\tan A + \tan B}{1 - \tan A \tan B}$

39. The accompanying figure shows a right triangle ABC superimposed over a unit circle in the xy-coordinate system. By considering similar triangles, show that

$$\sin\theta = \frac{BC}{AC} = \frac{\text{Opposite side}}{\text{Hypotenuse}}$$

$$\cos\theta = \frac{AB}{AC} = \frac{\text{Adjacent side}}{\text{Hypotenuse}}$$

$$\tan\theta = \frac{BC}{AB} = \frac{\text{Opposite side}}{\text{Adjacent side}}$$

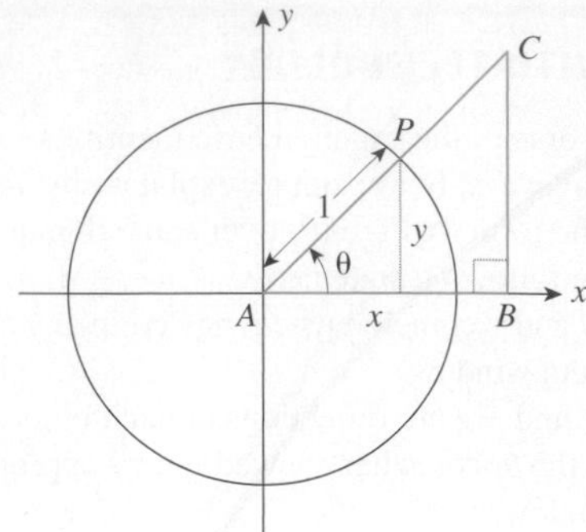

40. Refer to Exercise 39. Find the values of the six trigonometric functions of θ, where θ is the angle shown in the right triangle in the accompanying figure.

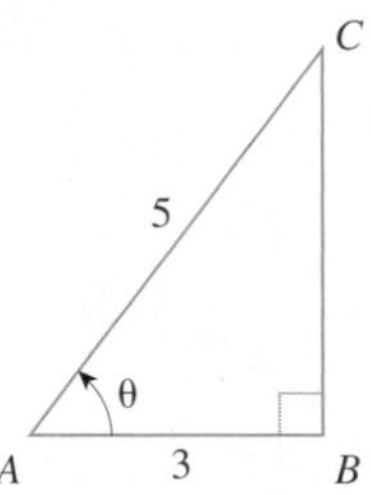

41. Refer to Exercise 39. Find the values of the six trigonometric functions of θ, where θ is the angle shown in the right triangle in the accompanying figure.

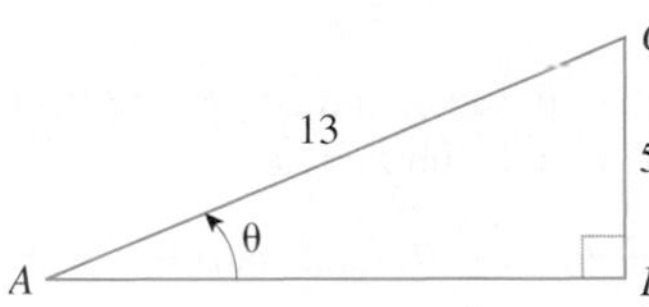

42. **Predator–Prey Population** The population of foxes (predators) in a certain region over a 2-yr period is estimated to be

$$P_1(t) = 400 + 50\sin\left(\frac{\pi t}{12}\right)$$

in month t, and the population of rabbits (prey) in the same region at time t is given by

$$P_2(t) = 3000 + 500\cos\left(\frac{\pi t}{12}\right)$$

Sketch the graphs of each of these two functions and explain the relationship between the sizes of the two populations.

43. **Blood Pressure** The arterial blood pressure of an individual in a state of relaxation is given by

$$P(t) = 100 + 20\sin 6t$$

where $P(t)$ is measured in mm of mercury (Hg) and t is the time in seconds.

a. Show that the individual's systolic pressure (maximum blood pressure) is 120 and his diastolic pressure (minimum blood pressure) is 80.

b. Find the values of t when the individual's blood pressure is highest and lowest.

In Exercises 44–48, determine whether the statement is true or false. If it is true, explain why it is true. If it is false, give an example to show why it is false.

44. $\sin(a - b) = -\sin(b - a)$

45. If $\sin\theta = -\dfrac{\sqrt{3}}{2}$ and $0 \leq \theta \leq 2\pi$, then $\theta = \dfrac{5\pi}{3}$.

46. If $\tan\theta = \dfrac{1}{3}$, then $\sin\theta = \dfrac{\sqrt{10}}{10}$.

47. $\cos 2\theta = 2\cos^2\theta - 1$

48. $\sin 3\theta = 3\sin\theta + 4\sin^3\theta$

12.2 Solutions to Self-Check Exercises

1.
$$\cos\left(-\frac{13\pi}{3}\right) = \cos\frac{13\pi}{3} = \cos\left(4\pi + \frac{\pi}{3}\right) = \cos\frac{\pi}{3} = \frac{1}{2}$$

2. Cos θ is negative for θ in Quadrants II and III. From Table 2, we see that the required values of θ are $3\pi/4$ and $5\pi/4$.

3. By comparison with the graph of $y = \cos x$, we obtain the accompanying graph of $y = 2\cos x$.

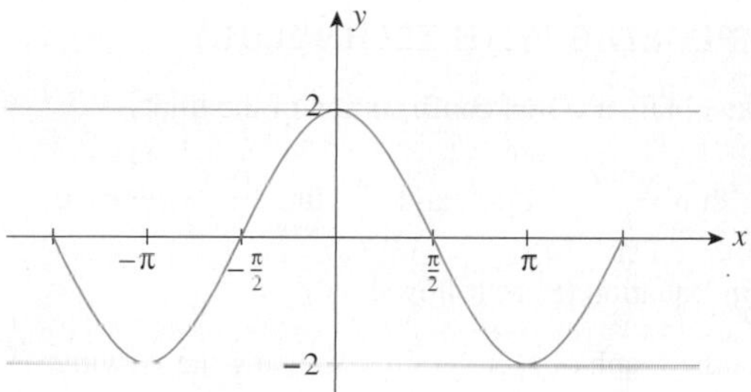

12.3 Differentiation of Trigonometric Functions

In this section we develop rules for differentiating trigonometric functions. Knowledge of the derivatives of the trigonometric functions helps us to study the properties of functions involving the trigonometric functions in much the same way we analyzed the algebraic, exponential, and logarithmic functions in our earlier work.

Derivatives of the Sine and Cosine Functions

First we develop the rule for differentiating the sine and cosine functions. Then, using these rules and the rules of differentiation, we derive the rules for differentiating the other trigonometric functions.

To derive the rule for differentiating the sine function, we need the following two results:

$$\lim_{h\to 0}\frac{\sin h}{h} = 1 \quad \text{and} \quad \lim_{h\to 0}\frac{\cos h - 1}{h} = 0 \tag{8}$$

The plausibility of the first limit can be seen by examining Table 4, which is constructed with the aid of a calculator.

TABLE 4

Values of $\frac{\sin h}{h}$ for Selected Values of h [in Radians] Approaching Zero				
h	± 0.5	± 0.1	± 0.01	± 0.001
$\frac{\sin h}{h}$	0.9588511	0.9983342	0.9999833	0.9999998

The second limit can be derived from the first using the appropriate trigonometric identity and the properties of limits (see Exercise 64).

EXPLORING WITH TECHNOLOGY

You can obtain a visual confirmation of the limits

$$\lim_{h\to 0}\frac{\sin h}{h}=1 \qquad \text{and} \qquad \lim_{h\to 0}\frac{\cos h-1}{h}=0$$

given in Equation (8) as follows:

1. Plot the graph of $f(x) = (\sin x)/x$, using the viewing window $[-1, 1] \times [0, 2]$. Then use ZOOM and TRACE to find the values of $f(x)$ for values of x close to $x = 0$. What happens when you try to evaluate $f(0)$? Explain your answer.
2. Plot the graph of $g(x) = (\cos x - 1)/x$, using the viewing window $[-1, 1] \times [-0.5, 0.5]$. Then use ZOOM and TRACE to show that $g(x)$ is close to zero if x is close to zero.

Caution: Convincing as these arguments are, they do not establish the stated results. As noted earlier, the proofs are given at the end of the section.

Now suppose $f(x) = \sin x$. Then

$$\begin{aligned}
f'(x) &= \lim_{h\to 0}\frac{f(x+h)-f(x)}{h}\\
&= \lim_{h\to 0}\frac{\sin(x+h)-\sin x}{h}\\
&= \lim_{h\to 0}\frac{\sin x\cos h+\cos x\sin h-\sin x}{h} && \text{Use the sine rule for sums.}\\
&= \lim_{h\to 0}\frac{\sin x(\cos h-1)+\cos x\sin h}{h}\\
&= \lim_{h\to 0}\sin x\left(\frac{\cos h-1}{h}\right)+\lim_{h\to 0}\cos x\left(\frac{\sin h}{h}\right)\\
&= \sin x\lim_{h\to 0}\left(\frac{\cos h-1}{h}\right)+\cos x\lim_{h\to 0}\left(\frac{\sin h}{h}\right)\\
&= (\sin x)(0)+(\cos x)(1) && \text{Use Equation (8).}\\
&= \cos x
\end{aligned}$$

With the help of the chain rule, this result can be generalized as follows: Suppose

$$y = h(x) = \sin f(x)$$

where $f(x)$ is a differentiable function of x. Then putting

$$u = f(x)$$

so that $y = \sin u$, we find, upon applying the chain rule,

$$h'(x) = \frac{dy}{du} \cdot \frac{du}{dx} = (\cos u)\frac{du}{dx} = [\cos f(x)]f'(x)$$

Derivatives of the Sine and Generalized Sine Functions

Rule 1a: $\frac{d}{dx}(\sin x) = \cos x$ **Rule 1b:** $\frac{d}{dx}[\sin f(x)] = [\cos f(x)]f'(x)$

Observe that Rule 1b reduces to Rule 1a when $f(x) = x$, as expected.

EXAMPLE 1 Differentiate each of the following functions:

a. $f(x) = x^2 \sin x$ **b.** $g(x) = \sin(2x + 1)$ **c.** $h(x) = (x + \sin x^2)^{10}$

Solution

a. Using the product rule followed by Rule 1a, we find

$$\begin{aligned} f'(x) &= 2x \sin x + x^2 \frac{d}{dx}(\sin x) \\ &= 2x \sin x + x^2 \cos x = x(2 \sin x + x \cos x) \end{aligned}$$

b. Using Rule 1b, we find

$$g'(x) = [\cos(2x + 1)]\frac{d}{dx}(2x + 1) = 2\cos(2x + 1)$$

c. We first use the general power rule followed by Rule 1b. We obtain

$$\begin{aligned} h'(x) &= 10(x + \sin x^2)^9 \frac{d}{dx}(x + \sin x^2) \\ &= 10(x + \sin x^2)^9\left[1 + \cos x^2 \frac{d}{dx}(x^2)\right] \\ &= 10(x + \sin x^2)^9(1 + 2x \cos x^2) \end{aligned}$$

To derive the rule for differentiating the cosine function, we make use of the following relationships between the cosine function and the sine function:

$$\cos x = \sin\left(\frac{\pi}{2} - x\right) \quad \text{and} \quad \sin x = \cos\left(\frac{\pi}{2} - x\right)$$

(see Table 3). Now, with the help of Rule 1b, we see that if $f(x) = \cos x$, then

$$\begin{aligned} f'(x) &= \frac{d}{dx}\cos x \\ &= \frac{d}{dx}\sin\left(\frac{\pi}{2} - x\right) \\ &= \cos\left(\frac{\pi}{2} - x\right)\frac{d}{dx}\left(\frac{\pi}{2} - x\right) \\ &= (\sin x)(-1) \\ &= -\sin x \end{aligned}$$

This result may be generalized immediately using the chain rule. In fact, if

$$y = h(x) = \cos f(x)$$

where $f(x)$ is a differentiable function of x, then with $u = f(x)$, we have $y = \cos u$ so that

$$\frac{dy}{dx} = \frac{dy}{du}\frac{du}{dx} = (-\sin u)\frac{du}{dx} = -[\sin f(x)]f'(x)$$

Derivatives of the Cosine and Generalized Cosine Functions

Rule 2a: $\frac{d}{dx}(\cos x) = -\sin x$

Rule 2b: $\frac{d}{dx}[\cos f(x)] = -[\sin f(x)]f'(x)$

Observe that Rule 2b reduces to Rule 2a when $f(x) = x$, as expected.

VIDEO

EXPLORE & DISCUSS

Derive Rule 2a, $\frac{d}{dx}(\cos x) = -\sin x$, directly from the definition of the derivative and the relationships given in Equation (8).

EXAMPLE 2 Find the derivative of each function:

a. $f(x) = \cos(2x^2 - 1)$ **b.** $g(x) = \sqrt{\cos 2x}$ **c.** $h(x) = e^{\sin 2x + \cos 3x}$

Solution

a. Using Rule 2b, we find

$$\begin{aligned} f'(x) &= -\sin(2x^2 - 1)\frac{d}{dx}(2x^2 - 1) \\ &= -[\sin(2x^2 - 1)]4x \\ &= -4x\sin(2x^2 - 1) \end{aligned}$$

b. We first rewrite $g(x)$ as $g(x) = (\cos 2x)^{1/2}$. Using the chain rule followed by Rule 2b, we find

$$\begin{aligned} g'(x) &= \frac{1}{2}(\cos 2x)^{-1/2}\frac{d}{dx}(\cos 2x) \\ &= \frac{1}{2}(\cos 2x)^{-1/2}(-\sin 2x)\frac{d}{dx}(2x) \\ &= \frac{1}{2}(\cos 2x)^{-1/2}(-\sin 2x)(2) \\ &= -\frac{\sin 2x}{\sqrt{\cos 2x}} \end{aligned}$$

c. Using the corollary to the chain rule for exponential functions, we find

$$\begin{aligned} h'(x) &= e^{\sin 2x + \cos 3x} \cdot \frac{d}{dx}(\sin 2x + \cos 3x) \\ &= e^{\sin 2x + \cos 3x}\left[(\cos 2x)\frac{d}{dx}(2x) - (\sin 3x)\frac{d}{dx}(3x)\right] \\ &= (2\cos 2x - 3\sin 3x)e^{\sin 2x + \cos 3x} \end{aligned}$$

■

Derivatives of the Other Trigonometric Functions

We are now in a position to find the rule for differentiating the tangent function. In fact, using the quotient rule, we see that if $f(x) = \tan x$, then

$$\begin{aligned} f'(x) &= \frac{d}{dx}\tan x = \frac{d}{dx}\left(\frac{\sin x}{\cos x}\right) \\ &= \frac{(\cos x)\dfrac{d}{dx}\sin x - (\sin x)\dfrac{d}{dx}\cos x}{\cos^2 x} \\ &= \frac{(\cos x)(\cos x) - (\sin x)(-\sin x)}{\cos^2 x} \\ &= \frac{\cos^2 x + \sin^2 x}{\cos^2 x} = \frac{1}{\cos^2 x} = \sec^2 x \end{aligned}$$

Once again, this result may be generalized with the help of the chain rule, and we are led to the next two rules.

Derivatives of the Tangent and the Generalized Tangent Functions

Rule 3a: $\dfrac{d}{dx}(\tan x) = \sec^2 x$

Rule 3b: $\dfrac{d}{dx}[\tan f(x)] = [\sec^2 f(x)]f'(x)$

The rules for differentiating the remaining three trigonometric functions are derived in a similar manner. We state the generalized versions of these rules. Here $f(x)$ is a differentiable function of x.

Derivatives of the Cosecant, Secant, and Cotangent Functions

Rule 4: $\dfrac{d}{dx}[\csc f(x)] = -[\csc f(x)][\cot f(x)]f'(x)$

Rule 5: $\dfrac{d}{dx}[\sec f(x)] = [\sec f(x)][\tan f(x)]f'(x)$

Rule 6: $\dfrac{d}{dx}[\cot f(x)] = -[\csc^2 f(x)]f'(x)$

Note As an aid to remembering the signs of the derivatives of the trigonometric functions, observe that those functions beginning with a "c" [$\cos x$, $\csc x$, $\cot x$, $\cos f(x)$, $\csc f(x)$, and $\cot f(x)$] have a minus sign attached to their derivatives.

EXAMPLE 3 Find an equation of the tangent line to the graph of the function $f(x) = \tan 2x$ at the point $(\pi/8, 1)$.

Solution The slope of the tangent line at any point on the graph of f is given by

$$f'(x) = 2\sec^2 2x$$

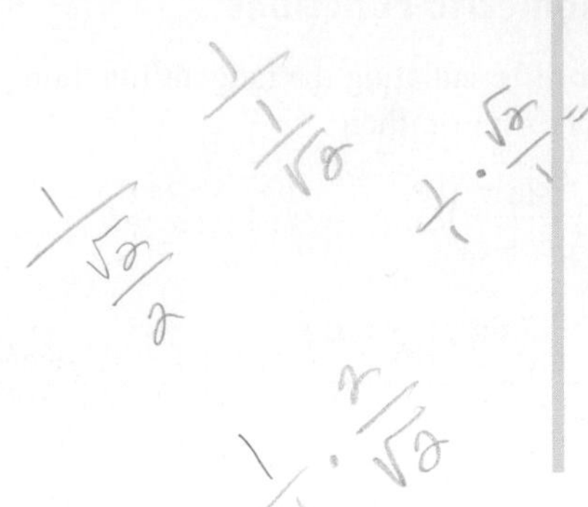

In particular, the slope of the tangent line at the point $(\pi/8, 1)$ is given by

$$f'\left(\frac{\pi}{8}\right) = 2\sec^2\frac{\pi}{4}$$
$$= 2(\sqrt{2})^2 = 4$$

Therefore, a required equation is given by

$$y - 1 = 4\left(x - \frac{\pi}{8}\right)$$
$$y = 4x + \left(1 - \frac{\pi}{2}\right)$$

The techniques developed in Chapter 4 may be used to study the properties of functions involving trigonometric functions, as the next examples show.

APPLIED EXAMPLE 4 Predator–Prey Population The owl population in a certain area is estimated to be

$$P_1(t) = 1000 + 100\sin\left(\frac{\pi t}{12}\right)$$

in month t, and the mouse population in the same area at time t is given by

$$P_2(t) = 20{,}000 + 4000\cos\left(\frac{\pi t}{12}\right)$$

Find the rate of change of each population when $t = 2$.

Solution The rate of change of the owl population at any time t is given by

$$P_1'(t) = 100\left[\cos\left(\frac{\pi t}{12}\right)\right]\left(\frac{\pi}{12}\right)$$
$$= \frac{25\pi}{3}\cos\left(\frac{\pi t}{12}\right)$$

and the rate of change of the mouse population at any time t is given by

$$P_2'(t) = 4000\left[-\sin\left(\frac{\pi t}{12}\right)\right]\left(\frac{\pi}{12}\right)$$
$$= -\frac{1000\pi}{3}\sin\left(\frac{\pi t}{12}\right)$$

In particular, when $t = 2$ the rate of change of the owl population is

$$P_1'(2) = \frac{25\pi}{3}\cos\frac{\pi}{6}$$
$$= \frac{25\pi}{3}\left(\frac{\sqrt{3}}{2}\right) \approx 22.7$$

That is, the predator population is increasing at the rate of approximately 22.7 owls per month, and the rate of change of the mouse population is

$$P_2'(2) = -\frac{1000\pi}{3} \sin \frac{\pi}{6}$$
$$= -\frac{1000\pi}{3}\left(\frac{1}{2}\right) \approx -523.6$$

That is, the prey population is decreasing at the rate of approximately 523.6 mice per month. ■

EXAMPLE 5 Sketch the graph of the function $y = f(x) = \sin^2 x$ on the interval $[0, \pi]$ by first obtaining the following information:

a. The intervals where f is increasing and where it is decreasing
b. The relative extrema of f
c. The concavity of f
d. The inflection points of f

Solution

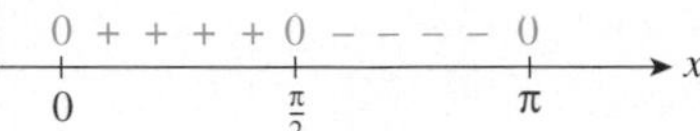

FIGURE 16
The sign diagram for f'

a. $f'(x) = 2 \sin x \cos x = \sin 2x$. Setting $f'(x) = 0$ gives $x = 0, \pi/2$, or π. The sign diagram for f' (Figure 16) shows that f is increasing on $(0, \pi/2)$ and decreasing on $(\pi/2, \pi)$.

b. From the result of part (a) and the following table

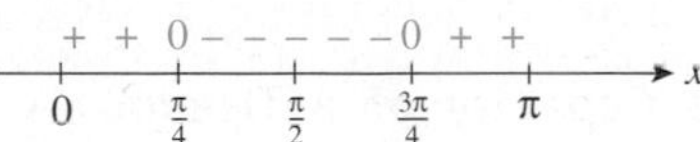

FIGURE 17
The sign diagram for f''

x	0	$\frac{\pi}{2}$	π
$f(x)$	0	1	0

we see that the end points $x = 0$ and $x = \pi$ yield the absolute minimum of f, while the absolute maximum of f is 1.

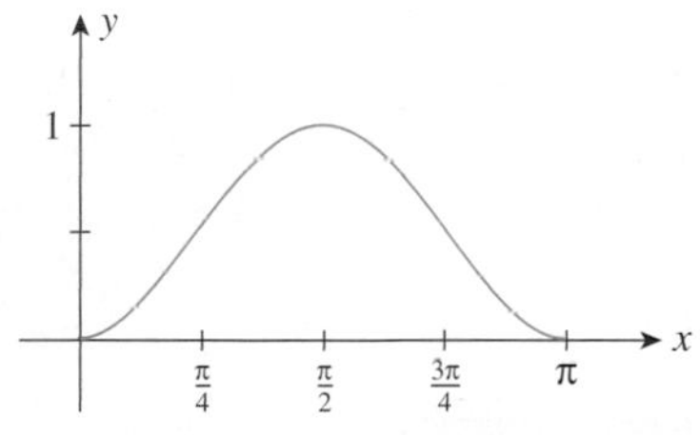

FIGURE 18
The graph of $y = \sin^2 x$

c. We compute $f''(x) = 2 \cos 2x$. Setting $f''(x) = 0$ gives $2x = \pi/2$, or $3\pi/2$; that is, $x = \pi/4$, or $3\pi/4$. From the sign diagram for f'', shown in Figure 17, we conclude that f is concave upward on the intervals $(0, \pi/4)$ and $(3\pi/4, \pi)$ and concave downward on the interval $(\pi/4, 3\pi/4)$.

d. From the results of part (c), we see that $(\pi/4, 1/2)$ and $(3\pi/4, 1/2)$ are inflection points of f.

Finally, the graph of f is sketched in Figure 18. ■

APPLIED EXAMPLE 6 Restaurant Revenue The revenue of McMenamy's Fish Shanty, located at a popular summer resort, is approximately

$$R(t) = 2\left(5 - 4 \cos \frac{\pi}{6} t\right) \qquad (0 \le t \le 12)$$

during the tth week ($t = 1$ corresponds to the first week of June), where R is measured in thousands of dollars. When is the weekly revenue increasing most rapidly?

Solution The revenue function R is increasing at the rate of

$$R'(t) = -8\left(-\sin \frac{\pi t}{6}\right)\left(\frac{\pi}{6}\right)$$
$$= \frac{4\pi}{3} \sin \frac{\pi t}{6}$$

thousand dollars per week. We want to maximize R'. Thus, we compute

$$R''(t) = \frac{4\pi}{3}\left(\cos\frac{\pi t}{6}\right)\left(\frac{\pi}{6}\right)$$
$$= \frac{2\pi^2}{9}\cos\frac{\pi t}{6}$$

Setting $R''(t) = 0$ gives

$$\cos\left(\frac{\pi t}{6}\right) = 0$$

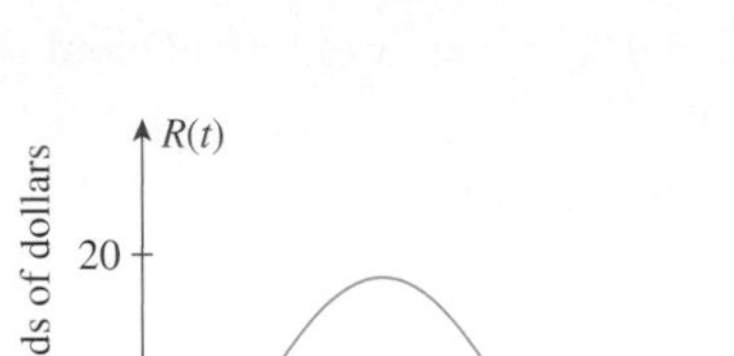

FIGURE 19

The graph of $R(t) = 2\left[5 - 4\cos\left(\frac{\pi}{6}t\right)\right]$

from which we see that $t = 3$ and $t = 9$ are critical numbers of $R'(t)$. Since $R'(t)$ is a continuous function on the closed interval [0, 12], the absolute maximum must occur at a critical number or at an endpoint of the interval. From the following table, we see that the weekly revenue is increasing most rapidly when $t = 3$—that is, in the third week of June (Figure 19).

t	0	3	9	12
$R'(t)$	0	$\frac{4\pi}{3}$	$-\frac{4\pi}{3}$	0

APPLIED EXAMPLE 7 Maximizing the Capacity of a Trough A trough with a trapezoidal cross section is to be constructed with a 1-foot base and sides that are 6 feet long (Figure 20). Find the angle of inclination θ that maximizes the capacity of the trough.

FIGURE 20

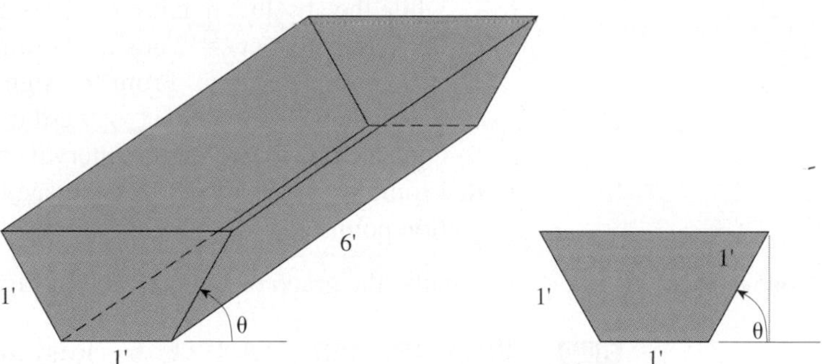

(a) We want to maximize the capacity of the trough with the given dimensions.

(b) A cross section of the trough has the shape of a trapezoid.

Solution The volume of the trough is the product of the area of its cross section and its length. Since the length is constant, it suffices to maximize the area of its cross section. But the cross section is a trapezoid with area given by one half the sum of the parallel sides times its height, or

$$A = \frac{1}{2}[1 + (1 + 2\cos\theta)]\sin\theta$$
$$= (1 + \cos\theta)\sin\theta$$

To find the absolute maximum of the continuous function A over the closed interval $[0, \pi/2]$, we first compute the derivative of A, obtaining

$$\begin{aligned} A' &= -\sin^2\theta + (1 + \cos\theta)\cos\theta \\ &= -\sin^2\theta + \cos\theta + \cos^2\theta \\ &= (\cos^2\theta - 1) + \cos\theta + \cos^2\theta \qquad \text{Use the identity } \sin^2\theta + \cos^2\theta = 1. \\ &= 2\cos^2\theta + \cos\theta - 1 \\ &= (2\cos\theta - 1)(\cos\theta + 1) \end{aligned}$$

Setting $A' = 0$ gives $\cos\theta = 1/2$, or $\cos\theta = -1$; that is, $\theta = \pi/3$ [the other values of θ lie outside the interval $(0, \pi/2)$]. Evaluating the function A at the critical number $\theta = \pi/3$ of A and at the endpoints of the interval, we have

θ	0	$\frac{\pi}{3}$	$\frac{\pi}{2}$
A	0	$\frac{3\sqrt{3}}{4}$	1

from which we see that the volume of the trough is maximized when $\theta = \pi/3$, or 60°. ■

12.3 Self-Check Exercises

1. Find the derivative of $f(x) = x\cos 2x$.
2. If $f(x) = x + \tan x$, find $f'(\pi/4)$.
3. Suppose the volume of air inhaled by a person during respiration is given by

$$V(t) = \frac{6}{5\pi}\left(1 - \sin\frac{\pi t}{2}\right) \qquad (t \geq 0)$$

liters at time t (in seconds). When is the volume of inhaled air at a maximum? What is the maximum volume?

Solutions to Self-Check Exercises 12.3 can be found on page 797.

12.3 Concept Questions

1. State the rules for differentiating $\sin x$ and $\cos x$.
2. State the rules for differentiating $\sin f(x)$ and $\cos f(x)$.
3. State the rules for differentiating $\tan x$ and $\tan f(x)$.
4. State the rules for differentiating $\csc f(x)$, $\sec f(x)$, and $\cot f(x)$.

12.3 Exercises

In Exercises 1–30, find the derivative of the function.

1. $f(x) = \cos 3x$
2. $f(x) = \sin 5x$
3. $f(x) = 2\cos \pi x$
4. $f(x) = \pi \sin 2x$
5. $f(x) = \sin(x^2 + 1)$
6. $f(x) = \cos \pi x^2$
7. $f(x) = \tan 2x^2$
8. $f(x) = \cot \sqrt{x}$
9. $f(x) = x\sin x$
10. $f(x) = x^2\cos x$
11. $f(x) = 2\sin 3x + 3\cos 2x$
12. $f(x) = 2\cot 2x + \sec 3x$
13. $f(x) = x^2\cos 2x$
14. $f(x) = \sqrt{x}\sin \pi x$
15. $f(x) = \sin\sqrt{x^2 - 1}$
16. $f(x) = \csc(x^2 + 1)$

17. $f(x) = e^x \sec x$

18. $f(x) = e^{-x} \csc x$

19. $f(x) = x \cos \dfrac{1}{x}$

20. $f(x) = x^2 \sin \dfrac{1}{x}$

21. $f(x) = \dfrac{x - \sin x}{1 + \cos x}$

22. $f(x) = \dfrac{\sin 2x}{1 + \cos 3x}$

23. $f(x) = \sqrt{\tan x}$

24. $f(x) = \sqrt{\cos x + \sin x}$

25. $f(x) = \dfrac{\sin x}{x}$

26. $f(x) = \dfrac{\cos x}{x^2 + 1}$

27. $f(x) = \tan^2 x$

28. $f(x) = \cot \sqrt{x}$

29. $f(x) = e^{\cot x}$

30. $f(x) = e^{\tan x + \sec x}$

31. Find the equation of the tangent line to the graph of the function $f(x) = \cot 2x$ at the point $(\pi/4, 0)$.

32. Find an equation of the tangent line to the graph of the function $f(x) = e^{\sec x}$ at the point $(\pi/4, e^{\sqrt{2}})$.

33. Determine the intervals where the function

$$f(x) = e^x \cos x \qquad (0 \le x \le 2\pi)$$

is increasing and where it is decreasing.

34. Determine the point(s) of inflection of the function

$$f(x) = x + \sin x \qquad (-2\pi \le x \le 2\pi)$$

In Exercises 35–38, sketch the graph of the function over the specified interval by obtaining the following information:

a. The intervals where f is increasing and where it is decreasing

b. The relative extrema of f

c. The concavity of f

d. The inflection points of f

35. $f(x) = \sin x + \cos x \quad (0 \le x \le 2\pi)$

36. $f(x) = x - \sin x \quad (0 \le x \le 2\pi)$

37. $f(x) = 2 \sin x + \sin 2x \quad (0 \le x \le 2\pi)$

38. $f(x) = e^x \cos x \quad (0 \le x \le 2\pi)$

39. a. Find the Taylor series about $x = 0$ for the function $f(x) = \sin x$.

b. Use the result of part (a) to show that

$$\lim_{x \to 0} \frac{\sin x}{x} = 1$$

40. a. Find the Taylor series about $x = 0$ for the function $f(x) = \cos x$.

b. Use the result of part (a) to show that

$$\lim_{x \to 0} \frac{\cos x - 1}{x} = 0$$

41. Predator-Prey Population The wolf population in a certain northern region is estimated to be

$$P_1(t) = 8000 + 1000 \sin\left(\frac{\pi t}{24}\right)$$

in month t, and the caribou population in the same region is given by

$$P_2(t) = 40{,}000 + 12{,}000 \cos\left(\frac{\pi t}{24}\right)$$

Find the rate of change of each population when $t = 12$.

42. Stock Prices The closing price (in dollars) per share of stock of Tempco Electronics on the tth day it was traded is approximated by

$$P(t) = 20 + 12 \sin \frac{\pi t}{30} - 6 \sin \frac{\pi t}{15} + 4 \sin \frac{\pi t}{10} - 3 \sin \frac{2\pi t}{15} \qquad (0 \le t \le 20)$$

where $t = 0$ corresponds to the time the stock was first listed in a major stock exchange. What was the rate of change of the stock's price at the close of the 15th day of trading? What was the closing price on that day?

43. Number of Hours of Daylight The number of hours of daylight on a particular day of the year in Boston is approximated by the function

$$f(t) = 3 \sin \frac{2\pi}{365}(t - 79) + 12$$

where $t = 0$ corresponds to January 1. Compute $f'(79)$ and interpret your result.

44. Water Level in a Harbor The water level (in feet) in Boston Harbor during a certain 24-hr period is approximated by the formula

$$H = 4.8 \sin \frac{\pi}{6}(t - 10) + 7.6 \qquad (0 \le t \le 24)$$

where $t = 0$ corresponds to 12 A.M. When is the water level rising and when is it falling? Find the relative extrema of H and interpret your results.

45. Average Daily Temperature The average daily temperature (in degrees Fahrenheit) at a tourist resort in Cameron Highlands is approximated by

$$T = 62 - 18 \cos \frac{2\pi(t - 23)}{365}$$

on the tth day ($t = 1$ corresponds to January 1). Which day was the warmest day of the year? The coldest day?

46. **Volume of Air Inhaled During Respiration** Suppose the volume of air inhaled by a person during respiration is given by

$$V(t) = \frac{6}{5\pi}\left(1 - \cos\frac{\pi t}{2}\right)$$

liters at time t (in seconds). When is the volume of inhaled air at a maximum? What is the maximum volume?

47. **Restaurant Revenue** The revenue of McMenamy's Fish Shanty located at a popular summer resort is approximately

$$R(t) = 2\left(5 - 4\cos\frac{\pi}{6}t\right) \qquad (0 \le t \le 12)$$

during the tth week ($t = 1$ corresponds to the first week of June), where R is measured in thousands of dollars. On what week does the restaurant realize the greatest revenue?

48. **Televising a Rocket Launch** A major network is televising the launching of a rocket. A camera tracking the liftoff of the rocket is located at point A, as shown in the accompanying figure, where ϕ denotes the angle of elevation of the camera at A. How fast is ϕ changing at the instant when the rocket is at a distance of 13,000 ft from the camera and this distance is increasing at the rate of 480 ft/sec?

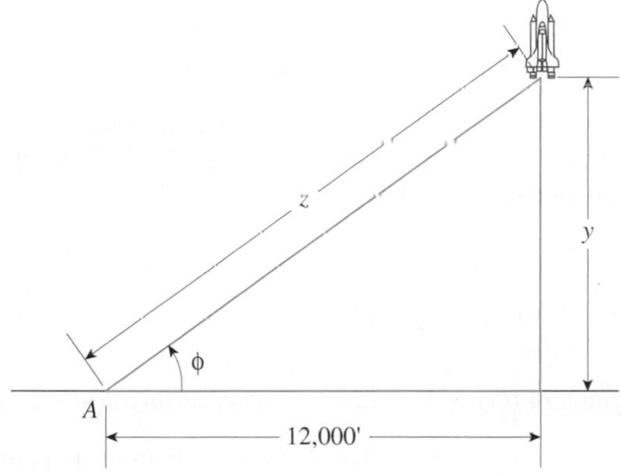

49. **Maximum Air Flow During Respiration** Refer to Exercise 46. Find the rate of flow of air in and out of the lungs. At what time is the rate of flow of air at a maximum? A minimum?

50. Refer to Exercise 60, page 231. Use the result of that exercise to find the rate at which the angle θ between the ladder and the ground is changing at the instant the bottom of the ladder is 12 ft from the wall.

51. **Tracking a Suspect** A police cruiser hunting for a suspect pulls over and stops at a point 20 ft from a straight wall. The flasher on top of the cruiser revolves at a constant rate of 90°/second, and the light beam casts a spot of light as it strikes the wall. How fast is the spot of light moving along the wall at a point 30 ft from the point on the wall closest to the cruiser?

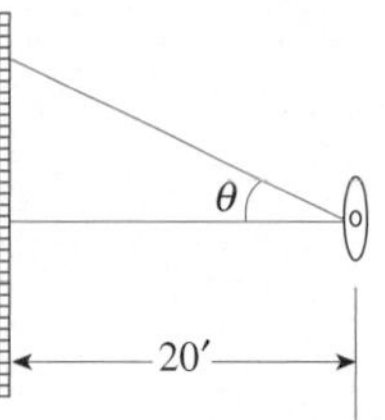

52. **Percentage Error in Measuring Height** From a point on level ground 150 ft from the base of a derrick, José measures the angle of elevation to the top of the derrick as 60°. If José's measurement is subject to an error of $\pm 1\%$, find the percentage error in the measured height of the derrick. (See the accompanying figure.)

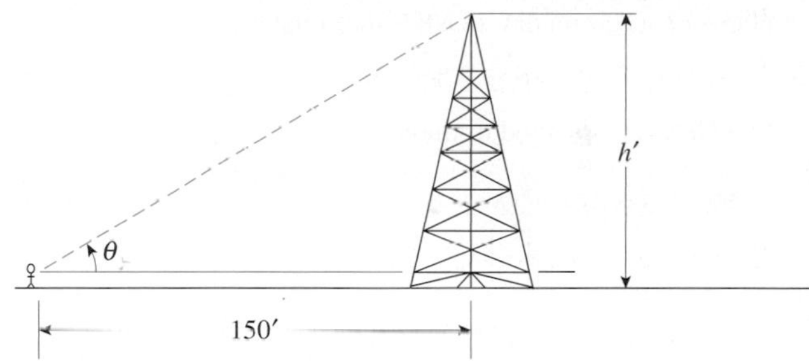

53. **Maximizing Drainage Capacity** The cross section of a drain is a trapezoid, as shown in the figure. The sides and the bottom of the trapezoid are each 5 ft long. Determine the angle θ so that the drain will have a maximal cross-sectional area.

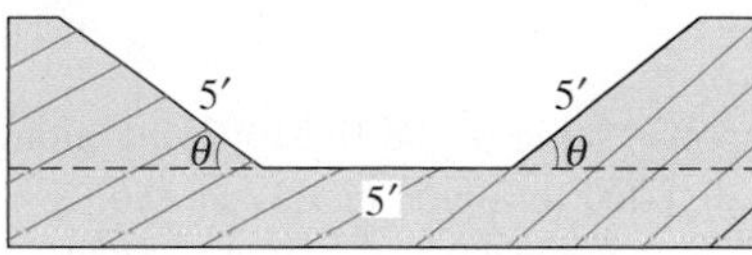

54. **Surface Area of a Honeycomb** The accompanying figure depicts a prism-shaped single cell in a honeycomb. The front end of the prism is a regular hexagon, and the back is formed by the sides of the cell coming together at a point. It can be shown that the surface area of a cell is given by

$$S(\theta) = 6ab + \frac{3}{2}b^2\left(\frac{\sqrt{3} - \cos\theta}{\sin\theta}\right) \qquad \left(0 < \theta < \frac{\pi}{2}\right)$$

where θ is the angle between one of the (three) upper surfaces and the altitude. Show that the surface area is minimized if $\cos\theta = 1/\sqrt{3}$, or $\theta \approx 54.7°$. Measurements of actual honeycombs have confirmed that this is, in fact, the angle found in beehives.

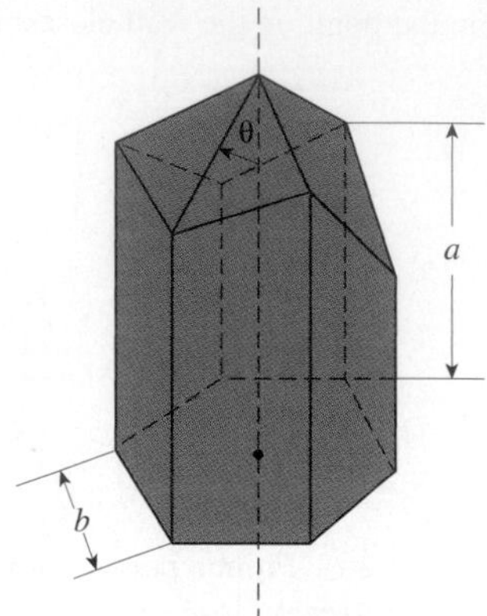

55. Finding the Position of a Planet As shown in the accompanying figure, the position of a planet that revolves about the Sun with an elliptical orbit can be located by calculating the central angle θ. Suppose the central angle sustained by a planet on a certain day satisfies the equation

$$\theta - 0.5 \sin \theta = 1$$

Use Newton's method to find θ.

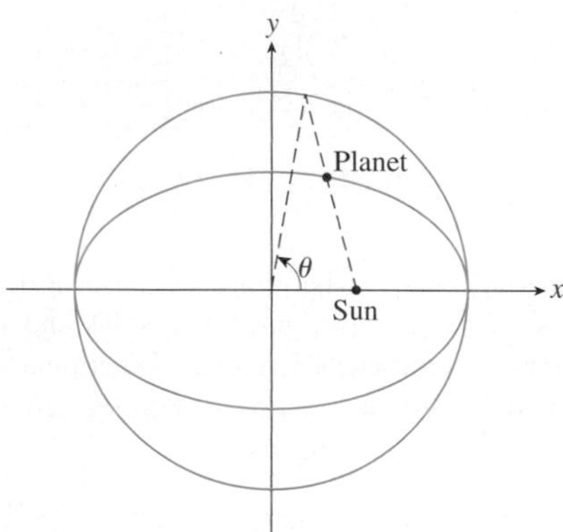

56. Flow of Blood Suppose some of the fluid flowing along a pipe of radius R is diverted to a pipe of smaller radius r attached to the former at an angle θ (see the figure). Such is the case when blood flowing along an artery is pumped into an arteriole. What should be the angle θ so that the energy loss due to friction in moving the fluid is minimal? Assume that the minimum exists and solve the problem using the following steps:

a. Use Poiseuille's Law, which states that the loss of energy due to friction in nonturbulent flow is proportional to the length of the path and inversely proportional to the fourth power of the radius, to show that the energy loss in moving the fluid from P to S via Q is

$$L = \frac{kd_1}{R^4} + \frac{kd_2}{r^4}$$

where k is a constant.

b. Suppose a and b are fixed. Find d_1 and d_2 in terms of a and b. Then use this result together with the result from part (a) to show that

$$E = k\left(\frac{a - b \cot \theta}{R^4} + \frac{b \csc \theta}{r^4}\right)$$

c. Using the technique of this section, show that E is minimized when

$$\theta = \cos^{-1} \frac{r^4}{R^4}$$

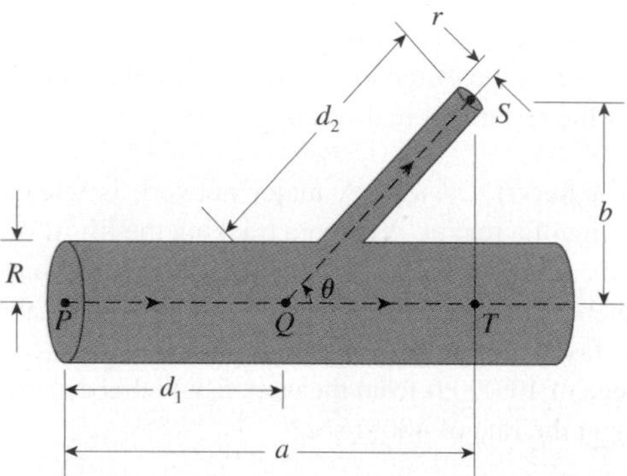

In Exercises 57–60, determine whether the statement is true or false. If it is true, explain why it is true. If it is false, give an example to show why it is false.

57. $\lim_{x \to 0} \frac{\sin x}{x} = 1.$
Hint: Use the definition of the derivative to study $f'(0)$ where $f(x) = \sin x$.

58. The graph of $f(x) = x - \sin x$ is nondecreasing on $(-\infty, \infty)$.

59. The function $f(x) = \cos x$ has a relative minimum at a point $x = a$ where $g(x) = \sin x$ has a relative maximum.

60. The graph of $f(x) = \sin x + \cos x$ is concave downward on the interval $\left(0, \frac{\pi}{2}\right)$.

61. Prove Rule 4:

If $h(x) = \csc f(x)$
then $h'(x) = -[\csc f(x)][\cot f(x)]f'(x)$
Hint: $\csc x = 1/\sin x$

62. Prove Rule 5:

If $h(x) = \sec f(x)$
then $h'(x) = [\sec f(x)][\tan f(x)]f'(x)$
Hint: $\sec x = 1/\cos x$

63. Prove Rule 6:

If $h(x) = \cot f(x)$
then $h'(x) = -[\csc^2 f(x)]f'(x)$
Hint: $\cot x = 1/\tan x = (\cos x)/(\sin x)$

64. Prove $\lim_{h \to 0} \frac{\cos h - 1}{h} = 0$

Hint: Multiply by $\frac{\cos h + 1}{\cos h + 1}$.

12.3 Solutions to Self-Check Exercises

1. $$\begin{aligned} f'(x) &= x\frac{d}{dx}\cos 2x + (\cos 2x)\frac{d}{dx}(x) \\ &= x(-\sin 2x)(2) + (\cos 2x)(1) \\ &= \cos 2x - 2x \sin 2x \end{aligned}$$

2. $f'(x) = 1 + \sec^2 x$. Therefore,

$$f'\left(\frac{\pi}{4}\right) = 1 + \sec^2\left(\frac{\pi}{4}\right) = 1 + (\sqrt{2})^2 = 3$$

3. We compute

$$V'(t) = \frac{6}{5\pi}\left(-\cos\frac{\pi t}{2}\right)\left(\frac{\pi}{2}\right) = -\frac{3}{5}\cos\frac{\pi t}{2}$$

Setting $V'(t) = 0$ and solving the resulting equation give $t = 1 + 2n$ $(n = 1, 2, \ldots)$ as critical numbers of V. The numbers $t = 3, 5, 7, \ldots$ give rise to the relative maxima of V, which coincide with the absolute maximum of V with value $12/(5\pi)$. Thus, the volume of inhaled air is at a maximum when $t = 3, 5, 7, \ldots,$ and its value is $12/(5\pi)$ liters.

USING TECHNOLOGY

Analyzing Trigonometric Functions

Graphing utilities can be used to analyze complicated trigonometric functions, as this example shows.

EXAMPLE 1 Let $f(x) = \frac{\sqrt{x}\cos 2x}{\sqrt{x^2 + \sin 3x}}$, where $0 < x \leq 2$.

a. Use a graphing utility to plot the graph of f in the viewing window $[0, 2] \times [-2, 1]$.

b. Use the numerical derivative operation of a graphing utility to find the rate of change of $f(x)$ at $x = 1$.

c. Find the inflection points of f.

d. Evaluate $\lim_{x \to 0^+} f(x)$. Explain why $f(0)$ is not defined. What happens if you use the evaluation function of the graphing utility to find $f(0)$?

e. Find the absolute maximum and absolute minimum values of f on the interval $[0.5, 2]$.

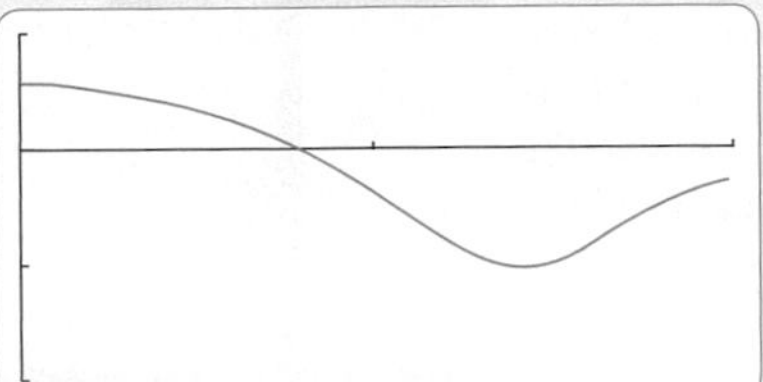

FIGURE T1

Solution

a. The graph of f in the viewing window $[0, 2] \times [-2, 1]$ is shown in Figure T1.

b. We find $f'(1) \approx -2.0628$.

c. The inflection points are $(1.0868, -0.5734)$ and $(1.9244, -0.5881)$. (*Note:* Move the cursor near $x = 1$ to locate the first inflection point and near $x = 1.5$ to locate the other.)

(continued)

d. Using **ZOOM-IN,** we find $\lim_{x \to 0^+} f(x) \approx 0.5773$. $f(0)$ is not defined because both the numerator and the denominator of $f(x)$ approach zero as x approaches zero. If we use the evaluation function of the graphing utility to find $f(0)$, the calculator will not give a value for y.

e. Using the operation to find the absolute minimum and the absolute maximum in the viewing window $[0.5, 2] \times [-2, 1]$, we find the absolute minimum value to be -1.0823 and the absolute maximum value to be 0.3421.

TECHNOLOGY EXERCISES

In Exercises 1–6, use the numerical derivative operation of a graphing utility to find the rate of change of $f(x)$ at the value of x. Give your answer accurate to four decimal places.

1. $f(x) = x^2\sqrt{1 + \sin^2 x}$; $x = 0.5$
2. $f(x) = \dfrac{1 + \cos x}{\sqrt{1 + \sin 3x}}$; $x = \dfrac{\pi}{4}$
3. $f(x) = \dfrac{x \tan x}{\sqrt{1 + x^2}}$; $x = 0.4$
4. $f(x) = e^{\cot x}(1 + x^2)^{3/2}$; $x = 1$
5. $f(x) = \cos[\csc(\sqrt{x} + 1)]$; $x = \dfrac{\pi}{3}$
6. $f(x) = \dfrac{\ln(\cos x)}{(1 + e^{-x})^{3/2}}$; $x = 0.8$

In Exercises 7–10, use a graphing utility to find the absolute maximum and the absolute minimum values of f in the given interval. Express your answers accurate to four decimal places.

7. $f(x) = \dfrac{\sin x}{x}$; $[1, 6]$
8. $f(x) = \dfrac{\cos x^2}{\sqrt{x}}$; $[1, 3]$
9. $f(x) = \dfrac{x \sec x}{1 + \cot x}$; $[0.5, 1]$
10. $f(x) = \dfrac{x + \cos x}{1 + 0.5 \sin x}$; $[0, 2]$

11. Refer to Exercise 42, page 794.
    a. Plot the graph of P, using the viewing window $[0, 20] \times [15, 35]$.
    b. What was the rate of change of the stock's price at the close of the 23rd day of trading?
    c. What was the closing price on that day?

12. Refer to Exercise 45, page 794.
    a. Plot the graph of T, using the viewing window $[0, 365] \times [20, 80]$.
    b. Using **TRACE** and **ZOOM** or using the graphing utility's operation for finding the absolute extrema of a function, verify the solution to the problem.

13. Refer to Exercise 46, page 795.
    a. Plot the graph of V, using the viewing window $[0, 12] \times [0, 2]$.
    b. Verify the solution to the problem using a graphing utility.

14. **Number of Hours of Daylight** The number of hours of daylight on a particular day of the year in Boston is approximated by the function

$$f(t) = 3 \sin \frac{2\pi}{365}(t - 79) + 12$$

where $t = 0$ corresponds to January 1.
    a. Sketch the graph of f on the interval $[0, 365]$.
    b. When does the longest day occur? When does the shortest day occur?

15. **Effect of an Earthquake on a Structure** To study the effect an earthquake has on a structure, engineers look at the way a beam bends when subjected to an Earth tremor. The equation

$$D = a - a\cos\left(\frac{\pi h}{2L}\right) \qquad (0 \le h \le L)$$

where L is the length of a beam and a is the maximum deflection from the vertical, has been used by engineers to calculate the deflection D at a point on the beam h feet from the ground (see the figure). Suppose a 10-ft vertical beam has a maximum deflection of $\frac{1}{2}$ ft when subjected to an external force. Using differentials, estimate the difference in the deflection between the point midway on the beam and the point $\frac{1}{10}$ ft above it.

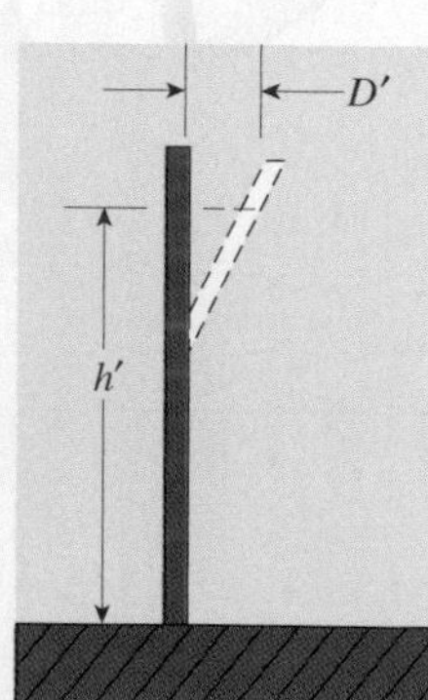

12.4 Integration of Trigonometric Functions

Integrating Trigonometric Functions

Each of the six equations pertaining to a rule of differentiation given in the last section may be integrated to yield a corresponding integration formula. For example, integrating the equation $f'(x) = \cos x$ of Rule 1 with respect to x yields

$$\int \cos x\,dx = \int f'(x)\,dx = f(x) + C = \sin x + C$$

where C is a constant of integration. The six rules of integration obtainable in this manner follow.

Trigonometric Integration Formulas

1. $\int \sin x\,dx = -\cos x + C$
2. $\int \cos x\,dx = \sin x + C$
3. $\int \sec^2 x\,dx = \tan x + C$
4. $\int \csc^2 x\,dx = -\cot x + C$
5. $\int \sec x \tan x\,dx = \sec x + C$
6. $\int \csc x \cot x\,dx = -\csc x + C$

EXAMPLE 1 Evaluate $\int \cos 3x\,dx$.

Solution Let's put $u = 3x$, so that $du = 3\,dx$ and $dx = \frac{1}{3}\,du$. Then

$$\begin{aligned}\int \cos 3x\,dx &= \frac{1}{3}\int \cos u\,du \\ &= \frac{1}{3}\sin u + C && \text{Use Rule 2.} \\ &= \frac{1}{3}\sin 3x + C\end{aligned}$$

■

EXAMPLE 2 Evaluate $\int \sec(2x + 1)\tan(2x + 1)\,dx$.

Solution Put $u = 2x + 1$ so that $du = 2\,dx$, or $dx = \frac{1}{2}\,du$. Then

$$\begin{aligned}\int \sec(2x + 1)\tan(2x + 1)\,dx &= \frac{1}{2}\int \sec u \tan u\,du \\ &= \frac{1}{2}\sec u + C \\ &= \frac{1}{2}\sec(2x + 1) + C\end{aligned}$$

■

EXAMPLE 3 Evaluate $\displaystyle\int \frac{\sin x\, dx}{1 + \cos x}$.

Solution Put $u = 1 + \cos x$ so that $du = -\sin x\, dx$, or $\sin x\, dx = -du$. Then

$$\begin{aligned} \int \frac{\sin x\, dx}{1 + \cos x} &= -\int \frac{du}{u} \\ &= -\ln|u| + C \\ &= -\ln|1 + \cos x| + C \end{aligned}$$

EXAMPLE 4 Evaluate $\displaystyle\int_0^{\pi/2} x \cos 2x\, dx$.

Solution We first integrate the indefinite integral

$$\int x \cos 2x\, dx$$

by parts with

$$u = x \qquad \text{and} \qquad dv = \cos 2x\, dx$$

so that

$$du = dx \qquad \text{and} \qquad v = \frac{1}{2} \sin 2x$$

Therefore;

$$\begin{aligned} \int x \cos 2x\, dx &= uv - \int v\, du \\ &= \frac{1}{2} x \sin 2x - \frac{1}{2} \int \sin 2x\, dx \\ &= \frac{1}{2} x \sin 2x + \frac{1}{4} \cos 2x + C \end{aligned}$$

We have used the method of substitution to evaluate the integral on the right. Therefore,

$$\begin{aligned} \int_0^{\pi/2} x \cos 2x\, dx &= \left. \frac{1}{2} x \sin 2x + \frac{1}{4} \cos 2x \right|_0^{\pi/2} \\ &= \left[\frac{1}{2}\left(\frac{\pi}{2}\right) \sin \pi + \frac{1}{4} \cos \pi\right] - \left(0 + \frac{1}{4} \cos 0\right) \\ &= -\frac{1}{4} - \frac{1}{4} = -\frac{1}{2} \end{aligned}$$

EXPLORE & DISCUSS

Refer to the Explore & Discuss question on page 455.

1. Show that if f is even and g is odd, then the function fg is odd.
2. Refer to Example 4 above, and use the result of part (1) to show that

$$\int_{-\pi/2}^{\pi/2} x \cos 2x\, dx = 0$$

3. What is the area of the region bounded by the graph of $f(x) = x \cos 2x$, the x-axis, and the lines $x = -\pi/2$ and $x = \pi/2$?

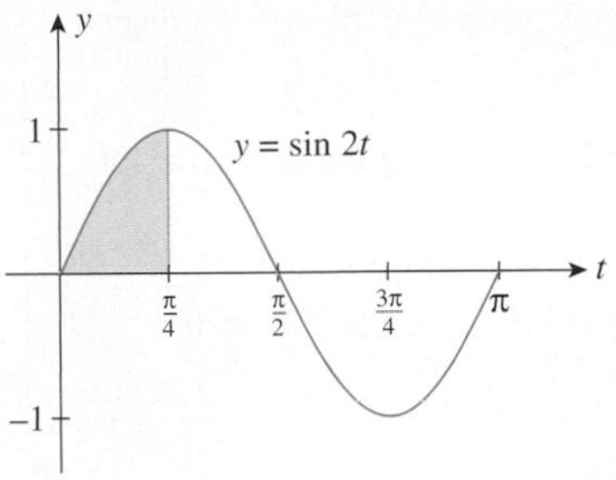

FIGURE 21
The area under the curve $y = \sin 2t$ from $t = 0$ to $t = \frac{\pi}{4}$

EXAMPLE 5 Find the area under the curve of $y = \sin 2t$ from $t = 0$ to $t = \pi/4$.

Solution The required area, shown in Figure 21, is given by

$$\int_0^{\pi/4} \sin 2t\, dt = -\frac{1}{2}\cos 2t \Big|_0^{\pi/4} = -\frac{1}{2}\cos\frac{\pi}{2} + \frac{1}{2}\cos 0 = \frac{1}{2}$$

or $\frac{1}{2}$ square unit. ■

APPLIED EXAMPLE 6 Stock Prices The weekly closing price of HAL Corporation stock in week t is approximated by the rule

$$f(t) = 30 + t\sin\frac{\pi}{6}t \qquad (0 \le t \le 15)$$

where $f(t)$ is the price (in dollars) per share. Find the average weekly closing price of the stock over the 15-week period.

Solution The average weekly closing price of the stock over the 15-week period in question is given by

$$\begin{aligned} A &= \frac{1}{15-0}\int_0^{15}\left(30 + t\sin\frac{\pi}{6}t\right)dt \\ &= \frac{1}{15}\int_0^{15} 30\, dt + \frac{1}{15}\int_0^{15} t\sin\frac{\pi}{6}t\, dt \\ &= 30 + \frac{1}{15}\int_0^{15} t\sin\frac{\pi}{6}t\, dt \end{aligned}$$

Integrating by parts with

$$u = t \qquad \text{and} \qquad dv = \sin\frac{\pi}{6}t\, dt$$

so that

$$du = dt \qquad \text{and} \qquad v = -\frac{6}{\pi}\cos\frac{\pi}{6}t$$

we have

$$\begin{aligned} A &= 30 + \frac{1}{15}\left(-\frac{6}{\pi}t\cos\frac{\pi}{6}t\Big|_0^{15} + \frac{6}{\pi}\int_0^{15}\cos\frac{\pi}{6}t\, dt\right) \\ &= 30 + \frac{1}{15}\left[-\left(\frac{6}{\pi}\right)(15)\cos\frac{15\pi}{6} + \frac{6}{\pi}(0)\cos 0 + \left(\frac{6}{\pi}\right)^2 \sin\frac{\pi}{6}t\Big|_0^{15}\right] \\ &= 30 + \frac{1}{15}\left[-\left(\frac{6}{\pi}\right)(15)(0) + \left(\frac{6}{\pi}\right)^2\sin\frac{15\pi}{6} - \left(\frac{6}{\pi}\right)^2\sin 0\right] \\ &= 30 + \frac{1}{15}\left(\frac{6}{\pi}\right)^2(1) \approx 30.24 \end{aligned}$$

or approximately $30.24 per share. ■

EXPLORE & DISCUSS

Without actually evaluating the integral, explain the following using the properties of the trigonometric functions:

1. $\int_0^{\pi/2} \sin x\, dx = \int_{10\pi}^{21\pi/2} \sin x\, dx$ **2.** $\int_{-3}^{3} (2 + \cos 3x - \sin 5x)\, dx \geq 0$

Additional Trigonometric Integration Formulas

The six integration formulas listed at the beginning of this section were immediate consequences of the six corresponding rules of differentiation of the previous section. We now complete the list of integration formulas for the basic trigonometric functions by giving the integration formulas for the tangent, secant, cosecant, and cotangent functions.

Additional Trigonometric Integration Formulas

7. $\int \tan x\, dx = -\ln|\cos x| + C$

8. $\int \sec x\, dx = \ln|\sec x + \tan x| + C$

9. $\int \csc x\, dx = \ln|\csc x - \cot x| + C$

10. $\int \cot x\, dx = \ln|\sin x| + C$

To prove Formula 7, we write

$$\int \tan x\, dx = \int \frac{\sin x}{\cos x}\, dx$$

Putting $u = \cos x$ so that $du = -\sin x\, dx$, we find

$$\begin{aligned} \int \tan x\, dx &= -\int \frac{du}{u} = -\ln|u| + C \\ &= -\ln|\cos x| + C \end{aligned}$$

as we set out to show.

12.4 Self-Check Exercises

1. Evaluate $\int_0^{\pi/2} \frac{\cos x\, dx}{1 + \sin x}$.

2. The rate of change of the population of foxes in a certain region in month t is

$$\frac{25\pi}{6} \cos\left(\frac{\pi t}{12}\right)$$

foxes per month. If the population at $t = 0$ is 400, find an expression giving the number of foxes in month t.

Solutions to Self-Check Exercises 12.4 can be found on page 804.

12.4 Concept Questions

1. State the rules for integrating $\sin x$, $\cos x$, and $\sec^2 x$.

2. State the rules for integrating $\csc^2 x$, $\sec x \tan x$, and $\csc x \cot x$.

12.4 Exercises

In Exercises 1–32, find or evaluate each integral.

1. $\int \sin 3x\, dx$
2. $\int \cos(x + \pi)\, dx$
3. $\int (3 \sin x + 4 \cos x)\, dx$
4. $\int (x^2 - \cos 2x)\, dx$
5. $\int \sec^2 2x\, dx$
6. $\int \csc^2 3x\, dx$
7. $\int x \cos x^2\, dx$
8. $\int x \sec x^2 \tan x^2\, dx$
9. $\int \csc \pi x \cot \pi x\, dx$
10. $\int \sec 2x \tan 2x\, dx$
11. $\int_{-\pi/2}^{\pi/2} (\sin x + \cos x)\, dx$
12. $\int_0^{\pi} (2 \cos x + 3 \sin x)\, dx$
13. $\int_0^{\pi/6} \tan 2x\, dx$
14. $\int_{\pi/8}^{\pi/4} \cot 2x\, dx$
15. $\int \sin^3 x \cos x\, dx$
16. $\int \cos^2 x \sin x\, dx$
17. $\int \sec \pi x\, dx$
18. $\int \csc(1 - x)\, dx$
19. $\int_0^{\pi/12} \sec 3x\, dx$
20. $\int_0^{\pi/8} \tan 2x\, dx$
21. $\int \sqrt{\cos x} \sin x\, dx$
22. $\int (\sin x)^{1/3} \cos x\, dx$
23. $\int \cos 3x \sqrt{1 - 2 \sin 3x}\, dx$
24. $\int \frac{\cos x\, dx}{\sqrt{1 + \sin x}}$
25. $\int \tan^3 x \sec^2 x\, dx$
26. $\int \sec^5 x \tan x\, dx$
27. $\int \csc^2 x(\cot x - 1)^3\, dx$
28. $\int \sec^2 x \sqrt{1 + \tan x}\, dx$
29. $\int_1^{e^{\pi}} \frac{\sin(\ln x)}{x}\, dx$
30. $\int_0^{\pi/2} \frac{\cos x}{1 + \sin x}\, dx$
31. $\int \sin(\ln x)\, dx$
Hint: Integrate by parts.
32. $\int_1^{e^{\pi/2}} \cos(\ln x)\, dx$
Hint: Integrate by parts.

In Exercises 33–36, find the area of the region under the graph of the function f from $x = a$ to $x = b$.

33. $f(x) = \cos \frac{x}{4}$; $a = 0, b = \pi$
34. $f(x) = x + \sin x$; $a = 0, b = \frac{\pi}{4}$
35. $f(x) = \tan x$; $a = 0, b = \frac{\pi}{4}$
36. $f(x) = \cot x + \csc x$; $a = \frac{\pi}{4}, b = \frac{\pi}{2}$
37. Find the area of the region bounded above by the graph of $y = x$ and below by the graph of $y = \sin x$ from $x = 0$ to $x = \pi$.
38. Find the area of the region bounded above by the graph of $y = \cos x$ and below by the graph of $y = x$ from $x = 0$ to $x = \pi/4$. Make a rough sketch of the region.
39. STOCK PRICES The weekly closing price of TMA Corporation stock in week t is approximated by the rule

$$f(t) = 80 + 3t \cos \frac{\pi t}{6} \qquad (0 \le t \le 15)$$

where $f(t)$ is the price (in dollars) per share. Find the average weekly closing price of the stock over the 15-wk period.

40. AVERAGE DAILY TEMPERATURE The average daily temperature in degrees Fahrenheit at a tourist resort in the Cameron Highlands is approximately

$$T = 62 - 18 \cos \frac{2\pi(t - 23)}{365}$$

on the tth day ($t = 1$ corresponds to January 1). What is the average temperature at the resort in the month of January?
Hint: Find the average value of T over the interval $[0, 31]$.

41. RESTAURANT REVENUE The revenue of McMenamy's Fish Shanty, located at a popular summer resort, is approximately

$$R(t) = 2\left(5 - 4 \cos \frac{\pi}{6} t\right) \qquad (0 \le t \le 12)$$

during the tth week ($t = 1$ corresponds to the first week of June), where R is measured in thousands of dollars. What is the total revenue realized by the restaurant over the 12-wk period starting June 1?

42. STOCK PRICES Refer to Exercise 42, Section 12.3. What was the average price per share of Tempco Electronics stock over the first 20 days that it was traded on the stock exchange?

43. VOLUME OF AIR INHALED DURING RESPIRATION Suppose that the rate of air flow in and out of a person's lungs during respiration is

$$R(t) = 0.6 \sin \frac{\pi t}{2}$$

liters per second, where t is the time in seconds. Find an expression for the volume of air in the person's lungs at any time t.

44. AVERAGE VOLUME OF INHALED AIR DURING RESPIRATION Refer to Exercise 43. What is the average volume of inspired air in the person's lungs during one respiratory cycle?
Hint: Find the average value of the function giving the volume of inhaled air (obtained in the solution to Exercise 43) over the interval $[0, 4]$.

45. GROWTH OF A FRUIT-FLY COLONY A biologist has determined that by varying the food supply available to a population of *Drosophila* (fruit flies) in a controlled experiment, the rate at which the population grows obeys the rule

$$\frac{dQ}{dt} = 0.0001(4 + 5\cos 2t)Q(400 - Q)$$

where Q denotes the population of *Drosophila* at any time t. If there were ten fruit flies initially, determine the number of fruit flies in the colony after 20 days.
Hint: Solve the separable differential equation using the following identity:

$$\frac{1}{Q(C - Q)} = \frac{1}{C}\left[\frac{1}{Q} + \frac{1}{C - Q}\right]$$

46. Use the sixth Taylor polynomial approximation of $(\sin x)/x$,

$$\frac{\sin x}{x} \approx 1 - \frac{x^2}{6} + \frac{x^4}{120} - \frac{x^6}{5040}$$

to obtain an approximation of the integral

$$\int_0^{\pi/2} \frac{\sin x}{x}\, dx$$

(See Exercise 39, Section 12.3.)

47. Use the twelfth Taylor polynomial approximation of $\cos x^2$,

$$\cos x^2 \approx 1 - \frac{x^4}{2} + \frac{x^8}{24} - \frac{x^{12}}{720}$$

to obtain an approximation of the Fresnel cosine integral

$$\int_0^x \cos t^2\, dt$$

(It is important in the study of the theory of diffraction.) In particular, evaluate

$$\int_0^1 \cos t^2\, dt$$

(See Exercise 40, Section 12.3.)

In Exercises 48–51, determine whether the statement is true or false. If it is true, explain why it is true. If it is false, give an example to show why it is false.

48. $\displaystyle\int_a^b \sin x\, dx = \int_{a+2\pi}^{b+2\pi} \sin x\, dx$

49. $\displaystyle\int_a^b \cos x\, dx = \int_a^{b+2\pi} \cos x\, dx$

50. $\displaystyle\int_{-\pi/2}^{\pi/2} \sqrt[3]{x} \sin 2x\, dx > 0$

51. $\displaystyle\int_{-\pi/2}^{\pi/2} |\sin x|\, dx = \int_{-\pi/2}^{\pi/2} |\cos x|\, dx$

12.4 Solutions to Self-Check Exercises

1. Let $u = 1 + \sin x$ so that $du = \cos x\, dx$. Then

$$\int \frac{\cos x\, dx}{1 + \sin x} = \int \frac{du}{u} = \ln|u| + C = \ln|1 + \sin x| + C$$

Therefore,

$$\int_0^{\pi/2} \frac{\cos x\, dx}{1 + \sin x} = \ln|1 + \sin x| \Big|_0^{\pi/2} = \ln 2$$

2. The population of foxes in month t is

$$P = \int \frac{25\pi}{6} \cos\left(\frac{\pi t}{12}\right) dt = \frac{25\pi}{6} \int \cos\left(\frac{\pi t}{12}\right) dt$$

Let $u = \frac{\pi t}{12}$ so that $du = \frac{\pi}{12}\, dt$. Therefore,

$$P = \left(\frac{25\pi}{6}\right)\left(\frac{12}{\pi}\right)\int \cos u\, du = 50 \sin u + C$$

$$= 50 \sin\left(\frac{\pi t}{12}\right) + C$$

To evaluate C, note that $P = 400$ when $t = 0$. That is,

$$400 = 50(0) + C, \qquad \text{or} \qquad C = 400$$

Therefore,

$$P(t) = 400 + 50 \sin\left(\frac{\pi t}{12}\right)$$

USING TECHNOLOGY

Evaluating Integrals of Trigonometric Functions

As we saw in Chapter 6, the numerical integration operation of a graphing utility can be used to evaluate definite integrals involving trigonometric functions.

EXAMPLE 1 Use a graphing utility to find the area of the region R that is completely enclosed by the graphs of the functions

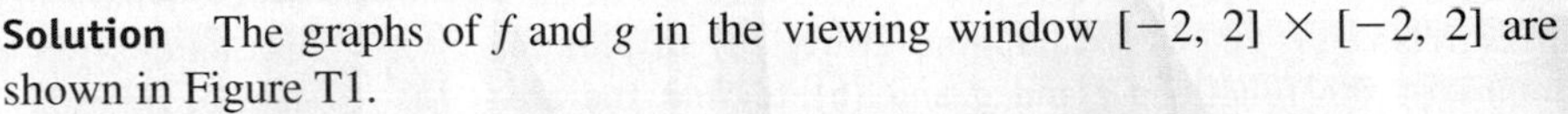

$$f(x) = \cos x + \sin 2x \qquad \text{and} \qquad g(x) = 1 - x^2$$

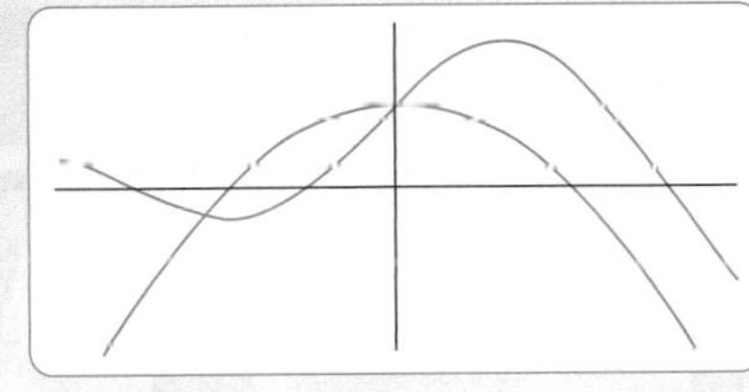

FIGURE T1

Solution The graphs of f and g in the viewing window $[-2, 2] \times [-2, 2]$ are shown in Figure T1.

Using the intersection operation of a graphing utility, we find that the x-coordinates of the points of intersection of the two graphs are ≈ -1.1554 and 0. Since the graph of g lies above that of f on the interval $[-1.1554, 0]$, we see that the area of R is given by

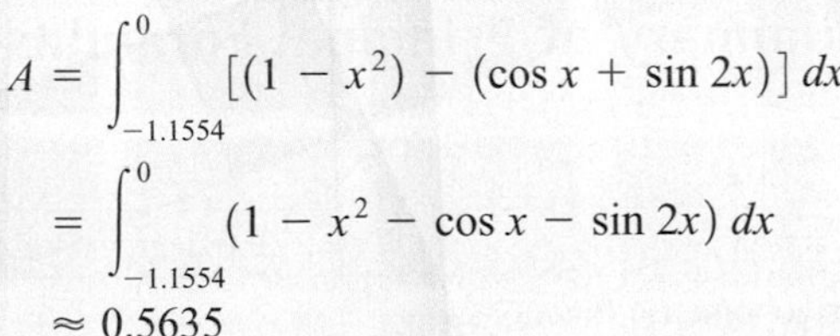

$$A = \int_{-1.1554}^{0} [(1 - x^2) - (\cos x + \sin 2x)]\, dx$$

$$= \int_{-1.1554}^{0} (1 - x^2 - \cos x - \sin 2x)\, dx$$

$$\approx 0.5635$$

or approximately 0.5635 square unit.

TECHNOLOGY EXERCISES

In Exercises 1–4, use a graphing utility to find the area of the region under the graph of f on the interval $[a, b]$. Express your answer accurate to four decimal places.

1. $f(x) = \sqrt{x} \sin 2x$; $[0, 1]$

2. $f(x) = \dfrac{x}{\sin x}$; $[1, 2]$

3. $f(x) = \dfrac{\cos x}{\sqrt{1 + x^2}}$; $[0, 1]$

4. $f(x) = (\cos x) \ln\left(\dfrac{1 + x}{1 - x}\right)$; $[0, \frac{1}{2}]$

In Exercises 5–12, use a graphing calculator to evaluate the definite integral. Express your answers accurate to four decimal places.

5. $\displaystyle\int_0^{0.3} (0.2 \cos x - 0.3 \sin 2x + 1.2 \tan 3x)\, dx$

(continued)

6. $\int_0^{1.1} \sqrt{1 + x^2} \sec x \, dx$

7. $\int_0^{2\pi} \frac{dx}{10 + 3 \cos x}$

8. $\int_0^{\pi/2} \sqrt{1 + \frac{1}{2} \sin^2 x} \, dx$

9. $\int_{0.5}^{1} \ln \sin x \, dx$

10. $\int_1^2 e^{\cot x} \, dx$

11. $\int_0^1 \cos x^2 \, dx$

12. $\int_0^3 \frac{\cos x}{(1 + \sin^2 x)^{3/2}} \, dx$

In Exercises 13–16, use a graphing utility (a) to plot the graphs of the functions f and g and (b) to find the area of the region enclosed by these graphs and the vertical lines $x = a$ and $x = b$. Express your answers accurate to four decimal places.

13. $f(x) = x \sin 2x$; $g(x) = \frac{1}{2}x^2 - 2$; $a = 0$, $b = 1$

14. $f(x) = \tan^2 x$; $g(x) = 5 - \sqrt{1 + \sqrt{x}}$; $a = 0$, $b = 1$

15. $f(x) = e^{\cos x}$; $g(x) = \ln \tan x$; $a = 0.2$, $b = 0.8$

16. $f(x) = \cot x$; $g(x) = \sqrt{1 + 2 \sin x}$; $a = 1$, $b = 2$

In Exercises 17–20, use a graphing utility (a) to plot the graphs of the functions f and g and (b) to find the area of the region completely enclosed by the graphs of these functions.

17. $f(x) = 2 \cos 3x$ and $g(x) = x^2$

18. $f(x) = \cos^2 x$ and $g(x) = e^{x^2} - 1$

19. $f(x) = \sin 2x$ and $g(x) = x - \sqrt{x}$

20. $f(x) = e^{\sin x}$ and $g(x) = 0.5x^2$

21. **Boston Harbor Water Level** The water level (in feet) in Boston Harbor during a certain 24-hr period is approximated by the formula

$$H = 4.8 \sin \frac{\pi}{6}(t - 10) + 7.6 \qquad (0 \le t \le 24)$$

where $t = 0$ corresponds to 12 A.M. What is the average water level in Boston Harbor over the 24-hr period on that day?

22. **Number of Daylight Hours in Chicago** The number of hours of daylight at any time t in Chicago is approximated by

$$L(t) = 2.8 \sin \frac{2\pi}{365}(t - 79) + 12$$

where t is measured in days and $t = 0$ corresponds to January 1. What is the daily average number of hours of daylight in Chicago over the year? Over the summer months from June 21 ($t = 171$) through September 20 ($t = 262$)?

CHAPTER 12 Summary of Principal Formulas and Terms

FORMULAS

1. Degree-radian conversion	360 degrees $= 2\pi$ radians
2. Derivatives of the trigonometric functions:	
Sine	$\frac{d}{dx}(\sin u) = \cos u \frac{du}{dx}$
Cosine	$\frac{d}{dx}(\cos u) = -\sin u \frac{du}{dx}$
Tangent	$\frac{d}{dx}(\tan u) = \sec^2 u \frac{du}{dx}$
Cosecant	$\frac{d}{dx}(\csc u) = -\csc u \cot u \frac{du}{dx}$
Secant	$\frac{d}{dx}(\sec u) = \sec u \tan u \frac{du}{dx}$
Cotangent	$\frac{d}{dx}(\cot u) = -\csc^2 u \frac{du}{dx}$

3. Integrals of the trigonometric functions:	
$\int \sin u\, du = -\cos u + C$	$\int \csc u \cot u\, du = -\csc u + C$
$\int \cos u\, du = \sin u + C$	$\int \tan u\, du = -\ln\|\cos u\| + C$
$\int \sec^2 u\, du = \tan u + C$	$\int \sec u\, du = \ln\|\sec u + \tan u\| + C$
$\int \csc^2 u\, du = -\cot u + C$	$\int \csc u\, du = \ln\|\csc u - \cot u\| + C$
$\int \sec u \tan u\, du = \sec u + C$	$\int \cot u\, du = \ln\|\sin u\| + C$

TERMS

angle (772)
initial ray (772)
terminal ray (772)
degree (773)
radian (773)
coterminal angles (774)
period of a trigonometric function (779)
predator–prey population model (781)
trigonometric identity (782)

CHAPTER 12 Concept Review Questions

Fill in the blanks.

1. **a.** The measure of the central angle subtended by an arc equal in length to the radius of the circle is called a ____.
   b. The angle subtended by one complete revolution is ____ radians or ____ degrees.

2. **a.** If x is the number of degrees, then $f(x) =$ ____ gives the number of radians.
   b. If x is the number of radians, then $g(x) =$ ____ gives the number of degrees.

3. If $P(x, y)$ is a point on the unit circle, then $\cos\theta =$ ____, $\sin\theta =$ ____, $\tan\theta =$ ____, $\csc\theta =$ ____, $\sec\theta =$ ____, and $\cot\theta =$ ____.

4. **a.** $\sin^2\theta +$ ____ $= 1$ **b.** $\tan^2\theta +$ ____ $= \sec^2\theta$
   c. $\cot^2\theta +$ ____ $= \csc^2\theta$ **d.** $\cos^2\theta = \frac{1}{2}(1 +$ ____$)$

5. **a.** $\sin^2\theta = \frac{1}{2}(1 -$ __$)$
   b. $\sin(A + B) = \sin A \cos B$ ____ $\cos A \sin B$
   c. $\sin 2A = 2$ **d.** $\cos 2A = \cos^2 A -$ ____

6. **a.** $\frac{d}{dx}[\sin f(x)] =$ ____ **b.** $\frac{d}{dx}[\cos f(x)] =$ ____
   c. $\frac{d}{dx}[\tan f(x)] =$ ____ **d.** $\frac{d}{dx}[\csc f(x)] =$ ____
   e. $\frac{d}{dx}[\sec f(x)] =$ ____ **f.** $\frac{d}{dx}[\cot f(x)] =$ ____

7. **a.** $\int \sin x\, dx =$ ____ **b.** $\int \cos x\, dx =$ ____
   c. $\int \sec^2 x\, dx =$ ____ **d.** $\int \csc^2 x\, dx =$ ____
   e. $\int \sec x \tan x\, dx =$ ____ **f.** $\int \csc x \cot x\, dx =$ ____

CHAPTER 12 Review Exercises

In Exercises 1–3, convert the angle to radian measure.

1. $120°$ 2. $450°$ 3. $-225°$

In Exercises 4–6, convert the angle to degree measure.

4. $\frac{11}{6}\pi$ radians 5. $-\frac{5}{2}\pi$ radians 6. $-\frac{7}{4}\pi$ radians

In Exercises 7 and 8, find all values of θ that satisfy the equation over the interval $[0, 2\pi]$.

7. $\cos\theta = \frac{1}{2}$ 8. $\cot\theta = -\sqrt{3}$

In Exercises 9–20, find the derivative of the function f.

9. $f(x) = \sin 3x$ 10. $f(x) = 2\cos\frac{x}{2}$

11. $f(x) = 2 \sin x - 3 \cos 2x$

12. $f(x) = \sec^2 \sqrt{x}$

13. $f(x) = e^{-x} \tan 3x$

14. $f(x) = (1 - \csc 2x)^2$

15. $f(x) = 4 \sin x \cos x$

16. $f(x) = \dfrac{\cos x}{1 - \cos x}$

17. $f(x) = \dfrac{1 - \tan x}{1 - \cot x}$

18. $f(x) = \ln(\cos^2 x)$

19. $f(x) = \sin(\sin x)$

20. $f(x) = e^{\sin x} \cos x$

21. Find the equation of the tangent line to the graph of the function $f(x) = \tan^2 x$ at the point $(\pi/4, 1)$.

22. Sketch the graph of the function $y = f(x) = \cos^2 x$ on the interval $[0, \pi]$ by obtaining the following information:
   a. The intervals where f is increasing and where it is decreasing
   b. The relative extrema of f
   c. The concavity of f
   d. The inflection points of f

In Exercises 23–32, evaluate the integral.

23. $\int \cos \frac{2}{3} x\, dx$

24. $\int \sin^2 2x\, dx$

25. $\int x \csc x^2 \cot x^2\, dx$

26. $\int x \cos x\, dx$

27. $\int \sin^2 x \cos x\, dx$

28. $\int e^x \sin e^x\, dx$

29. $\int \dfrac{\cos x}{\sin^2 x}\, dx$

30. $\int_0^{\pi/4} \tan x\, dx$

31. $\int_{\pi/6}^{\pi/2} \dfrac{\cos x\, dx}{1 - \cos^2 x}$

32. $\int_0^{\pi} \sin x(1 + \cos^3 x)\, dx$

33. Find the area of the region bounded by the curves $f(x) = \sin x$ and $f(x) = \cos x$ from $x = \pi/4$ to $x = (5/4)\pi$.

34. **Predator-Prey Population** The population of foxes in a certain region over a 2-yr period is estimated to be

$$P_1(t) = 400 + 50 \sin\left(\frac{\pi t}{12}\right)$$

in month t, and the population of rabbits in the same region at time t is given by

$$P_2(t) = 3000 + 500 \cos\left(\frac{\pi t}{12}\right)$$

Find the rate of change of the populations when $t = 4$.

35. **Hotel Occupancy Rate** The occupancy rate of a large hotel in Maui in month t is described by the function

$$R(t) = 60 + 37 \sin^2\left(\frac{\pi t}{12}\right) \qquad (0 \le t \le 12)$$

where $t = 0$ corresponds to the beginning of June. When is the occupancy rate highest? When is the occupancy rate increasing most rapidly?

36. **Hotel Occupancy Rate** Refer to Exercise 35. Determine the average occupancy rate of the hotel over a 1-yr period from June 1 through May 31 of the following year.
Hint: $\sin^2 x = (1 - \cos 2x)/2$.

CHAPTER 12 Before Moving On . . .

1. Sketch the graph of $f(x) = 3 \cos 2x - 1$ on $[0, \pi]$.

2. Verify the identity $\cos^2 x = \dfrac{1 + \cos 2x}{2}$.

3. Find the derivative of $f(x) = e^{-2x} \sin 3x$.

4. Evaluate $\int_0^{\pi/4} \cos^2 x\, dx$. *Hint*: Use the identity in Exercise 2.

5. Find the area of the region under the graph of $f(x) = \sin^2 x \cos x$ from $x = 0$ to $x = \pi/4$.

Appendix

A Inverse Functions

A.1 The Inverse of a Function

A.2 The Graphs of Inverse Functions

A.3 Functions That Have Inverses

A.4 Finding the Inverse of a Function

B Table

B.1 The Standard Normal Distribution

A.1 The Inverse of a Function

Consider the position function

$$s = f(t) = 4t^2 \qquad (0 \le t \le 30) \tag{1}$$

giving the position of the maglev at any time t in its domain [0, 30]. The graph of f is shown in Figure 1. Equation (1) enables us to compute algebraically the position of the maglev at any time t. Geometrically, we can find the position of the maglev at any given time t by following the path indicated in Figure 1.

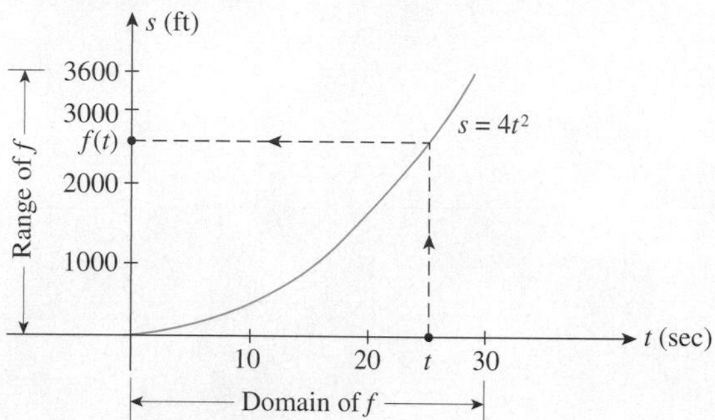

FIGURE 1
Each t in the domain of f is associated with the (unique) position $s = f(t)$ of the maglev.

Now consider the reverse problem: Knowing the position function of the maglev, can we find some way of obtaining the time it takes for the maglev to reach a given position? Geometrically, this problem is easily solved: Locate the point on the s-axis corresponding to the given position. Follow the path considered earlier but traced in the *opposite* direction. This path associates the given position s with the desired time t.

Algebraically, we can obtain a formula for the time t it takes for the maglev to get to the position s by solving (1) for t in terms of s. Thus,

$$t = \frac{1}{2}\sqrt{s}$$

(We reject the negative root because t lies in [0, 30].) Observe that the function g defined by

$$t = g(s) = \frac{1}{2}\sqrt{s}$$

has domain [0, 3600] (the range of f) and range [0, 30] (the domain of f) (Figure 2).

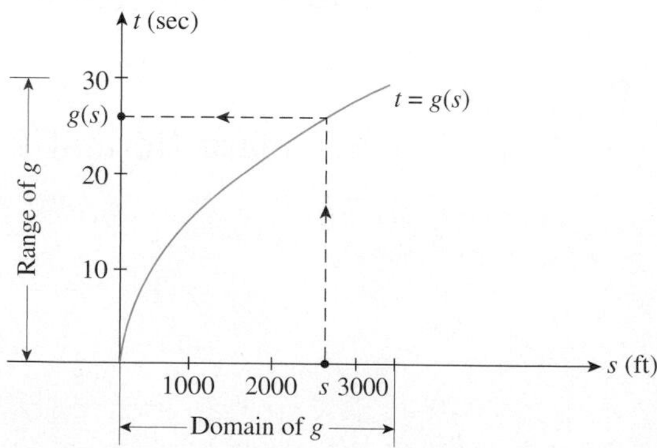

FIGURE 2
Each s in the domain of g is associated with the (unique) time $t = g(s)$.

The functions f and g have the following properties:

1. The domain of g is the range of f and vice versa.
2. $(g \circ f)t = g[f(t)] = \frac{1}{2}\sqrt{f(t)} = \frac{1}{2}\sqrt{4t^2} = t$ and $(f \circ g)(t) = f[g(t)] = 4[g(t)]^2 = 4\left(\frac{1}{2}\sqrt{t}\right)^2 = t$

In other words, one undoes what the other does. This is to be expected because f maps t onto $s = f(t)$ and g maps $s = f(t)$ back onto t.

The functions f and g are said to be *inverses* of each other. More generally, we have the following definition.

Inverse Functions

A function g is the inverse of the function f if

$$f[g(x)] = x \text{ for every } x \text{ in the domain of } g$$
$$g[f(x)] = x \text{ for every } x \text{ in the domain of } f$$

Equivalently, g is the inverse of f if the following condition is satisfied:

$$y = f(x) \quad \text{if and only if} \quad x = g(y)$$

for every x in the domain of f and for every y in its range.

Note The inverse of f is normally denoted by f^{-1} (read "f inverse"), and we will use this notation throughout the text.

Do not confuse $f^{-1}(x)$ with $[f(x)]^{-1} = \dfrac{1}{f(x)}$!

EXAMPLE 1 Show that the functions $f(x) = x^{1/3}$ and $g(x) = x^3$ are inverses of each other.

Solution First, observe that the domain and range of both f and g are $(-\infty, \infty)$. Therefore, both composite functions $f \circ g$ and $g \circ f$ are defined. Next, we compute

$$(f \circ g)(x) = f[g(x)] = [g(x)]^{1/3} = [x^3]^{1/3} = x$$

and

$$(g \circ f)(x) = g[f(x)] = [f(x)]^3 = (x^{1/3})^3 = x$$

Since $f[g(x)] = g[f(x)] = x$, we conclude that f and g are inverses of each other. In short, $f^{-1}(x) = x^3$. ■

Interpreting Our Results We can view f as a cube root extracting machine and g as a "cubing" machine. In this light, it is easy to see that one function does undo what the other does. So f and g are indeed inverses of each other.

A.2 The Graphs of Inverse Functions

The graphs of $f(x) = x^{1/3}$ and $f^{-1}(x) = x^3$ are shown in Figure 3.

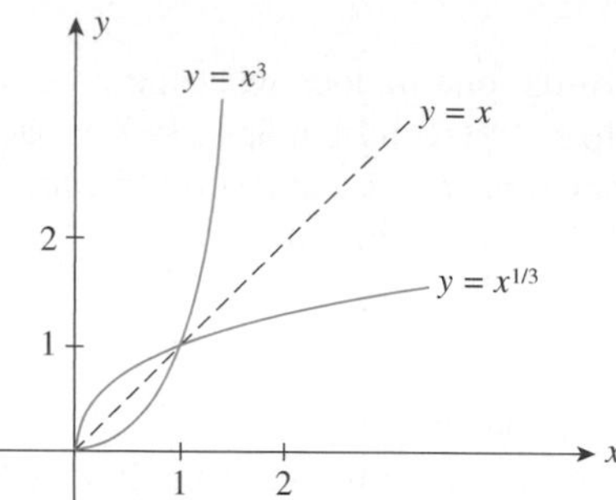

FIGURE 3
The functions $y = x^{1/3}$ and $f^{-1}(x) = x^3$ are inverses of each other. Their graphs are symmetric with respect to the line $y = x$.

They seem to suggest that the graphs of inverse functions are mirror images of each other with respect to the line $y = x$. This is true in general, a fact that we will not prove.

> **The Graphs of Inverse Functions**
> The graph of f^{-1} is the reflection of the graph of f with respect to the line $y = x$ and vice versa.

A.3 Functions That Have Inverses

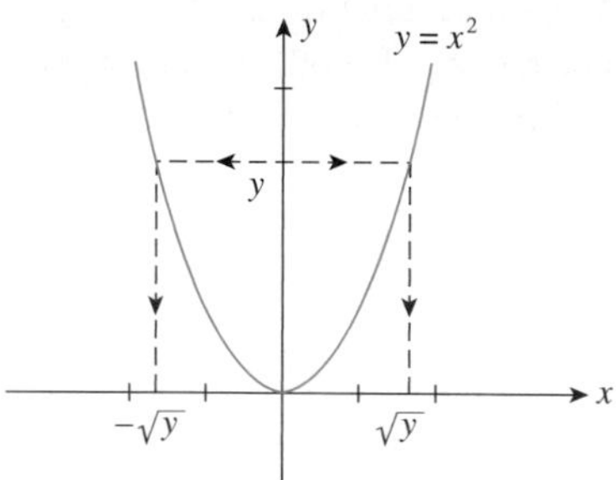

FIGURE 4
Each value of y is associated with two values of x.

Not every function has an inverse. Consider, for example, the function f defined by $y = x^2$ with domain $(-\infty, \infty)$ and range $[0, \infty)$. From the graph of f shown in Figure 4, you can see that each value of y in the range $[0, \infty)$ of f is associated with exactly *two* points $x = \pm\sqrt{y}$ (except for $y = 0$) in the domain $(-\infty, \infty)$ of f. This implies that f does not have an inverse because the uniqueness requirement of a function cannot be satisfied in this case. Observe that any horizontal line $y = c$ $(c > 0)$ intersects the graph of f at more than one point.

Next, consider the function g defined by the same rule as that of f, namely $y = x^2$, but with domain restricted to $[0, \infty)$. From the graph of g shown in Figure 5, you can see that each value of y in the range $[0, \infty)$ of g is mapped onto exactly *one* point $x = \sqrt{y}$ in the domain $[0, \infty)$ of g.

Thus, in this case, we can define the inverse function of g, from the range $[0, \infty)$ of g onto the domain $[0, \infty)$ of g. To find the rule for g^{-1}, we solve the equation $y = x^2$ for x in terms of y. Thus, $x = \sqrt{y}$ (because $x \geq 0$) and so $g^{-1}(y) = \sqrt{y}$, or,

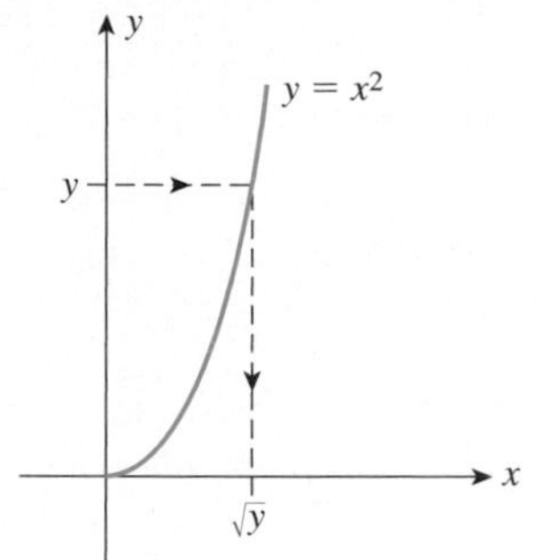

FIGURE 5
Each value of y is associated with exactly one value of x.

since y is a dummy variable, we can write $g^{-1}(x) = \sqrt{x}$. Also, observe that every horizontal line intersects the graph of g at no more than one point.

Why does g have an inverse but f does not? Observe that f takes on the same value twice; that is, there are two values of x that are mapped onto each value of y (except $y = 0$). On the other hand, g never takes on the same value more than once; that is, any two values of x have different images. The function g is said to be **one-to-one**.

One-to-One Function

A function f with domain D is **one-to-one** if no two points in D have the same image: that is, $f(x_1) \neq f(x_2)$ whenever $x_1 \neq x_2$.

Geometrically, a function is one-to-one if every horizontal line intersects its graph at no more than one point. This is called the **horizontal line test**.

The next theorem tells us when an inverse function exists.

THEOREM 1

The Existence of an Inverse Function

A function has an inverse if and only if it is one-to-one.

A.4 Finding the Inverse of a Function

Here is a summary of the steps for finding the inverse of a function (if it exists).

Guidelines for Finding the Inverse of a Function

1. Write $y = f(x)$.
2. Solve for x in terms of y (if possible).
3. Interchange x and y to obtain $y = f^{-1}(x)$.

EXAMPLE 1 Find the inverse of the function defined by $f(x) = \dfrac{1}{\sqrt{2x - 3}}$.

Solution To find the rule for this inverse, write

$$y = \frac{1}{\sqrt{2x - 3}}$$

and then solve the equation for x:

$$y^2 = \frac{1}{2x - 3} \qquad \text{Square both sides.}$$

$$2x - 3 = \frac{1}{y^2} \qquad \text{Take reciprocals.}$$

$$2x = \frac{1}{y^2} + 3 = \frac{3y^2 + 1}{y^2}$$

$$x = \frac{3y^2 + 1}{2y^2}$$

Finally, interchanging x and y, we obtain

$$y = \frac{3x^2 + 1}{2x^2}$$

giving the rule for f^{-1} as

$$f^{-1}(x) = \frac{3x^2 + 1}{2x^2}$$

The graphs of both f and f^{-1} are shown in Figure 6.

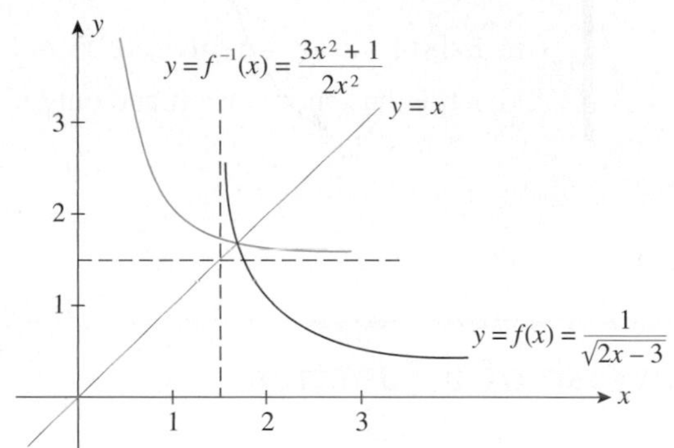

FIGURE 6
The graphs of f and f^{-1}. Notice that they are reflections of each other about the line $y = x$.

Appendix Exercises

In Exercises 1–6, show that f and g are inverses of each other by showing that $f[g(x)] = x$ and $g[f(x)] = x$.

1. $f(x) = \frac{1}{3}x^3$; $g(x) = \sqrt[3]{3x}$
2. $f(x) = \frac{1}{x}$; $g(x) = \frac{1}{x}$
3. $f(x) = 2x + 3$; $g(x) = \frac{x - 3}{2}$
4. $f(x) = x^2 + 1 (x \leq 0)$; $g(x) = -\sqrt{x - 1}$
5. $f(x) = 4(x + 1)^{2/3} (x \geq -1)$; $g(x) = \frac{1}{8}(x^{3/2} - 8)(x \geq 0)$
6. $f(x) = \frac{1 + x}{1 - x}$; $g(x) = \frac{x - 1}{x + 1}$

In Exercises 7–12, you are given the graph of a function f. Determine whether f is one-to-one.

7. 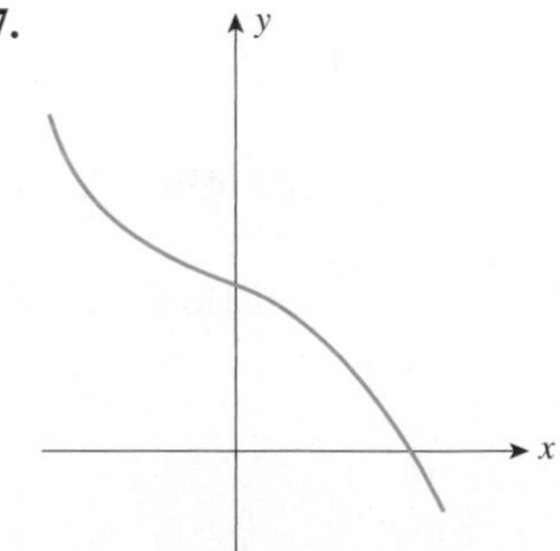

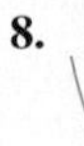

8. 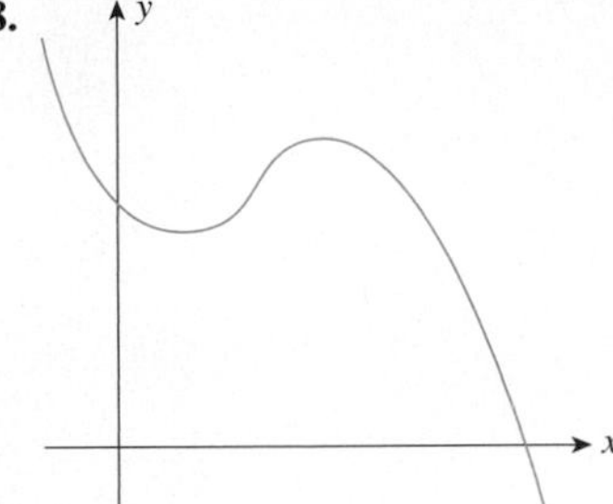

9. 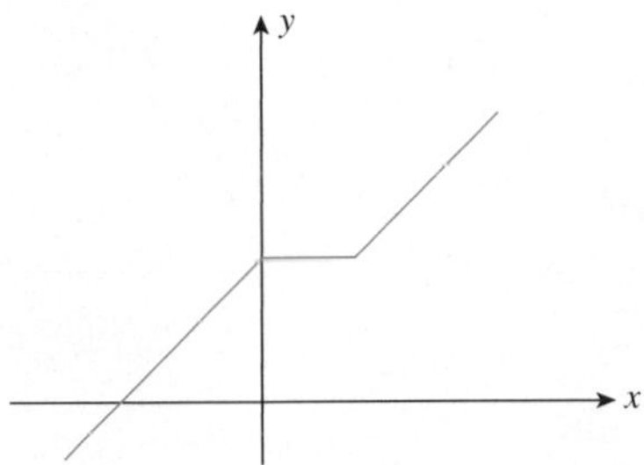

10. 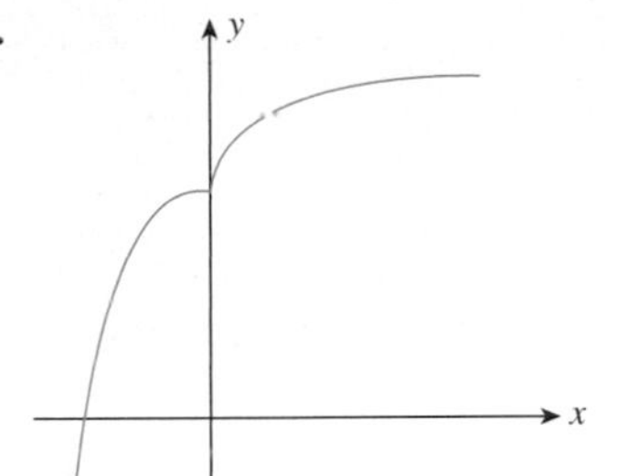

11. 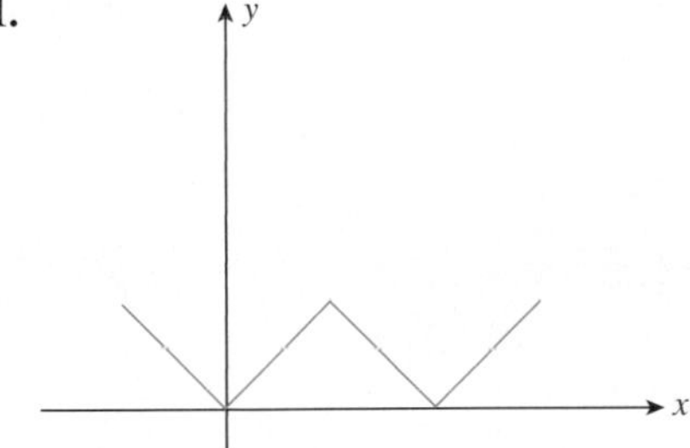

12. 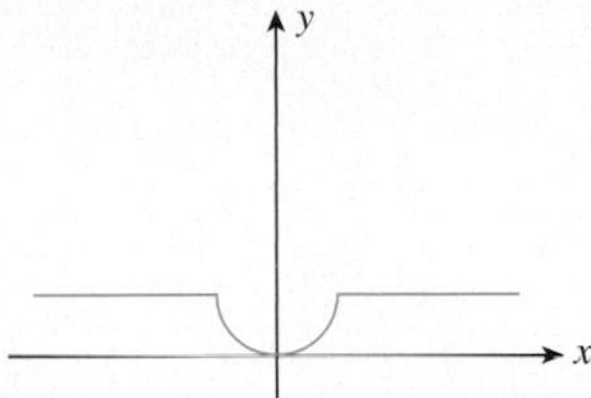

In Exercises 13–18, find the inverse of f. Then sketch the graphs of f and f^{-1} on the same set of axes.

13. $f(x) = 3x - 2$

14. $f(x) = x^2, x \leq 0$

15. $f(x) = x^3 + 1$

16. $f(x) = 2\sqrt{x + 3}$

17. $f(x) = \sqrt{9 - x^2}, x \geq 0$

18. $f(x) = x^{3/5} + 1$

19. A hot-air balloon rises vertically from the ground so that its height after t sec is $h = \frac{1}{2}t^2 + \frac{1}{2}t$ ft $(0 \leq t \leq 60)$.

a. Find the inverse of the function $f(t) = \frac{1}{2}t^2 + \frac{1}{2}t$ and explain what it represents.

b. Use the result of part (a) to find the time when the balloon is between an altitude of 120 and 210 ft.

20. Aging Population The population of Americans age 55 and over as a percent of the total population is approximated by the function

$$f(t) = 10.72(0.9t + 10)^{0.3} \qquad (0 \leq t \leq 20)$$

where t is measured in years and $t = 0$ corresponds to the year 2000.

a. Find the rule for f^{-1}.

b. Evaluate $f^{-1}(25)$ and interpret your result.

Source: U.S. Census Bureau

TABLE B.1

The Standard Normal Distribution

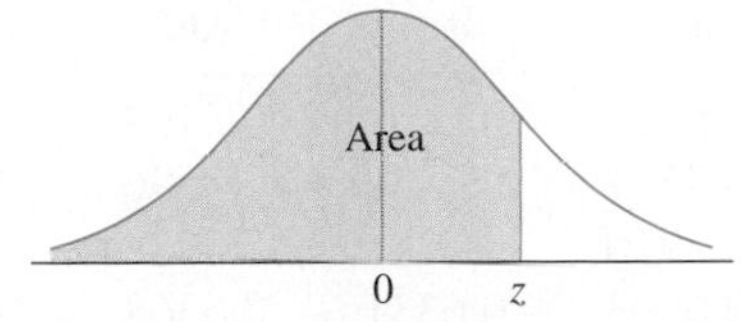

$F_z(z) = P[Z \leq z]$

z	0.00	0.01	0.02	0.03	0.04	0.05	0.06	0.07	0.08	0.09
−3.4	0.0003	0.0003	0.0003	0.0003	0.0003	0.0003	0.0003	0.0003	0.0003	0.0002
−3.3	0.0005	0.0005	0.0005	0.0004	0.0004	0.0004	0.0004	0.0004	0.0004	0.0003
−3.2	0.0007	0.0007	0.0006	0.0006	0.0006	0.0006	0.0006	0.0005	0.0005	0.0005
−3.1	0.0010	0.0009	0.0009	0.0009	0.0008	0.0008	0.0008	0.0008	0.0007	0.0007
−3.0	0.0013	0.0013	0.0013	0.0012	0.0012	0.0011	0.0011	0.0011	0.0010	0.0010
−2.9	0.0019	0.0018	0.0017	0.0017	0.0016	0.0016	0.0015	0.0015	0.0014	0.0014
−2.8	0.0026	0.0025	0.0024	0.0023	0.0023	0.0022	0.0021	0.0021	0.0020	0.0019
−2.7	0.0035	0.0034	0.0033	0.0032	0.0031	0.0030	0.0029	0.0028	0.0027	0.0026
−2.6	0.0047	0.0045	0.0044	0.0043	0.0041	0.0040	0.0039	0.0038	0.0037	0.0036
2.5	0.0062	0.0060	0.0059	0.0057	0.0055	0.0054	0.0052	0.0051	0.0049	0.0048
−2.4	0.0082	0.0080	0.0078	0.0075	0.0073	0.0071	0.0069	0.0068	0.0066	0.0064
−2.3	0.0107	0.0104	0.0102	0.0099	0.0096	0.0094	0.0091	0.0089	0.0087	0.0084
−2.2	0.0139	0.0136	0.0132	0.0129	0.0125	0.0122	0.0119	0.0116	0.0113	0.0110
−2.1	0.0179	0.0174	0.0170	0.0166	0.0162	0.0158	0.0154	0.0150	0.0146	0.0143
−2.0	0.0228	0.0222	0.0217	0.0212	0.0207	0.0202	0.0197	0.0192	0.0188	0.0183
−1.9	0.0287	0.0281	0.0274	0.0268	0.0262	0.0256	0.0250	0.0244	0.0239	0.0233
−1.8	0.0359	0.0352	0.0344	0.0336	0.0329	0.0322	0.0314	0.0307	0.0301	0.0294
−1.7	0.0446	0.0436	0.0427	0.0418	0.0409	0.0401	0.0392	0.0384	0.0375	0.0367
−1.6	0.0548	0.0537	0.0526	0.0516	0.0505	0.0495	0.0485	0.0475	0.0465	0.0455
−1.5	0.0668	0.0655	0.0643	0.0630	0.0618	0.0606	0.0594	0.0582	0.0571	0.0559
−1.4	0.0808	0.0793	0.0778	0.0764	0.0749	0.0735	0.0722	0.0708	0.0694	0.0681
−1.3	0.0968	0.0951	0.0934	0.0918	0.0901	0.0885	0.0869	0.0853	0.0838	0.0823
−1.2	0.1151	0.1131	0.1112	0.1093	0.1075	0.1056	0.1038	0.1020	0.1003	0.0985
−1.1	0.1357	0.1335	0.1314	0.1292	0.1271	0.1251	0.1230	0.1210	0.1190	0.1170
−1.0	0.1587	0.1562	0.1539	0.1515	0.1492	0.1469	0.1446	0.1423	0.1401	0.1379
−0.9	0.1841	0.1814	0.1788	0.1762	0.1736	0.1711	0.1685	0.1660	0.1635	0.1611
−0.8	0.2119	0.2090	0.2061	0.2033	0.2005	0.1977	0.1949	0.1922	0.1894	0.1867
−0.7	0.2420	0.2389	0.2358	0.2327	0.2296	0.2266	0.2236	0.2206	0.2177	0.2148
−0.6	0.2743	0.2709	0.2676	0.2643	0.2611	0.2578	0.2546	0.2514	0.2483	0.2451
−0.5	0.3085	0.3050	0.3015	0.2981	0.2946	0.2912	0.2877	0.2843	0.2810	0.2776
−0.4	0.3446	0.3409	0.3372	0.3336	0.3300	0.3264	0.3228	0.3192	0.3156	0.3121
−0.3	0.3821	0.3783	0.3745	0.3707	0.3669	0.3632	0.3594	0.3557	0.3520	0.3483
−0.2	0.4207	0.4168	0.4129	0.4090	0.4052	0.4013	0.3974	0.3936	0.3897	0.3859
−0.1	0.4602	0.4562	0.4522	0.4483	0.4443	0.4404	0.4364	0.4325	0.4286	0.4247
−0.0	0.5000	0.4960	0.4920	0.4880	0.4840	0.4801	0.4761	0.4721	0.4681	0.4641

TABLE B.1 The Standard Normal Distribution (*continued*)

$F_z(z) = P[Z \leq z]$

z	0.00	0.01	0.02	0.03	0.04	0.05	0.06	0.07	0.08	0.09
0.0	0.5000	0.5040	0.5080	0.5120	0.5160	0.5199	0.5239	0.5279	0.5319	0.5359
0.1	0.5398	0.5438	0.5478	0.5517	0.5557	0.5596	0.5636	0.5675	0.5714	0.5753
0.2	0.5793	0.5832	0.5871	0.5910	0.5948	0.5987	0.6026	0.6064	0.6103	0.6141
0.3	0.6179	0.6217	0.6255	0.6293	0.6331	0.6368	0.6406	0.6443	0.6480	0.6517
0.4	0.6554	0.6591	0.6628	0.6664	0.6700	0.6736	0.6772	0.6808	0.6844	0.6879
0.5	0.6915	0.6950	0.6985	0.7019	0.7054	0.7088	0.7123	0.7157	0.7190	0.7224
0.6	0.7257	0.7291	0.7324	0.7357	0.7389	0.7422	0.7454	0.7486	0.7517	0.7549
0.7	0.7580	0.7611	0.7642	0.7673	0.7704	0.7734	0.7764	0.7794	0.7823	0.7852
0.8	0.7881	0.7910	0.7939	0.7967	0.7995	0.8023	0.8051	0.8078	0.8106	0.8133
0.9	0.8159	0.8186	0.8212	0.8238	0.8264	0.8289	0.8315	0.8340	0.8365	0.8389
1.0	0.8413	0.8438	0.8461	0.8485	0.8508	0.8531	0.8554	0.8577	0.8599	0.8621
1.1	0.8643	0.8665	0.8686	0.8708	0.8729	0.8749	0.8770	0.8790	0.8810	0.8830
1.2	0.8849	0.8869	0.8888	0.8907	0.8925	0.8944	0.8962	0.8980	0.8997	0.9015
1.3	0.9032	0.9049	0.9066	0.9082	0.9099	0.9115	0.9131	0.9147	0.9162	0.9177
1.4	0.9192	0.9207	0.9222	0.9236	0.9251	0.9265	0.9278	0.9292	0.9306	0.9319
1.5	0.9332	0.9345	0.9357	0.9370	0.9382	0.9394	0.9406	0.9418	0.9429	0.9441
1.6	0.9452	0.9463	0.9474	0.9484	0.9495	0.9505	0.9515	0.9525	0.9535	0.9545
1.7	0.9554	0.9564	0.9573	0.9582	0.9591	0.9599	0.9608	0.9616	0.9625	0.9633
1.8	0.9641	0.9649	0.9656	0.9664	0.9671	0.9678	0.9686	0.9693	0.9699	0.9706
1.9	0.9713	0.9719	0.9726	0.9732	0.9738	0.9744	0.9750	0.9756	0.9761	0.9767
2.0	0.9772	0.9778	0.9783	0.9788	0.9793	0.9798	0.9803	0.9808	0.9812	0.9817
2.1	0.9821	0.9826	0.9830	0.9834	0.9838	0.9842	0.9846	0.9850	0.9854	0.9857
2.2	0.9861	0.9864	0.9868	0.9871	0.9875	0.9878	0.9881	0.9884	0.9887	0.9890
2.3	0.9893	0.9896	0.9898	0.9901	0.9904	0.9906	0.9909	0.9911	0.9913	0.9916
2.4	0.9918	0.9920	0.9922	0.9925	0.9927	0.9929	0.9931	0.9932	0.9934	0.9936
2.5	0.9938	0.9940	0.9951	0.9943	0.9945	0.9946	0.9948	0.9949	0.9951	0.9952
2.6	0.9953	0.9955	0.9956	0.9957	0.9959	0.9960	0.9961	0.9962	0.9963	0.9964
2.7	0.9965	0.9966	0.9967	0.9968	0.9969	0.9970	0.9971	0.9972	0.9973	0.9974
2.8	0.9974	0.9975	0.9976	0.9977	0.9977	0.9978	0.9979	0.9979	0.9980	0.9981
2.9	0.9981	0.9982	0.9982	0.9983	0.9984	0.9984	0.9985	0.9985	0.9986	0.9986
3.0	0.9987	0.9987	0.9987	0.9988	0.9988	0.9989	0.9989	0.9989	0.9990	0.9990
3.1	0.9990	0.9991	0.9991	0.9991	0.9992	0.9992	0.9992	0.9992	0.9993	0.9993
3.2	0.9993	0.9993	0.9994	0.9994	0.9994	0.9994	0.9994	0.9995	0.9995	0.9995
3.3	0.9995	0.9995	0.9995	0.9996	0.9996	0.9996	0.9996	0.9996	0.9996	0.9997
3.4	0.9997	0.9997	0.9997	0.9997	0.9997	0.9997	0.9997	0.9997	0.9997	0.9998

Answers to Odd-Numbered Exercises

CHAPTER 1

Exercises 1.1, page 9

1. False 3. False

5. 

7. 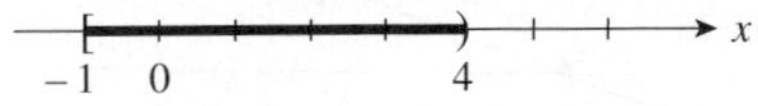

9. 

11. $(-\infty, 2)$ 13. $(-\infty, -5]$ 15. $(-4, 6)$

17. $(-\infty, -3) \cup (3, \infty)$ 19. $(-2, 3)$ 21. 4

23. 2 25. $5\sqrt{3}$ 27. $\pi + 1$ 29. 2

31. False 33. False 35. True 37. False

39. True 41. False 43. 9 45. 1 47. 4

49. 7 51. $\frac{1}{5}$ 53. 2 55. 2 57. 1

59. True 61. False 63. False 65. False

67. False 69. $\dfrac{1}{(xy)^2}$ 71. $\dfrac{1}{x^{5/6}}$ 73. $\dfrac{1}{(s+t)^3}$

75. $x^{13/3}$ 77. $\dfrac{1}{x^3}$ 79. x 81. $\dfrac{9}{x^2y^4}$ 83. $\dfrac{y^8}{x^{10}}$

85. $2x^{11/6}$ 87. $-2xy^2$ 89. $2x^{4/3}y^{1/2}$ 91. 2.828

93. 5.196 95. 31.62 97. 316.2 99. $\dfrac{3\sqrt{x}}{2x}$

101. $\dfrac{2\sqrt{3y}}{3}$ 103. $\dfrac{\sqrt[3]{x^2}}{x}$ 105. $\dfrac{2x}{3\sqrt{x}}$ 107. $\dfrac{2y}{\sqrt{2xy}}$

109. $\dfrac{xz}{y\sqrt[3]{xz^2}}$ 111. [362, 488.7] 113. \$12,300

115. \$32,000 117. $|x - 0.5| < 0.01$

119. Between 1000 and 4000 units 121. False 123. True

Exercises 1.2, page 21

1. $9x^2 + 3x + 1$ 3. $4y^2 + y + 8$ 5. $-x - 1$

7. $\frac{2}{3} + e - e^{-1}$ 9. $6\sqrt{2} + 8 + \frac{1}{2}\sqrt{x} - \frac{11}{4}\sqrt{y}$

11. $x^2 + 6x - 16$ 13. $a^2 + 10a + 25$ 15. $x^2 + 4xy + 4y^2$

17. $4x^2 - y^2$ 19. $-2x$ 21. $2t(2\sqrt{t} + 1)$

23. $2x^3(2x^2 - 6x - 3)$ 25. $7a^2(a^2 + 7ab - 6b^2)$

27. $e^{-x}(1 - x)$ 29. $\frac{1}{2}x^{-5/2}(4 - 3x)$ 31. $(2a + b)(3c - 2d)$

33. $(2a + b)(2a - b)$ 35. $-2(3x + 5)(2x - 1)$

37. $3(x - 4)(x + 2)$ 39. $2(3x - 5)(2x + 3)$

41. $(3x - 4y)(3x + 4y)$ 43. $(x^2 + 5)(x^4 - 5x^2 + 25)$

45. $x^3 - xy^2$ 47. $4(x - 1)(3x - 1)(2x + 2)^3$

49. $4(x - 1)(3x - 1)(2x + 2)^3$ 51. $2x(x^2 + 2)^2(5x^4 + 20x^2 + 17)$

53. -4 and 3 55. -1 and $\frac{1}{2}$ 57. 2 and 2

59. -2 and $\frac{3}{4}$ 61. $\frac{1}{2} + \frac{1}{4}\sqrt{10}$ and $\frac{1}{2} - \frac{1}{4}\sqrt{10}$

63. $-1 + \frac{1}{2}\sqrt{10}$ and $-1 - \frac{1}{2}\sqrt{10}$ 65. $\dfrac{x-1}{x-2}$

67. $\dfrac{3(2t+1)}{2t-1}$ 69. $-\dfrac{7}{(4x-1)^2}$ 71. -8

73. $\dfrac{3x-1}{2}$ 75. $\dfrac{t+20}{3t+2}$ 77. $-\dfrac{x(2x-13)}{(2x-1)(2x+5)}$

79. $-\dfrac{x+27}{(x-3)^2(x+3)}$ 81. $\dfrac{x+1}{x-1}$ 83. $\dfrac{4x^2+7}{\sqrt{2x^2+7}}$

85. $\dfrac{x-1}{x^2\sqrt{x+1}}$ 87. $\dfrac{x-1}{(2x+1)^{3/2}}$ 89. $\dfrac{\sqrt{3}+1}{2}$

91. $\dfrac{\sqrt{x}+\sqrt{y}}{x-y}$ 93. $\dfrac{(\sqrt{a}+\sqrt{b})^2}{a-b}$ 95. $\dfrac{x}{3\sqrt{x}}$

97. $-\dfrac{2}{3(1+\sqrt{3})}$ 99. $-\dfrac{x+1}{\sqrt{x+2}(1-\sqrt{x+2})}$

101. True 103. False

Exercises 1.3, page 28

1. (3, 3); Quadrant I 3. (2, −2); Quadrant IV

5. (−4, −6); Quadrant III 7. *A* 9. *E*, *F*, and *G*

11. *F* 13–19. See the accompanying figure.

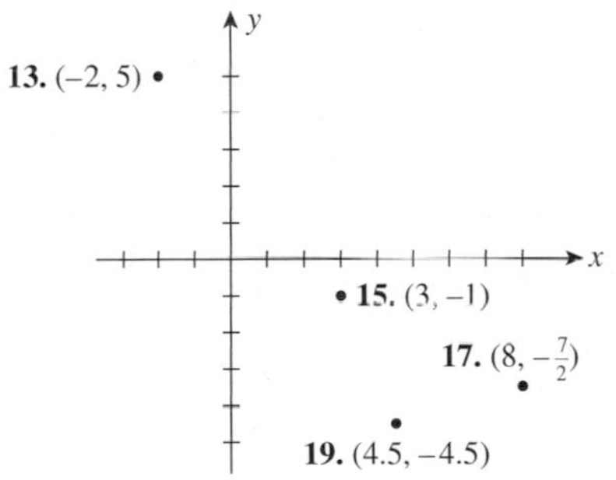

21. 5 23. $\sqrt{61}$ 25. (−8, −6) and (8, −6)

29. $(x - 2)^2 + (y + 3)^2 = 25$ 31. $x^2 + y^2 = 25$

33. $(x - 2)^2 + (y + 3)^2 = 34$ 35. 3400 mi 37. Route 1

39. Model C 41. $10\sqrt{13}t$; 72.1 mi 43. True 45. False

Exercises 1.4, page 40

1. e 3. a 5. f 7. $\frac{1}{2}$ 9. Not defined 11. 5

13. $\frac{5}{6}$ 15. $\dfrac{d-b}{c-a}$ $(a \neq c)$ 17. **a.** 4 **b.** −8

19. Parallel 21. Perpendicular 23. -5 25. $y = -3$

27. $y = 2x - 10$ 29. $y = 2$ 31. $y = 3x - 2$

33. $y = x + 1$ 35. $y = 3x + 4$ 37. $y = 5$

39. $y = \frac{1}{2}x$; $m = \frac{1}{2}$; $b = 0$ 41. $y = \frac{2}{3}x - 3$; $m = \frac{2}{3}$; $b = -3$

43. $y = -\frac{1}{2}x + \frac{7}{2}$; $m = -\frac{1}{2}$; $b = \frac{7}{2}$ 45. $y = \frac{1}{2}x + 3$

47. $y = -6$ 49. $y = b$ 51. $y = \frac{2}{3}x - \frac{2}{3}$ 53. $k = 8$

55.

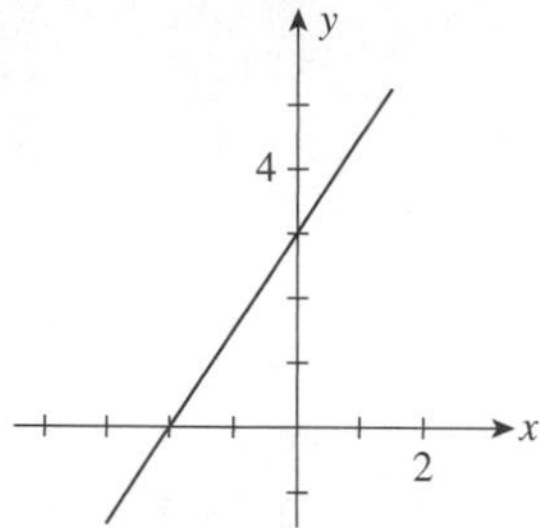

57.

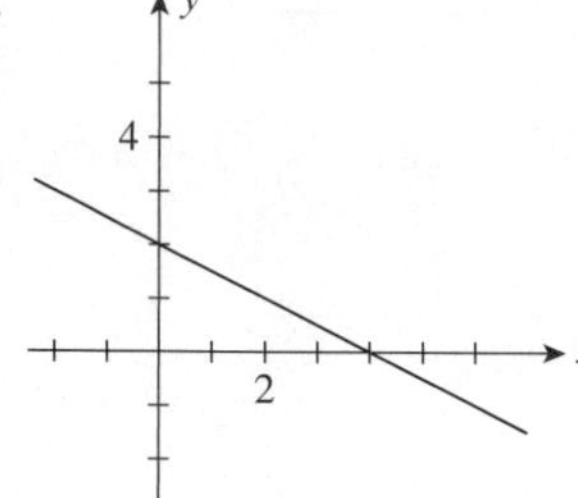

59.

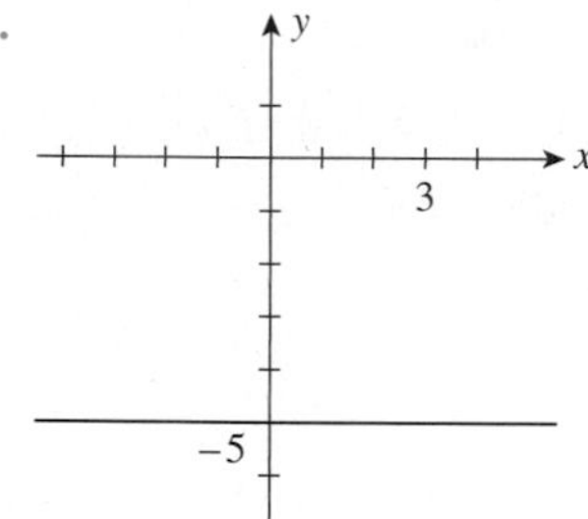

63. $y = -2x - 4$ 65. $y = \frac{1}{8}x - \frac{1}{2}$ 67. Yes

69. **a.**

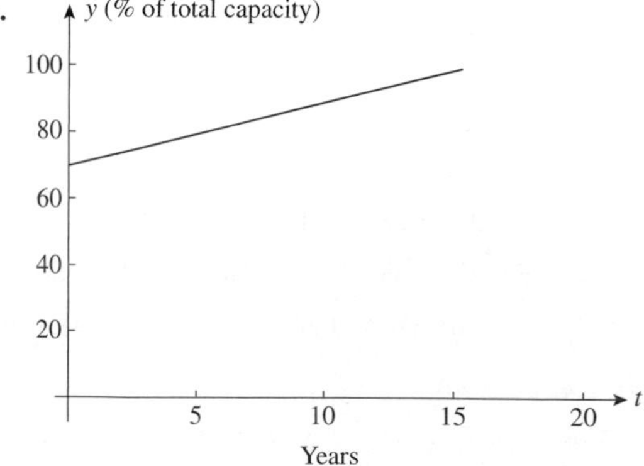

b. 1.9467; 70.082

c. The capacity utilization has been increasing by 1.9467% each year since 1990 when it stood at 70.082%.
d. Shortly after 2005

71. **a.** $y = 0.55x$ **b.** 2000 73. 89.6% of men's wages

75. **a.** and **b.**

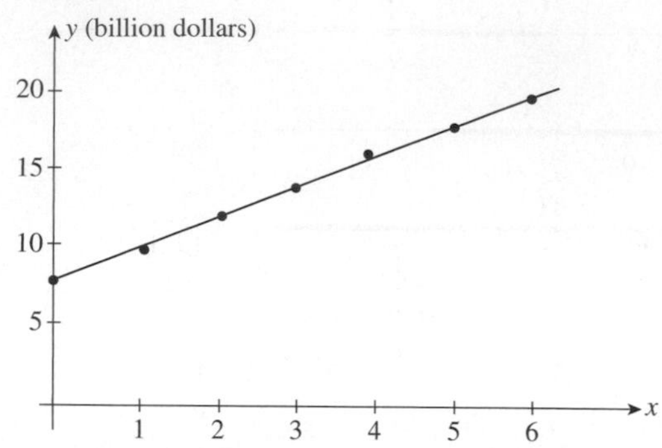

c. $y = 1.82x + 7.9$ **d.** $17 billion; same

77. True 79. False 81. True

Chapter 1 Concept Review, page 46

1. Ordered; abscissa (x-coordinate); ordinate (y-coordinate)

2. **a.** x-; y-; **b.** Third 3. $\sqrt{(c-a)^2 + (d-b)^2}$

4. $(x-a)^2 + (y-b)^2 = r^2$

5. **a.** $\dfrac{y_2 - y_1}{x_2 - x_1}$ **b.** Undefined **c.** 0 **d.** Positive

6. $m_1 = m_2$; $m_1 = -\dfrac{1}{m_2}$

7. **a.** $y - y_1 = m(x - x_1)$; point-slope form
b. $y = mx + b$ slope-intercept

8. **a.** $Ax + By + C = 0$ (A, B, not both zero) **b.** $-\dfrac{a}{b}$

Chapter 1 Review Exercises, page 47

1. $[-2, \infty)$ 2. $[-1, 2]$ 3. $(-\infty, -4) \cup (5, \infty)$

4. $(-\infty, -5) \cup (5, \infty)$ 5. 4 6. 1 7. $\pi - 6$

8. $8 - 3\sqrt{3}$ 9. $\frac{27}{8}$ 10. 25 11. $\frac{1}{144}$ 12. -32

13. $\frac{1}{4}$ 14. $3\sqrt[3]{3}$ 15. $4(x^2 + y)^2$ 16. $\dfrac{a^{15}}{b^{11}}$ 17. $\dfrac{2x}{3z}$

18. $-x^{1/2}$ 19. $6xy^7$ 20. $9x^2y^4$ 21. $-2\pi r^2(\pi r - 50)$

22. $2vw(v^2 + w^2 + u^2)$ 23. $(4 - x)(4 + x)$

24. $6t(2t - 3)(t + 1)$ 25. $-\frac{3}{4}$ and $\frac{1}{2}$ 26. -2 and $\frac{1}{3}$

27. 0, -3, 1 28. $\dfrac{\sqrt{2}}{2}$ and $-\dfrac{\sqrt{2}}{2}$ 29. $1 + \sqrt{6}$, $1 - \sqrt{6}$

30. $-2 + \dfrac{\sqrt{2}}{2}$; $-2 - \dfrac{\sqrt{2}}{2}$ 31. $\dfrac{180}{(t+6)^2}$

32. $\dfrac{15x^2 + 24x + 2}{4(x + 2)(3x^2 + 2)}$ 33. $\dfrac{78x^2 - 8x - 27}{3(2x^2 - 1)(3x - 1)}$

34. $\dfrac{2(x + 2)}{\sqrt{x + 1}}$ 35. $\dfrac{1}{\sqrt{x} + 1}$ 36. $\dfrac{x - \sqrt{x}}{2x}$ 37. 5

38. 2 39. $x = -2$ 40. $y = 4$ 41. $y = -\frac{1}{10}x + \frac{19}{5}$

42. $y = -\frac{4}{5}x + \frac{12}{5}$ 43. $y = \frac{5}{2}x + 9$ 44. $y = \frac{3}{4}x + \frac{11}{2}$

45. $y = -\frac{3}{4}x + \frac{9}{2}$ 46. $y = -\frac{3}{5}x + \frac{12}{5}$ 47. $y = 3x + 7$

48. $y = -\frac{3}{2}x - 7$

49.

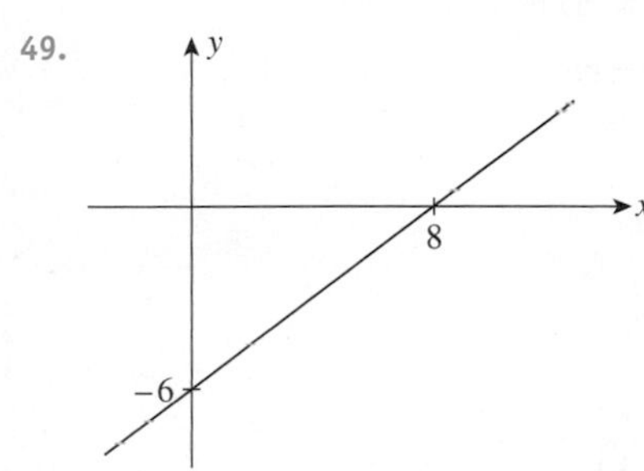

50.

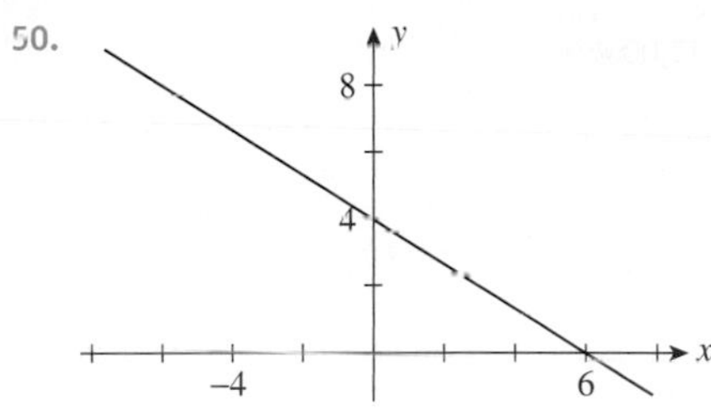

51. \$100 52. \$400

53. **a.** and **b.**

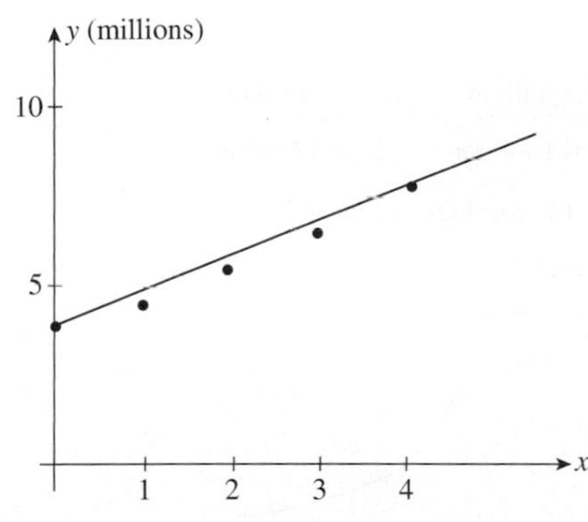

c. $y = 0.975x + 3.9$ **d.** 6,825,000; differ by 25,000

Chapter 1 Before Moving On, page 48

1. **a.** $\sqrt{3} + \sqrt{2} - \pi$ **b.** -3 2. **a.** $12x^5y$ **b.** $\dfrac{b^5}{a^3}$

3. **a.** $\dfrac{2x\sqrt{y}}{3y}$ **b.** $\dfrac{x(\sqrt{x} + 4)}{x - 16}$

4. **a.** $\dfrac{1 - 3x^2}{2\sqrt{x}(x^2 + 1)^2}$ **b.** $\dfrac{6\sqrt{x + 2}}{x + 2}$ 5. $\dfrac{x - y}{(\sqrt{x} - \sqrt{y})^2}$

6. **a.** $2x(3x + 2)(2x - 3)$ **b.** $(2b + 3c)(x - y)$

7. **a.** $x = -\frac{1}{4}$, or 1 **b.** $\dfrac{5 \pm \sqrt{13}}{6}$ 8. $4\sqrt{5}$

9. $y = \frac{7}{5}x - \frac{3}{5}$ 10. $y = -\frac{1}{3}x + \frac{4}{3}$

CHAPTER 2

Exercises 2.1, page 58

1. 21, −9, $5a + 6$, $-5a + 6$, $5a + 21$

3. −3, 6, $3a^2 - 6a - 3$, $3a^2 + 6a - 3$, $3x^2 - 6$

5. $2a + 2h + 5$, $-2a + 5$, $2a^2 + 5$, $2a - 4h + 5$, $4a - 2h + 5$

7. $\dfrac{8}{15}$, 0, $\dfrac{2a}{a^2 - 1}$, $\dfrac{2(2 + a)}{a^2 + 4a + 3}$, $\dfrac{2(t + 1)}{t(t + 2)}$

9. 8, $\dfrac{2a^2}{\sqrt{a - 1}}$, $\dfrac{2(x + 1)^2}{\sqrt{x}}$, $\dfrac{2(x - 1)^2}{\sqrt{x - 2}}$ 11. 5, 1, 1 13. $\frac{5}{2}$, 3, 3, 9

15. **a.** −2 **b.** (i) $x = 2$; (ii) $x = 1$ **c.** [0, 6] **d.** [−2, 6]

17. Yes 19. Yes 21. $(-\infty, \infty)$ 23. $(-\infty, 0) \cup (0, \infty)$

25. $(-\infty, \infty)$ 27. $(-\infty, 5]$ 29. $(-\infty, -1) \cup (-1, 1) \cup (1, \infty)$

31. $[-3, \infty)$ 33. $(-\infty, -2) \cup (-2, 1]$

35. **a.** $(-\infty, \infty)$
b. 6, 0, −4, −6, $-\frac{25}{4}$, −6, −4, 0
c.

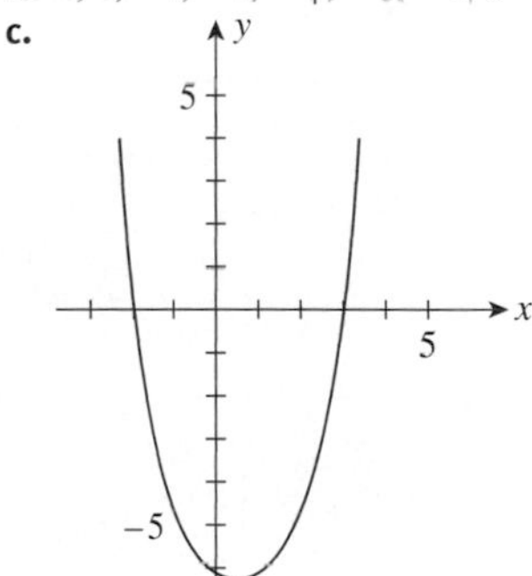

37.

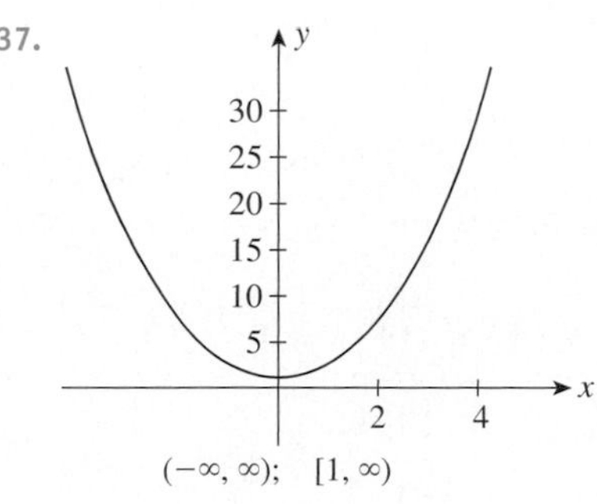

$(-\infty, \infty)$; $[1, \infty)$

39.

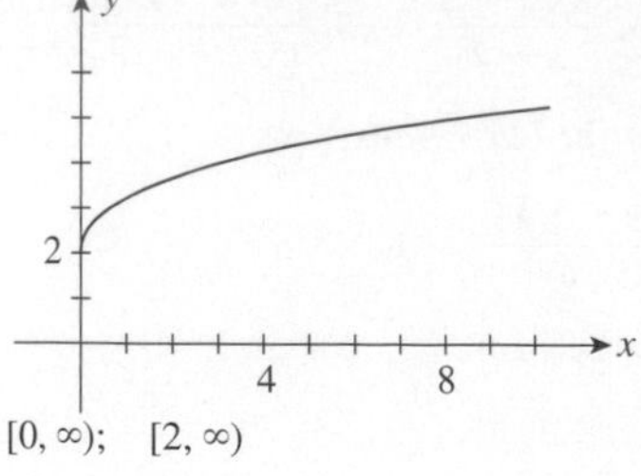

$[0, \infty)$; $[2, \infty)$

41.

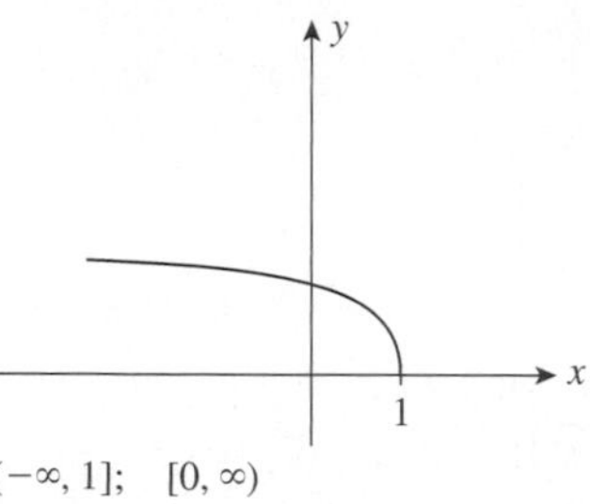

$(-\infty, 1]$; $[0, \infty)$

43.

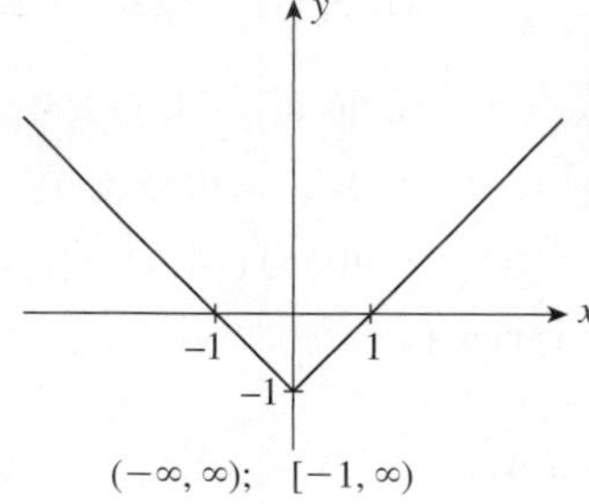

$(-\infty, \infty)$; $[-1, \infty)$

45.

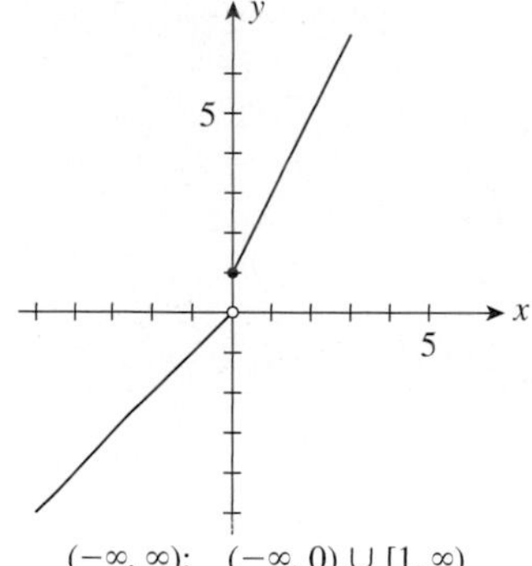

$(-\infty, \infty)$; $(-\infty, 0) \cup [1, \infty)$

47.

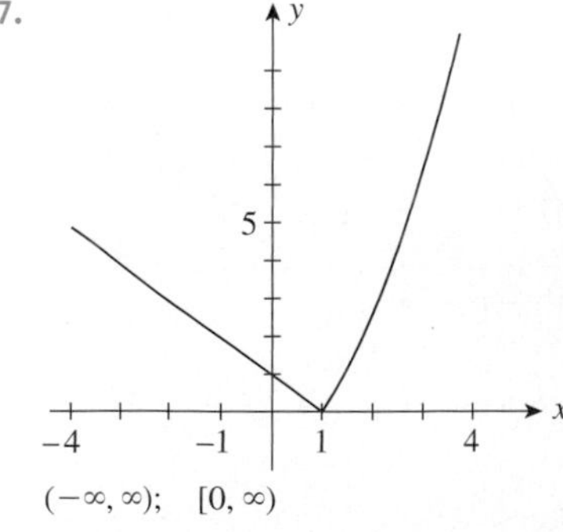

$(-\infty, \infty)$; $[0, \infty)$

49. Yes 51. No 53. Yes 55. Yes

57. 10π in. 59. 8

61. **a.** From 1985 to 1990 **b.** From 1990 on **c.** 1990; $3.5 billion

63. **a.** $f(t) = \begin{cases} 0.0185t + 0.58 & \text{if } 0 \le t \le 20 \\ 0.015t + 0.65 & \text{if } 20 < t \le 30 \end{cases}$

b. 0.0185/yr from 1960 through 1980; 0.015/yr from 1980 through 1990

c. 1983

65. **a.** $0.06x$ **b.** $12.00; $0.34 67. 160 mg

69. **a.** $f(t) = 7.5t + 20$ **b.** 65 million

71. **a.** $V = -12{,}000n + 120{,}000$

b.

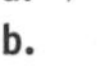

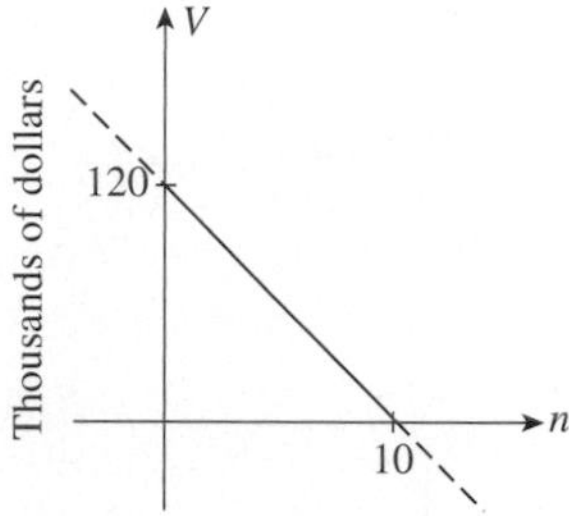

c. $48,000 **d.** $12,000/yr

73. $(0, \infty)$

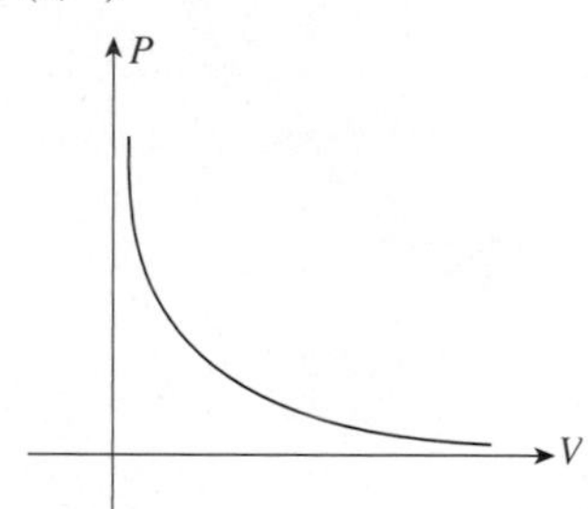

75. **a.** 3.6 million; 9.5 million **b.** 11.2 million

77. 0.77 79. **a.** 4.1 million **b.** 5.47 million

81. **a.** 130 tons/day; 100 tons/day; 40 tons/day

b.

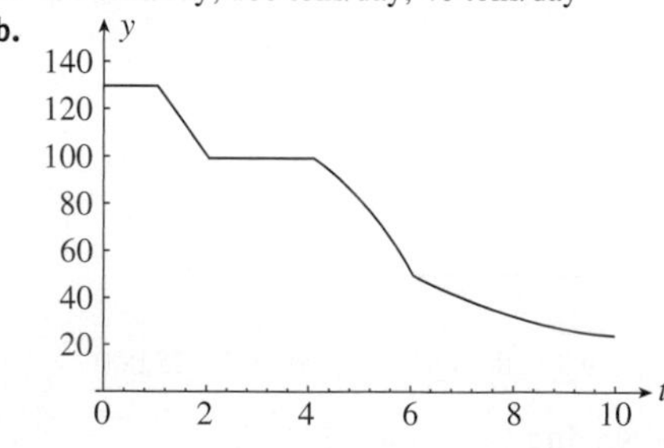

83. True 85. False

Using Technology Exercises 2.1, page 67

1.
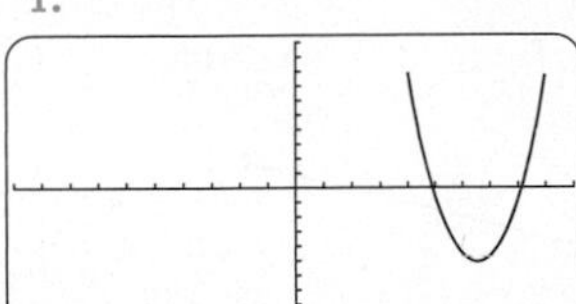

3.
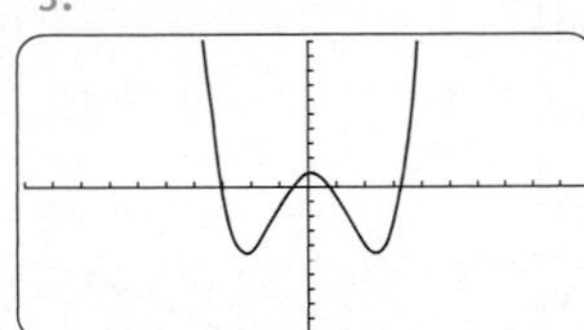

5. **a.**
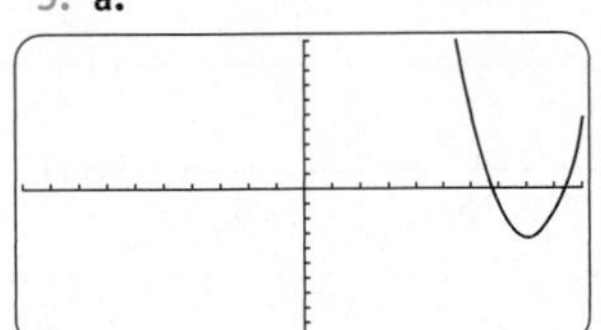

b.
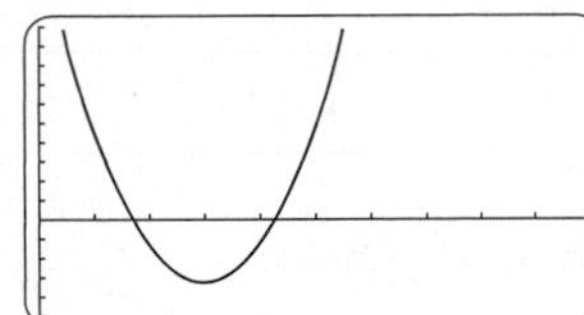

7. **a.**
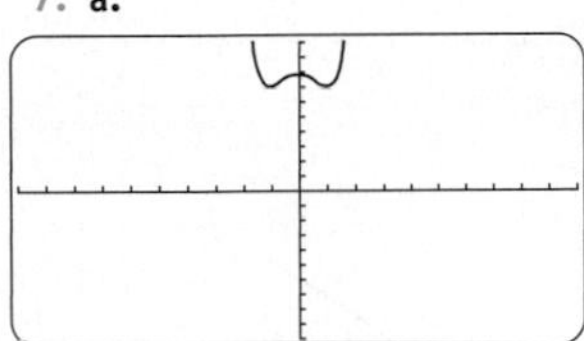

b.
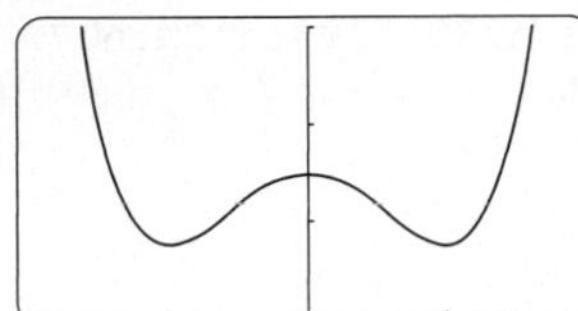

9. **a.**
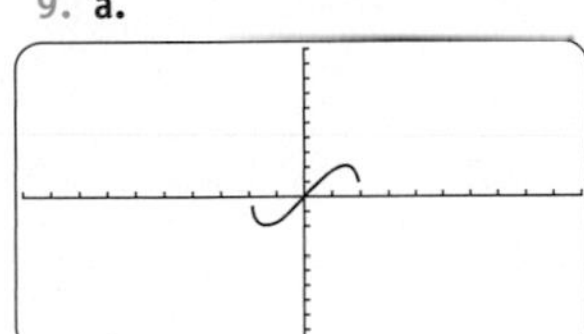

b.
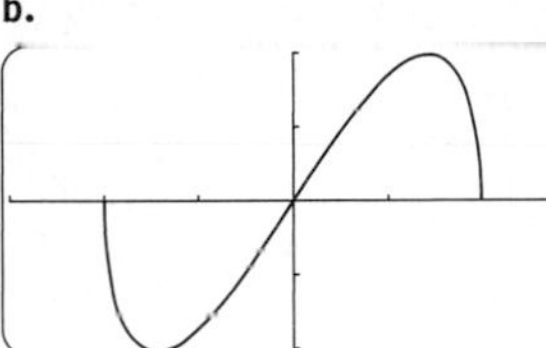

11.
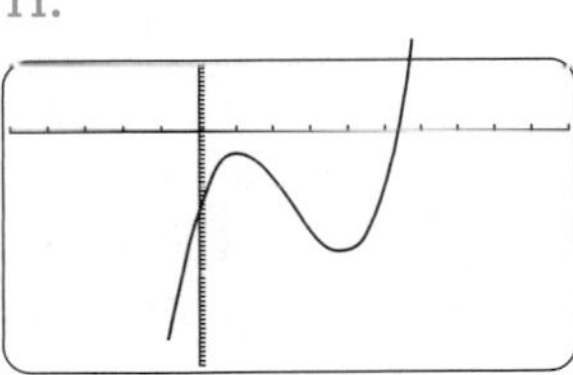

13.
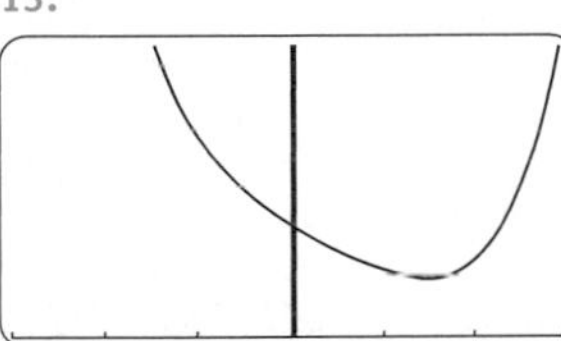

15.
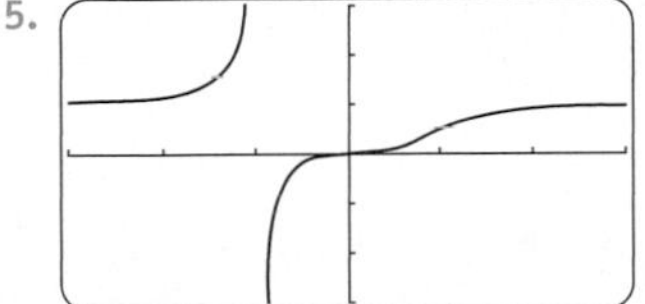

17. 18 19. 2 21. 18.5505 23. 4.1616

25. **a.**
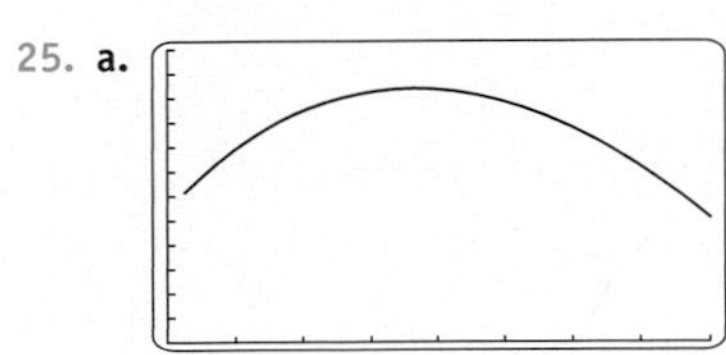

b. 9.41%; 8.71%

27. **a.**
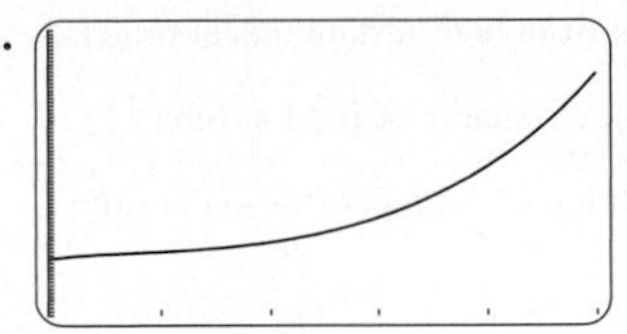

b. 44.7; 52.7; 129.2

Exercises 2.2, page 73

1. $f(x) + g(x) = x^3 + x^2 + 3$ 3. $f(x)g(x) = x^5 - 2x^3 + 5x^2 - 10$

5. $\dfrac{f(x)}{g(x)} = \dfrac{x^3 + 5}{x^2 - 2}$ 7. $\dfrac{f(x)g(x)}{h(x)} = \dfrac{x^5 - 2x^3 + 5x^2 - 10}{2x + 4}$

9. $f(x) + g(x) = x - 1 + \sqrt{x + 1}$

11. $f(x)g(x) = (x - 1)\sqrt{x + 1}$

13. $\dfrac{g(x)}{h(x)} = \dfrac{\sqrt{x + 1}}{2x^3 - 1}$ 15. $\dfrac{f(x)g(x)}{h(x)} = \dfrac{(x - 1)\sqrt{x + 1}}{2x^3 - 1}$

17. $\dfrac{f(x) - h(x)}{g(x)} = \dfrac{x - 2x^3}{\sqrt{x + 1}}$

19. $f(x) + g(x) = x^2 + \sqrt{x} + 3$; $f(x) - g(x) = x^2 - \sqrt{x} + 7$; $f(x)g(x) = (x^2 + 5)(\sqrt{x} - 2)$; $\dfrac{f(x)}{g(x)} = \dfrac{x^2 + 5}{\sqrt{x} - 2}$

21. $f(x) + g(x) = \dfrac{(x - 1)\sqrt{x + 3} + 1}{x - 1}$; $f(x) - g(x) = \dfrac{(x - 1)\sqrt{x + 3} - 1}{x - 1}$; $f(x)g(x) = \dfrac{\sqrt{x + 3}}{x - 1}$; $\dfrac{f(x)}{g(x)} = (x - 1)\sqrt{x + 3}$

23. $f(x) + g(x) = \dfrac{2(x^2 - 2)}{(x - 1)(x - 2)}$; $f(x) - g(x) = \dfrac{-2x}{(x - 1)(x - 2)}$; $f(x)g(x) = \dfrac{(x + 1)(x + 2)}{(x - 1)(x - 2)}$; $\dfrac{f(x)}{g(x)} = \dfrac{(x + 1)(x - 2)}{(x - 1)(x + 2)}$

25. $f(g(x)) = x^4 + x^2 + 1$; $g(f(x)) = (x^2 + x + 1)^2$

27. $f(g(x)) = \sqrt{x^2 - 1} + 1$; $g(f(x)) = x + 2\sqrt{x}$

29. $f(g(x)) = \dfrac{x}{x^2 + 1}$; $g(f(x)) = \dfrac{x^2 + 1}{x}$ 31. 49

33. $\dfrac{\sqrt{5}}{5}$ 35. $f(x) = 2x^3 + x^2 + 1$ and $g(x) = x^5$

37. $f(x) = x^2 - 1$ and $g(x) = \sqrt{x}$

39. $f(x) = x^2 - 1$ and $g(x) = \dfrac{1}{x}$

41. $f(x) = 3x^2 + 2$ and $g(x) = \dfrac{1}{x^{3/2}}$ 43. $3h$ 45. $-h(2a + h)$

47. $2a + h$ 49. $3a^2 + 3ah + h^2 - 1$ 51. $-\dfrac{1}{a(a + h)}$

53. The total revenue in dollars from both restaurants at time t

55. The value in dollars of Nancy's shares of IBM at time t

57. The carbon monoxide pollution in parts per million at time t

59. $C(x) = 0.6x + 12{,}100$

61. **a.** $f(t) = 267$; $g(t) = 2t^2 + 46t + 733$
b. $f(t) + g(t) = 2t^2 + 46t + 1000$ **c.** 1936 tons

63. **a.** $P(x) = -0.000003x^3 - 0.07x^2 + 300x - 100{,}000$
b. \$182,375

65. **a.** $3.5t^2 + 2.4t + 71.2$ **b.** 71,200; 109,900

67. **a.** 55%; 98.2% **b.** \$444,700; \$1,167,600

69. True **71.** False

Exercises 2.3, page 85

1. Yes; $y = -\frac{2}{3}x + 2$ **3.** Yes; $y = \frac{1}{2}x + 2$

5. Yes; $y = \frac{1}{2}x + \frac{9}{4}$ **7.** No **9.** Polynomial function; degree 6

11. Polynomial function; degree 6

13. Some other function **15.** $m = -1$; $b = 2$

17. **a.** $C(x) = 8x + 40{,}000$ **b.** $R(x) = 12x$
c. $P(x) = 4x - 40{,}000$ **d.** A loss of \$8000; a profit of \$8000

19. \$43,200 **21.** 104 mg **23.** \$128,000

25. 123,780,000 kWh; 175,820,000 kWh **27.** 0.7; 7.95

29. **a.** 320,000 **b.** 3,923,200 **31.** 582,650; 1,605,590

33. **a.**

b. 599.1 million

35. **a.** $R(x) = \dfrac{100x}{40 + x}$ **b.** 60% **37.** **b.** 2003

41. \$4770; \$6400; \$7560

43. **a.** \$16.4 billion; \$17.6 billion; \$18.3 billion; \$18.8 billion; \$19.3 billion

b.

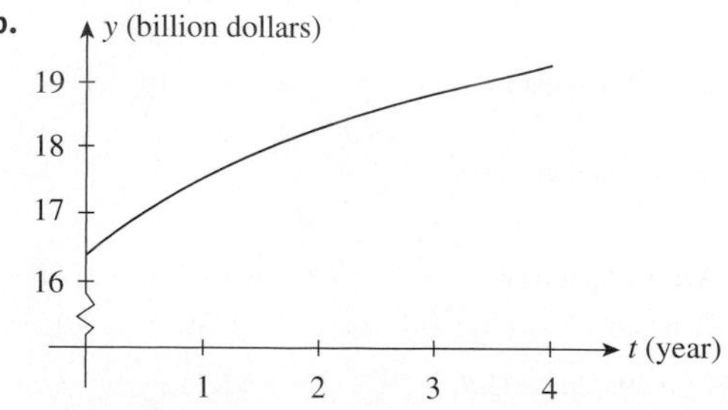

45. **a.**

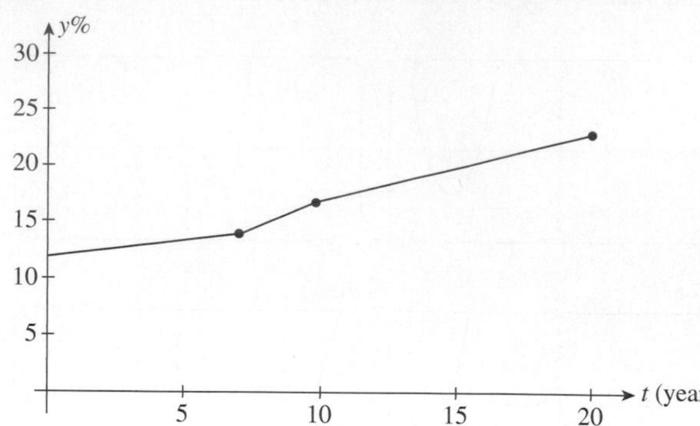

b. 13.43%; 20%

47. $\dfrac{110}{\frac{1}{2}t + 1} - 26\left(\dfrac{1}{4}t^2 - 1\right)^2 - 52$; \$32, \$6.71, \$3; the gap was closing.

49. **a.** 59.8%; 58.9%; 59.2%; 60.7%, 61.7%

b.

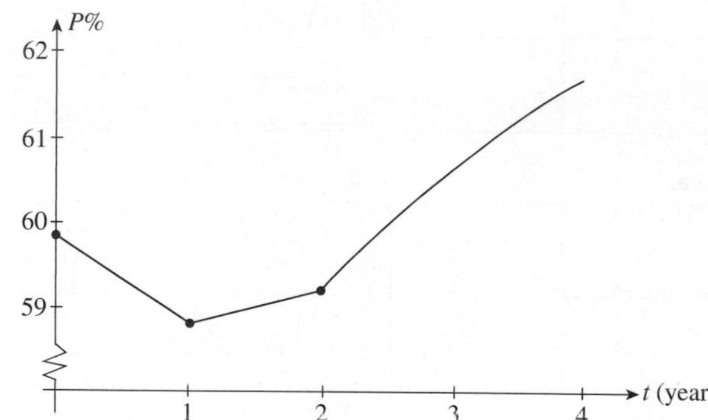

c. 60.66%

51. **a.**

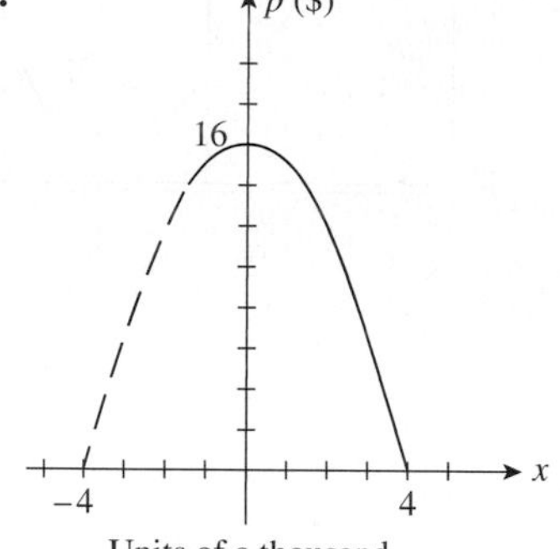

b. 3000 units

53. **a.**

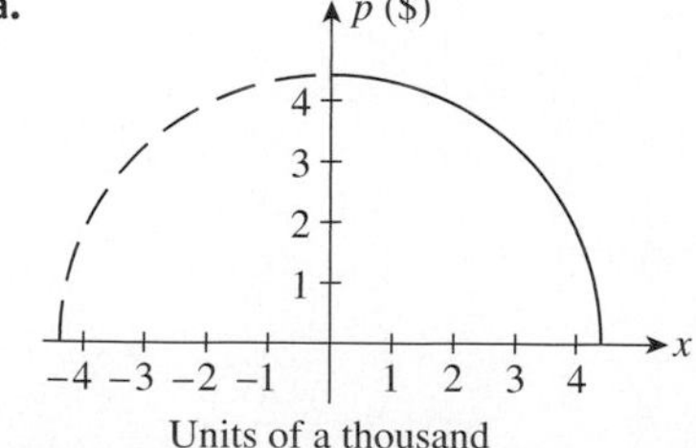

b. 3000

55. a.

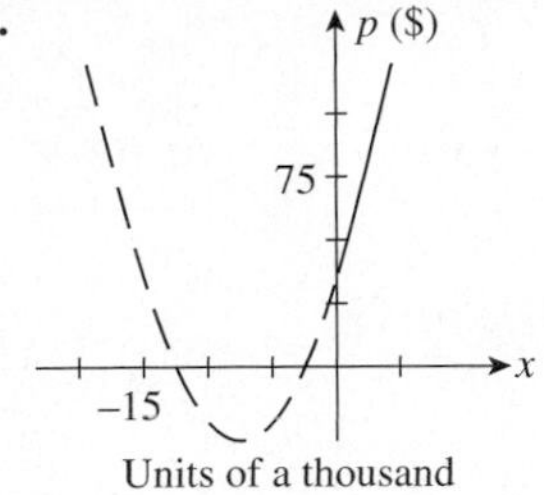

b. \$76

57. a.

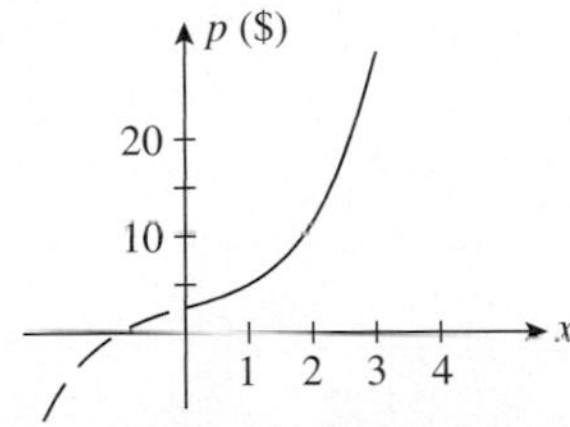

b. \$15

59. L_2; for each dollar decrease in the price of a clock radio, the additional quantity demanded of model B clock radios will be greater than that of model A clock radios.

61. a.

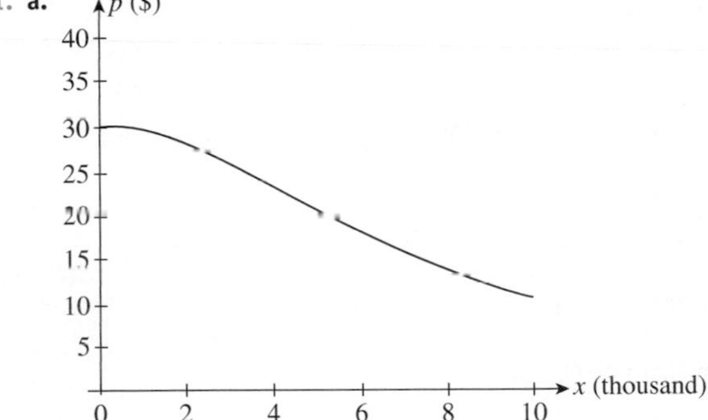

b. \$10

63. a.

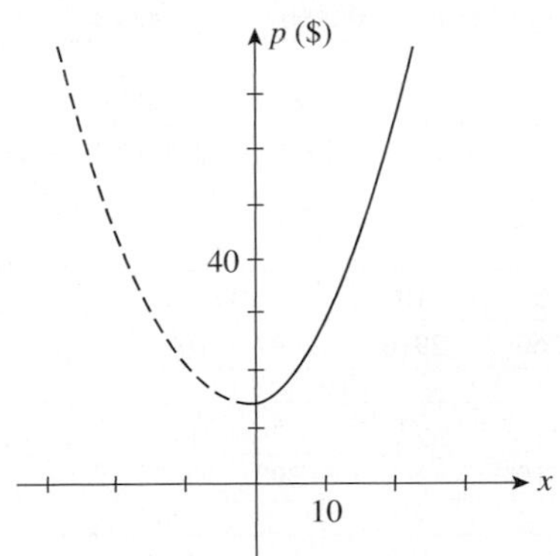

b. \$20

65. a. $\dfrac{b-d}{c-a}$; $\dfrac{bc-ad}{c-a}$

b. If the unit price is increased, then the equilibrium quantity decreases while the equilibrium price increases.

c. If the upper bound for the unit price of a commodity is lowered, then both the equilibrium quantity and the equilibrium price drop.

67. 2500; \$67.50 **69.** 11,000; \$3 **71.** 8000; \$80

73. $f(x) = 2x + \dfrac{500}{x}$; $x > 0$ **75.** $f(x) = 0.5x^2 + \dfrac{8}{x}$

77. $f(x) = (22 + x)(36 - 2x)$ bushels/acre

79. a. $P(x) = (10{,}000 + x)(5 - 0.0002x)$ **b.** \$60,800

81. False **83.** False

Using Technology Exercises 2.3, page 94

1. (−3.0414, 0.1503); (3.0414, 7.4497)

3. (−2.3371, 2.4117); (6.0514, −2.5015)

5. (−1.0219, −6.3461); (1.2414, −1.5931); (5.7805, 7.9391)

7. a.

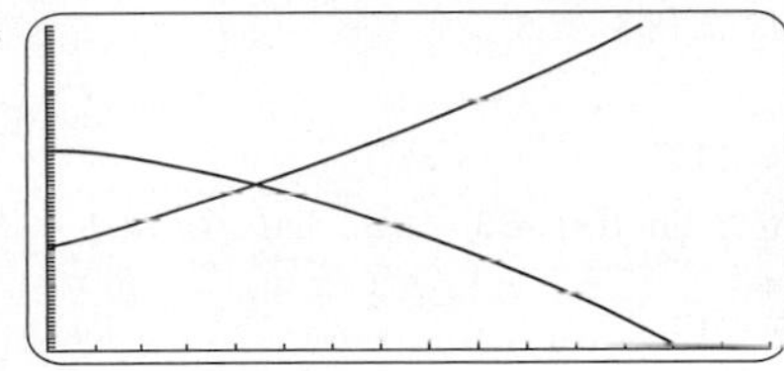

b. 438 wall clocks; \$40.92

9. a. $y = 0.1554t^2 + 0.5861t + 3.1607$

b.

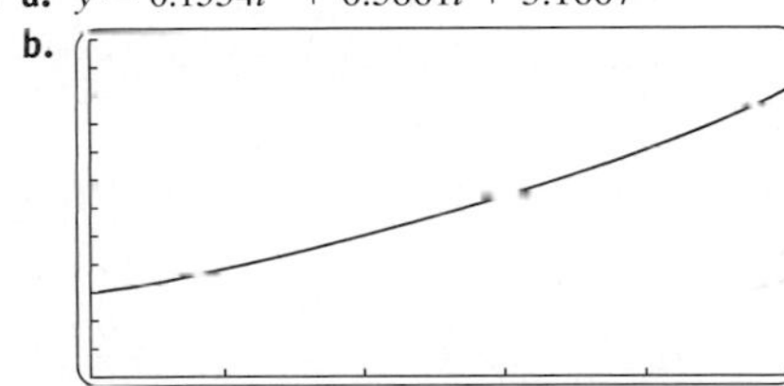

c. 3.16; 3.9; 4.95; 6.32; 7.99; 9.98

11. a. $y = -0.02028t^3 + 0.31393t^2 + 0.40873t + 0.66024$

b.

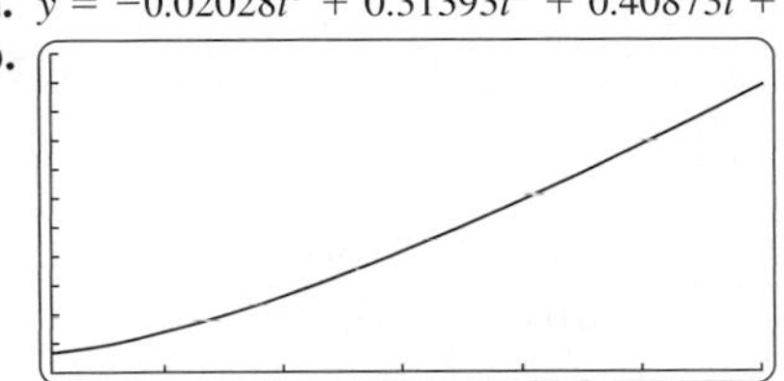

c. 0.66; 1.36; 2.57; 4.16; 6.02; 8.02; 10.03

13. a. $D = f(t) = 0.001532t^3 - 0.0588t^2 + 0.5208t + 2.55$

b.

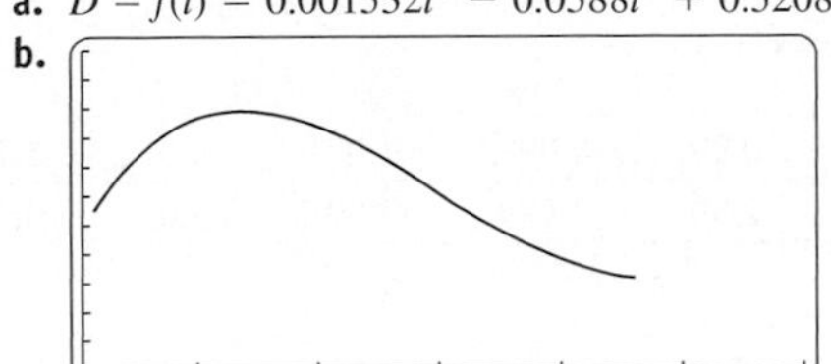

c. \$2.55 trillion; \$3.83988 trillion; \$3.41 trillion; \$1.702 trillion

15. a. $y = 0.05833t^3 - 0.325t^2 + 1.8881t + 5.07143$

b.

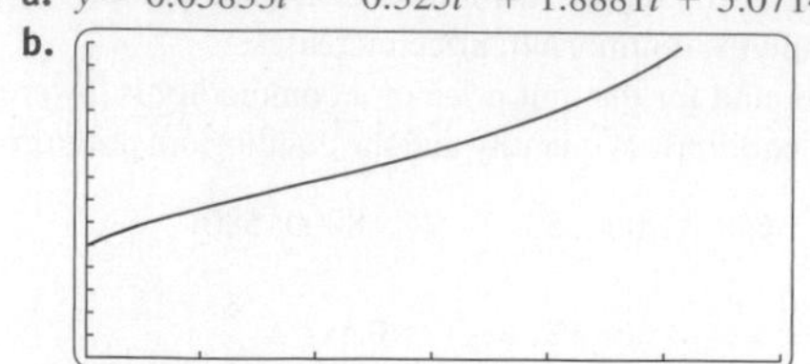

c. 6.7; 8.0; 9.4; 11.2; 13.7

17. a. $y = 0.0125t^4 - 0.01389t^3 + 0.55417t^2 + 0.53294t + 4.95238$ $(0 \le t \le 5)$

b.

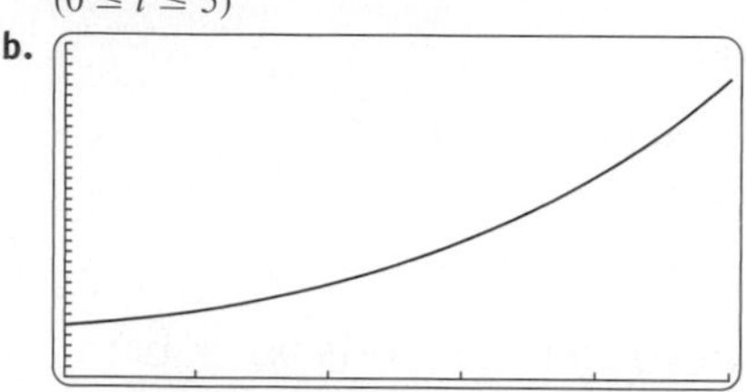

c. 5.0; 6.0; 8.3; 12.2; 18.3; 27.5

Exercises 2.4, page 111

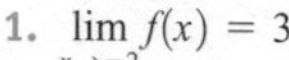

1. $\lim_{x \to -2} f(x) = 3$ **3.** $\lim_{x \to 3} f(x) = 3$ **5.** $\lim_{x \to -2} f(x) = 3$

7. The limit does not exist.

9.

x	1.9	1.99	1.999
$f(x)$	4.61	4.9601	4.9960

x	2.001	2.01	2.1
$f(x)$	5.004	5.0401	5.41

$\lim_{x \to 2} (x^2 + 1) = 5$

11.

x	−0.1	−0.01	−0.001
$f(x)$	−1	−1	−1

x	0.001	0.01	0.1
$f(x)$	1	1	1

The limit does not exist.

13.

x	0.9	0.99	0.999
$f(x)$	100	10,000	1,000,000

x	1.001	1.01	1.1
$f(x)$	1,000,000	10,000	100

The limit does not exist.

15.

x	0.9	0.99	0.999	1.001	1.01	1.1
$f(x)$	2.9	2.99	2.999	3.001	3.01	3.1

$\lim_{x \to 1} \dfrac{x^2 + x - 2}{x - 1} = 3$

17.

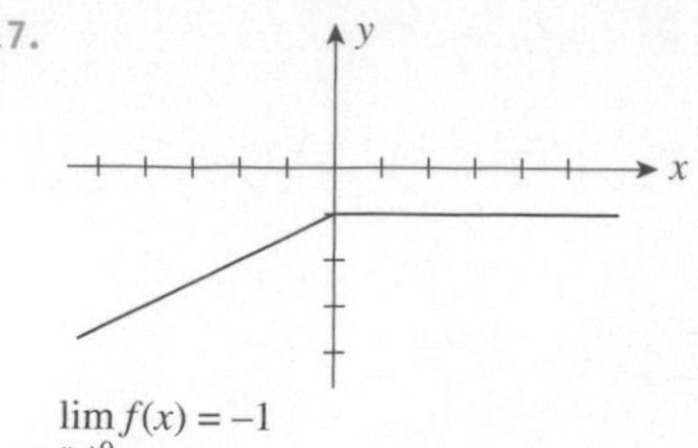

$\lim_{x \to 0} f(x) = -1$

19.

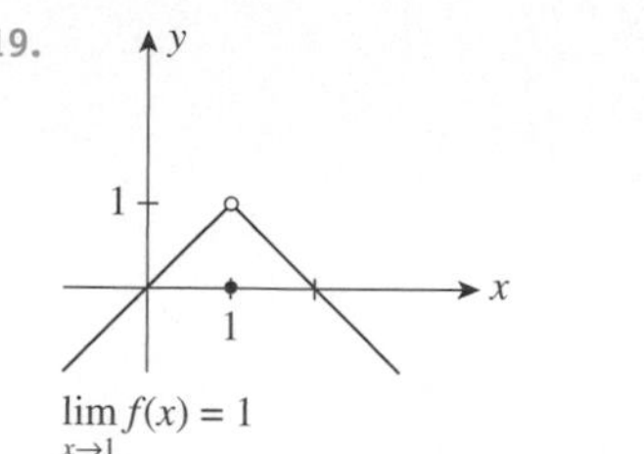

$\lim_{x \to 1} f(x) = 1$

21.

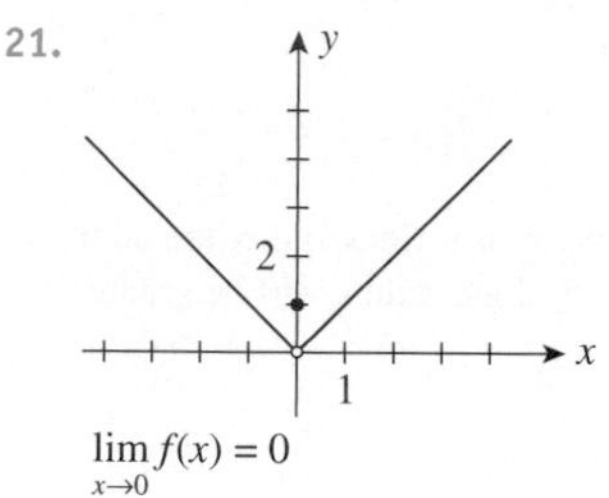

$\lim_{x \to 0} f(x) = 0$

23. 3 **25.** 3 **27.** −1 **29.** 2 **31.** −4 **33.** $\frac{5}{4}$

35. 2 **37.** $\sqrt{171} = 3\sqrt{19}$ **39.** $\frac{3}{2}$ **41.** −1 **43.** −6

45. 2 **47.** $\frac{1}{6}$ **49.** 2 **51.** −1 **53.** −10

55. The limit does not exist. **57.** $\frac{5}{3}$ **59.** $\frac{1}{2}$ **61.** $\frac{1}{3}$

63. $\lim_{x \to \infty} f(x) = \infty$; $\lim_{x \to -\infty} f(x) = \infty$ **65.** 0; 0

67. $\lim_{x \to \infty} f(x) = -\infty$; $\lim_{x \to -\infty} f(x) = -\infty$

69.

x	1	10	100	1000
$f(x)$	0.5	0.009901	0.0001	0.000001

x	−1	−10	−100	−1000
$f(x)$	0.5	0.009901	0.0001	0.000001

$\lim_{x \to \infty} f(x) = 0$ and $\lim_{x \to -\infty} f(x) = 0$

71.

x	1	5	10	100
$f(x)$	12	360	2910	2.99×10^6

x	1000	−1	−5
$f(x)$	2.999×10^9	6	−390

x	−10	−100	−1000
$f(x)$	−3090	-3.01×10^6	-3.0×10^9

$\lim_{x \to \infty} f(x) = \infty$ and $\lim_{x \to -\infty} f(x) = -\infty$

73. 3 **75.** 3 **77.** $\lim_{x \to -\infty} f(x) = -\infty$ **79.** 0

81. **a.** \$0.5 million; \$0.75 million; \$1,166,667; \$2 million; \$4.5 million; \$9.5 million
b. The limit does not exist; as the percent of pollutant to be removed approaches 100, the cost becomes astronomical.

83. \$2.20; the average cost of producing x DVDs will approach \$2.20/disc in the long run.

85. **a.** \$24 million; \$60 million; \$83.1 million **b.** \$120 million

87. **a.** 76.1¢/mi; 30.5¢/mi; 23¢/mi; 20.6¢/mi; 19.5¢/mi
b.

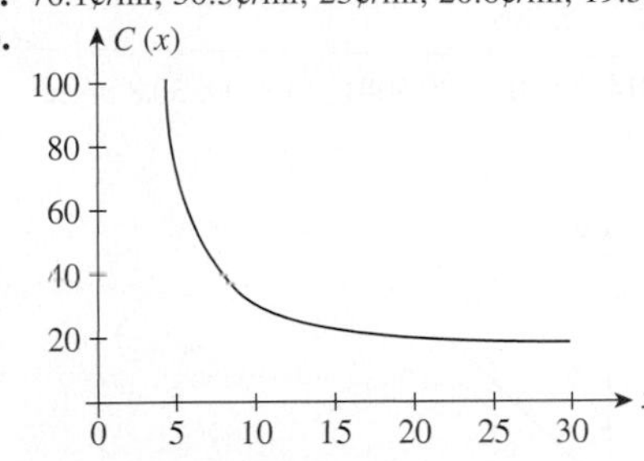

c. It approaches 17.8¢/mi.

89. False 91. True 93. True

95. a moles/liter/second 97. No

Using Technology Exercises 2.4, page 118

1. 5 3. 3 5. $\frac{2}{3}$ 7. $\frac{1}{2}$ 9. e^2

13. **a.**

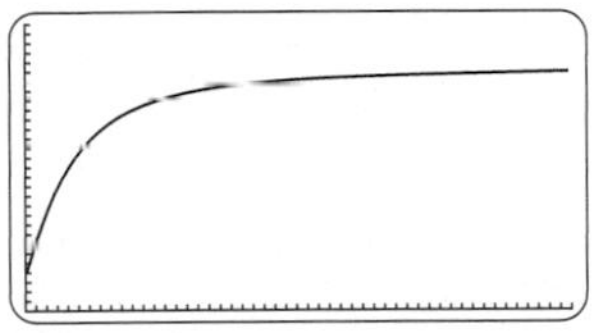

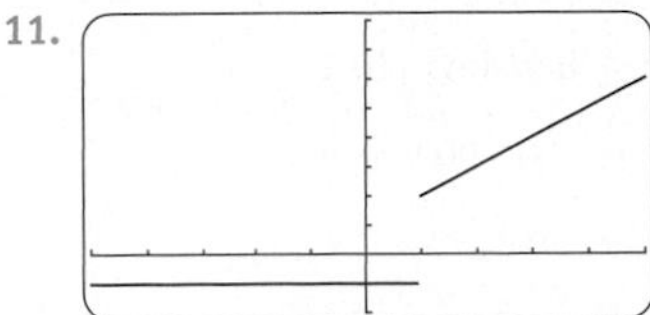

b. 25,000

Exercises 2.5, page 127

1. 3; 2; the limit does not exist.

3. The limit does not exist; 2; the limit does not exist.

5. 0; 2; the limit does not exist.

7. -2; 2; the limit does not exist. 9. True 11. True

13. False 15. True 17. False 19. True 21. 6

23. $-\frac{1}{4}$ 25. The limit does not exist. 27. -1 29. 0

31. -4 33. The limit does not exist. 35. 4 37. 0; 0

39. $x = 0$; conditions 2 and 3 41. Continuous everywhere

43. $x = 0$; condition 3 45. $(-\infty, \infty)$ 47. $(-\infty, \infty)$

49. $(-\infty, \frac{1}{2}) \cup (\frac{1}{2}, \infty)$ 51. $(-\infty, -2) \cup (-2, 1) \cup (1, \infty)$

53. $(-\infty, \infty)$ 55. $(-\infty, \infty)$ 57. -1 and 1 59. 1 and 2

61. f is discontinuous at $x = 1, 2, \ldots, 11$.

63. Michael makes progress toward solving the problem until $x = x_1$. Between $x = x_1$ and $x = x_2$, he makes no further progress. But at $x = x_2$ he suddenly achieves a breakthrough, and at $x = x_3$ he proceeds to complete the problem.

65. Conditions 2 and 3 are not satisfied at each of these points.

67.

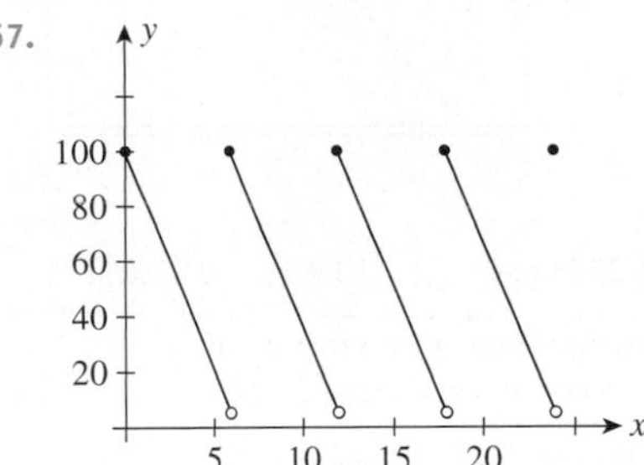

69.

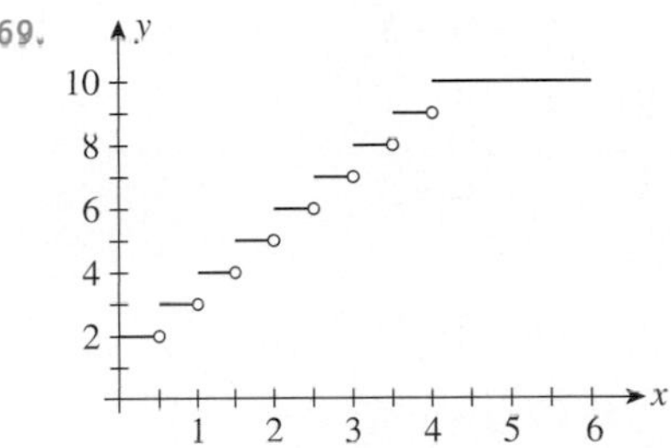

f is discontinuous at $x = \frac{1}{2}, 1, 1\frac{1}{2}, \ldots, 4$.

71. **a.** ∞; As the time taken to excite the tissue is made smaller and smaller, the strength of the electric current gets stronger and stronger.
b. b; As the time taken to excite the tissue is made larger and larger, the strength of the electric current gets smaller and smaller and approaches b.

73. 3 75. **a.** Yes **b.** No

77. **a.** f is a polynomial of degree 2. **b.** $f(1) = 3$ and $f(3) = -1$

79. **a.** f is a polynomial of degree 3. **b.** $f(-1) = -4$ and $f(1) = 4$

81. $x \approx 0.59$ 83. ≈ 1.34

85. **c.** $\frac{1}{2}$; $\frac{7}{2}$; Joan sees the ball on its way up $\frac{1}{2}$ sec after it was thrown and again $3\frac{1}{2}$ sec later.

87. False 89. False 91. False 93. False

95. **c.** $\pm\frac{\sqrt{2}}{2}$

Using Technology Exercises 2.5, page 134

1. $x = 0, 1$ 3. $x = 2$ 5. $x = 0, \frac{1}{2}$ 7. $x = -\frac{1}{2}, 2$

9. $x = -2, 1$

11.

13.

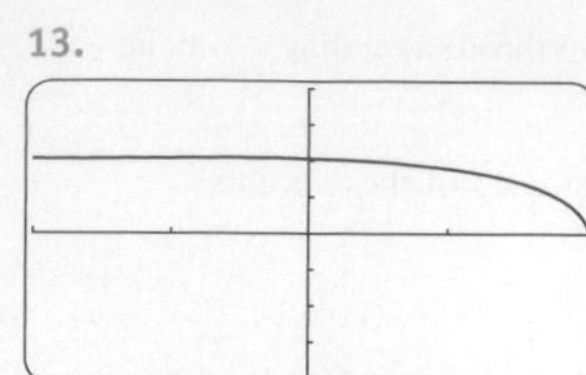

15.

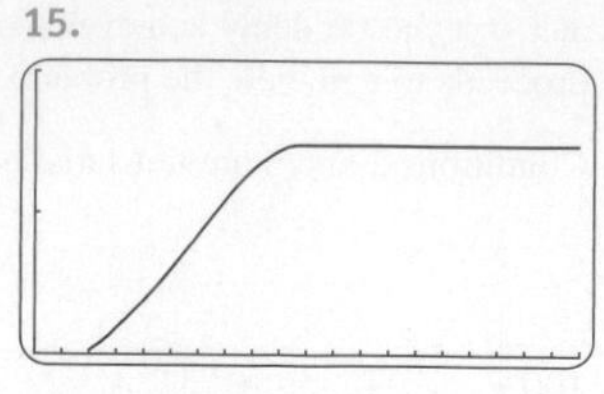

Exercises 2.6, page 147

1. 1.5 lb/mo; 0.58 lb/mo; 1.25 lb/mo **3.** 3.1%/hr; −21.2%/hr

5. a. Car A **b.** They are traveling at the same speed.
c. Car B **d.** Both cars covered the same distance.

7. a. P_2 **b.** P_1 **c.** Bactericide B; bactericide A

9. 0 **11.** 2 **13.** $6x$ **15.** $-2x + 3$ **17.** 2; $y = 2x + 7$

19. 6; $y = 6x - 3$ **21.** $\frac{1}{9}$; $y = \frac{1}{9}x - \frac{2}{3}$

23. a. $4x$ **b.** $y = 4x - 1$
c.

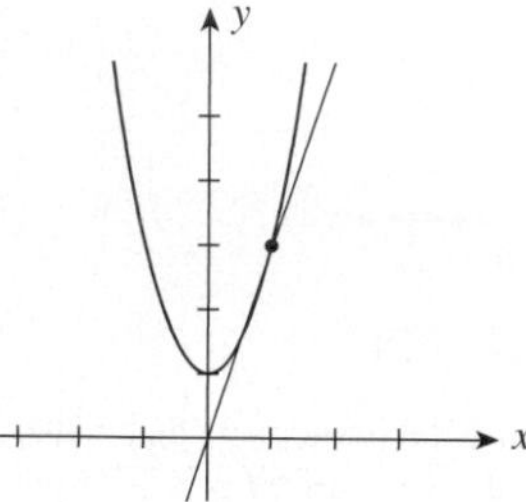

25. a. $2x - 2$ **b.** (1, 0)
c.

d. 0

27. a. 6; 5.5; 5.1 **b.** 5
c. The computations in part (a) show that as h approaches zero, the average velocity approaches the instantaneous velocity.

29. a. 130 ft/sec; 128.2 ft/sec; 128.02 ft/sec **b.** 128 ft/sec
c. The computations in part (a) show that as the time intervals over which the average velocity are computed become smaller and smaller, the average velocity approaches the instantaneous velocity of the car at $t = 20$.

31. a. 5 sec **b.** 80 ft/sec **c.** 160 ft/sec

33. a. $-\frac{1}{6}$ liter/atmosphere **b.** $-\frac{1}{4}$ liter/atmosphere

35. a. $-\frac{2}{3}x + 7$ **b.** \$333/quarter; −\$13,000/quarter

37. \$6 billion/yr; \$10 billion/yr

39. Average rate of change of the seal population over $[a, a + h]$; instantaneous rate of change of the seal population at $x = a$

41. Average rate of change of the country's industrial production over $[a, a + h]$; instantaneous rate of change of the country's industrial production at $x = a$

43. Average rate of change of atmospheric pressure over $[a, a + h]$; instantaneous rate of change of atmospheric pressure at $x = a$

45. a. Yes **b.** No **c.** No

47. a. Yes **b.** Yes **c.** No

49. a. No **b.** No **c.** No

51. 32.1, 30.939, 30.814, 30.8014, 30.8001, 30.8000; 30.8 ft/sec

53. False

55.

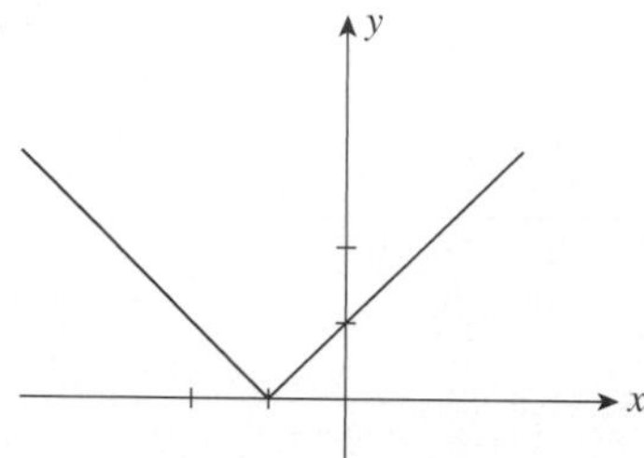

57. $a = 2, b = -1$

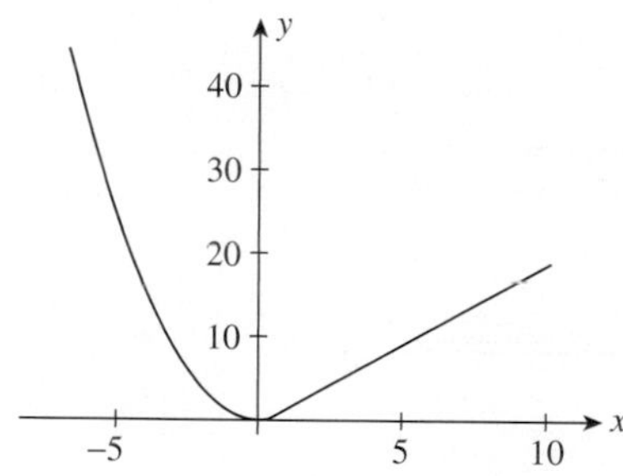

Using Technology Exercises 2.6, page 154

1. a. $y = 4x - 3$
b.

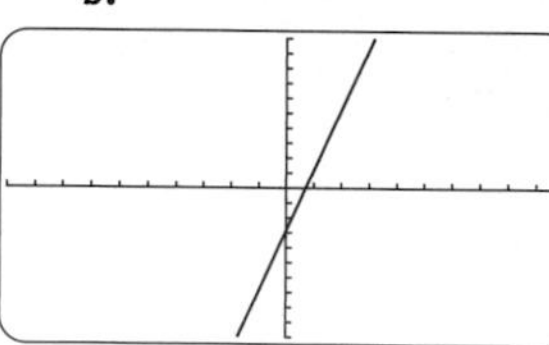

3. a. $y = 9x - 11$
b.

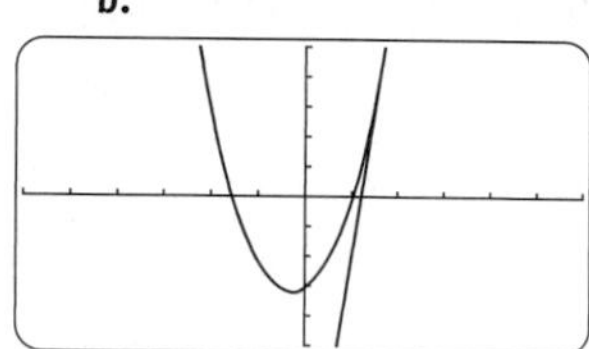

5. a. $y = \frac{1}{4}x + 1$
b.

7. a. 4
b. $y = 4x - 1$
c.

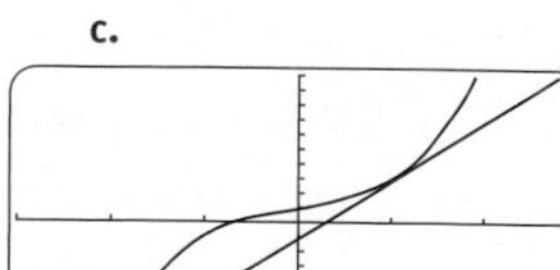

9. **a.** 0.75
b. $y = 0.75x - 1$
c.

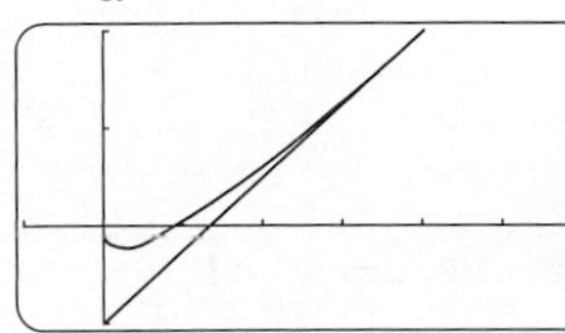

11. **a.** 4.02
b. $y = 4.02x - 3.57$
c.

13. **a.**

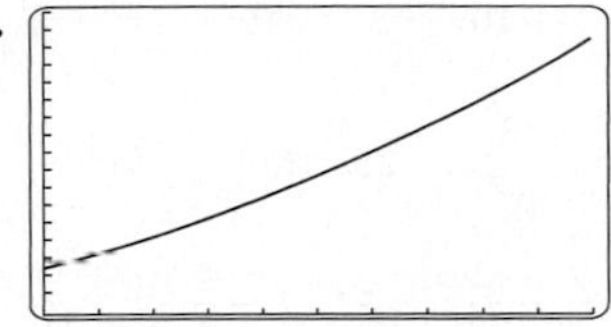

b. 41.22¢/mi **c.** 1.22¢/mi/yr

Chapter 2 Concept Review, page 155

1. Domain; range; B 2. Domain, $f(x)$; vertical, point

3. $f(x) \pm g(x)$; $f(x)g(x)$; $\dfrac{f(x)}{g(x)}$; $A \cap B$; $A \cap B$; 0

4. $g[f(x)]$; f; $f(x)$; g

5. **a.** $P(x) = a_0x^n + a_1x^{n-1} + \cdots + a_{n-1}x + a_n$ ($a_0 \neq 0$, n, a positive integer)
b. Linear; quadratic; cubic **c.** Quotient; polynomials
d. x^r (r, a real number)

6. L; $f(x)$; L; a

7. **a.** L^r **b.** $L \pm M$ **c.** LM **d.** $\dfrac{L}{M}$; $M \neq 0$

8. **a.** L; x **b.** M; negative; absolute

9. **a.** Right **b.** Left **c.** L; L

10. **a.** Continuous **b.** Discontinuous **c.** Every

11. **a.** a; a; $g(a)$; **b.** Everywhere **c.** $Q(x)$

12. **a.** $[a, b]$; $f(c) = M$ **b.** $f(x) = 0$; (a, b)

13. **a.** $f'(a)$ **b.** $y = f(a) + m(x - a)$

14. **a.** $\dfrac{f(a+h) - f(a)}{h}$ **b.** $\displaystyle\lim_{h\to 0}\frac{f(a+h) - f(a)}{h}$

Chapter 2 Review Exercises, page 156

1. **a.** $(-\infty, 9]$ **b.** $(-\infty, -1) \cup (-1, \frac{3}{2}) \cup (\frac{3}{2}, \infty)$

2. **a.** 0 **b.** $3a^2 + 17a + 20$ **c.** $12a^2 + 10a - 2$
d. $3a^2 + 6ah + 3h^2 + 5a + 5h - 2$

3. **a.**

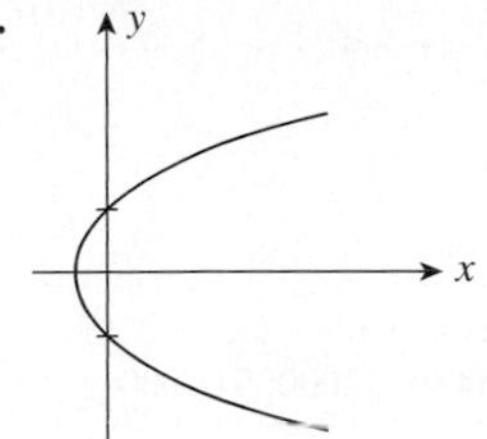

b. No **c.** Yes

4.

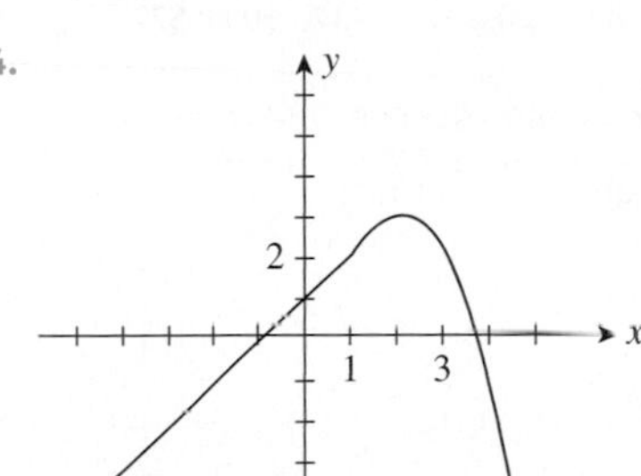

5. **a.** $\dfrac{2x+3}{x}$ **b.** $\dfrac{1}{x(2x+3)}$ **c.** $\dfrac{1}{2x+3}$ **d.** $\dfrac{2}{x} + 3$

6. -3 7. 2 8. -21 9. 0 10. -1

11. The limit does not exist. 12. 7 13. $\frac{9}{2}$ 14. 1 15. $\frac{1}{2}$

16. 1 17. 1 18. $\frac{3}{2}$ 19. The limit does not exist.

20.

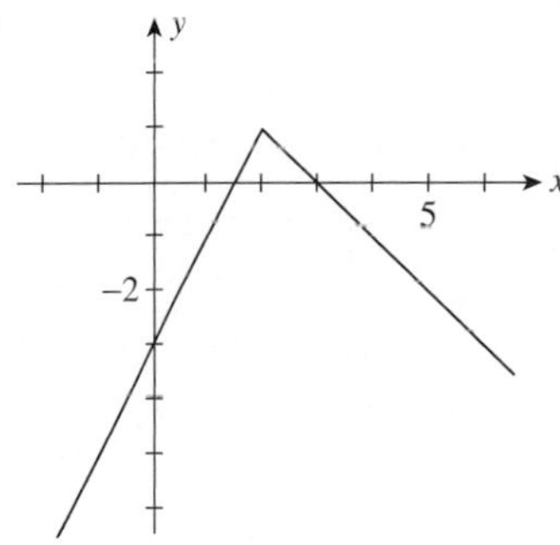

1; 1; 1

21.

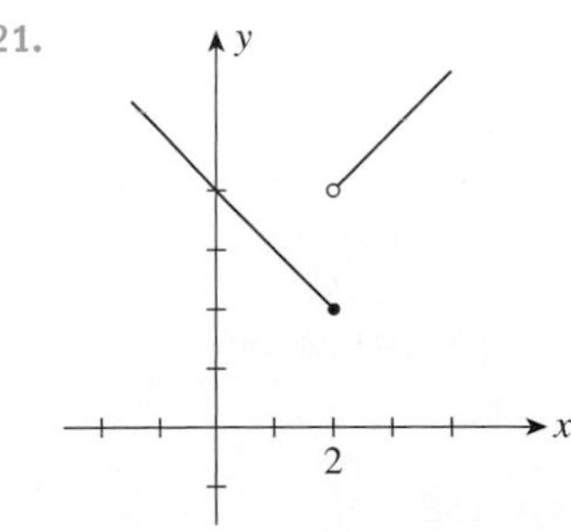

4; 2; the limit does not exist.

22. $x = 2$ 23. $x = -\frac{1}{2}, 1$ 24. $x = -1$ 25. $x = 0$

26. **a.** 3; 2.5; 2.1 **b.** 2 27. 3 28. $\dfrac{1}{x^2}$

29. $\frac{3}{2}$; $y = \frac{3}{2}x + 5$ **30.** -4; $y = -4x + 4$

31. a. Yes **b.** No **32.** 54,000

33. a. $S(t) = t + 2.4$ **b.** \$5.4 million

34. a. $C(x) = 6x + 30{,}000$ **b.** $R(x) = 10x$
c. $P(x) = 4x - 30{,}000$ **d.** (\$6000); \$2000; \$18,000

35. $\left(6, \frac{21}{2}\right)$ **36.** $P(x) = 8x - 20{,}000$ **37.** 6000; \$22

38. 117 mg **39.** \$400,000 **40.** \$45,000

41. 400; 800 **42.** 990; 2240

43.

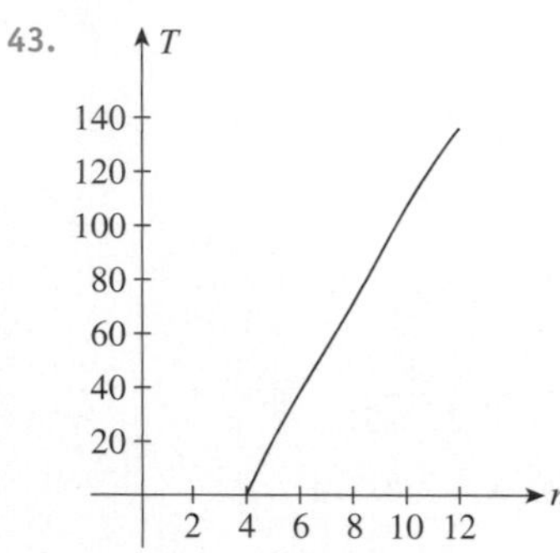

As the length of the list increases, the time taken to learn the list increases by a very large amount.

44. After $5\frac{1}{2}$ years **45.** 5000; \$20

46. a. πr^2 **b.** $2t$ **c.** $4\pi t^2$ **d.** 3600π ft²

47. $$C(x) = \begin{cases} 5 & \text{if } 1 \le x \le 100 \\ 9 & \text{if } 100 < x \le 200 \\ 12.50 & \text{if } 200 < x \le 300 \\ 15.00 & \text{if } 300 < x \le 400 \\ 7 + 0.02x & \text{if } x > 400 \end{cases}$$

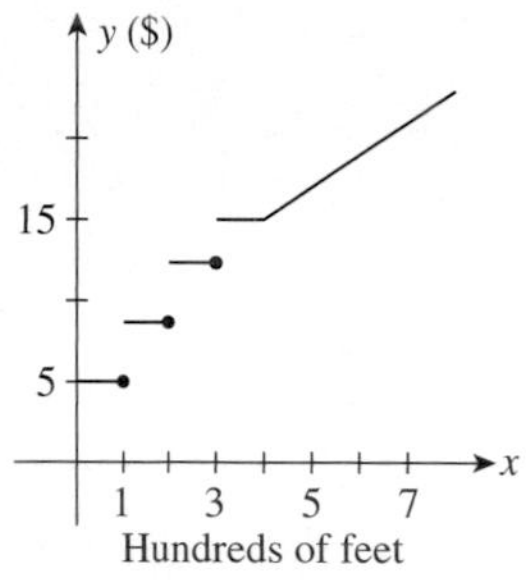

The function is discontinuous at $x = 100$, 200, and 300.

48. 20

Chapter 2 Before Moving On, page 158

1. a. 3 **b.** 2 **c.** $\frac{17}{4}$

2. a. $\frac{1}{x+1} + x^2 + 1$ **b.** $\frac{x^2+1}{x+1}$ **c.** $\frac{1}{x^2+2}$
d. $\frac{1}{(x+1)^2} + 1$

3. $108x^2 - 4x^3$ **4.** 2 **5. a.** 0 **b.** 1; no

6. -1; $y = -x$

CHAPTER 3

Exercises 3.1, page 167

1. 0 **3.** $5x^4$ **5.** $2.1x^{1.1}$ **7.** $6x$ **9.** $2\pi r$ **11.** $\frac{3}{x^{2/3}}$

13. $\frac{3}{2\sqrt{x}}$ **15.** $-84x^{-13}$ **17.** $10x - 3$ **19.** $-3x^2 + 4x$

21. $0.06x - 0.4$ **23.** $2x - 4 - \frac{3}{x^2}$ **25.** $16x^3 - 7.5x^{3/2}$

27. $-\frac{3}{x^2} - \frac{8}{x^3}$ **29.** $-\frac{16}{t^5} + \frac{9}{t^4} - \frac{2}{t^2}$ **31.** $2 - \frac{5}{2\sqrt{x}}$

33. $-\frac{4}{x^3} + \frac{1}{x^{4/3}}$ **35. a.** 20 **b.** -4 **c.** 20 **37.** 3

39. 11 **41.** $m = 5$; $y = 5x - 4$ **43.** $m = -2$; $y = -2x + 2$

45. a. (0, 0)
b.

47. a. $(-2, -7)$, $(2, 9)$ **b.** $y = 12x + 17$ and $y = 12x - 15$
c.

49. a. $(0, 0)$; $(1, -\frac{13}{12})$ **b.** $(0, 0)$; $(2, -\frac{8}{3})$; $(-1, -\frac{5}{12})$
c. $(0, 0)$; $(4, \frac{80}{3})$; $(-3, \frac{81}{4})$

51. a. $\frac{16\pi}{9}$ cm³/cm **b.** $\frac{25\pi}{4}$ cm³/cm

53. a. 16.3 million **b.** 14.3 million/yr **c.** 66.8 million
d. 11.7 million/yr

55. a. 49.6%; 41.13%; 36.87%; 34.11%
b. −5.55%/yr; −3.32%/yr

57. a. 51.5% **b.** 1.95%/yr

59. a. $120 - 30t$ **b.** 120 ft/sec **c.** 240 ft

61. a. 5%; 11.3%; 15.5% **b.** 0.63%/yr; 0.525%/yr

63. a. -0.9 thousand metric tons/yr; 20.3 thousand metric tons/yr
b. Yes

65. a. 15 pts/yr; 12.6 pts/yr; 0 pts/yr **b.** 10 pts/yr

67. a. $(0.0001)(\frac{5}{4})x^{1/4}$ **b.** \$0.00125/radio

69. a. $20\left(1 - \dfrac{1}{\sqrt{t}}\right)$ **b.** 50 mph; 30 mph; 33.43 mph
c. -8.28; 0; 5.86; at 6:30 a.m., the average velocity is decreasing at the rate of 8.28 mph/hr; at 7 a.m., it is unchanged; and at 8 a.m., it is increasing at the rate of 5.86 mph.

71. 32 turtles/yr; 428 turtles/yr; 3260 turtles

73. a. 12%; 23.9% **b.** 0.8%/yr; 1.1%/yr **75.** True

Using Technology Exercises 3.1, page 172

1. 1 **3.** 0.4226 **5.** 0.1613

7. a.

b. 3.4295 ppm; 105.4332 ppm

9. a.

b. Decreasing at the rate of 9 days/yr; increasing at the rate of 13 days/yr

11. a.

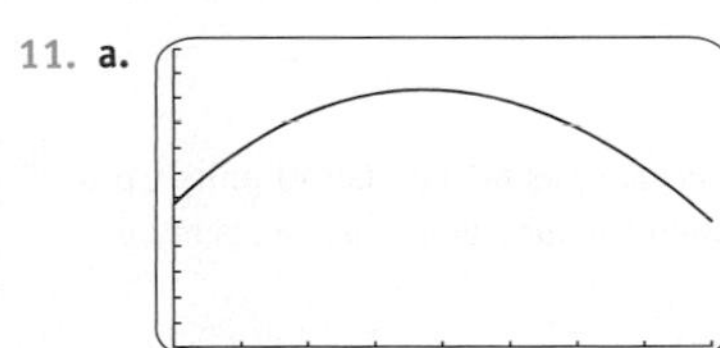

b. Increasing at the rate of 1.1557%/yr; decreasing at the rate of 0.2116%/yr

Exercises 3.2, page 180

1. $2x(2x) + (x^2 + 1)(2)$, or $6x^2 + 2$

3. $(t - 1)(2) + (2t + 1)(1)$, or $4t - 1$

5. $(3x + 1)(2x) + (x^2 - 2)(3)$, or $9x^2 + 2x - 6$

7. $(x^3 - 1)(1) + (x + 1)(3x^2)$, or $4x^3 + 3x^2 - 1$

9. $(w^3 - w^2 + w - 1)(2w) + (w^2 + 2)(3w^2 - 2w + 1)$, or $5w^4 - 4w^3 + 9w^2 - 6w + 2$

11. $(5x^2 + 1)(x^{-1/2}) + (2x^{1/2} - 1)(10x)$, or $\dfrac{25x^2 - 10x\sqrt{x} + 1}{\sqrt{x}}$

13. $\dfrac{(x^2 - 5x + 2)(x^2 + 2)}{x^2} + \dfrac{(x^2 - 2)(2x - 5)}{x}$, or $\dfrac{3x^4 - 10x^3 + 4}{x^2}$

15. $\dfrac{-1}{(x - 2)^2}$ **17.** $\dfrac{2x + 1 - (x - 1)(2)}{(2x + 1)^2}$, or $\dfrac{3}{(2x + 1)^2}$

19. $-\dfrac{2x}{(x^2 + 1)^2}$ **21.** $\dfrac{s^2 + 2s + 4}{(s + 1)^2}$

23. $\dfrac{(\frac{1}{2}x^{-1/2})[(x^2 + 1) - 4x^2]}{(x^2 + 1)^2}$, or $\dfrac{1 - 3x^2}{2\sqrt{x}(x^2 + 1)^2}$

25. $\dfrac{2x^3 + 2x^2 + 2x - 2x^3 - x^2 - 4x - 2}{(x^2 + x + 1)^2}$, or $\dfrac{x^2 - 2x - 2}{(x^2 + x + 1)^2}$

27. $\dfrac{(x - 2)(3x^2 + 2x + 1) - (x^3 + x^2 + x + 1)}{(x - 2)^2}$, or $\dfrac{2x^3 - 5x^2 - 4x - 3}{(x - 2)^2}$

29. $\dfrac{(x^2 - 4)(x^2 + 4)(2x + 8) - (x^2 + 8x - 4)(4x^3)}{(x^2 - 4)^2(x^2 + 4)^2}$, or $\dfrac{-2x^5 - 24x^4 + 16x^3 - 32x - 128}{(x^2 - 4)^2(x^2 + 4)^2}$

31. 8 **33.** -9 **35.** $2(3x^2 - x + 3)$; 10

37. $\dfrac{-3x^4 + 2x^2 - 1}{(x^4 - 2x^2 - 1)^2}$; $-\dfrac{1}{2}$ **39.** 60; $y = 60x - 102$

41. $-\frac{1}{2}$; $y = -\frac{1}{2}x + \frac{3}{2}$ **43.** $y = 7x - 5$ **45.** $(\frac{1}{3}, \frac{50}{27})$; $(1, 2)$

47. $(\frac{4}{3}, -\frac{770}{27})$; $(2, -30)$ **49.** $y = -\frac{1}{2}x + 1$; $y = 2x - \frac{3}{2}$

51. 0.125, 0.5, 2, 50; the cost of removing all of the pollutant is prohibitively high.

53. -5000/min; -1600/min; 7000; 4000

55. a. $\dfrac{180}{(t + 6)^2}$ **b.** 3.7; 2.2; 1.8; 1.1

c.

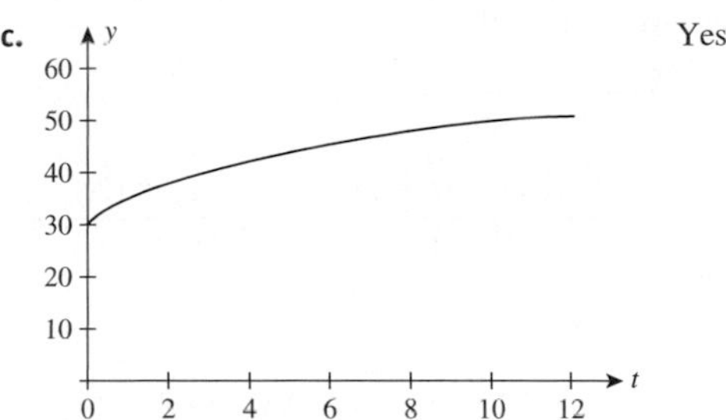

Yes

d. 50 words/min

57. Dropping at the rate of 0.0375 ppm/yr; dropping at the rate of 0.006 ppm/yr

59. False **61.** False

Using Technology Exercises 3.2, page 184

1. 0.8750 **3.** 0.0774 **5.** -0.5000 **7.** 87,322/yr

Exercises 3.3, page 192

1. $8(2x - 1)^3$ **3.** $10x(x^2 + 2)^4$

5. $3(2x - x^2)^2(2 - 2x)$, or $6x^2(1 - x)(2 - x)^2$ **7.** $\dfrac{-4}{(2x + 1)^3}$

9. $3x\sqrt{x^2-4}$ 11. $\dfrac{3}{2\sqrt{3x-2}}$ 13. $\dfrac{-2x}{3(1-x^2)^{2/3}}$

15. $-\dfrac{6}{(2x+3)^4}$ 17. $\dfrac{-1}{(2t-3)^{3/2}}$ 19. $-\dfrac{3(16x^3+1)}{2(4x^4+x)^{5/2}}$

21. $-2(3x^2+2x+1)^{-3}(6x+2)=-4(3x+1)(3x^2+2x+1)^{-3}$

23. $3(x^2+1)^2(2x)-2(x^3+1)(3x^2)$, or $6x(2x^2-x+1)$

25. $3(t^{-1}-t^{-2})^2(-t^{-2}+2t^{-3})$ 27. $\dfrac{1}{2\sqrt{x-1}}+\dfrac{1}{2\sqrt{x+1}}$

29. $2x^2(4)(3-4x)^3(-4)+(3-4x)^4(4x)$, or $(-12x)(4x-1)(3-4x)^3$

31. $8(x-1)^2(2x+1)^3+2(x-1)(2x+1)^4$, or $6(x-1)(2x-1)(2x+1)^3$

33. $3\left(\dfrac{x+3}{x-2}\right)^2\left[\dfrac{(x-2)(1)-(x+3)(1)}{(x-2)^2}\right]$, or $-\dfrac{15(x+3)^2}{(x-2)^4}$

35. $\dfrac{3}{2}\left(\dfrac{t}{2t+1}\right)^{1/2}\left[\dfrac{(2t+1)(1)-t(2)}{(2t+1)^2}\right]$, or $\dfrac{3t^{1/2}}{2(2t+1)^{5/2}}$

37. $\dfrac{1}{2}\left(\dfrac{u+1}{3u+2}\right)^{-1/2}\left[\dfrac{(3u+2)(1)-(u+1)(3)}{(3u+2)^2}\right]$, or $-\dfrac{1}{2\sqrt{u+1}\,(3u+2)^{3/2}}$

39. $\dfrac{(x^2-1)^4(2x)-x^2(4)(x^2-1)^3(2x)}{(x^2-1)^8}$, or $\dfrac{(-2x)(3x^2+1)}{(x^2-1)^5}$

41. $\dfrac{2x(x^2-1)^3(3x^2+1)^2[9(x^2-1)-4(3x^2+1)]}{(x^2-1)^8}$, or $-\dfrac{2x(3x^2+13)(3x^2+1)^2}{(x^2-1)^5}$

43. $\dfrac{(2x+1)^{-1/2}[(x^2-1)-(2x+1)(2x)]}{(x^2-1)^2}$, or $-\dfrac{3x^2+2x+1}{\sqrt{2x+1}(x^2-1)^2}$

45. $\dfrac{(t^2+1)^{1/2}(\frac{1}{2})(t+1)^{-1/2}(1)-(t+1)^{1/2}(\frac{1}{2})(t^2+1)^{-1/2}(2t)}{t^2+1}$, or $-\dfrac{t^2+2t-1}{2\sqrt{t+1}(t^2+1)^{3/2}}$

47. $4(3x+1)^3(3)(x^2-x+1)^3+(3x+1)^4(3)(x^2-x+1)^2(2x-1)$, or $3(3x+1)^3(x^2-x+1)^2(10x^2-5x+3)$

49. $\frac{4}{3}u^{1/3}$; $6x$; $8x(3x^2-1)^{1/3}$

51. $-\dfrac{2}{3u^{5/3}}$; $6x^2-1$; $-\dfrac{2(6x^2-1)}{3(2x^3-x+1)^{5/3}}$

53. $\frac{1}{2}u^{-1/2}-\frac{1}{2}u^{-3/2}$; $3x^2-1$; $\dfrac{(3x^2-1)(x^3-x-1)}{2(x^3-x)^{3/2}}$

55. -12 57. 6 59. No 61. $y=-33x+57$

63. $y=\frac{43}{5}x-\frac{54}{5}$

65. 0.333 million/wk; 0.305 million/wk; 16 million; 22.7 million

67. $\dfrac{6.87775}{(5+t)^{0.795}}$; 0.53%/yr; 64.9%

69. **a.** \$8.7 billion/yr **b.** \$92.3 billion

71. **a.** $0.027(0.2t^2+4t+64)^{-1/3}(0.1t+1)$ **b.** 0.0091 ppm

73. **a.** $0.03[3t^2(t-7)^4+t^3(4)(t-7)^3]$, or $0.21t^2(t-3)(t-7)^3$
b. 90.72; 0; -90.72; at 8 a.m. the level of nitrogen dioxide is increasing; at 10 a.m. the level stops increasing; at 11 a.m. the level is decreasing.

75.
$$300\left[\frac{(t+25)\frac{1}{2}(\frac{1}{2}t^2+2t+25)^{-1/2}(t+2)-(\frac{1}{2}t^2+2t+25)^{1/2}(1)}{(t+25)^2}\right],$$
or $\dfrac{3450t}{(t+25)^2\sqrt{\frac{1}{2}t^2+2t+25}}$; 2.9 beats/min², 0.7 beats/min², 0.2 beats/min², 179 beats/min

77. 160π ft²/sec 79. -27 mph/decade; 19 mph

81.
$$(1.42)\left[\frac{(3t^2+80t+550)(14t+140)-(7t^2+140t+700)(6t+80)}{(3t^2+80t+550)^2}\right],$$
or $\dfrac{1.42(140t^2+3500t+21{,}000)}{(3t^2+80t+550)^2}$; 31,312 jobs/yr

83. -400 wristwatches/(dollar price increase)

85. True 87. True

Using Technology Exercises 3.3, page 197

1. 0.5774 3. 0.9390 5. -4.9498

7. 10,146,200/decade; 7,810,520/decade

Exercises 3.4, page 208

1. **a.** $C(x)$ is always increasing because as the number of units x produced increases, the amount of money that must be spent on production also increases.
b. 4000

3. **a.** \$1.80; \$1.60 **b.** \$1.80; \$1.60

5. **a.** $100+\dfrac{200{,}000}{x}$ **b.** $-\dfrac{200{,}000}{x^2}$
c. $\bar{C}(x)$ approaches \$100 if the production level is very high.

7. $\dfrac{2000}{x}+2-0.0001x$; $-\dfrac{2000}{x^2}-0.0001$

9. **a.** $8000-200x$ **b.** 200, 0, -200 **c.** \$40

11. **a.** $-0.04x^2+600x-300{,}000$
b. $-0.08x+600$ **c.** 200; -40
d. The profit increases as production increases, peaking at 7500 units; beyond this level, profit falls.

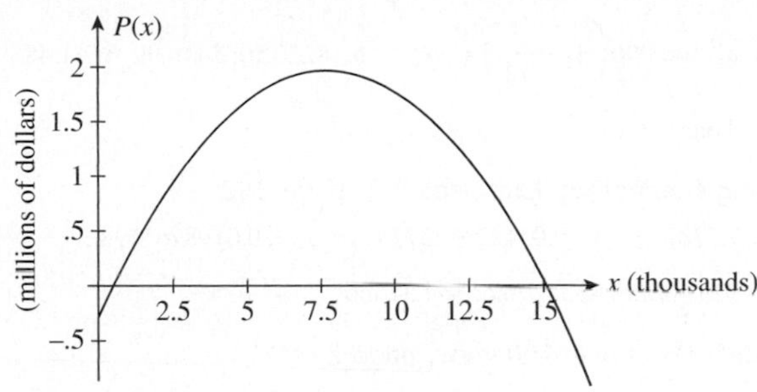

13. **a.** $600x - 0.05x^2$; $-0.000002x^3 - 0.02x^2 + 200x - 80{,}000$
b. $0.000006x^2 - 0.06x + 400$; $600 - 0.1x$; $-0.000006x^2 - 0.04x + 200$
c. 304, 400; 96
d.

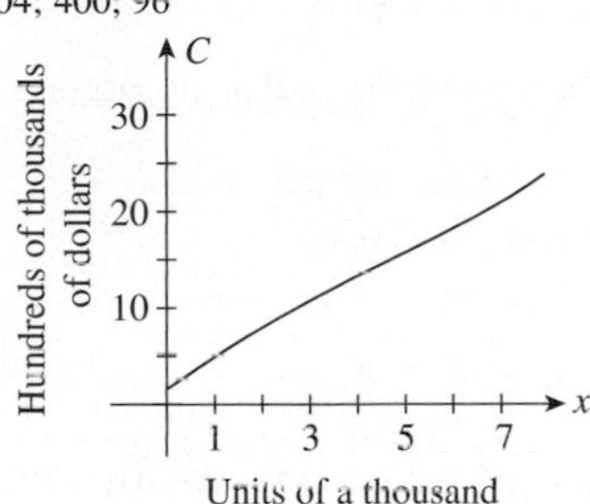

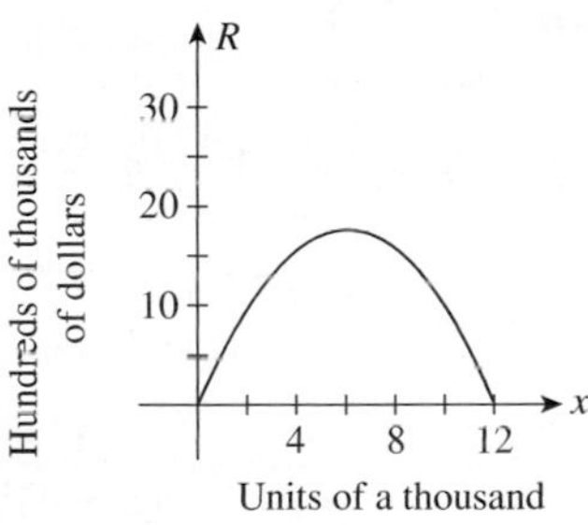

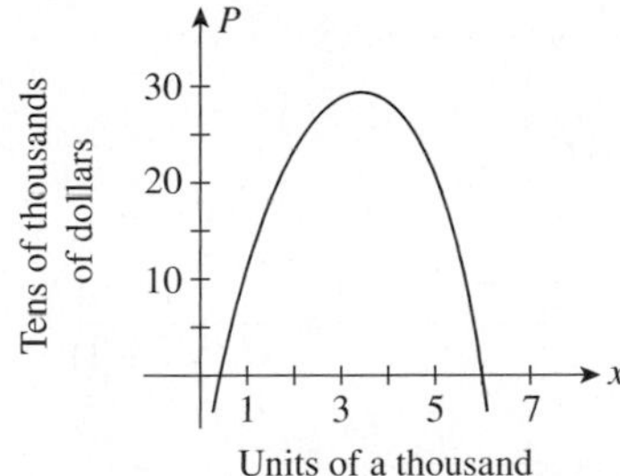

15. $0.000002x^2 - 0.03x + 400 + \dfrac{80{,}000}{x}$
a. $0.000004x - 0.03 - \dfrac{80{,}000}{x^2}$
b. -0.0132; 0.0092; the marginal average cost is negative (average cost is decreasing) when 5000 units are produced and positive (average cost is increasing) when 10,000 units are produced.

c.

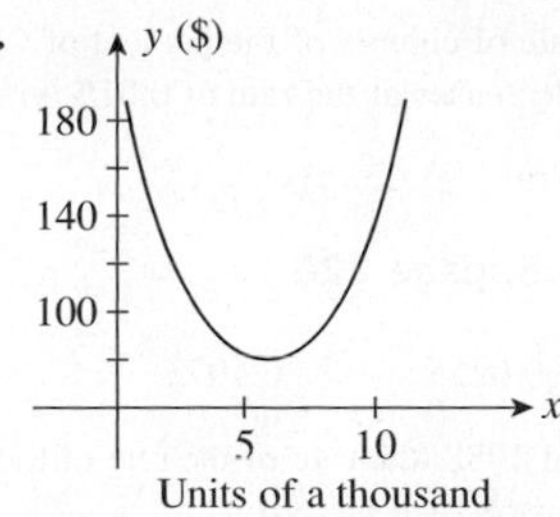

17. **a.** $\dfrac{50x}{0.01x^2 + 1}$ **b.** $\dfrac{50 - 0.5x^2}{(0.01x^2 + 1)^2}$
c. \$44,380; when the level of production is 2000 units, the revenue increases at the rate of \$44,380 per additional 1000 units produced.

19. \$1.21 billion/billion dollars 21. \$0.288 billion/billion dollars

23. $\frac{5}{3}$; elastic 25. 1; unitary 27. 0.104; inelastic

29. **a.** Inelastic; elastic **b.** When $p = 8.66$
c. Increase **d.** Increase

31. **a.** Inelastic **b.** Increase

33. $\dfrac{2p^2}{9 - p^2}$; for $p < \sqrt{3}$, demand is inelastic; for $p = \sqrt{3}$, demand is unitary; and for $p > \sqrt{3}$, demand is elastic.

35. True

Exercises 3.5, page 216

1. $8x - 2$; 8 3. $6x^2 - 6x$; $6(2x - 1)$

5. $4t^3 - 6t^2 + 12t - 3$; $12(t^2 - t + 1)$

7. $10x(x^2 + 2)^4$; $10(x^2 + 2)^3(9x^2 + 2)$

9. $6t(2t^2 - 1)(6t^2 - 1)$; $6(60t^4 - 24t^2 + 1)$

11. $14x(2x^2 + 2)^{5/2}$; $28(2x^2 + 2)^{3/2}(6x^2 + 1)$

13. $(x^2 + 1)(5x^2 + 1)$; $4x(5x^2 + 3)$

15. $\dfrac{1}{(2x + 1)^2}$; $-\dfrac{4}{(2x + 1)^3}$ 17. $\dfrac{2}{(s + 1)^2}$; $-\dfrac{4}{(s + 1)^3}$

19. $-\dfrac{3}{2(4 - 3u)^{1/2}}$; $-\dfrac{9}{4(4 - 3u)^{3/2}}$ 21. $72x - 24$ 23. $-\dfrac{6}{x^4}$

25. $\frac{81}{8}(3s - 2)^{-5/2}$ 27. $192(2x - 3)$ 29. 128 ft/sec; 32 ft/sec^2

31. **a.** and **b.**

t	0	1	2	3	4	5	6	7
$N'(t)$	0	2.7	4.8	6.3	7.2	7.5	7.2	6.3
$N''(t)$					0.6	0	−0.6	−1.2

33. 8.1 million; 0.204 million/yr; -0.03 million/yr^2. At the beginning of 1998, there were 8.1 million people receiving disability benefits; the number was increasing at the rate of 0.2 million/yr; the rate of the rate of change of the number of people was decreasing at the rate of 0.03 million people/yr^2.

35. **a.** $\frac{1}{4}t^3 - 3t^2 + 8t$ **b.** 0 ft/sec; 0 ft/sec; 0 ft/sec
c. $\frac{3}{4}t^2 - 6t + 8$ **d.** 8 ft/sec^2; -4 ft/sec^2; 8 ft/sec^2
e. 0 ft; 16 ft; 0 ft

37. $-0.01\%/\text{yr}^2$. The rate of the rate of change of the percent of Americans aged 55 and over decreases at the rate of $0.01\%/\text{yr}^2$.

39. False 41. True 43. True 45. $f(x) = x^{n+1/2}$

Using Technology Exercises 3.5, page 220

1. -18 3. 15.2762 5. -0.6255 7. 0.1973

9. -68.46214; at the beginning of 1988, the rate of the rate of the rate at which banks were failing was 68 banks/yr/yr/yr.

Exercises 3.6, page 229

1. a. $-\frac{1}{2}$ b. $-\frac{1}{2}$ 3. a. $-\dfrac{1}{x^2}$ b. $-\dfrac{y}{x}$

5. a. $2x - 1 + \dfrac{4}{x^2}$ b. $3x - 2 - \dfrac{y}{x}$

7. a. $\dfrac{1 - x^2}{(1 + x^2)^2}$ b. $-2y^2 + \dfrac{y}{x}$ 9. $-\dfrac{x}{y}$ 11. $\dfrac{x}{2y}$

13. $1 - \dfrac{y}{x}$ 15. $-\dfrac{y}{x}$ 17. $-\dfrac{\sqrt{y}}{\sqrt{x}}$ 19. $2\sqrt{x+y} - 1$

21. $-\dfrac{y^3}{x^3}$ 23. $\dfrac{2\sqrt{xy} - y}{x - 2\sqrt{xy}}$ 25. $\dfrac{6x - 3y - 1}{3x + 1}$

27. $\dfrac{2(2x - y^{3/2})}{3x\sqrt{y} - 4y}$ 29. $-\dfrac{2x^2 + 2xy + y^2}{x^2 + 2xy + 2y^2}$ 31. $y = 2$

33. $y = -\frac{3}{2}x + \frac{5}{2}$ 35. $\dfrac{2y}{x^2}$ 37. $\dfrac{2y(y - x)}{(2y - x)^3}$

39. a. $\dfrac{dV}{dt} = \pi r\left(r\dfrac{dh}{dt} + 2h\dfrac{dr}{dt}\right)$ b. 3.6π cu in./sec

41. Dropping at the rate of 111 tires/wk

43. Increasing at the rate of 44 ten packs/wk

45. Dropping at the rate of 3.7¢/carton/wk 47. 0.37; inelastic

49. 160π ft²/sec 51. 17 ft/sec 53. 7.69 ft/sec

57. 196.8 ft/sec 59. 9 ft/sec 61. 19.2 ft/sec 63. True

Exercises 3.7, page 237

1. $4x\,dx$ 3. $(3x^2 - 1)\,dx$ 5. $\dfrac{dx}{2\sqrt{x+1}}$ 7. $\dfrac{6x + 1}{2\sqrt{x}}\,dx$

9. $\dfrac{x^2 - 2}{x^2}\,dx$ 11. $\dfrac{-x^2 + 2x + 1}{(x^2 + 1)^2}\,dx$ 13. $\dfrac{6x - 1}{2\sqrt{3x^2 - x}}\,dx$

15. a. $2x\,dx$ b. 0.04 c. 0.0404

17. a. $-\dfrac{dx}{x^2}$ b. -0.05 c. -0.05263

19. 3.167 21. 7.0357 23. 1.983 25. 0.298

27. 2.50375 29. ± 8.64 cm³ 31. 18.85 ft³

33. It will drop by 40%. 35. 274 sec 37. 111,595

39. Decrease of \$1.33 41. $\pm$\$64,800

43. Decrease of 11 crimes/yr

45. a. $100{,}000\left(1 + \dfrac{r}{12}\right)^{119} dr$ b. \$220.50; \$440.99; \$661.48

47. True

Using Technology Exercises 3.7, page 242

1. 7.5787 3. 0.031220185778 5. -0.0198761598

7. \$26.60/mo; \$35.47/mo; \$44.34/mo 9. 625

Chapter 3 Concept Review, page 243

1. a. 0 b. nx^{n-1} c. $cf'(x)$ d. $f'(x) \pm g'(x)$

2. a. $f(x)g'(x) + g(x)f'(x)$ b. $\dfrac{g(x)f'(x) - f(x)g'(x)}{[g(x)]^2}$

3. a. $g'[f(x)]f'(x)$ b. $n[f(x)]^{n-1}f'(x)$

4. Marginal cost; marginal revenue; marginal profit; marginal average cost

5. a. $-p\dfrac{f'(p)}{f(p)}$ b. Elastic; unitary; inelastic

6. Both sides; $\dfrac{dy}{dx}$ 7. y; $\dfrac{dy}{dt}$; a 8. $-\dfrac{x}{y}$; $-\dfrac{y}{x}$

9. a. $x_2 - x_1$ b. $f(x + \Delta x) - f(x)$ 10. Δx; Δx; x; $f'(x)\,dx$

Chapter 3 Review Exercises, page 243

1. $15x^4 - 8x^3 + 6x - 2$ 2. $24x^5 + 8x^3 + 6x$

3. $\dfrac{6}{x^4} - \dfrac{3}{x^2}$ 4. $4t - 9t^2 + \frac{1}{2}t^{-3/2}$ 5. $-\dfrac{1}{t^{3/2}} - \dfrac{6}{t^{5/2}}$

6. $2x - \dfrac{2}{x^2}$ 7. $1 - \dfrac{2}{t^2} - \dfrac{6}{t^3}$ 8. $4s + \dfrac{4}{s^2} - \dfrac{1}{s^{3/2}}$

9. $2x + \dfrac{3}{x^{5/2}}$ 10. $\dfrac{(2x - 1)(1) - (x + 1)(2)}{(2x - 1)^2}$, or $-\dfrac{3}{(2x - 1)^2}$

11. $\dfrac{(2t^2 + 1)(2t) - t^2(4t)}{(2t^2 + 1)^2}$, or $\dfrac{2t}{(2t^2 + 1)^2}$

12. $\dfrac{(t^{1/2} + 1)\frac{1}{2}t^{-1/2} - t^{1/2}(\frac{1}{2}t^{-1/2})}{(t^{1/2} + 1)^2}$, or $\dfrac{1}{2\sqrt{t}(\sqrt{t} + 1)^2}$

13. $\dfrac{(x^{1/2} + 1)(\frac{1}{2}x^{-1/2}) - (x^{1/2} - 1)(\frac{1}{2}x^{-1/2})}{(x^{1/2} + 1)^2}$ or $\dfrac{1}{\sqrt{x}(\sqrt{x} + 1)^2}$

14. $\dfrac{(2t^2 + 1)(1) - t(4t)}{(2t^2 + 1)^2}$, or $\dfrac{1 - 2t^2}{(2t^2 + 1)^2}$

15. $\dfrac{(x^2 - 1)(4x^3 + 2x) - (x^4 + x^2)(2x)}{(x^2 - 1)^2}$, or $\dfrac{2x(x^4 - 2x^2 - 1)}{(x^2 - 1)^2}$

16. $3(4x + 1)(2x^2 + x)^2$ 17. $8(3x^3 - 2)^7(9x^2)$, or $72x^2(3x^3 - 2)^7$

18. $5(x^{1/2} + 2)^4 \cdot \dfrac{1}{2}x^{-1/2}$, or $\dfrac{5(\sqrt{x} + 2)^4}{2\sqrt{x}}$

19. $\dfrac{1}{2}(2t^2 + 1)^{-1/2}(4t)$, or $\dfrac{2t}{\sqrt{2t^2 + 1}}$

20. $\frac{1}{3}(1 - 2t^3)^{-2/3}(-6t^2)$, or $-2t^2(1 - 2t^3)^{-2/3}$

21. $-4(3t^2 - 2t + 5)^{-3}(3t - 1)$, or $-\dfrac{4(3t - 1)}{(3t^2 - 2t + 5)^3}$

22. $-\frac{3}{2}(2x^3 - 3x^2 + 1)^{-5/2}(6x^2 - 6x)$, or $-9x(x - 1)(2x^3 - 3x^2 + 1)^{-5/2}$

23. $2\left(x + \dfrac{1}{x}\right)\left(1 - \dfrac{1}{x^2}\right)$, or $\dfrac{2(x^2 + 1)(x^2 - 1)}{x^3}$

24. $\dfrac{(2x^2 + 1)^2(1) - (1 + x)2(2x^2 + 1)(4x)}{(2x^2 + 1)^4}$, or $-\dfrac{6x^2 + 8x - 1}{(2x^2 + 1)^3}$

25. $(t^2 + t)^4(4t) + 2t^2 \cdot 4(t^2 + t)^3(2t + 1)$, or $4t^2(5t + 3)(t^2 + t)^3$

26. $(2x + 1)^3 \cdot 2(x^2 + x)(2x + 1) + (x^2 + x)^2\, 3(2x + 1)^2(2)$, or $2(2x + 1)^2(x^2 + x)(7x^2 + 7x + 1)$

27. $x^{1/2} \cdot 3(x^2 - 1)^2(2x) + (x^2 - 1)^3 \cdot \dfrac{1}{2}x^{-1/2}$, or $\dfrac{(13x^2 - 1)(x^2 - 1)^2}{2\sqrt{x}}$

28. $\dfrac{(x^3 + 2)^{1/2}(1) - x \cdot \frac{1}{2}(x^3 + 2)^{-1/2} \cdot 3x^2}{x^3 + 2}$, or $\dfrac{4 - x^3}{2(x^3 + 2)^{3/2}}$

29. $\dfrac{(4x - 3)\frac{1}{2}(3x + 2)^{-1/2}(3) - (3x + 2)^{1/2}(4)}{(4x - 3)^2}$, or $\dfrac{12x + 25}{2\sqrt{3x + 2}(4x - 3)^2}$

30. $\dfrac{(t + 1)^3\frac{1}{2}(2t + 1)^{-1/2}(2) - (2t + 1)^{1/2} \cdot 3(t + 1)^2(1)}{(t + 1)^6}$, or $-\dfrac{5t + 2}{\sqrt{2t + 1}(t + 1)^4}$

31. $2(12x^2 - 9x + 2)$ 32. $-\dfrac{1}{4x^{3/2}} + \dfrac{3}{4x^{5/2}}$

33. $\dfrac{(t^2 + 4)^2(-2t) - (4 - t^2)2(t^2 + 4)(2t)}{(t^2 + 4)^4}$, or $\dfrac{2t(t^2 - 12)}{(t^2 + 4)^3}$

34. $2(15x^4 + 12x^2 + 6x + 1)$

35. $2(2x^2 + 1)^{-1/2} + 2x\left(-\dfrac{1}{2}\right)(2x^2 + 1)^{-3/2}(4x)$, or $\dfrac{2}{(2x^2 + 1)^{3/2}}$

36. $(t^2 + 1)(14t) + (7t^2 + 1)(2)(t^2 + 1)(2t)$, or $6t(t^2 + 1)(7t^2 + 3)$

37. $\dfrac{2x}{y}$ 38. $\dfrac{2x^2 - y}{x}$ 39. $-\dfrac{2x}{y^2 - 1}$ 40. $-\dfrac{x(1 + 2y^2)}{y(2x^2 + 1)}$

41. $\dfrac{x - 2y}{2x + y}$ 42. $\dfrac{4y - 6xy - 1}{3x^2 - 4x - 2}$ 43. $\dfrac{2(x^4 - 1)}{x^3}\,dx$

44. $-\dfrac{3x^2}{2(x^3 + 1)^{3/2}}\,dx$

45. **a.** $\dfrac{2x}{\sqrt{2(x^2 + 2)}}\,dx$ **b.** 0.1333 **c.** 0.1335; differ by 0.0002

46. 2.9926

47. **a.** $(2, -25)$ and $(-1, 14)$ **b.** $y = -4x - 17$; $y = -4x + 10$

48. **a.** $(-2, \frac{25}{3})$ and $(1, -\frac{13}{6})$ **b.** $y = -2x + \frac{13}{3}$; $y = -2x - \frac{1}{6}$

49. $y = -\dfrac{\sqrt{3}}{3}x + \dfrac{4}{3}\sqrt{3}$ 50. $y = 112x - 80$

51. $-\dfrac{48}{(2x - 1)^4}$; $(-\infty; \frac{1}{2}) \cup (\frac{1}{2}, \infty)$

52. **a.** $\frac{1}{3}$; inelastic **b.** 1; unitary **c.** 3; elastic

53. $\dfrac{25}{2(25 - \sqrt{p})}$; for $p > 156.25$, demand is elastic; for $p = 156.25$, demand is unitary; and for $p < 156.25$, demand is inelastic.

54. **a.** Inelastic **b.** Increase 55. **a.** Elastic **b.** Decrease

56. **a.** 15%; 31.99% **b.** 0.51%/yr; 1.04%/yr

57. 200 subscribers/wk

58. $-14.346(1 + t)^{-1.45}$; -2.92¢/min; 19.45¢/min

59. ≈ 75 yr; 0.07 yr/yr

60. **a.** \$2.20; \$2.20 **b.** $\dfrac{2500}{x} + 2.2$; $-\dfrac{2500}{x^2}$
c. $\lim\limits_{x \to \infty}\left(\dfrac{2500}{x} + 2.2\right) = 2.2$

61. **a.** $-0.02x^2 + 600x$ **b.** $-0.04x + 600$
c. 200; the sale of the 10,001st phone will bring a revenue of \$200.

62. **a.** $2000x - 0.04x^2$; $-0.000002x^3 - 0.02x^2 + 1000x - 120{,}000$; $0.000002x^2 - 0.02x + 1000 + \dfrac{120{,}000}{x}$
b. $0.000006x^2 - 0.04x + 1000$; $2000 - 0.08x$; $-0.000006x^2 - 0.04x + 1000$, $0.0000004x - 0.02 - \dfrac{120{,}000}{x^2}$
c. 934; 1760; 826
d. -0.0048; 0.010125; at a production level of 5000, the average cost is decreasing by 0.48¢/unit; at a production level of 8000, the average cost is increasing by 1.0125¢/unit.

Chapter 3 Before Moving On, page 245

1. $6x^2 - \dfrac{1}{x^{2/3}} - \dfrac{10}{3x^{5/3}}$ 2. $\dfrac{4x^2 - 1}{\sqrt{2x^2 - 1}}$ 3. $-\dfrac{2x^2 + 2x - 1}{(x^2 + x + 1)^2}$

4. $-\dfrac{1}{2(x + 1)^{3/2}}$; $\dfrac{3}{4(x + 1)^{5/2}}$; $-\dfrac{15}{8(x + 1)^{7/2}}$ 5. $\dfrac{-y^2 + 2xy - 3x^2}{2xy - x^2}$

6. **a.** $\dfrac{2x^2 + 5}{\sqrt{x^2 + 5}}\,dx$ **b.** 0.0433

CHAPTER 4

Exercises 4.1, page 258

1. Decreasing on $(-\infty, 0)$ and increasing on $(0, \infty)$
3. Increasing on $(-\infty, -1) \cup (1, \infty)$ and decreasing on $(-1, 1)$
5. Decreasing on $(-\infty, 0) \cup (2, \infty)$ and increasing on $(0, 2)$
7. Decreasing on $(-\infty, -1) \cup (1, \infty)$ and increasing on $(-1, 1)$
9. Increasing on $(20.2, 20.6) \cup (21.7, 21.8)$, constant on $(19.6, 20.2) \cup (20.6, 21.1)$, and decreasing on $(21.1, 21.7) \cup (21.8, 22.7)$
11. Increasing on $(-\infty, \infty)$

13. Decreasing on $(-\infty, \frac{3}{2})$ and increasing on $(\frac{3}{2}, \infty)$

15. Decreasing on $(-\infty, -\sqrt{3}/3) \cup (\sqrt{3}/3, \infty)$ and increasing on $(-\sqrt{3}/3, \sqrt{3}/3)$

17. Increasing on $(-\infty, -2) \cup (0, \infty)$ and decreasing on $(-2, 0)$

19. Increasing on $(-\infty, 3) \cup (3, \infty)$

21. Decreasing on $(-\infty, 0) \cup (0, 3)$ and increasing on $(3, \infty)$

23. Decreasing on $(-\infty, 2) \cup (2, \infty)$

25. Decreasing on $(-\infty, 1) \cup (1, \infty)$

27. Increasing on $(-\infty, 0) \cup (0, \infty)$ 29. Increasing on $(-1, \infty)$

31. Increasing on $(-4, 0)$; decreasing on $(0, 4)$

33. Increasing on $(-\infty, 0) \cup (0, \infty)$

35. Relative maximum: $f(0) = 1$; relative minima: $f(-1) = 0$ and $f(1) = 0$

37. Relative maximum: $f(-1) = 2$; relative minimum: $f(1) = -2$

39. Relative maximum: $f(1) = 3$; relative minimum: $f(2) = 2$

41. Relative minimum: $f(0) = 2$ 43. a 45. d

47. Relative minimum: $f(2) = -4$

49. Relative maximum: $f(3) = 15$ 51. None

53. Relative maximum: $g(0) = 4$; relative minimum: $g(2) = 0$

55. Relative maximum: $f(0) = 0$; relative minimum: $f(-1) = -\frac{1}{2}$ and $f(1) = -\frac{1}{2}$

57. Relative minimum: $F(3) = -5$; relative maximum: $F(-1) = \frac{17}{3}$

59. Relative minimum: $g(3) = -19$ 61. None

63. Relative maximum: $f(-3) = -4$; relative minimum: $f(3) = 8$

65. Relative maximum: $f(1) = \frac{1}{2}$; relative minimum: $f(-1) = -\frac{1}{2}$

67. Relative minimum: $f(1) = 0$

69. 

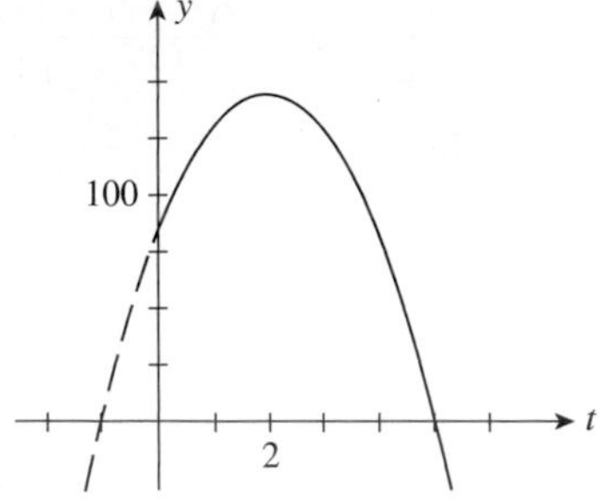

Rising in the time interval (0, 2); falling in the time interval (2, 5); when $t = 5$ sec

71. The percent of the U.S. population age 65 and over afflicted by the disease increases with age.

73. Rising on (0, 33) and descending on (33, T) for some positive number T.

75. f is decreasing on (0, 1) and increasing on (1, 4). The average speed decreases from 6 a.m. to 7 a.m. and then picks up from 7 a.m. to 10 a.m.

77. **a.** Increasing on (0, 6) **b.** Sales will be increasing.

79. **a.** Decreasing on (0, 21.4); increasing on (21.4, 30)
b. The percent of men 65 years and older in the workforce was decreasing from 1970 until mid-1991 and increasing from mid-1991 through 2000.

83. Spending was increasing from 2001 to 2006.

85. Increasing on (0, 1) and decreasing on (1, 4)

87. Increasing on (0, 4.5) and decreasing on (4.5, 11); the pollution is increasing from 7 a.m. to 11:30 a.m. and decreasing from 11:30 a.m. to 6 p.m.

89. **a.** $0.0021t^2 - 0.0061t + 0.1$
b. Decreasing on (0, 1.5) and increasing on (1.5, 15). The gap (shortage of nurses) was decreasing from 2000 to mid-2001 and is expected to be increasing from mid-2001 to 2015.
c. (1.5, 0.096). The gap was smallest ($\approx$ 96,000) in mid-2001.

91. True 93. True 95. False

99. **a.** $-2x$ if $x \neq 0$ **b.** No

Using Technology Exercises 4.1, page 226

1. **a.** f is decreasing on $(-\infty, -0.2934)$ and increasing on $(-0.2934, \infty)$.
b. Relative minimum: $f(-0.2934) = -2.5435$

3. **a.** f is increasing on $(-\infty, -1.6144) \cup (0.2390, \infty)$ and decreasing on $(-1.6144, 0.2390)$.
b. Relative maximum: $f(-1.6144) = 26.7991$; relative minimum: $f(0.2390) = 1.6733$

5. **a.** f is decreasing on $(-\infty, -1) \cup (0.33, \infty)$ and increasing on $(-1, 0.33)$.
b. Relative maximum: $f(0.33) = 1.11$; relative minimum: $f(-1) = -0.63$

7. **a.** f is decreasing on $(-1, -0.71)$ and increasing on $(-0.71, 1)$.
b. Relative minimum: $f(-0.71) = -1.41$

9. **a.**

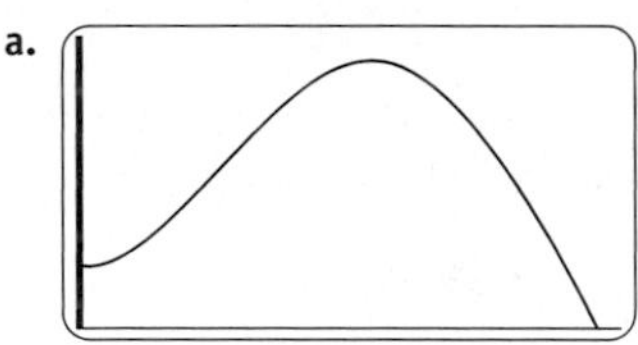

b. f is decreasing on $(0, 0.2398) \cup (6.8758, 12)$ and increasing on $(0.2398, 6.8758)$.
c. (6.8758, 200.14)

11. **a.**

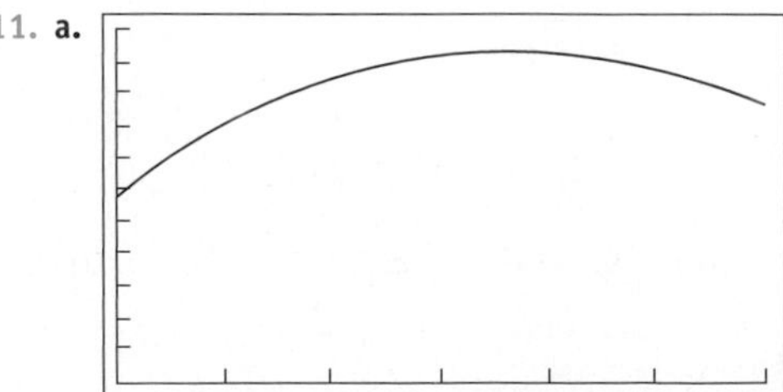

b. Increasing on (0, 3.6676) and decreasing on (3.6676, 6)

13. f is decreasing on (0, 1) and increasing on (1, 4); relative minimum: 32. The speed of the traffic flow drops from 6 a.m. to 7 a.m., reaching a low of 32 mph. Thereafter it increases till 10 a.m.

Exercises 4.2, page 276

1. Concave downward on $(-\infty, 0)$ and concave upward on $(0, \infty)$; inflection point: (0, 0)

3. Concave downward on $(-\infty, 0) \cup (0, \infty)$

5. Concave upward on $(-\infty, 0) \cup (1, \infty)$ and concave downward on (0, 1); inflection points: (0, 0) and $(1, -1)$

7. Concave downward on $(-\infty, -2) \cup (-2, 2) \cup (2, \infty)$

9. a **11.** b

13. **a.** $D_1'(t) > 0$, $D_2'(t) > 0$, $D_1''(t) > 0$, and $D_2''(t) < 0$ on (0, 12)
b. With or without the proposed promotional campaign, the deposits will increase; with the promotion, the deposits will increase at an increasing rate; without the promotion, the deposits will increase at a decreasing rate.

15. At the time t_0, corresponding to its t-coordinate, the restoration process is working at its peak.

21. Concave upward on $(-\infty, \infty)$

23. Concave downward on $(-\infty, 0)$; concave upward on $(0, \infty)$

25. Concave upward on $(-\infty, 0) \cup (3, \infty)$; concave downward on (0, 3)

27. Concave downward on $(-\infty, 0) \cup (0, \infty)$

29. Concave downward on $(-\infty, 4)$

31. Concave downward on $(-\infty, 2)$; concave upward on $(2, \infty)$

33. Concave upward on $(-\infty, -\sqrt{6}/3) \cup (\sqrt{6}/3, \infty)$; concave downward on $(-\sqrt{6}/3, \sqrt{6}/3)$

35. Concave downward on $(-\infty, 1)$; concave upward on $(1, \infty)$

37. Concave upward on $(-\infty, 0) \cup (0, \infty)$

39. Concave upward on $(-\infty, 2)$; concave downward on $(2, \infty)$

41. $(0, -2)$ **43.** $(1, -15)$ **45.** (0, 1) and $(\frac{2}{3}, \frac{11}{27})$

47. (0, 0) **49.** (1, 2) **51.** $(-\sqrt{3}/3, 3/2)$ and $(\sqrt{3}/3, 3/2)$

53. Relative maximum: $f(1) = 5$ **55.** None

57. Relative maximum: $f(-1) = -\frac{22}{3}$, relative minimum: $f(5) = -\frac{130}{5}$

59. Relative maximum: $g(-3) = -6$; relative minimum: $g(3) = 6$

61. None **63.** Relative minimum: $f(-2) = 12$

65. Relative maximum: $g(1) = \frac{1}{2}$; relative minimum: $g(-1) = -\frac{1}{2}$

67. Relative maximum: $f(0) = 0$; relative minimum: $f(\frac{4}{3}) = \frac{256}{27}$

69.

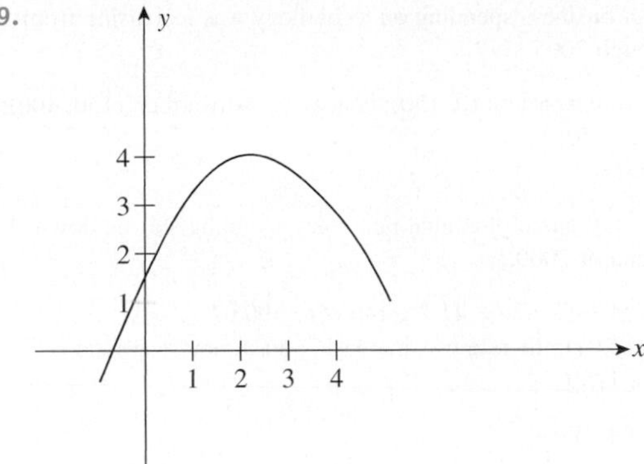

71.

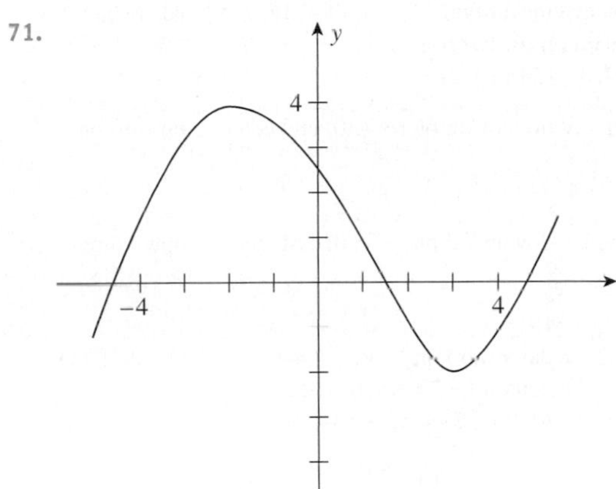

73.

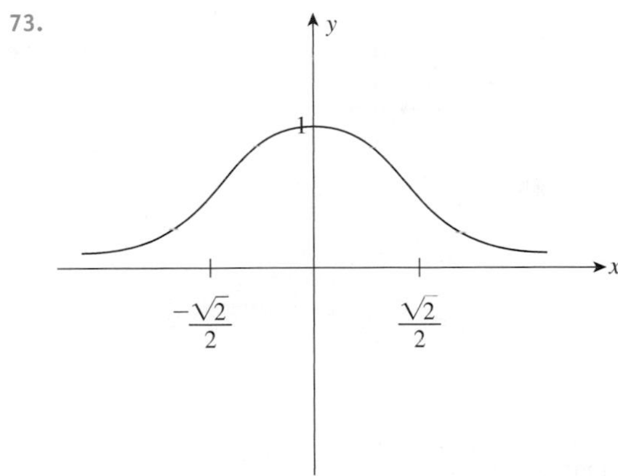

75. **a.** N is increasing on (0, 12).
b. $N''(t) < 0$ on (0, 6) and $N''(t) > 0$ on (6, 12)
c. The rate of growth of the number of help-wanted advertisements was decreasing over the first 6 mo of the year and increasing over the last 6 mo.

77. $f(t)$ increases at an increasing rate until the water level reaches the middle of the vase at which time (corresponding to the inflection point) $f(t)$ is increasing at the fastest rate. After that, $f(t)$ increases at a decreasing rate until the vase is filled.

79. **b.** The rate was increasing. **81.** 10 a.m.

83. The rate of business spending on technology was increasing from 2000 through 2005.

85. **a.** Concave upward on (0, 150); concave downward on (150, 400); (150, 28,550)
b. \$140,000

89. (1.9, 784.9); the rate of annual pharmacy spending slowed down near the end of 2000.

91. **a.** $74.925t^2 - 99.62t + 41.25$; $149.85t - 99.62$
c. (0.66, 12.91); the rate was increasing least rapidly around August 1999.

97. True 99. True

Using Technology Exercises 4.2, page 283

1. **a.** f is concave upward on $(-\infty, 0) \cup (1.1667, \infty)$ and concave downward on (0, 1.1667).
b. (1.1667, 1.1153); (0, 2)

3. **a.** f is concave downward on $(-\infty, 0)$ and concave upward on $(0, \infty)$.
b. (0, 2)

5. **a.** f is concave downward on $(-\infty, 0)$ and concave upward on $(0, \infty)$.
b. (0, 0)

7. **a.** f is concave downward on $(-\infty, -2.4495) \cup (0, 2.4495)$ and concave upward on $(-2.4495, 0) \cup (2.4495, \infty)$.
b. (2.4495, 0.3402); (−2.4495, −0.3402)

9. **a.**

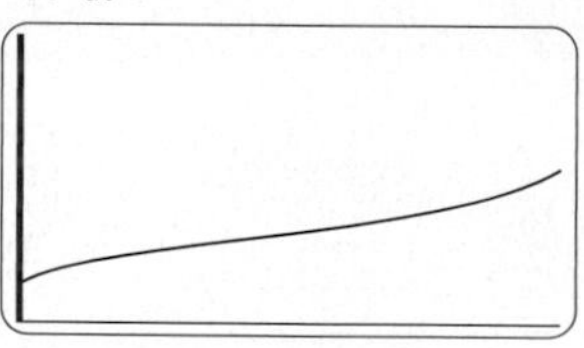

b. (5.5318, 35.9483)
c. $t = 5.5318$

11. **a.**

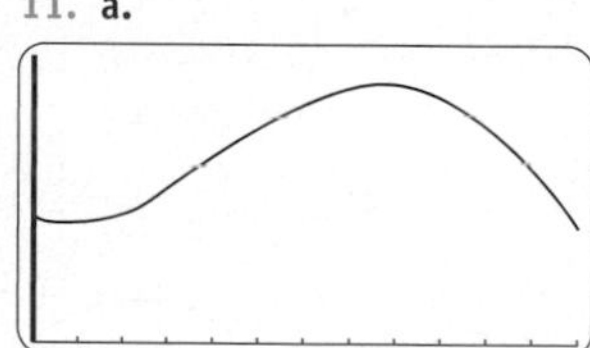

b. (3.9024, 77.0919)

13. **a.**

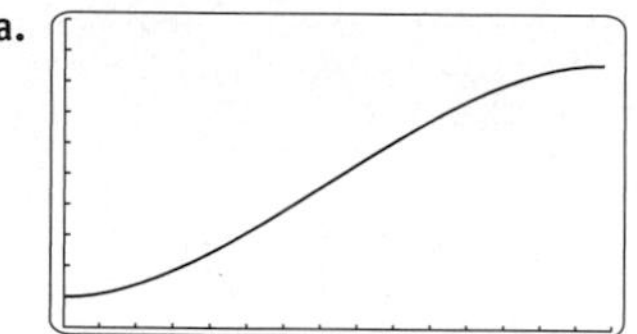

b. April 1993 ($t = 7.36$)

Exercises 4.3, page 293

1. Horizontal asymptote: $y = 0$

3. Horizontal asymptote: $y = 0$; vertical asymptote: $x = 0$

5. Horizontal asymptote: $y = 0$; vertical asymptotes: $x = -1$ and $x = 1$

7. Horizontal asymptote: $y = 3$; vertical asymptote: $x = 0$

9. Horizontal asymptotes: $y = 1$ and $y = -1$

11. Horizontal asymptote: $y = 0$; vertical asymptote: $x = 0$

13. Horizontal asymptote: $y = 0$; vertical asymptote: $x = 0$

15. Horizontal asymptote: $y = 1$; vertical asymptote: $x = -1$

17. None

19. Horizontal asymptote: $y = 1$; vertical asymptotes: $t = -3$ and $t = 3$

21. Horizontal asymptote: $y = 0$; vertical asymptotes: $x = -2$ and $x = 3$

23. Horizontal asymptote: $y = 2$; vertical asymptote: $t = 2$

25. Horizontal asymptote: $y = 1$; vertical asymptotes: $x = -2$ and $x = 2$

27. None 29. f is the derivative function of the function g.

31.

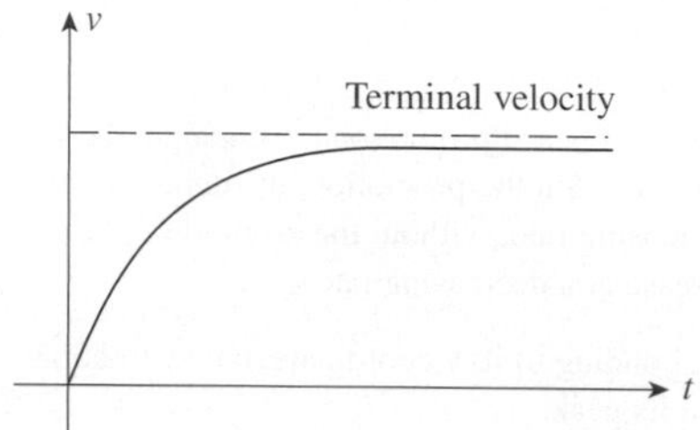

33.

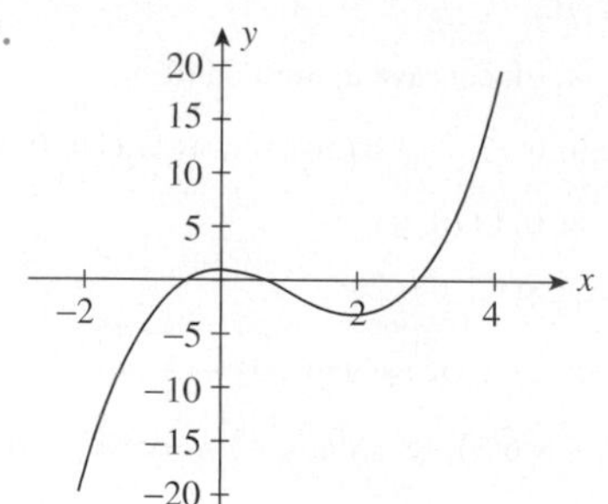

35.

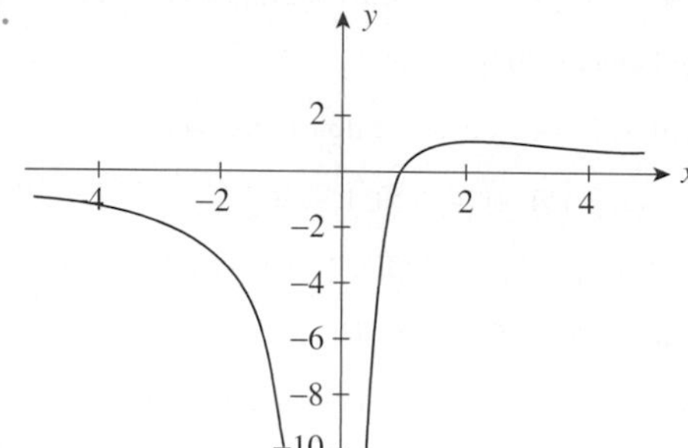

37.

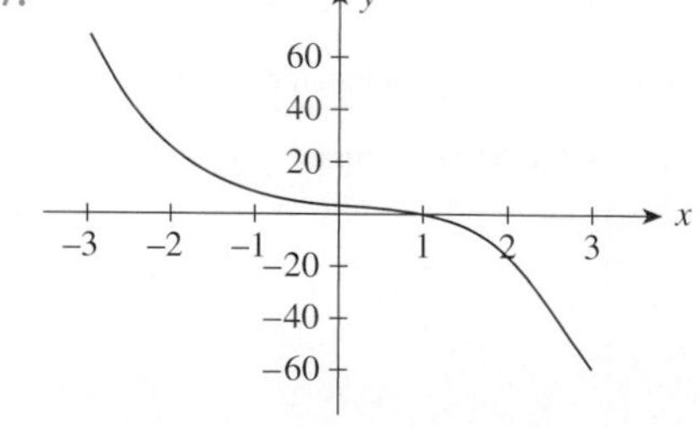

39.

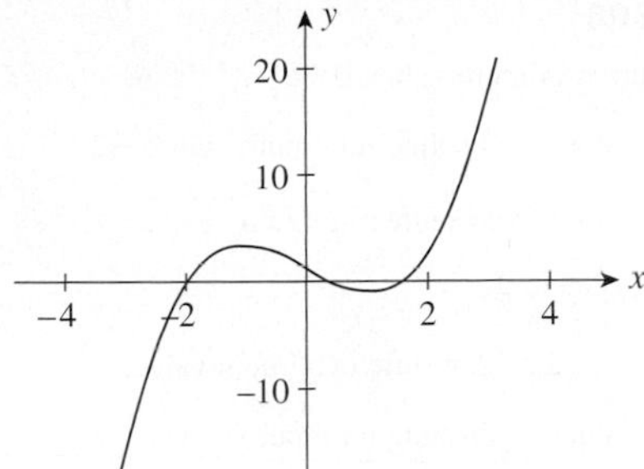

41.

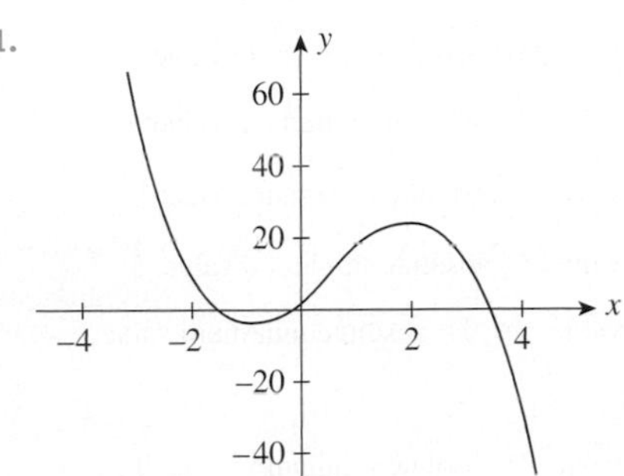

43.

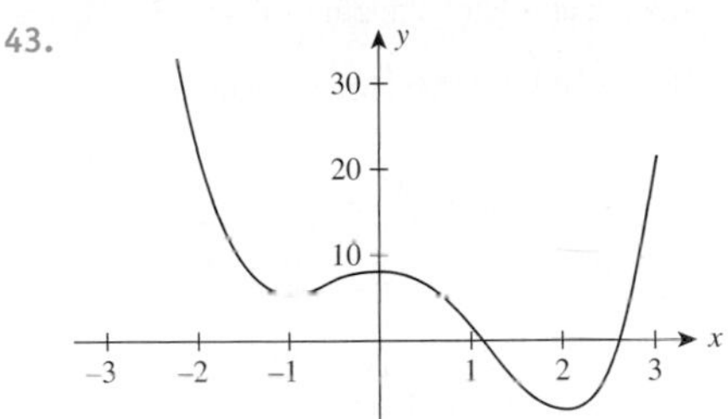

45.

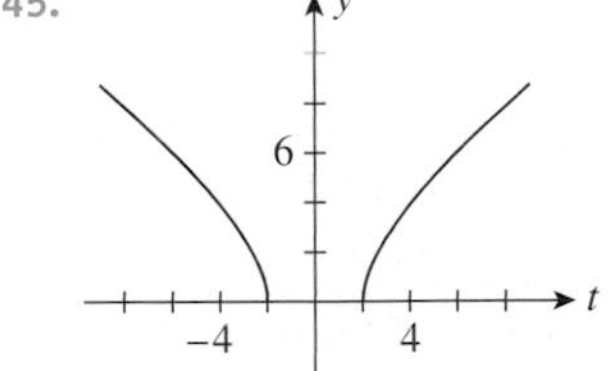

47.

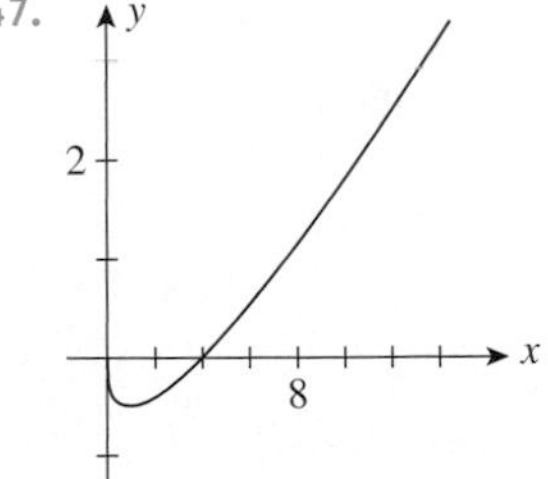

49.

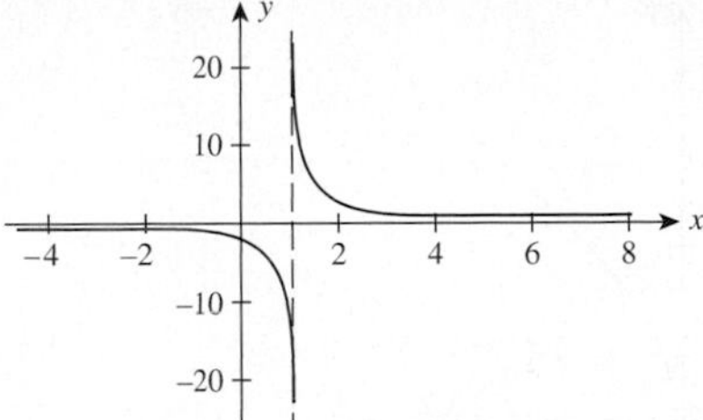

51.

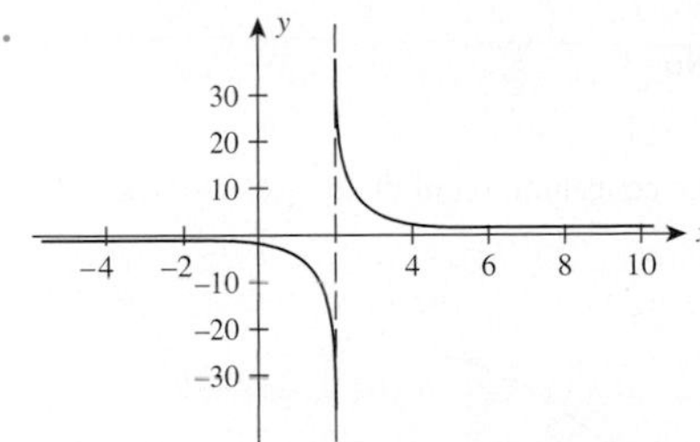

53.

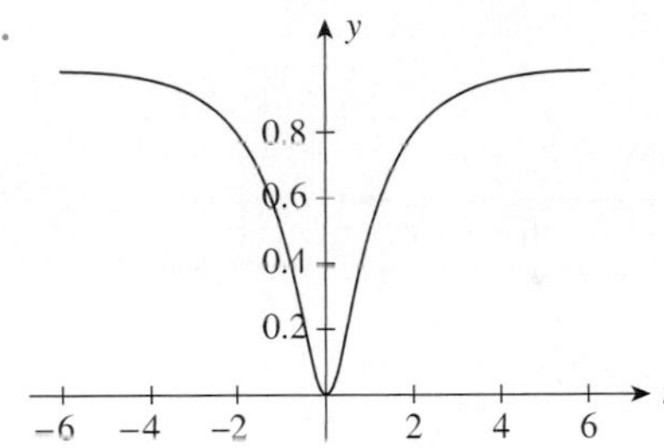

55.

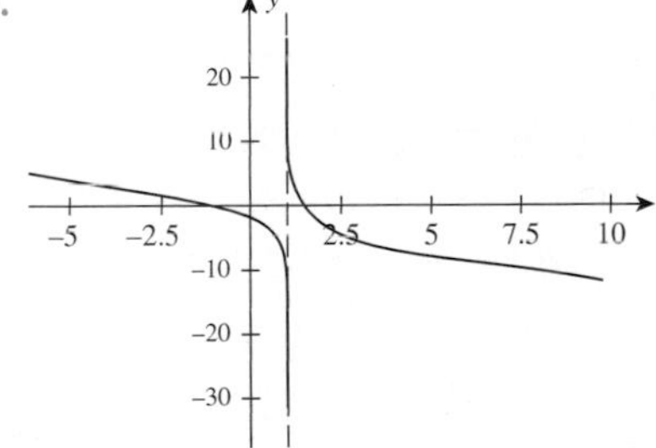

57.

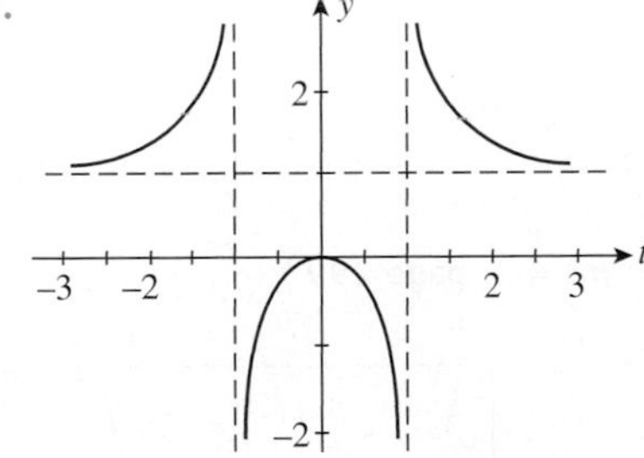

59.

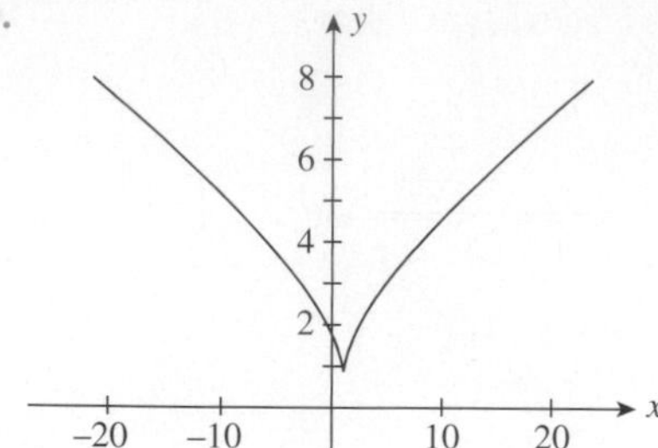

61. **a.** $x = 100$ **b.** No

63. **a.** $y = 0$
b. As time passes, the concentration of the drug decreases and approaches zero.

65.

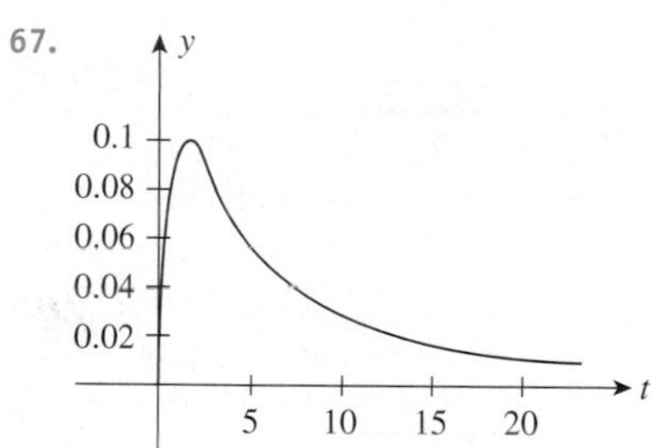

67.

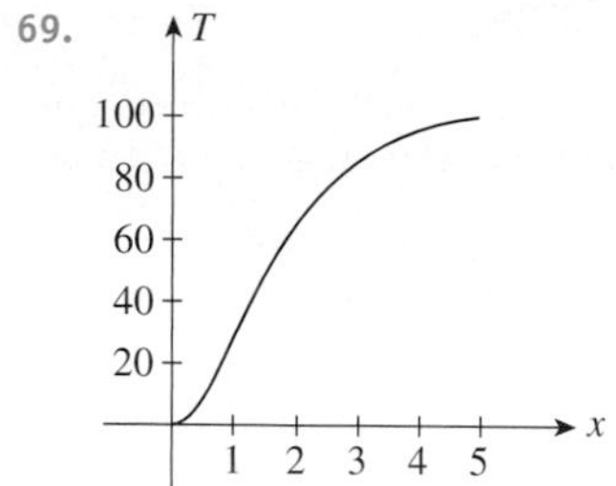

69.

Using Technology Exercises 4.3, page 299

1.

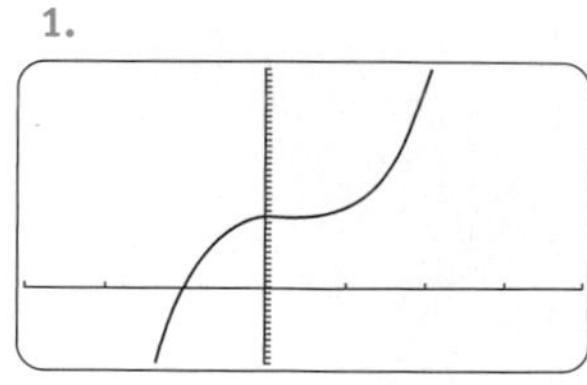

3.

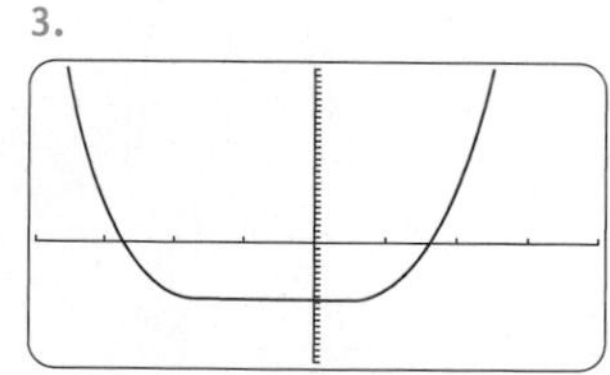

5. -0.9733; 2.3165, 4.6569 7. -1.1301; 2.9267

9. 1.5142

Exercises 4.4, page 308

1. None 3. Absolute minimum value: 0

5. Absolute maximum value: 3; absolute minimum value: -2

7. Absolute maximum value: 3; absolute minimum value: $-\frac{27}{16}$

9. Absolute minimum value: $-\frac{41}{8}$

11. No absolute extrema 13. Absolute maximum value: 1

15. Absolute maximum value: 5; absolute minimum value: -4

17. Absolute maximum value: 10; absolute minimum value: 1

19. Absolute maximum value: 19; absolute minimum value: -1

21. Absolute maximum value: 16; absolute minimum value: -1

23. Absolute maximum value: 3; absolute minimum value: $\frac{5}{3}$

25. Absolute maximum value: $\frac{37}{3}$; absolute minimum value: 5

27. Absolute maximum value ≈ 1.04; absolute minimum value: -1.5

29. No absolute extrema

31. Absolute maximum value: 1; absolute minimum value: 0

33. Absolute maximum value: 0; absolute minimum value: -3

35. Absolute maximum value: $\sqrt{2}/4 \approx 0.35$; absolute minimum value: $-\frac{1}{3}$

37. Absolute maximum value: $\sqrt{2}/2$; absolute minimum value: $-\sqrt{2}/2$

39. 144 ft 41. 17.72%

43. $f(6) = 3.60$, $f(0.5) = 1.13$; the number of nonfarm, full time, self-employed women over the time interval from 1963 to 1993 reached its highest level, 3.6 million, in 1993.

45. $3600 47. 6000 49. 3333

51. **a.** $0.0025x + 80 + \dfrac{10{,}000}{x}$ **b.** 2000 **c.** 2000 **d.** Same

53. 533

55. **a.** 2 days after the organic waste was dumped into the pond
b. 3.5 days after the organic waste was dumped into the pond

63. $52.79/sq ft

65. **a.** 2000; $105.8 billion **b.** 1995; $7.6 billion

69. False 71. False

75. **c.**

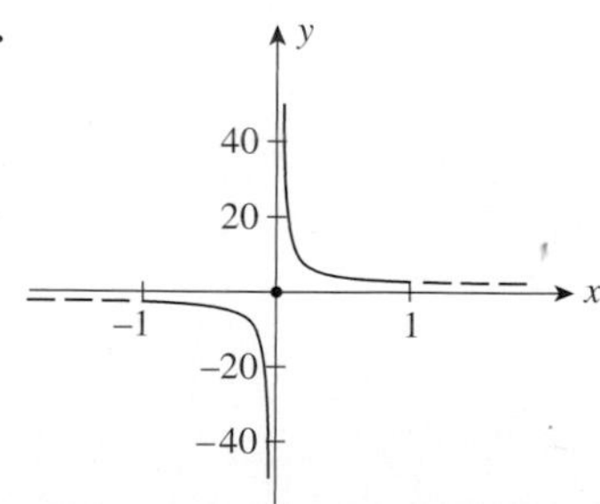

Using Technology Exercises 4.4, page 314

1. Absolute maximum value: 145.8985; absolute minimum value: -4.3834

3. Absolute maximum value: 16; absolute minimum value: -0.1257

5. Absolute maximum value: 2.8889; absolute minimum value: 0

7. **a.**

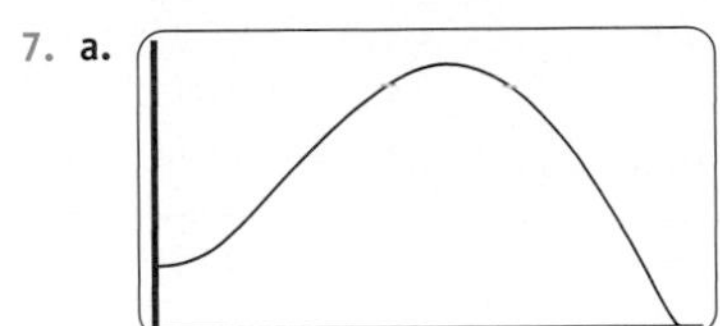

b. 200.1410 banks/yr

9. **a.**

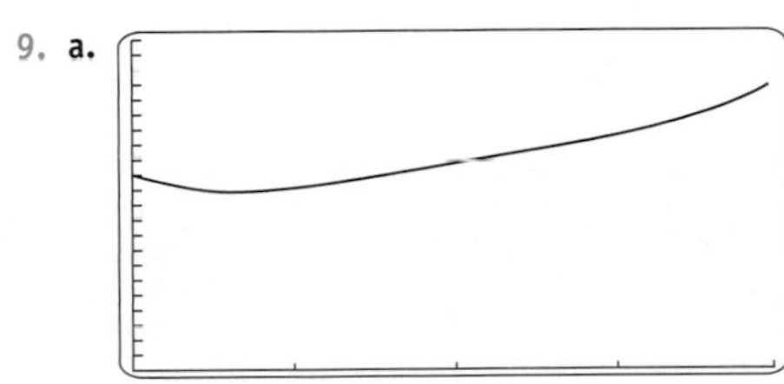

b. 21.51%

11. **b.** 1145

13. **a.**

b. 2029

Exercises 4.5, page 322

1. 750 yd × 1500 yd; 1,125,000 yd^2

3. $10\sqrt{2}$ ft × $40\sqrt{2}$ ft 5. $\frac{16}{3}$ in. × $\frac{16}{3}$ in. × $\frac{4}{3}$ in.

7. 5.04 in. × 5.04 in. × 5.04 in.

9. 18 in. × 18 in. × 36 in.; 11,664 in.^{3}

11. $r = \dfrac{36}{\pi}$ in.; $l = 36$ in.; $\dfrac{46{,}656}{\pi}$ in.^{3}

13. $\frac{2}{3}\sqrt[3]{9}$ ft × $\sqrt[3]{9}$ ft × $\frac{2}{5}\sqrt[3]{9}$ ft 15. 250; \$62,500; \$250

17. 85; \$28,900; \$340 19. $w \approx 13.86$ in.; $h \approx 19.60$ in.

21. $x = 2250$ ft 23. $x \approx 2.68$

25. 440 ft; 140 ft; 184,874 sq ft 27. 45, 44,445

Chapter 4 Concept Review, page 327

1. **a.** $f(x_1) < f(x_2)$ **b.** $f(x_1) > f(x_2)$

2. **a.** Increasing **b.** $f'(x) < 0$ **c.** Constant

3. **a.** $f(x) \le f(c)$ **b.** $f(x) \ge f(c)$

4. **a.** Domain; = 0; exist **b.** Critical number
   c. Relative extremum

5. **a.** $f'(x)$ **b.** > 0 **c.** Concavity
   d. Relative maximum; relative extremum

6. $\pm\infty$; $\pm\infty$ 7. 0; 0 8. b; b

9. **a.** $f(x) \le f(c)$; absolute maximum value
   b. $f(x) \ge f(c)$; open interval

10. Continuous; absolute; absolute

Chapter 4 Review Exercises, page 327

1. **a.** f is increasing on $(-\infty, 1) \cup (1, \infty)$ **b.** No relative extrema
   c. Concave down on $(-\infty, 1)$; concave up on $(1, \infty)$
   d. $(1, -\frac{17}{3})$

2. **a.** f is increasing on $(-\infty, 2) \cup (2, \infty)$ **b.** No relative extrema
   c. Concave down on $(-\infty, 2)$; concave up on $(2, \infty)$ **d.** $(2, 0)$

3. **a.** f is increasing on $(-1, 0) \cup (1, \infty)$ and decreasing on $(-\infty, -1) \cup (0, 1)$
   b. Relative maximum value: 0; relative minimum value: -1
   c. Concave up on $\left(-\infty, -\dfrac{\sqrt{3}}{3}\right) \cup \left(\dfrac{\sqrt{3}}{3}, \infty\right)$; concave down on $\left(-\dfrac{\sqrt{3}}{3}, \dfrac{\sqrt{3}}{3}\right)$
   d. $\left(-\dfrac{\sqrt{3}}{3}, -\dfrac{5}{9}\right)$; $\left(\dfrac{\sqrt{3}}{3}, -\dfrac{5}{9}\right)$

4. **a.** f is increasing on $(-\infty, -2) \cup (2, \infty)$ and decreasing on $(-2, 0) \cup (0, 2)$
   b. Relative maximum value: -4; relative minimum value: 4
   c. Concave down on $(-\infty, 0)$; concave up on $(0, \infty)$ **d.** None

5. **a.** f is increasing on $(-\infty, 0) \cup (2, \infty)$; decreasing on $(0, 1) \cup (1, 2)$
   b. Relative maximum value: 0; relative minimum value: 4
   c. Concave up on $(1, \infty)$; concave down on $(-\infty, 1)$ **d.** None

6. **a.** f is increasing on $(1, \infty)$ **b.** No relative extrema
   c. Concave down on $(1, \infty)$ **d.** None

7. **a.** f is decreasing on $(-\infty, 1) \cup (1, \infty)$ **b.** No relative extrema
   c. Concave down on $(-\infty, 1)$; concave up on $(1, \infty)$ **d.** $(1, 0)$

8. **a.** f is increasing on $(1, \infty)$ **b.** No relative extrema
   c. Concave down on $(1, \frac{4}{3})$; concave up on $(\frac{4}{3}, \infty)$
   d. $\left(\dfrac{4}{3}, \dfrac{4\sqrt{3}}{9}\right)$

9. **a.** f is increasing on $(-\infty, -1) \cup (-1, \infty)$
   b. No relative extrema
   c. Concave down on $(-1, \infty)$; concave up on $(-\infty, -1)$
   d. None

10. **a.** f is decreasing on $(-\infty, 0)$ and increasing on $(0, \infty)$
   b. Relative minimum value: -1
   c. Concave down on $\left(-\infty, -\dfrac{1}{\sqrt{3}}\right) \cup \left(\dfrac{1}{\sqrt{3}}, \infty\right)$; concave up on $\left(-\dfrac{1}{\sqrt{3}}, \dfrac{1}{\sqrt{3}}\right)$
   d. $\left(-\dfrac{1}{\sqrt{3}}, -\dfrac{3}{4}\right)$; $\left(\dfrac{1}{\sqrt{3}}, -\dfrac{3}{4}\right)$

11.

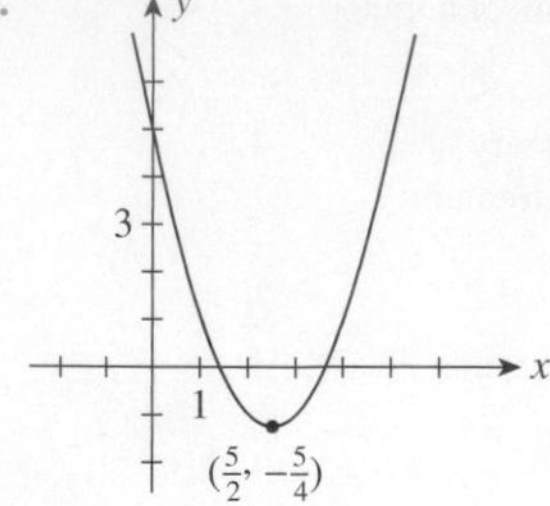

12.

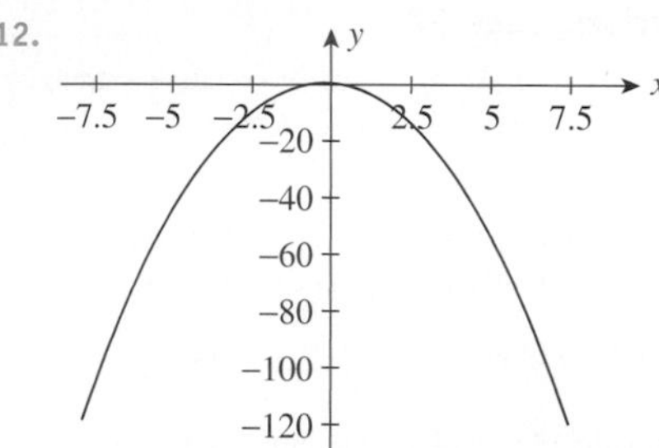

13.

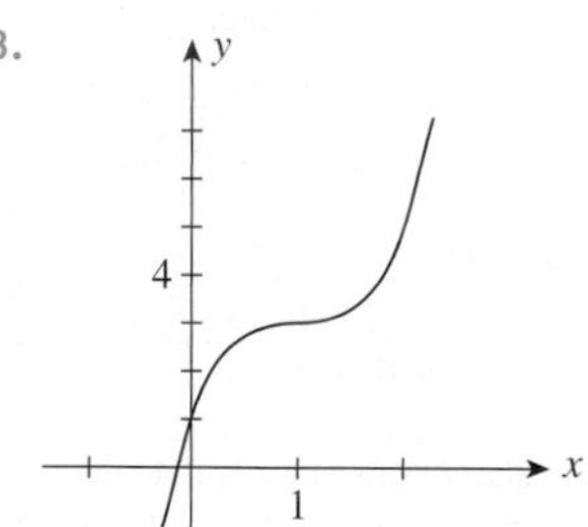

14.

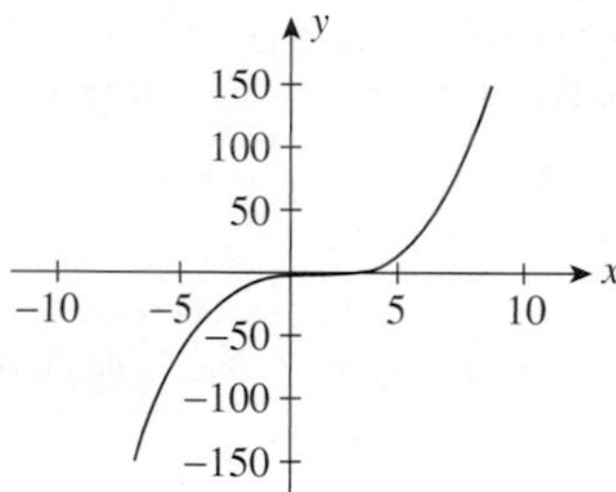

15.

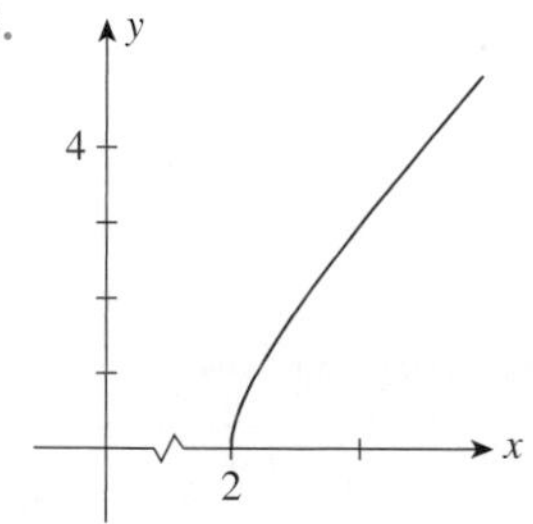

16.

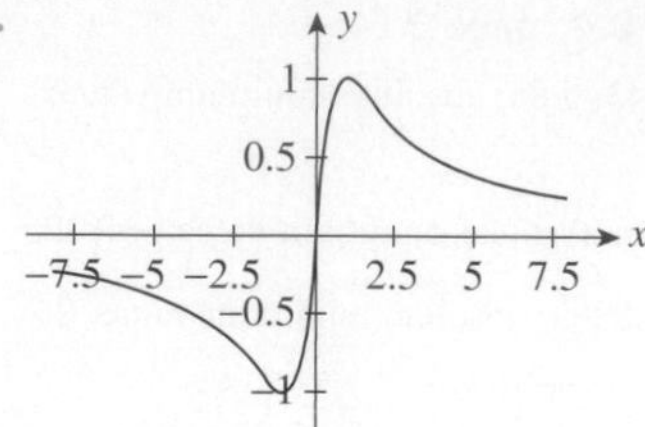

17.

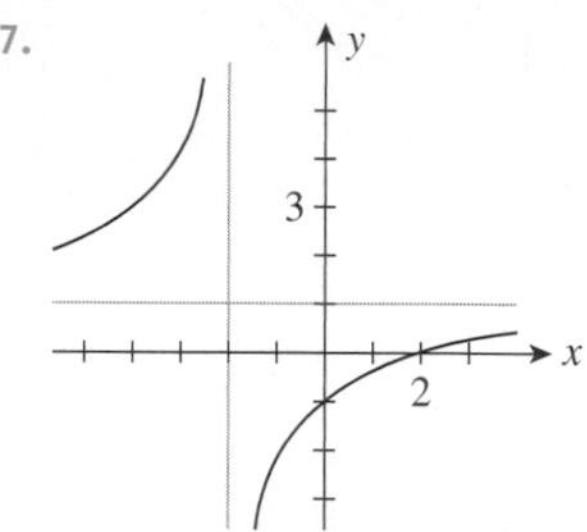

18.

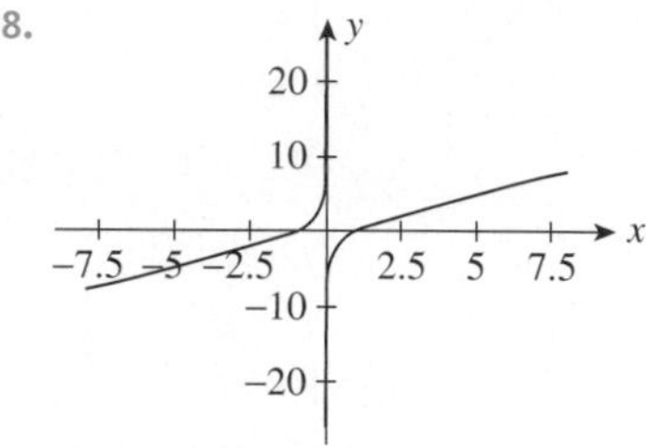

19. Vertical asymptote: $x = -\frac{3}{2}$; horizontal asymptote: $y = 0$

20. Horizontal asymptote: $y = 2$; vertical asymptote: $x = -1$

21. Vertical asymptotes: $x = -2$, $x = 4$, horizontal asymptote: $y = 0$

22. Horizontal asymptote: $y = 1$; vertical asymptote: $x = 1$

23. Absolute minimum value: $-\frac{25}{8}$

24. Absolute minimum value: 0

25. Absolute maximum value: 5; absolute minimum value: 0

26. Absolute maximum value: $\frac{5}{3}$; absolute minimum value: 1

27. Absolute maximum value: -16; absolute minimum value: -32

28. Absolute maximum value: $\frac{1}{2}$; absolute minimum value: 0

29. Absolute maximum value: $\frac{8}{3}$; absolute minimum value: 0

30. Absolute maximum value: $\frac{215}{9}$; absolute minimum value: 7

31. Absolute maximum value: $\frac{1}{2}$; absolute minimum value: $-\frac{1}{2}$

32. No absolute extrema **33.** \$4000

34. **c.** Online travel spending is expected to increase at an increasing rate over that period of time.

35. **a.** $16.25t + 24.625$; sales were increasing.
b. 16.25; the rate of sales was increasing from 2002 to 2005.

36. (100, 4600); sales increase rapidly until \$100,000 is spent on advertising; after that, any additional expenditure results in increased sales but at a slower rate of increase.

37. (266.67, 11,874.08); the rate of increase is lowest when 267 calculators are produced.

38. **a.** $I'(t) = -\dfrac{200t}{(t^2 + 10)^2}$

b. $I''(t) = \dfrac{-200(10 - 3t^2)}{(t^2 + 10)^3}$; concave up on $(\sqrt{10/3}, \infty)$; concave down on $(0, \sqrt{10/3})$

c.

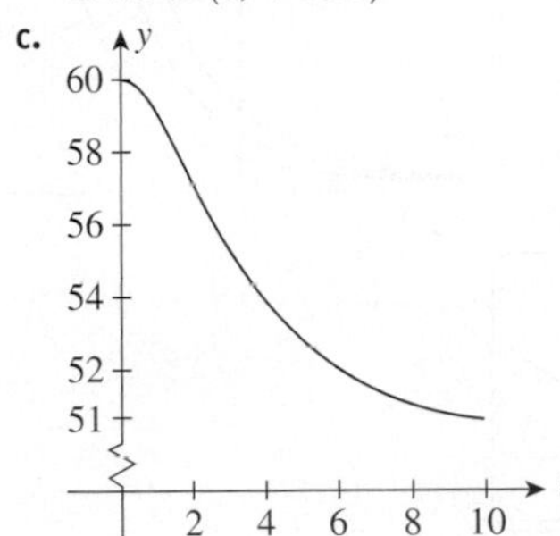

d. The rate of decline in the environmental quality of the wildlife was increasing the first 1.8 yr. After that time the rate of decline decreased.

39. 168 40. 3000

41. **a.** $0.001x + 100 + \dfrac{4000}{x}$ **b.** 2000

42. 10 A.M.

43. **a.** Decreasing on (0, 12.7); increasing on (12.7, 30)
b. (12.7, 7.9)
c. The percent of women 65 years and older in the workforce was decreasing from 1970 to Sept. 1982 and increasing from Sept. 1982 to 2000. It reached a minimum value of 7.9% in Sept. 1982.

45. 74.07 in.^{3} 46. Radius: 2 ft; height: 8 ft

47. 1 ft × 2 ft × 2 ft 48. 20,000 cases

49. If $a > 0, f$ is decreasing on $(-\infty, -\frac{b}{2a})$ and increasing on $(-\frac{b}{2a}, \infty)$; if $a < 0, f$ is increasing on $(-\infty, -\frac{b}{2a})$ and decreasing on $(-\frac{b}{2a}, \infty)$.

50. **a.** $f'(x) = 3x^2$ if $x \neq 0$ **b.** No

Chapter 4 Before Moving On, page 330

1. Decreasing on $(-\infty, 0) \cup (2, \infty)$; increasing on $(0, 1) \cup (1, 2)$

2. Rel. min: $(1, -10)$

3. Concave downward on $(-\infty, \frac{1}{4})$; concave upward on $(\frac{1}{4}, \infty)$; $(\frac{1}{4}, \frac{83}{96})$

4.

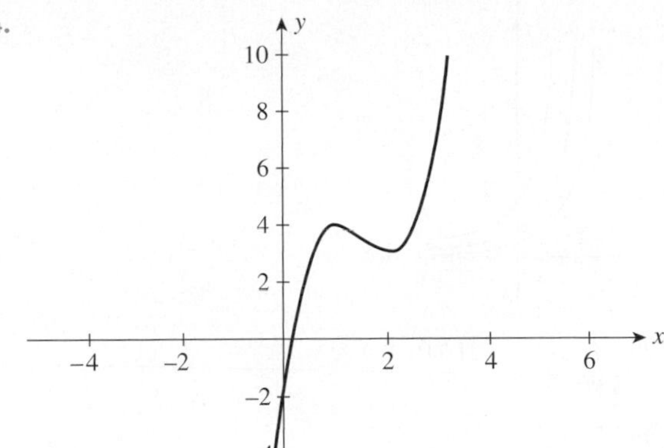

5. Abs. min. value: −5; abs. max value: 80

6. $r = h = \dfrac{1}{\sqrt[3]{\pi}}$ (ft)

CHAPTER 5

Exercises 5.1, page 335

1. **a.** 16 **b.** 27 3. **a.** 3 **b.** $\sqrt{5}$

5. **a.** −3 **b.** 8 7. **a.** 25 **b.** $4^{1.8}$

9. **a.** $4x^3$ **b.** $5xy^2\sqrt{x}$ 11. **a.** $\dfrac{2}{a^2}$ **b.** $\frac{1}{3}b^2$

13. **a.** $8x^9y^6$ **b.** $16x^4y^4z^6$ 15. **a.** $\dfrac{64x^6}{y^4}$ **b.** $(x - y)(x + y)$

17. 2 19. 3 21. 3 23. $\frac{5}{4}$ 25. 1 or 2

27. 29.

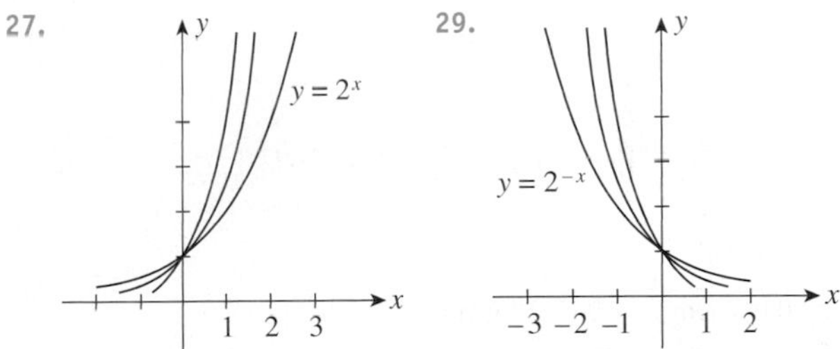

31. 33.

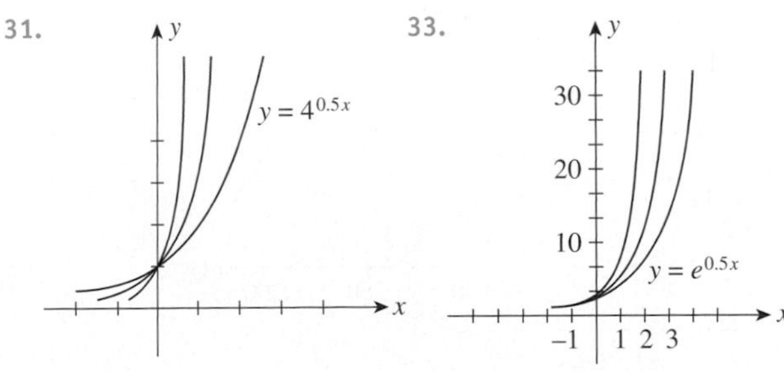

35.

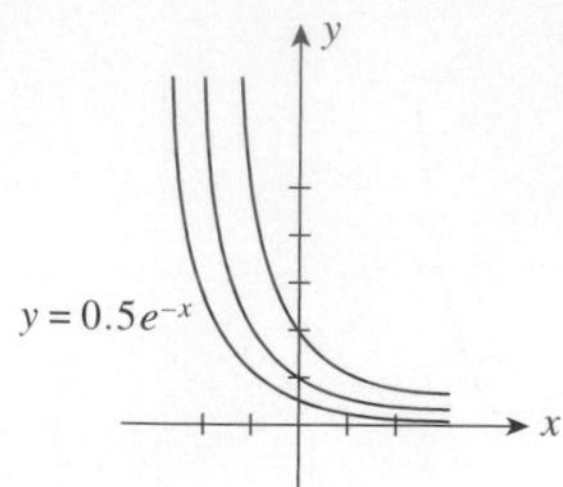

37. **a.** 26.3%; 24.67%; 21.71%; 19.72%

b.

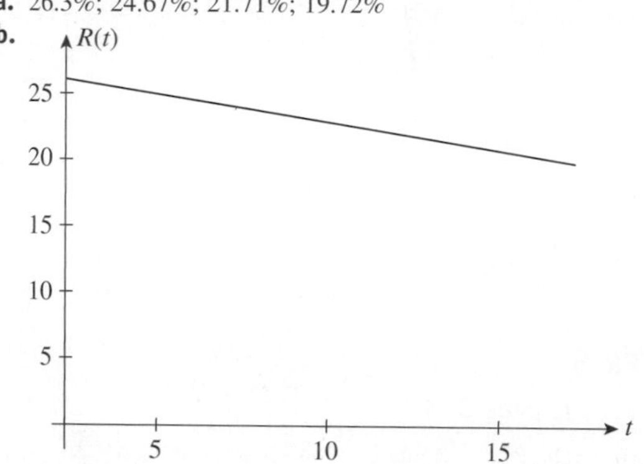

39. **a.**

Year	0	1	2	3	4	5
Web Addresses (billions)	0.45	0.80	1.41	2.49	4.39	7.76

b.

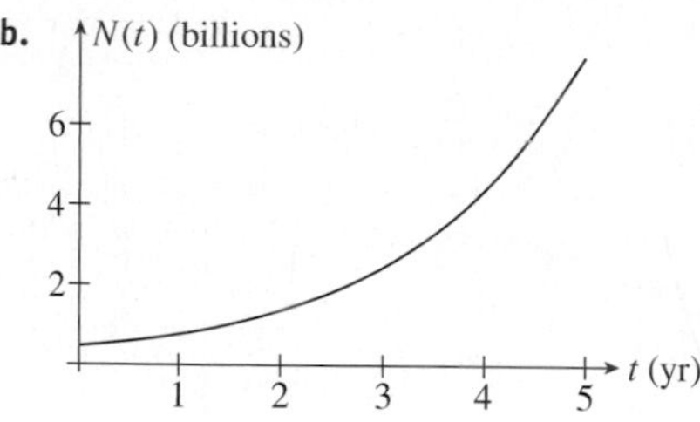

41. 34,210,000

43. **a.** 0.08 g/cm^3 **b.** 0.12 g/cm^3 **c.** 0.2 g/cm^3

d.

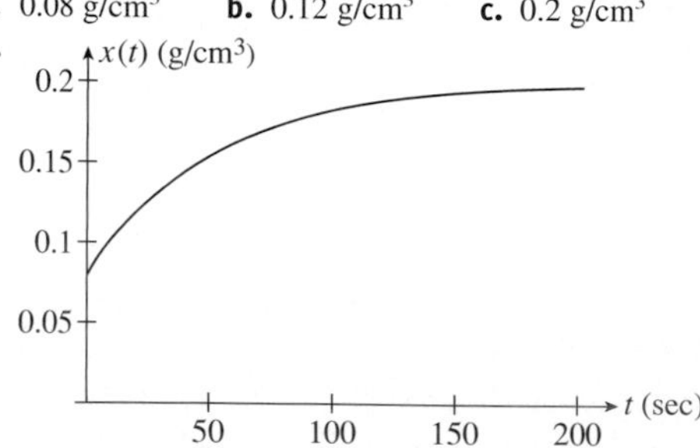

45. False 47. True

Using Technology Exercises 5.1, page 338

1.

3.

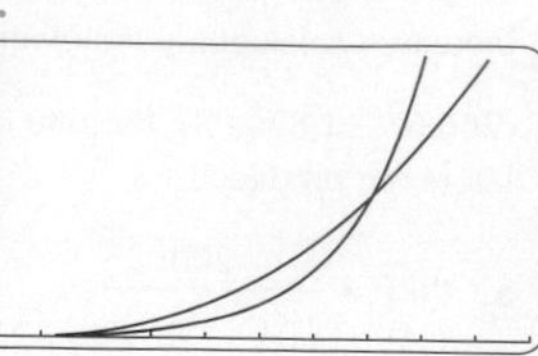

5.

7.

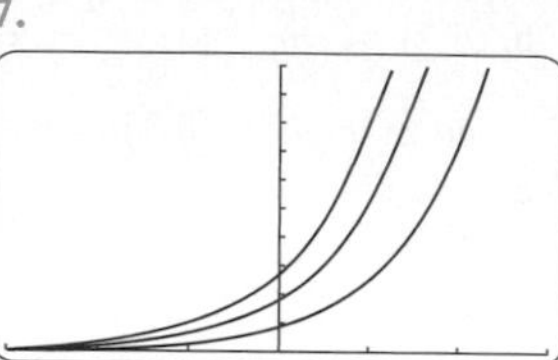

9.

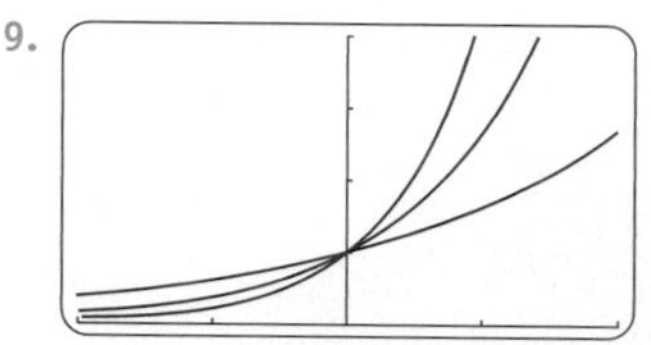

11. **a.**

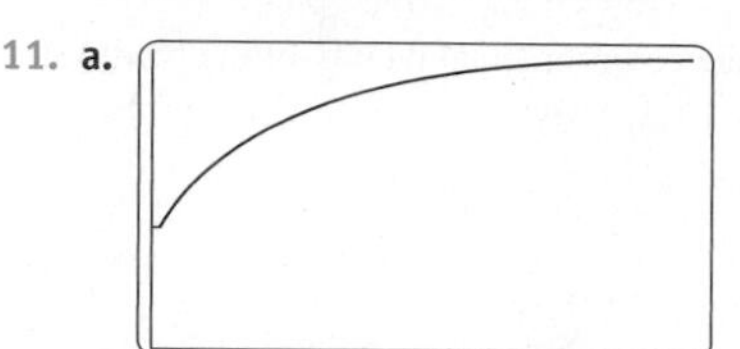

b. 0.08 g/cm^3 **c.** 0.12 g/cm^3 **d.** 0.2 g/cm^3

13. **a.**

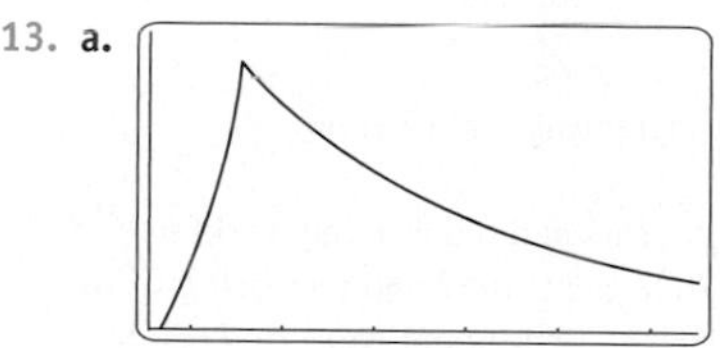

b. 20 sec **c.** 35.1 sec

Exercises 5.2, page 345

1. $\log_2 64 = 6$ 3. $\log_3 \frac{1}{9} = -2$ 5. $\log_{1/3} \frac{1}{3} = 1$

7. $\log_{32} 8 = \frac{3}{5}$ 9. $\log_{10} 0.001 = -3$ 11. 1.0792

13. 1.2042 15. 1.6813 17. $\ln a^2b^3$ 19. $\ln \dfrac{3\sqrt{xy}}{\sqrt[3]{z}}$

21. $\log x + 4\log(x + 1)$ 23. $\frac{1}{2}\log(x + 1) - \log(x^2 + 1)$

25. $\ln x - x^2$ 27. $-\frac{3}{2}\ln x - \frac{1}{2}\ln(1 + x^2)$

29.

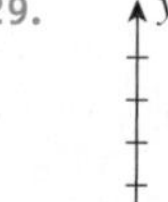

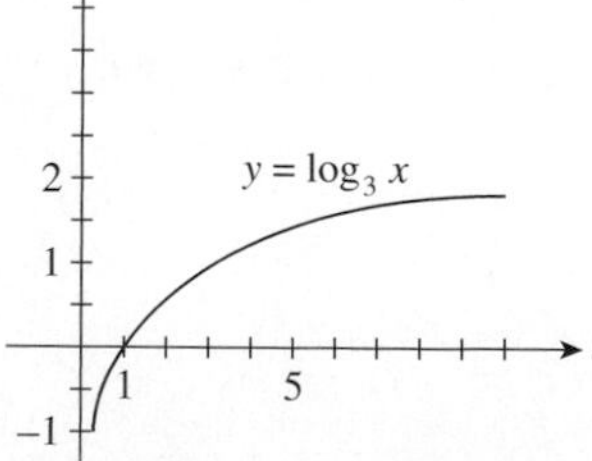

31.

33.

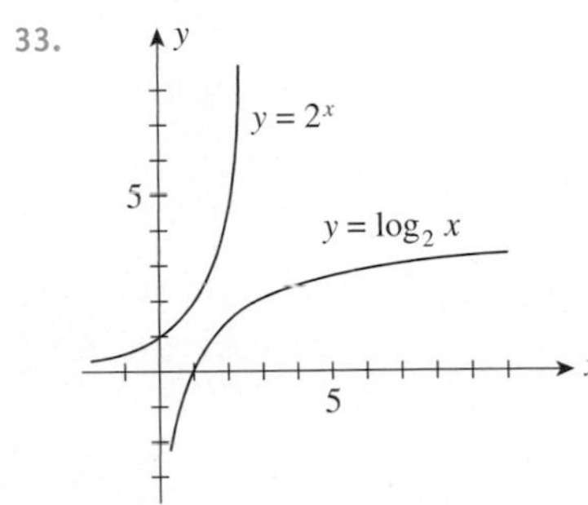

35. 5.1986 37. -0.0912 39. -8.0472 41. -4.9041

43. $-2\ln\left(\frac{A}{B}\right)$ 45. 105.7 mm

47. **a.** $10^3 I_0$ **b.** 100,000 times greater
c. 10,000,000 times greater

49. 27.40 yr 51. 6.44 yr 53. **a.** 9.12 sec **b.** 20.27 sec

55. False 57. True 59. **a.** ln 2

Exercises 5.3, page 357

1. \$4974.47 3. \$223,403.11

5. **a.** 10.25%/yr **b.** 9.31%/yr

7. **a.** \$29,227.61 **b.** \$29,137.83 9. \$6885.64

11. 13.59%/yr 13. 12.1%/yr 15. 14.87%/yr

17. 2.2 yr 19. 7.7 yr 21. 6.08%/yr 23. 2.06 yr

25. \$254,084.69 27. \$2.844 million 29. \$23,329.48

31. \$731,250 33. 91¢ 35. 12.75% 37. \$40,000

39. **a.** \$33,885.14 **b.** \$33,565.38

41. **a.** \$16,262.79 **b.** \$12,047.77 **c.** \$6611.96

45. Bank B 47. 9.531%

Exercises 5.4, page 365

1. $3e^{3x}$ 3. $-e^{-t}$ 5. $e^x + 1$ 7. $x^2e^x(x + 3)$

9. $\frac{2e^x(x-1)}{x^2}$ 11. $3(e^x - e^{-x})$ 13. $-\frac{1}{e^w}$ 15. $6e^{3x-1}$

17. $-2xe^{-x^2}$ 19. $\frac{3e^{-1/x}}{x^2}$ 21. $25e^x(e^x + 1)^{24}$ 23. $\frac{e^{\sqrt{x}}}{2\sqrt{x}}$

25. $e^{3x+2}(3x - 2)$ 27. $\frac{2e^x}{(e^x + 1)^2}$ 29. $2(8e^{-4x} + 9e^{3x})$

31. $6e^{3x}(3x + 2)$ 33. $y = 2x - 2$

35. f is increasing on $(-\infty, 0)$ and decreasing on $(0, \infty)$.

37. Concave downward on $(-\infty, 0)$; concave upward on $(0, \infty)$

39. $(1, e^{-2})$ 41. $y = e^{-1/2}(-\sqrt{2}x + 2)$; $y = e^{-1/2}(\sqrt{2}x + 2)$

43. Absolute maximum value: 1; absolute minimum value: e^{-1}

45. Absolute minimum value: -1; absolute maximum value: $2e^{-3/2}$

47.

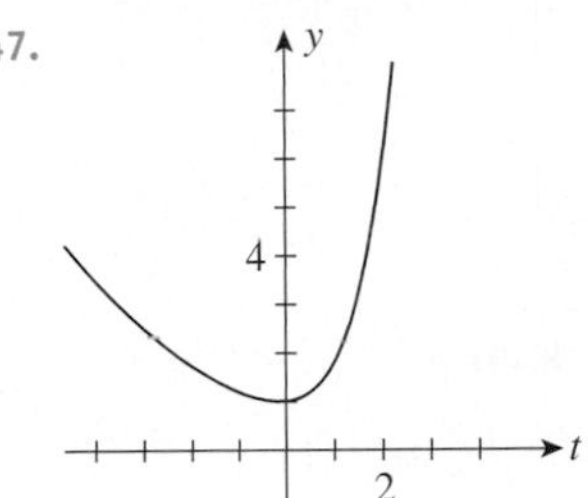

49.

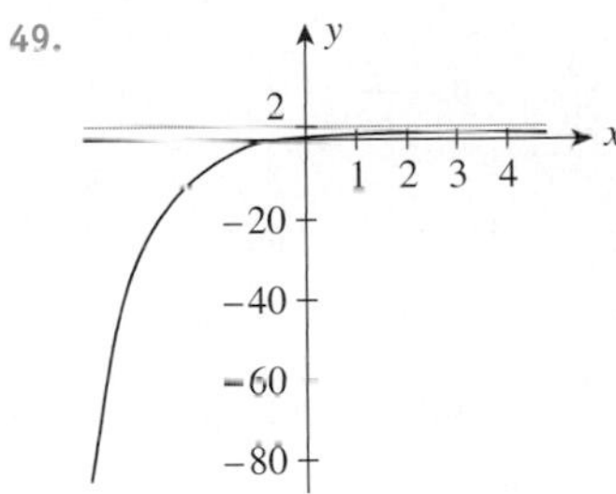

51. -0.1694, -0.1549, -0.1415; the percentage of the total population relocating was decreasing at the rate of 0.17%/yr in 1970, 0.15%/yr in 1980, and 0.14%/yr in 1990.

53. **a.** 70,000; 353,716 **b.** 37,800/decade; 191,000/decade

55. **a.** \$4.6 trillion; \$3.62 trillion
b. \$184 billion/yr; \$145 billion/yr

57. **a.** $-$\$6065/day; $-$\$3679/day; $-$\$2231/day; $-$\$1353/day
b. 2 days

59. **b.** 4505/yr; 273 cases/yr 61. 10,000; \$367,879

63. **a.** -1.68¢/case **b.** \$40.36/case

65. **a.** 45.6 kg **b.** 4.2 kg 67. 7.72 yr; \$160,207.69

69. 1.8; -0.11; -0.23; -0.13; the rate of change of the amount of oil used is 1.8 barrels/\$1000 of output/decade in 1965; it is decreasing at the rate of 0.11 barrel/\$1000 of output/decade in 1966; and so on.

71. **a.** \$12/unit **b.** Decreasing at the rate of \$7/wk **c.** \$8/unit

75. **a.** 0.05 g/cm^3/sec **b.** -0.01 g/cm^3/sec **c.** 20 sec
d. 0.90 g/cm^3

77. False 79. False

Using Technology Exercises 5.4, page 370

1. 5.4366 3. 12.3929 5. 0.1861

7. **a.** 50 **c.**

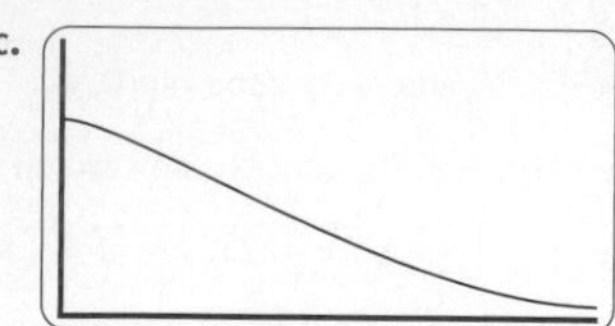

9. **a.**

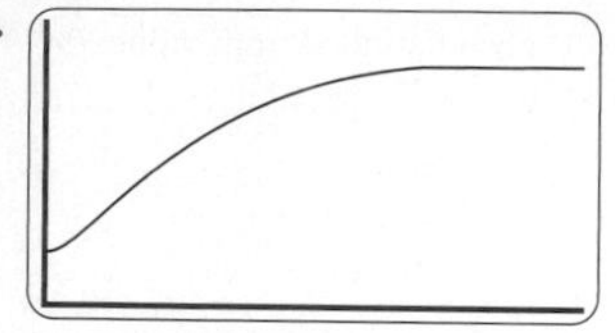

b. 4.2720 billion/half century

11. **a.** 153,024; 235,181 **b.** 634; 18,401

13. **a.** 69.63% **b.** 5.094%/decade

Exercises 5.5, page 376

1. $\frac{5}{x}$ 3. $\frac{1}{x+1}$ 5. $\frac{8}{x}$ 7. $\frac{1}{2x}$ 9. $\frac{-2}{x}$

11. $\frac{2(4x-3)}{4x^2-6x+3}$ 13. $\frac{1}{x(x+1)}$ 15. $x(1+2\ln x)$

17. $\frac{2(1-\ln x)}{x^2}$ 19. $\frac{3}{u-2}$ 21. $\frac{1}{2x\sqrt{\ln x}}$

23. $\frac{3(\ln x)^2}{x}$ 25. $\frac{3x^2}{x^3+1}$ 27. $\frac{(x\ln x+1)e^x}{x}$

29. $\frac{e^{2t}[2(t+1)\ln(t+1)+1]}{t+1}$ 31. $\frac{1-\ln x}{x^2}$ 33. $-\frac{1}{x^2}$

35. $\frac{2(2-x^2)}{(x^2+2)^2}$ 37. $(x+1)(5x+7)(x+2)^2$

39. $(x-1)(x+1)^2(x+3)^3(9x^2+14x-7)$

41. $\frac{(2x^2-1)^4(38x^2+40x+1)}{2(x+1)^{3/2}}$ 43. $3^x\ln 3$

45. $(x^2+1)^{x-1}[2x^2+(x^2+1)\ln(x^2+1)]$ 47. $y=x-1$

49. f is decreasing on $(-\infty, 0)$ and increasing on $(0, \infty)$.

51. Concave up: $(-\infty, -1) \cup (1, \infty)$; concave down: $(-1, 0) \cup (0, 1)$

53. $(-1, \ln 2)$ and $(1, \ln 2)$ 55. $y = 4x - 3$

57. Absolute minimum value: 1; absolute maximum value: $3 - \ln 3$

59. 0.0580%/kg; 0.0133%/kg

61. **a.** 78.82 million **b.** 3.95 million/yr

63. **b.**

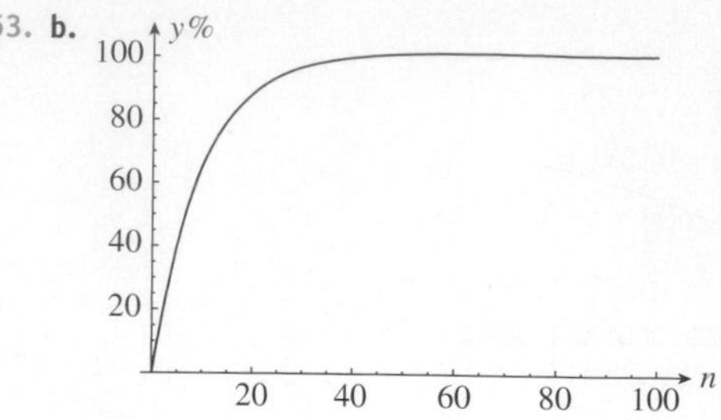

c. 100

65. **a.** 6 **b.** 276,310 I_0

67.

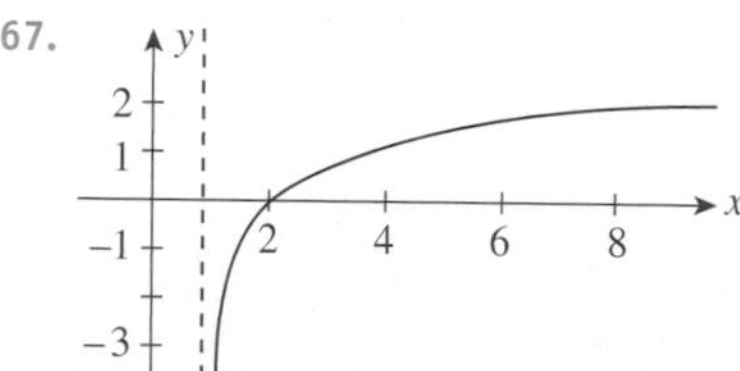

69. False

Exercises 5.6, page 386

1. **a.** 0.05 **b.** 400

c.

t	0	10	20	100	1000
Q	400	660	1087	59,365	2.07×10^{24}

3. **a.** $Q(t) = 100e^{0.035t}$ **b.** 266 min **c.** $Q(t) = 1000e^{0.035t}$

5. **a.** 54.93 yr **b.** 14.25 billion 7. 8.7 lb/in.^{2}; 0.0004 lb/in.^{2}/ft

9. $Q(t) = 100e^{-0.049t}$; 70.7 g; 3.451 g/day 11. 13,412 yr ago

13. $Q(t)$; 100; 10; 20; t

a. 60 words/min **b.** 107 words/min **c.** 136 words/min

15. $D(t)$; 2000; 1000; 50; t

a. 573 computers; 1177 computers; 1548 computers; 1925 computers

b. 2000 computers **c.** 46 computers/mo

17. **a.** 122.3 cm **b.** 14 cm/yr **c.** 200 cm

19. **a.** 86.1% **b.** 10.44%/yr **c.** 1970

21. 76.4 million 23. 1080; 280 25. 15 lb

27. **a.** $\dfrac{\ln \frac{b}{a}}{b-a}$ min **b.** $\dfrac{k}{b-a}$ g/cm^3 29. **b.** 5599 yr

Using Technology Exercises 5.6, page 390

1. **a.**

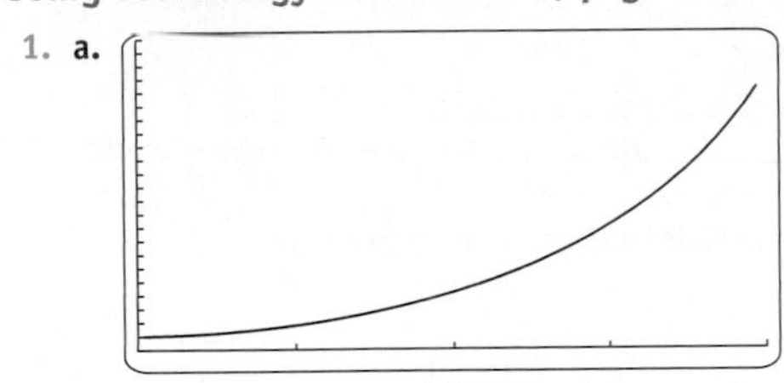

b. 12.146%/yr **c.** 9.474%/yr

3. **a.**

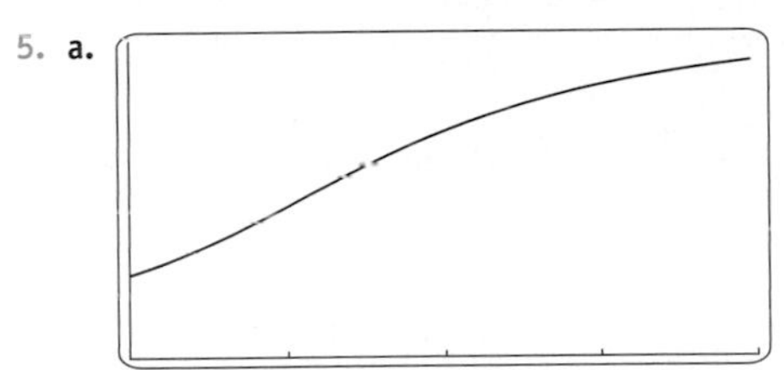

b. 666 million, 818.8 million **c.** 33.8 million/yr

5. **a.**

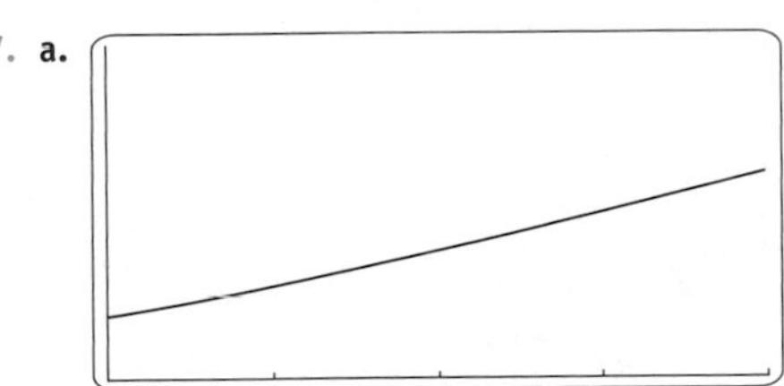

b. 86.12%/yr **c.** 10.44%/yr **d.** 1970

7. **a.**

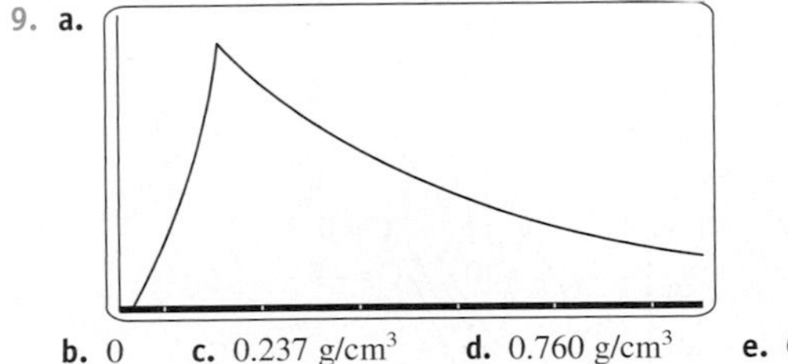

b. 325 million **c.** 76.84 million/decade

9. **a.**

b. 0 **c.** 0.237 g/cm^3 **d.** 0.760 g/cm^3 **e.** 0

Chapter 5 Concept Review, page 392

1. Power; 0; 1; exponential

2. **a.** $(-\infty, \infty)$; $(0, \infty)$ **b.** $(0, 1)$; $(-\infty, \infty)$

3. **a.** $(0, \infty)$; $(-\infty, \infty)$; $(1, 0)$ **b.** <1; >1 4. **a.** x **b.** x

5. Accumulated amount; principal; nominal interest rate; number of conversion periods; term

6. $\left(1 + \dfrac{r}{m}\right)^m - 1$ 7. Pe^{rt} 8. **a.** $e^{f(x)}f'(x)$ **b.** $\dfrac{f'(x)}{f(x)}$

9. **a.** Initially; growth **b.** Decay **c.** Time; one half

10. **a.** Horizontal asymptote: C
b. Horizontal asymptote; A, carrying capacity

Chapter 5 Review Exercises, page 393

1. **a.** and **b.**

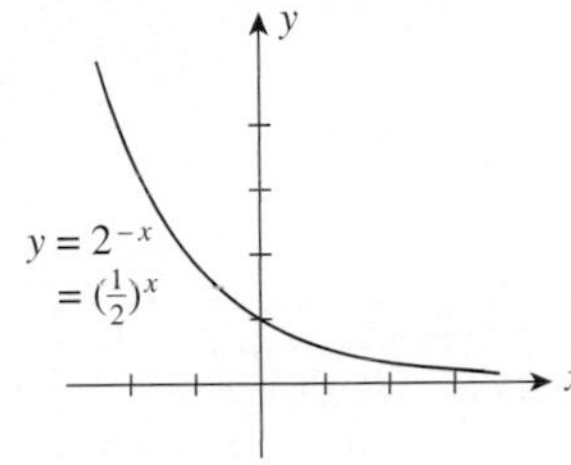

2. $\log_{2/3}\left(\frac{27}{8}\right) = -3$ 3. $\log_{16} 0.125 = -\frac{3}{4}$ 4. $\frac{15}{2}$ 5. $x = 2$

6. $x + y + z$ 7. $x + 2y - z$ 8. $y + 2z$

9.

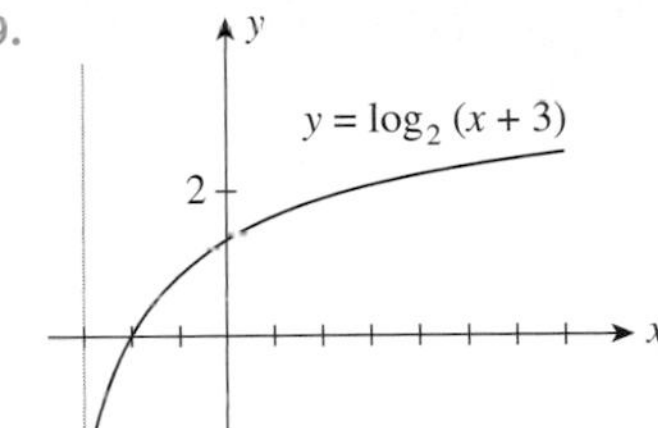

10.

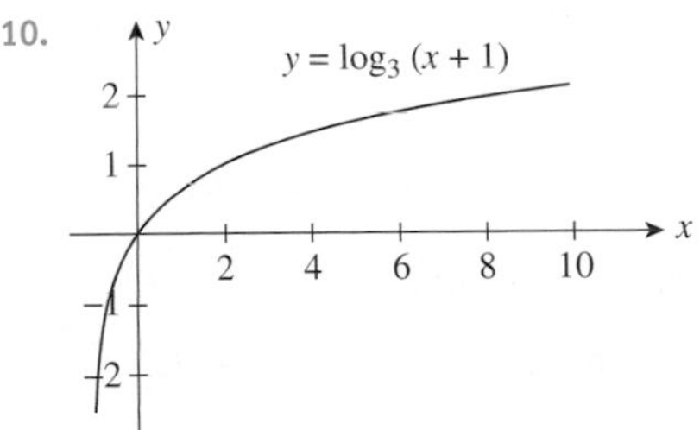

11. $(2x + 1)e^{2x}$ 12. $\dfrac{e^t}{2\sqrt{t}} + \sqrt{t}e^t + 1$ 13. $\dfrac{1 - 4t}{2\sqrt{t}e^{2t}}$

14. $\dfrac{e^x(x^2 + x + 1)}{\sqrt{1 + x^2}}$ 15. $\dfrac{2(e^{2x} + 2)}{(1 + e^{-2x})^2}$ 16. $4xe^{2x^2-1}$

17. $(1 - 2x^2)e^{-x^2}$ 18. $3e^{2x}(1 + e^{2x})^{1/2}$ 19. $(x + 1)^2e^x$

20. $\ln t + 1$ 21. $\dfrac{2xe^{x^2}}{e^{x^2} + 1}$ 22. $\dfrac{\ln x - 1}{(\ln x)^2}$

23. $\dfrac{x - x\ln x + 1}{x(x + 1)^2}$ 24. $(x + 2)e^x$ 25. $\dfrac{4e^{4x}}{e^{4x} + 3}$

26. $\dfrac{(r^3 - r^2 + r + 1)e^r}{(1 + r^2)^2}$ 27. $\dfrac{1 + e^x(1 - x\ln x)}{x(1 + e^x)^2}$

28. $\dfrac{(2x^2 + 2x^2 \cdot \ln x - 1)e^{x^2}}{x(1 + \ln x)^2}$ 29. $-\dfrac{9}{(3x + 1)^2}$ 30. $\dfrac{1}{x}$

31. 0 32. −2 33. $6x(x^2 + 2)^2(3x^3 + 2x + 1)$

34. $\dfrac{(4x^3 - 5x^2 + 2)(x^2 - 2)}{(x - 1)^2}$ 35. $y = -(2x - 3)e^{-2}$

36. $y = \dfrac{1}{e}$

37.

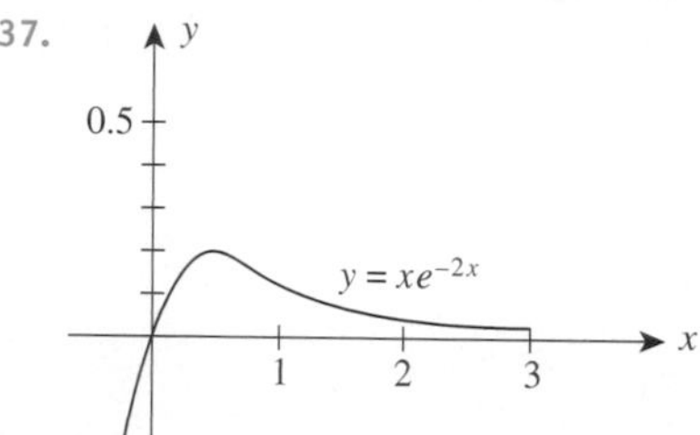

38.

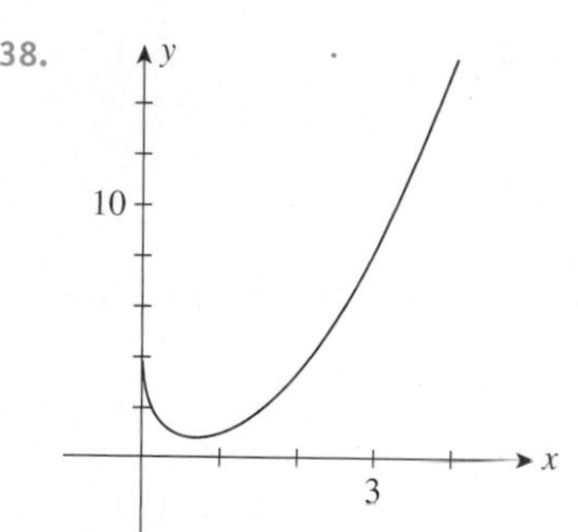

39. Absolute maximum value: $\dfrac{1}{e}$

40. Absolute maximum value: $\dfrac{\ln 2}{2}$; absolute minimum value: 0

41. 12%/yr 42. $80,000 43. 9.58 yr

44. **a.** $Q(t) = 2000e^{0.01831t}$ **b.** 161,992 45. 0.0004332

46.

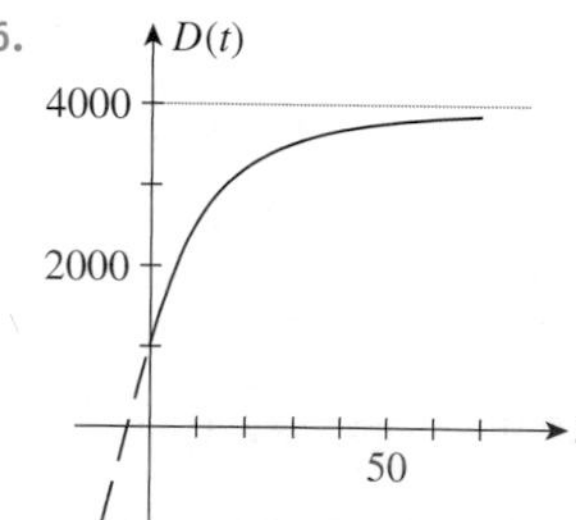

a. 1175, 2540, 3289 **b.** 4000

47. 970

48. **a.** 12.5/1000 live births; 9.3/1000 live births; 6.9/1000 live births
b.

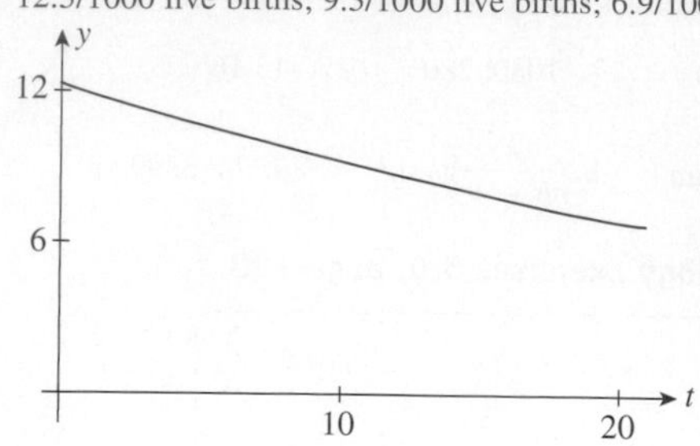

49. **a.** 0 g/cm³ **b.** 0.0361 g/cm³ **c.** 0.08 g/cm³
d.

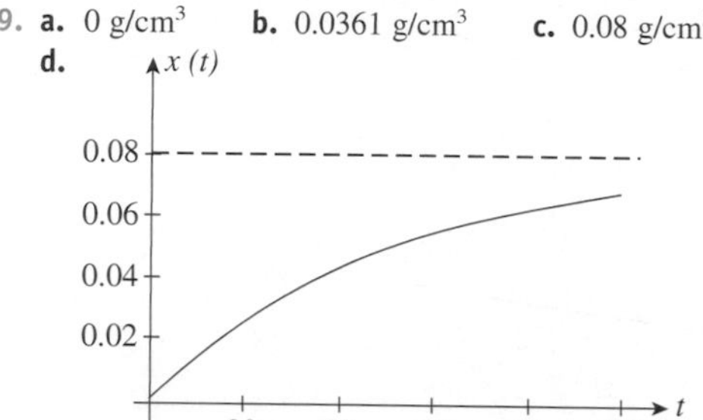

Chapter 5 Before Moving On, page 394

1. −0.9589 2. $4130.37 3. $\dfrac{e^{\sqrt{x}}}{2\sqrt{x}}$ 4. $1 + \ln 2$

5. $e^{2x}\left(\dfrac{4x^2\ln 3x + 4x - 1}{x^2}\right)$ 6. 8.7 min

CHAPTER 6

Exercises 6.1, page 404

5. **b.** $y = 2x + C$

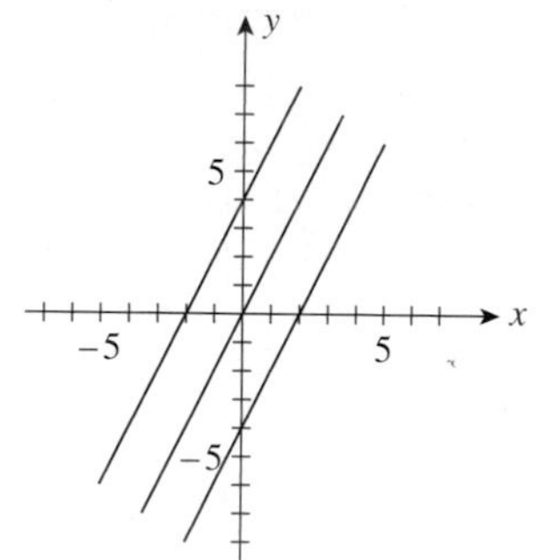

7. **b.** $y = \frac{1}{3}x^3 + C$

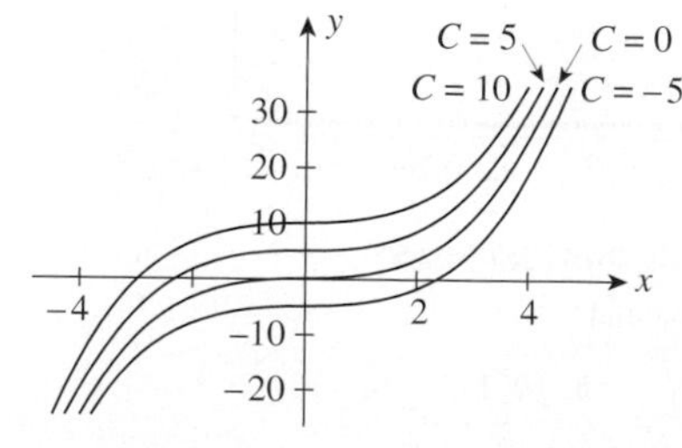

9. $6x + C$ **11.** $\frac{1}{4}x^4 + C$ **13.** $-\frac{1}{3x^3} + C$

15. $\frac{3}{5}x^{5/3} + C$ **17.** $-\frac{4}{x^{1/4}} + C$ **19.** $-\frac{2}{x} + C$

21. $\frac{2}{3}\pi t^{3/2} + C$ **23.** $3x - x^2 + C$

25. $\frac{1}{3}x^3 + \frac{1}{2}x^2 - \frac{1}{2x^2} + C$ **27.** $4e^x + C$

29. $x + \frac{1}{2}x^2 + e^x + C$ **31.** $x^4 + \frac{2}{x} - x + C$

33. $\frac{2}{7}x^{7/2} + \frac{4}{5}x^{5/2} - \frac{1}{2}x^2 + C$ **35.** $\frac{2}{3}x^{3/2} + 6\sqrt{x} + C$

37. $\frac{1}{9}u^3 + \frac{1}{3}u^2 - \frac{1}{3}u + C$ **39.** $\frac{2}{3}t^3 - \frac{3}{2}t^2 - 2t + C$

41. $\frac{1}{3}x^3 - 2x - \frac{1}{x} + C$ **43.** $\frac{1}{3}s^3 + s^2 + s + C$

45. $e^t + \frac{t^{e+1}}{e+1} + C$ **47.** $\frac{1}{2}x^2 + x - \ln|x| - \frac{1}{x} + C$

49. $\ln|x| + \frac{4}{\sqrt{x}} - \frac{1}{x} + C$ **51.** $x^2 + x + 1$

53. $x^3 + 2x^2 - x - 5$ **55.** $x - \frac{1}{x} + 2$ **57.** $x + \ln|x|$

59. $\sqrt{x}$ **61.** $e^x + \frac{1}{2}x^2 + 2$ **63.** Branch A

65. $s(t) = \frac{4}{3}t^{3/2}$ **67.** \$3370 **69.** 5000 units; \$34,000

71. a. $-t^3 + 6t^2 + 45t$ **b.** 212

73. a. $-0.125t^3 + 1.05t^2 + 2.45t + 1.5$ **b.** 24.375 million

75. a. $-1.493t^3 + 34.9t^2 + 279.5t + 2917$ **b.** \$9168

77. a. $3.133t^3 - 6.7t^2 + 14.07t + 36.7$ **b.** 103,201

79. a. $y = 4.096t^3 - 75.2797t^2 + 695.23t + 3142$ **b.** \$4264.11

81. 21,960 **83.** $1.0974t^3 - 0.0915t^4$

85. $-t^3 + 96t^2 + 120t$; 63,000 ft

87. a. $9.3e^{-0.02t}$ **b.** 7030 **c.** 6619

89. $\frac{1}{2}k(R^2 - r^2)$ **91.** $9\frac{7}{9}$ ft/sec²; 396 ft **93.** 0.924 ft/sec²

95. a. $\frac{2t}{t+4}$ **b.** $\frac{2}{5}$ in.; $\frac{2}{3}$ in. **97.** True **99.** True

Exercises 6.2, page 417

1. $\frac{1}{5}(4x+3)^5 + C$ **3.** $\frac{1}{3}(x^3 - 2x)^3 + C$

5. $-\frac{1}{2(2x^2+3)^2} + C$ **7.** $\frac{2}{3}(t^3+2)^{3/2} + C$

9. $\frac{1}{20}(x^2-1)^{10} + C$ **11.** $-\frac{1}{5}\ln|1 - x^5| + C$

13. $\ln(x-2)^2 + C$ **15.** $\frac{1}{2}\ln(0.3x^2 - 0.4x + 2) + C$

17. $\frac{1}{6}\ln|3x^2 - 1| + C$ **19.** $-\frac{1}{2}e^{-2x} + C$ **21.** $-e^{2-x} + C$

23. $-\frac{1}{2}e^{-x^2} + C$ **25.** $e^x + e^{-x} + C$ **27.** $\ln(1 + e^x) + C$

29. $2e^{\sqrt{x}} + C$ **31.** $-\frac{1}{6(e^{3x} + x^3)^2} + C$ **33.** $\frac{1}{8}(e^{2x} + 1)^4 + C$

35. $\frac{1}{2}(\ln 5x^2) + C$ **37.** $\ln|\ln x| + C$ **39.** $\frac{2}{3}(\ln x)^{3/2} + C$

41. $\frac{1}{2}e^{x^2} - \frac{1}{2}\ln(x^2 + 2) + C$

43. $\frac{2}{3}(\sqrt{x} - 1)^3 + 3(\sqrt{x} - 1)^2 + 8(\sqrt{x} - 1) + 4\ln|\sqrt{x} - 1| + C$

45. $\frac{(6x+1)(x-1)^6}{42} + C$

47. $5 + 4\sqrt{x} - x - 4\ln(1 + \sqrt{x}) + C$

49. $-\frac{1}{252}(1-v)^7(28v^2 + 7v + 1) + C$

51. $\frac{1}{2}[(2x-1)^5 + 5]$ **53.** $e^{-x^2+1} - 1$ **55.** 17,341,000

57. $21{,}000 - \frac{20{,}000}{\sqrt{1 + 0.2t}}$; 6858 **59.** $\frac{250}{\sqrt{16 + x^2}}$

61. $30(\sqrt{2t+4} - 2)$; $14{,}400\,\pi$ ft²

63. $\frac{65.8794}{1 + 2.449e^{-0.3377t}} + 0.3$; 56.22 in. **65.** $\frac{r}{a}(1 - e^{-at})$

Exercises 6.3, page 427

1. 4.27 sq units

3. a. 6 sq units

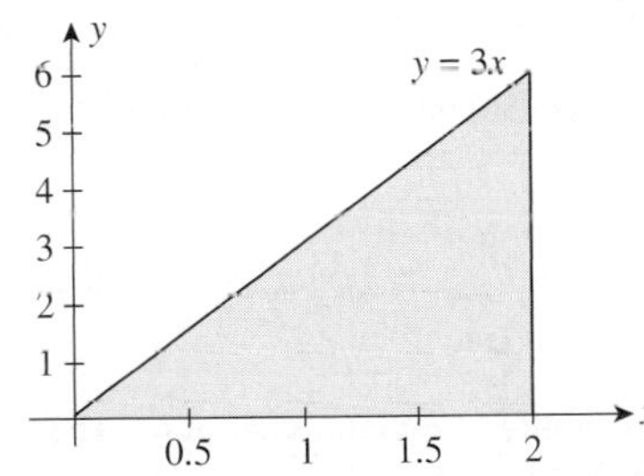

b. 4.5 sq units **c.** 5.25 sq units **d.** Yes

5. a. 4 sq units

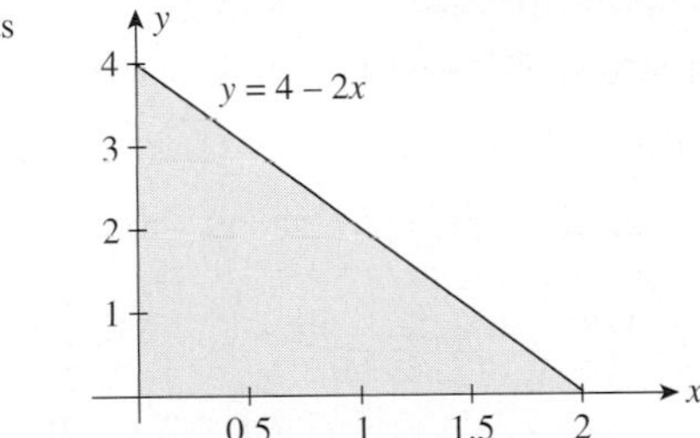

b. 4.8 sq units **c.** 4.4 sq units **d.** Yes

7. a. 18.5 sq units **b.** 18.64 sq units **c.** 18.66 sq units **d.** ≈18.7 sq units

9. a. 25 sq units **b.** 21.12 sq units **c.** 19.88 sq units **d.** ≈19.9 sq units

11. a. 0.0625 sq unit **b.** 0.16 sq unit **c.** 0.2025 sq unit **d.** ≈0.2 sq unit

13. 4.64 sq units **15.** 0.95 sq unit **17.** 9400 sq ft

Exercises 6.4, page 437

1. 6 sq units 3. 8 sq units 5. 12 sq units 7. 9 sq units

9. ln 2 sq units 11. $17\frac{1}{3}$ sq units 13. $18\frac{1}{4}$ sq units

15. $(e^2 - 1)$ sq units 17. 6 19. 14 21. $18\frac{2}{3}$ 23. $\frac{4}{3}$

25. 45 27. $\frac{7}{12}$ 29. ln 2 31. 56 33. $\frac{256}{15}$ 35. $\frac{2}{3}$

37. $2\frac{2}{3}$ 39. $19\frac{1}{2}$ 41. **a.** \$4100 **b.** \$900

43. **a.** \$2800 **b.** \$219.20 45. $10{,}133\frac{1}{3}$ ft

47. **a.** $0.2833t^3 - 1.936t^2 + 5t + 5.6$ **b.** 12.8% **c.** 5.2%

49. 49.7 million 51. $\frac{23}{15}$ sq units 53. False 55. False

Using Technology Exercises 6.4, page 440

1. 6.1787 3. 0.7873 5. −0.5888 7. 2.7044

9. 3.9973 11. 46%; 24% 13. 333,209 15. 903,213

Exercises 6.5, page 447

1. 10 3. $\frac{19}{15}$ 5. $32\frac{4}{15}$ 7. $\sqrt{3} - 1$ 9. $24\frac{1}{5}$

11. $\frac{32}{15}$ 13. $18\frac{2}{15}$ 15. $\frac{1}{2}(e^4 - 1)$ 17. $\frac{1}{2}e^2 + \frac{5}{6}$ 19. 0

21. 2 ln 4 23. $\frac{1}{3}(\ln 19 - \ln 3)$ 25. $2e^4 - 2e^2 - \ln 2$

27. $\frac{1}{2}(e^{-4} - e^{-8} - 1)$ 29. 5 31. $\frac{17}{3}$ 33. −1

35. $\frac{13}{6}$ 37. $\frac{1}{4}(e^4 - 1)$ 39. 120.3 billion metric tons

41. ≈\$2.24 million 43. \$40,339.50 45. 49.69 ft/sec

47. 1367 49. $\$6\frac{1}{3}$ million 51. 16 ft/sec

53. \$14.78 55. 80.7% 63. Property 5 65. 0

67. **a.** −1 **b.** 5 **c.** −13 69. True

71. False 73. True

Using Technology Exercises 6.5, page 451

1. 7.71667 3. 17.56487 5. 10,140 7. 60.45 mg/day

Exercises 6.6, page 458

1. 108 sq units 3. $\frac{2}{3}$ sq unit 5. $2\frac{2}{3}$ sq units

7. $1\frac{1}{2}$ sq units 9. 3 11. $3\frac{1}{3}$ 13. 27 15. $2(e^2 - e^{-1})$

17. $12\frac{2}{3}$ 19. $3\frac{1}{3}$ 21. $4\frac{3}{4}$ 23. 12 − ln 4

25. $e^2 - e - \ln 2$ 27. $2\frac{1}{2}$ 29. $7\frac{1}{3}$ 31. $\dfrac{3}{2}$

33. $e^3 - 4 + \dfrac{1}{e}$ 35. $20\frac{5}{6}$ 37. $\dfrac{1}{12}$ 39. $\dfrac{71}{6}$ 41. 18

43. S is the additional revenue that Odyssey Travel could realize by switching to the new agency; $S = \int_0^b [g(x) - f(x)]\,dx$

45. Shortfall $= \int_{2010}^{2050} [f(t) - g(t)]\,dt$

47. **a.** $A_2 - A_1$ **b.** The distance car 2 is ahead of car 1 after t sec

49. 30 ft/sec faster 51. 21,850 53. True

Using Technology Exercises 6.6, page 463

1. **a.**

b. 1074.2857

3. **a.**

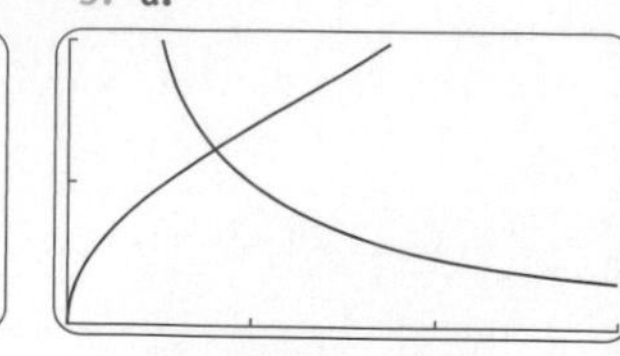

b. 0.9961

5. **a.**

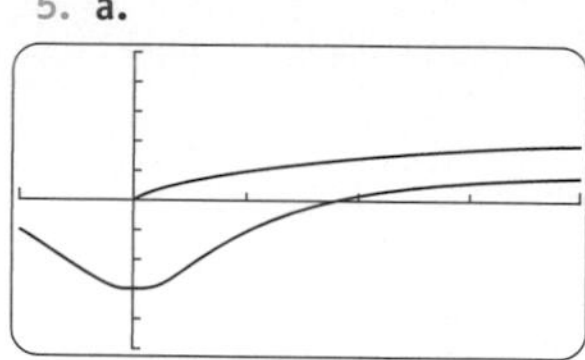

b. 5.4603

7. **a.**

b. 25.8549

9. **a.**

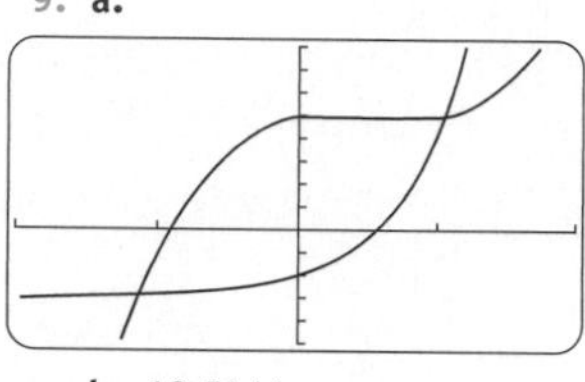

b. 10.5144

11. **a.**

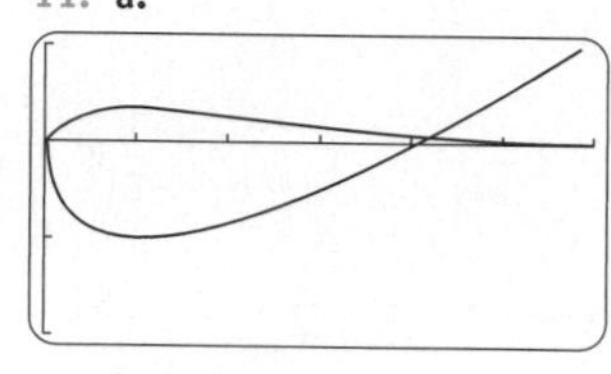

b. 3.5799

13. 207.43

Exercises 6.7, page 474

1. \$11,667 3. \$6667 5. \$11,667

7. Consumers' surplus: \$13,333; producers' surplus: \$11,667

9. \$824,200 11. \$148,239 13. \$52,203

15. \$76,615 17. \$111,869 19. \$20,964

21. **a.**

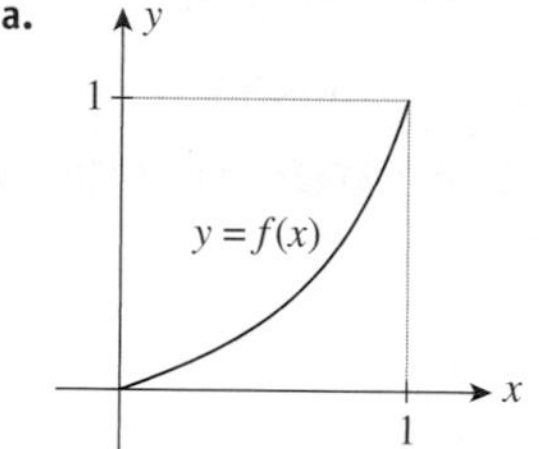

b. 0.175; 0.816

23. **a.**

b. 0.104; 0.504

Using Technology Exercises 6.7, page 476

1. Consumers' surplus: \$18,000,000; producers' surplus: \$11,700,000

3. Consumers' surplus: \$33,120; producers' surplus: \$2880

5. Investment A

Exercises 6.8, page 484

1. 3π cu units 3. $\dfrac{15\pi}{2}$ cu units 5. $\dfrac{4\pi}{3}$ cu units

7. $\dfrac{16\pi}{15}$ cu units 9. $\dfrac{\pi}{2}(e^2 - 1)$ cu units 11. $\dfrac{2\pi}{15}$ cu units

13. $\dfrac{136\pi}{15}$ cu units 15. $\dfrac{64\sqrt{2}\pi}{3}$ cu units

17. $\dfrac{\pi}{2}(e^2 - 2 + e^{-2})$ cu units 19. $\dfrac{\pi}{6}$ cu units

21. $\dfrac{6517\pi}{240}$ cu units 23. $\dfrac{64\sqrt{2}\pi}{3}$ cu units

25. $\dfrac{8(\sqrt{2} - 1)\pi}{3}$ cu units 27. $\dfrac{4\pi}{3}r^3$ cu units

29. $50{,}000\pi$ cu ft

Chapter 6 Concept Review, page 487

1. a. $F'(x) = f(x)$ b. $F(x) + C$

2. a. $c\int f(x)\,dx$ b. $\int f(x)\,dx \pm \int g(x)\,dx$

3. a. Unknown b. Function 4. $g'(x)\,dx$; $f(u)\,du$

5. a. $\int_a^b f(x)\,dx$ b. Minus

6. a. $F(b) - F(a)$; antiderivative b. $\int_a^b f'(x)\,dx$

7. a. $\dfrac{1}{b - a}\displaystyle\int_a^b f(x)\,dx$ b. Area; area 8. $\int_a^b [f(x) - g(x)]\,dx$

9. a. $\int_0^{\bar{x}} D(x)\,dx - \bar{p}\,\bar{x}$ b. $\bar{p}\,\bar{x} - \int_0^{\bar{x}} S(x)\,dx$

10. a. $e^{rT}\int_0^T R(t)e^{-rt}\,dt$ b. $\int_0^T R(t)e^{-rt}\,dt$

11. $A = \dfrac{mP}{r}(e^{rT} - 1)$ 12. $L = 2\int_0^1 [x - f(x)]\,dx$

13. region; $x = a$; $x = b$; $v = \pi\int_a^b [f(x)]^2\,dx$

Chapter 6 Review Exercises, page 487

1. $\frac{1}{4}x^4 + \frac{2}{3}x^3 - \frac{1}{2}x^2 + C$ 2. $\frac{1}{12}x^4 - \frac{2}{3}x^3 + 8x + C$

3. $\frac{1}{5}x^5 - \frac{1}{2}x^4 - \frac{1}{x} + C$ 4. $\frac{3}{4}x^{4/3} - \frac{2}{3}x^{3/2} + 4x + C$

5. $\frac{1}{2}x^4 + \frac{2}{5}x^{5/2} + C$ 6. $\frac{2}{7}x^{7/2} - \frac{1}{3}x^3 + \frac{2}{3}x^{3/2} - x + C$

7. $\frac{1}{3}x^3 - \frac{1}{2}x^2 + 2\ln|x| + 5x + C$ 8. $\frac{1}{3}(2x + 1)^{3/2} + C$

9. $\frac{3}{8}(3x^2 - 2x + 1)^{4/3} + C$ 10. $\dfrac{(x^3 + 2)^{11}}{33} + C$

11. $\frac{1}{2}\ln(x^2 - 2x + 5) + C$ 12. $-e^{-2x} + C$

13. $\frac{1}{2}e^{x^2+x+1} + C$ 14. $\dfrac{1}{e^{-x} + x} + C$ 15. $\frac{1}{6}(\ln x)^6 + C$

16. $(\ln x)^2 + C$ 17. $\dfrac{(11x^2 - 1)(x^2 + 1)^{11}}{264} + C$

18. $\frac{2}{15}(3x - 2)(x + 1)^{3/2} + C$ 19. $\frac{2}{3}(x + 4)\sqrt{x - 2} + C$

20. $2(x - 2)\sqrt{x + 1} + C$ 21. $\frac{1}{2}$ 22. -6 23. $5\frac{2}{3}$

24. 242 25. -80 26. $\frac{132}{5}$ 27. $\frac{1}{2}\ln 5$ 28. $\frac{1}{15}$

29. 4 30. $1 - \dfrac{1}{e^2}$ 31. $\dfrac{e - 1}{2(1 + e)}$ 32. $\frac{1}{2}$

33. $f(x) = x^3 - 2x^2 + x + 1$ 34. $f(x) = \sqrt{x^2 + 1}$

35. $f(x) = x + e^{-x} + 1$ 36. $f(x) = \frac{1}{2}(\ln x)^2 - 2$

37. -4.28 38. \$6740

39. a. $-0.015x^2 + 60x$; b. $p = -0.015x + 60$

40. $V(t) = 1900(t - 10)^2 + 10{,}000$; \$40,400

41. $3000t - 50{,}000(1 - e^{-0.04t})$; 16,939

42. $N(t) = 15{,}000\sqrt{1 + 0.4t} + 85{,}000$; 112,659

43. 26,027 44. \$3100 45. 37.7 million

46. 19.4 billion metric tons 47. 15 sq units

48. $\frac{1}{2}(e^4 - 1)$ sq units 49. $\frac{2}{3}$ sq units 50. $\frac{9}{2}$ sq units

51. $(e^2 - 3)$ sq units 52. $\frac{3}{10}$ sq units 53. $\frac{1}{2}$ sq units

54. 234,500 barrels 55. $\frac{1}{3}$ sq units 56. 26°F 57. \$270,000

58. Consumers' surplus: \$2083; producers' surplus: \$3333

59. \$197,652 60. \$98,973 61. \$505,696

62. a.

y
1.0
0.8
0.6
0.4
0.2
0.2 0.4 0.6 0.8 1.0
x

b. 0.1017; 0.3733 c. 0.315

63. 70,784 64. $\dfrac{2\pi}{3}$ cu units 65. $\dfrac{3\pi}{10}$ cu units

Chapter 6 Before Moving On, page 490

1. $\frac{1}{2}x^4 + \frac{2}{3}x^{3/2} + 2\ln|x| - 4\sqrt{x} + C$ 2. $e^x + \frac{1}{2}x^2 + 1$

3. $\sqrt{x^2 + 1} + C$ 4. $\frac{1}{3}(2\sqrt{2} - 1)$ 5. $\frac{9}{2}$ sq units

CHAPTER 7

Exercises 7.1, page 497

1. $\frac{1}{4}e^{2x}(2x - 1) + C$ 3. $4(x - 4)e^{x/4} + C$

5. $\frac{1}{2}e^{2x} - 2(x - 1)e^x + \frac{1}{3}x^3 + C$ 7. $xe^x + C$

9. $\dfrac{2(x+2)}{\sqrt{x+1}} + C$ 11. $\frac{2}{3}x(x-5)^{3/2} - \frac{4}{15}(x-5)^{5/2} + C$

13. $\dfrac{x^2}{4}(2\ln 2x - 1) + C$ 15. $\dfrac{x^4}{16}(4\ln x - 1) + C$

17. $\frac{2}{9}x^{3/2}(3\ln\sqrt{x} - 1) + C$ 19. $-\dfrac{1}{x}(\ln x + 1) + C$

21. $x(\ln x - 1) + C$ 23. $-(x^2 + 2x + 2)e^{-x} + C$

25. $\frac{1}{4}x^2[2(\ln x)^2 - 2\ln x + 1] + C$ 27. $2\ln 2 - 1$

29. $4\ln 4 - 3$ 31. $\frac{1}{4}(3e^4 + 1)$ 33. $-\frac{1}{2}xe^{-2x} - \frac{1}{4}e^{-2x} + \frac{13}{4}$

35. $5\ln 5 - 4$ 37. 1485 ft 39. 2.04 mg/mL

41. $-20e^{-0.1t}(t + 10) + 200$ 43. $131,324 45. 101,606

47. $(c_2 - c_1)\left[\dfrac{r_1}{r_2 - r_1} + \dfrac{1}{\ln r_1 - \ln r_2}\right] + c_2$ moles/L 49. True

Exercises 7.2, page 505

1. $\frac{2}{9}[2 + 3x - 2\ln|2 + 3x|] + C$

3. $\frac{3}{32}[(1 + 2x)^2 - 4(1 + 2x) + 2\ln|1 + 2x|] + C$

5. $2\left[\dfrac{x}{8}\left(\dfrac{9}{4} + 2x^2\right)\sqrt{\dfrac{9}{4} + x^2} - \dfrac{81}{128}\ln\left(x + \sqrt{\dfrac{9}{4} + x^2}\right)\right] + C$

7. $\ln\left(\dfrac{\sqrt{1+4x} - 1}{\sqrt{1+4x} + 1}\right) + C$ 9. $\frac{1}{2}\ln 3$ 11. $\dfrac{x}{9\sqrt{9 - x^2}} + C$

13. $\dfrac{x}{8}(2x^2 - 4)\sqrt{x^2 - 4} - 2\ln|x + \sqrt{x^2 - 4}| + C$

15. $\sqrt{4 - x^2} - 2\ln\left|\dfrac{2 + \sqrt{4 - x^2}}{x}\right| + C$

17. $\frac{1}{4}(2x - 1)e^{2x} + C$ 19. $\ln|\ln(1 + x)| + C$

21. $\dfrac{1}{9}\left[\dfrac{1}{1 + 3e^x} + \ln(1 + 3e^x)\right] + C$

23. $6[e^{(1/2)x} - \ln(1 + e^{(1/2)x})] + C$

25. $\frac{1}{9}(2 + 3\ln x - 2\ln|2 + 3\ln x|) + C$

27. $e - 2$ 29. $\dfrac{x^3}{9}(3\ln x - 1) + C$

31. $x[(\ln x)^3 - 3(\ln x)^2 + 6\ln x - 6] + C$ 33. ≈$2329

35. 27,136 37. 44; 49 39. 26,157 41. $418,444

Exercises 7.3, page 518

1. 2.7037; 2.6667; $2\frac{2}{3}$ 3. 0.2656; 0.2500; $\frac{1}{4}$

5. 0.6970; 0.6933; ≈0.6931 7. 0.5090; 0.5004; $\frac{1}{2}$

9. 5.2650; 5.3046; $\frac{16}{3}$ 11. 0.6336; 0.6321; ≈0.6321

13. 0.3837; 0.3863; ≈0.3863 15. 1.1170; 1.1114

17. 1.3973; 1.4052 19. 0.8806; 0.8818 21. 3.7757; 3.7625

23. **a.** 3.6 **b.** 0.0324 25. **a.** 0.013 **b.** 0.00043

27. **a.** 0.0078125 **b.** 0.0002848 29. 52.84 mi

31. 21.65 mpg 33. 17.1 million barrels 35. 1922.4 ft^3/sec

37. **a.** $51,558 **b.** $51,708 39. 474.77 million barrels

41. ≈50% 43. 6.42 L/min 45. False 47. True

Exercises 7.4, page 528

1. $\frac{2}{3}$ sq unit 3. 1 sq unit 5. 2 sq units 7. $\frac{2}{3}$ sq unit

9. $\frac{1}{2}e^4$ sq units 11. 1 sq unit 13. **a.** $\frac{2}{3}b^{3/2}$ 15. 1

17. 2 19. Divergent 21. $-\frac{1}{8}$ 23. 1 25. 1

27. $\frac{1}{2}$ 29. Divergent 31. −1 33. Divergent 35. 0

37. 0 39. Divergent 41. Convergent 43. $18,750

45. $\dfrac{10{,}000r + 4000}{r^2}$ dollars 47. True 49. False

51. **b.** $83,333

Chapter 7 Concept Review, page 531

1. Product: $uv - \int v\,du$; u; easy to integrate 2. $x^2 + 1$; $2x\,dx$; (27)

3. $\dfrac{\Delta x}{2}[f(x_0) + 2f(x_1) + \cdots + 2f(x_{n-1}) + f(x_n)]$; $\dfrac{M(b-a)^3}{12n^2}$

4. $\dfrac{\Delta x}{3}[f(x_0) + 4f(x_1) + 2f(x_2) + 4f(x_3) + 2f(x_4)$ $+ \cdots + 4f(x_{n-1}) + f(x_n)]$; even; $\dfrac{M(b-a)^5}{180n^4}$

5. $\lim_{a\to-\infty}\int_a^b f(x)\,dx$; $\lim_{b\to\infty}\int_a^b f(x)\,dx$; $\int_{-\infty}^c f(x)\,dx + \int_c^\infty f(x)\,dx$

Chapter 7 Review Exercises, page 531

1. $-2(1 + x)e^{-x} + C$ 2. $\frac{1}{16}(4x - 1)e^{4x} + C$

3. $x(\ln 5x - 1) + C$ 4. $4\ln 8 - \ln 2 - 3$ 5. $\frac{1}{4}(1 - 3e^{-2})$

6. $\frac{1}{4}(1 + 3e^4)$ 7. $2\sqrt{x}(\ln x - 2) + 2$

8. $-\frac{1}{3}xe^{-3x} - \frac{1}{9}e^{-3x} + \frac{1}{9}$

9. $\dfrac{1}{8}\left[3 + 2x - \dfrac{9}{3 + 2x} - 6\ln|3 + 2x|\right] + C$

10. $\frac{2}{3}(x - 3)\sqrt{2x + 3} + C$ 11. $\frac{1}{32}e^{4x}(8x^2 - 4x + 1) + C$

12. $-\dfrac{x}{25\sqrt{x^2 - 25}} + C$ 13. $\dfrac{1}{4}\dfrac{\sqrt{x^2 - 4}}{x} + C$

14. $\frac{1}{2}x^4(4\ln 2x - 1) + C$ 15. $\frac{1}{2}$ 16. $\frac{1}{3}$ 17. Divergent

18. 1 19. $\frac{1}{10}$ 20. 3 21. 0.8421; 0.8404

22. 1.491; 1.464 23. 2.2379; 2.1791 24. 8.1310; 8.041

25. **a.** 0.002604 **b.** 0.000033 26. $1,157,641 27. 48,761

28. $41,100 29. 274,000 ft^2; 278,667 ft^2 30. 7850 sq ft

31. $111,111

Chapter 7 Before Moving On, page 533

1. $\frac{1}{3}x^3 \ln x - \frac{1}{9}x^3 + C$ 2. $-\dfrac{\sqrt{8 + 2x^2}}{8x} + C$ 3. 6.3367

4. 3.00358 5. $\dfrac{1}{2e^2}$

CHAPTER 8

Exercises 8.1, page 542

1. $f(0, 0) = -4; f(1, 0) = -2; f(0, 1) = -1; f(1, 2) = 4;$ $f(2, -1) = -3$

3. $f(1, 2) = 7; f(2, 1) = 9; f(-1, 2) = 1; f(2, -1) = 1$

5. $g(1, 2) = 4 + 3\sqrt{2}; g(2, 1) = 8 + \sqrt{2}; g(0, 4) = 2; g(4, 9) = 56$

7. $h(1, e) = 1; h(e, 1) = -1; h(e, e) = 0$

9. $g(1, 1, 1) = e; g(1, 0, 1) = 1; g(-1, -1, -1) = -e$

11. All real values of x and y

13. All real values of u and v except those satisfying the equation $u = v$

15. All real values of r and s satisfying $rs \geq 0$

17. All real values of x and y satisfying $x + y > 5$

19.

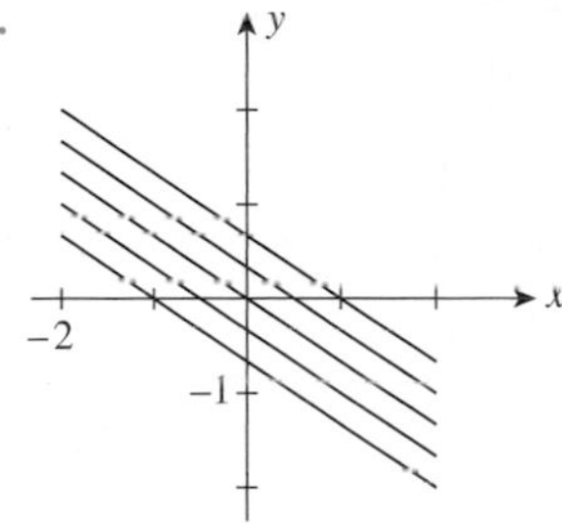

21.

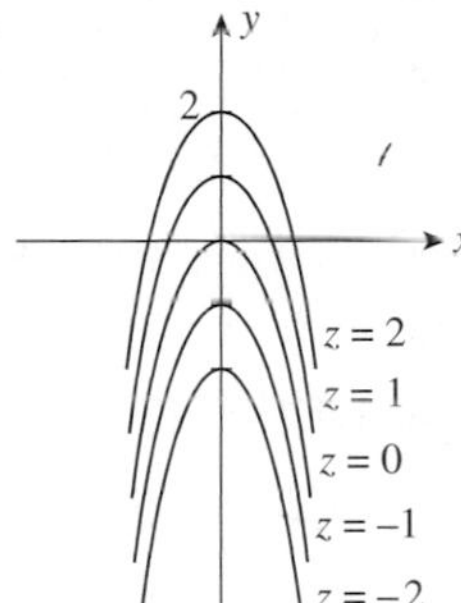

23.

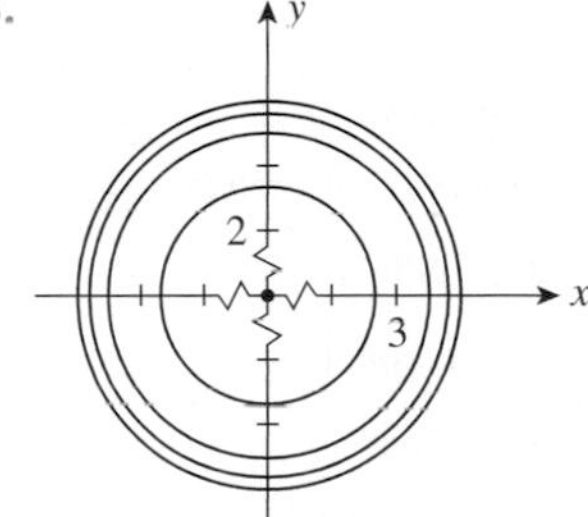

25. 9π ft³ 27. **a.** 24.69 **b.** 81 kg

29. **a.** $-\frac{1}{5}x^2 - \frac{1}{4}y^2 - \frac{1}{5}xy + 200x + 160y$
   b. The set of all points (x, y) satisfying $200 - \frac{1}{5}x - \frac{1}{10}y \geq 0$, $160 - \frac{1}{10}x - \frac{1}{4}y \geq 0$

31. **a.** $-0.005x^2 - 0.003y^2 - 0.002xy + 20x + 15y$
   b. The set of all ordered pairs (x, y) for which
   $20 - 0.005x - 0.001y \geq 0$
   $15 - 0.001x - 0.003y \geq 0$

33. **a.** The set of all ordered pairs (P, T), where P and T are positive numbers
   b. 11.10 L

35. \$7200 billion 37. 103

39. **a.** \$733.76; \$877.57 **b.** \$836.44

41. 40.28 times gravity 43. False 45. False

Exercises 8.2, page 555

1. 2; 3 3. $4x$; 4 5. $-\dfrac{4y}{x^3}; \dfrac{2}{x^2}$ 7. $\dfrac{2v}{(u + v)^2}; -\dfrac{2u}{(u + v)^2}$

9. $3(2s - t)(s^2 - st + t^2)^2; 3(2t - s)(s^2 - st + t^2)^2$

11. $\dfrac{4x}{3(x^2 + y^2)^{1/3}}; \dfrac{4y}{3(x^2 + y^2)^{1/3}}$ 13. $ye^{xy+1}; xe^{xy+1}$

15. $\ln y + \dfrac{y}{x}; \dfrac{x}{y} + \ln x$ 17. $e^u \ln v; \dfrac{e^u}{v}$

19. $yz + y^2 + 2xz; xz + 2xy + z^2; xy + 2yz + x^2$

21. $ste^{rst}; rte^{rst}; rse^{rst}$ 23. $f_x(1, 2) = 8; f_y(1, 2) = 5$

25. $f_x(2, 1) = 1; f_y(2, 1) = 3$ 27. $f_x(1, 2) = \frac{1}{2}; f_y(1, 2) = -\frac{1}{4}$

29. $f_x(1, 1) = e; f_y(1, 1) = e$

31. $f_x(1, 0, 2) = 0; f_y(1, 0, 2) = 8; f_z(1, 0, 2) = 0$

33. $f_{xx} = 2y; f_{xy} = 2x + 3y^2 = f_{yx}; f_{yy} = 6xy$

35. $f_{xx} = 2; f_{xy} = f_{yx} = -2; f_{yy} = 4$

37. $f_{xx} = \dfrac{y^2}{(x^2 + y^2)^{3/2}}; f_{xy} = f_{yx} = -\dfrac{xy}{(x^2 + y^2)^{3/2}};$
   $f_{yy} = \dfrac{x^2}{(x^2 + y^2)^{3/2}}$

39. $f_{xx} = \dfrac{1}{y^2}e^{-x/y}; f_{xy} = \dfrac{y - x}{y^3}e^{-x/y} = f_{yx};$
   $f_{yy} = \dfrac{x}{y^3}\left(\dfrac{x}{y} - 2\right)e^{-x/y}$

41. **a.** $f_x = 7.5; f_y = 40$ **b.** Yes

43. $p_x = 10$—at (0, 1), the price of land is changing at the rate of \$10/ft²/mile change to the right; $p_y = 0$—at (0, 1), the price of land is constant/mile change upward.

45. Complementary commodities

47. \$30/unit change in finished desks; −\$25/unit change in unfinished desks. The weekly revenue increases by \$30/unit for each additional finished desk produced (beyond 300) when the level of production of unfinished desks remains fixed at 250; the revenue decreases by \$25/unit when each additional unfinished desk (beyond 250) is produced and the level of production of finished desks remains fixed at 300.

49. **a.** ≈20°F **b.** ≈ −0.3°F

51. 0.039 L/degree; −0.015 L/mm of mercury. The volume increases by 0.039 L when the temperature increases by 1 degree (beyond 300 K) and the pressure is fixed at 800 mm of mercury. The volume decreases by 0.015 L when the pressure increases by 1 mm of mercury (beyond 800 mm) and the temperature is fixed at 300 K.

55. False 57. True

Using Technology Exercises 8.2, page 558

1. 1.3124; 0.4038 3. −1.8889; 0.7778 5. −0.3863; −0.8497

Exercises 8.3, page 567

1. (0, 0); relative maximum value: $f(0, 0) = 1$

3. (1, 2); saddle point: $f(1, 2) = 4$

5. (8, −6); relative minimum value: $f(8, -6) = -41$

7. (1, 2) and (2, 2); saddle point: $f(1, 2) = -1$; relative minimum value: $f(2, 2) = -2$

9. $(-\frac{1}{3}, \frac{11}{3})$ and (1, 5); saddle point: $f(-\frac{1}{3}, \frac{11}{3}) = -\frac{319}{27}$; relative minimum value: $f(1, 5) = -13$

11. (0, 0) and (1, 1); saddle point: $f(0, 0) = -2$; relative minimum value: $f(1, 1) = -3$

13. (2, 1); relative minimum value: $f(2, 1) = 6$

15. (0, 0); saddle point: $f(0, 0) = -1$

17. (0, 0); relative minimum value: $f(0, 0) = 1$

19. (0, 0); relative minimum value: $f(0, 0) = 0$

21. 200 finished units and 100 unfinished units; \$10,500

23. Price of land (\$200/ft^2) is highest at $(\frac{1}{2}, 1)$

25. (0, 1) gives desired location. 27. 10″ × 10″ × 5″; 500 in.^{3}

29. 30″ × 40″ × 10″; \$7200 31. False

Exercises 8.4, page 576

1. **a.** $y = 2.3x + 1.5$
b.

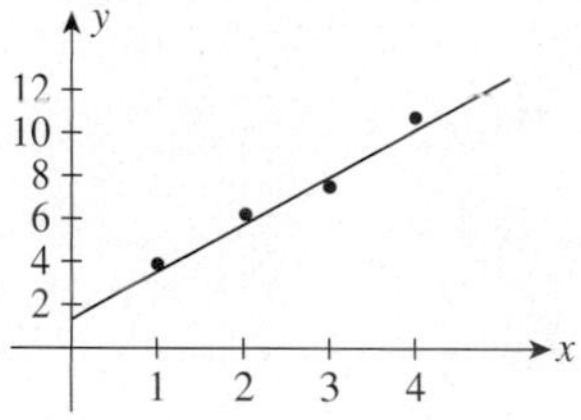

3. **a.** $y = -0.77x + 5.74$
b.

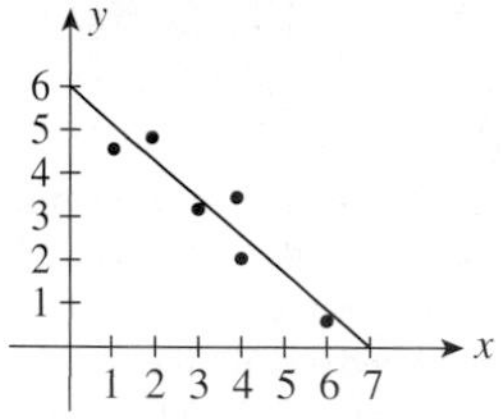

5. **a.** $y = 1.2x + 2$
b.

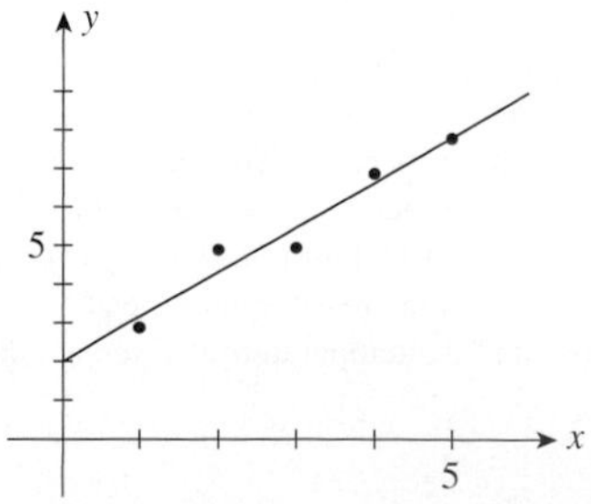

7. **a.** $y = 0.34x - 0.9$
b.

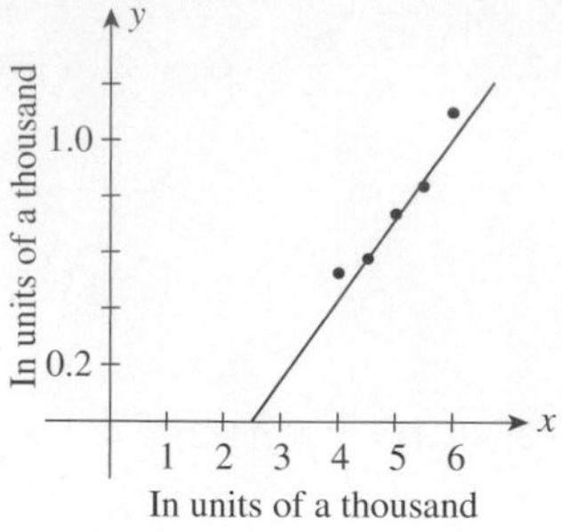

c. 1276 applications

9. **a.** $y = -2.8x + 440$
b.

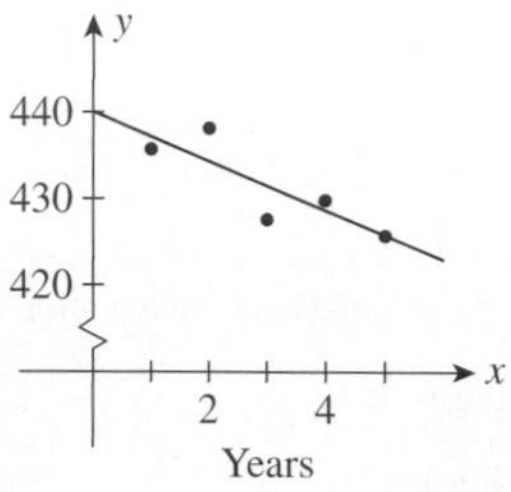

c. 420

11. **a.** $y = 2.8x + 17.6$ **b.** \$40,000,000

13. **a.** $y = 22.7x + 124.2$ **b.** \$260.4 billion

15. **a.** $y = 0.4x + 3.6$ **b.** \$0.4 billion/yr

17. **a.** $y = 1.76x + 7.9$ **b.** \$16.7 billion

19. **a.** $y = 2.43x + 78.88$ **b.** \$103,180

21. **a.** $y = 0.09x + 15.90$ **b.** 19.50 yr **c.** 18.6 yr

23. **a.** $y = 10x + 90.34$ **b.** 150.34 million; good estimate

25. False 27. True

Using Technology Exercises 8.4, page 580

1. $y = 2.3596x + 3.8639$ 3. $y = -1.1948x + 3.5525$

5. **a.** $y = 0.5471x + 1.1671$ **b.** \$5.54 billion

7. **a.** $y = 13.321x + 72.571$ **b.** 192 million tons

9. **a.** $y = 1.95x + 12.19$ **b.** \$23.89 billion

Exercises 8.5, page 590

1. Min. of $\frac{3}{4}$ at $(\frac{3}{4}, \frac{1}{4})$ 3. Max. of $-\frac{7}{4}$ at $(2, \frac{7}{2})$

5. Min. of 4 at $(\sqrt{2}, \sqrt{2}/2)$ and $(-\sqrt{2}, -\sqrt{2}/2)$

7. Max. of $-\frac{3}{4}$ at $(\frac{3}{2}, 1)$

9. Max. of $2\sqrt{3}$ at $(\sqrt{3}/3, -\sqrt{6})$ and $(\sqrt{3}/3, \sqrt{6})$

11. Max. of 8 at $(2\sqrt{2}, 2\sqrt{2})$ and $(-2\sqrt{2}, -2\sqrt{2})$; min. of −8 at $(2\sqrt{2}, -2\sqrt{2})$ and $(-2\sqrt{2}, 2\sqrt{2})$

13. Max.: $\frac{2\sqrt{3}}{9}$; min.: $-\frac{2\sqrt{3}}{9}$ 15. Min. of $\frac{18}{7}$ at $(\frac{9}{7}, \frac{6}{7}, \frac{3}{7})$

17. 140 finished and 60 unfinished units

19. $10\sqrt{2}$ ft × $40\sqrt{2}$ ft 21. $r = \frac{4}{3}\sqrt[3]{\frac{18}{\pi}}$ in.; $h = 2\sqrt[3]{\frac{18}{\pi}}$ in.

23. $\frac{2}{3}\sqrt[3]{9} \times \frac{2}{3}\sqrt[3]{9} \times \sqrt[3]{9}$

25. 1500 units on labor and 250 units of capital 27. False

Exercises 8.6, page 597

1. $2x\,dx + 2\,dy$ 3. $(4x - 3y + 4)\,dx - 3x\,dy$

5. $\frac{x}{\sqrt{x^2 + y^2}}\,dx + \frac{y}{\sqrt{x^2 + y^2}}\,dy$

7. $-\frac{5y}{(x - y)^2}\,dx + \frac{5x}{(x - y)^2}\,dy$

9. $(10x^4 + 3ye^{-3x})\,dx - e^{-3x}\,dy$

11. $\left(2xe^y + \frac{y}{x}\right)dx + (x^2e^y + \ln x)\,dy$

13. $y^2z^3\,dx + 2xyz^3\,dy + 3xy^2z^2\,dz$

15. $\frac{1}{y + z}\,dx - \frac{x}{(y + z)^2}\,dy - \frac{x}{(y + z)^2}\,dz$

17. $(yz + e^{yz})\,dx + xz(1 + e^{yz})\,dy + xy(1 + e^{yz})\,dz$

19. 0.04 21. −0.01 23. −0.10 25. −0.18

27. 0.06 29. −0.1401 31. An increase of \$19,250/mo

33. An increase of \$5000/mo 35. \$1.25 37. 38.4π cu cm

39. 0.55 ohm

Exercises 8.7, page 605

1. $\frac{7}{2}$ 3. 0 5. $4\frac{1}{2}$ 7. $(e^2 - 1)(1 - e^{-2})$ 9. 1

11. $\frac{2}{3}$ 13. $\frac{188}{3}$ 15. $\frac{84}{5}$ 17. $2\frac{2}{3}$ 19. 1 21. $\frac{1}{2}(3 - e)$

23. $\frac{1}{4}(e^4 - 1)$ 25. $\frac{2}{3}(e - 1)$ 27. False

Exercises 8.8, page 612

1. 42 cu units 3. 20 cu units 5. $\frac{64}{3}$ cu units

7. $\frac{1}{3}$ cu unit 9. 4 cu units 11. $3\frac{1}{3}$ cu units

13. $2(e^2 - 1)$ cu units 15. $\frac{2}{35}$ cu unit

17. 54 19. $\frac{1}{3}$ 21. 1

23. 43,329 25. 312,439 27. \$194/sq ft

29. True

Chapter 8 Concept Review, page 615

1. xy; ordered pair; real number; $f(x, y)$

2. Independent; dependent; value 3. $z = f(x, y)$; f; surface

4. $f(x, y) = c$; level curve; level curves; c

5. Fixed number; x 6. Slope; $(a, b, f(a, b))$; x; b

7. $\le$; (a, b); $\le$; domain

8. Domain; $f_x(a, b) = 0$ and $f_y(a, b) = 0$; exist; candidate

9. Scatter; minimizing; least-squares; normal

10. $g(x, y) = 0$; $f(x, y) + \lambda g(x, y)$; $F_x = 0$; $F_y = 0$; $F_\lambda = 0$; extrema

11. independent, $dx = \Delta x$, $dy = \Delta y$, $\frac{\partial f}{\partial x}\,dx + \frac{\partial f}{\partial y}\,dy$

12. Volume; solid 13. Iterated; $\int_3^5 \int_0^1 (2x + y^2)\,dx\,dy$

Chapter 8 Review Exercises, page 615

1. 0, 0, $\frac{1}{2}$; no 2. e, $\frac{e^2}{1 + \ln 2}$, $\frac{2e}{1 + \ln 2}$; no

3. 2, $-(e + 1)$, $-(e + 1)$

4. The set of all ordered pairs (u, v) such that $u \neq v$

5. The set of all ordered pairs (x, y) such that $y \neq -x$

6. The set of all ordered pairs (x, y) such that $x \le 1$ and $y \ge 0$

7. The set of all triplets (x, y, z) such that $z \ge 0$ and $x \neq 1$, $y \neq 1$, and $z \neq 1$

8. $2x + 3y = z$

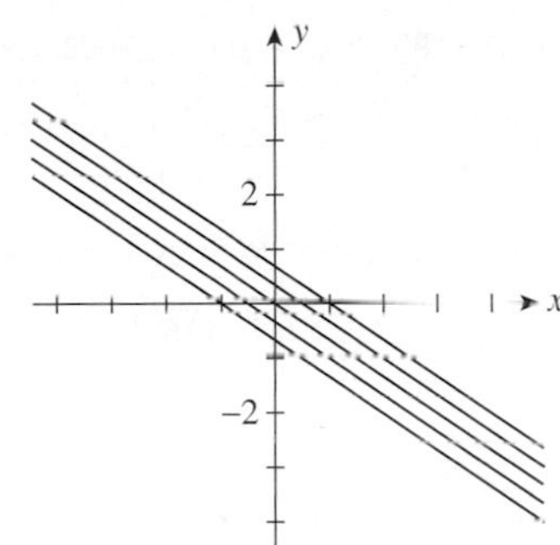

9. $z = y - x^2$

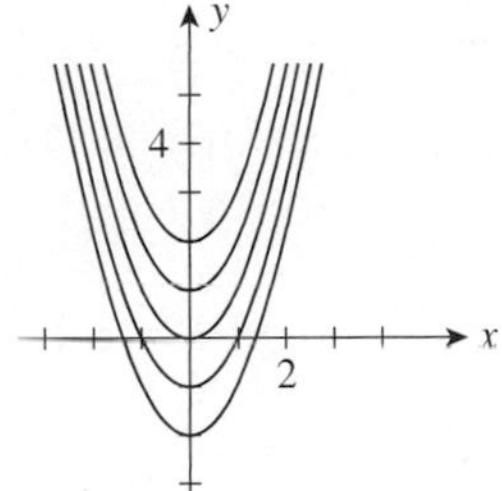

10. $z = \sqrt{x^2 + y^2}$

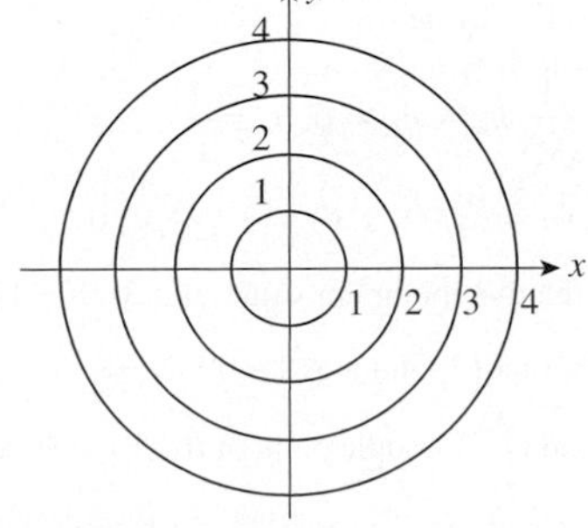

11. $z = e^{xy}$

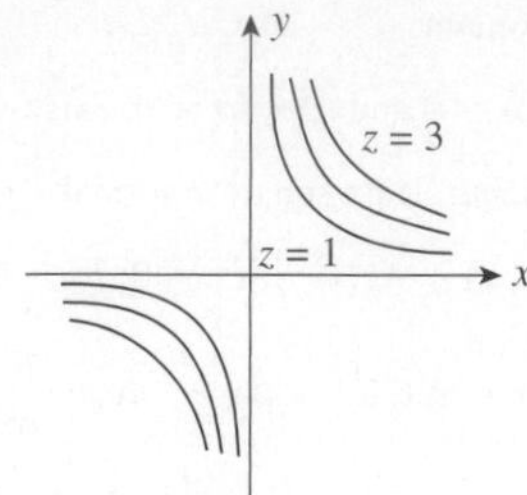

12. $f_x = 2xy^3 + 3y^2 + \dfrac{1}{y};\ f_y = 3x^2y^2 + 6xy - \dfrac{x}{y^2}$

13. $f_x = \sqrt{y} + \dfrac{y}{2\sqrt{x}};\ f_y = \dfrac{x}{2\sqrt{y}} + \sqrt{x}$

14. $f_u = \dfrac{v^2 - 2}{2\sqrt{uv^2 - 2u}};\ f_v = \dfrac{uv}{\sqrt{uv^2 - 2u}}$

15. $f_x = \dfrac{3y}{(y + 2x)^2};\ f_y = -\dfrac{3x}{(y + 2x)^2}$

16. $g_x = \dfrac{y(y^2 - x^2)}{(x^2 + y^2)^2};\ g_y = \dfrac{x(x^2 - y^2)}{(x^2 + y^2)^2}$

17. $h_x = 10y(2xy + 3y^2)^4;\ h_y = 10(x + 3y)(2xy + 3y^2)^4$

18. $f_x = \dfrac{e^y}{2(xe^y + 1)^{1/2}};\ f_y = \dfrac{xe^y}{2(xe^y + 1)^{1/2}}$

19. $f_x = 2x(1 + x^2 + y^2)e^{x^2+y^2};\ f_y = 2y(1 + x^2 + y^2)e^{x^2+y^2}$

20. $f_x = \dfrac{4x}{1 + 2x^2 + 4y^4};\ f_y = \dfrac{16y^3}{1 + 2x^2 + 4y^4}$

21. $f_x = \dfrac{2x}{x^2 + y^2};\ f_y = -\dfrac{2x^2}{y(x^2 + y^2)}$

22. $f_{xx} = 6x - 4y;\ f_{xy} = -4x = f_{yx};\ f_{yy} = 2$

23. $f_{xx} = 12x^2 + 4y^2;\ f_{xy} = 8xy = f_{yx};\ f_{yy} = 4x^2 - 12y^2$

24. $f_{xx} = 12(2x^2 + 3y^2)(10x^2 + 3y^2);\ f_{xy} = 144xy(2x^2 + 3y^2) = f_{yx};$
$f_{yy} = 18(2x^2 + 3y^2)(2x^2 + 15y^2)$

25. $g_{xx} = \dfrac{-2y^2}{(x + y^2)^3};\ g_{xy} = \dfrac{2y(x - y^2)}{(x + y^2)^3} = g_{yx};$
$g_{yy} = \dfrac{2x(3y^2 - x)}{(x + y^2)^3}$

26. $g_{xx} = 2(1 + 2x^2)e^{x^2+y^2};\ g_{xy} = 4xye^{x^2+y^2} = g_{yx};$
$g_{yy} = 2(1 + 2y^2)e^{x^2+y^2}$

27. $h_{ss} = -\dfrac{1}{s^2};\ h_{st} = h_{ts} = 0;\ h_{tt} = \dfrac{1}{t^2}$

28. $f_x(1, 1, 0) = 3;\ f_y(1, 1, 0) = 3;\ f_z(1, 1, 0) = -2$

29. (2, 3); relative minimum value: $f(2, 3) = -13$

30. (8, −2); saddle point at $f(8, -2) = -8$

31. (0, 0) and $(\frac{3}{2}, \frac{9}{4})$; saddle point at $f(0, 0) = 0$; relative minimum value: $f(\frac{3}{2}, \frac{9}{4}) = -\frac{27}{16}$

32. $(-\frac{1}{3}, \frac{13}{3})$, (3, 11); saddle point at $f(-\frac{1}{3}, \frac{13}{3}) = (-\frac{445}{27})$; relative minimum value: $f(3, 11) = -35$

33. (0, 0); relative minimum value: $f(0, 0) = 1$

34. (1, 1); relative minimum value: $f(1, 1) = \ln 2$

35. $f(\frac{12}{11}, \frac{20}{11}) = -\frac{32}{11}$ 36. $f(\frac{1}{22}, \frac{21}{22}) = \frac{179}{44}$

37. Relative maximum value: $f(5, -5) = 26$; relative minimum value: $f(-5, 5) = -24$

38. Relative maximum value: $f\left(\dfrac{\sqrt{2}}{2}, -\dfrac{\sqrt{2}}{2}\right) = e^{\sqrt{2}}$; relative minimum value: $f\left(-\dfrac{\sqrt{2}}{2}, \dfrac{\sqrt{2}}{2}\right) = e^{-\sqrt{2}}$

39. $45\,dx + 240\,dy$ 40. $2\,dx$ 41. 0.04 42. $\dfrac{227}{54}$

43. 48 44. $\frac{1}{2}(e^{-2} - 1)^2$ 45. $\frac{2}{63}$ 46. $\frac{1}{4}(3 - 2\ln 2)$

47. $\frac{34}{3}$ 48. $10\frac{4}{5}$ cu units 49. 3

50. **a.** $R(x, y) = -0.02x^2 - 0.2xy - 0.05y^2 + 80x + 60y$
b. The set of all points satisfying $0.02x + 0.1y \le 80$, $0.1x + 0.05y \le 60$, $x \ge 0$, $y \ge 0$
c. 15,300; the revenue realized from the sale of 100 16-speed and 300 10-speed electric blenders is $15,300.

51. Complementary 52. $490

53. **a.** $y = 8.2x + 361.2$ **b.** 7 hr 31 min

54. **a.** $y = 0.059x + 19.45$ **b.** 21.8 yr; same **c.** 21.2 yr

55. The company should spend $11,000 on advertising and employ 14 agents in order to maximize its revenue.

56. 337.5 yd × 900 yd 57. 75 units on labor; 25 units on capital

Chapter 8 Before Moving On, page 617

1. All real values of x and y satisfying $x \ge 0$, $x \ne 1$, $y \ge 0$, $y \ne 2$

2. $f_x = 2xy + ye^{xy};\ f_{xx} = 2y + y^2e^{xy};\ f_{xy} = 2x + (xy + 1)e^{xy} = f_{yx};$
$f_y = x^2 + xe^{xy};\ f_{yy} = x^2e^{xy}$

3. Rel. min.: (1, 1, −7) 4. $y = 2.04x + 2.88$ 5. $f(\frac{1}{2}, \frac{1}{2}) = \frac{5}{2}$

6. **a.** $(4x - y)\,dx - x\,dy$ **b.** −0.09 7. $\frac{1}{8}$

CHAPTER 9

Exercises 9.1, page 625

13. $y = 12x^2 - 2x$ 15. $y = \dfrac{1}{x}$ 17. $y = -\dfrac{e^x}{x} + \dfrac{1}{2}xe^x$

19. $\dfrac{dQ}{dt} = -kQ;\ Q(0) = Q_0$ 21. $\dfrac{dA}{dt} = k(C - A)$

23. $\dfrac{dC}{dt} = -kC;\ C(0) = C_0$ 29. True 31. False

33. False

Exercises 9.2, page 631

1. $\frac{1}{3}y^3 = \frac{1}{2}x^2 + x + C$ 3. $\frac{1}{3}y^3 = e^x + C$ 5. $y = ce^{2x}$

7. $-\dfrac{1}{y} = \dfrac{1}{2}x^2 + C$ 9. $y = -\frac{4}{3} + ce^{-6x}$

11. $y^3 = \frac{1}{3}x^3 + x + C$ 13. $y^{1/2} - x^{1/2} = C$

15. $\ln|y| = \frac{1}{2}(\ln x)^2 + C$ 17. $y^2 = 2x^2 + 2$

19. $y = 2 - e^{-x}$ 21. $y = \frac{2}{3} + \frac{1}{3}e^{(3/2)x^2}$ 23. $y = \sqrt{x^2 + 1}$

25. $\ln|y| = (x - 1)e^x$ 27. $y = \ln(x^3 + e)$

29. $y^2 = x^3 + 8$ 31. $Q(t) = Q_0e^{-kt}$ 33. $A = \dfrac{C}{k} - d_2e^{-kt}$

35. $S(t) = D - (D - S_0)e^{-kt}$ 37. True 39. True

41. False

Exercises 9.3, page 640

1. $y = y_0e^{-kt}$ 3. $Q(t) = 4.5e^{0.02t}$; 7.9 billion 5. $\frac{1}{2}$ in.

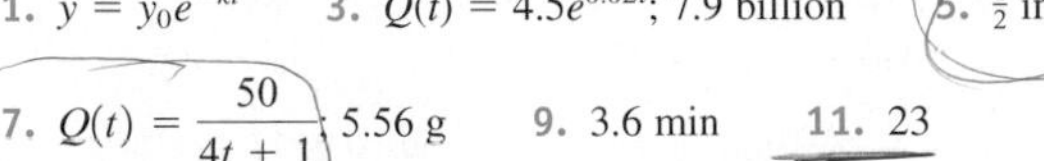
7. $Q(t) = \dfrac{50}{4t + 1}$; 5.56 g 9. 3.6 min 11. 23

13. 299.1 million 15. 395 17. $Q(t) = e^{C-(C-\ln Q_0)e^{-kt}}$

19. $x(t) = 40(1 - e^{-(3/20)t})$; 38 lb; 40 lb

Exercises 9.4, page 648

1. **a.** $y(1) = \frac{369}{128} \approx 2.8828$ **b.** $y(1) = \frac{70{,}993}{23{,}328} \approx 3.043$

3. **a.** $y(2) = \frac{51}{16} \approx 3.1875$ **b.** $y(2) = \frac{793}{243} \approx 3.2634$

5. **a.** $y(0.5) \approx 0.8324$ **b.** $y(0.5) \approx 0.8207$

7. **a.** $y(1.5) \approx 1.7831$ **b.** $y(1.5) \approx 1.7920$

9. **a.** $y(1) \approx 1.3390$ **b.** $y(1) \approx 1.3654$

11.

x	0.0	0.2	0.4	0.6	0.8	1
y_n	1	1	1.02	1.0608	1.1245	1.2144

13.

x	0.0	0.2	0.4	0.6	0.8	1
y_n	2	1.8	1.72	1.736	1.8288	1.9830

15.

x	0.0	0.1	0.2	0.3	0.4	0.5
y_n	1	1.1	1.211	1.3361	1.4787	1.6426

Chapter 9 Concept Review, page 649

1. **a.** differential equation **b.** satisfies

2. **a.** general **b.** particular

3. **a.** highest, unknown **b.** $f(x)g(y)$; not separable; separable
   c. separate, variable

4. **a.** kQ **b.** Q_0e^{kt} **c.** restricted

5. Approximate, actual, polygonal

Chapter 9 Review Exercises, page 650

4. $y = \dfrac{1}{x^2 + 1}$ 5. $y = (9x + 8)^{-1/3}$ 6. $\frac{1}{3}y^3 = \frac{1}{4}x^4 + x + C$

7. $y = 4 - Ce^{-2t}$ 8. $y = Ce^{(1/2)(\ln x)^2}$ 9. $y = -\dfrac{2}{2x^3 + 2x + 1}$

10. $y = 1 - 3e^{-(1/3)x^3}$ 11. $y = 3e^{-x^3/2}$ 12. $y = \dfrac{4}{(x^2 + 1)^2}$

13. **a.** $y(1) \approx 0.3849$ **b.** $y(1) \approx 0.4361$

14. **a.** $y(1) \approx 0.2219$ **b.** $y(1) = 0.2628$

15. **a.** $y(1) \approx 1.3258$ **b.** $y(1) \approx 1.4570$

16. **a.** $y(1) \approx 1.9083$ **b.** $y(1) = 2.1722$

17.

x	0.0	0.2	0.4	0.6	0.8	1
y_n	1	1	1.08	1.2528	1.5535	2.0506

18.

x	0	0.2	0.4	0.6	0.8	1
y_n	1	1.2	1.496	1.9756	2.8282	4.5560

19. **a.** $S = 50{,}000(0.8)^t$ **b.** \$16,384 20. $R = CS^k$

21. $A = \dfrac{P}{r}(e^{rt} - 1)$; \$342,549.50 22. $S = D - (D - S_0)e^{-kt}$

23. ≈7:30 P.M. 24. 105 words/min 25. **a.** 183

26. $x = 120(1 - e^{-0.1t})$; 120 lb

Chapter 9 Before Moving On, page 651

1. **b.** $y = 2x^2$ 2. $y = \dfrac{1}{xe^{x-1}}$ 3. 6.3 years 4. ≈1.7639

CHAPTER 10

Exercises 10.1, page 662

13. $k = \frac{1}{3}$ 15. $k = \frac{1}{8}$ 17. $k = \frac{3}{16}$ 19. $k = 2$

21. **a.** $\frac{1}{2}$ **b.** $\frac{5}{8}$ **c.** $\frac{7}{8}$ **d.** 0

23. **a.** $\frac{11}{16}$ **b.** $\frac{1}{2}$ **c.** $\frac{27}{32}$ **d.** 0

25. **a.** $\frac{1}{2}$ **b.** $\frac{1}{2}(2\sqrt{2} - 1)$ **c.** 0 **d.** $\frac{1}{2}$

27. **a.** 1 **b.** .14 29. **a.** .375 **b.** .75 **c.** .5 **d.** .875

35. $k = 2$ 37. $k = 12e^2$ 39. **a.** $\dfrac{1}{4}$ **b.** $\dfrac{1}{96}$

41. **a.** $\dfrac{1}{56}(2\sqrt{2} - 1)$ **b.** ≈.2575

43. **a.** .63 **b.** .30 **c.** .30 45. **a.** .10 **b.** .30

47. .9355 49. .4815; .7407 51. $\frac{1}{5}$ 53. ≈ .041

55. False

Using Technology Exercises 10.1, page 666

1.

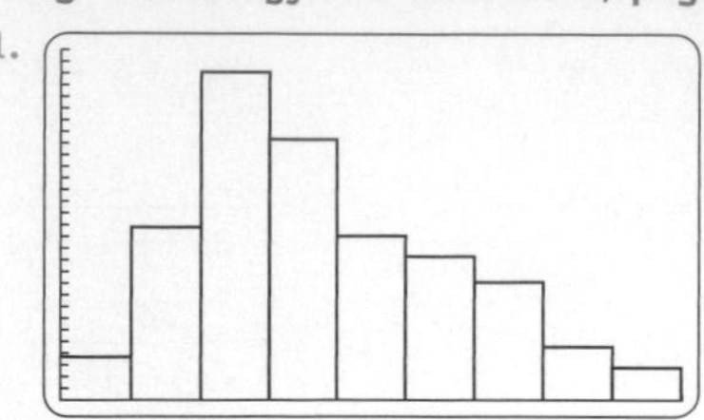

3. **a.**

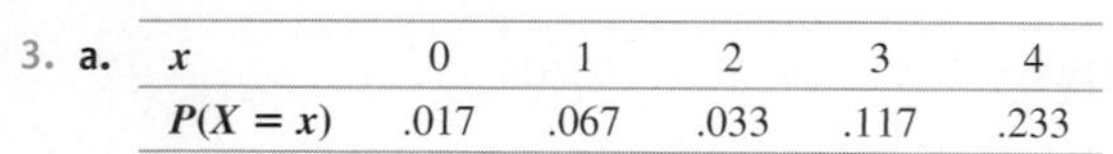

x	0	1	2	3	4
$P(X = x)$	.017	.067	.033	.117	.233

x	5	6	7	8	9	10
$P(X = x)$	.133	.167	.1	.05	.067	.017

b.

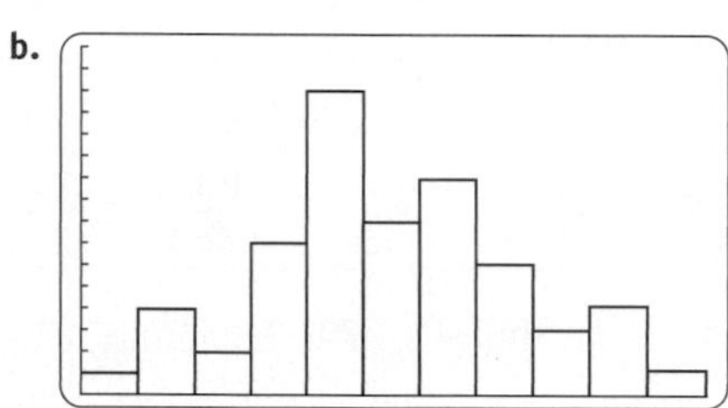

Exercises 10.2, page 678

1. $\mu = \frac{9}{2}$; $\text{Var}(x) = \frac{3}{4}$; $\sigma \approx 0.866$

3. $\mu = \frac{15}{4}$; $\text{Var}(x) = \frac{15}{16}$; $\sigma \approx 0.9682$

5. $\mu = 3$; $\text{Var}(x) = 0.8$; $\sigma \approx 0.8944$

7. $\mu = 2.3765$; $\text{Var}(x) \approx 2.3522$; $\sigma \approx 1.534$

9. $\mu = \frac{93}{35}$; $\text{Var}(x) \approx 0.7151$; $\sigma \approx 0.846$

11. $\mu = \frac{3}{2}$; $\text{Var}(x) = \frac{3}{4}$; $\sigma \approx \frac{1}{2}\sqrt{3}$ 13. $\mu = 4$; $\text{Var}(x) = 16$; $\sigma = 4$

15. 100 days 17. $3\frac{1}{3}$ min 19. 1.5 ft 21. 2500 lb/wk

23. $m = 5$ 25. $m \approx 2.52$ 27. $m = \frac{6}{5}$ 29. False

Using Technology Exercises 10.2, page 681

1. **a.**

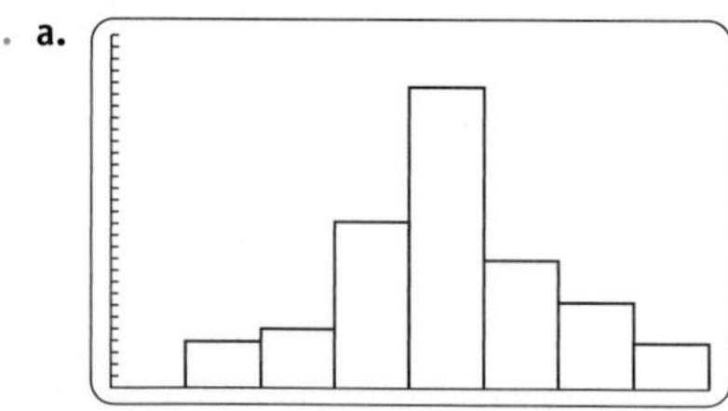

b. $\mu = 4$ and $\sigma = 1.40$

3. **a.** X gives the minimum age requirement for a regular driver's license.

b.

x	15	16	17	18	19	21
$P(X = x)$	.02	.30	.08	.56	.02	.02

c.

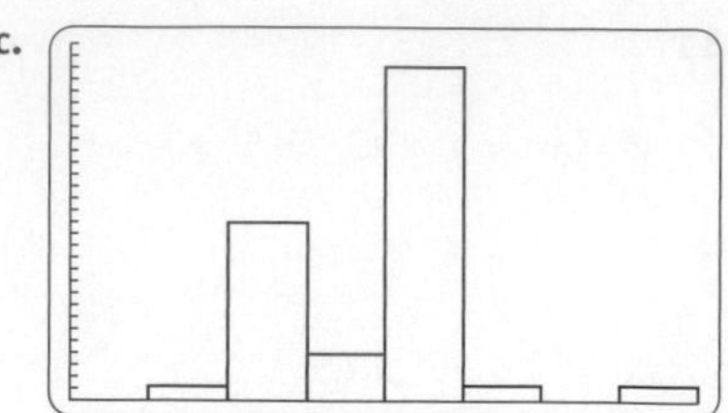

d. $\mu = 17.34$ and $\sigma = 1.11$

5. **a.** Let X denote the random variable that gives the weight of a carton of sugar.

b.

x	4.96	4.97	4.98	4.99	5.00	5.01
$P(X = x)$	$\frac{3}{30}$	$\frac{4}{30}$	$\frac{4}{30}$	$\frac{1}{30}$	$\frac{1}{30}$	$\frac{5}{30}$

x	5.02	5.03	5.04	5.05	5.06
$P(X = x)$	$\frac{3}{30}$	$\frac{3}{30}$	$\frac{4}{30}$	$\frac{1}{30}$	$\frac{1}{30}$

c. $\mu \approx 5.00$; $\text{Var}(X) \approx 0.0009$; $\sigma \approx 0.03$

Exercises 10.3, page 690

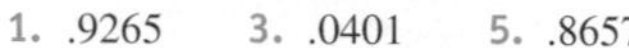

1. .9265 3. .0401 5. .8657

7. **a.**

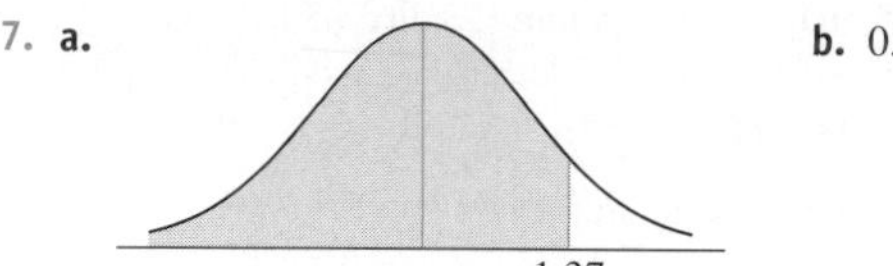

b. 0.9147

9. **a.**

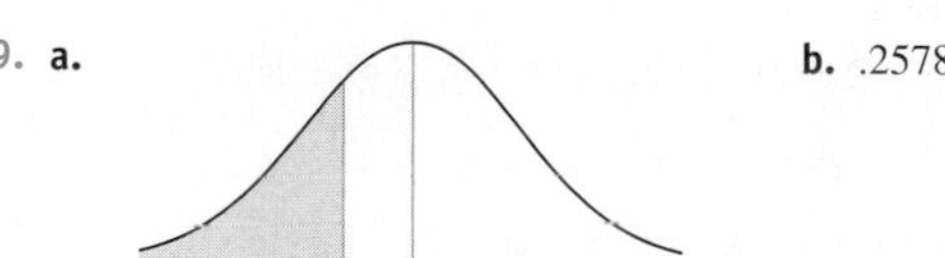

b. .2578

11. **a.**

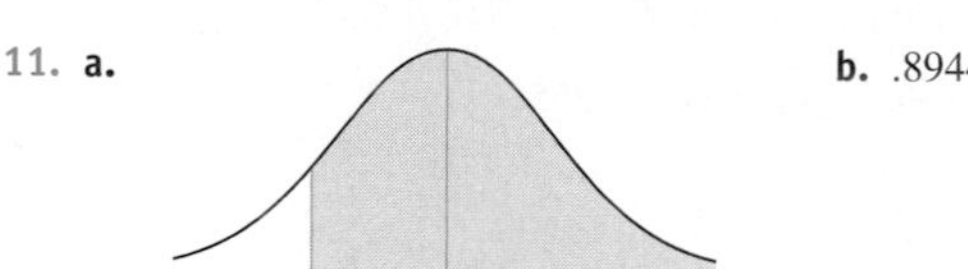

b. .8944

13. **a.**

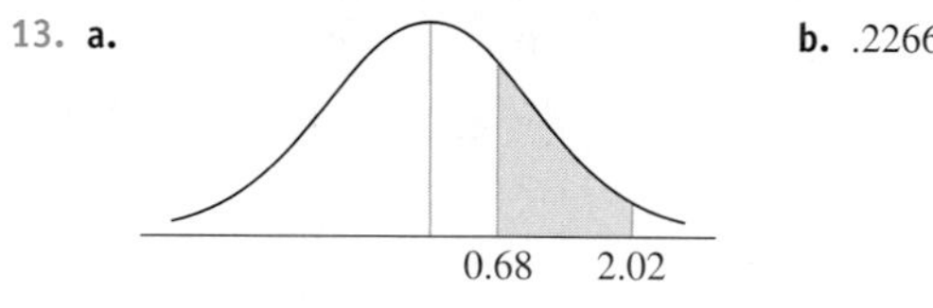

b. .2266

15. **a.** 1.23 **b.** −0.81 17. **a.** 1.9 **b.** −1.9

19. **a.** .9772 **b.** .9192 **c.** .7333

21. **a.** .2206 **b.** .2206 **c.** .3034

23. **a.** .0228 **b.** .0228 **c.** .4772 **d.** .7258

25. **a.** .0038 **b.** .0918 **c.** .4082 **d.** .2514

27. .6247 29. 0.62% 31. A: 80; B: 77; C: 73; D: 62; F: 54

Chapter 10 Concept Review, page 693

1. **a.** experiment **b.** outcomes, sample points
c. sample space **d.** 0, 1

2. **a.** random **b.** finite discrete **c.** 0, 1, 1

3. **a.** continuous **b.** nonnegative, 1, $\int_a^b f(x)\,dx$

4. nonnegative, 1, $\iint_R f(x, y)\,dA$

5. **a.** $(x_1 + x_2 + \cdots + x_n)/n$ **b.** central tendency, X
c. $x_1p_1 + x_2p_2 + \cdots + x_np_n$, X, probabilities
d. $\int_a^b x f(x)\,dx$, probability density, $[a, b]$

6. **a.** dispersion
b. $p_1(x_1 - \mu)^2 + p_2(x_2 - \mu)^2 + \cdots + p_n(x_n - \mu)^2$
c. $\sqrt{\text{Var}(X)}$

7. $\int_a^b (x - \mu)^2 f(x)\,dx$, $\sqrt{\text{Var}(X)}$

8. **a.** $x = \mu$ **b.** $x = \mu$ **c.** above, the x-axis **d.** 1
e. standard deviation, 2, 3

Chapter 10 Review Exercises, page 694

5. $\frac{1}{243}$ 6. $\frac{1}{6}$ 7. $\frac{3}{2}$ 8. 1.5

9. **a.** $\frac{4}{7}$ **b.** 0 **c.** $\frac{1}{3}$ 10. **a.** $\frac{1}{2}$ **b.** $\frac{1}{2}$ **c.** $\frac{3}{4}$

11. **a.** $\approx .52$ **b.** $\approx .65$ **c.** 0 12. **a.** $\frac{9}{10}$ **b.** $\frac{1}{4}$ **c.** $\frac{1}{2}$

13. $\mu = \frac{9}{7}$; $\text{Var}(x) \approx 2.083$; $\sigma \approx 1.44$

14. $\mu = \frac{50}{21}$; $\text{Var}(x) = 1.1882$; $\sigma \approx 1.0900$

15. $\mu = 0$; $\text{Var}(x) \approx \frac{7}{15}$; $\sigma \approx 0.6831$

16. $\mu = \frac{4}{3}$; $\text{Var}(x) \approx \frac{2}{9}$; $\sigma \approx 0.4714$ 17. .9875

18. .8925 19. .3049 20. .8294 22. **a.** $\frac{3}{16}$ **b.** $\frac{5}{6}$

23. **a.** .6915 **b.** .8944 **c.** .4681

24. **a.** .9525 **b.** .0475 **c.** .905

25. **a.** .22 **b.** .39 **c.** 4 days

26. **a.** .5507 **b.** .3012 **c.** .2210

Chapter 10 Before Moving On, page 695

1. $\frac{6}{7}$ 2. .2387 3. $\frac{13}{3}$, 5.42, 2.33

4. **a.** .9938 **b.** .8944 **c.** .9876

CHAPTER 11

Exercises 11.1, page 707

1. $P_1(x) = 1 - x$; $P_2(x) = 1 - x + \frac{1}{2}x^2$;
$P_3(x) = 1 - x + \frac{1}{2}x^2 - \frac{1}{6}x^3$

3. $P_1(x) = 1 - x$; $P_2(x) = 1 - x + x^2$; $P_3(x) = 1 - x + x^2 - x^3$

5. $P_1(x) = 1 - (x - 1)$; $P_2(x) = 1 - (x - 1) + (x - 1)^2$;
$P_3(x) = 1 - (x - 1) + (x - 1)^2 - (x - 1)^3$

7. $P_1(x) = 1 - \frac{1}{2}x$; $P_2(x) = 1 - \frac{1}{2}x - \frac{1}{8}x^2$;
$P_3(x) = 1 - \frac{1}{2}x - \frac{1}{8}x^2 - \frac{1}{16}x^3$

9. $P_1(x) = -x$; $P_2(x) = -x - \frac{1}{2}x^2$; $P_3(x) = -x - \frac{1}{2}x^2 - \frac{1}{3}x^3$

11. $P_2(x) = 16 + 32(x - 2) + 24(x - 2)^2$

13. $P_4(x) = (x - 1) - \frac{1}{2}(x - 1)^2 + \frac{1}{3}(x - 1)^3 - \frac{1}{4}(x - 1)^4$

15. $P_4(x) = e + e(x - 1) + \frac{1}{2}e(x - 1)^2 + \frac{1}{6}e(x - 1)^3 + \frac{1}{24}e(x - 1)^4$

17. $P_3(x) = 1 - \frac{1}{2}x - \frac{1}{8}x^2 - \frac{1}{16}x^3$

19. $P_3(x) = \frac{1}{3} - \frac{2}{9}x + \frac{4}{27}x^2 - \frac{8}{81}x^3$

21. $P_n(x) = 1 - x + x^2 - x^3 + \cdots + (-1)^n x^n$; 0.9091; 0.909090. . .

23. $P_4(x) = 1 - \frac{1}{2}x + \frac{1}{8}x^2 - \frac{1}{48}x^3 + \frac{1}{384}x^4$; 0.90484

25. $P_2(x) = 4 + \frac{1}{8}(x - 16) - \frac{1}{512}(x - 16)^2$; ≈ 3.94969

27. $P_3(x) = x - \frac{1}{2}x^2 + \frac{1}{3}x^3$; 0.109; 0.108

29. 2.01494375; 0.00000042

31. 1.248; 0.00493; 1.25 33. 0.095; 0.00033

35. **a.** 0.04167 **b.** 0.625 **c.** 0.04167 **d.** 0.007121

37. 0.47995 39. 48% 41. 2600 43. False 45. True

Exercises 11.2, page 716

1. 1, 2, 4, 8, 16 3. $0, \frac{1}{3}, \frac{2}{4}, \frac{3}{5}, \frac{4}{6}$ 5. $1, 1, \frac{4}{6}, \frac{8}{24}, \frac{16}{120}$

7. $e, \frac{e^2}{8}, \frac{e^3}{27}, \frac{e^4}{64}, \frac{e^5}{125}$ 9. $1, \frac{11}{9}, \frac{25}{19}, \frac{45}{33}, \frac{71}{51}$ 11. $a_n = 3n - 2$

13. $a_n = \frac{1}{n^3}$ 15. $a_n = \frac{2^{2n-1}}{5^{n-1}}$ 17. $a_n = \frac{(-1)^{n+1}}{2^{n-1}}$

19. $a_n = \frac{n}{(n + 1)(n + 2)}$ 21. $a_n = \frac{e^{n-1}}{(n - 1)!}$

23.

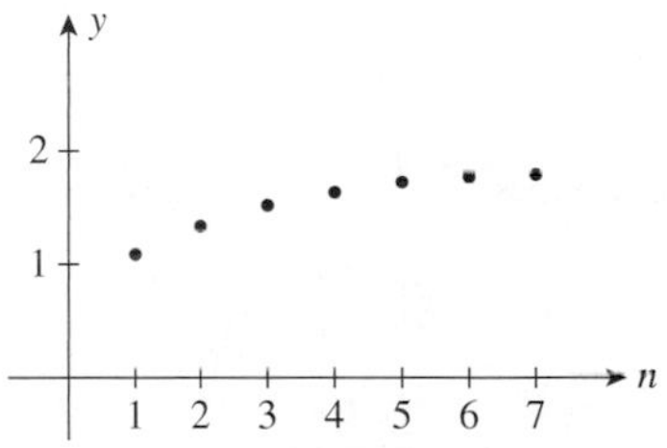

25.

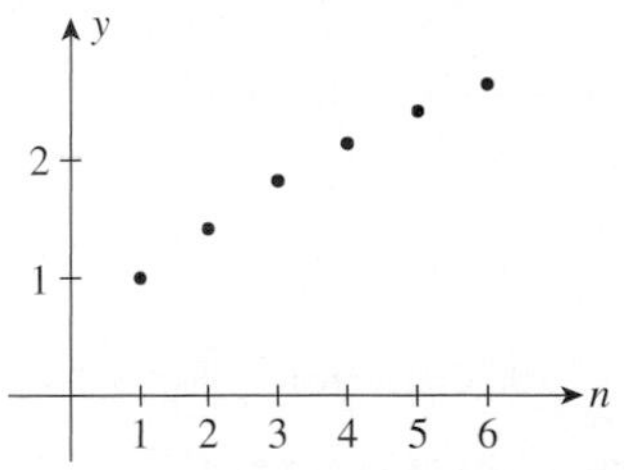

27.

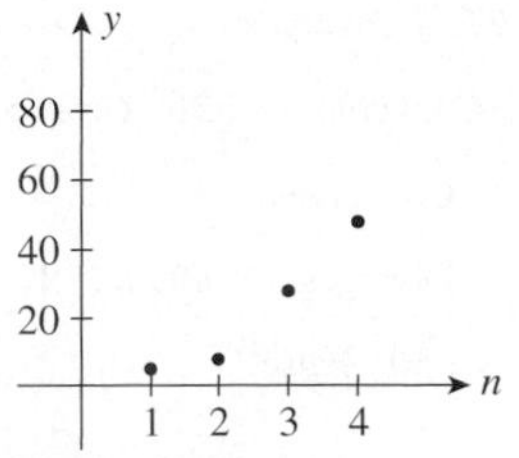

29.

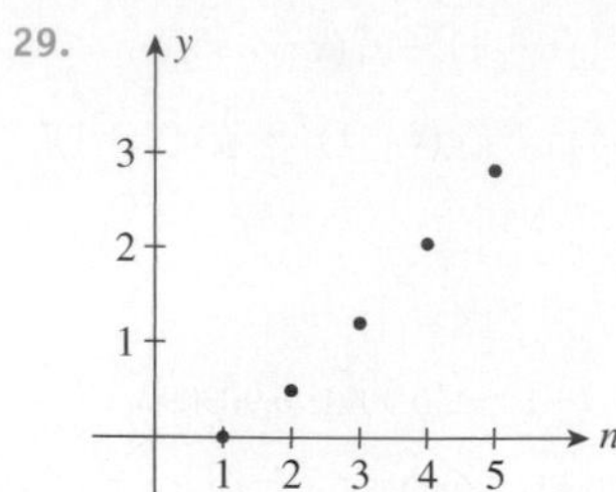

31. Converges; $\frac{1}{2}$ 33. Converges; 0 35. Converges; 1

37. Converges; 2 39. Converges; 2 41. Converges; 0

43. Converges; $\dfrac{\sqrt{2}}{2}$

45. **a.** $a_1 = 0.015$, $a_{10} = 0.140$, $a_{100} = 0.77939$, $a_{1000} = 0.999999727$
b. 1

47. **b.** $a_n = 100(1.01)^n$
c. $a_{24} = 126.97$. The accumulated amount at the end of 2 yr is \$126.97.

49. True 51. False

Exercises 11.3, page 728

1. $\lim_{N\to\infty} S_N$ does not exist; divergent

3. $S_N = \dfrac{1}{2} - \dfrac{1}{N+2}; \dfrac{1}{2}$ 5. Converges; $\frac{3}{2}$ 7. Diverges

9. Converges; $\frac{3}{5}$ 11. Converges; $\frac{1}{3}$ 13. Converges; 5.52

15. Converges; $\dfrac{\pi}{\pi + 3}$ 17. Converges; 13

19. Converges; $2\frac{3}{4}$ 21. Converges; $7\frac{1}{2}$ 23. Diverges

25. $\dfrac{e^3 - 2\pi e + \pi^2}{(\pi - e)(e^2 - \pi)}$ 27. $\frac{1}{3}$ 29. $\frac{404}{333}$

31. $-1 < x < 1; \dfrac{1}{1+x}$ 33. $\dfrac{1}{2} < x < \dfrac{3}{2}; \dfrac{2(x-1)}{3-2x}$

35. \$303 billion 37. $\frac{6}{11}$ 41. $\dfrac{P}{e^r - 1}$ 43. **b.** $\dfrac{1}{k} \ln \dfrac{S}{S - C}$ hr

45. False 47. True

Exercises 11.4, page 739

11. Diverges 13. Diverges 15. Converges

17. Converges 19. Converges 21. Converges

23. Converges 25. Converges 27. Converges

29. Diverges 31. Diverges 33. Converges 35. Diverges

37. Diverges 39. Converges 41. Converges

43. Diverges 45. Converges 47. Diverges 49. $p > 1$

51. $a = 1$ 55. True 57. False 59. False

Exercises 11.5, page 749

1. $R = 1; (0, 2)$ 3. $R = 1; (-1, 1)$ 5. $R = 4; (-4, 4)$

7. $R = \infty; (-\infty, \infty)$ 9. $R = 0; x = -2$ 11. $R = 1; (-4, -2)$

13. $R = \infty; (-\infty, \infty)$ 15. $R = 1; (-\frac{1}{2}, \frac{1}{2})$ 17. $R = 0; x = -1$

19. $R = 3; (0, 6)$ 21. $\sum_{n=0}^{\infty} (-1)^n (x-1)^n; R = 1; (0, 2)$

23. $\sum_{n=0}^{\infty} (-1)^n \dfrac{(x-2)^n}{3^{n+1}}; R = 3; (-1, 5)$

25. $\sum_{n=0}^{\infty} (-1)^{n+1} (x-2)^n; R = 1; (1, 3)$

27. $1 + \dfrac{1}{2}(x-1) + \sum_{n=2}^{\infty} (-1)^{n+1} \dfrac{1 \cdot 3 \cdot 5 \cdots (2n-3)}{n!\, 2^n} (x-1)^n$;
$R = 1; (0, 2)$

29. $\sum_{n=0}^{\infty} \dfrac{2^n}{n!} x^n; R = \infty; (-\infty, \infty)$

31. $\sum_{n=0}^{\infty} (-1)^n \dfrac{1 \cdot 3 \cdot 5 \cdots (2n-1)}{n!\, 2^n} x^n; R = 1; (-1, 1)$ 35. True

Exercises 11.6, page 757

1. $\sum_{n=0}^{\infty} (-1)^{n+1} (x-2)^n; (1, 3)$ 3. $\sum_{n=0}^{\infty} (-1)^n 3^n x^n; (-\frac{1}{3}, \frac{1}{3})$

5. $\sum_{n=0}^{\infty} \dfrac{3^n}{4^{n+1}} x^n; (-\frac{4}{3}, \frac{4}{3})$ 7. $\sum_{n=0}^{\infty} x^{2n}; (-1, 1)$

9. $\sum_{n=0}^{\infty} (-1)^n \dfrac{x^n}{n!}; (-\infty, \infty)$ 11. $\sum_{n=0}^{\infty} (-1)^n \dfrac{x^{2n+1}}{n!}; (-\infty, \infty)$

13. $f(x) = 1 + \dfrac{x^2}{2!} + \dfrac{x^4}{4!} + \dfrac{x^6}{6!} + \cdots + \dfrac{x^{2n}}{2n!} + \cdots; (-\infty, \infty)$

15. $\sum_{n=1}^{\infty} (-1)^{n-1} \dfrac{2^n x^n}{n}; (-\frac{1}{2}, \frac{1}{2}]$ 17. $\sum_{n=1}^{\infty} (-1)^{n+1} \dfrac{x^{2n}}{n}; (-1, 1)$

19. $(\ln 2)(x-2) + \sum_{n=1}^{\infty} (-1)^{n-1} \left(\dfrac{1}{n2^n}\right)(x-2)^n; (0, 4]$

21. $f'(x) = 1 - x + x^2 + \cdots + (-1)^n x^n + \cdots$

23. $f(x) = x - \dfrac{1}{2}x^2 + \dfrac{1}{3}x^3 + \cdots + \dfrac{(-1)^{n+1}}{n} x^n + \cdots$

25. 0.4812 27. 0.7475 29. 3.34 31. 15.85%

Exercises 11.7, page 764

1. 1.732051 3. 2.645751 5. 2.410142 7. 0.7709

9. 1.1219; 2.5745 11. 2.30278 13. 1.11634

15. 1.61803 17. 0.5671 19. **b.** 0.19356 21. **b.** 1.87939

23. 2.9365 25. 1.2785 27. 0.4263 29. 294 units/day

31. 8:39 P.M. 33. 26.82%/yr 35. 12%/yr 37. 10%/yr

39. 11,671 units; \$21.17/unit 41. 2.546

Chapter 11 Concept Review, page 768

1. **a.** $f(a) + f'(a)(x - a) + \cdots + \dfrac{f^{(n)}(a)}{n!}(x - a)^n$

 b. $f(a), f'(a), \dfrac{f^{(n)}(a)}{n!}$

2. **a.** $I, a, x, \dfrac{f^{(n+1)}(c)}{(n+1)!}(x - a)^{n+1}$

3. **a.** function, numbers, $f(n)$, nth term **b.** large, converge

4. $cA, A + B, AB, \dfrac{A}{B}, B \neq 0$

5. **a.** partial sums **b.** geometric, 1, 1

6. **a.** 0, diverges **b.** $cA + B$

7. **a.** convergent, divergent **b.** $\displaystyle\sum_{n=1}^{\infty} \frac{1}{n^p}, p > 1, p \leq 1$

8. converges, $a_n \geq b_n$

9. **a.** $\displaystyle\sum_{n=0}^{\infty} a_n(x - a)^n$ **b.** $x = a, x, (a - R, a + R)$, outside

10. **a.** $\dfrac{f^{(n)}(a)}{n!}$ **b.** $\displaystyle\lim_{N\to\infty} R_N(x) = 0$, the Taylor series

Chapter 11 Review Exercises, page 769

1. $f(x) = 1 - (x + 1) + (x + 1)^2 - (x + 1)^3 + (x + 1)^4$

2. $f(x) = e^{-1}\left[1 - (x - 1) + \dfrac{(x-1)^2}{2!} - \dfrac{(x-1)^3}{3!} + \dfrac{(x-1)^4}{4!}\right]$

3. $f(x) = x^2 - \frac{1}{12}x^4$ 4. $f(x) = 1 - 2x + 3x^2 - 4x^3 + 5x^4$

5. $f(x) = 2 + \frac{1}{12}(x - 8) - \frac{1}{288}(x - 8)^2$; 1.983

6. $f(x) = x - \frac{1}{3}x^3 + \frac{1}{10}x^5 - \frac{1}{42}x^7$; ≈ 0.74

7. 2.9992591; 8×10^{-11} 8. 0.909; 0.0001; 0.9090909. . .

9. 0.37 10. ≈ 0.199 11. Converges; $\frac{2}{3}$ 12. Diverges

13. Converges; 1 14. Converges; -1 15. 2 16. $\frac{3}{5}$

17. $\dfrac{1}{\sqrt{2} + 1}$ 18. $\dfrac{1}{e - 1}$ 19. $\frac{141}{99}$ 20. $\frac{3139}{999}$

21. Diverges 22. Diverges 23. Converges

24. Converges 25. $R = 1; (-1, 1)$ 26. $R = 1; (-1, 1)$

27. $R = 1; (0, 2)$ 28. $R = \dfrac{1}{e}; \left(2 - \dfrac{1}{e}, 2 + \dfrac{1}{e}\right)$

29. $f(x) = -1 - 2x - 4x^2 - 8x^3 - \cdots - 2^n x^n - \cdots; (-\frac{1}{2}, \frac{1}{2})$

30. $f(x) = \dfrac{1}{e} - \dfrac{1}{e}(x - 1) + \dfrac{1}{2!e}(x - 1)^2 - \cdots + \dfrac{(-1)^n}{n!e}(x - 1)^n + \cdots; (-\infty, \infty)$

31. $f(x) = 2x - 2x^2 + \frac{8}{3}x^3 - \cdots + \dfrac{(-1)^{n+1}2^n}{n}x^n + \cdots; (-\frac{1}{2}, \frac{1}{2}]$

32. $f(x) = x^2e^{-2x} = x^2 - 2x^3 + 2x^5 - \cdots + \dfrac{(-1)^n x^{n+2}}{n!} + \cdots;$ $(-\infty, \infty)$

33. 2.28943 34. ≈ 0.7549 35. (0.35173, 0.70346)

36. 2.65 sec 37. $212,372.20

38. **a.** $115 billion **b.** $90 billion 39. 31.08%

Chapter 11 Before Moving On, page 770

1. $P_1(x) = x$, $P_2(x) = x - x^2$, and $P_3(x) = x - x^2 + \dfrac{x^3}{2}$

2. **a.** $\dfrac{2}{3}$ **b.** 2 3. **a.** $\dfrac{25}{12}$ **b.** $\dfrac{1}{3}$

4. **a.** convergent **b.** divergent 5. $-\displaystyle\sum_{n=0}^{\infty} (x + 2)^n$, 1

CHAPTER 12

Exercises 12.1, page 775

1. $\dfrac{5\pi}{2}$ radians 3. $-\dfrac{3\pi}{2}$ radians

5. **a.** III **b.** III **c.** II **d.** I

7. $\dfrac{5\pi}{12}$ radians 9. $\dfrac{8\pi}{9}$ radians 11. $\dfrac{7\pi}{2}$ radians

13. 120° 15. −270° 17. 220°

19. 21.

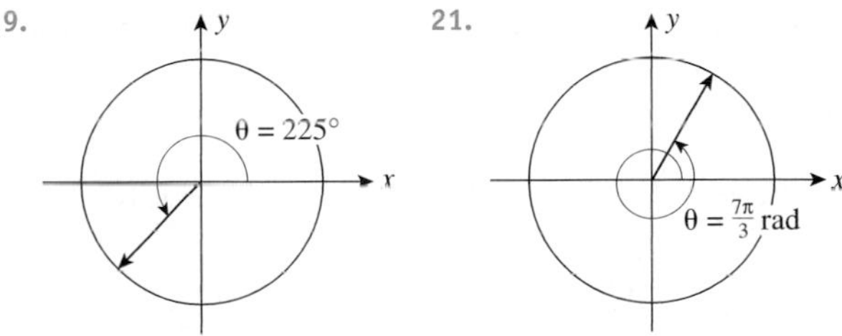

23. 150°, −210° 25. 315°, −45° 27. True 29. True

Exercises 12.2, page 783

1. 0 3. 1 5. $\dfrac{\sqrt{3}}{2}$ 7. $\dfrac{\sqrt{3}}{3}$ 9. -2.6131

11. $\sin\dfrac{\pi}{2} = 1$, $\cos\dfrac{\pi}{2} = 0$, $\tan\dfrac{\pi}{2}$ is undefined, $\csc\dfrac{\pi}{2} = 1$, $\sec\dfrac{\pi}{2}$ is undefined, $\cot\dfrac{\pi}{2} = 0$

13. $\sin\dfrac{5\pi}{3} = -\dfrac{\sqrt{3}}{2}$, $\cos\dfrac{5\pi}{3} = \dfrac{1}{2}$, $\tan\dfrac{5\pi}{3} = -\sqrt{3}$, $\csc\dfrac{5\pi}{3} = -\dfrac{2\sqrt{3}}{3}$, $\sec\dfrac{5\pi}{3} = 2$, $\cot\dfrac{5\pi}{3} = -\dfrac{\sqrt{3}}{3}$

15. $\dfrac{7\pi}{6}, \dfrac{11\pi}{6}$ 17. $\dfrac{5\pi}{6}, \dfrac{11\pi}{6}$ 19. π 21. $\dfrac{\pi}{3}, \dfrac{2\pi}{3}$

23.

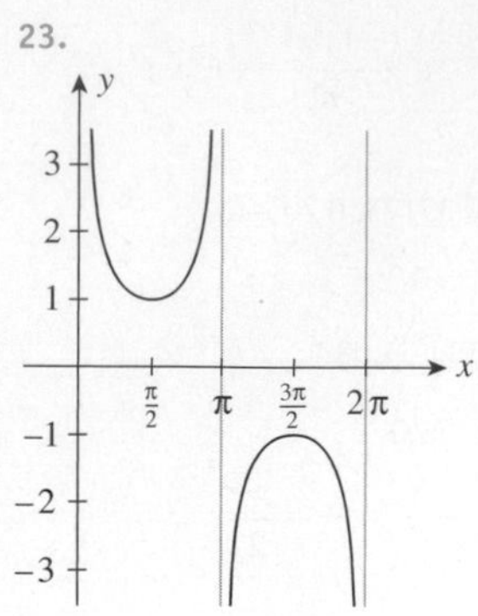

25.

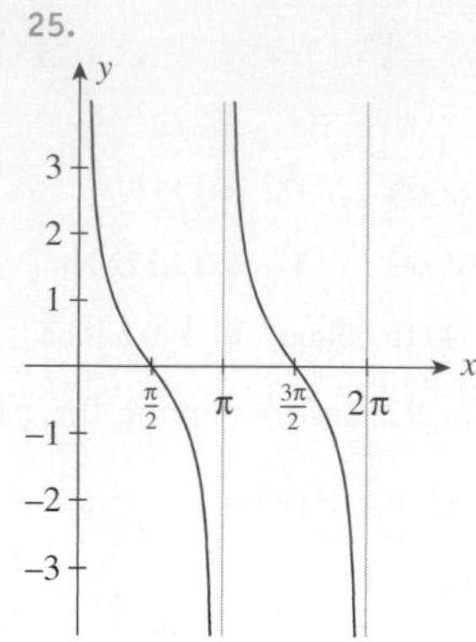

27.

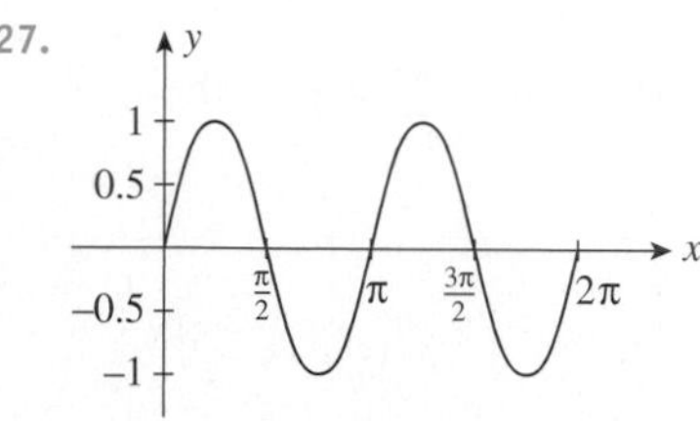

29.

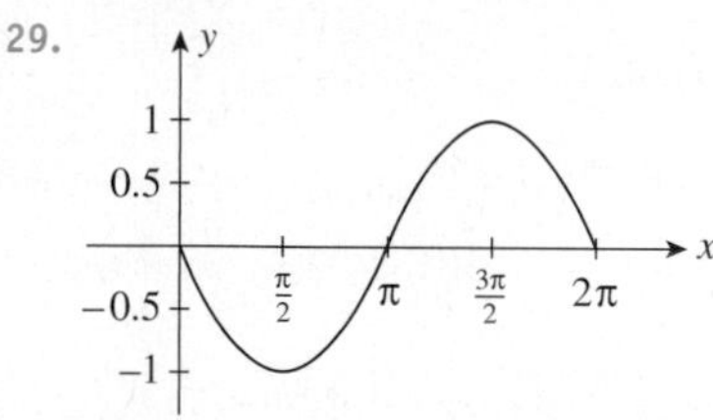

41. $\sin\theta = \frac{5}{13}$, $\cos\theta = \frac{12}{13}$, $\tan\theta = \frac{5}{12}$, $\csc\theta = \frac{13}{5}$, $\sec\theta = \frac{13}{12}$, $\cot\theta = \frac{12}{5}$

43. **b.** $\dfrac{\pi(4n+1)}{12}$ $(n = 0, 1, 2, \ldots)$; $\dfrac{\pi(4n+3)}{12}$ $(n = 0, 1, 2, \ldots)$

45. False 47. True

Exercises 12.3, page 793

1. $-3 \sin 3x$ 3. $-2\pi \sin \pi x$ 5. $2x \cos(x^2 + 1)$

7. $4x \sec^2 2x^2$ 9. $x \cos x + \sin x$ 11. $6(\cos 3x - \sin 2x)$

13. $2x(\cos 2x - x \sin 2x)$ 15. $\dfrac{x \cos\sqrt{x^2 - 1}}{\sqrt{x^2 - 1}}$

17. $e^x \sec x(1 + \tan x)$ 19. $\cos\dfrac{1}{x} + \dfrac{1}{x}\sin\dfrac{1}{x}$

21. $\dfrac{x \sin x}{(1 + \cos x)^2}$ 23. $\dfrac{\sec^2 x}{2\sqrt{\tan x}}$ 25. $\dfrac{x \cos x - \sin x}{x^2}$

27. $2 \tan x \sec^2 x$ 29. $-\csc^2 x \cdot e^{\cot x}$ 31. $y = -2x + \dfrac{\pi}{2}$

33. Increasing on $\left(0, \dfrac{\pi}{4}\right) \cup \left(\dfrac{5\pi}{4}, 2\pi\right)$; decreasing on $\left(\dfrac{\pi}{4}, \dfrac{5\pi}{4}\right)$

35. $f(x) = \sin x + \cos x$

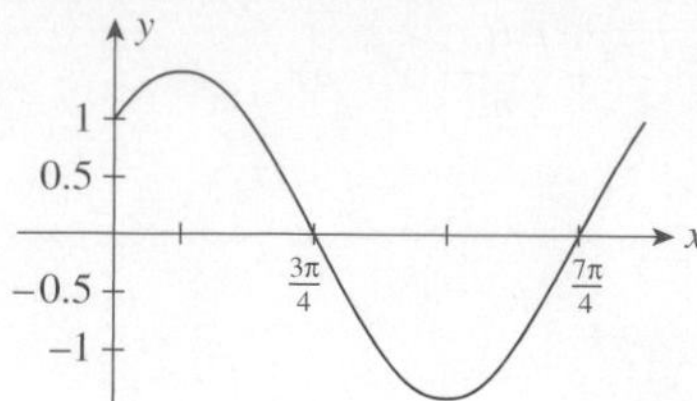

37. $f(x) = 2 \sin x + \sin 2x$

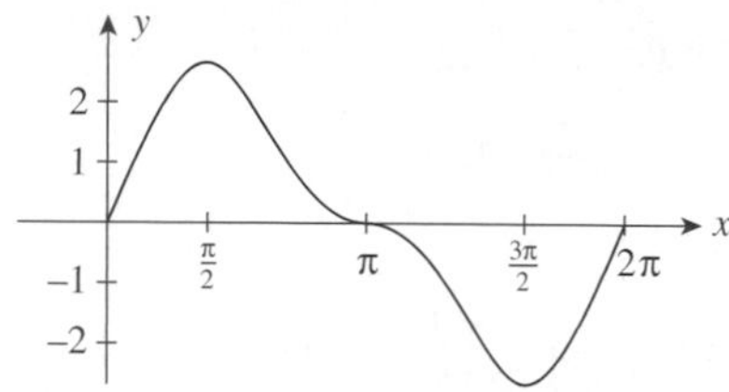

39. **a.** $f(x) = \sin x = x - \dfrac{x^3}{3!} + \dfrac{x^5}{5!} - \dfrac{x^7}{7!} + \cdots + (-1)^{n+1}\dfrac{x^{2n+1}}{(2n+1)!} + \cdots$

41. Zero wolves/mo; -1571 caribou/mo 43. 0.05

45. Warmest day is July 25; coldest day is January 23.

47. Sixth week

49. Maximum when $t = 1, 5, 9, 13, \ldots$; minimum when $t = 3, 7, 11, 15, \ldots$

51. 70.7 ft/sec 53. 60° 55. 1.4987 radians 57. True

59. False

Using Technology Exercises 12.3, page 798

1. 1.2038 3. 0.7762 5. -0.2368 7. 0.8415; -0.2172

9. 1.1271; 0.2013

11. **a.**

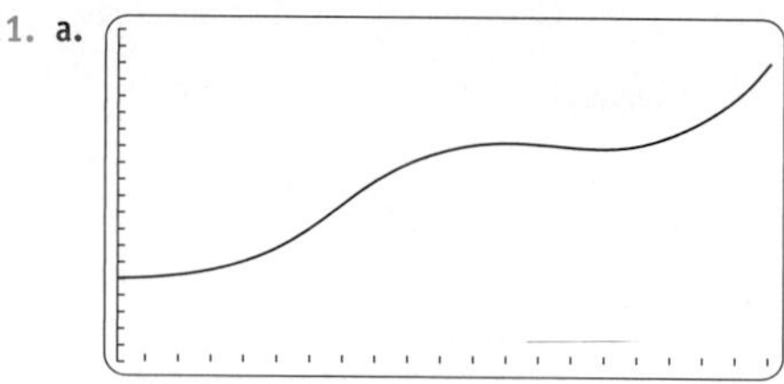

b. $\approx$\$0.63 **c.** $\approx$\$27.79

13. **a.**

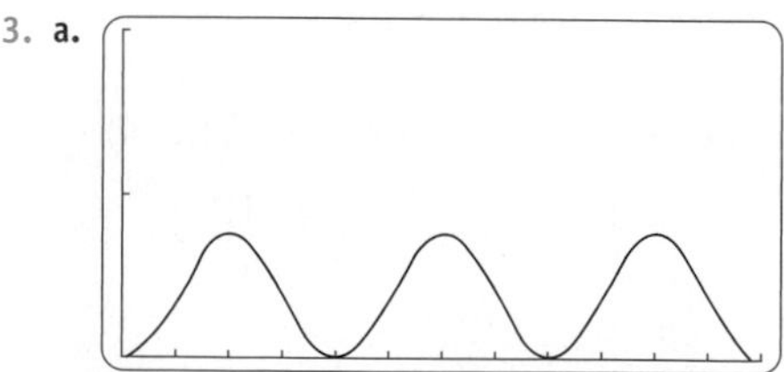

15. ≈ 0.006 ft

Exercises 12.4, page 803

1. $-\frac{1}{3}\cos 3x + C$ 3. $-3\cos x + 4\sin x + C$

5. $\frac{1}{2}\tan 2x + C$ 7. $\frac{1}{2}\sin x^2 + C$ 9. $-\dfrac{1}{\pi}\csc \pi x + C$

11. 2 13. $-\frac{1}{2}\ln\frac{1}{2}$ 15. $\frac{1}{4}\sin^4 x + C$

17. $\dfrac{1}{\pi}\ln|\sec \pi x + \tan \pi x| + C$ 19. $\frac{1}{3}\ln(1 + \sqrt{2})$

21. $-\frac{2}{3}(\cos x)^{3/2} + C$ 23. $-\frac{1}{9}(1 - 2\sin 3x)^{3/2} + C$

25. $\frac{1}{4}\tan^4 x + C$ 27. $-\frac{1}{4}(\cot x - 1)^4 + C$ 29. 2

31. $\frac{1}{2}x[\sin(\ln x) - \cos(\ln x)] + C$ 33. $2\sqrt{2}$ sq units

35. $\frac{1}{2}\ln 2$ sq units 37. $\frac{1}{2}(\pi^2 - 4)$ sq units 39. \$85

41. \$120,000 43. $\dfrac{1.2}{\pi}\left(1 - \cos\dfrac{\pi t}{2}\right)$ 45. 162 fruit flies

47. ≈ 0.9 49. True 51. True

Using Technology Exercises 12.4, page 805

1. 0.5419 3. 0.7544 5. 0.2231 7. 0.6587

9. -0.2032 11. 0.9045

13. a.

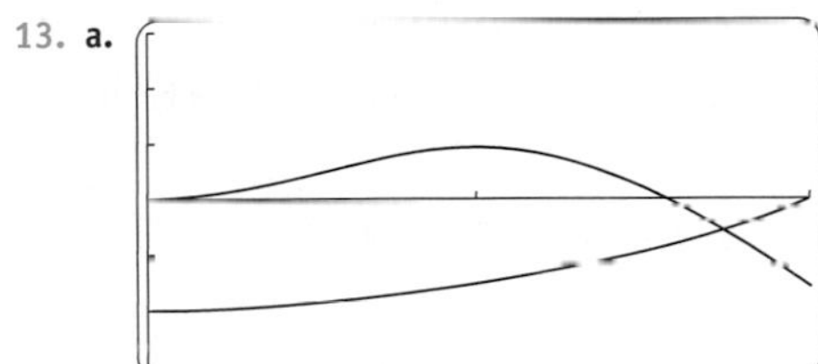

b. 2.2687 sq units

15. a.

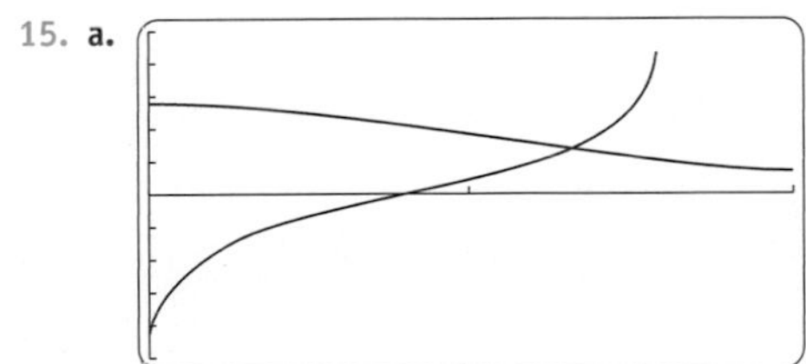

b. 1.8239 sq units

17. a.

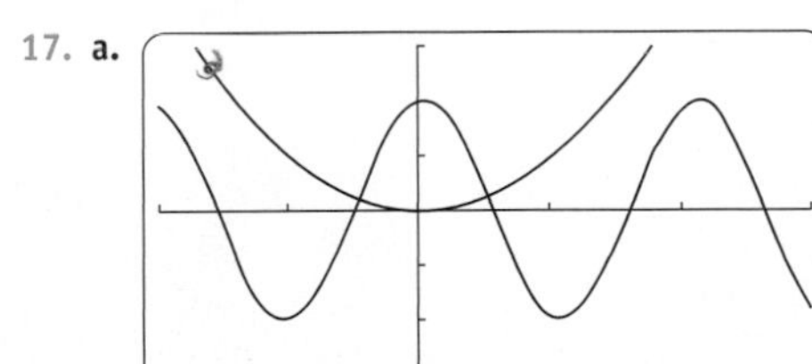

b. 1.2484 sq units

19. a.

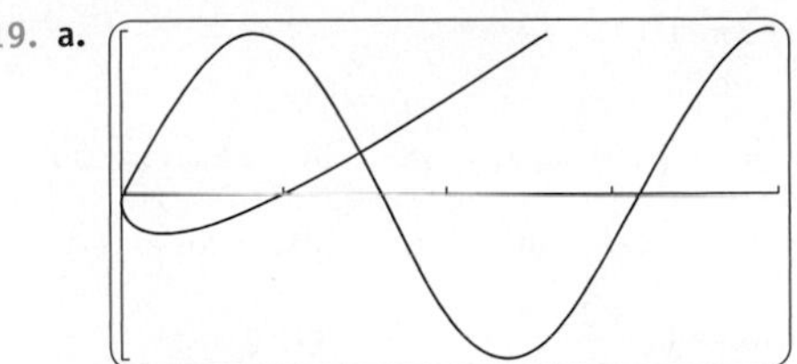

b. 1.0983 sq units

21. 7.6 ft

Chapter 12 Concept Review, page 807

1. a. radian b. 2π, 360 2. a. $\dfrac{\pi}{180}x$ b. $\dfrac{180}{\pi}x$

3. $x, y, \dfrac{y}{x}, \dfrac{1}{y}, \dfrac{1}{x}, \dfrac{x}{y}$ 4. a. $\cos^2\theta$ b. 1 c. 1 d. $\cos 2\theta$

5. a. $\cos 2\theta$ b. + c. $\sin A\cos A$ d. $\sin^2 A$

6. a. $[\cos f(x)]f'(x)$ b. $-[\sin f(x)]f'(x)$
c. $[\sec^2 f(x)]f'(x)$ d. $-[\csc f(x)][\cot f(x)]f'(x)$
e. $[\sec f(x)][\tan f(x)]f'(x)$ f. $-[\csc^2 f(x)]f'(x)$

7. a. $-\cos x + C$ b. $\sin x + C$ c. $\tan x + C$
d. $-\cot x + C$ e. $\sec x + C$ f. $-\csc x + C$

Chapter 12 Review Exercises, page 807

1. $\dfrac{2\pi}{3}$ radians 2. $\dfrac{5\pi}{2}$ radians 3. $-\dfrac{5\pi}{4}$ radians 4. 330°

5. $-450°$ 6. $-315°$ 7. $\dfrac{\pi}{3}$ or $\dfrac{5\pi}{3}$ 8. 150° or 330°

9. $3\cos 3x$ 10. $-\sin\dfrac{x}{2}$ 11. $2\cos x + 6\sin 2x$

12. $\dfrac{\sec^2\sqrt{x}\tan\sqrt{x}}{\sqrt{x}}$ 13. $e^{-x}(3\sec^2 3x - \tan 3x)$

14. $4(1 - \csc 2x)(\csc 2x\cot 2x)$ 15. $4\cos 2x$ or $4(\cos^2 x - \sin^2 x)$

16. $-\dfrac{\sin x}{(1 - \cos x)^2}$ 17. $\dfrac{(\cot x - 1)\sec^2 x - (1 - \tan x)\csc^2 x}{(1 - \cot x)^2}$

18. $-2\tan x$ 19. $\cos(\sin x)\cos x$ 20. $e^{\sin x}(-\sin x + \cos^2 x)$

21. $y = 4x + 1 - \pi$

22.

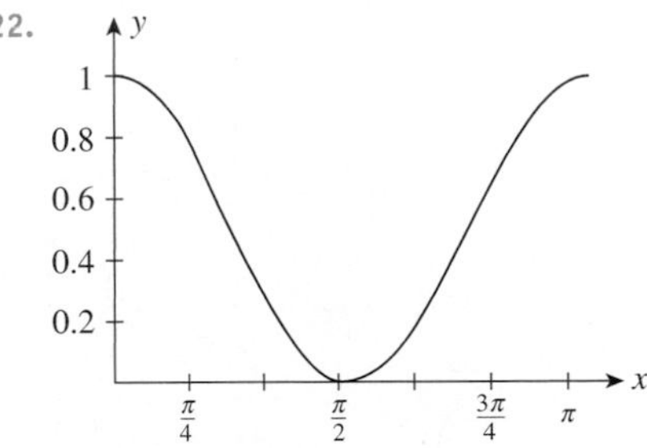

a. ↗ on $(\frac{\pi}{2}, \pi)$ and ↘ on $(0, \frac{\pi}{2})$ b. 0

c. Concave up on $\left(\dfrac{\pi}{4}, \dfrac{3\pi}{4}\right)$ and concave down on $\left(0, \dfrac{\pi}{4}\right) \cup \left(\dfrac{3\pi}{4}, \pi\right)$

d. $\left(\frac{\pi}{4}, \frac{1}{2}\right)$ and $\left(\frac{3\pi}{4}, \frac{1}{2}\right)$

23. $\frac{3}{2}\sin\frac{2}{3}x + C$ 24. $\frac{1}{2}x - \frac{1}{8}\sin 4x + C$ 25. $-\frac{1}{2}\csc x^2 + C$

26. $x\sin x + \cos x + C$ 27. $\frac{1}{3}\sin^3 x + C$ 28. $-\cos e^x + C$

29. $-\csc x + C$ 30. $-\ln\frac{\sqrt{2}}{2}$ 31. 1 32. 2

33. $2\sqrt{2}$ sq units

34. Fox population ↗ at the rate of ≈6.5 foxes/mo and the rabbit population ↘ at the rate of ≈113.4 rabbits/mo

35. December 1; September 1 36. 78.5%

Chapter 12 Before Moving On, page 808

1.

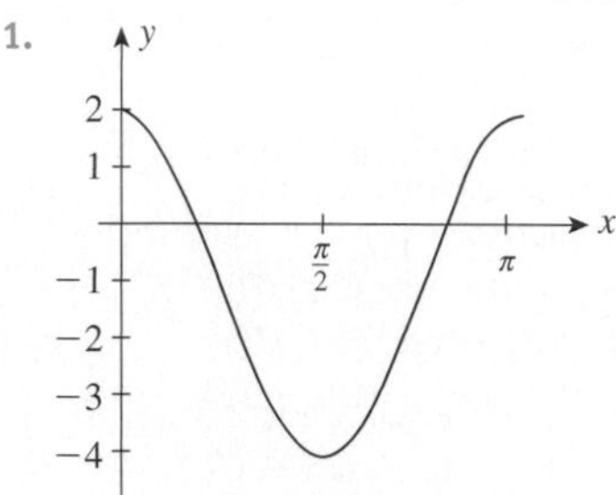

3. $(3\cos 3x - 2\sin 3x)e^{-2x}$ 4. $\frac{\pi + 2}{8}$ 5. $\frac{\sqrt{2}}{12}$ sq units

APPENDIX A

Exercises, page 814

7. Yes 9. No 11. No

13. $\frac{1}{3}(x + 2)$

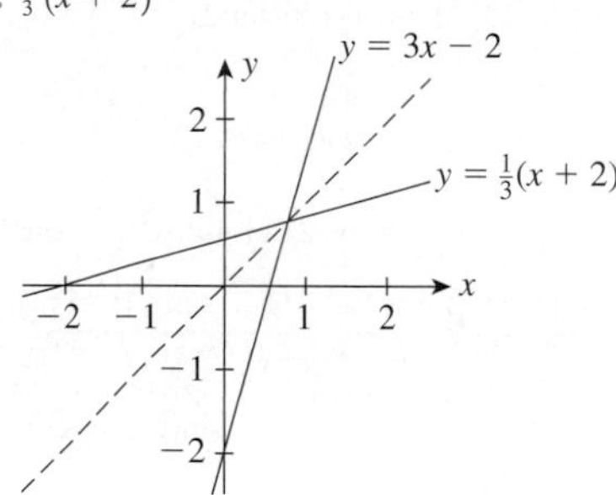

15. $\sqrt[3]{x - 1}$

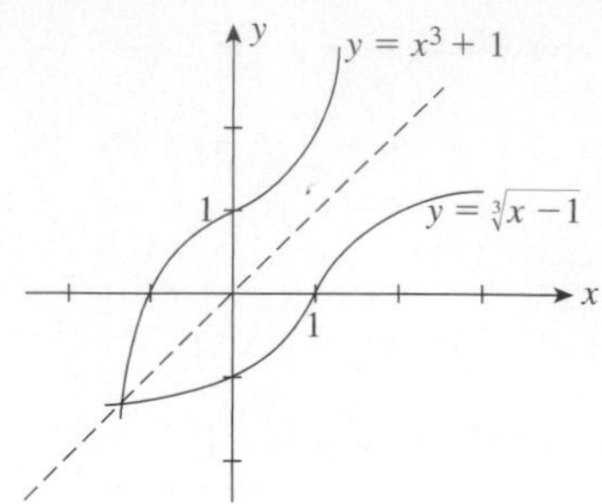

17. $\sqrt{9 - x^2}\ (x \geq 0)$

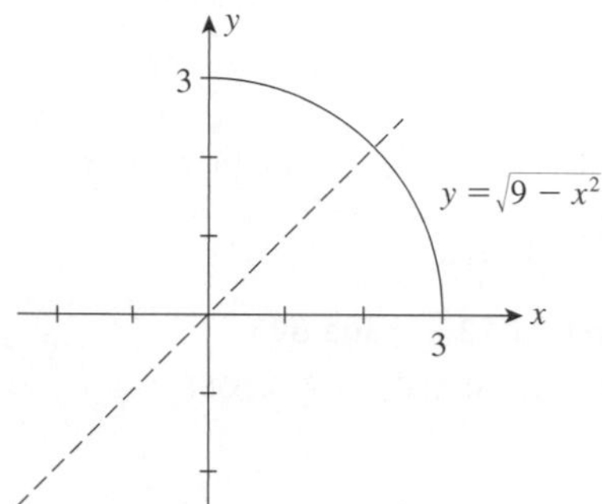

19. **a.** $-\frac{1}{2} + \frac{1}{2}\sqrt{1 + 8h}$; f^{-1} represents the time at which the balloon is at height h.
b. Between 15 and 20 sec

Index

Abscissa, 24
Absolute extrema, 300, 559
Absolute maximum value, 300, 559
Absolute minimum value, 300, 559
Absolute value, 5
Angle, 772
  coterminal, 774
  degree measure, 773
  initial ray, 772
  radian measure, 773
  standard position, 772
  terminal ray, 772
Annuity, 470
Antiderivative, 396
Antidifferentiation, 398
Area
  between curves, 452–457
  problem, 420
  under a curve, 420–424, 431–432, 444
Asymptotes, 106, 286–290
  horizontal, 106, 288
  vertical, 287
Average, 667
Average cost function, 200
Average rate of change, 139
Average value
  of an exponential density function, 670
  of a function of one variable, 444–446
  of a function of two variables, 609–611
Axes, 23

Base, 6, 332
Boyle's law, 62

Capital value of a perpetuity, 726
Carbon-14 dating, 382
Cartesian coordinate system, 23, 538
  abscissa, 24
  axes, 23
  ordered pair, 23
  ordinate, 24
  quadrants, 24
  three-dimensional, 538
Chain rule, 185–189
  for exponential functions, 362
  for logarithmic functions, 373
  for powers of functions, 187–189
Change of variable, 442–443
Closed interval, 3
Cobb–Douglas production function, 550
Common logarithm, 340
Comparison test, 737
Complementary commodities, 551
Composite function, 71, 185
Compound interest, 347–350. *See also* Interest
Concavity, 267–270
  intervals, 269
  test for, 269
Constant of integration, 398
Constrained extrema, 582
Constrained optimization, 583–584
Consumers' surplus, 465
Continuity, 120–124
  on an interval, 121
  at a number, 121
  of polynomial and rational functions, 123
Continuous compound interest, 354
Continuous function
  definition of, 121
  properties of, 123
Continuous random variable, 656, 669, 676
Contour map, 539
Convergence
  geometric series, 722
  of an improper integral, 523
  interval of, 742
  power series, 742
  radius of, 743
  of a sequence, 713
  of a series, 720, 741–743
Coordinate, 2, 24
Cosecant, 777, 789
Cosine, 777
  derivative of, 788
  graph of, 780
  integral of, 799
Cost function, 70, 198–200
Cotangent, 777, 789
Critical number, 254
Critical point, 561
Curve sketching, 285–292

Decay constant, 381
Decreasing function, 248–249
Definite integral, 424
  as area, 420–424
  geometric interpretation, 425–426
  as a limit of a sum, 424
  limits of integration, 424
  properties of, 441
Degree, 773
Degree of polynomial, 15
Demand
  curve, 82
  elastic, 206
  equation, 82
  function, 82
  inelastic, 206
  unitary, 206
Dependent variable, 51, 536
Depreciation, 61, 88
Derivative
  definition, 139
  first partial, 546
  higher-order, 212–214
  of an implicit function, 222
  instantaneous rate of change, 139
  notation, 139
  partial, 545–548
  as a rate of change, 143–146
  second, 212
  second-order partial, 552–554
Derivative rules, 160–164, 174–179
  chain, 186
  constant, 160
  constant multiple of a function, 162
  for exponential functions, 360, 362
  general power rule, 187
  for logarithmic functions, 371–372
  power rule, 161
  product rule, 174
  quotient rule, 175
  sum rule, 162
Deviations, 673
Difference quotient, 138
Differentiable function, 145
Differential equations, 620–623
  approximate solutions of, 642–647
  first-order, 627
  general solution, 622
  initial value problem, 629
  order, 627
  particular solution, 623
  separable, 627
  solution of, 621
Differentials, 232–237, 592–596
Discontinuous function, 121
Distance formula, 24
Divergence
  of a geometric series, 723
  of an improper integral, 523
  of a power series, 742–743
  of a sequence, 713

Domain
  of a function of one variable, 50
  of a function of two variables, 536
Double integral, 599–604

e, 334
Effective rate of interest, 351
Elasticity of demand, 206–208
Equation of a circle, 25
Equilibrium
  price, 82
  quantity, 82
Error analysis, 516–517
Error bound for polynomial approximation, 705
Euler's method, 642–647
Event, 654
Expected value, 667–671
Experiment, 654
Exponent, 6–7
Exponential decay, 381
Exponential density function, 660, 670
Exponential function, 332
  applications of, 380–385
  base *e*, 334
  derivative of, 360, 362
  graph of, 335
  indefinite integral of, 401
  properties of, 334
Exponential growth, 379–380

Factoring polynomials, 12–15
Finite interval, 3
First derivative test, 255
Fixed costs, 70
Function, 50
  algebraic operations on, 69
  average value of, 444–446, 609–611
  composite, 71
  continuous, 121
  cost, 70
  cubic, 80
  decreasing, 248
  demand, 82
  dependent variable, 51, 536
  differentiable, 145
  discontinuous, 120
  discrete probability, 654
  domain, 50, 536
  explicit representation, 221
  exponential, 332
  graph of, 53
  implicit representation, 221
  increasing, 248
  independent variable, 51, 536
  inverse, 344
  linear, 80
  logarithmic, 342
  marginal average cost, 200
  marginal cost, 199
  marginal profit, 203
  marginal revenue, 206
  piecewise-defined, 55
  polynomial, 80
  probability density, 656, 660
  profit, 70, 203
  quadratic, 80
  range, 50
  rational, 80
Functions of several variables, 536–542
  constrained relative extrema, 581–582
  critical points, 561
  dependent variable, 536
  domain, 536
  independent variable, 536
  maxima and minima, 559
  partial derivative of, 545–546
  saddle point, 560
  second derivative test, 562
Fundamental theorem of calculus, 429–430, 435–436
Future value, 352

Geometric series, 722–723
Gompertz growth curve, 389, 626, 641
Graph
  of an equation, 56
  of a function, 53, 538
Growth constant, 380

Half-life, 381
Half-open interval, 3
Harmonic series, 731
Higher-order derivative, 212–216
Histogram, 654
Horizontal asymptote, 106, 288
Horizontal line test, 813

Implicit differentiation, 221–225
Improper integral, 521–527
  convergent, 523
  divergent, 523
Income stream, 467–470
Increasing function, 248–249
Increment, 232–233
Indefinite integral, 398
  of a constant, 398
  of a constant multiple of a function, 399
  of the exponential function, 401
  power rule, 399
  sum rule, 400
Independent variable, 51, 536
Indeterminate form, 103
Inequalities, 4
Infinite interval, 3
Infinite sequence, 710–715
Infinite series, 718–724
Inflection point, 270
Initial value problem, 402, 629
Instantaneous rate of change, 139
Integral
  change of variable, 442–443
  of a constant, 398
  of a constant multiple of a function, 399
  definite, 424
  double, 599–605
  of the exponential function, 401
  improper, 521–527
  indefinite, 398
  notation, 398
  power rule for, 399
  properties of, 441
  sum rule for, 400
  tables, 500–501
Integrand, 398
Integration
  constant of, 398
  limits of, 424, 442–443
  by parts, 492–495
  rules, 398–401
  by substitution, 410–414
  See also Integral
Intercepts, 36
Interest
  compound, 349
  continuous compound, 354
  conversion period, 348
  effective rate, 351
  nominal rate, 348
  simple, 347
  true rate, 350
Intermediate value theorem, 124
Interval,
  closed, 3
  finite, 3
  half-open, 3
  infinite, 3
  open, 2
Interval of convergence, 742
Inventory control, 320–321
Inverse functions, 344, 810–815
  graphs, 812
  guidelines for finding, 813

horizontal line test, 813
one-to-one function, 813
Isotherms, 541
Iterated integral, 601

Joint probability density function, 660–662

Lagrange multipliers, 584
Laws of exponents, 7, 332
Laws of logarithms, 340
Learning curves, 383–384
Least common denominator, 19
Least-squares principle, 572
Level curves, 539–541
Limits, 97–109, 119–120
of a function, 100
at infinity, 107
of integration, 424, 442–443
left-hand, 119
properties of, 102
right-hand, 119
of a sequence, 713
Linear equation
general form, 39, 40
intercept form, 43
intercepts, 36
parallel lines, 33
perpendicular lines, 36
point-slope form, 35, 40
slope-intercept form, 37, 40
vertical lines, 34, 40
Line of complete equality, 472
Logarithmic differentiation, 374–376
Logarithmic functions, 342–344
derivative of, 371–372
graph of, 343
properties of, 343
Logarithms, 339–344
common, 340
laws of, 340
natural, 340
Logistic curve, 384–385
Logistic growth function, 384–385
Lorentz curves, 471–473

Maclaurin series, 742
Marginal
analysis, 197–204
average cost function, 200
cost function, 199
productivity, 550
productivity of money, 589
profit function, 203
revenue function, 203
Market equilibrium, 82
Mathematical model, 76–79
Maxima and minima
absolute, 300, 559
constrained, 582
relative, 253, 559
Method of bisection, 126
Method of integration by substitution, 410–414, 441–443
Method of Lagrange multipliers, 583
Method of least squares, 570–572
normal equations, 572
principle, 570
scatter diagram, 570
Minima. *See* Maxima and minima
Multiplier effect, 726

*n*th root, 6
Natural logarithm, 340
Net change, 432
Newton's method, 758–760
Nominal interest rate, 348
Normal curve, 682
Normal density function, 682
Normal distribution, 682
Normal equations, 572
Normal probability density function, 682
Number line, 2
Numerical integration, 506–516
error analysis, 516–517
Simpson's rule, 512
trapezoidal rule, 509

One-sided limits, 119
One-to-one function, 813
Open interval, 3
Optimization, 303–307, 315–321
Ordered pair, 23, 53
Ordered triple, 540
Ordinate, 24
Origin, 2
Outcome, 654

p-series, 735
Parabola, 55
Parallel lines, 33
Partial derivative
first-order, 546
second-order, 552
Partial sums, 720
Perpendicular lines, 36
Perpetuity, 526, 726
Point-slope form, 35
Poiseuille's law, 62, 796
Polynomial, 11
factoring, 12
function, 80
Power function, 81
Power rule
for differentiation, 187
for integration, 399
Power series, 741–743
Predator–prey population model, 781–782
Present value, 352
Principle of least squares, 570
Probability, 654–662
density function, 656, 660
of an event, 654
experiment, 654
normal distribution, 682
outcome, 654
random variable, 654
sample space, 654
Producer's surplus, 465
Product rule, 174
Profit function, 70, 203
Properties of inequalities, 4

Quadrant, 24
Quadratic formula, 16
Quadratic function, 80
Quotient Rule, 175

Radian, 773
Radicals, 8
Radioactive decay, 381–382
Radius of convergence, 743
Random variable, 654
continuous, 656–657, 669–670, 676
expected value, 667–671
finite discrete, 654
standard deviation of, 675
Range, 50
Rate of change, 138–139
average, 138
derivative as, 145
instantaneous, 139
Rational expression, 17
Rational function, 80
Rationalization, 8, 20
Real number line, 2
Regression line, 571
Related rates, 225–229
Relative maximum, 253, 559
test for, 255, 564
Relative minimum, 253, 559
test for, 255, 564
Remainder theorem, 704
Revenue function, 70, 202

Riemann sum, 423, 599
Roots of an equation, 15, 758

Saddle point, 560
Sample space, 654
Scatter diagram, 570
Scatter plot, 79, 570
Secant, 777, 789
Secant line, 137
Second derivative test, 274, 562
Second-order partial derivative, 552–554
Separation of variables, 627
Sequence, 710–715
  convergent, 713
  divergent, 713
  general terms, 710
  graphs of, 712
  infinite, 710
  limit, 713
  limit properties, 714
Series, 718–724
  comparison test, 737
  convergent, 720
  divergent, 720
  geometric, 722
  infinite, 718–723
  integral test, 732
  interval of convergence, 742
  Maclaurin, 742
  partial sums, 720
  power, 741–743
  properties of, 724–725
  sum, 742
  Taylor, 741–747
  test for divergence, 731, 751–755
Simplifying an algebraic expression, 12
Simpson's rule, 511–515
Sine, 777
  derivative of, 787
  graph of, 780
  integral of, 799
Slope, 32–34
  of a tangent line, 138
Slope-intercept form, 37
Standard deviation, 675
Standard normal curve, 683
Standard normal random variable, 686
Standard viewing window, 64
Stimulus response, 621
Substitute commodities, 551
Supply
  curve, 82
  equation, 82
  function, 82

Table of integrals, 500–501
Tangent, 777
  derivative of, 789
  graph of, 780
  line, 97, 137
Taylor polynomial, 698–706
  approximation, 700–702
  error bound, 741–747
Taylor series, 751–755
Total differential, 594
Trace, 539
Trapezoidal rule, 508–510
Trend line, 570
Triangle inequality, 5
Trigonometric functions, 777–782
  derivatives of, 785–789
  graphs of, 780
  integrals of, 799–800
  periods of, 779
Trigonometric identities, 782

Unit circle, 773
Unrestricted growth models, 620, 633–634

Variable costs, 70
Variance, 673
Velocity
  average, 98, 143
  instantaneous, 98, 143
Vertical asymptote, 287
Vertical-line test, 56
Volume of a solid, 606–607
Volume of a solid of revolution, 479
Volume of a solid under a surface, 607

Weber–Fechner law, 638

List of Applications (*continued*)

Percentage of females in the labor force, 371
Percentage of population relocating, 366
Politics, 62
Population density, 608, 610, 611, 613
Population growth, 114, 158, 170, 182, 184, 408, 418, 440, 461, 490, 640
Population growth in Clark County, 96, 282, 432
Population growth in the 21st century, 388, 391
Prison population, 95
Quality of environment, 263, 329
Recycling programs, 505
Registered vehicles in Massachusetts, 62
Rising median age, 62
Safe drivers, 68
SAT scores, 576
Scholarship funds, 726
Senior citizens, 438
Senior workforce, 309, 329
Single female-headed households with children, 438
Socially responsible funds, 193
Solar power, 86
Spending on medical devices, 86
Spread of rumor, 279, 641, 651
Student enrollment, 418
Thurstone learning models, 158, 194
Tracking with GPS, 336
Traffic studies, 170, 194
TV viewing patterns, 134, 193, 418
University admission eligibility, 688
U.S. Census, 489
U.S. nursing shortage, 263
U.S. population growth, 267
U.S. senior citizens, 448
Vacation trends, 708
Voter registration, 505
Waiting times, 667
Waste generation, 581
Working-age population, 89
Working mothers, 193, 218
World population growth, 267, 311, 370, 386, 640

LIFE SCIENCES

Absorption of drugs, 337, 339, 346, 368, 369, 389, 391, 394, 451
Administration of an IV solution, 130
Adult obesity, 244
Aids in Massachusetts, 451
Allometric growth, 626, 632
Amount of rainfall, 118, 409
Anticipated rise in Alzheimer's patients, 66, 219
Arterial blood flow, 598
Arteriosclerosis, 190, 194
Average weights and heights of infants, 147, 418
Birthrate of endangered species, 74
Birth weights of infants
Blood alcohol level, 367, 368
Blood flow in an artery, 408, 439
Blood pressure, 345
Body mass, 542
Brentano–Stevens law, 650
Cancer survivors, 62
Carbon-14 dating, 382, 387
Carbon monoxide in the air, 74, 173, 193, 408, 708
Cardiac output, 506, 513
Chemical decomposition, 640
Chemical reactions, 640, 642
Child obesity, 169
Clark's rule, 157
Concentration of a drug in an organ, 419
Concentration of a drug in the bloodstream, 114, 181, 262, 296, 368, 446, 497, 625
Concentration of glucose in the bloodstream, 389, 418, 625, 632
Conservation of species, 164, 170, 216
Contraction of the trachea during a cough, 304
Cricket chirping and temperature, 88
Crop yield, 148, 324, 371, 451
Crossing the finish line, 409
Diffusion, 498
Doomsday situation, 114
Drug dosages, 86, 181
Effect of bactericide, 148, 181
Effect of enzymes on chemical reactions, 296
Energy expended by a fish, 130, 312
Epidemic models, 368, 388, 394, 504, 637, 717
Eradication of polio, 367
Extinction situation, 370
Female life expectancy, 192
Flights of birds, 325
Flow of blood in an artery, 408, 796
Flu epidemic, 366, 394
Forensic science, 346
Forestry, 147, 261
Formaldehyde levels, 181
Friend's rule, 61
Gastric bypass surgeries, 407
Genetically modified crops, 406
Gompertz growth curve, 389, 626, 641
Groundfish population, 169, 173
Growth of a cancerous tumor, 60, 168, 238
Growth of bacteria, 150, 380, 386, 633, 640
Growth of fruit fly population, 106, 388, 391, 505, 641
Harbor cleanup, 62
Heights of children, 377, 408
Heights of trees, 346
Importance of time in treating heart attacks, 183
Index of environmental quality, 329
Length of a hospital stay, 694
Lengths of fish, 346, 387, 391
Lengths of infants, 520
Life span of a plant, 663, 679
Maximum air flow during respiration, 793, 795, 804
Measuring cardiac output, 520
Medical records, 690
Nuclear fallout, 387
Obese children in the U.S., 88
Obesity in America, 170, 217
Outpatient visits, 578
Over-100 population, 366
Oxygen content of a pond, 117, 131, 296, 311
Ozone pollution, 408
Period of a communications satellite, 242
Photosynthesis, 115
Poiseuille's law, 62, 542
Polio immunization, 367
Predator–prey population, 781, 784, 790, 794, 808
Preservation of species, 164
Prevalence of Alzheimer's patients, 62, 219, 261
Pulse rates, 194, 460
Radioactive decay, 381, 387, 389, 394, 625
Rate of growth of a tumor, 391
Reaction of a frog to a drug, 87
Reaction to a drug, 312
Residual drug in the bloodstream, 727, 729
Senior citizens' health care, 88
Serum cholesterol levels, 689, 755
Spam messages, 74
Speed of a chemical reaction, 115
Spread of contagious disease, 329
Spread of flu epidemic, 295, 368, 385, 388, 394, 637
Spread of HIV, 173
Storing radioactive waste, 324
Strain of vertebrae, 377
Surface area of a honeycomb, 795
Surface area of a horse, 239
Surface area of a lake, 532
Surface area of a single-celled organism, 61
Surface area of the Central Park reservoir, 519
Surface area of the human body, 408, 543, 556, 598
Surgeries in physicians' offices, 266
Testosterone use, 87
Time rate of growth of a tumor, 391
Toxic pollutants, 114
Transmission of disease, 717, 729
Unclogging arteries, 238
U.S. infant mortality rate, 394
Velocity of blood, 168, 311
Von Bertanlanffy growth function, 388
Waiting times, 664, 667
Walking versus running, 87
Water pollution, 186, 278
Weber–Fechner law, 378, 638
Weight of whales, 44
Weights of children, 368
Weiss's Law, 130
Whale population, 164, 448
Wind chill factor, 556
Yield of an apple orchard, 91
Yield of a wheat field, 635

GENERAL INTEREST

Acceleration of a car, 217, 408
Average highway speed of a vehicle, 170, 309
Blowing soap bubbles, 231
Boston Marathon, 259
Boyle's law, 62
Coast Guard patrol search mission, 231
Designing a grain silo, 324
Designing a Norman window, 91, 323
Driving costs, 81
Effect of an earthquake on a structure, 798
Effect of stopping on average speed, 168
Estimating the flow rate of a river, 519, 520
Expected snowfall, 664, 679
Fisk's law, 626, 631
Flight of a rocket, 165, 170, 261, 280, 306, 309, 408, 760
Flight path of a plane, 135
Force generated by a centrifuge, 544
IQs, 542, 691
Keeping with the traffic flow, 67
Lambert's law of absorption, 625, 640
Launching a fighter aircraft, 409
Lotteries, 475
Magnitude of earthquakes, 346, 378
Maximizing drainage capacity, 795
Mortgage payments, 239, 241, 242
Motion of a maglev, 97, 135, 214
Newton's law of cooling, 346, 390, 448, 625, 640, 641
Number of daylight hours, 794, 798, 806
Optimizing travel time, 29
Parking fees, 130
Postal regulations, 62, 323, 569, 591
Reaction time of a motorist, 679
Rings of Neptune, 236, 242
Spam messages, 74
Speedboat racing, 438
Stimulus response, 638
Stopping distance of a racing car, 169
Strength of a beam, 324
Televising a rocket launch, 795
Terminal velocity, 295
Tracking a criminal suspect, 795
Trial run of an attack submarine, 518
Turbo-charged engine performance, 461
Used car markup, 10
Velocity of a car, 143, 149, 405, 448
Velocity of a dragster, 497
Volume of a football, 484
VTOL aircraft, 217
Women's soccer, 282

BASIC RULES OF DIFFERENTIATION

1. $\frac{d}{dx}(c) = 0, \quad c$ a constant
2. $\frac{d}{dx}(u^n) = nu^{n-1}\frac{du}{dx}$
3. $\frac{d}{dx}(u \pm v) = \frac{du}{dx} \pm \frac{dv}{dx}$
4. $\frac{d}{dx}(cu) = c\frac{du}{dx}, \quad c$ a constant
5. $\frac{d}{dx}(uv) = u\frac{dv}{dx} + v\frac{du}{dx}$
6. $\frac{d}{dx}\left(\frac{u}{v}\right) = \frac{v\frac{du}{dx} - u\frac{dv}{dx}}{v^2}$
7. $\frac{d}{dx}(e^u) = e^u\frac{du}{dx}$
8. $\frac{d}{dx}(\ln u) = \frac{1}{u}\cdot\frac{du}{dx}$
9. $\frac{d}{dx}(\sin u) = \cos u\frac{du}{dx}$
10. $\frac{d}{dx}(\cos u) = -\sin u\frac{du}{dx}$
11. $\frac{d}{dx}(\tan u) = \sec^2 u\frac{du}{dx}$
12. $\frac{d}{dx}(\sec u) = \sec u \tan u\frac{du}{dx}$
13. $\frac{d}{dx}(\csc u) = -\csc u \cot u\frac{du}{dx}$
14. $\frac{d}{dx}(\cot u) = -\csc^2 u\frac{du}{dx}$

BASIC RULES OF INTEGRATION

1. $\int du = u + C$
2. $\int kf(u)\, du = k\int f(u)\, du, \quad k$ a constant
3. $\int [f(u) \pm g(u)]\, du = \int f(u)\, du \pm \int g(u)\, du$
4. $\int u^n\, du = \frac{u^{n+1}}{n+1} + C, \quad n \neq -1$
5. $\int e^u\, du = e^u + C$
6. $\int \frac{du}{u} = \ln|u| + C$
7. $\int \sin u\, du = -\cos u + C$
8. $\int \cos u\, du = \sin u + C$
9. $\int \sec^2 u\, du = \tan u + C$
10. $\int \csc^2 u\, du = -\cot u + C$
11. $\int \sec u \tan u\, du = \sec u + C$
12. $\int \csc u \cot u\, du = -\csc u + C$